The Encyclopedia

of the Biological Sciences

The Encyclopedia of the Biological Sciences

SECOND EDITION

EDITED BY

PETER GRAY

Andrey Avinoff Distinguished Professor of Biology
University of Pittsburgh, Pittsburgh, Pennsylvania

 **VAN NOSTRAND REINHOLD COMPANY**

New York   Cincinnati   Toronto   London   Melbourne

Van Nostrand Reinhold Company Regional Offices:
Cincinnati, New York, Chicago, Millbrae, Dallas
Van Nostrand Reinhold Company Foreign Offices:
London, Toronto, Melbourne

Copyright © 1970 by Litton Educational Publishing, Inc.

Library of Congress Catalog Card Number: 77-81348
ISBN: 0-442-15629-4

Published by Van Nostrand Reinhold Company
450 West 33rd Street, New York., N.Y. 10001
Published simultaneously in Canada by
Van Nostrand Reinhold Ltd.

15   14   13   12   11   10   9   8   7   6   5

INTRODUCTION

Purpose of this volume   This "Encyclopedia of the Biological Sciences" is intended to provide succinct and accurate information for biologists in those fields in which they are not themselves experts. And also to provide a comprehensive reference source for those who, like librarians, may be asked for information about biology even though this is not the field of their specialization. It should be emphasized that the term "biologist" is not intended to restrict the use of this encyclopedia to those who earn a living as professional biologists in universities, colleges, and research institutes. Every endeavor has been made to provide information which would be of value to teachers in high schools and their students, and to those nonprofessionals whose thirst for information goes beyond the popular accounts of general encyclopedias. The Editor realizes that it is impossible to please everyone, so that there may be professional biologists who are annoyed by the inclusion of such articles as NAMES, GROUP, GENDER, AND JUVENILE—information frequently sought from librarians—just as there may be those who will be slightly dismayed by the apparent technicality of some of the articles on more abstruse subjects. The joint aim of the Editor, and of the publishers, has been to provide the maximum amount of information for the greatest possible number of people.

Scope   This is an encyclopedia, not a dictionary. That is, it does not merely define the numerous subjects covered but describes and explains them. There are therefore no articles, except for a few biographies, which are less than 500 words in length and several are ten times as long. These 800 articles cover the broad field of the biological sciences as viewed by experts in their developmental, ecological, functional, genetic, structural, and taxonomic aspects. Numerous topics in the fields of biophysics and biochemistry have been included but the latter subject has been slanted to function rather than form. Readers whose primary interest is in the structure of chemicals involved in the life processes are referred to the Clark-Hawley *Encyclopedia of Chemistry* and the Williams-Lansford *Encyclopedia of Biochemistry* also published by Van Nostrand Reinhold.

No attempt has been made to touch on the applied biological sciences. The behavioral sciences have been included only as they apply to animals, not as animals apply to them. Thus there are articles on AGGRESSION, MATING DANCE, and the like, but none on the rat in any of its numerous applications to psychological research. The content of each individual contribution, and the method by which this content is presented, has been left entirely to the discretion of the contributor. The Editor has no love of uniformity for its own sake and is not convinced that any two articles on similar topics should be given identical treatment. Each contributor is a specialist and the Editor feels strongly that a specialist is best qualified to determine what is of importance in the subject which he is discussing, and what is the best method of making this importance apparent to the reader.

Alphabetization   The articles are arranged in alphabetical order, using the method of alphabetizing preferred by the U. S. Library of Congress and often referred to as the "encyclopedia method" from its frequent employment in such volumes. This method of alphabetizing involves the assumption that a space between two words is a mythical letter of the alphabet which precedes the letter A. Thus NEW YORK would precede NEWARK since the space after the w, in the first entry, precedes the a after the w in the second entry. Those who do not know that this system exists, or find it peculiar that it does, are entitled to an explanation of the purpose. Strict alphabetization by the conventional method results in main entries finding themselves jammed, out of context, between two subsidiary entries. Thus, were this rule not in effect, ASIA MINOR, to continue with geographical examples, would find itself sandwiched between ASIA, LABOR PROBLEMS and ASIA, MINORITIES. The use of the inverted title is necessary to bring similar topics together. Thus BIOLOGICAL STATIONS, INLAND and BIOLOGICAL STATIONS, MARINE should quite obviously be successive entries, and not separated as widely as would be the case were they alphabetized as INLAND BIOLOGICAL STATIONS and MARINE BIOLOGICAL STATIONS. The device of having a main entry BIOLOGICAL STATIONS with subordinate entries for INLAND and MARINE results in utter confusion as to what is a main, and what is a subordinate, entry.

Illustrations   The illustrations in this volume have been kept to a minimum, not only to conserve space, but also to avoid wasting the reader's time. "Popular" encyclopedias may have to illustrate, for example, an article on carnivora with a picture of a lion, tiger, and a cat but the Editor feels that biologists are more likely to be insulted than edified by this sort of thing. Illustrations, in fact, are used only when they tell a clearer story than could the words that they replace.

References   Almost all of the articles in this encyclopedia have references at the end which are designed both to refer the reader to a more copious source of information and, in some cases, to justify the views expressed by the contributor. This is the one place in which the Editor has insisted on uniformity and has ensured it by changing all references into a common pattern. This pattern is again that preferred by the U. S. Library of Congress. For books, the information is presented in this order: author's name, title, place, publisher, and date. The same form is followed for journal references save that the place and publisher of the journal, which is reduced to a standardized abbreviation, are not given unless confusion would result from the omission. The name of the journal is followed by the volume number in bold face type separated from the pagination by a colon. In the case of joint authorship, both authors are given, but where there are more than two authors, only the first is cited and followed by the phrase *et al.* Successive references are separated by a semi-colon. An example should make this clear.

> Hiatt, Robert W., "Directory of Hydrobiological Laboratories and Personnel in North America," 2nd ed., Honolulu, University of Hawaii Press, 1954; Toole, E. M., "Physiology of Seed Germination," Ann. Rev. Plant Physiol. 7:299, 1956.

Some contributors have preferred to cite the title of journal articles, as in the above example, and others have not. Some contributors have referred only to the first page of the article while others have given the complete pagination. The Editor has left decisions of this sort in the hands of the contributors.

The Editor would like to take this opportunity to point out one rather unfortunate result of the biologists' otherwise admirable view that all are members of one happy family, privy to each other's thoughts. Unfortunately a zoologist's reference to "Kukenthal's Handbuch" is just as meaningful to the average bacteriologist as is a social reference to "Uncle George" to one who is not of the family. The working zoologist will, of course, regard the following as hopelessly prolix:

> Kukenthal, W. and T. Krumbach, Eds. "Handbuch der Zoologie," Berlin, de Gruyter, 1923.

This form of reference is, however, more likely to be of use to most members of the biological family, and is the only one which is of the slightest value to a librarian assaulted with a demand for the volume.

Cross references Cross references are indicated in the body of the article by the device of setting the relevant word in small capitals. Thus a sentence reading, "the contribution of the GENE to our understanding of . . ." indicates that there is a separate article GENE in its appropriate alphabetic location. Cross references have only been inserted by the Editor in those places where he considers them to be germane to the subject of the reader's interest. To set such words as "invertebrate," "excretion," and "monocotyledon" in small capitals in every place where they occur in the text would create a typographic eyesore and serve to confuse, rather than clarify, the subject. The Editor has not hesitated to set in small capitals words that do not exactly correspond to the title of the article in those cases where confusion cannot possibly result. Thus, though the exact title of one of the articles is DICOTYLEDON, the Editor has set DICOTYLEDONS and DICOTYLEDONOUS since "dicotyledonous (see DICOTYLEDON)" appears to him a stupid pedantry.

Index It is difficult to keep the index of an encyclopedia, particularly a technical encyclopedia, within bounds. If every reference to every living form, and every technical term used to describe its functions, anatomy, embryology, and interaction with its environment, were to be transferred to an index entry, the index itself would become a second volume. The Editor has therefore tried to place in the index only those entries which, in his opinion, will be of value to the reader. For example, a cursory reference to the Diptera as part of a specific environment is omitted from the index, but a reference to the same flies as vectors in the life cycle of a protozoan parasite is included. Moreover, so far as possible, index entries refer back only to those places in the text where significant information will be found. The word "archegonium," for example, probably occurs in at least a hundred scattered places in this volume. A reader who comes across it, and who does not know its meaning, may refer to the index where he will find the word ARCHEGONIUM followed by a page reference. The small capitals indicate that a specific article is to be found commencing on the page indicated. A reader who finds the word "chelicera," in an article on one of the Arthropoda, and who is hazy as to its meaning, will find the word in ordinary type (since there is no special article of that name) in the index followed by page references which will lead him to those various articles on Arthropoda in which the appendage in question is identified.

Far more difficult decisions have had to be made on the subject of technical terms. The Editor has endeavored to make a reasonable number of index entries for technical terms, such as *pyrenoid* and *clone*, which are specifically biological. He has not indexed such terms as *clavate, geniculate* or *plumose* on the grounds that a person unacquainted with these words should seek them in a dictionary of the English language and not the index of an Encyclopedia of the Biological Sciences. It has, however, been extremely difficult to draw a clear line and the Editor fears that he may have committed more sins of commission and omission in the index than in any other part of the encyclopedia.

PETER GRAY

PREFACE TO THE SECOND EDITION

This edition, as was the first, is a cooperative effort among biologists the world over. I can only reiterate the admiration for their scholarship, courtesy, and cooperation that I expressed in the Preface to the First Edition.

Each contributor to the first edition received a questionnaire asking whether he considered that his article should be revised; if so, whether he would undertake the revision; if not, who should; and what articles he thought should be deleted or added. There was, naturally, a wide spread of opinion. Some articles in rapidly changing fields (SPACE BIOLOGY, PROTEIN SYNTHESIS) required to be completely rewritten while others of an equally fundamental, but less topically urgent, nature (SKULL, ANNELIDA) usually required no more than the alteration of a paragraph or two and the updating of the bibliography. Some topics omitted, either by accident (the coelenterates in general) or design (EXOBIOLOGY) from the first edition have been added as have also many new illustrations; to make way for these a number of topics, of which many contributors had expressed disapproval, have been omitted.

I am happy again to have the opportunity of acknowledging the continuous, friendly and skillful help of my publishers, represented in the compilation of the present volume by Mr. Charles S. Hutchinson, Jr., Mr. William K. Fallon, and Miss Karen Tobiassen. At this end of the line, Miss Bernadette Brown has cheerfully shouldered an immense number of those chores that a secretary enjoys no more than, but can usually do better than, an author.

PETER GRAY

Pittsburgh, 1969

PREFACE TO THE FIRST EDITION

*"The preface of a book affords the author an opportunity of speaking to his reader in a comparatively direct and personal manner, and of acquainting the prospective user of the book with the considerations which impelled the author to write it."**

The reasons which impelled me to edit this book are called Gessner G. Hawley and James B. Ross, respectively the Executive Editor of the Technical Editorial Department, and Manager of the College Department, of the Reinhold Book Division. They came into my office one day and asked me if I would like to edit an Encyclopedia of the Biological Sciences. I replied that I thought it would be fun and now, many months and about 5,000 letters later, I find that I was right. That is, I am confirmed in my opinion that biologists the world over are the most scholarly, the most courteous, and the most cooperative group with which any person could be privileged to associate.

I have been often asked, "What does the editor of an encyclopedia *do*?" The answer is quite simple. He does everything except write the articles signed by somebody else. He shoulders the whole responsibility, and bears the whole blame, for every sin of omission and commission which occurs in the next 1,000 pages of this book. With a view to placing myself in this position, I first of all sat down and broadly divided the biological sciences into their component parts, then these parts into lesser parts, and so on, as though I were classifying a phylum until I had a list of something over 800 topics which seemed to me to require articles. These articles had then to be weighted according to their importance and this weight expressed in thousands of words. I then, of course, went round and bothered all my friends but I am solely responsible for having taken, or rejected, their advice.

The next stage was to find contributors. The obvious persons for the more important articles were first approached and more than seventy per cent of them accepted. Most of those who were unable to accept (only one was unwilling) suggested to me the name of a contributor of almost equal eminence. Then came a long period of browsing through libraries, through book sellers' catalogues, through *Biological Abstracts*, and through the bibliographies in reference works in search of contributors for more obscure subjects. Finally, a list of the articles for which I had failed to find contributors was circulated to the contributors themselves who came up with all the necessary suggestions.

I was struck by a letter from one of the first contributors expressing the hope that I would give younger writers a break since he was "sick and tired of the compilations of middle-aged celebrities which are infesting the literature." It is unfortunately inevitable that the distribution curve of a group of distinguished biologists should peak rather sharply at "middle age" but I saw no reason why the ends of the curve should not be spread as widely as possible. Let it therefore be of record that the contributors to this encyclopedia range from the early twenties to the early nineties, with both groups being significantly more than 3 sigma away from the mean. In academic status, these persons range from a handful of graduate students to a considerably larger handful of Nobel prize winners,

---

*"The Bookman's Glossary." 3d ed. New York, Bowker [c. 1951.] Quoted by permission of the R. R. Bowker Co.

Fellows of the Royal Society, and Academicians of various National Academies. In academic affiliation, they range from graduate assistants to directors of internationally famous institutes. There is also a leavening of distinguished amateurs, as well as a newspaper editor and a librarian. These persons have, moreover, been drawn from every continent, or continental island, in the world and there is scarcely a major civilized country which is not represented by at least one contributor.

Let us return to the editor's duties. Having secured his contributors, he now waits for the contributions. When they arrive—I was pleasantly surprised that nearly fifteen percent arrived on schedule—the editor really has to start work. Each page of manuscript of each contribution has to be read carefully and marked up for the printer. Those contributions not written in English—some were received in French, German, Italian, Spanish, and Portuguese—have to be translated and the translation submitted to the contributor for his approval. The article is then sent to the publisher from whom, in due course, a galley proof arrives. The editor glances over this with a cursory eye, since it will be scrutinized for error by both the contributor and the publisher's proof reader. Then the page proofs arrive and the editor has again to get down to work preparing the index, writing the introduction, and writing the preface.

You will appreciate that all these functions of the editor do not occur in the clear chronological sequence in which I have presented them. There was a mad time in which I was performing almost all of these duties at once and, though I continue to assume complete responsibility, I am happy to have the opportunity to acknowledge the assistance of at least some of those who helped me to do so. Much of the correspondence had to be conducted by form letter and the results recorded on complicated forms and filing systems. The Central Printing Department, under the direction of Mr. Calvin Smith, of the University of Pittsburgh's Office of Administrative Services, reproduced all my material, and looked after problems of bulk mailing, and the like, in a manner which combined the promptness and efficiency of a commerical undertaking with the sympathetic understanding of a scholarly affiliate. Miss Hazel Johnson, and her assistants in the reference department of the University of Pittsburgh Library, have been invaluable in pursuing obscure references. Dr. Joan Eiger Gottlieb and Miss Letitia Langord have both shouldered more than a fair share of my other burdens while helping me to carry this one. Mrs. Leah Porter, assisted from time to time by Miss Annette Galluze, has survived two editions of a textbook, a small scholarly work, a large reference book, and now this encyclopedia, and has made to each of these the indispensable contribution which can only come from a first class secretary and indefatigable worker.

But my greatest debt, as I said at the beginning, is to those biologists the world over whose warm cooperation is the only thing that has made this volume possible.

PETER GRAY

CONTRIBUTORS

S. AARONSON, Queens College of the City University of New York, Flushing, New York. VITAMINS, WATER-SOLUBLE

D. P. ABBOTT, Stanford University, Stanford, California. UROCHORDATA

L. G. ABOOD, University of Illinois, Chicago, Illinois. MICROSOME

G. G. ABRIKOSSOV, Moscow University, Moscow, USSR. POGONOPHORA

C. J. ALEXOPOULOS, The University of Texas at Austin, Austin, Texas. FUNGI and MYXOMY-CETES

C. F. ALLEGRE, University of Northern Iowa, Cedar Falls, Iowa. SPOROZOA

R. W. ALRUTZ, Denison University, Granville, Ohio. ODONATA

E. ANDERSON, University of Massachusetts, Amherst, Massachusetts. PINEAL

W. ANDREW, Indiana University School of Medicine, Indianapolis, Indiana. INTEGUMENT

T. S. ARGYRIS, Syracuse University, Syracuse, New York. SKIN

C. A. ARNOLD, University of Michigan, Ann Arbor, Michigan. CYCADALES and PLANT KINGDOM

E. J. ARTHUR, University of Pittsburgh, Pittsburgh, Pennsylvania. PHEROMONE

G. ASBOE-HANSEN, University of Copenhagen, Copenhagen, Denmark. CONNECTIVE TISSUE

S. A. ASDELL, Cornell University, Ithaca, New York. OVULATION

W. R. ASHBY, Burden Neurological Institute, Bristol, England. HOMEOSTASIS

E. H. ASHTON, University of Birmingham Medical School, Birmingham, England. PRIMATES

C. R. AUSTIN, National Institute for Medical Research, London, England. EGG

C. BAEHNI, Conservatoire et Jardin Botanique, Geneva, Switzerland. EBENALES

S. BAGINSKI, Akademii Medycznej, Lodz, UI. Narutowicza, Poland. NEUROGLIA

H. B. BAKER, Havertown, Pennsylvania. MOLLUSCA

H. G. BAKER, University of California, Berkeley, California. ANGIOSPERMAE

W. BALAMUTH, University of California, Berkeley, California. CILIATA

E. BALL, North Carolina State College, School of Agriculture, Raleigh, North Carolina. MERISTEM

N. G. BALL, 4 Ennerdate Road, Kew Gardens, Surrey, England. RHIZOME

H. P. BANKS, Cornell University, Ithaca, New York. RHYNIOPHYTINA, SPHENOPHYTINA (part), TRIMEROPHYTINA and ZOSTEROPHYLLOPHYTINA

M. W. BANNAN, University of Toronto, Toronto, Canada. CAMBIUM

S. B. BARBER, Lehigh University, Bethlehem, Pennsylvania. SENSE ORGANS

R. BARER, The University, Sheffield, England. INTERFERENCE MICROSCOPY

S. B. BARKER, University of Alabama, Medical Center, Birmingham, Alabama. THYROID

J. L. BARNARD, Smithsonian Institute, Washington, D.C. AMPHIPODA

R. DONOSO-BARROS, Instituto Biologia Carlos Porter, Santiago, Chile. IXODOIDEA

G. W. BEADLE, University of Chicago, Chicago, Illinois. BIOCHEMICAL GENETICS

H. W. BEAMS, University of Iowa, Iowa City, Iowa. OVARY, ANIMAL

I. J. BENDET, University of Pittsburgh, Pittsburgh, Pennsylvania. DIFFUSION and SEDIMENTATION

L. L. BERGMAN, New York Medical College, New York, New York. GANGLION

A. BERNHARD, Pittsburgh Post Gazette, Pittsburgh, Pennsylvania. NAMES, GROUP, GENDER and JUVENILE(part)

S. S. BERRY, 1145 W. Highland Avenue, Redlands, California. AMPHINEURA

J. B. BEST, Walter Reed Army Institute of Research, Washington, D.C. THERMODYNAMICS OF CELLS

B. G. BIBBY, Eastman Dental Center, Rochester, New York. TOOTH

D. W. BIERHORST, New York State College of Agriculture, Cornell University, Ithaca, New York. SPHENOPHYTINA (SPHENOPSIDA) (part)

R. BIERI, Antioch College, Yellow Springs, Ohio. CHAETOGNATHA

J. BONNER, California Institute of Technology, Pasadena, California. NUCLEIC ACIDS (part)

H. A. BORTHWICK, U.S. Department of Agriculture, Beltsville, Maryland. PHOTOPERIODISM, PLANT

J. BOUILLON, Université Libre de Bruxelles, Brussels, Belgium. HYDROZOA

E. BOZLER, Ohio State University, Columbus, Ohio. MUSCLE

H. BRATTSTRÖM, University of Bergen, Bergen, Norway. ASCOTHORACIDA

D. C. BRAUNGART, Catholic University of America, Washington, D.C. ULTRASONICS

G. E. BRIGGS, University of Cambridge, Cambridge, England. OSMOSIS

H. BROCH, University of Oslo, Oslo, Norway. CIRRIPEDIA

A. F. BRODIE, University of Southern California, School of Medicine, Los Angeles, California. OXIDATIVE METABOLISM AND ENERGETICS OF BACTERIA

M. A. BROOKS, University of Minnesota, St. Paul, Minnesota. SYMBIOSIS

L. P. BROWER, Amherst College, Amherst, Massachusetts. SPECIATION

D. M. BROWN, Loma Linda University, Loma Linda, California. CORK

F. A. BROWN, JR., Northwestern University, Evanston, Illinois. ENDOGENOUS RHYTHMS

H. P. BROWN, University of Oklahoma, Norman, Oklahoma. FLAGELLA and SKELETON, INVERTEBRATE

J. M. A. BROWN, University of Auckland, Auckland, New Zealand. KREBS CYCLE

W. L. BROWN, JR., Cornell University, Ithaca, New York. ANT

R. B. BRUNSON, University of Montana, Missoula, Montana. GASTROTRICHA

R. BUCHSBAUM, University of Pittsburgh, Pittsburgh, Pennsylvania. INVERTEBRATES and TISSUE CULTURE, ANIMAL

M. BULJAN, Institut za Oceanografiju i Ribartstvo, Split, Yugoslavia. OCEAN

D. A. BURGH, R. D. No. 1, Box 86, Aliquippa, Pennsylvania. MICROSCOPE

B. L. BURTT, Royal Botanic Garden, Edinburgh, Scotland. TUBIFLORAE

P. H. CAHN, Long Island University, Brookville, New York. LATERAL LINE

T. W. M. CAMERON, McGill University, Montreal, Canada. PARASITISM

W. H. CAMP, University of Connecticut, Storrs, Connecticut. ERICALES

J. E. CANRIGHT, Arizona State University, Tempe, Arizona. RANALES

J. V. CARBONE, University of California Medical Center, San Francisco, California. DIGESTIVE SYSTEM

M. CARDENAS, University of San Simon, Cochabama, Bolivia. OPUNTIALES

F. M. CARPENTER, Harvard University, Cambridge, Massachusetts. ARTHROPODA and INSECTA

H. L. CARSON, Washington University, St. Louis, Missouri. VARIATION

J. D. CARTHY, Queen Mary College, London, England. COLONY, SOCIAL INSECTS, and TAXIS

R. V. CHAMBERLIN, University of Utah, Salt Lake City, Utah. CHILOPODA, DIPLOPODA, and SYMPHYLA

K. L. CHAMBERS, Oregon State University, Corvallis, Oregon. PLANT CYTOTAXONOMY

V. J. CHAPMAN, University of Auckland, Auckland, New Zealand. ALTERNATION OF GENER-
ATIONS and SALT MARSH

A. CHASE, Smithsonian Institute, Washington, D. C. GLUMALES

F. E. CHASE, Ontario Agriculture College, Guelph, Ontario, Canada. SOIL MICROORGANISMS

R. M. CHEW, University of Southern California, Los Angeles, California. CLIMAX and
WATER METABOLISM OF VERTEBRATES

B. G. CHITWOOD, 809 North Sheridan, Tacoma, Washington. NEMATA (part)

A. R. CHRISTOPHER, University of Pittsburgh, Pennsylvania. ELECTRON MICROSCOPE

T. CLAY, British Museum of Natural History, London, England. MALLOPHAGA

H. K. CLENCH, Carnegie Museum, Pittsburgh, Pennsylvania. LEPIDOPTERA

M. H. CLENCH, Carnegie Museum, Pittsburgh, Pennsylvania. PASSERIFORMES

W. J. CLENCH, Harvard University, Cambridge, Massachusetts. SHELL

R. R. CLOTHIER, Arizona State University, Tempe, Arizona. INSECTIVORA

P. E. CLOUD, JR., U.S. Geological Survey, Washington, D.C. BRACHIOPODA

M. E. CLUTTER, Yale University, New Haven, Connecticut. MORPHOGENESIS, PLANT (part)
and PITH

W. V. COLE, Kansas City College of Osteopathy and Surgery, Kansas City, Missouri.
NERVE ENDINGS

M. S. COLLINS, Howard University, Washington, D.C. ISOPTERA

A. COMFORT, University College, London, England. LONGEVITY

L. CONSTANCE, University of California, Berkeley, California. UMBELLALES

P. L. COOPER, Republic Aviation Division, Fairchild Hiller Corporation, Farmingdale
Long Island, New York. EXOBIOLOGY

R. G. COLODNY, University of Pittsburgh, Pittsburgh, Pennsylvania. HISTORY OF
BIOLOGY

G. COLOSI, Universitá de Firenze, Florence, Italy. ADAPTATION

G. W. COMITA, North Dakota State University, Fargo, North Dakota. COPEPODA

D. P. COSTELLO, University of North Carolina, Chapel Hill, North Carolina. LARVA, IN-
VERTEBRATE

R. S. COWAN, Smithsonian Institute, Washington, D.C. CENTROSPERMAE, RHOEDALES,
ROSALES, SALICALES, and URTICALES

E. L. CORE, West Virginia University, Morgantown, West Virginia. RHAMNALES

J. M. CRIBBINS, Lafayette College, Easton, Pennsylvania. SUCTORIA

J. F. CROW, University of Wisconsin, Madison, Wisconsin. GENETIC DRIFT

H. C. CURL, JR., Oregon State University, Corvallis, Oregon. PLANKTON(part)

W. A. DAILY, Fermentation Products Research, Lilly Research Laboratories, Eli Lilly
and Company, Indianapolis, Indiana. CYANOPHYCEAE

H. V. DALY, University of California, College of Agricultural Sciences, Berkeley, California. TAXONOMY(part)

H. DAMAS, Université de Liège, Liège, Belgium. CYCLOSTOMATA

R. DAUBENMIRE, Washington State University, Pullman, Washington. GRASSLAND

C. C. DAVIS, Case Western Reserve University, Cleveland, Ohio. HEART, INVERTEBRATE

D. D. DAVIS, Chicago Natural History Museum, Chicago, Illinois. MAMMALS

D. E. DAVIS, Pennsylvania State University, University Park, Pennsylvania. PREDATION

R. J. DAVIS, Idaho State College, Pocatello, Idaho. ARALES, PLANTAGINALES, and SANTA-
LALES

H. DAVSON, University College, London, England. SECRETION

A. B. DAWSON, Harvard University, Cambridge, Massachusetts. EPITHELIUM

E. DEICHMANN, Museum of Comparative Zoology, Harvard University, Cambridge, Massachusetts. ECHINODERMATA

J. DELACOUR, Los Angeles County Museum, Los Angeles, California. ANSERIFORMES and GALLIFORMES

R. K. DELL, Dominion Museum, Wellington, New Zealand. CEPHALOPODA

M. DEMEREC, Brookhaven National Laboratory, Upton, New York. GENE(part)

R. W. DEXTER, Kent State University, Kent, Ohio. BRANCHIOPODA

L. R. DICE, University of Michigan, Ann Arbor, Michigan. HUMAN ECOLOGY

H. J. DITTMER, University of New Mexico, Albuquerque, New Mexico. ROOT and STEMS, MODIFIED

T. DOBZHANSKY, Rockefeller University, New York. GENETICS (part)

E. O. DODSON, University of Ottawa, Ottawa, Ontario, Canada. EVOLUTION

R. I. DORFMAN, Stanford University, Palo Alto, California. STEROID HORMONES

A. DORIER, Institut de Zoologie, Université de Grenoble, Grenoble, France. NEMATO-MORPHA

M. S. DOTY, University of Hawaii, Honolulu, Hawaii. BENTHON

M. DOUDOROFF, University of California, Berkeley, California. BACTERIA

E. C. DOUGHERTY, Kaiser Research Foundation Institute, Richmond, California. NEMATA (part)

J. A. DOWNES, Entomology Research Institute, Department of Agriculture, Ottawa, Canada. MATING DANCE

L. C. DUNN, Columbia University, New York, New York. DEVELOPMENTAL GENETICS

M. DURCHON, Laboratoire de Biologie Animal. Université de Lille, Lille, Nord, France. HORMONE, INVERTEBRATE

H. L. EASTLICK, Washington State University, Pullman, Washington. FAT

T. H. EATON, University of Kansas, Museum of Natural History, Lawrence, Kansas. CHORDATA, PAEDOGENESIS, and TELEOSTEI

M. W. EDDY. Dickinson College, Carlisle, Pennsylvania. HAIR

G. F. EDMUNDS, JR., University of Utah, Salt Lake City, Utah. EPHEMEROPTERA

W. T. EDMONDSON, University of Washington, Seattle, Washington. ROTIFERA

H. ELIAS, University of Chicago Medical School, Chicago, Illinois. EXCRETORY ORGANS (part)

P. M. ELIAS, Department of Internal Medicine, University of California Affiliated Hospitals, San Francisco Medical Center, San Francisco, California. EXCRETORY ORGANS(part)

W. K. EMERSON, American Museum of Natural History, New York, New York. SCA-PHOPODA

R. O. ERICKSON, University of Pennsylvania, Philadelphia, Pennsylvania. GROWTH

F. G. EVANS, University of Michigan, Ann Arbor, Michigan. SKELETON, VERTEBRATE

R. L. EVANS, Mead Johnson and Company, Evansville, Indiana. ISLETS OF LANGERHANS

L. FAGE, Musee Nationale d'Histoire Naturelle, Paris, France. XIPHOSURA

J. E. FALK, Division of Plant Industry, Commonwealth Scientific & Industrial Research Organization, Canberra City, Australia. PORPHYRINS

J. FELDMANN, Institut Oceanographique, Université de Paris, Paris, France. PHAEO-PHYCEAE

G. FELSENFELD, Laboratory of Molecular Biology, National Institute of Arthritis and Metabolic Diseases, National Institutes of Health, Bethesda, Maryland. BLOOD, INVERTE-BRATE

S. J. FOLLEY, National Institute for Research in Dairying, University of Reading, Shinfield, Reading, England. LACTATION

R. H. FOOTE, Cornell University, Ithaca, New York. DIPTERA

D. FOREMAN, Case Western Reserve University, Cleveland, Ohio. HORMONE, ANIMAL

H. S. FOREST, State University College, Genesco, New York, PHILOSOPHY OF BIOLOGY

F. R. FOSBERG, Smithsonian Institute, Washington, D.C. CONSERVATION and RUBIALES

D. L. Fox, Scripps Institution of Oceanography, La Jolla, California. COLORATION OF ANI-
MALS

J. W. Fox, Carnegie Museum, Pittsburgh, Pennsylvania. STREPSIPTERA(part)

R. M. Fox, Carnegie Museum, Pittsburgh, Pennsylvania. STREPSIPTERA(part)

P. W. FRANK, University of Oregon, Eugene, Oregon. ECOLOGY

J. A. FREEBERG, Lehigh University, Bethlehem, Pennsylvania. LYCOPSIDA

H. F. FROLANDER, Oregon State University, Corvallis, Oregon. PLANKTON(part)

W. G. FRY, Marine Science Laboratories, University College of North Wales, Anglesey,
Wales. PORIFERA

C. GANS, State University of New York at Buffalo, Buffalo, New York. AMPHISBAENIA and
MORPHOLOGY

L. A. GARAY, Botanical Museum, Harvard University, Cambridge, Massachusetts.
ORCHIDALES (part)

E. D. GARDNER, Wayne State University, Detroit, Michigan. JOINT

R. GARRISON, Wheaton College, Norton, Massachusetts. LENTICEL, SHOOT, and STEM

C. L. GAZIN, U.S. National Museum, Washington, D.C. ARTIODACTYLA

T. A. GEISSMAN, University of California, Los Angeles, California. PLANT COLORATION

C. J. GEORGE, American College, Madurai, India. METAMERIC SEGMENTATION

L. J. GIER, Department of Biology, Missouri Southern College, Joplin, Missouri. MUSCI

E. M. GIFFORD, JR., University of California at Davis, Davis, California. PLANT ANATOMY

B. P. GLASS, Oklahoma State University, Stillwater, Oklahoma. RODENTIA

C. J. GOIN, University of Florida, Gainesville, Florida. AMPHIBIA

C. J. GOODNIGHT, Western Michigan University, Kalamazoo, Michigan. ARACHNIDA and
PHALANGIDA

R. E. GOODWIN, Colgate University, Hamilton, New York. CHARADRIFORMES

T. W. GOODWIN, University of Liverpool, P.O. Box 147, Liverpool, England. CAROTENOIDS
and FLAVONOIDS

ISABELLA GORDON, c/o British Museum of Natural History, London, England. CRUSTACEA

F. J. GOTTLIEB, University of Pittsburgh, Pittsburgh, Pennsylvania. GENE (part)

J. E. GOTTLIEB, Churchill Area Schools, Pittsburgh, Pennsylvania. FLOWER, GYMNO-
SPERMAE and SEED

A. GRAHAM, University of Reading, Berkshire, England. GASTROPODA

W. C. GRANT, JR, Williams College, Williamstown, Massachusetts. METAMORPHOSIS

P. P. GRASSÉ, Laboratoire d'évolution des êtres organisés, Université de Paris, Paris,
France. ANIMAL KINGDOM

J. C. GRAY, State University College at New Paltz, New Paltz, New York. HEART, VERTE-
BRATE

P. GRAY University of Pittsburgh, Pittsburgh, Pennsylvania. MICROTOMY

J. D. GREEN, University of California Medical Center, Los Angeles, California. PITUITARY

K. N. H. GREENIDGE, Dalhousie University, Halifax, Nova Scotia. WOOD

A. V. GRIMSTONE, University of Cambridge, England. MASTIGOPHORA

C. S. GROBBELAAR, 16 Hofmery Street, Stellenbosch, South Africa. HYRACOIDEA

A. GROLLMAN, Southwestern Medical College, Dallas, Texas. ADRENALS

TH. GROSPIETSCH, Max Planck Institute für Limnologie, Plon, Germany. RHIZOPODA

H. E. GRUEN, University of Saskatchewan, Saskatoon, Saskatchewan, Canada. SPORANGI-
OPHORE

A. B. GURNEY, U.S. National Museum, Washington, D.C. ORTHOPTERA and PSOCOPTERA

A. F. GUTTMACHER, Mt. Sinai Hospital, New York, New York. TWINNING

R. P. HALL, 10335 Cinnebar Avenue, Sun City, Arizona. PROTOZOA

B. A. HALL, State University College, Cortland, New York. SAPINDALES and SAR-
RACENIALES

B. W. HALSTEAD, World Life Research Institute, Colton, California. VENOMOUS FISHES

T. HALTENORTH, Zoologische Staatssammlung, Munich, Germany. DERMOPTERA, PHOLIDOTA and TUBULIDENTATA

A. HAM, University of Toronto, Toronto, Ontario, Canada. HISTOLOGY

H. L. HAMILTON, University of Virginia, Charlottesville, Virginia. FEATHER

M. J. HAMON, Université de Rennes, Rennes, France. NEMATOCYST

J. L. HANCOCK, Dryden Mains, Roslin, Midlothian, Scotland. SPERMATOZOA

C. HAND, Bodega Marine Laboratory, University of California, Bodega Bay, California. ANTHOZOA

W. J. HARMAN, Louisiana Polytechnic Institute, Ruston, Louisiana. SETA

O. HARTMAN, University of Southern California, Los Angeles, California. ANNELIDA

P. E. HARTMAN, Johns Hopkins University, Baltimore, Maryland. BACTERIAL GENETICS

F. T. HAXO, Scripps Institute of Oceanography, La Jolla, California. BIOLUMINESCENCE (part)

C. L. HAYWARD, Brigham Young University, Provo, Utah. BIOME

F. HARRISON, Southwestern Medical School at Dallas, University of Texas, Dallas, Texas. ELECTROPHYSIOLOGY

R. HEADSTROM, Worcester Museum of Natural History, Worcester, Massachusetts. NEST BUILDING

O. V. S. HEATH, Agricultural Research Council Unit of Flower Crop Physiology, Reading, England. STOMA

J. W. HEDGPETH, Oregon State University, Newport, Oregon. PYCNOGONIDA

C. B. HEISER, JR., Indiana University, Bloomington, Indiana. HYBRIDIZATION

J. HESLOP-HARRISON, University of Birmingham, Birmingham, England. APOMIXIS

M. J. HEUTS, University of Louvain, Louvain, Belgium. CAVE BIOLOGY

C. P. HICKMAN, DePauw University, Greencastle, Indiana. PIGMENTATION

W. C. OSMAN HILL, Yerkes Regional Primate Research Center, Emory University, Atlanta, Georgia. LEMUROIDEA

M. HINES, Johns Hopkins University, School of Medicine, Baltimore, Maryland. BRAIN

R. E. HODGES, Department of Internal Medicine, University of Iowa, Iowa City, Iowa. VITAMINS, FAT-SOLUBLE

C. C. HOFF, University of New Mexico, Albuquerque, New Mexico. PSEUDOSCORPIONIDA

N. HOTTON III, U.S. National Museum, Washington, D.C. REPTILIA

R. A. HOWARD, Harvard University, Cambridge, Massachusetts. MALVALES

L. HUSTED, University of Virginia, Charlottesville, Virginia. MITOSIS

J. HUTCHINSON, Royal Botanic Garden, Kent, England. DICOTYLEDONS and MONOCOTYLEDONS

K. E. HYLAND, University of Rhode Island, Kingston, Rhode Island. ACARINA

J. W. IRWIN, Massachusetts Eye and Ear Infirmary, Boston, Massachusetts. CIRCULATORY SYSTEM

T. IWAI, Kyoto University, Maizuru, Japan. PHARYNX

F. IWATA, Hokkaido University, Hokkaido, Japan. NEMERTEA

D. L. JAMESON, University of Houston, Houston, Texas. URODELA

D. C. JOHNSON, University of Kansas Medical Center, Kansas City, Kansas. GESTATION

M. T. JOLLIE, Northern Illinois University, DeKalb, Illinois. BEAK and FALCONIFORMES

E. R. JONES, University of Florida, Gainesville, Florida. TURBELLARIA

W. H. JONES, Southern University, Baton Rouge, Louisiana. CLADOCERA

T. G. KARLING, Naturhistoriska Riksmuseet, Stockholm, Sweden. KINORHYNCHA

A. KAWAMURA, JR., University of Tokyo, Tokyo, Japan. RICKETTSIA

R. H. KAY, Oxford University, Oxford, England. ELECTRONIC INSTRUMENTATION

C. KAYSER, Université de Strasbourg, Strasbourg, France. HIBERNATION

H. KENG, University of Singapore, Singapore. GNETALES

J. KEOSIAN, Rutgers, The State University of New Jersey, Newark, New Jersey. NEOBIO-GENESIS

A. K. KHUDAIRI, College of Liberal Arts, Northeastern University, Boston, Massachusetts. PHYTOCHROME

A. KILEJIAN, School of Veterinary Medicine, University of California, Davis, California. FLAME CELL

J. E. KINDRED, University of Virginia, Charlottsville, Virginia. BLOOD, VERTEBRATE

L. M. KLAUBER, 233 West Juniper Street, San Diego, California. VENOMOUS REPTILES

J. W. KNUDSEN, Pacific Lutheran University, Tacoma, Washington. DECAPODA

DOV KOLLER, The Hebrew University of Jerusalem, Jerusalem, Israel. GERMINATION

M. J. KOPAC, New York University, Washington Square College, New York, New York. MICROMANIPULATION

A. KOROS, School of Medicine, University of Pittsburgh, Pittsburgh, Pennsylvania. ABIO-GENESIS

V. E. KRAHL, University of Maryland School of Medicine, Baltimore, Maryland. LUNG

O. KRAUS, Natur-Museum und Forschungs-Institut, Frankfurt, Germany. PAUROPODA

E. G. KREBS, University of Washington, Seattle, Washington. BIOENERGETICS

M. LAGRECA, Instituto di Zoologia, Universitá de Napoli, Naples, Italy. MANTODEA

H. J. LAM, Rijksherbarium, Leiden, Holland. SPERMATOPHYTA

W. LANDAUER, Storrs Experiment Station, Storrs, Connecticut. PHENOCOPIES

I. LaRIVERS, Biological Society of Nevada, Reno, Nevada. HETEROPTERA

R. F. LAWRENCE, Natal Museum, Pietermaritzburg, Natal, South Africa. PEDIPALPI, SCOR-PIONES and SOLIFUGAE

W. W. LEATHEN, Carnegie-Mellon University, Pittsburgh, Pennsylvania. IRON BACTERIA and SULFUR BACTERIA

H. LEES, University of Manitoba, Winnipeg, Manitoba. BACTERIAL NUTRITION

A. LENGEROVÁ, Czechoslovak Academy of Sciences, Institute of Experimental Biology and Genetics, Prague, Czechoslovakia. SOMATIC CELL GENETICS

H. W. LEVI, Museum of Comparative Zoology, Harvard University, Cambridge, Massachusetts. ARANEAE(ARANEIDA)

M. LEVINE, Brookhaven National Laboratories, Upton, New York, BACTERIOPHAGE

J. C. LEWIN, Scripps Institute of Oceanography, La Jolla, California. BACILLARIOPHYCEAE

E. B. LEWIS, California Institute of Technology, Pasadena, California. ALLELISM and LINK-AGE

C. C. LI, University of Pittsburgh, Pittsburgh, Pennsylvania. POPULATION GENETICS

H. L. LI, University of Pennsylvania, Philadelphia, Pennsylvania. GINKGOALES

E. G. LINSLEY, College of Agricultural Science, University of California, Berkeley, California. TAXONOMY(part)

J. H. LOCHHEAD, University of Vermont, Burlington, Vermont. HEPATOPANCREAS

R. B. LOFTFIELD, School of Medicine, University of New Mexico, Albuquerque, New Mexico. PROTEIN BIOSYNTHESIS

G. H. LOWERY, JR., Louisiana State University, Baton Rouge, Louisiana. MIGRATION (part)

H. W. LISSMANN, University of Cambridge, Cambridge, England. ELECTRIC ORGANS

R. G. LUTFY, Ain Shams University, Cairo, Egypt. CONTRACTILE VACUOLE

E. S. LUTTRELL, University of Georgia, Athens, Georgia. ASCOMYCETES

R. H. MACARTHUR, Princeton University, Princeton, New Jersey. COMMUNITY

P. MAHESHWARI, University of Delhi, Delhi, India. CONIFERALES

P. G. MAHLBERG, Indiana University, Bloomington, Indiana. LATICIFERS

J. MAJOR, University of California at Davis, Davis, California. BIOCENOSE

W. E. MANNING, Bucknell University, Lewisburg, Pennsylvania. JUGLANDALES

S. M. MANTON, British Museum of Natural History, London, England. LOCOMOTION and ONYCHOPHORA

R. D. MANWELL, Syracuse University, Syracuse, New York. PLASMODIUM

A. S. MARRAZZI, Laboratories in Neuropsychiatry, Veteran's Hospital, Pittsburgh, Pennsylvania. IMPULSE CONDUCTION AND TRANSMISSION

R. E. F. MATTHEWS, University of Auckland, Auckland, New Zealand. ELECTROPHORESIS

H. A. MATZKE, University of Kansas, Lawrence, Kansas. NERVOUS SYSTEM

P. H. MAUER, The Jefferson Medical College of Philadelphia, Pennsylvania. IMMUNOLOGY

D. MAZIA, University of California, Berkeley, California. CELL

L. S. MCCLUNG, Indiana University, Bloomington, Indiana. ANAEROBIC BACTERIA

B. H. MCCONNAUGHEY, University of Oregon, Eugene, Oregon. MESOZOA

D. MCCONNELL, California Institute of Technology, Pasadena, California. NUCLEIC ACIDS (part)

C. J. MCCOY, JR., Carnegie Museum, Pittsburgh, Pennsylvania. ANURA

E. R. MCGILLIARD, University of Chattanooga, Chattanooga, Tennessee. LILIALES

T. M. MCMILLION, Geneva College, Beaver Falls, Pennsylvania. HOMOPTERA

R. MERTENS, Natur-Museum und Forschungs, Institut Senckenberg, Frankfurt, Germany. CROCODILIA

L. H. MILLENER, University of Auckland, Auckland, New Zealand. PHLOEM

E. V. MILLER, 7616 Homestead Road, Benzonia, Michigan. FRUITS

F. W. MILLER, Hartwick College, Oneonta, New York. ANTENNA

M. A. MILLER, University of California, Davis, California. ISOPODA

M. R. MILLER, University of California, San Francisco Medical Center, San Francisco, California. ENDOCRINE SYSTEM

J. MILLOT, Musée Nationale d'Histoire Naturelle, Paris, France. PALPIGRADA

H. B. MILLS, Illinois Natural History Survey, Urbana, Illinois. COLLEMBOLA

F. MOOG, Washington University, St. Louis, Missouri. REPRODUCTION

H. E. MOORE, JR., Cornell University, Ithaca, New York. PALMALES

H. J. MOROWITZ, Yale University, New Haven, Connecticut. BIOPHYSICS

J. H. MORRISON, Case Western Reserve, Cleveland, Ohio. MEIOSIS

J. E. MORROW, JR., University of Alaska, College, Alaska. COELACANTH

M. F. MOSELEY, University of California, Santa Barbara, California. VERTICILLATAE

H. W. MOSSMAN, University of Wisconsin, Madison, Wisconsin. OESTROUS CYCLE

J. M. MOULTON, Bowdoin College, Brunswick, Maine. LIMB

J. F. MUELLER, State University of New York, Upstate Medical Center, Syracuse, New York. TRYPANOSOMA

C. F. W. MUESEBECK, U.S. National Museum, Washington, D.C. HYMENOPTERA

J. H. MULLAHY, S. J., Loyola University of the South, New Orleans, Louisiana. RHODO-PHYCEAE

C. H. MULLER, University of California, Santa Barbara, California. FAGALES

E. G. MUNROE, Department of Agriculture, Ottawa, Canada. NEUROPTERA and TRICH-optera

D. M. MUNSHI, St. Xavier College, Bombay, India. SIPHONAPTERA

W. R. MURCHIE, The University of Michigan, Flint College, Flint, Michigan. OLIGO-chaeta

R. C. MURPHY, Indiana University, Bloomington, Indiana. SPLEEN

R. J. MYERS, Colgate University, Hamilton, New York. HIRUDINEA

A. V. NALBANDOV, University of Illinois, Urbana, Illinois. REPRODUCTIVE SYSTEM

A. E. NEEDHAM, Oxford University, England. REGENERATION, INVERTEBRATE

M. G. NETTING, Director, Carnegie Museum, Pittsburgh, Pennsylvania. ZOOGEOGRAPHY

R. J. NEWMAN, Louisiana State University, Baton Rouge, Louisiana. MIGRATION (part)

A. B. NOVIKOFF, Albert Einstein College of Medicine, New York. MITOCHONDRIA

E. C. OGDEN, New York State Museum, Albany, New York. NAJADALES

J. G. OGDEN III, Ohio Wesleyan University, Delaware, Ohio. CARBON DATING and PALY-NOLOGY

H. J. OOSTING, Duke University, Durham, North Carolina. SUCCESSION

A. I. OPARIN, Bach Institute of Biochemistry, Moscow, USSR. LIFE, ORIGIN OF

J. M. OPPENHEIMER, Bryn Mawr College, Bryn Mawr, Pennsylvania. MORPHOGENESIS, ANIMAL

J. L. OSCHMAN, Case Western Reserve University, Cleveland, Ohio. CHLOROPLAST and THEOPHRASTUS

J. D. PALMER, New York University, University Heights, Bronx, New York. BIOLOGICAL CLOCK

B. PARDUCZ, Biologisch Laboratoriums am Ungarisches Naturwissenschäftiche Museum, Budapest, Hungary. CILIA

F. M. PACKARD, International Committee on National Parks, 2144 P Street, N.W., Washington, D.C. NATURE RESERVES

A. M. PAPENHEIMER, JR., Harvard University, Cambridge, Massachusetts. TOXINS

C. R. PARTANEN, University of Pittsburgh, Pittsburgh, Pennsylvania. ENDOPOLYPLOIDY

A. D. PEACOCK, 2 Onslow Gardens, London S.W. 7, England. PARTHENOGENESIS

T. PEHRSON, University of Stockholm, Stockholm, Sweden. SKULL

R. W. PENNAK, University of Colorado, Boulder, Colorado. TARDIGRADA

T. PERRI, Universitá di Perugia, Rome, Italy. SYMMETRY

M. H. PETTIBONE, University of New Hampshire, Durham, New Hampshire. POLY-CHAETAE

D. R. PFOUTZ, Carnegie Library, Pittsburgh, Pennsylvania. NAMES, GROUP, GENDER and JUVENILE(part)

J. R. PLEASANTS, University of Notre Dame, Notre Dame, Indiana. GNOTOBIOTES

R. W. PILLSBURY, University of British Columbia, Vancouver, British Columbia, Canada. LITTORAL

R. B. PLATT, Emory University, Atlanta, Georgia. MICROCLIMATE

N. POLUNIN, "Biological Conservation," 1249 Avusy, Geneva, Switzerland. ARCTIC FLORA, ARCTIC VEGETATION, and TUNDRA

E. PONDER, Nassau Hospital, Long Island, New York. HEMOPOIESIS

E. J. POPHAM, University of Salford, Lancashire, England. DERMAPTERA

H. POPPER, Mt. Sinai Hospital, New York, New York. LIVER

D. M. POST, San Francisco State College, San Francisco, California. GENTIANALES (CON-TORTAE)

G. E. POTTER, A and M College of Texas, College Station, Texas. AMPHIOXUS and SCALES

G. W. PRESCOTT, University of Montana Biological Station, Big Fork, Montana. ALGAE

R. D. PRESTON, Astbury Department of Biophysics, University of Leeds, Leeds, England. CELL WALL

J. W. S. PRINGLE, University of Cambridge, Cambridge, England. FLIGHT

J. PROSKAUER, University of California, Berkeley, California. HEPATICAE

C. RALPH, University of Pittsburgh, Pittsburgh, Pennsylvania. CARBOHYDRATE METABOLISM

N. RASHEVSKY, University of Michigan, Ann Arbor, Michigan. MATHEMATICAL BIOLOGY

H. RASMUSSEN, Rockefeller Institute, New York, New York. PARATHYROIDS

R. L. RAUSCH, Arctic Health Research Center, College, Alaska. ARCTIC

J. F. REGER, The University of Tennessee, Memphis, Tennessee. NERVE CELL

G. K. REID, Florida Presbyterian College, St. Petersburg, Florida. LIMNOLOGY (FRESH-WATER)

R. H. REINHART, Miami University, Oxford, Ohio. DESMOSTYLIA and SIRENIA

R. REISER, Texas A and M University, College Station, Texas. PHOSPHOLIPIDS

C. L. REMINGTON, Yale University, New Haven, Connecticut. GYNANDROMORPH

O. E. REYNOLDS, National Aeronautic and Space Administration, Washington, D.C. SPACE BIOLOGY(part)

E. R. RICH, University of Miami, Coral Gables, Florida. POPULATIONS

A. G. RICHARDS, University of Minnesota, Minneapolis, Minnesota. CHITIN

O. W. RICHARDS, Pacific University, College of Optometry, Forest Grove, Oregon. MICROTOME

N. D. RICHMOND, Carnegie Museum, Pittsburgh, Pennsylvania. LACERTILIA

W. E. RICKER, Fisheries Research Board of Canada, Nanaimo, British Columbia. PLECOPTERA

R. RIEDL, University of North Carolina, Chapel Hill, North Carolina. HEMICHORDATA

R. A. ROBINSON, Johns Hopkins University, Baltimore, Maryland. CARTILAGE

T. W. ROBINSON, U.S. Department of the Interior, Water Resources Division, Menlo Park, California. TRANSPIRATION (EVAPOTRANSPIRATION)

J. A. ROPER, The University, Sheffield, England. FUNGAL GENETICS

J. L. ROSENBERG, University of Pittsburgh, Pittsburgh, Pennsylvania PHOTOSYNTHESIS

C. A. ROSS, Western Washington State College, Bellingham, Washington. FORAMINIFERA

D. M. ROSS, University of Alberta, Edmonton, Canada. CNIDARIA (COELENTERATA)

L. E. ROTH, Argonne National Laboratories, Argonne, Illinois. ORGANELLE

P. ROTHEMUND, Ohio State University, Columbus, Ohio. PURINES

A. S. ROUFFA, University of Illinois at Chicago Circle, Chicago, Illinois. PARENCHYMA, PLANT

D. RUDNICK, Albertus Magnus College, New Haven, Connecticut. ORGANIZER and ORGANOGENY

E. D. RUDOLPH, Ohio State University, Columbus, Ohio. LICHENES

R. RUGH, Radiological Research Laboratory, College of Physicians and Surgeons, Columbia University, New York, New York. RADIATION EFFECTS

O. RULON, Northwestern University, Evanston, Illinois. GRADIENT THEORY

L. S. RUSSEL, Royal Ontario Museum, Toronto, Canada. CARNIVORA

N. H. RUSSEL, Central State College, Edmond, Oklahoma. PARIENTALES

A. RYTER, Institut Pasteur, Paris, France. BACTERIAL NUCLEUS

M. R. J. SALTON, New York University School of Medicine, New York. BACTERIAL CELL

M. SANTER, Haverford College, Haverford, Pennsylvania. CHEMOLITHOTROPHIC BACTERIA

A. A. SAUNDERS, Box 141, Canaan, Connecticut. BIRD SONGS, FOREST, and PICIFORMES

D. E. SAVAGE, University of California, Berkeley, California. PERISSODACTYLA

B. SCHAEFFER, The American Museum of Natural History, New York, New York. CROSSOPTERYGII

W. E. SCHEVILL, Harvard University, Cambridge, Massachusetts. CETACEA

C. R. SCHROEDER, San Diego Zoological Gardens, San Diego, California. ZOOLOGICAL GARDENS

R. E. SCHULTES, Harvard University, Cambridge, Massachusetts. VENOMOUS PLANTS

C. SCHWEINFURTH, Botanical Museum, Harvard University, Cambridge, Massachusetts. ORCHIDALES(part)

J. P. SCOTT, P.O. Box 847, Bar Harbor, Maine. AGGRESSION

M. SEARS, Woods Hole Oceanography Institute, Woods Hole, Massachusetts. SIPHONOPHORA

S. J. SEGAL, Rockefeller Institute, New York, New York. SEX

R. B. SELANDER, University of Illinois, Urbana, Illinois. COLEOPTERA

J. T. SELF, University of Oklahoma, Norman, Oklahoma. PENTASTOMIDA

E. E. SELKURT, Indiana University, Medical Center, Indianapolis, Indiana. EXCRETION

C. E. SHAW, San Diego Zoological Garden, San Diego, California. VENOMOUS REPTILES (part)

S. SHOSTAK, University of Pittsburgh, Pittsburgh, Pennsylvania. MESOGLEA

H. SICHER, Loyola University, School of Dentistry, Chicago, Illinois. BONE

RICHARD SIEGLER, M.D., Cancer Research Institute, New England Deaconess, Boston, Massachusetts. THYMUS

N. W. SIMMONDS, Scottish Plant Breeding Station, Pentlandfield, Roslin, Scotland. ZINGIBERALES

F. S. SJØSTRAND, University of California, Los Angeles, California. ULTRASTRUCTURE

S. SKOWRON, Polish Academy of Sciences, Krakow, Poland. FERTILIZATION

W. R. SISTROM, Harvard University, Cambridge, Massachusetts. PHOTOSYNTHETIC BACTERIA

A. H. SMITH, University of Michigan, Ann Arbor, Michigan. BASIDIOMYCETES

H. M. SMITH, University of Illinois, Urbana, Illinois. OPHIDIA

K. M. SMITH, University of Texas, Austin, Texas. VIRUSES

W. J. SMITH, University of Pennsylvania, Philadelphia, Pennsylvania. BIOCOMMUNICATION

O. T. SOLBRIG, Gray Herbarium, Harvard University, Cambridge, Massachusetts. COMPOSITAE

T. M. SONNEBORN, Indiana University, Bloomington, Indiana. CONJUGATION and PLASMAGENE

J. D. SOULE, Allan Hancock Foundation, University of Southern California, Los Angeles, California. BRYOZOA

F. K. SPARROW, University of Michigan, Pellston, Michigan, PHYCOMYCETES

N. H. SPECTOR, Medical College of Virginia, Richmond, Virginia. NEUROPHYSIOLOGY

E. B. SPIESS, University of Illinois at Chicago Circle, Chicago, Illinois. DOMINANCE, DROSOPHILA, GENETICS, and HETEROSIS

S. SPRINGER, U. S. Fish and Wildlife Service, Washington, D.C. ELASMOBRANCHII

D. F. SQUIRES, Marine Sciences Research Center, State University of New York, Stony Brook, New York. CORALS

L. J. STANNARD, State Natural History Survey, State of Illinois, Urbana, Illinois. THYSANOPTERA

F. J. STARE, Harvard University, Cambridge, Massachusetts. NUTRITION

A. STARRETT, San Fernando Valley State College, Northridge, California. CHIROPTERA

W. T. STEARN, British Museum of Natural History, London, England. CLONE

G. L. STEBBINS, University of California at Davis, Davis, California. POLYPLOIDY

W. C. STEERE, New York Botanical Garden, Bronx, New York. BRYOPHYTA

T. A. STEEVES, University of Saskatchewan, Saskatoon, Saskatchewan. LEAF

A. C. STEPHEN, Royal Scottish Museum, Edinburgh, Scotland. ECHIURIDA, PRIAPULIDA, and SIPUNCULIDA

W. L. STERN, U. S. National Museum, Washington, D. C. XYLEM

W. STILES, University of Birmingham, Birmingham, England. TRACE ELEMENTS

H. ST. JOHN, University of Hawaii, Honolulu, Hawaii. PANDANALES

A. G. STOKEY, Mount Holyoke College, South Hadley, Massachusetts. ANTHERIDIUM and ARCHEGONIUM

L. STØRMER, Universitet i Delo, Oslo, Norway. TRILOBITA

B. STOWE, Yale University, New Haven, Connecticut. HORMONE, PLANT (part)

H. W. STUNKARD, The American Museum of Natural History, New York. PLATYHELMINTHES

M. SUGIMURA, Hokkaido University, Sapporo, Japan. LYMPHATIC TISSUE

I. M. Sussex, Yale University, New Haven, Connecticut. MORPHOGENESIS, PLANT (part) and TISSUE CULTURE, PLANT

C. P. Swanson, Johns Hopkins University, Baltimore, Maryland. CYTOGENETICS

B. M. Sweeney, Scripps Institute of Oceanography, La Jolla, California. BIOLUMINESCENCE (part)

J. M. Talbot, Office of the Surgeon General, Department of the Air Force, Washington, D.C. SPACE BIOLOGY (part)

J. H. Taylor, Florida State University, Tallahassee, Florida. NUCLEUS

T. T. Tchen, Wayne State University, Detroit, Michigan. STEROLS, BIOGENESIS OF

H. Thiel, Universität Hamburg, Hamburg, Germany. SCYPHOZOA

K. V. Thimann, Harvard University, Cambridge, Massachusetts. HORMONE, PLANT (part)

J. A. Thomson, University of Melbourne,Victoria,Australia. MARSUPIALIA and MONOTREMATA

C. S. Thornton, Michigan State University, East Lansing, Michigan. REGENERATION, VERTEBRATE

H. B. Tordoff, University of Michigan, Ann Arbor, Michigan. AVES

A. A. Buzzati-Traverso, Università de Pavia, Pavia, Italy. POLYMORPHISM

G. Trégouboff, Station Zoologique, Vellefranche-sur-Mer, France. ACTINOPODA

W. L. Tressler, U.S. Navy Hydrographic Office, Washington, D.C. OSTRACODA

W. B. Turrill, The Herbarium, Royal Botanical Garden, Kew, England, PLANT GEOGRAPHY

D. H. Valentine, University of Durham, Durham, England. PRIMULALES

R. C. von Borstel, Oak Ridge National Laboratory, Oak Ridge, Tennessee. GENETIC SUPPRESSION

D. von Wettstein, Institute of Genetics, University of Copenhagen, Copenhagen, Denmark. PLASTID

W. H. Wagner, Jr., University of Michigan, Ann Arbor, Michigan. FILICINEAE and FROND

'S. A. Waksman, Rutgers, The State University, New Brunswick, New Jersey. ACTINOMYCETES

W. F. Walker, Jr., Oberlin College, Oberlin, Ohio. MUSCULAR SYSTEM

G. J. Wallace, Michigan State University, East Lansing, Michigan. COLUMBIFORMES, CORACIFORMES, CUCILIFORMES, GRUIFORMES, PELCANIFORMES, and RATITES

H. L. Ward, University of Tennessee, Knoxville, Tennessee. ACANTHOCEPHALA

C. W. Wardlaw, 6 Robins Close, Bramhall, Stockport, Cheshire, England. PLANT EMBRYOLOGY

R. A. Wardle, University of Manitoba, Winnipeg, Manitoba. CESTODA

R. L. Watterson, University of Illinois, Urbana, Illinois. EMBRYOLOGY

E. R. Waygood, University of Manitoba, Winnipeg, Manitoba. RESPIRATION

E. C. Webb, University of Queensland, St. Lucia, Brisbane, Queensland, Australia. ENZYME

F. Weberling, University of Mainz, Mainz, West Germany. MYRTIFLORAE

T. E. Weier, University of California, Davis, California. PROTOPLASM

J. S. Weiner, University of Oxford, Oxford, England. ANTHROPOID

F. W. Went, University of Nevada, Reno, Nevada. DESERT

A. D. Whedon, Yale University School of Medicine, New Haven, Connecticut. HEMOCOEL

L. C. Wheeler, University of Southern California, Los Angeles, California. GERANIALES

M. J. D. White, University of Melbourne, Melbourne, Australia. CHROMOSOME

P. W. Whiting, University of Pennsylvania, Philadelphia, Pennsylvania. HABROBRACON

H. R. Whiteley, University of Washington, Seattle, Washington. FERMENTATION

W. Wieser, Zoologisches Institut der Universität, Vienna, Austria. ARCHIANNELIDA

R. J. WILLIAMS, University of Texas at Austin, Austin, Texas. BIOCHEMICAL INDIVIDUALITY

P. N. WITT, State University of New York, Syracuse, New York. WEB BUILDING

A. WOLFSON, Northwestern University, Evanston, Illinois. PHOTOPERIODISM, ANIMAL

J. J. WOLKEN, Biophysical Research Laboratory, Carnegie-Mellon University, Pittsburgh, Pennsylvania. EYE

A. E. WOOD, Rutgers, The State University, Newark, New Jersey. LAGOMORPHA

R. M. WOTTON, University of Nebraska, Lincoln, Nebraska. GOLGI COMPLEX

S. YAMAGUTI, 453 Tonodan Sakuragi-cho, Kamikho-ku, Kyoto, Japan. TREMATODA

C. M. YONGE, Whincroft, Helensburgh, Dumbartonshire, England. PELECYPODA

T. G. YUNCKER, DePauw University, Greencastle, Indiana. PIPERALES

R. ZANGERL, Chicago Museum of Natural History, Chicago, Illinois. CHELONIA

M. H. ZIMMERMANN, Harvard University, Maria Moors Cabot Foundation for Botanical Research, Harvard Forest, Petersham, Massachusetts. TRANSLOCATION

D. J. ZINN, University of Rhode Island, Kingston, Rhode Island. PSAMMON

C. E. ZOBELL, Scripps Institute of Oceanography, La Jolla, California. MARINE BACTERIA

The Encyclopedia
of the Biological Sciences

ABIOGENESIS

The doctrine of abiogenesis, or spontaneous generation, was the widely accepted belief in ancient times that living organisms are formed spontaneously from non-living matter.

Aristotle wrote in the fourth century B.C. of his observations that plants and animals sometimes arise from nonliving matter. He advocated, for example, that some animals are generated from decomposing meat, and that plant lice are formed from the dew on plants.

The theory expounded by Aristotle was generally accepted until about the middle of the seventeenth century. Francesco Redi, about 1665, was one of the first persons to present scientific evidence which opposed the theory of spontaneous generation. He showed that maggots which arise from putrid meat are actually the larval stages of flies, and that such larvae will never form spontaneously if the meat is placed in a container covered with a fine gauze; because the flies are thereby prevented from depositing eggs on the meat. Redi thus successfully refuted the theory that higher organisms arise from nonliving material. His experiments, however, did not refute the idea that microorganisms can arise spontaneously.

Antoni van Leeuwenhoek about 1683 observed with the aid of a simple microscope that microorganisms are everywhere. Throughout the seventeenth century there was great controversy concerning the abiogenesis of microorganisms, because evidence appeared both for and against such a theory. Joblot, for example, reported that hay infusions prepared with cold water and left standing for a few days contained many small organisms, but that similar infusions which had been boiled and placed in a closed vessel remained free of organisms for several days. If the containers were left uncovered, however, many organisms soon appeared in the liquid.

The experiments of Needham in 1749 seemed to support the concept of abiogenesis of microorganisms; for he reported that many different types of infusions gave rise to microorganisms, irrespective of whether the solutions were boiled or covered. By about 1765, however, Spallanzani showed that Needham's techniques were inadequate. Spallanzani presented evidence from many experiments that proper heating will prevent the formation of microorganisms in various types of plant and animal infusions. He reported also that the time required for heating the solutions to sterilize them varied considerably depending on the nature of the infusion. He also observed that the complete exclusion of air was required to effect such sterilization. Needham had used corks to seal his flasks, but Spallanzani used hermetically sealed vessels. Thus, Spallanzani proved that if an infusion is boiled sufficiently, and hermetically sealed, it will never give rise to organisms unless new air which is contaminated with microbes somehow comes in contact with the infusion.

Schwann in 1837 passed air into a boiled infusion through a coiled glass tube which he kept in a continuous flame. He showed that heated air added to a sterile infusion did not cause the formation of microorganisms, and thus confirmed Spallanzani's observations.

Louis Pasteur, about 1861 presented further evidence which refuted the concept of abiogenesis. He showed that microorganisms could be extracted from air by sucking large quantities of air through a tube containing a guncotton stopper. He dissolved the cotton in a mixture of alcohol and ether, examined the residue microscopically, and observed that it contained small spherical or oval structures, which resembled the spores of certain plants. Pasteur also repeated and confirmed Schwann's experiments. Both workers agreed that heated air introduced into a boiled infusion would not induce the formation of microorganisms.

Pasteur further demonstrated that he could maintain infusions sterile indefinitely even if they were exposed to the air, provided that he used a flask with a very thin neck bent downward. He deduced that microbes could not pass beyond the bend in such a flask, and thus could not enter the sterile solution. The infusion became contaminated with organisms however, if the neck of such a vessel were broken. Pasteur also showed that he could contaminate the infusion by tipping the flask slightly and thus bringing the solution in contact with the microorganisms of the air in the bent portion of the container. These experiments provided the best evidence of the role of air as a vehicle for the transmission of microorganisms.

Pasteur attempted to determine quantitatively the distribution of microbes in the air. He observed that microorganisms are unevenly distributed throughout the atmosphere. He exposed sterile infusions in sealed flasks to the air in various parts of France and immediately resealed the containers to prevent further contamination. He found, for example, that more of those flasks which he opened on a country road became contaminated than those flasks which he opened on the summit of a mountain.

John Tyndall, an English contemporary of Pasteur, observed that certain forms of living matter are more resistant to heat destruction than others. While working

1

with hay infusions, he observed that heating even for five hours was insufficient for sterilizing his solutions. He developed a method for sterilization by discontinuous heating which would destroy all bacteria in hay infusions. Tyndall's experiments thus further explained the appearance of living organisms in nonliving material, and furnished additional evidence refuting the concept of spontaneous generation.

At the present time there is no evidence to support the theory that organisms arise from nonliving matter in a short period of time. A concept of the stepwise formation of living material from nonliving substances over an extensive period of time differs from abiogenesis. The latter is referred to as archebiosis or archegenesis (SEE LIFE, ORIGIN OF).

The possibility of recurring biogenesis, or the concept of NEOBIOGENESIS is also distinct from abiogensis, and is discussed elsewhere in this volume

AURELIA KOROS

References

Koesian, J., "On the Origin of Life," Science 131: 479–482, 1960.
ibid, "Neobiogenesis," (in this volume).
ibid, "The Origin of Life," 2nd ed. New York, Reinhold, 1968.
Lederberg, J., "Exobiology: approaches to life beyond the Earth," Science 132: 393–400, 1960.
Oparin, I. A., "The Origin of Life," 4th ed. New York, Academic Press, 1957.
ibid, "Life, origin of," (in this volume).
Stanier, R. I., *et al.* "The Microbial World," Englewood Cliffs, N. J., Prentice-Hall, 1957.

ACANTHARIA see ACTINOPODA

ACANTHOCEPHALA

The Acanthocephala, or thorny-headed worms, are entirely parasitic, and occur in the intestines of all classes of vertebrates, chiefly fishes, birds and mammals. They are now classified as a separate phylum, although their taxonomic position has been uncertain until recently. Since the body becomes turgid and cylindrical after removal from the host, they have been placed as an appendix to the NEMATHELMINTHES. They also resemble the roundworms in the possession of a large pseudocoel. However, the acanthocephalan body is more or less flattened before it is removed from the intestine of the host, suggesting a relationship with the PLATYHELMINTHES. The absence of a digestive tract in both the acanthocephalans and the cestodes suggests an affinity, and in the embryological development of the acanthocephalans there is a larva with hooks which resembles the cestode hexacanth. The history of the taxonomy of the Acanthocephala has been outlined by Van Cleave (1948) who has shown clearly that this group should be recognized as a distinct phylum.

The worm is composed of a proboscis armed with sharp hooks, a neck, and a trunk or body proper. The proboscis is retractile and may be withdrawn into the proboscis sheath, which is contained in the body proper. The body proper is unsegmented and cylindrical or flattened in form. In some species delicate body spines

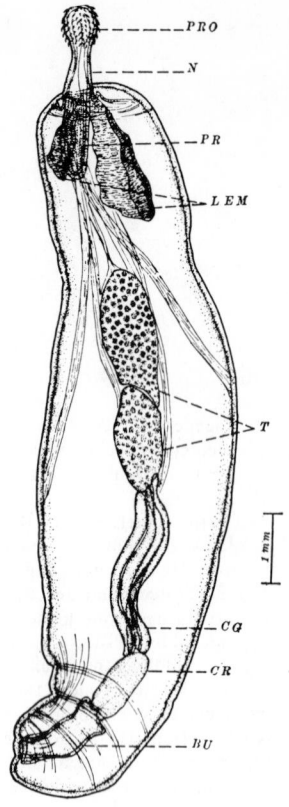

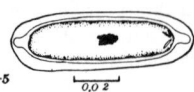

Fig. 1   *Above:* Adult male; *Below:* Embryo from a mature female; BU—bursa; CG—cement glands; CR—cement receptacle; LEM—lemnisci; N—neck) PR—proboscis receptacle; PRO—proboscis; T—testes.

are present. The body wall consists of an outer cuticula, a hypodermis, and an inner muscular layer. The syncytial hypodermis is penetrated by a branching lacunar system peculiar to the Acanthocephala. A pair of ribbon-shaped extensions of the hypodermis, the lemnisci, extend backward from the neck into the body cavity. The ligament, which consists of a strand of connective tissue enclosing the reproductive organs, extends from the proboscis sheath to the posterior extremity of the body. Digestive organs are completely lacking and nourishment is absorbed through the body wall. A pair of lateral longitudinal nerve branches, the retinacula, extend posteriorly from the cerebral ganglion, which is in the proboscis sheath. The sexes are always separate. Ovaries are present in the female only during the larval stage, and they dissociate very early into "ovarian balls" from which the eggs develop. The body cavity of the mature female is completely filled with these ovarian balls and eggs which have separated from them. The eggs develop into oval or spindle-shaped embryos

Meyer, A., Acanthocephala, *in Bronn's Klassen und Ordnungen des Tierreichs.* **4**: 582, 1932–33.
Hyman, Libbie H., "The Invertebrates," **3**: 1–53. New York, McGraw-Hill, 1951.

ACARINA

Mites and ticks which comprise the order Acarina are typical arachnoids. They may be separated from their close relatives the daddy longlegs (Order PHALANGIDA) because they lack a segmented abdomen, and from the spiders (Order ARANEIDA) because they lack a distinct division between cephalothorax and abdomen. Their mouth parts are set off from the rest of the body to produce a head-like structure, the gnathosoma. Characteristically the adult acarine possesses four pairs of legs but only three pairs are present in the larval stage. Antennae are never present.

The acarines are ubiquitous and diverse, and can be found in nearly every ecological niche—from the mountains to the ocean depths, and from the arctic circle to the equator. Free-living forms occur in all situations where vegetation can be found, some of the richest sources being moss, leaf litter, and tree holes. They may constitute 70 to 80 percent or more of the total arthropod population found in these habitats. Plant feeding forms have developed in one suborder (Trombidiformes). Although the number of species developing this habit is limited, the number of individuals developing on a single host plant may far exceed expectations. Some have developed an intimacy with other animals where they live as ectoparasites (Laelaptidae) or endoparasites (Cytoditidae).

It is estimated that there are more than 20,000 known species, and conservative estimates indicate several times this number are yet to be described. The more liberal estimates are that the number of mite species may approach that of the insects. Mites are small and will measure usually less than a millimeter in length, the smallest ones only a few microns long, and the ticks two centimeters or more when fully engorged.

Most species studied are oviparous, but in some ovoviviparity is found. In a few species the young are held within the body of the female until they complete their entire development, being released as adults (e.g. *Pyomotes ventricosus*). Parthenogenesis is known in several families.

Development proceeds in some (Trombiculidae) from the egg to a deutovum (after the shell has split to allow for expansion). The hexapod larva emerges and may actively feed. The larva becomes quiescent (now known as nymphochrysalis) and transforms within the larval skin into an octopod nymph. After feeding, the nymph becomes dormant (the imagochrysalis) and metamorphoses into the adult. In other species the larval stage is wanting or is passed within the egg with the result that the nymph is the first actively feeding stage. In some the nymphal stage is further divided into the protonymph, deutonymph, and tritonymph. Each of these might be feeding stages and accompanied by a molt. In still others (Acaridae) the second nymphal stage has become modified into a form, the hypopus, with well-developed claws and anal suckers which allow attachment to or penetration into a host. In this stage it can withstand dessication and the stage serves as one for dispersal.

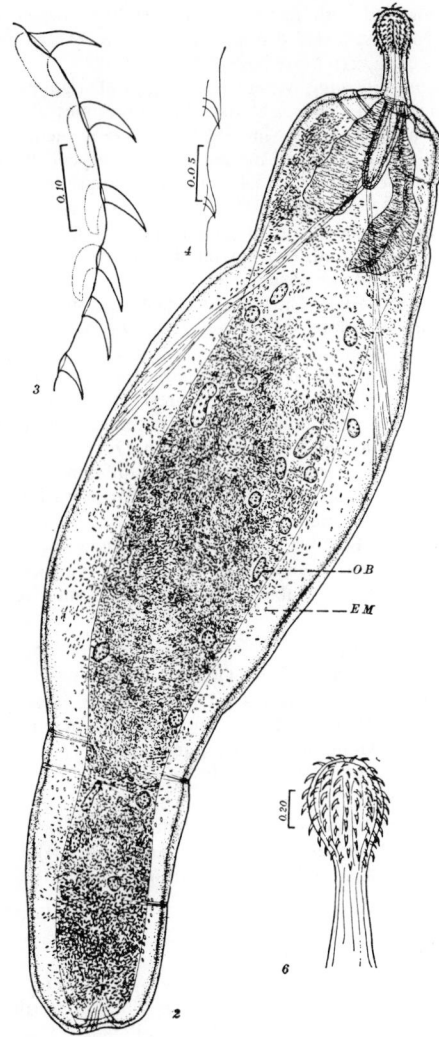

Fig. 2 *Center:* Adult female; *Upper left:* Proboscis hooks from a single longitudinal row; *Upper Center:* Body Spines; *Lower right:* Proboscis of specimen shown in Fig. 1 (above); OB—ovarian ball; EM—embryo.

covered with three or four sheaths before being discharged through the uterine bell. The ensheathed embryos are voided with the feces of the host and are capable of withstanding severe environmental conditions for many months. However, they do not develop further until they are ingested by a suitable intermediate host, which is usually an arthropod. If the infected arthropod is eaten by a suitable vertebrate host, the adult stage of the parasite develops. In some cases a second intermediate host is required for the completion of the life cycle.

HELEN L. WARD

References

Van Cleave, H. J., Expanding horizons in the recognition of a phylum. *J. Parasitol.* **34**: 1–20, 1948.

The integument may vary greatly in thickness and is clothed to varying degrees with setae of various types. In some the setae are so numerous and close-set as to give the mites a velvet appearance (Trombidiidae). In other forms the integument is equipped with one or more sclerotized plates and may possess also specialized sensory setae, eyes and pores.

The entire body is sac-like and there is no head as such. The portion of the body bearing the mouth parts is set off from the remainder of the body and is called the *gnathosoma*. The remainder of the body is termed the *idiosoma*. The gnathosoma bears the *chelicerae* which may be chelate or not, the *palps* (pedipalps) and the *hypostome*. The latter possesses a series of recurved teeth in the ticks and serves as a hold fast organ.

With exceptions the adult acarine has four pairs of segmented legs which possess from few to many setae of diverse types. The legs terminate typically in a pair of claws with which there is a pulvillus or other suctorial device. The number of claws may be reduced and the suctorial device may vary from a cup, to a comb-like structure, to a third claw.

Respiration may be accomplished through the thin integument, or through spiracular openings located at various positions on the body. In the ixodids spiracles are associated with coxa IV. Numerous unbranched tracheae arise from the atrium interior to the spiracle. In the mesostigmatids a peritrematal canal of varying lengths leads from the stigma. In the trombidiform mites the openings are associated with the gnathosoma.

Basically the acarine digestive tract is composed of a muscular pharynx, esophagus, stomach with its associated caeca, intestine, and rectum. Various modifications of the stomach and its caeca, and of the rectum exist. In parasitic forms the stomach is frequently small with a corresponding increase in size and complexity of the caeca. Expansion of the caeca in the ticks is enormous. Typically the trombidiform mites have lost the communication between the stomach and the hindgut. Here the rectum has developed an unpaired excretory organ and the anus has become the uropore. Other types of excretory organs are the primitive coxal glands of the ticks, and the excretory (Malpighian) tubules which empty into the hind gut.

There is great diversity in the systems of classification proposed and in current use. That proposed by Baker and Wharton (1952) remains in wide usage, with or without modification, especially among American workers. In this system the order is divided into five suborders as follows:

Onychopalpida: Mites with ambulacral claws on the pedipalps, with numerous setae on the hypostome, and more than one pair of stigmata. These are primitive mites. Some live under rocks and debris. One genus is apparently poisonous when swallowed.

Mesostigmata: Mites with one pair of stigmata located lateral to the legs and usually with an elongate peritreme, hypostome not developed for piercing, and lacking Haller's organ (a sensory pit). Dorsum of adult usually with one or more shields. Included are free-living (predaceous) and parasitic forms. Some are swift moving, and well sclerotized (Neoparasitidae) while others are sluggish and poorly sclerotized (Rhinonyssidae). Many beneficial and many harmful species are included here.

Ixodides: These possess one pair of lateral stigmata associated with a stigmal plate, hypostome fitted for piercing and with recurved teeth, Haller's organ on tarsus I present and distinct. All ticks are parasitic in all stages on vertebrate hosts.

Trombidiformes: Mites with one pair of stigmata associated with the gnathosoma but sometimes absent, palps usually highly modified, chelicerae generally modified for piercing, and anal suckers absent. Plant-feeding, free-living, and parasitic forms are to be found here. Many are parasites of invertebrates as well as vertebrates, and aquatic as well as terrestrial. Most are weakly sclerotized, fast moving and many are clothed with an abundance of setae.

Sarcoptiformes: These with tracheae opening through stigmata or on porose areas at various positions on body, or stigmata wanting, mouth parts with strong chelae and usually fitted for taking particulate matter; palps simple; anal suckers often present; and the coxae form conspicuous apodemes on the venter. Many are soil-inhabiting and heavily sclerotized (Oribatei), others are common as pests of stored foodstuffs and weakly sclerotized (Acaridia), while others are parasites of vertebrates and moderately sclerotized.

Van Der Hammen (1968) proposed a new system in which the order is elevated to the level of subclass and the five suborders above are rearranged into seven orders:

Subclass: ACARIDA
    Superorder: Anactinotrichida
        Order: Opilioacarida (= Onychopalpida, in part)
        Order: Holothyrida (= Onychopalpida, in part)
        Order: Gamasida (= Mesostigmata)
        Order: Ixodida (= Ixodides)
    Superorder: Actinotrichida
        Order: Actinedida (= Trombidiformes)
        Order: Oribatida (= Sarcoptiformes, in part)
        Order: Acaridida (= Sarcoptiformes, in part)

K. E. HYLAND

References

Grassé, P., "Traite de Zoologie," Tome VI. Onychophores, Tardigrades, Arthropodes Thilobitomorphes, Chelicerates. Paris. Masson et Cie, 1949.

Hughes, T. E., "Mites or the Acari," London. Athlone Press, 1959.

Baker, E. W. and G. W. Wharton, "An Introduction to Acarology," New York, Macmillan, 1952.

Baker, E. W., J. H. Camin, F. Cumliffe, T. A. Wolley, and C. E. Yunker., "Guide to the Families of Mites," Contribution No. 3, Institute of Acarology, College Park, University of Maryland, 1958.

Naegele, J. A., "Advances in Acarology," Vols. 1 and 2. Cornell, Comstock, 1963–1965.

Evans, G. O., J. G. Sheals and D. Macfarlane, "The Terrestrial Acari of the British Isles. An Introduction to their Morphology, Biology and Classification. Vol. 1. Introduction and Biology," pp. 219, figs. 1–216, London, British Museum (Nat. Hist.), 1961.

Van Der Hammen, L., "Introduction générale à la classification, la terminologie morphologique, l'ontogénèse et l'évolution des acariens," Acarologia, 10(3): 401–412, Paris, 1968.

ACOELOMATA

A division of the ANIMAL KINGDOM without specific taxonomic rank used to distinguish those invertebrate groups (e.g. PLATYHELMINTHES, etc.) which do not have a coelom from those which do.

ACTINOMYCETES

Nature. Actinomycetes are a group of branching unicellular organisms, which reproduce either by fission or by means of special spores or conidia. They usually form a mycelium which may be of a single kind, designated as substrate or vegetative, or of two kinds, substrate and aerial. They are closely related to the bacteria, and are frequently considered as higher, filamentous bacteria. Frequently they are looked upon as a separate group of organisms occupying a position between the fungi and the bacteria. Some even believe that antinomycetes are the prototypes from which both fungi and bacteria have been derived. Some forms of actinomycetes, such as certain members of the genus *Nocardia*, have their counterparts among the bacteria; others like species of *Streptomyces* and *Micromonospora*, have their counterparts among the fungi. The similarity in diameter between cells of bacteria and the mycelium and spores of the actinomycetes, as well as certain common chemical and biochemical properties, suggest that the actinomycetes should be classified with the bacteria. They are usually placed in a separate order, the *Actinomycetales*.

The actinomycetes comprise a number of genera and species, which vary greatly in their morphology, physiology, biochemical activities, role in natural processes, and practical utilization. They play an important part in the cycle of life in nature by bringing about the decomposition of complex plant and animal residues and the liberation of a continuous stream of available elements, notably carbon and nitrogen, essential for fresh plant growth. Some forms cause certain human and animal diseases; others cause certain plant diseases. The biochemical activities of some of the actinomycetes are now being utilized for the large-scale production of chemical substances such as antibiotics, vitamins, and certain potent enzyme systems. Many of these compounds have found extensive practical application in the control of infectious diseases of man, animals, and plants, in animal nutrition, and in the preservation of biological products, including virus preparations and human foodstuffs.

Characterization and occurrence. Ferdinand Cohn, in 1875, first observed an actinomycete in the concretions of a lachrymal duct; he named it *Streptothrix Foersteri*. Two years later, Harz studied an organism believed to be a causative agent of a cattle disease known as "lumpy jaw"; he designated it as *Actinomyces bovis*. Thus within two years, two different generic names were used to designate the same group of microorganisms. No pure culture was obtained in either case. Numerous other names soon followed, some of which, such as *Nocardia* and *Proactinomyces* were extensively used. Thaxter, in 1891, considered the causative agent of potato scab as an *Oospora*. The most prevalent of these names was the one proposed by Harz, *Actinomyces*; it came, in time, to be used for designating the group as a whole.

More recently, various attempts have been made to subdivide the actinomycetes into several genera. In

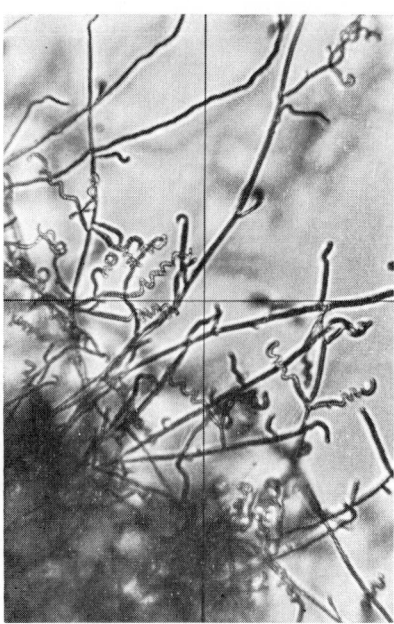

Fig. 1 Growth of an actinomycete belonging to the genus *Streptomyces*, in artificial culture.

1943, Waksman and Henrici recommended that four genera be recognized, *Actinomyces, Nocardia, Streptomyces,* and *Micromonospora*. The third genus was new and included the forms producing aerial mycelium. *Actinomyces* comprises the parasitic, microaerophilic forms which produce a substrate mycelium that breaks up readily into bacillary or coccoid elements. *Nocardia* includes the aerobic, partly acid-fast or non-acid-fast organisms which produce a fragmenting substrate mycelium; some are animal and human pathogens and others are saprophytes. *Streptomyces* produces a nonfragmenting substrate mycelium and an aerial mycelium forming sporophores that break up into chains of spores; the manner of sporulation of this group may serve as a basis for further subdivision. *Micromonospora* comprises organisms that form a nonfragmenting substrate mycelium and produces terminal spores on short sporophores. Several other genera were added later.

Actinomycetes are widely distributed in nature. They are found in vitually every natural substrate; in the air, in river and lake waters and bottoms, in foodstuffs, in various soils and manures, in oil deposits, and in the bodies of man, animals, and plants. Some genera favor one habitat and others favor another. Some substrates are ideal as permanent habitats, in which the actinomycates live and multiply; others represent only temporary habitats to which they are brought by water and air movements.

Isolation and identification. Most of the techniques used in the isolation and cultivation of bacteria and fungi also apply to actinomycetes. The isolation of these organisms from soils and other substrates is brought about by plating out such materials in proper dilutions on suitable agar or gelatin media. The plates are incubated at favorable temperatures for 2 to 7 days, and

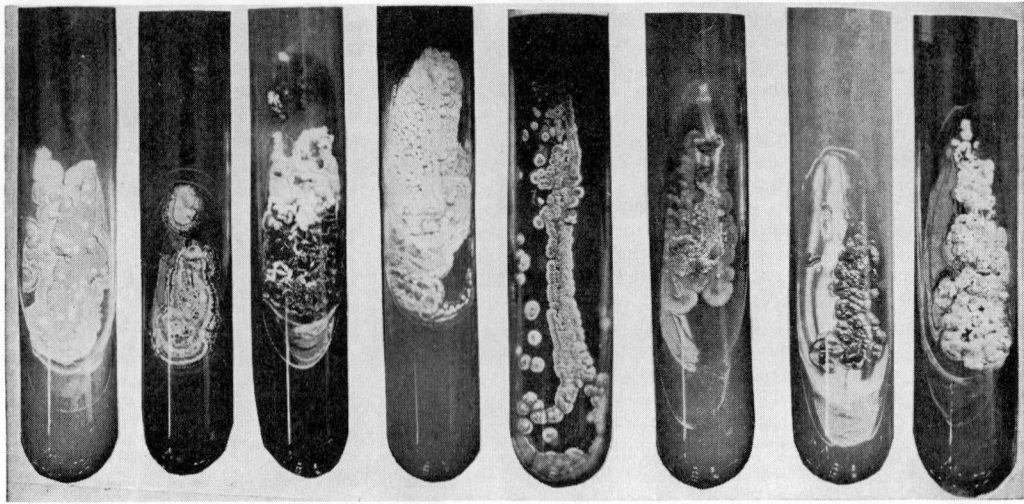

Fig. 2   A group of cultures of actinomycetes isolated from the soil (genus *Streptomyces*).

the colonies* picked and transferred to sterile liquid or solid media for further development. Actinomycete colonies can easily be distinguished on the plate from those of fungi and bacteria. They are compact, often leathery, giving a conical appearance and a dry surface, either colorless or colored red, yellow, blue, or green. They are often covered with cottony or powdery aerial mycelium, varying from white or gray through all the colors of the rainbow. If the colony is well developed and the aerial mycelium abundant, the surface spores can easily be picked with a sterile needle. If growth is limited, or the aerial mycelium not fully developed, sharp, razor-like needles are usually required to transfer a part of the growth to fresh media.

When grown in liquid culture, either in a stationary or in a submerged condition, the majority of actinomycetes form flakes or spherical compact masses, leaving the medium clear. The mass of growth can easily be removed by filtration through ordinary paper. When growth undergoes lysis, the cells disintegrate completely and a certain degree of turbidity occurs.

The great majority of actinomycetes are aerobic; very few are anaerobic; many are microaerophilic. To supply proper aeration, the organisms are grown on the surface of solid media, or in shallow liquid layers, or in a thoroughly aerated submerged condition. Temperatures of 25°–30°C are used for incubation of most actinomycetes. Pathogenic organisms require 37°C, and thermophiles thrive best at 50°–60°C.

The stable morphological properties of actinomycetes essential for characterization and classification include structure and subsequent changes in the substrate mycelium, production and nature of the aerial mycelium, nature of the sporulating branches of sporophores, and size, shape, and surface of the spores.

———

*A colony of an actinomycete is different from a bacterial colony. It represents a filamentous extension of the original cell or cells, spores and degradation products; it is not an accumulation of cells originating from one or more similar cells.

Among the physiological and cultural properties essential for characterization of actinomycetes, color of the substrate growth and of the aerial mycelium is most important; the formation of soluble pigments, both in synthetic and in organic media (melanins), is also significant. Among the other properties used for species identification are hydrolysis of proteins including gelatin and milk casein; hydrolysis of starch; inversion of sucrose; digestion of cellulose; reduction of nitrate; formation of hydrogen sulfide; utilization of sugars and related compounds, with and without the formation of acids; antagonistic activities and ability to produce antibiotics; and sensitivity to specific phages and to known antibiotics.

Nutrition. Actinomycetes are able to utilize a great variety of organic compounds as sources of energy. These include sugars and organic acids, starches, and certain hemicelluloses, proteins, polypeptides and amino acids, and nitrogenous bases. Some actinomycetes can also attack cellulose, fats, hydrocarbons, benzene ring compounds, and, to a more limited degree, lignin, tannin, and rubber. There is considerable selectivity in the utilization of these substances by different kinds of actinomycetes.

Actinomycetes are unable to fix nitrogen and have to depend, like the great majority of fungi and bacteria, upon fixed compounds of nitrogen for their cell synthesis. Proteins, peptones, and certain amino acids form the best sources of nitrogen for actinomycetes, followed by nitrates, ammonium salts, and urea. Nitrates are reduced to nitrites, and both are reduced to ammonia and assimilated for cell synthesis. The nitrogen content of the mycelium of actinomycetes varies between 6 and 15 per cent, depending upon the ratio of carbon to nitrogen in the substrate.

Among the minerals required for the nutrition of actinomycetes, are potassium, sodium, phosphorus, sulfur, and small amounts of calcium, magnesium, and iron. Traces of chlorine, cobalt, zinc, and certain other trace elements are often necessary, especially the synthesis of certain antibiotics (chlorotetracycline) or vitamins (B_{12}).

Under aerobic conditions, glucose is converted by actinomycetes mainly to cell material and CO_2. Under restricted aeration, lactic acid is also formed. Pyruvic acid is produced during the stages of most rapid growth. The metabolism of glucose by actinomycetes depends on the presence of phosphate, the optimal hydrogen-ion concentration for both glucose oxidation and disappearance of inorganic phosphate being about pH 7.

Actinomycetes can attack a large number of plant and animal proteins, which are hydrolyzed to amino acids, polypeptides, and ammonia. They also possess the unique capacity to decompose keratins. This was shown first in connection with the ability of certain pathogenic forms to attack the skin and horny portions of feet. When a soil is enriched with keratinized tissues, such as human hair and feathers, various forms will develop. The mechanism of keratin destruction by these organisms has not been determined. The biogenesis of antibiotics and vitamins by actinomycetes, problems of steroid oxidation, and enzyme formation, have raised many important questions bearing upon the metabolism of actinomycetes; most of these questions still remain unanswered.

Production of antibiotics. In recent years, interest in the actinomycetes has centered largely upon their ability to produce antibiotics, or chemical substances which possess antimicrobial activities. Preliminary observations on the ability of certain actinomycetes to inhibit the growth of bacteria and fungi were carried out by Müller in 1908, Gratia and Welsch (1924–1940), and a group of Russian investigators. The first crystalline antibiotic, actinomycin, was isolated in 1940. Streptothricin was isolated in 1942, streptomycin in 1943 in the laboratories of the Department of Microbiology of the New Jersey Agricultural Experiment Station, and numerous others followed. To date, more than 500 different antibiotics have been isolated from cultures of actinomycetes. Many of them have been obtained in the form of pure compounds, the chemical nature of which has been determined. A few have been synthesized. More than 50 of these antibiotics have found extensive practical application as chemotherapeutic agents. Of the total 2,400,000 pounds of antibiotics produced in the United States in 1955, valued at more than half a billion dollars, at least two thirds have been obtained from cultures of actinomycetes.

Numerous surveys have been carried out on the production of different antibiotics by actinomycetes, using certain screening procedures. Twenty to 50 per cent of all the cultures tested were found to possess antagonistic properties. The nature of the medium used for testing is of great importance. Most of the antagonistic organisms are active against gram-positive, including acid-fast, bacteria; fewer are active against gram-negative bacteria and fungi. The majority of the antibiotic-producing actinomycetes belong to the genus *Streptomyces*.

Certain general principles concerning the production of antibiotics by actinomycetes have been established: (a) different strains of an antibiotic-producing species may form different antibiotics that are not related chemically; (b) a single strain of an antibiotic-producing organism may form several chemically related antibiotic substances; (c) organisms producing the same antibiotic or closely related compounds are found in different soil regions throughout the world; (d) a change in the nutrition of the organism may result in a change in the nature of the antibiotic produced; (e) as a rule, antibiotic-forming organisms are resistant to the antibiotics they produce; this phenomenon is used to advantage in the isolation of fresh cultures capable of forming a given antibiotic, and in the selection, from a given culture, of more potent strains; not all antibiotic-forming organisms, however, behave in the same manner.

Other important biochemical reactions. Numerous actinomycetes, both producers and nonproducers of antibiotics, are able to form vitamin B_{12} if cobalt salts are added to the medium to serve as precursors. Steroid oxidation by actinomycetes has recently attracted considerable attention. Cultures of various species of *Nocardia* and *Streptomyces* are now being utilized to bring about desirable transformations of steroid hormones.

Among the other biochemical reactions carried out by actinomycetes that are of considerable theoretical and possible practical importance are production of enzymes, notably protease, keratinase, amylase, tyrosinase, and invertase; formation of yellow, red, blue, and black (melanin) pigments; formation of odors that may impart "earthy" smells to potable water, fish, milk, cacao, and other food products; decomposition of plant residues in soils and in composts, and the formation of humus; and oxidation of hydrocarbons.

Causative agents of disease. Two major diseases are now known to be caused by actinomycetes: actinomycosis, caused by anaerobic organisms, and nocardiosis, caused by aerobic organisms. Antibiotics have found extensive application in the treatment of these diseases.

Very few forms among the actinomycetes are capable of causing plant diseases. Except for two species—the Irish potato and the sugar beet—plants subject to infection by actinomycetes do not occupy a very prominent place in human economy.

SELMAN A. WAKSMAN

References

Waksman, S. A. "The Actinomycetes," New York, Ronald Press, 1967.
Waksman, S. A., "The Actinomycetes," Vols. 1–3. Baltimore, Williams and Wilkins, 1959–1962.

ACTINOPODA

The Actinopoda are a super class of the PROTOZOA containing the Heliozoa, Acantharia and Radiolaria. These three groups are combined in the Actinopoda because they have thin pseudopods and sometimes, axopods radiating round their body. Apart from this common character, they differ markedly among themselves, each possessing specific characteristics.

Class Heliozoa. The body, which may be spherical (Fig. 1), or a pedunculate oval in sessile forms (Fig. 2), is composed of ectoplasm and endoplasm but may be surrounded by a gelatinous or mucilaginous envelope of variable thickness. The body may be naked or protected by a skeleton. This last may be *heterogenous*, i.e., made from foreign inorganic fragments or *autogenous*, i.e. composed of small silicious, or rarely chitinous, spicules of various shapes embedded tangentially, but without specific pattern, in the mucilaginous layer.

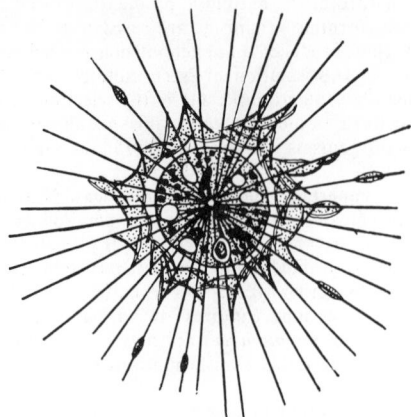

Fig. 1   Raphidisphrys pallida Schulze (Heliozoa).

Fig. 2   Actinolophus pedunculatus Schulze (Heliozoa).

Pseudopods and axopods radiate from the body and some forms may have prehensile lobopods of the amoeboid type. The ectoplasm, which often contains contractile vacuoles and symbiotic Zoochlorella, cannot be clearly distinguished from the endoplasm since the central capsule is absent. Some are uninucleate but others are multinucleate with numerous (20–500) small nuclei dispersed in the endoplasm. The single nucleus may be central or eccentric, and in this latter case the center of the body is occupied by the *centroplast*, on which the axopods are implanted. The role of the centroplast is not merely kinetic because it is involved, as a centrosome, in nuclear divisions. In those forms which lack centroplasts, the axopoda are fixed either to the membrane of the nucleus or terminate free in the cytoplasm. The peduncles of sessile forms are merely modified axopods adapted for attachment to the substrate. These forms are also motile and pass from place to place either by swimming or by rotating along the substrate with the aid of their flexible axopods.

Encystment is common and may be for protection, digestion, or reproduction. This last may be sexual, through binary division and budding, or sexual by pedogamy. This leads to the formation of physiologically anisogametic gametes, two being produced by the uninucleate forms and many by the multinucleate forms. The zygotes encyst after fusion. The classification of the helioza is artificial since numerous forms are still little known. They are divided into two orders but it is not yet possible to group the genera in well defined families.

1. *Actinophrydia* contains only a few primitive uninucleate and multinucleate forms, lacking skeletons and centroplasts, and having affinities with the fresh water amoebae.

2. *Centrohelidia* brings together all those forms having centroplasts but in which the body may be naked or protected by either a heterogenous or an autogenous skeleton. They may be free or sessile and live for the most part in fresh water or on humid soil. Some are marine, or, very rarely, pelagic.

Class Acantharia. These forms, mostly spherical (Fig. 3), are distinguished by their non-silicious skeleton made up in primitive forms of 10 supple diametric spicules or, in more advanced forms, by 20 radial spicules joined together in the center and reaching the exterior through points predetermined by Müller's law. In these last forms, the skeleton may be in the form of an incomplete armored shell, pierced (Fig. 4) or solid, formed by the anastomosis of lateral ramifications, which may become transformed into contiguous plates, of the spicules (*apophyses*). The body of these forms is clearly divided by the plasmatic membrane of the central capsule into ectoplasm and endoplasm. The periphery of the ectoplasm is differentiated into one or two gelatinous pellicules and into elastic fibers which surround, or radiate from, those places where they are attached to spicules by very contractile plasmatic fibers (*myonemes*), of which there are from two to 60 around each spicule (fig. 3). This

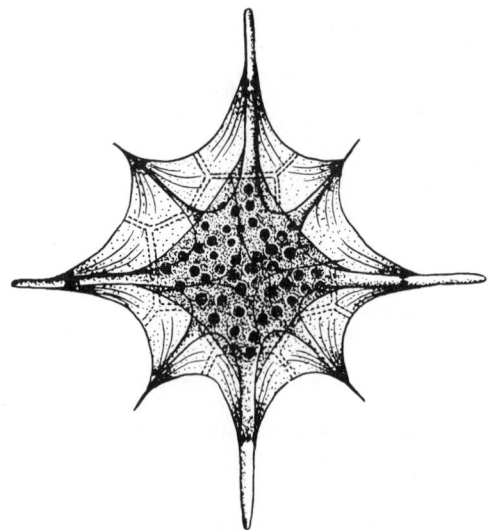

Fig. 3   Acanthostaurus purpurascens Hoeckel (Acantharia).

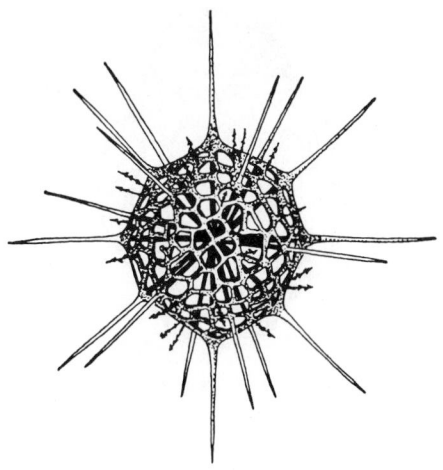

Fig. 4 Lychnaspis giltschi Hoeckel (Acantharia).

complex constitutes the hydrostatic apparatus, peculiar to Acantharia, thanks to which they can move vertically in the sea. The very young stages are uninucleate but the adults are always multinucleate, their numerous, small nuclei being arranged in two or three peripheral layers. The existence of a centroplast, connected to the axopods has been established. The endoplasm is often variably colored from pigments obtained through the assimilation of colored prey. Acantharia harbor symbiotic Zooxanthellae in the cytoplasm which facilitate, by their activity, the assimilation of the prey. They are also hosts to aberrant Dinoflagellate parasites. The more primitive forms in this group can reproduce by binary division but isosporogenesis, preceded by encystment, is more common. The isospores, uni- or biflagellate, do not appear to be gametes since their sexuality has not been established with certainty. The development of some forms from the uninucleate stage has been established. This group can be divided into four orders according to the shape of the central junction of the spicules, as well as by the numbers of the gelatinous pellicles and myonemes.

A few little known forms, including a sessile type from the Sea of Japan, are provided with a very large number of spicules which are not arranged to Müller's law. These have been assembled in the group Actinelia which appears to be closely related to some Heliozoa. All Acantharia are pelagic and marine.

Class Radiolaria. The classic definition of this group, which emphasizes the silicious nature of the skeleton, the central capsule of the protein membrane, and above all the axopods, is no longer accurate since there have recently been discovered a few forms which have both axopods and centroplasts. The body of Radiolaria consists of the ectoplasm, surrounding an undifferentiated gelatinous layer containing symbiotic peridinian Zooxanthellae, and the endoplasm clearly separated by the membrane of the central capsule, which is itself pierced either by a number of fine pores over all the surface or by three clearly localized holes. The members of the group may be naked, or protected by either a heterogenous or autogenous skeleton. This last is either discontinuous, being made up of isolated tangential or

radial spicules, which are never joined at the center, or continuous in the form of one or more complete or incomplete pierced shells of various shapes, resulting from the deposition of silica in the meshes of a superficial ectoplasmic net. The skeletal fabric is solid except in a single order in which its is hollow. Reticulated ectoplasmic pseudopods radiate from the body as do also, in some cases, axopods which may be, as in the Helioza, fixed to the nuclear membrane, or to the intranuclear or eccentric centroplasts, or may terminate without attachment in the endoplasm. The vegetative stages are uninucleate, the multinucleate forms occurring only as a preliminary to reproduction. This last takes place by binary division, or by isospores, generally biflagellate, of which the fate is unknown. The alleged sexual reproduction by anisospores derives, in fact, from the developmental cycles of various intraplasmic, or intranuclear, Dinoflagellate parasites. Little is known of the evolution of the Radiolaria and their present classification, based on the shape of the skeleton, is arbitrary. They are divided into three orders.

1. *Spumellaria.* The sub-order Collodaria contains the monocytic forms, either naked (Fig. 5), or having an autogenous skeleton formed of tangential spicules. There are no axopods, the membrane of the central capsule is simple, though with minute perforations, and the ectoplasm harbors Zooxanthellae. They reproduce by isosporogenesis, either in the monocytic stage, or through retardment, in their colonial stage (Fig. 6). The suborder Sphaerellaria contains those monocytic forms which are protected by a skeleton consisting of one or more concentric pierced shells, usually spherical (Fig. 7) or flattened, discoidal, or even irregular, ornamented with radial spines or spicules, attached to the shell. The membrane of the central capsule is simple, with minute pores, and Zooxanthellae are present. They have axopods and centroplasts and reproduce both by binary division and by isosporogenesis. The Order Nassellaria contains the oval monocytic forms, of which the skeleton is usually in the form tiaria or of a hat (Fig. 8). The central capsule has a single large orifice through which protrude the axopoda which are grouped in bundles and

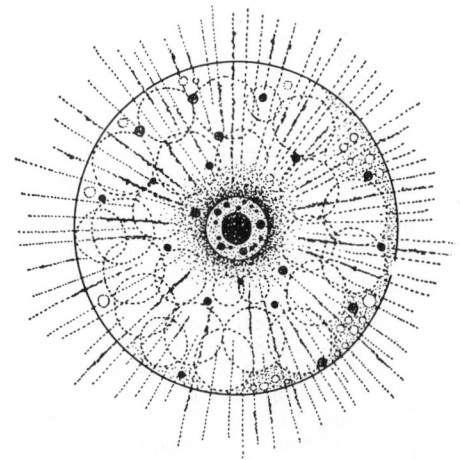

Fig. 5 Thalassicolla pellucida Hoeckel (Radiolaria Spumellaria Collodaria).

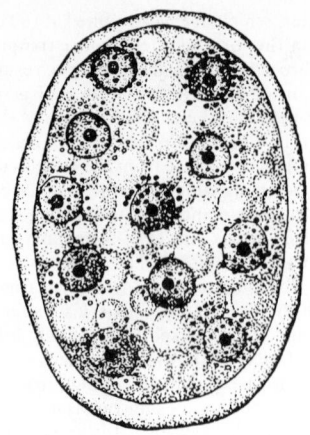

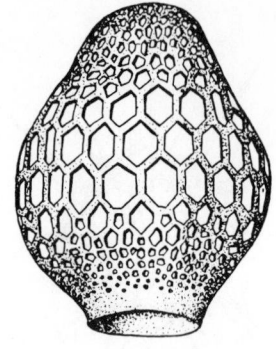

Fig. 8 Cyrtocalpis urceolus Haeckel (Radiolaria Nassellaria).

Fig. 6 Collozoum inerme Haeckel (Radiolaria Spumellaria Collodaria).

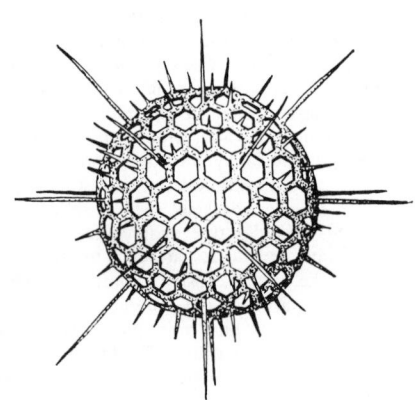

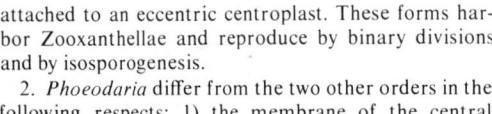

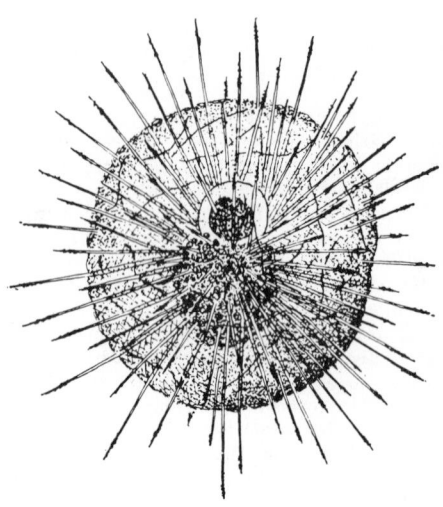

Fig. 7 Heliosphaera echinoides Haeckel (Radiolaria Spumellaria Sphaerellaria).

Fig. 9 Aulacantha scolymantha Haeckel (Radiolaria Phoeodaria).

attached to an eccentric centroplast. These forms harbor Zooxanthellae and reproduce by binary divisions and by isosporogenesis.

2. *Phoeodaria* differ from the two other orders in the following respects: 1) the membrane of the central capsule, made of two contiguous sheets, is pierced by three holes of peculiar shape—a basal (*astropyle*) and two lateral (*parapyles*); 2) the invariable presence of phoeodium, the residual detritus of digestion; 3) the skeleton is formed of hollow elements, filled with an amorphous silica combined with an organic silicate; 4) the absence of Zooxanthellae and axopods. Some are naked, and others have either a discontinuous heterogenous or autogenous skeleton which is sometimes continuous, which gives rise to variable shells. They multiply only by binary division, which follows the division of their nuclei, and are thus truly "polycaryous." The Radiolaria are without exception pelagic and marine.

G. TREGOUBOFF
(trans. from French)

References

Heliozoa: Wailes, G. H., "British Freshwater Rhizopoda and Heliozoa," London, Ray Society, 1921.
Acantharia: Schewiakoff, W., "Acantharia," Naples, Fauna e Flora del Golfo di Napoli, Monog. 37, 1926.
Radiolaria: Haeckel, E., "Report of the Radiolaria collected by H.M.S. Challenger during the years 1873–1876," *in* Tomson, C. W. "Challenger Reports—Zoology," Vol. 18. London, Ballantyne, Hanson, 1887.

ADANSON, MICHEL (1727–1806)

Michel Adanson, a French naturalist of Scottish descent, (the family name being originally Adamson), was born at Aix en Provence in 1727 and died in Paris in 1806. His two important works are *Histoire naturelle du Sénégal* (1757) and *Familles des Plantes* (1763), the latter of great importance in botanical nomenclature. Adanson, an opponent of Linnaeus and a vigorous advocate of a natural system of classification, was a man of very independent spirit and original views, long unjustly neglected owing to his adoption of uncouth native vernacular names as scientific generic names and his phonetic spelling of French. It is thus fitting that the remarkable if grotesque-looking Baobab Tree of tropical Africa should bear the generic name *Adansonia.* In 1748, at the age of twenty-one, he went as a naturalist to Senegal, West tropical Africa, where he spent four and a quarter years. Here he found the existing classifications of Tournefort and Linnaeus inadequate when applied to the tropical flora. These had been based on the shape of the corolla and the number of stamens and styles (or stigmas). Adanson accordingly began to make classifications based on other features and attempted ultimately to produce a natural classification by combining them. His views have received appreciative attention during recent years from students of epistemology and scientific methodology, notably from P. H. A. Sneath (*in J. Gen. Microbiol.* **17**: 195, 1957): the first taxonomist who conceived of the use of every feature impartially and with equal weight was Michel Adanson (1763). His views were far ahead of his time. He attacked the arbitrary methods of the day and insisted that every part of a plant should be used in making a (Préf. p. clv). He constructed in form all the classifications which he could, each one based on a single part of the plants, and on comparing them he found that some of the classifications divided up the plants in a manner which was almost identical with that of other classifications. These he chose as indicating the most natural groupings, which were therefore groupings based on correlating features of virtually equal weight. He also realized the nature of the dividing lines which were thus made (Préf. p. clxiv). Adanson's views were not highly esteemed, since his contemporaries considered that some features were more important than others, although they could never agree which features these should be. The ideal classification is the one with the greatest predictive value. Such a concept is based on giving every feature equal weight. Such classifications may conveniently be called 'Adansonian' (Sneath, *loc. cit*).

WILLIAM T. STEARN

Reference

Chevalier, A. J. B., "Michel Adanson, Voyageur, Naturaliste et Philosophe," Paris, 1934.

ADAPTATION

Adaptations are those details which result in suitable and convenient morphological and functional correlation between parts of an organ, between the organs of a living organism, between individuals of the same species or of different species, and, finally, between an organism and its organic environment. These adaptations consist of conformations, of structures, and of functions, particularly well adjusted to the role played by the organ in question on which they confer a high degree of effi-

ciency, or which are at least very advantageous either to the maintenance of the individual or to the perpetuation of the species.

The concept of adaptation differs in magnitude according to whether it is applied to an organism in its totality or in its details. It follows that any discussion can deal either with general phenomena of correlation and complementarity which exist, of necessity, at all levels of the organization between the elements of the organism, as well as those which are concerned with the congruence of the organism with its environment; but careful consideration must be given to those details which are particularly significant in the meshing of structures with each other and in the relation of the living organism with the exterior world. Finally, there is great difficulty in defining adaptation exactly by reason of the subjective element which must be used in judging the utilitarian value of the characters under consideration. In any case, speaking of the microcosm of adjustment, adaptations must be controlled in part by the necessity of fitting in to the general organization of the living organism itself, and in part by the varying conditions of existence. It is always necessary to bear in mind that organisms which differ in their structure may present strikingly analogous adaptations, whilst organisms belonging to the same systematic group may differ profoundly in those adaptations which correlate with the diversity of connections which they maintain with their environment.

Anatomical and physiological studies disclose internal adaptations which may concern the organism in its entirety or each of its organs without any relation to the external environment. In many cases, it is the structure and mechanical interaction of parts which make it possible for the organism or organ to function; in this case, the adaptations are primarily morphological. In many cases, mutually adapted parts develop in close proximity and thus result in the production of complex functional unities. Examples of this are the joints in the skeleton of vertebrates and arthropods; the connections between muscles and portions of the skeleton; the apparatus for the transmission of acoustical vibrations in aerial vertebrates; the eye and its adnexed structures, etc. The parts may also evolve separately but appear as though they had been molded together. Even though they remain separated, they may fit perfectly together and thus result in a true functional unity; examples of this type of adaptation are the "press studs" on the mantle of cephalopods; the buttons which immobilize the abdomen of male crabs; the complex which hooks together the wings of insects; the ridges and their corresponding grooves in the elytra of Coleoptera; the mouth parts of sucking arthropods; the ovipositors of Phasgonuridae; the stings of Hymenoptera; the jumping apparatus in Elateridae and collembola; the clutching legs of several arthropods; the operculum and peristome of prosobranchs; the cutting and crushing action of the upper teeth against the lower teeth in mammals; and the armature of the mandibules of chewing arthropods; etc.

Other internal adaptations are connected with the general plan of organization and function of a living organism. Examples of this are the correlation of the circulatory system with the mechanism for the absorption of substances and for the elimination of waste; the connections of the nervous system with receptor and effector organs. Other adaptations are predominantly functional; for example, hormonal integration; the mechanisms which regulate the successive functions in the length of the digestive tube; similar phenomena connect the function of the gonads with those of the genital ducts, etc.

Ecological and ethological studies provide data on adaptations in relation to a changing environment. In point of fact, there is no clear cut line of demarcation between interior adaptations and adaptations to the external environment, the latter being more or less closely linked to the former. As examples of this, there may be quoted the respiratory apparatus in relation to aquatic or aerial environment; the anaerobic respiration of intestinal parasites; the adaptation of the digestive system to the diet; protection against loss of internal heat and against drying; hibernation and estivation; those mechanisms which permit copulation between sexes. There are also several examples of an adaptive relationship between embryos, and newborn forms, with their parents, and there are numerous cases of adaptive relations between parasites and host. Cases in which two organisms are specifically adapted to each other provide remarkable examples of true adaptation ("lock and key" of copulatory apparatus, etc.).

On the other hand, some adaptations are solely or predominantly connected with the external environment. For example, the desert dwelling cacti, Opuntiacea, Euphorbiacea, and Asclepiades limit transpiration through the suppression of foliage and the development of a thick epidermis; the localization of stomata on the upper side of leaves spread on the surface of water; the pharyngeal pump of trematodes, cestodes, hirudinea, and cephalopods, etc.; color, and other types of mimicry (of which the actual value has been more often suggested than demonstrated); etc.

Apart from such obviously advantageous adaptations, there are those which are less useful and even some of which the organism could be deprived without damage. Examples of this are the hooking of hemielytrae with atrophied wings in some hemiptera; the respiratory tube of Nepa, etc. There must also be included among adaptations certain negative characteristics which favor an organism in a particular environment. Examples of this are the blind and colorless animals of subterranean caves, which can escape the sight of their enemies in the dark and can thus survive in competition with pigmented and eyed species.

True specialization occurs when the adaptation is pushed to the point that the organism by its structure, by its shape, or by its physiological exigencies, is strictly confined to a particular environment to which it is linked by an unbreakable chain, or else when some one of its organs can only be utilized for a restricted function. Specialization involves the greatest departure from the norm between the organism or organ under consideration, and the morphological and functional conditions which are usually found in the greater part of the representatives of branch, class or order to which it belongs. In addition, the idea of adaptation and of specialization carries in its train a judgment on the genesis of these conditions from more generalized former types living in a less specialized environment and possessing less differentiated structures. Indeed, all specialization, however advantageous it may be in specific circumstances carries always with it a limitation as to the use of the organ or as to the relations to the organism

with the external environment. Examples of this are the buccal apparatus of sucking animals; the jaws and tongues of ant-eaters and pangolins; the filter of the whalebone whale; the swimming paddles of whales, sirenians, ichthyosaurians, plesiosaurians, and of mosasaurians.

From the point of view of the regularity with which adaptations occur, there may be distinguished (A) *individual adaptations which are not heritable* and which consist of (i) modifications which follow and correct alterations which occur in the structure and function of organs (for example, the regulation of teratological conditions, and such things as atypical regenerations, cases like the hypertrophy of bone marrow in splenectomies and other functional compensations) or of the environment (for example, the increase in the surface area of the branchial tufts of tadpoles of frogs in hypooxygenated water); (ii) the phenomena of acclimatization and naturalization which are, in fact, the modifications which all representatives of a species undergo when they find themselves in a different environment from that in which they originated but in which they are forced to live; (B) *specific adaptations which are heritable* and which are found in all the representatives of a species, or of a genus, but which may be found repeated in varied systematic groups (i) morpho-physiological internal adaptations which concern the functioning of organs (for example, the intestinal siphon of sea urchins and of some polychaetes; the gizzard of birds, gasteropods and crustacea; the typhlosoles of oligochaetes; and the spiral valve of cyclostomes and elasmobranchs); (ii) ecological (or statistical adaptations) which repeat with a marked frequency among inhabitants of the same or analogous environments (the fins of fishes, ichthyosaurians, whales, sirenians, seals, etc.; the digging legs of the mole, of *Notoryctes typhlops*, of the mole cricket, etc.; the air sacs of birds and of insects; the pneumatophores of siphonophores and the swim bladders of fishes; the webbed feet of aquatic birds); (iii) adaptations which apply to specific relationships strictly limited to the interrelation of habitats and behavior (for example, sedentary life in the shelter of secreted or constructed tubes; hooks on the limbs of externally parasitic arthropods; adaptations required for copulation, for parental care, or for the role played by members of polymorphic societies, either symbiotic or parasitic).

The origin of adaptations may be explained differently according to views held with regard to the development of shape; (A) *teleological explanations:* (i) adaptations are due to a creation foreign to the world and represent the manifestation of a harmony preestablished by a superior intelligence (theism); (ii) they are due to intrinsic final causes which lead the organism to a predetermined form (Driesch); (B) *nonteleological explanations:* (i) factors in the external environment directly, or indirectly, induce, in some individuals, some kind of alterations in some of their germ cells; among the forms which develop from these natural selections there survive those in which the somatic alterations are favorable; these phenomena are repeated for several generations, the advantageous variations which are oriented by chance in the same direction accumulate and the adaptation is thus perfected (Darwinism, neodarwinism, mutationism); (ii) adaptability is a fundamental property of living things; the exterior environment produces heritable adaptive modifications in all individuals of the same species which are subjected to it

(Lamarckism, "nomogenesis" of Berg); (iii) the formation of shape in phylogeny, as well as in ontogeny, is due to an intrinsic mechanism, of which the process of development follows the law of differentiation with a division of physiological labor between the parts; these parts result necessarily coordinated and thus adapted one to the other; the harmony which results represents the chief condition which favors the incorporation of the organism with its exterior environment where it might encounter a suitable condition and utilize its advantageous characters in relation to this ("ologenesis" of Rosa).

Whatever may be the case, it must be emphasized that (A) the functioning of regulated, permanent and complicated mechanisms, as the living beings are, is possible only provided that the internal as well as the external conditions do but little change; life being indeed possible only when there is adequate reciprocity between organisms and their environment (Henderson); (B) general conditions of the biosphere have remained unaltered since the earliest known epochs (Vernadsky); (C) the original ecological environment represented by ocean waters, is the most constant and the most uniform of ecological environments (Henderson) and primordial organisms were of necessity adapted to this environment; (D) the interior environment of organisms has always retained certain constant fundamental characteristics both in animals (Macallum, Quinton) and in plants (Colosi), even when they have been separated from their original ecological environment to which they were adapted at the beginning, so that viability can only be assured when there is some method of adaptation of organisms and ecological environments however different they may be from the original; (E) the congruence between organisms and the exterior world has been maintained without interruption from the origination of life, notwithstanding the changes in the environment and the morpho-physiological transformations which are observed during phylogeny and demonstrated during ontogeny; (F) the new environments invaded, and the new characters appearing during phylogeny and ontogeny, must of necessity maintain the congruence between the organism and the exterior environment by means of adaptations; (G) the invasion of an environment, which differs from that previously inhabited, is dependent on the physio-morphological state being strictly linked to psychological behavior of the organisms at the moment of their migration which is not possible without preadaptation (Cuénot); (H) the interaction of organisms with the exterior world has remained fundamentally unchanged during the history of life; the vehicle with the aid of which there appear material exchanges between the two has been always represented by water (Colosi); it follows that the migration from aquatic to aerial environments requires very important adaptations: (I) new adaptations must be considered as compromises between the exigencies imposed by structural differentiations and by localizations and functional specializations on the one hand, and fundamental and general exigencies of the living body on the other hand; (J) even in a constant environment, the appearance of new forms, and new functional unities, requires a corresponding change in the relation with the environment, and the inauguration of new adaptations to this; (K) the more apparent the departure from the norm of a secondary environment is in relation to the environment inhabited by the greater part of the rep-

resentatives of a systematic group, the more rigorous must be the adaptations which permit the organism to live there; the more the relations of the organism with its environment are novel, the more the organism must show itself to be specifically adapted to them.

G. COLOSI

References

Allee, W. C., *et al.*, "Principles of animal ecology," Philadelphia, Saunders, 1949.

Caullery, M., "Le problème de l'evolution," Paris, Payot, 1931.

Colosi, G., "Gli organismi e il mondo esterno," Firenze, La Nuova Italia, 1945.

ibid, "La dottrina dell évoluzisme e le teorie évolutionistiche," Firenze, Le Monnier, 1945.

ibid, "Le correlazione nel mondo vivente," Boll. Labor. Zool. gen. e agraria Portici 33:1954.

Cuénot L., "L'adaptation," Paris, Doin, 1925.

ibid, "Invention et finalité en biologie," Paris, Flammarion, 1941.

Reiner, J., "The Organism as an Adaptive System," Englewood Cliffs, N.J., Prentice-Hall, 1968.

Srb, A. M., and Bruce Wallace, "Adaptation," 2nd ed., Englewood Cliffs, N.J., Prentice-Hall, 1964.

ADRENALS

The adrenal glands of the Eutheria are compound paired organs derived by the union of two embryologically and functionally distinct tissues which, because of their topographical relationship are referred to as cortex and medulla, respectively. The medullary or chromaphil tissue is part of the nervous system and is a homologue of a sympathetic ganglion. It forms epinepherine and norepinepherine, the mediators of sympathetic nervous activity, and participates with the remainder of the sympathetic nervous system in the maintenance of homeostasis, but may be extirpated without any obvious detriment to the organism. The adrenal cortex, on the other hand, is essential for life and its removal results in death within a matter of days.

The location and topographical arrangement of the components of the adrenal differ but the gland in all species occupies a position in relation to the anterior pole of the kidney. In some species, e.g., the AMPHIBIA, the adrenal is embedded in the kidney. In the ELASMOBRANCH fishes the chromaphil tissue is separate from the "cortical" tissue which constitutes the interrenal bodies.

The histological appearance of the adrenal also differs widely in various species. In the Eutheria the cortex is composed of an outer connective tissue capsule beneath which is a thin layer of cells arranged in whorls and designated as the zona glomerulosa. Immediately under this is a band of narrow cuboidal cells with dark staining nuclei designated as the zona intermedia. The greater part of the cortex is composed of cords of cells designated as the zona fasciculata. Between the latter and the medullary chromaphil tissue is the zona reticularis. During early life in certain species (e.g., the mouse) a transient zone (the X-zone) is also present. The significance of this zonation has not been established but some evidence indicates that the various zones are concerned with the elaboration of different hormones.

The adrenal cortex is concerned in many physiologic and metabolic functions as demonstrated by the effect of ablation of the glands or the administration of excessive amounts of its hormones. The most critical of the defects induced by adrenalectomy is the disturbance of electrolyte and water metabolism. If this be corrected the adrenalectomized dog may be kept alive for a month or more without recourse to hormonal therapy.

Adrenocortical insufficiency induced in the experimental animal or occurring spontaneously in man as the result of destruction of the gland by disease, results in a characteristic train of symptoms in which gastrointestinal disturbances, muscular weakness, lethargy, a lowered metabolic rate, an increased susceptibility to heat and cold, hypotension, anhydremia, hypoglycemia, hyponatremia and hyperkalemia are the most striking features. The loss of sodium chloride in the urine results in a decline in the concentration of sodium, chloride, and bicarbonate in the blood serum. With this loss of salt there is an increased loss of water leading to dehydration, a lessened blood volume, and a state of circulatory shock. The adrenal cortex is thus essential for the normal renal conservation of sodium and the excretion of potassium.

The adrenal cortex is concerned also with the metabolism of carbohydrate, fat and protein. The adrenalectomized animal tends to waste liver glycogen and to consume carbohydrate more rapidly than the normal. The rate of formation of glucose from protein is reduced with a resultant hypoglycemia. This hypoglycemia results from a disturbance in the catabolism of protein and interference with gluconeogenesis, and from an increased rate of utilization of carbohydrate.

Disturbance in function of many tissues and organs is observed in adrenal insufficiency or following the administration of excessive doses of the hormone. Some of these effects are secondary to the electrolyte and metabolic disturbances already described, but in addition the hormones also exert a permissive as well as a direct effect on many cellular functions.

Over thirty steroidal compounds have been isolated from adrenal cortical extracts. These are classified according to their principal actions into:

(1) the mineralocorticoids, desoxycorticosterone and aldosterone;

(2) the glucocorticoids, which include corticosterone, cortisone and hydrocortisone; and

(3) estrogenic, progestational and androgenic steroids. Aldosterone is believed to be the hormore responsible for the regulation of electrolyte and water balance. It is not available commercially, but synthetic analogues, desoxycorticosterone and fluorocortisone are available for inducing this effect. Corticosterone and hydrocortisone are believed to be the active principles responsible for the glucocorticoid functions of the adrenal. In addition to the latter, its dehydro-derivative, cortisone, and several synthetic derivatives, prednisone, prednisolone, traimcinolone, methyl prednisolone and dexamethasone are available for clinical use in a variety of disorders for their anti-inflammatory, antiallergic and metabolic effects.

The disease of the adrenal observed in the human include a deficient reduction of HORMONE due to atrophy of the gland or its destruction usually by tuberculosis which gives rise to a condition first described by Thomas Addison in 1855. It is characterized by gastro- intestinal disturbances, hypotension, pigmentation, *etc.* and is designated as Addison's disease. In certain congenital defects in which the enzymes necessary for the production of the glucocorticoids are lacking, the adrenal produces an abnormal androgenic hormone which gives rise to pseudohermaphroditism. The adrenal is also subject at times to the overproduction of glucocorticoids which gives rise to a condition designated as Cushing's syndrome, which resembles the iatrogenic disorder induced by prolonged administration of excessive amounts of these hormone glucocorticoids. An excessive production of aldosterone gives rise to a syndrome characterized by hypertension and periodic attacks of weakness. Tumors of the chromaphil tissue result in the production of excessive quantities of epinephrine and norepinephrine which induce hypertension and other symptoms of sympathetic overactivity.

ARTHUR GROLLMAN

References

Grollman, A., "The Adrenals," Baltimore, Williams & Wilkins, 1936.

Jones, I. C., "The Adrenal Cortex," Cambridge, Cambridge University Press, 1957.

Wolstenholme, G. E. W. and Cameron, M. P., eds., "The Human Adrenal Cortex" Boston, Little, Brown and Co., 1955.

AGASSIZ, JEAN LOUIS RODOLPHE (1807–1873) AND AGASSIZ, ALEXANDER EMANUEL (1835–1910)

This father and son are often confused with each other since both wrote extensively on the marine fauna of the United States. Both were born in Switzerland near Lake Neuchatel but the father was originally a physician having completed his doctorate in medicine at Munich in 1830. Several years before this, however, he had been asked by Professor Martius to describe the collection of fresh water fishes that he and Professor J. B. Spix had collected in Brazil. This description was published in 1829 and diverted Agassiz from medicine first into ichthyology and later into paleontology since he found it impossible to classify recent fishes without a study of their fossil ancestors. The mountains held more for him than fossils and in 1840 he published his *Études sur les Glaciers*, the first scientific study of the movements of glaciers and the results of these movements.

His ichthyological studies had made him dissatisfied with the confusion of zoological nomenclature and from 1842 to 1846 he produced the *Nomenclator Zoologicus* which is still a valuable reference tool. In 1846 he secured a grant from the King of Prussia which enabled him to accept J. A. Lowell's invitation to give a course of lectures on zoology at the Lowell Institute in Boston. So impressive were these lectures that he accepted the chair of zoology and geology in Harvard University in 1847 and remained in the New World till the end of his life.

His son accompanied him to Harvard and graduated from that institution in 1855. His original interests were engineering and chemistry. His exploitation of the copper deposits of Lake Superior made him an extremely wealthy man so that he was able to endow the Museum of Comparative Zoology at Harvard which had been built up by his father. In 1875 he surveyed copper deposits in Peru and Chile and made a great collection of Peruvian antiquities which he presented to the museum of which he became curator in 1874. His scholarly publications were subsequently divided about equally between marine ichthyology, marine zoology in general and the antiquities of Peru. His best known publications are his two-volume review of the Echini (1872-1874) and his two-volume account of the results of the dredging expeditions of the U.S. Steamer "Blake."

PETER GRAY

AGGRESSION

Aggression may be defined as the unprovoked attack of an animal against members of its own species. As such, it forms a part of agonistic behavior, the system of behavior patterns related to conflict between individuals. So defined, aggression rarely occurs except under unusual and abnormal conditions.

Fighting of any sort appears relatively late in evolutionary history. The first patterns of agonistic behavior are those of escape and passivity, which are seen even in protozoa. Social fighting is dependent upon the evolution of motor organs which can inflict injury, and of sense organs which enable the recognition of other individuals. Thus aggression is chiefly important in the higher phyla.

Fighting occurs in all the major classes of arthropods. Lobsters and crayfish fight during the breeding season. Spiders attack each other occasionally, although this is possibly only mistaken predation. Fighting occurs commonly in certain orders of insects, particularly the Orthoptera and Hymenoptera. In the latter it usually consists of an attack on a strange individual attempting to enter a colony, as in honey bees. Raids between different species of ants are a somewhat different phenomenon.

Most of our detailed studies of aggression have been made with vertebrates, and fighting of some sort has been recorded in all classes, even the amphibia, handicapped as they are by the lack of teeth and claws.

With a few exceptions, male vertebrates are more aggressive than females. This reflects a general tendency toward genetic control over basic social roles, in which females are specialized for the care of young and males are specialized for fighting and greater activity. There are also wide genetic differences in aggressive behavior between species and between larger taxonomic groups. For example, howling monkeys confine their fighting to vocal threats, whereas adult gibbons of either sex are highly aggressive. Many primates, like the rhesus monkey, may attack as a group, but ungulates such as sheep and goats typically fight only in individual combats.

Fighting may have different adaptive functions in different species. In some, females may fight for the defense of the young. Among the herd mammals, males during the rutting season attempt to round up all available females and attack other males which come near. Defense of a breeding territory is very common among birds and many fish, and males attack any stranger crossing the boundary.

Among vertebrates, fighting is normally organized into social relationships which reduce the amount of harmful aggression. Animals reared together generally develop persistent peaceful relations with each other. As they grow older they may develop a dominance-subordination relationship, in which one animal habitually threatens and the other habitually avoids the threat. In a well-organized society little fighting is evident.

Prolonged and destructive fighting chiefly occurs under conditions of social disorganization. This can be brought about by confining several adult males in close quarters in zoos or laboratories. Under natural conditions such social disruption rerely occurs except under conditions of excessive population growth, which forces strange individuals into close contact.

J. P. SCOTT

References

Carthy, J. D. and E. F. Ebling, eds., "Natural History of Aggression," New York, Academic Press, 1964.
Collias, W., "Aggressive behavior among vertebrate animals," Physiological Zoology, 17: 83–123, 1944.
Lorenz, K., "On Aggression," New York, Harcourt, 1966.
Scott, J. P., "Aggression," Chicago, University of Chicago Press, 1958.

ALDROVANI, ULISSI (1522–1605)

Bologna, Italy, was both benefactor and beneficiary of U. Aldrovani, naturalist. Of noble parentage he was originally set upon the path of commercial endeavors, but finding this distasteful, he changed of his own volition to the study of law and medicine.

On receiving his doctorate in medicine from the University of Bologna in 1553, he was appointed professor of philosophy and lecturer in botany. He held this position for six years at which time he was transferred to the chair of natural history. With the influence of this post he was able to establish a botanical garden, of which he was appointed director, in addition to being instrumental in founding the public museum of Bologna.

At about this same time he was made inspector of drugs, which resulted in the publication of *Antidotarii Bononiensis Epitome*, a work which became the model for many subsequent pharmacopoeias.

Aldrovani was chiefly noted for his patient and laborious research, the results of which may be found in his magnum opus of natural history, which is illustrated by well known artists of the time. While preparing this work he was supported by the senate of Bologna until his death in 1605.

The chief criticism which can be made of him is a poor critical judgement, a shortcoming which shows itself in the mixture of facts and fables, and in the confusion between the important and the trivial in his works. In spite of this, his major work is still of great scientific value to the naturalist.

LETITIA LANGORD

ALGAE

Algae (sea wrack) include marine and freshwater, chlorophyll-bearing organisms, diverse in size and form, ranging from microscopic unicells (0.5–2.5 μ in diameter) to the giant kelps of the ocean which may be 100 feet or more in length (*Macrocystis, Pelagophycus, Nereocytis*). In freshwater, algae are commonly known as "mosses," pond-scum, "frog-spittle," and "water-blooms." Although there are many microscopic algae in the sea, attached and free-floating, the more conspicuous are well-known as sea-weeds. Algae are differentiated from all other plants by a combination of characters, rather than by any single attribute. They are nonvascular, lack true tissues, except in the brown algae where tissue differentiation is approximated, and have reproductive organs, sexual and asexual, which are fundamentally one-celled. A negative characteristic is that algae lack root, stem and leaf organs.

Eight phyla (divisions) of the plant kingdom are represented: CHLOROPHYTA (green algae); CYANOPHYTA (blue-green); RHODOPHYTA (red algae); PHAEOPHYTA (brown algae); EUGLENOPHYTA (Euglenoids); CHRYSOPHYTA (yellow-green algae); PYRRHOPHYTA (Dinoflagellates); CHLOROMONADOPHYTA (Chloromonads), and conceivably a ninth Phylum, CRYPTOMONADOPHYTA. Some phyla are comprised entirely of motile, hence protozoan-like but autotrophic forms, whereas certain phyla include both motile and non-motile organisms. The Phaeophyta are all non-motile in the vegetative state but most of them have swimming reproductive cells. The Rhodophyta and Cyanophyta have no motile cells whatever. Phyla are differentiated primarily by: (1) chemical nature of the various pigments (in addition to chlorophyll); (2) the chemical composition and form of food reserve (starch, oil paramylum, floridean starch, leucosin, mannitol, glycogen?); (3) flagellation of motile cells; secondarily by such features as chemistry and structure of the cell wall, types of reproduction, and morphology of organs.

Occurrence. Although predominantly marine, fresh- or brackish-water inhabitants, many microscopic forms requiring only a film of moisture can exist on and in soil, on exposed rocks, or as epiphytes and endophytes of higher plants, especially in the tropics. In soil, Cyanophyta and diatoms have been recorded from a depth of several feet where they exist in a dormant condition or as saprophytes. Some Cyanophyta are able to fix nitrogen and various forms have been demonstrated to be capable of converting nitrates to nitrites. Thus they enter the nitrogen cycle and share in importance with bacteria in reconditioning tillable soils.

Cephaleuros is an example of a semi-parasitic genus on leaves of tropical plants, and on tea causes considerable economic loss. Among the Rhodophyta there are several genera which are specifically epiphytic or semi-parasitic on, especially, other red algae. A few species of both Chlorophyta and Cyanophyta combine with different forms of fungi in a parasitic or commensal association to produce Lichens. A smaller number of forms are to be found within cells or body cavities of animals, including Man and other mammals. *Chlorella* is a unicellular genus which occurs in Protozoa, *Hydra*, and Porifera. *Anabaeniolum* and *Simonsiella* are examples of blue-green algae occurring in the intestinal

tract of Man. Roots of *Cycas* (a primitive Gymnosperm) are invariably endophitized by *Anabaena*. A few forms live epizoically, such as the filamentous *Basicladia* on turtles (rarely on other shells), *Characium* on the carapace of microcrustacea, *Trichophilus* and *Cyanoderma* among the hair scales of the three-toed sloth.

Although generally restricted to an optimum type of habitat, some species are highly adaptive to environmental extremes. Especially Cyanophyta species may occur in water of average temperature or in nearly boiling hot springs (85°C.). Others are to be found in snow fields (red, yellow and green snow) and on glaciers. Pitting of the Greenland ice-cap is related to the accumulation of wind-distributed algae which maintain themselves in a film of water. Red snow is produced by unicellular Chlorophyta in which a red carotinoid pigment, haematochrome, masks the green color of chlorophyll.

All Phyla except the Phaeophyta (with questionable exceptions) are found in both fresh and salt water and most species are limited to one or the other environment. A few marine species, however, (*Enteromorpha*, e.g.) occur inland also. In brackish water are to be found overlaps while a few strictly halophytic species occur in brine lakes where NaCl is 17.55% or more.

The majority of fresh-water algae have a nearly world-wide distribution, being limited by optimum factors (light and water chemistry) in particular habitats. Most marine algal species, however, are restricted to particular coasts and certain latitudes or oceanic regions. Thus there are Pacific, Atlantic and Indian Ocean floras, and there are recognizable associations of tropical, neotropical, temperate and arctic species. Marine algae also show a vertical zonation related to light penetration and tide action. Freshwater algae (*Cladophora, Chara, Dichotomosiphon*) may grow at depths of from forty to sixty feet in middle latitudes. Marine algae, especially Rhodophyta, have been collected from as deep as 600 feet in the equatorial zone where solar rays are vertical, apparently making use of their red pigment (phycoeythrin) and phycocyanin to absorb blue rays for photosynthesis.

Aquatic algae which drift are known as phytoplankton, referred to as euplankton when they occur in open water, and as tychoplankton when they are entangled among mats of other algae and vegetation near shore. Organisms attached to the bottom, especially in deep water, are said to be benthic. Marine algae are mostly attached near, and on shore in the intertidal zone. They are well-adapted physiologically and structurally for living in turbulent water and for existing when exposed to drying at ebb tide. In fresh water a few species are confined to well-aerated, swiftly flowing water (*Monostroma, Lemanea*), but most occur in relatively quiet habitats. Basic waters which are well-supplied with electrolytes, phosphates and nitrates support the greatest numbers and abundance of species. Acid or soft water lakes are characteristically poor producers, although acid bogs are ideal habitats for luxuriant growths of desmids.

Reproduction. Reproduction is as variable as form among the algae but occurs in three general ways: (1) vegetative; (2) asexual; (3) sexual. The former includes fission (Cyanophyta), cell division by mitosis, fragmentation and proliferation. Asexual reproduction occurs by various kinds of spores, motile zoospores, or by various non-motile elements. Akinetes are spores

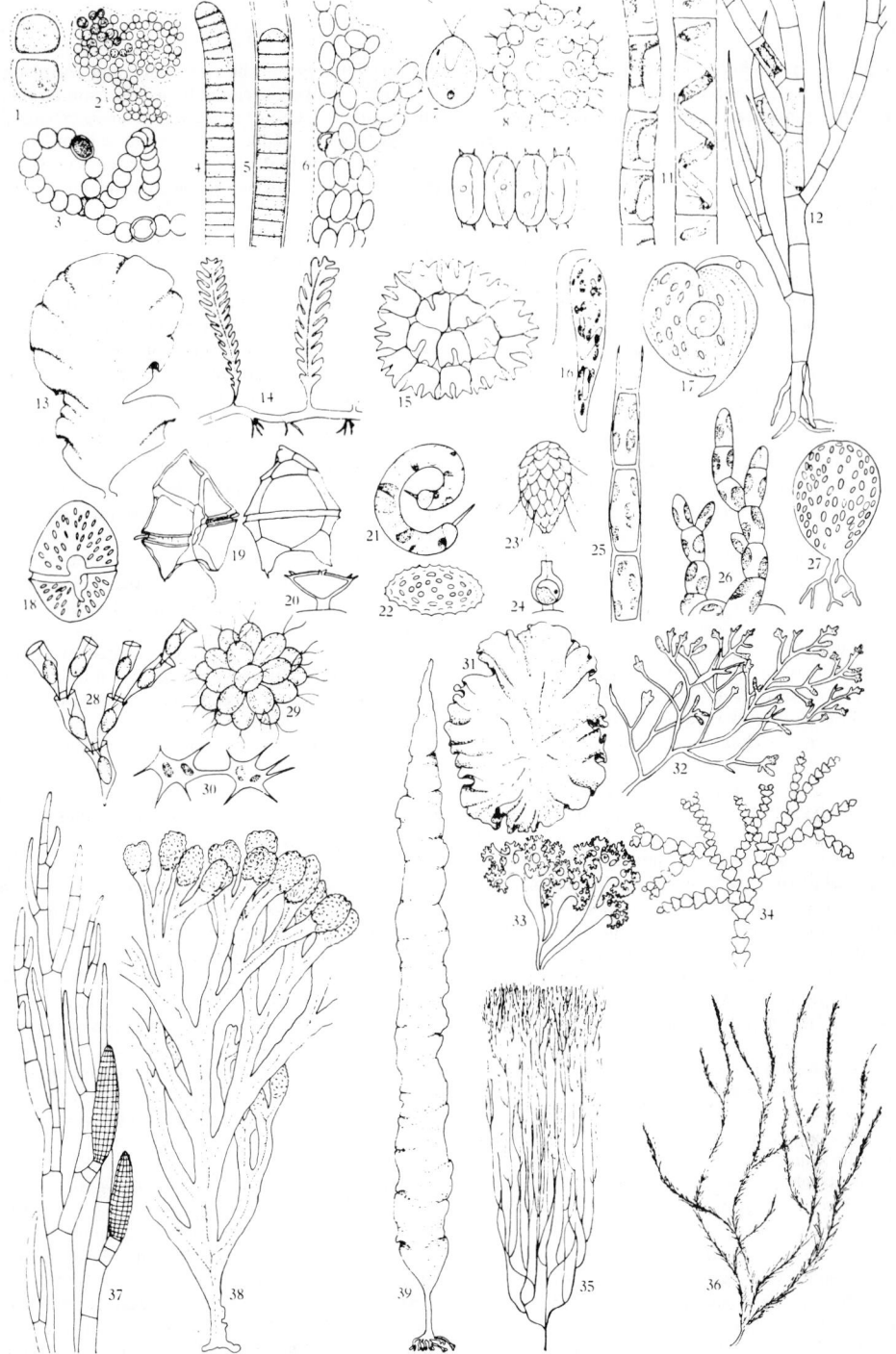

Figs. 1–39   Representatives of the Algal Phyla.

Figs. 1–6   Cyanophyta; 1, Chroococcus; 2, Microcystis; 3, Anabaena; 4, Oscilatoria; 5, Lyngbya; 6, Stigonema.

Figs. 7–15   Chlorophyta; 7, Chlamydomonas; 8, Eudorina; 9, Scenedesmus; 10, Ulothrix; 11, Spirogyra; 12, Stigeoclonium; 13, Ulva; 14, Caulerpa; 15, Pediastrum.

Figs. 16–17   Euglenophyta; 16, Euglena; 17, Phacus.

Figs. 18–20   Pyrrhophyta; 18, Gymnodinium; 19, Peridinium; 20, Tetradinium.

Figs. 21–30   Chrysophyta; 21, Ophiocytium; 22, Chlorallanthus; 23, Mallomonas; 24, Derepyxis; 25, Tribonema; 26, Phaeothamnion; 27, Botrydium; 28, Dinobryon; 29, Synura; 30, Chrysidiastrum.

Figs. 31–36   Rhodophyta; 31, Porphyra; 32, Gelidium; 33, Chondrus; 34, Bossea; 35, Polysiphonia; 36, Dasya.

Figs. 37–39   Phaeophyta; 37, Ectocarpus; 38, Fucus; 39, Laminaria.

formed by the enlargement of single vegetative cells with thickened walls. Aplanospores (walled) are produced many within a vegetative cell, each capable of regenerating a new plant. Endospores are bits of protoplasts cut out within a cell and liberated without having special walls.

In sexual reproduction (lacking in some phyla) there is a range of complexities both in respect to the differentiation of sex cells (male and female) and the cells (gametangia) producing them. Uniting sex cells having no detectable differences in respect to size and behavior are known as isogametes. When there is a slight amount of differentiation (size and/or behavior) sex cells are known as anisogametes, whereas heterogametes are those which are clearly recognizable as egg (large and non-motile) and sperm (small and motile) gametes. In the Rhodophyta the spermatia are non-motile and bring about fertilization by drifting to the female reproductive organ. Reproduction by egg and sperm (antherozoid) is referred to as oogamy. Gametes may be produced in unspecialized vegetative cells or in sex organs (male, antheridium; female, oogonium or carpogonium). Heterogamy is common among the "higher" algae but occurs also in the "lower" colonial and simple filamentous forms.

Union of gametes produces a zygote which may function in one of two ways, resulting in two general types of life histories. In most of the fresh-water algae the zygote develops a thick wall and enters a dormant period as a zygospore. Upon germination the zygospore nucleus (diploid) undergoes reduction division (meiosis) and produces four (usually) haploid spores (motile or non-motile) which in turn create a new haploid plant (gametophyte). The sporophyte phase thus exists only as a single-celled zygospore. In marine algae (Phaeophyta especially) the zygote germinates promptly and grows into a diploid plant that is similar to the plant that produced the gametes (isogeneratae) or into a plant different in size and shape (heterogeneratae). Certain Phaeophyta, *Fucus*, e.g. have no alternation of diploid and gametophyte generations, non-motile spores of the diploid generation metamorphosing directly into gametes.

In the Rhodophyta sexual reproduction is highly diversified, involving many specializations. Following fusion of the gametes the zygote behaves variously but in one of two general ways. There may be an immediate meiosis to form haploid spores (carpospores). Haploid carpospores develop haploid sexual plants. There may be a diploid growth and proliferation from the zygote all within the gametophyte, resulting in the formation of diploid carpospores. Diploid carpospores produce another diploid generation (tetrasporic) in which tetraspores are produced by meiosis, these in turn growing into haploid sexual plants.

Physiology. In their growth algae require the usual plant nutrients, but the optimum amounts and qualities vary from group to group and between species. In addition, vitamins (ascorbic acid, *e.g.*) and growth-promoting substances (metabolites from bacteria or from other algae) are essential to some. In certain instances growth promoters induce "water blooms" (Diatoms, *Microcystis, Anabaena, Aphanizomenon*). Trace elements may be equally important. In culture, freshwater algae respond to a medium composed of calcium nitrate, magnesium sulphate, sodium silicate, potassium hy-

pophosphate, magnesium sulphate and ferric citrate. Certain marine algae can be grown in a somewhat similar medium made up in sea water, using 50 cc. of soil extract to which sodium nitrate and sodium hypophosphate are added. Artificial sea water has been used successfully and soil extract in solution are often satisfactory.

Growth may occur by cell division throughout the thallus, at the base of a filament or frond, or at the apex by means of an apical cell or a meristem. Some (in the Phaeophyta) have a meristematic region at the base of colorless, terminal hairs (trichothallic growth). The multitudinous forms resulting from cell division are (in general) unicellular, colonial, or filamentous. Elaborate thalli are formed from the various uses of the filament as a structural element. The thallus may be entirely prostrate, prostrate with some erect development, or with erect growths only. Juxtaposition of filaments may form sheets or multiaxial fronds. There may be siphonous coenocytes or filamentous complexes with cortical cells variously disposed.

Economics. Among the several phyla of algae there are varied, both positive and negative economies. The positive values are far from exhausted. Economies vary according to phylum as well as between species because they are related to 1) specific substances contained within algal cells; 2) substances given off by them, or 3) in a few instances, to physiological habits and structure of the cells or the cell walls.

Most of the positive economies are related to the red and brown sea weeds. The Rhodophyta (e.g. *Porphyra, Rhodymenia, Gracilaria*) are used directly as food by especially Oriental and Polynesian peoples, although Dulse (*Rhodymenia*) is commonly eaten as a salty confection by north Europeans and in maritime North America. The copious mucilage characteristic of most red algae (*Chondrus, Gelidium,* e.g.) finds uses as food in comestibles (blanche, ice cream and other gelatinous deserts). Also this mucilage (carrageenin and agar-agar) find countless uses in industry and medicine. A few examples are: as coating for photographic film, clarification of liquors, cold creams, preparation of paints, 75,000 lbs. a year used in making dental impressions, preparation of petrol-agar laxative and other pharmaceutical products, bacteriological laboratories (250,000 lbs. used annually in the U.S. alone), preparation of stored meats.

Although a few of the brown algae are used as food by Man, these plants are fed mostly to cattle, sheep and chickens. In coastal countries kelps are harvested for soil conditioning since they are richly stored with salts. Because of the concentration of salts (leached through the ages from soil) brown algae (also *Phyllophora*, a red alga) are a major source of commercial and pharmaceutical iodine and potash. Also the kelps, especially *Laminaria* yield alginic acid and algin, the former important in the manufacture of rubber tires. The important uses of these substances has led to a thriving mechanical kelp-harvesting industry on the Pacific coast of North America.

The diatoms, both living and as dead silicious shells (diatomaceous earth) are of inestimable value. They, together with dinoflagellates in the plankton, form the bases of the food chain of aquatic animals and hence the foundation of all that this life means to Man. As diatomaceous deposits (Fuller's Earth) they are used in insulating materials, in metal polishes, for filters (Berk-feld) and as a clearing agent for solutions.

Blue-green algae cause economic loss far out of proportion to their small size. Because of their habit of floating high in the water and of forming sticky surface scums (*Microcystis, Aphanizomenon, Anabaena*) they ruin domestic water supply and recreation sites by the disagreeable tastes and odors produced. In "bloom" conditions these forms lead to the death of fish, either directly through oxygen depletion or indirectly by the production of poisonous substances resulting from their decay. Some blue-greens possess either endo- or exo-toxins (or both) which cause the death of domestic animals and water fowl that drink from bloom-bearing water. Fish death is also caused by dinoflagellates (*Gymnodinium* and the "red tide") especially in tropical and subtropical waters. Death of human beings may be caused by eating shell-fish that have subsisted on dino-flagellates (*Gonyaulax, Gymnodinium*) from which an alkaloid toxin has been ingested and stored in the digestive organs. The production of toxins and other antibiotics is a field that metits much research.

In various ways algae are of use in medical and biological research; for biological assays, cause of cancer, physiology of protoplasm. Sodium alginate derivatives have produced chemicals that have led to the elaboration of whole blood for transfusions in Man. As oxygenators and digesters many algae (fresh-water) are serviceable in the handling of sewage and wastes (space flights, e.g.).

G. W. PRESCOTT

References

Fogg, G. E., "The Metabolism of Algae," London, Methuen, 1953.
Fritsch, F. E., "The Structure and Reproduction of the Algae," Cambridge, University Press, Vol. I., 1935.
ibid, Vol. II., 1945.
Krauss, R. W., "Physiology of the Fresh-Water Algae," Ann. Rev. of Plant Physiol. 9: 207–244, 1958.
Lewin, R. A. (Ed.), "Physiology and Biochemistry of Algae," New York, Academic Press, 1962.
Prescott, G. W., "Algae of the Western Great Lakes Area," Dubuque, Iowa, Brown Co., 1962.
ibid., "The Algae," Boston, Houghton-Mifflin, 1968.
Pringsheim, E. G., "Pure Cultures of Algae, their Propagation and Maintenance," Cambridge, University Press, 1946.
Smith, G. M., "Marine Algae of the Monterey Peninsula," Stanford, University Press, 1944.
ibid., "The Fresh-Water Algae of the United States," New York, McGraw-Hill, 1950.
ibid, "Cryptogamic Botany," Vol. I. Algae and Fungi, New York, McGraw-Hill, 1955.
ibid., "Manual of Phycology," Waltham, Chronica Botanica, 1951.
Round, F. E., "Biology of the Algae," New York, St. Martins, 1965.
Taylor, W. R., "Marine Algae of the Northeastern Coast of North America," Ann Arbor, Univ. of Michigan Press, 1957.

arise from pre-existing ones by gene mutation. The diversity of alleles produced in this way is the ultimate basis of hereditary variation and evolution.

The different alleles of a given gene determine the degree to which the specific hereditary characteristic controlled by that gene is manifested. The particular allele which causes that characteristic to be expressed in a normal fashion is often referred to as the "wild-type" allele. Mutations of the wild-type allele result in "mutant" alleles, whose functioning in the development of the organism is generally impaired relative to that of the wild-type allele. If the genetic material is thought of as a coded set of instructions in chemical form (see, deoxyribose nucleic acid) for making a living cell, then the gene may be likened to an essential word in the code. In this analogy the wild-type allele corresponds to the correct word and the mutant alleles correspond to mistakes in the spelling of that word.

A gene or any of its alleles is said to occupy a fixed position or "locus" in the chromosome. In the body cells of higher organisms, including man, there are two chromosomes of each kind and hence two alleles of each kind of gene. Such organisms and their somatic cells are said to carry a "diploid" complement of alleles. In the formation of gametes, the members of each pair of alleles sort out or "segregate" so that each gamete comes to carry one member of each pair, or a "haploid" complement of alleles.

With respect to a given locus, a diploid individual may be "homozygous" if it has the same allele present twice, or "heterozygous" if two different alleles are present. Specifically, if A and a represent a pair of allelic genes then A/A and a/a represent the genetic constitutions, or "genotypes," of the two possible homozygotes, while A/a represents the genotype of the heterozygote. Frequently the appearance, or "phenotype," of A/a resembles that of A/A. In such cases A is said to be the dominant allele and a, the recessive allele. Another system of symbolizing alleles is in widespread use and is to be preferred whenever it is clear which member of a series is the wild-type allele. In this case the wild-type is designated by adding a $+$ sign as a superscript to the base symbol of the mutant gene. Thus a^+, or when no confusion will arise simply $+$, designates the wild-type allele of the recessive mutant a.

When more than two alleles of a gene are known the relationship is said to be one of multiple allelism. Symbolically, the multiple alleles are indicated by attaching distinguishing letters or numbers as superscripts to the base symbol. For example, a^2 represents another member of a multiple allelic series containing a^+ and a. The a/a^2 heterozygote may resemble the a or a^2 homozygote, it may be intermediate in phenotype or it may approach the wild-type, in which case the alleles are said to "complement" one another. The gametes produced by the a/a^2 heterozygote consist of equal numbers of a and a^2 types. Failure to observe recombination or crossing over between the a and a^2 gene constitutes evidence that they are alleles.

E. B. LEWIS

ALLELISM

Allelism is the relationship existing between alleles, which are the different forms of a GENE. New alleles

Reference

Pontecorvo, G., "Trends in Genetic Analysis," New York, Columbia University Press, 1958.

ALTERNATION OF GENERATIONS

The term was originally coined to describe the kind of life history found in the BRYOPHYTA and Vascular Cryptogams (PSILOPSIDA, SPHENOPSIDA, LYCOPSIDA, PTEROPSIDA). In all these groups the sporophytic diploid generation reproduces by means of spores. These germinate and grow into a gametophytic haploid generation which reproduces by means of male and female

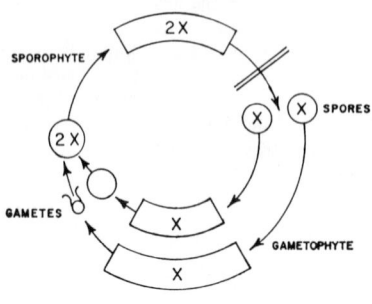

Fig. 1

organs. Fertilization of the female gamete produces again the diploid status of the sporophyte. Apart from certain cases where either the sporophyte or gametophyte can respectively give rise vegetatively to new comparable plants the two generations must succeed each other alternately (Fig. 1).

In these plants reduction division takes place at spore formation and there is an alternation, not only of morphologically unlike generations, but also of cytologically unlike generations (2x and x).

The term has since been applied to all members of the Plant Kingdom where it is relevant, and in the case of the land plants it is possible to trace a series in which at first the gametophyte (haploid) generation is the larger (or dominant) through a succession in which the gametophyte generation becomes successively more and more reduced, until in the flowering plants it is represented by only a few nuclei. In plants below the Bryophyta (ALGAE, FUNGI) alternation of generations has also been noted. Here, in many cases, however, the alternation is by no means regular. It is evident that the phenomenon emerged before the adoption of the land habitat, but that prior to the transmigration from water to land the alternation was not inevitably regular. In the algae there is regular alternation of morphologically similar generations, e.g. *Dictyota, Ulva*; of mor-

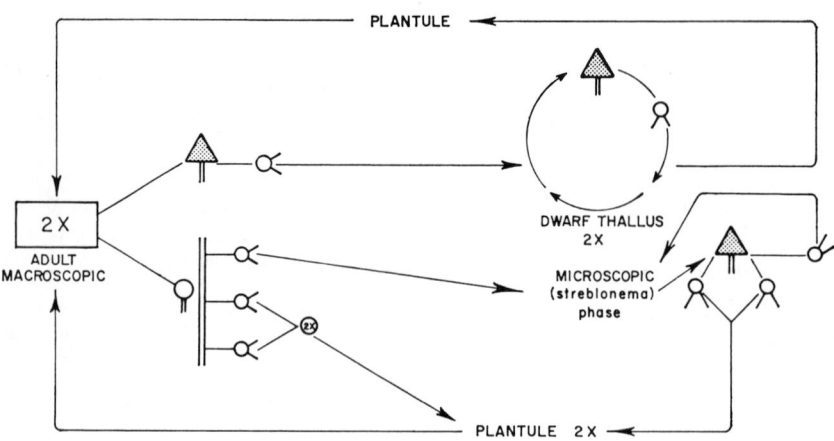

Fig. 2   Life cycle of *Asperococcus bullosus* (Phaeophyceae).

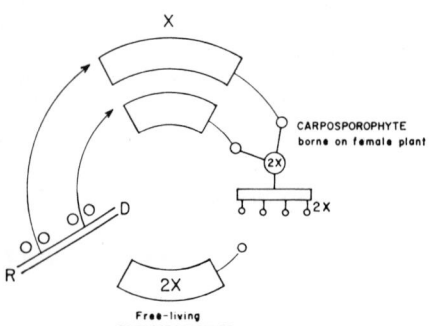

Fig. 3   Life cycle of typical red alga.

phologically dissimilar generations both of which may be macroscopic, e.g. *Cutleria, Halicystis, Galaxaura* or one macroscopic and the other microscopic, e.g. *Laminaria*. Many of the algae, however, possess an irregular alternation of generations (Fig. 2) and in some species, e.g. *Ectocarpus siliculosus*, the type of life cycle appears to vary from region to region. This irregularity is clearly associated to some extent with the relatively undifferentiated nature of the swarmers which can act as zoospores or gametes.

In the red algae (Rhodophyceae) the situation is more complex in that the majority have a cycle with two cytological generations (gametophyte (x), carposporophyte and tetrasporophyte (2x)), but with three morphological ones only two of which, however, are independent (male and female gametophytes belong to

	Sex plants only	Sex plants and alternation of generations			Spore producing plants only
Diploid only					Lomentaria rosea
Diploid and haploid		Galaxaura Asparagopsis armata	Polysiphonia	Phyllophora brodiaei Liagora tetrasporifera	
Haploid only	Batrachospermum	Naccaria wiggii			Ahnfeldtia plicata
	Heteromorphic	Isomorphic			
	DIMORPHIC (DIPHASIC)			MONOMORPHIC (MONOPHASIC)	

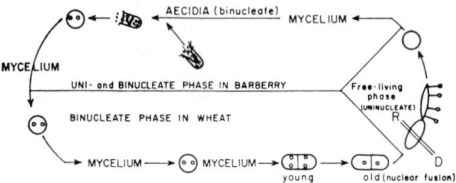

Fig. 4   Life cycle of *Puccinia graminis*.

the same generation) (Fig. 3). In some cases the separate generations have the same chromosome complement.

The various life cycle complexities in the Rhodophyceae are exemplified in the table above.

Life cycle complications also occur in the Fungi. In *Pyronema confluens* there is only one morphological generation which gives rise to sex organs that after fertilization produce binucleate threads (ascogenous hyphae) in which nuclear fusion takes place in the penultimate cell immediately prior to spore formation. Although the fungus is haploid, there is a parasitic binucleate phase which results in spore production. In the rust fungi, *Puccinia graminis* can be regarded as having three morphological and three cytological generations (Fig. 4), the binucleate phase being regarded as a separate cytological condition. Other rust fungi show modifications of this basic cycle.

Originally considerable debate took place as to whether the two generations (e.g. in the Bryophyta or Vascular Cryptogams) were completely distinct entities (antithetic theory) or whether, since one generation had arisen from the other, there was no such distinction (homologous theory). With our present day knowledge about chromosomes and the nature of the haploid and diploid nuclei, this apparent problem has now become one of historical interest only.

V. J. CHAPMAN

AMENTIFERAE

A taxonomically inacceptable botanical group used to draw together the catkin-bearing, DICOTYLEDONOUS, plants such as the Fagaceae, Betulaceae, etc.

AMPHIBIA

Amphibia is the name of the class of VERTEBRATES that stands midway between the fishes on the one hand and the REPTILES on the other. This class includes three living groups, the frogs, the salamanders, and the caecilians. Many are actually "amphibious"— that is, they spend part of their lives in water and part on land—although there are some amphibians that never come to land and there are others that reproduce on land and have no aquatic larval stage. The term Amphibia, as introduced by Linnaeus, originally included both the amphibians and the reptiles, but in present day usage it has been restricted to the former and reptiles are set apart as a separate class.

The amphibians have no unique characteristic, such as the feathers of the BIRDS or the mammary glands of the MAMMALS, that sets them off from all other classes. They are, however, readily distinguished from both the fishes (from which they originated) and the amniotes (to which they gave rise). Amphibians are poikilothermous animals with a moist or rather dry skin which lacks scales except for a few deeply imbedded scales in some of the caecilians. They are the first tetrapods, all of the modern forms except the caecilians having well developed limbs rather than fins. The eggs are essentially aquatic and lack both amnion and calcareous shell. They are often deposited in open water, but in many of the modern forms, life histories have become modified so that the eggs are deposited elsewhere. The egg usually hatches into a larva which later metamorphoses into the adult form; this double life history is characteristic of the amphibians, except for a few specialized forms that have succeeded in eliminating the larval stage, so that the young hatch out as essentially miniature replicas of the adults. The amphibians have well developed, multicellular integumentary glands; lungs are usually present, and are always simple in structure; and there are ten pairs of cranial nerves.

The amphibians can be distinguished from the fishes by the fact that they have developed the tetrapod limb; by the fact that the outermost layer of the epidermis dies and tends to become cornified, forming a stratum corneum; and by the fact that they have lost the bony operculum of the fishes. From the reptiles they differ in that they have not yet developed the amniote egg, well developed epidermal scales, or true claws. All the reptiles lack a larval stage and have direct development.

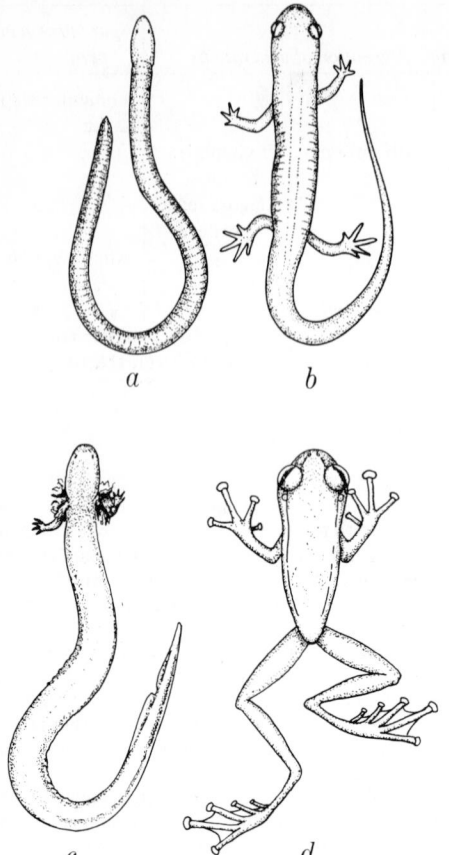

Fig. 1 The four types of living amphibians—a. caecilian; b. salamander; c. trachystome; d. frog.

The earliest known animal that can be difinitely assigned to the class Amphibia is *Ichthyostega,* found in freshwater beds of Greenland. These beds are of late Devonian or early Mississippian age, indicating that we should seek for the ancestor of the land vertebrates in early Devonian or earlier times.

Classification. *Subclass Lepospondyli.* These were all animals of modest size; many of them were eel-like forms in which the limbs were lost. There are six orders recognized in this subclass; first three, Aistopoda, Nectridia, and Microsauria, are extinct. The other three have living descendants.

Order Apoda. This order is known only from recent forms. However, Aistopoda and Nectridia were really specialized sidearms of amphibian evolution, while the microsaurs were more generalized; thus it seems locigal to assume some connection between the latter and the modern caecilians. The caecilians are long, slim, worm-like creatures without limbs or limb girdles, and with a very reduced tail. The eyes are minute, buried in the skin, and without lids. The more primitive genera have tiny dermal scales imbedded in the skin, apparently an heritage from the early scaled amphibians of the Carboniferous. They are primarily forest animals of the tropics; one form, *Typhlonectes,* is an aquatic river-dweller. There is but one family in this order, the Caecilidae.

Order Urodela. These animals, commonly known as salamanders, retain the tail throughout life instead of losing it at metamorphosis, as do the frogs. The head and trunk regions are distinct; most species have two pairs of legs and a poorly developed, primitive sternum. There is no cavum tympanum (middle ear cavity), and no tympanic membrane. Fertilization is either external or internal by means of spermatophores. Most salamanders are oviparous although a few have become ovoviviparous. The larvae closely resemble the adults in general form and have teeth in both jaws.

Suborder Cryptobranchoidea. These are the most primitive living salamanders and the only ones known to have external fertilization. The eggs are laid in gelatinous sacs. There are but two families in this suborder—the Asiatic land salamanders, Hynobidae, and the giant salamanders, Cryptobranchidae. The famous fossil cryptobranchid, *Andrias,* was first described as the mortal remains of a human sinner drowned by the Noachian Deluge and was known as *"Homo diluvii testis"* (man, witness of the flood). Fossil cryptobranchoids are now known as far back as the Upper Cretaceous.

Suborder Ambystomoidea. These are terrestrial, sturdily built, broad-headed salamanders. Fertilization is internal by means of the male spermatophore. In a typical situation the larva metamorphoses into a terrestrial adult which lives on land except when it retires to the water to breed. At least one species lays its eggs on land. Sometimes, though, because of environmental conditions, the aquatic larvae never metamorphose into morphological adults, although they become sexually mature. These forms, which are known as axolotls, stay in the water as unmetamorphosed but sexually mature individuals and breed as aquatic rather than as terrestrial salamanders. Such a condition is known as neoteny. There is but one family, the Ambystomatidae, in this suborder.

Suborder Salamandroidea. This is the dominant group of salamanders. Its members are more numerous, varied, and widely distributed than are those of any other suborder. Some are among the most terrestrial of all salamanders, but the newts and *Amphiuma* are quite aquatic. The European fire salamander *(Salamandra)* gave rise to the ancient legend that these animals are able to live in fire. It is easy to see how such a myth arose, for salamanders often live in crevices of fallen, partly rotten logs. When such a log is brought in and thrown on the fire, the animal, roused by the heat, immediately tries to escape. To an uncritical eye, it seems to emerge from the flame. Even at the present day objects that can withstand a great amount of heat are sometimes called salamanders. The often-heard words "newt" and "eft" come from the Anglo-Saxon "efete" or "evete," a word originally used for both lizards and salamanders. In medieval English this word became "ewt" and finally "newt" (an ewt). Since the only caudate amphibians found in England are rather aquatic members of this family, newt has become the common name for the more aquatic salamandrids. There are three families in this suborder, Salamandridae, Amphiumidae, and Plethodontidae.

Suborder Proteida. This suborder is made up of

aquatic salamanders that retain their gills throughout life and also have lungs. The only forms included are the European olm, *Proteus,* and the American mud-puppy or waterdog, *Necturus,* both in the family Proteidae.

Order Trachystomata. This small order is sharply distinct from all the urodeles. Its members are all aquatic creatures that retain their gills as adults; they have but one pair of limbs, the hind limbs and the pelvic girdle are entirely absent. The life history of these forms is poorly known, the mechanism of fertilization not yet determined. A single family, Sirenidae, contains but two living genera, *Siren* and *Pseudobranchus.*

Subclass Aspidospondyli. It is to this subclass that the great majority of both fossil and living amphibians belong. They differ from the Lepospondyli in that in them the centrum of the vertebra originated as several centers of ossification rather than simply as bone formed in a tube-like manner around the notochord. This subclass is divided into two superorders. The superorder Salientia contains the modern frogs and toads, and a few of their extinct relatives. The superorder Labyrinthodontia contains the great majority of the primitive extinct amphibians of the Paleozoic.

Superorder Salientia. This group goes further back in fossil history than any of the other groups of living amphibians. It contains three orders—Eonura, Proanura, and Anura.

Order Eoanura. The single family, Amphibamidae, comprises small-sized amphibians known only from the Pennsylvanian of North America. They did not look at all like modern frogs and lacked the long hind legs specialized for jumping.

Order Proanura. This order was created for a single specimen from the Lower Triassic. There is some evidence that this animal is nothing more than a metamorphosing tadpole of a true frog. If this should prove to be so, then the Order Proanura would have to be eliminated and the Order Anura carried back to the Lower Mesozoic.

Order Anura. We assign all of our living frogs and toads to this order. It is not entirely recent, for two families are known only from fossil forms. These amphibians are characterized by the absence of a tail in the adult and by having the hind limbs differentiated into four, rather than three segments.

Suborder Amphicoela. These, the most primitive of living frogs, have amphicoelous vertebrae. Like all true frogs, they lack tails, but they give evidence of their descent from tailed ancestors by retaining two tiny 'tail-wagging' muscles. The suborder has one living family, the Liopelmidae, and one fossil family, the Montsechobatrachidae, which was apparently widespread in the Upper Jurassic, being known from Europe, Africa, and North America.

Suborder Opisthocoela. This is another group of primitive frogs in which the tadpoles, and in some cases, the adults, have ribs. Such well known frogs as the European fire-bellied toad, *Bombina,* and the mid-wife toad, *Alytes,* belong here. Three families, Discoglossidae, Pipidae, and Rhynophyrnidae, are included in this suborder.

Suborder Anomocoela. This group is comprised of the "spade-foots," toad-like creatures standing between the two preceding suborders and the true toads. Most of them have procoelous vertebrae, but in some

the intervertebral disks do not fuse with the vertebrae but remain as separate structures. Most anomocoels lay their eggs in open water and pass through a tadpole stage. Two families are included in this suborder, Pelobatidae and Pelodytidae.

Suborder Procoela. This group of frogs is large, widespread, and very successful and comprises a bewildering variety of highly diverse forms. They are characterized by having a double condyle on the urostyle and uniformly procoelous vertebrae. None of them ever have any free ribs.

The frogs in this group are as diverse in life history as they are in structure. While a number of them still retain the aquatic egg—tadpole—adult form of life history, many have departed from this pattern. Modifications include laying on land and omitting a tadpole stage, carrying the eggs and young on the back in a specially developed pouch, having the larvae carried about in the vocal pouch of the male, and depositing the eggs in a nest above a stream so that the larvae, upon hatching, fall into the stream below. The aquatic tadpole of the toad *Bufo* is the polliwog of our ponds and streams. There are seven recent families: Leptodactylidae, Centrolenidae, Rhinodermatidae, Dendrobatidae, Hylidae, Bufonidae, and Atelopodidae, in this suborder. The family Paleobatrachidae is known only from the Miocene of Europe.

Suborder Diplasiocoela. This suborder comprises a large group of species characterized by having ten vertebrae usually including procoelous, amphicoelous, and acoelous vertebrae. It includes the so-called "true frogs" *Rana,* along with some less well-known forms. There are four families—Ranidae, Rhacophoridae, Microhylidae, and Phrynomeridae.

Superorder Labyrinthodontia. This superorder contains the great majority of the amphibians of the Paleozoic. Many of them were large forms; they were the dominant vertebrates of the Carboniferous swamps of the world. Paleontologists divide this group into several orders and suborders.

C. J. GOIN

References

Angel, F., "Vie et Moeurs des Amphibiens," Paris, Payot, 1947.
Goin, C. J. and O. B. Goin, "Introduction to Herpetology," San Francisco, W. H. Freeman, 1962.
Noble, G. K., "The Biology of the Amphibia," New York, McGraw-Hill, 1931.
Romer, A. S., "Vertebrate Paleontology," Chicago, University of Chicago Press, 1966.

AMPHINEURA

The Amphineura form a relatively small class of MOLLUSCA. They are decidedly primitive in structure, relationship being indicated on the one hand to the now nearly extinct Monoplacophora and more distantly on the other to some of the more archaic GASTROPODA, with which they were for a long time wrongly associated. Two decidedly divergent subclasses are here included, the Aplacophora or solenogastres, and the Polyplacophora (Crepipoda, Loricata) or chitons. These two groups agree in and are separated from other Mollusca by 1) the diffuse and probably archaic ladder-like central nervous system, comprising two ventral and two posteriorly continuous lateral cords connected by successive transverse commissures; 2) the generalized plan of the reproductive, alimentary, and renal systems; and 3) the lack of cephalic eyes and tentacles. Indeed primitive traits so dominate that in some respects these animals seem to have diverged less from the presumed molluscan archetype than even the remarkable primitive monoplacophoran, *Neopilina.*

Aplacophora. The Aplacophora are uncommon, aberrant, deep-water creatures, rarely to be found in collections. The whole organism is so worm-like that some authorities have denied their molluscan affiliation, but the present weight of authority is to rank them with the chitons (Polyplacophora) in the class Amphineura of that phylum. Principal support for this is to found in their possession of a simple radula in the buccal cavity, in the possession by some of a small posterior mantle-chamber containing a true ctenidium, and in a nervous system essentially similar to that of chitons. There is a (shell-less) mantle strengthened by calcareous spicules. The group falls into two sharply separated suborders: the one, including *Chaetoderma,* adapted for ooze-feeding; the other, typified by *Neomenia,* parasitic on arborescent coelenterates. In the former the sexes are separate and there is a single gonad as in chitons. In the latter the gonads are paired and the animals are reputed to be protandric, with a very short male phase.

Polyplacophora. The Polyplacophora, or chitons, are creeping, dorso-ventrally flattened animals, usually elliptic in outline, but occasionally (*Stenochiton, Crypttoplax*) elongate and worm-like. The foot, expanded into a broad sole, and the snout-like head are separated from the mantle by a narrow, ambient, pallial chamber containing the sometimes very long, repetitively laminate ctenidia. Protecting the animal dorsally is a segmented shell of overlapping transverse plates, termed "valves," which in the living forms are characteristically eight in number from the time of their first appearance in the embryo. These are held in place both by muscular attachments and by a peculiar peripheral structure, the tough and usually scaly or spinose *girdle,* which in some species is provided with tufts of movable spines or with spinose setae, often metamerically arranged. The valves are penetrated by numerous specialized sensory end-organs, the *aesthetes.* These are of two primary types, the smaller and more numerous *micraesthetes,* which are thought to be tactile in function, and the larger *megalaesthetes,* which may be light-sensitive, since in certain of the higher forms (*Tonicia, Acanthopleura,* etc.) some of them attain the level of true eyes, equipped with a retina, pigment-cup, and lens. The whole integument, inclusive of the mantle and shell, thus forms a protective and sensory shield, specialized through its coat-of-mail mechanics to enable its owner to traverse obstacles or angles of the rock without exposure of more vulnerable tissues, and even to curl up into an armored ball. In other respects the chiton organization is mainly noteworthy for its retention of the most simple, primitive, or generalized features, notably in the alimentary, renal, and reproductive systems. The gut is long and much wound, its opening median and posterior. Chitons are probably mainly herbivorous, but some animal matter (e.g., barnacles) is known to be eaten by certain species. At least one pair of teeth in each transverse series of the well-developed radula is powerfully cusped. The sexes are separate. The single median gonad opens through paired postero-lateral ducts. Fertilization is external but some species have been observed to brood their fry in the pallial chamber. The larva in the species investigated is a modified trochophore. Whether repetition of some structures reflects an earlier more truly segmented condition is still a moot point and students disagree therefore whether an annelid or a turbellarian origin is the more strongly suggested. In any event the importance of the chitons in any theory of molluscan origins is hardly to be overestimated.

Chitons are usually to be found on or under stones or other hard objects from the upper intertidal region to considerable depths, with their usual metropolis at or just below low tide. They are world-wide in distribution, but attain a maximum development in the waters of the west coast of the Americas and Australia. The Japanese fauna also is rich in chitons, but they are much more weakly represented in the Atlantic, the Mediterranean, and the polar seas. As fossils they are rarely abundant, but recognizable remnants occur well down into the Palaeozoic. Some kinds of chitons are very beautiful animals, exhibiting astonishing differences of color and pattern within the confines of a single species.

S. STILLMAN BERRY

AMPHIOXUS

Amphioxus (Branchiostoma, lancelet) belongs in the subphylum Cephalochorda. There are usually listed twenty-eight species of lancelets in this group, which is locally distributed over the world. There are four species on American shores: *Branchiostoma virginiae, B. floridae, B. bermudae, B. californiense.* It is a small, fishlike, marine animal whose average adult length does not exceed two or three inches on the average. There is one species along southern China and India that reaches six inches.

It is found in shore water and on sandy beaches of the subtemperate and tropical parts of the world. It burrows rapidly, head first in the sand by means of a vibratory action of the whole body and comes to rest with the anterior end exposed to the water. Usually at night and during the breeding season the animal leaves the burrow and swims about like a fish, then at dawn again burrows in the sand. Their food consists largely of sedimentation of organic matter and small animals taken principally while the animal is burrowed in the sand with the mouth exposed.

The body of the animal is laterally compressed and shaped like a small lance, the tail being the point. It does not have a distinct head. The mouth opens at the anterior portion of the ventral surface of the body by way of an oral hood (funnel). The mouth is surrounded by oral cirri, some tentacle-like structures. There is a low median fin along the dorsal side, continuing around the tail as the caudal fin and anteriorly about $\frac{1}{3}$ of the length of the body as the ventral median fin. There are

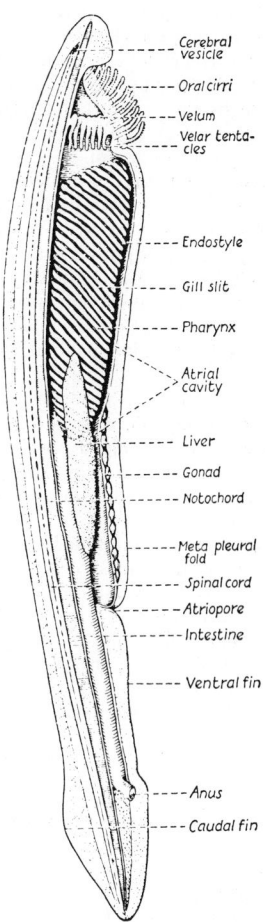

Fig. 1   Diagram of *Bronchiostoma* (Amphioxus) from the right side to show the structure.

they move the water in rotary fashion. At the posterior part of the mouth is a membranous velum to which are attached twelve velar tenacles, which are sensory and serve as a strainer to hold back coarser particles in the water. The mouth leads to the large, barrel-shaped pharynx. Gill slits are clefts in the lateral walls of the pharynx and they number from fifty to ninety pairs depending on the species. These slits open into the atrial cavity which surrounds the pharynx and other visceral organs except at the dorsal. In the floor of the pharynx is a ciliated groove, the hypobranchial groove with mucus secreting, glandular walls which constitute the endostyle. Food particles settle out in the hypobranchial groove and become entangled in the mucus and the mass is moved anteriorly by ciliary action. At the anterior end of the hypobranchial groove are two ciliated peribranchial grooves running obliquely dorsal up the sides of the pharynx to the hyperbranchial groove in the roof of the pharynx. The cilia here move the mass back to the intestine. The intestine is a straight tube leading to the anus. Digestion is carried on here, and there is a blind, finger-like diverticulum of the intestine, hepatic caecum, which extends anteriorly to lie on one side of the pharynx.

Respiration is completed as water passes through the gill clefts in the pharynx. There are branchial lamellae on the faces of the gill clefts and these are richly supplied with capillaries which absorb the oxygen from the water and give off carbon dioxide.

The circulation of the blood is carried on by a set of vessels, and there is no heart. The blood is forced along through the vessels by peristaltic contractions. Ventral to the pharynx is the ventral aorta which branches to afferent branchial arteries to the gills. Here the vessels branch into the capillaries of the branchial lamellae for aeration of the blood. These capillaries converge into the efferent branchial arteries which go from gills to dorsal aorta at dorsal side of pharynx. There are two branches of the dorsal aorta dorsal to the pharynx then at posterior end of the pharynx it converges into one and continues to the tail where it branches into caudal capillaries. These converge into the subintestinal vein which continues along ventral side of intestine to become hepatic portal in region of hepatic caecum, which it supplies and is collected in the hepatic vein which extends anteriorly as the ventral aorta with which we started.

This animal is dioecious and the mature individual possesses from 26 to 33 pairs of nodular gonads embedded in the body near the base of the metapleural folds. When germ cells mature, they break through the wall of the gonad into the atrial cavity and pass out through the atriopore with the water. Fertilization occurs in the water. Early summer is the breeding season and at that time the animals are quite active during the evenings and nights.

GEORGE E. POTTER

References

Weichert, C. K., "Anatomy of the Chordates," New York, McGraw-Hill, 1958.
Storer, T. I. and R. L. Usinger, "General Zoology," New York, McGraw-Hill, 1957.
Potter, G. E., "Textbook of Zoology," St. Louis, Mosby, 1947.

no clearly defined lateral fins, but there is a pair of metapleural folds, extending along the anterior two-thirds of the ventral surface of the body. On the ventral side just posterior to the metapleural folds is an opening, the atriopore, and beside the ventral margin at the left of the caudal fin is the anus. The segmental divisions or myotomes of the muscles are apparent on the body wall and the number of segments varies from fifty-eight to sixty-nine in different species. The myotomes on the two sides alternate with each other and adjacent ones are separated by myocomma or myosepta.

Internally one can see the notochord extending the length of the body as a rod of vacuolated cells which are filled with fluid to give it turgor. Immediately dorsal to this rod is the nerve cord which also runs the length of the body. The small canal inside the nerve cord is the neurocoele which extends the length of it, and is dilated at the anterior end to form the cerebral vesicle or rudimentary brain. There are dorsal sensory nerves going to the skin and alternating with these are ventral motor nerves going to the myotomes.

A current of water is carried into the mouth by ciliated bands on the inner surface of the oral hood, and

AMPHIPODA

An order of CRUSTACEA, characterized by 7 free pairs of thoracic appendages, an abdomen bearing 3 pairs of pleopods and 3 pairs of uropods, and the lack of a cara-pace. The body is laterally compressed and the append-ages are elongated, resulting in poor ambulation. In-stead, amphipods swim to and from positions among algae and crevices or remain in mud burrows and sef-constructed muddy parchment tubes. A few pelagic am-phipods are known to be carniverous but the benthic species feed on debris and detritus, in some cases using the setose mouthparts to sort out food particles, in others ingesting organic muds indiscriminately.

Amphipods are one of the most ubiquitous orders, ranging to 10,000 meters depth in hadal trenches and to 4000 meters altitude in moist environments of Indo-nesia. The largest suborder Gammaridea comprises 3200 species in 672 genera and 57 families. More than 230 species have evolved in Lake Baikal, another 400 species are in streams and other lakes and 90 species are terrestrial or "beachhoppers." The remainder are pro-fusely distributed in the sea. Algae literally swarm with herbivorus amphipods in size ranges 3 to 20 mm. Except for ostracods, the amphipods are the most abundant benthic crustacean on coastal shelves, rang-ing up to 4000 animals per square meter.

The first two pairs of legs are chelate or subchelate, better developed in males and useful for prehension primarily in copulatory amplexion. The numerous species of amphipods are monotonously similar in ap-pearance, unlike isopods, and classification is based on obscure and subtle criteria, largely mouthparts. Abys-sal and subterranean species lack eyes.

The suborder Hyperiidea is entirely pelagic, com-prising about 300 species. Hyperiideans differ from Gammarideans by the lack of a maxillipedal palp and pelagic adaptations such as oily bodies, enormous eyes and suspensory mechanisms of body and appendages. Many hyperiids suck tissues of medusae and salps and develop claws on the peraeopods for attachment.

The suborder Caprellidea with vestigial abdomen contains the "skeleton shrimps," with 200 species and the cyamid "whale-lice" with 30 species. Caprellids are abundant in epifaunal growths of hydroids. Their be-havior bears a remarkable resemblance to praying-mantises. Cyamids are epibionts in external orifices of cetaceans.

Amphipods are useful as food for many species of commercial and game fish, especially demersals. Ant-arctic seas teem with amphipods; fish, whale and seal stomachs often are gorged with them. One genus, *Chelura* is a minor wood borer in association with the isopod *Limnoria.*

Unlike most Crustacea, amphipods lack a larval stage. The eggs are laid commensurate with ecdysis, fertilized by the amplexing male and carried by the female under the thorax in a brood pouch formed by 4 pairs of soft plates. The young hatch in 9 to 30 days as miniature adults. Egg numbers range from 2 to 200 or more. Res-piratory and circulatory organs are thoracic, unlike Isopoda, where they are abdominal.

J. LAURENS BARNARD

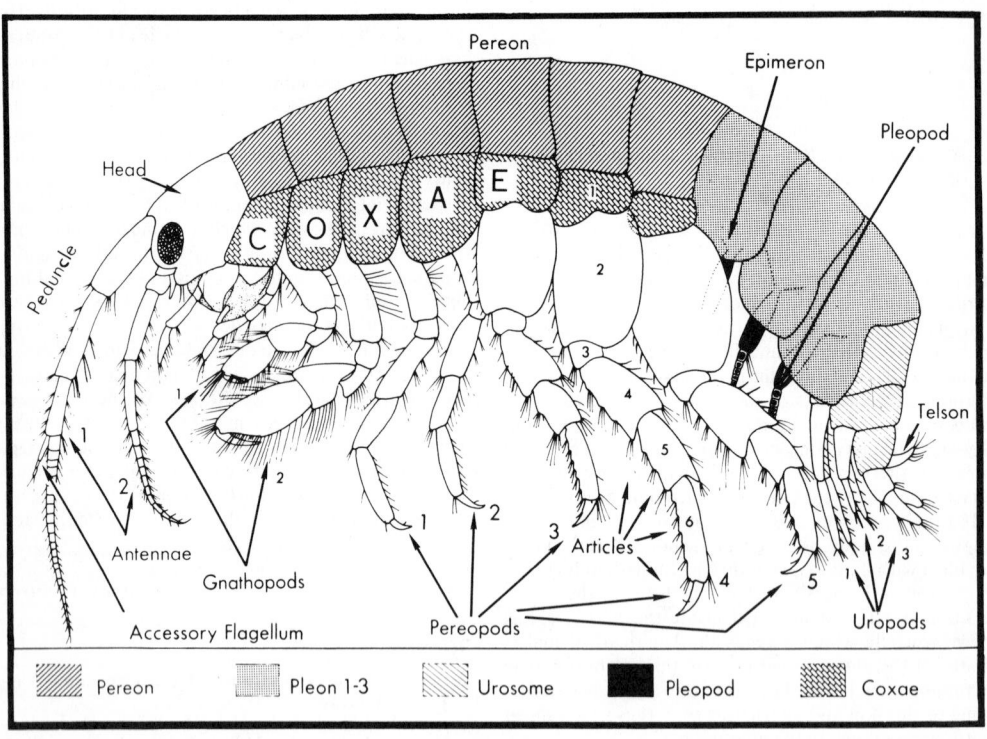

Fig. 1   Basic gammaridean.

References

Enequist, P., "Studies on the softbottom amphipods of the Skagerak," Zool. Bidr. Uppsala **28**, 1940.

Gurjanova, E., "Bokoplavy morei SSSR i sopredl'nyx vod (Amphipoda Gammaridea)," Opred po Faune SSSR, Izd. Zool. Inst. Akad. Nauk **41**, 1951.

Ortmann, A. E., "Crustacea. II. Malacostraca," *in* Brown, H. G. ed., "Ordungen und Klasse des Tierreichs," Vol. 5, Leipsig, Winter, 1901.

Stebbing, T. R. R., "Amphipoda I. Gammaridea," *in* Brown, H. G. ed. "Ordungen und Klasse des Tierreichs," Vol. 21, Leipsig, 1906.

AMPHISBAENIA

Members of a suborder of burrowing reptiles belonging to the order Squamata; all but one of its genera (*Bipes*) being completely limbless. Amphisbaenians may be differentiated from snakes and limbless lizards by their reduction of the right rather than the left lung, by the presence of a generally quite enlarged medial premaxillary tooth, by a heavily ossified and complexly reinforced skull, and a characteristic enlargement of the extracolumella. The presence of hemipenes, of a common head joint pattern, and of an egg tooth indicates that they belong with the snakes and lizards in the order Squamata.

The approximately 150 members of the suborder represent the only true burrowing species among the reptiles. They occur in Florida, Mexico, and the Antilles, in South America from Panama to Patagonia, in Africa from the Cape south to Somalia in the east and to Sierra Leone in the west as well as in North Africa, and in the Iberian Peninsula, and across the Arabian Peninsula from Persia to the Mediterranean and Asia Minor. Fossils are known back to the Eocene in North America, to the Pleistocene in South America, and to the Miocene in Africa and Europe. All fossils are clearly amphisbaenians and little is known of their affinities to other reptiles.

The various species feed on a variety of food animals which they crush and from which they bite pieces with their powerful interlocking dentition.

Amphisbaenians vary in length from 3 inches to a foot and a half and in girth from an eighth of an inch to more than one inch. They live in tunnels of their own making, whence their name: *Amphis*, both ways— *baenian*, to go. Progression is by lateral undulation and rectilinear locomotion and there are three major functional patterns for excavating the tunnel with the head and compressing the soil into its sides.

The eyes are highly reduced but light is perceived. The extracolumella is much enlarged and lies on the outside of the quadrato-mandibular joint whence cartilaginous and connective tissue connections attach it to the skin near the edges of the jaws. It is assumed that there connections serve as vibration receptors to permit the detection of prey within the soil.

Different species live at characteristic depths; those forms found closer to the surface are often recognizable by dense countershading or generally patterned pigmentation while deep ranging forms (1 to 2 meters) are often an unpigmented pink in life. Members of the pleurodont family Amphisbaenidae generally show

caudal autotomy with an intravertebral autotomy plane in one or two adjacent vertebrae, but no regeneration. Certain species have knobbed caudal tips that may acquire a coating of dirt, apparently deterring predators. The non-autotomizing caudal tip of the acrodont Trogonophidae facilitates their progression.

Within their range, the different species show a remarkable association with substrate rather than general environmental humidity. They burrow to levels where the soil is moist and can drink soil moisture by capillary movement between their lips. Different species vary widely in their dermal permeability; those found in humid areas show a relative water loss equivalant to that of amphibians, while species from xeric zones have a skin of very low permeability. In either environment, specimens are often found near river courses and in irrigated zones; they will move up to the surface during prolonged rains.

CARL GANS

ANAEROBIC BACTERIA

Anaerobic bacteria are those BACTERIA which, for growth, require a reduction or strict exclusion of the atmospheric oxygen from the environment. Since anaerobic bacteria commonly lack the enzyme catalase, the failure to grow in aerobic conditions may result from the inability to decompose toxic concentrations of the hydrogen peroxide produced in metabolic reactions. It would also appear that the enzymes of anaerobic bacteria require a reduced oxygen tension for optimum activity.

A wide variety of metabolic types exist among the anaerobic bacteria including groups which are extensively saccharolytic and others in which proteolytic activity predominates. In some species, both proteolytic and saccharolytic enzymes are produced and other species are relatively inert. The saccharolytic group includes species which ferment the complex polysaccharides starch, cellulose, or pectin as well as the simple carbohydrates. Some species, especially *Clostridium pasteurianum*, are active in the non-symbiotic fixation of atmospheric nitrogen. Certain bacteria, when grown anaerobically in the light are capable of carrying out a photosynthetic metabolism in which a variety of compounds (reduced sulfur compounds, alcohols, fatty acids, etc.) may serve as the oxidizable substrate which is dehydrogenated when accompanied by a simultaneous reduction of carbon dioxide. Some of these organisms will grow aerobically in the dark.

The anaerobic bacteria include both sporeforming and nonsporeforming types. In *Bergey's Manual of Determinative Bacteriology,* the most extensively used American taxonomic manual, the former are grouped as 93 species in the genus *Clostridium.* The nonsporeforming types, which have been less extensively studied, appear in several genera. Characteristics of taxonomic importance in the genus *Clostridium* include shape and position of spore, pigmentation of colonies, toxin production, ability to ferment cellulose and action on gelatin, coagulated albumin, and milk.

The genus *Clostridium* includes the etiological agents of botulism, tetanus, gas gangrene, and a variety of animal diseases including struck, lamb dysentery, pulpy kidney, and enterotoxemia. The pathogenicity of the clostridia depends largely on the production of potent toxins. Generally these are heat labile exotoxins and often several pharmacologically and immunologically different toxins may be produced by the same species. For example, one of the species, *Clostridium perfringens,* involved in food poisoning and gas gangrene in man and a variety of diseases in other animals, produces ten to twelve different toxins. Likewise, with the species producing the lethal botulinal toxin, six immunological types are known. In recent years, certain of the toxins have been purified to the point of crystalline purity and revealed to be high molecular weight proteins. The availability of the pure toxins has permitted more precise studies both of the chemical nature of the toxin and also of their properties as enzymes. Fortunately, from these pure or crude toxins, it is possible to produce efficient toxoids which have been used, particularly for tetanus and botulism, with notable success for prophylactic immunization.

In contrast to the above, some of the clostridia are of great benefit to man. The nitrogen fixing ability of certain of the species has been mentioned. Other species, as *Clostridium sporogenes,* are active in the decomposition of proteins and these no doubt are of great value in the cycle of nitrogen in nature. Others are important for their action on carbohydrate compounds. The mesophilic and thermophilic cellulose fermenting species are active in decomposition of plant remains. *Clostridium felsineum* and other species, by their action on pectin but not cellulose, are the agents of the anaerobic process of retting of flax and hemp. *Clostridium acetobutylicum* is an example of the species which are valuable due to the production of acetone and butyl alcohol by fermentation of the starch of cereal grains and other carbohydrates.

The anaerobic bacteria are widely distributed in nature. They may be isolated easily from rich soil, mud, poor quality raw milk, sewage, and the contents of the intestinal tract of man and other animals.

The first publication on the anaerobic bacteria appeared in 1861; the author was Louis Pasteur.

L. S. McCLUNG

References

Prévot, A. R., "Manual for the classification and determination of the anaerobic bacteria," Philadelphia, Lea and Febiger, 1966.
Smith, L. DS., "Introduction to the pathogenic anaerobes," Chicago, University Press, 1955.
McClung, L. S., "The anaerobic bacteria with special reference to the genus *Clostridium,*" Ann. Rev. Microbiol. **10**: 173–192, 1956.

ANGIOSPERMAE

The Angiospermae form a *class* of vascular plants which, together with the GYMNOSPERMAE, are usually grouped as the *division* SPERMATOPHYTA. The angiosperms (flowering plants) differ from gymnosperms in bearing their *ovules* (the potential seeds) within a hollow *ovary* (the ultimate FRUIT). The ovary forms part of the characteristic, relatively delicate and short-lived FLOWERS which may be borne singly or grouped into INFLORESCENCES. Within the flower, POLLEN grains (microspores) are produced in *anthers* which, with their supporting filaments form the *androecium.* The *gynoecium,* or pistil, consists of 1 or more separate or fused carpels which are usually differentiated into stigma, style and ovary.

Sporogenesis and gametophyte development are reduced further in the Angiospermae than in any other vascular plants. Nuclear division within the pollen grain produces a tube nucleus and a generative nucleus, the latter dividing again to produce 2 naked male gametes. Thus, the sexual generation from microspore mother cell to male gamete involves only 4 nuclear divisions. In the ovary, 1, 2 or 4 megaspores contribute to the development of a single embryo-sac (the female gametophyte). Most commonly the nucleus of 1 megaspore divides 3 times to produce an 8-nucleate embryo-sac. The subsequent fusion of 2 nuclei to produce the secondary nucleus reduces the embryo-sac to a 7-celled condition (without separating cellulose walls). One haploid nucleus, with associated cytoplasm, functions as the oosphere.

Pollination consists of the deposition of pollen grains on the receptive stigma through the agency of wind, water or animal vectors (to which the Angiospermae show particularly frequent and marked adaptations in the shape, coloration, odor and disposition of the flowers). The production of nectar in many flowers is directly related to visits by potential pollinators. Germination of the pollen grains produces a pollen-tube which penetrates the style to the ovary where the 2 male nuclei are released into the embryo-sac. Here double fertilization, which is unique to the Angiospermae, takes place, the fertilized oosphere producing an embryo and the fertilized secondary nucleus producing the nutritive tissue of the endosperm. The endosperm may be completely absorbed by the time the seeds are dispersed ("exalbuminous" seeds) or may be retained in part until germination ("albuminous" seeds). Food reserves in the seed largely consist of complex carbohydrates, proteins or fats, although in one case, *Simmondsia chinensis* (Link.) Schneid. (Buxaceae), waxes are stored and mobilized.

Self-incompatibility is widespread amongst angiosperms and is revealed by the failure of pollen from the same plant to germinate on the stigma or by the failure of the pollen-tube to grow successfully in the style and achieve fertilization. Outbreeding systems of this sort, depending upon the presence of the style, were not possible prior to the evolution of the angiospermous flower and may have played a significant part in the rise of this class of plants to its present dominant position. At the opposite extreme, various mechanisms are known promoting self-fertilization or even the formation of viable seed in the absence of fertilization of the oosphere (apomixis) as in *Taraxacum* (Dandelions) and *Poa* (Meadow Grasses).

Corresponding to the diversity of pollination systems is the great variety of dispersal mechanisms for fruits and seeds which may utilize the carrying power of water, wind and animals (by adhesion or passage through the gut). Seeds vary in size from the dust-like seed of the Orchidaceae, produced in vast numbers within each fruit, to the giant single seed in the Cocode-Mer (the fruit of *Lodoicea sechellarum* Labill., Palmaceae) which may weigh up to 9 lbs.

The Angiospermae contains annual, biennial and perennial taxa, the first two groups not being found at least amongst contemporary Gymnospermae. The perennials may be monocarpic (e.g. *Agave,* bamboos), growing vegetatively for a number of years and then dying after flowering, or polycarpic (flowering repeatedly). Vegetative reproduction by the detachment of usually somewhat specialized plant parts may be pronounced. Trees and shrubs are numerous but herbaceous plants are in the majority (and point another contrast with the Gymnospermae). The range of life-forms is immense, reaching from reduced, thallose types (as in the Podostemonaceae) to trees more than 300 feet in height (e.g. *Eucalyptus*). Ecologically, the Angiospermae range from desert-inhabiting xerophytes to submerged aquatics (although they have been distinctly less successful in colonizing marine rather than freshwater habitats). They dominate plant communities from the tropical forests to the arctic tundras with their numerical superiority threatened only in cool-temperate coniferous forest areas.

The diversity in size and morphology of the leaves of angiosperms is enormous. In most cases, a leaf base, a stalk-like petiole, and a flattened lamina may be distinguished. However, angiosperms characteristic of drought-ridden situations may show conspicuous reductions in leaf-surface or the development of succulence in the plant body. In some families, notably the Leguminosae, phyllodes (leaf-like petioles with no blades) may take the place of conventional foliage leaves. An ornamentation of hairs, spines or thorns may be found on the aerial parts of angiospermous plants.

In addition to the normal autotrophic way of life, partially and fully parasitic angiosperms are known, e.g. mistletoes and broomrapes *(Orobanche* spp.), as well as saprophytes (including a number of orchids). An extreme modification to saprophytic existence is shown by the entirely subterranean Australian orchids *Rhizanthella gardneri* R. S. Rogers and *Cryptanthemis slateri* Rupp. Insectivorous plants may utilize pitchers (e.g. *Nepenthes*), trigger-traps *(Dionaea* and *Utricularia*) or sticky leaves *(Drosera, Pinguicula,* etc.) in the capture of small insects, upon which proteolytic enzymes act subsequently. Many angiosperms (including all saprophytes) show an intimate association between their roots and soil-inhabiting fungi (mycorrhiza) which appears to be of nutritional assistance to them. Leguminous plants and some others develop root-nodules containing nitrogen fixing bacteria; a similar function may be performed by nodules developed in the leaves of tropical trees of the family Rubiaceae.

Traditionally, the classification of the Angiospermae has been based upon the morphology and disposition of the flowers, fruits and seeds. Avowedly "artificial" systems (such as the "sexual" system of LINNAEUS), and more nearly "natural" systems (such as those of Bentham and Hooker, Engler, Bessey, Hutchinson, Thorne, Takhtajan, and Cronquist), which come closer to a reflection of evolutionary relationships, are both based largely on floral characters. Nevertheless, in Hutchinson's system the Angiospermae are divided into predominantly herbaceous and woody lines (the Herbaceae and Lignosae, respectively). All of the more "natural" classifications recognize a division into two

series: DICOTYLEDONAE and MONOCOTYLEDONAE. Although the names for these series are founded on a difference in the numbers of embryonic leaves (cotyledons) in the emergent seedling, this apparently minor distinction is associated with a number of other morphological features and marks two very distinct groups of plants.

An increasing emphasis in taxonomic research is being put upon the utilization of information from other sources than macroscopic external morphology, for the structure and development of pollen grains and embryo-sacs, the anatomy of ROOT and SHOOT systems, and the analyses of chemical constituents are also providing information of value. Studies of the numbers and morphology of CHROMOSOMES and their behavior at MEIOSIS have been made with greater frequency on the Angiospermae than on other plant groups and have generally shown a good correlation with existing classifications while assisting sometimes in the elucidation of relationships. The range of known chromosome-numbers is from n = 2 in *Haplopappus gracilis* A. Gray and *Brachycome lineariloba* (DC.) Druce (Compositae) to n = 154 in *Morus nigra* L. (Moraceae). Limited numbers of serological studies have also been made but more notably successful amongst experimental techniques in clarifying relationships at or below the species level have been those which have involved comparative cultivations in uniform and varied environments and attempts at hybridization between suspected related taxa.

It has been estimated that over 200,000 species of angiosperms exist at the present day and roughly 80% of these appear to be Dicotyledons. Few of these species have any claim to be truly cosmopolitan, the nearest approach probably being *Phragmites communis* L. (Gramineae), the Common Rush. In general, there is a relatively sharp line of demarcation between tropical and extratropical taxa. Floristic comparisons between the angiosperms of the Old and New Worlds are at their greatest in northernmost regions and decrease southwards across the globe.

Anatomically, the STEMS of dicotyledons show basically different patterns from those of monocotyledons. In the former the vascular bundles are disposed in a circular arrangement; the stems of all woody and many herbaceous dicotyledons develop secondary vascular tissues from a continuous cambium (and, in addition, produce cork and phelloderm from a cork cambium in a more peripheral position). In monocotyledons, the vascular bundles are usually scattered throughout the ground tissue and there is a general absence of cambial activity (except in the peculiar type of secondary growth found in a few arborescent genera in the Liliaceae). The presence of open ended vessels in the XYLEM of angiosperms is characteristic except for certain members of the dicotyledonous order Ranales, and it has been suggested that the most likely common ancestor of monocotyledons and dicotyledons contained, in its wood, tracheids with scalariformly pitted ends, from which vessels have developed independently in the two groups.

The fossil record of seed plants is quite inadequate to demonstrate the phylogeny of the Angiospermae. Much of the material which is available consists of vegetative parts, particularly leaves, and almost none of the reproductive material appears to be more primitive than is found in some contemporary angiosperms. An origin from the Pteridospermae, seed-bearing plants with fern-like foliage, seems the most plausible suggestion, but origins in the Gymnospermae, Caytoniales and Bennettitales have also had their proponents. All must be regarded as highly speculative. Angiosperm fossils first become abundant in Middle Cretaceous deposits but, as these show a high stage of morphological specialization, it is likely that this class had a long history even then. Jurassic remains (particularly pollen grains) are apparently well substantiated and Axelrod has suggested that remains from the Late Triassic represent true Angiospermae, for he is inclined to place their origin in tropical conditions during Permo-Triassic time. If early angiosperm evolution should have taken place in upland regions, sufficiently well removed from the usual lowland basins of deposition, this could explain the paucity or absence of remains from the early days before the angiosperms rose to relative abundance.

In the framework of human economy, the Angiospermae occupy a position of the greatest importance, directly providing major sources of carbohydrates, proteins, fixed and essential oils, waxes, beverages, spices, drugs, rubber, textile and stuffing fibers, paper and lumber. Through the animals which graze upon them they also provide the greatest proportion of the protein in human diet.

H. G. BAKER

References

Willis, J. C., "A dictionary of the flowering plants and ferns," 7th ed., revised by H. K. Airy Shaw, Cambridge, The University Press, 1967.
Cronquist, A., "The evolution and classification of flowering plants," New York, Houghton Mifflin, 1968.
Baker, H. G., "Plants and civilization," Belmont, Wadsworth, 1965.

ANIMAL KINGDOM

No observer who considers the flora and fauna of any given region can avoid concluding that all seems disordered. There is no apparent correlation between the various plants which exist in any given region and the animals which populate it. This impression was that of almost every observer from ancient times to the middle ages. Aristotle alone, seeing in nature an expression of a real order, undertook a serious essay on animal systematics; but his efforts were not founded on research and, after him, there were long centuries during which the living world was considered as the result of independent and repetitive creations of a superior power.

It was not until the eighteenth century that Carl Linnaeus undertook a broad examination of the flora and fauna of the world by a rigorously analytical method. It is, however, a mistake to imagine that the work of this Swedish naturalist was no more than an effort at classification, an effort to produce a catalogue of the living things on our planet. His interests were quite different. He sustained the thesis that living things had been created by God according to a rigorous plan and it was to the disclosure of this plan that Linnaeus consecrated a life of uninterrupted work. Indeed, in his

pride, he went so far as to believe that his task, being a humble mortal, was confined to this relevation and to participation in the work of God.

However merited may have been this thesis, which was, after all, more closely allied to metaphysics than to science, one cannot deny to him the immense merit of having for the first time clearly enunciated the idea that living things were linked to each other and of having substituted the idea of an organized nature for that of a chaotic nature.

This idea was of extreme importance to the future of biological research. Without the systematics of Linnaeus (see TAXONOMY) champion though he was of the idea of "fixism," the theory of the evolutionists, invented by Lamarck and gloriously developed by Darwin and Wallace, would never have seen the light of day.

In fact, every research undertaken since the work of Linnaeus has shown, with daily increasing vigor, that GENETIC connections unite all species, be they animal or vegetable, which people the surface of the earth. The animal kingdom is inexplicable without the light of evolutionary ideas.

The fauna of today is no more than the residue of the defunct faunas of the past; contemporary species have left behind them immense cemeteries and are only the last links in a long chain of other, and now vanished, forms. It follows that no comprehension of nature can be obtained without a consideration of the evolution of animals through the ages. The study of the whole animal kingdom is dominated to great extent by this principle of chronology.

Transformations of living things are "historical" phenomena. The equilibrium of the moment carries in its train the concept of modifications to come. To ignore the perspective of time is to admit a lack of comprehension of life on the surface of our globe.

The overall view of evolution, taken chronologically, ignores episodic characters which are contemporary forms taken by themselves alone. The true progress of evolution continues, even though its actual cause may not be apparent, and even though hypotheses subsequently shown to be false may be brought to its support. The outstanding problem of transformation remains the origin of those great structural plans which have marked the essential stages in the development of living things.

No one knows if the first living things to appear on earth (see LIFE, ORIGIN OF and NEOBIOGENESES) were chemolithotrophic BACTERIA, or organisms with a perfect cellular structure. It is apparent that today life only appears on the surface of our globe as a structurally heterogenous substance and has never, for example, the intrinsic property of a determinate matter as luminescence is the property of a phosphore. Life only emerges when specific structures, made up of organic macromolecules of a specific structure, are arranged according to a specific architecture. ELECTRON MICROSPY, which permits great accuracy in the analysis of structure, has confirmed that the fundamental plan of all living things is the same, be they plants or animals, with the possibly transitory exception of the bacteria.

The first stage recognized by life is the *cellular stage.* Within the framework of the isolated CELL, EVOLUTION has shown itself to be extremely creative. The archtypes of unicellular animals, or PROTOZOA, are the Flagellates, the amebas, the Radiolaria, the Foraminifera, the Heliozoa, and the Ciliata. Certain protozoa which have become rigorously parasitic have given rise to two remarkable types: the Sporozoa and the Cnidosporidia.

The second great stage of evolution was the transformation of the unicellular organism to the multicellular state. In this latter stage, animals were able to increase in size since the number of their constituent units were no longer limited to one. The specialization which came to these groups of cells led to tissues, then to functional assemblages, or organs. Architectural variations, impossible to the unicellular state, were thus available to life and led to the colonization of environments forbidden to the unicellular. All in all, it was from this moment that there came the evolution of the animal kingdom.

From the Silurian epoch on, all the branches of the INVERTEBRATES and of the lower vertebrates (see CHORDATA) (cylostomes and fish) existed but they were restricted to water. The oldest amphibia and reptiles have been discovered in the upper Devonian and Pennsylvanian (upper carboniferous) deposits. Doubtful remnants of mammals have been discovered from the lower Jurassic; but this class stagnated more or less until the Eocene, an epoch in which there sprung forth a truly explosive evolution.

The remnants of the ancestors of invertebrates, lambent in precambrian deposits, have been subjected to the destructive action of metamorphic rocks.

Systematists and evolutionists, thus deprived of paleontological evidence, have fallen back as their sole source, on the findings of embryology and comparative anatomy to establish the classification of invertebrates and to trace "macroevolution" among them.

The systematic group here referred to as "Diploblastic" immediately above the Protozoa comprises three unquestionably different branches but possesses many anatomical and physiological characters in common. This group comprises the Sponges (see PORIFERA), the Cnidaria (see COELENTERATA), and the Ctenophores.

Until quite recently the Sponges were considered as those animals which provided the transition from Protozoa to the true multicellular forms. The possession of collared flagellated cells or *choanocytes* would seem to ally the sponges to the Choanoflagellata. The development of the Metazoa from the division of Choanoflagellates, imperfectly known as they are, is more than doubtful. The apparent similarity between choanoflagellates and the choanocytes is probably entirely fortuitous. Sponges have a reproductive cycle which is typically metazoan and not protozoan: for they produce eggs and typical spermatozoa. They are not aggregates of Protozoa and even less a disorganized cellular republic. Moreover, they have produced a marked cellular differentiation; foundation cells or *pinacocytes,* such contractile cells as *myoepithelium,* wandering cells (*amebocytes, archeocytes*), colored cells, sexual cells, scleroblastes which produce calcareous or silicious spines, spongioblasts which secrete fibers of spongin, and, finally, nervous cells. The existence of these latter, long denied, appears today certain. All this is far from the simplified concept which in the past was held of sponges. Moreover, these multicellular forms have demonstrated themselves incapable of building specific organs. They have, so to speak, remained in the stage of cell layers, variously folded and specialized.

The Coelenterates are animals of two cell layers; the external forming the structural foundation, the internal delimiting the digestive cavity. Between the two, there lies a jelly (MESOGLEA) without cellular structure. Differentiation in the Coelenterata is scarcely more developed than in the sponges. The nervous system, save in a few cases, is not aggregated into specialized centers; it is more often formed of localized networks on the interior surface of the superficial cellular layer. Nevertheless, some progress is seen in the differentiation of specific sensory organs and, in a few (Siphonophora), the rudiments of digestive glands and respiratory organs. Apart from this, the Coelenterata are characterized by specialized cells, the cnidoblasts, which are venomous mechanisms developed for the defense of the animal and for the capture of prey.

The CTENOPHORA would appear to be close relatives of the Coelenterata but they are separated from them by their fundamental bilateral symmetry and also by the presence of two more or less branched tentacles on which are localized peculiar cells or *colloblasts* which apparently serve to stick to the prey.

Evolution appears to have taken a decisive step forward when, in the course of development, there appeared a third germ layer, the mesoderm, which slipped between the ectoderm and the endoderm. New perspectives were thus available in organogenesis. At the same time that triploblastic animals developed, the anterior region gained the ascendancy and a head took form.

The third germ layer (see EMBRYOLOGY) did not behave in the same manner in all branches of animals; in some it remained diffuse, even though participating in organogenesis; this occurred in the Acoelomates. In others, it formed hollow, paired vesicles which often were repeated symmetrically from one part to another in a sagittal plane. These vesicles together formed the coelom and the animals which bore them are thus called the Coelomata.

The Acoelomata comprise two great branches, that of the PLATHELMINTHES in which the third germ layer forms an abundant intervisceral parenchyma; they have a digestive apparatus which lacks an anus but which shows clearly differentiated regions, and an excretory apparatus composed of protonephridia (little excretory ampules with a flame cell and an excretory tubule). The Platyhelminthes are divided into the classes Turbellaria, Temnocephla, Monogenia, Cestoda, and Trematodes. With the exception of the first, all lead a parasitic life and, for this reason, show an organic simplification correlated with their adaptation to a specific method of existence.

In the NEMATA (including the ROTIFERA) the mesoderm plays a less important role; it produces a parietal musculature and a parenchyma reduced to a few cells. Between the digestive tube and the body-wall, there is a vast cavity filled with a liquid in which float mobile cells. The digestive tube is open at both its ends, the excretory apparatus is represented by protonephridia. The NEMERTEA, covered with cilia, have a doubtful systematic position for there is an argument as to whether or not they possess a coelom. Modern biologists tend to the view that they lack one.

The Coelomata comprise a vast assemblage of animals, of which the blastopore (larval mouth) generally coincides with the adult mouth and frequently gives, by a median stricture, both mouth and anus; they are for this reason called Protostomians. Apart from this, their nervous system develops as a chain of ganglia situated under the digestive tube and as a completely anterior center (brain) situated above the digestive tube. This character has gained for them the name *Hyponeuria*. Coelomic sacs, nervous ganglia and excretory organs (metanephridia connecting with the coelom) are replicated along the length of their longitudinal axis. Partitions have developed between successive pairs of coelomic sacs and the body of the animal is thus segmented (or metameric).

The first branch of the Protostomia is the ANNELIDA comprising the Polychaeta, the Oligochaeta, and the Hirundinea. The SIPUNCULOIDEA and ECHIURIDA are branches of lesser importance allied to them. The branch of the MOLLUSCA is allied to the Polychaeta, as is attested by the similarity of the first embryonic stages and by the presence, in both groups, of a pelagic larval form, the trochophore (see LARVA, INVERTEBRATE). But the coelom of the Mollusca has a different origin; metameric segmentation, if it ever existed at all, has left only weak traces. Even in Neopilina, an extremely archaic form discovered a few years ago, the metamerism which is observed in the gills, the nephridia, the foot muscles and the coelom is probably not of the same type which is found in the Annelids. At least, this is the opinion of leading zoologists. The growth of Mollusca takes place in a manner quite different from that of Annelids. The development of organs along a dorso-ventral axis, perpendicular to the longitudinal axis, has profoundly altered the anatomy of Mollusca.

The ARTHROPOD branch, richer in species than any other, belongs among the hyponeural Protostomia. The resemblances and homologies with the Annelids are undeniable; and demonstrate an ancient, but nonetheless real, common origin of the two branches; the general structure of the body, the metamerism, the composition of segments, and the like, are signs of a very similar parentage.

The outstanding anatomical features of the Arthropods are the following; first, the fusion of specialized segments forming distinct regions (head, thorax, abdomen in the case of Insects); second, the development of an exoskeleton (see SKELETON, INVERTEBRATE) composed of hard plates of a glycopolysaccharide (chitin), beneath which there has developed a rich musculature with striped fibers; third, the acquisition of a pair of jointed appendages in each segment (locomotor, peribuccal, or sensorial according to the region to which they belong); fourth, the embryonic coelomic vesicles lose their individuality and produce a diffuse mesoderm from which is derived the muscles, the connective tissue, and the like.

Three major evolutionary lines have developed in the Arthropods, each corresponding to a sub-branch; first, the extinct Trilobites, with simple antennae, followed by biramous appendages, all of the same type; second, the Chelicerata, comprising the sections, Scorpions, Pseudoscorpions, Arachnida, etc. which, at the level of the mouth, carry a pair of uniseriate appendages (chelicera) unique among Arthropods; third, the Mandibulata or Antenulata, possessing a pair of uniseriate antennae in front of the mouth and a pair of biting appendages, the mandibles (of biramous origin). The CRUSTACEA, the MYRIAPODS, and the INSECTS are large classes in this third sub-branch.

The ONYCHOPHORA (or Peripatus), outstanding archaic survivals, demonstrate in themselves both an-

nelid and arthropodan characters, bearing witness to the relation of annelids and arthropods. They are most certainly not the ancestors of Arthropods but they give a good idea of what these might have been. The Tardigrada and the Pentastomida also belong at the base of the Arthropod tree.

To conclude the Protostomia, it is necessary to discuss briefly a probably heterogenous assemblage— the *Lophophorians*—which may be regarded either as a super-branch or as a branch. This group comprises the Endoprocta, the Phoronidia, the Ectoprocta, and the BRACHIOPODA; all are sessile, lack clearly defined head, and feed on small particles. A crown of tentacles, covered with cilia, the lophophore, rises round the mouth and carries the food particles towards the buccal orifice. The digestive tube is U-shaped so that the anus flanks the mouth. The coelom is not clearly segmented; in Phoronidea alone it occurs in three parts of which the front, or *prosome* is much reduced. A few anatomical characters, more in the larva than in the adult, suggest a relationship of the Lophophorians with the Annelids.

In many ways contrasted with the Protostomia are the Deuterostomia, in which the blastopore of the gastrula becomes the anus or marks the position of this opening. Two profoundly different groups compose the Deuterostomia; one consists of forms with a trimeric coelom (ECHINODERMATA, Stomocorda, and POGONOPHORA), while in the other the coelom is more or less metameric (Tunicates, Cephalochordates and Vertebrates).

The nervous system of each is very different. In the first group it remains more or less diffuse and does not fuse with the epithelium of the integument from which fact there derives the name neurepithelial Deuterostomia; in the second group the nervous system forms completely isolated centers, all situated above the digestive tube. They are thus called epinural Deuterostomia.

The Echinoderms, all sessile in origin show a pentaradial symmetry with a repetition of homologous organs (antimeres) round the apicobasal axis. There is no cephalization at all. In contrast, the coelom has evolved a peculiar method of development, linked with that of cavities appearing in the substance of the ectomesenchyme.

The Stomotochorda possess a very marked trimerism. The pharynx communicates with the exterior by lateral holes, which has given rise to the term pharyngotremata, and sends forward an unbranched diverticulum, the stomotocord, which does not appear to be homologous with the notochord of the Chordates. The Stomotochordates comprise the two classes, Enteropneusta (Balanoglossus) and the Pterobrachia (colonial, resembling Bryozoa).

It is possible that the Pogonophora, discovered quite recently, belong in the vicinity of the Stomotochorda. These animals, which resemble extremely thin, long, worms, inhibit the depths of the sea, living in their tubes. The trimerism is very marked. They do not possess a digestive tube and their anterior extremity carries one or more ciliated tentacles. Here also belong the little group of Chaetognatha which, by their embryology, are Deuterostomia but which differ greatly from all other Deuterostomia. The coelom is not trimeric in origin; the nervous system, with a double perioesophial collar, is in itself unique. As a matter of fact, no one knows where to place these planktonic glass-clear animals in the evolutionary scale.

The epineurian Deuterostomia had an extraordinary stroke of evolutionary good luck in their third branch, that of the vertebrates. All possess, at least during the larval or embryonic period, an elastic rod situated above the digestive tube. This is the dorsal cord, or notochord, in consequence of which these Deuterostomians are often called Chordates. All, at least during embryonic life, have pharynxes.

The Tunicates, without exception marine, have some forms which lead a planktonic life and others which are fixed to the substrate. The primitive organizational plan shows a dorsal nervous system and a notochord, but the digestive tube only remains in larvae and in adult Appendicularia. At no time do the Tunicates have a metameric coelom.

The Cephalochordata, of which AMPHIOXUS is the classic example, show the chordate plan to perfection, but their excretory apparatus, composed of protonephridia, is of a unique type, of which the origin remains inexplicable. Amphioxus shows the essential characters of the archtype of a vertebrate, even though it cannot be regarded as one but is rather a form arrested in its evolution by its peculiar mode of life with its body planted vertically in the sand and with fine particle feeding.

The Vertebrates are distinguished from Amphioxus by the brain, by the vertebral column molded on the notochord, and by the fusion of the pharynx with the walls of the body.

The evolution of Vertebrates is above all distinguished by the progressive development of the nervous system which went hand in hand with a more or less marked cephalization, involving the formation of, first a cartilaginous, and then a bony capsule (the skull) formed of parts developed from different sources. At the beginning, the brachial skeleton played no part in the constitution of the skull; this condition implies the absence of a lower jaw characteristic of the Agnathostomes or Cyclostomes, both living and fossil. In all other Vertebrates the branchial skeleton is incorporated with the skull, at least in its first two arches; the first adapted to new functions and became the lower jaw.

The evolution of the lower classes, cyclostomes and fish, took place entirely in an aquatic environment; the transition to a terrestrial environment became possible thanks to the coexistence of an aerial respiratory apparatus (lung), which developed from the swim bladder, and the pharyngeal gills, and to the substitution of the second by the first. With the conquest of terrestrial and aerial environments, the progress of the brain went rapidly. The existence of gill slits remained only in the embryo. The caudal region tended to lose its importance. The paired fins were replaced by members (limbs), of which the structural plan has scarcely varied from Amphibian to Man. All vertebrates since the Amphibia are four-legged. The branchial skeleton lost its respiratory function, partially disappeared and took part to a greater or lesser extent in the building up of the splanchnocranium; the skull, properly so-called, or neurocranium, drew to itself the first vertebrae and the first spinal nerves (occipital region). A new kidney, the metanephros, took the place of the primitive nephridia (coelomoducts).

The brain evolved in two ways; first, it increased in volume, primarily through an increase in the number of neurones; secondly, its parts became specialized and the neurones became divided into categories. From the Amphibia, which gave rise to them, there diverged two great evolutionary lines; the Sauropsidea and the Theropsidea. The first comprises the sauropsidian Reptiles and the Birds, the other the theropsidian Reptiles and the Mammals.

In both lines the modification of the lower jaw, the final development of the ear, and the progress of the brain went hand in hand; but in the Theropsidea there were not only more rapid transformations but they were more varied and allowed an advantage to forms with a relatively unspecialized mode of life. Independence in regard to the external environment came about through the maintenance of a constant internal environment (homeostasis) with the aid of regulating systems (constant temperature, constant chemical composition) which involved the cooperation of the nervous system and of endocrine glands.

The Birds, whose brains never reached more than a relatively low level of development, became morphologically and physiologically highly specialized but their behavior has never broken free of the bonds of automatism.

Mammals very early manifested a tendency to that diversification which has made it possible for them to penetrate the most varied environments, even returning, thanks to the acquisition of secondary adaptations, to an aquatic environment without abandoning pulmonary respiration. The evolutionary power of the Mammals derives from the fact that several stumps from its ancestral trees preserved a generalized structure unmarked by adaptations. The insectivore stump gave rise almost imperceptibly to the evolution of the Primates, of which some, escaping sterilizing specialization principally by an arboreal life, have freely developed their brain and have ended with the genesis of Man, last fruit of evolutionary creation.

In the course of this development, probably balanced among several twigs from the same branch but never clearly united in a single one, evolution has attained a new level just as important as the passage from unicellular to multicellular in producing a new organ, to wit a brain endowed with a neopallium (cortex) which gives to its possessor conceptual thought, the notions of morality and of liberty.

This material evolution has been paralleled by a social evolution, of which it is not possible to overestimate the importance. It is thanks to the strict twinning of the two evolutions that *Homo sapiens* has been able to appear.

PIERRE-P. GRASSÉ
(trans. from French)

References

Harmer, S. F. and A. E. Shipley eds., "The Cambridge Natural History," 10 vols. London, Macmillan, 1900–1906 [photoreprint, New York, Hafner, 1959].

Grasse, P-P. ed., "Traité de Zoologie," 17 vols. Paris, Masson, 1948–.

Bronn, H. G. ed., "Klassen und Ordnungen des Thierreichs," Leipzig, Winter, 1886–.

Krumbach, T. and W. Kükenthal eds., "Handbuch der Zoologie," Liepsig, de Gruyter, 1923–.

ANNELIDA

The phylum Annelida (sometimes Annulata) refers to a group of segmented wormlike, coelomate animals with bilateral SYMMETRY. It includes the orders POLYCHAETA (bristle worms), OLIGOCHAETA (earth and fresh-water worms) and HIRUDINEA (leeches). The SIPUNCULIDA (peanut worms), ECHIURIDA (tongue worms) and PRIA-PULIDA are other annelidlike marine worms which are more correctly assigned to separate phyla. Altogether they are sometimes called the *Vermes polymera* (with many segments) to distinguish them from the *Vermes oligomera* (with few segments) which include the tape-, flat-, thread- and ribbon-worms. Collectively, the Polychaeta and Oligochaeta have been called *Chateopoda*, since their representatives are character-ized by having segmentally arranged, lateral outgrowths of setal tufts or bristles, while the Oligochaeta and Hirudinea together have been called *Clitellata*, in reference to the girdle of clitellum, a cocoon-building organ in the anterior region of the body. The oligo-chaeta and Hirudinea are largely inhabitants of freshwater or moist earth, whereas the others are mainly marine.

Fundamentally, the annelid construction is that of an outer tube or body wall, separated from an inner tube mentary canal by a coelomic space. The mouth is at the anterior, and the anus at the posterior, end of the ali-mentary tract. A dorsal head or prostomium precedes the mouth and represents the episphere of the larval trochophore (fig. 17). It is followed by few to many rings or segments, similar to one another (*homonomous*) as in Oligochaeta, which may be secondarily divided (Hirudinea), or differentiated (heteronomous) to vary-ing degrees, as in many Polycheata, in which an an-terior thoracic, and a posterior abdominal, region may be identified. The anal pore, at or near the end of the body, follows a pre-anal growth zone in which the num-ber of segments is increased by proliferation.

The outer surface or epithelium is a thin to thick layer of cells; this may be overlain by secreted cuticle or other modifications, and variously adorned with papil-lae, spines, modified CILIA, or other structures function-ing in protection, food gathering, respiration or other ways. There is no skeleton. The setae or bristles are secretions of follicle cells within parapodial bases; they vary in numbers and kind according to species, and are replaced as required. In the Polychaeta they have their highest development and are the most diversified.

The epithelium is underlain by muscles consisting of circular, longitudinal and oblique series, and an inner peritoneum. They are accompanied by pigment bodies, connective tissue, and capillary vessels. The body is elongated by contraction of the circular muscles, and shortened by contraction of the longitudinal muscles; the oblique muscles give additional plasticity to move-ments.

The coelom develops as a split in the larval mesoderm. This comes to be longitudinally divided by transverse septa marking the characteristic segmental pattern, or the septa may be reduced or lacking. The coelom contains a watery or a colored fluid which functions not only to provide turgor pressure, giving rigidity to the soft body, but it may contain bodies which function in other ways.

The nervous system consists of a dorsal brain, cir-cumoesophageal connectives and a ventral nerve cord which may be longitudinally divided. In some Poly-chaeta it is accompanied by a giant axon system which is effective in greatly increasing the rate of conduction velocities, accounting for abrupt responses.

Polychaeta. The largest order, the Polychaeta (mean-ing many setae) are characterized for having segment-ally arranged, paired lateral parapodia provided with setae. They are to be distinguished from the Oli-gochaeta, which have few setae and no parapodia, and the Achaeta, without setae and parapodia. This order comprises 67 families; 1366 genera have been named of which 720 may be valid, and 10812 species of which 5341 may be valid (Hartman, 1959, p. 5). More genera and species are added as oceanographic investigations are pursued. Its members are widely distributed over the world in marine habitats, at all depths, but chiefly in less than 200 fathoms; fewer are in brackish to fresh-water and very few occur in moist earth. Certain groups such as eunicids and amphinomids, predominate in tropical waters; others such as bamboo worms and ampharetids are most numerous in colder sea bottoms. Some are pelagic (tomopterids and alciopids) but many others are pelagic in larval stages.

The general form (figs. 1–14) varies greatly, de-pending on whether habits are errantiate, sedentary or pelagic. Diversity in shape and habit are illustrated in some common group names as sea mouse, shield worm, proboscis worm, fringe worm, lug worm, bamboo worm, gold crown, feather duster. In size they vary from a fraction of a millimeter (sphaerodorids) to more than two meters (some enuicids). Color and pattern, re-sulting from both the presence of pigment and refrac-tion of light, vary from brilliant, to dull, to none. Lu-minescence may heighten the display of both, and may be the result of modified cells or associated organisms.

The outer surface of the body is unarmed, or pro-vided with cilia, stiff hairs, sensory papillae, scutes or other differentiated parts. Parapodia are fleshy, lateral outgrowths of the body wall, ornately developed or re-duced to low ridges. Fascicles of extensile, chitinized bristles or setae, and supporting rods or acicula, emerge from parapodia in upper or notopodial, and lower or neuropodial branches, or one or both may be lacking from few to many segments. Associated with these are other structures named for structure or function, as cirri, branchiae, tentacles or elytra.

Setae originate as secretions of epidermally invagina-ted cells formed in parapodia. They provide a re-markably accurate means of specific to family differen-tiation. Each is fully formed deep in the parapodium, from a cell enclosing a nucleus, a ciliary apparatus and a lacuna in which the seta develops. Composite setae are formed from two or more processes juxta-posed on one another and ornamentations are the re-sult of modifications of simple setae. The diversity of their structure is exemplified in their common names, as setae, paleae, lyre, uncini, platelets and brushes.

Branchiae or gills, when present, occur as epithelial outgrowths and are penetrated by a vascular loop. They may occur as simple filaments or tufts, or be much divided or spiralled. They may be stationary or evers-ible, and are frequently associated with parapodia. The tentacular crown of sabellids and serpulids functions both for respiration and food gathering. In ampharetids

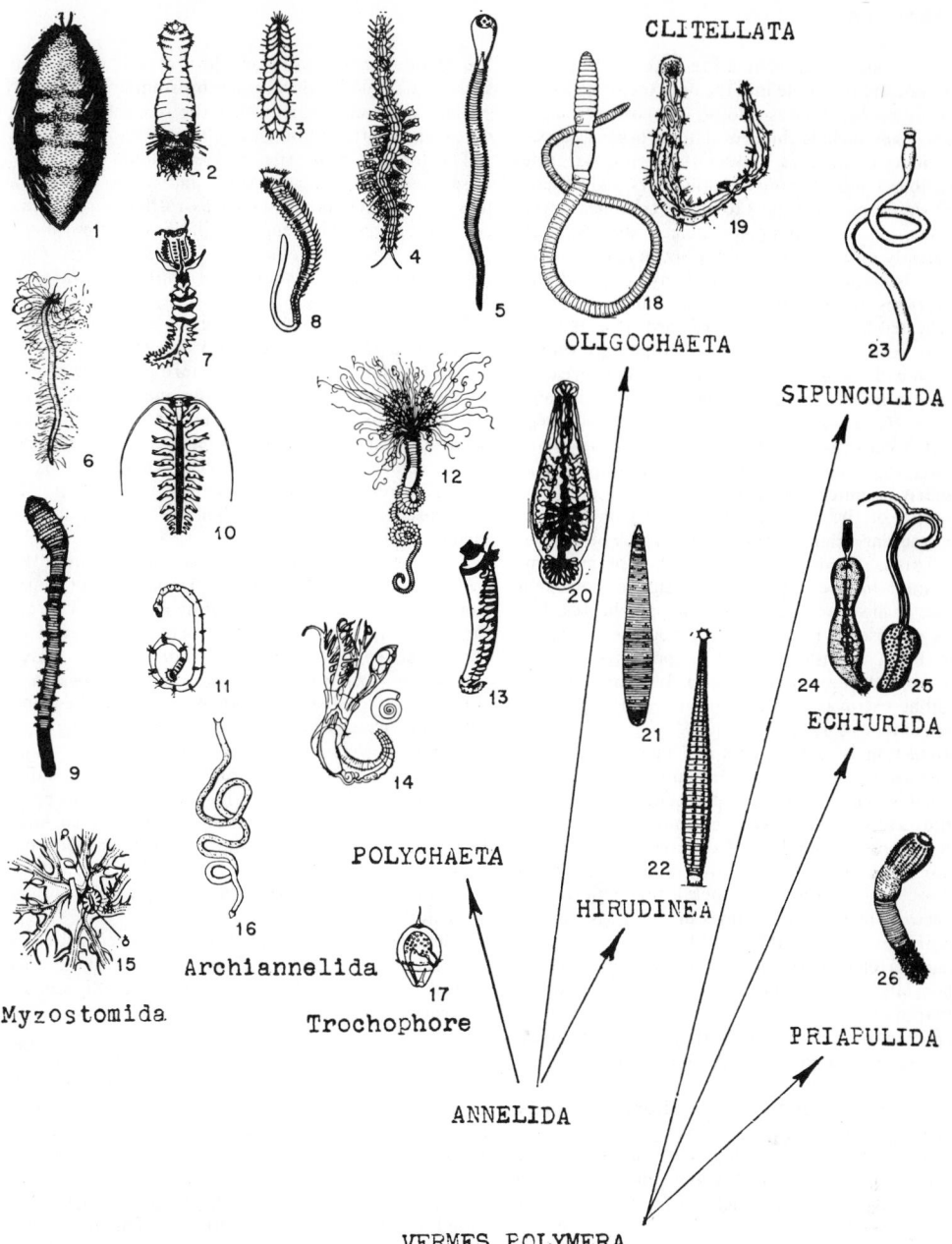

Figs. 1–17 POLYCHAETA. (Modified from Cambridge Natural History, 1896. Vol. II).

1. *Aphrodita*, sea mouse
2. *Sternaspis*, shield worm
3. *Lepidonotus*, scale worm
4. *Pionosyllis*, with attached embryos
5. *Glycera*, probiscis worm
6. *Cirratulus*, fringe worm
7. *Chaetopterus*, winged worm
8. *Sabellaria*, gold crown
9. *Arenicola*, lug worm
10. *Tomopteris*, glass worm
11. *Axiothella*, bamboo worm
12. *Terebella*, crested worm
13. *Pectinaria*, gold crown
14. *Spirorbis*, feather duster
15. *Myzostoma*, comatulid parasite

16. *Polygordius*, archiannelid worm
17. trochophore larva
18–19. OLIGOCHAETA
18. *Libyodrilus*, an earthworm with clitellum
19. *Aeolosoma*, an aquatic oligochaete, in fission
20–22. HIRUDINEA
20. *Glossiphonia*, a freshwater leech
21. *Hirudo*, medicinal leech
22. *Pontobdella*, a marine leech
23. SIPUNCULIDA
24–25. ECHIURIDA
24. *Echiurus*, a tongue worm
25. *Bonellia*, a tongue worm
26. PRIAPULIDA

and flabelligerids the branchiae are outgrowths of the peristomium and retractile into the oral aperature.

Sense organs function as photo-, chemo- and tango-receptors. They include the eyes, lateral organs, dorsal ciliated ridges or mounds, statocysts or otocysts, taste buds, various kinds of papillae, stiff hairs and others. Eyes are simple pigment spots to highly complex, lenticular organs, and best developed in pelagic species. They are frequently on the prostomium, but may occur also on the peristomium, sides of the body or pygidium. Lateral organs are usually segmentally arranged; they are innervated by the ventral nerve cord. Dorsal sense organs are derived from epidermal cells and are simple ciliated mounds or grooves; they attain their greatest development in errantiate polychaetes, especially amphinomids. Tango-receptors may be diffuse or segmental; they occur as stiff hairs or modified cirri (nephtyids). Statocysts function to maintain equilibrium and occur widely in some Sedentaria.

The color of the blood is clear, green or red, the change resulting from increased intensities of chlorocruorin. This is a red-green respiratory protein chemically similar to haemoglobin. It occurs dissolved in the blood, or in separate corpuscles in polychaetes. The presence or absence may have no genetic relationship. A heart body, consisting of loose, spongy, intravasal tissue is present in many Sedentaria. It if formed from an infolding extravasal epithelium and participates in excretion of food storage.

Reproduction is very diverse and most polychaetes are dioecious. Viviparity with hermaphoroditism (some syllids and serpulids) or with parthenogenesis (some cirratulids and syllids) are known. Epitoky, resulting in anatomical and morphological differentiation at maturity for a pelagic swarming, are common in nereids and syllids. Stolonization, in which a tail region gives rise to new individuals, sometimes in chains, is characteristic of some syllids. Ablation of the proventriculus may cause a change in the pattern of reproduction. (Durchon, 1952). Some hermaphoroditic populations show marked sexual variability, producing unbalanced sexual populations (Bacci, 1955); they are explained on a hypothesis of determination through multiple sexual genes.

Development frequently proceeds from a fertilized ovum, through spiral cleavage to a pelagic ciliated larva called a trochophore. In development, the entire body originates from a single cell. Metamorphosis results in the replacement of larval structures for those characteristic of the adult.

Oligochaeta. These are worms with segmentation indicated externally and internally. Setae are present in segmental series, without parapodia. Each individual is hermaphroditic, with both male and female gonads, few in number, situated in an anterior part of the body. Genital products are discharged through special ducts. A clitellum (girdle) is present in higher groups. Eggs are deposited in a cocoon where development proceeds; there is no larval stage. Most species are terrestrial; many inhabit freshwater and a few are marine. Some of the lumbricids are exceptionally peregrine, having adapted themselves to world-wide areas at the expense of endemic species. In size, oligochaetes vary from a millimeter long (*Chaetogaster*) to more than 2000 mm long (tropical earthworms). Number of segments varies from as few as 7 (*Aelosoma*) to

500–600 (large Australian earthworms). The body consists of a small, inconspicuous prostomium and many similar segments, from which the few setae occur in dorsal and ventrical fascicles.

The food of most species is vegetable or detrital; lesser numbers are carnivores, and a few are predators, on smaller animals such as rotifers, crustaceans and protozoans. The length of life has been estimated for some; *Eisenia foetida* may live 3–4 years, *Lumbricus terrestris* 5–6 years, and *Allolobophora longa* 5–10 years.

The classification of Oligochaeta is based on the kinds of reproduction, whether only sexual, or also by fission, the latter including the more primitive groups. Further, the position of the male and female gonads, the position of the male pore and funnel, and the relations of spermathecal pores in female, as related to the male copulatory pore, have been found the most reliable. A reliable source (Stephenson, 1930, p. 721) recognized 14 families with about 2400 species and 213 genera.

The largest group the Megascolecina, has more than 1390 species; they are terrestrial, small to very large, limited largely to the southern hemisphere. The next largest, the Lumbricina, with more than 350 species, are terrestrial, small to large earthworms. They include the common *Lumbricus, Eisenia* and *Allopophora* of the western world, and the glossoscolecids of the tropics and South America. The monilogastrids are terrestrial and best known in southeastern Asia; they are unique for having a multiple gizzard. Tubificids are small to moderately large, aquatic or intertidal, and widespread. The more primitive enchytraeids are small to large, terrestrial or equatic. The most primitive naids reproduce by fission. The branchiobellids, known for 9 genera and about 28 species are small, to 12 mm long, parasites of crayfishes.

Comprehensive accounts are to be consulted in Michaelsen (1928, Handbuch der Zoologie, Bd. 2, Lief. 2, Teil (8)) and Stephenson (1930, The Oligochaeta. Clarendon Press, Oxford, 978 pp., with extensive bibliography).

Hirudinea. Leeches differ from other annelids in that segments are lacking or secondarily divided, and parapodia with setae are absent. Suckers are present at posterior and anterior ends. The coelom is greatly reduced into a system of sinuses and lacunae. In size they vary from a few mm to several inches long. They are predatory or parasitic animals, provided with terminal suckers which serve for attachment, locomotion and feeding. Leeches occur on land, in fresh- and sea-water, as parasites of animals, or as carnivores or scavengers. The food is the blood or body fluids of animals.

The body typically consists of 34 segments, each divided into 2 to many rings. The body is differentiated into a head region; the first 10 segments are in front of the clitellum which includes segments 11 and 12. The middle region comprises segments 13 to 24, and the anal region segments 25 to 27. The caudal sucker is the most compact; it is deeply cupped, and includes segments 28 to 34. The oral sucker surrounds the mouth and usually has the main eyes. Sensory organs include eyes, papillae and tubercles.

The digestive tract is divided into a buccal chamber, pharynx, esophagus, crop, intestine and rectum. In

Hirudo the buccal chamber has 3 muscular jaws with teeth, and the salivary gland contains a non-coagulant, hirudin, to prevent clotting of blood.

Reproduction is hermaphroditic; gonopores are median, on the ventral face of the clitellum.

The order is recognized for 4 suborders, 7 families and 290 species (Scriban and Autrum, 1934, Handbuch Zool., vol. 2, Lief. 7 (8), p. 334). The largest group, the rhynchobellid, with more than 150 species, includes freshwater parasites. The Ichthyobdellids are parasitic on fishes. The gnathobdellids are terrestrial and amphibious; they include the hirudinids, most abundant in North America and Europe.

OLGA HARTMAN

References

Bacci, G., "La variabilita dei genotipi sessuali negli animali ermafroditi," Pubbl. Staz. Zool. Napoli, **26**: 110–137, 1955.
Durchon, M., "Rêcherches experimentales sur deux aspects de la reproduction chez les Annélides polychètes: l'épitoquie et la stolonisation," Ann. Sci. Nat. Zool. Paris, sér. 11, **14**: 117–206, 1952.
Hartman, O., "Catalogue of the Polychaetous annelids of the world, Los Angelos, Allan Hancock Found. Pub. Occas. Pap. no. 23, 1959.
Michaelsen, W., "Dritte Klasse der Vermes Polymera (Annelida) Clitellata-Gürtelwürmer," *In* Kukentham, W. and T. Krumbach, Eds., "Handbuch der Zoologie," **2**: 1–112, Berlin, de Gruyter, 1928.
Scriban, I. and H. Autrum, "Hirudinea," *In* Kukenthal, W. and T. Krumbach, Eds., **2**: 119–352. Berlin, de Gruyter, 1928.
Stephenson, J., "The Oligochaeta," Oxford, Clarendon Press, 1930.

ANSERIFORMES

The present order consists of three families: the Anatidae (Swans, Geese and Ducks) with 40 genera and 144 species; the Phoenicopteridae (Flamingoes) with 3 genera and 4 species; the Anhimidae (Screamers) with 2 genera and 3 species. They are all aquatic birds. Their skull is desmognathous and their bill holorhinous; their plumage is thick, waterproof, but devoid of pterylies, their bill is strong, fitted with lamellae on the sides except in the Screamers. Their anatomy is peculiar, but it indicates relationship with the Storks (Ciconiiformes) on one side, with the South American Cracidae (Galliformes) on the other; the Flamingoes approach the former, the Screamers the latter. The sexes are alike, or different, and in the last case, there can be an eclipse plumage in the males after the breeding season. The nest is often built on the ground or over water, in grass or reeds, under a bush, or in holes of trunks of cliff, between rocks, sometimes in large tree forks, or even in old nests of other birds. The eggs, 2 to 15, are plain colored. Chicks precocious and covered with thick down, capable of swimming soon after birth. Monogamous, the males help with the brood in numerous cases, but seldom incubate. Displays are elaborate as a rule.

Anseriformes are found in all parts of the world except Antarctica. Two species of ducks, the Mallard and the Muscovy (*Anas platyrhynchos* and *Cairina moschata*), and two of geese, the Greylag and Swan Goose (*Anser anser* and *Anser cygnoides*) have long been domesticated.

The three families of Anseriformes are easily distinguished:

The *Flamingoes* (Phoenicopteriformes) have very long legs, a thick bill curved at right angle, a very long neck and much white and pink in the plumage. They inhabit shallow salt lagoons on the sea shore and inland, sometimes at high altitudes (in the Andes and Central Africa). They are found in Europe, Western Asia, Africa, North and South America. They migrate a great deal and nest in large colonies on low banks and islets, building truncated cones of mud. They feed mostly on algae and small animal life.

The *Screamers* (Anhimidae), have a short, curved bill and large, unwebbed feet. Their plumage is mostly greyblack. The head is crested or adorned with a thin horn. They resemble geese in their aspect and habits, and are found in the low parts of South America where they are sedentary. They swim with ease despite their lack of webs; their toes are very long. They feed on vegetable matters and nest like geese. They fly and soar well, their wings being large and broad, armed with strong spurs; they readily perch on tree tops.

The *Waterfowl* (Anatidae) form the core of the order. They have moderately long to short legs and webbed feet, and their wings are usually capable of a fast and sustained flight; but many species are heavy on land while they all swim well. Their bill is broad, fleshy as is their tongue, with lamellae. All Waterfowl are closely related, but the subfamilies can be distinguished:

1. *Anseranatinae*, consisting of one very peculiar Australian genus, and species, the Magpie Goose, a primitive black and white bird with semi-palmated feet whose anatomical characteristics approach those of the Screamers.

2. *Anserinae*, which include the Whistling Ducks, the Swans, and the Geese. They are goose-like in shape and in posture, with a long neck; the tarsi are reticulated in front; a single annual molt; no important sexual difference in plumage, voice or display. There are 8 species of Whistling Ducks or Tree Ducks, all rather small and living in the tropics; one of Coscoroba, large and swanlike, from South America; 5 of Swans from the Northern Hemisphere, Australia and South America; 13 of Geese, all from the Northern Hemisphere.

3. *Anatinae*, a large group of 34 genera and 117 species distributed throughout the world. All are basically similar although some resemble Geese and others are divers, and they can only be sub-divided in tribes, which are linked by intermediate forms:

The *Shelducks* (Tadornini) are large, grazing birds with high legs, resembling geese, distributed rather locally throughout the world, but particularly abundant in the Southern Hemisphere (20 species);

The *Dabbling Ducks* (Anatini) are very numerous throughout the world (41 species), all more or less similar to the well-known Mallard and Teal in size, shape and habits;

The *Eiders* (Somateriini) are a group of 4 species of northern diving sea ducks feeding on marine animal life.

The *Pochards* (Aythyini) are another group of diving ducks, generally vegetable eaters, rounded in shape, living mostly on fresh water, composed of 15 species, 10 of which are northern, the other 5 found in South America, Africa, Madagascar, Australia and New Zealand.

The *Perching Ducks* (Cairinini) have a long tail and their legs are inserted forward; they perch readily, nest in tree holes and live in wooded countries. The Carolina and Mandarin Wood Ducks, and the Muscovy are the best known members of the tribe (12 species) which is represented in America, Africa, Asia and Australasia, particularly in the tropical parts. They vary in size from very small to large, and are mostly vegetable feeders.

The *Scoters, Golden-eye* and *Mergansers* (Mergini) also have long tails and nest in tree or rock holds, as a rule, but they are adapted to diving life and an animal diet. Some spend much of their time at sea.

The *Stiff-tails* (Oxyurini) are a very specialized tribe of fresh water divers, composed of 9 species distributed throughout North and South America, Southern Europe, Western Asia, Africa and Australia. They are thoroughly aquatic and feed on vegetable matters.

JEAN DELACOUR

References

Peters, J. L., "Check list of the birds of the world," vol. 1, Cambridge, Mass., Harvard Univ. Press, 1931.
Delacour, J. and E. Mayr, "The family Anatidae," Wilson Bull., 57: 3-55, 1951.
Delacour, J., "The waterfowl of the world," 3 vols. London, Country Life, 1954-1959.

ANT

Ants are INSECTS belonging to a single family Formicidae, coextensive with superfamily Formicoidea and containing perhaps 12,000 species. Superfamily Formicoidea is one of the major groups of Hymenoptera, falling nearest the solitary fossorial wasps of superfamily Scolioidea, family Tiphiidae, with which it apparently had common origin during the Cretaceous. Structurally, ants are usually distinguished from other Hymenoptera, many of which are wingless and superficially antlike, by the presence of a large gland, with bulla and orifice, on either side of the metathorax; the function of these metapleural glands, although they secrete to the outside, is unknown.

All modern ants are social, living in perennial colonies consisting, in the usual case, of two or more castes of females; winged males, like honeybee drones, appear at intervals and serve to fertilize the primary females. The two basic female castes, derived, as in other Hymenoptera, from fertilized (diploid) eggs, are the worker or "neuter" and the queen or true female. The worker is normally small and does not develop wings or flight muscles, and is specialized to perform most of the labor in the established colony; it is the form universally recognized as "the ant." The queen usually bears wings up to the time of the nuptial flight, but they are shed after mating. The queen functions as the primary reproductive; she founds the new colony alone, or is adopted by an established colony.

The male develops from unfertilized eggs. It mates with the female, often after a mass nuptial flight from the nest, and normally does not return to the colony. The individual ant develops through a typical holometabolous sequence: egg, grublike larva, pupa with or without cocoon, adult. The number of larval instars is not known certainly for any species, but may be 3 to 5. As with all true social insects, the generations overlap, and the adults rear the immatures, feeding the larvae by progressive provisioning. In congenial climates, two or more broods may be reared in a single season. Differentiation of the females into queen and worker castes, and subdivision of the worker caste into soldiers, minor workers, etc. (where it occurs) is determined by the type of food the larva is given, and to a lesser extent, in some cases, by the nourishment the embryo receives in the egg.

Ants vary by species from highly specialized monophagous predators to near-omnivores, seed harvesters, and even to one whole tribe, the Attini, in which all species cultivate and feed upon special fungi. Some, perhaps most, ants feed the youngest larvae and the nest queen primarily on unfertilized eggs laid by the workers. Many ants are fond of the excretions, called honeydew, of homopterous insects, and some of them, particularly subterranean genera such as *Acropyga* and *Acanthomyops,* are obligatory tenders of certain root-feeding plant lice.

Studies of communication and orientation in ants are hardly begun, and the very great differences in behavior between different taxa render generalizations hazardous. Communication among adults, and between adults and larvae, employs five known modalities: (1) olfactory, in which volatile pheromones are secreted by specific glands upon specific stimulus, and which elicit a specific response, such as trail-following or alarm behavior; (2) gustatory, particularly during food exchange, by regurgitation from the adult crop or social stomach, which has an elaborately modified valve in many groups, permitting prolonged storage of liquid food; (3) auditory, including stridulation, tapping and other sounds, transmitted most importantly through the substrate; (4) mechanical, including antennal stroking, body recoil and generalized jostling; various types of interindividual licking and grooming are probably combined with (1) and (2); (5) visual, as suggested by striking color patterns in some forms with well-developed eyes.

Foraging in ants apparently involves two main types of orientation: (1) by individual foraging fields, in which an individual worker tends to revisit the same restricted feeding area, becoming familiar with the routes to and from the nest by learning based on olfactory, visual, kinemetric, gravity-sensitive and other faculties; (2) by trail-following, in which foraging is induced and oriented by means of a chemical trail laid by appropriately stimulated workers between food source and nest. The mass foraging practiced by army-ants is apparently a special kind of trail-following.

Ants usually inhabit more or less permanent, definitely structured nests, excavated in the ground or in wood, or utilizing pre-existing cavities in plants or in rocks. Some ants live in apparently mutualistic relations with certain kinds of plants, offering some protection against insects and browsing mammals that fear their stings and bites, and receiving from the plant nectar from special nectaries, and often also a ready-made domicile on the plant. Some arboreal ants build nests of leaves joined by silk produced by their larvae (*Oecophylla*), and others build papery carton nests from masticated plant fibers (*Crematogaster*).

Ant colonies may contain as few as 8 or 10 adults, or

several millions, as in the African *Anomma* army-ants; probably most species average less than 1,000 workers per colony. The colonies, however, may be very numerous, especially in tropical forest or warm semidesert. Upwards of 150 species have been found nesting in a single square kilometer of tropical rain forest, whereas the same area in a deciduous forest in New England or Central Europe might harbor 25 or 30 species. Ants in general do not thrive in very cold climates or in cool, wet, poorly insolated forests. Their great abundance and relatively high activity rates make ants one of the most important animal influents in temperate and tropical climates.

Family Formicidae is divided into nine subfamilies: Myrmeciinae, Ponerinae, Pseudomyrmecinae, Dorylinae, Myrmicinae, Dolichoderinae, Formicinae, Leptanillinae, and a Cretaceous subfamily, Sphecomyrminae.

WILLIAM L. BROWN, JR.

Reference

Wheeler, W. M., "Ants," New York, Columbia University Press, 1910.

ANTENNA

The antenna, or feelers, of insects are never more than a single pair and are situated near the compound eyes. In the adult insect they are the first appendages of the head. In neither their segmentation and their musculature do they resemble the other insect appendages. Because of their innervation and the position of the nerve centers in the brain the antenna are considered as belonging to the preoral region of the head. Antennae are absent in the Protura and in many forms of the larvae of the Hymenoptera. The antenna is a useful distinguishing characteristic; for the spiders, mites and ticks have no antenna, the crayfish, lobster and crabs have two pairs, while the centipedes and the millipeds have a single pair.

The typical antenna is a many-jointed filament. Generally three parts may be distinguished: the scape is basal and attaches the antenna to the head, followed by a pedicel which is short and generally contains the sensory *organ of Johnston* and the flagellum which is usually long and compounded of many subsegments. The scape is set into a membranous area of the head known as the antennal socket. This articulation allows the antenna free motion in all directions. Movement is accomplished by muscles which are inserted in the base of the scape and arise on intercranial supports.

In form, the antennae are considerably varied. The flagellum shows the greatest difference in shape and size. Generally the antennal types are referred to in descriptive terminology as being either *filiform* (threadlike), *moniliform* (bead-like), *setaceous* (tapering), *geniculate* (bent), *capitate* (having a head), *lamellate* (leaf-like), *pectinate* (comb-like), *serrate* (saw-like) and *clavate* (clubbed). The total length of the antenna is subject to much variation. The long horned beetles (Cerambycidae) and the caddis flies (Trichoptera) have antenna considerably longer than the body while dragon flies (Anax) have antennae which are reduced to mere bristles. Generally the larval antenna bears very little

resemblance to that of the completely metamorphosed adult. The aquatic nymph of the May fly (Ephemera) has a very long many jointed antenna which is gill-like in appearance, while the adult has a much reduced antenna consisting of two short basal segments and a short, stiff bristle-like flagellum. In the plant lice (Aphis) the various instars have antenna with fewer segments in the flagellum than are found in the adult. The number of segments increases from two, with successive molts, until the adult number of four is attained. Active larvae, such as the entomophogous ground beetle (Carabid), have well developed antennae of four segments with conical sensory tubercles which are very similar to the antennal structures of the adult parent beetle.

Though antenna are homologous in all insects, they are by no means equivalent in function. They are tactile for grasshoppers, olfactory for beetles, moths, and flies, auditory for mosquitos and respiratory for aquatic beetles. At least in some cases the ants use their antennae for communicating with others of their own kind.

Antennal modifications, both as to structure and function, are also associated with sexual activity. In the Promethia and Cecropia moths the feathery, pectinate antennae of the male are much larger, more branched and segmented that those of the female. These sensory antennae of the male allow him to seek out the female by olfactory means. Beetles and aphids also have olfactory pits which are located on various antennal segments. The plumulose antennae of the male mosquito (Culex) is larger than that of the female of the species and is a highly developed organ of hearing. This antenna is used to locate the female through the use of sound vibrations; a type of mating call. In certain collembolans the antennae of the male are provided with hooks which are used for grasping the female antennae at copulation.

FORREST W. MILLER

References

Snodgrass. R. E., "Principles of Insect Morphology," New York, McGraw-Hill, 1935.
Folsom, J. W., "Entomology with Reference to its Ecological Aspects," 3rd ed. Philadelphia, Blakiston, 1913.
Miller, F. W., "Antenna of Habrobracon juglandis Ash," Penna. Academy of Science VII, 1933.

ANTENNATA

A group, without specific taxonomic rank, erected within the ARTHROPODA to contain those classes (e.g. INSECTA CHILOPODA, etc.) in which the most anterior appendage functions as an ANTENNA.

ANTHERIDIUM

The Antheridium is the structure which bears the sperms (microgametes) in THALLOPHYTES, BRYOPHYTES and PTERIDOPHYTES. In ALGAE it is usually unicellular, producing one to many sperms, and is more varied in form than in the higher groups; in *Oedogonium* a short cell of the filament produces two sperms (Fig. 1); in *Fucus* a sac-like structure in an elaborate fructification

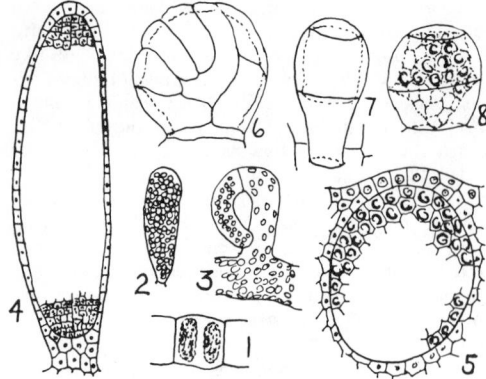

Figs. 1–8

bears many sperms (Fig. 2); in *Vaucheria* it is the tip of a slender curved branch (Fig. 3). In oogamous Fungi it is usually formed at the tip of a filament (hypha) which grows towards the oogonium and fuses with it; a pore forms at the place of contact through which the sperms pass to the egg or eggs.

In Bryophytes the antheridium is multicellular consisting of a stalk, and a jacket layer of cells enclosing a mass of spermatogenous cells each of which produces two minute biciliate sperms (*Mnium*, Fig. 4). When water comes in contact with the mature antheridium the tip is ruptured and the sperms are released in a mass from which they escape into the water and swim away.

In the Anthocerotales, and in some Pteridophytes (Lycopodiales, Equisetales, Marattiales) the antheridium develops below the surface, "embedded," (*Marattia* Fig. 5).

In most ferns the antheridium is superficial and more or less globular. In the more primitive ferns, e.g. *Osmunda*, the antheridium wall consists of many irregular cells enclosing a mass of spermatogenous cells each of which produces a coiled multiciliate sperm (Fig. 6). In higher ferns the antheridium wall consists of three cells: a basal cell (flat, funnel-shaped, or elongated), a ring cell, and a cap cell (Figs. 7, 8). Water is necessary for the dehiscence of the antheridium.

In heterosporous Pteridophytes a small antheridium (two in some forms) develops on the microgametophyte within the microspore wall.

ALMA G. STOKEY

References

Eames, A. J., "Morphology of Vascular Plants. Lower groups," New York, McGraw-Hill, 1936.
Smith, G. M., "Cryptogamic Botany" 2 vols., New York, McGraw-Hill, 1955.

ANTHOCEROTIDAE see HEPATICAE

ANTHOZOA

The Anthozoa are a wholly marine class of the phylum Coelenterata. They are solitary or colonial and may

or may not possess a skeleton. Such common marine animals as corals and sea anemones make up the bulk of the class though it also includes such organisms as sea fans, sea feathers, soft corals, organ pipe corals, sea pens, sea pansies, black corals and ceriantharians. It should be noted that the word coral is commonly used to include certain members of the class Hydrozoa as well as those noted above. As the word coral is used here, it excludes hydrozoan corals (see CORALS).

Anthozoans are basically radially symmetrical but this symmetry is usually modified such that the individuals are hexamerous, octamerous or polymerous or are biradially or radiobilaterally symmetrical. In the class no trace of the medusoid body form, so common elsewhere in the phylum, is to be found, and all individuals exist in the form of polyps. These polyps are generally cylindrical and more or less elongated with one end attached to some substrate and the other free. The free end is expanded into an oral disk which has the mouth at its center. The disk bears hollow tentacles which are arranged in a cycle or cycles around the mouth. The mouth is the primary body opening and no anus is present, though an anal pore exists at the aboral termination of the body in certain anthozoans. The mouth opens into a stomodaeal tube, usually termed the pharynx, which in turn opens into the general body cavity or coelenteron. The pharynx usually possesses one or more grooves along its length which are called siphonoglyphs. This groove is heavily ciliated and serves to pump water into the coelenteron, which in turn inflates the body of the animal. The pharynx ends as an open tube and is attached to the body wall by radially arranged logitudinal sheets of tissue, the mesenteries, which divide the coelenteron into a series of longitudinal compartments. In all the larger anthozoan polyps the mesenteries bear stomata which allow for the circulation of fluids between compartments. It will be noted that the Hydrozoa have no such longitudinal compart-

Fig. 1   Two plumose anemones, photographed from life. The stinging cells are concentrated in the feathery tentacles. (From Goodnight, Goodnight and Gray, "General Zoology," New York, Reinhold, 1964.)

ments and that the polyps of the Scyphozoa have four comparable compartments. Many Anthozoa have two sorts of mesenteries, complete ones which reach from the body wall to the pharynx and incomplete ones which do not reach the pharynx. The free edges of the mesenteries, which exist only below the pharynx on complete mesenteries, possess special structures, the filaments, which are richly provided with cilia, gland cells and nematocysts. The germ cells are derived from the endoderm, though the definitive gonad may become completely embedded in the mesoglea. Germ cells, at maturity, are either released freely into the environment through the mouth, or fertilization may be internal and the embryos and young may be retained for a period in the coelenteron.

Like other coelenterates the larva of this class is a planula. The planula may be planktonic but more often it exists only as a creeping, benthic stage in the life history. After a generally rather short free existence the planula attaches by its apparent anterior end and the mouth and tentacles are developed at the free end.

The mesoglea of the Anthozoa is cellular and is better developed as a third tissue layer than in either the Hydrozoa of Scyphozoa. The cellular components of the mesoglea resemble those of connective tissues. No elements of the nervous system are recognized as being present in the mesoglea. Nematocysts, the stinging capsules or cnidae of coelenterates, are also absent from the mesoglea though they are abundant in both ectodermal and endodermal tissue layer.

The Anthozoa, as a class, may be characterized as carnivores, though other feeding habits exist. Some sea anemones and corals feed only on small particulate material, including detritus, this material being first collected by the cilia of the tentacles and oral disk and subsequently swept into the mouth. The nutrition of other anthozoans, most notably the reef forming corals, depends in part on the presence of symbiotic algae in the endodermal tissues. It has been demonstrated that the algae produce certain substances, including sugars, which move from the alga into the tissues of the animal. The alga is not digested in this process. It seems rather to provide to the animal its excesses and perhaps in turn the animal provides some of the raw materials, such as carbon dioxide, which the plant needs in its own metabolism. More commonly, however, anthozoans capture living prey such as worms, molluscs, small fish, or plankters by stinging and paralyzing them with the nematocysts of the tentacles. The actions of the tentacles are such that captured prey is carried to the mouth which opens to receive the food object. Peristaltic waves move the food through the pharynx to the coelenteron where the mesenterial filaments are then brought into contact with the food object. Preliminary digestion, at least to the stage of breaking the food into small particles, is carried out within the coelenteron presumably as a result of enzymes released from the filaments and neighboring tissue. Essentially every endodermal cell is capable of the intake of particulate matter and as the food is broken into particles it is engulfed by these cells where final digestion occurs. Wandering cells move from the endoderm to the mesoglea and ectoderm, providing for the nutrition of those layers.

Only relatively restricted movements are possible for most anthozoans and these movements are largely concerned with the waving of tentacles and expansions and contractions of the cylindrical body. Water is pumped into the coelenteron, as noted earlier, and this acts as a fluid skeleton. The polyp body in general consists of circular endodermal muscles and to a greater or lesser extent depending on the particular animal, an outer, ectodermal sheath of longitudinal muscles. There also tend to be longitudinal muscles on the mesenteries which as they contract depress the oral disk and shorten the body. By closing the mouth a fixed volume can be maintained but considerable variation in shape, especially length and diameter, is possible as the circular and longitudinal muscles contract and relax. By unilateral or asymmetrical contractions of the longitudinal muscles, such movements as bending and twisting occur. Most solitary polyps, such as sea anemones, are also capable of creeping by means of coordinated movements of the muscles of the base which attaches them to the substrate. Such movement, called pedal locomotion, is exceedingly slow and is of utility in only short movements. One anemone is planktonic and floats base upward at the surface. In this animal, *Minyas,* the pedal disk secretes a frothy float of cuticular material which supports the individual. A few other anemones are capable of limited swimming. One, *Boloceroides,* swims when disturbed by movements of its tentacles, and another, *Actinostola,* swims in response to contacts by certain starfishes and the nudibranch, *Eolidea.* The latter, at least, is a known predator of anemones and the swimming of the anemone by a writhing of its body is interpreted as an escape response. Movement for most anthozoans is restricted to changes in shape however, since the great majority of species are firmly attached to various substrates and the class is best regarded as being sessile in habit.

There are two major sorts of anthozoan polyps, and the class is divided into two subclasses, the Alcyonaria or Octocorallia with octamerous symmetry and the Zoantharia or Hexacorallia which commonly has hexamerous symmetry. Alcyonarians are readily distinguished from other Anthozoa by their eight pinnately branched tentacles and their eight mesenteries. These animals are nearly all colonial and in some colonies there is a well developed dimorphism of the polyps with certain individuals in the colony, the siphonozooids, which are usually without tentacles, functioning to pump water into the animal and others, the autozooids with tentacles, serving to capture food for the colony. The form of colonies of alcyonarians varies from relatively isolated polyps connected by tubular stolons to massive and elaborate colonies which may be yards long or meters tall. The gorgonians, a group which includes the sea feathers and sea fans, have an axial skeleton of a black, horny material called gorgonin. Other skeletal structures in alcyonarians are the calcified axial rod of sea pens, the massive calcareous skeleton of the so-called blue coral, *Heliopora,* and the tubular skeleton of the organ pipe coral, *Tubipora.* All alcyonarians also possess calcareous spicules in their tissues.

The Zoantharia includes the remainder of the Anthozoa. Most commonly the tentacles and mesenteries are arranged in multiples of six, but many other patterns occur. However, the eight pinnate tentacles and eight mesenteries of the Alcyonaria are never met in the Zoantharia. Commonly the mesenteries of the Zoantharia are paired and usually there are both complete and incomplete mesenteries present. The subclass can be divided into a number of smaller groups or orders, the most notable of these being the Actiniaria or sea anemones and the Madreporaria, or true corals. Sea anemones exist in many sizes, shapes and colors, from minute individuals only a few millimeters long to giant anemones of tropical reefs over a meter in diameter. Sea anemones are not colonial and differ in this respect from most corals. They also lack the calcareous exoskeleton which characterizes corals. The remaining Zoantharia are more obscure forms such as the thorny corals (Antipatharia), the sea anemone-like Ceriantharia and the colonial anemones known as Zoanthidea. The skeleton of the Antipatharia resembles that of gorgonians and is fashioned into attractive jewelry which is sold as "black coral."

CADET HAND

Reference

Hyman, L. H., "The Invertebrates: Protozoa Through Ctenophora," New York, McGraw-Hill, 1940.

ANTHROPOID

The term "anthropoid" strictly means "manlike" or "of human form" and is often so used to denote the manlike resemblances of the apes—the gorilla, chimpanzee, orang and gibbon—collectively known as the "anthropoid apes." The term "anthropoid" has consequently come to be used as a substantive and as synonymous with "ape." This usage for scientific purposes is inexact and misleading since "an anthropoid" must be taken to refer, in terms of the agreed zoological classification, to a member of the Primate suborder, the anthropoidea, and this suborder contains in addition to present and extinct apes all the genera, past and present, of man and of monkeys. Monkeys, apes and men are classified together as "Anthropoidea" not merely because of their manlike resemblances but on the common possession of many fundamental traits in the skeleton and skull, teeth, brain, special senses and viscera. These characters taken together with the fossil evidence provide a basis for the belief in the common evolutionary origins of the Anthropoidea from earlier and simpler strata of primates which make up the sub order Prosimii (see table) represented today by tree-shrews, lemurs, lorises and tarsier.

The characters which convey the impression of human resemblance among monkeys and apes are briefly: the tendency to flattening of the face combined with enlargement of the brain case, the alertness of facial expression (arising largely from the forward setting of the eyes and the mobility of the musculature especially of the lips), the presence of a rather fixed and relatively small external ear and the frequent holding of the trunk in a near vertical position with, at the same time, the dextrous use of a free "hand" which, like the foot, is equipped with flattened nails. Among the Prosimians these characters, if not altogether absent, are in many cases only present incipiently. Thus like more primitive mammals, many prosimians retain some claws on hand or foot, the posture is on the whole pronograde, the orbits are not yet rotated completely forward and the muzzle shows only the beginnings of

recession; as in lower animals the nostrils are surrounded by a naked and moist glandular skin, the rhinarium, and the upper lip is fixed to the underlying gum. The prosimian hand shows on the whole a lesser degree of prehensility. Tarsius in some respects bridges the gap between the prosimians and anthropoidea, e.g. in the lack of the rhinarium and the development of the visual system.

The taxonomic relations of Anthropoidea as generally accepted are based on those proposed by G. G. Simpson (1945). In outline these are as follows:

Order: Primates
  Suborder: Prosimii
    Families: Living genera include treeshrews, lemurs, aye-aye, indris, lorises, galagos, tarsier. Extinct forms include treeshrews, ancestral lemurs and tarsioids.
  Suborder: Anthropoidea
    Superfamily: Ceboidea (New World Monkeys)
    Superfamily: Cercopithecoidea (Old World Monkeys)
    Superfamily: Hominoidea
      Family: Pongidae
        Subfamily: Hylobatinae (gibbons, including extinct genera)
        Subfamily: Dryopithecinea (extinct ancestral apes, including *Proconsul)*
        Subfamily: Ponginae (large apes-gorilla, chimpanzee & orang & extinct genera)
      Family: Oreopithecidae (extinct)
      Family: Hominidae (modern & fossil forms of Homo; Pithecanthropus & Australopithecus)

Among other important characters displayed by the Anthropoidea as a suborder, and in which they contrast with the Prosimii, may be noted the following: the body form tends to be less enlongated, due to a reduction in the number and size of the trunk vertebrae; the head also shows a progressive tendency to shorten and to become more globular; the snout region undergoes reduction, accompanied by a restriction in the olfactory cavities. The shortened face tends to bend downwards under the expanding neuro-cranium; this increased flexion of the basi-cranial axis is present even in the baboon, where there has been a secondary lengthening of the muzzle. With the cerebral expansion goes a forward displacement of the foramen magnum, the plane of which tends to look more downwards than backwards. This reaches a culmination in *Homo sapiens* where the foramen magnum is well forward on the base of the skull. In the anthropoid dentition the incisors are reduced to two, the pre-molars to three or two; the canines are on the whole enlarged and pointed except in the Hominidae where they have undergone secondary reduction (as testified by the very strong root of the human permanent canine and by its relatively late eruption). In the limbs, claws have given way to nails on all the digits (except in the marmosets). The digits retain the primitive formula with the middle digit as the longest. The feet have on the whole avoided the more extreme specialisations seen amongst some of the Prosimians. The Anthropoid brain has undergone expansion and elaboration of the cerebral hemispheres and of the cerebellum. In the highly convoluted brain certain sulci are characteristically present. In intrinsic structure the visual centres in particular have become expanded, while the olfactory centers are greatly reduced. These changes in the brain are correlated with the disapperance of the naked rhinarium and the dwindling of the olfactory apparatus and the simultaneous improvement in visual function, in the forward rotation of the orbit and the perfection of a diurnal type of retina and in the tactile sense.

The Anthropoidea may be looked on as a graded series with the lowest stratum represented by the monkeys. Those of the Old World are termed 'Catarrhine' because the nostrils are close together, those in the New World 'Platyrrhine' with the nostrils well spaced apart. The former share with the Hominidae and Pongidae the dental formula 2.1.2.3. (two incisors, one canine, two premolars and three molars). The latter has one more premolar, giving a formula 2.1.3.3. but the marmosets have 2.1.3.2. Most of the monkeys are arboreal specialists and certain platyrrhines have developed even the tail as a prehensile organ. Many of the Old World species descend frequently to the ground and some are entirely terrestrial. The gait is quadrupedal but a semi-erect position of the trunk is often assumed in climbing and in sitting. The fore-limbs are only slightly longer than the hind limbs and in some species are in fact shorter. In the pre-natal condition both man and ape approach these limb proportions, thus emphasizing the basic affinities of apes, monkeys and man. It is by a process of differential post-natal growth that the upper limbs of the great apes and the lower limbs of man attain their characteristic elongation (Schultz).

The anthropoid apes resemble the Hominidae more closely than they do the monkeys, as seen in the relative enlargement of the brain and its configuration, in many details of the skeleton (for example in the side to side broadening of the thorax, which reflects the general orthograde posture of the trunk in climbing and branch to branch swinging), in the absence of a tail, in details of the foot structure, especially in the size and musculature of the great toe, and in the absence of the ischial callosities or 'sitting pads' (though these are present in the gibbon which in this and some other respects shows affinity with monkeys). The evidence of comparative anatomy is sufficient for taxonomists to group Pongidae and Hominidae into the one superfamily—the Hominoidae. This carries the implication of a remote common ancestry for the two families and is supported by the evidence of palaeontology.

Superimposed on the structural similarities between modern apes and man are of course the divergences which reflect the separate evolutionary history and ecological specialisation of the two families. In trunk and limb structure the anthropoid apes are well adapted to their arboreal life and to their peculiar mode of locomotion by brachiation, that is their ability to swing by the arms from branch to branch. To facilitate this the upper limb as a whole is of course greatly lengthened (the percentage ratio of the upper to the lower limbs ranges from about 125 in the chimpanzee to 160 in the gibbon compared with about 80 in man) particularly in the radius and ulna; the thumb is greatly reduced while the fingers are lengthened.

The Hominidae have as their characteristic feature the adoption of an erect gait and posture with appropriate skeletal and visceral modifications for a wholly terrestrial life. This necessitates in the pelvis, amongst other things, a great broadening of the plate and crest of the ilium to provide the extensive attachments for the muscles concerned both with bi-pedal walking and for supporting the trunk upright; the hominid innominate bone contrasts strongly with that of the pongid which is narrow bladed, elongated and shallow notched, without backward extension of the iliac crest, and lacks a strong anterior inferior iliac spine. The hominid cranium as already noted is balanced on top of the vertebral column and so the foramen magnum and occipital condyles are placed well forward on the base. The more recent Hominidae and especially Homo sapiens are distinguished also by the possession of a greatly enlarged brain.

Basal Primates and primitive Prosimians are known from the Palaeocene and from the earliest Tertiary, but the first anthropoid remains are very scarce; some (see figure) come from the Eocene (Burma) and Oligocene (Egypt) and are represented by fragments of jaws and by teeth. *Parapithecus*, from the Lower Oligocene, is the most generalised and shows some tarsioid features in the teeth and the shape of the jaw but already in the Oligocene the differentiation of cercopithecoid had occurred if we can judge by the fragments of *Apidium*.

Other remains (see figure) carry the evolutionary sequence of the Old World monkeys through the Miocene and the Pliocene (e.g. *Mesopithecus*) into the Pleistocene when the material becomes relatively abundant and quite widespread (e.g. there were macaques in Northern Europe). The New World monkeys are believed by many to have had a separate New World tarsioid ancestry. Fossil records are very scanty; there are some remains from the Miocene of Patagonia and Colombia.

With the early separation of monkeys from the anthropoid evolutionary sequence the next phase is represented by the appearance of ancestral *Hominoidea*. They are of great importance as representing the stock from which by divergence the Pongidae and Hominidae evolved, leading eventually to modern apes and man. As such they would not be expected to show the specialisations of their descendants, particularly the large brain of Homo sapiens and the peculiarities of limb structure of brachiating apes and bi-pedal man. They should in fact have been sufficiently generalised to allow these developments to take place. *Dryopithecus* of the Miocene and Eocene has already made some advance in the pongid direction but still carries strong indications of the characters expected of the early hominoid ancestral stage. *Proconsul* in fact exhibits a limb structure which is largely that of a cercopithecoid with only a slight tendency in the direction of brachiaton but not sufficiently so to preclude the development of hominid limb proportions. The ancestral gibbons (e.g. *Limnopithecus* or *Pliopithecus*) likewise do not display the extreme lengthening of the upper limbs characteristic of the modern gibbon. In the skull, too, Proconsul displays an appropriate ancestral condition—the nasal region is cercopithecoid, the dentition is essentially pongid while the simian shelf, characteristic of recent apes, is absent as it is in the Hominidae.

It is possible that *Oreopithecus* represents an early Hominoid much nearer the hominid line. The pelvis is said to adumbrate the hominid type, the limb proportions, even though the upper limb shows some lengthen-

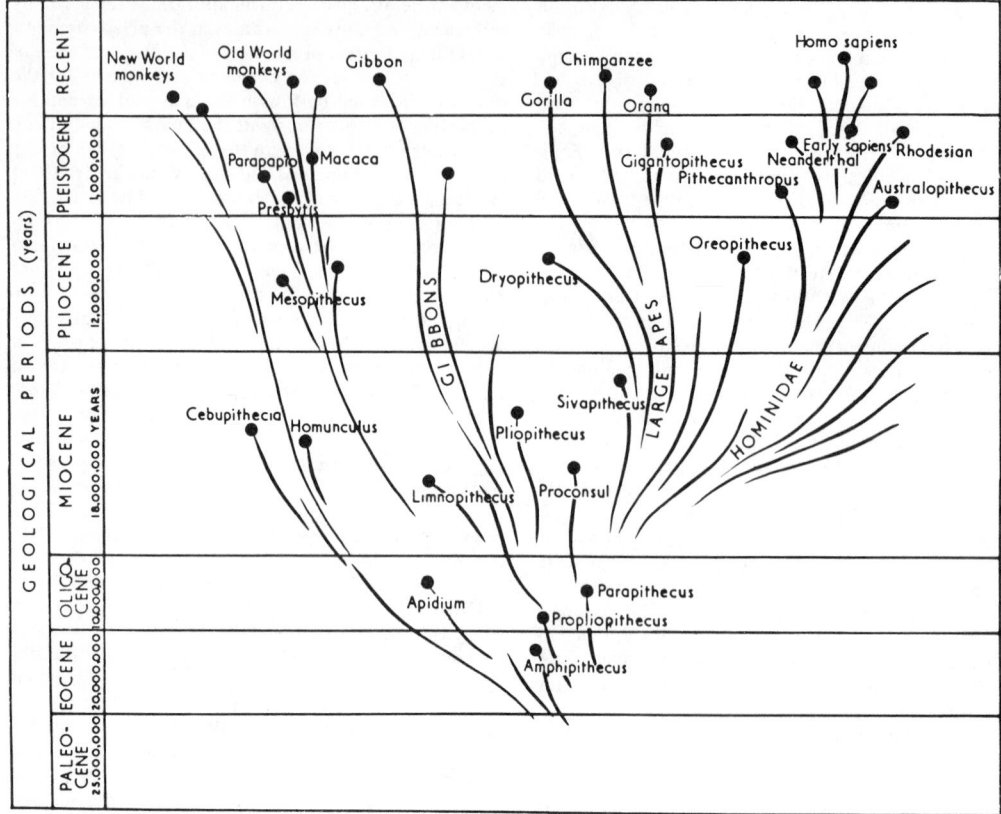

Fig. 1   Genealogical schema indicating the relationships of the Anthropoidea.

ing, are not specialised, and the canine is not projecting as in the pongids, though the molar teeth retain certain cercopithecoid traits. But only after a long gap, at the beginning of the Pleistocene, with the appearance of *Australopithecus* in South and East Africa do we encounter an undoubted early and primitive Hominid. In these creatures the upright posture is clearly in evidence as shown by the form of the pelvic bones, the position of the foramen magnum and the reduced attachment for neck musculature on the back of the skull; the canines and first pre-molars and the shape of the dental arcade are all essentially hominid; the primitiveness of this hominid resides in the relative smallness of brain size (barely larger than that of the gorilla but the body size was smaller), the relatively massive and chinless jaws and the continued presence of well developed bony ridges on the skull. The further sequence of hominid evolution concerns mainly the progressive enlargement of the brain capacity (*Pithecanthropus*) until the genus Homo (neanderthal and sapiens) emerges in the second half of the Pleistocene period.

J. S. WEINER

References

Buettner-Janusch, J., "Origins of Man," New York, Wiley, 1966.

Le Gros Clark, W. E., "The Antecedents of Man," Edinburgh, The Univ. Press, 1959.
Reynolds, V., "The Apes," New York, Dutton, 1967.
Schultz, A. H., "Characters common to higher Primates and characters specific for man," Quart. Rev. Biol., 11:259 & 425, 1936.
Simpson, G. G., "The Principles of Classification and Classification of Mammals," Bull. Amer. Mus. Nat. Hist., 85:1, 1945.

ANURA

The order Anura of the class Amphibia contains the forms commonly known as the frogs and toads (See AMPHIBIA). Anurans are highly modified tetrapod vertebrates characterized by a shortened vertebral column, hind legs adapted for leaping and the absence of a tail in adults. The anuran axial skeleton consists of five to nine presacral vertebrae. A long slender bone (urostyle) articulated with the sacrum represents the fused tail vertebrae. The pelvic girdle is elongate and forms a flexible union with the sacrum at about mid-point of the body. The powerful hind legs are made disproportionately large by elongation of the tarsal bones into a functional fourth segment. The radius and ulna are fused, as are the tibia and fibula. The skull is unusually large in proportion to the body, broad, and flattened, with a huge gape and dorsal orbits. The head is solidly united to the body without a flexible neck. A strong pectoral girdle composed of both bony and cartilaginous elements supports short but muscular front legs.

Anuran skin consists of a thick, heavily vascularized dermis richly supplied with mucus and poison glands, and an epidermal layer of variable thickness. The poison glands may be diffuse, or concentrated into discrete glandular protuberances. Some frogs produce exceedingly virulent alkaloid poisons. The epidermis is thin and fragile in aquatic forms and those adapted to humid conditions, and thick and heavily keratinized in desert species. Localized cornifications of the epidermis provide strengthened areas used in digging or grasping, or seasonally, for clasping during mating. The toes and fingers of anurans are usually adapted for locomotion in the medium in which they live by means of webs, adhesive pads, or sharp calluses.

Anurans have well-developed ears with a middle ear cavity, and usually an external tympanic membrane. The eyes are generally large, with movable eyelids, and vision is acute. Larval amphibians and adults of some permanently aquatic forms have a cutaneous sensory system of superficial neuromasts.

Sexual dimorphism is great, and commonly involves body size and proportions, color and pattern, and size of the legs. Males of most anuran species have well-developed vocal cords and one or two inflatable resonating chambers which communicate with the mouth. The incredibly loud and harmonically complicated calls of frogs are species-specific, and serve both to attract females to potential breeding sites and to reduce the possibility of accidental mis-mating in mixed breeding populations.

Mating typically occurs in the water, and fertilization is almost always external. The male frog clasps the female tightly from behind, and releases seminal fluid near the eggs as they are laid. Anuran eggs are enclosed in jelly envelopes which swell on contact with water and protect the egg until hatching time. The hatchling is an elongated embryo which soon develops into a bizarre, free-swimming larva. This larva, or tadpole, is characterized by the absence of legs, an extremely large head and short body, a long powerful tail, external gills housed inside a flap of skin, a long coiled gut and a complex of horny excrescences around the mouth which function as teeth. Metamorphosis, which occurs after a larval life of a few days to several years, involves resorption of the tail and growth of legs, atrophy of the gills and the change to air breathing, shortening of the gut and enlargement of the mouth to adult size and shape.

After metamorphosis the young frog usually invades an entirely different habitat, then rapidly grows to sexual maturity. Terrestrial forms characteristically breed annually, in response to seasonal rainfall and temperature cycles. Adult anurans are all predaceous, and eat a variety of invertebrate and vertebrate prey. The length of adult life is tremendously variable—very short in tiny species, relatively long in the larger forms.

Adult anurans as a group are very conservative in morphology and are all characterized by their specialization for leaping. As Robert Inger has paraphrased Gertrude Stein—"a frog is a frog, is a frog." The universality of the basic frog body plan has resulted in extensive parallelism and convergence in the sizable modern frog fauna (approximately 200 genera, 1800 species). Adult specializations include adaptation to arboreal, subterranean, terrestrial and aquatic life, but the few general adaptive types are frequently repeated in diverse phyletic lines. Anurans have escaped the evolutionary restrictions of their morphological conservatism in part by extreme variability in details of the reproductive cycle and larval adaptations. Breeding season, site of oviposition, larval form and ecology, and time of metamorphosis are among the life-history features subject to this adaptive variability. The most successful direction of this radiation has been toward freeing the anuran species from dependence on free-standing water during the life cycle. This is accomplished in a number of ways by different groups, among them the care of eggs and larvae by adults, post-metamorphic hatching of eggs and even direct development within the maternal oviducts.

The ancestry and relationships of modern frogs are obscure. The earliest known frog fossil, the Jurassic *Notobatrachus,* is a modern frog in all essential respects, and most later fossils can be placed in living families. Fossils annectant between the Anura and their presumed labyrinthodont ancestors are unknown. Construction of a phylogenetic scheme for modern frogs is hindered further by the extreme convergence and parallelism within the group. Students of anuran phylogeny have sought conservative and meaningful characters in details of the skeletal anatomy and larval morphology, but a definitive phylogeny of the Anura is far from possible at the present stage of our knowledge. The following arrangement of suborders and families is that employed by Goin and Goin ("Introduction to herpetology" San Francisco, W. H. Freeman, 1962).

Order ANURA
  Suborder AMPHICOELA
    Family Leiopelmidae   Tailed Frogs
  Suborder AGLOSSA
    Family Pipidae   Tongueless Frogs
  Suborder OPISTHOCOELA
    Family Discoglossidae   Fire-bellied Toads
    Family Rhinophrynidae   Mexican Burrowing Frog
  Suborder ANOMOCOELA
    Family Pelobatidae   Spadefoot Toads
    Family Pelodytidae   Slender Frog
  Suborder DIPLASIOCOELA
    Family Ranidae   True Frogs
    Family Rhacophoridae   Old-World Tree Frogs
    Family Microhylidae   Narrow-mouthed Toads
    Family Phrynomeridae   Narrow-mouthed Tree Frogs
  Suborder PROCOELA
    Family Pseudidae   Paradox Frogs
    Family Bufonidae   True Toads
    Family Atelopodidae   Variegated Toads
    Family Hylidae   Tree Frogs
    Family Leptodactylidae   Southern Frogs
    Family Centrolenidae   Glass Frogs

CLARENCE J. McCOY

References

Goin, C. J., "Amphibians, pioneers of terrestrial breeding habits," Smithsonian Report, **1959**: 427–455, 1960.

Inger, R. F., "The development of a phylogeny of frogs," Evolution, **21**: 369–384, 1967.

Jameson, D. L., "Evolutionary trends in the courtship and mating behavior of Salientia," Syst. Zool., **4**: 105–119, 1955.

Noble, G. K., "Biology of the Amphibia," New York, McGraw-Hill, 1931.

Orton, G. L., "The bearing of larval evolution on some problems in frog classification," Syst. Zool., **6**: 79–86, 1957.

Tihen, J. A., "Evolutionary trends in frogs," Amer. Zool., **5**: 309–318, 1965.

APATHY, ISTVAN (1863–1922)

This Hungarian zoologist graduated in medicine from the Univeristy of Budapest and held successively the chairs of zoology and comparative anatomy in Cluj and Szeged. He published considerable work, mostly as a result of studies at the Naples Marine Biological Station, on the structure of the nerve fiber and on the transmission of nerve impulses. He is best known, however, for his work "Die Mikrotechnik der Thierischen Morphologie," the second volume of which (1901) is the only comprehensive history of this subject.

APOMIXIS

At first employed by Winkler in 1908, the term apomixis refers to the replacement of sexual reproduction by any form of propagation which, by avoiding MEIOSIS and syngamy, permits the reproduction unchanged of a given genotype. In higher animals, apomixis necessarily depends upon PARTHENOGENESIS. In plants with an alternation of morphologically different generations (see ALTERNATION OF GENERATIONS), apomixis involves the elimination or modification of the normal cycle.

The alternation is wholly suppressed when one generation propagates itself by vegetative means (the gametophyte in mosses and liverworts, and the sporophyte in ferns and their allies and seedplants). Propagation may be by simple fragmentation of the individual, or by specialised accessory structures. In the flowering plants, vegetative reproduction may be regarded as a form of apomixis only when accompanied by a regression of sexual reproduction, as in the viviparous grasses,where flowers are replaced by bulbils. Vegetative reproduction may be accomplished through the seed, the embryo being formed directly by the outgrowth of a cell of the parent sporophyte. This process of *adventitious embryony* is one form of *agamospermy*, production of seeds without a sexual process.

In apomictic ferns and seed plants where the alternation of morphologically unlike generations is retained, the normal cycle of change in chromosome number, haploid in the gametophyte and diploid in the sporophyte, is suppressed. In general, this results from the correlation of two developmental aberrations, (a) the elimination or modification of the reduction divisions so that the gametophyte is diploid, and (b) the elimination or modification of the reduction divisions so that gametophyte is diploid, and (b) the subsequent development of the egg without fertilisation (parthenogenesis), or, in ferns, the formation of sporophyte through the outgrowth or other cells of the gametophyte than the egg (*apogamety*).

In certain apomictic ferns, the chromosomes fail to separate in the last mitotic division before meiosis, so that the reduction division merely restores the original diploid chromosome number in the spores. The new diploid sporophytes then arise apogametically from the diploid gametophyte. No parallel phenomenon is known in flowering plants. Here the reduction of CHROMOSOME number in the formation of the megaspores may be avoided by two principal means: either (a) the archesporial cell (or derivatives if it enters meiosis) is displaced by a neighbouring cell of the sporophyte, which forms an adventitious diploid embryo-sac (*apospory*) or (b) meiosis in the megaspore mother cell fails to result in a reduction of chromosome number, so that a diploid embryo-sac arises (*diplospory*). Where diplospory involves the failure of meiosis through the formation of a restitution nucleus, some GENE re-arrangement may be possible (*autosegregation*). Following apospory or diplospory, the apomictic cycle is completed by parthenogenetical development of the egg.

Agamospermy, whether arising from adventitious embryony or alter diplospory or apospory, may be autonomous, or may only follow pollination (*pseudogamy*). In pseudogamous apomicts, the pollen does not contribute genetically to the embryo, but is required to initiate the growth of the ovary, or to fertilise the indosperm nucleus.

Among vascular plants, apomixis is known in more than eighty genera. As a reproductive system, it offers the possibility of the indefinite propagation of specially favorably biotypes, which may be highly heterozygous or sexually sterile. Where apomixis is obligate, this advantage is bought at the expense of the long-term evolutionary flexibility which is the gift of sexuality, a sacrifice not necessarily made in groups where sexual and apomictic members co-exist, or when apomixis is facultative.

J. Heslop-Harrison

References

Gustafsson, Al, "Apomixis in higher plants," Lunds Univ. Arsskr. 42–43: 1946–1947;

Nygren, A., "Apoximis in the angiosperms," Bot. Rev. **20**: 577, 1954.

Stebbins, G. L., "Variation and evolution in plants," New York, Columbia Univ. Press, 1950.

Steil, W. N., "Apogamy, Apospory, and Parthenogenesis in the Pteridophyta," Bot. Rev. **17**: 90, 1951.

ARACHNIDA

Arachnida is a class of the phylum ARTHROPODA. It includes such well known forms as scorpions, spiders, mites, ticks, and harvestmen as well as a number of less common types. Except for a few aquatic mites, arach-

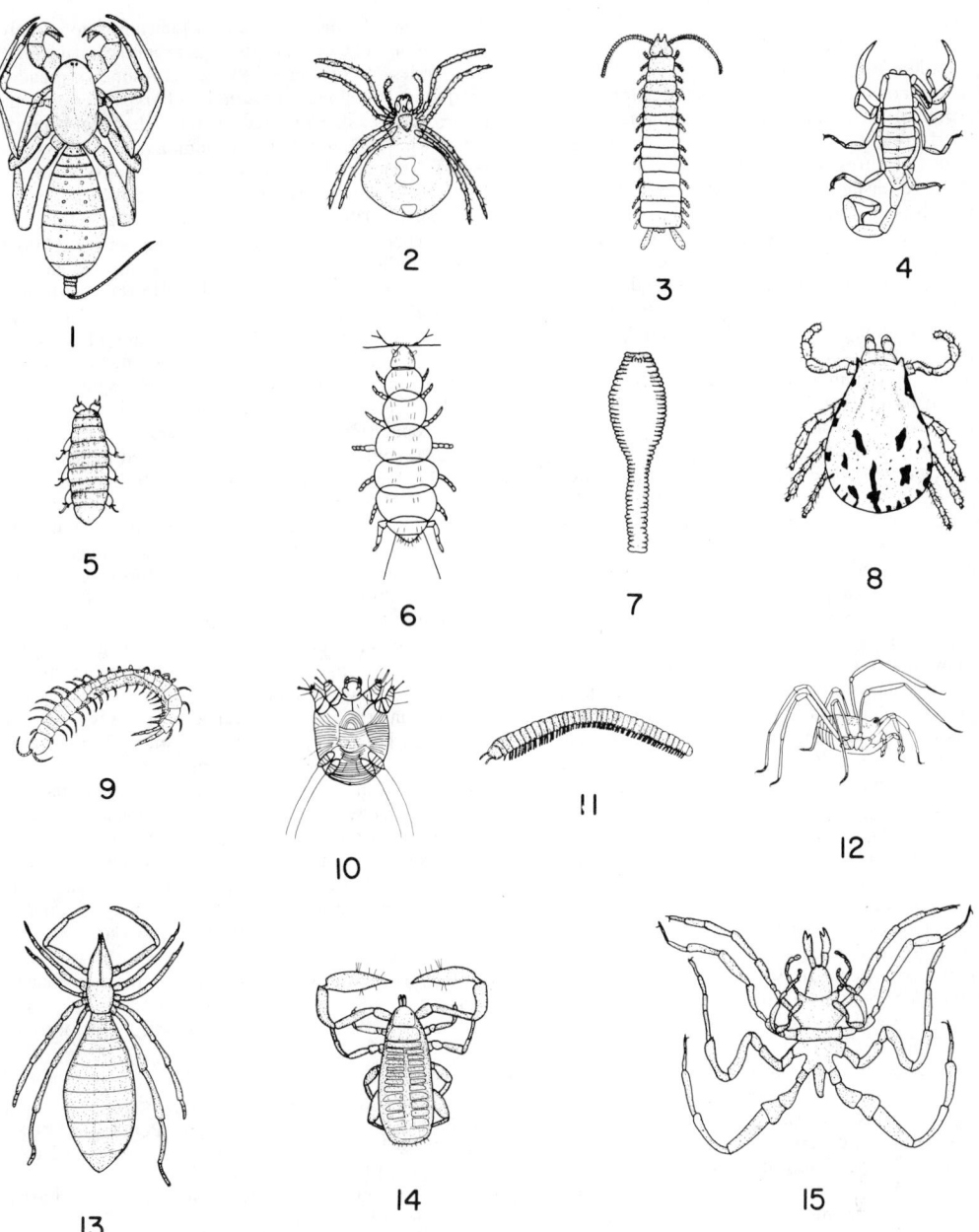

Fig. 1   Arachnids and related forms. 1, A whip scorpion. 2, Female black widow spider, *Latrodectus*, ventral view. 3, A symphylid. 4, A scorpion. 5, A tardigrade or water bear. 6, A pauropod. 7, A linguatulid or tongue worm. 8, A tick, *Dermacentor*. 9, A centipede. 10, The itch mite, *Sarcoptes*. 11, A millipede. 12, An opilionid. 13, A solpugid. 14, A pseudoscorpion. 15, A pycnogonid or sea spider. (From Goodnight, Goodnight and Gray, "General Zoology," New York, Reinhold, 1964.)

nids are terrestrial animals. On land they are found nearly everywhere, from deserts to rain forests. In the case of some small spiders, they have been collected thousands of feet in the air as they were wafted along on their silk strands.

In common with all arthropods, they have a chitinous exoskeleton and jointed appendages. The arachnid body is divided into two portions: a cephalothorax (combined head and thorax) and abdomen. The cephalothorax has about nine segments, but these are so modified that the exact number is difficult to determine. It is covered dorsally with a shield, the carapace, and usually has six pairs of appendages. The first pair of appendages are the chelicerae. In some arachnids these are chelate and used to hold and crush animals for food. In spiders they contain venom sacs, and the last segment is a fang that injects the venom.

The second pair of appendages is the palpi. The palpi or pedipalpi vary considerably among the different arachnid groups. In spiders and harvestmen they are elongate structures while in scorpions they are large, chelate, prehensile organs. The distal segments of the palpi of the spider are highly modified and act as sexual intromittent organs. Frequently the basal segment is modified into a structure for crushing or cutting the food. The remaining four pairs of appendages are the walking legs. At times, the legs are modified into tactile organs and among one group (Ricinulei) special copulatory organs are located on the third pair of legs. In some mites the number of appendages is reduced, and there may be only two or three pairs of legs.

The abdomen is composed of twelve segments which are variously modified. In some forms the abdomen is very elongate with free tergites and telson as in the whip scorpions. In others it may be shortened with indistinct segmentation. The sex organs discharge their products through a single opening on the ventral portion of the second abdominal segment. Frequently this opening is covered by a genital plate. The abdomen in some arachnids (mites and harvestmen) is broadly jointed to the cephalothorax, while in others such as spiders it is joined by a narrow pedicel.

Most arachnids are carnivorous. They lack jaws for chewing food, so with few exceptions, they predigest their food, then suck the juices into the mouth. The digestive tract of arachnids consists of three parts: the fore-gut, the mid-gut, and the hind-gut. The fore-gut and hind-gut are lined with chitin, and only the mid-gut has a digestive epithelium. The mouth leads into the foregut, the anterior portion of which is the pharynx. From this organ, a straight esophagus leads to the midgut. In some forms (spider, pedipalpi) the distal portion of the esophagus is enlarged and acts as a sucking organ to draw the food juices into the tract. The mid-gut has many diverticula in which digestion takes place; the hind-gut or rectum terminates in the anus.

Book lungs and tracheae are the respiratory organs of the arachnids. One or the other or both may be present. Arachnids have an open circulatory system with dorsal heart, arteries, veins, and extensive blood sinuses. Excretory organs also vary with some forms having Malpighian tubes, others coxal glands, and some both. The silk producing glands of spiders are believed to be modified nephridia. The nervous system consists largely of a massive ganglionic mass surrounding the anterior portion of the digestive tract and nerve cords. Arachnids possess many sense organs, including eyes and tactile hairs.

Arachnids are bisexual animals. Many have developed elaborate courtship patterns while others, such as many phalangids, simply mate without previous activity. Except for the viviparous scorpions, arachnids lay eggs either singly or in groups. These hatch into small individuals that closely resemble the adults except for certain special structures. By a variable number of molts, the adult structures and size gradually develop.

The nearest relatives of the arachnids are the king crabs (*Limulus*) of the class XIPHOSURA. These both probably arose from the extinct TRILOBITES, which were abundant in Cambrian Seas. The class Arachnida is divided into 16 orders, five of which are known only from fossils. The orders with living representatives are: Scorpiones (scorpions), Pseudoscorpiones (pseudoscorpions), Opiliones (phalangids, harvestmen), Acari (mites and ticks), Palpigradi (palpigrades), Schizomida (small whip-scorpions), Thelyphonida (tailed whip scorpions), Phrynichida (tailless whip scorpions), Araneae (spiders), Ricinulei (ricinulids), and Solifugae (wind scorpions). Mites and ticks, spiders, phalangids and pseudoscorpions are found over the entire United States. Other arachnids are found mostly in the southern United States, particularly in the Southwest. Ricinulids are almost entirely tropical, and only one species being known from the United States (Texas).

Arachnids are of great importance to man because of their numbers and predaceous habits. Spiders prey largely on insects and exercise considerable control of insect populations. Population studies have estimated 2,265,000 spiders per acre in undisturbed grassy areas in England and 14,000 per acre on herbs and shrubs of woodland of Illinois. Free living mites abound in the leaf mold of forests and play an important role in its conversion to humus. Many mites are parasitic in habit and are irritating pests to man and other forms. Chiggers, the larvae of one group of mites, cause irritating rashes on humans. Many mites and ticks are vectors of disease. Other mites are very destructive of crop plants. One mite "the red spider" is particularly troublesome to plant growers. All spiders possess poison which they use for capturing their prey. Fortunately only a few possess sufficiently powerful poison to affect man. One of these, however, the black widow, is dangerous to humans, its bite resulting in serious illness. The bite of another spider, *Loxosceles*, may result in severe local reactions. Scorpions too possess poison by which they kill prey. The poison gland in these forms is located at the end of the abdomen and the poison is delivered by means of a stinger. While the sting of any scorpion may be unpleasant only two small species of the southwestern United States are dangerous to man. However, the vast majority of arachnids live out their lives in their particular natural habitat unnoticed by man.

CLARENCE J. GOODNIGHT

References

Cloudsley-Thompson, J. L., "Spiders, Scorpions, Centipedes, and Mites," New York, Pergamon, 1958.
Gertsch, W. J., "American Spiders," Princeton, Van Nostrand, 1949.
Grassé, P-P., Ed., "Traite de Zoologie," vol. 6, Paris, Masson, 1949.
Lawrence, R. F., "The Biology of the Cryptic Fauna of Forests," Capetown, Balkerna, 1953.
Savory, T. H., "The Biology of Spiders," London, Sedgwick and Jackson, 1928.
Savory, T. H., "The Arachnida," New York, Academic Press, 1964.

ARALES (SPATHIFLORALES)

This order of plants belongs to the sub-class Coroliferae of the class MONOCOTYLEDONEAE. It is thought by some authorities to have evolved from the palms, but most others think it came from the lilies. Whatever the origin, the plants are peculiar in their floral arrangement. They are mostly herbs or climbers, often growing on other plants, with minute flowers usually on a thickened spadix, subtended by, and usually enclosed in, a single, usually large, herbaceous spathe.

The flowers are bisexual or unisexual, with an undifferentiated perianth, or these parts much reduced or entirely wanting. The ovary is usually superior, of 1–3 united carpels and develops into a berry or utricle. There are commonly 1–10 stamens.

There are two families in this order (Lemnaceae, Araceae). Those plants that are free-floating or submerged aquatics, with the plant body reduced to a thalloid-like structure without stems or leaves, and whose flowers are reduced to a single stamen or carpel, form the family Lemnaceae. Flowers, and especially fruit, are extremely rare. The vegetative stage is very plentiful and almost world wide. The plants are small or minute, oval, oblong, flat or globose. They reproduce mainly by budding, and in the fall minute bulblets are formed which sink to the bottom of the water and then rise again the next spring and vegetate. The flowers are usually enclosed in a membranous sheathing spathe, or this has been considered just a pouch in some plants.

The family is now usually considered to be composed of four genera (Lemna, Spirodela, Wolffia, Wolffiella). Plants with one root and green on the lower surface belong to Lemna. Those with several roots and red on the lower surface belong to Spirodela. Plants globular, without roots, and about 1 mm. or less in size belong to Wolffia. Plants strap-shaped, single or coherent in starlike masses, 6–8 mm. long, belong to Wolffiella.

The members of this family often cover fresh water ponds and streams in profusion, and furnish considerable food for wild fowl and water animals. They are the smallest and most simple of the flowering plants.

The other family (Araceae) of this order is composed of herbaceous or woody plants with rhizomes or tubers, and they mostly do not grow in water. Some of the plants are epiphytic with aerial roots and most of them are tropical. Most of them also have pungent sap and calcium oxalate crystals (raphides) in their tissues. The leaves vary from simple to compound and small to very large. The spadix, covered with inconspicuous flowers, is usually surrounded by a large, white or brightly colored spathe, making a very attractive structure (flower).

There are around 105 genera and 1500 species. They were divided by Engler into 8 subfamilies and 28 tribes. Eight genera are indigenous in the United States, but many are grown horticulturally in greenhouses and gardens. Jack-in-the-pulpit (Arisema triphyllum), skunk cabbage (Lysichiton americanum), calla lily (Calla palustris) and sweet flag (Acorus Calamus) are common examples of our native plants. The genera Zanthedeschia, Arum, Caladium, Helicodicerus, Scindapus and Amorphophallus are grown extensively for their showy "flowers," leaves, or as curiosities. Taro, the source of poi in the Hawaiian Islands and tropics, comes from the genera Colocasia and Alocasia. The large "fruits" of Monstera of the tropics are prized as a delicacy, while species of Pistia and Orontium are grown as ornamentals in pools or aquaria.

RAY J. DAVIS

Reference

Hutchison, J. "Families of Flowering Plants," vol. 2. London, Macmillan, 1934.

ARANEAE (ARANEIDA)

The order Araneae, spiders, belongs to the phylum Anthropoda, the subphylum Chelicerata, and the class Arachnida. In common with most other arachnids, spiders have the head and thorax fused into a cephalothorax and they possess six pairs of appendages: the chelicerae or jaws, the pedipalps, and four pairs of legs.

Spiders differ from most other arachnids in several characteristics. Their non-chelate chelicerae bear fangs, below the tips of which are the openings of poison glands. The second appendages, the pedipalps, are also non-chelate and leglike. Further distinctions are the slender pedicel or stalk connecting cephalothorax and abdomen, and the spinnerets, usually at the posterior end of the abdomen. The method of sperm transmission is also characteristic. A drop of sperm deposited on a special web made by the male is picked up and stored by a distal modification of the male palpus, the bulb. In mating, the sperm is transferred directly into the gonopore (or in higher spiders into a special structure, the epigynum, anterior of the gonopore) and is stored in seminal receptacles (spermathecae). Spiders usually have eight simple eyes, sometimes six or fewer. In the anterior median eyes (main, or direct eyes) the retina faces the cornea; in the others the retinal cells face the back. Respiration is by means of book lungs or tracheae or both. Species with an extensive tracheal system that takes oxygen directly to tissues have a reduced circulatory system: a dorsal heart in the abdomen, and an anterior aorta.

All spiders are predators, as are all other arachnids (except many mites). Digestion begins outside the digestive system as the spider injects enzymes into the prey. The soft tissues of the prey are liquified and sucked up by the pumping stomach of the spider. Or the prey may be chewed during digestion and the undigested portions discarded. Unique use is made of the silk for capturing prey: It is used as a trap or net for flying or alighting insects. The pedicel permits maximum movement of the abdomen for manipulating the spinnerets. Up to six kinds of silk glands producing different silks open through microscopic spigots on the spinnerets.

Irregular cobwebs made by cobweb weavers (combfooted spiders, Theridiidae) may have viscid trip threads around the periphery. An insect running against them will get stuck and break the thread. The breaking thread contracts and pulls the insect toward the center. The sheetweb weavers (Linyphiidae) construct silk sheets with irregular threads above; the spider hangs below. The insect trips on the irregular threads, calling attention to itself. The spider runs to the insect and pulls it through the sheet. Funnel weavers (Agelenidae) construct a tube-like retreat for the spider at the edge of a sheet; vibrations or an insect falling on the sheet summon the spider. The most complex are the orbwebs, independently evolved in two different families, the orbweavers, Araneidae (Argiopidae), and the Uloboridae. An orbweb consists of spokes radiating from a central hub and supporting a spiral of ensnaring threads, viscid in Araneidae, woolly in Uloboridae.

Other spiders hunt actively; they may run over the ground and catch insects encountered, sit in ambush on tree bark or flowers, or stalk prey. Jumping spiders (Salticidae) have excellent vision, among the best of invertebrate animals, permitting them to stalk potential prey and, when within 5 to 10 cm, jump on it. Wolf spiders (Lycosidae) take prey encountered when running over the ground and crab spiders (Thomisidae) are an example of a family whose members wait in ambush. Some crab spiders that sit on flowers can slowly change their color to match the background, making them less conspicuous to potential prey, often bees visiting the flowers.

Silk is also used to wrap subdued and trapped insects, as safety lines, as guidelines for males in finding sexual partners, for the construction of eggsacs and for ballooning. Males can determine from a silk dragline whether it was made by an adult, unmated female and follow the thread leading to its owner. Most spiders encase their eggs in a sac made of silk, often of several layers. Some (e.g. orbweavers, Araneidae) may hang up and abandon the egg sac, or the egg sac may be guarded (representatives in many families). Some may carry the egg sac around between the chelicerae (nursery web spiders, Pisauridae, and huntsman spiders, Heteropodidae) or attached to the spinnerets (wolf spiders, Lycosidae). Young wolf spiders are carried by the mother on her back for about a week before going off on their own.

Young spiders of most families at one time in their life history climb up vegetation and let out threads of silk. As the threads are caught by wind the little spiders let go of the plant and float off or "balloon," an important method of distribution for many spider species. The masses of threads are called gossamer.

The males in many groups have elaborate courtship displays, often species specific, which prevents their being mistaken for prey. The courtship of web weavers consists of plucking threads. In jumping spiders (Salticidae) and wolf spiders (Lycosidae) it is a visual display of hopping and waving first legs or palps.

Slit sense organs function for proprioception and some hearing; trichobothria, long hairs with a complicated articulation, are sense organs for hearing and vibrations. The sense organs of taste have not been identified. Experiments have demonstrated that jumping spiders (Salticidae) can differentiate colors.

The bite of some spiders is venomous to man. Black widows (*Latrodectus*, Theridiidae) have a neurotoxin, and brown spiders (*Loxosceles*, Loxoscelidae), widespread in the Americas, have a hemolytic poison. In Brazil, *Phoneutria* (Ctenidae) has a neurotoxin and *Lycosa raptoria* (Lycosidae) has a hemolytic poison. The funnelweb mygale, *Atrax* (Dipluridae) of Australia is much feared. Some of the 160 species of *Chiracanthium* (Clubionidae) have a venomous bite.

Female mygalomorph spiders (suborder Orthognatha), "tarantulas" in American usage, live up to 20 years; females of most true spiders (suborder Labidognatha) live only one or two seasons, except for some primitive groups whose members live 5 to 10 years. Males are short-lived and die soon after mating.

Of 30,000 species, only one is truly aquatic, though several live on ocean shores; all others are terrestrial, being found in many habitats on all continents except Antarctica.

HERBERT W. LEVI

References

Bonnet, P., "Bibliographia Araneorum," Toulouse, 1946–62.

Bristowe, W. S., "The World of Spiders," London, Collins, 1959.
Comstock, J. W., rev. by W. J. Gertsch., "The Spider Book," Ithaca, N. Y., Comstock Assoc., 1940.
Gertsch, W. J., "American Spiders," Princeton, Van Nostrand, 1949.
Kaestner, A., trans. and adap. by H. W. and L. R. Levi, "Invertebrate Zoology," Vol. 2, Arthropod Relatives, Chelicerates and Myriapods," New York, Wiley, Interscience, 1968.
Kaston, B. J., "Spiders of Connecticut," Bull. State Geol. Nat. Hist. Surv. 70, Hartford, Conn, 1948.
Kaston, B. J. and E. Kaston, "How to Know the Spiders," Dubuque, Iowa, Brown, 1953.
Levi, H. W. and L. R. Levi, "Spiders and Their Kin," New York, Golden Press. 1968.
Millot, J., "Araignées" *in* Grassé, "Traité de Zoologie," Vol. 6. Paris, Masson, 1949.

ARBER, AGNES (1879–1960)

English botanist and philosopher, was resident for most of her life in Cambridge, England. The erudition, clarity and insight of her works, "Water Plants" (1920), "Monocotyledons" (1925), "The Gramineae" (1934), "Herbals" (1912; 2nd ed. 1938) and her contributions to periodicals, mostly dealing with plant morphology, led to her election in 1946 as a Fellow of the Royal Society. "The Natural Philosophy of Plant Form" (1950), "The Mind and the Eye" (1954) and the "Manifold and the One" (1957); written with the same grace and scholarship. She was the wife of the palaeobotanist E. A. Newell Arber.

ARCHEGONIUM

The archegonium is the structure of the gametophyte in BRYOPHYTES, PTERIDOPHYTES and GYMNOSPERMS which bears the egg (megagamete). It is larger and more complex in Bryophytes than in the higher groups. The archegonium arises from the outgrowth of a single superficial cell, the archegonium initial. As it develops in Bryophytes there is formed an axial row (egg, ventral canal cell, neck canal cells) enclosed in a jacket layer of cells, of which the basal portion (venter) encloses egg and ventral canal cell, while the distal portion encloses neck canal cells, usually eight or more (*Marchantia*, Fig. 1). If water comes into contact with the archegonium when the egg is mature, the tip of the neck ruptures and there pours out a mucilaginous mass formed by the breaking down of the canal cells. Sperms which may be in the vicinity show a chemotactic response to this substance and move towards the open neck, enter the canal and pass to the egg. If a sperm enters the egg and fertilizes it, a wall is formed around the egg. The fertilized egg (zygote), the first cell of the sporophyte generation, develops within the growing venter (Fig. 2). At a certain stage varying with the type of liverwort, the venter is ruptured, and the spores which have developed are discharged.

In mosses the basal portion of the archegonium is relatively stout with a small cavity in the upper portion containing the egg and ventral canal cell (*Mnium*, Fig.

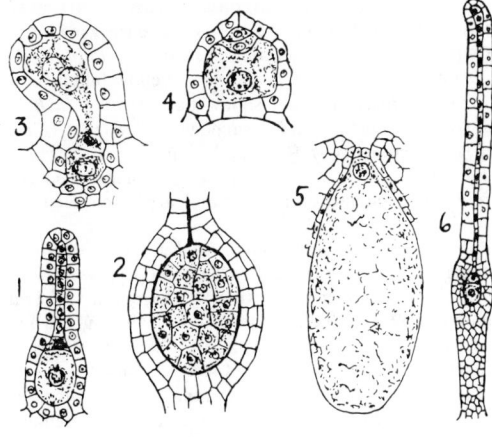

Figs. 1–6

6). There is a considerable development of the moss venter after fertilization, and a portion of it may be carried up on the elongating sporophyte as a Calyptra.

In one Bryophyte group (Anthocerotales), in all Pteridophytes, and in Gymnosperms, the venter of the archegonium is within the tissues of the gametophyte, "embedded," but most or all of the neck projects, above the surface.

In the Lycopodiales the neck is long with eight or more neck canal cells. In homosporous ferns the neck is short, usually curving towards the posterior end, and the neck canal has one binucleate cell (*Acrostichum*, Fig. 3). After fertilization and the egg divides and develops an embryo within the growing venter which is soon ruptured by the emergence of root, stem and leaf of the developing sporophyte.

In heterosporous Pteridophytes the gametophyte is retained partly or entirely within the megaspore wall; the archegonium neck is short usually containing one neck canal cell (*Marsilia,* Fig. 4).

In Gymnosperms the gametophyte with its archegonia develops within the megasporangium (ovule). The archegonium has a large venter and four small neck cells in a single tier, projecting into or in contact with the pollen chamber. The axial row is reduced to a large egg and a ventral canal cell. Fig. 5 shows a young archegonium of *Pinus* with the central cell which will form the egg and ventral canal cell or nucleus.

ALMA G. STOKEY

References

Eames, A. J., "Morphology of Vascular Plants Lower Groups," New York, McGraw-Hill, 1936.
Chamberlain, C. J., "Gymnosperms. Structure and Evolution," Chicago, The University Press, 1935.
Smith, G. M., "Cryptogamic Botany" 2 vols., New York, McGraw-Hill, 1955.

ganization once were considered to represent the most primitive annelids. However, it is now agreed that the order is not a natural one, that the organizational simplicity displayed by its members is the result of reduction and that the various genera can probably be derived from several, not necessarily closely related, families of polychaetes.

Archiannelids are distinctly segmented polychaetes in which the parapodia are either absent or very much reduced. SETAE are also absent or simple. Many forms show locomotion by means of the metachronous beat of ciliar fields, bands or circles. This mode of locomotion, absent in other adult polychaetes, points towards neotenic or paedo-genetic origin of at least some of the archiannelid types. The nervous system in most forms is situated in the epidermis. A muscular pharyngeal pouch is present in all families except the Polygordiidae. The excretory system is represented by protonephridia in *Dinophilus*, by metanephridia in the members of all other families. Development is direct in the Nerillidae and Dinophilidae, through a pelagic larva (Trochophora) in the other families.

Five families are recognized at present:

Polygordiidae. Many-segmented, long forms, with one pair of appendages at the anterior end. Nonciliar locomotion. Parapodia absent in all, setae absent in most species. The dorsal and ventral blood vessels are connected by segmentally arranged loops.

Saccocirridae. Like the former family but with simple cylindrical parapodia each of which carries a bundle of chisel-shaped setae. The nervous system lies under the epidermis and the longitudinal blood vessels are connected by only one pair of loops.

Protodrilidae. A ventral field and several incomplete bands of cilia may be present. The anterior end is equipped with one pair of appendages. Except in a few species of *Protodrilus* there are no parapodia or setae. The dorsal and ventral blood vessels are present in the anterior part of the body only.

Nerillidae. Body consisting of not more than nine segments. A ventral ciliar field and in some cases also several incomplete circles of cilia are present. In this family the parapodia are moderately well developed, each being supplemented by one cirrus and two bundles of setae.

Dinophilidae. Short animals without parapodia, anterior appendages or setae, but with incomplete circles of cilia. Circulatory system either completely absent (*Diurodrilus*) or very simple (*Dinophilus*).

Most archiannelids live in sandy substrates in the sea, but a few species also penetrate into brackish and even fresh water. Of the latter the most interesting is *Troglochaetus beranecki* which occurs in subterranean caves in Europe.

W. WIESER

Reference

Remane, A., "Archiannelida," *in* Tierivelt der Nord- und Ostsee, Tiel VIa: I-36, 1932.

ARCHIANNELIDA

An order in the class POLYCHAETA, consisting of several forms which on account of the simplicity of their or-

ARCTIC

The exact geographical extent of the Arctic is somewhat indefinite since there is considerable disagreement as to

the criteria that would serve to establish its southern limits. It can be defined, perhaps conservatively, as the areas of sea and land extending from the pole southward to the northern limits of the coniferous forest (taiga). The southern limits of the Arctic correspond roughly to the 10°C mean isotherm for the warmest month.

Climatically, the Arctic is characterized by low temperature and scant precipitation. The region near the sea coast is often foggy in the summer. The soil, permanently frozen to great depth, thaws but little during the summer months; since water cannot escape downward, and there is relatively little evaporation, the northern lowlands are largely covered by water. Inland the weather is cool but with less cloudiness and more evaporation.

The Polar Sea and its adjacent seas are largely covered by ice for much of the year. The pack-ice attains its maximum extent in late winter and, under the influence of off-shore winds and an influx of warmer waters, recedes gradually northward during the summer. Over a period of 2 or 3 months, the ice of the polar pack melts at the upper surface and the floes decrease in thickness. This loss is replaced during the winter, when new ice forms at the lower surface and snow accumulates. At all times the movements of the ice are influenced by the winds; leads open and, in winter, are soon covered by new ice. The closing of leads and other movements produce pressure ridges which account for the typically irregular surface of the pack-ice.

Some of the arctic islands, including Greenland, Ellesmere Island, Iceland, Svalbard, Franz Josef Land, Severnaia Zemlia, and Novaia Zemlia, are permanently covered largely or in part by ice. Icebergs result when large pieces break from the main masses and enter to the sea. Most of the arctic islands have no permanent surface ice, however.

The northern seas contain a remarkably rich invertebrate fauna, which in turn supports a variety of vertebrate animals. Among the marine fishes, the cods (Gadidae), eelpouts (Zoracidae), sculpins (Cottidae), and lumpsuckers and related forms (Cyclopteridae) are particularly prominent. Baleen whales, such as the bowhead (Balaena mysticetus) and the gray whale (Eschrichtius glaucus), migrate into the Arctic in spring and feed during the summer on the abundant nekton. During the warmer months the killer whales (Orcinus) occur in arctic waters, and the white whale (Delphinapterus leucas) is commonly present around the mouths of rivers. The narwhal (Monodon monoceros) is particularly common in the seas around Greenland. Pinnipeds of several species are abundant: walrus (Odobenus rosmarus) migrate into the Arctic Ocean in the spring, but few remain during the winter months; two species of seals, the bearded seal (Erignathus barbatus) and the ringed seal (Pusa hispida), live in close association with the pack-ice; harbor seals (Phoca vitulina) are found commonly around the mouths of rivers; the harp seal (Pagophilus groenlandicus) and the gray seal (Halichoerus grypus) enter arctic waters around Greenland and farther east. The pack-ice is also inhabited by the polar bear (Ursus maritimus), which feeds chiefly upon seals but which can subsist upon vegetation when summering on arctic islands, as individuals sometimes do. During the winter months arctic foxes (Alopex lagopus) are numerous on the pack-ice, where they feed upon the remains of seals killed by polar bears and upon fishes and other animals.

Many species of birds migrate, at least in part, over arctic waters. Particularly prominent are the king eider (Somateria spectabilis), common eider (S. mollissima), glaucous gull (Larus hyperboreus), yellow-billed loon (Gavia adamsii), and old squaw duck (Clangula hyemalis). Cliff-nesting birds, such as the murres (Uria spp.), guillemots (Cepphus spp.), common puffin (Fratercula arctica), and other species of alcids, feed entirely upon marine organisms and are often seen far from land.

The continental arctic lands of Eurasia and North America are predominantly low and relatively flat, with elevations increasing toward the south. Several mountain masses, such as the Brooks Range in Alaska, and the Verkhoiansk, Kolyma, and Anadyr Ranges in Siberia, lie at least partially in the arctic zone.

Floristically, the Arctic is characterized by a lack of arborescent vegetation in the north, but trees often are found at its southern limits where the tundra merges with the taiga. Several classifications of tundra have been defined, but none is universally accepted. The following divisions, beginning at the north, have been proposed: (1) arctic tundra, from which even shrubs are absent; (2) shrub tundra, in which arborescent willows (Salix), dwarf birch (Betula), Ledum, and other shrubs occur; (3) southern tundra, where trees are present mainly along watercourses; (4) forest tundra, an intermediate zone where tundra merges with taiga. The flora of the tundra consists essentially of perennials, many of which are species having holarctic distribution. Decumbent, mat-forming species are prominent, sedges of various kinds are abundant in wet situations, and lichens and mosses also are important.

The terrestrial fauna is remarkably uniform throughout the Arctic. Invertebrates abound, both in water and in the active zone above the permanently frozen ground. Although comparatively few species of insects occur, numerically they are abundant; Diptera of various species, particularly mosquitoes, are much in evidence inland. Few species of fishes, mostly salmonids, inhabit the fresh waters.

Great numbers of birds, representing many species, migrate to the Arctic in the spring for nesting. Many, including snowy owl (Nyctea scandiaca), Pomarine jaeger (Stercorarius pomarinus), snowbunting (Plectrophenax nivalis), longspur (Calcarius lapponicus), glaucous gull, herring gull (Larus argentatus), dunlin (Erolia alpina), red phalarope (Phalaropus fulicarius), willow ptarmigan (Lagopus lagopus), and rock ptarmigan (L. mutus), are distributionally holarctic.

The mammalian fauna of the Arctic is comprised largely of holarctic species, the distribution of which in most cases has been significantly influenced by the presence of the land bridge between Eurasia and North America during late Pleistocene Time. These include the arctic fox, red fox (Vulpes vulpes), wolf (Canis lupus), brown bear (Ursus arctos), ermine (Mustela erminea), mouse weasel (M. nivalis), wolverine (Gulo gulo), lynx (Felis lynx), reindeer (Rangifer tarandus), brown lemming (Lemmus sibiricus), collared lemming (Dicrostonyx torquatus), redbacked vole (Clethrionomys rutilus), tundra vole (Microtus oeconomus), arctic hare (Lepus timidus), and ground squirrel (Citellus undulatus). Other probably holarctic species are mountain sheep (Ovis nivicola), marmot (Marmota marmota), and narrow-skulled vole (Microtus gregalis). Other species are more restricted distributionally.

Of the terrestrial mammals, the arctic fox, the collared lemming, the arctic hares, the ermine, and the mouse weasel have white pelage in winter. In addition, the collared lemming grows specialized claws used in burrowing through snow. Cyclic behavior of microtine populations has its maximum development in the Arctic. Some species, such as the brown lemming, exhibit regular fluctuations in numbers, with a 3 to 4 year periodicity. These changes in density strongly influence the numbers of predatory birds and mammals that depend upon the rodents for food.

Several ethnic groups are represented among the human inhabitants of the Arctic. Prominent among these are Lapps, Yakuts, Lamuts, Yukaghirs, Chukchi, and Eskimos. The latter have the greatest geographic distribution, with settlements scattered from the tip of Siberia across North America to Greenland. These aboriginal peoples are primarily hunters or herders. The nomads of the Eurasian tundra are reindeer herders, but domesticated reindeer were not present in North America and Greenland until introduced comparatively recently by Europeans. Some groups of Eskimos in Alaska and Canada depend almost entirely upon the wild reindeer (caribou). Coast-dwelling people throughout the Arctic depend upon marine mammals for food, shelter, and clothing.

The Arctic regions are now being invaded and significantly modified by man. As a result of modern means of transportation and a rapidly growing world population, many important changes in the Arctic are to be expected within the forseeable future.

ROBERT L. RAUSCH

References

Berg, L. S. "Natural Regions of the U.S.S.R." New York, Macmillan, 1950.

Hultén, E. "Flora of Alaska and Neighboring Territories." Stanford, Stanford Univ. Press, 1968.

Lindroth, C. H., "The Faunal Connections between Europe and North America." New York, Wiley, 1957.

ARCTIC FLORA

Considering the harsh environment in which it ekes out its often precarious existence, the versatile arctic flora, at least in favorable situations for a brief period in summer, may be astonishingly attractive and even beautiful. Although akin to the alpine flora occurring at high elevations to the south, and showing comparable diversity due to varying geographical position and often still more marked and local changes due to habitat conditions, it forms a characteristic complex ranging around the top of the world. Precisely the same remarks apply to the plant communities comprising ARCTIC VEGETATION; and just as arctic plant taxa are often notoriously variable and difficult to classify, so are arctic vegetational eca, which, in favorable situations especially in the south, may be relatively luxuriant.

If, as seems desirable, the arctic lands are considered as confined to those lying north of the limit of arborescent growth (or north of the temperature-based Nordenskiöld Line where this lies farther north still), the total flora of the arctic lands and fresh waters is still very considerable, while the vegetation is widely diverse. Yet it tends to be in the sea that life abounds, and many large algae—including kelps several meters in length—are to be found in the waters surrounding truly arctic lands. Taking in general a rather conservative view of specific limits, the total known representation and characteristic habitats of the main plant groups in the Arctic may be indicated as follows, though undoubtedly numerous additions remain to be made especially in the lower groups (i.e. up to and including the mosses).

Bacteria. Although bacteria have been relatively little studied in the Arctic, it has long been known that a wide range of types occur—not only in the soil and air, but also in fresh and salt waters. Whereas in most respects the bacteria found in the Arctic appear to be similar to their counterparts living to the south, they tend to have lower temperature optima.

In soils, although the numbers of individuals are usually far smaller in arctic than in temperate regions, the types are much the same, and include non-symbiotic nitrogen-fixing, root-nodule, nitrifying, dentrifying, and both aerobic and anaerobic cellulose-decomposing types. They often go down as deeply as the soil itself. Actinomycetes also occur, and exhibit antibiotic activity.

In salt-water habitats in the Arctic, bacteria are often plentiful—especially in muddy bottom-deposits—and include putrefying, nitrifying, dentrifying, and both aerobic and anaerobic nitrogen-fixing types. They develop at temperatures down to at least $-7°C$. and may total many millions of individuals per gram of bottom mud.

In arctic sea water the density of bacteria tends to be low except where there is mixing with southern currents, where up to an estimated 500,000 or occasionally more individuals per cc. have been observed. Bacteria also occur in freshwater lakes and streams in the Arctic, in the pools of more or less fresh water that form in summer on the sea-ice, and in the atmosphere; indeed, as in other parts of the world, scarcely any possible habitat seems to be free of them.

The bacteria found in arctic air are common soil types but rather sparse, rarely exceeding a few individuals per cubic foot; however, they persist even above the North Pole—at least at low altitudes in summer. Surprisingly enough, the intestinal tracts of a wide range of arctic animals appear often to be sterile or nearly so.

Algae. In the Arctic, algae seem at least as diverse and ecologically important, and may be almost as numerous, as in many regions of more genial climate. Thus most of the major systematic groups of algae are well represented in the salt or fresh waters, where conditions tend to be more 'even' than in adjacent aerial habitats, and in addition numerous forms occur in the Arctic on damp soil, on damp rocks and the undersides of stones, in moist tufts of moss, on manured tracts, on other algae, and on ice-floes and even snow.

As regards numbers, although most arctic areas of land and sea are unexplored algologically, probably more than 2,000 different species of algae are already known from the Arctic, a considerable number being apparently limited thereto. Concentrations of the order of a million phytoplankters per liter have been observed in arctic seas near coasts, at the edge of melting ice, and where currents meet. As indications of the richness of the arctic algal flora it may be noted that over 500 species of freshwater diatoms (Bacillariophyceae) have already been reported from more or less arctic regions, and about 200 species of Dinophyceae (Peridinians) from arctic seas—where in general diatoms are more important. From the writer's inexpert collections made in one summer in the Canadian Eastern Arctic, some 530 species and many additional lower taxa of algae were identified, including 184 species of demids.

Even more than plants of most other groups, the smaller arctic algae tend to be extremely widespread, the same microscopic species turning up again and again in suitable habitats almost throughout the region, and in some cases occurring in fair numbers even in the vicinity of the North Pole. The larger types are often less widespread and more distinctly arctic; they may form luxuriant growths even where the temperature does not exceed 0°C. at any time of the year, algae being apparently better adapted than most other groups to growing at low temperatures. In some places benthic algae extend down to depths of at least 118 meters in arctic seas. Yet still more than in most other groups of plants does the algal flora vary with the often contrasting habitat conditions in the Arctic. Thus in sheltered situations on rocky coasts Brown (and to a lesser extent Red and Green) Algae are often plentiful, the Fucaceae sometimes forming thick mats, while below the influence of tides and grinding ice, stable rocky surfaces may support luxuriant 'beds' of Laminariaceae with thalli up to some 15 meters long.

In the open seas at 'peak' periods of the year, prolific phytoplanktonic maxima are often developed—particularly of diatoms and dinophyceae, which may render the water turbid as far down as light-penetration allows. Such phytoplanktonic algae are of course the main ultimate source of food of almost all animals.

On land the innumerable lakes and shallow tarns upheld by the permanently frozen substratum, and the streams and seepage waters especially of lasting duration, support a wide range of algae, and even small peaty pools and ephemeral puddles may teem with desmids, diatoms, or sometimes flagellates. The muddy shores of receding lakes and drying tarns may be strewn with colonial Cyanophyceae often several centimeters in diameter, while the beds of streams may support abundant mats or tassels of filamentous or other Green Algae. On persistently manured areas such as occur around bird-cliffs and native settlements, thalloid Green Algae may form soft mats, while other sizable Green Algae may occur in hot springs, brackish marshes, and run-off areas. Diatoms, particularly, may abound in the melt-water pools on the sea-ice in summer, forming considerable aggregations that render the aspect a dirty brown, while various algae may form a 'bloom' on the surface of the ice, or occur on the sides and undersurfaces of the floes. On land or sea-ice, considerable areas of snow or névé may have the surface colored by red-snow or other algae.

Fungi. These again are well represented practically throughout the arctic lands, though generally they are little in evidence and still so inadequately studied that it is scarcely possible even to guess at the order of number of the species occurring in truly arctic regions. But more than 850 being so far recorded from Greenland alone, it seems likely that the fungi will yet vie with the algae as the major group contributing the greatest number of species to the arctic flora.

The four main groups of fungi proper are all well represented in the Arctic, the majority of whose fungus denizens are widespread or at least known in cool-temperate regions, although some appear to be restricted to the boreal regions or even to be arctic endemics. Even these last are probably often circumpolar in range.

Many arctic fungi are parasitic especially on higher plants, while all groups of plants appear to be widely attacked by fungi in the Arctic, where the host-range tends to be wider than to the south. But although the species of fungi are probably at least twice as numerous as those of vascular plants in the Arctic, and may even outnumber these *plus* lichens *plus* bryophytes (which together comprise the main bulk of land vegetation in the most unfavorable situations), as actual components of the vegetation the fungi are relatively minor.

In accordance with their needs to be either saprophytic or parasitic, fungi tend to be more exacting in their habitat requirements than members of most other groups. Yet they are found in at least most land and freshwater habitats in the Arctic, including almost all areas of closed or not-too-sparse vegetation. Particularly productive of conspicuously fructifying Basidiomycetes are the damper areas of mossy tundra—especially where manured, as around bird-cliffs, scavengers' perches, nesting colonies, and foxes' earths. Animal excreta and dead bodies of animals and plants afford numerous micohabitats for fungi in the Arctic, where the slowness of decay favors the saprophytic forms. Arctic soils also support plentiful fungi, even if their numbers and activity tend to be less than to the south, while the air readily and probably universally transports fungal spores. Thus viable yeast and other fungal spores have been found in the atmosphere northwards practically to the North Pole, and the winds easily blow fragments of plants with resistant spores or perennial mycelium over the ice in winter and spring. Fungi also occur in arctic fresh waters, where the wide range of forms observed include many that attack living algae and some that grow on fishes, while there seems every likelihood that many others will be found in the sea. Even the algae occurring on the sea-ice in the Arctic can support some fungi, while other fungi on land are linked mycorrhizally to higher plants. Still others occur symbiotically in the bodies of lichens.

Lichens. These 'dual' organisms, consisting of photosynthesizing algal cells living symbiotically in a protective fungal 'envelope', tend to be far more prevalent in arctic lands than in most other regions. Frequently lichens carpet and may dominate considerable areas in the Far North, where the number of their species may greatly exceed that of the vascular plants. Thus against the 142 species of vascular plants now known from the Spitsbergen Archipelago, it was already estimated in 1938 that about 400 species of lichens had then been collected there, while from Greenland are known about 500 species of lichens, and from Novaya Zemlya 456, though doubtless more remain to be discovered in all arctic territories. In any case it seems that lichens, like the algae, fungi, and possibly also mosses, are each represented by more different species in the Arctic than are the vascular plants.

Lichens are extremely hardy organisms that may grow almost alone on rock surfaces in the most rigorously exposed situations, and become more and more prevalent as we go farther north in the northern hemisphere and the ranker competitors tend to disappear. Their habitats are almost exclusively on land, where they tend to grow on the drier surfaces. Thus the faces of rocks and stones and old bones, mossy tussocks

and bare or only partially vegetated earth (if not disturbed by solifluction etc.), and the bark of woody plants
or dead parts of herbaceous ones, are all apt to support a considerable range of crustaceous microlichens
or, in favorable situations, foliose or fruticose macro-
lichens. Yet lichens are extremely sensitive to chemical differences of the substratum or even air, and,
whereas many grow extremely slowly and may attain a
very great age, some are obviously favored by manuring and a moist atmosphere, so that out-sized specimens of nitrophilous species are often to be found
around bird-cliffs and scavengers' perches.

Whereas the vast majority of lichen species occurring
in the Arctic appear to be widespread both there and
(commonly at higher elevations) to the south, many
being circumpolar in range, there seem to be a fair number of arctic endemic lichens. Fruticose and foliose
lichens are widely important ecologically as 'fillers,'
especially in the drier types of vegetation in boreal regions, and they are also economically significant in composing much especially of the winter feed of caribou and
domesticated reindeer. Some can also be used, after
treatment, as emergency food for man.

Liverworts. Although they constitute a relatively
small group in terms of number of species, the liverworts (*see* HEPATICAE) are quite well represented in the
Arctic, where it is to be expected that some 300 species
may in time be found, and where some are apt to be of
considerable ecological significance not merely as indicators but also, very locally, as constituents of the
vegetation.

As regards floristic richness, whereas already nearly
100 species of liverworts are known from the Canadian
Eastern Arctic, it is almost certain that wider specialist
investigation would add more—as elsewhere probably
throughout the Arctic. So far at least fifty different
liverworts are known from Spitsbergen, more than
twice that number from the Soviet Arctic mainland,
and no fewer than thirty-nine from the small island of
Jan Mayen.

Liverworts occupy much the same habitats as mosses
(*see* below), apart from a tendency to avoid the more
persistently dry or otherwise inhospitable situations,
and are very often found intermixed or even intricately
interwoven with mosses or other plants. Individual
species tend to be rather exacting in their habitat requirements, but widespread and often circumpolar in
distribution. Thus in manured or otherwise disturbed
habitats, e.g. around settlements, Marchantiaceae
widely form characteristic mats.

Mosses. With their capacity for withstanding very
low temperatures, their common attachment to humid
conditions which widely prevail at least around the time
of melting of the snows, but on the other hand their
extraordinary ability to survive desiccation for prolonged periods if necessary, mosses (*see* MUSCI) are
plentiful and often ecologically important in the Arctic.
Although far more specialist investigation is needed,
the number of species of mosses occurring in the Arctic
is already known to be considerable, over 200 having
been reported from the Spitsbergen Archipelago, at
least 313 from the Canadian Eastern Arctic north of
the 60th parallel, and apparently the better part of 500
from Greenland.

Like other non-vascular plants, mosses lack true roots
and depend for water largely on absorption from any
surface film of moisture. Consequently they are less dependent than higher plants on the thawing of the substratum, being able to take advantage of microclimatic
rises in temperature which, with high insolation in the
arctic spring, may bring their dark (often reddish or
brownish) cushions and mats above freezing-point. Thus
in the manner of many other lowly plants, mosses are
apparently able to carry on their vital activities even
when the surrounding air may be well below freezing-
point—a faculty which may be expected to stand them
in particularly good stead in the Arctic.

The habitats of mosses in the Arctic are many and
various, including brackish marshes but apparently
nowhere of full sea-water salinity. Thus they can occupy
practically the complete range of terrestrial and freshwater habitats—from swiftly running water in which
long soft tassels may waver, to exposed rock faces on
which tiny hard tussocks eke out a seemingly precarious
existence. They are often abundant on rocky ledges, especially where water seeps. In lakes and tarns of a relatively stable nature, large and coarse types may form
luxuriant brown 'beds' sheltering a wide range of diatoms and other algae, and even streams and pools that
dry up in late summer often support some amphibious
mosses.

In the manner of lichens in the drier situations,
mosses tend to be by far the most important 'fillers' of
higher vegetation in mesophytic habitats, while in
damper places they may be subdominant to the virtual
exclusion of lichens and, except in very wet places, also
of algae. Sphagna (Bog-mosses) may even be dominant
in wet places, and other types may form luxuriant mats
in run-off areas below long-lasting banks of melting
snow or in manured situations such as perennial nesting-
grounds of wildfowl. The hillock tundras of marshy
lands also introduce a whole set of microhabitats for
mosses in arctic regions—ranging from the drier tus-
sock-tops and somewhat damper sides, to the aquatic
runnels where amphibious types flourish between the
hillocks. The intervening tracts of many of the less
dynamic 'polygon' soils often support a fine array of
mosses—sometimes forming extensive cushion-like investments—while on solifluction slopes mosses frequently show the advantage of quicker growth which
they have over lichens. On windswept or water-washed
sandy banks, mosses often act as effective stabilizers,
while on the surfaces of rocks especially in crevices a
wide range of mosses occur, and many grow among
detrital materials or on almost bare stony ground in exposed situations (including hill-tops). Other habitats include old bones, feces, late-snow areas, and the burrows
and lairs of mammals as well as the immediate environs
of owl-perches.

Being extremely sensitive to environmental conditions, the moss flora often changes entirely from one
small spot to the next. Yet quite a large proportion of
arctic species of mosses appear to be fully circumpolar
in range and also widely distributed to the south especially on mountains, few being endemic—as is to be
expected *inter alia* in view of the frequent occurrence
of moss spores in considerable numbers in the atmosphere. Many arctic moss denizens are so markedly depauperate when compared with their southern congeners as to be unrecognized at first by workers familiar only with areas to the south.

Pteridophytes. Of these, comprising the ferns and
their so-called allies (*see* PTERIDOPHYTA), 36 species, belonging to 6 different families, are now known from

truly arctic lands. All appear to be fully native. They comprise 7 Equisetaceae, 5 Lycopodiaceae, 3 Selaginellaceae, 2 Isoetaceae, 4 Ophioglossaceae, and 15 Polypodiaceae. Their habitats tend to be typical of the genera concerned, ranging from shallow fresh water to dryish heaths etc. for different Equisetaceae, being mostly heathy areas in the cases of the Lycopodiaceae and Selaginellaceae, muddy lake-bottoms for the Isoetaceae, turfy sands and open gravelly heaths for the Ophioglossaceae, and rock crevices, stony slopes, and damp mossy or humus-rich shady situations for the Polypodiaceae.

Gymnosperms. Of these only 6 species appear recognizable as arctic, including 5 Pinaceae (*see* GYMNO-SPERMAE) which exceed sufficiently the poleward forest limit that they usually form. In the Arctic these are found chiefly in rather dry rocky or sandy but usually sheltered situations, although some may occur in damper places by rivers or even in bogs. The remaining species is a member of the Cupressaceae, and it appears more properly to belong to the Arctic, in which it is fairly widespread as a low bush growing in dry sandy or rocky situations far beyond the tree-line.

Monocotyledons. Our latest accounting totals 226 species belonging to 11 families of MONOCOTYLEDONS in the Arctic, as follows: 2 Sparganiaceae, 1 Zosteraceae, 10 Potamogetonaceae, 2 Juncaginaceae, 92 Gramineae, 80 Cyperaceae, 2 Lemnaceae, 21 Juncaceae, 8 Liliaceae, 1 Iridaceae, and 7 Orchidaceae. With the exception of about 10 grasses that have evidently been introduced, all these appear to grow in the Arctic as natives. Although not a few occur in salt marshes and other saline habitats near sea-shores, only 1 (Zosteraceae) grows in the sea. Otherwise the habitats are orthodox land and freshwater ones, ranging from shallow lake-bottoms to high mountain-summits, which a few hardy Juncaceae and Gramineae often reach. In general the characteristic habitats of particular families and genera in the Arctic are closely comparable with those recognized to the south, where indeed the vast majority of monocotyledonous species known from the Arctic also occur; consequently these habitats need not be described here. But although many arctic monocotyledonous species are widespread in temperate regions, and at least 20 are now known to be fully circumpolar to the extent of occurring in all 10 sectors of the Arctic, a sprinkling of at least 10 species nevertheless appear to be endemic to the Arctic, or anyhow not yet to have been found to the south.

Although in the Arctic the cryptogamic groups tend to be more in evidence than in most other regions, the main bulk of arctic land vegetation, except in extremely unfavorable areas, is still made up of the present and following groups, which together comprise the Angiosperms. The angiospermous genus with the largest number of known and recognized arctic species is *Carex,* with 67, comprising the vast majority of arctic Cyperaceae.

Dicotyledons. Of these (*see* DICOTYLEDONS) our latest accounting of arctic species totals 624, belonging to 47 families as follows: 40 Salicaceae, 5 Betulaceae, 1 Urticaceae, 15 Polygonaceae, 3 Chenopodiaceae, 7 Portulacaceae, 50 Caryophyllaceae, 38 Ranunculaceae, 5 Papaveraceae, 1 Fumariaceae, 57 Cruciferae, 1 Droseraceae, 4 Crassulaceae, 31 Saxifragaceae, 41 Rosaceae, 29 Leguminosae, 4 Geraniaceae, 1 Linaceae, 1 Polygalaceae, 4 Callitrichaceae, 1 Empetraceae, 7 Vio-laceae, 1 Elaeagnaceae, 8 Onagraceae, 2 Haloragaceae, 1 Hippuridaceae, 11 Umbelliferae, 2 Cornaceae, 1 Diapensiaceae, 6 Pyrolaceae, 18 Ericaceae, 17 Primulaceae, 1 Plumbaginaceae, 12 Gentianaceae, 6 Polemoniaceae, 11 Boraginaceae, 5 Labiatae, 38 Scrophulariaceae, 1 Orobanchaceae, 6 Lentibulariaceae, 4 Plantaginaceae, 7 Rubiaceae, 1 Caprifoliaceae, 1 Adoxaceae, 1 Valerianaceae, 4 Campanulaceae, and 113 Compositae.

In the number of arctic representatives that may be accorded specific rank, the Compositae is thus the largest family among the vascular plants; but perhaps to an even greater extent than in the largest monocotyledonous family, the Gramineae, its numbers are swollen by apomictic and other narrow and even dubious species.

Except that they do not include any fully marine representative, the Dicotyledons occupy the same vast range of habitats in the Arctic as do the Monocotyledons. These include salt marshes and saline shores as well as shallow freshwater habitats and practically all others upwards to high mountain peaks, where a few depauperate Caryophyllaceae, Papaveraceae, and/or Cruciferae are commonly to be found.

Again with the Dicotyledons it can be said that the characteristic habitats of particular families and genera in the Arctic are closely comparable with those recognized to the south, where indeed the vast majority of species known from the Arctic also occur. Many of these are moreover widespread in temperate regions especially at the higher elevations, while at least 38 species of Dicotyledons are now known to be fully circumpolar in range, occurring in all 10 sectors of the Arctic; nevertheless some 30 (mostly rather narrow) species of Dicotyledons appear to be endemic to the Arctic. The dictotyledonous genus with the largest number of known and recognized arctic species is *Salix,* with 40.

The vascular plants, comprising the last four groups dealt with above, are thus represented in the arctic by a total of 892 species belonging to 230 genera placed in 66 different families. But although a considerable number of additions doubtless remain to be made to at least the first of these figures, especially in the southern tracts of the Arctic when they come to be more thoroughly explored, it is particularly among the lower cryptogams that, with due specialist investigation, by far the most numerous additions and striking range-extensions are to be expected. Each of their main groups—algae, fungi, lichens, and bryophytes—is probably represented by more species and lower taxa in the Arctic than are all the vascular groups together.

NICHOLAS POLUNIN

References

Polunin, N., W. R. Taylor, R. Ross, *et al.,* "The Cryptogamic Flora of the Arctic," Bot. Rev. **20**: 361–476, 1954.
Polunin, N., "Plant Sciences in the Arctic and Subarctic, vol. 1: Exploration, Taxonomy, and Phytogeography," Oxford, Clarendon, *in preparation.*
ibid., "Circumpolar Arctic Flora," Oxford, Clarendon, 1959.

ARCTIC VEGETATION

North of the scattered trees (subarctic 'taiga') or intermittent timbered tracts (in 'forest-tundra') that between them terminate the poleward limit of forests and still belong to the Subarctic, a zone of relatively luxuriant 'tundra' etc. is normally found. This tends to become thinner and poorer to the north, and extremely sparse as the northernmost lands are approached, while in the sea a similar tendency is apparent. Nevertheless relatively luxuriant plant communities may occur very locally at surprisingly high latitudes in the most favorable situations both on land and in the sea. For in spite of its tree-lessness and general dwarfishness, which give super-ficially an impression of monotonous sameness, the vegetation of arctic regions varies very markedly from place to place. This variation is often extreme in closely contiguous areas of different habitats, and, while commonly dependent on more or less drastic habitat changes, may even occur without the basis of any evi-dent marked difference in environmental conditions— suggesting that repeated readjustments to disturbance outweigh any tendency to equilibrium.

The striking variability of arctic vegetation, often from one tiny area to the next, depends *inter alia* on the absence of sufficient plant growth to control the physical forces of the environment, among which the climatic and geodynamic tend to be particularly strong. Against these the vegetation is relatively impotent, and so the struggle of arctic plants tends to be with the inimical forces of a harsh physical environment rather than with hostile competitors as is widely the case to the south, though there is still plentiful competition between plants in the more favorable arctic habitats.

With this commonly strong physical control, it is per-haps not surprising that the vegetation developed under similar habitat conditions in any particular climatic belt ranged around the top of the globe tends to look much the same in whatever sector it may lie. Accordingly the progressive depauperation from south to north allows general subdivision on the basis of vegetational physiog-nomy into three main belts: (1) the low-arctic belt (e.g. southern Greenland), in which the vegetation is con-tinuous over most areas that are not too exposed, (2) the middle-arctic belt (e.g. northern Baffin Island), in which the vegetation is still sufficient to be widely evident from a distance, covering most lowlands, and (3) the high-arctic belt (e.g. Spitsbergen), in which closed vegetation is limited to the most favorable lowland habitats and is rarely at all extensive.

Tundras. Arctic tundras are dealt with in the separ-ate article entitled "Tundra" (*q.v.*).

Scrub and heathlands. In lands of the low-arctic belt, a shaggy scrub of Willows (*Salix* spp.) and/or Birches (*Betula* spp.) is commonly developed on the most favored slopes, in damp depressions, and es-pecially along watercourses and the margins of lakes. This scrub is commonly around 60 cm. high, but tends to become lower and more restricted northwards until, about the center of the middle-arctic belt, it usually becomes very thin as well as limited in extent and stat-ure. On the other hand in the most sheltered situations in the extreme south, the Willows may be luxuriant and even exceed the height of a man, and the Birches in some places may become arborescent. The main domi-nants in different regions are the Dwarf Birch (*Betula nana* agg.) or Scrub Birch (*B. glandulosa* agg.), or such shrubby Willows as the Glaucous Willow (*Salix glauca* s.l.) or the Feltleaf Willow (*S. alaxensis* agg.). Often two or more such shrubs will dominate a mixed association, while in some places bushes of Green

Alder (*Alnus crispa* agg.) may be locally dominant. Such scrub at its best is so thickly tangled and pro-duces so much litter that few associated plants occur, apart from tall grasses such as Bluejoint (*Calamagrostis canadensis* agg.) and occasional straggling forbs. But where the dominants are less luxuriant, an extensive flora is often found, including a considerable variety of herbs and mosses. In dry situations there may be a well-developed subdominant layer of heathy plants such as Crowberry (*Empetrum nigrum* s.l.), with character-istic open patches of tall Cladonias, Stereocaulons, and other lichens, with or without Polytricha or other coarse mosses.

To the north such scrub thins out gradually, its most northerly expression about the northern limit of the middle-arctic belt being usually in the form of single or scarcely confluent, straggly birch or willow bushes that rarely exceed 50 cm. in height and are usually much lower, though often widely-spreading.

Heathlands are relatively widespread and various in the Arctic, though still commonly occupying only a very small proportion of the total land area. They are usually characterized by being dominated by members of the heath family (Ericaceae), or by heath-like plants such as, particularly, Crowberry, though sometimes broad-leafed plants such as Avens (*Dryas* spp.) are impor-tant. These arctic heathlands tend to be confined to the more favorable, sheltered situations that are snow-covered in winter—provided they are not too moist in summer. They are particularly characteristic of coarse-grained soils. In the low-arctic belt they are usually covered by a continuous thick sward of mixed woody and herbaceous plants, the main dominants being typi-cally 8–15 cm. high. These commonly include Crow-berry, Arctic Blueberry (*Vaccinium uliginosum* subsp. *alpinum*), Mountain Cranberry (*V. vitis-idaea* agg.), Arctic Bell-heather (*Cassiope tetragona*), Narrow-leafed Labrador-tea (*Ledum palustre* agg.), Dwarf or Scrub Birch, and various diminutive Willows. Often the dominants themselves are much mixed, and usually they are consolidated below by a layer of cryptogams in which mosses or lichens commonly subdominate ac-cording to whether the situation is relatively moist or dry, respectively. In the drier situations there may occur frequent gaps in the heath, which are actually dominated by lichens—particularly by 'Reindeer-moss' Cladonias that may form a sward 5 or more cm. high. On the other hand in depressions or behind ob-structions where snow drifts more deeply in winter, a characteristic dark sward dominated by Arctic Bell-heather usually develops, often with associated sedges (*Carex* spp.) and mosses at least where the soil is last-ingly damp.

In the middle-arctic belt, heathlands are usually somewhat lower in stature and more restricted in area than to the south, having the appearance of postcli-maxes developed in the most favorable situations. Of the above-mentioned dominants Mountain Cran-berry has usually disappeared, and even though the taller individuals of other plants may still exceed 20 cm. in height, the sward is usually only 5–10 cm. high. Al-though it may still be fairly dense, more often the 'heath' is of scattered ground-shrubs with interven-ing thin patches of Cetrarias, Alectorias, and other lichens.

In the high-arctic belt heathy plants are entirely ab-sent over considerable areas, the tracts that are popularly spoken of as 'heaths' being usually dominated by sedges or lichens or occasionally mosses. However, Crowberry and Arctic Blueberry are to be found in some regions, dominating limited heathy communities in unusually favorable situations, while Arctic Bell-heather is quite widespread, characteristically forming a dark tract where snow accumulates sufficiently to form a good protective covering in winter.

Fell-fields and barrens. These are types in which the evident vegetation occupies less than half of the area; and whereas the two categories are scarcely to be rigidly distinguished, it is usually those tracts that bear relatively few and scattered plants that are referred to as "barrens." Fell-fields typically have a surface of frost-shattered "detrital" material including much finer "soil," and usually support fairly numerous different species forming mixed communities, whereas barrens are apt to be characterized by one prominent size of particle and a single species of vascular plant, such as Mountain Avens (*Dryas octopetala* s.l.) or Purple Saxifrage (*Saxifraga oppositifolia* agg.). Such one-plant characterization is especially marked on barrens occupying the most exposed situations. Where sufficient moisture is present, these areas are commonly disturbed by all manner of frost-heaving and allied effects—such as solifluction on slopes and polygon-formation on the flatter terrain. The solifluction streaks and polygons are commonly accentuated by vegetation in one or more of many various ways, while on steep slopes and below weathering crags still more dynamic and often poorly vegetated 'screes' are common.

In low-arctic regions fell-fields, barrens, etc., are found chiefly in upland districts and in exposed areas near the coast—especially where the substratum is of porous material and the surface is consequently dry. Scattered tussocky Avens, hardy Sedges, and various tufted Saxifrages, Drabas (*Draba* spp.), Poppies (especially *Papaver radicatum* s.l.) and other herbs, may form a fell-field or 'half-barren,' sometimes including irregular but limited patches of matted vegetation. A fair number of cryptogams are often intermixed, though usually they are of poor growth. In the most unfavorable situations of all, such types may thin out to a stony barren supporting little more than diminutive crustaceous lichens and very occasional depauperate tussocks—particularly of Purple Saxifrage.

Although such poorly-vegetated areas tend to be more numerous and extensive in the middle-arctic belt than farther south, they do not normally occupy the general run of lowland terrain but are chiefly encountered in exposed situations. In most high-arctic regions, however, 'open' and often extremely sparse vegetation is the general rule, and so fell-field and barrens areas are widespread and plentiful. Commonly found in exposed situations are monotonous barrens of gravel supporting terricolous lichens of poor growth and only occasional diminutive herbs or tufts of Avens, Purple Saxifrage, or Arctic Poppy (*Papaver radicatum* s.l.), the general aspect being desolate in the extreme. Only in the less unfavorable situations may more mixed fell-fields occur, while on favorable banks there may be Grasses and some hardy but attractive forbs.

Seaside and other local types. In both low- and middle arctic belts there are usually scattered plants of Sea-purslane (*Arenaria peploides* agg.) and Sea Lungwort (*Mertensia maritima* agg.) on sandy maritime fore-

shores and, farther up, stabilizing beds of Lyme-grass (*Elymus arenarius* s.l.) which may be fairly tall and luxuriant. In the high-arctic belt, Lyme-grass is unknown and the other two are rare, so exposed sandy and shingly shores are liable to be barren around high-tide mark.

In sheltered and less well-drained seaside areas, muddy or sandy 'salt marshes' are common though usually of very limited extent in the Arctic. In low-arctic regions they typically consist of a dwarfish grassy sward dominated by Alkali-grasses (particularly the Creeping Alkali-grass, *Puccinellia phryganodes* agg.) or sometimes by Salt-marsh Sedge (*Carex salina* s.l.), with associated Low Chickweed (*Stellaria humifusa*), Scurvy-grass (*Cochlearia officinalis* s.l.), Pacific Silverweed (*Potentialla egedii* agg.), and other halophytes. Except for the usual absence farther north of the Silverweed, and the substitution of Salt-marsh Sedge by the possibly conspecific Hoppner Sedge (*Carex subspathacea*), the same plants generally play a similar role in middle- and high-arctic regions, though with progressive depauperation to the north.

The perennial manuring around bird-cliffs, where countless sea-birds nest every summer, leads very locally to unusually luxuriant vegetation even in exposed situations. For here unoccupied ledges may support coarse Grasses and rank Scurvy-grass, the rock faces being covered by lichens often of considerable size—even in the far north. The tops of bird-cliffs typically support a rich grassy sward at least in damp depressions, and, stretching back for 100 meters or so, a luxuriant and dense 'patchwork quilt' of mixed and variously-colored lichens and mosses—due apparently to manuring *inter alia* by scavengers and predators. The situation being usually very exposed, the adjacent unmanured cliffs and hinterland are apt to be contrastingly barren. Also striking are the grassy or flower-decked swards, often consolidated by aggrandized cryptogams, that develop in manured areas around human habitations, mammalian burrows and lairs, owl-perches, and the nesting grounds of geese, eiders, and other gregarious wildfowl.

A common and widespread type of local effect is that engendered by the drifting and late-melting of snow, which tends to take place similarly each year and to lead to characteristic vegetational zonation within the area of the drift. Owing to the later and later disappearance of the snow and consequent abbreviation of the growing-season as the center of such a snow-patch is approached, the vegetation tends to constitute a zoned series of subclimaxes. In the low- and middle-arctic belts the outermost zone of such snow-patches, which is well protected by snow in winter but does not have its growing-season markedly shortened by late melting, is commonly vegetated by a luxuriant mixed heath, or, in lastingly damp situations, by a thin willow scrub. In the far north, Arctic Bell-heather is commonly dominant just here, whereas to the south it usually forms a characteristic dark belt farther in, where the snow drifts sufficiently deeply for the growing-season to be appreciably shortened by late melting. Inside come zones which vary considerably in number and vegetation in different places and circumstances, but which typically include towards the outside a zone of Dwarf Willows—particularly the Herb-like Willow, *Salix herbacea*—and, farther in where the growing-season is too short for woody plants, a sparsely-vegetated zone with a considerable variety of bryophytes and open-soil herbs such as Mountain Sorrel (*Oxyria digyna*). Around the centers of the deeper drifts, where the snow melts only towards the end of summer or sometimes not at all, most herbs are unable to persist and, except in the run-off below, even cryptogams are little in evidence, though some small tufts of bryophytes and investments of algae are usually to be found, together with the diminutive grass *Phippsia algida* agg. (Frigid Phippsia). Owing to the local retardation and shortening of the growing-season in such late-snow areas, Saxifrages, Buttercups, Drabas, and other attractive forbs, are often to be seen still in flower just here at the onset of winter.

Seral types. Additional seral types on land include the marshy and boggy ones of hydroseres, various ones of xeroseres, and the 'flower-slopes' that probably belong to mesoseres. The hydrosere of arctic lakes and tarns usually has as its first aerial stage a 'reed-swamp' of aquatic Sedges (particularly the Water Sedge, *Carex aquatilis* agg.) and/or Cotton-grasses (particularly the Tall Cotton-grass, *Eriophorum angustifolium* agg.), although sometimes Common Mare's-tail (*Hippuris vulgaris* s.l.), or such coarse Grasses as Tawny Arctophila (*Arctophila fulva* agg.), may largely or wholly take their place. Any of these plants may form luxuriant 'beds' where the bottom is soft and not more than 40 cm. deep, projecting up to about a similar height above the surface of the water in the south but usually less in the far north. Commonly they are accompanied by aquatic mosses and numerous small algae. Behind stretches typically a marshy sedge-meadow with the same 'grassy' dominants and, in addition, *Arctagrostis* and lowly Willows. This in turn commonly merges into damp tundra. Alternatively, especially in the southernmost regions, boggy areas dominated by Bog-mosses (*Sphagnum* spp.) may be developed around pools in peaty tracts and be colonized by Baked-apple (*Rubus chamaemorus*) or heathy plants. On the other hand in the far north, many tarnside areas may remain uncolonized by higher plants, yet display their instability by supporting adjacent beds particularly of the attractive Scheuchzer's Cotton-grass (*Eriophorum scheuchzeri*).

Lithosere stages are abundant in the Arctic, where much of the terrain is of more or less bare rock that has, in many cases, been freed from glaciation only in relatively recent times. Yet it is evident that succession is proceeding, however slowly, at least in areas that are not too rigorously exposed or lastingly snow-covered or geodynamically disturbed. Thus rock and boulder faces are apt to be largely invested with crustaceous and foliose lichens, and to occupy considerable areas, while rock crevices often support higher life-forms, so that in time a moss-mat or mixed herbaceous community develops, and, ultimately, heathy vegetation in suitable situations. Screes, if not too active, may also be bound by hardy plants, and have dark strips of vegetation extending down their slopes. Also commonly stabilized by vegetation are inland sandy areas, though the psammosere may advance little beyond the pioneer stage of sand-binding grassy plants, mosses (such as *Polytrichum* spp.), or ground-shrubs (such as Crowberry and Alpine Bearberry, *Arctostaphylos alpina* agg.). On both rock and sand, a dense mat of the 'silvery' moss *Rhacomitrium lanuginosum* may cover substantial areas and apparently persist for many years, though in time it is usually colonized by lichens and grasses, etc.

The mesosere is represented by relatively short-lived communities in such favorable situations as alluvial deltas and the beds of receding tarns, and apparently by long-lived types on the earthy or gravelly 'flower-slopes' that form such a pleasing feature on steep south-facing inclines particularly in low-arctic lands. These flower-slopes develop under an unusually favorable combination of conditions of shelter, aspect, water, aeration, and soil, and support a large and various assembly of flowering forbs which typically grow much-mixed and in such profusion that the usual dominants are excluded.

Fresh and salt waters. Although Pondweeds (*Potamogeton* spp.) occur, and a few Buttercups (particularly *Ranunculus trichophyllus* agg.) and many more mosses grow as floating-leaf or more often submerged aquatics, such vegetation is relatively little developed in the Arctic. Yet in fresh waters a vast array of mostly microscopic algae—particularly diatoms and desmids but including some larger filamentous, etc., types—are to be found, most notably in reed-swamps and moss beds developed in shallow water in sheltered situations, or in algal etc. 'tassels' in streams. Thus from six samples taken in as many small vials from shallow pools in southern Baffin Island in July and August, 1936, no less than 179 different species and varieties of algae were determined—including some benthic types but many more desmids, which are particularly characteristic of the peaty pools—and in general the arctic freshwater algal flora appears to be very diverse, though not making much of a 'show' of vegetation. However, diatoms etc. may form a brownish encrustation on aquatic vegetation and color the beds of pools, as may investments of Blue-green Algae (Cyanophyceae) and filamentous types, and the waters of tarns may appear almost soupy (especially when viewed from aircraft) at 'peak' periods of phytoplanktonic activity in low-arctic regions, though in general the phytoplanktonic development in arctic lakes appears to be poor. Nevertheless some 200 species of algae have been recorded from freshwater plankton in the Arctic, about half of these being Green Algae.

In arctic seas there are apt to be phenomenal outbursts of diatoms when the sea-ice melts in early summer, but after this peak in density has receded and the population has come to a low ebb owing to marked stratification or incipient exhaustion of nutrients, Dinophyceae (Peridinians) may become dominant. For their nutritional requirements are lower than those of diatoms and their rate of growth is slower, so that they can continue propagation in impoverished waters, besides which they have the power of locomotion and consequently of adjustment to the level of the best available conditions. However, their daily increase under summer conditions is only 30 to 50 per cent, which is only about one-tenth that of diatoms. There are even suggestions that the diatom maxima may be preceded by vast numbers of unarmed flagellates which pass through the finest nets and are arbitrarily grouped as 'nannoplankton.' Almost all marine planktonic algae are unicellular, though these unicellular types may form chains or colonies. Of planktonic algae in arctic seas there have for some time been known about 470 species (including some 250 diatoms and 200 Dinophyceae), but to this total it is certain that many tiny nannoplanktonic forms, at least, remain to be added. Of the larger 'net plankton' alone, densities of the order of a million individuals per liter have been recorded in the Arctic; such densities exceed those found in most other parts of the world and frequently discolor the water markedly. However, there may be striking local as well as seasonal variations in these respects.

Because particularly of ice abrasion, and of the summer persistence of the ice-foot in the far north, the lower levels of sea-shores between tide-marks are relatively poorly vegetated. In sandy and muddy places they are largely barren, as may be the exposed parts even of rocky shores, though sheltered pools and the sides of rocks and boulders that are protected from ice-grinding may support quite luxuriant mats of Wracks (*Fucus* spp.) and other gregarious Brown Algae, often with admixture of Red and some Green Algae. Even below low-tide mark, the bases of thick ice-floes may scrape and abrade the upper surfaces of rocks and boulders, leaving only protected clefts etc. to be vegetated and have a darker color. Yet lower down where conditions tend in any case to be more 'even,' luxuriant 'beds' of large Laminariales commonly occur where there is a suitable bottom for attachment—providing vegetation that is incomparably more prolific than anything to be found on the adjacent land, and little if any less luxuriant than in comparable habitats in temperate or other regions. Gregarious species of *Laminaria*, often several meters long, and large ones also of *Alaria* and *Agarum*, commonly abound, with associated Red Algae such as species of *Lithothamnium* and *Lithophylum*, and often beset with epiphytes, while other Red Algae may extend down to a depth of at least 118 meters in some arctic seas. It has even been asserted that the richest algal vegetation in the Arctic may occur at depths where the temperature does not rise above 0°C. at any time of the year—possibly because low temperatures have a more repressive effect on respiration than on photosynthesis, allowing more accumulation of body-building materials than in warm habitats. Nevertheless the growth of benthic algae in the Arctic tends to be slower than in temperate regions, and development to take longer, so that, as with vascular plants on land, there are few functionally annual species. Nor is the long dark winter necessarily a period of quiescence for arctic marine algae, but, for many, one of active reproduction or vegetative propagation, even though they are covered with ice. Nevertheless there is little development of special reproductive features or forms in the Arctic, marine algae having a great capacity to adjust themselves to changed conditions without need of morphological modification, although the periodicity of their development may be changed considerably.

In waters of low salinity, such as occur near the heads of fiords into which large streams flow, in Hudson Bay, and off the coasts of Siberia near the mouths of great rivers, the benthic flora and vegetation tend to be far poorer than in waters of more normal marine salinity.

Cryophytic communities. The plant communities developing on snow and ice are in some ways akin to those found in ephemeral pools, which indeed often result from the melting of snow or ice and, initially at least, tend to harbor the same species. Thus phases of the common arctic and alpine Red-snow Alga, *Chlamydomonas (Sphaerella) nivalis*, are frequently abundant in pools of snow-water. The 'cryovegetation' is greatly influenced by the physical and chemical characteristics of the medium. Thus variations in salt-content and pH of the surface snow or ice and of any liquid water will

influence the composition and development of the community, and so will the nature of any near-by exposed rock or soil from which inorganic salts are obtained through windborne particles. Consequently the snow-fields and glaciers in the vicinity of acidic rocks, for example, are apt to support very different communities from those in limestone districts: in general, acidic environments support red or pink snows and basic ones yield the rarer green etc. snows. Cryophytic communities are often particularly well developed in the vicinity of 'bird-cliffs,' being presumably favored by manuring particles and foul stenches supplying nitrogenous and phosphatogenous substances.

The cryophytes may be usefully classified according to their preferred environments as growing (1) on ice —e.g. *Mesotaenium berggrenii*, (2) on snow and *névé*— e.g. *Chlamydomonas nivalis*, (3) on both snow and ice —e.g. *Cylindrocystis brebissonii*, and (4) occurring on snow and ice but only after transportation from their normal habitats, as in the case of various Blue-green Algae (including species of *Gloeocapsa*).

Although often a single cryophyte predominates in a particular community, sometimes giving it a distinctive color, usually others are present, the mixture sometimes involving a dozen or more species, often including fungi and bacteria. Dispersal appears to be mainly by wind. Whereas the color and texture produced vary with the organism and other circumstances, 'red snow' commonly appears in diffuse spots scattered over the surface, and often involving wide areas, though sometimes the colonizing is more uniform and extends to a depth of 3 or even 5 cm. The 'bloom' developed by organisms growing on ice may also extend for miles, as in the purplish-brown form on the largely snow-free glaciers of southern Alaska and Greenland. This is characterized by filaments of the alga *Ancyclonema nordenskioldii*, which form bunches up to 2 mm. in diameter on the surface of the ice or, chiefly, in small hollows formed by its melting around such dark bodies.

It is on the floating sea-ice, however, that the most plentiful cryovegetation commonly develops. Here diatoms, particularly, are often abundant in the superficial pools that result from summer melting, and especially at the bottoms of little holes where the ice melts around their dark, energy-absorbing bodies. Frequently diatoms form considerable aggregations that render the surface of the sea-ice brownish; they also occur on the sides and undersurfaces of the floes.

NICHOLAS POLUNIN

References

Polunin, N., "Botany of the Canadian Eastern Arctic, Part III: Vegetation and Ecology," Bull. Nat. Mus. Canada **104**, 1948.

Ibid., "Introduction to Plant Geography and some Related Sciences," London, Longmans, and New York, Barnes & Noble, 1960.

ARRHENIUS, SVANTE AUGUST (1850–1927)

The early training of this Swedish scientist was devoted entirely to physics and chemistry which he studied in Sweden, Germany, Russia, and Holland. In 1891 he assumed the chair of physics in the University of Stockholm and continued his, by then, world famous work in the theory of solutions and in the conductivity of electrolytes. Shortly after the turn of the century, he became deeply interested in the physical chemical problems associated with immunology and investigated particularly the physical and chemical relations between toxins and antitoxins. In consequence of a course of lectures he gave on this subject in the University of California in 1904, he published his "Immunochemistry" which was the first monograph in this field. He later involved himself in theories of the nature of the universe, and of the origin of life, and warmly espoused the theory that life was of universal distribution and that it passed from planet to planet in the form of spores.

PETER GRAY

ARTHROPODA

Arthropoda is a phylum of ANIMALS containing the ARACHNIDS, CRUSTACEANS, INSECTS and related forms. They stand apart from other INVERTEBRATES by having (1) bilateral symmetry, (2) an external skeleton, (3) segmented appendages, (4) a haemocoele and (5) a nervous system consisting of a dorsal brain and a pair of ventral nerve cords, which are more or less fused at ganglia. Of these traits, the most significant in respect to the history of the arthropods is the external skeleton, which is a cuticle secreted by the epidermal cells forming the inner part of the body wall. The cuticle itself is composed of several parallel layers and consists in part of chitin, a nitrogenous polysacharride; and of proteins, —sclerotin and cuticulin. These three substances are responsible for the rigidity and the resistance of the arthropod's integument. The formation of this skeleton in the very early arthropods apparently had a significant part in directing the evolution of this group of animals. It presumably made necessary the joints in their legs; it provided protection not only against predators in the aquatic environment but also against dessication and invasion by microorganisms in the terrestrial environment; it made possible the development of wings, which are simply extensions of part of the body wall; it limited the maximum size of terrestrial arthropods; and it made absolutely necessary some method of molting, a process which allowed the animal to change its size and form, as a new cuticle was being secreted to replace the old one.

The arthropods, along with the ONYCHOPHORA, presumably arose, in pre-Cambrian time, from simple, segmented worms. The presence of trilobites and of aquatic arachnids in early Cambrian deposits points clearly to the antiquity of the Arthropoda. Apart from the TRILOBITES, two divergent lines of arthropods developed early among the marine forms. One of these, the subphylum Chelicerata, consisting of the aquatic arachnids (Merostomata) and the terrestrial arachnids (Arachnida), evolved chelicerae (jaw-like structures) as the most anterior appendages; antennae were never formed. This subphylum first appears in the Cambrian rocks, where it is represented by king-crabs (Order XIPHOSURA). Members of the Order Eurypterida were probably also present in Cambrian times, although they are not known from deposits older than Ordovician age. From these aquatic forms or their ancestral stock, the true Arachnida arose, invading the terrestrial environment. The earliest record of these is that of the scorpions in the Silurian, followed later by the spiders, the harvestmen and other orders.

The second major divergent line of arthropods, the Mandibulata (Antennata), developed antennae as the most anterior appendages. This group was represented in the Cambrian period by various Crustacea. From this early stock or their ancestors there arose the terrestrial mandibulates, i. e., the myriapods, insects and related forms; these are first known from the strata deposited in the latter part of the Paleozoic Era. One of these groups, the insects, not only successfully occupied the land and fresh water but also achieved flight.

The geological record of the arthropods shows that, after originally arising in a marine environment, they invaded the terrestrial environment at least twice highly successfully, quite apart from minor invasions by some groups of Crustacea.

The several systems of organs in the arthropods show a similarity of structure throughout the diverse groups of the phyla, although some systems are highly modified in terrestrial types. The vascular system consists of a dorsal vessel comprised of a heart and of an aorta, which may be anterior as well as posterior to the heart. The latter, which is composed mainly of muscle cells, normally forces the blood anteriorly through an opening in the anterior part of an aorta into several incompletely isolated sinuses extending through the haemocoele. The blood is eventually drawn back into the heart again through lateral ostia. Among some arthropods, as in Malacostraca among the Crustacea and in Orthoptera among the insects, there are canals which lead from the ventral part of the heart to the sinuses or portions of a sinus; in this way the blood is conducted more directly to organs. The haemolymph is usually clear or pale yellow, and it contains many colorless blood cells, some of which do not circulate through the heart itself but remain restricted to limited areas of the haemocoele. Although respiratory pigments are absent in most arthropods, some (as the Malacostraca) have haemocyanin in the blood plasma; others (as some Entomostraca and a very few insects) have haemoglobin. The blood in the arthropods functions mainly as a vehicle of transportation of food materials and hormones as well as a place of storage of food and water; its respiratory function in most cases is limited to the part it plays in the diffusion of gases in the cellular level.

The excretory organs of arthropods are surprisingly diverse. In the arachnids there are two kinds of excretory structures; one of these consists of ducts which open on one or more of the pairs of legs, and the other of Malpighian tubules which open into the posterior part of the midgut. In the Crustacea there are ordinarily present a pair of ducts associated with the second antennae on the third somite or a pair on the maxillary segment. In some Crustacea there are other glands which seem to have an excretory function, and which may replace the usual ducts. In the myriapods and insects there are Malpighian tubules, which open into the alimentary canal at the beginning of the hind duct. It is notable that nephridia (characteristic of the annelids) are entirely unknown in the arthropods. The substances excreted differ in the two main lines of arthropods as well as in the crustaceans and insects; in the arachnids they are guanine, in the crustaceans they

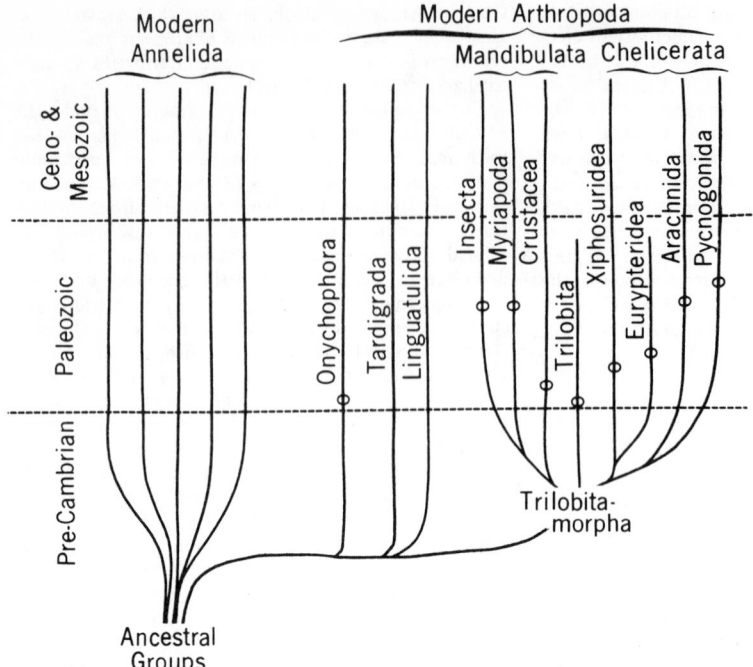

Fig. 1   Phylogenetic diagram to illustrate the probable relationship between Annelida and Arthropoda and the probable relationships among the arthropod classes. The time scale at the left is not proportional. Circles indicate the approximate age of the earliest fossil record for various classes. (From Fox and Fox, "An Introduction to Comparative Entomology," New York, Reinhold, 1964.)

are ammonia compounds and amines, and in the insects they are urates.

The eyes of arthropods, although occurring in diverse form, appear to have been derived from a basic unit consisting of a cluster of retinal cells covered on the outside by transparent cuticular material. The retinal cells of each eye are usually arranged in several or many groups called retinulae, surrounding a vertical rod termed the rhabdome. In the chelicerate arthropods, the most generalized condition of the eyes seems to be represented in the Xiphosura (e.g., *Limulus*). Here there are a number of units more or less united together. The eyes of true arachnids appear to have been derived from these compound eyes by separation of the units and loss of some of them. Among the Mandibulata, the myriapods have small groups of units placed together; the insects have simple eyes, consisting of a single unit each, and compound eyes, each of which consists of a number of units functioning as a single visual organ. The respiratory structures of arthropods are diversely modified in connection with the environments concerned. Aquatic arachnids possess a series of gills (gill books); in *Limulus* (Xiphosura) they are carried by the five pairs of posterior appendages. The gills consist of a series of leaf-like pads, which are attached to movable plates. Among the terrestrial arachnids, the scorpions have lung books, which appear to be modifications of the gill books, the leaf-like pads having been withdrawn into a cavity, the lung. In the spiders, several primitive families have only two pairs of lung books. In some others there is an anterior pair of lung books and a posterior pair of stigmata which open into

tracheae; and in still others, there are two pairs of stigmata both opening into tracheae. Some of the more highly specialized arachnids, such as the ticks and mites, have a tracheal mechanism which has probably arisen independently from that in the true spiders. Among the Mandibulata, the aquatic Crustacea have a series of blood gills which extend into a cavity (branchial cavity) formed by the arching of the carapace along the sides of the body. Some of the terrestrial isopods have air tubes which are formed by the invagination of the integument of the abdominal limbs. In insects and myriapods there are tracheae similarly formed by the invagination of the body wall; the opening to the tubes is controlled by a valve. The tracheae penetrate within the haemocoele and the fine tracheoles are closely associated with the cells of the various organs. The reproductive systems and reproductive processes in the arthropods show great variation. Although most of the arthropods have the sexes separate, there are some (chiefly the crustacean subclass Cirripedia) which are usually hermaphroditic. Parthenogenesis is common, especially among the insects, although in most cases normal sexual reproduction also occurs at intervals even in these parthenogenetic species. In most arthropods, the sperm cells are transferred to receptacles or cavities within the abdomen of the female, and the eggs are fertilized later as they are about to be laid. The immature stages of arthropods present a bewildering variety of forms. In some, the young are very much like the adults in appearance, such as among the Thysanura (Insecta), but in the great majority of arthropods the young are very different in appearance from the adult forms, such as the nauplius larva of crustaceans or the larval forms of insects.

The classification of the Arthropoda is necessarily complex, because of the large size of the group and because of the many phylogenetic lines which have been developed. The Onychophora comprise a group which shows some features of both arthropods and annelids; they are usually regarded as constituting a separate phylum. The Tardigrada and the Pentastomida, although placed by some zoologists as orders of the Arachnida, are only doubtfully assigned to the Arthropoda and are usually treated as separate subphyla of arthropods. Apart from these dubious groups, the phylum is generally considered to consist of four subphyla, as follows: Trilobitomorpha, consisting of the Trilobites and related extinct forms; Chelicerata, comprising the class Meristomata (with orders Xiphosura and Eurypterida) and the class Arachnida, with several orders; the Pycnogonida, marine forms which superficially resemble some arachnids; and the Mandibulata, which consists of the classes Crustacea, Myriapoda and Insecta, each of which includes a series of orders.

F. M. CARPENTER

References

Borradaile, L. A. and F. A. Potts, "The Invertebrata," 4th ed. revised by G. A. Kerkut, Cambridge, The University Press, 1961.
Carthy, J. D., "Behavior of Arthropods," San Francisco, Freeman, 1965.
Moore, R. C., Ed., "Treatise on Invertebrate Palaeontology," Part O, Arthropoda, 1. Trilobitomorpha, Lawrence, Kans., University of Kansas Press, 1959.
Part P, Arthropoda, 2. Chelicerata, 1955.
Part Q, Arthropoda 3, Ostracodes.
Part R, Arthropoda 4, Branchiopods, Malacostrans, Myriapods, Insects.
Snodgrass, R. E., "The Evolution of Arthropod Mechanisms," Washington, Smithsonian Miscellaneous Collections, vol. 138, no. 2, 1958.
Snodgrass, R. E., "Textbook of Arthropod Anatomy," Ithaca, Cornell University Press, 1952.

ARTIODACTYLA

The Artiodactyla are an order of MAMMALS comprising the even-toed ungulates (from Greek *artios*, even + *daktylos*, finger). The ordinal name was first proposed by Owen in 1847 although the modern concept of this order (as an ungulate suborder) was first recognized by De Blainville in 1816 under his "Ongulogrades à doigts pairs," based on osteological study.

The more readily observed features of the order are to be found in the structure of the feet, characterized as *paraxonic*, or having an axis of symmetry between the third and fourth digits, contrasting with the PERISSODACTYLS or odd-toed ungulates which are *mesaxonic*, in that the axis of symmetry is through the third digit. Correlated with the more even distribution of stresses between the third and fourth digits, the members of this order exhibit a highly distinctive astragalus that is distally as well as proximally wide and articulates broadly with the cuboid as well as with the navicular. The distal surface, moreover, somewhat resembles the tibial surface in its distinctly trochlear development.

A natural division of the living Artiodactyla occurs between the ruminant, or cud-chewing forms, and the non-ruminant or pig-like groups. The comparatively complex stomach of the artiodactyls reaches a full four-part division in the ruminants. In the first of these, the *rumen*, the partially chewed food is rotated and moistened with liquid from the *reticulum*. The regurgitated *bolus* after more thorough mastication descends to the *psalterium* where it loses much of its superfluous fluid and is moved to the *abomasum* for the normal digestive action. It may be noted, however, that there is a somewhat less distinctive separation of the reticulum from the rumen in the tragulids, whereas in the camelids there is the greatest development of the reticulum, providing unusual water storage in the stomach.

Although criteria of this nature cannot be extended to the fossil record, rather general, though not invariable correlation exists with tooth and foot structures. The ruminants have developed highly selenodont or grazing teeth and show greater lengthening of the metacarpals and metatarsals with marked reduction of second and fifth digits. The non-ruminants, while showing considerable variation in tooth structure from decidedly bunodont teeth observed in the suids to highly developed selenodonty in fossil oreodonts and cainotheres, have in general retained a shorter metacarpus and metatarsus with less reduction of the second and fifth digits.

Ruminants for the most part are also characterized by horns, but in the females they may be of smaller size or absent. In cervids the horns consist of bone which grows from the frontal with its integument or velvet and are shed annually. In the bovids, however, a pair of bony processes or horn cores grows from the frontals and is covered with a sheath of horny fibers developed from the epidermis. A similar horny development in the antilocaprids may branch and is shed annually. Giraffe horns differ in that they develop from separate ossifications that fuse with the skull and are covered with a hairy integument.

In a current classification of the Artiodactyla, as outlined by Simpson in 1945, three suborders are recognized: the Suiformes, Tylopoda and Ruminantia. Within the Suiformes (Palaeodonta, Suina, Ancodonta and Oreodonta) are included the many pig-like forms, from the primitive dichobunids to the more specialized entelodonts, anthracotheres, anoplotheres, cainotheres and oreodonts, as well as the pigs and peccaries. The Tylopoda comprise only the camelids and xiphodonts. Though functionally ruminant, the camels are placed in a separate suborder because of their otherwise distinctive characters and rather ancient lineage. The Ruminantia, an extremely large suborder, embraces the Tragulina, diverse in the fossil record but represented in the living fauna by only the chevrotains; and the Pecora with something like 225 or more genera, living and extinct, distributed between the cervids, lagomerycids, giraffids, antilocaprids and bovids.

Artiodactyla are first recorded in the early Eocene of both Europe and North America. *Protodichobune* of France and related *Hexacodus*, as well as the diacodexines of western United States, are the earliest known artiodactyls and represent beginnings of the primitive Dichobunidae which became more highly diversified in the later Eocene on both continents. Much of the European middle Eocene dichobunid material shows an early tendency towards selodonty not so evident until the later Eocene of North America and appears correlated with increasing aridity.

Following early dispersal of the dichobunids, possibly at the beginning of Eocene time, a faunal interchange involving artiodactyls between the Eastern and Western Hemispheres appears to have taken place at about the beginning of the Oligocene when the Old World anthracotheres appeared in North America, lasting there, however, only until Miocene time. This family in Africa is believed to have given rise to the hippopotami. The first peccaries in North America also appear in the Oligocene, possibly derived from European cebochoerids. The suids, however, remained Old World until more widely distributed by man. The Oreodonta, so abundant and enduring in the fossil record of North America, are not recorded elsewhere in the world. The camel family is peculiarly North American in its development from later Eocene time until land connections near the beginning of the Pleistocene permitted nearly world-wide distribution.

Among the Ruminantia, forms regarded as belonging in the Tragulina are recorded as early as Eocene time in Europe and North America, as well as Asia. It seems probable that the Old World Gelocidae includes the ancestry of the true tragulids, whereas the North American Tertiary protoceratids with weird premaxillary as well as frontal horns, are derived from the western Leptomerycidae. The Pecora are not truly distinctive until the Miocene and like the Old World tragulids may well have had their roots in the gelocids. The bovids did not reach the Western Hemisphere until the Pleistocene. The antilocaprids, however, are only North American in their fossil as well as recent distribution. Cervids were present in both hemispheres in the Miocene and Pliocene but apparently did not longer survive in the west and were reintroduced at about the beginning of the Pleistocene. The giraffe family showed marked diversification in the later Cenozoic of Eurasia, but survived only in Africa.

C. L. GAZIN

Reference

Simpson, G. G., "The principles of classification and the classification of the mammals," Bull. Amer. Mus. Nat. Hist., **85**, pp. i–xvi, 1–350 (143–162, 258–272), 1945.

ASCOMYCETES

The Ascomycetes with 25,000 to 35,000 species constitute the largest class of FUNGI, organisms usually included in the Division Eumycophyta of the PLANT KINGDOM. They are generally considered an intermediate group derived from the PHYCOMYCETES on the one hand and leading to the BASIDIOMYCETES on the other, although evidence on their origin is so incomplete that hypotheses of an independent derivation from ALGAE (Rhodophyta) can not be excluded. This highly variable assemblage of fungi is bound together by a single common character, the production of spores arising from sexual reproduction in an ascus. The ascus is a globose to cylindrical, sac-like cell which at its origin typically contains a pair of haploid nuclei. Fu-

sion of these nuclei, the essential feature of sexual reproduction, produces a single diploid nucleus which immediately undergoes two meiotic divisions, usually followed by one or more mitotic divisions. Cell walls cut out portions of cytoplasm around the resulting haploid nuclei and delimit the ascospores. Typically the ascus contains eight ascospores, although fewer spores may develop and multispored asci occur in some species. The ascospores are unicellular or multicellular and assume a variety of shapes, sizes, and colors. They are released by disintegration of the ascus wall or are discharged forcibly through a pore or split in its apex.

The ascospores germinate by the protrusion of cylindrical germ tubes which elongate to form branching, septate hypha each cell of which contains one to several haploid nuclei. The loose weft of hyaline or pigmented hyphae in or on the substrate constitutes the mycelium, which represents the assimilative phase. The Ascomycetes grow as parasites on plants or animals or as saprobes on almost every form of organic matter. Despite their dependence on an external supply of food, they possess remarkable synthetic abilities. Many are able to grow luxuriantly and produce a complex array of metabolic products if supplied only with water, inorganic compounds, and a single carbon source such as glucose; others have more complex nutritional requirements. Under some growth conditions, the mycelium may break up into individual cells which multiply by budding or fission. Forms classified among the yeasts occur only in this unicellular phase. Mycelium may be lacking also in an aberrant group of ectoparasites on insects in which the ascospore develops by division into a cellular thallus.

Hyphae of the mycelium may become aggregated to form tissues which often are so compacted that they resemble the parenchyma of seed plants. Tissues may also arise from three-dimensional division of a hyphal cell or group of cells. Such tissues occur in small globular or ellipsoidal sclerotia which function as resist- and resting bodies or in larger flat, pulvinate, or cylindrical stromata in or on which the reproductive organs develop. Most lichens are formed by Ascomycetes in which the mycelium develops into a histologically complex crustose, foliose, or fruticose thallus within which the parasitized algal cells are imprisoned.

Asexual reproduction may consist of the fragmentation of the mycelium or the budding of mycelial cells to produce individual dispersal cells. Usually the asexual reproductive bodies are conidia, which may be unicellular or multicellular and occur in a variety of distinctive forms characteristic of the various species. The conidia are abstricted from the tips of specialized hyphal branches, the conidiophores. The conidiophores may be distributed over the mycelium; aggregated on flat, tubercular, or columnar stromata; or borne in cavities in stromata. Many Ascomycetes are found commonly only in the conidial, or imperfect, stage; and the majority of fungi in the class Fungi Imperfecti are probably Ascomycetes in which the perfect, or sexual, stage is unknown.

The sexual organs may arise as branches from the mycelium or within a stroma. Typically, the female organ is an ascogonium composed of one or several enlarged cells and often terminating in a filamentous receptive organ, the trichogyne. Male organs are cells differentiated as antheridia. Plasmogamy is accomplished by fusion of the antheridium with the ascogonium or its trichogyne and migration of the antheridial nucleus into the ascogonium. The antheridial and ascogonial nuclei divide repeatedly by mitosis, and the daughter nuclei of each type migrate into ascogenous hyphae which grow out from the fertilized ascogonial cells. The ascogenous hyphae become divided into cells, each containing a pair of nuclei of opposite sex potential, and these produce asci in which the sexual process is completed by caryogamy. Often antheridia are lacking, and their function is taken by spermatia (microconidia), conidia, or mycelial cells. In some supposedly primitive species the sex organs are nearly identical gametangia which on fusing produce a single ascus. Such an ascus may be comparable to the germ sporangium arising from the zygote in certain Phycomycetes. In other Ascomycetes the sex organs are completely reduced, and fusions occur between morphologically undifferentiated cells. In some of these species fusions occur early in the life cycle, and the assimilative phase consists of a dicaryotic mycelium each cell of which contains a pair of nuclei and may develop into an ascus. The origins of the Basidiomycetes may be sought in these forms if it is assumed that the basidium could be derived from the ascus by the migration of the nuclei produced by meiosis into external spores budded out from the ascus wall. In some yeasts fusion of cells is followed immediately by nuclear fusion, and the assimilative phase consists of diploid cells which ultimately are transformed into asci, or there may be an alternation of diploid and haploid phases. Most Ascomycetes appear to be hermaphroditic, and many are self-fertile; others have inherited self-sterility factors, and mating of compatible strains derived from different ascospores is necessary for sexual reproduction.

The asci may be produced singly or in exposed layers on the mycelium; usually they are grouped in ascocarps. The ascocarp consists of an outer wall of compacted layers of hyphae originating from the stalk of the ascogonium or from adjacent mycelial cells and enclosing a centrum of filamentous or parenchymatous tissue within which the ascogenous hyphae proliferate and the asci arise. Globose or flask-shaped ascocarps surrounding the group of asci are perithecia; cup-shaped ascocarps bearing a palisade of exposed asci are apothecia. The perithecia or apothecia may be united in a stroma and form compound fruit bodies. In some species in which the ascogonia develop within a stroma no differentiated perithecial walls are formed, and the asci arise in cavities in the stroma. Ascocarps and stromata vary from black and carbonaceous to bright-colored and fleshy. Most ascocarps are minute, some being less than a tenth of a millimeter in diameter; others attain dimensions of several centimeters or in extreme cases up to forty centimeters.

E. S. LUTTRELL

References

Gäumann, E. A. (Trans. F. L. Wynd), "The Fungi. A Description of their Morphological Features and Evolutionary Development," New York, Hafner, 1952.

Bessey, E. A., "Morphology and Taxonomy of Fungi," Philadelphia, Blakiston, 1950.

ASCOTHORACIDA

Small, insufficiently known order of marine parasitic Entomostraca related to the Cirripedia. Body enclosed in a bivalve shell or, in most species, in a sac-like, coiled or branched mantle, in which ramifications of the intestine and ovaries enter. The mantle is connected with the head and the anterior part of the thorax and has a small, ventral opening.

The head is fused with the first thoracic segment. Antennae are missing, antennulae well developed. Mouth-parts, when not reduced, consist of three paired appendages and one unpaired structure within a conical sheath. The most primitive ectoparasites have six thoracic segments, each with a pair of jointed, biramous limbs. Abdomen with five segments and a furca. The endoparasites form a series showing gradual transformation into highly degenerate species in which only traces of segmentation remain. The thoracic feet undergo similar reduction in number and shape; like the furca they are lacking in the most reduced genera.

The internal anatomy is little known. The Ascothoracida are hermaphroditic or the sexes are separate. Some species have dwarf-males. The first larval stage is a nauplius differing from that of the Cirripedia in lacking frontal horns. Next is a metanauplius, and last a cypris resembling the adult of ectoparasitic species.

The Ascothoracida comprise four families: Synagogidae with *Synagoga* and *Ascothorax*, Lauridae with *Laura* and *Baccalaureus*, Petrarcidae with *Petrarca*, and Dendrogasteridae with *Ulophysema*, *Dendrogaster* and *Myriocladus*. *Synagoga* lives as an ectoparasite on Antipatharians or crinoids. All others are endoparasites: *Ascothorax* in ophiuroids, *Laura* and *Baccalaureus* in zoantharians, *Petrarca* in a madreporarian, *Dendrogaster* and *Myriocladus* in asteroids, and *Ulophysema* in irregular echinoids. Some have more than one host species.

The parasites affect the host in different ways. *Laura* and *Baccalaureus japonicus* cause gall-formation in the host. Other *Baccalaureus* species cause distortions in the zoanthids. *Baccalaureus* and *Ulophysema* make holes in their hosts, through which the larvae emerge, and *Ascothorax* and *Ulophysema* may cause complete obliteration of the hosts gonads. The way in which the Ascothoracida take up food is unknown. Probably they absorb the body fluids in their hosts through the mantle.

The larvae develop in the mantle cavity of the mother until the nauplius, metanauplius or cypris stage. Unpaired or paired dorsal hornlike processes in the adults stir up the larvae in the mantle. *Ulophysema* larvae seem to infect new hosts through their genital pores.

HANS BRATTSTRÖM

References

Brattström, Hans, "On the organization of the genus *Baccalaureus* (Ascothoracica) with description of a new South African species," Bertil Hanström. Zoological papers in honour of his sixty-fifth birthday. Lund 1956.

Krüger, Paul, "Ascothoracida," in Bronn, H. G. ed. "Klassen und Ordnungen des Tierreichs," Vol. 5 Leipzig, 1940.

Okada, yô K., "Les Cirripèdes ascothoraciques," Trav. Stat. Zool. Wimereux. **13**:Paris 1938.

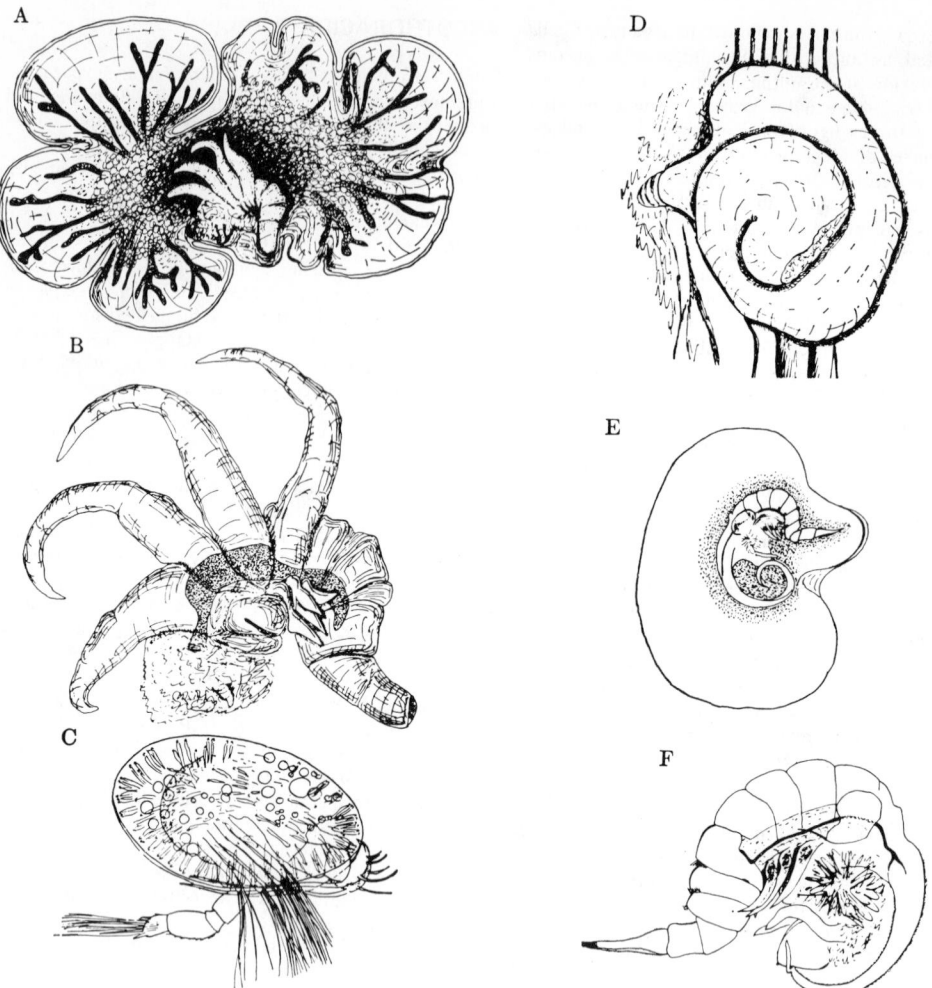

Fig. 1   (A–C) *Ulophysema öresundense* Brattström, approx. 1 cm long. (A) left half of mantle removed, (B) animal without mantle seem from the left, (C) Cypris larva, abt. 0.6 mm long.
(D–F) *Baccalaureus durbanensis* Brattström, approx. 1 cm diam. (D) from the right, (E) left half of mantle removed, (F) animal without mantle, from the right.

AVES

The Class Aves is the most easily recognized class of vertebrates (see CHORDATA) because all birds have FEATHERS . Despite this ease of recognition, birds are only weakly differentiated morphologically from REPTILES. No basic feature of the bird's internal anatomy serves by itself to separate birds from reptiles. Here, for once, we find that the layman's understanding of the Class fits the best scientific definition: warmblooded vertebrates with feathers.

One factor—FLIGHT—has been most important in the evolution of birds. No part of the bird's body, no aspect of its biology have been free from influence by natural selection for efficient flight. Even modern flightless birds, such as penguins and ostriches, show unmistakable signs of descent from flying ancestors. The

class has prospered as a result of exploitation of niches best filled by flying organisms.

The key to avian flight is feathers. Combining lightness, strength, and insulation qualities, feathers are remarkable derivatives of the scales of the reptilian ancestors of birds. At the time of the first appearance of birds in the fossil record, feathers had evolved to a stage comparable to feathers of modern birds. Special types of feathers serve special functions. The strong flight feathers of the wing and tail and the contour shaping feathers of the body are essential to flight. Hidden among the contour feathers may be downy feathers to provide more effective insulation; these are comparable in structure to the downy feathers of young precocial birds. The color and sometimes the shape of the feathers provide protection and may be used in courtship and other displays.

Feathers are composed of keratin and grow from papillae in the skin. The skin lacks sweat glands. In fact, the only skin gland is the oil gland, a bilobed structure located dorsally at the base of the tail. The oil gland's secretion is distributed by the bill in preening to feathers all over the body and is essential to maintenance in good condition of feathers and the covering of the bill, legs, and feet. Removal of the oil gland, for example, causes the plumage to become frazzled, disarranged, and easily soaked.

The feathers of birds are arranged in tracts on the body with intervening bare spaces. All birds molt at least once a year and some have an additional partial or complete molt. Ptarmigan (*Lagopus*) are said to molt three or four times per year. Coloration in feathers is produced by pigments, melanins and carotenoids being most important, by feather structure through light interference, and by a combination of pigment and structure.

Plumage coloration serves a variety of functions including reduction of wear, insulation and absorption of light and heat, concealment, and advertisement, especially in relation to breeding.

The most notable feature of the avian skeleton is the loss and fusion of bones. In the skull of adult birds most of the sutures between bones have disappeared so that the resultant structure combines great rigidity with light weight. Only three digits remain in the wing; these probably represent II, III, and IV of the original five and they have 1, 2, and 1 phalanges, respectively. The metacarpals are fused with the distal carpals into an important bone called the carpometacarpus, leaving only two free carpals in the wrist. Mobility within the wing is mostly confined to a single plane—jackknife fashion, producing the necessary rigidity in the extended wing. The proximal tarsal bones are fused with the tibia; distal tarsals and metatarsals are fused into the tarsometatarsus. Toes vary from four in most birds to two in the Ostrich (*Struthio*). The pelvis also shows much fusion, as does the thoracic cage. The importance of large pectoral muscles used in flight is reflected by the great development of the keeled sternum in flying birds. Other features of the skeleton which are adaptations to flight are the strong pectoral girdle with a robust coracoid, great fusion and reduction of tail vertebrae, and pneumaticity of many bones. Neck vertebrae in birds range from eight to 24, in contrast to seven in all but a few mammals.

The heavy, powerful muscles of flight are centrally located on the sternum, thus keeping weight near the center of gravity. Most muscles of the hind limb and the wing operate through long tendons from fleshy bellies lying near or within the main body mass, again centralizing weight. The heart of birds is four-chambered, with the right aortic arch persistent, rather than the left as in mammals. Red blood cells of birds are biconvex, oval, nucleated, and vary in number from roughly $1\frac{1}{2}$ million (Ostrich) to nearly 8 million (*Junco*) per cubic millimeter. Small birds as a rule have higher blood counts than large birds, and also a higher rate of heart beat—up to 700 beats per minute for a House Wren (*Troglodytes aedon*) compared to 220 for a Common Pigeon (*Columba livia*). Body temperatures vary between 100° F and 112° F; again small birds are high, large birds are lower.

Torpidity in birds has been reported recently for some goatsuckers, hummingbirds, and swifts. In at least some of these birds, torpidity is as profound as in hibernating mammals, with lowering of body temperature almost to the level of the surrounding air and great reduction in rates of respiration and heart beat.

The avian kidney is essentially like that of reptiles. Nitrogenous wastes are excreted as uric acid, rather than urea as in mammals. As a probable consequence of the ability of the avian kidney to excrete salt only in low concentration, some marine birds have evolved a remarkable accessory method of salt elimination—the "salt gland" located usually above the orbit. This gland is best developed in species which feed on marine invertebrates isotonic with sea water, or which drink sea water.

The salt gland functions only after salt has been ingested in concentrations beyond the excretory power of kidneys. It then produces a highly concentrated salt secretion which is discharged via the nares to the external surface of the bill.

The most notable feature of the avian reproductive system is the marked asymmetry. In most birds only the left ovary and oviduct are functional, the right ovary and oviduct degenerating in early embryonic development. Both testes are normally present and presumably both are functional, but again the left testis is usually considerably larger than the right.

The respiratory system is noteworthy for the system of air sacs which are extensions from the lungs to various parts of the body. These thin-walled air sacs, usually numbering from seven to nine, are not involved directly in oxygen-carbon dioxide transfer, but they serve to increase lung efficiency by providing the lungs with "unused" air on both inspiration and expiration. They also are important in regulation of body temperature through internal evaporation and radiation. Among other functions should be mentioned the cooling effect on the testes of the abdominal air sacs, which thus provide in a different way one of the functions of the mammalian scrotum.

The need for teeth in birds has been eliminated by the development of a muscular gizzard which serves to grind and macerate food. The gizzard enhances aerodynamic efficiency by substituting centralized weight for the heavy teeth, jaws, and jaw musculature. Many birds store food in the elastic esophagus and some have developed a dilation called the crop, which in breeding pigeons secretes "pigeon's milk," used as food for the young.

Birds are visually- and acoustically-oriented animals —as are humans, and their central nervous system reflects this. The eyes are relatively huge and are supported by a well-developed ring of bony plates in the sclera, as in all modern reptiles except snakes, crocodilians, and some forms with degenerate eyes. The pecten, a vascular comb-like structure extending into the vitreous chamber of the eye from the base of the optic nerve, occurs in birds and reptiles. Its function is unknown, even though it has been the subject of much speculation. It may have a nutritive function, and there is evidence that it may aid in detection of prey movement. The optic lobes of the brain are large, as is also the cerebellum, associated with coordination and equilibrium. Although little information is available, birds seem to have only poorly developed olfactory abilities, although there is evidence that kiwis

(*Apteryx*) and vultures (Cathartidae) may surpass most birds in this respect.

The earliest known avian fossil is *Archaeopteryx*, from Upper Jurassic limestone in Bavaria. *Archaeopteryx* might well be considered a reptile were it not for the impressions of well-developed feathers. It was toothed, showed less fusion of distal wing bones, separate, unfused pelvic bones, and unfused metatarsals. By the Cretaceous period, birds had radiated widely, so that the two marine orders found in chalk beds of Kansas were very different osteologically. *Hesperornis* fed on fish captured by swimming and diving in the shallow Cretaceous sea. *Ichthyornis* was a small flying bird which probably had gull-like habits. *Ichthyornis* was originally reported to be toothed on the basis of incorrect association of toothed jaws of a small mosasaur with the bird's skeleton. It remains for future discoveries to decide which Cretaceous birds were toothed. *Ichthyornis* and *Hesperornis* had typically avian tarsometatarsi, fused skull bones, and brains much like those of modern birds.

The fossil bird record from the Cenozoic contains many novelties, including some huge flightless terrestrial birds and an oceanic bird with an estimated 15 foot wingspread. These oddities are of interest, but the most important lesson of the Cenozoic record is that the great radiation of birds into modern orders occurred very early. By the Eocene, most modern orders were recognizable. Many Oligocene birds fit readily into modern families. This early radiation of birds makes interpretation of relationships of major groups extremely difficult. New discoveries of bird fossils from the Cretaceous and Paleocene can be expected; they shall, of course, be of crucial scientific interest.

The number of living species of birds can be given with confidence that it will err by no more than a few per cent from the final total. Approximately 8600 species are now recognized. Probably relatively few new species remain undiscovered, but differences of opinion among taxonomists as to what constitutes specific rank introduces a variable which makes a more precise figure impracticable. There is considerable disagreement as to the number of orders and families into which these 8600 species should be divided. One widely used classification, by Wetmore, employs 27 orders and about 170 families. Mayr and Amadon recognize 28 orders, while Stresemann uses 51 orders.

Only the extreme polar regions and the depths of the seas are unoccupied by birds. Still, despite the great powers of dispersal of most birds, ranges of individual species are ordinarily sharply delimited. Physical, physiological, and biological barriers serve to keep each species within its ecological tolerance. It should be remembered that one of the most formidable barriers to range expansion may be the presence in the newly-reached area of an already established species occupying, at least in part, the same ecological niche.

Ranges of birds vary from a few square miles—especially in some island forms—to most of the world. The Barn Owl (*Tyto alba*), Osprey (*Pandion haliaetus*), and the Common Gallinule (*Gallinula chloropus*) are species ranging over a major part of the land areas of the world.

Some species show conspicuous discontinuities in range. These may result from chance colonization of new remote areas; relicts through shrinkage of a larger continuous past distribution; or interruption of a con-

Fig. 1   Representatives of orders of birds. 1, Brown pelican. 2, Mallard duck. 3, Sandpiper. 4, Flicker, 5, Ostrich. 6, Golden eagle. 7, Cardinal. 8, Penguin. 9, Carolina parakeet. 10, Great horned owl. 11, Hummingbird. 12, Swallow. (From Goodnight, Goodnight and Gray, "General Zoology," New York, Reinhold, 1964.)

tinuous range by ecological change. The great Zoogeographic Regions of the world are admittedly arbitrary, yet each has its own distinctive avian fauna. Attempts to interpret the avifaunal history of these regions must of necessity be based primarily on living birds, because of the inadequacy of the avian fossil record. That this method is hazardous is amply proved by the much better mammal fossil record, which, for example, indicates a North American origin for horses and camels, neither of which left any modern descendants here. In spite of the certainty that speculations based on modern birds will be wrong to an undetermined degree, the analyses that have been made add to our understanding of modern avifaunas.

Bird migration has provoked more speculation than any other aspect of avian biology. Elaborate theories to account for the origin of migration have been proposed. Actually, migration of some sort appears to be the inevitable result of the combination of highly mobile animals, great seasonal fluctuations in food supply, and natural selection. The most interesting questions involve methods of orientation and navigation, environmental and physiological stimuli, and evolution of routes and destinations for individual species. It now seems clear that length of day is an important stimulus, operating through the hypothalamus which, through the pituitary gland, regulates gonadal activity and fat deposition. Patterns of migration, although under primary influence of photoperiod, are modified by weather, food supply, and other factors.

Some birds are sedentary; others migrate only a few miles; still others may travel many thousands of miles. Most tropical species are sedentary. In higher latitudes, longest migrations are characteristic of insectivorous species, while seed-eaters and most birds-of-prey usually make shorter migrations.

Much about avian navigation and orientation remains unexplained. However, present knowledge indicates that recognition of "home," the general region of birth, is learned rather than inherited; that the earth's magnetic field and Coriolis force are probably not used in orientation; that birds are able, with the help of an internal "clock," to determine position by use of the position of the sun; that birds may likewise be able to determine their position by stellar cues; and that birds are superbly able to see and remember topographic features and to use them in both local and long distance movements.

In recent years behavior has been studied intensively. Important progress has been made in description and analysis of behavioral components. There is good reason to think that behavior will become an important source of evidence of phylogenetic relationships of birds. The field is still young, however, and basic arguments about such things as instinct versus learning, and learning versus intelligence, are still unresolved. Much of the difficulty arises from very incomplete understanding of the operation of the central nervous system.

The breeding cycle can be conveniently described by starting with the establishment of breeding territory, an area defended during some part of the cycle. Territories may include all of the essentials required for nesting, including food; may exclude the feeding area; may be restricted to a mating area; or may be little larger than the nest itself, as in many colonial birds. Size of complete territories, including feeding areas, range from half an acre or less up to many acres in some birds-of-prey. Territorial behavior serves to insure adequate food and nesting cover, regulates population density, reduces interference in nesting activities from other members of the species, and serves as a focal point for formation and maintenance of the sexual bond between male and female.

Most territorial defense is accomplished by song and displays. Actual combat is relatively less common and ordinarily the owner of the territory wins encounters on his home ground. Vigor of territorial defense seems to decrease as the boundaries of the territory are approached.

Pairing in birds may involve little more than the moment of copulation, or it may last for life. A great many species remain paired for one breeding season. In some cases where birds remate in successive years, the attraction to the same territory seems to be more important than that to each other as individuals. Pairing is followed by courtship activities which serve to bring male and female into synchrony of breeding activity. Nest site selection and nest building are usually the female's responsibilities, but there are many exceptions. Copu-

lation is frequent during nest building, and in some species it continues through or even beyond the egg-laying period. Clutch size ranges from one egg, as in many seabirds, to about 20. Incubation may be performed by both sexes, but the female alone incubates in many species. In some, such as phalaropes (Phalaropodidae), males play the major role in incubation as in other nesting activities. Incubation periods range from 11 days in many passerines to 80 days in the Royal Albatross. Hatching of the eggs is synchronous when incubation starts after completion of the clutch, or asynchronous when incubation starts before the clutch is complete. Asynchronous hatching seems to be an effective means of regulating brood size to available food, the larger, first hatched young prospering at the expense of their younger siblings if food is scarce. Synchronous hatching, on the other hand, appears to be a more efficient means of exploiting food sources that are highly seasonal, that is, that are available in quantity for only brief periods. Synchronous hatching also serves to keep the brood together in precocial birds such as ducks, pheasants, and the like.

The nestling period may be as short as 8 days in altricial birds. Precocial birds remain in the next for only a few hours or days. At the other extreme, chicks of Royal Albatross (*Diomedea epomorphora*) may not fly until they are about 240 days old. In the remarkable megapodes (Megapodiidae) the young are wholly independent from the time of hatching, having developed through a long incubation period buried in mounds of sand, warm volcanic ash, or decaying vegetation.

One aspect of breeding biology of birds that deserves special mention is social or brood parasitism. In some cuckoos (Cuculidae), some cowbirds (Icteridae), one subfamily of weaver-finches (Viduinae, Ploceidae), honey-guides (Indicatoridae), and a few ducks, reproduction regularly involves laying eggs in nests of other birds which then rear the parasite's young. Considering the whole array of social parasites, practically all states occur from occasional placing of eggs in nests of other birds to obligate parasitism involving specialization of the parasite to a single host species. In the latter situation, remarkable similarities in egg size, shape, and color better fit the parasite to deception of its host. In the parasitic weaver-finches, the nestlings often have brightly marked mouth-linings which match those of the host species. The young of parasitic cuckoos may eject young or eggs of the host from the nest. Nestling honey-guides of the genus *Indicator* have sharp hooks on the tips of the bill which are lost by the time the young leave the nest. At least one species uses these hooks to kill host young by biting.

HARRISON B. TORDOFF

References

Grassé, Pierre P., *et al.* Oiseaux. Vol. 15 of *Grassé's* Traité de Zoologie. Paris, Masson and Co., 1950.
Marshall, A. J., "Biology and Comparative Physiology of Birds," 2 vols. New York, Academic Press, 1960–1961.
Streseman, E. *in* Krumbach, T. and W. Kukenthal, "Handbuch der Zoologie," vol. 7. Berlin, de Gruyter, 1927–34.
Van Tyne, J. and A. J. Berger, "Fundamentals of Ornithology," New York, Wiley, 1959.

AVICENNA (980–1037)

Ibn ben Sienna, known to the western world as Avicenna, was born the son of a tax collector in 980. Avicenna had already won recognition for his mental ability at the age of ten, by which age he had memorized the Koran. Noted primarily for his achievements in the field of medicine, he was for all purposes a self-taught physician. Before the age of sixteen he knew medical theory thoroughly, and according to his own account, had discovered new methods of treatment.

This "boy wonder" came under the protection of many patrons, some royal, and thus fell out of favor whenever a change of rule occurred. It was while under the patronage of a friend at Hycania, that his famous *Canon of Medicine* was begun. This five volume work, covering physiology, pathology, hygiene, treatment of disease and the composition and preparation of remedies, was the guide for medical studies in European universities from the twelfth to seventeenth century. The work's chief drawbacks seem to have been an excessive classification of bodily faculties and non-clarity in discrimination of diseases.

The precedence that the *Canon of Medicine* achieved over its predecessors derived, in all probability, from its improved methodology which earned for him his title of "Prince of Physicians."

Though his main fame rests on his medical achievements, there are also approximately 100 treatises attributed to him covering such subjects as theology, philology, mathematics, physics, astronomy and music.

Despite his love of knowledge, Avicenna's pursuit of the sensual pleasures of life was never satisfactorily subordinated, and was one of the chief contributing factors of his death in 1037.

LETITIA LANGORD

b

BACILLARIOPHYCEAE

Diatoms are microscopic, unicellular or colonial ALGAE constituting the Class Bacillariophyceae of the Phylum CHRYSOPHYTA. This phylum, which comprises also the CHRYSOPHYCEAE and the XANTHOPHYCEAE, is characterized by:
(1) the carotinoid pigments β-carotene and fucoxanthin, in addition to chlorophylls a and c,
(2) leucosin (a glucose polymer) and oil as storage products, and the absence of starch,
(3) a tendency to deposit silica in the cell walls.
In the diatoms, the cell protoplast is typically enclosed in a perforated siliceous wall (frustule) made up of two separate parts (thecae); hence the name "dia-tom" (= across + cut, Gr.). The frustule of a typical diatom is illustrated diagrammatically in Fig. 1.

Two subclasses are usually recognized: (1) the Pennales (oblong, feather- or boat-shaped diatoms), in which the valve exhibits two planes of symmetry, and (2) the Centrales (usually cylindrical or disc-shaped diatoms), in which the valve is radially symmetrical. There are exceptions to both of these generalizations: a few genera of pennate diatoms (e.g. *Gomphonema*) may exhibit only one plane of symmetry, and some centric diatoms (e.g. *Triceratium*) may be triangular or bilaterally symmetrical. Some authorities are of the opinion that the distinction between Pennales and Centrales may not be a natural one. Nevertheless, these subclasses are usually distinguished by certain associated features, viz:

Pennales	Centrales
1. Valves usually elongate, with striae more or less transverse to the apical axis. Spines absent.	1. Valves usually discoid or cylindrical, with radially disposed striae or concentric markings. Spines or protuberances often present.
2. Raphe usually present, often associated with ability to glide over substratum.	2. Raphe invariably absent: cells incapable of locomotion.
3. Usually one or two large plate-like or lobed chromatophores.	3. Usually numerous discoid chromatophores.
4. Resting spores absent.	4. Resting spores often present.
5. Gametes (when known) isogamous, not flagellate.	5. Gametes (when known) oogamous, with motile sperm.
6. Habitats: fresh, brackish, or salt water, usually on surfaces of plants or mud. Many soil species. Few planktonic.	6. Mostly planktonic, mostly marine. Never in soil.
7. Examples: *Navicula, Nitzschia.* (See fig. 2.)	7. Examples: *Coscinodiscus, Melosira.* (See fig. 3.)

Species of diatoms are distinguished from one another by the shape, structure, and ornamentation of the frustules. These may be most clearly seen after the organic matter has been destroyed by heat or acids, and the walls mounted in a medium of a suitable refractive index. Each silica wall or frustule consists essentially of two halves, each comprising a valve and a connecting band (see Fig. 1). When the protoplast divides by binary fission, two new silica walls are deposited on the new surfaces of the daughter cells, which are thus enclosed by an epitheca from the original parental frustule and a newly formed hypotheca (see Fig. 4). In some species without silicified girdles (e.g. *Navicula pelliculosa*) the cell size remains constant; in others which are more rigidly silicified (e.g. *Nitzschia frustulum*) this particular mode of cell division leads to the formation of smaller and smaller cells as the population multiplies. With a decrease in size, alterations of form may occur, so that the smaller cells are usually not geomet-

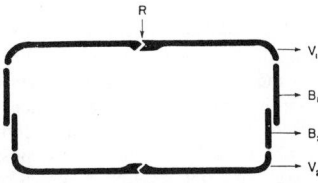

Fig 1 Diagrammatic section of a typical diatom frustule.

V_1, V_2 = Valves
B_1, B_2 = Connecting bands
$V_1 + B_1$ = Epitheca
$V_2 + B_2$ = Hypotheca
$B_1 + B_2$ = Girdle
R = Raphe

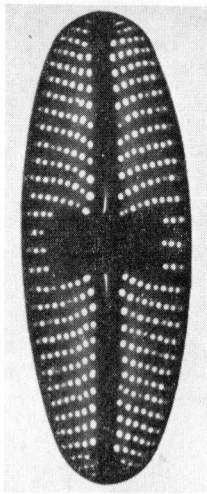

Fig 2   Silica valve of a pennate diatom. (Electron micrograph, ×7,550.)

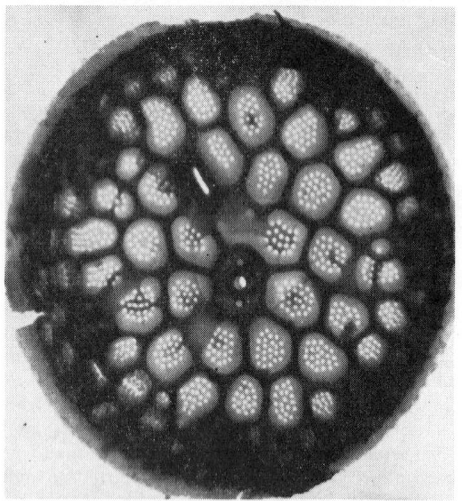

Fig 3   Silica valve of a centric diatom. (Electron micrograph, ×23,850.)

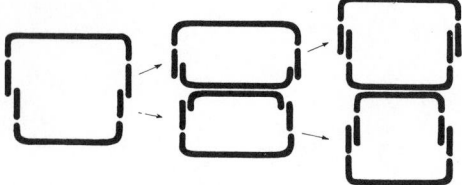

Fig 4   Diagram indicating the mode of division of a diatom cell showing the formation of new silica valves.

Many species of diatoms, both marine and fresh water, have been grown in the laboratory in chemically defined media. Some species (e.g. *Nitzschia closterium*) are auxotrophic, i.e. they require accessory growth factors such as cobalamin, thiamine, or both. Some species are able to grow heterotrophically in darkness using organic substrates such as glucose or lactate as an alternative to their usual mode of growth in light by photosynthesis. One littoral species (*Nitzschia putrida*) lacks chlorophyll, and is therefore obligately heterotrophic.

For virtually all diatoms, silicon is essential for the construction of the walls, and in the absence of soluble silicic acid or silicates the cells are unable to divide. Although amorphous silica is slowly soluble at the pH of natural waters, the silica walls of living diatoms are in some way protected against solution. After the death of the cells, the silica usually dissolves. In certain environments, however, the walls of planktonic diatoms may settle and accumulate on the bottom faster than they dissolve, thereby forming a diatomaceous ooze. With the passage of time and slow dehydration, the amorphous silica develops a microcrystalline structure and becomes much less readily soluble in water. After fossilization such a deposit is known as a diatomaceous earth or kieselguhr. In some places where considerable deposits occur (e.g. Lompoc, California) they are mined as a white, friable, highly porous rock, which is crushed, often calcined, and employed for insulation and filtration and as an abrasive.

Diatoms appeared rather late on the evolutionary scene; the earliest fossil records are from deposits of the late Cretaceous (e.g. Moreno shale in Central California). Centric types predominated during the early Tertiary, and pennate types became increasingly abundant in late Tertiary. Today diatoms frequently constitute the bulk of marine and fresh water phytoplankton. In all, about 10,000 species have been described.

JOYCE C. LEWIN

Reference

Fritsch, F. E., "The structure and reproduction of the algae," Vol. I. pp. 564–651. Cambridge, The University Press, 1935.

rically proportional to the larger ones, but are relatively wider. When the cells have reached a certain reduced size, auxosphore formation may take place. This is a process whereby the protoplast escapes from its rigid wall, expands by vacuolation, and then deposits a new, enlarged frustule, making possible a return of the cells to maximum size. In some Pennales auxosphore formation may also involve a sexual process.

Many diatoms secrete mucilage. The cells may be encapsulated, or embedded in a common gelatinous mass or in a simple or branched tube; they may be attached to one another in colonies by gelatinous threads or pads; or they may attach to a substrate by gelatinous stalks.

BACTERIA

The bacteria as a group are small microorganisms with a relatively primitive cellular organization. In *Bergey's Manual of Determinative Bacteriology* (7th Edition,

1957), they are treated as a single Class, *Schizomycetes,* of the Division *Protophyta* of the plant kingdom, but there is reason to believe that they comprise a rather heterogeneous collection of organisms of phylogenetically doubtful affinities, morphologically most closely resembling the blue-green ALGAE. Some types of bacteria appear to be, in fact, colorless counterparts or close relatives of certain blue-green algae. Although a few bacteria are PHOTOSYNTHETIC, these do not produce oxygen, as do the blue-green algae.

Bacterial morphology. Five major groups of bacteria have been recognized by some authors on the basis of morphology. These are the *eubacteria* (true bacteria), the *spirochetes,* the *gliding bacteria,* the *budding bacteria,* and the *pleuropneumonia group.* A sixth group, the *rickettsias,* have been sometimes classified as bacteria and at other times as viruses.

The eubacteria (Orders *Pseudomonadales, Chlamydobacteriales, Eubacteriales, Caryophanales* and *Actinomycetales*) include most of the known bacterial species. This major group is characterized by the possession of rigid cell walls and, in motile species, by flagellar motility. Cell division is always by binary transverse fission. The organisms are unicellular *(Pseudomonadales* and *Eubacteriales),* filamentous *(Chlamydobacteriales* and *Caryophanales),* or mycelial *(Actinomycetales).*

The cells and mycelia of the true bacteria differ from those of the higher protists (e.g. green algae, fungi and protozoa) in their generally smaller dimensions, in the absence of plastids such as mitochondria or chloroplasts, and in the apparently simpler organization of the nuclear apparatus and of the FLAGELLA in the motile types. In the unicellular forms, the cells may be spherical (coccus), rod-shaped (bacillus), curved rods (vibrio) or spiral (spirillum). The cells of the members of the Family *Caulobacteraceae* of the Order *Pseudomonadales* develop stalks, by which they become attached to a solid substrate. The diameter of eubacterial cells generally ranges from 0.2 to 2.0 microns (1 micron = 0.001 mm). The length of the rod-shaped and spiral cells varies for different species, usually being between 1 and 10 microns. The filaments of the filamentous forms, composed of many linearly arranged cells, may reach a length of hundreds of microns and are, in some instances, surrounded by a common sheath. A typical eubacterial cell is enclosed by a rigid cell wall which may be surrounded by a capsule or "slime layer." The cytoplasm is bounded by a morphologically distinguishable cell membrane, is not vacuolated and may contain inclusions such as globules of lipid, of sulfur (in the sulfur bacteria) or of polymetaphosphate (metachromatic granules). The nuclear bodies, often found in pairs or groups of four, are spherical or dumbbell-shaped. Bacterial flagella are usually less than 0.1 micron thick, arise in the cytoplasm and penetrate the cell wall and capsule. They are single-stranded, in contrast with the multi-stranded flagella of the higher protists and of the flagellated cells of the higher animals and plants. In elongate cells, the flagella may be located at the ends (polar) or around the periphery (peritrichous). In some eubacteria, the cell membrane is separable from the rigid cell wall, against which it is normally pressed by the turgor of the enclosed cell. In others, it appears to be structurally contiguous with the cell wall. Cell membranes of some species can be

isolated by enzymatic dissolution of the wall followed by the liberation of the internal contents by osmotic lysis. The cytochrome system and some associated enzymes of aerobic bacteria have been shown to be located in the cell membrane. In the photosynthetic bacteria, the photosynthetic apparatus is located in sub-microscopic "chromatophores" within the cell, that may be attached to or even evolved from the cell membrane.

Cell division is initiated by nuclear division. This is followed by the division of the cytoplasm with the formation of a plasma membrane. Next, a transverse cell wall, continuous with the outer cell wall, grows inward toward the center of the cell. Finally the transverse cell wall splits into two layers, leading to the separation of the daughter cells. In rapidly growing cultures of rod-shaped bacteria, the formation of walls and the separation of the cells lag behind nuclear division, so that the vegetative units may consist of more than one cell and contain several nuclei. After cell division has taken place, the cells may remain attached to each other in more or less characteristic patterns that reflect the plane of cell division and post-fission movement of the daughter cells. Thus, the spherical cells of different species may occur in pairs (diplococcus), chains (streptococcus), tetrads, cubical packets (sarcina) or irregular clusters. In the filamentous forms of the Order *Caryophanales,* the vegetative unit is always multicellular, reproduction occurring by transverse fission of the filaments. In *Chlamydobacteriales,* the filaments are reproduced by motile swarm cells that become detached from a parent filament.

Two types of resting cells occur in the unicellular eubacteria, cysts and endospores. Cysts are formed by the rounding up of vegetative cells that become surrounded by a heavy wall. These occur in some species of *Azotobacter,* and possibly in other eubacteria. Endospores are formed within the cells of certain eubacteria, being characteristic of the Family *Bacillaceae.* A single spore is developed per cell and liberated upon disintegration of the mother cell. Endospores are remarkably resistant to dehydration, toxic chemicals and heat. Those of some species can be boiled for an hour or longer without affecting their viability. It is because of the occurrence of such resistant cells that temperatures above 100° C. are required for sterilization of bacteriological media and surgical materials and for the preservation of foods in canning.

The mycelial eubacteria (actinomycetes) are immotile and resemble mycelial fungi. Their mycelium is nonseptate and has smaller dimensions and simpler organization than that of molds. Some actinomycetes reproduce by fragmentation of the mycelium into units that resemble true bacteria; others produce conidia for the purpose of dissemination. These somewhat resistant spores are budded off at the ends of aerial hyphae either singly or in characteristic chains.

The Spirochetes (Order *Spirochaetales*) are unicellular spiral-shaped motile bacteria and differ from the eubacteria in that they do not possess rigid cell walls and are hence more or less flexible. They exhibit a swimming motility, the exact nature of which is not entirely understood and, in some species, active flexing movements of the body. A bundle of very fine fibrils, spirally wound around the organism, has been demonstrated in several types of spirochetes and it appears likely that the movements may be effected by contractions of this bundle, resulting in undulations of the entire cell.

The Gliding Bacteria are flexible organisms that possess a gliding or creeping movement when in contact with solid surfaces. In this respect they closely resemble the blue-green algae, from which they differ in being nonphotosynthetic. Unlike the eubacteria, the gliding bacteria do not possess rigid cell walls and never show flagellar movement.

The unicellular rod-shaped myxobacteria (Order *Myxobacterales*) form thin, loose, creeping colonies and most of them produce resting cells known as microcysts. Each microcyst is formed by the shortening and rounding up of a vegetative cell. The microcysts are commonly formed in macroscopically visible fruiting bodies, whose shape is characteristic of the genus. The fruiting bodies are produced by the aggregation and cooperation of individual cells, much as are the fruiting bodies of the myxomycetes of the group *Acrasiae.*

There are a number of filamentous gliding bacteria (Order *Beggiatoales*) that resemble filamentous blue-green algae; to which they may be closely related.

The Budding Bacteria (Order *Hyphomicrobiales*) comprise two rarely encountered genera that reproduce by buds formed at the tips of fine threadlike extensions of the mother cell. At least one and possibly both of these genera may be closely related to the eubacteria.

The Pleuropneumonia Group (Order *Mycoplasmatales*) are very small microorganisms with irregular morphology and without rigid cell walls. Their small cells may develop into irregular spherical or filamentous "large bodies" that can break down and liberate small viable elements. Under special conditions, certain eubacteria have been found to lose their rigid cell walls and to produce vegetative forms, morphologically similar to the pleuropneumonia organisms. It is not known whether the naturally occurring pleuropneumonia and related organisms may not be, in fact, forms of true bacteria.

The Rickettsias are very small rod-shaped or spherical organisms that are obligate intracellular parasites of animals, particularly arthropods. It is uncertain whether they are very small bacteria or large viruses. Some species are pathogenic to man, causing such diseases as typhus, Rocky Mountain spotted fever and Q fever.

Bacterial nutrition and physiology. No other group of organisms includes as vast a variety of physiologically and nutritionally diverse types as does the group known as the "bacteria." The nutritional requirements of some species can be satisfied by entirely inorganic media. Other species, on the other hand, need complex mixtures of amino acids, PURINES, PYRIMIDINES and VITAMINS. In the extreme case of obligate parasites, the organisms are so dependent on the metabolism of the host that their *in vitro* cultivation has not yet been accomplished. Not only does there exist a whole spectrum of different metabolically and nutritionally distinct types of bacteria, but also many species are endowed with a remarkable versatility of their metabolic patterns, which permits them to adapt themselves to a great variety of environmental conditions. The adaptation, in many cases, is effected by the synthesis of special enzymes that are produced in response to the presence of specific substrates and are therefore spoken of as "adaptive" or "inducible" enzymes. The ability to form a variety of such inducible enzymes may make a given strain of bacteria remarkably omnivorous in its nutrition; some cultures are known to be capable of using any one of over a hundred different organic compounds as the sole source of carbon in the medium. Probably no naturally occurring organic compound exists that cannot be decomposed by some species of bacteria.

Many bacteria are obligately aerobic and require atmospheric oxygen for respiration. Others can carry out anaerobic respirations or fermentations and may be either obligately or facultatively anaerobic. A few are photosynthetic.

Although most of the bacterial enzymes are located in the cells, some species produce specific extracellular enzymes, particularly hydrolases that digest proteins, polysaccharides and other compounds of high molecular weight.

Bacterial Photosyntheses differ from green plant photosynthesis in that oxygen is never evolved. Some of the photosynthetic bacteria are, indeed, obligate anaerobes. The photosynthetic pigments, which include the bacterial chlorophylls and carotenoids, are contained in sub-microscopic chromatophores. The bacterial chlorophylls differ structurally in minor respects from the green plant chlorophylls and have absorption maxima in the infra-red region. The "green" and "purple sulfur bacteria" can grow autotrophically in the light, using H_2S, sulfur or thiosulfate for reducing CO_2 to cell material. For example the purple sulfur bacteria can oxidize H_2S to sulfur in the presence of light according to the following equation, in which (CH_2O) is an approximation of the state of cell carbon:

$$2H_2S + CO_2 \xrightarrow{\text{light}} 2S + (CH_2O) + H_2O$$

The elemental sulfur formed in the reaction may be stored as a reserve material in the form of globules within the cells and later oxidized to sulfuric acid with a concomitant reduction of more CO_2. The "nonsulfur purple bacteria" depend on organic substrates such as fatty acids, dicarboxylic acids, etc., in the nutrition. Both the organic compound and CO_2 may be used for the synthesis of cell material.

Aerobic Respiration as a source of energy is used by a great variety of bacteria. The respiratory enzymes are, at least in part, located in the cell membrane. The complement of cytochromes differs in various groups and these respiratory pigments are slightly different from those of the higher animals and plants. The Krebs tricarboxylic acid cycle has been demonstrated as a mechanism for the "terminal oxidation" of organic compounds in many heterotrophic species.

Some groups of heterotrophic bacteria are characterized by the fact that they carry out "incomplete oxidations" of organic compounds. For instance, the "vinegar bacteria" oxidize primary alcohols mainly to the corresponding acids. Ethanol, for example, is converted to acetic acid:

$$CH_3-CH_2OH + O_2 \rightarrow CH_3-COOH + H_2O$$

A number of bacteria can use the aerobic oxidation of inorganic materials as a source of energy for autotrophic growth. In fact, certain genera, such as *Nitrosomonas, Nitrobacter,* and *Thiobacillus* are said to be "obligately autotrophic," since they grow poorly or not at all in media that contain organic compounds. The "nitrifying bacteria" oxidize the nitrogen atom. Thus, *Nitrosomonas* and *Nitrosococcus* oxidize ammonia to nitrate:

$$2NH_3 + 3O_2 \rightarrow 2NO_2^- + 2H^+ + 2H_2O$$

Nitrobacter, on the other hand, oxidizes nitrite to nitrate:

$$2NO_2^- + O_2 \rightarrow 2NO_3^-$$

The "colorless sulfur bacteria" and members of the genus *Thiobacillus* obtain energy from the oxidation of the sulfur atom:

$$2H_2S + O_2 \rightarrow 2S + 2H_2O$$

$$2S + 2H_2O + 3O_2 \rightarrow 2SO_4^= + 4H^+$$

$$S_2O_3^= + H_2O + 2O_2 \rightarrow 2SO_4^= + 2H^+$$

Similarly, the "hydrogen bacteria" oxidize molecular hydrogen to water, while the "iron bacteria" oxidize ferrous to ferric ion.

Anaerobic Respiration is the oxidation of organic and, in some cases, inorganic compounds with inorganic oxidants other than molecular oxygen. This type of metabolism is characteristic of some obligately and facultatively anaerobic bacteria. The reduction of nitrate by different species leads to the formation of nitrite, nitrous oxide, molecular nitrogen or ammonia. Sulfate reduction gives rise to H_2S, while the reduction of CO_2 by the "methane fermenters" results in the evolution of CH_4 ("marsh gas" or "sewer gas").

Fermentation is the sole or auxiliary source of energy for a vast variety of obligately or facultatively anaerobic bacteria. Carbohydrates, hydroxyacids, dicarboxylic acids, amino acids, purines, pyrimidines and even fatty acids can be fermented. The useful substrates as well as the end-products of fermentation are characteristic of each bacterial species and the products include formic, acetic, propionic, butyric, lactic and succinic acids, ethyl, isopropyl and butyl alcohols, 2,3-butylene glycol, acetone, methane, molecular hydrogen and CO_2. The "homofermentative lactic acid bacteria," for example, ferment glucose with the formation of lactic acid almost exclusively. Members of the genus *Escherichia,* on the other hand, ferment glucose with the production of lactic, formic, acetic, and succinic acids; ethyl alcohol, hydrogen and CO_2.

Bacterial growth and reproduction. *Cell Multiplication* of the unicellular bacteria is, in most cases, accomplished by transverse fission and results in the doubling of the number of individuals with every cell generation. Hence, the growth of bacterial cultures in a favorable environment is exponential with time. The average "generation time" for a given culture depends on the properties of the organism and on the conditions of cultivation. Under optimal conditions for growth, many species have very short generation times, of the order of 20 to 30 minutes. A single cell of such a species can produce a progeny of billions of individuals in the course of half a day.

Gene Transfer (see BACTERIAL GENETICS) from one bacterial cell to another has been found to occur in a limited number of species. Three types of mechanisms have been discovered. In so-called "recombination," a copulation of two individuals is followed by a unidirectional transfer of genetic material from a "donor" to a "recipient" cell. The GENES are transferred in a linear order, as though arranged in a single linkage group, such as a CHROMOSOME. The number of genes that is transferred depends, at least in part, on the duration of contact between the cells. The recombinant nucleus incorporates genes from both parent cells. The resultant vegetative nuclei are normally haploid and the progeny gives phenotypic expression of the recombinant genotypes. Different clones of a single strain may act as donors or recipients respectively, but recipient clones may become donors after contact with donor cells.

Another type of gene transfer, known as "transformation" does not require physical contact between cells. In this process, heritable properties are transmitted by means of desoxyribonucleic acid molecules. A third mechanism for the transmission of genetic material called "transduction" resembles "transformation" in that contact between the donor and recipient cells is not required. The bacterial genes are transmitted through the agency of bacterial viruses known as BACTERIOPHAGES, that carry small portions of bacterial desoxyribonucleic acid into the cells which they infect.

Bacterial Variation may result not only from temporary phenotypic adjustment to environmental conditions but also from mutations such as those commonly observed in laboratory cultures, and from the subsequent genetic transfer of the mutant characters. Selective environmental pressures often lead to the rapid development of mutant or recombinant clones and to the virtual disappearance of some of the original characteristics of a given culture.

Bacterial taxonomy. The taxonomy of bacteria is in a far less developed state than that of the higher animals, plants and microorganisms. There is, in fact, no universally accepted system of classification, although "Bergey's Manual of Determinative Bacteriology" is the most complete and generally used compilation of the described types. The morphological simplicity of the organisms coupled with a wide variability in their readily observable characteristics, the lack of fossil remains and the absence of known sexual processes in most bacteria make it difficult, if not impossible, to construct a "natural" system of classification. Although recognizably related groups of bacteria occur in nature, and these have been classified as species, even the applicability of the "species concept" as developed for the higher organisms is of doubtful validity. The paucity of morphological features makes it necessary to utilize a variety of critera for differentiating between bacterial species, genera, and even higher categories. These criteria include gross physiological properties such as the nature of the substrates and products of metabolism, and more subtle ones such as symbiotic or parasitic relation to other organisms, toxigenic and immunological properties.

Bacterial ecology. The enormous variety of nutritional and physiological types encountered among the bacteria is a reflection of the multiplicity of ecological niches that these organisms occupy. The inhabitants of soil and water generally have world-wide distribution, being carried about by water or on dust particles by the wind or by water. Some highly specialized parasites, on the other hand, require specific hosts and, in some instances, are dependent on special mechanisms for transmission, such as insect vectors.

The Soil Bacteria, as well as those inhabiting surface waters and ocean muds are among the most important geochemical agents on the earth's surface. Together with the fungi, they are responsible for the bulk of the mineralization of organic matter and the evolution of CO_2 to the atmosphere. The "nitrifying bacteria" accomplish the oxidation of ammonia, that is liberated in the decomposition of organic materials, to nitrate. Similarly, the H_2S that is derived from sulfur-containing organic compounds is oxidized to sulfate by the "sulfur bacteria." The "denitrifying bacteria" reduce nitrates to molecular nitrogen, whereas the sulfate reducing species cause the local accumulation of sulfide. The "nitrogen fixing bacteria" play an essential role in the maintenance of the cycles of matter by reducing atmospheric nitrogen to amino nitrogen. Because of their ubiquity and their high metabolic and reproductive rates, the bacteria are eminently suited for their various roles in the turnover of the essential elements of the biosphere.

Symbiosis and Parasitism (see articles under these titles) are characteristic of many types of bacteria. The host-parasite relationships range from the mutualistic type through commensalism to pathogenesis. Well-known examples of symbiosis which is beneficial to the host are the root nodulation of leguminous plants and the rumen flora of ruminant animals. In the former case, the infection of the roots by bacteria of the genus *Rhizobium* makes it possible for the plant to utilize (fix) atmospheric nitrogen, which neither the plant nor the bacterium can do alone. In the latter case, the bacteria of the rumen digest and ferment foodstuffs that the host animal cannot itself utilize. The products of fermentation as well as the bacteria themselves are then used as nutrients by the animal. Examples of commensalism are the microflora of the skin, mouth and intestine of man.

Many diseases of animals and plants are caused by parasitic bacteria, which are in many cases highly host-specific. Among the diseases of man caused by bacteria are pneumonia, scarlet fever, diphtheria, epidemic meningitis, whooping cough, gonorrhoea, syphilis, tuberculosis, plague, tularemia, brucellosis, typhoid fever, dysentery, and cholera. Plant diseases caused by bacteria include leaf spots, fire blights, wilts, galls, tumors, and soft rots.

Exploitation of bacteria by man. Bacteria, along with other types of microorganisms, were used by man in many ways, even before their existence was suspected. The preservation of food and fodder by pickling and ensilage, and the preparation of cheese and other sour milk products depend on the growth and metabolism of the "lactic acid bacteria" which convert fermentable sugars to lactic acid. The manufacture of vinegar from alcoholic beverages, the retting of vegetable fibers and the disposal of sewage by the use of septic tanks or of the "activated sludge process" are other examples of bacteriological processes. Although, in modern times, pure cultures of bacteria are occasionally used for certain of the above processes, in all cases

the results can be obtained by the natural "enrichment" of the desired bacterial types from the mixed population of organisms present in the starting materials. Such "enrichment" requires that a suitable environment be provided for the predominance of the desired organisms.

The development of microbiological techniques and modern discoveries in the field of bacteriology have provided the tools for important new uses of pure cultures of bacteria. These uses include the industrial fermentation of carbohydrates with the formation of butyl alcohol and acetone, the production of a number of antibiotics, such as streptomycin, aureomycin and chloramphenicol, and the manufacture of certain enzymes and vitamins. Bacterial cultures are also used in the bio-assay of a large variety of vitamins and amino acids.

Pure cultures of bacteria have not only provided man with many useful products, but have served him as essential tools for the development of modern scientific knowledge. Thus, the science of biochemistry has depended greatly on studies with bacteria, while recent important developments in the fields of genetics and virology have concerned the genetic transfer and bacteriophage multiplication in these microorganisms.

M. Doudoroff

References

Stanier, R. Y., *et. al.*, "The Microbial World," 2nd ed. Englewood Cliffs, N. J., Prentice-Hall, 1963.

Thimann, K. V., "The Life of Bacteria," 2nd ed. New York, Macmillan, 1963.

Oginsky, E. L., and W. W. Umbreit, "An Introduction to Bacterial Physiology," 2nd ed. San Francisco, Freeman, 1959.

Wilson, G. S., and A. A. Miles, eds., "Topley and Wilson's Principles of Bacteriology and Immunity," Baltimore, Williams & Wilkins, 1955.

Breed, R. S., *et. al.,* "Bergey's Manual of Determinative Bacteriology," Baltimore, Williams & Wilkins, 1957.

BACTERIAL CELL

Bacterial cells vary enormously in size from about 0.3 μ in width or length, to 10–15 μ in length. There are only three principal shapes observed for bacterial cells. They are the spherical cells or cocci, rod-shaped or cylindrical cells and the spiral-shaped cells. Most BACTERIA possess an outer wall which confers mechanical rigidity upon the cell and is responsible for the cell's characteristic shape. Bacteria and plants thus possess a rigid wall as a major structural component of their cells and therefore differ from animal cells which are bounded by a more fragile membrane. The nature of the outer wall of bacteria is responsible for their differentiation by the Gram stain reaction (developed by Christian Gram) into two groups, the Gram-positive and the Gram-negative bacteria. There are many physiological and biochemical properties correlated with this division of the bacterial world.

Structurally, the bacterial cell may be subdivided into the following principal anatomical regions: (1) *surface appendages;* (2) *surface adherents;* (3) *surface layers;* (4) *cytoplasm, intracellular organelles, particles and granules;* (5) *special structures,* such as endospores and stalks.

1. Surface appendages. *Flagella.* Many of the rod and spiral-shaped bacteria and less frequently the cocci, possess thin, hair-like structures whose lengths are many times that of the cell. These appendages are called flagella and they are organs of locomotion. The numbers per cell may vary from a single flagellum to as many as 30 flagella. Bacterial flagella are about 120 Å in diameter and in the fixed state they appear as single-stranded, wave-like structures. Only one bacterium has been found to possess a flagellum made up of three spirally wound strands. The flagella extend through the surface layers and are attached to a basal granule embedded in the interior of the bacterial cell. It is possible to detach the flagella by agitating the cells in a "blendor" and if undamaged, the cells can re-form new flagella and resume their motility. Flagella have been isolated in a purified state by removing them from bacteria by shaking with small glass beads. The isolated flagella from the organism *Proteus vulgaris* have been chemically characterized as proteins (called flagellins). They possess a structure similar to that of other hair and muscle proteins of the keratin-myosinepidermis group.

Pili, Fimbriae. Some non-motile bacteria may be surrounded by several hundreds of short, hairlike structures obviously differing in form and function from the flagella. These have been called fimbriae and pili, the former term describing the appearance of the cell rather than the structure itself. The pili are responsible for the adhesion of bacterial cells to other cells such as red blood cells and bring about a flocculation of the latter, a phenomenon known as hemagglutination.

2. Surface adherents, *Capsules and slime layers.* Outside the rigid cell wall, some bacteria are surrounded by gelatinous material which may in some cases form a layer of up to 10 μ in thickness. The presence of a cytologically demonstrable capsule or loose slime layer probably depends on the viscosity and gel properties of the material produced by the bacterial cell. Mutant strains of capsule-forming bacteria have been found to produce chemically identical compounds which however, are of lower viscosity and are readily washed from the surface of the cell or diffuse into the growth medium.

Capsules may be simple uniform accumulations of one type of polymeric substance or in some instances they may be both chemically and structurally complex. The organism *Bacillus megaterium* produces a capsule composed of a polypeptide and a polysaccharide, the latter occurring in localized patches in the polypeptide capsular material. Another type of complex capsular structure showed the presence of a large number of banded fibrils in an amorphous matrix around the cells of a strain of *Escherichia coli.*

The majority of capsular and slime substances isolated and studied chemically have been found to be polysaccharides. Pneumococci produce a great variety of capsular polysaccharides containing different sugar constituents or similar sugars joined by a variety of linkages. A number of these polysaccharides contain

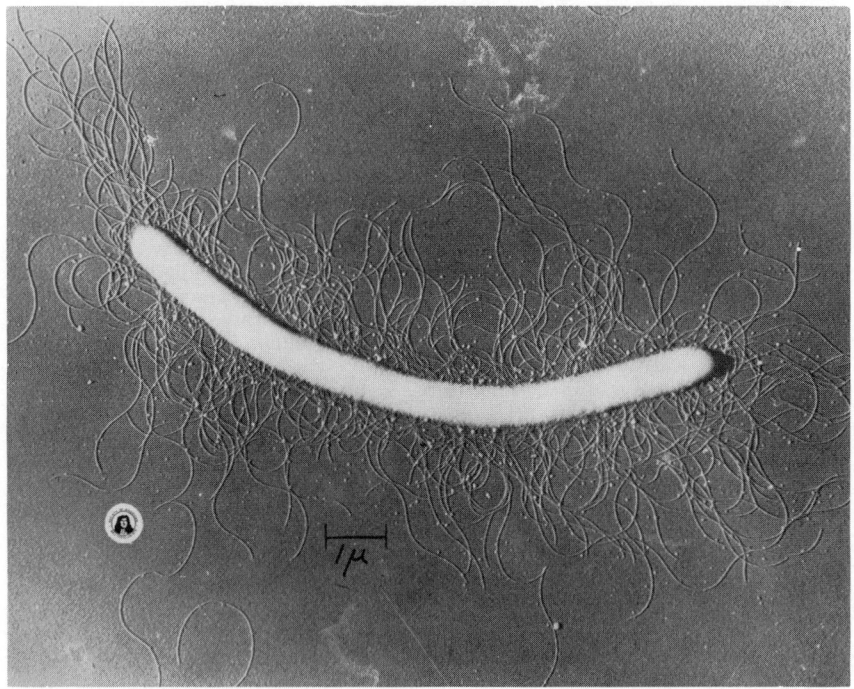

Fig 1   Electron micrograph of *Proteus vulgaris* showing flagella. (LS-258 Courtesy American Society for Microbiology.)

amino sugars. Ribitol phosphate has recently been found in a pneumococcal capsular polysaccharide. Some streptococci produce a capsule of hyaluronic acid, a polymer of glucosamine and glucuronic acid, which also occurs widely in animal cells. One of the most unusual capsular substances is the polypeptide composed of D-glutamic acid, isolated from the capsule of *Bacillus anthracis* (organism causing anthrax) and a strain of *Bacillus megaterium*. This material is unusual not only because it contains the D-isomer of an amino acid (most natural products apart from a number of bacterial compounds, contain the L-isomers) but also because of the linkage of the glutamic acid through its γ-carboxyl group. Such compounds have not been encountered in cells of higher plants and animals and seem to be unique to bacteria.

Many capsules can be removed with specific enzymes and this can frequently be achieved without affecting the viability of the bacterial cells. Cells from which the capsule has been removed enzymically can re-synthesize new capsular material. The capsules may have two possible functional advantages for the bacterial cell, one as a protection against phagocytosis where the organism is a pathogen, and a second function in providing a potential storage product. There is some evidence that capsular material may be reutilized by the cell.

There are several examples of surface materials that are not detectable as thick capsules or loose slime layers, but are firmly adherent to the outer cell surfaces. Their presence was first established by immunological reactions and later demonstrated by extraction from the cells. Certain streptococci are surrounded by a thin layer of a protein substance which can be removed from the cell by the proteolytic enzyme, trypsin. A number of the bacteria belonging to the group causing dysentery infections in man, possess a surface substance composed of an amino uronic acid.

3. Surface layers. Most bacteria possess two major surface layers, a rigid cell wall and an underlying protoplasmic membrane. It is conceivable that in some bacteria there may be a single, integrated structure performing the functions of wall and membrane. Thin sections of both Gram-positive and Gram-negative bacteria examined in the electron microscope have shown an external layer (wall) of 100–200 Å in thickness and a separate underlying layer (membrane) about 100 Å thick.

The cell wall constitutes the major structure of the bacterial cell and accounts for 20–40% of the weight of the cell, depending on the physiological age of the cells. As seen in the electron microscope, the walls of some bacteria are completely homogeneous in appearance, while others possess spherical macromolecules packed hexagonally in a multi-layered wall. Fine structure in the wall is encountered more frequently in the Gram-negative group of bacteria.

Mechanical disintegration, either by violent agitation with minute glass beads (0.13 mm diameter) or exposure to super-sonic vibrations has been used in the isolation of the cell walls. In general the wall is sufficiently rigid to withstand the effects of disintegration whereas the other cellular structures are broken down and can be readily separated from the walls.

Marked differences in cell-wall composition are found for Gram-positive and Gram-negative groups of bacteria. The entire walls of many Gram-positive bacteria may be accounted for by a class of chemical substances known as mucoids, including mucopeptides and mucopolysaccharides. Some contain in addition the teichoic acids (*teichos* = Greek for wall), polymers of ribitol or glycerol phosphates. The wall mucopeptides represent a new class of structural polymers confined to bacteria and blue-green algae. These substances contain glucosamine and a new amino sugar called muramic acid (3,O-carboxyethyl glucosamine) together with alanine, glutamic acid, lysine or diaminopimelic acid and in some, glycine, as the principal amino acids. Another unusual feature of the wall mucopeptides is the presence of a number of amino acids (alanine, glutamic acid and aspartic acid) as D-isomers. The walls of Gram-negative bacteria also contain mucopeptide components which account for only 10–20% of the wall in many strains. Although this component is only a small part of the whole wall, it is responsible for its mechanical rigidity, the rest of the wall being made up of more plastic protein-lipide-polysaccharide complexes. Lipide contents as high as 20% are thus a common feature of the walls of Gram-negative bacteria. It is believed that penicillin owes its bactericidal action to an interference in the biosynthesis of the mucopeptides of bacterial walls, thus explaining the susceptibility of both Gram-positive and Gram-negative bacteria and the greater sensitivity of the former group.

The walls of some Gram-positive bacteria are completely dissolved by the enzyme, lysozyme. This enzyme occurs abundantly in chick egg-white, but is widely found in animal tissues and secretions as well as in some plants and bacteria. The amino sugar backbone of the wall is broken down by the enzyme. If selective removal of the wall is performed by lysozyme treatment of the bacteria suspended in solutions of sucrose for osmotic stability, intact *bacterial protoplasts* are obtained.

Protoplasmic membrane. The relationship of the wall to the membrane has been established by selective enzymic removal of the wall and protoplast formation. Thus, with the rod-shaped organism *Bacillus megaterium*, the mechanical rigidity of the wall and its lack of biochemical function is demonstrated by the formation of spherical protoplasts capable of performing normal metabolic processes, protein and nucleic acid synthesis and growth. Treatment of Gram-negative bacteria with lysozyme, or growth in the presence of penicillin does give rise to spherical cells. These are called spheroplasts to distinguish them from protoplasts obtained by complete removal of the wall. Analogous protoplasts have not yet been obtained from Gram-negative bacteria as there is no suitable method of releasing the protoplast from the rest of the protein-lipid-polysaccharide part of the wall.

The functional properties of the protoplasmic membrane have been established by studying the isolated protoplasts. The membrane is the site of the cell's osmotic barrier and a number of enzymes are localized almost exclusively in the membrane. Cytochromes and various dehydrogenases are present in the membrane. The protoplast membrane can be obtained by osmotic lysis of protoplasts and it is chemically quite different from the cell wall. It is largely made up of protein and lipid, 60% and 20% respectively, and contains about 20% polysaccharide in addition. The chemical composition of bacterial membranes is in accord with classical concepts for the lipo-protein nature of membranes of higher plant and animal cells.

4. Cytoplasm, intracellular organelles, particles and granules. Thin sections of bacterial cells examined in the electron microscope have shown a fine granular cytoplasm without any evidence of a regular endoplasmic reticulum. Occasional sections have revealed series of parallel membranes within the cytoplasm, reminiscent of a reticulum, but this has not been a regular feature. What proportions of the cytoplasm and enzymes of bacterial cells are present in particulate form is not known. However, the cytoplasm contains soluble enzymes and cell solutes such as inorganic ions, and various amino acids, purines and pyrimidines and other substances constituting a 'metabolic pool.' Gram-positive bacteria generally contain a high concentration of inorganic ions and amino acids in the cytoplasm.

The principal intracellular structure demonstrable in bacteria is the *nucleus* (see BACTERIAL NUCLEUS). Its morphology and mode of replication differs from that found for the nucleus of higher organisms. The bacterial chromatin is not surrounded by a membrane nor is it organized into recognizable mitotic figures during replication. The nuclear body of *Bacillus megaterium* has been isolated by enzymic degradation of bacterial protoplasts. In addition to containing deoxyribonucleic acid (DNA) (see NUCLEIC ACIDS) it also contained ribonucleic acid (RNA) and protein.

Particles and granules. There is no doubt that bacteria possess the biochemical equivalents of mitochondria, but there are no structures within the bacterial cell showing the organized anatomy characteristic of plant and animal mitochondria. The enzyme systems associated with mitochondria are usually located in cytoplasmic particles of 100–200 Å size.

In the photosynthetic bacteria, the pigments are present in structures called *chromatophores*, the functional organelle corresponding to the chloroplast in higher plants. The chromatophores are hollow spherical particles about 300 Å in diameter and are made up of lipid, protein, carotenoids and the photosynthetic pigments. Sections of photosynthetic bacteria show the presence of a network of chromatophores, sometimes located around the periphery of the membrane and sometimes spreading throughout the cell.

The RNA of the cell occurs in the form of RNA-protein particles of about 100 Å diameter. These particles are called *ribosomes* and are active sites of protein synthesis in the bacterial cell.

There are many types of intracellular granules in bacteria. Among the most conspicuous types found are the *lipid granules* composed of a polymer of β-hydroxybutyric acid. Cells grown in media rich in acetate, sucrose and glycerol contain large numbers of butyrate granules and they may account for as much as 30% of the weight of the cell. This material is an energy storage product and is later utilized by the cell after exhaustion of the growth medium. Various *polysaccharide granules* have been observed and many bacteria accumulate a polyglucose, glycogen-like substances as a reserve material. Certain bacteria contain both butyrate and polysaccharide granules. *Volutin granules*

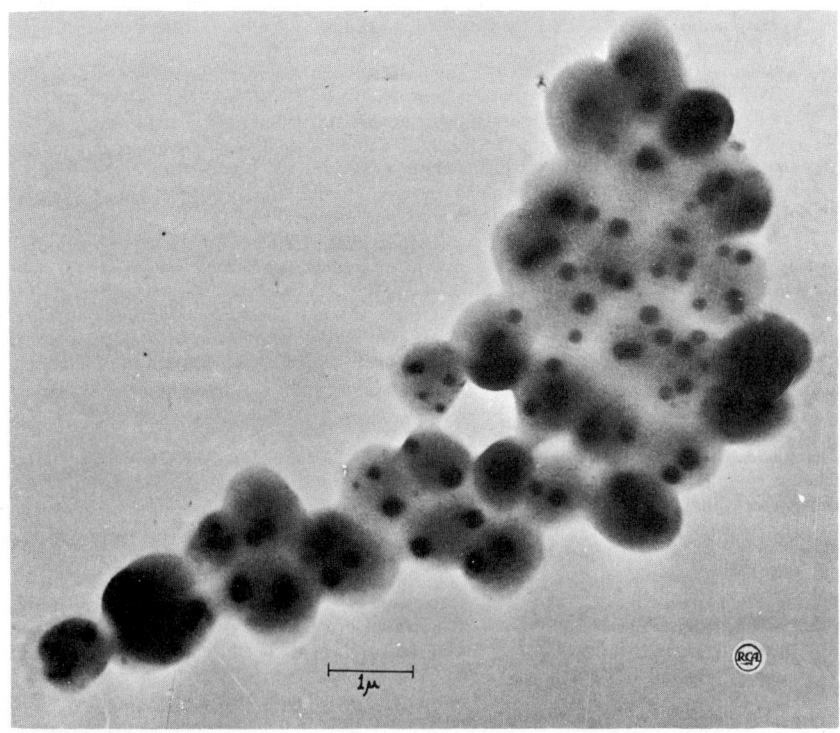

1μ

Fig 2   Electron micrograph of *Staphylococcus* showing nuclear material. (LS-89 Courtesy American Society for Microbiology.)

are present in a number of bacteria. Some of the sulphur bacteria contain granules of elemental sulfur.

5. The bacterial endospore is a specialized resistant structure formed within the cells of certain bacteria. The bacterial spore is much more resistant to heat and other physical and chemical agents than the parent vegetative cell. The spore possesses a multi-layered 'coat' or wall and its amino acid composition is more complex than that of the wall of the vegetative cell. Spores contain a large amount of dipicolinic acid, a substance so far only encountered in Nature in bacterial spores. This compound is responsible for the refractility of the bacterial spore and is also associated with its heat resistance. The bacterial spore contains fewer enzymes than the parent cell, but it is by no means biochemically inert and obviously carries all of the genetic information and enzymic constitution required for the synthesis of a normal vegetative cell.

6. Stalks. Members of the *Caulobacter* group produce a well-defined stalk-like structure which contains cytoplasm and is enclosed by a continuation of the wall. The stalks of the iron bacteria, *Gallionella*, on the other hand are excreted structures.

M. R. J. SALTON

References

Brinton, C. C., "Non-flagellar appendages of bacteria," Nature **183**: 782, 1959.
Gunsalus, I. C. and R. Y. Stainer, Eds., "The Bacteria," Vol. 1. New York, Academic Press, 1960.
Salton, M. R. J., "The properties of lysozyme and its action on microorganisms," Bact. Rev. **21**: 82, 1957.
Spooner, E. T. C. and B. A. D. Stocker, Eds., "Bacterial anatomy," (Symp. No. 6. of the Soc. Gen. Microbiol.) Cambridge, The University Press, 1956.
Weibull, C., "Bacterial Protoplasts," Ann. Rev. Microbiol. **12**: 1, 1958.

BACTERIAL GENETICS

Inactivation kinetics and GENETIC studies suggest that vegetative cells and endospores of some BACTERIA contain haploid nuclei but suggest that other strains may be POLYPLOID. Cytological studies by light and ELECTRON MICROSCOPY autoradiography, and isolation of nuclei from disrupted cells show that deoxyribonucleic acid (DNA) (see NUCLEIC ACIDS) is predominantly, but possibly not exclusively, located in nuclear bodies. The amount of DNA per nuclear body remains constant at various growth rates, although cell mass, the average number of nuclei per cell, and the content of DNA are functions of the growth rates. Studies on induced mutations indicate that the nuclear bodies bear much of the "genetic information" of the cell, and evidence cited below conclusively demonstrates that it is the DNA of the nucleus that contains this "information."

The DNA of bacteria is double-stranded, comprising the helical structure first pointed out by Watson and Crick. The PURINE adenine (A) content equals the PYRIMIDINE thymine (T) content; the purine guanine (G) content equals the pyrimidine cytosine (C) content. Minor quantities of other bases have been reported

in some bacterial DNA's and certain analogues for example, bromodeoxyuridine, are readily taken up from the growth medium and incorporated into the DNA's of some bacteria. The relative proportions of AT/GC vary from about 0.25 to 0.75 for different bacterial species, but DNA's of species classified together on morphological, biochemical and immunological grounds resemble one another in composition. Polynucleotide chains resembling bacterial DNA's have been synthesized *in vitro* by an enzyme system extracted from bacteria. The synthesis requires the presence of all four deoxynucleotide triphosphates and a DNA "primer." The synthesized DNA reflects closely the composition of the primer DNA, indicating that the system contains the high degree of specificity required of genetic material.

The bacterial species which have higher AT/GC ratios in their DNAs have slightly higher AU/GC ratios in their ribonucleic acids (RNAs). This correspondence may be even closer if only one of the several classes of bacterial RNAs (e.g. soluble, cell-membrane bound, ribosomal) varies with the DNA composition. DNA "molecules" extracted from a single bacterial species are quite homogenous in their molecular weights (either 6–7 or 14 × 10⁶). The melting points in heat denaturation and the densities for undenatured DNA molecules from a single strain are also quite homogeneous and quantitatively reflect the relative AT/GC contents characteristic of the total DNA of the species. At a given temperature, however, different molecules from a single species heat denature at slightly different rates or may be differentially inactivated by other treatments such as ultraviolet irradiation. DNA inactivated with ultraviolet light regains much of its biological activity when exposed to visible light *in vivo* or in the presence of a reactivating enzyme system and visible light *in vitro*.

P32-suicide and autoradiographic studies show DNA synthesis to be an essentially continuous process under steady-state conditions of growth. Procedures which artificially synchronize bacterial division, however, also induce step-wise bursts of DNA synthesis. As the DNA of *Escherichia coli* replicates *in vivo*, longitudinally-hybrid molecules are formed, comprised of one "old" unit (presumably a polynucleotide chain) is maintained at least through the next several replications ("semi-conservative replication"). In contrast, autoradiographic techniques demonstrate that the complete, presumably single, *E. coli* chromosome replicates in a more dispersive fashion than found on the level of the DNA molecules which, at least in part, constitute it.

Transformation. Genetically-active DNA may be extracted from bacteria and highly purified *in vitro* to contain less than 0.02% protein and no detectable uracil or serologically reactive substances such as capsular polysaccharides. During its extracellular existence the DNA is sensitive to degradation by the ENZYME, deoxyribonuclease. Genetic activities may be differentially inactivated by chemical and physical treatments of the cell-free DNA, and the DNA effectively subjected *in vitro* to chemical mutagens. DNA is taken up by only those cells in a population which are "competent." The development of competence requires protein synthesis. In some cases, but not in others, cell division and accessory factors such as calcium and serum albumin are required. Competent bacteria contain of the order of 50 sites to which DNA reversibly adsorbs. In high concentrations of DNA, there is a competition between DNA molecules for the DNA-adsorbing sites. A minimum molecular weight of about 10^6 is necessary for adsorption and uptake. DNA's from a wide variety of sources are equally well adsorbed and taken up by *Pneumococcus*, but *Hemophilus* spp. may preferentially take up DNA's from homologous and more closely-related bacteria than DNA's extracted from other ogranisms. The actual uptake of DNA from the adsorption sites is irreversible and occurs with about 1/30th the rate of loss of DNA from the site by dissociation into the external medium. Adsorption and uptake require no protein synthesis on the part of competent bacteria, and ultra violet-killed bacteria, in which DNA synthesis is presumably blocked, can nevertheless adsorb and take in DNA. Biologically active, irreversibly-bound DNA may be extracted quantitatively 10 minutes after uptake.

Kinetic experiments show that irreversible uptake of one DNA molecule is sufficient for transformation of one bacterium. In suitable conditions, the number of transformed clones produced by a DNA preparation approximates the number of nuclei present in the bacteria from which the DNA was extracted. The expression of transformed characteristics requires protein synthesis and in some cases takes place before the transformed bacteria begin dividing. However, the phenotypically-altered cells may still segregate off non-transformed (genetically unaltered) progeny for several divisions. The incorporation ("integration") of genetic factors into the replicating genome of the recipient bacteria involves replacement, in some of the daughter cells, of homologues originally present in the bacteria whose progeny they represent. Once the input DNA has been integrated into the replicating genome, a stably-altered clone of descendents is built up by vegetative multiplication. DNA extracted from this clone is again able to transform the same trait in still further bacterial recipients. Closely-linked loci are sometimes transformed together by a single DNA molecule ("joint or linked transformations"). Joint transformation is expected when it is realized that, for example, *Hemophilus* contains only about 200 molecules of DNA per cell while reasonable estimates of the number of gene loci in *Hemophilus* would be between 10–100 times this value. The cell receiving the DNA molecule gives rise to a mixed clone, including progeny of parental genotype as well as a single recombinant type. No clones mixed with respect to recombinant types appear, although just a portion of the genetic information in the input DNA fragment usually appears in the replicating genomes of bacteria comprising a CLONE of transformants. This indicates that the input DNA neither multiplies before integration nor is conserved and available for repeated interaction with additional chromosomes after the initial integration event. Integration of just a portion of the input DNA may account for the differential behavior of DNA upon re-extraction from certain transformed bacteria; in transformations between different, yet closely-related, bacterial "species," the number of recombinants is low although DNA uptake is normal. When the DNA, containing the same genetic markers, is again isolated from the transformed clone and re-tested on the homologous species, it transforms with greatly increased efficiency. Thus,

"effective pairing" (intimate synapsis) is implicated as underlying the increased activities found in intraspecies transformations.

Mutation. In bacteria dividing rapidly on adequate media under steady-state conditions, the rate of spontaneous mutation per hour is dependent upon the rate of multiplication. The rate of mutation per mutable unit per generation is the same at all temperatures. In contrast, under conditions severely limiting growth rate, the rate of mutation is the same per hour of growth regardless of the growth rate. Mutation also occurs at a slow rate during the stationary phase but may involve the same process as during growth since it exhibits the same temperature coefficient (Q10 about 2) as shown by growing cultures.

Point mutations involving single "sites" (presumably single nucleotide pairs in the DNA), and multisite mutations, involving several or many adjacent sites occur spontaneously. Some multisite mutations have been characterized as true deletions since the strains do not back mutate and only the entire mutation may be replaced during recombination.

Several chromosomally-localized mutator genes are known in bacteria; each induces a high frequency of mutations throughout a large portion of the genome. One of the mutator genes induces mutations only of the single-site type. Strains carrying these mutations can be induced to back-mutate with mutagenic agents such as 2-aminopurine (2AP), 5-bromodeoxyuridine (BD), and nitrous acid (NA), agents which cause "transition" mutations in the DNA, i.e. replacement of one base pair by the other so that purine substitutes for purine in one chain and pyrimidine for pyrimidine in the other chain. Transition mutations occur with lower frequency in stocks not possessing demonstrable mutator genes. There is a secondary mutagen specificity superimposed upon such basic changes in the DNA molecule; this is reflected in specific susceptibility of certain alleles to respond to only one or two of a number of mutagens. Of point mutations which revert spontaneously, some cannot be induced to higher levels of reversion by any of a wide variety of mutagens tested.

Genetic Markers. Since 10^9 bacteria may be rapidly grown vegetatively from a single cell and spread onto a single plate of chemically-defined medium, rare new phenotypes are readily detected on selective media. Also, in many cases, clues are supplied as to the biochemical nature of the chemical process underlying the change in phenotype. Markers most widely used in genetic tests include: (1) Relative resistance to antibiotics such as streptomycin (SM), penicillin (P), terramycin, etc., (2) Relative resistance to toxic agents such as azide, nitrofurans, sulfonamides, etc., (3) Relative resistance to physical agents such as ultraviolet light and heat, (4) Relative resistance to antimetabolites, such as amino acid and purine analogues, (5) Resistance, or partial resistance to biological agents such as bacteriophages and bacteriocins.

No non-controversial demonstration of locus-specific induced mutation has been made. A number of phenotypic changes have been shown to be due to mutations spontaneous origin and not to involve specific induction of the resistant phenotype by the agent under test, as widely proposed but insufficiently documented. Resistance to many agents is polygenically controlled (e.g. P resistance; low level SM resistance). Mutations to high level, single-step resistance, and even dependence occur in other cases (e.g. SM). Phage- and bacteriocin-resistant mutants lack the specific receptors in their cell walls necessary for adsorption of these agents.

Antimetabolite-resistance has been shown to be due to different mechanisms in different situations: (1) failure to take up the agent (e.g. in rhamnose resistance), (2) loss of an enzyme necessary for conversion of the inhibitor to an active substance (e.g. nitrofuran reductase; fluorouracil riboside kinase), (3) alteration in an enzyme which is normally sensitive to a natural compound and some of its analogues ("feedback inhibition") so that it is no longer inhibited (e.g. histidine and 2-thiazole alanine), (4) gain of an uninhibited or unrepressed enzyme system capable of over-production of a normal intermediate which competes with the antimetabolite (e.g. thienylalanine-resistant mutant excretes phenylalanine; some sulfonamide-resistant mutants excrete p-aminobenzoic acid).

Other widely utilized markers include: (1) Inability to ferment a sugar. Such mutants have been demonstrated to be relatively or completely deficient in active uptake of the substrate ("permease" mutants) or in one or several enzymes active in metabolizing the substrate, (2) Inability to synthesize a metabolite required for growth (e.g. amino acids, purines, pyrimidines, vitamins, hemin, chlorophyll in photosynthetic bacteria, etc). Many such mutants have been shown to be lacking adequate amounts of a specific enzyme, or of several enzymes, in a metabolic pathway, or to contain catalytically-defective enzyme proteins. In a few cases the genetic lesion appears to affect the control of necessary enzyme levels rather than more directly affecting the structure of the enzyme protein itself. (3) Ability to synthesize a cell constituent: A block in synthesis of a component unique to the cell wall, α,ϵ-diaminopimelic acid, leads in the presence of lysine to a specific inhibition of cell wall synthesis without affecting bacterial growth; under these conditions, the mutants "outgrow" their cell walls and lyse unless supplied with an adequate osmotic environment. "Dark" mutants of luminous bacteria have been described. The serological specificity of flagellar proteins in E. coli is under the control of two gene loci and their modifiers. For each of these loci, many isoalleles are known as well as mutants completely defective at one of the loci ("monophasic" strains). A number of other loci regulate the ability to produce either of the flagellar proteins. The genetic control of capsule synthesis in Pneumococcus resides in a number of genes regulating oligosaccharide synthesis. (4) Other alterations in colonial morphology reflect biochemical lesions of similar nature (e.g. smooth to rough mutations), the biosynthesis of pigments, the ability to undergo or complete spore formation, or the content of bacteriocinogenic or prophage elements (see "episomes," below).

Genetic fine structure. A large number of gene loci have been mapped on a single linear linkage group in Enteric bacteria. The gene loci comprise segments of the linkage group which control particular biochemical reactions (gene-enzyme relationship to Beadle and Tatum). The gene loci are complex segments, divisible both by mutation and by recombination. The component units of the gene locus ("sites" of mutation) presumably indicate nucleotide pairs of the Watson-Crick double helix. The gene represents a linear sequence of

about 100–1000 such pairs. Gene loci involving enzymes with related functions in many, but not in all, cases are closely-linked to one another and reflect in their linkage order the sequence of biochemical steps their respective enzymes carry out in the cytoplasm. The loci have discrete ends (do not overlap) and some appear nearly, or truly, adjacent to one another. Mutations resulting from changes at different sites of one locus are termed allelic. The mutants bearing these non-identical alleles are deficient in one enzyme activity. In some cases the enzyme protein appears to be totally lacking; other alleles govern the formation of catalytically defective proteins. Strains bearing alleles mutant at different sites of a single locus differ from one another in accessory respects (e.g. stability to back-mutation, response to suppressor mutations, temperature sensitivity, quantitative aspects of nutrition and of accumulation of biochemical intermediates, etc.). Even changes occurring at a single site, inseparable by recombination, may evidence such differences; changes at a single site may produce at least 3 distinguishable genotypes. Some gene loci appear to be constituted of single functional regions, as determined by the reconstitution of wild type phenotype in partial heterozygotes ("heterogenotes"). However, other loci are constituted of several (2 or 4) functional elements ("complementation units"). It seems probable that the enzymes determined by such genes are themselves complex, composed of a number of different polypeptide subunits, each of which is determined by a complementation unit. Enzyme reconstitution in mixed extracts has been demonstrated in vitro by Crawford and Yanofsky tryptophan synthetase) and Loper (imidazoleglycerol dehydrase), reflecting the in vivo complementation obtained in genetic tests (2 and 4 complementation units, respectively, in the two cases).

Infectious heredity. Temperate bacteriophages inject their DNA into bacteria. This DNA may replicate and elicit the synthesis of phage-specific proteins and, eventually, mature into infectious phage particles (the "vegetative" or "lytic" cycle). Alternatively, with "temperate" viruses, the DNA may remain in a non-infectious state ("prophage state") and divide more or less in synchrony with the chromosome of the bacterial host (lysogenic response). The prophage is associated with a specific location (or a few locations) on the bacterial chromosome and may be mapped as any other bacterial gene locus. The site of this "integration" is determined by just a portion of the total phage genome in conjunction with the specific gene locus (or loci) of the bacterial host. In the prophage state, the genetic material of the virus is presumably "naked"; it is noninfectious. However, each cell containing the prophage carries the hereditary potential to produce bacteriophage under suitable conditions. The prophage may mutate in any of a number of genes essential for its vegetative multiplication and maturation. Defective prophage may be genetically blocked in an early reaction, eliminating vegetative reproduction entirely, or in a late reaction, wherein replication occurs vegetatively but the phage genomes fail to mature into infective particles or obtain release from the cell.

Similar to defective prophages in their behaviors are elements allowing bacteria to undergo conjugation (F-elements, below) and elements engendering the hereditary potential for the lethal syntheses of proteins termed "bacteriocins." Just as are phages, bacteriocins are adsorbed to specific receptors on the surfaces of sensitive bacteria. Adsorption of a single bacteriocin particle, like the adsorption of protein coats isolated from phages, kills the sensitive bacteria.

The above elements are collectively termed "episomes" by Jacob and coworkers. They are capable of replication either as integrated elements at chromohal sites in synchrony with the bacterial host or in a vegetative state, out of snychrony with chromosomal replication. The presence of episomes universally appears to result in changes in the immunological constitutions of the host bacteria. Specific somatic (0) antigens (lipopolysaccharide-protein complexes) which are mainly present on the cell surface (cell wall) and are elicited by the episomes (episomal conversion). The most important haptenic groups of somatic antigens detected by serological or phage adsorption techniques appear to be oligosaccharides and are elicited by episomes when in both the vegetative and the integrated states. The protein component of one somatic antigen has shown to be a bacteriocin; another protein elicited by phage conversion has been demonstrated as the diphtheria toxin.

Episomal elements occasionally (10^{-6} per chromosomal division) incorporate genetic markers of the bacterium into their genetic continuity. In some cases this genetic material may be added to the preexistent episomal genome. However, in most cases it appears to replace other genetic material originally present in the accessory (episomal) genetic fragment. The classic examples are the defective, transducing derivatives of lambda prophage. Such phages exhibit "special transduction," that is they contain only genes known by other criteria to be normally closely-linked with the prophage locus at which lambda is localized (gene loci involved in galactose metabolism). They incorporate *gal* genes only during replication in the prophage state, not during vegetative multiplication. In contrast, other phages are capable of "general transduction." They incorporate various regions of the genome which may collectively encompass the entire bacterial genome. They carry out this incorporation during vegetative multiplication. They also convert serologically important somatic antigens. All transducing phages serve to transport ("transduce") relatively short pieces of genetic material from "donor" bacteria to "recipient" bacteria. The genetic constitution of the bacterial fragment they carry is characteristic of the last host upon which the phage was prepared. When the genetic fragment is injected by the phage particle into recipient bacteria, it may persist and function without replicating ("abortive transduction"), allowing the construction of partially heterozygous bacteria. In other cases, the input fragment undergoes rapid recombination with the recipient host chromosomal genes ("integration," "complete transduction"). In still other cases, it persists and multiplies in synchrony with the host chromosome ("heterogenote"), a situation which also allows examination of heterozygotes for restricted chromosomal regions. By growing up such heterogenotes a relatively pure clone of transducing particles may be obtained (HFT—"high frequency transducing" phage), enabling genetic analysis of the episome.

Conjugation. Recombination mediated by conjugation has been demonstrated to take place among and between various Escherichia, Shigella, and *Salmonella* strains and in *Pseudomonas* spp. In *E. coli* the process has been demonstrated to involve the following progressive sequence of events: (1) Random collision of donor (F^+ or *Hfr*) bacteria with recipient (F^-) bacteria. Recipients can mutate to a state incompetent in conjugation (F°). Donors, when F^+, contain a purportedly extrachromosomal element ("episome") which alters their surface properties, making the bacteria amenable to conjugation. However, during conjugation, only the F (fertility) element is transferred. In cases where the F-episome undergoes recombination with the host chromosome and becomes physically-linked with bacterial genes (F' element), it transfers those markers now linked with it. Both the extrachromosomal F^+ and F' elements are selectively and permanently removed from bacteria dividing in the presence of acridine compounds. A large number of positions on the chromosome is capable of mutation to an *Hfr* (high frequency recombination) condition in the presence of the F elements. In these cases, the F element becomes associated at the specific chromosomal location with a particular degree of reversibility and guides the transfer during conjugation of bacterial genes located on one side of it. When "integrated" into the chromosome in this fashion, the F element becomes insensitive to the presence of acridine dyes in the medium. (2) Formation of an intimate union and a fusion bridge between bacteria occurs after random collision. (3) The transfer of bacterial chromosome is unidirectional going from *Hfr* to F^-. Recombinants are formed only among the F^- bacteria entering the cross; P32 labelled DNA also is detectably transferred only from *Hfr* to F^- bacteria, not in the reverse direction. The transfer is slow and progressive, involving per minute only 10^5 nucleotide pairs of the total 10^7 nucleotide pairs contained in the haploid *E. coli* nucleus. Transfer requires the addition only of an exogenous energy source for the donor parent. Transfer is oriented in its starting point (origin = 0) and begins with genes adjacent to the F element attached to the *Hfr* site. The direction of sequential transfer of donor genes is also determined by the F element-*Hfr* complex and proceeds away from the element which itself enters the recombinant progeny with very low frequency. At intervals during the transfer process, some bacteria may spontaneously break apart. Also, they may be torn apart by agitation in a Waring blendor, may be killed with a drug or by infection with a virulent phage, or the transfer may be interrupted by allowing P32 decay to take place in labelled donor DNA before conjugation. Thus, most often, gene transfer involves only a portion of a haploid complement which, in each case, originates at 0 but ends at a variety of places along the chromosome. (4) The newly introduced genes function very shortly after entering the recipient bacterium but do not replicate *per se*. (5) "Integration" of the transferred genetic material into the replicating chromosome of the recipient bacterium takes place over a large number of divisions of an aberrantly-dividing cell by a series of successive (multiple) "matings" between the introduced genetic fragment and the bacterial chromosome. The foreign fragment appears to be conserved in this process. If certain *Het* mutations or elements are present on the *Hfr* chromosome and are transferred during conjugation, the input fragment is enabled to multiply along with the F^- chromosome in the recipient (partial diploidy).

PHILIP E. HARTMAN

References

Braun, W. D., "Bacterial Genetics," 2nd ed. New York, Saunders, 1965.

McElroy, W. D. and B. Glass, eds., "The Chemical Basis of Heredity," Baltimore, Johns Hopkins Press, 1957.

Hays, W., "Genetics of Bacteria and their Viruses," New York, Wiley, 1964.

Stent, G. S. and A. Adelberg, eds., "Papers on Bacterial Genetics," 2nd ed. Boston, Little, Brown, 1966.

BACTERIAL NUCLEUS

Chemical analysis reveals that bacteria contain deoxyribonucleic acid (DNA). Feulgen preparations and fluorescence microscopy show that the DNA is concentrated in certain structures which can also be recognized, unstained, during life because they are less dense than the cytoplasm that surrounds them. These structures have been loosely called CHROMOSOMES, NUCLEOIDS or NUCLEI but their variable shapes, uniform texture and lack of a containing membrane are more aptly indicated by the name "nucleoplasm" introduced by Kellenberger.

The nuclei of different bacteria are cast into slightly different patterns and in a given culture their configuration is liable to change with age owing to the accumulation of droplets of lipid, sulfur or other materials. Nevertheless, under defined conditions of cultivation the types of nuclei seen at different times after the start of a culture are entirely predictable. The variability of bacterial nuclei is one of shape only. It is known that a constant (average) amount of DNA is packed into the *spores* of *Bacillus* species and that this amount is half that found in the normally binucleate *vegetative* forms of these bacteria.

In several species, e.g. in gliding bacteria (genus *Vitreoscilla*) and in large forms from the gut of tadpoles the nucleus extends through the cell in the shape of a net. The behaviour of such nuclei is not well known. Perhaps they are not unit structures but aggregates of many independently duplicating subunits. Fortunately there are many species whose cells contain under ordinary condition of cultivation no more than a few neat and separately visible nuclei. In these species it is possible to follow the cycle of growth and division of the nuclei during life. Division is invariably direct and does not involve gross visible changes of the organisation of the nucleus. Just before it divides the nucleus is compact and stands out sharply from the cytoplasm under the phase contrast microscope. As it breaks in two its luminosity softens and its contours become indented and indistinct. Each half soon assumes again a compact shape in preparation for the next division. This visible alternation of different physical states is usually not clearly reflected in fixed and stained preparations and still needs to be reconciled with the repeatedly made observation that in growing bacteria DNA is made continuously and at a steady rate (See NUCLEIC ACIDS).

Electron micrographs of sectioned bacteria prepared

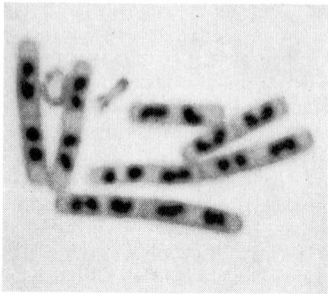

Fig 2   Dividing nuclei of *B. subtilis*. Fixed and stained preparation, magnified 3600×.

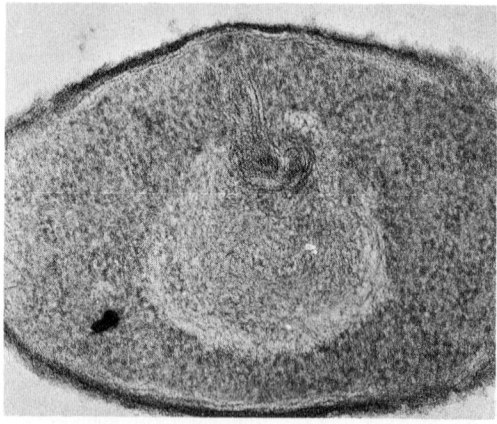

Fig 3   Electron micrograph of a section through a cell of *B. Subtilis*. The area of low density filled with fine fibrils represents a section of the nucleus. It is in contact with an extension of the membrane termed mesosome on which the point of attachment between the nucleus and the membrane is probably located. Magnification 75,000×.

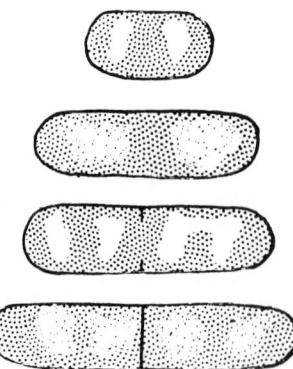

Fig 1   Diagram of the changes of shape and luminosity which dividing nuclei of bacteria undergo during life. Based on observations on *B. subtilis, B. cereus* and several other bacteria.

by the technique of Kellenberger show that the nuclei are not separated from the cytoplasm by a membrane and are composed entirely of bundles of fine fibers with diameters of 30–60 Å.

Genetic studies indicate that the genetic determinants are linearly distributed along a circular chromosome. These conclusions are entirely confirmed by direct observations. Two different techniques for chromosome spreading show that the chromosome is composed of a single DNA molecule, measuring about 1000 μ long and forming a closed ring. This giant molecule, one thousand times longer than the bacterium, is wound into a ball inside the cytoplasm. It is very probable that a certain organization exists in the arrangement of the DNA molecule as is suggested by the periodic structure found in the examination of thin sections. Little else is known, and no satisfying model has been proposed.

Genetic, morphological, and biochemical results indicate that the bacterial chromosome is probably attached to the membrane. In Gram-positive bacteria, this linkage generally seems to be located on a membrane extension, termed mesosome. The nature of this attachment is still quite unknown, but it seems to be indispensable for both nuclear expression and division.

C. F. ROBINOW
(revised by Antoinette Ryter)

References

Kellenberger, E., "10th Symposium Soc. Gen. Microbiol.," Cambridge, The University Press, 1960.
Kellenberger, E. and A. Ryter, in "Modern Developments in Electron Microscopy," Siegel, B. M. ed., New York, Academic Press, 1964.
Murray, R. G. E., in "The Bacteria," vol. 1, Stanier, R. Y. and I. C. Gunsalus, eds., New York, Academic Press, 1960.
Robinow, C. F. in "The Cell," vol. 3. Brachet, J. and A. E. Mirsky, eds., New York, Academic Press, 1961.

BACTERIAL NUTRITION

BACTERIA are aquatic microorganisms and are therefore cultured in aqueous media, on the surface of gels containing much water (e.g., 2% agar), or in the interior of such gels. Nutrients for the growth and maintenance of the bacteria are supplied in the medium and pass from the medium through the cell boundary into the cell proper. As primary energy sources bacteria may use light (PHOTOSYNTHETIC BACTERIA), the energy of an inorganic oxidation (chemosynthetic bacteria) (see CHEMOLITHOTROPIC BACTERIA), or the energy of an organic oxidation (heterotrophic bacteria). The study of bacterial nutrition involves investigations of how any given species of bacterium is best supplied with compounds essential for its growth and maintenance and investigations into environmental conditions under which such compounds are utilized with the greatest efficiency. The results of such investigations may be collated as follows:

Carbon sources. Autotrophic bacteria (chemosynthetic and photosynthetic bacteria in the autotrophic mode) require only carbon dioxide as a carbon source. In this mode they oxidize inorganic compounds and couple the oxidation with the reductive endergonic conversion of carbon dioxide to the level of carbohydrate, from which class of compounds they are able to elaborate all other cellular components. Heterotrophic bacteria are either not able to utilize carbon dioxide or, if they are, can obtain the necessary energy and reducing power only from the oxidation of some organic compound. *Hyphomicrobium vulgare*, for instance, appears to synthesize much of its cellular material from carbon dioxide while obtaining the necessary energy supply and reducing power from the oxidation of formic acid. Many heterotrophs are quite unexacting in their requirements for organic compounds, being able to utilize a wide variety of organic acids, sugars, amino acids, etc. Moreover, the formation of adaptive enzymes by bacteria allows certain organisms to utilize the most bizarre sources of carbon and energy such as cyanide and trichloroacetic acid. Gram positive organisms are generally more exacting than gram negative ones and may cease to grow unless certain organic compounds, particularly various amino acids, are incorporated in the growth medium. Such compounds enter the cells either by diffusion or by active transport against a concentration gradient; *Streptococcus faecalis* and *Staphylococcur aureus*, by means of such transport, may raise their internal concentration of glutamic acid to several hundred times that of the medium. Carbon compounds taken in by bacterial cells may be oxidized to yield energy and/or incorporated into the cell material; there may or may not be fragmentation of the compound prior to incorporation. The ratio between the amount of compound oxidized and the amount incorporated varies with the compound, the organism, and the age of the culture (young cultures often incorporate more carbon per atom carbon oxidized than do older cultures). In general, over appreciable periods of growth, aerobic organisms incorporate about one atom of carbon while oxidizing four to carbon dioxide. In ANAEROBIC species, where oxidation to carbon dioxide can never be complete and some 'end product' accumulates, the ratio of carbon incorporated to carbon converted to 'end product' is usually much less than 1:4. Over lengthy periods of growth chemosynthetic bacteria usually show an energy efficiency of 5–10%; i.e., 5–10% of the energy released by the inorganic oxidation is utilized in the endergonic reduction of carbon dioxide to cell material; in young cultures however the efficiency may be as high as 30%. This figure is similar to the maximum shown by photosynthetic bacteria in converting light energy into the chemical energy required for carbon dioxide reduction. Although carbon dioxide is obviously of prime importance in autotrophic metabolism, it is probably required by all bacteria in certain amounts, presumably for carrying out essential carboxylation reactions. In heterotrophic bacteria, the deleterious effect of removing carbon dioxide, or the beneficial effect of augmenting the supply is more noticeable in young cultures. Older (and thus denser) cultures generate enough carbon dioxide by metabolism to supply their metabolic needs.

Nitrogen supply. Many bacteria are capable of elaborating all their nitrogenous components from such simple sources as ammonium salts or nitrates but the more exacting species may require certain amino acids (see above) or other nitrogenous compounds. On the other hand, some anaerobes (e.g., *Clostridia*) and the aerobic *Azotobacter* spp. can 'fix' gaseous nitrogen and therefore do not require nitrogenous compounds in their growth media; *Rhizobium* spp. can fix nitrogen when they are enclosed in the root nodules of the appropriate legume. *Clostridia* may fix up to 20 mgm nitrogen per gram of glucose fermented, *Azotobacter* about 20 mgm per gram of glucose oxidized. Nitrogen fixation occurs only when active metabolism is proceeding and only when other sources of nitrogen such as ammonia are not available. About 10% of the dry weight of bacteria is nitrogen, and a satisfactory growth medium must be able to supply this level of nitrogen for the dry weight of organism expected to grow.

Mineral supply. Requirements for sulfur (see SULFUR BACTERIA) can usually be met by some inorganic sulfur source such as sulfate or thiosulphate; sulfite and sulfide may be toxic. Some exacting bacteria may require sulfur-containing amino acids. As far as is known, phosphate can always meet the phosphorus requirement of all bacteria. In addition to these 'bulk' minerals a satisfactory growth medium must also supply 'trace' elements. Small concentrations of iron and magnesium are probably required by all bacteria; in addition, such elements as copper, cobalt, manganese, zinc, molybdenum, may be required by certain organisms as cofactors for essential enzyme systems. The level of trace element needed in the growth medium is usually very small; bacteria that synthesize vitamin B_{12} are capable of performing the synthesis even when the level of cobalt in the medium is reduced to 10^{-6} p.p.m.

Growth factors. In common with animals some bacteria are not capable of synthesizing all those compounds essential for metabolism yet needed only in small quantities and they may require in the growth medium such compounds as biotin, thiamin, or other substances classed in animal nutrition as vitamins. The requirement for such growth factors by very exacting bacteria may be quite extensive and a proper determination of the requirements correspondingly difficult. As a consequence there remain a few species of bacteria that still cannot be grown on a precisely defined medium compounded from known chemical constituents; these bacteria require such complex additions to the medium as blood, yeast extract or soil extract to supply unidentified growth factors.

Energy supply. The energy supply essential for the performance of anabolic process is met in heterotrophic bacteria by the energy released during the oxidation of an organic compound. In such bacteria the carbon source and energy source many thus be identical. In bacteria living autotrophically the carbon source is carbon dioxide, the reducing power [H] necessary for its conversion to cell material is supplied by the oxidation of an inorganic donor (e.g., $H_2S \longrightarrow S° + 2[H]$) and the energy either by an inorganic oxidation (chemosynthetic bacteria) or light (photosynthetic bacteria). Inorganic oxidations known to be used by chemosynthetic bacteria include the following: ammonia $\longrightarrow$ nitrite, nitrite $\longrightarrow$ nitrate, hydrogen $\longrightarrow$ water, ferrous iron $\longrightarrow$ ferric iron, sulfide $\longrightarrow$ sulfur $\longrightarrow$ thiosulphate $\longrightarrow$ sulfate. Some chemosynthetic bacteria (all the hydrogen bacteria and one sulfur bacterium, *Thiobacillus novellus*) are facultative autotrophs capable of living heterotrophically when supplied with suitable organic compounds. The photosynthetic bacteria in the light obtain all their energy from the light and are strictly anaerobic, reducing carbon dioxide by means of hydro-

gen supplied by a hydrogen donor compound. The green sulfur bacteria use inorganic sulfur compounds or hydrogen gas as donors. The purple sulfur bacteria can, in addition, use organic donors. Both purple and green sulfur bacteria are strictly photosynthetic and will not grow in the dark. The purple non-sulfur bacteria include species not only capable of all the foregoing photosynthetic activities but also capable of growing aerobically in the dark, obtaining their energy supply from the aerobic oxidation of those hydrogen donors used anaerobically in the light.

Other factors. While the foregoing considerations cover most of those factors important in bacterial nutrition there remain other general points. It is usually necessary to control the pH of the medium, most bacteria have a pH optimum for growth on either side of which growth becomes increasingly scanty. The osmotic pressure of the medium should not be too high or too low (halophilic bacteria, however, may require a high osmotic pressure in the medium). Anaerobic bacteria demand as complete a removal as possible of oxygen from the medium, microaerophilic bacteria require a little, but not too much oxygen, while aerobic bacteria thrive on an oxygen supply and often grow better under conditions of forced aeration. Techniques of forced aeration, in which the culture is rapidly swirled in a vessel fitted with internal baffles while a stream of air is pumped in at the base of the vessel, have been developed in recent years. These techniques, which apparently facilitate diffusion of oxygen from the air bubbles to the bacterial surface, have sometimes resulted in yields of organism several times the maximum obtainable under conditions of normal sparger aeration.

HOWARD LEES

BACTERIOPHAGE

Most bacterial VIRUSES or bacteriophages (phages) are differentiated into head and tail. The tail parts and the head envelope consist of proteins; the envelope contains the DNA (see NUCLEIC ACIDS) of the phage. Infection of the host bacterium begins by attachment of the tail tip to specific sites on the bacterial cell wall. The envelope then acts as a "microsyringe" and injects the phage DNA, presumably through the core of the tail, into the host. Only a small amount of protein (less than 5 percent of the total), of unknown function, is injected along with the DNA; the rest of the protein shell remains outside the infected cell.

The essential feature of a phage infection is the injection of DNA. The DNA is the hereditary material of phage and phage infection may be thought of as genetic infection, resulting in a number of alternative interactions between host and virus.

Infection by *intemperate* or *virulent* phage results in cell lysis and liberation of newly formed infectious progeny particles after a characteristic latent period. The genetic material of the phage takes control of cell functions, forcing first the vegetative reproduction of phage DNA and later the production of proteins needed for transmission of the DNA to other cells. However, before synthesis of phage-precursor DNA is initiated, new proteins (and possibly some specific

RNA) required for this synthesis, must be formed. In the case of the coliphage T2, infected cells synthesize a new enzyme necessary for the formation of the unique base hydroxymethyl-cytosine instead of cytosine. Phage-precursor proteins are formed only after considerable DNA synthesis takes place; tail parts and head structures can be seen in electron micrographs of infected cells. The first mature particles are detected soon after the appearance of phage proteins and these accumulate until liberated.

Mutation and genetic recombination occur during multiplication. The clonal distribution of mutants in single cells is exponential, suggesting that replicas of newly formed phage DNA act as templates in subsequent replications. Mixed infections with differently mutated particles result in genetic recombination. Phage characters behave generally like Mendelian factors in a haploid organism and extensive linear linkage maps are known for a number of phages. Fine structure genetic analysis with particular phage mutants is contributing greatly to the understanding of genes and mutations on a molecular level. However, some features of phage crosses are unique and are still to be reconciled with classical genetic theory. For instance, the lack of significant correlation between reciprocal recombinants from single cells suggests that only one recombinant is formed in each recombinational act.

The larger phages contain double stranded DNA. Tracer experiments suggest that the DNA of phage T2 replicates in a semiconservative manner; that is, each strand maintains its structural integrity and each new doubled element consists of an old strand and a newly formed one. These findings may be interpreted according to the theory of the structure of DNA and its replication proposed by Watson and Crick. The small phage ϕX174, however, contains single-stranded DNA. It undergoes mutation and recombination, but its manner of replication seems not to be semiconservative.

Other interactions between host and virus occur following infection with *temperate* phage. The host may lyse and produce a yield of infectious particles or it may survive, multiply and give rise to genetically altered *lysogenic* progeny. In lysogenic bacteria, phage DNA persists in a nonvegetative form and, as *prophage*, becomes integrated into and replicates synchronously with the cell genetic material. As a consequence of prophage integration, these cells are immune to reinfection by the same phage, but have the genetic potential to produce the phage. Production of mature phage in lysogenic bacteria, by a shift of the prophage to the vegetative state, occurs spontaneously as a rare event, but can be induced on a large scale by ultraviolet irradiation and other treatments. The ability to become prophage and the specificity of immunity are controlled by specific phage genes. Temperate phage may mutate to virulent forms.

Incorporation of prophages into the bacterial genome is confirmed by the results of bacterial crosses. Prophages segregate as bacterial markers and are mapped at unique loci. A single cell may be lysogenic for many prophages, each at its own locus. A prophage contains as much essential DNA as a mature phage; the rate of inactivation of the two by the disintegration of P^{32} atoms is similar. How this DNA is placed on the bacterial chromosome is not at all clear.

Other cell properties besides immunity and phage production are controlled by phage genes. Every cell that becomes lysogenic shows these *conversions*. Among a list of many, the control of diphtheria toxin production and the control of some somatic antigens in *Salmonella* stand out.

Any *bacterial* character may be transferred from one cell to another by temperate phage. A fragment of bacterial DNA may be included in a phage during maturation and can become incorporated into a recipient cell genome, if the cell survives. This kind of genetic exchange, using the phage as a vehicle of transmission, is called *transduction.*

The properties of temperate phage demonstrate more intimate relationships between host and parasite than might be suspected from studies with virulent phage alone. The genetic interactions outlined above suggest the existence of genetic homologies between phage and host chromosomes. There is even some evidence for recombination between phage and host. By mutation, prophages may become defective so that a lysogenic bacterium loses the ability to produce mature phage, while maintaining its other properties. Except for the fact that its origin is known and that it may regain its ability to produce infective particles by back mutation, such mutant viral genetic material is indistinguishable from host genetic material. Indeed, a prophage can be conceived as a fragment of bacterial DNA capable of initiating its own vegetative multiplication and of controlling the production of proteins necessary for its transmission to another cell. Thus, phages, and possibly all viruses, may arise from normal constituents of cells.

MYRON LEVINE

References

Adams, M. H., "Bacteriophages," New York, Interscience Pub., 1959.

Stent, G. S., "Molecular Biology of Bacterial Viruses," San Francisco, Freeman, 1963.

BAER, KARL ERNST VON (1792–1876)

Few biologists have had such an impact on the science of their times as did von Baer on pre-Darwinian zoology. "Von Baer's Law," which states that animals repeat in their embryonic development the evolutionary development of their line, completely dominated comparative anatomy and comparative embryology practically to the present time, when the concepts of parallel evolution and adaptation to environment have eroded its foundations. The fact that his theories are no longer valid must not be allowed to cloud the magnificence of his discoveries which include the demonstration that the notochord was common to all vertebrates, that what are now known as "germ layers" are found in the development of all vertebrates and that the organs derived from them are similar in each instance and, possibly most important of all, that the human ovary contained a small spherical egg.

The author of these discoveries was born at Piep, in Astonia and had intended to become a physician until he was diverted from this by an interest in comparative anatomy. In 1817 he became chief of the

Zoological Museum at Konigsberg where all of his important discoveries were made. His published researches, besides those on which his fame rests, include geology and ichthyology.

BASIDIOMYCETES

Basidiomycetes are FUNGI in which the spores of the sexual stage are borne on basidia. There are well over 10,000 species belonging to this group distributed in over 600 genera. They include fungi with diverse living patterns: some are obligate parasites, some can live either as a parasite or saprophyte, and some only as saprophytes. There is some species of Basidiomycete which will grow on the remains of any given higher plant, and those parasitic on our crop plants can and have caused great economic loss.

The basidium, which gives the name to this group, is a single cell at least in its early stages, and is typically nothing more than the apical cell of a HYPHA or hyphal branch. Typically it contains two haploid NUCLEI in the young stage (as do other hyphal cells) but it is only in the basidium that the two haploid nuclei fuse to form a true diploid nucleus. Typically, reduction division also occurs in the basidium so that the four or more daughter nuclei found in it just prior to spore formation again are all haploid (with the n number of chromosomes). Because these nuclei migrate into the spore, the spore also has haploid nuclei in it when it is mature. Typically one nucleus migrates into each spore.

The spores, the actual reproductive bodies of the fungus, effect dispersal of the species by being carried away by air currents or animals such as insects or rodents, or to preserve it over periods unfavorable to growth. Different degrees of adaption are found among the numerous species to aid in these functions, and fungi may be said to actually compete with each other for survival on this basis.

When conditions are right the spore in its new location, starts to grow (GERMINATE). This growth is composed of threads mostly radiating outward from the position of the original spore, but becoming woven into a felt-like mass by the threads (hyphae) branching in all directions. The nuclei in a mycelium of this type are all of the same "sex," being daughters of the single nucleus which migrated into the spore. If a spore of opposite sex-potential falls on this mycelium or the mycelium of both grow until they meet, a change occurs. Hyphal fusion occurs between the two, and one nucleus of one sex-potential now finds its way into a cell where there is one of opposite sex-potential. However, instead of fusing to form a diploid nucleus, as one would expect, they do not fuse but divide in unison in such a way that each daughter cell of the new hypha contains a pair of nuclei each of opposite sex-potential. The resulting hyphae which contain the pair of nuclei (called a *dicaryon*) grow much more vigorously than the first mycelium and it is this stage which eventually produces the reproductive or fruiting stage. The mycelium of the first stage is termed the primary mycelium. The second type is called the secondary mycelium, and is the rapidly growing destructive stage. It digests its food by the process of excreting enzymes from the region of the hyphal tip, these then dissolve the food materials (sugars, etc.) and the dissolved food is absorbed by the hypha and either used there to produce further hyphal growth or is translocated to some other part of the mycelium or stored for future use in the production of the reproductive stage. There is no limit to the life of a single mycelium. It will continue to grow as long as food is available and dies only when that supply is exhausted. Some mycelia regularly live only a year or two, others live for hundreds of years. Although the mycelia of different Basidiomycetes are all genetically different, there is such similarity, morphologically, among species belonging to the major groups that species identifications in any group must be made from a study of the basidiospores and the specialized structures known as fruit bodies on which basidia are produced, i.e., the reproductive phase of the plant. As the study of the physiology of fungi comes into its own, however, and we learn more about the chemical activities of these organisms, we find that large numbers of species have peculiar physiological features which help to solve taxonomic problems when the usual morphological approach lacks effectiveness. The mycelia of each species of fungus are to some extent biochemical factories which perform some operation differently or make some chemical compounds which are different from those performed or made by any other species. The activities species have in common are used to group them into related groups. The commerical possibilities of using fungi generally to improve man's economy is an active field of investigation. The problem of the identification of Basidiomycetes, however, is still such an acute one that physiological and genetic studies on all but the common species are handicapped for lack of accurate species concepts. We still lack an adequate inventory of all of our fungous resources.

The classification of Basidiomycetes is based first of all on the morphology of the basidium. The simplest type of basidium is found in the largest number of species and at maturity (when the spores are mature) consists of a single cell, i.e., a hyphal tip with a cross wall at the base. But not every tip-cell in a fruit body is a potential basidium. Those destined to function as basidia are produced in a dense palisade on a special surface of the fruit body. Such a simple basidium is termed a homobasidium, and all species having this type are Homobasidiomycetes. The remaining species with various types of more complex basidia are termed Heterobasidiomycetes.

The classification of the Homobasidiomycetes is based in a large measure on combinations of basidiospore features in conjunction with the configuration of the basidium-bearing layer of tissue (termed the hymenophore) which is the functional tissue of the fruitbody. The palisade of basidia itself is termed the hymenium. The other tissues merely serve in a protective or supporting capacity. Groups at the rank of order are separated primarily on the configuration of the hymenophore.

The basic unit of structure of the fruiting body is the same as that of the mycelium, i.e., the hypha or thread. The genetic mechanism of the species, located on the chromosomes in the nuclei of the hyphae, determines, for each species, how these hyphae will organize themselves into a fruiting structure and hence determine what the features of the fruiting body will be. Although the fungous fruit-body is very simple, a surprising number of variations have evolved. Basically the fruit body is a device to place the hymenium in such a position that the basidiospores can be discharged effectively and be carried away by air currents.

Or, if the basidiospores are not forcibly discharged as in some orders, it has become modified in some way to facilitate spore dispersal. Basically it consists of a column of tissue designed to raise the hymenophore out of the material in which the mycelium is

Fig. 1 Fleshy fungi. 1, an Ascomycete, is the edible morel, *Morchella.* All others are Basidiomycetes. 2, a pore fungus, *Boletus*; 3, a meadow mushroom partly opened, *Agaricus*; 4, a milk mushroom, *Lactarius*; 5, an oyster mushroom, *Pleurotus*; 6, an inky cap, *Coprinus;* 7, a deadly poisonous mushroom, *Amanita;* 8, a coral fungus, *Clavaria;* 9, the bird's nest fungus, *Nidularia*; 10, a puff ball, *Lycoperdon.* (From Platt and Reid, "Bioscience," New York, Reinhold, 1967.)

growing. This structure is the stalk or *stipe*. At its top there develops a spreading structure of various shapes, flat, pulvinate, funnel-shaped etc., which shows some modification over the upper surface to function as a protective covering, and on the under side of which the hymenophore develops. This expanded stipe-apex is termed a pileus.

In the simplest Homobasidiomycetes the hymenophore is merely the smooth under surface of the pileus, and of course, is made up of the palisade of basidia. Hence a specialized distinct hymenophore is lacking. Species with this type of hymenophore are grouped in the *Thelephorales* and related orders.

On many fruit-bodies of Homobasidiomycetes, however, the underside of the pileus bears rather distinctive structures, such as a layer of tissue punctured by numerous minute holes called pores. The basidial palisade lines the wall of the pore so that when the spores are discharged they fall downward out of the pore where they can be carried away by air currents. Homobasidiomycetes with a poroid hymenophore belong in the *Polyporales* if the fruiting body is woody to fleshy-tough, and in the *Boletales* if the fruiting body is soft and decays readily. On other fruit-bodies the underside of the cap bears thin plates of tissue which extend from the stipe to the cap margin and have both sides or faces covered by the basidial palisade (hymenium). Such fungi belong to the true mushrooms or *Agaricales*. If the hymenophore of a Homobasidiomycete is in the form of teeth or needle like structures hanging down from the underside of the cap or other structure the fungus belongs in the *Hydnales*. In quite a different type of fruit-body we find the hymenium lining the smooth surface of upright clubs or branched structures as in the *Thelephorales*. Large fructifications of this type remind one of corals and the group has been named the *Clavariales* or coral-fungi. These fruit-bodies are fleshy to tough in consistency.

It is readily seen, then, that the fruit-body is a device to place the hymenium in such a position that the spores can be shed in such a manner that they will be readily carried to new locations and thus can perpetuate the species. The variations in the form of the hymenophore may be interpreted as ways of increasing the amount of basidium-bearing surface and hence, finally, the number of spores produced. In one group of Homobasidiomycetes, however, often referred to as the Gastromycetes, the basidium does not discharge the spores forcibly. In these fungi the spores are produced on basidia typically produced on completely enclosed hymenial layers which usually are convoluted layers forming the interior of the fruit body. When the basidium collapses the spores become free and eventually so many are produced that the central part of the fruit body is filled with them. In such species the outer layers of tissue usually show some type of modification to aid in spore dispersal. These consist of the outer layer opening by a pore so that spores can be forced out over a long period of time (2–4 weeks) by environmental agencies, or various modifications of elevating the spore sac have evolved in addition to the latter having a pore or peristome. In some Gastromycetes adaptation for insect dispersal of spores rather than air-dispersal has been reached. In these the spore-mass has a putrid odor and is elevated in a number of different ways. The odor attracts the insects who then fly to the fruiting-body to investigate and carry the spores

away with them. These fungi are the stinkhorns and belong in the order *Phallales*. A survey of fruiting-body types in the Homobasidiomycetes shows almost all the variations one can imagine. Some gastromycetes fruit underground and the fruiting body merely resembles a small potato. Some mushrooms lack stipes, especially in groups growing on stumps and trees. Some pore fungi and some of the *Thelephorales* are reduced to having a fruit body consisting of nothing more than the hymenophore. These, of course, use the stick or log in which the mycelium is growing as a support.

In the Heterobasidiomycetes there is less of a tendency to produce highly organized fruit bodies but nearly all the types outlined for the Homobasidiomycetes can be found though the number of species in each group is much fewer.

In the Heterobasidiomycetes we find more fungi with the parasitic mode of existence than in the Homobasidiomycetes and there are distinct evolutionary trends associated with this habit, such as a number of asexual spore forms inserted into the life cycle. In the Heterobasidiomycetes the morphology of the basidium furnishes the main features for classifying the species into orders. We have the *Tremellales* in which the mature basidium is divided lengthwise into four cells by two longitudinal walls at right angles to each other. Each of the four cells ultimately produces a basidiospore. In the *Dacrymycetales* the basidium is basically homobasidial in type but is shaped like a tuning fork, i.e., divided to near the base into two prongs with a basidiospore eventually produced on each prong. In the *Auriculariales* the basidium is totally different in shape, the part bearing the basidiospores being a filament divided into 4 cells by transverse walls much in the same manner that a vegetative hypha is divided by cross walls, only the total number of cells formed is typically four. The basidiospores are produced laterally on each of the four cells. In the *Uredinales* and *Ustilaginales* the basidia typically are somewhat like those of the *Auriculariales*, but no true fruit-body is formed though in some species spore-masses may simulate one. The species of these two orders are typically obligate parasites of green plants and often have accessory asexual-spore stages as a distinctive part of the life cycle. These fungi have caused such damage to crop plants that they have been studied intensively for the purpose of controlling them. The literature on them is very extensive. The spore stages, typically, are borne in pustules or lesions on the host plant and are termed sori (singular *sorus*). A classical example of a plant rust, as the *Uredinales* are called, is our common wheat rust with two spore stages on barberry and two on wheat. In the last two orders mentioned species identification is based largely on ability to infect certain hosts and not others, the manner in which the spores of various stages are borne and the details of the spore.

Anyone taking up the study of Basidiomycetes should pay particular attention to the details of the basidiospore. Such features as the color of a deposit of spores, the color of the spore wall under the microscope, any color changes it undergoes when treated with chemicals—particularly color changes, any thin spots (germ pores) in the wall and their position, and any projections on the spore or unevenness of the wall are very important along with the shape of the spore as seen both in profile and face view. Spore size has been given great prominence in species recognition in the past but is secondary to the other characters listed.

The features of the Basidiomycete fruit body of importance in arriving at species concepts involve the details of the hyphae in the various parts and especially the details of the surface (dermal) layers. The thickness of hyphal walls, incrustations on the wall, discontinuities in it, color and color changes produced by various chemical reagents such as potassium hydroxide, iodine, and ferric compounds (ferric chloride or sulphate) are all very important. The shape of the individual cells of the hyphae is also of diagnostic value. In some species the hyphal cells are almost perfectly cylindric whereas in most there is some degree of inflation at maturity—the cell being wider in the middle than at the ends. The ultimate development of this trend is the globose cell, and these are found in the tissues of certain species. The presence or absence of clamp connections at the cross walls is also a character of some importance. The clamp connection is a small hyphal outgrowth which grows outward and backward from the place in a dividing cell where the new cross wall is forming. It fuses with the hyphal cell back of the new cross wall and is interpreted as a device to keep the dividing nuclei of the parent cell separate. However, clamp connections do not occur in many Basidiomycetes.

The manner in which the hyphae are arranged in many parts of the fruit-body is an important feature. In the hymenophore, the hyphae of its context may be parallel in arrangement, interwoven, or show some more distinct pattern such as converging toward the central axis or diverging from it. The hyphae in the dermal layers of pileus and stipe may be modified in various ways, and all such modifications are important taxonomically. In fruit-bodies of many species we find additional morphological features to the essential ones already mentioned. In certain mushrooms a layer of sterile tissue at first covers the gill cavity and breaks away at maturity to allow the spores to be freely disseminated. The layer usually collapses on the stipe and its remnants are termed an annulus if sufficient to form a distinct ring. Some species have the young fruit body completely enveloped in a layer of sterile tissue called an outer veil, which may leave scales on the cap, or a cup at the base of the stipe, depending on its texture. In some species the two veils are intergrown, and in the mushroom group as a whole one can find species showing various degrees of veil development.

ALEXANDER H. SMITH

References

Corner, E. J. H., "A monograph of Clavaria and and allied genera," New York, Oxford, 1950.
Singer, R., "The Agaricales in modern taxonomy," Waltham, Mass., Chronica Botanica, 1951.

BATES, HENRY WALTER (1825–1892)

This English naturalist was destined for the mercery trade but developed such a love for nature study that, at the age of twenty-three, he joined Alfred Russel Wallace in what was intended to be a short collecting trip to the Amazon. Bates, however, became so enamored of the Amazonian tropical rainforest that it was eleven years before he returned to England and then only by reason of his broken health. He was relieved of his financial worries—for he had not sold as many specimens as he had hoped—in 1864 when he became assistant secretary of the Royal Geographical Society, a position which he retained until his death. He is most popularly known for his "The Naturalist on the Amazons" (1863) but his most important scientific contribution was his study of protective mimicry. He was long a friend of both Wallace and Darwin both of whom freely acknowledge in their writings the debt which they owe to his observations.

BAUHIN, CASPAR (GASPARD) (1560–1624)

Professor at Basel, the compiler of *Pinax theatri botanici* (1620), which was the most important and most used reference work on the scientific names of plants until Linnaeus's *Species Plantarum* (1753).

BAUHIN, JOHANN (JEAN) (1541–1613)

Brother of Caspar Bauhin, author with J. H. Cherler of *Historia Plantarum universalis* (1550–51).

BEACH see PSAMMON

BEAK

The term beak is generally thought of as applying to the bill of the bird. The term is used however for beak-like structures in other animals. The jaws of turtles have a horny sheath and are shaped much like the bill of a bird. Similarly there are beaked whales, beetles and even "beak-nosed" humans.

The beak of the bird is made up primarily of the premaxilla above, the dentary below and the sheaths of these bones. The external naris or nostril perforates the beak posteriorly. The sheath is produced by the skin overlying these bones. This skin generates eleiden (keratin) filled layers of cells which when compacted and dried form the hard, tough and shiny cover. The bill skin is like that of other parts of the body, its basal layer, lying next to the rich blood supply of the dermis, forms a stratum germinativum from which new cell layers are delaminated. These new layers are quickly converted to the horny layers of the sheath. Growth involves addition of new layers of cells below and the wearing off externally of the oldest layers. The form of the bill is achieved by differential growth rates and by proximal distal movement of the outer layers of the whole sheath. For example if the sheath is notched dorsally at the base of the bill, the damaged area moves distally as well as outward from the stratum germinativum until it is lost at the tip. The sheath grows fastest at the tip of the bill and along the margins, areas where the wear is greatest. The bill is comparable, in form, growth and relationship to the bony core, to the claw.

The beak may be enclosed entirely by the hard horny sheath or much of its surface may be covered by relatively soft, pliable skin with only a thin keratinized or waxy outer layer. In some birds only a terminal nail on the tip is hard and horny. The soft cover of the beak may contain highly developed tactile organs as in the probing beak of the snipe. The skin of the beak and that of the head are usually sharply demarcated. This is not the case in the cathartids and some other birds.

The form of the beak is extremely variable since it is a structure which is used for seizing food, for preparing food for swallowing, in maintaining the feather cover of the body, in defence, in courtship, and in nest building (from boring holes in trees to weaving fine fibers): in short the bill is the main tool of the bird.

Usually the form of the bill reflects the food securing habit of the species. The pelican has a very long bill with a large gular pouch for gulping in small fishes, the merganser has a long thin bill with marginal serrations for seizing small fish in flight through the water. The humming bird has a long thin bill for sipping nectar. Short, thick, hooked bills are observed in predators where they are used in tearing up prey or in seed and nut-eating species where they are used for hulling. Occasionally the bill shape can not be correlated with food habits as in the case of the toucans, some of the cuckoos, and the hornbills. Here the bill is greatly inflated and enlarged and in the hornbills may bear a large casque above. These birds feed on fruits but also prey on insects and small vertebrates. The large bill may increase their reach and does allow swallowing fairly large objects but otherwise appears to be a cumbersome appendage. The beak in these groups may have unappreciated functions which determine its size but more probably its size and inflation are related to other adaptive structures. This size and inflation is thus not in itself adaptive nor is it detrimental enough to cause selection against the species.

The bird beak can be viewed as a plastic and adaptive structure which makes possible easy handling of food materials while serving other "bird-required" functions. Its history can be assumed from comparisons with other classes of vertebrates and with the scale covered, toothed jaws of the Jurassic fossil, *Archaeopteryx* (including *Archaeornis*). Loss of teeth and their functional replacement by the sheath, as well as the reduction in depth of the snout and its transformation into a projecting beak, are thought of as weight reducing changes. More probable is the view that a bill is a more efficient tool for the way of life of the bird, any weight reduction achieved is coincidental.

The ancestry of the beak may be reflected by the retention of a subdivided bill sheath (figure) in palaeognaths (the large flightless forms like the Ostrich and the tinamous of Central and South America which can fly) and some Neognaths (the remainder of the birds). Of interest is the presence of a cere in the tinamous (or the Kiwi); this is a smooth skinned, featherless area surrounding the nostril which is marked off sharply from the rest of the beak by a fold. Ceres occur in most falconiforms (poorly defined in cathartids),

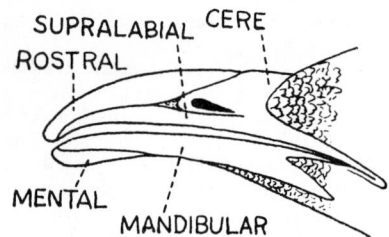

Fig. 1   *Crypturella*, a tinamou, probably represents the primitive style of subdivision of the bill sheath.

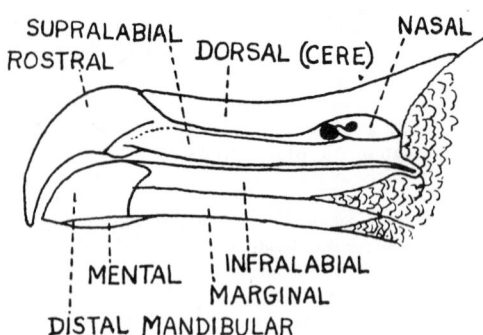

Fig. 2   *Phoebetria*, an albatross, shows an increased subdivision of the sheath related to lengthening of the nasal section of the bill with retention of hard labial margins.

owls, parrots, chickens and pigeons. In the latter the cere may be quite inflated and with an irregular surface. The cere may have a pad or operculum over the slit-like nostril. An operculum may be present when a cere is lacking.

The position and the shape of the nostril is quite consistent but in a few birds it is different. In the Kiwi the nostril is nearly terminal on the long flexible probing beak. It is well out on the side of the beak in the skuas or the Oil Bird, *Steatornis*. It is extended distally in a tube in the fulmars and shearwaters. The base of the beak and the area surrounding the nostril may be covered by small feathers (some of the grouse and parrots) or bristles (some parrots, falcons and hawks). The nostril may be vestigial (not functional) in many pelecaniforms. A hard sheath from the base of the bill grows forward over the nostril in the Sheathbill, *Chionis*.

The form and color of the beak may reflect internal states. In the Puffin or the White Pelican the beak acquires special outgrowths during the reproductive season. In the pelican this consists of a plate extending upward at a point two-thirds of the distance from the base of the bill. This plate does not occur in all individuals and is extremely variable in size and shape. In the Puffin a series of plates on the base of the bill are developed during the breeding season and then shed. In addition to these changes in growth the bill may undergo color changes. These may be sexual, seasonal, age or dietary. In most birds the bill is one unchanging color, usually brown or blackish, but red, orange or yellow bills crop up in many bird groups.

The bill may be strikingly multicolored as in the toucan.

MALCOLM T. JOLLIE

References

Grassé, P.-P., ed., "Traité de Zoologie: Anatomie, Systématique, Biologie," XV "Oiseaux," pp. 7–11, 418–419, Paris, Masson, 1950.

Stresemann, E., *in* "Handbuch der Zoologie: Eine Naturgeschichte der Stämme des Tierreichs," Kükenthal, W. and T. Krumbach, eds. VII "Sauropsida: Allgemeines. Reptilia. Aves," pp. 9–11, 41, 55–58, 326, 436–465, 470–484, Berlin und Leipzig, de Gruyter, 1934.

BEHRING, EMIL (1854–1917)

This German bacteriologist was born in Prussia at Lansdorf and studied at the University of Berlin. He was the first to demonstrate clearly that serum from infected animals contained a property which neutralized the poisons secreted by the infected agent and that immunity could be conferred by the injection of this material prior to infection. He was awarded the Nobel prize in 1901, both for his general studies on immunology and for his specific development of diphtheria antitoxin.

BENTHON

Benthon refers to the organisms living under water on or in the bottoms and shores. Perhaps all the plant, fungal, algal, animal and bacterial or precellular phyla of organisms are involved in physiological and physical roles. These roles are related to the distribution of the organisms as motile or sessile adult organisms on the surface, *epibioses*, in the surface, *endobioses*, or under objects, *hypobioses*, distinctions elaborated upon in the classical treatise on the bottoms of cooler oceans by Gislen, which work also contains a detailed history of the subject. There is no detailed treatise of the more tropical marine benthon of broad scope, and aside from some studies of lakes, less has been done on fresh water benthon than marine. In part, the great economic value of the marine and brackish benthon, and the larger generally more stable marine habitats may account for this.

The benthon may be distributed as individuals or as groups of species in one or more of the regions characteristically recognized (Figure 1) on the sea bottom or shores and in somewhat similar ways in fresh water when physiography of the bottom permits. As one traces vertical distributional limits of populations away from low water or low tide level, they become more and more variable and diffuse. With a few notable exceptions (*e.g.*, the poleward-equatorward distribution of reefs, river mussels and barnacles, where often annual temperature extremes coupled with time seem to be major factors) distribution on a broad geographic basis has been little investigated as yet. However, from local distributional and observational studies, the physiological, physical and ecological functions have become theoretically outlined.

Physiological roles of the benthon are in *primary production* (inorganic material and free energy converted into organic material and bound energy), *secondary production* (primary production shifted along the food chain) and *mineralization* (the return of organic material and bound energy to the inorganic and free-energy states). Measurement of benthon is usually in terms of numbers of individuals or mass of material (*e.g.*, chlorophyll, protein as calculated from nitrogen content, dry or wet weight). Such measurements of the benthonic *standing crop* are often in terms of the horizontal area under which they are found. When a change in the mass of, or energy in, the population per unit of area (or volume) with time is under consideration one is speaking of *productivity, i.e.*, the rates of production.

Primary productivity of the benthon is often in terms of the rates of production of oxygen or conversion of inorganic carbon to organic carbon because of the techniques involved. It is at times very importantly of silica structures or inorganic carbonate. Marine benthonic primary producers are almost entirely algae; largely of the phyla: Myxophyta; Chlorophyta; Phaeophyta and Rhodophyta. The benthic algae and flowering plants, which latter are insignificant below low tide level, are most significant for their production, along with the algal plankton, of the food for all the kinds of aquatic living things. Not generally known or understood is the role of algae as the major producers of limestone and coral reefs both directly through accumulation of inorganic salts from the sea, largely as "skeletal" calcium carbonate, and indirectly as the internal sources of food and animal waste disposal, the zooxanthellae, of the "reefbuilding" coelenterate corals. Except in calm water such as found in atoll lagoons and on lee atoll shores, it is almost always algae that produce and dominate the upper atoll reef surfaces, especially the exposed seaward edges where the rock-like red algal genus *Porolithon* is usually the dominant organism. Benthonic algae and plants forming extensive stands in the euphotic waters near shore are functional in accumulating a great mass of fertilizer elements and in producing a great mass of organic matter. This mass enables the higher standing crops of fish and other secondary producers found as one approaches shore from the open sea. In the sea it serves to stabilize the shore against erosion by binding particles (Myxophyta) and by reducing wave action (kelps) and it serves at times as the raw material of rather large industries (*e.g.*, suspensoids, iodine, fertilizer, human food, agar-agar). In size, the benthonic primary producers vary from microscopic unicellular organisms to those kelps weighing perhaps over 1000 pounds and reaching 100 feet in length. General information on the kinds may be had in the books by Smith (algae) or Fassett (flowering plants).

The benthonic secondary producers and mineralizers among the benthon belong for the most conspicuous part to invertebrate animal phyla, but others such as Fungi and those of even simpler organizational levels, especially the Bacteria, are extremely important. The principle role is in converting primary production into secondary production and thereby reducing or regulating the popu-

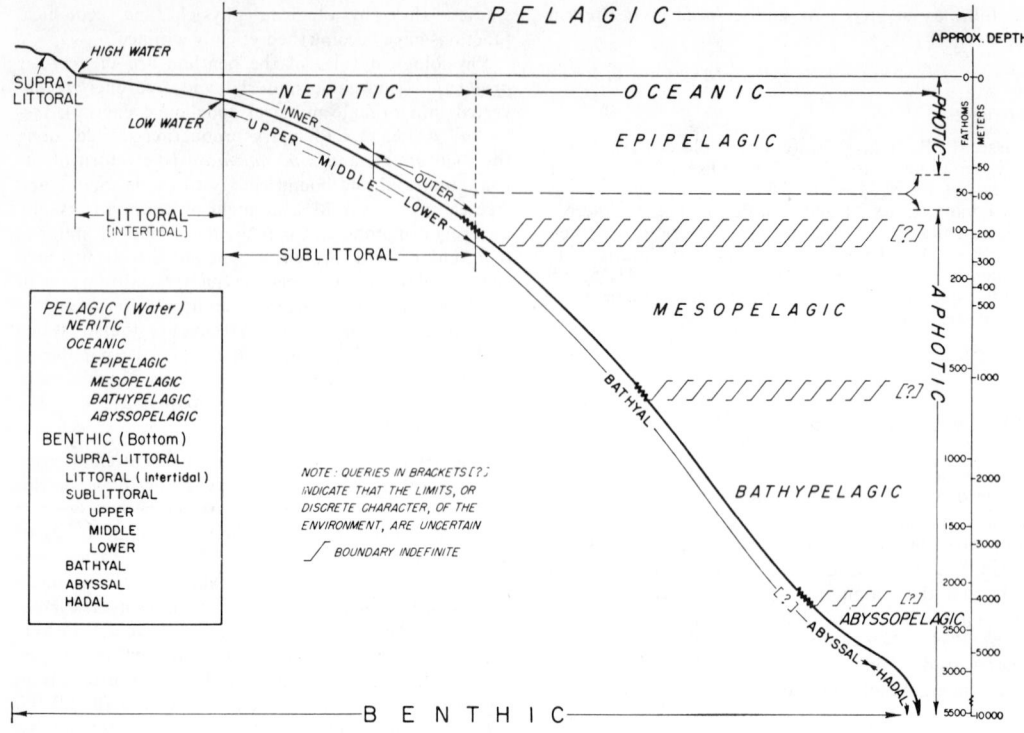

Fig. 1   Diagrammatic section of an ocean shore at right angles to the beach. The nomenclature and vertical and horizontal relationships of the gross ecological habitats most often recognized or used by marine biologists are illustrated. (Redrawn with some modification from Hedgpeth's figure on page 18 of his chapter "Classification of Environments" in volume I of memoir 67, of the Geological Society of America)

lation of primary producers. This is done as primary producers are grazed upon by *herbivores* and these in turn serve as the food of the other secondary producers, *carnivores*. When the populations are not quantified, this series of organisms through which the material and energy passes is referred to as a *food chain*; when it is quantified it may be represented as a *food pyramid*. In constructing food pyramids the coordinate largely-mineralizing action of the fungi and bacteria should not be overlooked.

According to their habits, the secondary producers may have physical roles in the environment. When they are limestone depositors (*e.g.,* coelenterate coral, foraminifera, gastropods) or depositors of other inorganic matter they may aggrade the surfaces whereon they dwell. Bottom dwellers (*e.g.,* coelenterate corals, worms, fungi) may be sessile and have free floating reproductive stages or achieve some distribution by growth. The motile benthon may burrow about in the bottom (worms, arthropods) or move in semi-permanent burrows (bivalve molluscs) or crawl on surfaces (echinoderms, gastropods, arthropods), sometimes swimming free briefly (crabs). In such cases their physical role is degrading or altering the habitat chiefly by reducing the particle size (*e.g.,* fish, holothurians and worms) and thus leading to its reconsolidation or transportation. The magnificent treatises of Hyman (for the invertebrates) and those

edited by Edmondson (for fresh water organisms) and Hedgpeth (for marine distribution and roles) should be consulted for the wealth of material they contain on the benthon as found in the different distributional areas distinguished (Figure 1) in fresh or marine waters.

The inefficiency of the benthonic secondary producers is ultimately represented in ecological schemes as mineralization. The metabolism of the primary producers is conservative of essential mineral materials (*e.g.,* phosphorus and fixed nitrogen): they do little or no mineralizing, except that if the rates of respiration exceed those of photosynthesis, organic carbon and energy are lost to the biological systems. The animals, herbivores less and carnivores more, return to the sea some of the food they ingest as degraded or mineralized matter. It appears the Bacteria and especially the Fungi are even more inefficient in converting nutrient material into more of their own substance. These organisms appear able to completely mineralize almost any organic matter in the benthonic habitat to which they have access. Organic matter bound in dense mud and rock is apparently not available to their often highly oxidative metabolism. Information on the functions of Bacteria is to be had in Zobell's work and "Bergey" is an exhaustive manual of the kinds. The marine fungi are little known except for nearshore forms.

MAXWELL S. DOTY

References

Breed, R.S., *et al.*, Eds., "Bergey's manual of determinative bacteriology," 7th ed. Baltimore, Williams and Wilkins, 1957.

Edmondson, W. T. Ed., "Ward & Whipple," Freshwater Biology. 2nd ed. New York, Wiley 1959.

Fassett, N. C., "A manual of aquatic plants," Madison, University of Wisconsin Press, 1960.

Gislen, T., "Epibioses of the Gullmar Fjord: I. Geomorphology and hydrography; II. Marine sociology," Kristinebergs Zoo. Sta. 1877–1927 (3) : 1–123; (4): 1–380, 1930.

Hedgpeth, J. W. Ed., "Treatise on marine ecology and paleoecology. Vol. 1, Ecology," Washington, The Geological Society of America memoir 67, 1957.

Hyman, L. H., "The invertebrates," Vols. 1–6, New York, McGraw-Hill, 1940–67.

Smith, G. M. Ed., "Manual of phycology," Waltham, Mass., Chronica Botanica, 1957.

Zobell, C. E., "Marine microbiology," Waltham, Mass., Chronica Botanica, 1946.

BERNARD, CLAUDE (1813–1878)

The debt which contemporary physiology owes to this pioneer is due to his failure in another field. Born in France near Villefranche he studied at Lyons where a brief provincial success as a playwright took him to Paris armed with the manuscript of another play. The critic to whom he presented this warmly suggested that he study medicine, advice which the world is the better for his having taken. By 1855 he had become a full professor at the College de France where he had already discovered the digestive function of the "pancreatic juice." Had he been provided with modern equipment, he would probably also have discovered the first hormone since his investigations on the glycogenic function of the liver led him to believe that it was the site where was produced some substance which concerned itself with the metabolism of sugar. He also discovered the nervous mechanism of vaso-dilation. He was thus an outstanding pioneer both in the fields of chemical and physical physiology and his greatest contribution to biology is probably his demonstration that these disciplines could be used to explain many life processes. Physiology would, of course, have developed without his aid but it might well have been delayed another half century but for the impetus he gave to it.

PETER GRAY

BERZELIUS, JÖNS JACOB (1779–1848)

Berzelius, the orphaned and penniless son of a Swedish priest, became one of the great scientific intellects of his period. While supporting himself as a teacher, he taught himself pharmacy and then while supporting himself as a pharmacist, studied medicine and eventually, in 1809, secured a doctorate in this field from Stockholm. An account of his contributions to chemistry must be sought in another place but the brilliance of these must not be permitted, as is often the case, to overshadow his fundamental work in the biological sciences. He was led directly to this work through an analysis of what we would now call organic compounds and, his interest being thus aroused, he passed on to develop new theories of the nature of life. He might well be described as the founder of the mechanist school since he firmly rejected every existing theory which dealt with a supernatural nature, either of the body or of the soul or of the functions of either. It was so clear to him that the processes of digestion, excretion, and secretion were as much chemical as were their inorganic counterparts that he refused to regard the organs involved in these processes as "living" and was therefore forced to the completely erroneous conclusion that life was resident in the nerves. He frankly admitted that he did not know how the nervous system worked and forcefully rejected suggestions that the transmission of impulses were "electrical" on the then very reasonable grounds that no adequate experimental evidence had been adduced in support of this theory. His insistence on the physical chemical basis of life led him into other errors for he believed, for example, that organogenesis was the result of the "crystallization" of organs rounds the ends of capillaries and he was, moreover, firmly convinced that lower animals developed by spontaneous generation. The inaccuracy of his theories, however, do not alter the fact that he exercised a profound influence on the development of contemporary biological thought.

PETER GRAY

BICHAT, MARIE FRANCOIS XAVIER (1771–1802)

The French physician Bichat's contribution to biology stemmed from close observations made while conducting post-mortems. Bichat, a classical biologist, performed his experiments with the naked eye, feeling deeper probe unnecessary. He was an extreme vitalist who felt that chemistry or physics could not aid in the understanding of life. Despite these handicaps, Bichat was the first to show that the organs of the body were a complex of simpler structures. He showed that each organ was composed of various types of tissue and that different organs might possess common tissues. For his identification of twenty-one types of tissue, he has been called the founder of histology. An accidental death at age thirty barred Bichat from seeing his work aid in the development of the cell theory of life propounded in 1839 by Theodore Schwann.

DOUGLAS G. MADIGAN

BIOCENOSE

A biocenose is a COMMUNITY of organisms. It consists of a number of different kinds of plants and animals living together in a particular habitat, or biotope. Such combinations of species or biotypes are frequently repeated in different places, and equivalence of ecological conditions and mutual relationships are therefore assumed.

The term biocenose thus describes in the first place concrete, empirical entities. The limits between entities may be sharp as when mutual reactions cause several organisms to react simultaneously, when a sharp break in environment exists, etc., or they may be continuous. A difference between two biocenoses can frequently be demonstrated without shedding light on the placement of a boundary between them.

The term biocenose was introduced by Mobius for an oyster bed and its associated organisms. The physical environment was not included. However, a biocenose and its biotope form a unit. Other, equivalent terms have been suggested from experience in various natural sciences for what is evidently this same unit. Thus, Tansley as a plant ecologist introduced the term ecosystem in 1935. Ecosystem is favored especially by English-speaking biologists, while biocenose-biotope is more used in continental Europe. "Ecosystem" has the advantage of suggesting an analogy with the systems of the physicists and chemists. Similar methods of study should apply. Sukachev as a plant sociologist has used the term biogeocenose, and as the dean of Russian phytosociologists his ideas carry great weight in the USSR. "Biogeocenose" suggests not only the existence of a biotic community but its relation to a particular habitat as well. The Russians follow Tansley in preferring not to use the term community. "Community" does have some anthropomorphic connotations of mutual "interests" if not of mutual aid. The term landscape is used for a synthetic unit, often composed of several biocenoses in several kinds of biotopes which cannot be ecologically combined but do have spatial contiguity. Foresters have used such terms as site, habitat, and environment for ecosystems. Clements from his preoccupation with plant succession and life forms coined the term biome for a biocenose whose vegetation was a particular formation. Jenny as a pedologist has used the term tessera, and he meant to suggest thereby the mosaic nature of the units making up a landscape.

It is evident that all these terms refer to the same kind of natural entity, namely an ecosystem whose biological part is a biocenose.

Because autotrophic plants are less mobile than animals, occupy the lowest trophic level, and are taxonomically less difficult than invertebrates, at least, plant sociology has been pursued more intensively than animal sociology. Since it has been shown that many animal communities coincide with, or even depend on, plant communities, the animal communities are frequently related directly to the plant communities. At any rate, the vegetation can often be considered a matrix for the animal groupings.

Considering a particular example of a biocenose, a forest stand with the animals living in it for example, we can define and describe this unit in various ways. These ways will reflect our interests. The following points of view give a reasonably complete picture: Composition, structure, physiology, ecology, genesis, history, chorology, and taxonomy.

Composition. The most objective and at the same time most exact description of a biocenose is given by the list of species contained.

The composition of the biocenose could be brought down to a more basic level by listing the chemical compounds and their amounts which are contained, or by using the chemical elements. Such treatments veil information on the organization of the biological units making up the cenose. A similar statement applies if energy instead of matter is used in description.

The biocenose can be described by trophic levels. The energies fixed and dissipated in various trophic levels within the biocenose are determinable amounts, expressible in universally recognized units of measurement.

The composition of a biocenose can be described by some system of life forms. If ecologically oriented, such a system can be most instructive—but only for ecologists. Taxonomists, floristic botanists, physiologists, geographers, foresters, etc. may not be able to use such a special classification. Most systems of life forms are based on a theory of relation between the organisms and their environments. They are an interpretation, not basic data which each investigator can use and interpret as suits his purposes.

The taxonomic system of organisms is the most highly developed, most consistent, most universally recognized, and aspires to correlate the most characteristics of organisms.

Structure. Structural studies of biocenoses lack universally applicable systems of nomenclature. The layers or stratification of the plant community, with the animals more or less confined to these layers, have been described. The morphology of soils has an abundant literature and it is in this part of the ecosystem that the most intensively studied animal communities live. Periodicity is another structural feature.

Physiology. The functioning of a biocenose can be studied once the composition and structure are known. Customarily such studies have been carried out in the laboratory. The results are then reproducible and interpretable in terms of more well-known physical and chemical processes. Their applicability to natural biocenoses is *a priori* unknown, and this forms a second problem after the purely physiological one has been designed, studied, and solved. By considering ideal or simplified laboratory biocenoses, at least general guides to the mechanisms obtaining in natural biocenoses have been laid down.

It is evident that study of the functioning of biocenoses must follow their description. This description will give not only the individual organisms whose functioning must be known, but the organisms with which they compete. Competition between organisms modifies not only the tolerance spans but even the position of environmental optima of individual organisms.

Ecology. While the physiology of biocenoses is concerned with their functioning, their ecology is a matter of relations to environment. The two are not identical. The ecologist requires not only data from the physiologist, including responses to the laboratory environment, but also description of the natural environment.

If functional relationships are to be set up between the biocenose as a whole, or between its members, and the factors of the environment, then mathematical logic demands that properties of the environment be not also measures of the organisms themselves. If the properties of the organisms can be considered dependent, then the factors of the environment can be defined to be independent. Six such factors, or groups of factors, have been recognized: Climate, soil parent material, relief, biota, fire, and time.

Climate on the regional level is independent of the biocenose in habiting a biotope or the biocenoses making up a landscape. Local climates are the regional climate as conditioned by relief. Microclimates are properties of the biocenose itself. Thus, a forest makes a stand climate which is different from the climate measured by the standard weather instruments. If the stand is cut, the micro-climate changes.

Soil parent material has been defined by Jenny as the state of the soil part of the ecosystem at time zero. The choice of time zero is the investigator's. In studies of soil formation and plant succession, representing two aspects of the dynamism of biocenoses, it has been customary to use the time of deposition of morainal material (Alps, Alaska) or of flood-borne debris (Mt. Shasta in California, the Swiss National Park), or of volcanic ejecta (West Africa, Krakatoa in the East Indies, St. Vincent in the West Indies, Craters of the Moon in Idaho, etc.), or of sand dunes (Australia, England, southern France) as time zero.

By a revolution in the factors of the environment, a new time zero will clearly ensue. In other words, the dynamic development of a cenose should take place under constancy of all factors of the environment except time. As a matter of fact cenoses do show a development under such conditions—see time below.

Relief produces local climates as modifications of the regional climate, and it is to such local climates that vegetation and animal aggregations respond. Relief conditions other features of the environment modifying the presence and abundance of plants and animals. Ground water, snow accumulation, and direct effects of the downslope component of gravity on steep slopes are examples.

The actual biocenose found in a particular biotope is dependent on the kinds of plants and animals available to colonize there. Identical biotopes in different parts of the world will therefore have different biocenoses occupying them. The vegetation and associated animals in those subtropical areas where precipitation comes in winter and summers are dry (the Mediterranean, California, Chile, the Cape, and southwest Australia) have quite different biocenoses because they lie in different floristic and faunistic regions. These differences are historically, not ecologically, determined.

To say that two biotopes have different biotic factors, it must be shown that the kinds of organisms, including their disseminules, which reach the two areas are different. Data are not at hand to demonstrate if differences in numbers of disseminules of a taxon are of importance. Plants, at least, have such a high rate of reproduction that the presence of one breeding pair is enough to saturate the biotope with this organism in a fraction of the time necessary for development of a cenose from time zero. However, for small biocenoses a long-lived organism such as a tree can be the dominant feature of that cenose's habitat.

Man is an independent biotic factor *par excellence*. He knows no historical limitations over most of the world's surface today. His actions are frequently taken without regard to the ecological conditions within a biotope.

Fire is usually connected with man's activities. This is not necessarily so, and the intensity and frequency of fires in some ecosystems are strictly independent in that the actual time when a fire will occur cannot be predicted from knowledge of all the other factors of the ecosystem ordinarily influencing plants and animals.

A given constellation of plants and animals, set in a particular climate, on a particular slope, on some kind of soil parent material will develop through several biocenoses to a more stable one. This autogenic development is known as plant succession and carries with it the animal communities as well.

In general, the rate of change decreases with time. Absolute stability is of course never reached, since the factors of the ecosystem will eventually change. At least in northern latitudes in postglacial time succession has had a time span within which to operate of roughly 5,000 years since hypsithermal time. Within this time changes in climate have occasioned changes in vegetation and associated animals at any one spot, and these vegetation changes have transgressed formation limits. Within time spans of greater length floras and faunas have changed by evolution, migration, and extinction. There are thus absolute time limits within which it makes sense to speak of plant and animal succession as autogenic processes.

The genesis of biocenoses is a problem of what happens with the passage of time. This autogenic succession in biocenoses can be considered their genesis. It is directional, predictable, and tends toward a more or less stable state. The "allogenic successions" of Tansley can be considered changes in the biocenose brought about by changes in the independent factors of the environment.

On the other hand the history of cenoses has no necessary direction of change. Climatic changes on a geological time scale conditioned by changes in the relationships of land and water, possibly by cosmic events, and opportunities for and obstacles to plant and animal migration as land connections are formed or broken follow no general rules.

The decipherment of such changes is a paleontological problem, but both botanists and zoologists have found that interpretation of the fossil record is surer if the cenoses in which the plants and animals occured in the past are considered. Thus, the presence of grass-eating herbivores in the Great Plains of North America in the Pliocene can be correlated with early grasses such as Stipa. The cenose which included primitive man in Europe is shown not only by the artifacts man left, the bones of the animals which lived in the vicinity and which formed part of his food, but also by the presence in the pollen record of his cereals, Plantago, and other plants of forest clearings.

The distribution chorology of biocenoses is little-known because the cenoses themselves have not been defined and described. Mapping can follow description and, preferably, ecological study.

In many cases a biocenose coincides in distribution

with a particular landscape, but the relationship is not necessarily reciprocal. A landscape can be formed of several biocenoses. Their integration into a landscape is a geographical problem, not a biocenotic one.

Classification. Many systems of classification of biocenoses have been suggested. An apparent one is based on physiognomy. Evolution from quite varied original stocks has produced organisms which not only look alike but may act alike in response to their environments. However, physiognomy is no necessary and sufficient guide to function. The kinds of organisms contained, then, are an objective description of a biocenose.

However, not all organisms are equally good indicators of the existence of a particular biocenose, of its ecology, its history, its successional stage, or its distribution. Dominant species in general contain an abundance of biotypes, all differing in indicator value. In plant sociology and in soil-dwelling animal communities, at least, constant and exclusive species provide good indicators. These are biotypes which always or only occur in one particular biocenose. The exclusiveness of occurrence can be absolute as with many relict species, or regional, or territorial only. Differential species which occur in only one of two similar biocenoses offer a further way to separate allied cenoses.

JACK MAJOR

References

Braun-Blanquet, J., Pflanzensoziologie., 2nd ed. Vienna, Springer, 1951.
Clements, F. E. and V. E. Shelford, "Bio-ecology," New York, Wiley, 1939.
Tischler, W. "Synökologie der Landtiere," Stuttgart, Fischer, 1955.

BIOCHEMICAL GENETICS

In contrast to classical GENETICS, which is concerned mainly with the manner in which GENES are transmitted from one generation to the next, either in limited lines of descent or in large populations of interbreeding individuals, biochemical genetics deals with the chemical nature of genes and the manner of their action in development and function. It is convenient to describe this branch of biology in terms of the answers to the following five questions.

(1) What are genes chemically?

(2) In what way are genetic specifications "written" in molecular terms?

(3) How are genes replicated, as they must be during cell division?

(4) How are genetic specifications "translated" during development and function?

(5) What is the molecular basis of mutation?

Although final answers cannot be given to these questions, it is possible to at least restate them in plausible molecular terms.

First of all there is convincing evidence that the primary genetic specifications of some viruses and of bacteria are carried in large molecules of deoxyribonu-

cleic acid (DNA). The evidence in bacterial viruses consists in showing that during the infection process, DNA enters the host cell whereas the protein components of the virus remain outside and may be removed without interfering with the process. That this is so was shown by Hershey and Chase,[1] who labeled DNA of viruses with radioactive phosphorous (^{32}P) or the protein components with radioactive sulfur (^{35}S). In this way it could be shown that DNA enters the host but that little protein does so. Later experiments make it clear that parental DNA is passed on to progeny viruses while the small amount of protein that enters the host during infection is not.

In bacteria it was shown by Avery et al.[2] in 1944 that one genetic type of pneumococcus bacterium can be transformed to that of another by treating the recipient with pure DNA from donor cells. In this way naked genes are evidently transferred and incorporated by replacement into the genome of the recipient.

Since chromosomes of higher plants and animals largely contain DNA and protein, it is a reasonable assumption that in them, too, DNA is the primary genetic material.

In 1953 Watson and Crick[3] proposed a detailed structure for native DNA that was tremendously exciting to biologists in that it suggested plausible answers to questions two to five as given above. In this structure, now abundantly demonstrated to be substantially correct, complementary but oppositely polarized polynucleotide chains are hydrogen bonded together and wound helically around a common axis. The complementarity is given by the fact that there are only two pairs of specifically hydrogen bonded nucleotides, those containing adenine and thymine or those characterized by guanine and cytosine. A four-unit segment of such a structure can be represented as follows:

where A, T, C and G represent adenine, thymine, cytosine and guanine, S indicates a deoxyribose sugar unit, and P stands for a phosphate group.

Genetic information is believed to be somehow encoded in sequence of nucleotides, perhaps in a simple four-symbol linear code. Since a modest sized bacterial virus contains DNA composed of about 200,000 such nucleotide pairs and the nucleus of a human cell something more than 10,000 times as much, there appears to be no shortage of DNA in which to encode the total genetic instructions of these organisms.

Replication is believed to be accomplished by separation of complementary neucleotide chains, each single chain retaining its integrity while serving as a template against which new complementary chains are built for the four nucleotide components. This can be represented as follows:

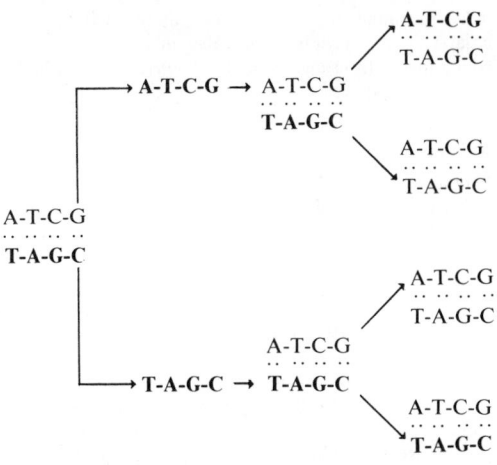

where original chains are indicated in bold face letters and new ones in their Roman counterparts. That this scheme of replication is correct is strongly suggested by labeling experiments with radioactive phosphorous or heavy nitrogen (^{15}N). In bacterial DNA Meselson and Stahl[4] have shown that the labels are distributed as indicated in the diagram, assuming bold face nucleotides to be labeled differently from those indicated in Roman type.

In the translation of the genetic information contained in DNA it is commonly assumed that RNA of corresponding information content is synthesized, perhaps by a template mechanism, carries information to the ribosomes of the cyto in the nucleus of cellular forms, that this RNA plasm where each kind of RNA molecule or its descendents serve as templates against which amino acids are properly ordered to form specific proteins, the amino acids being transported to their proper site on the template by specific carrier or soluble RNA molecules. The evidence for this view is incomplete. The postulated movement of RNA from nucleus to cytoplasm has not been clearly demonstrated. The coding system by which DNA information is related to amino acid sequence remains to be discovered. If it is a four-symbol linear code it may consist of non-overlapping quadruplets, each encoding a specific amino acid. There are 27 such quadruplets in a simple "comma-free transposable" code.[5]

Whatever the mechanism of translation, there is increasingly convincing evidence that specific segments of DNA, defined as genes in a functional sense, bear a one-to-one relation to poly-peptide chains of corresponding specificity. It is in this way that genes are presumed to determine the specificity of enzymatically active and other proteins.[6]

Mutation in terms of DNA is presumed to consist in alteration of nucleotide sequence either during the process of replication or at other times. Presumably occasional mistakes are made spontaneously as a result of improbable tautomeric forms of nucleotides at each instant of partner selection.[3] Base analogues are known to induce mutations, presumably by substituting A-T base pairs for G-C or vice versa.[7] Oxidation of amino groups with nitrites increases mutation fre-

quencies as do many other chemical mutagens. High energy radiation has been known since 1926 to be mutagenic. There is now convincing evidence that at least some radiation-induced mutations are produced by indirect processes.

Nucleotide substitutions or transpositions presumably lead to missense or nonsense coding. Assuming nucleotide quadruplets to encode single amino acids, a single unit substitution may result in a new quadruplet that encodes a different amino acid or one that does not encode any of the twenty amino acids.

GEORGE W. BEADLE

References

1) Hershey, A. D., and M. Chase, Jour. Gen. Physiol., **36**: 39–56, 1952.
2) Avery, O. T., et al. Jour. Exper. Med., **79**: 137–158, 1944.
3) Watson, J. D., and F. H. C. Crick, Nature, **171**: 737–738, 1953.
4) Meselson, M., and F. W. Stahl, Proc. Natl. Acad. Sci. U.S., **44**: 671–682, 1958.
5) Golomb, S. W., et al. Biol. Medd. Dan. Vid. Selzk., **23**: 1–34, 1958.
6) Beadle, G. W., Ann. Rev. Physiol., **22**: 45–74, 1960.
7) Freese, E., Proc. Natl. Acad. Sci. U.S., **45**: 622–633, 1959.

BIOCHEMICAL INDIVIDUALITY

Biochemical individuality is the possession of biochemical distinctiveness by individual members of a species, whether plant, animal or human. The primary interest in such distinctiveness has centered in the human family, and in the distinctiveness within animal species as it might throw light on human biochemistry.

While it has been known for centuries that bloodhounds, for example, can tell individuals apart even by the attenuated odors from their bodies left on a trail, the first scientific work which hinted at the existence of substantial biochemical distinctiveness in human specimens was the discovery of blood groups by Landsteiner about 1900.

A few years later Garrod noted what he called "inborn errors of metabolism"—rare instances where individuals gave evidence of being abnormal biochemically in that they were albinos (lack of ability to produce pigment in skin, hair and eyes), or excreted some unusual substance in the urine of feces. To Garrod these observations suggested the possibility that the biochemistry of all individuals might be distinctive.

About fifty years later serious attention to the phenomenon of biochemical individuality resulted in the publication of several artices and a book on this subject. These reported evidence indicating that every human being, including all those designated as "normal," possesses a distinctive metabolic pattern which encompasses everything chemical that takes place in his or her body. That these patterns, like the abnormalities discussed by Garrod, have genetic roots is indicated by the pioneer explorations of Beadle and Tatum in the field of biochemical genetics in which they established the fact that the potentiality for producing enzymes resides in the genes.

Biochemical individuality, which is genetically determined, is accompanied by and in a sense based upon anatomical individuality, which must also have a genetic origin. Substantial differences, often of large magnitude, exist between the digestive tracts, the muscular systems, the circulatory systems, the skeletal systems, the nervous systems, and the endocrine systems of so-called normal people. Similar distinctiveness is observed at the microscopic level, for example in the size, shape and distribution of neurons in the brain and in the morphological "blood pictures," i.e., the numbers of the different types of cells in the blood.

Individuality in the biochemical realm is exhibited with respect to (1) the composition of blood, tissues, urine, digestive juices, cerebrospinal fluid, etc.; (2) the enzyme levels in tissues and in body fluids, particularly the blood; (3) the pharmacological responses to numerous specific drugs; (4) the quantitative needs for specific nutrients—minerals, amino acids, vitamins—and in miscellaneous other ways including reactions of taste and smell and the effects of heat, cold, electricity, etc. Each individual must possess a highly distinctive pattern, since the differences between individuals with respect to the measurable items in a potentially long list are by no means trifling. Often a specific value derived from one "normal" individual of a group will be several times as large as that derived from another.

The implications of this individuality are extremely broad. For medicine they suggest that susceptibility to all disease—infective, metabolic, degenerative, mental or unclassifiable (including cancer) probably has its roots in biochemical individuality and that the differences in responses to drugs including alcohol, caffeine, nicotine, careinogens and morphine derivatives, which are well authenticated, have a sound and discoverable basis. It is a well-known fact that conditions which will produce disease in certain individuals will not do so in others. The basis for this observation has hitherto not been recognized; it doubtless has its roots in biochemical individuality—a development which has received little attention. People definitely possess what may be called "biochemical personalities" and it seems extremely likely that these are meaningful in connection with the numerous and increasing personality disorders and difficulties which afflict men and women in modern life.

Biochemical individuality offers a sound scientific basis for recognizing the existence in every individual of a unique make-up in the broadest sense. For centuries "individuality" has been written about and its place in the scheme of things has been discussed, but the knowledge of what individuality consists of and how it manifests itself have indeed been scanty. Only in recent years have we had a basis for understanding it in a definitive way.

The understanding of individuality is basic to the understanding of human behavior. It is not enough to know the ways in which all human beings respond alike to certain stimuli; it is fully as important to know also why different people confronted with about the same stimulus react very differently. Since biochemistry underlies many of our moods and our reactions to different types of stimulus, and since it is in the area of biochemistry that individuality is most definitively recognized, biochemistry merits inclusion as one of the most important of the so-called "behavioral sciences."

Because biochemical individuality points the way toward individuality in the broadest sense of the word, it has profound implications not only in medicine, psychiatry, and psychology but also in human relations, education, politics and even philosophy.

ROGER J. WILLIAMS

Reference

Williams, Roger J. *Biochemical Individuality*, New York, Wiley, 1956.

BIOCOMMUNICATION

A great many organisms have evolved means of being sources of information for other organisms whose responses can be important to them; even flowers may possess designs which "lead" bees to their nectaries. Animals commonly have markings or other visible features which identify their species or sex, and in addition most have conspicuously specialized activities apparently serving largely or solely to transmit information. Sounds, movements, postures, or excreted chemicals which have been evolutionarily specialized to function as signals are called "displays" by ethologists. While all activities, and indeed all detectable characteristics of an organism, are informative, it is events involving displays that are usually the main subjects of current research in biocommunication. The displays and their employment are, of course, not the whole story—for if animals did not also evolve the ability to respond appropriately to displays there could be no patterned communication.

Evolution of Patterns of Communication: As a species evolves, the nature of the communication pattern being produced is subject to several sources of influence. (a) The ecological characteristics of the populations of the species are established by the competitive influence of other species populations, the resources and limitations inherent in the total environment, and other causes not directly related to problems of communication. Yet these ecological characteristics set critical limitations on the type of communication pattern which can function most efficiently: Habitats determine the degree to which displays will be obstructed; the number of similar species present determines the extent to which innovations must be evolved to render the display repertoire of each species distinct from that of the others; seasonal changes in climates affect population structure and hence social interactions; and so forth. (b) The form of social organization of a population determines the sorts of behavioral interactions which occur, and hence the functions required of communication patterns. The evolutionary relationship is reciprocal, as complex social behavior cannot develop without the assistance of complex communication. (c) Finally, the evolution of any lineage is influenced by genetic opportunity, the sources of which include factors operating at random with respect to the direction taken by evolution. Because of this, some of the differences which develop between related evolutionary lineages are not functional differences, but are the products of chance.

In the evolution of behavior, there has often developed the ability to respond with innate selectivity to environmental features, and cases have been demonstrated in which a signal pattern has evolved which is capable of elicting a specific response. The red breast of a European Robin, for instance, while not in itself a display, is a persistent signal which can "release" threat-posturing, an aspect of aggressive behavior, from a male of this species by appearing within his territory. Aggressive behavior is forthcoming even when the red breast is not accompanied by the rest of the Robin, as has been shown experimentally by David Lack using stuffed decoys with the head, tail, or *all* of the remainder of the body removed. And yet the situation is not as simple as might be assumed, since the red breast must appear within a suitable context—e.g., a territorial male European Robin rarely attacks his mate, even though she is almost identical in appearance with an intruding rival. Thus the notion that a red breast "releases" aggression is true only within limits, and does not imply that an animal's responses to signals are invariably elicited by simple, innately specified rules. But the ethological concept of innate releasers does point out a fairly general feature of animal communication: Signals and responses are often to a considerable degree innate, their forms and usages are determined by genetic information while individual experience is capable of modifying only some aspects. This considerable degree of genetic control means that most members of the species will employ the species-typical pattern of communication when in the proper physiological state, even without opportunity for learning.

This considerable reliance on genetic information for the specification of communication patterns is violated by only one major exception, language, a sole prerogative of Man. The functions which must be subserved by animal communication are apparently sufficiently invariable that a relatively inflexible strategy can be used to cope with them. On the other hand, the inflexibility may often be required by the lack of opportunity for learning, the need to accomplish a social end within a short time of an individual's first encounter with the situation (e.g., pair formation in birds). As illustrated by the Robin, however, learning can make at least some responses more selective. This may be particularly true where social situations involve relatively prolonged contact among individuals; it seems generally true, for instance, that in colonies of vertebrate animals at least, a knowledge of individual characteristics enables animals to assign different weight to the same signal when given by different individuals. There is also some opportunity for the modification of signals themselves. For example, in bird "song," it is common for some aspects to be relatively invariable and species-typical, while other aspects vary from bird to bird and presumably function to permit individual recognition.

Functions. As has been indicated, a function of major importance in animal communication is to identify the communicator. An identifying signal may specify species, sex, which individual (in terms of both "who" and "where") and perhaps general physiological state (such as "immature," or "mature, and in breeding condition"). Some displays incorporate features yielding all of the identifying information, while some incorporate almost none. The latter are rare, however, since for most displays it is important to have

the signal labelled "pertinent to read" for any relevant member of the species; this enables displays produced by other sympatric species to be ignored, for they are differently labelled. Displays which serve in the initiation of pair formation, or in the initiation of copulation in species which do not form pair bonds, are customarily markedly distinctive among related, sympatric species. Examples are known in many kinds of animals, and are prominent in groups like the pond ducks in which relatively large numbers of closely related species live side by side. Much of the burden of preventing cross-species attempts to breed falls on these displays; they are critically important intrinsic isolating mechanisms protecting the species' gene pools.

Other functions of displays vary among species with different social patterns. In general, displays help individuals to attract or repel particular classes of other individuals, and to attain and maintain stable social relationships with particular individuals such as mates, offspring, or members of a dominance hierarchy. The functions may usually be achieved by making the behavior of the communicator more predictable to the recipient of a display, thus enhancing the probability of an appropriate response to that behavior. Infant animals often have repertoires of displays which are different from adult repertoires, and probably serve primarily to alert an adult to the infant's need of assistance.

Although there are many cases in which the functions of particular displays have been demonstrated from the responses of recipient animals, the study of responses is often very difficult. Recipients of displays may do a variety of things, or even nothing at all, and often make the same initial responses to different displays. Apparently they are making use of more information than is carried by the display itself—they are placing the display in a context of events which are coincident or immediately subsequent, and of information available from memory or genetic storage. For that matter, in the case of many displays of vertebrates (and at least some invertebrate displays) the communicator employs the display in a variety of circumstances, and may behave in a variety of ways during and after displaying. To a human observer, there is usually a common denominator in the usage of each display which makes it possible to predict the communicator's most likely subsequent behavior. Young gorillas, for instance, often conspicuously bow their heads and twist their necks so that their gaze is averted from the face of another gorilla they are approaching or being approached by, and then play in a mock fighting fashion. The same bowing of the head and presenting of the back of the neck has been seen when one gorilla is actually attacked by another and does not resist or flee, or attempt to play, but crouches and remains motionless. In both usages the movement and pose appear to be a display and seem to indicate that the communicator is not about to behave aggressively to the individual specified by the orientation of the head. The display is not, however, specifically a play invitation or a sign of submission; it has more than one social function. This is not unusual, and much of the breadth of animal communication may depend upon displays which function differently in different contexts.

Display Repertories. Most animals do not have a great many different displays; a common number for different species of birds or mammals is about twenty to forty, and many species have fewer. The actual number of displays a scientist may describe for a species is a matter of convenience and personal preference, since a repertoire can include various forms of grading. Thus two vocalizations in the repertoire of a bird or primate may intergrade with each other in some circumstances, and they and the intermediate forms may intergrade with some other vocalizations in the repertoire. Both may also vary continuously in loudness, harshness, and other properties. Repertoires with a wealth of graded possibilities are far from universal, but even in such cases it is usually possible to distinguish nodal points—display forms which are relatively constant and common, or which are the more often used extremes on certain of the continua of change. And while the existence of continuous gradations does imply at least a theoretical infinity of points along the continua, there are limitations (a) in the ability of receptors to distinguish small degrees of difference (limitations which are compounded by a noisy environment) and (b) in that no information is added which falls outside the possibilities of the continuum. An unspecialized intermediate in form between displays A and B usually carries no information outside that specified by A and B other than the degree of nearness to each. Thus even when gradation occurs, animals are limited in the range of communication open to them. Not only that, but the display repertoires of most species include considerable redundancy. Indeed, what an animal signals with its vocal repertoire it may largely recode into its visible signals.

Probably the size of display repertoires is set by the "cost" of errors of reception. In posturing, for instance, an animal has only so many stances it can assume, and only so many anatomical features it can employ to make one posture appear different from another, or from the posture of a closely related sympatric species. Beyond some level, the more postures it has, the more similar each must be, and the more easily confused they become. Some confusion can usually be afforded, particularly where contextual clues can add information to aid discrimination. But there are always limits beyond which the attempted communication is too inefficient for the species. Man has circumvented the problem with linguistic inventions, employing a repertoire of sounds in combinations to form a very large repertoire of discrete words, then combining the words according to a set of syntactical rules to provide an infinite range of possible utterances. This is an extraordinarily flexible means of communicating; no other known animal employs a pattern in which there are two levels of recombinable basic elements.

Display and Sensory Modalities. Displays exist for reception by a variety of senses: They may be visual, auditory, olfactory, or tactile. Different sorts predominate in different groups of animals: chemical signals in social insects, visual signals in fish and reptiles, and visual and auditory signals in birds. Mammals tend to make considerable use of all forms, although among different mammalian groups there tends to be considerable specialization. Even a mammal like a diurnal monkey, in which the evolution of vision and hearing has been at the expense of olfactory abilities, is usually found to have at least limited olfactory communication, but is never so dependent upon excreted chemicals as are some prosimians with their more prominent noses and better olfactory capabilities. Yet while the evolution of specialization on different sensory modalities is determined by many factors not related directly to communication, it is also true that signals appealing to different senses do have different adaptive values in different environments and for different social functions.

Persistence of signals, for instance, is highly desirable in some circumstances. Some visible signals, like the breeding plumage of a male songbird or the structure built of twigs by a male bowerbird, are relatively very persistent. Chemical signals can also be relatively permanent, particularly if they diffuse slowly. They are much used in marking. Various mammals have glands from which they secrete chemicals to mark their territories, and many hymenopterans lay chemical trails to food sources. Where evolution of chemical signals (called pheromones) has proceeded farthest, there has been a matching of such characteristics as the molecular size of the pheromone, the manner in which it is released, and of the response thresholds of the recipient. The alarm substances of ants, for instance, are released in a burst and diffuse rapidly, while the chemical trail laid by a foraging fire ant (Fig. 1) between a food source and the colony is continuously emitted by the communicator, and diffuses more slowly with the results that the trail is fairly persistent and has narrow boundaries. Such a trail does fade, however, and the fading time is probably adjusted in part to minimize the attraction of worker ants to food sources that have been exhausted.

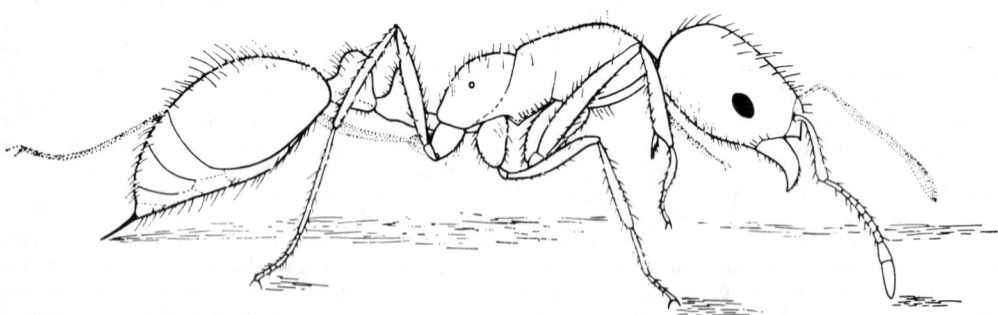

Fig. 1   The use of an olfactory social display in ants. A worker of the imported fire ant (*Solenopsis saevissima*) is shown laying a "recruitment" odor trail from left to right. The trail substance, which leads other workers to new food finds, is secreted by Dufour's gland in the abdomen and is released through the sting tip. (After E. O. Wilson.)

Natural environments provide various obstructions to the dissemination of signals, and chemicals and sounds both are relatively good at getting around these. Visible signals, which are much hindered by physical obstacles, can be to some degree modified to cope with this problem for long-distance signalling by the communicator's choice of a high, conspicuous place to stand while signalling, or by the evolution of flight displays which occur above the obscuring vegetation. Visible signals are easier to locate than are chemical and sound signals, and may often be advantageous when it is necessary to know which of many individuals is signalling. They can also function to indicate the relevant recipient by means which are directly analogous to pointing, as in the example of the oriented twisting of the head and neck in gorillas, described above. In some social birds such as gulls, and in diurnal primates like baboons which dwell in relatively open habitats, visible displays are often used in simultaneous combinations and may convey rather precisely specified and subtle messages. Yet visible displays have some important limitations even under conditions where they are not easily obstructed. Primary among these is that they interfere with other behavior of the communicator, and they require the recipient to watch (see Fig. 2).

Vocalizations (Fig. 3) are in many ways optimal vehicles for information which must be disseminated rapidly, over a long or short distance, with minimal interference of communicator behavior or pre-emption of recipient sensory modalities and yet still be complexly variable and be fairly readily located under most circumstances. Vocalizations can be loud or faint, and can be varied in such other parameters as pitch, duration, successive combination, rate of repetition, and clarity or harshness. The range of variation of some of these parameters can be held within limits set by habitat problems (e.g., low-pitched sounds suffer less

Fig. 2   The use of a visual display in birds. Hostile wing-flashing in the Olive-backed Thrush, *Catharus ustulatus*. (After W. C. Dilger.)

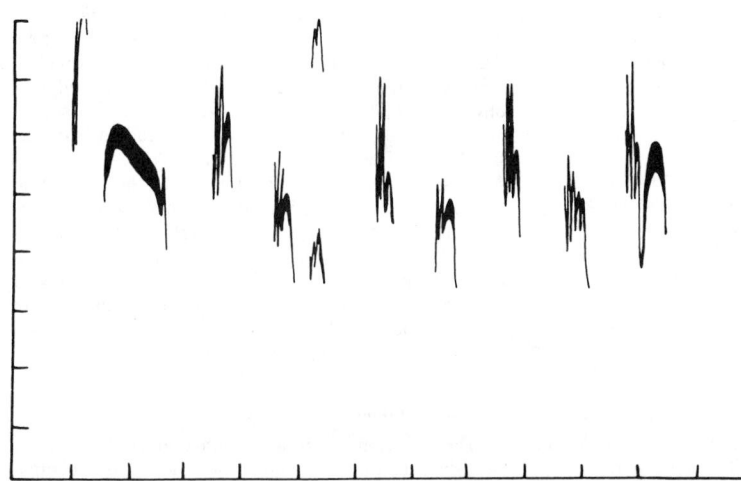

Fig. 3   A sonagram of a vocalization of an Eastern Kingbird (*Tyrannus tyrannus*). This call was uttered by a female kingbird as she flew from her nest and past her mate; he uttered a similar but more prolonged vocalization which is not shown. Sounds such as these are not accurately described by language symbols, and the illustration demonstrates a form of description in which measurable physical properties are portrayed. In this case, the duration of the sounds can be read from the horizontal axis, marked in tenths of seconds, and the frequency in kilocycles (now kiloHertz) per second from the vertical axis. (Original.)

Fig. 4 Tail-twining by two titi monkeys (*Callicebus moloch*). This signal is usually employed whenever two to four individuals of the same social group sit quietly side by side, awake or asleep. (After M. H. Moynihan.)

attenuation with distance, particularly in dense vegetation, than do high-pitched) while still leaving considerable opportunity for modifications encoding message alterations. They can be uttered while the communicator is otherwise occupied, and received by similarly active animals. Granted a recipient with binaural abilities, or even with the opportunity to move a single receptor while a signal is being repeated, vocalizations can be rather accurately located if the environment is not maximally noisy.

Tactile communication is by all odds the most limited form, yet is of very considerable importance in some cases. Among highly social species of birds and mammals, mutual (allo-) preening or grooming is common and apparently helps maintain social bonds between individuals. Some species have additional tactile displays for similar purposes—as in the tail-twining of titi monkeys (Fig. 4). Chimpanzees touch one another with their hands and even embrace in greetings. One

of the most famous examples of animal communication, the dancing of honey bees, occurs in the darkness of the hive, and workers follow the motions of the dancing bee tactually, apparently receiving information about the location of a foraging source, source of water, or a potential new location for a colony's hive.

W. JOHN SMITH

References

Lanyon, W. E., and W. N. Tavolga, eds., "Animal sounds and communication," A.I.B.S. Publ. 7, Washington, D.C., 1960.

Busnel, R.-G., ed., "Acoustic behaviour of animals," New York, Elsevier, 1963.

Sebeok, T. A., ed., "Animal communication. Techniques of study and results of research," University of Indiana Press, 1968.

BIOENERGETICS

Introduction and Historical

The subject of bioenergetics is concerned with energy transformations in biological processes and involves application of the basic laws of thermodynamics to living systems. In an absolute sense, it is impossible to exclude any problems of biological structure or function from energy considerations, but for many specialized topics the reader is referred elsewhere in this book. The important area of photosynthesis is treated in another section. No coverage of biological problems involving radiant energy will be attempted here. This section will deal primarily with the principles involved in bioenergetics and with those topics generally or traditionally classified under this heading. Illustrations will be drawn for the most part from examples worked out in higher animals.

The earliest studies in the field of bioenergetics dealt with the measurement of heat production by intact animals or organisms. This approach was necessitated by the general lack of knowledge concerning the individual aspects of the life process. As biochemical and biophysical research began to reveal these component parts, they too become systems to be studied from the standpoint of energy changes. Physiological processes such as muscle contraction came under the scrutiny of biologists thermodynamically oriented. Calculations were made of the osmotic work performed in the passage of compounds through membranes. The equilibrium positions and free energy changes in many of the individual steps of intermediary metabolism were studied. Innumerable examples could be cited showing a steadily increasing maturity of the subject.

General Metabolism in Animals

When an organic substance such as glucose is oxidized to carbon dioxide and water, the reaction is accompanied by the liberation of energy. If the reaction is carried out under conditions of constant temperature and pressure and without the performance of useful work, the energy is liberated in the form of heat and is denoted by the symbol, ΔH. This is called the change in "enthalpy" or "heat content."

Growing out of the pioneer work of Lavoisier and Laplace in the 18th century, it has come to be recognized that the heat produced in the combustion of a given substance in the laboratory can be related to the heat produced by an animal, when the same substance is oxidized within the body. Let us suppose that over a period of time the total metabolism of the body results in the oxidation of an amount of foodstuff having a total energy content of E_1, and that during this same period the total amount of respiratory gases and excreta resulting from the metabolism of this material has an energy content of E_2. The increment of energy ($E_1 - E_2 = \Delta E$) must be balanced, according to the first law of thermodynamics, by the heat given off (Q) minus any work (W) performed on the environment.

$$\Delta E = Q - W \qquad (1)$$

The experimental condition can be simplified by having the subject perform no work, and in this instance the total energy change is equal to the heat given off. For the measurement to be valid, it must be assumed that no significant structural changes occur within the animal body during the experiment, and to actually carry out this procedure requires the knowledge that the gases and excreta collected result exclusively from the oxidation of the foodstuff in question and that no other heat-producing process is in progress. In other words, the initial body state and the final body state must be identical, and any intermediate forms of energy-change must have been converted to heat, in order to have the measurement of body heat relate directly to the ΔH resulting from the degradation of the foodstuff.

These theoretical implications lie behind the design of the clinically useful test known as the determination of the basal metabolic rate (BMR), which is a measurement of the rate of heat production in a subject under what is defined as "basal conditions." For a human being these conditions are those in which the subject is supine, motionless, has not eaten for a period of about twelve hours, and is in a room at 20°. The heat produced under these conditions reflects the energy requirements for the maintenance and conduct of the total tissue processes fundamental to the life of the organism under the defined conditions. The basal metabolism usually comprises about one half of the total energy expenditure of an individual, but can amount to much less in a very active subject carrying out heavy work. The thyroid gland secretes a HORMONE, thyroxin, which is an important factor in the regulation of the metabolic rate.

Nowadays direct calorimetric studies on intact animals are seldom carried out, but instead an indirect method is used. In this method oxygen utilization, carbon dioxide output and urinary excretory products are measured; and the amounts of carbohydrate, fat, and protein oxidized during the period of the experiment are calculated from this data. A summation of the theoretical amounts of heat produced for each of the separate metabolic oxidations gives the total heat production of the animal. In clinical practice the method is further simplified in that oxygen consumption alone is measured and is used as an index of heat production.

Energy Changes in Intermediary Metabolism and Cellular Functions

In the preceding section the body was treated as though it were simply a bomb calorimeter in which oxidation reactions are carried out with the production of heat. The restricted conditions under which this condition is approximated were discussed. In reality the life process is much more complicated, and, as is well known, the living organism is carrying out synthetic reactions, growing, reproducing, and is also performing work. Even in those processes in which the final "energy product" does appear eventually as heat, the overall reaction involves many individual steps some of which are energy-yielding and some of which require energy. The goal in this section will be to look more closely at the manifold energy changes that occur in intermediary metabolism and other functions as they occur at the cellular level. To facilitate this discussion certain additional thermodynamic relationships will be reviewed.

The second law of thermodynamics. The heat liberated in a chemical reaction at constant temperature under conditions in which no work is done (ΔH) is not generally equal to the useful work energy (Δf) that can be derived from the reaction. The relationship is given in the equation,

$$\Delta F = \Delta H - T\Delta S \qquad (2)$$

in which T is the absolute temperature and ΔS is a quantity known as the entropy change. $T\Delta S$ represents energy that is not available for useful work. By convention ΔF is negative when there is a "release" of free energy and ΔH is negative when heat is given off by the reaction. In a reversible reaction, $A \rightleftharpoons B$, the equilibrium constant, K, which is equal to $\dfrac{[B]}{[A]}$, is related to ΔF by the equation,

$$\Delta F° = -RT \ln K \qquad (3)$$

Here $\Delta F°$ is the standard free energy change and obtains when unit concentrations ($1 M$) of A and B are involved. $\Delta F°$ is the maximal useful work that can be accomplished, when one mole of A is converted to one mole of B. From these relationships it can be seen that for those reactions in which ΔF is a large negative number (exergonic reactions) the equilibrium will be far to the right. These are the reactions that are capable of occurring spontaneously without having net energy put into the system. On the other hand for reactions having a large positive value of ΔF (endergonic reactions) the equilibrium position will be to the left and these reactions would not occur spontaneously.

High-energy phosphate compounds. Intermediary metabolism involves an intricate blending of exergonic and endergonic reactions, and inasmuch as many of these reactions implicate organic phosphate compounds, it is essential to have an understanding of "phosphate bond energy." Following the exposition on this subject by Fritz Lipmann, it has been customary to divide phosphate bonds into two groups called "high energy" or "energy-rich" phosphate bonds and ordinary or "low energy" phosphate bonds. Those in the former group include anhydric and N-P bonds, and in the latter group are found phosphate ester and glycosidic bonds. Some of these types are illustrated by considering the structure of the compound adenosine triphosphate (ATP), which plays an important role in bioenergetics. ATP has the following shorthand structure:

$$A-R-P \sim P \sim P$$

where A = the purine, adenine; R = the pentose, ribose; and P = phosphate. Hydrolysis of either of the two terminal phosphate bonds is accompanied by a large negative free energy change in the range of 5,000 to 8,000 calories. These bonds are in the high energy class and are depicted schematically using the symbol, ($\sim$). The bond between ribose and the other phosphate group is an ester bond, ($-$), and its hydrolysis results in the release of much less energy. It shall be noted here that the terminology used in reference to high energy or energy-rich phosphate bonds is often extended to non-phosphate-containing compounds with groups whose hydrolysis results in an energy release comparable to that which occurs in the splitting of ATP; e.g., thioester bonds are energy-rich bonds.

Synthesis of ATP in biological systems. In the over-all degradation of foodstuff a considerable portion of the energy released resides for a significant period of time in the form of compounds containing high energy phosphate bonds. These bonds may later be split with the resulting energy being used for biosynthetic reaction, cellular work of one form or another, or simply further heat production. It has become abundantly clear that the synthesis and utilization of energy-rich phosphate compounds constitutes a process which is of utmost significance in the functioning of the living cell. In this respect, ATP, used as an example above, constitutes one of the major compounds of importance. In the utilization of ATP the terminal phosphate group is commonly split off to yield adenosine diphosphate (ADP), inorganic phosphate (P), and energy. Resynthesis of ATP requires the reverse of this process as follows:

$$ADP + P + Energy \rightarrow ATP \qquad IV$$

Metabolism furnishes the energy indicated in Equation IV.

The coupling of metabolism to ATP synthesis can be illustrated by considering the utilization of glucose by cells. This sugar is broken down through a series of enzymatic reactions, which may terminate with the formation of two moles of lactic acid under anaerobic conditions (anaerobic glycolysis) or which may proceed to the production of six moles of carbon dioxide under aerobic conditions. ATP synthesis occurs under either condition, although by far the largest amount is formed when oxygen is present. In the breakdown of glucose as far as lactic acid the value of $\Delta F°$ for this reaction is approximately $-50,000$ calories. Assuming that all of this energy were available for the formation of ATP from ADP + P, it would be possible theoretically to synthesize 6–10 moles of ATP (on the basis of 5,000 to 8,000 calories required for each phosphoric anhydride bond synthesized). In reality the process is not this efficient and only two moles of ATP are found in the conversion of glucose to lactic acid. In the aerobic metabolism of glucose to carbon dioxide and water the value of $\Delta F°$ for the reaction is about $-690,000$ calories. Here, in theory, the order of 100 moles of ATP could be synthesized, but from what is known, it appears more probable that 30–35 moles are actually formed. Energy not utilized for ATP synthesis appears immediately as heat.

In the anaerobic metabolism of glucose the exact mechanism for ATP synthesis is well known and reference to the section on anaerobic glycolysis will show the actual enzymatic reactions in which ATP is formed. In aerobic metabolism, which involves the tricarboxylic acid cycle and the electron transport system, the details of the mechanism are less well understood. In the latter instance it is known that these reactions are carried out within subcellular units known as mitochondria, and that ATP synthesis from ADP + P or AMP + 2P is coupled to the oxidative process.

Under certain conditions it is possible to uncouple oxidative metabolism from ATP synthesis, and at this time all of the energy released in the oxidation reaction will appear as heat. No ATP will be formed, and as a consequence this substance is not available for many cellular functions in which it is required (see below). One hypothesis in reference to the mechanism of action of the hormone, thyroxin, is that it controls the coupling of ATP synthesis to oxidation metabolism. Thyrotoxicosis is viewed as a condition in which there is excess heat production and too little ATP synthesis.

Utilization of ATP in intermediary metabolism and cellular functions. *Unidirectional Steps in Metabolic Pathways.* In a phosphorylation reaction in which an energy-rich compound is utilized with the formation of an energy-poor compound, the value of $\Delta F°$ for the reaction is usually a large negative number and the position of the equilibrium is far to the right (Equation 3). Many such reactions occur within cells and are catalyzed by specific enzymes; these reactions are often irreversible from a physiological standpoint, and provide steps that serve to prevent the reversal of a given metabolic pathway. Specific examples of this type of reaction are the phosphokinase reactions in which the terminal phosphate of ATP is transferred to an alcohol group with the formation of a phosphate ester. In muscle glucose is phosphorylated by ATP in the presence of glucokinase to give glucose-6-phosphate. Since this tissue has no phosphatase which would catalyze the splitting of this compound, the direction of "flow" of glucose is always into the cell.

Biosynthetic Reactions. Most biosyntheses are endergonic reactions and would not occur spontaneously. It is clear that among the macromolecule found in living cells, the proteins, nucleic acids, and polysaccharides all represent structures whose syntheses requires energy. The simple and complex lipids cannot arise spontaneously from their component parts. Even many metabolites whose structures are comparatively simple require energy in their formation. It is beyond the scope of this section to present detailed mechanisms for these syntheses, but in principle it has been found that the source of energy for most of these processes is ATP. In these reactions ATP is involved in the activation of molecules taking part in the synthetic reactions.

Maintenance of the Intracellular Ionic Environment. Energy is required to maintain the relative concentration of ionic components on either side of cellular membranes. In the polarized state the interior of a cell is electronegative with respect to the outside solution. Intracellular [Na$^+$] is less than extracellular [Na$^+$], while the reverse is true for [K$^+$]. There is good evidence that ATP furnishes the energy for the "sodium pump" mechanism which is involved in the polarization of cells.

Muscular Contraction. The specific role of ATP itself in the actual process of muscle contraction has not been entirely clarified, but there is no doubt that the ultimate source of energy for the mechanical work of muscle is ATP formed as a result of muscle metabolism. Within muscle a storage form of high energy phosphate, creatine phosphate, serves to replenish the ATP rapidly when it is needed, but this in turn must ultimately be replenished through energy released in the breakdown of carbohydrate or other metabolites.

EDWIN G. KREBS

References

Lippmann, F., "Metabolic Generation and Utilization of Phosphate Bond Energy," *in* Nord, F. F. and C. H. Werkman, ed., "Advances in Enzymology," New York, Interscience, 1941.

Pardee, A. B., "Free Energy and Metabolism," *in* Greenberg, D. M., ed. Chemical Pathways of Metabolism," 3rd ed. New York, Academic Press, 1968.

Klotz, I.M., "Energy Changes in Biochemical Reactions," New York, Academic Press, 1967.

BIOLOGICAL CLOCK

Many biological processes, both at the cellular and multicellular level of organization, undergo *regularly* recurring quantitative and qualitative changes. A whole spectrum of period lengths is associated with these variable processes; however, we will be concerned with only one, the period of about a day. One of the most fascinating aspects of these 24-hour biological rhythms is that they will *persist* for some time when the obvious environmental time cues—day-night illumination and temperature cycles—are precluded. The persistence of these rhythms under constant conditions is attributed to the existence of a "biological clock" resident within each organism which continues to time organismic processes.

Attempts at understanding the nature of the clock mechanism are primarily arrived at by studying the overt rhythms under a variety of experimental conditions and assigning the properties of the rhythm to the ruling clock. Briefly, the following characteristics have been elucidated: (i) the clock is innate; (ii) the period of a rhythm usually, but not always, deviates slightly from 24 hours (these rhythms are called *circadian*: a Latin contraction signifying *about-a-day*); (iii) the period is either unaltered, or only slightly altered by temperature; (iv) metabolic inhibitors, drugs, and other direct chemical treatments have little influence on the length of the period; (v) the phase of the rhythms is controlled by illumination or temperature cycles (e.g., maxima can be made to come at any time of the 24-hour day in the laboratory by offering light at the desired time (see ENDOGENOUS RHYTHMS). In constructing a hypothesis concerning the mechanism of the biological horologe, all of these properties must be considered.

There are two basic hypotheses (into which all models fit) as to the nature of the biological clock: (i) it is an autonomous clock that generates its own time intervals, and (ii) it is a clock which contains no independent timing mechanism, but simply indicates time from information fed to it. By way of analogy we can contrast the spring-driven wind-up clock with the sundial. The former, via the *escapement*, allows the energy stored in the spring to "escape" at regular intervals; the latter is subordinate to an external time indicator, the sun, and merely signals the time of day to an onlooker. Clearly, the escapement clock is a time *measurer* while the latter is a time *signaler*. We will discuss the latter first.

Directly or indirectly, due to the rotation of the earth on its axis, most geophysical forces undergo primary 24-hour variations in intensity, vector, etc. Many of these rhythmic forces—such as geomagnetism, cosmic radiation, gravity, etc.,—pervade standard laboratory "constant" conditions. Therefore, an organism with a non-escapement-type clock capable of responding to the rhythmic changes in one or more of these pervasive geophysical forces could easily continue to time its activities in so-called constant conditions. It is postulated that this clock (called the responder), which always runs at precisely 24 hours, is coupled to a secondary unit (the mediator), the latter of which directly drives the manifest rhythms. Between the two is a coupling mechanism that can be disengaged and engaged, thus enabling the organism to adjust the phase of its rhythm to new photoperiods encountered during longitudinal migrations or new light regimes produced in the laboratory. It would be expected that when an organism was placed in unvarying light and temperature in the laboratory, it would attempt to adjust to this new "photoperiod" by eternally uncoupling and coupling the mediator and responder (i.e., rephase) to adjust to its new regime, and in the process display an overt circadian (i.e., other than exactly 24 hours) frequency. Thus the coupling mechanism acts as a variable frequency transformer producing circadian frequencies as artifacts of the unnatural constant conditions. Because the ultimate timing information originates outside the organism in its environment, the temperature independence of the period and also its insensitivity to chemical perturbations would be expected. The weakness of this hypothesis is that no geophysical force has been shown to act as the sought-after time giver. However, forces such as geomagnetism, cosmic radiation, and the geoelectrostatic field (all omnipresent geophysical forces with 24-hour periodicities) have been demonstrated to have heretofore unsuspected influences on living organisms.

The escapement-clock hypothesis differs in stating that the primary clock generates its own time intervals, i.e., it is a pacemaker. Some investigators envision this clock as simply possessing all the requisite properties necessary to drive temperature-independent, phase-labile rhythms. A more widely accepted model involves a clock complex quite similar to the nonescapement one described above; in this model a secondary unit is also coupled to the ultimate clock and driven by it. By experiment and supposition, the period of the primary clock is defined as being temperature-independent (the temperature-compensating mechanism has not yet been defined); and the phase may be instantly set by light-dark cycles. The secondary unit cannot be rephased by light perturbations, but is temperature-sensitive; it drives the overt rhythms manifested by organisms. How the two are conjoined is unknown, but a coupling mechanism is a necessary postulate. It is felt that organisms, during their phylogenetic development on an earth replete with 24-hour geophysical periodicities, have evolved these clocks capable of measuring a period of approximately the same length. In the normal habitat, this clock complex is entrained to day-night light and temperature cycles, but in laboratory constant conditions it runs at its own inherited circadian speed of *about* 24 hours. In the search for these clocks investigators have traced them down through nervous and hormonal systems to the individual cell where they are now thought to reside. The escapement clock mechanism has not been deciphered.

It would be expected that the escapement-type clock must have a biophysicochemical basis, though the evidence for this is meager. A variety of chemical inhibitors, stimulants, drugs, etc., have been administered to both plants and animals in an attempt to alter the phase or period of their rhythms. Though the amplitude of several rhythms has been reduced to zero by some of these treatments, only a few real alterations in the period have been described. Whether or not these chemical perturbations affected the primary clock, the coupler, or the secondary unit is not known.

The above formal clock models, while accounting for all the properties of biological rhythms, fail to establish a clearly recognizable relationship to cellular components and their functions. An attempt to overcome this difficulty has resulted in a phenomenological model developed around cycles in messenger RNA transcription. It is postulated that within every cell are hundreds of *chronons*, each of which is an unbroken polycistronic complex of DNA, 200 to 2000 cistrons in length. Messenger synthesis begins at one end of the chronon (with the initiator cistron) and proceeds sequentially along the chronon to the terminator cistron. The transcription rate of the entire chronon is prolonged because of numerous intercistronic (and interoperonic) events, which stop the transcription of one cistron and initiate transcription on the next one. It is thought that to accomplish this after the transcription of each cistron, the nascent mRNA passes out to the cytoplasm, directs polypeptide synthesis, and that certain of these products diffuse back into the nucleus, initiating messenger synthesis on the subsequent cistron of the chronon. This diffusion circuit consumes much more time than the transcription steps, thus essentially governing the rate of the transcription of the entire chronon. As a consequence, the duration of the total process is controlled primarily by diffusion effects, thereby engendering the system with the necessary approximate temperature independence. With the transcription of the terminator cistron, a period of cytoplasmic protein synthesis ensues—including the synthesis of a *cistron initiator substance*. The initiator substance accumulates and diffuses back into the nucleus where it reactivates the initiator cistron of the chronon, completing the feedback circuit and starting a new one. Because of the length of the chronon—which has been arrived at through natural selection—the completion of the entire loop takes approximately 24 hours. The obvious weakness in this model is that it does not presently provide an explanation for the phase-setting ability of light and temperature cycles. On the other hand, it has the advantage of being constructed from tangible cellular components,

and therefore, many of its postulated attributes should be directly testable.

JOHN D. PALMER

References

Brown, F. A., Jr., "A unified theory for biological rhythms," *in*. Aschoff, J., ed., "Circadian Clocks," Amsterdam, North-Holland Publishing, 1965.

Ehret, C. F. and E. Trucco, "Molecular models for the circadian clock. I. The chronon concept," J. Theoret. Biol., **15**: 240, 1967.

Pittendrigh, C. S. and V. G. Bruce, "Daily rhythms as coupled oscillator systems and their relation to thermoperiodism and photoperiodism," *in*. Withrow, ed., "Photoperiodism and related phenomena in plants and animals," Washington, D.C., Amer. Assoc. Adv. Sci., 1959.

BIOLUMINESCENCE

Bioluminescence is the production of light by living organisms. This light results from an oxidative reaction in which the light-emitting molecule is called luciferin and the ENZYME, luciferase.

Luminescent organisms are much more numerous in the marine environment than on land and are notably rare in fresh water. They are especially abundant in the deeper layers of the ocean. Species displaying luminescence are extremely diverse. Simplest among these are the BACTERIA, which account for the glow observed on decaying fish or meat. Several species, in particular *Achromobacter fischeri*, have been used extensively in laboratory studies of bioluminescence. The hyphae and fruiting bodies of a number of FUNGI, *e.g.* the basidiomycete *Armillaria mellea*, glow continuously. The unicellular plants and animals include many marine luminous species, particularly among dinoflagellates (see MASTIGOPHORA) and radiolarians (see ANTINOPODA). The occurrence in abundance of luminescent dinoflagellates, for example *Gonyaulax* and *Noctiluca*, is the most common cause of the occasional brilliant displays of "phosphorescence" in the sea.

The COELENTERATES include a large number of luminescent species, especially striking being the sea pansies and the sea pens. Examples of luminous species from other animal groups are the comb jelly *Mnemiopsis*, the polychaet worm *Achloë*, the clam *Pholas*, the millipede *Luminodesmus*, the crustacean *Cypridina*, and among the insects, the fireflies and the glowworms. Luminosity in fish is of common occurrence, but in many cases is due to symbiotic luminous bacteria. No luminous amphibians, reptiles, birds or mammals are known.

In bioluminescence, the light may be produced within the cell, or enzyme and substrate may be extruded and react outside the organism. Intracellular luminescence may be continuous, as in bacteria, or may be of very short duration, as in the flash of fireflies or dinoflagellates. In the single-celled protozoans, luminescence appears to arise from discrete granules distributed throughout the protoplasm. In multicellular animals, luminescence is often restricted to special light-emitting cells or photocytes which are distributed in characteristic patterns over the body. Luminous organs quite comparable in structure to the eye have evolved in some forms. The photophore of the euphausid *Meganyctiphanes norvegica* comprises a cornea, a lens, and a retina-like cup of photocytes, backed by a reflective material. In animals which show extracellular luninescence, the luminous secretion arises from special glands. The secretory cells within the photogenic gland of *Cypridina* contain distinguishable yellow granules of luciferin and colorless granules of luciferase, both of which dissolve when discharged into sea water. Luminescence in multicellular animals is under nervous control, and shows fatigue, facilitation and other phenomena characteristic of nervous responses in general.

Although the significance of luminescence in many organisms is not apparent, the ability to luminesce clearly serves a useful function in others. It can provide a means for recognition and attraction between members of the same species, as in some fireflies, the females of which flash in response to the spontaneous flashing of the males in the proximity. Luminous secretions in certain deep-sea squids are believed to aid in concealment from predators. The luminous organs of deep-water fishes may function as lures for attracting prey, or by illuminating the surroundings, facilitate the search for food.

All of the energy emitted in bioluminescence falls within the visible spectrum, typically giving blue or blue-green colors. The wave lengths of maximum emission for some of the luminescent forms mentioned above are as follows: *Achromobacter fischeri*, 495 mμ; *Armillaria mellea*, 525 mμ *Gonyaulax polyedra*, 478 mμ; *Cypridina hilgendorfii*, 495mμ; *Luminodesmus sequoiae*, 495 mμ; and the firefly *Photinus pyralis*, 567 mμ. In living animals, the color of bioluminescence may be modified by pigmented screens external to the photophores. Such screens probably account for the red luminescence in the shrimp *Lycoteuthis diadema*. The intensity of luminescence varies widely, that of the ctenophore *Mnemiopsis leidyi* being among the brightest recorded for marine animals (0.3 millilamberts).

Bioluminescence is a special case of chemiluminescence, since light emission results from the decay of a molecule brought to the excited state by a chemical reaction. The general reaction for bioluminescence may be written:

$$LH_2 + \tfrac{1}{2}O_2 \xrightarrow{\text{luciferase}} L^* + H_2O$$

$$L^* \longrightarrow L + light$$

where L stands for luciferin, and the asterisk denoted the excited state. Luciferin and luciferase are general terms, since the chemical nature of these substances is different in different organisms.

Best known from a biochemical point of view is the bacterial luciferin-luciferase system, Here, reduced flavin mononucleotide (FMNH), in the presence of the enzyme, oxygen, and an aliphatic aldehyde containing more than seven carbon atoms, yields the light-emitting complex. Only the enzyme need be of bacterial origin.

Both the luciferin and the luciferase of firefly have been crystallized. In this system, adenosine triphosphate (ATP) is necessary for the activation of luciferin

(LH_2), forming an adenyl-luciferin complex as follows:

$$LH_2 + ATP + enzyme \xrightarrow{\text{Mg}} LH_2\text{-AMP-enzyme}$$
$$+ \text{ pyrophosphate}$$

$$LH_2\text{-AMP-enzyme} + \tfrac{1}{2}O_2 \longrightarrow$$
$$L\text{-AMP-enzyme} + H_2O + light$$

Adenyl-oxyluciferin (L-AMP) strongly inhibits light production, having a stronger affinity for luciferase than has reduced luciferin. Coenzyme A reverses this inhibition, forming a complex with oxy-luciferin and freeing the luciferase.

Other luminescent systems which are understood to a more limited extent are those of *Cypridina, Gonyaulax* and *Luminodesmus*. All these systems differ biochemically from one another and from those of firefly and bacteria. The low bioluminescence detectable in algae and higher plants immediately following exposure to light is believed to represent a reversal of photochemical reactions involved in photosynthesis.

Luminescence has been used as a tool in a number of ways. The luminescence of bacteria provides a very sensitive method for the detection of oxygen. The firefly may be used as an assay for ATP, DPN or coenzyme A. Since the intensity of luminescence provides an instantaneous measure of the rate of the luminescent reaction, luminescent bacteria have been used in kinetic studies of temperature and pressure effects. In *Gonyaulax polyedra*, luminescence shows diurnal rhythmicity and so has been used in studies of physiological clocks.

FRANCIS T. HAXO
BEATRICE M. SWEENEY

References

Harvey, E. N., "Bioluminescence," New York, Academic Press, 1952.
Johnson, F. H. and Y. Haneda, "Bioluminescence," Princeton, The University Press, 1966.
McElroy, W. D. and B. L. Strehler, "Bioluminescence," Bact. Rev. **18**: 177–194, 1954.
Nicol, J. A. C., *in* Nicol, J. A. C., "The Biology of Marine Animals," London, Pitman, 1960.

BIOME

A biome is a large COMMUNITY of living organisms recognizable by having a peculiar life form of its dominant vegetation and an associated group of characteristic animals. The biome is under the control of a particular set of climatic conditions; and its extent, in general, coincides with a distinctive type of soil. The concept has been extended by Clements and Shelford 1939 to include certain marine communities, but this application has not been widely accepted.

The term "biome" was first proposed by F. E. Clements before a meeting of the Ecological Society of America in 1916. The concept of the biome is actually an extension of the plant ecologists' older term "plant formation" to include the animals, for it has long been

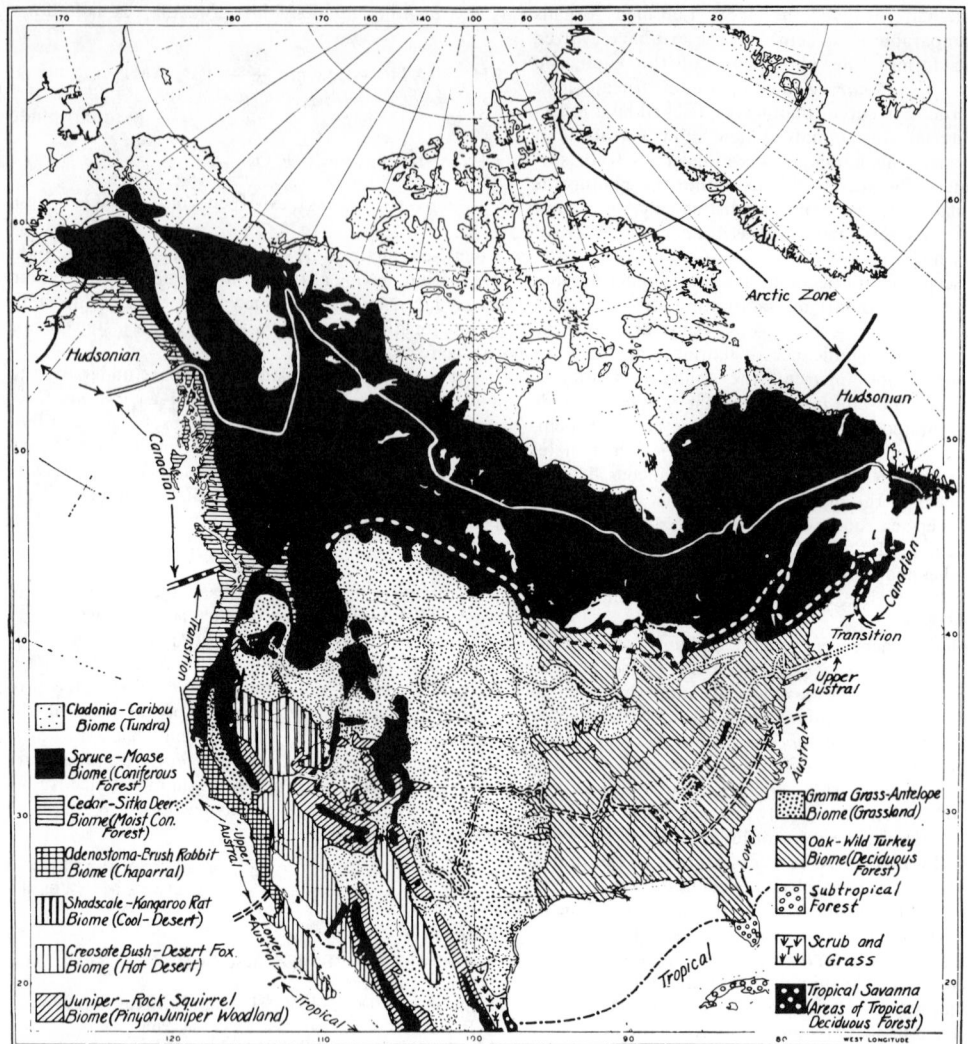

Fig. 1   Goode's Series of Base Maps, Henry M. Leppard, Editor. Prepared by V. E. Shelford, University of Illinois. Published by the University of Chicago Press, Chicago, Illinois. Copyright 1937 by the University of Chicago.

realized that while plants usually play a dominant role in land communities, animals nevertheless exert an important influence. The ecology of a community is an expression of both groups operating together.

It should be emphasized that biomes are large units often of the magnitude of the Great Plains grasslands of North America or the treeless arctic TUNDRA extending entirely across the continent. In a biome both the climax or mature stage and the developmental or successional stages of the community are included, with the climax stage being the basis for identification in the field. Biomes such as the arctic tundra and the northern coniferous forests which are relatively undisturbed by man may be easily recognized by the life form of their distinctive and dominant vegetation and by the presence of their larger and more characteristic animals.

Biomes of temperate climates have been so altered by man that today they show few of their pristine features and their primitive characteristics must be pieced together from historical information and from relatively small fragments of the original vegetation and animal life that remain today.

Owing to the irregularity of the topography and the concomitant variations in climate and soil, the biomes of western North America are smaller in area, more discontinuous, and less well defined. Tropical and subtropical biomes are even less well defined and understood.

There is never an abrupt transition from one biome to another but rather a zone of varying widths where there is a blending of the floral and faunal elements of the two. Such areas of transition are often called *eco-*

tones and are included on some maps of North American biomes (Pitelka 1941). A biome may be named for its predominant plant and animal such as the Grama Grass-Antelope Biome; or a term characterizing the life form of the dominant vegetation or a peculiar physical feature may be used such as Deciduous Forest Biome or Cool Desert Biome.

C. LYNN HAYWARD

References

Carpenter, J. R., "The Biome," Am. Midl. Nat., **21:** 75–91, 1939.
Pitelka, F. A., "Distribution of birds in relation to major biotic communities," Am. Midl. Nat., **25:** 113–137, 1941.
Shelford, V. E., "The relative merits of the life zone and biome concepts," Wilson Bulletin, 57: 248–253.

BIOPHYSICS

Biophysics is that branch of biology which attempts to formulate and study the properties of living systems in terms of the concepts of physics. This is not to imply that present day physics necessarily contains all of the principles necessary to account for all of the phenomena of biology. Indeed, should new physical theories be required to deal with some aspects of biology, the formulation of these theories would also be considered a part of biophysics.

Judging from the papers delivered at meetings of the Biophysical Society, contemporary biophysics may be divided into five overlapping categories:

1. Physiological Physics
2. Molecular Biology
3. Radiation Biology
4. Theoretical Biophysics
5. Instrumentation and Measurement

Physiological physics is the oldest branch of biophysics dating back to the mid 18 hundreds and the work of Carl Ludwig, Ernest von Brücke, Emil du Bois-Reymond and Hermann von Helmholtz. These workers resolved to "constitute physiology on a chemicophysical foundation." While the next 70 years of physiology did not necessarily stress the physical approach, the aim for physical explanation has always been a part of physiological doctrine.

The following fields indicate the scope of present day physiological physics and the relationship to physics:

(a) Hemodynamics, the study of blood flow, is a highly specialized case of hydrodynamics, involving the flow of a non-Newtonian fluid in complex elastic tubes.

(b) Conduction of electrical impulses by nerves is a problem of electro-dynamics. The complex character of the nerves also indicates that considerable knowledge of the molecular structure of nerves as well as a detailed understanding of electrolytes in solution is required to understand the physics of this problem.

(c) The conversion of chemical to mechanical energy by muscles and other organs of locomotion such as cilia poses problems in mechanics and thermodynamics. The most recent trend in this field has been to relate the contractile processes to detailed changes in the molecules constituting the fibrils.

(d) The problems of vision and photosynthesis are examples of systems involving the interaction of the electromagnetic radiation and matter. The development of the quantum theory of light has placed emphasis on such studies as quantum efficiency and threshold of perception. Molecular spectroscopy has played an important role in these fields.

(e) The study of hearing or perception of acoustical signals has long been a recognized branch of both the science of acoustics and sensory physiology. The classical fields of study have recently been augmented by the use of ultrasonics, both as a probe and as a perturbing agent in many studies.

(f) Electrocardiography and electroencephalography are important diagnostic tools for the physician. The distribution of electrical potentials in the body is currently being studied as a basic problem in physiology as well as mathematics and physics.

(g) Membrane phenomena in biology have been studied from a predominantly physical chemical point of view. One need only think of Planck's discussion of the diffusion potential, the Donnan equilibrium, Einstein's concern with osmotic phenomenon and Fricke's impedance measurements, to appreciate the essentially physical basis of membrane studies.

Modern molecular biology arises out of developments in four fields: genetics, biochemistry, physical chemistry of macromolecules and chemical physics. The advances in genetics provided the basic confidence in biology as an exact science. In addition genetic studies pointed to the importance of the material carriers of biological information. Biochemical studies revealed the nature of the organic molecules making up living systems. The composition of cells in terms of proteins, nucleic acids, lipids, etc., as well as the more detailed structure of these macromolecules and their constituent monomers are essential bases for studying the molecular physics of cellular systems. The nature of enzymatic activity and the existence of metabolic cycles and pathways also serve to provide the basis for further investigations. Studies on the size and shape of macromolecules by sedimentation, diffusion, electrophoresis, light scattering, dielectric dispersion and a variety of other methods have presented insight on the size and shape of biomacromolecules in solution. This knowledge is the essential basis for more detailed studies of molecular structure. The fourth class of studies which provides the basis for molecular biology deals with chemical physics. Based on quantum mechanical considerations these studies elucidate the nature of the chemical bond in terms of fundamental physical principles. These studies also make possible detailed considerations of the stereochemical details of chemical reactions.

Molecular biology poses two classes of questions analogous to the classical distinction between structure and function. The question of structure at this level is that of locating the atoms of a living system and specifying the relations between atoms. The question of function is a detailed examination of the molecular physics of the time dependent changes of these atomic systems.

The principal biophysical tools for the study of structure are microscopy, X-ray analysis and radioautography. Microscopy is carried out in the ordinary optical and phase MICROSCOPES, the ultraviolet and polarization microscopes, and in the ELECTRON MICROSCOPE. Optical and phase microscopy are useful for objects down to about 0.3μ in linear dimensions. Ultraviolet microscopy can examine somewhat smaller objects and is particularly useful with objects of distinctive ultraviolet absorption. Polarization microscopy is especially useful in looking at structures made of oriented macromolecules. Electron microscopy covers a wide range of sizes and is useful for objects ranging from several microns down to details near the limit of resolution of the instrument which appears to be around 7 Å. The necessity of using dried specimens in high vacuum imposes restrictions on the interpretation of results obtained by this method. Recently many advances have been made by combining electron microscopy with fractionating techniques, with radioautography and with the physical chemical methods of DNA chemistry.

The most detailed information about the position of atoms has come from X-ray studies. These investigations fall into two categories, X-ray scattering and X-ray diffraction. X-ray scattering gives information about the distribution of electrons and is particularly useful in structures that show some symmetry, such as spherical symmetry (Tomato bushystunt virus) or cylindrical symmetry (Tobacco mosaic virus). Research in this field has led to a convincing picture of how the protein and nucleic acid are arranged in these structures.

X-ray diffraction is in general carried out on crystals or fibres. The information obtained along with relevant chemical data on the structures enables the investigator to obtain a fairly precise picture of the positions of the atoms in the crystal or the fibre. Thus the protein α helix and pleated sheet and the DNA double helix structures have all been the result of X-ray diffraction studies.

Radioautography and the combination of radioautography with thin section techniques enables studies on the location and sites of synthesis of specific biological molecules. Since the isotopic tracer used can be introduced in a wide variety of compounds, these methods provide a link between biochemical and cytological information. The most successful isotope used to date has been tritium (H_3). It emits a short range beta particle and thus permits high resolution (the order of a micron). Almost all biological molecules contain hydrogen in relatively stable positions, so that the isotope can be used in many different kinds of studies. Of special note have been the recent detailed pictures of chromosomal replication made possible by the use of tritiated thymidine.

The studies of molecular process have largely involved the use of isotopes both stable and radioactive. Both kinetic aspects as well as the elucidation of biochemical pathways have been so carried out. A distinction has been drawn between stable structures and those undergoing turnover. Intensive efforts are now underway to obtain a detailed picture of the biosynthesis of the macromolecules, protein, deoxyribose nucleic acid and ribose nucleic acid.

Radiobiology has been associated with physics since the origin of the subject. Since early X-ray machines as well as radioactive isotopes were the possessions of physicists the historical association is an obvious one.

Contemporary radiobiology seems to concentrate in three directions.

1. The biological effects of various types of radiation on living systems.

2. The study of the detailed physical changes accompanying the interaction of radiation and biomatter.

3. The use of radiation as a perturbing agent to study biological systems.

The first phase of radiobiology is a phenomenological one to ascertain the quantitative relations between type and dose of radiation and specific biological effect. In addition to the intrinsic interest in this field research has been spurred by practical demands from radiology and health physics. Both of these disciplines require as detailed information as possible on the biological effects of radiation.

The second subdivision of radiobiology represents an attempt to provide rational physical and chemical explanations of the phenomena studied in the first subdivision. Pertinent questions involve the nature of the ionization and excitation processes and the distribution of these events in the irradiated material. It is necessary to distinguish between direct effect (ionization in the molecule under consideration) and indirect effect (chemical process resulting from the ionization of a neighboring molecule). This branch of radiobiology is rooted very firmly in physical analysis.

In radiation damage of living systems, a small amount of energy input causes a large biological effect. This fact in addition to the random nature of the distribution of ionization makes ionizing radiation a unique type of perturbation to study biological systems. The general philosophy behind this approach is as follows. Ionizing radiation represents an input and biological effect represents an output from a biological system. From a study of input-output relations plus some knowledge of the intermediate system, additional information about the intermediate system can be obtained. Using this approach as exemplified by target theory it has been possible to obtain information about the size and shape of the structures responsible for assayable biological functions.

Theoretical biology (see MATHEMATICAL BIOLOGY and THERMODYNAMICS OF LIVING MATTER) may be divided into four fields of interest: systems theory, application of computers, classical mathematical biology and the application of theoretical physics to biology. The recent developments of information theory, game theory, cybernetics, operations analysis and computer techniques have spurred biologists to consider the systems aspects of living organisms. Such diverse fields as cellular physiology (systems of inter-related chemical reactions) and neurophysiology (nerve networks) have been viewed from one or more of the system theory approaches. The field is still in its very formative stages.

Computer techniques, both analog and digital, have been applied to many biological problems. Many biological functions have been analogued by electrical elements and networks of those have been studied. Blood kinetics and the distribution of surface potentials on the human body are examples of problems that have been approached by analog methods.

Digital techniques have been used for solving specific problems as well as investigating general biological situations. The broader implications of computers and biology have been discussed by von Neumann in the monograph, "The Computer and the Brain." The computer field has opened the philosophical question of the relation between automata and living systems.

Classical MATHEMATICAL BIOLOGY is represented by the work of such investigators as A. Lotka and N. Rashevsky. The general approach is to formally state a number of postulates, about a given biological situation, in mathematical terminology and then to use the techniques of mathematics to deduce the consequences of the postulates and to compare them with observed data. This approach has been applied to many branches of biology ranging from ecology to cellular metabolism.

The application of theoretical physics to biology includes those attempts to apply kinetic theory, thermodynamics, statistical mechanics, quantum mechanics, electric field theory, mechanics and other formal physical disciplines to biological problems. While such a direct link between physics and biology has not been a very strong one in the past indications are, from the theory of membrane potentials and applications of irreversible thermodynamics, that studies of this type are being pursued more frequently.

Since precise measurement has always been important in many branches of physics, it was quite natural for this activity to carry over into biophysics. While it is quite impossible to catalogue all the contributions of biological instrumentation, particular note may be made of the use of electronic circuits in neuro- and muscle physiology and the use of radioactive counting equipment in almost every branch of biology.

Biophysics is thus clearly seen to be a borderline field bridging the gap between the biological and physical sciences. The approach of biophysics has been carried into almost every branch of biology and at the same time almost every branch of physics has contributed some of the working material. At present the combined approach appears to be a very fruitful one.

The ultimate in the approach of biophysics would be to write down an equation (if present theory is adequate, it would be Schrodinger's equation) and boundary conditions for a given living system. The solution of the equation subject to the boundary conditions would then predict the future behavior of the organism. In the absence of this idealized approach (and we now seem to be very far away) two general philosophical approaches are possible. We might start on the one hand from our knowledge of nuclei and electrons, proceed to the formation of chemical bonds and understand the physics of small molecules. We might next investigate the polymers of these small molecules and proceed in an attempt to conceptually synthesize a cell and an organism from its atomic constituents.

We might of course start from the other point of view, the phenomenological observations of living systems, and attempt to analyze each of the functions in terms of the active structures and the physical principles exhibited by the functions. Both approaches are part of the discipline of biophysics, and constitute the working assumptions of the field.

In concluding, such a broad field of biophysics can only be made clear by a specific example. To do this we shall present a biophysicist's view of a living cell. Our choice will be a hypothetical bacterial coccus 0.5μ in diameter. Such a cell has an area of 7.9×10^{-9} cm^2, a volume of 6.5×10^{-14} cm^3 and weighs 6.9×10^{-14} gms. Of this total weight 75% is water and 1.72×10^{-14} gms is dry weight. Of the dry weight 53% is protein, 16% carbohydrate, 15% ribonucleic acid, 10% lipid, 3% DNA and 3% small molecules. The constituent monomer and small molecular weight units are shown in Table 1 and the atomic composition is given in Table 2.

Structurally the cell consists of a polysaccharide-peptide cell wall, which is a rigid structure with a fairly

Table 1   Molecular Building Blocks of a Typical Small Bacterium

Alanine	18.54×10^6
Arginine	10.30×10^6
Aspartic Acid	12.36×10^6
Asparagine	8.24×10^6
Cysteine	4.12×10^6
Diaminopimelic Acid	2.06×10^6
Glutamic Acid	14.42×10^6
Glutamine	8.24×10^6
Glycine	16.48×10^6
Histidine	2.06×10^6
Isoleucine	10.30×10^6
Leucine	16.48×10^6
Lysine	16.48×10^6
Methionine	8.24×10^6
Phenylalanine	6.18×10^6
Proline	10.30×10^6
Serine	12.36×10^6
Threonine	10.30×10^6
Tryptophan	2.06×10^6
Tyrosine	4.12×10^6
Valine	12.36×10^6
Hexose	4.20×10^7
Ribose	1.83×10^7
Deoxy Ribose	3.93×10^6
Thymine	9.82×10^5
Adenine	5.70×10^6
Guanine	5.70×10^6
Cytosine	5.70×10^6
Uracil	4.72×10^6
Glycerin	5.7×10^6
Fatty Acid	1.14×10^7
Phosphate	2.79×10^7
Potassium	2.20×10^6
Sodium	4.40×10^6
Calcium	1.10×10^6
Magnesium	2.20×10^6
Manganese	1.10×10^6
Silicon	1.10×10^6
Iron	1.10×10^6
Assorted Small Molecules	8.50×10^6

Table 2   Atomic Composition of a Typical Small Bacterium

Carbon	173×10^7
Hydrogen	275×10^7
Nitrogen	37.1×10^7
Oxygen	75.0×10^7
Sulfur	1.66×10^7
Phosphorus	3.22×10^7
Potassium	$.220 \times 10^7$
Sodium	$.440 \times 10^7$
Calcium	$.110 \times 10^7$
Magnesium	$.220 \times 10^7$
Manganese	$.110 \times 10^7$
Silicon	$.110 \times 10^7$
Iron	$.110 \times 10^7$

high degree of crystalline order. The wall is approximately 200 Å thick. Within the wall is a protoplast membrane consisting of a bimolecular layer of lipoprotein and approximately 100 Å in thickness. The DNA is coiled up in a central structure about .25μ in diameter. The interior of the cell contains about 5,000 ribosome particles ranging from 100 Å to 200 Å in diameter. These particles are approximately half protein, half RNA. There are about 150,000 protein molecules in solution in the cell and most of these are enzymes. There is also some soluble RNA and about 850,000 small molecules, mostly metabolic intermediates and metabolic waste products. In addition there are about 5,000,000 ions.

The assigning of numbers to the above quantities stresses the quantitative aspect of the biophysicist's view. The molecules and particles in the cell are not randomly distributed. Interactions such as the Gibbs absorption effect, the Donnan equilibrium, the Debye-Hückel effect and the electrostatic interaction of charged structures, all serve to maintain a statistical order in spite of the ceaseless thermal motion. As a result certain molecules are concentrated near membranes (interfaces) and others are concentrated elsewhere. The cell in addition has a net negative charge of several hundred electron units. A number of computations can be made regarding the cell. These include the information content (entropy of information), the heat of formation, the effects of Brownian motion and other thermodynamic and kinetic parameters describing the cell.

The picture presented is essentially a static one, and the next step is a physical description of macromolecular synthesis, membrane growth, enzymatic action and cell replication. At this level the view becomes very incomplete since our knowledge of the processes is very incomplete. One branch of biophysics is thus actively engaged in research to learn more about these processes.

HAROLD J. MOROWITZ

References

Quastler, H. and H. J. Morowitz, Eds., "Proceedings of the First National Biophysics Conference," Yale, The University Press, 1959.
Oncley, J. L., et al., Eds., "Biophysical Science," New York, Wiley, 1959.

BIRD see AVES

BIRD SONGS

Birds are the most musical of wild animals. A bird song is a vocal performance, usually confined to the male and to a definite season of the year, that season including the time of mating. Singing is most abundant in spring, continued through the early summer, ceases abruptly during the time of post-nuptial molt, in late August or early September, and is renewed, to some extent, in the fall. Some species sing occasionally in the winter. In northern United States, Song Sparrows and Eastern Meadowlarks have been heard in every month of the year.

Birds also produce call-notes and alarm-notes that should not be confused with song. These are not seasonal and are used by both sexes. Female birds sometimes sing, but the song is weaker, less frequent, and likely to be only during the mating period.

Each species of bird has a distinctive song. When the song is known, the bird can be named accurately without being seen. In some birds, such as the tyrant flycatchers, the songs of individuals are alike. One phoebe's song is just like that of another. But in other species individuals vary. The song is recognizable as to species but individuals differ in details. Each individual may have one or more different songs. An individual Field Sparrow has but one song. A Song Sparrow has six or more. An Eastern Meadowlark has been known to sing fifty-three different songs in less than an hour. Every Meadowlark can sing any song it hears another Meadowlark sing, and the total number of Meadowlark songs on record is more than a thousand.

According to the territory theory (Howard, 1915) a male bird, arriving north in the spring migration, finds a place in the type of habitat that its species occupies, selects a perch, and sings from that perch regularly. The song is a warning to other males of its species to keep away from that territory. At the same time it is an invitation to a female bird to become his mate.

According to this theory, the male bird ceases singing when mating is established, and birds do not sing while migrating. This does not apply in many cases, at least with American birds. While the Hermit Thrush is silent while migrating, the Olive-backed Thrush is not. Most of our warblers sing on migration, and many sing during the nesting period, but have a different type of song at that time.

There are, in our passerine birds, two types of song—short songs, frequently repeated, and long-continued songs. Short songs vary in length from a fraction of a second in Henslow's Sparrow to eight seconds in the Winter Wren. Long-continued songs are indefinite in length; a series of varied phrases arranged in different orders. Such songs are to be found chiefly in the Mockingbird, Thrush, Shrike and Vireo families. Anyone who lives where there are trees, shrubs, and green lawns may hear such singing in the carolling of the Robin each early spring morning.

Birds that live in open areas, prairies or other grass lands, are likely to do most of their singing in flight. Larks, pipits, longspurs, the Lark Bunting and the Bobolink are such flight singers. Longspurs fly up silently, set their wings at an angle, and float back to the ground singing. Sprague's Pipit sings high in the air, circling around in rising and falling flight, silent when rising and singing when falling, but often so high in the air that we see the falling flight begin before the sound of the song reaches our ears. When one first hears this song, it is confusing to tell where it comes from.

A flight singer of the western mountains that is not well-known is Townsend's Solitaire. This bird sometimes hovers in one spot and sings continuously for as much as fifteen minutes. The song is clear, loud and gloriously melodious, and the bird one of our finest singers.

Bird songs are not always musical. The territory theory requires only that they be distinctive noises. Many songs are highly musical but others are not. The voices of shrikes are harsh. The Yellow-headed Blackbird squawks when it sings. The Clay-colored Sparrow has a song of harsh buzzes. Even the Mockingbird sings occasional notes that are mere noises.

The Yellow-breasted Chat is a long-continued singer. The song consists of squeaks, squawks, rattles, cackles, and retarded series of low whistles. At times it sings in flight, and the flight is as ludicrous as the song: the wings flapping like those of a big moth, the tail pumping up and down, and the legs dangling below. The entire performance is clown-like.

Birds that nest on the ground in woods, thickets or tall grass ordinarily sing simple songs from perches, but occasionally they sing what have been termed ecstatic flight songs. These songs generally begin with two or three chips, followed by a wild jumble of notes that, without other notes, would be unrecognizable. But in the midst of such songs there usually occur a few phrases of the normal song, then a short pause, as if the singer were catching its breath, then a few more jumbled notes as it drops back to the ground. Birds that sing such songs are the Ovenbird, Yellowthroat, and Mourning and Canada Warblers.

The Eastern and Western Meadowlarks, whose ordinary songs are quite unlike, sing flight songs that are like these ecstatic songs, and have a likeness to each other that suggests that such songs are primitive and originated before the two species became distinct. It may be that all ecstatic flight songs are such.

In one family, the tyrant flycatchers, a special form of song occurs that is of particular interest. This is the twilight song. That of the Wood Pewee has been subjected to a cooperative study practically throughout its range in the eastern half of the United States (Craig, 1943).

This song is to be heard from June to August in the very early morning hours. It begins in practical darkness and ends with the sunrise. It seems to have no relation to the nesting cycle. It consists of three phrases sung over and over in various orders. Two of these may be heard commonly in the daytime. These are numbered 1 and 2, and the third phrase, that belongs to the twilight song alone, is numbered 3. Phrase 2, "peeoh" has an air of finality and is least numerous in the twilight song. So the songs were studied in groups of phrases ending in phrase 2. The commonest of all of these groups was 3132.

The significance about this order of phrases is its similarity to man's music. One observer compared it to *Swanee River* and another to *Home, Sweet Home*. In this comparison a simple phrase of the pewee compares with a whole line of music in human tunes. The Stephen Foster songs are of this form. Many hymns are also like it. Even the great composers used it, as in the second half of Brahms' *Lullaby* and the singing portion of Beethoven's *Ninth Symphony*.

There are many other ways in which bird songs are like human music. The intervals in White-throated Sparrow songs are perfect fourths and major thirds. Ruby-crowned Kinglets frequently drop an exact octave between the first and second parts of its song. Wood Thrushes' phrases are frequently on the notes of the major tonic chord. Three-note phrases from Rossini's *Overture to William Tell* are practical duplicates of certain Wood Thrush phrases. Meadowlark songs occasionally are like phrases of human music, one such being a bit from the *Song of the Volga Boatmen*, and another like the first line of a hymn, the words being "There is a happy land."

While passerine birds are the ones we commonly think of as song birds, there are non-passerine birds well-known for their songs, and even named for them, as in the Bob-white, the Whip-poor-will, and the European Cuckoo. Other non-passerines substitute other sounds for song. The Woodpeckers drum, and the Woodcock and the Snipe produce sounds with special feathers of wings or tail. The Ruffed Grouse drums, using its wings to make the sound. These performances are just as seasonal as are true songs.

Bird songs differ in five characters—pitch, time, loudness, quality and phonetics. Pitch and time can be measured with reasonable accuracy. Loudness is relative. We can tell which part of a song is the loudest, but distance, wind and interfering objects make exact measurements impossible. Quality is indefinite. We can use terms like clear, sweet, harsh, strident, etc. but they are not exact. Phonetics can be represented by words, but they do not seem to sound the same to two different people. For example, two observers describe the Song Sparrow's call-note, one as 'chink' and the other as 'tsack.'

Early attempts were made to record songs by musical notation. This applies well in some cases, but not all, for not all birds sing on notes of the diatonic scale. The best known of these attempts is Mathews (1904).

A method has been devised by using horizontal lines to represent the notes. The vertical position of the lines represents pitch, and the horizontal length time. The heaviness of the lines represents loudness. The quality is written above the record, and the phonetics under each note. This was first made known in 1915, but is better represented today in a later publication (Saunders, 1951).

About 1932 Albert Brand, at Cornell, began making phonograph records of bird songs. By studying these records with a microscope, he worked out the vibration frequencies in passerine bird songs. He determined that they ranged from 1100 per second in the Catbird to 10225 in the Blackpoll Warbler. This latter is more than an octave higher than the highest note on the piano.

Records are now being made with tape recorders. Phonographic records of these are on the market. Some of them, however, ignore the fact that songs vary in most species, and a single record is treated as the song of the species.

There seems to be no reason, from the standpoint of biological necessity, why bird songs should be musical. One writer has stated that bird songs tend toward an ideal. Certainly the world is a better place to live in because there are bird songs. Why is man fond of music? How did he come to devise a system of musical notation or to write both simple songs and great symphonies? We have different tastes in music, some preferring popular and others classical. It seems to be the same with birds, for their music ranges from the sublime of the Hermit Thrush to the ridiculous of the Chat.

ARETAS A. SAUNDERS

References

Howard, H. Eliot, "Territory in Bird Life," New York, Dutton, 1920.
Craig, Wallace, "The Song of the Wood Pewee," New York State Museum Bulletin No. 334. 1943.
Mathews, F. Schuyler, "Field Book of Wild Birds and their Music," New York, Dover, 1966.
Saunders, Aretas A., "A Guide to Bird Songs," New York, Doubleday, 307 pp. 1951.

BLAINVILLE, HENRI MARIE DUCROTAY DE (1777–1850)

This French zoologist was born at Arques where he developed an interest in, and showed much talent for painting. To develop this talent, he went to Paris in 1796 where he fell under the spell of Cuvier, by whom he was trained in zoology and comparative anatomy. He joined the ranks of Lamarck and for this reason forfeited his friendship with Cuvier. In 1832, however, he succeeded the latter in the chair of comparative anatomy in the University of Paris. His major contribution to biology is his demonstration, through the scope of his writings, that all branches of the biological sciences were interconnected. He published major reference works in the fields of taxonomy, osteology, physiology, and a monumental "Faune Française" (1821–1830). It is as much to him as to Lamarck that the theory of the inheritance of acquired characters took such a firm hold on the minds of his contemporaries.

BLOOD, INVERTEBRATE

The blood of invertebrates, like that of vertebrates, carries oxygen to the tissues, removes carbon dioxide, and transports nutrients, hormones and the waste products of nitrogen metabolism. Invertebrate blood is generally less complex than vertebrate blood in composition, containing a smaller variety of proteins and of formed elements. Leucocytes, many with the amoeboid movement characteristic of phagocytes, constitute the only formed blood element in the majority of invertebrate species. The respiratory pigment, if it is present, is most often carried in solution in the plasma.

The relatively simple composition of invertebrate blood is compensated by the remarkable variety of methods for oxygen transport. The vertebrates all possess one class of respiratory pigment, the hemoglobins, which, despite variations in amino acid composition, are uniform in molecular weight and apparently identical in active site configuration in all vertebrates. An invertebrate animal, on the other hand, may make use of hemoglobin, chlorocruorin, hemerythrin or hemocyanin, or it may have no pigment of any kind. The mode of oxygen transport varies within phyla and even within classes, and within each group of pigments there is great variation in molecular size and shape. A classification of invertebrate blood according to the nature of the respiratory pigment thus transects the classical phylogeny.

The most striking example of this fact is hemoglobin, which is the dominant pigment of the annelids, and occurs also in certain arthropods (*Daphnia, Chironomus*), molluscs, echinoderms, nemathelminths and platyhelminthes. Invertebrate hemoglobins (erythrocruorins) vary in molecular weight from about 17,000 to 2,750,000, but all share with vertebrate hemoglobin the same prosthetic group, and the same molecular weight of protein per iron (17,000). The variations in the globin portions of the molecule have a profound effect upon physiological properties, so that while some invertebrate hemoglobins behave much like the vertebrate pig-

ment, others have half saturation pressures so low, or contribute such a small amount of oxygen-carrying capacity to the blood, that their role *in vivo* is not clear. In certain cases the hemoglobin is apparently used principally as an oxygen storage mechanism for temporary survival under anoxic conditions. Some species of *Chironomus* larvae have hemoglobin and others do not, so that in this case the pigment is clearly not essential to survival. Most insects have no respiratory pigment at all, but rely upon oxygen dissolved directly in the blood; a great many other invertebrates which do have pigments probably rely nonetheless upon dissolved oxygen as their principle source.

The only other well established oxygen carrying pigment employing a porphyrin prosthetic group is chlorocruorin, a green pigment limited to the polychaetes (*Sabella, Serpula*). It is closely related to the large molecular weight invertebrate hemoglobins in properties, differing in the substitution of a formyl group for a vinyl group in position 2 of the protoporphyrin molecule. The Fe to O_2 ratio, as in hemoglobin, is one.

The hemocyanins represent the only established departure from the use of iron in oxygen-carrying pigments. The hemocyanins are copper proteins, deep blue in color, found in a great variety of arthropods and molluscs. The hemocyanins range in molecular weight from about 400,000 (*Palinurus*) to 5,000,000 (*Helix*). In general, the hemocyanins of smallest molecular weight have been found in the crustaceans, those of intermediate size in the cephalopods and in the arachnoids, and the largest in the gastropods. These values are the upper limits of size for the species; the large molecules are capable of reversible dissociation into smaller subunits in a manner which depends upon the pH, ionic strength, protein concentration and nature of the ions in the medium. The subunits are all capable of carrying oxygen. The protein molecular weight per copper is 36,500 in the arthropods and 25,000 in molluscs. In fact, the smallest subunit found is about twice this size.

There are no heme or other porphyrin groups in hemocyanin. The copper appears to be bound directly to one or more amino acid side chains in the protein, and one oxygen molecule is bound for each two copper ions present. The oxygen-copper complex probably involves a pair of copper ions bridged by an oxygen molecule: Protein-Cu-O-O-Cu-Protein. It is well established that the copper of deoxygenated hemocyanin is in the cuprous state, while in oxyhemocyanin the copper possesses part cuprous and part cupric character, a chemical situation which would be expected to result in the intense blue color observed in the pigment.

The oxygenation curves (% oxygenated pigment vs partial pressure of oxygen) of the hemocyanins more or less resemble those of the hemoglobins. Detailed analysis suggests that there is considerable interaction among oxygen-binding sites, and the oxygenation curves vary greatly with conditions of ionic strength and acidity. In many cases, increasing acidity tends to decrease the affinity of hemocyanin for oxygen in a manner analogous to, but greater than, the Bohr effect in hemoglobin, but in others the effect is reversed (*Limulus, Helix, Busycon*) so that increasing acidity increases oxygen affinity. As in the case of invertebrate hemoglobins, the oxygen affinity of hemocyanins (*Homarus*) under physiological conditions, and their total oxygen capacity (*Limulus, Helix*) in the blood are in some species so small that it is unlikely they play a major role except in oxygen storage, but in the more active species (*Loligo*) hemocyanin is comparable in efficiency to vertebrate hemoglobin.

The brown respiratory pigment hemerythrin, though it is an iron protein, appears to resemble hemocyanin more closely than hemoglobin in its active site structure. There is no heme group in hemerythrin; fairly conclusive evidence exists that the iron is bound to protein sulfhydryl groups of cysteine, and again it is likely that the active metal ions are located in pairs. The earliest studies of hemerythrin indicated an $Fe:O_2$ ratio of 3:1, but more recent work suggests that one third of the iron atoms in the material used were attached to inactive sites, and that the proper combining ratio of $Fe:O_2$ is 2:1, the same ratio found in hemocyanin. The molecular weight of hemerythrin from *Sipunculus nudus* is 66,000. Hemerythrin has a limited distribution in nature. It is found in brachiopods, sipunculids, priapulids, and certain polychaetes (*Magelona*). There is considerable variation of the oxygenation curve and the pH dependence of oxygenation among the species.

Brief mention must be made of the blood pigment of *Ascidia* and other tunicates, which contains vanadium and has been called hemovanadin. The properties of this pigment have not been thoroughly investigated; there is some question concerning its ability to carry oxygen. The manganese containing pigment, pinnaglobin, obtained from the blood of *Pinna squamosa,* occupies a still more doubtful position in the list of oxygen-carrying pigments.

The ability of blood to transport carbon dioxide is related to the buffering capacity of the blood. In vertebrates there is a great variety of blood proteins to provide the buffering action; in invertebrates this function rests principally with the respiratory pigment, which is in most cases the principal blood protein. The invertebrate respiratory proteins in general possess adequate buffering capacity to fulfill this function. Since they are the principal blood proteins, the respiratory pigments, in these cases where they are dissolved in the blood, are also responsible for maintenance of osmotic balance of blood fluid with respect to the environment. In animals with closed circulatory systems they contribute to the counterbalancing of blood pressure, permitting retention of water within the circulatory system. In the invertebrates possessing hemoglobin a closed circulatory system is generally associated with hemoglobin of high molecular weight dissolved directly in the plasma, while an open system most often is accompanied by low molecular weight hemoglobin in corpuscles. The pigment hemerythrin is found only in corpuscles, while chlorocruorin and hemocyanin are found only in the directly dissolved form.

The blood of invertebrates provides a demonstration of the diverse ways in which an organism can solve its problems of adaptation and survival within a relatively limited framework of chemical and physiological structures.

GARY FELSENFELD

References

Prosser, C. L. and F. A. Brown, "Comparative Animal Physiology," 2nd ed., Philadelphia, Saunders, 1961.
Manwell, C., *Ann. Rev. Physiol.* **22**, 191 (1960).

BLOOD, VERTEBRATE

The blood of the vertebrates is a fundamental tissue in which the blood corpuscles are the cellular elements and the plasma is the intercellular matrix. The cells are specialized to carry out certain functions of about the same nature throughout the phylum, consequently information gained from the study of one group may be used as a basis for generalization, within certain limits. The cells are continually being worn out and replaced by new cells derived from the hemopoietic organs in the higher vertebrates and from hemocytoblasts in the blood stream in the lower vertebrates. According to their morphology and tinctorial reactions to standard dyes, the blood cells are classified into: (1) the erythrocytes or red blood cells, which contain a respiratory pigment, hemoglobin; (2) the leucocytes or white blood corpuscles which are motile and colorless; and (3) the thrombocytes or blood platelets, which are specialized functionally to participate in the clotting of the blood. The plasma contains in suspension, enzymes, hormones, and all the substances necessary for the metabolism of the tissues.

In all the vertebrates except mammals, practically all the erythrocytes in the circulating blood are nucleated. In mammals, the non-nucleated erythroplastid is derived from a nucleated ancestral cell, the erythroblast. Historically, as the erythrocytes were studied, methods of measurement were invented by which the number of cells per cu mm could be counted; the amount of hemoglobin per 100 cc of blood measured; and the volume of red blood cells per 100 cc of whole blood (Hematocrit) could be estimated. By combining these date certain information known as the mean corpuscular volume, the mean corpuscular hemoglobin, and the mean corpuscular hemoglobin concentration may be calculated and quantitative comparisons between species as regards hemoglobin function may be compared.

The leucocytes have been classified by their granular content, and the tinctorial reactions of these granules in special blood stains, such as Wright's or Giemsa's stain, which are widely used by hematologists. Because of their diverse contents, the leucocytes, do not lend themselves to mass study as do the red blood cells, although the count per cu mm is often a source of pertinent information regarding physiologic change. In this brief essay on vertebrate blood, the following general classification of the leucocytes will be based upon their appearance in blood smears stained with Wright's stain. The reader desiring more detailed information is referred to the classic descriptions by Jordan.

In this classification: (1) the *hemocytoblast,* found in the blood of the lower vertebrates, is a large or small cell, with a sieve-like nucleus in which the chromatin is distributed throughout the nucleus in the form of fine granules; one or more nucleoli are present; and the cytoplasm may be deeply or lightly basophilic and contain vacuoles, but it is nonspecific. (2) *Lymphocytes,* large, medium and small spherical cells, with a nucleus in which the chromatin is spread out in the form of blocks; a nucleolus may be present; and there is a rim of more or less basophilic cytoplasm which may contain azurophilic granules. (3) *Granulocytes,* in which the cytoplasm is full of fine granules which usually stain only faintly or not at all, the so-called *neutrophils or hetero-*

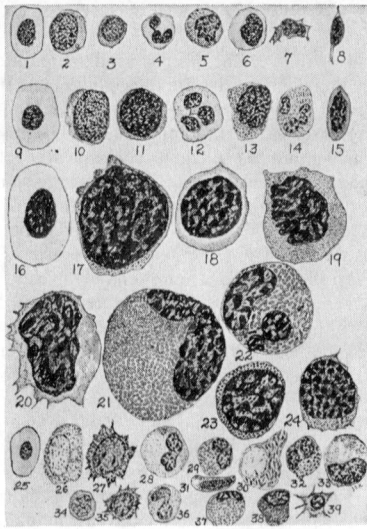

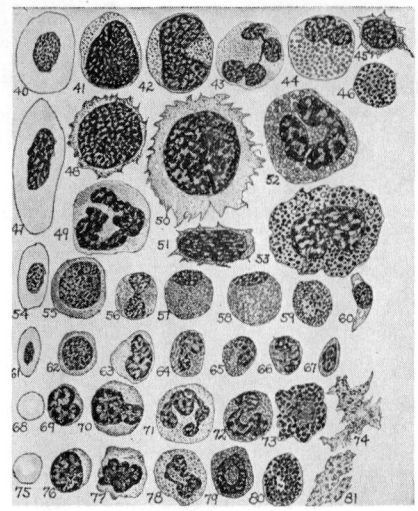

Fig. 1   Blood smears stained with Wright's stain of following Vertebrates: 1–8, Myxine glutinosa; 9–15, Mustelus canis; 16–24, Protopterus ethiopicus; 25–33, Amia calva; 34–39, Cyprinus carpio. All magnifications before reduction 1800 × except Cyprinus, 1300 × .

Fig. 2   40–46, Rana pipiens, 1800 ×; 47–53, Triturus viridescens, 1200 ×; 54–60, Terrapene carolina, 1500 ×; 61–67, Pigeon 1800 ×; 68–74, Opossum 1800 ×; 75–81, Rat 1800 ×. (Courtesy of Dr. Harvey E. Jordan).

phils; and in which the nucleus is lobulated and the chromatin stains deeply. (4) *Eosinophilic granulocytes* in which the cytoplasm is filled with coarse acidophilic granules which may be spherical or rod-shaped, and the nucleus is bilobed and stains deeply. (5) *Basophilic granulocytes,* in which the cytoplasm is filled with ir-regularly-shaped basophilic granules and the nucleus is usually darkly staining, and bilobed. (6) *Monocytes,* large cells in which there is a delicate faintly basophilic homogeneous cytoplasm containing a few azurophilic (metachromatic) granules, and in which the chromatin

of the large spherical or reniform nucleus is in the form of strings; a nucleolus may be present. These cells are usually counted in smears and the varieties recorded in percentages as the differential leucocyte count.

The *thrombocytes* are small oval or spherical cells present in the blood of the vertebrates except mammals. They often stain in such a way as to make discrimination between them and small lymphocytes a difficult matter. The nucleus is usually small and dense, and the cytoplasm forms a narrow rim around the nucleus. It may contain fine metachromatic granules. In the follow-

Table 1. Ranges of Quantitative Values of Cellular Components of the Blood From Typical Representatives of Som Orders of the Mammalia

RBC, red blood corpuscles per cu mm in millions; D, diameter of RBC, in μ; Hb, hemoglobin, g per 100 g whol blood; Ht, hematocrit, or volume RBC per 100 cc whole blood; WBC, number of leukocytes per cu mm in thousands Differential counts or percentage distribution as follows: L, lymphocytes; M, monocytes N, neutrophils or hetero phils; E, eosinophils; B, basophils; P, blood platelets, in 100,000's per cu mm (counts of cells from Jordan and Al britton, and recent literature; differential counts, data obtained by the writers).

Order	Animal	RBC	D	Hb	Ht	WBC	L	M	N	E	B	P
Marsupalia	Opossum	5–6	8–9	8–9	14–36	7–8	49–73	0–1	22–34	0–8	0–8	—
Insectivora	Hedgehog	9–11	7–9	—	—	5–6	14–30	5–6	56–77	7–13	2–5	—
Cheiroptera	Brown bat	7–14	5–7	—	—	1–7	30–44	3–5	8–20	—	0–5	—
Primates	Marmoset	5–7	5–8	12–13	42	5–13	21–47	5–8	31–55	0–2	0–1	1.7
Edentata	Anteater	—	8–13	—	—	—	63–95	0–2	12–26	Trace	Trace	—
Lagomorpha	Rabbit	5–7	6–8	12–13	37	6–14	51–91	3–6	11–27	0–4	Trace	1.7–11.1
Rodentia	Rat	6–7	5–7	13–14	46–49	7–13	69–82	Trace	18–22	0–2	0–5	2.0–10.0
Carnivora	Cat	6–8	5–7	8–12	28–41	15–21	13–26	2–8	69–86	3–11	0.2	1.6–7.6
Pinnepedia	Seal	6	7–8	—	48	—	9–22	3–5	47–53	17–22	3–4	—
Perissodactyla	Horse	7–11	4–7	11–15	27–47	6–14	7–44	0–7	44–83	2–8	0.5	2.4–5.6
Artiodactyla	Cow	5–10	4–7	11–14	33–48	5–12	54–72	4–8	14–50	4–11	0.4	5.4–9.7

ing brief assay of the formed elements of the blood of the vertebrates (except man), this classification will be referred to without further comment unless there is some peculiar cellular characteristic which deserves attention. In general the larger classification as given in *Biological Abstracts* will be followed. Space does not permit detailed systematic classification. In the two plates illustrating the article an attempt has been made to arrange in the same column the same type of cell in the sample from each group described. The numbers given in parentheses in the text will refer to the figures in these plates. Since the magnification of the cell in the drawings is given in the titles in the plates, no detailed reference to cell sizes will be given in the text.

Leptocardii. In Amphioxus, the representative of this order, the small lymphocyte-like cell is the only kind of cell found in the blood.

Marsipobranchii. The blood cells of the hagfish, *Myxine* are good examples of this order. The rbcs (1) vary in number from 120,000 to 130,000 per cu mm; hemoglobin, 4.0 to 6.7 gm per 100 cc blood. In the stained blood smear the following cells may be seen: hemocytoblasts (2), of which small lymphoid hemocytoblasts are the most numerous lymphocytes (3); not numerous, heterophils (4) with few granules; eosinophils with fine purplish-red granules (5); and a few monocytes (6). Basophils are absent. The thrombocytes (7) are characteristic, but in addition there are present peculiar non-granulated, elongated, narrow spindle cells (8). *Petromyzon* contains the same kinds of cells except spindle cells, but they are smaller; the hemoglobin value is about the same.

Elasmobranchii. The common saltwater dogfish, *Mustelus canis,* is used as a sample of this group. The rbcs are large (9) and the count ranges from 200,000 to 600,000 per cu mm; hemoglobin value 1.0. The number of leucocytes per cu mm ranges from 6–18,000. In sample blood smears the following cells 10–15 may be found: hemocytoblasts (10); large lymphocytes (11) and small lymphocytes (incidence of lymphocytes 63%); neutrophilic granulocytes, with a tendency of granules towards eosinophilia (12), incidence, 4%; young eosinophilic granulocytes (13), 14%). There are no basophilic granulocytes. Thrombocytes (15) form 18% of the differential count. All stages between hemocytoblasts and erythrocytes as well as transitional stages between small hemocytoblasts and thrombocytes may be found. The hematocrit is 5.5%. In the skates the conditions are about the same.

Dipneusta. This subclass includes the lungfishes of which *Protopterus ethiopicus* is used as an example. No counts, hemoglobin values nor hematocrits were available. The cells (16–24) of the blood are much larger than in any of the fishes. In the smears of blood the following cells have been described: erythrocytes (16) which may be traced to the differentiating and multiplying hemocytoblasts (17); smaller lymphocytes (18); large monocytes (20); large neutrophils (19) with fine eosinophilic granules; huge eosinophils with spherical granules (21); and smaller eosinophils full of coarse bacillary granules (22); basophils (23); and thrombocytes (24) which are often difficult to distinguish from small hemocytoblasts.

Pisces, Halecomorphi, *Amia calva,* the bowfin, is an example of this group. The cells (25–33) are small

and all stages of development are present in the blood. Counts of the rbcs range from 1.28 to 2.0 million. The following cells may be seen in the blood smear; erythrocytes (25); non-nucleated plastids; hemocytoblasts (26); lymphocytes (27), the most numerous cells in the blood; monocytes (33); neutrophils 28); eosinophils with spherical vesicular nucleus and bacillary granules (30); a few eosinophils with spherical granules and pycnotic nucleus; large numbers of cells with granules having an intermediate reaction (32) between basophilia and eosinophilia; and small thrombocytes (31).

Pisces, Isospondylii. In this suborder which includes the common marine and freshwater fishes and there is a wide fund of information concerning the counts of the erythrocytes, but not of the leucocytes. The most complete analysis has been given by Wintrobe for the rbcs of some species and a wide range of the values for all the measurable components has been presented from a low count of 700,000 per cu. mm. and hemoglobin of 4.0 in the wrymouth, to a high of 4.2 million and hemoglobin of 15.2 in the mackerel. The numbers of rbcs are associated with the ecological conditions of salinity, oxygen availability, tidal relations, rbc structure and numerous unknown factors. The carp, *Cyprinius carpio,* is a representative of this suborder. The rbcs range from 1.2 to 1.6 millions; hemoglobin 9.4–12.4; leucocytes 3200–4200. In the blood smear, the following cells may be seen (34–39): nucleated, ovoid erythrocytes with about the same characteristics as those of Amia; small hemocytoblasts (34); lymphocytes (35); neutrophils (36); eosinophils with spherules (37); eosinophils with bacillary granules (38); basophils; and thrombocytes (39). The differential count for the mackerel is as follows: lymphocytes, 54%; neutrophils, 33%; eosinophils, 3%; basophils, 4%; and monocytes 6%.

In the marine fish *Gobius,* the erythrocytes are smaller than in the other teleosts. In the so-called "bloodless fishes" of the Antarctic family Chaenichthyidae there are no erythrocytes and oxygen is carried in low concentration in physical suspension in the blood plasma. This condition limits its range much more than is the case of Arctic fishes or sedentary fishes with low rbc counts and low hemoglobin concentrations. Electrophoretic studies on teleost hemoglobin have shown that it does not migrate.

Amphibia. The blood cells of the Amphibia are gigantic when compared with those of the reptiles, birds and mammals; and they are larger than those of most fishes, (Plate II Figs. 40–53). The lowest rbcs counts were in *Necturus,* 20,000/cu mm. The erythrocytes of *Necturus* were also the largest (67 mu × 45 mu) except for *Amphiuma* (70 mu). Hemoglobin content varies from a low of 4.6 in *Necturus* to a high of 13.3 in *Cryptobranchus.* The hermatocrit varies from a low of 21% in *Necturus* to a high of 49% in *Cryptobranchus.* The values of these data in the Anura are between these two extremes.

In the Anura of which the frog, *Rana pipiens* is used as an example, the following cells may be seen in the blood smear (40–46): erythrocytes (40); small, medium and large lymphocytes (41); monocytes (42); neutrophilic granulocytes (43); eosinophils (44); basophils (45). The thrombocytes (46) resemble small lymphocytes and average 150,000/cu mm. In *Hyla,* the tree frog, the rbc count averages 674,000; leucocyte count,

29,000; hemocytoblasts 9%; neutrophilic myelocytes 2%; neutrophils 16%; thrombocytes 65%; eosinophils 8%; and basophils 3%.

The same types of cells are present in the blood of the Urodela, as represented by *Triturus viridescens* (Figs. 47–53). The rbc count ranges from 94,000–136,000. The following cells may be seen in the blood smear: erythrocytes (47) hemocytoblasts (48); lymphocytes (50); neutrophilic granulocytes (49); eosinophils with coarse granules (52); very large basophils (53); monocytes (52); and thrombocytes (51). In *Batrachoseps* more than 90% of the rbcs are non-nucleated plastids. Electrophoresis has revealed only one kind of hemoglobin in frog and toad.

Reptilia. Since the blood cells of the several orders of living reptiles, such as the Testudinata (turtles), Crocodilia (alligators and crocodiles), Lacertalia (lizards), and Ophidia (snakes), are much alike morphologically, the cells of the turtle, *Terrapene carolina* Figs. 54–60) are used as samples. The rbc counts range from 388,000 to 984,000; hemoglobin 11–14 grms; leucocyte count from 2,000–13,000. In the stained blood smear of the turtle the following cells may be seen (Figs. 54–60): erythrocytes (54); lymphocytes (55), 46%; heterophils (neutrophils) with small amphophilic granules (56), 4%; eosinophils with spheroidal granules (57), 12%; eosinophils with bacillary granules (58) 10%; basophils (59), 1%; and thrombocytes (60). Electrophoretic studies revealed three types of hemoglobin in turtles and one in snakes.

Aves. In the birds, blood formation does not take place in the circulating blood and the cells are mature elements. The cells are smaller and more numerous than in the reptiles. The erythrocytes are nucleated and the counts vary from 1.89 million in the turkey to 4.63 in the quail. The leucocyte counts range from 10,000 in the fowl to 21,000 in the Ostrich. In the stained smear of pigeon blood, chosen as a sample of avian blood, the following cells may be seen (Figs 61–67): erythrocytes (61); lymphocytes (62), 30–50%; monocytes (63), 1–2%; heterophils with bright red bacillary granules (64), 1–7%; eosinophils with spherical, dull red to purple granules (65), 1–7%; basophils (66), 1–7%; and thrombocytes (67) (28,000 to 47,000 per cu mm). Physiologic studies have shown that the oxygen capacity of the blood varies in birds living at different altitudes. Electrophoretic patterns of the hemoglobin show variation from three kinds in the chicken and duck to one kind in pigeon and robin.

Mammals. Except for the wild species, the blood of mammals, particularly domestic animals has been studied for many years and more is known about it than in all other groups together. There is no record of the blood of the Monotremes. As an example of the blood of the Marsupalia, the blood of the opossum, *Didelphys virginiana,* has been selected. As in all mammals, the erythrocyte has matured in the bone marrow and appears in the blood in the form of a small non-nucleated, biconcave disc, the erythroplastid (68). The rbc count ranges from 2.8–8.0 million; hemoglobin 8.2; hematocrit 23–36%; the leucocytes average 15,000. In a blood smear the following cells may be seen: lymphocytes (69), 62–73%; monocytes (70), 1.0%; neutrophils (71), 23–28%; eosinophils (72), 3%; and basophils (73), 1%. The thrombocytes of the lower classes have been re-

placed by blood platelets (74) which appear in smears as small irregularly shaped masses of clear material, the hyalomere, in which are embedded small masses of metachromatic granules, the chromomere. In all adult mammals the platelets are believed to be formed by fragmentation of megakaryocytes of the bone marrow. There are usually from 400,000 to 800,000 platelets per cu mm of blood.

In other marsupials, such as the kangaroo, wallaby and wombat, the rbc counts are higher and the leucocytes counts range around 10,000. The blood physiology is like that of man.

Many blood counts, morphological and histo-chemical studies have been made on the blood cells of the different members of the placental mammals. With few exceptions, individual and species variations in the counts of the rbcs run around 5 million/cu mm and there is more constancy in single animals than in members of the same species. By some the rbcs counts are regarded as genetic characteristics. The leucocyte counts are very labile, but usually range between 5,000 and 10,000/per cu mm. Platelets have same distribution as in marsupials. For convenience, the blood cells of the rat have been chosen to represent the kinds of cells present in the blood smears of most placental mammals (Figs. 75–81). The rbc count ranges from 6.0 to 10.0 million; hemoglobin, 14–15; hematocrit 41–47%; leucytes 4,000–10,000; and platelets 190,000 to 1.0 million. In the blood smear the following cells may be seen: erythroplastids (75); lymphocytes (76), 76%; neutrophils (77), 20%; eosinophils with ring nucleus (79); 1.4%; basophils (80) less than 1%; monocytes (78), 1–5%; and platelets (81). By modern methods of electrophoresis two different kinds of hemoglobin have been identified in the rat, sheep and buffalo.

The only deviations in shape and tinctorial reactions from the cells illustrated by the rat are the elliptical rbcs of the camels and llamas, the sickle-shaped rbcs of the deer; the very small rbcs of the goat; the presence of mucopolysaccharide bodies in the lymphocytes of the guinea pig; the extremely large spherical eosinophilic granules in the eosinophils of the horse (2.5 mμ in diameter against a usual diameter of 0.6 mμ; and the rod-shaped eosinophilic granules of the cat). The highest rbc counts have been recorded from the goat and llama.

There are many gaps in our knowledge of the morphology and physiology of the blood cells in the animals other than man, the domestic and laboratory animals. If the reader wishes to know more about the finer details of structure of the blood cells he may refer to the handbooks listed below.

JAMES E. KINDRED

References

Alder, A. and E. Huber, Folia Haematologica, **29**: 1923.

Albritton, E. C., "Standard Values in Blood," Philadelphia, Saunders, 1952.

Burnett, S. H., "Clinical Pathology of Domestic Animals," New York, Macmillan, 1917.

Jordan, H. E., "Comparative Hematology," (Section XII of "Handbook of Hematology," New York, Hoeber, 1938.

Kleineberger, C., "Die Blutmorphologie der Laboratriumstiere," Leipzig, Barth, 1927.

Wintrobe, M. M., Folia Haematologica **51**:1933.

BONE

Bone shares with the other mesodermal supporting tissues, CONNECTIVE TISSUE and CARTILAGE, some important characteristics. All these tissues consist of highly specialized cells, fibrocytes, chondrocytes, osteocytes, and a bulky intercellular substance. The latter, in turn, consists of collagenous fibrils or fibers and a cementing substance, mainly mucopolysaccharides. In addition, a variably high volume of water is also present. Bone, as its related tissues, dentin and cementum is also highly mineralized. The fibers or fibrils are identical in all these tissues. They consist of collagen with its characteristic periodicity of 640 Angstrom units. The mucopolysaccharides are different in these tissues in their chemical composition as well as in the degree of polymerization. The minerals are mainly a form of apatite present as submicroscopic hexagonal platelets.

The chemical properties of bone are as follows:

> organic components: about 30%
> inorganic components: about 45%
> water: about 25%

Most of the organic material—about 95%—is collagen. The rest is almost equally divided between mucopolysaccharides, elastin, and fat. The bulk of the inorganic substance—about 85%—are crystalline calcium phosphates. The rest is calcium carbonate with traces of other elements, e.g., magnesium and sulfur.

Physically bone is highly elastic and resistant both to pressure and tension. Tensile strength may rest on the fibrils of bone tissue, compression strength on its mineralized state.

The osteocytes, the bone cells, are characterized by their plum-stone shaped body and their numerous branching processes. The processes of adjacent osteocytes are continuous. Cell bodies are contained in bone lacunae, the processes in bone canaliculi. In dried preparations these spaces fill with air and appear black under the microscope (fig. 1.). Two other types of cells, though not permanently present, can be said to belong to bone tissue, namely, osteoblasts and osteoclasts. The former are involved in the formation of bone, the latter in its destruction.

Osteoclasts, if typically developed, are large, multinucleated cells, often forming a syncytium on a bony surface. Often they are found lying in shallow, bay like depressions, Howship's lacunae.

Though not all details of osteogenesis are known, it is indisputable that the osteoblasts are the bone forming cells. The first product of these cells is uncalcified and not easily stained. It is termed osteoid tissue. Its "calcification," also dependent on the activity of the osteoblasts, is not merely a precipitation of calcium salts, but also entails a change in the organic substance (mucopolysaccharides?), probably a depolymerization.

Where bone formation is fast, the osteoblasts line the surface as rather large cuboidal cells with an eccentrically located nucleus. Where bone formation is slow, the osteoblasts are much smaller and spindle-shaped or flat.

The bone destroying function of the osteoclasts also can hardly be questioned. It is important to realize that there is never in the living animal a decalcification of

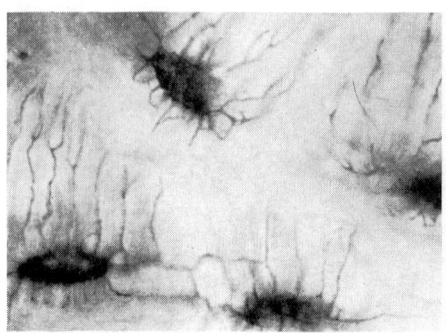

Fig. 1   Dried ground section through human bone. Note bone lacunae and canaliculi, (From Weinmann and Sicher, "Bone and Bones," Second edition, C. V. Mosby Co., 1955.)

bone tissue, but rather a total removal of all of its components. Hypothetically one assumes that the osteoclasts produce some proteolytic enzymes (collagenases?, hyaluronidase?) and some chelating agents, that split the calcium phosphates and make them soluble.

The apatite crystals are arranged in strict relation to the bands of collagen fibrils. It is still questionable whether one can speak of calcification of these fibrils or whether it is the binding substance between the fibrils that is mineralized.

The collagenous fibrils of the intercellular substance can be made visible by special stains, e.g. silver impregnation. The fibrils are found in bundles, crossing each other at acute angles. The most significant and functionally important feature of their arrangement is the fact that the course of the fibrils alternates regularly in adjacent layers of bone tissue. These layers, 4 to 12 μ thick, are known as lamellae. Adult bone, with the exception of the otic capsule, consists of lamellated or mature bone (figs. 2, 3, 4). The lamellae are separated by fine cementing lines. In dry preparations the alternating course of the fibrils results in the appearance of "striped" and "stippled" lamellae, in which the fibrils are either in the plane, or at right angle to the plane of sectioning.

The two types of bone tissue, spongy or cancellous and compact bone, differ not in the properties of their elements, cells and intercellular substance, but in the arrangements of the bony lamellae, the "construction units" of bone. Spongy bone consists of delicate bars, plates, or tubules forming a three dimensional lattice. The marrow spaces of spongy bone communicate throughout the spongiosa. The spongy trabeculae are composed of a few lamellae, arranged either parallel to each other or concentrically (fig. 2). The arrangement of the trabeculae in the spongiosa of bones follows the lines of force or stress (trajectorial arrangement).

In compact bone the lamellae are concentrically arranged around a specialized, elongated and narrowed marrow space containing blood vessels and some loose connective tissue. The long cylinders of the concentric lamellae, 5 to 20 in number, are called osteones or Haversian systems. The blood vessel containing spaces are the Haversian canals, surrounded by Haversian lamellae (fig. 3). The fibrils run in a longitudinal direction in

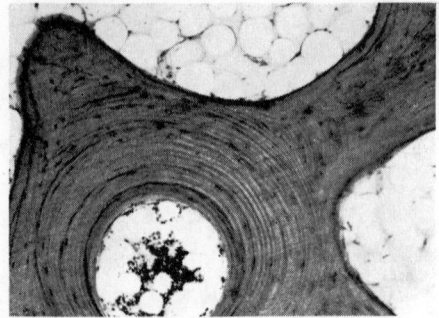

Fig. 2 Tubular spongiosa from a human rib. Note the concentric lamellae. (From Weinmann and Sicher, "Bone and Bones," Second edition, C. V. Mosby Co., 1955.)

Fig. 3 Cross section through a Haversian system. Dried ground section of human bone. (From Weinmann and Sicher, "Bone and Bones," Second edition, C. V. Mosby Co., 1955.)

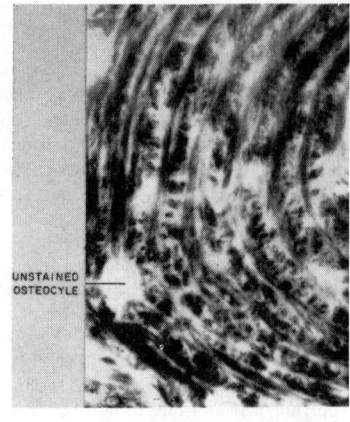

Fig. 4 Alternating course of fibrils in Haversian lamellae. Silver impregnation. (From Weinmann and Sicher, "Bone and Bones," Second edition, C. V. Mosby Co., 1955.)

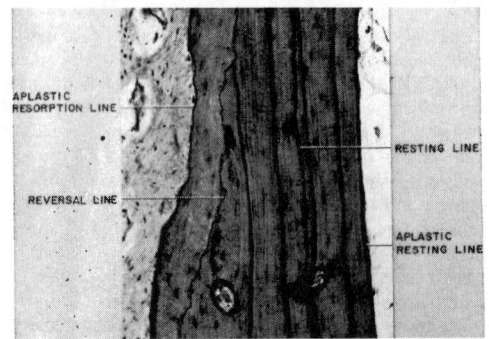

Fig. 5 Cementing lines from human mandible. (From Weinmann and Sicher, "Bone and Bones," Second Edition, C. V. Mosby Co., 1955.)

one, in a circular direction in the next lamella, and so on (fig. 4).

The blood vessels in the Haversian canals are connected to each other by crosslinks that are contained in short canals perforating the Haversian lamellae of two adjacent systems. Such canals are referred to as Volkmann canals. Similar canals, though sometimes much wider, allow communications of Haversian bloodvessels with those of the periosteum or the marrow.

The mode of transverse growth of tubular bones by periosteal apposition leads to the formation of lamellae running around the entire circumference of the shaft: circumferential lamellae.

Because of the fairly regular cylindrical shape of Haversian systems, irregular spaces exist between adjacent osteones. These spaces are filled by irregular interstitial lamellae, remnants of partially destroyed Haversian or circumferential lamellae.

Bone tissue is a shortlived tissue. Replacement of overaged bone occurs by osteoclastic destruction, resorption, of such parts and by osteoblastic formation of new bone. This is the reason why bone tissue is a highly active and dynamic tissue, though the bones of an animal are, for its entire lifetime, permanent organs. Bone tissue is said to be in constant flux. The internal reconstruction of bone tissue starts as a rule, by the differentiation of osteoclasts, either from undifferentiated cells of marrow spaces or in Haversian canals. It is important to realize that bone resorption, a destructive process, is always accompanied by proliferation of loose connective tissue, that is by a productive process. The cells of this proliferating tissue are at the same time the source of new osteoblasts and osteocytes. There is hardly any real evidence of a change of osteocytes into osteoblasts, or osteoclasts and vice versa, though this theory of cellular transmutability is widely expounded.

Whenever a bony surface, outer or inner, is at rest a strongly basophilic layer forms at its boundary with the surrounding connective tissue. On sectioning these layers appear as aplastic lines. They may be smooth when preceded by apposition: aplastic resting lines; or scalloped, when formed at the end of active resorption: aplastic resorption lines. If a new wave of bone apposition sets in, these lines persist. Separating layers of bone formed one after another, they are recognized as resting lines. If resorption is followed by a period of reparative apposition we speak of a reversal line (fig. 5). By an analysis of such lines much knowledge can be gained of the past history of any area of bone.

The bone tissue, first formed in the embryo and fetus, and later gradually destroyed and replaced by mature lamellated bone, is called immature or coarse fibrillar bone. It is always spongy bone and is characterized by the greater size and irregular shape of its osteocytes that rarely show extensions or processes. Furthermore, the fibrils and fibers of immature bone are irregularly arranged. It is interesting to note that this type of bone recurs in the adult wherever rapid production of bone tissue is essential, e.g. in the healing of fractures. It is further important that immature bone is much less radio dense than mature bone and therefore often not visualized in roentgenograms.

HARRY SICHER

References

Bourne, G. H., "The Biochemistry and Physiology of Bone," New York, Academic Press, 1956.
McLean, F. C., and M. R. Urist, "Bone," Chicago, The University Press, 1961.
Ponlot, R., "Le Radiocalcium dans l'étude des Os," Paris, Masson, 1960.
Stein, I., et al., "Living Bone," Philadelphia, Lippincott, 1955.
Weinmann, J. P., and H. Sicher, "Bone and Bones," 2nd ed., St. Louis, Mosby, 1955.

BOSE, JAGDISH CHANDRA (1858–1937)

Jagdish Chandra Bose is the most outstanding among the Indian pioneers of science to achieve international reputation in experimental science. His scientific career started with his remarkable researches on the properties of electrical waves. Then followed the work of his life—the similarity of response in the living and non-living—and later comparative physiological investigations on plant and animal tissues. His monumental work *Plant Responses* was published in 1906.

J. C. Bose was born on Nov. 30, 1858 in Mymensingh, East Bengal. In 1880 he graduated from Calcutta University and proceeded to England from where he obtained the B.Sc. degree of the London University. In 1897 he was invited to give a lecture at the Royal Institution in London and later at Paris, Berlin, etc. From 1900 onwards he continued his researches and through 1902–1919 he published six volumes and many papers in the 'physics of Biology.' The Optical Lever, the Crescograph, the Resonant Recorder stand outstanding among the devices he designed to record responses in plants. He was awarded the D.Sc. degree of the London University in 1896 and in 1902 the Société Française de Physique elected him to its Council. He made extensive lecture tours in America on two occasions in 1908 and 1914. He was elected a Fellow of the Royal Society of London in 1920 and a member of the Academy of Sciences, Vienna in 1928. He received the C.I.E. in 1902; in 1911 the C.S.I. and in 1917 he was knighted. In 1915 he retired as Professor Emeritus at the Presidency College, Calcutta. He opened the Research Institute at Cal-cutta in 1917 named after him and was himself closely associated with all its activities until his death on Nov. 23, 1937.

S. VENKETESWAREN

BRACHIOPODA

Brachiopods are a phylum of attached, lophophore-bearing, filter-feeding, exclusively marine invertebrates with a trochophore-like larval stage and a bilaterally symmetrical, bivalved shell. They are non-colonial but commonly gregarious in habit. The characteristically unequal valves are dorsal and ventral, rather than right and left as in a pelecypod; and, when calcareous, they consist of the stable mineral calcite, rather than its unstable polymorph aragonite as is usual among pelecypods. The sexes are separate but essentially identical, and the few available records indicate that fertilization and early development may take place either within the mantle cavity or outside the shell.

In their feeding structure, embryology, and soft individual anatomy brachiopods grossly resemble PHO-RONIDEA and BRYOZOA (Ectoprocta). Within themselves, they are sharply divided at the class or subphylum level into two very unequal groups. The larger of these groups, the Articulata or Pygocaulia, has an articulated calcareous shell, a blind intestine, simple musculature, and an apparently enterocoelous embryological development in which the mantle lobes are rotationally inverted after fixation of the larva. The smaller group, the Inarticulata or Gastrocaulia, has a non-articulated chitin-like or calcareo-phosphatic shell, an intestine that terminates in an anus, a more complex musculature, and an apparently schizocoelous embryological development in which the mantle lobes do not rotate after attachment.

In brachiopod classification below the level of these two main subdivisions, stress is laid upon the form and position of the pedicle opening and associated structures, structures having to do with muscle attachment, and both gross and minute external morphology and ornamentation. Including fossils, distinctions based on such criteria have resulted in the description of about 1400 genera, perhaps 25,000 species, and a number of distinctive and widely accepted familial and larger groupings. The status of the suprafamilial groupings and the basis for an ordinal classification are not agreed upon, however.

In keeping with their low organization and rudimentary circulatory and nervous systems, brachiopod responses are limited to gaping and closing of valves in reaction to feeding habits, light, or touch; extending and retracting mantle-edge setae and lophophore; functioning of the ciliary feeding system; liberation of the sexual products; and movement of or on the pedicle by means of which lifelong fixation is maintained.

Among the Articulata the pedicle normally protrudes through a posterior or apical foramen or gape in the pedicle valve and is ordinarily attached to a hard substrate; a few genera, however, are cemented by part or most of one valve, normally the pedicle valve. Living articulate brachiopods are known from the intertidal zone to abyssal depths and from tropical to polar waters, but are probably most common in waters less than 100 fathoms deep. Living species are mainly smallish, inconspicuous, smooth or radially ornamented, subovate to lenticular forms of pale or rarely brightly patterned colors. Some fossils, however, attained the size of a man's fist or larger, and displayed a variety of shapes from globular to ovate, alate, concavo-convex, ostreiform, conical, or even elaborately lobate; with straight or pointed beak ends; and with smooth, radially ornamented, concentrically corrugated, spiny, or spinulose ornamentation.

The pedicle of the Inarticulata (also excluding ce-

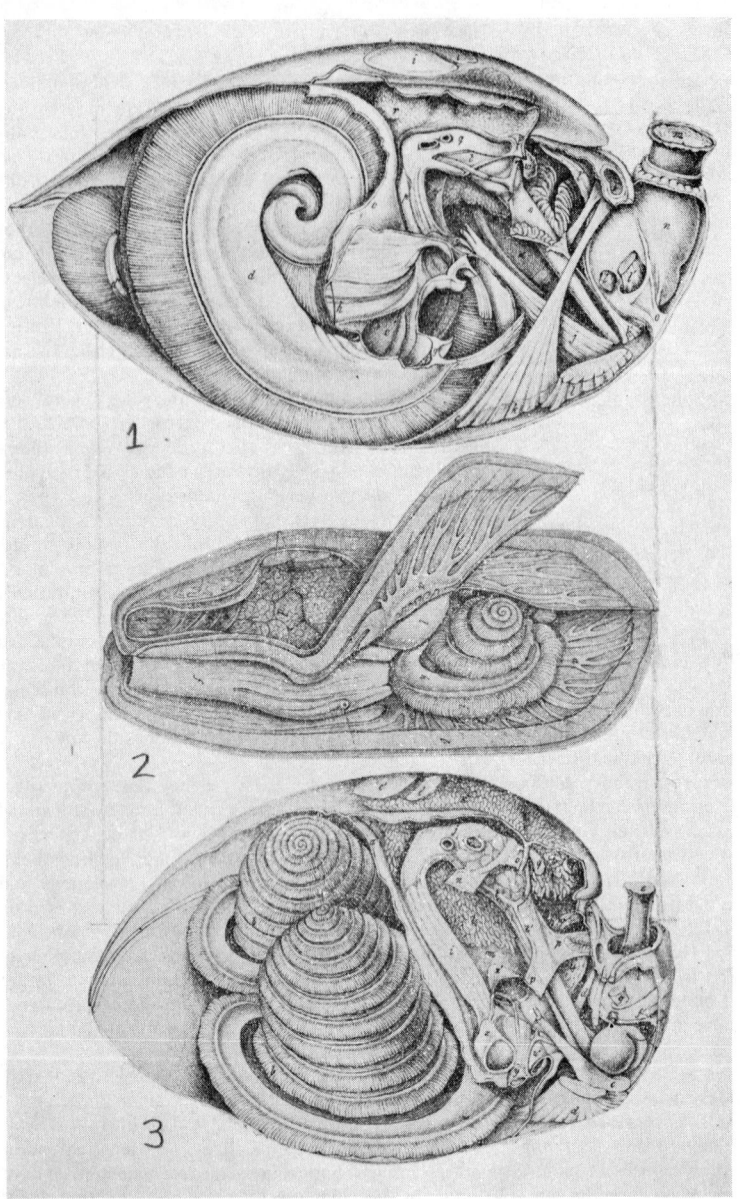

Fig. 1   (1) One of the Articulata, the terebratuloid *Magellania flavescens* (Lamarck). The mantle lobes of the excised brachial and pedicle valves preserve the general shape of the animal. The coiled lophophore occupies the anterior (left) half of the body space; the pedicle protrudes upward at the rear; the muscles and digestive, reproductive, and circulatory organs are between. Parts of the viscera have been removed so that other parts may be seen.
(2) One of the Inarticulata, a species of *Lingula*. The large, fleshy pedicle has been removed from the posterior (left). One of the paired spiral lophophore lobes is visible at the front (right) of the body space. The forked trunks of the pallial sinuses show up prominently on the turned-out mantle lobes.
(3) Another articulate, the rhynchonelloid *Hemithyris psittacea* (Gmelin). The spiral lophophore lobes occupy the front (left) half; the rest of the body is analogous to that in (1).

mented forms) either passes between the posteriorly gaping shells which are held together by the muscles only, or, in some discoid to conical species with highly inequivalved shells, it passes through a slot or perforation in one valve. Living inarticulates also encompass a wide zoogeographic range, except for members of the family Lingulidae which occur exclusively in shallow to intertidal, impure sediments of tropical to warm temperate waters, and whose ancestors apparently occupied a similar niche with little evolutionary change for the last half-billion years or so. The inarticulate shell is normally lingulate, discoidal, or conical, smooth, green or brown, commonly minute, and rarely more than a few centimeters in maximum dimension.

Brachiopods are only very locally abundant in existing seas, having undergone a persistent reduction in numbers and variety from their Paleozoic ascendancy to the present obscure handful of something like 75 living genera, including only 240 or so living species. They were among the most abundant and widespread marine organisms of the ancient shelf seas, however, and, in general contrast to present warm water reef communities, brachiopods were common associates of the Paleozoic tetracorallian reefs. As a rule the phylum is strictly poikilosmotic and stenohaline, but some of its members can tolerate fresh water for brief intervals, and the intertidal lingulids of Japan may survive exposure to brackish waters for days at a time.

Their former abundance, their wide variety of external and internal form and structure, and the fact that the calcitic, "chitinous" and calcareophosphatic shells survive indefinitely under chemical conditions normally found in a wide variety of sediments (in contrast to aragonitic shells) combine to give the brachiopods an importance for paleontologists far out of proportion to their place among living phyla. In the Paleozoic they are outclassed for stratigraphic correlation only by pelagic organisms, and, for a short while, by the fusulinid Foraminifera. In addition, their restricted living habits confer a large degree of paleoecologic utility, and their long duration and wide and systematic morphological variation provides excellent basis for long-range phylogenetic studies. They are rarely seen in aquaria or studied in a live state, however, although available at or near most marine biological stations.

PRESTON E. CLOUD, JR.

References

Hyman, Libbie H., Phylum Brachiopoda, in "The Invertebrates," v. 5: New York, McGraw-Hill, 1959.
Muir-Wood, H. M., "A history of the classification of the Phylum Brachiopoda," London, British Museum, 1955.
Thompson, J. Allan, "Brachiopod morphology and genera (Recent and Tertiary)," Wellington, New Zealand Board of Science and Art, Manual no. 7, 1927.
Hancock, A., "On the organization of the Brachiopoda," Phil. Trans. Roy. Soc. London, **148**: 791–869, 1858.

BRAIN

The animal body is made up of four types of tissues, 1) a lining-covering and secreting tissue, called EPITHE-LIUM, 2) a supporting tissue, known as CONNECTIVE TISSUE, 3) a tissue which is able to shorten, MUSCLE and 4) one, capable of conducting irritation. The latter is NERVOUS TISSUE which integrates the activity of the animal body. Nervous tissue is composed of nerve cells and their prolongations embedded among the supporting cells and fibers of the neuropil. Although nerve cells have many forms, they are cut to a single physiological pattern. Each nerve cell possesses a cell body and prolongations. The cell body contains one eccentrically placed NUCLEUS, poor in chromatin and a single nucleolus. In general the morphology of the prolongations which carry the stimulus to the nerve cell body are complex; whereas that of the prolongation which takes the stimulus on to another cell is simple. The length of these prolongations varies from a few micra to many centimeters (1).

The nerve cell bodies are arranged either, 1) in groups or 2) in layers. Nerve cells arranged in groups within the CNS (central nervous system) are known as *nuclei,* those outside the CNS, as ganglia. The nuclei might be classified as sensory, motor or adjustor groups of nerve cells. The GANGLIA are sensory ganglia found on cranial or spinal sensory nerves, and autonomic. The latter may belong either to sympathetic or to parasympathetic divisions of the autonomic division of the NERVOUS SYSTEM. The arrangement of nerve cell bodies in layers occurs in the roof of the midbrain, in the cortical surfaces of the cerebellum and in the cerebral hemispheres.

Nervous tissue, whether found in the peripheral or in the central nervous system is, like all other tissues supported, surrounded and held in place by connective tissue. Connective tissue, wherever found is composed of fibers and of cells. In the peripheral nervous system the fibrous supports are derived embryologically from two of the three embryological layers, the mesoderm and the ectoderm. In the central nervous system cells rather than fibers form the intimate support of nerve cells and their processes (2). The supporting cells share with nerve cells a common origin, the undifferentiated ectodermal cells which form the neural plate.

Outside the central nervous system the cell bodies of nerve cells within ganglia (sensory or autonomic) are supported by modified Schwann cells (3). Within the central nervous system the homologous supporting cell is called an oligodendroglia (see NEUROGLIA) cell. The processes of nerve cells, that is the axoplasm is surrounded by a tube of tissue or a sheath known as myelin, arranged in concentric layers (4). Although all the processes of nerve cells are myelinated some of the sheaths are so thin and contain so little lipid that it cannot reduce osmium tetroxide. Because of this finding many small nerve fibers have been called nonmyelinated. The name remains in spite of the fact that the electron microscope shows these fibers to be finely myelinated. The myelin sheath forms a continuous cylinder about the axoplasm of the prolongations of nerve cells which lie within the central nervous system, but is periodically completely interrupted in the peripheral nerves. This interruption is called the node of RANVIER. Again each nerve fiber in the peripheral nerves and its myelin sheath is completely encased in a thin structureless tube of tissue, the neurolemma sheath. These tubes of tissue characterize peripheral nerves, only; for they have never been seen covering the myelin of nerve fibers within the central nervous system.

Although nerve cells and their processes are arranged in definite orderly sequence of contact with one another, each nerve cell and its processes remain, in spite of their areas of contact with other nerve cells and their prolongations, a morphological entity. This independence is demonstrated by two findings, 1) each nerve cell develops from a single embryonic cell, the neuroblast and 2) when injured, degeneration is limited to the nerve cell suffering the injury. The multiple prolongations of the nerve cell body which pass the nerve impulse on to the cell body are called dendrites; whereas, the prolongation which relays the nerve impulse to another cell is known as an axon. Contact between an axon and another nerve cell may be made upon its dendrites, upon its cell body or upon its axon. The area of contact is the *synapse*, which varies in size from a simple loop to many square micra.

These orderly interrelationships of nerve cell to nerve cell occur relatively late in the history of development of the CNS. At the end of the third day after fertilization the human ovum emerges from the fallopian tube into the cavity of the uterus. Wandering freely within the lumen of that organ, the ovum continues the cellular division initiated by fertilization which had occurred in the distal end of the oviduct. After four days of freedom the fertilized egg comes to rest upon the maternal endomertium and corrodes its way to food and security. About two days after implantation, the dorsal and ventral surfaces are distinguished by the proliferation of a new layer, the endoderm (5). On the fourteenth day after fertilization, the cells lying laterally in the anterior part of the disc multiply, pile up, and form two lateral ridges. Posterior to these ridges lies the tissue from which the spinal cord develops; whereas, anterior to them the primordium of the encephalon forms (6).

On the fifteenth to sixteenth day (4 somite stage) the cellular proliferation continues to be active laterally and anteriorly. Thus the groove appears to deepen. The lips of the lateral ridges begin to make contact over the groove in the region of the 4th to 5th somite (7th to 9th somite stage, i.e. 21st to 22nd day) or in the region of the future upper cervical segments of the spinal cord. This process of closure once initiated continues anteriorly and posteriorly. Closure continues for another week (20 somite stage) leaving a small hole anteriorly in the future lamina terminalis. Shortly thereafter, generally only 24 hours later (29th day) this cephalic hole, the anterior neuropore closes. The closure of the posterior neuropore may occur immediately thereafter (29th to 30th day) or may be delayed for as long as five days (35th day). With the latter closure, the dorsoventral flexure disappears (6).

Later development proceeds by differential development in the walls, the roofplate and the floorplate which surround the slit-like central canal. This differential growth occurs as a thickening or a thinning or an evagination of the lateral walls of the roof or of floor plates (7). Some of these evaginations are hollow and remain so; others are solid growths or thickenings in the walls of the primitive CNS. At the anterior end of this irregular tube, a single vesicle appears as an evagination of the dorsal part of the lateral walls (7, 8). This single vesicle becomes two as the result of differential growth, for the tissue within the lateral walls grows more rapidly than does that on either side of the midline. These

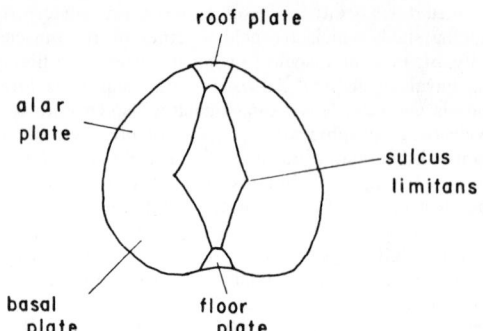

Fig. 1   Outline drawing of a cross section of the spinal cord of a 20 mm human embryo (thoracic level) showing the groove within the central canal known as the sulcus limitans of His. This sulcus divides the lateral wall into a dorsal or alar plate and a ventral or basal plate. × c 33.

evaginations are the primordium of the telencephalic vesicles.

A cross section of the spinal cord in a 20 mm. human embryo (fig. 1.) presents a narrow groove named by His, the sulcus limitans. Passing a plane through the side walls at the level of sulcus limitans divides each lateral wall into a dorsal or alar plate and a ventral or basal plate. The former contains the neuroblasts destined to become the sensory nuclei of the CNS, the later those which will form the motor nuclei. The lateral walls are joined by two thin plates of tissue, the roof plate and the floor plate. No nerve cells develop in either of these plates, although later the floor plate in particular becomes the bed through which commissural axons are to pass. The roof plate on the other hand may become thin and convoluted in the thalamus and in the posterior part of the rhombencephalon The connective tissue supporting the blood vessels follows closely the repetitive infolding of the single cell layer (ependyma cells) of these roof plates, forming the choroid plexus.

The early organization of the CNS is longitudinal and remains so in that part of the brain which lies caudal to the midbrain in spite of the transverse partitioning of the spinal cord and of peripheral nerves by the metameric development of surrounding vertebrae. Furthermore, the longitudinal organization of the cerebrum is, in primates, grossly interrupted by the subsequent overpowering growth of the dorsal evagination from the alar plate of the forebrain. In the hindbrain the subdivision of the alar plates which lies cephalad to the lateral recess begins a thickening which continues until the two parts of the hindbrain coalesce in the midline and become one, the future cerebellum (fig. 2). In lower forms the growth of these two suprasegmental regions is so limited that the longitudinal organization of the brain remains visible throughout life (fig. 3).

The CNS of all vertebrates is characterized by similar subdivisions which become visible in the developing neural tube shortly after closure of the neuropores (7,8). The cephalad separation between the spinal cord and the brain stem is marked by the most cephalad rootlets of the first cervical nerve. The brain itself falls naturally into two subdivisions, the *cerebrum* and the *rhomben-*

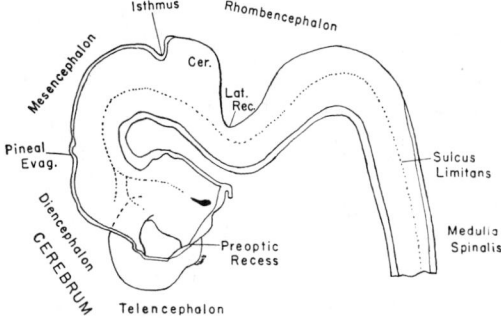

Fig. 2   Outline drawing of a midsagittal section of the brain of 14 mm human embryo belonging to the Carnegie Collection. The subdivisions of the brain are indicated. The sulcus limitans can be followed through the rhombenecphalon and the mesencephalon into the lateral wall of the third ventricle (diencephalon). × c 8. Used as fig. 14., in Hines, M. 1922. J. Comp. Neurol., **34**, 73–169.

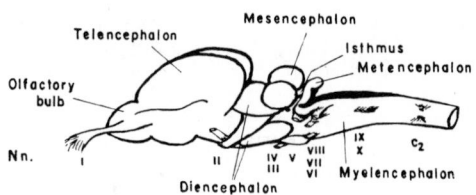

Fig. 3   Outline drawing of the lateral aspect of the brain of a frog (Rana pipiens) showing the main divisions of the brain. This brain has no midbrain flexure and no brachium points. The telencephalon does not cover the diencephalon and the cerebellum is only a small ridge-like projection at the anterior end of the rhombencephalon. × 2.

cephalon, separated by a region of slow growth, the *isthmus.* The cerebrum has two easily recognizable parts, the *prosencephalon* or forebrain and the *mesencephalon,* or midbrain. The prosencephalon is divided into two large divisions, the *telencephalon* (endbrain) and the *diencephalon (thalamus).* The former contains the *cortex cerebri* and the basal ganglia *(corpus striatum)* whereas the latter has three parts. The thalamus is separated in all vertebrates into *hypothalamus, epithalamus* and thalamus. The midstructure or thalamus has a dorsal part made up of many sensory nuclei and a ventral part containing only a few motor nuclei. The dorsal part varies greatly throughout the vertebrate phyla, the ventral part very little.

Caudad to the isthmus lies the rhombencephalon. The cerebellum and its attachment to the medulla oblongata, the pons, form the cephalad and dorsal division of the rhombencephalon, the *metencephalon.* The *medulla oblongata,* the ventral subdivision of rhombencephalon is known as the *myelencephalon,* (fig. 4). The accompanying table (table 1) will clarify the embryological derivatives of the early neural tube as well as the relation of these derivatives to the subdivisions of the adult human brain.

The generalization that sensory nuclei develop from neuroblasts located in the alar plate, and motor nuclei from those in the basal plate is true for the brain stem as far cephalad as the anterior boundary of the mesencephalon. Furthermore this generalization includes the site of the visceral motor nuclei in the brain stem and in the spinal cord (C_8 to L_2; S_2–S_4). Unequivocal visceral sensory nuclei seem to be lacking in the spinal cord, although present in the medulla oblongata. Besides the sensory and motor nuclei there are others, namely, adjustor nuclei. The major part of the CNS is composed of adjustor neurons. Their site of development is not limited to either the alar or basal plates. Indeed, the CNS adjusts the motor responses of the vertebrate body, be they visceral or somatic, to the variety of sensory stimuli brought to it over peripheral nerves.

Peripheral nerves originating in the body carry to the spinal cord the sensations which originate in the body, i.e. somaesthetic sensibility. Somaesthetic sensibility includes general cutaneous sensibility (pain, temperature touch and light pressure), general visceral sensibility (sensations from blood vessels, from visceral and parietal membranes and internal organs) and proprioceptive sensibility (joint, tendon and muscle sense). Stimulation of similar endings in the head reach the brain stem over cranial nerves. The cranial nerves, however, innervate specialized receptors such as those for sight and for hearing (distant receptors), those of smell and of taste (the chemical senses) and those of equilibrium and position of the head in space.

Adjustor neurones may be long or short, serving distant or local adjustment. Local adjustors provide for "simple reflexes" with their attendant inhibition of other motor activity; distant adjustor mechanisms provide for the complex motor responses which stem from activity of cephalic centers. These centers of control are found in the *cerebral cortex,* in the *cerebellum* and in the whole of the brain stem. The long ascending adjustors, which carry the results of stimulation of sensory endings of somaesthetic sensibility to cephalad centers, send collaterals or terminals into the generalized reticular formation of the brain stem and terminate in thalamic nuclei of sensory projection. Similarly all long adjustors stemming from nuclei of sensory reception in the brain stem follow this pattern of double termination, 1) in the reticular formation and 2) in the thalamic nuclei.

The cerebral cortex is subdivided into an old cortex, common to all vertebrates containing the *pyriform lobe* (uncus in man) and the *hippocampus* and a new cortex, which characterizes mammals, particularly the primates. The organization of the *neocortex* falls naturally into four lobes, named for the bones of the skull under which they lie. Each of the three posterior lobes, *parietal, occipital* and *temporal,* is organized respectively about a receptive area for somaesthetic sensibility, for sight and for hearing. The *frontal lobe* on the other hand is organized about a motor area, the site of origin of a part of a long descending system which enters the spinal cord without interruption in the brain stem, the cortico-spinal tract, and of other descending systems which do snyapse in the brain stem. The cortical tissue which surrounds each sensory projection area is called area of association. Just as the sensory receptive areas receive the terminals of neurones located in thalamic nuclei of sensory projection, so the associa-

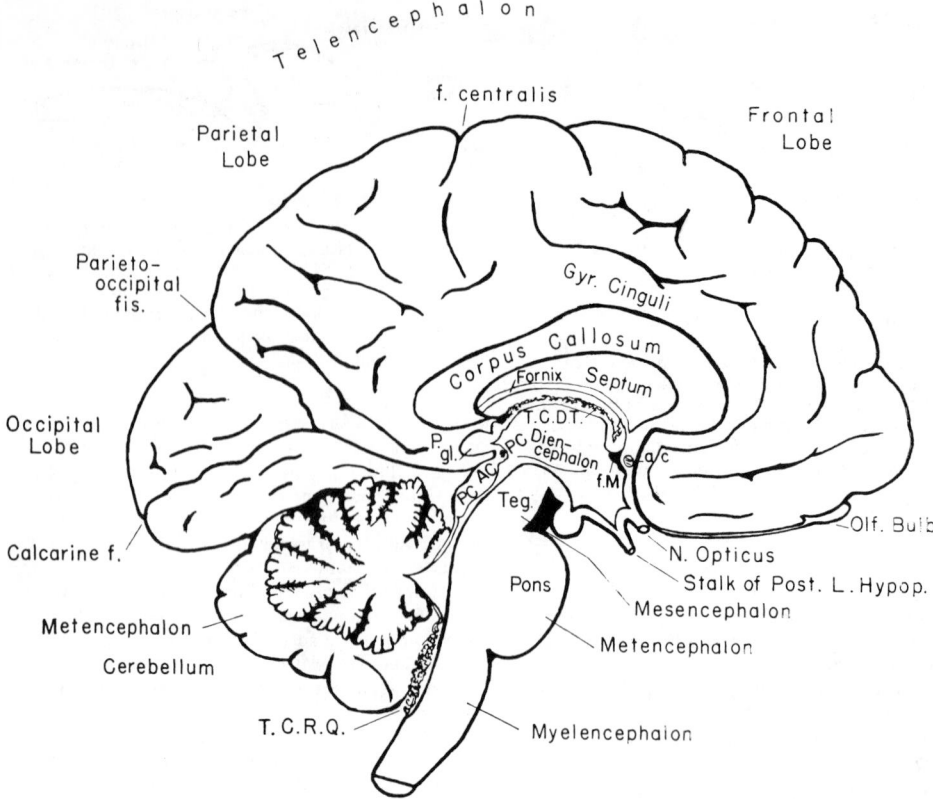

Fig. 4   Outline drawing of a midsagittal section of the brain of adult man, showing the five typical divisions. The telencephalon covers the greater part of the brain stem. Similarly the metencephalon (cerebellum and brachium points) dwarfs the myelencephalon.
Abbreviations used: T.C.D.C., tela chorioidea ventriculi tertii; T.C.M.Q., tela chorioidea ventriculi quarti; P.gl., pineal gland; A.C., anterior colliculus; P.C., posterior colliculus; teg., tegmentum of midbrain; p.c., posterior commissure; a.c., anterior commissure; f.M., foramen of Monro; c.m., corpus mamillare. × ¼.

tion areas receive terminals of neurones found in the thalamic nuclei of association. Within each of the sensory projection areas a topical localization is present, similar to that long assigned to the motor projection area of the frontal lobe (the *precentral gyrus* of primates). The post-central gyrus entertains a topical localization of somaesthetic sensibility, i.e. of general cutaneous sensibility and proprioceptive sensibility of a part of the body. For example, proprioceptive sensibility and general cutaneous sensibility of the thumb are lost or deleted when the thumb area is injured. The localization is topical, therefore, not functional. Similarly the retina is said to be projected upon the area striata of the occipital lobe and the organ of Corti upon Heschl's convolutions of the temporal lobe. Besides these well known systems, the cortical projection of olfactory sensations are now confined to the *pyriform cortex*. The site of cortical reception for taste and for the vestibular system has not been adequately delineated.

Besides the sensory projection nuclei of the dorsal thalamus two other varieties are present, namely, those of association and those of diffuse projection. The association thalamic nuclei project of the cerebral cortex.

The *nucleus lateralis posterior* projects to the parietal association areas and to the tip of the temporal lobe; the *pulvinar*, to the junctional regions of the three posterior lobes; the *nucleus anterior*, to the *gyrus cinguli* and the *nucleus medialis dorsalis*, to the prefrontal part of the frontal lobe.

Knowledge of the relationship of the thalamic nuclei of diffuse projection to the cerebral cortex and to the reticular formation was obtained by use of electronic recording devices. With the increase in number of investigators qualified to use this type of technique, the complexity of neural relationships multiplied. Certainly these methods have changed concepts of functional activity at every level of the central nervous system. New anatomical relationships have been sought to support the revolution in functional concepts. Many have been found. At the present time neuronal connections demonstrated by the evoked potential technique exist without benefit of anatomical substantiation or of functional interpretation. These findings are additional. The older well-known anatomical patterns remain.

The reticular formation, the central core of the brain stem receives direct fibers or collaterals from each of the sensory systems. From cell bodies within the

Table 1   Divisions of CNS and their Derivatives

		Basal Plate	Alar Plate	Floor Plate	Roof Plate
Cerebrum	Prosencephalon Telencephalon	None	Single telencephalic evagination, which becomes two	None	Telencephalon medium Preoptic recess Lamina terminalis— bed for commissures, ant., hippocampal and corpus callosum Angulus terminalis Tela chorioidea telencephali medii Paraphysial arch Velum transversum
	Diencephalon	Ventral thalamus Hypothalamus	Optic evagination Dorsal thalamus	Bed for optic chiasma, com. of corpus mamillare Postoptic recess Post. lobe of hypophysis	Tela chorioidea ventriculi tertii Pineal body Habenula commissure Post commissure
	Mesencephalon	Motor nuclei of Nn. III & IV Tegmentum	Cellular layers in colliculi	Bed for commissures	Bed for commissures
Isthmus					
Rhombencephalon	Metencephalon	None	Cerebellum Vestibular nuc. Cells destined for nuc. of pons & inf. olive	None	None
	Myelencephalon	Motor nuclei of cranial nerves V–XII Formatioreticularis	Cochlear nuc. Sensory nuc. of cranial nerves V, VII, IX & X	Commissures	Anterior medullary velum Posterior medullary velum Tele chorioidea ventriculi quarti
Medulla spinalis		Motor nuc. Adjustor nuc.	Sensory nuc. Adjustor nuc.	Bed for commissures ventral to central canal	Bed for commissures dorsal to central canal

reticular formation (cells are grouped or isolated) axons project directly to midline nuclei of the thalamus or synapse with cells in the midbrain tegmentum, the axones of which end in midline thalamic nuclei. This non-specificity is carried one step further, for these thalamic nuclei project diffusely to the upper layers of the cerebral cortex including the sensory projection areas as well as those of association. When these non-specific neurones in the reticular formation are activated with slow frequencies or are surgically severed the animal sleeps. In the former condition rapid frequency stimulation of the massa intermedia will awaken it, in the latter only momentary awakening is possible.

The descending systems modify the activity of spinal cord motor neurones. The system previously studied and evoked for control of all voluntary use of skeletal muscle was the corticospinal tract, originating in part in the motor cortex and terminating on cell bodies within the central gray matter of the spinal cord. All other descending systems synapse with at least one neurone before terminating in the gray matter of the spinal cord. The descending systems which lie within the generalized reticular formation can modify not only motor activity of the spinal neurones but sensory activity as well. The descending division of the reticular formation exerts generalized influences of a facilitory or of an inhibitory nature upon tone and upon movement. For example, movement is impossible in spite of intact corticospinal systems subsequent to large lesions in the reticular formation.

Indeed, sensory activity can be attenuated, enhanced, or inhibited by the activity of the reticular formation upon sensory endings, or upon synapses within sensory ascending systems. For example it is possible for attention given a new stimulus to attenuate the potentials of a cat's cochlea nucleus to click stimuli. Ascending or sensory conduction in the dorsal columns of the spinal cord can be inhibited by electrical stimulation of cortical loci, such as the sensory motor cortex. Again cephalic directed conduction in the reticular formation can be inhibited or facilitated by stimulation of cortreal loci. Many cortical fields seem to channel into the same system in the reticular formation because cephalic conduction over a definite pathway can be interfered with by stimulation of several cortical areas (10).

No similar non-specific projection has been found for the cerebellum. Phylogenetically that organ is subdivided in two regions, a small posterior region, the *flocculo-nodular lobe,* which receives the terminals of the vestibular neurones of the first and second orders and a large anterior division, the *corpus cerebelli* (11). The former (found in all vertebrates) is relatively small compared with the latter. The *corpus cerebelli* shows a remarkable topical localization as delimited by the evoked potential technique. For example general cutaneous sensibility and spinocerebellar fibers innervating a particular part of the body together with projection from the motor cortex or from the somaesthetic cortex for that same part are found to terminate in the same area of the cerebellum (12). On the other hand no evidence exists for a topical origin of efferent neurones. Those which stem from the anterior part of the corpus cerebelli inhibit tone, whereas those which stem from the lateral hemispheres facilitate it. Such fibers project to the reticular formation of the midbrain or medulla oblongata or terminate in the nucleus ventralis lateralis of the thalamus, which in turn projects to the motor cortex of the frontal lobe. This set of neurones together with the return corticopontile and pontocerebellar tracts forms one of the earliest reverberating circuits known to neurologists.

The recent explorations of the central nervous system reveal greater and more complex neuronal relationships. With feed-back circuits on a single neuron or between one region and another, together with evidence for greater control of motor activity and of sensory conduction by the cerebral cortex via the reticular formation, understanding of the contribution which the CNS makes to living, becomes little by little more satisfying.

MARION HINES

References

(1) Peale, T. L., "The Neuroanatomical Basis for Clinical Neurology," pp. 1–18. New York, McGraw-Hill, 1954.
(2) Penfield, W., "Neurologia: Normal and Pathological," *in* "Cytology and Cellular Pathology of the Nervous System," Vol. 2, pp. 421–480. New York, Hoeber, *2*, 1932.
(3) Young, J. Z., "The functional repair of nervous tissue," Physiol. Rev., **22**: 318–374, 1942.
(4) Fernandez-Moran, H., "The submicroscopic organization of vertebrate nerve fibers," Exp. Cell. Res. 3: 282–325, 1952.
(5) Heuser, C. H., and G. L. Streeter, "Development of Macaque Embryo," Contrib. Embry. No. 181. Carneg. Inst. Wash., **29**: 15–56, 1941.
(6) Streeter, G. L., "Developmental horizons in human embryos," Contrib. Embry. No 197. Carneg. Inst. Wash., **30**: 211–245, 1942.
(7) *ibid,* "Developmental horizons in human embryos," Contrib. Embry. No. 199. Carneg. Inst., Wash., **31**:27–63, 1945.
(8) Hamilton, W. J., *et al.* "Human Embryology," pp. 275–303 Baltimore, Williams and Wilkins, 1952.
(9) *ibid,* pp. 263–274.
(10) Henry Ford Hospital Symposium, "The Reticular Formation of the Brain," Boston, Little, Brown, 1957.
(11) Larsell, O., "The development of the cerebellum in man in relation to its comparative anatomy," J. Comp. Neurol., **87**:85–129.
(12) Hines, Marion, "The anatomic basis for a new concept of the functional activity of the cerebral cortex in primates," N. C. Med. Jour., **16**: 8 & 9, 1955.

BRANCHIOPODA (except Cladocera)

Branchiopods, often called phyllopods, belong to the phylum ARTHROPODA (exoskeleton of chitin and paired, jointed appendages), the class CRUSTACEA (two pairs of antennae, gill respiration, and usually a calcified covering), and the subclass Branchiopoda (leaf-like appendages on thorax that serve for both locomotion and respiration). They are found almost exclusively in temporary ponds of fresh-water, the only notable exception being the brine shrimp which occurs in Great Salt Lake, salt ponds, and in coastal salt-evaporating basins. Some species are also known to live at times in water of appreciable salinity, but always in bodies of inland waters; they are never found in marine waters. Phyllopods are sporadic in distribution and annual occurrence. The temporary pond habitat eliminates fishes as a regular inhabitant and predator which would devour these defenseless crustaceans, and permits their eggs to dry over a period of time which seems to be the chief stimulus needed for hatching in ordinary circumstances. Appendages are never found on the abdomen except in the Notostraca. Sexes are separate, and except for the brine shrimp, fertilization is necessary for development. Larval stages are nauplii. Four orders are recognized:

1. **Anostraca.** These are the fairy shrimps. There is no carapace; the body is elongated; the eyes are stalked. They swim upside down with their 11 to 17 pairs of thoracic appendages. They feed on plankton organisms and detritus and in turn are devoured by carnivorous insects and amphibians. The second pair of antennae on the males are distinctive and used as key characters for identification of species. These antennae of the male, plus a frontal organ or antennal appendages, in certain cases, are used as clasping organs. The female develops an ovisac at the junction of the thorax and the abdomen. After fertilization this egg sac becomes filled with numerous brown-shelled eggs which are either dropped in the water or remain in the sac at the death of the female when the water of the pond dries up or becomes too warm. Ordinarily the eggs are dried before they hatch. Drying and aging are stimuli for hatching. A total of 28 species is known in North America.

2. **Notostraca.** These are the tadpole shrimps, which have much of the body covered with a broad, shield-like carapace on the dorsal surface. These live on and in soft mud and frequently swim upside down.

Appendages are numerous (40–60), variable in number and size, and serve for both locomotion and respiration. Eyes are sessile. Some segments contain more than one chitinized ring; some rings contain more than one pair of appendages. Females carry ovisacs on the eleventh pair of appendages. Six species are known from North America, all of them found west of the Mississippi River.

3. **Conchostraca.** These are the clam shrimps or claw shrimps. They are usually more or less compressed and enclosed entirely with a bivalve shell. Eyes are sessile and close together. Trunk appendages number from 10 to 28 pairs, of which the first one or two are modified in the males. The second antennae as well as trunk appendages are used for swimming. Clam shrimps are found in warmer water and later in the season than the fairy shrimps. Species are determined by the growth rings on the shells, the structure of the head and its rostrum, the clasping organs of the male (modified first two trunk appendages), and the structure of the telson with its dorsal spines. Twenty-six species are known from North America.

4. **Cladocera.** These are the water fleas and are the subject of a separate article.

RALPH W. DEXTER

References

Edmondson, W. T. Ed., "Ward and Whipple's Fresh Water Biology," New York, Wiley, 1959.
Pennak, R. W., "Fresh-water invertebrates of the United States," New York, Ronald, 1953.

BREHM, ALFRED EDMUND (1829–1884)

This German naturalist was probably the best example in history of the self taught scholar. He was born in the little village of Renthendorf and died in the same house. His knowledge of animals was gained only in a very minor part from university study in Vienna and in major part from long travels in Egypt, Nubia, Sudan, Spain, Norway, Sweden, Lapland, Siberia, Russia, and in most other parts of the then civilized globe to which he could gain access. He undertook these journeys for the purpose of observing animals in their natural environment and in 1864 published the volume known universally as "Brehm's Tierleben." The "Life of Animals" was translated into most of the languages of Europe and is currently in print in many of them. The characteristic of this book which endeared it to so many readers was that the author based his writing on the personal observations of animals in the field. Most previous works of this type have been enlivened by uncritical reproduction of the tales of local neighbors. Brehm's accuracy set a standard which has been followed by every serious natural history since his time.

BROWN, ROBERT (1773–1858)

One of the greatest botanists of all time, was born on 21 December, 1778 at Montrose, Scotland, where his father was the Episcopalian minister. In 1789 he entered the University of Edinburgh to study medicine and there came under the influence of John Walker (1731–1803) the professor of natural history and began sedulously to collect and list the Scottish flora. He was commissioned in 1795 as an assistant-surgeon in the Fifeshire regiment of infantry. A recruiting mission in 1798 brought him to London and gave him the opportunity to study in the private herbarium and botanical library of Sir Joseph Banks (1743–1820) and to impress him with his ability and enthusiasm. No man possessed a greater influence on scientific investigation at this period than Sir Joseph; he was keenly interested in the development of Australia, then called New Holland, to which he had sailed with Cook, and he saw in Brown the ideal naturalist to accompany Matthew Flinders (1774–1814) on his voyage to explore the coast of New Holland. Brown immediately accepted the post when offered by Banks. Flinder's leaky ship M.M.S. *Investigator* sailed in July, 1801 and reached King George's sound, Western Australia in December, 1801. Thus Brown made the acquaintance of the Australian flora in one of its areas of greatest diversity and highest endemism; it confronted him immediately with problems of classification and morphology to which his solutions have had a far-reaching effect. The *Investigator* circumnavigated Australia, having unfortunately to leave the western coast unsurveyed owing to the increasing unseaworthiness of her timbers, and reached Port Jackson for the second time in June, 1803, there to be condemned. Flinders set off for England to get a new ship, was interned by the French at Mauritius and not released until 1810. Luckily Brown and the artist Franz Bauer stayed behind. They returned to England in 1805. Brown was then appointed librarian to the Linnean Society of London. In 1810 he became librarian to Sir Joseph Banks, who bequeathed him in 1820 the life-use of his house at Soho Square and his collections. In 1827 he agreed with the Trustees of the British Museum for the transfer of these collections to their keeping, with himself as keeper, a position he held until his death.

The most important of Brown's publications is his *Prodromus* Florae Novae Hollandiae (1810; facsimile, 1960) which sold so few copies that only the first volume was published. He followed it with a series of memoirs, many published as appendices to records of voyages and travels, e.g. by Flinders, Salt, Tucky, Ross, Parry, Denham and Clapperton, and Sturt, most of the others in the *Transactions* of the Linnean Society. The discoveries here recorded are so many and various, relating, among much else, to the morphology and pollination of the flower in Asclepiadaceae and Orchidaceae, the nature of the ovule, polyembryony, the gymnospermy of conifers and cycads, the nucleus of cells, the classification of families and genera, the statistical comparison of floras, that they cannot be concisely summarized. One of the most interesting outcomes of his work was his observation of the oscillatory motion of minute particles hence forward known as "Brownian movement."

WILLIAM T. STEARN

BRUNO, GIORDANO (1548–1600)

The speculations of this Renaissance philosopher on the relation between organic and inorganic matter were so far ahead of his time that only recently have biologists become aware of them. It was, in his day, sufficient that he dared to question Aristotle, a heresy that caused him to be chivvied all over Europe seeking a place in which he could both think and write with reasonable safety. He never found one and stupidly went to Venice, about 1592, where he was arrested on a charge of heresy and extradited to Rome. He spent seven years in prison before being excommunicated and burned at the stake in 1600. That part of Bruno's philosophy which bears on contemporary evolutionary thought dealt with his belief in a "a universality in which every particle of the universe is composed inseparably of the physical and the psychical." Bruno felt that this universality was not, however, static and though the concept of evolution was unthinkable to him, he nonetheless turned the minds of others in this direction. In fact, "the man is an expression of the innate psychical quality that Bruno thought existed as one with every particle of the universe. In short, the mind-matter-energy substance is seeking consciousness through evolution." The quotations in this this brief article are from McKinley, G. M., "Evolution: The Ages and Tomorrow" (New York, Ronald, 1956) which should be consulted for a more detailed examination of the impact of Bruno's philosophy.

PETER GRAY

BRYOPHYTA

For the past century or so, our concept of the division or phylum Bryophyta (the bryophytes) has comprised the mosses and the hepatics. Within the past few years, however, the intrinsic differences between the various groups of mosses and of liverworts have been considered to be large and significant enough to justify the establishment of two divisions or phyla, and still other liberties have been taken with the basic classification of the original Bryophyta (Steere, 1958). Nevertheless, when one looks at the various major groups of bryophytes, he finds certain fundamental and underlying similarities and relationships common to all groups, no matter how far apart they may appear to be, in our cross-sectional view of today, which represents only the evolutionary apices of widely divergent phylogenetic lines. The report of spores of a bryophytic nature from Lower Cambrian deposits of the eastern Baltic region makes one realize the enormous age of these plants and the time available for groups to evolve–and to become extinct.

All bryophytes have biflagellate sperms, multicellular sex organs with sterile one-layered coverings or jackets, and reasonably similar sporophytes permanently attached to the gametophytic plant and dependent upon it for part of its moisture, although it is normally green when developing and can produce all or most of its carbohydrates. Mosses and their allies are unique in the plant kingdom because the green vegetative plant is the haploid sexual generation, and the diploid spore-producing structure, although green throughout its development, is permanently epiphytic on it. The green sexual plants grow rapidly and may form extensive and often conspicuous masses of vegetation. Moreover, many species have developed some means of vegetative reproduction, so that a large colony of any one species may be only a single clone. Each sporophyte, however, results from a separate fertilization. Although the clonal nature of the sexual plants may tend to reduce the amount of outbreeding, it cannot prevent it, since free-swimming sperms may be splashed or washed in from some distance. As a consequence of the different nature and function of the haploid sexual plant and the diploid asexual plant that perches upon it, at least while it is producing spores, these two plant generations have evolved at different rates and in different directions.

To the botanist who reads one evolutionary story in the sexual plant and quite another in the sporophyte, mosses and their relatives are quite literally double-headed monsters. In the hepatics and in peat mosses, the sporangium is extremely conservative and rarely shows clear differences between closely related species. In the true mosses, on the other hand, one occasionally finds clear-cut and consistent differences between species in their spore capsules, occasionally even in genera in which the green plant may be highly conservative or unspecialized.

A brief resume of the major group of bryophytes will illustrate the foregoing remarks and set the stage for more detailed discussions later.

1. **The hornworts or Anthocerotes** form a highly isolated, natural and distinctive group of plants. Several peculiar or unique features characterize them: (1) the sporophyte is the only one among mosses and their relatives that continues to grow in length, forming new spore mother cells at the base while releasing mature spores at the apex, through the presence of a meristematic zone just above the absorbing foot. (2) The very large chloroplasts, sometimes only one to a cell, and the normal presence of a pyrenoid, are unique features in bryophytes. (3) The ANTHERIDIA and ARCHEGONIA immersed in the tissue of the thallose plant are entirely unlike those of other bryophytes, and resemble more those of ferns. (4) The presence of a columella around which the spores are produced, the presence of highly developed stomata and photosynthetic tissue in the walls of the capsule, the relative rarity of symbiotic fungi and especially the distinctive basic chromosome number, $n = 5$, among a whole constellation of other features, clearly demarcate the Anthocerotes from other bryophytes.

2. **The true hepatics or liverworts—Hepaticae or Hepatophyta** consist of several diverse evolutionary lines with a high degree of specialization. The vegetative plants vary from extremely simple to highly specialized cell masses or thalli, or consist of leafy stems, with almost all intermediate conditions. The sporophytes likewise present a wide range of variation, all the way from a simple spherical case full of spores to more complex structures consisting of an attachment organ or foot, a stalk—sometimes very long—and a spore case with specialized devices for

the release and distribution of spores. In contrast to the diversity of form and structure among hepatics, taken as a whole, we find a remarkable uniformity in the presence of elaters among the spores and, in late years, in the CHROMOSOME numbers. By far the great majority of hepatics so far investigated has the chromosome number, n = 9. In a very few genera we find the number, n = 10, and in somewhat more genera, n = 8. In a few species the chromosome number is complicated by the sex-determining mechanism (XO type), so that the different sexes will have different numbers. POLYPLOIDY is remarkably rare among hepatics, in comparison with most plants.

3. **Takakia lepidozioides** Hatt. & Inoue is a recently discovered new member of the bryophytes whose place in the over-all classification is still a mystery, since its structure is remarkably suggestive of an ALGA. Although close to the hepatics in some of its characteristics, it will undoubtedly end up eventually as a new class, evidence toward which is given by its chromosome number, n = 4, a number found nowhere else in bryophytes (Tatuno, 1959).

4. **Peat mosses or Sphagnobrya.** All belong to a single genus, *Sphagnum*, in which leafy plants arise from buds formed around the margin of a flat disc-shaped thallus that develops from the germinating spore. The fascicled branches, the complicated stem structure, the remarkable leaf consisting of a mesh of small green cells fencing in large bubble-like dead cells, the total lack both of rhizoids and of special means for vegetative reproduction are all unique features among bryophytes. Other noteworthy characters are the long-pedicelled spherical antheridia, more like those of hepatics than of mosses, and the development at the apex of the vegetative plant of a pseudopodium whose extension carries the sessile spore capsule aloft. Moreover, the sporophyte differs from that of true mosses in several fundamental ways, including its embryogeny. This group is quite separate from all other bryophytes, and its distinctiveness is further enhanced by its unique chromosome number, n = 19, which is varied only by rather rare cases of polyploidy.

5. **True mosses—Musci or Bryophyta** are the largest and most complex, as well as the most obvious, polyphyletic group of bryopnytes. The vegetative plants, without exception, have leaves and usually stems that range from very short to a length of several feet, especially in flowing water or when pendent from the branches of trees in rain forests. The sporophytes vary from thin-walled cases full of spores attached to the green vegetative plant by a simple absorbing organ to relatively specialized structures consisting of a highly developed absorbing organ, a long stalk with complex conducting tissues, and a highly specialized, highly photosynthetic capsule, with remarkable structures and mechanisms for the release and dispersal of spores. The usual occurrence of stomata on the capsule walls, the germination of the spores to produce protonemata, the invariably leafy stems, and many other features in common give a somewhat deceptive over-all appearance of homogeneity to the group that disappears only when careful attention is given to the component families, orders and classes. Overwhelmingly distinct and even unique modifications are to be found in groups not normally

thought of as being remarkable. The superficial lack of morphological diversity of the high degree found in hepatics has militated against detailed research on the morphology of true mosses. Several groups of mosses, as those centered about *Andreaea* and *Archidium*, are distinctive and of very uncertain phylogenetic position. Serious study of the mosses in future years will make necessary the reclassification, dismemberment and regrouping of many families and orders that represent present-day survivors of divergent phylogenetic lines.

In summary, our present-day Bryophyta must be considered as a constellation of divergent groups rather than as a homogeneous division or phylum. We would be well justified in recognizing a new group, the "Mosses and Moss Allies," just as a few decades ago we recognized the "Ferns and Fern Allies," then grouped together under the division Pteridophyta. Another decade or so may well see the complete breaking up of the Bryophyta and the reallocation of some of its constituent groups to other areas of the Plant Kingdom.

Ecology of bryophyta. Mosses and liverworts are small plants that would hardly be noticed as individuals. However, since they ordinarily grow in tufts, cushions or extensive mats, they contribute much, in the aggregate, to the green color of forests, mountains and moors, especially in foggy or rainy weather. The association of mosses and liverworts with moist habitats is no accident, since none of these plants has true roots, and they can, therefore, get little water from below the surface of the soil. They need abundant moisture not only for their ordinary vegetative growth but also for their normal reproduction. Although a few mosses are able to live on rocks exposed to the sun in deserts and may not receive water more than a few times a year, the great majority of them grow in moist places, and it is obvious even to the casual naturalist that the moister the climate, the more abundant and conspicuous are the mosses and liverworts.

In mountain forests where clouds hang low at night or where mist and rain are of almost continual occurrence, so that the air is saturated with moisture during at least part of each day, mosses and liverworts become so conspicuous that this type of vegetation is known to botanists and to geographers as a special type, the "mossy forest." Mossy forests of this type are found on the coastal ranges of the Pacific Northwest of North America, where mosses clothe the trees and hang from them in festoons, and may even become weeds in fruit orchards. In the wet, cool forests of tropical mountain peaks, mosses and hepatics are so abundant that the apparent diameter of tree trunks is doubled by their growth, and even the leaves of trees and shrubs may be covered with small specialized bryophytes. In arctic regions, where there are no forests, the abundance of mosses clothing rocks and ledges, and in bogs and marshes, is directly related to the amount of precipitation and humidity.

Mosses are very hardy plants and can resist extremely unfavorable conditions. As one ascends a high mountain, he finds mosses and liverworts covering the rocks and the ground long after the total disappearance of trees, and these plants which appear to be so delicate have been collected at altitudes as great as 18,000 feet above sea level in the Himalayas and in the Andes of South America. Likewise, as one goes north in Europe, Asia and North America, he finds that, in many tundra areas beyond the tree line, mosses may become the most conspicuous land plants as well as the most numerous in species, under favorable conditions. For this reason they are among the very few plants, otherwise mostly lichens, that flourish on the Antarctic Continent.

Bryophytes are unusually sensitive, among plants, to variations in the substratum upon which they live, not only in moisture, but in acidity, alkalinity, the presence of certain metallic ions, and other factors. Certain mosses and liverworts require an acidic environment, and some of them, as peat mosses, actually manufacture the acids themselves. Others require an alkaline or calcareous environment, and cannot tolerate acids. Still others grow best in neutral soils and have considerable tolerance both to acid and alkaline conditions. The very close relationship of bryophytes to the soils, rocks, and barks upon which they grow, because of their lack of roots, combined with their sensitivity to the environment, cause them to be excellent indicators of the nature of the environment—one can tell at a glance whether the substratum is acidic or basic, whether usually dry or wet, whether sunny or shaded, and many other combinations of conditions. Some mosses and hepatics are reported to be specific to rocks containing copper and other minerals, and a careful study may well reveal useful indicators for specific minerals.

W. C. STEERE

References

Steere, W. C., "Evolution and speciation in mosses," Amer. Nat. **92**: 5–20, 1958.

Tatuno, S., "Chromosomen von *Takakia lepidozioides* und eine Studie zur Evolution der Chromosomen der Bryophyten," Cytologia **24**: 138–147, 1959.

BRYOZOA

Bryozoa (C. G. Ehrenberg, 1831) and Polyzoa (J. B. Thompson, 1830) are terms that have been widely used to designate identical colonial animals.

The Bryozoans (= Polyzoa) include members of two phyla of sessile invertebrates, the Entoprocta and the Ectoprocta, which were formerly considered to constitute the Phylum Bryozoa. The term "bryozoan" is still used by some authorities to include both groups, while others equate it with the Ectoprocta alone. To avoid confusion, it seems best to drop both "Bryozoa" and "Polyzoa" as taxonomic terms, and retain them only in the vernacular sense, for both groups. The Entoprocta differ from the Ectoprocta in cleavage patterns, mode of coelom formation, and in some anatomical details. The phylum names refer to the location of the anus; in the Entoprocta both oral and anal openings lie within a circle of tentacles, whereas in the Ectoprocta the anus lies outside the tentacle circle.

Phylum Entoprocta Nitsche, 1869. The Entoprocta

are distinguished as having spiral determinate cleavage, pseudocoelomate body cavity formation, and primitive excretory organs, the protonephridia. Members of the phylum are sessile, solitary or colonial, stalked animals less than 2 mm tall, which occur primarily in shallow marine waters. There are no higher taxa, and only two families. The Loxosomatidae are solitary but form dense clusters of individuals epizoic on worms, worm tubules, ectoprocts, tunicates, sponges or crustaceans. The Pedicellinidae are colonial, stolonate forms, some of which are common fouling organisms.

The body of the entoproct consists of a stalk which is attached to the substratum at the base, and is topped by a bulbous calyx. The calyx contains the internal organs of the animal, and bears a circle of tentacles around the face of the calyx (the vestibule). The calyx and the stalk are covered by a thin cuticle. Internally, the pseudocoelom contains a cellular packing tissue. The "U"-shaped alimentary tract that occupies most of the calyx is subdivided into a mouth, esophagus, cardium, intestine and an anal aperture opening within the tentacle circle. The protonephridia of the Entoprocta occur as paired "flame bulbs" connected to a pore opening in the region of the oral aperture. The nervous system is very simple. It consists of a ganglion located near the stomach, from which a series of nerve fibrils extend to the tentacles, the stalk, and the wall of the calyx. Most species are dioecious, with two gonads lying above the stomach in the calyx. A single gonoduct opens in the center of the vestibule between the mouth and the anus. Eggs apparently are fertilized internally and covered with a membrane before leaving the gonopore. They then become attached to the vestibule, forming a larval "brood chamber." Cleavage is of the spiral determinate type, and results in the formation of free-swimming larvae. The free-swimming period terminates in the settling of the larva and the metamorphosis

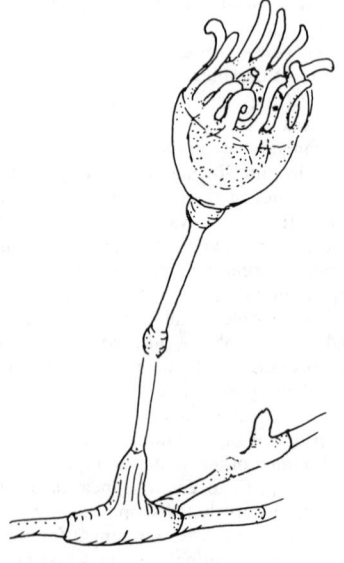

Fig. 1   Entoprocta.

into the initial individual of a colony. Colony formation then takes place by asexual budding, in some
forms from the calyx and in others from the stalk.
In solitary forms, budding produces new individuals
which detach and settle close to the parent. A foot
gland in the base of the stalk provides for attachment.

When one observes a living colony of entoprocts
under magnification, the feature of the calyx that
immediately catches the attention is the circle of
ciliated tentacles. The tentacle circle, when expanded,
creates water movements that bring food in the form
of protozoans and diatoms to the mouth opening. The
tentacle crown is very sensitive to tactile stimuli. The
tentacles cannot be withdrawn inside the calyx, but
roll inward upon themselves on the face of the calyx
like clenched fingers, when disturbed.

Phylum Ectoprocta Nitsche, 1869. The Ectoprocta
are characterized by having radial or biradial cleavage and true coelomate body cavity formation. Excretory organs are lacking in this phylum. The group
is predominantly marine, with a few freshwater
representatives.

The members of the phylum Ectoprocta form a
diverse group that is usually divided into two classes,
the Class Gymnolaemata and the Class Phylacto-
laemata. The latter includes the majority of the
fresh-water bryozoans, while the gymnolaemate
group is predominently marine. The Class Gymnolaemata is further divided into five major orders,
namely the Ctenostomata, the Cheilostomata, the
Cyclostomata, the Trepostomata and the Crypto-
stomata. Of this group, the last two named are known
only from the fossil record. The Order Ctenostomata
is divided into two suborders, the Stolonifera and the
Carnosa. The Cheilostomata likewise can be divided
into two main groups, the Anasca and the Ascophora.
The Cyclostomata have five living suborders, the
Articulata, the Tubuliporina, the Cancellata, the
Cerioporina and the Rectangulata.

Obviously there is a great diversity within this
phylum. However, all possess certain general characteristics in common. The ectoproct bryozoans are
small colonial, rarely solitary, aquatic animals that
are primarily marine in habitat, with only a comparatively few freshwater forms. They occur in great
abundance in present day oceans, occupying many
ecological niches from the intertidal zone to abyssal,
with probably the greatest abundance in the littoral
zone. Geographically the ectoprocts are found from
arctic and antarctic waters to the tropics, and the
group includes a surprising number of cosmopolitan
species in the ctenostome and cheilostome orders.
The colonies (called zoaria) are adherent to a wide
variety of substrata encrusting rock, mollusk shells,
algae, hulls of ships, dock pilings, in fact anything that
affords a satisfactory point of attachment is used.
Some species show a definite preference for certain
types of substrate. For example, the *Membranipora*
(cheilostome, anascans) are rarely found on anything
but the stipes, blades, or holdfasts of marine algae.
Infrequently they may be found on mollusk shells
or rock. The ectoproct colony presents a variety of
structure and form. It may be erect, or creeping, encrusting, or even boring into the shells of mollusks

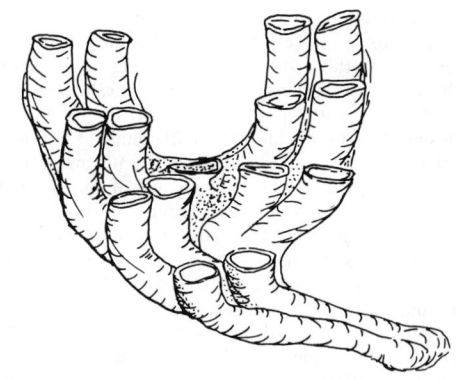

Fig. 3   Ectoprocta, Cyclostomata.

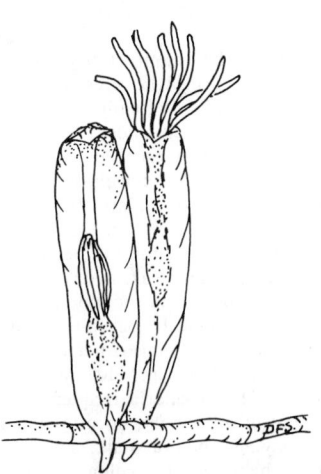

Fig. 2   Ectoprocta, Ctenostomata.

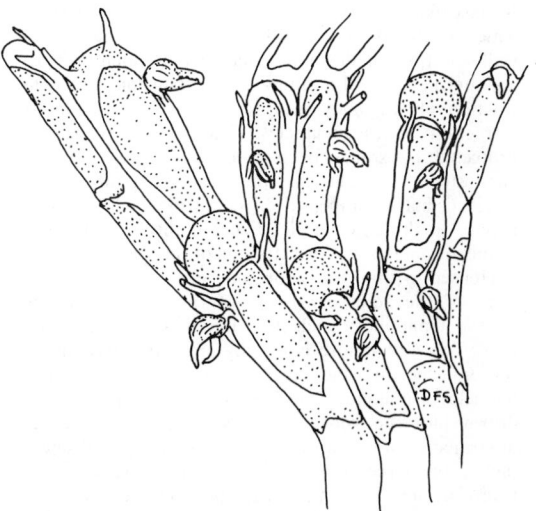

Fig. 4   Ectoprocta, Cheilostomata, Anasca.

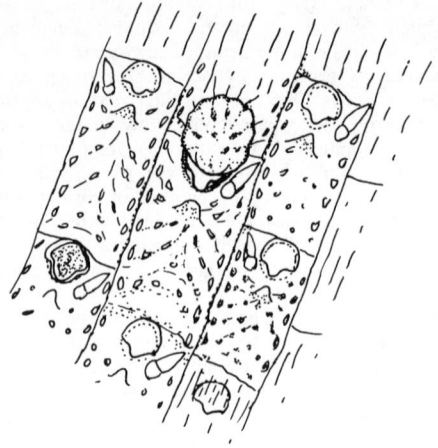

Fig. 5   Ectoprocta, Cheilostomata, Ascophora.

and brachiopods. The encrustations are unilaminar or multilaminar; by continued superposition, the latter may form masses of considerable size. The erect forms arise in tufts, fronds or branching tubes. Some of the erect forms construct an open meshed lattice-work of regular pattern.

Throughout the phylum, the individuals of a colony (zoarium) conform to a surprisingly uniform type of structure. Each zoarium is made up of many small individuals (zoids) which are rarely as much as a millimeter in length. The zoid is composed of a double-walled sac (the zooecium) which houses the lophophore, a protuberance bearing the tentacles and mouth, and the visceral mass or polypide. The ectoprocts are classified primarily according to the features of the zooecium. In the Ctenostomata the zooecium is noncalcareous, while the Cyclostomata and most Cheilostomata are calcareous. The cteno-stome and cyclostome body form is usually tubular with a terminal aperture. The cheilostome zooecium is box-like, usually with a subterminal aperture which is closed by a lid-like operculum. The frontal wall of the cheilostome anascans is membranous and uncalcified. This allows for expansion and con-traction of the body wall to facilitate extrusion or retraction of the tentacles. The frontal is usually protected by spines or a calcareous shield. In the cheilostome ascophorans the frontal membrane is calcified, but a compensation sac (asc) which opens by a pore to the exterior lies beneath the frontal mem-brane to compensate for extrusion of the tentacles.

Housed within the zooecium are the living por-tions of the zoid, consisting of the tentacles borne by the lophophore and the visceral mass, the poly-pide. The lophophore is usually circular, surrounding the oral aperture and bears a series of tentacles. The tentacles are hollow, possess cilia, and can be with-drawn into the zooecial cavity. They create water movements conveying food particles to the oral open-ing of the visceral mass. Within the zooecium is a true coelom in which the polypide or visceral mass is suspended. The polypide is provided with a tentacle sheath and a "U" shaped digestive tract. The

oral aperture or mouth is within the circle of tentacles as indicated above, while the anal aperture opens outside of the tentacle circle. The digestive tract is divisible into a number of regions. Leading from the mouth there is the pharynx and the esophagus which may or may not be sharply defined. The esophagus opens into the stomach that is divided into a cardia, a caecum and a pylorus. From the pylorus the rec-tum extends to the anal aperture that is outside of the tentacle ring. In the ctenostomes, some genera have the cardia modified into a grinding organ or gizzard.

Within the coelom the nerve ganglion is usually found in the region of the pharynx or esophagus. From the nerve ganglion a meshwork of fibrils run to the ten-tacles and the tentacle sheath, the musculature, and the digestive tract.

The muscles of the zooecium consist of the parietals, the aperturals and the retractors. The parietals help protrude the tentacle sheath by increasing hydro-static pressure within the zooecium. The aperturals open the oral aperture, while the retractors return the tentacles into the cavity of the zooecium.

Both sexes are usually combined in the same zoid, either simultaneously or in sequence. The reproduc-tive organs, while they may be found in various por-tions of the zooecial cavity, are usually arranged so that the testes are in the lower and the ovaries in the upper. Occasionally, the reproductive organs of each sex may occur singly in different individuals. Sperm are known to be extruded through the tentacle tips in some species. Generally, fertilization is internal, and the ova develops into the larval form in a special structure, the ooecium or ovicell. The cleavage is of the holoblastic radial or biradial type; the re-sulting larva escapes from the ovicells for a short free swimming period before settling on some suit-able substratum. Upon settling, the larval internal structure is broken down and the first zooecium of the new colony is formed. This initial individual is called the ancestrula, and gives rise to the first genera-tion of daughter zooecia by asexual budding. The buds of the first generation of daughter zooceia usually appear within a few hours after the formation of the ancestrula. Repeated budding from the daughter zooecia soon will produce a zoarium of consider-able size.

No mention of the ectoproct bryozoans would be complete without noting the presence of the avicu-laria and the vibracula. Actually, the avicularia and the vibracula are modified zoids, the result of a form of polymorphism that is especially prevalent among the Cheilostomata. The avicularia frequently re-semble a birds head, having a large beak-like mandible that can open and close. Numerous theories have been advanced to account for the presence and func-tion of the avicularia. There are two types of avicularia, the vicarious avicularia, that occupy a spot in a zoar-ium that normally would be occupied by a zooecium, and the adventitious avicularia that can occur almost anywhere on the colony, and do not take the place of a zooecium. The avicularia frequently show a sur-prising amount of movement, opening and closing the mandible, swaying upon a stalk, etc. Functionally, the avicularia help keep the colony clean and entrap and hold small animals that blunder onto the surface of the colony.

The vibracula, also modified zooecia, are frequently elongated bristle-like structures, capable of making sweeping movements over the colony. These sweeping movements probably serve to prevent debris from settling on the surface of the colony and also to create water currents to help in the feeding of the zooecia, to facilitate oxygenation and assist in the removal of the metabolic by-products.

JOHN D. SOULE

References

Harmer, S. F., "Polyzoa," *in* Harmer, S. F. and A. E. Shipley, Eds., "Cambridge Natural History," Vol. 2, London, Macmillan, 1910.

Hyman, L., "The Invertebrates," Vols. 3 and 5, New York, McGraw-Hill, 1951 and 1959.

BUFFON, GEORGE LOUIS LECLERC, CONTE DE (1707–1788)

This wealthy French nobleman was the first great encyclopedist of the sciences. He was born at Montbard in Burgundy and studied law at Dijon. His interest for this was, however, slight and he soon turned to physics and to mathematics as offering a better expression to his original and productive mind. His writings attracted the attention of Louis XV who appointed him keeper of the Jardin du Roi, probably with a view to providing a sinecure for a favorite. Buffon, however, threw himself enthusiastically into a study of the biological sciences and all his subsequent publications are in the field of "natural history," a subject which at that time included geology. His

monumental *Histoire Naturelle, Générale et Particulière* appeared in 44 quarto volumes over the period 1749–1804. Buffon was not, of course, the sole author of these volumes but acted more as editor of what would today be called an encyclopedia. This work attracted the attention of scholars all over Europe and successive volumes were eagerly awaited. This great work is often criticized today as being in parts superficial but in point of fact it far exceeded in accuracy any previously published work in this field. It was translated into many languages and many pirated editions, mostly of the anatomical and zoological portions, were still being published as late as the beginning of the present century.

PETER GRAY

BÜTSCHLI, OTTO (1848–1920)

This German zoologist was born at Frankfurt-am-Main and held the chair of zoology in Heidelberg for the last thirty years of his life. He was primarily a protozoologist and contributed the first three volumes on this subject to Bronn's *"Klassen und Ordnungun des Tierreichs."* He is, however, best remembered for his theory of the structure of protoplasm which held sway for many years. In Bütschli's view, the protoplasm was not, as it appeared under the optical microscope due to the artifacts of fixation, a fibrillar structure but was, in point of fact, a foam. It has often been said that he derived these views solely in virtue of the fact that he graduated in the fields of chemistry and mineralogy before turning his attention to zoology.

C

CAMBIUM

Cambium is the term given to meristems involved in diameter accretion in plants. Tissues derived from the cambium are termed secondary. Two types of cambium occur: vascular cambium concerned with addition to the wood (XYLEM) and bast (PHLOEM), and cork cambium (phellogen) which produces protective tissue (cork or phellem) making up a part of the bark. (See illustration to article STEM.)

The *vascular cambium* exists as a narrow strip between the xylem and phloem. In conifers and broad-leaved trees and shrubs it extends around the whole circumference in the form of a cylinder, but in many of the smaller dicotyledonous herbs it is reduced to small arcs within the vascular bundles.

When dormant the cambium in conifers is narrow, consisting of from 1 to 4 or more rows of cells, but during the growing season the zone expands to a width of several cells. Structurally the meristem consists of a single tier of cells, known as initials, which determine the basic pattern of the derived tissues, and immediate derivatives, termed mother cells, lying on either side. Growth entails division of the initials to produce the mother cells which redivide to yield cells that shortly mature into xylem and phloem elements. Width of the zone of division varies with the rate of growth. In fast growing trees the zone of cell production during the flush of spring growth expands to a width of 10–30 cells. At this time most active cell division takes place in the xylem mother cells, the rate of division among these surpassing that in the layer of initials. The width of the zone of dividing mother cells becomes much wider on the xylem than on the phloem side and hence the development of new xylem surpasses that of phloem. As growth slackens later in the growing season, the zone of dividing mother cells narrows, particularly on the xylem side, and the rate of production of new wood falls off. The development of xylem is thus usually characterized by a period of rapid growth through the spring, followed by slow and sometimes erratic accretion during the summer and early autumn. Phloem development generally begins somewhat later than that of the xylem. In some species it is produced at a more or less uniform rate through the growing season, whereas in others maximum development is coincident with the grand period of xylem growth. Often some of the last-formed phloem elements do not complete maturation until the following spring. The annual increments of phloem tend to vary less in width than those of the xylem, often being almost as wide in years of poor growth as in years of good development.

In broad-leaved trees various situations have been reported in the literature with respect to the timing of xylem and phloem development after cambial reactivation in the spring. A precedence of xylem production over that of phloem has been reported for maple and orange; a simultaneous beginning of xylem and phloem has been claimed for ash, grape, apple and pecan; and a delayed development of xylem has been observed in pear. Regardless of the relative times of xylem and phloem initiation, the annual production of xylem seems generally to exceed that of phloem.

On cessation of growth the meristematic zone becomes reduced to the layer of initials, one cell wide, and usually a very few rows of inactive mother cells. With respect to terminology it may be noted that some botanists prefer to restrict the term cambium to the uniseriate layer of initials, whereas others apply the term to the whole zone of cell production which includes the initials and the derived tissue mother cells.

The period of activity of the vascular cambium varies greatly. Among neighbouring trees of the same species the growing season may be as much as $1\frac{1}{2}$–2 months longer in some trees than in others. Duration of growth is also influenced by environmental factors such as light, rainfall and temperature. In cool temperate zones growth begins with the return of favourable temperatures, and ceases in the late summer or early autumn when soil moisture tends to become exhausted and day length shortens. Under these conditions the annual increments of wood are usually distinct and may be recognized as annual rings. In warm temperate and tropic zones, on the other hand, various conditions obtain. Some species in tropical rain forests exhibit continuous diameter growth without demarkation into annual rings. At other times rainfall and temperature may vary so as to induce rest periods of varying duration and occurring at different times of the year. When there is only one rest period per year a single growth ring is usually formed. This is sometimes clearly marked, in other cases ill-defined. If two or more rest periods occur during the year double or multiple rings generally result. The opposite extreme is met with in some desert areas where several years may pass without rainfall. Here the dry years are often marked by cambial inactivity. In cases where there are multiple rings on the one and, or only periodic activity on the other, ring counts clearly do not reveal the true age of the tree.

In temperate zones were growth is characterized by a definite periodicity, the inception of cambial activity seems related to bud opening. In most of the cases which have been studied, the resumption of cambial activity in the spring more or less coincided with opening of the leaf buds. From the point of initiation beneath the emergent leaves at the shoot and branch tips growth usually spreads down the branches and trunk to the roots. The stimulation of cambial activity is believed to be due to the movement of growth hormones originating in the buds.

Anatomically the initiating layer in the cambium of most plants consists of cells of two types. These are fusiform initials which are vertically elongate in form and give rise to the conducting and supporting elements of xylem and phloem (tracheids, vessel segments, sieve elements, fibres and vertical parenchyma), and short ray initials which are the source of the vascular rays extending radially through the xylem and phloem. Length of the fusiform initials ranges from 2–8 mm. in conifers down to as little as 0.2 mm. in certain broad-leaved trees. In species with very short fusiform initials neighbouring cells are usually in tangential alignment, an arrangement known as storied or stratified. Radial accretion entails circumferential expansion of the cambium which requires multiplication of the fusiform and ray initials. New fusiform initials are produced by a more or less transverse division of old initials in trees with a non-storied cambium, and by a radial longitudinal division in species with a storied cambium. Ray initials are formed periodically from fusiform initials by segmentation or the cutting off of small segments from the sides.

The vascular cambium is a general feature of the gymnosperms (cycads, ginkgo, conifers) and dicotyledonous angiosperms (flowering plants). In most of these plants there is a single cambial layer which lies between the xylem and phloem and by its activity adds to the xylem on the inside and the phloem on the outside. Occasionally additional cambial layers occur, usually in concentric arrangement, as in the root of the beet (*Beta vulgaris*). Another situation obtains in certain herbaceous dicotyledonous plants (*Chenopodium* and *Amaranthus*) where a cambium arises in the outer part of the stem beyond the vascular bundles. This additional cambium, sometimes referred to as an accessory cambium, is most active toward the inside (centripetally) where an embedding (conjunctive) tissue and groups of xylem elements are produced. Small amounts of phloem are sometimes laid down externally, but these become included on the inside with the conjunctive tissue and other vascular elements as a consequence of the development of new arcs of cambium. In certain arborescent monocotyledons (*Dracaena, Yucca, Alöe, Cordyline*) a cambium arises in the outer part of the stem and produces on the inner side a mass of parenchymatous conjunctive tissue within which vascular bundles differentiate.

A *cork cambium* (phellogen) generally arises in plants with extensive radial accretion. In stems and branches the phellogen usually forms underneath the epidermis, whereas in roots the site of origin is deeper, most often in the pericycle. The phellogen initials divide to produce more or less impermeable cork cells (phellem) externally and occasionally parenchyma cells (phelloderm)

internally. Collectively the phellogen and its derivatives are known as periderm. The first-formed phellogen sometimes continues producing cork cells for a number of years to develop a smooth bark, as in *Fagus, Prunus* and *Betula*. More often the first phellogen functions only briefly and is soon followed by others which originate in the phloem tissues within. The successive periderm layers and the intervening dead tissues form a bark which becomes furrowed or cracked, the rough bark of most evergreens and broad-leaved trees. The periderm functions quite efficiently to reduce water loss but heat transmission seems to be relatively unaffected.

M. W. BANNAN

References

Bailey, I. W., "Contributions to plant anatomy," New York, Ronald, 1954.

Bannan, M. W., "The vascular cambium and radial growth in *Thuja occidentalis L,*" Can J. Botany, 33: 113–138, 1955.

Büsgen, M. and E. Münch, "Structure and life of forest trees," New York, Wiley, 1929.

Esau, K., "Development and structure of phloem tissue," Bot. Rev. 16: 67–114, 1950.

Esau, K., "Anatomy of seed plants," New York, Wiley. *ibid,* "Plant anatomy," New York, Wiley, 1965.

Glock, W. S., "Tree growth II. Growth rings and climate," Bot. Rev. 21: 73–188, 1955.

Lier, F. G., "The origin and development of cork cambium cells in the stem of *Pelargonium hortorum,*" Amer. Jour. Bot. 42: 929–936, 1955.

Studhalter, R. A., "Tree growth I. Some historical chapters," Bot. Rev. 21: 1–72, 1955.

CARBOHYDRATE METABOLISM

Carbohydrate metabolism includes all the reactions undergone in the organism by the carbohydrates of the diet and by those formed in the organism from non-carbohydrate sources. Carbohydrates—literally, "hydrates of carbon"—are in reality the aldehyde or ketone derivatives of polyhydric alcohols. The starches and sugars are the chief members of the group. Carbohydrates have been classified in terms of the number of carbon atoms in the molecule. Thus, there are trioses ($C_3H_6O_3$), tetroses ($C_4H_8O_4$), pentoses ($C_5H_{10}O_5$), and hexoses ($C_6H_{12}O_6$). These simple sugars, or monosaccharides, are frequently found in polymerized states where two or more monosaccharide units are linked to one another through glycosidic linkages and are then called oligosaccharides. Examples of oligosaccharides are the two monosaccharide unit sucrose (a disaccharide) and the three unit one, raffinose (a trisaccharide). Large aggregates of monosaccharides are usually called polysaccharides. Starches and glycogen are common representatives of the latter.

Sources of carbohydrates. Of all foodstuffs, carbohydrates are used the most for energy sources by organisms. Pentoses and hexoses may be obtained in large quantity from animal and plant sources and, hence, are important food carbohydrates. Glucose, a hexose, is perhaps the most important of all carbohydrates. The disaccharide, sucrose, is found in all photosynthetic plants, and is probably the chief low-molecular weight carbohydrate in the natural diet of animals. Another disaccharide, lactose, occurs in the milk of mammals. The polysaccharides may be divided roughly into two categories: the "structural and "nutrient" polysaccharides. The "nutrient" polysaccharides, starches and glycogen, act as a metabolic reserve of monosaccharides for the vast majority of plants and animals. Starches occur abundantly in many plant tissues. Glycogen is the form of the important reserve of monosaccharides in carbohydrate metabolism of animals. However, the "structural" polysaccharides, such as cellulose of plants and chitin in the exoskeleton of arthropods may be carbohydrate sources for certain animals (e.g. certain snails, earthworms, symbiotic microorganisms in ruminants, and termites have cellulases; some snails have chitinases).

Digestion and absorption of carbohydrates. Extracellular digestion, as occurs under the influence of pancreatic and intestinal ENZYMES in the vertebrate gut, yields large amounts of monosaccharides from oligo-and polysaccharides. Assimilation of undigested disaccharides is known to occur, but it is the monosaccharides glucose, fructose, galactose, and mannose, that are most commonly absorbed. Differential rates of absorption of hexoses indicate some kind of active transport mechanism exists in the mucosa of the small intestine. The hormones insulin and thyroxin are known to influence the process of absorption. In the mammal, glucose must be considered as the form in which carbohydrate is transported through the blood from one part of the body to another. The liver is a site of regulation of blood glucose levels and can either remove excess glucose from the blood or add glucose to it. The conversion of glucose (or isomerized fructose) into liver glycogen may be summaried as follows:

$$\text{Glucose} \xrightarrow[\text{ATP}]{\text{Hexokinase}} \text{Glucose-6-phosphate} + \text{ADP}$$

(adenosinediphosphate)

$$\text{Glucose-6-phosphate} \xrightleftharpoons{\text{Phosphoglucomutase}}$$

Glucose-1-phosphate

$$\text{Glucose-1-phosphate} \xrightleftharpoons{\text{Phosphorylase}} \text{Glycogen} +$$

P (inorganic)

Glucose also may be supplied from other sources (gluconeogenesis), such as from amino acids and the glycerol portion of fat molecules.

In the cell the chemical energy of the glucose molecule may be liberated as heat or work. Perhaps twenty-five different enzymes participate in the process of glucose oxidation. These enzymes co-operate to bring about an organized chain of successive reactions to oxidize carbohydrates while retaining much of the free energy in a readily available form through the energy-rich adenosinetriphosphate (ATP).

Anaerobic utilization of glucose. The anaerobic degradation of glycogen to lactic acid by muscle cells (glycolysis), and the formation of alcohols or acids by microorganisms (fermentation), while occurring in widely differing organisms, serve to illustrate the remarkable similarity of the main line of anaerobic respiration through all life forms.

In fermentation the glucose molecule is phosphorylated first at one end and then at the other, then broken in two, then oxidized by a special mechanism which is coupled with a synthesis of ATP from ADP, then dehydrated and dephosphorylated to give pyruvic acid.

Glycolysis is essentially the same process as fermentation, except that only one mole of ATP is required per unit of glucose converted, an inorganic phosphate being required for the phosphorolytic cleavage of glycogen. The scheme below shows the events in anaerobic respiration.

(1a) Glucose + ATP (1b) Glycogen + phosphate

(1c)   Glucose-1-phosphate

(2)   ADP + Glucose-6-phosphate

(3)   Fructose-6-phosphate + ATP

(4)   Fructose-1,6-diphosphate + ADP

(5)   3-Phosphoglyceraldehyde +

Phosphodihydroxyacetone

(6) 3-Phosphoglyceraldehyde + P (inorganic) $\longrightarrow$ Phosphoglycerylphosphate

(7) Phosphoglycerylphosphate – 2H $\longrightarrow$ 1,3-Diphosphoglyceric acid

(8) 1,3-Diphosphoglyceric acid + ADP $\longrightarrow$ 3-Phosphoglyceric acid + ATP

(9) 3-Phosphoglyceric acid $\longrightarrow$ 2-Phosphoglyceric acid

(10) 2-Phosphoglyceric acid $\longrightarrow$ Phosphoenolpyruvic acid + H_2O

(11) Phosphoenolpyruvic acid + ADP $\longrightarrow$ Pyruvic acid + ATP

Yeast:

(12a) Pyruvic acid $\longrightarrow$ Acetaldehyde + CO_2

Acetaldehyde + 2H $\longrightarrow$ Alcohol

Muscle:

(12b) Pyruvic acid + 2H $\longrightarrow$ Lactic Acid

The coupling of oxidation with formation of energy-rich bonds is shown in reactions (6) to (8). The aldehyde group of the phosphoglyceraldehyde takes on a molecule of inorganic phosphate and is then enzymatically oxidized (by removal of hydrogens) to carboxyl with the phosphate still attached. The bond is thereby converted into a kind of anhydride link, which is energy-rich and unstable. Before it breaks down and loses energy, its phosphate is transferred to ADP forming an energy-rich (11,000 cal. per mole) ATP bond at the expense of triosephosphate oxidation. A similar reaction occurs when phosphoglyceric is converted to pyruvic acid (steps (10) and (11)).

Anaerobic decomposition of pyruvate ends in lactic acid formation not only in muscles but in some bacteria as well. Alcohol formation is characteristic of several bacteria and fungi, in addition to yeast. Formic, acetic, succinic, and propionic acids are end products in certain bacteria.

Aerobic utilization of pyruvic acid. Pyruvic acid is the focal point in the metabolism of several kinds of food-stuffs. Not only is it the hub toward which carbohydrate metabolism converges, but pyruvate is also produced in the utilization of fats and proteins. Only part of the energy in a molecule is liberated in pyruvic acid formation. Just as there are different pathways for the utilization of pyruvate under anaerobic conditions, there may exist a number of pathways for aerobic oxidation of pyruvate, but the one which is probably most generally utilized is KREBS TRICARBOXYLIC ACID CYCLE. Energy is liberated in a stepwise series of reactions involving dehydrogenations and decarboxylations in Krebs cycle with transfer of electrons from the substrate through flavoprotein and cytochrome carriers to the final acceptor, oxygen. In one revolution of the cycle, three molecules of carbon dioxide are liberated per molecule of pyruvate dissimilated and ten hydrogen atoms are started on the flavoprotein-cytochrome pathway to oxygen:

$$\underline{\text{Pyruvate } + \text{ 3 Water } + \text{ Oxalacetate}}$$
$$\downarrow \qquad\qquad\qquad \uparrow$$
$$\text{3 Carbon dioxide } + \text{ 10 Hydrogen } + \text{ Oxalacetate}$$
$$\downarrow$$
$$\text{Flavoprotein}$$
$$\downarrow$$
$$\text{Cytochrome}$$
$$\downarrow$$
$$\text{Oxygen}$$
$$\downarrow$$
$$\text{5 Water}$$

In anaerobic oxidation ATP formation occurs at the substrate level. In the aerobic phase the transfer of electrons through carrier systems to oxygen makes available rather large amounts of energy which also may be trapped as high-energy phosphate bonds of ATP. In fact, 36 of the 38 high-energy bonds formed in the conversion of a mole of glucose to carbon dioxide and water are formed at the electron transfer level.

Synthesis of carbohydrate. All the series of reactions from glycogen to pyruvate are reversible except the dephosporylation of phosphopyruvic acid. Energy therefore is required to synthesize glycogen. The synthesis of carbohydrates from fats may be accomplished through an acetoacetic acid link to pyruvic acid. Carbohydrate in excess over that which can be stored in the glycogen depots of animals is converted into fat by ways of synthesis of fragments from the Krebs cycle. Amino acids from protein metabolism may be transformed into carbohydrate through entry into the glycolytic pathway as pyruvate or as a component in Krebs cycle. In photosynthetic plants, the dark reduction of carbon dioxide apparently forms phosphoglyceric acid, which may subsequently form glucose and then starch.

C. L. RALPH

References

Buchanan, J. M., and A. B. Hastings, "The use of isotopically marked carbon in the study of intermediary metabolism," Physiol. Rev., 26: 120–155, 1946.
Pigman, W. W. "The carbohydrates; chemistry, biochemistry, physiology," New York, Academic Press, 1957.
Wood, H. G., "Significance of alternate pathways in the metabolism of glucose," Physiol. Rev., 35:841–859, 1955.
Greenberg, D. M., "Metabolic pathways," 3rd ed; Vol. I, New York, Academic Press, 1967.

CARBON DATING

Carbon dating combines the technology of nuclear physics with the materials of historians, archeologists, and geologists to provide a method for determining the absolute age of prehistoric objects and events. Discovered and developed by Dr. Willard F. Libby, the theory of radiocarbon dating is elegantly simple. The radioactive isotope Carbon-14 is produced in the upper atmosphere by the bombardment of Nitrogen-14 with neutrons derived from primary cosmic radiation. The radioactive carbon produced in this reaction is quickly converted to carbon dioxide and enters the biosphere by the uptake of carbon dioxide by photosynthetic organisms. Animals in turn incorporate radiocarbon into their tissues by eating the plants. As long as the animal or plant is alive, there is a continual exchange with atmospheric radiocarbon. This equilibrium is interrupted when the organism dies, and there is no further exchange with atmospheric radiocarbon. In approximately 5568 years, only half of the Carbon-14 originally present in the tissues will remain, and after another 5568 years, there will be only one-quarter of the amount of radiocarbon originally present. In most instances the organism will have long since decayed and the radiocarbon of its tissues will have been recycled into living organisms. Under favorable circumstances, however, the remains of an animal or a plant may be preserved as a fossil, or perhaps as a piece of charcoal. The ratio of the amount of Carbon-14 in the sample to the amount of Carbon-14 in modern wood or tissue indicates how long ago the organism died.

Production and distribution of radiocarbon. Nitrogen and Oxygen are the only elements in the atmosphere which are present in sufficient quantity to be appreciably affected by neutron bombardment. Although Oxygen is quite inert to both slow and fast neutrons, Nitrogen-14 is highly reactive.

Laboratory studies with neutrons of thermal velocity (slow neutrons) show that the reaction

$$N^{14} + n = C^{14} + H^1 \qquad (1)$$

is dominant, although other reactions can be demonstrated. Fast neutrons have effected the reaction

$$N^{14} + n = C^{12} + H^3. \qquad (2)$$

Both reactions occur in the upper atmosphere (30,000–50,000 feet) and the radioactive products, Carbon-14 and Tritium (H^3) occur in nature in amounts and concentrations which correspond closely to predicted values. The yield of Tritium from reaction (2) is of the order of one per cent of the yield of Carbon-14 from reaction (1).

A world-wide survey of radiocarbon in living organisms shows that Carbon-14 is distributed uniformly by atmospheric circulation. Of the earth's total radiocarbon inventory (about 81 metric tons), approximately 93 per cent is oceanic, in the form of carbonates, bicarbonates, and dissolved organic material. The remainder is found in the biosphere (4 percent) and the atmosphere (3 per cent).

Half-life of Carbon-14. The disintegration rate of radioactive substances is extraordinarily immutable and is independent of the chemical state, temperature, pressure, and other physical characteristics of the environment of the radioactive material. The disintegration of Carbon-14 takes place according to the reaction

$$C^{14} = \beta- + N^{14+}. \tag{3}$$

Determinations made by several different laboratories and methods have provided different estimates of the half-life of Carbon-14 (length of time in which a radioactive substance will lose one-half of its specific activity). The weighted average of these determinations is 5568 ± 30 years and is the standard value used by all radiocarbon laboratories.

Preparation and methods of sample measurement. The most serious problem in preparing samples for radiocarbon measurement is the possibility of contamination by younger carbon. The presence of intrusive rootlets or dissolved organic or inorganic carbon in a piece of fossil wood may completely invalidate the age of the sample. The replacement of 25 per cent of the carbon atoms in a 10,000 year old sample by modern carbon would cause an error of 50 per cent in the dating results. In a 40,000 year old sample only 5 per cent of modern carbon would cause a 50 per cent error.

After a thorough physical cleansing, the sample is tested with hydrochloric acid solution to insure that calcium carbonate has not been deposited secondarily by precipitation from ground water. A positive reaction necessitates digestion of the sample with hydrochloric acid until all evidence of carbonate is gone. In peat deposits, samples for radiocarbon dating are treated with potassium hydroxide to remove humic acids which may have been leached from older or younger layers in the deposit. Further fractionation is possible, such that lignin, cellulose, bone carbonate, and residues may be dated separately and compared for evidence of replacement of Carbon following deposition.

Once the samples have been cleaned and fractionated, they are burned to carbon dioxide in a sealed combustion apparatus. The sample is then purified to remove traces of Radon (a radioactive decay product of uranium), oxides of Nitrogen and Sulfur, and products of incomplete combustion. At the conclusion of the purification process, the sample may be converted to elemental carbon by passing the carbon dioxide over hot magnesium. The pure carbon is then washed and deposited as a thin film around the inside of a Geiger counter tube and the distintegrations of the Carbon-14 in the sample are counted electronically. An alternative procedure consists of passing the carbon dioxide into a counting chamber at high pressure and measuring the radioactivity of the sample in this way. In some laboratories, the carbon dioxide is converted to methane or acetylene and the radioactivity determined as for carbon dioxide. Another approach is to dissolve the sample (usually as acetylene or benzene) in a liquid scintillation counter cell and measure the radioactivity by photoelectric methods.

Published radiocarbon dates are the mean of two or more determinations of sample radioactivity which are alternated with determinations of background radiation. The net radioactivity of the sample is calculated by subtracting the background count from the total sample count. The age of the sample is the ratio of the sample count to the radioactivity of the modern reference sample (plus corrections) and is followed by an error figure which is one standard deviation of the counter error.

Sources of error. (1) There is no conclusive evidence that the cosmic ray flux or the Carbon-14 concentration in the atmosphere have remained constant in the past. Variations in cosmic ray intensity or atmospheric carbon dioxide could appreciably alter the amount of Carbon-14 available to plants and animals. (2) Biological systems are capable of discriminating between the isotopes of Carbon. Carbon-13/Carbon-12 measurements have shown that different organisms accumulate these isotopes in different proportions. Also, the Carbon-14 content of modern shell or bone material may differ significantly from the Carbon-14 content of the flesh of the organism that produced the shell or bone. (3). The burning of coal and petroleum products since the Industrial Revolution has resulted in the release of radioactively "dead" carbon dioxide into the atmosphere. Radiocarbon samples from industrial areas show Carbon-14 radio-activities appreciably lower than samples from non-industrial areas. (Suess effect). (4) Organisms, particularly aquatic plants, whose habitat is supplied mainly by water leached through ancient limestone may have less Carbon-14 in their tissues than samples from non-limestone areas. (5) Carbon-14 released into the atmosphere by atomic bomb explosions has increased the availability of this isotope to plants and animals. (6) Even the best made counters have a background radiation of their own. Also, counters must be shielded from the effects of cosmic rays and other stray radiation.

The effects of (1) and (2) on radiocarbon dating cannot be evaluated at the present time. (3) and (4) have the effect of making samples appear too "old" when compared with modern reference material. Living material from a hard water lake in New England contained only 77 per cent as much radiocarbon as modern wood. Because of the Suess effect (3), modern reference samples are derived from 19th century wood and corrected for radiocarbon decay since they were formed. (5) has an opposite effect, making samples appear too "young" when compared with reference standards. Carbon-14 from nuclear weapons testing reached a maximum in late 1963, nearly doubling the normal atmospheric inventory of Carbon-14. Following the weapons-testing ban, atmospheric levels have dropped down to approximately 67% above the modern reference standard.

The sensitivity of counting techniques (6) limits the maximum age of samples which can be determined by the radiocarbon method to about 40,000 years. Improvements in counting techniques may extend the maximum age to 60,000–70,000 years.

Significance of radiocarbon dating. A great many radiocarbon dates have been published since the first list of dates was presented by Dr. Libby in 1949. More than 80 laboratories in all parts of the world are investigating the age of materials, geochemistry, and other aspects of radiocarbon dating. As a direct result of the radiocarbon dating method, geologists now believe that many of the events of the last glaciation were synchronous in North America and Europe. Sequences of prehistoric cultures have been dated by Carbon-14, and in general, the dates confirm previous archeological interpretation. Ancient manuscripts, mummy cases, funerary boats, and other artifacts of organic origin are used to determine the age of cultural and historical objects.

J. Gordon Ogden, III

References

Deevey, E. S., Jr., R. F. Flint, J. G. Ogden, and I. Rouse, "Radiocarbon," v. 10, Amer. Jour. of Science, New Haven, Conn., 1968.
Libby, W. F., "Radiocarbon Dating," 2nd ed. Chicago, The University Press, 1955.

CARNIVORA

A well characterized order of placental MAMMALS distinguished by the flesh-eating habit and by structural features associated with this mode of life. The teeth are mostly pointed or blade-shaped for piercing and cutting, and the jaw motion is characteristically a simple opening and closing. The brain has a relatively large and distinctly convoluted cerebrum. The alimentary tract is simple, with a small coecum. The placenta is usually zonary.

The order is usually divided into three suborders, the Fissipeda or terrestrial carnivores, the Pinnipedia or marine carnivores, and the Creodonta, a miscellaneous group of extinct forms. The Fissipeda are mostly adapted for the capture, killing and devouring of living prey. The dentition may be reduced in the cheek series, but there is characteristically one tooth in each row which is distinctly blade-like, and functions with its opposing tooth as a shearing mechanism. These teeth are called carnassials, and in Fissipeda are always the fourth upper premolar and the first lower molar. The feet, except among the bears, are digitigrade. The claws are usually sharp and curved. In the skull the facial and cranial portions are of about equal length, and the sagittal crest and zygomatic arches are strong, to support powerful jaw muscles. The bony orbit is open behind. The clavicle is reduced or absent; radius and ulna, tibia and fibula are separate bones. In the carpus the scaphoid, lunare and centrale are usually fused. The manus normally has five toes and the pes has four, the hallux being reduced or absent.

The seven well-defined families of living Fissipeda may be grouped into two superfamilies, the dog-like Canoidea and the cat-like Feloidea. The Canoidea includes the *Canidae, Ursidae, Procyonidae* and *Mustelidae*. The Canidae, the dogs and their relatives, are mostly adapted to the capture of prey by swift pursuit. The head is long, with well developed jaws and teeth. The limbs are strong but not heavy; the claws are of moderate length. The dental formula is I$\frac{3}{3}$, C$\frac{1}{1}$, P$\frac{4}{4}$, M$\frac{2}{3}$ which is almost the basic placental formula. The carnassials are well developed. The clavicle is short and broad, and the head of the radius has little possibility of rotation. Unlike most other carnivores, the canids have a relatively long and twisted coecum.

The dog was probably the first animal to be domesti-

cated, and dogs occur on all continents including Australia, but it is not known whether one or more than one wild species were the progenitors. In any case, the extreme variation seen among domestic dogs is almost entirely the result of artificial selection. The wolves, close relatives of the dogs, are characteristic of the northern hemisphere. Jackals, found in Africa and southern Asia, are like small dogs with bushy tails; they range in packs and feed partly on carrion. Foxes differ from dogs in the reduction of the frontal sinuses, which gives the head a flatter profile; they are widespread in the northern hemisphere, including the Arctic. The fox-like *Octocyon* of South Africa is unique among placental mammals in having four molars.

Living members of the family Ursidae or bears are very distinctive, but their anatomy and fossil record show that they are relatively late offshoots of the dog family. Bears have an omnivorous diet, with less emphasis than in most other carnivores on the capture of prey. The teeth, with the same formula as in dogs, have the premolars and molars broad and blunt, with no functional carnassials. The skull is like that of dogs, but more massive. The body is heavy and the limbs powerful. There are four toes on each foot, and the claws are long but only moderately curved. The posture is plantigrade, and the motion is usually slow and deliberate. The tail is always short and the ears small. Bears are typical of the northern hemisphere, the brown bear in its various species being holarctic in distribution. The black bear of North America is smaller, and the spectacled bear of the Andes and the sun bear of India are still smaller. The polar bear is of large size but has a relatively small head and long neck; its fur remains white throughout the year. In late Cenozoic time the broad-faced bears, such as *Arctotherium*, inhabited North and South America; in these the characteristic ursid dentition was only partially developed.

The raccoons and their relatives make up the family Procyonidae. They are like small bears in body form and posture. There are only two molars in each tooth row, and the shearing nature of the carnassials is poorly developed. The tail is usually long, with characteristic ring-like markings. In addition to the familiar raccoon of North and South America there is the bassaris, which is like an elongated raccoon. Even more elongated and less raccoon-like is the coati-mundi of Central and South America, which has a long, turned-up nose. The panda of the southern Himalayas is of about the size and appearance of a cat; its relative, the giant panda, is large and bear-like, with black and white colouring.

The family Mustelidae includes the weasels, wolverines, badgers, skunks and otters. The molars are one above and one or two below, and the carnassials are moderately well developed. The true weasels are small mammals, but very ferocious; they have long necks and bodies, broad faces and short ears. Northern species, such as the ermine, change from brown to white in winter. Pole-cats are like weasels but larger, and keep the brown colour all year. The sable, marten and fisher are weasel-like, but are good climbers. The mink is an aquatic relative of the weasels. The wolverine is larger, and almost bear-like, with hairy feet for walking on ice and snow. Badgers have broad bodies and short limbs and are excellent burrowers. They are represented in Europe, North America, Asia and Africa. The skunks are confined to the Americas, and are only moderately good burrowers. They have a black and white pelage and a long bushy tail, and are noted for their ability to eject a highly offensive liquid from a pair of anal glands. Otters are mustelids modified for an aquatic mode of life, and a fish-eating habit. They have slender bodies and short limbs, with webbing between the toes. Otters are mostly northern in distribution but occur in India. The rare sea-otter is a true marine mammal, found in the North Pacific.

The Feloidea, or cat-like fissipeds, include the families Viverridae, Felidae and Hyaenidae. The viverrids are the least cat-like, with rather long skulls and moderate reduction of teeth. The body is slender, and the limbs rather short, usually with five toes on each foot. This family, of Old World distribution, is typified by the civet cats, the genets and the mongoose.

The Hyaenidae are carrion-feeders, and are represented by the spotted hyaena of Africa and the striped hyaena of Africa and Asia. Hyaenas have powerful jaws, with large carnassials, but the other cheek teeth are reduced. The strong jaw muscles are supported by a high sagittal crest and large zygomatic arches. The body form is characteristic, with heavy shoulders and long front legs, but with more slender hind quarters.

The Felidae seem the best adapted of all fissipeds for the flesh-eating mode of life. They have short, broad skulls, with reduced dentition, but with highly efficient carnassials. The limbs are powerful and the sharp, curved claws usually can be folded back. The head of the radius has some ability to rotate on the ulna. The pupils of the eyes can be contracted or expanded, permitting both day and night vision. Members of this family characteristically catch prey by stalking and sudden leaping attacks. The tiger of Asia and the lion of Africa and southern Asia are the largest of the cat family. The leopard, also of Africa and Asia, is spotted, and smaller than the lion but equally ferocious. In the Americas the mountain lion occurs on both continents, whereas the leopard-like jaguar is mainly South American. The lynxes are holarctic in distribution; they are of medium size, with ear tufts and short tails. The domestic cat was derived mostly if not entirely from the wildcat of Africa and southern Asia. The cheetah, of Africa and southern Asia, is spotted like a leopard, but is more dog-like in body form and in method of capturing prey.

The extinct machairodonts or sabre-toothed cats appeared in Oligocene time and reached maximum development in the Pleistocene epoch. They had the upper canines enormously elongated, and probably used them for stabbing their prey.

The suborder Pinnipedia, the marine Carnivora, is a relatively uniform group. Pinnipeds occur in all the oceans and some large inland seas. The head and skull suggest the dogs and bears, but the teeth are greatly simplified, with premolars and molars similar. The limbs are greatly modified as swimming organs. The upper limb segments are short, and the manus and pes are long, rather broad, and extensively webbed. Locomotion on land is awkward, but swimming is swift and graceful. The suborder is made up of three well-defined families, the Otariidae, the Odobenidae and the Phocidae.

The Otariidae or eared seals are also known as sea-lions. In addition to the presence of ears, they are distinguished by carrying the hind limbs turned forward when standing on land. Here belongs the California sea-lion, the trained "seal" of the circus. The fur seal of the North Pacific is also a sea lion. Other species are found on the coasts of Patagonia, Australia and South Africa.

The Odobenidae are the walruses, which lack external ears but hold the hind legs in the sea-lion fashion. The broad skull bears two enormous upper canines, which project far below the mandible, and are used in digging molluscs from the sea bed. The remaining teeth are simple peg-like structures, with rounded or flattened crowns. Walruses are found in the Arctic Ocean and the northern Atlantic and Pacific.

The Phocidae or true seals have head and teeth similar to those of the Otariidae, but have no external ears. The hind limbs are used only in swimming, and project backwards. Seals are found in all oceans but especially in the North Atlantic. The elephant seal of California, with proboscis-like nose, reaches a length of nearly 20 feet.

Most of the extinct families of Carnivora were formerly placed in the suborder Creodonta, but modern studies have largely discredited this as a natural group. The family Arctocyonidae, mostly Paleocene, was once considered to be the most primitive of carnivores and the link with the basic placental stock, but is now included in the Condylarthra, an order of primitive ungulates. The Mesonychidae, of the Paleocene and Eocene, had simplified molars with conical cusps. Many were small but the largest of all known carnivores, *Andrewsarchus,* from the Eocene of Mongolia, was a mesonychid. Some authorities also regard the mesonychids as primitive ungulates. The families Oxyaenidae and Hyaenodontidae were true carnivores, with shearing teeth, but in the Oxyaenidae the carnassials were the first upper molar and the second lower molar, while in the Hyaenodontidae the second upper molar and the third lower molar were the main shearing teeth. These families were characteristic of the Eocene epoch, but the Hyaenodontidae persisted into the Oligocene, and in southern Asia into the Pliocene.

The Miacidae of the Paleocene and Eocene had the last upper premolar and the first lower molar developed as carnassials, as in modern fissipeds. For this reason the miacids are now regarded as forming the superfamily Miacoidea, from which the Canoidea and Feloidea were derived. Miacids had the complete eutherian dentition, and were mostly small and dog-like.

Typical dogs appeared in the Oligocene fauna, as well as cats of the normal and sabre-tooth groups. Bear-like dogs of the Miocene gave rise to true bears by Pliocene time. Procyonids are known only as far back as the Miocene, but the Mustelidae existed in Oligocene tiime. Viverridae also were well represented in the Oligocene fauna. The Hyaenidae are not known before early Pliocene time and probably are related to the felids in much the same ways that bears are related to dogs. All three families of the Pinnipedia occur in Miocene marine faunas, but their earlier history is unknown.

L. S. RUSSELL

References

Flower, W. H., and R. Lydekker. "An introduction to the study of mammals living and extinct." London, Black, 1891.

Jayne, H., "Mammalian anatomy; a preparation for human and comparative anatomy. Part I. The skeleton of the cat, its muscular attachments, growth and variation compared with the skeleton of Man," Philadelphia, Lippincott, 1898.

Matthew, W. D., "The phylogeny of dogs," Jour. Mammalogy, **11**: 117–138, 1930.

Matthew W. D., "The phylogeny of Felidae," Bulletin Amer. Mus. Nat. Hist., **28**: 289–316, 1910.

Matthew, W. D., and W. Granger," A revision of the Lower Eocene Wasatch and Wind River faunas (Creodonta)." Bulletin Amer. Mus. Nat. Hist., **34**: 1–103, 1915.

Mivart, St. G. J., "The cat. An introduction to the study of backboned animals, especially mammals," London, Murray, 1881.

Mivart, St. G. J., "Dogs, jackals, wolves and foxes: a monograph of the Canidae," London, Porter, 1890.

Romer, A. S., "Vertebrate palentology." Chicago, University of Chicago Press, 1966.

Scheffer, V. B., "Seals, sea lions and walruses: a review of the Pinnipedia." Stanford, The University Press, 1958.

Simpson, G. G., "The principles of classification and a classification of mammals," Bull. Amer. Mus. Nat. Hist., **85**: 1–350, 1945.

CAROTENOIDS AND FLAVONOIDS

Carotenoids

1. Distribution. Carotenoids are lipid-soluble, yellow to orange-red pigments universally present in the photosynthetic tissues of higher plants, in algae, and in the photosynthetic bacteria. They are spasmodically distributed in flowers, fruit and roots of higher plants, in fungi and in bacteria. They are synthesized *de novo* in plants and protists. Carotenoids are also widely distributed in animals, especially marine invertebrates, where they tend to accumulate in the gonads and skin or feathers; all carotenoids found in animals are ultimately derived from plant or protist carotenoids.

In photosynthetic tissues of higher plants and in algae carotenoids are located in the chloroplasts where, in higher plants, they are confined to the grana. In the photosynthetic bacteria they are located in chromatophores which correspond to the grana of higher plants. In all cases they exist as lipoproteins in close association with the chlorophylls. In other tissues of higher plants and in fungi and bacteria they can exist as lipoproteins (carrot root, *Corynebact.* spp.) or as oil droplets (red palm, *Phycomyces*). Carotenoproteins are often differently coloured from the parent pigment: the colour of brown algae is largely due to a brown fucoxanthin-protein complex.

The leaf carotenoids of all higher plants are qualitatively similar, but variations from this pattern occur in some algae and in the photosynthetic bacteria. The individuality of higher plants with regard to carotenoids is demonstrated in flowers, fruit and roots, where numerous different and often highly specific pigments are found. Specific carotenoids often accumulate in reproductive regions of algae (*Ulva*). About 0.2% of the dry matter of leaves is carotenoid; highest values in non-photosynthetic regions are about 1% dry matter.

Carotenoids exist in animals as lipid droplets in specialized cells (xanthophores of trout skin), dissolved in body fat (locusts, cows), and as chromoproteins (lobster eggs and carapace).

2. Structure. Carotenoids are tetraterpenoids, consisting of 8 isoprenoid residues, and they can be re-

garded formally as being synthesized by the tail to tail condensation of two 20 C units, themselves produced by the head to tail condenstation of four isoprenoid units.

Hydrocarbon carotenoids are termed carotenes and the structure of the four from which almost all others are derived are as indicated, with that of β-carotene given in detail. New structures recently elucidated have acetylenic linkages between C-7 and C-8, whilst others have an allenic system (—C=C=C—) between C-6 and C-8.

β-Carotene

α-Carotene   γ-Carotene   Lycopene

$$2CH_3COSCoA \longrightarrow CH_3COCH_2COSCoA \xrightarrow{CH_3COSCoA} \begin{array}{l} CH_3C(OH)CH_2COSCoA \\ | \\ CH_2COOH \end{array}$$

Acetyl-CoA   Acetoacetyl-CoA   β-Hydroxy-β-methyl-glutaryl CoA

$$\begin{array}{l} CH_3C(OH)CH_2CH_2OPP \\ | \\ CH_2COOH \end{array} \longleftarrow \underset{2\,ADP \quad 2\,ATP}{} \begin{array}{l} CH_3C(OH)CH_2CH_2OH \\ | \\ CH_2OOH \end{array}$$

Mevalonyl 5 pyro-phosphate   Mevalonic acid

ATP⟍
A%P, Pᵢ⟋) → CO₂

$$\begin{array}{l} CH_3CCH_2CH_2OPP \\ \| \\ CH_2 \end{array} \longrightarrow \begin{array}{l} CH_3C{=}CHCH_2OPP \\ | \\ CH_3 \end{array}$$

Isopentenyl pyrophosphate   γγ-Dimethylallyl pyrophosphate

Oxygenated carotenoids (hydroxyl, oxo, epoxyl and carboxyl functions) are termed xanthophylls. Hydroxylated carotenoids occur in both esterified and free forms. The colour of carotenoids is due to the long series of conjugated double bonds in the molecules. Compounds in which the conjugation is interrupted (phytoene, 7, 8, 11, 12, 7′, 8′, 11′, 12′-octahydrolycopene) are often colourless and are more appropriately termed polyenes.

3. **Biosynthesis.** The common polymerizing unit for the biosynthesis of all terpenoids is isopentenyl pyrophosphate formed from acetyl-CoA via mevalonate (above). Polymerization is initiated by the isomerization of one molecule of isopentenyl pyrophospate to γγ-dimethylallyl pyrophosphate which acts as a starter on which further isopentenyl pyrophosphate molecules condense.

4. **Function.** (a) *Plants and protists.* In all photosynthetic organisms, except perhaps the blue-green and red algae. The light energy absorbed by carotenoids is utilized in photosynthesis. Carotenoid-absorbed light cannot initiate photosynthesis in the absence of chlorophylls. In algae and photosynthetic bacteria carotenoids also protect against chlorophyll-mediated photosenitization; carotenoid-less mutants are killed by simultaneous exposure to oxygen and light. A similar function for carotenoids in non-photosynthetic organisms has been suggested.

Carotenoids probably function as mediators in phototropic bending of seedlings and mould sporangia; phototaxis in photosynthetic bacteria and certain algae is mediated by carotenoids; in other algae the situation is confused. No convincing evidence exists for implicating carotenoids in sexual processes in plants and protists.

(b) *Animals.* Carotenoids frequently play an important part in the production of characteristic colour patterns, especially those concerned with sexual dichroism, in lower animals. Carotenoid accumulation in eggs is to enable the developing embryo to possess a correct colour pattern on hatching. Carotenoids are often important pigment components of display feathers. In mammals, carotenoids, *per se*, have no known function; however, β-carotene and those carotenoids with a structure in which one half of the β-carotene structure remains unaltered are converted into vitamin A as they pass across the intestinal wall; the mechanism of conversion is unknown. Vitamin A-active pigments (provitamins A) are the only dietary source of vitamin A in herbivores. In crustaceae astaxanthin (3,3′-dihydroxy-4,4′-dioxo-β-carotene) may be a vitamin A precursor. Many fish, particularly fresh-water fish, can convert carotenoids into vitamin A and vitamin A₂ (3,4-dehydrovitamin A).

Vitamin A

Flavonoids

1. **Distribution and structure.** Flavonoids are widely distributed plant products found in the cell-sap of all tissues but most frequently in flowers and fruit. They do not occur in protists or animals. They are C₆—C₃—C₆ compounds with each C₆ residue an hydroxylated aromatic ring, and they normally occur as water-soluble glycosides; the aglycones are insoluble in water. Flavonoids are classified according to the configuration and oxidation state of the C₃ connecting link; the major groups are on p. 147.

The algycones of anthocyanidins are termed anthocyanins. Flavonoids are pH indicators; anthocyanins are blue-purple in alkali and pinkish red in acid solutions; the pale yellow colours of flavones, flavonals, chalcones and aurones in acid or neutral solution are

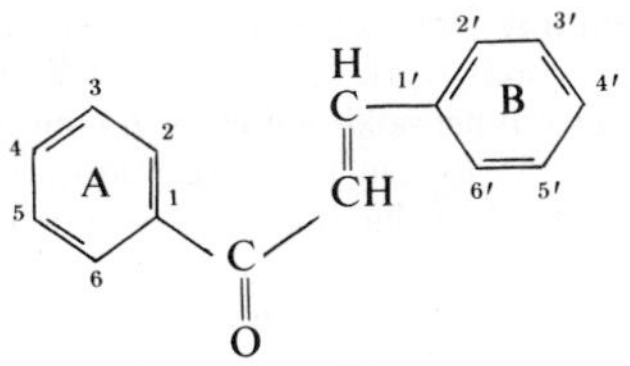

2′ 3′
H
C
A B 4′
CH
6′ 5′
C
O

Chalcones

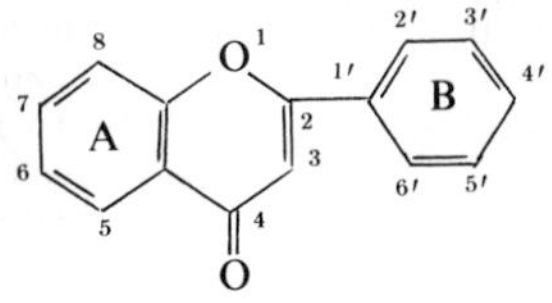

8 O 1 2′ 3′
7 1′ B 4′
A 2
6 3
5 6′ 5′
4
O

Flavones

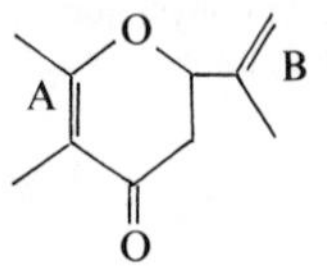

O B
A

O

Flavanones

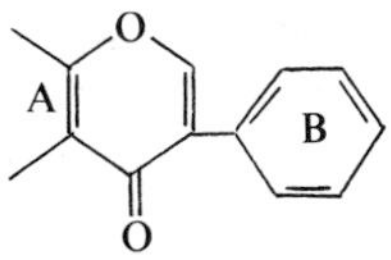

O
A B

O

Isoflavones

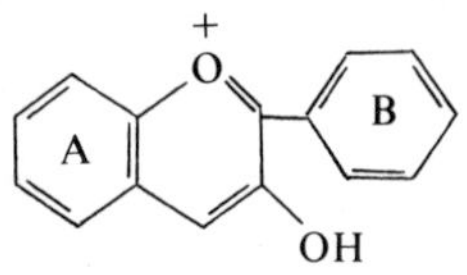

+
O
A B

OH

Anthocyanidins

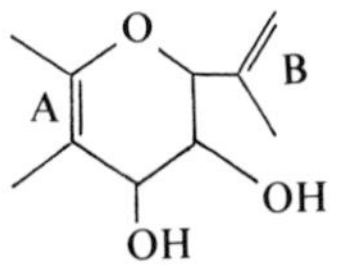

O B
A
OH
OH

Leucoanthocyanidins

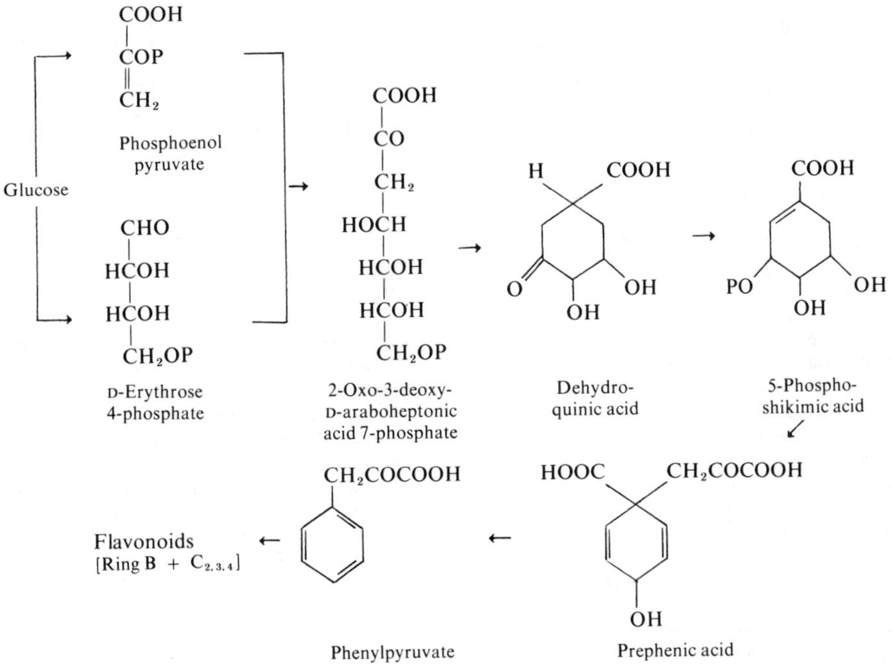

COOH
COP
CH₂

Phosphoenol
pyruvate

Glucose

CHO
HCOH
HCOH
CH₂OP

D-Erythrose
4-phosphate

COOH
CO
CH₂
HOCH
HCOH
HCOH
CH₂OP

2-Oxo-3-deoxy-
D-araboheptonic
acid 7-phosphate

H COOH
O
OH
OH

Dehydro-
quinic acid

COOH
PO
OH
OH

5-Phospho-
shikimic acid

Flavonoids
[Ring B + C₂,₃,₄]

CH₂COCOOH

Phenylpyruvate

HOOC CH₂COCOOH
OH

Prephenic acid

greatly intensified in alkali; flavonones are nearly colourless in cold alkali, but turn to yellow and red on heating owing to isomerization to the corresponding chalcones. Leucoanthocyanins are converted into anthocyanins by warming in acid solution.

2. Biosynthesis. Ring A arises from acetate by an unknown mechanism; ring B + $C_{2,3,4}$ from glucose via phosphoshikimic acid and (probably) phenylpyruvic acid.

Further details, mechanism of condensation of the two rings, the stage at which hydroxylations occur, the interconversion of various groups of flavonoids, are being actively investigated.

3. Contribution to flower colours. The blue and pink colours of flowers are due to anthocyanidins which can exist in very high concentration (30% of dry matter in some Viola spp., compare carotenoids). The pigment responsible for pink and blue hydrangea flowers is dephinidin (3,5,7,3′, 4,5′ - hexahydroxy - 2 - phenyl-benzopyrylium chloride). The colour differences are not due to pH variations in cell sap but to the formation of a stable aluminium complex in the blue flowers. Similar intergenus variations (cyanin in cornflowers and red roses) are probably due to related reactions.

Flavonoids other than anthocyanins impart yellow and orange colours to flowers and are generally present in white flowers. When associated with carotenoids, it is the small amounts of these pigments rather than the large amounts of flavonoids which determine the flower colour. The appearance of a yellow colour in white petals and the darkening of yellow petals on exposure to ammonia vapour demonstrates the presence of flavonoids.

A mixture of anthocyanins and carotenoids can yield brown (Wallflower) and scarlet (tulip) flowers.

4. Function. Apart from attracting insects no well-authenticated physiological or biochemical functions of flavonoids is known.

T. W. GOODWIN

Reference

Goodwin, T. W. ed. "Chemistry and Biochemistry of Plant Pigments," New York, Academic Press, 1965.

CARTILAGE

Most parts of the skeletons of vertebrates are formed first in cartilage and the ends of opposing bones at movable joints are covered by cartilage. The longitudinal growth of the long bones depends on epiphyseal cartilage proliferation with concurrent bone substitution. However, the increased girth of long bones depends on peripheral additions of bone under the periosteum without a cartilage precursor and certain flat bones are not preceded by a cartilage model. Pituitary growth hormone will cause an increased rate of cartilage cell proliferation in the epiphyseal apparatus as long as that apparatus is not completely replaced by bone. Other hormones increase or decrease the rate of cell proliferation and delay or hasten the bony closure of the epiphyseal apparatus.

Each cartilage cell lays down about itself a matrix composed largely of a protein (collagen) and a mucopolysaccharide (chondroitin sulfate). Both of these extracellular tissue components as well as the cells are highly hydrated so that for example the hyaline cartilage which covers the ends of the bones at the knee is by analysis 75-85% water.

Hyaline, fibrous, elastic, epiphyseal cartilage and the soft nuclear cartilage which forms an hydraulic cushion between vertebral bodies of the spine,—each vary in their collagen fibril/chondroitin sulfate ratio. This ratio increases with age. Elastic cartilage matrix contains an additional protein, elastin.

The elastic cartilage in the ears of young rabbits lose their elastic quality and become limp when crude papain is administered intravenously. This phenomenon is apparently caused by a loss, not of collagen, but of chondroitin sulfate and perhaps elastin from the extracellular matrix. If cortisone is subsequently administered the ears remain limp. Following the papain alone, the cartilage regains its normal elastic quality in 2–3 days. A similar phenomenon has been noted in epiphyseal cartilage.

S^{35} is incorporated into chondroitin sulfate in cartilage. The S^{35} appears in the cartilage cells first, then in the matrix and finally disappears.

Apparently, the cartilage cells produce chondroitin sulfate which then becomes extracellular and forms an association with the collagen fibrils in the matrix, but gradually passes into the vascular system to be replaced by the cartilage cells. Papain speeds the release of chondroitin sulfate A from the matrix and cortisone prevents its reaccumulation in the extracellular cartilage space.

The collagen fibrils of cartilage matrix are apparently introduced by the cells as "soluble" protofibrils of collagen, into the extracellular matrix space and these then become organized into fibrils. The active lathyrismic principle—β-glutamyl-amino-propionitrile or aminoacetonitrile—interferes with the proper bonding between collagen protofibrils, so that even though they do form fibrils these fibrils can be more readily solubilized: even after an extracellular matrix is fully formed, the administration of the active lathyrismic principle will make its collagen fibrils more soluble.

Severe Vitamin C deficiency, or scurvy, interferes with the formation of collagen protofibrils by the cells although once formed and delivered into the extracellular matrix space they appear to combine into collagen fibrils with normal internal bonding.

ROBERT A. ROBINSON

References

Weinman, J. P. and H. Sicher, "Bone and Bones," St. Louis, Mosby, 1955.
Eichelberger, L., et al. J. Bone and Joint Surgery **41-A**: 1127, 1959.
Happey, F., et al, "The Nature and Structure of Collagen," New York, Academic Press, 1953.
Dziewiatkowski, D. D. J. Exp. Med. **93**: 451, 1951.
ibid, loc. cit. **95**: 489, 1952.
Thomas, L. J. Exp. Med. **104**: 245, 1956.
Follis, R. H. and A. J. Tomimis, Proc. Soc. Exp. Biol. Med. **98**: 843, 1958.
Grillo, Ll and J. Gross, Proc. Soc. Exp. Biol. Med. **101**: 268, 1959.

CAVE BIOLOGY

Species confined to caves for the entire or partial individual life cycle (*troglobionts*) are known among PLATYHELMINTHES, ANNELIDA, ARTHROPODA, MULLUSCA (GASTERPODA, and CHORDATA (PISCES and AMPHIBIA). Such species are not known among Protista and Plants.

Morpho-physiologic traits. Most troglobiotic species show a convergent complex of traits, generally regressive, which are not as often encountered as such, and to such a degree, in other biota. For instance, out of the 40 known blind fresh water fish species 30 live in caves or artesian wells. Troglobiotic traits, usually subject to high phenotypic variation, can be described as follows:

1) Hypofunction of the body integuments: regression or loss of pigments (melanine in all taxa, guanine in fishes, carotenoids in Crustacea), weakness of the chitinous exoskeleton (in Arthropoda), regression or loss of scales (in Collembola and fishes), regression of glandular activity (in Urodela).

2) Regression or loss of the different parts of the optical system, photoreceptors (in flatworms), single and complex eyes (in Arthropoda), all parts of the vertebrate eye (cfr. Eigenmann[1]), optic nerves and optic lobes (in all taxa).

3) Elongation and slenderness of the body, particularly in Crustacea (the genera *Bathynella*, *Parabathynella*, *Stenasellus*, some Copepoda), Insects (most pronounced in the tribe *Trechitae*, Coleoptera) and Urodela. Elongation and slenderness of some body appendices, antennulae and antennae (in Arthropoda), legs (in Arthropoda and Urodela), chaetae (in Insects). Shortening or loss of other appendices, second pair of wings (in Coleoptera), barbels and fins (in *Caecobarbus geersti* Blgr, Belgian Congo, *Pimelodella transitoria* Ribeiro and *Typhlobagrus kronei* Ribeiro, Brasil).

4) Regression of the endocrine system, principally the thyroid which is lacking or obsolescent in several cave Urodela, and very small in *Caecobarbus geertsi*. The rudimentary thyroid of *Proteus anguinus* Laurenti (Dalmatia) is functional, but the tissues have lost their reactivity against the hormone. The hypophysis in *Caecobarbus geertsi* lacks thyreotropic cells. The same gland is very small in the blind *Amblyopsidae* (North America). Regression of endocrine functions in Crustacea is suggested since atrophy of the eye stalks often follows the disappearance of the eyes.

5) Relatively slow embryological development in all cases studied: 11–12 months in *Caecosphaeroma burgundum* Dolff (Isopoda, France); 40–80 days in the beetle *Speonomus longicornis* Saulcy (Bathysciinae, France); $1\frac{1}{2}$–2 months in *Amblyopsis spelaeus* De Kay (Pisces, Kentucky); 3 months, interovarially, in the Cuban cave Brotulids, *Lucifuga subterraneus* Poey and *Stygicola dentatus* Poey; 3 months in the viviparous *Proteus anguinus*. Slow larval and/or post-larval growth has been observed: in *Niphargus Virei* Chevr. and *Caecosphaeroma burgundum* (France) in conjunction with very slow succession of moltings; in *Speonomus longicornis* where no growth occurs during the larval stage, covering 5 to $8\frac{1}{2}$ months. Slow growth rate in *Caecobarub geertsi* has been deduced from scale readings. In the less regressed Characin *Anopthichtys jordani* Hubbs and Innes (Mexico) growth rates are more normal. In many cave Vertebrates and Invertebrates (Crustacea) rudimentation, especially of the eyes, is progressive during individual development. Strong negative allometry (which may be followed by a strong positive one, in relation to the sign of general growth acceleration) characterizes equally several body appendices in *Caecobarbus geertsi*, resulting in a rudimentation proportionate to the degree of positive allometry in its epigean relative *Barbus holotaenia* Blgr. Progressive depigmentation has been reported in other instances. Similar progressive rudimentation is known from several blind deep sea fishes.

6) Low standard metabolic rates as measured by oxygen consumption. Its value is 3 to 5 times lower in the following cave animals: *Caecobarbus geertsi* (compared to *Barbus conchonius* of similar size); *Niphargus Viréi* (versus *Gammarus pulex*); *Meta menarda* Latreille (versus *Araneus diadematus* Clerck, Arachnida). Indirect evidence is present from several other troglobionts.

7) Stenothermy, according to many occasional observations. Experimentally *Niphargus Virei* and *Asellus cavaticus* Schiödte (France) have been shown to die above 15° C.

8) Permanence of early ontogenic (neotenic) and palaeozoic (archaic) traits. Remarkable neotenics are: the polychaet *Troglochaetus Beranecki* Delachaux (Switzerland), with a high number of segments, the crustacean *Thermosbaena mirabilis* Monod (Tunisia), the cave Urodela. Many troglobionts have become taxonomically unclassifiable missing links through the accumulation of archaic traits, generally not considered as neoformations (*cfr*. Jeannel[2]). Famous examples are the genera *Bathynella* and *Parabathynella*, *Thermosbaena mirabilis* (Crustacea), *Uegitglanis zammaroi* Gianf. (Pisces, Italian Somalia). Single archaic features are the metameric extension of the lateral line system on the head in *Caecobarbus geertsi*, the ossification of branchial and hyoidal arches in *Proteus anguinus*, the

absence of lungs in the cave spider *Teleoma vernella* E.S., the atrophy of the tracheals in Coleoptera.

9) Very low fecundity, according to ecologic and experimental evidence. Quantitative data are available from *Caecosphaeroma burgundum* (7–8 eggs at a time as against 67–91 in the related *Sphaeroma serratum*), *Bathynella chappuisi* Delachaux (1 egg at a time), *Speonomus longicornis* (1 egg every 40–50 days). The loss of sexual periodicity in troglobionts is claimed on the basis of ecologic data.

10) Highly developed tactile and vibratory sensitivity in blind cave fishes according to laboratory observations. In the same animals and circumstances light sensitivity is always present. A great variation exists as to sign of phototactic behavior, ranging from negative over indifference to positive, depending on species, populations and individuals. In the regressive series of cave Characins: *Astyanax mexicanus* Filippi, *Anoptichtys jordani* Hubbs and Innes, *A. hubbsi* Alvarez, *A. antrobius* Alvarez, the sign of phototaxis is determined by whether the pineal complex is covered or not by pigments, more depigmented fish showing more positive tactism. Light sensitivity diminishes with age in *A. jordani*. Differential sensitivity to wave lengths is present in *Caecobarbus geertsi*. Modern studies have not shown the existence, on the anatomical or histological level, of neogenic compensatory improvements of sense organs in troglobionts.

Geo- and topographic distribution of cave forms. The majority of, and the most regressive, cave animals occur in the subtropical and warm temperate regions of the northern hemisphere. Highly regressive cave beetles are found at higher latitutes than the most regressive fishes. Some groups are more ubiquitous than others (*e.g.* cave Crustacea). Within the regions indicated, local conditions govern the development of cave life. Constant darkness, constant high humidity, constant temperature, absence of air currents, a calcareous substrate resulting in supernormal CO_2 concentrations in water and air, spatial isolation with its consequence of dependency on exogenous food resources and at least periodic food shortage, presumably constitute the most favorable complex of factors for troglobiotic colonization. Cave life is always missing from dry caves, basalt caves, and naturally or artificially excavated holes in other than calcareous rocks (mine galleries) except for ubiquitous forms such as *Niphargus spp.* In addition to the geographic distribution many other facts point to the importance of the temperature level for the occurrence of regressive forms. The most regressive cave beetles are found in caves with temperatures of 10–12° C (Europe and N. America). Carabid beetles with similar traits except somewhat shorter legs, have been found only in deeper litter layers of mountain forests in Central Africa, with similar constant temperatures, while they are not found in caves of the same region (with temperatures from 18–24° C).

Evolution of troglobionts. Very often populations of troglobionts are morphologically divergent from cave to cave within a single system. Sometimes intermediate variants connect subterranean with epigeic forms. In a few cases the genetic nature of this variation has been proved: in *Asellus aquaticus* and its cave populations (Postumia, Trïeste), in mine gallery populations of *Gammarus pulex* (Germany) and in the regressive series from *Astyanax mexicanus* to *Anoptichtys antrobius*.

Crosses within this last series yield F_1's with intermediate morphology and behavior. In F_2's increased variation and recombination types (pigmented blind fishes) are recorded, indicating chromosomal determination of the regressive traits. Experimental evidence as to the mechanism of troglobiotic evolution is lacking however. In addition to all the mentioned facts, theories concerning this evolution should consider the following aspects:

1) Classic albinotic mutants have played a negligible, if any role. 2) Regressive evolution is a family affair. For instance all 600 species of Bathysciinae (Carabidae) are regressive and cavernicolous, muscicolous or endogeic. Very few if any higher insects and vertebrates are real cave dwellers and regressive. 3) The relative rates of evolutionary loss of different organs is constant. In all groups pigment is lost before eyes, except in Siluroid and Brotulid fishes. 4) Morphologic evolution—leading to regression—and physiologic evolution—leading to obligatory cave dwelling—may or may not proceed at the same rate. As a consequence phylogenetically old and non-regressive troglobionts, as well as regressive animals in other biota (endogeum, litter, muddy stream beaches, hypolimnion of lakes, limnocrenes) exist. 5) Regressive evolution does not take place in cave animals which have much food at their disposal (guanobionts).

Theories on troglobiotic evolution can as yet only be evaluated by the number of facts they integrate and by their economy of secondary hypotheses. A single cause for evolution and actual occurrence of cave animals is invoked by lamarckist theories, namely disuse of organs of functions under influence of darkness or high humidity (Eigenmann, Fage). Many secondary, historical hypotheses are required to explain the absence of troglobionts in otherwise favorable conditions (Jeannel). Neodarwinist theories invoke usually two different causes, repeated and independent chance mutations and their accumulation by supposed relaxation of natural selection in caves (Lankaster, Hubbs, Kosswig, Emerson, Cuenot). These theories explain at most some morphological aspects of troglobiotic evolution. Many facts can be understood if this evolution proceeds through natural selection by the ecologic factors actually conditioning cave life and favoring low metabolism combined to genotypically low growth rates determining low food needs, internal instability and hypotely, viz. hypertely of highly positive allometric (including eyes) and highly negative allometric organs or functions, and if such selection would be limited by a preexisting variability of developmental and physiologic patterns shared by given taxa.

M. J. HEUTS

References

"Cave Vertebrates of America. A Study in Degenerative Evolution," Carnegie Inst. Washington Publ. 104, 1909.

Jeannel, R., "Les Fossiles vivants des cavernes," Paris, Nouvelle revue française, 1943.

Vandel, A., "Biospeleology: the Biology of Cavernicolous Animals," New York, Pergamon, 1967.

CELL

The generalization that the cell is the unit of biological structure and function has gained in validity since its formulation more than a century ago. The Cell Doctrine can be put in a more positive and self-defining form: cells are the elements or atoms of Life in the sense that they are the minimum biological units capable of maintaining and propagating themselves in an environment other than the interior of cells. The biological expressions of viruses and other parasitic reproductive entities presuppose the availability of all of the activities of a complete cell.

Cells are easily recognized operationally, for their material constitution, structural organization, and behavioral expressions have no counterparts in the natural inanimate objects on this planet. There is no difficulty in defining the variable features of cellular structure as modulations of a common plan. This is done by the sciences of *Histology*, in which it is seen that the variety of cell types is considerable but no more overwhelming than the periodic table of the atoms of Chemistry. The material organization of cells, as described by the science of *Biochemistry*, is even less variable. To an extraordinary extent, we find the same kinds of molecules and the same schemes of catalyzed reaction sequences in cells of different kinds. Indeed, the very great difficulties of accounting for variation in development and in evolution arise from the conservatism of cells.

Cells are rather large assemblages of matter. The smallest complete cells known, the PPLO (pleuropneumonia-like organisms of the genus *Mycoplasma*) have diameters of $0.2-0.2\mu$. It is unlikely that cells could be very much smaller because of the number of large molecules required for minimal function. A cell containing a few thousand enzyme molecules of molecular weight 10^4-10^5, the nucleic acids necessary for the coding of these molecules, a membrane, and a water content greater than 50 percent will already represent a body approaching microscopically visible dimensions. The upper range of cell size is limited only by problems of diffusion and by the number of sets of chromosomes present.

A cell is a strictly bounded system. It confines and maintains an internal aqueous continuum which is strikingly different in composition from the aqueous external environment, the boundary consisting of a "plasma membrane" about 100 Å thick, comprising only a few layers of protein and lipid molecules. Perhaps the most striking characteristic of the intracellular aqueous medium is its high content of potassium ions and its low content of sodium and chloride ions, the maintenance of which in a world rich in Na and poor in K demands the expenditure of energy for *osmotic work* (see OSMOSIS), and is a limiting factor in the environmental tolerances of organisms. The surface layer, which is responsible for regulating the internal medium, must be regarded as one of the important organs (or organelles) of the cell.

The 19th century notion of a PROTOPLASM, a complex substance or association of substances embodying the attributes of life, has given way to a greater stress on the cell as a self-contained system composed of distinct parts, the parts being associations of distinct molecules. The parts and the molecules of which they are composed may be isolated, and the several functions of cells may be studied in subcellular preparations. The analysis of the cell along these lines has proceeded rapidly, and now (1967) the major focus of attention is the integration and mutual regulation of the parts and functions.

Structure. Certain structural features seem to be

nearly universal for the cells of plants and animals and some microorganisms (e.g. fungi) and the term *eucaryotic* is often applied to their common structural plan. This plan is contrasted with the *procaryotic* plan characteristic of bacteria. Procaryotic cells lack internal membranes, including a nuclear membrane, and their genetic apparatus to consist of a single piece of DNA rather than multiple chromosomes made of DNA combined with proteins. The main features of eucaryotic cells are:

(1) The *cell membrane*, a double-layered structure about 100 Å thick in which the biochemical mechanisms of osmotic work—the so-called "carrier" or "permease" systems—seem to be located. An earlier view of this membrane as a rather stable semipermeable barrier equipped with "pumps" is now being supplemented by evidence that regions of it may invaginate, surround droplets of medium (in *pinocytosis*) or particles (*phagocytosis*), and carry these into the cell interior in vacuoles. Thus the membrane itself may be in a state of active turnover, and this may play a significant part in the transport of materials by cells.

(2) The NUCLEUS, which in many if not all cells is surrounded by a pore-bearing membrane whose basic structure appears to be similar to that of the cell membrane. The CHROMOSOMES within the nucleus are in an extremely extended state when the cell is not dividing, and cannot be resolved as the microscopic threads characteristic of the dividing cell. In this extended state the chromosomes carry out their major functions: the synthesis of duplicates of themselves to be distributed at division, and the control of the synthetic activities of the cell. This control, which is the ultimate expression of heredity, is now viewed as the determination of the character, amounts, and time of production of specific macromolecules. In molecular terms, the present theory is that the sequences of the nucleotide units of *deoxyribonucleic* acid (see NUCLEIC ACIDS) (which is found only in the chromosomes) embody a "code" which dictates the patterns of synthesis of other large molecules, and that genetic stability depends on the self-reproduction of the deoxyribonucleic acids.

The nucleoli, bodies associated with chromosomes and containing RNA, are thought to be the site of formation of the RNA of the ribosomes, which accounts for most of the RNA of the cell.

In a broad sense, it is evident that the nucleus is the administrative center of the cell, responsible for its development and for its long-term maintenance, but not for its short-term activities. Cells without nuclei survive for considerable periods of time, but have a limited capacity for growth and no capacity for reproduction.

(3) In the *cytoplasm*, the body of the cell, we observe a high degree of localization of function. The best-known functional particles (see ORGANELLE) are: (a) the MITOCHONDRIA, sac-like membranous bodies of micron dimensions, in which are localized the energy-yielding oxidations and phosphorylations of the cell. This represents a strict localization; mitochondria in isolation are capable of carrying out these energy-mobilizing reactions; (b) The *ribosomes*, particles about 150 Å in diameter composed of about half ribonucleic acid and half protein. These are thought to be the sites of protein synthesis, and, in the most influential current view, the ribonucleic acid of the ribosomes is thought to have received the genetic "information" from the deoxyribonucleic acid of the chromosomes. One view is that this ribonucleic acid is synthesized on the chromosomes and is transported to the cytoplasm. Protein synthesis *in vitro* has been achieved, using isolated ribosomes and additional soluble enzymes and intermediates; (c) The *endoplasmic reticulum* (or ergastoplasm) is a system based on double-layered membranes about 100 Å thick, organized into vesicles, and chains and networks of vesicles. The basic membranous structure resembles that of the cell membrane and the nuclear membrane, and it is possible that all of these are interconvertible. The endoplasmic reticulum commonly is associated with ribosomes (and vice versa), but the association is not obligatory; membranes without ribosomes and ribosomes without membranes are found. The membranes seem to contain enzymes capable of synthesizing cholesterol and other non-protein molecules. On the whole, the ribosome-reticulum association embodies the major biosyntheses taking place in the cytoplasm, and this is visually expressed in the high degree of organization of the system in cells that are very active in synthesis. (d) The *lysosomes* are a class of particles intermediate in size between the mitochondria and the ribosomes, in which a group of destructive (hydrolytic) enzymes seems to be confined. There is considerable speculation as to the meaning of these depots of destructive enzymes. (e) The so-called *soluble phase* of the cytoplasm consists of water, salts (chiefly organic salts of potassium), and "free" molecules, including a great many enzymes whose activity does not seem to depend on firm associations in particles.

Superimposed upon this general scheme of cytoplasmic organization, specialized structures may be found in highly differentiated cells. For example, 60–70 percent of the dry mass of a muscle fiber consists of a quasi-crystal-line array of filaments of two proteins, actin and myosin, which are not found in appreciable amounts in most other cells. A corresponding example from the plant world would be the highly organized PLASTIDS carrying the PHOTOSYNTHETIC pigments.

(4) *Centrioles* are self-reproducing bodies having rather unique functions in animal cells and those of certain lower plants; their existence in cells of higher plants is questioned. These unique bodies have two functions: (a) They determine the poles of dividing cells; chromosomes move to the centrioles and division takes place along the equator defined by two centrioles. (b) In the form of *basal granules* or *kinetosomes,* they determine the formation of motile fibrillar structures, cilia or flagella, at the surface of certain cells.

Biochemical organization. Probably all of the organic matter on the planet is found in cells or is the product of the activity of cells, ultimately deriving from photosynthesis. The fact that organic chemists are capable of imitating biological syntheses, or making complex carbon compounds not found in cells, does not alter this old generalization fundamentally, but merely damages its vitalistic implications.

For practical purposes, the rates of virtually all chemical reactions in the cell are governed by catalysts, and there is a specific catalyst, a protein which is an ENZYME, for each reaction. Thus the character of a cell is largely describable in terms of the specificity, the amounts, and the structural relationships of these enzymatic proteins, and it is reasonable to think that the genetic code in the chromosomes is, to a considerable extent, a set of specifications for the synthesis of enzymes.

The work of cells, including biosynthesis, the transport of material, movement, the generation of electrical potentials, and sometimes the production of light is carried out at the expense of the chemical potential energy of substrates, by oxidizing them. Only the net economics of biological oxidations can be compared with combustion. Biological oxidations take place at low temperatures and in a stepwise fashion, each step governed by enzymes. The unit oxidation step involves the removal of two hydrogen atoms (or two electrons), and these are carried to oxygen, the ultimate acceptor, by a stepwise "electron transport system," each step of which again involves enzymes.

This quantization of biological oxidations is intelligible in terms of one of the great generalizations of biochemistry: that the immediate source of energy for biological work is not oxidation, but the splitting of "high energy bonds" (see BIOENERGETICS). The most prominent of the high energy bonds is the P-O-P (pyrophosphate) bond, and the best known of the compounds transferring phosphate-bond energy is *adenosine triphosphate* (adenine-ribose-P-O-P-O-P, abbreviated *ATP*). The formation of the high energy P bonds depends on the *coupling* of oxidations and phosphorylations, (or oxidative phosphorylation) which can be represented by the reaction $XH_2 + C + ADP + PO_4 \rightarrow X + CH_2 + ATP$, where X is the reduced substrate, providing the energy, C is an acceptor for electrons (ultimately oxygen), ADP is adenosine diphosphate. Even the ultimate energetic reaction of the biological world, the capture of light energy in photosynthesis, has now been shown to involve the generation of high energy phosphate bonds at the expense of the quanta of light absorbed by chlorophyll. ATP is produced by *photosynthetic phosphorylation.*

The efficiency of oxidative phosphorylation is measured by the number of high energy P bonds produced per atom of oxygen reduced, and this so-called "P/O ratio" may be about 3. Since we can only form one high energy bond per oxidation event, it is obvious that cell oxidations require a stepwise mechanism for high efficiency, and this is embodied in the various pathways that have been described. In most kinds of cells, the enzymes for these complex pathways of oxidative phosphorylation are linked together in the mitochondria, which may be regarded as the centers for the conversion of "food" substrates to high energy compounds, especially ATP.

To perform useful work, the cell reacts ATP with the particular "working system" in such a way that high energy bonds are split and the working system is *activated*. The activated system can now proceed with its operations in a seemingly spontaneous way, having gained energy from the high energy bond. For example, a general formulation for the biosynthesis $X + X \rightarrow XX$ would be

(1) $2X + 2APPP \,(ATP) \rightarrow 2XP + 2APP$ or $2XPP + 2AP$

(2) $2XP \rightarrow XX + 2P$ or $2XPP \rightarrow XX + 2PP$ (inorganic)

A more obvious case of "work," the contraction of muscle, might be represented by:

Extended muscle + nATP $\rightarrow$
$\qquad$ contracted muscle + nADP + nP

where the nature of the activation step is not known and the changes in molecular configuration seen as contraction are still not understood.

Thus, the plan of biological energy transformations involves the sequence: coupling of oxidation to phosphorylation, storage in high energy bonds, activation of "working systems" by splitting high energy bonds, and the "spontaneous" transformations of working systems. Overall, this system is more efficient than heat engines, which oxidize fuel directly and exploit temperature differentials. The efficiency of muscle, for example, is reckoned at 40 percent.

The integration of biochemical events in the cell will be expressed by the interrelations of: transport of molecules across the surface, synthesis of enzymes, formation of intermediates for macromolecular syntheses, provision of information for specific syntheses and reproduction of that information, oxidations, phosphorylations and, of course, overt work or behavior. The solution of these problems of integration depends not only on simple kinetic considerations such as would apply to reactions in a test tube, but also on the structural organization of the cell. The following statements are examples of current views:

1. The selective transport of substances into and out of the cell depends on specific "carriers" or "permeases" in the surface. These are enzymes and may be derived from ribosomes. The energy for both the synthesis of the carriers and the transport itself will have come from ATP which was generated by oxidative phosphorylations in the mitochondria.

2. The specifications for the enzymes of the mitochondria, as well as for the proteins of the ribosomes, the carriers or permeases of the surface, and for structural proteins, derive from the nucleus. They may be encoded in the DNA of the chromosomes, "translated" into ribonucleic acid, which functions in the ribosomes to assemble amino acids (activated at the expense of ATP of mitochrondrial origin) in the correct sequence.

3. The ultimate basis of accurate cell reproduction may be the reproductive synthesis of chromosomal DNA. This involves the participation of the parental DNA as a model, but also involves, as shown above, the availability of nucleotides which are activated by transfer of high energy phosphate from ATP of mitochondrial origin, and which are synthesized by the participation of many enzymes. Many of these enzymes are synthesized by ribosomes which derived the information from the very DNA that is now being replicated. If these seem to be complex situations, it is because even a simple network of relationships seems complex compared to a linear causal sequence.

Development and reproduction. Cells tend to increase in mass and in number, while tending to maintain constancy of character. In nature, these tendencies are modified by insufficiency of nutrients, or of water, or by mutual interactions of cells in the development of multicellular organisms. Cellular increase is describable in terms of a reproductive cycle (sometimes termed the "cell cycle" which may be described by the following statements.

(1) Cells are "born" as individuals by the division of their "parent" cells. Division entails the provision of a complete nucleus to each daughter cell and the complete partitioning of the cytoplasm.

(2) In the differentiation of multicellular organisms, many of the cells which take over specialized functions

do not divide again. This is the result of organismal controls; if such cells are removed from the organism to cultures, they tend to lose much of their differentiation and to divide.

(3) Following the "birth" of a cell at division, it generally grows before dividing again. Typically, cells double in mass before division. Differentiating cells often increase to more than double their birth-mass.

(4) In addition to the general doubling in mass before division, there are certain processes which are specific preparations for division. One of these is the reproduction of the genetic material, measured as the doubling of the DNA content. Others are the reproduction of the centrioles in animal cells, the provision of the substances of which the division machinery (mitotic apparatus) will be made, and the storage of energy for the division process.

(5) In plant and animal cells, the division of the genetic material following its replication takes place by mitosis. The essential features of mitosis are the condensation of the already-doubled chromosomes into compact units, the separation of sister chromosomes, and their transport to the poles (which are determined by the centrioles in animal cells). In bacteria and some protozoa, the mitotic mechanism is not observed.

(6) A characteristic feature of cells that are dividing by MITOSIS is the mitotic apparatus, a body in which sister chromosomes seem to be engaged to sister poles by means of fibrils. Superficially, the movement of the chromosomes seems to involve the shortening of the fibers connecting them to the poles.

(7) The mechanisms of the division of the cell body following the separation of the chromosomes vary greatly from one kind of cell to the other, but in those cells which divide by mitosis the plane of division is almost always the equator defined by the polar axis of mitosis.

The capacity of the cell to grow following division seems to be limited by the nucleus, as though there were a nuclear "sphere of influence." The division of the nucleus doubles the capacity for growth, and this is independent of the division of the cytoplasm; a cell with two nuclei can grow to the same mass as would be attained by two cells each with one nucleus. Thus, division refreshes the capacity for growth, while the completion of growth, provided that it includes the specific preparations for division, makes further division possible.

Regulation. The relations of cells, as well as more complex biological systems, to the material world as a whole are usefully expressed in the concepts of regulation or control. Biological regulation has two faces: the maintenance of internal constancy in the face of a fluctuating outer world and the capability of rapid and adaptive changes in response to external changes. For example, cells maintain characteristic internal K^+ and Na^+ concentrations, quite different from those of the environment and relatively independent of the composition of the environment. Yet in the most responsive of cells, nerve cells, the response to stimuli and the ability to conduct impulses depends on a transitory reversal of the ion-regulating mechanisms. Another example is seen in the control of enzyme synthesis (see ENZYME INDUCTION, known in detail from studies on bacteria, where the level of enzyme production is kept relatively constant by genetic controls and by "negative feedback" effects of the enzyme products on enzyme formation, yet the cells can greatly increase the production of an enzyme adaptively in response to the supply of the substrate for that enzyme. Cellular regulations employ various devices, but one of the most widespread of these is inhibition and disinhibition. That is, the capabilities of cells are almost always in excess of their actual performance, and very commonly a so-called stimulation is explained as the removal of an inhibition. One example has already been mentioned: the release of potential enzyme production by the substrates, which is clearly adaptive.

A larger scale example is the fact that many of the cells of a differentiated organism do not divide or divide infrequently, a regulation whose failure is expressed as malignant growth. This can be shown to involve the inhibition of one or more of the specific preparations for division, and not the loss of the capacity to divide.

Altogether, our present insights into the cell—and hence into the nature of Life—lead us away from holistic conceptions implying vitalistic factors or an unassailable material intactness and toward analysis based on parts and their interactions, molecular events and their interactions, and intelligible principles of regulation and control. At the same time, the study of the cell is revealing remarkable capabilities of organized matter, such as the ability of certain molecules to store, transmit and reproduce information with extraordinary precision.

DANIEL MAZIA

References

Brachet, J. and Mirsky, A. E., eds. "The Cell." Vol. 1, 1959; vols. 2, 3, 5, 1961; vol. 4, 1960, New York, Academic Press.

Lowy, A. G. and Siekevitz, P., "Cell Structure and Function," New York, Holt, Rinehart and Winston, 2nd ed. 1968.

CELL WALL

In most green plants the cell wall consists of cellulose, pectic compounds and hemicelluloses, all linear polymers. Cellulose occurs as unbranched chains of β-glucose residues (10,000 or more) with a regular 1:4 link; except in a few cases (e.g. *Valonia, Cladophora*) it is closely associated with other sugar polymers, e.g. mannan, xylan. In parts, these chains lie parallel to each other and spaced a regular distance apart forming crystallites some 25–200 Å wide (dependent on species) and more than 600 Å long. In cellulose I, the normal cellulose of higher plants, this array corresponds to a space-lattice defined by the unit cell $a = 8.35$ Å; $b = 10.30$ Å (fibre axis); $c = 7.90$ Å; $\beta = 84°$. When cellulose I is dissolved and reprecipitated, or recovered from a highly swollen condition, it recrystallizes in a new form, cellulose II. Electron microscopically, naturally occurring celluloses are seen to consist of microfibrils about 100 Å wide (e.g. bacterial and wood cellulose) to 200 Å or more (e.g. *Valonia, Cladophora*) and about one half this thickness. These microfibrils are apparently smooth-surfaced, unbranched and endless, and consist of cellulose chains in parallel order. The crystallites form the central core of the microfibrils, clothed with chain molecules irregularly spaced presumably because of admixtures with non-glucose sugar chains. The "cortex" together with irregularly spaced interruptions in the crystalline core forms the paracrystalline fraction of cellulose.

Cellulose stains blue with chlorzinciodide or with 72% H_2SO_4 and aqueous iodine; is soluble in 72% H_2SO_4 or cuprammonium; is transparent to ultraviolet; and swells in alkali. It is usually birefringent, with $n_\gamma = 1.59$, $n_\alpha = 1.53$ in well oriented samples, and yields typical X-ray diagrams with characteristic spacings of 6.1 Å, 5.4 Å; 3.9 Å, 2.56 Å and others.

The basic component of the pectic compounds is polygalacturonic acid associated with an araban and a galactan. The acid is partially or completely methylated, or neutralized with Ca^{++} and/or Mg^{++}. These compounds are often fully soluble on prolonged boiling (ca. 12 hrs.) in water; insoluble in hot 0.5% ammonium oxalate after 50:50 alcohol, conc. HCl; they stain with methylene blue and ruthenium red but the reaction is not specific. The hemicelluloses are comparatively short chain compounds, sometimes branched, of hexoses (other than glucose) and pentoses, often mannose and xylose; these are extractable with alkali.

Additional substances occur in surface tissues and in lignified elements. Among the former are cutin and suberin, polymers of hydroxymonocarbonic acids. Lignin is a high molecular weight compound containing phenylpropane derivatives, extractable in hot 3% Na_2SO_3 after chlorination; the red coloration with phloroglucin and conc. HCl is not strictly specific.

There is evidence that wall polysaccharides are synthesized from nucleosyl diphosphate sugars (e.g. cellulose from either guanosine or uracyl diphosphate glucose). Cellulose microfibrils are almost certainly produced by end synthesis, possibly by granular enzyme complexes at or near the protoplasmic surface. Lignin is derived from phenylpropane derivatives such as coniferyl alcohol, through the agency of a peroxidase or a laccase. Cellulose forms the structural framework of the walls of cells in which it occurs; the other substances are deposited around it and together are classed as encrusting substances. Most fungi contain chitin (a polyacetylglucosamine) in place of cellulose, a substance known otherwise only in invertebrates. Cellulose occurs in the Oomycetes, and in some Chytridiaceae and Monoblepharidaceae. It is also found in some tunicates and has been discovered in mammalian skin. Among green plants, it has recently been shown that seaweeds belonging to the families Bryopsidaceae, Caulerpaceae, Udotaceae and Dichotomosiphonaceae contain a β-1,3-xylan in place of cellulose. This polysaccharide is crystalline in the wall, the microfibrils visible in the electron microscope consisting of hexagonally packed arrays of triple helices. Similarly, members of the Codiales (again green seaweeds) contain β-1, 4-linked mannan in place of cellulose.

In higher plants, the primary wall—the only wall layer present during the bulk of cell expansion—contains 2–12% cellulose by volume. This wall expands in area with no decrease in thickness and is pierced by plasmodesmata—intrusions of the cytoplasm—grouped together into pit fields. The microfibrils are narrow, tortuous and interwoven and tend in elongating cells toward an orientation transverse to the long axis of the cell. In some elongating cells, e.g. collenchyma cells, the microfibrils are oriented longitudinally. The secondary wall is deposited after most, at least, of the increase in cell dimensions has ceased. In wood tracheids and most fibers this is heavily lignified, lignification beginning often at corners toward the middle lamellae, after the wall has been thickened, and being most intense at the middle lamella region. These walls consist of three or more layers, usually an outer and inner lamella which are thin, and a central layer which is thick, in latewood tracheids. The run of the cellulose microfibrils in these is illustrated in Fig. 1. In some fibers, e.g. bamboo, lamellae with flat helices and lamellae with steep helices alternate more frequently than in Fig. 1, often 3 or 4 times. At bordered pits, microfibrils approaching the pit separate, pass round the border and come together at the far side. The separating membrane around the torus is pierced by pores ranging up to 2500 Å in diameter, compared with the intermicrofibrillar spaces in the rest of the wall, ranging around 40 Å. In vessels the microfibrils tend to lie transversely with considerable distortion due to the numerous pits. The walls of sieve tubes are primary except in some conifers. The sieve area, whether on a transverse wall or not, is pierced completely by pores which are filled with cytoplasmic material.

Little is known concerning the wall components of Pteridophytes and Bryophytes, though normal cellulose has been identified. In the green algae, cellulose in the form found in higher plants is common (see exceptions noted above). In the Cladophorales and in a few Siphonales (e.g. *Valonia, Siphoncladus*) the walls are thick and finely lamellated. In each lamella the microfibrils are thick, straight and parallel to, though rather frequently twisted round, each other. In adjacent lamellae the directions of the microfibrils lie at rather less than a right angle to each other, with occasional interposition of a lamella with a third orientation, lacking in most species of *Chaetomorpha*. These directions are inclined to the length of the cell, forming helices round the cell. In *Valonia* the three helices round the vesicles converge to two "poles," one at the region of attachment of rhizoids. Zoospores and gametes of *Chaetomorpha* and *Cladophora* develop this same wall structure soon after settling.

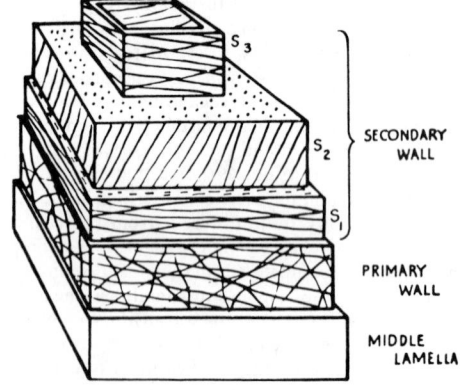

Fig. 1 Dissected model of wall of tracheid in *Pseudotsuga*. The lines on the model represent the run of microfibrils.

During cell growth it was early considered that the wall extends passively under turgor forces. Several phenomena suggest, however, that the wall grows (lack of bulk reorientation of microfibrils and increase in thickness of outer lamellae during cell elongation, presence of cytoplasmic constituents in walls, twisting of microfibrils round each other, difficulty in plasmolyzing growing cells etc.). The demonstration that in some elongating cells the microfibrils of inner lamellae tend to lie transversely, and of outer lamellae at random or even axially, has again been taken to imply passive extension. Both this process and metabolic processes may be involved. Growth hormones are considered to exercise their effect through the chemical bonding in the wall, possibly between a polysaccharide and a protein recently demonstrated to be present. Markers placed on the walls of some elongating cells separate not only longitudinally but also transversely; the top of the cell rotates with respect to the bottom. This spiral growth has been traced to interaction between turgor pressure and the mechanical properties of the walls.

Wall synthesis and growth can occur at localized regions (tips of hairs etc.) or over the whole surface (common). Theories involving passive extension assume that wall formation is in either case by apposition; if the wall also grows, intersussception also plays a part.

R. D. PRESTON

References

Roelofsen, P. A., "The Plant Cell Wall," Berlin, Borntraeger, 1959.
Frey-Wyssling, A., "Die Pflanzliche Zellwand," Berlin, Springer, 1959.
Preston, R. D., "The Molecular Architecture of the Plant Cell Wall," London, Chapman & Hall, 1952; Endeavour, 23: 152, 1964.

CENTROSPERMAE

An order of plants presumably derived from the Buttercup Order, including seven families; typically the flowers have both sepals and petals and these parts are attached to the floral axis beneath the ovary. The latter organ has a single seedbearing cavity. The embryo is generally coiled or at least curved (which characteristic was the basis for the old ordinal name Curvembryonae) and the seeds are attached by relatively long funiculi either to the base of the ovary or to a central projection from the ovary base. The order is in some systems of classification referred to as the Caryophyllales.

The species of the Chenopodiaceae or Goosefoot Family are predominantly herbaceous or shrubby and occur commonly on saline or alkaline soils. The leaves are simple, sometimes fleshy or scalelike, and nearly invariably alternate. The flowers are predominantly bisexual, minute and apetalous. The family includes some of the commonest weeds (Goosefoot or Lamb's Quarter, Red Sage, and Russian Thistle) but also some valuable food plants (beet and spinach).

The Amaranthaceae are mostly herbs with simple, alternate or opposite leaves and bisexual flowers which are usually arranged in dense clusters. Many of its members are noxious weeds, notably the pigweeds and tumbleweeds. It is a family with its best development in the tropics and subtropics.

The Nyctaginaceae, or Four O'Clock Family, are generally herbaceous in temperate zones but arborescent in the tropics. The leaves are usually opposite and simple. Many representatives are cultivated, their attractiveness due to the brilliant colors of the bracts which surround the otherwise insignificant flowers or flower-clusters. The best development of the family is in the southwestern United States; several species are grown for ornament (Mirabilis) and several are weeds of note.

While in temperate regions the Phytolaccaceae, or Pokeweed Family, is represented by one wide-ranging weedy species (Phytolacca americana L.), the bulk of the family is found in tropical America. The leaves are alternate, simple, and some parts of the plant, chiefly the roots and seeds, are poisonous. The flowers are usually bisexual, lack petals, and have a wide range in the number of flower parts.

The Carpet-weeds, the Aizoaceae, are remarkable for the great diversity of floral structure that is found in the family. The leaves are simple and either small and scale-like or fleshy. The flowers are apetalous; even though petals appear to be numerous in the Iceplant (Mesembryanthemum spp.), these are petaloid staminodia. The center for the family is in South Africa; weeds (Mollugo) and ornamentals occur in the warmer parts of North America.

The Purslane Family (Portulacaceae) is made up chiefly of herbs or infrequently shrubs; the leaves are simple, usually rather fleshy, alternate or opposite, and these often form a rosette at the base of the plant. The distinguishing characteristic of much of the family is the mode of fruit dehiscence: the capsule splits in an equatorial plane and the calyptrate top falls away to release many small seeds.

R. S. COWAN

References

Engler, A. and K., Prantl, Eds., "Die natürlichen Pflanzenfamilien," 2nd ed. Leipsig, Englmann, 1934.
Hutchinson, J., "The Families of Flowering Plants," 2nd ed. Oxford, Clarendon Press, 1959.

CEPHALOCHORDA see AMPHIOXUS

CEPHALOPODA

Cephalopods (Octopuses, squids, cuttlefish, Pearly Nautilus and their allies) are a class of the phylum MOLLUSCA and are the most highly organised invertebrate animals. The molluscan affinities of the group are demonstrated by the possession of a radula, the form of the ctenidia and the presence of a shell or the remains of a shell.

The Cephalopoda are bilaterally symmetrical mollusca with a definite head which is supposedly homologous with the foot of other members of the phylum. This head bears eight or ten tentacles (many in *Nautilus*). A mantle partly encloses the visceral mass leaving a mantle cavity, into which opens the anus, ink sac duct and the renal and genital apertures. Within this mantle cavity lies the single pair of gills (double pair in *Nautilus*). A funnel or siphon attached to the ventral surface of the head forms the exhalent aperture for the mantle cavity. The mantle edge bordering the mantle cavity is usually free, but may be closed by muscular action against the siphon and body. In the squids, the mantle edge is positioned against the siphon by a paired locking device, and in some groups may be permanently fused to the siphon.

The circle of arms or tentacles is diagnostic of the class. *Nautilus* has numerous arms without suckers but most other cephalopods have longitudinal rows of suckers on the arms. In the Decapoda chitinous rings strengthen the suckers, the free rims being serrated in varied patterns and sometime being modified to form hooks.

Most male cephalopods have one or more arms modified as secondary sexual organs. Such modification is known as a hectocotylus but it is not certain that all structures known by this name are homologous. Sperm are assembled into very complex spermatophores which are released into the mantle cavity of the male,

are transferred to the hectocotylus and during mating are deposited in the mantle cavity of the female. In many of the Octopoda the hectocotylus consists of a spoonshaped swelling at the tip of one arm, and in a few genera (*Argonauta* and *Tremoctopus*) the hectocotylus is detachable and may be left in the mantle cavity during coitus. The modified arm or arms in the Decapoda may bear modified suckers and additional subsidiary structures of varied form.

A muscular bulb-shaped buccal mass lies in the centre of the ring of arms, enclosing a pair of chitinous, beak-shaped jaws, within which lies the rather simple but typically molluscan radula. The alimentary canal is essentially a U-shaped tube extending from the mouth to the anus, which opens into the mantle cavity near the base of the siphon. The alimentary system can be divided into buccal mass, oesophagous, crop (sometimes absent), stomach, spiral caecum, intestine and rectum. Except in *Nautilus* two pairs of salivary glands open into the buccal mass or oesophagous. The posterior pair in some forms at least has been shown to produce a poisonous secretion. The most obvious accessory to the alimentary system is the mid-gut gland (otherwise known as "liver" or hepatopancreas) which is a very large saccular organ opening by paired ducts into the spiral caecum. An ink sac is closely associated with the mid-gut gland but opens by a separate duct, near or into the anus.

The sexes are always separate. The female reproduc-

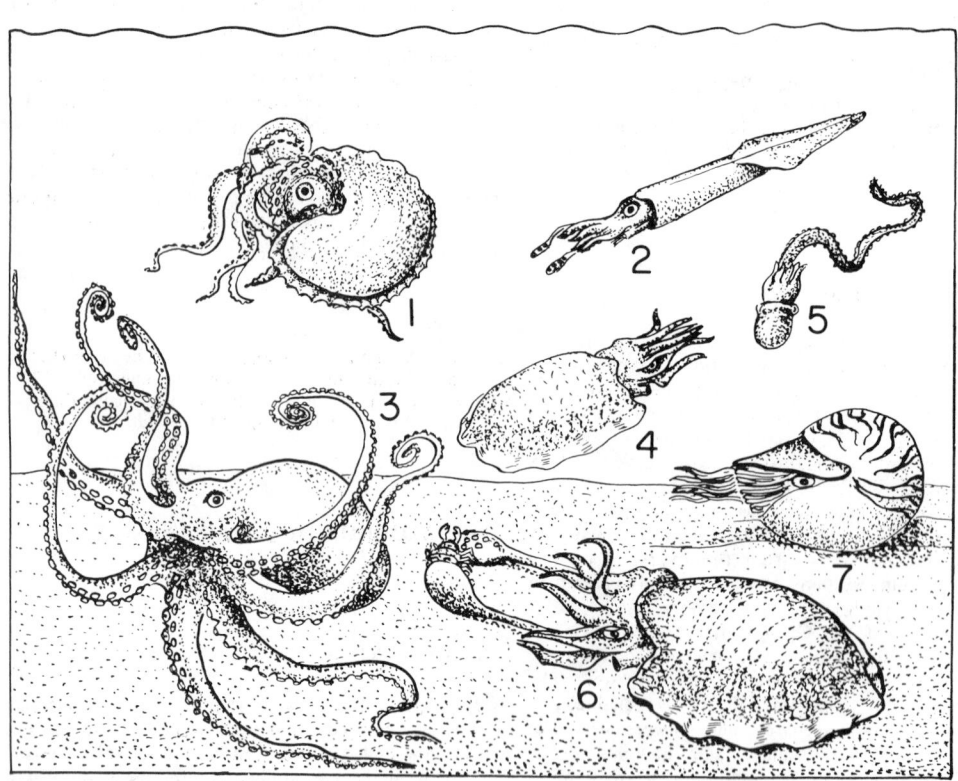

Fig. 1   Some members of the class Cephalopoda. 1, *Argonauta,* the paper nautilus; 2, *Loligo,* the squid; 3, *Octopus,* the octopus; 4, *Sepia,* the cuttlefish; 5, male argonaut; 6, *Architeuthis,* the giant squid; 7, *Nautilus,* the chambered nautilus. (From Goodnight, Goodnight and Gray, "General Zoology," New York, Reinhold, 1964.)

tive system is comparatively simple consisting essentially of a large ovary situated at, or near, the posterior extremity of the body giving rise to a single, or double oviduct which opens into the mantle cavity. Oviducal glands of varied form are usually associated with the oviduct. The male reproductive system is much more complex, some of the structures being concerned with the formation of the very complicated spermatophores.

The circulatory system consists of a complex system of closed vessels, radiating from, and returning to, a centrally situated heart system. Two branchial hearts, muscular thickenings at the base of each afferent branchial vessel pump blood to the gills. Re-oxygenated blood is returned from the gills by the efferent branchial vessels to a median systemic heart, consisting of two auricles and a ventricle. Three large vessels, an anterior, an abdominal and a genital aorta radiate from the ventricle branching in various ways to supply all parts of the body.

Four ganglia, cerebral, brachial, pedal and visceral are fused in the head to form the most complex brain of any invertebrate. Recently giant nerve fibres found in some squids have proved very useful material for physiological experimentation. Although they develop differently and although there are differences of structure there is very close analogy between the design of the Dibranchiate and the vertebrate eye, a remarkable case of parallel evolution.

The excretory system is comparatively simple, consisting of a pair of kidneys opening into the mantle cavity through short ureters.

Cephalopods are noted for rapid colour changes, brought about by contraction and expansion of numerous chromatophores imbedded the skin. Some squids produce vivid displays of bio-luminescence, some by utilising bacterial action, some by secretion and some by the most highly developed systems of complex photophores.

The size ranges from forms under two inches in length to species of the genus *Architeuthis* with bodies seven feet long and tentacular arms extending an additional fifty feet. All cephalopods are marine, members of the class filling most available niches in benthic, pelagic and bathypelagic habitats in all seas.

R. K. DELL

Reference

Lane, F. W., "Kingdom of the Octopus: The Life-History of the Cephalopoda." New York, Sheridan, 1958.

CESTODA

Definition. A class of *parasitic worms* within the zoological phylum PLATYHELMINTHES, familiarly termed tapeworms, *Bandwürmer, Bendelormer, etc.* The term Cestoda comes from the Greek *cestos*, a ribbon.

Features. Parasitic as adults, with few exceptions, in the small intestine of chordate animals—except cephalochordates and lampreys—; body depressed, dorso-ventrally flattened, ribbon-like, anchored to the intestinal wall by a spherical, ovate or brickshaped holdfast (*scolex*) provided with organs of adherence;

these may be shallow or trench like depressions of the dorsal and ventral surfaces provided with mobile margins, or may be in the form of four relatively deep suction cups (*suckers, acetabula*); either type of adherent structures may or may not be supplemented by an armature of hooks.

No trace of mouth or alimentary canal. Reproductive organs metamerised, as a linear series of sets of hermaphroditic *genitalia*; the area containing such a set is termed a *proglottis*; transverse grooves may demarcate the body into square or rectangular *segments*, usually but not invariably corresponding with the proglottides.

Each genitalium may be provided with a *uterine pore*, or escape of eggs may occur through a temporary split of the ventral surface, or a gravid segment may eventually leave the chain of segments, pass from the host and through decay liberate the eggs, or be swallowed by the intermediate host. A segmented tapeworm may be called a *strobila* (Gr. chain) and the progressive break-off of gravid segments may be called *apolysis.*

Forms with uterine pores do not shed ripe segments and are said to be *anapolytic.* Some forms, with uterine pores, may shed chains of egg-exhausted segments, and may be termed *pseudapolytic.*

Zoological interest. Cestoda offer a number of questions of profound biological interest. Is a tapeworm an individual or is it a colony of individuals? is the holdfast merely an anchor or has it a trophic function? to what extent do tapeworm structures, especially the skin, provide an exception to the Germ Layer Theory? what are their dietary and respiratory requirements? Until a method of keeping tapeworms alive outside their hosts for a reasonable length of time, or even of rearing them from egg to adult in laboratory media, has been established, our knowledge of tapeworm physiology must remain fragmentary and speculative.

Origin and phylogeny. The lack of fossils and the unlikelihood of fossil tapeworm material ever being available, make conclusions as to tapeworm origin and phylogeny purely speculative. Little help is provided by life-cycle studies, due to the great variability and complexity of embryonic development and life-cycle stages. Conclusions from their distribution in primitive host types are weakened by the wide choice of hosts shown by many types.

Two distinct types of tapeworm are found. On the one hand are the *difossate* forms—the orders *Caryophyllidea, Spathebothridea* and *Pseudophyllidea*—commonly regarded as primitive types—with saucer-like or trench-like bothria on the scolex—with uterine pores—with segmentation weak or lacking—with a life-cycle involving a free swimming embryo, the *coracidium* and a sequence of procercoid and plerocercoid larval stages requiring two intermediate hosts. On the other hand are the *tetrafossate* forms—comprising the remaining eight orders—commonly regarded as specialized types—with four suction cups on the scolex—without uterine pores—with pronounced segmentation and apolysis—with one larval stage, commonly bladder-like, requiring one intermediate host, a crustacean, arachnid, insect, or rarely another vertebrate.

It is the considered opinion of the writer that Cestoda came originally from a parasitic TURBELLARIAN

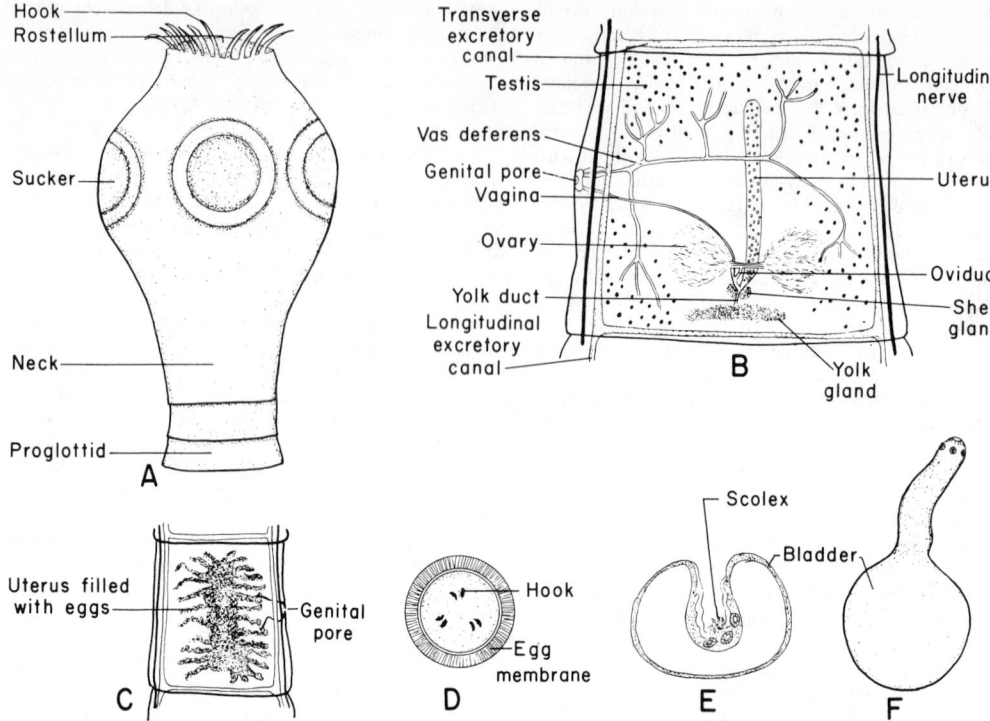

Fig. 1   The pork tapeworm, *Taenia solium*. A, scolex; B, mature proglottid; C, gravid proglottid; D, egg with the oncosphere (six-hooked embryo) inside; E, cysticercus (bladder worm) which encysts in muscle; F, cysticercus with scolex everted. (From Goodnight, Goodnight and Gray, "General Zoology," New York, Reinhold, 1964.)

stock which also gave rise to the present acoelous-alloiocoelous Rhabdocoelida. That this stock had already replaced the alimentary tract by a digestive syncytium. That the eggs were discharged into this syncytium and escaped through a temporary rupture of the segment wall, marking the site of the old mouth. Sexual reproduction was by linear budding of secondary zooids which broke away when mature. Premature autotomy provoked genitalial repition. Delay in the shedding of the zooids provoked segmentation, apolysis being a reminiscence of the former shedding of the budded zooid. A secondary tetra-radial symmetry arose through the development of mid-surficial and mid-marginal grooves, that invaded the apical end.

The oldest of Cestoda are the tetrafossate *Proteocephala*. Difossate forms, on the other hand, must be regarded as persistent, neotenic larval stages of tetra-fossates whose Palaeozoic and Mesozoic hosts are long extinct. Caryophyllidea and Spathobothridea may be larval of forms so ancient that tetra radial symmetry had not yet evolved. Haplobothriidae may be larval stages of Trypanorhyncha of long extinct sharks.

As regards individuality, tetrafossates are polyzootic—representing linear colonies of zooids. Difossates are monozootic; single individuals with genital metamerism. External segmentation in both tetrafossates and difossates presents an example of convergent evolution, of homoplasy rather than homology.

R. A. WARDLE

References

Hyman, H. L., "The Invertebrates," vol. 2. "Platyhelminthes and Rhynocephalia," New York, McGraw-Hill, 1951.
Wardle, R. A., and J. A. McLeod, "The Zoology of Tapeworms," New York, Harper, 1952.

CETACEA

The Cetacea are the whales and their smaller relatives, such as the porpoises. Among them are the largest animals ever known, some weighing perhaps 120 tons. They are completely aquatic MAMMALS, with many special modifications. The order Cetacea contains three suborders, the Archaeoceti, the Odontoceti, and the Mystacoceti (Mysticeti). The archaeocetes are all extinct, known as fossils from the older Tertiary rocks, and will not be considered further here. The earliest representatives of the other two suborders may be recognized in mid-Tertiary deposits. Since there is no vernacular equivalent for the technical term "cetacean", we will extend the term "whale", unless otherwise noted, to include all cetaceans—the big whales from 100-foot length down to about 35 or 30 feet, the killer whale-grampus-blackfish group down to about 15 feet, and below this, the porpoises (of seamen) or dolphins (of the literary world) to about 5 feet.

Living Cetacea. Whales live in all seas, from Arctic to Antarctic, and even in fresh water. A few are large, and are hunted commercially; the majority are small and are hunted only in a few regions, notably off Japan, Norway, and a few other places. Some species appear to be permanent residents where they are found; others migrate along paths still imperfectly known, though for a few commercial species, like *Megaptera*, the humpback, a part of the story has been learned. In any case, many species are cosmopolitan like *Orcinus* (the killer whale), or nearly so, like *Megaptera* or *Physeter* (sperm whale). Others have a discontinuous distribution, like *Eubalaena* (the right whales), of which several populations are separated, apparently by barriers of tropical waters. Others, like *Globicephala* (pothead, blackfish, pilot whale), have distinct forms separated by apparent marine barriers which we cannot detect, except to a degree by the distribution of the animals. This sort of discussion is somewhat undermined by the lamentable fact that many more species have been named than appear to exist. With this in mind, the count we are about to give is intentionally conservative.

Of all the many kinds of mammals that have returned to the ancestral sea, no group has done so as thoroughly as the Cetacea, which cannot long survive out of water. This is a small order, as zoological classifications run, comprising (Recent forms only) the suborders Mysticeti (baleen or whalebone whales) and Odontoceti (toothed whales); together, they muster about one hundred species in thirty-seven genera.Only six of these genera, embracing perhaps twelve species, are mysticetes. These two suborders are not particularly closely related, differing markedly in their anatomy and rather less in their biology, as far as it is known. Superficially, they are much alike, especially in their adaptive features—so much so that Herman Melville's excellent terse field diagnosis fits any living cetacean: "A whale is a spouting fish with a horizontal tail."

I cannot improve on this, but can offer a few subsidiary details for those who may want them. "Spouting" is the conspicuous exhalation through the blowholes (nostrils) on top of the head; these are paired (as in most vertebrates) in the mysticetes, but single externally in the odontocetes. All have a smooth and hairless "fish-shaped" body, the posterior part being especially elegantly streamlined; the head end is in some forms less so, especially in side view. Seen from above or below, all are impressively streamlined. The head runs into the body with no constriction for the neck, the fore limbs are modified into paddle-like flippers, and the hind limbs have vanished. The most spectacular and characteristic external modification is to the tail: fish-like, it tapers smoothly from the body into a slender, laterally compressed section, called "the small" by whalers, and "tail stock" and "caudal peduncle" by the learned (but all it is, is tail), at the end of which are two broad lateral extensions, the tail fins, ordinarily called flukes. This structure is the whale's propeller, and extraordinarily efficient it is, providing essentially silent propulsion and great maneuverability at speeds up to 20 and perhaps 25 knots. The laterally compressed tail ("small") has an especially sharp and thin upper (dorsal) edge extending forward into a ridge which dies out on the back; in many forms this ridge is produced into a conspicuous dorsal

fin, which varies from a mere lappet (as in the sperm whale) to the killer whale's fin, the height of which may be one fifth of the animal's length. In some forms unevenness in the caudal ridge may have the appearance of subsidiary dorsal fins. Since many whales lack a dorsal fin, it seems unlikely that this organ his any great hydrodynamic value; E. R. Gunther made the provocative suggestion that it is a response to turbulence caused by the whale's submerging after a blow (breath). All these control surfaces, flukes, flippers, and dorsal fins, have the cross-section of a hydrofoil. The flippers contribute to maneuverability, especially at slow speed. They range from less than one tenth the animal's length to about one third in the humpback whale.

The skin is very smooth, with a thin outer layer (the "blackskin" of the whalers). It is entirely hairless in the adult odontocetes (the young have a few hairs on the snout), and almost entirely so in the mysticetes, which retain a few sparse hairs along the mouth and snout. The lumps on the face of the humpback *Megaptera* and the bonnet and associated excrescences on the face of the right whale *Eubalaena* are probably related to sites of such hairs.

The innermost layer of the skin is the blubber, composed of tough fibrous tissue liberally larded with fat. This is the source of the oil for which men have long hunted whales. Its primary function is thermal, to conserve heat, and even in starvation it is often not drawn on as a food resource. The problem of heat conservation is important, for water is very heat-hungry; even in the warm seas a man's endurance is measured in hours. In the colder seas it is striking that whales can maintain their high body temperatures (between 95° and 100° F) even though their large-surfaced thin fins seem like huge radiators. The answer appears to lie in a double system of veins, one surrounding the arteries like a concentric conduit, providing a countercurrent heat exchanger, and the other a system of simple veins close under the skin, which are used to discharge excess heat. This arrangement is adequate in water, with its high heat conductivity, but usually fails in air, when a whale is stranded. There are no sweat glands to help, but heat stroke is a primary cause of death of stranded whales; cooling, and keeping moist, are the two most important points in the out-of-water transport of porpoises, as for aquaria.

The most obvious difference between the two kinds of whales is seen on looking into the mouth. The odontocetes have teeth, some species more than 200, some only 10 or 12 or less. Unlike most mammals, almost all species are isodont, that is, the teeth in all parts of the mouth are alike except in size (*Inia*, a freshwater form from Brazil, has the hinder teeth differentiated). Curiously, the teeth are evidently not of too great importance in feeding, for one often encounters well-fed whales of this group with teeth worn well down or even lost.

Although the mysticetes have teeth during foetal life, these disappear before birth, and instead there hangs from the roof of the mouth the whalebone or baleen (*mystax*, Greek: moustache) that gave them the name of Mystacoceti. Baleen is a horny epidermal growth that develops in the foetus as the teeth are resorbed. It grows down from both sides of the palate in a series of transverse plates, numbering, on one side, from about 140 in the Pacific gray whale (*Eschrichtius*) to well over 400 in the rorquals (*Megaptera*, humpback, and *Balaenoptera physalus*, finback). These plates are roughly triangular, with the outer side (toward the lips) smooth, but with the inner, longest side fringed with the frayed ends of their fibers. Not only the number, but the shape, size, and color are distinctive between species. Their function will be discussed under feeding.

The skelton shows normal mammalian characters, but with certain characteristic modifications, particularly in the skull, which in each suborder has its own striking divergences from the usual mammalian pattern. In each case the changes are related to the transfer of the nostrils (blowholes) from the foremost part of the head to the top. This change is of obvious fitness for aquatic life. In each suborder the cranium has been shortened and the bones of the "upper jaw" lengthened into a long rostrum or "beak"; the bones of the face appear as though they had slid back over the cranium, with the much reduced nasal bones up on top, right behind the bony nares. The odontocete skull is even more abnormal in the remarkable lateral asymmetry, which again is related to the nostrils. In the delphinids, for example, the right bony naris is the larger, and the structural mid-line of the skull is displaced to the left, especially dorsally. This asymmetry is less conspicuous, but still measurable, in the external features of the head. It has been suggested, with little plausibility, that this cephalic asymmetry is caused by dynamic inequalities in the pressure field around the head in swimming; this elaborate theory has yet to be clothed in supporting facts.

The neck is extremely shortened, the vertebrae being thin plates which in some whales are fused together. The rest of the axial skeleton, while still characteristic, is not so strikingly unusual. However, it may be noted that the entire post-cervical vertebral column is flexible, with the complete absence of a sacrum. The chief distinction in the thorax is the variable number of ribs in different species, one having only eight pair of ribs. Compared with land mammals, the sternum is reduced, especially in the mysticetes.

The fore limbs have a well developed scapula, a short, stocky humerus, radius and ulna much flattened, and the carpals flattened into a sort of mosaic. The metacarpals and phalanges are not distinguishable except by position, and the number of the latter is variable; most whales have five fingers in the flipper.

The hind limbs, entirely vanished externally, are represented by a pair of rod-like bones which presumably represent the two halves of the pelvis, near the ventral genital opening, and thus far from the backbone. Occasional teratological specimens have been found, notably humpback whales (*Megaptera*), in which other elements interpreted as relics of femur and tibia are to be seen, in some instances showing externally.

The internal organs and genitalia are of the basic mammalian pattern, but with a number of special features. The stomach is compartmented, there being from four to over a dozen sections in different species. The kidneys, too, are specialized to deal with sea water by the multiplication of renules.

Reproduction. The entire life of a whale, including birth, takes place in the water. While some, notably the Pacific gray whale (*Eschrichtius*), come into very shallow water for the birth of the calves, many appear to seek no especial place for this, birth taking place at sea. Normally one calf is born at a time, though cases of twins, and even of six foetuses, have been recorded. Gestation lasts about a year; for some species 9 months is recorded, for most from 10 to 12 months, and for the sperm whale (*Physeter*), 16 months. The calves are born quite large, their length in some species often surpassing 40% of their mothers'. Weaning is sometimes delayed until near the birth of the next calf. The interval between successive births may be a year in some kinds and as long as three years in others. There is a fairly regular mating schedule for different species in different regions.

Swimming and diving. Whales are among the most accomplished of swimmers, and may be fairly said to outswim the fishes. Not only do many of them catch and eat fish, but they more readily accept rapid pressure (depth) changes. The swimming speed of whales has been very much exaggerated by irresponsible reporters. Careful measurements have produced no figures higher than about 25 knots; even this is high for most species, and 20 knots (23 mph) would be a safer maximum. Higher speeds have turned out not to be based on reliable measurements. Almost as much nonsense has been written about the mechanics of whale swimming, including some elaborate accounts of spiral or semi-rotary swimming. Direct observations, including slow-motion moving pictures, do not support these torque theories. What actually happens has been satisfactorily explained by D. A. Parry in 1949. The basic propulsive motion is a simple up-and-down oscillation of the entire tail, the "hinge" being at the point opposite the pelvic remnants (in other words, at the root of the tail). A secondary and very important motion is the independent vertical tilting of the flukes and the last few vertebrae, to which they are attached. In this way large amplitudes and large angles of attack are employed for "low gear" operations, and small amplitudes and small angles of attack for "high gear" fast swimming. The tail beat is not particularly rapid; a 7-foot porpoise going about 18 knots had a tail-beat frequency of 3.4 cycles per second, and one at half that speed is recorded at about 2.5 cps. These beats would be even slower for large whales.

Whales spend almost all their time submerged, the breathing exposures being normally very short (less than a second for many species). The actual time submerged varies with and within the species. The longest well-attested dives are those of *Hyperoodon* (bottlenose whale), listed as 2 hours, and *Physeter* (sperm whale) with a maximum of 1½ hours. Porpoises have been timed to 6 minutes (*Stenella plagiodon*), but one would expect that all these numbers might be exceeded by individual animals. After these long dives, the whale usually takes a number of breaths at much shorter intervals ("having its spoutings out") while liquidating its oxygen debt. Several different rhythms have been described.

The depths reached by whales are known with any certainty only for the sperm whale (*Physeter*), whose 500 fathom dives are alluded to in connection with feeding, and for fin whales (*Balaenoptera*) which have taken manometers to over 160 fathoms. Here again there are many unverifiable statements of great depths, usually based on the amount of line taken out by harpooned whales (which undoubtedly did not go "straight down" all the way as the reporters seem to believe), or on bottom mud stuck to the whale (traditionally without precise indication of water depth).

One is often asked why whales do not get the bends (caisson disease) like human divers. The first reason is that they do not inhale compressed air, but take with them a lungful of atmospheric air obtained at the surface. Subsidiary reasons may include the special adaptations in their vascular system, especially the retia mirabilia, a greatly expanded, finely divided network of blood vessels, which may permit longer utilization of the oxygen in the inspired air. The lungs must be collapsed to at least some degree in even relatively shallow dives.

Food and feeding. Whales are carnivorous, eating both vertebrates and invertebrates (some litoral forms have been accused of vegetarianism, but the evidence is not conclusive). Evidently any whale will eat squid of some sort, although many of the odontocetes make a specialty of it, apparently to the virtual exclusion of any other food. Slippery though this food may be, many and sharp teeth are evidently not a requirement for catching it. Other odontocetes eat fish primarily, and one of the fish-eaters, the killer whale (*Orcinus*), is well known for its voracious incursions among seals and porpoises, not to mention the big whales. (There is no authenticated record of an actual attack on man.) The sperm whale *(Physeter)* is one of the squid specialists, evidently concentrating on the larger forms, including the giant squid which are often over 10 feet long. This prey lives deep in the ocean, at least during the day time, and in search of it the sperm whale dives to depths of 500 fathoms where the pressure is about 100 atmospheres. We know this from accidents that befall the whales, which occasionally get tangled in telegraph cables on the bottom, to be found later by the repair party. The mysticetes, with their special filters of baleen through which they strain the water, feed primarily on small animals, especially crustaceans. These are often as small as 2 mm (e.g., *Calanus*, one of the foods of *Eubalaena*, the right whale); the primary food of the rorquals in the Antarctic, the euphausids (called "krill" by the whalers), are about 40 mm long. Some of the rorquals also eat fish such as herring or small scombrids, usually not much over a foot in length.

Whereas the odontocetes seize individual victims in their mouths, the mysticetes feed by swimming through a mass of their prey, holding the mouth widely open. Water is taken in at the front and passes out the sides of the mouth, leaving the food animals entangled in the baleen fringes inside. When the mouth is shut, the high lower lips overlap the upper jaw, and the whale swallows the accumulation on the baleen, presumably with the aid of the tongue. There is an additional adaptation in the rorquals (fin whales, *Balaenoptera*, and humpback whales, *Megaptera*), which have a system of grooves and pleats on the throat and vicinity. These grooves permit a great expansion of the body surface and of the mouth cavity, so that more food-filled water can be taken in at once.

But how does the whale find its food?

Senses. Of the usual mammalian senses, that of smell has been reduced almost or quite to the vanishing point. The olfactory nerves are much reduced.

This is not unreasonable, since the whale's activity is not in the medium that it breathes; smells in the air would not be a very dependable clue to what is going on in the water. Taste is probably present, but not useful until after the food is caught. Eyesight is very useful, and is utilized by many whales both in water and air, but it is rare to be able to see more than, or indeed as far as, 100 feet through the water; usual ranges would be less than about one fourth or fifth of this, much less in turbid water. Moreover, it is night half the time. Whales often swim fast enough to be outdriving their headlights under normal conditions of visibility. But sound is available. It is well transmitted in sea water, where it travels at nearly one mile per second. It is available to whales, whose hearing is acute. Not only do they hear and utilize sounds that are faint for sensitive electronic gear, but they can localize a sound source very precisely, as experiments with captive porpoises (*Tursiops*) have shown. Moreover, similar experiments have also demonstrated that these porpoises, at least, can echo-locate objects in the water, including their prey. The sounds that they use for this purpose are sharp impulsive clicks of almost "white" frequency spectrum. Since all odontocetes whose sounds we know make such clicks, we venture to suppose that all may use them in echo-location. How the mysticetes find their food is even harder to explain. They do not appear to make clicking sounds like the odontocetes, but instead sonorous moans and screams of relatively low frequencies (less than about 1000 cps), which do not readily suggest echo-location; besides, we have not heard any sounds from them while they were feeding. There is a possibility that the "sinus hairs", a sort of sparse vestigial beard of a few dozen scattered hairs, may be detectors of some sort. Perhaps the mysticetes merely taste experimentally from time to time, sampling the water in this way.

Respiration. As one would expect, the breathing apparatus has been modified for being under water most of the time. The requirements include efficient and rapid exhalation, and a tight seal, not only to prevent escape of air, but also to keep the water out.

The first requirement is met by the situation of the nostrils (blowholes) on top of the head, and by musculature for rapid and wide opening of the blowhole as well as by thorough ventilation of the lungs (the whales' lungfuls are almost all tidal air, with only about 10% (*Tursiops*) to 21% (*Phocoena*) residual air; in man this unchanged residual air is about 85% of the lung capacity).

The second requirement is met by changing the external nasal passages (in the odontocetes) to a rather intricate system of fibrous plugs and membranous sacs, which are closed when at rest and which resist passage through them except when held open by the nasal muscles. The mysticete nose also has a collapse-seal, but lacks the other structures.

Another feature, especially developed in the odontocetes, is the intranarial larynx. The larynx projects through the back of the pharynx (beginning of the throat) into the bottom of the bony nares, and is held there by muscles. Thus breathing and swallowing are kept from interfering with each other.

Another odontocete feature related to the reorientation of the nose is the development of a fatty cushion (melon) on top of the rostrum in front of the nostrils. This is at a minimum in, for example, *Steno* and *Phocoena*, is more conspicuous in *Globicephala* (pothead or blackfish), and reaches its extreme development in *Physeter* (sperm whale) where the famous "case" (spermaceti organ) and associated structures constitute the most enormous nose on record.

Voice. Although for long supposed mute by scientists, whales have well-developed voices. Many fishermen and sailors have known this all along, but were discredited by the learned, especially after the latter ascertained that the cetacean larynx lacks vocal cords. Now, however, we know better again, and at least a dozen species of whales have been phonographically recorded. The odontocetes are apparently much more loquacious than the mysticetes, whose sonorous calls have been alluded to under "Senses." The odontocetes, besides uttering the clicks there spoken of, also squeal. These squeals are often called whistles by men who have not tried to whistle under water. The whales evidently use them mainly for communication, and the calls composed of these squeals, often mixed with clicks at varying repetition rates, are usually as distinct in different species as the bird and mammal calls familiar to us. Equally distinctive are some whale voices restricted entirely, as far as we know, to clicks, like those of *Physeter* (sperm whale) and *Grampus*.

It has been suggested by some that the fatty lump (melon) of the nose is part of the sound system, connected with limiting the sound field, thus helping provide directional sense.

All evidence so far points to the larynx as the source of the sound, although some students have suggested that parts of the nose, in odontocetes, may also be involved.

WILLIAM E. SCHEVILL

References

Gunther, E. R., "The habits of fin whales," London, "Discovery" Reports, **25:** 113–142, 1949.

Parry, D. A., "The anatomical basis of swimming in whales," Proc. Zool. Soc. London, **119:** 49–60, 1949.

Slijper, E. J., "Whales," New York, Hillary, 1962.

Tomilin, A. G., "Kitoobraznye [Cetacea]." *in* Zveri SSSR i prilezhashchikh stran [Mammals of the USSR and adjacent countries], vol. 9, Moscow, Akad. Nauk, SSSR, 1957.

CHAETOGNATHA

The Chaetognatha, a phylum of animals, are exclusively marine, with 16 to 17 genera and about 65 well-defined species. Commonly called arrowworms, *Sagitta* is the largest genus. The PLANKTONIC chaetognaths are widespread and very abundant in the sea, numbering second only to COPEPODS in the macroplankton of most seas. They occur from the surface to depths greater than 6,000 meters but are most abundant in the upper 200 to 300 meters. Some species migrate vertically with a diurnal period in response to light. The genus *Spadella* is benthic, one species, *Sp. schizoptera* having finger-like projections for temporary attachment to the bottom or eel grass. All other genera are planktonic with the possible exception of *Bathy-*

spadella. Amiskwia a reported fossil from the Middle Cambrian Burgess Shale is more likely a nemertean.

Although detrital particles and large diatoms are occasionally found in the digestive tract, they are essentially carnivorous feeding primarily on copepods, and rarely on Foraminifera and larval fish. They in turn are eaten by one another, by carnivorous crustaceans, by fish and in Arctic regions by whales.

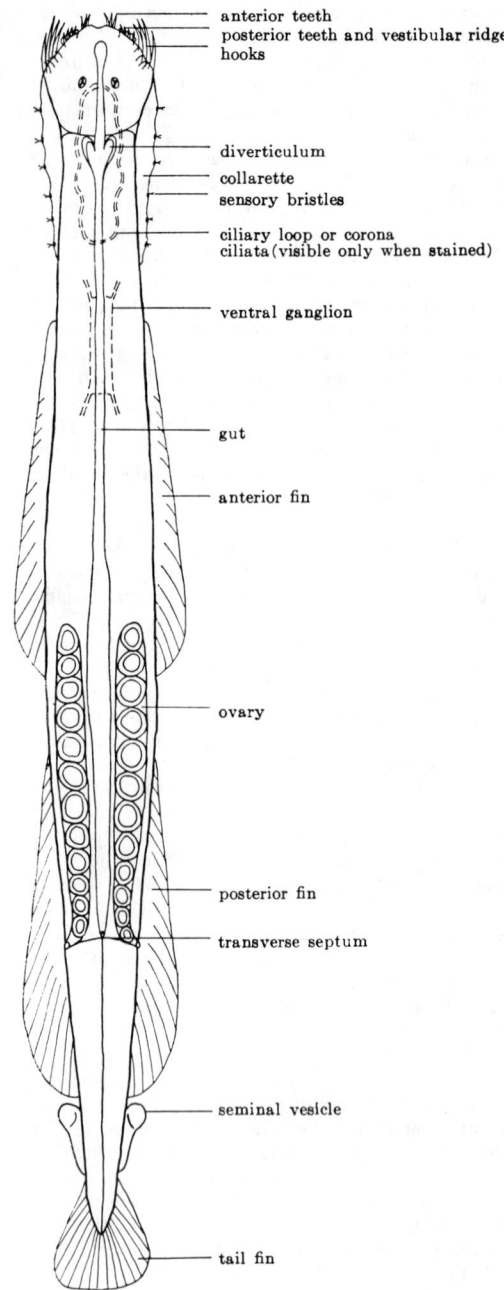

Fig. 1   Diagram of Sagitta. The ciliary loop is visible only after staining.

Length at maturity varies from five to 140 millimeters, most species falling between 12 and 25 millimeters. The anterior and ventral mouth is nearly circled by two lateral rows of chitinous hooks used to catch food and force it into the gut. In *Sagitta* there are two paired rows of shorter accessory teeth used for holding and cutting prey. Some genera have one paired row or no accessory teeth. The hooks and the mouth region are covered by a hood when swimming. The head, bearing two dorsal eyes, is separated from the trunk by a neck septum. The largest body segment, the trunk, is formed of two dorsal and two ventral longitudinal muscle bands with the straight tubular digestive tract suspended between. The body wall forms one or two paired lateral fins. Ovaries extend forward beside the gut from the posterior transverse septum which divides the trunk from the tail segment. Sperm tissue develops in the tail segment which is divided longitudinally in half by a thin dorsal-ventral septum. Seminal vesicles are conspicuous between the tail fin and posterior fins on mature specimens and are important in the systematics of species.

There are no specialized excretory, circulatory, or respiratory systems. Besides several ganglia in the head a conspicuous ventral nerve ganglion is present on the trunk. A dorsal ciliary loop or corona ciliata of varying length on the head (visible only by staining) and anterior trunk is of unknown function. Many species have a hispid appearance when alive due to clumps of projecting fibers presumed to be detectors of water motion.

Reproduction has been observed only in *Spadella*. Two mature individuals exchange sperm masses by breaking the seminal vesicle on the trunk region behind the head. Sperm migrate to the sperm receptacle at the transverse septum. Fertilization is internal, the zygote is expelled through the straight oviduct lying beside the ovary. *Pterosagitta* lays pelagic eggs in clumps of 100 to 300. Other species lay solitary eggs. *Eukrohnia hamata* is reported to brood its eggs in a marsupium formed by the lateral fins. Cleavage is total and equal. The mesoderm is enterocoelous and the mouth forms opposite the original blastopore. For these reasons the phylum is often grouped with the echinoderm-chordate superphylum. The reproductive tissues appear and differentiate unusually early in the embryology. Development is essentially direct. Length of life for different species is estimated from three months to two years.

Besides their importance as abundant carnivorous plankters, they also are frequently parasitized by protozoans and metazoans and may be important parasite vectors. In recent years they have been studied chiefly because of their use as indicators of water movement or "current indicators." They are particularly useful for this type of work because of their abundance and relatively large size.

ROBERT BIERI

References

Bieri, R., "The Distribution of planktonic chaetognatha in the Pacific and their relationship to the water masses," Limnol. Oceanog. **4:** 1–28, 1959.
Hyman, L. W., "The Invertebrates," vol. 5, New York, McGraw-Hill, 1959.

Tokioka, T., "Supplementary notes on the systematics of Chaetognatha," Publ. Seto Mar. Biol. Lab. 13: 231–242, 1965.

CHARADRIIFORMES

This diversiform avian order is taxonomically placed between the GRUIFORMES and COLUMBIFORMES. It is divided into three suborders, the Charadrii (shorebirds), Lari (jaegers, gulls, terns and skimmers) and Alcae (auks and allies). There are 16 families (second only to Passeriformes in number), 124 genera and 314 species. The order ranks sixth in the number of species that it comprises. Family relationships are not the same in all avian orders, e.g., the families of Passeriformes are much more closely related than are the families of other orders. Therefore, although 65 passerine families are recognized, the order probably represents a more compact taxonomic assemblage than does Charadriiformes with only 16 families. Three of the 16 families contain 72% of the total number of species. These are the Laridae (gulls and terns), the Scolopacidae (snipes, sandpipers and allies) and the Charadriidae (plovers) with 82, 82 and 63 species respectively.

Because of the diversity of the Charadriiformes, relatively few of the anatomical characters upon which the avian orders are based are constant throughout the group. Some of the less variable characters that delineate this order are the following: skull schizognathous (maxillo-palatines do not meet each other or the vomer; vomer small and pointed anteriorly) except in Thinocoridae (aegithognathous); nares usually schizorhinal (posterior edge of bony nostril cleft to or beyond the premaxillaries) but holorhinal in Dromadidae, Burhinidae, *Pluvianus* and Thinocoridae; lacrimals united with prefrontals; coracoids with a subclavicular process except in Alcidae; coraco-humeral groove distinct; furculum U-shaped; cervical vertebrae 15–16; dorsal vertebrae 5–8, opisthocoelous; flexor tendons Type I (flexor hallucis longus and flexor digitorum longus cross and are united by a vinculum); carotid arteries two except in Alcidae in which only the left carotid is present; syrinx tracheo-bronchial; oil gland tufted; aftershaft present; primaries 11 (one vestigial); secondaries 11; wing diastataxic (fifth secondary missing but its greater covert present) except in *Philohela*; rectrices 10–28, usually short; nidifugous (young leave nest shortly after hatching) or nearly so.

The phylogenetic relationships of the avian orders remain obscure because of the poor fossil record. Therefore, any attempt to arrange the orders in a phylogenetic sequence is largely speculative. Certain Charadriiform families show rather strong similarities to Gruiform families. The most notable example of this is found in the Jacanidae. In general appearance and habits these birds seem to be closely related to the Rallidae, but anatomically, although resembling the Rallidae in a few characters and unique in some respects, they appear to be closer to the Charadrii. The Burhinidae have some characters in common with the Gruiform family Otididae. The unusual Thinocoridae have been variously placed by taxonomists in Gruiformes, Charadriiformes and in a separate order.

With few exceptions the color of the plumage varies from black through various shades of brown and gray to white. Some Jacanidae are rather showy, having patches of maroon, yellow or green feathers. The neck and underparts of Ross's Gull, *Rhodostethia rosea*, are pink as are occasionally the underparts of the Roseate Tern, *Sterna dougalli*. A few species of Charadrii exhibit some metallic coloration. Striking flight patterns are rather common and the bill and feet are often brightly colored.

In most families there is little or no sexual dimorphism; in a few the male is slightly larger than the female. In the Jacanidae and Stercorariidae the female tends to be somewhat larger than the male, and in the Phalaropodidae and Rostratulidae the female is not only larger but also more brightly colored than the male.

The variation in the form of the bills reflects the diversity in feeding habits. In the seed-eating Thinocoridae the finch-like bill is short and stout and possesses operculate nostrils. Birds of this family also have a crop and well-developed gizzard. In some of the Scolopacidae the bill is extremely slender and long (over eight inches in some) in relation to the body length and often has the distal portion soft, flexible and richly supplied with nerve endings. These birds feed by probing into sand and mud. The strongly compressed bill of the Rynchopidae is unique in having the mandible much longer than the maxilla. In feeding, these birds fly low over the water and plough the surface with the mandible. In certain Alcidae the bills are strongly compressed, deep, multicolored and sculptured. The tip of the bill is hooked in the Stercorariidae.

Most representatives of this order are closely associated with large bodies of water. Many are littoral, some are more or less aquatic and a few are pelagic. The Jacanidae, Rostratulidae and some Laridae are marsh inhabitants. The Thinocoridae and certain species of Glareolidae are crepuscular while the Burhinidae with their large eyes and the Rynchopidae with their vertical pupils are partly nocturnal.

Many species exhibit strong colonialism. The most notable examples of this behavior are found in the Laridae and Alcidae, particularly the terns and murres. The eggs are laid on the ground (with or without nesting material) in most species, but some nest in shrubs or trees (*Gygis* and *Anous*), on floating platforms in marshes (Jacanidae, some Laridae and Recurvirostridae), in burrows (Dromadidae, some Alcidae) in rock crevices and on ledges (certain Laridae and Alcidae).

In most cases both sexes incubate and care for the young. However, in species in which the female is larger and more brightly colored, parental care may be restricted to the male. The Rostratulidae are believed to be polyandrous.

The Charadriidae, Scolopacidae, Recurvirostridae and Laridae are virtually world-wide in distribution. Three families are restricted to tropical and subtropical regions (Jacanidae, Rostratulidae and Dromadidae) and three families are largely polar (Chionidae, Stercorariidae and Alcidae). Two families are found only on the Old World (Dromadidae and Glareolidae) and one family is restricted to the New World (Thinocoridae). Ten families are represented in the United States by a total of 97 species.

Migratory tendencies are correlated with the latitude of the breeding range. Within a single family, species which breed in temperate or polar regions may be strongly migratory while tropical or subtropical species may be strictly non-migratory. The justly celebrated migration of the Arctic Tern, *Sterna paradisaea*, is undoubtdly one of the most remarkable feats in the animal kingdom. It has been estimated that some individuals of this species may cover a distance of 22,000 miles in flying from their Arctic breeding grounds to wintering grounds south of the Antarctic Circle and back again.

ROBERT E. GOODWIN

References

Baker, E. C. S., Fauna of British India: Birds. Vol. 6. London, Taylor & Francis, 1929.
Bent, A. C., "Life Histories of North American Diving Birds," Washington, Bull. 107, U.S. National Museum, 1919.
ibid., "Live Histories of North American Gulls and Terns," Washington, Bull. 113, U.S. National Museum, 1921.
ibid., "Life Histories of North American Shore Birds," Bull. 142, 146, Washington, U.S. National Museum, 1927–1929.
Low, G. C., "The Literature of the Charadriiformes from 1894 to 1928." London, Witherby, 1931.
Mayr, E. and D. Amadon, "A Classification of Recent Birds," New York, American Museum Novitates, no. 1496, 1951.
Murphy, R. C., "Oceanic Birds of South America," 2 Vols. New York, Macmillan, 1936.
Peters, J. L., "Check-list of Birds of the World. Part 2," Cambridge, Mass., Harvard University Press, 1934.
Ridgway, R., "Birds of North and Middle America," Part 8. Bull. 50, Washington, U.S. National Museum, 1919.
Witherby, H. F., *et al.,* "Handbook of British Birds," Vol. 4, 5. London, Witherby, 1940, 1941.

CHELONIA

Chelonia is an old, highly conservative order of the class REPTILIA comprising the turtles and tortoises. As a group the turtles are ancient, primitive reptiles that achieved, early in their evolutionary history, a morphological specialization so profound and eminently successful as to be without parallel among vertebrates: the turtle shell. The dorsal shell (*carapace*) is a complex of three morphological entities, the axial skelton, dermal (*thecal*) ossifications and epidermal, horny shields. Early in ontogeny the ribs come to lie within the thick dermis and the embryonic carapace disc thus formed grows at a notably faster rate than do the other parts of the embryo. As a result the ribs become displaced to the *outside* of shoulder girdle and pelvis. The ventral portion of the shell (*plastron*) comprises dermal parts of the shoulder girdle (*clavicles* and *interclavicle*) and probably *gastralia*, dermal ossifications corresponding to those of the carapace, and epidermal shields.

The phylogenetic origin of the turtle shell is not known. When the group first appears in the fossil record, in latest Triassic time, the shell is already completely formed. The effect of the acquisition of the shell by these reptiles was twofold: it became a highly successful adaptation to an animal of semiaquatic habits and rather sluggish disposition; it afforded protection against predators during adult life and, perhaps more importantly, retarded water loss during periods of adversity. On the other hand, the shell proved to be a phylogenetic straightjacket in the sense that it evidently limited the evolutionary potential of the order.

All turtles, living and fossil, have at least superficially similar shells, therefore similarity in overall habitus: an essentially rigid body with a flexible neck in which the number of vertebrae has been stabilized at eight, a head with a large mouth provided with horny beaks, limbs capable of raising the shell off the ground and/or propelling it in water, a tail whose major function is the housing of the copulatory mechanism.

Deeper insight into the phylogeny of the order reveals, however, a complex pattern and many biologically interesting phenomena. Probably as a result of the early acquisition of a shell the evolutionary history of the turtles is characterized by parallelisms and convergences. Close morphological similarity of parts in different turtles can never be relied upon as an indication of close relationship. Mandibles, for example, with broad masticatory surfaces of virtually identical design are found among scattered species of at least five families. The study of turtle relationships thus requires rather sophisticated comparative anatomical procedures.

Among Recent reptiles the turtles have by far the best fossil record, extending back to the top of the Triassic (Keuper, Germany). In spite of this the phylogenetic history of the order is not yet satisfactorily understood. *Eunothosaurus* of the Permian of South Africa has been widely suggested as an ancestor of the turtles. More recent analysis tends to question this view. Among the cotylosaurs the turtles show the greatest affinity to the Diadectomorpha.

Three major divisions are currently recognized: the Amphichelydia, a large, successful group of generally primitive, though widely diversified turtles that extend in time from the Triassic to the Pleistocene; the Pleurodira and the Cryptodira, which have evolved from the Amphichelydia probably in late Jurassic time. Pleurodires and cryptodires differ sharply by the mode of retraction of head and neck. In retracted position these organs form a horizontal S-curve in the pleurodires, a vertical S-curve in the Cryptodires. In the pleurodires, furthermore, the ventral pelvic bones coossify with the plastron.

It is of interest to note that the over-all evolutionary trends are much the same in all three groups; for example, the gradual simplification of the mosaic of the shell bones and shields. The pleurodires are now restricted to the southern hemisphere and all are aquatic. The cryptodires are clearly a more advanced group both in terms of ecological diversification and in speciation. While most of the cryptodires are semiaquatic freshwater turtles, some have become land turtles capable of surviving under dry, even severe desert conditions. Others have become highly specialized marine animals, with limbs modified as flippers and rudders.

Reduction of the thecal shell to mere vestiges and subsequent reconstitution of the bony shell by epithecal bone took place in two cryptodiran families: the soft-shelled turtles (Trionychidae) and the marine leatherback turtles (Dermochelyidae). In the latter the epithecal armour consists of a great number of small polygonal ossicles that have no relation to the underlying axial skelton. For this reason the dermochelyids were considered to be different from all other turtles (Athecae)—a view that is no longer accepted.

The most widespread and best known cryptodires are the testudinids. The family contains a great variety of pond and stream turtles (Emyinae) and true land tortoises (Testudininae), among which the genus *Testudo* with its many beautifully marked species is the most widely distributed. Members of this genus, furthermore, show a marked tendency toward giantism. At present the largest species are found on islands (Galapagos, Seychelles), but giant fossil species of *Testudo* are known from continental Europe and North America. While most testudinines have rigid, globular shells, there are a few noteworthy exceptions. In the African genus *Kinixys* a transversal hinge develops in the carapace (during ontogeny) enabling the animal to close its shell completely. The American box turtle (among others) is capable of the same feat, but the hinge is in the plastron. One testudinine, *Malacochersus*, is flat-shelled and has the bones reduced to narrow, thin, flexible bands. It lives in rocky situations where it habitually crawls into crevices between the rocks.

All turtles are oviparous. Internally fertilized eggs with parchment-like shells are carefully buried in favored situations. The clutch size varies from one to over two hundred. Turtle eggs are subjected to severe predation by small mammals (and more recently man) with apparently little effect on adult populations, except in a few cases where man has used the eggs as well as the adult animal for food.

RAINER ZANGERL

Reference

Romer, A. S., "The Osteology of the Reptiles," Chicago, The University Press, 1956.

CHEMOLITHOTROPHIC BACTERIA

The chemolithotrophic BACTERIA are a unique group of microorganisms since they can synthesize all their protoplasmic constituents from carbon dioxide, an inorganic nitrogen source, and other inorganic salts; they derive the energy for these syntheses from the oxidation of inorganic compounds. It is important to emphasize that these bacteria are not photosynthetic, since all their metabolism can take place in the dark. However, they are like photosynthetic organisms in that they use carbon dioxide as their sole carbon source. Although these bacteria can live in a completely inorganic environment, some of them are not obliged to do so. Our definition therefore includes both the obligate and facultative chemolithotrophs.

Discovery and types of chemolithotrophs. Historically the recognition of this kind of metabolism dates from the researches of S. Winogradsky in the latter part of the 19th century. He had observed a filamentous organism (*Beggiatoa*) in waters that contained a high concentration of hydrogen sulfide. These organisms are similar in structure and movement to the *Oscillatoria* (see ALGAE). In this environment Winogradsky noted that *Beggiatoa* was filled with sulfur granules which subsequently disappeared if the filament was placed in a non-sulfide medium; after the sulfur had disappeared, sulfate appeared in the surrounding fluid. Winogradsky proposed that *Beggiatoa* could oxidize $S^= \rightarrow S^0 \rightarrow SO_4^=$ providing energy to the cell. Winogradsky used these observations to define the chemolithotrophic mode of life. Winogradsky's discoveries stimulated interest in the so-called SULFUR BACTERIA and attempts were made to isolate more of these physiological types. In the course of subsequent investigations a new organism was isolated which is completely unlike the organisms studied by Winogradsky. This new type is a small rod-like bacterium subsequently called *Thiobacillus*. The chemolithotrophic nature of this bacterium was firmly established, since it grows very well in an inorganic medium with either sulfide or thiosulfate as the oxidizable substrate.

About the same time that the thiobacilli were discovered two other groups of chemolithotrophs were recognized. One group derives its energy from the oxidation of inorganic nitrogen compounds. The genus *Nitrosomonas* lives by oxidizing ammonia to nitrite ($NH_4^+ \rightarrow NO_2^-$) while *Nitrobacter* converts nitrite to nitrate ($NO_2 \rightarrow NO_3^-$). The other group of bacteria,

called *Hydrogenomonas*, lives by oxidizing hydrogen to water ($2H_2 + O_2 \rightarrow 2\ H_2O$). Other bacteria have been found to be able to live in an inorganic medium; they and the previously mentioned chemolithotrophs are listed in Table 1 which summarizes their characteristics. It may seem paradoxical that *Beggiatoa* is not included in Table 1. It seems clear, however, that *Beggiatoa* is not a true bacterium but rather a colorless counterpart of *Oscillatoria* of the *Myxophyceae*. Furthermore it is not certain that they can live in inorganic environments. It is evident (Table 1) that there is a fundamental structural similarity among all the recognized chemolithotrophic bacteria. They are rod-like bacteria, usually motile by means of a polar flagellum. They may be recovered from soil and aquatic environments while their presence in local areas can be considerably enhanced because of particularly favorable conditions. Thus, for example, it is extremely easy to isolate thiobacilli from water and surface mud where anaerobic conditions beneath the surface lead to the production of H_2S by bacteria.

Isolation of chemolithotrophs. The most fruitful procedure for isolating chemolithotrophs is the elective or enrichment culture technique. This involves placing a bit of soil or water into a liquid medium designed to promote the growth of one type of organism. Although the inoculum may only contain a few members of the desired bacterial group, they will have a selective ad-

Table 1   Characteristics of Chemolithotrophic Bacteria

Organism	Morphology	Energy Source	Obligate or Facultative	Final Electron Acceptor	Product of Oxidation
Thiobacillus thioparus *Thiobacillus thiooxidans* *Thiobacillus thiocyanoxidans* *Thiobacillus denitrificans*	small motile rods, 2–4 μ long, 1 μ wide; polar flagellum	$S^=, S_2O_3^=$ S_4O_6, S^0 SCN^-	obligate	O_2 O_2 or NO_3^-	$SO_4^= + H_2O$ $SO_4^= + N_2$
Thiobacillus novellus	non-motile rod	$S_3^= S_2O_3^=$	facultative	O_2	$SO_4^= + H_2O$
Nitrosomonas	small oval cells, 1.5 μ long, 1 μ wide, motile, polar flagellum	NH_4^+	obligate	O_2	NO_2^-
Nitrobacter	oval cells, 1 μ long, 0.8 μ wide	NO_2^-	obligate	O_2	NO_3^-
Hydrogenomonas	small rods, 1.5–2.0 μ long, 0.5 μ wide; motile, polar flagellum	H_2	facultative	O_2	H_2O
Ferrobacillus	small rod 0.6–1.0 μ wide, 1–1.6 μ long; motile; polar flagellum	Fe^{++}	obligate	O_2	Fe^{+++}
Desulfovibrio desulfuricans	0.5–1 μ wide, 1–5 μ long, curved rods	H_2	facultative	$SO_4^=$	H_2S

vantage in the medium and before long may constitute 90 per cent of the total bacterial count in the culture. Subsequent subcultures on the same medium solidified with agar (streak plate method) usually lead to the isolation of the bacterium in pure culture. Only after this procedure has been carried out can it be ascertained whether the organism in question is a chemolithotroph.

Biochemistry. The chemolithotrophic bacteria were considered, at one time, a bizarre form of life. Two findings were particularly responsible for this attitude. 1) In 1922 *Thiobacillus thiooxidans* was isolated and found to be able to convert elemental sulfur to sulfuric acid, like the other thiobacilli. However, the amount of acid produced was so great that the pH of the culture medium approached zero, while at the same time the bacteria could be recovered by subculture from this evironment. 2) In the latter part of the 19th century it had been reported that *Nitrosomonas* was not only unable to grow on organic material but was actually inhibited in its growth by simple organic molecules readily utilized by most nonchemolithotrophic bacteria. Later evidence disproved this latter finding and recent investigations have dispelled the notion of "peculiarity." Analysis of the chemical composition of *T. thiooxidans* and *Nitrosomonas* has demonstrated that they contain all the amino acids found in all other cells whether plant or animal, while many other organic compounds (i.e., vitamins) are also present. *T. thiooxidans* was found to contain phosphorylated sugars which are intermediates in the conversion of glucose to either lactic acid or ethanol and carbon dioxide in muscle tissue and yeast. Another important discovery was that this same bacterium contained adenosine triphosphate (ATP).

Carbon dioxide assimilation. The synthesis of cell material from carbon dioxide involves the creation of new carbon-to-carbon bonds between preexisting organic molecules in the cell and CO_2. CO_2 assimilation requires reducing power (electrons and protons) and energy in the form of ATP. In chemolithotrophic bacteria both of these components are derived from the oxidation of inorganic compounds. A third requirement for CO_2 assimilation is the continuous availability of a carbon dioxide acceptor molecule.

The pathway of carbon dioxide assimilation has been studied using $C^{14}O_2$. The experimental techniques used are almost indentical to those used to study CO_2 assimilation in photosynthetic organisms. The major pathway of carbon dioxide assimilation in chemolithotrophic bacteria turns out to be identical to the pathway in photosynthetic organisms. The details of this scheme are given in figure 1. In the first reaction carbon dioxide

is enzymatically combined with ribulose-1, 5-diphosphate, yielding two molecules of 3-phosphoglyceric acie; 3-phosphoglyceric acid is reduced to 3-phosphoglyceraldehyde, two molecules of which are used to make one molecule of hexose diphosphate (HDP). By a transketolase reaction and two transaldolase reactions involving HDP, a four carbon phosphorlyated sugar (erythrose-4-phosphate), and a seven carbon phosphorylated sugar (sedoheptulose-1, 7-diphosphate), a five carbon sugar is regenerated which is able to combine with a molecule of CO_2. In the scheme in Figure 1 six CO_2 molecules are shown to initially react, which leads to a net synthesis of a six carbon sugar and the regeneration of six molecules of a five carbon sugar.

Substrate metabolism and ATP synthesis. The thiobacilli oxidize thiosulfate to sulfate by a series of enzymatic steps, most of which are unknown. It is reasonably certain, however, that the initial step is:

$$2S_2O_3^= \rightarrow S_4O_6^= + 2e \ (\Delta F$$
$$= \text{about 10 k cal/mole})$$

The steps $\frac{1}{2} S_4O_6 + O_2 \rightarrow 2SO_4^=$ (ΔF = about 200 kcal/mole) obviously are the main energy-yielding reactions. Probably a portion of the energy (ATP) is derived from electron transport phosphorylation, as in all living systems. But an additional source of ATP would be available if there were substrate level phosphorylation. Such a possibility is suggested by recent experiments with thiobacilli.

The nitrifying bacteria *(Nitrosomonas* and *Nitrobacter)* carry out the following reactions:

Nitrosomonas $NH_4^+ + 1\frac{1}{2}O_2 \rightarrow NO_2^- + 2H^+ + H_2O$.

Nitrobacter $NO_2^- + 1\frac{1}{2}O_2 \rightarrow NO_3^-$

The primary step in ammonia oxidation is:

$$2NH_4^+ + O_2 \rightarrow 2NH_2OH + 2H^+$$

This step occurs with very little loss in free energy so that the subsequent reactions, hydroxylamine (NH_2OH) to nitrite, provide the energy for cell synthesis. Very little is known about the enzymatic reactions hydroxylamine to nitrite and nitrite to nitrate.

The distinguishing characteristic of the chemolithotrophic bacteria is their ability to carry out all their biosynthetic activities at the expense of the oxidation of inorganic compounds. While the synthesis of ATP is accomplished in this unusual way, the remaining reactions may be similar if not identical to those reactions in other living cells. This situation is analogous to the "light" and "dark" reactions in

6 CO_2 + 6 Ribulose-1,5-diphosphate ⟶ 12 3-phosphoglyceric acid (3-PGA)

12 (3-PGA) + 12 DPNH + 12 H$^+$ + 12 ATP ⟶ 12 3-phosphoglyceraldehyde (3-PGAld) + 12 ADP + 12 DPN$^+$ + 12 H_3PO_4

8 (3-PGAld) ⟶ 4 hexose diphosphate (HDP)

2 HDP + 2 (3-PGAld) ⟶ 2 pentose diphosphate + 2 erythrose-phosphate

2 HDP + 2 erythrose phosphate ⟶ 2 sedoheptulose-1,7-diphosphate (SDP) + 2 (3-PGAld)

2 SDP + 2 (3-PGAld) ⟶ 2 pentose diphosphate + 2 pentose diphosphate

Net 6 CO_2 + 12 DPNH + 12 H$^+$ + 12 ATP ⟶ 1 HDP + 12 ADP + 12 DPN$^+$ 12 H_3PO_4

Fig. 1

photosynthetic organisms. The "light" reaction results in the photochemical splitting of water, which is subsequently reformed providing the energy for biosynthetic reactions; these events are comparable to the oxidation of the inorganic substrate. The "dark" reactions in photosynthetic organisms are enzymatic steps present in all forms of life and similar to the carbon interconversions found in the chemolithotrophs.

Obligate and facultative chemolithotrophs. Few chemolithotrophs can grow on both inorganic and organic media. The inability to grow on a particular substrate is in many cases related to the inability of the molecule to penetrate the cell membrane or some structure within the membrane. It is evident that many bacterial cells transport organic molecules, like glucose, across their cell membrane by a mechanism (*permease*) which translocates the molecule from outside to the interior. If the cell lacks this permease no growth will occur on this substrate. Permeases are apparently specific proteins under genetic control and consequently may be absent or present depending on the genetic constitution of the cell. *Thiobacillus novellus* is capable of growth on both inorganic and organic media; growth on either medium following growth in the alternate one occurs in an adaptive manner. *T. novellus* is initially incapable of division on the organic substrate after growth on the inorganic medium, for it apparently lacks the "permease" for that compound; however, in a short time it can synthesize the transport mechanism. Obligate chemolithotrophs, of course, can not grow on the organic substrate and the reason might well be that they are genetically incapable of making that particular permease or for that matter any transport mechanism for any organic compounds. This hypothesis gains further support from the observation that cell-free extracts of obligate chemolithotrophs can rapidly metabolize organic molecules which are not utilized by the intact cell.

MELVIN SANTER

References

Fry, B. A. and J. L. Peel, eds., "Autotrophic microorganisms," Cambridge, The University Press, 1954.
Lees, H., "The Biochemistry of Autotrophic Bacteria," London, Butterworths, 1955.

CHILOPODA

Chilopods (centipeds) form a well-defined class of many-legged terrestrial ARTHROPODS in their total detailed anatomy more closely related to insects than to DIPLOPODS (millipeds) with which they were long associated under the name of myriapods. In chilopods there is a distinct head followed by a long body composed of many similar segments. Most of these segments bear a single pair of walking legs widely separated by a subrectangular sternite. At the end of the body are several segments on which the appendages are absent or abortive. The first pair of legs behind the head are modified into poison jaws which are fused at the base to form a median plate (*coxosternum*) and are characteristic of the class.

The integument of centipeds is a smooth, elastic chitinous membrane which, in contrast with that of diplopods, wholly lacks any deposit of calcium salts. In form the body is always long in proportion to the width and is more or less depressed and in some thread- or ribbon-like. The length ranges from about 3 mm. to near 275 mm. The number of pediferous segments in some groups is constant, being, e.g., always 15 in the two orders Lithobiida and Scutigerida, and 21 or 23 in the order termed Scolopendrida. In a third group (Geophilida), the number of segments varies greatly from species to species and often more moderately within many species themselves, the maximum known number being above 190.

The head is covered above by a flat-arched plate the borders of which are curved under and leave beneath a hollow space in which the mouth-parts lie. The anterior part or frontal plate of the head capsule, usually set off by a suture, bears a single pair of simple moniliform antennae in which the number of articles may be fixed or variable, ranging from a minimum of 14 to an exceedingly large number as possessed, e.g., by the common house centiped (Scutigera). Eyes, when present, are in the form of a convex group of independent ocelli which vary much in number. In the large order Geophilida eyes are never present. There is a labrum in the form of a median plate between two lateral pieces, but the median plate is sometimes abortive or absent. Mandibles are present as strong masticatory appendages behind which are two pairs of maxillae. The second maxillae may be fused at the base to form a plate homologous with the labium of insects.

Breathing is by means of air tubes or tracheae which open along the sides of the body excepting in the house centiped and its allies (order Scutigerida) in which they open on the mid-dorsal line. The tracheal systems are united by anastamoses as in the insects but not in the diplopods. The nervous system consists of a ventral nerve cord ending anteriorly in a circumoesophageal ring which presents a ganglion below one and one above (brain). The alimentary canal is a straight tube which receives anteriorly the secretion of the salivary glands and posteriorly that of the excretory (Malpighian) tubules. The reproductive organs lie above the alimentary tract and open toward the posterior end in a genital segment, never anteriorly as in millipeds. Different types of development furnish the basis for separating the Chilopoda into two subclasses. In the first type the young leave the egg with the full number of segments and legs (subclass Epimorpha). In the second type the young emerge with 7 pairs of legs and acquire the remainder in subsequent molts (subclass Anamorpha).

Centipeds are nocturnal in habit and generally conceal themselves during the daytime in retreats under leaves, wood, stones, etc. in damp localities where the females lay their eggs singly (Anamorpha) or in clusters which the mother guards until after hatching (Epimorpha). They are carnivorous and sometimes cannibalistic, killing their prey of such forms as slugs, earthworms and insects by means of their poison fangs. The bite of large forms is painful but rarely dangerous to man.

RALPH V. CHAMBERLIN

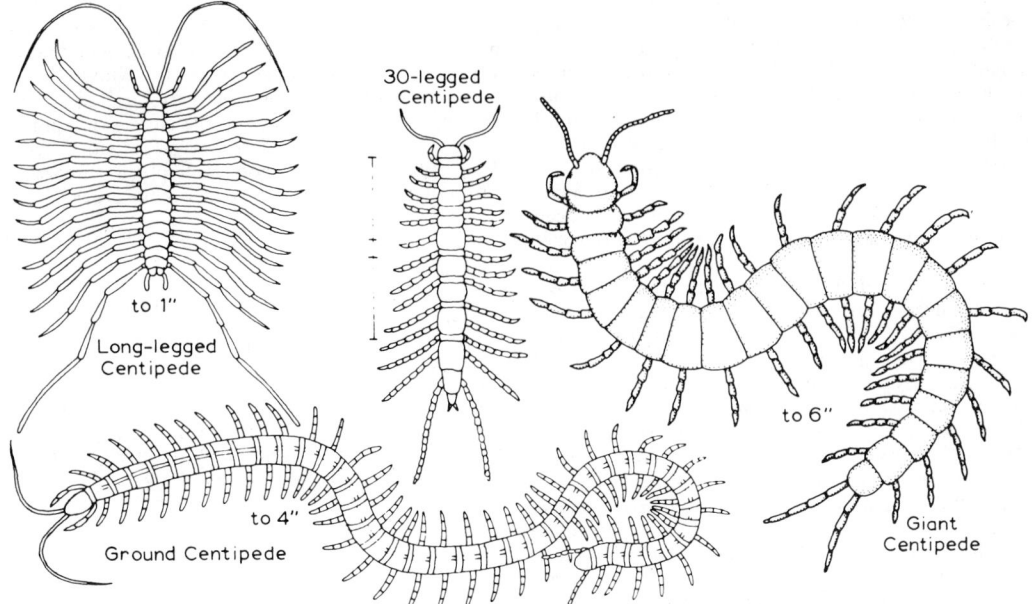

30-legged
Centipede

Long-legged
Centipede

to 1"

Ground Centipede

to 4"

to 6"

Giant
Centipede

(From Pimentel, "Invertebrate Identification Manual," New York, Reinhold, 1967.)

References

Attems, C., "Chilopoda," *in* Kükenthal, W. and T. Krumbach eds., "Handbuch der Zoologie," Vol. **4.** Berlin, de Gruyter, 1926.
Verhoeff, C. W., "Chilopoda," *in* Bronn, H. G. ed., "Klassen und Ordnungen des Tierreichs," Vol. **5.** Leipsig, Winter, 1923.

CHIROPTERA

This order contains the bats, whose modifications for true sustained flight make them unique in the class MAMMALIA. Flight is made possible by a membrane (patagium) which consists of extensions of the dorsal and ventral body integuments, joined together and enclosing the appendages which support the membrane. The radius of the highly specialized forelimb is elongate and is the main element of the forearm. The first digit is greatly reduced and bears a claw, while the elements of digits II–V are elongate and clawless (digit II has a claw in one group of bats), forming the main support for the membrane. In the more specialized families, the humerus has a secondary articulation with the scapula, adding rigidity to the joint and restricting limb movement to the antero-posterior plane. In all bats the femur is rotated so that the knee joint faces dorso-laterally, allowing the limb to participate in supporting the membrane and to function in suspending the animal while roosting. The sternum is usually keeled for attachment of the ventral flight muscles.

The order comprises two suborders, Megachiroptera (one family, approximately 40 genera, 155 species) and Microchiroptera (16 living, three fossil families; approximately 140 genera, 725 species), with distinguishing features which are related mainly to differences in degree of specialization for flight and feeding. Megachiropterans are generally larger than microchiropterans, the wingspan in the former ranging from eleven inches to just over five feet and in the latter from six to 30 inches. The features of the head in Megachiroptera are relatively unspecialized, whereas in Microchiroptera the ears are often enlarged or otherwise modified, the nose bears a fleshy leaf-like appendage in several families, and the facial region may be elongate or foreshortened in relation to specialized types of feeding. Most bats of the megachiropteran family Pteropidae consume chiefly fruits, some eat flowers, a few take insects as well, and others feed on nectar. The larger species often cause considerable damage to tropical fruit crops in the Old World. The members of most (twelve) microchiropteran families feed almost exclusively on insects, which are usually caught on the wing (the single species of family Mystacinidae, endemic to New Zealand, may also climb about in trees after its prey). Bats of family Megadermidae capture and eat lizards and other small vertebrates, as well as insects. *Noctilio leporinus* (Noctilionidae) eats small fish which it catches with greatly enlarged hind feet, as does the Mexican *Pizonyx vivesi* (Vespertilionidae). The true vampires (Desmodontidae) feed exclusively on the blood of living mammals. Fruit is eaten by most species of family Phyllostomidae, some feed on nectar and pollen, others are strictly insectivorous. The larger members of this family may also eat other bats and other small vertebrates. Along with the nectar-feeding habits of some pteropid and phyllostomid bats, pollenating mechanisms have evolved in certain trees. These chiropterogamous plants produce large flowers which are apparently attractive to bats and which bloom only at dusk when these mammals are active.

Although some bats commonly bear two, most species produce one offspring at a time. The young are born hairless, with their eyes closed, and they are usually carried by their mothers until they become too heavy, after which they are left at the roost until they learn to fly. Sexual maturity is usually attained in the second year, at least in temperate regions. Delayed fertilization occurs in some hibernating species, copulation taking place in the fall and gestation beginning the following spring. Longevity has been recorded at 19 years for captive *Pteropus* (Pteropidae) and over 20 years for marked wild individuals of several genera of Vespertilionidae. The relatively few predators of bats include chiefly predatory birds (the African bat hawk, *Machaerhamphus*, feeds mainly on bats), mammals and snakes. Bats may be solitary or they may roost together in colonies of from several individuals to more than a million (an estimated nine million freetail bats, *Tadarida*, inhabit Carlsbad Caverns, New Mexico). Different species may roost in any situations which provide shelter and satisfy their particular light tolerance requirements, most commonly in such places as trees, caves, mine tunnels, culverts and old buildings. Although most bats suspend themselves by the claws of the feet, some using the thumbs as well, those of families Thyropteridae (Neotropical) and Myzopodidae (endemic to Madagascar) attach themselves while roosting by means of four stalked adhesive discs located one at the base of each thumb and on the sole of each foot.

Orientation, obstacle avoidance and, in insectivorous species, food location are by means of ultrasonic echolocation, which is especially well developed in the Microchiroptera. The Megachiroptera apparently depend more on vision, although echolocation has been demonstrated in some species. Homing abilities may be quite well developed, with individuals able to return over considerable distances (180 miles in one known case) if removed from roosts. Some species are migratory, often travelling long distances between seasonal roosting sites (810 miles recorded in one instance). Non-migratory species living in temperate climates spend the winter in hibernation, with physiological activity reduced to a minimum. Some bats show similar reduction in body function while sleeping during the day, their body temperatures dropping to within a degree or two of ambient temperature.

Bats may be heavily infested with ectoparasites. Lice are extremely rare but mites, fleas and parasitic flies and true bugs (Hemiptera) are well represented, with seven families specific to Chiroptera. Vampires, particularly *Desmodus*, harbor and transmit several viruses (including that which causes paralytic rabies) which occasionally cause serious losses to livestock.

From available fossil evidence, it seems likely that the Chiroptera were derived from Cretaceous or Paleocene "insectivores." The oldest fossil bats are typical microchiropterans from the early and Middle Eocene, some of which apparently belong to extant families and genera. Although occurring in almost all tropical and temperate parts of the world, bats are most abundant in the tropics. Of the 17 living families, eight are restricted to the Old World (Pteropidae, Rhinopomidae, Megadermidae, Nycteridae, Myzopodidae—essentially tropical; Rhinolophidae, Hipposideridae, Mystacinidae—also in temperate regions), six are

entirely Neotropical (Natalidae, Thyropteridae, Furipteridae, Noctilionidae, Phyllostomidae, Desmodontidae) and three are common to both hemispheres (Emballonuridae—tropical; Molossidae—mostly tropical; Vespertilionidae—tropical and temperate). Some of the genera of the latter two families occur in both the Old and New Worlds, with *Myotis* (Vespertilionidae), the most widespread, occurring throughout most of the known range of the order.

ANDREW STARRETT

References

Allen, G. M., "Bats," Cambridge, Harvard University Press, 1939.
Grassé, P.-P., *et al*, "Ordre des Chiroptères," *in* Grassé, P.-P., "Traité de Zoologie," vol. 17. Paris, Masson, 1955;.
Griffin, D. R., "Listening in the dark," New Haven, Yale University Press, 1958.
Walter, E. P., "Mammals of the World," Vol. I, Baltimore, Johns Hopkins Press, 2nd ed. 1968.

CHITIN

Chitin is a high molecular weight polymer composed of N-acetylglucosamine residues joined together by β-glycosidic linkages between carbon atoms 1 and 4 (Fig. 1). Modern chemical terminology would call it a polymer of 2-acetamido-2 deoxy-αD glucopyranose. The molecular chains are long and unbranched. Freed from the protein with which it is normally associated, the chitin chains are parallel to one another and adjacent chains run in opposite directions—an arrangement that maximizes the number of interchain hydrogen bonds. In many respects chitin is similar to cellulose (see POLYSACCHARIDES) from which it differs by the presence of an acetylamine group on the second carbon atom.

Chitin is perhaps best known as a characteristic chemical component in the skeletons of arthropods. It is also found in setae, jaws, and gut lining of annelids, in the radula and dorsal shield of molluscs, in the perisarc of medusoid coelenterates, in the stalk wall of bryozoans, in the egg shells of nematodes and acanthocephalans, and less certainly in a few other animal groups. It is also the common constituent of the walls of most but not all groups of fungi.

As usually prepared for chemical study, chitin is a colorless solid which is chemically stable and is

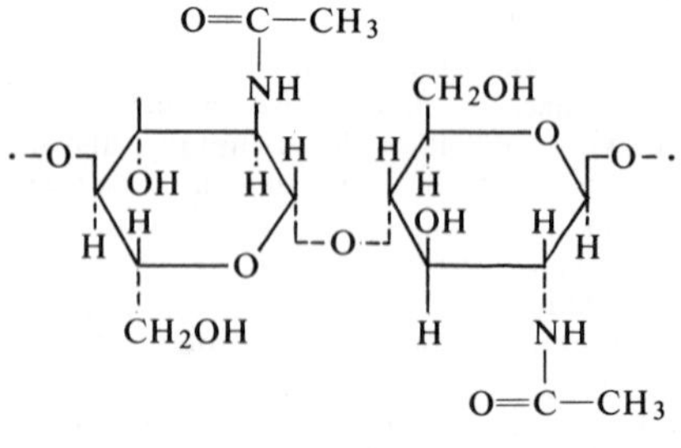

Fig. 1

insoluble and unaffected by most chemical reagents. Solutions in concentrated mineral acids quickly show degradation to short chain lengths, but some authors report that undegraded solutions can be made with lithium salts which are capable of strong hydration. Hot concentrated alkalies remove half of all of the acetyl sidegroups to give a product called chitosan which can be made to give a color reaction useful in identification of chitin in natural objects.

Birefringence and x-ray diffraction data show that chitin has crystalline areas, but there is not good agreement as to the details or dimensions of the unit cell of the crystal lattice. While chitin from diverse sources yields similar constituents on hydrolysis, x-ray diffraction studies reveal several crystallographic forms. According to Rudall, the commonest type is that found in arthropod skeletons, etc.; it has been termed α-chitin. Another type with water incorporated in the lattice is β-chitin found in the "pen" of squid. Another type is found in the perisarc of coelenterates. What type occurs in fungi remains to be determined.

Chitin is not found in nature in a pure condition. Rudall reports that β-chitin is associated with collagen whereas α-chitin is not. At least in the arthropod skeleton, α-chitin is always associated with proteins called arthropodins. On the average, chitin accounts for only $\frac{1}{4}$ to $\frac{1}{3}$d the dry weight of non-calcified arthropod cuticles.

It is now generally agreed that chitin is found in nature only linked to protein. The naturally occurring compound, then, is a glycoprotein from which the protein moiety is readily removed to yield what is called chitin. Purification is usually accomplished by prolonged heating in KOH or NaOH solutions followed by dilute acid and a strong oxidizing agent. Such treatment yields an ash-free product that seems to be pure chitin. Treatment of arthropod cuticles with solutions of Versene or lithium thiocyanate, however, results in the extraction of several chitin-containing glycoproteins. Analyses of these fractions indicate that chitin is bound to arthropodin by several kinds of bonds, including some covalent bonds, but the details of the bonding are not yet known. If, as some suspect, there is a small percentage of non-acetylated residues in the chains, these could readily be involved in cross-bonding. Purified chitin can be made to react with pure arthropodin, peptides, and amino acids (especially tyrosine), but it is not known whether these *in vitro* reactions are the same as the bondings in nature.

Other recent evidence has been interpreted as indicating that in the cuticle of arthropods the chitin and protein chains form an interpenetrating lattice with the protein chains at a right angle to the chitin chains. Both of these sets of chains are parallel to the surface of the cuticle. It is well known that arthropod cuticles become sclerotized in certain areas, that is, the originally soft cuticle becomes first elastic, then hard, and usually more or less dark. Some chemical information on the process of sclerotization has been obtained but we do not yet know whether or not linkages to chitin chains are involved. (See Hackman's review.)

The metabolic source of chitin may be glycogen since there are numerous reports of glycogen decrease concurrent with chitin synthesis. However, no proof exists. Furthermore, no detailed knowledge is available on the metabolic steps involved in chitin synthesis. Since the naturally occurring substance is a glycoprotein, one would expect that monomers are added *in situ* to a mixed lattice.

The decomposition of chitin in nature is accomplished by chitinases. The main source of these enzymes is various soil bacteria but chitinase activity has also been recorded for some fungi and even eelworms, earthworms and soil amoebae. The digestive juices of snails is a well-known source but it is uncertain whether the enzyme is produced by the snail itself or by associated bacteria. Chitinase is also present in the molting fluid of insects, but whereas the entire cuticle disintegrates in nature, only the soft cuticle is dispersed during molting.

A. GLENN RICHARDS

References

Hackman, R. H., "Biochemistry of the insect cuticle," Proc. 4th Int. Congress of Biochem., **12:** 48–62. 1959.
Richards, A. G., "The Integument of Arthropods," Minneapolis, The University Press, 1951.
Richards, A. G., "The cuticle of arthropods," Ergebnisse der Biologie, **20:** 1–26. 1958.
Rudall, K. M., "The distribution of collagen and chitin," Symp. Soc. Exp. Biol., **9:** 49–71. 1955.

CHLOROPHYCEAE

A class of grass-green ALGAE in the phylum Chlorophyta with the same pigmentation found in higher plants. Chlorophylls *a* and *b* predominate over the carotenes and xanthophylls. Food reserves are almost always starch. The photosynthetic pigments are localized in chromatophores of varied shape, form and position in the cell and often of use in separating genera. Species (*ca.* 5,500) of freshwater individuals far outnumber the marine ones.

The habitats of the Chlorophyceae are both marine and freshwater, the strictly aquatic species occurring in puddles, pools and ponds; lakes and marshes; in streams and rivers from those with little or no current to mountain torrents. Many are epiphytes on bark, leaves and wood (*Protococcus*), endophytes and epizoophytes, and some on ice and snow. The epiphytes and endophytes are mostly mere "lodgers," but sometimes are parasites (*Cephaleuros*, causing tea "rust"). The PLANKTON of ponds and pools and quiet, shallow areas of lakes and sluggish streams may be very abundant at times. Algae attached to some substrate by holdfast cells or jelly-like secretions may add to the plankton species after being torn from their attachment.

The body forms of green algae include single cells, irregular aggregates of cells, definite colonies varying from a few to thousands of cells, usually green and with or without flagella. There may be also branched or unbranched filaments, plates or tubes. The evolutionary tendencies in development of plant body are generally four: vegetation flagellation with cell division (*volvocine*), no vegetative motility but ability for cell division (*tetrasporine*), loss of both vegetative motility and ability to divide (*siphoneus*), and an indication of an amoeboid tendency in temporary stages of gametes and zoospores (*rhizopodal*).

Reproduction generally occurs both asexually and sexually. There may be fragmentation, or formation from the contents of a vegetative cell of such asexual structures as *zoospores, aplanospores* and *akinetes*, either of which may develop into a new plant. Sexual reproduction is diverse and often complex. Separate vegetative cells may develop into gametangia in which are produced male or female gametes. Upon union of the gametes a diploid zygote results. MEIOSIS occurs in the germination of the zygote, after a short or long period of dormancy, and a new plant directly or indirectly results. The male and female gametes may appear to be identical (*isogamy*), one may be somewhat larger than the other (*anisogamy*), or the differentiation may extend to the point where a small active SPERM and a large non-active EGG are produced (*oogamy*). Both gametes may escape and union occurs in the water, or the sperms (perhaps guided by some hormonal influence) swim to and fertilize the eggs *in situ*.

Since the green algae are able to photosynthesize, they furnish food for aquatic animals and thus indirectly for man. Soil algae are probably quite important as stabilizers and enrichers. Pilot plants are now established in several localities, and it has been found that the yield of algae is tremendous, often running to 20 or more tons of dry weight to the acre. The green algae thus represent a huge potential source of food for people.

The Chlorophyceae are separated into several orders, some of which will be briefly noted.

The *Volvocales* with haploid vegetative cells and flagella of equal length are either solitary or colonial. In simpler genera all cells of a colony may divide, forming daughter colonies (*Gonium*). In more complex genera a definite number of vegetative cells may produce daughter colonies (four to one-half of the cells in *Pleodorina*), or only a very few have this capacity (*Volvox*). Gametes may represent isogamy, anisogamy or oogamy.

Filamentous algae, branched or not, made up of uninucleate cells having parietal chloroplasts belong to the *Ulotrichales*. The branches may be free at the ends or united laterally forming discs (*Coleochaete*). Fragmentation or the production of asexual spores occurs, and gametic union ranges from isogamy to oogamy.

The *Ulvales* have uninucleate cells united into sheets or tubes. Most species are marine. Some species have a many-celled diploid sporophytic generation. Reproduction is similar to that in the *Ulotrichales*, although no oogamy occurs.

The *Oedogoniales* have uninucleate cells forming branched or unbranched filaments; the cells have reticulate chloroplasts and a unique division resulting in "apical caps." Most species are aquatic, with the exception of the small genus *Oedocladium*. Zoospores occur singly in vegetative cells and each has an anterior ring of flagella. Antheridia bearing sperms may occur on dwarf males (*nannandrous* species). If there are no dwarf males, the species is *macrandrous*. Dwarf male filaments grow from androspores, which are like zoospores but are incapable of germinating except in proximity to the oogonia. A single uninu-

cleate egg is produced in each oogonium. The zygote usually lies dormant for a year or more and then germinates meiotically into four zoospores. *Oedogonium* is the largest of the three genera, with *Bulbochaete* second.

The *Cladophorales* have multinucleate, cylindrical cells forming branched or unbranched filaments, usually attached. Akinetes and biflagellate zoospores and gametes are known. Species exhibit an alternation of haploid and diploid phases, but the common *Cladophora glomerata* has both generations diploid with gametic meiosis.

In the *Chlorococcales* the cells are solitary or united into non-filamentous colonies, uninucleate or multinucleate, not capable of division. Asexual reproduction is by zoospores or aplanospores; sexual reproduction usually isogamous.

The *Zygnematales* have solitary cells, or the cells are joined end to end into unbranched filaments. The chloroplasts are generally quite constant and distinctive; in the desmids they are varied both in number and in form. There are no flagellate cells. Gametic union occurs as a result of amoeboid movement of gametes through a special conjugation tube, formed between male and female gametangia. Zygotes have a wall highly ornamented with pits, lines, spines and reticulations; or unornamented and smooth. These are considered diagnostic for species. Upon germination the zygote undergoes meiosis and grows directly into a new plant. Common genera are *Spirogyra, Zygema, Mougeotia* and the desmids.

L. H. TIFFANY

References

Fritsch, F. E., "Growth and Reproduction in the Algae," Cambridge, The Univ. Press, 1952 and 1956.

Smith, G. M., "The Algae of the United States," New York, McGraw-Hill, 1950.

Tiffany, L. H., "Algae, the grass of many waters," Springfield, Illinois, Thomas, 1958.

CHLOROPLAST

Chloroplasts are cytoplasmic organelles (see also PLASTID) present in the green plants and are intimately involved in the process of PHOTOSYNTHESIS. They are characterized by the presence of colored pigments, including chlorophyll (green pigments) and carotenoids (yellow pigments). While there are several types of chlorophyll (chlorophylls *a* through *e*) all have the same basic porphyrin structure, with 4 pyrroles attached to a monionic magnesium nucleus. The carotenoids are of two main types, carotenes and xanthophylls. It is the pigments of the chloroplast that absorb the radiant energy of sunlight and transduce it into chemical energy. This energy is stored in the plant cell in the form of starch, fats, and proteins. Thus the chloroplasts and their biochemical machinery are the ultimate sources of all food materials, and of practically all organic substances on the earth.

Intact chloroplasts are capable of photosynthesis outside of the cell if all of the necessary cofactors are provided. Thus the chloroplast constitutes a complete photosynthetic unit. The three important reactions that intact chloroplasts are able to carry out are (1) the Hill reaction; (2) photosynthetic phosphorylation; and (3) the dark reaction, or carbon dioxide fixation. The immediate product of photosynthesis is a phosphate ester of a 3-carbon sugar, phosphoglyceraldehyde. Subsequent non-photosynthetic metabolic processes convert this molecule into compounds such as starch and lipid which may accumulate in the chloroplast in the form of starch grains or oil droplets. Protein granules may also be present.

The form and number of chloroplasts per cell is diverse, varying from a small number of large chloroplasts to a large number of small chloroplasts per cell. Algal chloroplasts occur in a variety of shapes, including discs, cups, rings, ribbons, flat plates, and spiral bands. Indeed, in many algae the shape of the chloroplast serves as the diagnostic character. Chloroplasts of higher plants are more uniform, usually being lens- or disc- shaped and 2–4 μ in diameter. A square millimeter of leaf has about 400,000 chloroplasts. In general, chloroplasts of plants grown in the shade are larger and contain more chlorophyll than those of plants grown in direct sunlight. In autumn the amount of chlorophyll decreases and the other pigments (carotenes and xanthophylls) become apparent.

The electron microscope reveals that the chloroplast has a double outer membrane. The inner structure consists of thin membranes that form flattened sacs or discs known as chloroplast lamellae (Fig. 1). It is thought that the lamellae are the surfaces upon which the chlorophyll molecules are disposed in a monomolecular layer. In some algae and in all higher plants the lamellae are stacked in localized regions forming the multi-laminate structures known to light microscopists as "grana." The grana were distinguished from the ground substance known as the "stroma." In the electron microscope the stroma appears as a matrix of low density. It is now known that the inner layer of the chloroplast lamellae has a granular organization comprised of tiny particles about 175 × 90 Å. These structures, called "quantasomes," are composed of hundreds of chlorophyll molecules and a cytochrome molecule, and are thought to have the capacity to trap light energy and begin its conversion to chemical energy.

In addition to the pigments, chloroplasts contain cytochromes, vitamins K and E, and metals such as Fe, Cu, Mn, and Zn. They also contain both RNA and DNA. The latter compound is concerned with a non-chromosomal genetic system that provides the information needed for the developing chloroplast to regulate the formation of its structure and molecular organization. Algal chloroplasts usually increase in number by simple division into two or more new chloroplasts (Fig. 2). In higher plants chloroplasts have a more complex developmental history. They originate from small particles known as proplastids. These structures are bounded by a double membrane and resemble mitochondria. The inner membranes of the proplastids invaginate into the matrix region to form numerous vesicular and lamellar units. These eventually form into a crystalline lattice of interconnected tubules called the prolamellar body. This structure seems to be a site of storage of membranes; it has been estimated that a single prolamellar body

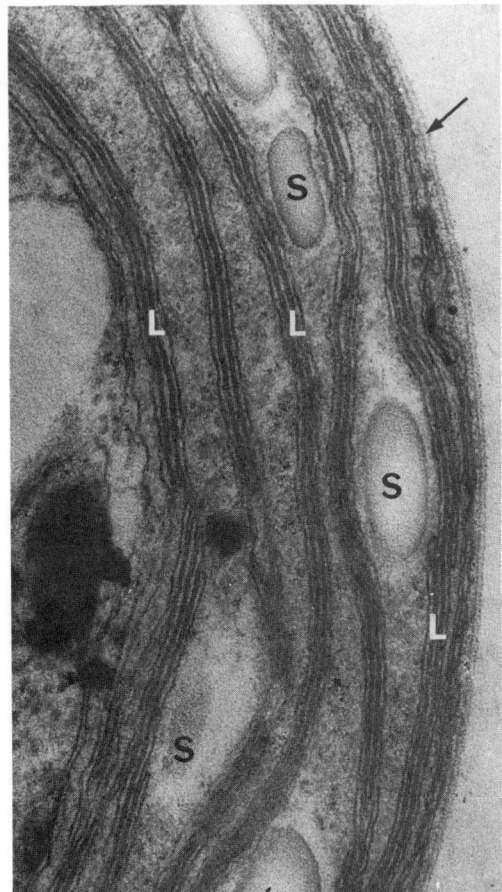

Fig. 1 A portion of the chloroplast of an alga (*Chlorella*) as seen in the electron microscope. The chloroplast lamellae (L) and starch grains (S) are embedded in a matrix of low density. A portion of the algal cell wall is also shown (arrow). Magnification 65,000.

less than 2 μ in diameter may have a surface area of 50–60 μ^2. These stored membranes are then available to differentiate into photosynthetic lamellae.

In addition to the photosynthetic lamellae, chloroplasts may also contain a number of inclusions. The most enigmatic is the pyrenoid, a dense proteinaceous structure that appears in the chloroplasts of many algae. In the light microscope these organelles have a refractile appearance, while the electron microscope (Fig. 3) shows that they are composed of a granular matrix that may be invaded by processes from the cell cytoplasm, by chloroplast lamellae, or even by extensions of the nucleus of the cell. In some diatoms the pyrenoid is surrounded by a membrane. Starch grains often form a sheath around the pyrenoid, as shown in Fig. 3. While this implies that the pyrenoid may function in starch synthesis, this has not been proven, and many algae without pyrenoids are able to form starch grains. In certain red algae the starch grains occur in the general cytoplasm of the cell outside of the chloroplasts. Some

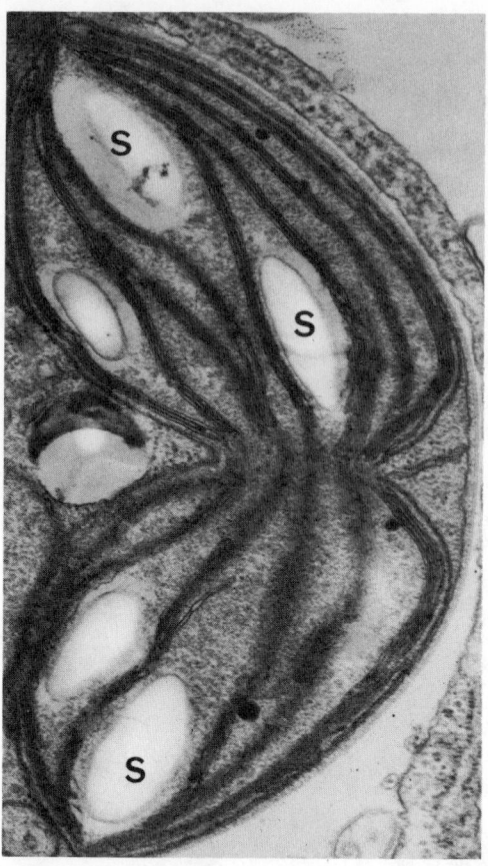

Fig. 2   Dividing chloroplast of an alga (*Chlorella*) as observed in the electron microscope. The chloroplast is pinching into 2 approximately equal parts, each of which contains some starch grains (S). Magnification 30,000.

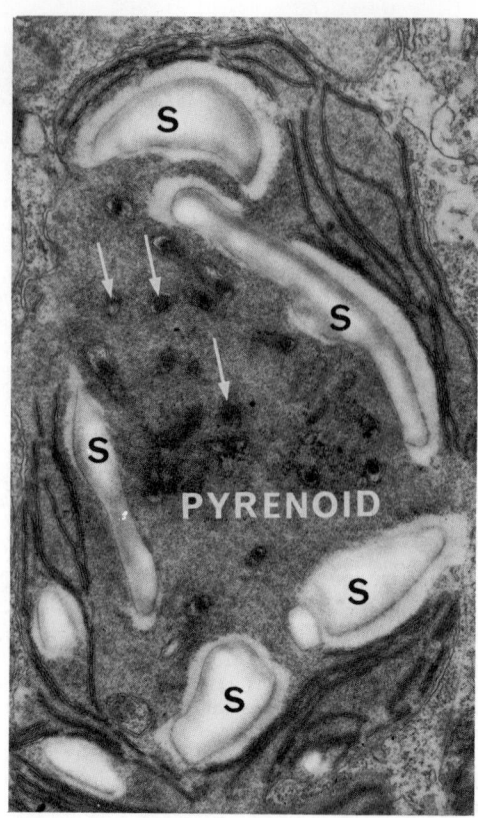

Fig. 3   An electron micrograph of the pyrenoid in the chloroplast of an alga (*Platymonas*). The pyrenoid is a granular structure and is nearly completely ensheathed by starch grains (S). Groups of cytoplasmic tubules (arrows) penetrate into the pyrenoid matrix. Magnification 12,000.

of these red algae have unusual particles coating the surfaces of the chloroplast lamellae. These granules, which have been termed "phycobilisomes," are thought to contain the accessory photosynthetic pigments phycocyanin and phycoerythrin. Another structure, recently discovered in chloroplasts of higher plants, is the "stromacenter," a proteinaceous aggregation of fine fibrils. Its function is unknown. Finally, the stigma or eyespot is present within the chloroplasts of many algae. The stigma usually takes the form of a dense aggregation of spherical pigment-containing granules. The stigma is no longer considered to be the photoreceptor in the phototaxic response of algae. Instead, it is thought to be an auxiliary structure that may aid in the orientation of motile algae, perhaps by periodically shading the actual photoreceptor structure as the cell rotates.

JAMES L. OSCHMAN

References

Brookhaven Symposia in Biology, Vol. 19, "Energy Conversion by the Photosynthetic Apparatus," U.S. Brookhaven National Laboratory, 1967.

Gunning, B.E.S., "The greening process in plastids. I. The structure of the prolamellar body," Protoplasma **60**: 111.

Frey-Wyssling, A., and K. Mühlethaler, "Ultrastructural Plant Cytology," New York, Elsevier, 1965.

Drum, R. W., and H. S. Pankratz, "Pyrenoids, Raphes, and other fine structure in diatoms," Amer. J. Bot. **51**: 405, 1964.

Goodwin, T. W., ed., "Biochemistry of Chloroplasts," New York, Academic Press, 1966.

Gantt, E., *et al.*, "Ultrastructure of *Porphyridium aerugineum*, a blue-green colored Rhodophytan, J. Phycol. 4: 65, 1968.

Rabinowitch, E. I., and Govindjee, "The Role of Chlorophyll in Photosynthesis," Scientific American, July, 1965.

CHORDATA

The Chordata are a major phylum of the ANIMAL KINGDOM, including the vertebrates and certain more simply built marine relatives. Characteristic features present in at least some stage of development of most members of the phylum include: (1) a notochord (whence the name of the phylum is derived), a stout yet flexible rod-like structure running down the back of the trunk in the position occupied by the backbone in typical adult vertebrates; (2) a hollow dorsal nerve cord lying above the notochord; (3) numerous gill slits, opening from the PHARYNGEAL region of the digestive tube to or toward the surface. No structure equivalent to the notochord is present in any other animals. Although longitudinal nerve trunks are present in various invertebrate groups, such trunks are typically solid rather than hollow and are never dorsal. Gills, primarily for water breathing, are present in a great variety of water-dwelling animals, but the chordates and some of their close relatives (Hemichorda) are unique in possessing a series of slits or pouches which open outward from the pharynx to the exterior, and the gills are borne on the bars that separate the slits from each other. A current of water taken in through the mouth passes outward through the slits, over the surface of the gills, where breathing takes place. But more important than this in the lower chordates is the fact that this mechanism is a food-gathering device. Primitive chordates are filter-feeders; food particles in the water are retained in the pharynx as the water is strained out through the gill slits, and then pass back into the intestine.

We shall here include in the Chordata three subphyla, the Cephalochorda (see AMPHIOXUS), URO-CHORDA, and Vertebrata. The last includes the vast bulk of members of the phylum, the first two a relatively small number of marine forms of much more lowly organization. Many workers also include in the chordates a further group of simple organisms, the HEMICHORDA, but these are here treated as a separate although closely related phylum, differing primarily in the absence of any structure clearly equivalent to the notochord.

Amphioxus. An early stage in chordate evolution, that of the subphylum Cephalochorda, is represented by the animal familiar to all students as *Amphioxus*. This small translucent marine creature, but an inch or two in length, has superficially the appearance of a minnow, but its structure is far more simple than that of any true vertebrate. There is little in the way of a skeleton; there is no major development of a head region and no jaws; no proper heart; almost nothing in the way of sense organs and no brain expansion; the sex organs are very simple in nature and the series of tiny kidney structures are built on a different plan from those of vertebrates. There is, however, a good series of segmental trunk muscles for swimming; for their attachment and support a highly developed notochord runs the entire length of the animal. Above the notochord is a typical hollow dorsal nerve cord. Here, in an adult, are two of the three basic characters of chordates well developed. Moreover the gill slits are emphatically present, for there may be half a hundred or so pairs of them, running down a good fraction of the body length (but enclosed in an outer chamber). *Amphioxus* is capable of free swimming, but most of its life is spent in burrows in the sand in shallow water with only the mouth region exposed. Here food is gathered by ciliary action and by gill filter-feeding. *Amphioxus* has certain specializations, primarily those of the pharynx, which debar it from a position on the line leading to the vertebrates. Although its simplicity may in part be due to the retention of larval rather than primitive adult char-acters, the general opinion is that it broadly represents an evolutionary stage antecedent to vertebrates.

Urochordates.—The Urochorda, or tunicates, are small simple marine animals few of which as adults show any obvious similarity to the higher members of the phylum. Their nature may best be understood by considering the structure of a solitary ascidian or sea squirt such as may be found in numbers attached to the rocks on many sea shores. Such an animal shows externally only a rather shapeless leathery covering or tunic (whence the popular group name), with an opening at the top, through which a current of water enters, and a second orifice at the side which serves as a water exit. If the animal be dissected, a digestive tract and other simple organs will be found in the lower part. Most of the interior, however, is occupied by a large barrel-shaped structure, pierced by numerous openings. The water taken in enters the "barrel" and is strained out through these openings; food particles are left inside and pass downward into the intestine. It is obvious that this structure is an enormous pharynx containing a complex framework of gill bars by which the animal gathers its food, and that the chamber surrounding it is equivalent to that of *Amphioxus*, the atrium.

No further chordate structures are to be found in the adult sea squirt. Many, however, develop from the egg into a larval form quite different from the adult and superficially resembling a miniature tadpole. In the expanded anterior end the pharynx is developing; in the tail is a typical notochord and above it a hollow nerve cord of true chordate type; there is even, at the front end, a brain-like structure and simple sense organs. The "tadpole" swims about briefly until, reaching a suitable spot, it settles down and attaches itself; the tail, with notochord and most of the larval nervous system, is resorbed and the saclike adult form is assumed.

The sessile sea squirts represent one major type of urochordates, many of which are colonial and others simple. A second major group is that of the salps, little floating or drifting animals which again may be simple or colonial. A few pelagic tunicates become adult while still retaining the tail.

Chordate Ancestry. The hemichordates are generally recognized as closely related to the chordates even if not included in the same phylum. They nearly always have paired gill slits and a dorsal nerve cord which, at least in its origin, is tubular. The development of hemichordates also shows strong affinities with the echinoderms. The tiny pterobranchs, simplest of hemichordates, are attached, sessile animals gathering food by a set of ciliated, branching arms. The most primitive echinoderms, like the crinoids (sea lilies) today, were likewise sessile and obtained food by ciliated bands running down to the mouth along a series of outstretched arms. Probably the common ancestors of echinoderms, hemichordates and chordates were sessile food-gatherers of this sort.

Some workers, on the other hand, have believed that the chordate ancestors were, like the vertebrates, active, free-swimming forms. In this view the sea squirt larva represents the chordate ancestor, and the sessile adult is degenerate, or, as an alternative, the adult sea squirt is a truly primitive chordate that, by means of a specially adapted larva, the function of which is dispersal, provided the basic patent for evolution of such active chordates as *Amphioxus* and fishes. In fact, however, there is nothing in the adult or larval stages of tunicates that indicates any relationship to either the sessile pterobranch hemichordates or sessile echinoderms, and there is ample evidence from embryonic and larval development that tunicates become sessile in a manner sharply contrasting with that of pterobranchs and echinoderms. Indeed *Amphioxus* shares some developmental features with hemichordates and others with vertebrates, which cannot be said of the urochordates. Probably, then, the latter are actually a side branch of early chordates, retaining from an *Amphioxus*-like ancestor (a) a free-swimming larva, and (b) some features of the complex pharyngeal basket and atrium of the adult. Yet, far back in what we may call the hemichordate stage of evolution, there was an echinoderm-like larva and a sessile, filter-feeding adult.

Vertebrates. Apart from the lowly forms already considered, all chordates are vertebrates, the backboned animals of the subphylum Vertebrata. Although many of the fishlike members of the group have the general body proportions of *Amphioxus,* even the most lowly of vertebrates are far more advanced in structure and more complexly built in almost every regard. The notochord is prominent in the embryo in every vertebrate, but in the adult it is typically supplanted by the backbone—the vertebral column to which the group owes its name. Except for traces of cartilage, *Amphioxus* has no further skeletal structures but every vertebrate has a complex internal skeleton formed, in the embryo, at least, of cartilage, and a majority have in addition bone, a skeletal material found in no other animal group. The digestive system is more complex in its subdivisions; a true liver is always present. The pharynx is less extensive; gill pouches persist even in embryonic land vertebrates but are reduced in number, usually to not more than half a dozen; well developed in adult fishes, they typically function in breathing alone, rather than in feeding as well. The kidneys are of a distinctive type found only in the vertebrates, their functional units quite different in nature from the varied types of nephridia found in invertebrates and even in *Amphioxus*. The sexes are always functionally separate, and in most forms a distinctive series of tubes and passages is developed for the transport of eggs and sperm. Specialized sense organs—nose, eye and ear—are always present. In the nervous system there is, in even the lowest vertebrates, a highly developed brain, with typical major components already established (although the cerebral hemispheres, dominant in higher forms, are little emphasized in lower classes).

We thus have, in vertebrates generally, a structural pattern far more complex than in their lowly chordate relatives. But there is a remarkable amount of structural and functional variation from fish to bird or man, and the problem of dividing the vertebrates into a series of classes within the subphylum arises. For some groups the problem is readily resolved. Obviously, the group of warm-blooded, hair- or fur-covered animals which nurse their young, to which we ourselves belong, can be readily set apart as a Class Mammalia. The birds, Class Aves, are equally distinct; so are the reptiles, as a Class Reptilia, and such lowly four-footed creatures as the frogs, toads and salamanders, living generally close to or in ponds and streams, are reasonably regarded as a distinct Class Amphibia. But what of fishes? By some writers all the lower water-dwelling vertebrates are "lumped" as a single class the Pisces. But if one seriously studies them it is seen that there is as much structural difference between one fish and another—between, say, a hagfish and a trout—as between a frog and a man. The fishes are best divided into some four classes. At the bottom are the lampreys and hagfishes, which lack such usual vertebrate structures as jaws and paired fins or limbs; they can be considered as forming a Class Agnatha (jawless). Next above is a group, now entirely extinct, but represented by many fossils, which we may term the Class Placodermi ("plated skin"); in these extinct fishes jaws and limbs were present, but present, so to speak, in an "experimental" stage. Higher come the sharks and their relatives, the Class Chondrichthyes or cartilaginous fishes, in which jaws and paired fins are well developed, but bone is absent and the skeleton is formed by cartilage alone. Finally, there are advanced bony fishes, the Osteichthyes, comprising most of the fish types with which we are familiar. We may, if we wish, consider these four classes as making up a Superclass Pisces, and it is not unreasonable to bracket similarly all the four-footed land dwellers as a Superclass Tetrapoda. Our classification of the vertebrates may thus stand:

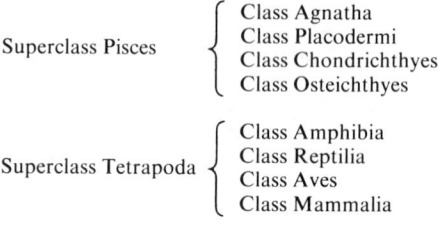

Subphylum Vertebrata

Superclass Pisces
{
Class Agnatha
Class Placodermi
Class Chondrichthyes
Class Osteichthyes
}

Superclass Tetrapoda
{
Class Amphibia
Class Reptilia
Class Aves
Class Mammalia
}

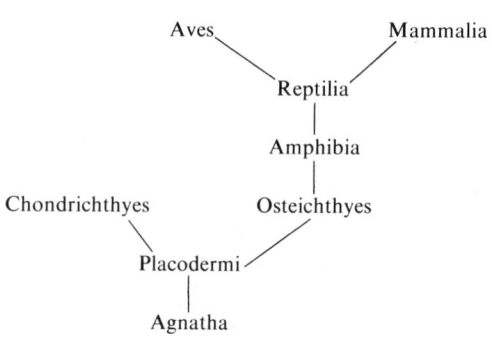

Jawless Vertebrates. Most lowly of living vertebrates and representatives of the Class Agnatha are the cyclostomes—the lampreys, some inhabiting fresh waters as well as salt, and their purely marine relatives, the hagfishes. These unattractive creatures resemble eels superficially; but the eels are highly developed, bony fishes, and cyclostomes, with not a bone in their bodies, represent in every regard a much lower stage. There is no trace of paired fins, nor are there biting jaws, one of the most significant structural and functional features of all other vertebrates. Cyclostomes prey on other fishes, but in place of jaws, have developed a peculiar protrusible rasping tongue-like organ, armed with sharp teeth, and in addition, in lampreys the round mouth (which gives these forms their group name) is an adhesive disk for attachment to the prey. The sense organs, too, are peculiar, for the ear structures are less developed than in other vertebrates, and the olfactory organs which in typical fishes form a pair of nasal pockets on the snout are here fused into a single deep pit which in the lamprey lies far back on the "forehead."

What is the evolutionary position of these curious creatures? Some of their characters may be ascribed to degeneration; others such as the rasping "tongue," are surely specializations. But the absence of jaws and of paired appendages are surely primitive features which justify placing the lampreys and hagfishes in a distinct class at the bottom of the vertebrate scale.

But these forms, as we know them, can hardly be the actual ancestral vertebrates. To cite the most obvious point, the lampreys and the hags prey on other higher fishes—which were not, of course, in existence at the beginning of vertebrate history.

Study of the fossil record has in recent decades given us a rather unexpected answer. There have been known for more than a century fossils coming from some of the oldest geologic periods (Ordovician to Devonian) termed ostracoderms ("shell-skinned"). These small forms have no particular resemblance in appearance to the recent cyclostomes, and, unlike them, had bony skeletons. But they show little or no trace of limb development; like the cyclostomes, they lack jaws and in many forms the nose and ear have the peculiar cyclostome characters. Although there is considerable variation among the ostracoderms, it seems clear that as a group they represent the actual ancestral vertebrates and that the feeble development of the skeleton in the living lampreys is not primitive but due to degeneration. The ostracoderms lack, as expected, the rasping "tongue." In many the gill region is highly developed and it seems clear they were filter-feeders, living like their lower chordate relatives on food particles strained out by the gill basket.

Placoderms. Somewhat later in geologic appearance than the ancestral ostracoderms was a varied series of extinct fishes, ranging in time from the later part of the Silurian to the early Permian, which are currently considered to constitute a Class Placodermi. As the name suggests, all are more or less armored by bony plates and scales on the surface of the body and bone was present in the internal skeleton in all well known types. Some placoderms, the acanthodians, or "spiny sharks," had a normal fish-like body shape; others, such as the arthrodires and their relatives, prominent in the life of the Devonian, were of grotesque appearance. As a group, the placoderms show structures well advanced over their ostracoderm predecessors and ancestors. Biting jaws are here developing, enabling these fishes to branch out into varied modes of existence and assume a dominant role in aquatic life. Limbs, too, are making their appearance in many cases in the form of small paired steering fins which aid greatly in fish locomotion. Most known placoderms are too late in time and too specialized in nature to be direct ancestors of later and higher fishes, but certainly the origins of late developments in fish history lie within the placoderm group.

Sharklike Fishes. Existing jaw-bearing fishes are sharply divided into two distinct groups, the sharks and their relatives, constituting the Class Chondrichthyes, and the Osteichthyes. The sharks, making up a modest fraction of the marine population, have a relatively uniform appearance—a fusiform body with an up-tilted tail, a skin protected by a shagreen of tiny tooth-like denticles, and a series of gill-slits opening directly on the surface. Internally, the intestine is a peculiar cigar-shaped structure traversed by a spiral food passage. In contrast with most fishes, the sharks produce large eggs with a plentiful yolk and with a protective shell. To facilitate fertilization of the eggs before they are laid, the males have a pair of finger-like "claspers"; in some forms the young are born alive. Distinctive, as the class name implies, is the fact that although there is a highly developed internal skeleton, there is not a trace of bone in the body. A corollary of the absence of bone is the fact that there is no formed skull in the usual sense of that term, the head skeleton consisting merely of the jaws, gill bars, and a cartilaginous braincase.

Closely related and derived from the sharks are the bottom-dwelling skates and rays; the body is much flattened, the tail reduced to a whiplike structure and swimming accomplished by undulation of the greatly enlarged pectoral (front) set of paired fins. The mouth parts too, are much modified, with the teeth arranged in plates suitable for crushing the shellfish which form the staple diet. A further group of Chondrichthyes is that of the chimaeras, relatively rare fishes of the deep seas which resemble the sharks in various basic features but differ in such respects as the presence of a flap of skin (an operculum) covering the gill openings and the fusion of the upper jaws, bordering the small mouth, with the braincase.

It was long assumed that the absence of bone in the shark skeleton was a primitive feature, and that this group of fishes preceded and was ancestral to the higher bony fishes. It now appears highly probable that this is not the case. The sharks appear in the latter part of the Devonian period, but all major groups of Osteichthyes had appeared well before this time; further, there are some Devonian fishes seemingly transitional between placoderms and sharks. It is probable that the sharks are descended from placoderms with bony skeletons and that the absence of bone is not a primitive feature but a degenerate retention of an embryonic condition; in all vertebrates internal skeletal structures are first formed in cartilage which is generally replaced by bone before maturity.

Higher bony fishes. Apart from those already mentioned, all fishes today belong to a final class, the Osteichthyes. As we have seen, the presence of bone, to which the name refers, is not distinctive of the class but is the retention of a skeletal feature abandoned by cyclostomes and shark-like forms. The Osteichthyes, however, have not merely retained bone, but have utilized it to advantage; this, combined with other progressive structures and functional adaptations, has made them the dominant modern fishes, both in numbers and in variety.

The bony fishes are an ancient group, appearing in the Devonian period, and rapidly assuming an outstanding position in fresh waters and, eventually, in the oceans. From their first appearance they were already clearly divisible into two major subclasses. Of the two, the Actinopterygii, or ray-finned fishes, are dominant in later periods and today. One of their many diagnostic characters is the fact that (as the name implies) the paired fins consist of little but rays supporting a web of skin. A few ray-finned forms of today, the sturgeons and garpike, for example, represent survivors of ancestral stages in the evolution of ray-finned forms; the overwhelming majority of ray-finned fishes, however belong to a final evolutionary stage appropriately termed the Teleostei. Here are included, among many other forms, every common food and game fish with which a reader might be familiar.

The second subclass of the bony fishes has been variably named (Choanichthyes, for example), but is perhaps best termed Sarcopterygii, or fleshy-finned fishes, because of the presence in the paired fins of a well developed lobe of flesh and bone. The presence of such paired fins is of great importance in the evolutionary picture, as giving rise to the potentiality of the development of the limbs of land vertebrates. However, this and other distinctive characters of the sarcopterygians availed them little in the world of fishes. For although this group was prominent in early fish history, only a very few representatives have survived to the present day. Three of these are the lungfishes of the Order Dipnoi, living in tropical regions of Australia, Africa and South America, where their possession of lungs aids them in surviving seasonal droughts. In many regards the lungfishes exhibit features comparable to those in primitive land animals, but specializations in teeth, jaws and skull show that the dipnoans, although related to the ancestry of tetrapods, cannot be direct ancestors. The true ancestors were ancient members of a second sarcopterygian group, the Crossopterygii. Of this group, there is a single survivor, *Latimeria*, only recently discovered in deep waters off the Comoro Islands in the Indian Ocean. The living form is a specialized one, however, for the typical crossopterygians from which the land vertebrates were derived were dwellers in fresh waters in the latter part of the Paleozoic era, particularly abundant in the Devonian period. The fins of these ancient crossopterygians, although small, had a pattern of a type expected in the limb of an ancestral land animal; many details of the skull and other skeletal features were comparable to those of early tetrapods; it is felt certain that, as in the living dipnoans, lungs were present (indeed, lungs may have been common in ancient fishes generally).

Amphibians. Lowest among living land vertebrates in their general organization are the forms included in the Class Amphibia. Commonest and most familiar are the members of the Order Anura, the frogs and toads. Also familiar are the Urodela, the newts and salamanders, mainly inhabitants of northern temperate regions. Little known except to the specialist are the rare tropical forms of the Order Apoda (or Gymnophiona). In life habits these forms merit the class name, for most of them seldom stray far from the water and spend much of their lives in that element; the typical mode of development is one in which the eggs are laid in the water in fishlike fashion and the young may long remain as tadpoles or comparable gill-breathing larvae before emerging from the water. But although the living amphibians as a group represent an ancestral tetrapod stage, no one of the three living types can be considered as typical tetrapod ancestors. The frogs, with a greatly

shortened body and long, peculiarly built hind legs, are specialized for a hopping gait. The salamanders and newts show much the general body proportions which we would expect in ancestral land vertebrates, but they are degenerate in their skeletons and possibly in other regards. The Apoda are small, limbless, worm-like tropical burrowers which are obviously highly specialized. The living amphibians are but the modified surviving remnants of a once great class that has been overwhelmed in importance by higher land types. In earlier times the amphibians played a much more important role. Most prominent of early amphibians were the members of a great group termed the Labyrinthodontia, of which we know a large array of fossil genera and species. This group appeared at the end of the Devonian and survived until the Triassic period. Some of the older forms were still very similar to the ancestral crossopterygians; some approached the ancestral reptiles closely in all determinable characters. While connecting links are few, it seems reasonably certain that the modest surviving groups of amphibians were derived, directly or indirectly, from the ancient labyrinthodonts.

Reptiles. Living reptiles pertain to four orders. (1) The Chelonia, the turtles—specialized in the development of a stout shell, both above and below the flattened body. (2) The Squamata, the "scaled reptiles," including the lizards and, as well, the snakes, derived from the lizards by loss of limbs and the development of a peculiarly flexible jaw apparatus, which enables them to swallow large prey. (3) The Order Rhynchocephalia, including but a single animal, *Sphenodon* of New Zealand, which is similar to the lizards in many ways but has a distinctive skull structure. (4) The Crocodilia, alligators and crocodiles and their relatives, amphibious predators whose most distinctive structural feature is a highly developed secondary palate in the roof of the mouth which enables them to breathe readily while nearly completely submerged. The diagnostic class feature of the reptiles, distinguishing them from the amphibians, lies in the mode of development. The typical amphibian's life history ties it to the water; the ancestral reptiles developed a type of egg with a protective shell and a large quantity of yolk. This can be laid on land, eliminating any necessary water-dwelling stage. As the embryo grows within the shell, various membranes develop around it which keep it moist, enable it to absorb the nutrient yolk, and allow it to breathe air through the porous shell. Because of this important reproductive improvement, the reptiles and the further classes derived from them are frequently termed the Amniotes, the name being derived from one of these embryonic membranes (the amnion).

Although the living reptiles are of interest in their own right, paleontology reveals the fact that here (as in the case of the amphibians) the surviving forms are but the remnants of a much greater group. Ancestral reptiles arose from the ancient labyrinthodont amphibians during the Carboniferous period and rapidly spread out into a variety of evolutionary lines. The turtles appear to have been an early side branch, as were such spectacular marine forms as the extinct ichthyosaurs and plesiosaurs. Of importance was the development of small reptiles whose skull was pierced by two openings in the temporal region (the diapsid condition). *Sphenodon* is a little-modified descendant of such ancient forms; the lizards and snakes come from the ancestral diapsid stock as the result of further skull modification. Other diapsid types tended toward a life as fast-running bipeds, and became the ancestors of the archosaurs, the "Ruling Reptiles," of which the crocodilians are the sole degenerate survivors. In the Mesozoic era, however, this group blossomed forth to produce the varied and spectacular types of dinosaurs that dominated the world for eons, and in addition the remarkable flying reptiles, the pterosaurs. A very early side branch from the stem reptiles produced a distinct line leading to the therapsids, the mammal-like reptiles whose remains are abundant in the Permian and Triassic periods, particularly in South Africa and Russia. As far as can be seen from the skeletons (to which, of course, our knowledge is confined), those forms bridge almost the entire gap between the reptiles and the mammals, the first of which appeared at about the end of the Triassic period. There is no indication in the therapsids of the increase in mental ability which was to play an important role in the eventual triumph of the mammals, but there is a vast improvement in four-footed locomotion, leading to mammalian conditions in this regard.

Birds. The Class Aves is today a very distinct vertebrate group, characterized by a long series of special features most of which are adaptations for a life in the air. Terrestrial locomotion is bipedal, the front limbs being transformed into wings containing the rudiments of three fingers. The body is compact, with an enormous breastbone (sternum) for the attachment of wing muscles. The skull (indeed, the whole skeleton) is lightly built, the teeth are lost and the bony tail reduced. The brain is highly developed (although in a fashion quite different from that of mammals). A high body temperature is maintained, in correlation with the need for continuous activity during flight. And last but not least of avian characters is the presence of feathers, believed to be derived from horny reptilian scales; these insulate the body and form the effective surfaces of wings and tail.

Birds are highly varied as to plumage and habits but, except for the ratites such as the ostrich, which have lost the ability to fly, are basically built on a uniform structural pattern. This pattern is clearly derivable from that of archosaurian reptiles, bipeds in which the front limbs were freed from use in terrestrial locomotion and hence were available for conversion into wings. A transitional fossil form is seen in *Archaeopteryx* of the Jurassic period. In this animal we are definitely dealing with a bird, for feather impressions are preserved in the three known specimens. But there is a long bony tail, there are three clawed toes in the "hand"; teeth are present. In fact, had not feathers been found, identification as a bird rather than a little dinosaur would have been far from certain.

Mammals. To this final class belongs a host of familiar animals, including man himself. There are numerous diagnostic features. As the class name implies, mammals nurse their young and in most cases bear them alive at an advanced stage of development, the embryo being nourished within the mother's uterus through a complex union of maternal and foetal tissues termed the placenta. As in birds, a high body temperature is maintained, giving the potentiality of continuous activity in any climatic condition; the presence of hair or fur is one of several factors which aid temperature maintenance. The brain is very highly developed,

notably the great expansion of the "gray matter" of the cerebral hemisphere—the seat of the highest brain centers and of learning ability. Present, too, are distinctive skeletal features.

Most primitive of living mammals are two forms found in the relative isolation of Australia, the duckbill (*Ornithorhynchus*) and the spiny anteater (*Echidna*), which differ from other mammals not only in specialized adaptations to their mode of life and certain primitive structural features but, most important, in that they still lay eggs in reptilian fashion. As a consequence, they are placed in a Subclass Prototheria, in contrast to a Subclass Theria for all other mammals combined. The Theria, again, are subdivided into two groups of unequal size. The Metatheria are the pouched mammals, the Marsupials. The opossums are common in the Americas, but it is in the Australian region that the marsupials flourish. Before the arrival of man no other mammals were able to reach Australia except certain types of rats and mice which had migrated down the East Indian chain. In the absence of competition the marsupials there radiated into a host of varied forms which paralleled to a great degree the adaptive types which evolved among the higher mammals in other continents. The marsupials bear their young alive. But in general the placental connections between mother and young are poor. The young are born at a very immature stage; in partial compensation, they enter a pouch on the mother's belly where the teats are placed, and growth continues here.

All other mammals are classed as the Eutheria, frequently termed the placentals. Here the placenta is more efficiently constructed, and embryonic development can proceed to a much later stage. On this placental level belong all other mammal groups of which only a fraction may be enumerated here: the little insectivores, such as the shrews which in many ways suggest the ancestral mammal types; the flying mammals, the bats; the primates, including lemurs, monkeys, apes and man; the edentates, such as sloths and armadillos; the whales; the great host of rodents; the hares and rabbits; the numerous and varied groups of hoofed mammals, or ungulates, and such related types as the elephants, sirenians and conies.

Mammalian evolution has been characterized by the development of intelligent activity, locomotor agility in the limb skeleton, and temperature regulation. Brain development is outstanding in mammals. The lengthening of individual development allows time for the proper elaboration of this complex organ; the nursing habit initiates family life and allows further time for mental training.

ALFRED S. ROMER
Revised by Theodore H. Eaton

References

Berrill, N. J., "The origin of vertebrates," Oxford, Clarendon, 1955.
Garstang, W., "The morphology of the Tunicata and its bearing on the phylogeny of the Chordata," Quart. J. Micros. Sci., 72:51, 1928.
Grassé, P.-P., ed., "Traité de Zoologie," Vol. 11, Paris, Masson, 1948.
Romer, A. S., "The Vertebrate Body," Philadelphia, Saunders, 1962.
ibid., "The Vertebrate Story," Chicago, The University Press, 1959.

CHROMOSOME

In all organisms above the level of the Bacteria and Blue-Green Algae the hereditary material of the nucleus is organized into thread-like bodies known as *chromosomes*. Whether this is also true of Bacteria, Blue-Green Algae and Viruses is still uncertain; geneticists who study the phenomena of heredity in these microorganisms speak of bacterial or viral "chromosomes," but until more critical microscopic studies have been carried out it is premature to assume that the organization of the hereditary material in these lower forms of life is really comparable with the situation in higher organisms.

In the nuclei of plants and animals the chromosomes float in a more or less viscous fluid, the *nuclear sap,* surrounded by the *nuclear membrane,* which separates the nucleus from the rest of the cell, or cytoplasm. The sperm and egg nuclei of one variety of the Horse Roundworm *Parascaris equorum* contain only a single chromosome, while certain secretory cells in insects contain at least 20,000 chromosomes and perhaps in some instances as many as 100,000.

In general, all the body cells of all individuals of a species contain the same number of chromosomes. This is known as the *somatic number* of that species. But there are some exceptions to this general principle. Thus the somatic number may be different in the two sexes; it may be different from one individual to another because of the existence in varying number of supernumerary chromosomes in addition to the "minimal" set; and in most higher animals and plants certain cells contain twice the somatic number, four times, eight times . . .etc.

During the formation of the sperm and ovum (in animals) and that of the pollen grains and ovules (in plants) the chromosome number is reduced to half the somatic number as a result of the process of MEIOSIS. The reduced number is known as the *haploid* number, in contrast to the *diploid* (somatic) number. By fusion of two haploid nuclei at fertilization the somatic number is restored. Thus in sexual reproduction every individual receives a haploid set from each parent. In the diploid set the chromosomes exist in homologous pairs, one member of each pair having been derived from the father, the other from the mother.

Many plant species possess more than two haploid sets, and are called *polyploids.* Individuals with 3 sets are *triploids,* those with 4 sets, *tetraploids,* etc. Odd-numbered polyploids (with 3, 5, 7 . . .sets) are usually incapable of normal sexual reproduction. Several very important crop plants such as wheat, cotton and tobacco are even-numbered polyploids.

Polyploidy is much rarer in animals than in plants, but a few animal species that reproduce by parthenogenesis are polyploids. In the Hymenoptera and a few other Arthropod groups the females are diploid, while the males, which arise by parthenogenesis, are haploid.

As a rule the chromosomes cannot be seen clearly during the interphase or resting stage of the cell and it is only during the process of cell division (MITOSIS) that they can be seen clearly. There are four main stages of mitosis, *prophase, metaphase, anaphase* and *telophase.* During prophase the chromosomes are visible within the nucleus as filaments with a somewhat irregular, "woolly" outline. Each filament is clearly double, i.e. it consists of two parallel strands or *chromatids* which have arisen by the biochemical duplication of the chromosome during the interphase. These parallel chromatids are united at one point, the *centromere.* Some chromosomes have the centromere near the midpoint, so that they have approximately equal limbs (metacentric type), while others have it close to one end, so that one limb is very much longer than the other

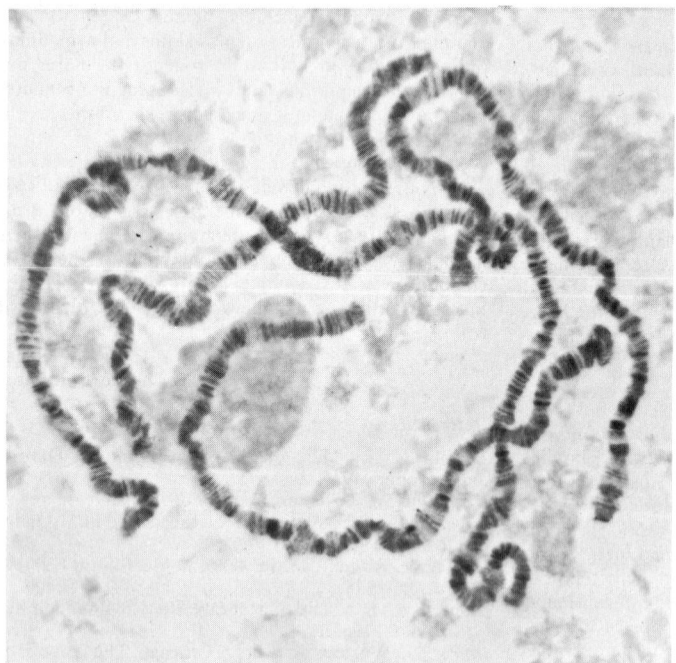

Fig. 1   Photomicrograph of giant polytene chromosomes in salivary gland cell of *Drosophila.* (From Wilson and Morrison, "Cytology," 2nd ed., New York, Reinhold, 1966.)

(acrocentric type), but apparently the centromere is never quite terminal, in natural chromosomes. Whereas other portions of the chromosome stain readily with basic dyes the centromere usually remains unstained.

During the prophase stage one can often see that certain regions of the chromosome are more condensed and thicker than the rest of the thread (*heterochromatic blocks*). Non-staining gaps or *constrictions* (not to be confused with the centromere) may also be seen. Some chromosomes have bladder-like structures or *nucleoli* attached to them, which may be very conspicuous during interphase and prophase, but shrink in size and usually become completely absorbed into the substance of the chromosome by metaphase. The nucleoli arise from special regions of the chromosomes, the *nucleolar organizers*. If there are several originally distinct nucleoli in a nucleus they often have a tendency to fuse. Thus many diploid nuclei, in which one pair of chromosomes carry nucleolar organizers, show a single large nucleolus that is double in origin.

The chromosomes become much more condensed during the latter part of prophase and by metaphase their outlines are smooth and they usually appear sausage-shaped, often with a waist where the centromere is. Actually each chromatid is a separate sausage-shaped body, and its structure is really a highly coiled helix whose gyres are in contact and closely pressed together.

At the beginning of metaphase the nuclear membrane disappears and a transparent gelatinous body, the *spindle* is formed, largely out of the nuclear sap. The chromosomes attach themselves by means of their centromeres to the equator of the spindle, mid-way between the two poles. In the case of long chromosomes their limbs may project out from the spindle into the cytoplasm, but many small chromosomes are entirely enveloped by the spindle substance at metaphase. At the poles of the spindle special structures (centrioles and asters) are present in most animal cells, but not in those of the higher plants.

At the beginning of the anaphase stage the centromeres, which have been functionally single up to this time, split and the daughter centromeres start moving in the direction of the poles, dragging after them the chromatids which are attached to them. The whole spindle also elongates so that the two groups of chromatids (called daughter chromosomes after they have separated) are pushed apart. The spindle now begins to break down and the groups of daughter chromosomes at the poles go through a series of changes (telophase) which are in many respects the opposite of those which occur in prophase. The nuclear membrane re-appears and eventually two interphase nuclei are formed. At about this time the cytoplasm of the cell is cut into two, either by the formation of a constriction, or by the growth of a cell wall.

Chemically, the chromosomes consist of two kinds of NUCLEIC ACIDS (desoxyribonucleic acid and ribonucleic acid) together with two main types of proteins, histones or protamines and a tryptophane rich protein. The exact way in which these are combined in the living chromosome is not yet understood.

The process of meiosis consists of two nuclear divisions in the course of which the chromosomes only undergo replication once. In the prophase of the first meiotic division the members of each pair of chromosomes become closely approximated, side by side. This process of *synapsis* seems to result from a highly specific force of attraction between homologous chromosome regions which operates at this stage. The double structures resulting from synapsis are called *bivalents*. When each chromosome has undergone replication, the bivalent consists of four parallel chromatids. At a slightly later stage the two chromosomes of each bivalent separate again but remain connected together at certain points (*chiasmata*) where two of the four strands—one derived from one chromosome and one from the other—have broken and rejoined "the other way round." This is the process of genetic crossing over which leads to intrachromosomal recombination.

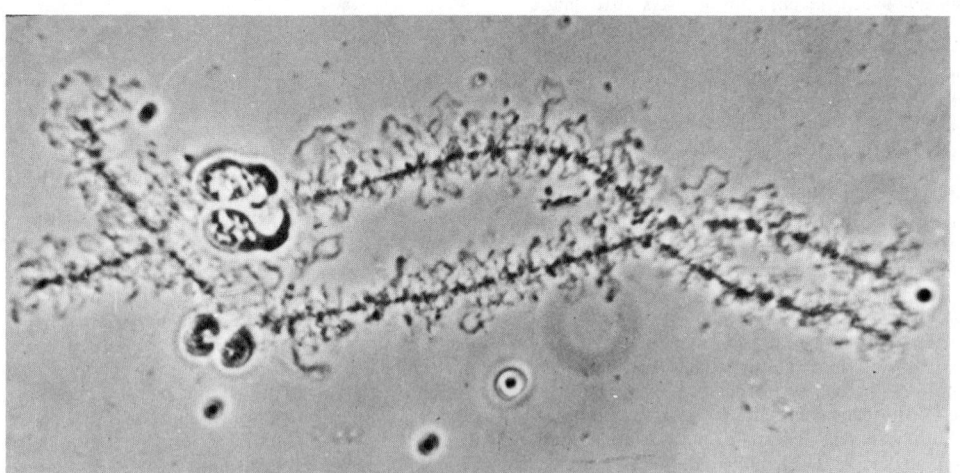

Fig. 2 Phase-contrast photomicrograph of a bivalent with homologous lampbrush chromosomes joined by two chiasmata in first meiotic prophase of the newt, *Triturus cristatus carnifex*. Note the numerous lateral loops associated with the central chromosomal axis of each lampbrush chromosome. (Courtesy of Dr. H. G. Callan, The University, St. Andrews, Scotland.)

At the first metaphase the bivalents arrange themselves on the spindle in a characteristic manner, with the two centromeres equidistant from the equatorial plane. At the first anaphase stage the centromeres do not divide as in an ordinary somatic division; instead each whole centromere goes to the nearest pole, dragging after it the chromatids attached to it. No duplication of chromatids takes place between the two meiotic divisions and at the second division the centromeres do split, each one taking a single chromatid to the pole.

In normal meiosis there is always at least one chiasma in each bivalent: but in the male *Drosophila* and some other insects no true chiasmata are formed, although they do occur in the females.

Giant chromosomes of a very special kind exist in certain somatic tissues of the two-winged flies (order Diptera). They usually attain their largest size in the nuclei of the larval salivary glands and are hence often called salivary gland chromosomes, although since they are also found in other tissues the general term *polytene chromosomes* is preferable. In these huge nuclei the chromosomes have undergone a kind of synapsis. They show cross striations, regions which stain intensely with basic dyes alternating with inter-band regions that stain little or not at all. In the haploid chromosome set of *Drosophila melanogaster* there are over 5000 stainable bands. Studies of stocks containing minute rearrangements such as *deletions* of a few bands have shown that the bands correspond in a general way to genetic loci. The study of chromosomal rearrangements such as *inversions* (where a section of a chromosome has been turned through 180° as in the sequence of letters ABCD*GFE*HIJ) is greatly facilitated by the existence of the polytene chromosomes.

Another type of giant chromosomes occurs in the oocyte nuclei of vertebrates with yolky eggs (sharks, birds and especially urodeles). These *lamp brush* chromosomes are in the mid-prophase of the first meiotic division. They show many stainable granules (*chromomers*) from which long lateral loops project.

In organisms with separate sexes there is frequently a difference between the chromosome sets of the male and female. A pair of microscopically indistinguishable chromosomes in one sex (called X-chromosomes) may be represented in the other sex by a pair of visibly different elements (X and Y). Usually the XX sex is the female and XY sex the male, but the situation is reversed in the Lepidoptera (Butterflies and Moths) and in the Birds, as well as in certain Fishes and Salamanders. The XY sex is called the *heterogametic* sex, since it produces two kinds of gametes (sperms or eggs) in equal numbers, while the XX sex is said to be *homogametic*, since it only produces one kind of gamete, as far as sex determination is concerned.

Both X and Y elements are known as *sex chromosomes*. The remaining chromosomes are referred to as *autosomes*. In certain species there is no Y; the number of chromosomes in the heterogametic sex is then an uneven one. Most of the common species of grasshoppers have 23 chromosomes in the male (eleven pairs of autosomes and an X) and 24 in the female (eleven pairs of autosomes and two X's). The heterogametic sex is said to be XO in such species, the symbol O designating simply the absence of a Y.

In *Drosophila melanogaster* the sex of the individual depends on the ratio of X-chromosomes to autosomes which it possesses, the former being female-determining, the latter male-determining. Thus an individual of this species with three sets of autosomes but only two X's has an X/Autosome ratio intermediate between those normally present in the male and female respectively, and is in consequence an intersex. The Y seems to have no direct effect on sex determination in Drosophila, since abnormal XI individuals, which occur occasionally, are males, although sterile and individuals with 2 X's and a Y are females. But in the Mexican Axolotl and the Silkworm (both with female heterogamety) the Y is an actively female-determining chromosome. In the human species there is also some evidence that the Y has an actively male-determining role, since individuals with a type of intersexuality known as Klinefelter's syndrome have been found to be XXY and individuals with a different kind of intersexuality (Turner's syndrome) are XO.

In some species the X and Y chromosomes can readily be distinguished under the microscope by differences of size or shape, but in others they are alike in appearance, although differing genetically. In a good many species of animals and in a few dioecious (bisexual) plants either the "X" or the "Y" is represented by two or more separate chromosomes. Thus in many species of Praying Mantids we find that the males have three sex chromosomes (X_1X_2Y) and the females four ($X_1X_1X_2X_2$). In most species of Spiders there are two different kinds of X-chromosomes but no Y, so that the males have X_1X_2 and the females $X_1X_1X_2X_2$.

M. J. D. WHITE

References

Swanson, C. P., "Cytology and Cytogenetics," Englewood Cliffs, N. J., Prentice-Hall, 1957.

Schrader, F., "Mitosis: the Movement of Chromosomes in Cell Division," New York, Columbia Univ. Press 1953.

White, M. J. D., "Animal Cytology and Evolution," Cambridge, Cambridge University Press, 1954.

CILIA

Permanent and highly vibratile ectoplasmic processes of certain cells, in every respect like the FLAGELLA save for the smaller size (about 0.2μ in diameter and 15μ in length. They are found in almost all groups of the animal kingdom, fulfilling varied and important functions, for which other motile structures are unsuitable in the life of the organism.

Electron micrographs reveal a fundamentally similar fine structure in the free shaft of all cilia: nine (sometimes double) longitudinally oriented peripheral filaments around a closely approximated central pair, all embedded in a structureless matrix. The external membrane enveloping the cilium is probably continuous with the cell membrane. The thickened proximal ends of the peripheral fibrils form the tube-shaped *basal granule*, generally considered as "self-reproducing" units and progenitors of new cilia. According to the "Lenhossék-Henneguy theory" (1898), the basal granules are centrioles or derivatives of the centriole. The ciliary rootlets often associated with the basal granule usually show a periodic cross-striation resembling those in collagen and fibrin. In structural details, basal granules and ciliary rootlets vary somewhat within the several systematic categories.

The work of the typical cilium is based on a paddle-stroke effect: it vibrates backwards and forwards, in regular intervals completing a cycle in about 1/25 sec. It is reasonable to suppose that, in this order of size, the productive effect is due not so much to the different velocities of the two phases, as—according to the Stoke's law of resistance—to the smaller shape-resistance during recovery.

The cilium is capable of spontaneous movement; in the bond of a ciliary field, however, their activity is strictly coordinated: the primary movement of individual cilia is modified to an axially directed paddle-stroke, and the succession of beats are exactly timed for the sake of the total effect of the ciliature as a whole. The spontaneous movement itself could be observed only after elimination of the coordinating forces, and it is noteworthy that in the phylogenetically most primitive groups this resembles the basic movement of the flagellum. Thus in the CILIATA, nearly all cilia circulate, when moving independently of metachronal coordination, in accordance with the cylindrical array of the nine peripheral fibrils around a wide-angled cone, at a constant angular velocity and in a counterclockwise direction (Fig. 1). This apolar, rotary movement is incapable of effecting a displacement of the ciliate body. Under natural conditions it is transformed by waves of excitation of endogenous origin, passing in regular intervals along the periphery of the cell. Cilia reached by a metachronal impulse strike out in an outstretched form from their momentary position in a plane perpendicular to the cell surface and directed towards their point of articulation. As the wave of activity passes away, the cilium becomes limp, bends out of the plane of beat to the right and swings, stiffening again from base to tip, counter-clockwise and parallel to the body surface until reached again by the next impulse. The recovery phase of the beat thus conforms to a segment of the primary apolar path. During normal forward swimming in a left-handed spiral, the endogenous impulses usually touch the cilium in a position pointing anteriad and slightly to the left therefore the beat is directed posteriorly and to the right (Fig. 2).

In ciliated Protozoa, the excitatory effects of external stimuli might temporarily transform or completely suppress the rhythm and direction of propagation of endogenous impulses at the stimulated region or on the whole body surface. Due to these new waves of coordi-

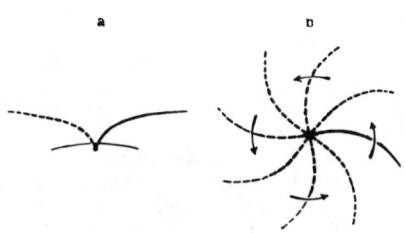

Fig. 1   The uncoordinated apolary movement of a single cilium from Paramecium, (a) side view, (b) face view. Direction of the funnel-shaped circling is marked by arrows.

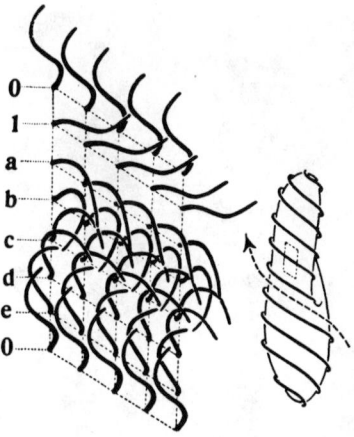

Fig. 2   Metachronal waves of Paramecium during normal forward swimming (right). Enlargement of the same showing a rhombic portion of the body surface with five ciliary rows in the successive stages of complete beating cycle: (0-a) effective stroke, (a-0) recovery stroke.

nation the amplitude, frequency and direction of the ciliary beats undergo modification, the latter simply owing to the fact that the frontline of the wave may reach and compel the leftward circulating cilium to strike in, respectively from, diverse positions. Accordingly, the diverse motor reactions of the Infusoria are always plastically suited to the existent situation of stimuli.—The material basis of the metachronal impulse-transmission is the cortical zone of the ectoplasm itself.

In metazoan epithelia, the ciliary activity is more stereotyped and adapted to special needs. Here, the autonomous movement of the single cilium is not rotatory but simply pendular. Accordingly, the movements occur in the identical plane during the two phases and the direction of the beat becomes more and more irreversible, towards the higher differentiated organisms. In an intact epithelium, the ciliary movement is also metachronally controlled but it seems to be synchronous in isolated ciliated cells. The ciliated epithelium of many Metazoa works under the secondary control of the nervous system, and this control is usually of an inhibitory nature.

As regards the essential mechanism of the ciliary movement, little definite can yet be said. Several considerations lead to the conclusion that the cilium is not moved by the cell. The seat of the movement lies obviously within the cilium itself and only refuelling occurs from within the cell. Numerous experiments demonstrate that the metabolism of the cilium has much in common with that of the muscle. In all, the evidence strongly suggests that the ciliary movement is based on contraction and that the contractile elements are represented by the peripheral fibrils. Unlike myofibrils, cilia obviously contract not synchronously but individually or in groups. The basic movement of the uncoordinated protozoan cilium may, for example, be conceived as follows: each fibril shortens successively and in a counterclockwise order so that the contraction of the

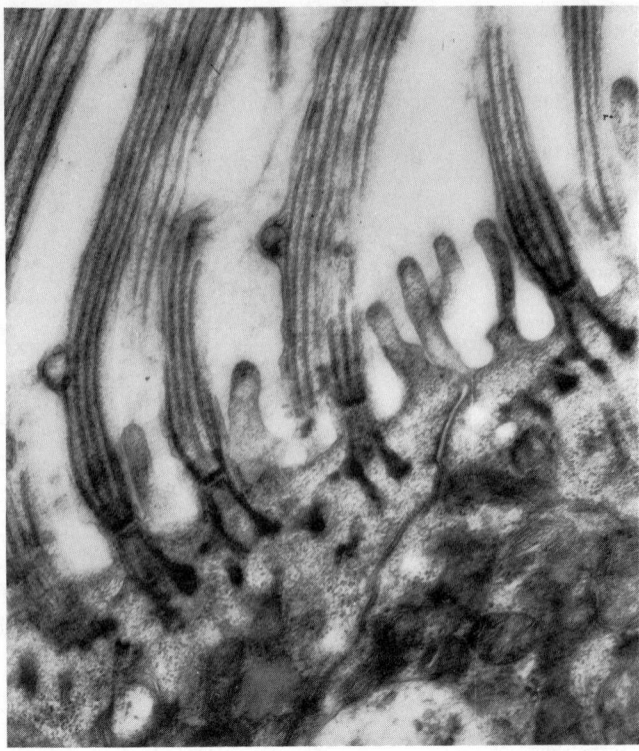

Fig. 3 Electron photomicrograph of cilia. The dark lobes at the base of each cilium are part of the nerve-like net which keeps the beat synchronized. (Courtesy Peter Satir.)

next fibril happens before the total relaxation of the preceding one. Accordingly, the activity of these fibrils is primarily spontaneously rhythmic, and the role of the impulses passing along the cell would only be the regulation of these spontaneous contractions to assure an adaptive behavior in the given conditions.

B. PARDUCZ

CILIATA

The ciliates comprise one of the four major subdivisions of the Phylum Protozoa (along with the flagellates, amoeboid forms and sporozoans). Its members form a homogeneous, sharply separated group with no clear linkages to other protozoa. Two exclusive and invariable group features are the presence of CILIA and nuclear dimorphism. Cilia are locomotor ORGANELLES whose basic structure is homologous to that of FLAGELLA, but which are recognized by their relatively short length, large number and arrangement into longitudinal and/or transverse rows on the body surface. The nuclear apparatus is unique in its differentiation into two distinct kinds of nuclei: a highly POLYPLOID macronucleus (without recognizable chromosomal organization) shown to be essential for continuous metabolic activity, and a typical diploid micronucleus equivalent to nuclei of most other organisms and providing the source of gametic nuclei during sexual reproduction. Individual ciliates may possess from one to many of each type of NUCLEI.

Knowledge of taxonomic relationships within the ciliates has advanced greatly during the past decades, stimulated especially by the findings and interpretations of Fauré-Fremiet. Whereas earlier taxonomic systems had attached much importance to superficial features associated with habitat (whether motile or sessile), mode of feeding, etc., the "new taxonomy" stresses basic morphology of the locomotor apparatus and developmental changes in the ciliature during ontogeny. The functional units of the locomotor apparatus are linear groups of cilia typically oriented in meridional rows. The individual cilia are anchored in subpellicular basal granules (*kinetosomes*), each row of which in turn is interconnected, or at least accompanied, by a fibril (*kinetodesmos*). Each of these functional units is termed a *kinety*; the aggregate of granules and fibrils comprises the *infraciliature*.

The evolutionary history of the ciliates apparently has involved modifications in an originally simple kinetal organization. The prototype, possibly resembling a form like *Prorodon*, presumably possessed numerous meridional rows of simple cilia converging toward an unspecialized, apical, anterior mouth (cytostome) opening at the surface of the body. Major evolutionary trends have involved: changes in number and orientation of kineties; differentiation of cilia into special compound oganelles (membranes, membranelles, cirri) resulting from condensation and duplication of kineties to form more efficient agents of locomotion and food-getting; secondary reduction of ciliation in certain regions of the body, or even complete loss of cilia in adult stages.

When due recognition is paid to the plasticity of ciliate organization and their extensive adaptations to diverse habitats, we may distinguish four coordinate groups (subclasses) amounting to several thousand species. The Subclass Holotrichia includes the least specialized types (seven orders), in which a uniform coating of simple cilia tends to accompany relatively simple axes of SYMMETRY. Variations are observable in the oral region, involving shifts of the cytostome to ventral or lateral locations and into depressions which may bear specialized ciliary organelles (long tactile cilia, membranes, small groups of membranelles). The sessile habit has been adopted by adult stages of some groups (e.g., chonotrichs, thigmotrichs), and special adaptations for attachment and food-gathering presumably account for some unusual morphological adaptations. Symbiotes are commonly encountered, and indeed three groups (astomes, apostomes, thigmotrichs) are exclusively symbiotic. In an evolutionary sense the key group of holotrichs are the hymenostomes (e.g., *Paramecium*, *Tetrahymena*), forms with a relatively prominent cytostome accompanied by accessory compound cilia. This appears to be a stem group from which other orders of holotrichs have evolved, and which also has provided the immediate source of at least one other subclass, the Spirotrichia.

The Subclass Spirotrichia marks the high point in specialized morphology among ciliates, and among protozoa in general. Axes of symmetry tend to be highly developed, yielding sharply polarized dorsoventral and antero-posterior axes. The oral ciliature typically includes both an undulating membrane and a prominent row of close-set membranelles leading toward the cytostome situated in a depression of the body. Simple body ciliation tends to become reduced, or to be replaced by compound ciliary organelles, culminating in the hypotrichs (e.g., *Euplotes*) in complete replacement by cirri. Of the six orders usually recog-

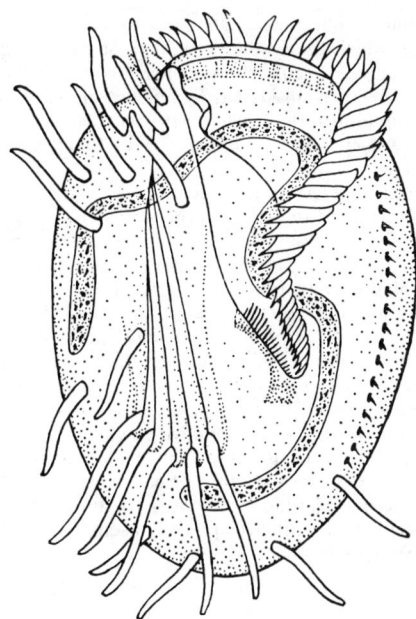

Fig. 2   Euplotes.

nized, the exclusively symbiotic Entodiniomorphida occupy the alimentary canal of herbivorous mammals, especially ruminants. The role of these highly specialized symbiotes in the nutrition of their economically important hosts has been extensively studied, without disclosing any evidence of an obligate mutual dependence as in the case of termite flagellates. The Tintinnida comprise another unusual order, characterized by reduced body ciliation and the presence of a shell-like covering (lorica); they are free-swimming, pelagic forms especially prevalent in oceanic waters.

The present account recognizes two additional subclasses—the Suctoria and the Peritrichia—whose members are predominantly sessile in their adult stages. Like the spirotrichs each of these groups shows clear evidence of derivation from separate holotrich stocks, with the Suctoria emerging at a more primitive level and in their modern expression comprising a widely separated group. No cilia or cytostome are present in the mature, sessile stages; instead suctorial tentacles extend from the body and function by ingesting the bodily contents of immobilized prey (exclusively other ciliates!). The ciliate nature and holotrich affinities of the group become evident during asexual reproduction, when migratory developmental stages appear as buds bearing rows of simple cilia. The Subclass Peritrichia also exhibits adaptations for attachment, but with less extreme modification of ciliate organization. Their most conspicuous feature is a crown of compound cilia enveloping the apical end of the body. The basal end bears either a stalk attaching the body to a fixed substrate or elaborates an adhesive, ciliated disc used in gliding over the surface of a host.

One still unresolved problem concerns the affinities of the Opalinata, a small group of symbiotic, astomatous, uniformly ciliated (?) forms which do not exhibit nuclear dimorphism. Most investigators regard the

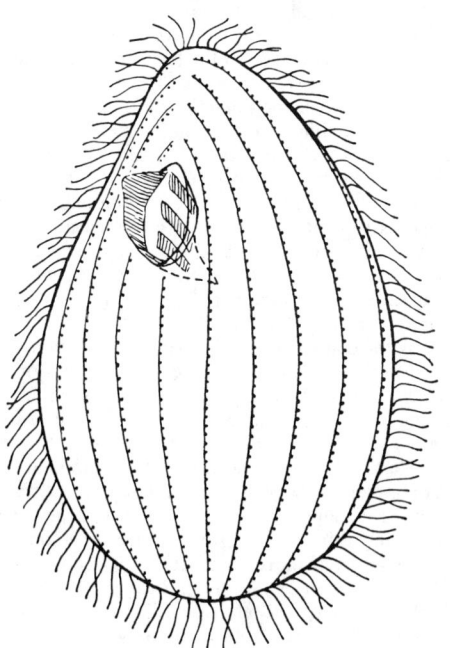

Fig. 1   Tetrahymena.

opalinids as aberrant ciliate survivors of ancient lineage, in which the symbiotic habit is somehow related to the unusual group features, including a peculiar life cycle. However, an increasing number of students are inclining toward the French view that these symbiotes of Anura actually may be closer to the zooflagellate assemblage.

Special interest attaches to the ciliates as the most highly evolved Protozoa, and through their pronounced expression of "animal" traits: bilateral symmetry, motility, mode of nutrition, biparental sexuality. Analyses of their nutritional requirements have disclosed striking similarities to patterns and metabolic pathways encountered in higher animals; many comparative biochemical studies now utilize representative ciliates like *Tetrahymena*. Important genetic studies have also been conducted with ciliates, especially *Paramecium*, owing to well-developed sexuality and the fact that the soma is transmitted along with germinal nuclei during sexual reproduction. Sonneborn and others have been exploiting a unique opportunity to study the relative roles of nuclear and cytosomal components in the inheritance of basic traits.

WILLIAM BALAMUTH

References

Corliss, J., "The Ciliated Protozoa," New York, Pergamon, 1961.
Honigberg, B., *et al.* "A revised classification of the Phylum Protozoa." J. Protozool., 11: 7–20, 1964.
Sonneborn, T., "Breeding systems, reproductive methods, and species problems in Protozoa." *In* "The Species Problem," E. Mayr, ed. A.A.A.S. publ., pp. 155–324, 1957.

CIRCULATORY SYSTEM

The circulation is intimately associated with all functions of living, and it is quite possible to view circulation as one of the integrating systems similar to the neural and hormonal systems. The circulation exists for the purpose of moving blood from the lungs to various tissues of the body. The circulation carries blood to the tissues so as to maintain the correct environment for cells. Claude Bernard called attention to "le milieu interieur," and Cannon used the word HOMEOSTASIS to maintain constancy of the environment of cells.

The Heart. The HEART is the pump of the circulatory system. This action may be watched when the chest is opened and the pericardium is removed. Contraction or systole appears to originate where the large veins enter the auricles. This contraction spreads quickly through the auricles and then to the ventricles. Relaxation or diastole follows.

The muscle of the heart (cardiac muscle) is made up of cross-striated fibers which are more capable of rapid contraction than smooth muscle but less capable than skeletal muscle. When a stimulus is applied to the heart, all or none of the cardiac fibers respond. This is known as the all-or-none law. Even though many factors contribute to the magnitude of the heart's contraction, the strength of the stimulus is not one of these factors. If the stimulus is too weak, the heart does not respond, but the heart responds maximally if the stimulus is

strong enough to induce contraction at all. During contraction the heart will not respond to another stimulus. This is known as the refractory period. During relaxation the excitability of cardiac muscle is below normal and will respond to a strong stimulus which causes a premature contraction. During this premature contraction the normal stimulus in time sequence naturally fails to cause contraction.

The heartbeat does not originate in the central nervous system because a heart excised from a living animal will continue to beat for a period of time. This leaves two possibilities: it may depend on transmission to cardiac muscle fibers of impulses from peripheral nervous tissue or it may depend on impulses made in the cardiac tissues. In the last theory only peripheral nerves and central connections would play a role in regulation.

Certain inorganic ions in correct portions are essential for the heartbeat. These include potassium, sodium, and calcium. The right amount of potassium is important. Too much or too little potassium can cause cessation of the hearbeat. Too little calcium or too much will also cause the heart to stop. Sodium is important in maintaining normal osmotic pressure, but even if osmotic pressure is maintained, an excessive loss of sodium will cause the heart to stop.

The human heart is divided into four chambers: the right auricle, the right ventricle, the left auricle and the left ventricle. In the normal heart the right and left portions of the heart are completely separated by walls. The right auricle and the right ventricle are connected by the atrioventricular orifice which is guarded by the tricuspid valve which consists of three triangular segments. The left auricle and left ventricle join through the left atrioventricular opening which is guarded by the mitral valve which consists of two triangular cusps. Venous blood enters the right auricle through the superior vena cava and the inferior vena cava. Then blood proceeds to the right ventricle to the pulmonary artery to the lungs. It returns from the lungs to the left auricle via the pulmonary veins. From the left auricle blood goes into the left ventricle and then into the aorta.

An understanding of the cardiac cycle has depended to great extent on two methods of study. Einthoven in 1903 described the string galvanometer which was very sensitive and quantitative. This invention led to a sudden increase in other studies of electrocardiography. This remains useful even though other principles such as the use of vacuum tube amplification have been replacing the string galvanometer. The electrocardiogram is simply a graphic tracing of the electric current produced by the contraction of cardiac muscle. The second method involves the study of the pressures within the heart chambers. Marcy pioneered this work by recording intraventricular pressures with the use of catheters threaded through large vessels into the various heart chambers and connecting a tambour type of manometer to record pressure. The introduction of the optical manometer allowed greater accuracy in recording intracardiac pressures. Electromanometers have proved even more satisfactory. By coordinating the pressure changes in the auricles and ventricles with the waves of the electrocardiogram, much can be learned of the events of the cardiac cycle.

The start of the P wave in the electrocardiogram indicates the beginning of excitation in the auricles.

Within a few hundredths of a second there is a rise in pressure in both the auricle and ventricle as the valves separating auricles and ventricles are open. When the contraction of the auricles begins to subside, the tricuspid and mitral valves close. The delay between closure of valves and ventricular contraction is due to delay of conduction in the auriculo-ventricular node and is represented by the P-2 interval on the electrocardiogram. The ventricles contract the intraventricular pressures rise. When the pressures are raised above the pressures in the aorta and pulmonary artery, their valves open and blood flows into the vessels. This is represented by the Q-RST part of the electrocardiogram. Gradually contraction ceases. As pressure in the vessels equals that in the ventricles, the valves of the vessels close. As tension lowers, ventricular pressure falls below that in auricles and the mitral and tricuspid valves open. During relaxation blood is stored in the auricles.

Cardiac output is the amount of blood expelled by one ventricle. Obviously, except for brief intervals, the output of the right and left ventricles must be equal. Therefore, determination of the output of either ventricle for a minute would give the volume of blood which flows through the entire systemic circulation in that minute. Two factors influencing cardiac output are stroke volume (output per heartbeat) and the heart rate. These two factors naturally depend on properties of the heart muscle and on extracardiac factors.

The Fick principle is used in determining cardiac output. It is as follows:

$$= \frac{\text{Cardiac output (liters per minute)}}{\text{Arteriovenous difference}} \quad \frac{\text{ml. of Oxygen used per minute}}{\text{of Oxygen in ml. per liter}}$$

Cardiac catheterization has enabled the taking of samples from the right auricle and ventricle. The consumption of oxygen can be determined by analysis of inspired and expired air.

Other methods such as inhaling foreign gases, the ballistocardiograph which records thrusts transmitted to the body as a result of impacts and recoils produced by the pushing of blood from the ventricles and the movement through the pulmonary and systemic systems, and roentgenographic methods have been used with some degree of success in studying cardiac output.

The arterial system. Blood flows from the heart into arteries. The large arteries normally are distensible and offer little resistance to the flow of blood. The more distal portions of the arterial system are called arterioles. At just what point any artery becomes an arteriole is not clear, but in the small arterioles the muscular walls are only slightly distensible so these arterioles offer a great resistance to blood flow.

Walls of arteries consist of highly distensible elastic tissue, muscle, and indistensible connective tissue. The arterioles have less elastic tissue. Thus a small change in pressure can markedly alter the potential volume of larger arteries. This is not so marked in arterioles.

Blood flow is opposed by frictional forces in all vessels. Frictional resistance to flow varies with the viscosity of the blood, diameters of the vessels, and the length of the vessels. The longer and narrower a vessel is, the greater is the frictional force. Resistance to flow occurs in all vessels, but it is greater in arterioles.

Flow resistance in arterioles depends to a great extent on the degree of contraction of the smooth muscle of their walls. Contraction appears to depend on a natural degree of tone, nervous control, and even chemical agents contained within the blood. The nerves which help control the diameter of arterioles are known as vasomotor. The effector neurons are called vasoconstrictor, whereas those which produce dilatation are called vasodilator.

Capillaries. The capillaries are the thin-walled vessels which connect arterioles to venules. In microcirculatory studies done by many investigators it has not been easy to determine just where the arteriole ends and the capillary begins or to know for certain just where the capillary ends and the venule commences. Smooth muscle cells do not end abruptly in the microcirculation. They tend to diminish until they disappear. One might wonder if a small smooth muscle cell is effective in contracting a blood vessel wall. In some preparations as in the circulation of the bat's wing, precapillary sphincters of smooth muscle have been described. Some feel that all of the exchange between blood and tissue occurs at the capillary, but other are certain that some exchange involves the venules. How material crosses the capillary wall is not fully understood. Fluids may be transferred according to the Starling hypothesis which suggests that the direction and rate of transfer between plasma in the capillaries and fluid in the tissue spaces rest on the osmotic pressure of plasma protein, on the hydrostatic pressure on each side of the capillary wall, and on the capillary wall which acts as a filtering membrane. Landis measured the hydrostatic pressure in capillaries of living frogs and humans. His data have given support to Starling's hypothesis. Additional support has been derived from the experiments of Pappenheimer and Soto-Rivera who performed perfusion experiments on amputated legs of mammals. The isolated limbs were perfused with blood. The movement of fluid from plasma to tissue fluid or vice versa was determined by gain or loss of weight of the limb. They found a number of values of arterial and venous pressures at which the leg remained at a constant weight. They found that the mean hydrostatic capillary pressure at any constant limb weight equalled all pressures opposing filtration.

The Starling hypothesis and supporting data led to the idea that water and water-soluble molecules pass the capillary wall through pores. It was felt that these pores were in the intracellular substance between the endothelial cells of the capillary wall. With the use of the electromicroscope some investigators have failed to find any intracellular substance between capillary endothelial cells and have found capillary endothelial cells overlapping each other. They favor the process of pinocytosis by which fluid is taken into small vesicles at the luminal surface and liberated into the perivascular tissue fluid. Others, with electron microscopy, have suggested that pores are present.

Only a portion of the available capillaries possess active blood flow at any one time. Only when a tissue or organ is in full activity does active blood flow occur in most capillaries of that specific tissue or organ.

There is considerable disagreement as to the mechanism concerned in the opening and closing of capillaries. Some feel that capillary walls contract actively, but others feel that the endothelial cells of the walls of capillaries have a certain elasticity and can contract passively on their contents. This issue is not settled, but most observers agree that blood flow in any set of capillaries shows some degree of intermittency.

To date, any extensive nerve supply to capillaries has not been satisfactorily demonstrated. It may exist, but for the present three forces appear to regulate capillaries. Hydrostatic pressure and colloid osmotic pressure play a role. An elastic force due to simple distention and the natural tone of the capillary walls may also help.

Arteriovenous anastomoses. In the skin direct connections between arterioles and venules have been demonstrated. These connecting vessels have been shown to contract actively and would appear to be under nervous control. They have been shown to contract during cold exposure and to dilate when heat is applied. This certainly suggests that in the skin these vessels are important in temperature control. These shunts have also been noted in the lung, kidney, intestines, spiral ligament of the inner ear, and the liver. What their functions are in these inner organs is not quite evident, but one might consider their importance as related to regulation of pressure.

Sinusoids. In the liver, connecting the portal venules and central venules, are vessels called sinusoids. In the lining walls of these vessels are cells which participate in phagocytosis. The splenic sinusoids appear to connect directly to splenic arterioles or capillaries and to splenic venules. The splenic sinusoids appear to have a function in the storing cells of the blood for varying periods of time.

The venous system. The flow of blood through venules and veins does not depend on hydrostatic pressure alone. In the extremities, valves in veins, and motion of the muscles of the extremities play a role. The large veins of the abdomen have no valves. The return of blood in this area would appear to depend on tissue fluid pressure. Strong movement of the diaphragm in respiration may aid venous flow by raising the pressure within the abdomen. The venules and veins are supplied by both vasoconstrictor and vasomotor nerves. Contraction and dilatation of the walls of veins and venules affect the flow of blood in venules.

Lymph vessels. The lymphatic system consists of a branching arrangement of closed vessels. The lymphatics have an endothelial lining. These vessels appear in most tissues, and the lymph capillaries are about as numerous as the blood capillaries. Thus the absorption area in lymphatics is as great as that of blood capillaries. Lymphatic capillaries join to form larger lymph vessels which reach a lymph gland to form a sinus. Later they branch to form a network of lymphatic capillaries within the gland. Again these smaller lymphatics merge into an efferent vessel. These vessels may enter other lymph nodes, go into the thoracic duct, or into one of the smaller collecting ducts to the large veins. The larger lymph vessels contain both elastic and muscular tissue. All of the lymphatics are equipped with valves which only permit flow centrally.

The flow of lymph in the extremities is variable and depends on activity of the particular limb.

Blood capillaries in many areas must be injured rather frequently. It is known that capillaries will permit proteins to pass their walls during periods of anoxia. There is evidence that lymphatic capillaries are very permeable even to microscopically visible particles. When the blood capillaries are permeable to protein because of damage to their walls, frequently normal fluid balance is quickly restored. It would appear that the lymphatics play an important role in the described condition.

For a full description of the lymphatic system the reader is referred to Drinker and Yoffey.[1]

The pulmonary circulation. This system deserves special mention as it is in the lungs that blood takes on oxygen which is so essential in life. In the pulmonary artery as in the aorta, the mean pressure depends on the relation of cardiac output to flow resistance of the pulmonary arterial tree. The pulmonary artery divides rapidly into terminal branches which have thinner walls than corresponding systemic blood vessels. This leads to a much lower flow resistance than in systemic vessels. Another factor of interest in the pulmonary circulation is the fact that many of the pulmonary vessels are in areas where the pressure is less than atmospheric. The pulmonary capillaries, however, are exposed to pressure of air in the lungs. These two factors have made it difficult to measure and interpret the pressures of the pulmonary circulation in both experimental animals and man.

Experiments by several investigators suggest that the pulmonary circulation can act as a blood reservoir.

During inspiration the intrapleural pressure falls to lower subatmospheric pressure. Experiments have showed that this results in a similar reduction in both plumonary arterial and aortic pressures. During inspiration there is a greater venous return to the right heart.

Hemodynamics. Here will be considered some of the basic principles underlying the flow of blood through the various blood vessels. A Newtonian fluid is one which contains no dispersed material. Here the flow rate is strictly proportional to the force causing it to flow. Blood is a mixture of cells in complex material known as plasma. Blood, therefore, is not a Newtonian fluid. The flow of blood does not occur until the necessary minimal force is applied.

When the velocity of flow of a liquid in a tube exceeds certain values, it does not flow smoothly and steadily, but it becomes turbulent with eddy currents. Turbulence occurs usually in hearts with organic disease of valves. Many question turbulence in blood vessels, but some think that turbulence occurs even in the microcirculation.

Luminar flow suggests that a thin line of plasma at the vessel wall barely flows, but within the vessel each successive concentric cylinder flows faster so that the fastest flow is at the center. In blood the erythrocytes are in the center and flow faster than plasma.

Problems. The heart has one important function— that is to pump blood, and the larger blood vessels appear to act only as conduits of blood. In the microcirculation occur the exchanges between blood and tissue.

Obviously, much profitable research remains to be done on the heart and larger blood vessels, but research on vessels of the microcirculation should be pushed vigorously if only because of the paucity of understanding. For a background in microcirculation the reader is referred to Krogh's book[2] and to the Microcirculatory Conferences.[3, 4, 5, 6]

Experiments have no doubt been limited because of the difficulties of approaching the microcirculation. Newer technics, however, are becoming available. With the use of the microscope, living preparations, and high speed photography, Bloch has showed that red cells do not rush through the small arterioles, capillaries, and venules in straight lines. Instead, any individual red cell will wander about a small vessel like a small pleasure sailing boat in any Cape Cod bay. Bloch has also recorded red cells bouncing off the wall at the branching of small arterioles like a basketball off a backboard. Rappaport, Bloch, and Irwin[7] have described an electromanometer which will better enable experimenters to describe full pressure curves within vessels of the microcirculation. This manometer must be improved. In fact, manometry must be developed so that the hydrostatic pressure and colloidal osmotic pressure of a vessel can be measured simultaneously. Then the capillary wall must be studied anatomically, physiologically, and chemically. With the development of many different types of artificial membranes, such studies become more feasible.

As such data are secured, it should be possible to learn a great deal more of the rheology of blood. The task at the moment appear formidable for here there is a complex mixture known as blood flowing through a set of channels, the diameters of which change frequently. One might view the arterial system as a cone in which the various diameters are constantly changing. The capillaries appear to be cylinders which can open or close in various parts of any organ at any one instance. The veins also are cones in which diameters also shift from moment to moment. New technics are constantly being developed in the areas of physics, chemistry, and biology. Many may prove applicable.

The nervous control of the microcirculation deserves more attention. Further studies are now possible. The new methods of electroneurophysiology allow one to stimulate single nerve fibers and nerve cells. With tape recorders responses can be recorded and interpreted rapidly and thoroughly with the use of electronic computers.

Television applied to visualization of living circulation may prove useful. It is true that television technics will not improve resolution since this depends on the objective of the microscope. One of the difficulties of studying living circulation, however, has been securing sufficient light without altering the tissue under study; and clear pictures can be secured with less light with the use of television. Colored television may also prove useful in identification of various substances in flowing blood.

Conclusion. Diseases of the circulatory system continue to play a major role in the death rate in spite of some understanding and new methods of therapy. Invariably, disease is not conquered until the cause is thoroughly understood. The circulation is complex and not completely understood as yet. As data increase, a better understanding of circulatory disease is inevitable. It should again be emphasized that one part of the circulation depends so much on the other parts. No one part should be unduly stressed; for example, in heart failure one should consider the microcirculation as well as the heart. Research is increasing,

and more knowledge of the circulatory system will result.

JOHN W. IIWIN

References

Drinker, C. K. and J. M. Yoffey, "Lymphatics, Lymph, and Lymphoid Tissue." Cambridge, Harvard University Press, 1941.

Krogh, A. "The Anatomy and Physiology of Capillaries." New Haven, Yale University Press, 1922.

The First Conference on Microcirculatory Physiology and Pathology." Anatomical Record 120: 1954.

"Vascular Patterns as Related to Function." Baltimore, Williams and Wilkins, 1955.

"Factors Regulating Blood Flow." American Physiological Society, 1958.

"The Microcirculation." Urbana, The University of Illinois Press, 1959.

Rappaport, M. B., et al. "A Manometer for Measuring Dynamic Pressures in the Microvascular System." J. Appl. Physiol. 14: 651–655, 1959.

CIRRIPEDIA

An order of CRUSTACEANS comprising (1) freeliving barnacles and acorn-shells, (2) *Acrothoracia* resident in madreporarians and snail-shells inhabited by hermit-crabs, and (3) parasitic *Rhizocephala*. The relationship between these types, which differ so greatly as adults, is shown by their larval development. Their nauplii have three longer or shorter spines on their carapace, viz. one posterior median spine, and one anterior lateral spine on each side. Between the nauplius stages, and the adult, a pupa or "cypris" stage is inserted, the latter designation suggesting its external similarity to an ostracod. The larval stages do not demonstrate a close relationship with any other order of the crustacea, the cirripedia in reality holding a rather isolated position within the class. Pupae of barnacles and acorn-shells fix themselves to a substratum by means of their first antennae, and their fixation is strengthened by a secretion from "cement glands," probably functionally transformed antennary glands of other crustacea. After fixation the anterior part of cephalon between the antennal region and the mouth is in barnacles prolonged into a "stalk," whereas in acorn-shells it grows in width remaining short in comparison' with cephalothorax. Most barnacles and acorn-shells develop a skeleton of calcareous plates in the mantle inside the shell shortly after the pupa has settled; the shell is then shed, leaving the animal in its final shape. Acorn-shells, and most of the barnacles, are hermaphrodites, but several barnacles have dwarf males serving as accessory males, and in a few cases the sexes are separate, although the males are also then dwarfs. In *Acrothoracica* the development and the sexual conditions are known only in one species, which has separate sexes and dwarf males; practically nothing is known about species living in madreporarians. In *Rhizocephala,* parasitic in decapod crustacea, the nauplius and pupa stages are similar to those of other cirripedia, but in some species the larva is hatched in the pupa stage. Most species have separate sexes, but in a couple of genera neither males nor male organs have been found, and here the ova evidently develop parthenogeneti-

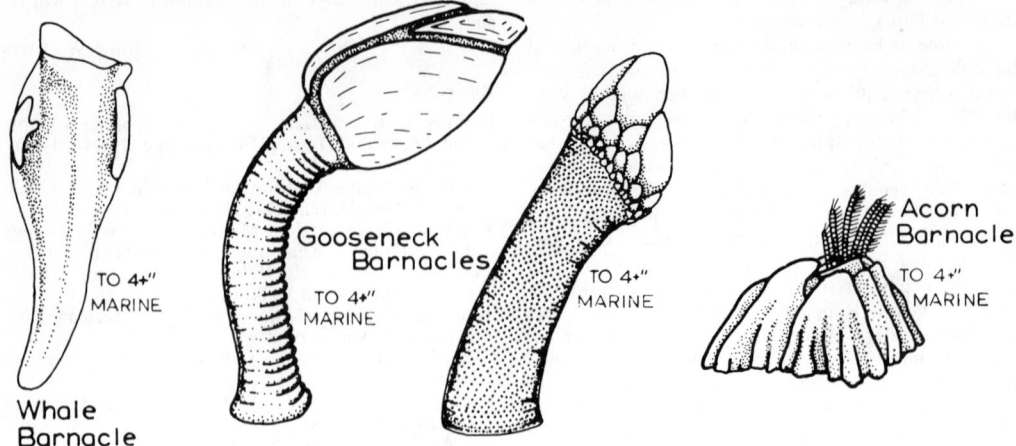

Whale Barnacle — TO 4+" MARINE

Gooseneck Barnacles — TO 4+" MARINE — TO 4+" MARINE

Acorn Barnacle — TO 4+" MARINE

(From Pimentel, "Invertebrate Identification Manual, "New York, Reinhold, 1967.)

cally. The males are dwarf and become sexually ripe in the pupa stage, which is their final phase of life. After having fixed itself at the base of a seta on its host, the female pupa sheds its organs of locomotion, together with their appurtenant body parts and the pupal shell. Only a group of embryonic cells (a *kentrogon*) penetrates into the interior of the host and develops a nutritive network (*rhizom*) within the body of its host. The female organs of reproduction break through the surface of the host as a tumor-like sac. The development from the pupa through a kentrogon makes it questionable, whether the adult stage ought to be considered as a complete individual.

HJALMAR BROCH

References

Darwin, C. "A monograph on the sub-class Cirripedia." London Ray Soc. 1851, Vol. I; 1854, Vol. II. (Reprint New York, Stechart, n.d.).
Krüger, P. 1940. Cirripedia, *In* Bronn's "Klassen und Ordnungen des Tierreichs." Vol. 5, Leipzig, Springer.

CLADOCERA

Cladocera, an order of the class CRUSTACEA, are composed of small organisms varying in length from around 15 mm. to .25 mm. The closest relatives of the Cladocera are members of the order Conchostraca, which show marked similarity to the former group in size and structure. In addition to the Cladocera the sub-class BRANCHIOPODA includes as the living representatives the orders Conchostraca, Notostraca, and the Anostraca. All orders of this sub-class are called *Phyllopods* because they possess leaf-like thoracic appendages.

The body of a typical cladoceran is generally short and compact with poorly defined segmentation. The head which is usually small and drawn out into a beaklike rostrum possesses a median compound eye and an ocellus. The thoracic appendages, four to six pairs, serve to strain food from water. Two pairs of antennae are located near the anterior end. Antennules, the first pair, are usually small and are primarily sen-

sory in nature. The antennae, the second pair, which serve as organs of locomotion, are biramous except in the female members of the genus *Holopedium*.

The unhinged bivaled shell which may cover the thoracic appendages provides for brood space for the eggs. In marine species and fresh water species of the division Gymnomera, the shell is reduced to a brood pouch only and does not cover the appendages.

Cladocera are cosmopolitan and are found chiefly in fresh water. *Evadne, Penilla,* and *Podon* are the only marine genera. A few members of the order have secondarily migrated into fresh water and then to brackish water, to which they have become adapted. Clado-

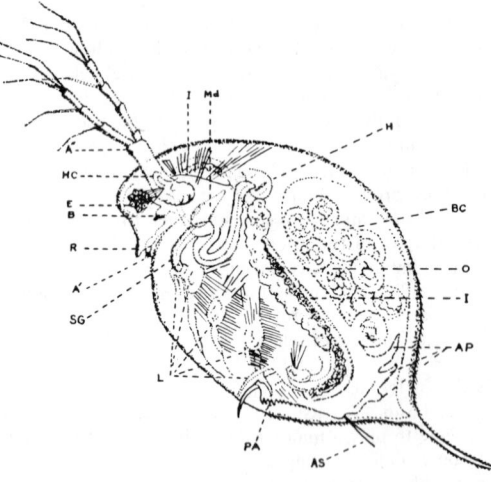

Fig. 1 *Daphnia longispina.* A', antennule; A", antenna; AP, abdominal processes; AS, abdominal setae; B, brain with optic ganglion and ocellus; BC, brood case with developing ova; E, eye, with three eye muscles of left side; H, heart with venous opening on side and exit in front; HC, hepatic cecum; I, intestine; L, legs; Md, mandible; O, ovary; PA, post abdomen with anal spines and terminal claw; R, rostrum or beak; SG, shell gland. (After Sars.)

cera occur in all types of lotic and lentic water-masses, but reproduce optimally in lentic shallow waters of protected bays. In such environments they employ their leaf-like legs to strain out and ingest the microscopic growth of plants as well as the detritus of the soil and digestible organic substances found in the bottom mud. Members of the division Gymnomera are predaceous and possess appendages that can be utilized to capture and hold the intended prey. This group comprises a major portion of the zoo plankton (see PLANKTON), and therefore is an important link in the food chain of aquatic environments. Its great reproductive capacity becomes evident when chemical and physical factors are favorable.

PARTHENOGENETIC reproduction alternates with syngamic reproduction. The *winter eggs* resulting from the latter can withstand adverse environmental conditions. They hatch after a period of rest, and when favorable conditions prevail. The newly hatched females then produce parthenogenetic eggs that in turn will give rise to females only, under favorable conditions, but may give rise to both males and females under adverse conditions. Later by syngamic reproduction winter eggs are again produced. These winter eggs are often provided with an extra shell called an *ephippium*. This pouch-like structure consists of two chitinous plates that covers the two or more enclosed eggs. The reproductive cycle in Cladocera in some instances is accompanied by a change in body form known as *cyclomorphosis*. The form of the head is usually affected. Confusion in taxonomy has résulted because of cyclomorphosis and a wide variety of names has been used. Recent taxonomic revisions on Caldocera are providing clarification in this respect, as well as information on the ecology and evolution.

Cladocera occur in a wide range of physiochemical habitats, thus they are widely distributed in space and time. Such distribution is undoubtedly enhanced by the resistant ephippial eggs, which are efficient means to insure wide zoogeographical dipersion. Few Cladocera are endemic to any particular region. Knowledge concerning factors that limit the distribution of cladocerans is incomplete although intensive investigations have been made in this regard. The use of shell-remains or exuviae, found on the bottom of bodies of water inhabited by Cladocera, is of great aid in studying the distribution of these organisms.

WOODROW H. JONES

References

Brooks, John L. "The Systematics of North American Daphnia." Mem. Conn. Acad. Arts and Sci. **13**: 1–180, 1957.
Ibid in Edmondson, W. T. ed., "Ward and Whipple's Fresh-Water Biology" New York, Wiley, 1959.
Frey, David G., "The Late-Glacial Cladoceran Fauna of a Small Lake." Arch. Hydrobiol. **54**: 209–275, 1958.
Jones, Woodrow H., "Cladocera of Oklahoma." Trans. Amer. Micros. Soc. 77: 243–257, 1958.
Wagler, E., "Crustacea (Krebstiere)" *In* Brohmer, P. *et al.*, ed., "Die Tierwelt Mitteleuropas," vol. 2, Lf. 2a, 1–224. Leipzig, Quelle and Meyer, 1937.
Kiser, R. W. A Revision of North American Species of the Cladoceran Genus *Daphnia*. Seattle, Washington, Edward, 1950.

CLIMAX

In the traditional sense, a climax is the relatively stable, self-perpetuating, terminal biotic COMMUNITY of a sequence of communities (= sere) developing in a particular location, such as a beech-sugar maple forest in Indiana, a *Stipa-Sporobolus-Bouteloua* grassland in Iowa, and a sagebrush community in Utah. As succession occurs in terrestrial environments, from the pioneer organisms of the first community to those of the climax, the dominant plants of each successive community gradually modify their physical environment, making it less favorable for themselves and more nearly optimum for the dominants of the next community, hence the succession of communities. The dominants of the climax, however, are not able to further modify the physical environment, and are able to live and reproduce indefinitely in the conditions they create, in equilibrium with the existing climate.

As a result of the sequence of physical changes, there is in the climax community a minimum ultraviolet light penetration, rate of evaporation and air movement, and a maximum relative humidity, soil moisture and humus. These physical conditions are created principally by the presence and biological activity of the dominant plants; the presence or absence of other organisms depends largely on their tolerance of these created conditions.

The stability of a climax is recognizable in a continual presence of young plants of the dominant species (i.e. maintained reproduction of the dominants), maintenance of the same average density of dominants, and probably an energy balance in which: (energy produced by green plants of the community) = (energy consumed in respiration by all organisms of the community). That is, there is no further accumulation of organic matter in the climax. In some seres the climax may have the greatest rate of energy production of all stages in the sere.

The specific type of climax that develops in a particular region depends on the climate of that area. Hence, climax vegetation types are good indicators of land climates. This dominant role of climate is apparent in the convergence of different seres to the same climax, as occurs in northern Indiana. Here, seres that begin on sandy, clayey or floodplain soils, or in ponds, all terminate in beech-sugar maple climax forest.

For a particular region, there is theoretically only one climatic climax. However, local conditions of soil, slope angle and exposure, drainage and air circulation may impede the development of the climatic climax and result in a different type of community of sufficient permanence to be called a climax, an edaphic (= local) climax. For example,[1] in southern Ontario, beech-maple forest occurs repeatedly on moderately level, moist but well drained areas, and is judged to be the climatic climax. However, oak-hickory forest occurs on warm dry south-facing slopes, hemlock-yellow birch on cool, north-facing slopes. These other types of forest may eventually give way to beech-sugar maple, but their stability in terms of human history is sufficient to rate them as edaphic climaxes. The moderate topography of northern Indiana has allowed convergence to a monoclimax situation, while the topography of Ontario is such as to maintain a poly-climax situation. In some situations the resistant nature of the rock that must be weathered to soil, or permanence of some other physical factor, may permanently prevent the development of the climax.

A climax community persists until the occurrence of some unusual change—natural cataclysm or human disturbance. Communities which persist in equilibrium with a continual disturbance are termed disclimaxes (=disturbance climaxes). Heavy grazing by cattle can maintain a mesquite-creosote bush-cactus disclimax where grassland climax would otherwise develop. The longleaf pine forests occupying much of the forested part of the coastal plain of southeastern United States is a fire dis-climax, in equilibrium with the frequently occurring fires. The speculated climatic climax of this region, a broad-leaved evergreen forest, such as in the "hammocks" of Florida, is very rare. The chestnut blight fungus, introduced through human agency, has killed off all large chestnut trees in the Appalachian mountains, converting certain areas of oak-chestnut climax to oak disclimax.

Climax communities are sometimes economically undesirable, in that valuable species of timber trees, game animals and fishes are often components of communities prior to the climax. In such instances human effort may be directed to preventing the development of the climax.

In the absence of unusual changes, a climax endures as long as the same climate prevails. As climate has changed in the past, climaxes have also changed, as can be reconstructed from pollen profiles. In Connecticut, beginning 10,000 to 12,000 years ago, there has been a sequence of climaxes[2]: pine-spruce (cold climate), spruce-pine-birch (cool), oak-beech (warm moist), oak-hickory (warm dry), to the present oak-hickory-chestnut-birch (cooler moist). Some relict examples of previous climaxes persist in locally favorable spots, as postclimaxes.

The life form and taxonomic identity of the dominant plants of climax communities are used as the bases for the biome type:biome:association system of classifying terrestrial communities. A biome type is an abstract grouping of all climaxes having dominants with a same particular life form, together with all the communities composing seres leading to these climaxes.

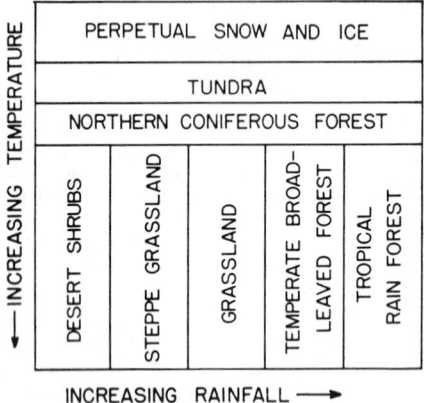

Fig. 1   Schematic representation of vegetation types in relation to macroclimates.

As shown in Figure 1, life form of the dominants: coniferous trees, deciduous trees, shrubs, grasses or some other major plant form, is a general climate indicator. Biomes, the geographically continuous subunits of biome types, are in turn divided into associations: abstract groupings of all climaxes (with their seres) having the same species of dominants. For example, the division of the North American deciduous forest biome into: beech-maple, oak-hickory, oak-chestnut, and other specific associations.

ROBERT M. CHEW

References

Odum, E. P., "Fundamentals of Ecology." 2nd ed. Philadelphia, Saunders, 1959.
Ibid "Ecology," New York, Holt, 1963.
Ooosting, H. J., "The Study of Plant Communities." 2nd ed. San Francisco, Freeman, 1956.

CLONE

Introduced in 1903 by Herbert John Webber as a new horticultural term and then spelled *clon* (cf. *Science* n.s. **18**: 501), the noun *clone* (together with the adjective *clonal*) has passed from horticulture and agriculture into botany, zoology and medicine without change of basic meaning but with wide divergence of application. Essentially a clone is an asexually produced population, of which all the members have been derived from one and the same progenitor exclusively by non-sexual (vegetative or parthenogenetic) multiplication. The word *clone* can thus be used to refer collectively to all the apple trees obtained by grafting from one tree, all the strawberry plants produced by runners from a single seedling, all the tapeworms arising from the segmentation of one, all the tissue (or group of cells) derived by simple mitotic division from one cell, and so on for an infinity of examples of vegetative descent from a common source. In 1903 Webber naturally did not foresee this extensive employment of his word but it has, nevertheless, come from his statement that '*clons* are groups of plants that are propagated by the use of any form of vegetative parts such as bulbs, tubers, cuttings, grafts etc. and are simply parts of the same individual.'

Implicit in this definition is the concept of the genetical identity of many physiologically autonomous individuals repeating exactly, as a consequence of vegetative descent, the genetical make-up, characters, behaviour and potentialities of their original progenitor. The introduction of such a convenient, short and distinctive term as *clone* has led to general recognition of its wide applicability. Its origin is a Greek word meaning a twig, spray or slip 'such as is broken off for propagation.' In the 19th century the lack of such a word, which would have helped to make evident the distinction between a sexually and an asexually produced population or between an individul as a physiologically independent unit and as the sum total of similarly independent members of a population forming the vegetative progeny of one individual, hampered discussion about the nature of the individual in biology. Nevertheless the clone-concept was clearly formulated as long ago as 1816 by the Italian pomolgogist Count

Giorgio Gallesio (1772–1839) in his *Teoria della Riproduzionne vegetale* (1816). Here (pp. 32–33) Gallesio said: 'I call an individual not only the plant which grows alone on its rootstock and enjoys the life established by nature for its kind; I call an individual also the collection of all the plants which originate from a single embryo and which consequently form but a single plant multiplied without changing, *** by grafting *** by layers or cuttings, *** and this one plant thus prolongs its own life as that of its kind, varying according to the places and conditions of its existence but carrying always in itself the principles of organization received as its conception, this being the one and only origin both of the individual which dies on the root with which it was born and that which renews its life for the millionth time through grafting and layering. This individual, though multiplied without end, will always carry in each of the numberless subdivisions of its being the same characters and the same general appearance, which it received at its birth'. Such a statement rested upon the empirical experience of gardeners through the ages who had learned that only by vegetative propagation could the special features of certain cultivated plants be maintained unchanged. Its scientific explanation had to await the development of cytology and genetics towards the end of the nineteenth century. The members (ramets) of a clone were then seen to owe their uniformity to their possession of identical sets of chromosomes reproduced by mitotic division.

The term *clone* was thus first applied to manmade populations of cultivated plants, resulting from propagation by grafting, budding, layering or rooting of cuttings, by division of rootstocks or rhizomes, by the separation of bulbs, bulbils or offsets. Many of these methods of propagation simply repeat in the garden processes of vegetative spread found among wild plants. Thus it is now recognized that some natural populations of great age and wide extent are really clones. A clone derived from a male member of a dioecious species will consist of only male plants; an example is the widely cultivated Lombardy poplar (*Populus nigra* var. *italica*). Similarly a clone from a female plant will be entirely female; an example is the Canadian pondweed (*Elodea canadensis*) extensively naturalized in Europe. Ramets of a clone derived from a self-sterile plant will not be interfertile; many examples of this are to be found among fruit trees. The ability to multiply asexually by the production of bulbils or young plants on the inflorescence, or of seeds by apomixis, has caused sterile forms differing in only a few minute characters to hold these constant over very long periods of time and has thus enabled them to spread over wide areas, thereby forming populations very difficult to classify in such genera as *Alchemilla, Crataegus, Hieracium, Taraxacum* etc. Thus an agamospermic Scandinavian dandelion, *Taraxacum gotlandicum*, on the basis of its ecology and isolated stations in Norway and Sweden would seem to have originated about 10,000 years ago (cf. Wendelbo in *Nytt Mag. Bot.* 7: 161–967; 1959). Calculated to be even older are certain apomictic clones of *Alchemilla* which may have originated 35,000 years ago (cf. Samuelsson in *Acta Phytog. Suecica* 16; 1943). Clones of great age may also be formed by ordinary vegetative growth, the young shoots or rhizomes pushing into new territory and forming plants which become isolated by the death of the older connecting growth. Thus the mile-long self-sterile colonies of box huckleberry, *Gaylussacia brachycera*, in the Losh Run ravines of Pennsylvania, with an average spread of about six inches a year, have taken about 5,000 to 10,000 years to reach their present size and seem to form part of one very much older clone (cf. Moldenke in *Wild Flower* 33: 4–8; 1957).

The use of the term *clone* by plant geneticists from 1903 onwards led animal geneticists between 1922 and 1929 to adopt it in place of the expressions 'strain' and 'pure line' as applied to the asexual progeny of a single individual. Thus N. S. Jennings when dealing with the genetics of the Protozoa (*Bibl. Genetica* 5: 105–126; 1929) defined a single clone as comprising the 'products of fission of a single individual' or, in other words, as 'all the individuals descended by uniparental reproduction from a single individual.'

Although the possibilities of variation within a population consisting of one clone are much less than those within an interbreeding population of many sexually produced individuals, nevertheless mutations taking place within them may result in some long-established clones being by no means homogeneous. Although the members of a particular clone may compete ecologically between themselves, they are usually isolated in space and physiologically independent, and may not be distinguished on this account from a population of sexually produced individuals. In cancer research the term *clone,* however, denotes a group of cells forming a tissue (a cell population) which has been created from one cell by mitotic division. A 'multiclonal mosaic' is thus a tissue compounded of cells derived from the multiplication of several original cells. The first use of the words *clone* and *clonal* in this manner may be by Hauschka in *J. Nat. Cancer Inst.* 14: 723 (1953) where he states that 'exact chromosome analyses of diverse mammalian tumors support the conclusion that most cancers are multiclonal mosaics of altered karyotypes.'

A further derivative of the term *clone* is *clonotype* introduced by W. T. Swingle in 1912 (*J. Washington Acad. Sci.* 2: 345) to designate 'a specimen taken from a vegetatively propagated part of the individual plant from which the type specimen was obtained.' Swingle (op cit. 2: 221; 1912) also proposed the term *merotype* to designate a specimen obtained from the original individual itself, in Swingle's words, 'a part of the individual organism that furnished the type specimen of a new species, such part usually containing organs homologous to those represented in the type specimen.'

In 1954 (*Cactus & Succ. J. Amer.* 26:82) P. C. Hutchison independently coined the same term *clonotype* to apply to 'an herbarium specimen prepared from the same plant from which the holotype was prepared or to herbarium specimens prepared from members of a clone which was vegetatively reproduced from the plant from which the holotype was prepared' or to the living plant itself.

The term *clone* should not be confused with *cline* introduced by Julian Huxley in 1938 (*Nature* 142: 219) to denote 'a gradual and continuous change in a character over a considerable area as a result of its adjustment to changing conditions, whether geographical (geoclines), ecological (ecoclines) or of others' (E. B. Ford, 1954).

WILLIAM T. STEARN

References

Stout, A. B. "The clone in plant life." J. New York Bot. Garden **30**: 25–37, 1929.
Stearn, W. T., "The use of the term 'clone'." J. R. Hort. Soc. London **74**: 41–47, 1949.

CNIDARIA

Cnidaria are a phylum of lower invertebrates constituting the main division of the old Phylum Coelenterata which also included the CTENOPHORA. The Cnidaria take their name from the *cnidae* (from the Greek for "nettle"), unique cell-organs of microscopic size, from which threads are discharged by an excitatory process triggered by external stimuli. The best known cnidarians are the hydras, various colonial zoophytes (class HYDROZOA), the jelly-fishes (class SCYPHOZOA), sea anemones and corals (class ANTHOZOA). The vast majority of cnidarians are marine and in that environment they are conspicuous in the fauna of every habitat. With approximately 10,000 species the Cnidaria rank as a major phylum.

The Cnidaria have been described as multicellular animals at the "tissue grade of construction." This means that in general they lack organs. Certain systems are well defined, e.g., the nervous system and the digestive epithelium, but there is no brain, no stomach, no liver, etc. Traditionally these animals have been described as "diploblastic," i.e., composed of two cell layers like a gastrula. The two cell layers, epidermis and endodermis, are separated by a noncellular fibrous layer, the mesoglea. This layer is not always acellular, however, and in the more advanced groups it may contain many cells, which are generally said to have moved in from the epithelial layers. In such cases it can be misleading to regard these animals as two-layered in any functional approach to structure.

Animals belonging to the Cnidaria are generally either polyps or medusae. *Hydra* is an example of a polyp. It has a cylindrical body with an adhesive base attached to the substratum and a ring of tentacles surrounding an aperture which serves as both a mouth and an anus. The single internal cavity is a *coelenteron*. Polyps commonly associate in colonies in which the individuals are joined together by common tissues and a common coelenteric cavity. The larger jelly-fish provide the most familiar example of the medusoid body. It is essentially umbrella-like with a single aperture, the mouth, placed in the center of the sub-umbrellar surface, often raised up on a manubrium. In these animals the *coelenteron* is more restricted, extending out to the periphery of the umbrella or bell only as a system of canals.

Some polyps reproduce sexually, producing eggs and sperm in single or separate individuals with direct development into new individuals after fertilization (e.g., most sea anemones). The colonial hydroids, however, frequently show an alternation of polyp and medusa in the life history of a species, the medusa being a sexual and the polyp an asexual phase. These sexual and asexual phases or generations in the life history do not always appear as free-living medusae. They frequently remain attached and in the more extreme cases, the medusoid body is hardly evident, the polyp producing eggs and sperms from so-called *gonophores*. Most cnidarians reproduce also asexually by budding off new individuals, usually in the polypoid phase. It is by such buds remaining joined together that colonies arise. In colonies the individuals are not always alike and may have specialized functions associated with the capture of food, ingestion, water circulation or reproduction. These polymorphic individuals take on the functions of organs in so-called higher animals. Thus the Cnidaria provide striking examples of POLYMORPHISM and especially in those species in which the polypoid form, basically sessile and asexual, alternates with the pelagic, locomotory and sexual medusoid form.

Some species are still known only as polyps and others only as medusae. In the past, before certain medusae were associated with their corresponding polyps, the two forms were given different systematic names. Thus two parallel systems of classification exist in certain hydrozoan orders.

Several types of cells can be recognized in the tissues of typical cnidarians. The epithelia lining the internal and covering the external surfaces are generally rich in gland cells but their bases are typically extended into contractile processes which form sheets of muscle. These cells are therefore known as "musculo-epithelial" cells. The so-called "muscle tails", which are simple in a hydra, can become fairly massive aggregates and take on the appearance of definitive muscles (e.g., the sphincters which close up the crown of a sea anemone) by growing out into the mesoglea.

Nerve cells with ramifying fibers underlie the epithelia. The entire system has been described as a nerve net, but the fibers which run in network fashion from the typical bipolar and multipolar nerve cell bodies do not fuse; they form junctions with one another of a typical synaptic type. Some signs of nerve fibers grouped in bundles and forming tracts or cords are found in jelly-fish and anemones but they lack nerve centers or trunks of any kind corresponding to central nervous systems of most other multicellular animals. Sensory cells exist in the epithelia and except for some jelly-fishes, which possess simple eyes (occasionally more complex) and balancing organs, these animals rely on scattered sensory cells, especially chemo- and mechano-receptors for information about the environment.

Interstitial cells which have developmental functions and differentiating capacities like mesenchyme cells in other animals, are found between the epithelial cells. They multiply and give rise to various other cell types, e.g., during regeneration, or in the development of gonads, and they give rise to the *cnidae* by a special developmental process.

The *cnidae*, or NEMATOCYSTS, are produced by special cells, *cnidoblasts* which develop from interstitial cells and migrate to their final positions in the epithelia. The fully developed *cnida* ends up as a capsule filling the interior of the cell which remains as a thin nucleated envelope. The final product is an extremely elaborate apparatus equipped with a sensory trigger and an involuted sac or more commonly a thread. This thread may be simple but more often it is complex, equipped with barbs and having characteristic shapes, lengths and patterns. Weill surveyed the *cnidae* occurring in 119 species in the phylum and he showed that they fall into about 18 classes according to these structural features. These classes follow certain systematic patterns in their distribution so that systematists frequently use the characteristics of the *cnidae* to identify species and to determine relationships of various groups with one another.

The *cnidae* are undoubtedly the major adaptive feature of the Cnidaria to which their success as a phylum of lowly invertebrates can be attributed. Food capture is based mostly on the use of *cnidae* to ensnare and often to poison other organisms which may be considerably larger. They enable the Cnidaria to be highly successful carnivores even though they are sessile or pelagic in their habits, not able to hunt for food. Most other sessile animals feed by setting up currents from which food particles are strained. *Cnidae* are also used for defense against marauders, for adhesion in certain locomotory acts or behavior patterns, as in *hydra* somersaulting or sea anemones forming associations with commensal partners such as hermit crabs.

Many features of the physiology of the *cnidae* are obscure. Their small size (usually less than 100 microns) presents problems for the investigator. Excitation causes the capsule to discharge the neatly coiled thread, a process which turns it inside out presumably by an intake of water creating a sudden increase of intracapsular pressure. The discharge is explosive and in the process the entire capsule is commonly thrown off altogether from the animal. Once used they cannot be used again but new *cnidoblasts* soon appear to replenish those lost.

In the discharge of the *cnida,* a sensory process on the surface, the cnidocil, is involved. Simple experiments showed that absolutely clean objects or food substances in solution by themselves would not bring about discharge. However, any object slightly contaminated by organic matter or preceded by food substances in solution caused discharge at once. This showed that a combination of chemical (sensitizing) and mechanical (exciting) stimuli was necessary for discharge. This has the advantage of avoiding indiscriminate firing.

Many *cnidae* deliver poisons which are highly toxic to such prey as crustaceans, and even fish. Men have been incapacitated and even killed by massive stings from certain Cnidaria, e.g., the Australian sea wasp (Cubomedusae) and the Portugese man-of-war (Siphonophora). The study of reactions to poisons of certain sea anemones by Charles Richet (Nobel Laureate, 1913) contributed much to the discovery of *anaphylaxis.* Attempts to identify the poisons of the *cnidae* have shown that they are proteins sometimes accompanied by painful histamine-releasing substances such as quaternary-ammonium or indole derivatives.

In general the Cnidaria are divided into the classes HYDROZOA, SCYPHOZOA and ANTHOZOA because: (a) polyp and medusa are both present in the HYDROZOA; (b) the medusoid phase is dominant in the SCYPHOZOA but with a reduced polypoid phase in which asexual reproduction occurs by strobilation; (c) the polyp alone occurs in the ANTHOZOA. Other distinguishing features of the classes are: (1) the hydrozoan *coelenteron* is not subdivided by partitions, the medusa has a shelf (*velum*) inside the rim and the gonads are located in the epidermis; (2) the scyphozoan *coelenteron* has special chambers (*gastric pockets*) with septa between, and gonads are endodermal; (3) the anthozoan *coelenteron*

is subdivided by vertical partitions (*mesenteries*), a stomodaeum is present and gonads are internal.

Studies on the physiology of Cnidaria have centered on the nervous system, the most primitive in existence. The hopes that it would show "elementary" features were disappointed; its basic physiological features were the same as those of nerves in higher animals. However, these studies have provided some particularly interesting examples of general phenomena which are not peculiar to the Cnidaria, e.g., neuromuscular facilitation in sea anemones, non-nervous conduction of excitation in siphonophores, and spontaneous rhythmic activity detected by mechanical and electrical systems of recording in *Hydra,* medusae and sea anemones. Some sea anemones display striking and complex behavior patterns in response to specific stimuli, e.g., certain sea anemones that "swim" when touched by some other species of invertebrates, and other sea anemones that live as commensals with crustaceans and molluscs.

The physiology of corals, including the deposition of calcium, are topics of cnidarian research of great general interest. There is disagreement still about the role of algal symbionts (found in zoantharian reef-building corals in vast numbers), in the nutrition or waste disposal or corals and possibly in the secretion of calcium carbonate.

Cnidaria interest ecologists largely as the basic organisms in the coral reef community in which so many other organisms make their homes, often with spectacular adaptations. Also the partnerships between hermit crabs and sea anemones are often cited as classic examples of commensal associations. Anemones involved in these partnerships, the genera *Adamsia* and *Calliactis,* gain transport from living on the shells but the advantages to the crabs are problematical, though defense, camouflage and food capture have been suggested. It is now known that some partnerships are established solely by the anemones climbing on shells and others by the crabs' activities in transferring anemones to their shells. Probably in most cases both the crab and the anemone participate, the crab by stroking, prodding or tapping the anemone in the manner specific to the species pair involved, and the anemone by reacting to the shell or to the crab's manipulations in its own specific way.

The phylogeny of the Cnidaria has been much debated. Their two-layered architecture has suggested that the group is at the gastrula stage of organization and therefore ancestral to the rest of the Metazoa. Impressions of medusae of Cambrian age and other evidence demonstrate the great antiquity of the phylum.

Most zoologists have regarded the HYDROZOA as the most primitive and closest to the ancestral Cnidaria and have discussed at length whether the polyp or the medusa is the prototype of the original cnidarian. If it is assumed that one of these hydrozoan forms was the original cnidarian, schemes for the subsequent evolution of the SCYPHOZOA and ANTHOZOA can be advanced without too much difficulty, although various details have been argued pro and con since the late nineteenth century.

Recent work has emphasized two aspects which did not enter into the older discussions on phylogeny. The types of *cnidae*, not unreasonably in view of their basic role in the group, are now regarded as indicators of relationships. This approach has completely changed the thinking of many systematists. The more problematical development has been the view advanced by Hadži that the flatworms are the most primitive true Metazoa and that the CTENOPHORA and the Cnidaria have been derived from them. This point of view has aroused interest but has gained few supporters. With so many problems to investigate in the known living and extinct Cnidaria, it is perhaps understandable that interest is focussed on them rather than on surmises about happenings that might have taken place far back in Pre-Cambrian times.

D. M. Ross

References

Hadži, J., "The evolution of the Metazoa," London, Pergamon Press, 1963.
Hyman, L. H., "The Invertebrata. Protozoa through Ctenophora," New York and London, McGraw-Hill, 1940.
Pantin, C. F. A., "The elementary nervous system," Proc. Roy. Soc. London B. **140**, 147–168.
Rees, W. J., ed., "The Cnidaria and their evolution," Symp. Zool. Soc. London, No. 16. pp. xviii × 449, 1966.
Ross, D. M., "Behavioural and ecological relationships between sea anemones and other invertebrates," Oceanogr. Mar. Biol. Ann. Rev. **5**, 291–316, 1967.
Weill, R., "Contributions à l'etude des cnidaires et leur nematocystes," Trav. Stn. Zool. Wimereux **10, 11**, 1934.

COELACANTH

Any of a group of primitive fishes of the subclass CROSSOPTERYGII, characterised chiefly by the presence of an ossified air-bladder, pediculate fins with well developed internal skeleton, fin rays superficially ossified, notochord un-ossified, and thick, bony scales.

Until recently, it was thought that Coelacanths became extinct at the end of the Cretaceous. However, in 1938, a living form, *Latimeria chalumnae,* was discovered off the mouth of the Chalumna River, South Africa, and was described and named by the South African ichthyologist, Prof. J. L. B. Smith. In 1952, a second specimen was discovered at the Comoro Islands north of Madagascar. Since then, about a dozen more have been brought to the attention of scientists. Their anatomy is being described in great detail in a series of monographs by the French scientist J. Millot, and his co-workers. Accounts of Comoro native fishermen indicate that they catch several Coelacanths every year.

The oldest known fossils of Coelacanths date from the Devonian, nearly 300 million years ago. Others have been recovered from rocks of subsequent periods through the Cretaceous, these last about 90 million years old. The fossils are more or less world wide in their distribution, being known from such diverse localities as England, New Jersey, Brazil. The earliest Coelacanths were marine, those of the Carboniferous lived in fresh water. Later forms reverted back to shallow seas. It has been suggested that the absence of their fossils after the Cretaceous may indicate that Coelacanths retreated to deep water, where sedimentation is slow and the probability of fossilization remote, but this seems unlikely.

Modern Coelacanths are known from relatively shallow water, about 300 fathoms or less, over a rough, rocky bottom. They appear to tolerate a wide range of temperature, and a salinity range between 35.10 and 35.25%. They are piscivorous, known to eat such forms as *Ruvettus, Promethichthys,* and *Myctophum.* Thus far, only a Nematode has been recorded as a parasite.

Contrary to the situation in the related Dipnoi and Rhipidistia, the Coelacanth air bladder never developed into a functional lung. There is no connection between the external nares and the mouth. The air bladder is encased in bony plates, a structure of unknown function which may possibly serve as a resonator, increasing the perceptibility of sound waves. The notochord is not ossified, forming, in *Latimeria*, a fibrous tube. The fins of Coelacanths are borne on muscular, scaly lobes rather than arising fan-like from the body as in teleosts. Each lobe is equipped with skeletal elements, which are not, however, homologous with the skeletal elements of the limbs of higher vertebrates.

Despite the popular conception of Coelacanths as "missing links" that may be "millions of years old," the living individuals are actually no older than other fishes of similar size—perhaps 15 or 20 years. As for the "missing link," this, too, is a misconception. Although

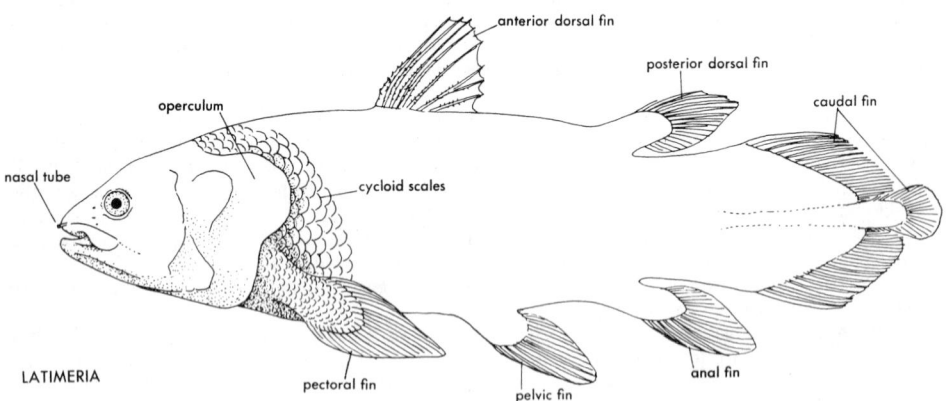

Fig. 1 The living fossil *Latimeria*, a coelacanth crossopterygian. (From Jollie, "Chordate Morphology," New York, Reinhold, 1962.)

ancient relatives of the Coelacanths, the Rhipidistia, gave origin to land animals in the middle or late Devonian, it is perhaps the most remarkable feature of the Coelacanths that throughout their long history they have never given rise to any other group.

JAMES E. MORROW JR.

References

Millot, J., "Le troisième coelacanthe" Naturaliste Malagache, 1er Suppl., 1954.
Millot, J. and J. Anthony, "Anatomie de *Latimeria chalumnae* Part I." Paris, Editions Centre National de la Recherche Scientifique, 1954.
Smith, J. L. B. Nature, 171:99–101, 1953.
Watson, D. M. S. Ann. Mag. Nat. Hist. 8: 320–327, 1926.
White, E. I., Ann. Rept. Smithsonian Inst. 1953: 351–360, 1954.

COELENTERATA see CNIDARIA

COLEOPTERA

The Coleoptera (*coleos,* sheath; *pteron,* wing) comprise an endopterygote (holometabolous) order of the ARTHROPOD class INSECTA. With about 275,000 species they constitute the largest order in the ANIMAL KINGDOM and nearly 40 per cent of all species of insects. All Coleoptera may be called beetles, but the Curculionoidea are frequently distinguished as weevils and the Stylopoidea as stylopids. In some classifications the latter group is given separate ordinal status, as Strepsiptera.

Early theories deriving the Coleoptera from blattoid or ORTHOPTEROID ancestors have proved unsatisfactory because of their failure to account for the endopterygote characters of the order. Within the division Endopterygota the Coleoptera most closely resemble and are regarded by many authorities as most closely related to the NEUROPTERA. Adults of both orders primitively possess trilobed male copulatory organs, tubular testes, polytrophic ovarioles, and relatively complete wing venation. In addition, there are a number of larval similarities, the most striking of which is the presence of lateral abdominal appendages in the neuropteran suborder Megaloptera and many aquatic adephagid beetles. Evidently both orders arose early in the evolution of the Endopterygota. However, inasmuch as most if not all of the characters which they share are likely primitive for the endopterygote line, the exact relationship of the two orders to each other is uncertain. Fossil remains of beetles are faily abundant but consist mainly of elytra and offer little evidence of phylogeny. The first fossils appear in the Permian; by Jurassic times most of the modern superfamilies and many of the modern families had evolved.

The most recent classification divides the Coleoptera into four suborders, 22 superfamilies, and 162 families. The suborder Myxophaga includes the Lepiceridae, Sphaeriidae, and Hydroscaphidae; these are all tiny, poorly-studied beetles usually placed in the Polyphaga, of which the suborder is perhaps an early branch.

The suborder Archostemata contains the Cupedidae, which retain the greatest number of primitive characters of any of the families of beetles, and the Micromalthidae. The suborder Adephaga includes several thousand species divided among 9 families, the more conspicuous of which are the Carabidae (including Cinindelidae), Dytiscidae, Haliplidae, and Gyrinidae. The remaining families of beetles are placed in the suborder Polyphaga. Within this suborder a distinction is made between the Haplogastra, in which the sternum of the second abdominal segment is represented by lateral plates, and the Cryptogastra, which lack the sternum entirely (rare exceptions). Haplogastran superfamilies are the Hydrophiloidea, Histeriodea, Staphylinoidea, Scarabaeodea, and Dascilloidea. Cryptogastran superfamilies are the Eucinetoidea, Byrrhoidea, Dryopoidea, Buprestoidea, Rhipiceroidea, Elateroidea, Cantharodea, Dermestoidea, Bostrychoidea, Cleroidea, Lymexyloidea, Cucujoidea (including Nitdulidae, Coccinellidae, etc., and the "Heteromera"), Chrysomeloidea (including Bruchidae and Cerambycidae), Curculionoidea, and Stylopoidea.

Few orders of insects compare with the Coleoptera in variety of structural adaptations. In most beetles the integument is heavily sclerotized and the tagmata are closely co-adapted, although in the Cantharoidea and some other specialized forms there has been an evolutionary return to a more flexible exoskeleton and looser body construction. The head, which is usually prognathous, lacks an epicranial suture; in most Curculionoidea it is produced into a distinct rostrum. Mouthparts are of the chewing type; the maxillary palpi have three to five segments and the labial palpi two or three. Ocelli are absent except in some Staphylinidae, Silphidae, and Dermestidae. Compound eyes are lacking only in some cavernicolous and parasitic

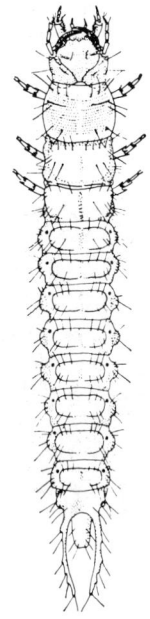

Fig. 1   Larva of *Laemostenus terricola* Herbst, family Carabidae (redrawn from Böving and Craighead).

forms. Antennae are usually 11-segmented; they vary greatly in form. The prothorax is free and has a large, undivided notum; primitively, notopleural and sterno-pleural sutures delimit the prothoracic pleurae, but the former suture is lacking in the Polyphaga and the latter in the Curculionoidea. The meso- and meta-thorax are fused; the mesocutellum is normally exposed. The most characteristic feature of adult beetles is the thickened fore wings or elytra, which ensheath the dorsum of the meso- and metathorax and abdomen. Elytra retain no venation; in most beetles they extend to the tip of the abdomen, although in the Staphy-linoidea and in some other forms they are truncate. The hind wings are folded beneath the elytra in repose, and many of their venational characteristics are adaptations for folding. As a rule, the radial and medial veins are obliterated basally and the apex has few if any veins; the presence of cell 2nd M-Cu (ob-longum) is characteristic of nearly all beetles except the Polyphaga. The abdomen normally and primitively lacks the first sternum; there are generally seven or eight pairs of abdominal spiracles. Male copulatory organs consist of a median lobe or aedeagus and paired lateral lobes which attach to a plate-like basal piece; many authorities doubt that these structures are homologous with the copulatory organs of the lower insects. Female copulatory organs consist of a membranous tube formed in part by the ninth abdominal segment and terminated by paired coxites and styli. Genital structures in both sexes are extensively used taxonomically. Malpighian tubules number four to six. There are three thoracic and eight abdominal ganglionic centers in the more primitive groups, with considerable concentration of centers among various Polyphaga.

Larvae are primitively campodeiform; grub-like forms have evolved in many groups of Polyphaga. As in other Endopterygota, larvae have six or fewer ocelli on each side of the head, no more than four segments in the antenna, and only one tarsal segment. The mouthparts are similar to those of adults. Thoracic legs are present in most beetles except the Curculionoidea and

many Buprestidae and Cerambycidae. In the Archostemata and Adephaga the tarsi generally have two claws; in the Myxophage and Palyphaga there is but a single claw, which is fused to the tarsal segment. Paired processes or urogomphi arise from the ninth abdominal segment in many larvae, and some aquatic species have respiratory filaments on other segments; otherwise, abdominal appendages are lacking. Pupae are exarate.

There is more ecological diversity in the Coleoptera than in any other order of animals. Within a single superfamily or family species often differ greatly in their habits and niche requirements, and this is often true also of larvae and adults of the same species. Most beetles are terrestrial, but the Dytiscidae, Hydrophilidae, Gyrinidae, Dryopidae, and several other families are aquatic, at least as larvae. Many beetles are found beneath bark, and it has been suggested that elytra evolved as an adaptation for life in this environment. A large number of species inhabit decaying material and other debris, feeding either on decomposing organic substances or on fungi. Larvae of the Archostemata, Lymexyloidea, Bostrichoidea, Cerambycidae, Buprestidae, and many Curculionoidea are wood borers, and many of them have symbiotic organisms which carry on the digestion of cellulose. As a general rule, the Chrysomelidae feed on the foilage or roots of vascular plants. The Dermestidae are the notable scavengers among the beetles. The Cantharoidea, Cleroidea, Histeriodea, and most Adephaga and Staphylinoidea are predacious on other insects, both as larvae and adults. Larvae of Meloidae, Rhipiphoridae, Stylopoidea, and some Staphylinidae parasitize other insects; adults of the Leptinidae are external parasites of mammals. Myrmecophily or termitophily occurs in several families, including the Paussidae, Pselaphidae, Staphylinidae, and Limulodidae. Regardless of their habits as larvae, a large number of beetles visit flowers as adults.

RICHARD B. SELANDER

References

Crowson, R. A., "The natural classification of the families of Coleoptera." London, Lloyd, 1955.
Jeannel, R., "Ordre des Coléoptères," in Grassé, P.-P., ed., "Traité de zoologie," vol. 9. Paris, Masson, 1949.

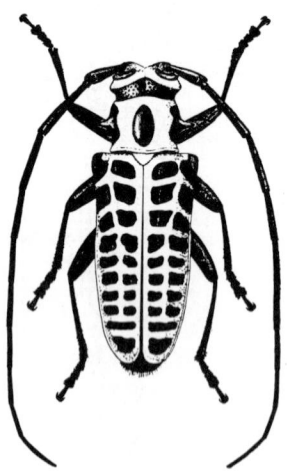

Fig. 2   Adult male of *Plectrodera scalator* (Fabricius), family Cerambycidae (after Knull).

COLLEMBOLA

An order of tiny INSECTS characterized by the absence of wings, gradual metamorphosis, six abdominal segments, not more than six antennal segments, and the presence of ventral abdominal appendages.

The wingless condition is considered primitive for Collembola apparently separated from the higher insects prior to the development of flight organs. The first abdominal segment bears ventrally a tube-like structure of uncertain function, called the *collophore*. A pair of appendages occurs ventrally on the third abdominal segment forming the *tenaculum*, a toothed structure which acts as a catch to hold the paired appendages of the fourth abdominal segment which form the *furcula*. When the tenaculum releases the fur-

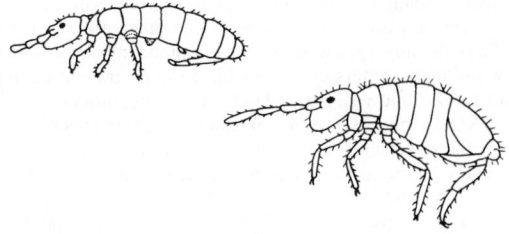

(From Pimentel, "Natural History," New York, Reinhold, 1963.)

cula, this structure springs back, catapulting the insect into the air. The furcula consists of a basal piece, the *manubrium,* and paired apical arms, the *dentes.* Each dens ends in a small hook, the *mucro.* Although in some groups the tenaculum and furcula are absent, these structures are so generally present that they have given the common name of springtails to the order. The antennae are basically four-segmented, but through secondary subdivision there may be five or six segments, and the apical one or two segments may be further subsegmented or even annulate. Each leg ends in a compound segment, the tibiotarsus, which bears apically a median claw, the *unguis,* and another ventrad to the unguis, the *unguiculus.* The mouthparts are typically of the chewing type but in some groups they have been reduced to piercing styli. They have the appearance of having been withdrawn into the head capsule, but embryological studies have shown that this condition has been brought about by an outgrowth of the genal area on either side, which growth joins with the labial and labral areas to surround all but the tips of the mouth appendages.

Collembola are separated into two suborders; the *Arthropleona* are elongate and the thoracic and all or most of the abdominal segments distinct, and the *Symphypleona* which have globular bodies with the thoracic and first four abdominal segments ankylosed.

These insects are often active at temperatures just above freezing and may appear in large numbers on snow. For this reason they are sometimes called snow-fleas. Most species live in leaf mould or decaying vegetation where they subsist on plant detritus, fungi, spores, pollen grains, and similar materials. A few appear to be carnivorous. They are covered with hairs or scales and, although they appear to be dependent on saturated atmospheres for existence, a few scaled forms, notably the genus *Willowsia,* can invade drier situations and may infest human dwellings. They are widely distributed, from the arctic to the antarctic, and some species are included in that small group of animals which are permanent residents of the Antarctic Continent. A few are modified for living on water surfaces and are found only in such situations. Soil inhabiting forms are often blind and pigmentless.

Collembola are considered by some authorities to form a group of arthropods below the level of true insects, and peculiarities in their embryonic development and adult morphology have been cited to substantiate this view.

HARLOW B. MILLS

References

Handschin, E., "Urinsekten oder Apterygoten," Die Tierwelt Deutschlands, Vol. 16. Jena, Fisher, 1929.

Maynard, E. A., "A monograph of the Collembola or springtails of New York State," Ithaca, Comstock, 1951.
Mills, H. B., "A monograph of the Collembola of Iowa," Ames, Collegiate Press, 1934.

COLONY

This term is applied to many different kinds of temporary and permanent groups of individual animals of the same species. These groups range from those found where individuals have merely gathered or settled in the same place as with the encrusting bryozoa (e.g. *Alcyonidium hirsutum*) through those in which the group is held together by behavioural bonds (SEE SOCIAL INSECTS), to those held by morphological links as are the polyps of CORALS. Among these last, the individuality of the members may be surrendered to an overall individuality of the colony; this occurs particularly among the SIPHONOPHORES.

The behaviour of the dispersive phase (larvae etc.) of sedentary animals may lead them to settle selectively in areas where adults are, or have recently been, either because they react to the presence, or traces, of adults, or because all members of the species select the same narrow range of substrata. Many marine invertebrates, e.g. barnacles (SEE CIRRIPEDIA) bryozoa, and so forth, have been shown to behave in this manner.

The term has also been applied to aggregations of animals which have come together because a particular place offers some optimal environmental condition, e.g. the gatherings of woodlice below the bark of fallen trees, where humidity and darkness combine in attractiveness.

The breeding colonies formed anually by birds (particularly sea-birds, like the gannet) and aquatic mammals (Alaska fur seal, elephant seal, etc.) are temporary, existing solely for breeding. It is not clear what advantage such colonies have. Social facilitation may speed up the process of breeding, for there is some evidence that in large sea-bird colonies egg-laying starts earlier and lasts for a shorter time than in small colonies; this may be expected to reduce the period during which the young are exposed to predators. The formation of such colonies does not necessarily occur on the most favourable sites, for others just as advantageous, or even more so, go unoccupied; the presence of a few birds seems to attract more. Land birds such as weaver finches may also be colonial nesters.

Trophallactic relationships are important in the cohesion of most colonies of social insects, solicitation for and acceptance of food is common among the members of the colony and forms the means whereby pheromone (ecto-hormones or sociohormones) are passed through the colony influencing the reproductive processes. Tactile stimuli from other members are also attractive potential cohesive forces. Such stimuli from the larvae are responsible for the rhythmic activation of army ant colonies (*Eciton* spp.) from their statary phase into their nomadic phase. Such social insect colonies are permanent though rupture, with the formation of new colonies, may occur, e.g. the swarming of honeybees, when a new queen leaves the hive accompanied by workers to found a new colony. The total colony behaviour is well shown by the complex architecture of termite and ant nests, structures produced by the mutual interpaly of individual behaviour patterns not by overall direction.

It is possible for most of the members of such insect colonies to live a solitary life, at least temporarily. However, the individuals of a coral colony, a *mille-porine* or a *stylasterine,* for example, are interdependent for each is specialised for one main function; the *gastrozooids* for feeding, the *dactylozooids* for collecting food and the *gonozooids* for reproduction. Their gastric cavities communicate with each other. The individual is incapable of carrying out functions other than that for which it is specialised and therefore cannot exist in isolation (unless one considers the free-living medusoid phase as another form of the individual). This loss of individuality is carried further in the floating colonies of Siphonophores, where loss of the individual's abilities has gone so far in submission to the total organisation of the colony, that the whole is now effectively the individual.

J. D. CARTHY

Reference

Allee, W. C., "Cooperation among animals," London, Pitman, 1951.

COLORATION OF ANIMALS

All opaque objects are seen by reflection of incident light from their surfaces, while transparent substances are viewed partly by reflection and in part by transmitted light, i.e., shining through them to the eye. White substances reflect, by simple random scattering, all components of the visible spectrum equally, while black objects absorb light of all wavelengths equally and entirely; they are therefore seen only by contrast with their surroundings. Various degrees of equivalent but incomplete absorption of light throughout the visible spectrum yield corresponding shades of grey. In a strict sense, however, black, white and grey are not defined among the true colors.

The manifestation of colors in animals depends upon the spectrally selective absorption of incident light before reflection of the residual components to the eye. But this fractionating process occurs at two distinct levels of morphology. The true *biochromes* owe their chemical or pigmentary colors to the absorption of one or more fractions of white light through the electron activities of their various special molecular types. The summation or integration of a set of electronic vibrational frequencies within any molecule evokes a resultant resonance throughout the whole molecule; the particular frequency of this resonance determines whether or not visible color is to be manifest. Should a compound's resonance frequency happen to match the wavelength of certain components of the visible light-spectrum, e.g., whether in the violet, blue, green, yellow, orange, red, or perhaps in more than one of these regions, then that portion of the incident light is quenched, being absorbed by internal frictional processes and converted into heat energy, while the trans-mitted beam of light, now lacking the absorbed component, appears to the eye as of complementary color. Thus the absorption of the *blue* component of incident white light confers upon the object the prominent color of *yellow,* and *vice versa.* These emergent colors are not spectrally pure, since they include other portions of the visible spectrum, save for the fractions absorbed; a yellow pigment, for example, will transmit all or much of the visible light save violet and blue. The perceived color of a substance is thus an integrated result, and depends upon the wavelength of the predominating or maximally transmitted fraction of visible light.

In contradistinction to the true biochromes, which express color arising from molecular resonance, there are the so-called *schemochromes*; these are physical or structural colors evoked by materials which, although without any pigmentary coloration, are disposed into extremely small, ultramicroscopic states of subdivision. Their architectural dimensions lie within the range of wavelengths of ordinary visible light, just as do the resonance-frequencies of colored molecules, i.e., within limits close to 0.4 and 0.7 μ, thus disintegrating the component light-rays usually expressed as lying between about 400 and 700 mμ, or 4000 to 7000 Ångstrom units (between approximately 1/62,500 and 1/35,700 inch).

In the animal world there are two outstanding classes of schemochromic expression, each of wide-spread occurrence: the iridescent colors and the scattered blues. The iridescent spectral colors arise neither from prismatic refraction of light, e.g., as in water droplets, producing a rainbow, nor commonly through the kind of diffraction effected by finely spaced gratings, although there are a few examples of this among certain insects viewed in a direct beam of light. Rather, the basis of these so-called changeable and "metallic" colors lies in the *interference* between light rays reflected from an object, i.e., through alternating reenforcement and quenching of the returning spectral components by respective synchrony or asynchrony of matching fractions returning from different reflecting strata within a translucent object such as a pearl.

The basic phenomenon, discussed by Sir Isaac Newton in the late 1600's and early 1700's, followed upon his studies of the iridescence of soap bubbles, thin layers of transparent glass, oil films, peacock feathers and the like, and was pursued much later through careful experiments, chiefly on colored lepidopteran wing-scales and beetle elytra, by the elder Lord Rayleigh in 1919, by his son the younger Lord Rayleigh in 1923 and by other investigators, notably C. W. Mason, who, between 1923 and 1929, published critical studies on the iridescent colors of feathers and of insect structures.

The manifestations of changeable iridescent colors depends upon the disposition of transparent or translucent materials in extremely thin layers, which themselves or the interstices between them do not exceed in thickness the wavelength of visible light to which reference has been made (ca. 0.4 to 0.7 μ). Moreover, the finely attenuated interstices between successive layers or films of solid or fluid material must contain or be constituted of some substance differing from the latter in refractive index, else the system will appear homogeneous, i.e., either transparent, white, or of some fortuitous pigmentary color involved in the material.

A few glass microscope slides, tightly pressed together between the (softened) jaws of a clamp will, through the imposed thinness of the air-spaces intervening between the glass surfaces, manifest Newton's interference zones. And this effect is greatly accentuated by having an absorbing layer of dark pigment at the bottom of the stack; of. the accentuation of light-interference colors by thin oil films lying on a black pavement, which serves as an absorbing screen, thus preventing the reflection of white light, which would obscure or weaken the observed effects. An oyster pearl involves very thin films of water between the concentric, attenuated layers of calcium carbonate. If the water be driven out of the interstitial layers of a pearl, or of other nacreous material, the object now assumes a "dead" whitish appearance, arising from mere random scattering of incident white light.

The brilliant colors of certain butterflies' wings, e.g., *Morpho menelaus*, predominantly blue but changeable with angle of incidence, arise from the presence, in the minute wing-scales, of ultra-thin air-layers between the keratinaceous lamellae. Substitution of the enclosed air by a fluid, such as carbon disulfide, or refractive index close to that of keratin (*ca.* 1.6) immediately renders the whole structure transparent down to the ventral surface, giving the wing the brown color of its underlying melanin deposits (see below). Evaporation of the foreign fluid now quickly restores the displaced air and thus the original coloration.

Unlike some biological products such as nacre or pearl, which enclose water, or certain lepidopteran or coleopteran wing-scales containing air in the interlamellar spaces, the iridescent feathers of many birds, such as the peacock, hummingbird, pigeon, turkey, birds of Paradise, and certain ducks, apparently do not enclose either gaseous or aqueous films between their thin keratinaceous laminations, but instead some substance of sufficiently different refractive index to manifest the interference colors, This applies also to the horny coverings of many scale-less beetles, wasps and flies, whose metallic lustre arises from the presence of much underlying melanin pigment, accompanied by highly specular reflectivity of the surfaces. The fine order of spacing between these multiple thin lamellae, and the cross-dimensions of the layers themselves, determine whether the predominant reflected color shall be of the relatively shorter or longer wavelengths. For example, a pile of twelve transparent plates, each of refractive index = 1.5, a thickness of 0.05 μ, and separated from its neighbor by an air-space of 0.15 μ, will appear violet at an incident angle of 60° and blue-violet at 90°, whereas if the dimensions of the lamellae and those of their interlamellar air-spaces should be the reverse of the above, the predominant colors at the respective angles will be green and yellow-green. In such manner, and provided with an underlying deposit of dark melanin, the integumentary surfaces, wing-covers and other structures of insects, as well as the feathers of some birds, many exhibit steel-blue, red, coppery, golden or even silvery metallic colors. Removal of the underlying pigment layer by chemical or mechanical means greatly diminishes or abolishes the chromatic effects, while substitution of the original melanin by black ink then restores the colorful appearance of the structure.

Changes may be induced in these iridescent interference-colors in many instances by great mechanical pressures, whereby the diminution in thickness in the interlamellar layers shifts the predominant color toward those of shorter wave-lengths (blue or violet), whereas increased thickening of laminations in structures possessing no internal air-spaces (e.g., the integument of certain scale-less beetles) by exposure to swelling agents such as steam, ammonia, alcohol or phenol yields changes of color in the opposite direction, from blue or blue-green through green to brassy or coppery hues. These changes are reversible. Certain chrysomelid beetles of the genus *Coptocycla* exhibit pronounced iridescent changes in their elytra by extensive dehydration of the interlamellar spaces, resulting in a general attenuation of the whole structure. Such changes occur on the death of the insect, and reversibly from golden through green, blue and violet to brownish orange when the living animal is disturbed.

The manifestation of *non-iridescent, structural blue colors*, discussed by John Tyndall in 1869, and subsequently referred to as Tyndall scattering, depends upon the presence of minute, ultramicroscopic micelles or other entities 0.6 μ or less in diameter and having a refractive index different from that of the containing medium. The longer rays of the spectrum pass through such systems, which therefore appear yellowish, reddish, brown or grey by *transmitted light*, while the blue colors of scattering are seen as the *reflected* portion (cf. the bluish colors of woodsmoke versus the ruddy brown *shadows* cast by it).

Blue eyes contain no pigment in the iris, which allows the green, yellow, orange and red rays to penetrate to the underlying dark, light-absorbing uvea, while the ultramicroscopic colloidal micelles in the iridial stroma reflect the scattered blue component. This scattering effect is precluded by the presence of dark melanin, which, in the stroma of the iris, confers any of several shades of brown or black. Minor quantities of pale yellow melanin, superimposed upon the blue colors of Tyndall scattering, are responsible for green eye-colors.

The conspicuous blue colors of skin areas arise from similar basic factors, i.e., a pale colloidal layer of epidermis overlying deposits of dark melanin pigment, e.g., in the corium. Examples of this are seen in the naked blue skin about the face and neck of turkeys, cassowaries, guinea fowl, certain psittacine birds and the like; also on the muzzle and buttocks of some baboons, notably the male *Mandrillus sphinx*, the face and scrotum of certain guenon monkeys, and integumental areas of numerous fishes and reptiles. The presence of super- or juxtaposed areas of rich blood supplies yields purple regions intermediately between the bright blue and the bright red patches; this is particularly noticeable in the naked haunches of the mandrill *M. sphinx*, and in the neck-skin of the double-wattled cassowary.

Turning from these examples of Tyndall blues evoked by solid-in-liquid or liquid-in-liquid colloidal systems, we find conspicuous instances of the same basic phenomenon in the feathers of countless species of birds, such as blue jays, some South American macaws and many others of the parrot group. In feathers, however, the blue-scattering entities are not solid or liquid, but are ultramicroscopic air-spaces, occurring as minute "bubbles" within the so-called alveolar or box-cells of pale keratinaceous material, lying beneath the horny sheath of the feather-barb and surrounding dorsally and in part laterally a mass of hollow medullary cells carrying dark granular melanin adsorbed to or embedded within their walls. The box-cells in the jay's feather-barbs are about 3 to 4 μ thick, and surround a central spheroid cavity 4 to 5 μ in diameter. The innumerable minute air-vesicles within the box-cells of some feathers are of the order of 100 to 250 $\mu\mu$ in diameter, separated one from another by horny walls only 15 to 20 μ thick.

The physical nature of these blue feather colors may be appreciated by (a) the fact that no blue pigment is recoverable by chemical means; (b) high mechanical pressures or a hammer-blow against the feather-barbs flattens the minute air-vesicles within, now rendering the area black from the quenching of Tyndall scattering; (c) replacement of the air within the barbs by a fluid of refractive index equal to that of feather keratin (ca. 1.6) renders the box-cells homogeneously translucent, and the feather therefore exhibits the dark or black appearance of the underlying melanin screen. This effect is readily reversible on evaporating or leaching out the foreign fluid.

Removal of the underlying black pigment causes the Tyndall blue feature to disappear or suffer great diminution, while restoration of a black under-surface restores the optical effect. The presence of a *yellow* pigment in the cuticle overlying the feather-barb renders its color *green* instead of blue, while mechanical or chemical removal of the yellow cuticle permits the expression of the Tyndall blue alone. The scattered blue colors of eyes or skin patches are readily imitable in some degree by placing cloudy, translucent (yellow-to-red transmitting) disks of agar gel upon black surfaces. Blue tattoo marks are made by injection of a black ink into the dermis underlying a white or very pale epidermal area.

Of the colored molecules displayed as true biochromes by animals, reference will be restricted here to those which impart conspicuous colors to the intact organisms or to products thereof, such as pelages, scales, feathers or certain secretions.

Carotenoids (Figure 1) are yellow, orange or red pigments of nearly universal distribution. Insoluble in water but readily soluble in various fat-solvents, they are characterized by definite crystalline habits and, in solution, by well defined spectral absorption bands in the violet, blue, and sometimes in the green regions. Closely similar carotenoids are resolvable by chromatographic separation, whether on paper or on columns of chemically inert powders, such as chalk, lime, magnesia or alumina. Synthesized *de novo* by many bacterial and fungal species, and by all green plants, carotenoids are thus acquired, directly or indirectly, by animals only through their food; hence the yellow

colors of body-fat in horses, cattle and many birds, and in the egg-yolk of countless birds, fishes and invertebrates. The hydrocarbon (or carotene) type is selectively absorbed by relatively few animals, e.g., the horse, and to a great extent by cattle, whereas the oxygenated or xanthophyllic class of carotenoid, e.g., hydroxyl, ketonic, or combinations of such derivatives, are selectively assimilated by the domestic fowl and innumerable other birds, fishes and invertebrates, while a few animals such as man and some frogs and cephalopods absorb and store both kinds equally readily; numerous other animals, including swine, goats, sheep, carnivorous mammals and raptorial birds, store no colored carotenoids but only varying supplies of vitamin A, derived metabolically from certain members of this class of compounds (notably from β-carotene), and important in vision, reproduction and antixerosis.

The feathers, eyes, naked facial or shank skin and in some instances the bills of many herbivorous, fructivorous or omnivorous birds exhibit brilliant pigmentation due to the presence of xanthophyllic carotenoids (e.g., toucans, roseate spoonbills, flamingos). Some green feathers, but not all, exhibit their color through the combined effects of Tyndall-blue scattering from the internal box-cells of the barbs and the xanthophyllic yellow outer cuticle through which the reflected light must pass (see above).

Marine animals are particularly rich in carotenoids, notably in astaxanthin or closely similar oxygenated, acidogenic derivatives; these appear in the integument of crustaceans, sea-stars, nudibranch mollusks, or in the chitinous carapace and the eggs of many crustaceans, wherein the chromogen often is conjugated chemically with protein to yield green, blue, purple, brownish, grey or other colored complexes. These chromoproteins, upon boiling or treatment with alcohol, acetone or other protein coagulants, immediately turn red or orange with the release of the carotenoid.

Quinone biochromes (Figures 2, 3) are of two classes in animals: the polyhydroxynaphthoquinones and the polyhydroxyanthraquinones. In each instance the oxidized $>$C=O or quinoid groups at p-positions on one of the phenyl rings are responsible for the color. The chemically reduced, corresponding hydroxy derivatives are without color. These quinones occur in countless plant species as yellow, orange, red or purple pigments, soluble in numerous organic solvents and have been used extensively as dyes for fabrics.

Chemical studies have revealed the existence of a number of different naturally-occurring, so-called echinochromes. These red, purple or sometimes green compounds are naphthoquinone derivatives encoun-

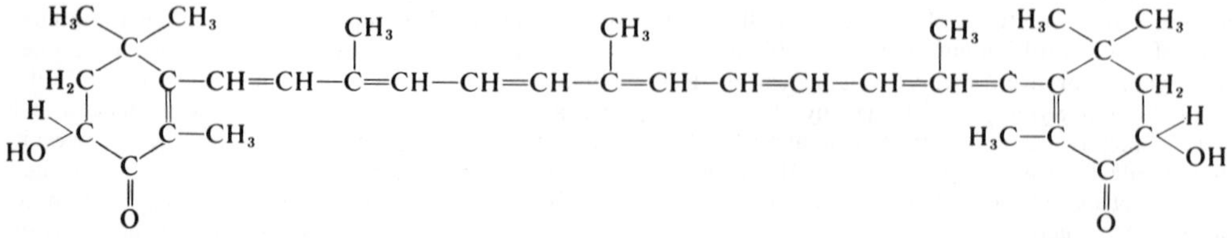

Fig. 1   Astaxanthin.

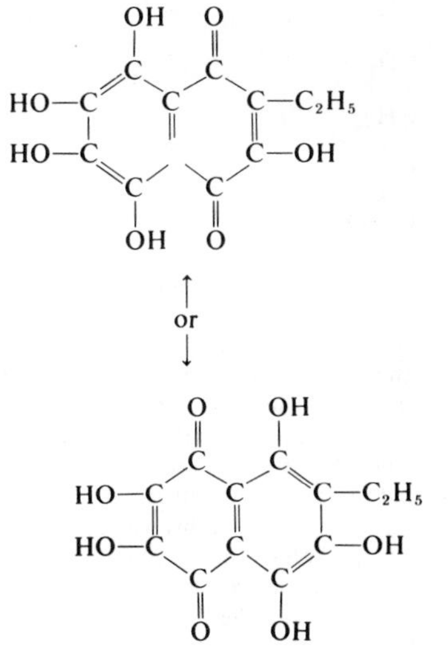

Fig. 2   Echinochrome.

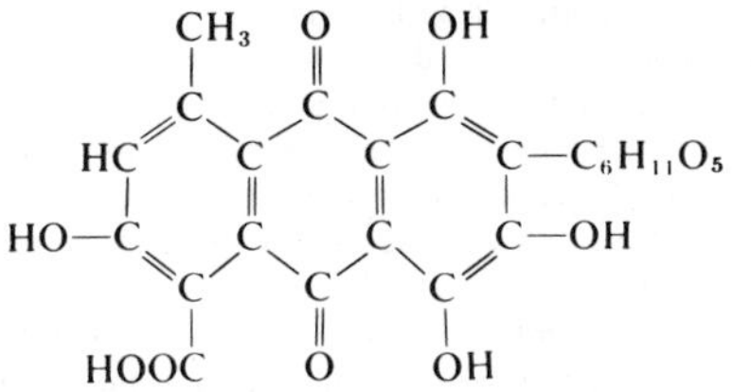

Fig. 3   Carminic acid.

tered in two echinoderm classes, the echinoids (sea-urchins and sand-dollars) and the crinoids (sea-lillies or feather-stars). Red masses of echinochrome can be seen in the lining of the gut and of the shell, in the body-fluid, and in some echinoid species within the egg-jelly. The calcareous shell and spines incorporate the red or purple pigments as well, presumably as insoluble calcium salts. Echinochromes have been recovered from certain fossil species of crinoid, as well as from the soft parts of a free-swimming species, *Antedon bifida*, presumably a particulate feeder.

The sea otter, *Enhydra lutris*, gradually develops, as a consequence of consuming large numbers of sea urchins, pink to purple colors in its teeth, skull and other skeletal parts due to the deposition of insoluble echinochrome salts.

Like the naphthoquinones, the related anthraquinone biochromes comprise a number of brightly colored compounds which have long found wide use in the dyeing industry. They exhibit color changes in acids and bases and are chemically reducible by certain reagents. Cochineal is recovered as the red potassium salt of carminic acid from the fat-body and eggs of scale-insect of the genus *Dactylopius*, while kermesic acid, another red dye popular in ancient times, is a chemically similar pigment extractable from the females of the kermes scale-insect *Lacanium ilicis*; and laccaic acid is a component of lac-lac, a solid excrescence covering the

female bodies of several species of lac insects. All of these insects are parasitic upon plants, whence they doubtless derive the anthraquinone pigments which, so far as we know at present, fulfill no essential physiological role in the insect's metabolism. Red anthraquinones are prominent pigments in crinoids of the genus *Comatula*, a particulate feeder which doubtless assimilates much marine plant material bearing the dye or its chemical precursors.

Tetrapyrroles (Figures 4, 5, 6) include both the closed ring class, or porphins, and the open chain variety or bilins, chemically and metabolically related to the porphins. The kinds and numbers of side-chains, as well as the relative degree of chemical unsaturation (double bonds), determine the character and functions of the endogenous tetrapyrroles. The so-called phaeophorbides, when binding a central atom of magnesium and bearing a molecule of the long, unsaturated alcohol phytol as a specifically placed side-chain to give the corresponding ester, constitute the photosynthesis-catalyzing chlorophylls of green plants. Similar quar-

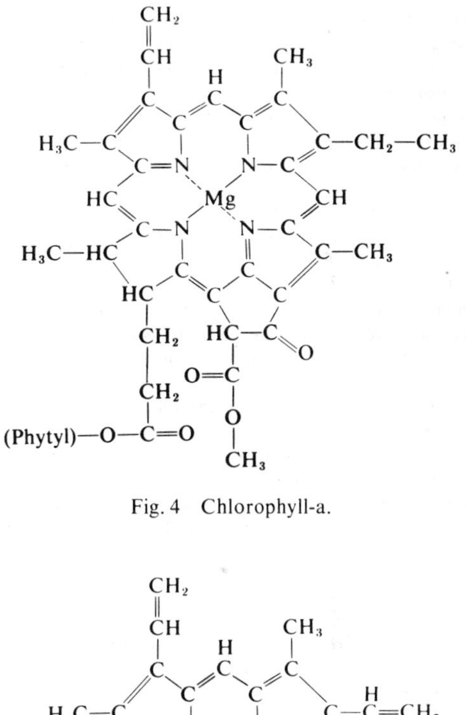

Fig. 4   Chlorophyll-a.

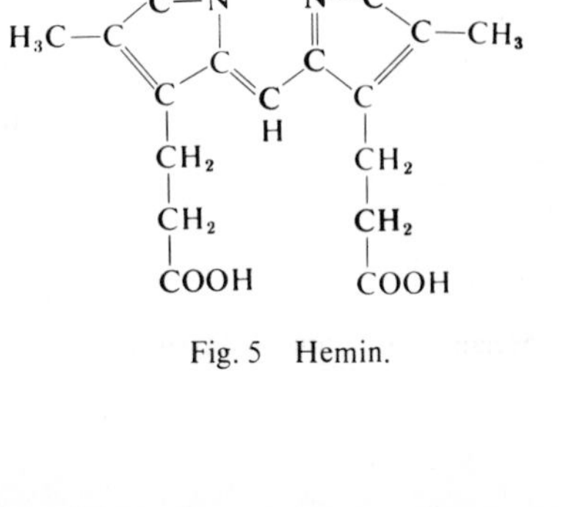

Fig. 5   Hemin.

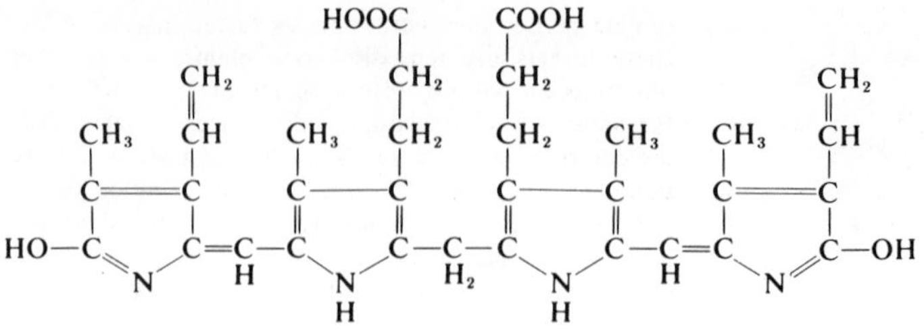

Fig. 6   Bilirubin.

tets or substituted pyrrole nuclei, linked each to its neighbor by the same unsaturated methene (—CH=) bonds, but lacking the isocyclic ring condensed in chlorophyll to one of the pyrrole groups, and binding, through the pyrrolic nitrogen atoms, a central atom of ferrous iron instead of magnesium, characterize the unesterified haems; these, when conjugated with a globin protein, constitute the familiar red pigment haemoglobin in the bloods of vertebrates and of many invertebrates, particularly among the polychaete worms and to a limited degree in certain gastropod mollusks and a few crustacean and holothurian species. The capacity of haemoglobin to combine reversibly with atmospheric oxygen and thus to serve the animal's respiratory needs varies considerably, and notably between vertebrate and invertebrate forms, in accordance with the type and molecular weight of the protein moiety involved. The red oxyhaemerythrin in the blood of certain sipunculid worms and the green chlorocruorin in that of chlorhaemid worms involve porphyrins conjugated with protein and iron, and serving as oxygen carriers (see below).

The haemoglobin present in the root-nodules of leguminous plants supporting vigorous populations of symbiotic *Rhizobium* bacteria, is associated with the extraordinary catalytic function of fixing atmospheric nitrogen.

Characteristic pink to bright red colors arise from the presence of haemoglobin and oxyhaemoglobin within the capillaries beneath the relatively thin skin of lips and other facial or specific body-surfaces of man and other primates, the wattles and combs of birds and the muscles of all vertebrates and of many invertebrates. Various porphyrins confer ruddy, brownish or greenish

colors to the shells of many gastropods and to the eggshells of numerous species of birds, while feathers of the African turaco, a fruit-eating bird, contain the red copper uroporphyrin salt turacin and a green oxidized derivative of this called turacoverdin.

Yellow, green, bluish or brown, open-chained nonmetallic tetrapyrroles, known collectively as bilins or bilichromes, occur in certain marine red and blue-green algae, wherein they are conjugated with proteins, and in the blood, skeletal parts, integument, excretory material, secretory products and eggshells of many animal forms.

Bilirubin ($C_{33}H_{36}O_6N_4$), red-brown when concentrated, confers orange or yellow colors to the blood plasma, sclera and skin of patients with jaundice, a condition involving disorders in the secretion of bile through the duct of the gall bladder. Its green oxidation product (actually a dehydrogenation derivative) biliverdin ($C_{33}H_{34}O_6N_4$) characterizes the color of bile and occurs also in the blue or blue-green eggshells of certain gulls and of some other wild birds, such as the emu, as well as in the bones and scales of several species of fish. Violet, red, blue or blue-green bilichromes confer their colors to the wings and integument of some insects, the integumentary tissues of some anemones and mollusks, the shells and inky secretions of some gastropods and the skeletons of some corals, e.g., helioporobilin, the blue bilichrome from the skeleton of the coral *Heliopora caerulea*. The bilichromes very probably are derived principally from the breakdown of haems or other porphyrins, but may play some role in the synthesis of new haemoglobin in vertebrates.

Of the *indole* biochromes, the *melanins* (Figure 7) are endogenous yellowish, orange, brown or black end-

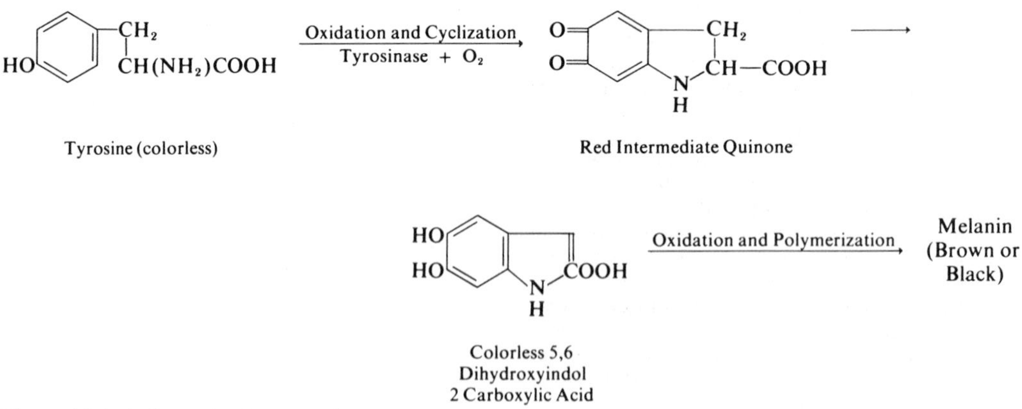

Tyrosine (colorless)

Red Intermediate Quinone

Colorless 5,6
Dihydroxyindol
2 Carboxylic Acid

Fig. 7   Melanin formation from tyrosine.

products of the oxidative metabolism of aromatic amino acids, principally tyrosine, accompanied by polymerization of the resulting derivatives. It is this class of compounds, relatively ill-defined in a stoichiometric sense, that controls the pigmentation of eyes, mammalian hair or fur, many feathers, scales of fishes and reptiles, ink secreted by cephalopods and by some fishes, and the integument itself of countless vertebrate and invertebrate animals.

Dark-skinned human races bear large, concentrated aggregates of so-called melanocytes in the corium, as well as dispersed melanin in the epidermis, while pale skin contains but little melanin, which may be yellowish, tan or other light brown hues rather than black. Many fishes, crustaceans, cephalopods and lizards are able to vary greatly their chromatic appearance by aggregating or dispersing the pigment microgranules within the melanophores of their skin.

The oxidation of tyrosine or similar phenolic compounds and their derivatives toward the ultimate production of melanin is catalyzed by a copper-containing enzyme tyrosinase. Copper is known to be a limiting factor in the ability of some animals to produce melanin.

Melanic pigmentation of the skin, resulting from "tanning" after exposure to unfiltered sunrays, doubtless affords to sensitive underlying tissues a degree of protection against injury from erythemogenic ultraviolet fractions, although most of the protection probably is effected by thickening of the outer, keratinized layers of skin.

The *indigoid* derivatives in this same general class (Figure 8), like the melanins, are among the excretory end-products of amino acid metabolism, but here tryptophan is involved, rather than tyrosine; moreover the indigoids are far more limited in their distribution among animals. And while the melanins are sombre, dark pigments, the indigoids exhibit blue, green, red or purple colors, and, for this reason, have sometimes been confused with the bilichromes. Indigo itself (indigo-blue or indigotin) occurs as glucocide derivatives in many plants, and was long employed as a dye, conspicuously by the ancient Britons, who smeared their naked skins with woad, a fermented blue paste of crushed cruciferous plants such as *Isatis tinctoria*. Caesar's soldiers encountered native British warriors so painted, while Pliny reported that their women similarly covered their nude bodies before participating in certain rites.

Indigoid biochromes are relatively inconspicuous, commonly of excretory occurrence and often of pathological significance in man and other mammals, but derivatives of the class found in purple secretions of certain marine gastropods, including *Murex, Purpura*

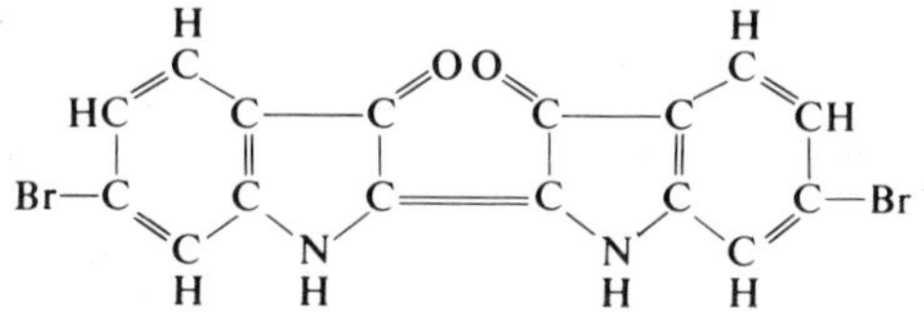

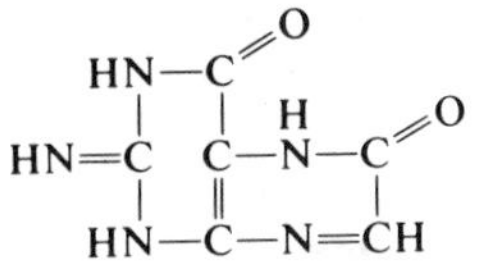

Fig. 9   Xanthopterin.

and *Mitra* spp., were known in Biblical and Phoenician times, and have been referred to as Purple of the Ancients or Tyrian Purple. The chief representative, 6-6'-dibromindigo, is derived from a colorless precursor in the pale hypobranchial or adrectal gland of these snails. When the gland is crushed and exposed to light, the red-violet dye is generated, presumably through the action of an accompanying enzyme. No physiological role has yet been assignable to dibromindigo or its precursors, all of which probably are of excretory character, somewhat reminiscent of the bilichromes aplysiopurpurin and aplysiorhodin, secreted from the ink-gland of the sea-slug *Aplysia*, save that the latter are in all probability endogenous products, while one thinks of dibromindigo as deriving more or less directly from some plant source.

The *pterins* (Figure 9) are close chemical endogenous relatives of the colorless purines, and were formerly classed among them. They are white, yellow, orange or red insoluble deposits encountered commonly in butterflies' wings, and give rise to the yellow body-colors of many wasps. Xanthopterin is a common example, appearing as a yellow pigment in the wings of *Gonepterix rhamni* and other lepidopterans, in the abdominal integument of *Vespa* and other wasps, and in human urine, whence its alternative name, uropterin. This compound has been shown to prevent nutritional anemia in young salmon and in rats.

Fluorescyanin, a blue-fluorescing pterin occurring as a chromoprotein in fishes' scales, exhibits interesting chemical and physiological properties, being easily reducible by dithionite (hyposulphite) as is riboflavin, and apparently contributing to the respiratory processes, e.g., oxygen consumption, in the living scales, besides exhibiting some similarities to vitamins B_1 and B_2 when administered to experimental animals.

Among miscellaneous pigments of currently unknown chemical structure, many could be mentioned, but only a few will be listed here.

Chromolipoids, sometimes mistaken for melanins because of their yellowish, reddish or brown colors and their insolubility in water, or for carotenoids because of their colors and fat-soluble properties, are found in the fat-droplets of lipid-rich tissues and certain other cells, as well as in eggs. They are commonly regarded as oxidized derivatives of fatty acids, phospholipids or allied compounds, and perhaps bear some relationship to stored fuel. They are differentiable from the carotenoids chiefly through general failure to exhibit discrete absorption bands in the visible spectrum, while their variable solubility in organic solvents, the exhibition, in some instances, of blue to green fluorescence, occasionally with an absorption band in the far violet, and their stability in dilute acids rather than precipitability therewith, all contrast them with the melanins.

Fig. 8   Dibromindigo.

The *haemocyanins* serve as oxygen-transporting agents in the blood of gastropod, cephalopod and some amphineuran mollusks and of crustaceans and certain other arthropods, such as the "horse-shoe crab," *Limulus polyphemus*, and some scorpions and spiders. This is a copper-protein complex, colorless in the reduced condition and blue when oxygenated. The reversible condition is believed to involve changes in the valency of the copper, i.e., between the cuprous (Cu^+) and cupric (Cu^{++}) states, unlike the condition of iron in haemoglobin, which remains in the ferrous (Fe^{++}) valency save when oxidized to the trivalent ferric state as in methaemoglobin under abnormal conditions.

The haemocyanins involve no porphyrin moiety as a link between the metallic ion and the protein; instead, the copper is bound in some other fashion, and to large protein molecules, conferring upon the whole biochrome very high molecular weights, e.g., from about 1.3×10^6 in *Limulus* to 2×10^6 in *Octopus*, or even 5×10^6 in *Helix*, the land snail, thus comparing with some of the invertebrate haemoglobins, which may reach molecular weights of 3×10^6, in contrast with vertebrate myoglobins (17,000) or haemoglobins (68,000). The oxygen-combining capacity of the haemocyanins is considerably lower than that of typical haemoglobins.

Haemerythrin, the iron-containing respiratory protein of some sipunculid worms and of the brachiopod *Lingula*, turns from colorless to red when oxygenated and involves a porphyrin, as does chlorocruorin, the green iron-protein oxygen carrier of some serpulimorphid worms. But the porphyrin of haemerythrin is not believed to be a haem, and that of chlorocruorin appears to differ from the protohaem of haemoglobins.

Vanadium chromogen refers to the pale green biochrome contained within the blood cells, or vanadocytes, of some ascidians or tunicates, which are primitive chordates. The vanadocytes may constitute 1% or slightly more of the total blood volume, and enclose, along with the vandadium-rich pigment, sulfuric acid in concentrations approaching 9%. Cytolyzed vanadocytes release the pigment, which immediately turns red or brownish in color. This so-called haemovanadin, when alkalized, separates as a blue precipitate and, on oxidation, as a dark blue-green, insoluble compound bearing 5 to 10% of vanadium. No conjugated protein seems to be present in the raw pigment after cytolysis. Nor have purines or porphyrins been detected in the chromogen, which is a dialyzable molecule involving a considerable proportion of pyrrole material, suggesting a possible conjugation with an open-chained bilin-like compound. The free trivalent vanadium ion (V^{+++}) in the cytolyzate, however, would suggest that any chemical bonds between the metal and the organic radicals within the cells must be rather tenuous. The vanadium biochrome, while exhibiting strong chemical reducing properties, seems not to be a respiratory catalyst, and its physiological role, if any, remains a challenging enigma.

DENIS L. FOX

Reference

Fox, D. L., "Animal Biochromes and Structural Colours," Cambridge, The University Press, 1953.

COLUMBIFORMES

The columbiform, or pigeon-like, birds consist of three families, the largest one including the familiar pigeons and doves. The smaller, less known families include some 16 species of Eurasian and African sand-grouse, and the extinct dodos and solitaire. All share in common the characteristics of a compact dense plumage, arranged into special feather tracts, with the feathers set loosely in the skin and easily shed. A well developed crop, and nesting habits that feature crude nest construction and usually two eggs are also characteristic of the columbiform birds.

The sand-grouse, family Pteroclidae, comprise 16 species and about 45 total forms (species and subspecies) of pigeon-like birds that in general appearance and habits resemble grouse. They lack the swollen operculum or cere that overlaps the nostrils in the true pigeons, and they have a short, feathered tarsus. Other more technical features set them somewhat apart from the pigeons, but in the main characteristics, as set forth above, their affinities seem to be with the columbiform birds.

The sand-grouse inhabit arid (desert) or semiarid (steppes) areas in southern Europe, central and southern Asia and suitable parts of Africa, including Madagascar. They are highly gregarious, nest on the ground, and have a dove-like gait, with toes well adapted (sometimes feathered) for walking in the sand. Certain species, particularly the Pallas' sand-grouse (*Syrrhaptes paradoxus*), have long been famous for their sporadic or irruptive migrations. Periodic but irregular build-ups in sand-grouse populations in the past have resulted in some spectacular irruptions, with the surplus individuals, like lemmings, spilling over into usually unoccupied range and becoming temporarily established but eventually dying out. The most famous, or best known, of these irruptions was in 1863 when large numbers invaded northwestern Europe, even nesting in Great Britain and Denmark, but, as always, they failed to establish themselves permanently.

Sand-grouse, as the well developed crop implies, feed largely on seeds and such vegetative material as can be found in the dry sands or grasslands they inhabit. Periodically during the day, usually in the morning and evening, they assemble in flocks and visit distant water holes to drink and bathe, usually drinking in the well known pigeon manner of drawing up water with the bill immersed instead of raising the head to swallow as other birds do. However, both drinking methods have been observed in these birds. For nesting the adults scrape out hollows in the sand and line the cavity skimpily with grasses before it receives the 2 or 3 eggs. Both parents share in incubation and care of the young, which are born in a precocial or advanced state, and, unlike pigeons, leave the nest soon after hatching.

The family Raphidae has been erected for the familiar dodos (2 species) and the less famous but similar solitaire. Though flightless, the dodos were essentially heavily built and clumsy terrestrial pigeons. A peculiar tuft of tail feathers is a well known feature, depicted in drawings of the extinct dodos. The raphids inhabited small islands off the coast of Africa, near Madagascar. One species of dodo inhabited the island of Mauritius, another was on Réunion, and a similar species, known as the solitaire, lived on Rodriquez.

The dodo was first discovered on Mauritius in 1598 and persisted there till 1681. It was said to be clumsy, stupid, and edible, so that it made an easily available source of food for visiting mariners and explorers; then the introduction of hogs onto the island spelled its final doom. The other dodo, on Reúnion, lasted till 1699, and the solitaire, on Rodriquez, survived still longer.

Aside from the extinct raphids and the somewhat restricted pteroclids, the columbiform birds are comprised of many species of widely distributed pigeons and doves (family Columbidae). Peters, in his Checklist of Birds, of the World, listed 302 living species (with 4 others, extinct) and 835 forms, but the taxonomic status of many of these forms was, and still is, open to question. Current taxonomic concepts (since Peters) have reduced the number of recognized species considerably. Distinctive features in the pigeons and doves (there is no real technical distinction between pigeons and doves) are the dense, soft plumage, often richly colored, with metallic reflections, especially on the neck; the peculiar fleshy cere or operculum overhanging the nostril; and the well developed, bilobed crop which produces a caseous substance, known as "pigeon's milk," for nourishing the young.

Pigeons are widely distributed over most of the world, but avoid the polar extremes and are lacking on many oceanic islands. They reach their highest development, however, in the Australasian region, which is presumed to be the center of origin, from which ancestral pigeon stock sent branches out to other parts of the world. In the Australasian region are found the gorgeously colored fruit pigeons, the highly ornamental crowned pigeons, and the tooth-billed pigeon. Thus three of the four groups (subfamilies) of columbids are largely restricted to their place of origin; members of the fourth sub-family, the Columbinae, have spread pretty much throughout the world, often aided in their dispersal by man.

Some pigeons, particularly the rock dove (*Columba livia*), have been closely associated with man for about 5000 years. About 200 domestic varieties or strains have been developed by man from the original rock dove or common pigeon, some primarily as meat producers, some purely for ornamental purposes. Others have been trained as carriers of messages, especially during times of war, and the training of homing and racing pigeons is still a popular pastime.

Among the most famous of the wild columbids is the fabulous passenger pigeon. Once said to number billions of individuals, perhaps more numerous than any landbird surviving today, the species was reduced, primarily by exploitation by man, to scattered flocks which soon disappeared completely. The last passenger pigeon died in captivity in the Cincinnati Zoo in 1914, having survived its last known wild relatives by a decade or more. The passenger pigeon has now been appropriately memorialized by a monument erected in its honor in Wisconsin, by a comprehensive book (Dr. A. W. Schorger's *The Passenger Pigeon*) detailing its known history completely, and by various other writings and tributes. But it is gone, and its plight is often evoked as a lesson to mankind of a valuable resource needlessly lost; but there is no assurance that history will not repeat itself, for several other species have departed since the passenger pigeon, and others are on their way.

GEORGE J. WALLACE

References
Gilliard, E. T., "Living Birds of the World," New York, Doubleday, 1958.
Peters, J. L., "A Check-list of the Birds of the World," Vol. 3, Cambridge, Harvard University Press, 1937.
Wallace, G. J., "An Introduction to Ornithology," New York, Macmillan, 2nd ed., 1963.

COMMUNITY

For the purpose of this article, a community is any collection of organisms living together at one time. Thus one may speak of the soil fauna community under a beech tree or the community of birds nesting in a certain marsh, or, with somewhat more interest, of the community of all living things in a given wood during the summer. This is somewhat vague due to wandering of animals, but for the most interesting problems this vagueness is unimportant.

Communities have been studied in several ways as follows:

Descriptive. A vast amount of research and argument has centered on the problem of how to describe a community so that other workers can recognize another example of the same type. Usually some (one or few) dominant or characteristic plant species are used in the description, but further agreement is lacking, and the workers of one region are usually dissatisfied with any classification proposed by workers familiar with other types of terrain. The trouble seems to be that no very clear problem is motivating these descriptions. Problems considered in later paragraphs have required descriptive studies but the measurements used are of other properties than those chosen arbitrarily by the plant and animal sociologists. Perhaps the most useful descriptive procedure is that used by Curtis (1959). This reference contains much other information on plant communities.

Communities as heat engines. The laws of thermodynamics impose the most fundamental bounds on patterns of energy flow through communities. The first law, the law of conservation of energy, assures that, for instance, the carnivore has no more energy available to it (at least in the long run) than its food, the herbivores, have available to them. The second law goes farther and asserts that energy can never be fully efficiently utilized by the organism, so that some is dissipated. Combining these, one can say that the rates of storage and utilization of energy decrease progressively from the plant community to the herbivore community, to the carnivores etc. This is often diagrammed as an "energy pyramid," the lowest layer which has length proportional to the rate of energy intake (or storage) of the plants; the second layer to the energy intake (or storage) of herbivores etc. The laws of thermodynamics then assure that an "energy pyramid" has indeed a tapering pyramid shape. Since energy transferred from organism to organism is always associated with nutrients, the rate of flow of nutrients through a community must fall off in the same general way so that a pyramid of the average rate of nutrient intake is a logical necessity. It is by no means necessary, however, that there should be a tapering pyramid of mass (at any given time) and there are, in fact, cases of stable communities with more mass at all times in the carnivores than in the herbivores. This does not, of course, contradict the deductions above since, by having a short life span, a small mass of organisms may utilize and store nutrients and energy at an enormous rate. Experimentally the most important conclusion which has been drawn from this type of study is that storage and utilization rates of plant communities (often called "net production" and "gross production," respectively) seem to be relatively independent of the general appearance of the plant community (e.g. phytoplankton or forest) and depend much more upon the general availability of nutrients, which have a fairly large scale geographic pattern. It seems that the efficiency (measured by the ratio of energy intake of predator divided by that of prey) is usually about 10% and may never exceed about 13%.

Chapter 3 of the book by the Odums (see references) with its bibliography gives the best account of this aspect of community study.

Communities as collections of interrelated individuals and species. If, instead of picturing energy and nutrients as flowing from plants to herbivores, etc, one is more specific and considers separately the paths from each green plant species to each herbivore species which eats that plant species, etc., one then gets a "food web" for the community. This may be made quantitative by stating for each species either the fraction of its energy or nutrients which goes to each predator or the fraction which comes from eacy prey species. More generally $p_{j\,i}$ will be the fraction of energy from species j going to species i. The nature of the food web controls the resistance to population fluctuation of the species in the community in at least two ways. If the $p_j i$ remain constant, a steady state population composition of the whole community will be approached for which the energy consumption P of the ith species is given by $\Sigma_j P_j\, p_j\, i$. The rate of approach to this steady state is more rapid as the equality (for different i's) of the $p_j i$ increases. This can most easily be seen by considering a temporary excess of energy in one species. As time progresses this excess will be distributed first over that species and its predators and then over that species, its predators and their predators, etc. and eventually over the whole community. While this may be interesting as a first approximation, it is known that predators almost universally switch their attention from temporarily rare food species to temporarily common ones. This activity increases the stability which the system has, for now it is easily seen that the temporary excess or defect of energy centered at one species will spread even more rapidly over the whole food web. A crude measure of this stability (often called ultrastability) is provided by giving the variety of alternative paths open to a small unit of energy beginning its trip through the community.

The number of species which inhabit an area varies greatly from place to place. A few acres of tropical forest may support well over a hundred bird species while, in temperate and polar regions the diversity of species in a given forest will be in the neighborhood of ten to thirty. The total density of bird individuals, however, may not vary particularly. It seems apparent that both history and the nature of the habitat are involved in the control of the number of species. That history (involving long continued or rapid species formation) is important is most easily seen by considering the exceptions to the general rule of increased diversity of species in the tropics. For instance, the flora of the coast regions of south west Australia and South Africa is comparable in diversity to that of the tropical forests. These regions have apparently suffered less by Pleistocene climatic fluctuations than have regions in comparable latitude and present day climates north of the equator which have relatively impoverished floras. There are a few very old lakes in the world (in the case of Lake Baikal probably extending back to the Cretaceous or at least Paleocene). These are scattered over many latitudes and yet nearly always have a very rich fauna in contrast with newer ones of comparable position. Finally, remote islands are impoverished in number of species compared to ones near the main continents. Thus history is clearly important. It is equally clear that history is not a complete answer, for nearby habitats differ in the number of species they support but not in correlation with the histories of these habitats but with the diversities of the habitats. Thus, for instance, the creosote bush desert of the southwestern U.S. has very few species and yet it is presumably no more recent than the adjacent succulent desert with a very rich flora and fauna. From the point of view of the animals, the succulent desert supports more species because it offers a greater variety of ways of life. At present it is difficult to make this more precise. One final point must be mentioned. Not only is the diversity of the habitat important, but so also is the predictability and constancy of the climate. Thus the tropical climates permit insects and even fruit to be available at all times of the year so that very specialized feeding habits may develop. In more severe climates, species must be more flexible in their requirements. The comparative importance of these causes of species diversity is unknown.

Given a number of species coexisting in a habitat, one may ask what controls their relative abundance. In species such as insects whose abundance fluctuates wildly, the abundance at any given time may vary exponentially with the length of time during which increase has been unimpeded. Therefore abundance of such species are likely to be lognormally distributed, and species which are common at one time and place are not necessarily common at other times or places even in the same habitat. At the other extreme, populations of birds, at least, seem remarkably stable and the relative abundance reflects the fashion in which the energy available to the bird community has been partitioned among the species. As a first approximation, the total rate of energy intake is independent of the number of species utilizing it (the fewer species of birds on Bermuda maintain as high a total population density as the many species in comparable habitats on the mainland); and an increase in the abundance of one species causes a roughly equivalent decrease in the abundance of others. Furthermore, adaptations seem to be "convex" in the sense that a bird which can make effective use of two conditions within its habitat can also make use of intermediate ones. These properties combine to predict that in an undisturbed habitat the abundance of the species may often be proportional to the lengths of the segments of a line cut at random points into the appropriate number of pieces.

The yearbook of general systems theory contains several articles of interest. The papers by Hutchinson (1959) and MacArthur (1960) may also be consulted.

Change in time, or succession. When the environment

changes, the species inhabiting it also change. An enumeration of the species and habitat preferences of each (really a part of environmental physiology) is sufficient to explain this part of succession. It becomes of pressing interest to the ecologist when a change in the community itself is responsible for that change in the environment which causes further change in the community, thus producing a continuing process with some internal control.

Most widely known is the succession leading from an essentially barren area through grass, shrubs and various types of forest until a more or less stable climax is reached. Basic to at least part of this succession is the shade cast by one species changing the environment under it so that another species is better able to grow there. The new species, being more tolerant of shade, has a thicker canopy of leaves and in turn casts a deeper shade. This process gives succession a direction as well as explaining the existence of change. (It should not be inferred that the shade is directly responsible for the change, however; in fact, competition for water in the presence of shade and soil changes seem often to play an important part.)

Less well known, but also interesting is the succession in the insects infesting stored grain. Here, it appears that the accumulated metabolic heat generated, combined with degradation of the grain, plays the same role as did the shade in forest succession. Succession of protozoa in cultures has often been described, but the precise nature of the chemical changes which direct the succession is not known.

It is a mathematical property of any replacement process such as succession that either a stable state will be reached (called a climax in the terminology of the ecologist) or else a cyclic series of states will develop. It is also a consequence of the mathematics that various initial states will approach the same climax or steady state. There seems to be nothing mysterious about this, and empirically it appears that the steady state, when it exists, may change continuously with at least the temperature and moisture of the environment. Of course unpredictable climatic fluctuations (such as hurricanes) may superimpose such erratic changes as nearly to obscure the underlying pattern, or prevent it from reaching a climax.

It is clear that in a steady state system energy utilization must balance its intake. The earlier succession stages may be in either direction from this, however. That is, in some early stages a hay infusion protozoan succession is dissipating energy by respiring faster than it takes in energy, while a forest succession is accumulating energy by taking energy in faster than it uses it. Early stages in this type of succession are often accompanied by large values of birth rate or number of seeds, while later stages have larger young or seeds rather than many.

Succession is usually studied by a sort of "ergodic hypothesis" that different places at one time may exhibit the same sequence that one place will in successive times. This is almost a necessity in studying succession taking many hundreds of years to reach a steady state. More recently, pollen analysis has made it possible to follow a single area through many centuries.

The articles on community changes are widely spread. The work of Watt (1947), Salisbury (1942), Baker (1950), Dimbleby (1957) and Whittaker (1951) all give different aspects for plants succession while Mar-

gelef's article in the Yearbook of general systems theory is a good general reference.

ROBERT H. MACARTHUR

References

Baker, F. S., "Principles of silviculture," New York, McGraw-Hill, 1950.
Curtis, J. T., "The vegetation of Wisconsin," Madison, The University Press, 1959.
Dimbley, G. W., "Pollen analysis of terrestrial soils," New Phytologist 56: 12-28.
Hutchinson, G. E., "Homage to Santa Rosalia, or, Why are there so many kinds of animals?" American Naturalist XCIII: 145-159, 1959.
MacArthur, R. H., "On the relative abundance of species," American Naturalist XCIV: 25-36, 1960.
Odum, E. P. and H. T. Odum., "Fundamentals of ecology," Philadelphia, Saunders, 1959.
Salisbury, E. J., "The reproductive capacity of plants." London, Bell, 1942.
Watt, A. S., "Pattern and process in the plant community," Jour. Ecol. 35: 1-227, 1947.
Whittaker, R. H., "A consideration of the climax theory," Ecological Monographs. 23: 41-78, 1951.
Yearbook of the Society for general systems research, 1958. Volume III. Published by the society at Mental Health Research Institute in Ann Arbor.

COMPOSITAE

The Compositae, or sunflower family (alternative name Asteraceae), is the largest family of dicotyledonous flowering plants (see ANGIOSPERMS). Although there are no precise estimates, figures of 1,000-1,500 genera and 20,000-25,000 species are usually given. The family is truly cosmopolitan, extending from the Arctic to the subantarctic islands (but absent from the Antarctic Continent) and from sea level to snow line of the highest mountains. Although Compositae have successfully invaded all types of habitats, perhaps with the exception of the aquatic (only a few species are truly aquatic), they are particularly conspicuous in dry areas and in montane regions. The family includes both small, isolated, well-marked genera and some of the largest, cosmopolitan, polymorphic complexes in the plant kingdom, with indistinct generic and specific lines, such as *Senecio*, the largest genus of flowering plants, with more than 1,000 species.

Compositae are characterized by having their flowers grouped into heads (*capitula*) which only rarely are single-flowered, a characteristic shared with other families (Proteaceae, some Umbelliferae, Leguminosae, etc.). The receptacle of the head is usually flat or slightly convex, occasionally conical, or more rarely concave. The surface of the receptacle often is paleaceous, with membranous scales, each subtending a floret. The capitulum is surrounded by a number of modified leaves, the involucral bracts, in one or more series, often armed with spines and/or hooks (particularly in Cardueae). The heads as a rule possess two types of florets: one or more rows of pistillate, ligulate, or filiform florets on the outside, and bisexual, tubular florets in the inside, but all florets can be either tubular or ligulate (in which case they usually are all bisexual). The florets have an inferior ovary, a single basifixed ovule with a single integument and with the funiculus oriented in an abaxial position which develops into a characteristic fruit; five anthers, united by their edges into a cylinder surrounding the style, a style bifid into two stigmatic branches, each of which is covered exteriorly by collecting hairs (which remove pollen from anthers) and interiorly by stigmatic hairs. The calyx is modified into a pappus, or absent. The florets are usually yellow, but they can be white, orange, red, blue or violet.

The flowering heads of Compositae are often very showy. Pollination is largely by insects, although ornithophily is also known. Anemophily is the rule among the ragweeds (*Ambrosia* spp.) and their allies, the pollen of which is among the most important causes of hay fever. Wind pollination has been accompanied by a series of special adaptations, such as free stamens, inconspicuous heads with reduced involucre, and reduction in the number of flowers per head.

The family Compositae has many genera and species that are rapidly evolving, with many weedy species, offering excellent examples of various evolutionary phenomena.

The most reduced chromosome number in flowering plants known to date (*n* = 2 in *Haplopappus gracilis*) as well as extremely high numbers occur in Compositae. The most frequent chromosome numbers, however, are in the range of 8–12, with a modal number of 9. Polyploidy and hybridization are frequent in many genera. Most Compositae are selfsterile.

Apomixis has been shown in 16 genera, and for eight others there are reports of triploids and unbalanced chromosome numbers, possible indication of apomixis. Although this is a rather low figure considering the size of the family, some of the apomictic genera such as *Hieracium* (hawkweeds), *Taraxacum* (dandelions), *Rudbeckia* (black-eyed Susan), *Erigeron* (flybean) and *Antennaria*, have many apomictic species.

The family is predominantly herbaceous or subshrubby, particularly in the temperate zones, but woody members, including large trees 20 m in height, are not rare. The Compositae are most likely primitively woody.

The Compositae form a distinct group, often treated as comprising the order Asterales. The relationships of Compositae to other families are not entirely clear. They have been considered allied to Campanulales on account of the connate anthers, milky juice, Compositae-like heads, and similar vegetative aspect of some species of Campanulaceae, particularly those of subfamily Lobelioideae. Calyceraceae and Dipsacaceae have also been cited as possible ancestral groups on account of their involucrate inflorescences and one-seeded fruits. Finally, an alliance to the Rubiales has also been considered because of the pappus-like calyx, the differentiation of the marginal flowers, and the tendency towards a 2-carpellate, uniovulate ovary, features which occur in different families and species of the Rubiales.

Classically, the Compositae have been subdivided into a number of tribes on the basis of such characters as differences in the styles of the bisexual flowers, the basal appendages of the stamens, the presence or absence of receptacular bracts, the shape of the corolla, and the type of pappus. The major tribal categories are fairly well agreed upon. Although different authors vary in according either tribal or subtribal status to some groups, most accept either 12 or 13 tribes.

Economically the family is important as the source of sunflower oil and seed, safflower oil, and wormwood, several garden crops, particularly lettuce, artichokes, chicory, endive, salsify, etc., as well as some of the most cherished garden flowers, such as *Chrysanthemum, Dahlia, Cosmos, Tagetes* (marigold), and *Ageratum*. Rubber is found in extractable quantities in guayule (*Parthenium argentatum* Gray) and kok-saghyz (*Taraxacum kok-saghyz* Rodin). Alkaloids have been reported from many Compositae, but are of little commercial importance. On the negative side, some of the most obnoxious garden and field weeds are members of this family.

OTTO T. SOLBRIG

References

Bentham, G., "Notes on the classification, history, and geographical distribution of Compositae," Journ. Linn. Soc. Bot. **13**: 335–577. 1873.

Lawrence, G. A. M., "Taxonomy of Vascular Plants," New York, MacMillan, 1951.

Solbrig, O. T., "The tribes of Compositae in the Southeastern United States," Jour. Arnold Arb. **44**: 436–461. 1963.

CONIFERALES

The conifers comprise 52 genera and about 570 species, usually grouped in 7 families. In contrast to CYCADS they are plants of the temperate region. A few also occur in the tropics but are then usually confined to the mountains. Of the forested areas of the earth more than a third are covered by Conifers. In fact the treeless TUNDRAS of the polar regions and the peninsula of India are the only extensive areas of the earth without some species of Conifers.

Included among the Conifers are not only the well known pines, firs and spruces but also the Redwood (*Sequoia sempervirens*) of the Californian coast ranges which is among the tallest and most long-lived trees of the world. *Sequoiadendron giganteum* of the Sierra Nevada mountains is less tall but much more massive. *Metasequoia* was long known as a fossil but about 16 years ago some Chinese botanists discovered living specimens in the Szechuan Province of China. Presently it is being cultivated in many parts of the world. Another new genus, described very recently from China, is *Cathaya*, a member of the Pinaceae.

There is no satisfactory classification of the group. Chamberlain (1934) recognized six families: Pinaceae, Taxodiaceae, Cupressaceae, Araucariaceae, Podocarpaceae and Taxaceae. Buchholz assigned *Cephalotaxus* to a separate family Cephalotaxaceae. Florin (1948) has produced some evidence in favour of giving the Taxaceae the rank of an order, Taxales, but this is a matter of opinion, for the only important point of distinction is terminal position of the ovule in the Taxaceae.

In addition to the above seven families there are also two other families consisting wholly of fossil members investigated largely by Florin. In the Voltziaceae the female cones have spirally arranged bract scales, which are sometimes forked, and axillary dorsiventral spur shoots or flowers consisting of 5–6 free or partially fused sterile scales and a few stalk-like sporophylls each with a terminal, erect or anatropous ovule. Among important genera belonging to this family are *Lebachia*, *Ernestrodendron* and *Voltizia*. The other family Palissyaceae has two genera, *Palisaya* and *Stachyotaxus*. In these the female cone comprises spirally arranged decurrent bracts and flattened, elongated, axillary seed scale complexes each having two or more short, stalk-like, biserial sporophylls with a single terminal orthotropous ovule.

Coniferous wood is easily recognized from that of the Cycadales and Gnetales but does not differ markedly from the wood of *Ginkgo*. In most conifers the tracheids bear a single row of bordered pits or sometimes two rows lying opposite to each other. In the Araucariaceae the bordered pits are in one, two or three rows on the radial walls but always in contact and therefore polygonal in outline; the pits of adjacent rows are alternate and not opposite. In *Taxus*, *Torreya* and *Cephalotaxus* the pitting is similar to that in pines but there are conspicuous spiral bands which make taxinean wood easily recognizable. Sporadically such bands also occur in *Pseudotsugy*, *Phyllocladus*, *Larix* and *Abies*. Bars of Sanio are present in all conifers except in the Araucariaceae but even here they have been seen in the cone axis.

The occurrence of resin canals in a characteristic feature of the wood and cortex and in the secondary xylem they may be produced as a response to wounding. *Taxus* has no resin canals, although resin cells are found here and there.

Florin (1951) has emphasized the value of epidermal characters in the identification of fossils. The stomata are always haplocheilic but may be mono- or amphicyclic. Of importance in identification are also the shape of the epidermal cells including the presence of papillae or hairs, the distribution and arrangement of the stomata on the upper and lower sides of the leaves, the longitudinal and transverse distance between adjacent stomata, the extent to which the guard cells are sunken, and so on.

The strobili are always monosporangiate although the plants may be dioecious or monoecious. Bisporangiate strobili have been seen only as an abnormality in *Abies, Juniperus, Larix, Picea, Pinus, Pseudotsuga and Sequoia*.

The male cones are rather small but those of *Araucaria* reach a length of 15 cm. In the Cupressaceae the microsporophylls are opposite or whorled but in the rest of the families the arrangement is spiral. In the Pinaceae and Podocarpaceae there are always two microsporangia borne on the abaxial side of the sporophyll; in the Taxodiaceae and Cupressaceae their number is variable. In the Araucariaceae there are several large, slender and pendent sporangia. In *Taxus* the sporophyll is peltate and bears 3–8 microsporangia on its lower surface.

The nature of the female cone has been a matter of controversy for many years and there was a keen debate as to whether the cone is simple (a flower) or compound (an inflorescence). After Florin's investigation of the fossil forms it now seems certain that the ovuliferous scale is really a shoot bearing both sterile leaves and sporophylls, all consolidated by reduction and adnation into one simple looking structure.

The development of the male gametophyte varies in different families. While the Taxodiaceae, Cupressaceae, Taxaceae and Cephalotaxaceae lack prothallial cells, the Pinaceae consistently show two and the Podocarpaceae and Araucariaceae have a larger number derived by secondary division. In some genera the body cell divides to give rise to two equal sperms; in other the inequality is evident from the beginning.

The female gametophyte shows prolonged free nuclear divisions. After wall formation a variable number of archegonia are differentiated. These may be organized into a complex as in the Cupressaceae or borne irregularly and even laterally as in *Sequoia*.

The pollen grains of the Pinaceae and Podocarpaceae are winged but the wings are perhaps of little value in keeping the pollen grains afloat in the air. More likely, as Doyle points out, the wings help to orient the pollen in the pollination drop so that the wings lie upward, and in ovules which are inverted or directed towards the axis the germinal side comes in close contact with nucellus.

In many Conifers the interval between pollination and fertilization is as long as 12 months. Possibly there are inhibitory substances in the ovule but this needs further study.

The embryogeny of conifers shows a wide variation and reference must be made to the numerous papers published by Buchholz and Doyle on this subject. One important feature is that except in *Sequoia* the first division and often several of the subsequent divisions are free nuclear. Another is the occurrence of simple as well as cleavage polyembryony. According to Doyle, the podocarps show the basal type in conifer embryogeny while the pines show a derived condition.

Conifers are among the most useful timber trees of the world. In addition they provide many kinds of resin, pitch, turpentine, and varnishes. Most of the newsprint of the world comes from coniferous woods, and the seeds of some of them are edible.

P. MAHESHWARI

References

Buchholz, J. T., "Gymnosperms," *In* Encyclopedia Britannica, 1946.
Chamberlain, C. J., "Gymnosperms, structure and evolution," Chicago, The University Press, 1935.
Doyle, J., "Aspects and problems of conifer embryology," Adv. Sci. **54**: 1–11, 1957.
Florin, R., "Evolution in Cordaites and Conifers." Acta Hort. Bergiani **15**(11): 1951.
Gaussen, H., "Les Gymnosperm actuelles et fossiles," Trav. Labor. Forest Toulouse 1941–1960.
Schnarf, K., "Embryologie der Gymnospermen," Hnd. Pflanzanat. **2**: 1933.

CONJUGATION

Conjugation denotes several processess that connect, couple or unite CHROMOSOMES, NUCLEI, CELLS or INDIVIDUALS. (1) Conjugation of chromosomes: the side by side pairing (*synapsis*) of homologous chromosomes during the first MEIOTIC division. (2) Conjugation of nuclei: (a) in general, *karyogamy*, the coming together and fusion of gamete nuclei; (b) as a special case in certain Fungi (Basidiomycetes), *dikaryotization*, the establishment throughout the plant of separate but adjacent haploid nuclei in pairs, one from each parent, without nuclear fusion until after repeated synchronous (*conjugate*) divisions and formation of fruiting bodies. (3) Conjugation of cells: (a) in general, *syngamy*, the union of sex cells (*gametes*); (b) the merging of a number of cells into a multinucleate plasmodium or syncytium. Union of organisms consisting of but a single cell or of a filament of equivalent cells becomes (4) conjugation of individuals: (a) in certain Algae (Zygnematales), the formation of connections between pairs of laterally juxtaposed cells leading to fusion of the connected cells and their nuclei; (b) in ciliated Protozoa, the temporary coupling of two individuals during meiosis and reciprocal cross-fertilization. In contrast to the latter denotation of temporary union, permanent or complete fusion of cells or cell-individuals is sometimes denominated copulation; but copulation also denotes temporary union of multicellular animals in the sexual act. Although this confusion in terminology remains unresolved, one distinctive technical meaning of conjugation is agreed upon, namely, the mating of ciliated Protozoa (4b above). With it alone the remainder of this article deals.

Conditions for conjugation. Depletion of food following a period of feeding is the commonest requirement for conjugation, but various additional special requirements such as particular temperatures or certain light-dark relations (appearing as rhythms of mating reactivity) may be exhibited by various species or strains. In at least some species, conjugation can be induced by adjusting the chemical constitution of the surrounding water (especially the amounts of potassium, magnesium and calcium). In the absence of such chemical requirements the occurrence of conjugation depends upon the meeting of individuals belonging to different and complementary physiological types, mating types. In some Ciliates there are just two mating types, in others up to fifteen or more. The usual system is that any two of the complementary mating types can conjugate with each other, individuals of the same mating type being unable to interbreed except under special chemical conditions. There are as a rule no visible differences between complementary mating types. However, in the Peritrichida, conjugation takes place only between a larger, stalked, sessile individual (the female) which looks like an ordinary vegetative individual, and a smaller, motile male which arises by special divisions. Visible differences between mates exist also in some Suctoria and a few other Ciliates.

The process of conjugation. Aside from minor variations in detail from species to species, the major events of conjugation follow the same pattern in all Ciliates in which the two mates are morphologically equivalent. The conjugants adhere in a position characteristic for each species (after the preliminary sex dance in Hypotrichida and after the preliminary agglutinative clumping in Paramecium). The diploid micronuclei then undergo the two standard meiotic divisons and one or more of the resulting haploid nuclei in each conjugant divides again into a male and a female gamete nucleus. One male nucleus of each conjugant migrates into the mate with whose female nucleus it unites. This nuclear transfer may or may not be accompanied by detectable transfer of cytoplasm; variation in this respect occurs even within a species. (In the extreme case, permanent cytoplasmic union may be effected

without loss of individual integrity of structure; this results in true-breeding doublet animals.) After reciprocal cross-fertilization, the diploid fusion nucleus (*synkaryon*) in each mate divides one or more times. Of the resulting diploid nuclei, some may disappear, but some develop into diploid micronuclei and others into polyploid macronuclei, the numbers varying with the species. Normally the mates separate during these post-fertilization events and soon begin their repeated cycles of growth and fission. Within a few fissions, the prezygotic macronuclei, or the pieces into which they disintegrated, disappear. Two important but relatively rare deviations from these events may occur: (a) failure of the male nuclei to pass into the mate, in which case one male nucleus unites with one female nucleus of the same animal (*cytogamy* or *double autogamy*); (b) the new macronuclei may arise from pieces of the old macronuclei (*macronuclear regeneration*).

In Ciliates such as Peritrichida, some Suctoria, and a few others, in which the two mates regularly differ in size, the pattern of events in conjugation differs in three important respects: (a) the two mates fuse completely and permanently into one; (b) each mate produces only one functional gamete nucleus; and consequently (c), each pair of mates forms only one fertilization nucleus.

The functions of conjugation. Like sexual reproduction in general, conjugation serves to bring together and recombine the genes of different individuals, the resulting varied combinations of hereditary traits providing an important source of materials upon which natural selection can operate. (The standard Mendelian rules of inheritance have been demonstrated to apply to all adequately studied Ciliates, such as Paramecium, Tetrahymena, and Euplotes.) This function of conjugation has as a corollary the delimitation of potentially common pools of genes: animals which can conjugate fruitfully can draw upon the same supply of genes. Remarkably, every well-studied taxonomically recognized species of Ciliate has turned out to consist of several, up to more than a dozen, sexually or genetically isolated varieties (*syngens*) which cannot interbreed or cannot interbreed fruitfully. Each syngen has its own characteristic geographical distribution and its own distinctive mating types; but thus far syngens have defied unique morphological definition. Syngens of the same taxonomic species may be considered as sibling species.

A second function of conjugation has emerged from recent genetic studies, namely, the production of a new kind of genetic variability which is independent of gene differences. Macronuclei, during their development from products of the fertilization nucleus, are uniquely sensitive to the conditions prevailing in the cytoplasm and the milieu. They may then be irreversibly differentiated so that they and their descendants can determine only one of two or more alternative traits in spite of possessing the genes for both or all alternative traits. (This new genetic principle of nuclear differentiation has subsequently been discovered also in Vertebrates as a feature of embryonic development.)

Conjugation has a third function in many, though perhaps not all, Ciliates: it initiates a new life cycle, as does fertilization in higher organisms. The progeny of an exconjugant go through a series of life cycle changes with the passage of successive fissions: immaturity (inability to conjugate), maturity (ability to conjugate), senescence (declining ability to conjugate fruitfully and declining vigor), and ultimate death after hundreds or thousands of fissions. These changes occur only in the absence of intervening fertilization; the occurrence of conjugation reinitiates a new life cycle (*rejuvenescence*), if the exconjugants survive. The probability of survival is high when conjugation occurs during maturity, but progressively declines to zero during advancing senescence. Survival of the species thus depends upon the periodic recurrence of conjugation at intervals which are not too long. The life cycle depicted here is typical for many Ciliates, although the extent of the various stages varies greatly from species to species. Some Ciliates appear to be able to live indefinitely without conjugation, although some of these (*e.g.*, Tetrahymena) become genetically dead by loss of their "germ plasm", the micronuclei, and loss of the capacity to conjugate.

T. M. SONNEBORN

Reference

Sonneborn, T. M., "Breeding Systems, Reproductive Methods, and Species Problems in Protozoa," *in* Mayr, "The Species Problem," pp. 155–324. Washington, American Association for the Advancement of Science, 1957.

CONNECTIVE TISSUE

Connective tissue is developed from the mesenchyme, fetal supporting tissue. Mesenchymology is the science of connective tissues.

Connective tissue is ubiquitous, constituting tendons, fasciae, ligaments, joint capsules, dermis, important parts of blood vessels and heart, the fibrillar framework of the organs, their capsules, loose interstitial tissue, cartilage, and bone. Fatty and slimy tissues are special types.

Fibrillar connective tissue consists of cells, fibrils, and an amorphous jelly-like ground substance. Hyaline cartilage owes its plastic-like consistency mainly to its ground-substance gel. Fibrous and elastic cartilage are characterized by the type and amount of the fibrils. In bony tissue inorganic salts are deposited in the ground substance.

The cavities of the eye, inner ear, joints, and the Graafian follicles are mesenchymal spaces filled with mucinous fluid of ground-substance type.

Ground substance. The amorphous ground substance contains acid mucopolysaccharides (A.M.P.), neutral sugar, proteins, inorganic salts, water, hormones, and various metabolic products on their way between blood and tissue cells.

The A.M.P. hyaluronic acid (h.a.) consists of N-acetyl-hyalo-biuronic acid polymerized in un-branched chains by glucosaminidic bonds. It is highly viscous and water-binding. It is depolymerized and hydrolyzed by bacterial and testicular hyaluronidase. H.a. forms complexes with proteins.

Chondroitin sulfuric acid is predominant in cartilage. It is insensitive to bacterial hyaluronidase. The molecule is smaller than h.a. Solutions are slightly viscous.

Chondroitin is isomeric with h.a., non-sulfuric, and contains galactosamine for the glucosamine moiety of h.a.

Heparin is a sulfomucopolysaccharide with anti-coagulation effect.

Heparitin sulfuric acid is heparin-monosulfuric acid which, contrary to heparin, contains acetyl.

Keratosulfate is the only A.M.P. without uronic acid.

Fibrils. *Collagen* is a fibrous protein occurring in white bundles with considerable tensile strength and low elasticity. The elementary fibril is un-branched and the diameter uniform. By electron microscopy and X-ray diffraction transverse bands appear at intervals of 640 Å, and, besides, several subunits are known. This periodicity is characteristic of collagen, as well as the content of hydroxyproline. The proline and glycine content is high. Neutral salt and acid soluble collagen are precursors of insoluble, fibrous collagen. Citrate soluble collagen is called procollagen. Monomeric collagen is synthesized in fibroblasts and polymerized extracellularly into collagen fibrils. Soluble collagen is a biologically active substance, whereas insoluble collagen has a very slow turnover.

Reticulin fibers are isotropic structures with close morphologic and chemical similarities to collagen. Their diameter is smaller, they show real branching, and the material is intimately bound to mucopolysaccharides. Unlike collagen, reticulin is demonstrable by silver stains and by the periodic acid-Schiff stain for polysaccharides.

Elastic fibers consist of an amorphous matrix substance and fibrils. They possess a considerable elasticity and tensile strength. Elastin is insoluble in acids or alkali. The proline and glycine content is high, but unlike collagen, elastin does not contain hydroxyproline. There is specific, unexplained, histochemical stainability with certain dyes, e.g. orcein.

Cells. *Fibroblasts* are spindle-shaped or rounded cells with an ovoid nucleus of loose chromatin structure and a non-granular cytoplasm. Their main function is production of collagen.

Chondrocytes are situated in lacunae in the intercellular substance of cartilage. If this is calcified the cells die.

Osteocytes are located in lacunae in the calcified intercellular substance of bone. They are nourished by a system of small canals.

Mast cells are large mesenchymal cells with a central nucleus. In the cytoplasm are globular granules staining metachromatically with toluidine blue and other basic dyes. After degranulation, the cytoplasm appears spongy. Mast cells synthesize A.M.P. of the ground substance type. Extracellular water controls the release of granules.

Mast cells contain and release *histamine*, and, in mice and rats, serotonin (5-hydroxytryptamine) as well.

The main function of *reticulo-endothelial cells* is phagocytosis and antibody-production.

Hormones control the functions of connective tissue. In myxedema mucin is predominant, in Cushing's disease fat overbalances mucin. A balance between fat and mucin seems to exist normally.

Besides stimulating thyroid function, thyrotrophin brings about mobilization of fat from the normal depots and deposition in muscles, liver, and kidneys, stimulates mast cellular synthesis of mucopolysaccharide, and produces exophthalmos. Thyroxine inhibits these effects.

In some animal species estrogens stimulate mucin production. Somatotrophin stimulates collagen formation.

Adrenal glucocorticoids inhibit connective tissue

regeneration, interfering with the functions of both mast cells and fibroblasts.

Histamine and serotonin stimulate connective tissue regeneration by producing edema.

Vitamin-C plays some role in the production of collagen in the fibroblasts. The tensile strength of healing wounds is decreased in scurvy. Mast cells and mucinous substances seem to be influenced too.

Functions. The considerable tensile strength of joint capsules and ligaments, the lubricating and shock-absorbing effect of synovia, the tension of the vitreous body, stuffing effect of fat, elasticity of cartilage; hardness of bone, etc. are important qualities. By its content of hyaluronic acid, connective tissue is a water depot.

Variations in the physical and chemical conditions of connective tissues influence the absorption of nutrients from intestine to blood, the passage of hormones from blood to tissue cells, the functions of the eyes and labyrinth, ovulation, etc.

Healing of wounds and bone fractures is primarily connective tissue repair. Inflammation and granulation-tissue formation is a tissue reaction to injury. Mucinous organization of tissue edema leads to fibrosis.

Infection by hyaluronidase-producing bacteria changes the viscous ground substance into a watery material enchancing spreading. Certain breakdown products counteract spreading, and some are believed to neutralize bacterial toxins.

By way of hyaluronidase the spermatozoon can penetrate the hyaluronate-containing mantle of the ovum.

Connective-tissue responses to tumor growth are similar to the inflammatory reaction. The interaction between tumor and host is reflected *int.al.* in accumulations of c.t. cells, mucin production, and fibrosis.

In *systemic c.t. diseases* lesions of the extracellular substances are primary. Characteristic tissue changes are edema, mucinosis, serous inflammation, necrosis, fibrosis, and fibrinoid deposition.

Drugs acting directly or indirectly on some point of connective-tissue regeneration are in clinical use. Therapy with hormones influencing the ground substance and mast cells bring about a prompt effect primarily because of the rapid turnover of hyaluronic acid. Cortisone brings quick relief to arthritic patients by reducing joint swelling. Thyroxine decreases the mucin content of myxedematous tissues. Desired effects may overlap untoward effects of a drug, involving changes also in healthy tissues or organs.

G. ASBOE-HANSEN

References

Asboe-Hansen, G., "Connective Tissue in Health and Disease," Copenhagen, Munksgaard, 1954.
Asboe-Hansen, G., "Endocrine Control of Connective Tissue," Am. J. Med. **26:** 470, 1959.
Randall, J. T., "Nature and Structure of Collagen," London, Butterworths, 1953.
Tunbridge, R. E., "Connective Tissue. C.I.O.M.S. Conference," Oxford, Blackwell, 1957.
Hall, D. A., "Internat. Review of Connective Tissue Research," New York, Academic Press, 1963.
Graumann, W. ed., "Handbuch der Histochemie II/2." Stuttgart, G. Fischer, 1964.
Jackson, S. F., Harkness, R. D., Partridge, S. M., and Tristram, G. R., ed., "Structure and Function of Connective and Skeletal Tissue," London, Butterworths, 1965.
Asboe-Hansen, G., "Hormones and Connective Tissue," Copenhagen, Munksgaard, 1966.

CONSERVATION

Conservation is essentially the preservation of man's environment in a condition to fulfill his needs for a healthy and satisfying life.

In his early hunting and food-gathering stages man's impact on the world around him was not serious in terms of the ability of the world to support him. If the food supply became depleted locally, it was usually possible to move a short distance to an undepleted area, leaving the old area to recover.

When man domesticated grazing animals the pressure on his environment increased, but he escaped the consequences by developing mobility and becoming nomadic. Competition for range lands led to a highly developed practice of warfare, which served to ease the pressure by thinning out both the herds and their masters.

The warfare led to the establishment of towns for defensive purposes, and, to maintain the resulting concentrations, agriculture became important. Cities grew, with great division of labor and with expanding dependent agricultural areas; these supplied food to and were defended by the cities. At about this period in the history of man, local scarcities began to develop, signs of over-use appeared, and man's impact on his environment became noticeable.

He had, by this time, developed a certain level of intelligence, begun to associate causes with effects, and in many ways to benefit from experience. When shortages developed it now was possible to associate some of them with their causes. Here and there measures were devised to alleviate some of these shortages. These measures took many forms, but taboos on the use of certain things were common. Certain prized items were reserved for royalty and the favored classes, or even the favored sex. Many of these customs were interwoven with religious beliefs. Game preserves and estates were set aside for the use of royalty and the nobility.

Some cultures developed patterns of frugality and discouraged waste, but most of them exploited their resources mercilessly, impoverishing their environments more and more. Some peoples developed habits of conquest, expanding their territories and taking what they needed from their more provident or more fortunate neighbors. Others developed commerce, bringing the things they no longer had in sufficient quantity from regions where there was still a sufficiency. Payment was in services, artisanship, or in goods which were still in adequate supply or were acquired elsewhere. Artisanship and commerce brought about the discovery and development of new resources. Ingenuity developed, mechanical advantage was employed to speed up the utilization of resources that could be converted to goods. Industrialization came into being, increasing many-fold the pressures on the environment and the consumption of resources. Short supply led to increased commerce, also to increased conquest. To a measure, in some cultures, it led to frugality and avoidance of waste.

Of course, this pattern of cultural evolution did not take place at the same rate, or in an identical manner, in different regions. Some cultures reached a high degree of industrialization while others were still in the hunting or nomadic stages. This made it possible for those with superior weapons, created by industry, to conquer their weaker neighbors who lacked the more effective organization and tools of war of the industrial cultures. By these conquests "new" lands were "opened up." These lands, only slightly affected by their former owners, were rich and the machines produced by industry brought about a great and sudden expansion of production of food and raw materials.

Heretofore, population had grown, along with more effective exploitation of resources, at a slow but increasing rate. Where this increase exceeded the augmented production, poverty became the prevalent condition or war was waged. The sudden expansion of production following the conquest of weaker peoples and the "development" of the resources of their lands gave to the aggressive dominant cultures the illusion that resources were limitless. The idea prevailed that all problems would be solved merely by learning how to develop these resources. The widespread poverty in the "backward," or later "underdeveloped" countries was regarded simply as the consequence of their lack of industrialization, to be remedied by education in industrial ways. Meanwhile, these poorer, unindustrialized regions were important sources of raw materials, as well as markets, for industry.

As this pattern developed, a scattering of people began to worry about the possible depletion of the resources needed for industry, about the increasing signs of loss in productivity of soils, and about the increasing ugliness that accompanied every manifestation of industry and its attendant commerce. It was no accident that this concern developed in the highly industrialized areas. The depletion and ugliness were most apparent there. Also, the profits of industry and commerce made possible the leisure and spare energy needed to notice and ponder on these disturbing phenomena, or at least to notice that all might not be well, even in the thriving industrial cultures.

As the people who worried about these symptoms became articulate and increasingly called attention to the prodigal exploitation and waste of resources and the deterioration of the once beautiful environment of man, they began to propose remedial action. A general word came into use for such action. This was "conservation."

As this word came into wider and more frequent use, the concept denoted by it underwent development, proliferation, and modification. Its devotees became more numerous and more influential. The meanings of the word likewise became more numerous, diverse, and hard to pin down. As the concept gained influence, it inevitably was exploited, distorted, and used for ends unsuspected by the innocent and well-meaning followers of the very worth-while "conservation movement" that had grown up. Some of these ends were, and are, the direct antithesis of the concept properly termed conservation. The camp followers of conservation were, of course, never more than an unscrupulous or deluded minority, but they have introduced much confusion into the popular idea of conservation. The current status of the conservation movement

is one of a powerful force, perhaps better a great confusion of forces. The resultant tends to develop in the public an increasing awareness of waste, destructive exploitation, and the deterioration of the environment, but no absolute certainty as to just what the proper objectives are. There are conflicts between the purely economic concepts of conservation for practical reasons and use for monetary gain and the aesthetic idea of conservation for continued enjoyment. There is great difference as to emphasis on immediate use of resources or their preservation for the use of future generations. The welfare of these future generations is indeed much discussed, but when a choice arises between immediate benefit and that of posterity, the present commonly wins out. This does not necessarily mean that careful-use principles are never followed, or that waste is always permitted, but most frequently wasteful and destructive practices prevail because they are cheaper. Conservation is often a matter of degree, or even only of lip service. However, at least in the United States, Western Europe, and certain dependencies of European countries and their Commonwealth associates, conservation is becoming increasingly important in thinking, in everyday practice, and in government policy.

It is maintained by some that conservation is strictly an economic matter, that the resources at stake are controlled by the laws of supply and demand, and that their values can be expressed in economic terms. There is no doubt that economics play a very important part in conservation, in both a positive and a negative way. However, large aspects of human values are strictly emotional. They can scarcely be measured in material terms, nor can a reliable money value be assigned to them. Even money, itself, often acquires an emotional value that may not be very closely related to its intrinsic value.

These facts introduce a serious unreliability into attempts to reduce conservation in certain fields to economics. Hence, many conservation value judgements may, and should, remain intuitive, and will be seemingly arbitrary rather than economically logical. This is especially true wherever the resources concerned are of aesthetic or sentimental significance, or where anything beyond the immediate future is concerned. Such value judgements may not be lightly dismissed as is shown by the fact that most of the values for which men will face death—such as love, patriotism, or honor—or for which they will face economic loss—such as friendship, dignity, sentiment, or, above all, beauty—cannot be measured in money.

Out of all the welter of trial and error, astuteness and blundering, altruism and selfishness that have characterized the development of the conservation idea, some of the better thinkers in the movement are beginning to forge a sound and rational philosophy of conservation. As this develops there are signs of the appearance of what may be called a conservation attitude, epitomized by one of the best thinkers and spokesmen of the movement, Luna B. Leopold. He urges us to "identify land-use and resource development measures which are financially sound investments, quite justified without a conservation label. We would then fortify our position and consolidate our ranks when there was need to maintain and protect some esthetic value, some piece of scenery, some wood lot full of ladyslippers, some stretch of white water or of wilderness, which could never be justified on strictly economic grounds. It will require self restraint which can come only with the development of a conservation attitude for us to fly the conservation banner only on those things whose value lies not in our pocketbook but only in our heart." In other words, we are perhaps approaching the point where we want to preserve our environment in a condition that will not only sustain life but also make possible a pleasant and rewarding life.

The really devoted conservationist commonly follows this path at a sacrifice of time, money, or both. He knows that the benefits from his activity will mostly accrue to future generations rather than to himself. It is probably not going too far to suggest that conservation is coming to be one of the higher motivations of man—comparable with loyalty to family and to native land, love of music and other forms of beauty, and even religious devotion.

Material resources are commonly classified as renewable and non-renewable, and the problems and approaches relating to their conservation are widely different.

Among non-renewable resources some can only be utilized by consumption, others by enjoyment as they are, without alteration. Between these two subclasses, again, conservation is accomplished by quite separate means. Natural resources that are consumed when used are ultimately all in limited supply, either great or small. When this supply is exhausted nothing can be done about continued use. Therefore, avoidance of waste and use for only essential purposes is about the only way they can be conserved. Resources that are only of value in an unaltered state, on the other hand, are conserved by protection and preservation.

Things that are of benefit through aesthetic or contemplative enjoyment, through educational or scientific value, or because they perform a protective function, often present some of the most difficult conservation problems. Their values are of a long-term nature. Economically the gain from their preservation may not equal that to be derived from destroying them. The debate may be in simple terms—is a magnificent tree of more value growing, as a thing of beauty, or cut down, as lumber? It may be more difficult, as whether a beautiful canyon is more valuable dammed, as a continuing source of electrical energy and irrigation water, or in its natural state, as a source of inspiration and outdoor enjoyment. All too frequently the alternative that can be put in economic terms wins out in such a conflict of interests, particularly since the economic values usually accrue to one or a few individuals, while the aesthetic values are distributed to large numbers scattered through long periods of time. The ones who have not yet been born are not in a position to make their interests known. And because of this dominance of economic considerations the world becomes steadily less beautiful. Since aesthetic and similar values can scarcely hold their own in the free market place, the way to conserve resources of this nature is to preserve them by arbitrary decision without regard to economic values.

Certain categories of resources, such as forests, water supplies, wild-life, soil fertility, and even pure air, have the property of replenishment or renewal if man-

aged properly and not over-used. These renewable resources present an entirely different picture, and it is in relation to them that the definition of conservation as "wise use" arose. Wise use is essentially a matter of utilizing the income and leaving the capital intact to continue producing income. For example, if water is used from a ground-water body or aquifer faster than it is replenished by infiltration, or if the catchment area is damaged so as to impede infiltration or increase runoff, a shortage will result. But if use does not exceed infiltration and the catchment area is protected so that its capacity to absorb water remains unimpaired, there is no reason why the aquifer should not yeild a supply of water indefinitely. Soil, if planted repeatedly to non-leguminous crops and not fertilized, soon loses its fertility and the crop yield decreases. If its surface is not protected, it erodes and agriculture may become impossible. However, if proper rotation with leguminous crops is practiced, fertilizer is applied, and the surface is protected from erosion, yields may continue indefinitely at a high level or may even improve. Wildlife, if over-exploited or deprived of a suitable habitat, will disappear. But if the habitat is maintained and only the increase harvested, the wildlife populations will remain steady.

Basically conservation is an aspect of applied ecology. Man behaves as a pioneer species, tending to increase rapidly and to destroy his habitat, to render it incapable of satisfying his requirements or unsuitable for him physiologically or psychologically. Conservation attempts to offset these destructive tendencies. Of course, if man continues to increase in numbers, especially geometrically, as at present, these efforts are doomed to failure. Therefore, one of the basic ecological necessities, and the *sine qua non* of long term conservation, is to halt this increase before the carrying capacity of the habitat is exceeded. This is necessary in order to preserve human society even at a level of bare existence. Most thoughtful conservationists do not, however, find the mere survival of the human race at a minimum level a satisfactory prospect. Hence, a very large proportion of the total conservation effort goes into the fight to preserve the features of the environment that, while perhaps not required for the continued existence of the human species, are what elevates human life above the strictly animal or physiological plane. The dullness and monotony of life in an environment completely saturated with human population are fearful to contemplate.

In western civilization, in spite of its emphasis on individualism, there has been for many years a growing realization that conservation is a matter for government concern. It cannot be left entirely to indivuals, even though individuals derive the benefits from it. Early government efforts were for the benefit of royalty and nobles in the form of royal game preserves. Louis XIV, of France, established royal forests in order to have timber and other materials available for his navy. Here scientific forest management saw some of its early development. Forestry reached a fairly high degree of accomplishment in the continental countries of Northern Europe relatively early, and is generally a national or community matter there. In most of these countries departments of waters and forests or the equivalent have the responsibility for conservation matters. Most governmental efforts of this sort in other parts of the world, except in the United States, at least prior to World War II, were introduced by European countries into their colonies or dependent countries. In some cases the mother country may have been even more concerned about a continued supply of raw materials from the colonies than about conservation at home.

The United States, after a history of profligate exploitation of resources seldom equalled in the history of the world, essentially started its conservation program in 1872, with the setting aside of the Yellowstone country as the world's first national park. Other national parks followed, each one only after a bitter struggle between a few visionaries who would later have been called conservationists, and those who saw a chance of individual economic gain through some form of exploitation of the areas. Only in 1916 was the National Park Service formed to look after these national treasures and the many others that have since been added to the national park system.

The U.S. Biological Survey was founded in 1885, to study the whole subject of wildlife. It became more and more strictly concerned with maintaining adequate populations of fish and game for fishing and hunting. It was transferred from Department of Agriculture to the Department of the Interior in 1939 and shortly afterward it was combined with the Bureau of Fisheries and its name was changed to the U. S. Fish and Wildlife Service.

Forest conservation in the United States came into being at the end of the 19th century largely through the efforts of Gifford Pinchot, advisor to President Theodore Roosevelt, governor of Pennsylvania, and first chief forester of the United States. The first national forests were set aside in 1891 by President Benjamin Harrison, and the U. S. Forest Service was created by President Theodore Roosevelt in 1901 to administer them.

In 1932, in a period of serious economic depression, and of severe wind and water erosion of soils in many areas, another President Roosevelt, Franklin D., made conservation history by setting up the Soil Conservation Service to cope with soil erosion and depletion on a nationwide scale. He also created the Civilian Conservation Corps. This was an organization to alleviate the widespread unemployment of the time by giving large numbers of unemployed young men the opportunity to work, for modest wages and subsistence, on various conservation projects on government lands. This was a temporary effort which ended when unemployment was reduced by preparations for World War II.

This war was a period of waste of resources on a scale not even imagined before. It might have been expected that this would be followed by a great upsurge of conservation measures to compensate for the loss. This took place on a disappointingly small scale, however. A small start has been made in the post-war years toward reducing water pollution, but probably the result has not even equalled the increase of pollution. An even more feeble beginning has been made toward watershed protection and reduction of floods at their sources. One comprehensive river-valley development, the Tennessee Valley Project, with considerable emphasis on conservation has been completed and others are being planned.

Some slight conservation activity has been evident in parts of the world where such things were unknown before. Some of this is spontaneous, more has been stimulated by aid and propaganda from the United States and European countries, and much of it has been started by certain agencies of the United Nations which have developed a substantial interest in conservation. Most post-war government projects in these countries have been so confused with development projects that it has been hard, sometimes, to distinguish conservation from stepped-up exploitation, the utilization of resources more effectively being the principal aim.

The development of private organizations devoted to various aspects of conservation has been one of the conspicuous features of the post-war period. Of course, some of these existed before World War II, especially in certain western European countries and the United States. But since then, there has been an enormous surge of such activity. Some of the U.S. organizations with their fields of interest are: The Wilderness Society, dedicated to the preservation of wilderness values and of large wilderness areas; the National Parks Association, to the protection and promotion of the national park system; The Nature Conservancy, engaged in preservation of natural areas; the National Audubon Society, basically ornithological, but interested in general preservation of nature, the Soil Conservation Society of America; the American Forestry Association; the National Wildlife Federation; the Wildlife Management Institute; The Trustees for Conservation, interested in conservation legislation; The Conservation Foundation and the Conservation and Research Foundation, both for promotion of research in conservation.

On the local government level, almost every state has one or more conservation departments, under one name or another. Organizations under country or community auspices exist in some places. There are also many local private bodies, some of them doing excellent work in the creation of a conservation consciousness and in the protection of environmental values.

Although most efforts in conservation have been on national or lower levels, conservation problems are not all confined by political boundaries, nor are natural features of interest only to the residents of the countries where they happen to occur. The Grand Canyon or Victoria Falls, for example, are just as highly valued elsewhere as in the countries that possess them. International conservation efforts have been slow in development, although an international migratory bird protection treaty has been in existence for many years and fur seals have likewise been internationally protected. In 1948 the International Union for the Protection of Nature was formed, an organization of governments and of conservation organizations. Its name was changed recently to the International Union for the Conservation of Nature and Natural Resources. Its headquarters is in Morge, Switzerland, and it has held several very effective general assemblies and technical conferences. Gradually obtaining more support, it hopefully portends a new era when conservation of the habitat of man will be considered a world-wide problem and a responsibility of the human race as a whole.   F. R. FOSBERG

CONTORTAE see GENTIANALES

CONTRACTILE VACUOLE

Contractile vacuoles are organelles present in many PROTOZOA and in the different cells of freshwater sponges (see PORIFERA). Usually the contractile vacuole fills up gradually with fluid (*diastole*), then collapses, discharging its contents to the outside (*systole*). Fresh-water RHIZOPODS and MASTIGOPHORA possess contractile vacuoles while marine forms only rarely do. Contractile vacuoles are completely absent in parasitic forms of these two groups as well as in the purely parasitic class SPOROZOA. Besides their presence in the fresh-water ciliates, contractile vacuoles are usually found in marine and frequently in parasitic CILIATES.

In its simplest form (many amoebae), the vacuole is a spherical vesicle of no fixed position. In Protozoa with constant form and firm pellicle (e.g. ciliates and flagellates), the contractile vacuoles lie at very definite positions and are of constant number. A more complex type (mainly in Euglenoidea) has one or usually more secondary vacuoles in which the fluid is first collected, later to pour into the main contractile vesicle. In a third or more advanced type, the main vesicle is fed by a varying number of either longitudinal or radial afferent canals.

The vesicle may be provided with a thick lipoid membrane which reduces OsO_4 and which has long been homologized with the Golgi apparatus of metazoan cells. Electron microscopy has revealed that, like the Golgi material, this lipoid substance may be formed of electron-opaque double membranes. The differentiation of a cortical lipoid layer is generally accompanied by a relatively fixed position of the contractile vacuole. The lipoid substance might be in the form of a ring and not a sphere, and it can form a cortex around both secondary and principal vacuoles, while in more complex vacuoles it surrounds the walls of the afferent canals.

The hydrostatic function attributed to the contractile vacuole is based on the fact that in a fresh-water protozoan the cytoplasm represents a medium denser than the surrounding water and that the role of the vacuole in water drainage prevents excessive dilution of the cytoplasm which would otherwise result from the continuous endosmosis. Facts which support the hydrostatic function of the vacuole are: general occurrence of the vacuole in fresh-water forms; its absence in many marine and parasitic species; injection of distilled water into amoebae increases the rate of pulsation and water output of the contractile vacuole; the rate of pulsation bears a reciprocal relation to the salinity of the outer medium in some marine and parasitic forms; appearance *de novo* of contractile vacuoles in some marine amoebae under decrease in salinity of the medium.

The frequency of pulsation is greater in freshwater than in marine and parasitic forms. A number of factors affect the pulsation frequency, thus it rises with rise of temperature within non-injurious limits, slows down with O_2 deficiency, and increases at low and high pH values. Different salts have different effects on the pulsation frequency as they have different precipitating and thus blocking effects on the outer pellicle of the animal. The size of the contractile vacuole is another important factor in determining the rate of water drainage and has been observed to change considerably under different conditions.

Other mechanisms for the regulation of water exchange do occur in many protozoa, e.g. through the change in permeability of the outer pellicle. But this does not exclude the primary role of the contractile vacuole in osmoregulation. The presence of the contractile vacuole in fresh-water flagellates and amoebae and its absence from related parasitic forms point clearly to its osmoregulatory function. If most ciliates, however, behave differently in this respect, it may be concluded that in this group excretion of water is not the only function of the contractile vacuole and that other excretory substances also leave the body via this route so that the contractile vacuole cannot be dispensed with under such conditions of life where endosmosis does not take place at all. In some ciliates, excretory crystals have been observed to dissolve close to the contractile vacuole and their substances are excreted with the latter's fluid. Ammonia was found in the contractile vacuole fluid of many forms.

Three theories were advanced to explain the working of the contractile vacuole: a) the osmotic theory postulates that the passage of water from cytoplasm into vacuole takes place by osmosis; b) the filtration theory claims that hydrostatic pressure forces water through the vacuolar membrane; and c) according to the secretion theory, water is secreted into the vacuole by the membrane. The latter theory is the most widely accepted and it conflicts with no available data.

R. G. LUTFY

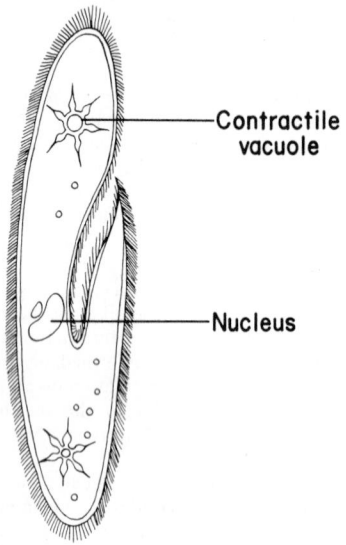

Fig. 1 Contractile vacuole in *Paramecium*. (From Goodnight, Goodnight and Gray, "General Zoology," New York, Reinhold, 1964.)

COPEPODA

The Copepoda constitute a subclass in the CRUSTACEA that are exclusively aquatic and are free-living, commensal, and parasitic. Nearly all are microscopic and found in all types of water bodies except in those of extreme salinity. This subclass contains at least four well established orders: Calanoida, Harpacticoida, Cyclopoida, and Monstrilloida, and a vast number of commensal and parasitic forms, poorly known, and variously included in three orders: Caligoida, Lernaeopodoida, and Notodelphoidea. In the Caligoida some authors include some or all of the parasitic copepods historically recognized in the order Lernaeopodoida, but the relations between all the orders and particularly those of the parasitic orders presently remain obscure.

In the free-living members the body is without carapace and is composed of two main regions: a cephalothorax and an abdomen. The cephalothorax consists of the head, without compound eyes, bearing five pairs of appendages, and a thoracic region with a maximum of seven pairs of appendages including a vestigial pair on the genital somite. Posterior to this is a limbless region of at most four somites. The anteriormost appendages are a pair of uniramous antennules of up to 25 segments used as a balancing organ in some, especially the calanoids. The antenna is second, being a biramous appendage having many segments (nine in some) in the exopod and generally more than one segment, usually two in the endopod. The third is a mandible, biramous, and may have four segments in both branches. Following these are the maxillule, biramous, and setigerous in many; the maxilla, uniramous and setigerous in calanoids but may have stout spines in other orders; the maxillipeds, uniramous and setigerous in some but strongly armed in others and may be a clawed prehensile organ. On each of the remaining segments of the cephalothorax there is a pair of biramous swimming legs, flattened and paddle-shaped having spines laterally and setae medially. The legs on segment six, which are in fact the fifth pair of legs, may be similar to the first four or only slightly modified, or progressively more modified in various species to being completely absent. Posterior to the cephalothorax there may be a marked constriction or a very slight one and even being absent depending upon the order in question. In members of the orders Calanoida, Cyclopoida and variously in the parasitic groups there is a marked constriction. In the Order Harpacticoida this constriction is much less evident and may be completely absent. The first segment of the abdomen consists of segment VI in the cyclopoids but is a fusion of segments VII and VIII in the calanoids. Segments VII and VIII are fused in both groups and bear the genital opening. Three segments follow the genital segment and terminally there are two rami, often elongate, bearing setae in a fan-like arrangement.

The Copepoda as a group have complicated life cycles. Eggs are produced. These hatch into a larval form, the nauplius. There may be as many as six nauplius instars but this number is often abbreviated in the parasitic forms. Some free-living cyclopoids have five. The first nauplius instar has three appendages. In each successive instar more appendages are added until there are seven in some, eight in others; the posterior two being the rudiments of the first and second swimming legs of the adult instar. The nauplii present a uniform morphology being more or less pear-shaped. The sixth nauplius stage, when present, molts into an instar that more nearly resembles the adult than do the earlier nauplii and is called a copepodid. There are six copepodid instars, the final one being the adult. In some of the parasitic forms the first two copepodids may be omitted along with all the nauplius stages so that the first appearing form is a copepodid III. Sexual dimorphism is evident in copepodid IV in the free-living forms. Where the sex ratios are known they approach 1:1 in all the sexual stages. Addition of swimming legs continues through copepodid IV, when a full complement is present but the most posterior are not fully developed. Sexual maturity occurs usually during the adult stage (copepodid VI) although some are precocious in stage V copepodid.

Eggs are extruded in a sac in the majority of Copepoda. Some calanoids and harpacticoids produce a large single sac attached to the ventral side of the genital segment, although *Calanus finmarchicus* (marine) and *Limnocalanus johanseni* (freshwater) extrude eggs without a sac.

The number of generations per year varies according to the species and the habitat. Some freshwater cyclopoids are known to produce a new generation every 10 to 14 days (*Eucyclops agilis*). Some marine cyclopoids (*Oithona similis*) are known to have three of four broods between March and September, having a longer life cycle ranging around 40 days. *Mesocyclops edax*, another freshwater cyclopoid, has two generations during the ice-free period from March to November in a temperate climate. Calanoids commonly have a life cycle requiring about 25 to 30 days, but one, *Diaptomus ashlandi*, is known to require the entire year for development from egg to adult. The amount of time spent in each instar varies but has been observed to be a few hours in *Eucyclops agilis*, and two or three days to as much as several months in calanoids.

Diurnal vertical migratory activity is prevalent throughout the group, and populations are known to occur in the range up to 700,000 per cubic meter in fresh-waters.

GABRIEL W. COMITA

Reference

Wilson, C. B., "The Copepods of the Woods Hole Region of Massachusetts," Washington, U.S. Nat. Mus., Bulletin 158, 1932.

CORACIIFORMES

The coraciiform birds comprise a surprisingly vast and interesting assortment of nearly 200 species that have been divided into ten different families. A striking feature that the ten families share in common is the syndactylous arrangement of the toes; that is, the anterior toes and sometimes the hind toe, are joined in various combinations, differing slightly in the different families. Large, long, or stout beaks and large heads are also characteristic. Many have brilliant plumages, with blues, greens, and shades of reddish brown predominating. All nest in holes or crevices of some sort, frequently in tunnels in banks. The eggs are usually white, often in large clutches, and the altricial young are slow to develop.

Largest and best known in the coraciiform families

are the kingfishers (Alcedinidae). J. L. Peters listed 88 species and 335 forms (species and subspecies). Though essentially world-wide in distribution, they are found mainly in the Old World and are largely tropical or sub-tropical in distribution. Only six species inhabit the New World and only one of these, the Belted Kingfisher, extends its range much north of Mexico.

Kingfishers are divisible into two rather distinct types: the wood kingfishers which inhabit woodlands in the Malayan and Australian regions and live on a varied diet of amphibians, reptiles and the larger invertebrates; and the true kingfishers which are sovereigns of the waterways and, as the name implies, feed largely on fish and other aquatic animals.

Kingfishers nest in tunnels in banks; these they excavate with great diligence, the male and female working in shifts. The sexes also share incubatory duties, some species changing over at regular 12-hour intervals with the female taking the night shift, some changing more frequently, and one, the large Ringed Kingfisher of tropical America, sitting for 24-hour periods. Kingfishers practice no sanitation at the nest which often becomes notoriously filthy; in sandy soils liquid excrement may be partially absorbed, but in clay banks it may accumulate and drain out the sloping entrance. After emerging from a visit to feed their young, the adults frequently plunge into the water and bathe. Fairly large broods of young, up to six or eight, are raised and remain in the deep tunnel, amid fish bones, scales, and feces, for nearly a month.

Closely allied to the kingfishers are the todies (Todidae, five diminutive monotypic species that inhabit the West Indies. They have somewhat flattened, weakly serrate, bright red bills, long tongues, and bright metallic plumages. Rictal bristles at the base of the bill are prominent, and the wings and tail are short. Todies are "flycatchers," waiting motionless on a perch and then darting out for approaching prey. Like other coraciiform birds they nest in holes which they excavate in banks.

Among the most curious of the coraciiform birds are the motmots (Momotidae) of Mexico, Central and South America. Eight species and about 45 forms are recognized. They are usually quiet sluggish birds, but at times they utter harsh "clack-clacking" notes. The most famous feature of the motmots is the long racket-tipped tail. The Blue-crowned Motmot (*Momotus momota*), for instance, is about a foot and a half in length, half of which is tail. The two central tail feathers are elongated with a bare subterminal portion preceding the expanded tip. Some claim the birds deliberately pluck the barbs from the subterminal portions of the tail; others maintain that the attachment of the barbs to the shaft is weak and that they drop off when the bird is preening its tail. Motmots exhibit various shades of turquoise, blue and green, presenting a handsome picture as they pearch nearly motionless over some woodland stream, slowly waving their tail like a pendulum.

In the Old World tropics occurs another fantastically plumaged family of birds known as the bee-eaters (Meropidae). Twenty-four species are known, and though primarily tropical, one species, the European Bee-eater (*Merops apiaster*) ranges into northern Europe. As in the motmots, the two central tail feathers in most, but not all, species are elongated, though not racket-tipped. Delicate pastel colors of blue and green predominate, with splashes of red, brown and yellow in some species. As the name implies, bee-eaters eat bees, often becoming a nuisance about apiaries, but they consume many other insects as well. Deep tunnels, up to ten feet in length, serve as nesting sites and sometimes as sleeping quarters for the adults in the non-nesting season. Bee-eaters are incined to be gregarious and often assemble in large colonies.

The Old World also harbors three families of birds known as rollers. The true rollers (11 or more species) belong to the family Coraciidae and occur on all Old World continents; but the ground rollers (Brachypteraciidae), which differ from the true rollers in several important skeletal features as well as having longer legs and more rounded wings, are confined to Madagascar. Still another roller (*Leptosoma discolor*) has so many peculiarities, that the single species is placed in a family by itself (Leptosomatidae). It is confined to Madagascar and adjacent islands.

Rollers get their name for their peculiar habit of turning somersaults or rolling over in the air during their courtship performances. They are noisy and somewhat quarrelsome, live mainly on large insects which they "hawk" from the air or ground, and nest in crevices and holes in trees or walls.

Perhaps the greatest oddity among the coraciiform birds is the well known Hoopoe (Family Upupidae) a single species well distributed over the warmer parts of Eurasia and Africa. A long decurved bill, an erectile crest of long black-tipped feathers, loud bars of black and white on wings and tail, and a fawn-colored body are among the features that make the Hoopoe unique. It is mainly terrestrial in habits, frequents open or semi-open places where it picks up or probes for a great variety of insects. It also nests in cavities, the female performing the duties of incubating while the male feeds her at the nest.

Six other species of hoopoes, called woodhoopoes, are considered sufficiently distinct (by some authorities) to merit placing them in a separate family (Phoeniculidae). They have elongate, rather than rounded, nostrils, a long graduated tail, and no crest. They occur in Africa.

Last but not least in this order of oddities are the hornbills (Bucerotidae). Forty-five species inhabit the warmer parts of Asia, Africa and islands of the Pacific and Indian oceans. Their greatest, or most conspicuous, anatomical peculiarity is the development of a horny casque, sometimes enormous, on the top of the bill. The union of the anterior toes for varying lengths in the different species gives them a broad-soled foot. Long eyelashes add still another peculiarity.

Hornbills are justly famous for their nesting habits. These vary considerably among the different species, especially as between the fruit-eaters and the insectivorous types. Both types, however, imprison the female in a cavity by plastering mast, mud and regurgitated fruit pits around the entrance until only the bill of the female can be extruded. The male then proceeds to feed the female during the long incubation period (from 28 to 40 days) and to feed both the female and the young during the nest period of the latter. The period of interment is about two months in the insectivorous species and four months in the frugivorous types. The sexes cooperate in the job of imprisonment, the female working from within and the male from without. The supposed function of this unique habit is to protect the female and young from predators during their long stay in the nest.

GEORGE J. WALLACE

References

Austin, O. L. Jr., "Birds of the World," New York, Golden Press, 1961.
Peters, J. L., "A Check-list of the Birds of the World," Vol. 5. Cambridge, Harvard University Press, 1945.
Skutch, A., "Kingfishers—Sovereigns of the Watercourses," Nature Magazine, **45**: 461–464, 500, 1952.
Van Tyne, J. V. and A. J. Berger, "Fundamentals of Ornithology," New York, Wiley, 1959.

CORALS

Living and extinct groups of marine organisms possessing hard skeletons, usually including only the COELENTERATA: Milleporina, Stylasterina, Spongiomorphida and Stromatoporoidea of the Hydrozoa; Scleractinia, Rugosa and Tabulata of the Zoantharia; Alcyonacea, Gorgonacea and Pennatulacea of the Octocorallia.

Most corals possess an ectodermally secreted corallum, a tube of calcite or aragonite surrounding the actiniform polyp, but in the Octocorallia the solid skeleton may be absent, replaced by a horny flexible axis or discrete calcareous spicules scattered through the tissues. A solitary polyp consists of ectoderm, mesoglea and endoderm forming a cylindrical body wall surmounted by an oral disc containing a mouth and one or more rows of tentacles. The internal cavity, the digestive sac, is more or less complicated by vertical, radial, pleat-like folds, the mesenteries. Asexual reproduction, which may produce discrete and separate polyps, may also result in the formation of complex compound polyps sharing a common digestive tract and bearing many mouths within a tentacular ring. Food taking is accomplished by the tentacles which are thickly studded with nematocysts or stinging cells characteristic of the Coelenterata. Mucous streams along the tentacles and oral disc may also convey entrapped organisms to the mouth. Reproduction is both sexual and asexual. Ripened eggs develop into ciliated planula larvae which are free for varying periods, finally attaching and growing into a polyp. Although sometimes resulting in discrete individuals, asexual reproduction more commonly produces compound or colonial coralla.

Scleractinia, the modern corals, are notable as the constructors of the massive calcareous structures of coral atolls, barrier and fringing reefs. Although they may be only a relatively small proportion of the biota of the coral reef community (calcareous algae may form up to 90 percent of the total reef mass), corals are conspicuous because of their brilliant coloration and the tremendous diversity of growth forms. Scleractinia form the interlocking framework of the reef providing a multitude of diverse habitats and controlling the accumulation of sediment. The reef formers are an ecologically similar but systematically diverse group of Scleractinia, the *hermatypic* corals, characterized by the presence of symbiotic unicellular dinoflagellate algae (zooxanthellae) in their endodermal tissues. These symbionts, in a relationship not yet fully understood, apparently stimulate the secretion of calcium carbonate, accelerating the growth rate of the corals. By providing an efficient mechanism for the absorption of waste products the hermatypic colonies may attain

Fig. 1   Coral reef. (Courtesy American Museum of Natural History.)

sizes not found in other colonial organisms; massive colonies 3 meters or more in height and breadth are not uncommon and may contain in excess of 30,000,000 polyps. The presence of zooxanthellae, although not obligatory for the coral, places the development of coral reefs under certain restrictions: radiant energy required for photosynthesis limits occurrences to 90 meters depth, usually less than 50 meters for active growth; and temperature requirements which limit growth to a range of temperatures of 16° to 36° C. although most active reef building occurs in the range of 23° to 25° C.

Through a propensity for asexual reproduction which results in colonial coralla, the Scleractinia are found in a bewildering array of highly adaptive forms. Many species are capable of assuming such shapes as a heavy dome, an encrusting mat, a stoutly branched form in the surf zone, or as foliaceous or delicately branched colonies in more protected areas. Other less plastic species may be specific for a certain ecological niche.

Largest and most luxurious reefs are found in the tropical regions of the western Pacific; lesser reefs occur in the eastern Pacific and western Atlantic, but they are almost absent from the eastern Atlantic. Because ecological requirements of hermatypic corals were apparently the same through the Tertiary, and possibly the Mesozoic, this group is one of the best indicators of past marine climates. The presence of reefs in rocks of these ages can usually be taken to indicate deposition under shallow water conditions in a tropical region.

The other group of Scleractinia, the ahermatypic corals, which lack symbiotic algae are distributed more widely from almost intertidally to depths of 6000 meters and through a temperature range of −2.0° to 36° C. They are found from the Antarctic to the Arctic on all types of substrate. Ahermatypic corals are usually simple forms with large polyps, most commonly solitary, but may be in simple dendroid colonies. Although individual colonies may be as large as shrubs in some deep water areas, the number of polyps is small in comparison to a similar hermatypic form. Because such colonies are formed only below wave base, the accumulation of dead material does not break up and ahermatypic corals may thus form thicket-like masses of considerable extent. Such occurrences are known from many localities along the western European continental shelf from Spain northward to Scandinavia.

Extinct Zoantharia include the Rugosa (= Tetracorallia), a solely Paleozoic Order ranging from the Ordovician to the Permian which are apparently not directly ancestral to the Triassic to Recent Scleractinia. With a corallum formed of calcite and possessing radial septa added in quadrants, the Rugosa contrast with the Scleractinia (=Hexacorallia) which have aragonitic skeletons and septa inserted radially in a hexameral fashion. Although most Rugosa were solitary in habit there were many colonial forms which were consituents of reefs during the Paleozoic, although not attaining the size of modern coralla. The Tabulata are a principally Paleozoic Order which differ from other Zoantharia by the almost complete absence of radial skeleton elements, having instead a dominance of transverse partitions of the corallite tubes, tabulae. Exclusively colonial, the Tabulata are important framework constituents of Paleozoic reefs.

Octocorallia differ from the Zoantharia by the invariate possession of eight pinnately branched tentacles on each polyp while a variable number are found in Zoantharian polyps. Included here are a diverse number of corals popularly known as the sea fans, sea whips, sea plumes and sea pens. Most lack a solid calcareous skeleton but are instead horny of chitinous, usually with calcareous spicules. Although common members of reef faunas, they range from the Antarctic to the Arctic and form shallow to abyssal depths. The Gorgonacea, most common as sea fans on the reefs of the Atlantic, also includes the precious coral of commerce. Most red or precious coral (*Corallium rubrum*), carved decoratively or utilized in jewelry, is obtained from the deeper waters of the Mediterranean by dredging. Despite competition from synthetics and Asian coral the industry continues, but is considerably reduced from its previous production. Gorgonian corals from Formosan and Japanese Seas are chiefly yellow or black varieties although some pink shades are also produced.

Stromatoporoidea, an extinct group ranging from the Cambrian through the middle Mesozoic, are a problematic group usually associated with the Hydrozoa. Occurring as large encrusting or massive bodies, they are important contributors to Paleozoic reefs. Another small group found exclusively in the Lower Mesozoic, the Spongiomorphida are found in the fossil coral reefs of Europe and North America. They are presumed to be Hydrozoa.

Among the living Hydrozoan corals are the Milleporina, which in the genus *Millepora* are important reef formers. This coral, also known as the fire coral, has a well deserved reputation for its nematocysts may inflict a sting not unlike that of nettles on an unwary diver. *Millepora* is most abundant in the surf zone of coral reefs and is usually a light shade of yellow. Closely related to the Milleporina are the Stylasterina. Although never abundant, these corals are widely distributed both in depth and temperature. Not uncommon on coral reefs, they are also frequently encountered on rocky bottoms at some depth and frequently may be the cause of "foul bottom" of trawlers whose nets are torn by the coralla.

DONALD F. SQUIRES

References

Bayer, F. M., et al., "Coelenterata," in Moore, R. C. Ed., "Treatise on Invertebrate Paleonthology," Part F, Kansas, Univ. Press. 1956.
Vaughan, T. W. and J. W. Wells., "Revision of the suborders, families, and genera of the Scleractinia," Geological Society of America, Special Paper 44, 1943.
Wells, J. W., "Coral Reefs," in Hedgpeth, J. W., Ed., "Treatise on Marine Ecology and Paleoecology," 1:609–632, 1957.

CORK

The spongy outer layers of bark of woody STEMS. Technically bark is composed of periderm only the outer layers of which are cork. It is formed by the deposition of suberin lamellae in the primary walls of the phellem cells. Since suberin is practically impermeable to water, drying and disintegration of protoplasts results. Cork is therefore a dead tissue. The peculiar qualities of cork are its relative imperviousness to water, its resilient structure and its relatively low rate of heat transfer. Commercial cork is produced by the cork oak, *Quercus suber* L., an evergreen of western Europe and northern Africa. About thirty years are required to grow the virgin crop which is so coarse and full of lenticular shafts that it is used mostly in the manufacture of linoleum, insulation and other ground cork products. The first growth cork is stripped without harming the tree while the cork cambium is still active. New cork initials form in the newly exposed tissues. Following a rapid generation of finely textured cork, it is harvested at nine year intervals, each harvest yielding a successively finer cork from which high grade sheet and bottle corks are cut.

Cork under certain conditions is produced in herbaceous plants as well. It is also formed in callus or wound tissues and in leaf abscission layers. In discussing cork formation, the terms cork, bark, and periderm are often used synonymously, but more specifically periderm is a collective term for the secondary tissue aggregate of phelloderm, phellogen (cork cambium) and phellem (cork). Ordinarily it is formed from cork cambium initials arising in any of the three primary tissues, epidermis, hypodermis or cortical parenchyma. Its point of origin is unique for each species. In wound areas phellogen may originate in practically any actively growing tissue. The cork cambium more frequently than not generates but a single layer of phelloderm to the inside while several layers of phellem are cut off to the outside. Cork growth is usually slower than vascular expansion from the true cambium, and accounts for the compensating deep fissures in the outer bark of some woody species. However, the texture of the surface may be greatly modified by the mode of periderm formation, three types of which are recognized. In one the phellogen arises in the epidermis or several layers beneath the cortex, producing a continuous uniform cylinder known as annular or ring cork. This bark appears uniform with ridges and fissures. (Ex. *Sambucus* and *Tilia*). In another type the phellogen originates in the hypodermis where the initials appear first beneath stomata. Lenticel formation is then the first prominent development. Subsequently the phellogen produces isolated tangential plates several cell layers thick which eventually connect into a loosely continuous but irregular cylinder. The cork thus formed is known as plate cork and because of inner pressure often proliferates off thin sheets of the smooth outer bark. (Ex. *Betula* and *Platanus*). A few woody stems produce a third type known as wing cork. The cork initials appear to form beneath rows of stomata and produce prodigious piles of cork in ridges or knife sharp wings superimposed on the smoother surface cork. (Ex. *Euonymous* and *Liquidambar*).

DONALD M. BROWN

References

Eames, A. J., and L. H. MacDaniels, "An Introduction to Plant Anatomy," New York, McGraw-Hill, 1947.
Esau, K., "Plant Anatomy," New York, Wiley, 1965.

CROCODILIA

The present species of the reptilian Order Crocodilia can be called "living fossils." During the Mesozoic representatives of this order were much more abundant (5 suborders with 15 families and about 100 genera), inhabiting the continents and seas. Of this past abundance only a few species, belonging to a single suborder with three families and eight genera, remain. These are the sole remaining representatives of a very old reptile stem, the Archaeosauria, to which moreover the Thecodontia, the "dinosaurs" (Saurischia, Ornithischia) and the pterodactyls (Pterosauria) belong. The Archaeosauria, from which birds stem also is characterized by the diapsid type of skull with two temporal openings in the posterior skull roof on either side (as in Rhynchocephalia). Characteristic of the Archaeosauria is the tendency to bipedal locomotion, wherein the fore part of the body is raised and the tail serves as a balance and support. Like many giant saurians, it is probable that the early crocodilian was bipedal, their much stronger hind limbs suggest this.

If one considers the form of recent crocodilians, they are similar to the lizards. Their head is elongated, the laterally compressed tail serves as an oar for movement in water. On the fore limb there are five, on the hind limb four toes; those of the hind limb are webbed. The seemingly small eye with the vertical slit pupil is somewhat raised and lies on a direct line with closeable nasal openings on the tip of the snout; also the ear opening has a skin valve for closure. With the exception of the dorsal aspect of the head, where the skin is attached to the underlying bone, the skin is covered with horny shields or plates, which on the back and in some species also on the belly, ossify and in this way form a dermal armor. The skull of the crocodile often has a greatly elongated snout (especially in the living genera, *Tomistoma* and *Gavialis*); its elements are bound fast one with another; the temporal region has two bony arches: an upper of postorbital and squamosal, a lower of jugal and quadratojugal. Further the skull is peculiar in having a secondary palate formed of the premaxilla, maxilla, palatine and pterygoid. In contrast to the other living reptiles, the pointed conical teeth (their greatest number, in recent species of *Gavialis*, is 27–29 on either side above and 25–26 below) are anchored in sockets. Other peculiarities of the skeleton include cervical ribs and bony abdominal ribs which are not attached to the vertebrae.

Peculiar is the short, flat tongue which is attached to the bottom of the mouth and is not extensible. It is arched upward along its hind margin; along with the fold hanging down from the hind margin of the palate, it can completely close off the throat. The nasal passage above the secondary palate opens along with the larynx behind this valve, allowing the crocodile to breath with open mouth while in the water with only the nares above the water surface. Very notable is the heart of the crocodile, which in contrast to that of other recent reptiles has a left and right ventricle connected only by a small opening, the foramen Panizzae. A urinary bladder is missing, the male genital organ is single, not double as in lizards and snakes. The anal opening is not a transverse slit as in the latter but rather longitudinal. Two large glands, producing a musk-like secretion, open to either side of the cloaca; also on the inner side of the lower jaw are two small, eversible glands.

The 7 meter total length which some species reach makes the crocodile the largest and strongest reptile of the present, although some of their prehistoric relatives reached substantially greater size. All of the existing crocodiles live an amphibious life, more aquatic than terrestrial. Their habitat are the large streams and inland lakes of the tropics and subtropics, only a few species occasionally enter the sea (especially *Crocodylus porosus* which is distributed from southern Asia to northern Australia). Crocodiles are essentially nocturnal; in the day time one often sees them with open mouth (serving for temperature regulation) sleeping on the bank, from which they can throw themselves, at the slighest noise, into the water. They feed on fish, frogs, waterbirds, and other vertebrates; the young feed on invertebrates as well. The female lays her hard-shelled eggs, the size and number varies with the species, in holes which she digs or in piles of vegetation, which stay warm because of fermentation and assist the development of the embryos. The female of *Crocodylus porosus* digs a mud hole next to the nest mound in which she lies and from time to time she splashes the nest with a sweep of her tail. The young of some species produce croaking calls just before hatching; whereupon the mother animal rakes the eggs out of the nest. Similar but louder calls are given by young crocodiles, while the adults of some species are capable of producing far-reaching roars.

The oldest crocodiles (Protosuchia) lived in the upper Trias and were closely related to the Thecodontia. The Mesosuchia of the Jurassic and Cretaceous had the upper temporal fenestra very large and the inner choanal opening further forward (between palatine and pterygoid). From these stem probably the unarmored Thalattosuchia of the Jurassic and Cretaceous, which were marine with very short, oar-like fore limbs, and a fin-bearing tail with a sharply decurved tip. The Sebecosuchia arose in South America in the Eocene. They had a high small snout and a short secondary palate without a pterygoid component. The suborder containing most species of crocodiles is the Eusuchia, which appeared in the Cretaceous period and to which, with the few exceptions of the surviving Eocene Mesosuchia and Sebecosuchia, all of the Tertiary and Quaternary species belong. Their secondary palate has involved the pterygoid in its fullest development. Along with two fossil families, three living ones can be distinguished. 1) Alligatoridae, in which the lower-jaw teeth lie inside the margin of the upper jaw and the fourth lower tooth fits into a socket inside the upper jaw. There are at present two species of *Alligator* in North America and China, two species of *Caiman* in Middle and South America, *Melanosuchus* with one and *Paleosuchus* with two species in South America. 2) Crocodylidae, in which the lower jaw teeth fit in notches between the upper teeth and the fourth mandibular tooth fits into a laterally open groove of the upper jaw. At present there is *Osteolaemus* with one species in Africa, *Crocodylus* with eleven species throughout the tropics, *Tomistoma* with one species in Malaya, Sumatra and Borneo. 3) Gavialidae with a long thin snout sharply set off from the cranium and without grooves or furrows to enclose the lower jaw teeth; the teeth fit between each other so that their points lie outside the jaw margin. At present there is *Gavialis* with one species in south Asia.

ROBERT MERTENS
(Trans. from German—Malcolm Jollie)

References

Romer, A. S., "Osteology of Reptiles," Chicago, The University Press, 1956.
Wermuth, H., "Systematik der Rezenten Krokodile," Mitt. zool. Mus. Berlin, **29**: 375–514, 1953.

CROSSOPTERYGII

The Crossopterygii comprise an order of lobefinned fishes assigned to the Subclass Sarcopterygii, which also includes the Order DIPNOI or lungfishes. The Sarcopterygii are distinguished from the other subclass of higher bony fishes, the Actinopterygii or ray-finned fishes, mainly by the structure of the paired fins. In the sarcopterygians, these fins have an internal, archipterygial skeleton covered proximally by lobes of muscle. The paired fins of the actinopterygians have a greatly reduced internal skeleton and usually no muscular lobes are developed.

At the time of their first known appearance in the Early Devonian, the Crossopterygii and the Dipnoi were already distinctive, and their common ancestry is unknown. The crossopterygian braincase is divided into two parts at the level of the trigeminal nerve. The palate is movably attached to the braincase. In the dipnoans, the braincase is a single unit to which the palate is firmly fused. Internal nostrils are present in some crossopterygians but probably do not occur in the dipnoans. The dermal bone pattern of the skull, the dentition, and the appendicular skeleton are also quite different in the two groups.

The separation of the Crossopterygii into two suborders, the Rhipidistia and the Coelacanthini, probably occurred early in the Devonian Period. Their common ancestry is clearly indicated by the structure of the braincase, the cranial dermal bone pattern, and the fins. The rhipidistian braincase is completely ossified and the dermal bones of the skull are in close articulation. Internal nostrils are usually present. The unrestricted notochord was surrounded by central ossifications in addition to the neural arches and short ribs or haemal arches. The dermal shoulder girdle includes an interclavicle and a small scapulocoracoid ossification with which the single proximal element of the pectoral fin skeleton articulates. The more distal elements of this skeleton exhibit a characteristic dichotomous arrangement. The pelvis is a small, single ossification; the pelvic fin skeleton is similar to that of the pectoral. The proximal radials of the two dorsal fins and the anal fin are in contact with the neural and haemal spines respectively. The caudal fin was either heterocercal or symmetrical and trifid.

The rhipidistians were world-wide, predaceous, fresh-water fishes ranging from the Lower Devonian

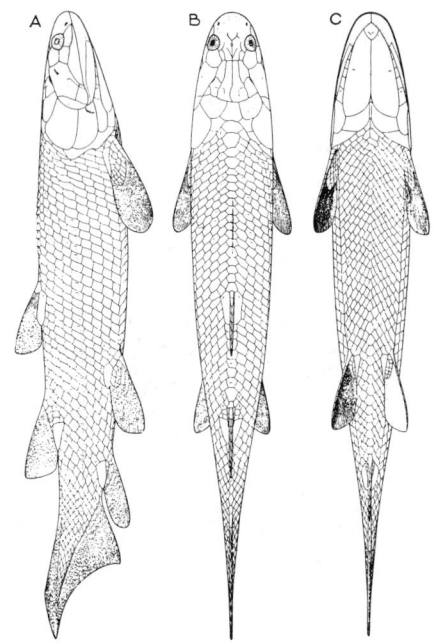

Fig. 1   *Osteolepsis macrolepidotus* Ag. Restorations in, A, lateral, B, dorsal, and C, ventral views. Approx ½ nat. size. (Erik Jarvik: "The Middle Devonian Osteolepid Fishes of Scotland.")

to the Lower Permian. They are of particular importance in vertebrate history because they gave rise to the Amphibia. This transition probably occurred in an environment subjected to seasonal droughts. The rhipidistians, which presumably had functional lungs, were able to leave the drying pools and move overland with their flexible paired fins in search of other bodies of water.

Although transitional forms between rhipidistian and amphibian are unknown, it is evident that most parts of the skeleton were modified. The braincase was fused into a single unit. There was a preorbital lengthening and a postorbital shortening of the entire skull, accompanied by changes in the dermal bone pattern. The modification of the hyomandibular into the tetrapod stapes and the gill arches into the hyoid apparatus and laryngeal skeleton was associated with a loss of the opercular bones. The vertebral structures surrounding the notochord remained essentially rhipidistian like in some early amphibians. In the shoulder girdle, the scapulocoracoid was enlarged and the dermal elements were reduced. The pelvis developed three ossification centers and expanded to join the vertebral column. The radical change from paired fins to limbs involved enlargement and elaboration of the proximal skeletal fin elements and particularly the formation of the carpus, tarsus and phalanges.

The coelacanths differ in many characters from the rhipidistians. Although the braincase retained the duplex condition, there was a marked reduction in ossification in the post-Devonian forms. The pattern of the dermal skull elements is very characteristic

although derived from that of the rhipidistians. The maxillary bone in the upper jaw is absent. The dentary bone of the lower jaw is reduced; the coronoid is enlarged. The gill region is mostly covered by a large, triangular opercular bone. The notochord is persistent; central ossifications are consistently absent. Neural and haemal ossifications are present, and pleural ribs occur in several related genera. The girdles are essentially similar to the rhipidistian type, and the internal skeletons of the lobate paired fins are modifications of the rhipidistian pattern. The rays of the anterior dorsal fin articulate directly with the basal plate, while those of the posterior dorsal and the anal fins are attached to a segmented internal skeleton. The candal fin is distinctive in having a small tuft of rays at the end of the notochord, which extends between the dorsal and ventral lobes.

During the course of their history, which extends from the Middle Devonian to the Recent, the coelacanths have inhabited a variety of freshwater and marine environments. In spite of this diversity in habitat, they apparently had only a very modest adaptive radiation that probably reached its climax in the Triassic. Modifications in the dentition indicate that some coelacanths were plankton feeders, while others were predaceous and a few perhaps lived on shelled invertebrates. Several Triassic genera had thoracic pelvic fins, a specialization found otherwise only among the teleosts.

Tertiary coelacanths are unknown, and there is a time gap of about 70 million years between the Late Cretaceous representatives and the living *Latimeria.* This genus, apparently restricted to the Comoro Archipelago northwest of Madagascar, resembles closely its Mesozoic ancestors in most respects.

BOBB SCHAEFFER

CRUSTACEA

Definition

Crustacea are a large and important class of the phylum ARTHROPODA comprising crabs, lobsters, shrimps, beach-hoppers, sow-bugs, barnacles, water-fleas and their allies. They exhibit such diversity of structure and habits that it is almost impossible to give a brief definition that will apply to all. They may be distinguished from the related classes such as INSECTA and ARACHNIDA in that the vast majority are aquatic, breathing by gills or by the general body surface, having two pairs of preoral ANTENNAE and at least three pairs of postoral appendages acting as jaws, the three corresponding segments being coalesced with the head. Some, like sow-bugs and various land crabs, are terrestrial with special organs for aerial respiration. The preoral appendages may be locomotor or prehensile, or may be absent, and some or all of the mouthparts may be suppressed. Certain extremely modified parasitic forms have, in the adult stage, lost almost all traces of Crustacean or even of Arthropodous structure, but their larval stages indicate their true affinities.

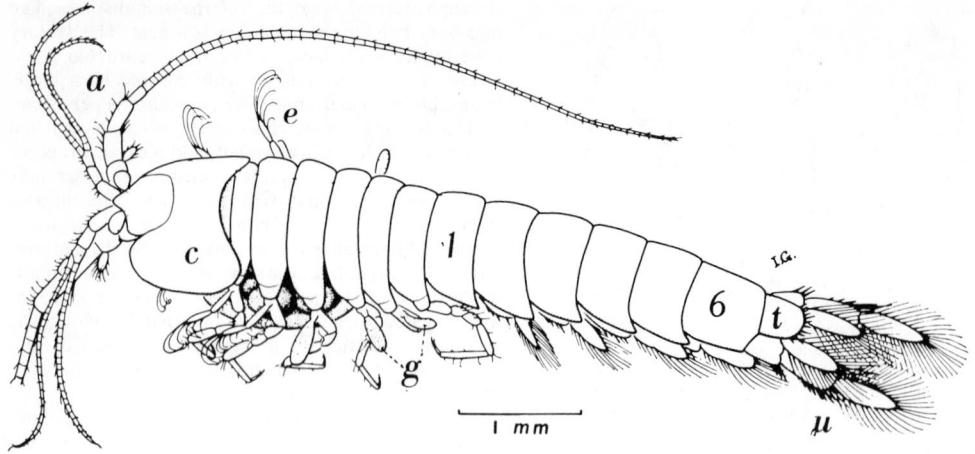

Fig. 1 *Spelaeogriphus*, a blind shrimp from a cave on Table Mountain (subclass Malacostraca) showing: carapace (*c*) with minute rostrum and a pair of oval eye-scales; seven free thoracic somites; abdominal somites (*l* to *6*); telson (*t*) and uropods (*u*) forming a tail-fan; biramous antennula (*a*); walking legs with exopod lash-like (*e*) on first three, and a vesicular gill (*g*) on next three, pairs; ova in a typical Peracaridan brood-pouch. After Gordon 1960.

Occurrence and Economic Importance

Crustacea are present in teeming multitudes in all OCEANS and seas and a few penetrate to the greatest known depths. They are abundant in fresh water and there is scarcely a ditch or pond that does not have at least some of the smaller forms. In recent years many new cavernicolous and "interstitial" forms have been discovered. The part played in the general economy of Nature by the edible crabs, lobsters and shrimps is small compared with that of the AMPHIPODA and ISOPODA which swarm in shallower waters, acting as scavengers and themselves forming the food of larger marine animals such as fishes. More important still are various pelagic shrimps and the minute COPEPODA which constitute an important part of the PLANKTON; they live on diatoms and other minute organisms, and provide food for fishes and even for gigantic whales.

General Morphology

The chitinous exoskeleton (see SKELETON, INVERTEBRATE) of the body consists typically of a series of segments (somites), which may be movably articulated or more or less fused. In no Crustacea are all the somites distinct. At least five coalesce to form the head or cephalon. The trunk throughout the subclass Malacostraca has eight thoracic, and six (rarely seven) abdominal, somites. But the terms thorax and abdomen are not equivalent in the other subclasses, where the number of somites varies greatly, exceeding forty in some Branchiopoda. The terminal part, on which the anus opens, is not a true somite as is clearly shown by embryology. All may bear true appendages save the terminal telson, which often has a pair of prongs, the caudal furca.

A structure present in the most diverse groups is the dorsal shield or *carapace* which arises as a fold of the integument from the posterior margin of the head. It may loosely envelop more or less of the trunk and limbs, form a bivalve shell, or a fleshy mantle. In many Malacostraca it fuses dorsally with some or all of the thoracic somites, and in Decapoda protects the gills on either side. A shell-fold is absent in some groups, secondarily reduced or vestigial in others.

Although they differ markedly in form and function, most Crustacean *limbs* are modifications of a fundamental type comprising a peduncle and two rami, exopod and endopod. The peduncle may have additional processes on its outer and inner margins, known as exites and endites. Some of the exites often act as gills and endites of appendages near the mouth often form jaw-processes or gnathobases. The shape of the rami is often, though not invariably, correlated with the function of the limb which may differ in the larval and adult phases.

The antennules differ from the rest since they are uniramous in the earliest larva and in many adults. In adult Malacostraca they are often biramous (Fig. 1) or even triramous; in CIRRIPEDIA they serve as organs of attachment. The antennae are biramous, natatory and sometimes also masticatory in the nauplius larva. In the adult they are chiefly sensory, but they may serve as organs of attachment in parasitic forms; the exopod may be a flattened scale or balancer, or may disappear.

The mandibles are biramous swimming organs with a spine-like jaw-process in the nauplius larve and this form and function are retained with little alteration in certain adults. In most, however, the exopod is lost, the endopod forms a palp or is lost, while the body of the mandible is derived from the peduncle. In parasitic forms with suctorial mouthparts they may form piercing lancets enclosed in a sheath formed by apposition of upper and lower lips. Maxillula and maxilla are nearly always foliaceous limbs with gnathobasic lobes or endites on the peduncle. The endopod is reduced to a palp or absent, an exopod and exite (epipodite) may be present.

The *trunk limbs* vary greatly in number. In Ostracoda, where the body is not distinctly segmented, there are never more than two, whereas in some BRANCHIOPODA there are over sixty, pairs. They are nearly all similar and foliaceous in Branchiopoda. Commonly one or two pairs may assist the mouthparts and are termed maxillipeds. In the subclass Malacostraca thoracic and abdominal limbs are sharply differentiated (Fig. 1). The thoracic endopods often act as walking legs, the exopods when present being lash-like or sometimes respiratory (*Spelaeogriphus*). In the DECAPODA three pairs are maxillipeds and one to three of the following pairs may end in pincers. The *abdominal limbs* are usually biramous and, in the more primitive forms, natatory. In Isopoda and Stomatopoda they have also assumed the respiratory function.

Sense organs. Eyes, when present, may be of two kinds. An unpaired median eye is usually present in the earliest larval stages and may be the only visual organ. It may persist in the adult along with paired eyes, or it may become vestigial or absent. The paired compound eyes are very similar to those of Insecta and may be sessile or set on movable stalks. Sensory hairs or setae on various parts of the body may respond to vibrations in the surrounding medium. Others associated with a chemical sense are found especially on the antennules and antennae. Balancing organs or statocysts may be present at the base of the antennular peduncle (Decapoda), in the endopod of the uropod (Mysidacea) or in the telson.

Internal Organs. In most Crustacea the *food canal* runs straight through the body, curving down anteriorly to the ventrally placed mouth. In a few instances it is twisted or actually coiled. It is divided into fore-, mid- and hind-gut, the first and last being lined by an infolding of the exoskeleton. In the more primitive forms there are spines and hairs on the lining of the fore-gut which help to triturate the food. In others there is a series of plates moved by muscles; this armature attains its greatest complexity in Decapoda forming a "gastric mill." Paired tubular outgrowths from the mid-gut secrete digestive juices and aid digestion and resorption; they may become more or less ramified and form a massive "liver" or hepatopancreas.

The *heart* lies in a peracardial sinus with which it communicates by means of valvular openings or ostia. In some primitive forms it is long and tubular, with a pair of ostia in each segment. But there is progressive abbreviation throughout the class, the heart being ultimately very short with only two or three ostia. In many small Crustacea there is no heart, the blood being driven hither and thither by movements of the alimentary canal, body and limbs.

The central *nervous system* consists of a supraoesophageal ganglionic mass or "brain" united by connectives round the oesophagus with a ventral chain. In primitive Branchiopoda this chain is ladder-like, the paired ganglia being united by double transverse bands. In higher forms the two halves of the chain are more or less fused and the ganglia tend to draw together longitudinally, ultimately coalescing into a single mass, as in crabs, and the "brain" becomes more complex.

The most important *excretory organs* are two pairs of glands situated at the bases of the antennae and the

maxillae respectively. The two pairs rarely function together, although one may replace the other during development. Various types of dermal glands are present on the body surface.

Certain pelagic and deep-sea Crustacea can *emit light* at will. Some pour out a luminous secretion from localised dermal glands or from their secretory organs. Others, like Euphausiacea and certain Decapoda, have complex *photophores* with a reflector and a condensing lens.

Reproduction and development. The sexes are separate, as a rule, and sex dimorphism may be marked. The male is often larger than the female but in parasitic forms a dwarf male may be attached to a large female. Certain limbs near the genital opening may be modified for sperm transference, and various appendages may be modified as claspers (antennule, antenna, mandibular palp, etc.). Hermaphroditism is the rule in Cirripedia, in some parasitic Isopoda and in a few Decapoda. PARTHENOGENESIS is frequent among Branchiopoda and Ostracoda (often alternating with sexual reproduction) and occasional in terrestrial Isopoda.

As a rule the eggs are carried by the female after extrusion though occasionally they are shed freely in the water or deposited on a substratum. They may be retained between the valves of the shell (OSTRACODA), in the mantle cavity (Cirripedia), in special egg cases attached to the genital somite (Copepoda). Among Malacostraca, the Peracarida have a ventral brood pouch formed by overlapping plates attached to the bases of the thoracic limbs (Fig. 1). In Pancarida the carapace enlarges to form a dorsal brood pouch reminiscent of that of certain parasitic Copepoda and of most Cladocera. In most Decapoda they are attached to the abdominal pleopods of the female.

The majority hatch from the egg in a form differing more or less markedly from the adult and pass through a series of free-swimming larval stages. There are many instances in which metamorphosis is suppressed and the newly-hatched young resemble the parent in general structure. Where the series of larval stages is most complete, the starting point is the nauplius larva. Typically, this has an oval unsegmented body, a median eye and three pairs of swimming appendages corresponding to antennules, antennae and mandibles. As development proceeds the body of the nauplius elongates and its posterior portion becomes segmented, new somites being added at each successive moult from a formative zone in front of the telson on which the anus opens. The limbs appear as buds on the ventral surface of each somite and become differentiated, like the segments that bear them, in regular order from before backwards. The dorsal covering of the body gradually extends backwards as a shell-fold. Paired eyes appear under the cuticle in the head region and only become stalked at a comparatively late stage. This plan is most closely followed in some Branchiopoda and Copepoda.

In many Crustacea the earlier stages are passed through in the egg and a more advanced larva is hatched. The gradual appearance of somites and appendages may be accelerated, so that comparatively great advances appear at a single moult. In the crab larva known as a zoea the posterior five or six thoracic somites are still undifferentiated when the abdominal ones are fully formed and may even bear limbs. Many of these larvae are specially adapted to a pelagic life, having long spines as organs of defense and flotation. Others, like the phyllosoma larva of spiny lobsters, are leaf-like and transparent.

Complete suppression of metamorphosis occurs in fresh water crayfishes, in river crabs and in a few marine crabs.

Classification

In 1806 Latreille divided the class Crustacea into Entomostraca and Malacostraca, but this is no longer tenable although it persists in some text-books. The "Entomostraca" is a heterogeneous assemblage of unrelated forms and is replaced by the following subclasses: Branchiopoda, Cephalocarida, Ostracoda, Copepoda, Mystacocarida and Cirripedia. The large subclass Malacostraca, on the other hand, is a natural group. It is subdivided into two Series: I. Leptostraca, having an adductor muscle connecting the two halves of the shell, seven abdominal somites and a caudal furca on the telson. II. Eumalacostraca, without an adductor muscle, and in the adult no caudal furca (except in the minute Bathynellacea) and only six abdominal somites. This series is further divided into Syncarida, Peracarida (including Isopoda and Amphipoda), Pancarida, Hoplocarida and Eucarida (including Decapoda).

Fossil remains are abundant in all geological strata and many of the chief groups were already differentiated before the beginning of the geological record as we know it. Shrimp-like forms that can be referred to the Malacostraca occur in Upper Devonian and Carboniferous rocks, and Decapoda first appear in the Trias. Except for the absence of a shell-fold and stalked eyes, the trilobites are not far removed from the hypothetical ancestor of the Crustacea.

ISABELLA GORDON

References

Calman, W. T., "Crustacea," in Lankester, E. R. Ed., "A Treatise on Zoology," London, Black, 1909.
Kukenthal, W. and T. Krumbach Eds., "Handbuch der Zoologie," Vol. 3. Leipzig, Oldenbourg, 1927.
Schmidt, W. L., "Crustaceans," Ann Arbor, University of Michigan Press, 1964.

CUCULIFORMES

The two families of the Cuculiformes appear at first glance not to have much in common, though the zygodactylous (yoke-toed) foot, a thin and tender skin, and some features in the development of the young are somewhat similar. Otherwise the gorgeously-plumaged plantain-eaters (Family Musophagidae) appear quite different from the soberly clad, but often handsome, cuckoos and allies (Family Cuculidae), and some authorities place them in separate orders.

The musophagids or plantain-eaters, also called touracos, are confined to Africa. About 20 species of these large (one to two or more feet in length), colorful, long-tailed, crested birds range over the forested parts of the continent south of the Sahara. They are strictly arboreal, climb about dexterously from branch to branch, and build a bulky nest of sticks placed in the trees. Male and female share in the incubation of the two or three eggs.

Chief interest of many biologists in the musophagids is their production of two color pigments not otherwise found in birds. One is turacin, a deep red pigment containing about seven per cent copper and said to be more or less soluble in water; the other is turacoverdin, a unique green pigment found only in the plantain-eaters (green colors in other birds are structural colors, usually due to the reflection of spectral blue from a basal yellow pigment). Experiments have not always borne out the theory of the solubility of turacin, which was based on the statement of a collector who found his hands stained with red when a bird slipped out of his clutches.

The other family, the Cuculidae, consists of an odd and varied assortment of 127 species of cuckoos, road-runners, anis, couas and coucals. They are mostly Old World, but the anis, road-runners, and several species of cuckoos occur in the Americas.

Considerable variability occurs, both in habits and appearance, among the various members of this family, but most of them have a long and more or less decurved bill, sombre though often harmonious plumages of gray, brown or black, a sagittate tongue with retroverted spines, and highly specialized nesting habits involving parasitism and social or communal nesting.

Most of the New World cuckoos build a crude platform of sticks for a nest and incubate their own eggs and rear their own young, but the Old World cuckoos, particularly the Common European Cuckoo (Cuculus canorus), are noted for the high development of brood parasitism. Many remarkable adaptations have been evolved by these cuckoos to make this type of nesting successful: they lay a small egg, to fit better and less conspicuously among the eggs of smaller host species; many egg colors and spotting combinations have been evolved, to blend harmoniously with eggs in the host nest (blue, white, spotted, even red); the incubation period is short, so that the cuckoo young will hatch first and get a "head start;" and the young cuckoo has a special saddle-shaped depression between the shoulders by means of which it "shoulders" the rightful occupants of the nest when they hatch and heaves them out of the nest.

In contrast to the Old World cuckoos, the neotropical anis have developed communal nesting habits. A large community nest is constructed and a number of females contribute eggs to it, depositing them in layers. The females take turns at incubation and the males guard the nest and females.

Another exceedingly interesting American species is the Road-runner (Geococcyx californianus), found in the arid regions of southwestern United States and Mexico. It is terrestrial in habit, a rapid runner, and feeds on a varied menu of grasshoppers, scorpions, lizards, poisonous snakes, and small birds and mammals. Early observers credited them with building corrals of spiny cacti and thorns around sleeping rattlesnakes and then attacking them in these artificial enclosures. Whether or not this is true, Road-runners do specialize to some extent on poisonous snakes which they dispatch with remarkable speed and dexterity by quick jabs in the back of the snake's neck.

Tropical species of cuculids are more of less sedentary, but those of higher latitudes undertake long migrations. The Black-billed and Yellow-billed Cuckoos (*Coccyzus erythropthalmus* and *C. americanus*) that nest in the northern states and Canada winter in South America, the latter even crossing the equator. Even more remarkable is the journey of the Bronzed Cuckoo (*Chalcites lucidus*) whose young are reared in foster flycatchers nests in New Zealand. According to Gilliard, the young cuckoos, unaccompanied by their parents, start out on a long trek across the water from New Zealand to Australia, some 1200 miles, then fly another 1000 miles to islands in the Pacific where they join their parents.

GEORGE J. WALLACE

References

Davis, D. E., "Social Nesting Habits of the Smooth-billed Ani," Auk, 57: 179–218, 1940.
Gilliard, E. T., "Living Birds of the World," New York, Doubleday, 1958.
Moreau, R. E., "A Contribution to the Biology of the Musophagiformes," Ibis: 639–671, 1938.

CUVIER, GEORGES LÉOPOLD CHRÉTIEN FRÉDÉRIC DAGOBERT (1769–1832)

Cuvier was the son of a retired military officer living on a small pension. He had the great good fortune to secure an appointment to the military academy at Stuttgart which was, at that time, primarily concerned with the training of civil servants, to which career Cuvier was destined. It is to the great fortune of biology that the French civil service of that period did not pay its subordinate officials, who worked for long periods in the hope of securing an appointment to the upper, salaried, ranks. Cuvier was therefore forced to secure a position as a tutor to a family living on the north coast of France, where he was brought, for the first time, in contact with an extensive marine fauna. He threw himself wholeheartedly into the investigation of this, not only dissecting and comparing the anatomy of fishes but also collecting, and studying exhaustively, the numerous invertebrates which he found left by the tides. He studied by drawing types—a method of study still used for elementary instruction in many contemporary schools—and some of his drawings so impressed the staff of the University of Paris that he was appointed there to the chair of comparative anatomy. He brought to this post a surprising knowledge of subhuman forms and a clear understanding of the relationships between many of these. He therefore approached human anatomy by the dissection of many of the great apes and thus developed an approach that was as revolutionary in its day as was its results on all subsequent teaching. In 1799 to 1805, he published "Leçons sur l'Anatomie Comparée" in which he not only drew attention to the structures but also discussed with great clarity the relation of these structures to each other in the different forms. This led to the formation of his "correlation theory" which pointed out that the parts of an animal were not only adapted to the environment but also to each other. Thus the teeth, stomach, and feet of herbivorous animals could only properly be used in connection with each other. He pointed out that the basis of any natural classification of animals should be this correlation of their parts, so that a competent scientist should be able, from an examination of one structure, to deduce the nature of others. This led Cuvier inevitably to the study of fossils and he was able, by applying his own principles, to turn the study of extinct vertebrates from an amiable game of guesswork to a definite science and to deduce the environment from the structures shown. The breadth of his knowledge and the clarity of his deductions were truly astonishing. For example, he removed hyrax from the rodents and placed it in association with the elephants solely on the basis of his studies of the comparative anatomy.

Cuvier's excursions into geology, whence he was inevitably led from his study of fossils, were not so fortunate in their results. It was obvious from his deductions as to the environment of fossil forms that vast changes had taken place in the climate of the past. He made, however, the mistaken assumption that these changes were a succession of catastrophes, of which the most recent was the biblical flood. It is extraordinary that the idea of evolution between the types he studied should have escaped so keen a mind but it must be pointed out that his principal source of geographical observation was the Swiss and French Alps, a region in which many cataclysmic changes have indeed taken place.

Armed now with a thorough knowledge of invertebrate and vertebrate anatomy and taxonomy, Cuvier turned to the greatest, and best known, work of his life the "Règne Animale" which appeared in 1817. He was by this time so strongly imbued with his catastrophic theory that he argues keenly in this work against any attempt to arrange living forms in progressive series but even in this he made a major contribution. It was customary in his day to speak of "lower" as against "higher" animals—a habit which still persists among the uneducated—and Cuvier points out, with some indignation, that to regard a mammal as higher than a bird is an argument which cannot be supported by any rational examination of structure and function.

It must not be thought that Cuvier's scientific contributions were all that he made. His early training as a civil servant stood him in good stead and he rose high in Napoleon's favor. The latter appointed him to the rank of Inspector General in the Department of Education and he undertook sweeping reforms in the educational system of France, founding new universities, both in that country and in the countries conquered by his master. So great, however, was Cuvier's reputation both as a scholar and an organizer, that he survived the fall of Napoleon and was ennobled by the Bourbons who continued him in all of his positions. He even survived the July revolution and was promoted to the peerage very shortly before he died in the cholera epidemic of 1832.

PETER GRAY

CYANOPHYCEAE

The Myxophyceae, or Cyanophyceae, generally known as the blue-green ALGAE, comprise the single class in the division Cyanophyta. They are one-celled to multicellular, coccoid to filamentous plants occurring in plankton or as globose, gelatinous masses and gelatinous or leathery strata on moist rocks, soil, wood and other materials. They are distinct from all other algae because of the usually prevailing blue-green pigmentation, the dispersal of photosynthetic pigments (c-phycocyanin, c-phycoerythrin, chlorophyll a, carotenes and xanthophylls) throughout the peripheral protoplasm, the formation of heterocysts and the presence of a primitive type nucleus. Other features, but not peculiar to them, are the production of glycogen, the abundance of gelatinous mucilage often as definite sheaths, and the absence of flagellated vegetative, asexual or sexual cells. Cell-division is by fission. Reproduction is chiefly by fragmentation, the formation of hormogonia, and in some genera the formation of non-flagellated spores.

The blue-green algae, often pioneers in barren habitats, are cosmopolitan and found most widely distributed in fresh-water and terrestrial environments. Some species are important in salt-water and occur in marine plankton, benthos, intertidal belts and marshes. Some of the planktonic forms (*Anacystis, Gomphosphaeria, Anabaena* and *Aphanizomenon* spp.) are responsible for the production of "water-blooms," the occurrence of myriads of microscopic plants in the water when accelerated growth by fragmentation occurs. *Trichodesmium erythraeum*, a marine filmentous species, is the causal agent of the red color in the Red Sea. Numerous species are the characteristic and most dominant organisms in hot springs, sometimes growing at a temperature of 85°C. Some grow in cold mountain streams and also in the Arctic and Antarctic regions.

Many Myxophyceae grow in association with other organisms. Besides true epiphytism (as *Entophysalis Lemaniae* growing on filaments of *Cladophora* sp.) some species grow within or between cells of other plants and animals. Symbiosis is exemplified by various species found living in higher plants (*Cycas* roots, *Azolla* leaves, *Gunnera* leafbases) and some in lower plants (lichen and liverwort thalli, diatom protoplasts and cells of some colorless algae).

The Myxophyceae resemble structurally many of the bacteria. A cell-membrane encloses an outer chromoplasm containing various pigments and granules. An inner centroplasm containing protein granules and chromatic granules resembles a nucleus but without a true nuclear membrane or nucleoli. Pseudovacuoles, apparently of a gaseous nature, are dispersed throughout the protoplasm of various planktonic species. Surrounding the cell membrane externally is the hyaline gelatinous sheath, chiefly pectin (hemicelluloses sometimes present) and often stratified. In some species the sheath material is hydrolized immediately upon extrusion from the protoplast. Often the sheaths are deeply pigmented and the coloration appears to be related to the surrounding environmental conditions particularly sunlight and the acidity or alkalinity of the medium. In many instances, the parasitization of the cells by fungi result in pigmentation.

Because of the rapid advances made in pure culture studies, the physiology and nutrition of these algae are much better known. Both coccoid and filamentous species have been grown in continuous culture in a single inorganic medium which indicates that organic nutrients or growth-promoting factors are not necessary for normal growth. It is known that some of these plants can utilize and fix atmospheric nitrogen.

These algae are grouped into eight families and are circumscribed here as follows:

Chroococcaceae. The plant body, uni- multi- cellular, of diverse shapes and sizes, free-floating or on various substrata; consists of spherical, discoid, pyriform, cylindrical or ovoid cells. The cells divide into two cells of equal size and soon become separated from each other by sheaths of gelatinous material. Reproduction is by fragmentation.

Chamaesiphonaceae. These plants, originally unicellular, grow eventually into strata or cushions from which filaments of cells penetrate the substratum. The solitary cells are basally attached to the substratum by a sheath of gelatinous material. Any of the solitary cells and any of the cells in the superficial layers of the cushions may enlarge and divide internally, wholly or in part, into numerous small cells (endospores), each of which is capable of growing as a solitary cell or as an endosporangium. Reproduction also takes place by fragmentation of the cushions.

Clastidiaceae. Plants of this family are elongate, epiphytic unicells contained in thin gelatinous sheaths. The entire protoplast divides into a uniseriate (rarely few-seriate in part) chain of rounded cells (endospores). Reproduction is by endospores.

Stigonemataceae. These plants are branched filaments consisting of uniseriate, biseriate or multiseriate trichomes. True branches arise where cell division occurs in a plane parallel with the axis of the trichome. Heterocysts are found in some genera and are terminal or intercalary. Reproduction is by hormogonia and occasionally by akinetes.

Nostocaceae. The plants are microscopic or macroscopic, globose, saccate, scalelike or in shapeless strata. The trichomes, composed of depressed spherical, cylindrical or barrel-shaped cells are unbranched, uniseriate and straight, spiralled or contorted. In some genera, the trichomes are somewhat parallel to each other. The sheaths are homogeneous, thin and often hydrolyzed, or become copious gelatinous hard or soft masses containing few to many trichomes. Heterocysts are terminal or intercalary and may be catenate. Reproduction is by akinetes and hormogonia.

Rivulariaceae. Here, the plants are globose, floating or attached, sometimes found as strata on rocks or woodwork in dripping, flowing or standing water. The uniseriate trichomes are attenuated from base to apex and in most species terminate in colorless hairlike cells. One or more trichomes are enclosed in a firm, homogeneous or lamellated sheath. A heterocyst developes from the basal cell in most species. Reproduction is by akinetes or hormogonia.

Scytonemataceae. In this family the plants occur as clumps, tufts or woolly strata. The uniseriate trichomes (usually one, but rarely several in a sheath) are falsely branched and in definite sheaths which

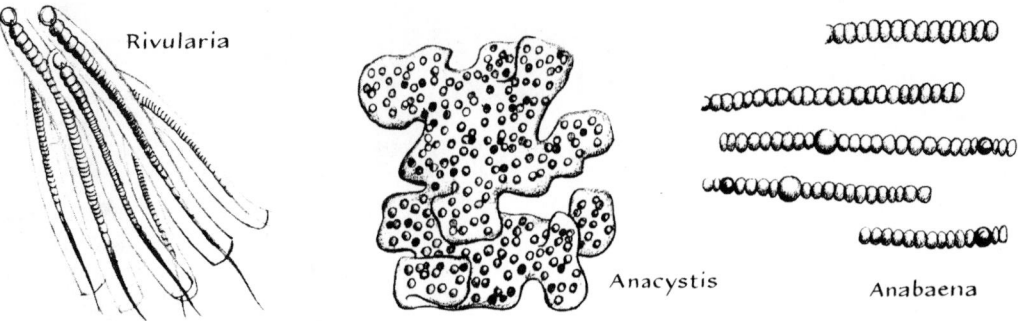

Fig. 1   Some blue-green algae. (Redrawn with permission from H. J. Walter in U.S. PHS Pub. No. 657.)

are hyaline or colored, homogeneous or lamellated. Branching often occurs at a heterocyst. Reproduction is by hormogonia or akinetes.

Oscillatoriaceae. In this family, the trichomes (one or more in a sheath) are unbranched, uniseriate, and the sheaths when visible are firm or gelatinous, homogeneous or lamellated, hyaline or colored. The filaments, branched or unbranched, occur singly or in strata. Reproduction is by hormogonia only.

WILLIAM A. DAILY

References

Drouet, F. and W. A. Daily, "Revision of the Coccoid Myxophyceae," Butler Univ. Bot. Stud. **12**: 1–218, 1956.
Fritsch, F. E., "The Structure and Reproduction of the Algae," Vol. 1, Cambridge, The University Press, 1935; [the same], Vol 2, 1945.
Smith, G. M. Ed., "Manual of Phycology," New York, Ronald, 1951.
Smith, G. M., "Fresh-water Algae of the United States," 2nd ed. New York, McGraw-Hill, 1950.

CYCADALES

Cycadales is a group of naked-seeded plants that grow in warm climates or in temperate climates in greenhouses. Ordinarily classified as GYMNOSPERMS, the cycads are only remotely related to the largest of the gymnosperm orders, the CONIFERALES. The cycads originated during the Paleozoic era presumably from seed-fern ancestors. From the vegetative standpoint they are closest of all seed plants to ferns. They became numerous and widely spread during the Mesozoic era, but declined rapidly during the latter half of the Cretaceous period when the angiosperms became dominant. Today 9 genera remain; *Cycas* in southeastern Asia, *Encephalartos* and *Stangeria* in Africa, *Bowenia* and *Macrozamia* in Australia, *Dioon* and *Ceratozamia* in Mexico, *Microcycas* in Cuba, and *Zamia* in Florida and South America.

The cycads have tuberous or columnar stems that are usually unbranched. A crown of large palmlike leaves of tough leathery texture spreads out from the summit. Most cycads are relatively small though one Australian species grows to a height of about 60 feet. In the tuberous forms the short thick stem remains mostly beneath the ground. Typically the plants are slow growers and specimens with trunks but 5 feet high have been estimated to be 1000 years old, but they usually grow faster where the ground is not too dry.

One who is inexperienced in distinguishing between plants that look much alike might have difficulty separating cycads from palms. In fact one cycad, *Cycas circinalis*, is the "sago palm" of the florist trade. Actually, however, the differences are great, of a degree comparable to those between a fish and a whale. In a cycad stem the tissues are arranged concentrically around a pith which is often quite wide that is surrounded in turn by wood, phloem, and cortex. In a palm stem there are hundreds of small separate vascular bundles that are scattered throughout a more or less uniformly constructed interior. The base of a palm leaf expands so as to nearly encircle the stem, whereas the cycad leaf is attached by a relatively narrow base. Palms produce true flowers and their seeds develop within fruits, but cycads bear their seeds exposed on the megasporophylls in compact cones.

The cycads are presumably wind pollinated, and they bear the pollen and seeds on different plants. The pollen is produced in small cones either borne singly or in clusters at the stem tip in the center of the leaf crown. The cycads and the ginkgo tree have the distinction of being the only living seed plants in which fertilization is accomplished by motile sperms. This is a primitive feature that apparently was retained from their remote sporebearing ancestors. In all other seed plants the sperms are conveyed to the egg cells through pollen tubes. In all genera except *Cycas* the seeds are borne in cones, but these are large cones that are often a foot long and weigh several pounds. The mature seeds may be nearly an inch long and have colored fleshy coats.

CHESTER A. ARNOLD

References

Arnold, C. A., "An Introduction to Paleobotany," New York, McGraw-Hill, 1947.
ibid, "Origin and relationships of the cycads," Phytomorphology **3**: 51–65.
ibid, "Fossil Plants," New York, Houghton, 1968.
Chamberlain, C. J., "The Living Cycads," Chicago, Univ. of Chicago Press, 1919.
ibid, "Gymnosperms. Structure and Evolution," Chicago, Univ. of Chicago Press, 1935.

Fig. 1   Characteristic cycads, primitive birds, flying reptiles, and small bipedal reptiles from the Jurassic. (From a painting by Charles R. Knight. Courtesy Field Museum of Natural History.)

Johnson, M. A., "On the shoot apex of the cycads," Torreya 44: 52–58, 1944.
Schuster, J., "Cycadaceae," in Engler, A. Ed., "Das Pflanzenreich," Vol. 4 Stuttgart, Englemann, 1922.
Seward, A. C., "Fossil Plants," Vol. 3 London, Cambridge Univ. Press, 1917.

CYCLOSTOMATA

This group includes the lampreys (Petromyzontoidea), the slime-eels and the hagfishes (Myxinoidea). All these aquatic bare-skinned, serpentiform animals are devoid of paired fins. The name comes from their sucker-shaped, jaw-less mouth. (The name Agnatha is also used). Cyclostomata is derived from the Greek kuklos (round) and stomax (mouth).

Biology. The lampreys are fresh water animals but certain species spend part of their life in the sea. The size of the adult varies from species to species from 10 cm to 1 m. They can be recognized by the following characters: round, sucker-shaped mouth, interiorly lined with horny denticles; relatively large eye; dorsal nasal opening behind which lies a transparent region of skin covering a "pineal eye"; seven pairs of oval branchial openings delimiting the sides of the head; dorsal fin divided into two more or less distinct regions; protocercus tailfin. At breeding time males and females gather in shallow, rapid, gravel-bottomed rivers. The animals fasten to the stones by means of their suckers, displace them and burrow a nest. Several dozen individuals may participate in this work. A very large number (up to several hundred thousand for the sea lamprey) of whitish eggs, from 0.8 to 1 mm in diameter are released into the nest. After about ten days a small larva hatches which immediately buries into the mud. This larva (ammocoetes) does not resemble the adult. Its minuscule eyes are hidden beneath the lateral muscles. Its mouth is horseshoe-

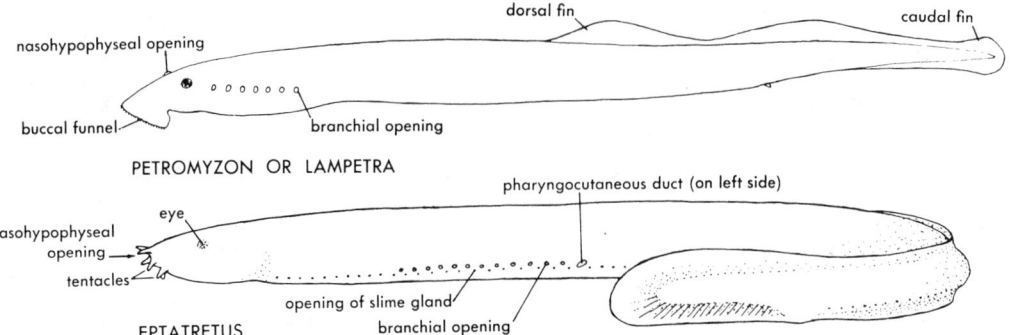

Fig. 1   Some external features of the lamprey, top, and hagfish, bottom. (From Jollie, "Chordate Morphology," New York, Reinhold, 1962.)

shaped, not sucker-shaped. This larva is microphage and like the Prochordates feeds on particles caught by its pharynx with the aid of the secretion of a specialized organ, the *endostyle*. The ammocoetes lives in the mud of brooks from three to four years. It reaches a length of 15 to 20 cm and then undergoes a metamorphosis which entails a partial destruction and a reorganization of the head region (muscles, skeleton, naso-hypophyseal apparatus, pharynx) and a remodelling of the intestine which becomes more complicated.

Afterwards the animals usually go down to the sea or to large lakes and there feed on fishes (chiefly Coregonae and Clupeidae). Great havoc was wrought in the fisheries of the American Great Lakes from 1921 on when sea lampreys which had been stopped until then in Lake Erie by Niagara Falls, reached Lake Ontario through the Welland Canal.

After a growing time estimated at several years the lampreys reach adult size and go back to the breeding sites. During this journey they stop feeding and their alimentary canal degenerates, and, in consequence, all die after the eggs have been laid.

Several forms, which certain authors consider as separate species, undergo metamorphosis and sexual maturation at the same time. In that way they acquire the structure of a predatory animal but do not feed between metamorphosis and breeding time. They remain buried in the mud of brooks.

On the contrary the hagfishes and the slime-eels never leave the sea. These are deep water animals (down to 4300 ft in the Gulf of Panama) which apparently live most of the time in the mud. Their reduced eyes are invisible and embryonic in structure. There is no trace of a pineal eye. The mouth is subterminal and surrounded with tentacles. When closed, it forms an anterior-posterior slit. An enormous horny rasp can be protruded from it. The branchial openings are displaced to the end of the first third of the animal. The number of visible openings varies from 15 (hagfish) to 1 (slime-eel) but in the latter case the visible opening is that of a duct common to five pairs of branchial pouches. On its left side the animal exhibits a supplementary opening behind the branchial oscules, i.e. the pharingo-cutaneous opening. One single round fin surrounds the posterior end of the animal.

The Myxinoidea are probably hunters by smell.

They feed on worms, and fishes such as the cod. They enter the branchial cavities of the fish and choke it by producing an enormous quantity of mucus. Thereafter they penetrate into the abdominal cavity and thence into the muscles by means of their buccal rasp. It seems that sick or immobilized fish are most often attacked. Cod attached to a hook may be completely emptied and reduced to a bag of skin.

The slime-eels and the hagfishes lay from 20 to 30 oblong rather large eggs (14 to 20 mm in length). These are enclosed in an orange or red horny shell bearing anchoring filaments. The hatched young has the same structure as the adult.

Geographical Distribution. The lampreys live in the lakes and rivers of North America (spreading southward roughly as far as New York and California), North Asia, whole Europe, Iceland, the south part of Greenland and the neighboring seas. Other species are found in the south point of America, the south extremity of Australia and New Zealand. The Myxinoidea have roughly the same geographical distribution: the west and east coasts of the U.S.A. and of Canada, the shores of Japan, of New Zealand and of the Cape of Good Hope. But they also exist in the Adriatic Sea and in the Gulf of Panama.

Morphology. The Cyclostomata are very simple (or simplified) and at the same time specialized vertebrates.

a) The traits of primitive vertebrates can be summarized as follows.

—Axial skeleton practically reduced to the notochord.
—Completely segmented mesoderm (at least in the lamprey) even in front of the otic capsules.
—Locomotory muscles divided into segments whose histological composition is less evoluted than normally. The first ones are situated behind the otic capsule and continue forward so far as to cover the eye.
—Straight alimentary canal exhibiting many branchial pouches (5 to 15).
—Segmental nerve roots from the otic capsule backward. In the lamprey the dorsal and ventral nerves are independent. Typical cranial nerves.
—Sympathetic system incompletely separated from the main nervous system.
—The three superior sense organs (nose, eye, ear) are present.

—Perfectly segmental renal rudiments. Pronephros in the cardiac region. It is functional during life in Myxinoidea but only during the larval period in lampreys. Long truncal mesonephros.

—Sex organs reduced to a single gonad hanging from the roof of the body cavity. Genital cells simply fall into the cavity and are carried outside through a genital pore independent of the urinary ducts.

b) *Specialized traits.* The single nasal opening leads into a single olfactory organ which presents two sensory areas. The monorhiny of Cyclostoma is thus perhaps an illusion. In Myxinoidea the nasal duct is continuous with a tube which opens out at the posterior end of the mouth, in front of the velum and the pharynx. The respiratory stream thus passes through the nose. On the contrary the posterior nasal duct of the lamprey is blind and ends above the branchial pouches whose beats periodically compress and dilate it. In ammocoetes a solid cordon replaces this duct and connects the olfactory to the hypophyseal rudiment. For the hypophysis does not originate from the buccal epithelium but from an anlage annexed to the nasal organ. This feature is unique in Vertebrates.

The eye is normal in adult lampreys. It is imperfectly developed in ammocoetes which live in the mud. It remains in an embryonic state and is deeply hidden under the tissues in Myxinoidea. The inner ear possesses only one semi-circular canal in Myxinoidea, two in lampreys which possess two vibratile sacs annexed to the ear. These sacs have no homologue in other vertebrates. The pineal eye of the lampreys is in fact twofold since it is composed of the superposed pineal and parapineal organs. The former is probably comparable to the pineal organ of Anura, the latter to parietal and parapineal eye of Lacertilia and Rhynchocephalia.

The feeding method of Cyclostoma is unique in vertebrates. The microphagy of the ammocoete which traps microscopic particles thanks to the mucus secreted by the endostyle recalls the way of living of Amphioxus and tunicates. The sucker of the lamprey, with its horny plates and its central tooth, the large horny rasp of Myxinoidea pulp the muscles of the fish which they attack. These techniques cannot be compared to the motion of the jaws by which the true fish capture their prey.

The buccal skeleton *(splanchnocranium)* of Cyclostomata is very different from that of the jawed vertebrates. This skeleton is membranous and elastic in the ammocoetes whose mouth is always gaping; in lampreys it is cartilaginous and constructed to permit the function of the sucker and the lingual piston; in Myxinoidea it is cartilaginous and built to support the large rasp. It is impossible to find in it any parts homologous to those of ordinary vertebrates and the three types can hardly be compared to each other. In lampreys, the buccal skeleton is continuous with a branchial basket situated outside and not inside the gills as is usual in branchial arches. In the adult lamprey a pericardial cartilaginous envelope is further added. This trait is absolutely unique in vertebrates.

In Myxinoidea the branchial arches are rudimentary and, in addition, topographically separated from the gills.

The cranial skeleton on the contrary is rather normal and primitive in its simplicity: it has parachordals, trabecules, otic capsules, nasal capsules but differs deeply in detail from lampreys and Myxinoidea.

The absence of paired fins is all the more remarkable as the arrangement of the nerves as well as that of the segmental muscles leads us to think that the ancestors of the present Cyclostomata have always been apodous.

The Cyclostomata contrast with all the other vertebrates in the form of the branchial pouches which are more or less hemispherical and hidden under the muscles. These pouches move backward during development. In the lamprey, the pouch which corresponds to the spiracle of elasmobranchia disappears. In Myxinoidea, the gills lie more or less at the midpoint of the body.

An endostyle is attached to the branchial pouches of the ammocoetes. It is comparable to that of Amphioxus but it is convoluted and much more voluminous. It disappears during metamorphosis but a part of its cells become transformed into thyroid follicles. By means of labelled iodine it has been shown that this organ already has a special metabolism in the larva.

The disappearance of the gall bladder and of the choledoch duct in the adult lamprey indicates a special physiology the details of which are not yet known.

Many venous sinuses surround the mouth and the gills of the lamprey. In Myxinoidea similar formations are scattered throughout the body. These sinuses play a role in the movements of the animal. In the lamprey for example, their swelling has an action antagonistic to the muscular contractions. The Myxinoidea are also remarkable for their accessory hearts (cardinal, portal, caudal heart). In both types the lymphatic system is not clearly separated from the venous system.

Systematic Position. The Cyclostomata are no doubt nearly related to the primitive vertebrates. Does their relatively simple anatomy prove their really primitive character or their degenerate state? The facts most often put forward in favour of the latter idea are the embryonic character of the eye of the hagfish, the poorly developed state of that of the ammocoete, the number of semi-circular canals in the inner ear of both types. These characters can probably be explained as adaptations to the conditions of life of these animals, i.e. in the ocean deeps or in the mud of brooks.

In the devonian deposits forms exist which have been authoritatively described by Stensio and like the lampreys, possess a dorsal nasal opening, a pineal eye, an inner ear with two semi-circular canals and a jaw-less mouth. These Cephalaspids, and other related forms, permit the assertion that the Agnatha and Cephalaspid have been formerly a successful group which is to be placed at the base of the vertebrates. But the fossil Agnatha possessed a bony shield and some of them had even rudiments of pectoral fins. Consequently Stensio thinks that the present forms have lost both their shields and their fins. The discovery of *Jamoytius* by White has shown that in the Devonian, forms existed which possessed a membranous skeleton. The present Cyclostomata probably come from similar forms and represent a very early separated and very early specialized branch.

The present Cyclostoma are thus primitive vertebrates which have never possessed either jaws or paired fins. Do they constitute a homogenous group? It seems that they do not. The relationships between the nose and the mouth, the structure of the splanchnocranium, the disposition of the nerve roots and of the parietal musculature and lastly the means of reproduction are too different to permit them to be considered as a single group. At the present time, zoologists distribute them among two super-orders: Myxinoidea divided into three genera and Petromysontoidea divided into eight genera.

H. DAMAS

References

Appelgate, V. C., "Natural history of the Sea Lamprey," Washington, U.S. Dept. Int. Fish. and Wild Life Service, Sp. Rpts. Fisheries No. 55, 1950.
Fontaine, M., *et al.* "Agnathes, formes actuelles," *In* Grassé, P. P. ed., "Traité de Zoologie," Vol. 13. Paris, Masson.
Marinelli, W. and A. Strenger, "Vergleichende Anatomie und Morphologie der Wirbeltiere," Vienna, Deuticke, 1954–1959.
Pietschmann, V., "Cyclostomata," *In* Kukenthal, W. and T. Krumbach eds., "Handbuch der Zoologie," Fasc. 6. Berlin, de Gruyter, 1929.

CYTOGENETICS

Cytogenetics, which is a natural fusion of the separately arising sciences of CYTOLOGY and GENETICS, rests upon the assumption that the CHROMOSOME is the principal vehicle of hereditary transmission. Its concern, therefore, is with the behavior of chromosomes in MITOSIS and MEIOSIS, and with a correlation of this behavior with the transmission of GENES from one cell to another through division and from one generation to the next through reproduction.

Cytogenetics became a recognized science in 1902–04 when Sutton and Boveri demonstrated a clear parallelism between the events taking place in cell division and recently re-discovered laws of Mendelian inheritance. These parallelisms include the following: (1) that the chromosomes consist of two homologous groups, one of maternal and the other of paternal origin, and that genes like-wise exist as pairs of factors of similar origin; (2) that gametes are "pure" in the sense that they contain only one chromosome and one gene of each kind; (3) that the chromosomes and the genes retain their identity throughout the life-cycle; (4) that each chromosome and each gene plays a definite part in development; (5) that the members of each pair of chromosomes and of each pair of genes segregate from each other during the reduction division; and (6) that one pair of chromosomes and one pair of genes segregate independently of other pairs of chromosomes and genes in the same cell. The abstract gene of Mendel was thus provided with a physical, visible basis.

Experimental proof of the Sutton-Boveri chromosome theory of inheritance came from several sources. A knowledge of meiosis, when the chromosomes are being reduced in number, indicated that gene segregation had a physical counterpart. Morgan's discovery of sex-linked inheritance and Bridges' study of the non-disjunction of X-chromosomes in *Drosophila melanogaster* showed conclusively that a particular gene is carried on a particular chromosome. Eleanor Carothers, through a study of the segregation of heteromorphic homologues in the grasshopper, provided a physical basis for the independent assortment of genes. By this time, Sturtevant had demonstrated the linkage of genes, and had devised a method for the calculation of gene distances, while Janssens had put forth the idea that the chiasmata observed between homologues in meiosis provided a chromosomal mechanism for crossing over. Cytological proof of genetical crossing over was provided by Stern in *Drosophilia*, and by Creighton and McClintock in maize, and it was now clear that an exchange of genes during crossing over was accompanied by an exchange of chromatin during the prophase of meiosis. Every feature of gene transmission had, therefore, a physical counterpart in chromosomal behavior.

The chromosomal theory of inheritance, as the fundamental basis of cytogenetics, has never been seriously challenged although ancillary forms of inheritance are known. Cytogenetics, however, has had its base strengthened and broadened by many other discoveries and applications. Two avenues of investigation, in particular, have been important in developing cytogenetics as an experimental science. The rediscovery of the giant chromosomes of *Drosophila*, with their distinctive banding pattern, opened the way for a detailed topological mapping of the chromosome, and for an accurate location of genes within the chromosome. More recently, variation in the morphology of these chromosomes in different tissues and at different stages of development, and a correlation of these findings with evidence of differential synthesis, offer an elegant cytological procedure for the study of gene action in cellular differentiation, thus merging cytogenetics with embryology. The second avenue was opened first by the discovery, mainly by Bridges and Sturtevant, of the spontaneous existence of deficiencies, duplications, translocations and inversions, and later by the radiations studies of Muller and Stadler which pointed the way to the artificial induction of both gene and chromosome changes. Thus variation as well as heredity was provided with a structural basis.

In another direction, cytogenetics tended to merge with studies dealing with the mechanisms of evolution. Such pioneer studies as Blakeslee's cytogenetic research on the evolution of the karyotype of the Jimson weed, *Datura*; the classic cytological work of Rosenburg on interspecific hybrids of *Drosera*; and Winge's investigation and clarification of polyploidy as an evolutionary factor in the plant kingdom were paralleled by comparable studies on population genetics by Fisher, Haldane and Wright. These studies have led to the concept of the species as a dynamic entity, its stability and evolutionary potential being determined by the variations in its genes and chromosomes.

On a more practical level, the principles of cytogenetics have been most successfully applied by the plant and animal breeder in the improvement of farm products. Hybrid corn has perhaps been the most spectacular achievement, but the same success applies equally well to the lesser known crop plants as wheat, rye, tomato and potato, and to beef and dairy cattle, poultry and hogs.

Medicine as well as agriculture has also benefited. The cytogenetics of man can only be considered in its infancy, but it is evident even now that many congenital abnormalities have their origin in an abnormal cytogenetic picture; e.g., mongolism has been shown to be due to the presence of an extra chromosome, while certain sexual aberrancies have been traced to variations in the number of sex chromosomes.

Modern cytogenetics has taken on a definitely chemical trend. A number of beautifully designed experiments have demonstrated with reasonable certainty that the deoxyribose nucleic acid (DNA) of the chromosome is the key molecule of the gene. Protein and ribose nucleic acid, also found in the chromosome, are necessary components in a pattern of reactions that bridge the gap between gene and character, but it is believed that the base pairs of DNA are the genetic alphabet out of which are constructed the language of heredity and variation.

CARL P. SWANSON

References

Babcock, E. B., "The Development of Fundamental Concepts in the Science of Genetics," Washington, Amer. Genetic Assn. 1950.

DeRobertis, E. D., et al., "Cell Biology," Philadelphia, Saunders, 1965.

Swanson, C. P., et al., "Cytogenetics," Englewood Cliffs, Prentice-Hall, 1967.

d

DARWIN, CHARLES (1809-1882)

Few men, other than the prophets of a new religion, have had a more profound effect on the thinking of their era than had this English naturalist. Indeed, much of the early opposition to his ideas came from those who, consciously or unconsciously, felt that his views were indeed a religion competing with an established order that brought comfort to its participants and profit to its institutions. Yet Darwin did not "invent" evolution, in the sense that it came to him as a flash of revelation. He merely arrived at certain views after long study and much thought, and expressed these views with clarity and force. Even the views were not particularly new. That some animals were related to others had been apparent to man since his first sentient days and had been openly discussed since the Reformation. Darwin's contribution was to present a plausible and logical reason as to how animals got that way.

He was born in 1809, the son of a country doctor and grandson of the great Erasmus Darwin. He received the usual conventional schooling of a middle class child of the period and found it irksome. His tastes were for the country and for nature, and he was removed from school at sixteen since his studies seemed to be of little benefit to him. His father wished him to follow in his footsteps as a physician and he accordingly went to Edinburgh to study medicine in 1825. The dislike of formal instruction which he had acquired at school stayed with him and he spent more time in the libraries than in the lectures. This distaste stayed with him until the end of his life and he differed many years later from T. H. Huxley whom he told that "there are no advantages and many disadvantages in lectures compared with reading." His main interest while at Edinburgh was in the long summer vacations which he spent in part rambling over the countryside and in part on the coasts of England, where he came in contact for the first time with marine zoology. It was here that Dr. Robert Edmund Grant seems to have persuaded him that a study of living things need not be confined to man. Gradually the influence of his father waned and his interest in medicine reached the vanishing point. But if not medicine as a profession, what then? The only two socially respectable fields for a man of his class were medicine and the church, and the idea of being a country clergyman appealed to him. He accordingly went to Cambridge with the idea at the back of his mind that he might take holy orders. His approach to divinity was, to put it mildly, lackadaisical but he managed to learn enough Latin and Greek, and to brush up on

Paley's "Evidences of Christianity" sufficiently to secure a bachelor's degree in 1831. It was while he was wondering what to do next that he received, through the good offices of some Cambridge dons who knew a first class potential scientist when they saw one, an offer to go as naturalist on the voyage of H.M.S. "Beagle". The story of this voyage, and of Darwin's gradually increasing comprehension of what he was seeing around him, has been often told and need not be repeated here. He came back from this voyage in 1831 with a vast collection of undigested data and with the idea of "natural selection" stirring in his mind. He was averse to exposing his ideas by publication until he had better support for them and it was not until 1858 that he was finally persuaded to present a paper to the Linnean Society, and that only because those of his friends who knew of his views felt that publication should take place before Wallace established claim as the originator of what subsequently came to be known as the theory of evolution. He decided that the limited space that could be afforded to him in the Journal of the Linnean Society was inadequate to present his ideas but he could not imagine who would be interested in a full length book. He prepared, however, a list of chapter heads and on the recommendation of the geologist Lyle, John Murray stepped in and provided a contract for the publication. He objected, however, to the title "An Abstract of an Essay on the Origin of Species and Varieties through Natural Selection" which he felt to be too vague. The printed volume therefore became "On the Origin of Species by Means of Natural Selection, or the Preservation of Favored Races in the Struggle for Life." This book immediately became a best seller and still sells well. That this was an unanticipated phenomenon may be judged by the fact that the first edition was only printed in 1,250 copies.

The immediate storm that then broke is again known to every biologist. Huxley, Gegenbaur, and most of the other great zoologists of his period accepted his ideas with enthusiasm. This was not the case with theologians who felt that they alone had the right to introduce new doctrine. Rarely has acrimony risen to such heights as in T. H. Huxley's reply to Bishop Wilberforce who had inquired which of Huxley's grandparents was to be considered a monkey. He replied that "a man has no reason to be ashamed of having an ape for his grandfather. If there were an ancestor whom I should feel shame in recalling, it would be a man, a man of restless and versatile intellect, who, not content with an equivocal success in his sphere of activity, plunges into scientific questions with which he has no real acquaintance, only to obscure them by an aimless rhetoric, and distract the attention of his hearers from the real point at issue by eloquent digressions, and skilled appeals to religious prejudice."

Darwin himself played little part in these battles. His health had never been good, though the fact that he describes an attack of tachycardia, while waiting for the "Beagle" to sail, and describes the eruption of a rash on his hands whenever he becomes nervous, would suggest that at least part of his troubles may have been psychosomatic even though there is some evidence that he contracted Chaga's disease while in South America. He purchased a country house at Downe, now a museum, and retired there, spending long periods in bed and rarely venturing beyond the confines of his gardens. His literary output was, nonetheless, prodigious in all the branches of the biological sciences.

He died in 1882, and was buried, as he would have wished, in Westminster Abbey in the company of the great men of England.

PETER GRAY

References

Darwin, C., "The Origin of Species" (facsimile of 1st ed. edited by E. Mayr) Cambridge, Harvard University Press, 1964.
Darwin, F. ed., "The Autobiography of Charles Darwin and Selected Letters," New York, Appleton, 1892 [reprint, New York, Dover, 1958].
Bates, M. and P. F. Humphrey, "Darwin Reader," New York, Scribner's, 1956.

DECAPODA

The Decapoda, the most highly organized CRUSTACEANS, constitute a large order (about 8,000 species), often grossly different in appearance, but basically similar in functional anatomy. The order includes macrurous forms (Penaeidea, Caridea, Astacura, Palinura: shrimp, prawns, lobster, crayfish, etc.) having large elongated abdomens which extend posteriorly, and compressed or cylindrical bodies; anomurous forms (Anomura: hermit crabs, porcellanid crabs, etc.) which have reduced abdomens that may bear uropods and show some evidence of spiraling, with cylindrical or depressed bodies and less than four pairs of thoracic legs for locomotion; and brachyurous forms (Brachyura: true crabs) with reduced abdomens flexed beneath the depressed cephalothorax, and usually with four pairs of thoracic legs for locomotion.

Decapods are characterized by having a well-developed carapace, stalked eyes, three pairs of thoracic maxillipeds, five pairs of walking legs (from which the order gets its name), thoracic limbs usually not all biramous, usually more than one series of gills, and with statocysts in most adults. The most closely related order is the Euphausiacea, whose members have no thoracic maxillipeds, all thoracic limbs biramous, one series of gills, and no statocysts.

General morphology: The body is covered with an exoskeleton containing chitin, which is often fortified with calcium carbonate. It may be rigid, firm, or extremely thin and pliable (between segments or on the abdomen of many Anomura). Of the three body divisions, the head and thorax are fused as a cephalothorax that is covered with a large, well-developed carapace which is fused to all thoracic somites; the abdomen is free and its somites rarely fused. The head is composed of five somites (some workers disagree) bearing the antennules, antennae, mandibles, first maxillae, and second maxillae. In addition a large pair of compound eyes is borne on movable stalks. The head appendages (not the eyes) are basically biramous and are variously used in sensory reception (chemical, tactile, vibration), feeding, and respiration (gill bailers). The thorax of the Decapoda is composed of eight somites characteristically bearing three pairs of usually biramous maxillipeds and five pairs of walking legs. The first pair of walking legs is often modified with chelae or pinching hands that are specialized for use in defense, feeding, and occasionally for mating by some species. Gills are thoracic and theoretically in four series: one on the coxae of the limbs (*podobranchiae*), two on the joint between limb and thorax proper (*arthrobranchiae*), and one series on the thoracic wall (*pleurobranchiae*); as a rule only part of these gills are actually developed, the number being different in various groups. Cephalothoracic invaginations (*apodomes*) of the ventral surface form an "endoskeleton" (see SKELETON, INVERTEBRATE) which serves to make the body rigid and allows a place for muscle attachment. The abdomen is composed of six somites, often with five pairs of biramous pleopods, a pair of uropods (except Brachyura and Lithodidae), and a terminal telson. Pleopods, often reduced in number throughout the order, serve for swimming in many species, are used to bear attached eggs, (except Penaeidea), or may be modified to transfer sperm.

The digestive tract is complete, equipped with a gastric mill in the cardiac stomach, and terminates with the anus at the telson. Digestive glands are often large and branched. The circulatory system is open, and consists of a sac-like heart in the thorax, arteries, and body sinuses. Plasma is clear. The respiratory pigment is hemocyanin. Amoeboid corpuscles are present. The nervous system is well developed, with a large supraesophageal ganglion, a subesophageal ganglion, and a double ventral nerve cord with paired ganglia in the somites (when not fused). Eyestalk hormones (sinus gland) may control color change, calcium metabolism, molting, eye pigment activity, etc. Other hormones may be produced by the commissure ganglion or gonads.

Food consists of both live and dead plant or animal tissue of microscopic or usually macroscopic size. Vision, chemoreception, or tactilreception, alone or together are used to obtain food. The specific use of the senses depends on the species involved, the food sought, and environmental conditions at the time of feeding.

Sexes are separate in most species. Sperm transfer is by paired pleopods, or if lacking, by possible contact transfer, or external fertilization (hermit crabs). Eggs are typically attached to pleopods of females until hatching. Direct development in a few genera (marine, freshwater and terrestrial); most with pelagic larval forms (see LARVA, INVERTEBRATE): nauplius (uncommon), protozoea, zoea, mysis, or postlarval stages (megalops, glaucothoe, etc.). The larval types present, their names, and the number of stages of each type vary greatly from group to group. Larvae filter feed or capture small organisms for food, molt between each stage, use various appendages for locomotion, are usually hyaline and may have chromophores. They usually are photo-positive becoming photo-negative in many groups before metamorphosis into the adult form.

Decapods molt (because growth is impeded by the exoskeleton, necessitating the molt and the production of a new, larger exoskeleton which will permit continued growth) by rupturing the exoskeleton dorsally between the carapace and abdomen and backing out of the opening. The tendons and the lining of the stomach and intestine are pulled from the animal as it molts. A new exoskeleton is secreted and hardening takes form days to weeks, depending on size and species. Autotomy of the walking legs (at a breaking plane by means of special muscles which sever the limb), is common. Regeneration of a new limb from the severed "stump" follows and requires two to three molts for completion. Autotomy is regarded as a defensive mutilation whereby an animal may escape an enemy, entrapment by foreign objects (rocks shifting), or fatal bleeding due to injury.

In habitat, most are marine; some are pelagic, but the majority are associated with various substrata ranging to great depths. Crayfish, a few crabs and others are freshwater dwellers; a few crabs and hermit crabs are terrestrial. A wide range of morphological and physiological adaptations to habitat is displayed.

Many decapods are of economic importance, directly supporting crab, shrimp, or lobster fisheries, and indirectly larvae or adults are used for food by other animals.

JENS W. KNUDSEN

References

Balss, H. *et. al. in* H. G. ed. Bronns Klassen and Ordnungen des Tierreichs," Vol. I. 7. Buch, Leipzig, Winter, 1940–1961.

Calman, W. T., *in* Lankester ed., "A Treatise on Zoology, Part VII, Apendiculata (3rd Fascicle, Crustacea)" London, Black, 1906 [reprint New York, Stechert].

Schmitt, W. L., "Crustaceans," Ann Arbor, The University of Michigan Press, 1965.

Waterman, T. H., "The Physiology of Crustacea," 2 vols. New York, Academic Press, 1960–1961.

THE DERMAPTERA

The Dermaptera is an order of insects known as "earwigs" because of the ear-like form of the hind wings. The ancestors of the Dermaptera probably arrived in Gondwanaland between the Permian and the early Jurassic. The main center of evolution seems to have been Africa and India, and the less specialized subfamilies and genera may well have been driven by selection pressure into South America, Australasia and probably Antarctica. While this process was still taking place, Gondwanaland disintegrated during the Cretaceous. This has probably given the more primitive genera and subfamilies a circumtropical distribution ranging from New Zealand and South America, whereas the more specialized subfamilies tend to occur in the southern continents with their genera restricted

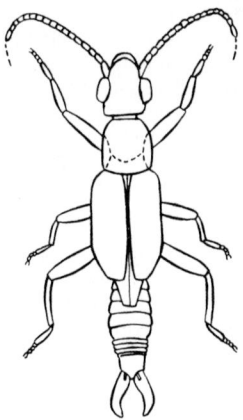

Fig. 1 Dermaptera. (From Fox and Fox, "An Introduction to Comparative Entomology," New York, Reinhold, 1964.)

to only one or two of them. The more specialized subfamilies and genera are restricted to one of the southern continents. Tertiary orogenesis formed land connections between the continents of the Northern and Southern Hemispheres, but only the more specialized subfamilies, such as the Labiinae and Forficulinae, have probably thus become widely established in the Northern Hemisphere, where their northern limit is determined by the occurrence of frozen soil in winter.

The fact that the Dermaptera possess such primitive insect features as vestiges of the eleventh and twelfth abdominal segments and paired penis lobes in some unspecialized subfamilies suggests that Dermaptera descended from a primitive stock of insects, which seems to have reached Gondwanaland after the Carboniferous period. Nevertheless, the specialized features characteristic of the order have been superimposed upon these primitive features and may be regarded as adaptations to life in litter and soil, under bark, in compressed vegetation and similar situations. These features may be summarized as follows:

Primitively, the Dermaptera are omnivorous, but some subfamilies have become predominantly carnivorous. For example, the Karschiellinae feed on ants, while the Chelisochidae are partial to leaf-mining insects which they dig out of plant tissues. The highly specialized Forficulidae shows trends to becoming secondarily omnivorous, and feed by tunnelling into their food.

The Dermaptera have, therefore, prognathous heads, and a complex neck, and are adapted to squeezing their way through the interstices of their environment by having the thoracic and abdominal segments sloping forwards to accommodate the large coxal muscles of the backwardly directed legs, which are used to push the insects forwards. The pronotum extends backwards as far as the wings. The front wings are short and sclerotized and when closed are held together by two medium rows of setae which act as "zip-fasteners." The membranous hind wings are normally folded like a fan under the front wings. Many earwigs, however, are apterous and only a few species are known to fly.

The cerci of the adults are sclerotized, comprise a single segment and show sexual dimorphism. The cerci are used to lever the insects backwards out of difficult situations, as organs of offense and defense, for measuring the height of the space in which the earwigs are enclosed and in some species for capturing the prey and holding it in front of the mouth for mastication. The insects mate end to end, the two sexes facing in different directions, but the posterior end of the abdomen of the males is twisted through 180° and in this position the cerci of the female support the male abdomen, while those of the male hold the female. After mating the females lay their elliptical eggs in a sheltered situation. Since the insects live in such microhabitats, the females need no ovipositor, though vestiges of this organ are to be found in the less specialized subfamilies. The females tend the eggs and very young larvae.

A natural classification of the order has been proposed by Popham (1965) in which four superfamilies, the Karschielloidea, the Pygidicranoidea, the Labioidea and the Forficuloidea are recognized. This classification differs substantially from that previously proposed by Verhoeff Zacher and Burrat at the beginning of the century. In particular the Arixenidae, previously regarded as a separate suborder, are seen as neotonous Labiids which have become adapted to external parasitism on cave bats of Java, while the Hemimerina are shown to have little affinities with the Dermaptera and accordingly are placed in a separate order.

E. J. POPHAM

References

Sharp, D., "Earwigs" *in* Harmer, S. F., and A. E. Shipley "The Cambridge Natural History," Vol 5. London, Longmans, Green, 1895.

Rehn J. G., A resume of one hundred years work on the Dermaptera. *Entomological News* **66**: 65–82, 1955.

Hincks W. D., A systematic monograph of the Dermaptera of the World based upon material in the British Museum (Natural History) Part 1, Pygidicranidae, sub-family Diplayinae, London, British Museum (Nat. Hist.), 1955.

Hincks, W. D., A systematic monograph of the Dermaptera of the World based upon material in the British Museum Natural History). Part 2, Pygidicranidae, excluding Diplayinae, London, British Museum (Nat. Hist.), 1959.

Popham, E. J., The anatom in relation to the feeding habits of *Forficula auricularia* and other Dermaptera. Proc. Zool. Soc. Lond., **133**: 251–300, 1959.

Popham, E. J., The functional morphology of the reproductive organs of the Common Earwig *Forficula auricularia*) and other Dermaptera with reference to the natural classification of the Order. Zool, J., **146**: 1–43, 1956.

DERMOPTERA

These gliding Lemurs are an order of MAMMALIA which have been distinct since Lower Paleocene times. The Greek name Dermoptera derives from "derma" (skin) and "pteron" (wing). Up to 70–80 million years ago, the order was connected with INSECTIVORES, Bats, (see CHIROPTERA) and LEMURS. But even in the Paleocene there was in North America an ancestral genus *Planetetherium* with great similarity to recent Gliding Le-

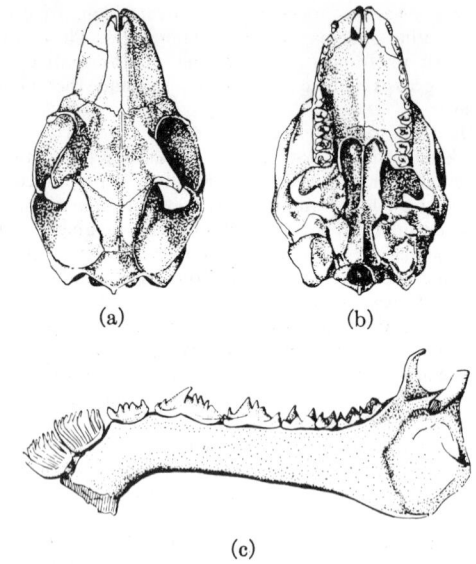

Fig. 1 Skull of the flying Lemur: (a) in view from above, (b) from below, (c) right part of mandible from the lingual side. (After Grassé, 1955.)

murs. Today the Kobegos, Cobegos, Korbugos or Kaguans—as they are called by the natives of various countries—are to be found only in the woods of tropical Asia.

Dermoptera are about the size of a domestic cat, with a total length from nose to tip of tail of about 60–75 cm, of which the body itself covers about 40–50 cm. The head of these nocturnal animals carries huge eyes, looking like those of a fox or dog, or even more like those of the Brown Lemur (*Lemur fulvus*) of Madagascar, which they also resemble in size, color, and the shape of the short, rounded ears. Behind the naked muzzle (*rhinarium*) there are a few short whiskers or *vibrissae*. The slender body has somewhat elongated limbs all of equal length. The comparatively long and round tail tapers gradually. The whole body is surrounded by a double furred, skin parachute (*patagium*), which extends between the neck and front paws as the *propatagium*, between the front paws and hind feet as pla-

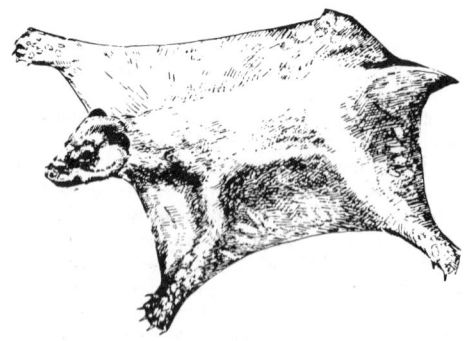

Fig. 2 Flying Lemur in flight. (After Grassé 1955.)

giopatagium, and between the hind feet and the tip of tail as *uropatagium*. The five fingers and toes are completely webbed except for the laterally compressed, narrow curved, and sharp claws. The first finger and toe are the shortest, the fifth the longest. All fingers and toes can be spread widely with the parachute between them. The whole upper surface of the animal is covered with a short, thick, excessively fine, silky, and chinchilla-like fur, in a varying mixture of grey and brown— with preponderance of the one or the other—with lighter parts on head and body, with a faint intermingling of pepper and salt, and with few or numerous irregular light or silver-grizzled small dots on the flanks. Thus the animals simulate lichens and mosses occurring on the tree trunks and branches where they sleep during the day. The silky, yellow to reddish brown, fur on the underside has a looser texture, its longest on the rump, and shorter and scantier on the parachute. The skin contains some scattered glands of alveolar and tubulus types, but there are no compact groups of these. The vulva is situated under the root of tail in a small, triangular pocket, capable of being closed. The well developed scrotum is postpenial in position, the pendulous penis carrying a baculum. Though only one young is born at a time, female carries four teats standing in two pairs near the armpits.

The flat and broad skull shows large, rounded eye-sockets or orbits, slightly frontally directed with the margin projected, and posteriorly opened to the temporal fossa. The bony palate is long, the bony ear, or *bulla tympanica*, is small and closely ankylosed with the mastoid process. The mandible is low, its mental part flattened, directed forward, and its ascendant branch or ramus is very short and low, There are 34 teeth with the formula 2.1.2.3./3.1.2.3 = 34. The small, multicuspid upper front teeth, or incisors, and the upper and lower canines, are displaced to the lateral part of the jaw, so that the frontal border of the upper jaw shows no teeth. The lower incisors are very wide, pectinate with 5–11 "teeth" standing closely together, but growing on a small base. The two outer incisors of the mandible, and the canines of the upper jaw, are inserted in the bone with two fangs. These extraordinary features are unique among mammals. It may be possible that the Dermoptera use the comb teeth to clean their fur. The molar teeth have several sharp cubs on a triangular base like the molars of the Insectivores—this is astonishing, because the Dermoptera are herbivorous (leaves, flowers, and fruits), as is shown by the structure of the digestive tract.

The chest is small, all the ribs being attached to the sternum. The number of vertebrae is remarkably variable. There are seven cervicals, 13–14 thoracics, 7–9 lumbars, 3–4 sacrals and 16–19 caudals. The forearm is longer than the upper-arm and in the forearm the distal end of the ulna is fused with the radius. In the hind-limb, the thigh and leg are equal in length, but the feeble fibula is closely lined with the strong tibia. The scapula and pelvis are small but the clavicle is sturdy.

The long tongue has no undertongue. The simple stomach opens into a gut 5–6 times as long as the body, with a large caecum; the large intestine is longer than the small intestine. The lobed liver contains a gall bladder. The two corned, or duplex, uterus carries a discoidal placenta only in loose connection with its wall.

At the beginning of embryonic development the placenta still shows a yolk sac.

The slightly folded hemispheres of the brain are small, the olfactory capsules and the nasal conchae (4 endo-, 3 ectoturbinalia) well developed. That would appear to mean that the Dermoptera have only an unimportant intellect, but a good sense of smell.

The solitary and nocturnal Dermoptera inhabit forests with high trees, where in day time they hang sleeping upside down and with all four feet together on a branch above, or rest curled up in forked branches. At night they awake and climb into the tops of trees where they eat leaves, flowers and fruits, putting them into their mouths with their hands. From time to time they utter terrific shrill cries. They do not voluntarily descend to the ground. If they wish to change trees, they push themselves off into the air, spreading the parachute to its extreme extent, and thus glide noiselessly as much as 130 m and more to the next tree, losing not more than one part in five in altitude in the flight. Most of the biology of these abstruse mammals is unknown. The breeding season seems to be unlimited throughout the year and duration of pregnancy about 60 days. One young is born at a time, and this is naked on the belly, upperside with finely patterned fur, with very large feet and claws and clinging to the mother's chest for a long time.

The recent Dermoptera all belong to the one family of Cynocephalidae Simpson, 1945 (=Galeopithecidae Gray, 1831) with the single genus *Cynocephalus* Boddaert 1768 (= *Galeopithecus* Pallas, 1780 and *Galeopterus* Thomas, 1908), including only two species, one with 16 subspecies. One species, *Cynocephalus variegatus*, (Audebert, 1799), the Malayan Gliding Lemur, is distributed from Tenasserim and southern Indo-China southward to Malaya, Sumatra, Java, Borneo and many small islands between. The body occupies 40–45 cm of its total length of 60–70 cm; the tail is 20–25 cm long.

The other species, the Philippine Gliding Lemur, *C. volans* (Linné, 1758) (=*philippinensis* Waterhouse, 1838), is somewhat bigger than *C. variegatus* with a total length of 65–75 cm (head and body 45–50 cm, tail 20–25 cm). Its first upper incisors are more reduced and the crests of the parietal bones of the skull stand closer together than in *C. variegatus*. The Philippine species varies in color and pattern following age and sex, but seems not to have developed distinct subspecies. It lives on the Philippine Islands Bohol, Samar, Leyte, Mindanao, Basilan, and the Tonkil group of the Sulu Islands.

THEODOR HALTENORTH

References

Walker, E., *et al.*, "Mammals of the World," Vol. I, Baltimore, Johns Hopkins, 1964.
Ward, P., "Flying Lemurs," Animals, London, **11**, 8, 364–365, 1968.

DESERT

A desert is defined to be an area in which less than twenty per cent of the ground surface is covered with permanent vegetation. According to this definition, there are a number of different types of deserts:

First, low rainfall deserts; second, cold deserts (polar and alpine areas); third, low nutrient deserts (due to excessive leaching of soil); fourth, man-made deserts (cities and industrial areas); fifth, deserts due to toxic conditions (near areas with volcanic activity, smelter areas); sixth, high salt deserts.

We will discuss here only the low rainfall deserts, because these are considered most typical.

The physical conditions of such deserts are:—low rainfall, high insolation, temperature extremes. Since watervapor in the atmosphere (which tends to act as a thermo regulator by absorption of radiation and by cloud formation), is present in small quantities only, sun radiation during day is at a maximum, causing a sharp temperature rise, and radiation into space at night is also high, causing rapid cooling during night. Since hardly any radiation energy is transformed into heat of evaporation, most radiation is transformed into sensible heat. The extremes in temperature also increase air movement and wind; frequent sand and dust storms are typical, especially in the flat deserts.

The lack of vegetation in the desert is due to insufficent rainfall, which limits plant growth. The plants occurring in such a desert have special adaptations, either to live with little water, or to utilize the available water to the greatest extent.

Desert plants can be divided into the following groups:

1. Succulents. These are plants which have reduced their transpiring surface to the extent that they lose very little water—Cacti and Crassulaceae. They have special adaptations to keep their stomata closed during day and absorb CO_2 especially during night. Thus they take up CO_2 at a time that water loss, through transpiration, is slight. A number of kinds of cacti can survive many months, or even years, without getting liquid water. When a rain comes, they are able to take up this water with hair roots developing within a few days, absorbing the surface water. The roots of cacti are usually very close to the surface.

2. Plants with deciduous leaves. The Ocotillo (Fouquieria) develops leaves within a few days after a rain, and sheds these leaves as soon as its water supply becomes limited. In this way the Ocotillo may develop leaves several times a year after proper rains.

3. Plants with deep root systems reaching the permanent water table (phreatophytes). Typical examples are mesquite (Prosopis), cottonwood (Populus), and palms (Washingtonia).

4. Annuals which develop only after heavy rains and continue to grow as long as there is water in the soil. These plants really are drouth-escaping, rather than drouth-resistant.

5. The real Xerophytes which have developed mechanisms with which they are able to withstand drouth. These plants, such as the creosote bush (Larrea), and the pygmy cedar (Peucephyllum), are able to grow in the driest deserts and can survive long periods of drouth. Ultimately, however, they still require small amounts of water, but they will be found in areas with precipitation of 50 mm or even less per year. It has been found that these plants are able to develop very high suction forces so that they are able to absorb water vapor from the desert air during night.

Most of the desert plants have developed special mechanisms by which they germinate only after heavy rains and not after a light rain. In this way, they can develop fully in response to a single rain of 25 or more millimeters, which wets the desert soil to a depth of 30–50 cm.

The mechanisms allowing germination only after a heavy rain are:

1. Presence of water-soluble germination inhibitors which must be leached out by a heavy rain.

2. Hard seededness. This condition means that the seed coat is so strong that the embryo cannot penetrate it. Only after a heavy rain which washed seeds together with gravel and sand down a wash, will the seed coat be scraped off and germination ensue.

3. Germination is often inhibited by salts which accumulate in desert soil. When these have been leached out by a heavy rain, germination can occur.

Both plants and animals have problems in the desert in connection with the very high temperatures especially at the ground surface. However, most desert animals live in burrows (rodents, reptiles, ants and other insects), and can escape the extreme temperatures by retreating in their burrows. But plants must be able to withstand soil surface temperatures of 60–70° C.

The water problem for desert animals is different from that of plants. In most deserts there are occasional water holes or springs, which are used by the larger animals as a source of water. They cannot live in a desert without surface water, like most of the Australian desert. A second group of animals (rabbits, turtles) lives off the vegetation which supplies the necessary water. A third group of animals is apparently able to live off metabolic water; that is to say, they are using water that is released by the burning of carbohydrates. These are ants and perhaps rats. The last group (snakes, birds of prey) comprises those which eat other animals and get a sufficient amount of liquid from their blood.

There is usually no regular soil formation in the desert. Weathering of the rocks leads to gravel and sand production which is washed down the slopes of the mountains and which accumulates at the foot of the mountains in the forms of alluvial fans of great porosity. Further decomposition of the gravel and sand produces salts and silt which ultimately are washed down in "basins." Water penetration in these silty basins is very poor and, therefore, results in poor plant growth.

Usually decomposition of rocks in desert areas proceeds at a faster rate than erosion can remove the debris and this results in an excessive amount of sand, gravel and rocks being washed down mountain sides during the infrequent, but often heavy, rains.

F. W. WENT

References

Shreve, F., "Vegetation of the Sonoran Desert" Washington, Carnegie Institute, 1951.
Jaeger, E. C., "Deserts Wild Flowers" Stanford, The University Press, 1942.
ibid. "Desert Wildlife" Stanford, The University Press, 1961.
Buxton, P., "Animal Life in Deserts" London, 1923.
Went, F. W., "The Ecology of Desert Plants" Sci. Amer. **192**: 68–75, 1955.

DESMIDS see ALGAE

DESMOSTYLIA

Four genera represent this aberrant extinct order, late Oligocene-late Miocene, of large, hippopotamus-like mammals with a circum-Pacific and Gulf coast (Florida) distribution. The order is divided into two families, the Desmostylidae (*Desmostylus, Cornwallius, Vanderhoofius*) and Paleoparadoxidae (*Paleoparadoxia*). The ancestral stock is believed to lie in an animal group which gave rise to the proboscideans, sirenians and perissodactyls during Paleocene time. A study of bone structure from the two better known genera and of the sedimentary environment in which all members have been found suggests these animals range from semi-amphibious forms capable of limited terrestrial movement (*Desmostylus*) to those almost completely amphibious (*Paleoparadoxia*) in brackish or salt water of coastal bays, lagoons and salt marsh swamps. Size range is about 9 feet for the estimated one-ton amphibious *Paleoparadoxia* to 12 feet for the semi-amphibious *Desmostylus*. All incisors and canines are procumbent, and according to genus, number 1, 2 or 4 pairs. Cheek teeth are brachyodont to hypsodont, replaced horizontally in adults from a large, ovate, globular dental capsule (*Desmostylus, Vanderhoofius*), composed of clusters of appressed, thick, peg-like columns of enamel with polydonty especially characteristic of the milk dentition. That the diet of the desmostylids is in doubt is shown by the group's being attributed to herbivores, omnivores and molluscivores.

ROY H.. REINHART

DEVELOPMENTAL GENETICS

Developmental GENETICS is primarily concerned with questions about the manner in which GENES control or modulate the processes of development in multicellular animals and plants. The essential theoretical problem is posed by considering that the set of genes in the first cell gives rise through mitosis to similar sets of genes in the descendant cells which nevertheless assume different forms and functions as development proceeds. Different genotypes as represented by different species or varieties show different patterns of development. How are such patterns related to the gene-differences involved? In more practical terms, by what mechanisms do gene-differences give rise to phenotype-differences? The question has to be asked in this way because at present we have no means of apprehending a gene except when by mutation it has assumed at least two ALLELIC forms which differ in their effects. Such questions inevitably pose larger ones as to how genes produce their effects and developmental genetics thus tends to become part of the more general field-physiological genetics.

Since at present we do not know the first effect or primary product produced by any gene, the observational data of developmental genetics consist of 1) the description of phenotypic differences which regularly accompany a known difference in genotype. This may be obtained by morphological (including histological and cytological), physiological, biochemical and other methods applied over as long a stretch of the life cycle as possible; 2) the results of experiments designed to test causal sequences suggested by differences as revealed by the descriptive study.

Before illustrating some results obtained by such methods some assumptions useful as guides in formulating questions and carrying out investigations should be pointed out.

First, that gene differences express themselves thru' products produced as a result of the vital activities of genes in metabolism and reproduction. The best evidence that the actions of a gene are specific is that each gene at each cell division produces a replica of itself out of materials available in the cell.

Second, some of the products of gene activity are able to effect action at a distance, both within the cell and upon other cells and tissues. This may occur through establishing gradients of concentrations of metabolites, through enzymes, hormones, inductors and similar agents.

Third, a common form of alteration of developmental processes may be deduced from the manner in which gene mutations affect metabolic processes. Mutant genes are known which affect one of the steps of a reaction sequence such as $a \rightarrow b \rightarrow c$, in which mutation of one gene may prevent $a \rightarrow b$, of another gene, $b \rightarrow c$, both resulting in defect of character or product c. Since such steps are catalysed, interference with production or action of enzymes will often be responsible for the altered phenotype.

Fourth, the effects of gene-controlled agents will be determined also by the competence of other cells or tissues to respond by changes in their forms or activities. Competence will be affected by the position of the responding cells in the temporal sequence of development. They may respond only during a limited period.

Fifth. Most effects of gene differences upon development will be removed by several steps from the primary gene products. This leads to the expectation that the effects of gene-differences, as deduced from phenotype-differences observed during development will usually be multiple, because of the diversity of response patterns which arise during differentiation. This expectation is realized in the rule that most gene mutations have pleiotropic or manifold effects. A second prediction, consequent upon this is that a phenotype will usually be the result of interaction among many pleiotropic genes, some tending to accentuate, others to diminish particular characters. The phenotype will thus represent the state of balance or equilibrium among gene-controlled reactions. The effects of mutation in any one gene is thus to be attributed to a disturbance of the normal developmental equilibrium which has been attained by natural selection through the establishment in the species of a harmoniously integrated system of genes, constituting the normal genotype.

Developmental genetics, as studied by the analysis of phenotypic differences during development has not yet been reduced to general laws or principles, comparable to those derived from analysis of the transmission mechanism of heredity. However, examination of mutant gene-differences in many animals and plants has revealed a general confirmation of the assumptions stated above. A few illustrative cases may be briefly cited.

Dwarfism is a striking difference which has appeared in many species as a result of single gene mutation. In one form known as chondroystrophy, studied in birds and mammals, including Man, the limbs are reduced in length relative to trunk length in heterozygotes while in homozygotes an associated syndrome of abnormalities, known as phokomelia, appears in limbs (reduced to stumps), eyes, heart, vascular system and other structures. As analysed in the domestic fowl, in which the whole period of development has been studied, the growth restriction in the limbs is determined (autonomous) when the first primordia are formed, probably as a consequence of a metabolic deficiency in the utilization of carbohydrate, which becomes limiting in different parts at different times depending on the growth intensities and consequent requirements of the parts. The effects of the gene on eye development, by contrast, are secondary and nonautonomous, imposed upon the rudiment by the defective character of the embryo as a whole. Thus a gene controlled difference in early metabolism reverberates through subsequent stages and systems, resulting in pleiotropic expression because of the interdependent, epigenetic character of development.

Other cases of dwarfism are known to result (in the mouse, for example) from gene-controlled alterations in the production of a growth hormone (pituitary dwarfism); some cases of dwarfism in plants are also traceable to genic effects on the production of a specific substance (giberellic acid).

In the mouse a number of gene differences are known which produce their effects by altering the dependence relations of one part on another which is known as induction. The alteration may be detected as incapacity of cells of one genotype to produce effective inductors, or of cells competent to given proper developmental responses to inductors. Such cases indicate that the integration of development by induction is controlled by the genotype.

In insects, a model of developmental control by genes was discovered when it was shown that the production of the brown component (ommochrome) of normal eye color involves at least two steps which may be separately altered by two different gene mutations; mutation in one gene causes failure of conversion of tryptophane to kynurenin, in another gene, conversion of kynurenin to 3-oxy kynurenin. Genic control of syntheses of pigments has been proved in numerous cases in flowering plants and in fungi.

In a few cases an identified gene has been shown to control the synthesis of a specific protein, and in a few other cases the specific activity of an enzyme has been found to be associated with a specific gene. These give clues to more general methods of control of developmental processes which are now being investigated.

L. C. DUNN

References

Hadorn, E., "Developmental Genetics and Lethal Factors" Methuen, London, and Wiley, New York, 1960.
Goldschmidt, R. B., "Theoretical Genetics" Univ. Calif. Press 1955.
Wagner, R. P., and H. K. Mitchell, "Genetics and Metabolism" 2nd ed. New York, Wiley, 1963.

DE VRIES, HUGO

Today mutation and recombination are recognized to be the principal sources of genetic variation. Toward the end of the nineteenth century such was not the case. Biologists had no good answer to the question: Where did the variation come from which natural selection acts upon?

The theory of mutation was developed from the observation of new forms in the evening primrose, *Oenothera,* by a Dutch botanist, Hugo De Vries. De Vries was deeply impressed by the sudden occurrence of radically different types of *Oenothera.* He postulated that new species are formed by sudden single mutations producing major changes. What De Vries actually observed in *Oenothera* was in effect recombination of linked complexes distal to chromosomal translocations and changes in chromosome number. Modern biology believes that speciation usually occurs by means of gradual accumulation of slight hereditary differences through isolation rather than by macromutation. These differences are caused by an array of different genes present in the gene pool due to mutation—not the gross mutation of De Vries, but minute changes in single genes. Due to different selective values under different environmental conditions new gene combinations are favored and eventually two populations differ so that they can not produce viable hybrid offspring.

Hugo De Vries was born February 16, 1848, at Haarlem, The Netherlands. Following his education at Leiden, Heidelberg, and Würzburg he was appointed lecturer at the University of Amsterdam in 1877, and became professor of plant physiology shortly thereafter. De Vries, Eric von Tschermak and Karl Correns independently verified and rediscovered Mendel's papers on heredity in 1900. In 1904 he lectured at the University of California. Following retirement in 1918 he continued active research, studying mutation and evolution in *Oenothera* at his home in Lunteren. Hugo De Vries died in Amsterdam, May 21, 1935. *Intrazellulare Pangenesis* (1889) and *Die Mutationstheorie* (1900) express his theories of evolution, and most of his other writing is collected in *Opera e Periodicus Collata* (1918–1927). Probably his greatest contribution to science was the demonstration that evolution could be studied experimentally.

ROBERT B. HELLING

DIATOMS see BACILLARIOPHYCEAE

DICOTYLEDONS

Flowering or seed plants (see SPERMATOPHYTA) constitute the largest division of the vegetable kingdom, and are primarily divided by all botanists into two very distinct groups, GYMNOSPERMS and ANGIOSPERMS. To the former belong the CONIFERS which have no true flowers and have naked ovules, and to the latter all those with more or less conspicuous flowers and with the ovules enclosed in a carpel or ovary.

Botanists also agree in dividing these true flowering plants into two principal groups, DICOTYLEDONS and MONOCOTYLEDONS, characterised by the number of seed-leaves or cotyledons as they are termed, Dicotyledons with two, and Monocotyledons with only one primary leaf. Cotyledons serve to protect the tender primary bud, the plumule, in the seed-stage and during germination, and to supply the young seedling with food material until it can fend for itself by means of its own roots. There are also other important differences between the two groups. In the stem of dicotyledonous plants the primary bundles are arranged in a circle or circles surrounding the pith. Owing to differences in the character of the bundles produced at the beginning and end of the growing season annual rings are usually discernible, especially in perennial plants such as trees and shrubs. Hence the approximate age of a dicotyledonous tree may be ascertained by counting the number of rings in a cross-section of the stem. On the other hand a cross-section of a monocotyledonous stem, such as that of a Palm tree, shows the bundles to be scattered and not arranged in rings. Frequently, but not always, the two groups may be recognised by the venation of the leaves, in the Dicotyledons these forming a network between the nerves, and in the Monocotyledons the nerves and veins run parallel with the margin and midrib.

Dicotyledons have been classified in various ways too numerous to record here in detail, and like the plants themselves their systematic arrangement has gradually been evolved from a small and crude beginning, like Jack's beanstalk, which was at first a tiny plant. In the very early herbals prominence was given to the habit of growth and their uses to mankind. THEOPHRASTUS (circa 370 B.C.) has the greatest claim to be called the 'father of botany', for he indicated the essential differences between Dicotyledons and Monocotyledons, and he also recognised that different plants grew in dissimilar habitats, such as woodland, marsh, lake, river, and other plant-associations. Subsequently writers of herbals greatly increased our knowledge of plants, but the first really important classification was made by the Swedish botanist CARL LINNAEUS, the 'father of modern botany' (hence there are *Linnean Societies* in various parts of the world). One of his most important works is his *Species Plantarum* in which he introduced the binomial system of nomenclature. Previous to Linnaeus a plant-name consisted of a short description. For example instead of calling the European field buttercup 'Ranunculus foliis ovatis serratis, scapo udo unifloro', he named it *Ranunculus bulbosus* L. In his sexual system he divided plants into twenty-four classes, determined mainly by the number, or some obvious character, of the stamens. Thus plants with one stamen were classed as *Monandria*, those with two as *Diandria*, and so on using the roman numerals up to *Dodecandria*, with twelve stamens.

More natural systems were soon forthcoming, beginning with the de Jussieus in Paris, the de Candolles in Switzerland, Bentham and Hooker in Great Britain, Baillon in France, and Endlicher and Engler & Prantl in Germany. These systems were written largely before the evolutionary views propounded by Darwin had taken root and whilst the dogma of the constancy of species still dominated biology.

In the Bentham and Hooker system the Dicotyledons were divided into several series mainly determined by one or only a few characters. Thus division *Polypetalae* had flowers with free petals, *Gamopetalae* with united petals, and *Apetalae* without petals. The *Polypetalae* were subdivided into smaller groups, those with a hypogynous type of flower, *Thalamiflorae* (examples *Ranunculaceae* and *Magnoliaceae*), those with a disk, *Disciflorae*, (examples *Rutaceae* and *Celastraceae*), those with a perigynous type of flower, *Calyciflorae* (examples *Rosaceae, Leguminosae* and *Myrtaceae*. The *Gamopetalae* were further subdivided into series, those with inferior ovary and those with superior ovary, and in a similar manner the *Apetalae*. These divisions did not result in a natural system because certain families in separate groups were clearly more or less related. For example *Caryophyllaceae* (*Polypetalae*) and *Gentianaceae* (*Gamopetalae*) are distinguishable by little more than the united petals of the latter family.

The German system of Engler and Prantl more or less reversed the sequence of the de Candolle plus Bentham and Hooker system. The groups without petals (*Apetalae*), some with catkins (*Amentiferae*) were placed first, those with free petals next, and finally those with united petals at the top of the scale, ending with the large Aster family *Compositae*.

From an evolutionary standpoint none of these systems was considered to be satisfactory. This was pointed out especially by Hallier in Germany, Bessey in the United States of America, and by Arber and Parkin and Hutchinson in Great Britain. The work of the last mentioned is the most recent attempt to provide an evolutionary or phylogenetic system (Hutchinson, *Families of Flowering Plants*, ed. 2, 1959: Oxford Press, England). The hypogynous flower with free sepals and petals, free stamens, and free carpels, is considered to be the most primitive type of present day flowering plants. Hutchinson recognises two main branches of the "family tree," one fundamentally woody, *Lignosae*, the other fundamentally herbaceous, *Herbaceae*. The *Lignosae* begin with the Magnolia family, *Magnoliaceae*, and terminate in an evolutionary sense with the Verbena family, *Verbenaceae*. In between these extremes in an ascending series are, to mention only a few of the larger families, *Annonaceae, Rosaceae, Leguminosae, Araliaceae, Moraceae, Urticaceae, Tiliaceae, Euphorbiaceae, Ericaceae, Myrtaceae, Rutaceae, Apocynaceae, Rubiaceae* and *Bignoniaceae*. Some of these families have flowers with free or united, or without, petals, and both superior and inferior ovaries.

The *Herbaceae* commence with the buttercup family, *Ranunculaceae* and the Hellebores, *Helleboraceae*, followed by the *Papaveraceae, Cruciferae, Caryophyllaceae, Gentianaceae, Saxifragaceae, Umbelliferae, Campanulaceae, Compositae, Solanaceae, Scrophulariaceae*, and end with the Dead-nettle family, *Labiatae*. Thus it will be seen that such families as *Araliaceae* and *Umbelliferae*, formerly placed in juxtaposition, are far apart, the former being regarded as having been derived from woody ancestors such as the stock of the *Rosaceae*, the latter from herbaceous ancestors, such as *Saxifragaceae*.

Certain families of the *Lignosae* contain many important economic plants. To *Rosaceae* belong the Apple, the Pear, the Plum, Apricot, Strawberry, Blackberry, Raspberry, Loquat, and others, besides many beautiful garden plants, Roses etc. Related to *Rosaceae* is the large group *Leguminosae*, often divided into three separate families, *Caesalpiniaceae*, *Mimosaceae*, and *Papilionaceae*; the first two of these contain many important timber trees, nearly all in the tropics, the last named food and forage plants in more temperate regions, such as the garden Pea, *Pisum sativum*, the Broad Bean, *Vicia faba*, Ground Nut, *Arachis hypopea*, Soy Bean, *Glycine max*, and the clovers, *Trifolium* spp. Dyes and insecticides are also obtained from these families, and many lovely garden plants, such as *Wisteria, Laburnum, Lupins*, etc.

The Oak family, *Fagaceae*, has been of great importance to man, the genera *Quercus* (Oak) and *Fagus* (Beech) providing valuable timbers, as also the related walnut family, *Juglandaceae*. *Sterculiaceae* provides us with Cocoa, *Theobroma cacaoa*, and the *Malvaceae* Cotton from the seeds of *Gossypium*. *Euphorbiaceae* contains the most valuable rubber plant, *Hevea brasiliensis*, and Castor Oil, *Ricinus communis*. The Tea family, *Theaceae*, is essential to comfort of many peoples, tea being the dried leaves of *Camellia chinensis*, the same genus containing some lovely flowering shrubs. The Grape vine, with all the delicious wines produced from its fruits, is *Vitis vinifera*, *Ampelidaceae*, to which family the lovely Virginia Creeper also belongs.

Especially important to mankind is the Rue family, *Rutaceae*, with the Orange, Lemon, Grapefruit, and Lime. Mahogany one of the most valuable timbers is *Swietenia mahogani*, of the large tropical family *Meliaceae;* to *Anacardiaceae* the Mango, *Magnifera indica*. The Ashes, valuable timber trees are species of *Fraxinus*, and the Olive, a valuable commodity in dry Mediterranean countries is *Olea europaea, Oleaceae*. *Apocynaceae* and *Asclepiadaceae* have very few plants of economic importance, but many are highly ornamental. *Rubiaceae* is a family of prime value for its contains Coffee, *Coffea arabica* and other spp., Quinine, *Cinchona* spp., whilst *Verbenaceae* provides Teak, *Tectona grandis*, the most important hard-wooded tree of Indo-Malaya.

The fundamentally herbaceous phylum, *Herbaceae*, is not so important from an economic point of view, though it provides many decorative plants of great value to the gardener and adorns fields and hedgerows. The Poppy family may be mentioned, the Opium Poppy being *Papaver somniferum*, with its significant specific name. The family *Cruciferae* has several food plants such as the Cabbage, *Brassica oleracea*, the Cauliflower, the Turnip, *Brassica rapa*, and Watercress, *Nasturtium*, with numerous lovely garden plants such as Wallflower, *Chieranthus cheiri*. *Gentianaceae* contains many beautiful rock-garden plants (*Gentiana*) as also *Primulaceae*, with many lovely species of *Primula*.

The more climax families of herbs are the Hemlocks, *Umbelliferae*, to which belong Celery, *Apium graveolens*, the Carrot, *Daucus carota*, Parsnip, *Peucedanum sativum*, and Parsley *Petroselinum crispum*. *Compositae*, the Aster or Daisy family, is the largest of all, but contains relatively few plants that are medicinal or edible. The most important are the Jerusalem Artichoke, *Helianthus tuberosus*, the Globe Artichoke *Cynara scolymus*, Chicory, *Cichorium intybus*, Lettuce, *Lactuca scariola*, Sunflower, *Helianthus annuus*, and very many decorative plants such as the *Chrysanthemum, Dahlia* and *Aster*.

And finally high up in the evolutionary scale is the Potato family, *Solanaceae*, the Potato, *Solanum tuberosum*, Tomato, *Lycopersicum esculentum*, and the medicinal plants *Atropa belladonna*, Henbane, *Hyoscyamus niger*, and of prime importance to both men and women the Tobacco plant, *Nicotiana tabacum*. Scented plants and some culinary herbs belong to the climax herbaceous family, *Labiatae*.

J. HUTCHINSON

References

Hutchinson, J., "Families of Flowering Plants" 2nd ed. Oxford, The University Press, 1959.
Engler, A. and K. Prantl., "Die natürlichen Pflanzenfamilien" Leipzig, Engelmann, 1897–1915.

DIFFUSION*

Diffusion, in a solution, is the movement of molecules from a region of high concentration to one of low concentration. This movement in position, more specifically translational diffusion, will occur until the entropy of the system is a maximum; that is, until such time as the molecules are completely dispersed.

The random movement of the solute molecules derives from their kinetic energy, and the net result of this Brownian motion under a concentration gradient is a displacement. The equation $D = \overline{x^2} / 2t$ expresses quantitatively the relationship between the diffusion rate, D the elapsed time, t, and the mean square displacement in the x direction. If the particles are large enough to be resolved by an optical microscope, then one can make use of this equation to determine their diffusion rate by observing individual particles in suspension and measuring their displacement in a given interval. Unfortunately, most biological macromolecules fall below the limit of resolution of the optical microscope, so that one is forced to study suspensions of such particles.

In 1855, Fick perceived that the principles governing the conduction of heat could be applied to diffusion. By analogy, his First Law states that the amount of material, dS, moving across an area, A, in the time, dt, under a concentration gradient, dc/dx, is governed by the relationship $dS = - DA (dc/dx) dt$, where D is the diffusion coefficient, usually expressed in cm^2sec^{-1}.

It is the above equation which constitutes the basis of the Porous Disc Method (Figure 1) for measuring diffusion coefficients. If two homogeneous solutions of known volume and concentration, usually solvent of concentration $c = 0$ and solution of concentration $c = c_0$, are separated by a sintered glass disc of thickness x, the diffusion coefficient can be determined by measuring the relative concentrations of both solutions at given

---

*Publication No. 78 of the Dept. of Biophysics of the University of Pittsburgh.

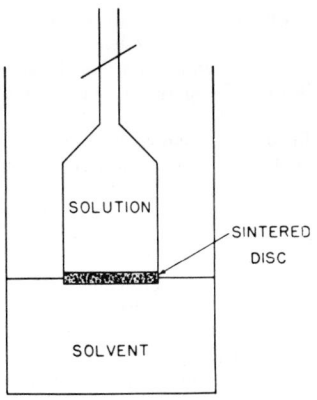

Fig. 1

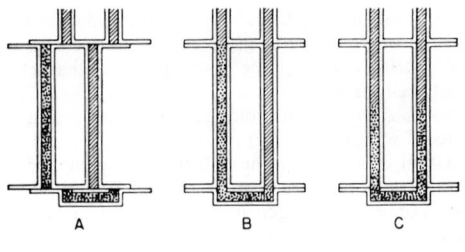

Fig. 2

time intervals where, A, the effective sum of all the pore areas, is a predetermined constant. The advantages of this technique are that it is relatively simple, the concentration gradient is stable against convection and mechanical movement, and only small amounts of the material are required for the determination. Its principle disadvantages are that the cell must first be calibrated with a standard, usually an inorganic salt whose diffusion properties may be considerably different from those of biological substances, and it is relatively insensitive to inhomogeneities in diffusion rates.

The most frequently employed method for carrying out diffusion experiments involves free diffusion at an interface. Generally, these are performed in a rectangular cross section Tiselius type electrophoresis cell (Figure 2), maintained at constant temperature in a refrigerated water bath.

By appropriate transverse motion of the three sections, the cell can be filled and interfaces can be formed between the lower solution and upper dialysate or buffer. These boundaries, formed when the limbs are brought into alignment, are established be-

tween the top and middle section of the left channel and between the middle and bottom section of the right channel (Figure 2b). Finally, by a procedure referred to as compensation, the boundaries are moved to the center of the mid section (Figure 2c), where they can be sharpened if necessary. The changes with time in the concentration gradient at the two interfaces can be followed by employing special optical systems, while photographs may be taken at suitable time intervals. These changes are illustrated diagrammatically in Figure 3.

Fick's Second Law, $dc/dt = D (d^2c/dx^2)$, relates the change of concentration with time as a function of the position in the cell. Assuming that the diffusion coefficient for the material under consideration is independent of concentration, and that none of the solute molecules at the original boundary reach the opposite end of the cell, Fick's Second Law may be expressed in the form

$$\frac{dc}{dx} = \frac{-c_0 e^{-x^2/4Dt}}{\sqrt{4\pi Dt}},$$

where x is the distance in cm from the original boundary, t is the elapsed time of diffusion in seconds, c is the concentration of the solute in g/ml, c_0 is the original concentration, and D is the diffusion coefficient in cm²/sec. The above equation reveals that: the concentration gradient is a function of D, x, and t; dc/dx is sym-

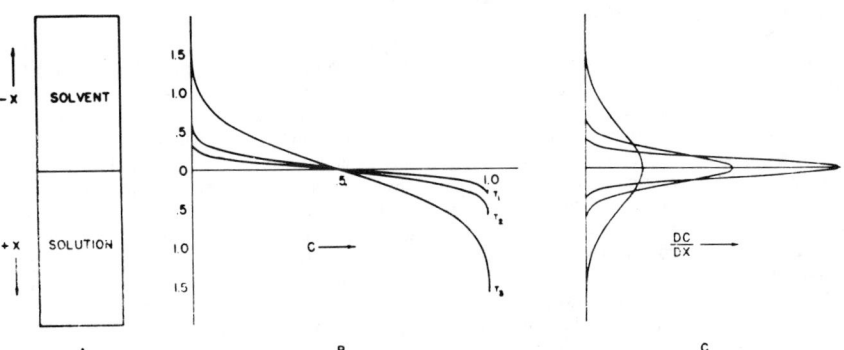

Fig. 3 (a)—Schematic representation of the center section of a Tiselius cell, at the start of a diffusion experiment. The biological material fills the bottom half, while buffer previously dialyzed against this solution occupies the top half. (b)—Graphic representation of the concentration, c, as a function of the linear position, x, in the cell, for the time intervals $t_1 = 10^5$, $t_2 = 3 \times 10^5$, and $t_3 = 2 \times 10^6$ seconds. The curves are calculated for the diffusion coefficient, $D_{20}^w = .672 \times 10^{-7}$ c.g.s. units, of T3 bacteriophage. (c)—The concentration gradient, dc/dx, as a function of the linear position in the cell at time intervals t_1, t_2 and t_3, for the bacteriophage T3.

metrical about $x = 0$, since x is in the exponent as x^2; and the curve is Gaussian or representative of a normal distribution, since the function e^{-x^2} is characteristically bell shaped.

To establish that a biological preparation is homogeneous with respect to rate of diffusion, the gradient equation indicates that the diffusion curve must be both symmetrical and Gaussian. A solution containing two populations of homogeneous material would not produce a Gaussian curve, although the curve should be symmetrical.

The gradient curve, such as Figure 3c, may be used for calculating the diffusion coefficient. Actually, the experimentally obtained curve relates the gradient in index of refraction to position in the cell. Since, however, the refractive index of a solution is a linear function of its concentration, according to the equation $n = n_0 + kc$, where n and n_0 are index of refraction of solution and solvent, respectively, c is concentration, and k is the specific refractive increment, one may employ the refractive index gradient in lieu of the concentration gradient. The derivative of the above equation, $dn/dx = k(dc/dx)$, indicates that the two are related simply by a proportionality constant.

While several different methods are available for calculating the diffusion coefficient, the choice generally is determined by the particular optical system employed to register the initial information. For gradient vs. position curves, as produced by the Philpot-Svensson Cylindrical Lens Method, usually either the Maximum Ordinate-Area or Maximum Ordinate Method are used. In the former, one measures the area, A, and maximum height, y_{max}, of the curve. These parameters, in combination with the time elapsed, t, and the magnification factor of the instrument, m, are related to the diffusion coefficient, D, by the equation $D = A^2/4\pi t m^2 y^2_{max}$. While this equation suggests that the diffusion coefficient can be calculated from a single observation or photographic exposure, this is true only if the boundary is infinitely sharp at the beginning of the experiment. Since, in fact, this is rarely the case, one eliminates this ambiguity by plotting $A^2/4\pi t m^2 y^2_{max}$ as a function of t, for several exposures, the resulting slope being D. Now, it does not matter where the curve intercepts the axis as it is the slope which determines D, and no time correction is necessary.

In the Maximum Ordinate Method, one measures the half width, z_i, of the Gaussian curve at its inflection point, since it is related to the diffusion coefficient by the equation $D = z_i^2/2m^2 t$. Again, one can plot $z_i^2/2m^2$ as a function of t, the slope of which is equal to D, in order to eliminate the time correction.

To systematize the diffusion values obtained by different investigators under various experimental conditions, it has become the practice to correct the diffusion coefficient to standard conditions; that is, the diffusion coefficient, D_{20}^w, that the particle would have in a solution whose viscosity was the same as that of water at $20°C$. This correction is made by using the equation

$$D_{20}^w = D_t \frac{293}{T_t} \cdot \frac{\eta_t}{\eta_t^w} \cdot \frac{\eta_t^w}{\eta_{20}^w},$$

where D_t is the measured diffusion coefficient at the temperature of the experiment, T_t is the absolute temperature at which D_t was measured, η_t/η_t^w is the relative viscosity of the buffer as measured at the temperature of the experiment, and η_t^w and η_{20}^w are the viscosity of water at the temperature t and $20°C$, respectively.

The diffusion coefficient's value derives from its association with the Einstein-Sutherland equation, $D = kT/f$, where k is the Boltzmann constant, T is the absolute temperature, and f is the frictional coefficient of the molecule. The latter coefficient is a function of the size, shape and hydration of the particle in solution and the viscosity, η, of the medium. For the simplest case, a particle which is spherical but large in comparison to the solvent molecules, Stokes' Law states that this coefficient of friction, f, is equal to $6\pi\eta r$, where r is the radius of the particle. Thus, one can estimate the size of spherical particles just from the diffusion coefficient. More sophisticated equations are available for characterization of prolate and oblate ellipsoids of revolution.

Of still greater significance, however, is the equation obtained by equating the value for the friction coefficient for diffusion, $f = kT/D$, with the value for the friction coefficient from sedimentation, $f = M(1 - Vd)$ $/Ns$. The result, $M = RTs/D(1 - Vd)$, is known as the Svedberg equation, where M is the molecular weight, R the gas constant per mole, s the sedimentation coefficient, V the partial specific volume, and d the density of the medium. Therefore, from diffusion and sedimentation studies one can determine the molecular weight of a biological material.

IRWIN J. BENDET

References

Edsall, J. T., and J. W. Mehl, "Translational Diffusion of Amino Acids and Proteins" "*in* Cohn, E. J., and J. T. Edsall, "Proteins, Amino Acids and Peptides," New York, Reinhold, 1943.

Alexander, A. E., and P. Johnson, "Translational Diffusion" *in* "Colloid Science" London, Oxford University Press, 1949.

DIGESTIVE SYSTEM

Introduction. The digestive system consists of specialized organs which function to propel, physically and chemically process, and absorb foodstuffs.

The esophagus. The esophagus develops from the primitive foregut, together with the pharynx and stomach. In the three-week embryo the esophagus is represented by a constriction between the pharynx and stomach. At the second month the esophagus undergoes rapid elongation. The development of the lung and pleural cavities forces the stomach into the abdomen. The septum transversum originates from cervical mesoderm and descends with the phrenic nerves. If there is delay in the descent of the stomach, the septum transversum may fuse and trap the stomach in part or in entirety in the chest. This results in a congenitally short esophagus. The esophagus and trachea become differentiated during the third and fourth weeks. The growth and fusion of the two lateral septa form the posterior wall of the trachea and the anterior wall of the esophagus. Defects in the development of these septa lead to tracheo-esophageal fistulae.

The adult esophagus is a muscular tube about 25 centimeters in length and functions as a conduit from the mouth to the stomach. The esophageal wall consists of the adventitia, muscularis, submucosa, and mucosa. The longitudinal and circular musculature of the upper third of the esophagus is striated, while that of the lower two-thirds is non-striated. The esophagus has a dual nerve supply. Parasympathetic fibers are supplied by the vagus nerve. Sympathetic fibers are supplied by the superior and inferior cervical ganglia, splanchnic nerves, and the celiac ganglion. The mucosa consists of stratified squamous epithelium, a layer of connective tissue, and a muscularis mucosa. The arterial blood supply arises from the inferior thyroid arteries, thoracic aorta, abdominal aorta, bronchial arteries, and left gastric artery. The venous drainage is to the subclavian, thyroid, azygos, hemiazygos, coronary, left gastric and splenic veins. The communication of systemic and portal venous systems through the esophageal plexus leads to the formation of esophageal varices when portal vein hypertension is present.

Esophageal dysfunction is manifested by dysphagia (difficult swallowing). Disorders of the esophagus include esophagitis, hiatus hernia, achalasia (megaesophagus) diverticula, benign tumors, malignant tumors, cysts, rupture, pellagra, scleroderma, and emotional disorders of swallowing.

The stomach. The stomach makes its appearance in the fourth week of uterine life as a fusiform dilatation of the foregut. The stomach undergoes further dilatation and by the sixth week its two curvatures can be recognized. The greater curvature grows more rapidly than the lesser curvature and is carried downward and to the left so that the left surface comes to lie anteriorly. The stomach functions as a reservoir for ingested food and converts the food into chyme by mechanical and chemical action.

The wall of the stomach consists of four coats. The serous layer is derived from the peritoneum and covers the surface of the stomach except for the attachments of the greater and lesser omentum and a small area near the cardia which is in contact with the diaphragm. The muscular coat consists of three sets of smooth muscle fibers. The longitudinal fibers are the most superficial. The circular fibers form a uniform layer over the whole extent of the stomach beneath the longitudinal fibers. At the pylorus these circular fibers aggregate into a circular ring which projects into the lumen—the pyloric valve. The oblique fibers are limited chiefly to the cardiac end of the stomach. The submucosa is a layer of loose areolar tissue. The mucosa consists of a multitude of glands which empty into the gastric pits. The epithelial lining, the gastric pits, and the free surface of the mucosa have the same structure throughout. Three areas of the gastric mucosa are recognized based on differences in the gastric glands. The glands in the cardia are either simple or compound tubular glands composed of cells with pale cytoplasm. They secrete mucus primarily. The glands of the fundus and body produce nearly all the enzymes and hydrochloric acid secreted in the stomach. They also produce some mucus. The glands consist of three cell types. The chief cell (zymogenic) secretes the enzyme pepsin, the parietal cell produces hydrochloric acid, and the mucous cells produce visible and soluble mucus. The parasympathetic nerve supply of the stomach is derived from the vagus nerves. The sympathetic nerve supply of the stomach is derived from the thoracic splanchnic nerves. The arterial blood supply of the stomach is derived from the celiac artery which branches into the left gastric artery, hepatic artery, and splenic artery. The venous drainage of the stomach is via the splenic, superior mesenteric, and ultimately, the portal vein.

Gastric secretion responds to psychic stimulation (cephalic phase), gastric stimulation (gastric phase), and intestinal stimulation (intestinal phase). The cephalic phase is mediated by the vagus nerves. The gastric phase of gastric secretion is mediated by a humoral substance or substances secreted by the antral mucosa in response to mechanical or chemical stimulation. The intestinal phase of gastric secretion is mediated by a humoral substance secreted by the mucosa of the small intestine. Gastric secretion and motility are inhibited by fat in the small intestine which stimulates the release of a humoral substance (enterogastrone) from the intestinal mucosa.

Gastric juice is colorless with a specific gravity of 1.002 to 1.004, and a pH of 0.9 to 1.0. The volume of secretions is between 2 and 3 liters per day. The three enzymes in gastric juice are pepsin (the principle proteolytic enzyme of gastric juice), rennin, and lipase. Rennin and lipase are not of major importance in the adult. Intrinsic factor is produced by the stomach and is essential for the absorption of cyanocobalamin (vitamin B_{12}). The electrolyte composition of gastric juice is as follows:

H+	Nat	K+	Cl–
20–100	20–100	5–25	90–155
mEq/liter			

Disturbances of gastric function are manifested by any one or combination of pain, anorexia, vomiting, weight loss, anemia. Disorders of the stomach include malignancies, peptic ulceration, inflammation, atrophy, and infection.

The small intestine. The duodenum, jejunum, and the major portion of the ileum are derived from the cephalic limb of the primitive gut loop. The remaining ileum and large intestine are derived from the caudal limb of the primitive gut loop. The duodenum serves to transport the chyme from the stomach to the jejunum. It begins at the pylorus of the stomach, has a horseshoe course, and ends at the ligament of Treitz. The duodenum is about 25 centimeters long and is the shortest, widest, and most fixed portion of the small intestine; it has no mesentery. Its secretions are alkaline, and the highly alkaline bile and pancreatic juices enter the gut in the second portion of the duodenum.

The jejunum begins at the ligament of Treitz and constitutes about 25% of the small intestine; the remainder of the small bowel is termed the ileum. The small bowel functions to propel the chyme, completes the processes of digestion and absorption, and produces hormones that regulate intestinal and pancreatic secretion, intestinal motility, and contraction of the gallbladder. The wall of the small intestine consists of four layers—the serosa, muscularis, submucosa, and mucosa. The muscular layer consists of two layers of smooth muscle—the external longitudinal and the internal circular layers. The inner surface of the small bowel is increased greatly by the formation of circular folds (valves of Kerking) and the intestinal villi. The villus consists of the lymphatic vessel (lacteal), blood vessels, epithelium, basement membrane, muscular tissue, and lymph tissue. The epithelium covering the villus continues into the intestinal glands (Glands of Lieberkuhn). The duodenal glands (Brunner) are limited to the duodenum and are found in the submucosal areolar tissue. They are small compound acinotubular glands opening by a single duct into the duodenal lumen. Three cell types are distinguished in the mucous membrane of the intestine. The columnar cells with a striated border, goblet cells and argentaffine cells occur along the free surface of the intestinal mucosa. The intestinal glands also contain large cells (Paneth) and increased numbers of argentaffine cells. The parasympathetic nerve supply of the small intestine is mediated by the vagus nerve and the sympathetic supply is derived from the celiac plexus. These fibers run to the myenteric plexus (Auerbach) located between the circular and longitudinal muscle layers and to the submucosal plexus (Meissner). The arterial blood supply is from the superior mesenteric artery. The superior mesenteric vein drains the small intestine and empties into the portal vein.

Duodenal secretions (Brunner's glands) are highly alkaline (pH 8.0) and mucoid. The secretions of the jejunum and ileum (Crypts of Lieberkuhn) are a cloudy fluid containing leucocytes, epithelial cells, and mucus (succus entericus). About three liters a day are secreted at a pH of 7.0 to 8.5. The intestinal juices contain the coenzyme, enterokinase, amylase, peptidases, arginase, phosphatase and enzymes that hydrolyze nucleic acids. Electrolyte composition of intestinal juice:

Na+	K+	HCO$_3$–	Cl–
80–150	2–10	20–40	90–130
mEq/liter			

Dysfunction of the small intestine is manifested by pain, diarrhea, and malnutrition. Disorders of the small intestine include malignancies (rare), arterial or venous thromboses, atrophy (Sprue), lipo-dystrophy (Whipple's Disease), non-specific inflammation, infection, and malrotation.

The large intestine. The large bowel is derived embryologically from the caudal limb of the primary gut loop. The gut rotates about the superior mesenteric artery as an axis. The caudal limb shifts to the left and cephalad. As growth continues, the gut herniates into the umbilical cord. At the fourth intrauterine month the gut returns to the abdominal cavity. The small intestine returns first and passes to the left, pushing the non-herniated descending colon to the left. As the remainder of the small bowel returns, the large bowel straightens out and is carried to the right and down. Thus, the cecum arrives in the right lower quadrant of the abdomen. The colon functions as a reservoir of intestinal contents, to absorb water and electrolytes, and to propel the feces to the anus for defecation.

The wall of the large intestine consists of four layers. However, it differs from the small bowel in having appendages on the external layer (appendices epiploicae). Also, the longitudinal muscle fibers do not form a continuous layer around the gut, but form three longitudinal bands (taeniae). The large bowel begins at the ileocaecal valve and consists of the cecum, ascending colon, transverse colon, descending colon, and sigmoid colon. It constitutes about 20% of the length of the bowel. The mucous membrane of the large intestine does not form folds except in the rectum. Being devoid of villi, it has a smooth surface. The glands of the large intestine attain a greater length than those in the small intestine and they differ from those in the small intestine by a preponderance of goblet cells. The free surface of the mucosa is lined with simple columnar epithelium with a thin striated border. The blood supply of the colon is from the superior and inferior mesenteric arteries and the hypogastric artery at the rectum. The veins correspond to the arteries and drain into the portal system. The nerves of the large bowel do not differ from those of the small bowel.

Dysfunction of the large vowel is manifested by cramping pain, diarrhea, constipation, hematochezia, and mucoid discharges. Disorders of the large bowel include ulcerative colitis, amebic colitis, polyps, carcinoma, megacolon (Hirshprung's disease) and functional disorders of motility.

The appendix. The vermiform appendix appears as an extension of the cecum at the third month of fetal life. It is highly variable in form and has no function in man. The epithelium of the appendix is similar to the large bowel but tends to undergo fibrosis with aging. The muscularis shows no deviation from the large bowel muscularis although the longitudinal fibers form a complete coat. The appendix has a rudimentary mesentery. The lamina propria contains more lymph tissue than the large bowel but the lymph tissue tends to diminish with age. The blood supply of the appendix is from the appendicular artery, a branch of the ileocolic artery. The appendix is frequently the site of infection and appendectomy is the usual treatment.

The rectum. The rectum is formed from the dorsal portion of the cloaca which has been separated from the urogenital sinus by the urorectal septum. At the same time the dorsal portion of the cloacal membrane is separated from the ventral portion and forms the anal plate which is later invaginated to form the proctodeum. The rectum and anal canal are separated until the end of the seventh week of intrauterine life when the anal membrane ruptures. Imperforate anus or atresia of the anus results from persistence of the anal membrane. The rectum functions to support and propel the fecal mass and is normally empty except during defecation. It is approximately 10–12 centimeters long and differs from the colon by having smoother walls and no appendices epiploicae. The rectum has a complete longitudinal muscle layer and the mucous membrane is thicker than that of the colon. There are three, and occasionally four, large transverse folds in the rectum (valves of Houston).

The anus. The anal canal is about 2.5 to 3.5 centimeters in length. The anus is the aperture by which the intestine opens externally and is contracted by sphincters. Anal closure is maintained by the external sphincter, levator ani muscles, and internal sphincter of the anus. The mucous membrane of the upper portion of the rectal canal presents a series of vertical folds known as the rectal columns (Morgagni) which contains smooth muscle longitudinally arranged. The columns are joined by arches of mucous membrane known as the rectal valves, and below the valves the anus is lined by modified skin. The blood supply of the rectum is from the superior, middle, and inferior hemorrhoidal arteries. Venous drainage corresponds to the arterial blood supply and the free anastomoses of the veins in the hemorrhoidal plexus provide a union of portal and systemic venous systems. The absence of valves in the superior hemorrhoidal veins, increased portal pressure, and increased intra-abdominal pressure lead to distension of the hemorrhoidal plexus and the development of hemorrhoids.

The parasympathetic innervation of the rectum is by the pelvic nerves from second, third, and fourth lumbar segments. The sympathetic nerve supply arises from the second and third lumbar segments and passes by the lumbar splanchnic nerves to the inferior mesenteric ganglion and thence to the bowel by the lumbar-colonic and the hypogastric nerves.

Dysfunction of the rectum and anus is manifested by pain, constipation, hematochezia, and diarrhea. Disorders of the rectum include malignancy, benign polyps, inflammation, fistulae, abscess formation, stenosis, hemorrhoids, and infection.

The liver. The hepatic diverticulum arises ventrally from the entodermal lining of the gut during the fourth week of embryonic life. As the closure of the foregut proceeds, the hepatic primordium is incorporated into the floor of that part of the developing intestinal tract which becomes the duodenum. The growth of the hepatic primordium proceeds between the two layers of splanchnic mesoderm which constitute the ventral mesentery of the gut. The continuing growth of the liver primordium spreads the two mesodermal layers over its surface. These investing layers become the capsule of the liver.

The gallbladder originates as a local dilatation of the hepatic diverticulum where the hepatic ducts become confluent. The gallbladder elongates very rapidly and its terminal portion becomes saccular. The narrow proximal portion becomes the cystic duct and the original area of the hepatic primordium becomes the common bile duct.

The liver functions as an accessory organ of digestion by secreting bile which is stored and concentrated by the gallbladder. The bile is introduced into the duodenum at appropriate times and is essential to efficient digestion of fat by the pancreatic and intestinal juices. Bile salts, sodium taurocholate and sodium glycocholate, aid emulsification of the intestinal fats and activate pancreatic lipase. In the absence of bile salts, pancreatic lipase has a weak lipolytic action. Bile is also necessary for fat absorption. Although the majority of fat in the intestine may be split without the presence of bile, absorption of fat is severly restricted. Bile is also necessary for the absorption of Vitamins D and K and carotene. The volume of bile approximates 500–800 ml. daily.

Electrolyte composition of the bile is:

Na+	K+	HCO$_3$—	Cl—
110–150	3–10	30–50	80–120
mEq/liter			

The metabolic functions of the liver include the conjugation and excretion of bilirubin, synthesis of cholesterol, deamination of amino acids, synthesis of albumin, fibrinogen, prothrombin, and accelerator clotting factors, glycogen storage, iron storage, urea formation, steroid conjugation, and many others.

The liver is the largest gland in the body and averages about 1500 grams in weight. The superior surface of the liver is divided by the falciform ligament into the right and left lobes. The right lobe is larger and shows on its posterior-inferior surface two smaller lobes, the caudate and quadrate lobes. The liver substance is composed of vascular units measuring from 1–2 mm. in diameter known as liver lobules. These lobules are supported by a fine reticulum network and, in part, separated by interlobular connective tissue. This connective tissue contains the terminal portions of the portal vein, hepatic artery, and bile ducts. The branches of the portal vein that encircle the lobule are known as the interlobular veins. These veins give off the hepatic capillaries (sinusoids). These capillaries anastomose freely but eventually unite to form the central veins which drain into the hepatic veins and inferior vena cava. The liver is a modified compound tubular gland. The liver cells are arranged in cords and plates between the hepatic sinusoids. The liver cells are polyhedral, and the surfaces not in contact with the hepatic sinusoids are enmeshed by a ring of bile capillaries. The reticulo-endothelial cells lining the capillaries are known as Kupffer cells. The blood supply of the liver consists of the hepatic artery and portal vein. The venous drainage is through the hepatic veins to the inferior vena cava. The portal vein is unique, beginning in the splanchnic capillaries and ending in the capillaries of the liver. Anastamoses between portal and systemic venous systems exist at the esophageal, rectal, retroperitoneal and diaphragmatic venous plexuses. The development of hepatic venous outflow resistance leads to portal venous hypertension which establishes functional anastamoses in these areas. The liver is innervated by the vagus nerve, and vagal stimulation increases bile flow. Bile flow may also be stimulated by secretin, bile salts and meat or fat in the diet. Epinephrine decreases bile flow.

Dysfunction of the liver is manifested by hyperbilirubinemia, anorexia, weight loss, abdominal pain, and nausea. Disorders of the liver include inflammation (hepatitis), malignancy (primary and metastatic), fibrosis (cirrhosis), infection, atrophy, and parasitic infestation.

The gallbladder. The gallbladder stores and concentrates the bile secreted from the liver. It is situated between the right and quadrate lobes on the visceral surface of the liver. It is pear-shaped and measures 7–10 cm. in length and 2.5–3.5 cm. in width at its widest part. The gallbladder empties into the cystic duct which enters the common bile duct. The common bile duct empties into the second portion of the duodenum. The wall of the gallbladder consists of four layers. The serosa is continuous with the peritoneum on the adjacent liver. The perimuscular coat is a layer of thick areolar tissue containing vessels and nerves. The muscular coat consists of interlacing bundles of smooth muscle forming an irregular network of transverse, longitudinal, and oblique fibers. The mucosal epithelium of the gallbladder is high columnar and contains no glands in the fundal area. There are mucus glands in the neck of the gallbladder. The blood supply of the gallbladder is from the cystic artery, a branch of the hepatic artery. The vagus nerve innervates the gallbladder but is felt to be of little importance in its contraction. Fat in the intestine or acid introduced into the duodenum causes a rapid emptying of the gallbladder. Cholecystokinin, an acid extract of the mucosa of the upper intestine, is considered to be the humoral substance responsible for gallbladder contraction.

Dysfunction of the gallbladder is manifested by pain and dyspepsia. Disorders of the gallbladder include stone formation, inflammation, gangrene, and carcinoma.

The pancreas. The pancreas arises from a dorsal and ventral primordium which develop independently and later fuse with each other. The dorsal pancreas arises from the dorsal wall of the duodenum and is located almost directly opposite the hepatic diverticulum. In its growth the dorsal pancreas pushes between the two layers of the splanchnic mesoderm which constitute the dorsal mesentery. The ventral pancreatic bud arises between the duodenum and hepatic diverticulum. As the hepatic diverticulum grows, it carries the ventral pancreas with it so that eventually the pancreatic bud appears to be a bud from the common duct. As the gut rotates, the ventral pancreas comes to lie to the right and posterior to the duodenum and extends into the dorsal mesentery where it meets and merges with the dorsal pancreas. The ventral bud gives rise to the head of the pancreas; the body and tail of the pancreas arise from the dorsal bud.

The pancreas has an exocrine and an endocrine function. The pancreatic juice is alkaline (pH 7.1–8.2). The proteolytic enzymes of the pancreatic juices include trypsin and chymotrypsin. Trypsin carries the digestion of protein beyond the peptone stage and is activated by the intestinal enzyme enterokinase. Chymotrypsin is activated by trypsin. The optimum pH for trypsin and chymotrypsin is 8.0. Pancreatic amylase is a very potent amylolytic enzyme which is active at pH 6.7 to 7.0. Chloride ion is essential for its action. Pancreatic lipase splits fats to glycerol and fatty acids. The electrolyte composition of pancreatic juice is:

Na+	K+	HCO$_3$—	Cl—
110–150	3–10	70–110	40–80
mEq/liter			

The volume of pancreatic juice is 500–1200 ml. per day.

The pancreas is tubulo-racemose in structure. The lobules are very loosely joined by areolar tissue and there is no distinct fibrous capsule about the gland. Scattered throughout the gland are small interlobular cell masses of varying size and shape (Islets of Langerhans). These structures have no ducts and secrete insulin.

The pancreas receives its blood supply chiefly from the splenic, superior mesenteric, and pancreaticoduodenal arteries. The blood is returned to the portal vein by means of the splenic and superior mesenteric veins. The pancreas is innervated by the vagus nerve and stimulation of the vagus promotes a viscous secretion of small volume which is rich in enzymes. The hormonal control of pancreatic exocrine secretion is mediated by secretin, a hormone derived from the duodenal mucosa. Secretin causes the secretion of water and inorganic substances from the pancreas. Pancreozymin, a duodenal extract, may play a role in pancreatic secretion. It causes the gland to discharge a secretion rich in enzymes and is similar to the vagus nerve in this respect.

Dysfunction of the pancreas is manifested by steatorrhea, weight loss, abdominal and back pain, and cachexia. Disorders of the pancreas include acute and chronic inflammation (pancreatitis), malignancy, and cystic disease.

JOHN V. CARBONE

DIPLOPODA

The diplopods, or millipeds, compose a distinct class of air-breathing (*tracheate*) ARTHROPODA. In them the head is distinct while the remaining segments form a continuous trunk or body. A distinctive feature, from which the name of the class derives, is that most of the ordinary segments of the body bear two pairs of legs due to the fusion of the primary segments in couples, each couple having a single dorsal plate but two sternites behind each of which a pair of legs are inserted close to the median line. One or two of the caudal segments and usually four of the most anterior segments remain single. In some forms the number of segments is nearly constant as, for example, in the large order Polydesmida in most of which the number is fixed at 20. In millipeds consisting of more than 32 segments, the number is variable. The length of millipeds ranges from about 2 mm. to the neighborhood of 300 mm. The body is typically cylindrical or hemi-cylindrical but may present a more flattened appearance due to lateral extension of the segments. In general, diplopods have the capacity and habit, when disturbed, of flexing the body into a spiral or, in broader forms, into a ball.

In some small forms (subclass Pselaphognatha) the body is soft, but in the vast majority of forms it presents a crustaceous shell, the integument consisting of a strong base of chitin stiffened with a rich deposit of calcium carbonate (subclass Chilognatha). However, growth is in successive stages at the end of each of which the shell is moulted, leaving the new integument soft and the animal exposed to danger unless it has found adequate concealment. With the exception of the crustaceous integument the only protection possessed by diplopods is in the possession by most forms of stink glands which open through pores along each side of the body. The secretion has offensive qualities sufficient to repel or kill insects.

The head bears a pair of relatively short antennae composed of 7 or 8 segments. In most the eyes are represented on each side of the head by from one to a group of many independent ocelli but they may be entirely absent as they are in the entire order Polydesmida. A pair of mandibles adapted to biting are always present. A first pair of maxillae are represented in adults only by rudiments while a second pair form a large complicated plate (*gnathochilarium*) homologus with, but more

Fig. 1 A typical diplopod, or milliped. (From Goodnight, Goodnight and Gray, "General Zoology," New York, Reinhold, 1964.)

highly evolved than, the labium of insects. Poison jaws are never present.

Breathing is by means of tracheae of which each double segment has two pairs opening adjacent to the sternites. Each tracheal system is independent, showing no anastomosis such as present in the systems of insects and chilopods. The reproductive organs are located anteriorly, their ducts opening behind the second legs or through the coxal joints of these. The eggs are laid in the ground in groups of from 25 to 50. The larva emerges from the egg with 3 pairs of legs. In the course of development double segments are added in groups of 1, 2, 3 or more. The life cycle lasts in different forms from 1 or 2 up to as many as 7 years.

Millipeds are inoffensive creatures living primarily upon decaying vegetable matter or, rarely, upon the rootlets or other delicate tissues of living plants. They are found generally under decaying wood or leaves but in dry seasons may descend to considerable depths in the soil.

R. V. CHAMBERLIN

References

Attems, C., "Diploda" in Kükenthal, W. and T. Krumbach ed., "Handbook der Zoologie," Vol. 4. Berlin, de Gruyter, 1923.

Verhoeff, K. W., "Diploda," in Bronn, H. G. ed., "Klassung und ordnungen des Tierreichs," Vol. 5. Leipzig, Winter, 1910–1914.

DIPTERA

Flies, or insects belonging to the order Diptera, are so-called because they have only one pair of wings. However, this order includes a number of kinds of two-winged INSECTS not usually known as flies. Thus, the midges (gall midges, biting and nonbiting midges), the mosquitoes, the so-called "gnats," most of the forms commonly called maggots (the larval stage of certain flies—see below), and the sheep ticks ("keds") also belong to the order.

The last insect census in 1948 listed approximately 85,000 species in the order, 16,700 of which occur in North America north of Mexico. All the species are grouped according to their physical and biological traits into smaller and smaller categories to facilitate identification and to demonstrate relationships. The first major grouping places all flies into one of two major classifications; the first contains all those flies with relatively long, threadlike antennae and whose adult emerges from its pupal case by splitting it lengthwise; the second group comprises all flies whose adults have only three visible antennal segments and escape from their pupal cases through a hole in one end. Further divisions in each group result in over 100 families, a few containing only one genus, a very few with up to 800 genera, and many with intermediate numbers of genera. Flies are worldwide in distribution and dwell in one or another of their developmental stages in every conceivable kind of habitat.

The flies belong to a group of insects, the Holometabola, the development of which takes place in a series of stages that have little resemblance to one another. The female mates and lays eggs—these hatch into legless larvae which carry within them the rudimentary tissue from which the adult fly will eventually grow. The larva is the feeding and growing stage and casts off its skin at intervals, eventually attaining full size for its species. At this time the larva stops eating and forms a pupa—a resting stage representing the developing adult. The adult eventually matures to complete the life cycle. Although there are minor variations in this mode of maturation, the same process regulates the development of every fly species.

Eggs. Eggs are laid in every kind of situation. The aquatic species deposit them beneath the water, directly on the water surface, or on vegetation near the water; flies parasitic on other insects lay their eggs directly on or in their intended prey, sometimes while both predator and prey are on the wing; the phytophagous flies deposit their eggs on or on stems, leaves, buds, twigs, fruit, and flowers. In many of the more highly developed species the eggs hatch inside the body of the female, and live larvae emerge from the mother fly.

Larvae. Many of the more primitive Diptera (mosquitoes and relatives) have an aquatic larva which spends its entire growth period beneath the water surface and has a well developed head with chewing mouthparts. These are usually very active swimmers, grow rapidly, and may be found in all kinds of water collections such as rain barrels, tree holes, rapidly running streams, boggy meadows, at the edges of lakes, large rivers, and even oceans. The larvae of the so-called higher flies are called maggots and usually have carrot-shaped bodies, large at the posterior end and tapering to a very small head with rasping mouth parts. Parasitic larvae live on or in their hosts and have lost much of their complicated external structure. Larvae of the plant-inhabiting Diptera occur in a variety of situations such as roots and fleshy underground plant parts, leaf mines, galls formed by the plant, or in fruit and seeds of various kinds. Some larvae are capable of living in malodorous accumulations of manure and decaying vegetation and in dead animals, while others exist on human or animal blood.

Pupae. Pupae may be formed in either of two ways—the fully fed larvae may cast its skin and form a shell of a different shape, or the larval skin may simply harden and become barrel-shaped. Pupae are found in a variety of situations, but most commonly occur in those places where the larvae have been feeding. Mosquitoes and their relatives with aquatic breeding places have pupae that actively swim about, but in most flies this period is spent without motion. Larvae living in or on plants may pupate on the plant tissue or drop to the ground to pupate. Most larvae living in decaying vegetation burrow into the earth before pupating.

Adults. In the tropics some flies may attain a wingspread of nearly three inches, but the great majority are not more than one-third to one-half inch in wing span, and a few are no larger than the head of a pin. In color they range from a uniform black, blue, brown, or yellow to every conceivable combination of hues, and the color patterns on the bodies, legs, or wings of an adult fly are often important clues to its identity. Flies live in all kinds of situations, often close to their breeding places. They usually fly only at certain times of the day, some preferring open sunshine, some the deep shade of the forest, and others only the dark of night. A few adult flies suck blood and have piercing mouth parts; these include the mosquitoes, biting midges, black flies and sandflies. The adults of most species have "lapping" mouth parts, sometimes provided with raspers, but none have the chewing type found in grasshoppers, beetles, and some other insects. Some flies, such as the sheep tick, are parasitic and have lost their wings, spending all their time on the animals whose blood they use for nourishment.

Importance. The flies constitute one of the most important orders of insects. A relatively few species cause the greatest amount of damage, while many are beneficial in one way or another. Many flies visit flowers and pollinate untold millions of plants; the larvae of a number of flies are scavengers and assist bacteria in destroying carcasses and decaying vegetable matter; still others have larvae that live in lakes, ponds, and streams and serve as abundant food for fish and other aquatic life. Certain flies prey on other, harmful insects, or by laying eggs that hatch into larvae that use them as food.

On the other hand, the harmful species have caused untold misery and economic loss throughout the world. Members of the order Diptera carry more disease to man than any other group of insects. Eighty-five or 90 of the 400 species of *Anopheles* mosquitoes over the world have been associated with well over one-half million cases of malaria every year. This disease has serious debilitating effects and occurs, sometimes undetected, on every continent and many islands of the world, resulting in untold economic loss. Yellow fever, once the scourge of nearly every tropical country except the Orient, is carried by *Aedes aegypti,* a mosquito with nearly world-wide distribution. Jungle yellow fever, so-called because it is a disease of monkeys living in tropical forests, is caused by the same virus but is carried by other species *Aedes* and by several *Haemagogus* mosquitoes that bite humans as they work out-of-doors. Dengue, or "breakbone fever," is caused by a virus transmitted to man by *Aedes aegypti* and *Aedes albopictus.* This disease is world-wide, and although cases are seldom fatal, they leave the patient physically exhausted for several weeks. Filariasis, primarily a disease of the Orient, Africa, and the Pacific Islands, affected a large number of our troops during World War II. Long-standing cases sometimes result in elephantiasis. This disease, caused by a minute worm, or filaria, is carried to man by species of *Culex, Mansonia, Aedes,* and *Anopheles*, and other forms are carried to animals by biting midges. A number of encephalitis viruses have been and are still being discovered. The effects on humans of many of these viruses are not known, but modern day methods of research have shown that a surprisingly large number of people throughout the world have been affected by one or another of them. Various mosquitoes have been incriminated in their transmission. A surprisingly large array of tropical diseases are carried by black flies (*oncocerciasis*), deer flies (*tularemia, loa loa*), tsetse flies (African sleeping sickness), sandflies of the genus *Phlebotomus* (*kala azar,* Oriental sore, pappataci fever, and Oroya fever).

A large number of disease organisms are transmitted by flies to domestic animals, thereby affecting the production of meat and dairy products. Some of these are anthrax, tularemia, botulism, and nagana, a form of sleeping sickness. The larvae of some flies are able to live on healthy animal tissue. Bots and screw worms are fly larvae that are able to exist in these situations; they cause considerable physical discomfort and damage to their animal hosts, and usually ruin potential hides by making holes for the escape of the adult flies. *Dermatobia hominis*, the human bot fly, hangs its eggs on a passing mosquito—when the mosquito lands on a warm-blooded animal to obtain a blood meal, the *Dermatobia* eggs hatch very rapidly and enter the mammal host. The so-called Congo floor maggot is a fly larva that sucks human blood.

Everyone is familiar with the unclean habits of the common house fly and its relative, the blow fly. Not only is the presence of these flies inconvenient and many times a signal of insanitary conditions, but they may carry disease-producing organisms mechanically on their body surfaces and hairs. These organisms cause typhoid fever, dysentery, diarrhoea, cholera, yaws, and trachoma.

Although flies that attack plants do not cause as much economic loss to man as do some of the other insects, they are nevertheless responsible for a large amount of annual loss. Fruit flies include the apple and cherry maggots; the Mediterranean fruit fly, which has occasioned great loss to growers in the Mediterranean and Africa, parts of South and Central America, and recently in Florida; the Oriental fruit fly and the melon fly in Hawaii and the Orient; the Queensland fruit fly in Australia; and the olive fly in the Near East. All these flies, and many others, have larvae that feed on the succulent flesh of many fruits. Leaf miners deface ornamental plants and reduce their potentialities for growth. The Hessian fly and several other midges have caused great losses to growers of grain crops in the United States and abroad by feeding on stems and seeds. Root maggots, burrowing in underground parts of plants, have seriously affected these crops all over the world.

RICHARD H. FOOTE

References

Curran, C. H., "The Families and Genera of North American Diptera," New York, American Museum of Natural History, 1934.
Felt, E. P., "Plant Galls and Gall Makers," New York, Hafner, 1940.
Needham, J. G. *et al.*, "Leaf Mining Insects," Baltimore, Williams and Wilkins, 1928.
Essig, E. O., "Insects of Western North America," New York, Macmillan, 1938.
West, L. S., "The Housefly," Ithaca, New York, Comstock, 1951.
Snodgrass, R. E., "The Metamorphosis of a Fly's Head," Smithsonian Miscellaneous Collections, Vol. 122, No. 2, 1953.
Usinger, R. L. ed., "Aquatic Insects of California," Berkeley, University of California Press, 1956.
Carpenter, S. J. and W. J. LaCasse, "Mosquitoes of North America," Berkeley, University of California Press, 1955.
Snodgrass, R. E., "The Anatomical Life of the Mosquito," Smithsonian Miscellaneous Collections, Vol. 139, No. 8, 1959.
Foote, R. H. and D. E. Cook, "Mosquitoes of Medical Importance," Washington, D. C., United States Department of Agriculture Handbook No. 152, 1959.

James, M. T., "The Flies That Cause Myiasis in Man," Washington, D. C., United States Department of Agriculture Miscellaneous Publication No. 631, 1947.

Oldroyd, H., "Natural History of Flies," New York, Norton, 1965.

DOHRN, ANTON (1840–1909)

The biological world owes an inextinguishable debt to this zoologist for having brought to fruition the concept that a "Marine Biological Station" was a natural and desirable facility. The world-wide existence of these stations today derives almost entirely from his indefatigable pursuit of what, in his day, appeared an extraordinary idea. His interest in zoology came from his entomologist father, C. A. Dohrn, who founded the *Entomologisch Zeitschrift* and who encouraged him to collect, and write papers on, beetles while he was still at school. He studied at the universities of Bonn, Jena, and Berlin, where he came under the influence of Gegenbaur and Haeckel. The latter, in particular, interested him in problems of evolution and phylogeny but Dohrn's view on the origin of the vertebrates espoused the cause of Darwin and thus differed sharply from his master. It was while pursuing his studies in phylogeny that he collected at Messina many larval stages of Crustacea. Messina, then a popular collecting ground and vacation resort for European zoologists, had few facilities and Dohrn developed the idea that a permanent marine station should be erected. He failed in an endeavor to collect funds for this purpose and was thus forced to the conclusion that such an institution would have to be self-supporting and that this could be brought about by associating it with a marine aquarium, the admissions to which would pay for the maintenance of scientific laboratories. In 1870 he presented this idea to the city council of Naples, who did not hesitate to condemn it as the impractical dreams of a visionary. Opposition, however, only served to encourage Dohrn and in 1874 he had built and opened to the public the marine aquarium and station which are his outstanding monument. The existence of this material monument should not, however, be allowed to overshadow the great scholarly contributions which he made to the origin of the vertebrates and to the embryology of the cartilaginous fish.

PETER GRAY

DOMINANCE

In 1865 Gregor MENDEL described an hereditary phenomenon which he called dominance: in the first hybrid generation following a cross between varieties of garden peas, Mendel noted that one variety trait always concealed, or dominated, the contrasting variety trait uniformly throughout the hybrid population. After inbreeding the hybrids, however, he found that the second generation displayed proportionately three times as many individuals with the dominant trait as those with the contrasting, or recessive, trait. Of those displaying the dominant trait in that generation, two-thirds were proved to possess a single dose while one-third possessed two doses of the dominant hereditary trait. Consequently we can now describe dominant genic effects in two ways 1) the GENE's effect is equivalent in single dose (heterozygote) and double dose (homozygote) and 2) the gene suppresses the action of its allele in production of the heterozygote's phenotype. Since the time of Mendel dominance has been found commonly though by no means universally in most sexually reproducing organisms. Some genes show no dominance or partial dominance with respect to their alleles; of such genes, the genic action may be additive or multiplicative in determining the phenotype. Dominance or lack of dominance describes interaction between alleles at a single genic locus.

Dominance of an allele at one locus may easily be modified by 1) environmental conditions, 2) interactions with other loci, or 3) substitutions of various other alleles at that locus which may in the new heterozygote change the dominance potency of the original allele tested. Examples of dominance modifications by these agencies may be found in the works of Goldschmidt (1955), Schmalhausen (1949), and Wagner and Mitchell (1955).

Certainly the suppressive action of one gene over its alleles is a growth phenomenon and must be analyzed from the basis of quantitative gene products. A dominant allele conceivably may act in production of a substance necessary for converting a substrate to a final phenotype produced. Possibly that allele may produce enough substance in single dose (heterozygote) to equal the amount produced by two doses (homozygote); conceivably also dominant genes may control velocities of reaction, thresholds for morphogenetic effects, or the conditioning of target tissues toward which morphogenetic substances may be directed. In any case the true dominant gene must either give equal effect in single and double dose or if excess gene substance occurs in the homozygote it must be ineffective in altering the biochemical reaction system compared with the heterozygote. Even dominance of antimorphs (genes with opposite or lower activity than their recessive alleles) could be explained on the basis of suppressor substance production.

The theories which have been proposed to help explain the evolution of dominance are neither mutually exclusive nor contradictory but are likely to be complementary or at least reasonably so in the light of current observation. R. A. Fisher has emphasized that in diploid organisms most mutations are recessive to the wild type. In DROSOPHILA especially this is true, and the so-called dominant mutants are not actually dominant in a strict sense but are rather only partially dominant to wild type since in most cases they are lethal in homozygous condition. Deleterious mutants in natural populations are several times more frequent as heterozygotes than as homozygotes; natural selection acts more efficiently on the more common than on the rare genotypes; consequently if genes are available which modify by reinforcing the dominance of normal alleles over newly arising mutants, those modifying genes in turn will be preserved by natural selection. The population may thus enjoy greater uniformity through the evolution of a system of genetic dominance modifiers. The evidence for the existence of such modifiers is incontrovertible. On the other hand alleles differ in their dominance potency, and it is also likely that selection could act to preserve those alleles which are potent enough to suppress directly the deleterious effects of newly arisen mutants at that locus. Haldane, Muller, Plunkett, Wright, and Schmalhausen feel that wild type alleles which possess a "factor of safety" by having a potent activity well above a minimum to keep normal phenotype production even in single dose might be selected far more rapidly and effectively than genes which modify dominance of wild type alleles. Since the evidence for both modifiers of dominance and "factor of safety" wild type alleles is clear, it is apparent that both theories may be correct in explaining the general evolution of dominance.

E. B. SPIESS

References

Fisher, R. A., "The Genetical Theory of Natural Selection," 2ed. New York, Dover, 1958.
Goldschmidt, Richard B., "Theoretical Genetics," Berkeley, University of California Press, 1955.
Schmalhausen, I. I., "Factors of Evolution," Philadelphia, Blakiston, 1949.
Wagner, R. P. and H. K. Mitchell, "Genetics and Metabolism," 2ed., New York, Wiley, 1964.

DRIESCH, HANS (1867–1889)

The father of experimental embryology was born in 1867 at Kreuznach. He studied in Hamburg, Freiburg, and Jena, and thus came under the influence both of Haeckel and Roux. In 1891 he performed the classic experiment, now known to every beginning student, of separating the blastomeres of an echinoderm egg and showing that an entire, but smaller, larva could be developed from each of the two halves. This led him to postulate a theory of "equipotentiality" as opposed to the theory of totipotentiality espoused by Roux. These experiments, and similar ones which he pursued at the Naples Marine Biological Station, led him to a vehement espousal of the views of the vitalists. His basic philosophy in this respect is that since living things can be regenerated out of severed parts, they cannot be machines for a part of a machine cannot reproduce the whole. This led him to borrow from Aristotle the word "entelechy," thought by its originator to mean the potentiality which is inherent in matter and to expand this to the new biological meaning of that which carries its purpose within itself. From this point on, he abandoned experimental biology in favor of abstract metaphysics outside the scope of this encyclopedia. Driesch travelled wipely in the course of his career, spending a year in Aberdeen as the Gifford lecturer, a year in Heidelberg, two years in Italian universities, and he travelled widely in China and the United States. The extent of his varied views can best be gathered from his book "The Science and Philosophy of the Organism" which has been translated into most of the languages of the scholarly world. The last years of his life were spent as professor of Philosophy at Leipzig where he died in 1889.

PETER GRAY

DROSOPHILA

The type genus of the family Drosophilidae (acalyptrate Diptera) was first described by Fallén in 1823. While the name *Drosophila* means "dew lover" these flies have been variously known as fruit flies, pomace flies, sour flies, and vinegar flies because they are generally attracted by fermenting fruit or vegetable products and especially by the associated yeasts in such decaying plant matter. Drosophila is a principal vector in transferring wild yeasts in nature; in ancient times the task of bringing yeasts from vineyard to vineyard was undoubtedly performed by these flies, an indispensable function for the alcoholic fermentation of grape juice.

Most species are general scavengers, fruit feeders, sap feeders, or fungus feeders. In the temperate zones a large number of the fruit feeders are introduced forms (*D. melanogaster*, for example) associated with man, and they prefer fruits of tropical origin. The general scavengers are also largely associated with man feeding often on decaying plant and animal refuse. Sap feeders and fungus feeders make up the bulk of temperate species, but they are not so specialized in their food habits that they will not be attracted by fermenting fruit. One rare food specialization in the genus is that of utilizing the pollen of flowers in certain tropical species; the larvae are not apparently dependent on yeasts or bacteria but are nourished by pollen in such flowers as *Datura, Ipomoea,* Malvaceae, and Cucurbitaceae.

It is easy to follow the main taxonomic characters of the genus by referring to the structures given in Figure 1 (adult *D. melanogaster*): arista plumose; vibrissae and ocellars present; three orbitals, lowermost proclinate, upper two reclinate, middle one smaller than the others; postverticals large and convergent; mesopleurae bare; sternopleural bristles present; anterior dorsocentral bristles well behind the mesothoracic suture; acrostichal hairs in six or more rows at the level of the anterior dorsocentrals; twice-broken costal vein; rudimentary auxiliary vein (subcosta); scutellum bare except for one or two pairs of bristles.

Out of 750 species catalogued by Patterson and Wheeler (1949) and Wheeler (1959), 262 are well known and have been assigned to eight subgenera by Sturtevant (1942), Patterson and Mainland (1944), and Wheeler (1949). The subgenera and some common species are listed in Table 1. The first six subgenera are largely tropical but the Ethiopian region is very poorly known. The subgenus Sophophora is predominantly Neotropical, though some of its members are cosmopolitan (*D. melanogaster* and *D. simulans*); members of the *obscura* group no doubt originated in the Orient and are now common throughout the Palaearctic and Nearctic with one species reported from South Africa. Finally the subgenus Drosophila includes many forms restricted to single

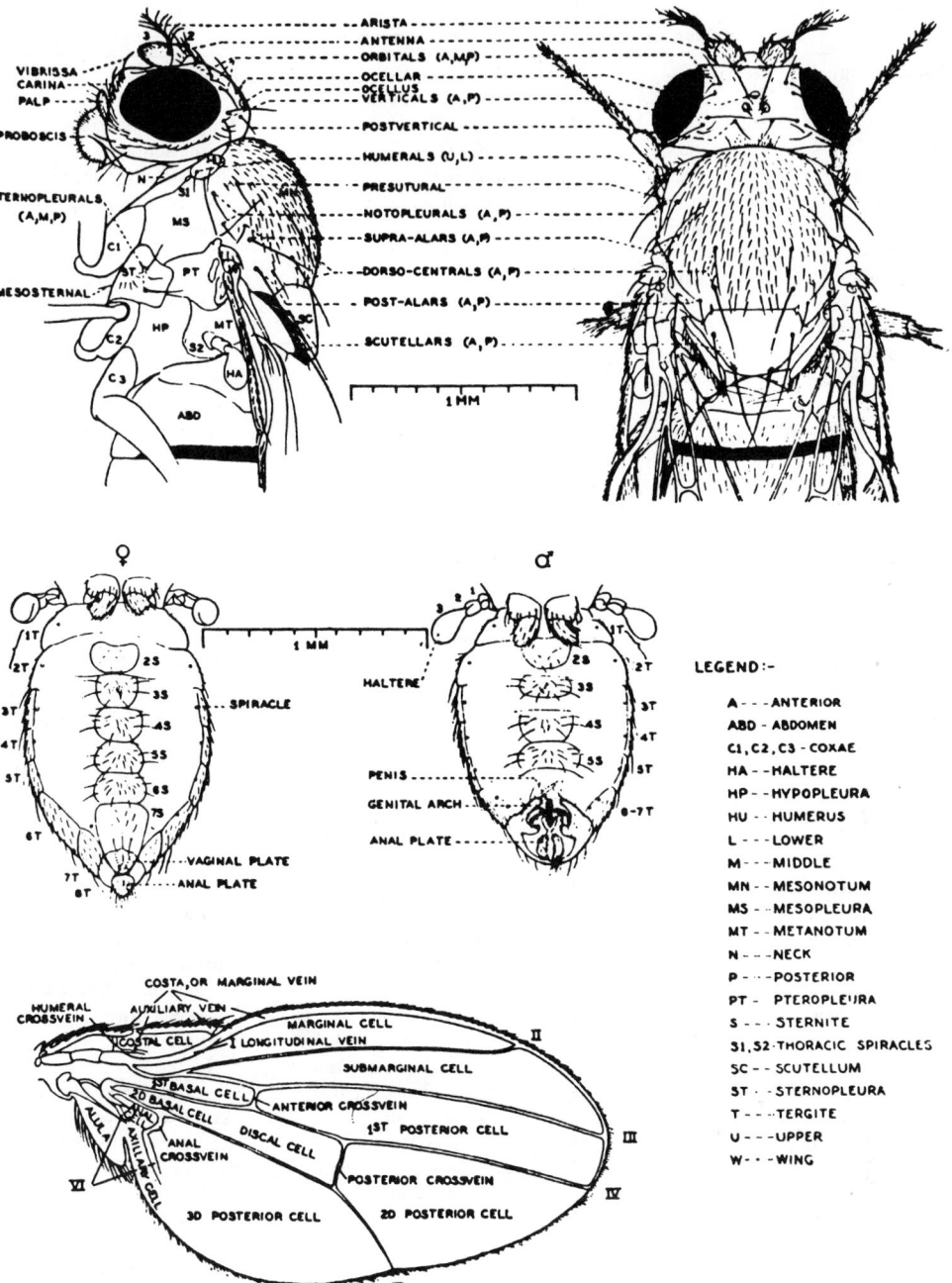

Fig. 1   External structure of *Drosophila melanogaster*. (From C. B. Bridges, Drosophila Information Service 9.)

regions; a few are cosmopolitan (*D. repleta, D. hydei, D. immigrans,* and *D. virilis*) but the vast majority are endemic or narrowly confined forms.

Genetics. After the year 1900 with the advent of genetics certain species of Drosophila came into central importance as advantageous organisms for demonstrating hereditary principles. They are unsurpassed today as laboratory tools in such widely diverse disciplines as evolutionary GENETICS, BIOCHEMICAL and DEVELOPMENTAL GENETICS, ECOLOGY, and EMBRYOLOGY. In the years 1902–1904, W. E. Castle and his students at the Bussey Institute of Harvard University found these flies easy to culture, and he subsequently drew them to the attention of T. H. MORGAN at Columbia University for his critical studies of Mendelian heredity. Morgan and his students, H. J. Muller, A. H. Sturtevant, and C.

Table 1   The Genus: Drosophila Fallén (from Wheeler, 1949)

Subgenera	Author	No. Spp.	Some common spp.	Remarks
1) Hirtodrosophila	Duda	26	duncani, grisea, cinerea, orbospiracula	Fungus-feeders
2) Pholadoris	Sturtevant	6	victoria	Tropical and Oriental
3) Dorsilopha	Sturtevant	1	busckii	Cosmopolitan
4) Phloridosa	Sturtevant	5	floricola	Flower-feeders Tropical
5) Siphlodora	Patterson and Mainland	3	sigmoides, flexa	Southern U.S.A. Mexico and West Indies
6) Sordophila	Wheeler	1	acanthoptera	Mexico
7) Sophophora	Sturtevant			Abdominal bands never broken or narrowed in mid-line
Species groups				
a. saltans		8	saltans	Mexico and Central American
b. willistoni		10	willistoni, equinoxialis, paulistorum, tropicalis, nebulosa	Southern Nearctic and Neotropical
c. melanogaster		18	melanogaster, simulans, ananassae	Sexcombs in males Cosmopolitan
d. obscura		22	obscura, subobscura, pseudoobscura, persimilis, affinis, athabasca, azteca	Nearctic and Palaearctic Sexcombs in males
e. alagitans		2		Mexico
f. nannoptera		1		Mexico
8) Drosophila	Fallén			Abdominal bands broken or narrowed in mid-line. 3–4 egg filaments (except melanica group)
Species groups				
a. quinaria		17	quinaria, transversa, palustris, occidentalis	Holarctic and Oriental
b. guttifera		1		Nearctic
c. pinicola		2		Nearctic
d. virilis		5	virilis, americana, novamexicana, montana	First sp. Cosmopolitan, others Nearctic
f. tripunctata		4		Nearctic and Neotropical
g. funebris		4	funebris (type sp. of genus) D. *funebris* Fabricius, macrospina	Cosmopolitan Nearctic
h. repleta		48	repleta, hydei, mulleri, mercatorum	Several endemic Holarctic and Neotropical
i. robusta		5	robusta, colorata	Holarctic and Orient
j. melanica		7	melanica, nigromelanca, melanissima	Nearctic and Orient
k. polychaeta		3		Endemics in Nearctic and Orient
l. carbonaria		1		Mexico and U.S.A.
m. cardini		9	similis, cardini, polymorpha	Neotropical
n. immigrans		18	immigrams	First sp. cosmopolitan, others endemics in Orient and Neotropical
o. macroptera		5		Nearctic and Neotropical
p. rubrifrons		5		Nearctic
q. annulimana		5	annulimana, gibberosa	Neotropical
r. melanderi		2		U.S.A. endemics
s. bizonata		3		Palaearctic and Orient
t. guaraní		7	guaraní, guarú, guaramunú, griseolineata	Neotropical except *subbadia* which is Nearctic
u. unclassified as to sp. group 9 ssp.				

B. Bridges, pooling their efforts from 1910 to 1915 brought out their classic work using *D. melanogaster* and formulating the GENE theory of herdity.

D. melanogaster, the most widely used species for genetic linkage studies, has numerous laboratory advantages: 1) ease of culture on yeasted nutritive media, 2) large numbers of progeny for good statistical treatment, 3) rapid life cycle (eleven days at 25°C.), 4) only four pairs of chromosomes, 5) giant salivary gland chromosomes with constant banding pattern for ease in cytological mapping of mutant loci and for comparison of break points in aberrations, 6) parallel with vertebrates in possession of chromosomal sex determination (male XY, female XX), and 7) lack of crossing over in males so that whole linkage groups may be held intact. The latter three advantages hold also for the entire genus.

Many basic genetic principles were established first in Drosophila and then in other organisms. From the time that T. H. Morgan found a white-eyed male in his otherwise red-eyed cultures, these flies have been exceedingly prominent in genetic history making. A few discoveries are listed below:

1) Morgan's white-eyed mutant was not inherited from father to son but from father through his daughters' lines and thereby to grandsons and granddaughters. This "criss-cross" mode of inheritance ("sex linkage") parallels the behavior of X and Y CHROMOSOMES: since the Y is typically found only in the male, Morgan postulated the location of the white-eye gene on the X, which is the father's contribution to his daughter's but not his son's chromosomal complement. This was the first parallel between a Mendelian factor and chromosomal behavior.

2) Linear arrangement of sex-linked factors and the construction of a linkage map with relative distances between genic loci were worked out by Sturtevant (1913) using crossing-over data. Subsequently linkage maps for autosomal loci from data collected by much the same methods were completed.

3) A sex determination hypothesis of chromosome balance, or the number of X chromosomes relative to the number of autosomes, was established by Bridges (1925) using the offspring of triploid females. A proper balance of equal dosage between X's and autosomes produced femaleness, a one to two ratio of X's to autosomes produced maleness, while all other ratios gave aberrant sexual types. This method of sex determination may be unique to Drosophila though in few other organisms can such a complex hypothesis be tested.

4) The concept of position effect first brought to

the attention of geneticists the fact that nonallelic genes are not completely independent but interact differently depending on their chromosome location. Sturtevant's "Bar eye" (1925), produced by unequal

crossing-over and consequently due to the duplication of a locus on the X chromosome, showed that "two genes lying in the same chromosome are more effective on development than are the same two genes when they lie in different chromosomes."

5) A major discovery that genetic material could be artificially altered by ionizing radiation was made by H. J. Muller (1927). The rise in spontaneous mutation rate was found to be a linear proportion with dosage in roentgens for *D. melanogaster*.

6) In 1933 the rediscovery of giant chromosomes in the salivary glands of Bibio (Diptera) by Heitz and Bauer followed by the emphasis on these chromosomes' crossbanding by Painter (1934) opened the field of cytological mapping. Bridges finally completed a map for about 500 loci in *D. melanogaster*, pinpointing mutant alleles to particular bands or groups of bands in the salivaries.

7) Beadle and Ephrussi (1936) pioneered in the techniques of biochemical genetics of *D. melanogaster* eye pigments by using transplantation of eye imaginal discs from one larva to another.

8) More recently the discovery of position pseudoalleles (for example Star-asteroid of E. B. Lewis) has elucidated the fine structure and functioning of Drosophila's genetic material, and the groundwork was laid for such studies in microorganisms.

Evolution. While the antiquity of Drosophila is poorly known (a few Drosophilids occur in Miocene Baltic amber), phylogenetics relationships can be inferred from comparisons of mitotic chromosomes, salivary bands, and morphological criteria. Karyotype evolution has probably started from a primitive complement of five pairs of rod-shaped (acrocentric) chromosomes plus a small dot pair. This primitive configuration is found in many species of wide distributions: *D. orbospiracula* (which lacks a Y chromosome in males), *D. subobscura, D. virilis, D. guttifera, D. palustris, D. pinicola, D. funebris, D. tripunctata, D. repleta,* and *D. macroptera.* With this basic number, centric fusions to form V-shaped (metacentric) chromosomes, subsequent translocations, pericentric inversions, and changes in amount of heterochromatin, which alter the chromosome shapes, have led to changes in chromosome number and shape without losing the basic number of chromosome arms in general. Homologous arms have been demonstrated in the Sophophora for example by Muller, Sturtevant, and Novitski, who designated the five basic arms A, B, C, D, E. By comparing similar mutants in *D. melanogaster* and *D. pseudoobscura,* the following homologies seem to hold:

	A	B	C	D	E	F	
melanogaster	X	IIL	IIR	IIIL	IIIR	IV (dot)	where L and R represent left and
pseudoobscura	XL	IV	III	XR	II	V (dot)	right arms of metacentrics

The chromosome number ranges from n = 3 (*D. saltans, D. willistoni,* and *D. busckii,* for example, up to n = 7 in the rare species *D. trispina* which has an extra pair of heterochromatic dots.

Many species exhibit chromosomal polymorphism, expecially in paracentric inversions. It is easy to ascertain the exact limits of the breakpoints in these aberrations by examining the salivaries of hybrids between any two inversion stocks. Phylogenies of overlapping inversions have been worked out in many species, the work of Dobzhansky on the *D. pseudoobscura* complex being a classic in extensive collecting and analysis. In some species inversions are distributed randomly over all the long chromosomes, while in others they are concentrated on a single chromosome. *D. willistoni* is an example of the first and ranks as cytologically the most polymorphic of all Drosophila, having fifty known inversions throughout its distribution area and in certain Amazonian localities having an average of nine inversion heterozygotes per individual. *D. pseudoobscura* and its sibling species, *D. persimilis*, on the other hand, are exceptionally variable on chromosome III, but chromosomes II and IV lack inversions generally while the X has two arrangements.

At first it was believed by Dobzhansky that the distribution of these chromosomal variants in *D. pseudoobscura* was a function of population size and the random effects of sampling. But examining natural populations over many years Dobzhansky found seasonal fluctuations in relative frequencies of certain common arrangements which regularly repeated the same pattern year after year. This discovery of seasonal cycles implied that the chromosomal frequency changes were adaptive. Consequently gene arrangements extracted from wild populations were allowed to compete in laboratory population cages at constant temperature. Changes in relative frequencies of the arrangements used were always determinate in outcome when chromosomes from a single wild locality were competing; and therefore the changes which in part imitated the changes in nature were certainly adaptive in nature: the zygotic combinations of gene arrangements had constant adaptive value in the laboratory. In most experimental populations both variants were preserved instead of one eliminating the other. Selection in the laboratory and presumably also in nature favored heterozygotes. This heterozygote superiority then provides a balance mechanism and a basis for maintenance of intrapopulational diversity.

The genus Drosophila capitalizes on inversion POLYMORPHISM for building up adaptive genetic complexes largely because it has overcome the loss of gametes which most organisms would suffer with crossing-over inside the inversion heterozygote loop: a single crossover within the loop produces 50% unbalanced gametes. No crossing-over occurs in Drosophila males, and in females a special spindle orientation removes deficient and duplicated chromatin to the polar bodies so that all eggs and sperm receive unrecombined inverted or original sequence chromosomes. In this manner inversions can build up mutant alleles or special genetic interactions which may have adaptive significance.

Some species do not display chromosomal polymorphism, for example *D. virilis* and *D. repleta*. The genetic architecture of such species has yet to be determined to find out how adaptive genetic complexes can be built up on chromosomes lacking crossover suppressors.

Species are isolated by a variety of mechanisms.

Sexual isolation and mating behavior demonstrated by "multiple choice" methods in the laboratory show very great differences between species. When opposite sexes come from different species the prevention of mating by incompatible mating behavior is perhaps the most important single isolating mechanism in the genus. Spieth has described mating behavior of several species and has found highly specific courtship reactions for nearly every species studied.

In addition it is common to find species isolated by more than one mechanism. For example, there may be mechanical differences in genitalia or incompatibility of foreign sperm in the female's vagina ("insemination reaction") following copulation between two different species. Hybrids may still form if a mating is "forced" by using an abundance of males, but usually one or both sexes is sterile at maturity due either to chromosomal or genetic incompatibility. If hybrids are by chance fertile as in the hybrid females from a cross between *D. pseudoobscura* x *D. persimilis*, they are much less viable than the parental species females but may nevertheless be backcrossed to either parent species. Backcross progenies are always very much weaker than the hybrid females; since these progenies contain various mixtures of chromosomes from the two species their decreased viability may be due in part to deleterious effects of recombination between the two parental genomes.

A few cases are known in which the transition from race to species is apparently taking place. Evidence for such "borderline cases" in which sub-species have almost become sexually isolated has been reported by Dobzhansky (1959) for six sub-species of *D. paulistorum,* or a cluster of species *in statu nascendi.* Darwin's opinion that "each species existed as a variety" may thereby be illustrated in Drosophila.

ELIOT B. SPIESS

References

Muller, H. J., "Bibliography on the Genetics of Drosophila," Edinburgh, Oliver and Boyd, 1939.

Herskowitz, I. H., "Bibliography on the Genetics of Drosophila," Part Two. Edinburgh, Commonwealth Agricultural Bureaux, 1950.

ibid., "Bibliography on the Genetics of Drosophila," Part Three. Bloomington, Ind. Indiana University Press, 1958.

ibid., "Bibliography on the Genetics of Drosophila," Part Four. New York, McGraw-Hill, 1963.

Strickberger, M. W., "Experiments in Genetics with Drosophila," New York, Wiley, 1962.

Lindsley, D. L. and E. H. Grell, "Genetic Variations of *Drosophila melanogaster*," Washington, D.C., Carnegie Institution of Washington, publ. 627, 1968.

Wheeler, M. R. (ed.), "Studies in Genetics of Drosophila," Austin, Texas, University of Texas publ. 6615, 1966.

DU BOIS-REYMOND, EMIL (1818–1896)

Du Bois-Reymond was a German biologist of Huguenot stock. At the University of Berlin, he wrote his graduation thesis on electric fish and from this work there developed a lifelong interest in the electrical properties of animal tissue. He invented new instruments and refined old ones to detect the passage of tiny electrical currents in nerves and muscles. Du Bois-Reymond showed that the nerve impulse was accompanied by a change in the electrical condition of the nerve. This contribution to biology did much to upset vitalism in one of its strongholds, for it showed that something as ethereal as the silent, unnoticeable nerve impulse which brings about motion could be reduced to inorganic terms.

DOUGLAS G. MADIGAN

EBENALES

Definition. Flowers sympetal, actinomorphic, showing diplostemony or triplostemony, exceptionally haplostemony, number of stamens as high as that of petals or higher, ovary generally inferior, septate, with axil placentation; ovule 1 or few in each locule, apotropous, tegument 1 or 2; woody plants with simple leaves.

Families included: *Ebenaceae, Sapotaceae, Sarcospermaceae, Styracaceae, Symplocaceae,* eventually *Lissocarpaceae (Hoplestigmataceae).*

Morphology. Trees or shrubs rarely thorny (*Bumelia, Argania*); *leaves* alternate; few exceptions as in *Euclea (Ebenac.)* and *Pouteria (Sapot.).* Stipules absent, except in some *Sapot.* and *Sarcosperm.,* generally caducous. Flowers solitary or inflorescence mostly few-flowered: racemes or umbels, the reduction being more advanced in the *Sapot.* and the *Styrac.* than in the other families. Calyx often deeply divided, sometimes (*Diospyros*) irregularly split when opening, often persistent and even accrescent (*Ebenac., Symploc.*). Corolla 4–12 lobes, the lobes divided nearly to the base (*Styrac.*) or united in a long tube (*Ebenac.,* many *Sapot.*). In many species, where the number of pieces varies, there is a tendency towards multiplication; in the same line, the more advanced genera (in the *Sapot.* at least) have more pieces than the primitive ones. Staminodes present (*Ebenac.* in ♀ fl., *Sapot.*) in many *Sapot.* accompanied by dorsal appendages of doubtful origin. Androecium of 1 to several verticils, the stamens (opening lengthwise or poricide) at the base of corolla (*Ebenac., Lissocarp.*) or attached higher up (*Sapot., Styrac.*), sometimes laterally united into fascicles (*Tridesmostemon, Symplocos* sect. *Ciporrima* and *Alstonia*). Ovary 2–5-celled (12 in some *Pouteria,* 16 in some *Ebenac.*) sometimes with secondary septa (*Ebenac.*) or incomplete partitions (*Styrac.*), very rarely 1—celled (*Pouteria,* sect. *Eremoluma,* some *Sarcosperma*). Ovules anatropous-apotropous ascending (*Sapot., Sarcosp.,* some *Styrac.*) or hanging (*Ebenac.,* some *Styrac., Symploc.*), in *Bruinsma (Styrac.)* ascending or handing in the same cell, 1—few in each locule (*Sapot.:* always 1, with the exception of *Diplöon* of doubtful position, *Ebenac.:* 1–2, *Symploc.:* 2–4, *Styrac.:* 1—few, *Sarcosp.:* 1, *Lissocarp.:* 2). Fruit a berry (*Sapot.,* some *Symploc.*) or drupaceous (some *Ebenac., Symploc.,* some *Styrac.*) or coriaceous and dry, a loculicidal capsule (*Ebenac., styrac.*). Seeds with small and roundish cicatrix or long and narrow, sometimes adhering (*Chelone-spermum*) by most of their surface to the placenta. Albumen present (*Ebenac.,* some *Sapot., Styrac., Symploc., Lissocarp.*) or absent (*Sarcosp.*).

Anatomy, vascularization. Stellate hairs (*Styrac.*) or bifid unicellular hairs (*Sapot., Sarcosp.*); secreting cells (*Ebenac., Symploc.*) or vessels (*Styrac.*); latex (*Sapot., Sarcosp.*); scalariform walls (some *Sapot., Styrac., Symploc.*); sometimes septate pith (*Halesia*). Presence of alkaloids (*Achras, Symploc.*), heterosids (*Vitellaria*). As far as we know, no attempt has been made until now to study the course of the vessels in the flower of the *Ebenales.* Although researches have been undertaken at the Conservatoire botanique of Geneva, they do not permit conclusions or generalisations at this stage.

Chromosome numbers. Little is known. The first to study them in the *Sapot.* seem to have been Brown and Clark (1940) giving 2 n = 24 for *Brumelia lanuginosa.* Since then, the same number has been found in the genera *Madhuca* and *Palaquium,* 20 in *Argania,* 48 in *Pouteria,* 26 and 52 in *Chrysophyllum,* 26 in *Achras.* This last number is highly interesting, because *Achras* has always been considered as rather advanced in evolution. But more counts are necessary before any conclusions can be drawn; one can only say that there is more than one line of evolution in this family. For the *Ebenac.* 2 n = 30 have been found with two notable exceptions in the genus *Diospyros*: *D. virginiana* (60, 90) and *D. kaki* (90) both being widely cultivated species. The *Styrac.* have two basic numbers: 16 (and 40) in *Styrax* and 24 (in *Halesa* and *Pterostyrax*).

Embryology. Since the time of Schnarf (1929–1933), only scattered information can be found in the literature concerning the embryology of the members of the *Ebenales* and probably much could be added. Copeland (1938), who has studied with great care *Styrax officinalis* var. *californica,* uses his findings for the clarification of the relationship between the *Styrac.,* the other members of the *Ebenales* and their possible ancestors (see below).

Systematics. For Engler (1897) the 3 orders *Primulades, Ebenales* and *Contortae* were parallel series. The *Ebenales* are more advanced than the *Ericales* on account of their always present sympetaly, the high number of stamens being considered, however as a back-step. Hallier (1912) thought that the *Sapot.* were forming an order by themselves. It was derived from the *Guttales* comprising among many other families as the *Ochnaceae,* the *Guttiferae,* the *Linaceae,* etc., the *Symploc.* In another branch, emanating from the

Guttales, the *Santalales* were included together with the *Santalaceae,* the *Ebenac.* and the *Styrac.*

Wettstein (1935) considers the order as a natural one, with the *Sapot.* somewhat separated from the other families on account of the presence of latex. But the relations with other orders (he places it between the *Bicornes* = *Ericales*) and the *Tubiflorae* are not clear: it stays more or less isolated. If any group of the *Dialypetaleae* had to be named as a distant relative, he would choose the *Geraniales-Celastrales* of the group *Euphorbiales* (in a wider sense). Copeland (1938) finds that the *Styrac.* are more closely related to the *Symploc.* than with the other families of the order. Embryologically, *Ebenac.* and *Styrac.* are "decidedly similar. In two particular details, the presence of a pedestal-like remnant of nucellus under the embryo sac and the delayed development of the zygote, there is a striking identity." This author also finds many points of resemblance between the *Ebenales,* as generally understood (i.e: in the sense of Engler, Wettstein, etc.) and the *Ericales*: the pubescence, the vascular bundles of the outer whorl of stamens associated with petal bundles, the hollow style, the poor development of ribs in the wall of pollen sac are among the points of reference. Looking further back among living plants, he finds that the *Theaceae* could be considered as a possible ancestor. Pulle (1952) made the *Ebenales* a derivation from the *Clusiales* (= *Guttiferales*) which is considered by him as one of the main centers of differentiation, itself derived from the *Ranunculales.* They are thus placed on the same level as the *Rutales,* the *Ericales,* the *Cistales* and the *Sapindales,* less advanced, however, and for obvious reasons, than, for instance, the *Brassicales,* the *Asterales* and the *Rubiales.*

Hutchinson (1959) (see DICOTYLEDONS) retains only 3 families in this order: *Ebenac., Sapot.* and *Sarcosperm.,* the *Lissocarp., Styrac.* and *Symploc.* forming another order, the *Styracales.* The *Ebenales* proper are placed at the end of one of the last branchlets of the phylogenic arrangement, related through the *Myrsinales* and the *Rhamnales* to the *Euphorbiales,* this conclusion being similar to that of Chadefaud and Emberger. The *Styracales,* on the other hand, are derived from the *Rosales,* and the *Cunoniales,* which in their turn, give birth to the *Araliales.* Emberger (Chadefaud and Emberger 1960) places the *Ebenales* (taken in the usual sense, but with the *Lissocarp.* being considered as related only and the *Hoplestigmataceae* excluded) in his group IV. Springing up from an as yet unknown ancestral stem, the *Ebenales,* as well as the *Rubiales, Ligustrales* (= *Oleaceae*), *Contortales* (= *Gentianales*) and *Tubiflorae* are considered as so many ends of parallel series. But, if more or less evident relationship could be detected between the *Rubiales* and some dialypetalous groups such as *Umbelliflores* and *Garryales* or between the *Ligustrales* and the *Rhamnales-Celastrales* complex, no such indications are given for the *Ebenales.* It is interesting to note in this connection that Alexnat using sero-diagnostic methods (1922), already arrived at the same conclusion viz. the isolation of the *Ebenac.*

From this incomplete survey of the classical and of the more modern views on the *Ebenales,* we shall retain two main facts: the *Ebenales,* as an order, are generally considered as a natural one; they are more closely related with some dialypetalous order than with any other gamopetalous order. It should be emphasized, however, that the naturalness of the order has not been accepted by all authors; it is believed that the lack of information we have, for example, on the structure of the ovules (in several families), on the vascularisation of the flowers (in all families) are serious drawbacks toward a good understanding of the *Ebenales.* Additional information would lead, it seems, to a new splitting, perhaps in the direction indicated by Hutchinson.

However divergent our opinions may be about the subject of supposed relationship, we should always carry in our mind the sound advice given by Pulle (1952: 142): "Daar men echter van het grote merendeel der levende vormgroepen niets weet omtrent hun relatieve ouderdom, berust de biologische stamboom voor 99% op hypothesen. Dit is natuurlijk voor hem, die zich dit goed realiseert."

CHARLES BAEHNI

References

Alexnat, W., "Sero-diagnostische Untersuchungen über die Verwandtschaftsverhältnisse innerhalb der Sympetalen," Bot. Archiv 1: 129–154, 1922.
Baehni, Ch., "Mémoires sur les Sapotacées. I. Système de classification," Candollea 7: 394–508, 1938.
ibid., "Mémoires sur les Sapotacees. II, "Le genre Pouteria," Candollea 9: 147–476.
Brown, W. L. and R. B. Clark, "The chromosome complement of Brumelia lanuginosa and its phylogenetic significance," Amer. J. Bot. 27: 237–238, 1940.
Chadefaud, M. and L. Emberger, "Traité de botanique (systématique). Tome 2: les végétaux vasculaires," Paris, Masson, 1960.
Copeland, H. F., "The Styrax of Northern California and the relationship of the Styracaceae," Amer. J. Bot. 25: 771–780, 1938.
Darlington, D. C. and A. P. Wylie, "Chromosome Atlas of flowering Plants," London, Allen and Unwin, 1955.
Engler, A., "Sapotaceae," in Engler, A. and H. Prantl, "Die natürlichen Pflanzenfamilien," 4: 126–153 Leipzig, Engelmann, 1897.
ibid., "Übersicht über die Unterabteilungen, Klassen, Reihen—der Embryophyta Siphonogama," loc. cit. 2: 340–357, 1897.
ibid., "Erläuterungen zu der Ubersicht," loc. cit. 358–380, 1897.
Gilg, E., "Hoplestigmataceae," in Engler, A, and H. Prantl, "Die natürlichen Pflanzenfamilien," 3: 349, 1908.
Gürke, M., "Ebenaceae," in Engler, A. and H. Prantl, "Die natürlichen Nat. Pflanzenfamilien," loc. cit., Nachtr. 4: 153–165, Leipzig, 1897.
ibid., Symplocaceae. loc. cit. 165–172.
ibid., Styracaceae. loc. cit. 172–180.
Hallier, H., "L'origine et le système phylétique des Angiospermes," Arch. néerl. Sci. exactes et nat. ser. 3, B, 1: 146–234, 1912.
Hiern, A. A., "A monograph of the Ebenaceae," Trans. Cambridge Phil. Soc. 12: 27–300, 1873.
Hutchinson, J., "The families of flowering Plants. I: Dicotyledons," 2nd ed. Oxford, Clarendon Press, 1959.
Lam, H. J., "On the system of the Sapotaceae with some remarks on taxonomical methods," Trav. bot. néerl. 36: 509–525, 1939.
Lam, H. J. and W. W. Varossieau, "Revision of the Sarcospermataceae," Blumea 183–200, 1938.

Maheshwari, P., "An Introduction to the Embryology of Angiosperms," New York, McGraw-Hill, 1950.

Parmentier, P., "Histologie comparée des Ebenacées," Ann. Univ. Lyon 6,2. Sep. 1–155, pl. 1–4, 1892.

Perkins, J., "Styracaceae," in Engler Pflanzenr. IV. 241: 1–111, 1907.

Pulle, A. A., "Compendium van de terminologie, nomenclatuur en systematiek der Zaatplanten," Utrecht, 1952.

Record, S. J., "American woods of the family Sapotaceae," Trop. woods 59: 21–51, 1939.

Schnarf, K., "Embryologie der Angiospermen," in Linsbauer, Handb. d. Pflanzen-anatomie, 10/2. Berlin 1929–1933.

Wettstein, R., "Handbuch der systematischen Botanik," (ed. 4) Leipzig u. Wien. 1935.

Woodburn, W. L., "Development of the embryo sac and endosperm in some seedless persimmons," Bull. Torr. bot. Cl. 38: 379–384, 1911.

ECHINODERMATA

The members of the phylum Echinodermata (the spiny-skinned) represent highly diversified, three-layered, marine animals, known to almost everybody who has visited the sea shore. In the temperate zone one finds the Sea-star and the Sea-urchin, the latter either represented by the ball-shaped form, covered by sharp spines, or the disk-shaped Sand-dollar. Under stones, in sponges or among sea-weed, one may find the Brittle-star with its slender, snakelike arms. In the tropical regions one is likely to find the Sea-cucumber, with its worm-shaped body, and with luck one may also pick up a Feather-star, the unattached relative of the stalked Sea-lillies. These five types exemplify the five classes of the recent echinoderms and excepting the Sea-cucumber, the five-rayed symmetry of the body is a striking feature of the adults, while the free-swimming larvae are bilateral.

Except for a few specialized plankton forms (as *Pelagothuria*), the echinoderms are bottom forms, with a calcareous, internal skeleton, recognizable by its optic properties. A unique feature is their Water vascular system, derived from parts of the coelomic sacs. Essentially it consists of a ring-canal, surrounding the gullet, with five, blind-ending canals, with a series of cylindrical outpocketings, the podia, which appear on the surface, clothed by the skin and with muscles and nerves. In some forms the podia have a terminal sucking disk and their chief function is ambulatory, in others no disk is present and the chief function is sensory and respiratory. In the Sea-cucumbers the podia around the mouth are modified into disks or are dendritic, digitate or feather-shaped, in all cases used for collecting food. In certain irregular Sea-urchins the tube feet around the mouth are lengthened and collect the food particles found on the surface of the mud beneath which these animals live. The water vascular system is in most cases connected with the outside by a canal (Stone canal), ending in a sieve plate (Madreporic plate), with ciliated pores through which water is taken in. In some Sea-cucumbers the Stone canal with its sieve plate opens into the body cavity, from which the liquid is drawn. In the Sea-lilies and Feather-stars several stone canals open into the body cavity, but a sieve plate is lacking.

Although far more complex than the two-layered Coelenterates with which the earlier zoologists lumped them, as Radiata, the echinoderms are still rather primitive forms. Their sensory organs are of the simplest type and they lack a higher central nerve-system as well as a well developed circulatory system. Food particles as well as waste products are transported mainly by ameboid cells, and the genital system is of the simplest type, with the sex products in most cases shed directly into the sea; rarely does one find adaptations for protection of the eggs and the young. The Sea-stars are chiefly carnivorous, while the majority of the Sea-urchins are vegetarians. The other echinoderms swallow mud or sand, or pick up food particles from the mud and sand wherein they live, or catch the dead or live food elements which drop down through the water. Echinoderms are found from extreme shallow water down to the greatest depth of the ocean. All are marine, with only a few capable of tolerating brackish water. None have become parasitic. In size they range from two feet in height (Sea-lilies) to a minimum of few cm. in length or diameter to 2–3 feet. In earlier periods larger sizes were reached; some Sea-lilies were known to reach a stem-height of 70 feet.

The general characteristics of the five recent classes are given in the key below. The group extends its range back into the palaeozoic period, when other, now extinct, classes also flourished.

Key

1. Primitive, archaic forms, attached by a jointed stem, at least during the early post-larval stage. Larva simple. Mouth and anus of adult upward directed; body small, cup-like, surrounded by articulated arms with lateral pinnae; nerve-band on upper side of arms superficially placed; food particles carried to the mouth by means of ciliae.
   1. Crinoidea (Sea-lilies and Feather-stars)
1. Less primitive forms, without stem; mouth downward directed, or more or less terminal.   2.
2. Body without arms.   3.
2. Body with arms.   4.
3. Simple larva; body worm-shaped; skeleton reduced to small plates or microscopical spicules; muscular system well developed; moves with the oral side forward by means of tube feet and/or muscle contractions.
   2. Holothurioidea (Sea-cucumbers)
3. Larva complex, with long arms, supported by a delicate skeleton; adult with strongly developed skeleton of plates which form a box-like test, with movable spines. Mouth either centrally placed with a complicated dental apparatus, or excentrically placed and lacking a dental apparatus, Anus on apex of test or pushed down in one inter-radius, thereby producing a secondary bilateral symmetry.
   5. Echinoidea (Sea-urchins)
4. Larva simple, though with short arms, but no skeleton; nerve-band superficial; body star-shaped, rarely with arms and disk set off from each other; moves chiefly by means of its tube feet.
   3. Asteroidea (Sea-stars)
4. Larva complex, resembles that of the Sea-urchins; nerve-bands not superficial; body sharply set off from

the slender, articulated arms; moves chiefly by means of muscular contractions.

4. Ophiuroidea (Brittle-stars)

1. The Sea-lilies of today represent a fraction of the hundreds of species which inhabited the sea of the Palaeozoic periods. About 80 species survive, all restricted to deeper water, hence their development is unknown. The Feather-stars represent a more modern offshoot which flourishes in the lagoons of the tropical seas, with some forms descending into deeper water or ranging into the Antarctic seas. In these forms a short-lasting, attached, stalked stage (the *Pentacrinus*-stage) has been observed, but later the animals become free, and develop a circle of jointed cirri around the area where the stem broke off. In forms which live on rocky bottom the cirri are few, short and strong, while in the mud-living forms they are numerous, long and delicate, enabling the animal to remain on the soft surface. Swimming is practiced by raising and lowering alternate arms. The Feather-stars are often found in large numbers, expanded like huge daises and feeding upon the organic matter which drops down through the water. With their firmly built skeleton and long geological history both groups are of importance for the palaeontologists, who in this case are able to study both the living and the extinct forms on the basis of equally well preserved material.

2. The Holothurioidea were originally classified

Fig. 1   Diagram of Feather-star, with the water vascular indicated in two arms, with the ring canal with the free stone canals, the dividing radial canals which extend into the side branches, the pinnulae, and end in clusters of usually three slender podia, the tentacles.

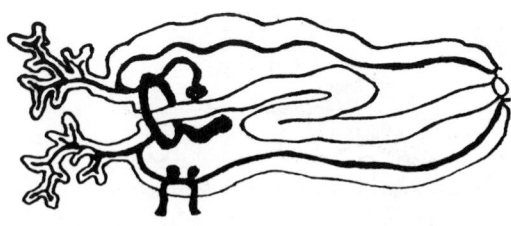

Fig. 2   Diagram of Sea-cucumber seen from the side, with the ring canal with the internal stone canal and madreporic "head" and a bladder, the Polian vesicle. Two radial canals are drawn, with the anterior podia, the tentacles, and a pair of ordinary tube feet, with sucking disk and an internal bladder, the ampulla.

among the worms, on account of their body-shape; later they were placed next to the Sea-urchins. The simple larva with its un-complex metamorphosis, the presence of only one set of gonads and the preservation in the adult animal of the bilateral symmetry, though the five radial canals of the water vascular system are present, justifies their position at the beginning of the more modern forms, although they in some respects are rather aberrant. The body is usually flexible, with the longitudinal and transversal muscles well developed, and the skeleton is reduced to small plates or microscopical spicules, often present in large numbers, and almost lacking in other forms. The tube feet may be few, or numerous, and in one group they are completely lacking, except for those around the mouth which help collect the food; in this extreme case, the radial canals have become reduced. The Sea-cucumbers are the only echinoderms which are commercially utilized to any extent. Certain large tropical forms are parboiled and smoked or dried after the sand-filled intestine has been removed, and sold in the East as Trepang or Bêche-de-Mer. The great power of regeneration which many species possess make them well suited for experimental laboratory work.

3. The Asteroidea have a larva slightly more complex than the Sea-cucumbers, with a more complicated metamorphosis. The adult is distinctly star-shaped, with the centrally placed mouth downward directed and the anus, if present, at the opposite pole, and the animals are able to move in any direction. In the more primitive forms, the number of tube feet is low, the skeleton is rather regular and firm, and the mouth is small, so only moderately-sized mollusks and other invertebrates can be swallowed, and the indigestible parts later ejected. In the more progressive forms, the number of tube feet is usually increased by staggering the rows, the skeleton is more loosely built and the entire stomach can be everted and wrapped around rather large bivalves and gastropods; when all the digestible parts have been absorbed, the stomach is withdrawn and the animal proceeds to its next victim. These swift-moving, aggressive Sea-stars do often great damage to commercial oyster and clam beds and many methods have been devised to curb their activities. Often they are swept up by tangles and used as fertilizer. Their power of regeneration is great, and some forms regularly break off arms which then reproduce the four missing arms and a new mouth.

4. The larva of the Ophiuroidea is extremely complex, with long, slender arms, supported by a fragile skeleton. In the adult the radial nerve-bands lie well

Fig. 3   Diagram of Sea-star with ring canal, with the stone canal with madreporic plate, and a Polian vesicle; two of the five radial canals are indicated with a pair of tube feet.

Fig. 4 Diagram of Brittle-star, with the ring canal, with the ventrally turned stone canal and madreporic plate. Two radial canals are indicated, with a pair of the slender podia, the tentacles.

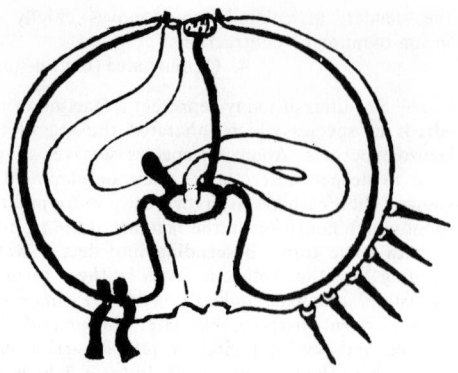

Fig. 5 Diagram of Sea-urchin, with the ring canal above the masticatory apparatus, with the stone canal and madreporic plate and a Polian vesicle. Two radial canals are drawn, with a pair of tube feet with their ampullae. To the right six spines are shown.

protected below the skin and the skeleton of the arms, essentially similar to that of the Sea-stars, has become strongly modified, with the elements from the opposite halves of the arm fusing into transversely placed, thin disks, resembling vertebrae, and connected by articulations and muscles. The tube feet lack a disk and act chiefly as sensory and feeding organs, while the animals move by twisting and bending the arms. An intestine and anus is lacking in the whole group; the animals feed on minute particles in the sand or mud which they sift by means of rows of spines around the mouth opening. They seem to be the most successful of all the recent echinoderms, often occurring in enormous numbers on soft muddy bottom, and they constitute an important food element in the diet of many bottom-living fishes.

5. The larva of the Echinoidea is almost identical with that of the Brittle-star, while the adults present the most extreme contrasts. The box-like body of the Sea-urchins is covered by movable spines which range from the extreme of long, lance-like rods to close-fitting disks which cover the upper side of the animal as a mosaic. Many forms use the spines of the underside as stilts upon which they move along. The tube feet with a large sucking disk are partly used for anchoring the animals to the substratum. As in the Sea-stars there is an increase in the number of tube feet in the five radial double rows, from the oldest, more archaic forms to the more recent types. The "regular" Sea-urchins are vegetarians, with a complicated masticatory apparatus with which they can cut up thick, leathery algae and scrape off the few mm high alga growth which encrusts many rocks. Certain species do also use their teeth for excavating holes in the rocks and live in these holes which they gradually enlarge as they themselves increase in size. In the flattened Sand-dollars the dental apparatus is present, but here it is used for sifting out the minute food particles which the animals find in the sand in which they burrow. In the fragile, egg-shaped Heart-urchins, which live in mud, below the surf zone, the dental apparatus had disappeared, and the food is gathered by the long tube feet around the here eccentrically placed mouth. In the two latter groups, the "irregular" Sea-urchins, the anus has shifted down, so the animals have become bilaterally symmetrical to a certain extent, and they have lost the ability to move in all directions and can only move forward. If a new direction is desired, the animal must turn around, pivoting on its spines, until its anterior end points in the new direction. The Sea-urchins were more numerous in previous earth periods

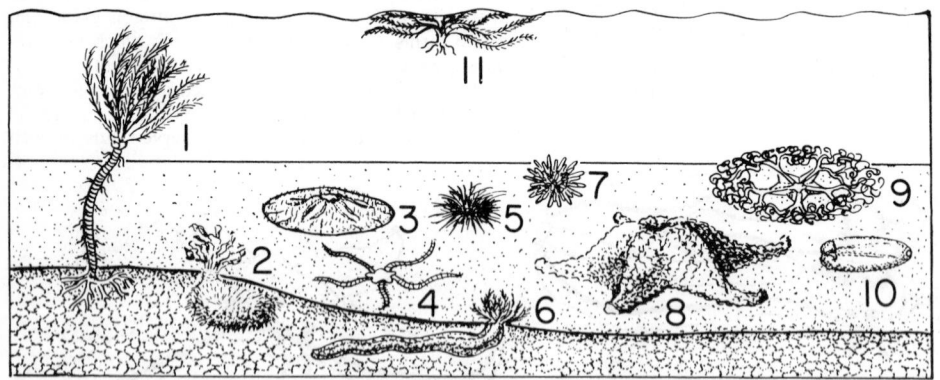

Fig. 6 Various echinoderms. 1, *Metacrinus*, a stalked crinoid; 2, *Thyone*, a Sea-cucumber; 3, *Echinarachnius*, a Sand-dollar; 4, *Ophiura*, a Brittle-star; 5, *Arbacea*, a Sea-urchin; 6, *Leptosynapta*, a Sea-cucumber; 7, the slate-pencil Sea-urchin; 8, *Oreaster*, a Starfish; 9, *Gorgonocephalus*, the Basket-star; 10, *Lovenia*, the Heart-urchin; 11, *Antedon*, the Feather-star. (From Goodnight, Goodnight and Gray, "General Zoology," New York, Reinhold, 1964.)

than in the present day's seas. With their well built skeletons they have been well preserved in the various strata, and next to the crinoids they constitute the most important group of the echinoderms, from the palaeontologists standpoint. For the experimental worker their eggs and sperm have supplied material for a large number of experiments; in many cases it has also been possible to produce hybrids between related species.

ELISABETH DEICHMANN

Reference

Hyman, L. H., "The Invertebrates, volume 4: Echinodermata," New York, McGraw-Hill, 1955.

ECHIURIDA

A group of marine animals, with about 100 species, allied to the ANNELID worms. Once placed in the obsolete "Gephyrea," now regarded as a separate phylum.

The body is divided into the long proboscis, used for gathering food, and the sac-like trunk, with highly glandular skin. Usually two hooked setae are present behind the mouth, and in two genera there are one or two rows of setae posteriorly. The alimentary canal is several times the length of the body and divided into distinct sections. Other internal structures are: anteriorly the nephridia, generally 1-4 pairs; posteriorly the long simple or branched anal trees carrying small funnels.

Mainly inhabitants of temperate and tropical seas from the shore to deep water.

In the Bonelliidae sex dimorphism is very marked, the males being minute and parasitic in, or on, the females.

A. C. STEPHEN

References

Kunkhenthal, W. and T. Krumbach, Eds., "Handbuch der Zoologie," Vol. 2, Hfte 2. Leipsig, de Gruyter, 1931.
Fisher, W. K., "Review of the Bonelliidae (Echiuroidea)," Ann. Mag. Nat. Hist. Series XI. **14**: 852–860, 1947.
ibid., "Echiuroid worms of the North Pacific Ocean," Proc. U.S. Nat. Mus. **96**: 215–292, 1946.

ECOLOGY

Ecology is commonly defined as the study of the interactions between organisms and their environment. The latter must be conceived as including everything that is not an intrinsic part of the organism or group under consideration, and thus includes living and non-living components. This primarily descriptive definition may be complemented by an alternate one that places more stress on the ecological point of view: ecology is the economics of organisms.

Ecological phenomena may be examined over the whole range of organisms and at a number of dif-

ferent levels of organization. As a result, specialization and fragmentation of the field have occurred. To some extent plant and animal ecology, and terrestrial as contrasted with aquatic ecology have developed separately. In part this has come about merely through historical forces. Real differences in emphasis do, however, exist. Animals generally are motile, whereas plants are not. On land plants are large and dominate the landscape of most animals. In water most plants are microscopic. Because different problems are amenable to solution at diverse levels of organization, other subcategories have arisen. One may profitably examine the reactions of individuals to various components of environment. So, for example, the exposure of a number of insects to some sort of environmental gradient, such as temperature or humidity, can establish limits of toleration and preferenda for these factors. This approach, termed autecology, can provide no information about group properties like spatial distribution, density or population growth rate, however. The characteristics of populations or larger units are the domain of synecology. One may choose to focus on a species population, or on the whole species complex that inhabits some particular area. Each approach suits certain problems. In gathering detailed data on a single population, complex interactions that affect this group only indirectly may be neglected. This defect is often more than compensated by the precision of the quantitative information that can be gathered in this way. To answer other questions, it may be essential to study the biota as a whole. The answers obtained at the present stage of the development of ecology are typically incomplete, but serve to illuminate other areas at different levels of organization.

Similar considerations apply to theoretical and laboratory approaches. In certain segments of ecology mathematical theory and laboratory experiments have played a significant role. The ultimate aim of such work is to provide a framework into which other features, observable only in natural situations, can be incorporated.

The major problems of ecology are necessarily so interwoven that progress in any one sub-category is limited by that of others. Consider some reasonably discrete area such as a forest surrounded by grassland, or a lake. This area contains an assemblage of diverse animals and plants, a community, as well as a set of physical and chemical features. Collectively the community and the inanimate matrix have been called an ecosystem. Ecology attempts to systematize description of such ecosystems, and to predict and possibly control the events that result from the presence of organisms within it. Specifically one might ask: How many plants and animals of which particular kinds occur in some area at a specified time? What factors cause them to exist in precisely these densities and proportions to each other? How do they modify the non-living components of the system?

The events with which ecology is concerned stem from the activities of organisms. All living things metabolize, and obtain energy either from sunlight or from chemical sources to maintain their activities and to grow. One may therefore regard an ecosystem as containing a set of coupled energy transformers (see BIOENERGETICS), which are mutually interdependent, and which store temporarily, and ultimately degrade into heat some of the incident light energy. Herbivores receive energy from photosynthetic plants, and are in turn eaten by carnivores. Eventually energy contained in all these organisms is transferred in part to decomposers. Many of the materials in the ecosystem participate in the energy transfers by forming part of the working substance of the transformers. Characteristically these materials, unlike the energy, need not become lost to the system, but may recirculate more or less indefinitely. Over a period of several years, it may be convenient to regard the physical and chemical climate and the properties of the species populations as constant. However, over longer time spans such as geological epochs, geological, climatological and evolutionary changes affect the ecosystems. In certain long-range problems the existence of these variables must therefore be considered explicitly; in the majority of ecological work they remain implicit. Ideally ecology should be able to explain the rates of transfer of energy and of various materials for every component of the ecosystem. In practice it is possible to answer detailed questions only about small parts. When considering the ecosystem as a whole, it is necessary so far to phrase questions in terms of aggregates, such as herbivores or carnivores, which may, in fact, contain many functionally different components.

Descriptive and functional questions about ecosystems are therefore asked in restricted ways. One may attempt to discover under what conditions some particular species can maintain itself. At the other extreme, one may want to estimate the total primary production rate, i.e. the amount of energy fixed by all photosynthetic organisms per unit of time, for the ecosystem. It is possible here only to indicate briefly a few selected areas of activity.

Autecological work has probably been emphasized more in the terrestrial plants than in other organisms. The interactions between moisture, vegetation and soil, in conjunction with the effects of temperature have far-reaching consequences visible on a geographic scale. It is possible to divide the terrestrial biota into large regions, the BIOMES, which differentiate between such diverse types of ecosystems as DESERTS, GRASSLANDS, TUNDRA or tropical rain FOREST. Although they are certainly not the sole causes of these divisions, temperature and moisture conditions are well enough correlated with the distribution of these biomes that the existence of these broad zones is evidence for the paramount influence of these factors. Detailed observation and experiment on the reactions of particular species to different régimes of temperature and moisture have therefore received much attention. Moreover, since plants themselves affect the soil and significantly modify the MICROCLIMATE, investigations can not confine themselves to simple cause and effect relationships, but must consider the possibilities of interaction and feedback. Other factors may, of course, be of overriding import under certain circumstances. Thus, for example plant species may be used as indicators of serpentine or gypsum substrates.

Among other organisms the autecological approach has often proved useful in correlating distribution with environment. By such methods it is sometimes possible to prognosticate the probability of success or failure of some species in a new area. However, by its nature, this mode of attack can not be applied simultaneously to all possible factors that may affect a species, and often can provide no answers to quantitative rather than qualitative questions.

Population ecology, which is primarily concerned with quantitative attributes of groups, has communications with both autecology and synecology in their strict sense. It is in this area that theory has developed most. Because population ecology restricts itself to simplified systems with a limited number of variables, it faces some of the same difficulties as autecology. However the questions answered are rather different. Most stress has been placed on biotic interrelations within and between species. The main attributes measured are statistical entities: density, birth and death rates, spatial distribution. Properties of single species populations, and of certain two species systems have been studied in considerable detail by means of mathematical models, laboratory experiment, and analysis of natural populations. The early models of single species growth and of predator-prey or parasite-host relationships, although useful in elucidating mechanisms of interactions, have not proved capable of giving adequate predictive estimates of future population size. The logistic curve, which depicts sigmoid growth of single species populations, is perhaps the best known example of these models. As more knowledge has accumulated about the effects and variability of population structure, in particular age distribution, the theoretical defects of these primitive mathematical theories have become increasingly clear. However they have had considerable value as stimuli for more penetrating study. Thus the simplified theory for predator-prey systems, which predicts cyclic fluctuations of density for both interacting species, has emphasized the dynamics of ecosystems and led to a number of important experiments and observations.

The theory of competition between species has possibly been more successful. A mathematical or logical model can be formulated whose assumptions are broadly consonant with the known behavior of populations. Such a model predicts that in an otherwise constant environment in which some resource required by both species is limiting, only one of the competing species can persist indefinitely. Many laboratory tests have demonstrated that the model is not entirely hypothetical. When two species are forced to compete under controlled conditions, one of the two does eventually die out. A large amount of circumstantial evidence suggests that interspecific competition is significant in modifying the distribution of animals and plants. However, crucial evidence from natural populations is extremely difficult to obtain, so that the degree to which competition acts as a major controlling force is a matter of controversy.

Some of the most valuable advances in population ecology have resulted from work with applied problems that demanded and made possible the gathering of large amounts of populational data. Commercial fisheries present an excellent example. Here the main desideratum is the maximizing of a sustained yield of certain size classes of fishes. To provide useful answers it is necessary (1) to develop methods of assessing the size of the population; (2) to determine birth, death and growth rates, and (3) to develop a theory of the effect of human predation on the dynamics of the exploited population. Other complicating factors, including economic considerations enter the picture, and the problem is not wholly solved. Nevertheless, the rather sophisticated models for population estimation and prediction have benefited population ecology as a whole, just as this discipline has provided the foundation on which the models could be erected.

In natural populations methods of measurement become so difficult that a major part of research effort has been expended on methodology. Taking a complete census even of a stand of trees is no simple task. For many populations the amount of labor required makes complete census prohibitive. More serious is the fact that in many animal populations the act of census so disturbs the population that its future behavior is affected. The purpose of census then becomes self-defeating. Various types of sampling methods have been developed to increase efficiency of making estimates, and because they may be the only feasible techniques for getting the desired information.

With sufficient knowledge of the distribution of the particular organisms to be measured, unbiased estimates with known variance can be obtained. However the sampling problem is typically distinct for every species, and encounters the same difficulties faced by the sociologist when he tries to sample characteristics of human populations. Most populations are not randomly dispersed, and different parts of a population may exhibit differential dispersion. In mammals, for example, females with young will not be distributed in the same way as are adult males. Methods of examining dispersion, usually by contrasting the observed distribution with a theoretical one like the Poisson, have received much attention. Studies of dispersion are of considerable interest in their own right, since they may provide clues to possible species interactions that might otherwise escape notice.

Consideration of the dispersion of several species with respect to each other leads logically to the examination of COMMUNITY structure. Communities composed of hundreds of species present a nexus of inter-relations; the possibilities for compensating mechanisms are certainly great. This means that causal relations, except for major climatic effects, are difficult to disentangle. If one uses the bottom community of a deep lake as an example, it is simple to delineate the effect of the seasonal temperature stratification that occurs in such an environment. The lower mass of water becomes sufficiently distinct from the upper that chemical exchanges are limited. Virtually anaerobic conditions therefore exist at times at the bottom. The consequences of this stagnation, and of the relative absence of light are great and largely predictable. However, aside from these dominating effects, the quantitative relations of the species to each other are affected by a great variety of other interacting factors whose total effect can probably not be ascertained by studying them singly. If this is true of a relatively simple community, such as the bottom fauna of lakes, it applies with added force to the more complex communities with greater numbers of species. It is not surprising therefore to find that generally community studies have not developed far beyond the descriptive stage.

The concept of SUCCESSION, developed quite early in the growth of the discipline, is one of the few major generalizations concerning communities. In a situation with a supply of energy and the materials necessary for life to exist, such as an exposed rock surface or a fallen log, a definite sequence of species can be observed over a period of time. The progression, for a particular set of physical and chemical factors, is consistent and fairly predictable. The first species that successfully colonize the area so modify their environment that other species can become established. The environment continues to change, new competitive and feeding relations appear, and conditions become unsuited to the needs of the pioneer colonizers, so that they disappear. The species that replace them are in turn succeeded by others. If a continuous supply of energy is available, stability ultimately increases. In areas that have been undisturbed for long periods, a grouping of organisms occurs that is capable of maintaining the *status quo:* production of various substances balances their consumption. This grouping is the CLIMAX community which, in theory, may persist indefinitely except for evolutionary changes or external disturbances. Since the rate of successional change may be not much greater than changes in physiographic and climatic conditions, it has been necessary to modify somewhat the idealized concept that has been described.

There is no doubt that a specific or concrete community is an internally compensated, interacting system rather than a mere passive aggregation of several species. This does not mean, however, that individual communities can necessarily be combined into groups of abstract community types each of which can be described objectively. Considerable debate has been aroused by the question whether one can objectively characterize, by species composition and other criteria, abstract community types. As it has become more apparent that different stands do not always conform to one or another description, alternate methods of describing communities have been advocated by various schools. While each method may have peculiar merits, none so far have completely solved this problem. Therefore it is usually impossible to determine any but arbitrary boundaries for a community.

Another question that has engendered heated controversy is what determines population balance in communities. One argument of considerable cogency is that a factor, to be effective in preventing a species from becoming indefinitely large in numbers, must change in severity disproportionately with density. Two questions have then to be answered. To what extent can this argument be applied in natural situations which are variable in time and space; and are there any factors which do not act in such a density-dependent manner. Only additional data can solve this problem which threatens to become largely semantic.

To gain understanding of the complete workings of a community, or of the ecosystem as a whole, would require quantitative information about all the interactions between species, as well as of their individual populational behavior. This is probably not humanly possible. Qualitatively, food relations and requirements have been worked out for many components. For economically important species some quantitative data exist. Such information can be formulated into food webs composed of many individual food chains that trace the progress of bits of energy from one species to the next. Such qualitative analyses are complemented by quantitative studies on trophic levels, such as photosynthetic plants or herbivores. Here interest has recently increased greatly. Knowledge of the efficiency with which energy is converted by a given group is obviously of practical as well as theoretical interest. As a rule of thumb, and merely to indicate the order of magnitude, the efficiency with which green plants fix solar energy is about 1 per cent. Animals have greater efficiencies of conversion; 10 per cent is a representative figure. These values are, of course, far below those physiologically attainable with energy available in optimal form and intensity.

The nature of ecological problems continually brings the science into contact with practical applications and decisions that must be made in an ecological context. Particularly as man progressively comes to be the major influence in virtually all ecosystems, more and more ecology is called on to provide policies for management and conservation in the broad sense. This is clearly desirable. However, with basic knowledge of complex ecosystems only at the stage where predictions are general rather than specific, and often qualitative rather than quantitative, applications sometimes rest on shaky foundations. Policy decisions, under these conditions, are somewhat delicate matters.

PETER W. FRANK

References

Allee, W. C., *et al.*, "Principles of animal ecology," Philadelphia, Saunders, 1949.
Andrewartha, H. G. and L. C. Birch, "The distribution and abundance of animals," The University of Chicago Press, 1954.
Clarke, G. L., "Elements of Ecology," New York, Wiley, 1956.
Daubenmire, R. F., "Plants and environment," New York, Wiley, 1959.
Odum, E. P., "Fundamentals of ecology," Philadelphia, Saunders, 1959.
Oosting, H. J., "Plant communities," San Francisco, Freeman, 1956.
Phillipson, G., "Biological Energetics," New York, St. Martin's, 1966.
Watt, K. E. F., ed., "Systems Analysis in Ecology," New York, Academic Press, 1966.

ECTOPROCTA see BRYOZOA

EGG

The egg (ovum, egg-cell, ovule, oosphere) is the female germ cell or gamete, and has as counterpart the male germ cell, the SPERMATOZOON. The terms germ cell and gamete, however, have a somewhat wider connotation than this implies for they can be used to refer to specialized conjugating cells even when, as in many simple organisms, the 'male' and 'female' forms are of about the same size and structure (*isogamy*); in these circumstances, and also when heterogamy is minimal, the terms egg and spermatozoon are inappropriate. On the other hand, the animal 'egg' as a descriptive name is commonly used not only for the gamete itself but also for the precursor cells (*gametocyte, oocyte*) and for the embryo at stages of development up to the blastocyst (*blastula*).

The male and female gametes are differentiated in representatives of all major groups of both plants and animals, and are especially adapted for a role that has come to be associated with multiplication. Union of male and female gametes (FERTILIZATION, *amphimixis*) forms the zygote and is the distinctive feature of sexual REPRODUCTION; it is followed by successive subdivisions of the zygote into many daughter cells. In Protista, this subdivision constitutes multiplication; in complex organisms, subdivision (cleavage) is the basis of embryogenesis, and multiplication of individuals depends on the production of numbers of eggs, or upon segregation of cell groups in the developing or adult organism, as in budding and polyembryony. The egg can also take part in asexual reproduction, giving rise to a fully viable embryo without the participation of the spermatozoon (PARTHENOGENESIS, APOMIXIS)—among plants, such a process is known in many species of Angiosperms; among animals, it is found in rotifers, some nematodes and echinoderms, but especially in insects.

In the great majority of organisms that form gametes, the essential and concluding act of fertilization is the fusion or syngamy of haploid nuclei, so that the egg, when fully prepared for syngamy, must contain only a haploid number of chromosomes. This state is achieved through meiosis and chromatin elimination. In the higher plants, chromosome reduction occurs in the sporophyte, so that the gametophyte, which bears the egg, is a haploid organism; in most Protista and Metazoa, on the other hand, it takes place during gametogenesis or in the early phases of fertilization. In some algae, meiosis in zygotic, immediately succeeding fertilization. When development is parthenogenetic, the new individual is haploid (rotifers and insects) or else diploidy is maintained, either by avoidance of meiosis (as in Angiosperms) or through the operation of one of the mechanisms of regulation to diploidy (as in most parthenogenetic animal species).

As a general rule, eggs are much larger than the tissue cells of the same organisms; in plants, eggs are seldom as big as in animals or their structure as complex, because the egg remains embedded in maternal tissues and does not need to be equipped for independent existence. Nevertheless, the eggs of some Gymnosperms are of a size that lies within the lower range of animal egg sizes. Most eggs are spherical or nearly so, but there are many other shapes to be seen, variations on the ellipsoidal or cylindrical forms.

The egg proper, or vitellus, consists, like other CELLS, of nucleus, cytoplasm and a limiting plasma (or permeability) membrane, but it is loaded with nutritive materials making up the yolk or deutoplasm. The nutrient stores can be carbohydrate, fat or protein in nature and vary greatly in relative amounts and in the physical forms assumed—fine particles, granules, rodlets, crystalloids, droplets, vesicles, etc. The deutoplasm does not account altogether for the large size of the vitellus, for the volumes of nucleus and cytoplasm also exceed those in tissue cells. Broadly speaking, however, the larger the vitellus the greater the proportion of deutoplasm. With most of the smaller eggs, the yolk is distributed throughout the cytoplasm, but, with larger proportions of yolk, there is an increasing tendency for this material to gather in a mass leaving part of the cytoplasm relatively free. The nucleus is found in the yolk-free region and marks the animal pole, the deutoplasmic region carrying the vegetal pole. The extreme condition is seen in the very large eggs of birds and reptiles, where the bulk of the cytoplasm is restricted to a small germinal disc lying on the surface of an immense yolk. A relatively large burden of yolk evidently interferes with cleavage of the egg after fertilization; as a result, while poorly yolked (*microlecithal*) and medium yolked (*medialecithal*) eggs undergo complete cleavage (holoblastic cleavage), heavily yolked (*megalecithal*) eggs become subdivided only at the animal pole, the rest of the vitellus containing the yolk remaining entire (*meroblastic cleavage*). The latter form is shown by the eggs of cephalopods, elasmobranches, teleosts, reptiles and birds.

Characteristically, eggs are enclosed within one or more coverings or investments, in addition to the plasma membrane, and of these there are many kinds. A common second layer is the thin vitelline membrane, which, in the eggs of echinoderms, *Branchiostoma*, and some fish and frogs, becomes modified after sperm entry and is raised from the surface as the fertilization membrane. Echinoderm eggs have, in addition, a much broader layer, the jelly coat, which rather resembles, in appearance though not in mode of formation, the gelatinous investments of various molluscs, amphibians and fish, and the mucinous or albuminous coats of the rabbit and marsupials. Another kind of coat that is particularly well developed in fish eggs is the moderately thick, resistant, radially striated zona radiata. Insect eggs emerge with a tough chitinous envelope distinguished as the chorion. Mammalian eggs are characterized by the possession of a transparent, mucoprotein membrane called the zona pellucida; it is notably thicker in the placental mammals than in monotremes or marsupials. Tough, horny egg capsules are formed about the eggs of selachians, chimeroids, molluscs and platodes. The eggs of monotremes, reptiles and birds have in common the possession of a wide layer of albumen (egg 'white'), a shell membrane and a calcareous shell.

Egg envelopes may be perforated, at or near one pole, by one or more narrow canals, the micropyles, which are known in the eggs of fish, insects, cephalopods, echinoderms, molluscs and nemertines. Micropyles exist, too, in the integumentary layer of cells surrounding the eggs of Spermatophyta. In general, micropyles are considered to provide the only means of access of the spermatozoon or pollen tube to the egg, though in nemertines and some sea urchins spermatozoa can apparently enter elsewhere as well.

Animal eggs differ enormously in size. Among the smallest are those of the bryozoan *Crisia*, of which the vitellus is only about 18μ in diameter. The smallest mammalian egg, with a vitelline diameter of 50μ, is that of the field vole *Microtus agrestis*; the clam *Spisula* has an egg of about the same size. Most plant eggs have diameters between 25 and 100μ. The diameters of the majority of mammalian eggs lie in the range of 70 to 140μ, and this is true too for the eggs of a number of echinoderms, tunicates, molluscs, polychaets, nemertines, platyhelminths, bryozoans, and coelenterates. Next come the eggs of the marsupial *Dasyurus* and the sea squirt *Amaroucium*, with a diameter of 250μ. The smallest fish eggs are about 400μ, and the smallest frog eggs 700μ. Some invertebrate eggs have a vitellus around 1400μ in diameter, as in the squid *Loligo*, the gasteropod *Busicon*, the starfish *Henricia*, and the crab *Libinia*. A vitelline diameter of 4000μ is found in the eggs of monotremes, the over-all size of which is about 16 mm. The largest frog egg, that of the marsupial frog *Gastrotheca*, has an over-all diameter of about 10 mm

and its vitellus is larger than that of the humming bird egg (6 mm), the smallest avian egg. Some reptile eggs are very big such as those of the python, but the largest vertebrate eggs are found among the sharks and ostriches, the vitelline diameter being about 80 mm and the over-all length 150 mm.

C. R. AUSTIN

References

Austin, C. R. and E. C. Amoroso, "The mammalian egg," Philadelphia, Davis, 1961.
Austin, C. R. and A. Walton, "Fertilisation." *In* Parkes, A. S. ed., "Marshall's Physiology of Reproduction," vol. 1, London, Longmans, Green, 1960.
Boyd, J. D. and Hamilton, W. J., "Cleavage, early development and implantation of the egg." *In* Parkes, A. S. ed., "Marshall's Physiology of Reproduction," vol. 2, London, Longmans, Green, 1952.
Costello, D. P., *et al.*, "Methods for obtaining and handling marine eggs and embryos," Woods Hole, Mass., Marine Biological Laboratory, 1957.
Romanoff, A. L., and A. J. Romanoff, "The Avian Egg," New York, Wiley, 1949.
Rothschild, Lord, "Fertilization," London, Methuen 1956.
Wilson, E. B., "The cell in development and heredity," New York, Macmillan, 1928.

EHRENBERG, CHRISTIAN GOTTFRIED (1795–1876)

This German naturalist, who held the chair of medicine in the University of Berlin, was early attracted to biology through the observations he made when accompanying von Minutoli in his 1820 expedition to Egypt. The scientific results of this expedition were reported by Humbold with whom Ehrenberg remained friends for the rest of his life, accompanying him on many of his expeditions. Ehrenberg's chief claim to fame, however, is in his study of the fossil Protozoa, for he was the first to demonstrate the existence of these forms. His huge "Mikrogeologie" (Leipzig 1854) was for many years the standard reference work in the field. He also named freshwater Protozoa and demonstrated for the first time that Noctiluca was the cause of much marine phosphorescence.

ELASMOBRANCHII

The sub-class of fish-like vertebrates which includes the sharks, skates, and rays, and which, with the chimaeroids of the sub-class **Holocephali,** make up the vertebrate class **Chondrichthyes.** The class as a whole differs from the more primitive cyclostomes and lampreys, regarded by some authorities as belonging to a separate vertebrate class, the **Agnatha,** in having well developed lower jaws, pectoral and pelvic girdles with paired appendages, a conus arteriosus with two or more series of valves, and a more highly developed cranium and visceral skeleton. Representatives of the Chondrichthyes and Agnatha lack true bone and thus differ from the higher vertebrates although the Cartilaginous skeletons of the Chondrichthyes may be stiffened by mineral deposits in the cartilage. In most of the Elasmobranchii, mineral deposits are sufficient to give the skeletal elements in adults great rigidity.

The elasmobranchs and chimaeroids are further characterized by the presence of placoid SCALES or dermal denticles consisting of a core of dentine over a pulp cavity, both of dermal origin, but overlaid with a hard enamel-like substance known as durodentine also of dermal origin. The denticles vary greatly in shape, size, and distribution. The denticles are densely distributed over the entire surface in most sharks, irregularly or sparsely distributed in the skates and rays, and reduced to a few patches or are absent from adult chimaeroids. The jaw teeth are similar in origin and in the general pattern of structure to the denticles. They are attached to bands or beds of connective tissue and are not imbedded in the jaws. The functional teeth may be in a single row as in the carcharhinid sharks, with several series in various stages of development in back of the functional row to move forward and replace lost or worn teeth of the functional row. In other sharks and rays several series may be functional at the same time.

Fertilization is internal and is effected by paired intromittent organs which arise from the inner margins of the pelvic fins of the males. The posterior part of the intestine is modified by various arrangements of fleshy flanges serving to increase the absorptive surfaces, forming the so-called spiral valve or scroll valve.

The elasmobranchs have 5 to 7 pairs of gills, each with a separate external slit-like opening, distinguishable vertebral centra, a notochord that is somewhat constricted segmentally or persists only between the vertebrae, and a cloaca. The chimaeroids, in contrast, have 4 pairs of gills with 4 pairs of gill clefts leading to a common branchial chamber on either side, each branchial chamber with a single external opening, no vertebral centra, no notable constrictions of the notochord, and an external urogenital aperture posterior to the anus.

Modern representatives of the sub-class Elasmobranchii may be grouped into two orders. The Selachii, sharks, are characterized by having the anterior edges of the pectoral fins unattached to the sides of the head and by having lateral gill slits, the upper portions of which extend above the plane of the insertion of the pectoral fins. The Batoidei, the skates and rays, are characterized by having the leading edges of the pectoral fins attached to the head, and by having the gill slits in a ventral position. This is an arrangement of convenience and other classifications placing greater weight variously on the structure of the vertebrae, the morphology of the male copulatory organs, differences in the spiral valve, or, giving greater consideration to the scanty evidence from paleontologic history, have considerable merit. All modern families of elasmobranchs appear to have existed by the beginning of the Tertiary period and the phylogeny is obscured by the comparatively poor fossil record. The origin of the sub-class is obscure and the early history of the group is based on fragments such as teeth, spines, and denticles. The diversity of such fragments occurring in Upper Silurian and Lower Devonian rocks suggests an even earlier origin.

The body fluids of elasmobranchs, unlike the body fluids of other vertebrates, contain 2 per cent or more

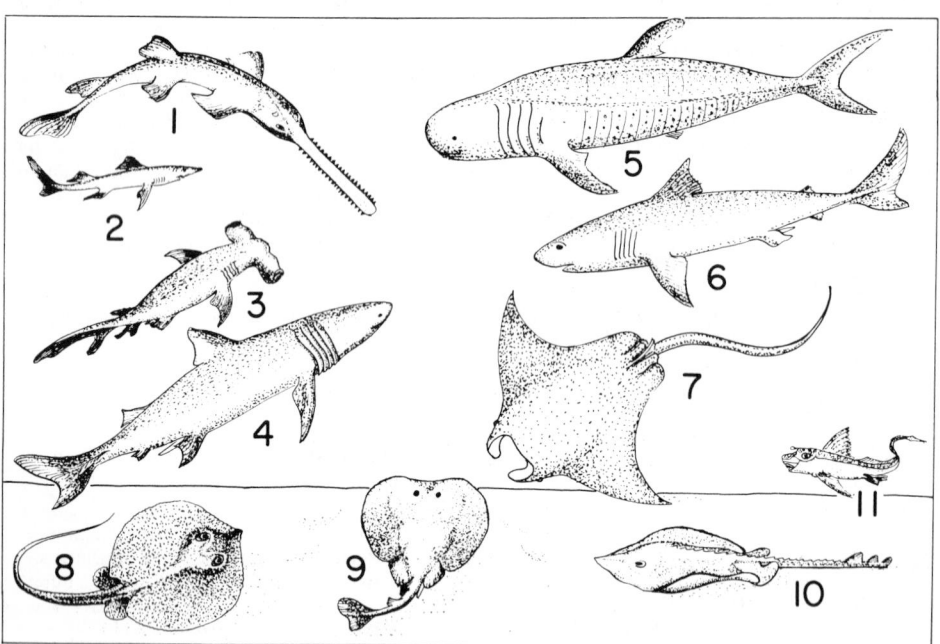

Fig. 1   Some examples of the class Chondrichthyes. 1, *Pristis*, the sawfish; 2, *Squalus*, the dogfish shark; 3, *Sphyrna*, the hammerhead shark; 4, *Cetorhinus*, the basking shark; 5, *Rhincodon*, the whale shark; 6, *Carcharodon*, the white shark; 7, *Manta*, the manta ray or devilfish; 8, *Dasyatis*, the stingray; 9, *Torpedo*, the electric ray; 10, *Raja*, a ray or skate; 11, *Chimaera*, a chimaera. (From Goodnight, Goodnight and Gray, "General Zoology," New York, Reinhold, 1964.)

urea as a normal constituent in those forms living in sea water. Presumably urea retention is regulated automatically and increased urea concentration tends, by raising the osmotic pressure, to make water available to the elasmobranch by direct absorption. The effects of this condition on the evolution of the modern elasmobranch, on its success in the invasion of fresh water habitats, and on mechanisms evolved in connection with development of the young appear to be important but are not well understood.

The comparatively static condition of evolutionary processes among elasmobranchs in recent periods in the geologic time scale suggests that basic patterns in evolutionary development of elasmobranchs have imposed limits on the adaptive capacity of members of the group. Elasmobranchs have no swim bladder and the specific gravity of most species is somewhat greater than sea water. Hence, species without special methods for remaining in hydrostatic balance tend to sink or are tied to a life near the bottom. The sand shark, *Carcharias taurus*, gathers gases in the stomach or perhaps gulps air which may be forcibly and noisily expelled. It has been suggested that these stomach gases may serve the same purpose as a swim bladder. Some deep water species such as *Dalatias licha* have large livers containing as much as 80 per cent hydrocarbon oils of sufficiently low specific gravity to assist in the maintenance of equilibrium.

Means to assist the elasmobranch in the maintenance of hydrostatic equilibrium are probably various and all of them may not be known, but most of the large sharks inhabiting midwater or surface layers maintain their positions above the bottom by constant forward motion. The paired fins of large sharks are stiff and have restricted capability for movement, functioning only as rudders or stabilizers when the shark is in forward motion. Associated with the need to maintain hydrostatic equilibrium, many of the larger sharks such as *Sphyrna, Eulamia, Isurus,* and *Carcharodon* apparently require forward motion not only to give a sufficient flow of water over the gills for adequate aeration, but also require the muscular activity of swimming to maintain sufficient circulation of the blood. Large species, in tropical surface waters, if stopped from forward motion, soon go into coma and die from respiratory failure unless adequately stimulated to continue swimming action. In comparison with the higher vertebrates, the relatively inefficient and poorly integrated nervous system of elasmobranchs, the sluggish circulatory system, and the paired appendages specialized for bottom life or for hydrostatic balance are all obviously inferior in performance. The resulting disadvantages to elasmobranchs in flexibility of response and in locomotion are less apparent, although probably important, to the smaller forms and to elasmobranchs inhabiting deeper and cooler water.

An extraordinary variety of reproductive patterns have developed within the Elasmobranchii, all tending to permit the production of small numbers of comparatively large young. Fertilization is internal in all species and all species, as far as known, produce either a leathery capsule for enclosing one or more fertilized eggs, or some homologous shell membrane of more or less functional significance but sometimes reduced and present for only a short period. Typically in the sharks, both oviducts are functional but in many rays only one serves in uterine development of the embryos. Two ovaries may be present but in many species only one is functional. Our knowledge of elasmobranchs is especially weak in the field of descriptive embryology but it is apparent that developmental methods vary greatly.

Some sharks, notably members of the families *Heterodontidae* and *Scyliorhinidae* as well as the skates of the family *Rajidae* lay eggs encased in leathery capsules that develop and hatch on the sea bottom. The nurse shark, *Ginglymostomia cirratum* may retain within the oviducts its encapsulated eggs until hatching occurs. The whale shark, *Rhincodon typus,* apparently lays encapsulated eggs which develop and hatch on the sea bottom.

Most other sharks and rays produce living young, nourished first during development of the embryo by a large yolk, but later by nutrient materials transferred from the lining of the oviducts of the mother to the embryo by various mechanisms. Typically the galeoid sharks develop a yolk-sac placenta while the rays receive nourishment by a secretion of the villi of the walls of the oviduct.

The young of the sand shark, *Carcharias taurus,* apparently develop in two stages, the first stage resulting in the hatching of a single embryo from a relatively small but leathery egg capsule in which many eggs (15 to 20) had been originally encased. The newly hatched embryo, less than 9 inches in length, then eats other eggs, and presumably other embryos as well, remaining within the oviduct until the supply of eggs from the ovary is exhausted. By this time the single embryo within either oviduct may have reached a length of 40 inches.

About 700 species of elasmobranchs are known and of these about 300 are sharks. Elasmobranchs range in size from the small deep water species, *Squaliolus,* which mature at a length of about 6 inches to the 45 foot whale shark, and from skates 5 inches across to giant Manta rays weighing more than a ton. The females in the species that have been studied average slightly larger than the males. Some of the larger species seem to be cosmopolitan in suitable habitats but many of the smaller forms, particularly those living at the bottom, have limited geographical ranges. Both sharks and rays are to be found from polar seas to the tropics and occur in depths at least as great as 1,500 fathoms. A comparatively large number of shallow water species enter fresh water. The bullshark, *Carcharhinus leucas,* of the Atlantic enters brackish and fresh water estuaries to give birth to its young and a species, *C. nicaraguensis,* closely allied to *C. leucas,* is said to be permanently established in Lake Nicaragua. In the absence of life history data it still appears uncertain that any species of elasmobranch maintains itself indefinitely in fresh water without recruitment from a marine stock.

Many of the shallow water species are migratory. Movements may sometimes be associated merely with seasonal temperature change but in some of the larger species, migrations appear to be related to travel to and from nursery grounds by gravid females. The young and the adult males and females of some of the migratory species such as *Eulamia milberti* occupy somewhat different geographical and vertical ranges only overlapping in part. Nursery grounds apart from the areas most frequented by adults are common to many of the larger predatory species. The adult males in the species studied appear to select a slightly cooler environment than the females.

The tendency to segregation by size and sex in many species of sharks may be the outgrowth of the need to protect young sharks from predation by larger sharks of their own species. This habit is well developed in Atlantic American carcharhinids and may be common in other and less known families. It is also common to many small species. The chief predators on small sharks and probably the only important predators are larger sharks. Habit patterns serving to protect newborn young appear to have developed in many species, some habits, such as fasting by the females when the young are born and immediately thereafter, perhaps being under endocrine control and made possible by the special ability of elasmobranchs to store energy in the liver as oil.

The largest sharks, the whale shark and the basking shark, and the largest rays, the devil rays or Mantas feed primarily on plankton and very small fishes. Their method is merely to swim through concentrations of small organisms, straining them out of the water with their specialized gill rakers. Most of the rays and some of the sharks are bottom feeders and those with heavy jaws and paved teeth usually feed on hard shelled mollusks and crustaceans. Crustaceans, small fishes, marine worms, and squids make up the bulk of the diet of the smaller sharks, skates and rays. The sawfishes (batoids) use their sawlike rostral processes, at least sometimes, to knock down and disable fishes for food. It seems probable that the sawsharks (selachians) have similar habits. Although many of the sharks have special feeding techniques peculiar to their species, most large sharks are omnivorous and opportunists in feeding. Tiger sharks are especially indiscriminate feeders and consume carrion, garbage, horseshoe crabs, and large conchs, as well as sea turtles, sea birds, other sharks, and even men in spite of a preference for fresh fish which they are usually unable to catch.

STEWART SPRINGER

References

Bigelow, H. B., and W. C. Schroeder, "Sharks," *In* "Fishes of the western North Atlantic," Memoir Sears Foundation for Marine Research, 1948.
ibid, "Sawfishes, guitarfishes, skates, and rays," *In* "Fishes of the western North Atlantic," Memoir Sears Foundation for Marine Research, 1953.
Daniel, J. F., "The elasmobranch fishes," Berkeley, Univ. of California Press, 1934.
Gilbert, P. W. *et al.*, "Sharks, Skates and Rays," Baltimore, Johns Hopkins Press, 1967.

ELECTRIC ORGANS

The only class of animals known to possess specific electric organs are the fishes. Within this class such organs must have evolved several times independently of one another. They cannot be derived from a common progenitor possessing these organs, because they are found only in a relatively small number of unrelated genera in which the organs occupy different positions in the body.

Electric fish include marine forms (Rajidae, Torpedinae, *Astroscopus*) and freshwater forms (Mormyridae, Gymnotidae, *Malapterurus*). In all cases which have been investigated embryology and innervation indicate that electric organs are derived from muscular tissue. The muscles which contribute to the formation of electric organs may be tail muscles, hypobranchial muscles, eye muscles etc. The substance of the electric organ is composed of units (electroplaques) often in regular series/parallel arrangement, each developing ontogenetically either from a single myoblast or from a syncitium of myoblasts. The electroplaques are supplied by cranial or spinal motor nerves, and the fibres terminate on one face of the plaque. In *Malapterurus* a single, branching fibre on each side innervates all electroplaques of that half of the electric organ. The arrival of a motor impulse leads to a discharge.

The total discharge varies greatly from species to species with regard to voltage, pulse shape and frequency. Amongst the strong electric fishes (*Electrophorus, Torpedo, Malapterurus*) the discharge is given off as a short train of 5 to 20 pulses each about 1 m sec. in duration. 550 volts have been measured in Electrophorus on open circuit. In species of weak electric fish the discharges may be only of a fraction of one volt, but in many such instances the discharges are emitted continuously throughout life and show a great regularity in frequency; this may vary between 50 and 1600 cycles per second and it is not affected by the state of excitation of the fish. Other types of weak electric fish (most Mormyridae and some Gymnotidae) have a slow and somewhat irregular basic discharge rate (1 to 6 per sec.); this frequency is considerably increased when the fish is excited. In such cases the pulse shape is essentially diphasic, and the duration of the discharge from the whole organ can be very brief (0.2 m sec.).

The strong electric organs function as offensive and defensive devices. The weak electric organs are used as part of a locating mechanism, and it has been suggested that these fish possess a specific electric sense. During each discharge the fish sets up in the water an electric field which resembles an electrical dipole. Objects with an electrical conductivity differing from that of water will distort the normal configuration of the field around the fish. It is thought that the fish can appreciate the distribution of electrical potential over its surface by means of special receptors, and can thereby locate objects in its vicinity. The theory is supported by morphological and physiological evidence. Probably the electric discharges have also a social significance in the life of these fishes.

Electrophorus is the only fish known to be able to produce weak searching discharges (from the caudal organ of Sachs), and strong discharges from the main organ. It is believed that in evolution the locating mechanism has preceded the strong discharge mechanism. In non-electric fish the functional equivalent of the discharge of an electroplaque is assumed to be represented either by the muscular action potential (*Electrophorus*) or by a motor endplate potential (*Torpedo, Raja*).

Work on single electroplaques of Electrophorus has shown that when one microelectrode is introduced into the cell and another placed on the outside a resting potential of -85 m V is recorded inside the cell. During the discharge there is a reversal of potential and the overshoot is about $+65$mV. The voltage of the discharge from the whole organ can be explained by the synchronised activity of about 6000 serial elements.

Various delay mechanisms along the pathways to the electroplaques have been demonstrated or postulated, but the whole mechanism of synchronisation is imperfectly understood.

It was formerly thought that only the innervated face of an electroplaque shows the phenomenon of reversed polarity during the discharge. Recently it has been shown however, that both faces can fire, one after the other, thereby producing diphasic spikes.

H. W. LISSMANN

References

Chagas, A., and A. Paes De Carvalho, eds., "Bioelectrogenesis," New York, Amer. Elsevier, 1961.

Grundfest, H., "The mechanisms of discharge of the electric organs in relation to general and comparative electrophysiology," Progress in Biophysics, 7, 1–85, 1957.

Keynes, R. D., "Electric organs," In the "Physiology of Fishes," Vol. 2., New York, Academic Press, 1957.

Lissmann, H. W., "On the function and evolution of electric organs in fish," J. Exper. Biol. 35, 156–191, 1958.

Lissmann, H. W. and Machin, K. E., "The mechanism of object location in *Gymnarchus niloticus* and similar fish," J. Exper. Biol. 35, 451–486, 1958.

ELECTRON MICROSCOPE

A device which forms highly magnified images of structures through the use of a focused beam of high-energy electrons.

History. It is generally agreed that electron microscopy had its beginning in 1924 when Louis de Broglie theorized that electrons had a wave nature. His formula $W = h/mv$, where h is Plank's constant, m is the mass of the particle and v is its velocity, gives the wavelength W of an electron. At an accelerating voltage of 60 KV, for example, the wavelength of the electron is only 0.05 Å, or 1/100,000 that of visible light. E. Schrödinger, in 1926, combined the work of de Broglie and W. R. Hamilton's analogy of 1830 between mechanics and optics. Later that year, H. Busch published a paper relating the work of de Broglie and Schrödinger to his work on the trajectories of electrons in magnetic fields and calling attention to the practical aspects of the study.

In 1927 Busch published a supplement to his paper in which he experimentally verified his theories through the use of a magnetic lens. In 1931 C. J. Davisson and C. J. Calbick performed similar work using an axially symmetrical magnetic field.

In 1932 E. Brüche and H. Johanssen, using a 300 KV accelerating potential, produced images of a heated oxide cathode. M. Knoll and E. Ruska, in that same year, constructed a 60 KV electron microscope using magnetic lenses of short focal length and succeeded in getting an image of the electron source within the evacuated column.

All of these early electron microscopes were only capable of forming an image of the heated electron-emitting source and were known as emission electron microscopes. Although this type of instrument is still available today, it has very limited usefulness in the field of biology. Of more general utility is the transmission microscope, first suggested by K. Rudenberg in 1931. In this instrument the electrons are sent through the specimen, and the resulting beam is focused to form an image of the "electron-shadow" of the specimen. Construction on this type of instrument was begun in 1932 by Knoll and Ruska and continues to the present. It was not until 1935, however, that E. Driest and H. O. Miller obtained results with the electron microscope that were better in resolution than the light microscope.

The basic principles of a new type of electron microscope were put forth by Knoll in 1935. This instrument utilized the electrons scattered and emitted secondarily from the surface of an electron-opaque specimen under electron bombardment to form its image. In 1938 von Ardenne described the first experimental instrument. Further studies were done by Zworykin and others in the United States in 1942. Development was discontinued until 1948 when Oatley in England began a program on the development of the instrument and its applications which is still under way. Current commercial examples of this instrument, called the scanning electron microscope, are capable of resolution approaching 100 Å.

Specimen preparation in the early days of electron microscopy consisted of using thick specimens that were heavily stained. These thick specimens would absorb electrons from the beam, heat up, and burn away, leaving behind only a stained skeleton. It was not until the work of Krause that it was demonstrated that ultrathin sections could better withstand the electron beam since they absorbed far fewer electrons, resulting in less heat damage.

In 1938 B. von Borries and Ruska designed an electron microscope for the Siemens and Halske company in Germany with a resolving power of 100 Å, suitable for laboratory use. Simultaneously, in Toronto, another practical microscope was being developed by A. Prebus and J. Hillier under the direction of E. F. Burton.

In 1939, at the Allegemeine Elektrizitats Research Institute, H. Mahl produced a high resolution instrument using electrostatic lenses. Also in that year Marton working for R. C. A. produced the first instrument using magnetic lenses with electronically regulated power supplies, which were designed by A. Vance. In 1940 R. C. A. announced their first commercial electron microscope with a resolving power of about 24 Å. Siemens also had a commercial model at this time with a resolving power of about 22 Å.

In 1941 Ardenne, Hillier, Vance, Ruska, Miller, and V. K. Zworykin were all working on different high accelerating-potential instruments. In 1943 two desk models were made, one electrostatic model by Bachman and Ramo and a magnetic one by Hillier and Vance. In 1944 Ardenne described an instrument using a high-strength objective lens that provided resolution down to 12 Å. In 1945 Hillier developed the axial symmetry of the magnetic lens still further and managed to get down to 10 Å. Since that time resolution in the 2–3 Å range has been recorded.

Description. The electron microscope can be broken down into four systems, electron optical, electronic, photographic, and vacuum.

The electron optical system, or column of an electron microscope, as in the light microscope, consists of an illuminating system and an image-forming system. (Fig. 1)

The illuminating system starts with a source of elec-

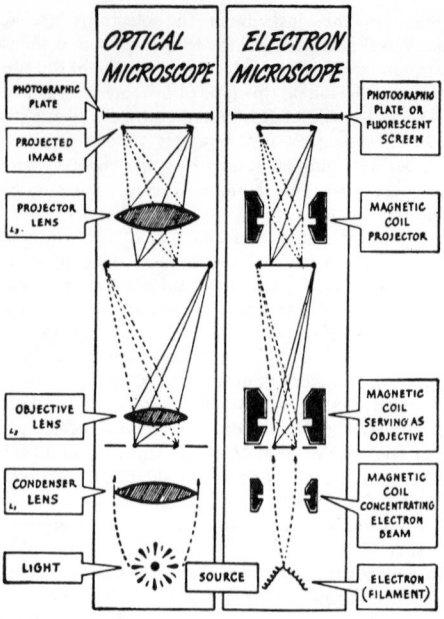

Fig. 1 Comparison of light and electron microscope

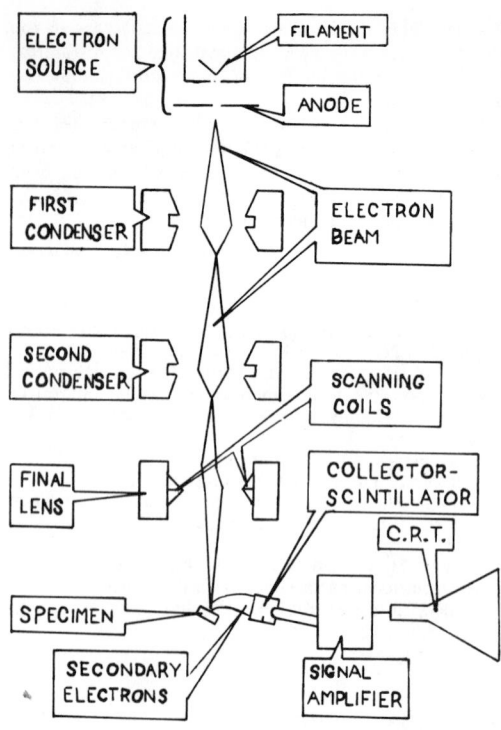

Fig. 2 Diagram of scanning electron microscope.

trons, usually heated tungsten wire maintained at a high negative electrical potential, (up to one million volts). The electrons that leave the hot wire are attracted to an anode several inches down the column that is maintained at ground potential. A small hole in the center of the anode cap allows the central portion of these electrons to pass through and become the beam of electrons that is used by the rest of the microscope. After the electrons leave the anode they enter the condensor lens field. This lens consists of a soft iron electromagnet, usually double, that focuses the beam down to a very fine spot at the plane of the specimen.

The image-forming system operates on the beam after it has passed through the specimen and some of the electrons on the beam have been scattered or absorbed by the specimen. The system starts with the objective lens, another circular, soft iron electromagnet that forms a magnified image of the effect that the specimen has had on the beam, forming a sort of electron density "shadow" of the specimen. This image then passes through the projector lens system, which may be either a single or a double lens. This lens, again another electromagnet, is usually variable in strength, like a zoom lens on a movie camera, and on some instruments can produce a final magnification of more than 500,000 diameters. The images that these electron lenses form are completely invisible to the operator of the microscope, so a fluorescent screen is placed below the projector lenses where the operator can see it through a glass viewing port for focusing.

The scanning electron microscope, (Fig. 2) has an illuminating system similar to that in the transmission microscope, except that it uses three condensor lenses in a row to focus the beam down to a very fine (100 Å) spot. The third condensor also contains a set of deflection coils to sweep the electron beam across the speci-

men as a scanning raster. The specimens used in this instrument are larger than those used in the transmission microscope and thick enough to be completely opaque to the beam. Reflected and scattered electrons are collected by a positively charged scintillation screen- photo-multiplier system that produces an electronic signal which modulates the intensity of an electron beam in a cathode ray tube. This beam scans synchronously with the one in the microscope column and produces an image of the surface texture of the specimen on the cathode ray tube screen.

The electronic circuitry backing up the electron optical components consists of a very finely regulated high-voltage power supply for the electron gun and a series of highly regulated low-voltage power supplies for the lenses. Any fluctuation of either the accelerating voltages or the lens currents will change the focal lengths of the lenses and will result in the blurring of the image on the screen, bringing about a loss of resolution. To attain a resolution of 10 Å, for example, the high-voltage circuits must be stabilized to one part in 10,000, the condensor lens current to one part in 1000, the objective lens to five parts in 100,000, and the intermediate and projector lenses to one part in 1000. These and even finer regulations are being achieved in present-day instruments through the use of solid-state circuitry.

A good vacuum is a necessity inside the column of an electron microscope since the mean free path of an electron in air at atmospheric pressure is on the order of about a millimeter. The column of a microscope is liable to be somewhat over a meter long and in order to insure that most of the electrons that leave the electron gun reach the specimen and the focusing screen, it is necessary to remove anything from their path that they might collide with, such as air molecules. This is achieved through the use of a high-speed oil diffusion pump backed up by a rotary mechanical pump.

Current practice is to maintain a pressure as low as 10^{-5} mm of mercury. Most contemporary instruments use systems of air locks on the specimen chambers and electron guns and cameras, so that the frequent changes of equipment in these areas do not necessitate the venting of the entire column to the atmosphere, which would slow down operation and greatly increase the contamination rate of the column.

The method for recording the electron images is photography. This is accomplished by fitting the microscope at some convenient level below the projector lens with a shutter, and a means for projecting the electron beam onto a photographic plate or film. This is usually achieved by placing a plate under the screen, swinging the screen out of the way and allowing the beam to strike the plate for the proper exposure, which can be either manually or electronically timed.

Some instruments also include 35 mm rollfilm cameras above the screen for low-magnification scanning of specimen grids and preliminary examinations. An ideal photographic emulsion for this purpose would be a thin, contrasty, very fine-grained emulsion, such as Kodak lantern slides, Kodak p426, Ilford E. M. plates and certain of the Gavaert emulsions.

The applications of the present-day electron microscope are wide. Transmission electron diffraction and reflection electron diffraction can be done on most units. In transmission diffraction the electrons are diffracted as they pass through the specimen giving rise to diffraction rings which can be measured and the material from which the rings originated can be identified. Either the entire specimen or an area as small as one micron can be studied in this way. With reflection diffraction thick specimens such as metal surfaces or bulk crystals can be studied. For the biologist minute cellular constituents can be identified. Reflection microscopy and dark field microscopy can similarly be done. Through stereo electron microscopy an accurate three-dimensional view of the specimen can be constructed. By inserting a target

in the path of the beam X-ray analyses of the specimen can be done. Particle size studies, surface analyses of solids, roughness tests and direct examination are but a few of the many applications. Through devices which heat, cool and stretch the specimen dynamic changes can be studied. Materials that cannot be sectioned can be replicated, specimens that are not electron-dense can be shadowed with a heavy metal and be rendered visible.

The scanning electron microscopes, which can produce images of the surfaces of specimens of extreme depth of field, even greater than the light microscope, have found some use in biology, but their greatest present use is in the field of semiconductor research and development.

The laboratory facilities necessary to the operation of an electron microscope must consist of the following three units: the preparation laboratory, the microscope room and the darkroom in which the photographic work is done. The preparation laboratory should be large enough to house a refrigerator, an oven, an evaporating unit, a microtomy table, a sink and a light microscope table in addition to the other routine materials found in any biological laboratory.

The size of the microscope room will depend upon the size of the unit purchased. Air conditioning is a most desirable feature since the dehumidified atmosphere will go a long way toward cutting down the time required for the microscope vacuum system to function. The room should be light-tight. The addition of an extension telephone in this room will assure that the operator retains his dark adaptation should contact with him be desired. A low-level room light switch close to the operator is convenient for changing specimens and plates.

The darkroom will contain the essentials of any darkroom, and size can be determined from any of the readily available darkroom guides. It is preferable to have the darkroom, or at least that portion of it where the negatives are developed, adjacent to the microscope room. In this way, the operator of the microscope can develop his negatives without losing dark adaptation. The enlarger used should be of the condenser type for maximum clarity and the optical system of the best quality.

Generally speaking, electron microscopes fall into four categories: the low-cost ($10,000–20,000) instrument with a resolution range of 25–50 Å for the range between light and electron microscopy; the intermediate range instrument ($25,000–35,000) with a resolution range of 10–20 Å, which is suitable for most work; and the high-powered ($40,000–60,000 instrument for specific research problems that require the ultimate in resolution below 5 Å. The lowest-range instruments make excellent "first" microscopes since the biologist familiar with the light microscope can make the transition to electron microscopy painlessly. As the desire to go to a better instrument arises, the first instrument can find a place as a screening and teaching microscope to allow the selection of the best samples for observation on the larger unit. Most of the microscopes in use today fall into the intermediate category and are suitable for all but the most critical work. The fourth category of microscopes is the scanning electron microscopes which range in price from $80,000–120,000 at this writing.

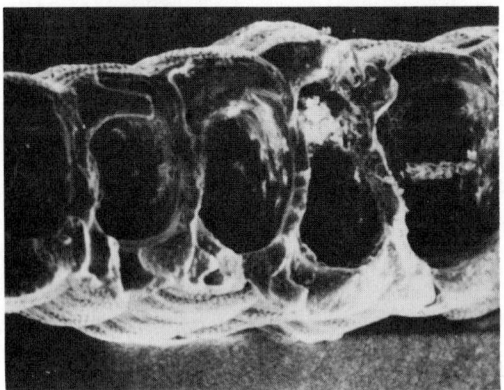

Fig. 3 Scanning electron micrograph showing a transverse section of a foraminiferan test (600 ×). (Courtesy of B. Akpati.)

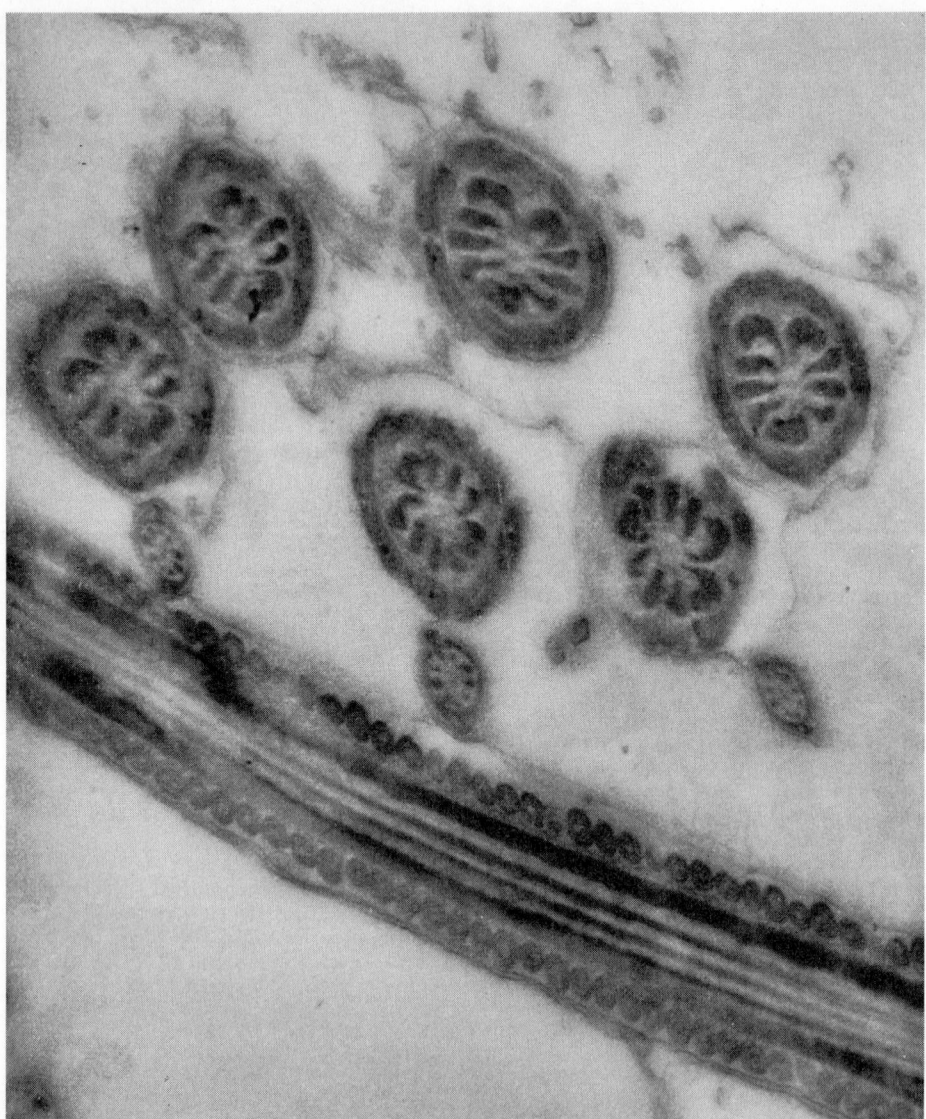

Fig. 4 Transmission electron micrograph showing cross sections of mouse sperm tails and one longitudinal section (magnification 45,000 ×).

Microscopes have progressed to the point where a skilled technician is no longer necessary for routine maintenance. Cleaning of the column and alignment can be carried out by most operators with a minimum of instruction from the manufacturer. Under average conditions an instrument should be operable 70–80 per cent of the time. Operation has been simplified to the point where the layman should be able to effectively operate his own instrument after only a few hours of instruction by the manufacturer.

ALAN R. CHRISTOPHER

References

Clark, G., "Encyclopedia of Microscopy," New York, Reinhold, 1961.

Hall, C., "Introduction to Electron Microscopy," 2nd ed., New York, McGraw-Hill, 1966.
Sjöstrand, F. S., "Electron Microscopy of Cells and Tissues," vol. 1, New York, Academic Press, 1967.
Wischnitzer, S., "Introduction to Electron Microscopy," New York, Pergamon Press, 1965.

ELECTRONIC INSTRUMENTATION

The modern biologist uses much electronic apparatus and needs to judge their appropriateness and limitations though, not necessarily, to understand circuit detail. Biological electronics is less peculiar in general principle than in detailed specification, particularly at the input, since wanted electrical signals are often extracted from minute, hardly accessible, regions buried within electrically inhospitable media, in the presence of man-made or biologically generated interference.

A generally uniform technique is used in electrophysiology to measure patterns of electrical change accompanying biological activity in diverse fields; the peripheral and central nervous systems, skeletal and smooth muscle and auditory, tactile, thermal and visual receptors.

Exploring electrodes detect potential changes (referred to some 'inert' part of the structure) ranging in amplitude from microvolts in, say, extracellularly recorded nonmedullated nerve impulses, to three or four orders of magnitude greater in, for example, intracellular investigation of single nerve or muscle cells. Even in given tissue, the potential detected depends on the nature of the exploring electrodes and the extent to which inactive tissue insulates, or tissue fluids shunt, the active region. Time course may be rapid—nerve cell spikes of duration a fraction of 1 msec. repeating at rates up to 1,000/sec. or more[1,2]; slow—the E.C.G. containing important components at 1 cycle/sec;[3] very protracted—mytilus muscle tension and membrane potential changes occupying 10–60 seconds;[4,5] or unvarying for still longer periods—the resting polarisation of the endolymph of the cochlea scala media, changing only when metabolic processes change.[6]

There varied signals are amplified sufficiently to attain a permanent record either by pen recording or photographing a cathode ray tube screen. The former requires appreciable current to work satisfactorily (depending on the frequency response needed), the latter tens of volts—the price paid for rapid response in a recording system being low sensitivity. The pen recorder or cathode ray tube are thus invariably preceded by electronic amplifiers.

Superimposed on the record are required: time signals to measure duration and time pattern of activity; calibration signals, of known amplitude and shape, applied to the input to assess both magnitude and fidelity of reproduction of the detected activity; electrical signals from transducers delineating non-electrical responses (length or tension change in muscle, for example).

Activity may be spontaneous or, more often since more easily interpreted, provoked by controlled external stimuli. If the latter, a display of magnitude, duration and time of application of the stimulus is necessary.

Amplification

Suitable instruments are manufactured commercially. An oscilloscope with amplification variable to 10^5 and frequency response uniform from d.c. to 500 kc/sec suffices, supplemented by an a.c. preamplifier of variable amplification (to 10^3), preferably with adjustable frequency response. This latter is set to the minimum band-width needed to amplify wanted signals with tolerable fidelity, yet maximally rejecting interference and noise outside the signal's frequency spectrum.

D.c. amplifiers are well understood and preferred over resistance-capacitance coupled amplifiers for low frequency phenomena.

The initial amplifier stages, and preferably all, are balanced symmetrically about ground to discriminate better the biological activity (arranged to occur out-of-phase at the two input terminals) from interference (which, by suitably locating the ground to the preparation, is made to affect each input in-phase, and at similar amplitude). In balanced amplifiers, in-phase amplification is heavily degenerated. Out-of-phase signals enjoy full gain. In difficult situations, the preparation earth is connected in a bridge network and the in-phase interference amplitudes more closely adjusted to equality.

D.c. drift is more easily minimised in balanced amplifiers than in single-sided, even-harmonic distortion largely cancels at the output automatically, while odd-harmonic distortion remains amenable to circuit cancellation. High fidelity reproduction is thus facilitated.

(In severe conditions the experiment is carried out in a fully screened 'box' with only the preparation and necessarily adjacent apparatus—including sometimes the biologist—inside. Such a screened room is not indispensable. It is always worthwhile to feed mains power into the experimental room via a 1:1 ratio isolation transformer, with secondary centre-tapped to ground and electrostatically screened from the primary. Mains borne interference is then much reduced.)

An input "cathode-follower" preamplifier is almost universal. Glass microelectrodes with tip diameters 0.5μ or less, commonly used to detect potentials close to or within cells, have high series resistance (10–100 Megohms). They must not be connected directly to an amplifier with low input resistance, or the signal amplitude at the amplifier would be much less than at the electrode tip, nor to one with appreciable input capacity, when only slow changes could be followed.

Cathode follower circuits have the necessary high input resistance, and low input capacity. The extra noise from a tube used like this is small and a quite negligible disadvantage far outweighed by the circuit's beneficial transforming action. The output impedance is low, allowing signals which would otherwise be severely attenuated and distorted to be fed along cable to a distance.

The price paid is a voltage gain of less than unity. The device is placed close to the preparation—the short lead to the electrode itself being screened by an encircling earthed, and/or cathode-connected, conducting shield.

The input tube must be selected for low grid current i_g (i.e. it must not exchange current from its own grid with the electrode and preparation); 10^{-10} a. can alter tissue excitability, given time, and in d.c. recording a change in electrode resistance, ΔR, will cause an apparent change, ΔRi_g, in recorded d.c. level. Grid currents $< 10^{-12}$ a. are possible with trouble and expense but a search among mass produced tubes is profitable. Some good audio amplifier tubes have $i_g \leq 10^{-11}$ a. when properly used. For a.c. measurements (grid isolated from electrode by condensers) i_g is of little importance.

To measure d.c. potential levels absolutely, rather than d.c. change, electrometer tubes may be necessary at the input. These have very low i and very high input impedance. True electrometer tubes are noisy, require careful circuit arrangement, and a lot of good biology is done without them. Their use should be limited to investigations where nothing else will do. However, the convenience of siting most of the electronics far from the preparation makes a cathode follower worth while even with low impedance electrodes. If extremely faithful reproduction of fast potential changes detected with high impedance electrodes is essential, positive feedback ("negative capacity") applied to the preamplifier input grid reduces the effective input capacity. Such circuits are somewhat noisy; initial stray capacity must still be kept small and adjustment of the positive feedback must somehow be monitored continuously.

Display

Given these amplifiers, all biologically generated signals from a few microvolts (a.c.) and ca. $100\mu v$. (d.c.) would yield cathode ray tube Y deflections of 2–3 cm (or, using power-amplifiers, equivalent pen recorder deflections, provided the frequency is low, < ca. 200 c.p.s.). The c.r.t. is versatile, easily adapted to control external circuits and much the best recording system, unless it is essential to monitor the permanent record immediately. It is the only satisfactory display for rapid signals.

Time base speeds variable from 10 cm/m sec to 1 mm/sec cover most needs. An extra convenience is automatically increased brightening during rapid Y deflections, particularly when photographing infrequent, fast Y signals on a slow time base.

A circuit 'expanding' a chosen postion of the time base within a single beam, or on the adjacent beam in a double-beam display, is useful; the slower trace showing steady or infrequent activity, while interesting events are singled out for examination in detail, more spread out in time.

Effective multiple-beam operation of a single beam c.r.t. can be achieved by electronic switching at the input, separating signals from each channel spatially in the Y axis. This fails when the switching time approaches the duration of the fastest signal displayed. A multiple gun tube is preferred. A short persistence blue 'photographic' phosphor is essential; after-glow screens aid visual inspection but the poorer photographic focus for a given actinic brightness cannot often be tolerated. Post deflection acceleration (P.D.A.) tubes should give the best ratio, beam brightness/deflection sensitivity, for a given sharpness of focus. However, in any amplification-display system. it is the overall signal/noise ratio and not sensitivity per se which is important and this is determined at the input and early amplifier stages.

Stimulation

(a) **Electrical stimulators** are designed to ease the application of predetermined patterns of stimuli when the biological preparation demands constant attention or the phenomenon is only briefly observable. The basic circuit elements are often conventional, variety arising in their interconnection to suit particular needs. A good stimulator incorporates two or more independent channels, providing rectangular current or voltage pulses, either positive- or negative-going, of independently variable amplitude (0–10 m.a. *or* 0–100v) and duration (20 μsec–500 msec). Stimuli are delivered singly, in pairs, or in 'tetanus' trains of controlled time duration (10 msec–10 sec) and pulse repetition rate (1/sec to 1,500/sec). Two single pulses (or a single pulse then a 'tetanus' train) can be successively applied with controllable delay between the two events. The two channels may feed separate stimulating electrodes or be mixed at a common electrode.

Stimulating pulses are often much larger than the electrical responses. Any residual stimulus artefact at the balanced amplifier outputs is further reduced by isolating the stimulating electrodes from earth by pulse transformers, or radio frequency links, inserted between the stimulator and the electrodes themselves.

(b) **Mechanical stimulators,** for investigating tactile sense organs, use piezoelectric or electromagnetic transducers operated by an electrical stimulator. The mechanical impulses follow more or less faithfully the electrical waveforms. Long duration stimuli or stimuli which need to do work against a considerable load are well applied by electromagnetic vibration generators which incorporate also a servo feedback circuit from a position detector. A voltage from the position detector is compared with a voltage corresponding with the required movement; any difference between these signals causes the vibrator to generate a force proportional to its magnitude. Error between the stimulus (mechanical displacement) actually present at any time and the desired stimulus is thus annulled. To effect this control rapidly enough against powerful opposing forces, as in muscle, necessitates the use of very powerful electromechanical vibrators and a very rigid mount for the mechanical stimulator and for the preparation.[7,8]

(c) **Acoustic stimulators,** using high quality loudspeakers or earphones, provide controllable patterns of sound stimuli, initiated as in (a), which are not always pure pulses ('clicks') but, frequently, pulsed a.f. carrier waves with controlled rate of onset, removal and duration, of constant audiofrequency and amplitude modulated, or of fixed amplitude and frequency-modulated.[9] The main problem is generation of an acoustic output free from harmonic distortion—particularly when the fundamental lies in an insensitive, the harmonic in a sensitive, region of the auditory system.

(d) **Optical stimulation** can be initiated as in (a), the difficulty here being to generate quantitatively repeatable light impulses by electrical control—most light sources which can be pulsed having rather unstable light output. Where the quantity of light per stimulus needs to be determined precisely, it is monitored photoelectrically. Alternatively, electrically initiated mechanical interruption of a constant standard lamp source is used. The shortest stimulus duration then possible is limited by inertia in shutters etc. to about 1 msec. Light amplitude control is always better effected by optical wedges than by current variation in the source—this would often concurrently change its colour spectrum.

(e) **Thermal stimulation** is difficult to control electronically, though infrared radiant heating is controllable as in (d). The very successful 'thermode' of Zotterman's school (hitherto the only pulsed 'cold' stimulator), operated by water streams, requires the recording apparatus to be timed with reference to the stimulator. Recent semiconductor developments promise success for a thermoelectric 'cold' stimulator, controlled electronically,[10] with the convenience which that implies. Microwave heating is directly controllable electronically, and the fact that energy penetration into the tissue varies from about 1 cm at 2,500 Mc/sec to 0.1 mm at 100,000 Mc/sec., is sometimes advantageous.

Time Marking

Crystal-controlled time signals (steep-sided voltage pulses) are easily generated with an accuracy 1 in 10^4 and better. With modern crystals, the primary standard frequency may be 10 Kc/sec (faster time marks are rarely needed in biology). This frequency is 'counted down' to submultiples by cold-cathode glow-discharge counting tubes. Reliability is all-important and these have a longer life than thermionic tubes with heaters and either divide properly or fail completely; any fault is immediately noticed. A central time marker can serve a whole laboratory, signals being 'piped' around from low output impedance circuits like the cathode follower.

Over-all Control of Stimulation and Recording

In the final display it is preferred that 1) the time marks will remain stationary on the time base from sweep to sweep. Therefore the time base repetition rate must be controlled by the standard time marker. Fixed submultiples of the available timing frequencies trigger each sweep. 2) That stimuli are applied once per sweep of the c.r.t. time base and are fixed stationary at selected spatial positions along that time base, whatever its repetition rate or velocity. Stimulators are therefore directly coupled to the time base waveform generator and triggered when this reaches selected voltage levels. Stimulus monitoring signals remain stationary on the c.r.t. from sweep to sweep; inter-stimulus 'delay' is the time appropriate to spatial separation along the time base, is varied by varying time base velocity, and enjoys the full linearity, accuracy and repeatability of the time base waveform.

This method of overall control, valuable in any circumstance, is essential if many successive c.r.t. sweeps are photographed, superimposed on one photographic frame, to increase the observing time and better discriminated weak signals from a noisy background.

(*Transistors*, rapidly increasing in importance because of their smallness and low power requirements, are well adapted to on-off applications, as in stimulators, counters etc. The high input impedance and stability required for amplifiers are available, except for input stages where noise is still a problem.)

Detectors of Non-electrical Activity

(a) Rapid *mechanical vibrations* are detected by the piezoelectric effect. Ceramics like barium titanate are particularly sensitive. Interruption of a light beam impinging upon a photocell works from d.c. to many kilocycles, imposing no load upon the vibrating system,[8] as does also the use of variable inductance transducers in which an electrical inductance is changed by mechanical displacement of a moving core within the inductor. This inductance is part of a tuned circuit, and the change in circuit tuning by the core displacement is converted into an electrical signal which follows both steady and alternating change in position of the core.[7] *Tension* is well measured by the RCA 5734 transducer—a small triode with a moveable anode. The anode pin is deflected via a cantilever upon which the muscle pulls. Signals of several volts for 10^{-2} mm pin deflection are usual. Natural frequency of the device itself is high but

limited by mechanical arrangements needed to convert large tension (or displacement) to small displacement of the anode pin. Sensitivity is varied electrically or mechanically—the jet force of a dragonfly nymph or the maximal tension of mammalian muscle being equally well measurable.

(b) *Pressure* and differential pressure, in gases (respiration) or liquids (circulation), are transduced into electrical responses by making the pressure-detecting diaphragm one plate of a condenser (diaphragm distortion varies the capacitance; electronic circuits convert this change into potential change for c.r.t. deflection) or by attaching the diaphragm to the 5734 transducer.

(c) *Light* output is detectable quantitatively by the electron-multiplier photocell—with sensitivity approaching the absolute sensitivity of the eye as a detector of weak light. The method has been exploited in observation of luminescent animals and in biochemical assay.

(d) *Sound* detection employing conventional air or underwater microphones is used in ornithology and marine biology.

(e) Remote recording of *temperature*, etc., is eased by potentiometer recorders. When balanced these absorb no current from the source (e.g., a resistance thermometer), and lead-length correction is abolished. Thermistors are often used for temperature measurement and its control, their high temperature coefficient of resistance and small mass giving increased sensitivity.

Microiontophoresis

Techniques have been perfected for introducing drugs or naturally occurring ionised substances accurately localised upon parts of a single cell membrane from within or without the cell. A solution of the substance, in ionised form, is contained within a glass micropipette of the size used for impulse recording, and controlled amounts of ion are ejected by applying an electrical potential gradient along the electrode in the appropriate direction; if the interior of the electrode is positive to the external environment near its tip then cation is ejected and vice versa.[11,12] The technique has had resounding success in elucidating chemical transmitter action at the neuromuscular junction[13] and at central nervous neuronal/neuronal synapses.[14] Although first developed a decade ago, there is little sign of its utility becoming less.[15,16]

Models and Computers

The central nervous system is in some respects analogous to electronic computers—though containing multitudes more computing elements than any machine and seemingly able to "program" itself. Since theoretical analysis of computer-like systems containing even few interconnected elements is unendurable—especially with threshold and safety factor linkages—relatively simple electronic models can be very illustrative.

Electronic computers themselves deal successfully with the successive approximation numerical integrations necessary in analyzing the behavior of systems, like the nerve membrane, where the numbers of parameters involved is large but the experimental observations sufficiently accurate to allow a reasonable hope of solution. A skilled biologist may program the machine to test the possibility of a new experiment.[17]

An essential to remember in constructing mathematical (or actual electronic) models is that the interconnections in real biological control systems are often very much more complicated in number and adaptability than even the most elaborate systems of the engineers—guided space rockets, for example—and in some cases it is unprofitable too facilely to equate the engineering analogue to the biology. Nevertheless there is a great utility, particularly for individual parts rather than the whole system, in establishing engineering analogues as an aid to understanding and a guide to future experiments[18, 19] but, at present, painstaking and critical experiment is still most needed, upon which bedrock the good theoretician can whet and the indifferent blunt his mathematical sword.

It is rare that either the physicist or the biologist makes, alone, a significant contribution to the other's field. Close experimental collaboration ensures suitable instrumentation for use in biology.

R. H. KAY

General References

Bures, J., *et al.* "Electrophysiological methods in Biological Research," Prague, Publishing House of the Czechoslovak Academy of Sciences, 1960.

Donaldson, P. E. K., "Electronic Apparatus for Biological Research," London, Butterworth, 1958.

Kay, R. H., "Experimental Biology," New York, Reinhold, 1964; Science Paperback 19, London, Chapman & Hall.

Nastuk, W. L., ed. "Physical Techniques in Biological Research: Electrophysiological Methods," Vol. 5 (1964), Vol. 6 (1963) New York, Academic Press.

Whitfield, I. C., "An introduction to electronics for physiological workers," London, MacMillan, 1953.

Text

(1) Hodgkin, A. L., Croonian Lecture "Ionic movements and electrical activity in giant nerve fibres," Proc. Roy. Soc. B. **148**:1, 1957.

(2) Phillips, C. G., "Actions of antidromic pyramidal volleys on single Betz cells in the cat," Q. J. expl. Physiol. **44**: 1, 1959.

(3) Suckling, E. E., "Some features of American electrocardiograph design," Electron. Engng. **24**: 243, 1952.

(4) Abbott, B. C. and J. Lowy, "Contraction in molluscan smooth muscle," J. Physiol. **141**: 385, 1958.

(5) Guttman, R. and S. M. Ross, "Effect of ions on smooth muscle response to cooling," J. gen. Physiol. **42**: 1, 1958.

(6) Davis, H., "Biophysics and physiology of the inner ear," Physiol. Rev. **37**:1, 1957.

(7) Matthews, P. B. C., "The differentiation of two types of fusimotor fibre by their effects on the dynamic response of muscle spindle primary endings," Q. J. expl. Physiol. **47**: 324, 1962.

(8) Machin, K. E. and J. W. S. Pringle, "The physiology of insect fibrillar muscle. Mechanical properties of beetle flight muscle," Proc. Roy. Soc. B. **151**:204, 1959.

(9) Whitfield, I. C. and E. F. Evans, "Responses of auditory cortical neurons to stimuli of changing frequency," J. Neurophysiol. **28**:655, 1965.

(10) Gordon, G. and R. H. Kay, "A thermoelectric cold stimulator," J. Physiol. **153**: 3P, 1960.

(11) Nastuk, W. L., "Membrane potential changes at a single end plate produced by transitory application

of acetylcholine with an electrically controlled microjet," Fed. Proc. **12**:102, 1953.

(12) Krnjevic, K., *et al.* "Determination of iontophoretic release of acetylcholine from micropipettes" J. Physiol. **165**:421, 1963.

(13) Katz, B., "The transmission of impulses from nerve to muscle and the subcellular unit of synaptic action," Croonian Lecture. Proc. Roy. Soc. B. **155**:455, 1962.

(14) Eccles, J. C., "The Physiology of Synapses," Berlin, Springer, 1964.

(15) Katz, B. and R. Miledi, "The timing of calcium action during neuromuscular transmission," J. Physiol. **189**:535, 1967.

(16) Miledi, R. and C. R. Slater, "The action of calcium on neuronal synapses in the squid," J. Physiol. **184**:473, 1966.

(17) Huxley, A. F., "Can a nerve propagate a subthreshold disturbance?," J. Physiol. **148**:80P, 1959.

(18) Grodins, F. S., "Control theory and biological systems," New York, Columbia University Press, 1963.

(19) Milhorn, H. T., "The application of control theory to physiological systems," Philadelphia, Saunders, 1966.

ELECTROPHORESIS

Electrophoresis concerns the migration of charged particles in an electric field and is a particular aspect of the more general phenomenon known as electrokinetics. A rather arbitrary distinction is sometimes based on the size of the migrating body, the term electrophoresis then being used for the motion of macromolecules and collidal particles and ionophoresis for the motion of small ionized molecules. Cataphoresis is an older term for electrophoresis. In the last twenty-five years electrophoresis has been developed into a most potent technique for the resolution of mixtures, the isolation, and the identification of a wide variety of substances of biological interest.

When charged particles or molecules are suspended or dissolved in a conducting medium and a potential gradient established through the medium, the particles or molecules will move towards the electrode of opposite charge. The rate of movement will depend among other things on (i) the net charge on the particle or molecule, (ii) the potential gradient and (iii) the resistance of the medium to movement of the particle, determined by the size and shape of the particle and the viscosity of the medium. Most charged substances of biological interest possess groups which ionize in the pH range 1–9. The degree of ionization of any group and hence the net charge on the molecule will depend on the pH of the conducting medium. For amphoteric molecules, possessing both positive and negative charges, there will be a pH value typical of the molecule known as the isoelectric point where the net charge is zero. Substances in a mixture which may migrate with the same mobility at any one pH will often separate at a different pH. Thus in studying unknown substances or mixtures it is usually advisable to carry out electrophoresis over as wide a range of pH values as is compatible with the solubility and stability of the material. This not only allows the greatest chance of resolving any possible mixture but may also give some indication of the nature of the ionizable groups on the molecules. Substances which do not themselves possess charged groups (e.g. hydrocarbons) may show migration in a field due to the preferential adsorption of charged groups of one type (e.g. OH^-) from the medium. Similarly, charged particles may have their net charge altered by the complexing or adsorption of ions from the medium.

A wide variety of experimental methods have been developed for applying the principle of electrophoresis to materials of biological interest. There are two main types of technique—the classical moving boundary method developed largely by Tiselius, in which the electrophoresis takes place in free solution, and zone electrophoresis in which the conducting fluid is supported in some stationary medium such as filter paper, cellulose powder or starch, or where electrophoresis takes place through a liquid density gradient made from some non-electrolyte. Electrophoresis in free solution is usually carried out in some form of U tube. The material to be studied, dissolved in an appropriate buffer solution, is placed in the lower part of the U tube and is then overlaid with buffer only. A direct current is then applied through the U tube and the particles in solution migrate according to their net charge. This migration gives rise to ascending boundaries in one limb of the tube and descending boundaries in the other. Each component of the mixture will migrate independently and provided the mobilities are sufficiently different, each will give rise to a separate boundary. At each boundary there is a change in refractive index and this change, which is proportional to the concentration of the component producing the boundary, may be detected by a suitable optical system of the schlieren or interferometric type. With the more commonly used schlieren system, electrophoretic analysis of a mixture gives rise to a pattern comprising a number of peaks, the area under each being proportional to the concentration of the corresponding component in the mixture. In contrast to zone electrophoresis, moving boundary electrophoresis is only suitable for use with large molecules (m. wt. > 10,000). It is most valuable in analysing complex mixtures of labile biological substances such as proteins. As a preparative method it suffers from the disadvantage that only the fastest and slowest migrating components may be isolated in a pure form and recovery of these is incomplete.

A great advantage of zone electrophoresis where the separation is carried out in some supporting medium is that complete physical separation of the components in a mixture can frequently be obtained. After electrophoresis spots or zones of colourless materials are located by treatment with some suitable reagent or by some other device, e.g. photography or visual examination in ultraviolet light for compounds which absorb or fluoresce in this region, or by bioassay for compounds which stimulate or inhibit growth. For separations on a relatively large scale columns or blocks of starch or purified cellulose fibres have been employed. For many purposes the most popular supporting medium has been filter paper. Acrylamide gel is becoming increasingly popular as a supporting medium for electrophoresis of macromolecules, used either in cylindrical tubes or as a thin layer on plates. The separations obtained are based partly on charge and partly on size and shape of the molecules. The combined use of two dimensional chromatography and electrophoresis on thin layer plates now provides effective micro procedures for the fractionation of such complex mixtures as peptides and oligonucleotides resulting from the partial degradation of proteins and nucleic acids. For example, a "fingerprint map" of the peptides resulting from tryptic digestion of a protein can now be obtained with as little as $50\mu g$ of protein digest.

R. E. F. MATTHEWS

Reference

Bier, M., ed., "Electrophoresis," New York, Academic Press, vol. 1 1959, vol. 2 1968.

ELECTROPHYSIOLOGY

Electrophysiology is the investigation and attempted explanation of the electrical properties of living cells and intracellular substances, and the correlation of these with physiological function.

This discipline arose in the early 18th century (Hoff). That muscle and nerve could be electrically stimulated and that certain eels and rays could duplicate these effects was well known before the Galvani-Volta discussions during and after 1791. Galvani demonstrated more clearly than the others that living tissues could produce electricity. Some aspects of electrophysiology, such as alternating current impedance (Schwan, 1959), thermal effects of imposed currents, and steady state membrane potentials can be studied in any cell. Variation of the membrane potential upon stimulation, however, has been investigated mostly in nervous tissue, muscle (striated, cardiac, smooth) and glands; these are the so-called excitable tissues. There is no cell whose membrane potential cannot be modified, so the term excitable is relative. In excitable cells a response may be obtained by light (retina), mechanical stimuli (cochlea, Pacinian corpuscle, stretch receptors, the axon itself), temperature variation (thermal receptors), chemicals, and electrical stimuli. The electrical variations may trigger additional events such as muscular contraction or glandular secretion. Electrophysiologists may therefore study (1) passive electrical properties, not necessarily of a linear nature (2) electrical membrane phenomena (3) electrical field effects of masses of cells (electrocardiography, electroencephalography) (4) responses to electrical stimulation, and (5) effects of drugs and other chemicals. It is not possible even in summary to discuss the many fields of study, so discussion must be limited to the basic mechanisms. Instruments and circuit diagrams can not be treated here.

Studies on many types of animal and plant cells have revealed surprising uniformity of electrical parameters. All cells have a physiologically active surface membrane of about 5 to 10 millimicrons in thickness across which a steady state potential difference of 50 to 100 millivolts is maintained, with the inside negative to the outside. One cm^2 of membrane has a resistance between 1,000 and 10,000 ohms, and a capacitance of about 1 microfarad. The resistivity is thus about 10^9 to 10^{10} ohm cm., the dielectric constant is approximately 5, there is an electrical phase angle of about 75°, and there is a potential gradient of about 100,000 volts per cm. Recently, a surface charge of about 2 to 4×10^{-8} coul per cm² has been measured in giant axons of invertebrates.

Conceptual advances have been linked with advances in instrumentation. Workers prior to Galvani used crude instruments which were insensitive and slowly reacting. Both Galvani and Volta used the electroscope for physiological experiments. They were not able to locate the site of production nor the magnitudes of biological voltages. Volta discovered that stimulation of a motor or sensory nerve elicits the proper function of the nerve. Nobili (1827) designed the astatic needle galvanometer and was able to measure the current of injury the same year that Ohm published his definition of resistance. Matteucci (1838) showed that a difference of potential exists between an excised frog's nerve and its damaged muscle, but it was du Bois-Reymond (1843) who described the decrease in injury currents (its negative variation) during contraction of muscle. He also proved the negative variation in nerve. These observations were important even though made on injured muscle. Although Helmholtz measured the speed of conduction of a nerve in 1852, and Bernstein by 1871 had demonstrated the presence of an action potential, the magnitude and time course were not known. Two additional advances were yet needed—faithfully responding equipment, and the use of intracellular electrodes. Sensitivity was increased with invention of the galvanometer by Sweigger in 1811 and the mirror galvanometer by Gauss and Weber in 1837. Speed of response was successively increased by Lippmann's capillary electrometer (1872), Einthoven's string galvanometer (1901), Matthews' magnetic oscillograph (1928), and the application of both vacuum tube amplification and the cathode ray oscilloscope by Gasser and his colleagues (1921 and later).

Many of the fundamentals of electrophysiology were established in the 19th century by du Bois-Reymond, Helmholtz, Gotch, and others, but recently there have been equally rapid advances. With better equipment, Gasser, Erlanger, and Bishop made many new discoveries concerning the properties of vertebrate nerve, but these were made with extracellular electrodes. Both Osterhout and Blinks inserted a capillary tube into algae to record from the sap vacuole. Hodgkin and Huxley (1939) and Cole (1940) used smaller capillaries to record intracellularly from the giant axon of the squid. They demonstrated the magnitude of the steady state membrane potential and showed for the first time that the negative internal potential does not merely fall to zero upon stimulation of the nerve but that the interior becomes positive. The positive overshoot of 30 to 50 millivolts has been shown to be a sodium concentration potential and follows closely the Nernst or Planck potential as calculated from the observed Na ratio (outside to inside) of 10 to 20:1. Graham and Gerard (1942) used true microelectrodes of the capillary type to record intracellularly the membrane potential of muscle. Many types of metallic microelectrodes have been described. However, the theoretical aspects of the use of microelectrodes are difficult, because the magnitude of the junction potentials of the electrode with cytoplasm of the cell is unknown. Gesteland et al. (1959) have given an excellent discussion of microelectrodes. A recent symposium (1968) was devoted entirely to the subject of "Bioelectrodes." In practice, the microelectrode is placed in the fluid near the external surface of the cell and a potential reading taken as a reference potential. The microelectrode is advanced through the membrane and there is a sudden jump to a more negative potential which is called the membrane potential. But the junction potential between the microelectrode and the extracellular fluid may not be the same as that between the same microelectrode and cytoplasm. By whatever magnitude the junction potential changes, the measured membrane potential will be in error. Since it is thermodynamically impossible to measure the junction potential alone, it is likewise impossible to measure the membrane potential alone. By making certain assumptions about ionic mobilities at the junction, the error of ascertaining the membrane potential can probably be reduced to a few millivolts. A pipette with less than 1 micron tip may have a resistance of 10 to 50 megohms. To prevent distortion of voltage changes occurring in a few microseconds, the input capacitance of the amplifier must be very small. With the recent development of compensated negative-capacity amplifiers of very high input resistance, amplifier distortion is not now a problem.

The steady state membrane potential. To maintain voltages and currents in a dissipative (resistive) system requires energy, and interference with metabolic energy production leads to loss of the membrane potential. After energy production has been blocked, the decline of membrane voltage may require many minutes or hours, and, if the block is removed, may be reversible. Such a dissipative system is obviously not in thermodynamic equilibrium. The nature of the coupling by which energy is utilized to maintain disequilibrium and thus an emf is currently being studied. It is accepted that metabolic energy is utilized to maintain the integrity and function of the membrane itself and to transport ions through the membrane against electrochemical gradients. Almost all ions which have been analyzed chemically in the cell have been found to differ in concentration inside and out. But to find a difference does not mean the ion is actively transported. It may be mechanically or electrically constrained. The constraint of even one type of ion will tend to cause redistribution of the freely diffusible ions by a Gibbs-Donnan mechanism. In a Donnan equilibrium, the electrochemical potential is the same on both sides and electrical energy can not be extracted from the system.

Bernstein in 1902 proposed that the membrane potential was a potassium concentration potential. Prior to this (1890), Planck had treated theoretically the potential difference at a diffusion boundary of aqueous phases with different electrolyte concentrations, and subsequently (1911) Donnan examined the case of phases separated by a membrane with at least one nondiffusible ion. The K concentration inside most cells is about 30 to 50 times that outside. The Planck equation for a single monovalent cation gives $E = RT/F \ln ([K]o/[K]i)$ and K ratios usually encountered give a calculated membrane potential of 90 to 100 m V. In most cases this is higher than the observed potential by several millivolts. For this and other reasons, a growing number of investigators believe K passively distributes itself in an electrochemical gradient set up by another mechanism (Lorente de No, Stampfli,

Eyring, Grundfest, Ling, and others). Every cell has been found to transport actively at least one ion, usually sodium, and many cells transport more. The transport of two ions may be coupled, so that interference with the transport of one will affect transport of the other. In one cell or another, almost every one of the light molecular weight ions has been shown to be transported actively, but only a limited number in any one cell. Since cells transport sodium actively toward the outside, many workers have identified the "sodium pump" with the source of emf. Eyring has shown that such an ion transport mechanism would be a source of emf and would maintain a membrane potential of the observed magnitude. The mechanism which serves as the ion "carrier" and the manner in which energy is injected into the carrier system remain to be demonstrated. Many believe that energy is contributed directly to the transport mechanism by adenosine triphosphate. The carrier is coupled to the ion, diffuses to the opposite membrane surface, unloads the ion, and diffuses back, to recycle. Enzymes necessary for the loading and unloading process are known to exist. Since the membrane is a double layer of lipid covered by a layer of protein on each surface, it has the correct properties for such activity. The low dielectric constant of the membrane would permit the carrier-sodium complex to be non-ionized, and a lipid soluble carrier could move freely through the membrane. Historically, Osterhout seems to have first suggested energy requiring active transport of ions.

Pumping out of sodium from the cell leaves the inside negatively charged, and there will be a tendency for positive ions to diffuse inward and negative ions outward. Such ionic conductances have been measured for many tissues by Hodgkin and Huxley, Curtis and Cole, Keynes and many others. For squid giant axon the K:Na:Cl conductances are in the ratio of 1:0.04:0.45. Since these are the ions which carry the majority of the current for nerve and muscle, Hodgkin and Huxley have developed a formula, based on one by Goldman, for the relation between membrane potential, ionic conductances, and ionic activities

$$E = \frac{RT}{F} \ln \frac{P_k[K]_o + P_{na}[Na]_o + P_{Cl}[Cl]_i}{P_k[K]_i + P_{na}[Na]_i + P_{Cl}[Cl]_o}$$

where P_k is membrane permeability (conductance) for K, and $[K]_i$ is inside K activity, etc. Although Goldman developed his theoretical treatment on the assumption of a constant electrical field within the membrane, and this was adopted in the Hodgkin and Huxley development, Harris has shown that the equations are correct for an arbitrary field distribution. Eyring has shown further that if the field is initially a series of potential jumps, ions diffusing through the membrane will accumulate in such a manner at the jumps as to produce a more nearly constant potential gradient.

For those ions moving under passive transport only, the fluxes of the various types may be different. These differences can be partly explained on a mechanical basis by assuming that the membrane has aqueous channels or pores of the same order of size as hydrated ions. Hydrated ions vary in size, and diffusion rates will differ in a constricted channel. In addition there may be fixed electrical charges near or in the pores which will repel ions of like charge and favor

those of unlike charge. This is discussed at length by Ussing, Teorell, Sollner, Mullins and others. Cole remarked that for a membrane resistivity of 10^9 ohm cm., the pores must constitute a very small fraction of the surface. In pores of ionic dimensions, diffusion of one ion type may modify the free diffusion of others.

Stimulation of the cell membrane. A membrane may respond primarily to chemical stimuli and be electrically inexcitable (electroplaque of the eel, subsynaptic membrane of neurons, muscle membrane of motor end plate region; see Grundfest), others may respond to many types of stimuli. The response may be a simple local graded change in membrane potential which is proportional to the stimulus and subsides without spread to the rest of the membrane. It may vary from a millisecond for nerve to many seconds in algae. A phylogenetically newer type of response is found in some membranes, but is always superimposed on the graded local response; this is a nongraded, all-or-none, propagated response of 70 to 130 mV which spreads to all areas of the membrane, and is the type which is used to transmit nerve impulses along the axon. There is a very large literature on electrical excitability of nerve and muscle (see Katz, 1939) and empirical formulae relating excitation to duration and strength of the stimulus have been developed by Nernst, Hill, Lapicque, Blair, Rashevsky, Monnier, and others; but the studies of Hodgkin and Huxley lead to more basic formulae. The latter workers in 1952 summarized studies on ionic movements in membranes. These studies, and similar ones by Cole, used the method of Marmont (1949) of electrically clamping the membrane and studying the resulting changes in ionic fluxes. If the inside of the membrane is made more positive, there is a surge of current into the membrane capacity requiring about 20 microseconds. Then an inward flow of sodium ions rises to a peak in a few hundred microseconds, but quickly falls again to a low value. This decrease occurring with membrane depolarization maintained is called sodium inactivation. About 0.5 msec after the initial voltage change, an outward current of potassium ions begins, rises to a peak more slowly than the sodium, and remains at a high level as long as depolarization is maintained. Since the rate of change of current during Na inactivation and during the rise of K conductance is proportional to the voltage change, the proportionality factor can be called an inductance. It amounts to as much as 0.4H in squid axon, but is temperature and voltage dependent. It increases three fold for a 10°C fall in temperature and decreases rapidly as membrane potential is increased. These changes depend on membrane potential only and are not affected by the current. Hodgkin and Huxley developed one formula for membrane current density at a specific membrane potential and another for the current associated with a propagated action potential. (See Hodgkin, 1964, and Katz, 1966.) Na conductance may be increased e-fold (2.7) by a membrane voltage change of 4 mV, while for K the figure is 5 to 6 mV. Using the formulae mentioned, almost all known functional properties of nerve can be calculated and predicted. It can be shown that weak stimuli which cause less than an 8 to 15 mV change in membrane potential will cause only a local graded response, but if the membrane changes by a little more the process is regenerative and complete depolarization with positive internal overshoot results. During an all-or-none response the membrane conductance may increase 40 fold, Na conductance may increase 500 fold and the total current may reach 3 mA/cm^2.

In a membrane (such as that of striated muscle underneath the motor end plate) which is not electrically excitable, membrane potential and conductance can be altered in a local graded manner by chemical transmitters (such as acetylcholine). If the membrane becomes sufficiently depolarized, it will electrically excite the surrounding muscle membrane, which is of a different nature, and a propagated all-or-none response spreads over the muscle fiber. Synaptic transmission of most neurons is of this type.

Conductance changes during alteration in membrane potential imply structural or charge changes in the membrane. Since each mV change in potential changes the voltage gradient of the membrane by some 1,000 to 2,000 V/cm., structure can be physically altered and fixed electrical charges modified. Membrane thickness must be unaltered for capacitance remains constant. Some workers (Nachmansohn) believe the initial event is chemical release (such as acetylcholine) which alters membrane structure, and conductance changes follow.

This discussion can only hint at the complexity of electrophysiology. Electrical studies have been made of heart, brain, smooth muscle, cornea, ureter, gastric mucosa, uterus, skin, ganglia, and salivary gland, among others. An adequate description of any one of these would exceed this one, but the basic electrical properties discussed here apply to most living systems.

FRANK HARRISON

References

Adrian, E. D., "The mechanism of nervous action," Philadelphia, Univ. Pennsylvania Press, 1932.

Bernstein, J., "Untersuchungen zur Thermodynamik der bioelektrischen Ströme," Pflüg. Arch. ges. Physiol., 92: 521–562, 1902.

Brazier, M. A. B., "The electrical activity of the nervous system," 2nd ed., New York, Macmillan, 1960.

Butler, J. A. V., ed., "Progress in biophysics and biophysical chemistry," vols. 1(1950), 2(1951), 3(1953), 6(1956) London, Pergamon, 1950–56.

Butler, J. A. V., ed., "Electrical phenomena at interfaces," New York, Macmillan, 1951.

Clarke, H. T., ed., "Ion transport across membranes," New York, Academic Press, 1954.

Cold Spring Harbor Symposia on Quantitative Biology, 17. The neuron. Cold Spring Harbor, New York, The Biological Laboratory, 1952.

Coster, H. G. L., and E. P. George, "A thermodynamic analysis of fluxes and flux-ratios in biological membranes," Biophys. J. 8: 457–469, 1968.

Davson, H. and J. F. Danielli, "Permeability of natural membranes," Cambridge, University press, 1952.

Eccles, J. C., "The physiology of nerve cells," Baltimore, Johns Hopkins Univ. Press, 1957.

Erlanger, J. and H. S. Gasser, "Electrical signs of nervous activity," Philadelphia, Univ. Pennsylvania Press, 1937.

Feder, W., ed., "Bioelectrodes," Ann. N.Y. Acad. Sci., 148: 1–287, 1968.

Gesteland, R. C., et al., "Comments on microelectrodes," Proc. IRE, 47: 1856–1862, 1959.

Goldman, D. E., "Potential, impedance and rectification in membranes," J. Gen. Physiol., 27: 37–60, 1943.

Grenell, R. G. and L. J. Mullins, eds., "Molecular structure and functional activity of nerve cells," Amer. Inst. Biol. Sci., Washington, 1956.

Glasser, O., "Medical physics," Chicago, The Year Book Publishers, 1944–1950.

Hecht, H. H., ed., "The electrophysiology of the heart," Ann. N. Y. Acad. Sci., 65: 653–1146, 1957.

Hodgkin, A. L., "The Conduction of the nerve impulse," Springfield, Thomas, 1964.

Hoff, H. E., "Galvani and the pre-Galvanian electrophysiologists," Annals of Science, 1: 157-172, 1936.

Johnson, F. H., et al., "The kinetic basis of molecular biology," New York, Wiley, 1954.

Katz, B., "Electric excitation of nerve," Oxford, Clarendon Press, 1939.

Katz, B., "Nerve, muscle, and synapse," New York, McGraw-Hill, 1966.

Krogh, A., "The active and passive exchanges of inorganic ions through the surfaces of living cells and through living membranes generally," Proc. Roy. Soc. B 133: 140–200, 1946.

De No, R. L., "A study of nerve physiology," Stud. Rockefeller Inst. Med. Res., 131: 1–496, 132: 1–548, 1947.

Murphy, Q. R., ed., "Metabolic aspects of transport across cell membranes," Madison, Univ. Wisconsin Press, 1957.

Nachmansohn, D., ed., "The physico-chemical mechanism of nerve activity," Ann. N. Y. Acad. Sci., 57: 375–602, 1946.

ibid., "Second conference on physicochemical mechanism of nerve activity and second conference on physicochemical mechanism of nerve activity and second conference on muscular contraction," Ann. New York Acad. Sci., 81: 215–510, 1959.

Nachmansohn, D., "Chemical and molecular basis of nerve activity," New York, Academic Press, 1959.

Nastuk, W. L. and A. L. Hodgkin, "The electrical activity of single muscle fibers," J. Cell. Comp. Physiol., 35: 39–74, 1950. Proc. Inst. Radio Engineers Biomedical Issue, November, 1959.

Shanes, A. M., ed., "Electrolytes in biological systems," Washington, Amer. Physiol. Soc., 1955.

Shanes, A. M., "Electrochemical aspects of physiological and pharmacological action in excitable cells. Part I. The resting cell and its alteration by extrinsic factors," Pharmacol. Rev., 10: 49–164. Part II. The action potential and excitation," Pharmacol. Rev., 10: 165–273, 1958.

Shedlovsky, T., ed., "Electrochemistry in biology and medicine," New York, Wiley, 1955.

Symposia of the Soc. for Exper. Biol., 1954, "No. 8, Active transport and secretion," New York, Academic Press, 1954.

Tasaki, I., "Nervous transmission," Springfield, Thomas, 1953.

EMBRYOLOGY

Embryology is the study of the embryo and its development. An embryo is a multicellular organism which arises most commonly by cleavage of and subsequent changes in a fertilized EGG. Development involves the emergence of new and varied differences (epigenesis) beyond those differences preexisting in the unfertilized and uncleaved egg (preformation). An embryo can also originate by cleavage of an unfertilized egg in those species characterized by natural PARTHENO-GENESIS and in those species whose eggs normally require FERTILIZATION but are capable of responding to application of parthenogenetic agents (*artificial parthenogenesis*). The role of the SPERMATOZOON in development of embryos appears to be limited to activation of the developmental potentialities of the egg and to contributing paternal GENES to the nucleus of the zygote, thereby making possible biparental inheritance. Since an embryo can develop in the absence of sperm in cases of parthenogenesis, the embryologist is more interested in the egg than in the sperm.

Although embryos are formed during the life cycles of both plants and animals, embryology, unless otherwise specified, is customarily considered to be a subdivision of zoology and hence deals with certain aspects of animal development. It is concerned primarily with those aspects of development which terminate in formation of the new individual as it appears at the time of hatching or birth. In some animals this new individual emerges in the form of a larva, and in such cases metamorphosis of the larva may be considered to be part of embryonic development. This sets a purely arbitrary limit to the period of development studied by the embryologist since many developmental processes occurring after hatching or birth or after metamorphosis are essentially no different than before. Such a restriction of the scope of embryology excludes the study of development of new individuals from asexual buds as well as the phenomena of REGENERATION. These and other aspects of development outside the realm of embryology proper belong to the broad field of developmental biology which has manifested considerable unity and prominence as a consequence of the developmental biology conference series of 1956.

Initially embryology was an observational science. Development of each kind of embryo was described and illustrated in great detail (*descriptive embryology*). Embryology was essentially a chronological study of the changing forms assumed by the embryo and its parts (*morphogenesis*) and of the visible changes in the cells constituting the tissues (*histogenesis*). As detailed morphological descriptions accumulated, the development of one species was carefully compared with that of others (*comparative embryology*) and certain common features in development were identified, although there were many differences in details which continue to bewilder the neophyte as well as the mature embryologist. The comparative aspect of embryology received a tremendous impetus as a consequence of Darwin's ideas about evolution and Haeckel's proclamation of the *biogenetic law* ("ontogeny recapitulates phylogeny"). Embryologists for many years after Haeckel examined embryos primarily to establish evidence of phylogenetic relationships. This at least gave a strong incentive for looking at embryos and many accurate observational data were collected which were later used to good advantage. Unfortunately at this time recapitulation was considered to be a sufficient cause for the succession of events found to be characteristic of embryonic development. Accordingly, no one sought for true causal factors in development until interest in Darwinism waned.

Interference with the normal course of development (*experimental embryology*) then emerged slowly as a new approach to embryology. The destruction of one blastomere of the two-cell stage of the frog's egg by Roux by pricking it with a heated needle resulted in development of only half an embryo from the remaining blastomere. This result seemingly verified Weismann's hypothesis that different blastomeres experience different developmental fates as a consequence of a differential distribution of nuclear material during cleavage. Shortly thereafter Driesch discovered that isolated blastomeres of the two- and four-cell stages of sea urchin eggs develop into entire larvae, a result directly opposed to Weismann's hypothesis. Such results led directly to the constriction experiments of Spemann on newt eggs which not only demonstrated the equivalence of the nuclei of blastomeres, but led eventually to the discovery of the phenomenal ORGANIZER capacity of the dorsal lip of the blastopore. Although the concept of embryonic induction first emerged from Spemann's earlier work on the relation of the optic vesicle to the lens, it surged to the forefront of biological attention with his discovery of the primary organization center of the amphibian gastrula. Once this latter discovery had been made, the way was open to explore the role of cellular interactions at all stages of development and in all kinds of embryos. This aspect of embryology continues today to intrigue embryologists and to challenge their skill in presenting questions to the embryo in such a way that greater insight into the nature of the role of cellular interactions can be gained. The length of the list of developmental transformations which depend upon cellular interactions continues to increase, but at the same time embryologists tend more and more to extend their analyses to biochemical and biophysical levels. Meanwhile the rapid developments in genetics have encouraged other embryologists to explore the role of genes in embryonic development, and the role of specific ooplasmic substances in development continues to receive attention. Embryology still seeks to discover what happens and where, but in addition it attempts to determine how and why it happens as it does.

The classical techniques of experimental embryology (extirpation, isolation, transplantation, centrifugation, parthenogenesis, hybridization, merogony, exposure of eggs and embryos to specific ions, etc.) have yielded rapidly to new techniques and approaches: transplantation of nuclei; dissociation of embryos or parts of embryos into individual cells and the study of their behavior during reaggregation; determination of nutritive requirements of tissues and organs; application of histochemical and cytochemical methods including the use of antibodies conjugated with fluorescein or with I^{131}; application of other immunological methods; modification of techniques of TISSUE CULTURE and organ culture for various purposes including insertion of millipore filters of known pore size between interacting tissues; attempts to induce differentiation of specific cell types by adding specific substrates to cultures of embryonic cells, by treating them with specific fractions of products of embryonic cells released into the culture media, or by treating them with specific fractions of products extracted from adult organs; controlling the genetic constitution of the gametes from which embryos originate, especially in insects and mammals; application of specific enzyme inhibitors and antimetabolites; use of radioactive isotopes as tracers and of ionizing radiations and radiomimetic substances; use of chromatography, autoradiography, electrophoresis and ultracentrifugation; use of electron microscopy, phase-contrast microscopy, polarization optics, microspectrography, microspectrophotometry; application of microrespirometry and microchemical methods; etc. As Oppenheimer has so aptly remarked, it is hardly possible to classify and neatly outline all the various methods currently used by embryologists to investigate all the varied methods used by embryos to achieve their ends of making highly varied adults! Actually many of these techniques have served to rekindle interest in descriptive embryology, but on a submicroscopic and subcellular level, rather than as tools for experimental embryology. Embryology is one of the most active fields of zoological research today in spite of the complexity of the problems with which it deals. Its dependence on progress in physics and chemistry for new approaches and new tools is clearly evident. Embryologists of the future must be trained adequately in chemistry, physics and mathematics if they are to take advantage of the ideas, as well as the methods, emerging from these fields and if they are not to be overwhelmed by the rapid progress made in their own field. At the same time they must constantly refocus their sights on the embryo itself and on the problems in developmental biology posed by the embryo if the myriad techniques and approaches at their disposal are to be used for the solution of fundamental problems of development.

It is helpful to review briefly certain stages in development which are characteristic of eggs of all animals. After these stages have been designated and characterized briefly, a general survey of the kinds of problems involved and the progress to date in the solution of these problems can be presented.

(1) *Preparatory phases:* development of eggs in the ovaries and sperm in the testes. (2) *Fertilization:* activation of the egg and all events leading up to union of the pronuclei derived from mature eggs and sperm. (3) *Cleavage and blastulation:* division of the activated egg into smaller and smaller cells called blastomeres and establishment of the special arrangement assumed by the blastomeres at the end of the cleavage period and prior to the onset of gastrulation, often around some sort of fluid- or jelly-filled cavity called the blastocoele. The spatial arrangement assumed by the first few blastomeres constitutes the cleavage pattern. (4) *Gastrulation* (or germ layer formation): individual or mass movements of blastomeres to form localized groups or layers called germ layers. The outermost cells constitute the ectoderm and the innermost ones the endoderm. A primitive digestive cavity, the archenteron, usually forms within the latter. In animals above a certain level of structural organization a third layer, the mesoderm, forms between ectoderm and endoderm. If a cavity arises within the mesoderm, it is called a coelom. (5) *Determination of fates of groups of cells* (establishment of organ-forming districts): the developmental capacities of small groups of contiguous cells become restricted until the prospective potency of the cells (what they are initially capable of doing) becomes restricted to their prospective fate (what they actually do in normal development). (6) *Organogenesis:* establishment of organ rudiments from organ-forming districts and emergence of the definitive gross form of the organs derived from these rudiments. Interaction and cooperation of re-stricted portions of two or more germ layers may be involved as well as the selective lodging of migratory cells (coaptation). (7) *Histogenesis:* cellular differentiation within those portions of germ layers involved in organ formation resulting in definitive histological structure of tissues and onset of the specific functional activities characteristic of each, including establishment of immune reactions. (8) *Growth:* attainment of definitive size of each part of the body; also includes the control of growth of each organ by its own products or by products of another organ of the same kind as well as by products of other kinds of organs (*hormone production*).

There is often considerable overlapping of these stages and one stage may merge imperceptibly into the next. Moreover, the sequence of stages is not necessarily the same in all animals. Ultimately embryology hopes to explain all these processes of development in terms of changes in populations of molecules making up the cells of the different tissues, but at the moment this goal is far from attainment.

Preparatory phases. These take place while the egg is developing in the ovary. It is during this period that the basic structure (organization) of the egg is somehow established under the direct influence of the maternal nucleus contained in the immature egg and under the indirect influence of maternal nuclei contained in the follicle or nurse cells which surround the immature egg and contribute substances to it. Polarity of the egg is established at this time. Bilaterality (dorsoventrality) is also frequently established then, at least in labile form. Apparently both polarity and bilaterality are properties of the egg cortex. This two-dimensional pattern in the egg cortex is frequently instrumental in causing specific kinds of cytoplasm to assume specific positions in the egg (segregation or localization of cytoplasmic or ooplasmic substances), and these different kinds of cytoplasm probably activate different genes after they have been separated into different blastomeres during cleavage, thereby initiating epigenetic aspects of development which emerge after fertilization. In other words, a system of coordinates is established during preparatory phases of development of the egg which initiates and controls in part all subsequent events in development. Too many embryologists accept the egg available to them at the time of fertilization or activation by parthenogenetic agents as the starting point for development of the embryo and consider the organization of the egg already established in the ovary to be preformed and beyond the boundaries of their area of endeavor. This preliminary stage in development may well prove to be the most important and critical stage of all, and yet it is relatively uninvestigated. Admittedly it is an extremely difficult stage to analyze. Interest has been rekindled in this period through application of immunological and histochemical techniques and electron microscopy, but this is only a beginning.

Fertilization. Fertilization involves the series of processes by which the spermatozoon initiates and participates in development of the egg. There is no convincing evidence that animal spermatozoa are attracted to eggs by chemotaxis. Certain interacting substances are produced by eggs and sperm of certain species and much is known about their chemical properties (fertilizin and antifertilizin of eggs and antifertilizin and egg-membrane lysins of sperm). Although the lysins obviously facilitate penetration of egg membranes by sperm, the significance of fertilizin and antifertilizin for fertilization remains uncertain; these substances may be important in the specificity of fertilization. There has been much interest of late in the acrosomal reaction of sperm of certain marine species as a consequence of which there is produced an acrosomal filament of characteristic length. Contact between the apex of this filament and the egg cortex may provide the stimulus (not necessarily mechanical) which activates the egg resulting in the rapid cortical changes, some of which are instrumental in establishing the block to polyspermy and in engulfment of part or all of the sperm with or without the aid of a fertilization cone. Many physical and chemical changes occur in the activated egg including sometimes marked redistributions of cytoplasmic substances. Meanwhile the male and female pronuclei are brought together by forces which are only partly understood and conditions are established which result in the first cleavage of the egg. Unfortunately in spite of the considerable amount of information available about events associated with fertilization, it is all too evident that much remains for future discovery if we are ever to distinguish between cause and effect. Since development of eggs can be initiated in the absence of sperm by an amazing variety of parthenogenetic agents, it has become increasingly apparent that the egg itself contains the basis for all the mechanisms essential for development and that no specific stimulus is necessary to put these mechanisms into operation. This realization has only served to reemphasize the tremendous importance of the developmental changes that occur during preparatory phases which culminate in production of a cell capable of developing into a new individual when given some slight provocation to do so.

Cleavage and blastulation. These are simultaneous rather than successive events; cleavage is considered as terminating upon establishment of the blastula. Eggs of each species characteristically exhibit a fairly definite cleavage pattern at least during the first 3 or 4 cleavages and in some species considerably longer. The locations of cleavage planes are largely the consequence of the positions assumed by the mitotic spindles and these, in turn, are controlled somehow by the organization of the cytoplasm including local differences in the egg cortex. Progressive changes in the organization of the cytoplasm with time have been demonstrated by delaying the time of onset of cleavage from the time when the first cleavage should occur to the time when the third cleavage should take place. In such cases the spindle forms in the vertical plane as it should do at the time of the third cleavage rather than in the horizontal plane as it should do at the time of the first cleavage. The rate of cleavage also appears to be controlled by the cytoplasm. At least the rate of cleavage is not changed in enucleated or nucleated eggs of one species when they are fertilized by sperm from another species characterized by a different rate of cleavage. In fact a cleavage of sorts terminating in establishment of a blastula-like stage can take place in enucleated eggs which have been activated parthenogenetically (*parthenogenetic merogones*). In some species the cleavage pattern can be modified experimentally without modification of subsequent development (*regulative eggs*);

in other species such modification of cleavage pattern results in modification of subsequent development (*mosaic eggs*). In eggs of the latter type specific cytoplasmic or ooplasmic substances are often separated by cleavage planes into separate blastomeres and the developmental fates of such cells frequently appear to be quite rigidly determined thereby with the result that cellular interactions are of little consequence for subsequent development. Several types of experiments have demonstrated the equivalence of nuclei during blastulation and early gastrulation (i.e., differentiation is not the consequence of differential divisions of nuclei during cleavage). Arrangement of blastomeres into a blastula somehow facilitates individual and mass migrations of cells characteristic of gastrulation.

Gastrulation. During this period of development blastomeres become strikingly rearranged in orderly ways relative to one another by so-called morphogenetic movements. Prospective mesoderm and prospective endoderm cells become displaced into the interior in a variety of ways in different species and eventually become completely enclosed by prospective ectoderm. Shifts in positions of individual cells or groups of cells have been followed and plotted in detail either as a consequence of visible differences in particles associated with the cytoplasm of different blastomeres or by tagging cells with vital stains or carbon particles or otherwise. The surface coat of some eggs and the hyaline layer of others seem to aid in coordinating movements of individual cells. Apparently these changes are due in large part to modified behavior of individual cells which becomes apparent at this time. At least dissociated cells obtained from gastrula stages not only reaggregate, but sort themselves out until endoderm comes to lie in the interior, ectoderm at the exterior, and mesoderm in between the other two layers. Gastrulation cannot take place in the absence of nuclei; parthenogenetic merogones fail to gastrulate. Moreover incompatibilities between maternal and paternal genes first make themselves felt at the time of gastrulation and may bring development to a standstill. Thus genetic factors located in the nuclei of blastomeres begin to manifest themselves at the time of gastrulation and in ever increasing manner control processes of development from this stage onward. Synthesis of new proteins apparently begins at this time and paternal genes begin to affect synthesis of proteins.

Gastrulation is a necessary prerequisite for subsequent stages of development since it brings cells into proper relationships relative to one another for them to interact and cooperate eventually in formation of organs. Gastrulation converts the two-dimensional system of coordinates established in the egg cortex at earlier stages into the three-dimensional pattern of the young embryo and hence is an extremely critical stage in morphogenesis. Evidence is accumulating that differentiation of nuclei begins to take place towards the end of gastrulation, at least in frog embryos.

Determination of fates of groups of cells. The mechanisms of determination appear to differ in mosaic and regulative eggs. In mosaic eggs the fates of many cells seem to be determined during cleavage by the kinds of cytoplasm isolated within them by formation of cleavage furrows. Current research on such eggs is concerned primarily with attempts to discover the nature of the cytoplasmic substances which exert such striking control over the fates of the blastomeres. Equally as important would seem to be the discovery of the factors which bring about the localization of these specific substances prior to cleavage which practically guarantees their incorporation into specific blastomeres.

In regulative eggs the fates of given groups of cells tend to be determined somewhat later (during late blastulation or gastrulation at the earliest) and to be decided primarily through interactions with adjacent cells (*embryonic inductions*). The demonstration of the general occurrence of the phenomena of induction has been one of the most momentous discoveries of twentieth century biology. The many studies devoted to embryonic induction attempt primarily to determine the nature of the influence which passes from inductors to competent cells. Is a transfer of specific substances involved? If so, are they in the form of small molecules which will pass through cellophane or large molecules which can pass only through the larger pores of millipore filters? Do such substances act at the surfaces of competent cells or do they actually enter those cells? Is cell-to-cell contact essential for embryonic induction? What part does intercellular matrix play in this process? In some instances ribonucleoproteins appear to be implicated. The question then arises whether the nucleic acid portion or the protein component plays the essential role. Thus the nature of the stimulus by which one group of cells affects another has not yet been answered unequivocally for any single inductive system. Even more important would seem to be the nature of the reactions released in competent cells by embryonic induction. The inductor does something to the cells which decides out of a given number of possible types of development which one will be actually realized. Doubtless this means in part setting in motion the synthetic machinery for production of specific kinds of proteins. Perhaps this is accomplished by differential activation of different sets of genes, thereby initiating one possible combination of gene-controlled processes. In spite of much effort devoted to problems of determination, solutions seem remote at the present time. We know only that the processes involved in determination eventually subdivide the organism into a complex of organ-forming districts each capable, upon isolation in appropriate surroundings, of developing into a given structure.

Organogenesis. A visible organ rudiment begins to emerge within each organ-forming district. Sometimes the rudiment is composed of cells derived from one germ layer only as in the case of the lens. More often it is composed of cells originating from two adjacent germ layers as in the case of the limb bud or the cornea of the eye. Sometimes migratory cells are incorporated into a specific organ rudiment as in the case of gonads where primordial germ cells originating outside the organ rudiment migrate into the latter and find conditions there favorable for their survival and continued development. Weiss has coined the term coaptation for this relationship. As a consequence of the original determination of each organ-forming district and of continued cellular interactions of varying degrees of complexity each organ rudiment progressively assumes its characteristic definitive form often as the result of seemingly complex morphogenetic processes. Surprisingly little is known of the physical forces involved in shaping a mass of cells into an organ of a specific shape. We must admit our ignorance even of the forces that roll the neural plate into a neural tube or a lens placode into a lens vesicle!

Histogenesis. The appearance of detectable differences between cells in different organ-forming districts as a consequence of the determination of the fates of these districts and/or continued cellular interaction constitutes differentiation. Initially our knowledge of differentiation was restricted to the gradual appearance of structural differences in cells. With the application of techniques for dissociation of cells, differences in behavior of cells arising from different cell strains were detectable. Cells of each cell strain were able to recognize each other in mixed populations of cells originating from different organs and aggregated selectively, whereupon they reformed the tissue or organ from which they originated. Moreover, the application of immunological techniques and the use of specific metabolic poisons and techniques for detecting enzymatic activities enabled the embryologist to detect emerging biochemical differences in specific cell strains long before visible histogenetic or morphogenetic differences became evident. Cellular differentiation on biochemical levels can therefore be equated to the synthesis of new macromolecules such as proteins. It is significant that once cell strains have become capable of such specific syntheses, they retain this capacity even though they may undergo "dedifferentiation" frequently when explanted *in vitro*. Thus the initial problem in the study of differentiation is one of establishing different mechanisms for the synthesis of different specific macromolecules, or, in other words, the study of differentiation resolves itself into a study of the acquisition and transmission of characteristic patterns of metabolic activity. Actually it has been demonstrated that the metabolic patterns of cells gradually change with successive stages of differentiation: new synthetic activities appear, previous synthetic activities disappear, and conspicuous shifts in relative synthetic activity occur. The progress of differentiation thus appears to depend on mechanisms for changing effective enzyme concentrations of cells. But, although considerable progress has been made in our understanding of differentiation on the biochemical level, we cannot lose sight of the fact that these changes must result ultimately in structural changes within the cells, and our understanding of differentiation of this level is slight indeed.

Growth. GROWTH in the narrow sense of the word is reproduction of the specific kind of cytoplasm characteristic of each cell type within the organism. In multicellular organisms it is difficult, if not impossible, to measure growth in this limited sense and our measurements of growth include increases in intracellular and extracellular products elaborated by cells. Embryologists are especially interested in mechanisms that regulate growth and that set upper limits to the size attained by organs. There is increasing evidence that certain organs at least emit products into the circulation that regulate their own growth by autoinhibition. Certain organs when transplanted to embryos stimulate growth of homologous host organs by emitting products into the circulatory system which become selectively incorporated into homologous organs of the host. It has been known for some time that endocrine organs emit specific products into the blood of embryos (HORMONES) which regulate growth of nonhomologous organs in specific ways. Heredity, of course, sets certain ultimate limits on growth in each kind of organism.

RAY L. WATTERSON

References

Willier, B. H., et al., eds., "Analysis of Development," Philadelphia, Saunders, 1955.
Waddington, C. H., "Principles of Embryology," New York, Macmillan, 1956.
McElroy, W. D. and B. Glass, eds., "The Chemical Basis of Development," Baltimore, Johns Hopkins Press, 1958.
Nickerson, W. J., ed., "Biochemistry of Morphogenesis," New York, Pergamon, 1959.
Balinsky, B. I., "An Introduction to Embryology," Philadelphia, Saunders, 1965.
Brachet, J., "The Biochemistry of Development," New York, Pergamon, 1960.

ENDOCRINE SYSTEM

Many processes occurring in living organisms are correlated one with the other or are regulated by two basic mechanisms; by the NERVOUS SYSTEM or by means of chemical regulators. The nervous system controls and regulates the activities of skeletal and smooth muscle and some of the exocrine glands. Chemical regulation is concerned primarily with the control of processes having to do with metabolism, growth, and reproduction. It must be kept in mind, however, that while the nervous system is effecting control over the musculature and certain glands, it mediates its effects by means of chemical substances. Thus, in the strict sense, all organismic regulation is effected by means of specific chemical agents.

Chemical regulating substances are defined or named according to the nature and site of origin, mode of transmission or dissemination, and the duration of their action. HORMONES, in the strict sense, originate in specifically differentiated and localized areas called endocrine organs or glands, and are secreted into and disseminated by the blood vascular system. Hormones may act on a wide variety of tissues (as in the case of growth hormone or insulin) or may stimulate only those tissues specifically differentiated to respond to these chemical mediators (as the thyroid epithelium reponds to thyrotrophic hormone, or the adrenal cortical cells to adrenocorticotrophic hormone). Hormones are usually effective in their action over a considerable period of time.

Chemical regulators that are similar to hormones in that they originate in localized anatomical regions, are blood borne, and have effects of considerable duration, but originate in specialized portions of the central nervous system are known as neurosecretory substances or neurosecretory types of neurohormones (i.e., vasopression and oxytocin). Chemical regulators or mediators originating from axonic terminals of both peripheral and central nervous system neurones

and which act by local diffusion on other neurones, nuscles, or glands for very brief periods of time are termed neurohumors. Neurohumors and neurosecretory substances may be classed together as neuroendocrine substances.

Products of metabolism, such as glucose and carbon dioxide that have regulatory effects on metabolic processes, but are widespread in origin are not considered endocrine in origin and are termed parahormones.

The endocrine system in vertebrate animals consists of a number of localized areas of specialized tissues that elaborate and disseminate substances which have specific regulatory effects. These include: 1) certain hypothalamic nuclei and the pars nervosa of the PITUITARY gland, 2) the pars intermedia and pars anterior portions of the pituitary gland, 3) the THYROID gland, 4) the PARATHYROID glands, 5) the pancreatic ISLETS OF LANGERHANS, 6) specialized portions of the gastrointestinal mucosa, 7) the ADRENAL glands, 8) the gonads (see OVARY) and 9) the PLACENTA.

While the regular chemicals secreted by all these various areas might be termed hormones, those substances produced by the hypothalamus are known as neurosecretory substances and those by certain portions of the gastrointestinal mucosa as gastrointestinal principles.

Other specialized areas, such as the pineal body, thymus, spleen, liver, and kidney cortex may exhibit endocrine-like mechanisms, but have not been determined to have clearly defined endocrine roles.

Hormones are characteristically effective in very minute quantities and probably act by stimulating or inhibiting specific intracellular enzyme systems, by modifying the permeability of cell membranes, or by other undetermined means. The tissue cells affected by hormones may be widespread as in the case of growth hormone, thyroxin, or insulin, or may be localized as in the case of the gonadotrophic, thyrotrophic, or adrenocorticotrophic hormones.

Regulation of the endocrine system itself is affected by several different mechanisms. This may be achieved by nervous or endocrine influences and may occur in a direct or indirect manner. In some cases the secretion of endocrine substances may be evoked by ingested food substances (secretagogues) as occurs in the elicitation of the production of the gastrointestinal principles, or by the blood level of circulating metabolites where the blood level of glucose in part regulates the secretion of insulin and the state of hydration of the blood regulates the secretion of vasopressin (antidiuretic hormone). Direct nervous control is exemplified by the evocation of epinephrine from the adrenal medulla upon stimulation by the sympathetic nerve fibers supplying this tissue. Indirect nervous regulation may occur when light stimulating the retina of certain birds or mammals, by neural pathways stimulates the production of hypothalamic neurosecretory substances which in turn affects gonadotrophin secretion by the anterior pituitary gland. Indirect endocrine regulation is seen in the control of gonadotrophic, thyrotrophic, or adrenocorticotrophic hormone secretion wherein the circulating levels of the sex hormones, thyroxin, or adrenal cortical steroids, in combination with the needs of the organism, either affect the cells of the anterior pituitary gland directly, or secondarily by effects on hypothalamic centers which in turn via neurosecretory substances regulate the output of pituitary trophic hormones.

The various components of the vertebrate endocrine system will now be considered in some detail.

The hypothalamus and pars nervosa of the pituitary gland. Certain hypothalamic areas (the supraoptic, tuberal, and paraventricular nuclei) serve not only as integrative centers for the autonomic nervous system, but also act as interrelating links between the nervous and endocrine systems. Neurosecretory principles or substances elaborated by modified neurones of the hypothalamus may be transported by the hypophysial-portal vascular system to the anterior pituitary gland and thus influence the secretion of anterior pituitary hormones, or may be disseminated by the systemic circulation and have more distantly removed effects.

Nervous influences from many different parts of the peripheral or central nervous system impinge upon the hypothalamus which thus serves as a focus of action for the nervous upon the endocrine system. Examples of this mechanism are the effects of suckling of the mammary glands which by direct nervous pathways stimulate the hypothalamus to produce neurosecretory substances which activate the anterior pituitary gland. The latter, in turn, stimulates the ovary to secrete estrogens and progesterone which aid in the maintenance of the lactational state of the mammary glands. The anterior pituitary also probably produces a mammotrophic substance that has a directly stimulating effect on the lactating breast.

The hypothalamus acts as a mediator in the regulation of secretion of the thyroid and adrenal glands. The circulating levels of thyroid hormone (*thyroxine*) and adrenal cortical hormones (certain steroids) may act upon hypothalamic centers which by the secretion of neurosecretory substances regulate the output of the appropriate trophic hormones by the anterior pituitary gland (in this case, thyrotrophic and adrenocorticotrophic hormones).

The pars nervosa portion of the pituitary gland has been included with the hypothalamus because it seems likely that the secretory principles that may be extracted from the pars nervosa are actually elaborated in the hypothalamus and are transported to and merely stored in the pars nervosa. From the pars nervosa two principles or substances have been isolated. These are polypeptides in nature and have been termed the pressor and the oxytocic principles.

The pressor principle has constrictive action on the smooth muscle of certain arteries, arterioles, and to some extent venules and capillaries, which results in an increase in blood pressure. The pressor principle also has an antidiuretic effect which is presumably achieved by stimulating the renal tubules to resorb water but not salt, and is thus known as the antidiuretic hormone. In the absence of antidiuretic hormone, the organism is incapable of renal water resorption and loses great quantities of water through the kidneys. This condition is known as diabetes insipidus.

The oxytocic princple acts directly and specifically upon the smooth muscle of the uterus, exciting it, causing it to contract, and increasing its tone.

The pars intermedia and pars anterior portions of the pituitary gland. While the pars nervosa is differentiated from the inferior portion of the diencephalon of the brain, both the pars intermedia and pars antereior portions of the pituitary gland are differentiated from an evagination of the stomatodeal ectoderm (*Rathke's pouch*). The pars intermedia usually lies between the pars nervosa and pars anterior and is composed primarily of basophilic cells. The secretory products of this portion of the pituitary are known as intermedin or melanophore expanding hormones. These hormones cause expansion of the pigment bearing cells of animals having these cellular elements (fish, amphibians, and reptiles). A darkening of an animal's body surface, or the expansion of pigment containing cells in other areas (peritoneal lining) is in part controlled by this endocrine mechanism. Although birds lack a pars intermedia, intermedin is extractable from the pars anterior. Melanophore expanding hormones are present in the pars intermedia of mammals, but the role of these hormones in these vertebrates is unknown.

In the absence of the pars intermedia in animals with numerous pigment cells (such as in amphibians and reptiles) the lack of melanophore expanding hormones leads to a constant state of contraction of the pigment containing cells and the animals appear very pale or colorless.

The pars anterior portion of the pituitary gland in most animals is made up of acidophilic, basophilic, and chromophobic cell types. In all vertebrate classes, the basophilic cells can be histologically differentiated into two distinct types. While both the basophilic cell types are periodic acid Schiff positive, the cytoplasmic granules of but one will stain with aldehyde fuchsin. At least in mammals, the aldehyde fuchsin staining basophil has been associated with thyrotrophic hormone production, and the non-aldehyde fuschsin staining basophil with follicle stimulating hormone production. The acidophilic cells stain differently with such dyes as azocarmine, orange G, or eosin, in various vertebrate representatives, and there has been no homologization of these cell types from one species to another. While it is likely that growth hormone originates from an acidophilic cell type, the origin of adrenocorticotrophic, interstitial cell stimulating, or luteotrophic hormone has not been established.

Six protein hormones have been isolated and characterized from the anterior pituitary. These are: 1) the follicle stimulating hormone (FSH), 2) the interstitial cell stimulating (ICSH) or luteinizing hormone (LH), 3) luteotrophic hormone (LTH) or prolactin, 4) thyrotrophic hormone (TTH), 5) adrenocorticotrophic hormone (ACTH), and 6) growth or somatotrophic hormone (STH).

FSH stimulates the growth of ovarian follicles and the seminiferous epithelium. ICSH acts upon the ovarium interstitium and the theca interna of the developing follicle and the interstitial or Leydig cell of the testis. Luteotrophic hormone or prolactin is involved in the process of ovulation and the growth and development of the corpus luteum. As a result of FSH and ICSH stimulation, the cells of the ovarian theca secrete estrogen. After ovulation of the ovum, LTH or prolactin induces the granulosa cells to secrete progesterone. ISCH stimulates the secretion of androgens by the testicular Leydig cells. The role of LTH in the male is not known.

Thyrotrophic hormone stimulates the thyroid gland to produce and secrete thyroxine. Adrenocorticotrophic hormone incites the formation and secretion of adrenal cortical hormones. Somatotropin or growth hormone promotes total body growth by exerting a major effect on the metabolism of protein, carbohydrate, water, and minerals.

In the absence of the pituitary gland, an animal shows many serious defects such as failure or cessation of growth, marked decline in the activity of the thyroid and adrenal glands and cessation of reproductive activity.

The thyroid gland. Derived from a ventral evagination of the primitive foregut in the region of the second branchial arch it is composed of irregularly shaped spheroidal follicles. The secretory epithelium of the thyroid follicle is one cell thick and surrounds a central mass of viscous albuminous material known as the colloid substance of the thyroid. Under the influence of the thyrotrophic hormone (from the anterior pituitary) the thyroid epithelium synthesizes thyroxin from iodine and tyrosine. Thyroxin (tetraiodothyronine) may be stored in the colloid as thyroglobulin. To be disseminated in the blood stream, thyroglobulin must be hydrolyzed, the thyroxin be passed back through the thyroid cell, and then be bound to a blood plasma globulin, in which form it is transported to the tissues of the organism.

TTH stimulates and increases the size and number of the thyroid follicle cells, increases the rate of thyroxin systhesis, the hydrolysis of thyroglobulin, and the rate of transfer of thyroxin from the thyroid gland to the blood stream.

A primary function of thyroxin is to increase the rate of tissue metabolism. A main site of action may be in the mitochondria where the formation of energy-rich phosphate bonds from the increased oxidation of substrates may be stimulated. Although the calorigenic effect of thyroxin has been well established in birds and mammals, a similar role has not been clearly established in the cold blooded vertebrates.

Thyroxin plays a fundamental role in the normal growth and development of the whole organism as well as a special role in the growth and development of the central nervous system. Thyroxin also is essential in the metamorphic changes that occur in the transformation from the larval to the adult state in many cold blooded vertebrate animals.

In the absence of the thyroid gland, or in conditions where there is insufficiency of iodine for the manufacture of an adequate amount of thyroxin, or where thyroxin production is blocked (in the presence of thiocyanates), the organism shows a number of abnormalities characterized as hypothyroidism. In hypothyroidism there is a decrease in metabolic rate, a decrease in the growth and development of the animal, and a loss of reproduction capacity. In most fishes and amphibians, larval metamorphosis will not occur in the absence of thyroid hormone. A severe deficiency of thyroid hormone in the human results in a condition known as myxedema.

An excess of thyroid hormone in mammals results in an abnormally high metabolic rate, increased utilization of protein, loss of weight, and weakness. In man, an excess of thyroid hormone due usually to an excessive output of thyrotrophic hormone leads to hyperthyroidism or Grave's disease.

The parathyroid glands. Are small (one or two pairs) organs composed of epithelial masses or cords of cells derived from the posterior halves of the third and fourth branchial arches. In adult animals they are found in the lateral anterior cervical regior, and in mammals are intimately associated with the thyroid gland.

There is no evidence that the anterior pituitary gland has a regulatory or trophic influence on the parathyroid glands. The secretory product of the parathyroid glands is a protein-like hormone known as parathormone. The function of this substance is the regulation of the calcium and phosphorus content of the blood. Parathormone may exert its regulating influence on the calcium and phosphorus levels of the blood by affecting the activity of bone forming cells (*osteoblasts*) or bone removing cells (*osteoclasts*), or by regulating phosphorus. excretion by the renal tubules, which indirectly affects the level of circulating calcium. The blood calcium or phosphorus level regulates the secretory activity of the parathyroid glands.

The normal function of both muscle and bone is dependent upon a proper ratio of calcium and phosphorus. In the absence of parathormone, the blood calcium declines and the phosphorus rises. Under these conditions an increased irritability of the nervous system ensues (*tetany*) which may result in muscular twitchings, cramps, or convulsions. An excess of parathormone produces a rise in blood calcium and a decrease in phosphorus. This may lead to the abnormal deposition of calcium in various organs and tissues, decrease the sensitivity of the nervous system, and produce demineralization and distortion of the skeletal system.

The pancreatic islets or islets of Langerhans. Derived from the dorsal pancreatic anlagen of the duodenal entoderm are relatively small masses or cords of epithelial cells found interspersed in the exocrine (or enzyme producing) tissue of the pancreas. These well vascularized islands of cells are composed usually of more numerous centrally located cells with small granules (the beta cells), and more peripherally located cells with coarser granules (alpha cells). The beta cells synthesize and secrete a protein hormone, insulin. It is postulated, but not proved, that the alpha cells produce a hormone with blood sugar increasing properties, known as the hyperglycemia producing factor (HGF) or glucagon.

The hormone insulin is essential for the proper utilization of glucose by most cells of the vertebrate organism. How insulin acts is not clear. It has been suggested that insulin is necessary for transfer of glucose across cell membranes, while another idea is that insulin is necessary for the intracellular phosphorylation of glucose. In the absence of insulin, most vertebrate organisms are unable to utilize glucose and must then oxidize fat as an energy source. The inability to utilize glucose increases the blood sugar, glucose is lost in the urine, large amounts of water are concommitantly voided and lost, the oxidation of fats leads to excessive production of anions, and the animal becomes acidotic. The metabolic abnormalities accruing from insulin deficiency result in a condition known as diabetes mellitus.

An excessive amount of insulin decreases the blood sugar level. In some types of animals, notably man and certain mammals, the central nervous system cannot function properly without adequate glucose, and insulin coma results. Other types of vertebrate animals (reptiles and birds) are apparently not as sensitive to low blood glucose levels, and while they may exhibit neuromuscular irritability, they do not succumb to the effects of insulin overdosage.

While a blood sugar elevating hormone (glucagon) has been extracted from the pancreas, its exact site of origin has not been unequivocally demonstrated. It probably acts by increasing the conversion of hepatic glycogen to glucose.

The blood glucose level is probably the main device for regulating insulin secretion (an increase in blood sugar level stimulates insulin secretion). A so-called *diabetogenic hormone* of anterior pituitary origin has been postulated. It has been suggested that this substance may act on the pancreatic islet alpha cell to stimulate glucagon secretion and thus elevate the blood sugar. It also has been argued that growth hormone has this property, and that an independent diabetogenic hormone does not exist. The status of both glucagon and diabetogenic hormone awaits further investigation.

Gastrointestinal hormones or principles. Originate from unidentified portions of the epithelium or glands of the mucus membrane of the stomach, or small intestine and are concerned with the regulation of certain digestive processes. These substances are evoked by the presence of ingested food or products of digestion ("secretogogues"), by gastrointestinal glandular secretions, or by direct nervous stimulation. Among these principles are secretin, a protein substance formed in the duodenum and released into the blood vascular system when acidic chyle from the stomach reaches the duodenum. Secretin is carried to the pancreas by the blood stream and there elicits the release of pancreatic digestive enzymes (amylase, lipase, and trypsinogen) which flow through the pancreatic ducts and into the duodenal lumen. Fat or fatty type food in particular excites the release of a duodenal hormone called cholecystokinin which stimulates release of bile from the gall bladder into the duodenum. Chyle (partially digested stomach contents) upon reaching the duodenum causes the release of another hormone, enterogastrone, from the duodenal mucosa, which inhibits the muscular activity and acid secretion of the stomach.

The adrenal glands. Are paired organs found in the posterior part of the abdominal cavity, usually in association with the kidneys. In fishes and amphibians they are contained within the kidney parenchyma, but in reptiles and birds are separate intra-abdominal organs. The adrenal gland is a composite of cells of two separate origins. From neural crest cells of ectodermal origin, the adrenal medulla of mammals, or the chromaffin tissue of sub-mammalian vertebrates is derived. This tissue is responsible for the elaboration and secretion of two hormones, epinephrine and norepinephrine, which are catecholamines. The adrenal medulla or chromaffin tissue is under direct nervous control, being innervated by pre-ganglionic sympathetic neurones. The primary effects of epinephrine are an acceleration of the heart rate and a vasodilation of the blood vessels of the skeletal muscle. Thus the supply of blood to the voluntary musculature is greatly increased. In addition to these effects, epinephrine also increases the blood sugar by stimulating hepatic glycogenolysis (breakdown of liver glycogen to glucose), inhibits gastrointestinal motility, relases the bronchial muscles, and stimulates the production of ACTH by the anterior pituitary gland.

Norepinephrine acts primarily on the peripheral blood vessels having a constrictive action and thereby limits the total blood volume and aids in the maintenance of an adequate blood pressure.

While the role of the adrenal medulla in normal physiology is not known, the above effects are important in mobilization of the resources of the organism for emergencies.

The adrenal cortex (of mammals) or the interrenal tissue (of submammalian vertebrates) is derived from the coelomic epithelium and is mesodermal in origin. In mammals, the adrenal cortex is usually differentiated into three zones or areas; the zonae glomerulosa, fasciculata, and reticularis. In a few avian species there is cytological differentiation of the interrenal tissue, but in practically all sub-mammalian species of vertebrates, no differentiation or zonation has been demonstrated. The adrenal cortex or interrenal tissue produces a variety of steroid hormones. These steroid hormones have three primary effects: 1) regulation of salt and water balance, 2) regulation of carbohydrate, fat, and protein metabolism, and 3) some effects similar to those of the steroids usually produced by the gonads.

While the type and quantity of the adrenal cortical steroids varies somewhat from class to class, or even from species to species, certain compounds have a fairly wide distribution and exert similar effects. Of these various steroid hormones, aldosterone is primarily effective in regulating salt and water balance by stimulating the renal tubules to resorb salt and water. Corticosterone and hycrocortisone aid in the mobilization of fat and protein from peripheral depots and influence in particular the conversion of protein to carbohydrate in the liver (gluconeogenesis).

In mammals at least, corticosterone and hydrocortisone are thought to originate in the zona fasciculata and their secretion regulated by ACTH of the anterior pituitary gland. The secretion of aldosterone, on the other hand, is thought to come from the zona glomerulosa, and is not dependent upon pituitary control. ACTH production is regulated by the effects of some of the adrenal cortical steroids acting either indirectly through the hypothalamus or perhaps directly upon the anterior pituitary gland cells.

In sub-mammalian species, where there is little or no cytological differentiation of the interrenal tissue, little is known concerning the regulation of steroid hormone secretion, other than hypophysectomy (removal of the pituitary gland) greatly decreases interrenal steroid secretion, or that the administration of ACTH stimulates the secretion of these compounds.

An insufficiency of adrenal cortical hormone results in abnormal loss of sodium and water with consequent dehydration and inability to maintain normal blood sugar levels (via conversion of protein to sugar in the liver) which leads to muscular weakness. An excess of adrenal cortical hormones results in abnormal retention of water and salt which produces edema (excessive amounts of extravascular fluid), overloading of the blood vascular system, and an increase in blood pressure.

The gonads. Besides producing the germ cells for

procreation of the species, also elaborate hormones which stimulate the growth and development of the sex accessories and the secondary sexual characteristics, and condition sexual behavior.

The growth and maturation of the sperm cells in the seminiferous tubules of the testis is dependent on both FSH from the anterior pituitary and testosterone from the testicular interstitial cell.

The interstitial or Leydig cells of the testis, interspersed between the seminiferous tubules, are dependent upon the secretion of ICSH by the anterior pituitary for their development and secretory activity. Under the influence of ICSH, the Leydig cells produce a male sex hormone (testosterone). The male sex hormone stimulates the growth and development of the epididymis, vas deferens, seminal vesicles, prostate, and bulbo-urethral or Cowper's glands. Likewise, the full growth and development of the penis, the male skeletal and muscular proportions, hair pattern, the enlargement of the larynx, and conditioning of male sexual behavior is dependent on male sex hormone.

The growth and maturation of the ovum is dependent on an orderly and properly proportioned sequence of FSH, ICSH (LH), and LTH secretion by the anterior pituitary gland. Under the influence of FSH and ICSH, the developing theca interna of the ovarian follicle secretes estrogen (female sex hormone). After ovulation of the ovum from the follicle, the follicular granulosa cells stimulated by LTH (prolactin), hypertrophy to form a corpus luteum, and secrete a steroid hormone, progesterone.

Estrogen (female sex hormone) initiates growth and development of the oviducts and the uterine endometrium (lining membrane of the uterus and its associated glands), as well as induces changes in the vaginal epithelium and in the mammary glands. The growth and development of the female breast, body proportions, hair pattern, and sexual behavior is dependent on female sex hormone.

Progesterone acts as an adjunctive agent to estrogen, furthering the development of the uterine endometrium and mammary glands and thereby prepares the uterus for the reception of a fertilized ovum and the breast for ultimate lactation.

Since implantation (embedding of the zygote in the uterine wall) is prerequisite for the continued stimulation of ovarian estrogen and progesterone, failure of fertilization and consequently of implantation, leads to cessation of ovarian hormone output. In the case of mammals with OESTRUS type reproductive cycles, ovulation without fertilization leads to regression of the sex accessories and the animal enters a quiescent phase of the reproductive cycle (anestrus). In mammals with a menstrual type of reproductive cycle (primates), failure of fertilization or implantation leads to the menstrual or bleeding phase of the reproductive cycle. Subsequent to the quiescent phases of the reproductive cycle, under the influence of anterior pituitary gland regulation, the estrus or the menstrual cycle begins anew. Thus, all the events regulating and ordering the reproductive cycles of vertebrate animals are controlled by the endocrine system.

In the absence of the gonads, not only will an animal be sterile, but also there will be deficiencies in the sexual attributes. If loss of the gonads occurs after puberty, the endocrine effects are not as pronounced as compared with an animal castrated before puberty. In a post-puberal castrate, there is considerable atrophy of the sex accessories, but only mild to moderate changes in the secondary sexual characteristics and behavior. In the pre-puberal castrate, on the other hand, there are marked changes, in that the sex accessories and secondary sexual characteristics fail to develop beyond a very rudimentary stage.

The placenta. If an ovum is fertilized and the zygote implanted in the wall of the uterus, the secretory activity of the anterior pituitary gland and the corpus luteum of the ovary continues for some time to maintain the oviduct or uterus in a condition appropriate and necessary for the growth and development of the embryo. Many mammalian species, however, by virtue of the endocrine secretory capacity of the placenta, may maintain pregnancy either without the anterior pituitary, or the ovary, or, in the case of complete placental autonomy, in the absence of both the anterior pituitary and the ovary.

Certain chorionic cells (epithelium of the chorion portion of the extra-embryonic membranes) are capable of elaborating and secreting estrogen, progesterone, and a protein anterior pituitary-like gonad stimulating hormone (chorionic gonadotrophin). The placenta, then either in conjunction with the anterior pituitary and the ovary, or with only the ovary, or completely by itself, is responsible for the maintenance of the oviduct or uterus in a condition necessary for the proper growth and development of the embryo and fetus. Gestation, or pregnancy, then, is in large part a hormonally regulated process.

MALCOLM R. MILLER

References

Turner, C. D., "General Endocrinology," 2nd ed., Philadelphia, Saunders, 1966.
Gorbman, A., ed., "Comparative Endocrinology," New York, Wiley, 1959.
Gorbman, A. and H. A. Bern, "Textbook of Comparative Endocrinology," New York, Wiley, 1962.
Williams, R. H., "Textbook of Endocrinology," 3rd ed., Philadelphia, Saunders, 1962.

ENDOGENOUS RHYTHMS

Endogenous rhythms comprise biological rhythms, with periods approximating the natural geophysical periods of the solar day, lunar day, month and year, which are observed to persist when such obvious biologically effective factors as light and temperature are held constant.

It is significant that these are the same periods as the extensive daily, lunar and annual changes in light, temperature, and the ocean tides, which have great importance to living things, evoking species-specific, adaptive, rhythmic patterns of response. In the solar day, some organisms are geared for diurnal activity, others for nocturnal, and still others for crepuscular. In the lunar-day cycles of the ocean tides, some animals feed on beaches uncovered at low tide, others only when covered by water at high tide. Once each synodic month solar and lunar "noons" become synchronized (new moon); of this average 29½-day period are reproductive activities of many marine animals and plants, locked to specific moon phases, and also the human menstrual cycle. Of annual frequency are numerous rhythms including ones of activity and dormancy, of reproductive cycles of innumerable plants and animals, and animal migrations.

Endogenous, or persistent, rhythmicity has been demonstrated to be widespread through the animal and plant kingdoms. Examples in plants are daily rhythms of cell-division and bioluminescence in unicellular forms, and of sleep-movement and growth rate in higher forms. Seeds exhibit an annual rhythm in capacity to germinate. Examples from animals are daily rhythms of skin color-change and lunar tidal rhythms of spontaneous running in crabs, daily rhythms in emergence-time of flies ready to leave their pupal cases, and daily rhythms of wakefulness or spontaneous activity and correlated underlying physiological phenomena in numerous kinds of animals including mammals. In all organisms in which they have been sought endogenous rhythms of basal metabolic fluctuations have been found displaying all the natural geophysical frequencies. It seems reasonable to postulate that endogenous rhythmicity is an universal attribute of living matter.

The persistent overt physiological rhythms possess certain properties in common, quite unorthodox in terms of usual biological phenomena. The frequency, or cycle duration, of the rhythms is essentially independent of the level of constant temperature at which the organisms are maintained and, furthermore, drugs or other chemicals known to alter substantially general metabolic rate have little or no influence on the rhythm frequency. In addition, the patterns of activity recurring daily in constant light and temperature can be abruptly shifted to bear any desired phase relationship to the outside day-night cycles simply by subjecting the organism to a few experimental daily cycles of appropriately adjusted light or temperature changes. The new phase relationships of the rhythm may then persist in continuing constant conditions. In fact, the recurring cycles very commonly appear to shift spontaneously from day to day to yield regular cycles a little longer or shorter than the solar day. The rate and direction of such spontaneous daily phase-shifting is related to the constant temperature or light level at which the organisms are held. It has also been a long-known fact that many organisms will, in response to artificial light-dark cycles, adopt "days" deviating from 24 hours by up to 4 to 6 hours but only as long as the light cycles continue to be imposed. When, however, the artificial light cycles deviate further from 24 hours, the organism commonly resumes its basic daily period, even in the presence of the continuing artificial cycles.

There is also a daily rhythm in the sensitivity to light and temperature as factors operating to shift phases of the persisting rhythms. There are reasons to postulate that these particular organismic rhythms are responsible for the apparent spontaneous phase-shifting even when the light and temperature are held constant, thereby giving rise to periods deviating from 24 hours, and responsible also for the inability of organisms to synchronize with artificial light-dark cycles when these deviate by more than a few percent from 24 hours.

There is adequate evidence that the same "clock-system" which times the periods of persistent physiological rhythms underlies a "time-sense" such as that exhibited by bees trained to feed at particular times of day. It also appears to comprise the chronometer which enables organisms as diverse as crustaceans, insects, and birds which normally use the sun or moon, and possibly the constellations, for celestial navigation, to correct continuously for the rotation of the earth relative to each of these types of heavenly bodies, to maintain straight compass courses.

There are two basic hypotheses as to the means by which the endogenous cycles are timed. One of these hypotheses depends upon the fact that living things, even in so-called constant conditions, exhibit exogenously timed metabolic rhythms containing all the natural geophysical frequencies. The pervasive, subtle geophysical factors immediately timing these rhythms are still unidentified, but it is known that their fluctuations are influenced by changes in atmospheric pressure and temperature, and other factors. These environmentally timed rhythms are postulated to be the primary "clocks" providing reference frequencies for timing the endogenous, adaptively phase-labile, physiological rhythms of approximately the same frequencies. In other words, the various endogenous rhythms are considered dependent upon continuing organismic responses to corresponding geophysical cycles.

The other hypothesis is that all living things possess inherited natural periods of oscillation which constitute independent internal clocks capable of measuring the various natural periods without any external source of information, and that these oscillations time the endogenous rhythms.

While it has been possible to establish that organisms, even in what has been commonly considered to be conditions constant for them, are still in a rhythmic geophysical environment to which they are sensitive, proof that organisms have, in addition, a fully autonomous complex of oscillations of all the observed endogenous rhythm periods must await experiments conducted safely away from all the highly pervasive geophysical rhythms of approximately the same periods.

FRANK A. BROWN, JR.

References

Brown, F. A., Jr., "Living Clocks," Science, 130: 1535, 1959.
Bunning, E., "The Physiological Clock," New York, Academic Press, 1964.
Cloudsley-Thompson, J. L., "Rhythmic Activity in Animal Physiology and Behavior," Academic Press, 1961.
Harker, J. E., "The Physiology of Diurnal Rhythms," Cambridge, The University Press, 1964.
Sollberger, A., "Biological Rhythm Research," New York, Elsevier, 1965.

ENDOPOLYPLOIDY

Endopolyploidy is a condition, found in many multicelled plants and animals, which is characterized by the presence of more than the basic chromosome number in some somatic cells. These increased chromosome numbers occur at euploid levels in a geometric progression, i.e., in a basically diploid ($2n$) organism, endopolyploid nuclei may be $4n$, $8n$, $16n$, etc., depending upon the extent to which the causal process had continued. In a strict sense, the term "endopolyploidy" makes reference to the mechanism by which the condition arose whereas the sometimes equivalent terms "somatic polyploidy" and "polysomaty" do not necessarily imply a causal mechanism.

The retention of the basic chromosome number, which is characteristic of most dividing somatic cells in an organism, is dependent upon the strict sequencing of the processes of replication, karyokinesis and cytokinesis, without deviation, in a cyclic pattern. Deviations from this sequencing, primarily the failure of karyokinesis to follow replication, may result ultimately in endopolyploidy if the nucleus undergoes another replication. Since in replication all existing chromosomal units are duplicated, and since the nucleus in question was already once duplicated, the result of the additional replication cycle is another doubling of the genome.

If the products of this and the preceding replication have not separated physically, the chromosomal condition is considered to be polytene (multistranded) rather than polyploid, although this distinction would be impossible to make in an interphase nucleus. However, extreme cases of polyteny, such as are seen for example is some Dipteran salivary gland cells, are notable exceptions.

The timing and visible manifestations of the separation of the replicated chromosomal units are the basis for some differences in terminology. However, the event that is directly responsible for the condition, however it be described and defined, is the occurrence of an additional replication cycle or cycles. Subsequently, these multiple chromosomal units may remain together (polyteny), they may separate visibly within the intact nuclear envelope, the chromosomes contracting during the process (endomitosis), or they may separate later during an otherwise normal karyokinesis. In the latter case, after one extra replication cycle, the chromosomes may either have separated very early in prophase, without any striking visible evidence, or they may enter karyokinesis with twice the usual number of chromatids, in which case they are called "diplochromosomes." These separate once sometime during the early part of karyokinesis, the result being twice the basic number of chromosomes, each consisting of two chromatids which then separate during a normal anaphase. The result, in a diploid organism, is two cells, each having a tetraploid chromosome number. In this latter case, in the absence of the events occurring within the nuclear envelope, the process has been called "endoreduplication" to distinguish it from the special case of "endomitosis," although the term "endoreduplication" refers more precisely to the mechanism by which the polyploid condition arises than to the morphological sequelae by which the prior occurrence of the additional replication cycle becomes evident. More than one additional replication cycle may occur prior to a karyokinesis. The chromosomes entering division would then have a larger number of chromatids (e.g., quadruplochromosomes). However, karyokinesis need never occur, in which case the distinction between polyteny

and polyploidy could not be made, and in the absence of an endomitosis, the former would have to be assumed. The condition would have to be revealed by indirect means or by induced divisions.

Indirect means for detecting the condition have included the counting of heterochromatic bodies, the estimation of nuclear volumes, or as a considerable refinement of the latter, microspectrophotometric estimation of the amount of DNA in individual nuclei. Based upon the assumption that the amount of DNA per completed chromatid set is constant, this technique reveals simply the number of completed chromatid sets that are present in a nucleus. Thus, it cannot be used to distinguish, for example, between a duplicated diploid and an unduplicated tetraploid, since both would have a "4C" amount of DNA.

Direct determinations of endopolyploidy can be made only by the analysis of dividing nuclei. Divisions of endopolyploid nuclei occur spontaneously in some systems. In others, in which the endopolyploids do not normally divide, karyokinesis can be induced by various means, e.g., as a response to wounding. In many plants it is possible to induce divisions in endopolyploid cells by the application of high concentrations of auxins.

Another mechanism by which somatic polyploidy can arise is through a normal sequencing of replication and karyokinesis, but with a subsequent failure of cytokinesis. This results either in a binucleate cell, or with the occurrence of nuclear fusion, or the formation of a "restitution nucleus," a single polyploid nucleus.

The occurrence of polyploidy in somatic cells is of widespread occurrence in both plants and animals. The question of how this phenomenon is to be viewed in light of morphogenesis and cell differentiation remains largely unanswered. However, it is perhaps best viewed as a type of cell differentiation in itself; the question then becomes one of its concomitance with or causality to other forms of differentiation.

CARL R. PARTANEN

References

D'Amato, F., "Polyploidy in the differentiation and function of tissues and cells in plants," Caryologia 4: 311–358, 1952.

Geitler, L., "Endomitose und endomitotische Polyploidisierung," Protoplasmatologia VI, C. Springer-Verlag OHG, Vienna, 1953.

Levan, A. and T. Hauschka, "Endomitotic reproduction mechanisms in ascites tumors of the mouse," J. Natl. Cancer Inst. 14: 1–43, 1953.

Partanen, C. R., "On the chromosomal basis for cellular differentiation," Amer. Jour. Bot. 52: 204–209, 1965.

ENZYME

An enzyme may be defined as "a protein with catalytic properties due to its power of specific activation." Enzymes are the directing and controlling catalysts which determine the particular chemical reactions which make up the complex metabolic pattern of a living organism. The fundamental characteristic of life, the ability to use chemical energy (for example, from food molecules) and to direct it into particular pathways of chemical synthesis, or to convert it into mechanical, electrical or other forms of energy, is primarily due to the possession of the right assortment of highly specific enzymes.

The study of enzymes is growing rapidly; several journals of enzymology exist, a number of large text-books have been written exclusively about enzymes, and in several countries institutes have been established specifically for enzyme studies. Several branches of research have developed; interest may be focussed on the reactions brought about by enzymes, on the quantitative aspects of the enzymatic process, the nature of the enzymes themselves, or on the mechanism of enzymatic catalysis. Although the topics are considerably inter-related, it will be convenient to consider them under separate headings.

Enzyme reactions. Enzymes catalyse a large number of chemical reactions, some of which have not so far been carried out by non-biological means. In many cases a separate enzyme is needed for each reaction, although some enzymes will catalyse a more or less narrow group of closely related reactions. This high specificity is one of the main features which distinguish biological from nonbiological catalysts.

In the most extreme cases the enzyme acts, so far as is known, on one substance only. In other cases, where the enzyme acts on a small group of closely related substances, it is often possible to recognise a common chemical structure which must be present in a substance if the enzyme is to act upon it. With the less specific enzymes this necessary structure is a relatively small part of the substrate molecule, e.g. an aldehyde or ester grouping, so that a large number of related substances can be attacked.

Nearly one thousand enzymes are known; 875 were listed and named in 1964 in the volume on "Enzyme Nomenclature." Apart from a very few old names (e.g., pepsin) the enzyme names in current use all end in "-ase." An International Commission was set up in 1956, to rationalize and standardize the nomenclature and classification of enzymes; their report was revised and adopted as "definitive recommendations" by the International Union of Biochemistry in 1964. These comprise a system of classification (described below), and a code number, systematic name, and trivial name for each well-established enzyme. Enzymes are classified into six main groups and 47 sub-groups according to the type of reaction catalysed, indicated by such names as dehydrogenase, oxidase, transaminase, phosphokinase, transacylase, synthetase and racemase. The names of individual enzymes include an indication of the chief substrate (or group of substrates) acted on, either by means of a prefix or a separate word: hence *lactate dehydrogenase, aspartate aminotransferase, phosphotransacetylase.* Enzymes which bring about hydrolysis usually have a trivial name which shows the name of the substrate plus "-ase", without further qualifications; e.g., *arylsulphatase* hydrolyses the phenolic ester link of aryl sulphates.

The six groups in the standard classification are *oxidoreductases, transferases, hydrolases, lyases, isomerases* and *ligases* (or *synthetases*).

The majority of enzymes either catalyse hydrolysis, of the type:

$$AB + H_2O \rightarrow AH + BOH$$

or transfer reactions, of the type:

$$AX + B \rightarrow A + BX$$

The main subdivisions of the hydrolases (hydrolytic enzymes) are: esterases, glycosidases, peptidases, deaminases, and enzymes hydrolysing acid anhydrides (such as the pyrophosphate linkage in adenosinetriphosphate). The main groups of transferring enzymes transfer the following radicals from one molecule to another: hydrogen (oxidoreductases), amino-groups (transaminases), phosphate (phosphokinases, pyrophosphatases, pyrophosphorylases), acyl groups (transacylases), glycosyl groups (transglycosylases), and methyl groups (transmethylases). Smaller groups of enzymes outside the two main sections catalyse changes of steric configuration (racemases) or the addition of molecules such as water, ammonia, or aldehydes across the double bonds of unsaturated molecules (lyases).

Despite the diversity of overall reactions, it can be argued that the number of reaction types catalysed by enzymes is very small. Hydrolysis can be regarded as transfer of a group to water as acceptor, so that hydrolases may be treated as a special type of transferase; and indeed, many hydrolytic enzymes will, under suitable conditions, bring about other transfer reactions. The structure in the substrate molecule actually acted on may be common to a large number of enzymes; for example, over 70 enzymes act on $-CO \cdot NH-$ links and over 80 enzymes oxidize $-CHOH-$ to $-CO-$ in various substrate molecules.

The metabolic processes of a living cell, both degradative processes (catabolism) which make chemical energy available, and the synthetic processes (anabolism) necessary for growth and reproduction, are brought about by interlinked systems of enzymes. In general, enzymes can be regarded as acting in chains, with the product of the action of one enzyme forming the substrate of the next. The high specificity and high catalytic activity of the enzymes making up such a chain are sufficient to explain its efficient working, with a minimum of side reactions. However, evidence is accumulating that there is in addition a physical positioning of the enzymes in the order of their reaction sequence. For example, many of the oxidative sequences take place within the mitochondrial membranes, and as D. E. Green has said in connexion with his own work on particulate enzyme preparations from mitochondria: "We are dealing with an organized mosaic of enzymes in which each of the large number of component enzymes is uniquely located to permit of efficient implementation of consecutive reaction sequences."

Enzyme kinetics. Since by definition enzymes are catalysts, speeding up a variety of chemical reactions, it is inevitable that a large proportion of the accumulated information about them should be of a quantitative nature. The methods used for the actual measurements are very diverse, because of the many chemical reactions whose rate is accelerated. The theoretical treatments and procedures used for the manipulation of such data are, however, more or less unique to enzyme kinetics; only a superficial account of these treatments can be given here. They are important in defining conditions under which enzymes can be used as tools in the laboratory, in understanding the mechanism of enzyme action, and also in experimental attempts to explain the action of vitamins, hormones and drugs in terms of an effect on particular enzymes.

The rate of an enzyme action is usually proportional to the amount of enzyme present. Under optimum conditions the rate is constant for a time as the reaction proceeds (i.e. the reaction is of *zero order*), but later falls off. At very low substrate concentrations the progress curve of the reaction (a plot of product produced against time) shows that the reaction is of the *first order*. Between these extreme cases progress curves are complex; consequently activity measurements are normally based on the tangent at the origin of a progress curve (*initial velocity*) or on the zero order portion obtained with high concentrations of substrate.

The most characteristic feature of enzyme kinetics is the dependence of the reaction velocity on substrate concentration, illustrated in Fig. 1.

A direct plot of initial velocity against concentration (a) is a rectangular hyperbola, which can be defined by the asymptotic limiting velocity (V) and the substrate concentration giving half the maximum velocity (*Michaelis Constant, K_m*); this curve is often described as a *Michaelis Curve*. (b) and (c) show improved methods of calculating the characteristic constants V and K_m from the same data. A logarithmic plot is symmetrical, and K_m is given by the point of inflexion; while if the reciprocals of velocity and substrate concentration are plotted against each other, as shown in the reciprocal plot (c), the reciprocals of V and K_m are given by the intercepts on the axes. There has been an increasing use of digital computers for obtaining V and K_m, by fitting the Michaelis Equation to the experimental data by the method of least squares.

A theoretical explanation for experimental observations of this type, with a mathematical treatment, was given by Michaelis and Menten in 1911. The saturation phenomenon, the existence of a limiting substrate concentration above which the enzymatic reaction velocity is sensibly constant, indicates a two-stage process, only the first of which is directly dependent on substrate. It can be formulated:

$$E + S \rightleftharpoons ES \qquad (1)$$

$$ES \rightarrow E + P \qquad (2)$$

ES represents a 'complex' of enzyme and substrate, formed in a reversible equilibrium (1), in which the substrate is 'activated' so that it is able to undergo reaction; equation (2) represents this reaction, in which the product P is formed and free enzyme is liberated; the latter is then able to undergo another catalytic cycle. At very low substrate concentrations, reaction (1) is important, but at sufficiently high concentrations the equilibrium is pushed over completely towards ES and the overall

reaction is then limited by the rate of reaction (2), and is independent of substrate concentration. If (1) is characterised by a dissociation constant K_s for ES, and k is the velocity constant of reaction (2), it may be shown that the velocity v at any substrate concentration s is given by

$$v = Vs/(K_s + s),$$

where $v = k.$[enzyme concentration]. This, the so-called Michaelis Equation, fits the experimental curves shown above. This fundamental equation has been modified and extended to cover many quantitative features of enzyme action, such as the action of activators and inhibitors. More precise treatments than that given above, which use the concept of a "steady state," give equations of the same form, but with a complex kinetic meaning for K_m, not equal to K_s.

Enzymes are readily inactivated by heating and by extremes of pH (i.e. by acid and alkali); this inactivation, due to denaturation of the enzyme protein, is normally irreversible. In addition, smaller changes of pH produce reversible changes in activity, which usually shows a pronounced pH optimum. (Fig. 2.)

The velocity of enzyme-catalyzed reactions increases with temperature, although usually less rapidly than the corresponding uncatalysed reaction. However, the effects are complicated by a much greater effect of temperature on the rate of spontaneous destruction of the enzyme, so that there appears to be an optimum temperature for the enzyme.

Purification of enzymes. The discovery of a new enzyme reaction is followed by attempts to isolate the enzyme responsible. Each purification is to a large extent empirical, but certain general methods have been widely used. They include salting-out with ammonium sulphate, precipitation by organic solvents such as alcohol or acetone, adsorption on gelatinous adsorbents such as calcium phosphate or alumina, and column chromatography, especially on ion-exchangers such as substituted celluloses, and on cross-linked dextrans such as

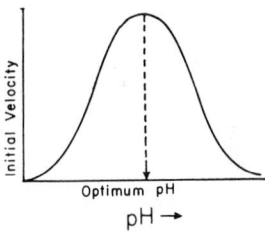

Fig. 2   Dependence of enzyme activity on pH.

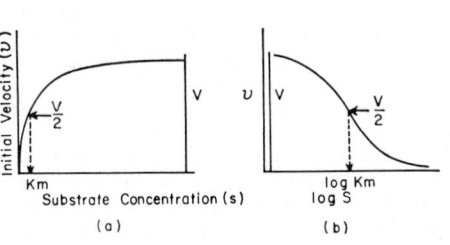

(a)

(b)

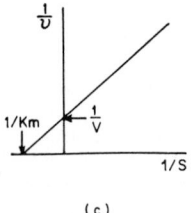

(c)

Fig. 1   Dependence of enzyme activity on substrate concentration.

Sephadex. All these methods can be used for *fractionations;* a series of protein fractions are obtained and tested, and the best fractions from the point of view of increase in specific activity and percentage recovery are combined for the next stage. To obtain a pure enzyme, the purification may have to be anything from ten-fold to several thousand-fold from the original tissue extract.

When a reasonable degree of purity has been obtained, the enzyme may often be *crystallized,* usually as micro-crystals. Since Sumner crystallized the first enzyme (urease) in 1926, altogether nearly 100 crystalline enzymes have been prepared. Unfortunately this is not necessarily an index of complete purity; the usual criteria of homogeneity used by the protein chemist must be applied.

All enzymes which have been obtained in a pure state have been found to be proteins, sometimes with a non-protein *prosthetic group* attached to the protein molecule; the molecular weight varies from 10,000 to 1,000,000 or more. In the last few years the structure of enzymes has been increasingly studied by the methods which have been so successfully applied to simpler proteins such as insulin. The sequence of aminoacids in a small number of enzymes of low molecular weight has been completely elucidated. In several other enzymes the sequence of important sections of the protein chain has been determined.

Evidence about the essential structure in the enzyme responsible for catalytic activity has been obtained by several methods. Inhibitors which prevent the activity of certain enzymes can sometimes be shown to react specifically with particular aminoacid residues. One group of enzymes is dependent on a free thiol group of a cysteine residue for its activity; in another group histidine plays an important role. The nerve gas di-iso-propyl phosphofluoridate (DFP) strongly inactivates a group of hydrolytic enzymes, including chymotrypsin, trypsin, cholinesterase and carboxylesterase. By using isotopically-labelled DFP it can be shown that the phosphate of the inhibitor remains firmly bound to the active centre of the enzyme, and by studying degradation products of the inhibited enzyme it is possible to determine an aminoacid sequence around the point attacked by the DFP. In this way it has been shown that all these enzymes contain somewhat similar sequences including a serine which reacts with DFP and which is presumably involved in the catalysis.

Mechanism of enzyme action. Many speculative accounts of the mechanism of catalysis by particular enzymes have been put forward, but it is probably true to say that in no case is the process really understood as yet. However, certain general features seem clear. Enzymes have an *active centre* (or several independent active centres on each molecule) which combines in a specific way with the substrate, perhaps because of close fitting of complementary molecular geometry in enzyme and·substrate. Stereochemical specificity, the preferential attack on one antipode of a racemic substrate, taken with other facts about enzyme specificity, suggest that the interaction at the active centre often involves attachment at 3 or more points. Electrostatic forces and hydrogen bonding are often important in enzyme-substrate interaction, but with many substrates such definite chemical forces cannot be invoked. The integrity of the tertiary structure of the whole protein molecule is generally necessary for enzyme action, and denaturation

inactivates the enzyme; this may be because the active centre is a pattern of groups extending transversely across two or more adjacent peptide chains, whose relative position is dependent on the intactness and configuration of the native protein. In any case, parts of the enzyme molecule beyond those actually involved in combination with substrate may be needed in order to produce the appropriate activation of the combined substrate molecule. In a few cases, however, it has been shown that the natural enzyme contains more than the minimum structure required for activity, and considerable portions of the protein molecule can be removed (chemically or enzymatically) without loss of activity.

In many cases there are probably additional intermediate stages between the initial combination of enzyme and substrate and the final liberation of the products of the reaction. When the reaction catalyzed is a transfer, the group transferred may be first accepted by the enzyme, either on to the protein itself or on to a prosthetic group, and then later handed on to the ultimate acceptor. Thus if E represents the enzyme, the general transferase reaction formulated earlier as

$$AX + B \rightarrow A + BX$$

may take place in the following stages

$$E + AX \rightleftharpoons E \cdot AX$$
$$E \cdot AX \rightarrow EX + A$$
$$EX + B \rightleftharpoons EX \cdot B$$
$$EX \cdot B \rightarrow E + BX$$

E·AX and EX·B represent the initial loose complexes formed with the substrates AX and B respectively, while EX represents a definite covalent binding of the −X radical on to the enzyme. Thus with the DFP-sensitive hydrolases mentioned earlier, which hydrolyse esters and peptides, it has been suggested that the enzyme is acylated with release of the alcohol or ammonia; the acyl-enzyme is unstable and hydrolysed very rapidly to give the other product, the acid. The specially reactive serine residue in the enzyme is the acyl-acceptor; a histidine residue takes some part in the hydrolysis of the acylated enzyme.

Rapid progress is taking place in this field, and it will probably not be long before it is possible to give a precise chemical account of enzymatic catalysis.

Comparative and developmental enzymology. Not all CELLS in the same organism contain the same pattern of enzymes, nor do similar cells from different species necessarily contain the same enzymes, either in quantity or precise characteristics. It must be admitted, however, that there are remarkable similarities in enzymatic pattern over a wide range of living organisms; morphological evolution has not been paralleled by any great increase in the number of enzymes. Nevertheless in many cases the enzyme proteins have undergone evolutionary modification, mainly small changes in the aminoacid sequences. Often when these differences cannot be detected in any other way, the enzymes can be distinguished by immunological cross-reactions; for example, ox and pig pepsin, or soy bean and jack bean urease, can be distinguished in this way.

Many enzymes exist in multiple forms (isoenzymes) in the same tissue. These isoenzymes have the same catalytic activity, but may differ slightly in aminoacid sequence or in conformation. The distribution pattern of isoenzymes is often characteristic of the particular tissue, and changes in the pattern have been used to study differentiation during development. These studies, together with those of enzyme content of mutants (see GENETICS), especially of molds and BACTERIA, can be interpreted in terms of the theory that enzymes are under close genetic control, one GENE being responsible for the production of one enzyme. A mutation which inactivates the gene will result in an enzyme deficiency which may cause a serious metabolic block. The fertilized ovum must obviously contain genes capable of producing all the enzymes and isoenzymes present in the tissues of the adult organism, but control mechanisms exist which stimulate or prevent the overt expression of the genes. It is hoped that elucidation of these control mechanisms will provide an insight into the important problems of embryological development, and work is actively proceeding in this field.

EDWIN C. WEBB

References

Dixon, M. and E. C. Webb, "Enzymes," 2nd ed. New York, Academic Press, 1964.
Dixon, M., "Multi-enzyme Systems," Cambridge, The University Press, 1949.
Laidler, K. J., "The Chemical Kinetics of Enzyme Action," Oxford, The University Press, 1958.
Boyer, P. D., H. Lardy and K. Myrbäck, eds., "The Enzymes," 2nd ed. (8 volumes) New York, Academic Press, 1959–1963.
Colowick, S. A. and N. O. Kaplan, eds., "Methods in Enzymology," 7 volumes, New York, Academic Press, 1955–64.
Wilkinson, J. H., "Isoenzymes," London, Spon, 1965.
"Enzyme Nomenclature:Recommendations (1964) of the International Union of Biochemistry," Amsterdam, Elsevier, 1965.

EPHEMEROPTERA

The Ephemeroptera or mayflies are the most primitive winged order of living INSECTS with no close relatives still extant. They are almost certainly derived either directly from ancestral lepismatoid THYSANURA or from the extinct order Palaeodictyoptera which was derived from the lepismatoid Thysanura. Whether the original winged insects would be classified as Palaeodictyoptera or Ephemeroptera is uncertain, but the terminal abdominal filament in addition to the cerci characterizes the Ephemeroptera and the Thysanura, but is not found in the Palaedictyoptera.

Adult mayflies ranges in wing expanse from about 3 to 75 mm. The antennae are inconspicuous and setiform. The compound eyes are small, simple and widely separated in both sexes of the Caenidae, Prosopistomatidae, and some Tricorythidae, and are simple in the females only of all other forms. The eyes of the males in many families have larger facets on the upper surfaces; these facets may gradually grade into the smaller facets below, may be separated by a clear line across the eye, or may be remotely separated off into a separate group on a turbinate stalk in the Baetidae and some Leptophlebiidae. The mouth parts are vestigial, but may be used to imbibe water.

The forewings vary in shape, but are usually somewhat triangular. The wing surface is fluted, with the veins alternately reaching the wing margin along the top of a ridge or the bottom of a groove. Midway in the length of the wing the principal concave or groove veins in the anterior half are interrupted by desclerotized spots called bullae. This is a specialization for vertical flight allowing the apical half of the wing to bend downward on the upstroke of the wing, but to retain full rigidity on the downstroke. The forewings of most genera have numerous crossveins, but the number is reduced in many groups, particularly in the small forms. In some forms (some Palingeniidae, *Behningia*, and the Oligoneuriinae) the concave veins lie adjacent to or under the convex veins so as to disrupt the fluting of the wings. These forms lack bullae on the veins and are capable of rapid flight. The hind wings are reduced in size or wanting and are often coupled to the forewings during flight.

The pro- and metathorax are reduced in size and may be partially consolidated into the enlarged mesothorax. The legs are moderately well developed in most forms but they are ill-adapted for walking. The forelegs of the male are specialized for grasping the female while copulating in flight. In some genera which have an adult life of only a few hours all legs of the female and the middle and hind legs of the male are obsolescent or vestigial.

The abdomen is composed of ten segments and bears two long cerci and a median terminal filament representing a modified eleventh segment. The females of some genera have paired genital openings, but in many these openings are fused. The male terminalia consist of paired styli of one to seven segments, and a pair of penes which may be separate or partly to completely fused into a single structure.

The mayflies have an incomplete metamorphosis; the winged stage which issues from the nymph is a subimago which must moult before becoming an adult or imago. This stage is easily indentifiable by the translucent wings with many setae along the hind margin. These setae persist in the adult in some small species. In the Oligoneuriidae the subimaginal pellicle is only partly shed, while some Polymitarcidae and Palingeniidae reproduce and die as subimagoes. The subimago stage lasts 24 to 48 hours in most species, but is much shorter in others, in *Ephoron*, only about 5 minutes. The adults are noted for their abbreviated life, but this is quite variable. Some genera, such as *Ephoron, Caenis, Prosopistoma* and others probably live only 1 or 2 hours, while the females of *Cloeon* and *Callibaetis* that retain the eggs in the body until they are ready to hatch, probably live for several weeks. Females of *Cloeon* have kept alive for over 50 days. Most species probably live 2 or 3 days.

Mating takes place in the air. The nuptial flight of some mayflies may take place in spectacular mass flights, but many others mate in small companies. Mating swarms of some species are oriented over specific markers such as rocks, clearings, streams, the "white water" of rapids, etc.

The eggs are laid or dropped on the surface of the water a few at a time or in a single or double cluster by most forms, but the females of *Baetis* descend beneath the water to lay the eggs on submerged objects.

In some, if not all, *Callibaetis* and *Cloeon* the eggs hatch within a few minutes after being deposited but for most species they hatch only after a few weeks to several months. The length of nymphal life varies considerably. Most species have an annual cycle with nymphal development taking 6 to 11 months, but many species produce several broods per year. The nymphal stage of *Parameletus columbiae* lasts only about two weeks, while *Hexagenia limbata* apparently spends two years in the nymphal stage.

Most of the nymphs are herbivores or scavengers, but a few are predators. They occupy a wide variety of fresh water and some species occur in brackish water. Within these habitats they are adapted for many niches. Most of the Ephemeroidea are burrowers in sand or silty bottoms; nymphs of the genus *Povilla* burrow in freshwater sponges or in wood. Some of the most extremely adapted mayfly nymphs are those burrowing in or living on shifting sand bottoms of rivers. The majority of the nymphs cling to rocks or vegetation or wander freely, crawling or swimming, along the bottom.

The subimago emerges from the nymph at the surface or the water for most kinds, but underwater emergence occurs in *Ephermera* and *Heptagenia*, and probably in others. Some of the Siphlonuridae nymphs crawl out of the water onto vegetation or stones.

The nymphal stages have setiform antennae of varying lengths and simple compound eyes. The mouth parts of most forms are of the typical chewing type, generally with three-segmented maxillary and labial palpi. Five segments persist in the maxillary palpi of most orthopteroids and Holometabola. The mandibles of most genera are robust with two teeth on the incisor lobe and well developed molar surfaces; the molar surfaces are absent and the teeth on the incisor lobe well developed in some Heptageniiae and Siphlonuridae that appear to be carnivores. The maxillae are highly variable but usually equipped with teeth on the laciniae. In *Ameletus* and *Metreletus* the apices are equipped with a bipectinate "diatom rake;" many Heptageniidae and Leptophlebiidae have hairs or spines probably of the same general use. The supposed carnivores have well developed sharp canines on the maxillae. The generalized three-segmented maxillary palpi are reduced to two or one segment or are wanting in some members of several phyletic lines; in the Ameletopsinae (Siphlonuridae) the palpi are threadlike and through secondary segmentation have 20 or more segments. The labium has become quite diversified in the many phyletic lines and its structural adaptations generally parallel the maxillae.

The legs are diversely adapted, usually with a single terminal claw, but this is lacking in a few sand dwellers. The legs of silt and sand dwellers are often extremely modified for burrowing or clinging to shifting sand surfaces.

The abdomen is ten-segmented and is moderately diverse in the various forms. It is usually cylindrical or sub-cylindrical, but it may be extremely depressed.

Gills arise from the pleurae of abdominal segments one to seven, but they may be missing from segments at either end of the series. The gill position is primitively lateral, but may be adaptively shifted to either dorsal or ventral positions. Not only are the gills modified for various current and oxygen conditions, they are often further modified as water circulation mechanisms, aids in swimming, protective devices against abrasion by silt and sand on other gills, adhesive holdfasts for clinging to stones, or for camouflage. Supplemental simple or branched gills arise from the mouth parts or the thorax in the Isonychiinae (Siphlonuridae) and Oligoneuriidae. In the Neotropical genus *Murphyella* these are the only

gills present; the abdominal gills are wanting. With their many functions, the gills show extremely diverse adaptive form.

GEORGE F. EDMUNDS, JR.

References

Berner, L., "The mayflies of Florida," Univ. Florida Studies, Bio. Sci. Series, 4(4): 1–267, 1950.
Burks, B. D., "The mayflies or Ephemeroptera of Illinois," Bull. Illinois Nat. Hist. Surv., 26 (Art. 1): 1–216. 1953.
Day, W. C., "Ephemeroptera" in "Aquatic Insects of California," Univ. Calif. Press, Berkeley and Los Angeles. pp. 79–105, 1956.
Edmunds, G. F., Jr., "Ephemeroptera" in "Fresh Water Biology," 2nd ed. New York, Wiley,, 1959.
Needham, J. G., J. R. Traver, and Yin-Chi Hsu. "The Biology of Mayflies," Ithaca, New York, Comstock, 1935.

EPITHELIUM

Epithelium constitutes one of four basic tissues that make up the animal body.

It most commonly appears as a cellular membrane which covers surfaces and lines cavitites but may also appear in the form of anastomosing cords or plates, especially in endocrine organs. Thus epithlium never presents a free margin but always exhibits cellular continuity unless injured. It is cellular, lacks a vascular supply and possesses a rich nerve supply. Its functions are diverse: protection, lubrication, absorption secretion, excretion, sensory reception, and even locomotion.

Epithelia are the oldest tissues, both phylogenetically and ontogenetically. For example, Hydra is essentially a two-layered, epithelial organism and blastulae and gastrulae exist as cellular layers. Epithelia are derived from all of the three germ layers.

When epithelia appear as membranes, they present free and attached surfaces. The free surfaces show many specializations, while the attached surfaces may be quite simple, separated from underlying connective tissue elements by a homogeneous layer, the basement membrane, possible of epithelial origin.

Epithelia are usually classified according to the arrangement of the cells in membranes and the form of component cells. Epithelia are described in terms of their appearance in vertical section. When there is only one layer an epithelium is simple. When there is more than one definitive layer, it is stratified. Simple epithelia are further classified according to the form of their component cells. The terms are self-explanatory; i.e., squamous, cuboidal, and columnar. Some simple squamous epithelia have been given special names according to their location: endothelium, lining blood vessels and lymphatics; mesothelium, lining the coelom and cavities derived from it.

In pseudostratified epithelium all the cells rest on the basement membrane but not all extend to the free surface. Stratified epithelia are ordinarily subdivided according to the differentiation of the cells of the superficial border. Thus we have cuboidal, columnar and squamous types. The squamous cells may become keratinized (cornified), especially on dry surfaces. The transition from one type of epithelium to another is frequently abrupt.

The epithelium of some organs, which are markedly distensible, such as the mammalian urinary bladder, is transitional. In the contracted state the stratification of the epithelium is very obvious but in full distention the number of layers is reduced and they are less distinct.

Stratified epithelia have a restricted phylogenetic distribution, being almost entirely limited to the vertebrates; not ordinarily found in protochordates or in invetebrates. However, the epidermis of an arrow worm, *Sagitta hexaptera,* has been described as stratified.

Epithelial cells are usually organized differently at their superfical and deep ends, a situation which confers on them the property of polarity. Polarity is probably fixed in part by the basic organization of the cytoplasm and this property is reflected by the relative positions occupied by such organelles as nucleus, mitochondria, Golgi-apparatus, centrioles (diplosomes) and accumulated secretory precursors.

At the free surface, the borders of some cells may appear as simple, structureless condensations, but, several kinds of processes may be present. The most numerous of these are microvilli; extremely short, delicate, uniformly shaped protoplasmic extensions, well resolved by the electron microscope. They are frequently associated with absorption, appearing as a uniform "striated border" on columnar intestinal epithelium, and as a less regular "brush border" on renal tubules. Other non-motile, long, irregular processes occur in male ducts of mammals. They are apparently concerned with the release of secretion.

Certain epithelial cells are transformed into sensory receptors (neuroepithelium). They possess either simple or multiple, non-motile processes (sensory bristles) and function in relation to hearing and equilibration (including the lateral-line system) and taste. They may be associated basally with a centriolar derivative. Neuroepithelia usually undergo regression if their nerve supply is interrupted.

The most notable differentiations of the free surface are the *motile* structures. Some are very long and slender and sharply localized, occurring either singly or in small groups. Each is a flagellum. The others which are shorter, of uniform length, and very numerous are cilia. They are widely dispersed on the cell tips, usually arranged according to a definitive plan. Both FLAGELLUM and CILIUM have basically the same structural pattern. Both are extensions of the protoplasm and their membranes are continuous with the plasma membrane. Microvilli, and cilia, or flagellum, may occur together on the same cell. Each motile filament contains two single, axial fibrils and nine double peripheral ones, all oriented in the long axis of each process.

Basally the cilia are associated with a row of granules (*basal bodies*), beneath the plasma membrane. These bodies are of centriolar origin and have been demonstrated to be specifically oriented in relation to the direction of the ciliary beat. Frequently, ciliary rootlets are present and continue from basal bodies far into the cytoplasm, often according to a definite pattern.

In many invertebrates the free surface of the epithelium is covered by a continuous layer of dense secreted material, a cuticla. This may remain thin and flexible as in worms or become greatly thickened and even calcified in the thick chitinous exo-skeleton of some arthropods.

In the enamel of the tooth, the secretion of each epithelial cell retains its individuality, so that with progressive secretory activity and coincident calcification long rods or columns are produced.

The lateral or contiguous surfaces of epithelial cells remain relatively simple. They are separated, or bound, by an exceedingly thin layer of cement, possibly a mucopolysaccharide. *Intercellular bridges*, which are most commonly found in stratified epithelium, are not, as the name implies and as the electron microscope reveals, direct cytoplasmic connections between cells but are closely apposed regional thickenings of adjacent cell membranes, forming attachment plaques. The tonofibrils, although they converge on these points, do not cross from cell to cell. *Terminal bars*, long regarded as collar-like sub-apical condensations of intercellular cement, appear in electron micrographs as dark double membranous thickenings. They apparently represent extensive regional elaborations, comparable to the smaller, disperse, plaques of deeper-lying cell surfaces.

Epithelia are subject to wear and tear, both physical and physiological, and are replaced by cell-division. In simple epithelia mitosis may occur at random throughout the tissue. In other cases, highly specialized cells may rarely or no longer divide and reserves of less differentiated cells remain capable of mitosis on the reception of a proper stimulus.

In the small intestine division is restricted to the tubular intestinal glands and the cells from this level migrate over the villi to replace the cells which are continuously lost at their tips. Leblond and Stevens have shown that in the rat such replacement may be completed every 36 hours. There are many variants of the pattern of normal replacement by division of regionally segregated cells accompanied by migration to a predetermined destination, to maintain or restore the cover. Even areas completely denuded by trauma are quickly covered by migration of neighboring cells; this early repair being later supplemented by mitosis.

In epithelia we have a series of living membranes, without gap or margin, which enable the organism to joust successfully with its environment—external and/or internal.

ALDEN B. DAWSON

References

Arey, L. B., "Wound healing," Physiol. Rev., **16**: 327–406, 1936.
Fawcett, D. W., "Structural specialization of the cell surface" *in* Palay, S. L., ed., "Frontiers in Cytology," New Haven, Yale University Press, 1958.
Fawcett, D. W. and K. R. Porter, "A study of the fine structure of ciliated epithelium," Jour. Morph. **94**: 221–282, 1954.
Leblond, C. P. and B. E. Walker, "Renewal of cell populations," Bio. Rev., **26**: 59–86, 1956.

ERICALES

The Heath Order, a group of DICOTYLEDONOUS angiosperms, widely known because of its horticulturally important members such as Azaleas, Rhododendrons, Heather and Blueberries. It is world-wide in distribution, but generally absent from areas with alkaline soils. Its affinities are with the Hypericales, Theales and Myrtales. The group is predominantly woody with many species in tropical regions, these usually being on mountain slopes or plateaus.

Biologically the group is closely tied to mycorrhizal fungi. This usually obligate association has permitted the evolution of non-green heterotrophic forms entirely dependent on their mycorrhizal associates. Such plants traditionally have been called "saprophytes"; physiologically they are parasites. On the whole the species have a fairly high light requirement, accounting for their individual abundance on rocky outcrops, on soils too sterile to support heavy forest, around ponds and bogs, on arctic tundra, as tropical epiphytes, in disturbed areas such as newly cut-over forest lands, and abandoned pastures. Those relatively few forms which regularly flower and fruit in dense shade usually depend to a considerable extent on their mycorrhizal associates for food, even if they possess green leaves. The group is predominantly evergreen, the leaves often being xeromorphic; these probably are characters of the group residual from its tropical and often epiphytic ancestors in which the evergreen, xeromorphic habit was fixed rather than a response to immediate ecological conditions. Tuberous "burls," sometimes a meter or more in diameter, are developed in a series of genera of both Ericaceae and Vacciniaceae; those of the Mediterranean *Erica arboraea* furnish the briar (bruyère) used in making tobacco pipes.

Cytogenetically the group is of considerable interest because of the high incidence of both auto-and allo-polyploids, close to half of the species being in one or the other category; the bulk of these are tetraploids, but 12-ploid material has been found in *Rhododendron*. A high degree of fertility in the offspring following crossings between homoploid species of the same genus appears to be the rule rather than the exception. This has been of considerable advantage to horticultural hybridists, but is the bane of classical taxonomists for the presence of natural populations with evidences of extensive genic introgression is all too frequent. The common occurrence of paired, simulative autopoly-ploids and highly segregative allopolyploids, these latter usually bridging the morphological gap between the parental species, merely adds to the taxonomist's problems. The Families of the Order are:

Ericaceae. World-wide in distribution, being forest trees sometimes to 30 meters or more (as certain Mexican and California spp. of *Arbutus*), shrubs, or minute prostrate shrublets (as the circumpolar *Cassiope hypnoides*). The most primitive genus is *Befaria*, its petals not united as in most other genera and with floral parts sometimes indefinite in number; it occurs from the equatorial Andes into Mexico, Cuba and Florida. Of the 1,500 species of the Family, over 600 are in *Rhododendron* and nearly 500 in *Erica*. *Rhododendron* occurs in Australia, New Guinea and Malaysia; its primary development is in SE. Asia with extensions into other temperate and arctic-alpine regions of the N. Hemisphere, less than 30 spp. being known in N. America. The plants commonly called Rhododendrons and Azaleas are placed in the same genus; the alternative is to split this polymorphic group into more than 30 poorly defined genera. *Erica*, which contains the true Heaths, is most abundant in S. Africa, extending northward into Europe. Heather is *Calluna*, a monotypic genus. Other characteristic genera are *Kalmia* (Mountain-Laurel and Lamb-kill), *Andromeda* (Bog-Rosemary), *Pieris* (Shrub-Andromeda), *Gaultheria* (over 100 species in

the mountainous areas circling the Pacific Basin; an outlier is *G. procumbens*, the Checkerberry of E. N. Amer.), *Pernettya* (about 25 spp. from Mexico to Patagonia and also in New Zealand and Tasmania; throughout its range the species often hybridize with homoploid spp. of *Gaultheria*), *Arctostaphylos* (about 50 spp., the Manzanitas of Mexico and W. U.S. being characteristic; *A. uvaursi*, the Bearberry, is circumpolar).

Vacciniaceae. A world-wide group, rare in Africa and Australia. It consists of coarse shrubs and creeping shrublets, many species being rhizomatous colony-formers; high-climbing woody vines are present in the Amer. tropics. The center of abundance is the Andean region of S. Amer., there dominated by members of the tribe *Thibaudieae*, these often being coarse plants with leathery, evergreen leaves and either brightly colored inflorescence bracts or large gaudily colored corollas; pollination in these is mainly by hummingbirds. The better known genera are *Vaccinium* (Blueberries, the most recent small-fruit crop to be extensively developed), *Gaylussacia* (the true Huckleberries; its center of concentration is Brazil), *Oxycoccus* (the true Cranberries).

Epacridaceae. Small trees and often xeromorphic shrubs, primarily of extratropical Australia, New Zealand and New Caledonia; *Leucopogon* extends into Indo-China and the Philippines, and *Lebatanthus* occurs in Patagonian S. America.

Pyrolaceae. Herbaceous and perennial evergreen plants of the N. Hemisphere, the group tending toward facultative parasitism. *Pyrola* (Shinleaf) has several species in NW. U.S. whose individuals either have leaves with chlorophyll or are leafless and essentially devoid of green, these sometimes being in adjacent colonies or even occurring as branches from the same rhizome. *Chimaphila* (Wintergreen or Pipsissewa, from its Amer-Indian name) inhabits dense forest areas; one species of E. N. Amer. has poorly developed chlorophyll in its leaves.

Monotropaceae. Parasitic plants of the N. Hemisphere with reduced leaves, the plants entirely lacking chlorophyll. *Monotropa* (Indian-pipes and Pinesap) is characteristic. Those genera with parietal placentae are suspect of being misplaced, perhaps being simulative developments from unrelated groups outside the Order.

Clethraceae. This family probably is related to the Saurauiaceae and Actinidiaceae; as a group these show some affinity with the Theales and Ericales, but apparently should be set apart.

Diapensiaceae. Recent anatomical studies indicate the group is not Ericalean; its relationships are not known.

Lennoaceae. Several parasitic, chlorophylless genera of SW. U.S. and Mexico. Their affinities may be with the Scrophulariales or Boraginales, but not with the Ericales.

W. H. CAMP

EVOLUTION

One of the most striking features of the world of life is its great diversity, comprising well over a million described species and an undetermined, but large, number of species still to be described. During post-Renaissance times, Linnaeus (1707–1778) and most of his contemporaries ascribed this to *special creation*, the idea that "there are just so many species as in the beginning the Infinite Being created" (Linnaeus). An alternative possibility, the gradual development or *evolution* of all species from one or a few original living forms, was considered by some biologists, such as Buffon (1707–1788), and it had been the subject of philosophical speculations since Empedocles (495–435 B.C.) and even before. In the absence of a reasonable hypothesis of the mechanism of evolution, however, none had been able to convince many of his contemporaries.

Lamarckism. The first major effort to deal with the origin of species on an evolutionary basis was that of Lamarck (1744–1829). He proposed the gradual development of species from similar predecessors on the basis of four principles: that all organisms tend to increase in size; that new organs arise because of new needs, and because of the activities resulting from such needs; that use of an organ results in its further development, while disuse results in its degeneration; and finally that changes produced by the action of these principles during the lifetime of an individual are inherited by its progeny (inheritance of acquired characters), so that the results are cumulative over long periods of time, thus resulting in the origin of species. After tentative statements as early as 1800, he developed his ideas fully in his *Philosophie Zoologique* (1809), and he fought for his ideas for the rest of his life. Yet he was unsuccessful, partly because of the opposition of Cuvier, the most influential zoologist of the time, but more because attempts to verify his basic theses, especially the inheritance of acquired characters, gave discordant or equivocal results.

Darwinism. Charles R. Darwin (1809–1882) was much more successful, because he developed a reasonable hypothesis on the mechanism of evolution, and because he assembled a great mass of data for the actual occurrence of evolution. His hypothesis can be summarized in five points. 1. All species tend to reproduce in numbers vastly in excess of those which can possibly survive (prodigality of nature), because reproduction is a geometric rather than an arithmetic process. 2. Nonetheless, adult populations in any region tend to remain fairly constant, which can only mean that there must be an enormous death rate in nature. This requires 3, that there must be a struggle for the means of survival, with the great majority of the contestants losing, for only thus could adult populations be restricted in the presence of a reproductive excess. 4. The competing individuals vary in many respects, for variation is one of the most universal characteristics of organisms; some of these variations may be neutral, but many will affect the ability of the individual to compete in the struggle for life. And 5, the result can only be *natural selection* of those organisms best fitted for their conditions of life, for in each generation relatively more of these will survive and leave progeny, which will generally inherit the characteristics of their parents. Thus species are gradually modified in the direction of better adaptation to their conditions of life.

Points 1 and 4 are observed facts. Point 2 is not true to the degree that Darwin thought, for natural populations are subject to great fluctuations. However, they always fall so far short of the reproductive potential that the point retains its logical force. Point 3 is a logical result of points 1 and 2, while point five is a logical result of points 1–4.

Darwin's evidence for the fact of evolution was drawn from such diverse fields as domestication, biogeography, taxonomy, comparative anatomy, comparative embryology, and paleontology. To these must be added comparative physiology, comparative biochemistry, and genetics. Each comprises a body of data which is readily understood on an evolutionary basis, but is anomalous or even contradictory on any other.

Modern Theories. The mechanics of evolution are now explained on the basis of mutation and selection, the predominant theory being called the *neo-Darwinian* theory, the *synthetic* theory, or even the *biological* theory. Inheritable variation is continually arising in all species by means of mutation, both gene mutation and chromosomal mutation. These are then reshuffled by the mechanism of sexual reproduction into new patterns of variability, which are put to the test of natural selection. How severe this may be is indicated by Blair's estimate that only 1 to 6 per cent of small mammals survive their first year. All of the conditions under which an organism lives act as selective agencies which permit more extensive survival and reproduction by the better adapted forms, while tending to restrict the less well adapted. A species is generally distributed over a wide range, with quite unlike conditions prevailing in various parts. Because of this, different populations of the same species are subject to different selective forces, and so the accumulation of mutations of selective value tends to make them first recognizable subspecies, then distinct species, and finally even genera and higher categories. For example the deer mouse, *Peromyscus maniculatus*, ranges over much of North America. Some of these mice never experience frost, while others must adapt to the Canadian winter; some live in marsh lands, and others in the deserts of the southwest; some live on prairies, and others in dense forests. The result has been the formation of a large number of geographically replacing subspecies. While this mechanism of formation of subspecies has been abundantly demonstrated, its extension to species and higher categories remains an unproven, though attractive, inference.

Natural selection was a logical inference for Darwin, but it was not actually demonstrated in nature. One of the major successes of modern evolutionary biology has been the demonstration of selection in nature and its duplication in controlled experiments. An example of selection in nature is afforded by the widespread use of DDT as an insecticide starting in 1945. At first, it seemed to be so effective an insecticide that some biologists feared a disastrous upset of the balance of nature. Campaigns of extermination of house flies were undertaken in many communities, and at first they seemed to be successful. A very small percentage of the flies, however, carried a mutant gene for resistance to DDT. Intensive poisoning campaigns provided strong selective pressure in favor of the resistant flies, and these increased rapidly, all but replacing the original form in many places. An excellent example of experimental selection is provided by Sukatchew's experiments on dandelions from the Crimea (temperate), Leningrad (sub-arctic), and Archangel (arctic). Mixed plots of the three varieties were sown at low and at high densities at Leningrad. In both cases, there was differential survival: in low density plantings, 60 per cent of the

Crimean plants, 96 per cent of the Lenigrad (local) plants, and 88 per cent of the Archangel plants survived; while at high densities, where selection should be more severe, 70 per cent of the Archangel plants, 11 per cent of the Leningrad plants, and only 1 per cent of the Crimean plants survived.

Thus, natural selection is now a well demonstrated phenomenon. Mathematical studies have indicated that population size and structure influence the effectiveness of selection. Medium size is optimal, for in very large populations changes in different parts may cancel out, while in very small populations selection may be swamped by genetic drift. That is, because the basic genetic processes (meiosis and fertilization) are random processes, sampling errors may cause changes in gene frequency without regard to adaptive value. One result is that permanently small populations are likely to be rather poorly adapted.

Population structure is also important. Total populations of most wild species are enormous, but these do not form continuous, interbreeding arrays. They are generally broken up into subspecies, each occupying a different territory and characterized by a somewhat different constellation of genes and gene-determined traits. Even the subspecies, however, may be very abundant and widespread, but each is typically restricted by its ecological requirements to limited parts of its total range. Thus the meadow mouse, *Microtus p. pennsylvanicus*, is spread over much of eastern North America. In this great range, however, it is excluded from forests and swamps, from river side and lake side, from cultivated land and urban communities. The result is a checkerboard of rather small meadow mouse communities, each somewhat isolated from the rest because of the unsuitability of the intervening territory. These local populations of a few hundred to a few thousand mice are the real breeding populations. Generally, their size is favorable for efficient action of natural selection. There is a limited amount of gene flow between such populations because of occasional migration of individuals. Further, wherever the territories of neighboring subspecies meet, there is a certain amount of gene-flow between them. The result is that neighboring subspecies tend to be most alike, while distant ones are likely to be strongly differentiated. Mathematical studies indicate that this type of population structure is especially favorable for evolution.

Dissenting views. While approbation of the neo-Darwinian theory has been general, it has not been universal. The most important dissenter has been Goldschmidt, whose views stemmed especially from studies on position effects and the theory of the gene. He conceived the gene not as a discrete particle but as an integrated part of a biochemical continuum (the chromosome). The entire genetic system he conceived as an integrated whole in which any considerable change (or cumulative series of changes) would probably be disharmonious, hence he did not consider it likely that accumulation of micromutations could result in neo-Darwinian speciation as a rule. For this purpose, a radical repatterning of the chromosomes to establish a new stable pattern was proposed. Such a *systemic mutation* would modify early embryonic processes, thus producing extensive changes throughout the adult organism. Goldschmidt believed that in this way new species might be formed in one or a few steps, then immediately put to the test of natural selection. The majority would fail this test and quickly become extinct, while the successful minority would then undergo refinement of adaptation by the neo-Darwinian process. While he regarded the systemic mutations as undemonstrated, there are some probable examples.

Goldschmidt was profoundly skeptical of many neo-Darwinian tenets, especially regarding the rate of evolution. Calculations of rates based upon weak selection pressures acting upon micromutations led to very slow rates, and he was doubtful that so slow a process could achieve the observed results (the actual world of life) even in the great reaches of geological time. It was partly to avoid this dilemma that he proposed the systemic mutations. Other accelerating mechanisms are also available: severe selection pressure, for example. Again, if two closely related groups are in competition, accentuation of their differences will tend to reduce competition between them and thus permit a larger total population. Such *character displacement* may be very rapid. Selection is directional and mutation random, but if predominant direction of mutation should happen to coincide with direction of selection, rapid evolution should result. It is probable that all of these accelerating mechanisms, including systemic mutation, have played roles in evolution.

Conclusion. Darwin attributed the great success of his theory to his having abstracted and reabstracted a great mass of material, selecting the most striking facts and conclusions, and to his habit of carefully noting anything which seemed opposed to his general conclusions. Quite as important was the fact that biology had accumulated a great mass of data which was chaotic because there was no general theory upon which it could be organized. Evolution quickly achieved a central position in biology because the theory of evolution by means of natural selection provided just such a theory: descent with modification is the basic fact of biology, playing much the same role in biology as do the laws of thermodynamics in physics.

EDWARD O. DODSON

References

Darwin, C. R., "On the Origin of Species by Means of Natural Selection, or the Preservation of Favoured Races in the Struggle for Life," New York, Modern Library Giants Series, 1859.
Dobzhansky, Th., "Genetics and the Origin of Species," New York, Columbia University Press, 1951.
Dodson, E. O., "Evolution: Process and Product," New York, Reinhold, 1960.
Goldschmidt, R. B., "Theoretical Genetics," Berkeley, Univ. of California Press, 1955.

EXCRETION

A universal phenomenon of living matter is production of metabolic wastes which need to be disposed of for efficient physiological function. Unicellular and simple multicellular organisms excrete by diffusion through cell membranes. In higher plants, excretion occurs mainly through the leaves. Excess carbon dioxide is excreted by this route, as is oxygen when produced in excess by photosynthesis. Solid wastes when produced are excreted through roots into the soil water.

Excretion in animals occurs through several organs:

(1) the kidneys which produce urine; (2) the skin by secretion of sweat; (3) the liver with biliary excretion; (4) the lungs which rid the body of carbon dioxide. In man, this amounts to 288 L./day (26 EQ. equal to 2.6 L. of HCl). (5) The intestine excretes bile residues and undigested food residues, modified by bacterial activity. Its epithelium can secrete certain inorganic constituents and water. In water-dwelling forms, gills function for gas, water and electrolyte exchange, and waste disposal.

All systems function together to maintain the constancy of the internal environment of the cells. In the sections which follow, emphasis will be placed on the role of kidney, liver, and skin, particularly as they function in man.

Kidney

Details of kidney structure should be reviewed before functional aspects are considered.

Nature and composition of the urine. About 1,250 ml. of urine are excreted in 24 hours by normal man, with specific gravities usually between 1.018 and 1.024 (extremes, 1.003–1.040). Flow ranges between 0.5 to 20 ml. per minute with extremes of dehydration and hydration. Maximum osmolar concentration is 1,400, compared to plasma osmolarity of 350 mOsm./L. In diabetes insipidus, characterized by inadequate *antidiuretic hormone* (ADH) production, volumes of 15–25 L. per day of dilute urine are formed.

In addition to substances listed in Table 1, there are trace amounts of purine bases and methylated purines, glucuronates, the pigments urochrome and urobilin, hippuric acid, and amino acids. In pathological states, other substances may appear: proteins (nephrosis); bile pigments and salts (biliary obstruction); and glucose, acetone, acetoacetic acid, and B-hydroxybutyric acid (diabetes mellitus).

The U/P ratios of the substances in Table 1 vary widely because of differential handling by the kidney. Quantitative knowledge of glomerular filtration, tubular reabsorption, and secretion of these requires an understanding of the concept of *renal plasma clearance.*

Excretion rates and plasma clearance. The rate at which a substance (X) is excreted in the urine is the product of its urinary concentration, Ux (mgm./ml.), and the volume or urine per minute, V. The rate of excretion (UxV) depends, among other factors, on the concentration of X in the plasma, Px (mgm./ml.). It is therefore reasonable to relate UxV to Px, and this is called the clearance ratio: $(Ux \cdot V)/Px$, or more generally, UV/P. This has the dimensions of volume, and is in reality the smallest volume from which the kidneys can obtain the amount of X excreted per minute. It must be understood that the kidneys do not usually clear the plasma completely of X, but clear a larger volume incompletely. The clearance is therefore not a real but a *virtual* volume. When substances are being cleared simultaneously each has its own clearance rate, depending on the amount reabsorbed from the glomerular filtrate or added by tubular secretion. The former will have the lower clearance, the latter the higher. Those cleared only by glomerular filtration will be intermediate and their clearance will in effect measure the rate of *glomerular filtration* in ml./min.

The best known substance which can be infused into the blood to provide a clearance equal to glomerular filtration rate is *inulin*, a polymer of fructose containing 32 hexose molecules (M.W., 5,200).

Strong evidence exists that it is neither reabsorbed nor secreted, that it is freely filterable, is not metabolized, and has no physiological influences. Its clearance in man is 120–130 ml./min. This is taken to be the glomerular filtration rate (GFR) or C_F (amount of plasma water filtered through glomeruli per minute). Besides inulin in the dog and other vertebrates, creatinine, thiosulphate, ferrocyanide, and mannitol also fulfill these requirements.

Knowing the glomerular filtration rate permits quantitation of the amount of any substance freely filtered (C_F, ml./min. X P_X (mgm./ml)). Subtracting from this one minute's excretion, $U_X V$, would give the amount reabsorbed in mgm./min. A classical example is the glucose mechanism. At normal plasma concentrations, none or a trace appears in the urine. When plasma glucose is elevated to about 180–200 mgm./%

Table 1 Composition of 24 Hr. Urine in the Normal Adult

	Amount		U/P*
Urea	6.0 –18.0 g.N		60.0
Creatinine	0.3 – 0.8 g.N		70.0
Ammonia	0.4 – 1.0 g.N		—
Uric acid	0.08– 0.2 g.N		20.0
Sodium	2.0 – 4.0 g.	(100.0–200.0 mEQ)	0.8– 1.5
Potassium	1.5 – 2.0 g.	(35.0– 50.0 mEQ)	10.0–15.0
Calcium	0.1 – 0.3 g.	(2.5– 7.5 mEQ)	—
Magnesium	0.1 – 0.2 g.	(8.0– 16.0 mEQ)	—
Chloride	4.0 – 8.0 g.	(100.0–250.0 mEQ)	0.8– 2.0
Bicarbonate	—	(0.0– 50.0 mEQ)	0.0– 2.0
Phosphate	0.7 – 1.6 g.P	(2.0– 50.0 mM)	25.0
Inorganic sulfate	0.6 – 1.8 g.S	(40.0–120.0 mEQ)	50.0
Organic sulfate	0.06– 0.2 g.S		—

(From White, Handler, Smith and Stetten: Principles of Biochemistry, McGraw-Hill, 1959.)

*U/P:ratio of urinary to plasma concentration.

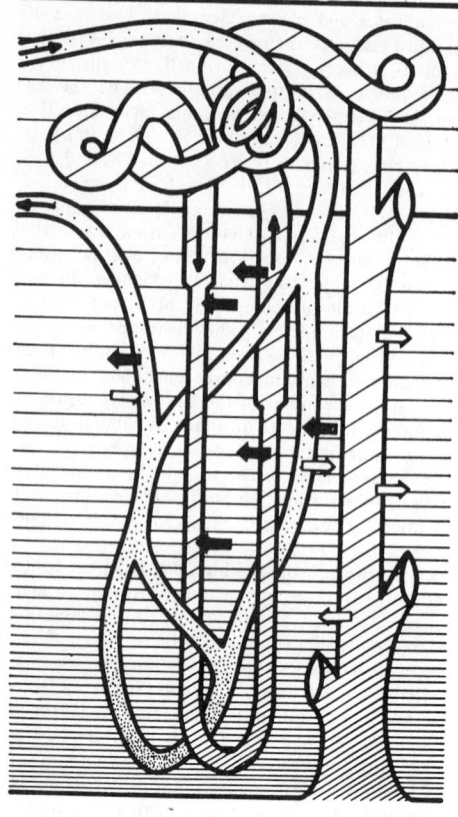

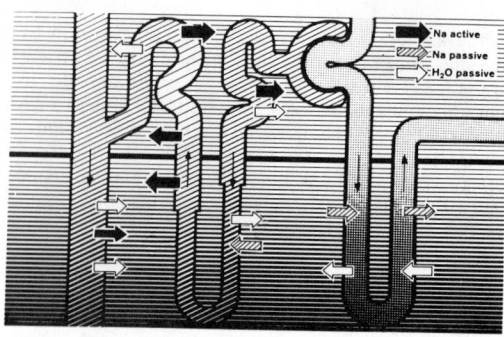

Fig. 1A

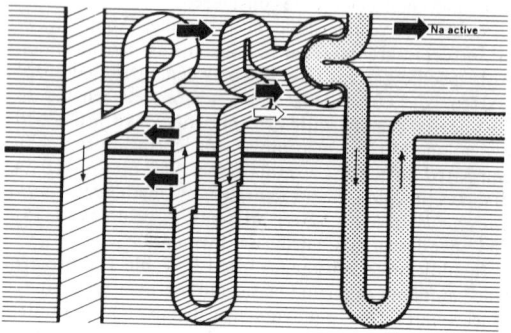

Fig. 1   Fig. 1B

Fig. 1   Nephron and related blood supply showing stratification of osmotic concentration from cortex to medulla (dounter-current system). White arrows: water transfer (passive). Dark arrows: transport of crystalloids (active or passive). Fig. 1A: Concentrating kidney. Fig. 1B: Diluting kidney (no ADH). (Courtesy of H. Wirz, Helvetica Physiologica et Pharmacologica Acta **11**: 20, 1953.)

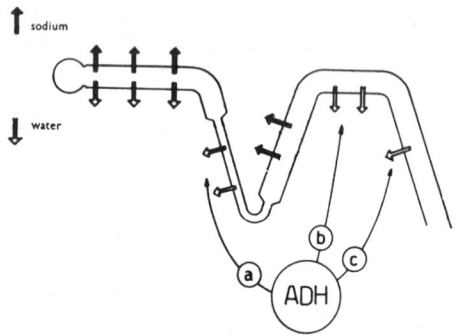

Fig. 2   Nephron showing sites of sodium and water reabsorption and ADH action. (a) descending limb of loop of Henle allowing tubular fluid to reach equilibrium with hypertonic extra-cellular fluid, thus starting the counter-current mechanism; (b) allowing water reabsorption to catch up with distal tubular reabsorption of sodium; (c) passive transfer of water to hypertonic surroundings, thus concentrating the urine in collecting tubule above isotonicity. (Courtesy of H. Wirz, *The Neurophypophysis*, H. Heller, Editor, Academic Press, Inc., New York, 1957, p. 165.)

(the "threshold"), the amount appearing in the urine begins to increase. As concentration is raised more, the nephrons become progressively saturated until the rate of reabsorption becomes constant and maximal. This indication of saturation of the transport system is referred to as the T_m, "tubular maximum," (here, T_{mG}). In humans T_{mG} has the value of 340 mgm./min. Absorption occurs in the proximal convoluted tubules.

Tubular Reabsorption of Organic Substances

Sugars. Xylose, fructose, and galactose when introduced into the blood are reabsorbed by the same transfer system as glucose, but much less completely. When glucose load is increased, their reabsorption is blocked (*competitive inhibition*). The glycoside, *phlorizin* (contained in the bark of apple, pear and other fruit trees) blocks reabsorption of all sugars with resulting increase in urinary excretion.

Amino acids. The total clearance is low in man (1–8 ml./min.). Much of the characterization of mechanisms has been done in the dog. Several (glycine, arginine, and lysine) demonstrate relatively poor reabsorption, with small T_m's Other amino acids are so efficiently reabsorbed that saturation is not achieved by concentrations which do not cause physiological disturbances.

Ascorbic acid. This has a T_m of $2.0 + 0.19$ mgm./1.73 SQ.M.S.A. in man (or ca 1.5 mgm./100 ml. of filtrate). In the dog, reabsorption is 0.5 mgm./100 ml. of filtrate. This represents a net figure, for secretion occurs under certain circumstances in the distal nephron. Reabsorption occurs proximally.

Urea. This major product of protein metabolism is filtered and reabsorbed to varying degrees (40 to 70%), inversely related to the rate of urine production. Reabsorption is largely a process of back-diffusion in man, dog, rabbit, and chicken, although active mechanisms operate in the kidneys of Elasmobranchii. Recent evidence suggests the possibility of active transport in the mammalian kidney. In amphibian (Anuran) kidneys, the tubules secrete urea.

Uric acid. In mammals this appears as a consequence of metabolism of purine bases. In most mammals, it is oxidized to allantoic acid, but not in primates or the Dalmatian coach dog. It has been generally assumed that a T_m characterizes its reabsorption, average 15 mgm./min. in man, a value so high compared to the amounts existing in the plasma that saturation should not occur normally. Under conditions of injections of an uricosuric drug (G-28315 (Sulfinpyrazone)) and vigorous mannitol diuresis, excreted urate/filtrated urate ratios up to 1.23 have been observed in man, demonstrating tubular secretion. A three-component system of filtration, reabsorption, and secretion is suggested, so that the amount excreted is the *net effect* of these operations. Definite evidence for tubular secretion has been found in birds and reptiles.

Creatine. This is a product of muscle metabolism which disappears from the urine of humans after adolescence. It is reabsorbed in concentrations below 0.5 mgm./%. At higher concentrations, reabsorption is incomplete and excretion is enhanced. No T_m has been demonstrated. At higher plasma levels, the creatine/C_F ratio becomes constant at 0.8

Tubular Secretion

Evidence has accumulated that the proximal tubule is the site of active secretion of some physiologically occurring substances as well as certain foreign substances when injected into the circulation.

Tubular Secretion of Foreign Substances

P-aminohippuric acid (PAH). At low plasma concentrations, this substance is almost completely cleared from the plasma by a combination of glomerular filtration and efficient tubular secretion. Hence, its clearance measures 90% of total plasma flow through the kidneys. In man, the clearance is 600–700 ml./min., and corrected for hematocrit, gives the *effective renal blood flow.*

When plasma levels are elevated in the range of 30 to 50 mgm./%, the excretory mechanism becomes saturated and a T_m can be discerned. The quantity of the substance filtered is $P_{PAH} \times C_F$; the total excreted in the urine is $U_{PAH} \times V$, with T_{PAH} designated as the tubular contribution. Then T, $T_{mPAH} = U_{PAH} \cdot V$ $-P_{PAH} \cdot G_F$. PAH in plasma is freely filterable. Some substances, e.g., diodrast and phenol red, are

bound to protein and a correction factor needs to be introduced. T_m for PAH in man is ca 77 mgm./min./1.73 M.S.A. In the dog, it is 19.1 mgm./min./SQ.M. S.A.

Others

Additional foreign substances, secreted by the *organic acid* transport system, are *diodrast, phenol red,* and *penicillin. Carinamide* and *benemid* block the secretion of the above. The blocking is competitive, for these substances are also secreted by the tubules. A separate *organic base* system secretes substances like tetraethyl- and tetramethylammonium, mepiperphenidol, and priscoline.

Tubular Secretion of Physiological Substances

Creatinine. This is derived from creatine in muscles. Exogenous creatinine is cleared by glomerular filtration plus tubular secretion in man; the T_m is small. It is also secreted by tubules of certain teleosts, alligator, chicken, goat, guinea pig, and rat. Glomerular filtration alone is the mechanism in the rabbit, sheep, seal, cat, frog and turtle. In the male dog, a small component may be secreted in the proximal tubule, although glomerular filtration is the dominant mechanism.

N-methylnicotinamide. This metabolic derivative of nicotinic acid has a clearance up to three times the C_F in dogs.

Ammonium. This is synthesized in the tubular epithelium from glutamine and other amino acids and secreted into the distal tubular urine.

Potassium. Its clearance is usually well below that of inulin in man and dog, suggesting fairly complete reabsorption. Under certain circumstances, e.g., giving large amounts of K salts of foreign anions, the clearance of K arises above C_F. It appears that a three-component system operates, with proximal reabsorption and distal secretion.

Hydrogen ions. The proximal and distal tubules are able to generate and secrete H^+ as a means of acidifying the urine by the aid of the enzyme, carbonic anhydrase:

$$CO_2 + H_2O \xrightleftharpoons{C.A.} H_2CO_3 \begin{matrix} \nearrow HCO_3 \\ \searrow H^+ \end{matrix}$$

Acid-base regulation. All three (NH^+, K^+, and H^+) are secreted into the distal tubular urine to be exchanged for Na (of Na_2HPO_4 and NaCl), conserving valuable base and ridding the body of metabolic acids. In the proximal, exchange of H^+ is for $NaHCO_3$.

Excretion of Electrolytes

Cations. Sodium. About 99% of the filtered sodium is reabsorbed by active mechanisms, 80–85% in the proximal tubule and 14–19% in the distal tubule and collecting duct. In man, ca 200 μEQ./min. are excreted. *Aldosterone,* secreted by the adrenal cortex, is necessary for efficient reabsorption.

The distal absorption is concerned with the acid-base regulatory mechanism. No T_m for Na can be discerned. Increased loading is followed by increase in total reabsorption, but with decreased efficiency, so that urinary excretion is increased.

Ca. Its excretion is complicated by the fact that a significant part of its plasma content is combined with plasma proteins. That which is filtered is in complexes poorly ionized; hence it is not handled as Ca^{++}. Urinary excretion is low (*ca* 8.5 μEQ./min.) suggesting efficient tubular reabsorption.

Mg. That not bound by plasma proteins is filtered and reabsorbed. About 5–6 mEQ./min. are excreted in man.

Anions. *Cl.* This renal mechanism is similar to sodium, being the chief "indifferent" anion that accompanies Na^+ through the kidney.

PO_4. In the plasma about 80% occurs as HPO_4^- and 20% as $H_2PO_4^-$, both usually combined with Na. The ultimate ratio of $H_2PO_4^-/HPO_4^-$ in urine is determined by the pH. PO_4 T_m in the dog is 0.10–0.15 mM/min., and in man, 0.13 mM./min. Excretion in man is 7–20 μM./min.

SO_4. This is actively reabsorbed in dog and man, and shows a well-defined, although small T_m . In the dog, it is 0.05 mM./min. With slight increases in plasma concentration, excretion is rapidly increased.

Excretion of water. Mechanisms which concern tubular reabsorption of water are intimately tied up with the handling of osmotic constituents, primarily sodium salts. Another factor is the action of anti-diuretic hormone (ADH), elaborated by the supra-optic and paraventricular nuclei of the hypothalamus. Finally the composite mechanisms are currently integrated in the light of a *counter-current diffusion multiplier system*, operating particularly in nephrons which project long loops of Henle and vasa recta into the papillary zones of the renal medulla, as exemplified by the golden hamster and kangaroo rat.

Since the fluid in the proximal convoluted tubule has been found to be isosmotic by direct puncture studies, it is assumed that water follows Na in on a passive osmotic obligatory basis. Evidence from microcryscopic methods shows the osmotic pressure in the cortex equal to that of the plasma, but that it is stratified in increasing concentrations proceeding from the cortex to the tip of the papillae where it is 3-4 times that of plasma.

Recently, this has been verified by micropuncture of the loops of Henle and the accompanying vasa recta, both arranged in the principle of a "hair-pin" counter-current system. Finally, the osmotic concentration in the collecting tubules has been found to parallel the concentration in the loops as the tubules pass from the cortex to their point of exit at the tip of the papillae.

The principle of the counter-current system as it applies to the medullary loop of Henle system is as follows: sodium, by an active mechanism, and chloride as the result of an electrochemical gradient thus established, are believed to be transported out of the relatively water impermeable ascending limb of the loop of Henle into the interstitium of the medulla until a gradient of *ca* 200 mOsm. Kg. H_2O has been established. This single effect is multiplied as the fluid of the thin descending limb comes into osmotic equilibrium with the interstitial fluid by the diffusion of water out (and probably by the diffusion of some NaCl into the descending limb), thus raising the os-

molarity of the fluid rounding the hair-pin loop into the ascending limb. The increased concentration here, also raising that in the interstitium, now favors further movement of fluid out of the descending limb, further increasing concentration and so on.

In this fashion, an increasing osmotic gradient is established in the direction of the tip of the papillae, and yet at no level is there a large osmotic difference between the luminal and interstitial fluid. The collecting ducts in the presence of ADH are believed to be water permeable and somewhat Na-impermeable (net transport small, although there may be diffusion into and active transport out). This results in diffusion of water out of the collecting ducts into the hyperosmotic medullary interstitium, and ultimately into the vasa recta to be carried away until the fluid in the ducts becomes corresponding concentrated.

The view is favored that ADH acts in a permissive fashion to let water diffuse out, perhaps altering the size of "pores" in the base of the cells of the tubular epithelium. In addition, the membrane of the opposite end of the cell presents an obstacle more readily passed by water than by hydrophilic solutes.

The active hormone is bound to a carrier neurosecretory material (NSM) as it is formed in the supra-optic and paraventricular nuclei of the hypothalamus. This substance is supposed to flow via the supra-optic-hypophyseal tract to the posterior pituitary lobe, which functions as a storage and a release center.

The change in effective osmotic pressure in the blood to the centers appears to be the effective stimulus, hypertonicity causing increased release, and urinary concentration. The probability exists that certain *volume receptors* respond to isotonic fluid expansion to inhibit ADH action and produce diuresis.

The active portion of NSM concerned with tubular water transfer is *arginine-vasopressin* in man, oxen, monkey, dog, rat, sheep, and camel. Lysinevasopressin is the active principle in the hog.

$$\overline{CYS \cdot TYR \cdot PHE \cdot GLU \cdot (NH_2) \cdot ASP(NH_2) \cdot CYS} \cdot$$
$$\overset{*}{PRO} \cdot A\overset{*}{R}G \cdot GLY \cdot (NH_2)$$

arginine vasopressin

(Lysine-vasopressin is identical except for lysine in place of arginine where starred.)

Liver

The liver as an organ of excretion must be qualified by the fact that a significant portion of its secretion, the bile, after temporary storage in the gall bladder, is released into the intestine, and returned to the liver by way of the portal vein to be re-excreted—the *enterohepatic circulation*. However, small amounts of bile salts escape in the feces. In addition, bile pigments form the chief pigment of the feces, (stercobilin) and some gets into the urine as *urobilinogen*.

Properties of bile. Bile is secreted at the rate of 500–1000 ml./day by the parenchymal cells of the liver. The composition of human bile appears in Table 2.

Bile salts. Cholesterol, synthesized in the liver, is the precursor of bile acids. Four have been isolated from human bile: cholic, deoxycholic, chenodeoxy-

Table 2   Composition of Human Bile (in per cent)

	Liver Bile	Gall-bladder Bile
Water	98.0	89.0
Total solids	2.0	11.0
Bile salts	0.7	6.0
Bile pigments	0.2	2.5
Cholesterol	0.06	0.4
Inorganic salts	0.7	0.8
pH	8.0–8.6	7.0–7.6

cholic, and lithocholic. Their structure is illustrated below using cholic acid, the most abundant in humans, as an example:

The four acids differ only by the number of —OH radicals at positions 3, 7, and 12 of the phenantrene nucleus.

These acids are coupled in amide linkage to the amino acids glycine and taurine:

$$H_2N \cdot CH_2 \cdot COOH \qquad H_2N \cdot CH_2 \cdot CH_2 \cdot SO_3H$$

glycine   taurine

$$\text{cholic acid} + CoA[1] + ATP \xrightarrow[Mg^{++}]{\text{enzyme}}$$

$$\text{cholyl CoA} + AMP[1] + PP[1]$$

$$\text{cholyl CoA} + \text{taurine} \xrightarrow{\text{enzyme}}$$

$$\text{taurocholic acid} + CoA$$

The products are glycocholic and taurocholic acid, the major representatives in human bile, in a ratio of 3:1.

$$C_{23}H_{26}(OH)_3 C \cdot N \cdot CH_2 \cdot COOH$$

glycocholic acid

$$C_{23}H_{26}(OH)_3 C \cdot N \cdot CH_2 \cdot CH_2 \cdot SO_3N$$

taurocholic acid

The conjugated acids are water-soluble, and their salts are powerful detergents. About 5 to 15 gm. are secreted per day, but most of its is returned to the liver via the enterohepatic circulation.

Bile pigments. When red blood cells break down, the hemoglobin is degraded into *choleglobin* by open-

[1]ATP—adenosine triphosphate; AMP—adenosine monophosphate; CoA—coenzyme A; PP—pyrophosate.

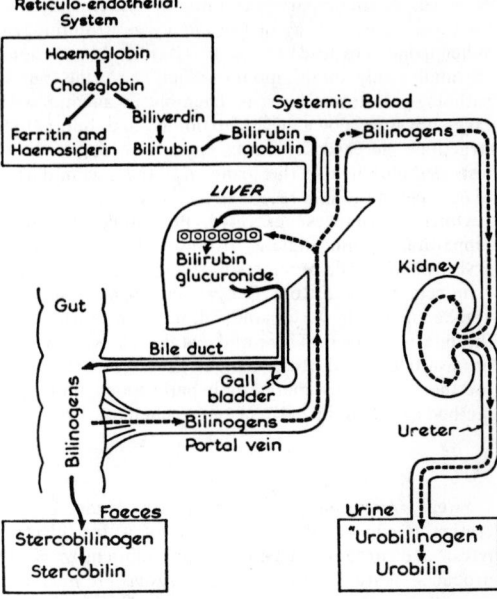

Fig. 3   Scheme of changes involved in the formation of bile. (Courtesy of G. H. Bell, *Textbook of Physiology and Biochemistry*, G. H. Bell, J. N. Davidson, H. Scarborough, Williams & Wilkins Co., Baltimore, 1959, p. 404.)

ing of the porphyrin ring. Next, the molecule loses both protein and iron, which are salvaged by the reticulo-endothelial system for new production of erythrocytes. The pigment which remains, *biliverdin*, consists of four pyrrole rings:

biliverdin

Bilirubin passes with plasma alpha globulin to the liver and is converted into water-soluble bilirubin mono- and diglucuronides. These are excreted in the bile and pass into the intestine where bacterial enzymes form colorless compounds known as *bilinogens*, most important of which are *mesobilirubinogen* and *stercobilinogen*. Undergoing autooxidation, stercobilinogen forms the chief pigment of feces, *stercobilin*, excreted in amounts of 40 to 280 mgm. per day.

Part of the bilinogen fraction is reabsorbed from the gut and passes back to the liver where it is re-excreted. A small part gets into the general circulation and leaves by way of the kidney as urobilinogen. When urine is exposed to air, it is oxidized to *urobilin*. Normally, only small amounts reach the urine, but in pathological states such as haemolytic anemia with excessive breakdown of hemoglobin, and biliary obstruction, increased amounts appear. Strong positive tests are obtained in the urine. Related accumulation of pigment in the skin results in jaundice.

Other substances excreted. Penicillin, p-amino-hippurate, bromosulfalein (bromosulfon-phthalein or BSP), phenol red, bromcresol green, fluorescin, phlorizin, cinchophen, Rose Bengal and eriocyanin. The uptake of dye is very rapid, although excretion into the bile is delayed. The rapid uptake of dyes such as bromsulfalein and Rose Bengal have been the basis for their use in estimates of hepatic blood flow by a method based on the indirect Fick principles.

Skin

Sweat formation. The skin as an organ of excretion is concerned with the loss of water, electrolytes, and organic substances predominantly waste products. Water is lost in part as *insensible perspiration* at the rate of 30 ml./hr. in man, $\frac{2}{3}$ of which is lost through the skin and $\frac{1}{3}$ through the respiratory tract. The origin of this water is the interstitial fluid of the dermis, which diffuses through the skin without wetting it. *Sensible perspiration* is supplied by some $2\frac{1}{2}$ million sweat glands in man, innervated by cholineric fibers of the sympathetic nervous system. There are two kinds of sweat glands: *apocrine*, in which the secretion end of the gland when filled is pinched off. (These occur largely in axillae and nipples); and *eccrine* (exocrine) glands, by far the most numerous covering the other areas of the skin. The surface film of skin is an emulsion of fatty and aqueous materials; the lipid component is derived mainly from *sebum*, the excretory product of sebaceous glands. Formation of sweat is not a simple filtration process, for the sweat glands may develop a pressure of 250 mm Hg., considerably higher than arterial blood pressure. The stimulus to secretion is a rise in blood temperature, activating centers in the hypothalamus, or reflex afferents, e.g., gustatory nerve stimulation.

Volume of sweat. The volume of production varies widely depending largely on environmental temperature and amount of physical activity. This is illustrated in Fig. 4. The volume lost may easily reach 1–2 L./hr. and with strenuous exercise in the sun, 4 L./hr.

Composition of sweat. Sweat is about 99% water, with a specific gravity of about 1.003. The pH is usually acid, from 5 to 7.5. Sweat is hypotonic, the solid content averaging about 680 mgm./%, of which about two-thirds is ash, and one-third organic substance.

Inorganic Content:

NaCl: Its concentration is very variable, but is usually in the range of 18-97 mEQ./L. The Na/Cl ratio averages 1:11. Men working in a hot environment, losing 12 L. of sweat per day with a concentration of NaCl of 100 mEQ./L., would lose 70 gm. of salt, 7 times the daily intake.

K: Its concentration averages 4.5 mEQ./L., and

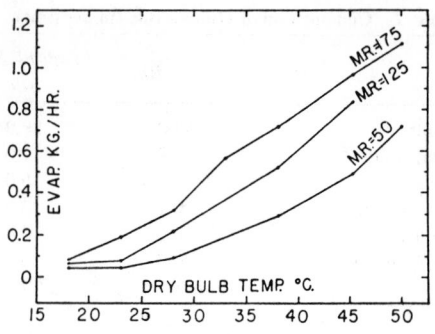

Fig. 4   Effects of increasing room temperature on rates of sweating at three different metabolic rates: at rest (M.R. 50 Cal./m²/hr.); walking on treadmill at 4.5 Km./hr. (M.R. 125 Cal./m²/hr.); and walking at 5.6 Km./hr. up a 2.5% grade (M.R. 175 Cal./m²/hr.). (Courtesy of S. Robinson, 1954.)

varies inversely with Na. The Na/K ratio is elevated to 15 in a hot environment without acclimatization; in about 5 days of adaptation it drops to 5, due to an increase in K concentration, and a 50% fall in Na. This is taken as an indication of increased adrenal cortical activity.

Other minerals: Ca ranges from 1-8 mgm./%, with a decrease in concentration as sweating proceeds. *Mg:* 0.04-0.4 mgm./%; *P:* 0.003 to 0.04 mgm./%; *SO₄:* 4-17 mgm./%; and traces of *Cu, Fe* and *Mn*.

Organic Content:

Nitrogen:

Urea N: 12–39 mgm./%; the sweat urea/plasma urea equals 1.92 ± 0.48, regardless of the concentration in the plasma or the rate of sweating.

NH₃–N: Its concentration in eccrine sweat ranges from 5-9 mgm./% (50–200 times the blood content).

Creatinine: Averages 0.4 mgm./% (0.1 to 1.3).

Uric acid: is less concentrated than in the blood (0 to 1.5 mgm./%).

Lactic acid: 4–40 mEQ. of acid per L. of sweat (4–40 times that in the blood).

Pyruvic acid: 0.1–0.8 mEQ./L. (1.3–10 times that of blood).

Miscellaneous organic: glucose in insignificant amounts; amino acids, vitamins, phenol, and histamine. *Drugs* such as nicotine, atabrine, morphine, sulfanilamide, and alcohol appear in sweat after administration.

EWALD E. SELKURT

References

Haslewood, G. A. D., "Recent Developments in our Knowledge of Bile Salts," Physiol. Rev. 35: 178–196, 1955.

Lorincz, A. L. and R. B. Stoughton, R. B., "Specific Metabolic Processes of Skin," Physiol. Rev. 38: 481–502, 1958.

Robinson, J. R., "Reflections on Renal Function," Springfield, Ill., Thomas, 1954.

Robinson, S. and A. H. Robinson, "Chemical Composition of the Sweat," Physiol. Rev. 34: 202–220, 1954.

Selkurt, E. E., "Kidney, Water and Electrolyte Metabolism," Ann. Rev. Physiol. **21**: 117–150, 1959.

ibid., "Physiology," Boston, Little, Brown, Ch. 23, 1966.

Smith, Homer W., "Principles of Renal Physiology," Oxford, The University Press, New York, 1956.

Sperber, I., "Secretion of Organic Anions in the Formation of Urine and Bile," Pharm. Rev. **11**: 109–134, 1959.

EXCRETORY ORGANS

Maintenance of stable conditions of form, structure and chemical composition within an organism (homeostasis) requires the elimination of substances which do not belong in its internal medium. Such substances include both normal and necessary constituents which are in *excess* of the quantity compatible with homeostasis and other ingested substances which are either useless or harmful. Finally, there are the end-products of metabolism which, if not removed, would have toxic effects. The use of the term "excretion" is reserved, by many, for the elimination of the nitrogenous end-products of protein metabolism, while elimination of water, salt and carbon dioxide and useless extraneous substances is not included in this term by the same authors. In this brief survey, however, we shall use the term excretion in its wider sense, including organs which eliminate any superfluous substance from the internal medium. Defecation of indigestible material, however, is not included because this material, having remained in the lumen of the gastrointestinal canal, i.e., in an inward extension of the outer world, never entered the internal milieu.

In freshwater animals elimination of water is an important function in itself, while in marine and desert dwellers, water serves as a vehicle in which waste materials can be carried to the exterior. Various mechanisms have been evolved by these animals to conserve this vital commodity. Water sufficient for many of the organism's needs derives partially from the catabolism of glucose to H_2O and CO_2. But most marine and xeric-adapted animals have evolved specialized means of conserving or creating additional H_2O. This can be, and is done in three different ways by various species: 1. active excretion of Na^+ by the *gills* in marine crustacea, teleosts, the *salt glands* in reptiles and marine birds, and *rectal gland* of insects; 2. transient decreases in glomerular filtration rate (GFR), whereby H_2O is retained temporarily at the expense of nitrogenous waste elimination, e.g., amphibians, reptiles and most birds, and 3. the production of a hypertonic urine, made possible by the development of structures in the kidney medulla capable of functioning as countercurrent multipliers, e.g., all mammals and most birds. (Table 1).

Mechanisms of Excretion

Diffusion through the surface of the body is the simplest method of excretion. It occurs, for example, among marine protozoans and in the Acanthocephala. In some teleosts, diffusion of waste occurs through the surface of the gills.

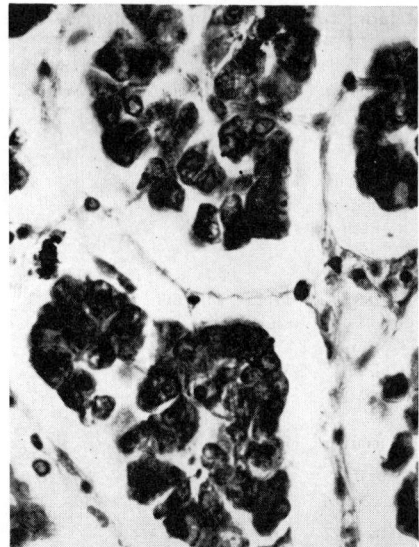

Fig. 1A

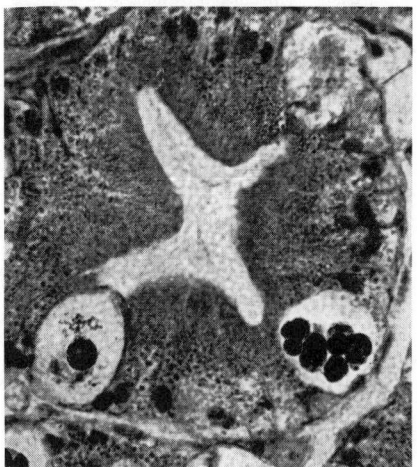

Fig. 1B

Fig. 1   A, liver of *Petromyzon marinus* (adult). Resemblance to liver is lost. Excreta are stored in very fine granules and diffuse in the cytoplasm, giving the cells a greenish and yellowish appearance. B, midintestinal gland of *Helix pomatia*. Excreta are stored in the form of fine granules and large concretions.

Active transport as a part of the process of urine formation occurs in virtually all animals. But in some species it is the primary mode of excretion, e.g., the gills of marine teleosts excrete Na^+ against a concentration gradient.

Phagocytosis by wandering cells occurs among the sponges and among the echinoderms. In these phyla the phagocytes, having picked up the excreta, expel them into the outer world, or they emigrate laden with excreta.

Storage excretion. In some hydroid coelenterates (Alcyconium), as well as in some crustaceans

Table 1   Relative Role of Kidney in Adaptation (Modified from Schmidt-Nielsen, B., 1964)

Phylum	Habitat	Excretory Organs	Excretory Mechanisms	Urine Osm. / Blood Osm.	Role of Kidney	Other Excr. Organs	Role
Coelenter.	marine	—	—	—	—	Body Surface	excr. nitr. waste
	fresh	—	—	—	—	Body Surface	excr. nitr. waste
Echinoder.	marine	—	—	—	—	Body Surface	excr. nitr. waste
Rotifera	marine	—	—	—	—	Body Surface	excr. nitr. waste
	fresh	Protonephr.	?	?	?	Body Surface	excr. nitr. waste
Gastrotr.	marine	—	—	—	—	Body Surface	excr. nitr. waste
Turbellar.	marine	—	—	—	—	Body Surface	excr. nitr. waste
	fresh	Proto = long diff. tubule	?	?hypotonic	ion + vol. + osmoreg.	Body Surface	excr. nitr. waste
Protozoa	marine	Contractile	?	?isotonic	ion + vol. reg.	Body Surface	excr. nitr. waste
	fresh	Vacuole	?	?	?	Body Surface	excr. nitr. waste
Kinoryncha	marine	Proto = short undiff. tubule	?	?isotonic	ion + vol. reg.	Body Surface	excr. nitr. waste
Nematoda	marine	Renette Cell	?	?isotonic	ion + vol. reg.	Body Surface	excr. nitr. waste
	fresh	Renette Cell + cuticle-lined tubule	?	hypotonic	ion + vol. + osmoreg.	Body Surface	excr. nitr. waste
Annelida	marine	Metanephr. = short undiff. tubule	?	isotonic	ion + vol. reg.	Body Surface	excr. nitr. waste
	fresh	Metanephr. = long diff. tubule	?	0.05–1.0	ion + vol. + osmoreg.	Body Surface	excr. nitr. waste
Crustacea	marine	Green Gland	filtr. + reabsorp.	1.0–1.6	ion + vol. + wk. osmoreg.	Gills	excrete nitr. + NaCl + osmoreg.
	fresh	Green Gland + nephr. canal	filtr. + reabsorp.	0.05	ion + vol. + osmoreg.	Gills	osmoreg. + uptake NaCl
Mollusca	marine	Bojanus Organ	filtr. + reabsorp.	isotonic	ion + vol. reg.	Gills	excr. nitr. waste
	fresh	Bojanus Organ	filtr. + reabsorp.	0.5–1.0	ion + vol. + osmoreg.	Gills	uptake NaCl + osmoreg.
Insecta	marine (larvae)	Malpigh. tub. + hindgut	secretion	0.5–10	ion + vol. + osmoreg. + excrete nitrog. waste	—	—
	fresh	Same + Rectal Gl.	secretion	.03	ion + vol. + osmoreg. + excrete nitrog. waste	Anal Papilla	osmoreg. + uptake NaCl
	terr.	Same + or − Peri-Rectal Mem.	secretion	.10–10+	ion + vol. + osmoreg. + excrete nitrog. waste	—	—
Cyclostom.	marine	glomer. + prox. tub.	filtr.-secr.	?isotonic	ion + vol. reg.	Gills	excr. nitr. waste
Elasmobr.	marine	glomer. + prox. + distal tubule	filtr.-secr.	.75–1.0	water excr.; ion + vol. reg.; urea conserv.	Rectal (Anal) Gl.	excr. NaCl
	fresh	glomer. + prox. + distal tubule	filtr.-secr.	0.10	ion + vol. + osmoreg.	Gills	excr. nitr. waste; osmoreg.; ?NaCl uptake
Teleostei	marine	glomer. or aglom. + prox. tubule	filtr.-secr. or secretion filtr.-reab. + secretion	.85–1.0	ion + vol. reg.	Gills	excr. nitr. waste; osmoreg.; excr. NaCl
	fresh	glomer. + prox. + distal tubule	filtr.-secr. or secretion filtr.-reab. + secretion	.10	ion + vol. + osmoreg.	Gills	excr. nitr. waste; osmoreg.; excr. NaCl
Amphibia	marine	glomer. + prox. + distal tubule	filtr.-secr. or secretion filtr.-reab. + secretion	.1–1.0	water excr.; ion + vol. reg.; urea conserv.	Skin	?

able 1 (continued)

Phylum	Habitat	Excretory Organs	Excretory Mechanisms	Urine Osm. / Blood Osm.	Role of Kidney	Other Excr. Organs	Role
	fresh	glomer. + prox. + distal tubule	filtr.-secr. or secretion filtr.-reab. + secretion	.1–1.0	ion + vol. + osmoreg.; excr. nitr. waste	Skin	osmoreg.; NaCl uptake
	terr.	glomer. + prox. + distal tubule	filtr.-secr. or secretion filtr.-reab. + secretion	.1–1.0	ion + vol. + osmoreg.; excr. nitr. waste	Skin	osmoreg.; NaCl uptake
Reptilia	marine	glomer. + prox. + distal tubule	filtr.-secr. or secretion filtr.-reab. + secretion	.1–1.0	ion + vol. reg.; excr. nitr. waste	Salt Gl.	osmoreg.; excr. NaCl
	fresh	glomer. + prox. + distal tubule	filtr.-secr. or secretion filtr.-reab. + secretion	.1–1.0	same + osmoreg.	Skin	osmoreg.; NaCl uptake
	terr.	glomer. + prox. + distal tubule	filtr.-secr. or secretion filtr.-reab. + secretion	.1–1.0	same + osmoreg.	Bladder	osmoreg. (= H_2O conserv.)
Aves	marine	same + Henle's Loop	filtr.-secr. or secretion filtr.-reab. + secretion	.1–4.5	same + osmoreg.	Salt Gl.	osmoreg.; excr. NaCl
	terr.	same + Henle's Loop	filtr.-secr. or secretion filtr.-reab. + secretion	.1–4.5	same + osmoreg.	—	—
Mammalia	marine	same + Henle's Loop	filtr.-secr. or secretion filtr.-reab. + secretion	.1–10	same + osmoreg.	—	—
	fresh	same + Henle's Loop	filtr.-secr. or secretion filtr.-reab. + secretion	.15–2	same + osmoreg.	—	—
	terr.	same + Henle's Loop	filtr.-secr. or secretion filtr.-reab. + secretion	.10–10	same + osmoreg.	Eccrine Glands	excr. NaCl; excr. nitr. waste

(Homarus), phagocytes fully loaded with excreta come to rest permanently within the body. This process is a primitive type of storage excretion. The liver of the Petromyzonidae (Fig. 1-A) while continuing its activity of temporary food storage and blood purification, must retain urea, bilirubin and biliverdin after the regression of its duct system, and thus becomes a storage place for waste products. In some Pulmonata, the midintestinal gland is loaded with similar excreta (Fig. 1-B). Guanine, a nitrogen-containing waste product, is deposited in crystalline form in special pigment cells in the skin of fishes, amphibians and reptiles, in the sclera of some fishes, in the tapetum lacidum of the eye of carnivores and in the irides of blue-eyed persons, serving in all these locations as a light reflector.

Organs of Excretion

Contractile vacuoles are bubbles of water within the cytoplasm of most freshwater and some marine protozoans and of freshwater sponges which rhythmically expand and collapse. Expansion (diastole) of contractile vacuoles may be a diffuse process of liquid accumulation (Amoeba), or the vacuole expands by contributions through "feeder canals" as in Ciliata (Fig. 2). In dinoflagellates, a permanently open bay receives liquid and conducts it to the outside. It is believed by many that contractile vacuoles eliminate only water.

Blind tubules embedded in the mesenchyme or in the coelom carry out excretion by an active process in which the excreta are secreted through epithelia into the tubular lumen. The tubules may empty directly through a bladder, or the intestine into the outer world.

Protonephridia are blind tubules provided with a terminal *flame cell.* The concavity of the flame cell represents the blind end of the tubular lumen. Toward the mesenchyme, the numerous slender processes of the flame cell reach far into the tissue spaces and probably actively collect fluid by pinocytosis, while

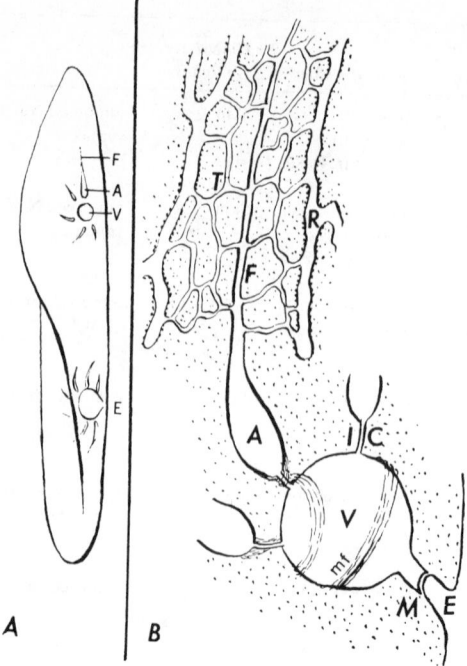

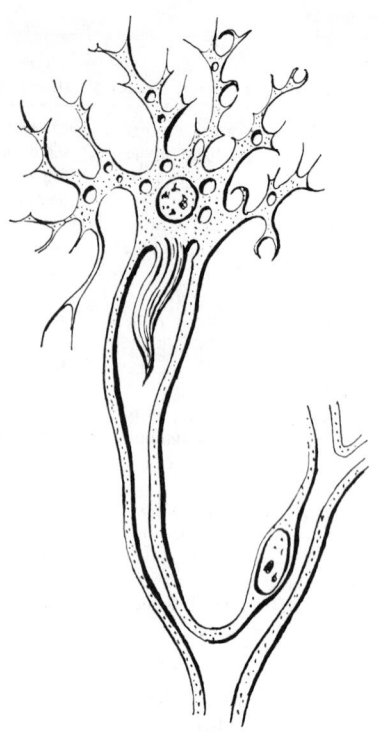

Fig. 2 Excretory apparatus of *Paramoecium caudatum*. A, Drawn from a living slipper animal at high power. B, A simplified diagram based on the electron-microscopic work of L. Schneider (J. Protozool. 7: 75–90, 1960). Legend: *A*, ampulla; *E*, excretory pore; *F*, feeder canal; *IC*, injector canal; *M*, closing membrane; *mf*, myofibrils; *T*, nephridial tubules; *R*, endoplasmic reticulum; *V*, contractile vacuole.

Fig. 3 Flame cell and protonephridial tubule of a tapeworm (composite based on several sources).

toward the tubular lumen of the flame cell, a bundle of long cilia project into the lumen, and cause a stream of liquid in the tubule by their flame-like waving. The protonephridial tubules of freshwater Rotifera and Gastrotricha are differentiated into an *anterior capillary* region and a *posterior glandular* region. In Turbellarians the degree of differentia-

tion also depends on habitat, marine forms having no excretory organs, freshwater species having very complex, convoluted protonephridia differentiated into a proximal *ampulla* and a *bladder*. The exclusively marine Kinoryncha and the Platyhelminths contain rudimentary protonephridia. Protonephridia empty singly or through common ducts to the outside.

Renette cells with a *lateral canal system* of variable length constitute the excretory organs of Nematodes, unlike any other in the animal kingdom. The terminal, excretory duct often is cuticle-lined in freshwater

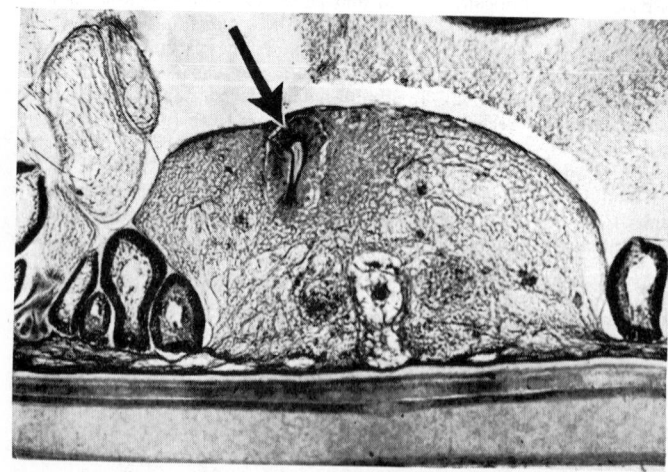

Fig. 4 Excretory canal of *Ascaris megalocephala*.

species, eventually emptying through a common, medial pore (Fig. 4). The length of the canal system and the presence of cuticle probably facilitate water excretion in freshwater forms.

Malpighian tubules (Fig. 5), the excretory units of Insecta, Myriapoda and Chilopoda, are blind tubules immersed in the coelom which open into the hindgut. They are lined by hemispherical epithelial cells with a brush border. In virtually all of these species, urine is *secreted* from the hemocoele into the blind tubules. Potassium is actively transported into the tubules against a large gradient, while other osmotically active substances passively follow. Within the posterior alimentary canal the *rectal epithelium* serves to concentrate the urine by actively reabsorbing electrolytes from the tubular fluid. In some Lepidoptera larvae and Coleoptera, the Malpighian tubules are applied closely to the rectal wall and the entire complex is surrounded by a *perirectal membrane*. Because these organisms excrete a dry pellet, it has been speculated that the urine is concentrated by countercurrent multiplication analogous to mammals (Kirschner, 1967).

Metanephridia (=nephridia) (Fig. 6) are tubules which possess a ciliated, funnel-like end that opens into the coelom. They are highly developed in the Annelida where, as a rule, a ciliated funnel, called a *nephrostome*, opens into the coelomic cavity of one segment, while the convoluted tubule runs through the next, emptying through a *nephridiopore* to the outside. The highly convoluted tubule of a metanephridium is composed of several cytologically specialized portions.

In contrast to the often numerous nephridia of Annelida, most Mollusca possess a single pair of nephridia, each opening from the pericardial cavity

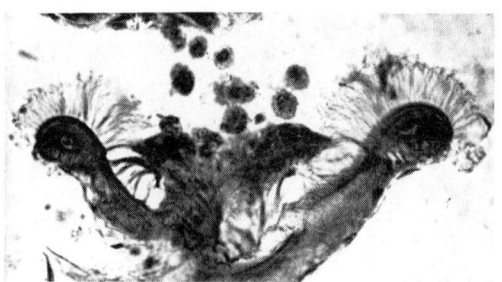

Fig. 6 Nephrostome of the earth worm *Lumbricus terrestris* (ventral side up).

through a small nephridiostome, into the *kidney* or *Bojanus organ*. These consist of an intensely convoluted nephridium and a thin-walled bladder lined by two cell types. Mollusca are the only invertebrates other than insects that can excrete a concentrated urine, presumably through reabsorption of water from the bladder. Interestingly, ultrafiltration of plasma across the heart wall occurs in a manner similar to that of glomerular vertebrates. Furthermore, urine formation is similarly subject to changes in blood pressure, urine composition being altered by reabsorption in proximal regions of the nephridium.

In the Crustacean *antennal-gland* urine formation also occurs by ultrafiltration of coelomic fluid into the *coelomosac*. By electron microscopy the *peritubular cells* of the coelomosac strikingly resemble the visceral epithelium of vertebrate glomeruli. Reabsorption of monovalent electrolytes, glucose and water occurs in the *labyrinth* or *green body* (proximal, convoluted portion of antennal gland). The elaboration of a hypotonic urine in freshwater forms occurs in the distal, *nephridial canal.*

The excretory system of Acrania (Amphioxus) consists of segmentally arranged tubules, receiving fluid from the coelom by way of nephrostomes and discharging it by a nephridiopore into the peribranchial cavity.

The nephron. Among vertebrates, we encounter three generations of kidneys, all derived from intermediate mesoderm; the *pronephros* (the persisting kidney or the Myxinoida = hagfishes), the *mesonephros* [persisting kidney of Anamniota = lampreys, fishes (except Myxinoids) and amphibians] and the *metanephros* (persisting kidney of Amniota = reptiles, birds and mammals).

In embryos, and in many larvae, the elements of the pronephros have an organization similar to that of the metanephridia, i.e., tubules which open with a ciliated nephrostome into the coelom, one pair per segment (Fig. 7A). But the tubules all empty into the continuous, longitudinal *pronephric (Wolffian) duct* which empties into the *cloaca.*

In older larvae with a functional pronephros, an arterial, capillary tuft located near the nephrostome, the *renal glomerulus*, serves to filter liquid from the blood into a pocket near the nephrostome. In 5-year-old "Amocoetes" larvae of Petromyzon (Fig. 7B) one very long glomerulus runs through many segments serving 50 or more tubules, each beginning with a long, ciliated neckpiece, the nephrostome, and emptying

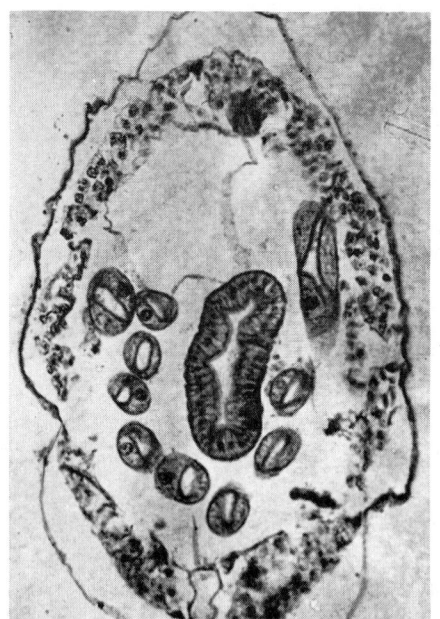

Fig. 5 Malpighian tubules surrounding the intestine of a mosquito *Anopheles quadrimaculatus.*

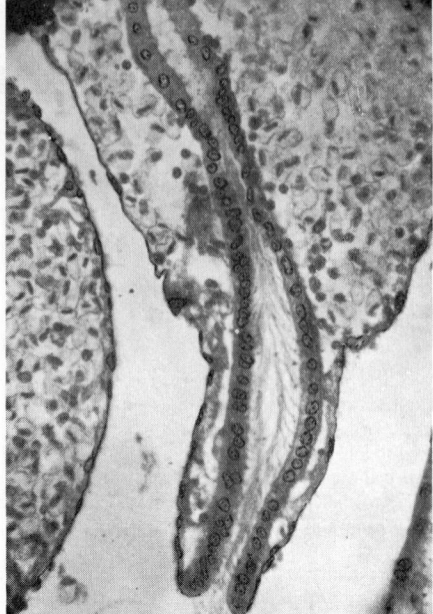

Fig. 7A

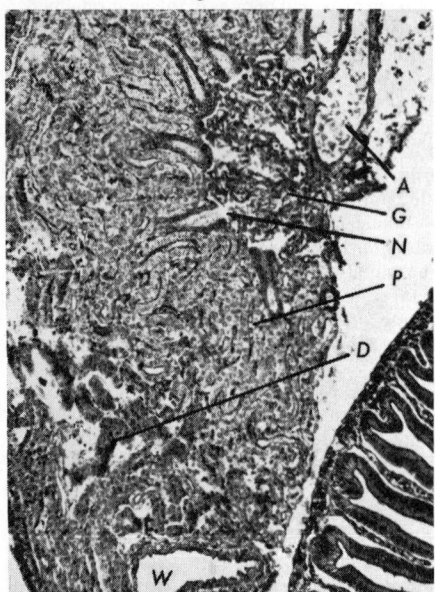

Fig. 7B

Fig. 7 Nephrostome and proximal part of pronephric tubule of "Amocoetes" larva 4 years old. B, pronephros of an "Amocoetes" larva of *Peteromyzon marinus*, 5 years old. *A*, afferent artery; *G*, glomerulus; *N*, nephrostome (7 nephrostomes appear in this section); *P*, proximal convolution; *D*, distal convolution; *W*, Wolffian duct.

into the pronephric duct which is lined with transitional epithelium.

In the kidney of adult vertebrates, however, nephrostomes persist rarely, if at all. Even the pronephros

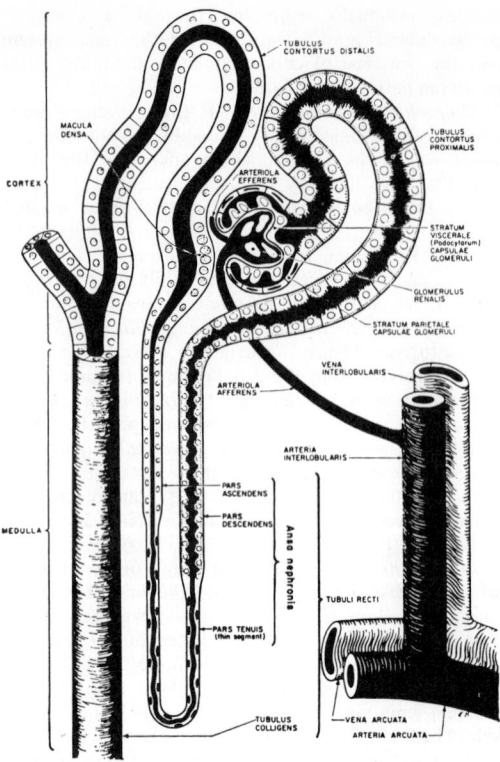

Fig. 8 Diagram of human nephron (from Elias and Pauly, Human Microanatomy, Chicago, 1960).

of adult Myxine does not possess open nephrostomes. Structurally, the nephra of all vertebrate kidneys are so similar to each other that the human nephron (Fig. 8) may serve as the paradigm.

The nephron is a twisted tubule lined by a simple epithelium showing varying structure along its course. The nephra of pronephroi and mesonephroi empty, one behind the other, into the Wolffian duct, while those of metanephroi empty into *collecting tubules* which unite into *papillary ducts*. These ducts discharge the definitive urine into extensions of the *ureter* called, according to their degree of branching *pelvis, calyces majores* and *calyces minores*. The ureter, in turn, empties into the urinary bladder which leads into the urethra.

Beginning at the exit, these passages have the following epithelial linings: *Urethra*—stratified squamous, stratified columnar and transitional epithelium; *bladder, ureter* and *calyces*—transitional epithelium; *papillary ducts* and *collecting tubules*—simple cuboidal epithelium; *distal convoluted tubule*—simple cuboidal epithelium with a few microvilli; *thin segment of ansa nephronis* (Henle's loop)—simple squamous epithelium with a few microvilli; *proximal convoluted tubule*—simple cuboidal epithelium with deep, basal indentations and rod-like mitochondria and with brush border consisting of tall, densely packed microvilli (Fig. 11); *neck piece*—(not present in mammals and Myxinoida)—clear, cuboidal cells with cilia. This piece is, perhaps, a remnant of the nephrostome.

The *blind end* of the nephron of adult vertebrates is enlarged into an invaginated bulb called the *glomerular capsule* (Bowman's capsule) lined by a parietal, simple squamous epithelium and by a specialized visceral epithelium which consists of large cells, often called podocytes, possessing long interdigitating processes (Figs. 9, 10).

Blood enters through the glomerular capsule at the *vascular pole* via the *arteriola afferens* and immediately ramifies into a tuft of anastomosing, arterial channels called the glomerulus. These vessels are lined by a specialized endothelium with many attenuated, cribriform portions through which the primary urine must filter (Fig. 9). Other cells of endothelial origin called the *mesangium*, occupy the root or stalk of the capillary tuft, but subserve an entirely different function—the "maintenance," replacement and "cleansing" of the basement membrane. The latter structure surrounds both endothelial cell types, and is the first filter which the plasma encounters. There is some evidence that the vascular channels, at least in amphibians, shift positions within the basement membrane. Thus, in these animals, the endothelium becomes a large, flat, branched sheet tunneled by shifting channels called the *lamina vasculosa glomeruli*.

Surrounding the basement membrane is the layer of podocytes. The podocytes ramify into interdigitating *pedicels*, which cover much of the surface of the basement membrane. Often a lateral, membranous extension of the podocytes, the *podocytic membrane*, can be seen covering the layer of pedicels (Fig. 9). The role of the podocyte in filtration and in variable states of anti-diuresis and hydration is the subject of considerable conjecture.

The *primary urine*, having traversed the endothelial fenestra, basement membrane, interpedicular spaces, and perhaps the podocytic membrane, is now in the *urinary space*, above the visceral epithelium. In composition and osmotic pressure it is very similar to serum, containing most of the body's waste products, as well as substances that the body needs such as glucose, amino acids, electrolytes and low molecular weight proteins. The human glomeruli filter three times the total body water each day. Fortunately, most of the water and all of the needed solutes are reabsorbed by the convoluted tubules, most reabsorption occurring in the proximal convoluted tubules (Fig. 11). The proximal convoluted tubule is not only the area of solute and water reabsorption, but also *secretes* additional waste products such as uric acid and creatinine into the tubule. On the other hand, the distal tubule is cytologically specialized to reabsorb sodium while remaining relatively impermeable to water. Thus, the production of a hypotonic urine in freshwater and hydrated species becomes possible.

In all mammals and most birds, long *nephric loops*

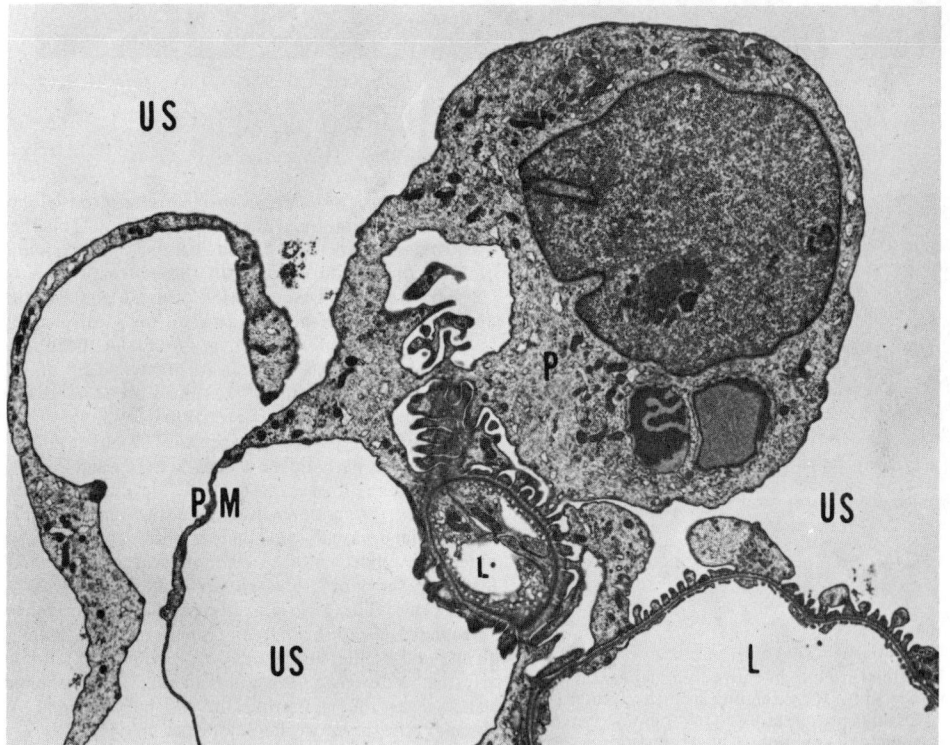

Fig. 9   Part of human glomerulus at high magnification: a podocyte (*P*) sits astride two capillary loops (*L*). The endothelium of the smaller vessel is cut tangentially, revealing the fenestrae through which the urine must filter. A tenuous podocytic membrane (*PM*) extends laterally through the urinary space (*US*). (Electron micrograph provided through courtesy of Dr. Ruth Bulger.)

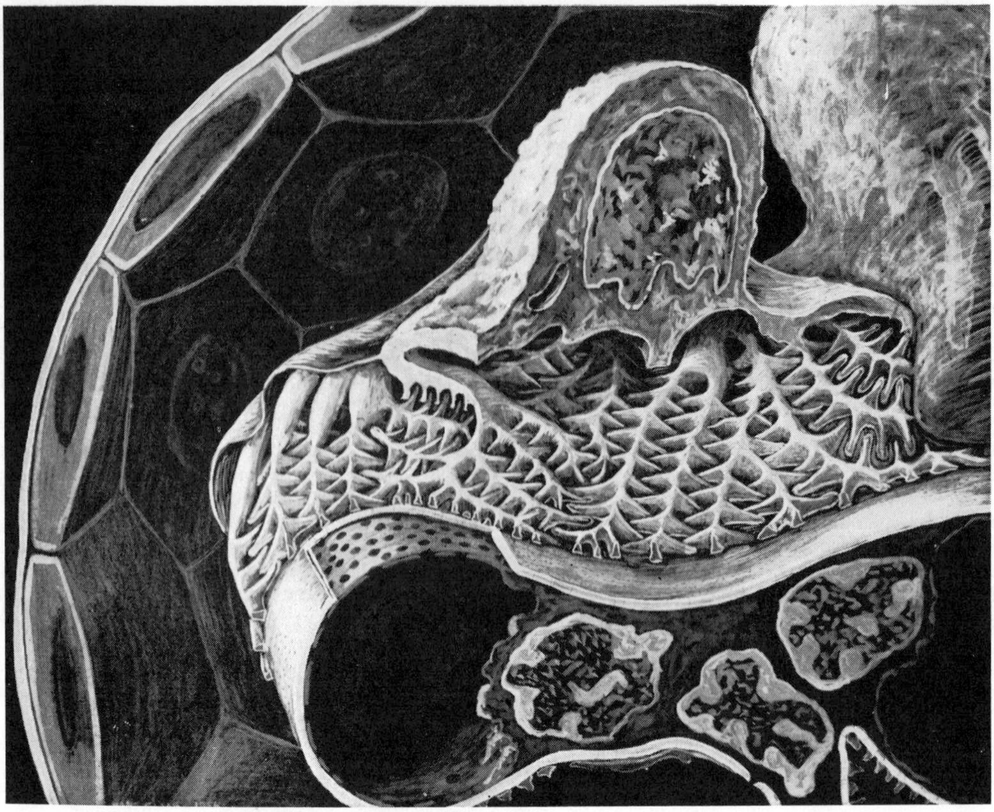

Fig. 10   Stereogram of a podocyte (Elias, Res. Serv. Med. **46**: 1–28, 1956).

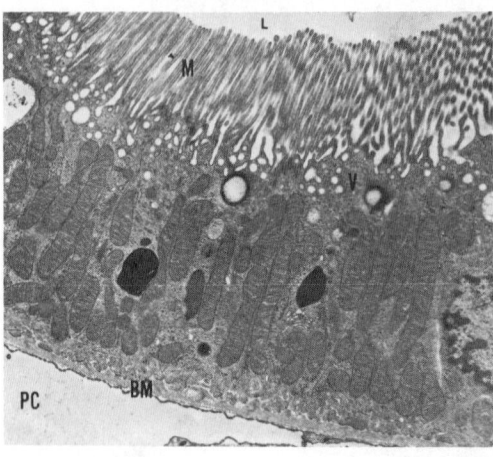

Fig. 11   Section through an entire human proximal tubule with microvilli (*M*) extending into the lumen (*L*). The apical cytoplasm contains many pinocytotic and lysozomal vesicles (*V*), while the basal portions contain deep interdigitating extensions of the plasmalemma of adjoining cells, largely occupied by mitochondria. A basement membrane (*BM*) and peritubular capillary (*PC*) surround the proximal tubule. (Courtesy of Dr. Bulger.)

(Henle's loops, ansa nephronsis) are interposed between the proximal and distal convolutions. Together with the *vasa recta* these loops comprise the *medulla* of the kidney. The function of these structures is to conserve water by concentration of urine and nitrogenous wastes. That these structures are ideally suited to serve as a *countercurrent multiplier* has been suggested by many, but the relative contribution of the ascending limb of Henle versus the vasa recta is still the subject of controversy (Lever, 1965).

In Myxine a pair of pronephric (Wolffian) ducts runs from the neck to the cloaca. Their renal corpuscles are some of the most highly developed among vertebrates. The glomeruli are richly branched, very thin laminae vasculosae glomerulorum. The vascular pole is extended into a long, mesentery-like band, fastening the glomerulus to the dorsolateral wall of the capsule. The arteriola afferens and its branches, surrounded by epitheloid cells, form a minute glomus at one end of the suspensory band, while the straight arteriola efferens, also surrounded by epitheloid cells, leaves the glomerulus at the opposite end. A vascular pole therefore, does not exist in Myxine.

Mesonephroi, wherever they are functional, are oval bodies containing the nephra which consist of large glomeruli in their capsules and highly convoluted tubules which empty into the *mesonephric duct* (Wolffian duct). This duct is identical with the pronephric

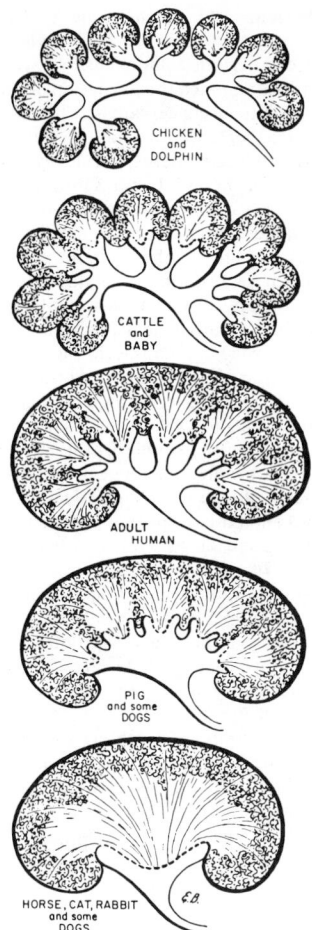

Fig. 12   Lobation of metanephroi in various higher vertebrates (adapted from Elias, Histology of Domestic Animals, Waltham, Mass. 1944).

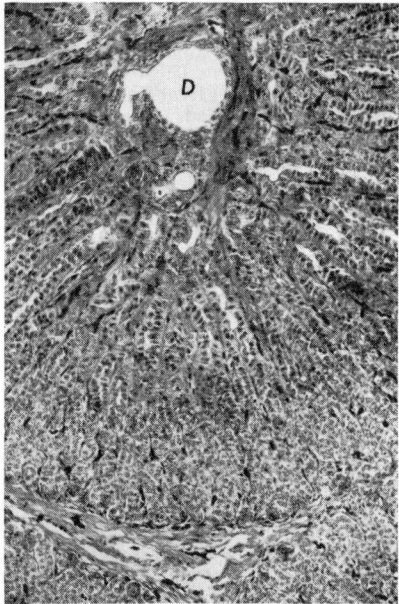

Fig. 13   Salt gland of the Herring gull *Larus argentatus* (photograph from slide prepared by Professor Knut Schmidt-Nielsen). The gland consists of distinct lobules, each containing a central duct (*D*) into which radiating tubules empty.

duct. In larvae with functioning mesonephroi and pronephroi, the latter is located cranial to the former, but both share a common duct. When the pronephros degenerates during metamorphosis, the middle and caudal portions of the duct remain.

In higher metanephroi, there are two innovations: the nephric loops (ansa nephronis) and the collecting tubules. The nephric loops are all crowded together in regions of the kidney forming the *medulla*, while the convoluted tubules and the renal corpuscles form the mottled *cortex*.

In birds and mammals, the convergence of the collecting tubules forms pyramidal shapes, each pyramid being surrounded by cortex. An individual pyramid together with the cortex that drains into it is known as a *renculus* or *lobe*. The number of renculi and the degree of their separation or fusion varies among the different kinds of birds and mammals. A few examples are shown in Fig. 12.

The eccrine *sweat glands*, found only in mammals and particularly well developed in the horse and in man, excrete water and solutes. While the most im-

portant of these solutes is sodium chloride, urea, uric acid and creatinine are other substances excreted by eccrine sweat glands in very small amounts.

Salt glands, i.e., compound tubular glands, found above the eyes in marine reptiles and birds, excrete large quantities of sodium chloride, thus permitting the animal to drink sea water (Fig. 13). The *anal glands* of sharks have the same function.

The *bladder* of most vertebrates serves only to store the final urine prior to excretion. But in amphibia and terrestrial reptiles, whose kidneys produce a hypotonic urine, it serves to conserve solutes and water.

The role of humoral influences in the formation of urine is a topic too complex to discuss here. In many invertebrates and all vertebrates neurohumoral antidiuretic and diuretic principles control the osmotic composition of body fluids by changing the urine composition. Volume regulation in vertebrates is subserved by the *juxtaglomerular apparatus (macula densa + juxtaglomerular cells* of afferent arteriole). These structures are sodium-sensitive and stimulate the release of aldosterone in response to tubular sodium concentrations.

PETER M. ELIAS AND HANS ELIAS

References

Andrew, W., "A Process of Elimination (The Excretory System)," *in* "Textbook of Comparative Histology," New York, Oxford, 1959, pp. 416–450.

Kirschner, L. B., "Comparative Physiology: Invertebrate Excretory Organs," Ann. Rev. Physiol., **29**: 196–196, 1967

Lever, A. F., "The Vasa Recta and Countercurrent

Multiplication," Acta med. Scand. Supp., **434**: 1–43, 1965.

Marshall, E. K., Jr., "The Comparative Physiology of the Kidney in Relation to Theories of Renal Secretion," *in* "The Kidney in Health and Disease," Philadelphia, Berglund and Medes, 1935, pp. 50–72.

Schmidt-Nielsen, B., "Organ Systems in Adaptation: The Excretory System," *in* "Adaptation to the Environment," Handbook of Physiology," **4**: 1964, pp. 215–243.

Schneider, L., "Elektronenmikroskopische Untersuchungen uber das Nephridialsystem von Paramaecium," J. Protozool, **7**: 75–90, 1960.

EXOBIOLOGY

At the present time a science of an unknown. Conjecture and theory fill all we know of life outside the earth. Attempting an understanding of this possibility man has presumed that life as we know it (a) cannot exist elsewhere in the universe, or (b) can exist but needs to be protected against a hostile environment.

The sun's radiation, galactic radiation, and electrical discharges offer a ready source of energy. If one were to follow the well known Miller-Urey pattern for the production of crude amino acids, a believable synthesis could be established. This approach employs an approximated primeval mixture of H_2, H_2O, CH_4 and NH_3 which is activated by an electrical discharge. Utilizing the residue of CO_2 and H_2O simple plant structures could have evolved. Primitive bacteria could have utilized the oxides present as a major source of oxidative energy. Once food was available, primeval animals would gradually have evolved to the structure complexes we see around us.

Exobiology is looked at with full recognition of the interactions that may prevail elsewhere. The "other" environments will undoubtedly possess a different radiation pattern, a gravitational field of a distinctive magnitude and magnetic fluxes. We would expect environmental characteristics such as gas mixtures, temperature and humidity to be unique to the point of even suggesting that perhaps we would have to revise many chemical concepts or design new ones. This last statement should not be construed as opting for a repeal of the current laws of chemistry and physics but rather the recognition of how little we know.

The approach today is to use the earth as the home base. From this sanctuary we attempt to duplicate the theorized environments of Mars, the Moon, Jupiter and others still to be decided on. Initially, we recognize the chronological onset of the important elements. Our universe appears to be 11 to 12 billion years old while our galaxy is about 7.5 billion years of age. The solar system assembled its array of elliptical eccentricities 5 billion years back in time. Hard on this figure at 4.75 billion years appears our little planet Earth. Primitive life currently is believed to have originated 3.2 billions of years past, while a primitive atmosphere prevailed that was capable of aiding in the formation of nucleic acids and protein. Why these two? To date, all references to any form of schema of what we call life has included without exception nucleic acids and protein. To retreat further back in biochemical time we must accept the prevalence of a reducing atmosphere, primarily hydrogen. Add to this, energy having both electrical and thermal potential and we find the beginning of primitive organic compounds.

Oparin has championed the formation of the colloidal mass of a sol into a coacervate. He likens this to a precellular arrangement. Fox has worked with polypeptides and produced microspherules which he calls proteinoid microspheres. These substances resemble micrococci. The two concepts present on the one hand a rather unstable coacervate more subject to extraneous influences, and on the other, a stable proteinoid.

Further theories resulting from investigations into the origin and duplication of basic units such as purines pyrimidines, amino acids, nucleotides, polynucleotides, nucleosides, monosaccharides, porphyrins, fatty acids, hydrocarbons and polypeptides offer much to ponder. For example, let us assume man can duplicate a chemical entity. Does it necessarily follow that he can produce even the simplest protein? An interesting point is that while self-replication had to follow initial synthesis, the point in sophistication that would permit of this should have been a polynucleotide. Conceding this then, one must search for a template capable of this action. In the earth are to be found particles of clay possessing imperfections in their lattice structure. These are repetitious self-selecting formations. Here in the dust we may have a primitive replicating tool able to store structure designs, align primitive compounds and stamp out a higher degree of biochemical sophistication.

So, here we have man stumbling for an awareness of how he came to be, and then trying to predict whether life, that elusive biochemical event, can be present elsewhere in the universe.

PHILIP L. COOPER

EYE

The eye is designed so that it can perceive light and transmit the sensation. This usually requires a focusing device, the lens, and a photoreceptor, the retina, containing photosensitive pigments. The image received on the retina is transferred through the neurons of the retina and the optic nerve to the brain. A variety of eyes and their structures are schematically illustrated in Fig. 1 A–C to permit comparison of the photoreceptors of the invertebrates and the vertebrates. In the vertebrate eye (C) the lens (a) projects an inverted image upon the retina. The diaphragm of the eye is the iris (i), through whose aperture the amount of light to the retina is controlled. The pupil of the eye opens in dim light to an extent governed by the activity of the retina. This adjustment admits more light through the lens. In man the lens also acts as a color filter, filtering out the ultra-violet light, which sharply cuts off the far edge of the violet region at about 400 mμ. The retina (r) is made up of receptor cells lying side by side to form a mosaic of light-sensitive elements. The receptors of most vertebrates are rods and cones, the cones being for bright light and color vision, and the rods for dim vision. Each is composed of an inner segment much like a nerve cell and a rod- or cone-shaped outer segment (os) which contains the photosensitive pigment. Each cone is usually

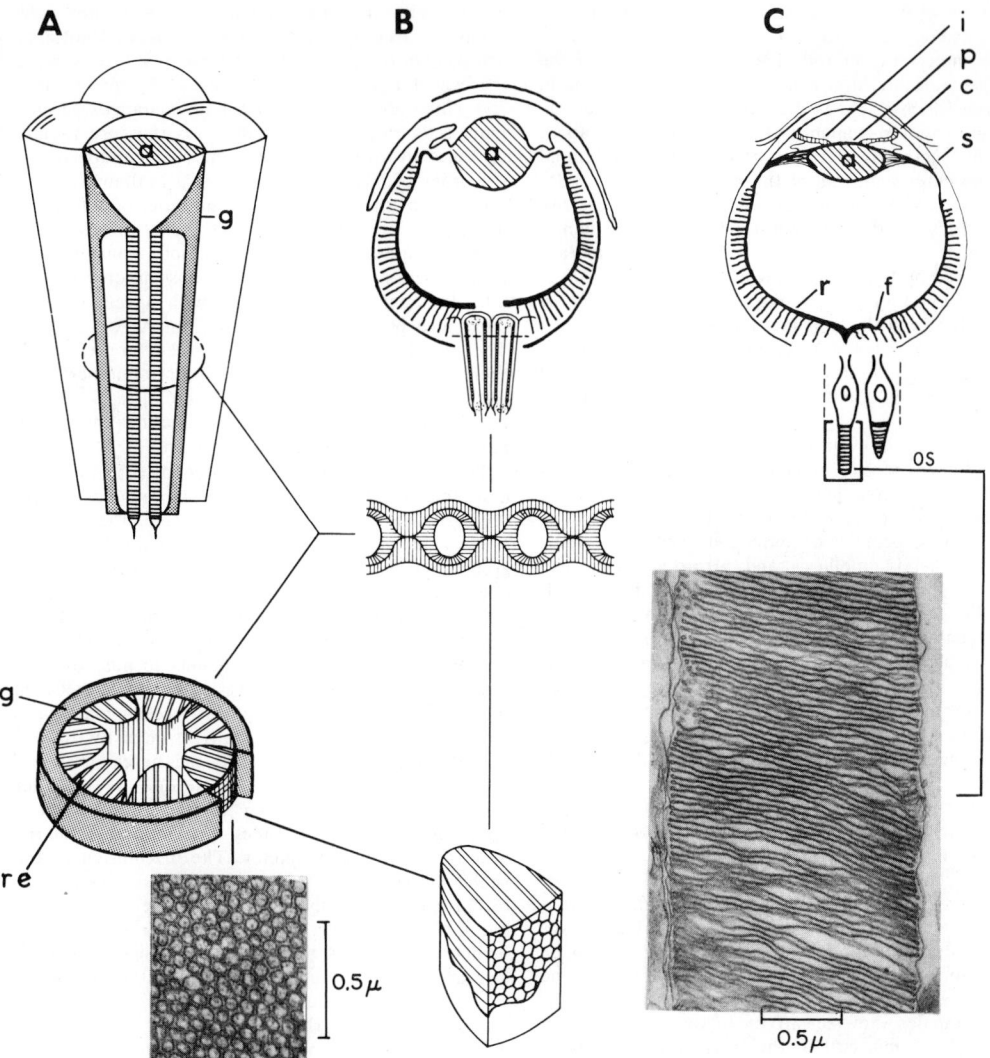

Fig. 1   Schematic representation of a variety of visual photoreceptors. A, Insect. B, Mollusc. C, Vertebrate. (a, lens; g, pigment-screening granules; re, retinal rod structure; i, iris; p, pupil; c, cornea; s, sclera; f, fovea; r, retina; os, outer segment.)

connected with the brain by a single fiber of the optic nerve, whereas clusters of rods are connected by single optic nerve fibers. Toward the center of the human retina there is a small depression, the fovea (f), which includes the fixation point of the eye, where vision is most acute, and contains only cones. Beyond the area of the fovea, rods begin and become more numerous as the distance from the fovea increases. The cones for bright light concentrate toward the center, and the rods for dim light concentrate toward the periphery of the retina. In man, the fovea and the region just around it, the *macula lutea*, are colored yellow; they contain a plant CAROTENOID (lutein, a xanthophyll) which absorbs light in the violet and blue regions of the spectrum.

The rods have their maximal sensitivity in the blue-green at about 500 mμ; the cone sensitivity is transferred toward the red, lying in the yellow-green at about 560 mμ. There are small shifts in these maxima depending on the animal species and its environment. All the rods and cones contain as their photosensitive pigments a chromophore, either retinal$_1$ or retinal$_2$ (the aldehydes of Vitamin A$_1$ or Vitamin A$_2$),* linked with a protein opsin. When extracted from the eye, these pigment-protein complexes are identified by their color and absorption spectra as rhodopsin (retinal$_1$ + rod opsin) or porphyropsin (retinal^2 + rod opsin) for the rods, and idopsin (retinal$_1$ + cone

*In the new chemical nomenclature, Vitamin A is *retinol;* Vitamin A aldehyde, or retinene, is *retinal.*

opsin) or cyanopsin (retinal$_2$ + cone opsin) for the cones. There are on the order of 1×10^6 to 10^9 rhodopsin molecules per rod. The outer segments of the rods are variable in size, but one of the largest is in the frog which is 6 μ in diameter and 60 μ in length. In man the outer segment is 1 μ in diameter and about 28 μ in length. By electron microscopy it was found that the outer segments of the retinal rods and cones are an arrangement of stacked plates or discs (re). These discs are double-membraned structures (lamellae) which are from 50 to 100 Å in thickness and are interspaced by lipids and aqueous proteins 200–400 Å thick. The visual pigment-complex is probably oriented on the surface of the lamellae. It is postulated that such an ordered or quasi-crystalline state is necessary for photoexcitation. It has been estimated that 8 quanta or less can bring about the experience of a visual sensation.

In the course of evolution, animals have used various kinds of devices for forming an image. These structures can be seen in the invertebrates, which possess the greatest variety of photoreceptor structures. Eyespots, sensory cells, ocelli and compound eyes have arisen among annelids, molluscs and arthropods, in each instance with differences in gross physical organization. For example, the stigma or eyespot described in flagellated algae such as *Chlamydomonas, Volvox, Euglena* and others is a photoreceptor for light perception. The eyespot in *Euglena* is an orange-red structure about 2×3 μ and consists of about 40–50 tightly packed grana on the order of 0.1 μ in diameter at the anterior part of the organism; this eyespot appears to be intimately linked to the flagellum at its base. The eyespot (receptor) plus the flagellum (effector) then act as a phototactic unit in directing the organism toward the light—a simple eye.

In the flatworm *Planaria*, the two eyes consist of *sensory cells* surrounded by screening pigment granules. The ends of the sensory cells continue as nerves which enter the brain. The pigment granules shade the sensory cells from light in all directions but one and so enable the animal to move away from the light. These granules show sensitivity to the ultra-violet as well as the visible part of the spectrum. The sensory cells are structurally similar to the retinal rods of the vertebrates; they consist of layers (lamellae). These retinal rods are about 5 μ in diameter, with a more variable length of approximately 35 μ. Each lamella is about 100 Å in thickness, as in the vertebrate rods. Studies of their visual pigment indicate that it too is a rhodopsin.

The compound eye and its structure (A) are of considerable interest in elucidating the visual apparatus. Compound eyes are restricted to the arthropods, among which are the insects and the crustacea. Many of these organisms exhibit orientation relative to the direction of polarized light, and such sensitivity to plane polarized light suggests the existence of a polarized light analyzer within the eye. The compound eye is particularly efficient in detecting movements in its total visual field. Structurally the compound eye consists of ommatidia; in the insect each ommatidium has a distal cone and lens (a) and a sheath (g) of pigment cells which extends throughout its entire length. The pigment sheath is not found in the white-eyed mutants. Each ommatidium consists of retinula cells, of which the differentiated retinal elements are the rhabdomeres (re). The rhabdomeres form the "light-trapping" area in which the visual process is initiated, and contain the photosensitive pigments. For example, the eye of *Drosophila* is composed of approximately 700 ommatidia;) each ommatidium contains seven retinula cells with seven rhabdomeres that are radially arranged to form a cylinder. The rhabdomeres are of the order of 60 μ in length and 1 μ in diameter. The thickness of the lamellae is on the order of 100 Å, the interspaces are from 200 to 400 Å. A photosensitive pigment has been isolated from both honeybee and housefly eyes, with a maximum absorption near 440 mμ. This peak was associated with a retinal$_1$ complex since both retinal$_1$ and Vitamin A$_1$ were found. The visual pigment within the rhabdomeres would therefore be a *rhodopsin*. In addition a number of insects show photosensitivity to the ultra-violet and possess color discrimination, indicating that other photosensitive pigments may be present.

The eyes of the mollusc cephalopods (B) show a remarkable resemblance in physical organization to those of the vertebrates. For example, the *Octopus* eye is a single lens eye, provided with a mechanism for accommodation. The lens, however, is formed out of two halves joined together; the photoreceptors of the retina are not inverted as in the vertebrate eye and are directly exposed to the incident light. In the *Octopus*, the retina is made up of rhabdoms formed by four radially arranged rhabdomeres. The space between the rhabdoms contains pigment-screening granules that migrate, depending on the light intensity, toward the crystalline cone or toward the basement membrane. Each retinal rod (rhabdomere) is of the order of 1 μ in diameter and 60 μ in length, and is made up of densely packed tubules. Each tubule is of the order of 500 Å in diameter. The orientation of the rhabdomeres in the rhabdom and their internal structure are schematically illustrated as viewed by longitudinal and cross-sectional cuts of the retina. It is to be noted in Fig. 1 A and B how the retina of the mollusc is similar to the ommatidium of the compound eye. The visual pigment extracted from the retinas of the molluscs is also a rhodopsin.

Within the retina of amphibians, reptiles and birds there are brightly colored oil globules ranging from 3 to 6 μ in diameter. They are not present in fishes and mammals. These oil globules are located near or between the inner and the outer cone segments. In the chicken retina the predominant colors are red, yellow, green and mixed colors. What function they have in the retina and in the visual process is still unknown, but the location of the globule between the inner cone segment and the light source indicates that they act as color filters for the retinal cones in a way suggestive of the color photographic process. However, there is no direct correlation between the color of the globules and the color discrimination of animals possessing them. The carotenoids lutein and zeaxanthin and a new carotenoid, galloxanthin, have been isolated from the chicken retina. These have been identified in the retina as astaxanthin for the pigment of the red globules, galloxanthin for the green globule pigment and lutein and/or zeaxanthin for the yellow globule pigment.

Recent studies indicate that color vision in vertebrates is probably mediated by three different light-sensitive pigments in different receptors; one for sensing blue, one for green and one for red. Three such spectral species have been identified in the cones by techniques in microspectrophotometry.

The photoreceptor must function in the conversion of the light energy to chemical and electrical energy. This requires a photosensitive pigment and an ordered arrangement within the photoreceptor. Electron microscopic and X-ray diffraction studies indicate a structure of ordered repeat units, lamellae; these are observed as plates, discs, tubes or rods. The lamellae are double-membraned systems of lipids and proteins. The photosensitive pigment complex *rhodopsins* are most probably oriented as monolayers on or within the lamellae. A variety of eyes have evolved which differ in their gross morphology, but their photoreceptor structures are similar in molecular organization and dimensions. A common pigment molecule, the carotenoid retinal is found in all functioning eyes.

JEROME J. WOLKEN

References

Duke-Elder, Sir Stewart, "The Eye in Evolution," St. Louis, Mosby, 1958.

Wald, George, "Molecular Basis of Excitation," Science **162**: 230–239, 1968.

Walls, G. L., "The Vertebrate Eye," Bloomfield Hills, Mich., Cranbrook Inst., 1942.

Wolken, J. J., "Vision," Springfield, Ill., Thomas, 1966.

f

FABRICIUS, JOHANN CHRISTIAN (1745–1808)

A Danish entomologist who studied under Linnaeus at Uppsala. He became one of this naturalist's most distinguished pupils and wrote many works on entomology, using the principles of Linnaen classification. His system of classification utilized mouth parts instead of wings and was retained for many years. He had little reputation in his own country where, at Kiel, he had the extraordinary appointment of professor of Natural History, Economy, and Finance. He spent much of his time abroad where his reputation was far greater than it was at home.

FAGALES

A DICOTYLEDENOUS order of nut-bearing trees or shrubs, leaves alternate, simple, stipulate, monoecious or rarely dioecious, the flowers borne in unisexual or bisexual catkins, in small clusters, or singly, apetalous, staminate flowers of fused or distinct sepals containing 1 to 40 stamens, pistillate flowers in groups of 2 or 3 in a cone-like catkin or this reduced, or borne singly or scattered on peduncles, epigynous or very obscurely so, the pistil borne on a bract or in an involucre. Two families are recognized:

1. Deciduous, monoecious trees or shrubs, staminate flowers in cymules of 3 forming groups in a pendulous catkin, pistillate flowers in groups of 2 or 3 in cone-like catkins, stigmas 2, the nutlets subtended by bracts but not surrounded basally by an involucre in the form of a cup or burr.—Betulaceae (Birch Family)

2. Deciduous or evergreen, monoecious or dioecious (in *Nothofagus*) trees or shrubs, staminate flowers singly disposed in a pendulous or erect catkin, pistillate flowers borne separately or upon the base of the staminate catkin, stigmas 3 to 6, pistils singly disposed or in groups of 2 or 3 enclosed in an involucre (or sometimes not), the involucre leaf-like, forming a cup, or a burr.—Fagaceae (Beech Family).

The Betulaceae consist of 6 genera:

1. *Betula* (Birch). Deciduous trees or shrubs, the pistillate catkin of thin, deciduous scales bearing compressed nutlets. About 40 species are recognized in the arctic and north temperate regions of both hemispheres, including some timber trees.

2. *Alnus* (Alder). Deciduous trees or shrubs, the pistillate catkin of thick, persistent scales, the nutlets usually winged. About 30 species occur in the north temperate regions of both hemispheres and south to Peru, a few valued as timber trees.

3. *Carpinus* (Hornbean). Deciduous trees or shrubs, staminate flowers without perianth consisting of 3 to 13 stamens subtended by a bract, pistillate flowers 2 to a bract and each subtended by 2 bractlets, the perianth of 6 to 10 teeth adnate to the pistil. About 25 species are recognized in the temperate regions of Europe, eastern Asia, the Himalayan region, and North America, south to Central America.

4. *Ostrya* (Hop-Hornbean). Deciduous trees, staminate flowers without perianth consisting of 3 to 14 stamens subtended by a bract, pistillate flowers on a slender upright catkin, paired in the axil of a deciduous bract, calyx adnate to the ovary, enclosed in a tubular involucre formed by the fusion of a bract and 2 bractlets, the fruit a nutlet enclosed in a bladder-like involucre and these grouped on a peduncle like a cone. Seven species occur in temperate regions of Europe, Asia, and North America, extending south into Central America.

5. *Ostryopsis*. Deciduous shrubs differing from *Ostrya* principally in the pistillate flowers being borne on very short spikes. Two species are known in China.

6. *Corylus* (Hazel). Deciduous shrubs or rarely trees, staminate flowers without perianth, 4 to 8 stamens per bract in a slender catkin, pistillate flowers in a small head-like cluster enclosed in bracts with the stigmas protruding, perianth adnate to the pistil, the fruit a nut wuth ligneous pericarp enclosed in a variously dissected often tubular involucre. About 15 species are known in the temperate regions of Europe, Asia, and North America, the nuts of some being esteemed as food.

The Fagaceae consist of 6 genera:

1. *Nothofagus*. Deciduous or evergreen trees or shrubs, dioecious, the staminate flowers solitary or in clusters of 3, calyx fused, 4 to 6 lobed, 8 to 40 stamens, pistillate flowers 1 to 3 in an involucre, nuts 3-angled, usually 3 in a lobed involucre with transverse lamellae on the outer surface. About 17 species are recognized in antarctic and south temperate regions of South America, Australia, and New Zealand.

2. *Fagus* (Beech). Deciduous trees, the staminate flowers numerous in head-like clusters, perianth 4 to 7 lobed, stamens 8 to 16, pistillate flowers usually paired in a 4-parted involucre of numerous fused bracts, nuts 1 or 2 in a 4-valved involucre, the bur-like

exterior covered with numerous free tips of bracts. Nine species are known in the temperate regions of the northern hemisphere, valued as timber trees, especially in Europe.

3. *Castanea* (Chestnut). Deciduous trees or shrubs, staminate flowers in erect catkins, calyx 6-parted, stamens 10 to 20, pistillate flowers borne upon the bases of the staminate catkins or rarely separately, usually 3 enclosed in a spiny involucre, the ovary 6-celled, the large nuts 1 to 3 or more in a prickly involucre. About 8 species are recognized in the temperate regions of the northern hemisphere. The large nuts of some species are important items of food and a few are important sources of timber. The most prominent American species (*C. dentata*) was virtually eliminated from a wide range by blight (*Endothia*) early in this century.

4. *Castanopsis* (Chinkapin). Evergreen trees or shrubs, leaves coriaceous, staminate flowers borne in erect simple or branched spikes, 10 to 12 stamens in a 5 or 6 parted calyx, pistillate flowers in separate short spikes or sometimes borne on the bases of the staminate spikes, 1 to 3 flowers in an involucre, ovary 3-celled, styles 3, nuts ripening the second year, 1 to 3 in an involucre covered by spines, tubercles, or short transverse ridges. About 30 species are recognized in southern and eastern Asia and in western North America, including timber trees.

5. *Lithocarpus.* Evergreen trees, leaves coriaceous, staminate flowers borne in erect simple or branched spikes, 10 to 12 stamens in a 5 or 6 parted calyx, pistillate flowers borne at the bases of the staminate spikes or separately, ovary 3-celled, styles 3, fruit a nut enclosed partially or almost wholly in a cup-like involucre the scales of which are free at the apices or connate into concentric rings. About 100 species are known in southern and eastern Asia, in Malaysia, and in western North America, including trees important for timber and tannin production.

6. *Quercus* (Oak). Deciduous or evergreen trees or shrubs, staminate flowers in pendulous catkins, stamens 4 to 12 in 4 to 7 parted calyces, pistillate flowers borne separately, solitary or several in short or elongate spikes, fruit a nut (acorn) enclosed at base or almost wholly in an involucre (cup) of many free scales or these fused into concentric rings. Some 400 to 500 species are recognized in temperate and sub-tropical regions of Europe, Asia, and North America, southward into North Africa, Malaysia, Central America, and reaching the southern hemisphere only in Colombia. These include dominant forest elements and important timber trees over much of this range.

The genus *Quercus* is not clearly delimited from *Lithocarpus* (whose fruit is also an acorn). Those species of *Quercus* with cup scales fused in concentric rings (subgenus *Cyclobalanus*) are separated from similar species of *Lithocarpus* only by the position of the staminate and pistillate flowers on different inflorescences. These groups of species occur only in Asia. The most widespread subgenus of *Quercus* is *Lepidobalanus*, the White Oaks, occurring throughout the bulk of the Eurasian and North American range of the genus. Two additional subgenera are peculiarly American, the subgenus *Erythrobalanus* (Red Oaks), ranging over almost the entire American area of the genus, and the subgenus *Chrysolepidae* (Intermediate

Oaks), confined to the California region of western North America.

Fossil representatives of both families of Fagales are known from the Cretaceous period onward, several genera being important forest constituents from early Tertiary time to the present. Present day species are readily recognizable in most fossils of the Pliocene deposits.

C. H. MULLER

References

Abbe, E. C., "Studies in the phylogeny of the Betulaceae," Bot. Gaz. 97: 1–67, 1935.
Camus, A. "Les Chênes, Monographie du genre Quercus." 3 vol. text, 3 vol. plates. Paris, Lechevalier, 1936–1954.
Langdon, L. M., "Ontogenetic and anatomical studies of the flower and fruit of the Fagaceae and Juglandaceae," Bot. Gaz. 101: 301–327, 1939.
Muller, C. H., "The Central American species of Quercus," Washington, U. S. Dept. Agric. Misc. Pub. 477: 1–216, 1942.
Rehder, A., "Manual of cultivated trees and shrubs," New York, Macmillan, 1927.
Trelease, W., "The American Oaks," Mem. Nat. Acad. Sci. 20: I–V, 1–255, 1924.

FALCONIFORMES

An order of diurnal birds of prey containing about 250 species of 90 genera separable into five main groups, usually designated as families. These are: Cathartidae, Sagittariidae, Accipitridae, Pandionidae and Falconidae. Characterization of this order is quite impossible in convincing anatomical terms although a combination of several features is generally offered (Stresemann, 1927–34). The most characteristic features are the hooked bill, the cere with the nostril centrally located, and the grasping type of foot which has the inner toe shorter than the outer, the middle toe longest and connected to the outer by a small web, and each toe armed with a long, curving claw. Each of these features finds some exception, and it is just these features that we generally think of as most plastic or open to adaptive modification and thus to convergent similarity.

The possibility of convergent evolution within the falconiforms finds support in the linking formerly of the hawks and owls in a common order. This association was based on the hooked bill and cere as well as the predatory habit. Fitzinger (1856) separated the hawks and owls as distinct orders, a schism which is supported by many anatomical features. However, the structural differences separating the owls from the typical hawks of the family Accipitridae are similar in degree to those separating the latter from the other "families" of the falconiforms.

The assumption that the several groups making up the falconiforms may share these physical features as a result of convergent evolution is inescapable. Fürbringer (1888) summarized much of the anatomical heterogeneity of this group and also discussed the relationship of these diurnal predators with other birds. Garrod, Gadow, Forbes and Beddard and more recent avian anatomists have also reviewed the anatomical details. Hudson (1948) has gone further in suggesting that the cathartids and *Sagittarius* should be separated as distinct orders. Since the evidence suggests a polyphyletic origin for these birds, each subgroup is best considered separately.

The Cathartidae or New World vulture group is small: there are six species in five genera. They range through temperate North America to the tip of South America. The largest flying birds of the New World belong in this group. The Andean and California Conders have wing spans of up to nine and one half feet and weigh as much as thirty pounds. All of the members of this group are vulturine in habit. The feet are not capable of strong flexion nor are they comparable in general form to those of the other falconiforms. This group differs from the typical hawks in almost every anatomical feature, being like them only in having a hooked bill and in their large size and soaring habit. This group extends back to the Eocene, at which time it appears to have been as distinctive as now. Anatomically, relationship with several other orders—pelicaniforms, ciconiiforms, charadriiforms—is suggested.

Sagittarius serpentarius is the only species belonging to the second "family." It is widely distributed over Africa in open savannah of scrub desert areas. In its nesting and general breeding behavior it is much like the accipitrids, however, it also resembles the Seriama of South America which is certainly not a hawk. This long-legged, crane-like "predator," feeds mainly on snakes, other reptiles and insects which it captures by stamping and grasping with the feet. Although not usually inclined to fly it occasionally soars in a hawk-like fashion.

There is no real fossil history for this group. The fossil form, *Amphiserpentarius*, is long-legged but does not resemble *Sagittarius* enough in detail to be identified with it. Anatomically, *Sagittarius* appears to be quite primitive with many superimposed specializations, explainable in terms of its crane-like form and predator habit. Its nearest relatives among living birds are members of several "primitive" orders, not including any of the other falconiforms.

The typical hawks, belonging to the family Accipitridae, form a large but well-defined group of species in which several of the larger genera or supergenera are cosmopolitan in distribution. This group extends back in time to the Eocene with apparently much the same anatomical detail (at least as regards the tarsometatarsus) as it has today. The wide range of adaptive modification observed in this group supports its great age. Although basically predators, the group includes species of such diverse habit as *Gypohierax angolensis* which eats the fruit of the Oil Palm, the species of the genus *Pernis* which consume the larvae of wasps and the large Old World vultures which feed on carrion.

The predation habits of a species tend to be quite fixed in any specific locality, but in widely separated areas may show quite different food preferences. Prey animals range from insects to sloths, monkeys, small deer and antelope. Interesting food habits include the crepuscular bat catching of *Machaeramphus alcinus,* of central Africa or Malaya, the snail feeding of the Everglade Kite, *Rostrhamus sociabilis,* or the taking of monkeys by *Stephanoaëtus, Harpia,* or *Pithecophaga* in the aequatorial forests around the world. Because of their predatory habit many species of this group are persecuted by man and have been greatly reduced in numbers. It will not be too many more years before some will become extinct.

In terms of their anatomy, the accipitrids, show little similarity to any other living birds. Their origin goes back perhaps into the Cretaceous, to the first radiation of bird types. There are some resemblances to the other "Pelargornithes" (of Fürbringer) but only in a very general way and the comparisons are no better with any one of these orders than another.

The single species *Pandion haliaetus,* the Osprey, makes up the Pandionidae, a distinct anatomical variant. This fish-eating species is nearly cosmopolitan in distribution (lacking in South America). Anatomically it agrees in some features with the accipitrids suggesting that it is a very early, or much specialized (or both) derivative.

The family Falconidae includes about eleven genera with nearly sixty species. The peregrine, known through its use in falconry, belongs here. The falcons show a wide range of feeding habits: fruit, carrion, insects or birds and mammals. They differ from the typical hawks in having a weak foot, less modified for clenching. The most hawk-like members are the Central and South American species of *Herpetotheres* and *Micrastur.* The latter is very much like *Accipiter* (the type genus of the Accipitridae) in its general appearance and in habit.

This group is without a clarifying fossil record but probably goes back at least to the early part of the Tertiary. They resemble in particular anatomical features the parrots, cuckoos and owls—an array of arboreal birds distinct from the water or shore bird with which the cathartid and *Sagittarius* appear to be allied.

MALCOLM T. JOLLIE

References

Fitzinger, L. J., "Uber das System und die Charakteristik der natürlichen Familien der Vögel," Sitzungber. K. Adad. d. Wiss. Nath.-nat. (Wien), **21**: 277–318, 1856.

Fürbringer, Max, "Untersuchungen zur Morphologie und Systematik der Vögel zugleich ein beitrag zur Anatomie der Stutz und Bewegungsorgane," K. Zool. Genoots. Bidr. tot de Dierkunde, **15** (2 parts):1–1751, 1888 (*also* Amsterdam, I. J. Van Holkema).

Hudson, George Elford, "Studies on the muscles of the pelvic appendage in birds II: the heterogeneous order Falconiformes," American Mid. Nat., **39**: 102–127, 1948.

Stresemann, Erwin, "Handbuch der Zoologie: eine Naturgeschichte der Stämme des Tierreiches," vol. 7, second half, "Sauropsida: Aves," Berlin und Leipzig, de Gruyter, 1927–1934.

FALLOPIO, GABRIELLO (1523–1562)

Fallopio, Italian anatomist, was born at Modena in 1523, where he later became a canon of the cathedral. Medicine seems to have been his dominating interest from early days, an interest which was strengthened and grew throughout his life until his early death in 1562.

After completing his medical studies at Ferrara, he became a teacher of anatomy. Prompted by various motives he continued this occupation at Pisa and later

Padua, where he worked with Vesalius, and where William Harvey came under his influence.

Individual achievements of especial interest are his discovery and discription of the chorda tympanani, the sphenoid sinus, the opening of the various tubes of the female human into the abdominal cavity, and the trigeminal, auditory and glossopharyngeal nerves. He also named the ovarian tubes, vagina, placenta, and the muscles of the forehead, occiput and tongue.

Fallopio's *Observationes anatomicae*, 1561, was his only treatise to be published during his lifetime, but his collected works, *Opera genuina omnia* were subsequently published at Venice in 1584.

LETITIA LANGORD

FAT

The white fat depots or fat organs. Many aspects of the origin, differentiation, physiology and other attributes of "adipose tissue" are still unknown or remain imperfectly understood. However, the increasing use of transplantation, electron and interference microscopy, radioisotopically labelled compounds, histochemical, biochemical and other techniques is leading to clarification of these problems.

"Adipose tissue" serves as energy reserves, shock absorbing tissue in the gluteal region and soles of the feet, packing material in the orbits of the eyes and in the joints, insulation against heat loss; also it contributes to body form. While small aggregations of fat cells are relatively widespread in the tissues and organs of higher vertebrates, the major fat depots or fat organs, as they may be called (since they are composed of a complex of tissues), are relatively restricted and are characteristic of each species. In mammals, the larger fat organs are found in the subcutaneous connective tissue, especially of the buttocks and abdominal wall, in the greater omentum and other mesenteries, about the kidneys and in the mediastinal, cervical, axillary and inguinal regions.

Two distinct types of fat cells occur in most mammals, namely a unilocular type which usually contains a single large vesicle of stored fat and is characteristic of the white fat organs, and secondly, a multilocular type which contains many droplets of lipid and makes up the brown fat bodies of hibernating and, to some extent, non-hibernating mammals.

Several investigators have postulated that the primitive fat organ is a syncytial structure composed of a meshwork of capillary sprouts and adjacent mesenchymal cells. The deposition of lipid within the reticulum converts the primitive fat organ into a fat lobule. This suggests that the fat lobule is a reticulo-endothelial type of organ. Droplets of lipid appear in the outstretched, conjoined stellate cells. As the lipid deposition continues, the cells withdraw their processes and round up. The fat droplets enlarge, coalesce, and eventually each cell contains a single vesicle of fat which is surrounded by a thin film of cytoplasm and the cell membrane. A network of silver staining (*argyrophilic*) reticular fibrils encircle each cell. The nucleus is pushed to the edge of the cell where it lies in a small pool of cytoplasm. MITOCHONDRIA and GOLGI BODIES are most numerous in the cytoplasm adjacent to the nucleus but may occur in the film of cytoplasm surrounding the lipid vesicle. MITOSIS occurs in developing fat organs and in immature fat cells but fully formed fat cells do not divide. Each fat lobule is separated from adjacent ones by areolar connective tissue, reticular, collagenous and elastic fibers. Fibroblasts, macrophages, mast and other types of cells are present in the areolar tissue. The lobules are highly vascularized and contain a rich complement of nerve fibers.

The lipide which is stored is dependent upon diet and is quite specific for each species. In general, the fat is composed of esters of glycerol and fatty acids (palmitic, stearic, oleic). The lipids are generally fluid at body temperature.

The fat bodies appear to contain a normal spectrum of enzymes, including those required for glycolysis and KREBS CYCLE oxidation. The lipides stored in the fat bodies are those present in the ingested food, those synthesized in fat bodies from carbohydrates and proteins, those resulting from the change of one fatty acid into another and those that are formed elsewhere in the body and subsequently transported to the depot.

The formation, deposition and mobilization of lipides in the fat organs takes place very rapidly. Instead of being passive storehouses of excess fuel, the fat bodies are exceedingly active metabolically and hence, play a fundamental role in organisms. The endocrine system particularly, and to some extent, the nervous system, controls the activities of the fat bodies.

The brown fat bodies or organs. The brown fat organs are histologically and physiologically different from white fat bodies. They are best developed in mammals which hibernate (see HIBERNATION) but also occur in non-hibernating species and in some primates. These bodies occur in the interscapular area, near the thymus, along the thoracic aorta, around the hilum of the kidney, in neck tissues and in small aggregations relatively widespread in the tissues of various rodents. The exact nature of the brown or yellowish color present in these bodies is unknown. The cells are much smaller than those of depot fat bodies. The cells are polygonal in shape with a central or slightly eccentric nucleus which is surrounded by granular, eosinophilic cytoplasm which characteristically contains a large number of lipide vesicles. Mitochondria and Golgi bodies occur in the cytoplasm. Brown fat bodies are highly lobulated and possess a glandlike appearance. The lobules are more highly vascular than white fat bodies and the network of argyrophilic, reticular fibers about each cell is more dense and ramified.

The brown fat bodies contain a wide complement of enzymes and are more active metabolically than the white fat depots. Hypophysectomy, adrenalectomy and nerve section cause more pronounced changes in brown fat bodies than occur in white fat organs.

H. L. EASTLICK

References

Greep, R. O. ed., "Histology," New York, McGraw-Hill, 1965.
Johansson, B., "Brown fat: A review," Metabolism. 8: 221–240, 1959.

FEATHER

Feathers, the specialized integumentary covering of birds (see AVES), are phylogenetic derivatives of reptilian SCALES. Transitional stages can be seen on the legs of birds where scales give way to feathers. Two principal types of feather are formed: the embryonic or down feather and the definitive or contour feather.

The down feather. Like scales, these arise from inductive action of a dermal condensation of mesenchyme upon the overlying epidermis. As a result, the epidermis grows outward to form a conical cylinder containing a central mesenchymal pulp. The outer portion of the epidermal cylinder remains smooth as a sheath, but the inner portion develops longitudinal ridges which lengthen from the tip of the feather towards its base. As many as fifteen such *barb ridges* appear around the inside of the cylinder. Along the sides of each barb ridge, cells become aligned in tangential rows to form *barbule plates.*

Nutrition is brought to the epidermal cylinder by the blood vessels of the pulp. The tip and outer parts of the epidermis, being farthest removed from the vascular supply, begin to cornify first. As growth diminishes in the feather, the pulp shrinks back and keratinization of the epidermal cylinder progresses towards its base. The timing of these events varies for different birds and for different tracts, but along the edge of the wing in the chick the feather germs first appear towards the close of the seventh day of incubation. Growth is most rapid between eleven and thirteen days, and cornification is completed by fifteen days. During the development of the feather, its base sinks into the surrounding skin. The epidermal pit, thus formed, is the feather *follicle.* At hatching, the down feather dries, its sheath splits, and the *barbs* are released as a whorl of slender branches radiating from a small circular base or *quill* which is set in the follicle. Along the basal two-thirds of each barb are two rows of tiny side branches, the *barbules.*

The contour feather. As the down feather is completed, there remains a small dermal papilla at its base with an epidermal covering. This is the germ of the juvenile contour feather. The dermal papilla acts as an organizer of the epidermis, again causing the latter to grow outward from the follicle as a cylinder. Mesenchymal pulp, proliferated from the dermal papilla, fills the epidermal cylinder and carries nutrition to it through many arborizations of a central artery which drain into venous sinuses. The feather cylinder is constricted and thickened at its base where it surrounds the dermal papilla. This region, the *collar,* is the proliferative zone. Above the collar there is growth, by cellular enlargement and elongation, and differentiation of specialized structures within the epidermis. The outer cells cornify to form a sheath; the inner cells of the cylinder become rearranged into longitudinal barb ridges. About fifty such ridges appear simultaneously within two-thirds of the circumference of the cylinder. The median ridge of the series marks the dorsal side of the feather. Additional ridges then appear on either side of the original set, so that the entire inner surface becomes grooved. The point where the circlet is completed lies diametrically opposite the dorsal side and thus is designated as ventral.

Two to four of the dorsalmost barb ridges then fuse at their bases, and the collar sends forth a triangular protrusion beneath them which proceeds upward with the growth of the cylinder as the *rhachis* or future shaft of the feather. Simultaneously, the bases of the other barb ridges become tilted towards the dorsal side of

the cylinder. New barb ridges continue to arise at the ventral side, with the same tangential slant as the other barbs. As the entire feather cylinder elongates by growth at its base, the base of each barb shifts progressively from its point of origin in the collar towards the rhachis and roots itself along the side of the rhachis. The vane of the definitive feather thus forms by the addition of a consecutive series of barbs, of ventral origin, to the rhachis, of dorsal origin, all parts proliferating from the circular base or collar.

Barbules develop from oblique rows of cells in the sides of the barb ridges. Specialized hooklets, or *barbicels,* may be formed on barbules, especially in regions of rapid growth. In the emerged feather, these hook the barbules of adjoining barbs together, thus forming a coherent air-foil as in flight feathers. When barbicels are absent, the barbs remain separate as down.

The apex of the feather cylinder cornifies and dries. Its tip becomes abraded, the sheath splits, and the barbs, formerly curved around the cylinder with their tips at the ventral side, are released to form the familiar nearly-flat vane. As growth of the feather nears completion, the pulp is resorbed towards the base of the cylinder until only a dermal papilla remains. The latter remains quiescent until the feather is shed, when it again becomes active and replaces the lost plume.

Environmental factors greatly influence the structure of feathers. Disturbances in nutrition or endocrine supply may temporarily prevent barbules or even barbs from forming, resulting in a defective band or *fault bar* across the vane. Sex hormones, in particular estrogens, drastically affect rates of growth, time of origin and differentiation of feather parts, and coloration, thus leading to the striking sexual dimorphism of plumage in many birds.

Color and pattern (see ANIMAL COLORATION) are due to lipochromes derived from the food of the bird, melanins produced by melanocytes of neural crest origin, optical effects due to the structure of the feather, or combinations of any of these factors. Melanocytes of two types produce either orange or black melanin, and their genetic constitution largely controls the patterns which they lay down. However, the feather organ provides the milieu in which the melanocytes differentiate, and controls their distribution and activities through its rates of growth, gradients of activity, responses to hormones, and other aspects of the chemical environment within the feather cylinder.

HOWARD L. HAMILTON

References

Hamilton, H. L., "Lillie's Development of the Chick," Chapt. 15, New York, Holt, 1952.
Rawles, M. E., "Skin and its derivatives," *in* Willier, B. H., *et al.* "Analysis of Development," Philadelphia, Saunders, 1955.
Rawles, M. E., "The integumentary system," *in* Marshall, A. J., "Biology and Comparative Physiology of Birds," vol. 1, Chapt. 6, pp. 189–240, New York, Academic Press, 1960.

FERMENTATION

The process of fermentation has been used by man from pre-historic times in the preparation of foods and beverages, but the causative agents of fermentation were not recognized until the middle of the nineteenth century. The end products resulting from the natural fermentation of glucose, namely alcohol and carbon dioxide (1), were identified by Gay-Lussac in 1810, but it was

$$C_6H_{12}O_6 \rightarrow 2CO_2 + 2C_2H_5OH \qquad (1)$$

thought that this process resulted from contact catalysis and the decay of animal or vegetable materials. This explanation was refuted by the work of Pasteur (1857) on the lactic acid fermentation. In the course of this investigation, Pasteur determined that fermentation was caused by living cells, that different microbial species caused different fermentations, that the nitrogenous materials present served only to support the growth of the cells, that lactic acid was produced when cells (removed from the fermentation mixture) were added to a sugar solution, and that the natural fermentation yielded both alcohol and lactic acid but that the amount of each could be altered by changes in pH. In later studies, Pasteur showed that the conversion of glucose to alcohol (1) was caused by yeast cells growing under anaerobic conditions, thus leading to the definition that fermentation was "life without air."

For the purposes of this discussion, fermentations will be defined not as "life without air," but as those energy-yielding reactions in which *organic* compounds act as both oxidizable substrates and oxidizing agents. Anaerobic reactions in which *inorganic* compounds are utilized as electron acceptors may be termed "anaerobic respirations," whereas reactions in which oxygen serves as a terminal electron acceptor are "RESPIRATIONS."

Almost any organic compound may be fermented provided it is neither too oxidized nor too reduced, since it must function as both electron donor and electron acceptor. In some fermentations, a compound is degraded via a series of reactions in which intermediates in the sequence act as electron donors and acceptors; in others, one molecule of the substrate may be oxidized while another molecule is reduced, or two different organic compounds may be degraded after a coupled oxidation-reduction reaction. These fermentations provide energy required for the growth of a variety of cells. In addition, many micro-organisms can carry out, in appropriate conditions, a number of fermentative reactions (e.g., oxidations, reductions, cleavages) which do not yield useful energy or do not yield sufficient energy for growth. The fermentations of carbon compounds and nitrogenous compounds discussed below will be concerned primarily with energy-yielding fermentations.

In view of the great variety of different compounds which may be fermented and the enzymatic capabilities of different micro-organisms, it is not surprising that many compounds important in industry (e.g., ethyl alcohol, butyl alcohol, acetone, 2,3-butylene glycol), in the production, preservation, and seasoning of food (e.g., lactic, citric and glutamic acids), and in medicine (e.g., vitamins are extracted from the yeast carrying out the alcoholic fermentation) may be produced most cheaply through microbial fermentations. In addition, fermentations continue to be important in the production of foods (e.g., the lactic and propionic acid fermentations in the making of cheeses), beverages (e.g., the alcoholic fermentations in the making of wine and beer), and in the leavening of breads (by the CO_2 produced in equation 1).

Fermentation of Carbon Compounds

A. Fermentation of carbohydrates to pyruvate. Carbohydrates are one of the chief sources of energy and cell components for the growth of plants, animals and micro-organisms. Oligosaccharides or disaccharides are first split by enzymes to the component sugars (glucose, fructose, mannose, or galactose). If a phosphorolytic enzyme mediates this reaction, sugar phosphates are produced; if the enzyme is hydrolytic, the resulting free sugars are phosphorylated at the expense of the "energy-rich" compound, adenosine-triphosphate (ATP). The subsequent metabolism of the phosphorylated sugars occurs by one of several pathways. In glycolysis (the only fermentative pathway found in plant and animal cells and the most common pathway in microorganisms), fructose-1-6-diphosphate is produced from hexose monophosphates by a second phosphorylation with ATP and is degraded to pyruvate by a sequence of enzymatic reactions outlined in Figure 1. In another pathway (discovered in the bacterium *Zymomonas lindneri*), glucose-6-phosphate is oxidized to phosphogluconic acid which is then split to phosphoglyceraldehyde and pyruvate. In a third pathway (found in heterofermentative lactic acid bacteria), phosphogluconic acid is oxidized and decarboxylated to yield ribulose phosphate which, in turn, undergoes cleavage to phosphoglyceraldehyde and ethanol. In all three pathways, phosphoglyceraldehyde is converted to pyruvate by the reactions shown in Figure 1 and is metabolized further to end products characteristic of each type of organism.

In glycolysis, 2 "energy-rich" bonds (indicated in Figure 1 as $\sim PO_3H_2$) are generated per molecule of triose oxidized, or 4 per molecule of fructose-1-6-diphosphate. These bonds are transferred to adenosine-di-phosphate (ADP), thereby yielding ATP which can be used as a source of energy for the synthesis of cell components (see Growth). Since the conversion of a hexose to its diphosphate derivative requires the expenditure of 2 molecules of ATP, glycolysis yields a net of 2 "energy-rich" bonds per molecule of hexose. The second and third pathways yield a net of 1 "energy-rich" bond—i.e., 2 "energy-rich" bonds arise from the metabolism of phosphoglyceraldehyde but 1 bond is required to produce glucose-6-phosphate. In the respiration of glucose (see Respiration), 30 "energy-rich" bonds could be generated per molecule of glucose, if all of the carbon were oxidized to carbon dioxide. This accounts for the observation, first made by Pasteur, that a larger number of yeast cells are produced for a given amount of glucose when cultures are grown under conditions of respiration rather than fermentation.

B. Metabolism of pyruvate. 1. *Lactic and alcoholic fermentations.* The reduction of pyruvate to lactate (2) may be carried out by a number of micro-

$$CH_3COCOOH + 2[H] \rightleftharpoons CH_3CHOHCOOH \quad (2)$$

organisms and also by mammalian muscle. In certain bacteria (e.g., homofermentative lactic acid bacteria), glucose is converted quantitatively to lactate; in others (heterofermentative lactic acid bacteria), ethanol and CO_2 are also produced (3). In the latter fermentation, ethanol arises by cleavage of ribulose phosphate (dis-

$$C_6H_{12}O_6 \rightarrow CH_3CHOHCOOH + \\ CH_3CH_2OH + CO_2 \quad (3)$$

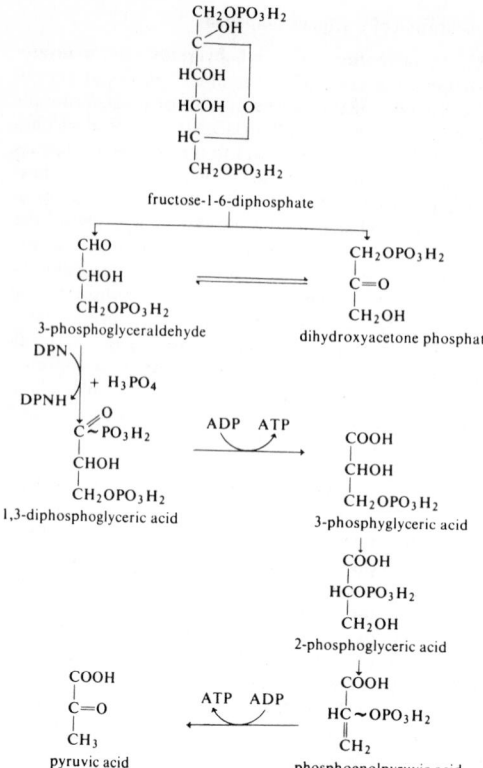

Fig. 1 Formation of pyruvate from fructose-1-6-diphosphate.

cussed above) but in the alcoholic fermentation of yeasts and certain bacteria (*Zymomonas lindneri*), ethanol arises from pyruvate. In this instance, pyruvate is decarboxylated to acetaldehyde (4a) and the latter is reduced to ethanol (4b). The reduction of acetaldehyde

$$CH_3COCOOH \rightarrow CH_3CHO + CO_2 \quad (4a)$$
$$CH_3CHO + 2[H] \rightarrow CH_3CH_2OH \quad (4b)$$

to ethanol, or of pyruvate to lactate (2), is linked to the oxidation of 3-phosphoglyceraldehyde to 1,3-disphosphoglycerate (Figure 1) and is mediated by the coenzyme diphosphopyridine nucleotide (DPN).

2. *Mixed acid fermentations.* The fermentation of pyruvate may yield more than one organic acid. For example, lactate and acetate are produced under certain conditions by the bacterium *Escherichia coli* (5a). This requires the oxidative decarboxylation of one molecule of pyruvate (5b) and reduction of another (5c). Certain bacteria (e.g., *Micrococcus lactilyticus*) do not form lactate from pyruvate, but carry out an oxidative decarboxylation in which the available hy-

$$2CH_3COCOOH + H_2O \rightarrow CH_3COOH +$$
$$CO_2 + CH_3CHOHCOOH \quad (5a)$$
$$CH_3COCOOH + H_2O \rightarrow$$
$$CH_3COOH + CO_2 + 2[H] \quad (5b)$$
$$CH_3COCOOH + 2[H] \rightarrow CH_3CHOHCOOH \quad (5c)$$

drogen is released as hydrogen gas (6). Others (e.g., *Clostridium butyricum*) metabolize pyruvate to acetate

$$CH_3COCOOH + H_2O \rightarrow$$
$$CH_3COOH + CO_2 + H_2 \quad (6)$$

and formate (7); the latter compound may be degraded to CO_2 and H_2 by other bacteria. Small amounts of

$$CH_3COCOOH + H_2O \rightarrow$$
$$CH_3COOH + HCOOH \quad (7)$$

succinate may also be formed in these fermentations (see section 5 below). The acetate produced in (5b) and (6) appears first not as free acetate but as the "energy-rich" Coenzyme A derivative, acetyl-Coenzyme A (acetyl-CoA). Thus, the energy of oxidation is conserved and is made available (via 8a and 8b) as ATP which may then be used for synthetic reactions. Re-

$$\text{acetyl-CoA} \underset{H_3PO_4}{\rightleftharpoons} \text{acetyl-phosphate} \underset{+ CoA}{\overset{ADP}{\rightleftharpoons}} \quad (8a)$$

$$\text{acetyl-CoA} + ADP \rightleftharpoons \text{acetate} + ATP + CoA \quad (8b)$$

action (8a) occurs only in bacteria and (8b) has been found in plant and animal cells and in many microorganisms.

3. *Butylene glycol fermentation.* In addition to variable amounts of formate, acetate, ethanol, CO_2 and H_2, some micro-organisms (e.g., *Aerobacter aerogenes*) produce 2,3-butylene glycol as the chief end product of glucose fermentation. This 4-carbon compound arises from the condensation of acetaldehyde (produced via 4a) with pyruvate to form α-acetolactate

$$(CH_3COHCOCH_3).$$
$$|$$
$$COOH$$

The latter is decarboxylated to acetoin ($CH_3CHOH\cdot CHOCH_3$) which is subsequently reduced to 2,3-butylene glycol ($CH_3CHOHCHOHCH_3$). Acetaldehyde may also be condensed with longer compounds of similar structure thereby yielding more complex acyloins.

4. *Butyric acid-butanol-acetone fermentations.* Certain anaerobic bacteria, notably species of clostridia, produce butyrate, butanol, acetone and isopropyl alcohol as the chief end products in the fermentation of glucose. The synthesis of butyrate requires the participation of the Coenzyme A derivatives of several organic acids (9a-9d). In the synthesis of butyrate, 2 molecules of acetyl-CoA produced from pyruvate via (5b) are condensed to form acetoacetyl-CoA (9a) which

$$2CH_3CO—CoA \rightleftharpoons CH_3COCH_2CO—CoA +$$
$$CoA + H_2O \quad (9a)$$
$$CH_3COCH_2CO—CoA + 2[H] \rightleftharpoons$$
$$CH_3CHOHCH_2CO—CoA \quad (9b)$$
$$CH_3CHOHCH_2CO—CoA \rightleftharpoons H_2O +$$
$$CH_3CH=CHCO—CoA \quad \text{or} \quad (9c)$$
$$CH_2=CHCH_2CO—CoA$$
$$CH_2CH=CHCO—CoA \quad \text{or}$$
$$CH_2=CHCH_2CO—CoA +$$
$$2[H] \rightleftharpoons CH_3CH_2CH_2CO—CoA \quad (9d)$$

is then reduced first to β-hydroxybutyryl-CoA (9b), then dehydrated to either crotonyl-CoA or vinylace-tyl-CoA (9c). The product of this reaction is reduced to butyryl-CoA (9d). Deacylation and reduction of the latter yields butanol. Acetoacetyl-CoA formed in (9a) may be deacylated and decarboxylated to acetone (10a) which may also be reduced to isopropyl alcohol (10b).

$$CH_3COCH_2COOH \rightarrow CH_3COCH_3 + CO_2 \qquad (10a)$$

$$CH_3COCH_3 + 2[H] \rightarrow CH_3CHOHCH_3 \qquad (10b)$$

5. *Propionic acid fermentation.* Two different path-ways have been described for the formation of pro-pionate. The acetate and CO_2 produced by propioni-bacteria from glucose (11) are formed via equation (5b) whereas propionate arises from a 4-carbon compound

$$3C_6H_{12}O_6 \rightarrow 4CH_3CH_2COOH +$$
$$2CH_3COOH + 2CO_2 + 2H_2O \qquad (11)$$

by the following reactions (the same reactions are re-sponsible for the metabolism of propionate by a va-riety of animal cells). Pyruvate is condensed with CO_2 to yield oxaloacetate which is converted to malate $\rightarrow$ fumarate $\rightarrow$ succinate through the Krebs tricarboxylic acid cycle (see Respiration). This pathway also ac-counts for the succinate found in the mixed acid fer-mentations (section 2 above). In propionate synthesis, succinate is isomerized to methylmalonate which is then decarboxylated to propionate and CO_2 (12). The latter reactions require the participation of the Co-

$$COOHCH_2CH_2COOH \rightleftharpoons$$
$$COOHCHCOOH \rightleftharpoons \qquad (12)$$
$$\underset{\displaystyle CH_3}{|}$$
$$CH_3CH_2COOH + CO_2$$

enzyme A derivatives of the acids rather than the free acids.

Another anaerobic bacterium (*Clostridium pro-pionicum*) is unable to carry out the sequence shown in equation (12) although propionate is formed from a variety of substrates. It has been postulated that pro-pionate is produced in this species by the dehydration of lactate (produced via 2) to acrylate and reduction of the latter to propionate (13).

$$CH_3CHOHCOOH \xrightarrow{\;-H_2O\;}$$
$$\qquad (13)$$
$$CH_2{=}CHCOOH \xrightarrow{\;+2[H]\;} CH_3CH_2COOH$$

C. *Fermentation of aldonic acids, polyhydric al-cohols and pentoses.* Bacteria adapted by growth on glucuronate, galacturonate, 2-ketogluconate and 5-ketogluconate are able to ferment these compounds to characteristic end products. Hexitols, glycerol and pentoses (ribose, xylose, arabinose and desoxyrobose) may also be fermented by a variety of micro-organisms. The pathways of fermentation of the latter group of compounds have not been completely established, but pentose fermentations are known to involve phos-phorylation, before or after isomerization, followed by cleavage to 2-carbon and 3-carbon compounds.

Fermentation of Nitrogenous Compounds

A. **Fermentation of amino acids.** 1. *Single amino acids.* About 20 species of anaerobic bacteria (chiefly clostridia and anaerobic micrococci) are known to fer-ment single amino acids. With the exception of pro-line, hydroxyproline and isoleucine, all of the common amino acids may be degraded by at least one species. As an example of this class of fermentation, the path-way for the fermentation of glutamate by the anaerobic bacterium, *Clostridium tetanomorphum*, will be out-lined (13a–13d). Glutamate is isomerized to β-methyl

$$COOHCHNH_2CH_2CH_2COOH \rightleftharpoons$$
$$COOHCHNH_2CHCOOH \quad (13a)$$
$$\underset{\displaystyle CH_3}{|}$$

$$COOHCHNH_2CHCOOH \rightleftharpoons$$
$$\underset{\displaystyle CH_3}{|}$$
$$\qquad (13b)$$
$$COOHCH{=}C{-}COOH + NH_3$$
$$\underset{\displaystyle CH_3}{|}$$

$$COOHCH{=}C{-}COOH + H_2O \rightleftharpoons$$
$$\underset{\displaystyle CH_3}{|}$$
$$\qquad (13c)$$
$$COOHCH_2COHCOOH$$
$$\underset{\displaystyle CH_3}{|}$$

$$COOHCH_2COHCOOH \rightleftharpoons$$
$$\underset{\displaystyle CH_3}{|} \qquad (13d)$$
$$CH_3COOH + CH_3COCOOH$$

aspartate (13a) which is deaminated to mesaconate (13b). Hydration of mesaconate yields citramalate (13c) which is cleaved to acetate and pyruvate, the lat-ter giving rise to butyrate by the mechanism outlined earlier (8a–8d).

2. *Pairs of amino acids.* Certain of the clostridia can-not ferment single amino acids, but are able to degrade appropriate pairs of amino acids (the "Stickland re-action"). One member of the pair is oxidized while another is reduced. For example, in the simultaneous fermentation of glycine and alanine (14a), alanine is oxidatively deaminated to pyruvate (14b) and pyruvate undergoes oxidative decarboxylation to acetate and CO_2 (14c). These reactions are coupled to the reduc-tive deamination of glycine (14d).

$$CH_2CHNH_2COOH + 2CH_2NH_2COOH +$$
$$2H_2O \rightarrow CH_3COOH + CO_2 + 3NH_3 \qquad (14a)$$

$$CH_3CHNH_2COOH + H_2O \rightarrow$$
$$CH_3COCOOH + NH_3 + 2[H] \qquad (14b)$$

$$CH_3COCOOH + H_2O \rightarrow$$
$$CH_3COOH + CO_2 + 2[H] \qquad (14c)$$

$$2CH_2NHCOOH + 4[H] \rightarrow$$
$$2CH_3COOH + 2NH_3 \qquad (14d)$$

B. Fermentation of heterocyclic and other compounds. Relatively few bacteria (for the most part clostridia and anaerobic micrococci) have been reported to ferment heterocyclic compounds. The variety of end products and differences in the intermediates formed in the fermentation of purines suggests that the degradation of these compounds by different bacteria does not proceed by the same pathway. Detailed studies on the degradation of xanthine by *Clostridium cylindrosporum* have shown that the following sequence is involved in this anaerobe: xanthine $\rightarrow$ 4-ureido-5-imidazole carboxylic acid $\rightarrow$ 4-amino-5-imidazole-carboxylic acid (plus NH_3 and CO_2) $\rightarrow$ 4-amino imidazole (plus CO_2) $\rightarrow$ 4-imidazolone (plus NH_3) $\rightarrow$ formiminoglycine $\rightarrow$ glycine $+$ NH_3 $+$ HCOOH. Pyrimidines are fermented by 3 bacterial species, but only *Zymobacterium oroticum* can use these compounds as a source of energy for growth. In addition, the following compounds have been reported to be fermented by bacteria: allantoin, nicotinic acid, creatinine, and ergothionine.

H. R. WHITELEY

References

Barker, H. A., "Bacterial Fermentations," New York, Wiley, 1956.

Stanier, R. Y., *et al.* "The Microbial World," Englewood Cliffs, New Jersey, Prentice-Hall, 1963.

Gunsalus, I. C., *et al.*, "Pathways of Carbohydrate Metabolism in Microorganisms," Bacteriol. Rev. **19**: 79–128, 1955.

Elsden, S. R. and J. L. Peel, "Metabolism of Carbohydrates and Related Compounds," Ann. Rev. Microbiol. **12**: 145–202, 1958.

FERTILIZATION

Fertilization is the union of two sexually differentiated cells, or gametes. In the Metazoa the fertilization process presents a comparatively uniform picture. SPERM fuses with the EGG (*syngamy*) and this includes both the union of the paternal nucleus with the maternal one (*karyogamy*) as well as the fusion of the cytoplasms of the gametes (*plasmogamy*). Fertilization has its developmental and genetic consequences. The fertilized egg begins to develop and the new individual originating therefrom is endowed with the hereditary material transmitted almost exclusively through the CHROMOSOMES of the pronuclei. Fertilization is preceded by two meiotic divisions of the germ cells in which reduction of chromosome number and the processes of crossing over take place. As a consequence, mature germ cells have the reduced (*haploid*) number of chromosomes restored to diploid condition during fertilization. The processes of meiosis and fertilization increase the genetic variability of the offsprings through the recombination of nuclear genes. The fusion of two gametes may be permanent or temporary (*conjugation*). Usually the gametes of the two sexes differ in size and shape (*anisogamy*), in some cases, however, they are morphologically undistinguishable (*isogamy*) but differ nevertheless in their biochemical and physiological properties. In certain unicellular organisms the meiotic divisions take place just after syngamy. As a special form of fertilization *autogamy* may be mentioned in which the processes of nuclear reorganization may occur without the conjugation of two individuals. Autogamy is therefore internal self-fertilization in which two daughter nuclei fuse within the same cell. The meaning of autogamy is not quite clear. Once the autogamy is passed no genetic changes at subsequent autogamies can appear, unless mutation takes place.

While considering the problems of fertilization reference should be made to the pre-fertilization behavior of the gametes i.e. the processes leading to the attachment of the sperm to the egg and to the maturation of the gametes. In all eggs which are incapable of natural PARTHENOGENESIS some block sooner or later sets in which inhibits their further development. This block can be established before the breakdown of the germinal vesicle in the oocyte, during meiotic divisions or after their completion. In all cases in which the eggs are fertilizable before the maturation is accomplished the removal of the block caused by sperm entrance leads to completion of maturation before karyogamy ensues.

Whereas in lower plants (FERNS, MOSSES) archegonia produce certain substances which attract spermatozoa, chemotaxis of this kind has not been established with certainty for animal sperm cells. Yet, in many, if not in all, animal species substances produced by the gametes have been found to exert definite influence on the germ cells predominantly of the opposite sex. In the opinion of many students of fertilization such substances facilitate this process or render it possible. The manner of operation of these interacting substances induced many workers to base the theory of fertilization on immunological analogies. On the other hand it has been suggested that fertilization consists in the activation of enzymes liberated through the action of sperm cell from the inhibitor-enzyme complex.

Two phases can be distinguished in fertilization. The first one includes the attachment of the spermatozoon to the surface of the egg and the reactions started thereby in the ovum itself which lead to development. The second phase occurs after the sperm has entered the egg and ends with the fusion or apposition of the pronuclei. The first phase can be naturally or artificially dissociated from the second, and in that case the activated egg develops without the participation of paternal chromosomes. This phenomenon is called *pseudogamy*. Pseudogamy may be artificially induced by heterologous insemination or by separation of homologous sperms shortly after insemination.

The first successful collision of the sperm with the egg releases a long series of reactions in the ovum which originate in its cortex and are the first sign of activation. To describe these changes sea urchin egg, a classical object, may serve as an example. The mature egg shed into the sea water is surrounded by jelly coat which gradually dissolves liberating one of the sperm-egg interacting substances (*fertilizin*) which is responsible for the often reversible agglutination of homologous spermatozoa. The outmost layer of the egg is covered by vitelline membrane under which cortical granules are located. The fertilizing sperm passes through the jelly coat and its reaction in the presence of the egg is the preduction of a filament from the acrosome of its head. The filament is anchored to the egg and the egg cytoplasm gradually engulfs the male germ cell. Under the influence of the sperm a local change in the cortical layer of the egg takes place which spreads from the point of sperm attachment. As found by Runnström this is accompanied by a change of color in the egg cortex. Immediately afterwards cortical granules undergo a breakdown and their substance is partly incorporated in the vitelline membrane which is lifted from the egg surface and now forms fertilization membrane. Under the fertilization membrane fluid accumulates in the perivitelline space, and on the surface of the egg a hyaline layer is formed. After the engulfment of the spermatozoon its centrosome provides the egg with the sperm aster and the female pronucleus migrates toward the male pronucleus. The fusion of the two pronuclei being completed synkaryon moves toward the center of the egg and the first mitotic figure ensues. The changes started in the cortex of the egg influence, in turn, the deeper parts of the egg cytoplasm. The influence of fertilization on the shape of the egg, on its viscosity, permeability, protein solubility and metabolism have been the subject of extensive researches.

Any theory of fertilization must account for the specificity of this process. As a rule the egg can be successfully fertilized only by homologous sperm, although the incompatibility in cross-fertilization is relative. Recently attempts were made to elaborate the views on the immunological nature of fertilization and the analogy between fertilization and transplantation, especially in connection with immunological tolerance (Hašek, Medawar). The results obtained do not preclude the possibility that the decrease in the reciprocal immunological reactivity of both parent animals (birds) improve the results of the remote crossing as regards the augmentation of the percentage and vitality in the progeny.

Normally only one sperm fertilizes the egg. If more than one sperm participate in fertilization (*polyspermy*) development is disturbed. In some species many sperms penetrate into the egg but only one provides the male pronucleus which fuses with the female one, while all others undergo degeneration. After fertilization the egg is very early protected against polyspermy but the mechanisms establishing this block have not yet been fully elucidated. Runnström suggests that there is a "multiple insurance" against polyspermy in which the jelly and hyaline layers play their part. In sea urchin egg the complete block reaches its full efficiency in 80 seconds, but a few seconds after sperm attachment the egg already acquires a certain degree of protection. It is not improbable that the egg may somehow choose which sperm attached to its surface should start the development and be engulfed. The process of fertilization in relation to the fate of sperms which do not take their part either in the fertilization proper or in the penetration into the egg cytoplasm was recently studied by Genin, Vojtiškova, Austin and others with quite contradictory results. On the one hand, evidence was presented showing that sperms penetrate into the somatic tissues of a female and even their genetic action was assumed, on the other hand the sperms found in the somatic tissues of a female are considered artifacts and there is evidence indicating that sperms not participating in fertilization become phagocytized in the lumen of the reproductive tract of the female.

S. SKOWRON

References

Austin, C. R., "Fertilization," Englewood Cliffs, N.J., Prentice-Hall, 1965.
Monroy, A., "Chemistry and Physiology of Fertilization," New York, Holt, 1965.
Rothschild, "Fertilization," London, Methuen, 1956.
Runnström, J., B. E. Hagström and P. Perlmann, "Fertilization," in Brachet, J. and A. E. Mirsky, eds. "The Cell," vol. 1, New York and London, Academic, 1959.
Tyler, A., "Gametogenesis, Fertilization and Parthogenesis," in Willier, B. H., P. A. Weiss and V. Hamburger, eds., "Analysis of Development," Philadelphia and London, Saunders, 1955.

FILICINEAE

The ferns are a major class of vascular plants (see SPERMATOPHYTA) represented as early as the Devonian, and with over 200 genera and 8,000–9,000 species living today. They are characterized by two free-living, perennial generations: the sporophyte, a diploid, spore-producing plant, which is usually conspicuous and long-lived and possesses stems, leaves, and roots; and the gametophyte, haploid, gamete-producing plant, inconspicuous and usually short-lived. No seeds nor pollen tubes are produced by existing Filicineae. Fern gametophytes are of three general types: the photosynthetic type, either terrestrial or epiphytic; the saprophytic type, subterranean; and the parasitic type, which derives nutrition from that deposited by the parental sporophyte in the spore. Fertilization depends upon free water, through which the sperms swim from the antheridium to the egg contained in the archegonium. Ferns with parasitic gametophytes have two types of spores, the microspores containing antheridia, and the megaspores containing archegonia; the embryo develops directly within the megaspore walls. In all ferns the zygote and embryo form in the venter of the archegonium and the embryo may be endoscopic or exoscopic.

Mature fern sporophytes generally have creeping rhizomes which die off in the older parts, but the stems may be more or less erect at the apex; and they form tall trunks in tree ferns. All stem and root tissues originate by primary growth except in the Ophioglossum group. The vascular tissues are composed of tracheids (rarely vessel elements) with scalariform pitting, and sieve cells. The stelar patterns are extremely diverse, ranging from simple protosteles to elaborate dictyosteles. The meristematic apices are protected usually by uniseriate or multiseriate hairs (paleae or ramenta), and typical bud-scales like those of seed plants are absent. Fern roots are formed in association with leaf bases. The leaves or FRONDS have circinate vernation and are usually divided. All the basic organs arise from the division products of single apical or marginal cells with few exceptions. The leaves or fronds are generally the most conspicuous part of the fern sporophyte; they function both as photosynthetic and as reproductive organs, but the two functions may be separated, producing a marked dimorphism between parts of fronds or between fronds.

The sporangia are borne upon the margins or abaxial

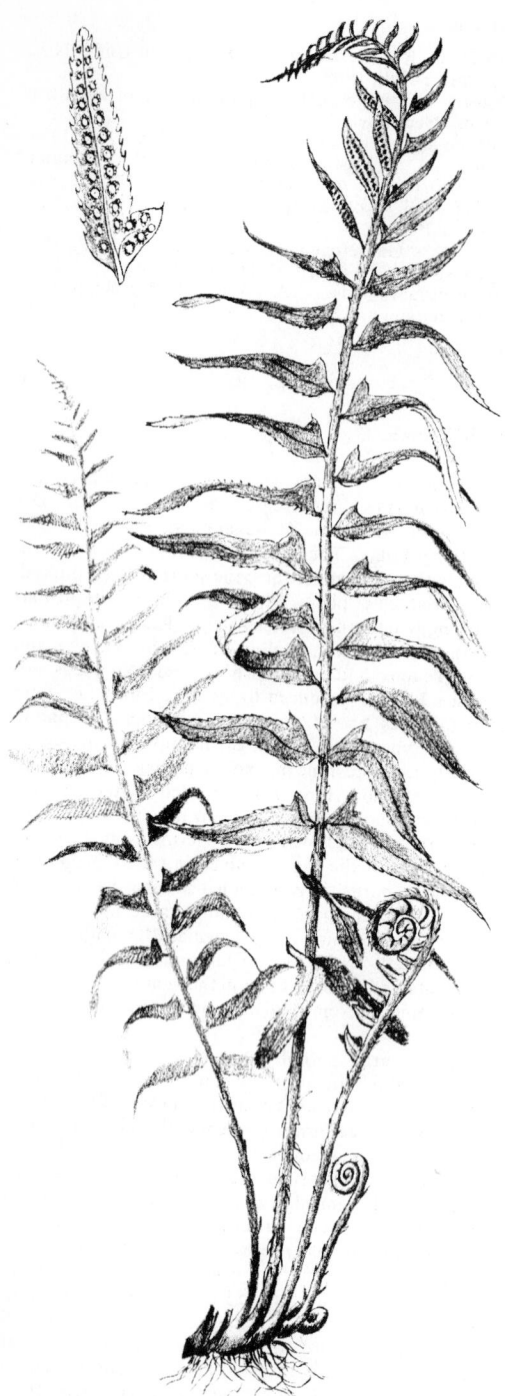

Fig. 1 The Christmas fern, *Polystichum acrosti-choides*; showing details of the rhizome, the coiled conditions of the young fronds, and the sori. (From Platt and Reid, "Bioscience," New York, Reinhold, 1967.)

surfaces of the lamina, either singly or in clusters, the grouped sporangia discrete (*sori*) or more or less fused together (*synangia*). Soriate sporangia mature simultaneously, successively, or in mixed order.

Much of fern classification is based upon sporangium and sorus structure. The sporangia fall into two general classes, but intermediate types occur. Eusporangia have many spores (ca. 500 or more), several initials, and massive walls. Leptosporangia have small numbers of spores (128, 64, 32, 16, or 4), single initials, and delicate walls. In heterosporous ferns, the sporangia are of two types, microsporangia and megasporangia. Leptosporangia typically possess an area or arc of mechanical cells, the annulus, which opens the sporangium. The sorus is commonly protected by the modified laminar margin or by special laminar outgrowths or indusia. The spores of ferns are either trilete or monolete, and may possess a perisporial covering on the exospore.

The most widespread type of autotrophic gametophyte is a photosynthetic, flat, ribbon- or heart-shaped prothallus, which grows from a meristematic notch. Antheridia are formed on the young growth and wings of the prothallus, and archegonia are produced on the thickened midrib or cushion of mature prothalli.

Vegetative origin of new sporophytes is by branching and fragmentation of rhizomes, lateral rhizome buds, and by leaf buds of various types. Some species reproduce apogamously, the sporophyte producing unreduced spores which germinate directly into diploid gametophytes with the ability to proliferate new sporophytes directly; both generations thus have the same chromosome number and there is no sexual fusion. Vegetative reproduction of the gametophyte occurs by fragmentation and, in certain tropical genera, by gemmae. The monoploid chromosome numbers of ferns are high, generally between 13 and 80, and most commonly between 25 and 60. Polyploidy is common, and diploid chromosomes may number well over a hundred (rarely over a thousand). Interspecific hybridization is widespread in many genera and families.

The majority of Filicineae flourish in tropical, wet regions, although some have adapted to arctic or xerophytic life. Habitats are mostly terrestrial in mesophytic forests, but include also open fields, rock outcrops, swamps and marshes, vernal pools, ponds, and boughs and trunks of trees. Epiphytes make up as much as one-third or more of the fern flora in certain tropical rain forests.

Although formerly classified with lycopods, articulates, and other pteridophytic plants on the basis of the life-cycle, the ferns are now generally believed to be more closely related to modern seed plants, especially on the basis of leaf structure and sporangial position. It is possible that ancient ferns were the prototypes of present-day gymnosperms and angiosperms. The most successful Filicineae are the "common ferns," with fully developed leptosporangia and sori. Parallel and convergent evolution has evidently been rampant in fern phylogeny, however, producing many problems of relationships which are still unsolved.

Some of the more important and distinctive living fern groups are the following: Marattia group (leaves stipulate, eusporangia, synangia, ca. 5 gen., 100 spp.); Ophioglossum group (leaf bases clasping, leaf with centrally attached fertile segment, eusporangia, 3 gen., 60 spp.); Osmunda group (leaves stipulate, sporangia intermediate, 3 gen. 20 spp.); Schizaea group (sporangia with apical annuli, 4 gen., 150 spp.); Gleichenia group (leaves usually falsely dichotomous and indeterminate, annulus oblique, 5 gen., 120 spp.); Filmy-fern or Hymenophyllum group (lamina filmy, one cell thick, sori involucrate, 20 gen., 500 spp.); Tree-fern group (giant ferns with well-developed trunks, ca. 2 families, 8 gen., 800 spp.); Common ferns or Polypodium group (typical leptosporangiate ferns of great diversity, ca. 5–10 families, 200 gen., 6500 spp.); Marsilea group (heterosporous, usually rooting in ponds and temporary pools, the leaves bearing sporocarps, 3 gen., 75 spp.); and the Salvinia group (much reduced, floating, heterosporous ferns, 2 gen., 15 spp.).

WARREN H. WAGNER, JR.

References

Bower, F. O., "The ferns," vols. 1–3. Cambridge, The University Press, 1923–1928.
Ching, R. C., "On natural classification of the family 'Polypodiaceae,' "Sunyatsenia 5: 201–268, 1940.
Copeland, Edwin B., "Genera Filicum," Waltham, Mass, Chronica Botanica, 1947.
Eames, A. J., "Morphology of vascular plants. Lower Groups," New York, McGraw-Hill, 1936.
Holttum, R. E., "A revised classification of leptosporangiate ferns," Jour. Linn. Soc. (Bot.) 53: 123–186, 1946.
Manton, Irene, "Problems of cytology and evolution in the Pteridophytes," Cambridge, The University Press, 1950.
Pichi-Sermolli, E. G., "The higher taxa of the Pteridophyta and their classification," Acta Univ. Upsaliensis 1958 (6): 70–90, 1958.
Verdoorn, F., ed., "Manual of Pteridology," The Hague, Nijhoff, 1938.

FLAGELLA

Flagella are thread-like projections of the cell surface which usually function either in locomotion or in the production of water currents. Bacterial flagella apparently vary in structure and function, usually consisting of several parallel protein fibrils with no enclosing membrane, ranging about 120–250 Å in diameter and a few micra in length. Although they are often important antigenically and otherwise of interest to microbiologists, they are not comparable to the flagella of eucaryotic cells and are not further considered in the following discussion.

Flagella of eucaryotic cells are essentially similar to CILIA, there being no absolute distinction between them either morphologically or physiologically. In general usage, the distinction is this: Cilia are typically shorter, present in greater numbers per cell (commonly there are only one or two flagella per cell), and coordinated in beat. Flagella serve primarily as locomotor organelles in most of the MASTIGOPHORA, zoospores and gametes of ALGAE and aquatic FUNGI, spermatozoids of BRYOPHYTES, PTERIDOPHYTES, and cycads (see GYMNOSPERMAE), and the spermatozoa of most metazoan animals. Flagella also play a vital role in PORIFERA and many CNIDARIA, in which they serve to produce water currents essential to respiration, food-getting, and internal transport.

A typical flagellum consists of an elongate *axoneme* enclosed by a *sheath*. In the axoneme, nine longitudinal tubular fibers form a cylinder which surrounds two additional longitudinal fibers, producing the now-famous "9 + 2 pattern." The two inner fibers or tubules are about 200 Å in diameter and 300 Å apart, and are often

shorter in length than the outer fibers. The outer fibers are typically double (presenting a "figure 8" appearance in cross section), with dimensions approximately 200 × 300–350 Å. Over-all dimensions of the flagellum exceed 2000 Å in diameter; length varies greatly—from a few micra to over 200 μ. The sheath is bounded peripherally by a membrane which is continuous with the cell membrane. The sheath may contain accessory structures, such as the *paraflagellar rods* of euglenoids which extend almost the entire length of the flagellum, lying alongside and parallel to the axoneme. Such structures may increase the diameter of the flagellum to more than a micron. The remaining material within the membrane, the flagellar plasm, exhibits diverse appearances: In some cases it appears to contain fibrillar elements oriented longitudinally or helically, in others it appears homogeneous or may present a striated aspect reminiscent of myosin fibrils. Fixation methods undoubtedly influence the appearance. Neither the chemical composition nor the functional roles of the various components have yet been elucidated.

Within the cell body at the base of each flagellum lies a *basal body* or *kinetosome*, presumably involved in both the formation and the function of the flagellum. Basal bodies may either be derived from or develop into *centrioles* (division centers), the two structures being homologous. The nine outer fibers of the flagellar axis are characteristically continuous with nine doublet or triplet microtubules of the basal body, but the two central fibers or microtublues of the axoneme end proximally within the flagellum.

Flagellar appendages known as *mastigonemes* or *flimmer* are characteristic of certain groups of phytoflagellates. Mastigonemes are filaments, about 100–150 Å in diameter and 1–4 μ in length, which extend laterally from the flagellar sheath. Among euglenoids, the mastigonemes are relatively long and lax, usually extending from only one side of the flagellum, although a shorter, denser fuzz of still finer fibrils occurs on both (all) sides, and the tip may bear a somewhat bottle-brush-like tuft of intermediate fibrils. In a number of groups with two unequal flagella, such as the chrysomonads (Chrysophyceae), the shorter flagellum is naked while the longer one bristles with stiff mastigonemes which extend outward like the bristles of a bottlebrush. The mode of formation of mastigonemes is no better understood than is that of flagella themselves, nor is it clear what advantage may be conferred by the flimsy mastigonemes of the plume-like euglenoid flagellum. The work of Jahn, et al., however, suggests that the rigid mastigonemes of such chrysomonads as *Ochromonas* may well be responsible for their rapid locomotion. Such work is also making it increasingly clear that the flagella of different taxonomic groups may utilize different basic mechanical principles in the production of propulsive force. Some serve to pull, whereas others push the cell body of the organism through the water.

HARLEY P. BROWN

References

Jahn, T. L., M. D. Landman and J. R. Fonseca, "The mechanism of locomotion of flagellates. II. Function of the mastigonemes of *Ochromonas*," J. Protozoology 11: 291–296, 1964.
Leedale, G. F., "Euglenoid Flagellates," Englewood Cliffs, N. J., Prentice-Hall, 1967.

Pitelka, D. R., "Electron-Microscopic Structure of Protozoa," New York, Macmillan, 1963.

Satir, P., "Structure and function in cilia and flagella," Protoplasmatologia 3E: 1–52, 1965.

FLAME CELL

Flame cells form part of the "excretory" organ, the protonephridium, in most but not all groups of acoelomate and pseudocoelomate animals. Although these cells vary in shape and may be uni- or multinucleate, they all share one common feature, the flame bulb. In certain species one cell may give rise to several flame bulbs. The latter are described as cup-shaped masses of protoplasm in which tufts of cilia (the flame) arise from a basal plate and project into the cavity of the cup. This cavity is continued by the protonephridial tubules. Electron micrographs of different sections through flame cells of a few species have shown the flame to be composed of 40–80 tightly packed cilia that are attached to the cytoplasm of the cell by electron-dense, conical rootlets passing through a basal plate. In cross section each cilium is hexagonal in shape and shows the conventional nine pairs of peripheral and two central fibrils enclosed within a membrane. The lumen around the flame appears to be wider in the area close to its origin from the cytoplasm and becomes narrower towards its free end. This lumen is surrounded by a tubular wall that contains rod-like supportive structures joined together by very thin membranes.

It has not been possible to ascribe a single definite physiological function for flame cells in all the groups of animals in which they occur. Based on experimental observations of the correlation between the tonicity of suspending media and flame-cell activity, in certain species they appear to function as osmoregulatory organs, while in other species this does not hold true. There seems to be no experimental proof for an excretory function. It has been suggested by several workers that the action of the cilia, in causing a distal movement of fluid within the lumen of the flame bulb, could produce a filtration pressure, and thus filtration could take place through the flame-cell membrane.

Some of the structural and functional aspects of flame cells are reviewed in the following references.

ARAXIE KILEJIAN

References

Hyman, L., "The Invertebrates," New York, McGraw-Hill, Vol. 1, 1940, Vol. 2, 1951.

Martin, A. W., "Comparative physiology (excretion)," Ann. Rev. Physiol. 20: 225–242, 1958.

von Bonsdorff, C. H. and A. Telkka, "The flagellar structure of the flame-cell in fish tapeworm (Diphyllobothrium latum)," Ztschr. Zellforsch. 70: 169–179, 1966.

FLIGHT

True flight, that is, the ability to move through the air for long periods, is found in only three groups of living animals, the birds (SEE AVES), bats (see CHIROPTERA) and INSECTS; in addition, it was well developed in the extinct pterodactyls. The ability to glide through the air at a more or less steep angle has been evolved independently by a wide variety of forms including all the main vertebrate groups. The large flat surfaces needed for gliding flight are formed in different ways in different animals. Thus, the frog Rhacophorus has very large webbed feet; the flying dragon (Draco volens) has a flexible membrane down each side of its body; the flying squirrel (Glaucomys) has a membrane of skin running between its wrists, body, and hind legs. All these animals merely prolong a jump from branch to branch or from the trees to the ground. The most remarkable of the pure gliding animals is the flying fish Exocoetus where the pectoral fins are enlarged into wings; here the glide is started with the tail continuing to swim in the water and may be greatly prolonged by use of the gradient of wind velocities near the surface of the sea.

True flight, by the birds, bats, pterodactyls and insects, always involves a definite WING and here again its construction is different in different animals. The wing surface of a bird is formed by FEATHERS borne on the forearm and the enlarged second digit. with the first and third digits present but reduced. The wing of a pterodactyl was a skin fold supported mainly on the enlarged fourth digit. The wing of a bat is again a fold of skin supported on the second to fifth digits of the hand and on the short hind legs and tail. The wings of insects are thin folds of exoskeletal cuticle and are not derived from segmental appendages. It is clear that aerial LOCOMOTION has evolved many times under different circumstances and that the similarities between flying animals are due to the strict physical requirements for an efficient flight system.

The flight of any heavier-than-air object is made possible by the aerodynamic reaction of the air flowing over the smooth wing surfaces. In a steady downward glide this reaction, which is nearly at right-angles to the airflow, can have a forward and upward component—the thrust and the lift—(Fig. 1), which balance respectively the drag and the weight of the animal. The wing is pulling the body forward through the air at the expense of the loss of potential energy. In flapping flight, whether by birds, bats or insects, active movement of the wings relative to the body, particularly on the forward and downward stroke, produces a greater aerodynamic force than that from a stationary wing; the direction of the force may also be changed. The backward and upward stroke of the wings takes place with an upward twist of the leading edge; its contribution to the flight varies in different animals and under different conditions. In the rapid flight of most birds, bats and insects, the upward twist is slight and the upstroke generates lift at the expense of increased drag. At the start of flight in many small birds, the upward twist is greater and the wing may be flexed so that the upstroke is a recovery stroke with reduced aerodynamic effect. In humming birds and in the smaller fast-flapping insects such as the bees, wasps and flies, the wing-tip turns right over on the upstroke so that lift is again generated. It is more correct in the hovering flyers to speak of the forward and backward strokes of the wings, since they move more nearly in a horizontal plane with the body tipped upwards at a steep angle. In the smallest insects, where the wing may be less than 0.1 mm long, special aerodynamic effects begin to be important; in many of

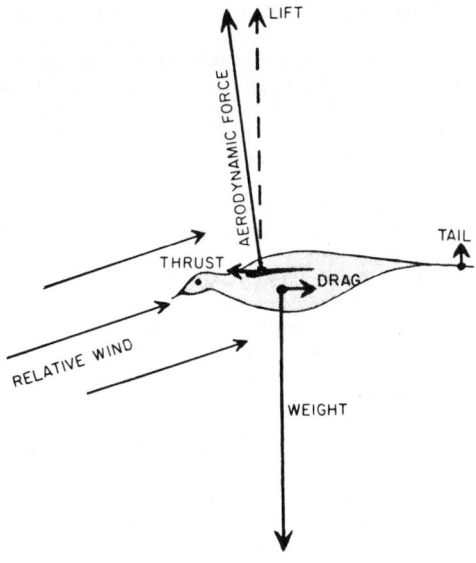

Fig. 1   The force acting on a gliding bird.

that of a helicopter than that of a conventional aero-plane, reaches its highest perfection in the flies (see DIPTERA), in which the second pair of wings is modified to form a gyroscopic sense organ detecting rotations in the three planes of space and controlling balance entirely through the single pair of wings. Other senses such as sight or hearing are usually also well developed in flying animals. Sight is particularly important for birds, bees and wasps (see HYMENOPTERA), none of which can fly blind; hearing, including the detection of echoes of sounds produced by themselves, is important for bats, and a well developed sense of hearing is also found in the nocturnal owls and moths (see LEPIDOP-TERA).

In other aspects of physiology the flight of birds and bats has involved merely the refinement and adaptation of mechanisms already developed for terrestrial loco-motion; respiratory metabolism, for example, is higher in flight than in any other type of movement. In insects there have been some special developments of the muscular system and the high frequencies of wing beat of bees, flies and beetles (see COLEOPTERA) have been made possible only by the evolution of a peculiar type of flight muscle with rhythmic mechanical prop-erties. The contraction of the wing muscles of a small midge at 1,000 times per second is the fastest mechani-cal event known in the animal kingdom.

Small insects and even small birds are still at the mercy of air currents and indeed may use these, as in the case of locust (see ORTHOPTERA) and greenfly (see HEMIPTERA) swarms, to achieve dispersal over a con-siderable distance. Larger birds and insects such as the butterflies use their ability to fly to perform controlled migrations, sometimes involving hundreds or thou-sands of miles of flight. The navigational problems of bird and insect migration are not yet fully solved but appear to involve the preception of direction from the sun. The use of other senses is not, however, completely excluded.

J. W. S. PRINGLE

References

Allan, G. M., "Bats," Cambridge, Harvard Univ. Press, 1940.

Brown, R. H. J., "The Natural History of Bird Flight," Cambridge, Univ. Press, 1960.

Graham, R. R., "Safety devices in wings of birds," *British Birds*, **24**: 1930.

Gray, J., "How Animals Move," New York, Penguin, 1964.

Hertel, H., "Structure, Form and Movement," New York, Reinhold, 1966.

Horton-Smith, C., "The Flight of Birds," London, Witherby, 1938.

Matthews, G. V. P., "Bird Navigation," Cambridge, Univ. Press, 1968.

Pringle, J. W. S., "Insect Flight," Cambridge, Univ. Press, 1957.

Storer, J. H., "The flight of birds," Cranbrook Insti-tute of Science, Bulletin No. 28, 1948.

these insects the wings consist merely of a rod with pro-jecting hairs. It is possible also that the rapid accelera-tions involved in the flight of small insects may intro-duce aerodynamic principles different from those in-volved in gliding or slow flapping.

Soaring flight has been perfected by many large birds and by a few butterflies, and may have been im-portant for the largest of the pterodactyls which had a wing span of twenty feet. All surviving flying animals flap their wings at some stage in their aerial locomo-tion but birds such as the vulture, eagle, and albatross do this only in order to get in the air and then hold their wings stationary. Soaring flight is made possible by the fact that energy is extracted from the motion of the air. The vultures and eagles, rely on thermal upcurrents produced by the heating of the ground by the sun; gulls and other sea birds use the deflected airflow over cliffs. The albatross and other birds of the open sea use the upcurrents from waves and gradient of wind velocities near the surface of the sea. Prolonged soaring flight is possible only when the upward component of the air motion is equal to or greater than the downward com-ponent of the glide, or when the gradient of horizontal wind velocities is sufficiently great to enable the bird to gain energy in moving from one airstream to another.

To be useful to an animal, flight must not merely in-volve motion through the air but there must also be an accurate control of balance and direction. The variety of the flight apparatus in birds, bats and different types of insects stems mainly from these considerations. A bird can vary the span and angle of twist of its wings and can also separate its primary feathers to produce the increased lift of a slotted wing at low speeds. Its tail and its feet are used in balance and in braking. A bat has somewhat similar control of the wing membrane stretched from its fingers to its legs and tail. Insects have no tails or other auxiliary aerofoil surfaces and rely for balance and control on the differential twisting of their two pairs of wings at different phases of the stroke. This mechanism, which more closely resembles

FLOWER

A flower is a short branch bearing appendages spe-cialized for the sexual reproduction of ANGIOSPER-

MOUS plants. Albeit an unaesthetic definition, this is the only technically comprehensive one. A typical flower consists of floral primordia in a close (*cyclic*) pattern on a central axis (*receptacle*). The outer cycle of appendages is called the *calyx* and typically consists of scale-like, protective leaves called *sepals*. Next as the bud opens are one or more cycles termed the *corolla*, composed often of brightly colored *petals*. These *accessory* appendages—calyx and corolla—(not directly concerned with sexual reproduction) collectively form the *perianth*. In consecutive order within the perianth are the male structures or *stamens* (collectively, the *androecium*), and the female parts or *pistils* (together, the *gynoecium*). These two are the essential or sexual parts of the flower. A typical stamen consists of a slender stalk (*filament*) and a terminal, saccular region called the *anther*. The anther is composed of two or more *pollen sacs (microsporangia)* and a *connective region*. The pistils consist of a basal ovary containing one or more ovules (*megasporangia* plus integuments), an elongate style and a terminal *stigma*.

The flower bud, then, is generally homologous with the GYMNOSPERMOUS "cone," and is merely a highly modified apex. The foliar nature of the floral appendages is easily demonstrated in favorable genera. The perianth parts are leafy both in gross appearance and vascularization. Ontogenetic evidence is provided by *Opuntia*, in which there is a visible gradation from vegetative leaves to sepals and petals in the floral bud. In *Lilium*, the sepals and petals are indistinguishable and are called *tepals*. The stamens are highly modified sporophylls, the petiole and midrib being represented by the elongated filament and the blade reduced to the connective between pollen sacs. Evidence for this may be seen in primitive ranalian genera (*Degeneria, Magnolia*) in which the stamen is in fact a broad, three-veined sporophyll bearing two pairs of elongate pollen sacs embedded near the midrib. In some plants (*Mentzelia, Castalia*) there is a gradual transition within the flower from stamens to petals; this tendency has been exploited in the successful breeding of prized, double-petalled varieties of *Rosa* and *Camellia*.

The relation of the pistil to a leaf can be demonstrated in the ranalian genera *Drimys* and *Degeneria* in which the pistil consists of a *conduplicate* (folded) leaf which encloses the ovules centrally. In these genera the margins of the sporophyll are closely adherent, but not fused. There is no style; the stigmatic surface is marginal along the unfused edges of the ovary. A fine developmental series illustrating the evolution of the typical flask shaped pistil is preserved in other members of the Ranales. A single ovule-bearing leaf of this kind is called a *carpel*. The ovule—containing chamber is the *locule* or cell. If the pistil consists of only one such leaf it is a simple pistil, if it is composed of two or more fused together, it is a compound pistil. In most compound pistils the number of fused carpels is indicated in cross section by the number of locules, but occasionally fusion is complete to the edges and only a single locule is enclosed. External characteristics of such a compound pistil will usually reveal the number of incorporated parts. The stigma may be lobed, cleft or divided (*Lilium*); the styles may be separated (*Malus*) into the component number; or the outer part of the ovary wall may be ridged. The floral parts are typically in multiples of 3 in the monocots and 4 or 5 in the dicots.

Although there is disagreement among some botanists over the form of the primitive flower, most evidence points to a type not unlike that of the present day *Magnolia*. The salient features are a rather conspicuous, strobilus-like organ containing a large and variable number of perianth parts—the sepals and petals indistinguishable or only texturally different, and numerous stamens and simple pistils arranged in a spiral fashion around an elongated receptacle. Floral evolution from this basic type has proceeded along many diverse lines.

a. from *complete* flowers—all four cycles or series of parts present (*Ranunculus, Magnolia*); to *incomplete* types—one or more of the characteristic series missing (*Salix, Ilex*). In some apetalous—incomplete flowers (*Anenome, Clematis*) the sepals are large and brightly colored, assuming the function of the missing cycle.

b. from *perfect, monoclinous,* or *bisexual* flowers —both of the essential cycles present (*Cornus, Nyssa*); to *imperfect, diclinous* or *unisexual*—only one of the essential series present (*Fagus, Betula*). Imperfect flowers may be either *pistillate* (possessing pistils, etc., but not stamens) or *staminate* (possessing stamens etc., but not pistils), and *monoecious*-pistillate and staminate flowers on the same plant (*Betula, Quercus*) or *dioecious*—pistillate and staminate flowers on separate plants (*Ilex, Populus*). Some species produce perfect and imperfect flowers (*polygamy*) on the same plant (*Fraxinus, Acer*). A few flowers have no sexual parts and are therefore *sterile* (cultivated *Hydrangea,* and ray flowers of *Helianthus*).

c. from *hypogyny*—perianth and androecium attachments below and free from the ovaries (*Lilium, Pisum*); through *perigyny*—perianth and androecium partially fused into a floral tube surrounding, but not attached to, the gynoecium (*Malus, Saxifraga*); to *epigyny*—floral tube united to or fused with the ovary wall so that the other floral parts appear to arise from the top of the ovary (*Cucurbita, Fuchsia*). In hypogynous and perigynous flowers the ovary is said to be *superior*; in epigynous ones, the ovary is *inferior*.

d. from *distinct* floral parts—individual floral appendages and cycles of appendages separate or free from one another (*Fragaria, Magnolia*); to *coalescent* parts—members of the same series united partly or completely into a tube, e.g., *synsepaly* (*Nicotiana*), *synpetaly* (*Convolvulus*), *synadelphy* or coalescence of stamens (*Lupinus*), and syncarpy or coalescence of carpels into a *compound* pistil (*Viola*); or *adnation* of parts—union of members of different cycles e.g., stamens to pistils (*Asclepias*).

e. from actinomorphic or radially symmetrical flowers (*Opuntia*) to zygomorphic or asymmetrical flowers (*Antirrhinum*).

It is important to recognize the differences between a flower (a single reproductive unit) and an inflorescence (several flowers arranged in order). Thus, the "flower" of the Compositae is, in reality a cluster of many, small flowers arranged on a common receptacle or *head*. The marginal or ray flowers are often sterile, consisting solely of large attractive petals while the minute disc flowers are usually incomplete but perfect. In *Poinsettia, Cornus, Bougainvillea,* the small reproductive flowers of the head are not fringed by sterile, attractive ones, but are subtended by modified leaves (*bracts*) called the *involucre,* often highly colored and functioning as the analog of petals. A very large, single bract called the *spathe* or hood protects the columnar floral head (*spadix*) of *Arisaema* and other Arales. Most inflorescences are rather more open, the individual flowers better spaced and more casually distinguishable.

The largest known flowers (4–5 feet in diameter) are found in certain species of *Rafflesia*, a Malayan genus of root parasites. The smallest flowers are found among the aquatic *Wolffia* and the terrestrial grasses.

Flowers are highly specialized for sexual reproduction. The first step in this process is *pollination*—the transfer of the microspore (*pollen grain*) from the microsporangium (*pollen sac*) where it is produced to the glandular receptive stigma where it germinates. Here it forms the male gametophyte (*pollen tube*, tube nucleus, and 2 sperm nuclei) and digests its way through the style to the ovary. The female gametophyte (*embryo sac*) is retained within the original megasporangial wall (*nucellus*) and these, together with one or more integumental outgrowths of the ovary, constitute the ripe ovule. An opening called the micropyle is created at the point where the enveloping integuments come together. There may be one (*Zea*) or many (*Antirrhinum*) ovules within the ovary. These are attached by a short stalk called the funiculus to a part of the ovary called the *placenta*. The placentation is usually *parietal* (along the walls of a simple ovary—(*Dicentra*); *axile* (along the central axis of a compound ovary—*Fuchsia*); or *free central* (on the axis of a simple ovary—*Primula*). The details of fertilization are outlined in the article SEED.

Although many angiosperm flowers are bisexual, self fertilization or inbreeding is the exception, rather than the rule. Cross fertilization is necessary in introducing genetic variation and is important in evolution. These advantages of cross pollination are reflected in the modifications developed by many flowers to favor the process. Cross pollination may occur between flowers on the same plant or, ideally, between flowers on different plants.

Adaptations favoring cross pollination include:

1. unisexual flowers and dioecism
2. *protandry* (maturation of pollen before stigma is receptive) and *protogyny* (readiness of stigma before pollen is matured)
3. location of stamens high above or far below the stigma of the same flower
4. sterility of flowers to their own pollen
5. construction, coloration, and odor of flower such that pollinating animals are attracted to the stigma first (depositing pollen from another flower there) and the anthers next (picking up pollen as they pass out).

Some flowers are regularly self pollinated, the floral envelope remaining closed or tightly appressed (*Antirrhinum, Pisum*) while others have both open cross-pollinated flowers and cleistogamus self-pollinated ones to insure at least some successful fertilization and seed production (*Viola*).

The most effective and common agents of cross pollination are wind, and animals (mostly insects). Although some aquatic plants produce submerged water pollinated flowers (*Ceratophyllum*), most bear long stalked aerial flowers which are pollinated in more conventional ways (*Castalia*). Many angiosperms and all gymnosperms are wind pollinated (*anemophilous*). Among the former, the "amentiferae" (catkin bearing families) are conspicuous among the tree forms and the grasses among the herbaceous types. Nearly all wind pollinated flowers are characterized by small size, reduction or elimination of the perianth, and the production of large amounts of dry powdery pollen. The stigmas of these are usually feathery or hairy and hence adapted for filtering the air-borne pollen.

Most flowering plants are insect pollinated (*entomophilous*), and are indeed so strikingly adapted for this in some cases that the plant species and animal pollinator appear to have evolved in intimate correlation. The *Yucca* flower, for example, is methodically pollinated only by the *Pronuba* moth which lays its eggs among the young ovules; some, but not all, of which are consumed later by the emerging larvae. Insects are the most common of the animal pollinators, but a few plants are pollinated by snails and slugs (*Calla, Aspidista*), humming and other birds (*Lonicera, Aquilegia*) and bats. Animals visit flowers for food in the form of pollen and nectar. Many of the floral structures already discussed may be interpreted in terms of their function in attracting pollinators. Bright colors and perfumed odors of flowers are definitely attractive to certain insects, particularly bees. The fetid odors of skunk-cabbage (*Symplocarpus*) and carrion flower (*Smilax herbacea*) suit the olfactory preferences of carrion fly pollinators, if not humans. Sugary exudates are produced by secretory cells called *nectaries* located in various parts of flowers. Nectar is familiarly collected by bees and other insects and converted to honey. Flowers are usually constructed so that splashes of color on the petals, the position of the nectary, or the symmetry of the flower serve to position the insect as it enters so that cross pollination is favored, the stigma being touched first.

Complexity in floral evolution reaches elaborate levels in some families. In the ORCHIDALES, for example, (see fig.) there are three sepals and three petals, but one of the petals forms a lip which is very distinctive in shape or color, and may be saclike (*Cypripedium*), tubular (*Cattleya*), lobulate and toothed (*Oncidium*), or spurred (*Habenaria*). These perianth parts attach to an inferior ovary. The stamens and pistils are united into a gynandrium or column composed of one anther terminal on the column or of two lateral anthers midway along it. The pollen is imbedded in two to eight waxy masses called *pollinia*. An insect entering such a flower passes the viscid stigma first, depositing pollinia from a flower previously visited, collects nectar at the base of the specialized petal, and then brushes by the anthers on its way out. The success of this elaborate scheme is evidenced by the fact that the orchids are the second largest order of angiosperms. Several other bizarre flower types found in successful groups are illustrated in the accompanying figure.

The mechanisms governing the shift from vegetative to reproductive meristems (the first visible stage in flowering) have received much attention from plant physiologists since the classic observations of Garner and Allard in 1920 that flowering in Maryland Mammoth Tobacco is controlled by PHOTOPERIOD—the relative length of night and day. Production of a flowering hormone (*florigen*) has been postulated. When this substance accumulates at the meristematic regions of the plant to threshold quantity, the entire physiology of the growing point is altered. It loses its capacity for unlimited growth, its dominance over lateral buds, and

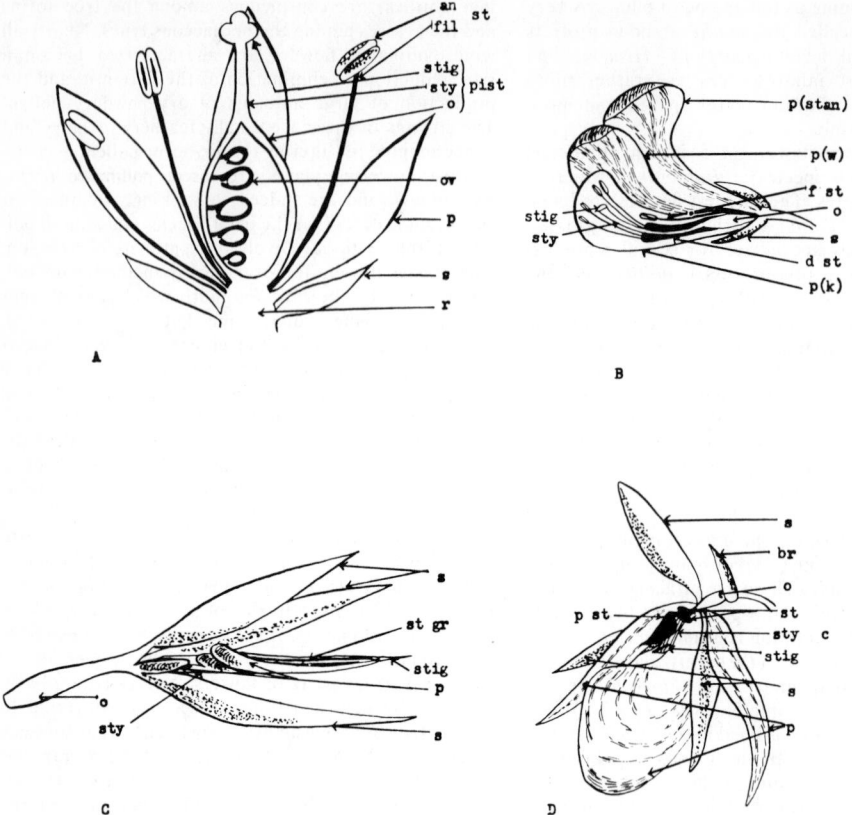

Schematic drawings of A. an actinomorphic, hypogynous flower, B. the zygomorphic flower of *Cercis* (Legumi-nosae), C. a single flower of *Strelitzia* (Musaseae), D. a flower of *Cypripedium* (Orchidaceae). br—bract; p—petal, p(stan)—standard (uppermost petal in flower), p(w)—wing (2 lateral petals in flower), p(k)—keel (2 fused, lowermost petals in flower); s—sepal; st—stamen, st gr—stamens in groove, f st—free stamen, d st—diadelphous stamens, p st—petaloid stamen; o—ovary; ov—ovules; sty—style; stig—stigma; an—anther; fil—filament; r—receptacle; c—column (formed by union of style, stigma and stamens).

its histological and phyllotactic patterns. The entire apical meristem converts to a floral meristem, the peripheral region producing the appendages of a single flower or developing the axillary branches and flowers of an inflorescence.

JOAN EIGER GOTTLIEB

References

Bailey, I. W., "Contributions to Plant Anatomy," New York, Ronald, 1954.
Pool, R. J., "Flowers and Flowering Plants," New York, McGraw-Hill, 1941.
Wetmore, R. H., E. M. Gifford, Jr., and M. C. Green, "Development of Vegetative and Floral Buds" *in* "Photoperiodism and Related Phenomena in Plants and Animals," Washington, A.A.A.S., **1959**: 255–273.

FOL, HERMANN (1845–1892)

This Swiss zoologist has variously been claimed as French, since he was born in Paris, and German in

virtue of his association with Gegenbaur and Haeckel. His chief claim to fame is that he is the first man to have seen, or at least to have recorded that he had seen, the penetration of an egg by a sperm. Before this observation, it was well known that eggs became fertilized but the mechanism of fertilization was a matter of speculation and argument. Fol was a wealthy man who used his own yacht, the "Aster," for marine collecting. The yacht vanished in the Mediterranean in 1892 on one of the expeditions without any survivor to leave a record of the circumstances.

FORAMINIFERIDA

The Order Foraminiferida of the Sarcodina (PRO-TOZOA) includes about 20,000 known species placed in nearly 100 families, 17 superfamilies, and 5 sub-orders. These unicellular animals are comparatively large, ranging in size from 0.01 mm to over 190 mm but are generally 0.1–1.0 mm. They have chitinous, agglutinated, or calcareous tests and anastomosing threadlike pseudopodia. Most Foraminiferida are

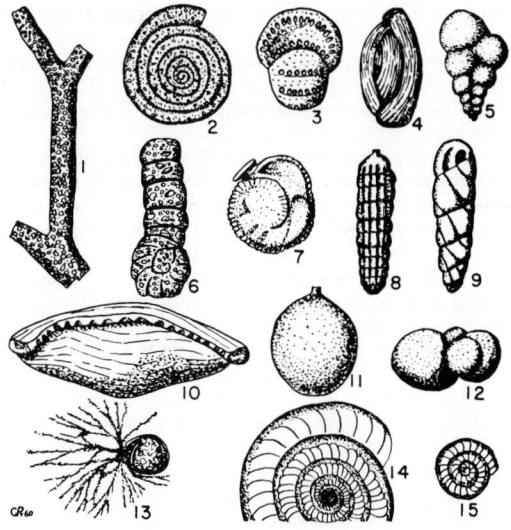

Figs. 1–15 1, *Rhabdammina*, Silurian to Recent (× 5); 2, *Involutina*, Silurian to Recent (× 45); 3, *Bradyina*, Pennsylvanian to Permian (× 4); 4, *Quinqueloculina*, Jurassic to Recent (× 25); 5, *Heterohelix*, Cretaceous to Oligocene (× 25); 6, *Ammomarginulina*, Jurassic to Recent (× 15); 7, *Siphonina*, Cretaceous to Recent (× 25); 8, *Siphogenerina*, Eocene to Recent (× 10ca); 9, *Turrilina*, Jurassic to Recent (× 30); 10, *Triticites*, Pennsylvanian to Permian (× 5); 11, *Ellipsobulimina*, Cretaceous to Pliocene (× 20); 12, *Globigerina*, Cretaceous to Recent (× 15ca); 13, *Lieberkühnia*, Recent, showing pseudopodia, marine and freshwater species (× 15); 14, *Nummulites*, Paleocene to Recent, microspheric individual (× 5); 15, *Nummulites*, Paleocene to Recent, megalospheric individual (× 5). (Figs. 1–13 redrawn with permission from Cushman, "Foraminifera," Cambridge, Harvard University Press, 1955. Figs. 14 and 15 redrawn with permission from Glaessner, "Principles of Micropalaeontology," New York, Wiley, 1948.)

marine, a few are adapted to brackish water or to salt lakes, and one primitive, chitinous suborder (Allogromiina) has freshwater representatives (Fig. 13). Some species live attached, most are free-living and benthonic, and a few are pelagic, and these occur in great abundance.

The cytoplasm of the organism is differentiated into an ectoplasm which fills the aperture or coats the outer surface of the test and an endoplasm which contains the nucleus or nuclei and fills the test. The ectoplasm and endoplasm are connected by one or more apertures and/or by tiny pores in the test.

Foraminiferida differ from other orders of the Sarcodina such as the Amoebida, Heliozoa, and Radiolaria (see ACTINOPODA) in secreting a shell or test. These tests vary in composition and complexity, the simplest are chitinous (Suborder Allogromiina, Fig. 13), others incorporate foreign particles such as sand grains, sponge spicules, or small shell fragments to form agglutinated tests (Suborder Textulariina, Figs. 1, 2, 6), and others secrete calcium carbonate in various ways to strengthen their tests. Three types of calcareous tests are common; an imperforate test which is formed by a mixture of organic material and fine

calcium carbonate particles and which appears porcellaneous (Suborder Miliolina, Fig. 4), a perforate test of nearly clear calcite which if thin may appear hyaline and which may have a branching canal system within the test (Suborder Rotaliina, Figs. 4, 5, 7, 8, 9, 11, 12, 14, and 15), and a very finely granular calcareous perforate test (extinct suborder Fusulinina, Figs. 3, 10). The tests have diverse shapes including tubular, spiral, spherical, and trochoid representatives. The aperture of a test is variously modified and may be multiple (Fig. 3) or single. Single apertures may be toothed, flanged (Figs. 7, 9), or ornamented in numerous ways.

The test of the animal grows either by continuous addition of material at the aperture (single-chambered test) or by periodic addition of chambers (multichambered test). In multi-chambered Foraminiferida the individual begins life swimming freely without a test. The initial test, the proloculus, is nearly spherical and, as the animal grows, it secretes successively larger chambers over the aperture of the preceding chamber. Each species is usually dimorphic having two forms of tests, microspheric and megalospheric (Figs. 14, 15). Microspheric individuals have a minute proloculus and a large test at maturity. Megalospheric individuals have a large proloculus and a small test at maturity. This dimorphism is the result of a complicated reproductive cycle in which sexual and asexual generations alternate. The minute proloculus of the microspheric generation results from conjugation of gametes to form a zygote. The large proloculus of the megalospheric generation results from individual agametes having 2n chromosomes due to incomplete subdivision of the mother nucleus.

Bottom-dwelling Foraminiferida live in the upper few centimeters of the substrate, in cavities of other organisms such as sponges, or on surfaces of other organisms. These benthonic forms may be divided into two faunal associations; cold- and warm-water faunas. The cold-water faunas are distributed in the polar regions and in the deeper colder waters of the high temperate regions. Generally the number of species is few, although individuals may be abundant. The warm-water faunas are subdivided into a Mediterranean fauna extending from the eastern Atlantic Ocean into the northwestern Indian Ocean, an Indo-Pacific fauna common throughout the Indian Ocean and the western and central Pacific Ocean, and a West Indian fauna distributed in the western and northern Atlantic Ocean. Many species are restricted to definite depth and temperature zones, bottom sediments, and salinity, although a few have wide tolerance to these conditions.

Pelagic Foraminiferida appear in only four families and are distributed by warm-water currents. Their tests settle to the ocean floors (Fig. 12) and accumulations of these calcareous tests on the ocean floors are commonly termed *Globigerina* ooze.

Foraminiferida have a long and complex geologic history. A few primitive genera are reported from the Cambrian but in general Foraminiferida are rare until about the close of the Ordovician and these are mostly agglutinated forms. The Mississippian, Pennsylvanian, and Permian Foraminiferida had abundant genera with calcareous tests and these have proven to be excellent guide fossils for correlation of strata of these periods. In the Triassic, Jurassic, and Cretaceous most modern families of calcareous perforate Foraminiferida arose. Families having pelagic species began in the early Cretaceous and became abundant by middle Cretaceous time. In the Cenozoic many new genera of Foraminiferida appeared and commonly persist in Recent seas. In a few superfamilies, genera having extremely large sizes evolved, became highly specialized, and are important limestone formers. These include the Fusulinacea (Pennsylvanian and Permian, Fig. 10) and the Orbitoidacea, Miliolacea, and Rotaliacea (Cretaceous and Cenozoic, Figs. 14, 15).

CHARLES A. ROSS

References

Cushman, J. A., "Foraminifera," Cambridge, Harvard University Press, 4th ed., 1948.
Glaessner, M. F., "Principles of Micropaleontology," New York, Wiley, 1948.
Loeblich, A. R., Jr., and Helen Tappan, "Sarcodina; Treatise on Invertebrate Paleontology," Pt. C, Protista 2, vols. 1 and 2, Lawrence, Univ. Kansas Press, 1964.

FOREST

A large part of the land areas of the earth are or were covered naturally by forest. Areas not originally covered by forest are ones too dry or wet for tree growth. Forests extend northward to timber line, above which there is perpetual snow and no growing season. Such a timber line is longitudinal. Where mountains are high enough to have perpetual snow on their tops, there is an altitudinal timber line.

Areas that are not naturally forested are such as deserts, plains, prairies and swamps. In the United States the Appalachian mountains on the east and the Rocky Mountains on the west are forested, but between them lies an area of treeless plains and prairies.

In tropical regions, especially where the trade winds produce a heavy rainfall, the forest is dense, and composed of many species of broad-leaved trees. These trees are evergreen because there is no winter season.

Farther north forests consist of deciduous, broad-leaved trees and needle-leaved evergreens. These make up varied types of forest. One factor that determines the kind of forest is the dryness or wetness of the soil. Different kinds of trees differ in the character of their root systems. Some trees have long tap roots that penetrate deep into the ground to reach water. Such trees grow in dry and often rocky soil.

In soils that are not rocky nor especially dry, trees with diversified root systems that spread out in all directions, grow. In wet soils roots cannot go deep into the soil because they need air. So trees that grow in such places have superficial roots that grow out close to the surface. A common type of wet soil forest in the United States consists of Pin Oak, Swamp White Oak, Red Maple, Elm and Black Ash.

A character of trees that has much to do with the kind of forest is tolerance of shade. A tolerant tree is one that will grow well under the shade of other trees. An intolerant tree will die if its crown is not up in the sunlight. In a forest of intolerant trees, the individual trees crowd each other for light space, and some of them get overtopped and ultimately die. Foresters classify the trees in such a forest as dominant, intermediate and suppressed, and they remove the inter-

mediate trees to give the dominant ones more room to grow.

A forest of intolerant trees is one-aged, and often pure, that is, all of one species. Forests of tolerant trees are likely to be of all ages and often of mixed species.

There are times when the geological formations affect the tree growth. In the Allegheny State Park, New York, the forest on the higher hills is a mixture of Sugar Maple, Beech, Yellow Birch and Hemlock. Part way up the hills there is a layer of rock known as the Salamanca Conglomerate. This layer is horizontal and outcrops at certain places on the hillsides. Rain water on the upper hills soaks in till it reaches the rock layer, and then follows it to the outcrop and forms a line of springs. The trees above this layer are maple and beech only, but at the line of springs, hemlocks form small groves, and from there down the hills, hemlock and birch occur with the maple and beech.

Another factor that has to do with the distribution of tree life is that of exposure. In hilly country slopes facing south get abundant sunlight and have dry soils. Slopes facing north are cool and moist. Slopes facing east or west, and level areas, are intermediate. The distribution of the different oak species varies with these differences in exposure. Chestnut Oak grows on the south slopes, and Red Oak on the north ones. On east and west slopes and level areas, Black and White Oaks grow, and formerly Chestnut grew with them.

In the Rocky Mountain country trees are practically all evergreens. In places east of the continental divide, the main forests are composed of Douglas Fir, Lodgepole Pine and Engelmann Spruce. Each species forms areas of pure forest. The Douglas Fir grows on steep slopes. The Lodgepole Pine grows on level areas, and the Engelmann Spruce grows in wet soil areas along the brooks.

But higher up in the mountains, just below timber line, grow the deep-rooted Limber and Whitebark Pines. These trees are small and stunted. Due to heavy winds from the west they are misshapen, leaning toward the east, sometimes to such an extent that they are almost lying on the ground. Such a forest is of no commercial value, but is is important for watershed protection.

When forests have been removed by cutting or fire, and the land then left to nature, a temporary forest comes in. Such a forest may consist of Aspen and the wild Red Cherry. The Aspen seeds are cottony and may be blown a long distance by the wind. The cherry seeds are undoubtedly brought by birds that are particularly fond of the wild red cherry. The oak hickory type of forest may spread to open areas through the Gray Squirrel, that has a habit of burying nuts and large seeds.

At one time in the past there was a glacier age, and the northern United States was covered with snow and ice. Then the glaciers receded, and apparently that recession is still going on. This is indicated by a northern movement of certain species of mammals and birds. Forests are moving northward as the glaciers recede. In places in the Glacier National Park one may observe this. Following a trail from a glacier, down the stream of water that melts from it, one notes an occasional willow bush, then more of them until the old bed of the glacier becomes a willow thicket. Then we may note a small spruce tree growing among the willows. Then several more appear until the willows are gone, and we are in a forest of spruce trees, where once there was a glacier. The timber line is moving northward.

ARETAS A. SAUNDERS

References

Sargent, C. S., "Manual of the Trees of North America," Boston, Houghton-Mifflin, 1905.
Gordon, R. B., "The Primeval Forest Types of Southwestern New York," Albany, New York State Museum Bulletin No. 321, 1940.

FRIES, ELIAS MAGNUS (1794–1878)

This distinguished Swedish botanist was born at Uppsala and died in the same town. His life shows a curious dichotomy of interests between exact descriptive science and philosophical speculation. In the first capacity, his observations, drawings, and publications on the fungi placed the taxonomy of this group for the first time on a scholarly basis. The main outlines of his classification of this group, and of the lichens which he associated with them, are followed to the present day. He did not permit this dedication to descriptive work to interfere with his philosophical speculations. Indeed, it was probably his firm devotion to the latter which made the former acceptable at Uppsala, at that time without distinction in the field of science. Even before Haeckel, he regarded biology as a "supernatural science" and pointed out that its study should more properly be a branch of theology than of physics and chemistry. This happy phrase permitted theologians, at that time forming the majority of learned men, to turn their attention to botany, to the great advantage of that science through the next fifty years.

FROND

The word frond (Lat., *frons, frondis*) refers to the leaves of ferns, but it has been applied to leaves of CYCADS and PALMS and fern-like organs of ALGAE and LICHENS. Typical fronds or fern leaves are extremely variable in form. The average frond is composed of three recognizable parts or areas, the petiole or stipe, the midrib or rachis, and the blade or lamina. It is pinnately organized, whether the lamina is undivided (simple) or divided (compound). The frond is, however, most commonly divided into leaflets or pinnae. Certain fronds do lack a petiole and are thus sessile, and rarely (as in Schizaea) the true lamina may be reduced or missing. The pinnate frond architecture may be modified into palmate (i.e., with all pinnae arising from a common center) or falsely dichotomous (the frond and/or frond parts appearing to fork by equal division of the leaf meristem). Although the once-divided or pinnate condition is average, roughly 15–20 per cent of ferns have simple or merely lobed leaves, and 25–35 per cent are decompound, the pinnae divided into pinnules (twice-pinnate)

or the lamina more divided (thrice-pinnate, etc.). The venation patterns vary from free veins to reticulate veins to complex-reticulate with veinlets included in the areoles. However, free venation is by far the most common. The growth of most fronds originates in a single apical cell, and is largely acropetal rather than intercalary as in leaves of flowering plants. In growth the fronds and frond parts display curling or circinate venation, appearing as fiddleheads or crosiers. Fronds of a given species may be monomorphic or dimorphic, the latter differentiated into fertile and sterile forms. The sporangia of ferns are borne either singly or in clusters, and either marginally or adaxially on the lamina.

The phylogenetic origin of the frond bears upon the interpretation of leaves and reproductive organs of seed plants. The earliest vascular plants probably had leafless aerial stem systems. Two views have been adduced to explain the origin of fronds from stems. The Enation Theory regards the leaf as a lateral protuberance or appendage of the cortical, photosynthetic area of the stem, the simple unvascularized lamina or enation being the earliest state, the extension of the stem stele laterally into the lamina as a midrib more advanced (the microphyll stage), and the origin of lateral venation from the midrib into the lamina the most advanced (the megaphyll stage). The Telome Theory considers the leaf a more or less fused and flattened cluster of stems, the ultimate veins of the frond being "telomes" and the intercalary veins "mesomes." The origin of the megaphyll or frond is seen by telomists as a process of planation (orientation of stems into one plane), webbing (formation of lamina between stems), and overtopping (emphasis in growth of alternate branches of a dichotomous system to produce a pinnate organization). Very simple leaves (e.g., Asteroxylon, Lycopodium, Psilotum) are regarded as much reduced branch systems.

WARREN H. WAGNER, JR.

References

Foster, A. S. and E. M. Gifford, Jr., "Comparative morphology of vascular plants," San Francisco, Freeman, 1959.
Lam, H. J., "Classification and the new morphology," Acta Biotheoretica, 8: 107-154, 1948.
Troll, W., "Vergleichende Morphologie der höheren Pflanzen. I. Vegetationsorgane. Part II. Morphologie des Blattes," Berlin, Borntraeger, 1938.
Wagner, W. H., Jr., "Types of foliar dichotomy in living ferns," Amer. Jour. Bot. 39: 578-592, 1952.
Zimmerman, W., "Die Phylogenie der Pflanzen," Stuttgart, Fischer, 1958.

FRUITS

Botanically speaking a fruit is the ripened ovary, or the ovary with adjoining parts; i.e., it is the seedbearing organ. Fruits may be classified on the basis of structure and development as follows:

Berry. A true berry consists of a fleshy fruit, derived entirely from the ovary of a flower and its contents. Usually many seeds are embedded in the flesh. Common examples are the tomato, grape, gooseberry, and currant.

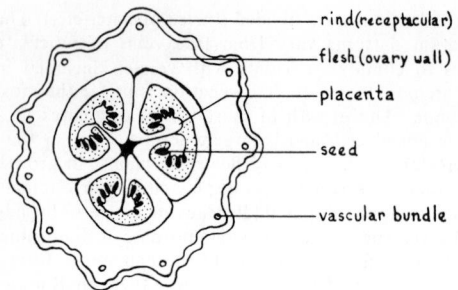

Fig. 1   A pepo type cucumber fruit in cross section. (Reproduced by permission from Kessler, G. M. "Fruits, Vegetables and Flowers" Burgess Publishing Co., 1954.)

Hesperidium. The hesperidium is a berry-like fruit which is represented by citrus fruits (orange, lemon, grapefruit, tangerine). It differs from a true berry in having a leathery rind of ovary tissue containing oil ducts, and many membraneous, juice-filled sacs in place of solid flesh.

Pepo. A pepo is represented by the cucumber, squash and pumpkin. These fruits resemble berries to a certain extent. The hard outer covering originates from the receptacle of the flower. (In the case of the hesperidium the rind arises from ovary tissue).

The banana resembles a pepo but the recepticular skin is softer and separates readily from the ovary.

Drupe. A fleshy fruit with a thin, edible, outer skin derived from the ovary, is called a drupe. A layer of edible flesh of varying thickness lies beneath the skin. Within this is the stone or pit, which is actually a hard inner wall of the ovary. Enclosed within the pit is the seed. The cherry, peach and plum are typical drupes. They are also called stone fruits. The raspberry consists of a cluster of small, individual drupes, or drupelets. Botanically speaking, it is *not* a berry.

Aggregate. An aggregate fruit is one which is formed from numerous carpels of one flower. The fruit therefore consists of a cluster of small, individual fruitlets. Examples are the raspberry, blackberry, and strawberry. The fruitlets of the raspberry and blackberry are actually small drupes. In the strawberry, the seed-like achenes are fruitlets, embedded in a fleshy, edible floral receptacle.

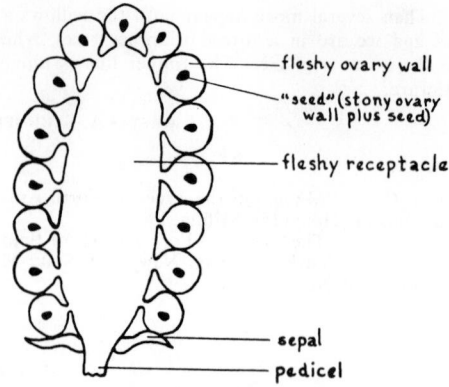

Fig. 3   A longitudinal section of a blackberry, an aggregate fruit. (Reproduced by permission from Kessler, G. M. "Fruits, Vegetables and Flowers" Burgess Publishing Co., 1954.)

Multiple. A multiple fruit is one which is formed from individual ovaries of several flowers. Fruits of mulberry, fig, and pineapple constitute common examples. In the pineapple, portions of the flower stalk, sepals, petals, and ovaries of many flowers are fleshy and edible, and all are so tightly compressed together that they appear fused to each other.

Pome. Pomes are fleshy fruits consisting of a thin skin and outer zone of edible flesh. Common examples are the apple and pear. The fleshy portion beneath the skin is ovary tissue. The core in the center consists of a number of seed-containing, leathery little compartments called carpels. These are derived from the inner ovary wall.

Legume. The main characteristic of a legume is the shell-like pod containing a number of relatively large seeds. Peas and lima beans are typical legumes. The pod which has developed from a single ovary, dries out as it matures, splits into two halves and releases the seeds.

Capsule. A capsule is somewhat like a legume but differs in that it consists of more than one seed com-

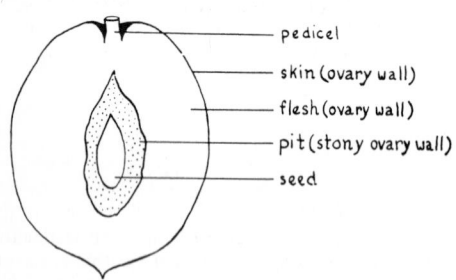

Fig. 2   A longitudinal section of a peach, a drupe type fruit. (Reproduced by permission from Kessler, G. M. "Fruits, Vegetables and Flowers" Burgess Publishing Co., 1954.)

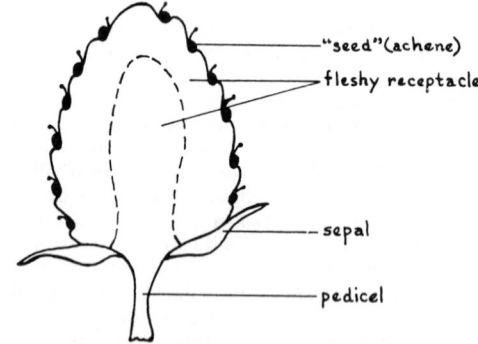

Fig. 4   A longitudinal section of a strawberry, an aggregate fruit. (Reproduced by permission from Kessler, G. M. "Fruit, Vegetables and Flowers" Burgess Publishing Co., 1954.)

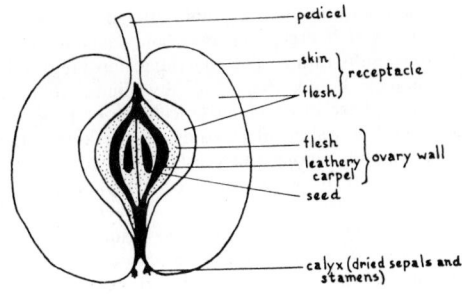

Fig. 5 A longitudinal section of an apple, a pome fruit. (Reproduced by permission from Kessler, G. M. "Fruits, Vegetables and Flowers" Burgess Publishing Co., 1954.)

partment and splits along more than two lines when ripe. The fruit of okra is a familiar example.

Caryopsis. The kernel of sweet corn is a kind of fruit called a caryopsis. The more or less horny outer coat is the ovary wall. This is firmly attached to the seed coat of a single seed. The remaining portion, i.e., the endosperm and embryo, comprises the seed in this instance.

Nut. A nut is defined as a hard, dry, single-seeded fruit, partly or entirely enclosed in a husk, which remains with the fruit as it ripens. Common examples are chestnuts and filberts. Although the term "nut" is popularly applied to many hardshelled fruits that may be stored dry, many of these are *not* true nuts. The peanut, for example, is not a nut but a legume. The almond is actually the pit of a drupe-like fruit and the Brazil nut is actually a seed.

The popular conception of a fruit is based largely on edibility so that the definition stipulates the more or less succulent product of a perennial or woody plant, consisting of the ripened seeds and adjacent or surrounding tissues, or the latter alone. Horticulturists likewise tend to adhere to this interpretation. Thus legumes, capsules, caryopses and nuts are not thought

of as fruits by the layman. Furthermore, a popular distinction is made between fruits and vegetables, although technically there is no valid distinction providing that the item in question is a product of the ripened ovary and/or related parts. Thus if the edible portion of the plant is the leaf, petiole, stem or root, it is definitely a vegetable. From the popular standpoint a fruit is more frequently eaten raw as a dessert, and it possesses a characteristic aroma and flavor due to the presence of various organic esters. A vegetable is ordinarily eaten cooked, or when raw as a salad or relish. It is the product of a herbaceous plant, rarely of a shrub or tree.

The ripening of a fruit (horticulturally speaking) involves, among other changes, the reduction of acidity and an increase in sugars, so that ultimately a "pleasantly tart" flavor is developed. Fruits like the apple, pear and banana have a "carbohydrate reserve" in the "mature green" stage. Because of this they may be harvested at this stage and permitted to ripen in storage. Other fruits, like citrus, raspberries, cherries, etc., do not develop a carbohydrate reserve and must therefore be ripened on the tree if they ripen at all. The softening that accompanies certain fruits during ripening is caused by conversion of pectic substances in the cell wall from the insoluble to the soluble form.

Many fruits have been found to undergo a postharvest "climacteric" or peak in respiration. During this respiratory peak the fruit has its best flavor and aroma. Maximum quantities of carbon dioxide and ethylene are evolved at this time, the latter gas exerting an accelerating effect on the ripening of preclimacteric fruits stored in close proximity. Following the climacteric the respiratory rate declines and the fruits are said to be in senescence. The efficient preservation of fruits by cold storage involves the storing of the fruits prior to the development of the climacteric.

ERSTON V. MILLER

References

Bailey, Liberty Hyde., "Cyclopedia of American Horticulture," Vols. 1–6, New York, MacMillan, 1944.
Ulrich, R., "La Vie des Fruits," Paris, Masson, 1952.
Kessler, George M, "Fruits, Vegetables and Flowers," Minneapolis, Burgess, 1954.
Von Loesecke, Harry W., "Bananas: Chemistry, Physiology, Technology," New York, Interscience, 1949.
Smock, Robert M. and A. M. Neubert., "Apples and Apple Products," New York, Interscience, 1950.
Webber, J. J. and L. D. Batchelor, "The Citrus Industry. Vol. 1. History, Botany and Breeding," Berkeley, Univ. of Calif. Press, 1943.

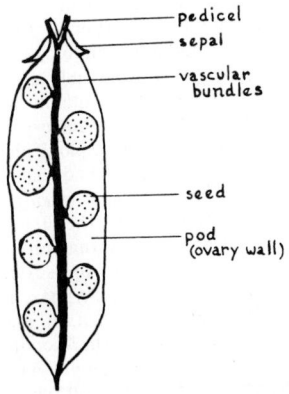

Fig. 6 An opened pod of a pea, a legume type fruit. (Reproduced by permission from Kessler, G. M. "Fruits, Vegetables and Flowers" Burgess Publishing Co., 1954.)

FUNGI

A group of organisms devoid of chlorophyll, generally included for convenience in the Division Mycota of the Plant Kingdom. The fungi, unable to manufacture their own food obtain their energy *saprobically* by breaking down organic matter in dead plant and animal bodies, or *parasitically* by infecting living organisms and thus often causing disease. Molds, mildews, yeasts, rusts, smuts, mushrooms, puffballs, and earthstars are common examples of fungi. The limits of the group are difficult to define. The more usual limits drawn amound the fungi, more often than not include the true slime molds (Myxomycetes). Accordingly, as we shall consider them here, the fungi include the Myxomycetes, "Phycomycetes," Ascomycetes, Basidiomycetes, and Deuteromycetes (see section on classification in this article). The origin and affinities of the fungi are obscure. Some mycologists still defend the theory of a monophyletic or polyphyletic origin from one or more groups of algae, but the majority view perhaps places the origin of most fungal groups among the protozoan flagellates and concedes that the Oomycetes may have originated from the Siphonales.

Occurrence. Many fungi are of almost universal occurrence having been reported from all parts of the world which have been investigated with any degree of thoroughness. Others are limited in their distribution and may be confined to tropical or temperate regions as the case may be. The fungus population of the extreme arctic or antarctic regions is probably limited, but we do not have sufficient data on which to base sound conclusions. Aerobiological studies over the Canadian arctic show that fungal spores are not rare in the air above these regions.

Fungi differ greatly in habitat. Some are strictly aquatic, fresh water or marine, others amphibious, still others soil inhabiting. Many complete their entire life cycle on or in the tissues of higher plants which they parasitize. If they are obligately parasitic (see section on biology of this article) they are unable to grow on any substratum other than the host they parasitize and their distribution is therefore limited by the distribution of their host.

Morphology. The great structural variation exhibited by the fungi makes it extremely difficult to discuss the morphology of the entire group without danger of misleading the uninitiated reader. The vast majority of the fungi are filamentous organisms, their bodies being constructed of microscopic filaments, the *hyphae*. A hypha, the unit of structure, consists, with few exceptions, of a tubular wall the characteristic component of which is chitin, in most forms, but cellulose in the Oomycetes. In at least two fungi both cellulose and chitin have been found in the wall.

The hyphal tube is filled or lined internally with a layer of protoplasm which may be continuous (Phycomycetes), or which may be divided into uninucleate, binucleate, or multinucleate compartments by means of cross-walls, the *septa*, as in the higher fungi. Hyphae grow apically, branch, rebranch, and anastomose, and form an extensive network, the *mycelium* which spreads in all directions over or within the substratum —living or non-living—which the fungus is utilizing for food. The hyphae of a fungus are often organized into tissues to which we give the general name *plectenchyma*. Such a tissue, composed of loosely woven hyphae, which to a great extent retain their individuality, we call *prosenchyma*, whereas a tissue in which the hyphae are so tightly woven as to have lost their individuality is called *pseudoparenchyma*, in reference to the parenchyma tissue of higher plants which it resembles. Thus, although the hyphae themselves appear to be relatively simple in structure, they are capable of giving rise to very complex tissues of which fruiting bodies, for example, are usually constructed. Not all fungi possess mycelium. The assimilative phase of the Myxomycetes, for example, is a *plasmo-*

dium, a multinucleate, protoplasmic mass devoid of cell walls which creeps in amoeboid fashion over or within the substratum. The chytridiaceous fungi (lower Phycomycetes) too, have no mycelium, their assimilative phase often being unicellular or being represented by a system of rhizoids. In the higher fungi again, we find, among the yeasts, many forms which are unicellular or which produce a *pseudomycelium* composed of short chains of cells. Nevertheless, generally speaking, the mycelium is the assimilative phase of a fungus, often invisible to the casual observer, as contrasted to the propagative phase represented by the *spores*, borne in many fungi by conspicuous spore-bearing organs such as the morels, the mushrooms, and the puffballs.

The spores of fungi are extremely varied. In size they range from a few microns to over a millimeter in length; in color, from colorless to black through a great variety of shades and hues; in shape, from spherical or oval to elongated, needle-shaped, curved, star-shaped, helical, etc. Spores may be unicellular or multicellular; thin or thick-walled; smooth, spiny, or variously sculptured; motile or non-motile; water-, air-, or insect-borne. In short, spore characteristics have reached the zenith of variation in the fungi. It should be noted also that a single fungus may produce more than one kind of spore, some rusts producing as many as four distinct types in regular sequence. Spore-bearing structures are almost as varied as the spores themselves. They range from undifferentiated microscopic hyphae through the relatively simple sporangia of the Phycomycetes or the more intricate ones of the Myxomycetes, to the complex fruiting bodies of such Basidiomycetes as the puffballs which may reach a diameter of five feet.

Life Cycle. In simple terms, the life cycle of a fungus may be thought of as an alternation of the assimilative phase with the spore bearing phase, one giving rise to the other; the mycelium producing spores, either directly or through spore-bearing structures of varying complexity, the spores germinating into hyphae which combine to form the mycelium. In more technical terms, the life cycle of a fungus must be viewed, in the light of the life cycles of other organisms, as consisting of a haploid phase alternating with a diploid phase. This is true of all fungi capable of reproducing sexually.

With the notable exception of the Myxomycetes, in which the plasmodium is diploid, and the Oomycetes in which the hyphae are possibly diploid, the assimilative phase of most fungi is mainly haploid, *i.e.* contains haploid nuclei. As in other organisms with similar life cycle patterns, the diploid phase begins with nuclear fusion (*karyogamy*) and ends with reduction in the number of chromosomes (*meiosis*) giving way to the haploid phase. Typically, the fungus life cycle may also be divided into an asexual phase and a sexual phase. The hyphae give rise to asexual spores which germinate and again produce hyphae. This is the asexual cycle which may repeat itself indefinitely under certain conditions. Under other conditions, the same hyphae give rise to sex cells (*gametes*) which unite and initiate the sexual phase. Special types of spores, produced as a result of karyogamy and meiosis, germinate and again give rise to hyphae which repeat the asexual and sexual phases of the life cycle under the appropriate conditions. In a large number of fungi which parasitize plants, the asexual phase is prevalent during the growing season producing many generations of asexual spores which serve to propagate and spread the fungus. With the approach of autumn, the fungus shifts into the sexual phase which is often characterized by structures resistant to winter temperatures thus carrying the organism over unfavorable conditions. In such organisms, therefore, the asexual phase is most important for the propagation of the species, whereas the sexual phase concerns itself mainly with the survival of the species.

Sexual reproduction in the fungi, as in all sexually reproducing organisms, is divided into three distinct phases: plasmogamy, karyogamy, and meiosis. We have already defined the last two terms. *Plasmogamy* is the process of bringing two nuclei into close proximity within a single protoplast in preparation for karyogamy. This usually involves the fusion of two protoplasts. Fungi employ a variety of methods in accomplishing this important task.

In the Myxomycetes (q.v.) plasmogamy takes place by the union of two swarm-cells or two myxamoebae derived from swarm cells by loss of flagella. The fusing cells are identical morphologically. Some authors hesitate to call such a process sexual and prefer the term *karyallagy* when morphological sexual differentiation is lacking. Karyogamy follows plasmogamy closely in the Myxomycetes and the resulting zygote grows into a diploid plasmodium.

In the "Phycomycetes" the fusing protoplasts may be *isoplanogametes* (equal, flagellated, sex cells), *anisoplanogametes* (unequal), or *aplanogametes* (nonmotile). In one group (Monoblepharidales) the male gamete is flagellated, but the female gamete is nonmotile. Typically, karyogamy follows plasmogamy immediately and the resulting zygote becomes transformed into a resting spore or a resting sporangium in which meiosis takes place.

In the Ascomycetes plasmogamy is usually accomplished by the transfer of nuclei from an *antheridium* (male sex organ) to the ascogonium (female sex organ); from a *spermatium* to a specialized female receptive structure; or from one simple, undifferentiated hypha into another. After plasmogamy, a migration of nuclei occurs through the *ascogenous hyphae* into the *ascus mother cell* where karyogamy and meiosis occur. The ascus mother cell then becomes transformed into the *ascus* (a sac-like structure characteristic of the Ascomycetes) in which typically 8 ascospores develop by free cell formation.

In most Basidiomycetes sex organs are no longer produced, plasmogamy usually taking place by the fusion of two cells in the haploid, uninucleate hyphae. This results in a binucleate cell which gives rise to a binucleate mycelium which in turn produces the *basidia*. Karyogamy and meiosis occur in the basidium which produces usually 4 *basidiospores.*

In the Deuteromycetes (Fungi Imperfecti) i.e. those fungi in which no sexual life cycle has been discovered, sexual reproduction seems to have been replaced by a *parasexual cycle* in which protoplasmic fusion, nuclear fusion, and chromosome reduction occur haphazardly and unpredictably rather than in an orderly sequence.

A discussion of sexuality in the fungi must not ignore the matter of compatibility. In *homothallic* fungi every thallus is sexually self-fertile and all copulatory cells are compatible. *Heterothallic* fungi, on the other hand, consist of at least two mating types (bipolar), the copulatory organs produced by a single thallus necessitating the presence of those of another thallus of the opposite mating type before plasmogamy may be accomplished. Thus a single thallus of a heterothallic fungus is always sexually self-sterile even though it may produce functional male and female organs. In many Basidiomycetes the situation becomes even more complex, a single basidium in tetrapolar species producing basidiospores of 4 mating types.

Biology. Inasmuch as all fungi are devoid of photosynthetic pigments they depend on other organisms for elaborated food. Most fungi are either saprobes or facultative parasites, *i.e.* are able to parasitize organisms susceptible to their attack, but are also able to live on dead organic matter. Both of these groups may be grown on artificial media and studied conveniently in the laboratory. It is from the study of such fungi that most of our knowledge about fungus physiology has been obtained. A small group of fungi— including the rusts, the downy mildews, and the powdery mildews—are obligately parasitic. These occur in nature only as parasites on living organisms and have defied all attempts to grow them in artificial culture.

Parasites in many instances have developed special organs which adapt them to their mode of life. Such are the *haustoria* for example, which are special "sucking" organs produced by hyphae. They penetrate into the cells of the host and come in intimate contact with the protoplasm. All obligate parasites as well as some facultative ones produce haustoria.

There are, of course, all degrees of parasitism from the very virulent type, which kills the host, to a very mild type which may be harmless to the host as a whole and, according to some biologists, even beneficial. Such mild, "balanced" parasitism is sometimes termed *symbiosis* and is exemplified by the lichens, and by mycorrhiza. In the lichens, the parasitized plants are algae which the fungus mycelium surrounds and often penetrates and from which it obtains nourishment and water. This type of parasitism is very mild, however, most of the algal cells surviving and appearing to be in a healthy state. Furthermore, the lichens apparently are able to survive under conditions where the algal host by itself is unable to live. Thus the fungus affords some protection to its host and extends its range.

Mycorrhiza is an association of a fungus with the roots of a higher plant. There is little doubt that here too we have a parasitic relationship which is in delicate balance. The fungus surrounds the fine rootlets of the plant with a hyphal mat and individual hyphae penetrate into the host cells of the outer layers. The penetration stops here, however. The fungus undoubtedly obtains some carbohydrates from the plant. It appears also that in many instances there is a transfer of nutrients from the soil to the plant through the fungal hyphae, that hyphae are digested by the plant cells eventually, and that nourishment is obtained from them. Even though the mycorrhizal relationship has been extensively studied, our knowledge of the physiology involved is rather meager. Plants which lack root haris, such as the conifers, are said to be particularly benefitted by mycorrhiza, infected plants invariably growing significantly better than non-infected ones under experimental conditions.

Physiology. Most of the information we possess on the physiology of the fungi has been obtained in the laboratory under highly artificial, controlled conditions with saprobic or facultatively parasitic fungi growing in pure culture. Whether fungi behave the same way in nature we can only conjecture. The nutrition of fungi appears to be very similar to that of other heterotrophic organisms. Most fungi prefer glucose to other carbohydrates as a source of carbon, but are able to utilize fructose, maltose, and sucrose almost as well, and other sugars to a lesser degree. Organic sources of nitrogen appear to be most favorable, with ammonium and nitrate salts next in that order. A very small number of species have been reported to be able to fix atmospheric nitrogen. Of the mineral elements, phosphorus, potassium, and magnesium are essential to fungus nutrition as are the so called micro-elements such as manganese, iron, zinc, boron, molybdenum, etc., traces of which appear to be essential for growth and reproduction. Many fungi also require an external source of various vitamins, particularly thiamin and biotin.

The temperature relationships of many fungi have also been investigated. The optimum temperature for growth and reproduction for most fungi appears to fall in the range of 20°–26° C. When the temperature exceeds 30° C most fungi stop growing, but some are thermophilic and thrive at temperatures of 40°–50° C or above.

Light is not necessary for the growth of fungi, but is required by many species for sporulation. It is interesting to note that some fungi will sporulate more abundantly if grown under alternate conditions of light and darkness than in continuous darkness or continuous light. Ultra-violet irradiation is also known to stimulate sporulation in some species.

Classification. As mentioned in the introduction, mycologists in the past placed the fungi into four classes (Myxomycetes, Phycomycetes, Ascomycetes, Basidiomycetes), and one so-called form class, the Deuteromycetes or Fungi Imperfecti. Three of these classes are homogenous and appear to be good natural groupings. The Phycomycetes, on the other hand, have long been recognized as a heterogenous group. The modern tendency, therefore, is to replace the old class Phycomycetes with five classes: Chytridiomycetes, Hyphocytridiomycetes, Plasmodiophoromycetes, Oomycetes, and Zygomycetes, following Sparrow's suggestion of 1958.

Importance of fungi to man. Fungi play a tremendous role in the constant changes which take place in nature inasmuch as they are responsible for much of the disintegration of organic matter particularly of plant origin. This is due to their ability to produce specific enzymes which act on cellulose, lignin, and other substances found in plant tissues. In the same way they destroy fabrics, leather, stored food and other consumers' goods. Fungi are responsible for the majority of known plant diseases such as late blight of potatoes, various downy mildews and powdery mildews, Dutch Elm disease, oak wilt, apple scab, and chestnut blight which has destroyed the chestnut forests of this country. Black stem rust of cereals, bunt of wheat, and corn smut claim millions of dollars worth of crops annually. White pine blister rust threatens our white pine forests. Fungi also cause many important diseases of animals and of man himself. Candidiasis appears to be increasing in importance since the widespread use of antibiotics which destroy beneficial as well as harmful bacteria in the human body. Aspergillosis, and, coccidioidiomycosis, are serious, often fatal deep infections. Skin infections such as athlete's foot and ring worm have been difficult to control until the discovery of some effective antifungals.

Not all fungi are destructive. Man has learned to harness a number of molds for his own use. The ergot fungus—the same one that causes ergotism—produces important pharmaceutical alkaloids; the yeasts are the basis of the baking and brewing industries; citric and many other organic acids are manufactured with the aid of fungi; so are penicillin, cortisone, some vitamins, and some enzymes. For the manufacture of roquefort, camembert, gorgonzola, and blue cheese members of the fungus genus *Penicillium* are indispensable. The mushroom growing industry based on a single species is constantly on the increase. In science too, the fungi are proving to be very important biological tools for fundamental investigations. Gibberellin, a product of a parasitic fungus is important in the study of growth regulating mechanisms in plants. As a research tool in biochemical genetics the ascomycete *Neurospora crassa* has yielded invaluable information on metabolic pathways in living organisms.

C. J. ALEXOPOULOS

References

Ainsworth, G. C. and A. S. Sussman, "The Fungi," 3 vol. New York, Academic Press, 1965, 1966, 1968.
Alexopoulos, C. J., "Introductory Mycology," 2nd ed. New York, Wiley, 1962.
Bessey, E. A., "Morphology and Taxonomy of Fungi," Philadelphia, Blakiston, 1950.
Cochrane, V. W., "Physiology of Fungi," New York, Wiley, 1959.
Cooney D. G. and R. Emerson, "Thermophilic Fungi," San Francisco, Freeman, 1964.
Hawker, L. E., "The Physiology of Reproduction in Fungi," Cambridge, The University Press, 1957.
Ingold, C. T., "Spore Liberation," Oxford, Clarendon Press, 1965.
Smith, G. M., "Cryptogamic Botany Vol. I. Algae and Fungi," New York, McGraw-Hill, 1955.
Sparrow, F. K., "Interrelationships and Phylogeny of the Aquatic Phycomycetes," Mycologia 50: 797–813, 1958.

FUNGAL GENETICS

The catalytic stage in the study of fungal genetics came through the work of Beadle and Tatum who, using *Neurospora crassa*, showed the ease with which nutritionally defined mutants could be obtained. This was a step towards a union of far-reaching consequence, that of genetics, microbiology and biochemistry. The fungi as a group offer many advantages for genetical work. In most species the nuclei are haploid throughout the major part of the life cycle; species with a sexual stage have a short generation time; test of large numbers of vegetative or sexual spores is simple and selective techniques are available to sieve for infrequent types arising by mutation or rare events of recombination; collectively the fungi enjoy a variety of life cycle features—haploidy, diploidy, homokaryosis, heterokaryosis, and dikaryosis—which, individually or comparatively, permit the study of extra-chromosomal inheritance, gene action and interaction, gene action and cellular organization; ascus analysis has confirmed in an absolute, as opposed to a statistical, way the fundamental laws of genetics, and it provides a valuable tool for the detailed study of recombination. In a complementary way, the application of genetical techniques to fungi has revealed details of their life cycles which would otherwise have remained undetected.

Fungal mutants. Mutations affecting morphology, growth rate, conidial color, ascospore pigmentation, uni- versus multi-nucleate conidia, resistance to inhibitors, ability to grow on various sugars and so on have been analyzed. In almost every case the difference from wild type is due to mutation in a single gene. The largest and most important class of mutants is the nutritionals. Many fungi grow on a simple medium of a sugar and inorganic salts with, in some cases, a vitamin supplement. Nutritional mutants are readily obtained following mutagenic treatment of conidia or other haploid, preferably uninucleate, cells. Mutants requiring a wide range of amino acids, vitamins, purines and pyrimidines have been described. Mutagenic treatment, especially with high-energy radiation, produces the expected chromosomal aberrations; of these, translocations have been analyzed in detail in *Aspergillus nidulans*.

The formal genetics of fungi. A number of species have now been subjected to a thorough genetic analysis which is equalled in few higher organisms. For instance, genes have been mapped on all seven chromosomes of *Neurospora crassa* and all eight chromosomes of *Aspergillus nidulans*. Tetrad analysis and Mendelian analysis based on random sexual spores have been the main analytical approaches, but the parasexual cycle is being used increasingly in some species.

Meiosis in fungi appears to be standard; the expected results of rare meiotic errors are found in, for instance, disomic ($n + 1$) ascospores of *Neurospora crassa* and diploid ascospores of *Aspergillus nidulans*.

Various mating type systems have been described. They range from the system determined by multiple alleles of two genes in certain Basidiomycetes, through the single gene (A/a) system of *Neurospora* and certain yeasts, to the homothallic situation obtaining in *Aspergillus nidulans*. *Neurospora tetrasperma* is a secondary homothallic species; it has the single gene (A/a) mating type system, but the ascospores usually carry one nucleus of each mating type so that single spore cultures are self-fertile.

Crosses in which there is a genetically determined sexual difference offer no technical difficulties; for instance, in *Neurospora crassa*, a culture of either mating type (A or a) forms protoperithecia and can be regarded as the female parent which is fertilized by applying conidia of the opposite mating type. Homothallic species require a different approach. For example, in *Aspergillus nidulans*, a heterokaryon carrying nuclei of types a and b is made. a and b differ in nutritional requirements and usually also in visible properties; the heterokaryon is usually maintained in

"balanced" condition on the simple medium which permits growth of neither component type alone. Sexual spores are carried in the perithecia which contain as many as 10^5 asci each; a perithecium is usually initiated by two nuclei only, which undergo conjugate divisions before fusion. Heterokaryons form three classes of perithecia ($a + a$, $b + b$, $a + b$) which are morphologically indistinguishable. Only ascospores from the $a + b$ type are of use in analysis. The products of hybrid meiosis are obtained either by sampling from individual perithecia or by plating ascospores of pooled perithecia on a selective medium which permits growth of neither parental type. In the case of selective plating, the resulting segregants are then analyzed for the segregation of unselected markers.

Tetrad analysis. Each ascus contains the products of meiosis of a single diploid nucleus. Tetrad analysis has demonstrated the Mendelian laws in an exact way. The two alleles of a gene usually segregate 2:2 in each ascus; exceptions, rare or very rare, are described as instances of gene conversion. From crosses involving two linked genes a proportion of the asci contain, individually, both parental classes as well as the two reciprocal crossover types. This shows that crossing-over is, at least in a gross sense, reciprocal and that it occurs at the 4-strand stage. In some species, e.g., *Neurospora crassa* and *Bombardia lunata*, the spores are linearly arranged in the ascus and their order reflects precisely the events of the first and second meiotic divisions. First and second meiotic division segregations of the alleles of a gene are determined by dissecting out the spores in linear order; the frequency of second division segregation asci is twice the frequency of recombination between the gene and its centromere. Species of *Aspergillus* and most yeasts have spherical asci with randomly arranged spores. Determination of gene-centromere recombination frequencies can be made only when at least three genes are segregating.

Fine genetic analysis in fungi. By crossing nutritional mutants and plating the spores on selective media it is a simple matter to select rare recombinants occurring with frequencies as low as 1 in 10^6 spores. This achieves a fine analysis of the genetic material, and it is plausible to suppose that mutational sites only a few nucleotide pairs apart have been resolved in this way. Crosses between strains carrying independently arising mutations of the same gene almost always yield some wild type recombinants. As a result of such analyses in fungi and other organisms the earlier concept of the gene has broken down. The gene is seen as a segment of a chromosome larger than either the unit of mutation or recombination. Fine structure analysis in fungi has also revealed the clustering of exchanges, termed "negative interference," over short chromosome segments.

Heterokaryosis. Heterokaryosis has played a major role in fungal genetics. The earliest work was concerned largely with tests of allelism. The most recent physiological studies are on complementation in heterokaryons formed between strains both carrying mutations in the same gene though presumably at different mutational sites in that gene. Interallelic complementation may occur when the relevant enzyme is a multimer. Each mutant allele determines a defective polypeptide which gives an inactive homomultimer; differently defective polypeptides, determined by different mutant alleles of one gene, may form a heteromultimer with some enzyme activity.

The conidia formed by a heterokaryon may combine the genotype of one component with either cytoplasmic elements of the other or with a mixture of cytoplasmic elements. Such a system has considerable potential for the study of nucleocytoplasmic interactions and of extra-chromosomal heredity. The possibilities are being exploited in the filamentous fungi and through the transitory heterokaryotic stage of heterothallic yeasts.

In some species, heterokaryons respond to an environmental change by an adaptive change of nuclear ratio while in *Neurospora crassa* such adaptive change does not occur. This process has been studied in *Aspergillus nidulans* through the use of heterokaryons carrying nuclear types which were visually distinguishable in the hyphae. Each hyphal tip has only 50–100 nuclei and, in a heterokaryon, there is variation in nuclear ratio among the hyphal tips. In this species, adaptation, and even normal maintenance of a balanced heterokaryon, depends on selective growth of tips with advantageous ratios. In *Neurospora crassa* the nuclei are in rapid circulation and each growing tip has sufficient nuclei to prevent significant inter-hyphal variation.

Heterokaryosis provides a means for the sheltering of recessive mutations; this, and the scope for heterokaryon adaptation in some species, might suggest a role for heterokaryosis in nature. Some factors limiting this possible role are already known. There are barriers, probably gene-determined, to the ready formation of heterokaryons between different isolates of one species. Such barriers would also reduce or prevent gene flow between isolates of those species in which heterokaryosis is an essential feature of the sexual cycle.

Parasexual processes and the genetics of certain asexual fungi. In heterokaryons of some filamentous fungi there are rare but regular fusions of pairs of haploid nuclei in the mycelium. The resulting diploid nuclei are relatively stable at mitosis and give rare vegetative spores with diploid nuclei. When the parent haploid nuclei carry appropriate and different genetic markers the spores with diploid nuclei can be selected out and strains with heterozygous diploid nuclei prepared.

Strains with diploid nuclei show vegetative segregation through rare "accidents" at mitosis. Two processes giving rise to segregation are well understood. These are: mitotic crossing-over, which brings about recombination, and haploidization, in which the haploid condition is restored and chromosomes are reassorted without crossing-over. Thus, recombination is achieved outside the standard sexual cycle. These events (heterokaryosis, nuclear fusion in the mycelium, mitotic crossing-over, haploidization) were worked out in *Aspergillus nidulans* and were termed by Pontecorvo the "parasexual cycle."

The elements of the parasexual cycle have been looked for in many perfect and imperfect species with positive results in most cases. Some or all elements have been found in species of various genera including *Aspergillus, Penicillium, Ustilago, Fusarium, Coprinus, Emericellopsis* and *Cochliobolus*. The cycle offers a valuable supplement to the sexual cycle in perfect species; in imperfect species, it opens the way to formal genetic analysis, to aspects of phytopathology and to the planned breeding of some industrially important organisms.

J. A. ROPER

References

Fincham, J. R. S. and P. Day, "Fungal Genetics," Oxford, Blackwell, 2nd ed. 1965.

Ainsworth, G. C. and A. S. Sussman eds., "The Fungi," New York, Academic Press, 1966, vol. 2, Chapters by R. H. Davis, S. Emerson, K. Esser, J. L. Jinks, J. R. Raper and J. A. Roper.

g

GALEN (130 A.D.–200 A.D.)

Physician, anatomist, physiologist and philosopher; all these terms can be applied in their fullest sense to Galen, born in 130 A.D. at Pergamum in Asia Minor. After two years of medical study, begun in 146, he roamed widely in search of knowledge, finally settling in Rome in 164 where he aroused the animosity of his fellow physicians by unsparing attacks upon the medical sects, methodists, dogmatists, pneumatists and empirics, who were prevalent in Rome at this time.

The author of some 500 treatises ranging from philosophy to drama, Galen still remains best known for his numerous medical works, of which, unfortunately, the authenticity of over one third is doubtful.

Readily regarded as the founder of experimental physiology, Galen is also considered one of the most distinguished physicians of ancient times, second only to Hippocrates. An avid dissector, his investigations on the anatomy of apes and lower animals were full and accurate, and result in a greatly improved understanding of the human body.

As a physiologist his work was revolutionary. For example, one of his greatest contributions was a demonstration that blood, not air, is contained in the arteries, a discovery completely contrary to the Alexandrian school. Yet he also propounded the unfortunate theory that the septum of the heart was pierced by a minute opening, through which some blood was supposed to have access into the left ventricle from the right.

A firm monotheist, Galen felt that God's purposes could be found quite exactly by an examination of His works. This was undoubtedly one of the main reasons for his popularity in following ages.

If it were not for Latin translations from Arabic, Galen would probably have never been known, as his works were lost to Western Europe after the breakup of Rome. As it was, by the fifteenth century, translations were completed and consequently studied in medical schools throughout Europe until the early nineteenth century.

LETITIA LANGORD

GALLIFORMES

There are about 275 species in the order Galliformes, commonly known as Game Birds. They form a well defined and fairly isolated group. They all are basically terrestrial in habits and usually feed on the ground. The majority of them roost on trees at night, but are heavy on their wings, which are short, curved, fitting closely on the body, enabling them to a fast, powerful, but usually not long sustained flight, with a few exceptions. They have strong legs, with four toes and heavy nails. The tarsi are often armed with spurs.

The skull is schizognathous, their bill, holorhinous, without a bony partition between the nostrils; bill short and thick, the upper mandible convex and overhanging the maxilla. Breast bone with a high keel and two pairs of deep terminal notches. Tail short to very long, but usually well developed, of 8 to 32 feathers, rounded or pointed, with long coverts. Plumage soft and ample. Sexes either alike or different. Nest simple, often built on the ground; eggs 2 to 15, plain or mottled. Chicks precocious and born well covered with down, some having already well developed wing feathers; they leave the nest soon after hatching.

The smaller species are generally monogamous, pairs being formed and maintained, and the male helping rear the brood, although the female alone hatches the eggs. Their displays and calls are not highly specialized. The larger species, however, at least among the Grouse, Pheasants and Turkeys, are usually polygamous. The cock, who has a brilliant plumage and emits striking calls, whistles and screams, displays elaborately in his territory, so attracting the females, and mating takes place more or less casually. The hens sit and rear the broods.

Most species of game birds have a compressed tail and their display is therefore lateral. Those which have a flat tail: most of the Grouse, Lophophorus, several Polyplectron, Arguisianus, Pavo and Apropavo have a frontal display of great splendor including the spreading of the tail and of the wings.

Galliformes are found all over the world except on some small islands and in the Antarctic.

Three of the most useful domestic birds belong to this order: The Chicken, derived from the Red Junglefowl (*Gallus gallus*); the Guineafowl (*Numidea meleagris*), and the Turkey (*Meleagris gallopavo*)—other species have long been kept in captivity and raised regularly under artificial conditions, such as the Peafowl (*Pavo*), the Game Pheasants (*Phasianus*), the Golden Pheasant (*Chrysolophus pictus*) and the Silver Pheasant (*Lophura nychthemera*). But they have not yet changed in size or in general characteristics, although mutations have occurred. All are of great importance to man as food and game.

Galliformes are easily subdivided into three families. The Phasianidae have the posterior notches of the sternum very large, and the hallux inserted higher than the other toes. They are cosmopolitan. The Cracidae and the Megapodiidae have much less developed sternum notches and the hallux is at the same level as the other toes.

The Cracidae (Curassows and Guans) have weaker legs and nest on trees, leading a more arboreal life. They possess two carotids and their uropygial gland is feathered. They inhabit the American tropics.

The Megapondiidae (Mound Builders) have extremely strong legs and feet, only one carotid and a naked uropygial gland. They build large mounds of vegetation or sand in which their eggs are buried and incubated by the heat produced by decomposition or by sun rays. They are proper to Australasia and some islands in the Pacific Ocean.

The very peculiar Hoatzin (*Opisthocomus*) from the South American Tropics, has often been included in the Galliformes. It may resemble some common ancestor of the Game birds, but it is too different in many characteristics and in its living habits to be so closely associated with them. It is better considered as representing a separate order. It has some affinities with the African Touracous (Musophagidae).

A—**Phasianidae.** Most of the popular game birds belong to this large family, which can be divided into four subfamilies.

I *Grouse* (*Tetraoninae*). Game brids of the northern hemisphere, adapted to cold climates. Bill short, the base covered with hair-like feathers as well as the legs and sometimes the toes, which are pectinate, enabling them to walk on snow. The small Ptarmingans are partridge like and live in the far north or at high altitudes. The larger Grouse live in forests or grassy plains.

II *Quails, Partridges and Pheasants* (*Phasianinae*). The largest group of Galliformes. There is no important difference between the tiny, short-tailed Quails and the huge, long-tailed Pheasants such as the Argus and Peafowls. For practical purposes, the larger species, the males which are brilliantly colored, are called Pheasants, and they all live in Asia and Indonesia with the exception of the Congo Peacock. The numerous Quails, Partridges, Francolins and allied genera extend throughout the world, the American Quails being fairly distinct on account of their sharp, serrated bill.

III *Guineafowls* (*Numidinae*). An African family characterized by the short tail, a non-glossy, striped, spotted or speckled plumage where grey is dominant; head and neck more or less bald, often adorned with horns and wattles. They differ structurally from Pheasants in lacking a protuberance on the second metacarpian of the wing. The males often have a tarsal spur. No elaborate display. Strong and disagreeable voice.

IV *Turkeys* (*Meleagridinae*). Two north and central American species of large size; sexes similar in plumage, but the male much larger, and spur-legged. Head and neck naked and covered with ornamental, erectile wattles. Plumage highly metallic, and greatly specialized frontal display.

B—**Cracidae.** A Central and South American family of large tropical birds. Sexes alike with the exception of the Curassows of the genus *Crax*. The latter, and other allied genera are large, and rather Turkey-like in appearance, mostly glossy black; bill adorned with highly colored basal wattles or ridges, head crested. Display fairly elaborate, and whistling calls.

The Guans have shorter legs and are more arboreal in habits. All nest on trees, laying few eggs of very large size; chicks much developed at birth.

C—**Mound builders** (**Megapodiidae**). These very strong-legged game birds are found on islands from the Nicobars, the Philippines east to the Mariannas, south to Celebes, the Moluccas, New Guinea, the Solomons, the New Hebrides and Australia. They differ from all other games birds in their nesting habits, the eggs being hatched by the heat developed in the mound, which is built and attended by the male. The chicks are hatched with well developed wings and are completely independent from birth. The sexes are alike in plumage and the

Fig. 1 Grouse. (Courtesy U.S. Soil Conservation Service.)

male is polygamous. The display is simple and the voice low.

JEAN DELACOUR

Reference

Delacour, J., "The Pheasants of the World," London, Country Life, 1951.

GALTON, SIR FRANCIS (1822–1911)

This towering intellect of the 19th century would have become better known had his discoveries not been overshadowed by those of his cousin, Charles Darwin. He first devoted himself to meteorology and was the first to establish the existence of, and a theory to account for, anticyclones. He then turned his attention to heredity and published a series of five books on this subject, of which the best known is "Inquiries into Human Faculty and Its Development," in which he foreshadowed many of the discoveries of contemporary psychology. It was in this work that he demonstrated the principles of composite photography in an endeavor to establish a basis for heritable human types. In later life he became passionately devoted to the theory that the human race could be improved by selective breeding and coined the word "eugenics" to describe this process. His principal contributions to evolution were his introduction of exact measurements and statistical analysis to data and in his firm opposition to the then universally accepted theory of the inheritance of acquired characters.

GANGLION

An accumulation of nerve cell bodies, usually circumscribed. Plural: ganglia, ganglions. Primarily nodes in the peripheral nervous system; then: certain nuclei in the central NERVOUS SYSTEM (basal ganglia, interpeduncular ganglion). Also all NERVE CELL aggregates in invertebrates. (However, "ganglion cell" refers to any nerve cell.) The discussion follows for mammals, especially man.

Peripheral Nervous System

(1) Nodular or weblike enlargements of nerves containing no cells, sometimes termed ganglia, are *gangliform swellings*.

(2) *Cerebrospinal or "sensory" ganglia* contain cell bodies of peripheral afferent nerve fibers. Located in each intervertebral foramen at the beginning of a spinal nerve where dorsal and ventral roots join (spinal or dorsal root ganglia). Also in the skull as swellings of afferent cranial nerves near foramen of exit (cranial ganglia). Olfactory nerves (I): Cells in the mucosa of the nasal roof correspond to a profoundly modified sensory ganglion, serving as sense organ. Trigeminal nerve (V): Gasserian or trigeminal ganglion lies in the dura of the middle cranial fossa in Meckel's cave. Facial nerve (VII): Genicular ganglion within the tem-

poral bone. Cochlear nerve (VIII): Spiral ganglion within the bony axis and spiral shelf of cochlea. Vestibular nerve (VIII): Scarpa's vestibular ganglion in the internal auditory meatus of the temporal bone. Glossopharyngeal nerve (IX): Superior and inferior ganglion, closely linked, in and below the jugular foramen. Vagus nerve (X): Superior and inferior (or jugular and nodose) ganglion, located as for IX.

(3) *Autonomic ganglia: Sympathetic trunk ganglia* paired and roughly segmental, lie in the sympathetic trunk (ganglionated chain) on the costal heads, along the lumbar spine and presacrally. The cervical portion (Fig. 1) of the trunk has only an inferior, a middle (occasionally missing) and a superior cervical ganglion with variable segmental relationships. The fused first thoracic and inferior cervical ganglia are known as stellate ganglion. Collateral (prevertebral) sympathetic ganglia lie near the midline in visceral nerve plexuses mostly preaortic (coeliac, mesenteric ganglia, etc.). Ectopic smaller ones are incorporated in visceral or periarterial nerve plexuses. The suprarenal medullary cells are homologues of collateral sympathetic ganglia.

Parasympathetic ganglia: Numerous peripherally scattered cells near or within the viscus to be supplied. Anatomically defined ganglia, cell aggregates, found as ciliary ganglion in the orbit; as pterygopalatine ganglion in the lateral wall of the nasal cavity; as otic ganglion below the foramen ovale of the skull; as submandibular ganglion below the floor of the oral cavity. Numerous cardiac and pulmonary ganglia about the roots of heart and lungs are probably intermingled with

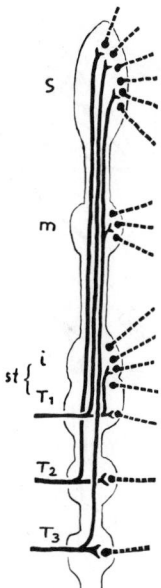

Fig. 1   Schema of cervical sympathetic trunk and ganglia. White rami communicantes on left, gray rami and other post-ganglionics on right, supplying sympathetic components to upper spinal nerves, heart, cervical and cephalic viscera and vessels, pilo- and sudomotor end organs. Note upper limit of preganglionic outflow at T1. s = superior cervical ganglion; m = middle, i = inferior cervical ganglion, st = stellate ganglion.

sympathetic collateral ganglia. Cells in the intestinal submucous (Meissner's) and myenteric (Auerbach's) plexus are considered extensive parasympathetic ganglia, so are the aggregates near the uterine and vesical cervix, those near other pelvic viscera as well as scattered nerve cells in various glands.

Connections and functions. Ganglia in the central nervous system are integral parts of intracerebral pathways to be considered with the brain.

Sensory ganglion cells have a peripheral process (afferent fiber from somatic or visceral area) and a central one that makes up the sensory nerve root to transmit impulses to the central nervous system. (Fig. 2)

Autonomic ganglion cells send their axones (postganglionic fibers) to cardiac and smooth muscles and to glandular epithelium. Stimulation occurs by preganglionic fibers which arise from certain restricted areas of the central nervous system and divide to contact synaptically several ganglion cells, thus ensuring spread of the impulse. Only the synapsing fibers permit to denote the ganglion as sympathetic or parasympathetic, while fibers passing through form additional "roots." (Fig. 3)

Sympathetic trunk ganglia (Fig. 2) receive their preganglionics from the lateral horns of the spinal cord, segments T1 through L2. The fibers pass through

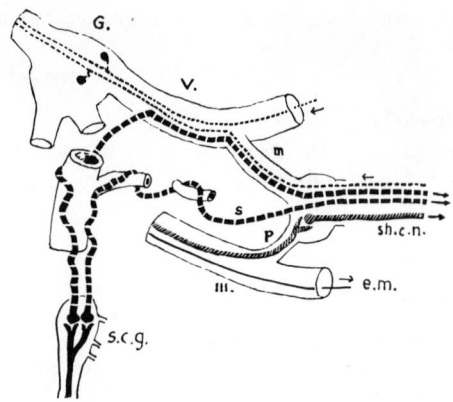

Fig. 3   Schema of connections of a parasympathetic (ciliary) ganglion. e.m. = somatic motor fiber to eye muscle, G. = Gasserian ganglion, m = mixed (sensory and symp.) root of ciliary ganglion, p = parasympathetic root, s = sympathetic root, s.c.g. = superior cervical ganglion. sh.c.n. = short ciliary nerves to eye globe, carrying symp.vasomotors and pupillodilators, parasympathetic pupilloconstrictors and sensory fibers, III. = oculomotor nerve, V. = trigeminal nerve. Character of lines as in Fig. 2. Note periarterial plexus.

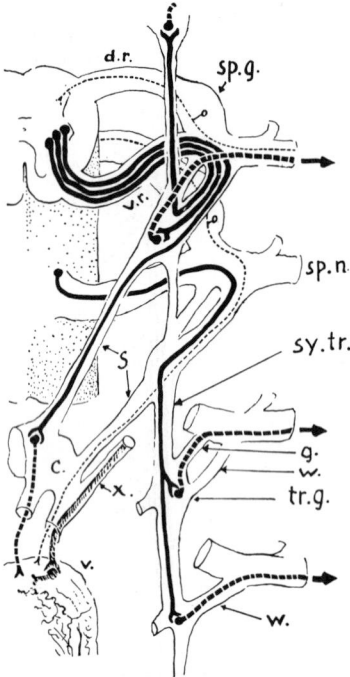

Fig. 2   Diagram of some connections of sympathetic and spinal ganglia. c. = collateral ganglion, d.r. = dorsal spinal root, v.r. = ventral spinal root, g. = gray ramus communicans, w. = white ramus communicans, S. = splanchnic nerves, sp.g. = spinal ganglion, sp.n. = spinal nerve, sy.tr. = sympathetic trunk, tr.g. = trunk ganglion, v. = viscus, X. = parasympathetic branch of vagus nerve. Heavy broken lines = symp.postganglionics, heavy solid = symp.preganglionics, thin broken = afferents, shaded = parasympathetic.

the respective segmental anterior roots and go as white rami communicantes to one of the trunk ganglia at the same level or, when ascending or descending they form the sympathetic trunk and synapse in a more distant level (usually at a ratio of 1 pre- to ± 30 postganglionic neurones). Thus all levels of the body receive sympathetics from a restricted outflow. Postganglionic fibers join the spinal nerve as gray ramus communicans to reach the periphery (vascular musculature, arrectores pilorum and sweat glands). Other preganglionics pass via white rami communicantes through trunk ganglia without synapses to continue as splanchnic nerves to collateral ganglia. Their postganglionics reach viscera to innervate smooth musculature, to act as vasomotor nerves and to assist in regulating secretion. A small group of splanchnic nerves forms synaptic endings around the suprarenal medullary cells, stimulating adrenaline production.

Parasympathetic ganglia receive their preganglionics from autonomic components of cranial nerves (III for the ciliary ganglion, VII for the pterygopalatine and submandibular ganglion, IX for the otic and X for most other visceral areas). Another parasympathetic outflow issues from the sacral spinal nerves 2 and 3 to synapse in the ganglia of the lower colon and pelvic viscera. The ratio of synapses is 1 : ± 10. Most autonomic ganglia are embedded in or connected to a dense nervous plexus (perivascular or perivisceral) containing pre- and postganglionic sympathetic and parasympathetic fibers as well as visceral afferents frequently defying detailed morphological and experimental analysis.

Variations in shape, grouping and connections of ganglia are numerous. Not all visceral ganglia can be definitely denoted as sympathetic or parasympathetic.

Structure. Ganglia in the central nervous system vary in regard to density of cell population, size, form and pigmentation of cells, neuroglia distribution and

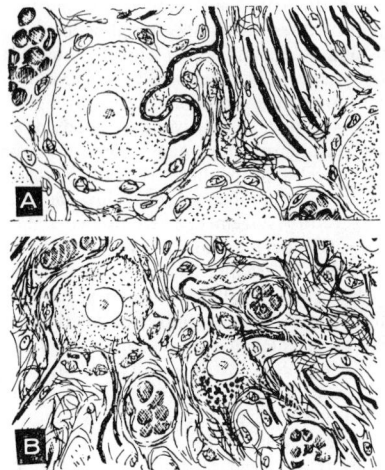

Fig. 4   Microscopical appearance of a spinal (A) and a sympathetic trunk ganglion (B) at about 400× magnification. Note numerous blood vessels and all-pervading connective tissue framework forming sheaths for vessels, neurones and their processes. A, upper right corner: myelinated nerve fibers. B: Left border of left cell is hugged by a preganglionic terminal.

vascular pattern. The ganglion cells are multipolar in type.

In peripheral ganglia (Fig. 4), sensory and autonomic, each cell is closely surrounded by an inner capsule of satellite cells which correspond to and are continuous with the neurolemma of the nerve fibers and by an outer connective tissue capsule, carrying blood vessels, continuous with the endoneurium of nerves. More than one cell may share a capsule; in the smaller ganglia capsulation may be poorly defined. Larger ganglia are invested by a tough connective tissue envelope extending as epineurium onto the nerve. The cells are intimately contacted by capillaries which abound in autonomic (synaptic) ganglia, and are fairly numerous in sensory (nonsynaptic) ganglia. Nerve fiber bundles have a scanty capillary bed. Intracellular accumulation of lipochrome granules occur with age without known pathological significance. In man, degenerate cell forms and reduction in cell number becomes conspicuous from the third decade onward.

Sensory ganglion cells, 15–100 μ in diameter, are derived from embryonic bipolar neurones and retain this appearance in the cochlear and vestibular ganglia. In the other cerebrospinal ganglia the two processes gradually fuse into one which splits T-wise and acquires a myelin sheath shortly after leaving the "pseudounipolar" cell. The process often coils within the capsule or runs straight outside to split into a central (axonal) and a peripheral fiber. The latter conducts toward the cell (hence considered dendritic) but is of axonal morphology. The chromidial (Nissl) substance is finely granulated throughout the cell body, the nucleus centrally located, vesicular with one or two nucleoli; neurofibrils and mitochondria are numerous, a Golgi apparatus present.

Autonomic ganglion cells are multipolar and 20–50 μ in diameter. Preganglionic axon terminals make synaptic contacts by forming spirals and arabesques on the cell body and dendrites within the capsule and on the numerous dendrites which extend extracapsularly. The single (postganglionic) axon pierces the capsule, may acquire a thin myelin sheath but usually remains naked. Neurocytological features are similar to sensory ganglion cells, but the chromidial substance frequently concentrates peripherally or centrally, the nucleus tends to be placed somewhat off center. Bi- and multinucleate cells occur regularly in certain rodents, occasionally in man.

Functional considerations. The anatomical and functional integrity of afferent nerve fibers and roots depends on the life of their sensory ganglion cells. Experimental destruction of autonomic ganglia, however, tends to cause temporary disturbances only, since most viscera cannot be completely denervated anatomically due to plexus formation and the presence of intramural ganglia. But cholinergic and anticholinergic drugs exert a profound effect by disturbing the metabolism of acetylcholine which is considered the chemical transmitter of impulses in all autonomic ganglia. Nicotine applied there topically has a blocking effect.

LOUIS L. BERGMANN

References

Bloom, W. and Fawcett D. W., "A textbook of histology," Philadelphia, Saunders, 1962.
Penfield, W., "Cytology and cellular pathology of the nervous system," Vol. 1. New York, Hoeber, 1932. (Articles by de Castro, F.: (A) Ganglia of the cranial and spinal nerves, normal and pathological, Vol. 1, pp. 91–143; (B) Sympathetic ganglia, normal and pathological, pp. 317–380.)

GASTROPODA

Gastropoda is the name of a class of Mollusca, equivalent to Lamellibranchia, Cephalopoda and Amphineura. Gastropods are often called univalves because their shell is in one piece in contrast to the two valves of lamellibranchs, and also, popularly, snails and slugs after the most familiar types. Most are marine and littoral (limpets, winkles, cowries, whelks, cones), a few pelagic (sea butterflies) and many are freshwater or terrestrial.

The head bears 1–2 pairs of tentacles and a pair of eyes. The foot has a flat sole for creeping, a process lubricated by mucus from special glands, and the visceral mass has a deep mantle cavity posteriorly. During development the visceral mass is twisted on the head-foot, partly by muscular action, partly by growth, through an angle up to 180° in a counterclockwise direction, a process known as torsion, which results in its right side lying on the left and vice versa, and in the mantle cavity being rotated to the anterior end, over the head. Head and foot may then be withdrawn into its shelter and, in many, a posterior pedal lobe secretes a cuticular plate, the operculum, which closes its mouth and that of the shell. This, secreted by the mantle or covering of the visceral mass, has necessarily the shape of that part of the body. Internally, torsion twists structures such as gut and nerves which run between head and visceral mass, the nervous system then being called streptoneurous. Gastropods are also distinguished by an asymmetrical helical spiral coiling of the visceral mass, its (post-torsional) left side hypertrophying and its right shortening. It may, with the shell, become secondarily conical, as in limpets, or a loose spiral as in *Vermetus*.

Within the gastropods may be distinguished a basal group Prosobranchia, showing the full effects of torsion internally and externally and possessing a well developed helical shell. Prosobranchs exhibit two grades, a lower with a double set of pallial structures (Diotocardia, Aspidobranchia or Archaeogastropoda), and a higher in which most of the pallial organs on the right have disappeared (Monotocardia, Pectinibranchia or Mesogastropoda with Neogastropoda). From a monotocardian ancestry evolutionary lines lead to two orders, Opisthobranchia and Pulmonata. The latter mostly retain the prosobranch facies (though terrestrial slugs are shell-less), but have lost streptoneury. A lower grade of pulmonate is represented by the Basommatophora, mainly inhabitants of freshwater though breathing air, and a higher grade by the Stylommatophora, the terrestrial snails and slugs. Opisthobranchs have followed an evolutionary trend towards reduction and loss of the shell, eversion of the mantle cavity, diminution of the visceral mass and reversal of torsion (detorsion) so that streptoneury is again lost. Lower opisthobranchs (Tectibranchia) merge in many ways with prosobranchs, retaining a shell; higher opisthobranchs (Nudibranchia or sea slugs) have none.

The mantle cavity contains gills (ctenidia) and associated sense organs (osphradia) and into it discharge the gut, kidneys and single gonad. Diotocardians typically have two ctenidia, two osphradia and right and left kidney openings, the former also serving as reproductive aperture. Cilia on the ctenidia ventilate the cavity, sucking water in laterally and expelling it medianly after it has washed over excretory openings and anus. In monotocardians the right gill and osphradium are lost and only the left kidney persists as an excretory organ, the right being incorporated in the reproductive duct. In these animals the respiratory current is a transverse one from left to right. In a few prosobranchs like the limpet *Patella* ctenidia are replaced by gills round the mantle edge. Some others (*e.g. Crepidula*, the slipper limpet; the freshwater snails *Viviparus, Bithynia*) have evolved a mechanism which collects for food particles sieved from the respiratory current. In tectibranchs the everting cavity exposes the gill and reduces the need for ventilation and in nudibranchs it is lost, the ctenidium gone and respiration cutaneous. In pulmonates the cavity becomes a lung, the gill lost, the roof vascularized and its mouth narrowed to prevent desiccation. In adverse—hot, cold, dry—circumstances pulmonates may aestivate or hibernate, withdrawing into their shell and sealing the entrance with a calcareous epiphragm secreted by the mantle edge.

The gut contains a radula and jaws. The radula is used for raking small particles together (rhipidoglossan type), scratching and raking (taenioglossan), scraping surfaces (docoglossan, pulmonate), tearing and pulling (stenoglossan), grasping (ptenoglossan) or piercing (toxoglossan). It is lubricated by saliva. The prosobranch oesophagus is glandular (oesophageal gland, gland of Leiblein) and leads to a complex stomach where food and enzymes from a digestive gland are mixed and digestible and indigestible matter sorted. An elaborate intestine moulds faecal pellets. In opisthobranchs and pulmonates oesophageal glands are absent and the stomach is functionally replaced by an expanded oesophagus, sometimes (tectibranchs) armed for crushing food. Primitive gastropods are microphagous herbivores but many higher prosobranchs (whelks, cones) and opisthobranchs (dorids, eolids) are carnivores; most pulmonates are herbivorous.

The excretory system consists of two kidneys connected to the pericardial cavity internally and the mantle cavity externally. The right invariably acts as genital duct (additionally to excretory activity in diotocardians only). In some diotocardians the left kidney is unlike the right and may be solely osmoregulatory. In other groups it is the only excretory organ and may show differential regional activity. The gonad is single, ovary or testis in prosobranchs, an ovotestis in other orders. Its duct leads to the right kidney or its vestige, which may retain a pericardial connexion. In diotocardians gametes are broadcast and fertilization external; in others a penis is developed and fertilization is internal. The female (section of the) duct is elaborated to provide a bursa copulatrix, a receptaculum seminis and glands providing the eggs with a covering, food albumen and a capsule or jelly mass in which many may be laid attached to a suitable substratum. In pulmonates, especially slugs, copulatory behavior may be elaborate with cross fertilization. In diotocardians the egg develops to a trochophore larva before settling, in most other marine gastropods to a larva provided with ciliated cephalic lobes (veliger). Freshwater and terrestrial gastropods hatch as miniatures of their parents.

A. GRAHAM

Reference

Hyman L. H., "The Invertebrates," Vol. 6, "Mollusca I," New York, McGraw-Hill, 1967.

GASTROTRICHA

Gastrotrichs are microscopic, multicellular, invertebrate animals. They are characterized by a muscular pharynx, ventral rows of cilia, and, in most species, a pair of adhesive glands. The species range in size from 65 to 500 μ in length. They can be found in most aquatic habitats, swimming along the bottom, around the roots of *Lemna*, on the basal leaves of *Nuphar* and among algae and moss.

Historically, the Gastrotricha were considered as a class of the Trochelminthes, later as a distinct phylum. In her book on the Invertebrata Dr. Libbie Hyman categorized the gastrotrichs as a class of the phylum Aschelminthes. Although many modern-day biologists follow Dr. Hyman's classification, some still consider the Gastrotricha as a distinct phylum. The approximately 400 species of the Gastrotricha fit into two orders: Macrodasyoidea and Chaetonotoidea. The members of the former are strictly marine, the latter are primarily fresh-water. The genera of fresh-water gastrotrichs most often found are *Chaetonotus, Lepidodermella* and *Ichthidium.*

Gastrotrichs are pseudocoelomate animals with distinct integumentary, muscular, excretory, nervous,

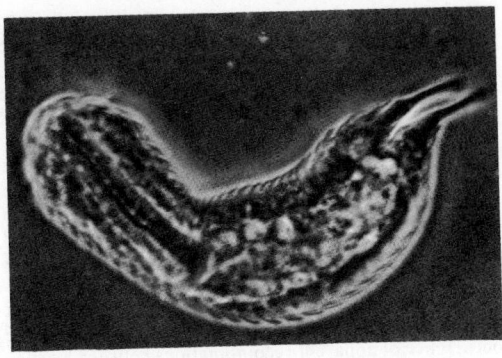

Fig. 1 Gastrotrich. (Courtesy James L. Oschman.)

digestive and reproductive systems. The integumentary system consists of a syncytial hypodermis and a secreted cuticular, outer layer. The cuticle is modified in various species to form scales, spines, or a combination of the two. There is very little variation in the number or size of the spines within a particular species. The muscular system is made up of about six pairs of longitudinal muscles. Contractions of these muscles produce a shortening of the body or a turning action. Excretion is accomplished by two flame bulbs which act in conjunction with two protonephridial tubules. The tubules may also act as an osmoregulator. Two lateral nerve strands connected to an anterior saddle shaped brain compose the nervous system. Ciliary tufts on the head appear to be tactile in function. The digestive system includes a mouth, pharynx, straight intestine, short rectum and an anus. Oral bristles occur in some species. Food varies from bacteria in some species to algae in others.

Although most marine species of gastrotrichs are monoecious, only parthenogenetic females of freshwater forms are known. There is some evidence of "maleness," however; and in some species there are as yet undescribed structures. The average number of eggs per female is 3.7; the range is from one to five; and four is the number produced by most females. Two types of eggs are produced. Tachyblastic eggs are those that begin cleavage immediately after oviposition and hatch within 35 to 50 hours. Opsiblastic eggs have a heavier covering and must have a period of dormancy before hatching. The latter eggs can withstand periods of freezing and drying. The newly hatched young are like the adult and grow at the rate of four μ per hour. Gastrotrichs in cultures have lived to a maximum of 22 days.

ROYAL BRUCE BRUNSON

Reference

Brunson, R. B., "Gastrotricha," *in* Edmonson, W. T., ed. "Ward & Whipple's Fresh-water Biology," 2nd ed. New York, Wiley, 1959.

GEGENBAUR, CARL (1826–1903)

Though Gegenbaur's contributions to zoology are of great importance, it is as a teacher that he should be best remembered. It is safe to say that the vast majority of distinguished zoologists of the late 19th century had either studied with him or had come directly under the influence of his towering intellect and masterful personality. He was born at Würzburg and attended the university there during the period in which, to the great good fortune of science, Kölliker joined the faculty. The papers which he wrote on coelenterata, following his first trip to the ocean with his master, attracted such attention that he was offered the chair at Jena in 1855, the most important of his students there being Haeckel who only in the later years of his life escaped from the dominant personality of his master. Gegenbaur wrote innumerable papers on vertebrate and invertebrate anatomy but his outstanding discovery was that all eggs of vertebrates were unicellular. Before his time it had been accepted that small invertebrate eggs and the like might be a single cell but it was presumed that the eggs of birds must be a large cellular mass. In connection with his destruction of the theory that the yolk granules in large eggs were cells, he brought to the fore the fact that the important components of the cell were the nucleus and the cytoplasm, and that the extent and nature of the inclusions were secondary, rather than fundamental, components of the system. His major published work was on the comparative anatomy of vertebrates, in which he opposed strongly the then current view that the skull was composed of modified vertebrae and developed the idea that it had derived as bony plates superimposed on the cartilaginous cranium of the elasmobranch. In 1872 he was attracted to the University of Heidelberg by the better facilities for his research there available and died in 1903 some years after his enforced retirement on the score of ill health.

GENE

In early studies of Mendelian inheritance the term gene, the name given to the unit of heredity, could not be very precisely defined but meant something that segregates in the germ cells and is in some way connected with a particular effect on the organism that contains it (Morgan, Sturtevant, Muller, and Bridges: *The Mechanism of Mendelian Heredity*, 1915). Through the brilliant work with Drosophila done by T. H. Morgan and his associates, vague concepts of the gene and of the mechanisms of heredity became clearer and more definite. Their work provided evidence that genes are located on CHROMOSOMES and, moreover, are arranged there in a linear order. They discovered that a certain gene is found regularly at a certain place on a chromosome ("gene locus" or, in brief, "locus"). It became well established that changes in genes, called "mutations," occurring with a very low frequency, give rise to various different forms of the same gene, called "alleles." Thus a locus on a chromosome may be occupied by any one of its alleles, either the normal ("wild-type") or a mutant form. Hence the gene was operationally defined as a unit of mutation, or as a unit of physiological function or as a unit of structure within the chromosome which could not be.

There is ample experimental evidence that the basic genetic mechanism is fundamentally the same in all living organisms. Even in bacterial viruses (bacteriophages), the lowest known form of life, the genes segregate as they do in higher forms and are linked together in linear order. Every living organism possesses a number of different gene loci. It has been determined that the genomes of the RNA bacteriophages consist of only 3 functional genes. The estimates of the number of different gene loci in bacteria vary between 1,000 and 13,300; in Drosophila, between 3,000 and 13,300; and in man and the other mammals, between 2 and 1,000 times as many as in Drosophila. As a rule, a full complement of genes is present in every cell of the body and their functional information units exercise primary control over the intricate and well balanced system of biochemical reactions operating in the living cell.

Since genes are present in every living cell, it is evident that mutations may occur in any living cell. In general, the frequency of mutations at any one gene locus is very low, between 1×10^{-10} and 1×10^{-5}, and can be detected only by observing a very large number of individuals. However, a culture of small bacteria (see BACTERIAL GENETICS), Escherichia coli, grown in 100 ml of broth contains about 2×10^{11} cells and the human body about 6×10^{13} cells. During the development of such large populations of cells, it is fairly certain that many of the different loci represented in the organism will undergo mutation in some cell or cells and that mutant genes will therefore be present at many loci somewhere within the population.

Our knowledge about the biological structure of genetic mechanisms has been derived primarily by observing the relative numbers of various genotypes among the offspring of parents that differ in two or more genes. The principle is very simple. When two genes are located in different chromosomes they will segregate freely, but when two genes are located in the same chromosome (i.e., are linked) the parental genotypes will be more numerous than the recombinant genotypes. Recombination involving two linked loci is accomplished through a process known as crossing over, which involves the interchange of homologous regions between two members of a pair of chromosomes. The frequency of crosssing over is a function of the linear distance between two genes. This distance relationship forms a basis for the construction of "linkage maps," delineating the genetic structure of chromosomes. Crossing over occurs regularly during meiosis, and occasionally also during mitosis.

Especially sensitive methods of detecting genetic variants, either mutants or recombinants, have been developed in studies with microorganisms—bacteriophages, bacteria, and fungi—because large numbers of these individuals can easily be observed. In a bacterial population certain types of variants can be detected if they are present with a frequency of 1×10^{-11}.

A requisite for biological studies of the structure of genetic material is the presence in the same cell of two homologous chromosomes that differ with respect to certain "marker" genes; for only under such circumstances can crossing over be observed. In bacteriophages, which are haploid (i.e., have each gene locus present only once), such a condition is attained when one bacterium is infected by two phage particles that differ in genetic constitution. During their multiplication in the bacterial cell crossing over occurs between the chromosomes of the different phages. In bacteria, which usually are haploid, conditions that permit crossing over can be brought about in several ways. The most useful of these for a study of gene structure is the process known as transduction (see BACTERIAL GENETICS), in which phage serves as a vector in the transfer of small fragments of bacterial chromosomes from one bacterium to another.

With the demonstration that X irradiation, and other physical and chemical agents, could increase the rate of occurrence of mutations it became possible to accumulate mutant genes in a variety of experimental organisms. As induced mutants were recovered and mapped Drosophila, a number of loci, such as "lozenge-eye" and "white-eye" accumulated numerous different alleles. In the early 1940's C. P. Oliver reported the occurrence of apparent recombinations between lozenge alleles. These apparent alleles were termed "positional pseudoalleles" by E. B. Lewis and explained as functionally related "subloci" (i.e., smaller genetic units very closely linked to one another). In this interpretation recombination between pseudoalleles is "intergenic," that is, it occurs between genetic loci. Similar observations of intraallelic recombination in microorganisms led to numerous studies of gene fine structure and to a more precise definition of the operational gene elements. Benzer, as a consequence of his fine structure analysis of the rII region of the T4 bacteriophage proposed the name "cistron" for the functional gene and describes it as that unit of genetic information which controls the synthesis of one polypeptide chain and within which no functional complementation occurs between mutant alleles in heterozygotes or in heterocaryotes. He further proposed the terms "muton" for the smallest unit of the gene within which mutation can occur and "recon" for the smallest unit of recombination. With the establishment of DNA (and RNA) (see NUCLEIC ACIDS) as the chemical substance of the gene, these units have been expressed in terms of the nucleic acid molecule. The cistron varies in size but is usually about 400–600 nucleotide pairs while the muton represents one nucleotide base pair and the recon the distance between adjacent nucleotide bases along the DNA molecule. The fine structure map of the microbial cistron indicates that recombination is possible between adjacent bases within the cistron and that this represents true intragenic recombination. There are obvious differences between the pseudoallelic locus of Lewis and the cistron of Benzer when viewed in terms of intragenic recombination. However, on the bases of function and mutation, there is amazing uniformity among the genetic systems of all organisms. Indeed, the recombination difference between the pseudoallelic loci and the cistrons may represent structural and physiological differences in the chromosomes and not real differences in the gene. There is ample evidence that the basis of function of the gene in the biosynthetic pathways (see PROTEIN BIOSYNTHESIS) is common for all organisms.

The number of sites per gene locus is large, and may amount to thousands. In several phage and Salmonella loci, as many as a hundred sites have already been identified. Mutants resulting from changes at different sites of the same locus, or even at a single site, may differ from one another in several respects (i.e., spontaneous and induced mutability, reaction to mutagens, temperature sensitivity, nutritional requirements, complementary relations). As a rule, however, such mutants have one feature in common: they are all changed with regard to one specific function, controlled by that particular locus. In some or perhaps most cases, control of the function is effected through an enzyme, whose structure and consequently in a large measure character and specificity are determined by the gene.

Within a locus the mutational sites are not distributed at random; instead, sites representing similar alleles tend to be grouped together. Such arrangements are particularly notable in the case of sites of noncomplementary alleles. In Salmonella, and presumably in bacteria generally, there is a marked tendency for genes controlling related biochemical reactions to be located next to one another, in the same order as the biosynthetic sequence of these reactions. This kind of arrangement makes it possible to determine, with a reasonable degree of accuracy, the limits of certain gene loci, and to analyze areas near the boundary line between two adjacent loci. The evidence suggests a sharp division between adjacent loci.

M. DEMEREC
F. J. GOTTLIEB

References

Carlson, E. A., "The Gene: A Critical History," Philadelphia, Saunders, 1966.

Demerec, M. and P. E. Hartman, "Complex loci in microorganisms," Ann. Rev. Microbiol. 13:377–406, 1959.

Dunn, L. C., "A Short History of Genetics," New York, McGraw-Hill, 1965.

Pontecorvo, G., "Trends in Genetic Analysis," New York, Columbia Univ. Press, 1958.

Sturtevant, A. H., "A History of Genetics," New York, Harper and Row, 1965.

GENETIC DRIFT

Genetic drift is the irregular change in gene frequencies in a population from generation to generation as a result of random processes. It is best called random genetic drift, since in physical sciences drift is often used for directional processes. Random drift as an evolutionary influence has been emphasized in the mathematical theory of Sewall Wright, and it is therefore called the Sewall Wright effect.

The process of population reproduction may be thought of as drawing a sample of genes from the parents to constitute the progeny generation. If the population number is small, there may be considerable fluctuations in gene frequencies due to sampling accidents. The results of such fluctuations are cumulative, and the eventual result is that some alleles may become lost from the population or fixed by the loss of alternative alleles.

The process of random fixation is equivalent to inbreeding in that homozygosity is increased. The number of heterozygous loci is decreased by a fraction $1/2N$ each generation, where N is the population number. This relationship is strictly true only when an individual offspring has an equal chance of having come from any parent. If the parents are not equally fertile, the rate of decrease is greater, being given by $(1 + V/2)/4N$ for a population of stable size N, where V is the variance of the number of progeny per parent.

The change in gene frequencies is shown by the diagram below. The abscissa is the frequency of the gene ranging from 0 (loss) to 1 (fixation). The ordinate is the probability of this frequency after t generations. As can be seen, with successive generations the probability distribution becomes flatter. When the number of generations is twice the population number ($t = 2N$), one gene frequency is as likely as another. At this time the probability of any intermediate gene frequency decreases by $1/2N$ each generation while the probability of fixation and loss each increase by $1/4N$.

These results are for a strictly neutral gene. In a very small population the effects of random drift may predominate even if there is a selective advantage of one allele over another. On the other hand, if the selective advantage is strong, chance effects are unimportant. The critical point is when the difference in fitness between two types is roughly equal to $1/2N$. If the selective difference is much less than this, random drift is predominant; if greater than this, selection is the important factor.

Random genetic drift due to finite population size is likely to be important, in any but the smallest populations, only for genes of very minute effect, such as those determining small changes in quantitative traits. But such genes of minute effect are probably in the long run the ones of greatest evolutionary importance. It should be emphasized that random gene frequency drift can occur because of randomly fluctuating selection intensities as well as because of small population number.

Random gene frequency drift plays a central part in the theory of evolution as put forth by Sewall Wright. In this theory the population structure is critical. Optimum evolutionary opportunity arises when a population is divided into numerous subgroups, almost but not entirely isolated from one another. Within each subgroup some random gene frequency drift occurs. Among these subpopulations a particularly favorable gene combination might arise which can then spread through the whole population. Thus, there is selection not only between individuals, but also between subpopulations. In this population structure, Wright be-

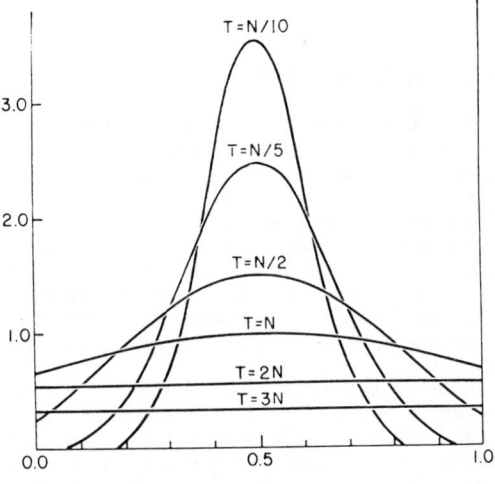

Fig. 1

lieves, lies the greatest opportunity for the emergence of novel types.

The alternative view, emphasized especially by R. A. Fisher, is that the optimum population is a very large one where the potential genetic variability is greatest and where random fluctuations are minimal. In such a population a genotype with even a minute selective advantage will ultimately succeed and meticulous evolutionary adjustments can occur.

These two views have been the subject of much discussion by evolutionary theorists. It is possible that both have their place—that a large population is optimum for steady improvement of adaptation, such as adjustment of an eye for better focusing, whereas the Wright model is more likely to lead to real novelty, such as new structures or new uses for old.

The mathematical theory of random fluctuation of gene frequencies is due to Fisher and especially to Wright. More recent mathematical advances have been made by M. Kimura. The process of gene frequency change in an evolving population is treated mathematically as a stochastic process. The probability distribution of the gene frequency is given approximately by the Fokker-Planck equation

$$\frac{d\phi}{dt} = \frac{1}{2} \frac{\partial^2}{\partial p^2} (\phi V) - \frac{\partial}{\partial p} (\phi m)$$

where p is the frequency of the gene, $\phi(p,t)$ is the probability that this gene has frequency p at time t, V is the variance in gene frequency change per generation due to random fluctuations, and m is the mean change in gene frequency due to the systematic forces of mutation, selection, and migration. The graph referred to above shows the solutions for the special case, $m = 0$, $V = p(1 - p)/2N$.

JAMES F. CROW

Reference

Wright, S., "Classification of the factors in evolution," Cold Spring Harbor Symposia, **20**: 16–24D.

GENETIC SUPPRESSION

Genetic suppression is the reversal of a mutant phenotype by a nonallelic mutation called a suppressor. A suppressor can be within the original gene (intragenic suppression) or without (extragenic suppression). It can be mutant independent but gene dependent (single-locus suppression). It can be mutant dependent but gene independent (nonsense and missense suppression) and may or may not be associated with resistance to an antibiotic such as streptomycin (antibiotic-induced suppression). Finally, it can be mutant dependent and gene dependent.

Intragenic suppression is of two types: (1) Consider a case where a mutation causes a replacement of an amino acid in a protein which destroys the function of the protein; a second mutation which replaces an amino acid somewhere else in the protein could restore some or all of the function. (2) The "reading-frame" of a RNA molecule could become out of translating register by the loss of one or two nucleotides (nucleotides are translated into amino acids in groups of three nucleotides called "codons"); a gain of one or loss of two in the first case, or a gain of two or loss of one in the second case could put the reading-frame in register again, even if the second mutation were some distance away from the original mutation though still in the same gene.

Extragenic suppression occurs when the gene that reverses the mutant phenotype is somewhere else in the genome. The gene dependent, mutant independent suppression (all mutants of the same gene, but only that gene, responding) is often related to the opening of an alternate metabolic pathway, the substitution of a new gene product for the nonfunctional gene product, or the decomposition of a toxic product accumulated by the first mutant.

Nonsense suppression or "super-suppression" is another case of extragenic suppression where the original "nonsense" mutation calls for a "stop" during translation and a protein fragment is produced. The suppressor that calls for a "start" again is most likely to be a transfer RNA molecule with its "anticodon" region altered to complement the nonsense codon. The nonsense mutations can be one of three types, an "amber" mutation, with the codon being UAG (for uracil, adenine, and guanine), an "ocher" mutation (UAA), or a "topaz" mutation (UGA). Different suppressors have been implicated for each type, though some ocher suppressors can also suppress amber mutations. In each case an amino acid (not necessarily the original one) is emplaced by the transfer RNA whose mother gene has been mutated in a way so that the anticodon region either efficiently or inefficiently reads the nonsense codon.

Missense suppression can only be inefficient. It is like nonsense suppression except that the original altered codon emplaces the wrong amino acid. A mutation of a transfer RNA gene causes it to put in the right one again. It would have to be inefficient because it is misreading the same codon in "natural" positions in all other "messenger RNA's" from all other genes.

Streptomycin-resistant mutations in bacteria appear to be mutations of the ribosome-forming region of the bacterial chromosome. Some of these mutants will suppress nonsense or missense mutants elsewhere in the genome if streptomycin is added to the medium. The phenomenon is not restricted to the action of streptomycin but is apparently limited to antibiotics of the aminoglycoside group.

In general, the types of suppression discussed above are fairly well understood. Other types of suppression do exist, and undoubtedly more will be discovered. For example, the gene dependent, mutant dependent type of suppressor is probably common. It is likely that the suppressors may have unique actions upon the gene products of the genes they suppress to render each gene product functional again.

R. C. VON BORSTEL

Reference

Gorini, L. and J. R. Beckwith, Ann. Rev. Microbiol., **20**: 401–422, 1966.

GENETICS

A branch of biology defined by Bateson (1905) as "the elucidation of the phenomena of heredity and variation." An old proverb has it that "like begets like," and it is *heredity* which makes children resemble their parents. And yet children are never exactly like their parents, nor are siblings, brothers and sisters, or any two living individuals, exactly alike. This is VARIATION. In another respect, every kind of organism, from virus to man, reproduces itself. New individuals arise from old ones, by converting materials taken from the environment, food in the broadest meaning of that word, into living bodies. The essence of heredity is, then, *self-reproduction* and resemblance among relatives in lines of descent. Indeed, some biologists believe that self-reproduction, and hence heredity, is the main characteristic distinguishing the living from the non-living. The ORIGIN OF LIFE would then be the origin of self-reproduction.

Phenotype and genotype. Johannsen (1911) proposed to distinguish the *phenotype*, the appearance of the organism, and its *genotype*, the sum total of the heredity received by the organism from its parent(s). The phenotype is changing as life and development proceed; thus, a man's phenotype is obviously different in the embryo, infant, child, adult, and senescent person. The genotype is said to be constant, or nearly so, during the entire life span. This should however be understood as meaning that the components of the genotype, GENES, reproduce themselves, by synthesizing their own copies from the non-genic materials, (ultimately from food) during growth and aging. An adult person does not have the same genes he had as an infant; what he does have are true copies of the same genes.

The development and growth of the organism, from conception to birth to maturity to death, are resultant of the interactions of the organism's genotype with its environment. What the genotype determines is, then, the *norm of reaction* of the organism, the way the organism responds to its environment via biochemical and physiological growth pathways to produce the phenotype. The phenotype is necessarily a product of interaction of the genotype with the sequence of the environments which individual organisms meet during their lifetime.

A lot of polemics among biologists, psychologists, sociologists, and others arose because of the difficulty of deciding which characters or traits of the organism are due to "*nature*" (= genotype) and which to "*nurture*" (= environment). This is a fallacious way to state the problem, since all characters and traits are always due to interaction of nature with nurture. To be meaningful, the "nature-nurture problem" must be stated in a different way. We observe that different human persons, as well as individuals of any other living species, differ in many traits—size, color, behavior etc. Now, it can validly be asked to what extent the *observed differences* between organisms are due to the *difference in their genotypes* and to the *differences in their environments*. With enough observational and experimental evidence, the problem so stated is susceptible of solution. It should, however, be remembered that the question must be asked and answered separately for each character and trait. For example, in man individual differences in blood types appear to be due almost entirely to genotypic differences; variations in skin or hair colors are contributed to mainly by heredity but also by environment; variations in intelligence have about equal genetic and environmental components; those in temperaments are more environmental than genetic; while the languages that people learn to speak are due to their environments and presumably not at all to their heredities. Since the genotype determines a set of biochemical pathways in growth and aging, it does not act in a vacuum but continually depends on a set of environmental factors for production of a phenotype. It is often possible, for example, to alleviate an hereditary condition (diabetes) when the gene's primary action is found (lowering of insulin) by the environmental change (administration of insulin). Another example is to be found in the alleviation of mental retardation in children with phenylketonuria by regulation of the diet, minimizing phenylalanine ingestion. These environmental "corrections" are analogous to "overcoming a genetic block in a biochemical pathway" (see primary gene action in section **Genetics and development**). It is obvious that with these environmental remedies, the condition is still hereditary because there has been no change in the genotype.

Gene theory. Heredity is particulate, i.e., the genotype is the aggregate of entities called genes, which are to some extent (but not wholly) independent in hereditary transmission, in function, and in capacity to undergo change (mutation). The evidence for the existence of genes was obtained first by Gregor Mendel (1822–1884), who published in 1866 the results of his studies on the heredity in crosses between different varieties in peas. His findings were essentially that the heredities of the varieties crossed do not fuse, or contaminate each other in the hybrid, but instead segregate when the hybrid forms its sex cells (the *first law of Mendel*). Furthermore, the segregating units are recombined and transmitted partly or wholly independently of each other (the *second law of Mendel*). The segregation and recombination take place regardless of whether the traits contributed by the parental varieties are *dominant* (manifesting themselves in the phenotype of the first generation hybrids) or *recessive* (suppressed in the first generation hybrids), or intermediate (the phenotype of the hybrid is a compromise between those of the parents). Furthermore, from his progeny ratios Mendel deduced that the appearance of the recessive phenotype implied a "two dose" condition (receiving each dose from one parent), or *homozygosity* (as it was later called by Bateson); while the dominant phenotype could result from "two doses" (homozygous for the dominant) or from a "single dose" (*heterozygosity*) or, receiving different doses from each parent. Dominance was said to be complete, then, when the single and double dose conditions were alike in phenotype. The dominant and recessive members of the segregating pair ($A = a$) were called *alleles*.

The work of Mendel was not appreciated until it was rediscovered in 1900, independently by Correns in Germany, deVries in Holland, and Tschermak in Austria. Their joint conclusions clearly showed that the idea held until then, that heredity is transmitted by fusable parental "bloods," was wrong. The segregating units of heredity were termed *genes* in 1909 by Johannsen in Denmark. It was gradually shown that gene transmission accounts for heredity of all kinds, including that of quantitative traits which do not show clear segregations in discrete classes (see POLY-GENES). It was also shown that the physical carriers of most genes are the CHROMOSOMES, although some self-reproducing units, PLASMOGENES exist also in the cell cytoplasm. T. H. Morgan (1866–1945) and his collaborators were able to show that certain genes, the inheritance of which was known from crossing experiments, are borne in certain chromosomes visible in microscopic preparations (see CYTOGENETICS). Furthermore, they showed that the genes within a chromosome are arranged in a single linear file, and constructed *genetic maps* (or linkage maps) of chromosomes of a species of fly, *Drosophila melanogaster*, which they selected as material highly favorable for genetic experiments (see DROSOPHILA, LINKAGE). Subsequently genetic chromosome maps, and also *cytological maps*, showing the location of certain genes in the microscopically visible chromosome, were constructed for some other animal and plant species (especially for maize, or Indian corn).

The basic mechanisms of the transmission of heredity, genes and chromosomes, operate with remarkable uniformity in all living organisms. Mendel's laws, although discovered originally in peas and elaborated in Drosophila fly, have early been shown to apply as well to man. Human traits, both differences between persons of normal health and numerous diseases, malformations, and constitutional weaknesses, differences observable in the appearance of persons (skin, eye, and hair colors and the like) as well as physiological and biochemical differences (e.g., blood types and inborn errors of metabolism) are inherited through genes. It was more difficult to demonstrate that the hereditary materials in simplest organisms, such as bacteria and viruses, also consist of genes. The difficulty was that sexual reproduction, and phenomena like parasexuality (cf. FUNGUS GENETICS) and transduction (cf. BACTERIAL GENETICS), have been discovered only recently, thus yielding the possibility of analysis of the heredity by means of what amounts to crosses of strains and varieties, paralleling the classical methods used by Mendel, Morgan, and others.

Individuality. Mendel's laws of segregation and of independent assortment of genes explain the origin of genetic differences between parents and children and between siblings. Suppose that two varieties differing in two genes, *A* (*a*) and *B* (*b*) genes either with genotypes of *AABB* in one variety and *aabb* in the other or with *AAbb* in one and *aaBB* in the other, are crossed. The progeny will be hybrid, *heterozygous*, for these two genes, *AaBb*. Four kinds of sex cells will be produced; *AB*, *aB*, *Ab*, *ab*. Nine genotypes will arise in the second generation of hybrids, four of them *homozygous* —*AABB*, *aaBB*, *AAbb*, *aabb*, and five heterozygous—*AaBB*, *AaBa*, *Aabb*, *AABb* and *aaBb*. If the varieties crossed differ in three genes, the heterozygous hybrid (*AaBbCc*) will form eight kinds of sex cells, and there will appear 27 genotypes in the second generation of hybrids, 8 of them homozygous and 19 heterozygous. And generally, hybrids heterozygous for n genes will be potentially capable to form 2^n kinds of sex cells, and to produce 3^n genotypes in the second and further generations, 2^n of them homozygous.

It is evident that with parents heterozygous for large numbers of genes, the potentially possible genotypes outnumber by far the living individuals in which they could be realized. It is not known for how many genes a person, or an individual of any sexually reproducing species is usually heterozygous, but a conservative estimate will be a hundred or more. Mankind is now approaching a population of 3 billion persons, which must have arisen from somewhat fewer than 6 billion sex cells. This latter number lies between 2^{22} and 2^{23}. The number of possible human genetic endowments is therefore vastly greater than the number of people living or having ever lived. Although children receive their genes from their parents, they never have all the genes which their parents had (since a sex cell contains only one-half of the genes which the parent possessed). The probability of two siblings obtaining exactly the same sets of genes from their parents is therefore negligible (excepting identical twins, who arise from a single fertilized egg cell). More than that, the probability of any two persons anywhere having the same genotype is infinitesimally small (identical twins again excepted). The genetic endowment of any one person is most probably unique, unprecedented in the past and unlikely to reappear in the future.

It must be emphasized that the vast amount of variation implied by these results of genetic recombination is based entirely on the conservative *single pair of alleles* (*A-a*, *B-b*, etc.) at each genic *locus*. Mutation of each gene may produce diverse "states," or *multiple alleles*, of that gene (*A*, *A′*, *a′*, *a* . . .) which may be responsible for still greater levels of variation. Furthermore, it must be kept in mind that phenotypic variation may involve considerable genic interaction; that is to say, genes do not act independently of each other during development (see *epistasis and* HETEROSIS, for example); they can be compared with players in a symphony orchestra, rather than with soloists. The same gene may act differently depending upon what other genes a given genetic endowment contains. Hybrid progenies often contain individual genotypes producing traits which none of the varieties crossed seem to possess. Crossing, hybridization, is therefore a powerful method for creating new varieties. It is utilized on a large scale in animal and plant breeding, to produce new and useful strains (cf. HYBRIDIZATION).

Mutation. Heredity is a conservative force, since it makes the progeny resemble their parents. If heredity were perfect, all organisms would carry the same genotype and evolution would not occur. The conservatism of heredity is, however, opposed by a factor of change, *mutation*, discovered around 1900 by deVries, working on the evening primrose, *Oenothera lamarckiana*. Mutations are sudden changes in hereditary materials. DeVries believed that mutational changes must be drastic as well as sudden, but it has been amply demonstrated that the visible effects of mutations vary all the way from alterations so slight that they are detectable only with the aid of refined statistical methods, to radical upsets which cause death (*lethal* mutations).

All characters or traits of the organisms are subject to mutational change. Mutations alter visible structures, such as size, shape and coloration of the body and its parts, physiological and developmental processes, chemical composition, behavior, adult characteristics and those of immature stages. Many types of mutation are known. Probably the most widespread and important types are changes in the constituent parts of single genes (*gene mutation*). Others, (*chromosomal mutations*) are due to absence (*deficienty*, or *deletion*) or multiplication (*duplication*, or *repeat*) of sections of chromosomes or of whole chromosomes or chromosome sets (cf. POLYPLOIDY). Still other mutations are caused by rearrangements (*inversion* or *transplantation*) of genes within the chromosomes (cf. POSITION EFFECT).

The frequencies of origin of mutations vary within very wide limits, not only in different organisms but for different genes of the same species as well. On the average, specific mutations are rare events. The few existing estimates of mutation rates of human genes are mostly of the order of 1:100,000 (or 10^{-5}, i.e., about one sex cell in 100,000 carries a newly arisen mutation of a given kind). A human sex cell contains, however, many genes—10,000 to 20,000 is a conservative estimate. If many or all of these genes mutate at the rates indicated above, then it would follow that an appreciable proportion of the sex cells, perhaps as high as 10 per cent or higher, carry one or more mutations newly arisen in the individual who produced these sex cells.

Much effort has been devoted to elucidation of the sources of origin of mutations, particularly gene mutations. Most mutations arise "spontaneously," as single genotypically altered individuals among masses of unchanged representatives of a strain living under apparently normal conditions. They differ from their unchanged siblings usually in a single mutated gene. Muller working with Drosophila showed in 1926, and Stadler working with cereals soon thereafter, that the frequency of mutations is enhanced in the progeny of parents treated with X-rays. Many of the mutations thus induced by X-rays resemble those arising spontaneously, although it is still an open question whether the irradiation merely increases the frequency of all kinds of mutation in proportion to their spontaneous frequencies.

Timofeff-Ressovsky, Muller and others found that all kinds of ionizing (or high-energy, or penetrating) radiations, from softest X-rays to gamma rays of radium and presumably to cosmic rays, are mutation-inducing (*mutagenic*). The frequency of gene mutations induced is directly proportional to the amount of radiation applied (as measured in so-called r-units). There was some doubt whether very small amounts of radiation are still mutagenic, but the existing evidence seems to be overwhelmingly in favor of the view that there is no "threshold" or "safe" or "permissible" dose of radiation that would produce no mutations. Nevertheless, only a fraction of the "spontaneous" mutants that arise are produced by the "background" radiation omnipresent in nature; for Drosophila this fraction is below 1 per cent; it is unknown for man but is suspected of being appreciably higher.

Apart from ionizing radiations, certain wavelengths of ultraviolet, but not of visible light, and apparently not the infra-red and longer wavelengths, are mutagenic. High temperature increases, and low temperature decreases, the mutation rates, at least in Drosophila. Some chemical substances are undoubtedly mutagenic; more work in the field of chemical mutagenesis is needed, since, in contrast to radiation, the mutagenicity of chemicals varies from organisms to organism and with the method of treatment applied.

Genetic loads. (cf. POPULATION GENETICS) A great majority of mutations arising in any organism are more or less deleterious, at least in homozygous condition. This may seem surprising since the process of mutation is assumed to be the mainspring of evolutionary change. The harmfulness or usefulness of a genetic change is, however, contingent on the environment in which the organism lives. Consider, for example, mutants which confer resistance to certain drugs or antibiotics on some microorganisms, or insecticide resistance on some insects. These mutations are highly useful to their possessors in environments containing the respective drugs, antibiotics, and insecticides, but apparently detrimental in the environments in which these microorganisms or insects normally live.

Mutations occur however without regard to whether they may be useful where and when they occur. The problem that arises is what happens to the mutants after they enter the populations of plants, animals, or people. Since genes do not fuse but segregate, the frequency of a gene variant which has no influence on the fitness of its carriers tends to be preserved constant indefinitely in large populations. If, however, the gene variant arising by mutation confers upon its carriers a lower, or a higher, fitness in some environment than the parental form has, then natural selection will tend to decrease the frequency of the harmful, or increase the frequency of the useful mutants. But natural selection is not perfectly efficient, and if the mutant does not kill or sterilize its carriers outright, several or many generations may elapse between the origin of the mutant and its final elimination from the population.

We have, thus, two mutually opposed processes—mutation which in every generation supplies a number of harmful genetic variants, and selection which eliminates some of them. The opposing forces will tend eventually to establish a *genetic equilibrium*, at which the numbers of the mutants produced and eliminated per generation will be approximately equal. The equilibrium frequencies of harmful mutant genes will be the greater the more frequently they arise by mutation, and the less are the detrimental effects which these mutants produce in their carriers.

Are the theoretically expected *genetic loads* of harmful mutants actually found in living populations? Indeed they are, in populations of sexually reproducing and outbred organisms (i.e., a majority of higher plants, most animals, and man). Chetverikov showed in 1927 that many individuals taken from natural populations of Drosophila flies, though they are themselves healthy and vigorous, contain a variety of harmful recessive mutants concealed in heterozygous condition. A mating of a female and a male which happen to be both heterozygous for the same recessive mutant (*Aa*, where *a* is the recessive and *A* the normal dominant gene) produces, however, a fourth of the progeny homozygous for the recessive, *aa*, and thus showing the deleterious effects of the mutation. Subsequent work in Drosophila has shown that the genetic loads are heavy indeed. Few if any of the apparently "normal" flies in nature are free of concealed deleterious mutants; one half or more of the individuals are heterozygous for one or more recessive lethal genes (which would kill the animal if they were homozygous), and the rest carry recessive mutants which would cause a loss of vigor, sterility, and various structural and physiological abnormalities.

The genetic loads in human populations are not as well known as in Drosophila, but without doubt a lot of human misery is caused by defective genes, hereditary diseases, malformations, and debilities of various kinds. Dominant detrimental conditions are relatively easy to detect, because they are usually present in at least one of the parents and grandparents of the afflicted person; but the recessives may not be recognized at all as due to heredity, they often "skip" one or several generations, and appear suddenly among the offspring of quite healthy persons.

Detrimental effects of inbreeding; classical and balance theories of population structure. The presence of genetic loads composed of deleterious recessive mutants can be inferred from a fact known for centuries or even millennia, that mating of close relatives (*inbreeding*, incest) in normally outbred species often results in defective or weak progeny. Intercrossing the weakened inbred strains leads, on the contrary, to *hybrid vigor*, or HETEROSIS, in the progeny. Both inbreeding degeneration and heterosis are, however, weak or absent in normally inbred and self-fertilized organisms (as, for example, in wheat and some other cereals, in which seeds normally arise by union of female and male generative elements produced on the same individual or in the same flower). All this is expected if many or most individuals of normally outbred forms are heterozygous for different recessive mutants. The chances that two individuals who mate will carry the same concealed recessives are greater if these individuals are close relatives than if they are taken at random from a population.

Since all or a great majority of individuals in normally outbred species (including man) carry a part of the population's genetic load of concealed detrimental recessives, the question naturally arises, what an individual would be like who would be free of this load? Would he be some superman or a Drosophila superfly? The *classical theory* of population structure assumes that there is one state for each gene which is normal and beneficial, and one or more mutant states all of which are more or less detrimental. If so, the ideal condition of a population would be genetic uniformity and homozygosis for the normal states of all genes. It is however, possible that in normally outbred species, high fitness is a product of heterozygosity for many genes (such as *AaBbCc* . . . etc.), all of which would be detrimental when homozygous (i.e., *AABBCC* and *aabbcc* would be low in vigor). According to the *balance theory* of population structure, the optimum states of a population require presence of genetic variety, or *polymorphism*, in many genes. It is possible that the population structure in some forms of life is closer to that envisaged by the classical theory, in others by the balance theory, and in still others intermediate. This is one of the many unsettled problems of modern genetics.

Nevertheless the total extent of genetic polymorphism in natural populations including mankind is of considerable interest because it is of such a vast magnitude that any attempt to account for its maintenance in terms of mutation and selection equilibria has so far been met with difficulty. Recent studies in Drosophila (Lewontin and Hubby in 1966) and in mankind (Harris, 1966; Livingstone, 1967) in which large samples systematically random for genic loci with isoallelic (isozyme) or hemoglobin variants have estimated that

about 40% of loci may be polymorphic showing two or more alleles and that the average individual has about 12% heterozygosity if all his genic loci are considered. This is a conservative estimate for Drosophila, since all the biases in the estimate conspire to lower it compared with the probable true value. We can only say that our genetic reservoir is indeed vast, but we cannot yet say that any agreement on the mechanisms which provide this immense supply of genic variation is forthcoming.

Genetics and development. Between the genotype found in a fertilized egg and the characteristics of the adult body there is interposed a long and complex chain of developmental processes. These processes, which lead to a realization of the potentialities of a given genotype in certain environments, are included in the fields of DEVELOPMENTAL GENETICS and BIOCHEMICAL GENETICS. Early strides in fundamental concepts for these fields were made in the 1930's and 1940's especially by Beadle, Ephrussi, Tatum, and Lederberg; though the first authentic statement of primary gene action was that of Garrod, a British physician studying the genetic basis of alcaptonuria, in 1909, when he proposed that the recessive homozygote condition was due to a mutant gene's production of a defective enzyme in a biochemical pathway with the normal allele being responsible for normal enzyme.

The chemical machinery of development can be visualized as a network of chemical reactions, many of which are facilitated by specific catalysts or enzymes. When mutants of the bread mold (Neurospora), or of a species of bacteria, are compared with the original type, it is often discovered that the changes stem principally from a blockage of some one chemical step in the reaction network, which may then have ramifying consequences in other reaction chains. The block may be supposed to be due to absence of a facilitating enzyme, and in some instances this has actually been demonstrated. By further extrapolation, it was surmised that each enzyme is produced by just one gene, and that each gene produces just one enzyme.

Such simplification was revised during the 1950's when genic loci were shown to be complex, or containing substructure, the mutant sites "within a gene" representing the allelic states which as a *set of subloci linked* in the region known as the genic *locus* came to be called a *cistron* by Benzer. The cistron concept was a necessary antecedent to working out the mechanisms of primary gene action and development. Briefly it refers to the functional genic unit (the Morgan-Mendel locus) with responsibility for production of an enzyme in a biochemical pathway (the Garrod-Beadle concept) which includes in linear order two or more mutant sites, recombinable by crossing-over. All such mutant sites are most likely to produce "mutant phenotype" (non-complementing) in heterozygotes because they produce "defective," or mutant, enzyme. (See also *pseudoalleles* of Lewis).

With perfection of methods for analyzing proteins chemically especially by electrophoresis ("fingerprinting"), Pauling and, following, Ingram, were able to work out the sequence of amino acids in such complex proteins as human hemoglobin. The sequence of amino acids in the genetic variant sicklecell hemoglobin, which produces sever anemia in the homozygote, was shown to differ from normal hemoglobin in the position #6 where, out of 146 amino acids in the beta chain, a glutamic acid in normal is replaced by a valine in sickle-cell hemoglobin. In another hemoglobin variant (*C*), the same position is occupied by lysine. These amino acid substitutions result in their respective hemoglobin molecules' bearing a net positive charge, accounting for the electrophoretic differences in these three polypeptides. All three variants are Mendelian alleles and are detectable in heterozygotes.

The next major step was to show that these amino acid substitutions in the protein gene product were the direct result of the colinearity of cistron mutant sites in the chromosome. Yanofsky was able to establish that the mutants of the tryptophane synthetase locus of *Escherichia coli* occur in the same order and in the same relative positions as the amino acid sequence of that enzyme's polypeptide primary structure. Consequently the one gene-one enzyme hypothesis has been modified to be more nearly accurate by the concept of one cistron-one polypeptide.

Finally the gene's specificity, or "genetic information," for the amino acid sequence of the polypeptide, its primary product, has been found to reside in the remarkable class of substances, the NUCLEIC ACIDS. The chromosomes are composed of *deoxyribonucleic acid* (DNA) and protein (histones and residual high molecular weight proteins). Watson and Crick proposed in 1954 that the molecular structure of DNA was composed of a double spiral helical chain, the links of which are longitudinally deoxyribose plus phosphoric acid while the crosslinkages transversely binding the two chains together are two purines plus two pyrimidines (adenine, guanine, cytosine, and thymine, or A, G, C, and T respectively). Their finding that the relative proportions of these four substances are such that A = T and C = G, indicating that they must be paired in crosslinkages with hydrogen bonding holding the two helices together. Countless words and sentences can be represented by means of different sequences of our 26-letter alphabet, which can of course also be symbolized by only two signs, the dot and dash of the Morse code; similarly, the differences between countless genes may be specified by sequences of the four genetic "letters," the nucleotides in pairs: A-T, C-G.

Finally the vehicle for "transcribing" this information from DNA was found to be RNA (ribonucleic acid, or specifically messenger, mRNA), a single-stranded and short-lived molecule which is synthesized presumably in colinearity with one strand of the DNA molecule complementing its nucleotides sequentially. From the cell nucleus the mRNA migrates to the cytoplasm where in the neighborhood of ribosomal RNA (rRNA), the small transfer RNA molecules (or soluble, sRNA) are matched again in complementary fashion to the nucleotides of mRNA. Each sRNA molecule in its active site is a double helix with a specific trinucleotide unpaired (for matching to its complementary mRNA triplet) at one end while an amino acid is attached off a free terminal C-C-A chain at the opposite end. According to present theory the sRNA's trinucleotide is specifically associated with a particular amino acid at opposite ends of the same molecule. As the trinucleotide of sRNA fits the mRNA specification at one end, its amino acid is unbound from it at the other to join a chain of amino acids in peptide linkage in construction of a polypeptide. Thus the sRNA is often known as the adapter, or more pertinently, the "translator" for the process of putting genetic information

from the "nucleic acid language of DNA" to the "peptide language" of the protein product. Each mutant site in the complex genic locus, therefore, may be responsible for the linear arrangement of amino acids in sequence in the polypeptide product. The genetic *code* worked out especially by Crick and Nirenberg was found to consist of triplet nucleotides in the mRNA.

This concept of primary gene action is very largely useful, and it constitutes the function of *structural genes*. Needless to say, however, the complexities of gene direction to development cannot all be accounted for only in terms of structural genes. There must be controlling mechanisms which inform the structural gene when to produce its polypeptide during the course of development. At the chromosomal level there have been demonstrated *regulator* genes and *operator* genes (Jacob and Monod in 1966). The former may respond to extracellular substrates in producing products which specifically act on *operator* loci that may control the production of mRNA or "reading" of the message at one end of a structural gene. Such mechanisms in general may provide the feedback necessary for "turning a gene off and on." Further levels of interaction between gene products (enzymes) and their possible role in regulation of gene function (nucleocytoplasmic interactions) are beyond the scope of this article. The vast problems of cellular differentiation and determination of both nuclei and cell types all indicate a far more complex series of gene-cytoplasm-environmental interactions than was ever imagined two decades ago. Finally we know many phenotypic effects which result not from simple genic action but only from whole chromosome (or large linkage group) action, namely in *sex determination* of organisms with heterologous chromosomal mechanisms (such as the *XX* female and *XY* male both in man and in Drosophila), and also as seen in chromosomal anomalies such as mongolism (Down's syndrome) in man with the addition of an extra chromosome (aneuploidy). In short, gene action, both in terms of single loci and in multiple sets of linked loci or independently assorting sets of loci, is still far from described with any completeness, though progress in this field is now so rapid that an optimistic view of the future seems justified.

THEODOSIUS DOBZHANSKY
(revised by Eliot B. Spiess)

References

Braun, W., "Bacterial Genetics," 2nd ed. Philadelphia, Saunders, 1965.

Dobzhansky, Th., "Genetics and the Origin of Species," 3rd ed. New York, Columbia Univ. Press, 1951.

ibid., "Mankind Evolving," New Haven, Yale Univ. Press, 1962.

Dunn, L. C., "A Short History of Genetics," New York, McGraw-Hill, 1965.

Fincham, J. R. S., "Genetic Complementation," 2nd ed. New York, Benjamin, 1966.

Hadorn, E., "Developmental Genetics and Lethal Factors," New York, Wiley, 1961.

Roman, H. L., ed., "Annual Review of Genetics," Vol. 1, Palo Alto, Calif., Annual Reviews, 1967.

Sinnott, E. W., *et al.*, "Principles of Genetics," 5th ed. New York, Macmillan, 1958.

Srb, A., *et al.*, "General Genetics," 2nd ed. San Francisco, Freeman, 1967.

Stern, C., "Principles of Human Genetics," 2nd ed., San Francisco, Freeman, 1960.

GENTIANALES (CONTORTAE)

The order Gentianales traditionally includes the following families of DICOTYLEDONOUS flowering plants: Apocynaceae (Dogbane), Asclepiadaceae (Milkweed), Gentianaceae, Loganiaceae, Menyanthaceae (Buckbean), and Oleaceae (Olive). The small obscure genus *Desfontainia* which had been unsuccessfully assigned to various families, usually Loganiaceae, is now regarded as a monogeneric family. Whether it should be retained in this order is doubtful since its 5-loculed ovary contrasts markedly with the almost exclusively bicarpellate condition of the other families. Although it is generally conceded that the remaining six families are related, their great diversity of form as well as paucity of unifying characters has led some authors to treat them as two or three smaller orders. However, there has been a singular lack of agreement among proposed realignments except that the Apocynaceae and Asclepiadaceae, which formerly comprised a single family, are consistently assigned to the same order.

The alternate descriptive name Contortae, referring to the curious convolute and twisted arrangement of the unopened corolla, is not really definitive of the order since neither the Loganiaceae, Menyanthaceae or Oleaceae conform whereas some members of the closely related order Polemoniales have this type of aestivation. The most consistent feature of Gentianales is the bicarpellary superior ovary, but again this is a feature shared by most Polemoniales. In fact, there are no reliable criteria for separating these two orders as they are presently constituted. Presumed differences appearing in most taxonomic keys are overemphasized. For example, leaf arrangement in Gentianales is by no means exclusively opposite nor is this type of phyllotaxy rare in Polemoniales. Incidentally, with the exception of Menyanthaceae and *Fraxinus* (ash) in the Oleaceae, leaves in Gentianales are simple and usually entire.

Both the anatomy and floral morphology indicate a moderately high phylogenetic position for the order. Vessel members are of the advanced type with simple perforations. Anomalous or specialized features include vestured pits, laticiferous tubes (Apocynaceae and Aslepiadaceae), and both internal and included phloem (Apocynaceae, Asclepiadaceae, Gentianaceae, and part of the Loganiaceae). The flowers show a decided tendency for fusion of parts and for the production of various types of appendages and ornate glands. Thus, corollas are typically sympetalous with alternate epipetalous stamens, although in a few genera, e.g. *Fraxinus* and *Swertia* (Gentianaceae), the petals are almost distinct and the stamens are free. Rarely are flowers obscurely zygomorphic, and only in some species of *Fraxinus* do apetalous or unisexual flowers occur.

Floral morphology in Asclepiadaceae deserves special mention because of its complex organization and the intricate pollinating mechanism, which correspond surprisingly close to the structure of orchid flowers. In both groups the pistil and stamens are united into a central column, the gynostemium. In Asclepiadaceae fusion actually involves only the stylar region, the enclosing staminal tube being free of the two ovaries which, as in the Apocynaceae, are not joined. The whorl of five bract-like anthers comprising the gynostegium cover the enlarged stigmatic head except for the alternate stigmatic grooves. The most con-

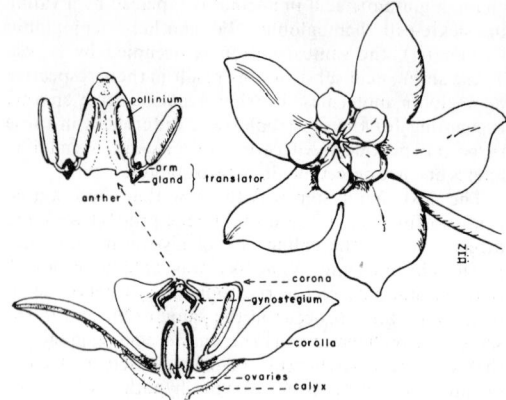

Fig. 1   Wax-flower, *Hoya* sp. (Asclepiadaceae).

spicuous and ornate feature of the flower, at least in the more highly evolved forms, is the corona, a term which is commonly used collectively in this family for both staminal and corolline appendages. In many cases (see Fig. 1) it would be difficult to make a more precise distinction due to the fusion of corolla, appendages and gynostemium into a complex unit. The unique feature of the pollinating mechanism is a small non-cellular device, the translator, which lies between anthers and collects the contents of the adjacent pollen sacs. The larger subfamily, Cyanchoideae, is characterized by a wishbone-shaped translator each arm of which is attached to a pollinium, consisting of the agglutinated mass of pollen from a single pollen sac. When the translator hooks onto an insect's leg the pollinia are withdrawn from the pollen sacs and are transferred intact to the stigmatic surface of another flower. The Periplocoideae differ slightly in that the pollen, which adheres merely in tetrads, is shed from adjacent pollen sacs into a single spoon- or horn-shaped translator which attaches to the insect by means of a basal adhesive disk as in orchids.

Gentianales are more numerous and widely distributed than is generally realized. Whereas members of the Gentianaceae and its segregate Menyanthaceae are familiar herbs of middle and higher latitudes of both hemispheres, the other families are predominantly tropical or subtropical where they consist of lianas, shrubs and small trees. Although the woody habit prevails in these groups, only a few genera, such as *Fraxinus* and *Aspidosperma* (Apocynaceae) attain sufficient size for lumber. Among the diverse forms of Asclepiadaceae the most distinctive is found in the carrionflowers of South African deserts which closely resemble xerophytic euphorbs or cacti. Epiphytes also occur in this family, and a few Gentianaceae are non-green saprophytes. Menyanthaceae are a small family of aquatic or marsh plants.

The Oleaceae is the only family of appreciable commercial importance. For centuries olive has been utilized for its oil as well as a food. The hardness and fine grain of ash wood makes it especially valued as handles for implements. Several common ornamentals such as lilac, privet, forsythia, jasmine, etc. belong to this family. The other families are represented in gardens or as house plants by gentians, oleander, periwinkle, wax plant, carrion-flowers, buddleja, and others. During World War II many of the latex producing plants (Apocynaceae and Asclepidaceae) were investigated as possible sources of rubber, but only one species, *Cryptostegia grandiflora*, has proved to be of commercial value. Many species are known to have been used as home remedies for stomach ailments, fevers, nervous disorders, and as heart stimulants. However, only very recently with the rediscovery of the remarkable effects of reserpine from *Rauvolfia* in reducing hypertension has this attribute begun to be extensively exploited. Related alkaloids derived from latex of other members of the Apocynaceae also promise to be effective in treating mental diseases. Another alkaloid, strychnine from a member of the Loganiaceae has been utilized for a long time both as a poison and in small doses as a drug. Primitive peoples recognized the poisonous effect of extracts from many plants of this group, using them as fish poisons and as a coating for arrow-heads. When these plants occur in pastures they are hazardous to livestock.

DOUGLAS M. POST

Reference

Lawrence, G. H. M., "Taxonomy of Vascular Plants," New York, Macmillan, 1951.

GERANIALES

Trees, shrubs, woody vines or herbs. Leaves simple or compound, alternate, opposite or whorled, stipulate or exstipulate. Flowers solitary, or in racemes, cymes or mixed inflorescences; mostly actinomorphic, with perianth and stamens hypogynous, mostly 5- or 4-merous, complete and bisexual or variously reduced even to unisexual naked flowers; sepals free, or united below, sometimes 0; petals usually free, sometimes 0; stamens commonly twice as many as the petals and in 2 whorls, or the outer whorl missing, anthers usually longitudinally dehiscent; hypogynous disk often present; pistil usually of 2-5 usually at least basally fused carpels with axile placentation, each carpel unilocular; ovules usually 1-2 per carpel, pendulous and with the raphe ventral when the micropyle directed distally, or erect and raphe dorsal when micropyle directed proximally.

Distinguished from the SAPINDALES which have ovules, when pendulous, with raphe dorsal and micropyle directed proximally, or when erect raphe ventral and micropyle directed distally. The Geraniales and Sapindales have stamens twice as many as the sepals and in 2 whorls, or in 1 whorl and opposite the sepals, while the Rhamnales have the stamens the same number as the sepals and alternating with them.

Oxalidaceae: Oxalis Family

Genera 7, spp. ca. 1,000; mostly pantropic, decreasing into the temperate zone; Oxalis the principal genus.

Herbs, with oxalic acid or oxalates. Leaves exstipulate, usually palmately compound. Flowers bisexual, actinomorphic, pentamerous: stamens 10. Fruit a capsule.

Various species of Oxalis often cultivated as ornamentals.

Geraniaceae: Geranium Family

Genera 11, spp. ca. 850; of temperate and subtropical regions; principal genera: Geranium, Pelargonium and Erodium.

Herbs usually. Leaves stipulate. Flowers bisexual, actinomorphic, usually pentamerous. Fruit a regma or schizocarp, styles often elongate and persistent on the segments of the fruit.

Pelargonium and Geranium often cultivated for flowers and essential oils.

Tropaeolaceae: Nasturtium Family

Tropaeolum, the only genus, of ca. 80 spp.; South American, mainly Andean, a few Mexican.

Semi-succulent herbs with watery acrid sap. Roots often tuberous. Leaves usually simple, exstipulate, blades commonly peltate. Flowers zygomorphic, bisexual; sepals petaloid, the uppermost spurred; petals clawed; stamens 8. Fruit a schizocarp.

Tropaeolum majus and other spp. cult. for flowers.

Linaceae: Flax Family

Genera 9, spp. 200; cosmopolitan—temperate; Linum, flax, the principal genus.

Herbs. Leaves stipulate or exstipulate entire, simple. Flowers actinomorphic, bisexual; calyx persistent in fruit; petals fugacious, usually clawed. Fruit a capsule or drupe.

Linum usitatissimum, common flax, yields flax fiber and linseed oil.

Erythroxylaceae: Coca Family

Genera 3, spp. ca. 200; mainly of the American tropics, a few African; Erythroxylum the principal genus.

Trees and shrubs with prismatic crystals in the parenchyma. Leaves simple, stipulate. Flowers actinomorphic, bisexual. Fruit a berry or drupe of the one developed carpel.

The alkaloid cocaine derived from the leaves of 2 or 3 Andean spp., especially E. coca.

Zygophyllaceae: Caltrops Family

Genera 27, 200 spp., mainly pantropic, especially Mediterranean region, Africa and Australia; principal genera: Zygophyllum, Larrea and Tribulus.

Shrubs, herbs or trees; often xerophytic. Leaves evenpinnate, stipulate, opposite. Flowers usually actinomorphic, bisexual. Fruit a capsule, schizocarp, or drupe.

Lignum vitae, a very hard heavy wood used for bearings and other wear-resistant parts is from *Guaiacum officinale* and *G. sanctum;* an antioxidant used for retarding rancidification of edible fats is extracted from *Larrea divaricata.*

Cneoraceae: Cneorum Family

Including only Cneorum, 3 spp. of the Mediterranean region and the Canary Islands.

Shrubs, somewhat resinous. Leaves alternate, simple, exstipulate. Flowers actinomorphic, bisexual, 3- or 4-merous throughout. Fruit a drupe with separable segments.

Rutaceae: Rue Family

Genera 140; spp. 1,300; widespread in tropical and temperate regions especially South Africa and Australia; important genera: Citrus, Diosma, Xanthoxylum.

Trees, shrubs and herbs with glands containing essential oils especially in leaves and fruit. Leaves exstipulate, simple or palmately or pinnately compound. Flowers bisexual, actinomorphic. Fruit usually a soft-rinded berry.

Citrus fruits are borne by the genus Citrus.

Simaroubaceae: Quassia Family

Genera 32, spp. 200; mostly pantropical, a few in temperate regions; important genera: Simaba, Ailanthus, Picramnia.

Trees or shrubs usually with a bitter principle in bark and wood; devoid of clear oil glands. Leaves pinnately compound; exstipulate. Flowers usually unisexual and dioecious, actinomorphic. Fruit a capsule or schizocarp, or cluster of samaras.

Various bitters used as tonics derived from the family.

Burseraceae: Bursera or Torchwood Family

Genera 20, spp. 500–600; tropics and hot deserts especially of tropical America and northeastern Africa; principal genera: Bursera, Commiphora, Canarinum and Protium.

Large trees to shrubs, with essential oils and aromatic resins. Leaves deciduous, exstipulate, pinnate or decompound. Flowers unisexual or bisexual, actinomorphic. Fruit 1–5 seeded, berry-like or a capsule.

Frankincense from *Boswellia carteri,* and myrrh from Commiphora spp.

Meliaceae: Mahogany Family

Genera 50, spp. ca. 800; pantropical; principal genera: Cedrela, Trichilia and Guarea.

Trees or shrubs. Leaves usually once or more pinnate, exstipulate. Flowers mostly bisexual, actinomorphic, pentamerous; filaments usually united into a tube. Fruit a berry, capsule or rarely a drupe.

Mahogany lumber from *Swietenia mahagoni,* and African mahogany from *Khaya senegalensis* of tropical Africa; West Indian cedar, for cigar boxes, from *Cedrela odorate.* The Chinaberry tree, *Melia azedarach,* is often cultivated in warm parts of the U.S.

Akaniaceae: Akania Family

One monotypic genus of eastern Australia.

Trees. Leaves alternate, odd pinnate. Flowers bisexual, actinomorphic, 5-merous except stamens usually 8. Fruit a capsule.

Malpighiaceae: Malpighia Family

Genera 60, spp. 850; mostly of American tropics and subtropics; principal genera: Banisteria, Malpighia and Hiraea.

Mostly woody vines. Hairs parallel to the surface on which borne, attached by their middles. Leaves simple, stipulate. Flowers bisexual, actinomorphic; sepals with external nectaries; petals conspicuously clawed. Fruit a samara, schizocarp, capsule, berry or drupe.

Banisteria caapi yields an hallucinant.

Trigoniaceae: Trigonia Family

Genera 3, spp. ca. 30; tropical America.

Trees or climbing shrubs. Leaves simple, stipulate. Flowers bisexual, obliquely zygomorphic; largest petal often saccate at base; filaments united. Fruit a capsule.

Vochysiaceae: Vochysia Family

Genera 5, spp. ca. 100; tropical America and 1 West Africa; principal genus: Vochysia.

Trees, shrubs or woody vines with resinous sap. Leaves simple; stipules small or 0. Flowers bisexual, obliquely zygomorphic; the largest sepal often saccate at base; petals mostly 1–3; 1 fertile stamen, with staminodia, filaments free. Fruit a samara or capsule.

Tremandraceae: Tetratheca Family

Genera 3, spp. ca. 24; Australia.

Shrubs or subshrubs. Leaves simple, exstipulate. Flowers bisexual, actinomorphic; anthers opening by an apical pore. Fruit a capsule.

Polygalaceae: Milkwort Family

Genera 10, spp. 700; mostly tropical to warm temperate; both hemispheres; principal genera: Polygala, Muraltia and Bredemeyera.

Shrubs, herbs or woody vines. Leaves simple, exstipulate. Flowers superficially simulating pea flowers: bisexual, zygomorphic, sepals 5, 2 often petaloid; petals 3; stamens usually 8, filaments fused below and adnate with the petals, anthers opening by apical pores. Fruit usually a capsule.

The dried root of *Polygala senega* is the official drug Radix Senegae.

Dichapetalaceae: Dichapetalum Family

Genera 4, spp. ca. 250; tropical, Dichapetalum, the principal genus, mainly of Africa and Madagascar, a few in tropical South America; the other three genera mostly tropical American.

Small trees, shrubs, or often woody vines. Leaves simple, stipulate. Flowers often borne on the petioles, mostly bisexual, actinomorphic or zygomorphic; fruit a drupe.

Often notably poisonous.

Euphorbiaceae: Spurge Family

Genera ca. 280, spp. ca. 7,000, most abundant in tropics and subtropics, but also occurring in temperate regions; principal genera: Acalypha, Euphorbia, Croton and Phyllanthus.

Trees, shrubs, herbs and vines; laticiferous cells or vessels present in many genera and widely distributed in the tissues. Stems often cactoid. Leaves usually simple, often stipulate. Flowers unisexual, usually actinomorphic; sepals sometimes absent; petals often absent; stamens 1 to indefinitely numerous; carpels mostly 3. Fruit a capsule. Seeds often carunculate, with abundant oily endosperm.

Economic products: Rubber from *Hevea brasiliensis*, other spp. of Hevea and *Manihot glaziovii*; starch (cassava and tapioca) from *Manihot esculenta*; tung oil from Aleurites especially *A. fordii*; castor oil from *Ricinus communis*; croton oil from *Croton tiglium*. Various used medicinally in addition to the two preceding: *Euphorbia hirta, Croton eleuteria* (cascarilla bark), *Euphorbia resinifera* (euphorbium). Many members poisonous: Phytotoxins known in Ricinus, Jatropha, Manihot; blistering and/or skin dissolving substances in *Hippomane mancinella, Excoecaria agalloch* and *Euphorbia marginata;* dangerous allergens common, e.g., Ricinus.

Many cultivated as ornamentals, e.g., *Euphorbia pulcherrima,* the poinsettia; *Euphorbia marginata,* snow-on-the-mountain; *Codiaeum variegatum* "croton;" *Acalypha wilkesiana,* chenille plant; and various cactoid African species of Euphorbia.

Daphniphyllaceae: Daphniphyllum Family

Including only Daphniphyllum with ca. 30 spp. of Eastern Asia and adjacent islands.

Small trees or shrubs; with clusters of calcium oxalate crystals in all organs. Leaves simple, exstipulate. Flowers dioecious, apetalous, often asepalous, pistillate with 5–10 staminodia, carpels 2; fruit a 1-seeded drupe.

Callitrichaceae: Water Starwort Family

Including only Callitriche, a cosmopolitan genus of ca. 30 spp., of terrestrial or more often aquatic, annual or perennial herbs.

Stems slender, delicate. Leaves, simple, exstipulate. Flowers monoecious, actinomorphic, naked, consisting of either a single stamen or a single pistil. Fruit a schizocarp.

LOUIS C. WHEELER

Reference

Engler, A., and K. Prantl, "Die natürlichen Pflanzenfamilien," 2nd ed. vol. 19a, 1931, Englemann, Leipzig. 19bI, 1940, Englemann, Leipzig. (Reprinted 1960, Duncker und Homboldt, Berlin). 19c, 1931, Englemann, Leipzig.

GERMINATION

Definition. The term *germination* signifies the sequence of processes by which an omnipotent body that is produced by the plant for the specific function of reproduction and dispersal (as part of a "dispersal unit") commences growth, after becoming cytoplasmically detached from the plant which produced it.

Germination leads to the eventual growth of new organisms (gametophyte, sporophyte, haploid or diploid thallus, mycelium or plasmodium) from (1) individual cells of unicellular or multicellular dispersal units, generically known as SPORES, (2) embryos in the SEEDS of the SPERMATOPHYTA, whether formed by gametic fusion or without it (e.g. by APOMIXIS or asexual polyembryony), (3) dispersal units comprising of tissues which were formed as vegetative outgrowths (e.g. bulbils, gemmae). Stricter definitions exclude the latter category.

Processes and requirements of germination. The growth which follows germination is considered as the only reliable indication that germination had occurred. Indeed, a controversy still exists as to whether germination is more than the first unmeasurable phase of growth.

Information is scant regarding the nature of the processes which intervene between the start of germination and the growth manifested at its termination. Nevertheless, preparations for subsequent growth are doubtlessly concerned in germination. Some of these, involving the reversal of arrangements which the dispersal unit had undergone preparatory to its dispersal, are fairly obvious. Dispersal generally takes place through environments unsuitable for growth. Preparations for dispersal therefore involve reduction of actual growth, provision for the requirements of subsequent growth and protection of the living contents of the dispersal unit in transit through the unsuitable environment. With few exceptions (e.g. vivipary) dispersal units neither grow beyond a specific stage of development, nor do they germinate, while attached to the plant which bore them, though they may appear as complete in anatomy and morphology then as they are when they eventually do germinate. The correlative mechanism by which growth is thus halted still requires elucidation. When germination occurs promptly upon release, it is likely that a chemical or physical inhibition to growth has been exerted by the bearer-plant. However, the dispersal units generally undergo partial desiccation while still plant-borne, i.e., while still inhibited. Dehydrated dispersal units never germinate unless rehydrated. Removal from the inhibitory influence of the bearer-plant and hydration are therefore prerequisites for germination. Other environmental conditions are required for germination, because they are essential for the operation of those processes which are preparatory for growth: an optimal supply of oxygen, and a temperature suited to the requirements of the various partial processes. A supply of building materials, available energy, and the ENZYMES and growthregulators (see HORMONE, PLANT) needed for the organized synthesis of cellular components, are also required for the same purpose. These supplies are either provided in the dispersal unit in a more or less inactive form, and have to be activated, mobilized and distributed, or they have to be synthesized *de novo* before growth can occur. Finally, before they can start growing, the cell(s) or tissue(s) which take part in germination have to release themselves from their enclosing protective structures, by means of internal pressure or chemical action.

Regulation of germination. It is a common phenomenon that germination may not occur even though the external conditions which are known to be suitable for subsequent growth are supplied. In such cases, there are at least four possibilities: (1) some internal requirement for germination (energy source, building material, enzyme or hormone) is lacking, (2) some external condition (e.g. moisture, oxygen) is denied access to its site of action, (3) an essential biochemical or hormonal system is blocked by an internal inhibitor, (4) germination is inhibited by mechanical restriction of growth which follows it.

The first possibility may explain the requirement for soluble carbohydrate for the germination of microspores in Spermatophyta, or the requirement for mycorhizal association for the germination of seeds in ORCHIDACEAE. Examples illustrating the other three possibilities follow.

These inhibitions to germination are usually removed under some specific combination of conditions, mostly environmental. Germination is thus regulated both internally and by the environment. Germination has been most extensively studied in the seeds of the Spermatophyta, and the following discussion will be confined to these plants.

The regulation of germination in seeds. Strictly speaking, germination in seeds is a function of the embryo. The latter, however, is nearly always part of a more or less complex dispersal unit, which may consist of several additional structures that enclose and accompany the embryo (e.g. testa, pericarp, perianth, floral bracts and receptacle). Some of these structures are clearly related to some mechanism which controls the "dispersal in space." Others are likewise involved in the regulation of germination. Hence, in speaking of germination the entire dispersal unit has to be considered.

Several well-defined, though not always as well-understood, mechanisms which regulate germination have been recognized. These operate (1) by control of water entry, (2) by control of gas exchange, (3) by mechanical restriction of embryo growth, (4) by reacting to specific temperature conditions, (5) by chemical control, (6) by reacting to specific conditions of illumination.

Control of water entry occurs when one of the structures enclosing the embryo is water-impermeable, and thus effectively prevents germination unless it is structurally modified (by microorganisms, weathering, etc.), or unless it has an aperture which opens up to admit water under some specific environmental condition (e.g. by hygroscopic movement).

Control of gas exchange is likewise due to selective permeability of embryo envelopes, and brings about a requirement for higher oxygen tension for germination. The selective permeability of such structures may be modified by the environment (temperature, possibly also light) or completely removed by weathering, or the activity of micro-organisms.

Lignified or horny enveloping structures may be responsible for restricting embryo expansion. This barrier to germination may also be overcome through weathering or the activity of microorganisms.

The means by which temperature exerts control over germination are not clear. When germination occurs within the entire range of temperatures favorable to subsequent growth, the regualtion is straightforward. However, the response of germination to temperature often takes forms not falling into this cateogry: (1) The range of temperatures in which germination occurs may be a lot narrower, and may or may not fall within the range favorable to growth. (2) Germination may occur by exposure to "seasonal thermoperiod" (i.e. one or more long exposures to near-freezing temperatures, alternating with long exposures to higher temperatures). This is usually the case with temperate-zone plants, the buds of which usually undergo a seasonal dormancy that is also overcome by a similar seasonal thermoperiodicity. The fact that the plant growth-substance gibberellin may substitute for this requirement may offer a lead to the elucidation to this mechanism. (3) Germination may be promoted by exposure to a "diurnal thermoperiod" (diurnally alternating temperatures). Again there is no explanation to the elucidation to this mechanism. However, the importance of diurnal thermoperiodicity as a factor in growth is increasingly recognized. Also, the presence in plants of an autonomous rhythm of a most basic physiological nature is suspected. These may tie up with the mechanism of germination response to diurnal thermoperiodism.

Chemical control of germination operates when some part of the dispersal unit contains non-toxic chemicals which are inhibitory to some partial process essential to germination (or growth). Germination is thus inhibited until the part of the dispersal unit which contains the inhibitor is shed, until the inhibitor becomes innocuous by chemical modification, or until its concentration is sufficiently reduced by evaporation (if volatile) or leaching (if water-soluble). Although many of these inhibitors have already been identified, their mode and site of action are yet unknown.

The control of germination by illumination is achieved through inhibition or promotion by specific conditions of illumination. The response is usually strongly modified, sometimes even reversed, by temperature. Decoating the embryo may also strongly modify the response. Most of the effects of light on germination are additional manifestations of the universal phenomenon of the morphogenetic effect of light, similar to etiolation and photoperiodism. This possibility is supported not only by the existence of *short-day seeds* and *long-day seeds*, but also by the striking resemblance between some of the partial processes in the various phenomena. This applies particularly to the *dark process* and the *low light intensity process*: A short illumination with low light intensity cancels the effects of darkness in the promotion or inhibition of the physiological process. The similarity is carried farther by the action spectrum of the *low light intensity process*: Low intensities of red light (6700 Å) inhibit the "dark process." Low intensities of far-red light (7300 Å), if applied immediately after the red light, will cancel its effects completely, and this effect may in turn be cancelled by a re-application of the red light, etc.

The responses to light are doubtlessly due to the existence of a phytochrome in the embryonic tissues. In many seeds the capacity for dark-germination is associated with presence of active phytochrome (P_{FR}) in the dry seed. Increase in light-requirement as temperature becomes more supraoptimal is attributable to thermal inactivation of P_{FR}. Suppression of light-requirement by incubation at near-freezing temperatures is attributable to fixation of P_{FR}, or of its products, in a thermo-stable form. The metabolic pathway which starts with P_{FR} and ends in germination may involve biosynthesis of gibberellin. Exogenous gibberellin often satisfies requirement for light, and inhibitors of gibberellin biosynthesis may, under certain conditions, inhibit P_{FR}-induced germination. This inhibition is reversed by exogenous gibberellin.

The various germination-regulating mechanisms may occur singly or in combinations. In the latter case they may either be independent or interdependent in operation.

Regulation of germination is sometimes modified with age, inasmuch as controls which are evident at time of dispersal (or harvest) are gradually relaxed. This phenomenon is known as *after–ripening*.

Induced dormancy. When only part of the environmental requirements for germination are supplied, so that germination does not take place, a state of dormancy is often eventually induced: the seed remains alive, but becomes incapable of germinating even when the missing requirements are supplied. This induced dormancy may be overcome by the supply of some new specific environmental condition not hitherto required in the non-dormant dispersal unit. This dormancy is thus induced by the active formation of new germination-regulating mechanisms.

The biological significance of regulated germination. The regulation of germination has some far-reaching implications for the survival of the species. Thus, inhibitors in the pulp of fleshy FRUITS effectively prevent germination within the fruit. The presence of water-impermeable coats ensures that only a fraction of any seed crop becomes permeable, imbibe and germinate at a time. The *dispersal in time* which is thus achieved is analogous to the *dispersal in space* achieved by means of the dispersal mechanisms, and provides insurance against annihilation of the entire generation by chance adverse environmental conditions (drought, disease, fire, etc.).

Water soluble inhibitors are often found in dispersal units of arid-zone plants. These have been shown to serve as rain-gages for germination, since the amount and duration of rainfall required to leach away the inhibitors will usually moisten the soil sufficiently for seedling establishment.

A requirement for specific photoperiods is a selective mechanism which restricts germination to a season suitable for the start of the life cycle. Similarly, a requirement for a specific temperature, or a seasonal or diurnal temperature fluctuation, may also be a selective mechanism which restricts germination to climates, habitats and soil depths most suitable for starting the life cycle. It is thus obvious that a requirement for seasonal thermoperiodicity in seeds of temperate-zone plants serves to minimize the danger of seedling mortality through freezing, by permitting germination to occur only after winter.

In summary, the regulation of germination increases the survival potential of the species, because, either through dispersal in time or by restricting germination of the species to the suitable niche in the environment, it provides insurance against the uncertainties of the environment. Since they endow the species with increased survival potential, the germination regulating

mechanisms are hereditary properties of high selective value in evolution. To the farmer, on the other hand, they have always been a nuisance and they have consequently had low selective value in cultivation. This is doubtlessly the reason why most cultivated species can rarely become established without continuous reseeding, while most wild species can seldom be eradicated by any single killing.

DOV KOLLER

References

Amen, R. D., "A Model of Seed Dormancy," Bot. Rev. **34**: 1–31, 1968.
Crocker, W., and L. V. Barton, "Physiology of Seeds," New York, Ronald, 1953.
Koller, D., "The Regulation of Germination in Seeds," Bull. Res. Council, Israel. **5D**: 85–108, 1955.
Koller, D., *et al.*, "Seed Germination," Ann. Rev. Pl. Physiol. **13**: 437–464, 1962.
Mayer, A. M., and A. Poljakoff, "Germination of Seeds," New York, Pergamon, 1963.
Toole, E. M., *et al.* "Physiology of Seed Germination," Ann. Rev. Plant Physiol. **7**: 299–319, 1956.

GESNER, ("GESSNER," "GESNERUS"), KONRAD VON. (1516–1565)

One of the most important "universal scholars" of the Renaissance, best known to biologists for his monumental *Historia Animalium* (4 vols. folio, 1551–1558) and his *Catalogus plantarium* (1542). His father, a Swiss furrier, fell in the battle of Kappel (1531) but the genius of young Gesner was already so apparent that he had little difficulty in securing patrons to further his education. At the age of 21, being then most interested in languages, he assumed the chair of Greek at the University of Lausanne. Four years later he became Professor of Natural History at Zurich, having in the interval studied medicine at Montpellier and taken his doctorate in this field at Basel. In the next 24 years he wrote 20 important books, mostly in Latin and Greek, though he was equally fluent in Arabic and Hebrew. His catalogue of plants is in Latin, Greek, German and French. He also compiled, in Latin, Greek and Hebrew, a catalogue *Bibliotheka Universalis* (1545) of every writer and every writing then of record and in 1555 published a philological study of 130 languages known to him. His love of nature extended to scenic beauty and he is as well known to athletes as the father of mountaineering as he is to biologists as the father of zoology. He was ennobled in 1564 and died of the plague in 1565, having refused to abandon his patients when the disease struck Zurich.

PETER GRAY

GESTATION

In the evolutionary history of the animal kingdom there has been a tendency to limit the number of EGGS shed by females with a concomitant increase in parental care of the offspring. The latter, called *nurture*, has prenatal or uterine, and postnatal or lactating phases. In viviparous MAMMALS, *gestation*, with implantation of embryonic tissue into the maternal uterine endometrium, represents the highest degree of mother-fetal association. Although no intermixing of maternal and fetal blood occurs, the outermost fetal membrane, the *chorion*, comes into close contact with maternal tissue for exchange of nutriments, respiratory gases, and wastes. Prior to implantation, uterine secretions called embryotroph sustain the developing fetus.

Maintenance of gestation, or *pregnancy*, is dependent upon the HORMONES of the ovaries, placenta, and pituitary gland. The *luteinizing hormone* from the pituitary is responsible for ovulation of the mature egg, and transformation of the ovarian follicle into a *corpus luteum*. This structure secretes a steroid hormone, *progesterone*, which causes development of a secretory type of uterine lining which is necessary for implantation of the fetus. The estrogens, also steroidal, are produced by the ovary and play an important role in the actual process of implantation as well as maintenance of the pregnancy. Continued function of the corpus luteum requires the *luteotrophic hormone* of the pituitary, as well as other gonadotropins. In humans the highly developed fetal placenta produces a luteotrophin, *chorionic gonadotrophin*, which very early in pregnancy replaces the need for a pituitary luteotrophic complex. The presence of this hormone makes possible the pregnancy test for humans. While chorionic gonadotrophin is unique to primates, some type of luteotrophin is produced by the placenta in most animals, particularly in the later stages of preganacy. In some animals such as the goat, cow or rabbit, presence of the corpus luteum for the entire length of gestation is essential, while in the guinea pig, cat and human among others, ovariectomy can safely be performed during the last half of pregnancy. Thus there appears to be an evolution of placental independence beginning with the ability to produce a luteotrophin which replaces the need for a pituitary source. In the highest forms, the placenta has acquired the ability to produce the ovarian steroids as well, and achieves complete autonomy.

Termination of pregnancy, or *abortion*, can occur spontaneously or can be induced. Surgical abortion has been practiced for many years and is now used in various parts of the world as a method of population control. There is current widespread use of *intrauterine contraceptive devices* which appear to prevent implantation, but not fertilization of the egg, therefore acting as abortifacients. The complexities of the endocrine interrelationships in gestation account for many natural abortions, and interference with these delicate balances by pharmacologic methods is currently under investigation as a possible means of birth control.

Gestation normally ends with *parturition*, which consists of three phases: relaxation of the entire birth canal; muscular contractions of the uterus and abdomen, with expulsion of the fetus; and finally, delivery of fetal membranes and placenta. The physiologic initiation of the process is unknown but certainly involves both nervous and endocrine systems. A hormone from the posterior pituitary, *oxytocin*, which in conjunction with estrogen brings about contractions of the pregnant uterus, is released when active *labor* begins. This compound is used clinically to aid delivery.

The *gestation period* is a genetically determined character. Although minor variations of a few days are common, the time between conception and parturition is species, and to some extent strain, specific. In some animals, such as the American Marten or the iactating mouse, implantation can be delayed, which causes a wider variation in length of pregnancy even though the actual period of development is always the same. Generally, smaller animals with large litters have shorter gestation periods. This is also true of animals which live in protection of burrows or caves and those that give birth to very immature young. The breeding season (see Oestrous cycle) appears to be correlated with the length of gestation, births occurring at a time of optimum food availability. The horse, for example, with an eleven month period, breeds in spring so as to have a foal the following spring. In the case of the raccoon breeding takes place in later winter and birth occurs 65 days later, again in spring.

Average gestation periods in days for some common animals with variations given in parentheses are: alpaca 240; ant-eater 190; armadillo 150; domestic ass 365 (15); baboon 186; American badger 183; brown bat 55 (5); bear 208 (20); American bison 270; camel, dromedary 318 (20); domestic cat 57; cow 283 (10); cheetah 92; chimpanzee 245 (10); chinchilla 115 (5); coyote 65; deer 225(20); dog 63; dolphin 276; elephant, African 640 (40); elephant, India 630 (100); elk 240; ermine 65 (10); ferret 42; fox 55 (5); gibbon 210; giraffe 430 (10); goat 150 (30); guinea-pig 63; hamster, common 21; hamster, golden 16; hippopotamus 240 (12); horse 335 (15); human 280 (25); hyena 91; jaguar 100; kangaroo 39; leopard 95; lion 108 (4); lynx monkey, rhesus 165 (15); American marten 210–280; mink 42; mouse 20; orangutan 218; pig 115; porcupine 112; porpoise 183; rabbit 31; raccoon 65; rat 21; reindeer 220 (15); rhinoceros 530 (20); sable 250; sea lion 342; seal 245; sheep 150 (6); shrew 18; tapir 390; tiger 106; walrus 330; weasel 35; whale, blue 305; whale, sperm 365; wolf 63; zebra 365.

DONALD C. JOHNSON

Reference

Asdell, S. A., "Patterns of Mammalian Reproduction," Ithaca, N.Y., Cornell Univ. Press, 1964.

GINKGOALES

Ginkgoales, an order of the GYMNOSPERMS, contain only one living species, *Ginkgo biloba* Linn., the maidenhair tree, native to eastern China and now widely planted as a shade tree. The tree is deciduous and with the habit of a conifer. The twigs are long shoots bearing many spur branches. The long shoots have a small pith and cortex and a very large volume of wood while the spur branches have a relatively thick pith and large cortex. The long shoots continue the growth at the end and bear alternate leaves. The spur branches bear a cluster of 3–5 leaves terminally. The leaves are fan-shaped, generally notched in the center and with open dichotomous venation.

The species is dioecious. The microsporangiate strobilus is a loose pendulous catkin-like structure of many spirally arranged microsporophylls on a stalk-like portion. The ovuliferous strobilus consists of a stalk bearing terminally two erect naked ovules. The ovule is subtended basally by a collar-like rim, a structure generally interpreted as a reduced and vestigial sporophyll.

The microgametophyte is endosporic, and when shed consists of two prothallial cells, a generative cell and a tube cell. Fertilization is effected by wind. In the nucellus of the ovule, the grain produces a haustorial pollen tube and the generative cell divides forming the stalk cell and the body cell. The latter produces two large multiflagellated sperms.

The development of the megagametophyte begins at the time of pollination of the ovule. The functional megaspore enlarges, accompanied by an extended development of free nuclear divisions followed by the centripetal development of cell walls. There are two or three archegonia situated at the micropylar end. The mature archegonium consists of four neck cells and a large egg.

Fertilization and embryogeny may occur either on the tree or after the ovule has fallen to the ground. After a series of about eight free nuclear divisions of the zygote, centripetal wall formation begins and the young embryo becomes cellular. The lower end develops into the shoot apex and cotyledons and the cells immediately behind this portion the primary root. There are two or occasionally three cotyledons. Growth of the embryo continues through the winter and germination occurs in late spring without any apparent dormancy.

Fossil remains, mostly leaves, of *Ginkgo* and allied genera such as *Baiera, Czekanowskia, Ginkgoites,* etc. occur abundantly in rocks of the Triassic and Jurassic Periods, during which time, these plants were apparently world-wide in distribution. Following the period of its vigorous development in the Mesozoic Era, *Ginkgo* declined progressively in its distribution until it is confined only to some parts of China.

In reproductive structures, there are many points of similarities between *Ginkgo* and the living cycads. In habit and many features of the sporophyte, *Ginkgo* resembles closely the conifers. Ginkgoales, linked by a series of many fossil forms, are generally regarded as to be closely related to the extinct order Cordaitales and Palaeozoic Coniferales. They may have developed from a separate but at present unknown group in the seed ferns.

H. L. LI

GLUMALES

Of all plants grasses are the most important to man. All our breadstuffs, wheat, corn, oats, rye, barley and rice, and sugarcane as well, are grasses. Bamboos are grasses and so are Kentucky bluegrass and creeping bent of our lawns, and timothy and redtop of our meadows. If plants so unlike as timothy, corn, bluegrass and bamboos are all grasses what is it that characterizes a grass? It is the structure of the plant.

All grasses have stems (culms) with solid joints (nodes) and 2-ranked leaves, one at each joint on opposite sides of the stem. The leaves consist of two parts, the sheath, which fits around the culm like a tube, usually split on one side, and the blade, mostly long and narrow. No other plant family has just this structure.

Clover and alfalfa, built on a very different plan, are not grasses. The seedheads of grasses are still more distinctive. The minute flowers are borne on tiny branchlets, often several crowded together, always 2-ranked as are the leaves.

Being wind-pollinated grass flowers need no fragrance, no gay colors, nor honey, to attract insects to pollinate them. The flower consists of a single pistil with one ovule, two styles, each with a feathery stigma, and three (rarely 1 or 6) stamens. (Fig. 1). Only 2, rarely 3, delicate little scales (lodicules) remain of the floral envelope, the calyx and corolla, of other flowers. The usually minute grass flowers are borne singly or 2 to many together in spikelets, which are really little flowering branchlets.

The axis of the spikelet, the rachilla, is jointed as is the culm (stem) of a grass, and the lemmas (specialized leaves reduced to blade-like sheaths) are 2-ranked as are the leaves.

Annuals, wheat, barley, rye, oats, rice, corn and many other grasses flower every year. But some perennial grasses, which spread by specialized underground stems (rhizomes or rootstocks), may cover extensive areas, especially in brackish marshes, without flowering regularly. Bamboos flower mostly at intervals of few to many years.

The root, stem and leaves constitute the vegetative parts of the plant. These are more uniform and characteristic in grasses than in most other plant families. From the stem and leaves of a plant one can readily decide whether or not it is a grass. The only plants that may reasonably be mistaken for grasses are the sedges. In these the culms, commonly 3-sided, are solid, not jointed, and the leaves always 3-ranked.

Specialization in the grasses takes place mostly in the spikelets. By its vegetative characters a given plant is shown to be a grass, but it is the spikelets and their arrangement, which indicate the kind of grass it is. The spikelets of a common brome (*Bromus commutatus* Schrad.) are shown (Fig. 2), the 2 glumes at base, the florets, lemma, palea, and enclosed flower together, borne on opposite sides of the jointed rachilla, the enclosed flower concealed. The palea with 2 nerves, its back to the rachilla, subtends and usually surrounds the

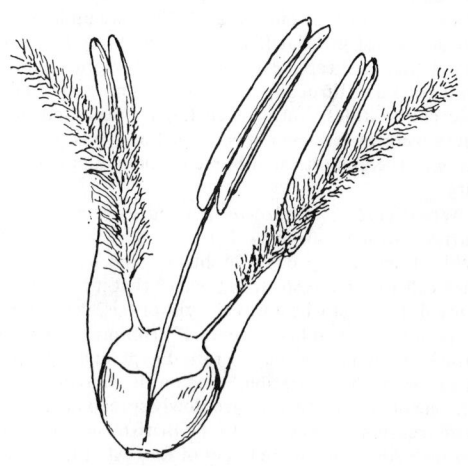

Fig. 1   Grass flower.

Fig. 2   Spikelet. *Bromus commutatus* Schrad.

flower. The glumes bear no flowers and are without paleas. This simple fundamental floral structure is subject to all manner of modifications, but every organ found in the most highly specialized spikelet is to be interpreted as an elaboration or reduction of some part of this structure. The floret is the unit of the spikelet; the spikelet is the unit of the inflorescence.

These true grasses seem to have appeared on the earth during Upper Cretaceous time, as their earliest fossil representatives have been found in formations laid down in this period. In the Eocene there was a notable expansion of the grass family, and in the Miocene it was well on its way to becoming one of the dominant types of plant life. Little Eohippus, of the Eocene, the ancestor of all the horses, and his descendents in the Oligocene, which have left their fossil remains in our Western States, had teeth for eating twigs and bark. During the Miocene our Great Plains were uplifted and became a vast grassland. The little browsing horse, no larger than a sheep, developed teeth for grazing, and, on a grass diet through many generations, increased in size and swiftness until, when the Ice Age appeared, there were at least ten species of the genus, some as large as the domesticated horse of today and one even larger.

Wheat (*Triticum aestivum* L.) and its related genus, barley (*Hordeum vulgare* L.) have been the basis of civilization. At the dawn of history the beginning of such cultivation was so far in the past that it had become a myth. In Egypt wheat was the gift of Isis, in Greece of Ceres. Our breakfast "cereals" commemorate the Greek myth to this day. From Egypt and adjoining Asia, cradle of civilization based on the cultivation and grazing of grasses, this culture slowly spread in all directions, reaching from China to the British Isles and down through Abyssinia to the tribes of East Africa, a culture based upon the economic foundation of grain fields and herds of grazing animals.

None of the cultivated races of wheat are known in the wild state. A wild form of emmer (*Triticum dicoccum* Schrank) was discovered in 1906 on Mount Hermon in Palestine, and later in Moab, by Aaron Aaronsohn, and named *Triticum dicoccoides* Koern. by Kornicke, the renowned student of Old World grains. In 1910 it was found again in western Persia, in the Zagros Mountains. It seems fairly certain that this form is the ancestor of cultivated wheat. In the related emmer the axis of the head breaks up, the grain remaining enclosed in the chaff. In cultivated wheat, *Triticum aestivum* L. the axis does not break up and the grain is readily freed from the chaff in threshing. This character must have been developed and fixed by selection, yet so long ago was it accomplished that the wheat found in the earliest known graves in Egypt is free from the chaff. Breasted, the historian, states that the stomachs of mummies in these graves contain the chaff of barley, which, adhering to the grain, was present in barley-bread, but no chaff of wheat has been found.

Barley was also cultivated in the New Stone Age, for it is found in Egyptian pottery jars dating from 4000 B.C. and also in the remains of Swiss lake dwellings.

Wheat reached China long before the Christian era, but rice is the more widely cultivated grain in eastern Asia. Rice (*Oryza sativa* L.) was developed in the dim past, being cultivated before 3000 B.C. In an ancient Chinese ceremony five kinds of seed were planted, rice by the emperor himself, the other four by princes of his family. Of the five, esteemed as the greatest gifts to man, four are grasses, rice, wheat, sorghum, and millet. The fifth is a legume, soya or soy bean, a common food of the Chinese today.

Rye (*Secale cereale* L.) came into cultivation far later than wheat, barley and rice. It seems to have originated in a region further north, somewhere in the steppes of eastern Europe or Asia, and, unlike the other grains, it will run wild and persist under favorable conditions.

The common oat, *Avena sativa* L., was known to the ancient Greeks as a weed in grain fields, and it seems to have been cultivated in middle Europe during the Bronze Age.

Indian corn or Maize, *Zea mays* L. the most highly specialized grass in the world, and the one with the greatest number of uses, originated in America where its cultivation formed a second center of civilization. Like wheat its cultivation began so far back in antiquity that its origin is veiled in myth. To the American Indian, maize was a gift of the gods. One of the legends is familiar to us—the one which relates how Hiawatha prayed that the lives of his people might not depend on hunting and fishing and how, in answer to his prayer, came Mondamin, with whom Hiawatha wrestled mightily, whom he slew and buried, and from whose grave, carefully tended according to Mondamin's instructions, sprang maize, a never failing food for the people.

While Eurasia had wheat, barley, rice and other grains America had but one. When white men arrived maize was cultivated from Central America south to Peru and north to Quebec. The Inca, Maya, Aztec and Pueblo civilizations were based upon it, and it was cultivated by the North American Indians over much of what is now the United States. The hungry Pilgrim Fathers, we are told, found a buried hoard of Indian corn during their first terrible winter in the New World and thankfully appropriated it. The Indians taught the Pilgrims how to plant maize, or corn, as it was called by the English settlers, fertilizing it by burying two fish in each hill.

Maize has never been found growing wild, and it is singularly unadapted to maintaining itself without cultivation. There are wild species of grass related to each of the Old World grains, from which the cultivated forms have probably been derived, but maize (*Zea mays* L.) is the only known species of its genus. The genus most nearly related to it is *Euchlaena* Schrad. to which belongs teosinte, a native annual of Mexico, occasionally cultivated for forage.

Maize is the most highly specialized grass in the world and it was the American Indian who, by selection through thousands of years before the coming of the white man, produced this marvel of plant-breeding.

The Indians of the Great Lakes region had, besides maize, the grain of an annual aquatic grass, *Zizania aquatica* L. or wild rice. Down to today Indian women have gathered the grain, first by going about in canoes and tying together the heads of as many of the plants as could be gathered together in the arms. These tied heads were left to ripen, when the women returned and, holding the tied heads over the canoes, beat out the grain. From two to three thousand bushels a year have been gathered in this way. Today Indian rice is an expensive dainty, served with game on the table of the epicure. In China the young shoots of a perennial species of *Zizania* are used as a pot herb.

All the grains, to which man owes his civilization, are annual grasses, that is, the plant bears one crop of seed and dies. Perennial plants live over the winter or the dry season by means of underground parts that remain alive but dormant. Perennial plants, perpetuated vegetatively by rhizomes or bulbs bear fewer and smaller seeds than do annuals which depend upon seed for survival. Primitive women, gathering seeds for the food supply, naturally took the larger and more plentiful seeds of annuals. These being short-lived, produce seed within a few months after planting, while perennials seldom bear seed the first year.

Grasses were the basis of civilization in prehistoric time, and they are largely the basis of advanced civilization today. The United States census reports show that the grass family, maize, hay, wheat, barley, oats, and rye together, not counting sugarcane, rice, millets, crops of grass seed, and forage, accounts for more wealth than all other crops, cotton, tobacco, fruits and the rest. Agricultural statistics do not give the value of pasture, but it must reach an enormous figure, though pasture includes clovers and alfalfa, which are not grasses. A large part of the value of dairy products and of beef and mutton must be credited to grasses. The proportional value of the grass family in agriculture is about the same throughout the world, rice and sugarcane being the most important in tropical regions.

Besides our daily bread, wheat bread or corn pone, knackbrod or bannocks, swartzbrod or macaroni, rice or cakes of millet or sorghum, the grains furnish other important food products. Maize, the one native American grain, is a host in itself, giving us delicious sweet corn, pop corn, corn flakes, cornstarch, hominy, glucose, corn syrup and a palatable oil besides. This oil, "Mazola" is obtained from the germ in the kernel of corn, a bushel of corn yielding about a pound of oil. As a by-product the germ (the oil extracted) yields a rubber substitute, the red-rubber now in common use as erasers, rings for fruit jars, sponges, soap dishes, and bath mats. Dextrin obtained from cornstarch has replaced gum arabic as the basis of mucilage. According to Slosson "more than a hundred different commercial products are made from corn, not counting cob pipes." Cornstalks, formerly a waste product of huge proportions, are now coming into use as a source of cellulose and promise to be of especial value in the production of paper and of wall board. A corncob stone, called maizolith, has been developed. It can be worked and polished and used for such pruposes as are now supplied by hard rubber and bakelite.

Corn is also a source of alcohol. As Slosson further says "This was one of the earliest misuses to which corn was put, and before the war put a stop to it 34,000,000 bushels went to the making of whiskey in the United States every year, not counting the moonshiners' output. . . .The output of alcohol, denatured for industrial purposes is more than three times what it was before the war."

Rye and barley are used extensively in making fermented and distilled beverages, and in the Orient a wine is made from rice. Much of commercial vinegar is made from malt liquor, the alcohol being converted into acetic acid (the acid of vinegar) by means of ferments.

The juice extracted from the stems of sugarcane is concentrated until the sugar (sucrose) crystallizes and can be separated from the molasses. Formerly the sugar was only partly extracted from the juice and the molasses, still rich in sugar, was an important by-product. The bagasse, or crushed cane, from which the juice has been extracted, is now being used in the manufacture of wall boards.

Besides supplying us with our daily bread, the grasses, by providing a large part of the forage of grazing animals, indirectly supply us with dairy products, beef and mutton, wool, leather, and horsepower. And, since hogs and poultry are fed largely on corn, ham and eggs are also secondary products of the grasses.

The range today is the modern equivalent of the grasslands of our remote nomad ancestors. It is unfenced public land upon which the cattle or sheep of several stockmen graze in common, the cattle being separated by a yearly roundup according to their brands, the calves being branded with the mark borne by the cows they claim as mothers. Our Western States were once immensely rich in good range land, but the best of the land has now been settled and brought under cultivation. But, even so, the acreage upon which livestock is grazed exceeds that under cultivation.

The range lands lie almost entirely west of the 100th meridian and comprise the vast semiarid region, with a rainfall of less than twenty inches. This land, covered with the hardy buffalo grass and grama grasses, the wheat grasses, bromes, porcupine grasses, and numerous other native species, affords excellent grazing.

Until the end of the last century the Federal Government allowed stockmen uncontrolled use of the public domain. As a result rolling hills knee-deep in grass were reduced to bare knobs, deeply gullied, their fine soil eroded and blown over the land in blinding dust storms. Vast natural pastures of grama grass were despoiled of the palatable and valuable forage and given over to weeds or left denuded and subject to erosion. When land is overstocked the animals devour the good forage so completely that no plants are allowed to seed and the spiny and woody plants, avoided by the cattle, are all that are left to replenish the range.

When we read of the wars of the Hebrews and the neighboring tribes in the light of the history of our western range lands we realize that overgrazing changed the Promised Land of plenty to a land of want. One of the great achievements of the United States Department of Agriculture has been the study of grazing problems, and the working out of a system of controlled use of grazing lands. Permits are issued to stockmen limiting the stock to the number which the range can bear without injury, and so timed as to permit the plants to set seed. The wars and invasions of ancient nomads were due to their ignorance of range management, as are many people today.

The largest of the grasses are the bamboos, all perennials with woody, usually hollow, culms and often with extensively creeping rhizomes. Only one genus, *Arundinaria,* of two species, is native to the United States. *Arundinaria gigantea* (Walt.) Muhl. grows to 8 meters tall, forming extensive canebrakes in low woods, river banks and moist ground from southern Missouri to Ohio and south to Texas and Florida.

A second species, *A. tecta* (Walt.) Muhl., usually not more than 2 meters tall, forms colonies in swampy woods and sandy margins of streams on the Coastal Plain from southern Maryland to Alabama and Mississippi. Both species furnish forage for cattle and hogs which are turned into the canebrakes in spring.

Several species of bamboos are cultivated for ornament in parks and gardens in the United States, particularly from China, Japan, India and Java, mostly in gardens of Florida and California.

Sedges (Cyperaceae) differ from grasses in having mostly solid, often 3-angled, stems (culms) and closed sheaths. The inflorescence consists of 1–3 flowered spikelets borne in heads, spikes or umbels, rarely solitary. The flowers have 1–3 stamens and a sessile or stipitate 1-celled ovary; fruit a 1-celled, lens-shaped achene. A very large family divided into 5 tribes: 1. Cariceae, 2. Scirpeae, 3. Rhyncosporeae, 4. Cypereae, 5. Sclerieae. The largest genus, Carex, includes some 533 species in North America. (K. K. Mackenzie, North American Flora 18: 504 pp. 1935). The sedges are abundant throughout the United States, mostly in moist or wet ground.

The papyrus of Egypt, the source of early papermaking, is a sedge, *Cyperus papyrus* L., the great bulrush, which grew abundantly along the Nile and the other rivers. It was this bulrush of which the mother of the baby Moses, "when she could no longer hide him" made "for him an ark of bulrushes and daubed it with slime and pitch and put the child therein . . .and laid it in the flags by the river's brink," where Pharaoh's daughter saw it and rescued the child—to be the great lawgiver. It was the papyrus of which the earliest paper was made, the pith of the culms being cut in thin strips vertically, the slices placed side by side, water added, then beaten with a wooden instrument until smooth, then pressed and dried in the sun. The ancient records of Egypt were written on such paper.

Cyperus papyrus is cultivated in parks in California, and perhaps elsewhere. It forms a tall handsome clump with great tassels of slender drooping bronze spikes.

AGNES CHASE

References

Arber, Agnes, "The Gramineae, a Study of Cereal, Bamboo and Grass," Cambridge, The University Press, 1934.

Barnard, C., ed., "Grasses and Grasslands," New York, St. Martin's, 1964.

Chase, Agnes, "First Book of Grasses," 3rd ed. Washington, Smithsonian Institution, 1964.

Elias, M. K., "Tertiary Prairie Grasses and Other Herbs from the High Plains," Geological Society of America. Special Papers No. 41. Published by the Society, 1942.

U. S. Dept. Agriculture, "Grass, the Yearbook of Agriculture 1949," Washington, U.S. Gov. Printing Office, 1950.

Hitchcock, A. S., "Manual of Grasses of the United States," 2nd ed. revised by Agnes Chase, Washington, U. S. Gov. Printing Office, 1950.

Sampson, A. W., "Range and Pasture Management, "Principles and Practice," New York, Wiley, 1923.

Sears, Paul B., "Deserts on the March," Norman, Oklahoma, University of Oklahoma Press, 1935.

Weaver, H. L. and E. W. Albertson, "Grasslands of the Great Plains, their nature and use," Lincoln, Nebraska, Johnsen, 1954.

GNETALES

Gnetales, the most advanced order of the Gymnosperms, comprise only three genera. 1. *Ephedra,* with about 30 species, mostly low, profusely branched shrubs or shrublets possessing scaly, opposite leaves, occurs in warm desert-steppe areas of western North and South America, the Mediterranean region, and temperate and subtropical Asia. 2. *Welwitschia,* having only one species, *W. bainesii* Carr. (syn. *W. mirabilis* Hook. f.), is a most curious plant with a thick turnip-shaped rhizomatous stem and a pair of huge, opposite leaves which persist and grow slowly throughout the life of the plant and can reach several feet in length. It is confined to desert regions of southwest Africa. 3. *Gnetum,* with about 35 species, mostly woody climbers except for a few species which are shrubs or small trees, all bear opposite, reticulate-veined leaves. Its range is humid tropical regions of the world: the lower Amazon, west Africa and Malesia. The general appearance and habit of these three genera are so different as not to suggest any relationship.

These three genera were once grouped together in a single family Gnetaceae, based essentially on the following features which are unique among the gymnosperms. (1) True vessels are present in the secondary xylem. (2) The ovules are enclosed within two or three envelopes, only the inner envelope representing the integument, the upper portion of which is prolonged into a long and narrow tubular structure known as micropylar tube. (The outer and middle envelopes are probably the fused bracts and bracteoles.)

Because of the heterogeneous nature of the order, several taxonomic treatments have been proposed. The group has been split into two families (Ephedraceae and Gnetaceae, with *Welwitschia* and *Gnetum* remaining in the latter), three families, or even three separate orders, each thus comprising a monogeneric family.

The origin of the Gnetales and their relationship to other Gymnosperms, fossil and living, remains an unsolved phylogenetic puzzle. Dubious *Ephedra-* and *Welwitschia-*like fossil pollens were reported in Permian deposits. Only in recent years, however, pollens with gnetalian affinities, those possessing a varying number of ridges and occasionally a furrow as in *Welwitschia* from Cretaceous sediments in Nigeria and Venezuela, and those *Ephedra* pollens from Cretaceous formations in Long Island, New York, have been accurately described. Several authors emphasize the resemblances between *Ephedra* and Cordaitales (a group of fossil Gymnosperms related to the Coniferales), while others point out the similarities between *Gnetum* and Bennettitales (another group of fossil Gymnosperms related to the Cycadales).

The Gnetales, especially *Gnetum,* also possess some strong angiospermous features—habit, vessels, the development of male and female gametophytes, and so forth. But the fact that there are resemblances to Angiosperms does not mean that the Gnetales are evolving toward Angiosperms or that they have given rise to any Angiosperms. The Gnetales are fundamentally different from the Angiosperms in these two characters: the former bear *naked* ovules in strobili; and pollination in them is an ovular, rather than carpellary, function (no true style or stigma being present). It can only be stated that the Gnetales have evolved parallel to and nearly as far as the Angiosperms.

HSUAN KENG

References

Chamberlain, C. J., "Gymnosperms, structure and evolution," Chicago, Chicago Univ. Press, 1935.
Eames, A. J., "Relationships of the Ephedrales," Phytomorph. **2:** 79–100, 1952.
Maheshwari, P., and V. Vasil, "Gnetum" (Bot. Monog. no. 1), New Delhi: Council of Sci. and Ind. Res. 1961.
Pearson, H.H.W., "Gnetales," Cambridge, The University Press, 1929.
Steeves, M. W., and E. S. Barghoorn, "The pollen of Ephedra," J. Arnold Arb. **40:**221–259, 1959.

GNOTOBIOTES

The gnotobiote is an animal or plant associated with a known microbiota. It is an experimental tool designed to bring new precision to any biological, microbiological, or medical investigation in which the interaction of host and microbial associates could be an important variable. The number of associated microbial species can vary from none (in the case of germfree or axenic animals) to as many as can be consistently identified as present through use of available microbiological techniques. The germfree animal provides baseline information about the animal as such, its potentialities and its limitations. The animal harboring defined microbial associates provides information about the interaction of host and microbes and of microbes with microbes, under simplified and controlled conditions.

Animal gnotobiotes have been or are presently being used (1) to study problems in immunology ("natural" resistance; roles of the thymus, spleen, and lymph nodes; effects of antigen-free diet; effects of antibiotics); oncology (spontaneous tumors in germfree rats; vertically transmitted leukemia virus in "germfree" mice; virus etiology of tumors; possible tumor vaccines); gerontology (germfree rats and mice outliving controls; germfree males outliving females); tooth decay (necessity of bacteria for decay; vaccine against caries); nutrition (determination of exact requirements; unexpected role of intestinal flora in cholesterol and mineral metabolism); radiobiology (pure radiation sickness without infection; protective effect of slow mucosal turnover in germfree animals; protective effect of endotoxin); virology, parasitology, and bacteriology (unexpected antagonisms and collaborations among viruses, bacteria, and endoparasites; variations in the disease process according to the combination present); space exploration (hazards of flora change in men or animals confined in space capsule, hazards first brought to light by early studies with gnotobiotes).

The word *gnotobiote* itself (the "g" is silent) is derived from the Greek prefix *gnotos* (known) and the ecological term *biota* which is used to summarize the flora and fauna of a region. With such a derivation it could be applied to an isolated region or to a pure culture of bacteria, but the word has been limited by its users as well as its proponents (2, 3) to an animal (or plant) associated with a known microbiota. In theory, any plant or animal could be called a gnotobiote if we could specify, at a moment in time, all the various forms of life with which it is associated. However, present methods for identifying microbial species are neither instantaneous nor exhaustive. Only while an animal harbors a simple microbiota and is protected from invasions of new bacterial species is it capable of having its microbiota defined. In practice, therefore, the gnotobiote has resulted, not from a breakthrough in bacteriological identification, or from the development of decontamination methods, but from a breakthrough in isolation procedures for higher organisms, and from the discovery that animals and plants pass the earliest stages of their lives as gnotobiotes. What gnotobiotic technology does is to maintain throughout life, by means of artificial barriers, the gnotobiotic condition of the unborn mammal, of the unhatched bird, reptile, fish, or insect, and of the unsprouted plant seed.

It had been hoped that all the important species of laboratory animals would prove to be germfree before birth, rather than merely limited to a few associates. This hope was first shaken by the discovery that dogs may be infested with worms before birth; however, it proved possible to prevent this uterine infestation. More recently, all the strains of mice used for germfree research were found to harbor latent virus while still *in utero* (4). The virus could be activated to produce leukemia in these otherwise germfree mice by exposing the mice to repeated small doses of X-irradiation. Efforts are underway to eliminate the viral associate(s) so that mice which are strictly germfree as well as gnotobiotic may be made available for special types of experiment. Similar screening of germfree rats has revealed no evidence of latent viral associates. Other species of gnotobiote have not yet been so exhaustively tested. Whatever the prenatal condition of the animal, it has proved possible to maintain this condition indefinitely by use of metal, plastic, or glass enclosures and by developing techniques which allow passage of air, food, water, and supplies into the system without accompanying microbes. Pure cultures of microbes may be taken into the system to add to the existing microbiota and thus produce a gnotobiote of the type desired for experiment.

Almost as soon as the pure culture of bacteria became possible, attempts were made to rear some higher organisms in "pure culture." As early as 1885, the unsuccessful attempt of Duclaux to grow germfree peas prompted a statement by Louis Pasteur (5) that he would have liked to try the rearing of germfree animals on the preconceived notion that under these conditions life would become impossible. Yet by 1895, Nuttal and Thierfelder (6) had shown that rearing of guinea pigs without microbes was technically possible and biologically promising, even though their animals did not grow during the period of several weeks in which they were kept germfree. Sporadic attempts to rear germfree chickens, goats, frogs, and more guinea pigs continued through the first quarter of the century, with considerable improvement in the quality of the animals as better diets and techniques were developed. By the early 1920's simpler animals, worms and crustaceans, had even been reared through successive generations (7).

It was only at the close of the 1920's, however, that laboratories in three different countries, the United States, Sweden, and Japan, began to approach the germfree animal not as a scientific curiosity but as an experimental tool whose development could bring new precision to a wide range of biological and microbiological problems. In particular, the Lobund Laboratory, founded in 1930 by J. A. Reyniers at the University of Notre Dame (Notre Dame, Indiana, U.S.A.), has carried on since that time a continuous and expanding program centered on the germfree animal. By the 1950's technical and nutritional problems had been sufficiently solved to permit rearing of germfree rats and mice through successive generations as a colony operation (8). At about the same time, widespread experimental use of gnotobiotic animals was further facilitated by the development of plastic isolation barriers and chemical germicides (9) to replace the expensive and cumbersome steel barriers and steam sterilization used until then. This development encouraged commercial production of germfree rats and mice and thus made both equipment and animals available to many types of laboratory. Plastic equipment made easier the rearing of larger gnotobiotes, such as dogs, cats, swine, sheep, goats, calves, and burros. The availability of gnotobiotes also made possible an important new development in animal care: the "disease-free," "pathogen-free," or "specific-pathogen-free" animal colony (8). This is derived from gnotobiotes and maintained within stringent but not impervious barriers. The animal so reared is a more dependable and reproducible laboratory animal, much less subject to intercurrent infection than the conventional animal.

Besides this expansion of gnotobiotic research in the United States, an expansion occurred in Europe centering on the laboratory of B. E. Gustafsson (10) in Stockholm, Sweden. In Japan, a number of different laboratories used gnotobiotic animals at different times, the longest continuous operation being that of M. Miyakawa at Nagoya (10). Exchange of information is carried on in the United States by the Association for Gnotobiotics and on the international level by means of symposia and workshops. It is important that users of gnotobiotes understand not only the techniques involved but also the characteristics of the gnotobiotes (11, 12), as they have been discovered by laboratories already using them. The germfree animal, for example, has a less well-developed lymphoid apparatus and lower levels of gamma globulin than its conventional counterpart. The intestinal mucosa shows characteristic differences between germfree and conventional animals. In addition, germfree rats, mice, rabbits, and guinea pigs display a cecal enlargement which may occasionally reach pathological proportions. Association of germfree animals with selected species of bacteria may alter these germfree characteristics to almost any extent desired. Except for the anomaly of cecal enlargement, which occurs only in rodents and rabbits, germfree animals generally show a state of health and function equal or superior to that of their conventional controls.

Details of the most common technique for maintaining germfree and other gnotobiotic animals may be briefly described (9). Arm-length rubber gloves are sealed to the wall of a clear flexible plastic envelope (the isolator) whose size and shape are determined by the range of a man's reach. Openings in the envelope are made to transmit air through filters of fine glass fiber. A cylinder of rigid plastic is inserted through and sealed to the wall of the envelope to serve as an entry port for supplies; its openings are normally closed with tight-fitting inner and outer caps made of flexible plastic. All interior surfaces are sterilized with a spray of 2% peracetic acid containing 0.1% sodium alkyl-aryl sulfonate as detergent. Supplies which require autoclaving are placed inside a perforated steel cylinder which is wrapped in fine glass fiber and is closed at the end with autoclavable plastic film. After autoclaving of the cylinder, the plastic-covered opening of the cylinder is sealed to the opening of the entry port with a plastic sleeve and plastic tape. After 20 minutes exposure of the entry port to peracetic acid spray, the inside cap is removed, the plastic cover of the supply cylinder is punctured, and the supplies are drawn into the isolator. Animals are transferred from one isolator to another by sealing entry ports together and sterilizing the resulting corridor with peracetic acid. At regular intervals samples of waste or tissues are brought out of the isolator and subjected to a battery of microbiological tests to verify gnotobiotic status (8).

The original germfree mammals of any line must be delivered into the isolator by aseptic caesarian section. Those which normally depend on nursing for their nutrition must be hand-fed sterilized milk substitutes until weaning (8). After weaning they are maintained within their isolators by normal animal husbandry, except that their diets must be specially formulated to compensate for sterilization losses and for the absence of nutrients normally supplied or digested by those microbes which have been excluded from the system. These first generation animals then reproduce within the system and nurse their own young. Oviparous species of animals—birds, fish, reptiles, etc.—are obtained as gnotobiotes by passing their eggs into an isolator through a germicidal bath or spray, and completing their incubation within the system.

The possible uses of gnotobiotic animals are almost unlimited for increasing the precision of animal experiments. Without the use of gnotobiotic animals at some point in an extensive study of health or disease, it becomes difficult to exclude a possible role of the "normal" flora in the process being studied. The gnotobiote is increasingly justifying its existence by revealing unexpected interactions of host and microbiota; it also permits study of subtle and long-term influences of the microbiota, providing an assessment of the positive role of microorganisms in maintaining health. By simplifying an often intolerably complex microbial situation in the conventional experimental animal, the gnotobiote offers hope of new insights into the basic mechanisms of health and disease.

JULIAN R. PLEASANTS

References

1. Pollard, M., "Germfree animals and biological research," Science **145**:247–251, 1964.
2. Reyniers, J. A., *et al.*, "The need for a unified terminology in germ-free life studies," Lobund Rept. **2**:151–162, 1949.
3. Ward, T. G., and P. C. Trexler, "Gnotobiotics: a new discipline in biological and medical research," Persp. Biol. & Med. **1**:447–456, 1958.
4. Pollard, M., "Viral status of germfree mice," Nat. Cancer Inst. Monograph No. 20:167–172, 1965.
5. Pasteur, L., "Observations relative à la Note précédente de M. Duclaux," Compt. rend. Acad. Sci. **100**:68, 1885.
6. Nuttal, G. H. F., and H. Thierfelder, "Thiersches Leben ohne Bakterien im Verdauungskanal," Z. physiol. chem, Hoppe-Seyler's **21**: 109–121, 1895–96.
7. Dougherty, E. C., ed. "Axenic culture of invertebrate metazoa," Ann. N. Y. Acad. Sci. **77**: 25–406, 1959.
8. Reyniers, J. A., ed. "Germfree vertebrates: present status," Ann. N.Y. Acad. Sci. **78**: 1–400, 1959.
9. Trexler, P. C., and L. I. Reynolds, "Flexible film apparatus for the rearing and use of germfree animals," Appl. Microbiol. **5**: 406–412, 1957.
10. Tunevall, G., ed. "Recent Progress in Microbiology," Stockholm, Almqvist & Wiksell, 1959.
11. Luckey, T. D., "Germfree Life and Gnotobiology," New York, Academic Press, 1963.
12. Coates, M. E., ed. "The Germfree Animal in Research," London, Academic Press, 1967.

GOLGI, CAMILLO (1843–1926)

This Italian pathologist was born at Cortona, Italy on July 9, 1843. He studied at the University of Pavia where he later became Professor of General Pathology and Histology. While doing research on malarial parasites, he discovered three new species. Golgi's major field of research was, however, neuroanatomy. Perhaps his greatest contribution to that field was the demonstration of the nervous system as interlaced rather than connected in a complete network. The Golgi method for staining nerve cells, the sensory Organs of Golgi, Golgi bodies, etc. are monuments to his intensive research. He shared the 1906 Nobel Prize for Medicine with Santiago Ramón y Cajal for their pioneering work in neuroanatomy. He became the physician at the Home for Incurables in the village of Abiategrasso and was a frequent contributor to many Italian journals. His collected writings were published in three volumes in 1903. Camillo Golgi died at Pavia, Italy, on January 21, 1926.

MARTIN J. NATHAN

GOLGI COMPLEX

The Golgi (apparatus) complex is considered to be a cytoplasmic organelle occurring in almost every type of vertebrate cell; as well as in many kinds of invertebrate cells. In plant cells, especially among those of the lower forms, an equivalent structure is believed to be present. In 1898 Camillo Golgi observed in nerve cells of certain vertebrates this entity which subsequently has become associated with his name. In suitably preserved material the Golgi substance may be demonstrated by the classical, cytological techniques, as a complex, fenestrated membrane about 0.2μ in thickness. In certain cells it has been reported correctly as filamentous in structure. Some investigators, especially those using silver impregnation, have maintained that a canalicular appearance represents the true form of the complex. When living cells are examined in the presence of supravital dyes like methylene blue and neutral red, the Golgi complex appears as a single, or as a series of vesicles. Such vesicles usually consist of a clear, chromophobic center surrounded by a limiting membrane with a cap-like structure on one side. Phase contrast illumination has been successful in revealing this element in the cytosome without the aid of coloring agents. No detectable, morphological differences have been noted between Golgi complexes found in sex cells compared with those present in somatic cells. Compression of the Golgi material indicates that it possesses a certain amount of elasticity. Entire Golgi complexes have been removed from living cells and found to retain their original shape following this procedure. Ultracentrifugation of intact cells stratifies the Golgi material above the rest of the cytoplasmic constituents at a density of about 1.09 to 1.13. Chemically it has been found to contain lipids and proteins. Enzymes and other substances aggregate on the surface of Golgi membranes, indicating that they may function as an interface for the condensation of such substances. In general an increase in cellular metabolism brings about a concomitant enlargement of the Golgi substance. Its participation in the formation of secretory products in gland cells, and the formation of the acrosome in male sex cells are representative of some of its activities. At present its full rôle in the function of the cell is not completely understood. Examination at high magnifications, under the electron microscope, of ultra-thin sections of cells which have been suitably preserved with osmium tetroxide reveal areas in the cytoplasm containing X-ray dense membranes, vesicles and granules. These membranes are arranged usually as parallel lamellae separated by about 50 to 200 Å. Frequently such membranes are observed joined to each other at their extremities, suggesting a laterally compressed sphere. In close association with these membranous structures are irregularly shaped vesicles and vacuoles which possess a limiting cuticle. Some of these are elongate giving the appearance of a tubular structure. These lamella-form structures and vesicles represent the submicroscopic morphology of those areas in the cytoplasm which correspond to the classical Golgi material.

ROBERT M. WOTTON

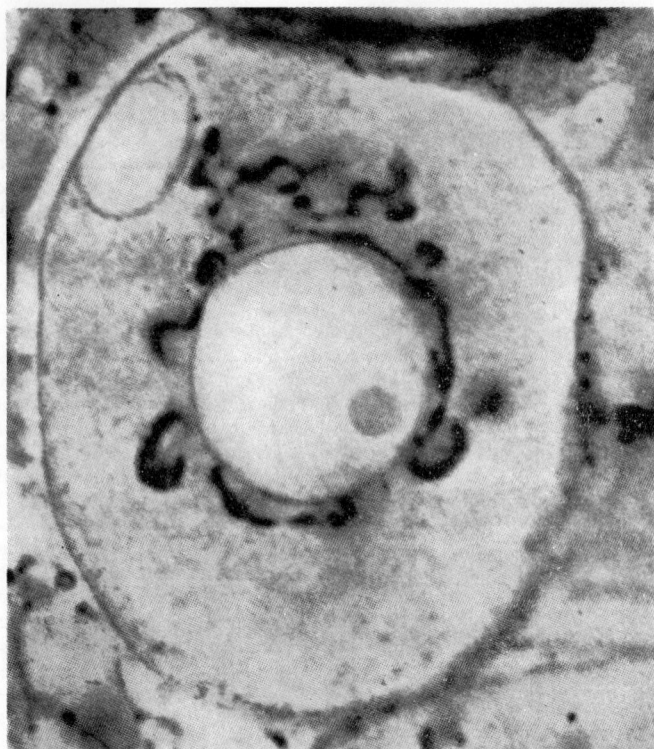

Fig. 1 Photomicrograph of Golgi complex in the neuron of a normal albino rat. Not filamentous structure of the anastomosing strands of the complex. (From Bourne, G. H., "Mitochondria and the Golgi Complex," *in* Bourne, G. H., ed., "Cytology and Cell Physiology," 2nd. Ed., London, Oxford University Press, 1951.)

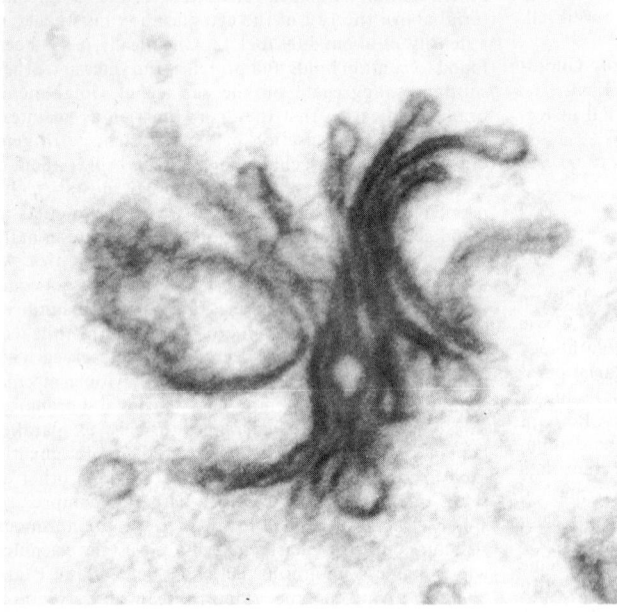

Fig. 2 Electron micrograph of Golgi complex in pea root cell. Note vesicular swellings at ends of the lamellae. (Courtesy Mr. Gordon Spink and Mrs. June Mack, Electron Microscope Laboratory, Biology Research Center, Michigan State University, East Lansing, Michigan.)

References

Dalton, A. J., and Marie D. Felix, "A comparative study of the Golgi complex," Jour. Biophys. and Biochem. Cyt., Supplement, **2**: 79, 1956.

Haguenau, F., and W. Bernhard, "L'appareil de Golgi dans les cellules normales et cancereuses de vértebrés," Arch. Anat. Microsc. Morph. Exper., **44**: 27–55, 1955.

Pollister, A. W., and Priscilla F. Pollister, "The structure of the Golgi apparatus," Intern. Rev. Cyt. **6**: 85–106, 1957.

GRADIENT THEORY

The pattern of development and the processes of integration between parts in living organisms have long been of interest to biologists. One of the more important areas of experimental embryology has dealt with factors which permit cells of identical heredity to differentiate into dissimilar tissue. The *gradient theory* has been of great value in offering a possible mechanism for integration and differentiation. The vigorous work of C. M. Child for over a period of fifty years has caused it to be accepted as one of the important biological generalizations.

Child pointed out that the first pattern to become apparent in the echinoderm egg is *polarity*, a differential along an axis. Polarity is obviously a gradation of materials but it is also a gradation of activity. Gradation of many materials and activities contributes to the total picture. There is a gradient in carbohydrate and protein metabolisms with carbohydrate being greater at the animal and protein greater at the vegetal regions. The animal pole is electronegative to, and more alkaline than, the vegetal pole. The animal pole shows more rapid janus green reduction and greater indophenol oxidase (cytochrome oxidase) activity than the vegetal pole. Differences in rate of cell division, permeability, susceptibility to certain poisons, yolk distribution, etc. all contribute to (or are the result of) the phenomenon of polarity.

Child held to the view that a single gradient of activity is most important as a controlling factor in the early development of echinoderms. This gradient is a gradation in rate and intensity of physiological reactions with the greatest activity centered at the animal pole. He did not deny the existence of concentration gradients but believed that factors of rate and intensity are more important in determining form and proportion than specific qualitative factors which are undoubtedly concerned with cellular differentiation.

With the animal (apical) region possessing greater physiological activity it is able to *dominate* or control regions of less activity which grade away from it. Such dominance may be brought about by more effective competition for oxygen, substrate, etc. or by control through chemical or physico-chemical transmission. Dominance may grade from strong to weak depending upon the system and environment. Agents such as KCN (in weak solution) inhibit the apical end more than the basal; dominance is thereby decreased, the field of control is diminished and structures near the apical region are differentiated on a smaller scale. The *scale of organization* is said to be decreased.

Dominance (and scale of organization) may be increased by the stimulation of apical more than basal regions (thereby steepening the gradient field) or by inhibiting the basal more than apical regions. The young embryos of echinoderms may be caused to undergo *ectodermization* or *entodermization* by strengthening or weakening dominance and thereby increasing or decreasing the scale of organization in the gradient system.

With the development of the blastula a second gradient appears in the echinoderm larva. This gradation (as shown by indophenol oxidase activity and janus green reduction) extends from the ventral to the dorsal region and, aside from giving bilaterality to the embryo,

exerts control over differentiating structures in this plane. This gradient-field may be steep or flat, depending upon intrinsic and extrinsic factors. Experiments have been performed which cause *ventralization* or *dorsalization* by strengthening or weakening activities that grade from ventral to dorsal areas.

The third gradient to appear in the echinoderm embryo is the invagination field at the base of the older blastula. This is followed by other gradients or fields as development becomes more and more advanced. It has been postulated that all of the gradient fields from major (polar) down to minor (those concerned with organs, tissues, cells) begin first as quantitative, dynamic differentials which become qualitative as differentiation proceeds.

Experimentation has indicated that the apical region of a developing egg dominates but does not *direct* the morphogenesis of more basal regions. When basal regions are freed from this dominance, as in the case of differential inhibition or the surgical removal of the apical end, vegetal structures (gut, mesenchyme, etc.) tend to grow larger than normal. The dominant apical end tends to hold down or inhibit the basal region from developing to its full capacity but it does not direct its development. There are also inhibitory influences of basal on apical regions as shown by surgical and chemical methods. When these influences are removed the degree of apical development is much increased but, again, not directed.

With increasing distance (spatial or physiological) from a focus of high activity, a condition is reached when cells or protoplasms yield partially or not at all to the controlling influences of the apical region. This region, removed from apical dominance, is a region of *physiological isolation* and with proper activation may develop a secondary gradient system. The region of invagination of the late echinoderm blastula is such a graded field or system. As development goes on other minor fields become isolated and differentiate in the direction of arms, stomodaeum, etc.

Gradient fields have been found in developmental processes throughout the animal and plant kingdoms. They are not always as clearly defined as in the echinoderm egg and larva and they may be largely obscured by such factors as yolk, prefertilization differentiation, etc. The significance, however, of graded differentials in early embryonic development seems well established.

In lower organisms (and even in some higher forms) gradient systems may persist into the adult and provide interesting correlating mechanisms for not only the maintenance of form but in regeneration and reconstitution. For many years Child and his students have experimented with the flatworm, *Dugesia dorotocephala*, and have shown the importance of gradient systems in integrations and correlations in the development and maintenance of body form and proportion.

Three main physiological (activity) gradients may be demonstrated in this planarian. The most apparent and most easily demonstrated is the *axial gradient*. It has been shown that the anterior (head) end of the animal is the most active and that this activity grades posteriorly to a region behind the mouth where the anterior zooid ends. The presence of this graded system has been correlated with such phenomena as *differential susceptibility, differential recovery, differential tolerance,* etc. where it is shown that anterior regions are more susceptible to lethal concentrations of certain poisons (KCN) but will recover and acclimate more readily to certain non-lethal concentrations than posterior regions. It has also been shown that anterior regions will regenerate a head more rapidly and more completely than posterior regions.

Gradients of planarians, not as obvious as the axial, are the *medio-lateral* and the *ventro-dorsal* whose presence may be demonstrated and functions noted.

The head is the dominant region of the axial gradient in planarians. It controls to a remarkable extent the type and degree of differentiation posterior to it. It appears to exert its dominance by competitive and transmissive forces. If the head is removed from a large normal planarian, fission of the posterior zooids from the anterior zooid usually results if the posterior zooids are able to gain purchase on a rough surface. Fissioning seems to be the result of a decrease of dominance and the physiological isolation of the fission area from the control of anterior regions.

When the head is removed from a planarian of the above named species the posterior body region immediately begins to form a new head at its anterior cut surface. Migrating primitive cells (amoebocytes) move toward the region of the missing head where they mobilize and differentiate. The newly developing head early begins to exert dominance over posterior regions and the field of *reconstitution* is established. Under experimental conditions the new head may be caused to vary in size, differentiation, and in the strength of its dominance and thus modify the scale of organization of the field (gradient system) posterior to it. If the head is grossly inhibited (as by nonlethal concentrations of KCN) the scale of organization is greatly decreased and the reconstituting structures (i.e. pharynx) will differentiate far anterior in the animal.

As indicated above the highest point in the gradient in planarians gives rise to the head—more specifically, the *cerebral ganglia*. If the central region of the head (cerebral ganglia) be removed and transplanted to a region low in the gradient system the transplant will not only become grafted to the posterior tissue but will grow at the expense of this region and assume the role of an *organizer*. The graft reorganizes the tissue of the host (even though the host may be of a different species) with reference to itself. This means that the polarity of the host cells may be changed and that new structures or organs (i.e. pharynges) may be caused to form in the host tissue in the new gradient field induced by the graft.

New gradient fields may be induced in old planarian tissue without transplants. If a small region of the planarian some distance from the head (or if the head as the center of dominance is removed) is continuously irritated by chemical or mechanical means this region will shown an increase in oxidative metabolism. Under certain conditions this irritated region may be activated sufficiently to form a head which may then organize a new gradient field. This new pattern may eventually take complete control from the old organization. Again, with the reconstitution of old tissue into a new gradient system quantitative changes precede qualitative and the region highest in the physiological activity of the gradient system assumes dominance over the differentiating field.

Child found abundant material for many of his fundamental assumptions in the hydroids. It was noted that the region about the hypostome had the greatest physiological activity and dominance. It was found that the scale of organization in the reconstitution of hydranths could be increased or decreased by physical and chemical means. By manipulating oxygen and carbon dioxide the polarity in a piece of stem could be reversed (i.e. the end of the stem receiving the most oxygen reconstituted into a hydranth). One of Child's most suggestive experiments was that of allowing a piece of *Corymorpha* to lie untouched at the bottom of a dish. In time a new hydranth appeared, but on the uppermost side. This was the old lateral side and at right angle to the old polarity of the piece. This experiment showed the origin of a new gradient across the piece with the region in contact with the dish receiving less oxygen and more subject to the narcotizing action of carbon dioxide and the region opposite, with a greater oxygen supply, becoming activated to the highest point in the new gradient system.

In the common brown hydra, *Pelmatohydra oligatis*, gradient factors are much in evidence. Levels down the body show a graded decrease in rate and number of tentacle regeneration. Under normal conditions buds appear only at the budding zone—a region of physiological isolation partially removed from the dominance of the apical region.

In cell aggregates of sponges and hydroids a quantitative gradient precedes differentiation. Gradients, degrees of dominance, and physiological isolation may have much to do with the development of segmentation in the higher animals. Gradients throughout the plant and animal kingdoms provide mechanisms of nuclei to behave differently. In different regions of a quantitatively graded system, similar nuclei may well be influenced by different cytosomes. In response to the different cytoplasms different genes may be brought into play or nuclei change in some other manner which would give the mechanisms for further qualitative differentiation.

OLIN RULON

Reference

Child, C. M., "Problems and Patterns of Development," Chicago, The University Press, 1941.

GRASSES see GLUMALES

GRASSLAND

Grassland is a major physiognomic category of vegetation in which the dominant plants are herbaceous perennials and usually members of the grass family. Large tracts of such vegetation lie in the rain-shadows of mountains in temperate latitudes, with desert on the drier side and forest on the wetter side, covering approximately 10% of the land surface in the aggregate. The word *steppe* is rather widely used to distinguish this type of grassland from other kinds of vegetation of similar life form but with very different ecologies. In South America *pampa* is the equivalent term.

Among those grasslands excluded from the steppe category are grass-dominated portions of arctic and alpine tundras where heat is inadequate for plants with a shrub or tree form. Neither does steppe include island-like grassed areas to be found (below the alpine zone) in most forest regions of the globe. These small areas are consequences of local soil conditions, such as excessive lime content, drouth too extreme for woody plants, deficiency of one or more nutrients, inadequate aeration, repeated burning, etc. When such enclaves support lush vegetation, and especially if they are rich in dicot herbs, they are commonly designated as *meadows*. Grasslands on mineral soil that is saturated at one or more seasons, but is not covered with peat, are called *marshes*. If peat covers the wet soil and Sphagnum is absent, vegetation dominated by grasses, sedges, cattail, etc. is called *fen*.

In addition to the above there are two physiognomically distinctive types of vegetation which contain a conspicuous grass component that from a functional standpoint dominates the dispersed shrub or tree phase. In the tropics there are extensive carpets of herbaceous vegetation supporting scattered trees or tall shrubs, that are usually excluded from even the concept of grassland and referred to as *savannas*. Some of the latter owe their character to deficient rainfall in relation to heat and are thus the tropical equivalents of steppe, but much of tropical savanna seems a result of forest burning, or of soil conditions inimical to closed forest. In temperate latitudes are *shrub-steppes* in which medium or low shrubs are conspicuously dotted over an herbaceous matrix, the sagebrush-grass communities in the rain-shadow of the Cascade Mountains providing an excellent example. Some have grouped this vegetation with steppe, and others place it with desert, with the hybrid character of both vegetation and climate providing some justification for either viewpoint.

In North America the term *prairie* has been applied to several of the above grassland types, but the word has been used in so many senses (for salt marsh, shrub-steppe, and even forest) that it is highly ambiguous.

Grasses of the steppe category, to which the remainder of this article refers, are mainly perennials that vary greatly in height but die back to the ground annually. The dominant species may be rhizomatous so that a continuous or broken sod is formed, or they may be caespitose, forming "bunchgrass" or "tussock grassland." The genus Carex is common but usually inconspicuous. Dicot herbs (i.e., forbs) are of minor importance at the drier edge of grassland, but toward the wetter edge they become conspicuous and may even exceed graminoids in dry-weight production. Shrubs are either dwarfed, shorter than the herbs and interspersed among among them, or aggregated into thickets confined to relatively moist slopes and ravines. Trees commonly form continuous forest on subirrigated recent alluvium in the least xerophytic margin of grassland, dwindling to narrow riparian strips of very limited taxonomic variety in more arid regions, and often disappearing almost completely in the very dry parts.

Grassland climates are characterized by precipitation low enough in relation to temperature that at one season all readily available soil moisture is exhausted in the rooting horizon, which for grasses is usually 0.5–2m. At another season this rooting horizon must be kept moist for a number of consecutive weeks without an excess to penetrate into the permanently dry subjacent soil layer. Such an environment differs from that of forest in the presence of the dry subsoil which prevents deeply-rooted woody plants from gaining access to moisture near the water table. It differs from desert in the greater duration of wetness in the rooting horizon.

As a consequence of low precipitation plant nutrients are not lost from the soils by leaching, and thus are conserved as they are released by normal processes of soil weathering and mineralization of humus. Calcium becomes especially abundant, normally accumulating in a layer near the limit of rooting and annual moisture penetration.

Owing to the numerous short-lived roots of grasses, the soils supporting them tend to become darkened with humus to a depth that varies directly with precipitation. As a consequence of the rich supply of nutrients, high humus content, and excellent soil structure, grassland soils are very fertile and the herbage in turn has high forage value.

Fire of lightning or aboriginal cause has frequently run over grasslands, probably since their inception. In broad view, fire does no more than consume the cured forage after its photosynthetic value is ended, and most plants regenerate rapidly from underground organs. But whereas grasslands are essentially unaffected by burning, fires starting in them tend to erode the edges of contiguous forest regions, commonly resulting in a border of fire-induced grassland to the leeward of areas where climate is the primary cause of herb dominance.

Steppes originated when orogeny in mid- or late-Cenozoic time created pronounced rain-shadows. This new vegetation was of immediate significance for animals in that it made possible the evolution of large, hooved mammals possessing teeth especially adapted for grinding herbage that is harsh by virtue of its high silica content. For modern man the importance of grassland (all types) can be inferred from the fact that the U.S.A. has twice as much land devoted to grazing as to other crops. This is at once indicative of the importance of grasslands as an indirect source of high-protein foods, and of the inherent inefficiency of converting the primary production of herbage into usable animal protein.

Along the least dry edges of steppes the land is most valuable for the production of grain crops annually without irrigation. Under lower rainfall cropping becomes possible only by fallowing on alternate years, and in the most arid steppes grazing is the best land use. The main problem in managing steppe for grazing is to regulate the time and intensity of animal use so as to harvest the highest percentage of the forage without reducing the productive capacity of the ecosystem. About 100–400g dry shoot material are produced per square meter per year, only about half of which can be harvested without damaging the vegetation. A generally low degree of success in management has led to the widespread replacement of original herbs of high forage value with others, native or exotic, that are not acceptable to the animals or are not very available owing to low stature or short growing seasons. Shrubs may figure prominently in this secondary vegtation, and annuals nearly always become conspicuous.

R. Daubenmire

References

Hanson, H. C., "Ecology of grassland." Bot. Rev. **4**: 51–82, 1938.

Malin, J. C., "The grassland of North America," Magnolia, Mass., Smith, Peter, 1967.

Shantz, H. L., "Grassland and desert shrub," *in* "Atlas of American Agriculture, National vegetation," U.S. Dept. Agric. Bur. Agric. Econ, 1924.

Thornthwaite, C. W., "Grassland climates," Internat. Grassl. Congr. State College, Pa. Washington, Proc. 6th 1952.

Weaver, J. E., and F. W. Albertson, "Grasslands of the Great Plains." Lincoln, Neb., Johnsen, 1965.

GRAY, ASA (1810–1888)

The most important American botanist of the 19th century and one of the most prolific taxonomic botanists of all time. His "Manual of the Botany of the Northern United States," the first edition of which appeared in 1847, is still in print and is likely to remain so for so long as anyone is interested in the identification of North American plants. Gray was by training a physician, having graduated as an M.D. in 1831, but it was only four years after this that his first contributions to botany appeared, and by 1836 he had published his first textbook. From then on he never looked back and continued to throw the weight of his learning into the description of the innumerable plant species which were discovered as the exploration of the United States continued. In 1842 he accepted the chair of Natural History at Harvard University where botany was, at that time, for all intents and purposes nonexistent. He developed the Herbarium, the Library, and the Botanical Gardens, contributed the greater portion of the profits from his books, and bequeathed to the University the royalties on the latter. He was an ardent exponent of Darwinism and took immediate issue with Darwin's many opponents in the United States. In the course of his services at Harvard, he trained two generations of American botanists so that his influence on this science was overwhelming.

GRAY, JOHN EDWARD (1800–1875)

This indefatigable and quarrelsome English naturalist became early interested in botany from the fact that his father, and grandfather, before him had written widely in this field. He had intended to become a physician but soon lost interest in his studies, in part from the ill health which was to plague him throughout his life. He then determined to devote his life to botany and in 1821 published a two-volume "Natural Arrangements of British Plants" based in great part on the collections in the British Museum in London. Even at that early age, however, his opinionated truculence had made many enemies for him and in 1822 his application for membership in the Linnean Society was rejected. Outraged at this insult from his fellow botanists he turned his attention to zoology and was in 1824 made an assistant in that subject in the British Museum. The collections at that time were poor in quantity and ill-arranged. He flung himself so enthusiastically into the task of expanding and arranging the collection that in 1839 he became Keeper of Zoology. Within a few years he had raised the standards of the institution until it rivaled the already famous museums in Paris and Berlin.

He was one of the most voluminous writers in history and a bibliography published by his family after his death shows 1162 entries, of which well over 200 are books in stiff covers. The most important of these are the Catalogs of the Collections of the British Museum in the compilation of which, with the assistance of his brother, George Robert Gray, whom he made Keeper of Birds, he probably attached his name to more genera and species than any man except Linnaeus. No subject was, however, too controversial to receive his enthusiastic attention and his writings include the collection of postage stamps, the emancipation of slaves, and a considerable amount on the decimal system including a long argument with the government of France, which he unsuccessfully endeavored to persuade to standardize the meter at forty inches in the interest of interchangeability.

Even though his recurrent illnesses resulted in partial paralysis in 1870 he did not retire from the keepership until 1874 one year before his death. Six years later the British Museum, no longer able to house the stupendous collections which John Edward Gray had brought to them, removed these collections to what is now the British Museum (Natural History) in South Kensington.

He was the direct ancestor of the compiler of this encyclopedia.

PETER GRAY

GROWTH

In its simplest aspects, growth means a gradual increase in the size of an organism, or in the number of members of a population of organisms. However, the growth of an organism involved changes in its form, the epigenetic formation of new structures, the differentiation of cells and tissues, biochemical changes, changes in its metabolism, and other complexities. Cases can be cited in which development occurs with no increase, or a decrease in size. Similar complexities are involved in the growth of a population: changes in birth and death rates, changes in the age distribution of its members, etc. Although many authors have attempted to make definitions of growth which will take account of its complexities, it would seem best to regard the term as representing an essentially undefined, intuitive concept, and in a specific context to deal specifically with the aspects of the phenomenon under consideration, such as increase in weight or numbers, cellular differentiation, metamorphosis, etc., in the appropriate terms. Since growth is a complex and poorly understood phenomenon, and since its details vary greatly from one organism to another, this article is limited to its metric aspects, that is to the biometry of growth.

Growth Curves

When some measurable property of an organism, organ, or population is plotted against time, the resulting graph is termed a growth curve. There are thus growth curves in terms of length, weight, dry weight, nitrogen content, number of cells or organisms, etc.

The growth curve of an individual can only be had in terms of harmless measurements such as length or fresh weight. Growth curves in terms of other measurements which require a destructive analysis, can only be obtained by the statistical procedure of sampling a population of similar organisms at various ages. Growth curves, of whatever sort, are often said to be sigmoid in form, with an early acceleration phase when the rate of increase in size or number is increasing, and a later deceleration phase, of decreasing rate, leading to a maximum value. Actually, growth curves take a variety of forms, and are rarely symmetrically sigmoid (Fig. 1–4). In a population of micro- or larger organisms, the acceleration phase may be closely exponential (v.i.) for a considerable period (Fig. 1), and may sometimes be preceded by a lag phase, when the relative rate (v.i.) of increase in number is zero or very low. In the growth in height of many plants with apical meristems, a prolonged linear phase, when the absolute rate of elongation is more or less constant, may be interposed between the acceleration and deceleration phases (Fig. 2).

In higher animals which are viviparous or oviparous, the acceleration phase may be limited to the prenatal,

and perhaps the early postnatal periods, with a prolonged phase of gradual deceleration (Fig. 3). In addition, there may be inflections of the growth curve at the time of birth (or hatching), weaning, or puberty. Inflections of the growth curve are particularly pronounced in the case of organisms which undergo metamorphosis, such as the insects and the amphibia (Fig. 4), and the growth curve may be said to consist of a succession of sigmoid phases. The variety of forms of growth curves is emphasized when it is realized that little is known about the growth of many groups of organisms.

Growth Rates

Absolute rate. As in the preceding paragraph, it is frequently useful to speak of the rate of increase of the measured property of the growing system, at various times, under various conditions, or for various parts of the growing system. If X is the property measured and t is time, the rate (or absolute rate) or increase at a given instant is properly the first derivative, dX/dt. This may be estimated empirically by finding the slope of

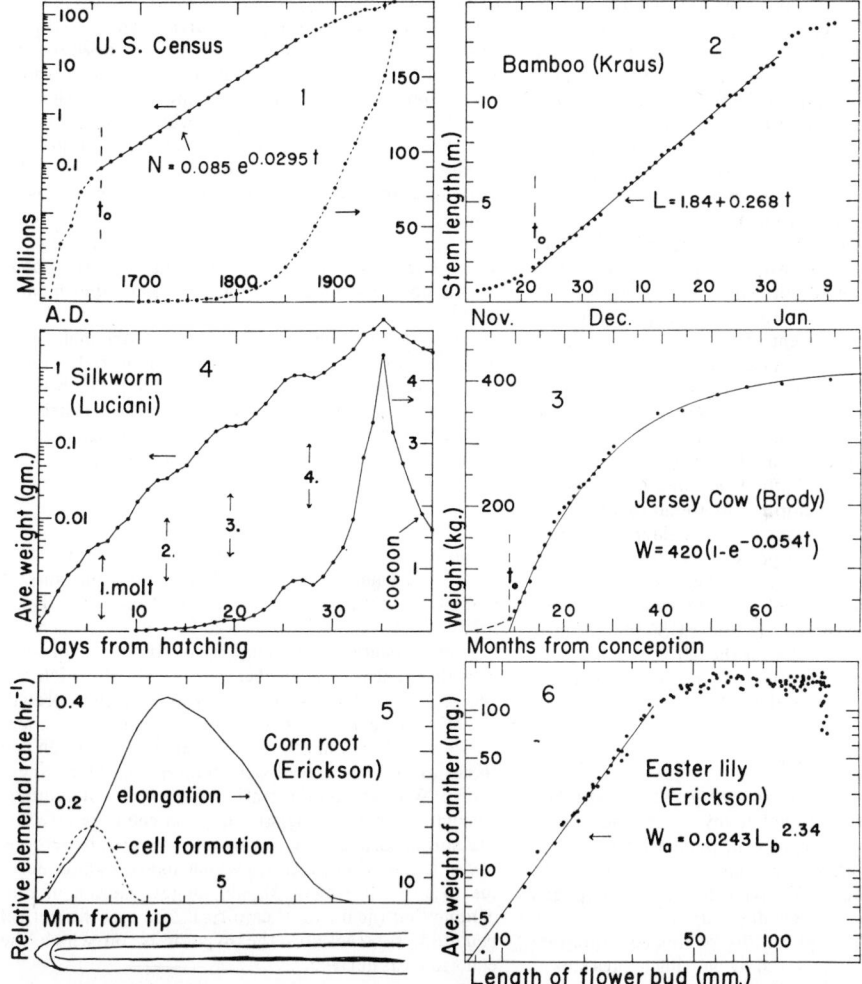

Fig. 1–6.

the growth curve at a particular t or X, by the devices of numerical calculus, or, if the growth curve is sufficiently nearly linear over an interval of measurement $(t_2 - t_1)$, by forming the quotient $(X_2 - X_1)/(t_2 - t_1)$.

Relative rate. It is frequently useful to express the rate of growth in proportion to the size of the growing system at a given instant. The resulting expression, $(1/X)(dX/dt)$, has been variously termed a relative growth rate, specific growth rate, growth coefficient, efficiency index, or when multiplied by 100, percentage growth rate, and has the dimensions of $(time)^{-1}$. Since $(1/X) \cdot (dX/dt) = d \ln X/dt$, it is often useful to plot the logarithm of X against t, or X vs. t on semi-log paper as in Figs. 1 and 4. Estimates of relative rates may be made from such a plot by determining its slope at various points. If the plot is linear, or the interval of measurement short, the formula $(\ln X_2 - \ln X_1)/(t_2 - t_1)$ may be used. Erroneous formulas have sometimes been used. For example Minot's formula, $(X_2 - X_1)/X_1(t_2 - t_1)$, overestimates the instantaneous relative rate in continuous growth, because of the division by X_1. It is appropriate, however, to discontinuous growth, as in a population where reproduction occurs at the intervals $(t_2 - t_1)$.

Relative elemental rate. For the study of small parts of a growing system, Richards and Kavanagh have proposed that it be referred to a system of X, Y and Z coordinates, and that an expression of the form $\partial/\partial X$ $(\partial X/\partial t) + \partial/\partial Y (\partial Y/\partial t) + \partial/\partial Z (\partial Z/\partial t)$ be estimated for each element of volume of the system $\Delta V = \partial X \cdot \partial Y \cdot \partial Z$. It has the dimension of $(time)^{-1}$. This method has had a simple application in the analysis of apical growth in length of a root, where relative elemental rates of increase in length and in cell number were worked out for various portions of the root (Fig. 5).

Allometric coefficient. Huxley has proposed another method of studying the growth of organs or parts of an organism, which consists of plotting the logarithm of some measurement of the organ or part, Y, against the logarithm of a measurement of the whole, the remainder, or another part, X. The slope of the resulting curve at any point, $d \log Y/d \log X$, is variously termed the allometric, heterogonic or heterouxetic coefficient, and other terms have been used. The allometric coefficient may be considered as the ration of the relative growth rates, $(d\ln Y/dt) / (d\ln X/dt)$, and is dimensionless. In some cases it is empirically found that the allometric coefficient is constant for a considerable period of time, in which case the power function $Y = Y_1 X^k$, or $\log Y = \log Y_1 + k \log X$, can be fitted, where k is the allometric coefficient and Y_1 is the value of Y when $X = 1$ (Fig. 6). When $k = 1$, growth of the organ is said to be isometric with that of the body. When $k > 1$, or $k < 1$, the terms positive and negative allometry respectively are sometimes applied.

Theoretical Growth Equations

Exponential equation. By assuming that the absolute rate of growth is proportional at any time to the size, X, of the organism, or number in the population $(dX/dt = rX)$, or that the relative rate is constant, $(1/X)(dX/dt) = d \ln X/dt = r$, the simple exponential or "compound interest" equation may be derived: $X = X_0 e^{rt}$, or $\ln X = \ln X_0 + rt$. The applicability of this equation to a given set of data may be tested by plotting logarithms of the measurements against time, or plotting the measure-

ments against time on semi-log paper, and determining whether the resulting graph is linear (Fig. 1). It is frequently found that the early acceleration phase of growth of an organism or a population is approximately exponential. In the "chemostat" of Novick and Szilard, bacteria may be grown exponentially for long periods of time. Curiously, exponential growth is sometimes termed logarithmic growth.

Monomolecular equation. The fact that many organisms grow to a fairly definite mature size has led some authors to analogize growth with a first-order chemical reaction in which the reactant is limited in amount. It is assumed that the rate of growth is proportional at any time to the difference between mature size, X_f, and the size at that time, X, $dX/dt = k(X_f - X)$. On integration, the equation, $X = X_f(1 - e^{-kt})$, or $\ln (X_f - X) = \ln X_f - kt$, is obtained, where k is analogous to the chemical rate constant. This equation often fits the later deceleration phase of an asymmetrical growth curve satisfactorily (Fig. 3).

Logistic equation. By assuming that the absolute rate of growth is proportional to the size, X, of the organism or population, and to the difference between final size, X_f, and the size at a given time, that is, $dX/dt = k'X(X_f - X)$, the following equation may be derived, $X = X_f / (1 + e^{-kt})$, or $\ln [X / (X_f - X)] = kt$. It is symmetrical about the point corresponding to $X_f/2$, and asymptotic to $X = 0$ and $X = X_f$. The assumptions made in deriving this equation are analogous to those which apply to an autocatalyzed chemical reaction, as Robertson has pointed out. Modifications of this equation have been devised to fit asymmetrical growth curves. Hutchinson has suggested that a time lag be introduced into this equation, so that the rate of growth is made to depend upon the size at time $(t - \tau)$ as well as at time t, $dX_t /dt = k'X_t [X_f - X_{(t-\tau)}]$. Integration of this equation yields curves which may oscillate about X_f, as do the sizes of some natural and experimental populations.

Gompertz equation. This equation was originally used in actuarial science for fitting of human mortality data. Its use as a growth equation follows from the observation that in mammals, the relative rate of growth decreases with time. If it is assumed that the logarithm of the relative rate decreases linearly with time,

$$\ln (d \ln X/dt) = a - kt,$$

integration yields the Gompertz equation, $X = Ae^{-ce^{-kt}}$, where A, c and k are arbitrary constants. It will often fit asymmetrically sigmoid curves satisfactorily.

Other equations. A number of other theoretical growth equations have been suggested, of which the recent proposal of Weiss and Kavanau that growth might be accounted for in terms of specific substances which promote, and others which antagonize organ and tissue growth in a feedback manner, might be mentioned. Winsor's demonstration that a series of numbers calculated from the logistic equation could be fitted by the normal probability integral, the inverse hyperbolic tangent, and other functions which appear to have nothing to do with growth, should suggest caution in concluding that the fitting of empirical data by a theoretical equation, substantiates the hypotheses on which the equation was derived.

RALPH O. ERICKSON

Reference

Clark, W. E. LeGros and P. B. Medawar, "Essays on Growth and Form," Oxford, Clarendon Press, 1945.

GRUIFORMES

The Gruiformes include perhaps the most heterogeneous assemblage of forms to be found among the birds. Sometimes the order has been referred to as a group of misfits that seem not to belong anywhere else. Twelve families of living birds currently make up the Gruiformes. They are loosely called "marsh birds," an appropriate designation for most of them, but some species are as aquatic as ducks, while still others are field, forest, or even desert types. Hence, included families are quite variable in both habits and structure, so that few ordinal characteristics can be defined for the group as a whole.

The classification of the Gruiformes is further complicated by an abundance of fossil material that seems to belong here. Dr. Wetmore, in going over these fossil-rich deposits from Patagonia in southern South America, has sorted out the gruiform types and tentatively placed them in ten different families. Typical of these is *Phororhacos,* a crested, large-billed, long-legged, cursorial giant which, with its close allies *Brontornis* and *Andrewsornis,* roamed the Patagonian uplands during the Miocene and Oligocene. Presumed to be raptorial in habit, they resembled, and possibly were quite closely related to, the Secretary-bird (*Sagittarius serpentarius*) of Africa. Five families of Phororhacos-like birds have been described by Wetmore; two others seem more closely related to the cariamas (see below) and three others are grouped elsewhere in this odd gruiform assemblage.

Among the twelve families of living gruiform birds, eight are small groups, limited to three or less species in each. Four of the families are composed of a single species, another has two members, and the remaining three have three species apiece. These restricted families are dismissed rather briefly below, whereas the four larger and more widely distributed families are described in more detail.

Among the three-member families are the roatelos or monias (Mesoenatidae), which are terrestrial, flightless or nearly flightless, forest birds of Madagascar. By contrast, the Plain-wanderer or Collared Hemipode (*Pedionomus torquatus*), a plump, virtually tailless, quail-like bird of Australia, is a bird of the open plains and is the sole representative of the Pedionomidae family.

Of more interest to Americans in these limited gruiform families is the Limpkin (Aramidae). A single species of these large rail-like birds survives, but its five subspecies are rather widely distributed over Central and South America. Presumed to be North American in origin, a single race (*Aramus guaruana pictus*) survives in the swamps of southern Georgia and Florida, where its chief food supply of snails is in jeopardy by the perennial threat of drainage for agricultural and real estate developments. Other members of the species have retreated to Central and South America. South America is also the home of three species of trumpeters (Psophiidae), which are noisy, hump-backed, guinea-like birds inhabiting the humid forests of Venezuela, the Guianas, and the Amazon valley.

The sun-grebes or finfeet (Heliornithidae), so-called because they dive and swim like grebes and have broad scallops or lobes on the toes, are *pantropical* in distribution, that is, its three species are widely scattered in the tropics of both the Old and the New World, with one species in South America, one in Africa, and one in southern Asia. By contrast the Kagu (Rhynochetidae), a crested, apparently flightless, heron-like bird, is limited to the island of New Caledonia where it faces an uncertain future. The Sun-bittern (Eurypygidae), of Central and South America, constitutes another one-member family, which in some respects resembles the Kagu. Both have spectacular courtship displays or dances. Two species of cariamas (Cariamidae) in South America wind up this queer assortment of one to three member families of gruiform birds.

Among the larger families of wider distribution are the button or bustard quails (Turnicidae). These include about 15 species that occur over much of the warmer parts of the Old World from Southern Spain and Portugal through Africa, southern Asia, Australia and nearby islands. Button quails are plump, concealingly colored, brown or grayish terrestrial birds of the grasslands or open forests. They are classic examples of polyandry, which is infrequent among birds. The females are larger, more brightly colored and take the initiative in courtship. They may mate with several different males, leaving each with a clutch of about four eggs to be incubated and cared for without help from the females.

Perhaps the most spectacular and best known of the gruiform birds are the cranes (Gruidae) which include 14 species of tall, long-legged, long-necked, handsome birds widely distributed over all continents except South America. North America has only two species, the Sandhill Crane (*Grus canadensis*), composed of five geographic races or subspecies, and the Whooping Crane (*Grus americana*), a monotypic species currently threatened with extinction. Great interest, stimulated by publicity through frequent press releases, magazine articles and books, has attended the Whooping Crane's dramatic fight for survival. During the past decade the total population of these large white cranes has fluctuated between two and three dozen or more individuals, which breed in the remote and as yet relatively unmolested wilderness of northwestern Canada (Wood Buffalo Park) and winter on the Aransas Wildlife Refuge in Texas. In spite of more or less complete protection on their breeding and wintering grounds, civilization is rapidly encroaching on both areas; lumbering and mining operations are dangerously close to the breeding grounds and the Aransas Wildlife Refuge is not large enough to permit any expansion of the winter range.

Currently the Sandhill Cranes are in a much better, but still precarious, position. The Arctic breeding race, formerly known as the Little Brown Crane, is quite abundant (250 thousand estimated), but the Cuban and Florida forms are rare, and the interior race in the Great Lakes states and westward is being watched with great interest. Zoos may help to perpetuate some of the spectacular Old World species, as they make very attractive and often easily cared for exhibits.

Largest of the gruiform families is the Rallidae, which consists of about 130 species of rails, gallinules and coots widely distributed over most of the world. Most of these are rails, which are usually secretive, largely nocturnal, concealingly colored birds which skulk about in the dense marsh vegetation, pushing their laterally compressed bodies ("thin as a rail") between the closely packed stems of rushes and cattails. They seldom take flight except during migration, are more often heard than seen, and usually have to be flushed out with dogs. Railhunting is popular with some sportsmen, but most species—some hardly larger than sparrows—offer only scanty morsels for the table. Some rails, in other parts of the world, are flightless and limited to small islands which when taken over by man has resulted in the extermination or near extermination of many species. Rails usually nest in marshes, building a loose platform of rushes over or near the water, and laying a large clutch of a dozen or more surprisingly large eggs, but there are many variations in egg number, nest structure and nest site among the many rail species.

Gallinules are larger, colorful relatives of the rails; usually they sport a bright red bill. The Purple Gallinule (*Porphyrula martinica*) in particular is handsomely marked with bright blue and purplish reflections. Gallinules often emerge from hiding in the dense marsh vegetation and swim or paddle about in open pools or along the edge of the marsh. Their toes are greatly elongated, as an adaptation for paddling over floating vegetation. Coots are even more aquatic in habit and swim about like ducks in open water. Often they occur in large rafts on the open lakes during migration. They are poor divers, however, and sometimes consort with diving ducks, depending on the latter to do the diving; then the coots promptly steal bits of succulent vegetation brought to the surface by the ducks.

One of the most unique of the coots is the Horned Coot (*Fulica cornuta*) which sports a decorative top knot or horn and resides on cold mountain lakes in the high Andes. Vegetation for nests is scarce, so the birds heap up huge islands of stones, up to a ton or more of them, for nesting sites, using plant material only for the lining of the nest. Another extremely interesting rallid is the large flightless "Takahe" (*Notornis hochstetteri*) of New Zealand. Discovered in 1849, only four specimens were known up to 1898 when the species was presumed to be extinct. Some 50 years later, however, tentatively identified tracks, and then actual specimens and nests were discovered on the mountains of South Island.

The last of the gruiform families, the Otididae, known as bustards, have departed from the ways of their marsh relatives. These are large, heavybodied, ostrich-like birds that inhabit grasslands, savannas, and open plains. Though strong fliers when launched, they prefer running or crouching to escape danger; perhaps because of this, as well as their reputed high palatability, they have been exterminated over most of Europe, but occur sparingly over much of Africa, southern and central Asia, and Australia. Largest of the 23 species is the Great Bustard (*Otis tarda*), which attains a weight of about 30 pounds and is one of the heaviest of flying birds. Males perform colorful courtship ceremonies which consist of short flights, posturing and showy displays.

GEORGE J. WALLACE

References

McNulty, Faith, "The Whooping Crane," New York, Dutton, 1966.
Van Tyne, J. V. and A. J. Berger, "Fundamentals of Ornithology," New York, Wiley, 1959.
Walkinshaw, L. H., "The Sandhill Cranes," Michigan, Cranbrook, 1949.
Wetmore, A., "A Check-list of Fossil Birds of North America," Smithsonian Misc. Coll., **99**: 1–81, 1940.
ibid., "A Revised Classification for the Birds of the World," Smithsonian Misc. Coll. 117 (4) : 1–22, 1951.

GYMNOSPERMAE

A class of higher vascular plants generally distinguished by the production of SEEDS in a superficial, exposed position on the sporophylls. The literal meaning of gymnosperm, in fact, is "naked seed" and serves to contrast these plants with the ANGIOSPERMAE (flowering plants) in which the seeds develop within a folded, sutured sporophyll or ovary.

The group is an old and extremely diversified one, both morphologically and geographically. It probably arose from FILICINEAN stock during the Devonian and within the group may be traced the evolution of the seed habit and the ancestry of the flowering plants. Thus, if the fossil record is examined carefully, even the character of "naked seededness" breaks down in some of the transitional forms.

According to Chamberlain (1935), the Gymnospermae may be divided into two distinct and parallel evolutionary lines. The first of these is the "Cycadophytes" with pinnate, fern-like foliage, short, stout, sparsely branched trunks, and a vascular cyclinder composed of scanty amounts of vascular tissue in narrow, radiating strips, surrounded by massive parenchymatous areas in the form of wide rays, a large pith, and an extensive cortex. The only extant group of this line is the modern CYCADALES (cycads), but a large assemblage of extinct forms belongs here, including the Cycadofilicales, (seed ferns) which, as their names implies, establish a liaison between the gymnosperm and filicinean lines, and the Bennettitales, whose bisporangiate strobili strongly bring to mind the structure of a primitive *Magnolia*-type FLOWER.

The Cycadofilicales were an alliance of primitive gymnosperms whose fossil history extends from the late Devonian to the Triassic. They were mostly small, undergrowth trees with stout trunks up to 50 feet in height bearing a crown of compound, large, feathery leaves at the summit. The late Paleozoic was erroneously called the "Age of Ferns" because of the many fern-like leaf fragments of these seed plant in the fossil record of this era.

Directly descended from the Cycadofilicales was the Bennettitales—a diverse assemblage of plants which ranged from the Permian through the Comanchean. It is the only gymnospermous group in which the reproductive leaves were grouped into *bisporangiate* strobili bearing sterile leaves at the base and mega- and microsporophylls arranged in a spiral pattern around a central axis or receptacle, identical to the arrangement of these parts in the angiospermous flower.

Fig. 1   Gymnosperm leaves: left, scale-like, the eastern white cedar; right, broad-leaved, the maidenhair or Ginkgo. The former has staminate cones, the latter a mature fruit. (Courtesy U.S. Forest Service.)

Thus the Bennettitales either resemble the Angiosperms as a result of parallel evolution or represent the true ancestral stock of the flowering plants. It remains, however, to uncover transitional forms which show stages in the development of a true flower, e.g. the reduction in number of pollen sacs to four per microsporophyll (there were many per sporophyll in the Bennettitales) and the enclosure of the ovules by the megasporophylls (the sporophylls merely subtend the ovules in the typical gymnospermous way in the Bennettitales). Despite these gaps in the fossil record, it is noteworthy that the Bennettitales were a large and successful group during the Jurassic, the earliest period from which true angiospermous fossils are known.

The Caytoniales (Permian through Jurassic), an offshoot from the early bennetitalean line, had no strobili, but the seeds were enclosed within rolled up sporophylls. Thus they do not really fall into the gymnosperm alliance as defined earlier but they do illustrate that the potential for a typical angiospermous condition was present in the line.

The Nilssoniales (Upper Permian to Jurassic) were an assemblage of mesozoic cycadophytes which were specialized as xerophytes.

The Cycadales (Jurassic to present) are the only extant remains of this experimental gymnospermous branch. They are thus distinguished from all other living gymposperms by their large, pinnate leaves and stem structure. There are only 9 genera and 60 species of them left, but fortunately, many are cultivated in conservatories and botanical gardens throughout the world and some are grown as ornamentals in the warmer parts of the United States and Europe. *Zamia* is a small cycad native to Florida and South America; *Dioon* and *Ceratozamia* are endemics of Mexico; *Microcycas* is found very locally in Cuba; *Cycas* is the most abundant and widespread of the cycads, occurring from Australia to Japan and over to Africa;

Macrozamia, tallest of the cycads, and *Bowenia* are endemics of Australia; and *Encephalartos* and *Stangeria* are rare endemics of Africa. All cycads are dioecious and bear cones or strobili which are remarkable for their extremely large size—up to $\frac{2}{3}$ of a meter long for the female cone of *Macrozamia*. The cycads share with *Ginkgo* the last flagellated sperm cells to be found in the plant kingdom. In all other gymnosperms and all angiosperms the male gametes are reduced to unflagellated nuclei which are ejected into the egg cytoplasm directly by the pollen tube. The flagellated sperms of *Dioon* are the largest known, measuring over 300 microns in length and visible to the naked eye.

The other line, the "Coniferophytes" is believed to be an early offshoot of the Cycadofilicales distinguished by simple leaves, tall, branched trunks, and the dense vascular anatomy of a good lumber tree. Included here are the extinct Cordaitales, Voltziales, etc., and the extant Coniferales, Ginkgoales, and Gnetales.

The Cordaitales (Mississippian to Early Jurassic) represent an early split from the cycadofilicalean complex in which the leaves were long, simple, and linear and the wood dense.

The Ginkgoales appeared in the Upper Permian and are represented today by one species—*Ginkgo biloba*—the Maidenhair tree. It was originally native to the temperate zone of Asia, but is now known only in cultivation, which saved it from extinction 2,000 years ago. It grows to be a 150 foot deciduous tree with broad fan shaped leaves borne on stubby spur branches (short shoots), as well as on the main stems (long shoots), and is remarkably resistant to parasites and diseases. Like the cycads, *Ginkgo* is dioecious and has motile sperm, but the strobili in *Ginkgo* are very small reduced structures.

The Gnetales are another small group, but one which is equally distinctive. It has no fossil history and there are three genera and 45 species extant. *Gnetum* is a broad leaved plant found in tropical Africa and the East Indies; *Ephedra* is a leafless desert shrub widely distributed in the arid regions of the world; and *Welwitzschia* is a bizarre rarity of the rocky barren deserts of southwestern Africa where its entire habitat has been set aside as a botanical reserve. The group as a whole is notable for its net-veined leaves, vessels in the wood, and degenerate, spiked strobili, all of which have been cited as evidence for a relationship with the angiosperms. Most botanists, however, believe that this is a very highly specialized dead-end group.

The Coniferales are the dominant and conspicuous gymnosperms and are represented in the present day flora by some 50 genera and about 500 species, widely distributed in all parts of the world, but most numerous in the temperate and sub-arctic regions where they form extensive forests encircling the globe. Their adaptation to colder regions (by virtue of their small-surfaced leaves) has undoubtedly been an important survival factor. By contrast, the large leaved cycads, being restricted to the tropics where they evolved, are on the threshold of extinction in these areas as a result of intense competition from the angiosperms.

The conifers arose in the Pennsylvanian as a group of branched tree forms having dense solid wood consisting of tracheids with *borded pits*, and simple tightly clustered leaves. The southern hemisphere conifers have rather broad linear leaves and the northern hemisphere ones possess xerophytic needle or scale leaves. The conifers produce two kinds of strobili, in either a monoecious or dioecious habit and the cones may be either the aggregated dense type of the pine, or the simple fleshy kind found in the yews.

This group has many members of great economic worth, as well as some of botanical interest. *Sequoiadendron gigantea* (the Big Tree or Giant Sequoia of the California forests) is renowned for its enormous size and longevity. The stunted, gnarled, slow growing *Pinus aristata* (Bristlecone Pine) is believed to be the oldest known plant; specimens more than 4,000 years old are still alive in California. *Metasequoia glyptostroboides* (Dawn Redwood), long known from its fossil remains, was discovered alive in the remote interior of China in 1944 and together with *Ginkgo biloba*, is one of the most popularly cultivated "living fossils" today.

The reproductive process is a variable one in the gymnosperms, and the separate articles on the individual groups should be consulted for specific details. Basically the vegetative sporophyte tree or shrub produces megasporangial (seed bearing) and/or microsporangial (pollen bearing) strobili (compact branches with fertile appendages). The microsporangial strobili vary in length, from over half a meter (*Cycas*) to a centimeter or less (*Pinus*). They may be single and terminal in position (*Zamia, Cycas*) to several and axillary (*Pinus, Macrozamia*). The microsporophylls may be hard, thick structures (*Zamia*), flexible scales (*Pinus*), or soft flexuous filaments (*Ginkgo*), and may bear as many as a thousand microsporangia (pollen sacs) each (*Cycas*) or as few as two (*Pinus*).

Many microsporocytes develop within each pollen sac and meiosis results in large numbers of dustlike pollen grains. Anyone who has parked a car under a mature pine tree at pollen shedding time in the spring can appreciate the magnitude of this process. In some genera the microspores are provided with wing-like outgrowths of the wall which make them even more easily air borne. The pollen nucleus divides mitotically to form a trinucleate immature male gametophyte, (one prothallial, one generative, and one tube nucleus) and in this condition it is shed and carried by the wind to the young open megasporangiate cone and finally to the micropylar canals of the ovules. Here the male gametophyte completes its development. The prothallial nucleus degenerates, the tube nucleus prompts the growth of the pollen tube, and the generative nucleus produces a stalk cell (possibly the remains of the antheridium) and a body cell (subsequently divides to form two sperm). The pollen tube digests its way through the megasporangial wall to the archegonia, ruptures, releases its gametes, and fertilization occurs. The sperm may be flagellated cells (cycads and *Ginkgo*) or more typically, are non-motile nuclei.

In contrast to the megasporangial strobilus of the cycads (which is basically similar to that of the microsporangial, i.e. a collection of sporophylls), the seed-bearing cone of the conifers is a very complex organ which has been interpreted by Florin to be a compound structure, representing an *inflorescence*, or tight cluster of fertile branches. Thus, each component (termed an *ovuliferous scale*) is not a sporophyll at all, but a highly modified, reduced lateral shoot consisting of a bract (representing the vegetative part of the shoot) and an axillary ovule-bearing appendage (representing the reduced sporophylls) which is sometimes free of the bract as in *Pseudotsuga* and *Abies* and sometimes fused to it as in *Pinus*. One or two ovules are formed at the base. In some genera, no megasporangial strobili are produced. In *Taxus* and *Podocarpus* for example, the ovule is terminal on a short lateral shoot (probably equal to a single cone scale complex of Pine). A fleshy and often brightly colored *avil* grows up and around the ripening ovule. In *Ginkgo*, the ovuliferous structure consists of a stalk bearing two or three ovules, each subtended by a *collar*. The stalk and collar have been variously interpreted as a cone axis and reduced sporophylls respectively, but this is by no means clear. The strobili of the Gnetales (both male and female) are very peculiar structures in which sterile appendages, fusion of parts etc. are very complicating features.

A single megasporocyte is formed in each ovule, and by meiosis four haploid megaspores are produced, three of which degenerate and are resorbed. This is typically the stage of the ovule at pollination time. The cone scales are fleshy and are apart at their outer margins to receive the pollen. Following the transfer of the microspores, the cone scales close tightly and harden, and the megaspore develops into an endosporic free-nuclear gametophyte. During the spring of the second year, the gametophyte becomes cellular, and several archegonia differentiate at the micropylar end. The male gametophyte or pollen tube also completes its development during this period and in late spring of the second year fertilization occurs. The zygote develops first as a free nuclear stage consisting of as many as 1,028 nuclei in *Dioon* or as few as 4 in *Pinus*. Following wall formation, three tiers of cells can be distinguished—a lower *pro-embryo*, a middle *suspensor* and an upper *rosette*. A rapid elongation of the suspensor cells forces the embryo into the center of the female gametophyte which is then partly digested as the embryo grows to full size.

If more than one egg is fertilized, the first to get buried in the female gametophyte survives. The others are resorbed. Most conifer embryos have several cotyledons but other gymnosperm embryos have two cotyledons. The integument hardens as a seed coat and the ripe ovule is ready to be shed. There are minor exceptions to this latter. In *Ginkgo* and many of the cycads, the outer part of the integument becomes very enlarged and malodorously fleshy, while the inner part becomes hard and stony, so that the structure resembles a plum-like fruit, although it is really a highly developed seed.

One of the distinctive features of gymnosperm seed production is the time lapse between pollination and fertilization. In pine, for example, a whole year (from one spring to the next) elapses from the time the pollen grains lands at the micropyle and the time the sperm nucleus fuses with the egg. During this first year the female and male gametophytes develop, the cone scales close tightly, harden and enlarge. During the second year fertilization occurs and the seeds ripen. In the spring of the third year the cone scales open to release the seeds. Another interesting feature is the development of a free-nuclear stage in the young female gametophyte and young embryo of nearly all gymnosperms.

JOAN EIGER GOTTLIEB

References

Arnold, C. A., "Origin and relationships of the cycads," Phytomorphology 3: 51–65, 1953.

Chamberlain, C. J., *The Living Cycads*, Chicago, The University of Chicago Press, 1919.

ibid., Gymnosperms: Structure and Evolution, Chicago, The University of Chicago Press, 1935.

Coulter, J. M. and C. J. Chamberlain, *Morphology of the Gymnosperms*, Chicago, The University of Chicago Press, 1917.

Delevoryas, T., "Morphology and Evolution of Fossil Plants," New York, Holt, Rinehart and Winston, 1962.

Eames, A. J., "Relationships of the Ephedrales," Phytomorphology 2: 79–100.

Esau, K., *Plant Anatomy*, New York, Wiley, 1965.

Florin, R., "Evolution in Cordaites and Conifers," Acta Horti Bergiana 15: 285–388, 1951.

Foster, A. S. and E. M. Gifford, Jr., *Comparative Morphology of Vascular Plants*, San Francisco, Freeman, 1959.

Johansen, D. A., *Plant Embryology*, Waltham, Chronica Botanica, 1950.

Seward, A. C., *Fossil Plants. Vol. 3*, London, The Cambridge University Press, 1917.

ibid., "The story of the Maidenhair tree," Science Progress 32: 420–440, 1938.

GYNANDROMORPH

An animal in which are combined normally distinctive female and male characteristics (literally: *female-male form*). Gynandromorphs have some parts purely female and some purely male and are distinguished from *intersexes* (which are wholly or in part intermediate in sexual characters) and *hermaphrodites* (which have in the same individual fully functional male and female reproductive systems). Hermaphroditism is the normal condition in many snails, earthworms, leeches, flatworms, flowering plants, and some other groups. On the other hand, gynandromorphs and intersexes are always rare and usually sterile. Intersexuality often results from hybridization between species or strongly differentiated subspecies, whereas gynandromorphs are usually unrelated to hybridization.

In many cases the gynandromorph is a nearly or quite perfect bilateral mosaic, half male and half female, but in some groups of animals there are smaller sectors of cells having characters of one sex in larger fields of cells of opposite sexuality. Gynandromorphs are most easily recognized, of course, in forms with conspicuous sexual dimorphism. Many gynanders have no doubt been missed in species with superficial similarity of males and females.

Gynandromorphs in *Drosophila* have been shown to be produced by loss of one X-chromosome in an early blastomere of a female embryo, the presumption being that the cells descending mitotically from that blastomere were all male in character. In *Drosophila melanogaster* the females have two X-chromosomes and the males one X and one Y, the Y being mostly inert and maleness being determined by the single dose of X in diploid individuals; cells having one X and no Y (XO) are like those which are XY in being phenotypically male. The same mechanism of gynandromorphism has been found in fungus gnats of the genus *Sciara*.

In the Gypsy Moth (*Lymantria*) and many other Lepidoptera, a similar system operates, but in this order it is the male that normally is homogametic (XX) and the female heterogametic (XY). Here the zygote of a future gynander is at first male, and the loss of an X during one or more embryonic divisions produces the XO (male) cells of a gynandrous individual.

At least one of the Lepidoptera, the Silkworm (*Bombyx*), is somewhat different from *Lymantria* in requiring the Y chromosome for femaleness. Both XX and XO silkworms are male, and XY are female. Some *Bombyx* gynanders should therefore be expected to come from female (XY) zygotes which lose the Y during cleavage and produce XO (male) cells. But certain gynanders of *Bombyx* have been shown to result from fertilization by two separate sperm (X) of two egg pronuclei (one carrying an X and one a Y), there developing into a single embryo with half the body male (XX) and half female (XY).

In butterflies and moths many gynandromorphs have some patches of femaleness on the male side. An explanation is that there is a physiological tendency in these individuals for loss of one X, with the X being eliminated at various times during embryonic mitosis, each resulting new XO cell producing a cluster of daughter cells expressing female characters.

Most, if not all, groups of Arthropoda produce gynandromorphs, and these are especially well known in Crustacea, spiders, ants, and bees, in addition to the groups already considered.

Among birds and mammals the secondary sexual characters are controlled by complexes of sex hormones. Whatever the sex hormones present, they are distributed similarly to all parts of the body, thus preventing localized discrete areas of some purely male and some purely female characteristics. Nevertheless, rodents have been reported to be gynandrous in having the reproductive system male in one half and female in the other, and bilateral gynandromorphs of sexually dimorphic birds are known. All are exceedingly rare. True gynandromorphs of man are apparently unknown, although many pseudohermaphrodites have been reported.

CHARLES L. REMINGTON

References

Goldschmidt, R. B., "Intersexuality and development," *American Naturalist* **72**: 228–242, 1938.

Morgan, T. H. and C. B. Bridges, "Contributions to the genetics of *Drosophila melanogaster*. I. The origin of gynandromorphs," *Carnegie Inst. Publ.* 278, 1919.

h

HABROBRACON

Habrobracon juglandis (Ashmead), now known to taxonomists as *Bracon hebetor* Say, is a small (3 mm.) wasp parasitic on the catepillars of moths infesting stored cereals. As with the Hymenoptera in general, females are diploid developing from fertilized eggs, males haploid from unfertilized. The eggs are scattered over the caterpillars which have been paralyzed by stinging. This external parasitism together with the short life cycle, 10 days at 30° C, makes the insect ideal material for embryological study and for investigation of radiation-induced mutations. X-radiation work has been carried out intensively and it was with this insect that dominant lethals caused by neutrons were first shown.

Male haploidy makes it possible to distinguish induction of dominant lethals in sperm from sperm inactivation. Sperm with dominant lethals are active and penetrate the eggs which therefore fail to develop. Inactivated sperm do not penetrate the eggs which therefore develop parthenogenetically producing males. With increasing treatment there is first decreasing fecundity followed by increasing fecundity.

Because of male haploidy visible mutations even if recessive are readily detected. Mutant traits include eye and body colors and structural changes in eyes, wings, antennae and legs. The insect has been used for genetics class work and extensive linkage tests have been made. These have shown the great majority of loci to segregate independently. From the few linkages obtained maps have been constructed with intervals totalling 615. However, many of the intervals are so long that true map distance must be much greater. Although chromosome number is not large, 10 haploid, and chromosomes are very small, the entire genome as measured in crossover units is probably very extensive.

In species with haploid males, sex determination has been an outstanding problem. If the haploid is male then it would be expected that the diploid should likewise be male for the ratio of sex genes is not different in the two cases. Diploid males were in fact found in Habrobracon as soon as gene markers were obtained by mutation. These males are biparental showing dominant traits inherited from both parents. They are of low viability, a large number of them dying in the egg stage. They are also near sterile and their few daughters are triploid. Their sperm are diploid for no reduction occurs in spermatogenesis. Sex in Habrobracon is determined by a series of alleles (xa, xb, xc, etc.), the homozygotes and azygotes (haploids) being male, the various compounds being female. The locus of this series of sex alleles is between *glass* (eyes) and *fused* (antennae) on the linkage map. In three-allele crosses (xa/xb × xc for example) all fertilized eggs produce females. In two-allele crosses (xa/xb × xa for example) half the fertilized eggs are female-producing, half male-producing. Low viability of these diploid males appears if their number is compared with that of their sisters. The problem of sex determination appears to have been solved for Habrobracon, but there is evidence that this solution cannot be applied to Hymenoptera in general.

Genetic mosaics occur not uncommonly in Habrobracon. These may be gynanders, male-female mosaics, which by use of genetic markers have been shown to result in most cases from binucleate eggs with one nucleus fertilized. Study of gynanders indicates that the brain rather than external sensory organs or gonads determines the type of reproductive reactions which differ markedly in the two sexes.

Haploid male mosaics develop not infrequently from unfertilized binucleate eggs when the mother is heterozygous for one or more genes. In some of the mutant eye colors of these mosaics there is a complementary effect when substances from one type of tissue diffuse through and produce wild-type black pigment near the border of the other type of tissue. Many of these mosaic males were found with feminized genitalia and it was these "gynandroids" that gave the first suggestion by analogy with the eye colors leading to the theory of multiple complementary sex alleles.

The single word Habrobracon has been used extensively in the experimental literature to indicate *H. juglandis* (Ashmead) (= *Bracon hebetor* Say). There are a number of synonyms resulting in some confusion with other species of the genus. Experimental work has been carried out on *H. brevicornis* (Wasmael) (= *Bracon brevicornis* Wesmael). It was shown that haploid chromosome number is ten here also and that diploid males occur.

P. W. WHITING

References

Whiting, P. W., "Reproductive reactions of sex mosaics of a parasitic wasp, *Habrobracon juglandis,*" J. Comp. Psychol. 14: 345–363, 1932.

Whiting, P. W., "Multiple alleles in complementary sex determination of Habrobracon," Genetics 8: 365–382, 1943.

Whiting, A. R., "Effects of X-rays on hatchability and on chromosomes of Habrobracon eggs treated in first meiotic prophase and metaphase," Am. Naturalist 79: 193–227, 1945.

HAECKEL, ERNST HEINRICH (1834–1919)

It is really difficult to account for the amazing influence which Haeckel had on biological thought in the late 19th and early 20th centuries. It is true that he had a prodigious literary output but very few of his works have survived and many of them were subject to heavy criticism in his own time. He is probably best known for what he called "The Fundamental Biogenetic Law" which is better known in English as the "Theory of Recapitulation." This states in effect that every animal reproduces in its own life history the evolutionary series through which it itself has passed in its development from a protozoan. Thus the egg represented a protozoan, the blastula represents the sponge, the invaginated gastrula represents the coelenterate, and so on. Such obvious flaws as the facts, known even in Haeckel's time, that a large number of blastulae, including those of the coelenterata and man, do not invaginate, did not prevent the enthusiastic acceptance of this theory which probably did far more harm than good to the study of embryology. For nearly a hundred years almost every new embryological fact discovered was examined to see how it fitted into this theory and quite startlingly little thought was given to examining the structure of the theory itself. This theory, coupled with a belief in the sanctity of germ layers, led to such imbecilities as lengthy papers endeavoring to find, within the framework of these theories, an "explanation" of the perfectly obvious fact that many organs received contributions from several germ layers and that the same germ layers do not always contribute the same structures in different animals. The awkward fact that the notochord is of endodermal origin, while the germ layer theory requires that it be mesodermal, led to a mass of writings which future students of the history of biology will study as an example of the wasted effort which results from endeavoring to fit facts to theories. Another of Haeckel's less felicitous efforts was his theory of general morphology developed in a book of that name. To quote Erik Nordenskiöld ("The History of Biology," New York, Tudor, 1949), "first comes an assertion that every natural object possesses three qualities: *matter, form,* and *energy.* In connection with this idea natural sciences divided into three disciplines: *chemistry, morphology,* and *physics.* Then the knowledge of inorganic nature is divided into mineralogy, hydrology, and meteorology; and biology is divided into zoology, protistology, and botany. Thus we have here four threefold groups, all extremely ill-founded." To have produced two fallacious, and widely accepted, theories in one lifetime is a startling background for one often thought by philosophers to be a founding father of contemporary biology. Haeckel's morphological work was far happier than his philosophical. His great monograph on the Radiolaria, and his enormous account of the same group in the Challenger Reports, are still the standard reference works in this field, even though the system of classification he applied is artificial. His 1872 monograph on the calcareous sponges was also of great value and the system of primary classification that he employed, based on the structure of the canal systems, is still followed by most morphologists. His monograph on the group Monera, which he established to contain living forms lacking nuclei, has been rather unjustly attacked on the ground that most of the forms in question have subsequently had nuclei observed in them. The microscope was far from perfect in Haeckel's time and he can scarcely be blamed for not observing what he could not see. Many students of Haeckel's relation to evolution take an ambivalent view of his contribution. He was the warmest exponent of Darwin's cause in continental Europe but the often unwise fervor of his espousal was, as much as anything else, the cause of the sharp cleft which developed between the more conventional exponents of Protestant theology and the naturalists of the period. His book "Die Perigenesis der Plastidule" set fire to the oil which others were endeavoring to cast on these troubled waters by, to oversimplify his arguments, explaining that the soul was the physical energy which caused the molecules of living matter to have the atomic structure which he attributed to them. Again, to quote Nordenskiöld, ". . . indeed, the whole plastidula theory sounds like a romance; in producing it Haeckel had abandoned himself entirely to romantic natural philosophy and there he remained for the rest of his life."

PETER GRAY

HAIR

Hair, fur and wool have a common origin and they stem from the SCALES of the age of reptiles, they in turn from fish scales. Many snake scales have a pattern similar to human hair scales.

Fur consists of as many as six types of fibers in one area: (a) *guard hair* is longest, the tip end is pointed, enlarges and gets narrower in the basil one-third end, the second one-third slender and having a characteristic pattern depending on the species and the basil one-third having a scale pattern which becomes more primitive toward the base; (b) *long fur hair;* (c) *fur hair;* (d) *short fur hair;* (e) *wool fur hair* and (f) *wool.* Many animals possessing fur may have some of the above characteristics in b, c and d.

Animals with *wool* have the fibers of approximately the same diameter and usually have prominent scale pattern characteristics for the species. There may be a double refractive index, lustrous or nonlustrous. The prominent sharp edged scales result in cloth fabric which is par excellent for resisting extremes in abrasion.

Hair is characterized as being cylindrical with variations as flattened (*kinky*), irregular flattening (*pubic hair*), varying in diameter from .005+ inches to one-fourth inch as in walrus whiskers.

The horn on the rhinoceros is a hair mat plated and the longest was 38 inches on a female.

Hair usually consists of three portions—the medulla, the cortex and the scales. The central portion, the *medulla,* consists of cubic cells usually with pigment cells and in some animals air spaces. These spaces may be filled with a fluid which, if removed, may result in grey hair or white hair.

The second area, the *cortex,* consists of spindle shaped cells interspersed with pigment cells occasionally, in some races always present. In many individuals microscopic branch canals, *canalculi,* which may ramify through the cortex and can be demonstrated by placing in a high vacuum and releasing turpentine phenol indophenol to demonstrate their presence.

On the outside is an overlapping layer os *scales,* ranging from 1 to 10 up to 1 to 16. These scales may form a regular or a very irregular pattern. The inheritance of a sulphate quotient from the parents is a large factor in the individuals pattern which is a principle characteristic in the fourteen items of identification of an individual from one hair.

Preparation of hair for microscopic examination: clean hair with ether to remove dirt and grease, and place in compression clamp—(1) ⅛ inch 1 × 3 brass plate; (2) 1 × 3 hard fiber board; (3) 10 mm 1 × 3 Dow thermoplastic; (4) 2 mm thermoplastic; (5) hair; (6) 2 mm thermoplastic; (7) 10 mm thermoplastic; (8) 1 × 3 hard fiber board; (9) 3—U clamps to tighten. Place in a drying oven and hold for fifteen minutes at 90° C. Remove the 2 mm thermoplastic and the impression of the scale pattern will be pressed in the plastic. The sides may be trimmed and the strips cemented to 1 × 3 glass slides.

MILTON W. EDDY

References

Bachrach, Max, "Fur," Englewood Cliffs, N. J., Prentice-Hall, 1930.
Eddy, M. W., "Hair Classification," Proc. Penna. Acad. Sci. **12:** 19–26.
Hausman, L. A., "The Microscopic Identification of Commercial Fur Hairs," Scientific Monthly, **15:** 70–78.
Locht, T., "Atlas der Menschlechen und Tierischen Haare," 306 p., illus. Leipzig, Schops, 1938.
Lynn, A. G. and B. F. Short, eds., "Biology of the Skin and Hair Growth," New York, Amer. Elsevier, 1965.

HALES, STEPHEN (1677–1761)

Stephen Hales was born into the rather unfortunate status of a younger son of a younger son of a minor member of the nobility. In accordance with the convention of the time, he therefore went to Cambridge and studied for holy orders which he received shortly before being appointed to the perpetual curacy of the village of Teddington in Middlesex. While at Cambridge he had mixed his theological studies with quite intensive work in chemistry, physics, and botany, the last of which was to remain his life interest. However, unlike all his contemporaries, his primary interest was in function and not in form, and his careful experimental methods, and his impeccable deductions from his well controlled experiments, have rarely been surpassed in quality by any botanist. His primary interest was the question of plant growth. He carefully measured the water and solid absorption of plants and came to the conclusion, in the absence of all direct evidence, that they must be capable of synthesizing material from other sources. Since the only other source available to them was air, it was a logical deduction that air penetrated to the leaves and Hales, by what can only be regarded as a stroke of intuitive genius, suggested that air interacting with light, was the source from which plants derived food not available to them through their roots. The results of these investigations published in his famous "Vegetable Staticks" earned for him a fellowship in the Royal Society. His subsequent investigations into blood, though of some interest to his contemporaries, produced little new to science. In addition to his scientific studies, Hales was a humanitarian of note and his concern with the inmates of prisons led him to invent a "ventilator" (a word which originated with him) for the purpose of inducing drafts of air in jails, hospitals, and ships holds. It was also his humanitarian interest in the lack of food for the poor which led him to discover the use of sulfur dioxide as a fumigant to prevent the destruction of stored wheat by weevils and to experiment on the preservation of meat by various methods of salting.

PETER GRAY

HALLER, ALBRECT VON (1708–1777)

This Swiss anatomist-physiologist did a vast amount of work in many areas. Haller published numerous poems, novels and medical tracts. While a professor at the University of Gottingen, he edited a monthly journal, the *Gottingishe gelehrte Anzeiger,* which he contributed 12,000 articles to in a period of seventeen years. Haller's greatest contribution to biology is found in his method of physiological experiment. He showed that irritability, the ability to contract when touched, is a property of many tissues. Haller distinguished this reaction from normal muscular action. He also proved all sensation to be channeled through the nerves. Haller's eight-volume work *Elementa Physiologiae* is the first modern physiological study.

DOUGLAS G. MADIGAN

HARRISON, ROSS GRANVILLE (1870–1959)

Two American Zoologists have been voted and awarded the Nobel Prize for their discoveries in the field of genetics. A third, Ross Granville Harrison was voted the prize by the Nobel Committee in 1917 for solving the problem of nerve outgrowth. Due to World War I, the Institute made no awards in Physiology and Medicine from 1914–1918, and so the award was never made.

Born in 1870, he combined the older traditional ideas with the later and more sophisticated ones. His scientific interest began during his undergraduate years at The Johns Hopkins. It continued in his graduate studies with W. K. Brooks and later in his work with Moritz Nussbaum at Bonn, Germany. He then became interested in the problem of nerve outgrowth which he analyzed by transplanting limbs. This did not give a crucial answer so he invented the method of tissue culture which provided the critical experiment leading to the solution of the problem. This contribution provided the foundation for a new outlook in neurology for the neurone was shown to be both the morphological and the physiological unit of the nervous system. The tissue culture method has since become vital to the study of disease. Polio vaccine was made possible by its application. By this approach Harrison changed biological thought. He brought to his field the idea of the relation of the whole to the part and yet separated the two into meaningful philosophical and logical entities. Harrison's later studies on the asymmetry problem involved experiments upon the ear, limb and eye, all of which follow much the same rules of development. These studies constitute the broad foundation for macromolecular research.

He came to Yale from The Hopkins Medical School in 1907 and built the outstanding department of Zoology in the world during his years of service. It attracted not only students from America but scholars from abroad. His Chairmanship of the Department and Directorship of the Osborn Zoological Laboratory, which he built, marked the high point in productive zoological research in this country and in science at Yale.

Harrison retired in 1938 and in 1939 was awarded a D.Sc. from Yale. His retirement was transitory for he immediately assumed the responsibilities of Chairman of the National Research Council which he held throughout the years of World War II. With Dr. Jewett, then President of the National Academy of Sciences, he served the country in many useful ways on national policy including that of the Manhattan project which produced atomic fusion and laid the groundwork for atomic fission.

Harrison's academic interest never flagged. He returned to his laboratory after his retirement from the Research Council and delivered the Silliman lectures at Yale in 1949. He substantially completed a comprehensive study of the limb in which he brought together his old and new work to explain in greater detail the processes involved in the attainment of asymmetry.

He died September 30th, 1959, after a wonderfully fruitful life. His approach to the problems of science has influenced biological thought in as great a way as did that of Darwin.

J. S. Nicholas

HARVEY, WILLIAM (1578–1657)

The unvarying statement that "Harvey discovered the circulation of the blood" is unjust both to the nature of the discovery and to the talents of the discoverer. No one supposed in Harvey's time, or indeed for many centuries before him, that the blood was stationary. There was a fairly general misapprehension that it ebbed randomly backwards and forwards after being developed from food in the liver, carrying a "natural spirit" to the lungs and a "vital spirit" from them. This vague concept of "circulation," moreover, depended on the belief that the ventricles were divided only by a perforate septum and that the sucking action caused by their expansion regulated the movement of the fluid in the great vessels. Harvey's own teacher Fabricius, among others, had demonstrated that the ventricles did not communicate and had also found valves, the function of which he did not suspect, in the veins of the arms. To this welter of half-truths, Harvey applied the methods of accurate observation and logical deduction. He dissected every animal, vertebrate or invertebrate, on which he could lay his hands and observed the beating of their hearts through a lens. He studied the pulse of man and animals

in health and disease and thus found that contraction, not expansion, was the heart's most important function. By 1616 his students at the Royal College of Surgeons were learning that the flow of blood was not aimless and in 1628 he published *Exercitatis anatomica de motu cordis et sanguinis* ("An anatomical exercise [discussion] on the motion of the heart and blood"). This advanced the thesis that a constant volume of blood circulated continuously in a closed system, being impelled by the contraction of one ventricle to the lungs and, on its return via the atrium, being impelled by the other ventricle round the body. This thesis, as startling in its day as the discovery of isotopes in ours, was based on such a wealth of anatomical and experimental data, of mathematical demonstration, and of logical analysis, that only halfhearted attacks were made on it. This is the more remarkable in that Harvey failed either to demonstrate, or ever postulate, the existence of capillaries.

It was fortunate for Harvey that he was the son of eminently respectable, and wealthy, parents who not only supported him through school and university (Cambridge) but also sent him to Italy for four years of post-doctoral study. He returned from Padua in 1603 sufficiently endowed with learning, recommendations and money to secure an appointment to the Royal College of Surgeons and to marry the daughter of Queen Elizabeth's physician. He speedily built up a practice at court and in 1618 was appointed physician to James I. In all this he showed great wisdom for his revolutionary theories would have received short shrift without such powerful protection. Indeed he probably did more than any man of his time to establish that observation and experiment were as respectable as classical scholarship.

On the outbreak of the revolution, he accompanied Charles I to Oxford where he continued those studies in reproduction which had replaced his interest in circulation. The fall of this city to the insurgents caused his return to London where, now protected more by his reputation than his wealth, he continued his research. He was 73 when, in 1651, he published his second great work *Exercitationes generatione*. This laid the foundation for modern embryology, particularly in its elucidation of fertilization and the role of the egg. Harvey did not, as is popularly supposed, write "*omne ex ovo*" (everything [comes] from an egg). This remark decorates the title page and is some unknown artist's summary of Harvey's more scholarly *omnia omnino animalia, etiam vivipara, atque hominem adeo ipsum, ex ovo progigni* ("absolutely every animal, including [those which are] viviperous, and even very man himself is brought forth of an egg").

Harvey died peacefully in his 80th year, leaving his fortune to the Royal College of Surgeons to endow an annual lecture which is still offered in his honor.

Peter Gray

HEART, INVERTEBRATE

Nearly all hearts are modified blood vessels. In some animals where there are vessels but no special heart, some of the vessels themselves may perform the function of propulsion through peristaltic waves of contraction. Examples are to be found among NEMERTEANS, ANNELID worms, and the invertebrate chordate *Amphioxus.*

In the earthworm there are five pairs of enlarged lateral vessels lying in segments 7 to 11. These serve as hearts, although the dorsal and ventral vessels also are contractile. The POLYCHAETE *Arenicola* has one pair of lateral hearts, each consisting of an auricle and a ventricle.

Among MOLLUSCA blood passes through veins to the auricles from the gills and the nephridia. The heart typically consists of a single muscular ventricle into which open two thin-walled auricles, but the heart of tetrabranchiate cephalopods has four auricles, and that of most gastropods, where the gill and nephridium on one side of the body disappear during development, has only a single auricle. The molluscan heart lies in a remnant of the coelom, the pericardial cavity, which lies in the vicinity of the rectum in PELECYPODS. The ventricle partially or completely surrounds the digestive tract in a few GASTROPODS and most of the pelecypods. Frequently the only arteries leaving the heart are the anterior and posterior aortae, carrying blood respectively in the direction of the head and of the visceral mass. Accessory branchial hearts drive the blood into the gills in the dibranchiate cephalopods and in many pelecypods, while gastropods have an enlargement on the aorta that serves as an accessory heart.

Simple bulbular hearts occur in the BRACHIOPODA, but little can be said about them other than that they are small contractile bulbs near the stomach.

Among ARTHROPODA and ONYCHOPHORA the primitive heart is a long contractile vessel suspended by alary ligaments or muscles within a pericardial sinus that lies dorsal to the digestive tract. Blood from the pericardial sinus passes into the heart during diastole through segmentally arranged ostia. The ostia are guarded by valves, and the blood therefore can pass through the openings only in one direction. Systole is a peristaltic wave of contraction of the tubular heart, while diastole is accomplished by contraction of the elastic alary ligaments or muscles. Hearts of Onychophora and of the Anostraca (see CRUSTACEA) extend nearly the entire length of the body, but in other forms they are more limited. For example, the heart is confined in the insects to the abdominal region, while in *Limulus* it lies both in the prosoma and in the opisthosoma, but it does not extend all the way forward in the former, nor all the way back in the latter. It is confined to the preabdomen in *Scorpio* and in spiders, though it is shorter and more concentrated in the latter. The heart has become short and very concentrated among decapod crustaceans, so that its tubular character is no longer clear, and its three pairs of ostia are not obviously segmentally arranged. Some of the smaller Crustacea (cyclopoid copepods and cladocerans), on the other hand, have a greatly reduced heart which bears only a single pair of ostia. In still other entomostracans, mostly small forms, the heart is lacking entirely.

Blood leaves the hearts of those arthropods that have more primitive circulatory systems through a single anterior artery. This may be short (CLADOCERANS, cyclopoid COPEPODS and anostracans), or it may be longer (INSECTA, where it extends from the abdominal heart through the thorax to the head). The number of arteries leaving the heart may be greater, as in the hearts of the scorpion and *Limulus,* which have paired lateral arteries and a posterior artery in addition to the anterior aorta. In decapod crustaceans typically there are five arteries supplying the anterior regions of the body, one artery supplying the posterior dorsal region, and one passing ventrally from the posterior region of the heart.

Most insects have accessory hearts in the form of special muscle fibers in the blood sinuses, whose contractions drive blood into the antennae and the legs.

Blood vessels occur in the hemichord *Balanoglossus,* but they are not lined with endothelium. In the proboscis region the dorsal vessel is enlarged and has a contractile lower wall which propels blood into the excretory organ. The urochord heart is a simple muscular enlargement lying near the stomach. Intercellular channels without endothelial linings leave each end of the heart. One channel carries blood towards the gill slits and the other to the viscera. The heart periodically reverses direction, so that the blood is driven first to the gill slits, then to the viscera. Similar reversals of heart beat have been observed in many pupal insects.

The physiology of invertebrate hearts is poorly understood. In *Limulus* removal of the cardiac ganglion stops the heart, although the heart muscle still maintains its contractility. Weak solutions of acetylcholine, which stimulate synapses and neuro-muscular junctions, accelerate the intact heart, and therefore it is clear that the heart of *Limulus* is neurogenic. Most crustaceans also are thought to have neurogenic hearts, although certain primitive forms such as the fairy shrimps appear to have myogenic hearts.

The hearts of molluscs are inhibited rather than accelerated by acetylcholine, and therefore are to be considered as myogenic. There is a difference of opinion concerning the origin of the beat in insect hearts. In some forms it has been demonstrated that the heart is myogenic, for the heart continues to beat in the absence of all connections with ganglion cells. In other insects ganglion cells associated with the heart are present and pharmaceutical evidence suggests that they are neurogenic.

Most hearts are innervated, and the nerves to them may have either inhibitory or acceleratory functions, or both. In many insects there are both inhibitor and accelerator nerve fibers supplying the heart, and earthworm hearts are inhibited by electrical stimulation of the ventral nerve cord. Stimulation of the visceral nerve in the clam retards the heart rate, whereas in some snails and chitons stimulation of the cardiac nerves increases it. Much more work is needed for a clearer understanding of these matters.

CHARLES C. DAVIS

References

Borradaile, L. A., et al. "The Invertebrata," 4th ed. Cambridge, The University Press, 1961.
Kükenthal, W. and T. Krumbach, "Handbuch der Zoologie," Berlin. Walter de Gruyter, 1923–1960.
Meglitsch, P. A., "The Invertebrates," New York, Oxford University Press, 1967.
Parker, T. J., et al. "A Textbook of Zoology," 2 vols. 6th ed. London, Macmillan, 1949.
Prosser, C. L. et al. "Comparative Animal Physiology," Philadelphia, Saunders, 1961.

HEART, VERTEBRATE

The heart is an integral part of the circulatory system. It is a hollow organ composed of muscle. The muscle is of the cross-striated type and is made up of cells that are relatively long and somewhat connected to one another to form an apparent syncytium. Histologically this is known as cardiac muscle. It differs from the skeletal muscle (also cross-striated) in that the nuclei are centrally located and the myofibrillae on which the cross-striations are located, are thicker and less numerous.

The heart function is to pump the blood, which enters it, out through the arteries. The muscle of the heart has its own blood supply and drainage and is not supplied by the blood that passes through it.

Anatomy. In the *fish and larval amphibians* the heart is a more or less straight muscular tube consisting of 4 chambers; posteriorly the sinus venosus receives the venous blood from several large veins. Proceeding anteriorly, next is the atrium followed by the ventricle and then the conus arteriosus. Only venous blood is propelled through the heart.

In the adult amphibians the atrium is divided into two cavities, the left and right atria (auricles). The interatrial septum has some perforations in it. The sinus venosus which receives the venous blood from large veins empties into the right atrium. The left atrium receives the arterial blood from the lungs which it has been oxygenated. When both atria contract, valves between sinus venosus and atrium close and the blood is forced into the ventricle. Very little mixing of the venous and arterial blood takes place here. When the ventricle contracts, valves prevent the blood from returning to the atria and thus it is forced out of the conus arteriosus which is guarded by valves. The blood then continues into the bulbus arteriosus (derived from the posterior part of the ventral aorta). The bulbus arteriosus has a longitudinal flap-like valve along its wall. This directs the blood either into the systemic arteries or the pulmonary arteries (carry venous blood).

In the reptiles the heart is similar to that of the amphibia with the exception that (1) the interatrial septum is not perforated, and (2) the ventricle in some is in the process of dividing into left and right ventricle. The interventricular septum is incomplete. The conus arteriosus has become divided into three parts: (1) the pulmonary artery, (2) the left aorta, (3) the right aorta.

In birds there are four chambers to the heart. Two atria and two ventricles. The interatrial and interventricular septa are complete. The sinus venosus has disappeared except for the valves at the entrance of the large veins to the right atrium. As in the reptiles the left ventricle is larger than the right one. There is no conus arteriosus, the pulmonary artery arises from the right ventricle and the aorta from the left ventricle. These two arteries lie next to each other and come from the ventricles in the region of the bases of the atria. The aorta arches dorsally to the right and becomes the dorsal aorta.

In the mammals there are the four chambers to the heart as in the birds. The pulmonary artery and the aorta arise in the same way. However the arch of the aorta in the mammals goes dorsally to the left.

As in the vertebrates previously mentioned there are valves (semilunar) between the right atrium and right ventricle and similar valves between the left atrium and

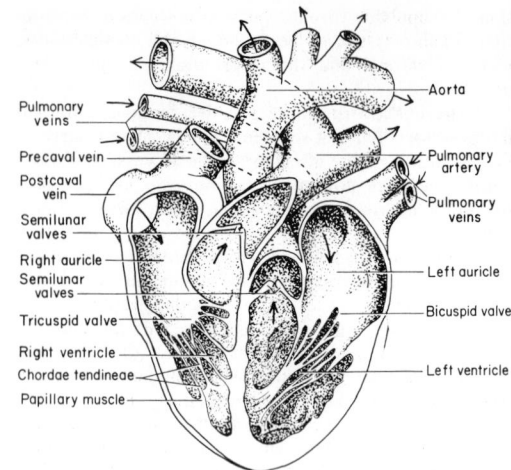

Fig. 1   Frontal section of the mammalian heart showing the chambers, valves, and principal blood vessels. (From Goodnight, Goodnight and Gray, "General Zoology," New York, Reinhold, 1964.)

left ventricle. Also there are valves between the ventricles and the respective arteries that lead from them.

Embryology. The heart arises from left and right primoria ventral to the embryonic foregut. It is primarily mesodermal in origin. These primordia form two endothelial tubes which connect anteriorly with the developing ventral aortae and posteriorly with left and right vitelline veins. The mesoderm just lateral to the endothelial tubes thickens and forms the epimyocardium, which differentiates into the heart muscle and the visceral pericardium. The paired endothelial tubes fuse into a single median one. This tissue forms the endocardium of the heart. From this point on through differential growth the various chambers of the heart are formed and the development proceeds according to the type of vertebrate the embryo belongs to.

Function. In the fish, amphibians, and reptiles the contraction starts in the thin muscle wall of the sinus venosus and proceeds, by means of an internal conduction system, over the atria and then the ventricles. This conduction system can be either differentiated from cells similar to those that form the muscle cells or nerve cells with their processes (see conduction system below).

In birds and mammals the sinus venosus is absent. At the junction of the great veins with the right atrium there is the sino-atrial node (modified muscle cells) which has branches extending throughout the atrial walls. They do not continue into the ventricular walls. At the junction of the atria and ventricles there is a node —A-V node—(node of Tawara) and the bundle of His which connects with it. The former node is mainly located in the septal wall of the right atrium and the latter branches beneath the endocardium of the right ventricle and fans out through the walls of both ventricles. These branches are known as Purkinje fibers.

There is an extrinsic innervation to the heart from the parasympathetic and sympathetic systems, in the vertebrates. Stimulation of the vagus nerves slows down or inhibits the heart action while stimulation of the sympathetic nerves accelerates the heart activity.

The heart of the vertebrates is called a myogenic

type. This means the initiation of the contraction of the heart is within the heart and not primarily stimulated by nerves or hormones from outside it.

The contractile elements of the heart muscle lie within the muscle cells and are called the myofibrillae. These are composed of chains of long protein molecules (polypeptide chains). The proteins localized in the myofibrillae are actin and myosin. During contraction the myofibrillae remain straight and do not coil, twist or accordion pleat. It is thought that the changes are internal ones.

The energy necessary for the contraction of the myofibrillae may be derived from two sources. One, the formation of lactic acid from glycogen (this is called glycolysis). Two, regular oxidative processes utilizing free oxygen. These two processes each result in the production of high energy phosphate compounds. Chemical energy is potentially established in compounds such as adenosine triphosphate (ATP) and phosphocreatine. The process of setting up this potential energy involves chemical processes utilizing ATP, ADP (adenosine diphosphate) and AMP (adenosine monophosphate).

J. C. GRAY

References

Willier, B. H. *et al.* eds. "Analysis of Development," Philadelphia, Saunders, 1955.
Davson, H., "A Textbook of General Physiology," 3rd ed. Boston, Little Brown, 1964.
Fruton, J. S. and S. Simmonds, "General Biochemistry," 2nd ed. New York, Wiley, 1958.
Maximow, A. A. and W. Bloom, "A Textbook of Histology," 8th ed. Philadelphia, Saunders, 1962.
Neal, H. V. and H. W. Rand, "Comparative Anatomy," Philadelphia, Blakiston, 1936.

HELIOZOA see ACTINOPODA

HELMONT, JOANNES BAPTISTA VAN (1577–1644)

Van Helmont is best known for his classical experiment with a willow branch. A five pound branch was placed in 200 pounds of dry soil which, in turn, was watered with rain water or distilled water for five years. At the end of this time the weight of the willow twig had increased to 169 pounds three ounces and the dry weight of the soil remained the same. Van Helmont erroneously concluded that the material of the plant had been formed from water. The experiment is remembered primarily because it represents an early attempt to employ the balance in determining the source of plant constituents and even though his conclusions were wrong, his observations were carefully recorded. One wonders, now, what would be said of Van Helmont today if, as suggested by Palladin, the conclusion from the experiment had been that "the greater part of the material of plants does not come from the soil."

Van Helmont's early training was in medicine but he was also a very excellent chemist, so much so that he received many attractive offers from princes for his services. He preferred to carry on his experiments in his own private laboratory. He possessed a considerable interest in the physiology of both plants and animals. The breadth of his interests is indicated by the fact that at an early age he studied philosophy and theology and became engrossed in mysticism.

Actually Van Helmont was an alchemist. He believed in spontaneous generation and was the author of a recipe for the production of mice from meal. But great credit should be given to Van Helmont for stepping across the gap between alchemy and chemistry. He seems to have done this in his very carefully planned experiment with the willow twig. He did not subscribe to the Aristotelian doctrine that matter is composed of air, earth, fire and water. He showed that carbon dioxide can be produced by treating limestone or ashes with acids, from burning coal and from fermentation of beer and wine. It was Van Helmont who introduced the word "gas," no doubt because he was aware that carbon dioxide is produced during alcoholic fermentation.

ERSTON V. MILLER

HEMICHORDATA

Three classes of primitive Deuterostomia, viz. *Enteropneusta, Pterobranchia* and POGONOPHORA, are combined to form the phylum Hemichordata. They all possess a very primitive nervous system situated at the base of the epidermis. There is no brain proper although a concentration of nervous tissue can be observed near the anterior end. The principal nervous strands are middorsal and midventral. One of the main characteristic features of the phylum lies in the formation and arrangement of the coeloms. In the anterior region of the body there is always a single coelom which communicates with the exterrior through one or two excretory ducts. The second coelom is paired and does not contain any organs. The third coelom is also paired and in it are embedded the paired gonads. The mode of development in the three classes referred to this phylum is also—as far as we know—very similar.

Enteropneusta ("acorn worms"): Characterized by the transformation of the two anterior body divisions into a burrowing organ, consisting of proboscis and collar. The species measure 50 to 500 mm in length and 2 to 20 mm in width. They occur on muddy and sandy bottoms of the oceans from high water level down to great depths. Many build complicated burrows into the substrate or they construct systems of tubes glued together from sediment and mucus. Some species move freely on deep water substrates. It has been known for some time that the enteropneusts eat large amounts of sediment. The water is ejected through numerous gill slits in the pharynx. More recently it was learned that some (perhaps many) species produce a current of mucus on the body surface and direct it into the gut where the food particles that attach to the mucus are digested. Ingestion of material through micropinocytosis also in the skin is to be taken into consideration. .

Reproduction is sexual. The majority of the more primitive forms perfer deep or cold water, have few eggs which are rich in yolk, and undergo direct, benthic development. The more specialized forms prefer shallow or warm water, have numerous eggs which are poor in yolk, and possess complicated planktonic larvae, the Tornarians. The latter are characterized by the development of richly folded bands of cilia by means of which small plankton is propelled into the gut.

Three families are distinguished, viz. *Harrimanidae, Spengelidae* and *Ptychoderidae,* mainly on the basis of increasing differentiation of their inner anatomy. The characters involved include the structure of pieces of skeleton between proboscis and collar, and of the dorsal nervous cord, the state of formation of lateral septa in the trunk, of hepatic sacculations and of lateral extension of the trunk, the genital ridges. Nevertheless, the whole group is very uniform.

Pterobranchia. Characterized by the differentiation of the first body division into a large suction disc and by tentaculate arms which originate at the dorsal surface of the second body division (region of the first and second coelom). The body measures one-third to a few mm in length. Most of the species live on hard bottom between 100 and 66 m depth. In the West Pacific and Indian Ocean a few species penetrate up to low water level. Two orders are distinguished: firstly the *Cephalodiscidea,* represented by solitary individuals, forming aggregations housed in a common secreted encasement (up to a height of 25 cm). One genus from Japanese waters, *Atubaria,* is free-living. Secondly, the *Rhabdopleuridea,* represented by individuals which form true colonies in which the members are in organic continuity, each enclosed in a chitinous tube. They probably feed exclusively on plankton and detritus, the food being transported to the mouth opening by a mucus and water current propelled by the cilia of the tentaculate arms. In the *Cephalodiscidea* the water is ejected through one pair of gill slits which is absent in the *Rhabdopleuridea.*

Reproduction is mainly asexual through budding. In the *Rhabdopleuridea* the buds remain with the colony while in the *Cephalodiscidea* they are set free. Sexual development is unknown in the *Rhabdopleuridea,* and the little that is known of it in the *Cephalodiscidea* points towards a simple (probably short-lived) larva.

The orders *Cephalodiscidea* and *Rhabdopleuridea* are distinguished by the number of arms and by the structure of the colony. Each order contains only a single family, the *Rhabdopleuridae* only a single genus.

Pogonophora. Characterized by the complete reduction of the intestine and by the differentiation of the tentacles which originate on the dorsal surface of the first body division. The species measure 50 to 200 mm in length but sometimes only 0.3 mm in width. They live in mud, down to the greatest oceanic depths. Several species occupy a depth range from 8000 to 150 m. *Siboglinum* reaches up to 23 m. Most of the species are known from the West Pacific. All the animals are solitary and live in thin chitinous tubes. They probably feed on plankton and detritus which is accumulated by means of the ciliar mechanism of the tentacles. If there are numerous tentacles they bunch up to form a tube within which food undergoes external digestion and is then resorbed. A similar effect is produced in *Siboglinum* by the corkscrew coiling of the single tentacle. More recently ultra-structure studies have shown tentacles covered with microvilli; and there is ample evidence for a passage of material (micropinocytosis) from exterior into the epidermal cells. The circulatory system which—as in all *Hemichordata*—is well developed provides for the distribution of nutrients through the elongated body.

In some species, embryos are known to be produced sexually and to undergo direct development. In other species the existence of simple larvae can be surmised.

The class *Pogonophora* is represented by two orders, *Athecanephria* and *Thecanephria* which are distinguished by features concerning the execretory ducts (coelomoducts), the spermatophores and by the presence or absence of a pericard. There are five families, distinguished, amongst others, by differences in the structure of the tentacular apparatus.

There is no full agreement as to the taxonomic position of the *Pogonophora.* Undoubtedly the closest relatives are *Enteropneusta* and *Pterobranchia,* but it is possible to put more weight on the specific features of the Pogonophora and to give them a more isolated status than is implied by their inclusion in the phylum *Hemichordata.*

Planctosphaera is a rare and peculiar larva, related to the Tornaria, which probably belongs to an unknown class of *Hemichordata.*

Xenoturbella, a small, worm-like animal from marine mud is considered by some investigators to be a relative of turbellarians, by others it is referred to the hemichordates.

RUPERT RIEDL

References

Burdon-Jones, C., "*Enteropneusta*" *in* Kükenthal, W. and T. Krumbach eds. "Handbuch der Zoologie," Vol. 3. Berlin, de Gruyter, 1956.

Hyman, L. H., "The Invertebrates," New York, Mc-Graw-Hill, 1940.

Ivanov, A. V., "Classe des Pogonophores" *in* Grassé, P.-P., "Traité de Zoologie," Vol. 5. Paris, Masson, 1960.

Van der Horst, C. J., "*Hemichordata*" *in* Bronn, H. G. ed. "Klassen und Ordnungen des Tierreichs," Vol. 4. Leipzig, Winter, 1939.

Norrevang, A., "Structures and functions of the tentacle and pinnules of Siboglinum ekmani Jägersten (Pogonophora)," Sarsia **21,** 1965.

Riedl, R., "Hemichordata" *in* "Handbuch der Biologie," Vol. 6, Konstanz, Athenaion, 1961.

Westblad, E., "*Xenoturbella bocki* n.g.n.sp. a peculiar primitive turbellarian type," Ark Zool. **1,** 1949.

HEMOCOEL

Two types of body cavities exist in triploblastic animals: the hemocoel and the COELOM. The body interior of the more primitive INVERTEBRATES is usually occupied by mesenchymal parenchyma. Normally the hemocoel is an extensive sinusoidal cavity devoid of peritoneal lining, with extensions to the appendages. Its fluid content containing corpuscles of mesodermal origin is circulated by contractions of a dorsal tubular vessel, the heart and aorta, thus forming an open circulation. In its simplest form it is limited to the phyla ONYCHOPHORA, MYRIAPODA and ARTHROPODA. With modifications for respiration it becomes more complicated in the CRUSTACEA, ARACHNIDA and MOLLUSCA.

The coelom, on the other hand, is formed from the fusion of the coela of the embryonic mesodermal somites into lateral cavities. It is the definitive peritoneal cavity of the vertebrates and of some invertebrates. It is never a part of the circulatory system.

In the tracheate Onychophora, Myriapoda and Arthropoda the hemocoelic circulation has no respiratory function: the tracheae of ectodermal origin connecting the body cells directly with the atmosphere. But in the Crustacea, Arachnida and Mollusca respiration is added by the development of gills, gill books or lung-like areas of the mantle. Such respiratory pigments as hemocyanin occur but no hemoglobin, except in Chironomus.

While the embryological steps in the development of the hemocoel vary in different groups it is best known in the insects, corroborating the findings on other groups. The most extensive recent work has been done upon the Orthoptera, notably by Wiesmann (1926) and Thomas (1936) on the "walking stick," (Carausius); by O. Nelson (1934) on *Melanoplus differentialis,* and by Roonwal (1936) on *Locusta migratoria.*

During blastokinesis in the locust and the walking stick the ventral nerve cord forms, and dorsolateral to it the mesodermal bands arise, grow forward and begin to segment. At about three days the somites form coela, and the resulting somatopleural and splanchnopleural layers grow dorsad. Above the nerve cord an epineural space, by some thought to be the remains of the blastocoele, appears beneath the yolk and splanchnopleure. This space is the anlage of the definitive body cavity. At this stage numerous earlier embryologists, from Heider in 1889 to Hirschler in 1909, claimed that openings formed in the thin medial coelomic walls to connect the lateral coela with the hemocoel, thus producing a mixocoel. Since 1926, however, the investigations of Wiemann, Paterson and others have shown this to be untrue: that the coelomic walls are gradually converted into fat body, muscle and blood cells, allowing the two cavities to unite as the definitive hemocoel. This is now the accepted concept.

From the first formation of the coela there appears at the growing dorsal junctions of the newly separated somatic and splanchnic mesodermal layers on each side single rows of large-nucleated cells, the cardioblasts. Growth carries these toward the mid-dorsal portion of the hemocoel, the cardiac sinus. Approaching one another the medial faces of these cardioblasts become concave, forming opposing grooves. The ultimate union of these, enclosing blood cells of the sinus, forms the tubular heart in the abdomen. Soon the dorsal portions of the accompanying somatic mesoderm thin out and expand to form the alary muscles stretched between the heart and body wall, while the ventral portions become the dorsal or pericardial diaphragm. Between these layers remain some pericardial cells. This mode of heart formation was first described by Claus about 1863 for the fairy shrimp (Branchipus) and has since proven true in all groups. The aorta, lying in the thorax, forms similarly but from the dorsal antennary coelomic pouches, its anterior part becoming the blood-distributing pulsatile vesicle in the head. Thus the cavity of the whole dorsal vessel is a part of the hemocoel, and the heart's muscular walls are formed from the cardioblasts and mesoderm.

The blood from the large sinuses of the hemocoel, usually separated by horizontal fenestrated diaphragms, enters the posterior segmental heart chambers through ostia, and is pumped by successive contractions through the aorta to the head and back again to the hemocoel. While the aorta has not regular ostia and is but slightly pulsatile, aortic diverticula with dorsal pulsatile organs exist in winged insects at least. As early as 1917 Brocher described these briefly in beetles, dragonflies and lepidopters. The most extensive investigation was done on the Odonata by Whedon in 1938. Blood forced into the wings is returned to the aorta by way of these diverticula. On the stalk, also, between each pulsatile organ and the aorta a pair of ostia draw blood from the visceral hemocoel. Each ostium is surrounded by a phagocytic gland-like mass, the cells of which have been seen to engulf carbon particles injected into the body cavity. These are probably similar to the ostial glands of the heart as seen by Zawarsin in 1911.

Of the remaining phyla especial interest attaches to the Onychophora as forms morphologically intermediate between the coelomate Annelida and the hemocoelic Arthropoda. Its more primitive development has been widely studied since the work of Balfour and Sedgwick in the 1880s. In the Myriapoda hemocoelic development is practically identical with that of insects. In the Arachnida the coelom remains dominant somewhat longer, but is finally reduced to the arthropod condition, varying mainly in the addition of gill and lung books for respiration. In the Crustacea there is great variation between the simple condition in the Branchiopoda and the gill and blood vessel possessing Malacostraca. Of the Mollusca the gastropods seem remarkable for an extensive hemocoel with heart, arteries and mantle-lung combined in a most elaborate closed circulatory system.

Thus the developmental history of the body cavity has everywhere involved embryonic competition between hemocoel and coelom, apparently with no direct evolutionary sequence, but often with the respiratory function forced upon the hemocoelic circulation by adaption to environment.

ARTHUR D. WHEDON

References

Borradaile, L. A. and F. A. Potts, "The Invertebrata," 4th ed. New York, McGraw-Hill, 1961.
Johannsen, O. A. and F. H. Butt, "Embryology of Insects and Myriapods," New York, McGraw-Hill, 1941.

HEMOPOIESIS

In the Embryo

Hemopoiesis in the human embryo occurs in three stages.

1. **Pre-hepatic period.** The first blood cells appear in "blood islands" in the walls of the yolk sac about the third week. They are mesodermal and arranged in little groups. The peripheral cells differentiate to form the endothelium of blood vessels, and the central cells become the first blood cells. They are carried into the circulation when the heart begins to beat. Later, their nuclei become smaller and their cytoplasm more refractile; at this stage hemoglobin appears. The cell is then called a primitive red cell and multiplies by mitosis.

This first generation of primitive blood cells is succeeded by a second generation of nucleated red cells similar to the erythroblasts found in the adult bone marrow. These cells multiply by MITOSIS; after they have acquired hemoglobin their nuclei degenerate, and they become typical erythrocytes.

2. **Hepatic period.** From the third to the ninth month, the LIVER is a hemopoietic organ. The formation of red cells of the second generation occurs in various hemopoietic tissues, first the liver, then the SPLEEN, and then the bone marrow.

Granulocytes appear about this time, and at the fourth month become numerous. Megakaryocytes appear, but not lymphocytes. Lymph glands develop at the fifth month; they develop rapidly, and soon the number of circulating lymphocytes is double the number of granulocytes.

3. **Lympho-medullary period.** After the fifth month, hemopoiesis begins in the marrow, spleen, and lymph glands, which at birth take on the specialization which they have in the adult. Nevertheless, totipotential tissue (hemocytoblasts) is found in the bone marrow, the liver capillaries, the splenic sinusoids, the lymph glands, the serous membranes, and in free histiocytes. These cells constitute the hemopoietic section of the reticuloendothelial system. The second section again starts from the reticuloendothelial cell but forms fibroblasts and cells which are not blood cells; the earliest cell of this kind is the hemohistioblast. It is important to remember that the reticuloendothelial system, under conditions of stress and disease, may form cells which are either hemocytoblasts or hemohistioblasts. The adult liver, spleen, and other tissues may form blood cells if the marrow cannot do so; this is called extra-medullary hemopoiesis.

Hemopoietic Functions of the Reticuloendothelial System

The reticuloendothelial system has a metabolic function, a phagocytic function, and a cytopoietic function. Only the last of these concerns us here.

In the adult, hemocytoblasts are found in the bone marrow, spleen, and lymph glands. They normally give rise to cells of the erythrocytic, granulocytic, and thrombocytic series; if situated in lymph glands, they give rise to the cells of the lymphocytic series. The monocytic or histiocytic series and the plasma cell series are differently conceived by different investigators. Probably everyone would agree with the scheme shown in Table 1, which represents the *monophyletic theory*. Some hematologists, however, go so far as to speak of the hemocytoblast and the hemohistioblast as mythical cells and derive the cells of all six series from the reticuloendothelial cells. This is the *polyphyletic theory*, which has several variations. The disagreement between the monophyletists and the polyphyletists was once almost violent, but nowadays few hematologists would care to resume the quarrel. Almost everyone, however, recognizes what is meant by differentiation and maturation.

1. **Differentiation.** Differentiation is "the appearance of different properties in cells which are initially equivalent" (Weiss). As it differentiates, a cell acquires characteristics different from those of the cells from which it originates and loses some of its original potentiality. A hemocytoblast is an undifferentiated cell which can differentiate into an erythroblast, a lymphoblast, etc., but once this step is taken the cell begins its maturation. In losing its totipotentiality, it acquires characteristics which are irreversible; it can mature, but what it can never do is to turn back to its original state and develop in another direction.

2. **Maturation.** We speculate on the causes of differentiation, but we have many observations bearing on maturation. Some are morphological and some are biochemical. Most of the latter are concerned with the quantity of nucleic acid in the nucleus and in the cytoplasm; the morphological characteristics are that young cells are larger than mature cells, that the nucleo-cytoplasmic ratio is greater, that the chromatin is clearer and finer, that there are one or more nucleoli, and that the cytoplasm is usually basophil and does not possess specific granules. As the cell matures, the nucleus becomes smaller while the cytoplasm becomes less basophil and acquires the specific properties of the series to which the cell belongs. Normally there is a "law

Table 1

Red Cell Series	Granulocytic Series	Thrombocyte Series
Hemocytoblast	Hemocytoblast	Hemocytoblast
Pronormoblast	Myeloblast	Megakaryoblast
Polychromatophil (Normoblast)	Promyelocyte	Megakaryocyte
Acidophil Normoblast	Myelocyte	Thrombocyte
Reticulocyte	Metamyelocyte	
Erythrocyte	Polymorph	

(There is a similar series for eosinophil and basophil granulocytes)

Lymphocytic Series	Histiocytic Series (Monocytic Series)	Plasmocyte Series
Hemocytoblast	Reticulo-endothelial Cell or Hemocytoblast	Reticulo-endothelial Cell or Hemocytoblast
Lymphoblast	Histioblast or Monoblast	Histioblast
Lymphocyte	Histiocyte or Monocyte	Plasmocyte or Plasma Cell

of synchronism," a young nucleus corresponding to a young cytoplasm, and a mature nucleus to a mature cytoplasm. Under abnormal conditions, all kinds of asynchronism are observed.

3. Passage of mature cells into the circulation. The passage of mature cells into the circulation is not understood. White cells supposedly leave the hemopoietic tissues when they exhibit amoeboid movement and have acquired enzymes which cause temporary breaches in capillary walls. Some think that red cells, which are not amoeboid, develop extra-vascularly and enter the circulation in the same way as they do in the embryo. Others think that red cells are surrounded by a viscous substance which becomes less viscous as the red cells mature. These ideas are only speculations.

ERIC PONDER

Reference

Bessis, M., "Cytology of the Blood and Blood-forming Organs," *trans.* Ponder, E., New York, Grune and Stratton, 1956.

HENLE, FREIDRICH GUSTAV JACOB (1809–1885)

This German physician was born near Nuremberg and studied at Bonn, where he became one of the more distinguished disciples of Johannes Müller. He had the misfortune to hold liberal views, at that time very unpopular in Berlin, and was arrested and condemned for treason. Through the intercession of various scientists, he was, however, pardoned but was forced to retire to Switzerland, where he held the chair of anatomy in Zurich. His particular field of research was Epithelium and he was the first to describe the various types found. His three volume "Handbuch der Systematischen Anatomie des Menschen" was primarily a text book of histology which he approached, for the first time, from the cytological rather than from the organismic aspects. His name is best known from Henle's loop in the kidney.

HENNEGUY, LOUIS-FELIX (1850–1928)

This French biologist was born in Paris and died in the same city. He studied medicine at Montpellier and was a student of Claude Bernard in 1876. He contributed numerous papers on comparative anatomy and embryology, particularly that of the fish, to the literature. He is, however, best known for his contributions to microtechnique and for his joint authorship, with Bolles Lee, of the French edition of the latter's classic treatise.

HEPATICAE

The name "Hepaticae" is obsolete. By it were meant the following members of the Class *Bryopsida*.

Hornworts (Subclass *Anthocerotidae*, Order *Anthocerotales*)

Bryopsida with a spindle-shaped sporophyte built up through the activity of an intercalary meristem at the juncture of foot and capsule. The photosynthetic cells of both gametophyte and sporophyte basically contain a single laminate chloroplast with a compound central pyrenoid. The survival of a simple pyrenoid as found in the ancestral green algae has been shown in one member, closing this cytological gap. By dencentralization of the pyrenoid material and subdivision of the chloroplast certain members (e.g., some species of *Megaceros*) approach the chloroplast condition of most other green land plants. The gametophyte is usually flat, thalloid, and usually "webbed," i.e., its apical growing points dichotomize, but the forks tend to adhere laterally. It is anchored by simple rhizoids. Stoma-like pores on the lower side lead into chambers which may be invaded by blue-green algal hormogones. If the blue-green algal colony established inside the thallus happens to be one capable of nitrogen fixation, a symbiotic system is established. The archegonia are almost completely embedded in the thallus. The stalked antheridia are formed internally and become exposed by rupture of the overlying tissues. The sporophyte jacket is highly photosynthetic and in two of the five genera possesses stomata in the epidermis. In the center of the capsule runs a narrow columella of elongate sterile cells, surrounded by the sporogenous tissue. The latter gives rise to spore mother cells and compound elaters, with or without spiral thickening. Maturation is from the tip downwards, and the capsule dehisces gradually into two valves showing a hygroscopic twisting. Affinities: Possibly reduced derivatives of the Psilopsida.

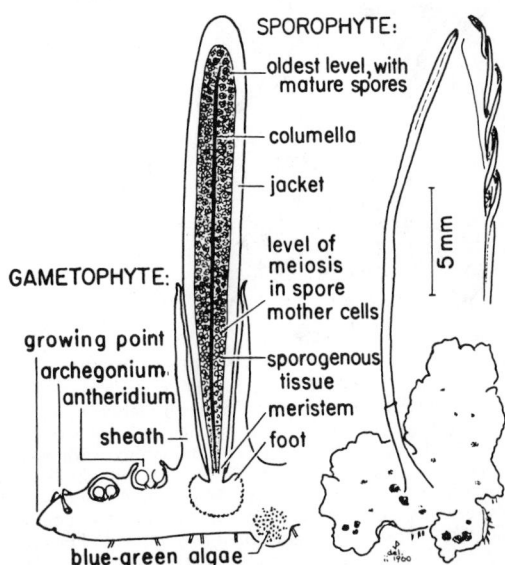

Fig. 1 HORNWORTS. Left: Semidiagrammatic vertical section through a gametophyte bearing a mature but not yet dehisced sporophyte. Right: *Phaeoceros laevis* supbsp. *carolinianus*. A gametophyte with antheridial cavities and bearing a sporophyte, shown indehisced and during dehiscence.

406   HEPATICAE

Distribution: Worldwide, but requiring a relatively temperate microclimate. Acidic substrates; mostly disturbed soil surfaces, some species on bark.

Liverworts (Subclass *Hepaticidae*)

Excluding the hornworts (see above). In ancient times the liverworts shared the appellation "lichen" and were employed in the treatment of skin disorders. The name "hepatica" or "liverwort," suggested by the overlapping broad thalli and texture of some of the larger Marchantiales, they began to share only in the late Middle Ages, and with it acquired some internal medicinal use.

Bryopsida with a sporophyte built up by over-all differentiation rather than by the activity of a meristem. The gametophyte is highly variable, ranging from a leafy axis to a flattened thalloid structure. The rhizoids, when present, are unicellular. Mucilages and essential oils may be produced in abundance, and the latter account for the characteristic scent of certain species (e.g., *Conocephalum conicum*). In many cases the gametophyte is associated with fungal endophyte (*mycorhiza*). Asexual reproduction by gemmae is common. The antheridia and archegonia are superficial in origin. Both sporophyte and gametophyte are highly

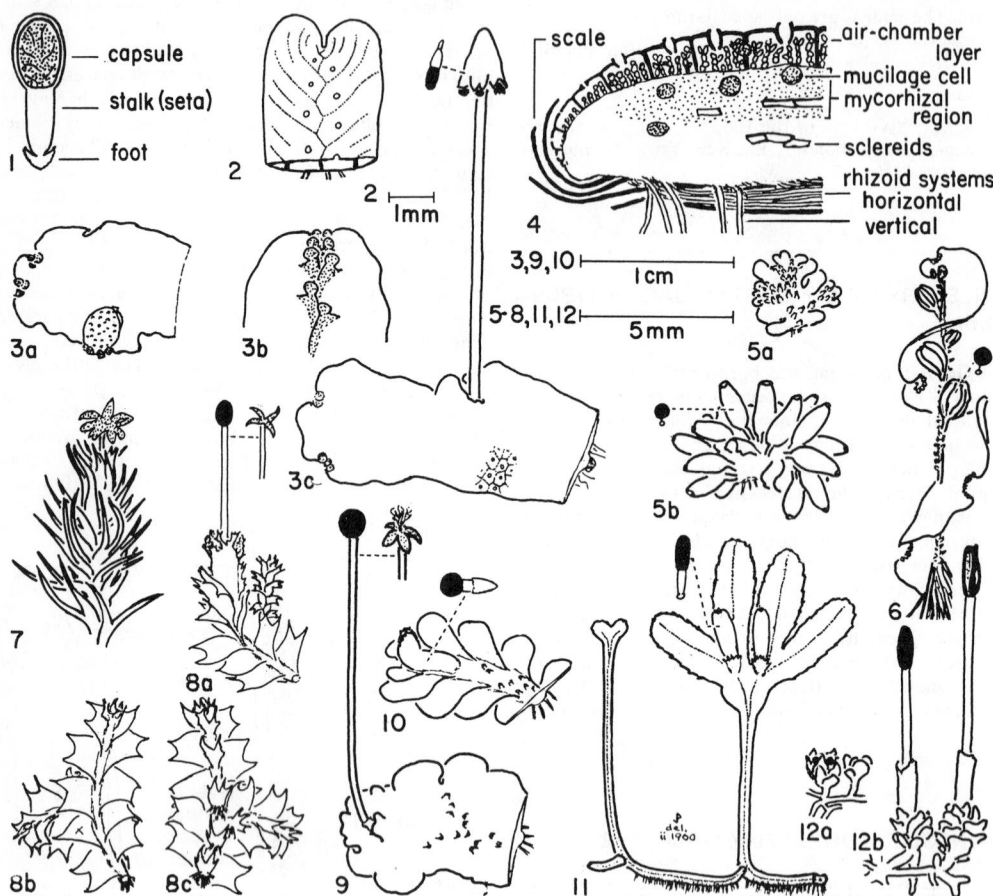

Fig. 2   LIVERWORTS. 1. Semidiagrammatic longitudinal section through typical sporophyte. *Marchantiales:* 2. *Cyathodium spruceanum.* Thallus largely composed of metameric air-chambers in two ranks. 3. *Conocephalum conicum.* a. Male plant, with sessile antheridial receptacle. b. Tip of thallus, from below. Scales in two ranks. c. Female plant, bearing archegoniophore with sporophytes. Air-chambers indicated in part. 4. *Marchantia chenopoda.* Semidiagrammatic vertical longitudinal section through tip of gametophyte. *Sphaerocarpales:* 5. *Sphaerocarpos stipitatus.* a. Male plant. b. Female plant with bottles containing sporophytes. 6. *Riella affinis.* *Jungermanniales:* 7. *Herberta juniperina.* Extreme tip of female plant bearing a sporophyte with irregularly dehisced capsule. 8. *Lophocolea cuspidata.* a. Fertile plant, with a small adventitious male branch, and terminal "perianth" containing a mature sporophyte shown with indehisced and dehisced capsule. b. Sterile plant, from above. The second lobe of the leaf marked "x" has been replaced by the branch on the left. c. Same plant, from below, showing the third row of reduced leaves. 9. *Pellia epiphylla.* The "thalloid" gametophyte bears scattered antheridial cavities and a mature sporophyte, shown with indehisced and dehisced capsule. 10. *Noteroclada confluens.* Closely allied to *Pellia,* but "leafy." With mature but not yet elongated sporophyte. 11. *Symphyogyna stipitata.* Filmy fern habit. The conducting strand is indicated. Female, with mature but unelongated sporophytes. *Calobryales:* 12. *Haplomitrium hookeri.* a. Male plant. b. Female plant with sporophytes with indehisced and dehisced capsules.

photosynthetic but lack stomata. The sporophyte is usually organized into foot, stalk (*seta*), and capsule. The phase of elongation of the stalk cells is frequently long deferred but very rapid, resulting in the mature capsule being lifted out from the protective structures. The sporogenous tissue of the capsule gives rise to spore mother cells, and, with few exceptions (e.g. *Riccia*), sterile cells. The latter in most cases are single celled elaters, with spirally thickened walls, dead and water-filled at maturity, and capable of hygroscopic twisting movements. In most cases the jacket of the mature capsule splits on drying along preformed lines of weakness, exposing the mass of spores and elaters.

Orders with Round Spore Mother Cells:

Marchantiales. The gametophyte is thalloid, metamerically segmented and of remarkable complexity. The photosynthetic upper part is organized into air-chambers opening through small pores which prevent waterlogging. On the lower surface, and protecting the growing point, are scales. In addition to vertical anchoring rhizoids, there is a system of horizontally running rhizoids held in place by the scales. It provides the main water conducting system. Male and female sex organs are either scattered along the midrib, or united into receptacles which are sometimes raised into the air. Many genera (but not *Marchantia*) show an explosive discharge of the antheridia, resulting in a spray of spermatids that liberate sperms after landing on a moist surface. There are about 35 genera, presenting many possible combinations of a series of variable characters. It is believed that the majority derived by reduction from the most complex forms such as *Marchantia*. Included are such extremes as *Cyathodium*, in which the thallus consists almost entirely of a layer of air-chambers, and, on the other hand, *Monoselenium*, in which there remains not a trace of the air-chambers.

Sphaerocarpales. A group of three genera with leafy axes, sex organs in individual protecting bottles, and fairly simple (probably reduced) sporophytes. *Sphaerocarpos* historically was the first plant in which sex-chromosomes were demonstrated. The highly specialized *Riella* is the only bryophyte successfully adapted to complete its life cycle under water.

Orders with Spore Mother Cells Four-lobed Priot to Meiosis:

Jungermanniales. The suborder Jungermanniinae ("leafy liverworsts") comprises the bulk of the liverworts. Although common throughout, a high proportion of the genera are epiphytes on leaves and bark of tropical forest trees. An apical cell with three cutting faces initiates a highly regular segmentation pattern. The axis bears three rows of typically bilobed leafy appendages. *Herberta*, probably primitive, is a robust, erect, "mossy" plant with an equal development of the leaves. In most cases the plants are delicate, with a prostrate axis having two lateral rows of leaves fully developed, and the third row, on the lower side, reduced. Branching of the axis is never by dichotomy. Commonly a branch replaces one of the two lobes of a leaf, another method is the formation of adventitious branches endogenous in origin. The terminal archegonia and developing sporophytes are typically surrounded by a "perianth" formed by fusion of the last three leaves of the axis. Germination of spores and gemmae often results at first in a juvenile or protonemal phase. The suborder Metzgeriineae represents a heterogeneous assemblage of forms with broadened axes. Several series show detailed transitions between a leafy and a thalloid habit. Most have probably been derived from Jungermanniineae, although a few (e.g., *Blasia*) may be archaic. Among specialized genera *Symphyogyna* has a complex internal conducting system, and some of its species closely resemble filmy ferns. *Cryptothallus* is the only bryophyte completely devoid of chlorophyll. It lives at the expense of its mycorhizal fungus.

Calobryales. The three genera are the only bryophytes lacking rhizoids. The small gametophytes are erect, leafy, and have subterranean rhizomatous axes with mycorhiza.

Affinities: It is as yet not clear if the erect leafy or the flattened thalloid habit is primitive here, or indeed if the liverworts are a monophyletic group. The majority share a most peculiar haploid karyotype of 8 + 1 chromosomes, probably derived from a 4 + 1 chromosome complement as is characteristic of most hornworts.

J. PROSKAUER

References

Müller, K., "Die Lebermoose" *in* Rabenhorst's "Kryptogamen-Flora," Vol. 6. 3rd ed. Leipzig. Akad. Verlagsges., 1951–1958.
Parihar, N. S., "Bryophyta," 5th ed. Allahabad, India, Central Book Depot, 1965.
Schuster, R. M., "The Hepaticae and Anthocerotae of North America," Vol. 1. New York, Columbia Univ. Press, 1966.

HEPATOPANCREAS

Hepatopancreas is a term applied to invertebrate digestive glands believed to function like both the liver and the pancreas of vertebrates. Present usage largely restricts the term to certain organs in malacostracan Crustacea and in Xiphosura. More rarely the term is employed for digestive diverticula in some other Crustacea, Arachnida, Gastropoda, and Pelecypoda. It seems not to have been applied to appropriate organs in other invertebrates, such as starfishes.

Among Malacostraca, the hepatopancreas has been studied intensively only in crayfish, crabs, and other Decapoda. In them it occurs on each side as a very large, soft organ, an outgrowth from the short midgut. From the gut a single duct enters each hepatopancreas and shortly divides, often into three main branches, leading ultimately to numerous blind tubules, comprising the bulk of the organ. The tubules are grouped in lobes, covered with a thin sheet of connective tissue.

A single tubule may be several millimeters long and consists of one layer of low columnar cells. The apical cells are embryonic, the others successively absorptive, glandular, fibrillar, and degenerative. Just external to the epithelial cells is a network of very thin muscle fibers. The circular fibers and some of the longitudinal ones are striated. Between the tubules there are connective tissue storage cells and many small blood spaces.

The gland cells are either apocrine or holocrine, and their secretion ultimately enters the stomach. Rhythmic waves of secretory activity may occur after each meal. Ingredients reported from the secretions of various decapods include a weak acid (perhaps NaH_2PO_4), brown and yellow carotenoid pigments, a bile salt (of a taurocholic acid) which helps to emulsify fats, an activator for tryptase, and a variety of digestive enzymes. Among the last are cathepsin, tryptase, collagenase, carboxypeptidase, aminopeptidase, dipeptidase, lipase, amylase, maltase, sucrase, and lactase, as well as enzymes acting on hemicellulose, amygdalin, salicin, arbutin, coniferin, phloridzin, and raffinose. Not all these enzymes are present in any one decapod, and it is possible that some of the carbohydrases are of bacterial origin. Certain substances, such as dialyzable dyes, may appear in the tubule cells or in the secretion if they have been introduced into the blood. After massive injections of such substances, the hepatopancreas becomes the main route for their excretion.

Digested food almost all passes from the decapod stomach into the hepatopancreas, there to be taken up by the absorptive cells. Active absorption by these cells has been demonstrated for fats, ferrous iron, and certain dyes. In the same cells there is considerable storage of food reserves, some of which accumulate also in the gland cells and in the connective tissue storage cells. Ultimate utilization of most of the stored substances is for secretion of a new exoskeleton, or in some other phase of the molting cycle. The amounts stored therefore vary with the stages of this cycle. Substances found so to vary in the hepatopancreas of at least some decapods include glycogen, mucopolysaccharide, protein, fermentable glucoprotein, glucosamine, neutral fat, lecithin, esters of cholesterol, other unsaponifiable lipids, unsaturated fatty acids, nonfermentable reducing substances, calcium phosphate, magnesium phosphate, sodium, potassium, copper, and chloride. Also found to vary with the molting cycle are alkaline phosphatase, polyphenoloxidase, riboflavin, and some of the digestive enzymes. At least some of these changes appear to be under hormonal control. Other pigments reported in the hepatopancreas, besides the flavoproteins, include carotenoids and a special pterine. Other enzymes found include oxydoreductase, peroxidase, and all the enzymes required for the breakdown of purines through urea to ammonia. Evidently the hepatopancreas is an important site for the degradation of these nitrogenous compounds. It probably also functions in regulating the blood sugar level and in protecting the body against toxins. Various dyes and other toxic substances when fed to a decapod enter the hepatopancreas but fail to reach the blood. Presumably, cells of the hepatopancreas either fail to absorb these substances, or actively detoxify them.

In the Xiphosura, or horseshoe crabs, the hepatopancreas is a very extensive, multilobed structure, which on each side opens by two ducts into the stomach-intestine. All the lobes are surrounded by both circular and longitudinal muscle fibers. Gland cells within the lobes secrete amylase, lipase, proteinase, and carboxypeptidase, which exert their digestive action in the stomach-intestine. Absorptive cells having a striated border are present in the same lobes. These cells not only absorb the products of digestion but also contain an intracellular dipeptidase. Storage of fat, glycogen, and protein

occurs in connective tissue cells between the lobules. The gland may also sometimes function in excretion, at least of calcium phosphate.

In arachnids a number of intestinal diverticula are present to which the term hepatopancreas sometimes is applied. These diverticula secrete digestive enzymes, are active in absorption, store food reserves, and may take part in nitrogenous excretion.

In gastropod molluscs there is a midgut gland likewise sometimes spoken of as the hepatopancreas. From this gland a secretion flows into the stomach and crop, containing enzymes which digest fats and carbohydrates, as well as proteins if the animal is carnivorous. Other ingredients in the secretions of some species include a substance helping to emulsify fats, a nutrient necessary for the growth of cellulase-producing bacteria, a derivative of a heme-pigment, and some calcium secreted by specific calciferous cells. In herbivorous species, and in at least some of the carnivores, phagocytosis of food particles occurs within the gland. Digestion of protein is intracellular in the herbivores, and there may also be intracellular digestion of fats and carbohydrates. Products of extracellular digestion are presumed to be absorbed in the gland, by columnar cells with a striated border. Other functions of the gland include storage of food reserves, synthesis of fatty acids, and purine synthesis and degradation.

In pelecypod molluscs there is a digestive gland which sometimes has been termed the hepatopancreas. Some controversial evidence indicates that this gland secretes digestive enzymes into the stomach. The gland also functions in absorption of digested food, phagocytosis of food particles, intracellular digestion, storage of fat, and catabolism of purines.

J. H. LOCHHEAD

References

Von Buddenbrock, W., "Vergleichende Physiologie," Vol. 3: "Ernährung, Wasserhaushalt und Mineralhaushalt der Tiere," Basel, Birkhäuser, 1956.

Maloeuf, N. S. R., "Physiology of the alimentary tract of arthropods," Riv. Biol. 24: 1–62, 1938.

Waterman, T. H., ed., "The Physiology of Crustacea," Vol. 1: "Metabolism and Growth," New York, Academic Press, 1960.

Wilbur, K. M. and C. M. Yonge, eds. "Physiology of Mollusca," Vol. 2. New York, Academic Press, 1966.

HERTWIG, OSCAR (1849–1922)
HERTWIG, RICHARD H. (1850–1872)

These brothers were both born at Friedberg. The elder became professor of anatomy first, both studied at Jena under the direction of Haeckel, and secured degrees in medicine from Bonn. The elder successively held the chairs of anatomy at Jena and Berlin, while the younger held the chairs of zoology successively at Königsberg and Bonn. They cooperated closely in many zoological and anatomical monographs, the most important being in the field of embryology. They together enunciated the "theory of the coelom," the comparative anatomy of which they studied extensively and which contributed so much to the early understanding of comparative embryology. The elder devoted much of his time to studying fertilization and described for the first time the fusion of the nuclei in this process. He was very unsympathetic to Roux's theory that early embryonic development represented a struggle for existence between the cells, pointing out that if this struggle took place it should be visible, but was not. He held the view that each of the metameres fitted into an allotted pattern and he thus, though in ill-defined terms, foresaw the development of field theories. The younger brother, who is the author of the well known "Lehrbuch der Zoologie" also studied the cytology of the nucleus but devoted most of his time to this structure in the Protozoa and particularly investigated the role that it played in vegetative reproduction.

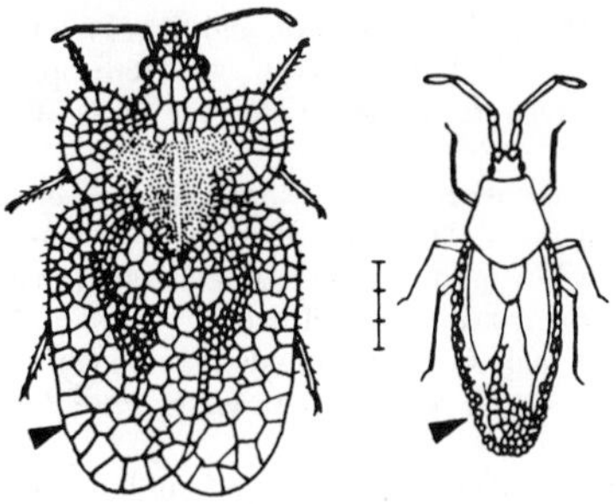

Fig. 1 Lace bugs. (From Pimentel, "Invertebrate Identification Manual," New York, Reinhold, 1967.)

HETEROPTERA or TRUE BUGS

This distinctive group of nearly 30,000 insects has evolved its discernible characteristics over a long period of time. From the lower Permian Period come fossil representatives of primitive heteropterans which show, essentially, the beginnings of what we call the modern Heteroptera. This was approximately 225 million years ago. An antecedent type, the Protohemiptera, connects Heteroptera with that great reservoir group of Pennsylvania Period insects known as the Paleodictyoptera. These preceded the true Heteroptera by some 20 millions of years and seem to have been ancestral to the majority of living insects.

Two major structural features quickly define these diverse types—the thin or wide sharp-pointed beaks which signify piercing-and-sucking insects, and the bipartitioned nature of the forewings which distinguishes them from their very close relatives, the Homoptera. These forewings, called *hemelytra,* have the anterior two-thirds or *corium* variably thickened and the posterior one-third membranous and changeably clear. The growth pattern is one of gradual metamorphosis, meaning that the young resemble the adults except for the usual differences in size and minor variations in proportion.

The majority of Heteroptera are terrestrial, but a relatively large segment of the suborder has adapted itself to the aquatic environment, with a wide spectrum of intermediate types living around the edges of the water habitat. This is not true aquatic existence, however, as occurs in many other insects in the sense that gills extract oxygen from the water and the animal can stay permanently submerged without an air film or bubble. Methods of respiring under water vary from the simple procedure of contacting the surface at necessary intervals for a fresh supply of air to the development of plastrons of remarkably fine and dense hairs which hold a thin layer of air to the body surface and through which oxygen dialyzes from the water in sufficient quantity to meet the usually lethargic insects' requirements.

Of more than passing interest is the fact that the only insect inhabitants of the high seas are surface striders belonging to this aquatic segment—although here, the insect lives upon the surface of the sea and not beneath it, much as the terrestrial species walk about on land.

Size range is considerable, varying from less than 2 mm (1/16 of an inch) to more than 100 mm (4 inches) in length, the giants of the group being water-dwelling species of the family Belostomatidae. In some places, these have quasi-economic overtones in that they have acquired the name of "toe-biters."

Feeding habits run the gamut from herbivorous to carnivorous—the lack of biting or chewing mouthparts makes the saprophagous habit a very rare one with Heteroptera. Plant feeders include an array of economic species of several families such as the chinch, squash, box elder, stink, plant and lace, bugs. These and other phytophagous species feed on a great variety of agricultural, range and miscellaneous plants.

Other bugs are predaceous, and several families account for a good deal of the natural control of insect-against-insect, without which our problem of keeping damaging species in check would be much more difficult. The most important of these belong to the large family Reduviidae, appropriately named "assassin bugs." In the early days of the electric street light, some species became known as "kissing bugs" because of their propensity for alighting on faces shining under the lights.

All true bugs, regardless of feeding habits, inject a salivary secretion into the tissues being fed upon. In the case of herbivorous species, this contains an enzyme which begins starch digestion. In predaceous types, the saliva contains a venom and an anticoagulant, the former for subduing its prey, the latter for keeping the blood in a fluid state, of particular importance in the bedbugs (Cimicidae) and in the group of reduviids (subfamily Triatominae) parasitic upon warm-blooded hosts. Bedbugs have had a long association with their hosts, losing the poison so that their bite is virtually painless, but the triatomines have a toxicity capable of killing their insect prey and causing intense and lingering pain in man. The latter bugs have the added economic impact of being vectors for certain types of human trypanosomiases or protozoan-caused diseases such as Chagas' disease.

As in the case of the beetles or Coleoptera, glands for the production of odoriferous fluids are common in Heteroptera. However, unlike beetles, the odor emanating from bugs is much less offensive, even when concentrated, and when diluted often has a pleasing fragrance. Studies indicate that such fluids function both to attract mates and to dissuade predators. Insectivorous vertebrates such as birds and lizards, which might otherwise feed unrestrictedly on Heteroptera, have been observed to avoid or retreat from odoriferous species.

Like many other insects, some bugs produce sounds by various stridulatory devices such as beak-against-sternum, tibia-against-abdomen, wing-against-abdomen and legs-against-labrum. However, there is little evidence that bugs hear the sounds they produce, with the exception of the water boatmen (family Corixidae), which have a tympanum.

A unique method of obtaining food has arisen in one group of reduviid bugs. Some members of the subfamily Holoptilinae possess a gland called a *trichome* which attracts ants by its fragrance and flavor, and then paralyzes them so that the bug can feed at its leisure. Man has not been the first to effectively use poisoned bait!

Omnivorous man has utilized most kinds of insects as food at one developmental stage or another, and Heteroptera are no exception. Species which can be collected in numbers and which are non-odoriferous are included here, with such large types as the belostomatid water bugs being popular in some parts of the world.

IRA LA RIVERS

References

Blatchley, W. S., "Heteroptera or True Bugs of Eastern North America," Indianapolis, Nature Publ. Co., 1926.
Essig, E. O., "Insects and Mites of Western North America," New York, Macmillan, 1958.
Miller, N. C. E., "The Biology of the Heteroptera," Londons Leonard Hill, 1956.
Parshley, H. M., "A Bibliography of the North American Hemiptera-Heteroptera," Northampton, Mass., Smith College, 1925.
Ross, H. H., "A textbook of Entomology," 3rd ed. New York, Wiley, 1965.
Van Duzee, E. P., "Catalogue of the Hemiptera of America north of Mexico," Univ. Calif. Publ. Entomology **2:** 1–902, 1917.

HETEROSIS

A group of phenomena including increased vigor, or stimulation to growth, size, yield, and resistance to disease has long been associated with the hybrid offspring from variety crosses and in many cases species crosses. Mankind has benefited by using the vigor of the mule and the yield of modern hybrid maize for example. Modern breeding practices have capitalized on these phenomena in breeding poultry, swine, beef and milk cattle, silkworms, sugar beets, sorghum, grasses, tomatoes, and numerous other crop plants and ornamentals.

The term *heterosis* was coined by George Harrison Shull in 1914 to include hybrid vigor as the manifest effect of a "developmental stimulation resulting from the union of [genetically] different gametes." It also includes as part of the definition the decrease in vigor in generations subsequent to the inbreeding of the hybrids. Shull together with Edward Murray East developed the heterosis concept in genetic terms through their work on hybrid corn from about 1905 to 1930

which led them to conclude that heterosis resulted from the increased heterozygosity in hybrids as compared with their parents. While most inbred varieties of corn showed extremely low vigor and yield, hybrid seed from crossing these inbreds grew into plants two or three times as productive as either parent. In turn when the hybrids were self-pollinated or inbred closely this high yield fell off until after a few generations of inbreeding the low vigor characteristic of the highly inbred parent was again attained.

Inbreeding has only one essential genetic effect, namely to increase homozygosity. That inbreeding does not have an intrinsic detrimental effect is evident from the fact that many self-pollinating plants such as legumes and cereals in which strains are highly vigorous though homozygous suffer nothing on inbreeding. For these species homozygosity of the genotype is the normal adaptive state.

Most cross-fertilizing species on the other hand possess a vast store of genetic variability in the form of deleterious recessive genes or genic complexes hidden in the heterozygous condition. Inbreeding increases the homozygous combinations of such deleterious genetic effects and thereby results in lowering of viability and other "fitness" traits. Crossing inbreds therefore in part restores the level of heterozygosity comparable with the normal adaptive state of the wild genotype and the increase in vigor results.

To East and Shull the genetic basis of heterosis was dependent on heterozygosis of alleles, implying an interaction between alleles (A_1 and A_2) in the heterozygote (A_1A_2) of a stimulatory nature. The term "overdominance" has been used to describe this interaction. With the knowledge that most deleterious mutants in wild populations are recessive it was postulated by A. B. Bruce (1910) and by D. F. Jones (1917) that heterosis resulted from the combined action of favorable dominant factors. If the vigor can be principally produced by dominants, however, it should be possible to fix the vigor in homozygous lines from inbreeding the hybrid generation; so far this possibility has not been realized. Jones points out that linkage would tend to reduce the probability of multiple dominant recovery, and since many loci are likely to be involved, the task of combining all dominant vigor loci into a single line seems to be infinitely arduous though possible of attainment. A resolution of the two hypotheses (dominance of vigor genes vs. overdominance) has come from a population analysis: J. F. Crow (1948) and F. G. Brieger (1950) have pointed out that in large cross-fertilizing populations the replacement of deleterious recessives with dominant increased vigor genes would not increase the average vigor more than a few percent, far below the observed increase in vigor. Specific interaction from overdominance at a small number of loci could explain the magnitude of increase more easily. In small populations, on the other hand, where inbreeding or random drift may establish a large proportion of homozygous recessives, the gain in vigor by crossing may be considerable if dominant vigor genes replace the deleterious recessives.

Among most crossbreeding populations genetic heterozygosity commonly confers upon individuals superiority in fitness characters such as viability, fertility, rate of development, rate of survival, and resistance to environmental fluctuations (HOMEOSTASIS). Adaptive superiority of heterozgotes in natural populations had been demonstrated in such organisms as Drosophila, mice, land snails, and man; and it accounts largely for the preservation of genetic variability (or balanced polymorphism) in those populations (see POPULATION GENETICS). The genetic structure of a natural population then is largely balanced because heterozygotes have highest net fitness (*euheterosis* of Dobzhansky); thus natural selection acts largely to preserve the balance, or the total genetic variability, and only in part to affect gene substitution. Apparently natural populations capitalize on a fundamental genic interaction (overdominance, or "luxuriance") if that interaction benefits the total net fitness of the population.

E. B. SPIESS

References

Crow, F., "Alternative hypotheses of hybrid vigor," Genetics **33**: 477–487, 1948.
Dobzhansky, Th., "A review of some fundamental concepts and problems of population genetics," Cold Spring Harbor Symposium **20**: 1–15, 1955.
Gowen, J. W. ed., "Heterosis," Ames, Iowa State College Press, 1952.
Shull, G. H., "What is 'heterosis'?" Genetics **33**: 439–446, 1948.

HIBERNATION

Definition. Very different states are called "hibernation." It is necessary to separate these states: the hibernation of poikilothermic animals is called by Eisentraut *winter rigidity.* The hibernation of the badger, raccoon, the bear, is *dormancy* or *carnivorean lethargy.* A man cooled artificially down to 30°C is in a hypothermic stuporous state, but does not hibernate. True hibernation is observed in only a small number of mammals and in some birds (Jaeger).

We shall define true hibernation as a state of seasonal, spontaneously reversible hypothermia, together with a state of sleep—not of stupor or narcosis; it is necessary to add the spontaneous periodic arousals which may be frequent (hamster) or rare (dormice, ground squirrels).

Which species hibernate? Hibernators are mostly Rodents: in the family *Sciuridae* the woodchuck or ground hog (*Marmota monax, M. marmota, M. bobak* etc.); in the genus *Citellus* (ground squirrel), *Citellus citellus, C. tridecimlineatus, C. columbiensis,* etc.); in the genus *Gliridae* (dormice), *Glis glis, Eliomys quercinus, Muscardinus avellanarius, Dryomys nitedula.* There are *Zapodidae, Sicistidae, Cricetidae* (hamsters) which are true hibernators.

Among the Insectivora, the best known is the hedgehog (*Erinaceus europaeus*); the tenrec of Madagascar and the hedgehog of Aethiopia also hibernate.

Among the bats, the Microchiroptera of the Northern hemisphere are all true hibernators (*Rhinolophus, Myotis, Macrotus, Plecotus* etc.) The Macrochiroptera of tropical regions do not hibernate. Speaking of Chiroptera. Boulière said that hibernation is a special adaptation of animals of tropical origin to temperate climates. We think that this conclusion is to be extended to all hibernators.

Geographic distribution of hibernators. All hibernators quoted above are found in the Northern hemisphere except the tenrec. In the regions where hibernators are found, the average winter temperature is low; this is not the case for the tenrec: the average temperature seldom falls below 15°C in its living area. The tenrec, when transported from Madagascar to Paris, hibernates in the cool days in June–July. It does not seem to tolerate well temperatures below +10°C.

The state of Monotremata, small marsupials and prosimians from June to September in the Southern hemisphere is not described precisely enough to decide whether it is dormancy or true hibernation.

Dormancy. The dormancy of the bear (*Euractos americanus, Ursus arctos*) is now well known. Only the female "hibernates." She brings forth her cubs in January, during the dormancy, and suckles them in this state. The hypothermia is not very deep (central temperature 29° – 34°C); the respiratory frequency is 12 cycles/minute, against 30 cycles/minute in the normothermic state. The oxygen consumption is reduced by only 40%. The animal is in a stuporous state.

The skunk, the badger, the raccoon show similar states. Dormancy depends much more on climatic conditions than does true hibernation. The seasonal rhythm is less developed than in true hibernators.

The aestivation of certain ground squirrels, hibernators living in steppes, is near to dormancy: it is a state of drowsiness, in summer when the green vegetation has disappeared. It may be interrupted by a period of rain.

Seasonal rhythm of hibernators. When a hamster is placed, in May–June, in a refrigerator, it does not enter hibernation. When a ground squirrel is placed in December at 15–20°C, it enters hibernation. These two examples are extreme cases. In the field, by trapping methods, hibernators are caught mostly during the fair season; in winter, they remain in their burrows. In these conditions of living, the effect of the "internal clock" and that of the seasonal rhythm fit together. In captivity, one sees in general that an environmental temperature of +20°C in winter, prevents the occurrence of hibernation.

A stay in a refrigerator, regulated at +5°C, in summer, may lead to hibernation in the captive ground squirrel, and to states of very deep hypothermia in the garden dormouse. But these hypothermic states are temporary, lasting about 12–14 hours per day. In these artificial conditions, these small hibernators behave like bats: they are poikilothermic during a large part of the day. This effect of exposure to cold in summer is rather exceptional.

Hibernation in autumn is entered into progressively. The daily number of sleeping hours increases and the diurnal rhythm becomes exaggerated. Then the phases of activity become more and more reduced, the temperature, during sleep, falls lower and lower; at a certain moment, it reaches a critical value (10°C for instance) and the animal hibernates.

At the beginning of hibernation (October), the arousals are more frequent. They become rarer and rarer in November. Then during some weeks, the animal goes through the deepest hibernation. Afterwards, the arousals become more and more frequent, the phases of uninterrupted sleep become shorter and, finally, the hibernator enters the state of normothermic spring activity. This cycle is observed in animals kept in a sound proof refrigerator, lit for 5 minutes per day only, at constant temperature during six months. These facts speak surely in favour of an internal rhythm.

Autumn behavior of hibernators. The accentuation of the diurnal temperature rhythm is not the only manifestation of the entrance into hibernation. Many hibernators become very fat, as Aristotle knew; he said of the common dormouse: "It retires in tree-holes and becomes fat." Other hibernators gather in their burrows considerable reserves: a 400-gram hamster stores up to 10–15 kg of corn.

Many hibernators build new burrows for the winter. The Alpine woodchuck goes down, from 2,000 m to 1,500 m usually. Bats show true migrations.

The non-migrating hibernators improve their burrows. The hamster builds living and storage chambers. It has been known since 1551 (Gesner) with what care the marmot closes the entrance of its den.

Seasonal cycle of endocrines in hibernators. Complete sexual rest is a rule in hibernating hibernators. The different behavior of the female bear was at the origin of the distinction between dormancy and true hibernation.

The thyroid, the adrenal cortex, the anterior pituitary gland involute. This is proved by histological study, the evaluation of circulating hormones and the results of homograftings.

In the endocrine pancreas, the insulin (hypoglycaemic factor)-secreting cells increase in number as compared to the glucagon (hyperglycaemic factor)-secreting cells.

The adrenal medullary gland undergoes no change; the parathyroids are hyperactive. The brown fat (hibernating gland) is hyperactive; its once controversial endocrine role is now well established (Hook and Barron).

The modification of the endocrine precedes the entrance into hibernation and takes place even if the hibernations are prevented from hibernating. It is also found, at least under some aspects, in animals which, as the badger, do not hibernate in the strict sense.

Temperature regulation. The cyclic evolution of hibernation—in constant conditions of temperature, noise and illumination—proves that it is a regulation. The hibernator "sleeps," that is, it can always be wakened by a sensory, auditory, tactile or thermal stimulation.

Comparing the oxygen consumption of the hibernating ground squirrel at $+5°C$ and at $+2°C$ (environmental temperature), one finds that at $+5°C$ it uses about 20 ml O_2/kg/h, and twice as much at $+2°C$. *It regulates its temperature.* But there is a true poikilothermia between $+5°$ and $+10°C$.

There are many exceptions to this rule: many hibernators die in hypothermia at temperatures around and below $0°C$.

Homiothermia of hibernators in summer. All hibernators have, in summer, an imperfect thermoregulation. They behave as very young homiotherms, uncompletely developed at their birth, like young rats; they can very well increase their exchanges in a cold environment, but they cannot protect themselves against heat loss; the 2 kg-woodchuck behaves as a shaven rabbit. In autumn, their heat production in cold environments decreases parallely to the involution of their anterior pituitary, thyroid and gonads.

Hibernators show a thermoregulation similar to that of the animals adapted to tropical climates.

If on one hand all hibernators are poor homiotherms, on the other hand many poor homiotherms as the sloth are not hibernators. The sloth is found only in tropical climates.

The bats behave also in summer like poikilothermic animals: their oxygen consumption between $+2°$ and $+28°C$ increases parallel to the environmental temperature.

Hibernators being easily overstrained by cold, it should be noted that true hibernators weighing more than 5 kg are not found. Since a large animal weighing more than 10 kg withstands very well cold environments, one can conclude that true hibernation is necessarily related to the small size of the hibernating species.

Hypothermia. An adult rat artificially cooled dies by respiratory and heat failure when its temperature has fallen down to about $+15°C$. The temperature of an awake hibernator, in summer, artifically cooled, may be lowered down to $+3°C$ from where the hibernator may warm up again and survive.

A young rat or mouse can tolerate much deeper hypothermias than adults: but these young mammals cannot warm up without external help.

Studying the resistance to asphyxia, one finds again the same facts: young homiotherms and awake hibernators show an abnormal resistance to asphyxia.

The nervous system of hibernators works—*in vivo* and *in vitro*—at temperatures inconsistent with the functioning of the nervous system of homiotherms. Hibernation may thus be traced back to two peculiarities: 1) a special type of functioning of the nervous system—tolerance to cold and asphyxia—present during summer and winter; 2) a seasonal cycle of the endocrine glands such that the animals are overstrained by cold without too much resistance.

Arousal of hibernators. There is an apparent contradiction between this conclusion and the fact that, spontaneously, the hibernator periodically interrupts its sleep to become normothermic again for about 12 hours, and then lets itself cool again. During arousal the animal makes an enormous thermogenetical exertion; in the ground squirrel, the heart, which frequency was 3–5 beats/minute during hibernation, reaches a frequency of 300–400 beats/minute around 20 minutes after the beginning of the arousal. The oxygen consumption, which was 20 ml/kg/h reaches 3,500 ml after 30–40 minutes. We could show that the number of hours of arousal does not reach 10% of the total duration of hibernation (6 months), but that the energy expenditure during this short fraction of time is 90% of the total expenditure!

We still are uncertain what causes the hibernator to awake periodically. But we know that the mechanism of arousal is a perfectly integrated process, that the experimenter tries to imitate when he "resuscitates" a cooled rat: he warms up the heart locally so that it "precedes" in its activity the energetic needs of the other tissues. The hibernator short-circuits its circulation at the level of the renal arteries. It provides for the circulation of its brain substance and heart muscle and temporarily sacrifices the remainder of the body. Only by this way can the heart realize the enormous work necessary without yielding.

Conclusion. Hibernation appears as an extremely complex physiological process needing, for its realization, a special internal adaptation (nervous system, endocrine glands). It is related to special climatic conditions. Hibernation has both external and internal origins. It appears by no means as a mere thermoregulatory insufficiency, but as a higher degree of evolution. If the features of the thermoregulation lead to classifying hibernators among the animals of tropical origin, it seems that they have later, by staying in cold temperate regions, acquired new adaptations which have led them to an original solution of life in unfavourable climatic conditions. This view derives from the simultaneous occurrence of very primitive features (resistance to asphyxia) and of more recent adaptations such as the seasonal homiothermia.

CHARLES KAYSER

References

Eisentraut, M., "Der Winterschlaf mit seinen ökologischen und physiologischen Begleiterscheinungen," Jena, Fischer.

Kalabukhov, I. N., "Hibernation in the animal kingdom" (in Russian). State University M. Gorki Publ. Charkov, 1956.

Kayser, C., "Physiology of Natural Hibernation," New York, Pergamon, 1961.

Lyman, C. P. and Chatfield, P. O., "Physiology of hibernation in mammals," Physiol. Rev., 1955, 35: 403–425.

HIRUDINEA

The Hirudinea (leeches) are all soft bodied, highly extensible ANNELIDS which generally lack setae. The body is dorso-ventrally flattened or cylindrical in outline and has a cup-like sucking disc at each end. The posterior sucker is most prominent and serves primarily for anchorage. Leeches are all hermaphroditic.

Three taxonomic orders established on the basis of the structural organization of the anterior alimentary tract include (1) Rhynchobdellae in which the pharynx is specialized into a protrusible proboscis with attendant sheath; (2) Gnathobdellae in which the pharynx contains three longitudinal folds armed with muscular jaws with or without chitinoid teeth; and (3) Pharyngobdellae in which the pharynx has only three longitudinal folds (lacks jaws and proboscis).

Leech bodies with one exception (Acanthobdella) are composed of 33 complete segments and one partial, the *prostomium*. Each typical body segment may be subdivided into as few as two or as many as fourteen *annuli*. The number of annuli for a typical segment of a given species is always the same. Anteriorly and posteriorly in a given species the annuli per segment become reduced so that the first and last somites equal one annulus. Sensory end organs are prominently located on the middle annulus and usually less so on the adjoining annuli of a somite. The *clitellum* always occupies segments X, XI, and XII but is more apparent in the breeding season.

As a result of the pronounced development of a connective tissue (*botryoidal tissue*) the coelom is reduced to lacunae. A closed blood vascular system is present in the Rhynchobdellae but is represented only by the lacunae in others.

The alimentary canal is complete with mouth, pharynx, oesophagus, anterior and posterior stomach, rectum and anus. In many species more or less extensive caeca are attached to the stomach for food storage. Common foods include other annelids, planaria, insect larvae, crustacea, molluscs, invertebrate and vertebrate eggs, and blood. Following one good meal they can survive for several months.

The central nervous system is ventral in position, composed of 34 ganglia, some of which are partially fused. Numbers 2 to 6 form an anterior ganglionic mass while numbers 28 to 34 form an anal ganglionic mass. All leeches are especially sensitive to the slightest changes in environmental stimuli as heat, light, chemicals, shock waves, etc. Special pigmented light-sensitive eye spots on the cephalic segments are common in many species.

Male and female sexual systems open externally on the ventral body surface with the male aperture always anterior to the female. The female system is composed of a pair of ovarian sacs joined to the exterior by one or more ducts. The male sex organs consist of from four to many pairs of testes joined to a common duct on their respective body side which after uniting with its counterpart may end in a cirrus or bursa.

Spermatozoa are transferred by means of spermatophores in Rhynchobdellae and Pharyngobdellae but by a special penial organ in Gnathobdellae. Spermatophores may be attached to the clitellar area where special conductal tissue leads the spermatozoa to the ovarian sacs, Piscolidae; or they may be attached anywhere on the body surface and hypodermically inject the spermatozoa into the body sinuses, Placobdella.

Self fertilization is highly improbable. Fertilized eggs are deposited externally and develop within cocoons. The latter may be attached to objects in the environment, Pharyngobdellae; or brooded on the ventral body surface, Rhynchobdellae.

Leeches are world wide in distribution. Some live in salt water, a few in terrestrial areas and all others in fresh water. About 75% of all species are temporary ectoparasites, a very few are permanent parasites, and the remainder are predators and scavengers.

RAYMOND J. MYERS

References

Harding, W. A. and J. P. Moore, "Hirudinea" in "Fauna of British India," Sewell, R. B. ed. pt. 4. London, Taylor and Francis, 1927.

Mann, K. H., "Leeches (Hirudinea), Their Structure, Physiology, Ecology and Embryology," New York, Pergamon, 1962.

HIS, WILHELM, SR. (1831–1904)
HIS, WILHELM, JR. (1863–1889)

Both father and son were comparative embryologists of some note, working at the University of Leipzig, at which the father held the chair of anatomy. The father is principally distinguished by his insistence on the theory, which had little acceptance in his day, that the course of embryology was influenced by the physiology and activity of the cells rather than by the more mechanical foldings and migrations which were at that time thought to be the basis of organogenesis. He never put his theories to experimental proof but they were, of course, the basis from which much later work derived. The son was the discoverer of the fasciae in the heart which bear his name.

HISTOLOGY

The word "histology," being derived from *histos* and *logia*, means the science or study of the tissues. The word "tissue" was derived from the French *tissu*, which means weave or texture, and was introduced into the language of anatomy toward the end of the eighteenth century by Bichat, a brilliant young French anatomist. As he dissected bodies, Bichat observed that they were composed of layers and structures of different textures. He wrote a book describing these, classifying more than twenty.

Bichat, however, did not make his classification of the tissues from their microscope study; indeed he distrusted the microscope. Bichat, moreover, did not coin the term histology, which was introduced 17 years after Bichat's death by A. F. J. K. Mayer. Since by that time many studies had been made on the microscopic structure of various parts of the body it is interesting that despite the existence of microscopic anatomy as a branch of science a new term seemed to be required, and that this term has gained increasing acceptance throughout the succeeding years.

There would seem to be two reasons for the general acceptance of the term histology when it was introduced. First, as various parts of the body were studied with the microscope it became obvious that there were certain patterns of microscopic structure that were repeated over and over again in various parts of the body and in the make-up of certain organs. The concept that these patterns of microscopic structure were the building blocks of most parts and organs of the body probably recalled Bichat's observations and led to their being termed the *tissues* of the body. Secondly, embryology was being studied more and more with the microscope and it became possible to trace the formation of the tissues, classified by microscopic study, from the germ layers of the embryo, and study the way in which they became assembled together during development to form the various structures and organs of the body. It should be noted that the tissue concept, stressed by the term histology, greatly simplified the study and understanding of microscopic anatomy, for with the knowledge that everything in the body is built of tissues, the understanding of the complicated microscopic structure seen in organs was greatly simplified.

For a considerable time histology was regarded as a somewhat different subject from microscopic anatomy. Since a study of the tissues was such a useful prelude to the study of the microscopic structure of the parts of the body, and in particular the organ systems, it was not unusual in the past for textbooks dealing with these fields to be entitled Textbooks of Histology and Microscopic Anatomy. Roughly the first half of such books would deal with the tissues and the second half with the microscopic anatomy of the organ systems.

Since histology means the science of the tissues it is a broader term than microscopic anatomy and in due course the term histology came to include the subject matter of microscopic anatomy. However this created a problem because since histology came to be used to include microscopic anatomy it led to a widespread impression to the effect that histology includes no more than microscopic anatomy. The broad term histology came to include more and more as follows.

First, the physiologist found the microscope as helpful as did the anatomist and soon it was found that structure at the microscopic level was often ever-changing in relation to its state of function. It was realized, moreover, that the great variety of microscopic pictures seen in various tissues and organs represented highly specialized arrangements for the performance of the particular functions performed in that site. In other words at the microscopic level it was wasteful of time and effort to attempt to study structure or function apart from each other and it was realized that the study of one facilitated the study of the other. With this realization histology became somewhat of a hybrid science with anatomy and physiology as its parents. One result of this was that in some countries histology came to be taught in physiology departments. In this country (U.S.), however, it is almost universally taught in anatomy departments.

In modern times the scope of histology has continued to expand. The employment of the electron microscope, the development of radioautography and other histochemical methods, immunofluorescence microscopy and many other techniques have pushed histological studies into the subcellular level where histology, along with many other disciplines with which it integrates at that level, has led to great advances in understanding many of the processes on which life depends. Modern histology has thus become increasingly concerned with activities, and for this reason it is a much better term in this day and age than microscopic anatomy.

Histologic methods. For tissue to be studied effectively with the MICROSCOPE requires that a tissue preparation be very thin. Thin preparations are commonly obtained by cutting tissue into thin slices (see MICROTOME, MICROTOMY), from three to ten microns in thickness. Tissue may be made suitable for being sliced this thinly by freezing it, or what is more common, by immersing it first in a chemical solution that coagulates its proteins (a fixative); secondly, in a series of dehydrating solutions; thirdly, in a paraffin solvent that is soluble in the dehydrating solution; and then finally, in melted paraffin. When the latter hardens the block of tissue can be easily sliced into thin slices on a special machine called a microtome, which has a very sharp knife and an accurate and delicate advance mechanism which permits successive slices of the same thickness to be cut. Slices are mounted on glass slides, the paraffin is dissolved, the paraffin solvent is removed, and the slices are again hydrated. A thin slice of tissue mounted on a glass slide is called a *section*.

In sections, or in other types of sufficiently thin preparations, such as teased preparations or smears, lack of contrast can be overcome in two ways. First, although the components of tissue do not differ greatly in optical density they alter the phase of light that passes through them to different extents. The PHASE MICROSCOPE exploits this and translates differences in the phase of the light waves coming through different tissue components, into differences in brightness and darkness. More commonly, however, contrast is obtained by staining sections and these, of course, can be studied with the ordinary light microscope. A time-honored method is that of immersing a hydrated section in a basic dye of one color (hematoxylin is commonly used), and then later into an acid dye of another color (eosin is commonly used). The stained section is then mounted in a medium of suitable refractive index and a coverslip is placed over it to make it into a permanent preparation.

The resolving power of the light microscope is unalterably limited by the wave length of light and this means that effective magnifications of only about one thousand times is possible. Vastly greater resolution can be obtained with the electron microscope and this is now used in histology to elucidate the *fine structure* of tissues.

In the ELECTRON MICROSCOPE a beam of electrons in a vacuum is directed through a thin slice of tissue. The lenses, which bend the electron beam just as glass lenses bend a beam of light, are magnetic fields and their strength can be regulated to give different magnifications. The electron beam penetrates poorly, hence sections for electron microscopy must be particularly thin, a millionth of an inch or less. Glutaraldehyde followed by osmium tetroxide is commonly used as a fixative and epoxy resins as embedding medium. The edge of a piece of fractured glass provides a good knife and special microtomes with mechanical or thermal advances permit the short advances necessary between successive sections of such extraordinary thinness. Contrast is obtained because electrons penetrate different tissue components differently and the contrast can be enhanced by the use of certain metallic solutions which act to make some parts of tissue components more electron dense than others.

Two other modern histologic technics, histochemistry and radioautography, which are described elsewhere, will only be mentioned here. Histochemistry is discussed elsewhere. Radioautography involves the use of sections of tissue obtained from animals given radioactive isotopes which are specifically incorporated into products in certain tissues. Sections cut from the tissues of the animal are taken to the dark room where they are either coated with photographic emulsion or covered tightly with a strip of film. They are left in the dark long enough for the emission from such radioactive material as is present in certain tissue components in the section to cause ionization in the emulsion above sites where it is present (beta emitters, which give short tracks, are most suitable for radioautography). After a suitable time the coated sections are developed and fixed; they may then be brought into the light and examined under the microscope. Dark "grains" are observed here and there in the emulsion that overlies the section, these will be directly above or very close to the sites where the radioactive material is present.

Another method now widely used hinges on homogenizing tissue and separating out different fractions from it by means of differential centrifugation. Different fractions so-obtained can be collected in pellets and sectioned for electron microscopy.

The components of tissues. Three basic structural elements enter into the composition of the tissues. They are:

1. Cells: these are living entities, minute in size and jelly-like in consistency.

2. Intercellular substances: these are non-living substances made by certain kinds of cells and as their name implies they lie between cells. Some are soft but many are firm and strong.

3. Fluids: Two important ones, blood and lymph, are confined to systems of tubes (blood and lymph vessels) in which they circulate. Another important fluid, tissue fluid, bathes the cells of the body; this requires that it occupy any minute spaces that exist in the intercellular substances that separate blood capillaries from cells and through which diffusion of oxygen, food and waste products occurs.

Intercellular substances are responsible for the body having form. The animal body is best regarded as an edifice of intercellular substance in which billions of jelly-like cells, of many different kinds and families, live for short or long periods as residents, exchanging their products with one another chiefly by means of the great fluid transportation system which is provided by blood circulating continuously through all parts of the body.

The Four Basic Tissues

Epithelial tissue (epithelium). EPITHELIAL tissue is specialized to protect, absorb and secrete. It consists only of cells. To protect, epithelial cells are arranged into sheets that cover and line all body surfaces that are exposed to, or connect with, the outside world, for example, the outer layer of the skin, and the lining of the digestive, respiratory and genito-urinary systems. In some sites epithelial membranes perform an absorptive function as well as a protective one, as in the small intestine. In this and as in parts of the respiratory system, the lining cells also perform a secretory function. While secretory and absorptive cells are both protective, they are different from one another, and different still from cells primarily specialized for protection.

The secretory capacity of a membrane is limited. Secretory capacity is increased at many points along epithelial membranes because of cords of cells from the membrane invading, during development, the underlying connective tissue and then branching. Lumens develop in the cords which thereupon become the ducts of glands. The ends of the cords open up and otherwise develop into tubular or alveolar secretory units and the secretion from the cells in these reaches the surface through the duct which delivers the secretion to the site from which the gland developed. These are termed *exocrine glands* because they deliver secretion through a duct to a surface which is *outside* the body. In some sites glands form similarly but the cord of cells, connecting the secretory parts of the gland to the surface, disappears; hence, cords and/or alveoli of secretory cells are left buried in the connective tissue as *endocrine glands* because they deliver their secretions *into* the body, by way of the blood stream. Endocrine glands secrete hormones.

Covering and lining epithelial membranes may be composed of a single layer of cells (simple epithelium) or of more than one layer (stratified epithelium). Simple epithelium is found where there is not much wear and tear or where diffusion or absorption occurs through the membrane, as in the intestine. Statified epithelium is found where there is more wear and tear and no need for absorption or secretion. If simple epithelium is composed of flattened squamous cells, it is called *simple squamous epithelium*; if of cuboidal cells, *simple cuboidal epithelium*, and if of columnar cells, *simple columnar epithelium*. This is sometimes ciliated, that is little hair-like projections extend from the free surface of the cells. Ciliated cells are commonly interspersed with mucous secretory cells (goblet cells), and the cilia beat to move the mucus along the free surface of the membrane.

For wear and tear epithelial membranes are commonly stratified. In a very common kind the deepest layer of cells are columnar, but near the surface the cells become squamous. On wet surfaces the superficial cells remain alive but on dry surfaces the outermost layers of cells die as they become converted into a tough protein called *keratin*. Such epithelium constitutes the outer layer of the skin. Keratin withstands wear and is relatively impervious to water and bacteria.

Epithelial tissue has no blood supply. Moreover, epithelial membranes or glands are relatively delicate and must rest on connective tissue. The latter carries the blood vessels and capillaries close to epithelial membranes or glands and this arrangement permits diffusion of food and waste substances back and forth from the capillaries to the cells of the epithelial membranes or glands.

Connective tissue. CONNECTIVE TISSUE consists of cells and intercellular substances) it "connects" by means of the latter. The cells of connective tissue are of many types. They are all formed from the mesenchymal cells of the embryo, and they exist in postnatal life in various stages of differentiation along several different pathways. Those that form intercellular substances are named in relation to the particular kind they form; *fibroblasts* make the fibers of ordinary connective tissue; *chondroblasts*, the intercellular substance of cartilage; and *osteoblasts*, the intercellular substance of bone. Other cells of connective tissue perform other functions. Some, the *fat cells*, store fat. Others are concerned with the production of the cells of the blood. Others serve as large phagocytes—macrophages and reticulo-endothelial cells. Still others, plasma cells, produce antibodies. Another kind, mast cells, make heparin, an anticoagulant, and histamine, which invokes inflammatory responses.

The intercellular substances of connective tissue are fibrous or amorphous. The former consist first, of the white fibers of *collagen*, a tough protein which yields gelatin on boiling; secondly, *elastin*, which is in the form of yellow fibers or membranes, and which is very impervious to change and hence generally very long-lasting; and thirdly, *reticular fibers* which are delicate and form lacy networks to give intimate support to cells. The amorphous intercellular substances exist as firm or soft gels. They are mucopolysaccharides and contain much water which permits diffusion to occur through them. Fibrous intercellular substances are commonly immersed in the amorphous type.

If the intercellular substance content of connective tissue is not very substantial and cells fairly prominent, the tissue is termed loose connective tissue; this is elastic and permits some movement, while holding parts of the body together. Dense connective tissues are characterized by a great abundance of intercellular substance and relatively few cells as in tendons and aponeuroses; here the intercellular substance is commonly collagen and this arranged in bundles with fibroblasts lying between the bundles.

The intercellular substance of cartilage is chiefly a firm mucopolysaccharide gel, chondroitin sulfuric acid, and collagenic fibers are embedded in it. The cells that make the intercellular substances are termed *chondroblasts* and after they make it they lie in little lacunae in it as *chondrocytes*. The bound water in the intercellular substance of cartilage is sometimes replaced with calcium salts. If this change happens the cartilage is said to be calcified. After calcification occurs diffusion can no longer take place through the bound water of the intercellular substance and the chondrocytes die. The intercellular substance of bone, which is made by osteoblasts, is collagen embedded in an amorphous gel. Calcification normally occurs in bone, but the *osteoblasts*, which have become embedded in the intercellular substance as *osteocytes*, do not die because minute canaliculi connect the lacunae in which they lie with some surface where there are capillaries. The canaliculi permit some diffusion to occur through the calcified intercellular substance of bone, but only for short distances, this is why calcified bone can live while calcified cartilage cannot and it also explains why dense bone is arranged in *Haversian systems*, which consist of concentric rings of bone (of a limited number) around, a capillary in an Haversian canal.

One type of connective tissue is termed *hemopoietic tissue* because it produces the cells of the blood. There are two subdivisions of this, *myeloid*, which occupies the marrow of bones, and *lymphathic*. The latter is distributed in nodules in the loose connective tissue under many wet epithelial membranes, in little acorn-shaped nodes, arranged along the lymphatic channels, which drain off excess tissue fluid from many sites in the body, and in two larger organs, the thymus gland and the spleen.

The hemopoietic tissues have a loose coarse framework of collagenic fibers and a finer framework of reticular fibers. Primitive mesenchymal cells, called *reticular cells* are scattered along the reticular fibers. These give rise to *fixed macrophages*, called *reticulo-endothelial cells*, and probably also to free cells. In myeloid tissue the free stem cells give rise to three lines of cell differentiation with the end products being erythrocytes (red blood cells), granular leucocytes and megakaryocytes (which produce platelets) respectively. Monocytes are also formed in myeloid tissue. In lymphatic tissue lymphoblasts and plasmoblasts form lymphocytes and plasma cells respectively.

Connective tissue carries the blood vessels. The capillary beds of connective tissue provide the tissue fluid, which bathes all tissue and which permits diffusion from capillaries to cells. The connective tissues, moreover, provide the medium in which the chief reactions of the body occur in response to invading disease agents; this is known as *inflammation* and is associated, at sites where it occurs, with a dilatation of blood vessels, exudation of plasma, migration of leucocytes from the capillaries into the tissues, phagocytosis of disease organisms by leucocytes and macrophages, production of antibodies by plasma cells that develop in response to the foreign antigen, and many other phenomena.

The spleen is a great depot of lymphatic tissue through which blood percolates in close association with phagocytic cells (called either fixed macrophages or reticulo-endothelial cells). These, together with similar cells in myeloid tissue, and others in certain other sites, remove worn-out erythrocytes and other debris from the circulation. The hemoglobin of the erythrocytes is broken down to an iron pigment (which slowly dissolves with the iron going back to the bone marrow to be used in new erythrocytes), and to an iron-free pigment (which is essentially bilirubin or bile pigment and which goes to the liver to be secreted in bile).

Muscular tissue. The cells of muscular tissue are called MUSCLE fibers. They are highly specialized for contractility. For this purpose the cells (fibers) are relatively long and narrow, so that contraction in their long axes is effective, and they contain longitudinally disposed contractile fibrils called *myofibrils*.

There are three kinds of muscular tissue, *smooth* (involuntary), *striated* (voluntary, skeletal), and *cardiac*.

Smooth muscle is so-named because its fibers show no cross striations as do the other two types. Smooth muscle fibers have elongated tapered shapes and vary from about 20 microns to 0.5 millimeters in length. They are termed involuntary fibers because they are under the control of the autonomic nervous system. They are generally arranged into sheets with the individual fibers in the sheets adhering to each other by a little glue-like intercellular substance that is produced by the fibers themselves. The fibers contain longitudinally-disposed myofibrils embedded in an amorphous material termed *sarcoplasm*. The nuclei are single, centrally-disposed and elongated. They appear partly folded in contracted fibers. Smooth muscle is adapted to maintaining different states of contraction, that is, different degrees of tonus. Hence, sheets of smooth muscle fibers surrounding a lumen, as occurs in arteries and arterioles can, by maintaining different degrees of tonus, keep the size of the lumen small or large at the dictates of the autonomic nervous system. Smooth muscle fibers can also contract actively, as occurs when a wave of peristalsis sweeps down the intestine. In general, smooth muscle is slow and sluggish as compared to striated.

Striated muscle fibers are larger, measuring from 1 to 40 millimeters in length and from 10 to 40 microns in diameter. They are multinucleated. They contain myofibrils that run longitudinally and which have sarcoplasm between them.

With the light microscope the fibers exhibit cross striations. With the lower powers each fiber appears as if it were composed of individual discs much like the coins in a roll that a bank teller unwraps when he needs more dimes or quarters. The discs, however, alternate between being light and dark; the light ones are termed I discs, and the dark ones, A discs. On closer inspection, a dark line can be seen bisecting each I disc; this is termed the Z line. With the higher powers, it can be seen that the alternate light and dark discs do not cross the whole fiber but are due to alternating light and dark segments of the myofibrils. With the electron microscope it can be seen that the Z line also does not cross the entire fiber but exists only in myofibrils; in other words, none of the striations cross the sarcoplasm between myofibrils. The reason for the striations' seeming to involve the whole fiber is that the light and dark segments of parallel myofibrils are always in register with one another. The portion of one myofibril between two Z lines is termed a *sarcomere*; this is a contractile unit.

With the electron microscope the myofibrils are seen to be composed of still smaller thread-like elements termed *myofilaments*. These do not continue from one Z line to another. Huxley and his associates have shown that there are two sets. The filaments of one set are attached to the Z line and extend from it, from both ends of the sarcomere, toward its middle, but do not reach it. Those of the second set are more centrally disposed and do not reach the Z line, but

interdigitate with those of the first set, that do. In contraction the filaments slide between one another to shorten the sarcomere.

The impulse for contraction comes from nerve impulses (see IMPULSE TRANSMISSION) that reach motor-end plates, which are attached to the double (cell) membrane that surrounds each muscle fiber, and which is called its *sarcolemma*. A continuation of the double membrane in the form of tubules extends from the sarcolemma throughout the sarcoplasm of the whole fiber to reach each sarcomere and conducts the impulse for contraction to it.

Striated muscle fibers are harnessed by means of connective tissue sheaths. A delicate one called the *endomysium*, which contains capillaries, surrounds each fiber. Bundles of fibers are surrounded by stronger sheaths of perimysium, and whole muscles are surrounded by an epimysium. The sheaths and the ends of fibers are firmly attached to dense connective tissue structures such as tendons, ligaments, aponeuroses, and the covering membranes of bones and cartilage.

Cardiac muscle is striated but under the control of the autonomic nervous system. In contrast to ordinary striated muscle, the fibers of cardiac muscle branch. The electron microscope has shown that cardiac muscle is not a syncytium as was once believed and that the dark lines, termed *intercalated discs*, which are found along the branching fibers, show two cell membranes and hence represent boundaries between two cells. In some sites the outer layers of the membranes fuse to form tight junctions. The impulse-conducting system of the heart is composed of a special kind of cardiac muscle fibers; characteristically these contain much glycogen in their middle parts and hence are paler than ordinary fibers.

Nervous tissue. Nervous tissue is highly specialized for irritability and conductivity. It provides a means whereby stimuli originating either within or without the body can induce a response either in muscle cells or in gland cells. Nervous tissue is the link between sites where stimuli arise and the tissues in which responses to stimuli can occur.

Nervous tissue is arranged two ways. First, it comprises the BRAIN and spinal cord. These are continuous with one another and constitute the *central nervous system*, the nervous tissue here is soft and delicate. Protection is afforded first by a wrapping of two membranes, the *pia mater* and the *arachnoid*, between which there is a cushion of cerebrospinal fluid. Outside is the surrounding bone of the cranium and vertebral column, the cavities of these are lined by another membrane, the *dura mater*. Secondly, nervous tissue comprises the cord-like nerves, a set of which leaves the cranium, as cranial nerves, and the vertebral column, as spinal nerves, on each side of the body. This system of nerves that flow from the central nervous system out through its bony encasement to reach all parts of the body, constitutes the *peripheral nervous system* and the tissue of which these nerves are composed is tougher than that of the central nervous system.

Nerve cells are called *neurons*. To conduct over distances they must be long; this is effected by each having a cell body and long cytoplasmic processes, called *nerve fibers*. Those that extend from the cell body to the source of stimulus are termed *dendrites*, and those that pass to the sites where the impulse is to be conducted are called *axons*. A neuron may have many dendrites and these branch. The axon is single, but may branch.

Most nerves of the peripheral nervous system consist of many nerve fibers each with a special wrapping. Most have a coating of non-living lipoid material called *myelin* and outside this a delicate cellular membrane termed the neurolemma or sheath of Schwann. Outside this is a delicate wrapping of connective tissue, the endoneurium. Some fibers have very little myelin and are said to be non-myelinated. Bundles of fibers, with their individual wrappings, are enclosed in stronger connnective tissue sheaths of perineurium, and bundles of these, that comprise an individual nerve, by a further sheath of connective tissue, the epineurium. Because of their connective tissue components, the nervous tissue of the peripheral nervous system is tougher than that of the central nervous system.

The fibers in nerves have two different functions. Some conduct impulses into the central nervous system and others away from the central nervous system to gland and muscle cells. The former are termed *afferent*, the latter *efferent*, fibers. The cell bodies of the former are in little connective tissue bodies, just outside the bony encasement of the central nervous system, termed GANGLIA. From here the axons of the afferent fibers pass into the spinal cord or brain. The cell bodies from which the efferent fibers of nerves arise are in the brain and spinal cord (the only exception to this are the postganglionic fibers of the autonomic nervous system).

Neurons connect with each other by synapses; these are sites of contact between an axon from one neuron and the cell body or dendrites of another. A wave of excitation sweeping along an axon to a synapse stimulates a wave of excitation in the next neuron, and by this means waves of excitation can sweep along chains of neurons.

The central nervous system consists essentially of chains of neurons supported by a glue-like cellular supporting tissue of nervous origin called NEUROGLIA. On gross inspection the tissue here is either gray or white and is referred to as gray or white matter. Much of the brain is covered with gray matter, with its interior being mostly white, while the spinal cord is white outside but has a more or less H-shaped column of gray matter in its more central part. Gray matter is gray because it lacks myelin; it consists chiefly of the bodies of nerve cells, naked nerve fibers, and an abundance of neuroglia cells which have delicate processes which support the bodies and naked fibers of the nerve cells. White matter consists of myelinated nerve fibers supported by neuroglia cells. The myelinated fibers of the central nervous system lack neurolemma and connective tissue.

Sensory endings are specialized to be affected by particular kinds of stimuli, for example, heat or touch. When a nerve ending, sensitive to touch, is touched, a wave of excitation is set up in that fiber and conducted to the spinal cord. There, by means of a synapse, another fiber is stimulated and this conducts an impulse toward the brain and by chains, usually of several neurons, the impulse stimulates cell areas in the brain which give rise to the sensation of touch and localizes its origin in consciousness. From here an impulse may stimulate an efferent system which carries the impulse

down the spinal cord and out through efferent neurons in a nerve to move the muscles of the part.

The autonomic nervous system is a functional division of nervous tissue comprising the efferent system that controls the smooth muscle and glands of the body. It operates outside consciousness. It has two divisions, the thoraco-lumbar outflow of fibers, which constitutes the sympathetic system, and the craniospinal outflow, which constitutes the parasympathetic system. In each, efferent fibers leave the central nervous system and terminate in ganglia that are variously disposed in the body. Here the cell bodies of second neurons are located. From these postganglionic fibers travel to smooth muscle and gland cells. Commonly the muscles and glands have fibers from both the sympathetic and parasympathetic divisions, and these fibers are generally antagonistic in their functions.

The histology of the organs is dealt with elsewhere in this volume under their individual names.

ARTHUR W. HAM

References

Bloom, W., and Fawcett, D. W., "Textbook of Histology," 9th ed., Philadelphia, Saunders, 1968.

Copenhaver, W. M., "Bailey's Textbook of Histology," 15th ed., Baltimore, Williams & Wilkins, 1964.

Greep, R. O., ed., "Textbook of Histology," 2nd ed., New York, McGraw-Hill (Blakiston Div.), 1966.

Ham, A. W., "Histology," 6th ed., Philadelphia, Lippincott, 1969.

HISTORY OF BIOLOGY

This volume is a monument to the achievement of thousands of investigators. It records in a systematic manner the most fundamental accomplishments in the quest for understanding of living forms and processes. Embedded in nature, *homo sapiens* has been engaged in this effort since long before recorded history. Ancient plant breeders and herdsmen secured a nutritional base for the human family ages before the first calendars, astronomical models or number systems were elaborated. Knowledge of the nutritive healing or toxic properties of substances of botanical origin were recorded in the age of the pyramid builders and codified in Babylonian clay tablets fifteen centuries before the era of the Hippocratic school. The legendary hanging gardens of Babylon may represent an ancient attempt to study the plant world or secure a supply of drugs for Babylonian pharmacologists.

With the rise of Hellenic civilization, the heir of the ancient Near Eastern culture, biological science, entered its first theoretical stage. The encyclopedic Aristotle, whose vast system was permeated with biological imagery, created the foundations of zoology and began the unending labors of taxonomists. He was a meticulous observer, particularly of marine life. He grasped the central importance of the problem of development and was the first embryologist. By a profound intuitive leap, he attempted to construct a theoretical account of the functioning of all living things. His *Scala Naturae* was a premature vision of the unity of life, lacking as it did a sense of progressive change or any idea of the magnitude of biological time. Aristotle's successor in the Lyceum, Theophrastus, began the careful study of the plant kingdom, describing over five hundred species gathered throughout the Macedonian Empire.

Physiology and anatomy were studied with skill and energy by Galen, but his excessive moralizing and uncritical teleological bias tended to stultify his advances over Aristotle, particularly in neurophysiology.

Biological knowledge, like other aspects of Hellenic science, was submerged in the successive tides of Roman imperialism, Christian mysticism, and barbarian invasion. The profound insights of Lucretius and the naive enthusiasms of Pliny mark the end of ancient biology.

Islamic scholars preserved, enhanced, and eventually were instrumental in transmitting Aristotelian and Galenic texts to the nascent science of the Latin West. After long incubation, this biological tradition was incorporated into the European curriculum, particularly that of the medical schools, as sacrosanct orthodoxy.

The intellectual and sociological ferment of the Renaissance, which culminated in the scientific revolution of the seventeenth century, was reflected in a renewed vigor of investigation by scholars probing the living world. The burning of canonical medical texts by Paracelsus was more symbolic of the future than the burning of Servetus by Calvin.

In 1543, the year in which the astronomical treatise of Copernicus was published, Andreas Vesalius produced his revolutionary anatomical work, *De Humani Corporis Fabrica Libri Septem*. Thus Ptolemy and Galen were stripped of that unquestioned authority which had ossified both astronomy and biology.

The revolt against tradition was transformed into a creative scientific impulse by merging with a methodological revolution compounded of a reliance on experiment and observation, the creation of a new mathematical language, and the construction of new instruments: telescope and microscope, thermometer and barometer. As Galileo's glass revealed the cosmic realities of the boundlessly large, the lenses of Anthony Van Leuwenhoek brought into view the equally amazing world of the vanishingly small. The advances in both branches of science were thereafter linked to the constant improvement of optical instruments and observational techniques. The work of Hooke, Malpighi, Swammerdam, and a host of lesser workers gave to biology the stimulus that Newtonian mechanics contributed to the older physics and astronomy. The triumphant mechanism and experimentalism of the epoch found its greatest biological embodiment in the work of William Harvey, whose *Exercitatis Anatomica de Motu cordis et Sanguinis* was memorable not merely because of the demonstration that the blood was pumped in a closed circuit, but because its mode of *discussion* was as rigorous as Newton's *Principia*.

Borelli, Jung, and Descartes also applied principles of mechanics to the behavior of animals, and there was early sketched the outlines of the vitalist-mechanist disputes of the life sciences. In this setting, the old Aristotelian problems of taxonomy and embryology were taken up again, particularly in the work of John Ray, Fabrizzi, and Harvey.

Eighteenth-century biology was highlighted by the embryological disputes between the preformationists and the successful counter theory of the epigenesists, the monumental work in natural history of Buffon, the adoption of the Linnean system of classification, the clear statement of the *problem* of the relationships and mutability of species: work of Buffon, Maupertuis, Cuvier, Erasmus Darwin, and Lamarck, experimental attacks on the problem of spontaneous generation (Redi and Spallanzani), the beginnings of systematic physiology of plant and animal, embodying discoveries in physics and chemistry: Hales, Ingenhousz, Lavoisier, Laplace, Haller, the recognition of bioelectricity (Galvani and Volta).

The incredible labors of Bichat laid down the foundations of histology which, aided by improved microscopy, maintained a link between physiology and anatomy. Comparative anatomy, in establishing the homology of organs (Cuvier and Owen), provided important empirical evidence for the emerging evolutionary ideas. The concurrent development of historicism by a brilliant school descended from classical scholarship and the formulation of philosophies of becoming (Hegelians) reinforced this trend of European thought. Applied biology became a powerful social force with the introduction of inoculation by Jenner.

Developments in astronomical theory, particularly the Kant-Laplace nebular hypothesis, and the rapid developments in geology (Hutton and Smith *inter alia*) established the principle of uniformitarianism as descriptive of earth processes. This hypothesis implied that the earth was immensely old. Geological theories contributed along with other currents of Enlightenment thought, including an aggressive materialism (La Mettrie) to the undermining of scriptural authority in matters becoming of prime importance to biologists and other scientists.

The nineteenth century, in retrospect, appears as the period of maturation of all of the major divisions of the sciences. The epoch was characterized by an increasing specialization, professional training, and interaction between the various scientific disciplines. All of these elements were reflected in the very rapid growth of biology and can be partially illustrated by the careers of the pupils of the German physiologist, Johannes Müller (1801–1858).

Schwann and Virchow (with Schleiden) developed the elements of cell theory, a conceptual advance which was analogous to the importance of atomic theory in chemistry; Schwann also isolated pepsin, thus contributing to the foundations of biochemistry. Helmholtz, the last of the universal geniuses, mobilized the resources of electrodynamics and thermodynamics to establish the foundations of biophysics, with particular reference to vision, hearing, speed of transmission of nerve impulses, and the conservation of energy in vital processes. Du Bois-Reymond, through the invention of new instruments, was able to interpret the currents in nerve and muscle tissue in terms of the inorganic environment, thus undermining vitalism, as his colleagues had demolished the mystical element of *Naturphilosophie*.

The culmination of one line of biological thought in the theory of Charles Darwin requires little commentary here. Among the many advances which contributed toward making Darwinism a plausible hypothesis in the specific formulation by the great Englishman was the development of comparative embryology by Kurt von Baer and his followers, particularly Peltier, Kölliker, Pringsheim, and Fol. The repercussions of Darwinism indicate that the publication of *On the Origin of Species by Means of Natural Selection, or the Preservation of Favoured Races in the Struggle for Life*, was a turning point not only in biology, but in many other domains of human thought.

Another fundamental biological conceptual leap is represented by the work of Claude Bernard, pupil of Magendie. Starting with problems of the chemistry of digestion, but expanding his horizon to include more comprehensive problems of physical and chemical regulation, Bernard was led to the theory of the constancy of the *milieu interieur* as a primary property of a living system.

With more detailed understanding of the integrative functions of the nervous system (Ramon y Cajal to Sherrington, Pavlov, and Cannon), homeostatic principles emerged as the cornerstone of biochemistry and biophysics.

These developments depended, of course, on the constant progress of both biochemistry and microscopy. The former may be traced back to the work of Berzelius and his pupils. Pasteur's work in many phases of microbiology did not begin *de ovo*, but continued the line of discovery of Bradley ("living contagion"), Agostino Bassi (pupil of Volta and Spallanzani), and Davaine. Pasteur, like Jenner before him, was not only a scientist of genius, but also one of the greatest benefactors of mankind.

These biochemical insights led with great speed, considering the complexity of the issues, to the development of the science of immunology (Richet, von Behring, Landsteiner, Metchnikov).

Evolutionary theory, with the long delayed incorporation of the Mendelian laws (De Vries, *et al.*), coupled with the refinement of ideas introduced by Weismann, led in the twentieth century to a general genetic philosophy; and the continuous input of evidence from biochemistry created a theory which infuses most branches of biological work and links problems of evolution, development, and regulation in one vast synoptic biological science. Here also, the contemporary insights of Information Theory (Shannon, *et al.*) intersect with older ideas from thermodynamics.

Molecular biology, in the widest sense of the term, with its supreme instrument, the electron microscope, is a rapidly expanding frontier wherein enzymology, the study of endocrine systems, the chemistry and architecture of nucleic acids, proteins, and steroids, metabolic pathways, and homeostatic processes, is revealing the essential relations of form and function. At the level of cellular chemical processes (active sites, neural thresholds, etc.), quantum phenomena have their effects and the fine grain of biological activity its causal explanation. With the new tools of radioactive isotopes, crystallography, and computers, many of the problems of molecular biology are yielding to mathematical analysis, with the extension of the ideas advanced by D'Arcy Wentworth Thompson. Extraordinary advances in this field have been recorded by Linus Pauling and his school.

With the work of T. H. Morgan, Müller, and others in genetics and the discovery that a virus is a complex of nucleotides encased in protein, the ancient issue of the origin of life could be discussed in terms of modern physics and chemistry. Here the work of Oparin, Urey, and Miller links biology to geology, cosmology, and astrophysics. These theories suggest the existence of primeval seas and a reducing atmosphere where nutrient broths brewed from inorganic reactions and catalyzed by radiant energy brought forth the original macromolecules whose stochastic combinations formed the lowest rung of the ladder of life.

The epoch of space travel now beginning promises further developments in biology. The studies of the effects of weightlessness and cosmic radiation are but prologue to a direct examination of the possibility and perhaps the nature of life forms on at least the closer planets of the solar system; the data expected from lunar exploration are apt to clarify still obscure aspects of the earth's history.

The next chapters in the history of biology can hardly be suggested, but it appears certain that biology is entering a golden age of revolutionary advance comparable in scope to the reorganization of physics after Planck and Einstein. In this era biology will be more than a grateful borrower of ideas. A hundred years ago Darwin's intuition about the age of the earth proved to be more prescient than the dogmatism of Lord Kelvin. The well confirmed theories of living systems now stand in no need of a crude reduction to physics and chemistry. On the contrary, biological facts are the parameters to which large-scale physical theory must adjust.

ROBERT G. COLODNY

References

Barnett, S. A., ed. "A Century of Darwin," Cambridge, Harvard University Press, 1958.

Beck, W. S., "Modern Science and the Nature of Life," New York, Doubleday, 1961.

Dunn, L. C., "A Short History of Genetics," New York, McGraw-Hill, 1965.

Ehrensvärd, G., "Life: Origin and Development," Chicago, University of Chicago Press, 1960.

Eisley, L., "Darwin's Century, Evolution and the Men Who Discovered It," New York, Doubleday, 1961.

Gabriel, M. and S. Fogel, eds. "Great Experiments in Biology," Englewood Cliffs, N. J., Prentice-Hall, 1955.

Granit, R., "Charles Scott Sherrington, a Biography of the Neurophysiologist," New York, Doubleday, 1967.

Graubard, M., "Circulation and Respiration, Evolution of an Idea," New York, Harcourt, Brace & World, 1964.

Mendelsohn, E., "Biological Sciences in the Nineteenth Century: Some Problems and Sources," History of Science, Vol. III, 1964.

ibid. "Physical Models and Physiological Concepts: Explanation in Nineteenth Century Biology," *in* Cohen, R. S., ed. "Studies in the Philosophy of Science," New York, Humanities Press, 1965.

Ritchie, A. D., "Studies in the History and Methods of Science," Edinburgh, The University Press, 1958.

Schrödinger, E., "What is Life?" Cambridge, The University Press, 1945.

Sherrington, C. S., "Man on His Nature," 2nd ed. New York, Doubleday, 1953.

Sigerest, H., "On the History of Medicine," New York, M.D. Publications, Inc., 1960.

Sirks, M. J. and C. Zirkle, "The Evolution of Biology," New York, Ronald, 1964.

Taton, R., ed. "Histoire Générale des Sciences," 4 vols., Paris, Presses Universitaires, de France, 1957–1961.

Taylor, G. R., "The Science of Life, A Picture History of Biology," New York, McGraw-Hill, 1963.

Thompson, W. D., "On Growth and Form," Cambridge, The University Press, 1917 and 1940. Abridged edition, Bonner, J. T., ed. Cambridge, The University Press, 1961.

Toulmin, S. and J. Goodfield, "The Architecture of Matter, The Physics, Chemistry and Physiology of Matter, both Animate and Inanimate as it has evolved since the beginnings of Science," New York, Harper & Row, 1962.

Verdoorn, F., ed. "A Short History of Plant Sciences," New York, Ronald, 1942.

von Bertalanffy, L., "Modern Theories of Development, An Introduction to Theoretical Biology," adapted by Woodger, J. H., New York, Harper & Bros., 1962. Original edition, Oxford, Oxford University Press, 1933.

Weyl, H., "Philosophy, Mathematics and Natural Science," Princeton, Princeton University Press, 1949.

HOFMEISTER, WILHELM FRIEDRICH BENEDICT (1824–1877)

Hofmeister was one of the very few self-taught naturalists ever to hold a chair in the German University. After the equivalent of a high school education in Leipzig he set up as a music seller in that town. His attention was, however, greatly attracted to botany. He studied, in particular, the fertilization of the lower plants and demonstrated beyond question that the embryo was developed from the ovum and not, as had previously been thought, from the pollen tube. He published the main facts of his investigation in 1851 in a book astonishingly modern for its period, in which he drew attention to the fact that the alternation of generations common in lower plants must point to a common origin for them. He indulged in no philosophical speculation but merely drew attention to similarities that had escaped his contemporaries, and which became explicable only with the subsequent pulication of Darwin's work. The production of so superbly scholarly a work by a book seller attracted wide attention in Germany, and in 1863 he was nominated to the chair of botany in Heidelberg and nine years later to the same chair in Tubingen. In addition to his work on the fertilization of plants, he contributed much to the distribution of the conifers.

HOMEOSTASIS

Definition: Homeostasis is said to be shown by a physiological system if, given a moderate disturbance that tends to displace the system from its normal values, the parts so react and interact that the harmful effects of the disturbance are much diminished. Here are some examples.

When a man is much chilled, the cooling stimulates a mechanism in the base of the brain that sets him shivering. The muscular activity generates heat, which *opposes* the chilling.

Glucose is required in the blood to provide energy by oxidation. If the amount of glucose in the blood should fall to an unusually low level, the fall stimulates the suprarenal glands to secrete epinephrine. When epinephrine arrives at the liver, it makes the liver convert some of its store of glycogen into glucose, which is passed into the blood. So the fall is opposed.

Sudden hemorrhage causes a sharp fall in blood pressure. The fall, however, causes the arterioles to constrict, thus lessening the amount of fall.

The word was coined by the physiologist Walter B. Cannon (1) to refer to a phenomen widely demonstrable in physiological systems. The significance of the phenomenon, however, can be appreciated adequately only if it is seen in its proper place in the general theory of systems; for phenomena that are demonstrably isomorphic with physiological homeostasis occur in systems far removed from the physiological.

The drive to stability. When *any* dynamic system drives forward in time, so that its state changes but not the laws that it obeys, it tends towards some subset of the possible states, i.e. toward some equilibrium (2). In this way every dynamic system (whether a watch, or a dog, or a species, or a digital computer) acts *selectively* in that it tends toward states that are peculiarly resistant to the change-inducing action of the dynamic laws. The reason is the fundamental asymmetry: when resistant, the system tends to stay as it is (i.e. to stay resistant), but when vulnerable it tends to change (i.e. to move from being vulnerable). From one point of view this statement is truistic; it is not trivial however, for it provides the link explaining how a dynamic drive can be essentially "blind" and yet act so as to cause the appearance of highly selected forms. There is no incompatibility, for the forms are necessarily those characterised by an unusual power of survival against the forces of the drive.

Such a process could be exemplified in a digital computer that is given a definite program (any program) and then left to follow the laws (physical and programmatic) that govern it. In this example, however, the result is apt to be trivial. The power of this selective process, of this asymmetry, can be seen adequately only when it occurs on a scale vastly larger than that in a computer.

The most notable example known to us is that in which a certain planet weighing 10^{22} tons was kept under constant dynamic laws for 5×10^9 years. The forms that have developed in *this* system are showing a truly astonishing tenacity (thus the mammals are outlasting the Alps). But the basic process is much more general than the biologist's "natural selection" —resistant forms will develop in *any* dynamic system that changes its state but not its laws. When such forms develop, they are bound to show the essential features of homeostasis; for to react to a disturbance in such a way as to diminish it (rather than to aggravate it) is a powerful factor for continued existence.

Claude Bernard (3) probably appreciated this fact in its generality when he wrote "Tous les mechanismes vitaux, quelques variés qu'ils soient, n'ont toujours qu'un but, celui de maintenir l'unité des conditions de la vie dans le milieu intérieur." But he spoke essentially as a physiologist.

Homeostasis in physiology. In the century that followed Bernard, the physiologists studied the parts that make up the mammalian body, and found a vast number of facts, of which the three examples given above are a very small sample. Then, in 1932, Cannon showed that a single theme ran through them all—the partial processes often formed circular chains of cause and effect, and the effect of the whole circuit was almost always to evoke a change that tended to *neutralise* the initial disturbance. The neutralisation he called "homeostasis."

The concept of homeostasis can thus be arrived at by two very different routes. On the one hand, the mathematician, considering *any* large-scale process, predicts that there will appear forms that show it—that whatever the form's parts, they will be so related that disturbance will evoke a response that acts as neutraliser. On the other hand, the physiologist has found the concept as an induction from a great number of empirical facts. Had he known nothing of evolution he could still have discovered it as a uniformity over the living organisms of today. The concept of homeostasis thus has both an abstract theoretical foundation and a wealth of actual examples.

Homeostasis generalized. Cannon defined the word originally with reference only to activities in the vegetative, physiological, and internal systems of the living organism, chiefly the mammalian. But he did not exclude the possibility that similar principles might hold more generally:

"It seems not impossible that the means employed by the more highly evolved mammals . . . may present some general principles for the establishment, regulation, and control of steady states, that would be suggestive for other kinds of organisation—even social and industrial—which suffer from the distressing perturbations. . . . Are there not general principles of stabilization?"

Sommerhoff (4), in 1950, produced what is perhaps the most general formulation, showing that all the biological concepts that use ideas of integration, coordination, and goal-seeking are variations on the basic theme that a disturbance (the *coenetic variable*) threatens to defeat the occurrence of some goal (the *focal condition*); the integrated regulator, however, so reacts as to neutralise the effects of the disturbance, thereby bringing the focal condition to achievement.

The present author, independently in 1952 (5), (following a preliminary observation of 1940 (6)), identified essentially the same phenomenon and showed its relationship to the well known physical phenomenon of "stability."

These two approaches were later (2) found to be closely related to the work of Shannon (7) on the transmission of information in the presence of noise. We now know that Sommerhoff's "directive correlation" and Ashby's "stability" are *homologous* with the communication engineer's concept of the elimination of noise from a corrupted message. The homology at first passed unnoticed because the proportions are so different: in Shannon's case one thinks naturally a great deal of message, with a small amount of noise driving it from its true values; in the biological system the "noise" becomes the major variant (for it includes all the disturbances that may drive the organism from its normal values), while the "message" is of extremely low variability (in the homeostatic cases); for what is "transmitted" is almost a constant (e.g. "the temperature should be 37°C," or "the blood sugar should be 120 mgm per 100 ml."). Thus, the concept of "homeostasis" is abstractly *identical* with that of the minimisation of error in a message.

From this identity follows the fact that Shannon's Tenth Theorem must apply; and so we reach the law (of requisite variety (2)) that is beginning to be recognized as fundamental in the theory of homeostasis: *the power of any regulator to achieve homeostasis cannot exceed its capacity as a channel for communication.*

Theory of homeostasis. The study of homeostasis in general is thus essentially co-extensive with the study of stable dynamic systems or processes.

When the system is simple the theory is correspondingly simple, and a great deal is known today about linear systems (in which effects are proportional to causes). The properties of feedback are well known, and calculable in the simpler cases, the whole knowledge being summed up in modern "servo-mechanism theory." The non-linear types are fairly well understood in the simpler cases. Thus the general theory of homeostasis, as stability, is well developed in those cases describable as a simple mechanism.

The machines we handle in everyday life however—the watch, the typewriter, the automobile—are really extremely simple types, quite unsuitable in their degeneracy to act as models for "machines in general." Only the digital computer, the brain, and the living organism can as yet make any claim to adequate generality. In these really complex types all sorts of complex stabilities may occur, and these stabilities may be of far more interest than those degenerate forms of stability seen in the everyday machine. The going of a Markov chain to equilibrium, for instance, corresponds on the one hand to a process as trivial as that of an insect coming to rest on the first place that is of the right color, and on the other to the extremely complex and interesting process by which the infant slowly makes its way to the state described as the "well educated adult." Both of these processes are forms of "going to equilibrium" and both will show the properties of homeostasis, the first so trivially as to be hardly detectable, the latter in so complex and rich a form as to be hardly recognizable. The study of these complex homeostases is a task for the future.

Conclusion. The concept of "homeostasis" may thus justly be considered as the most fundamental in biology. It is related essentially to "survival," for what survives in a vigorous world must be homeostatic in its reactions; and the ability to behave homeostatically enormously increases a system's chance of survival. The modern logic of mechanism has shown its importance in the general theory of dynamic systems; Cannon showed its importance in the theory of physiological mechanisms; there remains the task of showing its applications to the theory of learning, for the brain is wholly subject to the truism that what *persists* in it must react to the disturbances in the homeostatic nammer.

W. Ross Ashby

References

(1) Cannon, Walter B., "The wisdom of the body," New York, Norton, 1939.
(2) Ashby, W. Ross, "An introduction to cybernetics," New York, Wiley, 1956.
(3) Bernard, Claude, "Leçons sur les phénomènes de la vie," Paris, Baillière, 1878.
(4) Sommerhoff, G., "Analytical biology," New York, Oxford, 1950.

(5) Ashby, W. Ross, "Design for a brain," 2nd ed., New York, Wiley, 1960.
(6) Ashby, W. Ross, "Adaptiveness and equilibrium," J. Ment. Sci., **86**: 478, 1940.
(7) Shannon, C. E., and Weaver, W., "The mathematical theory of communication," Urbana, Univ. Illinois Press, 1949.

HOMOPTERA

The Homoptera is an order of terrestrial INSECTS which includes cicadas, spittle bugs, leafhoppers, treehoppers, aphids, scale insects, and some other forms. There are four membranous wings in the winged forms, the front and rear wings being of similar structure. The wings are typically roofed over the body when at rest. The mouthparts are of the piercing-sucking type, the beak being well back on the ventral side of the head, in some forms appearing to be between the coxae of the front legs. The metamorphosis is gradual, but variable in the aphids and coccids. Homoptera puncture plants and feed on the sap. Many of them are destructive to plants and some transmit plant diseases.

The Order Homoptera was formerly included in the Hemiptera as a suborder, along with the suborder Heteroptera, the true bugs. About 20,000 species of Homoptera have been described. Members of the order are known from the Permian to the recent.

The wings of Homoptera show reduction in the usual vein pattern and modification of the wing cells. In the treehoppers, leafhoppers, and froghoppers the front wings are almost leathery, but not hemipteran in form. The wings of these forms often show color patterns. There are wingless forms among the aphids and coccids, and male coccids have two wings.

Notable forms of Homoptera include the periodical cicada (Magicicada septendecim), commonly called the seventeen-year locust in the United States, which occurs in many broods across the eastern and central states. The nymphs of the periodical cicada grow thirteen or seventeen years (for different broods) in the ground where they feed by sucking juices from tree roots. The final instar crawls from the ground to an elevated place, as on a tree trunk, and the adult emerges. The eggs are deposited in woody twigs by a chisel-like ovipositor. The male cicada sings by a vibrating organ in the ventral portion of the metathorax and abdomen; the female is silent.

The harvest fly (Tibicen) is a large black and green cicada of the eastern and central United States. The nymphs develop in the ground in one year. The adults emerge from June to September, with the male singing in the trees on hot days.

Spittle bugs are best known in the nymph stage when they form frothy masses of fluid on herbaceous plants. Adult spittle bugs are the attractive brownish "froghoppers" of field and garden. Both the nymphs and adults suck juices from many ornamental plants, including Campanula and Gaillardia. The fluid for the froth is secreted by glands of the posterior abdominal and anal region.

The treehoppers occur on a variety of herbaceous and woody plants and are notable for the bizarre form of the head and the prothorax. The prothorax is elongated posteriorly above the abdomen and may bear horns and spines in some species. Several forms

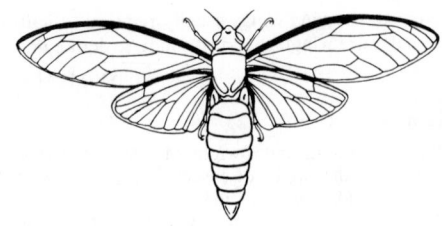

Fig. 1   Hemiptera, family Cicadidae. (From Fox and Fox, "An Introduction to Comparative Entomology," New York, Reinhold, 1964.)

of treehoppers show marked protective resemblance and are rarely seen except when they happen to move, or are captured in a net. They have wide distribution but are not abundant in numbers and do little damage to the host plants.

Leafhoppers are attractive and often multicolored insects about one-fourth inch long, abundant in grassy and weedy areas. The have strong jumping legs and can walk sideways as well as forward. They come to lights in large numbers during summer nights.

The large lantern flies are largely tropical and oriental in distribution. North American forms of lantern flies are small. They are not luminous. Many of them have an anterior median elongation of the head. The antennae are attached to the cheek below the eyes. A common form in eastern North America is pale green in color, powdered with white, and superficially resembling a small moth.

The plant lice, or aphids, are among the best known insects because of the damage they produce in a variety of cultivated plants including roses, maize, grapes and composites. Different species attack various succulent parts of plants from roots to leaves, flowers and fruits. There are species of aphids which require both primary and secondary host plant species to complete the life cycle. Typical aphids pass the winter in a winter egg stage from which the stem mother hatches in the spring. She produces wingless agamic forms which reproduce parthenogenetically, followed by winged agamic forms. The last winged generation on the secondary plant host is composed of winged males and oviparous females. These females deposit the winter egg from which the parthenogenetic stem mother form develops the following spring.

The phylloxerids differ from the aphids in several wing characteristics and in reproductive forms. The males and sexually perfect females are wingless; both the parthenogenetic and the sexually perfect females are oviparous. The group includes the adegids which feed on conifers. A two year life cycle of some of these involves two genera of conifers in succession, as spruce and pine, with several generations of adegids on each. The genus Phylloxera includes many species which infest forest trees. A notable form feeds on the wild grape of the eastern United States and is highly injurious to the European grape, notably in France and California.

The scale insects of the family Coccidae include mealy-bugs, bark lice, and such scales as the San Jose scale of fruit trees, and the oyster-shell scale which attacks a variety of woody plants. The best known scale insects are noxious, but a few scale insects are the source of useful products. Shellac and cochineal dye are produced by scale insects. Most adult male coccids

are winged, the hind wings usually reduced to halteres;
the females are wingless and the body scalelike in form.

T. M. McMillion

References

Borror, Donald J., and Dwight M. Delong, "An intro-
duction to the study of insects," New York, Holt,
Rinehart and Winston, 1964.
Comstock, J. H., "An introduction to entomology,"
Ithaca, New York, Comstock, 1940.
Marlatt, C. L., "The periodical cicada," Washington,
U. S. Department of Agriculature, Bureau of Ento-
mology, Bulletin 71, 1907.
Van Duzee, E. P., "Catalogue of the Hemiptera of
America north of Mexico excepting the Aphididae,
Coccidae, and Aleurodidae," Berkeley, University
of California Publications in Entomology, Vol. 2,
1917.

HOOKE, ROBERT (1635–1703)

This English physicist did scientific work in many
areas. However, it is believed that Hooke's achievements would have been more dramatic if his attentions
were not so diverse. While Hooke originated much,
he perfected little. His contributions to physics include:
an imperfect undulatory theory of light, work with
thermal expansion, an attempt to observe paralaxes of
the stars systematically, and the expression of a relationship known as elasticity. In biology, Hooke is best
known for his microscope studies. His *Micrographia*,
while written for another purpose, contained scattered,
original biological studies. Hooke showed the compound nature of the eye of a fly. He observed the metamorphosis of a mosquito from the larva stage. From
this study, Hooke cast doubts on the theory of spontaneous generation; however, he dropped this idea, and it
remained for Redi to disprove the theory. Hooke
also made studies on cell structure. But once again, it
remained for others like Grew and Malpighi to elaborate on the theory. While Hooke's primary concentrations were not in the field of biology, he was one of the
greatest biological researchers of his time.

Douglas G. Madigan

HOOKER, SIR JOSEPH DALTON (1817–1911)

Sir Joseph acquired from his father, Sir William, a
taste for botany. He undertook many collecting expeditions, mostly to the Far East, and in 1855 was appointed Assistant Director of the Royal Botanic
Gardens at Kew and, ten years later, succeeded his
father as Director. He was an early friend and sup-
porter of Darwin and was one of those who encouraged
the latter to publish the "Origin of Species." He is
best remembered for his "Flora of the British Isles,"
a work on which all subsequent floras of that region
have been based.

HOOKER, SIR WILLIAM JACKSON (1785–1865)

Sir William Hooker was one of the most prolific writers of the age of taxonomic botany. He early developed
an interest in plants from the extensive greenhouses
maintained by his father, Joseph Hooker, a wealthy
member of the socially prominent family. The son
made expeditions to Iceland, France, and Switzerland
before settling on his own estate in Suffolk, where he
devoted himself to developing a very large herbarium
which attracted the attention of most of the great botanists of his period. In 1820 he was drawn from his
country retirement to assume the chair of botany in
the University of Glasgow. He was indefatigable in
drawing the attention of the government to the necessity of including botanists on voyages of exploration
and the gratitude of the individuals so appointed led
to tremendous additions to his herbarium. In 1841 he
accepted the appointment of Director of the Royal
Botanic Gardens at Kew. His collections formed the
basis of the vast herbarium now there and under his
leadership, the Gardens were greatly enlarged.

HORMONE, ANIMAL

A hormone is a chemical secreted by an endocrine
(ductless) gland whose products are released into the
circulating fluid. Hormones are regulators of physiological processes by chemical means. Hormones produced by one species usually show similar activity in
other species. This is true even though there may be
small species differences in the chemical composition
of the hormones produced. The hormones showing
greatest species specificity are proteins or conjugated
proteins. However, highly purified preparations of
even these hormones are active in a variety of species.

The chemical nature of hormones varies widely.
There are steroids (estrogen, progesterone, aldosterone, cortisone), amino acids (thyroxine and similar
analogues), polypeptides (vasopressin), proteins of low
molecular weight (insulin) and conjugated proteins
(follicle stimulating hormone is a glycoprotein). The
polypeptide and protein nature of many hormones has
made them hard to isolate and purify. As mentioned
above, these hormones are the ones which show species differences in chemical composition. In contrast
to the difficulty in isolation of protein substances, the
greater ease of isolation and synthesis of steroidal and
amino-acid hormones has made these products generally available for medical purposes.

Hormones are chemicals which help regulate over-
all physiological processes such as metabolism, growth,
reproduction, metamorphosis, molting, pigmentation,
and electrolytic and osmotic balance. The regulation
of such functions of the endocrine system is accomplished either directly or indirectly. This has been
demonstrated in studies of the control of the vertebrate
pituitary gland and the control of the molting glands
of insects. Immediate effects of the central nervous
system on hormonal secretions have been shown in
the regulation of pigmentation changes in crustacea

and lower vertebrates. Similarly, direct neural control as well as chemical changes in the organs themselves stimulate the production and release of gastric hormones (pancreozymin, gastrin) which, in turn, cause the release of digestive enzymes. Neural control may operate either by direct action of nerve impulses on hormone-producing cells, as in the digestive tract or more commonly indirectly, by formation of substances by cells of the central nervous system. Neurosecretory substances may be transported to a target gland by movement whithin the axons of the nerve cells. Alternatively, the neurosecretory substances may be transported to the target gland by the circulatory system. In the latter case, the nervous system itself becomes an endocrine gland. In *either* case, the adaptive efficiency is obvious: the nervous system receives the stimulus of the ever-changing environment and coordinates metabolic adjustments within the organism to give adaptive responses to that environment.

Hormones often act on target cells as complex chemical macromolecules formed by combinations with various molecules in plasma or within the target cells themselves. Thus, thyroxine appears in the plasma bound to plasma proteins and estrogen has been found bound to proteins of liver and uterine cells. Hormones also may form associations with enzymes and thereby affect the rate of reaction or even the substrate utilized. This appears to be true for the effects of some steroids on some amino acid dehydrogenases.

The metabolic effects of hormones may be conveniently divided into two major types of activity: those which affect *rates* of cellular enzyme systems and, therefore, help maintain homeostasis in their target organs and in the organism; and those which affect the *kinds* of enzymes and products formed and, therefore, are temporarily anti-homeostatic (in some cases, developmental) to their target cells and often to the organism. Not all hormones can be classified into these two categories because research on their cellular effects has not progressed far enough.

Hormones which affect *rate* (homeostatic hormones) have their most obvious effects on cell permeability, mitochondrial activity, glycolysis, amino acid and fatty acid metabolism. These effects may be direct or mediated by effects on a variety of enzyme systems. For example, anti-diuretic hormone appears to affect membrane permeability directly while aldosterone affects only after it affects several systems (including RNA synthesis) first. These hormones also probably affect the synthesis of messenger RNA (mRNA) from DNA by affecting the activity of various RNA polymerases, although this has not been shown for all of them as yet. The homeostatic hormones do not change the species of mRNA produced since it is the rate, not the kind, of cell activity that is changed when they are removed. Thus, homeostatic hormones appear to increase the amount of the same species of mRNA, which in turn can increase the synthesis of enzymes and these increase the rate of various reactions. It would appear, under these circumstances, that mRNA synthesized under the stimulus of homeostatic hormones is long-lived because there is no immediate decrease in cell activity after the removal of the glands producing these hormones. Some of the hormones which affect rates of cell activity are thyroid hormone, corticosteroids, aldosterone, anti-diuretic hormone,

oxytocin, calcitonin, parathyroid hormone, gastric hormones, insulin, growth hormone, and glucagon. Some examples of their actions at the cellular level are given below.

Phosphorylation of glucose is influenced by insulin and the growth hormone. Insulin accelerates phosphorylation while growth hormone depresses it. These hormones also act at the cell membrane on active transport mechanisms. Insulin increases glucose uptake by cells, growth hormone decreases it. Both hormones also affect other enzymes in carbohydrate pathways, for insulin increases the store of glycogen in the liver while growth hormone decreases it.

The breakdown and utilization of glycogen are increased in liver and muscle by epinephrine. This hormone increases the concentration of active posphorylase. Glucagon, a hormone formed by the alpha cells of the pancreas, has a similar effect but acts on liver but not muscle phosphorylase. Adrenal cortical hormones increase liver glycogen by increasing the synthesis of glucose from other sources such as protein. This is correlated with an increased excretion of nitrogen from amino acids.

Thyroid hormone causes swelling of the mitochondria and probably, therefore, can affect all the oxidative enzymes and substrates present in them. Concentrations of thyroid hormone which stimulate growth in one tissue will depress or even cause regression of another. Thus, limb formation in amphibians is stimulated at the same time that the tail regresses.

As indicated above, processes which affect carbohydrate metabolism also affect fat and protein metabolism. Thyroid increases the utilization of fats for energy as do corticosteroids, epinephrine, and growth hormone. The synthesis of fats is increased by insulin, while in the absence of this hormone fats are used metabolically for energy and as a source for increased cholesterol formation. These processes are independent of the effects of insulin on glucose utilization.

Adrenal cortical hormones can increase the breakdown of proteins by increasing the amounts of intracellular peptidase formed. Thyroid hormone increases incorporation of amino acids into new proteins presumably by affecting the rate of synthesis of ribonucleic acids from diphosphate nucleotides. Insulin modifies the translation of messenger RNA at the ribosomal level and therefore affects the rate of protein synthesis.

Electrolyte balance and osmotic equilibrium are related processes in living organisms and are also affected by the homeostatic hormones. In mammals, sodium reabsorption and potassium excretion in the tubules of the kidney are increased by aldosterone and deoxycorticosterone. Aldosterone is the most effective and its action has been cited above. Calcitonin decreases phosphorus excretion in the urine, while parathyroid hormone increases phosphorus loss. Again, neither of these hormones appears to act initially on the cell membrane. The loss of glucose in the urine of the diabetic shows that insulin indirectly affects osmotic balance by affecting glucose utilization.

Some of these hormones affect the resting potential or threshold potential of irritable tissues so they must also, therefore, affect cell permeability to ions. Oxytocin, a principle isolated from the posterior pituitary, increases contraction in smooth muscle by decreasing the threshold. Muscular weakness is a common symptom of insufficiency of the adrenal cortex.

The foregoing discussion is not exhaustive but is intended not only to point out the complex actions of these hormones on cell enzyme systems but also to indicate the complicated interactions that may occur between hormones. It is also apparent that permeability changes caused by hormones can affect rates of biochemical reactions when either metabolites or ion concentrations have been rate limiting.

Anti-homeostatic hormones are those which not only affect the rates of cellular activity but especially affect the *kinds* of activity possible in the cells of their target organ. These hormones temporarily antagonize normal feedback systems, temporarily upset homeostasis, and ultimately force a new level of activity to restore it. They *reprogram* the cells of their target organs so that new enzymes and products are produced. Such extensive changes at the cellular level may also produce vast changes in the entire organism and cause organismic development. The overall changes at puberty are an upset of the existing homeostasis and a new homeostatic balance is not reached until sexual maturity. Such reprogramming necessitates basic changes in the synthesis of mRNA from DNA. These hormones, therefore, have their greatest effects on translation of the genetic code of the cells, and one hormone, ecdysone, has been shown to change the patterns of puffing of the salivary chromosomes of insects. Some investigators suggest that steroid hormones in this group may affect nuclear proteins (e.g. histones) rather than the chromosomes themselves. Testosterone has been found to increase the synthesis of messenger RNA in prostates from castrates and it has been suggested that it acts on nuclear proteins masking chromosome regions. It has also been suggested that some anti-homeostatic protein hormones may affect operator gene sites which control the synthesis of mRNA by other regions of the chromosomes as suggested by the classical model of Jacob and Monod. No direct evidence of such action by hormones is yet available.

New research should clarify the methods by which these hormones affect the synthesis of mRNA. However, it is obvious that the anti-homeostatic hormones have more profound effects at the genetic level of target cells than do hormones which primarily affect rate. Because new enzymes and products do not appear in target cells unless the stimulating hormone is present, and quickly disappear when that hormone is removed, it seems that mRNA synthesized under stimulus of anti-homeostatic hormones may be relatively less stable than other RNA in systems from multicellular animals. Evidence from the treatment of prostate glands of castrate rats with testosterone indicates that some messenger RNA may be very short-lived.

In order to reprogram the target cells (and sometimes the entire organism as well), the anti-homeostatic hormones must modify enzyme systems which yield energy to be used in the synthesis of RNA. Therefore, these hormones will also affect the rate of existing systems which yield high-energy phosphorus compounds such as ATP. Thus, testosterone affects the concentration of succinic dehydrogenase in the accessory glands and estrogen activates isocitric dehydrogenase in the placenta. Similarly, one may expect anti-homeostatic hormones to affect cell permeability both to ions and to metabolites. Several adenohypophysial hormones have been found to increase the uptake of amino acids by cells of their target organs. Estrogen increases both irritability of the central nervous system and utilization of calcium by osteocytes synthesizing bone at the epiphysial junction. The molt and metamorphosis in insects also involves changes in both permeability and metabolism which are controlled by anti-homeostatic hormones. Some of the anti-homeostatic hormones are trophic hormones of the anterior pituitary, estrogen, testosterone, progesterone, corpus cardiacum hormone, ecdysone, and the hypothalamic factors affecting secretion of the pituitary of mammals. The gross effects of the animal hormones on the organism are presented below.

Table 1.   Animal Hormones

Hormone	Found in	Where Secreted	Stimulus for Production	Gross Effect of Hormone
Growth hormone (STH)	Vertebrates	Adenohypophysis (anterior pituitary)	Hypothalamus secretions carried via pituitary portal blood vessels regulate secretion	Stimulates growth, antagonizes action of insulin
Adrenocortico-trophic hormone (ACTH)	Vertebrates	Adenohypophysis	Same as STH	Stimulates secretion by adrenal cortex
Follicle stimulating hormone (FSH)	Vertebrates	Adenohypophysis	Same as STH	Stimulates growth of ovarian follicles, estrogen secretion
Luteinizing hormone (LH)	Vertebrates	Adenohypophysis	Same as STH	Stimulates ovulation, and corpus luteum growth, estrogen secretion
Luteotrophic hormone (LTH)	Vertebrates	Adenohypophysis	Same as STH	Stimulates secretion by corpus luteum

Table 1 (continued)

Hormone	Found in	Where Secreted	Stimulus for Production	Gross Effect of Hormone
Thyrotrophic hormone (TSH)	Vertebrates	Adenohypophysis	Same as STH	Stimulates secretion by thyroid
Insulin	Vertebrates	Islets of Langerhans of pancreas, β-cells	Blood sugar changes	Increases blood sugar
Glucagon	Vertebrates	Islets of Langerhans, α-cells	Blood sugar changes	Decreases blood sugar
Thyroid hormone	Vertebrates	Thyroid gland	Thyrotrophic hormone	Increases metabolic rate
Calcitonin (Thyrocalcitonin)	Vertebrates	Thyroid gland (Ultimobranchial cells)	High blood levels Ca^{++}	Lowers blood Ca^{++}
Corticosterones	Vertebrates	Adrenal cortex	ACTH	Antagonizes action of insulin on glucose metabolism
Aldosterone	Vertebrates	Adrenal cortex	ACTH	Electrolyte balance
Epinephrine	Vertebrates	Adrenal medulla	Stress stimuli	Glucose metabolism, vasoconstrictor
Estrogen	Vertebrates	Ovary, follicle	FSH, LH	Growth of uterus
Progesterone	Vertebrates	Ovary, corpus luteum	LH, LTH	Maintenance of pregnancy
Testosterone	Vertebrates	Interstitial cells of testis	LH	Growth and secretion male accessory glands
Parathyroid hormone	Vertebrates	Parathyroid glands	Low blood Ca^{++}	Increases blood Ca^{++}, decreases blood phosphorus
Anti-diuretic hormone (vasopressin)	Vertebrates	Hypothalamus via axons to posterior pituitary (neuropophysis)	Osmotic changes cause neurosecretory cells to form it	Increases water reabsorption in tubules of kidney
Oxytocic hormone	Vertebrates	Hypothalamus via axons to neuropophysis	Unknown	Causes contraction of smooth muscle, probably helps terminate pregnancy
Intermedin	Vertebrates	Intermediate lobe of pituitary gland	Neurosecretory cells?	Causes chromatophore changes, produces lightening and darkening of body color
Gastrin	Vertebrates	Stomach	Neural and chemical changes in stomach	Causes secretion of stomach enzymes
Secretin	Vertebrates	Duodenum	Neural and chemical changes in duodenum	Causes secretion of pancreatic HCO_3^-
Pancreozymin	Vertebrates	Duodenum	Same as secretin	Causes secretion of pancreatic enzymes
Enterogastrone	Vertebrates	Duodenum	Same as secretin	Inhibits gastric secretion
Cholecystokinin	Vertebrates	Duodenum	Chemical changes in duodenum (fats)	Causes gall bladder contraction
Corpus cardiacum hormone	Insects	Nervous system via axons to corpus cardiacum	Environmental and internal changes produce neural secretion	Trophic hormone to prothoracic gland, affects molting

Table 1. (continued)

Hormone	Found in	Where Secreted	Stimulus for Production	Gross Effect of Hormone
Ecdysone	Insects	Prothoracic gland	Trophic hormone or corpus cardiacum	Controls molting, stimulates meta-morphosis
Juvenile hormone	Insects	Corpus allatum gland	Neural stimuli cause secretion	Promotes develop-ment of larval characters; causes ovarian develop-ment in some insects
Sinus gland hormone	Crustacea	Neurosecretory cells via axons to sinus gland	Internal and external environmental changes	Inhibits molting, also stimulates ovarian growth
Molting hormone	Crustacea	Y-gland	"Trophic" hormone from neurosecre-tory cells	Increases molting rate
Chromatophore lightening and darkening hormones	Crustacea	Eyestalk glands and Postcommissure organs	Neural stimuli cause secretion directly	Cause lightening or darkening of body in response to environment
Androgenic gland	Crustacea	Gland attached to vas deferens	Unknown	Causes development of testis and male characteristics

References

Pincus, G., and K. V. Thimann, "The Hormones," New York, Academic Press, 1955, 5 vols.

Gorbman, A., and H. A. Bern, "A Textbook of Comparative Endocrinology," New York, Wiley, 1962.

Turner, C. D., "General Endocrinology," Philadelphia, Saunders, 1966.

Barrington, E. J. W., "An Introduction to General and Comparative Endocrinology," Oxford, Clarendon Press, 1963.

Krahl, M. E., "The Action of Insulin on Cells," New York, Academic Press, 1961.

Gaillard, P. H., R. V. Talmage and A. M. Budy, "The Parathyroid Glands," Chicago, Chicago University Press, 1965.

Tata, J. R. in "Progress in Nucleic Acid Research and Molecular Biology," Vol. 5, 191–250 (1966).

HORMONE, INVERTEBRATE

The existence of hormones has been definitely established in most invertebrates (Platyhelminthes, Nemertea, Annelida, Mollusca, Arthropoda and Prochordata), and they probably also occur in Coelenterata and Echinodermata.

Those active principles of which the effects have been demonstrated by experimental methods (excisions, grafts, injections of extracts and parabiosis) are produced either by neurosecretory cells of the central nervous system, in which case they are called ergones; or from endocrine glands, which may or may not be of neural origin, in which case they are called hormones.

Ergones may act directly on an effector organ or on

an endocrine gland; the latter, in its turn, may act on an effector organ or on a second endocrine gland which produces the active hormone.

The development of the activity of neurosecretory cells, and of endocrine glands, is modulated by a stimulus received by the nervous system. These stimuli may be of external origin [light, temperature, humidity, volatile chemicals (pheromones)] or produced internally (hormonal activity, nutrition, variations of the composition of the internal environment). Both ergones and hormones are liberated in the internal environment (blood, coelomic fluid).

Neurosecretory Cells

These can be seen throughout the whole central nervous system; they are most frequently concentrated in the ganglia, particularly the cerebral, subesophageal, and some in the nerve chain. Prolongations of the cells (axons) serve as the distribution route for the neurosecretory products which are then liberated in the internal environment through a neurohemal organ (for example, the sinus gland and the post-commissural organs of Crustacea).

In Platyhelminthes and Nemertea neurosecretory cells have been demonstrated in both the anterior ganglia and the nerve chain. In Annelida they are principally localized in the cerebral ganglian. In Pelecypod molluscs neurosecretory cells are practically confined to the cerebral, visceral and pedal ganglia but in gastropods occur in most ganglia.

In arthropods these cells are largely grouped into the protocerebral portion of the supraesophageal ganglion, in the tritocerebrum, in the subesophageal ganglion, and in some ganglia of the ventral nerve chain. Insects and crustacea have been the subject of most studies. In insects the main groups of neurosecretory cells are found in the protocerebrum (*pars intercerebralis*), in the tritocerebrum, in the subesophageal ganglion and also in the thoracic and abdominal ganglion. In crustaceans, they occur in the brain, the optic ganglion, the X-organ and the abdominal ganglia.

Neurosecretory cells have been described from the cerebral ganglion of Urochordates. They may occur in coelenterates but more certainly in the echinoderms (Ophiuridea, Asteridea) in the radial nerves and in the circumoral nerve ring.

Endocrine Organs

a. Of neural origin. The *corpora cardiaca* of insects are apparently neural in function and the araneid organs of Schneider are probably homologous with them as are also the cerebral glands of Chilopoda.

b. Of non-neural origin. The gonads of Polychaetes (*Nereis, Arenicola*) probably show a feedback of endocrine origin; in the Mollusca, the gonads of Gastropoda, and the posterior salivary and the optical glands of Cephalopoda, are endocrine organs. In arthropods the endocrine glands are well differentiated. In insects they are the *corpus allatum*, the molting gland, the apical tissue of the embryonic gonads and the ovary; in Crustacea they are the molt-

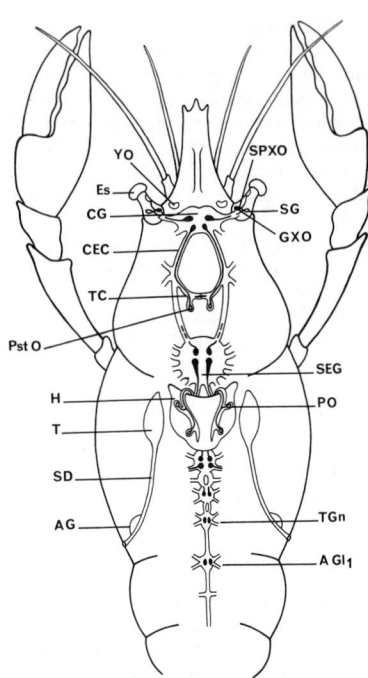

Fig. 2   Endocrine system of a male Crustacean. In solid black, situation of neurosecretory cells and their axons. AG, androgenous gland; AG1₁, first abdominal ganglion; CEC, circumesophageal commissure; CG, cerebroid ganglion; Es, optic stalk; GXO, ganglionic organ X; H, heart; PO, pericardial organ; PstO, ostcommissural organ; SD, sperm duct; SEG, subesophageal ganglion; SG, sinus gland; SPXO, sensory pore of the X organ; T, testis; TC, tritocerebral commissure; TGn, last thoracic ganglion; YO, Y organ. (After I.R. Gabadorn.)

ing glands (Y-organ), the androgenous glands and the ovaries.

Two examples of the anatomical interrelation existing between neurosecretory cells and endocrine organs are shown in Figures 1 and 2.

Endocrine Control

The functions in invertebrates that are under endocrine control are of the same nature as those in the vertebrates; they may be grouped in four categories: 1) growth, maturation and regeneration, 2) reproduction, 3) metabolism and homeostasis, 4) adaptation to external factors.

1) Growth, maturation, regeneration. a) *Annelida.* In Polychaeta (*Nereis, Nephtys*) posterior regeneration is stimulated by an ergone of cerebral origin. In Oligochaeta (*Lumbricus, Eisenia, Allolobophora*) posterior regeneration and diapause are, on the contrary, inhibited by a cerebral ergone.

b) *Arthropoda.* In arthropods with a rigid integument growth occurs only at each molt, which involves the development of a new integument and the shedding of the old one. This mechanism is subject to endocrine control.

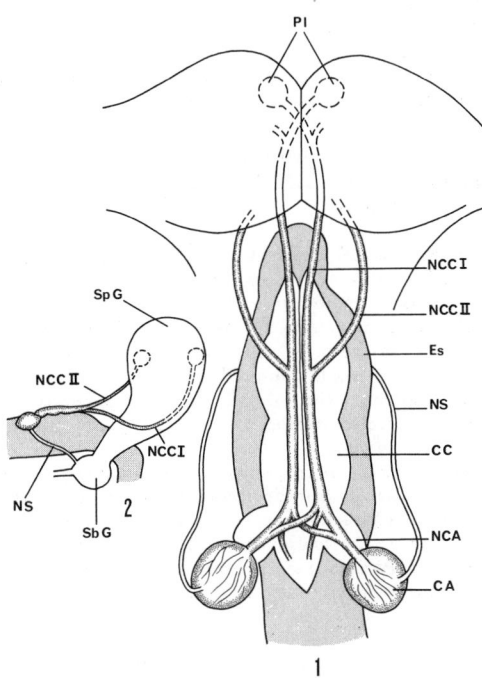

Fig. 1   Diagrammatic representation of the brain and retrocerebral complex of an insect (Periplaeta), 1, dorsal aspect, 2, lateral aspect. CA, *corpus allatum*; CC, *corpus cardiacum*; Es, esophagus; NCA, nervus corporis allati; NCC I and II, *nervis corporis carciaci* I and II; NS, *nervus allato subesophagealis*; PI, *pars intercerebralis*; SbG, sub-esophageal ganglion; SpG, supra-esophageal ganglion. (After I.R. Hagadorn.)

α) *Insecta*. In higher insects growth occurs in three phases; larva, nymph and adult (imago). Each molt is subject to endocrine control; the neurosecretory cells of the *pars intercerebralis* secrete an ergone (ecdysiotrophine) which acts on the molting gland. This in its turn produces the molting hormone, ecdysone, which stimulates the formation of a new cuticle. In the course of larval and nymphal molts, a second hormone is secreted by the *corpus allatum*. This is the juvenile hormone of which the presence inhibits the development of adult characters. The production of this juvenile hormone is discontinued at the time of the imago molt. The diagram below summarizes the endocrine control of growth in insects.

they also control the differentiation of external sexual characters (clitellum, etc).

c) *Mollusca*. Gametogenesis is subject to an endocrine control of neurosecretory origin in Gastropoda (*Patella, Calyptraea*) and Pelecypoda (*Mytilus*). In Cephalopods the optic gland, of which the function is inhibited by the central nervous system until the moment of reproduction, produces a hormone that inaugurates sexual maturity.

The differentiation of the genital tract is determined in gastropods either by a cerebral ergone (penis, uterus of *Calyptraea*) or by a hormone produced by the gonad (hermaphrodite gland, ovispermiduct of slugs).

d) *Insecta*. The process of previtellogenesis and

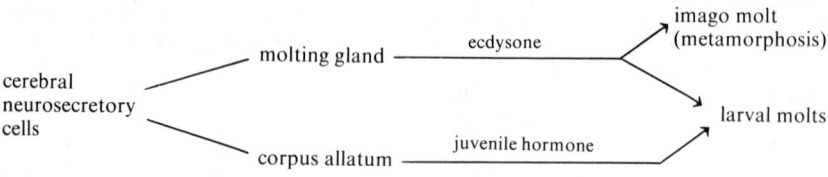

A diapause (discontinuance of development or of activity) can occur in the life history in either the egg, the larva, the nymph or the adult. The embryonic diapause is determined by a neurosecretion from the subesophageal ganglion which acts on the egg before it is laid. Larval and nymphal diapauses are produced by blockage of the cerebral neurosecretory activity or by the absence of ecdysiotrophine. The adult diapause is induced by a blockage of the activity of the *corpus allatum*.

β) *Crustacea*. The molts are under endocrine control.

vitellogenesis are subject to endocrine control. Ergones of cerebral origin act on the *corpus allatum*, a hormone from which inaugurates the formation of yolk in the ovules. The same mechanism is observed in the development and function of female accessory glands and also stimulates the production of certain sexual attractants (pheromones). Possibly sexual behavior is influenced by ergones.

The mechanism of sex differentiation in insects (*Lampyridae*) has just been elucidated. It can be diagramatically represented as follows:

neurosecretory cells ⟶ corpus allatum ⟶ apical tissue of the undifferentiated gonad $\xrightarrow{\text{androgenous hormone}}$ male differentiation

The molting gland (Y-organ) secretes the molting hormone. The neurosecretory cells of the *medulla terminalis* produce an ergone inhibitory to molting. The interactions of these two systems are poorly understood.

The regeneration of appendages is partly controlled by the molting hormone.

γ) *Chilopoda*. The cerebral gland produces a hormone that inhibits molting.

2) **Reproduction.** a) *Nemertea*. Cerebral ergones inhibit genital maturation.

b) *Annelida*. Cerebral ergones regulate genital maturation; they are inhibitory in *Nereis* and stimulatory in *Arenicola*, leeches and Oligochaeta. In these last

Female sexual differentiation takes place without the help of this mechanism; the female sex is autodifferentiated. Females can be inverted into males exposing very young ovaries to the action of androgenous hormone. In the same way, a young male deprived of his *corpus allatum* is feminized.

e) *Crustacea*. Genital maturation is inhibited by ergones secreted by the neurosecretory cells of the *medulla terminalis*.

Male genital differentiation, the maintenance of testicular activity and the development of male secondary sexual characters are all dependent on a hormone produced by the androgenous gland. In the Isopoda (*Anilocra, Porcellio*) male sexual differentiation occurs according to the following diagram:

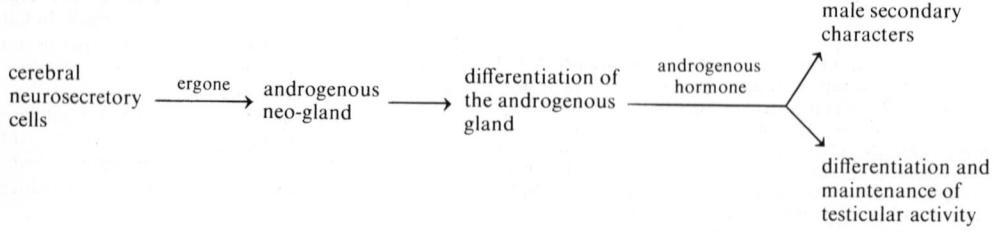

In protandrous hermaphrodite Crustacea, sexual inversion in the female direction occurs through the degeneration of the androgenous gland.

The differentiation of sexual characters (oostegites, ovigerus setae) is dependent upon a hormone of ovarian origin.

f) *Echinodermata.* Neurosecretory cells of the radial nerve are the source which controls the emission of gametes and of meiosis in the ovules of starfish.

3) **Metabolism and homeostasis.** a) *Nemertea.* Neurosecretory cells of the cerebroid ganglion and the cerebral organs exercise an effect on the regulation of osmotic equilibrium.

b) *Annelida.* Protein metabolism, oxygen consumption and osmoregulation are all controlled by cerebral ergones.

c) *Mollusca.* In Gastropods (*Lymnea*) water equilibrium depends on ergones liberated by the pleural ganglia.

d) *Insecta.* Water balance of insects appears to be under endocrine control. Neurosecretory cells of the *pars intercerebralis-corpora allata* produce an antidiuretic factor; the brain and the *corpora cardiaca*, as well as the thoraco-abdominal ganglion, produce an anti-diuretic factor. Protein metabolism depends either on the neurosecretory cells of the *pars intercerebralis* or on the combined action of the *pars intercerebralis* and *corpus allatum.*

Sugar metabolism (trehalose, glucose) is regulated by a hyperglycemic hormone produced by the *corpora cardiaca* and a hypoglycemic hormone secreted by the *corpora allata.*

Metabolism of other substances appears to be influenced by the endocrine system, fats by the *corpora allata,* phosphorus by the *corpora allata* and *corpora cardiaca* and nitrogenous substances by the *pars intercerebralis, corpora cardiaca* and *allata* and ovaries.

Cardiac rhythm is speeded up by a substance liberated by the pedicardial cells after they have been activated by the *corpora cardiaca.*

Cyclic motor activity (cockroaches) is regulated by the activity of the neurosecretory cells of the subesophageal ganglion.

e) *Crustacea.* Glucide metabolism is regulated by the neurosecretory cells of the eyestalk which secrete a hyperglycemic hormone; that of calcium is subject to a control of an endocrine from the molt gland (Y-organ) which acts on the hepatopancreas. Cardiac rhythm is regulated by neurosecretory cells of the thoracic ganglion.

4) **Adaptation to external factors.** a) *Annelida.* In leeches, chromatic adaptation is dependent on a cerebral ergone which controls the dispersion of pigment in the chromatophores.

b) *Mollusca.* The expansion and contraction of the chromatophores in Cephalopoda are controlled in part by neural centers (brain) and in part by substances (5-hydroxytryptamine, tyramine) liberated by the posterior salivary glands.

c) *Insecta.* Physiological color changes are controlled by the neurosecretory cells of the tritocerebrum (stick insects) or those of the brain (the larva of *Corethra*) which produce the dispersion of pigment. The formation of a hypodermic pigment (stick insects, crickets) is inaugurated by the juvenile hormone (*corpora allata*).

d) *Crustacea.* The dispersion and concentration of chromatophore pigments are regulated by ergones derived from the brain, from the eyestalk and from the ventral nerve chain.

Adaptation of the eyes to light intensity results from the migration of distal retinal pigment under the control of two ergones secreted by the brain and eyestalks (light-adapting ergone, dark-adapting ergone).

Isolation and Mode of Action of Active Principles

1) **Ergones.** Histochemical reactions show that the ergones are proteins or polypeptides. In insects (*Bombyx*), some authors consider the cerebral ergones to be a peptide and others believe it to be a sterol.

2) **Hormones.** The molting hormone has been isolated by Karlson and its chemical identity established. Ecdysone is a steroid (sterol nucleus, lateral chain of 5 hydroxyles; ecdysone, 5 hydroxyles [ecdysone] or 6 hydroxyles [ecdysterone]). It acts at the chromosomal level and controls gene activity by modifying the concentration of certain ions (Na-K). The juvenile hormone has been isolated by Williams and purified by Roller, *et al.* in 1967. It is a sesquiterpene.

M. DURCHON
(trans. from French)

References

Carlisle, D. B. and F. G. W. Knowles, "Endocrine control in Crustaceans," Cambridge, The University Press, 1959.

Durchon, M., "L'endocrinologie des Vers et des Mollusques," Paris, Masson, 1967.

Gabe, M., "Neurosecretion," New York, Pergamon, 1965.

Gersch, M., "Vergleichende Endokrinologie der Wirbellosentiere," Leipzig, Geest und Portig, 1964.

Gilbert, K. I., "Physiology of growth and development: endocrine aspects," *in* Rockstein, M., ed., "Physiology of Insects," Vol. 1, New York, Academic Press, 1964.

Hagadorn, I. R., "Neuroendocrine mechanisms in Invertebrates," *in* Martini, L. and W. F. Ganong, eds., "Neuroendocrinology," Vol. 2, New York, Academic Press, 1967.

Joly, P., "L'endocrinologie des Insectes," Paris, Masson, 1968.

HORMONE, PLANT

A plant hormone or *phytohormone* is a natural compound produced by one part of a plant which when it moves to another place in the plant controls growth or physiological functions there. Such substances act at very low concentrations (almost always less than one thousandth molar and often in the millionth to billionth molar range).

In common with other higher organisms, plants have a number of different types of hormones which regulate many aspects of growth, development, and metabolism. Since plants lack nervous systems, the hormonal control system is essential for communication between cells. These hormonal carriers of chemical messages vary both chemically and functionally. The best known plant hormones are those belonging to the auxin class, but knowledge of other hormone groups such as the gibberellins, cytokinins, dormins, anthesins, and ethylene is advancing rapidly.

The *auxins* are defined as organic substances which promote irreversible elongation of plant cells. This has been considered the most critical property of an auxin, but like all hormones the activity of an auxin is manifold and to a certain extent overlaps those of other hormones. Thus, auxins also are noted for their inhibition of bud elongation and for decreasing the growth of roots. In addition, they can promote seedless fruit formation and initiate roots on plant cuttings. Inhibitory effects on leaf shedding and the dropping of fruits also are observed and have natural significance as well as practical applications to agriculture. One of the most useful properties of auxins to man has been their inhibitory and even toxic effect at biologically high but chemically low concentrations, which has led to their use as weed killers in agriculture and as defoliating agents in chemical warfare.

The best established natural auxin is the substance indole-3-acetic acid, and all natural auxins appear to be closely related to this compound. Many synthetic chemicals also have auxin activity. Among the most useful of these in agriculture have been "2,4-D" (2, 4,dichlorophenoxyacetic acid), naphthalene-1-acetic acid, and 2,3,6-trichlorobenzoic acid. The relation between molecular structure and activity of these compounds is relatively well understood. Auxin activity requires an acid group about 5.5 angstrom units away from a partially positive center which is located in a planar source of π electrons. In addition, the shape of the rest of the molecule is important.

Natural auxins are formed in fruits, seeds, pollen, growing points, young leaves, and especially in developing buds. They travel away from the site of synthesis by a special polarized transporting system which requires metabolic energy and which moves auxin in only one direction. This direction is always from the tips of the shoots and roots of the plant towards the morphological base, and is thus anatomically determined. In its normal basal movement the auxin stimulates the cells below the tip to elongate and sometimes to divide also. Thus, growth and differentiation within the plant can be regulated by the production of auxin.

Tropistic movements of plants also are under auxin control. In responses to gravity (geotropism), tilting of the plant shoot causes the auxin transport system to redirect some of the auxin flow to the lower side of the shoot. As the concentration of auxin builds up on the lower side, the growing cells there elongate more than the cells on the upper side which have become auxin deficient. As a consequence, the shoot bends upward from the curvature induced by the differential growth. Roots grow down in a gravitational field rather than up, but fundamentally the same mechanism is responsible. The difference is that roots are much more sensitive to auxin, and as a consequence the buildup of auxin on the lower surface of a root reaches the inhibitory level and thus reduces the growth rate rather than promoting growth. Hence, the upper part of the root grows faster than the lower, and the root curves downwards.

A similar redistribution of auxin also plays a part in the most sensitive tropistic responses to light (phototropism). But at higher levels of illumination it probably is destruction of auxin by the light on the illumi-

nated side which is responsible for the difference in auxin levels between the two sides of the shoot and the consequent bending toward light. Auxin destruction also occurs naturally by means of endogenous oxidative enzymes, and these may cause the termination of growth of certain plant parts by removing the growth stimulating substances.

The *gibberellins* are plant growth substances too, but have novel properties of their own. At least twenty-four natural gibberellins are now recognized, but as there are problems in producing them commercially, at present only one, gibberellic acid, is readily available. All natural gibberellins appear to be derived from a tetracyclic diterpene named kaurene. Difficulties of synthesis have thus far prevented the production of any synthetic gibberellins, but undeniable gibberellin activity by two simpler molecules, helminthosporol and phaseolic acid, makes progress in this direction likely.

Gibberellins have some unique properties. They can return to normal the growth of many genetic dwarf mutants of higher plants, and there is good evidence that the genetic defects are in the system which synthesizes the natural endogenous gibberellins in these plants. In addition, gibberellins stimulate flowering in many plants which have a rosette morphology and which normally produce a tall flowering shoot or "bolt" at flowering time. These plants normally need a series of short night periods, or in the case of biennials, a cold period during the winter. Gibberellins bypass these needs and apparently duplicate the action of natural gibberellins whose formation is initiated by the light period or cold period normally required.

Gibberellins also replace the light requirement for germination of many seeds, and like auxins can produce parthenocarpic seed-free fruits. But they do not inhibit lateral bud development, and they inhibit rather than stimulate rooting of cuttings. Their movement within plants is in all directions and exhibits no polarity. Gibberellins are produced by germinating embryos and stimulate the release of enzymes in the endosperm which then liberates the food substances required by the growing seedling.

A number of substances have been discovered which inhibit the formation of gibberellins by the plant and thus dwarf normal plants. Chlorinated derivatives of quaternary ammonium and phosphonium compounds are particularly effective, and some of these such as chlorocholine chloride are in horticultural use. Other materials, known as morphactins, appear to act as dwarfing agents by a direct competition with gibberellins at their site of action.

The *cytokinins* were discovered when plant cells in culture enlarged but refused to divide in the presence of auxin alone. Cell division was found to be stimulated by a variety of natural extracts whose activity is now known to be due to various derivatives of the purine adenine. The best established natural cytokinin is known by the trivial name zeatin and is an N^6 hydroxyisopentenylamino purine. Many synthetic analogs have similar activity and also are N^6 aminopurines, such as kinetin (N^6 furfurylamino purine) and N^6 benzylamino purine. A number of phenyl urea compounds also appear to have cytokinin activity under certain conditions.

Cytokinins are found naturally in maturing seeds. Like auxin they show a directional polar movement within plant tissues, but counteract the effect of auxin by promoting bud growth. Leaf expansion and seed germination are often promoted by cytokinins, but perhaps only as a secondary effect of their cell division stimulation. Little is known yet of their natural regulatory roles. They can preserve the green color of detached leaves, apparently by controlling protein synthesis.

The recently established *dormins* are presently best known for their inhibitory action on certain aspects of plant development, notably their presence in dormant buds of trees and other species which have an arrested period of development. Dormins may also act to cause the abscission layer of the leaf stalk to form when leaves are shed. The only natural dormin identified is abscisic acid, a sesquiterpenoid hydroxy-cyclohexenone acid which has been detected in a wide range of plants. Synthetic substances with similar activity have been reported but are not as yet widely available.

It has been known since the 1930's that flowering in plants is regulated by the formation of a hormone in the leaves which then moves to the apical buds and causes them to differentiate into flower buds. This plant hormone was originally named florigen, but this term has been dropped in favor of the more classically correct term *anthesin*. Despite many efforts, the chemical nature of anthesin has never been established, nor are any synthetic compounds known which will cause flowering of all groups of plants. Clearly the gibberellins interact with anthesin, particularly in those plants known as long-day varieties, but gibberellins have not yet been implicated in the flowering of short-day plants.

It comes as a surprise to most persons that the simple gas *ethylene* is a hormone in plant tissues. It has long been recognized as being produced by ripening fruits, and that its production by one fruit during its ripening can initiate the ripening of others. Thus, the tale of one bad apple in the barrel causing the others to spoil has firm foundation, although any spoilage comes after ripening and is not directly due to ethylene. Fruit ripening can be caused by ethylene at concentrations of a part per million or even less. It has found commercial utility in the fumigation of the holds of banana boats a suitable time before docking, guaranteeing that fruit picked green will have started ripening before sale. Ethylene appears to strongly inhibit geotropic responses of plants, and begins to be formed when auxin concentrations reach a certain limit in plant tissues. Ethylene can be a factor in air pollution damage to plants, as toxic effects are sometimes shown at parts per billion concentrations.

Other hormones, particularly wound and root growth stimulating substances, have been investigated in higher plants, but are as yet little understood. The site of action of all these hormones is one of the most active areas of current research, but as yet few definite conclusions can be stated. Good evidence has been presented that gibberellins act by promoting the synthesis of specific enzymes and that this may result from a locus of action on the DNA of chromosomes. Cytokinins have been shown to enter into the molecule of transfer RNA which is involved in the incorporation of specific amino acids into protein, while the auxin indoleacetic acid has been found to bind with some macromolecules whose natural role is uncertain. These pieces of evidence suggest that plant hormones regulate via a control of protein synthesis. However, some responses to hormones seem to take place too rapidly to be explained this way. Other modes of action, such as effects on the permeability of membranes of cells, or of organelles within the cell, are still possible.

K. V. THIMANN and BRUCE B. STOWE

References

Went, F. W., and K. V. Thimann, "Phytohormones," New York, MacMillan, 1937.
Audus, L. J., "Plant Growth Substances," London, Leonard Hill, 2nd ed., 1959.
Leopold, A. C., "Plant Growth Substances," London, York, McGraw-Hill, 1964.
Setterfield, G., and F. Wightman, "Biochemistry and Physiology of Plant Growth Substances," Ottawa, Runge Press, 1968.

HUMAN ECOLOGY

Human Ecology is that branch of science which considers the interrelations between man and his physical and biotic environment. Two major subdivisions, personal ecology and community ecology, may be recognized.

Each person must constantly adjust his internal physiology to the changing conditions of his immediate environment in order to remain alive. He acquires energy from his food and loses energy in radiated heat, muscular effort, metabolic processes, excretion, growth, and in other ways. Regulatory mechanisms within the body operate to maintain the body temperature at its optimum level. Other physiologic mechanisms regulate the concentration within the body tissues of water, oxygen, carbon dioxide, sodium chloride, sugar, and all the other chemicals essential for life. Through the process of acclimation each individual is able to adjust his physiology so that he can exist in a wide range of climates. The invasion of the body by disease organisms is resisted by antibodies, phagocytes, and other special mechanisms. The sense organs, nervous system, and locomotor apparatus enable each individual to secure food, avoid enemies, find friends, and reach or remain in situations which are suitable for his survival. A dynamic equilibrium thus is maintained between the internal physiology of every individual and the conditions of his habitat.

Each human ecologic community consists not only of men, women, and children, but includes also numerous species of other animals and plants. Some of these associated species supply man with food and other essential materials, either directly or through a food chain. Herbivores may devour or damage the food-producing plants, but man may in turn utilize some of these herbivores for food. Carnivores also may prey upon the herbivores. Disease organisms and parasites attack the plants, herbivores, carnivores, other parasites, and even man himself. Saprophytes and scavengers convert dead organisms into soil materials which can be reused by the plants. The food and energy relations within each community thus form a complex web of life. Furthermore, the physiography, soils, climate, and other features of the physical habitat interact with and control the community in many ways. Each ecologic community and its habitat consequently constitutes an interacting system, which is called an ecosystem. The essential feature of an ecosystem is that it contains regulatory mechanisms which maintain a dynamic balance among the several member species of the community and also between the community as a whole and the constantly changing conditions of the physical environment.

Man is a social animal and he owes much of his success as a species to the efficiency by which through cooperative efforts he is able to control his environment and to secure the food and other materials needed for his survival. Through division of labor within his social groups, invention of tools and technological processes, domestication of plants and animals, and development of social, economic, and political institutions, man has within recent millenia attained a high level of civilization in many parts of the world. This has been made possible by the transfer of culture from generation to generation and from people to people by means of language, writing, printing, and more recently by the telephone, radio, and television.

Culturally acquired modes of behavior greatly aid man in adjusting to his personal habitat. Through the use of clothing, fire, shelters, and now air conditioning man is able to thrive in climates that otherwise would be inhospitable or barely tolerable for him. Through cooperation with his fellows, modern man is able to produce, store, transport, and process food and other materials; to construct dwellings, factories, public buildings, highways, railroads, ships, and airplanes; to maintain public order; and to provide elaborate institutions for communication, sanitation and health, power supply, education, and research.

No human society, however, no matter what its size or complexity of organization, is ever self sufficient. Instead, each human social group always operates as a part of a larger biologic unit, the ecologic community. From the other species of animals and plants which are associated with him as members of the same ecologic communities, man obtains all his food and much of the materials he requires for clothing, fuel, construction of buildings, and other essential purposes. Man's domestic animals and plants thus supply much food and many other useful materials to modern man. A considerable amount of human food is also obtained from wild species. Many human populations, for example, depend heavily on food fishes and on wild game for their supply of protein. Much of our lumber and of wood for fuel is obtained from forests which were not planted by man. Disease organisms and parasites often damage domestic plants and animals and those wild species which are of value to man. Weeds and other wild species likewise compete with or damage man's crops and domestic animals. The particular species and strains of plants and animals that live in each geographic area consequently affects man's welfare in many ways. In turn, man seriously affects the existence of all those species of plants and animals which happen to be associated with him.

At a primitive stage of culture man is only one of the numerous species of animals and plants that together constitute a natural ecologic community. In such a community the natural regulatory mechanisms operate very much as they do in communities of which man is not a member. Any species of animal which becomes too abundant locally for its food supply will be subject to starvation and this in time will reduce its numbers to the carrying capacity of the habitat. The activities of herbivores, carnivores, diseases, and parasites likewise tend to prevent those species on which they feed from becoming overly abundant. Competition, migration, dormancy, social cooperation, and other agencies also serve as regulatory mechanisms to keep the populations of the member species of each community adjusted to each other and to the resources of the habitat.

In those ecosystems in which man is a dominant member, however, the effectiveness of many of the natural regulatory mechanisms has been reduced or destroyed. Starvation among men has been largely eliminated by increases in food production, food storage, and efficient food distribution. Large predaceous animals have been reduced in abundance or locally extirpated. Wars among men have decreased in frequency though unfortunately not in destructiveness. Diseases and parasites have increasingly been brought under control by improvements in sanitation and in medical practices. Because of the ineffectiveness of most of the natural regulatory mechanisms in human ecosystems, man must himself take the responsibility for developing special mechanisms for maintaining stability in those modified communities in which he lives and in those which produce the food and other organic materials that civilization requires.

Among the special regulatory mechanisms which operate in a human society are public opinion, punishment, rewards, competition, and supply and demand. Social cooperation also operates as a regulatory mechanism by mitigating the harsher results of competition and by binding the society together. The democratic process may also be considered to be a regulatory mechanism which operates to give consideration to the needs of every member of the society. Through the operation of these and other social, economic, and political mechanisms, each social group of men usually is kept in a working balance with the changing conditions of its habitat. The mechanisms which regulate human communities, however, are by no means perfect. Constant efforts are needed to improve the existing regulatory mechanisms in every human community and to devise new methods for maintaining community stability.

Those ecosystems of which man is an important member occur in many sizes and exhibit a wide range of complexity in the organization of their social institutions. No satisfactory classification has yet been made of human ecosystems. Important units in any such classification, however, will probably be those ecosystems which have their centers of human influence respectively in camps, homesteads, villages, towns, and cities.

Human communities of every type may be assumed to be at most times in process either of progressive or degenerative evolution. Progressive evolution is occurring at the present time in many of the major human communities of the world. Contributing to this evolution are improvements in the productivity of the land and in the utilization of mineral resources. Technological advances have increased the efficiency with which many kinds of goods are manufactured and distributed. Important improvements have been and are being made in those regulatory mechanisms which control the social, economic, and political interrelations within and between human societies. On the other hand, retrogression has in the past occurred in numerous communities that at one time had attained high levels of civilization. The causes of retrogression in human communities are not well known, but over-utilization of natural resources, war, and inadequate ecologic regulatory mechanisms may be contributing factors.

Human ecology evidently is closely related to many other branches of natural and social science. The ecology of individual persons involves physiology and psychology as well as physiography and climatology. Community ecology overlaps in part anthropology, botany, demography, geology, history, political science, sociology, and zoology. Most of the branches of applied ecology, including medicine, hygiene, parasitology, agriculture, animal husbandry, forestry, game and fish management, and conservation, also involve human ecology.

The science of human ecology, however, is as yet relatively undeveloped. Its basic concepts are not well recognized nor agreed upon. Quantitative methods for measuring the relations between human communities and the physical and biotic features of their habitats are still in process of development. Little is known about the operation of the regulatory mechanisms which keep human societies in adjustment with the resources of their habitats. Nevertheless, no other subdivision of science has the breadth of view which is required for full consideration of the complex interrelations of human beings with the biotic and physical features of their environments.

LEE R. DICE

References

Cannon, W. B., "The wisdom of the body," New York, Norton, 1963.
Dice, L. R., "Man's nature and nature's man, the ecology of human communities," Ann Arbor, University of Michigan Press, 1955.

HUMBOLT, ALEXANDER VON (1769–1859)

Humbolt was one of the very great men of his day and one of the greater intellects since the Renaissance. The somewhat impecunious son of a noble family, he studied mining and worked as an employee of the Prussian Department of Mines. In 1799 a fortunate death placed him in the position of a huge inheritance which enabled him to indulge that passion for science which distinguished his life. He could think of no better use for his fortune than to equip an expedition to South America which left in the year 1799. His five years in that country were spent in traversing the subcontinent in every direction and on his return to Paris, he published the accounts of his expedition which made him world famous. The King of Prussia recognized the value to mankind of his former employee and appointed him to a well paid sinecure as Chamberlain which enabled Humbolt to reject all subsequent offers of university employment and to devote himself entirely to research. His principal contribution to the Biological Sciences was

what amounted, for all intents and purposes, to the invention of plant geography. He regarded vegetation less from the Linnaean taxonomic aspect than from the aspect of the mutual interaction of the plant with its environment. He was also interested in the contributions made by plants to the total landscape. Actually, his botanical contributions are among the least important of his great career for he established the principles of studying terrestrial magnetism which is still employed and invented the method of depicting climates by isothermal lines. He lived to a great old age, respected and beloved of all and showered with honors from every government and great scientific body.

Reference

Kellner, L., "Alexander von Humbolt," New York, Oxford, 1963.

HUXLEY, THOMAS HENRY (1825–1895)

Thomas Henry Huxley was not only a great biologist but certainly the most outstanding science educator that Britain has yet produced. He was the seventh son of an impecunious, but learned, schoolmaster and his formal education was confined to two dubious years of instruction between the ages of eight and ten. From that age on, Huxley acquired his learning solely by reading. He acquired sufficient knowledge to be able to join his elder brother James as a medical student at Charing Cross Hospital and at the age of twenty, he secured his license to practice medicine. Since he had not available to him the capital at that time necessary for the purchase of a "practice," he secured an appointment as a naval surgeon and commenced to practice at Hasler Hospital. His obvious love of scholarship, and his keen mind, had however attracted the attention of the Arctic explorer, Sir John Richardson, who secured for him an appointment to H.M.S. "Rattlesnake," which was under orders to explore the waters north of Australia. The waters of those parts teemed then, as now, with plankton but at that time no satisfactory method for its preservation was known. His detailed studies of the Medusae, and of the planktonic Urochordates, were in an almost virgin field but the harvest which he reaped is proof of his genius. His most important publication was on the Medusae, in the course of which he demonstrated not only the relationship between the medusoid and hydroid forms and thus founded the class "Hydromedusae," but he also studied the anatomy and realized the significance of the outer and inner layers. This led him to name the phylum Coelenterata. The value of this contribution was immediately realized and at the age of twenty-five, he was elected a fellow of the Royal Society and in the following year received the royal medal of the society and was elected to the Council. Huxley was still nominally a naval surgeon and was ordered three years later on active service. Since this would have terminated Huxley's scientific research, he immediately resigned and would again have been penniless but for the fortunate circumstance that a lectureship fell vacant at what was then the School of Mines. One year later, 1855, the Royal School of Chemistry (now the Royal College of Science) was joined to the

School of Mines and Huxley remained in the chair of natural history and zoology in that institution for the rest of his life. The publication of the "Origin of Species" by Darwin in 1859 was the most momentous event in Huxley's career. He threw himself enthusiastically into the propagation of the views of evolution. He criticized many of Darwin's minor conclusions but felt the whole thesis to be of such value that he wished to assist in the revolution of scientific thinking which he was sure it would produce. He joined Darwin on the dangerous ground that man could not be excluded from a theory of evolution even though the latter was the product of a human brain. This occasioned his famous quarrel with Richard Owen which was the most famous scientific controversy of its day. Huxley felt that paleontological evidence should be brought to the support of Darwin's theory and for the next thirty years, devoted himself to the study of fossil vertebrates. In the course of this, he discovered and pointed out the affinity between the birds and reptiles and founded the class Sauropsidia for the reception of both. During the whole of this period, Huxley was extremely active in promoting scientific education, not only in the college where he taught but in the schools of London. He is responsible for introducing at the Royal College of Science the method of instruction still employed there which he called the "tandem" method, by which only subject is studied at a time and that intensively. He pioneered the introduction of basic science into the elementary schools of his time but, curiously enough for a free thinker, insisted on the retention of bible instruction. His reason for this is given in his essays, where he says that he was "seriously perplexed to know by what practical measures the religious feeling, which is the essential basis of conduct, was to be kept up in the present utterly chaotic state of opinion in these matters, without its use." When the Royal College of Science was moved to its present location in South Kensington, Huxley was made "Dean" of the College. Such a title at that time was a tribute only to his high scholarship and popularity among the faculty and did not involve any administrative duties. In 1885 his health completely broke down and he died ten years later after a long and painful illness. His "Collected Letters" and his "Collected Essays" are a neglected part of his contribution and should be more widely known among all interested in the history, philosophy and teaching of science.

PETER GRAY

Reference

Bibby, C., "T. H. Huxley," New York, Horizon, 1959.

HYBRIDIZATION

The mule is generally acknowledged to be the first hybrid recognized as such, and the ancient Babylonians are known to have artificially pollinated plants by bringing pollen bearing branches of date palms to female trees of the species. It was not until 1694, however, that Camerarius was to produce scientific proof that sex existed in plants by carrying out hybridization experiments. LINNAEUS and his students from 1748 to 1760 produced several treatises upon hybridization. Linnaeus actually made hybrids in plants and was perhaps the first to suggest that many of our cultivated plants arose by hybridization as well as that new ones could be created by such a process. Kölreuter, however, is credited with making the first systematic study of species hybrids (1763–1766). In the early part of the nineteenth century Thomas Andrew Knight and William Herbert of England made some of the first applications of hybridization to plant improvement, and to this day hybridization has continued to be the principal tool of the plant and animal breeder. In 1881 W. O. Focke brought together a summary of the then known knowledge on hybridization in his "Pflanzenmischlinge." J. P. Lotsy in 1916 ascribed all evolution to hybridization, and although today hybridization is considered an important evolutionary factor, Lotsy's conclusions are generally unacceptable.

The precise definition of hybrid holds some difficulty since the word has been used both to designate crosses of dissimilar individuals of the same species and crosses of individuals belonging to different species. Stebbins has proposed an "evolutionary definition" to apply to crossing between individuals belonging to separate populations which through isolation have acquired different adaptive norms. Any attempt to use sterility as a criterion of hybrids is of limited use, since the degree of fertility of hybrids shows great variation. Hybrids between members of the same species having the same chromosome number are generally fertile, whereas hybrids between distinct species may be either as fertile as the parents, show various degrees of impairment of fertility, or be completely sterile.

Since the pioneering work of Edgar Anderson in 1936 there has been increasing recognition of the great number of natural hybrids in the plant kingdom, and it is now apparent that interspecific hybrids are quite common in higher plants. Many groups of woody plants, the Orchidaceae, the Compositae and certain other families contain a great number of hybrids. A large part of the so-called species problem in plants results from the frequency of natural hybridization, much of which stems from disturbance of the environment by man which allows previously isolated species to come together. Man at the same time has created intermediate habitats and new habitats in which the hybrids may survive and spread. One of the most common results of natural hybridization in plants in introgression which may be defined as the infiltration of germ plasm from one species into another as the result of continual backcrossing of hybrids and hybrid derivatives to the parental species. Since backcrosses may resemble their recurrent parents very closely, ordinary taxonomic techniques are not always adequate to indicate introgression, and a number of special biometrical methods have been designed by Anderson to detect introgression, and more recently biochemical methods have been used by Alston and others. The introduction of experimental methods into taxonomy has been particularly important in the acceptance of the vast numbers of hybrids now reported. Although introgression occurs most frequently between species which produce fertile hybrids, examples of introgression are known from groups in which the hybrids are nearly sterile (*Elymus* and *Helianthus*). Although continued introgression between two species might be expected to lead to an amal-gamation (secondary speciation), such rarely seems to happen.

Hybrids between species of animals, while not nearly so common as in plants, are not as once supposed and well documented hybrids are now known for several groups, particularly in amphibia, fishes, and birds. One of the most effective barriers in preventing animal hybrids is the strong development of ethological isolation. In addition, the fact that development in most animals is much more complex than in plants tends to prohibit the formation of hybrid individuals.

The primary evolutionary role of hybridization is its potential for the enrichment of variability. Genetic recombination, along with mutation and natural selection, are the major factors in evolution, and inter-specific hybridization offers far greater possibilities for recombination than are possible within a species. Although many of the resulting hybrid combinations are ill adapted, some may prove superior to their parental types, particularly in a new environment. Hybrid vigor, so important in breeding, may also give hybrids an advantage over their parents in nature. The doubling of CHROMOSOMES in a hybrid leading to the production of a new fertile allopolyploid has been an important mode of SPECIATION in plants. It is also possible that interspecific hybridization may give rise to a new species without chromosome doubling through the stabilization of hybrid derivatives. Harlan Lewis and Carl Epling have postulated such an origin for *Delphinium gypsophilum.* As shown by R. E. Cleland hybridization has had a unique role in one group of *Oenothera* where structural changes of the chromosomes combined with a system of balanced lethals and self pollination has lead to the development of complex hybrids which have been extremely successful. All in all, therefore, it appears that hybridization has been most important in the evolution of plants, although it may have occupied a more minor role in animals.

CHARLES B. HEISER, JR.

References

Anderson, Edgar, "Introgressive Hybridization," New York, Wiley, 1949.
Heiser, C. B., Jr., "Natural hybridization with particular reference to introgression," Bot. Rev. **15**: 645–687, 1949.
Stebbins, G. L., Jr., "The role of hybridization in evolution," Proc. Amer. Phil. Soc. **103**:231–251, 1959.
Zirkle, Conway, "The Beginnings of Plant Hybridizations," Morris Arboretum Monographs 1. Philadelphia, Univ. Pennsylvania Press, 1935.

HYDROZOA

Hydrozoa are CNIDARIA in which the mouth orifice opens directly into the gastric cavity without the intervention of a pharynx and in which the gastric cavity is not divided by partitions of septa similar to those found in the Anthozoa and the Scyphozoa. The Hydrozoa are divided into several subclasses, the Hydroid-Hydromedusae, the Hydrocorallina and the Siphonophora.

Hydroids and Hydromedusae

These are the simplest of the Hydrozoa, and they include organisms commonly occurring in two forms; one, a fixed asexual form, the polyp, and the other a free asexual form, the medusa. Occasionally only one of these two phases exists as, for example, in *Hydra,* in which there is only the polyp, or the Trachymedusae, in which the polyp is lacking.

Polyps. Polyps usually form colonies by budding but are occasionally solitary (*Hydra, Tubularia*). A hydroid polyp is basically composed of: (1) A hypostome in the form of a dome at the summit of which the mouth orifice opens and which, in most cases, is surrounded by a crown or whorl of tentacles. The hypostome, as well as the tentacles, plays a basic role in predation. (2) A gastric column, which may or may not bear tentacles according to the species, and at the level of which the *medusoid elements* are usually differentiated by budding. These elements will consequently be liberated or, if they do not reach the final stage of development, will remain fixed to the polyp as gonophores (Fig. 1). The various stages of the ingestion of prey occur in this zone. (3) A sphincter, which is a restricted zone lacking tentacles, interposed between the gastric column and the stolon (Fig. 1). The sphincter is an extremely muscular region which prevents the passage of too large pieces of prey from the gastric cavity of the column to that of the stolon. (4) The stolon system, in the form of a network of diblastic tubes surrounding the gastric cavity of the stolon. This network is covered with a largely chitinous sheath called the perisac. The stolon system is the most important and durable structure in a hydroid colony, since it not only serves to attach the colony firmly to its support but is also the principal site of growth. Moreover, it is at the stolon level, in numerous species, that the gonophores and medusary buds develop. Finally, in the winter or in other ecologically difficult conditions, only the stolon tissues survive and then serve as the basis for the development of a new colony. The stolon may be creeping (creeping and spreading colonies) or upright with the colonies resembling plants (arborescents, cymous, and other types of colony). Some polyps, as has been indicated, are solitary in which case they have neither sphincter nor stolon but are attached by an adhesive glandular disc (*Hydra*) or by the production of an anchorage system (*Branchiocerianthus*). (5) Various types of tentacles can be distinguished in the hydroids according to the way of distribution of their nematocysts. In the most primitive forms, the tentacles are arranged uniformly around the hypostome and the gastric column. In more specialized forms the arrangement is either in several circles or in either of two types of crown, one around the hypostome and the other around the column. Other species show only a ring around the hypostome, though this ring may be incomplete (*Lar*). Lastly there are forms that lack tentacles during their entire life (*Protohydra, Limnocnida*). The perisac, which, as we have seen, completely surrounds the stolon system, terminates at the level of the Athecates at the level of the sphincter (glymnoblastic). In Thecate hydroids, on the contrary, it is continued beyond the sphincter and appears as a complete theca surrounding the polyp, (the hydrotheca), or around the buds of medusae, or gonophores, (the gonotheca). Around those specialized polyps that

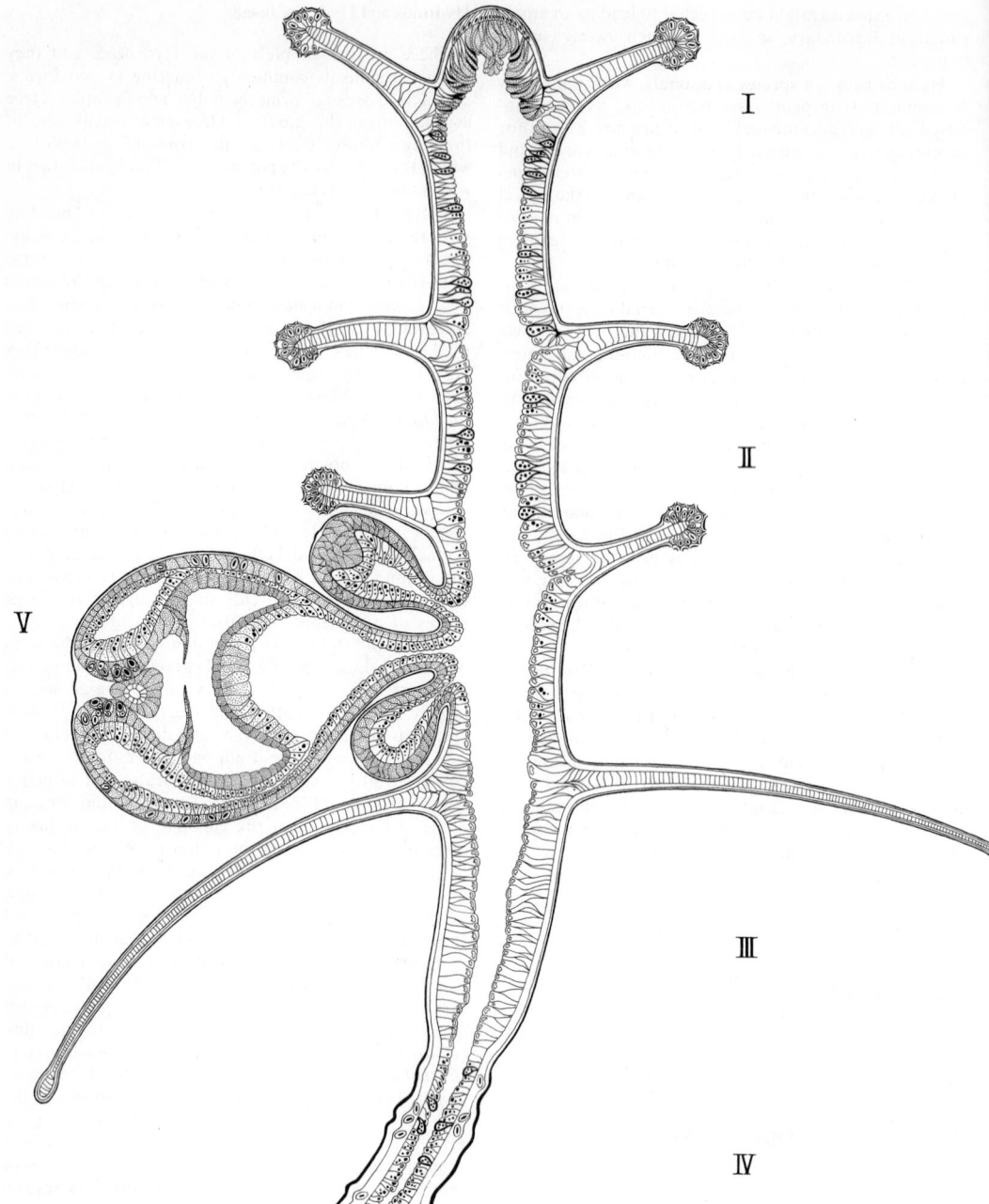

Fig. 1   Diagrammatic longitudinal section of a polyp of *Dipurena ophiogaster* (Athecate Hydroid). I, hypostome; II, gastric column; III, sphincter; IV, stolen; V, medusary bud. (Original.)

protect the colony (the dactylozoids), the perisarc forms a dactylotheca. Hydroid colonies are very often polymorphic; that is to say the individuals of which they are composed occur in several forms and we have already discussed two of these: the polyps and the medusae (or gonophores at various stages or regression). In addition there are sexual or gonozoid polyps, protective individuals either furnished with

nematocysts or dactylozoids or which may lack nematocysts and be modified into chitinous spines, the acanthozoids.

These various types of polyps are best seen in colonies of *Hydractinia echinata*, a Thecate hydroid living on molluscan shells inhabited by hermit crabs.

The wall of the polyp body is diblastic, being formed of two clearly distinct epithelia, an external protective one, the ectoderm, and an inner digestive one, the endoderm. These two cellular layers are separated by a non-cellular supporting lamella, the mesoglea.

The ectoderm which forms the exterior coat consists principally of epithelio-muscular cells, so called because they combine the roles of covering and muscular cells (smooth longitudinal fibers). Among these epithelio-muscular cells are other elements varying according to the species, the best known being the stinging cells, or cnidocysts, interstitial cells, nervous cells, senosry cells and glandular cells.

The interstitial cells are cellular elements retaining the embryonic state and are found in certain types of Hydrozoans intercalated between the epithelial cells. Their primary function is the replacement of cnidocysts, the initiation of asexual budding, the formation of sexual elements and the replacement of various types of other cells. The endoderm, or internal layer, is similarly formed of epithial muscular cells (smooth, circular fibers) that are phagocytic and absorb and digest the nutritive particles derived from the breakdown of the ingested prey. Interspersed between these absorbent cells can be found various types of glandular cells of which the secretions aid the ingestion of the prey and in its disintegration in the gastric cavity of the column. The endodermal ephitelium, as well as the ectodermal, contains nervous and sensory elements.

Hydroid polyps are voracious animals and feed principally on small crustacea, on worms and on larvae. They frequently ingest prey of a size greater than their own. The prey, harpooned and anesthetized by the cnidocysts arming the tentacles, are carried by these to the level of the gaping mouth and then engulfed slowly into the digestive cavity where they fragment in proportion to their penetration.

The hydroids are principally marine littoral forms attached to rocks, to algae, to other animals (Mollusca, Crustacea, Tunicates, etc.) or to the interior of other organisms (sponges, *Tubularia ceratogyne, Dipurena halterata*; and mussels, *Eugymnanthea*). Littoral species attached to a suitable substrate are frequently found at greater depths; for example, *Tubularia, Stylactis* and certain plumularians that are usually littoral can sometimes be found as deep as 1000 meters. Certain species fixed to movable bases inhabit extreme depths, such as *Branchiocerianthus imperator* living at 5000 meters, and *Aglaophenia galatheae* down to 7000 meters. Most hydroid colonies are of small size (from several millimeters to 20 centimeters in height; however, certain *Plumulariidea* and *Branchiocerianthus imperator* cited above can obtain a length of 2 meters. Occasionally Hydroid colonies become pelagic as, for example, *Pelagohydra mirabilis* and *Margelopsis.*

Some Hydroids are ectocommensals of other animals. For example, *Hydrichtys* living on fish, *Eugymnanthea,* cited above, living on the mantel and the palps of mussels; some become parasites, such as *Polypodium hydriforme,* which during a part of its existence lives free and during the other is parasitic in the eggs of sturgeon.

The Hydromedusae. The Hydromedusae fundamentally have a radial symmetry. They are generally considered to be polyps adapted to a pelagic life in view of their sexual reproduction from which the name gonozoid is sometimes applied to them. The body of the Hydromedusae, or umbrella, is usually in the form of a bell, a mushroom or sometimes a disc. The convex aboral surface of this umbrella carries the name exumbrellar surface, the lower concave area being called the subumbrella. The free border of the exumbrella is provided with tentacles and sensory organs (eyes or ocelli, static organs or statocysts, modified tentacles or cordyli) and frequently shows a thickening filled with young cnidocysts known as the nettle ring. From the center of the subumbrella there hangs, as the "clapper" of the bell, a cylindrical or quandrangular tubular structure called the manubrium, which is homologous with the hypostome of polyps. This manubrium can be long or short, simple or lobed, and with or without oral tentacles.

The orifice of the subumbrellar cavity is diminished by a muscular horizontal diaphragm, the velum. This velum differentiates the Hydromedusae, or Craspedote medusae, from the Scyphozoa that lack it and which are accordingly known as the Acraspedote medusae.

The greater part of the umbrella volume is occupied by a gelatinous mass that is homologous to the mesoglea of the hydroids and that contains the canals of the gastrovascular system. The latter consists in the gastric cavity of the manubrium, opening to the exterior by the mouth, and of which the proximal part or gastric pouch is prolonged across and through the mesoglea by the radial gastrovascular canals which are primitively four in number but may be more numerous. These radial canals join the gastric cavity to the circular canal which extends around the whole periphery of the umbrella. Centripetal canalicules arise from the circular canal and penetrate the tentacles when these are hollow. A single-layered membrane, the gastrodermal lamella links the radial canals to one another and like these connects the gastric cavity to the circular canal.

Medusae, just as polyps, have a diblastic structure. The exumbrella, umbrella, the subumbrella and the manubrium are limited on the outside by the ectodermal layer. The gastric cavity is limited in the interior by the endoderm; the gastrovascular canals and the gastrodermal lamella are endodermic. The velum is composed of two ectodermic epithelia, one of exumbrellar origin, and the other subumbrellar.

The distinction between the different groups of Hydromedusae is principally based on the structure of the sense organs which are of various types:

(1) The eyes or ocelli which are best developed in the Anthomedusae. They are also rarely found in some Leptomedusae, notably in the Laodiceidae, Mitrocomidae and Aequoridae.

(2) The statocysts, organs of equilibrium and orientation, are absent in the Anthomedusae. They may be arranged in two categories; those exclusively ectodermal, found in the Leptomedusae, and those of an ecto-endodermal origin that are found in the Limnomedusae, Actinulidae, Trachymedusae and Narcomedusae.

(3) The *cordyli.* These are ecto-endodermal sense organs in the form of clubs lacking both statoliths and nematocysts. They are found implanted on the umbrella side of medusae of the family Laodiceidae. Their function is doubtful.

The Hydromedusae have a complex system of radial and circular muscular contractile fibers. Some of these muscles are striated; they are the first such fibers to evolve in the animal kingdom. They are circular and are situated in the subumbrellar ectoderm and in the velum. The contraction of these striated fibers results in

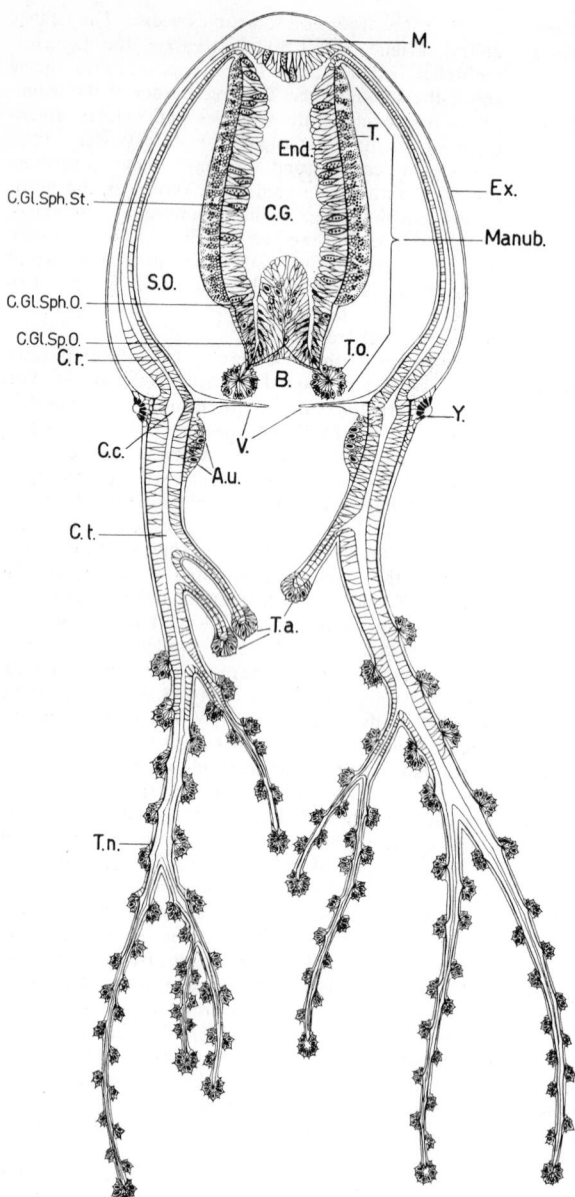

Fig. 2 Diagrammatic section of a medusa of *Cladomena radiatum* (Athomedusae). A.u., stinging ring; B., mouth; C.c., circular canal; C.G., gastric cavity; C.Gl.Sp.O., oral glandular cells; C.Gl.Sph.O., spherular oral glandular cell; C.Gl.Sph.St., Stomatic cellular glandular cell; C.r., radial canal; C. t., tentacular canal; End., endoderm; Ex., exumbrella; M., mesogloea; Maub., manubrium; S.O., subumbrella cavity; T., testicle; T.a., adhesive tentacle; T.n., stinging tentacle; T.o., oral tentacle; V., velum; Y., eye.

the expulsion of a part of the water contained in the subumbrellar cavity through the restricted aperture of the velum. In consequence of this the medusa moves with the umbrella forwards. These muscles work against the elasticity of the mesoglea which causes the dilation of the subumbrellar cavity. These alternate contractions and dilations are responsible for the jerky locomotion characteristic of medusae. Such movements allow the medusa either to remain at the surface or to migrate vertically, while water currents and winds are responsible for horizontal motion. In correlation with the complexity, the concentration of the sense organs, and the umbrellar and tentacular movements the nervous cells of the Hydromedusae are condensed into a marginal coordinating center. This coordinating center is usually found in the form of two nerve rings situated at the base of the velum and separated one from the other by the velary mesoglean lamella. As well as these two nerve rings, there are a subectodermal, manubrial, tentacular and subumbrellar nerve plexus. Certain elements of these plexi are more concentrated to the level of the radial canals and form true radial nerves, which connect with the central nerve rings, themselves interjoined.

Certain Hydromedusae reach great depths such as *Solmissus* (5000 meters). Some medusae are attached during the greater part of their existence such as, for example, *Eleutheria, Cladonema, Gonionemus*. The

Hydromedusae are generally of small size, the majority of them being less than a centimeter long, though there are a few that reach 5 centimeters, and the largest of all (*Aequorea*) can attain a size of 20 centimeters.

The Hydromedusae, as the polyps, are essentially carnivorous feeding on planktonic crustacea, Chaetognata, and planktonic larvae belonging to various groups.

Reproduction. The genital elements are generally of ectodermal origin, the sexes usually being separate, save in a few hermaphroditic forms (*Eleutheria* and certain *Hydra*). In the greatest number of medusae the eggs are laid directly into the external environment though a few types have incubatory chambers (*Eleutheria*). The fertile eggs divide and give rise to a ciliated larva, the planula, which after a certain length of free life becomes attached to a suitable support and gives rise to a new colony of Hydroids. Some forms (*Tubularia*) produce an intermediate larvae, the actinula.

In many hydroids only imperfect medusae develop. These remain attached to the colony forming more or less degenerate gonophores from which the larvae are produced.

The Trachymedusae, Narcomedusae and Actinulida have a direct development without alternation of phase, the larvae developing directly into organisms identical in form to those from which they themselves arose.

Hydroid polyps can give rise to various types of buds:

(1) Growth buds that produce new polyps, which may remain separate (*Hydra*), or form colonies.

(2) Medusary buds giving rise to the free sexual phase, the medusae, or to reduced sessile gonophores.

(3) Propagatory buds, or frustules of various types, ensuring both the propagation and the direct dessemination of the species.

(4) Resistant buds capable of surviving environmental conditions unfavorable to the species.

All of these various types of buds arise from an evagination of the diblastic wall of the hydroid trunk.

Certain Hydromedusae can also reproduce by medusary budding (*Limnocnida, Eleutheria*).

Hydroids and Hydromedusae have strong powers of regeneration and of reorganization from fragments of the two layers of which they are formed. In some cases normal colonies can be reconstituted from part of a single layer, the endoderm in the case of *Hydra* and the ectoderm in *Cordylophora* for instance.

Normal polyps can even be redeveloped from dissociated tissues passed through a sieve.

Hydrocorallina

Hydrocorallina are hydrozoans of which the skeleton, or perisac, is thickened and impregnated with calcarious salts. In the course of the development of a Hydrocoralline colony the initial stolons die but their calcareous skeletons remain to serve as a substrate for other individuals which will die in their turn and so on, to build an encrusting or branching coralline colony of constantly increasing thickness but of which the superficial layer alone is alive.

The surface of the calcareous mass, covered by ectoderm common to the whole colony, is pierced with apertures corresponding to the holes occupied by the polyps. These apertures are polymorphic and one can distinguish those of large size, the gastropores con-

taining the gastrozoids or feeding polyps, and the holes with small apertures containing the dactylozoids or defensive organs the dactylopores. All the polyps of a single colony are joined to each other by the ramifying gastrovascular canals.

In the Stylasteridae the base of the hole occupied by the gastrozoids has a central projection, the "style," from which the name Stylasteridae is derived. The feeding polyps of Stylasteridae bear solid filiform tentacles which are lacking from the dactylozoids. The sexual elements of hydrocorallines are carried in capsules (not medusoids) visible as swellings. In primitive forms the dactylozoids are scattered at random among the gastrozoids; in more specialized forms there is the tendency for them to be grouped around the gastrozoids wither in a circular or a stellate formation. In the Milleporidae the gastrozoids and dactylozoids lack a style and they possess capitate tentacles. Dactylozoids form a circle around the gastrozoids. Colonies are monoecious, that is to say some are male and others female. The sexual elements are developed in the bodies of imperfect medusoids which become liberated but die a short distance from the colonies from which they issue after having emitted the gametes. The subclass Hydrocorallinae is today strongly controverted, the two orders which used to constitute it being incorporated in the Hydroid-Hydromedusae.

Stylasteridae and Milleporidae live principally in warm waters where their polyps contribute to the formation of coral reefs, though a few forms exist in temperate seas. The Millepora are powerful stinging organisms to the extent that they have been called "fire corals."

Siphonophora

The Siphonophora are Hydrozoans forming beautiful and fragile pelagic swimming or floating colonies living in warm seas and made up of modified medusae and polyps. Extremely polymorphic, the colonies of Siphonophora are fundamentally made up of an organ of locomotion corresponding to a modified craspedote medusa and of a diblastic stolon of gastrovascular origin inserted in the ventral gutter of the umbrella of this medusa and budding off groups of polypoid zoids. These last form a secondary association, the cormidium. Each cormidium typically consisting of a gastrozoid polyp without tentacles "the siphon" at the base of which is inserted a dactylozoid or "fishing filament," of two gonozoids (one male and one female), of several cytozoids or excretory polyps and of an aspidozoid or protective element.

The flotation apparatus can be complex, being made up with several medusae of locomotion or swimming bells (nectophores) and of a float or pneumatophore similarly homologous with a rudimentary medusa and provided with a gas chamber rich in nitrogen, oxygen and argon.

The contraction of the pneumatophore allows colonies to submerge into deeper water, particularly in times of bad weather, and the contractions of the swimming bells enable them to move laterally. The fishing filaments are frequently very long, several meters in the case of *Physalia*, and form in their totality a kind of net in which sometimes relatively huge prey (fishes, for example) become entangled, immobilized and then

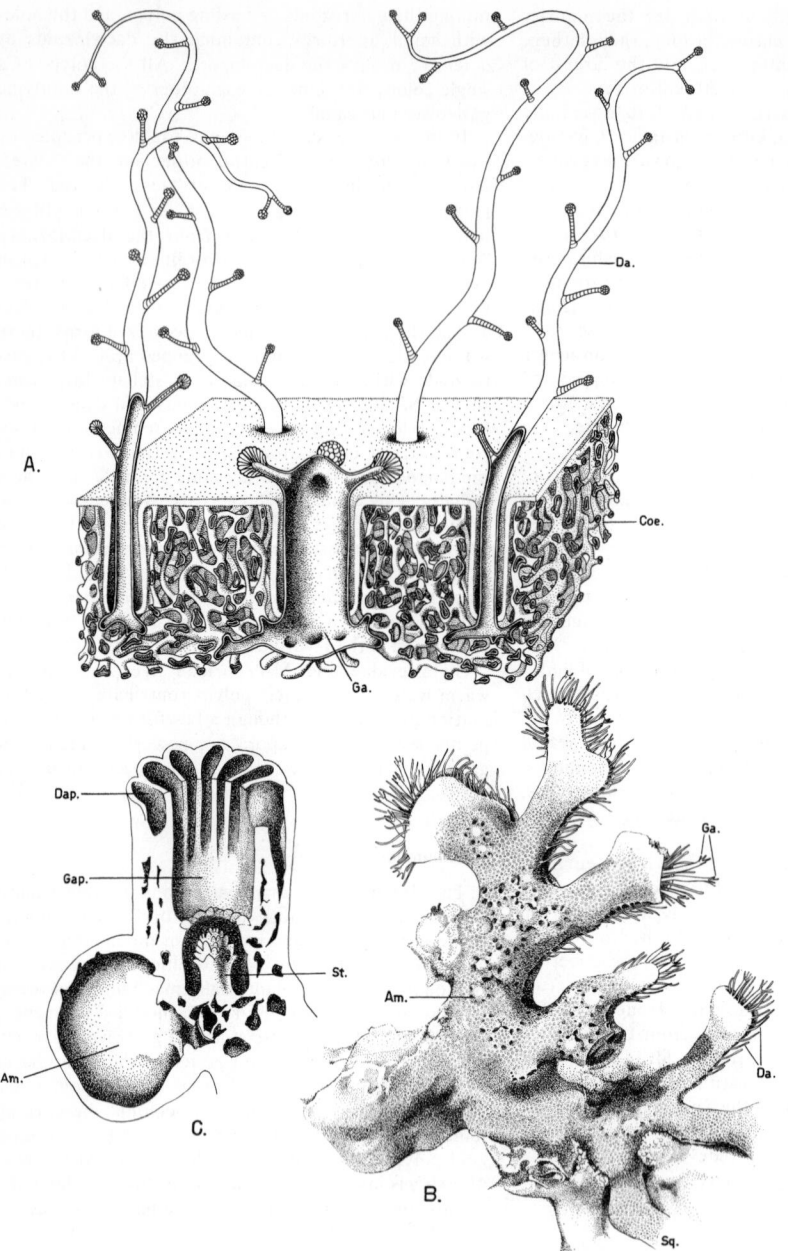

Fig. 3   A. Fragment of a colony of *Millepora nodosa* (Hydrocorallina, Milleporidae) showing a group of ex-pounded dactylozoids (Da.) surrounding a partially retracted gasterozoid (Ga.). Coe., coenosarc. B. Diagram of a colonly of Stylasteridae (Hydrocoralinae) belonging to the genus *Distochopora*. Am., sexual ampule; Da., dactyl-ozoid; Ga., gasterozoid; Sq., skeleton. C. Section of a skeleton of the housing and of an ampule of *Stylaster*, show-ing the gasteropore (Gap.) and the dactylopores that surround it (Dap.). At the bottom of the gastropore the char-acteristic stylets of the Stylateridae can be distinguished. Am., sexual ampule.

seized and led to the level of the gastrozoids or nutritive polyps.

There is a remarkable case of commensalism be-tween *Physalia* and fishes of the genus *Komeus*, which can swim with impunity among the fishing fila-

ments of *Physalia* the wounds from which are extremely dangerous even for man.

Many Siphonophora are luminescent.

Completely developed eormidia become detached from the colony and form "eudoxids," organs of distri-

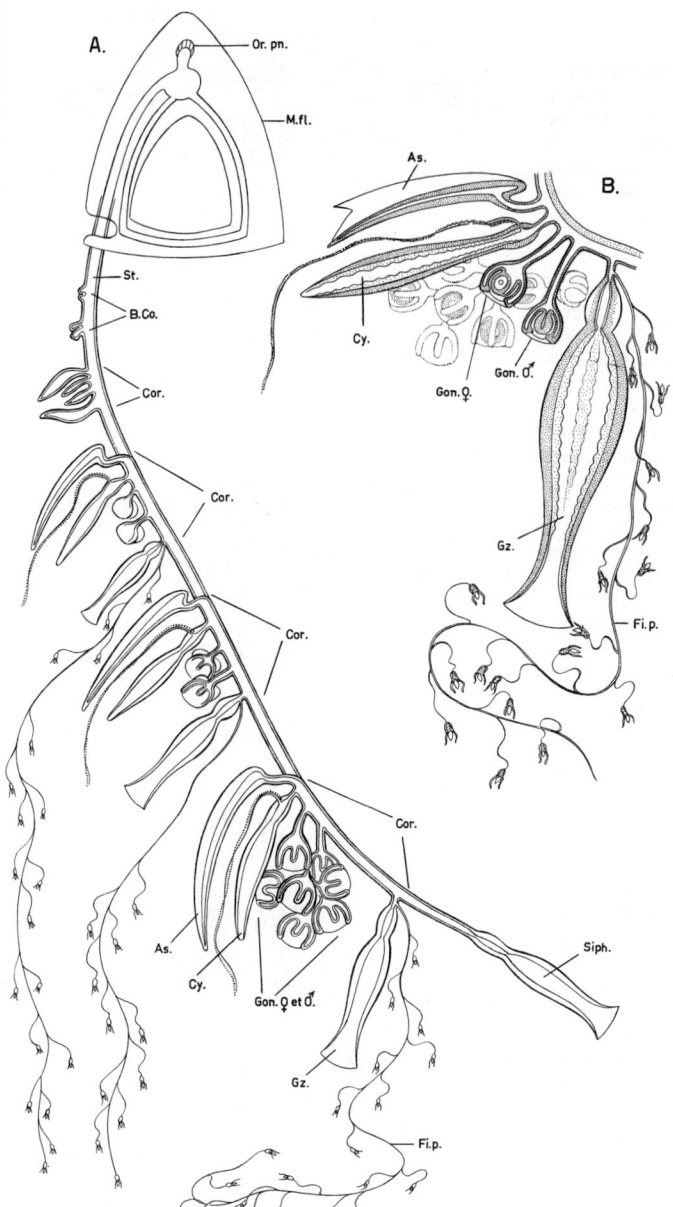

Fig. 4 A. Diagram showing the general structure of a Siphonophore. B. Diagram of a cormidian. As., aspidozoid; B.Co., bud of the stolen forming cormidia; Cor., cormidia at various stages of development; Cy., cystozoid; Fi.p., fishing filament; Gon., gonozoid; Gz., gastrozoid; M.fl., flotation medusa; Or.pn., flotation organ; Siph., primary gasterozoid; st., two-walled stolon supporting the cormidia.

bution giving rise to sexual elements. Certain Siphonophora are of large size, in particular *Physalia* of which the float may attain a size of 30 centimeters, and the fishing filaments, as we have seen above, several meters. Siphonophora are incapable of regeneration. They are divided in two groups. The Calycophora of which the flotation organs are exclusively composed of swimming bells and the Physophora in which the flotation organ is composed of floating bell or pneumatophore associated or not with swimming bells.

The Chondrophoridae (*Vellela* and *Porpita*) were previously included in the Physophora but are now placed with the Hydroids.

JEAN BOUILLON

References

Grassé, P.-P., *et al.,* "Précis de Zoologie. Vol. 1. Invertébrés," Paris, Masson, 1961.
Hyman, L. H., "The Invertebrates. Vol. 1. Protozoa through Ctenophora," New York, McGraw-Hill, 1940.

HYMENOPTERA

Hymenoptera is the order of INSECTS which contains the sawflies, bees, wasps, ants, ichneumon-flies, chalcid-flies, gall wasps and related forms. More than 110,000 different species have been named and described and many thousands are still unnamed and undescribed. They vary greatly in size and body shape. Some wasps and certain ichneumon-flies attain a length of about two inches, while some minute chalcid-flies that develop as parasites in the eggs of other insects are less than half a millimeter long. Most species have four membranous wings, the anterior pair always the larger, and the venation is more or less reduced; in some small parasitic species the wings are veinless. Winglessness, usually accompanied by certain structural modifications of the thorax, is not uncommon. The workers of ants are always wingless, as are the females of some of the solitary wasps (families Mutillidae and Thynnidae) and those of many small parasitic species. The winged and wingless individuals of the same species usually differ so widely that their association may not be determinable except through rearing or observation of mating. The compound eyes of the Hymenoptera are usually moderately large to very large, and there are normally three simple eyes, or ocelli, on the fronto-vertex. The mouthparts are usually developed for chewing but may be modified for sucking or lapping. The antennae are extremely varied in form and length. Sometimes they are long and slender and composed of many segments, as in many ichneumon-flies; but more commonly they are relatively short and not more than 15-segmented; the bees and wasps regularly have 12 segments in the female antennae and 13 in those of the male. Often the antennae are elbowed, with the basal segment, or scape, very long, and sometimes they are more of less thickened apically, especially in the females; rarely they are strikingly branched or broadly pectinate as in the males of certain chalcid-flies and sawflies. The legs are usually slender, although occasionally the hind femora are conspicuously enlarged and sometimes armed with one or more teeth; and in many bees the hind legs are modified to aid in collecting and carrying pollen. The tarsi are normally 5-segmented. Usually the ovipositor is well developed. In sawflies it is in the form of a saw-like structure with which the female saws tiny slits in the leaves or stems of the host plants for the deposition of its eggs; whereas in the higher groups, as in the bees and wasps, it is modified into a sting. In certain parasitic species, especially among the ichneumon-flies, it is often very long, sometimes much longer than the body. Hymenoptera are usually more or less hairy. They may be thickly covered with long hair, as many bees and wasps, or they may be sparsely hairy or only more or less pubescent. The surface of the body is often sculptured, and the extent and precise character of the sculpturing are important aids in identification.

The more primitive Hymenoptera belong to the suborder Symphyta, the members of which are characterized by having the abdomen broadly joined to the thorax. Included are the sawflies, horntails and wood wasps, all phytophagous, and a single parasitic group, the Orussidae, the comparatively few species of which parsitize certain kinds of wood-inhabiting beetle larvae. The sawflies, which comprise the greater part of the Symphyta, usually have caterpillar-like larvae, pro-

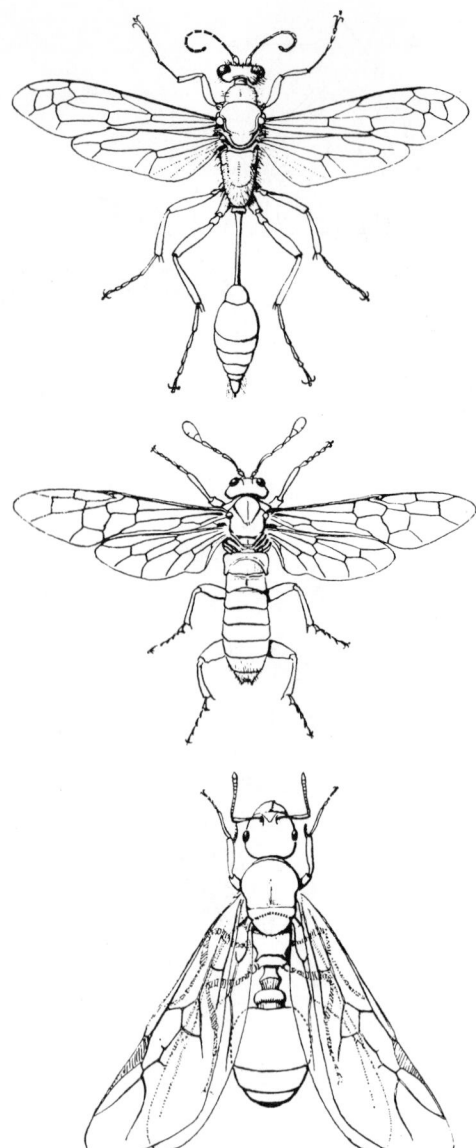

Fig. 1   Wasp (top), sawfly (middle), ant (bottom).

vided with legs; most of them feed openly on foliage or develop as leaf miners; and some, like the pine and spruce sawflies of the genera *Diprion* and *Neodiprion*, are destructive pests. In contrast with the Symphyta the members of the suborder Apocrita, which includes all other Hymenoptera, have the abdomen more or less constricted at the base; and the larvae are without legs and are grub-like or maggot-like in general appearance. This suborder is much larger than the Symphyta in number of species. Two major divisions are commonly recognized: the Aculeata, or stinging Hymenoptera, consisting principally of the bees, wasps and ants; and the Parasitica, composed largely of parasitic forms like the ichneumon-flies and chalcid-flies, but including also some plant feeders, like the gall wasps, most of which are tiny insects that are seldom seen in the adult state and are best known from the galls they produce, each type of gall being characteristic of the particular species responsible for it. To a large extent the Apocrita are beneficial, many bees as pollinators or plants, many wasps as predators on different kinds of injurious insects, and ichneumon-flies and chalcid-flies as parasites of other insects.

Various groups of Hymenoptera have attracted interest because of their specialization for social life (see SOCIAL INSECTS). All the ants are social; the colonies of some species contain only a few individuals, those of others many thousands. Each colony is founded by a queen and when normally developed contains a varying number of sterile workers and young. The young are helpless larvae that must be fed by the workers until fully grown and ready to transform to adults. Although most wasps are solitary in nesting habits some are social. These belong to the family Vespidae and include the hornets and yellow-jackets which build large paper nests, the former usually in trees or on buildings and always above ground, the latter in the soil; a single nest may contain thousands of individuals. Certain bees, expecially the bumble bees and honey-bees, also are social. Unlike other social Hymenoptera the queen honeybees cannot found new colonies alone but must be accompanied by numerous workers from the old colony.

PARTHENOGENESIS, or reproduction without fertilization, and polyembryony, the development of more than one individual from a single egg, occur in the Hymenoptera, the former commonly and in most of the major groups of the order, the latter rarely and only in certain genera of Chalcidoidea, Braconidae and Proctotrupoidea. In parthenogenetic reproduction the progeny are usually males but occasional species regularly produce females parthenogenetically, and in these species males are either very rare or do not occur at all (as in *Apanteles carpatus*, a world-wide parasite of the larvae of clothes moths). The phenomenon of polyembryony is most strikingly exhibited by *Copidosoma truncatellum*, a chalcidoid parasite of certain cutworms; as many as 2000 individuals may result from the deposition of a single egg in the body of the host caterpillar.

C. F. W. MUESEBECK

References

Bischoff, H., "Biologie der Hymenopheren," Berlin, Springer, 1927.
Brues, C. T. *et al.*, "Classification of Insects," Bul. Mus. Comp. Zool., **108**: 621–685, 1954.
Muesebeck, C. W. F. *et al.*, "Hymenoptera of America North of Mexico," U.S. Dept. Agriculture, Agricultural Monog. No. 2, 1951.

HYRACOIDEA

Hyracoidea ("Dassies") are moderate-sized mammals with 1 upper and 2 lower pairs of enlarged incisors. The upper incisors are long, curved, growing downward from persistent pulps, not chisel-shaped, but prismatic in section, and with no enamel coating on hinder surface; the lower incisors are straight, somewhat procumbent and gouge-shaped. There is a considerable space between these front teeth and the cheek teeth, which, in pattern, are not unlike those of *Rhinoceros*. The tail and ears are short. There are 4 functional fingers and 3 toes bearing flattened nails. The skull and dentition are of the ungulate rather than the rodent type with the palate well-developed and normal and a straight ascending ramus of the mandible. The stomach is horse-like, with a pair of conical-pointed caeca, quite unique in mammals. There is no gall-bladder, and the brain is of the ungulate type. The testes are abdominal and the placenta zonary. Dentition: 1, 0, 4, 3/2, 0, 4, 3.

According to Scott, the Hyracoidea are derived from Old World forms of the American Paleocene and Lower Eocene Condylarthra, an order regarded as a connecting link between clawed and hoofed mammals, and therefore ancestral to the ungulate orders. Cuvier, who first examined their internal structure and dental characters, regarded them as related to *Rhinoceros*. Subsequently Milne-Edwards and Huxley disproved this view, and demonstrated that the Hyracoidea in reality occupied a very isolate position, with only a general affinity to the Ungulates.

Fossil remains of the dassie have been found in Northern Egypt (the Fayum). Regardless of the little variation observed in the two genera, and the few species and geographical variations referred to below, the dassie has changed but little throughout geological times. *Limulus* (king crab) of the Pacific, the *Trigonia* of Australia and several other forms not listed here have not departed in a single important feature from their type in the Secondary, the Carboniferous or even the Cambrian. In this connection Pierre Teilhard de Chardin remarks: "While certain regions of the animal world were completely renewing themselves, others therefore remained strictly stationary. This is a curious fact. But what is even more disturbing is that the immobilized types which we find in nature are not only final twig-ends, species squeezed into a sort of morphological blind alley. The nautilus of the Indian Ocean, or the Syrian rock-badger (the dassie—author), or the tarsier of Malaysia or the *Cryptoproctus* and the lemuroids of Madagascar might, if known in their fossil state, quite easily assume the role of genealogical intermediates. Now all of them have remained living around us, unchanged for an immense period." (Pierre Teilhard de Chardin p. 10).

The order consists of a single family the Procaviidae with two genera, *Procavia* and *Dendrohyrax*, differing from each other in length of the crowns of the molars and that of the 3 molars in upper jaw compared with that of the 4 premolars. *P. capensis* with dorsal dark spot has about 16 geographic variations, found throughout South, Central and North Africa, South Arabia and Syria—the subspecies (Arabia, Syria) is the coney of the Bible (Ps. 104:18). *D. brucei* (the yellow-spotted dassie), with creamy to buffy white dorsal spot, has 8 subspecies ranging from Northern Transvaal to Somaliland; *D. arboreus* (the tree or bush dassie), with one subspecies, is found from Eastern Cape Province, northwards to Belgian Congo. This species is tree-loving and has a whitish dorsal spot. The length of the adult is 15–18 inches from tip of nose to root of tail and weighs from 6–8 lbs. The body is thickset and strong. The loose skin is covered with soft, moderately long brownish, yellowish to greyish brown fur. Two or three young are born after a gestation period of $7\frac{1}{2}$ months.

Little is known about the habits of the tree dassie that frequents forest regions, save that it subsists on leaves and makes its lair in the hollowed-out trunks of trees. The common dassie and the yellow-spotted dassie are agile and frequent rocky slopes and cliffs among piled-up boulders; they have soft "tacky" pads on the sole of the foot, and underside of the toes, giving them a safe foothold on the smoothest of rocks. They graze during daytime; their natural food consists of leaves of dwarfish shrubs, bark, berries, fruits and certain grasses. While grazing one or more always remain as scouts at look-out posts, and when they give the alarm the whole troop dash for the nearest cover—they are extremely wary, since they have no method of defence. The drink water only when it is available. The water necessary for digestion and absorption of food and metabolism is evidently produced in the body itself by endogenous carbohydrate and fat oxidation. Dassies are much infested with fleas, with a variety of lice and internal helminths. In recent years, many have contracted bubonic plague from their fleas, and died off. Their natural enemies are leopards and eagles.

Dassie colonies, often very large, are in the habit of selecting a particular spot in their rocky habitat to which all or most of the members repair for the purpose of urinating. The result is the formation on the rocky surface selected of deposits of a dark brown gluey substance. By reason of its contents of herbal extracts, medicinal properties have been ascribed to it, and it is listed in the Cape Pharmacopeia as "klip sweet." As an article of export, it is known in Europe as "hyraceum". In the early days inhabitants of the northwestern arid areas in the Cape Province, living hundreds of miles from the nearest towns, had, in case of serious illness, to forbare professional medical treatment and rely on their little stock of patent medicines or homemade remedies. Of the latter, a much valued article was an ointment made of "klip sweet" mixed with goats fat and administered as a sure cure for malignant ulcers and inflammation.

As a result of the reduction in numbers of the dassies natural enemies such as the South African lynx, jackals and certain of our large birds of prey, they have con-

siderably increased in numbers during the past thirty-forty years, and colonies have established themselves in rocky shelters in the open country adjoining their rocky strongholds. The result is that they are destroying much valuable pasture, and have become a definite menace to sheep and cattle farmers and methods had to be devised to check their prodigious increase.

Jack Russel terriers that are known to be excellent hunters and energetic eradicators of foxes and other mammalian vermin in Great Britain and other countries may prove to be very useful to check the dassie plague in South Africa. This finding is based on the results of field experiments undertaken by the Department of Nature Conservation of the Cape Province on the farm Skilpaddop belonging to Mr. A.G. Vermaak, near Glenconnor in the Uitenhage district about 50 miles from Port Elizabeth. With two Jack Russel terriers and five other terriers of another breed 423 dassies were destroyed within 43 hours. The subterranean burrows where they took refuge when pursued were scattered in a rather undulating terrain covered with dense brushwood. Of this number 123 were females, each with 2–3 embryos, so that at least 900 animals were accounted for in a very short time. Further experimentation is in progress particularly with the object of determining the modifications of technique to be applied in rocky and stone-covered localities.

C. S. GROBBELAAR

References

Scott, W. B., "History of Land Mammals in the Western Hemisphere," New York, Macmillan, 1937.

African Wildlife (Official Journal of the Wild Life Protection Society of South Africa.) Vol. 1, No. 1, p. 64; No. 2, p. 99; No. 3, p. 83; No. 4, p. 83. Distributors: Central News Agency, (Ltd.), cor. Rissik and Commissioner Streets, Johannesburg, S.-Afr.

Ellerman, J. R., "Die Taksonomie van die Soogdiere van die Unie van Suid-Africa," In: Annale van die Universiteit Stellenborch, Jaarg 30, Reeks A. No 1 (1954) p. 81.

Roberts, Austin, "The Mammals of South-Africa," distributed by the Central News Agency, South-Africa (p. 252.) (1951).

de Chardin, P. T., "The Vision of the Past," (trans. J. M. Cohen) London, Collins, 1966.

IMMUNOLOGY

Immunology in a restricted sense deals with the procedures used and the mechanisms involved whereby a host establishes resistance to disease (immune state) after a specific exposure to a foreign infectious agent (antigen). In a broader sense, immunology has become concerned with the response of a host to foreign macromolecules and hypersensitive biological phenomena of altered tissue reactivity such as allergies, acquired tolerance to and rejection of foreign tissues and autoimmune diseases.

Active immunization can be produced by injecting the host (intramuscularly, intravenously or intraperitoneally) with the antigen (usually proteins, polysaccharides, viruses or bacteria) incorporated in various vehicles (alum precipitate, water in oil emulsions i.e., Freund's adjuvant). The specific serum proteins produced in response to the injections are proteins with molecular weights of 150,000–1,000,000, sedimentation contents of 7S–19S and electrophoretic mobilities of gamma and beta globulins. They are heterogenous with respect to their specificities and combination with the antigen. In addition, other globulins (myeloma proteins, Bence Jones proteins) are found which do have some antibody activity and also share some structural features with antibodies. This entire group of proteins is termed immunoglobulins. Structurally the immunoglobulins are built of heavy (H) and light (L) chains having molecular weights of 20,000 and 50,000 respectively and held together by —S—S—bonds.

The transfer of immune serum either from a homologous species (i.e., measles therapy) or a heterologous species (i.e., horse antitoxin), is known as passive immunization. The active, in contrast to the passive state, is long lived, and challenge later in life with the same antigen results in a rapid synthesis of more antibody (secondary or anamnestic response) compared with the lower levels of antibody produced after a single exposure to antigen (primary response).

In the primary response, 19S antibody is formed first, followed by 7S antibody. In the secondary response 7S antibody forms the largest part of antibody produced. Exceptions exist depending on the host and the nature of antigen injected.

Much of what we know about the mechanisms by which antibodies are formed has been learned by the use of radioisotopes, fluorescent and electron microscopy and tissue culture techniques (immobilization of flagella in single drop techniques, hemolysis of red cells in presence of antibody secreted from lymphoid cells (Jerne)). Following the intravenous injection of a radio-labelled serum protein antigen, there is a rapid equilibration and distribution of the antigen throughout the body of the host. Although radiolabel from the antigen is found in many tissues and organs, it is believed that only a small fraction of the injected material stimulates antibody formation. (The role of persisting antigenic material in antibody formation is still unsettled). There is then a latent period (*induction phase*) during which time the antigen is catabolized at the same rate as homologous serum proten. A sudden increase in the rate of disappearance of the radiolabelled antigen indicates that newly formed antibody has entered the circulation, formed soluble complexes with the circulating antigen and is being rapidly removed (*immune elimination*). After the complete disappearance of circulating antigen (antigen-antibody complexes), free antibody appears in the serum and is catabolized at the same rate as the host's normal gamma globulin. The antibody is formed *de novo* from the available amino acid pool by the reticuloendothelial and lymphoid tissues, particularly the plasma cells. Cellular studies have indicated that under the influence of antigens, multi-potential and uncommitted stem cells differentiate and form the 19S antibody. Upon further differentiation 7S antibody is formed. Of major importance for the development of an adequate supply of immunocompetent lymphoid cells is the presence of the thymus gland throughout the life of the host.

Chemically, the immune state is recognized by the specific *in vitro* reactivity of the antiserums with the antigen (antigen-antibody reaction). With soluble antigens precipitation (precipitin reaction) results. Different zones of precipitation, depending on the amount of antigen added, are, antibody excess, equivalence zone and antigen excess (Fig. 1). With large cellular or particulate antigens (erythrocytes and bacteria), the reaction is termed agglutination.

Antigens have many combining sites and are termed multivalent. The valency is a function of the molecular weight of the antigens. 7S antibodies are bivalent and 19S antibodies have 5–6 antigen "combining sites." Antibodies against a single antigen therefore exhibit a spectrum of non-homogeneous reactivities. The reactive sites of antibody molecules are not all directed against the same specific sites in the antigen nor are the antibody combining sites against an antigenic site all equal (i.e., some sites may have specificity against 3, 4, 5, 6 or more amino acid or sugar residues).

The same basic mechanism of formation of an insoluble framework (lattice) of aggregates of antigen and antibody is involved in both the precipitin and agglu-

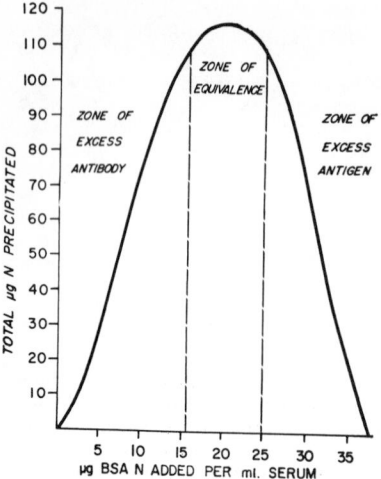

Fig. 1

tination reactions. The reactions are the results of specific chemical interactions between complementary configurations arranged on both antigen and antibody. The reactive sites of the antigen are believed to be small, consisting of several amino acids, nucleotides or sugar residues measuring about 34 Å × 12 Å × 8 Å (area of 700 Å²). Studies with chemically modified antigens, haptens (simple chemical substances coupled to proteins or polysaccharides) and antibodies have indicated that van der Waals forces, hydrogen bonds and coulombic forces are involved in the very specific immunological interactions. The sensitivity of the precipitin reaction (5 μg antibody N per ml) has been increased by allowing it to take place in gellified media such as agar (Oudin, Ochterlony and immunoelectrophoretic techniques) or by coating bacteria or erythrocytes with the soluble antigen before the reaction with antiserum.

In the presence of complement, red blood cells agglutinated or sensitized with macroglobulin antibody will lyse. Complement (C') consists of at least 9 proteins present in the globulin fraction of fresh serum designated $C'1-C'9$. The presence of these proteins acting in a concerted fashion leads to the lysis of sensitized erythrocytes.

The mechanism by which the various components act, possibly through enzymic action, in lysing sensitized red cells (EA) has been presently formulated as follows:

$$E^+A \rightleftharpoons EA$$
$$EA + C'1 \xrightarrow[Ca^{++}]{} EAC'1a$$
$$EAC'1a + C'4 \xrightarrow[Mg^{++}]{} EAC'1a,4$$
$$EAC'1a4 + C'2 \rightarrow EAC'1a4,2a$$
$$EAC'1a,42a + C'3 \rightarrow EAC'1a,4,2a,3a$$
$$EAC'1a,4,2a,3a + C'5,C'6,C'7 \rightarrow$$
$$EAC'1a,4,2a,3a,5,6,7a$$
$$EAC'1a,4,2a,5,6,7a + C'8,C'9 \rightarrow E^* \text{ (activated cell)}$$
$$E^* \rightarrow \text{ghost}^+ \text{ hemoglobin}$$
$$E = \text{sheep erythrocyte}$$
$$E^* = \text{injured erythrocyte}$$

A = rabbit antibody directed against sheep erythrocyte.

Complement can also interact with soluble antigen-antibody aggregates and cause them to increase in weight, sometimes precipitate, and be more readily phagocytized. The forces involved in the reaction of C' with antigen-antibody systems are also believed to be specific but weak forces. Complement has been implicated in the defense mechanisms of the body as well as in some hypersensitive reactions.

There are two basic biological manifestations of the immune reaction: (a) immunity to infectious agents (b) specific hypersensitivity. Hypersensitivity, or the heightened response to an agent, can be divided into anaphylactic, allergic and bacterial. Anaphylaxis, which can be produced by either active or passive sensitization, is a laboratory tool for studying the fundamental nature of hypersensitivity. The amounts of antigen and antibody involved, as well as the nature and source of the antibody, govern the extent of the reaction. Intravenous injection of antigen into an appropriately sensitized guinea pig usually leads to the release of pharmacologic agents such as histamine from cells and, within minutes, to shock and death. If the animal recovers, it may be refractory to another injection (desensitized). A piece of smooth muscle (intestine or uterus) from the same guinea pig placed in vitro will contract when challenged with antigen (Schultz-Dale reaction). If antibody is injected intradermally and antigen plus Evans blue dye are injected intravenously after a suitable latent period of three hours, the release of histamine alters the permeability of the small venules and blueing of the antibody site occurs within minutes (passive cutaneous anaphylaxis). These three reactions require a latent period for fixation of the appropriate antibody to cells and can occur with either precipitating or non-precipitating antibody. The property of binding to a specific site is related to the "Fc" portion of the H chain of the antibody. Bovine, chicken, horse and sheep antisera do not give good reactions.

When the challenging injection of antigen is given intradermally, within hours (4–24) edema and erythema may ensue, followed by tissue necrosis (Arthus reaction). This reaction requires more antibody (non-skin-fixing,) and antigen, but fixation of complement is not necessary. The lesion is caused by the deposition of antigen-antibody precipitates followed by the invasion of polymorphonuclear leukocytes.

The principal difference between allergy (atopy) and anaphylaxis is that allergy, which in the context presented here applies mostly to humans, may be entirely dependent upon a unique kind of non-precipitating antibody (reagin), which has an ability to fix to cells, is heat labile and does not pass the placental barrier. The most recent classification of this antibody is IgE. When serum from an allergic individual (hay fever) is injected intradermally into a normal person and the specific antigen (allergen) injected into the same site 24 hours later, typical skin reactions (wheal and flare) develop (Prausnitz—Kustner reaction). Immunization of an allergic individual with allergen results in the production of "blocking" antibodies (IgG or IgM), which have the properties of reacting with the allergen and protecting the sensitive individual.

"Cellular or delayed hypersensitivity" phenomena are complex and not as well understood as are "immediate hypersensitivity" reactions. The classical example of this type of hypersensitivity is observed 48–72 hours (delayed) after skin testing a sensitive individual (one who has previously been exposed to tubercle bacillus) with tuberculin or purified protein derivative (PPD) (local erythematous lesions). This delayed hypersensitivity is not transferable with serum. However, if the buffy coat of blood (leucocytes) or lymph node suspension is transferred, the recipient becomes sensitized. Delayed hypersensitivity has also been produced following contact with simple chemical allergens and by injection of small amounts of purified protein, alone, or as an antigen-antibody precipitate with the inclusion of the myco-bacteria in Freund's adjuvant. This transfer of the sensitivity is species-specific in contrast to anaphylactic sensitivity. In man, cell-free extracts of dead human leucocytes from sensitive donors (transfer factor) have also been effective in the above *in vitro* transfer reaction.

Exudate cells (peritoneal, spleen) of sensitive animals are markedly inhibited *in vitro* from migrating out of capillary tubes in the presence of antigen. In contrast, no inhibition occurs when cells are obtained from non-hypersensitive (delayed) animals that are producing circulating antibody to the same antigen. Recent studies indicate that delayed reactions might involve the reaction of sensitized cells containing on their surface antibody-like molecules directed against the specific antigen (migration inhibitory factor).

The homograft reaction, i.e., the sloughing of tissues several days following the transfer within the same species, and the heterograft reaction between two different species involve immunological reactions possibly of the delayed type. (The anamnestic response in this reaction is termed "second set response.") In most situations, antibody against the transferred tissue has not been demonstrated to be the major factor causing rejection whereas leukocytes from the sensitive donor are effective. (Injections of anti-lymphocytic serum can prolong the life of homografts). When animals are injected *in utero* with live splenic cells or other tissues from another adult, and subsequently, in adult life, presented with a graft from the same donor of the cells, the graft will not slough. The phenomenon is termed "actively acquired tolerance." Another form of acquired tolerance (immunological unresponsiveness), or inhibition of antibody formation, can be produced by injections of foreign serum protein into neonatal animals, or following injections of either small (low dose) or very large amounts (high dose) or proteins. This form of tolerance can be prolonged indefinitely by periodic injections of antigen during adult life. When a small amount of pneumococcal polysaccharide is injected into adult mice, they become protected against a subsequent challenge with virulent pneumococci (immunity). However, if a large dose of polysaccharide is injected, the mice do not produce antibody and are not protected ("immunological paralysis").

Ordinarily, immune responses are made only against foreign determinants and materials because of acquired immunological tolerance to autologous proteins. However, there are instances where responses against the host's autologous antigens have been implicated (nervous tissues, lens, uvea, testes, thyroid, adrenal). Autosensitization to these tissues, which leads to various inflammatory reactions, may, among other reasons, be initiated by the breakdown of the normal protective barrier "isolating" the tissue within the host. Although antibody can be detected against the tissue antigens, the autoimmune response resembles the "delayed" and homograft rejection reactions in many respects, including the transferability of the reaction with living cells but not with serum from sensitive donors. Histologically the reaction resembles the tuberculin reaction (accumulation of lymphocytic or histiocyte cells).

Two other groups or diseases in which antibody (gamma globulin) has been implicated as the etiologic agent are (a) acquired hemolytic anemia, thrombocytopenic purpura and lupus erythematosis and (b) rehumatiod arthritis, glomerulonephritis and thyroiditis. Antibody has been suggested because positive precipitation, complement fixation and agglutination tests have been observed as well as the localization of gamma globulin molecules by fluorescent antibody techniques. However, the precise role of these circulating factors in the disease process is neither completely understood nor proven.

PAUL H. MAURER

References

1. Lawrence, H. S., ed., "Cellular and Humoral Aspects of the Hypersensitive States," New York, Hoeber–Harper, 1959.
2. Kabat, E. A. and M. M. Mayer, "Experimental Immunochemistry," 2nd ed, Springfield, Ill., C. Thomas, 1961.
3. Humphrey, J. H. and R. G. White, "Immunology for Students of Medicine," 2nd ed, Oxford, Blackwell, 1964.
4. Dixon, F. J. and J. H. Humphrey, eds., "Advances in Immunology," Vols. 1–8 New York, Academic Press, 1961–1968.
5. Samter, M. and H. L. Alexander, eds., "Immunological Diseases," Boston, Little Brown, 1965.
6. Boyd, Wm. C., "Fundamental of Immunology," 4th ed., New York, Interscience (Wiley), 1967.

IMPULSE CONDUCTION AND TRANSMISSION

Communication is a vital and universal need of every organism from the most complicated and multicellular down to the unicellular. The problem is two-fold. It involves communication along and within continuous portions of the cell and this is termed conduction. It also involves communication across protoplasmic gaps, i.e., from one cell to another, termed transmission. Both processes make use of properties inherent in the protoplasm of all cells.

Among the fundamental properties of protoplasm are conduction as an active process accomplished by the release of stored potential energy, converting it into the kinetic energy of the moving impulse, which is therefore the result of an active process. An equally important passive conduction can be achieved through the process of simple diffusion of chemicals, which then constitute messengers. The specialization of these properties (most developed in nerve cells) has evolved, on the one hand, the self-propagated wave of membrane collapse and restoration which is the conducted nerve, muscle or gland cell impulse, and, on the other hand, the diffusion of the chemicals—secreted by cells—across intercellular gaps, constituting chemical or "humoral" transmission.

The possibilities for conduction are facilitated by the creation and storage of potential energy in the form of a concentration battery produced at the cell membranes by the unequal distribution of ions, particularly of the potassium concentration which is about 27 times higher on the inside than on the outside of the cell and of the sodium, whose concentration gradient is in the opposite direction because its concentration is 10 times higher outside than inside. The concentration of the anion, Cl^- is about 14 times higher outside than in. These uneven distributions result from the differential diffusion through selectively permeable membranes, which thereby become polarized. The potential from this battery can be recorded between an electrode on the outside and another on the inside of the cell, effecting connections to the two sides of the cell membrane and reading the voltage which is known as the membrane or resting potential. The measured potential is less than the value calculated (by the Nernst equation) for the potential that would be developed by a simple concentration battery formed by the unequal distribution of these ions. The reduction of the potential appears to be accomplished by a special back-pumping, against their concentration gradients, of the sodium and potassium ions, which goes on simultaneously with and subtracting from the diffusion determined by the membrane semipermeability. The resultant modified potential is the source of the current that eddies into the point of dropped (or even reversed) potential resulting at the site of stimulation, which experiences a transient loss of its restricted permeability permitting temporary free diffusion of ions and a resulting depolarization. The transient free diffusion of ions across the cell membrane carries the current into the cell. The eddy currents in turn act as stimuli to the surrounding membrane, which is then, in its turn, depolarized and draws current from more regions adjacent to the new stimulus sites. These new currents in turn stimulate more distant regions and so on and so on. Meanwhile, the original sites of stimulation and depolarization are repolarizing or recovering. Fig. 1 illustrates these events in a diagramatic and simplified fashion, showing the change of resting, polarized to active, depolarized region and the influence of the subsequent eddy current in creating a progression of the active region from left to right tailed by region of restored membrane.

The latter is temporarily more stable than normal and therefore resistant to re-stimulation, or refractory, for a short period.

The eddy of currents in this way spreads into a self-propagated wave, which is the conducted impulse whose minute electrical accompaniment can, after suitable amplification, be recorded as the action potential. The growth of electrical charge preceding and culminating in the membrane breakdown, causing the action potential, is evident as the local potential from which the action potential abruptly emerges. The explosive change from a graded disturbance (the accumulation of charge) to the full-fledged self-propagated disturbance, that is the impulse, marks the fact that the latter is a yes or no response or an "all or none phenomenon." It is for this reason that computer analogs of nerve networks are built of binary systems.

The chemistry underlying the conducted impulse is

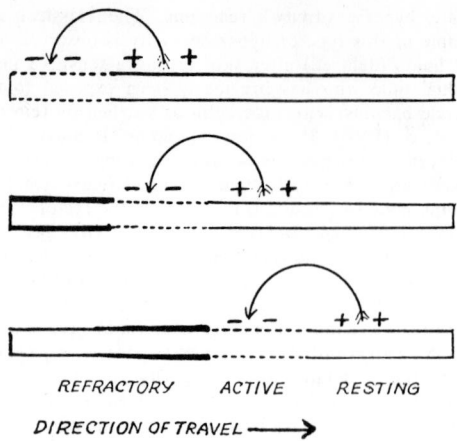

REFRACTORY   ACTIVE   RESTING

DIRECTION OF TRAVEL ⟶

Fig. 1 Spread of active (depolarized) area according to the membrane hypothesis. The local currents cause a repair of the active region and a breakdown of the resting surface beyond.

that of the energy supplying processes which polarize membranes by segregating ions into unequal concentrations on the two sides of the membrane and of the ionic pumps that partially curtail the concentration differences and bring about an observed resting potential at variance with the one expected from calculation. The renewal of the potential energy mechanism, as distinguished from the sodium pumps, which presumably operate steadily, is accomplished intermittently by oxidative restorative processes, which are characteristically delayed so that a so-called oxygen debt is built up to be payed off during periods of diminished or no call for impulse conduction.

The passage of an impulse across an intercellular gap we have defined as transmission. Examples of these are interneuronal or synaptic and neuroeffector or junctional transmission. The eddy currents and electrical fields effects generated here by the arrival of an impulse at the protoplasmic discontinuity, are sufficiently attenuated by the passage through the inert conductor offered by the non-polarized ionic population of the gap (which therefore does not regenerate impulses in the self-propagating manner of a polarized conductor) that they appear generally inadequate *per se* to effect transmission.

A passive conduction mechanism, therefore, becomes an important step in the transmission process at synapses and neuroeffector junctions. This step starts with the liberation by the nerve impulse of chemical messengers from the ends of the nerve (Fig. 2). The messengers diffuse across the microscopic interspace of about 200 angstroms and then exercise their specific effects on the cell membrane of the postsynaptic or postjunctional cells. They activate these, generating within them impulses that are conducted throughout the cells in the same manner that conduction took place in the cells originating the messages. Thus the impulse has been relayed from one cell to another by the intervention of a chemical diffusion step which transmits it across the gap. This describes transmission of an excitatory impulse. An

SYNAPTIC ACTIONS   (EXCITATORY ⇌ INHIBITORY)

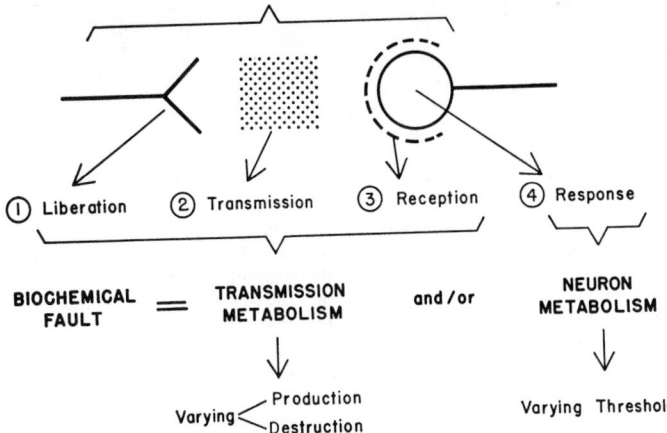

① Liberation   ② Transmission   ③ Reception   ④ Response

BIOCHEMICAL = **TRANSMISSION**   and/or   **NEURON**
FAULT   **METABOLISM**   **METABOLISM**

Varying ⟨Production / Destruction⟩   Varying Threshold   Fig. 2   Potential factors in disturbed synaptic equilibrium.

excitatory messenger chemical that has been identified is acetylcholine; and there may be others.

The influence of some cells upon their neighbors is inhibitory. In such cases inhibitory messenger chemicals have been identified, e.g., noradrenaline, adrenaline, serotonin and histamine (in order of increasing effectiveness); and, again, there may be others. These messengers, like the excitatory ones, combine with discrete and specific areas or receptors occupying a small fraction of the surface of the responding cells. The receptors are key spots, which initiate, according to their nature, an excitatory or depolarizing process, or an inhibitory or hyperpolarizing process. Some regard the excitatory action of acetylcholine in terms of "punching a hole" in the polarized membrane into which ions can freely diffuse and carry the eddy currents that constitute the first segment of the self-propagating mechanism described above.

The chemical transmitter, excitor or inhibitor, is terminated by the chemical breakdown of the transmitter accelerated by enzymes strategically located at the gaps where transmission is taking place. This clears the "switchboard" for further messages.

It seems probable that the simultaneous operation of varying numbers of excitatory and inhibitory fibers, sending messages into the synaptic switchboard, offers the possibility of a finer regulation through a "check and balance" system. An equilibrium between excitatory and inhibitory influences is diagrammed in Fig. 2, which summarizes the synaptic mechanisms and suggests the possible derangement or dysequilibrium that could occur spontaneously in disease or could be induced experimentally. The points where equilibrium is susceptible to change are both the sites vulnerable to disease and the targets for corrective measures or therapy.

The depolarization process at the receptor is recordable as a local potential (synaptic and motor end plate) from which, again abruptly, rises the much larger action potential indicating that the self-propagated process, that will invade the whole membrane, has been initiated. A detailed analysis of suitable electrical records can, therefore, identify the complete sequence of events that have been described as conduction, relay or transmission and post-transmission conduction.

Individual components can be exhibited with greater clarity or can be isolated by, so to speak, dissecting the various steps through the use of appropriate chemicals, which will enhance or block certain steps, and by the use of polarizing and depolarizing currents to abet or to resist the influences of biological polarization phenomena. Study of the phenomena can likewise be carried on at the level of energy storage and consumption by making use of the usual techniques in the analysis of metabolism.

It takes considerably longer to describe the events in impulse conduction than for them to take place. Thus conduction, at its slowest, goes on at the rate of 0.6 meters per second or about 30 yards a minute, and, at its fastest, at a rate of 120 meters per second or better than four miles a minute. It is this "flash-like," split-second speed that makes possible the arrival of impulses from a stubbed toe to the brain and back in time to remove the toe before further injury can take place. This is true despite the relatively longer delays in traversing synaptic switchboards with pauses of half to several microseconds.

The faster impulses are also larger and are conducted in suitably larger and better insulated "cables." These are the large A fibers, which are sheathed with a prominent lipid or myelin layer. Small thinly myelinated C fibers conduct the slow, low amplitude impulses and the intermediate, moderately myelinated B fibers conduct the intermediate size impulses that travel at intermediate speeds. A nerve trunk can conduct the several varieties of waves simultaneously and because of their varying amplitude and speeds, they can be readily, separately identified in the record obtained by "tapping in on the line" with suitable electrodes and amplifying the minute potential changes. The record will display in sequence, the faster and larger waves from the A fibers, the waves of intermediate speed and amplitude from B fibers and lastly the slow, low amplitude waves from C fibers.

The stream of conducted signals is both amplitude and frequency coded to indicate magnitude of environmental change or stimulus. Thus increased signal strength is accomplished by the increase in the number of fibers participating. Within the natural frequency range of particular fibers, a mechano-transducer action of sensory receptors also converts increasing stimulus strength to increasing frequency of impulses generated in the individual fibers of the sensory nerve connected to the receptor.

Although a minor degree of interaction can take place between the electrical fields of adjacent fibers, the neural impulse conduction system is essentially a fairly well insulated system suited to discrete communication. The development of widespread of patterns of activity reflects the functioning of interconnections at the synaptic level. It should not be surprising, since transmission is effected by specific chemicals, to find that, under special circumstances of great stress, a major depot of such chemicals (adrenaline and noradrenaline) located in the medulla of the adrenal gland can pour these synaptic inhibitory chemicals into the blood stream. Distributed in this way to all synapses, the adrenal medullary secretion can supply a cut-off influence limiting the massive discharge of impulses that stress is apt to initiate, and prevent it from becoming so excessive that it is detrimental. In this way the neurohumoral transmission mechanism also affords a means of chemical homeostatic regulation.

The vulnerability of impulse transmission to chemical influences may not always serve homeostasis; for here also is where the action of poisons like mescaline and lysergic acid diethylamide takes place. They produce in man a temporary mental derangement or psychosis. Experiments show that these substances inhibit impulse transmission at cerebral synapses and that tranquilizers prevent this effect. Such experiments in disturbed impulse transmission are developing the basis for understanding of chemically induced psychosis and the manner of action of tranquilizers. A sufficient parallelism appears to exist between the experimental laboratory findings and data in clinical psychosis to suggest that the latter may, also, sometimes be a disturbance of impulse transmission in the brain and that tranquilizers tend to restore synaptic equilibrium.

It seems an inescapable observation that the highly elaborated methods of impulse handling by complex organisms are foreshadowed in the mechanisms found in the unicellular organism or in any single cell.

AMEDEO S. MARRAZZI

Reference

Marrazzi, A. S., "Messengers of the Nervous System," Scientific American **196**: 87, 1957.

INGENHOUSZ, JAN (1730–1799)

This Dutch physician and plant physiologist in 1779 published his oxygen-carbon dioxide experiments clarifying the previous work of Hales and Priestley. Ingenhousz showed that in light, green plants take up carbon dioxide and give off oxygen. However, in the dark, plants, like animals, absorb oxygen and give off carbon dioxide. This experiment was the first demonstration of the function of sunlight in the life processes of green plants. Ingenhousz felt that in the long run, oxygen and carbon dioxide were in balance, neither to be overproduced nor consumed. The technical details

of these processes are only now, a century and a half later, being determined.

DOUGLAS G. MADIGAN

INSECTA

Insecta is a class of antennate ARTHROPODS which possess six legs and fourteen post-cephalic segments and which develop without adding body segments in post-embryonic stages. The term has often been used more broadly to include all hexapodous arthropods but it is employed here in the restricted sense. The hexapod groups thus eliminated as insects are the COLLEMBOLA, which have nine post-cephalic segments; and the Protura, which have fifteen such segments, some being added during post-embryonic development.

The insects comprise the largest class in the animal kingdom, more than three quarters of a million species already being formally described. They have become established in a great variety of environments. Most of them are strictly terrestrial, although many spend at least part of their lives in fresh water. Some breed in hot springs at temperatures as high as 52°C. and others in cold regions where a temperature of 4°C. is their optimum. Although not many insects have become established in the oceans, a few are entirely submarine, passing all of their stages beneath the surface of the sea.

The closest living relatives of the insects appear to be members of the myriopod group, comprising the subclass SYMPHYLA; the true insects probably arose from the ancestral stock of the Symphyla at least as far back as Devonian times. All available evidence indicates that these first insects were wingless and not very different in general appearance from those now included in the order Thysanura. The winged insects, constituting the subclass Pterygota, apparently arose from this Thysanuran-like stock also in Devonian times, although the earliest remains of winged insects are of Upper Carboniferous age. The wings themselves are clearly not homologous with any appendages of other arthropods. The evidence from fossils and from the structure of primitive winged insects now living indicates that wings arose as lateral outgrowths (*paranota*) of the dorsal plates of the thoracic segments. Many palaeozoic insects possessed small membranous lobes on the first thoracic segment (*prothorax*), but functional wings were present only on the meso- and metathorax. The primitive winged insects, comprising the division Palaeoptera, were unable to fold their wings back over the abdomen in resting position. Most of the palaeopterous orders existed only in late Palaeozoic times, but two living orders, Ephemerida and Odonata, are representatives of this ancient stock. From the Palaeoptera there arose, probably in the early Carboniferous, the large division of insects known as the Neoptera, characterized by having a wing articulation which enabled the wings to be folded back over the abdomen when the insect was at rest. The ability to flex the wings in this manner undoubtedly opened up new environments to the adult insects; they were able to inhabit terrestrial areas covered with dense foliage, to hide under stones and logs, and even to burrow underground. The early Neoptera, which were similar to the existing stone-flies (Perlaria), gave rise to a multiplicity of other neopterous types, represented today by such orders as the ORTHOPTERA and the HEMIPTERA. In all of these the immature forms (nymphs) resemble the adults at least superficially, and they inhabit the same general environment as the adults and feed on the same sort of food. Such Neoptera comprise the section Exopterygota. As more specialized derivatives of this early neopterous stock, there evolved the Endopterygota, which include the DIPTERA, LEPIDOPTERA, HYMENOPTERA, COLEOPTERA and related orders. In these the young do not resemble the adults, and they typically occupy very different environments from those occupied by the adults and feed upon different sorts of food. This endopterygote line of the Neoptera is the dominant group of insects at the present time, comprising about 88% of all living species. The time of origin of the Endopterygota is not certain, but the presence of Neuroptera in Upper Carboniferous strata shows that they must have evolved at least by middle Carboniferous time.

Adult insects differ markedly from the other classes of the Mandibulata by their tagmosis. The first six somites of the insect body are fused to form the sharply defined head, which is connected by a short cervix to the next tagma, the thorax. This is composed of three somites, which, although not nearly as completely fused as those of the head, are combined to form a compact unit containing mainly the muscles operating the legs and wings. The remaining segments of the body form the abdomen, which contains most of the viscera. The terminal abdominal segments are often abbreviated or vestigial in adults, although the full complement is developed in the embryo or immature stages.

The insect head has undergone many modifications; however, there are almost always three sets of conspicuous structures present. One of these is the pair of ANTENNAE, corresponding to the first antennae of Crustacea and developed from the first postoral somite. In their simplest form, the antennae consist of a series of tapering segments; modifications of these segments result in serrate, feathered or clubbed antennae, which appear to have developed in connection with specific environments or habits. The antennae are mainly the site of chemical and mechanical receptors and almost always have present, in the second segment (pedicel) from the head, an elaborate sense organ (Johnson's organ), which is highly receptive to such stimuli as vibrations. With the exception of the Entotrophi, true insects do not have intrinsic muscles in antennal segments beyond the pedicel. The eyes are also conspicuous structures on the insect head. The compound eyes, composed of a cluster of cuticular lenses and groups of retinal cells (each visual unit being termed an ommatidium), are the main photoreceptors of insects. They are capable of producing a mosaic image, and, at least in some insects, are sensitive to different colors of light. In addition to compound eyes, most insects possess three simple eyes, or ocelli, each of which consists of a single lens and one group of retinal cells. The ocelli probably do not form images. The mouth-parts (trophi) are usually the most conspicuous structures on the insect's head, apart from the antennae. Of the four mouth-parts generally recognized, three are the homologues of paired appendages present in other arthropods: the mandibles, the maxillae and the labium. The mandibles are typically hard, strong, triangular structures, often with tooth-like projections. They are ordinarily used for cutting or crushing food. In many insects, however, the mandibles are modified to a very different shape, appearing, as in the Hemiptera and some Diptera, as long, slender stylets; in other insects, as in the higher Diptera, they are apparently absent or vestigial. The maxillae are mainly concerned with the manipulation of food as it enters the mouth; each maxilla consists of a basal plate to whch are attached one or two lobes and a segmented palpus; all parts of the maxillae, especially the palpus, have sense organs distributed over their surfaces. In some insects, such as the HEMIPTERA, the maxillae are as markedly modified as the mandibles, forming slender stylets. The labium, although a median structure, is actually formed of the fusion of the two second maxillae as they occur in other arthropods. It apparently functions as the floor of the preoral cavity, with its several processes manipulating the food. As in the case of the maxillae, chemical sense organs are common on its surface. In addition to these parts there are two other structures involved with the mouth that are commonly considered part of the trophi. One of these is the upper lip, or labrum, which is a flat plate hinged to the head directly in front of the base of the mandibles. The other is the hypopharynx, a median structure which arises near the base of the labium and which resembles a tongue-like lobe; in some insects, such as certain of the Diptera, the hypopharynx becomes a piercing stylet.

The thorax is the locomotor center of the insect's body. Functionally, it is a thick-walled box, with numerous internal supports, that houses the muscles operating the locomotor appendages. The first thoracic segment, the prothorax, is usually the smallest in most insects, since it bears only legs. The meso- and metathoracic segments, the second and third respectively, are markedly modified in the Pterygota from the more generalized condition in the Apterygota; the chief differences are the increased sclerotization of the pleural regions and the development of internal struts formed by invaginations of the body wall. The three pairs of legs are primitive walking limbs and consist in most insects of six segments; from the base outward there are the coxa, trochanter, femur, tibia, tarsus and pretarsus. In various insects different pairs, expecially the front pair and the hind pair, have become specialized for particular functions. The front legs are often modified for prehensile purposes by the elongation of the coxae and the formation of spines on the tibiae and femora. The hind legs are often modified for jumping by the elongation of the tibiae and the increase in diameter of the femora. In many aquatic insects, such as the water beetle, all legs may be modified to form swimming appendages. The pretarsus and tarsus of most insects possess glandular structures which secrete substances enabling the insects to adhere to surfaces. Bristles and spines are often also developed in this connection.

The wings are the most remarkable structures which insects possess. Unlike wings of flying vertebrates (birds, bats and extinct flying reptiles), those of insects are not modified appendages that previously existed, but are new structures. Their basic up-and-down movement is caused by the dorsoventral muscles and the dorsal-longitudinal muscles of the thoracic segments.

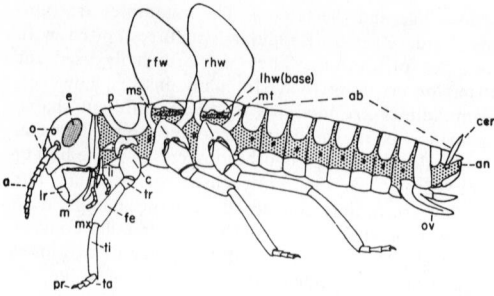

Fig. 1   Lateral view of winged insect (schematic). a, antenna; ab, abdomen; an, anus; c, coxa; cer, cercus; e, compound eye; fe, femur; lhw, left hind wing (base); li, labium; lr, labrum; m, mandible; mo, mesothorax; mt, metathorax; mx maxilla; o, ocelli; ov, ovipositor; p, prothorax; pr, pretarsus; ta, tarsus; ti, tibia; tr, trochanter. (After Grandi)

In addition, moreover, there is a series of special muscles responsible for the individual movements of the wings. Peculiarities of the wing muscles (not clearly understood) enable some insects to make as many as 300 wings strokes per second. The horizontal force which causes the insect to move in flight is produced by the twisting of the wings in the process of movement. In their ontogenetic development, the wings arise as lateral outgrowths of the body wall; in the exopterygote Neoptera, they appear as external pads which gradually increase in size as the insects molt and grow; in the endopterygote Neoptera, the pads develop under the integument during the larval stages and are not exposed on the exterior surface until the pupal stage is reached. The mature wing consists typically of a thin membrane (actually two contiguous membranes) supported by a series of cuticular ribs, termed veins. The general pattern of arrangement of the veins is peculiar (within certain limits) to the various families and orders, the venation thus supplying one convenient basis for insect classification. In some orders, as Orthoptera, the front wings have become thickened, forming leathery covers (*tegmina*); in others, the covers (*elytra*) are even thicker and are convex, as in the COLEOPTERA. In almost all orders of the Pterygota, some species have secondarily lost their wings, becoming completely flightless; in the case of a few orders, mostly parasitic types (as Siphonaptera and Anoplura), all species are wingless.

The insect abdomen is devoid of true appendages except for the eleventh segment, which may bear a pair of long segmented processes termed the cerci. The embryos of many insects do possess a pair of rudimentary appendages on their abdominal segments; some of these are retained in a reduced form in thysanuran adults. Many female insects possess a well-developed external ovipositor, composed of six slender plates which combine to form a tubular process; by this device eggs may be deposited precisely in special locations.

The integument of insects is fundamentally like that of all other arthropods, having an outer cuticular layer secreted by an inner layer of epidermal cells. The body is often covered with a variety of cuticular outgrowths, forming a coat of hairs, spines or scales. The integument usually contains pigments which are responsible for the coloration of the insect's body and wings; in addition, the structure of scales may produce physical colors (i.e., iridescence) by interference, as in the Morpho butterflies.

The internal anatomy of insects differs from that of other arthropods only in respect to a few organic systems. The fore and hind parts of the gut are derived from ectodermal tissue and bear a cuticular lining; the middle part, formed from endodermal tissue, lacks the cuticle. Although partial digestion of food may be extraoral, resulting from the deposition of saliva on food, digestion takes place mainly in the posterior part of the fore gut (*crop*) and in the midgut. Excretion is accomplished chiefly by the Malpighian tubules, which lie in the haemocoel and lead to the anterior end of the hind gut; the urates are eliminated with the feces. The vascular system is like that of other arthropods, consisting of a dorsal heart and tubular aorta; these move the blood anteriorly into the haemocoel, through which it gradually moves posteriorly and back into the heart by lateral openings (ostia). The nervous system, also typical of that of other arthropods, consists of a dorsal brain, composed of three pairs of fused ganglia; the subesophageal ganglion, also composed of three fused small ganglia; and a pair of ventral nerve cords, which fuse at the several thoracic and abdominal ganglia. Associated with the brain are neurosecretory cells which secrete a hormone that activates other endocrine glands, controlling molting and the assumption of the adult form. The reproductive system has undergone an extensive evolution within the insects. In the more primitive forms, the gonads are segmentally arranged, but in most insects the egg tubes are clustered into two ovaries and the sperm tubes into two testes. The paired gonaducts fuse to form a common duct before terminating at the gonapore, on the eighth or ninth abdominal segment. Fertilization is internal, the sperm cells often being transferred in a gelatinous sac (*spermatophore*); they eventually reach the spermathecal sac of the female and are stored there until the time when eggs are laid. Parthenogenesis is common in certain families of insects (such as the Aphidae); in virtually all such cases, so far as is known, normal sexual reproduction takes place at intervals. By far the majority of insects are oviparous, but many, as some aphids and roaches, are viviparous. The life history beyond the egg stage differs in the exopterygote and the endopterygote insects. In the former, as represented by the ORTHOPTERA and HEMIPTERA, the embryo gives rise to a nymphal form which resembles in general features the adult insect, with the exception of size and degree of wing development. After a number of molts, varying from two to more than twenty, the adult stage is reached, the wings developing gradually. In the endopterygote insects the embryo produces a larval form which bears no resemblance to the adult insect. After a series of molts the larva passes into a quiescent pupal stage, during which extensive alteration of tissues takes place; the imaginal form appears after the next molt.

Insects feed on a great variety of foods. Most are phytophagous but many are predacious, preying on small arthropods, chiefly other insects. Still others are true parasites, such as the fleas and sucking lice,

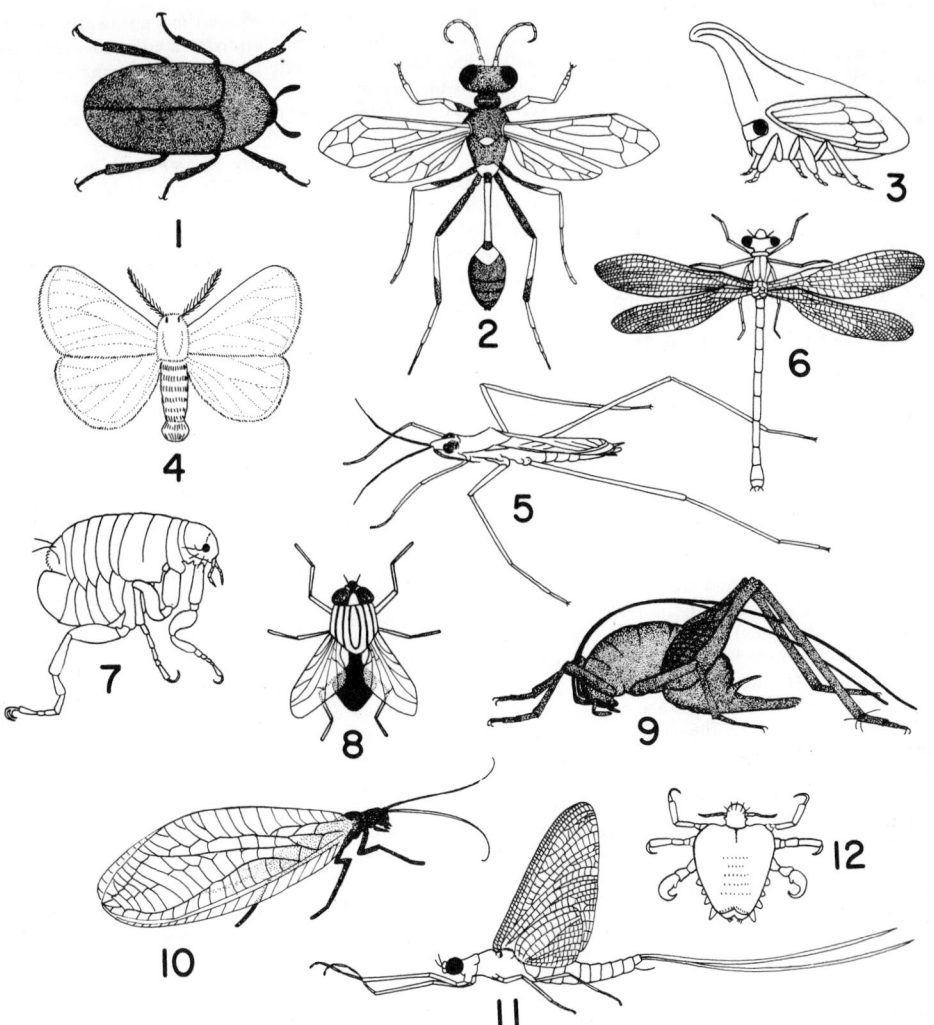

Fig. 2   Some representative insects. 1, Order Coleoptera, carpet beetle; 2, Order Hymenoptera, mud dauber wasp; 3, Order Hemiptera, suborder Homoptera, treehopper; 4, Order Lepidoptera, the brown-tail moth; 5, Order Hemiptera, suborder Heteroptera, water strider; 6, Order Odonata, damselfly; 7, Order Siphonaptera, common flea; 8, Order Diptera, housefly; 9, Order Orthoptera, cave cricket; 10, Order Neuroptera, lacewing; 11, Order Ephemeroptera, mayfly; 12, Order Anoplura, crab louse. (From Goodnight, Goodnight and Gray, "General Zoology," New York, Reinhold, 1964.)

which feed on the blood of mammals. Some of these parasites are of medical importance because of the transmission of pathogenic organisms (as the Protozoa which cause malaria).

The insects are ordinarily divided into two subclasses, the Apterygota and the Pterygota. Within the Apterygota there are two orders recognized: the Entotrophi (Diplura) and the Thysanura. The subclass Pterygota is divided into two major divisions: the Palaeoptera and the Neoptera. The Palaeoptera include two living orders, the Ephemerdia and the Odonata. The Neoptera are divided into two sections, one section the Exopterygota and the other the Endopterygota. The following living orders, each the subject of a separate article, are included in the Exopterygota:

Plecoptera (Perlaria), Blattaria, Orthoptera, Dermaptera, Embiodea, Isoptera, Mallaphaga, Corrodentia, Thysanoptera, Hemiptera and Anoplura. The Endopterygota include the following orders: Neuroptera, Mecoptera, Trichoptera, Lepidoptera, Diptera, Siphonaptera, Coleoptera, Strepsiptera and Hymenoptera.

F. M. CARPENTER

References

Brues, C. T., A. L. Melander and F. M. Carpenter, "Classification of Insects," Bulletin 108, Cambridge, Mass., Museum of Comparative Zoology, 1954.
Imms, A. D., "General Textbook of Entomology," 9th ed., New York, Barnes and Noble, 1964.

INSECTIVORA

Order of mammals consisting of twelve extinct and eight living families, including the solenodons (Solenodontidae) of the West Indies, the tenrecs (Tenrecidae) of Madagascar, the potomogales (Potomogalidae) of western equatorial Africa, the golden moles (Chrysochloridae) of southern Africa, the elephant shrews (Macroscelididae) of Africa, the hedgehogs (Erinaceidae) of Europe, Asia, and Africa, the shrews (Soricidae) of North and South America, Europe, Asia, and Africa, although mainly of the Northern Hemisphere, the moles (Talpidae) confined to the Northern Hemisphere. Generally small, terrestrial, plantigrade, placental mammals with primitive, sharp cusped teeth, a relatively simple brain without convolutions, a long snout, and a primitive skull. The order includes a number of groups of mammals of doubtful affinities, many of the members showing no close relationships except that all are primitive placental mammals. Some eat a wide variety of animal and vegetable foods, but most are insect or other invertebrate feeders, except the potomogales which feed on fish.

The solenodons (or alimiquis) resemble a long-snouted rat about the size of a red squirrel, and are rare inhabitants of Cuba and Haiti. They are mainly nocturnal, live in burrows and caves, and have been nearly exterminated by the introduced mongoose and domestic cats and dogs. The tenrecs are a group of strange insectivores confined to the island of Madagascar. Some are shrew-like in appearance and habits, some are mole-like or rat-like, and others resemble hedgehogs, but none are closely related to any other group of mammals. One, the common tenrec, is the largest of all the insectivores, having a head and body length of twelve to sixteen inches. The potomogales (or otter shrews) look something like a small, long-snouted otter. The flat-sided tail that makes up almost half of the animal's two-foot length is used to propel the animal through the water at a fair rate of speed. The potomogales live in the water-courses of the forested portions of central and west Africa, where they feed on fresh-water crustaceans and small fish. The golden moles are the African counterparts of the true moles of the Northern Hemisphere and behave like them. The golden moles are burrowers with greatly enlarged claws on their forefeet, cylindrical bodies, pointed muzzles, only a trace of a tail, and have a full, soft fur usually with a brilliant metallic luster varying from golden bronze to violet. The golden moles are found only in southern Africa.

The elephant shrews of Africa consist of many species that occupy almost all types of country from the deserts of the Sahara to the forested regions of the Congo and elsewhere. Most species have a body length of four or five inches and a tail of about the same length. The largest species has a body length of less than a foot. Unlike most insectivores, elephant shrews are most active during daylight hours.

Hedgehogs are widespread throughout most of the Old World except Australia and Madagascar. Most are covered with barbless quills and are from four or five inches up to eight or ten inches in length including a short stubby tail. When attacked, the hedgehog rolls up in a ball and tucks its head and feet in so that a sphere of bristling spines is presented to its enemy. In countries where the winters are cold, the hedgehog hibernates. Closely related to the hedgehogs and placed in the same family are the gymnures, small rat-like animals, and the moonrats, of Asia. These lack the spines, and the moonrat is reported to have an odor like an onion.

The true shrews include the smallest mammals in the world, although some African species are almost the size of a house rat. Most are mouse-like with a long, pointed head, minute eyes, and short rounded ears. They feed almost continuously and the smaller ones consume two to three times their own weight of insects and other small animal life every twenty-four hours. The major groups are: the red-toothed or long-tailed shrews (mostly *Sorex*) with many species in the Northern Hemisphere in both the Old and New World; the short-tailed shrews, of which the American short-tailed shrew, common in the eastern United States, has a poisonous bite; and the white-toothed shrews, one of which, the Etruscan shrew of the Mediterranean region of southern Europe, is the smallest mammal, weighing less than two grams, less than the weight of a dime.

The moles are mostly burrowing insectivores with soft thick fur, tiny eyes and ears, and rather large forefeet. Included in this family are the shrew moles, somewhat intermediate in physical traits between the shrews and moles, the desmans, aquatic members of the mole family, and the true moles. The true moles are remarkable diggers capable of excavating over a hundred yards of tunnel in twenty-four hours, although the mole itself is only five or six inches long. Its presence is usually made known by the mounds of earth and ridges of broken earth made by the mole's tunneling. Moles are highly beneficial, feeding on insects and worms, and working the soil. They also have a fine fur that is of importance in the fur trade.

The order Insectivora represents a primitive placental stock and includes the ancestors of all the placental mammals. The oldest forms were contemporary with the dinosaurs of the Age of Reptiles, and many of the modern representatives differ little in mode of life and general appearance from these early forms.

RONALD R. CLOTHIER

INTEGUMENT

The integument may be defined as the covering of the animal body and in this sense includes the outer layer or layers of all organisms, VERTEBRATE and INVERTEBRATE. A more specific term SKIN is used to describe the integument of the vertebrates as such.

The integument is located at the surface of an organism, which may be compared to a frontier at which the organism and its environment are in contact. The modifications of the surface, and hence of the integument, are primarily for purposes of defense and awareness of the outer world.

Even the one-celled animals may possess a strikingly modified surface layer or integument. Thus, while the pellicle, or integument, of PROTOZOA may be homogeneous, it often presents specific patterns consisting of longitudinal or spiral striations, polygonal depressions bounded by ridges, or other types of architecture. In many Protozoa, as in Paramecium, organs of defense, the trichocysts, are developed in the integument.

Among the sponges (see PORIFERA), the surface shows flattened cells which may be spoken of as an epidermis. In many of these animals, however, the epidermis is syncytial.

Among the COELENTERATES, where true tissues first make their appearance, many types of specialization of integument are seen. The epidermis may be syncytial or cellular. Conspicuous and characteristic defensive and offensive structures, the nematocysts, explosive, piercing organs, are present in many parts of the epidermis, particularly on the tentacles and in the oral region.

In a closely related phylum, the CTENOPHORA, peculiar structures occur in the epidermis which are known as lasso cells and which send out long, adhesive threads which serve to entangle the prey.

In the several phyla of "worms" there are many types of integument even among the members of one phylum, the flatworms (see PLATYHELMINTHES). The integument of some types, as of the free living planarians, is soft, coated with mucus, and bearing cilia; while in other forms, as in the trematodes, the integument is covered with a hard cuticle. The roundworms show a hard cuticular covering with many modifications in the forms of spines, tubercles, and other appendages. Among the annelids a thin cuticle overlies the epidermis. This cuticle has been shown to be chemically of albuminoid nature but non-chitinous, i.e., not the same as the covering material of the arthropods.

Other phyla of invertebrates are especially characterized by the manner in which the soft underlying tissues are protected from harm. Thus, in the MOLLUSCS, the integument proper remains soft, covered with mucus, and usually having cilia, while a new structure, the hard shell outside of the integument, serves as the protective mechanism. In the echinoderms, on the other hand, the integument is strengthened by the presence of many firm, calcareous plates which everywhere are overlain by a thin, living epidermis. In this phylum the appendages of the integument are of many and varied types including spines, pincers, and others.

The great phylum of the ARTHROPODA has, as one of its characteristic features, an *exoskeleton*, a very hard integument covering the entire body and becoming softer and more pliable only at the joints between segments of the body or of parts of the appendages. In this phylum the integument serves also as a site of attachment, both origin and insertion, for the striated muscles. Here, unlike the case with the echinoderms, the hardened portion of the skeleton is outside of the epidermis (hypodermis). The hardened, non-cellular portion again is called the cuticle, but here it may be many times thicker than the layer of epidermal cells and may be divided into horizontal layers which vary in their staining capacity and chemical composition.

Among the vertebrates, the integument is called *skin* and consists of two layers, an outer cellular layer, the epidermis, and an inner connective tissue layer, the dermis. With a few exceptions the dermis lacks blood vessels, its nutritive elements coming to it from the dermis. While the beginnings of a dermis and epidermis are seen in the lower chordates, it is first in the cyclostome fishes that stratified epithelium is found composing the epidermis. The epidermis of these fishes and of other fishes differs from that of higher vertebrates particularly in the variety of cells found in this layer, for various types of secretory cells, as well as those with only the covering function, occur in the epidermis in the fishes and also in larval AMPHIBIANS. Such gland cells may be either mucous or serous and form the slimy covering of these animals. Sensory cells may also be present within the epidermis. In some bony fishes there are still remnants of a genuine cuticular layer such as seen among the invertebrates.

Another type of cell also seen in the epidermis of fishes and amphibians is the chromatophore. Such cells may be found both in the epidermis and dermis and lend the varied colors which characterize different species of these animals. Color changes depend to a large extent on changes in shape of such cells.

Important changes occur in the skin of adult amphibians as compared to their larvae and to fishes: first, the superficial layer of epidermis often is cornified, a single layer of cells generally undergoing this change prior to being "shed"; second, the secretory function is usually taken over by special glands which dip down from the epidermis, thus giving to the epidermis proper the rather uniform appearance found in all of the higher vertebrates. The scales of fishes are dermal structures of varying composition, some being composed of bone-like substance, others of materials resembling greatly those found in the teeth. It is thought indeed that the teeth of elasmobranchs, which represent the beginnings of teeth of higher vertebrates, were first developed from scales of the integument.

In REPTILES, in contrast to amphibians, scales are found very frequently in the integument. Here, however, the surface of the skin generally is dry and the epidermis, not needing to be covered with mucus, often is heavily cornified and is involved with the dermis in the formation of the scales. The integument of the birds is marked particularly by the presence of the distinctive feathers, which are chiefly epidermal derivatives. Scales also may be found in the integument of birds, particularly on the legs and feet. The integument of the bird is characterized by a general lack of glands, the only gland which is present being the so-called "preen" gland which resembles the sebaceous glands of a mammal.

In man and other MAMMALS, a covering of hairs is a distinctive feature. These are entirely epidermal in origin. The hairs form a light soft cover, very well adapted to conserve body heat.

The epidermis in mammals, while it varies considerably in different species and on different parts of the body, tends to show a rather constant pattern of differentiation if it reaches a particular degree of thickness. It becomes divided into three general layers, (1) a stratum germinativum, or germinal layer, in which the living and dividing cells are found; (2) a stratum granulosum, a layer in which large granules of keratohyalin, are present, and (3) a stratum corneum, a layer of dead and cornified cells from the surface of which a constant slow desquamation is occurring. From the basal layer of the integument going toward the surface a process of flattening occurs so that while the lower cells are cuboidal or even columnar, the cells above them are polyhedral and beyond that become flattened or squamous. The polyhedral cells often are marked by the presence of *intercellular bridges* which appear like

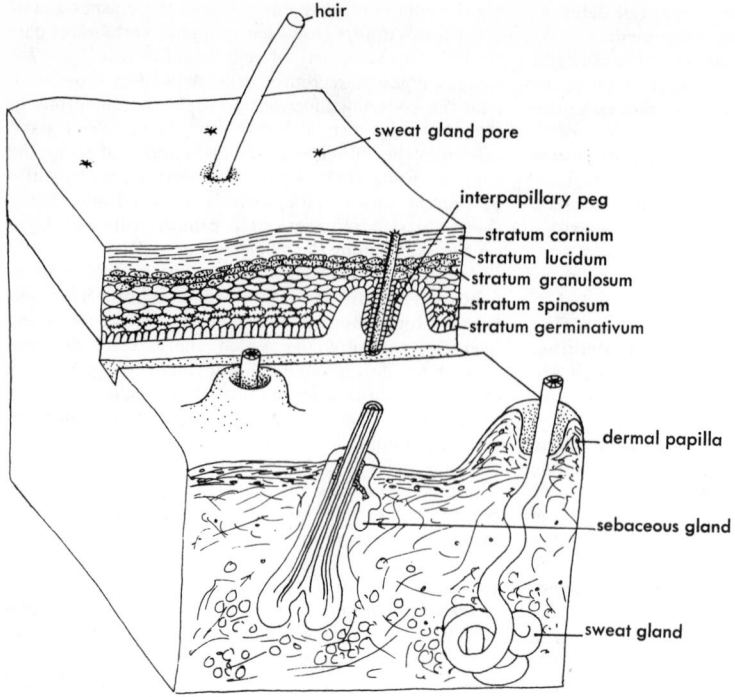

Fig. 1 A stereodiagram of human skin. The outer layer of epidermis is several times thicker than it should be in order to show its layering. The hair follicle is proportionally reduced in order to keep the follicle within the limits of the diagram; the diameter of the hair is actually somewhat greater than the thickness of the epidermis.

radiating spines and give to this layer the name of "the prickle cell layer."

While the normal epidermis of mammals does not present the varied cellular aspect seen in the skin of lower vertebrates, there are present several types of cells in addition to the ordinary epidermal ones. These include melanoblasts, melanocytes (clear cells), Langerhans (aurophilic) cells and, according to some, lymphocytes. Recent studies with the electron microscope are giving new criteria for distinguishing these cell types. The cells of Langerhans, considered at times in the past to represent artifacts, are now considered as "high level clear cells" and have been demonstrated to contain distinctive "granules," elongated bodies with rounded ends and a striated line running down the center.

In the dermis of the mammalian skin there are two types of glands present, namely, the sweat glands which form a secretion, which has a function of temperature control and of some excretion, and the sebaceous glands, which form the sebum, an oily material for lubrication of the hairs and the surface of the skin.

In the dermis, and often extending into the epidermis of mammals, are sensory endings of nerves. These may be in the form of simple naked terminations of fibers, which generally are the type which penetrate the epidermis, or of special types of receptors, which include those for pressure, touch, pain, and temperature. The more specialized sense organs in the integument are characterized by the presence of connective tissue sheaths, by specific forms, and by characteristic patterns of termination of the nerve fibers.

While the integument as viewed throughout the animal kingdom presents a tremendous variety of structure and adaptation, its general functions are seen to be those of protection and sensation.

WARREN ANDREW

References

Andrew, W., "Textbook of Comparative Histology," New York, Oxford University Press, 1959.
Andrew, W., "Some features of fine structure in cells of human dermis and basal layer of epidermis in relation to age," Anat. Rec., 157: 206.
Zelickson, A. S., "Electron Microscopy of Skin and Mucous Membrane," Springfield, Ill., Thomas, 1963.

INTERFERENCE MICROSCOPY

Definition. The interference MICROSCOPE is an instrument for observing transparent structures such as living cells and for making quantitative measurements of optical path differences through them. Since the optical path difference or phase change produced by a biological tissue can be related to its dry mass per unit area, interference microscopy is an important method of quantitative cytology.

Basic principles. The interference microscope was largely developed in an attempt to overcome certain deficiencies in the PHASE CONTRAST MICROSCOPE. In the latter the light waves incident on the object and those diffracted by it are separated and made to traverse unequal paths so that their phase relationships are altered before they recombine. The image of a transparent object can be made to appear with variations in light intensity corresponding to variations in optical path (i.e. the product of refractive index and thickness) through different parts of the object. However, the inability to separate the incident and diffracted waves completely results in the appearance of a bright halo around dark object details, the accentuation of discontinuities and the lack of any simple relationship between optical path and image intensity, so that it is not possible to measure optical path differences. In other interference microscopes interference is produced between the wave transmitted by the object and another wave, which is derived from the same light source but does not pass through the object. These waves add algebraically so that the appearance of the image will depend on their relative phase and amplitude. If the trough of one coincides with the peak of the other, maximum darkness results. If the peaks coincide there will be maximum brightness.

The image intensity can be varied between these extremes if some means of controlling the relative phase of the two beams is provided. Such a calibrated control will also allow the optical path difference through the object to be measured. The simplest method is to set the phase control to produce maximum blackness, first at a clear part of the field near the specimen and then at a point on the specimen itself. The difference between the two settings is proportional to the phase change introduced by the specimen. Variation of the phase control thus makes it possible to set any selected part of the object to maximum blackness, giving variable contrast. If white light is used spectacular colour changes can be produced. Two commercial instruments will now be described.

The Dyson microscope. This was manufactured by Vickers instruments, Ltd., of York, England. Although it is now obsolescent, its principle of operation is instructive. The object O is placed between a slide S and a cover slip (Fig. 1). The slide is in oil immersion con-

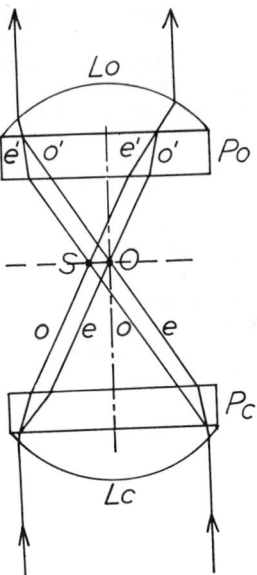

Fig. 2   Smith microscope.

tact below with a glass plate P_1 and above with a plate P_2 cemented to a glass block B. The upper surface of P_1 and both surfaces of P_2 are semi-reflecting. The upper surface of B constitutes a fully reflecting spherical mirror with its center at O. There is a small clear area on the optical axis near I. A small fully reflecting spot is located at R. A typical ray, a, from the condenser is split into two rays b, c; b passes through the object O and takes the path indicated to I; c is reflected downwards at first and then takes a different path to I. Whereas b passes through the object, c passes lateral to it provided that the object is small. These waves can interfere at I, where an image is formed at unit magnification and can be magnified by means of an ordinary microscope objective L. The plates P_1 and P_2 are very slightly wedge shaped so that when P_1 is moved horizontally by the calibrated micrometer screw S_1 the phase difference between the two beams is altered.

The Smith microscope. This is manufactured by Vickers Instruments of York, England, and instruments based on similar principles are available from several manufacturers. It uses double refraction to produce two interfering beams. Two similar doubly refracting plates of calcite, P_c and P_0 are cemented to the front lenses of both the condenser L_c and objective L_0 (Fig. 2). The condenser is illuminated by plane-polarised light. On entering the plate P_c each ray is split into an ordinary ray o and an extraordinary ray e. The e rays are focused on the object O but the o rays come to a focus at S, at some distance from the optical axis. Without going into the optical details, which are somewhat complex, it can be seen that these two sets of rays are recombined by P_0 and pass through the objective together. Interference can occur if these waves are brought to a common plane of polarization by means of an analyzer placed above the objective. It is possible to vary the phase difference between these waves very simply with a birefringent compensator as used in ordinary polarizing microscopes.

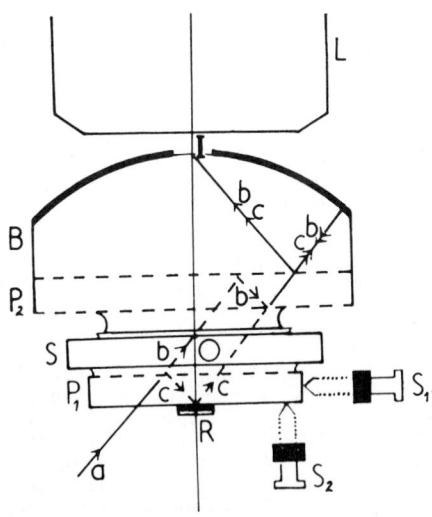

Fig. 1   Dyson microscope

Applications. The interference microscope possesses few advantages over the simpler phase contrast microscope for observational work. The supposed defects of the latter in fact result in increased contrast of intracellular details, whereas the interference contrast image is often disappointingly "flat." The main use of the interference microscope is for quantitative measurements, particularly of dry mass. This depends upon the fact that the refractive index of a solution varies linearly with the concentration C (in gm per 100 ml) thus $n = n_0 + \alpha C$ where n_0 is the refractive index of the solvent and α a characteristic constant for the solute. The average value of α for protoplasmic constituents is .0018. The optical path difference ϕ through any part of a cell is defined by $\phi = (n - n_0)t$, where n and n_0 are now the refractive indices of the cell and of the surrounding medium and t the cell thickness. On combining these relationships we obtain $\phi = \alpha C t$ or $\phi A = \alpha A t C$, where A is the projected cell area. But $ATC/100$ is the product of volume and concentration, i.e. the dry mass M, so that $\phi A = 100\ \alpha M$ or finally $M/A = \alpha/100\alpha = \phi/.18$. Thus the dry mass per unit area at any point is proportional to the optical path difference at that point. The total dry mass of the cell can in principle be obtained by integration over the whole projected area of the cell.

R. BARER

References

Barer, R., "Phase contrast and interference microscopy in cytology," *in* Pollister, A. W., ed., "Physical Techniques in Biological Research," Vol. 3A, 2nd ed., New York, Academic Press, 1966.

Barer, R., "Phase, Interference and Polarizing Microscopy," *in* Mellors, R. C., ed., "Analytical Cytology," 2nd ed., New York, McGraw-Hill, 1959.

Davies, H. G., "Microscope Interferometry," *in* Danielli, J. F., ed., "General Cytochemical Methods," Vol. 1, New York, Academic Press, 1958.

Hale, A. J., "The Interference Microscope in Biological Research," Edinburgh, Livingstone, 1958.

Ross, R. F. K., "Phase Contrast and Interference Microscopy for Cell Biologists," London, Arnold, 1967.

INVERTEBRATES

Invertebrate (=non-vertebrate) is a name given to any of the animals without backbones as contrasted with the vertebrates: FISH, AMPHIBIANS, REPTILES, birds (see AVES), and MAMMALS, all of which have a vertebral column. The invertebrates are grouped into a number of phyla, or major body plans, 30 in the list below, though some zoologists recognize fewer phyla by grouping. Zoologists differ in the number of phyla they recognize primarily because there is no precise definition of a phylum. A phylum is generally thought of as a very old major branch of evolutionary descent leading to a reasonably homogeneous group of animals bearing a unique combination of distinctive characteristics. Thus, the invertebrates do not make up a natural grouping of animals and include all of the animal phyla, even some of the members of the phylum CHORDATA, to which the vertebrates belong. Nevertheless, the term, invertebrate, persists because it distinguishes between animals closely related to man from those not so re-

lated, however much they vary from each other. And this distinction is recognized in the literature.

Following is a list of the commonly recognized phyla, an estimate of the numbers of species, and their common names where there are any. The PROTOZOA, 30,000 species, one-celled organisms or colonies of cells in which there is no differentiation among the cells; PORIFERA, 4,500 species, sponges; MESOZOA, 7 species; COELENTERATA or cnidaria, 9,000 species, the polyps, jellyfishes, anemones, corals, etc.; CTENOPHORA, 90 species, sea gooseberries and sea walnuts; PLATYHELMINTHES, 9,000 species, free-living flatworms, flukes, and tapeworms; NEMERTEA, 650 species, ribbon worms; Nematoda or NEMATA, 10,500 species, thread worms; ROTATORIA, 1,200 species, rotifers; GASTROTRICHA, 150 species, gastrotrichs; GNATHOSTOMULIDA, 45 species, gnathostomulids; KINORHYNCHA, 50 species; kinorhynchs; PRIAPULIDA, 6 species, priapulids; NEMATOMORPHA, 100 species, horsehair worms; ACANTHOCEPHALA, 400 species, spiny-headed worms; ENTOPROCTA, 60 species, entoprocts; CHAETOGNATHA, 30 species, arrow worms; POGONOPHORA, 22 species, beard worms; PHORONIDEA, 15 species, phoronids; ECTOPROCTA, 3,300 species, bryozoans or moss animals; BRACHIOPODA, 250 species, lamps shells or brachiopods; SIPUNCULOIDEA, 250 species, peanut worms; ECHIUROIDEA, 60 species, echiuroids; MOLLUSCA, 40,000 species, chitons, clams, snails, squids and octopuses, etc.; ANNELIDA, 6,000 species, segmented worms: earthworms, leeches, clam worms, etc.; TARDIGRADA, 180 species, bear animalcules; ONYCHOPHORA, 65 species, peripatuses or velvet worms; ARTHROPODA, 65,000 species plus about 750,000 to 850,000 or more insects, non-insect arthropods include spiders, scorpions, ticks, mites, centipedes, millipedes, crustaceans, etc.; ECHINODERMATA, 5,000 species, starfishes, serpent stars, sea urchins, sea cucumbers, and sea lilies; HEMICHORDATA, 100 species, acorn worms; CHORDATA, tunicates, and amphioxuses or sea lancets (1,500 species), and vertebrates (fishes, 20,000 species, amphibia and reptiles, 6,000 species, birds, 9,000 species, and mammals, 3,200 species). All of these numbers of species are approximate and serve merely to show the relative magnitude and diversity of each group.

RALPH BUCHSBAUM

References

Borradaile, L. A. and F. A. Potts, Rev. by G. A. Kercut, "The Invertebrata," Cambridge, The University Press, 1961.

Buchsbaum, R., "Animals Without Backbones," Chicago, The University Press, 1948.

Buchsbaum, R. and L. Milne, "The Lower Animals," New York, Doubleday, 1960.

Chandler, A. C. and C. F. Read, "Introduction to Parasitology," New York, Wiley, 1961.

Edmondson, W. T., ed., Ward, H. B. and G. C. Whipple, "Fresh-water Biology," 2nd ed., New York, Wiley, 1959.

Harmer, S. F. and A. E. Shipley, eds., "The Cambridge Natural History," London, Macmillan, 1895-1909.

Hyman, L. H., "The Invertebrates," Vol. I, 1940. "Protozoa through Ctenophora," Vol. II, 1951. "Platyhelminthes and Rynchocoela," Vol. III, 1951. "Acanthocephala, Aschelminthes, and Entoprocta," Vol. IV, 1955. "Enchinodermata," Vol. V "Smaller Coelomate Groups 1959, Vol. VI "Mollusca I" 1967, New York, McGraw-Hill.

Lancaster, R., ed., "A Treatise on Zoology," London, Black, 1900.

Pennak, R. W., "Fresh-Water Invertebrates of the United States," New York, Ronald, 1953.

Prosser, C. L., and F. A. Brown, "Comparative Animal Physiology," Philadelphia, Saunders, 1961.

Ramsey, J. A., "Physiological Approach to the Lower Animals," Cambridge, The University Press, 1952.

Ricketts, E. F. and J. Calvin, Rev. by J. Hedgepeth, "Between Pacific Tides," Stanford, University Press, 1952.

Yonge, C. M., "The Sea Shore," London, Collins, 1949.

IRON BACTERIA

Iron BACTERIA do not belong to any uniform morphologic or physiologic group of microorganisms. As originally used by Winogradsky (ca. 1888), the term was intended to denote those bacteria that oxidized iron for energy and used carbon dioxide as the sole source of carbon. Later research demonstrated that the bacteria described by Winogradsky obtained energy from organic sources and not through the oxidation of iron. Such microorganisms withdraw iron from their environment and deposit it in the form of hydrated ferric hydroxide, either within or without cellular structures. Though usually referred to as iron bacteria, rightfully, these organisms should be designated as iron-depositing bacteria. Recent research, however, has demonstrated the existence of bacteria which conform to Winogradsky's original premise. These are truly the iron bacteria, inorganic ferrous iron being oxidized to the ferric state for energy and carbon dioxide serving as the sole source of carbon.

Morphologically, the iron-depositing bacteria are placed in these Families: Chlamydobacteriaceae, Crenotrichaceae, and Caulobacteriaceae. One Family, Siderocapsaceae, consists of nine genera of iron-depositing bacteria and the only genus of the true iron bacteria, Ferrobacillus.

The Chlamydobacteriaceae comprise three genera: Sphaerotilus, Leptothrix, and Toxothrix. Positions of the species of these genera are uncertain, and it has been recommended that these genera be combined under the heading Sphaerotilus. The microorganisms are in sheaths often impregnated with ferric and/or manganese oxides. Reproduction is by swarm cells which are motile by means of a tuft of polar flagella. The genera are widely distributed in iron-bearing waters. Some species thrive in large tassels on substrates covered by running water, while others grow in quiescent pools. All species require organic matter for growth and may become noxious in sewage-purification plants and polluted streams.

The Crenotrichaceae Family is composed of two genera, Crenothrix and Clonothrix. Reproduction is by means of non-motile conidia. Sheaths containing iron or manganese oxides are thick and may show false branching. These organisms have never been grown in pure culture or on artificial media. Crenothrix is widespread in nature, being found in stagnant or running waters containing organic matter and iron salts. It is harmless to man, but frequently becomes troublesome in water pipes and water supplies. Often, growths of the microorganism may completely block the flow of water through pipes. Clonothrix was originally described as a blue-green alga, but failure to find pigments prompted placing the organism among the bacteria. Sheaths, encrusted with iron or manganese, may reach a thickness of 20μ, or more. The microorganism is usually found in rivers and streams with gravelly, manganese-bearing bottoms. It may cause technical difficulties in pipelines and treatment plants, and may occur in dark brown masses large enough to be seen readily in tap water.

The Caulobacteriaceae comprise two genera, Gallionella and Siderophacus. The Gallionella consists of round, or kidney-shaped, cells placed at the end of a stalk with the long axis of the cell transverse with the long axis of the stalk. Slender, twisted stalks secreted by the cells are composed of ferric hydroxide. The spiral twist of the stalks is the result of a rotary motion of the cells. The bacterial cells are from 0.5μ wide to 1.5μ long, but stalks 200μ long, or more, may be formed. Siderophacus cells are larger, 0.6μ by 3.0μ, and the stalks are shorter, $15–30\mu$. The stalks are not twisted, but horn-shaped. Both genera are found in iron-bearing waters and may float in irregular flocs.

In older studies, Gallionella stalks were described as the microorganism, the minute cells at the tip having been dislodged or at least overlooked. The stalks are attached separately in great numbers to solid surfaces. Often, tubercles heavily encrusted with ferric compounds are formed in old pipelines. There is some evidence that Gallionella may utilize iron autotrophically.

The Siderocapsaceae Family is divided into ten genera, nine of which are composed of iron-depositing bacteria. These are separated on a morphological basis. The cells are spherical, ellipsoidal, or bacilliform and are embedded in a thick mucilaginous capsule in which iron or manganese may be deposited. The cells do not form sheaths, but are free-living in surface films, or attached to the surface of submerged objects. The microorganisms form deposits of iron or manganese compounds.

One genus, Ferrobacillus (especially the type species, *ferrooxidans*), fulfills Winogradsky's original concept of iron bacteria. These bacteria are bacilliform and are free-living, neither sheaths nor mucilaginous capsules being formed. The microorganisms are chemoautotrophs, requiring only inorganic compounds for complete metabolic processes. Indeed, organic matter is detrimental to the growth of the organism. Iron is oxidized from the ferrous to the ferric state for energy, and carbon dioxide of the atmosphere serves as the sole source of carbon. The cells are actively motile, and are from 0.5μ to 1.0μ in width and from 1.0μ to 1.5μ in length. The microorganisms thrive in acid environments (pH 2.0–4.5), and are indigenous to bituminous coal regions. The high acidities encountered in some bituminous-coal-mine effluents may be attributed, in part, to the activity of this bacterium. In addition, the yellow ferrous iron of waters bearing acid mine drainage is oxidized rapidly to the colorless ferric state, confining the unsightly appearance of such streams to a relatively small area. In this respect, the organism is of distinct benefit in nature.

The Genus Ferrobacillus of the Family Siderocapsaceae includes one species, *F. sulfooxidans*, which oxidizes elemental sulfur as well as ferrous iron.

WILLIAM W. LEATHEN

References

Breed, Robert, S., *et al.*, "Bergey's Manual of Determinative Bacteriology," 7th Ed., Baltimore, Williams and Wilkins, 1957.

Ellis, David, "Iron Bacteria," New York, Stokes, ca. 1913.

Kinsel, N. A., "New Sulfur Oxidizing Iron Bacterium: Ferrobacillus Sulfooxidans Sp. N.," J. Bacteriol., 80: 628–32, 1960.

Leathen, W. W., *et al.*, "*Ferrobacillus ferrooxidans*, A Chemosynthetic Autotrophic Bacterium," J. Bacteriol., 72: 700–4, 1956.

Pringsheim, E. G., "Iron Bacteria," Biological Rev., 24: 200–45, 1949.

Starkey, R. L., "Transformations of Iron by Bacteria in Water," Jour. AM. W. W. A., 37: 963–84, 1945.

Temple, K. L. and A. R. Colmer, "The Autotrophic Oxidation of Iron by a New Bacterium, *Thiobacillus ferrooxidans*," J. Bacteriol., 62: 605–11, 1951.

ISLETS OF LANGERHANS

The islets of Langerhans first described in 1869 by Paul Langerhans are located throughout the parenchyma of the pancreas. The islet volume has been estimated to be one one-hundredth of the pancreas and in the adult man the islets number from 200,000 to 1,800,000 being slightly more concentrated in the tail than in the head or body of the pancreas.

newborn infant, however, the islet tissue consists of approximately an equal number of alpha and beta cells.

The beta cells of the islets produce the hypoglycemic hormone, insulin. Insulin facilitates the utilization of glucose and upon injection results in a decrease of blood glucose concentration with an increase in the formation of the products derived from glucose.

The first stable preparation of insulin was obtained by Banting and Best in 1921. It was later purified in crystalline form by Abel and his colleagues in 1926. Insulin was the first HORMONE to be demonstrated as a protein and has a minimum molecular weight of 6,000. The entire amino acid sequence of insulin has been established by Sanger and can be represented as follows for bovine insulin.

Insulin is composed of two polypeptide chains held together by the disulfide bonds of three cystine residues. One chain (glycyl) has the N-terminal residue of glycine while the other chain yields phenylalanine as the N-terminal amino acid. Insulins from various species (beef, pig, sheep, horse and whale) have been determined and found to have very similar molecular structures differing only in three amino acid residues of position 8, 9 and 10 in the glycyl chain.

It is now generally believed by most workers that the alpha cell is the site of origin of glucagon, the hyperglycemic-glycogenolytic factor. Glucagon stimulates hepatic glycogenolysis by increasing the concentration of active phosphorylase in the liver to give an overall effect of increasing blood sugar concentration.

The islets are composed of internal secretory cells grouped together in irregular rounded masses measuring 120 to 240 microns in diameter. They are delimited from the acini by a thin reticular membrane and are provided with a rich blood supply.

Four types of islet cells have been described: the alpha, beta, gamma and delta cells. The alpha, beta and delta cells are granular while the gamma type are nongranular and it has been suggested that they are precursors of the alpha cells. The delta cells have thus far been seen only in the human (Bloom, 1931) and have not been well defined. The delta cells give a distinct blue color with the Mallory-azan stain which stains the alpha cells red and the beta cells brown-orange. The granules of the alpha cells are insoluble in alcohol while the beta cell granules are alcohol-soluble.

In the adult approximately 20 per cent of the islet cells are represented by the alpha cells while the beta cells constitute the bulk of the remaining cells. In the

Glucagon was obtained in crystalline form from extracts of pancreas by Staub, Sinn and Behrens in 1953. The same workers have completely established the structure of this hormone. It is a straight polypeptide chain composed of 29 amino acids and has a molecular weight of 3482. The following is the structure of glucagon:

R. L. Evans

ISOPODA

The Isopoda comprise an order of the class CRUSTACEA, subclass Malacostraca, division Peracarida. As in other malacostracans, the number of somites in the isopod body and its three subdivisions are standardized, with five coalesced segments fused with the first thoracic somite forming the head (cephalon), seven free somites in the thorax (pereon), and six in the abdomen (pleon). In various groups, the pleonites show varying degrees of fusion with each other and with the terminal telson (not a true somite). Isopods are generally dorsoventrally depressed which distinguishes them from the laterally compressed Amphipoda (a closely related order).

Of eight suborders of Isopoda, only one is terrestrial, the Oniscoidea, represented by the sow bugs, pill-bugs and slaters. The aquatic suborders are Cymothoidea, Gnathidea, Anthuridea, Idotheoidea, Aselloidea, Phreatoicidea and Bopyridea. In size, isopods range from a few millimeters to about 36 centimeters (*Bathynomus giganteus*).

The isopod head bears a pair of sessile compound eyes. Its appendages are two pairs of antennae, one pair of mandibles, two pairs of maxillae, and one pair of maxillipeds (representing the coalesced first thoracic segment). An upper lip (labrum plus clypeus) precedes the mouthparts.

The pereon bears seven pairs of walking legs, all nearly alike, although the first is often modified into a subchelate gnathopod. In females, thin plates (oöstegites) from the bases of several pairs of legs overlap beneath the sternites to form a marsupium.

The pleon bears ventrally five pairs of biramous, lamellar pleopods and one pair of uropods. The pleopods are branchial, sometimes also natatory, or adapted for aerial respiration in the Oniscoidea. The uropods vary among the suborders being lateral, biramous and fan-like (Cymothoidea and Gnathidea); lateral and superior with the outer branch arching over the telson (Anthuridea); laterally attached but folding together, valve-like, under the pleon (Idotheoidea); terminal, biramous and styliform (Aselloidea, Phreatoicidea and Oniscoidea); or terminal and simple, when present (Bopyroidea).

Isopods of economic importance include the marine wood-borers, especially the "gribble" *(Limnoria)*, and the fish parasites (some members of several suborders, notably in the Cymothoidea); crustacean ectoparasites (the Bopyroidea); and the garden pests (e.g. the common *Porcellios* and *Armadillidium vulgare*). On the credit side, they are important links in food chains, and many serve as useful scavengers.

The Oniscoidea doubtless evolved from marine isopods that emigrated from sea to land. Their success in invading terrestrial habitats may be correlated with the degree of specialization of the pleopods for aerial respiration and adaptations for water regulation involving the integument, cutaneous glands, excretory organs, etc. The primitive Ligiidae represent early transitional stages as their essentially branchial pleopods and moisture requirements limit them to damp littoral or riparian habitats. More advanced families, with ramifying pseudotracheae ("white bodies") in the exopodites of two or more pairs of pleopods and better water regulation, can inhabit drier environments. At best, however, terrestrial isopods are im-

perfectly adapted to land life as none can long survive exposure to unsaturated air. All require ecological niches with high microhumidity or available water.

MILTON A. MILLER

References

Edney, E. B., "Woodlice and the land habitat," Biol. Rev., **29**: 185–219, 1954.
Richardson, H., "A monograph on the isopods of North America," Washington, U.S. Nat. Mus., Bull. **54**: i–liii + 727, 1905.
Vandel, A., "Essai sur l'origine, l'evolution et la classification des Oniscoidea (Isopodes terrestres)," Bull. Biol. France et Belgique, Suppl. 30: 1–136, 1943.
Van Name, W. G., "The American land and freshwater isopod Crustacea," New York, Bull. Amer. Mus. Nat. Hist., **71**: iii–vii + 535, 1936.

ISOPTERA

The Isoptera (termites) are morphologically primitive social INSECTS related to roaches, apparently diverging from a common stock prior to, or during, the Mesozoic. All have gradual metamorphosis, chewing mouthparts, and moniliform antennae shorter than the body. The thorax is broadly joined to the abdomen. Winged forms have two pairs of membranous wings of equal length provided with a breakage suture close to the base of each, facilitating shedding after colonizing flights. The family Mastotermitidae differs from other termite families in that members of this group have 5-jointed tarsi, well developed genitalia, and the hind wings have an anal lobe, while the others have 4- or imperfectly 5-jointed tarsi, lack genitalia in both sexes, and hind wings lack anal lobes. Short segmented abdominal cerci are present. The terminal abdominal sternites show sexual dimorphism, with males having 9 distinct sternites, females only 7. Emerson, 1955, recognized 2071 species in 168 genera. All live in colonies composed of different types of individuals with different functions (castes). Males and females are represented in each caste. Two or three fixed castes are present, depending on species:

1) *Reproductives* (1st form or Imagoes and 2nd form or supplementary sexuals); imagoes are pigmented, sclerotized, winged insects with well developed compound eyes and ocelli, phylogenetically the oldest caste and the one showing closest affinities to roaches; secondary sexuals resemble enlarged, more pigmented nymphs

2) *Soldiers*, specialized for defensive functions, with unsclerotized bodies, poorly developed or no wings and eyes, and heavily sclerotized heads with large mandibles (some species), or modifications enabling them to release noxious glandular secretions, or heads so shaped as to serve as plugs, barring entry to nest galleries; usually sterile; and

3) *Workers*, functioning in construction and maintenance of nests, care of eggs, young, and other castes; usually lightly pigmented, unsclerotized, lacking eyes, wings and ability to reproduce.

Termites of the primitive families (Mastotermitidae, Hodotermitidae, Kalotermitidae and probably Rhinotermitidae) lack adult workers, the older nymphs assuming these functions. Caste determination depends

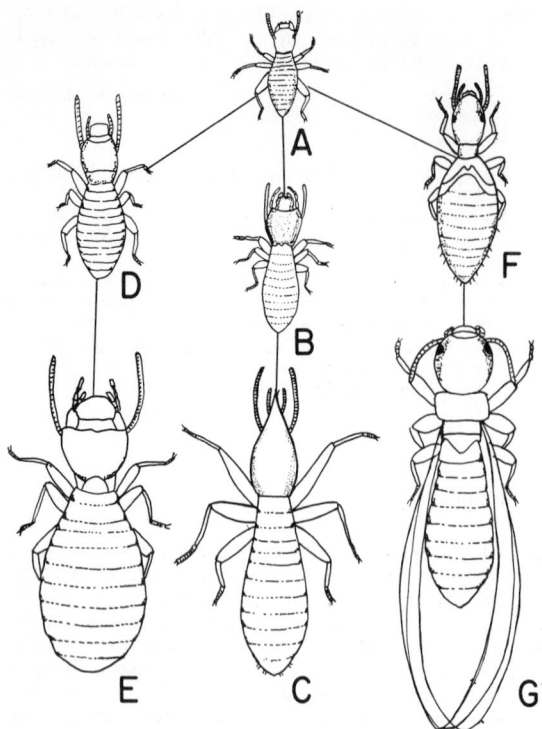

Fig. 1   Castes in a colony of *Nasutitermes*. A, Undifferentiated nymph; B, young soldier; C, adult soldier; D, young worker; E, adult worker; F, young winged reproductive; G, adult winged reproductive. (From Goodnight, Goodnight and Gray, "General Zoology," New York, Reinhold, 1964.)

upon secretion of pheromones, some of which inhibit development of adults of like caste. Induction of development of female reproductives by males has been established by Lüscher. Complex interactions between pheromones, hormones, nutrients and social behavior in caste determination are being recognized.

Colony founding follows swarming of imagoes from large colonies. Flights occur at night (species with light-colored imagoes) or during the day, usually after a rain (dark imagoes). These emerge from nest exits made by workers or nymphs, fly out, alight, pair, and shed the wings. Female behavior invokes a following response from the male. The female moves about until a suitable site for nest excavation is located, the male following closely ("tandem" behavior). Both make the excavation, enclose themselves, and later copulate. After egglaying begins, they share care of eggs and young. The first brood develops into workers and one or more soldiers (species where colony development has been observed). Nymphs cooperate in raising successive groups. The primary reproductives may live for many years, copulating at intervals. Queens of some tropical species become so swollen as to be unable to walk. Rate of oviposition increases in such physogastric queens to an estimated 8,000 per day, maintained for many years. Colonies of specialized genera may contain millions of individuals, but primitive genera have colonies of several hundred or fewer.

Nests may be simple excavations in wood or soil, or may be elaborate raised or buried shelters of characteristic shape and composition. Some subterranean nests are provided with chimneys and pores, probably facilitating ventilation, and those on trees may have rain-shedding devices. Some species build nests of chewed wood fragments, carton. Covered runways are utilized in extending workings from soil to wood, or from one structure to another, in some termite species, while others are capable of diurnal foraging above ground without protection.

Termites feed on cellulose-containing materials. Some species grow fungi in special gardens, eating the hyphae. Primitive termites and the roach *Cryptocercus punctulatus* have an intestinal fauna of flagellated protozoans, some of which effect most of the digestion of cellulose. There are also other types of protozoans in the intestine. All species have a rich intestinal bacterial flora.

Nests of the most specialized termites frequently harbor insects of several other orders, many of which have evolved glandular structures secreting substances consumed by termites. These termitophiles eat eggs or young termites, or are fed by workers, which also groom and carry them about. Seevers, 1957, showed that this association is a longstanding one, and that the two groups have evolved reciprocally.

Enemies of termites include ants, some of which live exclusively on termites, and have been described as important factors shaping the evolution of termite nest-building. Predacious winged insects consume large numbers of imagoes during daytime flights. Toads, lizards, and birds prey extensively on them, and a number of mammals ("anteaters") show remarkable adaptive developments associated with breaking into nests and licking up termites. Man is an agent of destruction, where cellulose-seeking activities of termites take them into fences and buildings, foraging activities conflict with agricultural operations, or mounds interfere with maintenance of airplane runways.

Termites vary considerably with respect to temperature and moisture requirements. Nearly all species are limited to areas south of the 49° F annual isotherm line, and the majority are tropical. Some of the more primitive genera contain species capable of living in dry wood such as furniture inside houses, while subterranean species so far studied are limited to a high and constant environmental moisture supply.

Termites are significant agents in the maintenance of tropical floras, aerating soil and releasing nutrients stored in wood to it. In such communities they occupy an ecological position comparable to that held by earthworms in temperate regions.

MARGARET S. COLLINS

References

Emerson, A. E., "Geographical Origins and Dispersions of Termite Genera," Fieldiana: Zool., Vol. 37, 1955.

Kofoid, C. A., et al., "Termites and Termite Control," 2nd Ed., Berkeley, University of California Press, 1946.

Snyder, Thomas E., "Our Enemy the Termite," Ithaca, New York, Comstock, 1948.

IXODOIDEA

The ticks constitute the super family IXODOIDEA. They are large ACARINA parasitic on the skin of vertebrates and characterized by the presence of spiracles on the third and fourth coxa, and by the chelicerae bearing pseudochelae.

They are grouped in four families; Argasidae, Ixodidae, Spaelaeorhinchidae and Nuttallielidae. The principle parts of the animal are the body (soma), a buccal apparatus (capitulum) and the legs. The form of the body varies. In Argasidae it is elliptical in section with rounded extremities. In the other families it is indented anteriorally to enclose the capitulum. In Ixodidae and Spaelaeorhinchidae the body bears a dorsal shield, which is rudimentary in Nuttallielidae and lacking in the Argasidae. In Ixodidae the regions not covered by the shield show longitudinal and marginal indentations. The anus is located in the ventral region and may show anterior or posterior grooves. In several species there are ventral plates which are classified, in their relation to the anus, as preanal, postanal or adanal. In Argasidae the dorsal and ventral surfaces are separated by a line of suture characteristic of the genus Argas. The capitulum is made up of the rostrum and the palps. The rostrum covers the piercing apparatus which is composed of the curved chelicherae enclosed in a sheath, and the hypostome, a toothed organ which fastens the tick to its host. In all the families the capitulum is visible from above except in the adult of Argasidae in which it is recurved in a dorsal fold (camerostome).

The limbs show the following segments: coxa, trochanter, femur, tibia, protarsus, and tarsus. The tarsus terminates in hooks which are aggregated in the Ixodidae into an adhesive organ (ambulacrum). The hooks may be lacking in adults of the family Spaelaeorhinchidae.

The digestive apparatus commences with the mouth which opens on the dorsal side of the hypostome and which runs under the chelicerae to open into a dilated pharynx which is continued by a narrow esophagus which ends in the stomach. From this point originate the blind intestinal diverticula which show lumbricoid movement and which terminate at the end of the body without connection to the exterior. A pair of salivary glands open into the esophagus.

Ticks are blood-suckers. In order to make the puncture the palps select the zone, then the chelicerae pierce the skin to permit the introduction of the hypostome which is secured in place by the recurved teeth. The continued suction of blood is assured by the anticoagulant saliva. The ingestion of blood takes place slowly in Ixodidae and rapidly in Argasidae. In Ixodidae only the females are blood-sucking; the males attach themselves without eating. In Argasidae blood sucking is common to both sexes. The excretory apparatus does not terminate in an anus but in the excretory vesicle. Respiration is by trachea connected with the spiracles in the coxa.

Reproduction in the majority of species is sexual but, nevertheless, there exist examples of parthenogenesis which in some forms is obligatory and in others facultative at times of overproduction of females.

The voluminous ovary is continued as an oviduct which opens in the genital apertures situated in the two first coxae. Copulation takes place on the ground in Argasidae and on the host in Ixodidae. The male introduces the hypostome into the female genital aperture which is dilated with the chelicerae to facilitate the entry of the spermatophore. The male dies a short time after fertilization. Once fertilization is completed, the ovary grows progressively as the digestive tube atrophies. When the gravid female has reached its largest size, it abandons the host in order to deposit its eggs in cracks in the ground under leaves and stones. In Ixodoidea the female oviposits continually until the time of its death. In Argasidae the process is not continuous and death intervenes after a number of ovipositions. Under the best conditions of temperature and humidity the eggs hatch in 4 to 5 weeks, though under unfavorable conditions this period may be prolonged several months.

A hexapod larva hatches from the egg and seeks a suitable host which it parasitizes. There the larva becomes gorged with blood and drops to the ground from which it passes to a permanent host where it changes into the octopod instar. This instar then seeks a new host and continues in this manner until in the course of 8 changes it reaches the adult form.

The geographical distribution of the Ixodoidea is extremely wide and they are found in extensive areas of the world. Africa is the center of the origin and it possesses the largest number of individuals and genera. Several genera, as Ixodes and Haemophysalis, are of universal distribution. Dermacentor is more common in the Arctic region and Amblyomma in the neotropical. Riphicentor, Riphicephalus, Hyalomma, Aponoma, and Margaropus are typically tropical. Argasidae are of wholly tropical distribution. Nuttallielidae are peculiar to Africa, and Spaelaeorhinchidae to neotropical regions.

The ecology of ticks is dependent on the particular habits of the host. Argasidae nourish themselves sporadically on the blood of many species of vertebrates to which they are very damaging since they are vectors for numerous microorganisms. Among the genera which comprise this family Argas prefers birds, Otobius the ears of domestic mammals, Antricola prefers bats while Ornithodores attacks in turn reptiles, birds and mammals.

Ixodidae are well adapted to the parasitic life and live permanently on their hosts save when they abandon them during the reproduction cycle. The species of the genus Boophilus pass the whole of this cycle on the same host, a phenomenon also observed in Margaropus and some Dermacentor. *Riphicephalus bursa* passes the larvae and nymphal stages in one host but transfers as an adult to another. Haemophysalis, Amblyomma and Ixodes alternate their developmental stages between different hosts.

From the epidemiological point of view those species which live permanently on one host cannot spread disease in the course of their development. Those which pass from host to host are the most likely cause of the spread of infections and this in direct relation to the number of hosts utilized.

The exclusive adaptation to a single host is relative and only appears to exist as a predilection. It is thus that the Ixodidae can adapt their capacities to new hosts. Boophilus, which was imported with cattle, has adapted itself in South America to deer and indigenous marsupials. Amblyomma and Dermacentor, normally parasites on forest animals, have transferred to domestic animals. Riphicephalus, which normally parasitizes dogs has often been found on cattle.

Spaelaerhinchidae, very little known forms, are cavernicolous and found in the ears of bats. Practically nothing is known of the parasitic activities of the Nuttallielidae.

The economic and medical importance of ticks derives both from their parasitic activities and from their transmission of pathogenic microorganisms between man and animals. In regard to the first, massive parasitism produces anemia, causes local irritation with toxic and allergic symptoms, which include what has been described as a neurointoxication of the body known as "tick paralysis."

With the regard to diseases transmitted, the members of the family Ixodidae transmit Rickettsiasis and Piroplasmosis which are of extreme importance in vast areas of the world. The Argasidae for their part, are responsible for considerable areas of endemic Borrelias (recurrent fever). These diseases are maintained within the Ixodoidea and the mechanism of infection is congenital through the eggs. It has also been observed that copulation and even canabalism can be factors which maintain microorganisms in populations of ticks. Other parasites like *Hepatozoon canis, Pasteurella tularensis,* and the larval stages of helminths, are also transmitted by ticks. A wide variety of transmissions has been demonstrated experimentally.

R. DONOSO BARROS
(translated from Spanish)

References

Cooley, R. A. and Kohls, G. M., "The Argasidae of North America, Central America, and Cuba," Notre Dame, Ind., The University Press, 1944.

Boero, J. J., "Las garrapata de la Republica Argentina," Buenos Aires, Imprenta Universitans, 1957.

j

JOHANNSEN, WILHELM LUDWIG (1857–1927)

This Danish botanist, geneticist, and creator of the terms, "gene," "genotype," "phenotype," and "pure line" was born at Copenhagen, Denmark on February 3, 1857. Johannsen's life was greatly influenced by his father, a Danish army officer characterized by a fondness for order and punctuality, and by his mother who cherished a strong interest in both plants and animals. Johannsen himself felt that his father's sense for concrete objects and fine nuances, combined with his mother's appreciation for nature was the inheritance responsible for his becoming a scientist. He began his education at one of the finest schools in Copenhagen but, in 1868, the Johannsen family moved to Helsingoer, a small but cosmopolitan town, near the Danish coast. This cosmopolitan atmosphere was responsible for Johannsen's proficiency in languages. He passed the preliminary examination which would have allowed him to enter the University of Copenhagen, but his father's modest salary and the fact that his brother was studying engineering, prevented the realization of this plan, and he became an apprentice apothecary. After spending a year in Germany as a pharmaceutical assistant, he returned to Denmark, in 1879, and, within one year, completed his pharmaceutical examination with high honors. With the aid of a small inheritance, he was able, for a short time, to continue his studies, especially in botany and in analytical chemistry. One of his teachers at the University of Copenhagen was the famous botanist, Eugenius Warming. Johannsen was then appointed, in 1881, as assistant in the chemistry department of the newly founded Carlsberg Laboratorium in Copenhagen. The head of this department was the brilliant chemist, Johan Kjehdahl. During his six years at the Laboratorium, Johannsen studied, among other things, the anatomy of barley and wheat grains. It was at Carlsberg that Johannsen first began his work on variability in barley and on dormancy in plants. In 1892, at the age of thirty-five and still with no university degree, Johannsen was appointed lecturer in botany and plant physiology at the Royal Veterinary and Agricultural College at Copenhagen. It was here that he first developed a method for waking plants from their normal winter dormancy by etherization. Johannsen's interest then turned to the problem of variation in self-fertilized plants, especially barley (*Hordeum vulgare*) and the Princess Bean (*Phaseolus vulgaris*). He isolated and maintained strains of these organisms and studied the effects of variation and selection, and, in 1903, created the term "pure line." It was also during this year that his classical work, "On Inheritance in Populations and Pure Lines," appeared. In this work he formulated a new interpretation of Galton's Law of Regression. In 1905, he was appointed Professor Ordinarius in plant physiology at the University of Copenhagen; in 1910, he was named doctor honoris causa in medicine, and in 1917, became Rector of the University. The last few decades of his life were spent in the fruitful capacity of both author and critic and his esteem and renown increased with the years. In 1909, Johannsen's greatest work, "Elemente der exakten Erblichkeitslehre" was published and it has enjoyed several subsequent editions. At first Johannsen was opposed to the chromosomal theory of heredity but eventually became reconciled to the usage of the term "gene" to designate a concrete, localized unit of inheritance. Wilhelm Ludwig Johannsen, pioneer in genetics and botany, died on November 11, 1927 at Copenhagen, Denmark.

MARTIN J. NATHAN

JOINT

In everyday language the term joint means a place at which two things are joined together. In anatomical usage, a joint may be described as the connection between any of the rigid component parts of the skeleton, whether bones or cartilages. Joints vary widely in structure and arrangement and are often specialized for particular functions. However, they have certain common structural and functional features. On the basis of their most characteristic structural features, they may be classified into three main types: fibrous, cartilaginous, and synovial.

Fibrous joints. The bones of fibrous joints (sometimes called *synarthroses*) are united by fibrous tissue. *Sutures,* which are tightly united, and *syndesmoses,* which are more loosely united, comprise two types of fibrous joints. The joint between a tooth and the bone of its socket is termed a *gomphosis* and is sometimes classed as a third type of fibrous joint. Little if any movement occurs at fibrous joints.

Cartilaginous joints. The bones of cartilaginous joints (sometimes called *synchondroses*) are united either by hyaline cartilage or by fibrocartilage.

Hyaline cartilage joints. These may be called primary cartilaginous joints, owing to the fact that the hyaline cartilage that joins the bones represents a persistence of a part of the embryonic cartilaginous skeleton. Little if any movement occurs at such joints. The cartilage serves as a growth zone for one or both of the bones that it joins. Most hyaline cartilaginous joints are obliterated, that is, are replaced by bone when growth ceases. An epiphyseal plate of a long bone is an example of a hyaline cartilaginous joint.

Fibrocartilaginous joints. These joints are sometimes called secondary cartilaginous joints and are sometimes called *amphiarthroses*. The skeletal elements are united by fibrocartilage during some phase of their existence. The fibrocartilage is usually separated from the bones by thin plates of hyaline cartilage. The intervertebral discs between the bodies of the vertebrae are examples of fibro-cartilaginous joints. The discs act as resilient pads, and a series of such discs permits a rather considerable range of movement in a vertebral column.

Synovial joints. Synovial joints, which are often termed *diarthrodial* joints, possess a cavity and are specialized to permit more or less free movement. The articular surfaces of their component bones are covered with cartilage, which is either hyaline or fibrous in type. The bones are united by a fibrous *capsule* and by *ligaments*. The inner surface of the capsule is lined by a vascular connective tissue, the *synovial membrane*, which produces the synovial fluid that fills the joint cavity and lubricates the joint. The cavity is sometimes partially or completely subdivided by fibrous or fibrocartilaginous *discs* or *menisci*.

Synovial joints are classified according to the shapes of the articular surfaces of the constituent bones. These shapes determine the types of movement and are partly responsible for determining the range of movement. The most common types of synovial joints are plane, hinge, and condyloid. Less common are ball-and-socket, pivot, and saddle.

The movements that occur or can be produced at joints include (1) gliding or slipping movements, (2) angular movements about a side-to-side axis (flexion and extension), or about an anteroposterior or ventrodorsal axis (abduction and adduction), and (3) rotary movements about a longitudinal axis (medial and lateral rotation). Whether one, several, or all of the movements occur at a particular joint depends upon the shape and ligamentous arrangement of that joint. The range of movement is limited by muscles, by ligaments and capsule, and by the shapes of the bones.

The lubricating mechanisms of synovial joints are such that the effects of friction during movement are minimized. Were this not the case, articular cartilage would be quickly worn away. Lubrication appears to be most efficient when the bearing surfaces are incongruous, when a viscous lubricating fluid is present, and when the articular surfaces are moving at certain rates of speed with respect to each other.

Synovial fluid, which is particularly important in lubrication, is produced by synovial membrane. This connective tissue lines the inner surface of the capsule but does not cover articular cartilage. Its most characteristic morphological feature is a capillary network adjacent to the joint cavity. Synovial fluid is a sticky fluid, much like egg white in consistency. Its viscosity is due to the presence of a nonsulfated mucopolysaccharide known as hyaluronic acid.

Adult articular cartilage is a resilient and elastic tissue, which serves as a bearing surface for synovial joints. It is, however, an avascular, nerveless, and relatively acellular tissue which has lost most of its power of growth and repair. Its nutrition seems to depend upon synovial fluid and upon diffusion from capillaries in underlying bone and at the periphery of the joint.

In most joints, the capsule and ligaments are composed of bundles of collagenous fibers. The capsule and ligaments serve obvious mechanical functions. They also contain sensory nerve endings which are very sensitive to position and movement and which are chiefly responsible for the peripheral contribution to the kinesthetic sense. In addition, the endings and their central connections are important in the reflex control of posture and locomotion. The capsule and ligaments are also plentifully supplied with pain endings, and the blood vessels in joints are supplied with sympathetic fibers.

ERNEST D. GARDNER

References

Gardner, E., "Physiology of Movable Joints," Physiological Reviews, **30**: 127, 1950.
Gardner, E., Chapter 3 *in* "Anatomy," by Gardner, Gray, and O'Rahilly, Saunders, Philadelphia, 1963.

JUGLANDALES

A DICOTYLEDONOUS order consisting of two families of important temperate or subtropical trees with leaves mostly deciduous, alternate or rarely opposite, pinnately compound, resin-dotted (*lepidote*), aromatic; the inconspicuous FLOWERS mostly unisexual in clustered or solitary catkins or elongate spikes, apetalous, four-sepalled (or asepalous); filaments short, anthers longitudinally dehiscent; pistil bicarpellate, with two styles or style branches, the ovary incompletely one-celled, but only one-ovuled; FRUIT one-seeded, the seed lacking endosperm with straight embryo.

The principal family, Juglandaceae, with six genera and sixty species produces valuable wood and nuts, especially the Persian, Circassian or English walnut (*Juglans regia*), black walnut (*J. nigra*), pecan (*Carya illinoensis*), and shag-bark hickory (*C. ovata*). The leaves are estipulate; the trees monoecious or rarely dioecious; the flowers, one to a bract, have the primary bract and two secondary bracts partly or completely fused with the ovary or staminate receptacle; the stamens are 3–100; the ovary is one-loculed, but 2- to 4- (to 8-) celled at base, inferior; the ovule is orthotropous, erect at top of the incomplete partition (placentation modified axile, but superficially appearing basal), and has one integument; the fruit is a nut (Juglans, Carya, Alfaroa) or a two- to three-winged nut or nutlet (Pterocarya, Platycarya, Engelhardia; the nut is enclosed in an adherent dehiscent or indehiscent leathery or fibrous husk, the whole resembling a drupe but with the husk ("exocarp") developed from the involucre and perianth not from the pericarp; the wings of the winged fruits are also derived from the adherent involucre (in *Alfaroa* true husk and wings are absent, the involucre being minute); cotyledons are four-lobed, epigaeous or hypogaeous in germination. In primitive members, *Alfaroa* and *Engelhardia chrysolepis,* the pistillate and staminate catkins are combined into a terminal panicle, the flowers having sepals; in the most advanced genus, *Carya,* the pistillate catkin is reduced to a few-flowered terminal spike, and the staminate catkins are lateral, the flowers typically lacking sepals. *Juglans,* walnut or nogal, occurring from eastern Canada to Argentina, in West Indies, and in eastern Asia, has 20 species; *Carya,* hickory and pecan, occurring from eastern Canada to eastern Mexico and in eastern Asia has 19 species; *Pterocarya* has 6 species in Caucasian mountains and eastern Asia; *Platycarya* has 1 species in eastern Asia; *Engelhardia* has 9 species in Mexico, Central America, eastern Asia and Malay Archipelago; *Alfaroa* has 5 species in Mexico, Central America and Colombia.

In the probably ancestral Rhoipteleaceae, with one species of *Rhoiptelea* from China, stipules are present; the flowers, arranged in a terminal panicle of "spikes," are three to a primary bract, the central flower perfect with six stamens, the lateral two pistillate but abortive; the ovary is completely two-celled, superior; the ovule in the fertile locule is half anatropous, unlobed, and has two integuments; the small fruit is two-winged because of outgrowths from the ovary wall.

The order is considered by many botanists primitive and related to families of the Amentiferae, by others advanced with reduced flowers and related to the Sapindales. Thorne reduces the order to a suborder, Juglandineae, under Rutales (Sapindales).

WAYNE E. MANNING

References

Manning, W. E., "The morphology of the flowers of the Juglandaceae. I The inflorescence," Amer. J. Bot. 25:407–419, 1938.

ibid., "The morphology of the flowers of the Juglandaceae. II The pistillate flowers and fruit," Amer. J. Bot. 27:839–852, 1940.

ibid., "The morphology of the flowers of the Juglandaceae. III The staminate flowers," Amer. J. Bot. 35:606–621, 1948.

LeRoy, J. F., "Etude sur les Juglandaceae," Memoires Mus. Nat. Hist. Nat., Nov. Ser., Serie B, Bot., 6: 1–246, 1955.

Thorne, R. F., "Synopsis of a putatively phylogenetic classification of the flowering plants," Aliso, 6: 57–66, 1968.

Whitehead, D. R., "Pollen morphology in the Juglandaceae. II. Survey of the Family," Jour. Arnold Arb., 46: 369–410, 1965.

JUSSIEU, ANTOINE DE (1686–1758)

JUSSIEU, BERNARD DE (1699–1777)

JUSSIEU, JOSEPH DE (1704–1779)

These three brothers, and their descendants noted below, were outstanding among French botanists of the 18th and 19th centuries. All three were born at Lyons to Christoph de Jussieu, who had already acquired some repute in the scholarly world by his contributions to contemporary pharmacy. All three brothers studied medicine at the University of Montpellier but subsequently devoted most of their writing to the then young science of botany. The elder brother became director of the Jardin des Plantes in Paris in 1708 but continued in the practice of medicine, being known for his devotion to the cause of the poor. He brought out a new edition of his predecessor, de Tournefort's "Institutiones rea Herbariae" and edited various other works of botany.

The second brother early abandoned the practice of medicine and became an assistant to his brother in Paris in 1722. He also edited new editions of de Tournefort's works but his principal contribution was probably his discovery that the hydrozoan Coelenterata and their relatives belonged indeed in the animal kingdom and not, as was firmly held up to his time, in the plant kingdom.

The youngest brother was more attracted to exploration than to botany and took part in geographical expeditions to South America. He was, however, a diligent collector of plant seeds and sent to his two elder brothers many South American plants not previously cultivated in Europe.

JUSSIEU, ANTOINE LAURENT DE (1748–1836)

This de Jussieu was the nephew of the three preceding scholars who trained him early botany and medicine. His remarkable brain enabled him to take every advantage of this opportunity and in 1789 published his "Genera Plantarum Secundum Ordines Naturales Disposita, Juxta Methodum in Horto Regio Parisiensi Exartum," which expanded and amplified a system of classification proposed by his uncle, Bernard and which is, in point of fact, the basis of modern plant classification. In spite of the royal favors shown to his uncles, he survived the revolution, the authors of which had so high an opinion of his scholarship and integrity that they placed him in charge of the whole hospital system of Paris. In 1793 he was charged with the task of organizing the Museum of Natural History which he set up on its present footing, and for which he was able to salvage a vast library from those of the monasteries and convents broken up by the revolutionaries.

JUSSIEU, ADRIEN LAURENT HENRI DE (1797–1853)

This was the son of the subject of the last entry and is frequently confused with him. His connection with medicine was rather nominal for the thesis with which he secured his doctorate in this field was on the classification of the Euphorbiaceae. He published many monographs on plant families. His text book of botany received universal acceptance and was translated into most of the principal languages of Europe.

K

KINORHYNCHA

The Kinorhyncha (Echinodera) is a class of invertebrates belonging to the phylum NEMATHELMINTHES
(Aschelminthes). Their nearest relatives are according
to Lang the Priapulida and Acanthocephala. They have
a cuticular body surface, a pseudocoelomate body, a
straight digestive tract with the mouth at the anterior
and the anus at or near the posterior end. The pharynx
is highly differentiated. Specific characters are the
superficial segmentation into 13 (14) *zonites* (= segments), the segmentally arranged musculature and
nervous system and the protrusible mouth cone
(= head) covered with circlets of *scalids* (= strong
spines).

The adult Kinorhyncha are less than 1 mm in length,
yellowish or brownish in color, sometimes with eye-
spots on the brain. They cannot swim but move on the
substratum by alternated protrusion and withdrawing
of the head. Cuticular plates of varying structure in the
main groups of the Kinorhyncha close over the retracted head (figures). The cuticular body wall bears
spines and hairs in varying arrangements and is in the
main part of the body divided into an arched *tergal* (=
dorsal) plate and two flattened *sternal* (= ventral) plates.
A pair of adhesive tubes occurs in front part of the
ventral surface.

The musculature is split up into fibers, arranged in
longitudinal, diagonal and dorsoventral groups. In the
head there are also some circular fibers. All muscle
fibers are nucleated and most of them are cross-striated.

A mid-ventral cord with a ganglion in each zonite
springs from the circular brain.

The main parts of the digestive tract are the buccal
cavity, the pharynx, the oesophagus with salivary
glands, the intestine and the end gut. The epithelium of
all these parts lacks cilia. A circulatory system and a
blood fluid are lacking.

The sexes are separate, generally with only small
differences. The gonads are a pair of elongate sacs
opening laterally on the thirteenth zonite. The fertilization is internal. The adult develops through different
juvenile stages separated by molts, in which the entire
cuticle is shed.

The Kinorhyncha have been taken in mud and sand
and among algae, most of them on the European coasts.

Highly specialized Kinorhyncha have been found in
littoral subsoil water (cf. Higgins, 1968). Blake, Chitwood and Higgins have studied the N. American species of Kinorhyncha.

TOR G. KARLING

References

Zelinka, C., "Monographie der Echinodera," Leipzig,
Engelmann, 1928.
Nyholm, K. G., "Studies in the Echinoderida," Arkiv
f. zool. **39**, 1947.
Lang, Karl, "Die Entwicklung des Eies von Priapulus
caudatus," Arkiv f. zool. **5,** 1953.

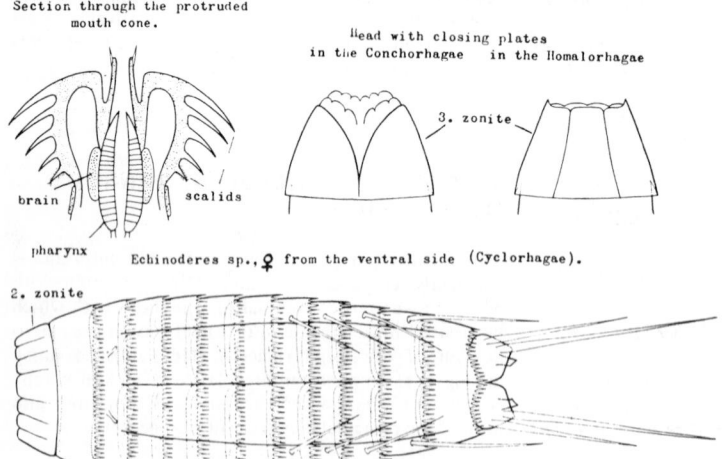

Section through the protruded mouth cone.

Head with closing plates
in the Conchorhagae   in the Homalorhagae

3. zonite

brain   scalids

pharynx

Echinoderes sp., ♀ from the ventral side (Cyclorhagae).

2. zonite

Fig. 1

Higgins, Robert P., "The Homalorhagid Kinorhyncha of Northeastern U.S.," Trans. Amer. Microsc. Soc. **84**, 1965.

ibid., "Taxonomy and postembryonic development of the Cryptorhagae, a new suborder for the mesopsammic kinorhynch genus Cateria," Trans. Amer. Microsc. Soc. **87**, 1968.

KOCH, ROBERT (1843–1910)

As a medical student at the University of Gottingen Koch's ambition was to become an explorer. After graduating from medical college, in 1866 at the age of twenty-three, Koch married Emmy Fraatz who insisted he give up his dreams and become a respectable practitioner. In 1872 Koch was appointed district physician of the small east Prussian town of Wollstein. A common disease of both people and livestock in this rural area was anthrax. Koch, carefully applying what was later to become familiar to every student of microbiology as Koch's postulates, was able to demonstrate that anthrax was caused by a bacillus. This was the first time that a specific microorganism had been shown to be the etiological agent for a specific disease. The hitherto unknown physician became prominent and in 1880, four years after publishing his paper on anthrax, he was invited by the German government to Berlin and made an Associate of the Imperial Health Office. During his early years in Berlin, Koch developed a technique for isolating bacteria in pure culture and the use of solid media for culturing bacteria, two methods which are still in use today. These discoveries of Koch's enabled his contemporaries to uncover the causative organisms of many diseases which, heretofore, had remained obscure in mixed cultures. On March 24, 1882, Koch announced, before the Physiological Society in Berlin that he had isolated the organism causing tuberculosis. In 1885, he was appointed Director of the Hygienic Institute at the University of Berlin. While at this post, Koch traveled widely. He went to India to investigate the cholera epidemics and was able to show cholera was caused by the comma bacillus and transmitted by ingestion of contaminated food or water. During other expeditions he contributed to the etiological origin of such diseases as bubonic plague and malaria. During this time he also developed a substance effective against tuberculosis, (although later he discovered it was not as successful as he had hoped), and announced his findings in 1890 at the Tenth International Medical Congress in Berlin. For this and his research with the tubercle bacilli he received the Nobel Prize in 1905. In 1891 he resigned from the Hygienic Institute and became head of the Institute for Infectious Disorders. He held this post until his retirement in 1904. He died on May 21, 1910. Koch must also be noted for his improvements on staining methods which enabled more accurate descriptions of the bacteria he observed and his discovery that steam was a more effective destructive agent of organisms and spores than the chemicals used at that time. But his most important contribution was that of establishing set principles which conclusively proved certain organisms were the cause of a particular disease.

RICHARD M. CRIBBS

KOELREUTER, JOSEPH GOTTLIEB (1733–1806)

This German botanist, had he been as acute an observer as Mendel, might well have preceded the latter in the establishment of the laws of inheritance. He conducted many experiments on the cross fertilization of plants and noticed most carefully the relation of the characters of the hybrids to the parent. Moreover, by crossing hybrids he noticed the reversion to the parent type but did not detect—or indeed apparently look for—any mathematical relationship in his experiments. His careful investigations into the mechanics of fertilization in plants established for the first time the necessity of insects to many of them but his work was, in general, very vague and mixed up with mystical and alchemical speculations. He believed, for example, that the actual fertilization of plants was due to the mixing of the oily secretions he thought he observed on pollen with the secretion he observed on the pistils. He was completely ignored in his day and his work was only rediscovered after the publication of Mendel's results had raised interest in the hybridization of plants.

KÖLLIKER, RUDOLPH ALBERT (1817–1905)

This Swiss zoologist, born in Zurich, might well be called the father of modern histology. His "Mikroskopisch Anatomie und Gewebelehre des Menschen," published in two volumes in Leipzig from 1850 to 1854, was in the form followed by almost every text book of this subject published since his time. His interests were not, however, confined to the vertebrates and his investigation of the embryology of the cephalopod mollusks deserves special mention. He was also a pioneer in the development of the theory that spermatozoa were actually true sex cells and not, as was often believed at the time, a protozoan parasite of seminal fluid. In spite of his renown in research, he thought of himself primarily as a teacher and from 1847 until 1902, held the chair of zoology in Würzberg. His resignation from his chair at the age of 85 was universally regretted and he maintained active contact with his hundreds of students in all parts of Europe until his death three years later.

KOWALEWSKY, ALEXANDER (1840–1901)

Kowalewsky was one of a large band of Russian zoologists who studied in Germany before returning to teach in their native land. The greater part of his career was spent in Leningrad (then St. Petersburg) where he held the chair of zoology. He is principally remembered for his work on the urochordata and cephalochordata. His embryological studies in both groups convinced him of their relationship with each other and with the vertebrates. He was also notable as a pioneer in the development of the germ layer theory and in the philosophical speculations, now no longer accepted, which derived from it.

KREBS CYCLE

By 1937 it was known that small amounts of succinic, fumaric, malic, oxaloacetic, α-ketoglutaric, and citric acids have a catalytic effect on aerobic oxidation in animal tissues. The stimulations are abolished by malonate, a known competitive inhibitor of succinic dehydrogenase. Krebs suggested that if each acid participates in a cyclic series of reactions leading to the complete oxidation of pyruvate, the catalytic effects are more easily explained than by previous theories. Pyruvate, a 3C product of glycolysis, reacts with oxaloacetate (4C) to produce a 7C acid from which, via 6C, 5C and 4C acids, oxaloacetate is re-formed.

Evidence for the cycle included (a) α-ketoglutarate and succinate accumulation during malonate inhibitions (b) citrate accumulation when pyruvate and oxaloacetate were incubated anaerobically with tissue and (c) the complete oxidation of pyruvate in the presence of trace amounts of the other acids. Equally convincing was the explanation of an RQ of unity for aerobic carbohydrate oxidation by the combination of the "Krebs cycle" with the glycolytic pathway.

Subsequent research revealed that the cycle operates in many aerobic organisms (plant, animal and bacterial) where it functions as: (i) a major pathway of terminal oxidation, not only for carbohydrates, but probably also for proteins and fats; (ii) the main source of energy for other enzyme pathways; (iii) a source of carbon skeletons for synthesis.

The modern version (fig. 1) is strikingly similar to Krebs' original hypothesis except for the removal of the 7C acid reaction and the addition, in plants and bacteria, of the 'glyoxylate cycle.'

Krebs Cycle Reactions

(1) *Pyruvic oxidase*, which catalyzes the oxidative decarboxylation of pyruvate to yield acetyl ~ S·CoA,

$$CH_3 \cdot CO \cdot COOH + HS \cdot CoA + DPN \rightarrow$$
$$CH_3 \cdot CO\,S \cdot CoA + CO_2 + DPNH_2$$

is composed of at least three enzymes requiring TPP, Mg^{++} and lipoic acid as additional cofactors. The number of decarboxylation reactions involving both pyruvate and TPP suggests that a common initial stage is the formation of an aldehyde-TPP complex. The reaction mechanism is obscure but it has been suggested that the thiazole ring of TPP opens hydrolytically to form a sulphydryl compound (R·SH) which reacts with the carboxyl of pyruvate:

$$R \cdot SH + CH_3 \cdot CO \cdot COOH \xrightarrow{Mg^{++}}$$
$$R \cdot S \cdot CHOH \cdot CH_3 + CO_2$$

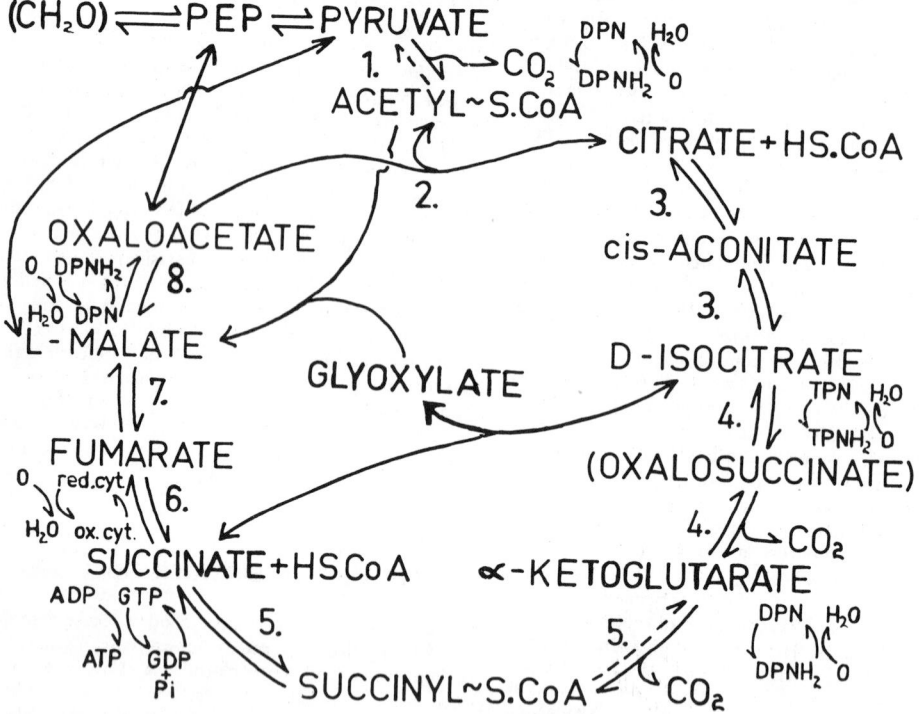

Fig. 1 The Krebs cycle (also called the Citric acid cycle or Tricarboxylic acid cycle). The numbers beside reactions refer to the enzymes discussed in "The Krebs cycle reactions." The over-all equations (ignoring phosphate esterification) are:

Glycolysis $C_6H_{12}O_6 + O_2 \rightarrow 2CH_3 \cdot CO \cdot COOH + 2H_2O$
Krebs cycle $2(CH_3CO \cdot COOH + HOOC \cdot CH_2 \cdot CO \cdot COOH + 3H_2O$
$\rightarrow 3CO_2 + 5H_2O + HOOC \cdot CH_2 \cdot CO \cdot COOH)$

Sum $C_6H_{12}O_6 + 6O_2 + 6H_2O \rightarrow 6CO_2 + 12H_2O$

Pyruvic dehydrogenase catalyzes this reaction and a subsequent acetyl transfer to oxidized lipoate, producing 6-S-acetylhydrolipoate:

$R \cdot S \cdot CHOH \cdot CH_3$

$$+ \; CH_2 \cdot CH_2 \cdot \underset{|}{CH} \cdot (CH_2)_4 \cdot COOH \rightarrow$$
$$\underset{S\text{------------}S}{}$$

$$R \cdot SH + CH_2 \cdot CH_2 \cdot \underset{|}{CH} \cdot (CH_2)_4 \cdot COOH$$
$$\underset{SH \qquad\qquad S \cdot CO \cdot CH_3}{}$$

Experiments with a mutant strain of *Escherichia coli* suggest that lipoate and TPP combine to form lipothiamide diphosphate (LTPP); however, TPP can be removed and restored to the enzyme whilst lipoate remains firmly bound. In *Streptococcus faecalis*, a specific protein, requiring ATP and a divalent metal ion, binds lipoate to the dehydrogenase (which is inactive after enzymatic hydrolysis removing lipoate). All reactions, apparently involving free lipoate, are probably due to a disulphide interchange between the free and protein-bound forms.

Thioltransacetylase A then transfers the acetyl group to HS·CoA producing acetyl ~ S·CoA and reduced lipoate:

$$CH_2 \cdot CH_2 \cdot \underset{|}{CH} \cdot (CH_2)_4 \cdot COOH + HS \cdot CoA \rightarrow$$
$$\underset{SH \qquad\qquad S \cdot CO \cdot CH_3}{}$$

$$CH_3CO \; \sim \; S \cdot CoA$$

$$+ \; CH_2 \cdot CH_2 \cdot \underset{|}{CH}(CH_2)_4 \cdot COOH$$
$$\underset{SH \qquad\qquad SH}{}$$

The thiolester bond of acetyl ~ S·CoA has a free energy of hydrolysis ($-8{,}200$ cal/mole at pH 7 and 25°C) comparable with that of the terminal phosphate of ATP ($-8{,}900$ cal/mole at pH 7.5 and 25°C). This "bond energy" may be used for synthetic reactions or be transferred to ADP.

The third known enzyme of the pyruvic oxidase complex, *lipoic dehydrogenase*, regenerates oxidized lipoate by a hydrogen transfer to DPN:

$$CH_2 \cdot CH_2 \cdot \underset{|}{CH} \cdot (CH_2)_4 \cdot COOH + DPN \rightleftharpoons$$
$$\underset{SH \qquad\quad SH}{}$$

$$CH_2 \cdot CH_2 \cdot \underset{|}{CH} \cdot (CH_2)_4 \cdot COOH + DPNH_2$$
$$\underset{S\text{------------}S}{}$$

In all organisms oxidizing added acetate (including fluoroacetate) an alternative mechanism for generating acetyl coenzyme A must exist. One suggested mechanism is:

$$acetyl + ATP \rightleftharpoons acetyl\text{-}P + ADP$$

$$acetyl\text{-}P + HS \cdot CoA \rightleftharpoons acetyl \sim S \cdot CoA + H_3PO_4$$

However, since acetyl phosphate is metabolised only in bacteria and not in animal or plant cells, a more probable sequence is:

$$acetate + ATP \rightleftharpoons acetyl\text{-}AMP + PP$$

$$acetyl\text{-}AMP + HS \cdot CoA \rightleftharpoons$$

$$acetyl \sim S \cdot CoA + AMP$$

(2) *Condensing enzyme (oxaloacetic transacetase)* was first crystallized by Stern and Ochoa who established that citrate, not *cis*-aconitate, is the product of oxaloacetate reaction with acetyl ~ S·CoA:

$$\overset{\alpha}{CO} \cdot COOH$$
$$\underset{|}{} \qquad\qquad + \; CH_3CO \; \sim \; S \cdot CoA + H_2O \rightleftharpoons$$
$$\underset{\beta}{CH_2 \cdot COOH}$$

$$CH_2 \cdot COOH$$
$$\underset{|}{C(OH) \cdot COOH} + HS \cdot CoA$$
$$\underset{|}{CH_2 \cdot COOH}$$

Since such a trimolecular reaction is unlikely, Ochoa postulates that citryl coenzyme A is formed first, with a subsequent hydrolysis of the thioester. Unlike its other reactions, acetyl ~ S·CoA combines here via the methyl and not the carboxyl group. The reaction is freely reversible.

(3) *Aconitase* dehydrates citrate to form *cis*-aconitate and then replaces the H_2O molecule in a reverse position forming D-isocitrate:

$$
\begin{array}{ccc}
CH_2 \cdot COOH & & CH_2 \cdot COOH \\
| & \xrightarrow{-H_2O} & | \\
C(OH) \cdot COOH & \rightleftharpoons & C \cdot COOH \\
| & \xleftarrow{+H_2O} & \| \\
CH_2 \cdot COOH & & CH \cdot COOH
\end{array}
\quad
\begin{array}{c}
\xrightarrow{+H_2O} \\
\xleftarrow{-H_2O}
\end{array}
$$

$$CH_2 \cdot COOH$$
$$\underset{|}{CH \cdot COOH}$$
$$CHOH \cdot COOH$$

For full activity the enzyme requires molar equivalents of Fe^{++} and an SH compound and exhibits a rigid stereospecificity for *cis*-aconitate and D-isocitrate. The competitive inhibition of both reactions by fluorocitrate and the failure to obtain two enzymes (except in *Aspergillus*) suggests that one enzyme catalyzes both reactions. Citrate produced from *cis*-aconitate in deuterated water contains deuterium but does not when isocitrate is the initial substrate (i.e. excluding OH exchange). An enzyme bound intermediate has been postulated:

$$citrate \rightleftharpoons (intermediate) \rightleftharpoons isocitrate$$
$$\Updownarrow$$
$$cis\text{-}aconitate$$

At equilibrium there is approximately 89% citrate, 7% isocitrate and 4% *cis*-aconitate.

(4) *Isocitric dehydrogenase* oxidatively decarboxylates isocitrate to form α-ketoglutarate. Oxalosuccinate is probably an enzyme bound intermediate

$$
\begin{array}{ccc}
CH_2 \cdot COOH & & CH_2 \cdot COOH \\
| & \xrightarrow{-2H} & | \\
CH \cdot COOH & \rightleftharpoons & CH \cdot COOH \\
| & \xleftarrow{+2H} & | \\
CHOH \cdot COOH & & CO \cdot COOH
\end{array}
\quad \xrightarrow{Mn^{++}}
$$

$$CH_2 \cdot COOH$$
$$\underset{|}{CH_2} \qquad\qquad + \; CO_2$$
$$CO \cdot COOH$$

The more widely distributed enzyme uses TPN as hydrogen acceptor, acts reversibly and also decarboxylates or reduces added oxalosuccinate. Moyle found that Mn^{++} is required for decarboxylation but not for dehydrogenation. This suggests a reaction sequence:

$$\text{Enzyme} + \text{isocitrate} \underset{k_2}{\overset{k_1}{\rightleftharpoons}} \text{E} \cdot \text{iC} \overset{\text{TPN}}{\rightleftharpoons}$$
$$\text{(E.)} \qquad \text{(iC)}$$

$$\text{E-oxalosuccinate} \overset{Mn^{++}}{\rightleftharpoons} \alpha\text{-ketoglutarate}$$
$$k_6 \Updownarrow k_5 \qquad\qquad + CO_2 + E$$

$$\text{E} + \text{oxalosuccinate}$$

In the absence of Mn^{++}, oxalosuccinate is slowly produced from isocitrate and TPN (k_5 has a low value), but with Mn^{++} α-ketoglutarate production is so rapid that only vanishingly low amounts of oxalosuccinate accumulate. The DPN-enzyme has no oxalosuccinate carboxylase activity, requires adenylic acid and does not reductively carboxylate α-ketoglutarate under conditions in which the TPN-enzyme acts reversibly. Possibly a substrate inhibition of the reverse reaction prevents its demonstration. Each enzyme shows rigid specificity for its coenzyme and both may be found together (e.g. in yeast and heart muscle).

Since citrate is symmetrical, ^{14}C in carboxyl-labeled acetate should appear in both carboxyls of α-ketoglutarate

$$CH_3 \cdot {}^*Co \sim S \cdot CoA$$
$$+ \qquad \longrightarrow$$
$$\text{oxaloacetate}$$

$$\begin{array}{c} CH_2 \cdot {}^*COOH \\ | \\ C(OH) \cdot COOH \\ | \\ CH_2COOH \end{array} \rightleftharpoons \left(\begin{array}{c} CH_2COOH \\ | \\ C(OH) \cdot COOH \\ | \\ CH_2 \cdot {}^*COOH \end{array} \right) \rightarrow$$

$$\begin{array}{c} CH_2 \cdot {}^*COOH \\ | \quad\gamma \\ CH_2 \qquad\qquad + CO_2 \\ | \quad\alpha \\ CO \cdot {}^*COOH \end{array}$$

In fact, only the γ-carboxyl is labeled. Similarly, β-carboxyl-labeled oxaloacetate (pyruvate + $^{14}CO_2$) yields only α-carboxyl-labeled α-ketoglutarate

$$CH_3 \cdot {}^*CO \sim S \cdot CoA$$
$$+ \qquad\qquad \longrightarrow \begin{array}{c} CH_2 \cdot {}^*COOH \\ | \\ CH_2 \qquad + CO_2 \\ | \\ CO \cdot COOH \end{array}$$
$$\text{oxaloacetate}$$

$$\begin{array}{c} CH_3 \cdot CO \sim S \cdot CoA \\ + \\ CO \cdot COOH \\ | \\ CH_2 \cdot {}^*COOH \end{array} \rightleftharpoons \begin{array}{c} CH_2 \cdot COOH \\ | \\ CH_2 \qquad + CO_2 \\ | \\ CO \cdot {}^*COOH \end{array}$$

Ogsten's 3-point attachment theory explains how an assymmetrical enzyme may impose steric orientation on a symmetrical substrate. Figure 1 illustrates this diagrammatically; only one $-CH_2 \cdot COOH$ of citrate can combine with the reactive site. Similar models could be made for each 3-point combination theoretically possible.

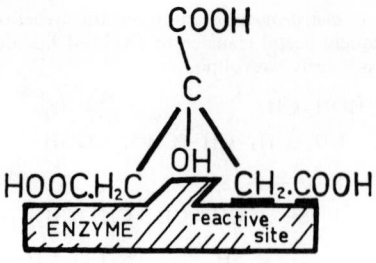

Fig. 2

(5) *α-Ketoglutarate oxidase* produces succinate from α-ketoglutarate by oxidative decarboxylation. Two distinct enzyme systems have been separated: (i) The first closely resembles, and has a similar reaction sequence to, pyruvic oxidase giving succinyl $\sim$ S·CoA in the presence of TPP, HS·CoA, lipoic acid, Mg^{++} and DPN

$$\begin{array}{c} CH_2 \cdot COOH \\ | \\ CH_2 \qquad + HS \cdot CoA + DPN \overset{TPP, Mg^{++}}{\underset{lipoate}{\rightleftharpoons}} \\ | \\ CO \cdot COOH \end{array}$$

$$\begin{array}{c} CH_2 \cdot COOH \\ | \qquad\qquad\qquad + CO_2 + DPNH_2 \\ CH_2CO \sim S \cdot CoA \end{array}$$

(ii) In the second, the succinyl $\sim$ S·CoA is either hydrolysed by a specific deacylase

$$\begin{array}{c} CH_2 \cdot COOH \\ | \qquad\qquad + H_2O \rightarrow \\ CH_2CO \sim S \cdot CoA \end{array}$$

$$\begin{array}{c} CH_2 \cdot COOH \\ | \qquad\qquad + HS \cdot CoA \\ CH_2 \cdot COOH \end{array}$$

or reacts with GDP in the presence of succinyl $\sim$ S·CoA synthetase

$$\begin{array}{c} CH_2 \cdot COOH \\ | \qquad\qquad + GDP \overset{Mg^{++}}{\rightleftharpoons} \\ CH_2CO \sim S \cdot CoA + H_3PO_4 \end{array}$$

$$\begin{array}{c} CH_2COOH \\ | \qquad\qquad + GTP + HS \cdot CoA \\ CH_2COOH \end{array}$$

The relationship between these two enzymes within tissues is not clear. The deacylase "wastes" the potentially useful thioester bond energy. The synthetase, by a coupled kinase reaction producing ATP, is the only example of substrate level phosphorylation in the Krebs cycle.

$$GTP + ADP \rightleftharpoons GDP + ATP$$

Arsenite inhibits the over-all reaction, as it does pyruvic oxidation, presumably by combining with the SH groups of lipoate.

(6) *Succinic dehydrogenase* has FAD and Fe^{++} as prosthetic groups and is activated by Pi to catalyse the dehydrogenation of succinate to fumarate. The hydrogen is transferred to oxygen through the cytochrome oxidase system.

$$\begin{array}{c}CH_2\cdot COOH \\ | \\ CH_2\cdot COOH\end{array} \underset{+2H}{\overset{-2H}{\rightleftharpoons}} \begin{array}{c}CH\cdot COOH \\ \| \\ CH\cdot COOH\end{array}$$

Malonate competitively inhibits the dehydrogenase and some part of the oxidase complex is inhibited by CO_2.

(7) *Fumarase*, the second cycle enzyme to be crystallized, hydrates fumarate to produce L-malate

$$\begin{array}{c}CH\cdot COOH \\ \| \\ CH\cdot COOH\end{array} + H_2O \rightleftharpoons \begin{array}{c}CH_2\cdot COOH \\ | \\ CHOH\cdot COOH\end{array}$$

The reaction kinetics are complex but, in general, malate predominates in the equilibrium mixture. The substrate stereoisomers, D-malate and maleate, and high concentrations of Pi are among the many known competitive inhibitors of the enzyme. Low concentrations of Pi activate the system.

(8) *Malic dehydrogenase* re-forms oxaloacetate which can then react with another molecule of acetyl ~ S·CoA. DPN is the normal hydrogen acceptor although several extracted dehydrogenases can react slowly with TPN.

$$\begin{array}{c}CH_2\cdot COOH \\ | \\ CHOH\cdot COOH\end{array} + DPN \rightleftharpoons \begin{array}{c}CH_2\cdot COOH \\ | \\ CO\cdot COOH\end{array} + DPNH_2$$

In vitro, with the purified enzyme, a pH of 10 is required before dehydrogenation proceeds at all rapidly; at neutrality the equilibrium favours malate production. Nevertheless, with coupled reactions which oxidise $DPNH_2$ or remove oxaloacetate, the oxidation can be readily demonstrated. Cellular organisation probably permits rapid oxidation *in vivo*.

Of the four carbon atoms in the original oxaloacetate only two (2 and 3) remain in the re-formed molecule; number 1 (α-carboxyl) and 4 (β-carboxyl) are removed as CO_2 by isocitric dehydrogenase and α-ketoglutaric oxidase respectively.

Ancillary Enzymes

The *glyoxylate cycle* first suggested by Kornberg and Krebs utilizes some of the Krebs cycle enzymes in addition to isocitritase and malic synthetase. Labelling experiments with ^{14}C and enzyme extractions have indicated its presence in plants (e.g., castor bean) and bacteria but not yet in animals.

Isocitritase, an aldolase type of enzyme, splits isocitrate directly to succinate and glyoxylate

$$\begin{array}{c}CH_2\cdot *COOH \\ | \\ CH\cdot *COOH \\ | \\ CHOH\cdot COOH\end{array} \overset{Mg^{++}}{\rightleftharpoons} \begin{array}{c}CH_2\cdot *COOH \\ | \\ CH_2\cdot *COOH \\ + \\ OHC\cdot COOH\end{array}$$

Malic synthetase catalyses the condensation of glyoxylate and acetyl ~ S·CoA to form malate:

$$\begin{array}{c}CH_3\cdot *CO\,S\cdot CoA \\ + \\ CHO\cdot COOH\end{array} + H_2O \overset{Mg^{++}}{\longrightarrow}$$

$$\begin{array}{c}CH_2\cdot *COOH \\ | \\ CHOH\cdot COOH\end{array} + HS\cdot CoA$$

The glyoxylate cycle reactions are:
acetyl ~ S·CoA + oxaloacetate $\rightleftharpoons$ isocitrate
(condensing enzyme + aconitase)
isocitrate $\rightleftharpoons$ glyoxylate + succinate (isocitritase)
glyoxylate + acetyl ~ S·CoA $\rightleftharpoons$ malate
(malic synthetase)
malate + DPN $\rightleftharpoons$ oxaloacetate + $DPNH_2$
(malic dehydrogenase)

sum: 2 acetate + DPN $\rightleftharpoons$ succinate + $DPNH_2$

β-*Carboxylation.* Several enzymes, frequently found with the Krebs cycle enzymes, carboxylate α-keto acids:

Oxaloacetic decarboxylase reversibly carboxylates phosphoenolpyruvate, the immediate precursor of pyruvate in the glycolytic pathway, with GDP as phosphate acceptor

$$\begin{array}{c}CH_2 \\ \| \\ P \sim OC\cdot COOH\end{array} + GDP + CO_2 \overset{Mn^{++}}{\rightleftharpoons}$$

$$\begin{array}{c}CH_2\cdot COOH \\ | \\ CO\cdot COOH\end{array} + GTP$$

ITP can replace GTP but the apparent participation of ATP is due to a linked kinase reaction.

PEP carboxylase also produces oxaloacetate but liberates Pi.

$$\begin{array}{c}CH_2 \\ \| \\ P \sim OC\cdot COOH\end{array} + H_2CO_3 \overset{Mg^{++}\,SH}{\longrightarrow}$$

$$\begin{array}{c}CH_2\cdot COOH \\ | \\ CO\cdot COOH\end{array} + H_3PO_4$$

This enzyme was the first isolated carboxylase known to be inhibited by CO_2 at levels higher than those for maximum reaction rates and, accordingly, is thought to be concerned in malate accumulation in Crassulacean plants.

Malic enzyme carboxylates pyruvate in the presence of $TPNH_2$, producing malate:

$$\begin{array}{c}CH_3 \\ | \\ CO\cdot COOH\end{array} + CO_2 + TPNH_2 \rightleftharpoons$$

$$\begin{array}{c}CH_2\cdot COOH \\ | \\ CHOH\cdot COOH\end{array} + TPN$$

Oxaloacetate is also decarboxylated but not reduced (cf. isocitric dehydrogenase).

At equilibrium, oxidative decarboxylation predominates except at high CO_2 concentrations or when linked reactions are present. This, together with the malic dehydrogenase equilibrium position, suggests a possible alternative source of oxaloacetate and pyruvate for entry into the cycle.

malate $\rightarrow$ pyruvate + CO_2........ malic enzyme

PEP + $CO_2 \rightarrow$ oxaloacetate........

oxaloacetic or PEP carboxylase

Oxaloacetate *in vitro* is very labile and, *in vivo*, might not be re-formed if earlier cycle acids contribute their C-skeletons to synthetic reactions. Each of these

β-carboxylases explains a possible resynthesis which would allow continuous operation of the Krebs cycle.

Krebs Cycle Functions

Oxidative Phosphorylation. The esterification of Pi at both substrate (see α-ketoglutaric oxidase) and coenzyme levels of hydrogen (or electron) transfer traps energy produced by oxidation and

$$ADP + Pi + \epsilon \rightleftharpoons ATP$$

permits its transfer to synthetic reactions. Theoretically, hydrogen transfer via TPN or DPN to oxygen, could synthesise 6 $\sim$P but experimentally determined ratios show an overall efficiency of 50% i.e. 3 $\sim$P.

Substrate	Enzyme	Coenzyme	$\sim$P trapped
Pyruvate	(1)	DPN	3
isocitrate	(4)	TPN	3
⁻ketoglutarate	(5)	DPN	4
			(1 substrate)
succinate	(6)	cytochrome	2
malate	(8)	DPN	3
		total	15

With the 5 $\sim$P produced glycolytically, 20 $\sim$P are produced per molecule of triose oxidised (i.e. hexose = 40 $\sim$P $-$ 2 $\sim$P required for initial phosphorylation = 38 $\sim$P).

Oxidation. Coenzyme linked reactions permit the continuous oxidation of cycle acids (and therefore carbohydrate) in the presence of oxygen

substrate (2H) $\rightarrow$ DPN(2H$^+$ + 2e) $\rightarrow$

flavin (2H$^+$ + 2e) $\rightarrow$ cytochrome(e) $\rightarrow$ O$_2$(H$_2$O)

Many enzyme pathways are known by which amino-compounds may be converted to cycle acids and thus be completely oxidised. A *few* examples are:

glutarate, glutamine, ketoglutaramate, ornithine $\rightarrow$

α-ketoglutarate

aspartate, arginosuccinate, tyrosine $\rightarrow$ fumarate

γ-aminobutyrate $\rightarrow$ succinate

aspartate $\rightarrow$ oxaloacetate

alanine $\rightarrow$ pyruvate

Fats may also be completely oxidised via the cycle

fat $\rightarrow$ glycerol $\rightarrow$ triose P $\rightarrow$

pyruvate $\rightarrow$ CO$_2$ + H$_2$O

fatty acid $\rightarrow$ acetyl $\sim$ S·CoA $\rightarrow$

HS·CoA + CO$_2$ + H$_2$O

Where the fatty acid is the sole C-source, the glyoxylate cycle can explain the formation of oxaloacetate for use in the condensing enzyme reaction.

Synthetic Reactions. Amino-acids. Glutamic dehydrogenase reductively aminates α-ketoglutarate to form glutamate:

CH$_2$·COOH
|
CH$_2$ + DPNH$_2$ + NH$_3$ $\rightleftharpoons$
|
CO·COOH

CH$_2$·COOH
|
CH$_2$ + DPN + H$_2$O
|
CH·NH$_2$·COOH

Transaminases may then transfer the amino-group to other α-keto acids e.g.

glutamate + pyruvate $\rightleftharpoons$

alanine + α-ketoglutarate

glutamate + oxaloacetate $\rightleftharpoons$

aspartate + α-ketoglutarate

Other amino-compounds may be synthesised by further reactions.

Carbohydrate production from fats cannot be explained via Krebs cycle reactions alone since for every acetyl S·CoA molecule consumed 2CO$_2$ are released. However, the glyoxylate cycle enables 4-C acids to be synthesised from which PEP can be produced

fatty acid $\rightarrow$ 2 acetyl $\approx$ S·CoA $\rightarrow$ succinate $\rightarrow$

oxaloacetate $\rightarrow$ PEP $\rightarrow$ carbohydrate

Other known synthetic reactions include β-ketoadipate formation from succinyl $\sim$ S·CoA plus acetyl $\sim$ S·CoA and pyrrole ring structures from succinyl $\sim$ S·CoA plus glycine.

Intracellular Location of Cycle Enzymes

Most mitochrondrial preparations can completely oxidise the cycle acids in the presence of pyruvate. However, most of the aconitase and isocitric dehydrogenase of the cell appears in the soluble cytoplasmic fraction. Recent evidence indicates that, in the intact cell, the Krebs cycle reactions may result from the combined activities of several cytoplasmic structures. In only a few tissues are all the cycle enzymes exclusively located in the mitochondria.

J. M. A. BROWN

References

Dixon, M. and E. C. Webb, "Enzymes," 2nd ed., New York, Academic Press, 1964.
Krebs, H. A., "Chemical pathways of metabolism," New York, Academic Press, 1954.
Krebs, H. A. and H. L. Kornberg, "Energy transformation in living matter," New York, Springer, 1964.

LACERTILIA

A suborder of the order Squamata of the class REPTILIA, popularly known as lizards. Most closely allied to snakes, the other suborder of Squamata. Both are the most recent of reptiles, and are not known from fossil remains until the Jurassic and are not common as fossils until late Cretaceous.

Lizards are considered the more primitive members of the order. In most groups they still retail both temporal and postorbital arches. Uusually, they still show a parietal foramen. Mandibles united by suture in contrast to snakes in which the mandibles are separate or loosely attached by ligament.

Lacertilians, like other reptiles are covered with a usually scaly skin that lacks sweat, oil and mucus glands. They are poikilothermic and are adapted to spending their entire life on land. Fertilization is internal and various species are either oviparous or ovoviviparous. Like the other suborder, Serpentes, lizards have a transverse anal opening and paired copulatory organs. They differ from snakes in that the mandibles are united by suture; at least vestiges of both pectoral and pelvic girdles are present.

Lacertilian teeth are either acrodont or pleurodont and may be variously modified from simple peg-like teeth to molariform or trilobate. In most forms the teeth are all more or less alike, although in some there are marked differences between the anterior and posterior teeth.

Typically, lizards have scales, frequently with osteoderms; four limbs with five toes each; a nonforked protrusible tongue; movable eyelids and functional ears. Exceptions to the typical pattern are frequent and occur in several families. Lizards are world wide in distribution except for the most extreme cold areas; most numerous in forms and individuals in warmer climates. They have reached such remote islands as the Marshall and Hawaiian Islands of the Pacific, and they are common throughout the West Indies. Approximately 300 genera of recent lizards are known. These are presently grouped into 16 families. Classification here used follows Romer, 1956.

Families of Recent Lacertilia

IGUANIDAE—the Americas—Madagascar and Fiji Isl.—One of the largest and most widespread families of lizards. It includes such familiar genera as *Anolis* the so called "American Chameleon" most common pet lizard in United States. *Phrynosoma,* the bizarre horned lizards of Western United States. *Amblyrhyn-*chus,* the Marine Iguana of the Galapagos Isl., the only marine lizard, unique in its adaptations for feeding on algae in the surf. *Iguana,* large lizards of South and Central America, by many considered a delicacy.

AGAMIDAE—Old World parallels of the Iguanidae to which they are closely related, includes *Agama,* widespread in Africa, Southwest Asia and Southeastern Europe. *Drasco,* the remarkable "flying" lizards of Southern Asia; that glide from tree to tree. *Moloch,* of Australia which parallels in form and habit the North American horned lizards.

CHAMELEONTIDAE—the true chameleons—mostly African but with representatives in Europe and Southwest Asia, also Madagascar. These lizards are highly specialized for an arboreal life with prehensile tail, grasping feet in which the digits are in two opposing groups, turreted eyes and greatly protrusible tongue. One of the most grotesque families of lizards.

GEKKONIDAE—includes the Eublepharidae and Uroplatidae of some authors. Nocturnal, exceedingly abundant in tropics, usually with specially modified toes that enable them to climb apparently smooth surfaces or to run across ceilings. Unique among lizards in that some members of the genus *Gekko* have distinctive calls.

PYGOPODIDAE—An Australian group of small snake-like forms—lacking front limbs and hind limbs reduced to flaps.

XANTUSIIDAE—A small family of four genera restricted to Southwestern United States, Central America and Cuba.

TEIIDAE—A large family in the New World and in appearance and habit parallels the Lacertidae of the Old World. This family is extremely varied in size and form, from elongate almost limbless, snakelike members to heavy-bodied lizards such as the Tegu (*Tupinambis*) of South America. Two of the largest genera *Cnemidophorus* and *Ameiva* are the familiar "race runners" of the American tropics.

SCINCIDAE—Widespread in both New and Old Worlds characterized by rounded smooth overlapping scales that contain osteoderms. In body form they range through all gradations from forms with strong limbs to those that have mere vestiges of limbs. Some of the genera in this group are remarkable for their great geographic range and large number of species. Noteworthy are: *Eumeces, Lygosoma,* and *Leiolopisma,* which occur in Asia, North America and the Pacific Islands.

LACERTIDAE—Common lizards of the Old World, paralleling in form the New World Teiids. The genus

Lacerta includes the common Wall lizards of Southern Europe.

HELODERMATIDAE—This family with but one genus *Heloderma* (Gila Monster) is noteworthy as they are the only venomous lizards in the world. They have grooved teeth, the venom is produced in a modified and enlarged submental gland. These are large, heavy-bodies lizards, with rounded bead-like scales, with osteoderms. The genus is restricted to the Southwest United States and Mexico.

ANGUIDAE—This relatively small family contains both limbed and limbless forms of which the most familiar are the "Slow Worm" (*Anguis*) of Europe and the "Glass Snake" (*Ophisarus*) of Southeastern United States. Both of these are limbless and snake-like in form although both retain vestiges of the pectoral and pelvic girdles. Externally, they are readily distinguished from snakes by their movable eyelids and visible ear openings.

ANNIELLIDAE—This family contains but a single genus, *Anniella*, restricted to Southern California and part of Baja California. Limbless, wormlike, no visible ears, but with eyelids and is ovoviviparous.

XENOSAURIDAE—Another family with but a single genus, *Xenosarus*, of Mexico. Although Shinisaurus of East Asia may be related, both have limbs and are related to the Anguidae but differ in having poorly developed osteoderms.

VARANIDAE—A single genus of large lizards that range over Africa, Australia and the East Indies, contains the largest of all living lizards, *Varanus komodoensis*, of Komodo Island in the Dutch East Indies which may attain a length of over ten feet.

LANTHANOTIDAE—Based on the genus *Lanthanotus*, formerly included in the family Helodermatidae. A relatively rare form of the East Indies.

AMPHISBAENIDAE—A large family of limbless lizards in which the skull and associated musculature are highly specialized for burrowing. A large number of species, widely distributed in the tropics of America and Africa. Only one species in United States; *Rhineura floridana*, restricted to Florida. The resemblance of this family to earthworms is increased by their small size and their scales arranged in rings around the body, giving them a segmented appearance.

NEIL D. RICHMOND

References

Romer, A. S., "Osteology of the Reptiles," Chicago, Univ. of Chicago Press, 1956.
Brehm, A., "Brehm's Tierleben, 2nd vol.," Leipzig, Bibliographisches Institut, 1930.
Smith, Hobart M., "Handbook of Lizards," Ithaca, Comstock, 1946.

LACTATION

The possession of mammary glands and hence the ability to secrete milk is, by definition, the most distinctive characteristic of mammals. In MONOTREMES the glands are relatively primitive, consisting of collections of tubes lined with milk-secreting cells; the milk exudes along hairs from which it is sucked by the young. In higher MAMMALS the mammary gland is a compound tubulo-alveloar gland based on an arborescent system of excretory ducts communicating with the exterior by a teat or nipple. The glands occur in pairs on the ventral aspect of the body and the number characteristic of a given species is broadly related to the number of young normally born at one time. Thus in the mouse, rat and rabbit, there are usually five or six pairs, while in the guinea-pig, aquatic mammals, primates and man, there is only one pair. Among ruminants, the goat and sheep have two glands and the ox four. Where there are only one or two pairs of mammae they may be thoracic as in the bat, elephant, primates and man, or inguinal as in the guinea-pig, ruminants and the whale. In MARSUPIALS the glands are grouped inside the pouch so that the young can each remain attached to a teat throughout the suckling period.

The mammary glands are modified skin glands, but it is not known whether they are derived from sweat or sebaceous glands. During the embryonic and fetal stages, the growth of the mammary rudiments is largely ahormonal, though in some forms (mouse, rat) androgen from the fetal gonads causes the mammary bud of the male to develop differently from that of the female, so that the males are born without nipples. In most species mammary development from birth to puberty, in the female, consists of slow extension of the mammary duct tree with relatively little growth of alveoli (milk secreting tissue); full lobulo-alveolar development occurs during pregnancy. This is brought about by the ovarian hormones, *estrogen* and *progesterone*, in co-operation with certain anterior-pituitary, and probably placental, hormones. Recent studies on the rat indicate that estrogen together with pituitary growth hormone (*somatotropin*) are responsible for growth of the duct system, while progesterone and the pituitary (or placental) lactogenic hormone, *prolactin*, are additionally needed for lobulo-alveolar development; hormones of the adrenal cortex (and hence, pituitary *adrenocorticotropin*) are also involved in both phases. This probably represents the broad picture in most mammals though species may differ in detail.

Although characteristic milk constituents (see below) are found in mammary tissue during late pregnancy, showing that milk synthesis has already begun, copious lactation is normally only initiated after parturition. The hormonal mechanism involved has not yet been fully elucidated, though it is known that a prominent part is played by prolactin and the pituitary-adrenal system. The release of the optimal constellation of pituitary hormones necessary for lactogenesis may be governed by alterations in the body levels of estrogen and progesterone and also perhaps by neuroendocrine influences arising from the suckling stimulus. As regards the maintenance of lactation, somatotropin appears to be important along with prolactin and adrenocorticotropin, and perhaps also *thyrotropin*. In most species the suckling stimulus appears to cause reflex secretion of prolactin, and possibly the rest of these galactopoietic pituitary hormones. In primates the separate identities of prolactin and somatotropin are not yet certain. *In vitro* studies indicate that *insulin* is necessary for mammary function.

The discharge of milk from the alveoli during suckling or milking is known to involve a neuroendocrine reflex, the milk-ejection reflex, which is evoked by stimulation of sensory nerve endings in the teat or nipple, or by conditional stimuli. Afferent impulses reach the brain and evoke release of a hypothalamic hormone, *oxytocin,* from the neurohypophysis into the blood stream. Oxytocin reaching the mammary gland causes contraction of myoepithelial cells thus squeezing the tenaciously held alveolar milk down into the larger ducts, sinuses, or, in ruminants, gland cisterns. The milk is thus made available to the suckling or milker. In domestic ruminants bred for high milk yield, an appreciable proportion of the milk present in the udder just before milking is contained in large spaces, the gland cisterns, into which the teats open. This milk can be drawn from the cisterns without the intervention of the milk-ejection reflex. Soviet physiologists believe that purely neural reflexes, involving efferent innervation of motor elements of the udder, are also important in milk discharge. Reflex changes in the tonus of smooth muscle fibres in the walls of the large ducts and cisterns, initiated by the response of baroreceptors to increases in intramammary pressure, are believed to favour the flow of alveolar milk into the cisterns not only during milking, but also during the milking interval. The gland cisterns of ruminants thus accommodate considerable amounts of milk before the intramammary pressure rises high enough to stop milk secretion. In the sheep and goat completely denervated udder-halves have recently been found to secrete virtually normal quantities of milk. It therefore appears that in these animals the milk-ejection reflex may not be so important for the maintenance of lactation as in other species not provided with capacious gland cisterns.

Milk contains three principal constituents, two of which, casein and lactose, are normally not found elsewhere; the third constituent, milk fat, is also characteristic in that in most species the milk glycerides, unlike body fats, contain appreciable amounts of fatty acids with fewer than 16 carbon atoms. A considerable proportion of milk fatty acids, mainly long-chain acids, enters the mammary gland from the blood triglycerides, while the rest are synthesised in the gland from small molecules derived from blood glucose, or in ruminants, blood acetate and β-hydroxybutyrate. The principal precursor of lactose is believed to be blood glucose, which a series of enzymes transforms into a nucleotide, uridine diphosphogalactose; another enzyme, lactose synthetase, couples the galactose residue with glucose, giving lactose. Casein, and the other principal milk proteins, α-lactalbumin and β-lactoglobulin, appear to be mainly derived from blood amino acids. These are assembled into protein molecules in the mammary gland. Ultrastructural studies indicate that all three classes of principal milk constituent are synthesised by the same type of alveolar epithelial cell. Fat droplets form in the ergastoplasm and vacuoles containing protein granules in the Golgi appartus; both move to the apical zone of the cell where the vacuoles empty their protein into the lumen and the fat droplets are pinched off with a plasmalemma envelope forming the fat globule membrane.

S. J. FOLLEY

References

Folley, S. J., In "Marshall's Physiology of Reproduction," 3rd ed., A. S. Parkes, ed., vol. 2, chap. 20, London, Longmans, Green, 1952.

ibid., "The Physiology and Biochemistry of Lactation," Edinburgh and London, Oliver & Boyd, 1956.
Kon, S. K. and A. T. Cowie, eds., "Milk: The Mammary Gland and its Secretion," New York, Academic Press, 1961.
Zaks, M. G., "The Motor Apparatus of the Mammary Gland," Edinburgh and London, Oliver & Boyd, 1962.

LAGOMORPHA

An order of mammals including the Leporidae (rabbits and hares), Ochotonidae (pikas) and an extinct family from the Paleocene of Mongolia, the Eurymylidae. The lagomorphs were formerly included among the RODENTS, but it is now generally agreed that the two orders have nothing in common. The ordinal relationships of the lagomorphs are uncertain, but they probably have a distant relationship to the ungulate stock (Wood, 1957).

All lagomorphs possess two pair of upper incisors, the second pair being very small, and a single pair of lowers. The incisors (especially the lowers) are relatively short, and do not form efficient chisels as do those of the rodents, since the enamel surrounds the tooth, instead of being limited to the anterior face, although it is thicker on the anterior face. Histologically, the incisor enamel has a single layer, instead of two as in the rodents. There are two or three upper and two lower premolars, whereas no known rodent has more than two upper and one lower. The pattern of the upper cheek teeth is very peculiar, and there has been no success in homologizing the cusps with those of other mammals. A peculiar set of nutritive foramina penetrate the maxilla, reducing it to a network in the leporids and forming a large fenestra in the ochotonids. Similar foramina are present in much of the rest of the skull, especially in leporids. The incisive foramina are very long, and the palate behind them is very short. There are large supraorbital processes of the frontal. The lower jaws are solidly united at the symphysis, permitting no motion between them. The glenoid fossa is short anteroposteriorly and the jaws are essentially restricted to vertical and transverse movements. There is no entepicondylar foramen on the humerus, and the motion at the elbow is limited to the anteroposterior plane. The tibia and fibula are fused in all forms where they are known, and the fibula articulates with the calcaneum, as in the artiodactyls. There is a spiral valve in the caecum. No os penis is present, and the scrotum is pre-penial (Petrides, 1950).

It has been frequently stated that the embryological similarities between the rodents and lagomorphs justify their being kept in the same order, but Hartman (1925) considered that the similarities had been greatly overemphasized, and that the embryologic patterns are actually so different as to support their ordinal separation. The few serological data available suggest that lagomorphs are no more closely related to the rodents than to the cow, raccoon or man (Moody, Cochran and Drugg, 1949).

The earliest known lagomorphs occur in the upper Paleocene of Mongolia. While they are fragmentary, the evidence suggests that the order was as distinct from the rodents then as now, and that it ties in with the basic ungulate stock (Condylarthra). The leporids and ochotonids have been traced back to the lower Oligocene. In the upper Eocene, the two families appear to have been little, if any, differentiated.

The Leporidae have a broad Holarctic distribution, and have spread also into Africa and South America. They were introduced into Australia by the white man, and rapidly spread to become a nuisance, which has recently been reduced by disease (mixomastosis).

The hares (*Lepus*) are characterized by long hind legs, giving great leaping ability and high speeds (up to 38 m.p.h.). They have long ears. They are primarily dwellers in open country, escaping their enemies by their speed and dodging ability. The rabbits (esp. old world *Oryctolagus* and new world *Sylvilagus*) have shorter hind legs, shorter ears, and live in wooded, brush, or mixed areas. They escape predators primarily by dodging or by remaining frozen in concealment. Leporids have litters that vary from averages of two to five, in different forms, although a litter of 18 has been reported in *Oryctolagus*. There are often 3 or 4 litters a year. The young reach sexual maturity at about three months of age, but usually do not produce young their first year. The gestation period ranges from about 28 days in *Sylvilagus* to about 43 in jack-rabbits.

Hares and rabbits are subject to great ranges in numbers, periods of high fertility leading to a maximum, followed by a great increase of predators, disease and parasites (e.g., up to hundreds of ticks per individual, largely on the ears). This results in a rapid decrease in the population. While these cycles are repetitive, they do not appear to have any fixed periods.

The northern hares (snowshoe hare in North America) have a brown summer coat and a white winter one, the old hair being lost rapidly in spring and fall as the new hair comes in, often giving a splotched appearance.

The ochotonids are small animals, about 20 cm. long, with short hind legs, short rounded ears, and practically no tail. They are most unusual for placental mammals in that there is a cloaca in both sexes (Duke, 1951). The male has no scrotum. They are essentially confined to elevations above the tree line in mountains of Europe, north Asia and North America, although they get down into the zone of coniferous forests. During the middle Tertiary, their range was broader, as they have been found in considerable numbers in plains deposits in both the old and new worlds. One species in Manchuria still inhabits the high steppes, making burrows through the ground in open country (Loukashkin, 1940). Other forms make nests under boulders, dodging between and under rocks, and running or making short leaps, to escape predators. The voice is a high shrill whistle, commonest at daybreak and sunset. It often is used as a warning call. Hay is cut and stored near the nests as a reserve food supply for the winter. The litters are smaller than in the leporids, two apparently being a common size. The gestation period is about 30 days, and there are two litters a year.

All leporids, apparently, exhibit the unique habit of refection. Soft fecal pellets are produced by the caecum, being egested in the early hours of daylight. They are taken by the animal directly from the anus and are swallowed whole. They are retained in the stomach for some hours (4 in *Sylvilagus*, 9 in *Oryctolagus*). Hard fecal pellets are produced by the colon, and are not eaten. Refection starts before weaning is complete, at about the time

the young begin to feed for themselves. There is some evidence that the young eat the maternal soft pellets, resulting in a transfer of the intestinal flora and fauna from mother to young (Lechtleitner, 1957). Refection seems to be a necessary part of leporid nutrition, to permit adequate time for bacterial digestion of plant tissues, and is analogous to cud-chewing among the ruminants. It is not known whether the ochotonids practise refection.

ALBERT E. WOOD

References

Duke, Kenneth L., "The external genitalia of the pika, *Ochotona princeps*," Jour. Mammalogy, 32: 169–173, 1951.
Hartman, C. G., "On some characters of taxonomic value appertaining to the egg and ovary of rabbits," Jour. Mammalogy, 6: 114–121, 1925.
Lechtleitner, R. R., "Reingestion in the black-tailed jack rabbit," Jour. Mammalogy, 38: 481–485, 1957.
Loukashkin, A. S., "On the pikas of north Manchuria," Jour. Mammalogy, 21: 402–405, 1940.
Moody, P. A., V. A. Cochran and H. Drugg, "Serological evidence on lagomorph relationships," Evolution, 3: 25–33, 1949.
Petrides, G. A., "A fundamental sex difference between lagomorphs and other placental mammals," Evolution, 4: 99, 1950.
Wood, A. E. "What, if anything, is a rabbit?" Evolution, 11: 417–425, 1957.

LAMARCK, JEAN BAPTISTE ANTOINE DE MONET (1744–1829)

The evolutionary theories of the Chevalier De Lamarck, the outstanding evolutionist before Darwin, are now generally repudiated, though he is more renowned today than at any time during his life. His theories gained recognition after his death only to be eclipsed by the theories of evolution by mutation and by natural selection. Although Lamarck's inheritance of acquired characteristics is today generally held untenable by all but the Lysenkoists, he is recognized nevertheless for having helped in the development of modern evolutionary theory. His greatest work in zoology was a classification scheme based on a single line of descent from primitive organisms to the most advanced.

Jean Baptiste Lamarck was born August 1, 1744, at Bazantin, Picardy, the son of Pierre de Monet, a French lord. As his father wished him to enter the priesthood he was sent to the Jesuit college at Amiens. After the death of his father in 1760, he immediately left college to join the French army then engaged against the Germans in the Seven Years' War. Arriving on the eve of the battle of Fissingshausen, he supposedly was commissioned an officer that very day for bravery. After five years of peacetime garrison duty, he left the military service.

Lamarck went to Paris and began the study of medicine, supporting himself with clerical work. He became intensely absorbed in botany and abandoned thoughts of a medical career in order to devote his full attention to botanical studies. In 1778 he published *Flore Fran-*

caise and as a consequence was elected to the Academy of Sciences.

Accompanied by Buffon's son, he travelled through Europe under commission of the king of France collecting rare plants. Later he contributed two volumes to *Encyclopédie Méthodique* and then received an appointment at the *Jardin du Roi.* In 1793 the *Jardin du Roi* was reorganized by the National Assembly as the *Muséum d'Histoire Naturelle* whereupon he was forced to take a chair in invertebrate zoology, though not familiar with zoology at the time of his appointment; nevertheless he devoted the rest of his life to zoology. The *Philosophie Zoologique,* a volume which expressed his views on evolution, physiology and psychology, was published in 1809. Toward the end of his life he published *Histoire naturelle des animaux sans vertébres* (1815–1822) in seven volumes.

Poor all his life and discredited by his poor attempts at meteorology and at chemistry where he disputed Lavoisier, Lamarck died December 18, 1829, having been totally blind for ten years.

Lamarck's work on classification is probably his best although he is known chiefly for his evolutionary theories. He introduced the term *Invertebrata* to differentiate the lower from the higher animals, a distinction whose basis had previously been the lack of red blood. Though Lamarck's evolutionary theories seem naive today to the biologist they appeal to the layman's "common sense" reasoning. According to Lamarck spontaneous generation of simple forms of life was followed by development of more complex organisms which was governed by four principles of evolution: 1. The environment governs the natural tendency of an organism to increase in size. 2. Because of new physical needs new structures arise and old structures are modified. 3. Use or disuse causes variation in a structure. 4. Acquired characters are transmitted to following generations. This latter concept has been overwhelmed by modern genetics which demonstrates no positive evidence for the inheritance of traits acquired by action of the environment upon the individual.

Much of Lamarck's classification scheme remains valid today. Although his evolutionary theories are considered invalid they provided a stepping stone for the work of Darwin and later evolutionists.

ROBERT HELLING

LANKESTER, SIR EDWIN RAY (1847–1929)

This English zoologist is chiefly known for his studies of the coelom. He was the first to point out the taxonomic value of this structure in showing the relationships, or lack of relationships, between various invertebrate groups. He held the chair of zoology at University College, London, from 1874 to 1890, then went to Oxford for seven years before becoming Director of the British Museum (Natural History). He edited an eight volume "Treatise of Zoology," to which contributions were made by the leading specialists of the day and which is still a standard reference work in morphology and taxonomy. This work is frequently, by reason of the editor's association, referred to as the "Oxford Natural History" to the great confusion of bibliographers.

LARVA, INVERTEBRATE

A larva is the immature form of an animal which is unlike its parents and must pass through some sort of metamorphosis before assuming the characteristics of the adult. The term is usually applied to stages later than the early embryo. Larvae may be either free-living (most common), parasitic (rare), or be contained within embryonic capsules, membranes, cases or brood spaces of the mother. The primary purpose of the larval stage or stages is to increase the possibility for the distribution or dissemination of the species. Distribution of larvae must be brought about by locomotor organs, currents and tides, or, in the parasitic types such as the glochidium, by the movements of its temporary host. Another method for the dissemination of a species occurs in those forms where non-motile larvae, such as sporocysts and rediae, undergo polyembryony to produce a great number of offspring. Such a broad definition extends far beyond the usual limited definition in terms of free-swimming, immature animals. Where there is no metamorphosis, but merely direct development of mature forms from immature, the post-embryonic stage is called a "juvenile."

Some remarkable larvae are found as early stages in the life-histories of many invertebrates which, as adults, live in the littoral or benthic zones of the ocean. These larval forms include many which drift freely in the upper waters as plankton before they eventually settle down and are transformed into sessile creatures. Here the question arises as to whether the planktonic larval stages represent primitive and ancestral adult types (the view of early workers) or specialized forms highly adapted for a particular type of larval existence (interpretation of Garstang and others). F. R. Lillie (1895) explained the special features of cleavage as adaptations to produce the specialized structures of the larvae. Garstang and others extended this concept and interpreted the manner in which many special larval features are completely lost during metamorphosis as evidence that they are larval adaptations, of temporary significance, only. The structures of the adult arise in certain cases from a very few relatively undeveloped rudiments of the larva. During this process larval structures may be thrown aside so violently as to warrant calling the transformation a cataclysmic metamorphosis (as in Polygordius). Locomotor organs, sensory eye-spots, notochord and perhaps much of the nervous system can be discarded when a motile tadpole (such as that of the tunicate) becomes a sessile adult.

Garstang described as competing selective advantages (1) the need of the larva to grow up rapidly into the adult stage so as to be able to reproduce the species, and (2) the need to remain as a floating organism as long as possible in order to further distribute the species. Furthermore, natural selection could be acting as potently on the young stages as on the adults. In fact, the developmental stages must be as much adapted to their particular mode of life as the adult, in order that they may survive. The young, varying as much as or even more than the adults, could be modified in completely different directions, especially since the larvae and adults may inhabit completely different environmental areas. There is, then, an evolution of development as well as an evolution of the adult. If any one of the stages of larval adaptation should by chance become the point of depar-

Table 1

Phylum	Group	Larva	Type Genus or Group
Mesozoa		Infusiform larva (swarmer)	Dicyema
		Wagener's larva	Microcyema
Porifera		Amphiblastula	Sycandra
		Parenchymula	Clathrina
		Gemmule larva (asexual)	Mycale
Cnidaria	Hydrozoa	Planula	Many Hydrozoa and Actinozoa
		Actinula	Tubularia
		Conaria, rataria	Velella
		Physonectid larva	Physonectae
	Scyphozoa	Planula, edwardsia larva, halcampoides stage	Actiniaria
		Scyphistoma (scyphula or hydratuba), ephyra (ephyrula), strobila	Aurelia
	Actinozoa	Cerianthid larva, cerinula	Cerianthus
		Arachnactis larva	Arachnactis
		Cerianthulid larva	Cerianthula
		Semper's larva; zoanthella and zoanthina	Zoanthidae
Ctenophora		Beroid larva	Beroë
		Cydippid larva	Cydippidae
Platyhelminthes	Turbellaria: Polycladida	Müller's larva (Mülleria)	Planocera
		Götte's larva	Stylochus
	Trematoda	Miracidium, sporocyst, redia, cercaria	Fasciola
	Cestoda	Cysticercus	Taenia
		Cysticercoid	Dipyllidium
		Coracidium, procercoid, plerocercoid (sparganum)	Diphyllobothrium
		Coenurus	Multiceps
		Hydatid cyst	Echinococcus
Rhynchocoela	Heteronemertini	Pilidium	Cerebratulus
		Desor's larva	Lineus
Annelida	Polychaeta	Trochophore (trochosphere), meta-trochophore, nectochaeta	Nereis
		Mitraria	Owenia
		Trochophore, polytrochula	Phyllodoce
Echiurida		Trochophore	Urechis
Sipunculida		Trochophore	Phascolosoma
Phoronida		Actinotroch	Phoronis
Ectoprocta		Cyphonautes (from non-brooded eggs)	Membranipora (Electra)
		Non-cyphonautes, trochophore-like (from brooded eggs)	Bugula
Entoprocta		Trochophore-like	Pedicellina, Barentsia
		Rotifer-like	Loxosoma
Brachiopoda		Trochophore-like; cilia on tentacles of lophophore	Terebratulina
		Lingulid	Lingula
		Discinid	Glottidia
Acanthocephala		Acanthor, acanthella	Macrocanthorhynchus
Mollusca	Gastropoda	Ctenophore stage, trochophore, veliger	Patella
		Trochophore, veliger	Buccinum
	Amphineura	Trochophore, veliger	Chaetopleura
	Bivalvia	Trochophore, veliger	Mercenaria
		Glochidium	Unio, Anodonta

Table 1   (continued)

Phylum	Group	Larva	Type Genus or Group
Echinodermata	Asteroidea	Dipleurula, bipinnaria, brachiolaria	Asterias
	Echinoidea	Dipleurula, echinopluteus (pluteus)	Echinus, Arbacia
	Ophiuroidea	Dipleurula, ophiopluteus	Ophiothrix, Ophioderma
	Holothuroidea	Dipleurula, auricularia, pupa, pentactula	Synapta
		Dipleurula, metadoliolaria, pentactula	Thyone
	Crinoidea	Dipleurula, pentacrinoid	Antedon
Pogonophora		Larva	Oligobrachia
Arthropoda	Crustacea	Nauplius, metanauplius	Cyclops
		Protozoea	Squilla
		Zoea, megalops	Carcinus
		Metazoaea, mysis (schizopod larva)	Decapoda
		Phyllosoma	Palinurus
		Nauplius, cypris, pupa	Lepas
		Erichthus (pseudozoea), erichthoidina, alima	Stomatopod
		Nauplius, cypris, kentrogon	Sacculina
	Insecta	Nymph	Romalea
		Caterpillar, pupa	Pieris
		Grub, pupa	Melolontha
		Larva, pupa	Drosophila
	Trilobita	Protaspis	Trilobita
	Merostomata	Trilobite larva	Limulus
Protochordata	Tunicata	Tadpole (appendicularia)	Amaroucium
Hemichordata	Enteropneusta	Tornaria	Balanoglossus
		Non-tornaria (ciliated larva)	Saccoglossus
	Pterobranchia	Tornaria-like	Cephalodiscus

ture for the development of a new mature form capable of reproduction ("paedomorphosis"), the possibilities for the evolution of extremely diverse types become infinitely more pronounced than if evolution occurred only by increments added to the existing adult stages, as postulated by the rigid adherents to the recapitulation theory. In fact, one could scarcely take Haeckel's recapitulation theory seriously without eliminating larval adaptations from the direct line of evolutionary descent. Only in those forms that show a direct development could ontogeny possibly recapitulate phylogeny.

Table 1 lists the larvae, insofar as they are known, for the various animal phyla. Omitted from the table are many groups which have no larval stages but which undergo direct development. These include the Turbellaria Acoela, Rhabdocoela and Tricladida, the Chaetognatha, Rotifera, Oliogochaeta, Cephalopoda, Cephalochorda, and Aschelminthes. In addition to these, within many groups listed as having typical larvae there are species with telescoped development or other special embryonic adaptations, enabling them to omit the larval stage.

The phylum Aschelminthes, which includes the Nematoda, presents an interesting situation. Hyman (and others) state that there are no nematode larvae, but that the young are juveniles. However, the transition from microfilaria to adult in Wuchereria, and from rhabditiform to filariform "larvae," etc., in the hookworms, with intervening molts, suggests a process sufficiently akin to metamorphosis to justify retaining the older terminology.

Space limitations make it impossible to describe more than a few of these larval types. Those selected represent a series with increasing complexity up to the trochophore, the most widespread in occurrence of all larvae.

The free-swimming amphiblastula of calcareous sponges is ovoid, with anterior hemisphere consisting of flagellated micromeres (to direct the forward movement) and posterior hemisphere of larger, non-flagellated, granular macromeres. Subsequently the macromeres overgrow the micromeres, by a process of embolic gastrulation, producing a blastopore, which becomes the region of attachment. The parenchymula of the Demospongia is a flagellated larva with a solid interior cell mass. Instead of invaginating, the flagellated cells are said to migrate into the interior during metamorphosis to become choanocytes.

The planula of the Cnidaria is a stereogastrula with ciliated ectoderm and solid endoderm which has arisen, in different forms, by unipolar ingression, multipolar ingression or by delamination. This polarized, mouthless larva is at first free-living, then attaches by its side or anterior end and develops a gastrovascular cavity in its endodermal mass.

In some Hydrozoa there is a larva intermediate between the planula and the attachment stage. This is a creeping form, the actinula, which is essentially a planula with blunt aboral tentacle buds which later lengthen. Buds of the oral tentacles develop after the gastral cavity has differentiated internally and broken through to the exterior to form the mouth.

In the Scyphozoa, the planula is free-swimming for four or five days. After attachment the hollow planula elongates, broadens, endodermal ridges (taeniolae) form in the gastral cavity, and the mouth opening develops, along with four tentacle buds. This larva with its flat, open oral disc is a scyphistoma (also called a scyphula or hydratuba). Extra tentacles develop between the original four. Then a constriction around the scyphistoma, separating the oral disc from the rest of the body, leads to the formation of the ephyra larva. This process is usually repeated (polydisc strobilization) varying numbers of times, producing a series of ephyrae. The scyphistoma tentacles are resorbed and replaced by the lobes of the ephyrae, making the strobilized larvae resemble a pile of saucers.

The free-living larva of the polyclads was discovered by Johannes Muller. After epibolic gastrulation, a two-layered, flattened, oval embryo with ciliated ectoderm results (Planocera). There is a stomodaeum opening into a broad gastral sac, lined with small endodermal cells. At this time eight ciliated lobes (characteristic of Müller's larva) connected by a continuous band of especially long cilia appear below the equator, projecting posteriorly. At the aboral pole a plate of small cells covers a ganglionic mass. An apical sensory tuft projects at this point, and there may be a caudal tuft at the opposite pole. The mouth shifts ventrally. During metamorphosis growth in length is coupled with gradual reduction of the lobes, eyes differentiate and the external shape characteristic of a typical polyclad results.

Some species of Stylochus produce a larva which has only four lobes, called Götte's larva.

There is a superficial resemblance, at least, between a Müller's larva and a pilidium larva of a nemertean. The eight lobes suggest, also, an affinity with the ctenophores.

The larva of Cerebratulus and certain other Heteronemertini is called the pilidium (J. Müller, 1847). Its external shape resembles a Greek helmet with spike and ear-lappets, the spike being an apical tuft of long cilia, the ear-lappets (oral lobes) lateral ciliated appendages on each side of the mouth. The helmet encloses a primitive digestive system consisting of large mouth, roomy oesophagus (foregut) and smaller blind intestine (midgut). A blastocoele is present, in which larval muscle bands develop. Invaginations from the ectoderm, originally called the prostomial and metastomial discs but now recognized as pairs of cephalic, cerebral and trunk discs and unpaired dorsal and proboscis discs, grow around the digestive tract, finally fuse together, and form the skin and muscular body wall of the future nemertean. When metamorphosis is sufficiently advanced, the juvenile worm frees itself from the investing helmet. This metamorphosis is a classic example of the formation of the adult from a small portion of the larva, the remainder, functional only during the pelagic larval existence, being cast off.

In Lineus the larva is a ciliated postgastrula lacking apical plate and spike, lappets and ciliated oral band, which remains inside its egg membrane. During metamorphosis four or more ingrowths give rise to the definitive nemertean body wall. Subsequently the external cell layer is stripped off. This reduced, sedentary pilidium is called the "larva of Desor," after its discoverer, E. Desor (1848).

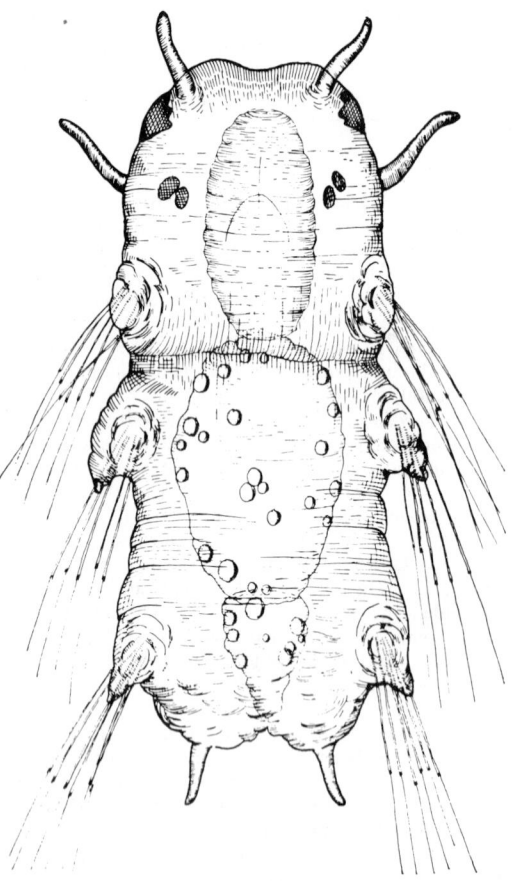

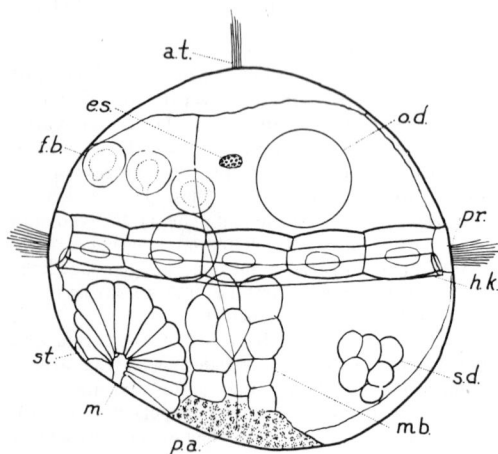

Fig. 1   Nereis trochophore (20 hours) viewed from left. a.t., apical tuft; e.s., eye-spot; f.b., frontal body; h.k., head-kidney; m., mouth; m.b., mesoblast band; o.d., oil droplet in endodermal cell; p.a., anal pigment area; pr., trototroch; s.d., somatoblast derivatives (anlage of seta sac); st., stomodaeum. (Published with permission of the *Journal of Experimental Zoology*.)

Fig. 2   Nereis nectochaeta.

The annelidan or molluscan trochophore varies considerably from species to species. Typically ovoid, it may be nearly spherical in shape, as in Nereis, or possess the double cone shape of a toy top, as in Hydroides. The most significant characteristic is the presence of the prototroch (which gives the larva its name), consisting of one to three equatorial bands of preoral cilia, separating the pretrochal from the post-trochal hemisphere. Certain trochophores develop secondary ciliated bands, such as a metatroch (or paratroch) in the post-trochal hemisphere and a telotroch near the anus. At the animal pole is an apical tuft of one or more stiff cilia. Below this is a thickened ectodermal ganglionic plate, the apical sense organ. The Nereis trochophore has a single ciliated equatorial prototrochal band. In the pre-trochal region there is a pair of red eye-spots, symmetrically placed, and anterior to them are five spherical "frontal bodies," of unknown function, symmetrically arranged in an arc. The mouth, which opens into a blind tube, the stomodaeum, lies in the median ventral line about halfway between the prototroch and the margin of the area of greenish-black anal pigment which marks the posterior extremity. The proctodaeum, extending up into the four solid endodermal cells which fill the interior of the larva, is likewise blind. There is no body cavity. Toward the fortieth hour of development, a granular reddish-brown pigment appears in the cells adjacent to the twelve large prototrochal cells. This pigment parallels the position of the head-kidneys. Seta

sacs (derived from the ventral plate) make their appearance during early metamorphosis as elongation of the larva begins. This metamorphosing larva is called the metatrochophore by Thorson. Setae are protruded from the two anterior pairs of sacs at about the fortieth hour. Segments arise from the post-trochal region as elongation proceeds, their musculature being supplied by the mesoblast bands derived from the teloblasts. The swimming-crawling larva with setae is called a nectochaeta.

The Hydroides trochophore differs from that of Nereis in several respects, including the presence of metatroch and telotroch. It has a single eye-spot of red pigment, located on the right side (Sabellaria has one, only, on the left). The digestive system is a complete tube, consisting of mouth, oesophagus, stomach, intestine and anus, enabling this larva to feed on diatoms. There is a blastocoele but no true coelom at this stage. Larval muscles are well developed as circular bands under the prototroch and metatroch, and as longitudinal strands between the apical plate and stomach. A pair of larval kidneys (typical protonephridia) extends from the oesophagus to near the anus. There is likewise an anal vesicle (specialized in Hydroides), which displaces the anus dorsally.

Some annelid larvae (including those of Arenicola, Arica and some terebellids) are often uniformly ciliated (atrochal) and are poor swimmers. In Arenicola the atrochal larvae remain within the jelly mass and do not

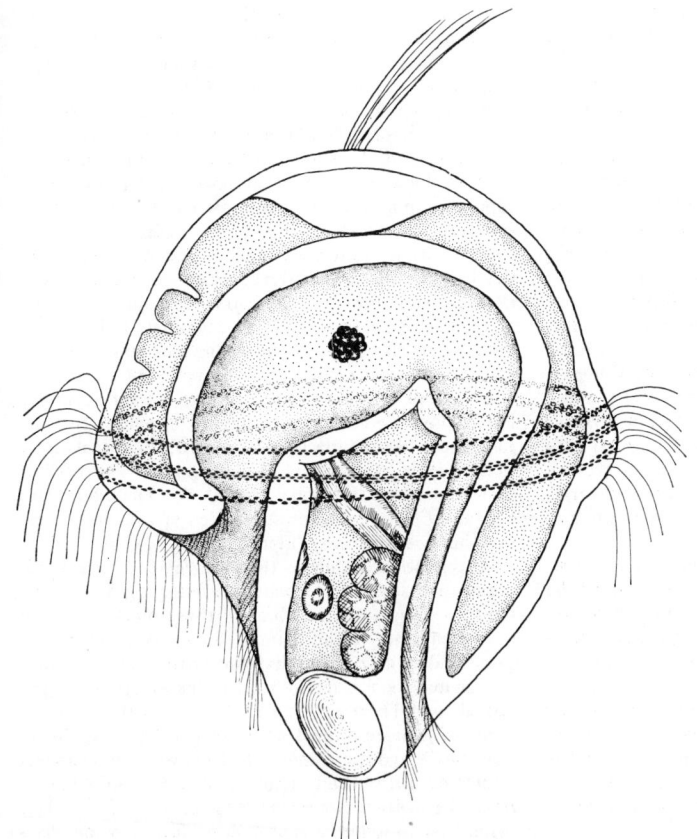

Fig. 3   Hydroides trochophore
(6-day). See text for description.

hatch until three to five pairs of setigerous segments have developed.

DONALD P. COSTELLO

References

Hyman, L. H., "The Invertebrates," Vols. 1–6, New York, McGraw-Hill, 1940–1967.
Richards, A., "Outline of Comparative Embryology," New York, Wiley, 1931.
Thorson, G., "Reproduction and Larval Development of Danish Marine Bottom Invertebrates," Copenhagen, Reitzel, 1946.

LATERAL LINE

This system of sense organs, restricted to cyclostomes, fish and amphibians, was discovered about the middle of the seventeenth century (Stenonis, 1664) but its sensory nature was not recognized until much later (Jacobsen, 1813; Leydig, 1850). The essential functional units are discrete, localized sensory areas, composed of receptor and supporting cells, known as sensory hillocks, or neuromasts. These neuromasts are typical hair cells, and show a strong structural resemblance to auditory hair cells. The lateral line is ontogenetically, phylogenetically and functionally related to the inner ear. Highly specialized head neuromasts called ampullae of Lorenzini are found in sharks and rays, and function as receptors of weak electrical stimuli (in the microvolt range). The function of the regular lateral line organs is still not completely understood. They are hydrodynamic detectors, sensitive to water turbulence phenomena (Cahn, 1967), but few quantitative studies have been carried out.

The presence of the lateral line system is associated with orientation in an aquatic habitat. It is persistent during the entire lifetime of fishes, but variously restricted in the amphibians by such factors as metamorphosis, and the permanent or temporary adoption of a terrestrial existence. Normally, lateral line organs, and the associated nerves and ganglia, are lost when amphibians metamorphose and become permanently terrestrial, but not all metamorphosed amphibians leave the water or lose their lateral line system (i.e., *Xenopus, Pipa, Bombina*), nor do all metamorphosed aquatic amphibia retain these organs (i.e., *Pseudis*). In *Triturus* the organs regress and are covered by epidermal cells during the terrestrial red eft phase but again became functional on the return to water. There are other variations in this water-lateral line relationship which cannot be cited for lack of space.

Ordinarily, the individual sense organs are distributed in rows which are usually arranged in groups, on the head and on the trunk. The main rows of the head are the supraorbital, infraorbital, pre-opercular, and mandibular, with various accessory rows in some animals. On the trunk, a single row is most common, located at or near the level of the external margin of the horizontal myoseptum. This may be supplemented by additional rows dorsal and/or ventral to the main row. The lateralis nerves parallel the rows and give off small branches to the individual organs. Afferent and efferent components are present.

In its simplest form the system may consist of neuro-masts which occur separately and occupy a superficial position in the epidermis in direct contact with the environment, or may lie at the bottom of shallow basins or deep pits. The ampullae of Lorenzini are located in pits. In other cases the neuromasts are spaced along open grooves of varying lengths and depths, or the grooves may be completely closed except at certain intervals where they remain open to provide pores which connect the canals with the exterior. Various combinations of superficial organs and systems of canals may be present. (Lowenstein, 1957; Wright, 1951).

The lateral line arises as a thickening of the ectoderm on either side of the head, in association with the developing ear, and differentiates into pre-and post-auditory placodes. From these centers cells extend or migrate in definitive lines to produce the patterns of rows just described. At intervals along these developing lines local sensory areas differentiate into supporting and sensory cells. The head organs derived from the preauditory placode are innervated by way of the seventh cranial nerve, while those originating from the postauditory placode are supplied by fibers associated with the ninth and tenth cranial nerves. Contributions from the eighth cranial nerve will also appear.

The cells that form the respective line-ganglia differentiate *in situ* and are closely associated with ganglia of the 7th, 9th and 10th nerves. Lateral line nerve connections do not appear necessary for the differentiation of the primary organs or their subsequent budding to produce additional organs. The central connections are in the dorsal medulla in close association with the auditory nuciei. Structurally the lateral line neuromasts are remarkably similar whether they belong to cyclostome, fish or amphibian. Under regular microscopy they usually appear pear-shaped in vertical section, with centrally located club-shaped sense cells, found in the narrower superficial half of the organ. The tapered free surface of the sense cell bears a hair-like process, with an elongated cone of gelatinous material called a cupula, capping the hairs, reminiscent of the inner ear cupulae. The proximal ends of the neuromasts are in contact with nerve terminations, so that deformation of the hairs initiates the nerve action potential. Under the electron microscope the hairs were found to consist of a bindle of stereocilia with an asymmetrically located kinocilium. Successive organs along a canal show alternate positions of kinocilium relative to the stereocilia, at either the anterior or posterior end of the neuromast. Electrophysiological studies have shown that this arrangement of the cilia imparts a directional sensitivity to the organ. Stimuli that displace the hairs in a direction from the stereocilia toward the kinocilium are excitatory, whereas when stimulated in a reverse direction they are inhibitory (Flock, 1967).

All investigators agree that the regular lateral line organs are sensitive to water turbulence phenomena, but they disagree as to which of these phenomena are most significant to the animals. Low frequency propagated underwater sound and non-propagated or slowly propagated local water perturbations all stimulate the lateral line. There is some evidence that the response to sound is more a function of the trunk organs (Suckling, 1967), and the response to local water movements a function of the head organs (Schwartz, 1967), but no broad distinctions can be made as yet. If the lateral line is sensitive to sound, it is probably affected by the physical parameters such as particle velocity or particle displacement rather than the pressure component of the sound wave (Harris and van Bergeijk, 1962). The lateral line appears sensitive to all kinds of local water movements such as those generated by swimming movements of other fish, moving objects in the water, water currents, surface waves, and even to the changes produced in the flow pattern around an individual when he swims near a motionless object in the water. There is much need for quantitative experiments with these stimuli using behavioral and neurophysiological techniques. Behavioral studies have been hampered by the fact that total extirpation of the lateral line has been unsuccessful, because the system is so diffuse. Dijkgraaf (1967) and Schwartz (1967) demonstrated that local elimination caused localized loss of sensitivity to hydrodynamic cues, but much further work is needed. There has also been very little work done on the regulation of lateral line function by the central nervous system and on the extent of functional overlap between lateral line and ear.

PHYLLIS H. CAHN

References

Cahn, P. H., ed., "Lateral Line Detectors," Bloomington, Indiana, Indiana University Press, 1967.

Dijkgraff, S., "Biological Significance of the Lateral Line Organs," in "Lateral Line Detectors," (see above), 1967.

Flock, A., "Ultrastructure and Function in the Lateral Line Organs," in "Lateral Line Detectors," (see above), 1967.

Harris, G., and W. Van Bergeijk, "Evidence that the Lateral Line Organ Responds to Near-Field Displacements of Sound Sources in Water," J. Acoust. Soc. Am., 34: 1831, 1962.

Jacobsen, 1813, see Wright 1951.

Leydig, 1950, see Wright 1951.

Lowenstein, O., "The Acoustico-Lateralis System," in M. E., Brown, ed., "The Physiology of Fishes, 1957.

Schwartz, E., "Analysis of Surface Wave Perception in Some Teleosts," in "Lateral Line Detectors," (see above), 1967.

Stenonis, 1664, see Wright, 1951.

Suckling, E. E., "Electrophysiological Studies on the Trunk Lateral Line System of Various Marine and Freshwater Teleosts," in "Lateral Line Detectors," (see above), 1967.

Wright, M., "The Lateral Line System of Sense Organs," Quart. Rev. Biol. 26: 264, 1951.

LATICIFERS

The laticifer can be defined as a particular type of living plant cell containing a specialized vacuolar content, termed *latex*, which may be conspicuously colored. Two types of laticifers have been described: 1) an articulated form consisting of individual latex-containing cells irregularly distributed in the plant body (*Parthenium*, Compositae), or occurring as series of superimposed cells forming vessel-like structures of variable length (*Achras*, Sapotaceae). Several separated series are usually present in the primary body, each originating during the development of the embryo. Additional series may be differentiated from deriva-tives of vascular cambium, perpetuating the laticifer system in the secondary body (*Hevea*, Euphorbiaceae). Anastomoses may develop between the several series of laticifers (*Cichorium*, Compositae), while in other plants no anastomoses are present (*Allium*, Liliaceae). 2) a non-articulated form consisting of a single extremely elongated cell. A small number of laticifer initials (approximately 4–46) are described as being formed during development of the embryo, no additional initials arise from the vascular cambium. As plant develops each laticifer commences to grow bidirectionally, the tip of each cell intrusively penetrating between adjacent cells until occupying a position in the vicinity of the meristem in the apex of shoot and possibly root. The number of initials is quite constant in a given species (approximately twenty-eight in *Nerium oleander* L.), all arising during formation of cotyledons, the series of initials forming a ring at the level of the cotyledonary node. The initials occupy a position in the zone of future procambium, usually becoming distinguishable before these elements. During subsequent growth the laticifers invade various tissues of the axis obscuring their initial position of origin. In some genera (*Nerium*, Apocynaceae; *Euphorbia*, Euphorbiaceae; *Asclepias*, Asclepiadaceae) growing tips of laticifers possess a capacity to bifurcate forming a system of branches which permeates most of the plant body. Laticifers initials in certain genera (*Eucommia*, Eucommiaceae, *Cannabis*, Cannabinaceae) reportedly do not bifurcate. During growth of laticifer repetitive nuclear divisions result in formation of a multinucleated protoplast.

The chemical composition of latex, constituting cell sap, is variable in different genera as well as between species of a genus. It may contain various substances including: carbohydrates (sugar, Cichorieae; starch, *Euphorbia*); protein (*Ficus callosa*, Moraceae); alkaloids (*Papaver somniferum*, Papaveraceae); tannins (*Musa*, Musaceae), enzymes (papin, *Carica papaya*, Caricaceae), as well as organic acids, salts, fats, sterols, and mucilages. Various terpene derivatives may be dispersed in latex, rubber being most commonly identified. Rubber, present in many genera, usually constitutes less than five percent of latex, although in *Hevea*, it may be as high as fifty percent. In some plants latex may be clear (*Nerium oleander)*, milky (*Euphorbia*, *Asclepias*) or highly colored by carotinoid-containing plastids (*Argemone*, *Chelidonium*, *Sanguinaria*, Papaveraceae).

Laticifers are widely but sporadically distributed throughout plant kingdom, being present in certain fungi (*Peziza*, Ascomycetes; *Lactarius*, Basidiomycetes); certain ferns (*Regnellidium*, Marsileaceae); and, monocotyledoneae and dicotyledoneae (Cabombaceae, Cactaceae, Convolvulaceae, Lobeliaceae) in addition to previous references.

Function(s) of laticifer system remains obscure having been described as a storage reservoir, food conductive system, protective, excretory and secretory system.

The concept of laticifers represents a broad interpretation, since the two types are not homologous ontogenetically. Further investigations are necessary to determine possible phylogenetic relationships between the articulated and non-articulated types, as well as other cell types.

PAUL G. MAHLBERG

References

Chauveaud, M. G., "Récherches embryogeniques sur l'appareil laticifere des Euphorbiacées, Urticacées, Apocynées et Asclepiadées," Ann. Sci. Nat. Bot., ser. 7, **14**: 1–161, 1891.

Bonner, J., and A. W. Galston, "The physiology and biochemistry of rubber formation in plants," Bot. Rev. **13**: 543–596, 1947.

LEAF

A leaf is a lateral appendage borne upon an axis, or stem, and the stem and leaves together constitute a shoot. A leaf is typically a flat and expanded organ, primarily concerned with PHOTOSYNTHESIS; but frequent and drastic deviations in both morphology and function occur. Although leaves are treated as distinct organs in classical morphology, the full biological significance of the leaf emerges only when it is considered as part of the SHOOT system, having its origin together with the stem in the shoot apical MERISTEM. In general usage, and even in some technical treatments, the term leaf has been used for flattened appendages in many groups of plants, and particularly for those of certain algae and of the sexual stages of mosses, as well as for organs of the vascular plants. It seems wise, however, to restrict the use of the term to the sporophytic generation of the Tracheophyta (vascular plants) in order that some degree of homology may underlie the concept of the leaf. With rare exceptions, leaves are present on the sporophytic axes of all groups of vascular plants, including, according to modern interpretations, the living members of the primitive Psilopsida. Certain extinct members of this group, known only from fossils, were truly leafless. It is not, however, certain that leaves in all vascular plants have had a common evolutionary origin. The leaves of the Pteropsida (ferns and seed plants) are called megaphylls and are held to have arisen through modification of parts of the shoot system so that they are equivalent in one sense to lateral shoots. In the other groups, on the other hand, the leaf, or microphyll, is thought to represent an enation (outgrowth) from the stem which has gradually become vascularized in the course of its evolution. Whether or not this view is correct, leaves of all vascular plants show certain similarities in origin and development.

The traditional representation of a leaf depicts a stem-like stalk, or petiole, attached to the stem and surmounted by a single, flattened blade or lamina. In many cases, however, the petiole is lacking and the lamina abuts directly, or nearly so, upon the stem, a condition described as sessile. The shapes of the lamina are numerous (Fig. 1a) and are described and named in many textbooks of general botany or of taxonomy. In some species leaf shape is relatively constant; but in other cases it varies widely even on a single plant, a fact which causes considerable difficulty to taxonomic botanists, and especially to paleobotanists who must attempt to identify leaf impressions. Particularly fine examples of such leaf variability are to be seen in species which have distinctive juvenile leaves. Ferns ordinarily produce in the sporeling stages a series of progressively more complex leaves until the typical adult form is attained (Fig. 1b). Leaves with a single lamina are simple. Many leaves do not have a single lamina; but rather consist of several leaflets borne upon an axis. Such leaves are said to be compound, and are pinnately or palmately so depending upon whether the leaflets are disposed featherlike along the axis or arise at or near a common point. The leaflets themselves may also be compound. Some leaves, especially those of monocotyledons, have bases which ensheath the stem, forming distinctive leaf sheaths. In some cases the presence of stipules, small, usually paired, lobes at or near the point of attachment of the leaf to the stem, complicates the structure of the leaf.

The leaf is composed of tissues which are fundamentally similar to those found in stem and root; but their organization is ordinarily very different. An epidermis, with a well-developed cuticle in most cases, covers both blade and petiole. Stomata are characteristically present, with a larger number typically being found in the lower or abaxial epidermis of the lamina than in the upper or adaxial. The fundamental or ground tissue of the lamina is designated as mesophyll and frequently shows a separation into an adaxial palisade layer and an abaxial spongy layer. The palisade layer consists of one or more tiers of columnar cells, containing many chloroplasts, and is generally recognized as the major photosynthetic tissue. Although the palisade layer does contain well developed intercellular spaces, these are much more extensively developed in the spongy layer which is adjacent to the numerous stomata of the lower epidermis. The extensive intercellular spaces of the mesophyll facilitate the process of gas exchange which is essential for efficient photosynthesis. In some leaves, the mesophyll is not differentiated into two regions, and in others there are palisade layers beneath both the upper and lower surfaces. In the petiole, the fundamental tissue shows none of the specializations found in the lamina and closely resembles the cortex of the stem. The vascular system, consisting of xylem and phloem, is continuous with that of the stem, and extends through the petiole in the form of one or several bundles. In the lamina it expands greatly by repeated ramifications to all parts of the blade in either a pinnate or a palmate pattern. Monocotyledons usually exhibit a venation pattern in which the main veins approach a parallel condition. So extensive is the ramification of the vascular system in the lamina that no part of the mesophyll is more than a few cells distant from a vein ending. The ultimate veinlets commonly consist of xylem only and are sometimes associated with thick walled sclereids.

The leaf has its origin from the apical meristem of the shoot, along with the tissues of the stem. It arises as an outgrowth at or near the base of the apical cone or mound, often beginning as a rather broad protuberance called a leaf buttress. As soon as the outgrowth is distinctly visible it is referred to as a leaf primordium. Early growth of the primordium is primarily at its apex and the outgrowth increases in length for some time after its inception. In monocotyledons the apical phase is very brief. In ferns, wh:ch have a distinct apical initial cell, this phase is of long duration, a matter of several years in some slow-growing species. Whatever its duration, apical growth ultimately ceases so that the leaf, in contrast to the shoot as a whole, is determinate. Almost as soon as it is recognizable, the leaf primordium begins to show evidences of a dorsiventrality which ultimately is expressed in the flatness of the leaf and its distinct upper and lower surfaces. At an early stage, lateral growth

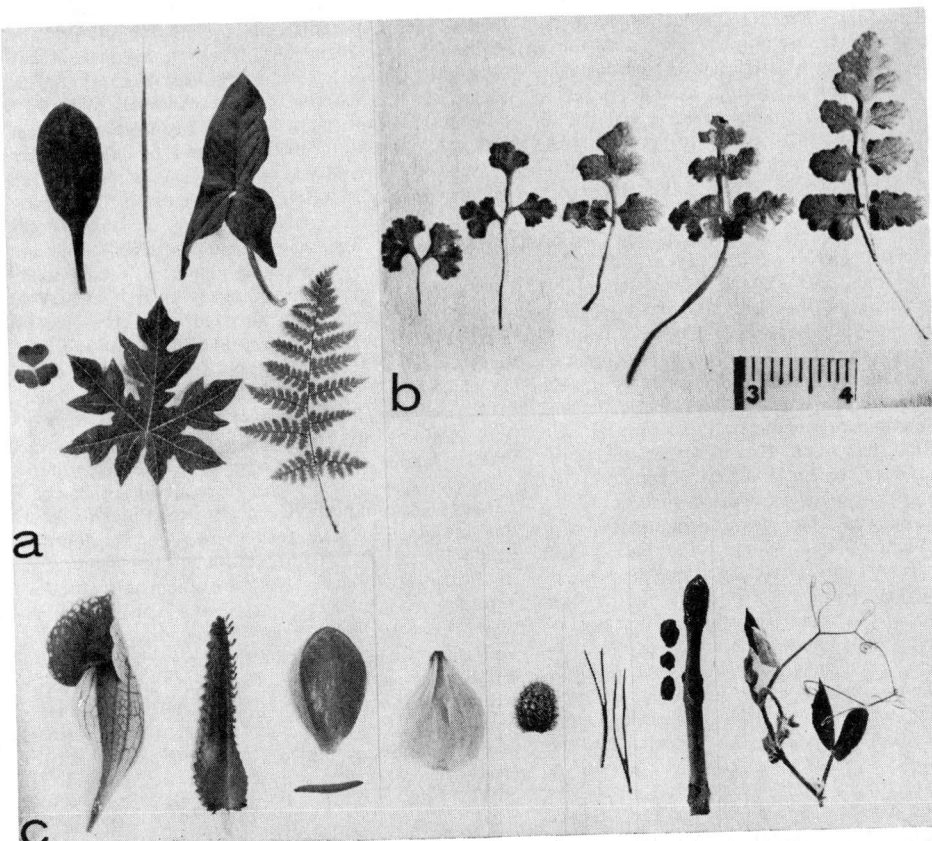

Fig. 1   (a) Variation in leaf form. Upper row, left to right: obovate leaf of *Euphorbia;* linear grass leaf; sagittate leaf of *Philodendron.* Lower row, left to right—palmately compound leaf of *Oxalis;* palmately lobed leaf of *Carica;* pinnately compound leaf of *Dryopteris.* × 0.15 (b) Ontogenetic variation in leaf shape of *Pteridium aquilinum.* From left to right, the first five sporeling leaves are illustrated. × 0.6 (c) Functional modifications of leaves. Left to right: insectivorous leaf of *Sarracenia;* asexual reproductive leaf of *Kalanchoe;* water-storing, succulent leaf of *Bryophyllum* with cross section below; food-storing and bud-protecting leaves of *Allium* bulb, foliar spines of *Mammillaria;* reduced-surface, needle sclerophyll leaves of *Pinus;* protective winter bud of *Hippocastanum;* foliar tendrils of *Lathyrus.* × 0.15.

begins in the leaf by means of marginal meristems, and it is to these that the typical broad expanse of lamina is to be attributed. If these meristems are general in distribution, a single lamina is produced resulting in a simple leaf. If they arise in restricted regions only, a compound leaf is the result. Marginal growth, too, is of limited duration; and ultimately ceases, to be replaced by intercalary growth. In monocotyledons, intercalary growth occurs at a very early stage and most of the leaf is produced in this fashion by means of an intercalary meristem at its base. In other plants, intercalary growth is responsible for a considerable enlargement of the lamina, for the development of the petiole, and for the formation of much of the axis in pinnately compound leaves. The final phase of development is one of expansion and maturation involving final enlargement and differentiation of cells. This phase occurs, in many perennial plants, at the beginning of the growing season, the leaves having passed the period of dormancy in a partially developed state within a terminal or an axillary bud. Within such a bud the leaves are variously folded

or rolled in a pattern ordinarily characteristic of the species.

The development of the leaf has, in recent years, been the subject of extensive experimental investigation, prompted in part by the importance of the organ itself, and in part by the fact that a discrete organ provided favorable material for such study. Experimental surgery involving cuts made with fine knives in the region of the shoot apex, has been especially revealing as to the nature of causal factors underlying leaf development. Isolation of an incipient leaf position from the apical meristem by a deep incision has, in potato, lead to the formation of a radially symmetrical, or centric, leaf instead of a normal dorsiventral one. In certain ferns, such isolated presumptive leaf positions, and even very young but distinctly visible primordia, frequently develop as shoots rather than as leaves. This evidence indicates that a leaf acquires its dorsiventrality, and even its determinate growth pattern, as a result of the conditions to which it is exposed at the shoot apex, not the least of these conditions being its one-sided relationship to the shoot apical meristem. Another approach to the same problem has made use of the techniques of sterile nutrient culture. Leaves of several ferns and flowering plants excised as young primordia, have developed to maturity in sterile culture, demonstrating that the development of the leaf is largely controlled within the leaf itself once the fundamental pattern has been established in the process of leaf determination. In at least one fern, however, it has been possible to excise leaf primordia prior to irreversible determination; and, as might be expected, such primordia give rise to shoots rather than to leaves. Although the basic pattern of leaf development appears to be internally controlled after determination has become irreversible, such features as size, shape and even the production of sporangia, in the case of certain ferns, can be influenced by specific treatment in culture. It seems very likely that in the next few years a combination of surgical and cultural techniques will lead to a much fuller understanding of the control of leaf development, and consequently to a better grasp of morphogenetic processes in general.

The leaf, in spite of a high degree of internal control in its development, is part of an integrated shoot system, and some of the most interesting aspects of leaf development relate to this integration. Phyllotaxis, the arrangement of leaves on the stem, has long been a subject of considerable interest and has, in recent years, been subjected to experimental analysis. An individual node may bear one, two or several leaves, and these arrangements are referred to respectively as alternate, opposite and whorled. In the alternate arrangement, the greatest interest is attached to the helical pattern in which the leaves, in the order of formation, are distributed along the axis. The final position of a leaf on the mature axis depends upon its relative position of initiation at the shoot apex, with some adjustments during expansion and maturation. A great deal of attention has consequently been focused upon phenomena in the shoot apex. The best evidence at present suggests that a field concept is the most plausible interpretation. Such a concept postulates, around the center of the apical meristem and each developing primordium, an inhibitory field, presumably of a chemical nature, within which no new primordium can occur. Such a pattern in a growing apex could lead to the placing of new primordia in an orderly fashion; but it leaves unexplained the factors which actually initiate leaf development in those areas which are not inhibited. It is also evident that the leaf, during its development, exerts a considerable influence upon the development of the stem, at least in seed plants and ferns. In the ferns it has been repeatedly demonstrated that the formation of a parenchymatous leaf gap in the vascular system is dependent upon the primordium with which it is associated. A similar claim for seed plants is less widely accepted; but the close association of the mature vascular system of the stem with its leaves in these plants makes a developmental relationship almost a certainty. The microphylls of the lower vascular plants do not seem to exert a corresponding influence. At a somewhat grosser level of overall expansion of the shoot, it is apparent that auxins from the developing leaves may play an important role; and the involvement of other growth factors as well is highly probable.

From the functional point of view, the leaf may be regarded as typically an organ of photosynthesis, and external morphology as well as internal anatomy show a high degree of specialization in relation to this function. It has long been recognized, however, that the leaf may be highly modified in its morphology (Fig. 1c); and, in such cases, its function is correspondingly altered. A few examples of such modified leaves may be considered. On underground axes and in inflorescences leaves are often reduced to mere scales whose function in any capacity is doubtful. In some cases, particularly in xerophytes, such reduction occurs on aerial shoots with the stems assuming the photosynthetic function. In species which develop dormant buds, cataphylls or bud scales, in which the lamina is greatly reduced or entirely suppressed, are often formed as protective structures. Tendrils, thorns, absorbing leaves and storage leaves (in bulbs) represent other modified forms in which the photosynthetic function has been largely or entirely lost. The modification of leaves as specialized trapping and digesting organs in insectivorous plants should also be mentioned. One of the more remarkable aspects of such modifications is the occurrence of leaves of more than one type on the same plant. In such instances divergent pathways of development are followed by primordia which are presumably equivalent at initiation. Such a phenomenon poses interesting morphogenetic questions which may be analysed experimentally. From the evolutionary point of view, the various parts of the angiosperm flower (sepals, petals, stamens and carpels) are held to be of a foliar nature; but this opinion is by no means universal. Support for it derives principally from a study of members of the presumed primitive order Ranales in which the foliar origin of these organs seems quite evident.

TAYLOR A. STEEVES

References

Esau, K., "Plant Anatomy," New York, Wiley, 1965.
Wardlaw, C. W., "Phylogeny and Morphogenesis, London, MacMillan, 1952.

LEEUWENHOEK, ANTHONY VAN (1632–1723)

One of the most extraordinary self taught scholars of all time. His sole "education" was as an apprentice to the cloth trade. He early abandoned this pursuit and, having acquired some local repute as a naturalist, was given a sinecure by the city fathers of his native town of Delft, which enabled him to devote the rest of his life to his hobbies. In the course of his life he contributed 112 papers to the Philosophical Transactions of the Royal Society of London and 26 papers to the Memoirs of the Paris Academy of Sciences. He was elected a fellow of the Royal Society in 1680 and of the Paris Academy of Sciences in 1697.

These almost incredible feats derived first from his powers of accurate observation, and second from his outstanding manipulative ability which enabled him to grind small lenses with a magnification and resolution far in excess of the compound microscopes of his day. Little is known of his instruments since he was of a secretive and jealous disposition and neither sold nor gave his better instruments to anyone. The few preserved in museums are those which he himself

disposed of as inferior. He had no fixed plan of study—
indeed he could not study since he knew no other language than Dutch—but passed with lightning rapidity
from one subject to another, making major contributions to each. By 1674 he had accurately described
numerous Protozoa and Bacteria as well as the erythrocytes of fish, amphibia and mammals. Three years
later, he published a description of the spermatozoa
of many mammals and two years after that, demonstrated for the first time the nature and life history of
yeast. He described the microscopic structure of dicotyledonous and monocotyledonous plant stems and contributed an accurate account of the embryology of the
fresh water mussel Unio. He was continuously vehement in his attack on the theory of spontaneous generation and published in accurate detail the life histories
of the grain weevil, the flea, the aphis, and the ant. He
was the discoverer of the phylum Rotatoria and
pursued his investigations of these to the point of dis-
covering their ability to withstand drought. He discovered, without knowing it, the Mendelian dominant
hair colors in rabbits and described in great detail the
structure of bone and teeth. He lived to a ripe old age,
craggy and argumentative to the last.

PETER GRAY

Reference

Dobell, C., "Anton van Leeuwenkoek and his Little
Animals," New York, Russell, 1958. [reprint, New
York, Dover, 1962].

Fig. 1

LEMUROIDEA

A suborder of the mammalian order PRIMATES comprising all the more primitive forms not included under
the designation of monkeys or apes (simians). In this
sense the term is synonymous with prosimians and
includes, besides the lemurs proper of Madagascar,
such bizarre forms as the lorises, pottos and bush-
babies (or galagos).

All prosimians agree in retaining the naked, moist
skin at the tip of the snout, associated with slit-like
nostrils. In these features they contrast with all the
higher Primates which have lost these primitive attributes insofar as the nostrils are rounded and hairy
to their margins or for a short distance within. The
teeth are peculiar, the lower front teeth in particular
being modified into a comb-like structure, their crowns
being horizontally disposed and forwardly directed.
This specialization is associated with the cleaning of
the dense woolly fur with which most prosimians are
covered. Correlated with this, a further toilet adaptation is the modification of the nail on the second toe of
the hind foot into a claw-like structure, all the other
toes being provided with normal flat nails, as in monkeys. The thumb and big toe are large and widely
separable from the other digits, rendering the hands
and feet capable of great grasping power—an important
adaptation to a life among the trees. As most prosimians are nocturnal in habit, the eyes are large and
directed forwards, while the ears are frequently large
and mobile. Smell also continues to play a large part
in the animals' appreciation of its environment.

In the skull the eye-sockets have complete bony
margins, but deeply they are not shut off by bone from
the space occupied by the temporal masticating muscles. The tail is usually long and well furred, but never
prehensile; a few species (e.g. the lorises) are tailless.
Palms and soles are naked and covered with modified,
ridged skin, but instead of being completely ridged as
in monkeys, the markings are confined to small islands.

In the brain the parts concerned with smell are large,
but the cerebral hemispheres usually rather smooth
and where convolutions are present their design resembles that in carnivores rather than that of monkeys.
The cerebellum is not covered by backward growth of
the cerebrum.

A curious character common to all lemuroids is
the persistence of a structure called sublingua. This is
a pointed or serrated horny plate beneath the fleshy
tongue. Its presence is correlated with the modified
lower front teeth and it seems to act as a cleaner for
them.

In the alimentary system the stomach is usually
simple, but shows some complications in a few specialized leaf-eating species. The intestine is long and
its hinder part often sacculated and held in a fixed loop.
In the female the uterus is of the lower mammalian
type (*uterus duplex*), whilst the foetal membranes are
also of a primitive character, the placenta being diffuse
and showing little intimacy of connection between maternal and foetal components. There is a distinct
breeding season; between seasons the reproductive
organs, in both sexes, remain quiescent.

As regards geographical distribution all the typical
lemurs are confined to the island of Madagscar, where
they are represented by a wealth of widely varied forms
which comprise no less than $\frac{2}{5}$ of the total mammalian
fauna of the country. Represented sparingly in the
tropical parts of the African mainland, and in parts of
Asia are the forms collectively grouped as lorisoids.
These are derived from an early offshoot of the ancestral prosimians and have specialized in one of two
ways. In the Asiatic Slender lorises and Slow lorises
and the African Pottos (*Perodicticus* and *Arctocebus*)

the limbs are specialized for very powerful grasping and a slow deliberate hand-over-hand movement. These phenomena are correlated with anatomical specializations in the muscles and blood vessels of the limbs. By contrast the African Bush-babies are adapted for rapid movements especially in the elongation of the hind-limbs, which endows them with leaping powers of a high order. Bush-babies also have large, modified ears capable of rapid folding, enabling them to avoid damage during rapid leaps at night.

Of the Madagascar lemurs the beautiful Ring-tailed *L. catta* is one of the best known. It differs from its fellows in being a rock-dweller and in its diurnal activity. In one species (*L. macaco*) there is a well marked sexual difference, the males being entirely black and the females brown with white ear-tufts. The dwarf lemurs form a subfamily ranging in size from a mouse to a large rat; they are among the most primitive of the group. The peculiar leaf-eating species of the family Indriidae are delicate and rarely seen, as they thrive badly in captivity, in contrast to the other members of the family which are, under suitable conditions, surprisingly long-lived and fecund. Perhaps the most bizarre form is the Aye-Aye (*Daubentonia*), a nocturnal form with chisel-like front teeth, a round face and elongated and attenuated fingers. Originally mistaken for a rodent, its position has been long debated, but it is now regarded as an aberrant offshoot from the same stock as the Indriidae.

Many other species including some giant forms thrived in Madagascar until comparatively recent times, but seem to have been killed off by the invading tribes from Malaya. They are known only from subfossil bones preserved in the marshes.

In geological time the lemuroids date back to the dawn of the Tertiary epoch, being especially numerous during the Eocene, when they were widespread both in North America and Europe.

W. C. OSMAN HILL

References

Attenborough, D., "Zoo quest to Madagascar," London, Butterworths, 1961.
Von Boetticher, H., "Die Halbaffen und Kobold-makis," *in* "Die neue Brehm," Bucherei, Wittenberg, 1958.
Hill, W. C. Osman, "Primates / Strepsirhini," Edinburgh, Edinburgh University Press, 1953.
Jolly, A., "Lemur Behaviour; a Madagascar field study," Chicago, Chicago University Press, 1966.
Jouffroy, F. K., "La musculature des membres chez les lemuriens," Mammalia, Paris, **26:** 323, 1962.
Petter, J. J., "The lemurs of Madagascar," *in* "Primate Behaviour," ed. I De Vore, New York, Holt, Rinehart & Winston, 1965.

LENTICEL

A *lenticel* is a porous region in the bark of woody STEMS and woody ROOTS through which an exchange of gases between the interior of the plant and the external atmosphere occurs. In surface view lenticels appear as rounded, elliptical, or elongate zones, slightly elevated from the external cork layer. In size they range from structures just visible to the naked eye to regions sev-

eral centimeters in length. Some remain a fixed size, others become divided into smaller units by the formation of intervening areas of cork, and still others become larger as the circumference of the stem increases. These regions are prominent in twigs and in the smooth surfaces of older branches.

Lenticels form a part of the periderm. Structurally each region is composed of a zone of loosely arranged cells in which there is a relatively large amount of intercellular space. The cells of this region are termed *complementary cells* and are characterized at maturity by the lack of a protoplast and by thin, unsuberized walls. Some lenticels are composed of a homogeneous group of cells, others have narrow bands of more compact cells (*closing layers*) alternating with complementary tissue.

The development of lenticels is associated with secondary growth. Young shoots with primary tissues only lack them. Just before or just as stem elongation becomes completed, the first lenticels are initiated, usually in angiosperms from cells located directly below the stomata of the epidermis. As divisions take place among the parenchyma cells to form the complementary tissue, a phellogen originates. Subsequently, by the activity of the phellogen a lenticel increases in size and protrudes at the surface of the stem. The loss of cells by weathering is compensated for by the differentiation of additional cells from the phellogen. Later in development, if other periderm layers form internally, new lenticels arise in regions opposite the vascular rays, thereby facilitating the aeration of these parenchymatous patches.

RHODA GARRISON

Reference

Esau, K., "Plant Anatomy," New York, Wiley, 1965.

LEPIDOPTERA

An order of INSECTS comprising the butterflies and moths, numbering an estimated 200,000 species. They are most closely allied to the TRICHOPTERA (caddis flies), somewhat more remotely to the MECOPTERA, NEUROPTERA, DIPTERA and SIPHONAPTERA, these orders forming the so-called Mecopteroid orders or Panorpoid complex.

Lepidoptera are holometabolous, i.e., having structurally and functionally different larval, pupal and adult stages; adults typically have four similar wings, long multiannulate antennae and (a major ordinal distinction) wings and body covered with flat scales, overlapping like shingles on a roof, colored and forming a mosaic pattern; venation of the wings is surprisingly primitive, with nearly all the principal ancestral radiating veins and very few cross-veins; the mouthparts are formed for sucking (major ordinal distinction), with mandibles absent, the galea of the maxillae united into a long tube (proboscis), coiled like a watchspring beneath the head when not in use; in certain groups which do not feed as adults this tube may be reduced even to complete absence; in wingspread adults range from about 2 millimeters ($\frac{1}{8}$ inch) to 30 centimeters (12 inches) or more. The larvae (caterpillars, "worms") are elongate, usually nearly cylindrical, may be nearly naked or variously adorned with setae, spines (often venomous), "hair," tubercles, tufts; they have mouthparts formed for chewing, three pairs of thoracic legs and a variable number of pairs of prolegs or false legs on some of the abdominal segments.

Classification

The order is divided into two suborders of vastly unequal size, each of these in turn into a series of superfamilies, the latter containing from one to a dozen or more families each. Higher classification within the group—below suborder down to (and often including) family—is still in a very imperfect state. The suborders and the more important families in each are as follows:

Suborder *Homoneura*. Fore and hind wings similar in length and shape, with substantially the same number and arrangement of the veins in each. Only a few rather obscure families are known, among them the Hepialidae (ghost moths) and the Micropterygidae (the only Lepidoptera with functional mandibles as adults).

Suborder *Heteroneura*. Hind wing usually distinctly smaller than fore wing, and of different shape, with fewer veins: typically 8, compared to 10 or 12. Over 100 families are recognised, of which these are the most important: Cossidae (carpenter moths), Tineidae* (tineids), Psychidae* (bag worm moths), Tortricidae* (leaf rollers and others), Eucleidae (slug caterpillar moths), Zygaenidae (burnet moths), Pyralidae* (pyralids), Geometridae* (geometers, measuring-or inch-worm moths), Bombycidae (true silk worm moths), Saturniidae (giant silk worm moths), Sphingidae (hawk, sphinx, hummingbird moths), Noctuidae* (noctuids, owlet moths); Hesperiidae (skippers), Papilionidae (swallowtails), Pieridae (whites and sulphurs), Lycaenidae (blues, coppers, hairstreaks), Nymphalidae (nymphalids).

The last five families are butterflies; the remainder are moths. Families marked with an asterisk (*) contain especially many or important economic pest species. In total number of species the three families, Pyralidae, Geometridae and Noctuidae, probably outnumbered all the other families combined.

Very small moths (most of which are also primitive) require special techniques of preparation and can be studied only under a microscope. This has led to a widespread, but always informal, division of the order into *Microlepidoptera* or "micros" (these families of very small moths) and Macrolepidoptera or "macros" (all other moths, and all butterflies).

Though taxonomically only a rather minor part of the order, the butterflies are by far the most popular: most frequently collected, most widely known, most thoroughly studied. Yet to distinguish them from the moths is by no means easy. "Rough and ready" characters—bright colors, clubbed antennae, day flying habits—are not particularly good, even in temperate regions where they work best. In tropical regions all are practically worthless. The most reliable traits are the *combination* of (a) clubbed antennae, and (b) absence of a frenulum, which are almost universally valid, though neither is sufficient alone. The frenulum is a single spine (males) or group of spines (females) arising from the base of the hind wing, caught in a small chitinous loop (males) or fan of scales (females) on the fore wing, its purpose being to synchronize the motion of fore and hind wings in flight. It is found in most moths and absent from all butterflies, save for males of a single Australian species.

Distribution and Fossil History

Lepidoptera are virtually world-wide in distribution, from the shores of the Arctic Ocean to near Antarctica. At least a few exist on virtually every vegetated islet, no matter how far from the mainland. Indeed, migrating butterflies have been seen hundreds of miles at sea, far from any land at all. They have managed to adapt to an extremely wide range of terrestrial habitats, and even some aquatic ones, though even the best adapted among the latter must leave the water for their adult stage. The greatest numbers of species are found in tropical rainforests around the world. As one leaves these regions for areas of lower and more variable temperature and less rainfall the number falls off rapidly; yet even over the arctic tundra, over mountain rockslides far above timberline, and in all but the most sterile stretches of deserts, some butterflies and moths occur.

Lepidoptera, however, do not lend themselves well to fossilization, and knowledge of their geologic history is fragmentary. The earliest known Lepidoptera are from the early Tertiary, but they probably arose much earlier. One famous species is the butterfly, *Prodryas persephone*, perfectly preserved with wings outspread and pattern clearly visible, almost like a cabinet specimen. Other Lepidoptera, chiefly "micros," have been found in amber.

Importance

Lepidoptera affect the lives of men in many ways. Three of these deserve special attention. (1) *Economic importance*. They are among the most important insects from an economic standpoint, due chiefly to the ravages of certain larvae on crops, forests, stored products and clothing. This, however, is more than offset by the beneficial activities of the adults in the cross-pollination of flowering plants. (2) *Esthetic importance*. It is an unfortunate man who has not seen, and a poor one who cannot appreciate, the anticipatory thrill of the first bright blue fleck of a spring azure exploring a reawakening wood, or the poetic saddness of a lone, frayed sulphur wandering among the browning weeds of an autumn field. There are, among the moths and butterflies, colors and patterns to suit every taste, from the bold and brilliant colors of tropical American *Morpho* and *Agrias* and New Guinean *Troides* to the subtle shade contrasts and pattern intricacies of noctuids and satyrids; and the "yin and yang" symbol of Korea, important in certain oriental philosophies, may well owe its origin to the similar mark on several noctuid moths of the region. (3) *Scientific importance*. The Lepidoptera have provided the source material for many significant advances in a wide variety of biological disciplines. From the beginning they have served importantly in the study of evolution, the most recent contribution of note being in the phenomenon of industrial melanism (the rapid increase of melanistic forms of some species in industrial regions, coincident with the growth of industry there), and one of the few instances of observed, documented and studied evolutionary change was in a colony of a species of butterfly in England. In the field of physiology notable contributions have been made, using Lepidoptera, in research on insect metamorphosis, relations of development to temperature, pigment chemistry. Study of mimicry and protective resemblance began with the Lepidoptera and they have undoubtedly contributed more to it than any other group; and in the still infant investigation of the relationships between numbers of individuals and numbers of species in biological universes they have played a leading role. Though they have also contributed to knowledge of genetics and of zoogeography, their capacities so far have not been at all fully utilized in these fields.

General

A butterfly or moth passes through four states in its lifetime: egg, larva, pupa, adult; as different from one another in their functions as in their appearance. In brief, these functions are: the *egg* has both a passive function of being properly located for the beginning of larval life and an active function of embryological development; the *larva* is concerned almost entirely with growth and hence feeding is its major activity; the *pupa* has a primary function of converting the insect from larval to adult structure and often a subsidiary function of carrying it through periods of extreme dryness; and finally, the *adult* stage is concerned chiefly with procreation. Each of these stages may now be examined in more detail.

Egg. The egg is usually fertilized (from the stored sperm supply in the female) shortly before laying, so it begins independent life in a very early embryological stage; events, however, proceed rapidly and often in less than ten days a small, but fully formed, larva emerges. Eggs of Lepidoptera are of various shapes (subspherical, elongate-ovoid, depressed, and many modifications of these), and may be smooth or variously, sometimes intricately, sculptured. Those laid singly are usually merely attached to the appropriate surface with an adhesive; those laid in masses often have a protective secretion deposited over them, or may be mixed or covered with scales drawn from a terminal tuft on the female's abdomen.

Larva. Most larvae feed exposed, chiefly at night, on the leaves of various plants. Larvae of many "micros" make tunnels or mines within the leaves, entirely between the upper and lower epidermal layers. Boring in stalks, roots or branches, often in hard wood, is a specialty of some groups. Other larval foods may include bark, developing seeds and fruit, dead leaves, dead animal matter, fungi. Especially interesting are the few scattered groups that are carnivorous. Larvae of the Epipyropidae live attached to adult fulgorid or cicadid Homoptera, feeding chiefly on their waxy secretions. Larvae of the American lycaenid *Feniseca* and its Old World relatives feed on woolly aphids; two American pyralids (*Chalcoela*) parasitize the larvae of *Polistes* wasps. The relationships of lycaenid larvae to ANTS are many and diverse, ranging from sometimes elaborate mutual benefit arrangements to strictly unilateral predation on the part of the caterpillar.

Larvae of many of the exposed feeders have no protection beyond their own integument; others, such as the Psychidae, some tineids and others, construct individual portable sacks or cases in which they live; still others, such as the tent caterpillar, construct communal nests in which they rest, leaving the nest to feed. An ever present threat to larvae, from which these devices offer little protection, is that of parasitism by various DIPTERA (especially Larvaevoridae) and HYMENOPTERA (especially Ichneumonidae, Brachonidae, Chalcididae).

Pupa. The transformation from the larval to the pupal state in the sack or case bearing forms is done within that structure. Most other Lepidoptera at this stage construct a shelter or cocoon, usually attached firmly to the substrate, within which metamorphosis occurs. These cocoons may be elaborate or simple; composed solely of silk or mixed with other materials: larval setae in some, supplementary secretions in others, often one or more leaves. Those pupating in the ground or in the larval tunnel usually line the area of pupation with silk. An exit may be prepared in the cocoon for escape of the adult, or the adult may have special structures designed to force an opening. Most butterflies have abandoned the construction of a cocoon altogether, save for the silk strand or girdle which many use to hold them against the substrate.

The pupa serves two major functions. Its primary purpose is to enable the radical change from a crawling, chewing, sexless "worm" to a flying, liquid-sucking (or completely non-feeding), sexually mature adult. This is accomplished within the pupa by the extensive dissolution (histolysis) of certain internal larval organs, the resulting fluid then reforming (histogenesis) about various organ buds (histoblasts) to produce the structures of the adult insect, the whole process often taking only ten days, or even less. The second function is due to the nearly moisture-tight nature of the pupal integument, or its ready capacity to become so. This makes the pupa an excellent resting stage in areas where growth periods alternate with periods when growth would be impossible or hazardous because of aridity, such as circumpolar regions with extremely cold, drying winters as well as warmer regions with an extreme dry season. Elsewhere there are either continuously favorable growing seasons or, as in many temperate regions, the unfavorable season is not particularly beset with problems of dryness, and resting can be accomplished in any stage.

Adult. Procreation is the principal function of this stage, and comprises three basic steps: the sexes must find and recognize one another; they must mate; and the female must deposit her eggs.

In those species with extremely short adult lives—an hour or so in some kinds, up to a few days in others—these activities must be compressed into that brief period. This requires many special and often curious adaptations. The adults, for example, do not feed and have lost their mouthparts: their relatively small energy needs are met by fat stored in the body during the larval stage. Finding and identifying one another are generally accomplished by scent (which works best in the still air of night, probably the reason that all these short-lived forms are moths). In some species the female does not actually lay her eggs but scatters them broadcast, leaving the emerging larvae to seek out their own food. In other species the female deposits her eggs in one single mass. Female psychids mostly die in their sacks, with the eggs still enclosed in their bodies.

Most moths and all butterflies live longer lives as adults, generally about three to six weeks (some butterflies as much as several months). Finding, identification and mate selection are often, especially in butterflies, accomplished visually, aided by complex and still little-known devices of territorialism and courtship rituals. Egg laying is a protracted and exacting task. The female typically seeks out particular kinds of plants, lays her eggs singly on particular parts of these plants, and spaces the eggs so that the future larvae will find enough to eat.

These activities and the greater time they require have several consequences. The chances of predatory attack are greater, so protective (cryptic) coloration and mimicry are generally well developed, the former especially among the moths, which rest by day, the latter chiefly among butterflies, which are active by day. The variety of patterns and devices, and the perfection of the resemblances, have long been renowned. The greater and more prolonged activity also requires more energy than could possibly be stored in the body, so these species all need a continuously available source of energy food. Some feed on the sugar secretions of aphids, some on the juices of ripe or rotting fruit, some on the sap of wounded trees. Others feed on the liquid components of carrion or excrement. Most of them, however, depend on the nectar of flowers. Energy is made available and used most efficiently at rather high and fairly uniform temperatures, and these moths and butterflies maintain such temperatures during their active hours by two different kinds of behavioral thermoregulation. The moths produce body heat by muscular activity (myothermy), partly by their more rapid wing beats in normal flight and partly by "shivering"—a rapid vibration of their wings while perched. Their woolly insulating body vestibule aids in containing this heat. Butterflies, like reptiles, absorb heat from the sun (heliothermy), chiefly by exposing their wings so that blood circulating in the wing-veins can gain in temperature. By such methods these moths and butterflies are apparently able to maintain a high and constant body temperature through a wide range of different air temperatures.

HARRY K. CLENCH

References

Brues, C. T., A. L. Melander and F. M. Carpenter, 1954, "Classification of Insects," (Revised edition). Bull. Mus. Comp. Zool. vol. 108 (Lepidoptera, pp. 20–21, 226–305, revised by C. L. Remington). Contains conspectus of higher classification, keys to families and subfamilies (world-wide), and a very lengthy bibliography, well selected.

Klots, A. B., 1951, "A Field Guide to the Butterflies," Boston, Houghton Mifflin. (40 plates in color and black-and-white). Specifically deals with the butterflies of eastern North America, but also contains general information on collecting and preparing specimens for study.

LEUCKART, KARL GEORG FRIEDRICH RUDOLPH (1822–1898)

Leuckart is one of the most underrated pioneers in the history of zoology. He was trained in the University of Gottingen, at that time famous for its school of anatomy. Leuckart, however, became more interested in invertebrate zoology and for the first time brought to bear on this subject that insistence on detailed structural investigation which had previously been the mark of the vertebrate anatomist. In consequence of this, he established the phyla, coelenterata, echinodermata, vermis, and arthropoda as basic to the animal kingdom and was thus the first man to recognize the close relation between the worms and the mollusca. He was able, moreover, to show clearly that the resemblance between the coelenterata and the echinodermata, based on their radial symmetry, was quite fortuitous and that the coelenterata were more closely allied to the sponges. His studies on the siphonophora led him to the realization that polymorphism within a single species was possible, which opened up an entirely new philosophical concept to his zoological colleagues of the period. He then turned his attention to parasitology and worked out in accurate detail the life histories of the human tapeworm and the sheep liver fluke, neither of which before his time had been clearly understood. He also investigated trichina and showed the manner of its passage from the pig to the human. Finally, he discovered the presence of the micropyle in the insect egg and thus demonstrated that direct contact between sperm and ova was necessary for fertilization. Any one of his three major discoveries should have been sufficient to assure his fame and his relative neglect by historians is probably due to the fact that these discoveries were so instantly accepted, being based on impeccable observation, that they are regarded today as something that was always known.

LEYDIG, FRANZ (1821–1905)

A German cytologist, remarkable in his day for having paid as much attention to the tissues of invertebrates as to those of vertebrates. His "Lehrbuch der Histologie" (1856) must be regarded as the starting point of comparative histology. It was also the first to include bone among the connective tissues.

LICHENS

Lichens are plant organisms, really dual organisms, composed of two Thallophyte partners, a fungus and an alga. About 23,000 species have been named. The biological relationship between the partners is termed symbiosis, or more correctly consortism or mutualism. Each partner can be called a symbiont: the alga the phycobiont, the fungus the mycobiont. The scientific name given to a lichen, according to the "International Code of Botanical Nomenclature," applies to the fungus as well as the lichen, and this is logical as the fungus usually determines the morphology of the lichen thallus.

The mycobionts are predominantly sac fungi (*Ascomycetes*) and the majority of these form disc-shaped structures, apothecia, in which the asci and ascospores are found. A number of them form either flask-shaped structures, perithecia, or more elaborate tissue structures, stroma, in which the asci are found. A few mycobionts are club fungi (*Basidiomycetes*) and these forms are restricted to tropical regions. Some obviously ascomycetous lichens do not normally form asci nor ascocarps. The phycobionts are mostly green algae (*Chlorophyta*) or sometimes blue-green algae (*Cyanophyta*). The algae grow and reproduce asexually in the lichen. They are usually found in a layer in the thallus, the algal layer, just beneath the fungal covering layer, the upper cortex, and subtended by a loosely compacted fungal layer, the medulla.

Lichens may be divided into groups based upon their thallus growth habit. The crustose ones are closely attached or embedded in their substrates and usually lack a lower cortex fungal layer. The foliose ones are usually flat, circular or lobed and grow loosely or only centrally attached to their substrates; they have a lower cortex and often specialized filamentous attaching structures called rhizinae. The fruticose ones are upright or pendent and attached to the substrates only at their bases. They may be ribbon-shaped, ropey, or shrubby in form.

Sexual reproduction is morphologically similar to that found in non-lichenized Ascomycetes although for lichens many cytological details are lacking. The ascogonia frequently have trichogynes and there are conidia formed in flask-shaped structures (pycnidia) that probably function as spermatia and provide the second nucleus that finally takes part in the karyogamy within the ascus. The conidia do not ordinarily germinate, while the ascospores germinate readily.

Asci can be found in fruiting lichens in various stages of development. The inoperculate asci may have from one to hundreds of ascospores, the usual number being eight. Ascospores vary in their morphology and may be one to multicelled. The characteristics of the ascospores as well as of the fruiting structures are important in lichen identification and classification.

Asexual reproduction takes place either by fragmentation of portions of the lichen thallus or by dispersal of specialized outgrowths of the thallus containing fungal hyphae and algal cells in the form of minute spherical bodies called soredia or larger coralloid outgrowths called isidia.

The symbionts may be cultured separately on synthetic media in the laboratory. The phycobionts will grow on mineral media but grow better with organic additives. Some blue-green phycobionts have been found to fix nitrogen. The mycobionts require organic media. The phycobionts resemble free-living algae and some have been equated with these forms. The mycobionts in almost all cases are not known in the free-living state. They are very slow-growing fungi in artificial culture. Attempts to resynthesize a lichen from its isolated symbionts have been successful only in one or two cases. Usually an abnormal growth results which does not resemble the original thallus. In the successful cases the synthesis has been achieved by using starved symbionts and by manipulation of the environment particularly by drying, or alternate wetting and drying. It is thought that in nature the fungal symbiont must contact a suitable phycobiont soon after ascospore germination or it perishes. The mutualistic relationship in lichens is thought to be of variable intensity with different members of this polyphyletic group. In general, the mycobiont which has intimate contact with, or even penetrates the phycobiont cells with haustoria, derives organic nutrients and vitamins from it. In turn, the mycobiont accumulates mineral nutrients such as salts

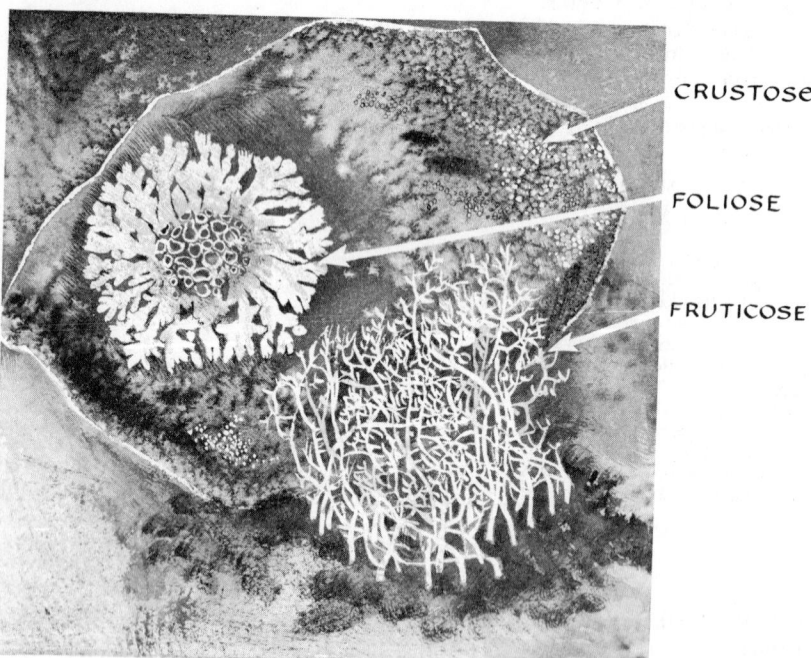

CRUSTOSE

FOLIOSE

FRUTICOSE

Fig. 1   Morphological types of lichens. (From Platt and Reid, "Bioscience," New York, Reinhold, 1967.)

or nitrogen and phosphorous and makes them available for phycobiont growth. The phycobiont may also benefit at times by the increased CO_2 concentration and the shading effect of the fungal layer when light is at a photosynthetic inhibitory intensity. Phycobionts are stimulated in their cultural growth by organic additives and in their photosynthetic rate by ascorbic acid, and these may be available in the lichen from the mycobiont, under certain conditions.

The lichens accumulate many distinctive organic substances. Most of these are aromatic compounds, the most characteristic being depsides and depsidones. Other aromatic compounds are quinones, pulvic acid derivates, xanthone derivatives, and dibenzofuran derivatives. Various nonaromatic acids, triterpenoids, and polyalcohols are also found. The lichen substances have been used to a considerable extent in the taxonomy of some lichen groups.

Lichens are widespread in their distribution, being found on every continent and in most habitats. They are particularly notable for being able to grow in unfavorable habitats. They are the major vegetation of the Antarctic continent, and in the Arctic the reindeer mosses (*Cladonia* spp.) are the principal winter forage of reindeer and caribou. Because the thalli of these plants act as mineral traps there has been interest in them as radioactive fallout accumulators and as members of the arctic food webs. In general, lichens grow very slowly and survive for long periods of time. In this connection, they have been used in estimating exposure times for rocks in areas of glacial retreat and there is some lichenometric literature. As pioneer plants they have importance as initiators of rock degradation and may be important in soil formation. They are very sensitive to air pollutants such as sulfur dioxide and disappear in industrial areas.

Economically lichens are not very important. They are used as natural dyes and one of these, litmus, is used as a pH indicator. Some are used as perfume bases. A few have antibiotic substances and are used in folk medicine.

EMANUEL D. RUDOLPH

References

Abbayes, H. des, "Traité de Lichénologie," Paris, Lechevalier, 1951.

Ahmadjian, V., "The Lichen Symbiosis," Waltham, Mass., Blaisdell, 1967.

ibid, "A Guide to the Algae Occurring as Lichen Symbionts: Isolation, Culture, Cultural Physiology and Identification," Phycologia, 6: 127–160, 1967.

Alvin, K. L. and K. A. Kershaw, "The Observer's Book of Lichens," New York, Warne, 1963.

Asahina, Y. and S. Shibata, "Chemistry of Lichen Substances," Tokyo, Japan Society for the Promotion of Science, 1954.

Beschel, R. E., "Dating Rock Surfaces by Lichen Growth and its Application to Glaciology and Physiography (Lichenometry)," in "Geology of the Arctic," Toronto, University Press, 1961.

Culberson, W. L., "A Guide to the Literature on the Lichen Flora and Vegetation of the United States," Beltsville, U. S. Dept. of Agr., Plant Industry Station, Special Publ. 7, 1955.

Duncan, U. K., "A Guide to the Study of Lichens," London, Buncle, 1959.

Fink, B., "The Lichen Flora of the United States," Ann Arbor, Univ. Michigan Press, 1935 (reprinted 1960).

Hale, M. E., Jr., "Lichen Handbook," Washington, D. C., Smithsonian Institution, 1961.

ibid, "The Biology of Lichens," London, Arnold, 1967.

Lamb, I. M., "Lichens," Scientific American 201 (4) October pp. 144–156, 1959.

Llano, G. A., "Economic Uses of Lichens," Smithsonian Annual Report 1950 pp. 385–422, 1951.

Ozenda, P., "Lichens," Handbuch der Pflanzenanatomie, Vol. 6, Berlin, Gebruder Borntraeger, 1963.

Rudolph, E. D., "Lichen Distribution," American Geographical Society Antarctic Map Folio 5, pp. 9–11, maps, 1967.

Smith, A. L., "Lichens," Cambridge, Cambridge University Press, 1921.

Smith, D. C., "The Biology of Lichen Thalli," Biological Reviews, 37: 537–570, 1962.

Vartia, K. O., "On Antibiotic Effects of Lichens and Lichen Substances," Annales Medicinae Experimentalis et Biologiae Fenniae, Supplement 7, pp. 1–82, 1950.

LIEBERKUEHN, JOHANN NATHANAEL (1711–1756)

This German physician, best remembered for the glands in the interstinal mucosa which bear his name, taught in Vienna, London, and Paris before settling in Berlin. He made significant contributions to the design of microscopes and was one of the very early pioneers in the application of this instrument to the study of anatomy. The reflecting condenser which bears his name did not go out of use until the early part of the present century.

LIFE, ORIGIN OF (see also Neobiogenesis)

The question of the origin of life is one of the fundamental problems confronting natural scientists today. It is of key importance for the rational understanding of the surrounding world.

Life is represented on Earth by a multitude of different organisms. Direct observation and experience show convincingly that the only way by which any living creature comes into being at the present time is by being produced from others like itself. This is true of highly organized plants and animals as well as of the most primitive bacteria, protozoa and algae. Everywhere in nature there are long series of generations of living things, but the origins of these series lie buried in the past and are not directly perceptible.

Charles Darwin and later evolutionists gave a materialist explanation of the method by which modern higher organisms have developed from lower ones. The mechanism of this gradual progressive development was convincingly confirmed by the study of fossilized remains, preserved in the earth's crust, of living things which inhabited the planet in the remote past.

All attempts to demonstrate direct emergence of even the most primitive living things from inorganic matter, either under natural conditions or in a laboratory, have proved futile. It is now known that life on earth could have come into being only as a result of successive historic development—a prolonged and onesidedly directed process of gradual complication of primitively evolved organic substances and complex systems formed in this process. The very large amount of factual material accumulated at present allows us to draw a picture of the successive stages in the process underlying the origin of life and even to reproduce some of them in the laboratory.

It is now the prevailing hypothesis of cosmogonists that the Earth, as well as other planets of the solar system, was formed from gases and dust particles of the primeval cloud which once surrounded the sun. Investigation of the chemical composition of interstellar aggregates of dust and gases shows that they contain the simplest hydrocarbon—methane. However, the detailed physico-chemical analysis carried out by H. C. Urey, respecting the phenomena that are presumed to have occurred during the formation of planets, showed that methane could not be retained in the region of earth's formation, since this occurred too close to the sun. The methane was continually escaping and accumulating on the surface of large planets, where it is found at present.

The main form of carbon precipitation on the then emerging Earth was metallic carbides and graphite. It is from them that hydrocarbons were formed as a simple chemical reaction, (i.e., abiogenically) during the formation of the earth's crust, more particularly when carbides reacted with the hydrated rocks in the deep layers of the crust.

The formation of the earth's crust that began at the earliest periods cannot be considered complete even now. It may, therefore, be possible to discover today the same processes as those of the initial hydrocarbon formation in the deep strata of the crust. Indeed, many geological findings seem to justify these hopes. They show that the process of the abiogenic formation of hydrocarbons is still taking place, though on a very limited scale. In particular, geologists have succeeded in finding hydrocarbons in the basic rocks of the foundation that have no connection with living organisms or products of their decomposition; these compounds have been discovered in clefts in granite as hydrocarbon gases (methane, ethane, propane), as well as slight traces of liquid hydrocarbons which could only have originated abiogenically.

Based on a study of the abundances of isotopes in the earth's crust, the age of the Earth is estimated at about five billion years. The first living organisms came into being about two billion years ago. Thus there was no life on Earth during the greater part of its existence. During this period complex chemical transformations of the initially appearing hydrocarbons took place. Following the formation of the earth, the hydrocarbons could not escape into the interplanetary space since they were retained by the force of gravity and large quantities were accumulated in the atmosphere.

The atmosphere of lifeless Earth was basically different from that of today in its chemically reducing character. This is proved not only by general theoretical considerations but by much direct geological data. The overwhelming preponderance of free oxygen in the present-day atmosphere was undoubtedly formed, and continues to be formed, biogenetically as a result of the activity of green plants. If the Earth again became devoid of life the free oxygen would very soon disappear from its atmosphere due to absorption by the inorganic materials on the earth's surface.

Hydrocarbons released into the initial earth's atmosphere reacted with water vapors, ammonia, hydrogen sulfide and other gases of the reducing atmosphere. Short-wave ultraviolet light and silent and spark electric discharges, intensified these reactions. A number of scientific papers have been published recently which describe the laboratory reproduction of those conditions which could have existed in the primeval atmosphere of the Earth. This work shows that even such a relatively inert gas as methane can serve as the original substance for the formation of organic compounds.

An example is the well known experiment of S. L. Miller, who obtained amino acids—the substances forming the basic components of the protein molecule —by passing silent electric discharges through a mixture of methane, hydrogen, ammonia and water vapor. Analogous syntheses of amino acids were carried out by T. E. Pavovskaya and A. G. Passynskii using ultraviolet. Other investigations have stated the possibility of forming the precursors of nucleotides, porphyrins, etc., under such conditions (T. Oró, C. Ponnamperuma).

With their gradual complication organic substances lost their gaseous character and precipitated from the atmosphere to the Earth's original water-envelope (hydrosphere). There the main bulk of organic compounds was accumulated, and it was there that principal processes in the formation of compounds of high molecular weight, and their conversion into multimolecular systems essential to the emergence of life, took place.

Contemporary literature is replete with reports of investigations demonstrating the way in which the gradual polymerization and condensation of organic substances proceeded (T. Akabori, G. Shramm). As a result of this process, in waters of the primeval ocean, protein-like substances, polynucleotides, porphyrins and other complex organic compounds emerged in a purely abiogenic, (chemical) way. Thus, the waters became a so-called "nutrient broth" of rather high concentration. If only half of the carbon existing now on the earth's surface had been present in the form of dissolved organic compounds, the concentration in the waters of the primeval ocean would have been between 0.1 and 1.0 per cent.

The present hydrosphere of the Earth contains a small quantity of organic matter arising mainly biogenetically, as products either of activity of living organisms or of decomposition of their dead bodies. However, waters of contemporary seas and oceans cannot now serve as a place where, under natural conditions, we can observe and investigate processes of gradual complication of organic substances analogous to those that led to the emergence of life. This is prevented, in the first place, by the abundance of free oxygen in these waters and secondly, by the ubiquitous distribution of existing organisms.

Organic substances originating in this way at the present time are soon destroyed by these living organisms. A prolonged evolution of organic substances in nature was possible only in the absence of living organisms on the then lifeless Earth. Therefore, we are unable to observe it directly in nature and have to judge the processes prior to the appearance of life by drawing analogy with the phenomena we can produce in the laboratory [see NEOBIOGENESIS for a contrasting opinion. (Ed.)]

The characteristic feature of life is that it is not just disseminated in space but that it is represented by very complex individual systems distinctly separated from the external world. These organisms interact directly with the environment, and their internal composition is well adapted to their continued existence, self-renewal and self-reproduction.

The present data do not permit the direct formation of such living systems in the primary solution of organic substances by simple self-assembly. We do not observe this process in nature and are not able to create it artificially either for the whole cell or for its separate organelles. This gap between the "primary soup" and the most primitive living creatures can be filled by the processes of natural evolution and by gradual development and improvement of prebiological systems, which are rather simpler than organisms. These systems are able to form spontaneously in the "primary soup" and sediment from it in the form of separate individual units which interact chemically and energetically with the environment. Not only is it possible to envisage the formation of self-assembling systems, but they may be experimentally obtained in different ways. In particular, it is necessary to recall Fox's microspheres, or coacervate drops, which are most suitable, although not unique, models for the reconstruction of some phenomena which have taken place in the "primary soup."

For their formation it is not necessary to have in solutions polymers with systematized intramolecular organization, like modern proteins or nucleic acids. According to experiments of the laboratory of the Institute of Biochemistry, Academy of Sciences of the USSR, coacervate drops are formed by a simple mixture of solutions of nonspecifically or monotonously constructed polypeptides and polynucleotides. The most important factor is the size of the molecules, and as soon as a certain level of polymerization would occur in the primitive primary solution, coacervate drops would result.

Thus, the formation of coacervates in the "primary soup" is the direct result of the formation of high-molecular-weight primitive polymers. The improvement of intra-molecular structures of such polymers during further evolution should continue, not in the solution but in the polymolecular systems.

Polymers in coacervate drops are in a very concentrated condition (50% and more); this concentration is preserved even if drops are isolated from actively diluted solution. At the same time drops are able to adsorb selectively from the external "solution" different low-molecular-weight substances. If any of these substances are able to stimulate catalytically the reactions which proceed in the drop, these drops convert to open systems, specifically interacting with the environmental medium.

In the experiments in the laboratory, it became possible to incorporate different simple and complex catalysts into the coacervate drops and to induce oxidative-reductive reactions with synthesis and destruction of polymers. In such a way some primitive schemes of metabolism in coacervates were reproduced. In these schemes not one, but many different reactions were involved and a more active growth and relatively slow destruction of drops was obtained.

Open polymolecular systems with a primitive metabolism, like our models, could easily form in the "primary soup," owing to the coacervation of polymers and the incorporation of different organic and inorganic catalysts from the external medium into the drops. Such

systems (which we call protobionts) might have assembled and grown in "primary soup" conditions and then might have broken apart under mechanical actions (for instance, blows of waves) in a way not unlike the division of emulsion drops under shaking.

The protobionts produced by breakup of grown protobionts retain to a certain extent the character of their interaction with the external medium. An important factor in conserving the composition of protobionts which expanded as a result of synthesis of new polymer molecules should have been the reduplication of the polynuclotides in the droplets based on complementarity. However, the main factor was that the relative reaction rates and coordination of the reactions in the protobionts remained constant. This was possible because the protobionts retained a high concentration of the simplest inorganic or organic catalysts during growth by selectively adsorbing them from the environment.

Of course, this phenomenon cannot in the least be compared in its constancy and accuracy with the self-reproduction of even the most simple organisms now existing. All sorts of changes and shifts in the protobionts could have occurred during their growth and disintegration, especially if the environmental conditions changed. However, all this taken together would inevitably lead to the appearance of a peculiar type of prebiological "natural selection" which predetermined the further evolution of protobionts toward the formation of primary living things.

The possibility of such "prebiological selection" has been demonstrated in the model experiments with coacervates. For this purpose, drops were used with a complex of catalysts, which allowed active syntheses of polymers and increase of the whole system. On the contrary, the catalytic complex of other drops was more primitive. We found that the drops of the first group grew more actively than those of the second group.

Only in this sense could prebiological selection be understood. But even this is enough to understand how the organization of growing and reproducing protobionts could gradually improve.

Only dynamically stable systems, where breakdown and synthesis were in balance, could give rise to stable, permanently repeating chains and cycles of reactions. Such stable systems could even multiply with the aid of substances entering from the external medium. In consequence there was a constantly repeating neoformation of this or that substance or structure. It is from this permanent repetition of interacted reactions, co-ordinated into a single network of metabolism, that there developed the specific ability of living things to reproduce. Polynucleotides played an exceptionally important role in this organization.

In contemporary organisms, synthesis of proteins is carried out by an exceedingly complicated and highly developed mechanism by means of which the amino acids are successively strung together along a polypeptide chain in just the sequence required by the specific, strictly ordered association of mononucleotide residues in the molecules of DNA and RNA (129). A mechanism of this sort, of course, could have arisen only during prolonged evolution of living systems. However, even at far earlier stages of development, polynucleotides in the protobionts could have exerted a certain effect on the polymerization of amino acids which occurred in the systems. The intramolecular structure of the very earliest polynucleotides was still very primitive and changed markedly during growth of the protobionts. Any change appearing in this way could be partially perpetuated in the expanding system as a result of complementarity and simultaneously affect the order of amino acid residues in the polypeptides synthesized within the system.

If the arrangement of amino acid residues produced in this manner was favorable from the point of view of increasing the catalytic activity of the polypeptides, the system which created it would be superior in respect to rapid growth and multiplication. In the opposite case it would be destroyed by natural selection. Thus the structure of protein-like polypeptides and also that of the polynucleotides which controlled their synthesis would gradually become more ordered and better adapted to the functions which the polymers carried out in the expanding and multiplying biological systems.

At the stage of evolution, where matter devloped the capacity to reproduce itself, natural selection acquired its biological sense. Under the control of natural selection in the subsequent development of organisms the further complication and improvement of metabolism, characteristic of life, took place.

Natural selection must have eradicated all these intermediate systems in which the organization of metabolism was imperfect. But comparative investigation of metabolism in most primitive organisms available today makes it possible for us to judge how a new order of chemical transformations, so characteristic of life in highly organized plants and animals, gradually came into being.

Data derived from comparative biochemistry convincingly show that some forms of metabolic systems arose at the very beginning of life and are inherent in all organisms without exception, while others evolved much later as a superstructure to early existing systems.

Only those organic substances that had arisen in an abiogenic way could serve as the source of nutrition for earth's initial organisms. The process of organic nutrition underlies life itself and is essential to all living things.

The absence of free oxygen in the primary terrestrial atmosphere and hydrosphere necessitated the anaerobic character of energy metabolism in primitive organisms. Comparative biochemistry clearly demonstrates that anaerobic metabolism is the precursor of the energetics of all modern organisms.

During the development of life the reservoir of abiogenically formed organic substances gradually became exhausted as the development of life went on at a greater rate than the formation of these substances. Such a change gave an advantage to those organisms which had acquired the ability to absorb light and to build de novo organic substances from inorganic carbon compounds.

Instead of the previous slow abiogenic method for the formation of organic substances, there appeared a new biological process in the synthesis of these substances —photosynthesis. The emergence of photosynthesis completely changed the living process on Earth. Some organisms began to build essential organic compounds themselves, while others retained the former ways of nutrition, using the organic substances which were created biogenically. On this basis living organisms assumed the characters which divide them into the animal and plant kingdoms.

The emergence of photosynthesis not only provided an abundance of organic substances but also resulted in the appearance of free gaseous oxygen previously absent from the Earth's surface. This changed the whole character of all chemical processes and enabled the majority of living organisms to increase significantly the level of their energy metabolism. Having built a superstructure of new systems of oxygen respiration over the former anaerobic metabolism, these organisms could use energy previously lacking in organic substances.

The evolutionary development of the structure of living things was closely linked to the improvement of metabolism. Unicellular and then multicellular organisms developed at an ever-increasing pace. We can establish now the way this development took place by studying fossilized remnants of those living things that once inhabited this planet.

A. I. OPARIN

References

Oparin, A. I., "The Origin of Life," 4th ed., New York, Academic Press, 1957.

Oparin, A. I., "The Chemical Origin of Life," Thomas, Springfield, Ill., 1964.

Fox, T. W., ed., "Origin of Prebiological Systems," New York, Academic Press, 1965.

Oparin, A. I., "The Origin of Life and the Origin of Enzymes," in Nord, F. F., ed., "Advances in Enzymology," Vol. XXVII, 1965.

LILIALES

This order of monocotyledonous plants contains families considered to be typical of the monocotyledons as a whole. Plants are generally herbaceous with subterranean perennating organs, e.g., bulbs, corms, rhizomes. Some are vines and a few, e.g., Dracaena, and Cordyline, are trees in which the axis is thickened by peripheral cambial activity. Leaves, which are usually simple in shape, are borne in a basal rosette or along an elongated stem. Roots are frequently fleshy and contain starch.

Flowers are borne singly or more usually in inflorescences. They are actinomorphic and trimerous, with a biseriate perianth not usually differentiated into calyx and corolla. There are 3 or 6 stamens and a compound 3-carpellate ovary. The ovary is superior or inferior and previously its position was used to differentiate families within the order. Recent redefinitions have paid less attention to ovary position as a taxonomic criterion and emphasized the type of inforesence as a more reliable indication of affinity.

The approximately 400 genera and 8,000 species of the order are distributed among 9 to about 17 families by different authorities. Several of these families are small, consisting of one or a few genera. Others, such as the Liliaceae with over 200 genera, are among the largest in the plant kingdom. The following are the most important families of the order.

Liliaceae. The lily family is by far the most important in the order. In the United States there are well over one hundred native species and they are well distributed in about thirty genera. They range in size from the petite grape hyacinth a few inches high with flowers 3-4 mm. to superb lilies four to six feet tall and bearing flowers several inches in diameter. Though the leaves vary somewhat in size and shape, all are simple with parallel veins. The flowers show characteristic monocotyledonous features. The petals and sepals are alike forming a six parted perianth. The six stamens opposite these enclose a pistil with a three parted ovary; the stigma suggests this tri-part character.

The bulbs of *Allium* and *Lilium* and rhizomes of other genera are used as food, as are the young shoots of *Asparagus.* Colchicine, used in the treatment of gout and as a mitotic inhibitor, is obtained from the European *Colchicum autumnale.* Fiber used for nets and fishing-lines in tropical Africa is obtained from the leaves of various species of *Sansevieria.*

Amaryllidaceae. Plants in this family are the well-known ornamentals *Narcissus* and the equally striking century plant, *Agave.* The leaves of many are succulent. The distinction of this family from the Liliaceae is debatable. Separation has commonly been made on the basis of the inferior ovary in the Amaryllidaceae. However, in most other respects the two families are quite similar. In the genus *Narcissus* floral structure is complex, a corona formed from the inner face of the fused perianth segments constitutes the trumpet of jonquils and daffodils.

Plants of the genus *Agave* are used extensively in tropical regions as sources of fiber for producing sisal hemp, and a sugary liquid which is fermented and distilled to produce the alcoholic beverage, pulque. The subterranean stem of a few plants in this family are used as food.

Iridaceae. Irises are represented by more than 1,000 species of world wide distribution but they are centered particularly in central America and subtropical South Africa. The type genus, *Iris,* is characterized by a massive development of the perianth segments, stigmas and styles giving the common "flag" effect. Each of the three stigmas and stamens forms a separate pollination system and the flower is highly specialized in this respect.

Economically the family is unimportant save for the dye saffron which is extracted from the dried styles of the European crocus, *C. sativus.* Approximately 5,000 styles are required to yield one ounce of the dye. Rhizomes of *Iris florentina* and other species are the source of Orris used as a scent in perfume and dentifrice.

Juncaceae. The rushes which may be easily mistaken for grasses or sedges on casual inspection are distinguished from these by the presence in the rushes of regular flowers with small, but distinct petals.

Many plants of the order Liliales have become highly prized as garden ornamentals, or are well-known and easily recognized members of North American plant associations. Among these are the following.

The Easter Lily, which is a native of China, belongs to the genus *Lilium.* There are a dozen native species of true lilies and many beautiful horticultural varieties. In the native species, though the flowers may be smaller than the Easter Lily the flowers are all attractive, and in some species there are many flowers. The large size of greenhouse flowers in the Easter Lily is due to forcing and removal of some buds.

The Canada Lily, resembles the Easter Lily, but is yellow and nods its head. The Superb or Turk's Cap Lily, the Tiger Lily and others have reflexed petals and sepals. In these red and orange or red-orange prevails with conspicuous speckling. In all of these there is a tall, leafy flower stalk. There are two species of *Hemorocallis* and many varieties, all known as *Day Lilies.* These are very popular garden plants. They are fibrous rooted, the leaves basal and linear, the flowers yellow to orange or red.

Tulips belong to the lily family but are not natives of the U. S. The flowers are solitary, waxy in texture and erect. Some varieties have crenulated sepal and petal margins, and in others, infected with virus the flowers are mottled.

Wild Hyacinth (Camassia), is represented by one native species in the eastern United States. The flowers are lavender, and smaller than the cultivated ones, as are the leaves.

Trout Lily (Erythronium), often called Dogtooth Lily or miscalled dogtooth violet is found in rich woods. The yellow flowered species is best known, and has a wide distribution but there is a species with white flowers. An *Erythronium* of the west with yellow flowers is known as the Glacier Lily. The leaves are speckled and otherwise similar to the eastern one only much larger, up to eight or ten inches long. The leaves and flowers of the Trout Lilies of the South are much smaller. The single flower springs from the ground between two leaves in all species.

Grape-Hyacinth (Muscari) was introduced from Europe many years ago and is a favorite in gardens. The leaves which are borne in a basal rosette are a few mm. broad and quite long. In the center of a cluster of leaves, there is a single flower stalk topped with a few very small bluish, lavender or purple flowers.

Yucca. Species of this plant, sometimes called *Spanish Bayonet,* are abundant in desert areas of the United States. The leaves, all basal in most species, are long, thick and stiff. The flowering stem, which may be several feet tall, bears a great many pendant bell-shaped white flowers. One species is widely distributed over the United States.

Indian Cucumber Root (Medeola) is found in rich woods. One's attention is gained by the whorl or whorls of leaves. The plant is from 8 to 18 inches tall. The 2-3 flowers are small, have slender petals a few mm. long, greenish yellow in color, above the top whorl of leaves.

Solomon's Seals, the true *(Polygonatum)* and the false *(Smilacina).* In both genera there is a leafy stalk bearing oval leaves at regular intervals. In *Polygonatum,* the very small greenish-yellow flowers are in the axils of the leaves, one or two suspended by slender threadlike pedicels. In *Smilacina* many tiny white flowers are arranged in a large terminal group or panicle. Both species are common over a large part of the United States. The individual plants vary in height from 16 inches to six feet.

Onions (Allium) represented by a small number of species in areas over the country. From the subterranean bulb the leaves emerge as narrow tubular structures attached to the button-like stem. The flowers are small, and are borne at the top of the flowering stalk. In some species they are white, in a few pink or yellowish. Many species are easily located by noting the characteristic odor of onions.

Trillium is both the generic and the common name. Some species are also called Wake Robin. This is definitely a spring flower, and found in many of our states. Some species will dominate the area, being several feet tall, others as the species usually referred to as "Wakerobin" may be less than one foot tall and of infrequent occurrence. Wherever you find them, they can be recognized—a slender stalk topped by a whorl of three leaves and above these a terminal flower. In some species the flower is sessile and in others it is carried on a slender stem or pedicel. The "Tri" in Trillium suggests the number of leaves, the number of sepals and petals, the three parted stigma at the top of the pistil, and the three celled ovary at its base. Only the stamens differ. There are six of these. The sepals and petals vary with the species, in form, color and in position. The sepals are usually green and often smaller than the petals. The petals, in different species, are distinctive, as to size, shape and color. The colors are white, pink, maroon, and greenish brown.

Smilax or Greenbriar (*Smilax*) includes about a dozen species. Some are herbaceous vines, delicate or coarse, some are woody. Coarse thorns on some species help to make areas they frequent difficult to penetrate. Southern *Smilax* is one of the more delicate ones. The flowers are very small and often overlooked. The fruit is a berry and those species with red ones are most attractive even though the tangle of vine and thorns may be objectionable.

ELEANOR MCGILLIARD

LIMB

This article deals with the paired pectoral and pelvic appendages of vertebrates. Thus restricted, a *limb* is a pectoral appendage (pectoral fin, foreleg, wing, flipper, arm) or a pelvic appendage (pelvic fin, hind leg, flipper, leg) of familiar vertebrates.

Limbs are lacking in many of the most primitive class of VERTEBRATES, the Agnatha, and are variable in presence, form and numbers in another vertebrate class, the extinct Placodermi. They are present in all other vertebrate classes (CHONDRICHTHYES, OSTEICHTHYES, AMPHIBIA, REPTILIA, AVES, MAMMALIA), although some of these have lost the limbs as external appendages through adaptive evolution. Tetrapods are those vertebrates with appendages adapted to locomotion on land. Both pairs of limbs have disappeared in snakes, caecilian amphibians, and some lizards, and one pair has disappeared in some salamanders and in whales. Even in cases where limbs are lacking in the adult, they may be present as buds in the embryo as in caecilians. In several cases, the limbs have become greatly modified in proportion to body size as among salamanders, birds, extinct flying reptiles, and some marsupial mammals.

Limbs are paired and are bilaterally symmetrical; i.e., each limb of a pair is a mirror image of the other. In the strict sense, the limb includes not only the external appendage; it includes also the pectoral or pelvic girdle, which comprises part of the appendicular skeleton, and its musculature. Rudiments of limbs occur in the form of girdle parts in some adult vertebrates with missing limbs as in the python and whales.

Embryologically, limbs arise from the flank of the vertebrate embryo (limb field) in the form of limb buds or short ridges extending over several body segments; the anterior (pectoral) pair usually appears first. The limb bud is composed of a core of loosely arranged mesoderm covered by ectodermal epithelium. Beneath the ectoderm lies a basement membrane; these two together form the epidermis of the skin. Closely associated mesodermal cells will form the dermis of the skin. Integumentary derivatives of the limbs of various vertebrates include glands, hairs, scales, feathers, hooves, nails, claws and skin folds useful in flight.

The mesodermal core of the limb bud is derived from the ventral tips of myotomes in elasmobranchs and from mesenchyme detaching from the somatic mesoderm in most vertebrates. The detached mesoderm or mesenchyme early arranges itself within the limb bud into three primordia—a central and proximal concentration which is the primordium of the limb skeleton formed in or from cartilage, and a dorsal and a ventral concentration which are primordia of the dorsal and ventral muscle masses. From these muscle masses are differentiated the appendicular muscles proper, reflecting the ancient organization of appendicular muscles in fishes. From the muscle masses will differentiate the flexors and extensors, supinators and pronators, rotators, levators and depressors, abductors and adductors of the tetrapod appendicular musculature.

In every case among vertebrates, musculature other than that derived from the limb bud also attaches to the shoulder girdle. Most of this is axial musculature: serratus ventralis, e.g., of mammals. But in addition, some musculature homologous to branchial or gill musculature of fishes also acts upon the shoulder region in higher vertebrates: trapezius, e.g., of mammals. With the evolution of terrestrial locomotion, the shoulder girdle has become progressively freed from the body wall into which it is thoroughly incorporated in fishes, thus effectively increasing its mobility; the axial and branchial musculature have followed outward migration of the girdle.

In most vertebrates, some skeletal parts derived directly from mesenchyme rather than from cartilage, are present in the shoulder or pectoral girdle. These dermal or membrane bone elements are the cleithrum, clavicle and interclavicle, the first two being paired. In fishes, only the first two occur, and in modern bony fishes, the cleithrum alone is present. On the tetrapod line (amphibians, reptiles, birds and mammals) the dermal shoulder girdle is progressively reduced; the cleithrum is lost in birds and mammals, and in most mammals only a clavicle (collar bone) remains. The "wishbone" of birds is comprised of two clavicles fused together.

The hip or pelvic girdle is comprised only of endochondral bone in all cases in which it is ossified: a posterior ischium and anterior pubis in fishes, with a dorsal ilium being added in tetrapods. In some salamanders a single mid-ventral ypsiloid cartilage assists in hydrostatic control, and in marsupial mammals a pair of marsupial bones assists in support of the pouch.

Basically in all vertebrates, the endochondral shoulder girdle consists of a scapular (dorsal to the shoulder joint) and a coracoid (ventral to the shoulder joint) region. The coracoid remains in shouldered vertebrates except marsupial and eutherian mammals, in which it is reduced to a nubbin (coracoid process) on the bony scapula. The scapula is expanded in these latter groups to accommodate musculature historically arising from the coracoid region.

The internal limb skeleton beyond the girdle consists entirely of cartilage or endochondral bone elements; the presence of a cartilaginous or largely cartilaginous limb skeleton is a specialized characteristic of sharks and salamanders. In the tetrapod pectoral appendage these elements are humerus (upper arm or brachium), radius and ulna (forearm or antibrachium), and the carpals (wrist), metacarpals (palm), and phalanges (digits) which together comprise the manus of the forelimb. In the pelvic appendage, the skeletal elements distal to the girdle are the femur, tibia and fibula, tarsals, metatarsals and phalanges.

In most fishes, the Chondrichythyes and Actinoptergii, the tetrapod arrangement of the limb skeleton is lacking; instead a series of basal and radial skeletal elements occurs distal to the girdle, with stiffening connective tissue rays supporting the fin membrane. The paired appendages especially the pelvic, are highly variable in position in teleosts.

The origin of paired appendages (fins) occurred among the Agnatha and Placodermi in a manner still subject to much debate. The tetrapod appendage can be derived from the lobetype fin of rhipidistian fishes (CROSSOPTERYGII) now extinct; the first tetrapods (labyrinthodont amphibians) arose from this group of fishes.

JAMES M. MOULTON

References

Romer, A. S., "Vertebrate Paleontology," 3rd ed., Chicago, The University Press, 1966.
Romer, A. S., "The Vertebrate Story," 4th ed., Chicago, The University Press, 1959.

LIMNOLOGY (FRESH WATERS)

The study of fresh waters, or limnology, is generally said to have begun with the work of F. A. Forel on Lake Geneva in the latter part of the nineteenth century. It was furthered significantly in North America by E. A. Birge and C. Juday in Wisconsin. As a science, limnology strives toward understanding the dynamic interrelationships between organism and environment, and energy utilization in organic production within aquatic ecosystems. Since the total economy of each of the various types of freshwater systems is dependent upon certain effects of light, heat, hydrodynamics, dissolved substances derived from the substrate and organisms, and of organisms themselves, it is evident that thorough investigations of aquatic ecosystems must draw from geology, physics, chemistry, and biology. Recent years have seen the development of limnological specialists in those fields. Information derived from both specialized and general studies of lakes, streams, and other waters has resulted in the accumulation of a great, but as yet incomplete, knowledge of structural and functional aspects of freshwater ecosystems.

Research on lakes and lake dynamics has exceeded that on other types of fresh water. It is therefore possible to recognize certain distinctive features of lakes and to erect schemes of classification which, although

far from being rigid, are useful descriptive instruments. Lakes may be categorized according to carbon dioxide-carbonate relationships and relative concentrations as soft water or hard water; and on the basis of vertical seasonal mixing of their waters. Obviously any number of gradations exists among these types.

Organic productivity—determined as oxygen production, carbon dioxide assimilation (using C^{14}), chlorophyll concentration, or organic carbon production—is a useful measure in designating lake types. Many inorganic and biological factors enter into productivity; these include: the ionic content of the waters and its contribution of materials for plant growth (much as fertilizer in agriculture); the content of oxygen and carbon dioxide; photosynthesis as the mechanism by which green plants ("producers"— mainly plankton algae) transform solar energy into chemical energy; several levels of "consumers" (herbivores → carnivore A → carnivore B, etc.) involved in food webs in which energy is transferred and the ecosystem maintained; death and decomposition of living organisms which release inorganic substances available for recycling through living plants and animals. All of these processes are regulated to some degree by light and temperature and thus are subject to diurnal and seasonal variation.

Eutrophic, or highly productive, lakes occur throughout the world, although usually on soluble substrates rich in inorganic ions. These lakes are characterized by considerable concentrations of calcium, bound carbon dioxide, and high specific conductance; pH values range from ±7 to about 9. An abundant plankton, often exhibiting "blooms," and high turbidity are typical of eutrophic lakes. The lakes are shallow, generally less than 12 meters, and during summer stratification the upper, warmer layer of circulating water, the epilimnion, is deeper than the bottom, cooler, non-circulating layer, the hypolimnion. Between the epilimnion and hypolimnion lies a zone in which the vertical temperature gradient decreases rapidly with depth; this is the metalimnion, or thermocline, region. An upper, lighted zone often corresponding to the epilimnion is a region of high primary productivity, and is therefore rich in oxygen and low in carbon dioxide; this is the trophogenic zone. The unlighted zone is predominantly one of decomposition in which carbon dioxide is abundant and in which oxygen may become depleted; it is the tropholytic zone. Gross productivity of dry organic matter in a typical eutrophic lake is on the order of 2 gms/M²/day.

Oligotrophic, or weakly productive, lakes are also distributed widely, although normally on less soluble substrates. This type is therefore characterized by low ionic content, low specific conductance, and low carbonate and calcium concentrations. The waters are acid, the pH being less than 7. These are deep lakes, typically more than 25 or 30 meters. Plankton is scarce, and "blooms" are unknown. Gross primary productivity amounts to about 1 gm/M²/day. The hypolimnion is usually deeper than the epilimnion and, as a result of low productivity, frequently contains oxygen sufficient to support populations of whitefish and trout. The Great Lakes are oligotrophic lakes.

Lakes and ponds are not permanent but rather follow a pathway of "succession" leading eventually to filling-in and the development of some type of terrestrial community. In the early stages the lake usually passes through the oligotrophic type, and as a result of increase in the nutrient store, decomposition, and sedimentation, reaches a eutrophic stage. This process of enrichment is termed eutrophication, and typically is long-termed. In recent years, however, man has become an agent to the acceleration of the process. This has come about through industrial and municipal pollution of lakes by inorganic and organic substances, particularly those containing nitrogen and phosphorus, and through agricultural practices that cause crop fertilizers to be introduced into nearby waters. Eutrophication in lakes is highly relevant in water supplies, fisheries, natural area management, and other aspects of human concern.

Various types of lakes may be recognized on the basis of morphology or chemical composition. Bog lakes, found notably on glaciated terrain, may be acid or basic depending upon the chemical nature of the substrate, and are usually rather productive. Fishes in such lakes include catfish, sunfishes, and perch. In arid regions, where precipitation is less than evaporation, lakes of high concentrations of various salts such as chloride, sulfate, or carbonate occupy closed basins.

Reservoir lakes present many unique conditions, and research on such impoundments has not kept pace with the rapidly increasing numbers and varieties of these waters. Release of hypolimnial water through deep intakes in summer often has the effect of decreasing both temperature and oxygen content of the stream below the dam, while drawdown from the warm surface region may increase the temperature of downstream waters; either or both may have serious effects on the biota of communities below the impoundment. Within the reservoir, inflowing waters and the method of drawdown create unusual conditions of temperature, density, and oxygen.

Land-water interchange and current are the fundamental factors which distinguish lake and stream communities. Streams are also characterized by the considerable variety of physical, chemical, and biological conditions presented along the stream course. Throughout the longitudinal extent of a stream of significant length, certain generally distinctive regions are recognizable. The head-water region of mountain brook characteristics is marked by clear, cool, rapidly flowing water, often seasonally intermittent. The scouring action of high current velocity results in a rocky stream bed. Organisms inhabiting this region typically possess some type of holdfast structure or are decidedly streamlined as adaptations to the environment. In the lower stream course, the velocity is reduced and the waters are usually very turbid and warm. The stream bed is frequently deep in sediments and harbors an abundant fauna. Dissolved oxygen is normally less than in the upper regions, due to increased turbidity and organic decomposition. Other chemical features such as pH and ionic content may change along the stream gradient, thereby limiting certain biological elements to particular zones.

A number of classification schemes based on such dimensions as size, velocity, temperature, chemical composition, nature of the stream bed, and permanence have been proposed. These prove inadequate for general usage due simply to the highly variable nature of the properties of streams.

GEORGE K. REID

References

Frey, D. G., ed., "Limnology in North America," Madison, Univ. Wisconsin Press, 1963.

Goldman, C. R., ed., "Primary Productivity in Aquatic Environments," Berkeley, Univ. California Press, 1966.

Hutchinson, G. E., "A Treatise on Limnology, vol. 1, Geography, Physics and Chemistry," New York, Wiley, 1957.

Hutchinson, G. E., "A Treatise on Limnology, vol. 2, Introduction to Lake Biology and the Limnoplankton," New York, Wiley, 1967.

Macan, T. T., "Freshwater Ecology," New York, Wiley, 1963.

Reid, G. K., "Ecology of Inland Waters and Estuaries," New York, Reinhold, 1961.

Ruttner, F., "Fundamentals of Limnology," 2nd ed., Toronto, Univ. Toronto Press, 1963.

LINKAGE

Linkage is the type of inheritance exhibited by GENES located in the same chromosome. It is characterized by the tendency for such genes to remain together in passing from one generation to the next. By contrast, genes located in different chromosomes assort independently of one another—Mendel's second law.

Genes located in the same chromosome are said to be linked and to constitute a linkage group. In a given organism the number of linkage groups that can be identified through breeding experiments is invariably equal to or less than the total number of pairs of chromosomes of that organism.

The strength of linkage between two pairs of genes varies inversely with the amount of recombination or crossing over between them. To illustrate, consider a diploid individual who is a carrier or "heterozygote" for two defective genes a and b, located in the same chromosome. Let a be derived from one parent and b from the other. The genetic constitution of such a heterozygote is symbolized: aB/Ab, where A and B designate the normal alleles of the genes in question, and the / sign is used to separate alleles carried in homologous chromosomes; that is, one chromosome in this case carries the alleles a and B, while the homologous chromosome carries A and b. Four classes of gametes may be produced by such a heterozygote. Two of the classes, aB and Ab, are referred to as parental types while the remaining two, AB and ab, are the recombinant, or "crossover" types. The latter result from physical interchanges of the homologous chromosomes in the region bounded by the a and b pairs of genes. The process resulting in such interchanges is known as crossing over. The farther the genes a and b are apart in the chromosome, the more likely that crossing over can occur; therefore, the greater the percentage of the crossover types. The latter is known as the percentage of recombination. It may vary between 0% and 50% depending upon the particular pair of genes in question. Another quantity, the percentage of crossing over, is a more accurate measure of relative distance in the chromosome. It takes into account the possibility of occurrence of multiple crossing between a given pair of genes. If only two pairs of genes are involved in a cross the occurrence of multiple crossing over goes undetected and the percentage of recombination underestimates the true amount of crossing over. However, for a relatively closely linked pair of genes (showing less than about 15% recombination), the occurrence of multiple crossing over is generally rare or absent and the percentage of recombination therefore closely approximates the percentage of crossing over. With a sufficient number of such pairs of genes it is possible to construct the so-called "linkage map" in which the position, or "loci," of the genes in a chromosome is shown in terms of a standard unit of genetic recombination; namely, one per cent of crossing over.

In the above example, the aB/Ab heterozygote has the normal alleles in opposite chromosomes and is called the "trans-heterozygote." In this case linkage is said to be in the "repulsion" phase. The ab/AB heterozygote has the mutant alleles in the same chromosome and is the "cis-heterozygote." In this case linkage is said to be in the "repulsion" phase. The ab/AB heterozygote has the mutant alleles in the same chromosome and is the "cis-heterozygote." In this case linkage is in the "coupling" phase. The strength of linkage, however, is the same, other conditions being equal, whether linkage is measured in coupling or in repulsion.

E. B. LEWIS

References

Morgan, T. H. *et al.*, "The Mechanism of Mendelian Heredity," New York, Holt, 1915.
Pontecorvo, G., "Trends in Genetic Analysis," New York, Columbia University Press, 1958.
Sturtevant, A. H., "The linear arrangement of six sex-linked factors in Drosophila as shown by their mode of association," Journ. Exp. Zool. **14**: 43–59, 1913.

LINNAEUS, CARL (1707–1778)

Described in 1754 by an English contemporary, William Watson, as 'the most compleat naturalist the world has seen,' Carl Linnaeus dominated the study of natural history for the greater part of the 18th century by his numerous widely used publications, of which the most important (*Systema Naturae, Genera Plantarum, Philosophia botanica* and *Species Plantarum*) went into several editions or reprints, and by his effective teaching at Uppsala, Sweden, of students from many lands who caught his enthusiasm and later, as professors themselves, communicated his methods of classification, description and nomenclature to their students in turn. The first edition (1753) of Linnaeus's *Species Plantarum* has been since 1905 the internationally accepted starting point for botanical nomenclature, because in this work the binomial system (see below) was applied for the first time to all known plants. Likewise the first volume (1758) of the tenth edition of his *Systema Naturae* has been taken as the official starting point for zoological nomenclature. These and associated works of Linnaeus have thus become reference books of lasting importance. They deal with the classification of all the plants and animals then known.

In the absence of a practical and generally acceptable 'natural system' of classification for plants Linnaeus assembled their genera into admittedly artificial classes (*Monandria, Diandria,* etc.) and orders (*Monogynia, Digynia,* etc.). He accepted the constancy of species, taking the view in 1737 that 'species are as many as the Supreme Being produced diverse forms at the beginning' but recognizing the existence of varieties which had arisen from these later in time as 'the work of Nature in a sportive mood.' Thus Linnaeus has come to be regarded primarily as the inventor and advocate of an artificial system of botanical classification long superseded and as a firm believer in the original creation and fixity of species whose views Darwin and Wallace had to overthrow, his sole memorable achievement being the introduction of binomial nomenclature. Although right in the main, such an assessment is very incomplete; it covers only part of Linnaeus's activities and interests as a naturalist, physician and professor, requires modification to include his later somewhat evolutionary albeit farfetched views on the origin of species through hybridization and ignores his significant contribution to the development of taxonomic descriptions. Essentially Linnaeus was an 18th century biological encyclopedist surveying a field so wide that his efforts could be neither deep nor adequate. His systematic and comprehensive listing of plants and animals, together with his convenient method of naming them, came at the right time to provide a basis for the expansion of the biological sciences brought about by European exploration and collecting in the tropics and elsewhere; his work thereby influenced not only his chosen taxonomic field but also studies of which he knew nothing or merely touched upon in dissertations, but to which Linnaean methods have been extended, and also such activities as flowerpainting. It met the needs of the period.

Thus, in the words of Loren Eiseley, 'Linnaeus shares, with the Comte de Buffon *** the distinction of being a phenomenon rather than a man. This achievement, though it demands great energies and unusual ability, is, in reality, dependent upon the psychological attitudes of a given period. The genius must receive extraordinary support and co-operation in intellectual circles. Linnaeus wrote and flourished in a time when the educated public had become fascinated with the word, the delight in sheer naming, whereby the many new plants in gardens from overseas, the shells and butterflies and other natural history objects in collections, the wild flowers gathered on country walks, could by means of their names be referred to a place in the system of knowledge erected by Linnaeus. Had Linnaeus lived a century earlier he would have lacked the vital stimulus provided by Camerarius's proof of sexuality in plants and by the friendship of Artedi, as well as the examples upon which to improve provided by Roy, Morison and Tournefort. Had he lived a century later he might have been eclipsed by such workers of high industry, ability and insight as A. P. de Candolle and George Bentham. The works of Linnaeus must be consulted with their 18th century background kept in mind.

The surname Linnaeus which Carl made so renowned was coined by his clergyman father Nils (Nicolaus) Ingemarsson (1674–1733), son of a farmer Ingemar Bengtsson (1633–93). It referred, as also did the name of the family property Linnegård and the surname Tiliander adopted by his uncles, to a large and old linden tree (Tilia), which survived into the 20th century. Swedish peasant families did not then possess family names handed down from generation to generation; a youth of peasant origin registering at a university in the 17th and early 18th century had to invent a surname for himself, usually of Latin form. The name Linnaeus was thus latinized from the start but, when Carl Linnaeus was ennobled in 1761, he derived from it the title of von Linné. Both names are from the Swedish *linn*, a now obsolete variant of *lind*.

Carl Linnaeus was born on 13 May (Swedish Old Style), 23 May (New Style) 1707 at Råshut, Småland, southern Sweden, where his father Nils was then curate. In 1709 Nils Linnaeus moved to Stenbrohult as pastor and laid out a garden wherein his little son Carl soon acquired a similar love of flowers. The boy was sent to school at Växjo, then in 1727 enrolled as a medical student at the University of Lund, changing in 1728 to the University of Uppsala. At both he first suffered from his poverty; at both he soon received support and co-operation from perspicacious men of learning. Neither university then offered much official teaching; in both, however, were good libraries and Linnaeus largely instructed himself with their aid. In 1729 he met Petrus Artedi (1705-35), a student two years older than himself with much the same interests but greater learning and a temperament which complemented his own, being, as described by Linnaeus, 'more seriously minded, more attentive to details, slower in observation and in everything he did, but on the whole more accurate.' For five years he and Linnaeus worked together on a survey of the whole living world, developing methods of diagnosis, description and nomenclature and dividing their labours so that Artedi took upon himself the study of fishes and reptiles, which repelled Linnaeus, and the dull-looking botanical family *Umbelliferae*, while Linnaeus applied the same methods to the more aesthetically pleasing birds, insects and plants in general; both students looked to medicine for their future livelihood. During these years of joint endeavour Linnaeus drafted the first versions of most of his later works.

The most important happening of this time for Linnaeus was a journey made in 1732 into the wilds of Lapland, which provided the material for his *Flora Lapponica* (1737) and several months of that mental isolation and solitude invaluable to a maturing creative intellect.

In 1735 Linnaeus left Sweden for Holland, quickly obtained the degree of doctor of medicine at the little University of Harderwijk, then went on to Leyden, where he was befriended by the Dutch scholar J. F. Gronovius (1686-1762) and his Scottish friend Isaac Lawson, who together printed the first edition of Linnaeus's *Systema Naturae* (1735) at their own expense. He then became physician to a wealthy banker, George Clifford, and superintendent of the latter's garden and menagerie at Hartecamp between Leyden and Haarlem. Meanwhile Artedi had also come to Holland and was working on fishes in Amsterdam. Unfortunately he fell into a canal on a dark night and was drowned. This tragedy deprived zoology of one who might have proved its most brilliant 18th century exponent. By virtue of an agreement made between the two students in Uppsala that if one died the other

would take over his work, it now became Linnaeus's task to edit his dead friend's remarkable monograph on fishes, *Ichthyologia* (1738), and to catalogue and classify the whole living world as then known, which Linnaeus achieved in 1758.

In 1738 Linnaeus returned to Sweden, having published during his residence in Holland his *Systema Naturae* (1735), *Fundamenta botanica* (1736), *Genera Plantarum* (1737), *Flora Lapponica* (1737), *Hortus Cliffortianus* (1738), and *Classes Plantarum* (1738). From 1738 to 1741 he practiced as a physician. In 1741, through the influence of a powerful friend, Count Tessin, he achieved his ambition of becoming a professor at Uppsala; in 1742 he exchanged his chair of medicine for the more congenial one of botany then held by Rosén. The period 1741-70 was an arduous and successful one of university teaching and administration, of travel within Sweden, and of writing his *Flora Suecica* (1745), *Fauna Suecica* (1746), *Philosophia botanica* (1751), *Species Plantarum* (1753), *Systema Naturae*, 12th edition (1758-59), numerous dissertations reprinted in his *Amoenitates academicae*, and other works. His health failed in 1772 and he died on 10 January 1778 at the age of seventy. His only son Carl did not long survive him, dying in November 1783. In 1784 Linnaeus's widow Sara Lisa (1716-1806) sold his collections of botanical, geological and zoological specimens and his library to a young Englishman, James Edward Smith (1759-1828), who later sold the geological specimens but carefully preserved the rest. This was purchased in 1829 from Smith's widow by the Linnean Society of London and remains in the Society's keeping.

Linnaeus, like other naturalists of his time, used essentially the same methods of classification, diagnosis and naming for both plants and animals and even attempted to apply them to minerals and diseases. His major divisions within each of the three kingdoms of Nature were Classes and Orders. Thus in his *Systema Naturae* of 1737 he divided the Animal Kingdom (*Regnum animale*) into six classes: *Quadrupedia* later called *Mammalis* (mammals), *Aves* (birds), *Amphibia* (reptiles and amphibians), *Pisces* (fish), *Insecta* (insects), *Vermes* (other invertebrates). These were further divided into orders. Thus the *Quadrupedia* in 1737 consisted of five orders (increased to eight in 1758) defined by dental characters: *Anthropomorpha*, later called *Primates* (the primates), *Ferae* (carnivores, insectivores, bats), *Glires* (rodents), *Jumenta* later divided into *Bruta* and *Belluae* (horses, hippopotamuses, elephants, pigs) and *Pecora* (ruminants); to these he later added the *Cete* (narwhal, whales, dolphins). His classification of the Plant Kingdom was more artificial, the orders being defined by the number and disposition of the stamens (e.g. *Monandria*, one stamen; *Diandria*, two stamens, etc.) and these in turn divided into classes by the number of styles or stigmas (e.g. *Monogynia*, one style or stigma; *Digynia*, two styles or stigmas, etc.). 'The prurient mind of Linnaeus, so visibly exhibited in his method of describing bivalve shells,' to quote S. F. Gray, led him to describe this 'sexual system' of classification metaphorically with the stamens as husbands and the styles as wives, leading in the *Compositae* to such extraordinary situations as 'necessary polygamy' and 'superfluous polygamy,' which aroused an interest denied to more prosaically expressed works. Basically, however, this system was arithmetical and mechanical. It sometimes placed far apart genera which resembled one another in almost everything except the number of their stamens and sometimes brought together genera agreeing only in this. It conflicted so severely with Linnaeus's keen perception of affinities based on general resemblance, particularly in habit, that he himself ignored it to the extent of putting species with different numbers of stamens in the same genus, thereby defeating on occasion the main purpose of his system as an aid to diagnosis. Linnaeus himself frankly regarded it as an artificial system, needed then because 'from want of knowledge of matters still hidden' it had not yet been possible, and might never be possible, to construct a natural system giving insight into the true nature of plants. His orders and classes were simply headings under which, for want of anything better, were to be placed the genera, the fundamental units of his taxonomy. Tournefort had defined genera in 1694 and 1703; Linnaeus in 1737 much improved the method and style of generic description and also reformed generic names, admittedly in a manner more to his own taste than to that of his contemporaries and with unconcern for the value of his predecessors' work, by making them accord with rules of his own devising set out in his *Critica botanica*; thus he got rid of all two-word generic names such as *Herba Paris* and *Corona solis* and some cumbersomely long ones as *Pseudo-dictamnus* and *Eupatoriophalacron*, barbarous names as *Lilac, Bihai* and *Gale*, and names ending in -*oides* as *Echioides* and *Jasminoides*, but he violently disturbed traditional nomenclature by such changes. Within the genus he at first distinguished the species by several-word diagnostic names (polynomials) such as *Lithospermum seminibus laevibus, corollis calycem vix superantibus, foliis lanceolatis, Lithospermum seminibus rugosis, corollis vix calycem superantibus* and *Lithospermum seminibus laevibus, corollis calycem multoties superantibus*, these being carefully drafted so as to form a key to the species within the genus when studied side by side. In 1753 he supplemented these specific names by providing alternative two-word (binomials), such as *Lithospermum officinale, L. arvense* and *L. purpuro caeruleum*, consisting of the generic name (e.g. *Lithospermum*) followed by a specific epithet or trivial name (e.g. *officinale*). This method of naming species by two terms is called the binomial system. Linnaeus himself first used it when citing books, treating the surname of the author as a generic name and a word taken from the title of the book as a specific epithet, e.g. *Bauh. pin.* referring to Bauhin's *Pinax* and *Bauh. prodr.* referring to Bauhin's *Prodromis*, and later applied the same system to plants and animals, for which it indeed had good precedent in vernacular nomenclature and the incidental binomials in the Latin nomenclature of earlier authors. Linnaeus established it as the most convenient system for international scientific use by coining such names for all the organisms known to him (roughly 4,400 species of animals, 7,700 species of plants), publishing them in comprehensive well-organized textbooks, and there linking them with descriptions, diagnoses or illustrations by which the application of such names could be determined and kept stable. He used two systems, polynomial and binomial, concurrently. His followers abandoned the polynomial system when increase in the number of known organisms made

impossible the drafting of convenient adequately diagnostic names for them.

As B. Daydon Jackson wrote in 1911, Linnaeus 'made many mistakes; but the honour due to him for having first enunciated the principles for defining genera and species, and his uniform use of specific names is enduring. His style is terse and laconic; he methodically treated of each organ in its proper turn, and had a special term for each, the meaning of which did not vary. The reader cannot doubt the author's intention; his sentences are business-like and to the point. The omission of the verb in his descriptions was an innovation, and gave an abruptness to his language which was foreign to the writing of his time; but it probably by its succinctness added to the popularity of his works.' These contributions to the methods of descriptive taxonomy provided systematic biology in particular and indeed biology generally with valuable tools for recording information that have made Linnaeus's influence permanent.

WILLIAM T. STEARN

References

Gourlie, N., "The prince of Botanists: Carl Linnaeus," London, 1953.
Hagberg, K., "Carl Linnaeus," London, 1952.
Jackson, B. D., "Linnaeus (afterwards Carl von Linné), the story of his Life," London, 1923.
Stearn, W. T., "An introduction to the Species Plantarum and cognate botanical works of Carl Linnaeus (prefixed to Ray Society facsimile of Linnaeus, Species Plantarum, vol. 1)," 1957.
Stearn, W. T., "The background of Linnaeus's contributions to the nomenclature and methods of systematic biology," Systematic Zool. 8: 4–22, 1959.
Stockholm, Kungl, svenska vetensakademien. 1909. Carl von Linnés Bedeutung als Naturforscher und Arzt (Jena).
Svenson, H. K., "On the descriptive method of Linnaeus," Rhodora 47: 273–302, 363–388, 1945.

LITTORAL

This is the *shore*, or where a body of water meets the land. It therefore corresponds almost exactly to *strand* except that this term applies merely to shores of unstable substrates such as sand, gravel, etc. *Littoral* is either the shore itself or the life found there, and may be used as an adjective or noun.

There are two different usages in practice, limnological and marine, with two further distinctions in marine usage. Limnologically the littoral includes everything from as high on shore as ordinary waves of calm weather seasons, or capillarity, produce unusual dampness of the ground (substrate), down below water level to the limit of rooted plants with floating or emergent leaves or flowers. Some difficulty appears in very shallow lakes where this would make the whole bottom "littoral." In such cases limits are best set by the particular case and ought to be stated in written accounts.

In marine use, ecologists now generally restrict *littoral* to the region between tide lines, and include the *spray zone* which is the region above highest tides, wetted from time to time by spray of ordinary surf. Without the spray zone, littoral means *intertidal*, the region from the level of highest spring tides down to the level of lowest normal low water of spring tides.

Oceanographers and some geologists call *littoral* the whole region from high tide level down to approximately the compensation point. This usually means the whole continental shelf down to between 200 and 400 meters.

Zoogeographers now tend to call this simply the *shelf region* or *shallow waters*. Oceanographers divide their littoral into *eulittoral* (intertidal) and *sublittoral* (subtidal). Biologists prefer to make littoral and intertidal synonymous, adding the spray zone for ecological reasons.

Littoral then applies only to the region exposed from time to time to the atmosphere. Within this region, it refers only to those organisms actually living upon, attached to, or buried in the ground or substrate. Organisms primarily suspended in the water which returns to cover the littoral are not littoral but pelagic. Littoral organisms are benthic. A few species are intermediate ecologically; these include certain fishes and crustaceans always found just at the water line, following it up and down with the tides. A few others like the Californian grunion (*Leuresthes*) and the Arctic capelin (*Mallotus*) lay eggs in littoral sands. The midshipman fish (*Porichthys notatus*) of the north-east Pacific lays eggs under intertidal rocks and stays by them until hatching and freeing of the attached larvae. Many of these species are really benthic, but some like the sand lances (*Ammodytes*) are actually pelagic most of the time, swimming in schools like smelt.

The lower limit of the littoral is not necessarily the *zero* or 0.0 level of tide tables. This level is the *tide datum* and in practice varies from country to country. Tide datum of the British Admiralty, Canadian Hydrographic Service and many others is the level of *lowest normal low water* or *lowest normal spring tides. Normal* is not defined by the Canadians. Datum of the U.S. Coast and Geodetic Survey tables is *mean lower low water*. There is a difference of about 2 feet (0.6 meter) between the two data, and this is also the region of the *subtidal fringe*. Where the tidal range is less than 15–16 feet (3 meters) the difference in datum-levels is correspondingly less. Nevertheless because the "fringe" is the zone where intertidal and subtidal species ranges meet and overlap, and are at their respective limits, it is quite important to know the datum of the tide tables in use. If there are "minus" tides shown nearly every month, as in most U.S. tables, it is probable that the datum is set biologically too high.

Tables showing only rare sub-zero tides are in the Americas usually based on British or Canadian datum levels and their indicated zero level is also very close to the true limit of the littoral. The subtidal (sublittoral, infralittoral) fringe is the zone which on calm days during a spring tide low water is exposed to atmosphere for a short time but still slightly washed by gentle ripples. During windy weather or in surf it is either continually washed and exposed or not exposed at all, during these low tides.

The marine littoral is biologically and ecologically divisible into three or four zones. On rock or stable boulder shores there are: *spray zone; littorine zone; balanoid zone; subtidal fringe*. Names are from the conspicuous life forms of each, respectively: black lichen and algae; *Littorina*-like snails; acorn (*Balanus*-like) barnacles; serpulid worms, reef corals and numerous subtidal algae.

On sandy, or other strand shores, the zones are not always distinct, due to the percolating and retention of water in the beach substrate. The spray zone has strand plants, mangroves and accompanying bushes in the tropics, and various herbs including forms like *Salicornia, Glaux, Abronia* and sedges, with spiders and insects and various crustaceans. The subtidal fringe has eelgrasses like *Zostera* and *Phyllospadix*, animals like *Macoma, Cardium* and other very active clams and dense populations of burrowing polychaete worms, tube-forming and free living. At various levels begin populations of other burrowing animals such as clams, burrowing shrimps, mole crabs and so on. The upward limits of these depend much on texture of the beach material and its ability to hold water, also on exposure to wave action and currents which shift the beach materials more or less depending upon many physical factors.

The above ecological zonation of the littoral into levels recognizable by the life-form of the most conspicuous species is apparently world-wide, varying from tropics to temperate regions by presence or absence of certain organisms like reef corals, mangroves and the like.

R. W. PILLSBURY

Reference

Hedgpeth, J. W., "Classification of marine environments," Geol. Soc. Amer., (Mem. 67) **1**: 17–28, 1957.

LIVER

The blood returning from the intestine to the heart is shunted through a capillary system, the hepatic sinusoids, which are surrounded by epithelial cells arranged in plate forms. These plates cross each other in space at different angles, to permit the greatest possible contact between the blood and these polygonal epithelial cells. The resulting spongelike organ located under the diaphragm and covered by the connective tissue capsule of Glisson is the largest organ of the body. Under normal circumstances, the major part of its blood, between 66 and 75%, comes from the portal vein which drains the splanchnic capillaries, particularly those of intestine, pancreas and spleen. Approximately $\frac{1}{4}$ to $\frac{1}{3}$ of the hepatic blood comes from the hepatic artery originating from the aorta at the celiac axis. Both hepatic artery and portal vein enter at the hilus of the liver and divide in a dichotomic fashion into parallel running branches. They are surrounded by ramified extensions of Glisson's capsule. The hepatic artery sends branches to the capillary plexus of the portal tracts whereas the bulk of its blood is released into the sinusoids parenchyma as does the portal vein which forms by confluence of superior mesenteric, inferior mesenteric and splenic vein, and receives additional internal radicles from the portal capillary plexus. The sinusoids are blood capillaries characterized by great permeability for serum proteins and by modifications of their star-like endothelial cells, the Kupffer cells, which are parts of the reticulo-endothelial system. The cytoplasmic extensions of the Kupffer cells form the sinusoidal wall and leave small stomata open through which macromolecular substances pass into a tissue space between liver cell plates and sinusoids. Tissue fluid is drained towards the lymphatics in either the central canal or, in the human, mainly the portal tract. Arterial and venous *blood*, mixed to a varying degree flows towards the tributaries of the hepatic veins which combine to larger veins into which frequently small branches enter at almost right angles. The largest branches enter into the vena cave inferior behind the liver. Vascular sphincter mechanisms in various locations regulate hepatic blood flow and thus function. The portal tracts and the central canals around the tributaries of the hepatic veins cross each other in space and are throughout the liver about 0.3 mm. apart. The direction of the blood flow from the portal tracts to the central canals produces the concentric arrangement of the liver cell plates characterizing the liver lobule which conventionally is considered the structural unit of the liver.

The liver forms bile which is released into slits between the liver cells, the bile canaliculi, which are arranged in a chicken-wire-like fashion. They are drained by small tubes with an independent cuboidal epithelial lining, the ductules or cholangioles. Under normal circumstances hardly any are found within the lobule, the majority being in the periportal zone or in the portal tract. Under abnormal circumstances, they increase in number either by sprouting from preformed ductules or by transformation from liver cells and are then found deep within the lobule. The ductules continue into the bile ducts located in the portal tracts which unite in dichotomic fashion to finally form the common hepatic duct which leaves the liver where hepatic artery and portal vein enter it. It combines with the cystic duct draining the gallbladder, which concentrates bile by water reabsorption to form the common duct running toward the duodenum. This entrance is controlled by the choledochoduodenal sphincter of Oddi. Bile is produced at an almost constant rate but released from the biliary system in human beings and many animals only if food appears in the duodenum. As a result of this or other mechanisms, the sphincter of Oddi relaxes and the gallbladder contracts. This serves proper utilization of bile which while being partly an excretory product is a secretion essential in intestinal digestion and absorption.

In the liver, several fluid currents exist. Blood and some tissue fluid flow towards the central canal while bile and most of the tissue fluid (at least in the human) is flowing towards the portal canal. The normal liver consists of approximately 60% hexagonal epithelial cells, 30% littoral endothelial or Kupffer cells, and about 2% each of bile duct cells, connective tissue and blood vessels. The hexagonal cells have three types of borders. Where they are in contact with each other, the border is straight indicating a limited, if any, exchange of substance between individual cells. The border towards the tissue space is elongated by narrow extensions of the space between neighboring liver cells and particularly by the formation of irregularly shaped finger-like projections in the form of microvilli. This tremendous elongation of the border of the liver cells and the preferential location of enzymes in this location reflects structurally the extensive exchange of substances between liver cells, tissue space and blood. Much shorter is the border towards the bile canaliculus which is also thrown into microvilli which are far more regular and disappear upon impairment of biliary secretion. Preferential accumulation of ATPase in these villi indicates the intensity of the metabolic processes in bile secretion.

The NUCLEUS is normally vesicular and has conspicuous nucleoli. It varies considerably under normal and pathologic conditions, the majority being tetraploid in adult rodents. Binucleated cells increase in regeneration. The cytoplasm normally contains many and relatively large mitochondria in which the enzymes of the energy metabolism can be demonstrated. In addition, smaller bodies, the lysosomes, presumably the side of various hydrolytic enzymes, can be demonstrated as well as dark bodies related to pigment. An extensive endoplasmic reticulum flanked by fine Palade granules of ribonucleoprotein character, correspond to the microsomes and is considered the site of synthesis of proteins including serum proteins and of steroids as well as of detoxification. The soluble fraction of the cytoplasm, the hyaloplasm, corresponding to the supernatant fluid in cytochemical analysis contains proteins and enzymes and cofactors related to carbohydrate metabolism and activation of amino acids and nucleic acids. In addition, in the normal liver, glycogen and few fat droplets are found as well as some ferritin crystals which, under abnormal circumstances, become hemosiderin deposits giving histochemical iron reaction.

The main functions of the liver cells are: 1) secretion of substances into the blood stream of which the serum proteins particularly albumin, alpha globulin, the proteins concerned with blood coagulation and some blood enzyme, for instance esterase, as well as serum cholesterol and blood glucose are probably the most important; in contrast to all other tissues which utilize but do not form blood glucose, the liver cells are the main source of the blood glucose because of a specific phosphatase system; 2) storage of various metabolites particularly glycogen, proteins, fat and vitamins; 3) transformation of various compounds into each other, e.g. fats into carbohydrates and vice versa; 4) detoxification mainly by oxidation or conjugation, the latter mainly for better solubility and urinary excretion; 5) formation of the bile into which bile pigment is transmitted by conjugation and bile acids and cholesterol by transformation.

The Kupffer cells have a cytoplasm of varying and irregular outlines and ameboid extensions. They contain few MITOCHONDRIA but varying inclusions. Their function concerns besides that of other endothelial cells particularly phagocytosis of circulating exogenous and endogenous macromolecular or corpuscular elements including bacteria as well as of hepatocellular breakdown-products; they are active in transformation of blood pigment to bile pigment. In contrast to other reticuloendothelial cells, they seem to form little, if any, serum gamma globulin or antibodies under normal circumstances but seem to do so in disease.

The liver, as a whole, because of its strategic situation near the right heart and because of its sheer bulk, influences circulating blood volume, as well as electrolyte and water metabolism.

The main diffuse pathologic reactions of the liver are:

(1) cytoplasmic disorganization and necrosis of the liver cells which, if extensive and diffuse enough, accounts for insufficiency of most of the crucial functions of the liver; (2) fatty metamorphosis of the liver cells which, if not complicated by other injury, interferes with few hepatic functions; (3) arrest of bile flow, as a result of extrahepatic biliary obstruction or intrahepatic cholestasis (so-called cholangiolitis); (4) inflammatory reaction involving Kupffer cells and portal tract; (5) fibrosis which, as such, causes little functional disturbance of the liver but sometimes interferes with portal blood flow to produce portal hypertension; (6) cirrhosis, characterized by regenerative nodules and septa dividing the parenchyma resulting in portal hypertension, diversion of blood from the liver, and damage of hepatic cells with subsequent disturbance of hepatic function; (7) cancer either primary from the hepatic cells or from bile duct cells or metastatic from other organs.

HANS POPPER

References

Brauer, R. W., ed., "Liver Function. A Symposium on Approaches to the Quantitative Description of Liver Function," Washington, D.C., Am. Inst. of Biol. Sci., 1958.

Child, C. G. III, "Liver and Portal Hypertension," Philadelphia, Saunders, 1966.

Elias, H., "A re-examination of the structure of the mammalian liver. I. Parenchymal architecture," Am. J. Anat., **84**: 311, 1949.

Elias, H., "A re-examination of the structure of the mammalian liver. II. The hepatic lobule and its relation to the vascular and biliary systems," Am. J. Anat., **85**: 379, 1949.

Fawcett, D. W., "Observations on the cytology and electron microscopy of hepatic cells," J. Nat. Cancer Inst., **15**: 1475, 1955, supplement.

Hartman, F. W., et al., eds., "Hepatitis Frontiers," Boston, Little, Brown, 1957.

Popper, H. and F. Schaffner, "Liver: Structure and Function," New York, McGraw-Hill, 1957.

LOCOMOTION

Perhaps the most outstanding characteristic of animals is their capacity for moving about, although some of them are sessile for part or all of their life histories. Locomotion is used in search of food, suitable living conditions and mates, and serves also for the distribution of species and the prevention of overcrowding.

The mechanisms whereby locomotion is achieved are not passive, as is the distribution of plant seeds, but active, involving the utilization of the animal's energy in the contraction of minute fibrils which are situated within muscle cells or in cilia or flagella. Muscle cells are sometimes isolated, but more often they are elaborated into sheets or bundles or into compact muscles. Increase in the tension of muscles or in actual contraction is an active process, utilizing energy and performing work, but muscular relaxation is passive, and muscle fibres elongate again only when subjected to an outside extensor force. This may be supplied by antagonistic muscles, working via a hydrostatic or a rigid skeleton, as in an actinian, an earthworm or a vertebrate, by fluid pressure, as in the extension of some echinoderm tube feet, or by the elasticity of a notochord or of collagen fibres in mesogloea or connective tissue. The design of an animal's body is correlated with provisions for the extension of relaxed muscles just as much as with the means for delivering propulsive thrusts.

Interest in the locomotory mechanisms of animals centers about (1) their nature, which varies according to the absolute sizes of the animals, the principal modes of locomotion being few, (2) the physiological and mechanical principles involved which have led to the most detailed studies of neurosensory and muscular physiology, biochemistry, applied mechanics, anatomy and cytology, and (3) the evolution and perfection of the various modes of moving about, which are bound up with the evolution of many of the most conspicuous characteristics of the trunk regions of phyla, classes and orders of animals.

Speed of movement from place to place has been of importance in escaping from predators or in the pursuit of prey at all size levels, except for animals which rely upon other methods of protection, such as concealment or a fecundity and rapidity of growth which makes good a large scale predation by other organisms. In the larger animals the most speedy running and swimming have been independently evolved many times; there is little difference between the maximum speeds shown by mammals of various orders, although their body sizes may differ. The cheetah is the fleetest of animals, but it can sprint at 70–75 mph for only ¼ minute; the horse can attain 44 mph for 300 yards and can run at 20 mph for 20 minutes. Inertia is of greater importance to large animals than it is to small ones. Both a porpoise and a whale can reach 18 knots; the whale is about 1000 times the weight of the porpoise and the tail beat takes about ten times the duration of that of the porpoise. In small animals, such as myriapods, inertia may be negligible; a 6000 fold range in body weight in diplopods is associated with similar pace durations, and thus the speeds achieved are not equal but bear a linear relationship to the body length. However, in fast-running centipedes, where the propulsive backstroke may be of very short duration, >0.01 sec., inertia is of importance, and the larger and the longer-legged species step more slowly than do the smaller ones.

Elastic storage of energy is important to the mammals and to fast moving insects although not so important to the slower moving arthropods. Much of the bulk of an elephant is made up by elastic tissue and short muscles and the limbs of a horse possess much elastic tissue. In flying insects special elastic cuticular ligaments are associated with the rapidly moving flight muscles, which give 330 wing beats per sec. in a blow fly, and the elasticity of sclerotized cuticle is also of importance when deformations are small.

In some phyla locomotory movements are so slow as to be unobservable to the eye. Speeded up cinematography demonstrates the slow dilatation of the leading edge of the actinian foot and the contraction at the "heel" and body which displaces enteric fluid, forcing it into the "toe." Such slowness is mechanically advantageous when body deformations are great because the hydrostatic pressure in the enteron is not appreciably raised thereby. Actinian muscle fibres can contract by 4–500%, in contrast to fast moving skeletal muscle in which the shortening is 20% or less, and it is mechanically advantageous for a hydrostatic-muscular system to work slowly. Faster movements can be ef-

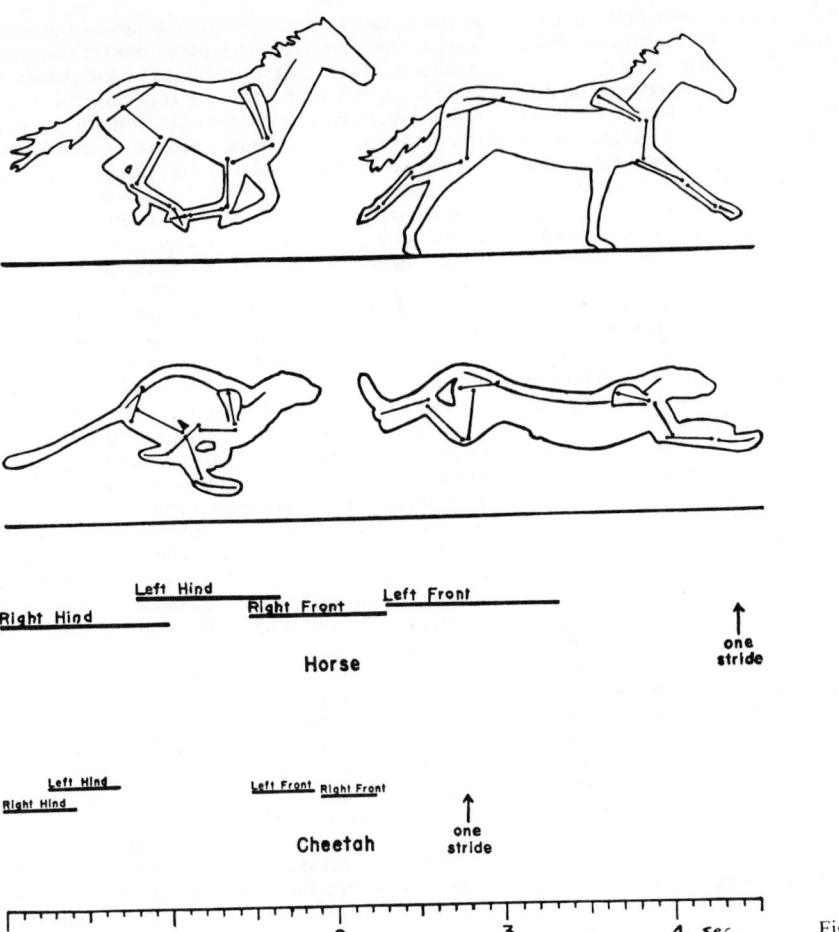

Fig. 1

fected by Actiniaria, but with less efficiency. Slow locomotory movements involving extreme deformations of the body, or of its parts, are similarly advantageous in other phyla, for example to the muscular-haemocoelic system of the molluscan foot.

Fairly slow movements can be found even in characteristically fast-moving animals, such as mammals and myriapods, which are both equipped with striated muscles attached to articulated skeletons. Slow strong movements used in burrowing and faster weaker movements used in running are mutually exclusive in the same animal. The anatomical necessities for the perfection of these two opposed habits are in many respects opposite, so that an animal showing specializations which fit it to execute the former type of movements leave it all equipped to perform the latter, and vice versa. For example, two short and wide extrinsic limb muscles in a iuliform diplopod provide the characteristic strong leg swing, and operate from a heavy rigid trunk skeleton, but 33 extrinsic limb muscles, many of them very long, mediate the rapid swing of each leg of the fleet centipede *Scutigera*, their number compensating for the weakness of quickly contracting muscles. Much economy in other muscles is needed to divert so much musculation to the legs, and the trunk exoskeleton is very light and flexible and ill suited to a pushing habit. In the larger, fleeter vertebrates leg muscles are always short and tendons and bones are long in contrast to the slowly moving or burrowing species. Throughout the animal kingdom perfection of one type of locomotory mechanism limits the animal's capacity for other achievements.

Each type of locomotory mechanism is usually efficient only within a certain range of body size. The relative velocity of movement, that is the distance traversed in 1 second over the body length, frequently decreases with increase in size of animals of similar type. Flagellates cover from tenths to 1 mm per sec., and their speeds increase rapidly as their body sizes increase from 100 to 300 μ. Between 300 and 500 μ increase of speed in small and beyond 500 μ there is no further increase in speed. In the Metazoa the larger ciliated animals are very slow movers, and cilia alone are effective swimming organs at body sizes of less than 0.5 mm. Above that level swimming by means of many pairs of limbs results in faster movement. But copepods below 0.1 mm have velocities of less than those of the ciliated creatures. Fishes swimming by body undulations have much higher velocities than Crustacea, but fish below a 10 mm body length swim more slowly than copepods. Millipedes of the iuliform type burrow by strong head on pushing into the soil, by utilizing the motive force of their many pairs of legs, but again these animals are efficient only within a certain size range. The larger species exert a progressively smaller pushing force relative to their size, and the larger species are the least efficient pushers because so much more of the force put out by their legs is utilized in carrying the body weight, the force roughly increasing in terms of squares and the weight in cubes. Very small species also are not efficient pushers because the force they can exert is small in relation to the size of soil particles and it is then more useful to move between particles than to push them aside. Fairly large diplopods of similar weight to small terrestrial vertebrates can push (and pull) much more strongly, but as size increases the latter outstrip the diplopods, showing that the tetrapod type of organization, with the four pairs of large limbs deriving their musculature from many myotomes, is more serviceable when the body weight is above the order of 30 g than is the diplopod organization with very many pairs of small limbs.

Arthropod swimming mechanisms utilizing the metachronal movement of a long series of paddlelike limbs are unsuitable for organisms of the size of vertebrates because the swimming they produce is too slow, and we can appreciate the reason why so many of the larger and some of the smaller Crustacea have evolved mechanisms for intermittent fast jumping through the water, such as seen in Malacostraca which flap the abdomen under the thorax and Copepoda which jump by the simultaneous movement of the four pairs of thoracic swimming feet. Only in animals swimming by jet propulsion is there an enormous range in body size, which extends from medusae of <1 mm to giant squids with a length of 20 m.

Some Metazoa move slowly over the substratum by waves of muscular contraction passing over the body surface, as seen in some planarians or the foot of gastropods, and nematodes move through viscous media by undulations passing along their whole bodies. A few land planarians show pseudo-metameric segmentation of their musculature at the anterior end of the body. The evolution and the ontogeny of metameric segmentation appears to be based upon the mesoderm, and its origin was probably associated with the usefulness of segmental muscles. Segmentation promotes the localization of muscular action and this allows the exploitation of metachronal patterns of movement, which, with some exceptions such as the movements of cephalopods, provide more rapid locomotion than that of unsegmented animals. Such metamerism, which has doubtless been evolved several times (chordates, annelids and anthropods, and the arms of echinoderms) leads to locomotory mechanisms which are dependent upon i) trunk movements, or ii) limb movements, or iii) a combination of trunk and limb movements.

Waves of unilateral contraction of segmental trunk muscles pass along the body from before backwards in fish, snakes and certain limbless arthropods (ex. larvae of *Anisopus*). The outward and backward thrusts from the convex body surfaces drive the animals forwards, the tail end of the fish being the organ which exerts the greatest pressure on the water as well as steering. The tail fin is not essential to the forward movement. Somewhat similar waves pass along the bodies of crawling and swimming polychaetes but from behind forwards; alone they would cause backward locomotion, but the

waves in fact enhance the effectiveness of the backward stroke of the parapodia, permitting the successive paddles to diverge on the propulsive backstroke and to move through a wide arc before they converge on the forward recovery stroke. If the metachronal waves were transmitted in the opposite direction, as in fish, successive parapodia would converge during the backstroke, an unsuitable disposition for swimming.

Arthropods primarily utilize their intrinsic and extrinsic limb musculature for locomotion. Their trunk musculature seldom contributes towards the direct locomotory thrust, and is concerned with maintaining the rigidity of the body as a base for limb action. Undulations of the trunk are deleterious to speed in arthropods, and many are the structural features which reduce or eliminate their occurrence. Extreme undulations in the horizontal plane appearing in centipedes running at speed do so only when the controlling mechanism breaks down and further increase in speed is impossible. Fish-like body undulations are apparent and useful in quadrupedal walking in urodeles and in some reptiles, but are eliminated in birds and mammals. Trunk movements of some fleet mammals (ex. cheetah) do contribute to speed, but they take place largely in the vertical and not in the horizontal plane, and have been perfected along with the structure and use of the four pentadactyle limbs.

Increase in the length of limbs elongates the stride and leads to speedy running in all animals, but projecting limbs hinder burrowing and such shelter seeking habits. The length and the structure of legs is correlated in great detail with habits in vertebrates and invertebrates. When legs are long and many, little choice of gait is possible, and stumbling must be avoided even when several successive legs overlap during their forward stroke, as they must do if their propulsive stroke is to be practicable. The mechanical advantages to arthropods resultant upon reduction in their leg number to four or three pairs has led to the independent evolution of hexapodous animals in certain Crustacea and Arachnida, besides the pterygote insects and the several groups of apterygotes. Rarely is the number of legs used for walking reduced to two pairs, as in Protura, and the gait here is the same as that of a walking horse.

The gaits of annelids, arthropods and vertebrates are becoming well known, and certain ranges of gaits, large or small, are practicable to each animal. A rapid propulsive backstroke and a slower recovery forward stroke leads to speed, and a slow backstroke followed by a quick recovery swing leads to strong movements. In the many-legged arthropods the latter type of gait results in the simultaneous pushing by very many legs against the ground, some 9/10ths of the total number of legs of a diplopod (some 80–140) may push at once in contrast to a 42-legged fleet centipede which may have but two legs on each side of the body in contact with the ground at one moment. Legs, like machines, work most easily when their loading is even and not greatly changing from moment to moment. The evolution of segment numbers in arthropods and the choice of gaits is controlled by such factors. Paired legs usually move in similar phase in gaits providing strong movements and in opposite phase in those providing fast movements for mechanical reasons. The hexapods need to use their paired legs in opposite phase in all gaits in order to obtain stability at all moments; jumping gaits result from the use of paired legs in the same phase in hexapods.

Other phase relationships exist between the paired legs and these are also dictated by mechanical needs. The exact phase difference between sucessive legs in annelids and arthropods is also of great importance to the animal and is determined by mechanical needs; the phase difference is readily changeable and is correlated with the gait in vertebrates and invertebrates, and the apparent direction of transmission of the waves of invertebrate limb movements (the visible effect of the phase difference between successive legs) can be reversed by the same animal when the phase difference passes from just less to just more than 0.5 of a pace. Animals whose gaits show predominantly small or large phase differences between successive legs, such as 0.2 or 0.8 of a pace, may never alter the gait sufficiently to cause an apparent reversal of the direction of the wave. There appears to be no phylogenetic significance in the direction of transmission of metachronal waves of limb movements, closely related animals can show opposite conditions and all can be related to functional needs and habits.

Many invertebrates and Enteropneusta move on the surface of or through the substratum by body movements in which sections of the body shorten, widen and grip the substratum or widen a burrow. Each zone of thickening usually passes along the body from before backwards, elongating sections or segments paying out in front and moving up behind each zone of thickening. Annelids and burrowing geophilomorph centipedes use the same principles of movements, but the secondary specializations which permit such shape changes in the scute-bearing arthropods are remarkable.

There are many different methods of swimming, jumping and flying to be found in both vertebrates and invertebrates besides the details of the gaits and the structures used in walking and running which cannot be considered in a short article. Neither is there space to devote to the generation of locomotory rhythms, both forward and backward, in the central nervous system, or to the coordination of locomotory movements by proprioceptor and exteroceptor sense organs, or the automatic compensation for mechanical damage by alteration of the normal rhythms, etc. The size of animals; the form of the trunk region; the number and shape of the segments; the shape and rigidity of scutes and the positions of arthrodial membranes and tendons; the shapes of apodemes and endoskeleton; the nature of joints, be they ball and socket, hinge or pivot, incompressible or loose (all are found in both vertebrates and invertebrates); the presence of limbs and their number; the length and number of the muscles; and so on are all intimately associated with locomotion, and these characters have doubtless been evolved along with the perfection of the locomotory mechanisms.

S. M. MANTON

References

Gray, J. et al., (Studies in Animal Locomotion, etc.) J. exp. Biol. (numerous papers 1933–present).
Howell, A. B., "Speed in Animals," Chicago, Univ. of Chicago Press, 1944.
Manton, S. M., (The Evolution of Arthropodan Locomotory Mechanisms, etc.) J. Linn. Soc. (Zool). (numerous papers 1950–present).
Marey, E. J., "Movement," transl. E. Pritchard, London, 1895.
Zenkevich, L. A., "The Evolution of Animal Locomotion," J. Morph. 77, 1, 1945.

LOEB, JACQUES (1859–1924)

Jacques Loeb, a German American scientist, was born at Mayen Germany on April 7, 1859. He obtained his medical degree from the University of Strasbourg in 1884 and then became assistant professor in physiology at the universities of Wurzburg and at Strasbourg. After that he worked at Naples biological station before becoming professor of biology and physiology at Bryn Mawr in the United States. He later joined the faculty of the University of Chicago and then the University of California. He became head of the department of experimental biology at the Rockefeller Institute for Medical Research where he worked until 1924 at which time he was persuaded to go to Bermuda on a short vacation, and there he died of a heart attack.

His significant work resulted from his great responsibility to discover truth, and in his experiments he was constantly trying to explain life. The directions of his experimental work were: brain physiology, tropisms and regeneration, antagonistic salt action, duration of life and colloidal behavior. In his studies on the dynamic or chemo-dynamic theory of living processes he produced larvae from unfertilized eggs of the sea urchin by stimulating them with different concentrations of sugar or salt in the sea water containing the eggs. He also produced tadpoles from unfertilized frogs' eggs. In his research in the field of chain reflexes he was among the first to determine the magnitude of the smallest particle showing all the phenomena of life. From this research Loeb stated that length of life is conditioned by chemical reaction.

Other achievements include the Journal of General Physiology which he founded in collaboration with W. J. V. Osterhout and a series of monographs on experimental biology which he wrote in collaboration with T. H. Morgan and W. J. V. Osterhout. Thus Jacques Loeb was a true leader and pioneer in biology.

BETTY WALL

LONGEVITY

In organisms which undergo AGING the potential life is limited by the rate at which mortaility increases with age, and longevity is usually expressed for comparison as the typical performance under good conditions (as against the ideal *physiological* longevity, or *specific age*, under hypothetically optimal conditions). In published records it also commonly means the highest age recorded for the species. For scientific comparison of life-spans the modal age of adult death and the last decile survival are often the best measures of "average" and extreme longevity: in populations e.g. of small wild birds, where mortality is effectively age-independent, it may be necessary to use the half-life and the recorded maximum.

Natural life-spans of organisms range from 2–3 days in some rotifers to > 150 years in tortoises and 3 or 4 ×

10^3 years in the Bristle-cone Pine (*Pinus aristata*).
Many assumed animal ages in literature and reference
books are imaginary. The longest-lived mammal is
Man, the highest plausible human record being 115–120
years: higher claims are undocumented. Longevity in
mammals is roughly correlated with size, more closely
with relative brain weight, and most closely and inversely with net reproductive rate, the slow-breeding
and partially poikilothermal bats living much longer
than similar-sized rodents: within a mammalian species
large breeds are usually shorter-lived than small. Life-
spans are considerably longer in birds than in mammals
of comparable size, and in flying than in terrestial
birds, and they increase steadily down the vertebrate
scale, becoming partially temperature-dependent in
poikilotherms. In large, slow-growing reptiles and fish
longevity as well as growth seems to be virtually indeterminate, as in the longer-lived trees. Both among
plants and fish, "annual" species exist for which reproduction is normally fatal. In sub-vertebrate phyla
there is a similar wide range of life-spans, the closest
correlate of longevity being always continued growth or
cell replacement. Sea anemones in captivity appear
capable of remaining indefinitely vigorous and have no
ascertainable "life span"—contrary to common belief,
by contrast, the individual lives of protozoa such as suctorians, where parent and progeny differ visibly, is apparently determinate as in higher animals. The highest
plausible records for typical species are given in the
Table.

**Table of maximum reliably-reported ages for representa-
tive species and groups, in years.**

Mammals. Man 115–120, Indian elephant 77+:
largest ungulates etc. 40–50 (horse 46, hippopotamus,
rhinoceros 49, zebra 38+): large primates 30–40
(chimpanzee 39+, gibbon 32+): large ungulates
and carnivores 20–35 (domestic cow 30+, lion 30–
35): domestic cat 27–31, dog 24+: smaller ungulates,
seals 20–25 (sheep 16–20, goat 20+): largest rodents
10–20 (agouti 15, rabbit 15): small chiroptera 10–15:
medium to small rodents 3–10 (guineapig 7, *Peromyscus* 7–8, *Apodemus* 6, white rat 4+, *Micromys*
4, *Mus* 3¼): shrews not more than 2 (*Blarina* 1½,
Sorex 1¼).

Birds. Large raptorial birds probably 100—(Vulture
?117): many other large flying species probably 60+
(*Bubo bubo* 68, Grey parrot ?73, Golden eagle ?80+,
Cockatoo ?70, ?85, domestic goose 47, ?80), Pelicans,
cranes, geese 40–55+: gulls, corvids, columbiformes
30–45 (Herring gull 41, ?49; domestic dove 42, domestic pigeon 35): Emus, ostriches 30–40: small
finches, waders, parakeets etc. 10–30 (chaffinch 29,
oyster-catcher 27, arctic tern 27, starling 19, swallow
9½).

Reptiles, amphibia. Larger chelonia 100, 150 or more
(*Testudo sumeirii* 152+, *T. graeca, Terrapene carolina, Emys* all 100+): crocodiles and alligators 50–
60, *Megalobatrachus* 50+; many snakes, lizards and
amphibia 25–30+ (*Sphenodon* 28+, *Anguis* 33, *Bufo
bufo* 36, *Triton* 35): smaller frogs 10–20 (*Hyla* 16,
Xenopus 15+, *Rana* 12+).

Fish. Sturgeon 82+, *Hippoglossus* 60–70, *Silurus* 60,
Anguilla 55. Aquarium species roughly in propor-
tion to size—small live-bearers can reach 5 years
(*Lebistes, Gambusia*).

Arthropods. Termite primaries estimated 40–60, *Homarus* ?50, *Avicularia* sp 20, *Lasius niger* ♀ 19, *Blaps*
imago 10+, *Ctenolepisma* all stages 7, *Apis melli-
fica* ♀ 5. Many others have long and variable larval
lives.

Other. *Margaritana, Megalonaias* (Mollusca) 50–60
+, *Cereus pedunculatus* (Coelent.) 80–90, *Actinia*
60–70. Platyhelminth parasites 25–50, *Loa* (Nematoda) 15, *Wuchereria* (Nemotoda) 17. *Allolobophora*
(Annelida) 5–10. *Suberites* (Porif.) 15.

Life-spans which differ between the sexes most commonly do so in favour of the female, regardless of which
sex is homogametic. Environmental factors which affect
longevity (other than by simply raising the early mortality) include temperature in all poikilotherms, nutritional restriction of growth rate, and probably low-level
radiation. The life-span of rats is said to be doubled by
allowing them to grow after keeping them artificially
juvenile by calorie restriction. Another important influence is heterosis, hybrids living commonly up to
twice as long as highly inbred strains.

ALEX COMFORT

References

Comfort, A., "Ageing—The biology of senescence,"
New York, Rinehart, 1963.
Ciba Foundation, "Colloquium on Aging No. 4: animal
life spans," London, Churchill, 1959.

LUDWIG, HUBERT (1852–1913)

A German anatomist who held the chair of comparative anatomy at Bonn for many years. He published
much on marine invertebrates, especially the echinoderms, but his chief claim to fame is the relentless
manner in which biographers have confused him with
the subject of the next entry.

LUDWIG, KARL FRIEDRICH WILHELM (1816–1895)

This German physician, born in Kassel, in 1816, was
one of the first early physiologists to devote a great
deal of his time and attention to the development of
apparatus designed to record physiological activities.
The kymograph, of which he published the first description in 1846, is in use at the present time. He
taught anatomy—since physiology was at that time an
insignificant science—in Mahrburg and Zurich but in
1855 was given the opportunity of developing a new
chair of physiology in the University of Vienna, which
he retained for ten years before going to Leipzig in
the same capacity. In addition to his work on the heart,
which led to the production of the kymograph, he
published much on the mechanism of glands and of the
kidney. His *Lehrbuch der Physiologie des Menschen*,
of which the first edition was published in 1856 in
Leipzig, remained a standard text book in the medical
schools of the world for many years after his death in
1895.

LUNG

General. The lungs are large, paired, spongy organs which occupy the two thoracic pleural cavities. Each pleural space is lined by a serous epithelium, the parietal pleura, which is continuous at the root of the lung (hilus) with the visceral pleura—the membranous investment of the lung. (See Fig. 1). The lung completely fills its space so that visceral and parietal pleurae are virtually in contact, separated only by a capillary film of serous fluid. As the thoracic cavity increases in volume upon inspiration the tendency for a subatmospheric pressure to develop in the closed pleural cavity is offset by the entrance of atmospheric air (at 15 lb/in²) through the tracheobronchial tree. Thus the lungs are pressed against the moving thoracic walls and, in a deep inspiration, may increase in volume from a residual capacity of about 1 liter of air to a total capacity of 5–6 liters. As lungs expand, their rich elastic fiber framework is placed under a tension which exerts a circumferential traction upon the walls of pulmonary vessels and airways. With the expiratory decrease in thoracic volume the lungs recoil elastically and the contained gases are forced out through the airways to the atmosphere.

Embryology. The human lung begins its development in the fourth week of embryonic life as a ventral laryngo-tracheal outgrowth from the endodermal epithelium of the pharyngeal floor. This diverticulum soon bifurcates and the resulting primary bronchial buds elongate and produce lateral buds—2 on the right and 1 on the left. Through subsequent dichotomous and monopodial branchings, 5 original buds provide the airways of the 5 definite lobes of the lungs. At birth some 17 generations of airways are present. Postnatally, additional ramifications increase the number to about 21 families of branches. The endodermal tubes ramify within a mass of undifferentiated mesenchyme on either side to provide the epithelial lining of the respiratory tree, while the supporting structures (cartilage, fibrous connective tissue) and smooth muscle fibers differentiate from the investing mesenchyme. In the final ⅓ of gestation the cuboidal epithelium of the terminal air spaces (alveoli) becomes greatly attenuated and there is a marked proliferation of the pulmonary capillaries. These lie immediately beneath the alveolar epithelium to provide a minimal blood-air barrier at the respiratory surface. Thus, a premature infant, born subsequent to the seventh month may possess sufficient respiratory surface to permit survival.

The respiratory tree. The airways are analogous to arborizations of a deciduous tree. The trachea represents the trunk of the tree and the bronchi, its primary, secondary, etc. branches. The bronchi enter the lung at its hilus and produce lobar branches which, in turn, divide into segmental bronchi to supply discrete areas of lung tissue termed segments. There are usually 10 such bronchopulmonary segments in the right lung and 8 in the left.

The walls of the trachea and extrapulmonary bronchi are supported by U-shaped cartilages. The cartilages are invested by an elastic fibrous membrane which spans the intervals between them and completes the tracheal wall dorsally. Transverse bands of smooth muscle extend between the ends of the cartilages. The lining is a pseudostratified ciliated epithelium with goblet cells, and rests upon an elastic lamina. Mucous glands situated between cartilages empty their secretions onto the lining of the airway via their secretory ducts. Intrapulmonary bronchi are encircles by irregular plates and spirals of cartilage embedded in the fibrous coat. Internal to this is a muscular layer consisting of spiralling bands of smooth muscle. Followed peripherally, the ciliated epithelium becomes simple columnar with goblet cells, and rests upon a lamina propria containing a diffuse network of elastic fibers.

Successive generations of bronchi progressively diminish in caliber until the diameter is 1 mm. or less, whereupon the branches are called bronchioles. The walls of bronchioles lack cartilages and mucous glands; their EPITHELIUM narrows to non-ciliated cuboidal and rests on a thin elastic lamina propria. Bronchioles have a relative abundance of smooth musculature, supported by connective tissue. A terminal bronchiole still exhibits a continuous cuboidal epithelium, but produces branches whose walls bear occasional outpocketings (alveoli) lined by true squamous respiratory epithelium. Accordingly, these are termed respiratory bronchioles, of which there may be a short series. All subsequent subdivisions are lined by true respiratory epithelium.

Next appear elongated passages—the alveolar ducts —which open on all sides into a close succession of respiratory spaces called alveoli. Adjacent alveoli share the frame which separates their entrances into the alveolar duct and they also share a common wall or alveolar septum. The frame of each entrance-way has a core of elastic and collagenous fibers and some smooth muscle, covered by simple squamous epithelium.

Each hall-like alveolar duct terminates in one or more rotunda-like alveolar sacs and each sac bears from 2 to 5 or 6 terminal alveoli. (See Fig. 2.) The myriad ramifications, the minuteness of the finer

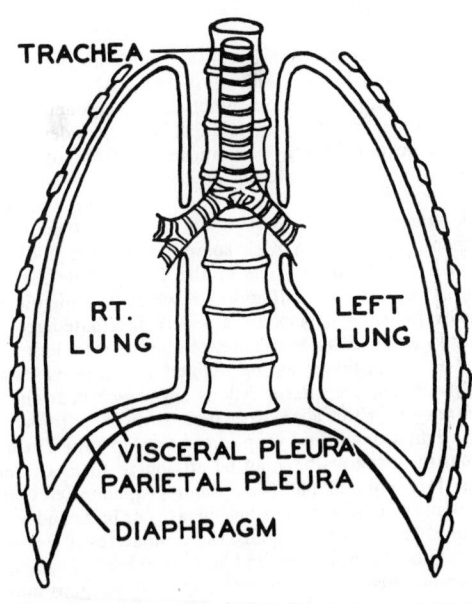

Fig. 1   Diagram showing continuity of visceral and parietal pleura at hilus of lung.

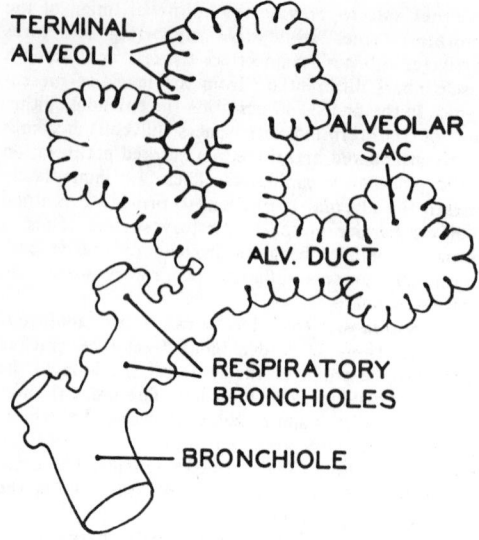

Fig. 2   Diagram of terminal ramifications of respiratory tree.

branches, and their intimate relationships render difficult the estimation of the number of pulmonary alveoli and their combined respiratory surface area. The best estimates place the number of alveoli between 300 and 400 millions, with a combined surface area of about 100m². All except subpleural alveoli share walls or interalveolar septa with others. The septa contain a rich network of pulmonary capillaries supported on a framework of elastic and collagenous fibers. The alveolar surface (long argued to be imperfectly covered by epithelium) has been proved electron microscopically to have a continuous squamous epithelial lining the cells of which send greatly attenuated cytoplasmic extensions across the capillaries. These attenuations may be as thin as 0.2μ or less in man. The air-blood barrier in respiratory areas, therefore, consists only of alveolar epithelium, capillary endothelium, their respective basement membranes and a tenuous—sometimes absent—tissue space.

Interalveolar communications or pores in some of the interstices of the capillary net permit the passage of gases (also exudates, cells, bacteria) between adjacent alveoli. Recently discovered bronchiole-alveolar communications provide a second, important means of collateral ventilation in instances of collapse or blockage of smaller peripheral airways.

The pulmonary circulation. Right and left pulmonary arteries convey blood from the right side of the heart to the lungs. Each enters the hilus of a lung and produces branches which, in general, parallel the course and branching of the bronchi. At the respiratory bronchioles and beyond, pulmonary arterioles give off short precapillaries which supply the rich capillary networks about the alveoli. Here, the delicate blood-air barrier permits the exchange of CO_2 in the blood for O_2 in the alveolar air. The blood of alveolar capillary networks flows into postcapillaries, venules and, finally, pulmonary veins. Whereas pulmonary arteries and their branches usually travel through the

central portions of lobes, sublobes and lobules of lung tissue, veins enter the connective tissue septa which partially or completely separate adjacent lobular subdivisions of the lung and unite into progressively larger channels as the hilus is approached. Finally, the oxygenated blood is returned to the left atrium of the heart by 2 right and 2 left pulmonary veins, thence to the left ventricle for distribution via the aorta through the systemic circulation.

Pulmonary lymphatics. Lymph vessels are abundant throughout the lung wherever loose connective tissue is found; i.e., beneath the pleura, in interlobular septa, and in periarterial and peribronchial connective tissue. In the latter location there is a rich network of valve-bearing lymphatics which extends toward bronchial lymph nodes within the lung and to hilar nodes in its root. Lymphoid tissue occurs in small masses along the smaller airways, but principally as patches or nodules in the walls of bronchioles and bronchi. The masses contain deposits of variable quantities of carbon, dust particles, etc., plus lymphocytes and pigment-laden "dust cells."

Some of the dust and particulate matter which reaches the terminal airways is removed via the lymphatic pathway; most, however, is ingested by wandering alveolar phagocytes. This phagocytosed debris plus particles trapped in mucus are propelled by ciliary action toward the larynx.

VERNON E. KRAHL

References

von Hayek, H., "Die Menschliche Lunge," Berlin, Springer, 1953.
Krahl, V. E. ed. & trans., "The Human Lung," New York, Hafner, 1960.

LYCOPSIDA

The plant sub-phylum Lycopsida contains the single class Lycopodineae. The class consists of the five orders, Lycopodiales, Selaginellales, Lepidodendrales, Isoetales, and Pleuromeiales. The Lepidodendrales and Pleuromeiales are composed entirely of fossil members while the present day lycopods are confined only to four genera: *Selaginella* having approximately 700 species, *Isoetes* with about 60 species and two genera, *Lycopodium* with some 200 species and the monotypic *Phylloglossum* in the Lycopodiales. With the exception of *Phylloglossum* which is endemic to Australia, the lycopods are a widely distributed group, largely tropical, but with many temperate members.

The Lycopsida may be defined as those vascular plants which are microphyllous and which bear single sporangia on the adaxial surface of their sporophylls. The definition of the Lycopsida as a group quite distinct from the PTEROPSIDA may be attributed to E. C. Jeffrey. He pointed out that, though siphonosteles are present in supporting the large leaves of ferns and seed plants, while in the lycopods, the siphonestele if present is related to branches and not to leaves. Thus, the Pterosida display both branch gaps and leaf gaps in the vascular tissue of the stele, while the Lycopsida evidence branch gaps only. Megaphyllous plants are those plants which exhibit leaf gaps, whereas microphyllous plants have no leaf gaps.

At present, many botanists agree that this distinction between the microphyll of the Lycopsida and the megaphyll is a valid one and, from a phylogenetic standpoint, regard microphylls as originating simply as vascularized outgrowths from the stem, and megaphylls as resulting from the condensation of an entire branch system.

Jeffery included the present Psilopsida and Sphenopsida as a part of his Lycopsida. These, however, are now regarded as being quite distinct from the Lycopsida on various morphological grounds. The most important is that the sporangia are terminal on the branch in the Psilopsida, while they are borne on sporangiophores and reflexed toward the axis of the plant in the Spenopsida. Thus, the Lycopsida is the only group which is truly microphyllous and which bears sporangia adaxially on the leaf surface.

Since microphylls are so different from megaphylls in their relation to the vascular tissue of the stele, one would expect that there would be certain obvious developmental features which would set the Lycopsida apart from other groups. In *Lycopodium* and *Selaginella*, studies made thus far by various workers indicate that the vascular tissue of the stele differentiates to a higher level in the stem apex than in the youngest leaf primordia having recognizable vascular tissue. In some members of the group where the apex is hyperboloid in shape the vascular tissue differentiates to a higher level than the point of initiation of even the youngest visable leaf primordia. This suggests a degree of independence of stelar differentiation and leaf differentiation which is not to be found in megaphyllous plants where vascular differentiation occurs in relation to leaves and never develops in the stem without being associated with leaf development. Thus, there is a possibility that microphylls might be regarded as a "developmental afterthought" on the part of the plant. This is by no means proven, however, and more evidence of a developmental nature is certainly needed in respect to this problem.

The fossil history of the lycopods is extensive, the oldest, known lycopod-like fossil, *Baragwanathia longifolia*, being of Silurian age. The lycopods reached their climax in the late Paleozoic where they were represented by many diverse forms and were dominant members of the Carboniferous coal swamp flora. The best known of these are the genera *Lepidodendron* and *Sigillaria*. These were tree-like forms reaching a height of well over 100 ft. in some cases with trunks up to 1½ ft. in diameter. They branched dichotomously at the top, and also at the base into a supporting structure known as *Stigmaria*. The trunks possessed only a small core of vascular tissue including secondary tissue. The greatest portion of the trunk was periderm which served to support the plant. Occuring with these larger forms there were many smaller lycopods, presumably ancestral to the modern genera. By the end of the Paleozoic, the Lepidodendrales and Pleuromeiales had disappeared and only the small herbacious lycopods remained.

The class Lycopodineae can be divided into two distinct groups on the basis of the presence or absence of a ligule, a small outgrowth from the adaxial surface of the leaf. In the Lycopodiales which are homosporous the ligule is absent, while the other orders are hetero-

sporous and ligulate. In the heterosporous orders, the development of both microgametophyte and megagametophyte is endosporal. The small microspores produce a male gametophyte and single antheridium within the microspore wall, and similarly the megaspore produces a female gametophyte within the larger megaspore wall with several archegonia protruding from the tri-radiate opening of the spore at maturity. Fertilization in extant members of the group ordinarily occurs on the ground with water providing a medium for the transfer of sperm from antheridium to archegonium. Usually 4 to 8 megaspores are produced in the megasporangium but in some fossil lepidodendrids apparently the number was reduced to one. The whole megasporangium with its single megaspore was enclosed by the megasporophyll, and retained in this enclosed fashion on the parent plant for at least a part of its development. These, though not seeds, represented a rather remarkable approach to the seed habit, complete with a slit-like micropyle which presumably sufficed for the entrance of wind-blown microspores.

In the homosporous Lycopodiales, the gametopyte undergoes a considerable exosporal development, with the prothallus achieving a size of several centimeters in some species. As far as is known, all lycopodium prothalli are capable of producing chlorophyll and carrying on an independent existence. Many do not do so in nature, however, and live for periods of up to ten years buried beneath the soil surface, existing by virtue of an association with an endophytic fungus. All degrees of endophytism occur naturally among prothalli of the genus. Those of many tropical species and our temperate *L. inundatum* are green photosynthetic, relatively short-lived prothalli, and moderately infested with an endophyte, while such subterranean forms as *L. complanatum* or *L. clavatum* lack chlorophyll, are heavily infested with an endophyte and takes 6–8 years to become sexually mature.

JOHN A. FREEBERG

References

Foster, A. S. and E. M. Gifford, "Comparative Morphology of Vascular Plants," San Francisco, Freeman, 1959.
Arnold, C. A. "An Introduction to Paleobotany," New York, McGraw-Hill, 1947.

LYMPHATIC TISSUE

Definition. Lymphatic tissue may be defined as a tissue composed of two consitutents, a sponge-like framework or stroma of fixed cells (reticular cells) supported on reticular fibrils and free cells (mainly lymphocytes) in the mesh of the stroma, although there are some confusions regarding the use of this definition. By Hellman, the term "lymphatic tissue" has been limited to a tissue, whose reticular cells derive from the mesenchyme, in which secondary nodules appear. In addition to the above definition, the term "lymphoid tissue" has been given to other similar tissues, e.g., the thymus, lymph infiltration, etc. But, according to Kihara, in the lymphatic system, subendothelial lymph-infiltration lacking the appearance of secondary nodules becomes transformed to a lymph node with secondary nodules, in the course of development. There is therefore some uncertainty where to draw the line between lymphatic and lymphoid tissues.

Morphology and distribution. The stroma, and the lymphocytes in the meshes of the stroma, occur in different proportions in various parts of the lymphatic tissue, so that one may distinguish (1) loose lymphatic tissue, consisting predominantly of stroma; (2) dense lymphatic tissue, in which free cells predominate; (3) nodular lymphatic tissue, containing "secondary nodules," which are dense accumulations of free cells in dense lymphatic tissue. Such accumulations are usually 300–400 μ in diameter and show various structures under various physiological conditions. Ehrich distinguished: (1) solid secondary nodules, consisting of densely crowded small lymphocytes; (2) Flemming's secondary nodules, composed of a light center with accumulations of large and medium lymphocytes, and a peripheral dark zone consisting of small lymphocytes; (3) a transition form, which is a transitional stage from Flemming's type to diffuse lymphatic tissues; and (4) pseudo-secondary nodules, large nodular masses, sometimes over 1 mm in diameter, in which venules may be present.

The three tissue arrangements mentioned above may give rise to three types of organ structure: (1) *Lymph nodes* are discrete organs covered by a capsule of dense collagenous fibers. They are always located along the course of lymphatic vessels and are scattered in large numbers, usually in groups at constant locations. The dense lymphatic tissue forms an outer cortical part, in which most often the secondary nodules appear and an inner medullar part, the tissue of which forms medullary cords. The loose lymphatic tissue occurs as sinuses both under the capsule and among the medullary cords. Many afferent lymphatic vessels, which supply lymph to the node, pierce the capsule on its convex surface. From here the lymph passes through the sinuses of both the cortex and medulla and then flows out from the node by way of one efferent vessel at the hilus, where blood vessels enter and leave the organ (Fig. 1). (2) *Lymphatic nodules* are a mass of dense lymphatic tissue, without sinuses in which there is usually a secondary nodule. They do not form an organ, but occur as solitary or aggregated nodules in the subepithelial tissue of the digestive, urogenital, respiratory, and other tracts, and along blood and lymphatic vessels. Examples are the tonsils, Peyer's patches of the intestine, the white pulp of the spleen, etc. (3) *Lymph-infiltrations* are small diffuse masses of dense lymphatic tissue with a vague boundary. They are found more extensively than the lymphatic nodules, though in almost identical locations. Under various physiological and pathological conditions, these structures may be changed in quantity and probably also in quality.

Lymphatic tissue may therefore be classified into the following groups and sub-groups: (1) Lymphatic tissue related with lymphatic vessels occurring as (a) lymph nodes, (b) subepithelial lymphatic nodules, with lymph-infiltrations of the digestive, urogenital and respiratory tracts, (c) subendothelial lymphatic nodules and lymph-infiltrations of lymphatic vessels. The thymus and bursa of Fabricius may be included in group (b), as most of the reticular cells are of endodermal origin. (2) Lymphatic tissues associated with blood vessels

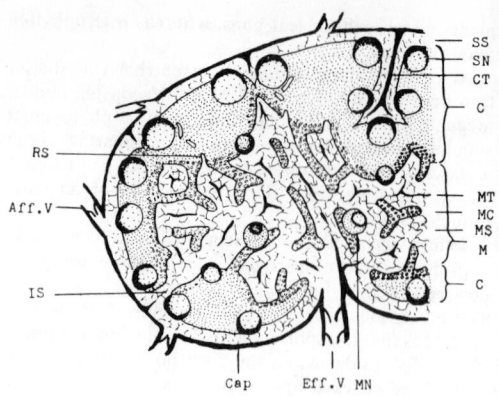

Fig. 1 Lymph node, C, cortex; M, medulla; Cap, capsule; CT, cortical trabecula; MT, medullary trabecula; MC, medullary cord; MN, medullary nodule; SN, secondary nodule; SS, subcapsular sinus; IS, intermediate sinus; MS, medullary sinus; RS, reticular fiberless sinus; Aff. V, afferent lymphatic vessel; Eff. V, efferent lymphatic vessel.

which appear as lymphatic nodules, or lymph-infiltrations, in the spleen, bone marrow, venules of the pancreas, lung, and probably also in the hemal node.

From origin of the reticular cells, lymphatic tissue may also be divided into two groups, epithelial (thymus, bursa of Fabricius) and mesenchymal (lymph node, spleen and others).

With the help of electron microscopy, structural differences between two similar types of cells are even more clearly manifested than by light microscopy: e.g., plasma cells, which are rich in granular endoplasmic reticulum, and lymphocytes, that rarely include it; epithelial reticular cells showing desmosomes with

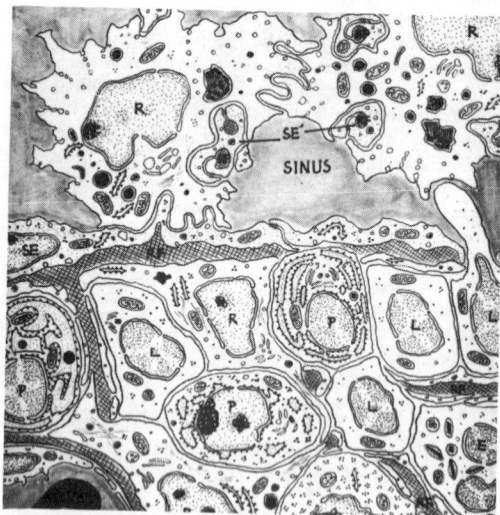

Fig. 2 Fine structure of medulla of lymph node. E, eosinophil leucocyte; L, lymphocyte; P, plasma cell; R, reticular cell; RF, reticular fiber; SE, sinus endothelial cell; SE', process of SE.

tonofibrils and mesenchymal reticular cells that do not show these. It is clearly proven that reticular fibers are extracellular components (Fig. 2).

Functions. (1) The production of lymphocytes is the first function of the lymphatic tissue. Flemming suggested that the secondary nodule is the principal productive field of lymphocytes ("germinal center"). Hellman, however, asserted that the nodule is a reactive field against some toxic agents ("reaction center"). In general, however, it is usually accepted that lymphocytes are produced in both the dense and nodular lymphatic tissues.

In the embryonic period, lymphocytes first appear in the epithelial lymphatic tissue and then are in early life transferred into mesenchymal lymphatic tissue and proliferate there in adult life.

(2) The second function is to serve as a defence mechanism against bacteria and the toxins produced by them. Active defence is represented by phagocytosis of reticular cells. The fixed reticular cells, and free macrophages originating from them, ingest bacteria, and other toxic foreign materials which have invaded the lymphatic tissue and render them harmless. Antibody formation is a passive defence function of the tissue, and there is little doubt that antibodies are produced by plasma cells in the lymphatic tissue. Moreover, there are some evidences that large baso-philic cells in the typical germinal center may contribute to this function.

It is believed that small lympocytes are immunologically competent cells; after antigenic stimuli, they transform to blast cells which change to lymphocytes and plasma cells, producing cellular and serum antibodies respectively. Transformation into other cell types is a possible activity of lymphocytes but there is some dispute because of the cytological confusion between lymphocytes and other similar mononucleated cells.

The thymus and bursa of Fabricius produce immunologically competent cells and are the central organ controlling antibody formation, because the neonatal thymectomized mice become severely lymphopenic with diminished serum antibody and fail to reject skin grafts from foreign strain. Thymectomy and/or bursectomy prove that the thymus and bursa have separate functions in birds; the former relates to homograft immunity and the latter to serum antibody.

(3) Questionable functions of the lymphatic tissue, especially the reticular cells, include participation of these cells in the metabolism of lipids, vitamin A, et al., and the transformation of lymphatic cells into other cell types.

M. SUGIMURA

MAGENDIE, FRANCOIS (1783–1855)

This French physician-physiologist made three contributions to biology. First, he experimented with the actions of certain drugs on the human system, introducing the medicinal use of strychnine and morphine as well as compounds containing iodine and bromine. Second, he showed that it was impossible to sustain life without nitrogen containing foodstuffs, i.e. protein, and thus laid the groundwork for the science of nutrition. Finally, he was obsessed with experimentation, even to the extent that he gained the reputation as a vivisector. Significantly, he gave rise to experimental physiology; and this study, in the hands of his more analytic followers and especially Claude Bernard, grew in importance. Magendie's experiments primarily concerned themselves with the nervous system. He was the first to deal with cerebrospinal fluid. In 1825, he showed that anterior nerve roots were motor, carrying impulses to the muscles to be enacted into motion, and that posterior nerve roots were sensory, carrying impulses to the brain to be interpreted as sensation.

DOUGLAS G. MADIGAN

MALLOPHAGA

The Mallophaga or biting lice, insects parasitic on birds and mammals, belong to the order Phthiraptera which also includes the Anoplura or sucking lice. Amongst the free-living insects they are most nearly related to the PSOCOPTERA or book lice.

They live mainly among the feathers or hair of their host, though some species live inside the quills of the wing feathers or in the throat pouch of pelicans and cormorants. The bird Mallophaga feed on the feathers and the blood and other tissue fluids of their host; some species live entirely on the fluids and the usual chewing mandibles have become adapted to piercing. The mammal Mallophaga probably never feed on hair but live on tissue fluids, skin debris and secretions. The whole of the life cycle (including three nymphal stages) is spent on the host, the eggs being attached to the hair or feathers or laid inside the quills.

There are about 2,800 named species of Mallophaga, but many others are still undescribed. Probably all kinds of birds have Mallophaga and most are parasitised by more than one species. The mammal Mallophaga are less universal; they comprise fewer genera

and species and are found on the marsupials, primates, rodents, land carnivores, hyraxes, ungulates and on the elephant, if the elephant louse, *Haematomyzus* (sometimes placed with the ANOPLURA) is included among the Mallophaga. The size of the parasite population on individual hosts varies considerably; preening and scratching keep down the numbers, and birds with damaged bills are often heavily infested. The Mallophaga do no damage to their hosts unless present in large numbers, when the irritation may cause the host to damage itself by scratching, and in the case of birds may interfere with egg production and fattening in poultry. The common dog louse, *Trichodectes canis*, acts as an intermediate host in the life cycle of one of the tapeworms of the dog.

Some Mallophaga at least show host specificity, being unable to survive on any but their own or related hosts; this close adaptation to the host has been brought about by the extreme isolation in which the population of each host species must live, as normally birds and mammals of different species do not come into close enough contact for transfers of lice to take place. The present distribution of the Mallophaga suggests that they became parasitic during the early stages of the evolution of their hosts and evolved with them so that in general, related hosts are parasitised by related Mallophaga. A genus of Mallophaga may be restricted to one group of birds or mammals, and a species may be found on only one species of host or on a group of related hosts. However, there may have been at all stages of their evolution, transference of lice between hosts of different species, and this was probably more common at a time when the Mallophaga were still partly free-living or before they had developed any degree of host specificity.

Thus, it is frequently possible to place a species of bird or mammal in its true systematic position by the Mallophaga which parasitise it. However, owing to secondary infestations and other factors, not least the limitations of our knowledge, this correlation is not invariably possible.

THERESA CLAY

MALPIGHI, MARCELLO (1628–1694)

Malpighi, often considered to be the father of microscopic anatomy and of embryology, was born near Bologna in 1628. He began the study of medicine at 21, receiving his doctorate from the University of Bologna

four years later. He spent the major part of his life occupying various chairs at universities throughout Italy
until shortly before his death, when he became private
physician to Pope Innocent XII. During these last three
years he drew up an account of his varying labours and
achievements, which was presented to the Royal Society
of London and published in 1696 two years after his
death.

Malpighi's accomplishments and contributions are
varied and important. Since he dissected live animals
for observational purposes, he was the first to see Harvey's correctly inferred theory of capillary circulation,
the results of which he published in two letters to his
contemporary, Borelli. In addition, these letters also
contained the first account of the vesicular structure of
the human lung, and thus made a theory of respiration
possible for the first time.

His demonstration of the structure of the secreting
glands, in which he maintained that the secretion was
formed in terminal acini standing in open communication with the ducts, was of equal importance.

Among other achievements was the discovery of the
vascular coils in the cortex of the kidney, which still
bear his name.

LETITIA LANGORD

Reference

Adelmann, H. B., "Marcello Malpighi and the Evolution of Embryology," 5 vols. Ithaca, N.Y., Cornell
University Press, 1966.

MALVALES

Malvales is an order of flowering plants of the class
DICOTYLEDONAE comprised mainly of woody plants of
pantropical and subtropical distribution. The flowers
are usually bisexual, actinomorphic and cyclic with
mostly five-parted perianths, valvate calyces, numerous
stamens in one or more whorls and ovaries multicarpel-
late, the placentae united at the axis. The presence of
stellate pubescence and mucilage-producing parenchymatous cells is typical. Classification of taxa within the
order is commonly based on the nature of the pubescence on the foliage and seeds, persistence of stipules,
aestivation of the perianth, presence or absence of an
epicalyx, number of anther locules, ornamentation of
pollen grains, fusion of parts and nature of dehiscence
of carpels.

The order, in its generally accepted form, was first
proposed by Eichler under the name Columniferae,
having reference to the united staminal filaments. The
order was renamed Malvales by Engler who typified it
with the family Malvaceae.

Most taxonomic publications recognize seven families within the order: Bombacaceae, Chlaenaceae,
Elaeocarpaceae, Malvaceae, Scytopetalaceae, Sterculiaceae and Tiliaceae. Bessey expanded the order to
include the Balanopsidaceae, Ulmaceae, Moraceae
and Urticaceae, a step which is unacceptable to most
specialists. Hutchinson redefined the Malvales to include only the Malvaceae and accepted the order Tilia-
les as comprising the Tiliaceae (including therein the
Elaeocarpaceae), Bombacaceae, Sterculiaceae, and

Scytopetalaceae, as well as the monotypic families
Dirachmaceae and Peridiscaceae. Hutchinson excluded
the Chlaenaceae (as the Sarcolaenaceae) placing this
family near the Theaceae, a position supported by anatomical evidence.

The Malvaceae, the typifying family, is the largest
(80 genera, 1500 species), most widely distributed, most
diversified and specialized. The largest genera are
Sphaeralcea, Hibiscus, Sida and Abutilon. The Malvaceae are of economic importance for cotton (single
celled hairs developed on the seed coat) and the oil and
pulp obtained from the seeds of Gossypium spp.; the
edible fruit okra (Hibiscus esculentus), and commercial
fibers obtained from the bark of species of Sida, Abutilon, Hibiscus, Urena and Thespesia. Many ornamental
species of Hibiscus, Althaea, Callirhoe and Malva are
cultivated.

The Bombacaceae (22 genera, 140 species) occur primarily in the American tropics and include a number
of massive trees with large buttressed roots in the genera Adansonia (baobab), Ceiba (kapok) and Bombax
(cotton tree). Balsa wood (Ochroma lagopus) is commercially significant for its very light soft wood composed of abundant parenchyma and thin walled fibers.
Hairs developed from the wall of the fruit surround the
seeds in species of Bombax and Ceiba (kapok) and are
economically important. Species of Adansonia, Bombax, Chorisia and Pachira are ornamental trees of
tropical areas.

The family Sterculiaceae (50 genera, 750 species)
comprised of herbs, shrubs and trees is pantropical
and subtropical. Alkaloids used in beverages are obtained from the American cocoa tree (Theobroma cacao) also the source of chocolate and cocoa butter, and
from the African cola nut tree (Cola nitida). Cultivated
as ornamentals are species of Dombeya, Fremontia,
Firmiana and Sterculia.

The Tiliaceae (40 genera, 400 species) are mostly of
tropical distribution with Tilia (American basswood or
linden, English limetree) a common and conspicuous
genus of shade and timber trees in temperate areas.
Phloem fibers from species of Corchorus, Grewia and
Triumfetta are the bast of commerce, i.e. jute etc.

The four families Malvaceae, Bombacaceae, Sterculiaceae and Tiliaceae are closely related on floral and
anatomical characters; the Malvaceae are the most specialized. The presence of mucilage in cells, cavities and
canals is characteristic. The nodal anatomy and the vascular structure of the petiole are basically similar. Pericyclic fibers are present and the phloem is characteristically arranged in triangular areas as seen in transverse
sections of the stems, broadest near the xylem and
stratified tangentially into alternating fibrous and non-
fibrous bands. The pubescence is mostly stellate or
peltate with simple hairs also present.

The Elaeocarpaceae (7 genera, 125 species) are of
pantropical distribution. Most of the species are trees.
The family has been united with either the Malvaceae
or Tiliaceae but the anatomical characteristics of simple hairs, lack of triangular phloem strands and of mucilage cavities and canals seem sufficient to warrant separation as a family. Members of the family are of little
economic importance, although a few species of Elaeo-
carpus and Muntingia are cultivated as ornamentals.

The Scytopetalaceae (4 genera, about 25 species) are
limited to west tropical Africa. The family shows some
similarities anatomically to the Tiliaceae but also im-

portant differences. It is poorly known and of no recorded economic importance.

RICHARD A. HOWARD

References

Hutchinson, J., "The Dicotyledons," London, Oxford Univ. Press, 1959.
Lawrence, G. H. M., "Taxonomy of Vascular Plants," New York, Macmillan, 1951.

MAMMALS

The mammals (class Mammalia) constitute one of the great classes of the Vertebrata (see CHORDATA). Mammals are distinguished from all other vertebrates by the fact that the young are nourished with milk produced by special mammary glands of the mother. The body is typically covered with HAIR, although hair is often secondarily scanty and in certain CETACEANS the hair coat may be reduced to a few transitory bristles present only in the fetus. The complex mandible of the REPTILIA has been reduced to a single pair of bones, the dentaries, which articulate directly with the SKULL instead of indirectly through a quadrate bone. There are three auditory ossicles (malleus, incus, and stapes) in the middle ear, derived from reptilian jaw elements, and typically a cartilaginous external ear. The left aortic arch persists, and red blood cells are non-nucleated. The BRAIN is characterized by great development of the cerebral cortex, often referred to as the "neopallium." Except in the monotremes, the young are born alive after a relatively long period of gestation within the uterus of the mother.

About 3200 species of living mammals, representing about 950 genera, are known. About 2,000 genera of extinct mammals, thus more than twice the number of living forms, are known and additional discoveries are constantly being made. Mammals are worldwide in distribution, including all the seas, and the class is probably at the peak of its phylogenetic development today.

Man himself is a mammal, and most of the domestic animals on which he depends for flesh, fiber, companionship, and transportation (except in the most advanced technological cultures), are mammals. Most of these were domesticated so long ago that their origins are obscure or even unknown. Some, the camels for example, no longer exist in the original wild state.

Origin and Early Evolution

Mammals arose from therapsid reptiles in the Late Triassic. They evolved slowly during the Jurassic and Cretaceous, but did not become a dominant group until the Tertiary. The Therapsida were characterized by progressive trends in the mammalian direction throughout their history. This evolution culminated in the Ictidosauria, a group of advanced therapsids that bridge the gap between reptile and mammal. Most authorities believe that mammals had a polyphyletic origin, and many suspect that each of the mammalian subclasses was derived separately from a distinct therapsid stock. Mesozoic mammals were very small and their remains are extremely rare and fragmentary, but specimens of

the mandible and posterior skull region are known (e.g., *Diarthrognathus*) that are almost exactly intermediate between the reptilian and mammalian condition and can be assigned to either the Reptilia or the Mammalia. Actually there is no way of knowing when other equally distinctive mammalian features appeared, such as hair and milk glands, since these are not preserved in the fossil record. It is highly improbable that they all appeared at once, and hence there would have been no moment at which reptile became mammal.

At least three, and probably four, subclasses are represented in the first radiation of the mammalian stock, which took place in Late Triassic times. Only two of these survived into the Tertiary, and only one, the Theria, is of any importance in the over-all history of the Mammalia. The Prototheria (monotremes) are not known from fossils until the Late Tertiary, but must have been represented already in the Triassic radiation. The Allotheria, the multituberculates, survived into the Early Eocene before becoming extinct. The multituberculates were small mammals with very peculiar dentition, apparently adapted to a rodent-like life. About 30 genera are known. The Theria were represented in this earliest radiation by the Pantotheria, a group that was ancestral to modern marsupial and placental stocks. The Pantotheria were divided into two orders, the Symmetrodonta and the Pantotheria proper, differing from each other in details of tooth structure. Two additional groups of uncertain affinities, the Tricondonta and Docodonta, are represented. The docodonts are interesting for the fact that the quadrate and articular still form a subsidiary part of the jaw suspension.

Prototheria

The Prototheria, represented by a single order, the MONOTREMATA, are so distinctive that some students have argued that they should not be classed as mammals. Living monotremes are confined to Australia and New Guinea, where they are represented by three genera: *Ornithorhynchus*, the platypus, and *Tachyglossus* and *Zaglossus*, the echidnas or spiny anteaters. Fossil monotremes, not generically distinct from the recent forms, are known only from the Pleistocene of Australia. It is evident that the Monotremata were never very abundant or widespread.

All known monotremes are highly specialized, but basically their structure is remarkably primitive. Two typically mammalian features, hair and milk glands, are present and it is almost exclusively because of these that monotremes are regarded as mammals. Teeth are absent in the adult. Various reptilian characters persist in the skull, and the shoulder girdle is extraordinarily reptilian, with an interclavicle, large coracoids, and the scapular spine absent or barely indicated. The rectum and urogenital system open into a common cloaca, and monotremes reproduce by laying eggs that are hatched outside the mother's body.

Classification of Therian Mammals

The monotremes and Australian MARSUPIALS were unknown at the time Linnaeus was preparing the various editions of his *Systema naturae*, and consequently they do not figure in his classification. It was not until 1834 that the French anatomist Blainville recognized the gulf separating the monotremes, marsupials, and

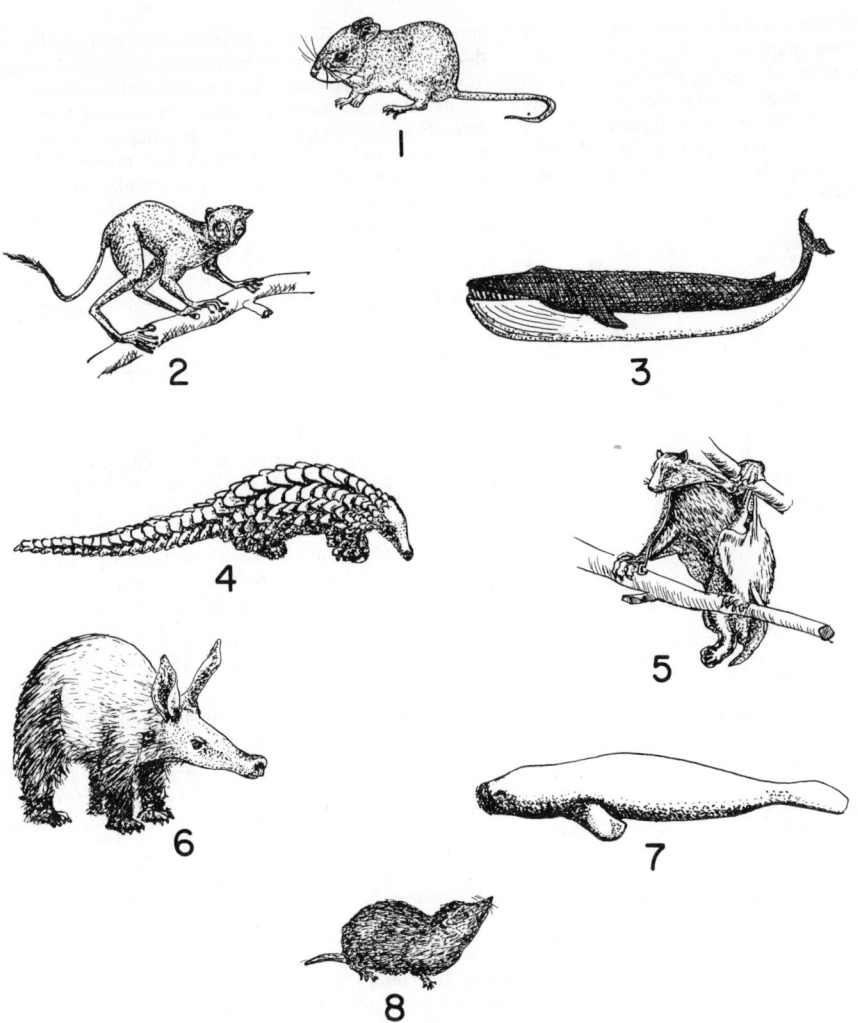

Fig. 1   Some representative mammals. 1, *Peromyscus,* the deer mouse; 2, *Tarsius,* the tarsier; 3, *Physeter,* the sperm whale; 4, *Manis,* the pangolin or scaly anteater; 5, *Cynocephalus,* the flying lemur; 6, *Orycteropus,* the aardvark; 7, *Dugong,* the sea cow; 8, *Blarina,* the short-tailed shrew. (From Goodnight, Goodnight and Gray, "General Zoology," New York, Reinhold, 1964.)

placentals and proposed a classification expressing these distinctions. In 1872 the American zoologist Gill formally proposed placing the marsupials and placentals together in one subclass (the Theria of modern classifications), the monotremes in another (the Prototheria). Gill's division is, in essence, the one in use today.

The Theria are divided into three infraclasses: the Pantotheria, Metatheria, and Eutheria. The Pantotheria are divided into two orders, the Symmetrodonta and the Pantotheria proper, both of which became extinct during the Cretaceous. In these the teeth are basically similar to the teeth of later mammals, and the pantotheres are generally regarded as the direct ancestors of the Metatheria and Eutheria.

The Metatheria includes a single order, the Marsupialia, characterized by a pouch (marsupium) in the female in which the young are sheltered for a time after birth. A special pair of bones, the marsupial bones, attached to the anterior border of the pelvis in both sexes, supports the marsupium. The typical marsupial dentition has 3 premolars and 4 molars in each jaw, the reverse of the typical placental formula. Various attempts have been made to divide the marsupials into suborders on the basis of the dentition (Diprotodontia vs. Polyprotodontia) or the condition of the second and third toes of the hind foot (Didactyla vs. Syndactyla). The resulting groupings are obviously unnatural and have been abandoned by modern students, who divide the order into six superfamilies: Didelphoidea, Boryhaenoidea (extinct), Dasyuroidea, Perameloidea, Caenolestoidea, and Phalangeroidea. There are about 60 genera of living marsupials, all but about a dozen confined to the Australian region. Nearly a hundred extinct genera are known. Marsupials persisted in Europe until the middle of the Tertiary. Fossil marsupials are unknown from Asia, and there is no evidence that they ever reached Africa or Madagascar. The main marsupial radiations were in South America and Australia.

The Eutheria, the placental mammals, embraces the vast majority of living mammals. The placentals were dominant throughout the Cenozoic, 95% of all known genera of Cenozoic mammals being placentals and only 5% marsupials. In the Eutheria the young are born in a relatively advanced state of development. The primitive dental formula is 3-1-4-3. The infraclass is divided into 26 orders, ten of which are extinct. These are:

Insectivora
    shrews, moles, hedgehogs, etc.
Dermoptera
    flying lemurs
Chiroptera
    bats
Primates
    lemurs, monkeys, apes, man
Tillodontia (extinct)
Taeniodontia (extinct)
Edentata
    sloths, anteaters, armadillos
Pholidota
    scaly anteaters
Lagomorpha
    rabbits, hares
Rodentia
    rodents
Cetacea
    whales
Carnivora
    dogs, bears, weasels, cats, etc.
Condylarthra (extinct)
Litopterna (extinct)
Notoungulata (extinct)
Astrapotheria (extinct)
Tubulidentata
    aardvarks
Pantodonta (extinct)
Dinocerata (extinct)
Pyrotheria (extinct)
Proboscidea
    elephants
Embrithopoda (extinct)
Hyracoidea
    hyraxes
Sirenia
    sea cows
Perissodactyla
    odd-toed ungulates (horses, tapirs, rhinoceroses)
Artiodactyla
    even-toed ungulates (pigs, cattle, deer, etc.)

The Eutheria are probably at the peak of their development today, nearly a thousand genera of living eutherians being known. They inhabit all the continents and all the seas. The largest order is the Rodentia, with about 350 living genera, followed by the Chiroptera and Carnivora, with about 120 living genera each. Thus these three orders account for more than half the living mammals. About 60 genera of living primates are known.

Dentition

Individual TEETH or groups of teeth often became specialized in various fishes, amphibians, and reptiles, but only in mammals has the dentition become subdivided into fixed functional units: incisors, canines,

premolars, and molars. A trend in this direction is clearly evident among the mammal-like reptiles. Aside from its biological significance, this differentiation of the dentition places an important tool in the hands of students of mammals, particularly paleontologists who often have only the teeth of extinct mammals with which to work. Since each tooth in the placental dentition is homologous throughout the Eutheria, it has an identifiable position in relation to the whole row of teeth. Each tooth may then be assigned a number and the dentition represented by a formula. The expanded formula for the full placental dentition, for example, is:

$$I \frac{1\ 2\ 3}{1\ 2\ 3}\ C \frac{1}{1}\ P \frac{1\ 2\ 3\ 4}{1\ 2\ 3\ 4}\ M \frac{1\ 2\ 3}{1\ 2\ 3}\ X2 = 44.$$

The formula gives the figures for one half of each jaw, and therefore the total must be multiplied by 2 to give the full complement of 44 teeth. Among living placentals the full dentition occurs only in a few insectivores, the mole for example. In almost all cases various teeth have been lost in connection with the adaptive radiation of mammals, and such losses are easily indicated in the formula. The formula for man, for example, is:

$$I \frac{1\ 2\ .}{1\ 2\ .}\ C \frac{1}{1}\ P \frac{.\ .\ 3\ 4}{.\ .\ 3\ 4}\ M \frac{1\ 2\ 3}{1\ 2\ 3} = 32,$$

the dot indicating that the third incisor and the first and second premolars in each jaw have been lost.

Usually it is sufficient to indicate the total number of each kind of tooth in each jaw, and a simplified formula may then be used. The formula for man is commonly written:

$$I \frac{2}{2}\ C \frac{1}{1}\ P \frac{2}{2}\ M \frac{3}{3}.$$

The milk or deciduous dentition is indicated by using lower case letters instead of capitals. The formula for the milk dentition of man is written:

$$i \frac{2}{2}\ c \frac{1}{1}\ p \frac{2}{2} = 20.$$

Not only do the teeth represent homologous units throughout, but the individual cusps on the premolar and molar teeth are homologous from species to species and from tooth to tooth in the toothrow, making it possible to identify each cusp wherever it appears among the Eutheria. In primitive placentals there are three main cusps, arranged in the form of a triangle, on each molar tooth. On the upper molars the base of the triangle is toward the lips, on the lower molars it is toward the tongue. In the Cope-Osborn *tritubercular theory* of mammalian dentition it is assumed that this cusp arrangement represents the original therian pattern from which all later patterns were derived by modification. Behind the primary triangle the lower molar has a basin-like heel (the talonid), into which the innermost cusp of the upper molar fits. In many more advanced mammals a fourth cusp appears on the lingual side of the upper molar, in a new area called the talon, producing a square tooth adapted to crushing. In 1888 the American paleontologist H. F. Osborn proposed the nomenclature for these cusps that has been generally

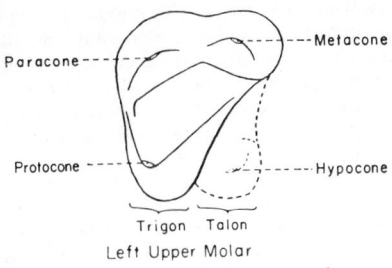

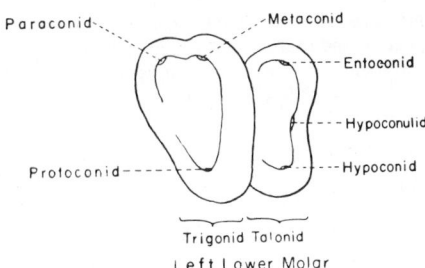

Fig. 2 Cusp nomenclature of mammalian molar teeth.

used since. (Fig. 2) Osborn's names are:

Upper Molar

Trigon

 Anterio-internal: Protocone
 Antero-external: Paracone
 Postero-external: Metacone

Talon

 Postero-internal: Hypocone

Lower Molar

Trigonid

 Antero-external: Protoconid
 Antero-internal: Paraconid
 Postero-internal: Metaconid

Talonid

 Postero-external: Hypoconid
 Postero-medial: Hypoconulid
 Postero-internal: Entoconid

In use the trigon on the upper tooth shears past the trigonid on the lower tooth to produce a cutting action. In addition, the biting action of the protocone of the trigon into the basin-like talonid produces a crushing action. With the adaptive radiation of the mammals this primitive mechanism underwent enormous modification. In carnivores the cutting blade tends to be greatly enlarged and the crushing function eliminated. In the primates the crushing function predominates. In many rodents and hoofed mammals the cheek teeth are tall prismatic structures that grind like millstones; such teeth may be so modified that the original cusps can no longer be identified. In many whales and porpoises the cheek teeth are simple conical structures, the upper and lower teeth interdigitating to form an effective grasping mechanism, but useless for mastication. In some mammals, notably the anteaters, the teeth have disappeared completely and the task of mastication has been taken over by a gizzard-like portion of the stomach.

So distinctive are the teeth of eutherian mammals that it has been said that if these creatures were all extinct and known only from their teeth, our classification of them would scarcely differ from the one in use today.

Adaptive Radiation

The first eutherian mammals were small creatures, very similar to small opossums in appearance and prob-

ably also in behavior. These early mammals evidently played a very minor role among the vast numbers of reptiles that populated the earth during the Cretaceous. They were unspecialized, feeding largely on insects and other small animals, the feet provided with five generalized digits, and the body terminating in a long heavy tail. They were probably largely nocturnal.

This primitive mammalian architecture has been adapted to exploiting a wide variety of ecological situations during the evolution of the Mammalia, often involving radical modifications of the basic mammalian structural plan. The bats fly in the air and their structure is superficially birdlike, the cetaceans are thoroughly aquatic and their body form resembles that of fishes. Efficient digging machines have been developed repeatedly and independently: moles, marsupial moles, and pocket gophers, for example. Perissodactyls and artiodactyls are essentially machines for gathering and metabolizing leaves and grasses on the one hand, and for speedy locomotion on the other. Carnivores are designed for preying on other vertebrates. Specialization for termite feeding appeared independently in the anteaters, pangolins, and aardvarks. A few mammals—insectivores and some of the small rodents—are still generalized in structure and occupy essentially the same ecological situation as their Cretaceous forebears.

It is evident that these major ecological types coincide for the most part with the orders of mammals. Furthermore, families, subfamilies, and even genera tend to represent adaptive types to increasingly narrow ecological situations. In other words, the features characterizing taxa are usually adaptive, and the taxonomic hierarchy is also an adaptive hierarchy. This is precisely what would be expected under evolution by natural selection.

The adaptive radiation of the Australian marsupials provides one of the most remarkable examples of parallelism known. From at least the beginning of the Tertiary the marsupials were isolated in the Australian region from the more progressive placental mammals that were evolving elsewhere. Here the marsupials underwent an adaptive radiation of their own, and in filling the available ecological niches they produced ecological counter-parts of many of the placentals found elsewhere in the world. In some cases the morphological similarity between the marsupial and placental types is extremely close: the marsupial "wolf" vs. eutherian canids, the marsupial "mole" vs. eutherian moles, marsupial "cats" vs. eutherian cats and mustelids, wombats, phalangers, and marsupial "mice" vs. eutherian rodents. Kangaroos are the ecological equivalents of eutherian artiodactyls, although there is little morphological resemblance. No truly flying bat-like marsupial has ever appeared, and there are no completely aquatic marsupials, although the water opossum (*Chironectes*) of South America is otter-like in its habits.

Geographic Distribution

The evolution of therian mammals was primarily a Teritiary phenomenon, and the geographic distribution of living mammals does not agree completely with the pattern of older groups such as the amphibians and reptiles. The distribution of living mammals points quite clearly to the northern Eurasian-North American land mass as a center of dispersal. Penetration into the southern peninsular land masses (South America, Africa, southeastern Asia, Australia) evidently took place at different times, depending upon the date and duration of their connections with the great northern land mass. In most instances there was more than one wave of migration into an area, not always from the same source, and the resulting histories of some faunas is very complex. Nothing in the present or past distribution of mammals requires direct connection between any of the southern peninsular land masses.

Eurasia and North America. The fossil record shows that many orders and families of mammals arose in North America and Asia during the Tertiary. These include the edentates, the rodents, the primates, carnivores, persissodactyls, and artiodactyls. The American and Asiatic land masses were connected across the Bering Sea at various times during the Tertiary, and at such times mammals were able to move in both directions across the land bridge. As a result of this intermingling, the American and Eurasian mammalian faunas are very similar. Among living North American mammals only three, the opossum, the armadillo, and the porcupine, originated in South America.

South America. The South American continent was probably unconnected with North America at the beginning of the Tertiary, and remained so until latest Pliocene, at which time the Panama land bridge was re-established. The indications are that the elements of the earliest South American mammalian fauna reached the continent from North America in the Late Cretaceous. These included marsupials, edentates, and an assortment of archaic hoofed mammals including condylarths, litopterns, notoungulates, astrapotheres, and pyrotheres. The primates, hystricomorph rodents, and procyonid carnivores reached South America by way of the West Indies at various times during the Tertiary. Near the end of the Tertiary the Panama land bridge was re-established and a wide assortment of modern mammalian types entered South America from the north. These included shrews, rabbits, squirrels, field mice, carnivores, elephants, horses, tapirs, peccaries, camels, and deer. These progressive types greatly enriched the South American fauna, exterminating many, but not all, of the original inhabitants and themselves occupying most of the ecological situations. Some of the invading groups, notably the elephants and horses, became extinct before the end of the Tertiary.

Thus the existing mammalian fauna of South America is an exceedingly complex mixture composed of three major elements: the vestiges of an original fauna of archaic mammals, a complex of later arrivals, and a very recent wave of immigrants. All were probably ultimately derived from North America.

Africa. Like South America, Africa is today a peninsula of the great northern land mass, but unlike South America it was not isolated as an island throughout the Tertiary and consequently the history of its mammalian fauna is quite different. The most important single factor in the history of the African fauna is the Sahara Desert. Africa north of the Sahara has an essentially Eurasian mammalian fauna. The desert has acted as a filter rather than an absolute barrier to the main part of Africa, for a few species of mammals managed to cross it in both directions during the Tertiary. Nevertheless the species that reached trans-Saharan Africa before the desert appeared evolved in almost complete isolation and the fauna has a very characteristic stamp. On the one hand there are mammals unknown elsewhere, or known only as fossils: golden moles, aardvarks, the okapi, and elephant shrews, for example. On the other hand, forms that evolved elsewhere relatively late in the Tertiary, after the Sahara Desert was established, are absent in trans-Sahara Africa: deer, sheep, goats, and bears, for example. Marsupials, although present on the northern land mass before the Sahara appeared, are absent in Africa both fossil and recent and seem never to have reached the continent. In general, the African fauna has a somewhat archaic cast, and it has been compared with a typical Miocene mammalian fauna.

Madagascar. Although separated from Africa by less than 250 miles of water, the island of Madagascar has a mammalian fauna that is only very remotely African. The island has been separated from the mainland since the very beginning of the Tertiary, and a few new forms were introduced after the Mozambique Channel was established. The result is that the Madagascar mammalian fauna resembles an unbalanced Eocene fauna. The most conspicuous elements are generalized insectivores, generalized lemurids, and numerous viverrids, the most primitive of living carnivores. The rodent fauna is impoverished, canids and felids are completely absent, and there are no ungulates except a pig that is obviously a recent entrant from Africa. Marsupials are completely absent from the Madagascan fauna.

Australia. The Australian region, which includes New Guinea, has been separated from the rest of the world since the beginning of the Tertiary. The original seeding of this area was with monotremes and marsupials, and the marsupials form the over-whelmingly dominant element in the modern fauna. A few species of rats and mice reached Australia some time during the Tertiary, along with many species of bats. The presence of marsupials in Australia and South America, along with a few other common primitive elements, has led some zoologists to speculate that these two land masses were once connected by a transoceanic land bridge. Others have postulated a connection via Antarctica. Most modern students have abandoned these hypotheses in favor of immigration from the north.

D. DWIGHT DAVIS

References

Flower, W. H. and R. Lydekker, "Introduction to the Study of Mammals, Living and Extinct," London, Black, 1891.

Weber, M. W. C., "Die Säugetiere," Jena, Fischer, 1927–28.

Bourliere, F., "Mammals of the World, their Life and Habits," New York, Knopf, 1955.

Simpson, G. G., "The Principles of Classification and a Classification of Mammals," Bull. Amer. Mus. Nat. Hist., **85**; 1945.
Hall, E. R. and K. R. Kelson, "The Mammals of North America," New York, Ronald, 1959.
Cabrera, A. and J. Yepes, Mamiferos sudamericanos Buenos Aires, Camp. Argent. Edit. 1940.
Matthews, L. H., "British Mammals," London, Collins, 1952.
Troughton, E., "Furred Animals of Australia," New York, Scribner, 1947.
Anderson, S. and J. K. Jones, "Recent Mammals of the World," New York, Ronald, 1967.

MANTODEA

Order of Polyneopterous insects of the Blattopteroid group (ISOPTERA, Zoraptera, Blattodea and Mantodea), by some classed with the Blattodea as suborder of the Dictyoptera. It includes medium or large-sized terrestrial species ("preying mantis") with an elongated body, which are termophilous and predacious. The head is triangular, hypognathous, very mobile, with three ocelli (larger in males) and with large eyes sometimes conical or furnished with a lateral grain or with a spine at the apex; antennae shorter than the body, normally setaceous and sometimes serrate or pectinate; the vertex between the antennae may be conical. Mouthparts of the biting type.

The prothorax is very long, and mobile on the mesothorax, the pronotum widens above the coxae, in some instances very noticeably and is divided by a transverse groove into prozona and metazona; propleura developed inside the prozona, divided into anapleurite in the upper part (with a very large anepisternum and a small anepimeron) and katapleurite in the lower part, and having a precoxal bridge anterior to the episternum. Mesothorax and metathorax normally and equally developed with pleura constituted by lateropleurite, episternum and epimeron; the sterna have also the laterosternite.

Forelegs are raptatorial, very mobile, having elongated prismatic coxae, frequently as long as the femora; these are strong, flattened and have in the lower margins two rows of spines, external and internal; the tibia is strong, shorter than the femur, having a hook apically and also an external and internal row of spines ventrally; the insect catches its prey by seizing it between the spines of the tibia and the femur. Second and third pair of legs unmodified. All the legs have a trochantin and the tarsi are 5-segmented. Fore wings are tegmina, more or less hardened; hind wings are membranous, often coloured and with the anal lobe very enlarged. Many species may be brachypterous, micropterous or apterous in both sexes or more especially in the female.

The abdomen is elongated with 11 segments; the first urosternite is reduced in size or even absent; in the female the modified 8th and 9th urosternite form the sclerites of the large genital chamber to which is connected the ovipositor which consists of three pairs of valves; many-segmented cerci are present in both sexes; the subgenital plate of the male (9th urosternite) bears two stili.

The alimentary canal is a little convoluted or straight with a distinct crop and a gizzard scarcely developed;

there is a mandible gland; about a hundred Malpighian tubules. Two pair of thoracic stigmi and eight abdominal. The nervous system has three ganglions in the thorax and seven abdominal. The reproductive system has paired gonads and ducts; well developed accessory glands in both sexes; panoistic ovarioles. The number of chromosomes (2 n) is often 28; the heterochromosomes in the male are of the type x-0, or x1, x2, y; in the female of the type x,x or x1, x1, x2, x2.

The eggs are laid in groups contained in a large and spongy ootheca of the consistency of parchment, very variable in shape, the majority globose, always attached to the substratum.

During and after copulation the female often devours the male beginning with its head: this does not however prevent the mating from being completed. Cases of parthenogenesis are extremely rare (*Miomantis savignyi* and *Brunneria borealis*).

Mantids are heterometabolous (Pauro- or pseudometabolous). Many species are lively coloured and homochromy and mimicry is frequent.

About 2,000 species are known; the majority are found in tropical regions and these are divided into groups of 12 families, not as yet defined. Therefore, fossil species known are very few and the earliest fossils date back only to the Miocene and Baltic amber.

MARCELLO LA GRECA

References

Giglio-Tos, E., "Mantidae," *in* "Das Tierreich," Berlin, de Gruyter, 1927.
Chopard, L., "La biologie des Orthoptères," Paris, LeChevalier, 1938.

MARINE BACTERIA

The sea is the natural habitat of many kinds of BACTERIA. Morphologically marine bacteria are much like their better known terrestrial or freshwater counterparts. Although movements of water, air, and certain animals carry terrestrial bacteria into the sea, most of the bacteria found in the ocean appear to be specifically indigenous to the marine environment, except in coastal waters.

Upon initial isolation from the sea, most marine bacteria grow best or only in nutrient media having a salt content of from 2 to 4 per cent. The average salinity of sea water is approximately 3.5 per cent, about 68 per cent of which is accounted for by sodium chloride. The growth-promoting property of natural sea water is attributed to many minerals and trace elements as well as to certain organic accessory growth factors.

Most marine bacteria are thermosensitive and psychotolerant. Not only do they tolerate refrigeration temperatures; many species grow and are otherwise physiologically active at near 0°C. By water volume and bottom area, approximately 80 per cent of the sea is always colder than 3°C. Sea water of average salinity freezes at around – 1.9°C. Excepting a few anomalous thermophilic species found in the sea, nearly all marine bacteria grow best between 15 and 25°C. A good many species are killed by 10 minutes' exposure to temperatures exceeding 25°C. Some of the more thermosensitive species are killed by being flooded or mixed with molten nutrient agar cooled to 42° C as in routine plating procedures.

The temperature tolerance of bacteria is influenced by the hydrostatic pressure, which increases with water depth by about 0.1 atmosphere per meter. In general, the temperature tolerance of bacteria is increased by compression, provided that the increased pressure is not injurious. A good many bacteria, taken from depths of 7000 to 10,000 meters, grow only when compressed to 700 to 1000 atm. Such bacteria are characterized as being *barophilic*. In this respect deep-sea bacteria differ from surface-dwelling organisms, most of which are slowly killed by such pressures.

On the average, marine bacteria are somewhat smaller than those living in soil, sewage, and other terrestrial materials. In general, marine bacteria grow more slowly and they form smaller colonies than their terrestrial counterparts. Facultative anaerobes predominate in the sea, there being relatively few strict aerobes. As a class, marine bacteria are actively proteolytic, more than half of the species being able to liquefy gelatin. Likewise more than half of the species produce pigments when grown in appropriate nutrient media. Non-sporulating, gram-negative rods predominate. Although both agar-liquefying and bioluminescent bacteria are commonly encountered in marine materials, only a small fraction of the bacteria isolated from the sea have the ability either to digest agar or to generate light.

Cultural procedures reveal the presence of from nil to millions of viable bacteria per gram of sea water or marine mud. Throughout the euphotic zone (i.e., the topmost hundred or so meters penetrated by sufficient light for photosynthesis by green plants), the bacterial population of sea water ranges roughly from 10 to 10^6 per ml, the mean density being about 10^3 bacteria per ml. Bacteria have been found in the sea at all latitudes, in all seasons, and at all distances from land wherever water samples from the euphotic zone have been examined. The largest bacterial populations are found associated with phytoplankton. It is not uncommon to find 10^6 viable bacteria per ml of water during or immediately following a plankton bloom. In certain polluted estuaries, bays, harbors, and areas around sewage outfalls, from 10^6 to 10^5 bacteria have been found per ml of water or bottom ooze, but such conditions are not typically marine.

Below the euphotic zone the bacterial population decreases with depth. Although appreciable numbers of bacteria have been recovered from all depths, at many oeanic stations sometimes a liter or more of water must be examined to detect any bacteria. In such water masses, the low organic content and not depth or the resultant hydrostatic pressure limits the bacterial population. From 10^3 to 10^8 viable bacteria per gram wet weight have been found in many samples of bottom sediments taken from depths exceeding 10,000 meters.

Based upon the results from analyzing several hundred samples, it seems that the mean bacterial population of the topmost two or three centimeters of bottom sediments is about 10^5 viable cells per gram wet weight. The abundance as well as the kinds of bacteria in such sediments is primarily a function of the organic content; secondarily of particle size, predators, pH, oxygen tension, and other environmental factors. Although the organic content of shallow bottoms is generally higher

than that on the floor of deep trenches, there seems to be little correlation between the bacterial population of marine sediments and water depth. The distribution of bacteria in bottom sediments is even more sporadic than in water. Throughout extensive areas of red clays and other highly oxidized pelagic sediments, the bacterial population is very low. In most sediments, the bacterial population falls off rapidly below the mud-water interface with increasing core depth. However, a few living bacteria have been found in deep-sea deposits buried under several meters of bottom deposits. Most of these deeply buried bacteria are strictly anaerobic sulfate reducers, which utilize either organic matter or molecular hydrogen as an energy source.

By virtue of their action on organic matter and certain inorganic substances as well, bacteria influence chemical and physicochemical conditions in marine sediments. For example, the gradual diminution in the amounts of organic carbon and sulfate in bottom deposits with age or depth of burial is attributable to bacterial activities. The precipitation of calcium carbonate in shallow tropical seas has been attributed in part to bacteria which tend to increase the pH of sea water. This they may do by reducing either nitrates or sulfates, by forming ammonia from nitrogenous compounds, and by the oxidation of organic acids. Other kinds of bacteria may tend to decrease the pH under certain environmental conditions by producing either carbon dioxide or organic acids, by oxidizing ammonium to nitrate, and by oxidizing either hydrogen sulfide or elemental sulfur to sulfate. Bacteria are believed to be the principal dynamic agents affecting the Eh or redox potential of marine sediments. This they may do by consuming free or dissolved oxygen, by producing hydrogen sulfide, and by destroying organic substances having a high electron-escaping capacity.

Besides being important as geochemical agents in the sea, bacteria influence the well-being of marine plants and animals in many ways. A few varieties cause infectious diseases. In localized environments, bacteria may vitiate the water by consuming all of the free oxygen and by producing hydrogen sulfide or toxic amines. Their beneficial activities, however, greatly exceed their harmful activities. Their most important beneficial function is the decomposition of waste organic matter and the conservation of carbon. Bacteria attack and mineralize virtually all kinds and classes of organic materials—sugars, starches, celluloses, proteins, lipids, hydrocarbons, carotenoids, chitins, keratins, etc. They quickly decompose the organic remains of both plants and animals and they are able to utilize dissolved and colloidal compounds of carbon. So effective are bacteria in decomposing organic materials that the sea has been characterized as being the world's largest and most efficient septic tank. In the vicinity of solid surfaces, which favor bacterial activity in dilute nutrient solutions, bacteria or their enzymes continue to act on organic matter in sea water until its content is reduced to less than 1 mg per liter. Ordinarily the organic carbon content of sea water ranges from 1 to 10 mg per liter.

On the average, approximately two-thirds of the carbon in organic compounds assimilated by marine bacteria is liberated as carbon dioxide and the remaining one-third is converted into bacterial cell substance or protoplasm. The conversion efficiency for various bacteria utilizing various kinds of organic matter ranges from 5 to 75 per cent. The conversion of soluble, colloidal, and waste organic materials into bacterial cell substance is of greater importance in the economy of the sea, because these bacterial cells constitute an important source of food for many kinds of marine animals. Quantitatively the standing crop of bacteria—the bacterial biomass—ranges from 0.001 to nearly 500 mg per cubic meter of sea water, the average throughout the euphotic zone being about 1 mg per cubic meter. There may be from 10 to 100 bacterial crops (cell divisions) per year, meaning an annual yeild of from 10 to 100 mg of nutritious bacterial protoplasm per cubic meter of water.

Besides forming bacterial cell substances and producing carbon dioxide from the organic remains of plants and animals, bacteria liberate sulfate or hydrogen sulfide, ammonium, phosphate, and certain other substances. The ammonium and phosphate are important as plant nutrients. Bacteria which liberate ammonium from proteins or amino acids are spoken of as ammonifiers. Many species of marine bacteria are endowed with this ability. There are also many species which produce the enzyme urease, which acts upon urea in aquatic environments to yield ammonium. Ammonium as such may be assimilated by photosynthetic plants or it may be oxidized by nitrifying bacteria to nitrite and subsequently to nitrate, another plant nutrient. Although nitrogen-fixing bacteria have been found in the sea, the extent to which they contribute to the fertility of sea water is not known.

Under certain conditions (perhaps highly reducing sediments where large quantities of organic matter are being rapidly buried) there appears to be a tendency for anaerobic bacteria selectively to reduce the oxygen, sulfur, phosphorus, and nitrogen content or organic compounds, leaving a residue that is relatively richer in hydrogen and carbon—more petroleum-like in chemical composition. Methane is a common product of bacterial activity and small quantities of more complex petroleum hydrocarbons are produced by marine bacteria. There are several ways in which bacteria may contribute to the origin of oil, but whether they play an essential part in the process has yet to be proved.

Most marine bacteria are heterotrophs dependent upon organic matter to satisfy their carbon and energy requirements. However, the sea is also the home of numerous species of autotrophic bacteria, including both photosynthetic and chemosynthetic varieties whose carbon requirements are satisfied by either bicarbonate or carbon dioxide. Photosynthetic autotrophs, dependent upon radiant energy, are confined to shallow coastal waters penetrated by sunlight. These are the purple sulfur and green bacteria having bacteriochlorophyll and certain accessory photosynthetic pigments. They differ from higher green plants by not liberating free oxygen during the photosynthetic process; instead the oxygen is consumed in the oxidation of hydrogen sulfide and/or elemental sulfur. Among the chemosynthetic autotrophs are those that obtain their energy by oxidizing either molecular hydrogen, hydrogen sulfide, elemental sulfur, methane, ammonium, or nitritte. Such bacteria are widely distributed in the sea, but not in great abundance.

Large numbers of saprophytic bacteria are ordinarily associated with fin fish, shellfish, and other marine animals. As in terrestrial animals, the intestinal tracts of marine animals have a typical microflora. *Escherichia coli* is generally absent except in the gut of

animals inhabiting polluted coastal waters. Despite the large numbers of *E. coli* and allied coliform bacteria discharged into the sea with sewage, such bacteria of sanitary significance are rarely found in the sea except near the source of terrigenous or human contamination. These as well as other freshwater or terrestrial bacteria entering the sea rapidly disappear as a result of flocculation and sedimentation. Death is attributed to predation, starvation,and the germicidal properties of natural sea water. A good many species of marine bacteria as well as certain phytoplankton produce potent antibiotic substances.

The bacterial flora found in the gut of marine animals appears to depend in part upon the animal species and in part upon the food habits of the animal. Surface slime on the integuments and gills of fish is usually a veritable bacterial garden. Predominating are species of *Achromobacter, Pseudomonas, Flavobacterium, Proteus, Bacillus, Serratia,* and *Micrococcus.* The muscle tissue of living and recently killed fish is generally sterile. The bacterial population of improperly handled dead fish increases rapidly at a rate that is fluenced primarily by handling and temperature. Spoilage is best controlled by careful handling to minimize the contamination of fish flesh, fast freezing, refrigeration, and by the use of appropriate antibiotics.

CLAUDE E. ZOBELL

References

Wood, E. J. F., "Marine Microbial Ecology," New York, Reinhold, 1965.
Zobell, C. E., "Marine Microbiology," Waltham, Mass., Chronica Bontanica, 1946.

MARSUPIALIA

Three groups, or infraclasses, are recognised within the subclass Theria in which all present day MAMMALS except the MONOTREMATA have been placed. The infraclass Metatheria includes only the order Marsupialia ("pouched" mammals) while the placental (higher) mammals are grouped in the infraclass Eutheria. The mammals of the third infraclass, the Pantotheria, are known only from fossil remains of Jurassic age, but they had a bunodont dentition with separate molar cusps closely resembling that of some marsupials and of the eutherian INSECTIVORA. For this reason most authors consider that the pantotheres were, or were derived from, the common ancestral stock of the marsupials and at least some of the placentals. Features common to these two groups, the Metatheria and Eutheria, include viviparity and the presence of a placenta, mammary glands provided with teats, the absence of separate coracoids and an interclavicle and the development of epiphyses in most of the bones.

The classification of marsupials was long based on a distinction between the Polyprotodontia, including the American Didelphidae and Caenolestidae, as well as the Australasian Peramelidae and Dasyuridae, all of which have at least 4 upper incisors, and the Diprotodontia. In the latter, to which the Phalangeridae of the Australian region are assigned, the upper incisors never number more than 3 and there is no lower canine. Further division of these groups was often based on the fusion of the second and third pedal digits in the syndactylous peramelids and phalangerids. This latter character was, in other classifications, made of primary importance to distinguish the Didactyla from the Syndactyla. Modern workers, following Simpson, prefer to recognise 5 super-families: Didelphoidea, Caenolestoidea, Perameloidea, Dasyuroidea and Phalangeroidea.

The marsupials and placentals appear contemporaneously in the Upper Cretaceous fossil record. The early Tertiary didelphoid marsupials were apparently almost world wide in distribution, although there is no fossil evidence of their existence in South-East Asia. Only in South America and in Australasia did extensive adaptive radiations develop; the continued dominance of the marsupial fauna in the latter region has apparently been due to limitation there of competition with placental mammals, which are represented only by murid rodents, bats, and the more recently introduced dingo (native dog).

Much confusion has been occasioned by the adoption of the name 'opossum', properly used for the American didelphoids, for certain of the Australian phalangeroids. In the following account, in accordance with a suggestion made by Troughton, 'opossum' will be used exclusively for American forms, 'possum' for Australian species of similar form.

Anatomy. The cranium of the marsupial is small relative to that of a placental of the same size. The palate of the adult is usually fenestrated posteriorly (except in the marsupial ant-eater *Myrmecobius*) and the jugal is well developed, extending posteriorly from its articulation with the lachrymal and maxilla to the glenoid fossa. The nasal bones of the marsupial are narrower in front than behind; the reverse is normally true of the eutherians. The angular process of the mandible of the marsupials is characteristically turned inwards, the pigmy honey possum *Tarsipes* is in this regard an exception and resembles the placental mammals.

The auditory region of the marsupial skull is formed chiefly by the alisphenoid which may be expanded to form a bulla which is often incomplete. The tympanic forms a tubular auditory canal in the phalangeroids; in the dasyuroids and and perameloids this canal is incomplete.

As in the monotremes, the brain of the marsupial has no corpus callosum, but despite the absence of this tract, the connection between the neopallial cortex of the two hemispheres is better developed than that in the prototherian group. The dentition of the marsupials is highly variable. In the carnivorous and insectivorous groups the basic pattern is

$$\text{i. } \tfrac{5}{4}; \text{ c. } \tfrac{1}{1}; \text{ pm. } \tfrac{3}{3}; \text{ m. } \tfrac{4}{4}$$

and is never derived from the placental pattern.

$$\text{i. } \tfrac{3}{3}; \text{ c. } \tfrac{1}{1}; \text{ pm. } \tfrac{4}{4}; \text{ m. } \tfrac{3}{5}.$$

The only milk teeth found in the marsupials are the posterior upper and lower premolars. The deciduous tooth resembles a molar in form, the tooth which replaces it is a typical premolar. In the diprotodont superfamily Phalangeroidea the middle lower incisors are enlarged and procumbent for grazing or leaf-chewing and much variation in dentition is encountered. Amongst the kangaroos (Macropodidae) the usual dental formula is

$$\text{i. } \tfrac{3}{1}; \text{ c. } \tfrac{1}{0}; \text{ pm. } \tfrac{2}{2}; \text{ m. } \tfrac{4}{4},$$

there being a marked diastema between the cropping incisors and the grinding molars.

The marsupial vertebral column and limb girdles have the same basic structure as those of the eutherian mammals. Seven cervical vertebrae, free except in the marsupial "mole" *Notorcytes* in which the second and third vertebrae are fused, are succeeded behind by 13 (11 in *Phascolarctos*, 15 in *Vombatus* and *Notoryctes*) rib-bearing dorsal and 4-7 lumbar vertebrae. Two to 6 vertebrae usually participate in the formation of the sacrum, while the number of caudal vertebrae is extremely variable.

In the pectoral girdle the coracoid is reduced to a process of the scapula and the interclavicle is entirely lost. The clavicles remain prominent except in the Perameloidea, where they are absent. The pelvic girdle carries, as in the monotremes, a pair of epipubic ("marsupial") bones which help to support the body wall in all species except *Thylacinus* in which they are represented by cartilaginous vestiges.

The foot-pads of the marsupials are usually finely ridged or striated in arboreal forms and granular in ground dwelling species. The hand is pentadactyle except in the bandicoot *Chaeropus*, in which it has only three digits. Grasping in arboreal forms is usually achieved by opposing the two inner digits to the three outer ones. In the fossorial *Notoryctes* two digits of the manus bear very enlarged flattened digging claws. The foot is variable in form, the hallux being relatively well developed only in the arboreal didelphoids and phalangeroids in which it is opposable to the other digits. The water opossum *Chironectes*, alone among the marsupials, has strongly webbed hind feet. The second and third digits are fused together to form a combing organ in the superfamilies Perameloidea and Phalangeroidea, but it is now considered that the condition evolved separately in the two groups. The perameloids have retained a presumably primitive polyprotodont dentition. The phalangeroids are specialised herbivorous diprotodonts, in which the brain is unique in possessing the fasciculus aberrans, a bundle of fibers derived from the dorsal portion of the ventral commissure connecting the cerebral hemispheres. Elongation of the hind foot, suppression of the first digit and use of the tail as a supporting organ have developed convergently in the didactylous dasyuroids specialised for saltatory locomotion as well as in the syndactylous kangaroos and wallabies (Macropodidae) of similar habit. A trend towards functional monodactylism is apparent in the bandicoots, where the fourth digit of the hind foot has become relatively enlarged, and, although the forelimbs are used alternately, the hind limbs are used together in saltatory fashion.

The respiratory and vascular systems of the marsupials closely resemble those of the placentals. At birth the lungs consist merely of several air sacs which gradually acquire the usual structure.

The endocrine glands of the marsupials have been little studied. The adrenal bodies, at least in the unstressed animal, appear to be unnecessary for maintenance of life over quite long periods in *Didelphis*. This is not true of the quokka (*Setonix*, Macropodidae) in which bilateral adrenalectomy is followed by death in 2 days.

The thymus is a persistent, sometimes complexly lobed, adult structure in a number of species.

Many of the herbivorous diprotodonts have a large sacculated stomach and well developed caecum. It has recently been shown that digestion in the quokka (*Setonix*) is of the ruminant type and is dependent on symbiotic micro-organisms, while regurgitation and further mastication of previously swallowed food has been observed in other wallabies. The caecum is of variable length; in the koala (*Phascolarctos*) it is 8–9 feet long. The stomach of insectivorous and carnivorous species is usually a simple sac, while the intestine is shorter in these forms than in the herbivores. The caecum is absent in the Dasyuridae and in the pigmy honey possum (*Tarsipes*) which has a unique globular diverticulum connected by a narrow duct to the stomach. The gall bladder is present in all marsupials.

The body temperature in the Metatheria, although a little lower than that of most placentals, is well regulated, lying within the range 31–36°C. There is some variation between species (e.g. *Didelphis virginiana* 34.2–34.5°C.; *Phascogale tapoatafa*, 32.0–33.5°C.).

The male and female are usually similar in size and colouration. Amongst dasyuroid marsupials the female is sometimes larger than the male, while in the brush-tailed possum (*Trichosurus*) a reddish-brown saddle across the shoulders is often found in sexually mature males. The males of several species of kangaroo are more often reddish in colour than the females. This is especially true of the red kangaroo (*Megaleia rufa* in which the male is red-brown above; the female is blue-grey.

The testes descend into a scrotum in all marsupials except the mole-like *Notoryctes* in which they remain abdominal. In the wombats (Vombatidae) the testes occupy the scrotum only during the breeding season, when spermatogenesis takes place. Many marsupials (e.g., *Trichosurus, Pseudocheirus, Myrmecobius*) show such a seasonal cycle of spermatogenesis, although not necessarily in all individuals. The scrotum is always prepenial in position and is not homologous with that of the placental mammals placed behind the penis. There are no vesiculae seminales. The prostate is well developed and surrounds the urethra. There are 2–3 pairs of bulbo-urethral (Cowper's) glands opening to the urogenital sinus. One to 2 pairs of anal glands surround the rectum and often at least one pair produces a strongly odoriferous secretion. The penis is erectile and often has a divided glans. In the bandicoots (Peramelidae) the penis serves, as in the monotremes, only for the transport of the seminal fluid. In other families a more advanced condition in which the urogenital sinus opens at the base of the penis and not into the rectum foreshadows the development in the Macropodidae of a penial structure similar to that of the placentals with both urine and semen passing along the same duct. A common sphincter encircles the rectal and urinary orifices. In the males of *Notoryctes, Thylacinus* and *Chironectes* a vestigial pouch is found.

The female usually possesses a pouch (marsupium) in which the young, born at a very early stage of development, are reared until able to fend for themselves. The pouch may open either to the anterior or less commonly to the posterior (Peramelidae, and certain Dasyuridae); in some forms it is absent or much reduced (*Marmosa, Philander, Myrmecobius* and some Dasyuridae). A sphincter derived from the postero-lateral part of the musculus panniculus carnosus may or may not completely encircle the pouch. Unlike that of the monotreme *Tachyglossus*, the marsupial pouch is a permanent structure although during anoestrous it may become reduced in size. At oestrous, and during lactation, numerous sebaceous glands become active and dark pigment may develop in the surrounding hair.

The number of mammae is highest in the didelphoids where they are usually arranged in pairs with a single additional unpaired mamma in the mid-ventral line. The American opossums *Monodelphis* (9–27) and *Didelphis* (5–17) have the largest number. The dasyuroids commonly have 2–4 pairs of mammae, bandicoots (Peramelidae) 3–4 pairs, phalangers 1–2 pairs, macropods 2 pairs, wombats and *Notoryctes* but a single pair. Since the young are permanently attached to the teats during their early pouch life, the number of teats determines the number of offspring raised although excess are often born. Milk is in many cases force fed to the young until they have developed suctorial powers. The compressor muscle responsible is homologous with the cremaster muscle of the male and passes over or around the mammary gland to an insertion on the epipubic bone.

The uteri remain separate in all marsupials and open on separate papillae into vaginal chambers. Development of the ureters precedes the development of the Mullerian ducts and as the former lie in a medial position in the marsupials, the Mullerian ducts are prevented from fusing to form a median vagina such as found in the placental mammals. The paired lateral vaginae form median culs-de-sac and loop backwards around the ureters to form together a posterior vaginal sinus or pass separately into the urogenital sinus. The paired vaginal culs-de-sac are thus initially kept apart, the mesial walls of the culs-de-sac forming a septum which may be resorbed, or ruptured at parturition, to leave a common median vaginal chamber. Primitively, birth probably occurred as it does in the macropod *Potorous* by way of the lateral vaginae, but in other existing marsupials a median birth route, the pseudovaginal canal, passes between the ureters to connect the vaginal culs-de-sac and the urogenital sinus. In most macropods (e.g. *Setonix, Thylogale, Macropus robustus*) the pseudovaginal canal remains open after parturition. The septum between the culs-de-sac is usually temporarily resorbed during gestation in the Didelphoidea, Caenolestoidea and Peramelidae. In the latter group, the pseudovaginal canal is especially long and closes rapidly after parturition, often enclosing some of the foetal membranes. Many of the Phalangeroidea (e.g. *Trichosurus, Pseudocheirus, Petaurus*) show coalescence of the culs-de-sac before sexual maturity is reached, while in others (*Vombatus, Phascolarctos*), as in the Dasyuroidea, separation is maintained. Both the phalangeroids and dasyuroids have a short pseudovaginal canal which closes after parturition. A clitoris, frequently divided, is present in all marsupials.

Reproduction. Ovulation in marsupials is at least in most cases spontaneous. The ovum is intermediate in size between that of the monotremes and the eutherian mammals. In *Dasyurus* and *Didelphis* it is known that a large number of ova are liberated from each ovary, in the latter species an average of 11, with a maximum from a single ovary of 44, but there is a high intrauterine mortality and in addition many foetuses fail to reach the pouch after birth. At least some of the Macropodidae (e.g. *Setonix*) are monovular. The opossum (*Didelphis virginiana*), normally in anoestrous between October and December, may breed twice, or rarely three times, in the year. Most Australian marsupials enter an anoestrous period during the late spring and early summer, the period at which the young usually become independent of the mother. Some species, as for example the red kangaroo (*Megaleia rufa*) and the brush-tailed possum (*Trichosurus*), breed throughout the year. In the latter species, however, there are two peaks of reproductive activity: March—May (southern autumn) and September–October (southern spring).

The didelphoid, macropod and phalangerid marsupials are all polyoestrous, 28 days elapsing between successive ovulations except in *Didelphis azarae* in which the oestrous cycle is completed in 7 days. There is some evidence that *Dasyurus* and *Phascogale* (Dasyuridae) are *monoestrous*. Delayed blastocyst implantation occurs regularly in a number of macropod marsupials, and possibly also in at least one phalangerid; it is best known in the quokka, *Setonix*. In this species a postpartum copulation results in the development to the 100-cell stage of a blastocyst which then becomes quiescent during lactation. If the early pouch foetus is destroyed lactation stops and implantation of this blastocyst takes place in about 20% of cases.

A placenta of the eutherian chorio-allantoic type has so far been found in only two groups of marsupials—the Perameloidea (*Perameles, Isoodon*) and the Phalangeroidea (*Vombatus, Phascolarctos*). In the didelphoids, macropods and some phalangerids, the allantois does not make contact with the chorion (yolk-sac or chorio-vitelline placentation), whereas in *Dasyurus* there is an intermediate condition in which the allantois meets the chorion but does not contribute to the formation of the placenta.

Gestation is of short duration, lasting in *Didelphis virginiana* 13 days, in *Setonix* (Macropodidae) 27 days, and in *Trichosurus* (Phalangeridae) 17½ days. Compared with those of the placentals the young are born at a very early stage of development. The forelimbs, their muscles and nerve centres and the sense of smell are precociously developed. The forelimbs usually carry enlarged claws which may be temporary, deciduous structures (*Didelphis, Perameles*) or prematurely developed but permanent (*Setonix, Trichosurus*). The young are born while the mother rests on her back in a reclining position with the cloacal aperture raised towards the pouch. Most females lick the hair in a line between the cloacal aperture and the pouch, working systematically from the cloaca forwards. The pouch is also frequently licked out before the birth. The young make their way to the pouch along the dampened track in the hair without help from the mother. In *Didelphis* the journey takes 16½ seconds, in *Megaleia* the young may require up to 30 minutes to reach the pouch and become attached to a teat. The young remain continuously attached to the nipple for the first weeks of pouch life. The offspring leave the pouch when about 3 months of age in *Dasyurus*, 3½ months in *Didelphis* and 5½ months in *Setonix*. The young may for some time enter and leave the pouch freely, a number of didelphoids and phalangeroids carry the young on the back. In the koala (*Phascolarctos*) the young eat a special exudate from the mother's rectum during weaning, and probably symbiotic microorganisms are thus transferred from mother to offspring.

Cytology. The Dasyuridae, often considered a primitive family, and the Peramelidae all have a diploid complement of 12 + XX (female) or 12 + XY (male), while all Didelphidae show 2N = 20 + XX or XY. A striking feature of the Peramelidae is an apparently unique mechanism of dosage compensation involving suppression in certain somatic tissues, but perhaps not all, of one X in the female soma, and the Y in that of the male. Germ line cells thus have the constitution 12 + XX and 12 + XY, somatic cells in affected tissues show 12 + XO in both sexes.

Amongst the Phalangeridae diploid numbers range from 14 to 22. Certain populations of the greater glider (*Schoinobates volans*) have a variable complement of supernumerary chromosomes in addition to the basic diploid set of 20 + XX or XY.

The Macropodidae show 2N = 10 to 2N = 24. Three species are known to possess multiple sex chromosome systems. In the male of both *Potorous tridactylus* ($2N = 10 = XY_1Y_2$) and *Protemnodon (Wallabia) bicolor* ($2N = 8 = XY_1Y_2$) three chromosomes are involved in sex-determination so that a trivalent is observed at meiosis. A different sex chromosome system occurs in the hare-wallaby (*Lagorchestes conspicillatus*) where the female has $X_1X_1X_2X_2$ and the male X_1X_2Y; at meiosis the male complement forms a chain trivalent orienting so that X_1 and X_2 normally move to the same pole.

Distribution. Two main centres of marsupial evolution are recognisable. In South America the group was well represented in the Tertiary and has now largely been replaced by eutherian mammals; in Australia the marsupials were dominant until the arrival of man and the introduction of the placental mammals which are now replacing many indigenous species.

The South American marsupials included the highly specialised carnivores of the extinct super-family Borhyaenoidea which appeared during the Miocene. *Borhyaena* was a fast moving dog-like animal resembling the existing Tasmanian "marsupial wolf," *Thylacinus*. *Thylacosmilus*, a larger more heavily built animal, bore a remarkable resemblance to the Old World sabre-toothed tiger. A second South American superfamily, the Caenolestoidea, was represented during the Tertiary by many species inhabiting the Andean region of South America; 3 genera are still represented in Ecuador and Peru. The living caenolestoids, (e.g. *Caenolestes*), three genera of small rat-like arboreal carnivores, have a reduced pouch. The enlargement of the procumbent lower middle incisors and the reduction or suppression of the other lower incisors has led to a remarkable, probably convergent, resemblance to the Australian diprotodont phalangeroids. The third American superfamily, the Didelphoidea, including forms which may have been ancestral to the borhyaenoids and caenolestoids, was probably world wide in distribution in the early Tertiary, although no fossil marsupials are known from South-East Asia. The existing forms, found in North, Central and South America, are nocturnal and usually arboreal; few are insectivorous, most are omnivorous. The opossum, *Didelphis virginiana*, like the Australian brush-tailed possum (*Trichosurus vulpecula*), is one of the few marsupials which has increased in number and extended its range despite disturbance and destruction by man. *Metachirops* constructs spheroidal twig and leaf nests like those of the Australian ring-tailed possum *Pseudocheirus*. *Philander* and *Marmosa* are also arboreal genera. Some species of *Marmosa* have thickened tails in which fat is stored as a reserve food supply: a close parallel is seen in the Australian dasyurid "pouched mouse" *Sminthopsis crassicaudata*. *Monodelphis* includes didelphoids which are largely ground dwelling. The two species of *Chironectes* found in tropical South America are fresh water animals with webbed hind feet. These water opossums ("yapok") feed mostly on aquatic insects and crustacea and live in burrows close to the water.

There is no fossil record of the entry of the ancestor, or ancestors, of the Australian marsupials into that region, although they almost certainly entered by way of South-East Asia. No representatives are now found west of Wallace's Line, in the Old World tropics. Many of the highly specialised marsupials are restricted to the Australian mainland and Tasmania. There appears to have been considerable faunal interchange with New Guinea facilitated by the existence of land connections as recently as the late Pleistocene. The arboreal genus *Phalanger* ("cuscus"), ranges west to Celebes and east to the Solomon Islands, the limits of marsupial distribution in the Australasian region. No marsupials reached New Zealand until introduced there by man.

The superfamily Dasyuroidea includes polyprotodont insectivorous and carnivorous animals without a syndactylous hind foot. Some are competent climbers, but most are terrestrial. The group includes the predatory "native cats" (*Dasyurops, Dasyurus*) and many rat-like carnivores and insectivores (*Antechinus, Murexia, Phascogale*) as well as the smaller marsupial mice (*Sminthopsis*) and the saltatory sand dwelling marsupial counter-part of the jerboa, *Antechinomys*. Two large dasyuroids confined to Tasmania are the Tasmanian "devil" (*Sarcophilus*), a heavily built carnivore, and the fast moving marsupial wolf (*Thylacinus*), which is now very rare. The marsupial mole (*Notoryctes*), a blind insectivorous subterranean species, resembling the true mole, which lives in central and western Australian deserts, and the West Australian marsupial anteater (*Myrmecobius*, the "numbat"), are apparently phylogenetically old members of this group.

The Perameloidea (Bandicoots) have polyprotodont dentition like that of the dasyuroids, but have in addition syndactylous hind feet. Most are of the size of a rabbit and have long pointed snouts used for obtaining insects and roots from the characteristic conical workings. Many species are also scavengers. The tail is short except in the long-eared rabbit-bandicoots of the genus *Macrotis*. *Isodon* has a shorter nose than *Perameles* and lacks the dark brown transverse bars which mark the rump of several species of the latter genus; both are more common in Tasmania, where there are no introduced foxes, than on the mainland. The pig-footed bandicoot, *Chaeropus*, is a rare arid inland genus. The peramelids are well represented in New Guinea.

The Phalangeroidea have both diprotodont dentition and syndactylous hind feet. The two main families, the possums (Phalangeridae) and kangaroos (Macropodidae) are widely represented throughout the Australian region. The wombats (family Vombatidae) are heavily built terrestrial animals with short tails. They construct large tunnels in which they rest during the day, and feed on roots and other plant material. The eucalypt leaf eating koala (*Phascolarctos*) is an arboreal phalangeroid with a vestigial tail which has apparently evolved from a terrestrial wombat-like stock and regained the arboreal habit. A large, possibly carnivorous, diprotodont (*Thylacoleo*) and heavily built giant herbivores (*Nototherium, Diprotodon*) became extinct during the Pleistocene. New finds of late Tertiary fossil material from South Australia are at present being investigated. The Australian arboreal opossum-like marsupials are mainly leaf, fruit and flower eaters, but many are partly insectivorous and the brush-tailed possum (*Trichosurus*) takes small birds, eggs, and sometimes carrion in addition. The ring-tailed possum (*Pseudocheirus*) is widespread in Australia and New Guinea. The cuscuses (*Phalanger*) do not extend into southern Australia. The small black and white striped possums of Queensland and New Guinea (*Dactylopsila*) have an elongated 4th digit on the hand, which, with specialised procumbent upper and lower incisors, are used for gouging insects from wood. The tiny pigmy possums (*Dromicia, Tarsipes*) are of mouse like proportions and feed on nectar and insects. The gliding habit has probably evolved three times amongst the phalangers; in each case skin flaps which stretch from the fore to the hind limb have been developed, and the tail is long haired. The smallest glider, *Acrobates*, is only about 3 inches long. *Schoinobates*, capable of gliding well over 100 yards between trees, has a body length of about 18 inches. *Petaurus* is a glider of intermediate size.

The recent capture of a single adult male of *Burramys*, a genus previously known from abundant bones in cave deposits dated about 10,000 years BP, has confirmed the phalangeroid affinities of this animal.

The Macropodidae comprise the kangaroos, wallabies and rat-kangaroos. All are saltatory, with well developed hind limbs and a tail used as a prop. During slow movement the forelimbs are employed; bipedalism is adopted in faster locomotion. The large kangaroos (e.g. *Macropus*) are capable of leaping 40 feet in a single bound when in fast motion; the largest stand up to 8 feet in height, but considerably larger forms occurred during the Pleistocene. Kangaroos often move in herds and are amongst the few diurnal marsupials. The wallabies (*Wallabia, Thylogale*) and rock-wallabies (*Petrogale*) are generally smaller than the kangaroos. The rat-kangaroos (*Potorous, Bettongia*) and the West Australian quokka (*Setonix*) are about the size of a hare. The tree kangaroos (*Dendrolagus*) of Queensland and New Guinea have become arboreal in their mode of life; the tail is used as a balancing organ and prop, but is not markedly prehensile.

J. A. THOMSON

References

Bodenheimer, F. S. and W. W. Weisbach ed., "Monographiae Biological VIII. Biogeography and Ecology in Australia," The Hague, Junk, 1959.

Marlow, B., "Marsupials of Australia," Brisbane, Jacaranda Press, 1962.

Troughton, E. Le G., "Furred Animals of Australia," 5th ed., Sydney, Angus and Robertson, 1954.

Wood-Jones, F., "The Mammals of South Australia," 5th ed., Sydney, Angus and Robertson, Fauna of South Australia, British Science Guild (S.A.), Government Printer, Adelaide, 1923–25.

MASTIGOPHORA

The Mastigophora (or Flagellata) are a class of PRO-
TOZOA, characterized by the presence of one or more
FLAGELLA and a single nucleus (or, if more than one are
present, nuclei of only one kind). The former feature
distinguishes them from the RHIZOPODA and SPOROZOA
and the latter sets them apart from the CILIATA, in
which there are two kinds of nuclei. The group includes both pigmented, photosynthetic flagellates,
which, together with certain morphologically similar
but non-pigmented organisms, constitute the *phytoflagellates* (see ALGAE), and colorless forms which lack
obvious pigmented relatives and which are grouped
together as the *zooflagellates*. Both of these are heterogeneous, probably polyphyletic assemblages.

Most of the organelles of the flagellates—centrioles,
flagella, plastids, pyrenoids, MITOCHONDRIA, GOLGI
BODIES, etc.—are comparable with those of the cells
of higher plants and animals, and evidence is accumulating that the similarities extend to the submicroscopic
level. Centrioles play an important role in the organization of many flagellates, not only in the production of
the mitotic apparatus but in acting as foci about which
there may be produced a whole range of other organelles made of fibrous proteins—flagella, axostyles,
parabasal filaments, flagellar bands etc.

The phytoflagellates usually possess a pellicle or cell
wall and seldom have more than two flagella. The
various orders are distinguished chiefly on the basis
of the colour of their plastids, the type of reserves,
the nature of the cell wall or pellicle and the number
and arrangement of the flagella. The more important
orders are the *Phytomonadina*, with two (rarely four)
equal flagella, green plastids, cellulose cell walls and
food reserves in the form of starch; the *Euglenoidina*,
in which the plastids are green, carbohydrate reserves
take the form of paramylum, the flagella (one or two
in number) emerge through an anterior gullet and
there is a complex, spirally organized pellicle; the
Chrysomonadina, which have yellow–brown plastids,
leucosin reserves and often a silicified cell wall; the
Cryptomonadina, small forms in which there is a single
plastid (the colour of which is variable), an anterior
groove or gullet in which two flagella arise, and starch
reserves; and the *Dinoflagellata*, with two flagella (one
directed backwards, the other transversely or spirally,
both usually running in grooves), numerous plastids,
the colour of which varies in different species, and
frequently an armour of cellulose plates. There are
also a number of less important groups, many of them
with silicified or calcified cell walls.

Certain evolutionary tendencies can be discerned
in all of these groups. *Firstly*, loss of plastids seems
to have occurred frequently, resulting in colourless
heterotrophs. Aplastidic forms of *Euglena* can be
produced experimentally by treatment with streptomycin and other methods and are indistinguishable
from the naturally occurring genus, *Astasia*. The
survival of these colourless forms depends on the fact
that many of the pigmented flagellates lead a partly
heterotrophic existence, even in light: indeed, a wholly
autotrophic existence may well be exceptional. *Secondly*, colonial forms are often produced. These range
in complexity from small, temporary aggregates
of non-motile cells embedded in a gelatinous matrix—
the palmelloid form, which may be assumed by

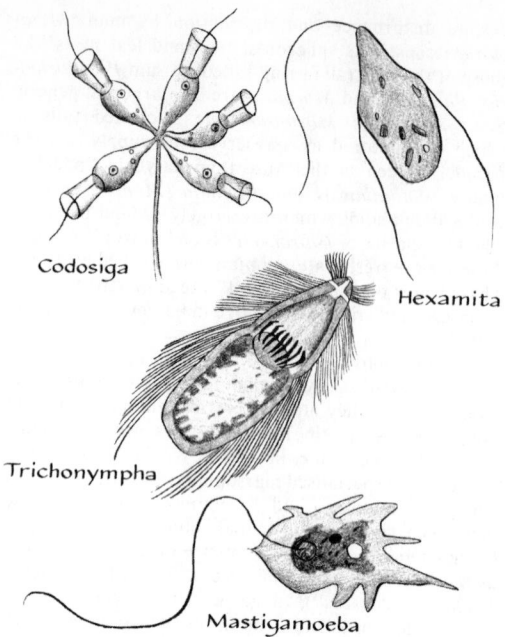

Codosiga

Hexamita

Trichonympha

Mastigamoeba

Fig. 1 Some representative flagellates. (Highly enlarged.) (From Platt and Reid, "Bioscience," New
York, Reinhold, 1967.)

otherwise free-swimming organisms, such as *Chlamydomonas*—to motile colonies of hundreds of cells
which may be differentiated into somatic and generative
types (e.g. *Volvox*). It is highly probable that colonial
phytoflagellates similar to these gave rise to the various
groups of algae: colonial phytomonads, for example,
to the *Chlorophyceae*, colonial chrysomonads to the
Chrysophyceae. These flagellates, in fact, are often
classified with the algae. *Thirdly*, the flagella may be
lost and an amoeboid form assumed. This is often accompanied by the adoption of a phagotrophic method
of feeding. (Both the amoeboid habit and phagotrophy
may also occur, however, in forms which retain their
flagella.) The clear derivation of amoeboid forms from
flagellates is one reason for the recent suggestion that
the Mastigophora and Rhizopoda should be grouped
together as the Rhizoflagellata.

The zooflagellates are as diverse as the phytoflagellates. They include firstly a large number of small
forms with a single flagellum, of which the trypanosomes are probably the most important. These are
parasitic, usually with a complex life-cycle comprising
a number of morphologically distinct forms which
develop in different hosts. The choanoflagellates, in
which the flagellum is surrounded by a collar, are of
interest in view of their probable affinity with the
sponges. A number of these simpler zooflagellates can
assume an amoeboid habit under certain conditions
and in some instances this transformation can be
brought about experimentally. The *Diplomonadina*
is a small group in which the nucleus and cytoplasmic
organelles are doubled, yielding bilaterally symmetrical organisms. The *Polymastigina*, which are mostly
parasitic, range in complexity from small forms with
four flagella and a single set of cytoplasmic organelles,

through types in which the organelles are enlarged but not multiplied, to forms of very great complexity in which there may be thousands of flagella and numerous and varied organelles. These latter forms represent one of the peaks of protozoan evolution. Colonial forms also occur among the *Polymastigina*. The *Opalinida* are parasitic forms, uniformly covered with short flagella. They are now usually regarded as flagellates, though they are somewhat distant from the other groups.

Flagellates commonly reproduce by longitudinal division. Their mitotic processes are diverse. Sexual processes are at present known only in some of the phytoflagellates and complex polymastigotes: they may take highly unusual forms in the latter.

It is reasonable to suggest that the ancestral group of Protozoa was probably composed of forms not greatly dissimilar from some of the simpler pigmented flagellates which exist today. Such organisms provide a plausible starting point for both the plant and animal kingdoms, giving rise on the one hand to the algae and higher plants, and on the other to the colourless flagellates, the amoebae and, presumably, the rest of the Protozoa.

Flagellates are important to man as the causative agents of a number of diseases (notably sleeping sickness, Chaga's disease and the various forms of leishmaniasis, all caused by TRYPANOSOMES), and as components of the phytoplankton.

A. V. GRIMSTONE

References

Grassé, P.-P., "Traité de Zoologie," Vol. 1. Paris, Masson, 1952.

Kudo, R. R., "Protozoology," 5th ed., Springfield, Ill., Thomas, 1966.

MATHEMATICAL BIOLOGY

In its broadest meaning, mathematical biology includes all applications of mathematics to biology. It encompasses therefore biometry. We shall use here the words in a restricted sense, meaning the science of developing mathematical and physicomathematical theories to explain different biological phenomena, as well as establishing general mathematical principles of biology, akin to the mathematical principles of physics. When the emphasis is on the physical aspects of biology, we speak of mathematical biophysics.

Mathematical biology stands in the same relation to experimental biology as mathematical physics to experimental physics. Its purpose is, on one hand, to explain and understand the ultimate processes of different biological phenomena, processes which may sometimes not be directly observable; on the other hand, as in mathematical physics, the theory is expected to lead to the prediction of phenomena or relations, not yet observed experimentally and thus stimulate the discovery of such phenomena.

As a coherent, systematic science, mathematical biology is still young and dates back to the work of Alfred J. Lotka in 1925, and N. Rashevsky of 1927, as well as Vito Volterra of 1931. Lotka and Volterra restricted themselves, however, to a mathematical theory of interaction of species and of some problems of ecology. Contemporary mathematical biology covers practically any branch of biology.

Earlier sporadic attempts at creating mathematical theories of biological phenomena were not lacking. No less a person than the famous mathematician Leonard Euler developed in the eighteenth century a mathematical theory of some aspects of cardiovascular phenomena. His work, written in Latin, and unpublished until the very end of the last century, remained until very recently unknown. Quite independently it was "rediscovered" at the turn of the century by Otto Frank and developed in a series of papers. We may also mention here the work of Hoorweg and that of Weiss, made in the nineteenth century on the theory of nerve excitation. The greatest development came, however, in the last forty years. In the *Bulletin of Mathematical Biophysics* founded by N. Rashevsky 30 years ago, over 900 papers have been published on subjects covering such a wide range as cardiovascular phenomena (Roston, Cope, Karreman); respiratory phenomena (Defares, Wise, and Landahl); kidney functions (Stibitz); electrocardiography (Martinek and Yeh, Plonsey); color vision (Landahl); pharmacological phenomena (Landahl, Segre, Berman); biochemical reaction rates (Hearon, Bartholomay); blood-tissue exchange (Morales and Smith, Schmidt, Macey); active transport (Patlak); struggle for life (Rescigno and Richardson); theory of some endocrine phenomena (Danziger and Elmergreen, Rashevsky); parasitism and symbiosis (Rapoport and Foster); and theory of tracer experiments (Hart), to mention only a very few. In 1960 the *Journal of Theoretical Biology* was founded by J. F. Danielli, in which numerous papers on similar and other subjects have been published. As the field grows, a specialization into different subfields begins, and recently the *Journal of Mathematical Psychology* (edited by Richard C. Atkinson) has been established. Quite recently still another journal, *Mathematical Biosciences,* began its publication under the editorship of Richard Bellman.

The mathematical methods used in mathematical biology range from some very elementary ones to such highly specialized branches as group theory, topology, theory of categories, theory of sets and theory of relations. The mathematics used depends largely on the subject matter treated.

Many biological phenomena depend on various processes of diffusion of metabolites. Mathematical theory of such processes requires the solution of partial differential equations for systems with sometimes very complicated boundaries. Except for the simplest cases, the biological application of which is limited, the exact solutions are either unknown or are very complicated. On the other hand, essential properties of a cell do not depend on details of its shape. Nor does the experimenter know the exact distribution of concentrations of diffusing substances. He is restricted to the knowledge of *average values*. With this in mind, an approximate method of handling diffusion problems was developed (Rashevsky). It is relatively crude and very simple, but it applies to almost any situation. In this approximation the time variable diffusion fields are described by *ordinary* differential equations, while stationary states are described by algebraic equations. If the over-all length of a cylindroid system is $2a$; its over-all diameter $2b$; if $\bar{c}$ denotes the average concentration of a metabolite produced at a rate q per unit volume in the system; if c_0 denotes the concentration of the metabolite at an infinite distance from the cell; if D_i and D_e denote correspondingly the diffusion coefficient of the metabolite inside the system and in the external medium; if h is the permeability of the boundary and if Λ denotes the following expression,

$$
\begin{aligned}
\Lambda = {} & ab(2D_i D_e + 2bD_i h + ahD_e) \\
& \cdot (2D_i D_e + 2bD_i h + bhD_e) \\
& \cdot \{3hD_i D_e[2(2D_i D_e + 2bD_i h + ahD_e)a \\
& + (2D_i D_e + 2bD_i h + bhD_e)b\}^{-1}
\end{aligned} \tag{1}
$$

then the differential equation which governs the variation of $\bar{c}$ with time is

$$
d\bar{c}/dt = q - (\bar{c} - c_0)/\Lambda. \tag{2}
$$

Integrating (2) and denoting by C a constant of integration, we find

$$
\bar{c} = c_0 + \Lambda q + C\Lambda \exp(-t/\Lambda). \tag{3}
$$

As t tends to infinity, $\bar{c}$ tends assymptotically to

$$
\bar{c} = c_0 + \Lambda q. \tag{4}
$$

In the above equations q may be any function of $\bar{c}$. It also can be either positive (produced metabolite) or negative (consumed metabolite). When $a = b$, the system approximates a sphere, for which the exact solution is known. The general form of the exact and approximate solutions is the same. They differ only by practically insignificant numerical coefficients.

Because of q being in general a function of $\bar{c}$, the steady state solution (4) may be under certain conditions infinite, which means that no steady state exists. Consider a spheroid cell, that is, $a = b$, let α denote a positive constant, and let $q = \alpha\bar{c}$. This means an autocatalytic production of a substance in the cell. Also consider that $D_e = h = \infty$ which corresponds to a thorough mixing of the external medium. (These assumptions are made for simplicity only and do not entail any loss of generality; the results, though more complicated, hold for $a \neq b$, and a finite D_e.) We now find from (1) and (4)

$$
\bar{c} = \frac{c_0}{1 - \dfrac{a^2 \alpha}{9D_i}} \tag{5}
$$

When $a > \sqrt{9D_i/\alpha}$, $\bar{c}$ becomes infinite. For still larger values of a, the concentration c becomes negative, which is physically impossible. Thus certain types of reactions in the cell lead to instabilities and are impossible.

Equation (2) shows that the rate of change of the concentration of a substance in the cell depends not only on its intrinsic biochemical rate q, but also on the constant Λ, that is, on the size and shape of the cell, as well as on the diffusion coefficients and permeabilities. This point has been developed in detail by J. Z. Hearon, and also used by Rashevsky to show the effect of the histological structure of endocrine glands on possible periodic fluctuations of the blood level of thyroid and the thyrotropic hormones.

If several metabolites are involved, equation (3) holds for each of them. Each of the corresponding q_i's may be a function of the $\bar{c}_i$'s. We then have for the steady state a system of algebraic equations which determine the $\bar{c}_i$'s.

When applied to a system of reactions which involve the consumption of glucose, glycolysis, and oxidation of lactic acid, the above method leads to the following equation, which gives the relative rate y of oxygen consumption as a function of the external oxygen concentration x:

$$
x = \zeta y + \xi y/(1 - y), \tag{6}
$$

where ζ and ξ are coefficients which depend on the size and shape of the system, and the diffusion coefficients and permeabilities of the metabolites involved (Landahl). Equation 6 describes well a number of available observations.

The approximate method can also be applied to diffusion fields in infinitely extending systems. It has been applied to the theory of spread of populations (Landahl). This problem has also been investigated, for special cases, by exact methods (Skellam, Kerner, Barakat). Landahl successfully applied the method to spread of rumors. Both the approximate and the exact methods have been used in the theory of diffusion of metabolites to and from capillaries (Schmidt, Macey). They have also been applied to vibrations of walls of blood vessels (Rashevsky). More exact mathematical techniques have been applied to the theory of pulse waves (Frank) and their reflections at arterial branchings (Karreman, Jacobs). The approximate method has been applied to the theory of experiments involving use of radioactive tracers (Landahl; Stevenson, Sangren, and Sheppard).

Diffusing substances exert forces on the solvent in the direction of the diffusion. Those forces are approximately proportional to the concentration gradients of diffusing substances. It has been shown (Rashevsky, Landahl) that under certain conditions those "diffusion drag forces" produce a division of a system (cell) into two parts. Average sizes of cells, computed on the assumption that diffusion drag forces are responsible for cell division, are correct. Some other gross features of cell division are correctly described quantitatively by the theory. It fails, however, in explaining mitosis. Several suggestions have been made but a mathematical theory of mitosis is still lacking.

A mathematical study of interaction of diffusion fields with suspended particles in the cell shows that asymmetries may arise, which result in a polarity of the cell. This polarity is automatically reestablished when destroyed by external factors (self-regulation). Light is thrown on some embryological phenomena (Rashevsky).

Mathematical theories of growth of cells and of metazoans have been developed.

A mathematical theory of mechanical interaction of several diffusing substances has been developed. It explains certain types of active transport. Other aspects of this phenomenon have also been studied (Patlak), including so-called apparent active transport (Bierman).

Mathematical theories of excitation and conduction

in peripheral nerves assume the existence in the nerve fiber of two "factors," an excitatory ϵ and an inhibitory j (Rashevsky, Hill, Monnier). Both are supposed to move toward the cathode of a current applied to the fiber. Their rates of accumulation at the cathode are assumed to be governed by the equations

$$d\epsilon/dt = KI(t) - k(\epsilon - \epsilon_0);$$
$$dj/dt = MI(t) - m(j - j_0),$$
(7)

where K, M, k, and m are constants, ϵ_0 and j_0 are the amounts of ϵ and j in the absence of a current, and $I(t)$ is the current as a function of time. It is assumed that excitation occurs when $\epsilon > j$. Hence in the absence of a current we must have $\epsilon_0 < j_0$. With the further assumption that $m << k$, the theory leads to an excellent agreement with experiments on excitation by constant currents, alternating currents, break excitation at the anode, and non-excitability by slowly rising currents. To account for other phenomena, more complicated equations have been assumed (Hodgkin and Huxley). A physical interpretation of equations (7) has been suggested (Karreman) considering diffusion processes in the membrane and using the approximate method. Mathematical theories of nerve conduction have been developed, leading to equations that are in general agreement with experiments. Considerations of molecular fluctuations of solutes in the nerve fiber lead to expressions for natural fluctuations of excitation thresholds (Landahl). The derived expressions agree with observations.

Recently Robert Rosen has shown a connection between the two factor theory, general properties of neural nets and biochemical automata, as well as the connection between the two factor theory and some genetic regulatory mechanisms in cells with the theory of Türings "morphogens," and with sequential machines.

A theory of the central nervous system has been developed by considering the interactions of very large numbers of excitatory and inhibitory neurons. The interaction of individual neurons at synapses is a discontinuous process, governed by the all-or-none law. The appropriate tool for the description of such interactions is Boolean algebra (McCulloch and Pitts). In most psychophysiological phenomena we deal, however, not with a few but with large numbers of neurons. It is shown (Rashevsky) that in this case the discontinuous interactions average out into a "pseudocontinuous" process, which is governed by simple linear differential equations. Those equations describe the interaction between units, each of which is composed of a large population of neurons. Those units are designated as neuroelements. If a particular configuration of neuroelements is assumed, the differential equations can be used to describe the properties of the configuration. The theory proceeds by assuming certain configurations, sometimes ad hoc, and by deriving their properties. In this manner a theory of reaction times, of psychophysical judgments (Landahl), of discrimination of intensities (Householder), of learning, and of a number of other phenomena has been developed. In many instances the quantitative conclusions of the theory are in very good agreement with experimental data. This does not prove that the postulated configurations are actually present in the brain, but it shows their value as working hypotheses.

On the same basis a theory of perception of Gestalts and of perception of abstract relations has been developed (Rashevsky). Different approaches to the Gestalt problem have also been made (McCulloch and Pitts; Culbertson). A theory of perception of geometric patterns and of their aesthetic evaluation has been developed. Its conclusions agree with experiments.

A theory of some of the above phenomena, e.g. learning, has been developed from purely formal assumptions, without involving a structural model of the brain (Mosteller and Bush; Rapoport). A general theory of configurations consisting of very large numbers of randomly arranged neurons has been developed (Rapoport, Shimbel). Mathematical theories of vision, some of a biochemical nature (Hecht, Kamiya), others based on the above neurobiophysical approach (Landahl, Greene), have been developed.

In connection with a theory of interactions of secretions of the anterior pituitary gland and the thyroid gland, a theory of periodic catatonia has been developed (Danziger and Elmergreen) which suggests some interesting practical conclusions. The theory was generalized to include the effects of histological structure. This effect in case of any number n of interacting glands was also studied (Rashevsky).

Methods of mathematical biology have also been applied to social phenomena by considering the psychophysiological interactions of individuals in a social group (Rashevsky).

In the last twenty-five years, the formulation of general mathematical principles of biology has been attempted. The first principle deals with the problem of organic form. While in principle the explanation of the shapes of different unicellular and multicellular organisms should be derivable from considerations of physical interactions between individual cells and cell groups, the complexity of such an approach defies any such attempt. Moreover, an *exact* mathematical description of any given organic form, though in principle possible, would be meaningless because no two organisms are quite alike. It has been shown (Rashevsky) that a practically useful quantitative description of the form of a plant or animal can be made by specifying approximate sizes of its different parts. For a plant this may be the specification of the average diameter and height of the trunk, the average diameter and the average length of primary branches, the average number of primary branches, etc. For a quadruped we may specify the approximate length and width of the body (trunk), the length and average diameter of the extremities.

It is further shown that those specifications are determined by the requirement that the organism be able to perform prescribed physiological and mechanical functions. This leads to the formulation of the principle of optimal design: For a set of specified physiological functions the geometric design of an organism is optimal with respect to economy of material and expenditure of energy.

The following are some conclusions drawn from this principle. For quadrupeds, the average width of the body should vary as the $3/2$ power of the length. This is found to be approximately true. The rate of respiration in homeoterms should vary approximately as the inverse cubic root of the mass of the animal. This is found to hold approximately in the range from rat to horse. The pulse rate should vary in a similar manner. This is found to hold in the same range. The principle permits a calculation of the size of the aorta, of arteries and capillaries in mammals (Cohn). The calculations are found to be in agreement with available data. A theoretical evaluation of the orders of magnitudes of such important parameters as the size of the aorta, the stroke volume, the average blood pressure, the systolic and diastolic pressures, the peripheral resistance, the elasticity of the aorta and the time interval between heartbeats can be made on the basis of this principle (Rashevsky). The theoretically evaluated values are found to be in agreement with observed ones.

The second general principle emphasizes the importance of relational aspects of biology. The physicochemical models of different individual phenomena, as well as the principle of optimal design, emphasize the quantitative aspects. Relational aspects are of equal importance. Considering such properties or manifestations of organisms as sensitivity and response to stimulation, food intake, digestion, excretion of indigestible products, absorption, assimilation, etc., we notice that in one form or another they are present in all organisms. A given property in a lower organism is, however, quite different both quantitatively and physiocochemically from the same property in a higher organism. But the basic relations between those properties within an organism are the same for all organisms. Moreover, to a simple property, e.g., general sensitivity, in a lower organism, there corresponds a complex set of properties in a higher organism such as sensitivity to light of different wavelengths, to sounds of different pitches, to heat, to touch, etc.

An organism may be described as a set of basic biological properties, in which certain relations are defined. The above-mentioned well-known facts lead then to the following general principle:

The set of properties of any higher organism can be mapped in a many-to-one way on the set of properties of a lower organism, in such a manner that the basic relations between the elements within the set are preserved under the mapping.

The principle leads to a number of verifiable conclusions. One of them is that in higher animals emotional disturbances may produce gastrointestinal disturbances. This is not only a well-known fact, but a fact the existence of which cannot be deduced from any *model*, unless the model is designed to explain the fact, which must then be considered as given. It is, however, deducible from the principle of mapping. Another conclusion is that some animals possess glands the secretions of which are used to catch food (e.g., spider). Among other things the principle leads us to the expectation that *some* unicellular organisms must be able to produce antibodies to appropriate antigens.

In the earlier formation of the principle of mapping, the organism was described not by a set, but by either a topological space or a topological complex (Rashevsky). The principle, in this formulation, requires that the topological spaces or complexes which correspond to each organism be obtained from one basic space or complex by the same multiparametric transformation. The different formulations are in many respects equivalent and the choice may be determined by mathematical convenience. One way of representing an organism by a topological complex is to represent it by a directed graph which gives the "flow diagram" or "organization chart" of the biological functions of the organism (Rashevsky). The points of such a graph may represent either the different biological functions (Rashevsky) or the different organs which perform those functions (Rosen). The graph-theoretical representation of organisms has been elaborated by Rosen and his pupils (Bramson, Demetrius). By including into his theory the problem of restitution of damaged or lost organs, Rosen arrives at his theory of (M,R)-systems, which leads to biologically important conclusions. Comorosan and Platica applied it to some regulatory mechanisms in the cell. An organ may be considered as having certain inputs and certain outputs. There is in general a mapping between the elements of the input and those of the output; e.g., the set of amino acids which are produced as a result of digestion and which form the output of a digestive organ, stand in a definite correspondence to the set of proteins that form the input of the organ. In mathematical terms, there exists a mapping between the two sets. Thus an organism may itself be considered as a set of mappings.

The consideration of the organism as a set of mappings leads to a representation of organisms in terms of the theory of categories which considers *mappings of mappings* (Rosen). It leads to a number of mathematical consequences which are verifiable experimentally.

The development of this new branch of mathematical biology, now called *relational biology*, leads us away from mere *physicomathematical models* of biological phenomena, which were originally the starting point of mathematical biology. The question as to whether the relational aspects of biology are realizable through physical systems depends on the validity of Church's Thesis, an unproven surmise in pure mathematics (Rosen). A close connection between biological phenomena and sequential machines also exists (Rosen).

Relational biology does not supersede the model-building biology, nor is it a substitute for the latter. Purely physiochemical models of biological phenomena implicitly assume that all biological phenomena can be *explained* in terms of physical mechanisms or can be represented by such mechanisms. This is actually not an assumption, because the only way we can perceive and study an organism is through its physical manifestations. From the possibility of representing any biological phenomenon by a physical model it does not follow that it is possible to *deduce the existence of biological organisms* from the postulates and laws of physics (Rashevsky). Similarly from the circumstance that it is possible to explain many sociological phenomena in terms of the behavior of individuals as biological units, it does not follow that the existence of societies, human or animal, can be deduced from known biological principles. Therefore, quite recently a conceptual superstructure has been proposed (Rashevsky) called *organismic sets* of which biology and sociology are two different aspects of the same class of phenomena. A society is a set of individuals, plus the products of their activities and their interactions. A multicellular organism is a set of cells plus the products of their activities. A simple collection of individuals not doing anything and not interacting is not a society. Similarly a suspension of noninteracting bacteria is not an organism. A unicellular organism may be considered as a set of genes plus the products of their activities, because everything in a cell is a direct or indirect result of the activities of its genes.

An organismic set is defined as a set of elements which are endowed with certain activities and induce different relations between the elements. Several postulates are made about the properties of such sets, postulates which apply equally to organismic sets that represent unicellulars, multicellulars, or societies. One of such postulates is the existence of gradual division of labor and specialization. The only completely specialized set possible is shown to be that of genes, which are indeed so specialized according to Beadles principle of "one gene—one enzyme." Another postulate is that dispersed elements of an organismic set will aggregate if as a result of such aggregation certain relations which characterize the organismic set are realized (postulate of relational forces). A theorem is then proven that already specialized elements will never aggregate, thus making impossible a spontaneous appearance of organisms at present. Interesting implications as to the origin of life on earth or conceivably in other parts of the cosmos are drawn. A unicellular organism may be considered as an organismic set of order zero. An organismic set of order n is one whose elements are organismic sets of order $n - 1$. It is then shown that any organismic set or order $n > 0$ is mortal. This accounts for the mortality of multicellular organisms as well as for the decay of social organizations and of whole civilizations (Rashevsky). Those ideas appear to be quite revolutionary, especially the postulate of relational forces. It has, however, been shown that the aggregation of physical particles due to attractive forces is only a special case of relational forces (Rashevsky). Two or several physical objects aggregate when as a result of aggregation the potential energy of the system decreases. But the potential energy is a function of the distances between the objects. Any function is a special type of relation, but not every relation is a function. Thus the interplay between biology and physics remains even though it now appears in an entirely different light.

N. RASHEVSKY

References

Rashevsky, N., "Mathematical Biophysics: Physicomathematical Foundations of Biology," 3rd ed., 2 vols., New York, Dover, 1960.

Rashevsky, N., "Mathematical Biology of Social Behavior," revised ed., Chicago, University of Chicago Press, 1959.

"Physico mathematical Aspects of Biology." Proceedings of the International School of Physics "Enrico Fermi," Varenna, Italy. Course 16, Director, N. Rashevsky. New York and London, Academic Press, 1962.

Rosen, R., "Optimality Principles in Biology," London, Butterworths, 1967.

Also numerous papers in the Bulletin of Mathematical Biophysics and the Journal of Theoretical Biology.

MATING DANCE

The term "mating dance" has not been precisely defined and may be used for many of the repeated or rhythmical displays and movements associated with mating. Such movements are exhibited by animals of many different groups and have a variety of functions.

Perhaps the most usual function is stimulatory; to attract the attention of the female and to bring her into a state of readiness for mating. Cases in point are the display movements of peacocks and the rapid wing vibrations and posturing of the males of many flies of the families Trypetidae and Drosophilidae. Sometimes, as in the great crested grebe, the display is mutual. Similarly in the stickleback a succession of displays, swimming movements and body quivering by the male evoke complementary responses by the female, and the whole forms a chain of instinctive actions that lead the pair through the various stages from sexual recognition to the entering of the nest and finally to the discharge of eggs and sperm. In the sage grouse display and promenade of a number of males leads to the establishment of a dominant male who will then be accepted as mate by the majority of the attendant females. The male salticid spider approaches the female with elaborate movements of the conspicuously colored legs; this suppresses the attack that is otherwise the immediate response of the female to a potential prey and allows mating to take place in safety. In the scorpions and pseudoscorpions the male first secretes a structure onto which he deposits the seminal fluid and then, while the female is held by the great claws of the pedipalps, the pair engage in a to-and-fro promenade that brings the female above the sperm mass in position for taking it into the genital opening.

The characteristics of these movements are often sharply distinct in closely allied species, and commonly these differences are the principal means by which mating between the members of a species is ensured. In other instances this important result is achieved more indirectly as will be seen below. The movements themselves are usually derived from those already in use in other situations, but they have become ritualized, that is stabilized and perhaps elaborated, to serve as the signals or stimuli for these various functions.

Perhaps the most widely-known examples of the mating dance, and the ones to which the term itself is most regularly applied, are the aggregations or "swarms" of mayflies, midges and other insects in sustained and rhythmically repeated flight. These swarms are a familiar sight, especially in the evening, at the margins of lakes, along roads, above outstanding bushes, etc. They consist principally of males, in to-and-fro or up-and-down flight at the chosen place, and females that fly into them are quickly captured and the mating pair drift to the ground.

The nature of these swarms has sometimes been misinterpreted. It is often believed that they are formed and held together by the mutual responses of the individuals of a species and are examples of true gregarious behavior; that the dancing flight is stimulatory and prepares the male for copulation; and that the females are attracted from a distance by the sound of the great numbers of flying males. None of these ideas is well founded. It can be shown that the dancing flight is a response by the individual insect to the landmarks already noted; thus a "swarm" may consist of a single individual only and the great size often observed results merely from the large population and the relative scarcity of appropriate landmarks. Related species are often adapted to different landmarks or fly at different heights or at different times, and the swarm, consequently, consists usually of one species only. The females seem to reach the swarm by responding to the same landmark, rather than by any positive attraction. Once within it they are perceived, not indeed as females of the species but only as small flying objects, by the specialized sense organs of the males—the plumose antennae of mosquitoes and midges, which function as auditory organs, or the peculiar large-facetted upper part of the eye of the mayflies and black flies. Thus the site functions as the specific place of assembly, and the males are held there in the rhythmical dance ready to capture an incoming female. Apparently, therefore, the essential feature of the "mating dance" is nothing more than its location; and the single males of horse flies or warble flies hovering at their "waiting-stations" are merely the limiting case found in scarce or pugnacious species. Some observers of the honey bee believe that the meeting of queen and drone on the mating flight likewise depends on an instinctive response to a landmark of a particular kind.

This aerial dance has become a central feature of the adult life of certain insects. Male Empidae (Diptera) capture prey and carry it during the dance, and then transfer it to the female who feeds on it during mating. Some forms enclose the prey in a ball of white silk. In the course of evolution the visual stimulus of the dancing ball has evidently become dominant in certain species, and the process has been ritualized—the males no longer hunt for prey but offer the silken container with an inedible fragment only, or even empty. In the mayflies the adult does not feed, and the life span is at most a few days; the mating flight and the subsequent egg-laying are the main activities. A few kinds are unfitted even to come to rest, the legs being aborted; they emerge directly from the water to take part in one brief flight, and then die.

J. A. DOWNES

References

Downes, J. A., "Assembly and mating in the biting Nematocera," Proceedings Tenth International Congress of Entomology, Montreal, 1956, 2: 425–434, 1958.

Richards, O. W., "Sexual selection and allied problems in the insects," Biological Reviews, 2: 298–364, 1927.

Tinbergen, N., "The Study of Instinct," Oxford, Clarendon Press, 1951.

ibid. "The origin and evolution of courtship and threat display," in Huxley, J. et al., Eds., "Evolution as a process," London, Allen and Unwin, 1954.

ibid. "Animal Behavior," Morristown, N.J., Silver, 1965.

Wendt, H., "Sex Life of Animals," New York, Simon and Schuster, 1965.

MAUPERTUIS, PIERRE DE (1698–1759)

This eighteenth century Frenchman was an unsung scientific genius; he possessed a great knowledge of military science, mathematics, physics, and biology. He was the first man on the continent to expound the ideas of Newton. Maupertuis gained recognition for his Lapland expedition; but his greatest contribution to science lies in the realm of biology, for he was the first person to produce a coherent theory of evolution. He supplied this theory with an adequate genetic base and foresaw mutation. Maupertuis also applied the Principle of Least Action to biology, stating that the internal milieu of the body will tend to remain stable. This principle was formally presented a century later by Claude Bernard. Maupertuis rejected the idea of fixity of species claiming that the species present are but a small part of all those a blind destiny has produced. Through fantastic insight, Maupertuis felt that heredity was attributable to particles derived from both the mother and father. These particles through affinity existed in pairs with either particle having the capacity to dominate. Thus, Maupertuis accurately predicted, one hundred years in advance, the theories of Darwin and Mendel.

DOUGLAS G. MADIGAN

MEIOSIS

The word, meiosis (G. meiōn, less), literally means "to make smaller." It is in this same sense that the term is used in biology to designate the cellular process in sexually reproducing organisms which leads to a reduction in the number of CHROMOSOMES of a cell. Because the sexual process results in production of a fertilized egg or zygote, all somatic (G. sōma, body) cells of the organism derived through MITOTIC division of the zygote contain two chromosome sets or a diploid complement of chromosomes. The chromosomes of one set, called a genome, are of paternal, the other of maternal, origin. For every paternal chromosome in the diploid nucleus there is, in most instances, a corresponding maternal chromosome which is similar in basic form, structure and genetic function. Maternal and paternal chromosomes showing such correspondence are referred to as homologous chromosomes and are, on this basis, considered to be mates or homologues of each other.

In plants meiosis is associated with the production of spore cells (sporogenesis) carrying the reduced or haploid number of chromosomes. The spore cells undergo mitosis to form a haploid body, the gametophyte, from which arises the functional germ cell or gamete. Meiosis in animals, however, plays a more direct role in germ cell formation (gametogenesis), the end products of the meiotic process being either the sperm or egg cell containing half the somatic number of chromosomes.

Meiosis typically consists of two successive nuclear divisions. The first meiotic division involves (1) the pairing of homologous chromosomes to form paired or bivalent chromosome units, (2) the resolution of each homologue of the bivalent or pair into two half chromosomes or sister chromatids, and (3) the separation of homologous chromosomes to opposite poles of the cell. The two resulting daughter cells each contain half the number of chromosomes characteristic of the cell from which they were derived. The second meiotic division involves the longitudinal separation and distribution of the sister chromatids of each chromosome to their respective poles. Because the nucleus of the developing germ cell undergoes two divisions during meiosis, four cells are formed, one or all of which may function as gametes. Meiosis is regarded as being essentially mitotic in character on the grounds that the morphological changes and movements of the chromosomes are basically similar in the two processes. There is, however, only *one* chromosome replication during the two divisions of meiosis, namely, that which is associated with the first division. Since the factors or genes controlling the production of hereditary characters are associated with the chromosomes, the distribution of homologous chromosomes into different daughter nuclei during meiosis results in segregation of the genetic material of the parent organism. The genes segregated out in each of the parent organisms are recombined by union of male and female gametes with the result that genetic variation is accomplished.

First Meiotic Division

Meiosis is conventionally subdivided into a series of stages based on the changes in morphology and behavior of chromosomes during the process. The initial or *prophase* stage of meiosis is made up of five separate stages which are, in order of their sequence, leptonema, zygonema, pachynema, diplonema and diakenesis.

Leptonema (G. *leptos,* **slender,** + *nema,* **thread).** This stage is associated with the initiation of the meiotic process and the beginning of enlargement of the nucleus of the developing germ cell. The chromosomes, which are present in the *diploid number,* appear as long, slender threads called *chromonemata.* Each chromosome thread or *chromonema* is differentiated along its length into a series of bead-like structures, the *chromomeres,* which are relatively constant in size, position and number for a given chromosome of the complement.

Zygonema (G. *zygōsis,* **a joining).** In this stage the homologous chromosomes undergo pairing or *synapsis* along their length to form bivalent chromosome units. Pairing is usually initiated between one or more chromomeres in the two homologues of the bivalent followed by a more intimate chromomere-to-chromomere association which brings the homologous chromosomes in opposition throughout their entire length. The nucleus of the cell at this stage contains paired units corresponding to the haploid number.

Pachynema (G. *pachys,* **thick).** During this stage the chromosomes undergo longitudinal contraction with the result that the individual members of each bivalent appear markedly shorter and thicker.

Diplonema (G. *diplous,* **double).** The homologues of each bivalent, which in the previous pachynema appeared single, are now visibly double, each homologue consisting of two half units or *sister chromatids.* Each chromosome pair, therefore, is made up of four chromatids and, in this condition, is called a *tetrad* chromosome. There is a marked tendency for paired chromosomes to separate from each other during diplonema, however, homologues frequently remain in contact at one or more points along their length. The individual points of contact produce a characteristic X configuration, hence, are called *chiasma* (plural, chiasmata). Each chiasma represents the region where two chromatids of the tetrad are in the process of exchanging corresponding segments with each other. As the result of chiasmata formation two of the chromatids of the tetrad become structurally reorganized in that each contains segments derived from the other. Hence, each

homologue now consists of two chromatids, one in its original form, the other made up of an original and an exchanged component. The cytological manifestation of chiasmata is considered equivalent to genetic crossing over since the genes associated with the exchanged segments become a permanent part of the genetic makeup of the reorganized chromatid.

Diakinesis (G. *dia,* **through, in different directions,** + *kinēsis,* **movement).** In this stage the homologues continue to move apart from each other and their contraction or shortening becomes accentuated. The gradual separation of the homologues results in *terminalization* or displacement of the chiasmata toward the terminal parts of the chromosome pair. Once terminalization is complete, the homologous chromosomes are found associated only at their ends where they are still connected by the unresolved chiasmata.

By the end of the first prophase of meiosis replicated homologous chromosomes have paired, exchanged chromatid segments and initiated their longitudinal separation. The stages in the meiotic process following prophase are largely classified in terms of chromosome movement. These stages are as follows:

Metaphase I. At this stage the paired chromosome units move into the plane of the equator of the cell with their *kinetochores* or *centromeres* oriented toward opposite poles. Each homologue possesses a single kinetochore to which its component sister chromatids are associated. Active repulsion of homologous kinetochores is initiated and progresses to the extent of almost completely separating the homologues.

Anaphase I. This stage is characterized by movement of each homologue of the pair to opposite poles of the cell. The homologous maternal and paternal chromosomes distributed to opposite poles may differ in genetic composition as the result of chiasmata formation between chromatids during the previous prophase. Since whole chromosomes are separated in anaphase the two cells resulting from the first meiotic division each have a haploid number of chromosomes.

Second Meiotic Division

Interkinesis. This stage represents the interphase between the first and the second divisions of meiosis. In certain organisms interkinesis as a stage does not exist, the telophase chromosomes immediately reorienting themselves in the plane of the cell's equator and passing directly into metaphase of the second meiotic division.

Prophase II—Metaphase II. In the second meiotic prophase sister chromatids are generally condensed in the form of rods or threads which are separated from each other except at their point of attachment to a common kinetochore. The second metaphase is initiated by orientation of the chromosomes on the equatorial plane. This is followed by cleavage of the kinetochores and longitudinal separation of sister chromatids.

Anaphase II. In this stage sister chromatids, now called chromosomes, are moved to their respective poles in the cell. Each of the two daughter nuclei have a complete chromosome set (genome) corresponding to the haploid number.

The first meiotic division results in a reduction in chromosome number and is, on this basis, referred to as a *reductional* division. The second meiotic division

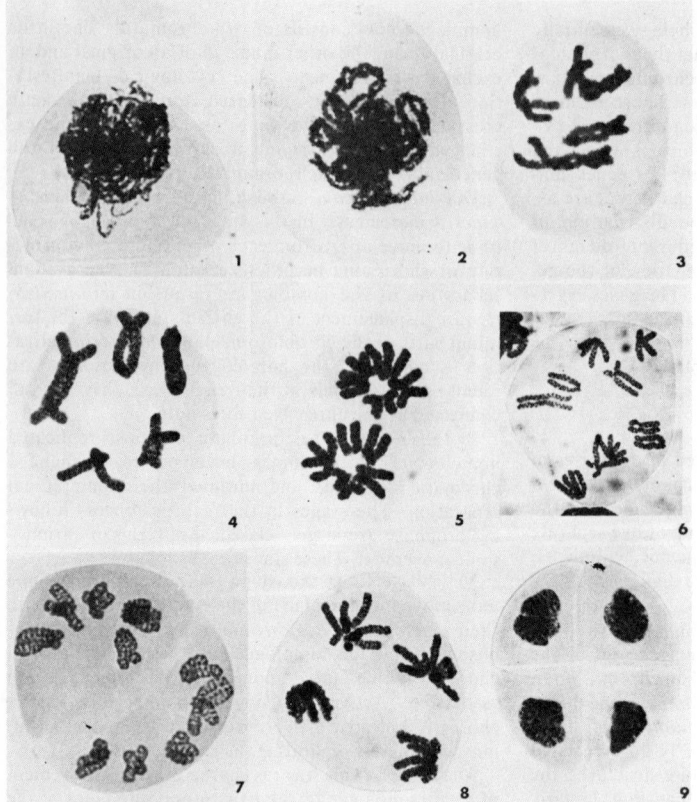

Fig. 1   States of meiosis in *Trillium erectum*. 1, pachytene; 2, diplotene; 3, diakinesis; 4, first metaphase; 5,6, first anaphase; 7, second metaphase; 8, second anaphase; 9, second telophase. (Courtesy A. H. Sparrow, Brookhaven National Laboratory.)

is called an *equational* division since separation of sister chromatids results in equal numbers of chromosomes being distributed to two daughter nuclei. From the genetic point of view, this division would not be equational if exchange of segments has occurred between chromatids during the first meiotic prophase to produce sister chromatids which are genetically different from each other.

JOHN H. MORRISON

References

DeRobertis, E., W. W. Nowinski, and F. A. Saez, "General Cytology," 2nd ed., Philadelphia, Saunders, 1954.

Sharp, L. W., "Fundamentals of Cytology," 1st ed., New York, McGraw-Hill, 1943.

Swanson, C. P., "Cytology and Cytogenetics," Englewood Cliffs, N. J., Prentice-Hall, 1957.

Wilson, G. B., "Cell Division and the Mitotic Cycle," New York, Reinhold, 1966.

MENDEL, GREGOR JOHANN (1822–1884)

It has been said that a man's greatness can be measured by the mount of adverse criticism he arouses during his lifetime. By this yardstick Gregor Johann Mendel would be doomed to obscurity, for he suffered the worse fate of having no significance at all attached to his experiments—experiments which added an entire realm to the world of biology. Rediscovered independently in 1900 by DeVries, Correns and Tschermak, these experiments have laid the foundation for the science of genetics.

Born July 22, 1822, at Heinzendorf, Silesia, in the Austro-Hungarian Empire, the second child of Anton Mendel, a poor peasant and veteran of the Napoleonic Wars, Mendel learned the methods of fruit-culture and gardening, and here he also came to fear poverty and hardship of a peasant's life. Showing marked ability in the village elementary school, he went on to school at Leipnik and in 1834 to the gymnasium at Troppau. Four years later his parents were forced to discontinue his support due to a series of misfortunes, and at the age of 16 he was obliged to tutor other students in order to finance his schooling. After completing his studies at Troppau, he attempted to continue his education at Olmütz Philosophical Institute but because of lack of resources, and illness brought on by worry regarding the future, he returned to his home. Here he recuperated and a year later returned to complete the two year course at Olmütz, supported in part by a portion of the dowry of a younger sister.

The rigors of his early life influenced his decision to embrace the leisure and security of monastic life. He entered the Königskloster at Brünn in 1843, and in

1847 was ordained a priest. He completed his theological studies and accepted a position as substitute teacher of mathematics and Greek. The next year he took the examination for a teaching certificate and failed. In 1851 he was sent by the cloister to the University of Vienna where he studied the sciences. Again he returned to teaching, this time as a supply teacher of physics and natural history at Brunn. Again he attempted to gain his teacher's certificate and failed. He continued as a supply teacher and now, in 1856, began a series of experiments on peas which was to become the subject of a paper published in *Verhandlungen des Naturforschenden Vereins*—"Versuche uber Pflanzen-hybriden" (Experiments in Plant-hybridization, Brünn, 1866).

Mendel's interest in the production and crossing of hybrids resulted from his observation of the regularity of color change following artificial fertilization of one species of flower by another. Three criteria were used in selecting the plant with which to study this phenomenon: The plants had to possess constant differentiating characters, the hybrids had to be capable of protection from foreign pollen, and the hybrids had to be fertile. Twenty-two varieties of pea (*Pisum*) were selected, with seven pairs of easily differentiated traits (length, shape, color, for example).

By following one pair of traits at a time, Mendel demonstrated that the hybrids consistently resembled one of the parents, and that following self-pollination of these hybrids the offspring were in a 3:1 ratio of the trait shown by the hybrid, or dominant trait, to the recessive trait which was similar to that of the second original parent. The recessive always bred true. Two-thirds of the dominant progeny produced mixed offspring in exactly the same ratio of three dominants to one recessive, and the remaining $\frac{1}{3}$ of the dominant progeny bred true.

From the results of these experiments Mendel postulated that each inherited trait was determined by a pair of differentiating factors. These factors (later named "genes" by Johansen) must separate, or segregate, and independently assort in each generation. Presumably their paired nature depends on their biparental origin, but as each individual becomes a parent in turn, it hands on only one member of the pair to its offspring. One factor is dominant over the other in an unlike pair. If the parental characters are designated a and A, the hybrid is Aa and the offspring of the hybrid cross are in a ratio AA:2Aa:aa. This ratio is constant regardless of other pairs of factors. (Each of Mendel's characteristics showed complete dominance and was later found to exist in a different linkage group.)

Although the journal in which Mendel's paper appeared was a relatively unimportant one, it was exchanged with more than 120 institutions. Probably the real reason that no significance was attached to this paper was not owing to its publication in an obscure journal but because the attention of biologists was focused on Darwin and on the evolutionary doctrine maintained in *The Origin of Species* published a short time before in 1859. Although Mendel himself realized that his work had provided an understanding of heredity which Darwin had been unable to explain, to others his work showed the existence of stable characters which seemed to contradict Darwin's basic variability of species. He carried on an active correspondence

with Karl Nägeli, one of the foremost botanists of the day, but even Nägeli attached little importance to Mendel's results, believing this experiment to be an isolated case, and mistrusting the constancy of the original characters.

Mendel's most important work was done with peas but he also used several other varieties of plants. Records of his extensive experiments with bees were unfortunately destroyed. A second major contribution of Mendel was his proof that, contrary to earlier work, only a single pollen grain was necessary for plant fertilization. Mendel's experiments were successful because of careful planning, the study of only one characteristic at a time, and because he was able to generalize mathematically the constant nature of his results. He was familiar with the literature and his own papers were very carefully written.

Gregor Mendel was appointed abbot of the monastery in 1868 and eventually had to give up his experiments in hybridization, although he kept an interest in scientific ideas because he maintained meterological records until his death. He became involved in a struggle with the government over what he believed to be a discriminatory tax law against religious institutions. Widespread racial strife then troubling Austria and chronic nephritis further strained his health and he became sullen and morose. Following his death January 6, 1884, he was mourned not as a great scientist but as a teacher and friend. An obscure priest during his lifetime, Gregor Johann Mendel is today recognized as a great scientist, the "father of genetics."

ROBERT HELLING

Reference

Iltis, Hugo, "Life of Mendel," New York, Hafner, 1966.

MERISTEM

Meristems are localized areas of cells in the plant body which have the potential for producing new cells, tissues, or organs in characteristically morphogenetical ways. Thus, meristems are regarded as such when they are producing new cells during periods of growth, and also when they are in periods of dormancy. Meristematic cells cannot be cytologically characterized, since they vary from approximately isodiametric, thin-walled, densely-cytoplasmic ones, to highly vacuolate, greatly elongated, thick-walled ones. The walls may be symmetrically or asymmetrically thickened. With certain exceptions, meristematic cells lack intercellular air spaces. Also, with some exceptions, they appear to have a generally higher rate of metabolism than mature living plant tissues. Although the prime function of meristems is the formation of new cells, secondarily they appear to function in the determination of the differentiation of the tissues that they have produced; examples are the maturing tissues subjacent to the SHOOT and ROOT apices, the maturing secondary XYLEM external to, and the maturing secondary PHLOEM internal to the vascular CAMBIUM. Conversely, the meristem may itself be determined by the mature parts of the plant body, as in the formation of a floral meristem from a vegetative shoot apex under the influence of leaves of short-day or of long-day plants that have been given the critical photoperiods.

Meristems may be classified in various ways, and all such systems are unsatisfactory because they do not cover all exceptional cases. For an understanding of morphogenesis, a classification of meristems on the basis of mode of development is helpful. Indeterminate meristems are those that may continue activity over a long period of time without maturing, ageing, or ceasing. Growth is potentially unlimited as to extent and time. They constitute a prime example of the open system of growth. The shoot apex is a circular meristematic region which produces leaves and primary stem, and of course, maintains itself. Its diameter may vary from several millimeters in the cycads (see CYCADALES), a primitive group (Foster, 1943) and in certain Cactaceae (see OPUNTIALES), a specialized group (Boke, 1941) down to 50μ–300μ in most ANGIOSPERMS (Popham, 1951). Its form may be a dome, a flat expanse, or a concavity. Functional, though not necessarily homologous, shoot apices are found in certain ALGAE, all liverworts, (see HEPATICAE), mosses (see MUSCI), vascular plants. Those of the less advanced plants are composed of cells not in discrete layers, while those of more advanced ones may be layered. Certain primitive groups have shoot apices with a definite apical cell. The root apex is a circular meristematic region that produces in the forward direction the root cap, and subjacently, the primary tissues of the root (Guttenberg, 1940, 1941). Root apices usually have a layered structure. The oustanding advance in the study of structure of both shoot and root apex is the recognition of various zones which may have certain cytological characteristics, e.g., the central zone, which is low in DNA, and has been described as having low rates of cell division. To other described zones have been attributed the formation of various parts of the shoot or root, and to those which may be peripherally or basally located, the majority of cell divisions.

The lateral meristems are those located at a certain distance beneath the surface along the organs such as stems and roots. Vascular cambium, a prime example of this category of meristem, arises in GYMNOSPERMS and DICOTYLEDONS in the residual parenchyma between the differentiated primary xylem and primary phloem of the primary vascular tissue, and in the parenchyma of LEAF gaps between the bundles. It constitutes a tube surrounding the cylinder of wood, and consists of long, slender, pointed, fusiform initials, and almost isodiametric ray initials. Theoretically, this meristem consists of a single layer of initials of both types, although many authors have thought it expedient to consider a cambial zone as proper terminology, since derivative cells may divide far more frequently than do the cells of the cambium proper. Cambial cells show various patterns of thick walls, are among the most vacuolate of all living plant cells, and may show the lowest cyto-nuclear ratio yet measured. Cell divisions are preponderantly radial, producing derivatives which mature into wood or phloem, but a minor number are tangential. These latter produce cells which add to the tangential dimension of the cambial tube, and permit the meristem to keep pace with the expanding woody cylinder. Such divisions are vertically oriented in stratified cambia (e.g., Robinia) and oblique in non-stratified ones (e.g., Pinus), where the ends glide past each other during their elongation. During the increase of girth in the cambium, new ray initials may be derived from certain fusiform initials (Bailey, 1954). Thus, while new initials of both types are continually being formed, it should also be emphasized that some of them are being displaced and cease to function.

Cork cambium arises by a regenerative process usually in an outer layer of a plant axis or organ. The locale may be epidermis, any layer of cortex, pericycle (in roots), secondary phloem, and in certain dicots, in the secondary xylem. When it arises in the outer parts of an axis, it often occurs first in connection with the origin of lenticels and spreads by subsequent periclinal divisions into surrounding parenchyma. In one instance it is known to arise in cortex beneath living hairs (Lier, 1955). It finally achieves a tube-like form and surrounds the axis. Cells of cork cambium are generally thin-walled, have a narrow radial dimension, are only slightly elongated in the direction of the axis bearing them, and are without intercellular spaces except under the lenticels. After the originating division, one derivative cell becomes the phellogen (cork cambium); if the inner one, the outer one becomes cork: if the outer one, the inner one becomes phelloderm (a kind of parenchyma that functions in a cortex-like fashion). The first complete tube of phellogen may persist for the life of the organ in certain kinds of woody dicots. In others, it is replaced by the regeneration of new phellogens beneath it. The latter usually extend only partially around the axis, and later unite laterally with other similar new phellogens. In both categories of plants the formation of new phellogens may be hastened or even initiated by injury to or removal of the outer phellogen. Certain categories of injury (e.g., fungal infection) have been shown to stimulate the phellogen to produce a rough cork layer.

Determinate meristems go through a definite cycle of activity culminating in the production of a mature organ in which growth process the meristem is all, or mostly, used up. A prime example is the floral apex. It may be derived by transformation of a vegetative shoot apex, or from specific flower buds in an inflorescence. Much recent work indicates that the structure of the floral apex may be little or no different from that of the vegetative shoot apex (Boke, 1947). After production of the flower parts, the remaining apical meristem normally grows no further; production of tissues and organs is normally never resumed. Closely related to the floral meristem is the inflorescence meristem, as seen in Compositae and certain other families. Here a great distinction obtains, compared to the vegetative shoot apex, in that there is a thin meristematic mantle surrounding an extensive parenchymatous core (Popham and Chan, 1952). From the former the specific floral meristems are produced.

A second category of determinate meristem is the intercalary one. This meristematic region occurs at the bases of internodes and of young leaves, especially in the grasses, the horsetails, and Ephedra. These meristematic cells are usually traversed by protoxylem (Stafford, 1948), or by the lacunae left when the latter are stretched and destroyed (Buchholz, 1920), or xylem may be absent and water transport apparently occurs through living parenchyma cells (Sharman, 1942). After varying numbers of cells have been added to the base of the leaf or stem by the activity of such a meristem, it matures along with xylem that connects it with subjacent tissues.

In a less specific sense, all maturing plant organs and tissues are determinate meristems in that they pass through early stages of frequent cell divisions, intermediate stages of less frequent ones, and final stages of few or none. The final stages are only comparably so, for natural or artificial cell-division stimuli have been found for a wide variety of mature, living tissues. Among the natural stimuli to cell division may be listed the gall- and tumor-producing plant and animal parasites. In some instances, such a stimulus will cause the production of a new determinate meristem that yields a highly differentiated structure (e.g., an insect-gall). In others, a stimulus may so change the metabolism of the tissues of the host that a continuously growing tumor, which is, in a real sense, an indeterminate meristem (e.g., crown-gall), arises. Artificial cell-division stimuli have been widely utilized in the procedures of PLANT TISSUE CULTURES. Many kinds of mature plant tissues have been caused to proliferate indefinitely as to time and extent in sterile culture (Gautheret, 1959).

ERNEST BALL

References

Bailey, I. W., "Contributions to plant anatomy," Ronald, New York, 1954.
Boke, N. H., Amer. Jour. Bot. **28**: 656–664, 1954.
ibid. **34**: 433–439, 1947.
Buchholz, M., Flora **114**: 119–186, 1920.
Foster, A. S., Amer. Jour. Bot. **30**: 56–73, 1943.
Gautheret, R. J., "La culture des tissus végétaux," Paris, Masson, 1959.
Guttenberg, H. von, *in* Linsbauer, K., "Handbuch der Pflanzenanatomie," Bd. 8, Lf. 39, 1940 and Lf. 41, 1941.
Lier, F. G., Amer. Jour. Bot. **42**: 929–936, 1955.
Popham, R. A., Ohio Jour. Sci. **51**: 249–270, 1951.
Popham, R. A. and A. P. Chan, Amer. Jour. Bot. **39**: 329–339, 1952.
Sharman, B. C., Ann. Bot. N. S. **6**: 245–282, 1942.
Stafford, H. A., Amer. Jour. Bot. **35**: 706–715, 1948.

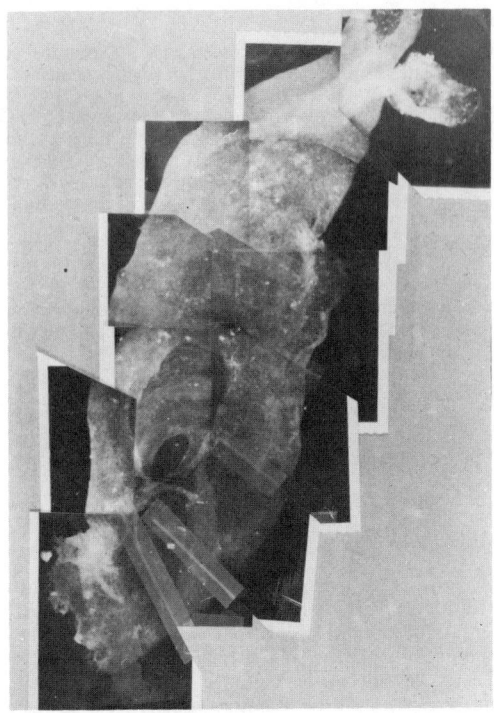

Fig. 1   Isolated mesoglea of *Hydra viridis* (500 ×) illustrates the retention of the form of the animal by the mesoglea in the absence of cells. Preparation stained by fluorescine labelled anti-*Hydra viridis* serum prepared in rabbits. Differences in the intensity of staining indicate different concentrations of the antigens along the length of the mesoglea. The hole in the mesoglea occurs at the budding region. The tentacles can be seen at the top.

MESOGLEA

Mesoglea is the connective tissue of sponges and coelenterates. It is found between the cell layers composing the remainder of the body walls of these organisms. It is cellular except in the hydrozoans. In the scyphozoan jellyfish it is a highly hydrated loose connective tissue and in the anthozoan mesentaries it can be a tough fibrous connective tissue. Intermediate grades of connective tissue are also found. Skeletal elements are often produced or found in the mesoglea of sponges, and anthozoans, but are rare among the hydrozoans, and absent from the scyphozoans. Hard spicules impregnate the axes of some corals.

Mesoglea contains proteins resembling vertebrate collagen by way of x-ray diffraction pattern (Chapman, 1966), and the presence of hydroxyproline, hydroxylysine, and unusually large amounts of proline, and glycine (Piez and Gross, 1959). An acid-soluble protein extracted from the mesoglea of the sea anemone (actinocol) can be reprecipitated in the segment-long-spacing (SLS) configuration. Moreover, the dimensions and banding pattern of this protein, as discerned in electronmicrographs, corresponds exactly to that of calf-skin collagen reprecipitated in this configuration (Nordwig and Hayduk, 1967). Mesogleal collagen differs from typical vertebrate collagen, however, in being complexed with considerably more carbohydrate, and often containing appreciable amounts of sulfur-containing and aromatic amino acids. The presence of elastin-like proteins in the mesoglea of medusae has also been reported (Bouillon and Vandermeerssche, 1956).

Mesoglea may function in several ways. In sponges it may act as an intercellular cementing substance (Humphreys, 1963). In anthozoans mesoglea seems to be the primary supporting structure, and in hydro- and scyphomedusae aids in flotation since it is less dense than sea water. The fibers lacing through the mesoglea of these medusae seem to function in support since they are oriented along lines of stress within the umbrella. In the freshwater polyp, *Hydra*, the mesoglea may provide a substratum for moving cells (Shostak, Patel, and Burnett, 1965).

STANLEY SHOSTAK

References

Bouillon, J. and G. Vandermeerssche, "Structure et nature de la mesoglee de hydro—et scyphomeduses," Ann. Soc. Zool. Belg. **87**: 9–25, 1956.

Chapman, G., "The structure and functions of the mesoglea," in Rees, W. J., "The Cnidaria and their Evolution," New York, Academic Press, 1966.

Humphreys, T., "Chemical dissolution and *in vitro* reconstruction of sponge cell adhesions," Devel. Biol. **8**: 27–47, 1963.

Nordwig, A. and U. Hayduk, "A contribution to the evolution of collagen," J. Mol. Biol. **26**, 351–352, 1967.

Piez, K. A. and J. Gross, "The amino acid composition and morphology of some invertebrate and vertebrate collagens," Biochem. Biophys. Acta **34**: 24–39, 1959.

Shostak, S., *et al.*, "The role of mesoglea in mass cell movement in *Hydra*," Devel. Biol. **12**: 434–450, 1965.

MESOZOA

Mesozoa are a group of small worm-like organisms parasitic in various marine invertebrates. The body consists of a single layer of ciliated cells, more or less constant in number and arrangement in any given species, enveloping the reproductive cells. Authorities differ respecting their affinities, usually treating them as a small phylum of doubtful position, or as a class of, or appendix to, the flatworms. There are two orders, the **Orthonectida** and the **Dicyemida.**

The orthonectids occur as multinucleate amoeboid syncytia in the tissues and body spaces of various invertebrates, often damaging their hosts. The syncytia eventually give rise, from agametes, to free swimming sexual adults (Fig. 1), the female larger than the male. The fertilized egg cells develop in the female into ciliated larvae which infect new host individuals where they disaggregate, their germinal cells giving rise to new syncytia.

The dicyemids occur in the renal organs of CEPHALOPODS. The reproductive cells and developing larvae are contained in a single long axial cell which is enveloped by the ciliated somatic cells (Fig. 2). There are two reproductive phases. During the *nematogen* phase, vermiform larvae, similar to their parents, are formed from agametes (axoblasts). These escape from the parent organism, attach themselves to the renal organ of their host, thus increasing the number within an infected host. This is succeeded by the *rhombogen* phase during which a few agametes give rise to *infusorigens* (very reduced hermaphroditic individuals) which stay in the axial cell of the rhombogen, producing egg and sperm cells. The fertilized eggs develop into *infusoriforms,* which, when fully developed escape from the cephalopod. Their fate is unknown.

In some species the first nematogens found in very young cephalopods have three (or two) axial cells in linear series. These are termed stem nematogens. They give rise to and are replaced by ordinary nematogens as the population increases.

The mesozoa are of phylogenetic interest because of the simplicity of their structure. If primitive, they would indicate that the first metazoans had a solid

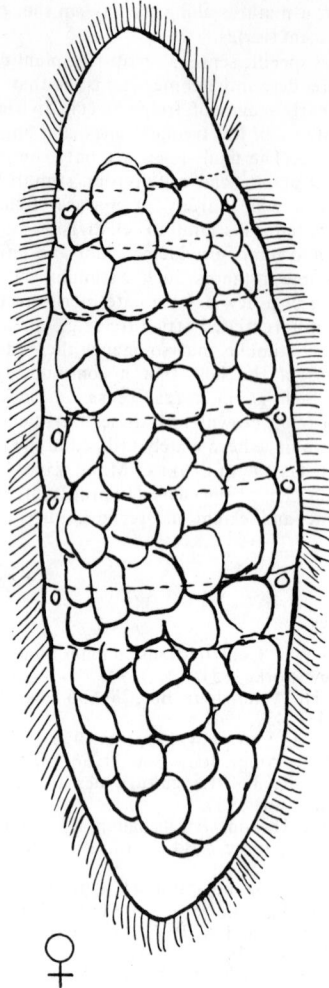

♀

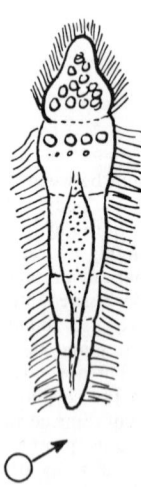

♂

Fig. 1　*Rhopalura granosa* Atkins.

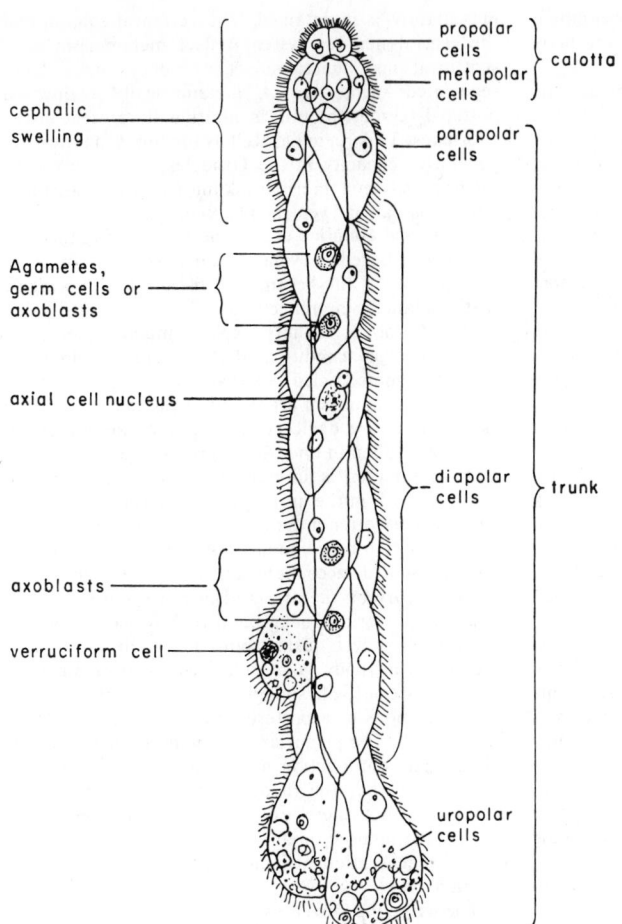

propolar
cells

metapolar
cells

} calotta

cephalic
swelling

parapolar
cells

Agametes,
germ cells or
axoblasts

axial cell nucleus

diapolar
cells

} trunk

axoblasts

verruciform cell

uropolar
cells

Fig. 2

rather than a hollow construction, the first step being the relegation of reproductive cells to the interior.

B. H. McConnaughey

References

Hyman, Libbie H., "The Invertebrates," Vol. I, New York, McGraw-Hill, 1940.
Stunkard, Horace W., "The life history and systematic relations of the Mesozoa," Quart. Rev. Biol. 29: 230–244, 1954.

METAMERIC SEGMENTATION

Among the various phya of animals, the ANNELIDA ARTHROPODA and CHORDATA alone show a serial division of their bodies into a number of units—the so-called metameres which show considerable resemblance to one another. The protozoans, sponges, coelenterates, molluscs, echinoderms and some others manifest no such demarcation of their bodies into metameres.

The three phyla of metamerically segmented animals constitute a large section of the animal population of the world, amounting to about 525,000 species, while the rest of the animal kingdom includes only 100,000 species. Even in terms of protoplasmic mass their share in the animal world is impressively high. To what extent their success in a competitive existence is due to their bodies being metamerically segmented is difficult to assess, though there seems some reason to suppose that their peculiar organisation might have been a contributory factor in their differential survival.

Metamerism in annelids. There is, as should be expected, considerable variation in the extent as well as in the intensity of metamerism as it occurs in the three phyla of animals. The generalised type of metamerism is met with in the annelids. In an annelid like the earthworm the individual segments behind the head, are marked out from one another by external grooves as well as by internal partitions which are pierced by the alimentary canal and the nerve cord. Though the longitudinal muscles of a worm form a continuous series, the fibers of a segment terminate at the intersegmental junctions. The circular muscles which run outside the longitudinal muscles are also grouped segment-wise, those of one segment not encroaching on another segment. The contraction of the musculature is consequently successive, beginning from one segment followed by that of the next and then of the third and so on. Thus the body wall consisting of the outer cuticle and the circular and longitudinal muscles participate in the serial division of the body into metameres.

Among the internal organs the nephridia are arranged serially, a pair or more in each segment. The blood-vascular system is metameric insofar as in each segment a set of afferent and efferent vessels conduct or take away blood either to or from the main longitudinal vessels. The gonads, especially the male element, show serial repetition in many cases. The nervous system consisting of the brain and the ventral ganglionated cord shows a ganglionic enlargement in each segment from which nerves supplying the respective segments originate. The innervation to that extent is independent in each segment—which in terms of nervous function means a localization of a reflex or relay center in the segmental ganglion.

The only organs of the body which do not take part in the metameric scheme are those derived from the endoderm, though some would regard the alimentary pouches also so disposed. Thus within the frame-work of the integrated body, each metamere, constituted and functioning as a unit of the body corporate, possesses a limited individuality characterised by separate musculature, nephridia, afferent and efferent blood vessels, a ganglion and, in some forms, gonads.

Among the POLYCHAETE worms a feature closely associated with metamerism is their asexual reproduction. In forms like Nereis, any one of the segments situated towards the middle region of a mature individual, in favorable circumstances, might be transformed into a head segment which in due course separates from the parent worm together with a number of succeeding segments and swims off as a daughter individual. The break up thus of a free-swimming Nereis into two or sometimes into more daughter individuals illustrates a very remarkable feature associated with metamerism.

Metamerism in arthropods. In the arthropods the external cuticle is stiff and hard through chitinisation. As a result the extension of the body segments, which is possible in the annelids, occurs only to a limited extent in the arthropods, though the intersegmental region of the cuticle remains less chitinised. An arthropod has a definite number of segments characteristic of its order. On the whole an arthropod has fewer segments than an annelid. The intersegmental septa are entirely absent in the arthropods. The circular muscles are degenerate in the arthropods and consequently the compression of the body segments is possible only to a limited extent. In annelids like Nereis and the earthworm the tissue at the region of a transection, when one is made, becomes embryonic. This embryonic tissue cannot only heal the cut, but also regenerate a segment. In this way regeneration of a few segments is possible. This faculty of adult tissue to revert to embryonic condition is confined to the segments of the appendages in the arthropods. On ultimate analysis the fundamental difference between metamerism in annelids on the one hand, and arthropods and chordates on the other, is the determinism in the number of segments in the latter two phyla.

Metamerism in chordates. In the adult chordate metamerism occurs somewhat masked, though more pronounced in the embryo which possesses a segmented nervous system and myotomes. In Amphioxus and fishes alone the primitive metamerism of the axial musculature is maintained. The vertebral column and the central nervous system reflect metamerism both structural and functional. The kidneys arise from segmented primordia. A metamerically segmented postanal region, the tail, is additionally present in the chordates. This region as well as the limbs manifest regenerative capacity in the Urodeles. In the tetrapods the primarily metameric trunk muscles are divided into dorsal epaxial, and ventral hypaxial, series.

Origin and evolution of metamerism. Several theories have been offered to explain the origin of metamerism. Protagonists of the "corm," or fission, theory suggest that metamerism reflects an incompletely divided chain of zooids. Others explain metamerism as a further evolved condition of the pseudo-metamerism present in some triclad PLANARIANS. Still others maintain that metamerism arose first in the muscular tissue as a locomotory device and other mesodermal structures have fallen in line. A last group, however, uphold that metamerism is an extension of the septa-forming trait met with in the anthozoan Cerianthus. Since evolution is very often a multiphased phenomenon, a simple explanation based on any one of the theories would not satisfactorily account for the origin of metamerism.

Any theory on the origin of metamerism cannot ignore the planarian background. It has been shown by susceptibility tests using poisonous substances that the planarian body is so polarized as to manifest an antero-posterior physiological gradient field. There are indications of a postero-anterior reverse polarization as well. The planarians also manifest a linear type of asexual reproduction and remarkable regenerative ability. Whether the theory of pseudometamerism is accepted or not, there is no doubt that metamerism has been superimposed on an already polarised body capable of regeneration and linear type of asexual reproduction.

The wriggling movements of aquatic dipterous larvae involving simultaneous contractions of segments from the anterior end backwards and from the posterior end forwards, are indicative of a locomotory polarization corresponding to the two physiological gradient fields present in planarians and annelids.

Incidentally it may be mentioned in this connection that a school of zoologists hold the view that the progressive formation of proglottids in a CESTODE is a form of metamerism. Since proglottids are reproductive "segments," their formation is a projection of the potentiality for asexual reproduction. It is doubtful if a proglottid possesses the morphological and physiological features of a metamere.

There are two opposing views on the origin of metamerism in the chordates. Some are of the view that metamerism in the chordates has arisen independently of that in the annelids and arthropods and, in support of this, point out that the dorsal position and hollow nature of the chordate central nervous system is in contrast to the ventral solid cord of annelids and arthropods. Recent evidence on the nature and arrangement of the segmental nerves in some annelids and arthropods indicates similarity between these groups and chordates. In annelids and arthropods the dorsal nerves are composed exclusively of motor fibers, while the ventral ones contain both motor and sensory fibers. In the chordates the dorsal nerves are sensory and the ventral ones motor in the trunk region, whereas in the head the ventral nerves are exclusively motor and the dorsal ones include both motor and sensory fibers. If this reverse parallelism between annelids and arthropods on the one hand, and the chordates on the other, is not a strange conincidence, the annelid-arthropod and chordate stems are genetically related. The chordate ventral side in that case is equivalent to the dorsal side of the annelid-arthropod and *vice versa*.

In summing up it should be emphasized that the morphological pattern known as metamerism contains a corresponding compartmentalized physiological set-up, primarily perhaps intended for locomotive purposes. Metamerically segmented animals carry also a heritage probably from planarian ancestry of an antero-posterior as well as a postero-anterior polarization, and a mechanism for linear asexual reproduction and regeneration. This heritage is fully illustrated in the annelids, but only to a limited extent in the arthropods and chordates.

C. J. GEORGE

References

Child, C. M., "Problems and patterns of development," Chicago, Univ. Chicago Press, 1942.
George, C. J. and P. T. Muthe, "Is the occurrence of dorsal and ventral nerves arising from the segmental ganglia in annelids and arthropods a feature associated with metamerism?" J. Anim. Mophol. Physiol. India (1955).
Lang, A., "Beitrage zu einer Trophocoltheorie," Jena Ztschr., Naturwiss. 38: 1904.
Marcus, E., "On the evolution of animal phyla," Quart. Rev. of Biol. 1958.

METAMORPHOSIS

The term metamorphosis is generally used in biology to denote a normal process in the life history of an organism which results in radical alteration of its structure within a relatively short period of time. Gorbman (1959) interprets metamorphosis as "a rapid differentiation of adult characters after a relatively prolonged period of slow or arrested differentiation in a larva" while Etkin (1956) describes the process as "a definitely delimited period in postembryonic development during which marked developmental changes in non-reproductive structures occur." However, the term has frequently been applied to local changes at the tissue level (i.e., ovulational metamorphosis) and in recent years to factors related to the alteration of physiological, behavioral and biochemical patterns as well. Because of its wide application and because of the existence of a confusing array of transitional types, a precise definition as to what constitutes an actual state of metamorphosis seems unwarranted. Metamorphosis is reasonably restricted to patterns of change that are intrinsic in nature and thus moderated by genetic and physiological factors, rather than environmental or pathological ones.

Invertebrates. The most striking examples of metamorphosis among the invertebrates occur in the certain insects which undergo complete or holometabolous metamorphosis. In these groups the egg develops into a larva (maggot, caterpillar, grub, etc.) which grows successively through a regular number of larval molts

with little change of form. A new larval cuticle is secreted by epidermal cells as the old cuticle is shed. In the giant silkworm *Hyalophora cecropia*, four larval molts are followed by a fifth molt which produces the pupa or chrysalis characterized by a heavy, dark cuticle. Most of the larval tissues are destroyed at this time and replaced by new tissues. A scaly cuticle is secreted at the sixth or terminal molt, resulting in the formation of the imago (adult). This process is characteristic of the Diptera, Hymenoptera, Coleoptera, and Lepidoptera but may vary in detail.

Gradual or hemimetabolous metamorphosis is characteristic of such groups as the Orthoptera, Isoptera and Hemiptera. Changes during the life cycle are more moderate than in holometabolous forms. No pupal stage occurs and the larva (nymph) resembles the adult except for its smaller size and the absence of wings. In the case of the grasshopper or the blood-sucking bug *Rhodnius prolixis*, the larvae pass through five nymphal stages, each one followed by a molt. The winged adult emerges following the fifth and terminal molt.

Metamorphosis in insects is controlled by hormones. In the hemimetabolous *Rhodnius*, neurosecretory centers in the brain produce a substance which induces the prothoracic gland of the thorax to produce a growth and differentiating hormone (GDH) which conditions molting and differentiation to the adult stage. Though GDH is present, it fails to elicit final metamorphosis during the first four nymphal molts because of the presence of another hormone secreted by a pair of small glands, the corpora allata, located in the posterior head region. This substance has been called the "juvenile hormone" because it prevents the differentiation of adult characters. While both juvenile hormone and GDH are essential for proper molting during the nymphal stages, the greatly decreased activity of the corpora allata during the terminal molt allows the prothoracic gland secretion to produce its full effect. The factors controlling holometabolous metamorphosis are similar to those described for *Rhodnius*. In *Hyalophora*, decrease in the production of juvenile hormone during the last larval stage results in the formation of pupal cuticle. The pupae are maintained in a state of reduced activity or diapause until reactivation of the brain-prothoracic gland mechanism releases GDH, which initiates the final steps in the metamorphic process.

Many groups of echinoderms display a pronounced metamorphosis which involves an alteration in their basic plan of symmetry. For example, the common sea star *Asterias* produces a bilaterally symmetrical, free-swimming larva (Bipinnaria). This form persists for several weeks until the larva attaches itself to some object such as a rock by a temporary sucker which develops anteriorly. As adult organs undergo progressive and rapid differentiation in the posterior areas, the sucker diminishes in size. At the end of this process the young, radially symmetrical sea star frees itself and crawls away.

A tadpole-like larval form possessing the common chordate characteristics is produced by many tunicates (subphylum Urochordata). At first the larva swims downward away from the light, but later displays positive phototaxis and swims toward the surface where it attaches itself to a substratum below the low tide level. Here it undergoes rapid transformation into the adult form, losing all major chordate features except the pharyngeal slits. As an adult it bears little resemblance to its free-swimming larva. In addition to the few examples described above, metamorphic phenomena are found in many other invertebrate groups such as the Crustacea, Mollusca, and Cnidaria.

Vertebrates. Among the vertebrates, distinct forms of metamorphosis are found in the Cyclostomata, Amphibia, and a few Osteichthyes.

The embryos of frogs develop in jelly membranes, and at the time of emergence possess external gills and U-shaped suckers for attachment to water plants, etc. During larval development the tail broadens, the gills expand, the cornea of the eye becomes transparent, and horny teeth are produced. By the time the larva is feeding with its newly acquired teeth, the suckers are being absorbed and a gill chamber is being formed by the extension of an opercular fold of skin over the external gills. The resultant tadpole of most frogs and toads continues to grow in size for periods of weeks or months, depending on the species. At the end of this time, feeding ceases and primary metamorphosis ensues.

In the woodfrog *Rana sylvatica,* for example, some of the major changes associated with the metamorphic process are: resorption of the tail, emergence of limbs, loss of horny teeth and widening of the mouth, loss of gills and the development of functional lungs, growth of the eyes, and over-all changes in bodily proportions. The time necessary for complete transformation varies with the species; the European frog, *Rana temporiana,* four months; bullfrogs, *Rana catesbiana,* from one to over two years; many toads from four to six weeks.

As early as 1912, Gundernatsch produced precocious metamorphosis in tadpoles by feeding them with thyroid gland material. Extension of this work by numerous investigators has demonstrated that the hormone thyroxin produced by the thyroid gland is the effective factor in the initiation of the metamorphic process. Metamorphosis fails to occur following removal of the gland during embryonic or larval development, and the giant tadpoles which result usually die. Metamorphosis follows the implantation of thyroid primordium into thyroidectomized tadpoles if the transplant undergoes differentiation.

Iodine is essential to the synthesis of thyroxin, and treatment with this substance alone can induce precocious metamorphosis in tadpoles. Conversely, metamorphosis will fail to occur in the absence of sufficient iodine. The hypophysis (pituitary gland) secretes thyrotropic hormone, TSH, which conditions the proper development and functioning of the thyroid, thus bringing metamorphosis under the control of a pituitary-thyroid axis. Hypophysectomized tadpoles fail to undergo metamorphosis because of the absence of TSH, while metamorphosis can be induced in individuals from which both thyroid and pituitary have been removed by thyroxin alone. Thus thyroxin is the immediate causal agent of metamorphosis, so that in normal frogs the activity of the thyroid gland is closely correlated with the metamorphic process. Unknown factors stimulate the pituitary-thyroid mechanism just prior to metamorphosis. This activity increases and reaches a peak during the final stages of the process, after which the thyroid returns to its relatively inactive, premetamorphic condition.

The pattern of metamorphosis varies considerably among the amphibians. The larvae of urodeles (newts, salamanders, etc.) posses a well-developed tail, four limbs and general body proportions that characterize the adult. Consequently, metamorphic changes are less pronounced than in the Anura, the major effect being the loss of the external gills. In certain tropical toads metamorphosis is greatly reduced as development proceeds rapidly from the egg to the young individual without the appearance of a free-swimming, tadpole stage. Arrested metamorphosis occurs in some salamanders. The Mexican Axolotl (a form of *Amblystoma tigrinum*) and the mud puppy *Necturus maculosus* are examples. In these neotonous forms the animals grow and become sexually mature, although some larval features such as the external gills are retained.

Second Metamorphosis. A second metamorphosis occurs in some organisms which undergo radical change during their adult life. Some of these changes may be essentially the reverse of those which took place during the original or primary metamorphosis. These changes may be slight and transitory, as in the case of adult amphibians which migrate to breeding ponds for a few days every year. In other animals, however, the changes may be profound and of a permanent nature, and it is here that events associated with second metamorphosis are most clearly defined. The newt *Diemictylus viridescens* has an aquatic larva with external gills, flattened tail and green pigmentation. During primary metamorphosis it transforms into the orange, terrestrial *eft* stage which has a rough skin and inhabits wooded areas. After a period of growth which may last for several years, the eft undergoes a second metamorphosis to the aquatic adult phase which develops the smooth skin, keeled tail and pigmentation of the larval form. The external gills, however, are not reconstituted. Physiological and biochemical change are also features of second metamorphosis. The visual pigment rhodopsin, characteristic of land animals and marine fish, is replaced in the aquatic adult by the porphyropsin of fresh-water fish and amphibian larvae. The terrestrial eft secretes nitrogen primarily as urea. Following second metamorphosis, the percentage of nitrogen excreted as ammonia rises, showing a significant shift towards the larval condition. The thyroid activity of the adult becomes greatly reduced as compared to that of the eft. It has been shown that the series of events associated with second metamorphosis in *Diemictylus* are triggered by a prolactin-like substance originating in the anterior pituitary.

The sea lamprey *Petromyzon marinus* and the eel *Anguilla* provide additional examples of second metamorphosis. The lamprey metamorphoses from a small, blind ammocoete larva living in the sand of fresh-water streams to a free-swimming form which migrates downstream to the sea. Following this phase which lasts several years, second metamorphosis ensues, and the now mature adult returns to fresh-water streams to spawn. The visual system of lampreys migrating towards the sea contains rhodopsin, while that of animals migrating towards the streams contains porphyropsin. Following hatching, the leptocephalus larva of Anguilla leaves the Sargasso Sea and migrates to the shores of Europe and North America, where it undergoes primary metamorphosis into the typical fresh-water form. European eels pass through a distinct second metamorphosis before returning to the spawning grounds in the Sargasso Sea. Changes occur in the pigment and digestive tract, and the eye doubles in diameter in preparation of its life as a deep-sea creature. The fresh-water porphyropsin is replaced at this time by a special form of rhodopsin.

The patterns of development and differentiation associated with metamorphosis have evolved by the process of natural selection. The fact that many larval forms, such as the caterpillar and tadpole, are adapted to modes of life which differ radically from those of their adults suggests the operation of selection pressures which have reduced intraspecific competition for food and living space between juveniles and adults of the same species. In addition, where distinct form phases occur in the life cycle of a species, neotonous evolution is possible. Arrested metamorphosis in sexually mature individuals or the retention of larval characteristics by adults following metamorphosis may produce evolutionary changes of considerable magnitude within relatively short periods of time.

WILLIAM C. GRANT JR.

References

Etkin, W., "Metamorphosis" in Willier, B. H. et al., Eds., "Analysis of Development," Philadelphia, Saunders, 1955.
Wald, G., "The Significance of Vertebrate Metamorphosis," Science **128**: 1481–1490, 1958.
Gorbman, A., et al., "Comparative Endocrinology," New York, Wiley, 1959.

METCHNIKOFF, ILIA ILICH (1845–1916)

This Russian biologist, best known for his observations on phagocytosis, a word and a theory which he himself invented, was originally trained as an invertebrate zoologist. He was born in Kharkov and went to the University of Tiessen, where he studied zoology under Siebold and Leuckart. His deep interest in invertebrates took him first to Naples and then later to Odessa where Kovalevsky was at that time presiding over his famous school of invertebrate embryology. Under Kovalevsky's guidance, Metchnikoff went to Messina then, as now, a rich collecting ground for invertebrate larval forms, and fell under the influence of C. Claus, with whom he subsequently studied in Vienna. His major works at this time were on echinoderm larvae, the transparent nature of which permitted him to make those observations that led him to postulate phagocytosis as a function of major importance in all animal life. In 1886 he was called to Odessa for the purpose of founding there a School of Bacteriology but political difficulties intervened and in 1888 he joined the staff of the Pasteur Institute in Paris. Here he continued his researches on immunology and made the first definitive experiments on the transmission of syphilis between humans. In 1908 he shared, with Paul Ehrlich, the Nobel prize for his studies in immunology. His death in 1916 in Paris deprived the world not only of a fine research worker, but also of a colorful figure in the scientific life of his period.

LA METTRIE, JULIEN OFFROY DE (1709–1751)

This French physician-metaphysician is significant to the study of the history of biology because of his mechanistic philosophy of man. Le Mettrie gained more notoriety as a translator of Boerhaave and a polemicist than as a practicing physician, although his *Practical Medicine* was a fairly comprehensive medical tract. For his medical satire *Penelope*, La Mettrie was banished from the medical profession as well as his native France. While in Holland, La Mettrie wrote his metaphysical *Man a Machine*. In this work, he described man's existence in terms of seventeenth century Newtonian mechanics, claiming all human processes were mechanical or acquired through the mechanical process of learning. He believed that there was a continuum between man and the lower animals and that life was a process of adaptation. La Mettrie felt, "The human body is a machine which winds its own springs." For this "heretical" work, La Mettrie was forced to move to tolerant Prussia where he served in the court of Frederick the Great.

DOUGLAS G. MADIGAN

Reference

Mettrie, J. O. de la, "Man a Machine," London, Opencourt, 1912.

MICROCLIMATE

The concept of microclimate recognizes the environment of smaller areas in contrast to that of larger ones, or at times, of particular organisms in contrast to that of the generalized environment of communities. Other terms interchangeably used for this concept are *microenvironment* and *bioclimate*.

In America, the term *microclimate* is often restricted to atmospheric conditions above the surface of the ground, while the more comprehensive term, *microenvironment*, is used to include the entire environmental complex below the surface as well as above. In Europe, however, both terms have the same connotation as that in America for microenvironment. The term *bioclimate* simply restricts the concept to living things. The choice of terms is usually dependent upon one's background and the orientation of the problem he is working with. Thus, in its broadest sense, the concept implies that the functional environment of a specific organism or community of organisms is not the macroenvironment, but is that which is composed of the continuously changing complex of all conditions and influences which are in direct and indirect contact with and are interacting with that organism or community. Since these terms are relative, they may be used to designate a large area, as a pine stand many acres in extent, or a very small area, as that occupied by a lichen within that stand. Or it may designate the climate of a city, as being different from the surrounding country-side, or of a garden within the city.

Although well-developed in the fields of meteorology and agriculture in certain regions in Europe for the most of this century, the concept was not developed in America until after World War II. Subsequently, it has had an extraordinary rapid and fruitful development, being influential in the establishment of many other concepts relating to the interrelationships between living organisms and their environment.

Some of the basic considerations and safeguards which consequently have now come into common use, but which were not well-understood or generally accepted a decade or more ago include: (1) the environment of a particular organism at a particular time is not the same as that some distance removed either by millimeters or by miles; (2) data cannot be transposed in time any better than in space; (3) environmental effects observed on an individual cannot be applied without question to the population or the species; (4) data obtained in a greenhouse under unnatural conditions cannot be used for interpretation of the same organism growing in its natural environment; and (5) vertical as well as horizontal gradients are of far greater significance than point measurements of environmental conditions.

Microclimatic studies have now become a part of all biological fields which attempt to relate life to its environment, as well as a part of various physical sciences such as meteorology, engineering and construction. As shown above, microclimatic conditions vary in time through diurnal, seasonal and other cycles, and in space through horizontal and vertical gradients. Climatic conditions within the pore spaces of the soil are different from that at its surface, and the climate within a grass stand is radically different from that a few inches above the stand. Beetle larvae in a rotting log live in a world far apart from that of the adult winging its way through the tree tops. Recognition of these conditions has stimulated and tied together, through their environmental aspects, the interrelated fields of evolution, genetics, physiology, taxonomy and ecology.

Applied fields, such as forestry, agriculture and wildlife management, have found the principles involved essential in the understanding of such diverse areas as disease control, seedling establishment and livestock management. The interests of industrial hygiene are centered on air pollution of areas where people congregate and work, such as factories, office buildings, mines and subways, and on a broader scale, the atmosphere of whole cities where smog is the extreme consequence. In the field of communicable disease, a vast body of knowledge has developed on the microclimate of disease producing and transmitting organisms in efforts to develop ecologically oriented control measures. The government has established large microclimate programs dealing with such fields as weather, soils, and the preservation and use of military materials in extreme climatic areas as the tropics and the arctic.

The study of microclimate and microenvironment has progressed within the last fifteen years from being purely descriptive to being primarily experimental. Clear cut concepts and highly specialized instrumentation have made it possible to establish laboratories in the natural environment and also be create controlled environments within the laboratory, the latter ranging from small temperature control chambers to large phytotrons or climate-controlled greenhouses.

ROBERT B. PLATT

References

Geiger, R. (trans. Milroy N. Stewart, *et al.*), "The climate near the ground," Cambridge, Harvard University Press, 1957.

Platt, Robert B. and John Griffiths, "Environmental Measurement and Interpretation," New York, Reinhold, 1964.

Tromp, S. W. (ed.), "Biometeorology," Proceedings of The Second International Bioclimatological Congress, Pergamon Press, 1962.

Tromp, S. W., "Medical Biometeorology," New York, Elsevier, 1963.

Wadsworth, R. M. (ed.), "The Measurement of Environmental Factors in Terrestrial Ecology," British Ecological Society Symposium Number Eight, Oxford & Edinburgh, Blackwell Scientific Publications, 1968.

Wang, Jen-Yu, "Agricultural Meteorology," Milwaukee, Wisconsin, Pacemaker Press, 1963.

MICROMANIPULATION

Micromanipulation, microsurgery, or micrurgy (*mikros*, minute; *ourgos*, working) deals with the instrumentation, techniques and applications of microdissection, microvivisection, microisolation, and microinjection. Many mechanical operations can be performed under high magnification by micrurgy when properly instrumented. Applications of micrurgy are many, especially in cancer research, embryology, cytogenetics, cytochemistry, microchemistry, biophysics, colloid technology, subminiature assemblies, and metallurgy.

The applications of micrurgy to the study of living cells include: single-cell isolation, especially bacteria, spores, and ascites tumor cells; microdissection of virus inclusion bodies in plant and animal cells; microdissection of the neuromotor system in ciliated protozoa; measuring adhesiveness of cells; enucleating cells; stretching and cutting chromosomes; dissecting polytene chromosomes; production of translocations in chromosomes; microinjecting pH and redox indicators, salt solutions, enzymes, drugs, and oils into cells; bioelectrical studies involving action and membrane potentials; electrical resistance of protoplasm; properties of extraneous coats; mechanisms of cell division; cytochemistry with the isolation of subcellular structures; transplantation of subcellular structures. Subcellular components that have been successfully transplanted from cell to cell include: micronuclei of ciliated protozoa; normal and irradiated cytoplasm; nuclei of amoebas; nuclei of differentiated frog cells into cytoplasm of unfertilized eggs; inclusion bodies of viral origin; and normal or malignant nucleoli usually with attached nucleolar chromosomes.

Non-biological applications include: measuring consistency and elasticity of colloidal systems; demonstration of elasticity of metallic crystals and of synthetic or natural fibers; manufacture and assembly of fine mechanical devices and electronic components such as diodes and transistors; orienting fine particles and preparing specimens for electron microscopy; isolating and determining chemical properties of rare elements such as Plutonium.

The basic instruments required for micrurgy are: (1) micromanipulators or micropositioners which reduce the crude hand movements to delicate fine motion of the microtools, (2) microtools, such as needles, pipettes, hooks, loops, electrodes, scalpels or forceps, and (3) microscopes and accessories for recording optical data.

Functionally, the micromanipulator is a micropositioner with which one can place the tip of a microneedle or micropipette into any position (three axes in space) covered by the optical chain. At any instant, the tip of a microneedle or micropipette must be either where it is needed or out of the way. Over 200 different types of micromanipulators have been described since Schmidt (1859) published a description of his "Microscopic Dissector." Of these, perhaps 10 have reached the status of commercial production. These instruments range from simple rack and pinion assemblies to massive, accurately fitted ball-bearing slides actuated by precise feed screws or other driving mechanisms. Many commendable adaptations of this principle have been engineered in the excellent fine adjustments incorporated into the modern research microscope. Through the inclusion of spring-loaded ball-bearing slides and various actuating mechanisms, the movements are smooth and without backlash.

Although precision positioning of microtools can be accomplished with feed screw type micromanipulators there is, however, a serious drawback in such instruments. They are inconvenient to operate, especially if a microtool needs to be moved diagonally. Undoubtedly, this handicap was responsible for the development of several varieties of lever or joystick actuated micromanipulators. With these instruments, the tip of a microtool can be positioned fairly rapidly—an advantage if the object is motile. Movements may be transmitted from the control lever to the slides by pneumatic or hydraulic means; in others, the movements are directly transmitted through mechanical couplings. One lever controlled micromanipulator generates motion for the microtools through the expansion of spring-loaded electrically heated wires.

The Leitz micromanipulator, one of the more recent commercially available instruments, consists of three ball-bearing slides mounted at right angles to one another to produce movements in three directions in space. A single joystick controls all horizontal fine motions. Vertical movement is provided by a coarse and fine adjustment operated by turning two coaxial knobs. There are also two horizontal coarse controls for preliminary positioning of the microtools. Movement is transmitted from the joystick to the two horizontal slides through a ball-sphere segment. By raising or lowering the eccentrically mounted ball segment, the ratio between lever movement and needle motion can be continuously varied from 16:1 (coarse) to 800:1 (fine).

A new instrument has been designed that combines the precision of feed screw controls with the convenience of joystick controls. This can be achieved by actuating the feed screw drives by servo motor mechanisms. Such drives can be operated by push button controls. Furthermore, through added electronic feedback devices, such micropositioners can be programmed for performing certain operations automatically.

Each micrurgical problem generally requires special instrumentation. Obviously, the construction of a microtransistor presents entirely different requirements from those needed to transplant a nucleus, a nucleolus, or even a chromosome. The transplantation of a nucleus from one amoeba to another requires one approach. The transplantation to a nucleus from a frog embryonic cell into an unfertilized egg needs an entirely different procedure. Micrurgical investigations on cells in tissue culture present strikingly different problems from those encountered with sea urchin eggs.

The microtools, such as needles, hooks, loops, and pipettes, can be fabricated by hand using a simple gas microburner and glass rods or tubing. Glass is the best substance for making microtools since this is the only material that consistently has ample rigidity even when reduced to micro or even submicro dimensions. Several mechanical devices, usually with electric heating elements, are available for pulling micropipettes. Micropipettes with special tips or shapes can be constructed in the microforge. Microforges are optical-mechanical devices for controlling the position of needles or pipettes in the field of a low power microscope by a simple micromanipulator. With these devices, microhooks and microloops can be made; micropipettes can be bent near the tip and their minute openings may be beautifully fire polished. The tip of a microneedle may be submicroscopic in size ($<0.2\ \mu$). Micropipettes are frequently used with functional tips $0.5\ \mu$ in diameter. Microelectrodes with tips of the order of $1\ \mu$ are routinely made and used in biophysical laboratories.

The glass microtools are mounted in metal holders which, in turn, are held by clamps on the micropositioners. The better instruments provide coarse adjustments for the preliminary centering and orientation of the microtools in the field of the microscope.

Any conventional research microscope equipped with a good mechanical stage and a long working distance substage condenser will suffice for most micrurgical studies. Ordinarily, biological specimens such as living cells, suspended in a fluid medium, are mounted on cover glasses and supported by a box shaped device, the moist chamber. The moist chamber which prevents excessive evaporation of the preparation is positioned by the mechanical stage. With conventional microscopes, the cells are suspended in a hanging drop from the bottom surface of a cover glass. With inverted microscopes, however, lying drops are used so that the cells can rest on the upper surface of the cover glass.

The basic micrurgical procedures include: setting up micromanipulator, microscope and light source; preparation of moist chamber; centering of microtools in optical field; proper vertical placement of microtools in relation to depth of hanging drop or height of lying drop; preparation and placement of cells or other material on cover glasses and subsequent mounting on the moist chamber; practice in manipulating needles or pipettes in the field of the microscope by familiarizing oneself with all the adjustments provided by the micromanipulators; practice in manipulating the microinjectors; and skill in making microtools.

Many parts of a cell such as nucleoli, chromosomes, and mitochondria can be transplanted into another cell with a micropipette by microinjection techniques. The removal of a nucleolus from a nucleus and its insertion into the cytoplasm of the same or different cell presents special and rather difficult technical problems. Specifically, the orifice of the micropipette must be correctly positioned towards the nucleolus. Then, suction must be applied carefully to start pulling the nucleolus, usually with its chromosome(s) into the orifice of the micropipette. At the right instant, the suction must be increased vigorously in order to dislodge and to extract the nucleolus from its site while avoiding the entry of extraneous material into the micropipette.

Such microsurgery requires exceptionally fine micrurgical and microinjection equipment and skill. The proper positioning of the tip of a micropipette towards the nucleolus with minimal injury to the cell can be done only with superb micromanipulators. The transplantation procedures and facilities should be designed so that the entire operation can be performed in a matter of seconds; otherwise there is the danger of overexposing and thus damaging the nucleolus.

Spherical cells can be conveniently and harmlessly held during removal and implantation of a nucleolus by a modification of the microelastimeter. The "cell-holding micropipettes" are made from thick-walled tubing. These micropipettes may have an orifice diameter of 10–15 μ, approaching $\frac{1}{4}$ to $\frac{1}{3}$ the diameter of the cell to be held. The tips of such micropipettes are fire polished to avoid tearing the cell surfaces. The cell is picked up with the cell-holding micropipette by bringing the tip against the surface of the cell and applying sufficient negative pressure with the microinjector to aspirate a small segment of the cell into the micropipette. Such cells are moved into position with a micromanipulator. The "transplanting micropipettes" are made from thin-walled tubing. The orifice has a diameter of approximately 2 μ.

Transplantation of subcellular structures by microinjection techniques can be executed properly only if volume displacment in the microinjector is under control. Heretofore, such controls were provided by fine pistons (0.005 inch in diameter) actuated by complex, carefully fitted, worm gear-feed screw mechanisms. A new method of controlling minute volume changes was recently developed by employing the principle of the differential piston. Two pistons, about 1 mm in diameter, with one piston slightly larger than the other are mounted on a common carrier bracket. Each piston enters the same volume chamber but from opposite ends. As one piston moves into the chamber, the other piston moves out. Accordingly, the volume displaced per unit length of travel will be the difference in volume displaced by the two pistons. The piston actuator is a slightly modified coarse-fine focusing mechanism as used on microscopes. This unit can be driven by a servo motor, thereby establishing a programmed or semiautomatic microinjector, so that different rates of aspiration needed for removing nucleoli from a nucleus can be automatically regulated.

Four Leitz micromanipulators, each equipped with a microinjector, are used for transplanting nucleoli. The micropositioners are grouped around the microscope. Each one carries a micropipette, a micropipette holder, and a micropositioner clamp. Two microinjectors, mounted behind the micromanipulators are connected to the micropipette holders by polyethylene tubing. These microinjects consist of glass syringes, the spring loaded pistons being moved by micrometer screws. A modified micrometer with a small piston provides the necessary fine volumetric control. The micropipettes are large, with fire-polished orifices and are used for holding the cells. In front of the micromanipulators are mounted the two microinjectors of the differential

piston type. These are also connected to micropipette holders by polyethylene tubing and are used for transplanting nucleoli and other subcellular structures. These microinjectors include a large piston moved by a micrometer screw for coarse volume adjustments in addition to the differential piston assembly.

A new procedure was devised to permit the transplantation of a nucleolus into the nucleus. A nucleolus is placed behind one of two anaphase sets of chromosomes in a cell undergoing mitosis. Both donor and host cells are held by large micropipettes which function as microsuction cups and are moved into place with micromanipulators. The donor cell should be in late interphase or early prophase with the nucleolus prominently displayed. By this time, the chromosomes begin to condense. The host cell should be in the mid-anaphase stage of mitosis. The nucleolus with its attached chromosome(s) is gently extracted by a fine micropipette, using the differential piston microinjector. Then the micropipette is removed from the donor cell with the nucleolar complex inside. This micropipette is inserted into the host cell so that the nucleolus can be deposited just behind one of the anaphase sets of chromosomes. The nucleolus is then gently expelled and the micropipette is carefully removed. This is an extremely delicate operation. During telophase when the new nuclear membrane is formed, the transplanted nucleolus will be enclosed and thereby adopted by the new nucleus.

What are the limitations of micrurgy? As far as delicacy of micropositioning is concerned, there is no limit. Micromanipulators are now available that permit micropositioning of a microneedle within the limits of resolution of the light microscope. Microinjections can now be performed on a quantitative basis, using either the micropiston or differential piston principles. Volumes of the order of 10–100 micromicroliters can be routinely measured.

There are certain practical limitations in the size of the micropipette. If the micropipette is too small, great difficulties are encountered in causing the flow of liquid through the orifice. Generally the rate of flow varies as the 4th power of the radius of the orifice, other factors being constant. The second limitation in micropipette size is determined by the toughness of the extraneous coats which universally cover all cells. Obviously, a superfine micropipette would lack the strength needed to penetrate a tough cell wall. For this reason, micrurgy on plant cells has been seriously restricted. There are many animal cells with extraneous coats too tough to permit entry of a micropipette (*Urechis* eggs, for example).

Since the minimum size of a micropipette is limited for practical considerations to a diameter of 0.5 μ, this will also limit the minimum size of a cell. Based on the author's experience, the minimum *cell/micropipette* (*c/m*) ratio, based on diameters, should be not less than 10. Accordingly, the minimum size of a cell, for a 0.5 μ micropipette would be 5 μ. Such a *c/m* ratio would obviously exclude bacteria since in these cells the diameter is rarely greater than 1 μ. The second limitation that more or less excludes bacteria is the cell wall which unfortunately is a tough structure. Whether this factor can be overcome by the use of lysozymes needs to be investigated. It is unfortunate that microinjections into bacteria cannot, as yet, be performed. There are many problems open for investigation such as the obvious injection of viral DNA or RNA into bacteria. In this connection, the bacterial viruses have solved, and quite admirably, the problem of microinjecting nucleic acid into the interior of a bacterial cell! This is probably one of the most elegant examples of nature's microinjection techniques.

The transplantation of nucleoprotein-rich subcellular structures, along with the induction of various cell changes, can add new horizons to the study of somatic cell genetics. No longer is one dependent on the chance inclusion into cells of, or infection by, some exogenous subcellular particle. One may now deliberately take out a selected structure from one cell and place it precisely into another cell. The field of virus-cell interactions can be approached through new experiments since inclusion bodies or structures associated with viruses can also be transplanted into uninfected cells. The interaction of viruses and cells, especially where striking cellular effects may be induced such as interference with normal processes of differentiation, could very well be one of the most fruitful areas of future work. Much of the success of survival and propagation of the cells, following transplantation, will be enhanced by more precise procedures as well as the most modern instrumentation.

M. J. KOPAC

References

El-Badry, H., "Micromanipulators and Micromanipulation," New York, Academic Press, 1963.
Chambers, R. and M. J. Kopac, "Micrurgical technique for the study of cellular phenomena," in Jones, R. M., Ed., "McClung's Handbook of Microscopical Technique," 3rd ed., New York, Hoeber, 1950.
Kopac, M. J., "Cytochemical micrurgy," Internat. Rev. Cytol., 4: 1–29, 1955.
Kopac, M. J., "Micrurgical studies on living cells," in Brachet, J. and A. Mirsky, Eds., "The Cell," Vol. 1, New York, Academic Press, 1959.

MICROSCOPE

The word "microscope" comes from a Greek word that means "to view small." Before the microscope could come into existance lenses had to be developed. Early knowledge of lenses is very obscure, but we do know that the Chinese porcelain vases had figures on them wearing glasses. These vases date back to 1000 B.C. The Assyrian lens of 700 B.C. was found to be double convex. Sir John Layard discovered a plano-convex lens of rock crystal at the Nineveh excavation. The Greeks and the Romans wrote about the burning properties of glass lenses as well as the use of glass globules filled with water to aid in seeing small objects. Rodger Bacon wrote about the magnification power of lenses to aid elderly persons in reading. After his book "Opus Majus" was published in 1266 A.D. nearly two hundred years passed before little if anything was written discussing advancements in optics. In 1542 the first references about telescopes appeared and these were written by Nicholaus Copernicus of Poland, who discovered that the sun is the center of our universe and that other universes exist.

In Middleburg, Holland, about 1590, the story of the compound microscope begins. Zaccharios Janssen developed, by accident, a microscope which was one inch in diameter and six feet long. This discovery started the parade of inventors and improvers of the microscope, too numerous to mention in their entirety in this article.

Galileo in 1610 introduced a microscope similar to Janssen's but with one major improvement. The Galileo microscope was focusable by means of screw threads on the body and mount. Fontana, Drebbell and Kepler also made microscopes in this early period.

The early microscope lenses had been formed on the end of glass rods by heating and pressing into a given shape. Campani was the first to design and grind lenses which had curves that could be reproduced. Anthony Von Leuuwenhoek studied all types of specimens under a simple microscope with ground lenses of his own design. He made in excess of one hundred microscopes with single lenses, some of which were $\frac{1}{8}$" in diameter and having a very strong curve. With these instruments he discovered bacteria and the existence of corpuscles in the blood.

Robert Hooke in the last quarter of the 1600's made a compound microscope that was easy to use. Hooke used a doublet eyepiece from Christian Huyghens to make up the telescope portion of his compound microscope. The Royal Microscopical Society published Robert Hooke's "Micrographia" which describes tissues, blood vessels, textiles, papers, sugars and salt crystals. Another Englishman by the name of John Dolland in about 1750 improved lenses by devising an achromatic system compounded from both hard and soft glass. In 1759 Dolland succeeded in making an achromatic lens, however, this type of lens was not used in a microscope until 1825. In 1854 the first steps towards standardization of the microscope took place with the × designation for magnification based on 250 mm. for a given magnification. In 1873 Ernst Abbe from the University of Jena wrote "Theory of Microscope Image Formation" and "Sub-Stage Illumination Apparatus." If the theories discussed in Abbe's book had not been followed, today's microscopes would form very poor inferior images.

Lenses. The use of simple lenses with high curves for high magnification presents problems that make their use impractical. As the curve of the lens increases, the focal length becomes smaller and more difficult to use. An example of this is the Leeuwenhoek microscope in which one's eye had to be placed on top of the lens in order to see the magnified specimen. A microscope is nothing more than an objective lens magnifying a specimen, the image of which is being viewed by a simple telescope at some convenient distance from the objective lens (Figure 1).

The aberration of lenses is always a problem to the designers of lenses. The two most common problems are color aberration (chromatic) and distortion of shape aberration (spheric). If a beam of white light passes from a medium of one density into a medium of another density such as air to glass, the beam is bent. As white light is made up of different colors each wavelength bends a different amount thus we have a spread of the colors called dispersion (Figure 2). The higher the index of refraction the greater the dispersion. Figures 3 and 4 show the reasons that aberrations appear in lenses. In order to correct spherical aberration one way is to place a diaphragm so as to cut off the thin edges of the lens which bend the light more (Figure 5). To correct

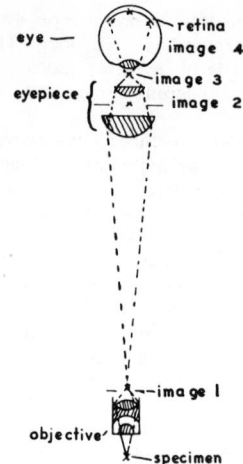

Fig. 1

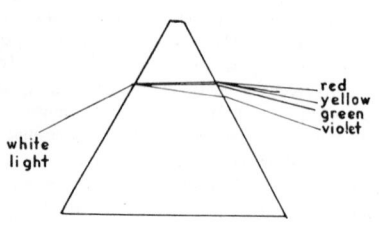

Fig. 2   Bending illustrates refraction: spreading illustrates dispersion.

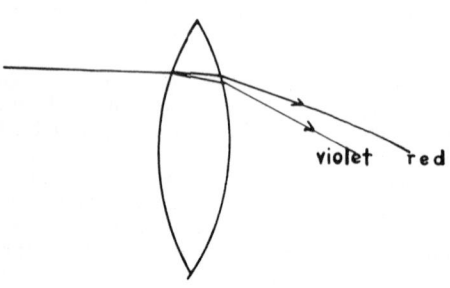

Fig. 3   Chromatic aberration.

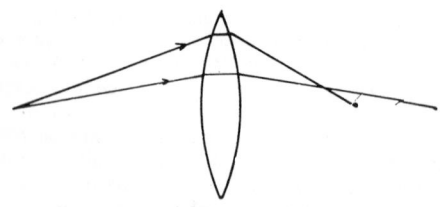

Fig. 4   Spherical aberration.

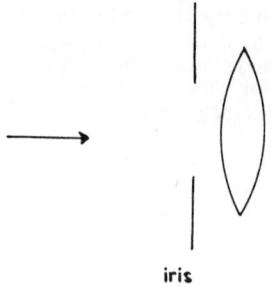

iris

Fig. 5

chromatic aberrations glass must be selected for its index of refraction and dispersion, then combined to have as many colors as possible come to focus at the same point.

The objective is the most important single factor of the microscope, for through its power to resolve minute structure we see small objects crisp and clear. Magnification, although it is of secondary importance to the resolving power of the objective, is absolutely essential. In order to have high resolution, a lens must be the best possible compromise on the correction of the various aberrations such as spherical, coma, and chromatic. The matter of resolution in actual practice is based not only on the correction of the lenses, but on the ability of the user to use the proper ocular, condenser, illumination, and controlling of the aperture diaphragms, and field diaphragms to their optimum settings for the objective being used. It is not always possible to achieve the resolutions of a given lens even after proper manipulation because the detail is obscured in the background.

Resolution, angular aperture and wavelength of light being used are related. The angle of the cone of light is dependent on the refractive index of the medium between the objective front lens and the glass cover slip over the specimen (Figure 6). Present day objectives have clearly marked on their mounts the relationship between the angular cone and refracture index of the working medium, expressed in terms of numerical aperture (N.A.).

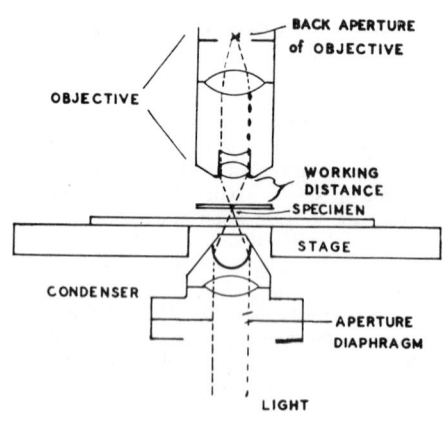

Fig. 6

$$N.A. = i \sin \theta$$

θ is one half the angle of the cone of light entering the objective, i being the refractive index of the medium in which the objective is working. From this formula theoretically one can see the N.A. in air can never be greater than 1 because i would represent air with a refractive index of 1.

The relationship of N.A. and resolving power is as follows:

$$R = N.A./\lambda$$

λ is wavelength of light and R is the number of lines being separated by the lens.

There are three main classes of objective achromatic, fluorite, and apochromatic: the achromatics are corrected for two colors (red and green) chromatically and one color spherically. To obtain more in correction substituting of the sandless glasses such as fluorite gives a little better correction than the achromats. The apochromatic objectives produce the best images, those finest in color corrector and clarity. Apochromatic objectives are corrected for three colors chromatically and two spherically. To achieve the added value in the apochromatic objectives compensating eyepieces must be used to correct for the color spread and a corrected condenser of a numerical aperture equal to or greater than the objective aperture.

When using the microscope it is very important to consider the working distance, depth of field and cover of glass thickness. The working distance is the amount of free space from the front of the objective to the cover glass and as the power and the N.A. increases the working distance becomes smaller. Therefore when a counting chamber with its normal cover glass is used magnifications up to about $43\times$ and N.A.'s up to about .66 can be used to view specimens. If an objective of a higher N.A. and magnification is used the working distance would not be great enough.

Depth of focus is the amount of thickness of the specimen that is in focus at one time. This amount decreases as magnification and N.A. increases. In photomicrography this is very important unless the specimen must be thin enough or several layers of the specimen will be out of focus, obscuring detail of the portion that is in sharp focus. High power dry objectives with correction collars to compensate for cover glass thickness should be used with cover glasses of .18 mm. to have optimum correction of the spherical aberration.

Eyepieces. The eyepiece acts as the telescope of the system merely to remagnify the primary image of the objective, thus eyepiece magnification adds nothing to the resolution of the system. The practical rule is that one should not normally magnify more than $1000\times$ the N.A. More magnification is normally called empty. However, in some instances to aid in counting, measuring, or drawing, ease of use can be achieved by using a higher power eyepiece. Many companies are now making eyepieces with high eye relief so that persons wearing glasses can continue to wear them while using the microscope.

Condensers. The condenser is commonly the most missused part of the microscope. A condenser has one optimum setting as to height and aperture diaphragm setting which controls the N.A. for each object. The condenser height and aperture diaphragm should not be used to control the intensity of the illumination. If the aperture diaphragm is open too far the specimen is flooded with light and detail is washed out. If closed too far the N.A. is lost and detail is again lost and defraction becomes quite apparent.

Illuminators. The illuminator is a very essential part of the microscope. If one is to achieve the maximum from the microscope in terms of resolution, clarity and ease on the eyes of the user, a good illuminator must be employed. An illuminator must have a condensing system large enough to accommodate the apertures of the microscope and an iris diaphragm, sometimes referred to as a field stop. The condensing system and field stop must be focusable. To achieve the ultimate in illumination filters and adjustable mirrors in the illuminator help to adjust for maximum light through the system.

In the early days illuminators were used to give critical illumination. The source of light was broad such as the sun or white clouds or oil wick lamps. The source was focused so it would appear in the field of view (Figure 7). Of course, the disadvantages of this system were size of the source and unevenness of illumination. The introduction of the concentrated coil filament lamps made the use of critical illumination impractical due to the fact that the coil source was small and did not fill the aperture. Kohler solved this problem and the system is named after him. In Kohler illumination the filament of the lamp is focused on the condenser iris (often called the aperture stop) and the field iris is focused so it will be in focus at the same place as the specimen on view through the microscope (Figure 8).

The use of the filters will make it possible to change the visual appearance of the specimen. The use of a green filter for photo-micrography will, in many cases, increase the contrast and produce an excellent photograph.

The proper way to control the intensity of illumination is to use neutral density filters. These filters control the intensity without changing the color balance at the source, e.g., if a 50% filter is used and the source emits 20% blue, 50% green, 30% red at the 100% level when the 50% filter is placed in the system, only 50% of the original light comes through the filter. This 50% is still composed of 20% blue, 50% green and 30% red. We have cut down only the total intensity, not the balance.

There are many types of microscopes. Basically they are all similar, that is to say, they have one or more objectives, eyepieces, condenser and some mechanical means to focus and manipulate the specimens. In order to change from one type of microscope to another many times only accessories have to be added.

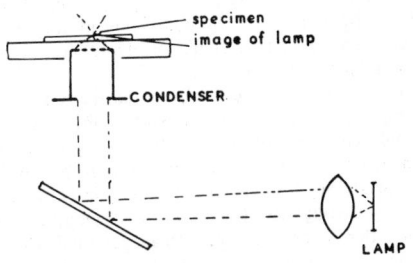

Fig. 7 Critical illumination.

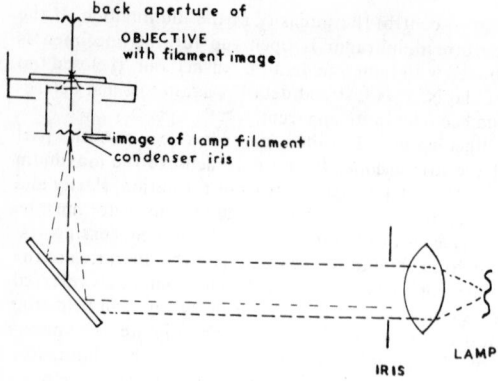

Fig. 8   Kohler illumination.

The phase contrast method of microscopy was introduced in 1935 by Professor F. Zernike. Phase contrast is most useful in research where other than the normal approach is necessary. The main advantage lies in the fact that contrast can be greatly enhanced in living, unstained material where, normally, little or no contrast exists with bright field microscope technique. Direct observation of living material, such as yeast, bacteria and fungi have made possible rapid, positive identification of specific organisms as well as observation of phenomenons within the cells that have gone unnoticed for many years. Phase contrast relies on the combining of light waves and is dependent on amplitude and phase of these waves.

Dark field microscopy consists of a condenser with an opaque central stop which will allow no direct light to enter the objectives. The circular cone of light is focused on the specimen which becomes very bright on a black background. Dark field shows objects that are too small to be seen by the bright field method, and is often used in the identification of spirakytes and colloidal materials.

Polarizing microscopes, in addition to having the normal components, have a polarizer and an analyzer. The light enters the polarizer where it is forced to vibrate in one direction. The light then passes through the condenser and objective to the analyzer which is set with the direction of vibration 90° to the polarizer; consequently no light is allowed to pass to the eyepiece. When a crystal is placed in the system at the specimen plane, the crystal, in many cases, rotates the plane of polarization. The crystals that rotate the plane of polarization become either colored or white on a dark background. Crystals are classified by the way that they look under polarized light. Other properties of crystals that can be studied by the polarizing microscopes are optical sign, extinction angle, birefringence, refractive index and pleochroism.

The stereoscopic microscope is basically two microscopes, one for each eye. These two microscopes are mounted conveniently in a single housing for easy use. Each eye views the same specimen from a different angle producing a stereoscopic, three dimentional view. The view through this type of microscope gives the user greater depth, erect image and larger fields than with the compound microscope. These instruments are used is dissection, genetics and small part assemblies.

There are many universities, companies and individuals that have advanced the microscope to its present form. It is certainly one of the most important factors in our knowledge of bacterias and other medical sciences. Even industry, in its quest for quality today, is investigating more closely the minute workings of their products, and with the advent of space exploration the need for miniaturization has reemphasized the part the microscope plays in studying and manipulating the small.

DONALD A. BURGH

References

Spitta, E. J., "Microscopy," 3rd ed., London, John Murray, 1920.
Gage, S. H., "The Microscope," 17th ed., Ithaca, Comstock, 1943.
Gray, P., "Handbook of Basic Microtechnique," 3rd. ed., New York, McGraw-Hill, 1964.
Birchon, D., "Optical Microscope Technique," London, Newnes, 1961.

MICROSOME

Within the cytoplasm of most CELLS is found a fibrillar network or cytoskeleton which has been given the term *hyaloplasm*. The hyaloplasm, which is normally invisible by light microscopy and is optically non-refractive, becomes birefringent after the cells are centrifuged; a process resulting in the parallel orientation of minute fibrous proteins comprising the fibrillar structure. Long before the advent of the ELECTRON MICROSCOPE this cytoplasmic reticulum was believed to not only serve as a structural framework for the cell, but to play a functional role in the many activities of the cell.

The term *microsome* refers to small RNA-containing granules which are derivable after cellular fractionation in sucrose media and high speed centrifugation. Actually the so-called microsomal fraction obtained in this manner consists of two separate components: spherical or oval vesicles with a limiting membrane to which are attached numerous small spherical particles, 50–200 mμ in diameter. Electron microscopy of the intact cell readily reveals the *endoplasmic reticulum* as a vacuolar system consisting of canaliculi and cisternae oriented in a parallel concentric fashion throughout the cytoplasm. During homogenation the endoplasmic reticulum disintegrates into the spherical or oval vesicles which are separated by high speed centrifugation along with recognizable fragments of the reticulum itself.

On the basis of morphological observations in glandular cells, the microsomes appear to be involved in the formation of specific proteins, a conclusion reached by Garnier and Bouin in 1897, when they coined the word *ergastoplasm* in reference to the vacuolar system. In view of the fact that electron microscopal data are based exclusively on one type of fixative (osmium tetraoxide), while observations on living cells with anoptral microscopy failed to reveal a reticulum until after exposure to osmium, it has been contended by some that the reticulum is an artifact. Nonetheless, the microsomes separated by differential centrifugation assume definite biochemical identity with at least some morphological specificity in their isolated state. In addition to comprising as much as $\frac{1}{5}$ of the total mass of some cells, of which about 50% (dry weight basis) is lipid, the microsomes contain most of the cytoplasmic RNA and a number of ENZYMES. Such enzymes as esterases, arylsulfatase, DPN-cytochrome C reductase, ATP-ase, as well as other phosphatases, adenylic kinase, adenylic deaminase, in addition to thromboplastic activity are presumed to be microsomal in origin.

Studies in plants and lower animal organisms have revealed the cytoplasmic fibrils to be in a state of continual reorganization and reconstitution as determined by the functional requirements of the cell. Within higher organisms there are numerous fibrils which are known to respond during functional activity: myofibrils, neurofibrils, epithelial fibrils, and CILIA. Such activity usually consists of visible contraction or motility involving rearrangements in polypetide chains. There may be polypeptide rearrangements associated with most cytoplasmic fibrillar activity involving energy-releasing mechanisms comparable to the ATP-ase system of myofibrils, i.e., actomyosin. Although it has been possible to produce beadlike threads by extruding microsomal suspensions into 10^{-3} M $MgCl_2$, such threads are incapable of contraction upon the addition of ATP.

Of particular significance is the fact that such artificial fibers exhibit a high degree of organization as well as birefringence of flow. Because of the colloidal nature of proteins, polysaccharides, and other cytoplasmic constituents, aggregation into beaded fibrils or meshworks may readily occur. Such cellular physical properties as fluidity, plasticity, and elasticity are probably attributable to reversible formation of such structures. To what extent the term microsome is to be applied to the basic component of such fibrils is purely a matter of definition. In the present stage of our knowledge of biocolloids and subcellular morphology, it would, perhaps, be unpragmatic to restrict the term to granules containing RNA, but rather attempt to define it in terms of functional morphology.

L. G. ABOOD

References

Roberts, R. B., ed., "Microsomal Particles and Protein Synthesis," New York, Pergamon Press, 1958.
Bracket, J., "Biochemical Cytology," New York, Academic Press, 1957.
Frey-Wyssling, A., "Submicroscopic Morphology," New York, Elsevier, 1953.

MICROTOME

Microtomes are machines for cutting thin sections of material for examination with a microscope. During the 18th and 19th centuries, they were called cutting engines. Chevalier introduced the term microtome in 1839. The microtome includes an adjustable means for holding the specimen, the knife, for advancing the knife or the specimen as needed for the desired thickness of section and means to move the specimen across the knife, or the knife across the specimen to cut the specimen. Automation can include a motor drive.

Many arrangements have been tried, but only a few have met the practical test of use. Several classifications are possible. Ordinary microtomes cut sections of tis-

sues, softer materials, and plastics from 1μ to 50μ or so, massive microtomes are designed to section hard materials or are made large for slicing whole organs, and ultramicrotomes section materials thinner than 1μ as required in electron microscopy. Microtomes are often called rotary or sliding according to the main motion of sectioning; or in accordance with the nature of the specimen embedding material, e.g., paraffin, frozen section, etc.

Kinds of microtomes. A simple classification of microtomes separates those with fixed knives from instruments with moving knives (Figure 1).

I. Microtomes with a movable knife
  A. With vertical advance for section thickness control
    1. Knife held by hand during cutting (Hand or table models)
    2. Knife pivoted at one end with the other end free or supported on a slideway
    3. Knife and knife holder pivots mounted to form a parallelogram with unequal arms to make a slicing stroke
    4. Knife holder moves on a horizontal track (Adams model)
  B. Advance for section thickness control obtained by moving specimen holder up an incline (Capanema-Rivet model)
II. Microtomes with a stationary knife
  C. With knife edge vertical
    5. Specimen is moved on a vertical track (Minot model)
    6. Specimen is moved on a pivot about a horizontal axis and the section is cut on an arc (Pfeiffer model, Rocking model)
    7. Specimen is mounted on a vertical parallelogram so that flat sections are cut (Flatcutting, Rocking model)
    8. Specimen rotates around a horizontal axis parallel to the knife edge (Peeling, or veneer model)
  D. With knife edge horizontal
    9. Specimen holder moves on horizontal ways
    10. Specimen holder rotates on a vertical axis

Valentine devised an adjustable double bladed knife in 1839 for making tissue slices of a predetermined thickness.

The hand microtome, A1 in Fig. 1, is a flat circular plate with one end of a tube fitted into a hole in the center of the plate and a screw at the other end of the tube to force a specimen placed in the tube beyond the surface of the plate. The section is cut with a razor held against the surface of the plate so as to cut off the part of the specimen extending above the plate. A more convenient, although less portable arrangement, adds a clamp for holding the microtome onto a table, A2 in Figure 1. Sometimes a clamp is placed in the cylinder to hold the material during the cutting process; otherwise, the specimen is wedged into the tube with pith, cork, or other easily cut support.

Many specimens are too soft for so simple a procedure and must be supported during cutting by a suitable embedding medium, e.g. paraffin, celloidin, methacrylate, epoxy resin, etc. Another method freezes the material on the specimen holder with carbon dioxide and the sections are cut when the specimen has thawed to a cuttable hardness.

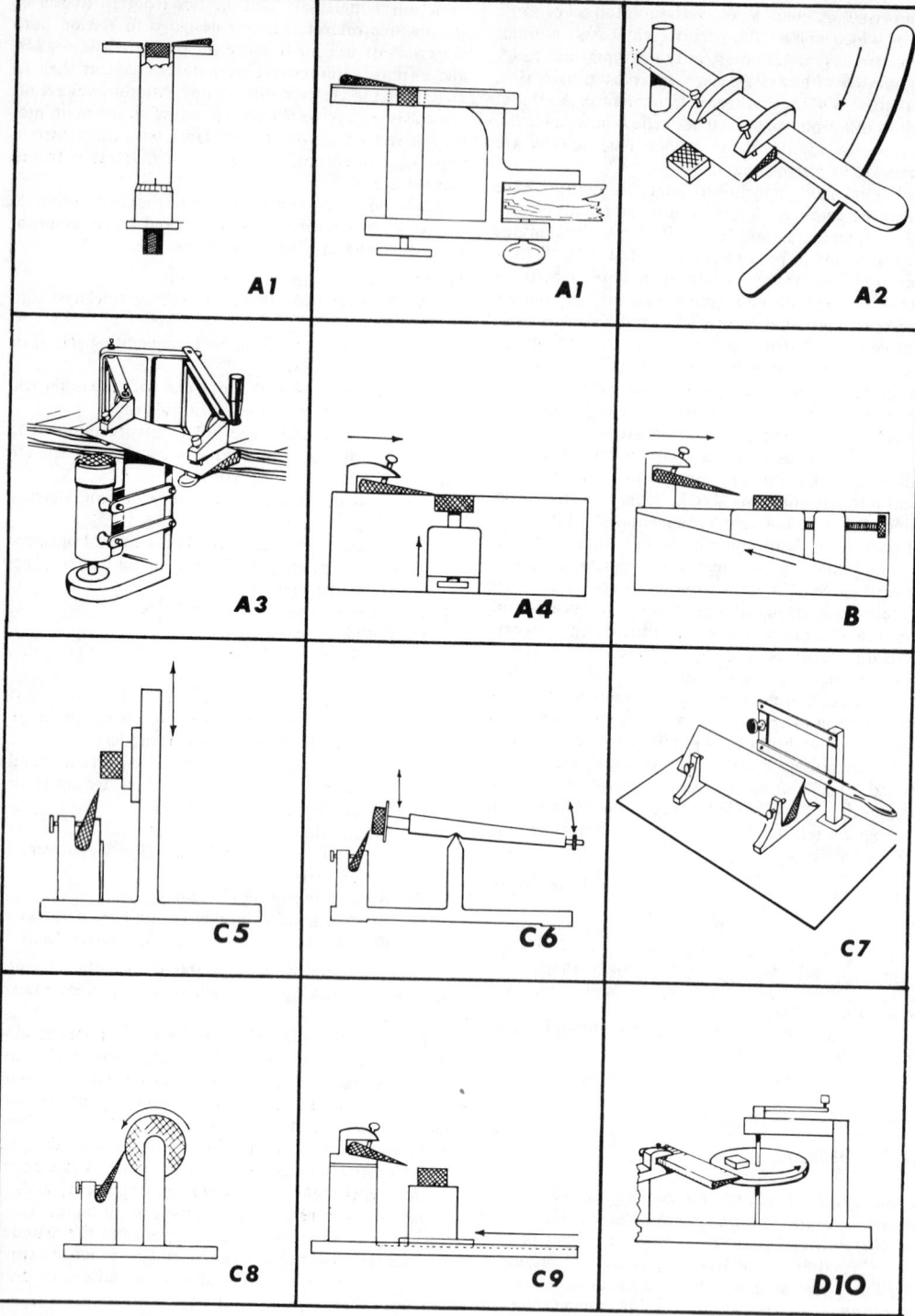

Fig. 1   Diagrams of the kinds of microtomes; A1–4, B, with movable knife; C5–10 with stationary knife. (Courtesy of American Optical Co.)

(For C9 read D9)

Sections of paraffin embedded material stick to each other, forming a ribbon of sections that is easy to handle and mount (Figure 3). To obtain a ribbon the knife should be fixed and the specimen cut by moving across the knife edge. Fairly rapid cutting is possible with designs like C5 and C6 of Figure 1.

Rotary microtomes. The first rotary microtome (1883, Adam Pfeiffer, John Hopkins University) carried the specimen on the periphery of a wheel and the sections were cut as the specimen passed across the knife. The knife was drawn a predetermined amount toward the specimen before the next cut. Shortly thereafter a rotary microtome was designed (1887, C. S. Minot, Harvard University) with a vertical motion for the specimen and a fine screw for the control of the section thickness. The precision of the screw feed and the smooth advance of the inclined plane were combined in a rotary microtome (1910, H. N. Ott, Spencer Lens Company) that is in general use a half century later (Figure 2).

A slicing cut is not practicable with a rotary microtome as the knife edge has to be mounted at right angles to the direction of the cut in order to form a ribbon of sections. While it is possible to hold the knife at an angle, the gain is slight due to the short stroke of the mechanism.

Sliding microtomes. The sliding microtome permits setting the knife for the best slicing stroke and is useful for delicate and less easily sectioned materials. The designs A4, B, and D9 of Figure 1 are used in sliding microtomes. The specimen advance for section thickness may be direct by a screw feed, or the screw may push the specimen holder up an incline. The latter method was introduced about the middle of the nineteenth century to provide a smooth slow movement. Microtomes based on the D9 design can be made very rigid and are capable of cutting harder specimens than possible with the other designs. The sledge microtome is powered by a drive wheel and gears, or other mechanism, to deliver the power needed for cutting the harder materials.

Soft metals can be surfaced with a sliding microtome. Delicate materials are usually embedded in celloidin and sectioned with a slicing stroke on a sliding microtome.

The microtome knife. The quality of the section is determined to a large extent by the nature of the cutting edge. Knives for biomedical work usually are made of special steel in the shape of an old-fashioned razor. The slides of the blade are flat or nearly so, of about 15° included angle and of sufficient width for adequate strength to support the edge against the strain during cutting. The cutting edge is formed along the thinner part of the wedge, by alternately grinding two narrow facts (±0.3mm) until they meet at a line, figure 3, with an included angle of about 28° between the plane facets. A smaller angle (±20°) is preferable for cutting soft materials; a greater angle for harder materials. An angle of 28–32° is a good compromise for commercially available knives for biomedical use.

As steel has structure, it is not possible to sharpen the edge to such a perfect geometrical line and there will be a fine irregular structure at the actual edge which should be polished until it appears smooth when examined by reflected light at 100–200 times magnification. Microtome knives can be sharpened against a plate glass surface with graded abrasives and a corundum polishing powder. When the edge is perfectly formed,

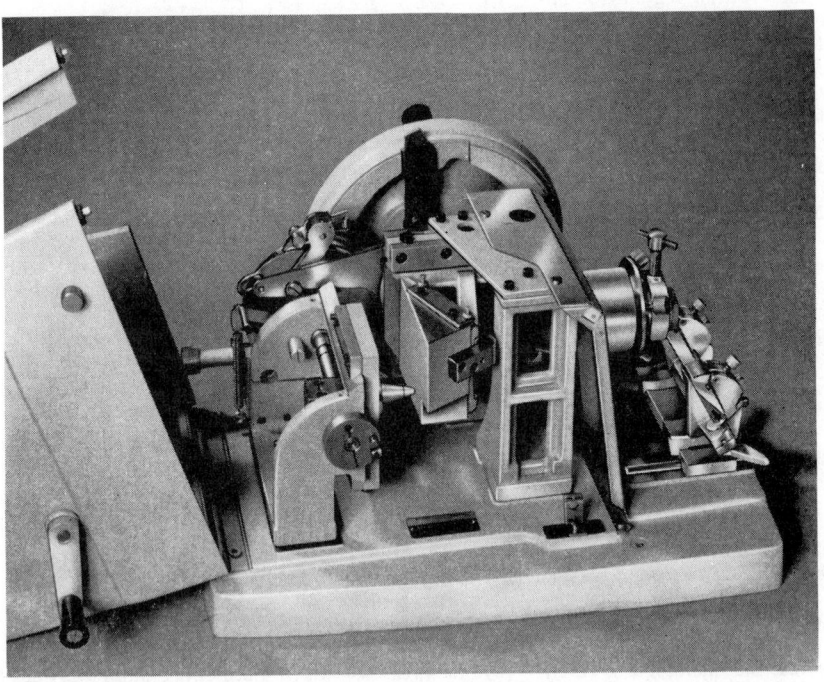

Fig. 2   A rotary microtome mechanism combining a feed screw drive and an inclined plane for the control of the section thickness. (Courtesy of American Optical Co.)

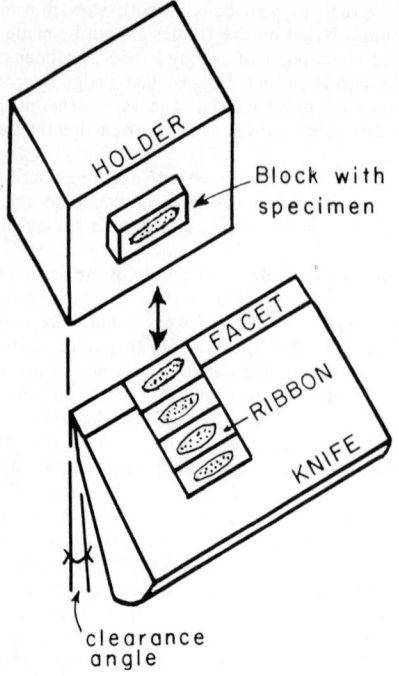

Block with specimen

clearance angle

Fig. 3 Diagram of microtome knife and specimen showing the formation of a ribbon of sections of paraffin embedded material.

it should never be unsharpened (rounded) by stropping. A slight wire edge from incorrect honing may be improved somewhat by stropping.

The proper knife angle in the microtome is essential for cutting sections. The knife must be tilted enough to provide clearance between the surface of the specimen and the nearest facet, figure 3 (clearance angle). Too little angle will produce sections of irregular thickness and too great an angle will only scrape off the outside of the block.

How hard material can be sectioned depends on the skill of the operator, the nature of the specimen and the strength of the knife edge and the microtome. Ordinary steel microtome knives can section materials to about 18 Brinnell hardness, depending somewhat on the brittleness of the specimen and the section thinness required. The thicker safety razor blades and the wallboard type knife blade can be used in a suitable holder for cutting hard materials instead of the more expensive microtome knives. For some applications the blades can be used as purchased, for others, the blades can be sharpened as necessary.

Successful sectioning requires 1) properly prepared material, hardened or softened as necessary or embedded in supporting material of closely similar hardness and rigidity, 2) a sharp knife (free from nicks and other defects) set at the correct angle for cutting the particular specimen 3) a proper microtome of the size and type needed and 4) a skilled operator. Unless very old, damaged, or mistreated, the microtome is rarely the reason for poor sectioning. An experienced operator will obtain sections of far better quality and sometimes can section material beyond the possibilities of the un-

trained operator, but no technician should be expected to section improperly prepared material or use inadequate equipment.

Frozen sections. Sections of fresh tissue for enzyme studies and chemical examination are usually made from frozen tissue to avoid adverse effects of fixing solutions and extractions by solvents. Carbon dioxide gas is used to freeze the tissue on the specimen holder of the microtome and sections are made when the block has thawed to the proper denseness. Microtomes of types A2 and A3 are used. (Figure 1). Considerable skill is needed to make sections less than 10μ thick.

Mounting a rotary microtome in a *Cryostat* cabinet with provision for sectioning at a constant coldness makes possible rapid preparations at thicknesses to 1μ with very little distortion. A section flattener is usually mounted at the knife edge. Soft tissues are cut at -5 to $-10°$ C, hard and fatty tissues at -20 to $-30°$ C. The sections can be brought out to room temperature for staining etc., or treated and examined in the cold chamber. Mechanical refrigeration is built into the Cryostat.

OSCAR W. RICHARDS

References

Gray, F. and P. Gray, "Annotated bibliography of works in Latin, alphabet languages on biological microtechnique," Dubuque, Iowa, Brown, 1956.
Richards, O. W., "The effective use and proper care of the microtome," Buffalo, American Optical Company, 1959.
Steedman, H. F., "Section cutting in Microscopy," Springfield, Ill., Thomas, 1960.

MICROTOMY

This term is commonly, but inaccurately, applied to the whole art and science of the preparation of objects for microscopic examination. It is, indeed, so used in the titles of two of the leading reference works in the field. This article, however, concerns itself entirely with what microtomy strictly should be; that is, cutting thin sections of objects, and particularly of biological materials. Another article deals with the apparatus (MICROTOME) with which this is done and in this place is discussed solely the methods used and the reasons for them.

There are three prerequisites to cutting thin sections. These are, a sharp knife, material of the proper consistency, and some method of supporting the latter.

Many plant materials either in the fresh condition (e.g., carrot) or preserved in alcohol (e.g., plant stems) are of a consistency which can readily be cut. If, therefore, they be fitted in some holder, they can be cut with a sharp razor held in the hand to produce what are known as "hand sections." Materials of a consistency, but not of a shape, to cut readily (e.g., leaves) may be cut as hand sections by holding them between two pieces of carrot. Hand sections are, however, rarely cut and it is usual to prepare materials for sectioning on a microtome.

In those cases in which speed of operation overrides quality of result, tissues may be hardened to a cuttable condition by simple freezing. Thus biopsy tissue may be taken directly from the operating theatre to the laboratory where it is placed on the metal table of a freezing microtome, hardened to a suitably frozen consistency by directing a jet of carbon dioxide on the underside of the metal table and then sectioned by drawing a knife across it. These frozen sections, though adequate for immediate diagnostic purposes, are rarely used for research. Almost all sections which are intended for permanent preservation are cut from tissues which have previously been embedded.

The process of embedding consists first in impregnating the tissues with some material which is itself of a consistency suitable for cutting and then casting the impregnated tissue into a block. The three materials most commonly used for embedding are gelatin, nitrocellulose and paraffin. Gelatin embedding is only employed in those cases where the solvents used in the nitrocellulose and paraffin methods would remove from the tissues some substance, such as fat, which it is desired to retain. In this method, the objects are soaked successively in 1%, 2%, 4%, and 8% gelatin which is maintained in a molten condition on a water bath and to which has been added some preservative such as phenol. The thoroughly impregnated object is now cast in a rectangular mold which is chilled to set the gelatin. The block is then transferred to the stage of a freezing microtome and the gelatin hardened to a suitable consistency by the freezing method. The sections are removed from the knife as they are cut to a 1% dilution of formaldehyde which hardens the gelatin to the point where it may be conveniently handled for subsequent staining and mounting.

Nitrocellulose embedding is used with two classes of objects. First, botanical, or more rarely zoological, specimens in which delicate tissues are so loosely arranged and widely spaced that their mutual relationships would be distorted were the specimen to be embedded in paraffin. Thus, for example, the partially opened bud of a flower in which it is desired to preserve in the section the spatial relationships of petals, pistil and stamen, are commonly embedded in celloidin. The specimen, after suitable fixation, is dehydrated in alcohol, being usually passed through a succession of increasingly strong solutions, and the alcohol then replaced first with a mixture of equal parts of alcohol and ethyl ether and then by pure ethyl ether. The nitrocellulose, often referred to by the brand names of Celloidin (Schering) or Parlodion (Mallinkrodt), is first "swollen" in absolute alcohol and then dissolved in a mixture of equal parts of absolute alcohol and ether. The specimen is usually placed in a one-half percent solution which is allowed to concentrate to about a 15% solution by slow evaporation; or the specimen is transferred from one-half per cent solution to a 1% solution and then from a 1% solution to a 2% solution, etc., until it is impregnated in a solution of the required strength.

The nitrocellulose is converted from a thick syrup to the crisp condition necessary for cutting by the action of chloroform. In the case of large specimens, a rectangular container containing the specimen covered with the nitrocellulose solution is placed under a bell jar in an atmosphere of chloroform maintained by the evaporation of this solvent from small vessels. Minute specimens, such as invertebrate larvae, are drawn up an eye dropper type pipette, from which a drop of syrup containing the specimen is squeezed into liquid chloroform. These minute specimens are then transferred from the chloroform directly to paraffin and embedded as described later. Large specimens are commonly taken from chloroform to oil of cedarwood which both renders the block transparent for easier orientation of its content and improves the cutting consistency. These large nitrocellulose blocks are usually cut on a "sliding microtome."

Sections of histological and cytological interest are usually cut by the paraffin method. Properly fixed specimens, after washing to remove the fixative, are dehydrated in a graded series of alcohols. Animal tissues are commonly taken through a series of ethanol but plant tissues are more commonly dehydrated in butanol or butanol-ethanol mixtures. Schedules for this complex process will be found in the references appended to this article. In either case the perfectly dehydrated specimen is "cleared" in xylene or benzene before being transferred to a bath of molten paraffin held slightly above its melting point in a thermostatically controlled oven. The paraffin is often mixed with small quantities of additives like rubber and bayberry wax which improve its consistency under the knife. Two or three changes of paraffin are necessary to make certain that all the solvent is removed. The wax impregnated specimen is then cast in a rectangular block which is rapidly chilled to insure a microcrystalline structure. The hardness of the resultant block is dependent in part on room temperatures and in part on the melting point of the wax itself. Waxes with a melting point as low as 52°C. are used in some cold European laboratories but the range 56° to 58° is more commonly employed in the United States.

The block containing the specimen is then trimmed so that its opposite faces are parallel and mounted on a microtome suitable for paraffin sectioning. Rotary microtomes are most commonly employed as these produce a "ribbon" of sections since successive sections stick to each other. The ribbons are then flattened on warm water and attached to glass slides with some adhesive, usually egg albumen or starch. These slides, after they are dry, are treated with xylene or benzene to remove the wax, then with absolute alcohol to remove the solvent, and then run down a graded series of alcohols to water whence they are transferred to the necessary STAINS, discussed elsewhere. The stained sections are then again dehydrated, cleared, and mounted under a coverslip in balsam.

PETER GRAY

References

Gray, P., "Handbook of Basic Microtechnique," 3rd ed., New York, McGraw-Hill, 1964.
Gray, P., "The Microtomist's Formulary and Guide," Philadelphia, Blakiston, 1954.

MIGRATION

Many phenomena are called migration: the exodus of whole faunas from one continent to another in prehistoric time and the daily passage of flocking birds to and from roosts; shifts of human population and rhythmic responses of marine invertebrates to tides; the suicidal cyclic outbreaks of lemmings and the equally suicidal emigrations of some insects; irregular southward influxes of boreal birds and sudden permanent range expansions, as that of the armadillo; the return of eels to the sea and of Pacific salmon in from the sea, to spawn at their birthplaces and die there. These activities, with no common feature but mass and movement, fail to satisfy a more stringent definition of migration as a semiannual alternation between definite summer and winter homes. Animals of several classes show at least a semblance of such alternation: butterflies, moths, lady-bugs, and dragonflies; various crustaceans and marine fish; toads and salamanders; bats, ungulates, seals, and whales; and, most clearly of all, a majority of nonseden-tary birds.

The reasons for migration. Migration permits mi-grants to exploit advantages offered by particular places at particular times. Sea birds, sea turtles, and seals must come to land to breed; most amphibians must go to water. Elk, caribou, cetaceans, and certain so-called *weather migrants* among birds must seasonally change range to find food or avoid extreme tempera-ture. A great many other migratory birds seem subject to all these motivations at once. In the nesting season they require increased food, especially soft kinds to feed their young. In summer the northern latitudes proffer a large supply, together with longer days in which to gather it, less crowded living space, and fewer nest predators. Though migration is perilous, though counts of migrants seen against the moon sug-gest that only half the birds leaving in fall return in spring, the habit promotes a larger world population than would otherwise be possible. The evolutionary steps by which migrants now wintering in the tropics adjusted to this opportunity are uncertain. Some say bird migration is as old as bird flight, others that it has arisen independently many times. The alternation of cold and warm world climate, which periodically made the North tropical during the period birds have existed, doubtless accelerated the process; but whether the northward or the southward flight of long-distance migrants today is the return to their ancestral home remains debatable.

During development, much migration became tem-porally stereotyped. The selective pressures that brought it into being no longer serve as the immediate "causes" that now set it in motion. Many birds whose biologic reason for migrating is alimental head south-ward as soon as the crisis of rearing young is over, while the North still abounds with food. Possessing, as their orienting ability has proven, a keen sense of time, birds might seem to require no signal when to start moving except the lapse of time. Such a simple reaction is sug-gested by the passage of some across the equator, from a zone of decreasing to a zone of increasing day-length and by the punctual arrival of others at breeding places. But the issue has been complicated by a series of labora-tory experiments—using in this country seed-eaters that do not winter in the tropics—in which readiness to migrate has been manipulated by control of light. Re-sults indicate that birds respond not to the entire pas-sage of time but to the amount of daylight between the phases of their annual cycle. The tests show that re-crudescence of the gonads and deposition of fat as fuel are common features of the premigratory state in spring. Once considered the basis of readiness to mi-grate, these developments are now widely regarded as mere concomitants. The whole process seems to be con-trolled by the anterior pituitary.

Even when physically ready to move, most birds await environmental signals before choosing the day or night of their departure. In spring the impetus typically comes with tropical air surging northward; in fall with polar air pouring southward. These situations almost always pair potentially critical influences—rising tem-perature with northward air flow, falling temperature with southward air flow. Some analysts stress the role of assisting winds, which permit travel with the least expense in energy. They point to the fact that any drop in temperature stimulates southward migration, even when the decrease is not enough to prevent the birds from remaining where they are. The regard cold in au-tumn and warmth in spring merely as clues that alert the birds to favorable air flows. Others argue that tem-perature affects migration directly. They seek support in the phenomenon of reverse migration, in which birds faced with unseasonable temperatures about-face and go southward in autumn or northward in spring.

The 24-hour rhythms of migration. The hours when migration takes place and resumes are rather uniform from date to date, though not the same for all species. In this respect the birds fall into two great groups that cut across taxonomic lines—daytime and nighttime mi-grants. Proven American nocturnal migrants on the basis of calls heard overhead or the numbers killed at night at lighthouses, TV towers, and ceilometers are most of the rails, the thrushes of the genus *Hylocichla,* all the vireos, the numerous American wood warblers, the tanagers, and the Indigo Bunting (*Passerina cyanea*). Conspicuous diurnal migrants are most herons, vultures and hawks, pigeons, swallows, and the blackbirds. In addition, some birds, such as ducks and geese and many shore birds, travel about equally by day and by night. Many common species still cannot be classified with certainty. The smoother air flows after the sun has set and the need to spend the day feeding explain why so many small birds migrate after dark.

The varying amount of hopping about of migrants in cages suggests which periods are normally spent in migrating. In many day migrants, migrational behavior shows up only as an increase in the year-around morning peak of activity; and as seen by radar and ordinary ob-servation most recognizable migration of exclusively day migrants takes place before noon. Usually, caged night migrants begin migratory restlessness after a brief sleeping pause in the first hour of darkness, and the ac-tivity culminates before midnight then gradually fades. The counting of flight calls and radar echoes of birds abroad, as well as data in this country from observers watching migration against the moon has indicated hourly changes in the volume of flight similar to typical activity patterns in caged night migrants. But the fre-quency of flight calls heard in America has generally varied in a dissimilar way and sometimes even in-versely, with the peak period in the hours just before dawn.

Migration routes and migration ranges. The much-used term *migration route* suggests that birds tend to move in streams along fixed, restricted pathways re-sembling human highways. Some species do this very thing, e.g., birds of the seacoast that travel coastwise to stay near tidewater habitat and soaring birds that exploit the thermal updrafts along mountain ridges. In-deed some authors still picture nearly all migration as of the preferred lane type. The criteria available for mapping bird movements in the past—the visible behavior of low-level diurnally migrating birds, daytime observations of grounded nocturnal migrants, and banding recoveries—are all tinged with biases that foster such a view. The more recent use of telescopes and radar to watch high-level migration by night and day has as yet produced little evidence that flight along set linear lines is the rule. The growing impression is that the bulk of bird migration is basically a broad-front movement.

During the period when banding records and observations of transients seemed a revelation of the whole truth, the so-called routes of a majority of North American migrants were plotted on maps. What was usually portrayed was actually the entire assumed migration range. These maps remain the chief source of information about the migratory distribution of individual species. As depicted, the areas of transient occurrence of a majority of the long-distance passerine migrants taper southward funnelwise from a broad northern breeding range and cross the Gulf of Mexico. Most exclusively western species travel overland in fall and do not go beyond Guatemala, while more than 50 eastern species venture, wholly or in part, all the way to South America.

Species with mapped migrations of special interest include the following: the Arctic Tern (*Sterna paradisea*), the distance champion, which flies from within the Arctic Circle to within the Antarctic Circle, sometimes by way of Africa; the eastern race of the American Golden Plover (*Pluvialis dominica dominica*) supposed to pass up the Mississippi Valley in spring but to return to South America by a direct flight from Labrador over 2500 miles of open sea; the Bobolink (*Dolichonyx oryzivorus*), which journeys extremely far for a songbird, from the northern United States and Canada to southern Brazil; and the Red-eyed Vireo (*Vireo olivaceus*), one population of which is said to retrace in migration the east-west trend of its recent range extension into the state of Washington, before heading southward.

Not all bird migration is even basically northward in spring and southward in autumn; it can be more universally described in terms of climate. Most migratory birds that breed in south temperate regions move in the same directions in the same periods as northern breeders but do so under reversed seasonal conditions. In mountainous regions, some birds satisfy their needs by altitudinal migrations up and down the slope from one life-zone to another. In the Old World many fall migrants head westward rather than southward, seeking sanctuary from the rigorous winter of Eastern Europe in the British Isles, where warm ocean currents provide more equable temperature.

Standard direction, drift, and lines of influence. A concept quite out of harmony with the idea of narrow fixed migration routes, yet explaining many of their apparent manifestations, has had great influence in Europe. It holds that each migrant inherits a tendency to proceed in a set direction appropriate for its population of the species. This *standard* or *preferred direction* leads the bird directly to its proper winter home or to an intermediate way-station, where a new standard direction may take over, Since each member of a population starts migrating in the same direction, regardless of its own exact location, the initial flow of birds is as wide as the summer range.

Certain kinds of birds, however, are reluctant to fly over certain kinds of terrain. Confronted with a formidable expanse of unfavorable habitat, they may veer temporarily from the standard direction to continue in a concentrated stream along the borders of the barrier. Boundaries with such effect are the *Leitlinien* of German authors—literally "leading lines" but more aptly translated simply as *lines of influence.* They apparently exert the most influence when their departure from the standard direction is least drastic, when the weather is bad, and when the birds are flying low or against the wind or in small numbers. Under extreme conditions, they may cause retromigration, in which a local migration stream becomes completely turned around. Radar observers feel that lines of influence are wholly disregarded at night. Those who have watched migrants crossing before the moon are not so certain though they would agree that the effects are considerably diminished.

A bird that keeps its body aligned in a standard direction will not actually move in that direction if a wind with a cross component is blowing. Then it may maintain its original heading and drift laterally or may turn to "drift" downwind. Whichever it does, the result is displacement. Thus drift helps to account for many puzzling features of migration: the straying of migrants from the eastern United States to the Far West and of migrants of the West to the eastern states; the occurrence of American vagrants, even small ones, in Europe; the diversion of flights of Continental migrants to Britain; and fall concentrations of migratory land birds along the western shores of the North Atlantic during northwest winds.

No other lines of influence have such marked effect as the borders of the sea. Birds sometimes hesitate to cross bodies of water so narrow that they can see the other side. Yet, beyond any doubt, some species of small land birds traverse hundreds of miles of open ocean as part of normal migrations, and evidence is strong that numbers of others do likewise. A strikingly clear-cut example is *Chalcites lucidus lucidus*, a race of Bronzed Cuckoo that summers in New Zealand and winters in the Solomon Islands, more than 2000 miles away. The most detailed studies of transmarine migration concern three great appurtenances of the oceans, each with different relationships of land and water—the North Sea, the Mediterranean, and the Gulf of Mexico. The majority opinion is that vast numbers of land birds regularly advance across these seas on a broad front. Only 15 years ago, it could be persuasively argued that nearly all spring migrants detour the Gulf of Mexico by overland or Caribbean routes. The several large flights of migrants that have been noted over the Gulf since then are only part of the accumulating evidence that has made this view less and less tenable. The possibility that these flights consist largely of birds blown back from the northern shore has been eliminated; but the extent to which drift from the shores of the southern half of the Gulf may be involved remains undetermined.

Concentrative factors. Most active migration is an unobtrusive process, invisible to the naked eye, taking place under cover of darkness or at high elevation. Moreover, the total number of migratory birds in comparison with the total volume of the air space through which they move is small. Passerines on their breeding grounds, migrants and nonmigrants together, spaced by their territorial demands, seldom seem lavishly abundant. If the entire continental breeding population of migratory birds were to fly southward on the same night, retaining their original spacial separation, the spectacle would not be very impressive. Thus, the only way migration can take on an appearance of profusion is through the bunching of birds. The factors that may contribute to such a result are various; the banking up of birds along lones of influence, the flocking habits of certain species, and the descent of migrants from an extensive area of sky into small areas of preferred habitat, such as islands in the sea and ecologic "islands" formed by oases in the desert or parks in urban districts. But the most sepctacular concentrations involve action of the weather.

In America in spring, mirgrants propelled northward in warm tropical air are likely to collide eventually with cold fronts moving southward, accompanied by rain, turbulence, cloudy skies, and adverse winds. The vanguard of advancing migration tends to stop along the forward edge of the cold air while other birds continue to pile in from the warm sector. The result is a damming up of migratory flow, the telescoping of migration into a dense grounded "wave" along a broad front. In the interior of the continent, the process seldom lasts very long on any one line. Continuing to edge southward, the front precipitates birds over an area of considerable depth. If it advances out over the Gulf of Mexico, however, the birds arriving from the southern Gulf and encountering the unfavorable conditions cannot descend at once but must remain aloft until they reach land. There, if the effort has exhausted them, they all tend to jam into the coastal woods. Since in such circumstances the ratio of compression may approach 100 to 1, every bush and tree may teem with migrants. Concentrations caused by the damming up of migration on a broad front have been called *arrested waves*. In the Gulf states most waves are of this type and so are associated with bad weather. But when favorable conditions again develop the arrested wave may take to the air en masse and surge forward as an *onrushing wave* that passes over some districts without alighting but deluges others with birds.

In fall, when the roles of cold and warm air are reversed, meteorological obstacles to migration are less effective and true migration waves do not develop in such obvious form. Thought adult migrants have been joined by their young and many more birds are on the wing, ground concentrations are less frequent and the drama of the spring movement is lacking.

Homing and navigation. Banding has shown that the end results of migration are often astonishingly precise. Year after year as long as they live, some individuals alternate between the same exact breeding place and the same exact wintering place. The many hypotheses that have sought to explain how these birds find their way have variously implicated all the known senses and several others besides. One writer even suggested that birds may *smell* their way to the goal. Homing performances and migratory orientation now seem to be, in large part, applications of the same faculty, though once there were determined attempts to divorce them—to represent the former as a product of random visual search, the latter as the following of landmarks known to the adult from previous experience or revealed to the immature through racial memory.

Homing experiments, most prolifically pursued with racing pigeons but carried out with numerous kinds of wild birds as well, have yielded returns to the loft or nesting site that far exceed the predictable results achievable by search patterns. The directions of departure from release points in unfamiliar territory have frequently shown approximately correct initial orientation by nearly all the birds even when released singly. On the other hand, the subjects of homing trials are far from infallible. Their records in terms of initial choice of direction, homing speed, and eventual attainment of the goal vary confusingly with the direction and distance of displacement, the region, the season, the weather, the temperature, the prior experience of the birds, and the stock employed. And careful efforts to duplicate the conditions of previous trials do not always succeed in duplicating results. The most startling homing performance on record is the return of a Manx Shearwater *(Puffinus puffinus)* 3200 miles from Boston to an island off the coast of Wales in twelve and a half days.

At any rate, a generally acceptable modern hypothesis of bird navigation must account for an ability to home quickly over long distances from strange territory. The specifications call for either a direct stimulus emanating from the goal (inconceivable in view of the distances sometimes involved) or a grid of perceptible and intersecting gradients of force that enables the bird to assess its present latitude and longitude in comparison with the latitude and longitude where it wants to be. Several proposed systems are based upon pairs of factors capable of producing grids: magnetic intensity and dip (1882 Viguier); east-west and north-south gradients of gravity (1946, Ising); Coriolis force and the vertical component of the earth's magnetic field (1947, Yeagley); the arc angle of the sun and the inclination of the plane of the arc (1951, Matthews); the altitude and azimuth of the sun (1952, Kramer); components of the star pattern (1957, Sauer).

The postulated navigational clues all exist. The difficulty is that their use by animals demands either unknown senses or incredible refinements of known senses. The celestial hypotheses (Matthews, Kramer, Sauer) further require that birds possess an internal chronometer. The existence of such a mechanism has been convincingly demonstrated by Kramer and his associates through tests in which caged birds were trained to choose the receptacle containing hidden food by reference to the position of the sun. A similar solar awareness had already been shown in bees, which locate objects in their immediate environment in much the same way.

The importance of a celestial element in bird navigation is now generally conceded, but doubt has arisen that the sun suffices as the only factor in daytime. The stellar navigation theory, impressively supported by the reactions of Old-World warblers to changing star patterns in a planetarium, has so far not been subjected to much criticism. With shifts of the star pattern simulating transportation to Russia and to the mid-Atlantic, birds with no previous experience in the wild reoriented quickly and appropriately. Yet the innate ability to navitate that such evidence implies has two very puzzling aspects: how the birds derive the necessary information from the continually moving stars of the real sky and why, if the faculty is general among birds, so many young ones should go astray under natural conditions. Banding recoveries of migrants of other species released in fall after actual transportation to localities off their normal route suggest that adults are able to readjust for the displacement and to fiy to their regular winter quarters. The displaced immatures, on the other hand, seem to maintain the standard direction that would have led them to the right goal, had they started out at the right place. Thus they tend to fly parallel to the normal track and to end up in abnormal wintering places.

ROBERT J. NEWMAN
GEORGE H. LOWERY, JR.

References

Lockley, R. M., "Animal Navigation," New York, Hart, 1967.
Thévenin, R., "Animal Migration," New York, Walker 1963.

MIRBEL, CHARLES-FRANCOIS BRISSEAU (1776–1854)

This French botanist, born in Paris, was one of the very few early workers to devote his attention to the structure of plant tissues in an era when taxonomy was dominant. He was the major exponent of the theory that all living organisms are composed of cells and demonstrated the cellular nature of the moss plant to the satisfaction of his contemporaries. It should perhaps be mentioned that his botanical success in part derived from his failure as a politician since he took up the study of plants only in middle age to relieve the boredom induced by his banishment from active politics.

MITOCHONDRIA

Mitochondria are cytoplasmic particles found in all cells, with the possible exception of BACTERIA and CYANOPHYTES. The name was coined in 1897 by the German cytologist, BENDA. It derives from the shapes which these particles frequently display: threads and granules.

Mitochondria are easily visible in living cells, particularly when examined by phase contrast microscopy. They can readily be stained in sections of properly fixed cells and tissues, both by classical cytological methods and by newer cytochemical techniques for oxidative enzyme activities. They have a characteristic structure in thin sections examined with the electron microscope. A multitude of enzymes, some of paramount importance, are localized in these cytoplasmic organelles. Because this includes enzymes involved in converting food energy to a form which the cell can utilize, mitochondria are often referred to as the cell's "furnaces." Mitochondria are usually highly plastic structures which change their form in response to physiological and pathological conditions.

Cells in tissue culture provide the best means of studying mitochondria in living cells. The mitochondria are constantly moving and changing shape in the cyto-

plasm. They may fragment, generally transversely, and several may fuse together, end to end. The addition of chemical agents to the culture medium provokes characteristic changes in movement and structure. During cell division the mitochondria which happen to be on either side of the cleavage plane are usually carried into the respective daughter cells. No signs of mitochondrial division are seen. Yet, in some way not understood, the mitochondrial number remains fairly constant in a given cell type.

The number has been estimated both in stained sections and in dispersed tissue suspensions. It is estimated that an average liver cell has 1,000 mitochondria, a sea urchin egg 15–30,000 and the "giant" ameba, about 500,000. In a few cells such as developing sperm cells of some species, the mitochondria may fuse into 4 or 5 large bodies, or even into a single body (the "nebenkern") which in one species attains a length of over 200 μ.

The size, and frequently the shape, of mitochondria are characteristic of a given cell type. In most cells, they are about 2–5 μ long, but in the pancreas they may have a length of 10–12 μ.

In some cells, the mitochondria appear to be anchored in specialized areas, where the energy they release is probably used in specialized function. For example, they are concentrated in areas of nerve, muscle, and kidney cells where impulse conduction, contraction and molecular transport, respectively, occur in specialized fashion. In other cells, where the energy they produce is presumably used in more general fashion, they are randomly distributed in the cytoplasm through which they probably move about.

The electron microscope reveals an external membrane around each mitochondrion; its existence had previously been deduced by biochemists from permeability experiments. Within the mitochondrion, there is an apparent chamber into which project many baffles or *cristae*. The most generally accepted hypothesis is that some of the mitochondrial enzymes are in solution within the chamber and that others, including those of oxidative phosphorylation, are integrated, in specific fashion, within the lipoprotein structure of the cristae. Oxidative phosphorylation involves many enzymes catalyzing the Krebs tricarboxylic acid cycle and electron transport, via the cytochrome system, to oxygen. Through their action, some of the food-stuff's energy is trapped in chemical bonds of adenosine triphosphate (ATP), the chief medium of energy exchange in cells.

Mitochondrial enzymes also catalyze a wide array of other chemical reactions. These are involved in such activities as the concentration of ions and the synthesis of proteins.

Many agents are known which uncouple phosphorylation from oxidation. Some of these, like the hormone, thyroxin, produce a marked change in mitochondrial structure. This may involve rounding of the mitochondria, loss of material from the matrix and tearing of the cristae. This is the kind of change commonly called "cloudy swelling" by pathologists. The electron microscope is helping to elucidate the relation of mitochondria to the intracellular fat droplets which develop in many physiological or pathological situations, to protein accumulations, and to many formed products of cell metabolism. In all instances thus far studied, the mitochondria do not *transform* into the other structures. The role of the mitochondria in the development of such cell structures is probably a metabolic one.

Among the important problems awaiting solution are the origin of new mitochondria, their interrelations with other cell structures, and the role—if any—which these organelles play in cell heredity.

ALEX B. NOVIKOFF

Reference

Brachet, J. and A. E. Mirsky, Ed., "The Cell," vol. 2, New York, Academic Press, 1961.

MITOSIS

Mitosis, the division of the NUCLEUS which results in daughter nuclei being quantitatively and qualitatively alike and with the same genetic constitution as the nucleus from which they arose, is the mechanism by which lineal heredity has been established. The desoxyribose NUCLEIC ACID component of the CHROMOSOMES, the genetic component, is present in duplicate in the resting nucleus. The amount of nucleic acid is doubled immediately before the onset of division and the subsequent events permit the separation of the products of replication to daughter nuclei in a confining cytoplasmic mass. Each longitudinal half-chromosome (*chromatid*) carries the identical nucleic acids of the chromosome before replication occurred. The chromatids of each chromosome are shortened, separated from each other and moved to daughter nuclei before the division of the cytoplasm (*cytokinesis*) is accomplished. Each chromosome in the resting nucleus of the solid type is a cylindrical spiral as it was at the end of the previous division. In the resting nucleus of the vesicular type the spiral may be relaxed and somewhat extended. The chromatids of each chromosome become identically coiled as prophase begins to form a spiral the gyres of which are numerous and of small diameter. As a spiral is assumed the gyres of the old spiral, which were assumed at the previous prophase and have persisted through the resting stage, become fewer in number and increase in diameter in an irregular fashion so that they become increasingly difficult to recognize as part of a previously cylindrical spiral. As the gyres of the old spiral (*relic*) decreases in number the gyres of the newly formed spiral also, but in a more orderly manner, decrease in number and increase in diameter until a constant number of gyres is found in the chromatids of each chromosome at metaphase. Any gyres of the old spiral that now remain result in chromosomes being variously curved or bent. Since the chromosome coiling sequence is initiated by many gyres which then decrease in number, the chromosome becomes more "loosely" coiled as division progresses and is usually uncoiled at the end of the prophase of the next mitosis and after the new coiling of that mitosis has begun. The dissolution of the nuclear membrane and nucleolus is accompanied by the dehydration and transformation of the karyolymph and a cytoplasmic element to form the spindle. The centromere of each chromosome moves to a position midway between the two poles of the spindle (*congression*) at a speed somewhat slower than the movement at anaphase which follows. The

centromere from the standpoint of behavior has been single. It becomes effectively two-parted and the halves separate from each other as the spindle becomes longer and more slender. They are directed and move toward opposite poles thereby pulling the chromatids (now chromosomes) behind them. When the movement of the chromosomes led by their centromeres toward the poles ceases, the chromosomes shorten, often by one-third, as the gyres continue to decrease in number and increase in diameter as seen in the prophase period. A nuclear membrane forms at the surface of the chromosomes assembled compactly at each pole. This membrane becomes inflated as newly formed karyolymph and nucleoi are resolved within and the division of the nucleus is completed.

A better understanding of chromosome structure and behavior during mitosis has been delayed not so much by the low resolving power of the visible light microscope as by the inadequateness of many methods of fixation and subsequent treatment. The relationally coiled chromatids so often described at anaphase, as well as the chromatic chromonemata embedded in the achromatic matrix are examples of observations based on material that is poorly preserved. The electron microscope will be especially useful when adequate fixation can be predicted.

The term mitosis now refers almost exclusively to the process by which a nucleus divides equationally. It is seldom applied to the two divisions that comprise meiosis.

LADLEY HUSTED

References

Darlington, C. D. and L. F. LaCour, "The Handling of Chromosomes," London, Allen and Unwin, 1960.
Swanson, C. P., "Cytology and Cytogenetics," Englewood Cliffs, N.J., Prentice-Hall, 1957.
Wilson, C. B., "Cell Division and the Mitotic Cycle," New York, Reinhold, 1966.

MOLLUSCA

This polymorphous phylum, the second largest in species, traditionally was divided into 5 classes Figure 1: [A] Polyplacophora or AMPHINEURA (chitons only). [B] SCAPHOPODA (tusk-shells). [C] GASTROPODA (snails, etc.). [D] PELECYPODA or Bivalvia (clams, etc.). [E] CEPHALOPODA (squids, etc.). In addition, the shell-less, worm-like Solenogastres, if they be mollusks, form at least a distinct class. The fossil shells of the class (?) Conularida remotely resemble those of living pteropods. The mainly Eopaleozoic Monoplacophora, recently considered a 7th (or 8th) class, include the living *Neopilina*, which was dredged from Pacific depths during the last decade. Finally, Ihering divided Gastropoda, which are most diversified in living forms, into 3 classes, which correspond to Prosobranchia, Euthyneura (Opisthobranchia and Pulmonata) and Pteropoda.

Mollusca are closely allied to the ANNELIDA, since their micromeres are formed by spiral cleavage. Commonly they develop trochophore-like or veliger larvae (see LARVAE, INVERTEBRATE), which may be free swimming in marine species. In all stages, they lack jointed

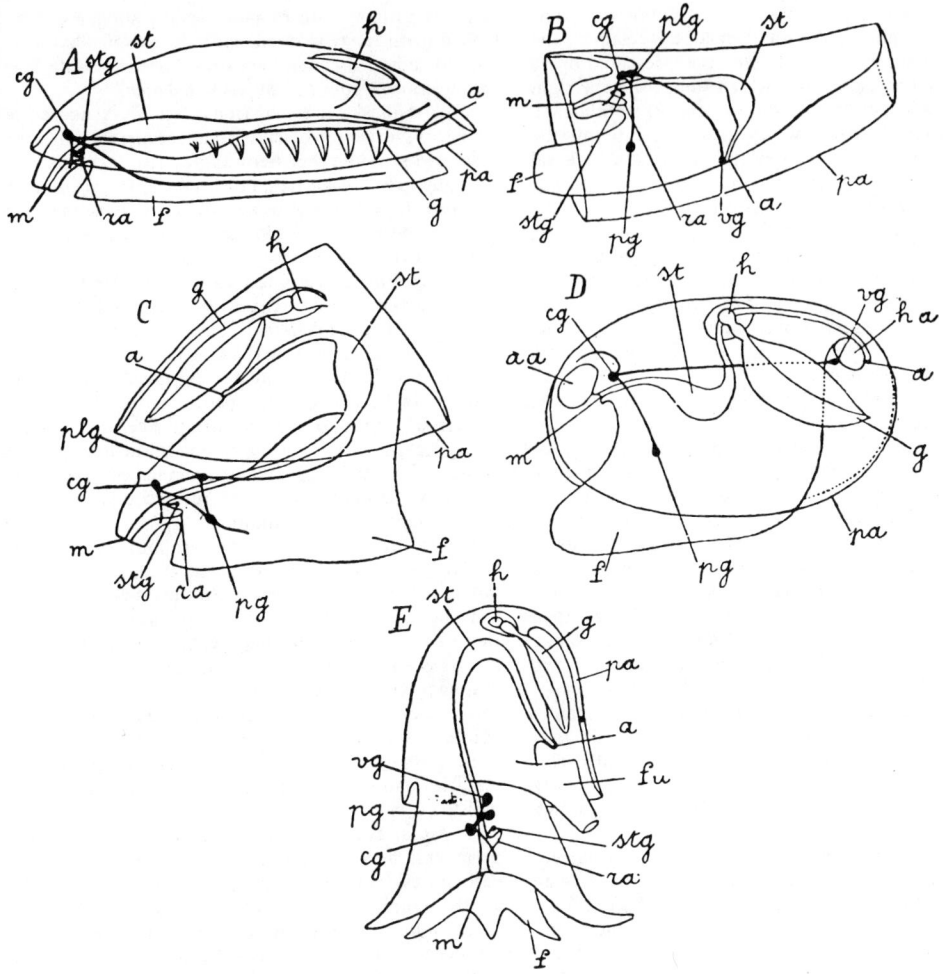

Fig. 1 Diagrams of five molluscan classes, from left sides. Symbol explanations are given in the text between brackets [], except stomach [st]. (From Pelseneer.)

appendages, and their bodies are unsegmented al- though, in the cephalopod *Nautilus* and in *Neopilina*, at least two metameres appear to be represented. Gen- erally they develop a univalve shell, 2 dorsally hinged valves (Pelecypoda), or 8 dorsally articulate plates (chitons). The underlying skin forms the mantle [pa], which usually is demarcated even when it invests an "internal" shell, or in naked slugs. Each shell or valve is fashioned by the free, mantle circumference [D:pa]; it grows by addition of successive rings; and differences in rates of growth along diverse arcs of its periphery determine its shape and sculpture. But, when first added, each new zone consists largely of organic ma- terials; its stiffness and hardness come later from de- positions of mineral salts, largely $CaCO_3$ in the form of calcite or aragonite, from body fluids through the surface of the mantle. Because $CaCO_3$ can be reab- sorbed or added at any time during life, mollusks maintain the balance of such salts in their blood better than do most animals.

A SHELL commonly has 3 layers: an outer, often horny and pigmented epiostracum; the middle layer, which ordinarily contains criss-crossing prisms of calcite; and the pearl-forming, usually more aragonitic nacre, which is added in laminae by the mantle surface. Thus, non-marginal incisions may be patched only by new nacre.

In many univalves, the animal fuses to its shell only in a small area, on which the faster, retractor muscles insert. Inside broadly conic [Cf. C] shells, this attach- ment may widen into a submarginal ring, or it may be subdivided, especially in slugs. In *Neopilina*, retractors occur in pairs. Also in bivalves and polyplacophores, retractors and adductors [D:aa & ha] may insert on several "scars," which often are paired, and/or on sub- peripheral zones.

Most mollusks retain varying degrees of bilateral symmetry, but in Gastropoda, the dextral side (left in sinistral animals) above the foot becomes much re- duced. Since this also involves the mantle periphery, the shell and the enveloped visceral mass commonly are spiral. Even when the shell is reduced [C] or lacking, viscera generally remain noticeably asymmetric. However, in most snails, a dextral liver-lobe is retained, and in Pulmonata, the right retractor is represented by an inconspicuous muscle-band, which runs down the right side of the hindgut. In many, bivalvular Prosobranchia, an operculum develops, which also may coil spirally, almost in one plane, but dextral shells have sinistral trap-doors. When shell apertures are not circular, opercles may develop almost concentrically.

The mantle collar usually is the edge of a fold [E:pa], which covers or surrounds a pallial cavity [A & C:pa]. Although any external region may develop "pseudo-branchs" and/or respire; "true" ctenidia [g], each usually with a sensory osphradium, commonly are inside the mantle cavity, which ordinarily loses the gill when it becomes a lung.

Molluscan epithelia, especially outside the mantle, generally are rich in mucigenous cells, although in some snails, they are most abundant in special crypts or tubes, like pedal "glands." These provide the stream of slime along which a flat sole [C:f] glides, primarily by the beating of cilia on its surface. Locomotion may be assisted by muscular waves, in which transverse strips are lifted and carried forward, and then pressed down and pushed caudad. Many pelecypods have a burrowing foot [D:f], which can be swollen terminally and then retracted, so as to pull the clam jerkily along. Some, like scallops, swim dorsad by rapid adduction of two valves, working against the hinge ligament. Cephalopods are propelled by jets of water from the mantle cavity through a funnelform foot [E:fu], but ordinarily develop circumoral tentacles ["f"], which usually bear suckers, and may be utilized when crawling. Many squids and some Gastropoda have fin-like flaps, which serve mainly to direct their course.

As in Arthropoda, body cavities are largely haemocoeles, and the "true" coelome usually is reduced to pericardial [h] and gonadic coela; from the coelomoducts, the kidneys and parts of the gonaducts supposedly are developed. The generally colorless blood may be bluish due to haemocyanin. Rarely, it is red and contains haemoglobin, which may be throughout the plasma or localized in corpuscles. Usually, blood from pallial veins is pumped into arteries and through haemal cavities by a heart [h], with primitively paired atria ("auricles") and a ventricle. In chitons and pelecypods, mouth [m] and anus [a] generally are at opposite ends, but in narrow univalves, they are juxtaposed by the shell aperture. In naked cephalopods, the anus remains near the oral end, but in gastropods with limpet-like or no shells, it often is deflected partly or completely caudad, as is also true in *Neopilina*. In limpets, the shell apex rarely is open, but in tusk-shells [B], it discharges faeces, ova and sperm.

Although absent from Pelecypoda and some Gastropoda and Solenogastres, the toothed tongue or radula [ra] is characteristic of mollusks. Its basement membrane bears denticles, which constitute longitudinal and "transverse" rows. Since the often cuspid blade of each tooth curves dorsocaudad and the ribbon commonly is drawn back and forth over "cartilages" by muscles, the radula rasps off food and helps ingestion. Transverse rows are laid down progressively from the caudal end in a radular pouch, and are shed from the oral end. In new rows, each tooth is softly pliable, but, while passing forward, acquires stiffly hard, imperious surfaces, although, if the enamel be broken or etched by acids, the flimsy core remains readily stainable. Gastropod teeth range in numbers from over 50,000 usable ones to some which function singly as a convolute fang for injection of poison, that has killed men.

Most mollusks are oviparous, but every degree of viviparity, except placental, may occur. Primitively, sexes are separate, but hermaphroditism appears sporadically in Solenogastres, Pelecypoda and prosobranch Gastropoda. It is characteristic of Euthyneura, in which ova and sperm may develop in one ovotestis, and self-fertilization seems common, although male organs may mature before female ones, and rarely vice versa. In many marine forms, fertilization is external, but most gastropods and cephalopods have copulatory organs. In Prosobranchia, a verge (lacking in terrestrial Helicinidae!) usually is permanently extruded although retractile. In Euthyneura, the penis commonly is an introvert, which may or may not sheath a verge or develop a vergic capsule. In some cephalopods, the hectocotylus is shed, and travels in search of a female. Noteworthily, accessory genitalia tend to be most complicated and multiversant in hermaphrodites.

In primitive mollusks [A], ganglion cells may be widely dispersed through scalariform nervous systems, not unlike those of flat worms. At the other extreme, in Cephalopoda [E] and some Pulmonata, the concentration of nucleate centers into cephalic ganglia [cg, pg, plg, stg & vg] exceeds that in other animals, but peripheral ganglia also are present, as in visceral systems of vertebrates. Thus, most mollusks develop heads, with paired tentacles, and eyes having univerted retinae. In contrast, pelecypods are acephalous, but may have tentacles and eyes on mantle edges; and chitons form indistinct, eyeless heads, but develop shell-eyes. However, pallial eyes also occur in some slugs. Origins of peripheral nerves seem far less conservative than in vertebrates, but nevertheless have been used analogously to support molluscan "homologies."

All molluscan classes are limited to salty water except Pelecypoda, which are always aquatic, and Gastropoda. Some marine gastropods, e.g., Heteropoda and Pteropoda, and many Cephalopoda are pelagic. Most other mollusks are bottom dwellers, although many marines have planktonic larvae. Adult scaphopods are sedentary; numerous pelecypods and some marine gastropods become sessile. Many gastropods and cephalopods are carnivorous and aglossate gastropods commonly are parasites, as are unionid larvae; adult pelecypods usually are planktonivores; but perhaps the majority of mollusks are omnivorous, with a predilection towards smaller, softer plants and animals, or products and organisms of decay. Adult mollusks range from about 1 mm. to 17 meters long.

On discontinuous continents and islands, land distances of a kilometer may form zoögeographic barriers, which are crossed only adventitiously and sporadically by proverbially slow snails or aquatic gill-breathers. Correlated with this, the few major groups of inland mollusks probably almost equal the marines in species and other minor categories.

Among fresh-water pelecypods, the Uniones constitute a specialized group, but the few other families and genera are less divergent. In Prosobranchia, 3 families form isolated groups, but the other families and genera, although including most freshwater species, seem to have been separated from marine counterparts mainly because of habitat.

The pulmoniferous Gastropoda include several sporadic families and genera of Prosobranchia and all Pulmonata, although in the last, some aquatic species fill enlarged lungs with water, or pseudobranchs may be developed. In fact, even terrestrial snails live actively only in the presence of water or high humidity, although desert species may be abundant because of aestivation. During such periods of reduced metabolism, they seal themselves into their shells by means of opercula, or by temporary formation of mucous (also calcareous) epiphragms. They may survive for years, primarily because most mollusks easily absorb $CaCO_3$ from their shells, and thus buffer the CO_2 of their blood. Correspondingly, most drought-resistant species or individuals are calcophiles.

The shells of most Mollusca readily become fossils, and apparently all the classes with tests were represented near the beginning of known paleontologic time. Marine fossils are abundant, but records of land forms are scantier, and their simple shells render identification with living groups difficult. The earliest fossil referred to the Pulmonata is Carboniferous, but, since prosobranch superfamilies, which include living pulmoniferous genera, are dated from the Ordovician, probably primitive Gastropoda were breathing air in the Eopaleozoic. Devonian unionoid shells are known, but the time of the first fresh-water pelecypod remains conjectural.

H. BURRINGTON BAKER

References

Abbott, R. Tucker, "American sea-shells," Princeton, van Nostrand, 1954.
Bronn, H. G., ed., "Klassen und Ordnungen des Tier-Reichs" Vol. 3, Mollusca, Leipsig, Winter, (various dates).
Hyman, L., "The Invertebrates" Vol. 6, "Mollusca I," New York, McGraw-Hill, 1967.
Pelseneer, P., "Mollusca," in Lankester, E. R., ed., "A treatise on zoology," London, Black, 1906.
Morton, J. E., "Molluscs," London, Hutchinson, 1958.
Pilsbry, H. A., "Land Mollusca of North America North of Mexico," Philadelphia, Academy of Natural Sciences, 1939-1948.
Thiele, Johannes, "Handbuch der systematischen Weichtierkunde," Jena, Fischer, 1929-1935.

MONOCOTYLEDONS

As noted under DICOTYLEDONS, all botanists divide the true flowering plants, apart from GYMNOSPERMS, into two main groups or phyla, i.e. DICOTYLEDONS, with two seed leaves, and Monocotyledons, with only one seed-leaf. THEOPHRASTUS (370 B.C.) wrote a *Historia Plantarum* in which he indicated the essential difference between these two groups of plants, and he was perhaps the first ecologist, for he gave accounts of woodland, marsh, lake river and other plant-associations. Nearly two thousand years later John Ray in Britain also called attention to the importance of the embryo and the presence of one or two cotyledons.

The origin of Monocotyledons has long been of great interest to botanists. Some have considered them to be polyphyletic (derived from several different stocks), others monophyletic (derived from a single source).

The latter view has more universal acceptance at the present day. Only a few families of Monocotyledons show close affinity with those of Dicotyledons; otherwise they have evolved on more or less parallel lines. The flowering-rush family, *Butomaceae*, and the Alisma family, *Alismataceae*, are related to the Hellebores and Buttercups (*Helleboraceae* and *Ranunculaceae* respectively), the carpels being free (an ancient character) in all of these families. Hutchinson's new system (see DICOTYLEDONS) for the Monocotyledons, therefore, begins with these apocarpus families, in contrast with the system of Bentham & Hooker, in which the orchids were placed nearly at the beginning, and the grasses at the end, and that of Engler, which began with the grasses and finished with the orchids.

Besides the apocarpous character, the more primitive families have also a separate *green calyx* and a *coloured corolla*, as in the related families mentioned above. On this account Hutchinson divides Monocotyledons into three main groups, 1, **Calyciferae**, with distinct calyx and corolla; 2, **Corolliferae**, with the calyx and corolla largely merged into one whorl, as in the lilies; and 3, **Glumiflorae**, in which the perianth has become much (sometimes completely) reduced, as in sedges (*Cyperaceae*) and grasses (*Gramineae*). This classification has gained a great measure of support amongst present day taxonomists.

In the first and most primitive group, *Calyciferae*, besides having free carpels, *Butomaceae* have retained that which appears to be another ancient character, i.e. the ovules are *scattered* over the inner surface of the carpels and not confined to the placentas, a feature they share with *Cabombaceae* among the early Dicotyledons. Related derived families have much reduced flowers and have become aquatic, some such as the wrack-or eel-grass family, *Zosteraceae*, having become adapted to life in the sea.

A large family which has retained a distinct calyx and corolla is *Commelinaceae*, to which the Virginian Spiderwort, *Tradescantia virginiana* belongs. Most of this family, however, is found in the tropics and favors moist habitats. Related to it are three small families which may be regarded as equivalent to the *Compositae* amongst the Dicotyledons, for their flowers are collected in a similar way into heads. A single British example of one of these families, *Eriocaulaceae*, namely *Eriocaulon septangulare*, is of exceptional interest because it is found only on the coasts of the islands to the west of Scotland and on the west coast of Ireland, and also in North America;—naturally the subject of much speculation as to its mode of dispersal.

Of the large tropical family *Bromeliaceae*, many are epiphytic and a few of economic importance, such as the pineapple, *Ananas sativus*. This family was until fairly recently considered to be confined to tropical and subtropical America until *Pitcairnia felicina* was discovered on the opposite atlantic coast of Guinea, in tropical Africa. Climax families in this alliance are the Ginger family, *Zingiberaceae*, and, of very great economic importance, the banana family, *Musaceae*. In the more advanced or highly evolved of the families the stamens are reduced to one, as in most orchids.

The family *Liliaceae* is basic for the *Corolliferae*, in which the two original whorls of perianth leaves are more or less alike and petaloid, often merged into one whorl. Many of this family possess a more recently evolved type of root-system, namely a *bulb*. Those with an ordinary kind of rootstock include the Aloe, and Asparagus, lily of the valley, *Convalaria majalis*, and, not least of all the Aspidistra, made famous in verse and song at any rate in western parts of the old world. The lily, *Lilium*, tulip, *Tulipa*, bluebell, *Scilla*, and the meadow saffron, *Colchicum*, are familiar bulbous examples of this large and attractive family.

Closely related through *Aspidistra* is the arum family, *Araceae*. In all this family the minute flowers are arranged on a fleshy axis, the spadix, which is usually more or less enclosed by a large often coloured bract, the spathe (really a modified leaf). The skunk cabbage, *Lysichitum*, is a very primitive example of this family with the spathe only loosely wrapped around the stalk of the spadix and clearly revealing its origin.

The Amaryllis family, *Amaryllidaceae*, is clearly a step higher up than the *Liliaceae* in the evolutionary scale. In this family the rootstock is always a bulb and the inflorescence an umbel (a parallel to the hemlock family in the Dicotyledons); it also retains the six stamens of the *Liliaceae*, but its ovary has largely become inferior. An inferior ovary is generally regarded as having been derived from a superior ovary, hence the ovary of the more highly evolved family *Orchidaceae* is always inferior. The onion and the leek are important garden crops. Next in order comes the Iris family, *Iridaceae*, with many very beautiful decorative plants but very few of economic importance. In this the stamens are reduced to three. To a closely related family, *Agavaceae*, belong the remarkable Dragon tree of the Canary Islands, *Dracaena draco*, and several valuable fibre plants such as New Zealand hemp, *Phormium tenax*, and sisal hemp, *Agave sisalana* and *A. amaniensis*, much cultivated in tropical East Africa.

Palms represent a complete climax development both in habit and structure. Many are trees which give the vegetation of tropical regions such a different look from that of more temperate zones. Their fibres are much used, and their fruits are vital to the life of the inhabitants of certain more or less desert regions; examples are the date palm, *Phoenix dactylifera*, the coco-nut, *Cocos nucifera*, and the oil palm, *Elaeis guineensis*.

Also related to the Amaryllis family is the haemodorum family, *Haemodoraceae*. These are found mainly in the southern hemisphere, and they are especially interesting because they seem to be something of a connecting link with the orchids. In the related *Apostasiaceae*, for example, the stamens are sometimes reduced to two, as they are in the more primitive members of the orchid family.

The peak or climax of floral structure in the *Coroliferae* or petaloid monocotyledons, is reached in the orchid family. Some of them grow on the ground (terrestrial orchids), but the majority (perhaps 3 to 1) are epilhytic and grow on trees and bushes or rocks. There is great economy in the number of stamens, 2–1, and the pollen, granular in the more primitive group, becomes waxy and in masses in the more advanced.

The most important family of the *Glumiflorae* is the grass family, *Gramineae*. They are perhaps the most highly evolved and the most successful of the monocotyledons, for they grow where other vegetation is almost non-existent. Without grasses mankind would be but a shadow of itself, as grasses are almost the sole food for cattle, sheep and other animals which in turn provide food for the human race.

The much-reduced flowers of grasses are pollinated by the wind, and grass-pollen is the chief cause of hayfever in many countries. Grasses are easily recognized from sedges (*Cyperaceae*) by their cylindric jointed stems which are closed at the nodes but otherwise hollow; and at the junction with the leaf sheath and blade there is usually a membranous rim or hairy outgrowth, the ligule. Probably the most primitive tribe of grasses is that to which the Bamboo belongs, the *Bambuseae*, some of which are woody and of economic value. It is significant that in one subtribe of this group there are still retained 6 stamens as in more primitive monocotyledons. In most other grasses the number of stamens is reduced to 3. The most important food grasses are wheat, *Triticum*, barley, *Hordeum*, oats, *Avena*, rice, *Oryza*, maize or indian corn, *Zea mays*, and various kinds of millet. Some grasses are troublesome weeds, especially couch grass, *Agropyrum repens*. On the other hand hay often owes its fragrance to the sweet vernal grass, *Anthoxanthum odoratum*.

J. Hutchinson

Reference

Hutchinson, J., "Families of Flowering Plants" 2nd ed., Oxford, Clarendon Press, 1959.

MONOTREMATA

The monotremes form a distinctive order of MAMMALS, represented today by two families (Ornithorhynchidae—platypus; Tachyglossidae—echidnas) in which the oviducts are separate along their length, and together with the ureters, enter a urogenital sinus which passes back to form with the rectum a cloaca with a single external orifice. The order has been grouped in the subclass Prototheria on account of the egg laying habit found neither in the early mammals of the subclass Allotheria nor in the more advanced subclass Theria.

The monotremes retain many of the structural features of their reptilian ancestors. Unlike other mammals, they possess separate pre- and post-frontal ossifications and the glenoid cavity is formed entirely by the squamosal. The pectoral girdle shows a persistent interclavicle (episternum) with a precoracoid, coracoid and scapula on each side. There is no scapula spine as in the higher mammals in which the precoracoid is lost and coracoid reduced to a process of the scapula. The pelvic girdle carries epipubic bones which, as in some reptiles and marsupials, act as supporting structures for the ventral body wall. The brain is well developed compared with that of reptiles, but the corpus callosum, a bundle of nerve fibres which connects the neopallial cortex of the cerebral hemispheres of higher mammals, is absent in the monotremes.

The egg of both the platypus and echidna has a flexible shell and is heavily yolked; early development follows a reptilian pattern.

In the male monotreme, the testes are abdominal and do not descend into a scrotum. The penis, which carries only semen and has a divided glans, is provided with a prepuce as in higher mammals. Of the accessory glands, only bulbo-urethral (Cowper's) glands are present. The males only have a spur on the inside of each hind leg. A toxic secretion is formed by femoral (crural) glands, active only during the breeding season, which drain through the spurs. In the platypus the secretion contains a powerful blood coagulant and may cause death in small animals and extreme discomfort in man. In the echidnas the spurs and femoral glands are less well developed.

Like other true mammals, the monotremes possess hair and suckle their young. Body temperature varies widely between 25–36°C., being often lower in the morning than the afternoon, although regulation appears to be largely independent of the external temperature. Seasonal changes in body temperature have been observed, and the lowest temperatures are encountered during "hibernation"—rest periods of reduced activity which have been observed to last from 6–120 days in the echidna, but which have not been fully investigated.

It is usually considered that the monotremes arose in the early Mesozoic from either a basic mammalian stock or from a mammal-like reptile which may already have diverged from the evolutionary line, or lines, leading to the Theria. Nothing is known of the former distribution of the group and the earliest fossils recognisable with certainty come from Australian Pleistocene deposits. Some authors have thus been led to suggest that the monotremes have always been confined to the Australian region and arose there in the late Mesozoic from marsupial ancestors, a conclusion which has also been supported on the basis of anatomical similarities.

Despite their fundamentally primitive structure, both the platypus and echidna show many specialisations related to their mode of life.

The platypus (*Ornithorhynchus anatinus*) is found in streams and lakes of all altitudes in Eastern Australia from Cape York to Tasmania. The body reaches a total length of 2 feet. The head is flattened, the skull and lower jaw are projected forwards to form a duck-like bill with a highly sensitive rubbery covering, the tail is beaver-like and the feet are webbed but retain powerful claws used in burrowing. The hair is dense and velvety in texture, brown above and grey beneath, with a variable suffusion of yellow or reddish-brown. There is no ear pinna, the eyes and ears are covered by a fold of skin while the animal is swimming. Vocalisation is limited, the only sound uttered is a scolding growl.

The platypus feeds during the early morning and evening. Worms, crustaceans and insects are taken underwater and masticated at the surface. Horny tooth plates replace the deciduous teeth of the juvenile. Both males and females dig burrows in the banks of streams or lakes; these are normally open above water at each end. In the breeding season (July–September) the female alone constructs a long burrow, sometimes up to 60 feet in length, off which a nesting chamber is formed. Grass and leaves for the nest are carried between the body and the tail folded under it. The female lays 1–3 eggs and incubates continuously until they hatch (7–10 days). The young first enter the water at about 4 months of age.

The echidnas are specialised diurnal anteaters with strong digging claws and a long muzzle with a small terminal mouth. The genus *Tachyglossus* is represented by two species in Australia and Tasmania, while *Zaglossus*, with both 3-toed and 5-toed species, is confined to New Guinea. The body is about 20 inches in length and

rounded in form, the tail a mere stump. Teeth are absent; horny ridges on the tongue grind the insect food against the bony ridges of the palate. The sense of smell is extremely well developed. Strong backwardly projecting spines project through the brown under hair and render the animal almost immune to attack from predatory animals. Although the echidna digs itself into the ground when alarmed, it does not form tunnels. A single egg is laid usually between the end of June and the end of September and is incubated in a shallow transitory pouch formed by a thin flap of abdominal skin during the breeding season. This pouch also serves for the transport of the offspring until a body weight of about 400 g is reached some weeks after hatching.

Chromosome numbers and DNA values suggest that the monotremes are more reasonably regarded as extreme mammalian forms rather than as closely related to the birds and reptiles. The male appears to be heterogametic but the sex-determining mechanism is still not known in detail. Somatic cells of the female platypus have 54 chromosomes, those of the female echidna have 64. Corresponding counts of male somatic complements are 53 and 63 respectively.

J. A. THOMSON

References

Burrell, H., "The Platypus," Sydney, Australia, Angus and Robertson, 1927.
Griffiths, M., *Comp. Biochem. Physiol.* **11**:383–392, 1965.
Griffiths, M., "Echidnas," London, New York, Pergamon Press, 1968.

MORGAN, THOMAS HUNT (1866–1945)

Awarded the Darwin Medal in 1924, the Nobel Prize in Medicine in 1933 and the Copley Medal of the Royal Society in 1939 for his pioneering research in genetics, Morgan formulated many principles which have served as working hypotheses for subsequent investigations of hereditary mechanisms. He was born September 25, 1866 at Lexington, Kentucky, the son of the American consul to Italy and later, a soldier in the Confederate Army. He received his undergraduate degree in 1886 at the Kentucky College of Agriculture and Mechanical Arts (now the University of Kentucky). He then entered Johns Hopkins University in the fall of 1866 and studied embryology under William Brooks. He was graduated with a Ph. D. Degree in 1890 and in 1891 was appointed Associate Professor of Biology at Bryn Mawr. Here he remained until 1904 teaching and doing research in embryology and it was at this time he married Lillian V. Sampson, one of his students. In the same year he was made Professor of Experimental Zoology at Columbia, a post he held until 1928, when he went to California Institute of Technology to organize a Division of Biology. He remained here, with frequent summer trips to Woods Hole until he died at Pasadena, California, December 4, 1945.

Morgan's first interests were in embryology and differentiation, later turning to genetics as a basis for solving the current problems in adaptation and evolution. His major contributions to genetics were published while he was at Columbia. Using rats and mice as his first laboratory animals, Morgan quickly recognized the advantages of the fruit fly, *Drosophila melanogaster* for studies related to variation and heredity. With him in his laboratories at Columbia were additional significant contributors such as A. H. Sturtevant, C. B. Bridges, H. J. Muller, and others. The results of the research of Morgan and his co-workers led to a series of genetic publications which were the most important to appear since the work of Gregor Mendel. Space does not permit the naming of all his papers and books on regeneration, evolution and genetics but among the more classical are, *The Mechanisms of Mendelian Heredity* in 1915 along with Sturtevant, Muller and Bridges which postulates an explanation of inheritance on the basis of the chromosome theory. A later edition (1923) contains many revisions indicating the rapidity with which knowledge was gathered during this eight year period. In 1916, appeared *Sex-Linked Inheritance* in *Drosophila* which gives understanding and significance to linkage from his work with *Drosophila*. *Theory of the Gene* published in 1917 still serves as a practical foundation for basic research. Beginning in 1888, during his graduate studies, Morgan's summers were spent at the Woods Hole Biological Laboratory where he carried on research with both *Drosophila* and marine organisms. Toward the end of his stay at Columbia and while at the California Institute his active research with *Drosophila* diminished and most of his time was spent at Woods Hole and the marine laboratory he created at the California Institute of Technology investigating regeneration and development of marine forms. At the time of his death in 1945, he was still actively engaged in research.

RICHARD M. CRIBBS

MORPHOGENESIS, ANIMAL

"The Lord . . . formed thee from the womb" is an expression of one of the prophets. Morphogenesis is the acquisition of form by the developing organism; it is the process of becoming a behaving being. The process, particularly in early stages, follows a fairly uniform pattern in most many-celled animals, and morphogenesis follows some predictable rules. This article will discuss some selected aspects of it in many-celled animals, primarily vertebrates.

Morphogenesis begins when the two gametes, the egg and the spermatozoan, fuse to form the zygote, the fertilized egg. The egg itself is relatively structureless at the time that it is fertilized; it is a complex consisting of cytoplasm and nucleus. The latter contains the chromosomes, which in turn contain deoxyribonucleic acid; this is the substance that conveys genetic information from one generation to the next. The unfertilized egg contains the genes for the new organism provided by the mother; at fertilization, the spermatozoan introduces those from the father. In most many-celled organisms the egg does not develop unless fertilized by a spermatozoan. The spermatozoan has a double role at fertilization. It not only transfers to the egg the genetic contribution from the father; it also activates the egg to develop. The unfertilized egg and the spermatozoan themselves are essentially end-products that have lost their power to divide. By uniting they form a new and temporarily unspecialized unit, the zygote, which will divide and which will eventually give rise to all the cells of the body and to new reproductive elements, eggs or spermatozoa. The sex of the new being is one of the features fixed at the time of fertilization, by the chromosome complement resulting from fusion of the egg and the spermatozoan. The large problem with which we are faced in the study of morphogenesis is the way in which the many million specialized cells that make up the adult are derived from the zygote.

The fertilized egg divides by mitosis to form many cells. These then move according to specific patterns until they are in the positions which they occupy in the adult organism. Once they have attained these positions, they differentiate, that is, they specialize and develop the structure that enables them to carry out their particular functions in the adult: A nerve cell becomes specialized to conduct the wave of electrochemical change that is the nerve impulse; rod and cone cells in the eye specialize to become receptive to light; body muscle cells become striated and are then able to contract to move the body or its parts. At the same time that the cells are differentiating, they become grouped into tissues, assemblages of like cells, and tissues assemble with other tissues to become organs. The organs assume their own particular form; for instance, the kidney of a man is bean-shaped, and that of a whale is lobular. The organs grow as they change their shape, and the organism as a whole also changes its shape and grows, until morphogenesis is complete.

Since the amphibian egg has received much study, the description of early morphogenesis to follow will be undertaken using this form as an example. Such a description serves only as an illustration; frogs and salamanders, which belong to the group Amphibia, develop somewhat differently from each other, but their variation from the pattern is minor as compared to that of birds and mammals, for instance, which vary not only from amphibians in their development, but also from each other. The morphogenesis of invertebrate animals differs even more from the pattern described here, but fertilization, cell division, cell movement, and cell differentiation are still essential features of it.

In the amphibian, after the spermatozoan has entered the egg, the nuclei of egg and spermatazoan fuse, and considerable activity then ensues which results in rearrangement of many of the materials contained in the egg. This shows grossly in the reaction of the egg to gravity. The egg rotates, so that its dark pole is uppermost and its pale yolkier pole down. There is little further change in the polar axis until later, in amphibians a matter of several days.

Once oriented, the egg divides by mitosis to form many cells; during early morphogenesis, the cell division is often called cleavage or segmentation. It divides the one-celled zygote into many cells, but it does not involve growth; the total amount of protoplasm has not yet increased. In the amphibian, the segmentation continues until there are about 1000 cells, which, although they vary in yolk-content and pigment marking, are still quite similar. This multicellular aggregate is sometimes called a blastula. It is a rounded ball of cells which becomes hollow; the cells at the top of the egg have divided more rapidly and are smaller than the yolky cells which are larger because they are found at the bottom of the blastula and inside it. These will later form the main portion of the alimentary canal. The more rapidly dividing cells will form most of the rest of the embryo after undergoing the movements about to be described.

The movements of the cells are highly patterned (Figs. 1–6) and organized; they are called the morphogenetic movements. They first become evident in the amphibian egg when a slight shift in the polar axis begins to occur. This indicates the beginning of a process,

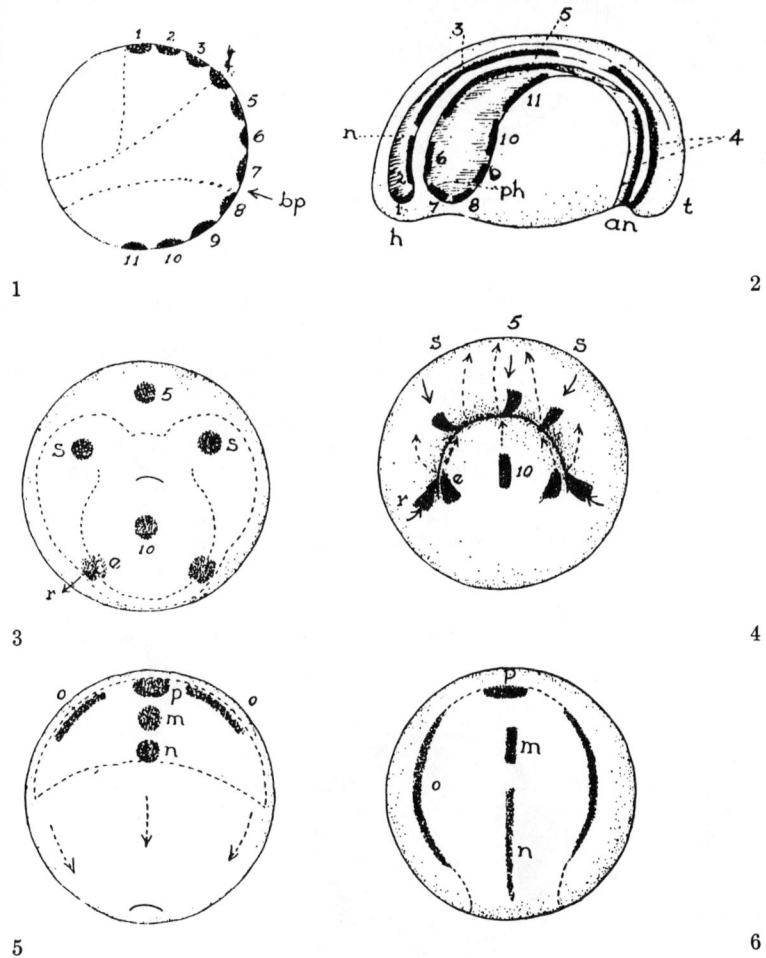

Figs. 1–6   Vital-staining experiments on urodele gastrulae.

Fig. 1   Eleven marks (1–11) placed in the median line of a late blastula. Lateral view. The upper arrow indicates the border in invagination; *bp* = point of origin of blastopore. The dotted lines indicate the borders between the main prospective areas.

Fig. 2   The same embryo in early tail-bud stage, with position of marks 1–11. *An*, anus; *h*, head; *n*, neural tube; *ph*, pharynx; *t*, tail.

Fig. 3   Marking of prospective mesoderm and entoderm in the early gastrula. Ventral view. The dotted lines indicate the borders between prospective areas. Marks 5 (prospective notochord) and 10 (median entoderm) as in Fig. 1; *e*, mark on the outer edge of entoderm field; *r*, mark on ventrolateral part of marginal zone (prospective lateral mesoderm); *s*, mark on anterior somite material.

Fig. 4   The same embryo as in Fig. 3, in middle gastrula stage. Note the changes in shape and in position of the marks. The solid arrows indicate the direction of past movements on the surface; the dotted arrows indicate the movements after invagination. Designations as in Fig. 3.

Fig. 5   Marking in the prospective medullary-plate area (dotted lines) of the early gastrula stage; *m, n,* marks in the median line; *o*, lateral marks; *p*, mark on the animal pole. The arrows indicate the directions of movements.

Fig. 6   The same embryo as in Fig. 5, in medullary-plate stage.

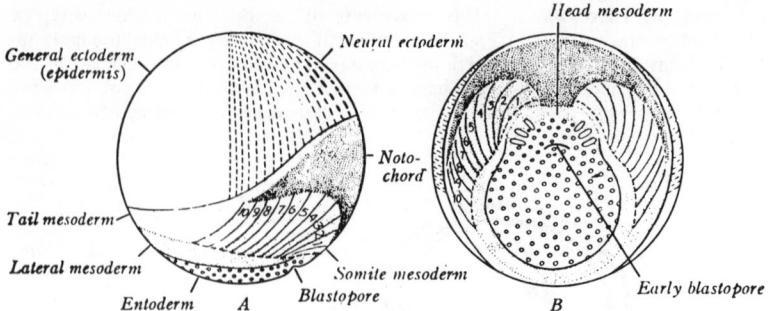

Fig. 7   Maps of prospective parts of embryos of tailed amphibians at the beginning of gastrulation. *A*, side view; *B*, view from vegetal pole.

often called gastrulation, which is first made evident by the appearance, near the lower pole of the egg, of an indentation called the blastopore. At first this is simply a rough series of cracks appearing near the lower pole as cells from the outside of the egg begin to move towards the inside. Eventually a distinct lip is formed as the cells move inwards; the blastopore lip becomes a transverse line, then crescentic, then horseshoe-shaped. As cells turn in around the whole circumference of the egg it becomes a complete circle. The outside cells push towards the blastopore, then turn in, establishing the embryonic axis.

Among the first cells to move inwards are those which will form the chorda-mesoderm, the middle layer of the embryo. In the midline, this later forms a

long bar, the notochord, around which later the vertebrae are formed. The mesoderm on either side of it will form muscle, skeleton, and other internal tissues and organs, with the exception of the alimentary canal and its derivatives, formed by the yolky cells (entoderm). The cells remaining on the outside, after the inward migration of chorda-mesoderm, are ectoderm, which will form the skin, and the nervous system and sense organs.

As a result of the shift in position of the chorda-mesoderm cells, the yolk-containing cells, more passive than the rest, become displaced, and since they are heavier than the others, the whole embryo rotates (compare the position of the yolky cells, indicated by small open circles, in Figures 8B and 8C in the accom-

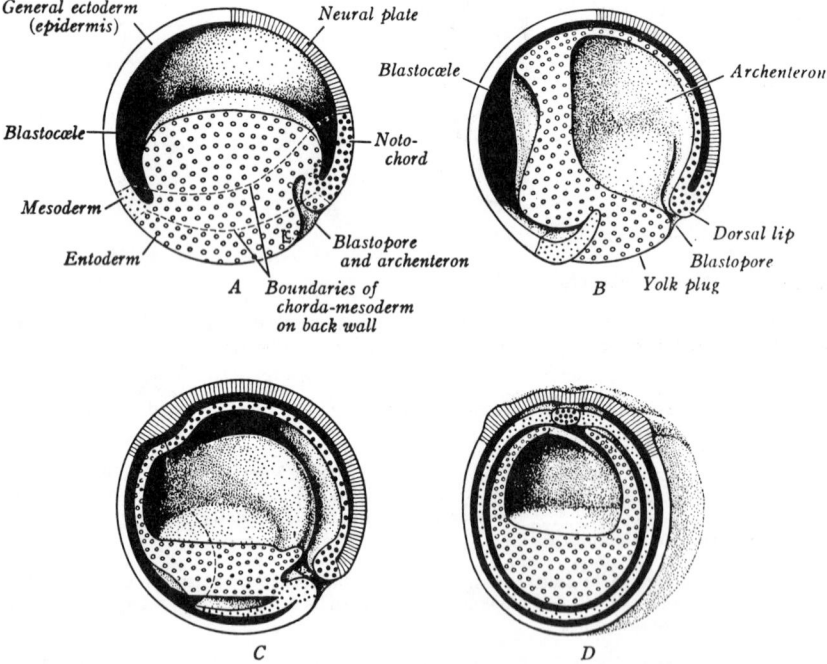

Fig. 8   Stereograms of gastrulation in tailed amphibians. *A–C*, early to late stages, showing the movements of areas differentially marked on the cut surfaces of longitudinal hemisections. *D*, caudal half of stage *C*, shown by a transverse hemisection.

panying diagrams). The region where the blastopore first began to form, originally inferior, in its new location after the rotation now marks the hindmost part of the embryo, where the tail will form after the yolk is completely covered by the pigmented cells that grow over it. At about this time, the embryo loses its spherical form and becomes ovate, its long axis representing the head-tail axis of the developing organism.

After the egg has undergone repeated cell divisions, and after the cells have moved into their position, many of the cells still resemble one another, but their capacity for the formation of diverse cell types has become very much delimited. By the time head and tail ends of the embryo are visibly distinguishable from one another, these structures are not transformable one to the other, although it is possible, in some larvae, to interchange the two structures by grafting; they maintain their integrity, if grafted at this stage, as head in tail region, or vice versa.

It is during the course of gastrulation that the ultimate fate of many of the cells is fixed, and in the amphibian the movement of cells through the blastopore lip is of importance in determining this fate. The materials which were originally on the outside of the egg have assumed new values as they have migrated through the blastopore lip into the inside of the egg, as first demonstrated by Spemann (1924).

Once the chorda-mesoderm cells pass through the blastopore lip into the interior, they come to lie under the part of the ectoderm that will later change its shape and form nervous system. Actually, the chorda-mesoderm cells influence the overlying cells to differentiate in this direction; they form only body skin, not nervous system, if they are not underlain by chorda-mesoderm. In other words, during the time that the chorda-mesoderm is accomplishing its own destiny, it is also influencing the surface material under which it has grown and compelling it to form brain and spinal cord. The imposition of the change in the overlying cells is called induction.

Furthermore, as the material from the blastopore lip moves inwards, it also develops the specific property of organizing head, trunk, and tail. Spemann showed that when the early blastopore lip is grafted to another embryo of the same age, not only do the cells that turn in induce a new nervous system; a new embryonic axis is induced with characteristic structures of head, trunk, or tail. Sometimes a whole new embryo is formed. Accordingly, Spemann called the first part of the blastopore lip to be formed the organizer.

At the same time that induction is taking place, a part of the inductive property seems to be passed on, and we find that we have inductions of the second or third order. For instance, in the amphibian, head mesoderm induces the nervous system to form a cup which will become the retina of the eye (the optic cup). During its early growth, the optic cup touches the outside embryonic ectoderm, and influences it to form the lens of the eye instead of skin. If this contact is experimentally prevented the outer layer forms skin and the eye is lensless, which renders it useless for sight. The contact of the optic cup is a critical agent necessary for the normal morphogenesis of the eye. Somewhat later, the lens, after it has been formed by the ectoderm and has sunk below it, influences the ectoderm which covers it to become transparent and form part of the cornea. The embryonic tissues resemble the runners in a relay race. As the baton is passed, a new runner is activated. He runs at his own pace, and in due time passes the baton to another runner. In the embryo the passing of the baton signifies that one tissue or cellular layer is changing the *quality* of another.

The process of morphogenesis is now well on its way, for the pattern of regional disposition, the movement from the outside to the inside has been completed, the axis of the future organism has been laid down, and the quality for forming diverse structures has been imparted by induction. The limitation of the capacity of cells imposes a rigidity and in its way controls the resulting form. By cellular differentiation the tissues are produced, and by tissue combinations are secured the organs that play such a necessary role in the organism as a whole.

These are the fundamentals in the development of form—the rest is continued growth with not only the limitation in the capacity of the cells compelling them in the main to duplication without change but also a limit on the size and shape of the structure which is produced. It is this important regulation which stops the growth process; this regulation is as important in morphogenesis as the factor of induction.

Many attempts have been made to pinpoint biochemically specific agents responsible for embryonic induction but none have yet been completely successful; in the case of some inductive systems, nucleoproteins, nucleic acids, or nucleotides seem to be implicated but it is not clear exactly how. It is probable that no single agent is responsible for the induction of all embryonic structures. Furthermore, some embryonic structures, even within the amphibian egg, differentiate in the absence of inductive influences; this is especially true of intestine and other entodermal structures. In the eggs of some invertebrates, the fate of all the cells seems fixed as early as during segmentation stages. Thus cellular differentiation during morphogenesis may be influenced by other than the inductive processes described above.

While the study of induction continues, experimental embryologists also investigate the control processes of morphogenesis from other points of view. Some experimental embryologists plan their experiments so that the embryo or a part of it is compelled to complete its morphogenesis in a way other than that normally maintained. A constant is varied so that the investigator can study the provoked variation which proceeds along a line determined by the embryo; in some cases an embryo which normally forms a structure in one way can form it in another. This is called embryonic regulation. If the embryo does not regulate, it may die; organs or body parts may be doubled, or they may be missing. Experiments in the early days of experimental embryology showed that factors controlling morphogenesis reside in the genes, in the cytoplasm, in interactions between genes and cytoplasm, and in interactions between cells. These determinants are modified in each individual by environmental factors—chemical and physical—which limit, regulate, and influence the various qualities which involve an embryo's manifestation of its capacities both to conform to or to diverge and deviate from its plan of organization. Under optimal conditions they determine not only the characteristics of the genus and species to which the individual organism belongs, but also the characteristics of its constituent parts: in man, the contour of arms, legs, and body, the color of the eyes and hair, and so forth.

The study of such determinants is carried out at both cellular and subcellular levels. At the cellular level, experiments are being performed to ascertain the factors governing the morphogenetic movements which are so important for early morphogenesis. Groups of cells may be removed from the embryo and disaggregated, mingled with other cell types, allowed to reaggregate and to differentiate in tissue culture. They sort themselves out according to type, exhibiting numerous specific affinities and disaffinities. This suggests that considerable cellular autonomy may be responsible for the cell movements that occur during morphogenesis. But what organizes the movements into orderly sequences in time and space still remains to be elucidated. Ultimately, of course, the difference between one cell and another, during this phase of morphogenesis as well as in other phases of the life cycle, is determined by the genes and influenced by various environmental factors.

Many biochemical studies are being carried out which illuminate the differences between one part of an embryo and another in terms of specific biochemical and metabolic systems. These analyses are extremely important, in that they describe the morphogenetic components and activities in far more precise and accurate terms than is possible by any other mode of study. Furthermore, they permit some approaches to the problems of how genes act in development. It has been shown, for instance, that different forms of certain enzymes (isozymes) are present in different organs during development, and the particular type of enzyme present has been shown to be genetically determined. Geneticists have made great progress in the elucidation of genetic control systems in microbial organisms; the applicability of their models to an explanation of morphogenesis in multicellular organisms is being widely discussed, and attaining an understanding of development in terms of biochemical genetics is the next great challenge to investigators of morphogenesis.

J. S. NICHOLAS
(Revised by JANE M. OPPENHEIMER)

References

Balinsky, B. I., "An Introduction to Embryology," 2nd ed., Philadelphia, Saunders, 1965.
R. L., DeHaan, and H. Ursprung, eds., "Organogenesis," New York, Holt, Rinehart and Winston, 1965.
Weber, R., ed., "The Biochemistry of Animal Development," New York, Academic Press, 2 vols., 1965, 1967.

MORPHOGENESIS, PLANT

During the development of a plant from a one celled zygote to a mature multicellular organism new cells produced in meristematic areas become differentiated from one another, and in most organisms become incorporated into tissues and organs of distinctive form. The descriptive account of the origin and development of this form and the analysis of the factors, both internal and external, by which it is regulated constitute the study of morphogenesis.

In most multicellular plants new cells are produced throughout the life of the plant. In these plants morphogenetic processes occur continuously. In unicellular and colonial plants the mature state is characterized by an absence of growth and cell division. These plants, like most animals, complete their morphogenetic development during early embryonic growth. The morphogenetically active parts of the plant are the meristems. In vascular plants these are the terminal root apex and shoot apex, the lateral vascular cambium and cork cambium, and the transitory leaf meristems. These meristems are small in size, and frequently inaccessible so that direct experimental studies are not always possible.

Because morphogenesis is concerned with all aspects of ontogenetic development no single technique or method is adequate to obtain a full understanding of the processes involved. Of the various experimental methods which have been employed two have so far yielded much more information than others. These are: surgical operations by which developing portions of the plant are removed or partially isolated by incisions and the resulting developmental changes studied; and tissue and organ cultures in which detached aseptic portions of the plant are grown in sterile nutrient media of controlled composition. Physiological studies have included the effect of externally applied plant hormones in meristematic areas, and ontogenetic changes in chemical composition. These studies have, however, been hampered by the small size of plant meristems. Genetic studies, in which the development of plants differing by known genes can be compared, have begun to show the relationship between genes and the development of tissues and organs. Natural or synthetic chimeras, in which some or all of the cells in one or more of the apical meristem layers are polyploid, are useful in elucidating the fate of chromosomally "tagged" cells in the organism.

Of the various morphogenetic processes which have been identified four are of fundamental importance in regulating the orderly development of the individual. These are *polarity, symmetry, differentiation,* and *correlation.* Each of these is an intrinsic property of the plant, but each of them is subject to external control so that in its development a plant is molded by the interaction of intrinsic and extrinsic forces.

Polarity. Most multicellular plants (and animals) exhibit a tendency towards polarity in that they are constructed on an axiate pattern. In vascular plants the shoot-root axis is established early in embryogeny and persist throughout growth. Without polarity growth would be equal along all axes and a spherical structure would result. Polarity is, therefore, the expression of preferred directions of growth.

That the polarity exhibited by the organism is preserved in detached portions of the organism was shown by the German botanist Vöchting in 1878 who cut willow twigs and found that, regardless of size or orientation, buds tended to develop at the morphological apical end, and roots at the morphological basal end of the twig. Similar behavior is found in root cuttings of *Taraxacum,* but in other species small cuttings may appear to lose their original polarity. This has been taken to indicate that polarity is a property of the organism and that individual cells do not possess a polarity of their own. Experiments with fertilized eggs of the brown alga *Fucus* and with 1-celled spores of ferns have shown that the axis of polarity is determined by external stimuli such as gradients of light, temperature, acidity, gravity, or nutrilites, and that once determined the axis is irreversibly fixed, or is reversible only with great difficulty. Similar results were obtained using germinating gemmae of *Marchantia* and embryos of *Selaginella*. Coinciding with the shoot-root axis is the polar distribution of auxin, the plant hormone, within the plant. In general auxin is transported only in the basipetal direction. Its transport can proceed against a concentration gradient and can be prevented or reversed temporarily only by anaesthetics or by massive external appications of hormones. Whether organism polarity is dependent on the physiological polarity of auxin flow, or whether this is merely another manifestation of a deeper-seated polarity is not clear.

Symmetry. Each organ possesses a characteristic symmetry as the result of the particular polar forces which operate during its development. The radial symmetry of a typical stem and the dorsiventral symmetry of a typical leaf are commonplaces of observation, but the stems of some plants are flattened and the leaves of others are rounded. Surgical operations in which the stem apex was isolated from the surrounding tissues by means of a series of vertical incisions were carried out in the fern *Dryopteris,* and in water-lily, lupin, potato, and some other flowering plants. These have shown that regardless of the shape of the tissue panel left by the incisions the apex formed a radially symmetrical stem after the operation. Similar results were obtained if the apex was bisected (Fig. 1a) or cut into quadrants (Fig. 1b) or into smaller pieces (Fig. 1c, d). Experiments with leaves have, however, revealed a different situation. If a leaf primordium, which has recently emerged on the apical flank, was isolated from the shoot apex it continued to grow as a dorsiventral leaf. If, however, the isolation was carried out at an earlier time, or even before the leaf appeared, the structure which developed was not dorsiventral but radially symmetrical. In potato the operation produced radial structures no larger than adjacent leaves,

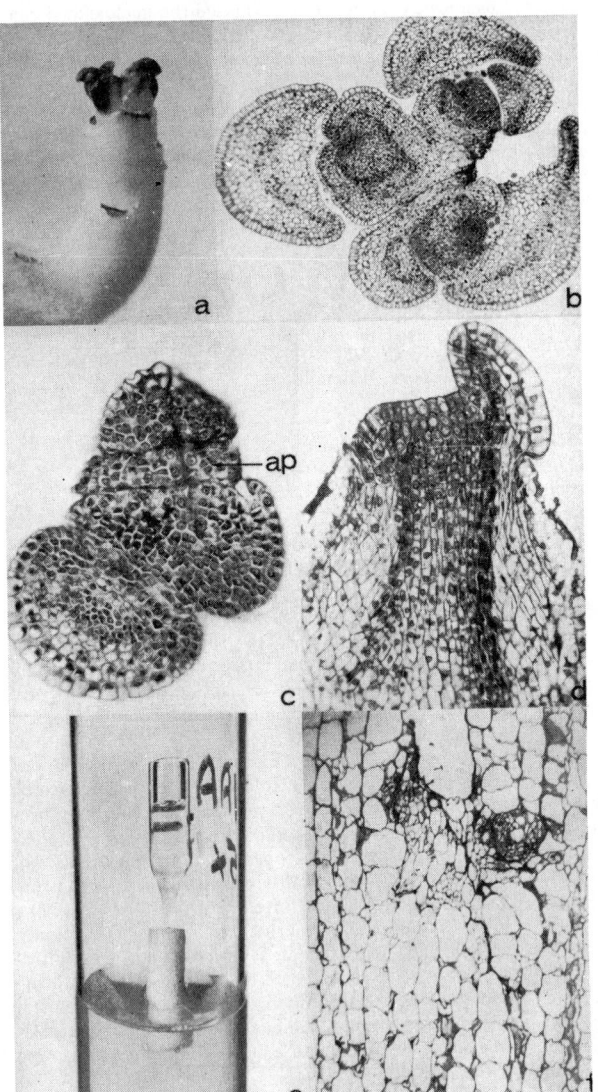

Fig. 1   a–f. *a.* Bisected shoot apex of potato regenerating a new shoot from each half. × 10. *b.* Potato apex cut into quadrants, three of which have regenerated as shoots. The fourth failed to grow. × 56.5. *c.* Potato apex at time of operation to produce small apical panel (ap) by two series of incisions at right angles. After the operation all apical tissue except that in the panel was excised. × 131.5. *d.* Shoot apex regenerated following operation shown in c. Note new formation of provascular tissue below apex. × 131.5. *e.* cultured tobacco pith with inserted glass pipette containing IAA. × .95. *f.* Section of tissue from e showing base of the pipette insertion and adjacent zone of dividing cells. Among these are lignified vascular elements. × 16.

but in *Dryopteris* the radial organs themselves bore leaf primordia and functioned as shoots. In the fern *Osmunda cinnamomea* similar results were obtained when leaves of different ages were excised and grown in sterile culture, the youngest were converted into shoots, the older continued to develop as leaves.

Although the stem and its attached leaves differ in their symmetry they together form the shoot which has its own characteristic symmetry resulting from the regular disposition of leaves along the stem. The pattern of leaf arrangement, called phyllotaxis, has been the subject of extensive experimentation. Small surgical incisions made in the leaf-forming flanks of the apical mound have indicated that the regular phyllotactic leaf sequence is not fixed but, on the contrary, may be easily modified. The position at which a new leaf emerges appears to be determined by the relative distance from the tip of the shoot apex and adjacent developing leaves. These areas operate as though surrounded by zones of inhibition within which new leaves cannot be initiated. New leaf primordia appear only in those areas cleared of inhibition, and the sequential clearing of inhibition zones in a regular manner during growth results in the regular phyllotactic pattern characteristic of the shoot. The postulated inhibition zones which surround newly emergent organs have been elaborated into a growth center theory which has been valuable in interpreting many of the observed morphogenetic phenomena at the shoot apex.

Differentiation. The growing organism produces in its meristems cells which are more or less similar. These cells enlarge and undergo specialization of function, wall structure and contents. Such mature, functional cells are stated to be differentiated, in contrast to the undifferentiated cells of the meristems. The word differentiation has also been used to describe differences in gross external form which are more properly described as polarity or symmetry changes.

Cellular differentiation has been studied experimentally in a number of cases e.g., in celery petioles more collenchyma was differentiated in those plants subjected to wind action. This result can be explained only by assuming that some of the meristematic cells destined to differentiate as parenchyma in the normal petiole underwent a change of developmental fate, becoming collenchyma. The result cannot be explained by migration of collenchyma cells into the experimentally stimulated petioles because, unlike animals in which morphogenetic migration of cells is an important aspect of embryonic development, cell movements in plants either do not occur or are limited to intrusive growth of adjacent cell tips.

Dramatic results have been achieved in the study of differentiation of one cell type—the lignified xylem element. When vascular strands in the stem of *Coleus* were severed, continuity was restored by the transformation of some of the pith parenchyma cells into xylem-like cells. This involved wall lignification, in some cases end-wall perforation, and protoplasmic death. The direction of strand regeneration followed strictly the polar direction of auxin transport, and exogenously supplied auxin resulted in a larger number of regenerated xylem strands. In tissue culture, tobacco pith was shown also to differentiate xylem-like cells when supplied with auxin through glass pipettes (Fig. 1e, f). In other experiments callus cultures of lilac

and other species which normally did not produce xylem did so in response to grafted buds. It was concluded, therefore, that the differentiation of xylem in the shoot is caused by apically produced auxin.

Correlation. Overall control of growth is brought about by a series of growth correlations in the organism. Lateral buds may remain in a dormant state unless the shoot tip is removed, removal of a newly mature leaf may accelerate the development of the next, removal of the cotyledons may inhibit root and stem development but not that of the leaves. These examples of correlations indicate the pervasiveness of the phenomenon. Considerable evidence has accumulated showing that many of the correlated phenomena are dependent on auxin distribution in the plant. In tissue cultures of tobacco pith little growth occurred if indole-3-acetic acid alone was added to the sugar-inorganic element nutrient medium, but when kinetin was supplied in varying amounts in addition to indole-3-acetic acid callus growth, bud growth, or root growth could be independently stimulated. It was postulated that a balance exists between auxin and the nucleic acid constituents in the plant, and that the state of the balance may well determine the kind and amount of growth which occurs.

Such a hypothesis could be a valuable aid in explanations of normal and abnormal, or tumerous growth, and should permit a deeper insight into the phenomena underlying morphogenesis.

IAN M. SUSSEX
MARY E. CLUTTER

References

Sinnott, E. W., "Plant morphogenesis," New York, McGraw-Hill, 1960.
Wardlaw, C. W., "Phylogeny and morphogenesis," New York, Wiley, 1955.

MORPHOLOGY

Morphology is the study of shape in biological systems; it deals with structures and their spatial relations. In more classical European usage morphology refers to the synthetic and experimental view of structure, while anatomy refers to the analytical and descriptive approaches. In the older American usage anatomy referred primarily to the description of gross structure in adults, while morphology included embryological and histological concepts.

Morphological research can be done on a variety of levels of organization which are perhaps best defined by the kinds of tools applied to them. Analysis may proceed on the gross level studied with the aid of stereoscopic dissecting microscopes or without magnification. This level is still very poorly known for many organisms. The intermediate level often referred to as histology involves compound microscopes, phase interference optics, and polarized light microscopy. The ultrastructural level is analyzed by the use of transmitting and scanning electron microscopes. New histochemical and electron probe techniques permit the characterization of the chemical nature of the component parts even at this level.

A selectionist view of morphology implies an understanding of population structure and the genetics of populations. Description hence has to be statistical. A species does not possess one ideal or "normal" condition with each departure from this referred to as a deviant, aberration or abnormality. Populations are generally polymorphic and include several distinct structural modes each at a different frequency.

Description represents a major task of morphology. It is an absolute precursor for comparative statements. Classical descriptions of form proceeded on a purely verbal level often coupled with elegant illustrations but with minimal comparison even to simple geometrical shapes. In more recent studies there has been a gradual shift toward the use of measurements and their combination into ratios, as well as to more complex statistical comparisons. A variety of techniques in stereometry and morphometry attempt to derive information about a three-dimensional structure from its surface appearance in section. The often extensive numerical manipulations underlying these techniques have been greatly facilitated by the availability of computers. More recent studies attempt to have the computer scan and determine the primary parameters as well. While many aspects of this work are still experimental, its long-range utility is unquestioned.

The description of adult conditions is often coupled with the study of ontogeny, the change of shape during the growth of a single individual. Very important here is the study of allometry, of differential growth rates in different parts of an organism. Such rates may appear to follow mathematical relations, but the proportions of the organism at each particular size are actually established by the adaptive demands posed by the environment.

The third task of morphology is that of comparison of similar structures in different organisms. The concept of homology implies that structures found in two species are parallel in terms of such parameters as embryology, position and innervation. The term and its evolutionary corollaries have recently been subject to extensive discussion and multiple re-definitions.

An organism's utilization or loading of a structure, the so-called form-function complex, has also received renewed attention under such names as biomechanics or functional morphology. While the concepts involved bear some relation to the old definitions of analogy, much of the recent work has utilized the tools of physiology to understand the workings, the action and the mechanism of structural systems, and from this to derive some of the relative selective advantages of different conformations of the organism. Functional morphology is now providing results leading to important conclusions about the process of animal evolution.

CARL GANS

References

de Beer, Gavin, "Embryos and ancestors," 3rd Ed., London, Oxford University Press, 1962.
Dullemeijer, P., ed., "De Anatomia Functionali," Folia Biotheoretica, **6:** 1–50, 1966.
Evans, F. Gaynor, ed., "Studies on the anatomy and function of bone and joints," New York, Springer-Verlag, 1966.
Mayer, Edmund, "Introduction to dynamic morphology," New York, Academic Press, 1963.
Weibel, Ewald R. and Hans Elias, eds., "Quantitative methods in morphology," Berlin, Springer-Verlag, 1967.

MÜLLER, FERDINAND JACOB HEINRICH (1825–1896)

This German botanist early emigrated to Australia and became Director of the Botanical Gardens at Melbourne, which he raised to international fame. He published much on Australian flora.

MÜLLER, FRITZ (1821–1897)

Fritz Müller, a student of, but no relation to, Johannes Müller, was an enthusiastic exponent of the theory that "ontogeny repeats phylogeny." He was born in Erfurt but emigrated to Brazil shortly after receiving his degree in medicine from Berlin. He accepted Darwin's views and determined to see how far they could be applied to a small group—the decapod crustacea—when studied in extreme detail. He described the life histories of many of these forms and pointed out in extensive detail how each larval stage demonstrated first, an ancestral conditon, and second, how this ancestral condition had become modified to the exigencies of the environment of the larva in question. His enthusiasm frequently outstripped his intelligence and many of his diagrammatic illustrations owe more to the theory they illustrate than to the objects which they were supposed to represent. His views were, however, ardently supported by Haeckel, who alternated between abusing his detractors and revising his illustrations.

MÜLLER, JOHANN

This Swiss botanist, from 1870 to the time of his death, was Director of the Herbarium at Geneva. His preoccupation with the flora of Latin America and his first name have frequently caused him to be confused with other Müllers.

MÜLLER, JOHANNES PETER (1801–1858)

The life of this great scholar presents many curious anomalies. He was of weak physique and apparently endeavored to compensate for this by driving himself into exertions far beyond those of a normal man. In his mind he was what would nowadays be called hopelessly neurotic and alternated between presenting what were revolutionary scientific ideas while fighting tenaciously to prevent any change in the structure of the University of Berlin, of which he was warden at a time when the students were in open revolt against the conservatism of that institution. He is popularly supposed to have committed suicide though this fact is based on supposition derived from the failure of the physicians of his period to find any rational cause for his unexpected death. He originally studied medicine at Bonn but went from there to Berlin, where he fell under the spell of Rudolphi whom he succeeded in the chair of anatomy and physiology in 1833. His earliest academic passion was physiology and he was the first exponent of the experimental method in this science in Germany. His outrage that mammals succumbed so freely in the course of his experiments led him to experiment first with reptiles and finally, a startling innovation in his day, with amphibia. The widespread use of the frog, and many of the experiments performed on this animal in elementary classes today, derive directly from his discoveries. He joined to his physiological experiments a love of metaphysics and his insistence that the latter was subject to experimental investigation led him to discover the role of the ganglia in the functions of the brain. He also established the existence of many other sense organs, such as the tactile, and demonstrated the principle of irritability—that is, that these sense organs respond to specific stimuli in a specific way. He also skirted round, but did not discover, the basis of sensory perception through his experiments showing that electrical irritation to the eye caused an apparent sensation of light. Many of his experiments in sensory perception were performed on himself and led to the first of the many complete breakdowns which were to punctuate his career. On his recovery from this, he threw himself into human histology and physiology, and in 1833 published his *Handbuch der Physiologie des Menschen* which was the standard text in this field for half a century. He threw himself also, with notably less success, into studies of invertebrate anatomy and embryology, and in the collection of marine forms. In the course of the latter, a ship-wreck which cost the life of his closest friend, and almost his own, still further shattered his mental and physical health. There is little doubt, however, that the most outstanding of all his contributions to science was his introduction of "academic freedom" as applied to the relation of student to master. He utterly refused to permit his views to influence those of his students who were always instructed that the only worthwhile view was that which they themselves formed in consequence of their own experiments. He was ruthless in demanding that these experiments followed logical scientific method, but insisted that no useful results could be derived from them save by the conductor of the experiment. In consequence of this staggering new approach, almost all of his students became famous and it should be pointed out that Henle, Virchow, Helmholz, Reichert and Lieberkuhn, were among those he trained. When the contributions of these and their own students are considered, it will be realized how outstanding was the influence that Johannes Müller exercised on the biological sciences in the last century.

PETER GRAY

MÜLLER, OTTO FRIEDRICH (1730–1784)

This Danish zoologist was primarily trained in theology and law and spent his life as a civil servant and mi-

nor diplomat. He was, however, deeply interested in invertebrate zoology and is the author of a considerable number of books in this field, mostly published posthumously. He was a pioneer in the study of protozoa and published, for the first time, accurate descriptions of many ciliates and attempted a classification of the entire group.

MUSCI

Musci belong to the phylum BRYOPHYTA, characterized as a group of photoautotrophic land plants without specialized water conducting cells (*Atracheata*) and having heterogametes produced within multicellular sex organs with an outer jacket of sterile cells. In this phylum we also have HEPATICAE and Anthocerotae which may be distinguished from Musci by their dorso-ventrally differentiated gametophytes.

In a typical life cycle, characteristic of most Musci or mosses, the spore develops into multi-cellular branched filaments called *protonema*. Buds arising from the protonema develop into the leafy gametophyte. True LEAVES, STEMS, and ROOTS are lacking but leaf-like organs are attached along the stem-like structure from the base of which arise numerous multicellular rhizoids. The gametophytes give rise to the multicellular ANTHERIDIA producing motile sperm and to the ARCHEGONIA, each with a single egg. Upon maturation of the egg, a row of cells (the neck canal cells) dissolve forming a passage down the neck to the egg. With a light rain, or heavy dew, the antheridium ruptures, releasing the sperm which swim along a gradient of chemicals released from the neck canal cells until they reach the egg. Following fertilization, the egg divides to produce the sporophyte, consisting of a capsule supported by a seta, the base of which (the foot) is embedded in the gametophyte tissue. Reduction division occurs in certain cells inside the capsule, forming tetraspores. Capsules of most mosses are surmounted by a peristome, usually made up of 16 hygroscopic teeth. Upon maturity of the spores, the teeth open and allow dispersal of the spores.

The variations in the peristome teeth are so constant and so closely correlated with the systematic groups that they serve as valuable aids for identification and classification.

In addition to the usual methods of reproduction outlined above, mosses also propagate vegetatively. Many species develop brood bodies or propagulae and bulbils or gemmae. Some will propagate from broken "leaves" or "stems" while others proliferate by branching from normal plants. In some species, the protonema is persistent, allowing growth over a long period of time. Proliferation by branching from existing plants results in mats or cushions in some kinds of mosses and perennial layers in others.

There are about 660 recognized genera of mosses made up of nearly 14,000 species ranging from little more than an inconspicuous mass of protonema with sporophytes (*Buxbaumia*) to large fern-like plants (Hylacomiaceae), pendant forms (Meteoriaceae, Neckeriaceae, Pterobryaceae), and floating forms (Fontinalaceae, *Leptodictyum laxirete*) with "stems" 4 decimeters or more in length.

Rhizoids range from very few to enough to form masses of felt around the bases of some kinds of mosses. The "stems" vary from a minute strand of undifferentiated tissue to some organized with evidence of a primitive stele as seen in section or by stripping off the cortical cells. Some "stems" are more or less covered with thread-like or leaf-like outgrowths (paraphyllia) among the "leaves" as in *Thuidium.*

The "leaves" are mostly one cell in thickness with a single midrib (costa), short double midrib, or none (ecostate). They are mostly more or less ovate in shape and range in length from about 0.3 mm. in *Amblystegiella* to 16 or more mm. in certain others as *Leiomela.* The "leaf" cells range from tiny opaque cells (in *Andreaea*) to large clear circular cells ±40 μ in diameter in others and much elongated cells in others. Cell walls may be thin or thick, smooth or rough (papillose) or irregular in outline (erose). Some "leaves" have an extra lamina (*Fissidens*) or several laminae, a few to several cells high arising from the costa (Polytrichaceae). The cells in the lower extremes of the "leaves" (alars) may be modified.

Fossil evidences of mosses are few, probably because of the lack of hard parts to be preserved. Most of those found have modern counterparts.

From the economic standpoint, mosses are sometimes used by florists for decorations. A greater economic use is of *Sphagnum,* which is used for mulch, for wrapping plants for shipping, and as an absorptive layer in poultry houses. It has been suggested for surgical dressings because of its natural antiseptic (acid) quality and its ability to absorb large quantities of liquids. The conservation value of mosses in soaking up water and holding soil cannot be over-emphasized. Mosses growing on rocks are also of value for their soil building properties.

Mosses may be found on exposed rock or soil from polar region to polar region. Many species grow on bark or branches of trees and others in fresh water. Mosses are not found in marine habitats. Some mosses will grow under a wide variety of environmental conditions but others will grow only in a very narrow range of light, temperature, pH, and moisture of soil and atmosphere. This makes some of them nearly ideal organisms for micro-environment studies.

Musci may be divided into three natural orders: Sphagnales, Andreales, and Bryales. Sphagnales is made up of a single very complex genus *Sphagnum* with many puzzling species. The members of this order are recognized by their spreading recurved whorled branches and gray green color. The ecostate "leaves" are made up of large hyaline cells separated by narrow chlorophyllose cells. The operculate black spherical capsule is raised on a pseudopodium. It has no peristome.

Andreales also has a single genus *Andreaea* but has fewer, usually distinct, species. They are small fragile black plants growing on acid rocks at high altitudes or latitudes. The "leaf" cells are small, thickwalled, and opaque. The capsules are on pseudopodia but open by four longitudinal lateral slits.

All the other mosses belong to the Bryales or Eubryales with a capsule more or less elevated on a seta and usually having a single or double peristome made up of 4 to 64 teeth although a few of them are cleistocarpous and split open upon maturing.

There are yet many taxonomic problems among mosses, mostly depending upon morphological characters not yet studied and correlated. Most of these problems will be solved by amateur students of mosses: housewives, engineers, doctors, teachers, and others desiring a stimulating hobby. The study of mosses can be done by anyone, anywhere, with little special equipment and furnishes entertainment and relaxation with a promise of valuable contributions to this field for anyone with an inquiring mind, an interest in nature, and a few hours leisure time.

L. J. GIER

References

Conard, Henry S., "How to know the mosses and liverworts," Dubuque, Iowa, Brown, 1956.
Grout, A. J., "Moss Flora of North America," 3 vols., Chicago, Museum of Natural History, 1928–1940.
van der Wijk, R., Ed., "Index Muscorum," 5 vols., Utrecht, International Bureau for Plant Taxonomy and Nomenclature, 1959–.
"The Bryologist," published quarterly by the American Bryological Society, Duke University, Durham, North Carolina.

MUSCLE

Muscles consist of many fibers held together by connective tissue. Their structure and function vary widely in different organs and animals. On the basis of structure they are divided into smooth and striated muscle. The former are usually found in organs which carry out only sluggish movements. The latter are capable of rapid contractions. Some invertebrates have only smooth muscles, but in arthropods all muscles, also those of the viscerae, are striated. Smooth muscles are adapted to many specialized functions. They should not be considered to be primitive; both types of muscles are found side by side even in the lowest forms, such as coelenterates. The obliquely striated muscles of molluscs and worms are a special type of cross-striated muscles in which the sarcomeres are arranged in a steep helix around the fiber axis.

Structure

Smooth muscle. One type of smooth muscle has spindle-shaped fibers which are 4 to 7 μ thick and rarely longer than 0.2 mm. Microscopically they are nearly homogeneous except for a single nucleus, but the electron microscope has shown that they are filled with longitudinal filaments about 80 Å thick (Fig. 1A). This type of muscle is found in the walls of the viscerae and blood vessels of vertebrates and in some invertebrates. Another type of smooth muscle, observed only in invertebrates, contains large fibers with microscopically visible fibrils, often located near the surface of the fibers (Fig. 1B).

Striated muscle. Striated muscle fibers contain fibrils which are about 1 to 2 μ thick. The fibers are made up of alternating bands with different optical properties. The A (anisotropic) bands have a higher refractive index and much greater birefringence than the I (isotropic) bands. The middle of the A bands appears light and is called the H band. In the middle of the I bands is

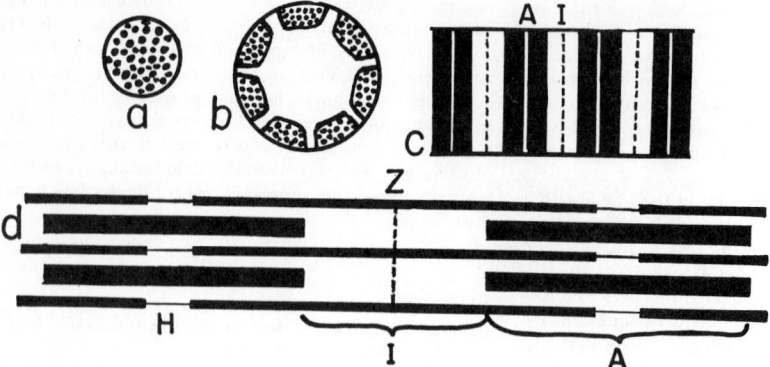

Fig. 1   Diagrammatic representation of structure of smooth and striated muscle. a and b, cross-sections of two common types of smooth muscles. Dots are submicroscopic filaments in cross-section. c, longitudinal section of striated muscle as seen in a microscope. d, the same at higher magnification as seen in an electron microscope.

the Z band, a fine network which connects all myofibrils of a fiber transversely and extends to the outer wall of the fibers, the sarcolemma. The Z band maintains the A and I segments of all fibrils at nearly the same level, thereby giving the whole fiber its striated appearance (Fig. 1C). It may also help to transmit tension from the fibrils to the outside.

The fibers contain a system of tubules which is most extensive in the most rapid muscles with the quickest contraction. It consists of transverse tubules which are open to the outside and the sarcoplasmic reticulum which has junctions with the transverse tubules and extends between the myofibrils. With the ELECTRON MICROSCOPE it has been found that the myofibrils of striated muscle contain a regular array of two types of filaments. Those with a thickness of 130 Å extend through the A bands and have, when isolated, except in the middle region protrusions which are formed by the heads of the myosin molecules. Thinner filaments pass through the I and part of the A bands, but are interrupted in the middle of the A bands, thereby forming the H bands (Fig. 1D). The protrusions of thick filaments form cross bridges with the thin filaments in the region of overlap between the filaments.

At rest most muscles are elastic bodies with rubber-like extensibility. However, the tension of extended muscle is largely, perhaps entirely, due to connective tissue and the supporting structures of the muscle fibers, such as the sarcolemma. Some smooth muscles are plastic within the physiological range of lengths. An external force extends them slowly, as if opposed by a viscous resistance. This resistance probably is caused by a sliding movement between two types of contractile elements. Sliding between thick and thin filaments is responsible for changes in length of striated muscle, whether passive or due to contraction.

Function

Muscles differ greatly in the speed of their contraction. In vertebrates the response to a single stimulus lasts about 1/40 second in external eye muscles, 0.1 to 0.3 second in most other skeletal muscles, a few seconds in smooth muscles. There are important differences also between visceral muscles, such as those of the heart and intestine, and skeletal muscle. In the former, contractions are initiated by a spontaneous, slow depolarization of some of the muscle fibers which leads up to a conducted response. These muscles are composed of many small fibers, but contractions normally are conducted over the whole muscle, as if it were a single giant fiber, and therefore follow the all-or-none law.

In skeletal muscles and some smooth muscles contractions are initiated by motor nerve impulses. In these muscles the contraction induced by a nerve impulse generally spreads from the neuromyal junction and thereby activates the whole muscle fiber. Conduction is associated with a brief action potential similar to that of nerve fibers. In some skeletal muscle fibers a nerve impulse activates only the region near the junction. These fibers have multiple motor innervation. Many visceral muscles of vertebrates and some skeletal muscles of arthropods are also supplied with inhibitory nerve fibers which diminish responsiveness. Some chemical agents induce long continued activity which may not be associated with repetitive action potentials and then are called contractures.

In vertebrate skeletal muscle usually more than one, sometimes more than 1000 muscle fibers, are innervated by a single motor nerve fiber. Because normally all these fibers respond at the same time, the motor neuron with all the muscle fibers it innervates is called a motor unit. During normal contractions the motor units discharge repetitively at frequencies varying from 5 to 50 per second. Because the individual responses fuse, a sustained contraction, called tetanus, is produced. Such contractions are stronger than single responses because the mechanical effects of successive responses are added up (summation) if relaxation after each response is incomplete. The strength of contraction increases with the frequency of discharge. Gradation of contraction is also due to the fact that in weak contractions only a small number of motor units participates and that their number increases with the height of contraction.

Because a single response is automatically followed by relaxation, a sustained contracted state can be produced only by repeated activation and, therefore, continuous energy expenditure. In some muscles such a state may persist under normal conditions for a long time and is then called tonus. Tonus is generally controlled by the central nervous system through motor nerve fibers, in some smooth muscles also through inhibitory nerve fibers. In slowly contracting muscles, particularly some smooth muscles, tonic contractions can be maintained by a small energy expenditure because it requires infrequent activation of the muscle fibers. Tonus does not represent a distinct physiological mechanism different from that of phasic muscular activity, but is a descriptive term for a sustained, involuntary contraction.

Contraction does not require the presence of oxygen. As in other tissues, energy liberated by the burning of carbohydrate and fat is utilized as phosphate bond energy. In skeletal muscles carbohydrate is the chief source of energy, as shown by the rise in the respiratory quotient during exercise. The active heart, however, can take up large amounts of fatty acids and ketone bodies from blood and always has a respiratory quotient below 1. If oxygen supply is insufficient lactic acid is produced in muscles during activity. In humans this is true only in severe exercise which cannot be sustained for more than an hour. During recovery from exercise O_2 consumption is increased for a period of time which increases with the severity of the exercise.

The mechanical efficiency of contraction eff $= W/(W + H)$, where W is mechanical work and H heat, can be as high as 40%, but depends greatly on speed of shortening and other factors. The total energy $(W + H)$ liberated depends on mechanical conditions such as tension and the extent of shortening.

Relaxation is essentially a passive process. Heat production during relaxation is accounted for by the conversion of mechanical energy stored in the muscle during contraction into heat. In smooth muscles with slow relaxation the process is governed by the viscous properties of the contractile elements.

Initiation of contraction. The energy for muscular contraction is furnished by the breakdown of adenosinetriphosphate (ATP) into adenosinediphosphate and inorganic phosphate. However, in a brief contraction the amount of ATP in a muscle does not diminish measurably, while inorganic P increases and phosphorcreatin disappears, demonstrating that ATP is turned over many times during a single contraction due to rapid resynthesis by the Lohman reaction.

ATP also is essential for maintaining the soft condition of the contractile elements during relaxation. Muscles which have been stored in glycerol solution are made soft and translucent again by ATP in the presence of Mg. A state closely resembling normal relaxation is produced in such muscles by the combined action of ATP, Mg^{++} and an agent such as a strong chelating substance which lowers the concentration of Ca^{++} below $10^{-6}M$. An extract from striated muscle, containing vesicles derived from the sarcoplasmic reticulum, acts as such an agent. A contraction of the relaxed preparation is produced by raising the concentration of Ca^{++} to about $10^{-5}M$. It has been concluded that the activity of the muscle is controlled by the binding and release of Ca inside the fibers and that this control is the function of the sarcoplasmic reticulum.

Under normal conditions contractions are triggered by depolarization of the cell surface such as it occurs during the conduction of an impulse. Depolarization by isosmotic KCl solution or electric current also causes contraction. A special mechanism is required to transmit activity from the surface to the inside of the muscle fibers. The fact that in large skeletal muscles heat production reaches its maximum within a few milliseconds shows that activity is conducted rapidly from the surface to the interior of the fibers, faster than can be accounted for by diffusion of any material. This function is probably carried out by the transverse tubules, which are invaginations from the surface, and are responsible for the large electric capacity of the muscle fibers. During conduction of an impulse they probably carry most of the outward electric current, and are thereby activated. How activity is transmitted to the sarcoplasmic reticulum and causes release of Ca^{++} is not understood.

Mechanism of contraction. Selective extraction of muscle has shown that the thick filaments are composed mainly of myosin, which makes up about 40% of the total muscle protein and has a molecular weight of about 500,000. Its molecules have a head and a tail. The thin filaments consist largely of actin and are made up of two helically round beaded strands, the beads representing molecules of actin G with a molecular weight of about 70,000. However, crude actin solutions also contain other, recently discovered proteins, tropomyosin, troposin and α- and β-actinin, which are not essential for contraction, but have important regulatory functions.

Shortening of striated muscle is brought about by a force which draws the thick filaments in between the thin filaments, their lengths remaining unchanged. During contraction the cross-bridges, which probably are the active centers of the contractile mechanism, attach themselves to points along the thin filaments which are progressively closer to the Z-bands, thereby producing a sliding movement. In agreement with this assumption the tension produced by a fiber is proportional to the overlap between the two types of filaments within the normal range of lengths. One of the basic unsolved questions is whether shortening takes place in very small steps due to chemical or electric forces acting over distances of only a few Angstroms or whether each step consists of a larger shortening in which protein chains in the cross-bridges coil as a result of a change in entropy or the formation of various types of bonds.

EMIL BOZLER

References

1. Gergely, J., ed., "The relaxing factor of muscle," Federation Proceedings 23: 885, 1964.
2. Huxley, H. E., "Electron microscope studies on the structure of natural and synthetic protein filaments from striated muscle," J. Mol. Biol. 7: 281, 1963.
3. Reichel H., "Muskelphysiologie," Berlin, Springer, 1960.
4. Horvath, S. M., ed., "Muscle as a Tissue," New York, McGraw-Hill, 1962.

MUSCULAR SYSTEM

In dealing with the vertebrate muscular system one is concerned primarily with the striated or skeletal muscles of the body, which comprise about 40% of the body weight in man. These muscles attach by *tendons* or *aponeuroses* upon connective tissue septa or parts of the skeleton, and they are arranged in antagonistic groups so that one muscle, or group of muscles, will move a part in one direction, and another in the opposite direction. Various sets of terms describe the more common antagonistic actions. *Abduction* is the movement of an organ, such as a limb, away from some point of reference, usually the ventral midline; *adduction,* movement toward the reference point. *Flexion* is the movement of a distal segment of an appendage toward a more proximal one, such as the movement of the forearm toward the upper arm. It also describes a ventral bending of the head or trunk. *Extension* is movement in the opposite direction. Flexion and extension are sometimes used to describe fore and aft movements of the entire appendage at the shoulder and hip. But their usage has been inconsistent with respect to the arm, so it seems best to describe forward movement of the entire appendage in quadrupeds as *protraction* and backwards movement as *retraction.*

Muscles acting upon bones do so in the manner of forces upon a lever system. Most of the lever systems in the body are of the third order, i.e., the point of pull of the muscles is closer to the fulcrum than the point of application of the force being acted against. Since the muscle arm of such a lever is shorter than the force arm, the muscles must generate a greater pull than the force to be overcome. This is inefficient from the point of view of power, but makes for a compact system, with the muscles close to the bones, and for relatively extensive and rapid movements at the end of the lever.

Adaptations of vertebrate muscles for power, or for speed of movement, have entailed changes in the ratio of the muscle and force arms through shifts in the attachments of muscles and changes in the relative lengths of the bones upon which the muscles attach. For example, the biceps brachii of a softshell turtle (*Trionyx*), which uses its forearm extensively in swimming, inserts close to the wrist, thereby giving a relatively longer muscle arm than that found in most turtles. Adaptations for power, or for extent and speed of movement, are also seen in the arrangement of the fibers within muscles. Power of contraction is a function of the number of fibers a muscle contains. Extent and speed of movement is related to the length of the fibers, for fibers can shorten about one half their resting length. Some muscles, the interossei between the metacarpals for example, consist of many short fibers having a bipinnate arrangement; they extend diagonally from the periphery of the muscle to attach onto a central tendon that penetrates far into the muscle. Such a muscle can exert a more powerful pull than one of equal mass consisting of fewer, longer and parallel fibers. On the other hand, muscles, such as the biceps brachii, consisting of relatively long, parallel fibers can bring about a more rapid or extensive contraction than a comparable sized muscle with more but shorter fibers.

The individual vertebrate muscles are exceedingly numerous, diverse and complex. They can only be discussed in an article of this scope by generalizing even though this runs the risk of oversimplifying. At the outset, a study of the muscular system is facilitated if the system is sub-divided into groups of muscles whose homology can be established in widely divergent species. Two main groups of muscles are commonly recognized—(1) *somatic* or *parietal* and (2) *visceral.* Somatic muscles are those of the outer tube of the body,

i.e., the body wall and appendages. All develop embryonically in the lower vertebrate from the myotomes. In higher vertebrates, certain of them (muscles of the appendages and tongue) develop from condensations of mesenchyme *in situ,* nevertheless these muscles are considered to be of myotonic origin on the basis of their innervation and phylogeny. Visceral muscles are those of the gut tube, and all develop from the lateral plate mesoderm. The only ones of these that are associated with the skeleton are those in the pharynx wall that primitively attached onto the visceral arches. These form a distinct group of branchiomeric or branchial muscles.

The somatic musculature in turn can be subdivided into axial and appendicular groups. Axial muscles are those located close to the longitudinal axis of the body, e.g., the trunk and tail muscles, the epibranchial and hypobranchial muscles located respectively dorsal and ventral to the gill region, and the extrinsic muscles of the eyeball. Appendicular muscles, *sensu strictu,* are only those girdle or appendage muscles that develop from mesenchyme that early in development is located within the limb bud. In lower vertebrates this mesenchyme is derived from myotomic buds, but such an origin has not been demonstrated in higher vertebrates. Certain of these appendicular muscles, the latissimus dorsi for example, spread from the appendage back onto the trunk, but most remain within the confines of the limb. Muscles, such as the rhomboideus and trapezius, which in the adult extend from the trunk to the girdle, are not appendicular muscles in the sense defined. These muscles develop embryonically respectively from certain trunk myotomes and branchial premuscular masses, and they extend secondarily to the girdle. They are, strictly speaking, modified trunk and branchial muscles.

Trunk and tail muscles form the bulk of the muscular system in fishes, and they are the important locomotor muscles. They take the form of segmental myotomes or myomeres which, in all fishes higher in the evolutionary scale than cyclostomes, are divided into dorsal (epaxial) and ventral (hypaxial) portions. These, and their derivatives in terrestrial vertebrates, are innervated respectively by the dorsal and ventral rami of spinal nerves. Muscle fibers within the myomeres extend primarily longitudinally, attaching onto the myosepta located between successive myomeres. The myomeres are usually complexly folded and have a zig-zag appearance on the body surface. Waves of contraction, affecting first one side of the body and then the other side, pass caudad along the myomeres causing undulatory movements of the trunk and tail. As regions of the body move from side to side, they also incline along their vertical axis in such a way that their thrust is directed somewhat posteriorly and ventrally as well as laterally. The effect is analogous to that of a sculling oar. The foldings of the myomeres may be related to these complexities in movement.

Tetrapods propel themselves by thrusts of the appendages against the ground, not by thrusts of the trunk against water. Appendicular muscles become increasingly important as one ascends the evolutionary scale among terrestrial vertebrates, and the trunk musculature less important in locomotion. Urodeles retain myomeres, and fish-like undulations of the trunk assist the appendages in locomotion. But the segmentation is

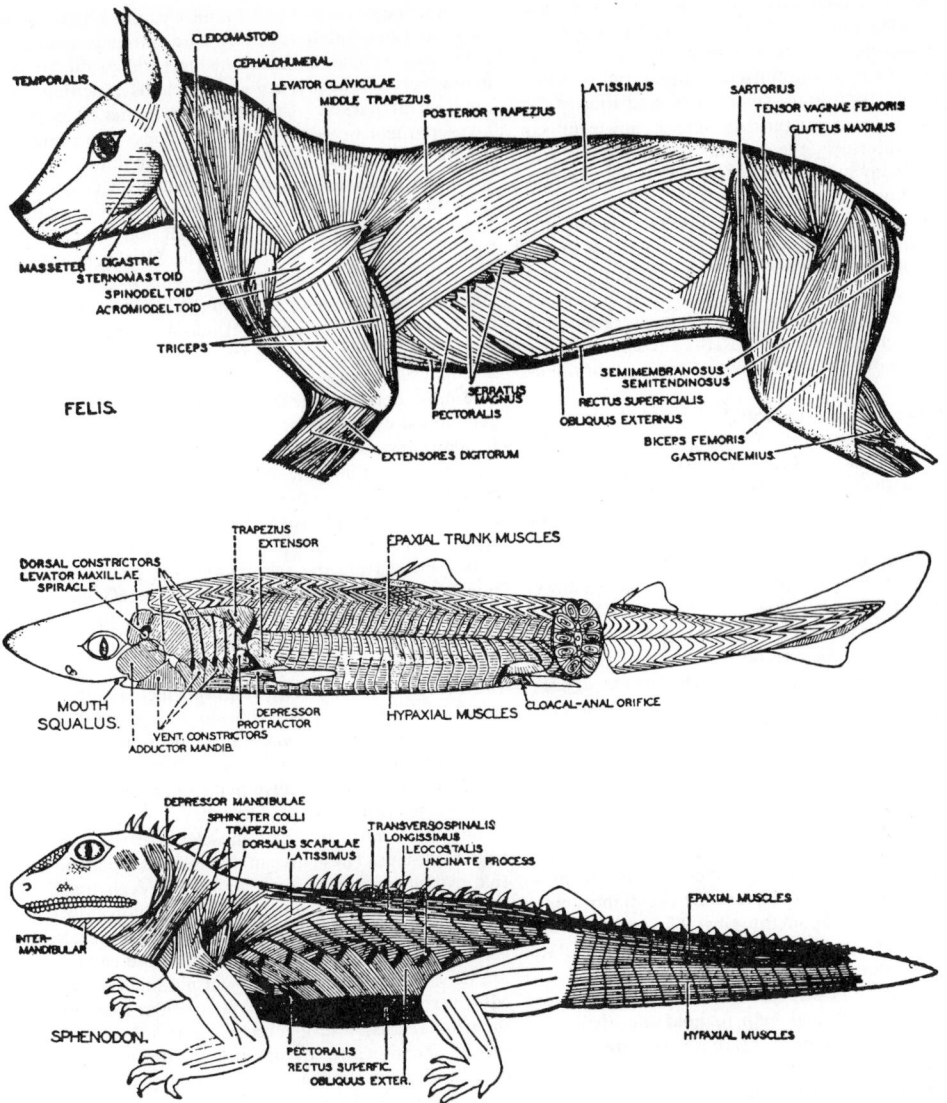

Fig. 1   Lateral views of the musculature of selected vertebrates: *Squalus* (the dogfish), *Sphenodon* (a reptile), *Felis* (the cat). Certain of the muscles illustrated are given different names in the text: *Squalus:* depressor = adductor, extensor = abductor, trapezius = cucullaris; *Sphenodon:* ileocostalis = iliocostalis; *Felix:* cepholohumeral = anterior trapezius plus clavodeltoid, levator claviculae = levator scapulae ventralis, rectus superficialis = rectus abdominis, serratus magnus = serratus ventralis. (Courtesy Neal and Rand, "Chordate Anatomy," Philadelphia, The Blakiston Company.)

neither as obvious nor are the myomeres as complex as in fishes. In other tetrapods segmentation is lost in most parts of the trunk (the intercostal muscles being a notable exception) by the fusion of successive myomeres.

The epaxial musculature remains powerful in most terrestrial vertebrates. In primitive forms, such as urodeles, it consists primarily of a single thick longitudinal bundle, the dorsalis trunci, occupying the area between the neural spines and transverse processes of the vertebrae. The deeper fibers of this bundle interlace the neural arches and spines, and are sometimes considered to

represent a distinct interspinalis. In typical reptiles, the epaxial musculature is more complex, for it has cleaved into three distinct, longitudinal bundles. A medial and deep group, the transversospinalis complex, binds successive vertebrae together. Two other groups, the longissimus dorsi and iliocostalis, lie lateral to this, and consist of fibers extending over several segments. The epaxial musculature is reduced in birds, correlated with the fusion of many of the trunk vertebrae, but the reptilian divisions persist and are elaborated further in mammals. Mammals hold their trunks off of the ground, and the epaxial musculature plays an important role in resisting tension stresses along the vertebral column. The reptilian transversospinalis complex is represented by a median multifidus spinae and spinalis, and the more lateral longissimus dorsi and iliocostalis have partially fused posteriorly into a powerful sacrospinalis. Anteriorly the epaxial muscles have subdivided into many muscles concerned with the complex movement of the head and neck.

The hypaxial musculature is reduced in tetrapods. Part of it, the subvertebral muscles, consists of longitudinal muscles situated ventral to the vertebral transverse processes and lateral to the centra. In mammals the group consists of the longus colli, quadratus lumborum and psoas minor, and it assists the epaxial muscles in the support and movement of the vertebral column.

Most of the hypaxial musculature is represented by broad, thin layers of muscle forming the lateroventral part of the trunk wall. Tetrapods usually have three such layers with different fiber directions. Mammals are typical, having an external oblique, internal oblique and transversus abdominis in the abdominal region and external intercostals, internal intercostals and a reduced transversus thoracis in the thorax. Reptiles have a more complex pattern for the superficial layer has become subdivided. The rectus abdominis muscles, a pair of longitudinal muscles located on each side of the midventral line, are closely associated with the ventral portion of the oblique muscles and develop from them embryonically. All of these muscles help to form the body wall, support the trunk viscera, and most of them are also involved in respiratory movements.

Other muscles, derived from the lateroventral part of the hypaxial musculature, are adapted for more specialized functions. The mammalian diaphragm, serratus dorsalis, scalenes, and transversus costarum are primarily respiratory muscles. Except for the diaphragm, these muscles all act upon the ribs. The mammalian rhomboideus, serratus ventralis, and levator scapulae ventralis, all of which extend from the trunk to various parts of the pectoral girdle, transfer body weight to the girdle and appendage, and help to hold the girdle in place. Homologues for most of these are present in lower tetrapods, but these species do not hold their bodies off of the ground as much of the time as do mammals, and these muscles are not as highly developed.

The epibranchial and hypobranchial groups of somatic muscles are essentially continuations of the epaxial and hypaxial trunk musculature dorsal and ventral to the gill region. In most vertebrates the epibranchial musculature is not separated from the epaxial muscles and is described along with them. But the hypobranchial musculature is a more clearly defined group, for it is separated from the hypaxial trunk muscles by the pectoral girdle. The myotomic derivation of the hypobranchial muscles is not as clear in mammals as in lower vertebrates, but their innervation by the ventral rami of certain cervical nerves and by the hypoglossal nerve, which is considered to be serially homologous to the ventral rami of spinal nerves, is evidence of their somatic nature and their close relationship to the hypaxial trunk muscles.

In all jawed vertebrates, the hypobranchial musculature can be divided at the level of the hyoid arch into prehyoid and posthyoid divisions. The posthyoid mus-

cles of fish consist of a pair of muscles, the rectus cervicis (sometimes subdivided into coracoarcuals and coracohyoids), extending from the pectoral girdle to the hyoid arch, and deeper coracobranchials attaching onto the ventral parts of the branchial arches. Traces of myosepta can be seen in these muscles. The prehyoid group consists of a pair of coracomandibulars extending from the mandibular symphysis posteriorly to attach onto the surface of the rectus cervicis. All of these muscles support the floor of the mouth and pharynx and help to open the jaws and expand the gill pouches.

Terrestrial vertebrates lose the gills of their piscine ancestors together with much of the brachial apparatus, including the coracobranchial muscles. Other parts of the hypobranchial muscles are retained and elaborated in connection with the evolution of a mobile tongue. The most primitive prehyoid muscle of tetrapods is a geniohyoid extending between the hyoid arch and chin. In addition most tetrapods have a number of small muscles that extend into the tongue from adjacent parts; the common ones in mammals being a genioglossus, hyoglossus, and styloglossus. These, and also the intrinsic musculature of the tongue, have evolved by the separation of fibers from the geniohyoid. The posthyoid musculature of primitive tetrapods, such as urodeles, consists simply of a rectus cervicis, but this has become subdivided among higher tetrapods into separate muscles extending from the girdle, sternum, and larynx to the hyoid. In mammals, the group includes the omohyoid, sternohyoid, sternothyroid and thyrohyoid. All of the hypobranchial musculature in higher tetrapods is concerned with manipulation of the tongue and the anterior and posterior movements of the hyoid apparatus in various phases of deglutition.

The most anterior axial muscles are the extrinsic ocular muscles. These are small, ribbon-shaped muscles that extend from the orbital wall to the eyeball and are responsible for its movements. They develop embryonically in all vertebrates from three pairs of mesenchymal condensations, which are often called head cavities and are believed by many to represent modified myotomes. Most vertebrates have six extrinsic ocular muscles. The superior rectus, inferior rectus, anterior (or medial) rectus, and the inferior oblique are derived from the first head cavity and are all innervated by the oculomotor nerve. The superior oblique arises from the second condensation, and is supplied by the trochlear nerve; the posterior (or lateral) rectus develops from the last head cavity, and is supplied by the abducens nerve. In addition, certain terrestrial vertebrates have a retractor bulbi supplied by the abducens, and mammals typically have an elevator of the upper eyelid, the levator palpebrae superioris, supplied by the oculomotor nerve. Variable slips acting on the nictitating membrane are present in many amniotes.

In contrast to most of the axial musculature, the appendicular musculature of fishes is relatively weak. The paired fins are basically stabilizing keels, but they also assist the fish in turning and in other delicate maneuvers. Typically each fin has a single dorsal abductor arising from the girdle and trunk musculature and inserting on the fin base, and a comparable ventral adductor. Certain fibers in these muscles arise sufficiently anteriorly and posteriorly so that they can also protract and retract the fin.

During the evolution of tetrapods the appendicular

muscles become increasingly important in locomotion and in holding the trunk off of the ground, and the muscles have become correspondingly complex and powerful. However, on the basis of their embryonic origin from dorsal and ventral premuscular masses within the limb bud, the numerous appendicular muscles of tetrapods can be sorted into dorsal and ventral groups homologous to the corresponding fish muscles.

The upper arm and leg of primitive terrestrial vertebrates, such as urodeles, project laterally from the trunk and move back and forth in the horizontal plane. The forearm and leg extend down to the ground at right angles to the proximal part of the limb. When at rest, the animal's belly is upon the ground, but when the animal moves, powerful adductor muscles, such as the pectoralis and supracoracoideus of the pectoral appendage, pull on the proximal segments of the limb and raise the body slightly off of the ground. The ventral appendicular muscles, which also spread onto the anterior surface of the limb are, in general, responsible for the first movements of the limb: adducting and protracting the proximal segment and flexing the distal parts. The dorsal musculature, which also spreads onto the posterior surface of the limb, then retracts, abducts and extends the limb.

In the line of evolution leading to mammals, the appendages rotate beneath the body. The trunk is held off of the ground by pillars of bone braced by muscles. Ventral adductor muscles are less important, and many of them have migrated from their primitive position dorsally where they now assist in protracting and retracting the limb. A classic example is the derivation of the mammalian supraspinatus and infraspinatus, which lie dorsal to the shoulder joint on the lateral surface of the scapula, from the ventral supracoracoideus of primitive tetrapods. The appendicular musculature is also more powerful, for movements of the trunk no longer assist in locomotion.

Branchiomeric muscles are less conspicuous than somatic muscles in most vertebrates, but they constitute a very important group. In jawed fishes the group is subdivided into mandibular muscles, which act upon the first visceral arch and are innervated by the trigeminal nerve, hyoid muscles, which act upon the second arch and are innervated by the facial nerve, and branchial muscles proper, which act upon the remaining arches. The glossopharyngeal nerve supplies the muscles of the third arch, and the vagus those of the remaining arches.

A typical branchial arch has constrictors, interbranchials, adductors and interarcuals, all of which compress the gill pouches. The pouches are expanded primarily by coracobranchials, a part of the hypobranchial musculature. In addition, each arch has a dorsal levator, but the levators usually have united to form a cucullaris, most of whose fibers insert on the pectoral girdle rather than on the arches. The first arch forms, or helps to form, the jaws in most fishes, and the hyoid arch is often involved in supporting the jaws. As might be expected, the mandibular muscles, and to some extent those of the hyoid arch, have departed from the typical pattern and are specialized for the support and movement of the jaws.

Many changes occur in the branchiomeric musculature of terrestrial vertebrates correlated with the loss of the gills and the reduction and modification of the visceral arches. Most of the mandibular musculature remains associated with the jaws, and forms the muscles of mastication. In mammals there are the temporalis, masseter, pterygoids, anterior belly of the digastric and mylohyoid. A part of this musculature, the tensor tympani, has followed a part of the mandibular arch into the middle ear, and part, the tensor palati, is associated with the soft palate. Only a few hyoid muscles, such as the mammalian stylohyoid, posterior belly of the digastric, and stapedius, remain associated with the hyoid arch and its derivatives. It will be recalled that intrusive hypobranchial muscles act on the hyoid apparatus. Most of the hyoid musculature has spread out beneath the skin to form the muscles of expression over the face and the platysma in the neck region. It may be added, parenthetically, that the remaining important integumentary muscle of mammals, the panniculus carnosus, is a derivative of the pectoral musculature. Most of the musculature of the remaining arches is lost, although some of it forms the intrinsic laryngeal muscles and certain pharyngeal muscles. The cucullaris is a notable exception, for it expands and cleaves to form the trapezius and sternocleidomastoid complexes of muscles which act upon the shoulder and head. It is of interest that the mammalian motor nerve to these, the spinal accessory, is homologous to part of the vagus of fishes.

WARREN F. WALKER, JR.

References

Edgeworth, F. H., "The Cranial Muscles of the Vertebrates," Cambridge, The University Press, 1935.
Romer, A. S., "The Vertebrate Body," Philadelphia, Saunders, 1962.

MYRIAPODA

The Myriapoda were at one time considered to be a class of the Arthropoda though they had little in common save the presence of more appendages than were to be found in any other class. They are no longer regarded as a natural group and are represented in this encyclopedia by articles on CHILOPODA, DIPLODA, SYMPHYLA, and PAUROPODA.

MYRTIFLORAE

According to Engler and Diels this order of the DICOTYLEDONEAE-Archichlamydeae comprises 23 families to be placed into 4 suborders:

Thymelaeineae: Geissolomataceae, Penaeaceae, Oliniaceae, Thymelaeaceae, Elaeagnaceae;

Myrtineae: Lythraceae, Heteropyxidaceae, Sonneratiaceae, Crypteroniaceae, Punicaceae, Lecythidaceae, Rhizophoraceae, Nyssaceae, Alangiaceae, Combretaceae, Myrtaceae, Melastomataceae, Hydrocaryaceae, Onagraceae, Haloragaceae;

Hippuridineae: Hippuridaceae, Thelygonaceae;

Cynomoriineae: Cynomoriaceae.

Typically the families of this order have cyclic, actinomorphic flowers, which show a transition from perigyny (in the primitive) to epigyny (in the advanced members). Style commonly 1. The leaves are mostly opposite, simple, often coriaceaous. The vascular bundles of many families have intraxylary phloem.

The **Thymelaeineae** (often classified as a separate order) contain mostly shrubs with estipulate leaves, perigynous (often apetalous) flowers with few-ovulate carpels. The Geissolomataceae, Penaeaceae, Oliniaceae, small African families, bear bisexual isomerous flowers. The Thymelaeaceae often have unisexual flowers with petaloid sepals, while the petals are scalelike, 4–12 (or absent, *Daphne, Edgeworthia*); ovaries often pseudo-monomerous. Possibly the Elaeagnaceae (*Elaegnus, Shepherdia, Hippophäe*) are not related to them, their gynoecium being monerous; their fruits are enveloped by a fleshy hypanthium. Lepidote or stellate hairs are common.

The **Myrtineae** commonly have rudimentary stipules and inferior ovaries with many-ovulate carpels, but the Lythraceae still have free ovaries. They comprise many herbs (*Peplis, Rotala*), but also trees (*Lagerstroemia, Lafoensia*) and shrubs (*Lawsonia, Heimia*); flowers sometimes zygomorphic (*Cuphea, Lythrum*).—The Nyssaeceae and Alangiaceae, small families with uniovulate ovaries, are possibly related to the Cornaceae. —The essentially tropical Myrtaceae (and the closely related Heteropyxidaceae) are distinguished by glandular-punctuate leaves. Stamens (fig. I, II, III), often fasciculated and colored. About 80 genera, some of them producing spices (*Eugenia caryophyllata*, cloves; *Pimenta*, allspice) and aromatic oils (*Eucalyptus*). Some Australian Eucalyptus-trees grow 350 feet tall. The allied Punicaceae have no glands, their inferior gynoecium consists of 2–3 whorls of carpels (fig. IV), small trees (pomegranate).

Lecythidaceae (incl. Barringtoniaceae and Asteranthaceae). The stamens are united at their bases, often forming rings or helmlike structures (fig. VI). The fruit is often a woody pyxix (fig. V) sometimes with edible seeds (*Bertholletia*, Brazil-nut). Trees without intraxylary phloem.

Several genera of the Rhizophoraceae (*Rhizophora, Ceriops, Brubuiera*), Sonneratiaceae (*Sonneratia*) and Combretaceae (*Laguncularia, Lumnitzera, Conocarpus*) habitate in the mangroves of the tropical shores, forming specific adaptations as prop roots (Rhizophoraceae) or aerating roots (*Sonneratia, Bruguiera, Laguncularia*). Many Rhizophoraceae develop viviparous fruits. The Crypteroniaceae are closely related to the Sonneraticeae. The Combretaceae have uniloculate ovaries producing only one seed; that of *Terminalia catappa* is edible. *Quisqualis indica* serves as ornamental shrub.

The pantropical (especially American) Melastomataceae are characterized by estipulate, commonly opposite leaves with 3–9 palmate veins, ± parallel running and transversely anastomosing. Connectives of the stamens often with specific appendages (fig. VII). The typically 4–5-merous, actinomorphic flowers (androecium sometimes zygomorphic) are perigynous or epigynous, mostly having 2 whorls of stamens, 4–14 carpels. 150 genera, 4000 species, mostly herbs and shrubs (*Miconia* (900), *Leandra, Clidemia*).—*Medinilla (magnifica), Centradenia, Tibouchina, Heterocentron, Rhexia* are ornamentals. 10 species of *Rhexia* occur along the Gulf and the Atlantic coast. 1 *Tetrazygia* in Florida.

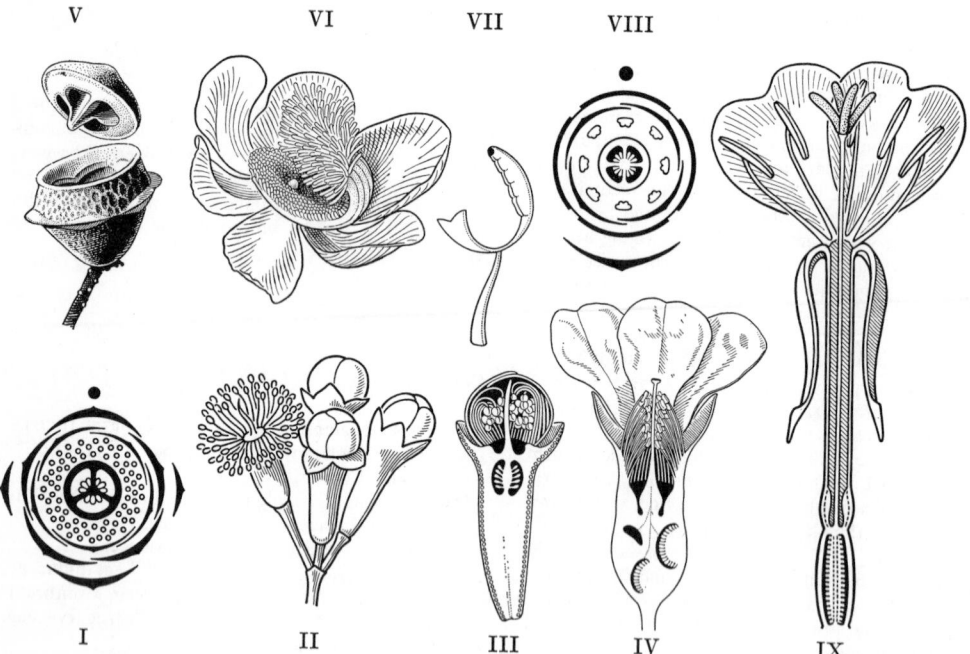

Fig. 1   I, *Myrtus communis*, floral diagram; II, III, *Eugenia caryophyllata*, flowers and vertical section of a flower; IV, *Punica granatum*, flower, vertical section; V, *Couroupita guianensis*, flower; VI, *Lecythis elliptica*, pyxis with lid; VII, *Centradenia inaequilateralis*, stamen; VIII, IX, *Oenothera*, diagram and vertical section of a flower.

Onagraceae. Typically the flowers have the diagram of fig. VIII and are distinguished by the tubular or cup-like epigynous hypanthium (fig. IX, hyp.). Leaves mostly alternate, deciduous, stipules ±. The family is distributed especially in (temperate) America. Small trees (*Hauya*), shrubs (*Fuchsia*), herbs (*Oenothera, Epilobium, Jussieua, Ludwigia, Gaura, Godetia, Circaea, Lopezia*), many of them garden flowers.— The unigeneric Hydrocaryacea (*Trapa*, water-chestnut) differ by the absence of the inner whorl of stamens and a half-inferior (2-loculed) ovary, which develops to an edible drupe, whose persistent endocarp bears 2–4 large horns. Floating annuals with rosulate leaves and inflated petioles.

The Haloragaceae commonly have unisexual flowers, 1–4 loculed ovaries with as many styles; each locule contains a single ovule; waterplants with pinnate leaves (*Myriophyllum, Proserpinaca*), herbs, subshrubs. *Gunnera* constitutes a separate family.—The flowers of *Hippuris vulgaris*, Hippuridaceae, have no perianth, 1 stamen, an unilocular uniovulate gynoecium; aquatic plants with whorled linear leaves. This family and the Thelygonaceae are of doubtful position, as are the Cynomoriineae (Cynomoriaceae, root parasites with some resemblance to the Balanophoraceae).

FOCKO WEBERLING

References

Engler, A. and K. Prantl, "Die natürlichen Pflanzenfamilien," Leipzig, Engelmann, 1st ed. 1889–1902, 2nd ed. since 1924 (Engler and Harms, H.).
Engler, A. and L. Diels, "Syllabus der Pflanzenfamilien," 11th ed., Berlin, Engelmann, 1936.
Lawrence, G. H. M., "Taxonomy of Vascular Plants," New York, Macmillan, 1951.

MYXOMYCETES (MYCETOZOA, SLIME MOLDS)

A very specialized group of organisms classified by some with the FUNGI as Myxomycetes, by others with the PROTOZOA as Mycetozoa. Martin (1960) argues convincingly for the fungal relationships of the Myxomycetes, and his views are accepted by many biologists.

The Myxomycetes are characterized by a free-living, slowly creeping, non-cellular, multinucleate, jelly-like assimilative stage, the plasmodium, which gives rise to a fixed, propagative stage of intricately constructed, often brilliantly colored, generally minute fruiting bodies on or in which spores with cellulose walls are produced.

Most of the 400 odd known species of Myxomycetes are cosmopolitan in their distribution, but a few seem to be confined to the tropics and a few others appear to be strictly temperate zone forms. The slime molds may be found in moist shady places in almost any woodland where leaf mold and decaying wood abound. The fruiting bodies are formed on dead leaves, sticks, decaying wood or bark, fungus sporophores, various portions of living plants, soil, or indeed on any substratum on which the plasmodium creeps just before fruiting. Nevertheless, some species fruit preferentially on certain substrata. Fruiting appears to be seasonal, at least with some species.

A typical fruiting body consists of a cellophane-like, horny, or lime-encrusted base (*hypothallus*), a stalk, and a spore case (*sporangium*). The sporangial wall (*peridium*) may be persistent or early evanescent. The spores inside the sporangium are usually enmeshed in a mass or network of threads (*capillitium*) which in many species is elastic and upon rupture of the peridium expands, elevating the spores, thus aiding in their dissemination. Any of these accessory parts of the propagative stage (hypothallus, stalk, peridium, capillitium) may be absent, their presence or absence, their color and structure when present, serving as important characters on the basis of which the various taxa are differentiated.

The class Myxomycetes consists of two subclasses, the Ceratiomyxomycetidae (Exosporeae), a small subclass with only 3 species, and the sub-class Myxogastromycetidae (Myxogastres, Endosporeae), to which all other species belong.

The life cycles of members of the two sub-classes differ in some important details. Only the general life-cycle pattern of the Myxogastromycetidae will be summarized below: Under favorable conditions the spores germinate releasing 1 to 4 uninucleate, amoeboid cells (*myxamoebae*). These divide by extranuclear mitosis and form large populations. In the presence of free water myxamoebae become anteriorly flagellated swarm cells. These cells fuse in pairs to form zygotes or withdraw their flagella and fuse as myxamoebae depending on the species. Some myxomycetes are heterothallic and fusion takes place only between cells of opposite mating types (Collins, 1963; Henney, 1967). Nuclear fusion follows and each zygote, by successive, synchronous, intranuclear divisions, grows into a plasmodium. Zygotes may also coalesce with other zygotes and other plasmodia of the same species thus growing by accretion as well as internal growth. In nature, plasmodia spread under the bark or in the wood of decaying logs, or under moist decaying leaves on the forest floor. In response to stimuli which are not well understood, they emerge, come to rest, and give rise to fruiting bodies in which spores are formed. The position of meiosis in the life cycle is controversial but it appears that in some species meiosis takes place just prior to spore formation whereas in others it occurs in the young spores. If so, 3 of the resulting 4 nuclei disintegrate leaving the spores uninucleate and haploid (Aldrich, 1967.)

Laboratory experiments indicate that such factors as temperature, pH, moisture, and food supply, all play major roles in inducing fruiting. Pigmented plasmodia also require light before they fruit, some of their pigments acting as photoreceptors under acid conditions which are known to favor the fruiting process. A shift in oxidases from ascorbic acid oxidases to cytochrome oxidases appears to take place at the time of sporulation.

Several types of plasmodia are recognized (Alexopoulos, 1960), but the best known is the phaneroplasmodium which, when well developed, consists of a network of many-branched, vein-like strands, one edge of which spreads into a fan-shaped sheet of protoplasm, the so-called advancing fan. Each strand is differentiated into a central channel, in which the protoplasm flows, and a gelified outer region which confines the protoplasmic stream. There are no cell walls, the plasmodium being a multinucleate, diploid, acellular, protoplasmic mass. Streaming in a plasmodial vein is periodically reversible, flowing in one direction for 50 to 90 seconds, coming to a gradual stop, and resuming flow, generally in the opposite direction. Streaming in a plasmodium may attain a speed of 1.35 mm a second, the greatest velocity of protoplasm ever recorded. The motive force in a plasmodial strand is thought to be generated through the interaction of a contractile protein (*myxomyosin*) with ATP (adenosine triphosphate) both of which are known to occur in the plasmodium of *Physarum polycephalum*, one of the Myxomycetes.

Myxamoebae, swarm-cells, and plasmodia of Myxomycetes normally obtain their nutrition by ingesting and digesting solid food in the form of bacteria, yeasts, fungus spores and hyphae, etc. In the laboratory, myxomycete plasmodia are fed rolled oats on which most species seem to thrive, at least in crude culture. It was not until 1939 that undoubted pure cultures of myxomycete plasmodia were obtained (Cohen, 1939), but of the 400 odd known species of Myxomycetes no more than 30 or 40 have been induced to complete their entire life cycle in artificial culture under any circumstances. *Physarum polycephalum*, which has been the subject of most physiological and biochemical research (Rusch, 1968), *Physarum flavicomum*, and *Physarum rigidum* are the only Myxomycetes whose plasmodia have been grown in axenic culture in chemically defined, liquid media.

Because of their unusual life cycle, the Myxomycetes, at the borderline between the plant and animal kingdoms, are of great interest biologically. They have become extremely important experimental tools in the study of protoplasmic structures, of protoplasmic streaming, and of the physiology of reproduction.

CONST. J. ALEXOPOULOS

References

Aldrich, H. C., Mycologia 59:127, 1967.
Alexopoulos, C. J., Mycologia 52:1, 1960.
ibid. "Introductory Mycology," 2nd ed., New York, Wiley, 1962.
ibid. Bot. Rev. 29:1, 1963.
ibid. in "The Fungi," (G. C. Ainsworth and A. S. Sussman, eds.) Vol. II:211, New York, Academic Press, 1966.
Cohen, A. L., Bot. Gaz. 101:243, 1939.
ibid. op. cit. 103:205, 1941.
Collins, O. R., Am. J. Bot. 50:477, 1963.
Gray, W. D. and C. J. Alexopoulos, "The Biology of the Myxomycetes," New York, Ronald Press, 1968.
Henney, M. R., Mycologia 59:637, 1967.
Lister, A., "Mycetozoa," 3d ed., London, Brit. Mus. (Nat. Hist.), 1925.
Martin, G. W., "The Myxomycetes," N. Am. Flora, 1, pt. 1, New York Botanical Garden, 1949.
ibid. Mycologia 59:119, 1960.
Martin, G. W. and C. J. Alexopoulos, "Monograph of the Myxomycetes," Iowa City, University of Iowa Press, (In Press)
Rusch, H. P., *in* "Advances in Cell Biology," (D. M. Prescott, ed.), Vol I, New York, Appleton-Century-Crofts (In Press).

NAJADALES

An unnatural order of herbaceous MONOCOTYLEDONES of doubtful phylogenetic position. Variously treated to include from one to eight families. Here treated in the broad sense (known also as *Helobiae* or *Fluviales*) although grouping the taxa into three or even six orders has much in its favor. The aquatic habitat, simplicity of their flowers and absence (or nearly so) of endosperm are the characters grouping the families in a unit. The fruit is usually an achene. Vegetative and reproductive parts exhibit a wide range of variation. Of little economic value except as food for water birds and cover protection for aquatic animals.

The *Najadaceae* include a single genus, *Najas*, of world-wide distribution. The approximately 40 species are submersed aquatics with linear leaves on much-branched stems. The plants are monoecious or dioecious; the flowers are simple and sessile in the branch axils.

The *Potamogetonaceae* (*Zosteraceae*) include eight or nine genera of which the best known are *Potamogeton*, with nearly 100 species of fresh or brackish waters, and *Zostera* (six species), *Ruppia* (one species) and *Zannichellia* (two species) of saline waters. Considered to comprise several families by some phylogenists. Recent studies indicate that the flowers are unisexual and lack a perianth even though, as in *Potamogeton*, the pistils and stamens with subtending bracts may be grouped to simulate a bisexual flower with a calyx.

The *Aponogetonaceae* include a single genus with some 15 species mostly of the Eastern Hemisphere.

The *Scheuchzeriaceae (Juncaginaceae)* include *Scheuchzeria palustris*, a marsh herb of the northern hemisphere, and *Triglochin* with a dozen species of north temperate regions. *Lilaea*, with a single species, is often placed here but probably is better placed in a distinct family: *Lilaeaceae*.

The *Alismaceae (Alismataceae)* differ from other families in the order (except the *Butomaceae* which often are treated as not comprising a separate family) by having bisexual flowers and a perianth with both calyx and corolla. Nectar may be produced and pollination effected by insects. They have been considered to be among the most primitive monocots. However, recent studies indicate that other families in this order may be more primitive. Of the 14 genera, the best known are *Sagittaria* and *Alisma*. Both are seen in aquaria. The fleshy tubers of *Sagittaria* are an important food for aquatic animals and were used by early man in America and in Eurasia.

The *Butomaceae* are variously treated to include from one to six genera, depending on the characters used for separating from the *Alismaceae*. *Butomus* has but one species, *B. umbellatus,* an attractive plant grown for ornament in water gardens. A native of Eurasia, it has become a serious weed in marshes of northeastern North America, crowding-out plants of more value as food to wildlife.

The *Hydrocharitaceae* (here treated as including the *Vallisneriaceae* and *Elodeaceae*) are a family of 16 genera and nearly 100 species of fresh and marine waters in the warmer parts of the world. The most familiar are *Elodea (Anacharis)* and *Vallisneria,* being common in aquaria. *Elodea* is native to America only but *E. canadensis* has now become a weed in some of the inland waters of Europe. It is dioecious and apparently only the pistillate plant is aggressive in Europe as the staminate plant is rarely found there. It reproduces freely by fragmentation. *Vallisneria* produces the female flowers singly on long scapes which extend to the surface of the water. The male flowers are clustered in a spathe near the roots; their pedicels break at maturity allowing the flowers to float to the surface where the pollen is released and floats to the pistils. After pollination, the ovary is drawn below the surface by the coiling of the slender elongated scape. *Stratiotes* (one species) has a rosette of stiff leaves which at pollination time lift the unisexual flowers to the surface of the water, becoming submerged again for maturation of the fruit. *Hydrocharis* (one species) floats on quiet water and is dioecious. The flowers are large, showy and produce nectar. Pollination is by insects. One of the two species of *Thalassia* occurs in such abundance in the Gulf of Mexico that great quantities are washed ashore and gathered for fertilizer.

EUGENE C. OGDEN

References

Rendle, A. B., "The Classification of Flowering Plants," 2nd ed. Vol. 1, Cambridge, The University Press, 1953.
Lawrence, G. H. M., "Taxonomy of Vascular Plants," New York, Macmillan, 1951.

NAMES, GROUP, GENDER AND JUVENILE

Collective nouns for groups along with names for the male, female, neuter and young evolved from prehistoric language. Primitive men used them to describe animals they saw and hunted. The Choctaw word *lukoli* signified a drove, herd, cluster, bunch or flock. The Osage *tse* indicated a buffalo cow; *wadsuta tonga* a bull.

Today a variety of terms designate assemblages of animals that feed, travel or live together. *Herd*, an ancient word, denotes an assemblage of cattle, as well as of antelopes, buffalo, cranes, deer, elephants, seals, swans, swine, whales, and wrens. *Flock* refers more specifically to a gathering of birds as well as to sheep or goats and sometimes to camels or lice. Young produced at one birth are a *litter*, while birds hatched together are a *brood, covey, hatch, clutch*, or *peep*. Other specific terms are: sloth of bears, rag of colts, skulk of foxes, mob of kangaroos, kindle of kittens, leap of leopards, stud of mares, nest of mice, pride of lions.

Names for genders and young vary with localities and frequently suggest size or age. An animal is better described as a knobber, pricket, brocket, stag, buck, doe, or fawn rather than just a deer. *Cock* and *hen* are used for birds; *boar* and *sow* for domestic swine; *colt* for a young horse; *heifer* for a female calf, and *vixen* for a female fox. Less well known usages are *cock* or *hen* for lobsters; *boar* for a male skunk; *sow* for a female raccoon; *colt* for a young crane. The young quail, turkey, green turtle, walrus, beaver, and monkey may be called a *squealer, poult, chicken, cub, kit*, and *infant*, respectively.

With the domestication of animals and growth of farming, this terminology was further developed. Refinements in vocabulary appeared as hunting evolved from the means of earning a living into a sport for the nobility. Knowledge of the special jargon of the hunt with its stilted, formalized nomenclature was essential.

Although origins often are buried in the history of language and hidden in folklore, it is known that many words were invented or derived from Norman-French. Certain terms developed from animal habits, cries, or characteristics, while others imitated older words. Anglo-Saxon words were absorbed by the expanding English vocabulary along with contributions from Latin, Teutonic, and other languages. Incorrectly transcribed letters and printers' errors often were responsible for variations in form and meaning. Even though archaic, some words are still encountered as colloquialisms while others survive in polite conversation.

Group names or terms applied to companies of animals or birds were popular in medieval "courtesy books," listing terms considered proper for usage by gentlemen. Such lists often were entitled "Terms of Venery." Examples appear in the fifteenth century "Boke of St. Albans," and the eighteenth century "Lore of the Chase."

In 1951 there appeared a flurry of discussions of such group names in the press of England and the United States. To the list of group names applied to animals and birds, several writers at that time suggested such jocosities as "a pomposity of bishops" and the like.

The lists which follow are not complete but cover the terms most commonly found in contemporary literature.

Andrew Bernhard
Daniel R. Pfoutz

Reference

Hare, C. E., "The Language of Field Sports," London, Country Life, 1949.

Group Names of Mammals: Antelope herd; **Ape** shrewdness; **Ass** pace or drove; **Badger** cete; **Bear** sloth or sleuth; **Beaver** family or colony; **Bloodhound** sute; **Boar** sounder; **Buffalo** troop, herd or gang; **Camel** flock or train; **Caribou** herd; **Cat, domestic** cluster or clowder; **Cat, wild** dout or destruction; **Cattle** drove, herd, mob (Australian) or drift; **Chamois** herd; **Colt** rag or rake; **Cur** cowardice; **Deer** herd; **Dog** (see also hound, Greyhound, Bloodhound) kennel, pack (dogs pursuing an animal), gang; **Donkey** (see ass); **Eland** herd; **Elephant** herd; **Elk** gang or herd; **Ferret** business or fesynes; **Fox** cloud, skulk, or troop; **Gelding** brace; **Giraffe** herd; **Goat** flock, trip, herd or tribe; **Greyhound** brace (two) leash (three); **Hare** huske, down, drove or trip; **Hartebeest** herd; **Horse** haras, stable, remuda, stud, herd, string, field, set or team; **Hound** mute; **Impala** couple (2 running); **Jackrabbit** husk; **Kangaroo** troop or mob; **Kine** drove; **Leopard** leap or lepe; **Lion** pride, sowse, sault troop or flock; **Lynx** (see Bob-cat); **Mare** stud; **Marten** richesse or richness; **Mouse** nest; **Mole** labor; **Monkey** troop or cartload; **Moose** herd; **Mule** barren; **Muskox** herd; **Ox** team, yoke, drove or herd; **Pig** (see swine); **Porpoise** school, crowd, herd, shoal or gam; **Reindeer** herd; **Roedeer** bevy; **Rhinoceros** crash; **Seal** pod, herd or trip; **Sheep** flock, hirsel, drove, trip, pack or hurtle; **Squirrel** dray; **Swine** sounder (wild), drift, herd or trip; **Walrus** pod; **Weasel** pack; **Whale** school, gam, mob, pod or herd; **Whelp** litter; **Wolf** rout, route or pack; **Zebra** herd.

Group Names of Birds: Bittern siege or sedge; **Chicken** flock or run; **Chough** chattering; **Coot** fleet or pod; **Cormorant** flight; **Crane** herd or siege; **Crow** murder; **Curlew** herd; **Dove** flight, flock or dule; **Duck Hawk** (Peregrine Falcon) flight, leash or cast; **Duck** paddling, bed, brace, flock, flight or raft; **Dunbird** rush; **Eagle** convocation; **Goose** gaggle (on water), flock (on land), skein (in flight), covert; **Goldfinch** charm; **Grouse** pack or brood; **Gull** colony; **Hawk** cast; **Hen** brood; **Heron** siege or sege; **Jay** band; **Lapwing** desert; **Lark** exaltation, flight or ascension; **Magpie** tiding; **Mallard** flush, sord or sute; **Nightingale** watch; **Partridge** covey; **Peacock** muster, ostentation or pride; **Pheasants** nye, brood (young with parent) or nide; **Pigeon** flock or flight; **Plover** stand, congregation, flock or flight; **Quail** covey or bevy; **Rook** building; **Shelldrake** dropping; **Snipe** wisp or walk; **Sparrow** host; **Starling** chattering or murmuration; **Stork** mustering; **Swallow** flight; **Swan** herd, team, bank, wedge or bevy; **Swift** flock; **Teal** spring; **Turkey** rafter; **Turtle dove** dule; **Widgeon** bunch, company, flight or knob; **Wildfowl** fall or flight; **Woodpecker** descent; **Wren** herd.

Other Group Names: Ant nest, army or colony; **Bee** swarm, cluster or nest; **Eel** swarm or bed; **Fish** school, shoal, haul, draught, run or catch; **Fly** business, hatch, grist, swarm or cloud; **Fly, May** swarm; **Frog** army; **Gnats** swarm or cloud; **Grasshopper,** cloud; **Herring** army, glean or cran (a measure for fresh Herring); **Hornet** nest; **Jellyfish** smuck; **Locust** swarm, cloud or plague; **Louse** flock; **Mackerel,** pack, school or shoal; **Oyster** bed; **Perch** pack, school or shoal; **Sardine** family; **Shad** run; **Shark** school or shoal; **Smelt** quantity or run; **Snake** bed (a group of young) or knot; **Termite** colony, nest, swarm or brood; **Toad** nest or knot; **Trout** hover; **Turtle** bale; **Wasp** nest.

Gender Names: Alligator m. bull; **Ant** f. queen; **Ass** m. jack, jackass f. jenny, jennyass, she-ass or jennet;

Bear m. boar or he-bear f. sow or she-bear; **Bee** m. drone f. queen or queenbee; **Bobcat** m. tom f. lioness; **Buffalo** m. bull or ox f. cow; **Camel** m. bull f. cow; **Canary** m. cock f. hen; **Caribou** m. bull, stag or hart f. cow or doe; **Cat, domestic** m. tom, tomcat, gib, gibcat, boarcat or ramcat f. tabby, grimalkin, malkin, pussy or queen; **Cattle** m. bull, ox, steer, beef, cow-brute, cow-creature, gentleman-cow, male-cow, seed-ox or Jonathan f. cow, ox, beef, milch cow, mulley, cow-beast or she-cow; **Chicken** m. rooster, cock, stag, game-chicken, he-biddy, crower, chanticleer or game cock f. hen, partlet or biddy; **Cougar** m. tom or lion f. lioness, she-lion or pantheress; **Coyote** m. dog f. bitch; **Deer** m. buck or stag f. doe; **Dog** m. dog f. bitch; **Duck Hawk** m. tercel, tiercel, tarcel, tarsel, tassel, tercelet, jack, falcon-gent or tercel gentle f. falcon, falcon-gentle, slight falcon; **Duck** m. drake, stag f. duck; **Eland** m. bull f. cow; **Ferret** m. dog, buck, jack or hob f. bitch, doe or jill; **Fish** m. cock or milter f. hen; **Fox** m. fox, dog-fox, stag, reynard or renard f. vixen, bitch or she-fox; **Giraffe** m. bull f. cow; **Goat** m. buck, billy, billie, billy-goat or he-goat f. she-goat, nanny, nannie or nanny-goat; **Goose** m. gander, stag f. goose or dame; **Grouse** m. cock f. hen; **Guinea Pig** m. boar; **Hare** m. buck or jack-hare f. doe or puss; **Hartebeest** m. bull f. cow; **Horse** m. stallion, stag, horse, stud, stot, stable-horse, sire or rig f. mare or dam; **Impala** m. ram f. ewe; **Kangaroo** m. buck f. doe; **Kudu** m. bull f. cow; **Leopard** m. leopard f. leopardess; **Lion** m. lion, tom f. lioness or she-lion; **Lobster** m. cock f. hen; **Manatee** m. bull f. cow; **Merganser** m. drake; **Mink** m. boar f. sow; **Moose** m. bull f. cow; **Mule** m. stallion or jackass f. she-ass or mare; **Muskox** m. bull or musk bull f. cow or musk cow; **Otter** m. dog f. bitch; **Ostrich** m. cock f. hen; **Owl** f. jenny howlet; **Ox** m. ox, beef, steer or bullock f. cow or beef; **Partridge** m. cock f. hen; **Peacock** m. peacock f. peahen; **Pheasants** m. cock or rooster f. hen; **Pigeon Hawk** m. blue hawk or jack-merlin; **Pigeon** m. cock f. hen; **Prairie Falcon** m. American Lanneret f. American Lanner; **Quail** m. cock f. hen; **Rabbit** m. buck f. doe; **Red Deer** m. hart or stag f. hind; **Reindeer** m. buck f. doe; **Roedeer** m. roebuck f. doedeer or doe; **Robin** m. cock; **Salmon** m. kipper; **Sandpiper** m. ruff f. ree or reeve; **Sea Lion** m. bull f. cow; **Seal** m. bull f. cow; **Shad** m. buck shad; **Sheep** m. buck, ram, male-sheep, mutton or tup f. ewe or dam; **Skunk** m. boar; **Sparrow** m. cock f. hen; **Sparrow-hawk** m. musket; **Swan** m. cob f. pen; **Swine** m. boar, hog, pig, porker or male-hog f. sow, boar (wild) hog, pig or porker; **Terrapin** m. bull f. cow; **Tiger** m. tiger f. tigress; **Turkey** m. gobbler, tom f. hen; **Termite** m. king f. queen; **Walrus** m. bull f. cow; **Weasel** m. boar f. sow; **Whale** m. bull f. cow; **Wolf** m. dog-wolf or dog; f. bitch, dam or she-wolf; **Woodchuck** m. he-chuck f. she-chuck; **Woodcock** m. cock f. hen; **Wren** f. jenny or jennywren; **Yak** m. bull or ox f. cow **Zebra** m. stallion f. mare; **Zebu** m. ox.

Juvenile and Juvenile Group Names: Ant antling; **Antelope** fawn, kid or yearling; **Bear** cub *(group* litter); **Beaver** kit or kitten; **Bobcat** kitten or cub; **Buffalo** calf, yearling or spike-bull; **Camel** calf or colt; **Canary** chick **Chicken** chick, chicken, poult, cockerel or pullet; **Caribou** calf or fawn; **Cat, domestic** kitten, kit, kitling or kitty *(group* litter); **Cat, wild** kitten, kit, kitling or kitty *(group* litter); **Cattle** calf, hog stirk, stot; (m. bull-calf or f. heifer); **Chimpanzee** infant; **Clam** littleneck; **Coal-fish** parr; **Cod** codling, scrod or sprag; **Cicada**

nymph; **Condor** chick; **Cougar** kitten or cub; **Cow** calf (m. bull; f. heifer); **Coyote** cub, pup or puppy; **Deer** fawn; **Dove** pigeon or squab; **Duck Hawk** eyas, nestling or hack bird; **Duck** duckling or flapper; **Eagle** eaglet; **Eel** fry or elver; **Eland** calf; **Elephant** calf; **Elk** calf; **Fish** fry, fingerling, minnow or spawn; **Fly** grub or maggot; **Fly, Bott** bot or bott; **Fly, Caddis** caddis worm; **Fly, Chalcid** joint worm; **Fly, Crane** leather jacket or meadow maggot; **Fly, Damsel** naiad; **Fly, Dobson** hellgrammite, dobson, crawler, hell-devil, hell-diver, coniption bug or arnly; **Fly, Dragon** naiad; **Fly, May** naiad; **Giraffe** calf; **Goat** kid; **Goose** gosling; **Grouse** chick, poult, squealer or cheeper; **Hare** kitten, pussy, puss or leveret; **Hartebeest** calf; **Herring** sardine, sprat or brit; **Horse** colt, foal, stot, stag, filly hog-colt, youngster, yearling or hogget; **Kangaroo** joey; **Kudu** calf; **Leopard** cub; **Lion** whelp, cub or lionet; **Louse** nit; **Lobster** chicken; **Mackerel** spike; **Manatee** calf; **Marten** cub or kitten; **Merganser** yearling; **Mink** kit or cub; **Monkey** suckling, yearling or infant; **Mosquito** larva, flapper, wriggler or wiggler; **Moth, Bombycid** silkworm; **Moth, Clothes** larva; **Moth Geometrid** inch worm, measuring worm or looper; **Moth, Hawk** tobacco worm; **Moth, Mulberry** silkworm; **Moth, Psyche** bagworm; **Moth, Sphinx** horn worm; **Moth, Silkworm** silkworm; **Moth, Tussock** tussock or caterpillar; **Musk-ox** calf; **Muskrat** kit; **Otter** pup, kitten, whelp or cub; **Ostrich** chick; **Owl** owlet or howlet; **Ox** stot or steer; **Oyster** set seed, spat or brood; **Partridge** squealer or squeaker; **Pelican** chick or nestling; **Penguin** fledgling or chick; **Petrel** chick or nestling; **Pheasant** chick or poult; **Pigeon** squab, nestling or squealer; **Pike** jack; **Quail** chick or squealer; **Rabbit** kitten or bunny; **Raccoon** kit or cub; **Red Deer** calf, fawn, brocket, spay, staggard, hart or hind; **Reindeer** fawn; **Roedeer** fawn or kid; **Rhinoceros** calf; **Salmon** parr, smolt, grilse, graul, jerkin, fingerling or alevin; **Sea Elephant** pup; **Sea Lion** pup; **Seal** whelp, pup, cub, bachelor, belamer or half bull; **Shark** cub; **Sheep** lamb, lambkin, hog, tag, teg, shearhog, shearling or hogget; **Skunk** kitten; **Squirrel** dray; **Swan** cygnet; **Swine** shoat, trotter, pig, shote, piglet, pigling, piggy, piggie, farrow, grice, gilt, squeaker (wild), hogget (wild) or hogsteer (wild) *(group* litter) **Termite** nymph; **Tiger** whelp or cub; **Tiger Salamander** axolotl; **Toad** tadpole; **Turkey** chick or poult; **Trout** fingerling, fry or alevin; **Turtle** chicken; **Walrus** cub; **Weasel** kit; **Whale** calf; **Wolf** cub or pup; **Woodchuck** kit or cub; **Zebra** colt or foal; **Zebu** foal.

NATURE RESERVES

Human survival on the earth depends not solely on control of atomic weapons and the advance of cultural mores and ideologies, but with equal force on the manner in which man utilizes and protects the natural resources that support him. Former civilizations, as great in their time as is ours today, are now swept by drifting sands because their peoples ravaged the natural verdure that was the foundation of their strength. As they wasted their natural wealth, they waged violent wars, seeking *lebensraum* by stealing their neighbors' more fertile lands.

The modern world has been as rapacious as were the

ancients. In all history no nation has been more waste-fully profligate with its resources than has America. In the brief span of three centuries she leveled the verdant forests of vast regions, pillaged wildlife, tore the cover from her priceless soil until the silt polluted her rivers, and committed crimes against nature that following generations will pay dearly to expiate.

Some regions, such as parts of western Europe, have practiced wise farming and forestry practices for cen-turies, and in less industrialized countries natural re-sources remain in abundance; but man still is wasting the wealth of the earth. The industrial revolution of the 19th century aggravated demands for conversion of soils and forests into agricultural crops, lumber and other commodities; slaughter of wildlife for food and sport together with competition for its habitat by domestic livestock threatened native species every-where; and misuse of land accelerated erosion, com-pounding destruction of topsoil and watercourses. At the same time, human populations began to expand explosively, and frontiers of arable land became scarce. A new concept gradually became a recognized aspect of modern civilization, the tenet of CONSERVATION, that every land and water area must be devoted to its wisest use to ensure perpetuation of renewable resources and to avoid depletion of those that are not renewable. Translation of this ideal into practice on the land is an imperative of the continuance of human culture.

During the past few decades it has been learned what must be done to ensure perpetuation of these re-sources; but in spite of truly astonishing progress, this knowledge is not yet being applied on a sufficiently comprehensive scale. Forests still are cut far more rapidly than they are replaced; overgrazing and out-moded farm practices are widespread. Despite im-proved laws to curb slaughter of wildlife, violations are frequent and many species continue to decrease. Nat-ural resources must be utilized to serve the economic needs of increasing numbers of people, but other human requirements are dependent on them also. The rapid spread of urbanization and modern technology through-out the world is creating social tensions and an arti-ficiality from which surcease must be available if man is to enjoy the full richness of life. Personal associa-tion with a natural environment not only contributes to physical and mental health but also provides that tranquillity which is essential to creative thought. The advance of science is achieved not solely in labora-tories, but more fundamentally by the searching mind which requires a measure of solitude and freedom from distraction for its inspiration. The highest use of certain resources is their preservation undisturbed and unmodified to safeguard assets which, once altered, cannot be restored. Science relies, too, on the natural associations of plants and animals and other features of the earth for the materials for its research into the mysteries of the origin and evolution of life and natural processes. The extinction of even a single species is a detriment to science. The natural world provides the conditions for correlating and equating investigations into methods of deriving human benefit from its re-sources. The beauty of the outdoors, the patterns of nature, the ways of wildlife, the magnificent order of the environment in which man lives are assets which warrant inviolate protection for the enjoyment of man and the realization of his full stature.

In earlier times, royal preserves were decreed in many parts of the world, sometimes to protect valuable resources but more usually as hunting grounds for the nobility; they were not dedicated to public use and en-joyment, although they did serve to prevent the extir-pation of characteristic forest types and species of wild-life, and many of them have become national parks or nature reserves in recent years. The first modern nature reserve was established in the Forest of Fontainebleau, in France, in 1853, and stimulated the concept of pro-tecting nature for its own sake especially in Europe and its colonial empires. In 1864, President Abraham Lincoln established a new pattern of public land ad-ministration when he signed an Act of Congress which ceded the incomparable Yosemite Valley and the Mariposa Grove of *Sequoia gigantea* to the State of California for "public use and preservation," which was returned to the Federal Government in 1905 for inclu-sion in Yosemite National Park. Establishment of Yellowstone National Park, in 1872, set the stage for large reservations to safeguard resources and to pre-serve representative examples of natural ecological en-vironment. This first national park, and the system of American parks and monuments that grew from it, in-spired other nations to undertake similar programs.

By 1960, forty-eight nations had set aside some 500,000 square miles as national parks or equivalent reserves, and more than seventy countries had under-taken programs of protecting ecological associations, wildlife, and its habitat. Japan has a superb series of national parks containing some of the finest scenery in the world comprising four per cent of its land area. The national and provincial park systems of Canada include vast expanses of undisturbed wilderness and abundant wildlife. Some of the African parks and re-serves are the largest yet established. Kruger National Park in the Union of South Africa and Kafue National Park in North Rhodesia cover 8,000 square miles each, while the Kalahari Gemsbuck National Park in Cape Province and Bechuanaland protects 17,000 square miles; millions of wild animals depend on these and other reserves for their perpetuation. A number of Asian and Australian parks and reserves are of special value in protecting endangered species, such as the Great Indian, Sumatran and Javan species of rhinoc-eros, the Asiatic Lion, Neotornis, and other rare forms. South America possesses some of the most striking scenery on earth, including mountain ranges, fiords, jungles, cloud forests and deserts, inhabited by myriad kinds of wildlife, and some of the finest terrain has been reserved there.

Many governments have modeled administrative agencies for managing these reserves along the lines laid down by the United States in 1916 when its Na-tional Park Service was created to "conserve the scenery and the natural and historic objects and the wildlife . . . and to provide for the enjoyment of the same in such manner and by such means as will leave them unimpaired for the enjoyment of future genera-tions." In other countries, administration of the re-serves is sometimes less definitive, often a collateral responsibility of the forestry or game departments, while in some instances little more has been done yet than to designate the boundaries of the reserved areas. The concept of protecting vanishing natural assets has gained vigorous momentum throughout much of the world.

Growth of the national park idea into a national

policy in the United States was slow. It was not until 1890 that three new parks were set aside; but thereafter the system expanded more rapidly. Vast regions, mainly in the West and in the territories, were reserved by congressional and presidential action, and to the system were added outstanding places of historical and archaeological interest. By 1959, twenty-nine national parks encompassed 13,459,637 acres and the entire system of 180 areas included 22,928,333 acres. Scenically, they are the nation's treasures of natural beauty, superb exhibits of the natural heritage such as the Grand Canyon, the towering ranges of forests of Sequoia, Rocky Mountain and other mountain parks, the rugged coastline and rain forests of the Olympic peninsula, miles of cactus-clad deserts, with myriad animals of many species. National parks are established by Congress, and it requires congressional approval to modify their boundaries or to alter natural conditions in them. National monuments are authorized by the Antiquities Act of 1906, which empowers the President, as well as Congress, to reserve public land to protect objects of historic, prehistoric and scientific importance. Thirty-three of the eighty-three national monuments are natural areas, some of them vast in size, such as Katmai which protects 2,697,590 acres of volcanic mountains and conifer forests, and Glacier Bay, a 2,274,248-acre fiord region of glaciers and peaks, both the Alaskan habitat of several kinds of bears and other wildlife. Other monuments are smaller, preserving particular ecological niches and species of flora and fauna, unusual geological features, or prehistoric Indian relics.

The philosophy laid down in the National Park Act is applied equally to all of the areas in the system. Exploitation of their natural resources is prohibited, except that grazing permits existing when the respective parks or monuments were established are respected, and the mining laws apply to four areas. Constant efforts are made to secure authorization for hydroelectric projects in some of them, to open some or all of them to mining or grazing, or otherwise to violate the protection given them; but such threats have been resisted successfully. All wildlife is protected from hunting, but the Secretary of the Interior has authority to reduce excessive local populations of animals such as elk and bison. The objective of administration is to perpetuate natural conditions and behavior patterns with as little interference as possible. Accommodating some 60,000,000 annual visitors requires facilities, but under the Mission 66 program inaugurated in 1956 to rehabilitate the system and to improve interpretive services, physical structures are concentrated insofar as possible in restricted areas, leaving some ninety-five per cent of the parks and monuments essentially in a wilderness state. The national park system is not regarded as complete and a number of additions have been proposed. The respective states and counties also have initiated park and reserve programs, some of their reservations being nearly the equivalent of national parks, such as the Adirondacks Forest Preserve in New York, the Porcupine Forest in Michigan, and the extensive state, regional and county park systems of California.

The United States Fish and Wildlife Service administers a large program of wildlife refuges. The first was the three-acre Pelican Island in Florida, reserved by Theodore Roosevelt in 1903. Threatened extinction of egrets, spoonbills and other birds pillaged by plume hunters at the turn of the century, and severe declines

in the national populations of ducks, geese, and swans, as well as of some of the larger mammals and the fur-bearers, prompted additional reserves for their protection and management. Effectiveness of federal protection of avian species migrating across Canada, the United States and Mexico was greatly increased by ratification of treaties with Great Britain and Mexico which assigned to the Federal Government a measure of responsibility for the management of these species, which heretofore had been solely the responsibility of the states except in national parks and monuments. In 1959, the Service administered 275 national refuges, covering 17,319,745 acres, including twenty game refuges and ranges for bison, pronghorn antelope, elk, deer, and other animals; and 255 migratory bird refuges, primarily to safeguard and restore nesting, feeding and wintering grounds of migrating waterfowl along the major flyways. These refuges have contributed significantly toward the recovery of many endangered or declining forms of American wildlife and have stimulated public interest in wildlife as a matter of national concern. Sportsmen have supported the use of funds derived from duck stamp licenses and a proportion of the taxes on ammunition and sporting goods for the program, and the growth of the refuge system and of Federal responsibility for the welfare of wildlife reflects the maturation of a large body of public sentiment interested in wildlife in its own right as a source of personal enjoyment and scientific study.

Late in the 19th century, Presidents Harrison and Cleveland established large forest reserves in the western states to protect watersheds and to safeguard public timber and grazing land from further ruinous exploitation. President Theodore Roosevelt increased the number of such reserves, redesignated as national forests, and created the U.S. Forest Service to administer them. The first Chief Forester was Gifford Pinchot, who successfully introduced sound forestry practices into the United States. The 160,582,000 acres of national forests represent most of the commercially used forests and much of the livestock ranges in Federal ownership, operated for the benefit of private enterprise under careful regulation. Within the national forests, 13,985,072 acres have been especially dedicated to preserve their untouched wilderness character. The twelve Wilderness Areas and a number of the forty-two Primitive Areas and three Roadless Areas encompass more than 100,000 acres each, some covering more than 1,000,000 acres. The twenty-six Wild Areas, each less than 100,000 acres, preserve areas of especial scientific importance. Since one of the purposes of wilderness preservation is to enable people to travel for periods of time by their own efforts without mechanized aids, roads are not constructed in these areas, timber is not harvested (except to a controlled degree in the three Roadless Areas), nor are other commercial uses authorized except as required by statute. Hunting is permitted, since the states retain jurisdiction over wildlife.

In addition to nature reserves administered by federal state and local governments, a large number of parks and wildlife sanctuaries are managed by citizens' organizations and by private individuals. The National Audubon Society, the Nature Conservancy and other related associations safeguard natural areas of considerable size and scientific significance, and these organizations have stimulated public appreciation of nature and of the conservation program throughout the country.

Recognition of the imperative need to protect natural resources for economic and social benefits developed during the same period that the biological, geological and related sciences expanded out of museum laboratories into ecological research in the natural environment. Biologists especially were among the first to realize the basic material for their research required preservation of unmodified ecological associations and wildlife habitat, so that these scientists provided the strongest impetus for establishing nature reserves on an international basis. They were able to present information about the features concerned and to provide justification for official action to reserve them.

As parks, refuges and reserves were set aside in many countries and protective laws and regulations were promulgated, the need for international programs and agreements became apparent. The Migratory Bird Treaty between the United States and Great Britain (1918) and the similar Convention for the Protection of Migratory Birds and Game Mammals between the United States and Mexico (1937) gave added protection to many avian species and stimulated expansion of refuge programs in North America. The Convention on Nature Protection and Wild Life Preservation in the Western Hemisphere (1940) provided a uniform framework for national park and reserve programs in the American Republics and marshalled support for such action there. The London Convention for the Protection of African Fauna and Flora (1933) led to establishment of vast parks and wildlife reserves in many parts of Africa and set forth criteria and principles for their administration.

The United Nations and its specialized agencies are taking increased interest in perpetuation of natural resources and preservation of endangered wildlife. The Food and Agricultural Organization, while primarily concerned with food production and efficient land-use practices, recognizes the economic significance of protective measures. UNESCO has a direct interest in the educational, scientific and cultural benefits to be realized from a strong and broad program of reserves. At the request of the Secretary-General, in 1958 the Economic and Social Council approved the establishment of a register of national parks and equivalent reserves of the member nations as an official activity of the United Nations. Correlating the work of administrators, technicians and scientists of all nations in this field is the International Union for Conservation of Nature and Natural Resources, established in 1946, which holds regular international conferences to provide forums for international discussions and recommendations.

FRED M. PACKARD

References

Anon., "Derniers refuges: atlas commente des reserves naturelles dans le monde," Brussels, International Union for the Conservation of Nature, 1957.

Anon., "The Position of Nature Protection throughout the World in 1950," Brussels, International Union for the Conservation of Nature, 1951.

Anon., "Addendum to the Position of Nature Protection," Brussels, International Union for the Conservation of Nature, 1954.

Butcher, D., "Exploring our National Parks and Monuments," Boston, Houghton Mifflin, 1963.

Butcher, D., "Seeing America's Wildlife in our National Refuges," New York, Devin-Adair, 1955.

Clawson, M. and B. Held, "The Federal Lands, their Use and Management," Baltimore, John Hopkins Press, 1957.

Pinchot, G., "Breaking New Ground," New York, Harcourt Brace, 1947.

Shankland, R., "Steve Mather of the National Parks," New York, Knopf, 1951.

Tilden, F., "The National Parks, What They Mean to You and Me," New York, Knopf, 1954.

Lieber, R., "America's Natural Wealth," New York, Harpers, 1942.

NEMATA

The phylum of eelworms, or nemas ("nematodes")—organisms also variously ranked in a class or order (commonly Nematoda). Nearest relatives, though remote, appear to be members of the phylum GASTROTRICHA. The Nemata and Gastrotricha are here placed, with other phyla of somewhat similar level of construction, in the Subkingdom Scolecida (syn. Amera).

Nemas are the most successful worm-like organisms, in numbers both of species and of individuals, and are adapted to life in nearly all accessible habitats, an important exception being to a free-swimming existence in the PLANKTON. Their practical importance stems particularly from their role as parasites of domestic plants and animals and of man himself; as agricultural pests they compete with ARTHROPODS in destructiveness.

A widespread misconception, even among specialists, is that nemas as a group are largely similar in form and organization. While, to be sure, a threadlike shape is typical for a number of groups, there are important exceptions. So varied in structure are they, in fact, that few generalizations can be made that are valid for all members of the phylum or for members of the major included taxa. The following discussion of morphology is based on typical nemas; limitations of space preclude a consideration of exceptions and variations.

The typical nema is an unsegmented, thin, vermiform organism covered by a cuticle of secreted, layered scleroprotein with an extremely delicate outer lipoid or lipoprotein membrane and possibly penetrated by submicroscopic pore canals presumed to secrete the lipoid layer. The outer-most protein layer may undergo tanning. Externally the cuticle is commonly striated transversely, or annulated, sometimes with additional longitudinal striae, or ridges (alae). It may bear sensory, or non-sensory, immovable setae. The cuticle is usually molted four times in development, one or more molts sometimes being cast inside the egg shell.

Beneath the cuticle the ectodermal epithelium (epidermis) forms a complete tube, which may be cellular or syncytial, typically with two median and two lateral enlargements projecting into the body cavity as longitudinal chords and consisting, if cellular, of uninucleate or multinucleate cells. Special uninucleate hypodermal gland cells may be present in the lateral chords. Some nemas possess three caudal glands with a single terminal pore, through which a sticky protein thread, serving for attachment, is emitted. The excretory system, where occurring, apparently develops from the ventral chord and ends in a ventral pore opening through the cuticle in the anterior part of the body. Lateral excretory canals, if present, are usually confined to the lateral chords. Such canals occur only in the Class Secernentea.

Beneath the epidermis, longitudinal, spindle-shaped muscle cells are arranged in four bands. Individual cells each have a sarcoplasmic innervation process to either the dorsal or ventral nerves situated in the corresponding chords. The muscle cells are uninucleate and few (*platymyarian, meromyarian*) or many (*celomyarian, polymyarian*). The position and number of the unstriated, ribbon-like muscle fibrils in relation to the muscle cells and chords depend on the age and evolutionary development of the particular nema. Adjoining muscle cells act as a unit, causing an eel-like movement. Special somato-esophageal, somato-intestinal, rectal, and copulatory muscles are also attached to the body wall.

Mesenchymatous cells in the body cavity may tend to grow together over the somatic muscles, as well as over the digestive tract and gonads, so as to support these structures. In no case does the body cavity originate as a splitting of a single layer; it is a pseudocelome.

The digestive tract is complete, being a rather straight tube with an anterior, terminal, round, or subtriangular oral opening and posterior, ventral anus leading from rectum or cloaca. The mouth is usually surrounded by labial structures and sensory organs. The mouth cavity, or stoma, is basically a rather short, cylindroid tube leading to the foregut or esophagus (= pharynx). In the stoma of all nemic parasites of plants and of some of those of animals is a hollow hypodermic needle serving to penetrate cells and to extract their contents. The esophagus, a syncytial muscular tube, has a triradiate lumen, with one ray directed ventrad. Its walls usually contain one dorsal and two subventral esophageal glands. There is an esophago-sympathetic nervous system. Various modifications permit the esophagus to serve as a pumping organ for liquids or as a particle-swallowing organ.

The mid-gut, or intestine, is a straight tube, one cell thick; sometimes its cells are multinucleate. Stored

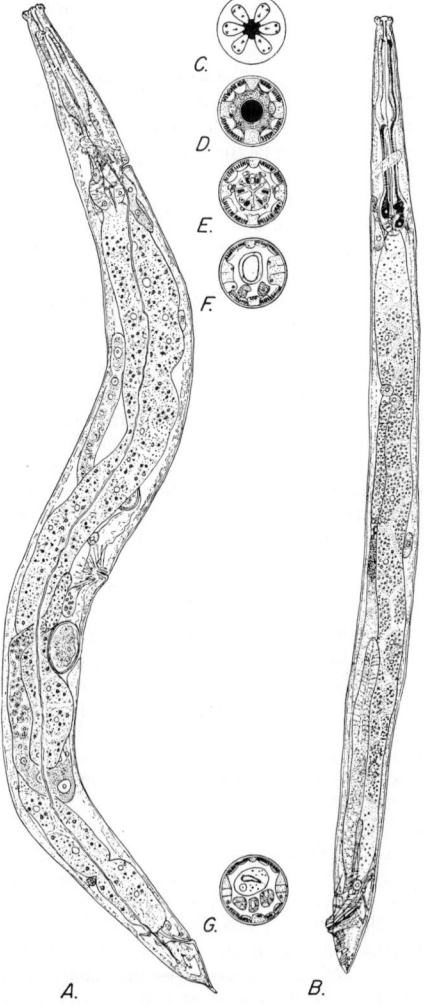

Fig. 1 *Pelodera strongyloides. A*, Female, lateral view—two pseudocelomocytes showing, one between vulva and intestine, second between posterior part of intestine and body wall, single dorso-rectal suspensory cell, dorso-rectal muscle, rectum, rectal glands, excretory system with anterior and posterior lateral canals, sinus and subventral gland cells, anterior and posterior flexed ovaries opening into corresponding uteri, one sperm at junction of posterior ovary and uterus, one fertilized egg in third cleavage, small seminal receptacle just anterior to latter egg, vulva with musculature and part of fertilized egg in anterior uterus; nerve ring, nerve cells, one dorso-esophageal amphidial gland cell, and phasmids. *B*, Male, lateral view—caudal alae with enclosed tacto-genital papillae, two pseudocelomocytes showing, one pair, at anterior flexure of single testis and one, unpaired, dorsal to intestine in mid-region; single dorso-esophageal suspensory cell at base of esophagus; cloaca with two subventral (or four) and one dorsal rectal gland, spicules and gubernaculum with muscles, dorso-rectal muscle cell; dorsal and subventral esophageal glands (in black); growth zone of single testis, with seminal vesicle, vas deferens with one (of pair) of lateral out-pocketings which serve as cement or copulatory glands. *C–G*, Cross sections of male: *C*, Cephalic region, *en face*, showing oral openings, lips, papillae and amphids; *D*, Stomatal level, showing stoma, papillary nerves, amphidial sensillae, somatic musculature, and hypodermis; *E*, Region of forepart of esophagus showing amphidial nerves, esophageal radii and musculature in addition to structures shown in *C*; *F*, Anterior intestinal region showing subventral excretory gland cells, lateral chords and lateral excretory canals; *G*, Preanal region of male showvas deferens, with a pair of copulatory glands and posterior lateral excretory tubes. After Chitwood, B. G. & Chitwood, M. B. (1937, 1950). An Introduction to Nematology, Sec. 1, fig. 3. Reproduced with permission of publisher.

proteins and lipoids and solid waste materials are common as microscopically demonstrable, usually granular inclusions within the cytoplasm.

The hind-gut (*rectum* in the female, *cloaca* in the male) is a dorsoventrally flattened, short tube set off from the intestine by a circular sphincter muscle and terminating postero-ventrally; a few large, flat epithelial cells comprise the wall. Rectal glands may open into the lumen anteriorly. In the male, there are usually two spicules (sometimes one, rarely none)—sclerotized cuticular modifications of the dorsal cloacal wall projecting into the cloaca and acting as sperm conducting (non-intromittent) organs; an additional grooved structure (*gubernaculum*) is a spicule-guiding organ. Sperm enter the cloaca ventrally. Both esophagus and rectum (or cloaca), being ectodermal invaginations, are lined with cuticle, which should theoretically be molted, although in many forms the poststomatal and esophageal lining apparently are resorbed.

Basically the reproductive system consists of two tubular, opposed gonads. In female development, uterine epithelial tissue between the gonads meets ectodermal epithelial cells, which grow from the ventral chord, forming the cuticularly lined vagina. In male development, a long ventral epithelial tube, the vas deferens, grows posteriad to meet a similar outgrowth from the hind-gut. The epithelium of uterus and vas deferens is one cell thick and is continuous with that of ovaries and testes, respectively. Gonadel epithelium (of testis or ovary) is a delicate sheath surrounding the developing germ cells. Musculature is commonly confined to those parts of the gonoducts near the genital opening and is arranged as a somewhat spiral layer, one cell in thickness. Reduction from two to one ovary is common; multiplicity of ovaries, exceptional. In some groups the number can be used for taxonomic characterization at certain levels. Suppression of the posterior testis is characteristic of many of the major taxa, but two testes are basically primitive, as indicated by the two primordial germ cells in the larvae of all forms.

The generative cells of a gonad may form a central cord, one cell thick, all definitive germ cells arising in sequence from a single primordial germ cell; or, exceptionally, the latter may proliferate, resulting in an elongate tube, with definitive germ cells arising along the wall over an extended region. Dilatations of the gonoducts form such functional structures as seminal vesicles and receptables (or *spermathecae*). Oöcytes, during their growth, acquire all the stored food necessary for the formation of the vermiform larvae and are homolecithal, no nurse cells entering into the formation of the eggs. Although, commonly, spermatozoa are grossly ameboid, they usually reveal vestiges of flagellate structure—unquestionably the ancestral condition. (In fact, a few nemas are known to have elongate spermatozoa.) They are introduced into the female by the male at copulation, after which they migrate up the uterus where they may be temporarily stored in a modified region of the uterus or in a seminal receptacle. After fertilization, the zygote first secretes a chitinous egg shell and then, inside that, forms the lipoidal vitelline membrane. Thereafter, an external protein layer may be deposited on the outside of the chitinous shell. This latter layer, when present, assumes many, often bizarre patterns and shapes.

The nervous system, of ectodermal origin, consists centrally of: 1) an anterior circum-esophageal commissure (*nerve ring*), with directly and indirectly attached ganglia; and 2) longitudinal nerves, situated in the four longitudinal chords. Anteriad, six sensory nerves (two subdorsal, two lateral, and two subventral) proceed from the nerve ring through their corresponding ganglia of bipolar neurones in the body cavity to the cephalic tactile organs (*papillae*). In addition, a pair of dorsolateral amphidial nerves enter the corresponding amphidial glands, in which they terminate as presumed chemo-receptors, the amphidial sensillae; the amphids, in a comprehensive sense, each consist of gland, sensilla, and lateral (or dorsolateral) external pores, simple in some groups, but associated with elaborate cuticular ornamentation in others. Posteriorly, four major nerves—dorsal, lateral (2), and ventral —are situated in the corresponding chords, sometimes with four additional submedian nerves likewise situated in the hypodermis. The chief nerve of the body is the partially double, ganglionated ventral trunk, serving both a motor and a sensory associational function. This and the dorsal nerve give off repetitive groups of motor nerve fibrils, which are met in the body cavity by the protoplasmic "innervation processes" of the somatic musculature. The ventral nerve also connects with the rectal commissure and, through the epidermis', with the median caudal nerve and paired lateral caudal nerves. Nerve cell nuclei have not been reported in the dorsal or submedian nerves.

So-called "somatic," tactile, sensory organs are largely confined to subventral (and ventral) genito-papillary organs in the male and to a single pair of specialized, presumably tactile organs, the deirids (*cervical papillae*), in both sexes of one sub-phylum (Secernentea). In the other sub-phylum (Adenophorea) there may be numerous, usually lateral or sublateral, tactile organs *(papillae*, or *setae)* over the entire length of the body, in addition to the paired genito-sensory organs, and one (rarely a paired) series of ventral, preanal, presumedly tactile, or tacto-glandular, organs (*supplementary organs*) is commonly present.

A pair of lateral, caudal presumed chemo-receptors, the phasmids, similar to, but simpler than, the amphids, are characteristic of one Class (Secernentea). Like the amphids, they have laterally situated pores, each leading into a phasmidial gland cell, which ensheaths a simple sensilla. Unlike the amphids, their neurones are situated nearby, the axones running anteriorly through the lateral caudal nerves and hypodermal commissures to the ventral nerve.

Evidence of photoperception has been offered by B. G. Chitwood and D. G. Murphy (1964) in the case of a marine monhysterid, *Diplolaimella* (*Diplolaimita*) *schneideri*. Other nemas with such organs occur sporadically in the Adenophorea. These may be in the form of paired pigment cups with lenses in the body cavity near the nerve ring, as cited, or in the subdorsal sectors of the esophagus, as in the enoplid, *Leptosomatum*; or take the form of paired colored crystalloids in the body cavity near the nerve ring as in the desmoscolecid, *Tricoma.*

Nemic development follows a determinative pattern so far as known, with a peculiar type of bilateral cleavage. Nuclear constancy, or eutely, for parts of the body is characteristic generally, some of the smaller and possibly more primitive nemas often having more parts of the body characterized by such constancy than the larger forms. Regeneration is unknown for nemas.

An astonishing variety of patterns of sexual reproduction is known among free-living forms—including typical gonochorism, pseudogamic gonochorism, self-fertilizing hermaphroditism (both syngamic and pseudogamic), and parthenogenesis; more than one of these may be combined in a single species.

The physiology of nemas has been relatively little explored in general, although certain aspects have received intensive investigation. Thus the cultivation of certain free-living forms (especially rhabditids) has been extensively carried out, and this work has led to their exhaustive study in axenic culture. *Ascaris* has been the subject of considerable biochemical investigation.

The physiology of nemas is only now receiving the attention it deserves. In the past 10 years articles on the biochemistry of nemic membranes, enzymology, and respiration have become an integral part of control investigations.

L. De Coninck, A. G. Chabaud, M. Maurice Ritter and Jean Theodorides (1965) conservatively recognize the Orders Araeolaimida (Suborders Araeolaimina and Tripyloidina), Monhysterida, Desmodorida (Desmodorina and Draconematina), Chromadorida (Chromadorina and Cyatholaimina), Enoplida (Enoplina and Oncholaimina) and Dorylaimida (Dorylaimina and Alaimina) in the Adenophorea and Rhabditida, Strongylida, Ascaridida, and Spirurida (Camallanina and Spirurina) in the Secernentea. Though many additional orders will soon come to be recognized, this work is the most comprehensive covering the Nemata at the present time. Classification of the phylum is undergoing such rapid revision that a listing of the lower taxa included in each of the orders had best await a less transitory period.

Phylogenetic relationships of nemas to other organisms and of groups within the phylum to one another have been, and remain, controversial. A relationship to gastrotrichs is now generally accepted, but other affinities are variously disputed. As a more general question one might well ask what feature (or features) has made the nemas so conspicuously successful. The writers regard the method of nemic locomotion to be an attribute of particular importance in this connection. The abandonment of vibratile cilia (*pecilokonts*) and the assumption of serpentine movement have clearly yielded organisms more versatilely mobile than, say, the gastrotrichs, which have continued to rely on ciliary locomotion. Quite likely, with this evolutionary improvement in nemic mobility, the primitive nema was able to invade a wide variety of habitats more effectively than its ancestors and their closely related contemporaries and thus to originate a considerable number of evolutionary lines.

Because of the extremely rapid development of nematology during recent years, all reference works are now out of date. Readers interested in information not contained in the works cited in the next paragraph should consult "Biological Abstracts," "Helminthological Abstracts," and "Zoological Record."

B. G. CHITWOOD
ELLSWORTH C. DOUGHERTY

References

Chitwood, B. G., *et al.* "An Introduction to Nematology," Sec. I. "Anatomy," Baltimore, Monumental Printing, 1950.

Chitwood, B. G. and D. G. Murphy, "Observations on two Marine Monhysterids—their Classification, Cultivation, and Behavior," Tr. Amer. Microsc. Soc. 83:311–329, 1964.
Christie, J. R., "Plant Nematodes, their Bionomics and Control," Gainesville, Univ. Florida Press, 1959.
DeConick, L., *et al.* "Classes des Nematodes," Traite de Zoologie, Vol. 4, Masson, Paris, 1965.
Goodey, J. B., "Soil and Freshwater Nematodes," New York, Wiley, 1963.
Hyman, L. H., "The Invertebrates", vols. 3 and 5, New York, McGraw-Hill, 1959.
Jenkins, W. R. and D. P. Taylor, "Plant Nematology," New York, Reinhold, 1967.
Thorne, G., "Principles of Nematology," New York, McGraw-Hill, 1961.

NEMATOCYST

Nematocysts or *cnidocysts* or *cnidae* are intracellular structures peculiar to the Cnidaria among COELENTERATA. They develop in ectodermal cells, the *nematoblasts* or *cnidoblasts*, in between other kinds of cells, such as sensorial, glandular and epithelial proper.

They consist of an ovoid rigid *capsule*, containing a long invaginated thread, the *nematocyst tube*. The capsule is capped by a stopper or *operculum*, adjacent to the surface of the ectoderm, which closes the orifice of the invaginated tube. Just below the stopper the tube is wider and forms the *proximal shaft*, shorter than the capsule; its end tapers to the long, thin *terminal tube*, coiled in the intracapsular space and bathed in fluid. Attached inside the tube's wall is a series of spirally arranged barbs, differing in size according to the diameter of the tube. When evagination occurs, the barbs stick out, harpoon-like, on the outer surface and help to attach the thread on bodies coming into contact with the animal.

Nematocysts vary greatly in size and shape and have been classed in numerous complex groups. The length ranges from 5 μ to 250 μ; and the evaginated thread ranges from 50 μ to 1 mm. The ellipsoid capsule is more or less elongated and curved, and the tube barbs most diversely disposed. Each species has two or more sorts of cnidae, which are of taxonomic value.

Cnidae are mostly localized on tentacles where they are often grouped in stinging buds (Hydrozoa, etc.) Actinians produce nematocysts all over their body.

The mechanism of the explosion of cnidae in natural conditions is still uncertain. The operculum is first ruptured, apparently as a result of contractions in surrounding tissues. The evagination of the tube follows instantaneously, due to the internal pressure developed in the ripe cnidocyst, increased by endosmosis of water after the stopper is torn. Nerve action is not thought to be a part of the process.

Cnidogenesis takes place in the deeper parts of the ectoderm, usually in nematoblasts, but sometimes in the protoplasm of a syncytial layer (e.g. Actinian acontial filaments). The cnidocyst appears as a dense secretion body, often enclosed in a vacuole as it grows. This body then differentiates into the capsule and its content. Later, the nematoblast migrates to the surface of the epithelium, with its cnida ready to explode.

The substances constituting the capsule wall and tube

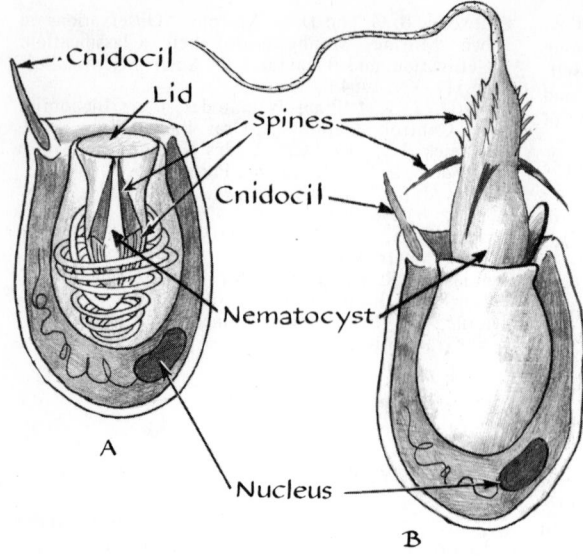

A

B

Fig. 1   Cnidoblast of hydra. (From Platt and Reid, "Bioscience," New York, Reinhold, 1967.)

differ according to species and the kinds of cnidae. Most seem to be protein bound polysaccharides; the protein has certain points in common with keratin. The intracapsular fluid has an albumin content with a variable phenolic compound which accounts for the poisonous effects of nematocyst stings.

Structures resembling cnidae occur in the spores of Cnidosporidia and in the body of the Peridinian Flagellate *Polykrikos*.

MARYVONNE J. HAMON

References

Hyman, L., "The Invertebrates," Vol. I, New York, McGraw-Hill, 1940.
Picken, L. E. R. and R. J. Skaer, "A Review of Researches on Nematocysts," *in* Rees, W. J., ed., "The Cnidaria and their Evolution," New York, Academic Press, 1966.

NEMATODA

The name most commonly used at present for the taxon (usually phylum or class) of nemas (= nematodes), or eelworms. It, and its earlier variant Nematoidea, were originally applied to both nemas and horse-hair worms (NEMATOMORPHA). This historically unambiguous name NEMATA is preferred in this volume.

B. G. CHITWOOD
ELLSWORTH C. DOUGHERTY

NEMATOMORPHA

The Nematomorpha, which are part of the NEMATHELMINTHES, include two orders, the Gordioidea and the Nectonematoidea.

Gordioidea. These are worms from several centimeters to more than a meter long, having the appearance of the large horse hair or violin string, and which are found in springs, in running and stagnant waters, particularly in mountainous country. The sexes may be distinguished by the form of the posterior extremity which is generally bilobed in the males and either entire, or trilobed (*Paragordius*) in the female. The body of the Gordioidea is covered by a thin cuticle formed of two layers. These are a thin external layer, usually covered with prominences (*areolae*) of varying forms and also with papillae, hairs, tubercles, and the like, and an internal thicker layer formed of matted fibrils.

Underneath the cuticle lies the epidermis, the outer layer of which is composed of a single layer of cells, or of a syncitium. Under this, is a muscular layer made up of fusiform fibers with a central sarcoplasm containing polyploid nuclei and peripheral myofibrils. This muscular layer is interrupted along the whole length of the ventral surface by a vertical partition which joins the nerve cord to the epidermis.

The central part of the body contains much interstitial cellular material, forming a mesenchymatous tissue in which the following longitudinal cavities are found: a ventral cavity containing the digestive system which is a straight tube, not functional in the adult and which ends in the cloaca; two dorsolateral gonadial cavities filled with gametes and opening to the exterior by two short ducts which end at the cloaca, the orifice of which is terminal in the female and subterminal in the male.

There are no specific respiratory, circulatory or excretory organs. Copulation, which occurs in a tangled skein of male and female individuals (truly a "gordian knot") is ended by the emission of a spermatophore. The egg masses are produced as short rods.

The so-called "echinoderoid" larva (55 to 145 μ long) is transparent and bears at its anterior end three crowns of chitinous spicules, and a rectractile proboscis supported by stylets. This free-swimming aquatic larva becomes a juvenile parasite in the hemocoel cavity of insects, particularly the Coleoptera (*Dystiscus, Carabids*); Orthoptera (*Decticus, Tettigonia*); in larval Trichoptera (*Stenophylax*) and Odonata; or, more rarely, in Myriapods (*Lithobius*) or Arachnids. In this environment, the larva increases in size and develops its organ systems. As soon as it has reached the adult condition, the parasite leaves its host by boring a hole through the tegument, usually in the vicinity of the anus.

When the echinoderoid larva is taken into an animal unsuitable for its development, (e.g., the larvae of Ephemeroptera or Chironomids) it encysts and passes to a dormant state. The larva of *Gordius aquaticus* Duj., after a brief active period, can also encyst in water and, in this condition, is capable of surviving desiccation. The fact that most of the hosts of the Gordioidea are terrestrial arthropods raises the question as to how the aquatic echinoderoid larva gains access to the body of its terrestrial host. If, as is usually the case, this latter is of carnivorous habit, it becomes infected by eating a vector host (Ephemerid, Chironomid), developed from an aquatic larva which contained the encysted form of the parasite. Those hosts which have a herbivorous or mixed diet become infected by eating, along with their food, larvae in the dormant condition or cysts lying out of the water.

The Gordioidea are divided into two families. The Gordiidae, the cuticle of which is smooth or bears flattened areolae, are essentially confined to the genus *Gordius* Linneus 1766, which has a vast geographical distribution. The Chordodidae have a roughened cuticle which may carry a complex pattern formed from several types of areolae. The principal genera, which are divided among four sub-families, are: *Chordodes* Creplin 1847, in which the posterior extremity of the male is solid, is represented by numerous species found principally in tropical regions; *Chordodiolus* Heinze 1952, lacks areolae on the cuticle but has paired cuticular bands joined together by transverse anastomoses. *C. echinatus* is found in Lake Nyassa; *Gordionus* Miller 1927, *Parachordodes Camerano* 1897, and *Paragordionus* Heinze 1935, are widely distributed in Europe and Asia; *Paragordius* Camerano 1897, in which the posterior extremity of the female is trilobed, occurs in the south of Europe and in America.

Nectonematoidea. The Nectonematoidea are represented by the single genus *Nectonema* Verrill 1879. These are filiform planktonic worms (from 10 to 845 mm. long) which occur in the Atlantic, the Mediterranean, and the Pacific. They resemble Nematodes in certain characteristics, particularly the possession of a thin, single-layered cuticle, though this may show rows of cuticular "hairs"; by the possession of muscular cells, of which the contractile portion is oriented towards the exterior, and by an unpaired male genital system. But the morphology of the larva, and in particular the retractile anterior extremity armed with hooks, recalls that of the Gordioidea, as does the life history which includes a polymorphic parasitism among various decapod crustacea (*Leander, Anapaqurus, Portunus*).

A. DORIER
(Trans. from French)

Reference

Hyman, L. H., "The Invertebrates," Vol. 3, New York, McGraw-Hill, 1951.

NEMERTEA

The body is slender, soft and contractile, being usually 2–60 cm long and 1–10 mm wide. The head is in general broader than the portion following and flattened dorsoventrally. The body is cylindrical in the foregut region and becomes flattened and broader in the intestinal region.

The epithelium consists of ciliated, glandular and sensory cells. The basement membrane between the epithelium and the body musculature is a homogeneous hyaline layer of connective tissue. In the paleo-hoplobdellonemerteans the body musculature is composed of outer circular and inner longitudinal muscle layers, while in the heteronemerteans it is composed of three layers (outer longitudinal, circular, inner longitudinal). The cutis demarcated in the outer longitudinal muscle layer of the heteronemerteans is divided into an outer layer of cutis glands and an inner layer of connective tissue. In the cephalic region there are usually found the cephalic glands. The submuscular glands are found in the foregut region of the hoplonemerteans. The parenchyma inside the body musculature is thicker in the Enopla than in the Anopla.

The nemerteans are provided with a long eversible proboscis for immobilizing their prey. The proboscis lying in the rhynchocoel surrounded by its sheath attaches anteriorly to the cephalic tissue and posteriorly to the end of its sheath by a bundle of retracter muscles. The epithelium of the proboscis facing the inner lumen contains numerous glandular cells. The muscle layers of the proboscis are the same as those of the body wall. One or two muscle crosses are found in the heteronemerteans, and the proboscidial nerves are usually constant in number in the hoplonemertean species. In the Monostylifera of the Hoplonemertea, the armature of the proboscis consists of a central stylet and a base usually conical in shape, while two or more pouches containing accessory stylets are arranged around the base. In the Polystylifera it consists of a small number of short stylets lying on a sickle-shaped base, together with several pouches of accessory stylets. The proboscis sheath has an outer longitudinal and an inner circular muscle layer. The rhynchodaeum has a thick epithelium and feeble musculature.

The alimentary canal is a straight tube divided into foregut and intestine. In the Enopla the foregut is usually differentiated into oesophagus, stomach and pylorus, while in the Anopla the pylorus is wanting. In the Anopla the mouth is situated behind the brain, while in the Enopla the oesophagus running under the

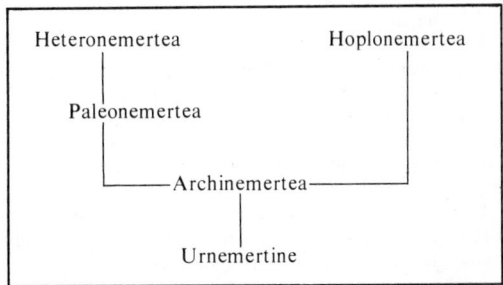

Fig. 1   Phylogenetic tree illustrated by Iwata.

brain opens in general into the rhynchodaeum. In the Cephalothricidae the mouth is situated far behind the brain. The intestine is generally provided with an intestinal diverticulum. In the Hoplonemertea the intestinal caecum is extended anteriorly under the pylorus. The rectum opens at the posterior end of the body.

In the cephalic region the blood circulates in a pair of cephalic blood lacunae, from which two lateral and one dorsal blood vessels arise, usually with numerous anastomoses. The dorsal blood vessel is wanting in the Cephalothricidae, Tubulanidae and Carinomidae of the Paleonemertea, but in the foregut region the Tubulanidae have a pair of rhynchocoel blood vessels and the Carinomidae have a pair of dorsolateral blood vessels in addition to the rhynchocoel blood vessels. In the foregut region the dorsal blood vessel runs posteriorly in the wall of the proboscis sheath.

The nephridia lying in the foregut region are branching tubes, lying in close proximity to the lateral blood vessels. One or more efferent ducts derived from the main tube of each nephridium open externally on the dorsolateral wall of the body. The excretory organs of the Cephalothricidae consist of a number of individual nephrostomes.

The brain is composed of a pair of dorsal and ventral ganglia with a thin dorsal and a thick ventral commissure. The lateral nerves originate from the ventral ganglia. They are found outside the body musculature in the Paleonemertea except the Cephalothriciae in which they are situated in the longitudinal muscle layer. In the Heteronemertea they are situated on the circular muscle layer lying between the outer and the inner longitudinal muscle layers, while in the Enopla they are situated inside the body musculature. The dorsal nerve extends posteriorly from the dorsal commissure of the brain and two oesophageal and two proboscidial nerves originate from the ventral ganglia or the ventral commissure.

As sense organs, the ocelli, the cerebral sense organs, the frontal sense organs and the lateral sense organs are recognizable. The cerebral sense organ is composed of sensory and glandular cells and opens anteriorly on the lateroventral wall of the head through a ciliated canal, while posteriorly it connects with the dorsal ganglion. Depressed areas, known as sensory grooves, in which the epithelium lacks pigment and gland cells, presumably with a chemo-tactile function, occur on the head and neck of the nemerteans. They are directly connected with the pair of canals leading inward to the cerebral sense organs.

The nemerteans are dioecious except for a few genera which are monoecious. The gonads are sac-shaped in shape and arranged metamerically along the lateral side of the intestine.

The nemerteans are littoral and marine, except for a few terrestrial genera. They live under stones and among algae, burrowing in mud or sand. Pelagic and bathypelagic nemerteans float at the surface or in the abyssal regions of the ocean. A few species are commensal in pelecypods, tunicates or sponges and parasite in crabs.

Embryologically the nemerteans are divided into the direct and the indirect types. The pilidium, Desor's and Iwata's larvae belonging to the indirect type are characteristic of the Heteronemertea. In these cases the adult epithelium is formed secondarily by five amniotic invaginations of the larval epithelium.

For the last ten years (1950–1959) Iwata has studied the comparative embryology of the nemerteans and has concluded that the Cephalothricidae should be separated from the Paleonemertea as a new order, Archinemertea, from the fact that the position of the lateral nerves in the body wall is definite in a post-larval stage of development. The Bdellonemertea accordingly must be settled in the Hoplonemertea as a new suborder, Bdellonemertoidea.

Fumio Iwata

References

Coe, W. R., "Biology of the nemerteans of the Atlantic coast of North America," Trans. Connect. Acad., 35: 129–328. 1943.

Iwata, F., "Studies on the comparative embryology of nemerteans with special reference to their interrelationships," Publ. Akkeshi Mar. Biol. State. 10:1–51, 1960.

Hyman, L., "The Invertebrates," vol. 2 "Platyhelminthes and Phynchocoela," New York, McGraw-Hill, 1951.

NEOBIOGENESIS

The theory of EVOLUTION holds that all living things are interrelated by common descent from an original case of successful biogenesis. Plausible explantions of how this may have taken place have been developed in the past three or four decades by Haldane[2], Oparin (see LIFE, ORIGIN OF), and Bernal[3]. All are agreed that abiotic synthesis of organic compounds, first simple then complex, preceded the origin of life on earth. This view has been supported recently through the work of Urey, Miller, Fox, Abelson, Oró, and others. Many have come to accept the following outline of evolution: inorganic compounds → simple organic compounds → complex organic compounds (including metalloorganic catalysts) → isolated systems of correlated reactions of increasing degree of complexity and interdependence → living things—as an expected and inexorable outcome of the evolution of matter. With modifications this explanation can be marshaled in support of the hypothesis that neobiogenesis has been a continuing possibility since life first originated. The rejection of this possibility on various grounds is not entirely justified.

Arguments against Recurring Biogenesis

Thermodynamic considerations. The applicability of the second law of thermodynamics to organic evolution has been critically examined and postulated by Blum[4]. There can be little argument against the view that evolution proceeds irreversibly or that the flow of this entire event in time is irreproducible on this planet. The whole spectrum of the origination of life from inorganic beginnings could have occurred in only one period of the earth's history. It is unsound to argue that this whole episode can be repeated exactly over and over again in time. The precise conditions for a repetition of the occurrence are no longer available and probably never again will be available. To assume from this, however, that life can no longer originate de novo is unwarranted, for it is not necessary to postulate the origin of life de novo from *inorganic* beginnings. The immediate prerequisite for the origin of life is the presence of highly complex organic molecules involved in related chain reactions within isolated colloidal systems. Such a state is supposed to have existed for a considerable period of time long ago in the prebiotic history of the earth. During this period of time, which may have had a long duration, self-replicating systems of much the same kind may have arisen repeatedly wherever and as long as conditions permitted.

In fact, the immediate prerequisites for the origin of life have since never ceased to exist. Living things have perpetuated them throughout biological time. True, evolution has changed and diversified the chemical basis of life, but this merely means that neobiogenesis at any given period might establish primitive forms of life in keeping with the level of structure and organization of molecules and pathways characteristic of the period, and differing from previous and future times. Put another way, recurring biogenesis is part of the fabric of evolution and not a repetition of any part of its history. What is ruled out on thermodynamic grounds is not the origin of life *de novo* throughout time, but rather the origination of life throughout time *identical* to the *first* instance.

Competition against existing organisms. Another objection to the occurrence of repetitive neobiogenesis concedes, for the sake of argument, that it is a possibility but proceeds to claim that it would be impossible for a primitive living thing thus evolved to survive in the face of fierce competition from the organisms already present and adjusted to the environment. The assumption that an organism, simply because it is newly arisen, will have no adaptive features and will meet insuperable competition in any place and at any time is unwarranted and untenable. The argument has validity only if it can be established as certain that all organisms of neobiogenetic origin would meet with overpowering competition. This reasoning is based on an exaggeration of the concept of the struggle for existence and ignores interspecific compatibility and aid. As long as it is conceivable that a newly arisen organism may be compatible or symbiotic with, or indeed parasitic upon one of the existing species of organisms, the argument is invalid. The original organisms had to have a metabolism independent of the existence of other organisms. Since then it has been possible for organisms which lack various functions (for example, forms without the complete metabolism necessary for independent existence) to arise and lead a parasitic or symbiotic existence. If a unit possessing an incomplete metabolism should arise out of the variety of organic compounds existing today, it would be destined to be destroyed in any local environment devoid of living things. In the presence of cells or organisms possessing the requisite complementary metabolism, such a neobiogenetically evolved form would have a chance of survival. Indeed, for such forms, the presence of other organisms instead of posing the threat of certain extinction through fierce competition, becomes the *sine qua non* of their survival. Thus, a wider variety of simple organisms can originate through neobiogenesis today than was the case when living forms first evolved; the first organisms perforce, had to be of more specific and limited metabolic scope.

Biochemical similarities among all organisms. Another line of argument against repetitive neobiogenesis points to the similarity in the chemistry of all organisms. This subject has many aspects. For example, all naturally occurring amino acids, regardless of source, are of the L form, with notably few exceptions. It is argued that since D and L forms are mutually antagonistic or require different enzymes, the first organisms could incorporate one or the other, but not both, into their metabolism. The evidence is interpreted as showing that chance favored the L form. Repeated neobiogensis, it is argued, would establish organisms which by chance incorporated the D form, and the notable absence of this form is taken to mean a lack of successful instances of neobiogenesis since the period when living organisms first evolved. Another aspect of this argument is based on the identity of, or great similarity among the organic compounds and biochemical pathways found in most forms throughout the living world.

The foregoing observations are taken to mean that the original instance of biogenesis occurred through chance incorporation of specific stereoisomers and specific types of organic compounds. All subsequent organisms, it is claimed, arose by descent from the original form and thus were compelled to utilize the same compounds. Mutation, based on the existing substances, established an increasing variation in compounds, pathways, and species, which are all interrelated, however, through common descent. Organisms descending from separate forms of neobiogenetic origin it is argued, would be expected to show greater variation, some of these forms having been established, by chance, with opposite antipodes, different organic compounds, and different pathways.

All these observations overlook the probability that after living things came into existence, the types of organic compounds available from the environment underwent a radical change. Two factors contributed to this change. First, the rate of synthesis of organic compounds, *abiotically*, from inorganic compounds diminished as conditions necessary for this type of synthesis changed. Secondly, the *selective* synthesis of organic compounds by living things increased at a rate proportional to the increase in mass of organisms, probably an exponential increase in the beginning. This would result eventually, for all practical purposes, in the exhaustion of biologically rejected organic compounds in the environment and their replacement by organic compounds synthesized by living things. Most significantly, racemic mixtures of biologically important optically active substances were replaced by the isomers characteristic of the first successfully established organisms. Subsequent instances of neobiogenesis would have to occur within such a milieu, and having as a basis not only existing organic molecules but also existing reacting systems—that is biochemical pathways—would exhibit common metabolic aspects. Such organisms would consequently contain many compounds, biochemical pathways, and exhibit reactions characteristic of pre-existing organisms.

The time factor. This is another argument leveled against the possibility of recurring biogenesis. It will be conceded that from an inorganic atmosphere the origin of organic substances of sufficient complexity and concentration to support the establishment of the first living things might have taken many millions of years. The sterility which existed before the origin of organisms put no premium on time. But in the presence of a complex organic milieu, on the other hand, the time required for the transition from highly complex lifeless systems to metabolizing replicating systems (of the nature of primitive living things) is greatly reduced. Given the proper combination of substances and circumstances, neobiogenesis actually may take only a relatively short time.

A complexity of organic compounds and reacting systems exists today almost everywhere. For example, cells undergoing cytolysis release into their environment (which may already be rich in organic substances) globules of colloidal material, chromatin substance (deoxyribonucleoprotein), mitochondria, microsomes, compounds in different stages of reaction with one another, compounds undergoing sequential reactions still in progress, and so on. It is conceivable that out of such surroundings and under specific conditions, a metabolizing system can arise which has the attributes of life. This is meant to point to the possibility of the existence locally, at times, of circumstances capable of supporting neobiogenesis in a manner similar to that proposed for the first instance from the original mixture of complex organic interacting substances.

Spontaneous generation and the Pasteur myth. All of the arguments against neobiogenesis have a root in the almost unshakable belief that Pasteur disproved the possibility of the spontaneous generation of any living thing "once and for all." Pasteur used fermentable mixtures and, by ingeniously devised apparatus and method, showed that if such mixtures were protected from contact with particles in the air, fermentation would not take place. Even air minus the particles, or *preheated air* containing particles, would not cause fermentation. The only valid conclusion that can be drawn from Pasteur's experiments is that the air contains minute organisms which can cause fermentation, or the corollary conclusion that fermentable mixtures will not ferment in the absence of organisms. His experiments also permit the inference that his predecessors who claimed to have proved the possibility of spontaneous generation may have employed faulty techniques. But Pasteur's experiments do not constitute an unequivocal refutation of neobiogenesis—the origin *de novo* of living things. Today, if a competent scientist versed in modern biochemistry, virology, and genetics were to test this hypothesis seriously, about the only component he would use in common with Pasteur is water. It is one thing to prove the need of the presence of organisms to cause fermentation in a simple food mixture, but an entirely different problem merely to devise, and then conduct, a valid experiment to *test* the possibility of neobiogenesis. In more recent years, with the technological, biochemical, and philosophical advances that have been made, allowing deeper penetration into the problem, the ever-recurring question once again may be raised, but at a more sophisticated level.

Suggested Modification of the Monophyletic Theory of Evolution.

Repetitive neobiogenesis, as suggested above, would establish organisms similar in metabolism to known forms. The suggestion that repetitive neobiogenesis may be expected to establish bizarre forms of life different from the form of life as we know it may have a place only in science fiction. It may very well be that life as we know it—that is, the complex interdependent metabolic reactions supported in a structure we rec-

ognize as protoplasm—is the only form that matter can eventually take in its evolution toward the origin of organisms. Indeed, it is possible, even though it appears improbable, that life can exist only with the specific isomers which we find associated with it, and that the presence of these isomers was not the result of random choice but of biochemical necessity.

One may well ask what is gained by proposing the repetitive origin through time of organisms based on a structure and metabolism similar to pre-existing organisms. The answer, of course, is that it does not matter how similar the results of neobiogenesis are to pre-existing organisms. But the idea that neobiogenesis is possible, and may have been taking place ever since life first occurred, does matter. Specifically, it would appear more plausible to accept present-day viruses as units of recent and present origin than to suppose that they descended through two billion or so years relatively unchanged. Throughout time, VIRUSES either evolved into higher organisms or were eliminated in the process of evolution, being ever re-established through neobiogenesis. The similarity between microorganisms and higher forms in many aspects of their metabolism has led to the belief that such metabolic pathways have had a long and persistent history extending back almost to the dawn of life. However, this similarity is a strong argument for viewing these microorganisms as recent derivatives of higher forms. A virus can be regarded as a small fragment of "genetic information" of a higher species and must depend on it or similar forms for metabolic complementation or supplementation. BACTERIA, too, may be viewed in the same light except that they show different degrees of evolution, resulting in differing morphological and biochemical development. They, also, are destined for further evolution or elimination, while progenitors, already arisen, continuously evolve into the newer bacteria.

The discontinuities in the paleontological evidence are explained away by the contention that some forms are not subject to fossilization, while many that are, did not encounter the conditions favorable for fossilization, that some fossils have been permanently lost through natural processes like erosion, and, finally, the contention that many discoveries have yet to be made. Some of the discontinuities, however, can be viewed as the result of separate cases of neobiogenesis. The same may be said of the discontinuities in the taxonomic arrangement of existing organisms. The difficulty of placing viruses, bacteria, certain "algae", sponges, and others, into a fitting place in any taxonomic scheme based on a monophyletic hypothesis may stem from the possibility that the discontinuities are real and represent the existence of separate lines of descent from independent instances of neobiogenesis at different times in the history of earth down to the present.

JOHN KEOSIAN

References

1. Keosian, J., "The Origin of Life," 2nd ed., New York, Reinhold, 1968.
2. Haldane, J. B. S., The origin of life. Rationalist Annual. 1929.
3. Bernal, J. D., "The physical basis of life," London, Routledge and Kegan Paul, 1951.
4. Blum, H. F., "Time's arrow and evolution," Princeton Univ. Press, 1955.

NERVE CELL

Our concepts about neurons are based upon information drawn from a wide variety of animal species and technical procedures, and include morphological, physiological and biochemical data. No attempt can be made here to present either a survey of all the possible neuronal cell types, nor can descriptions of the techniques used to arrive at our present understanding of neurons be included. For extensive bibliographies on neuronal morphology, physiology and biochemistry reference may be made to Beams, *et al.* (1 and 2), De Robertis (3), Eccles (4), Hodgkin (5), McIlwain (6), Nachmansohn (7), Palay, *et al.* (8) and two symposia on nerve cells (9, 10). The following account is based on current knowledge of the fine structure, physiology and biochemistry of neurons from select mammalian, amphibian and invertebrate species which have been the most widely investigated.

The morphological characteristics which distinguish neurons are Nissl substance (granular endoplasmic reticulum), neurofibrillae, and cell membrane specializations at synapses and junctional contacts. Other constituents of the cell such as lysosomes, Golgi complex, mitochondria, microtubules and the nucleus are common to other cell types and the reader is referred to the section in this encyclopedia dealing with the CELL for a more complete discussion of these elements.

Neurons are relatively large cells due to their extensive dendritic and axonal extensions. They therefore have a relatively large protein synthetic activity and it has been repeatedly demonstrated that this activity takes place in the perikaryon, the resulting products of which stream into the cellular expansions. The morphological substrate for this synthetic activity resides in the basophilic Nissl substance which consists of scattered masses of granular endoplasmic reticulum of a relatively highly oriented type and is known to contain a high concentration of RNA. The cisternal membranes of the Nissl bodies are 60 to 70 Å thick and have situated on their non-luminal surface RNA particles 150 Å in diameter. Neurons are further biochemically characterized by an almost exclusive utilization of glucose metabolism as an energy source (6).

The neurofibrillar component of neurons first observed by Remak in 1843, and since studied by a variety of methods, consists of long 50 to 100 Å thick filaments dispersed in the cytoplasmic matrix between masses of Nissl substance, Golgi bodies, mitochondria and vesicles. These appear to traverse the perikaryon in all directions and planes, although loose bundles of filaments sometimes are seen, predominantly oriented in one direction. They are found in axons as single, 100 Å thick, longitudinally oriented filaments. Such filaments extend from the axon-hillock into the axonal processes where they are found scattered along the length of axons intermingled with varying numbers of microtubules, mitochondria and vesicles.

The neuronal cell membrane is 50 to 100 Å thick. As determined by studies of myelin and retinal rod cell membranes it appears to consist of a double bimolecular lipid leaflet with protein absorbed, or bound, at the polar surface of the leaflets. The exact molecular configuration and complete chemical constitution is unknown, although such things as Na and K ions, acetylcholine, and acetylcholinesterase have been demonstrated to be associated with neuronal cell membranes

functioning in the process of neuronal excitation and transmission. It is within the structural framework of this semi-permeable cell membrane that the molecular events leading to neuronal activity and impulse transmission occur. These events include membrane permeability changes which allow rapid increases in the inward flux of Na ions during the rising phase of the action potential followed by an increased outward flux of K ions during its descending phase. Nachmansohn (7) has postulated that acetylcholine combines with a receptor protein in the cell membrane which causes a change in the molecular configuration of the receptor protein such as to alter "pore size" and/or "pore charge" thus triggering ionic movements during the action potential. For convincing arguments against such a mechanism reference may be made to Eccles (4) and Hodgkin (5). For a complete discussion of nerve cell activity on a chemical and molecular basis and discussions on the Hodgkin-Huxley theory of nerve impulse conduction see Eccles (4), Hodgkin (5) and Nachmansohn (7).

At synapses and junctional contacts structural modifications in the cell membrane occur which include localized thickenings and increased electron density of pre- and postsynaptic and junctional membranes. These areas are usually 150 to 400 millimicra in length and three-dimensional in extent. They represent differentiated patches in the synaptic surface, the function of which is unknown; the close association of 200 to 300 Å vesicles with these areas had led some investigators to conclude that they represent regions where vesicles and their contents are transported to the intersynaptic space by a process of membrane fusion. A variety of morphological specializations of both chemically and electrically mediated synapses have now been described. For reference to these the reader should consult Eccles (4).

On the presynaptic side of chemically mediated synaptic and junctional contacts there is an accumulation of 200 to 300 Å, membrane-limited vesicles. Vesicles of this size are found elsewhere in the neuron in the perikaryon, particularly in the region of the Golgi complex and in axonal and dendritic expansions but it is at the synaptic regions they are the most heavily concentrated. It has been repeatedly speculated that: 1) these vesicles are involved in synaptic and junctional transmission by serving as either precursors of, or carriers of, acetylcholine; and 2) protein-bound, inactive acetylcholine is transferred from the presynaptic to the intersynaptic space by a process of vesicular and cell membrane fusion. Such a process of cell membrane and vesicular fusion is known to occur in a variety of cells and appears to be the basis for transport mechanisms. Vesicles at synapses are often seen in contact with synaptic and junctional membranes. The most recent biochemical evidence (3) indicates that synaptic vesicles do in fact contain a relatively high concentration of acetylcholine.

The Golgi complex of neurons is similar to that of other cell types in its fine structure. It is circumferentially disposed about the nucleus as individual crescent-shaped bodies which are made up of highly oriented, closely packed, flattened, cisternal membranes. The individual cisternae of each Golgi body present dilatations of various dimensions at the extreme ends of the crescent. The cisternal membranes are 100 to 150 Å thick and the lumina of the cisternae have an average width of 200 Å in their undilated portions. Numerous 200 to 700 Å vesicles are found near the Golgi membranes, particularly at the dilated extremes of the closed cisternal membranes, and some of the membranes of the vesicles are seen to be continuous with the membranes of the cisternae; this leads one to suspect that vesicles are being formed from Golgi membranes. Such an event, if true, lends support to suggestions which have been made that the Golgi substance functions in a secretory manner. As in the case of synaptic vesicles, irrevocable proof for such a function must await cytochemical studies determining the specific chemical constitution of Golgi membranes and cisternal contents; such determinations must progress beyond what is already known, for some cell types, about the phospholipid, polysaccharide and phosphatase content of Golgi substance.

Mitochondria of nerve cells are not different from those of other cell types, either morphologically or, presumably, in function and do not, therefore, warrant extensive discussion. It should be pointed out, however, that they are always found heavily concentrated presynaptically and junctionally; they are known to function in a variety of energy-converting mechanisms concerned with synaptic and junctional transmission.

JAMES F. REGER

References

1. Beams, H. W., *et al.,* "A correlated study on spinal ganglion cells and associated nerve fibers with the light and electron microscopes," J. Comp. Neurology, **96:** 249–282, 1952.
2. Beams, H. W., *et al.,* "Studies on the neurons of the grasshopper, with special reference to the golgi bodies, mitochondria and neurofibrillae," La Cellule, **105:**293–304, 1953.
3. De Robertis, E., "Ultrastructure and Cytochemistry of the Synaptic Region," Science, **156:** 907–914, 1967.
4. Eccles, J. C., "The Physiology of Synapses," New York, Academic Press, 1964.
5. Hodgkin, A. L., "The Conduction of the Nervous Impulse," Liverpool, Liverpool University Press, 1964.
6. McIlwain, H., "Biochemistry and the Central Nervous System," Boston, Little, Brown, 1966.
7. Nachmansohn, D., "Chemical and Molecular Basis of Nerve Activity," New York, Academic Press, 1959.
8. Palay, S. L. and G. E. Palade, "The fine structure of neurons," J. Biophys. Biochem. Cytol., **1:**69–88, 1955.
9. Symposium on "The submicroscopic organization and function of nerve cells," Exp. Cell Research, Supplement **5:**1–644, 1958.
10. Symposium on "Current Problems in Electro-Biology," Ann. N. Y. Acad. Sci., **94:**339–654, 1961.

NERVE ENDINGS

The motor-nerve ending begins its embryonic development at thirteen weeks, with the appearance of Doyere's eminence (see below), the future junction of the epilemmal and hypolemmal axons. The nerve ending completes its organization and is probably functional between the fourteenth and twenty-fourth weeks. Studies in the rat show that differentiation of the adult forms of the motor-nerve ending is not complete until maturity is reached. There is evidence also that the endings of motor nerves associated with striated muscle are specialized morphologically, both phylogenetically and within species. It appears that nerve endings are modified to meet special functional demands.

The ending of a motor nerve associated with striated muscle is termed the *motor ending, motor end plate,* or *myoneural junction.* In the mammal the motor nerve divides into single axons. As the axon approaches the muscle it loses its myelin sheath and continues to the surface of the muscle fiber. Here the nerve ending is formed, appearing as a flat, oval plate on surface view, and thin and tapering on side view (Figs. 1 and 2). The point of contact between the epilemmal and hypolemmal axon is slightly raised into an eminence, *Doyere's hillock.* Electron micrographs have demonstrated that the sarcolemma divides in the region of the motor ending, sending a thin layer over the surface of the ending and an inner layer between the motor endplate and the underlying sarcoplasm. The motor ending is encapsulated by the sarcolemma and should be considered intralemmal. The Bielschowsky silver-staining technique reveals a delicate network of apparently nonmedullated nerve fibrils, the *periterminal network*, stretching from the neurofibrillar branches to make intimate contact with the myofibrils. The motor endings are associated with a granular substance, *Kühne's granules,* located around the nerve elements. It has been found that the motor endings in animals exhibiting finely coordinated muscular activity contain a larger amount of Kühne's granules and a more complex plasmodesmata (nerve net) which

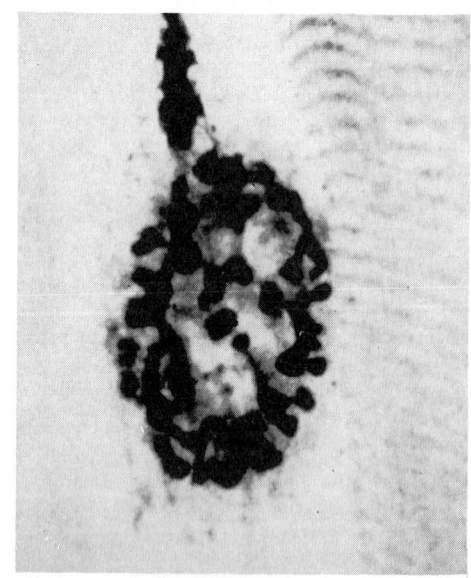

Fig. 1   Motor ending in rat, surface view. Gold chloride technique × 450 and enlarged. Motor nerve at top terminating in nerve net. Kuhne's granules located between meshes of net.

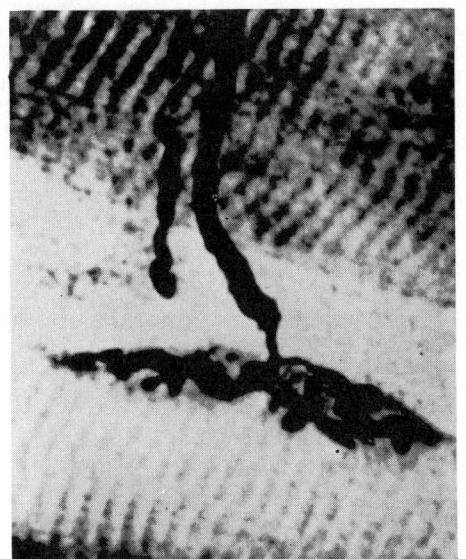

Fig. 2 Motor ending in rat, side view. Gold chloride technique × 450 and enlarged. Motor nerve at top terminating in motor ending. The sarcolemma and the nuclear elements are not demonstrated by this staining method.

makes possible a relatively greater innervation of the muscle fiber.

Cellular elements are associated with the motor endings in mammals and in all the animals of the subphylum Vertebrata. By means of the Bielschowsky technique, nuclear details can be demonstrated. Nuclei associated with the cells of the motor endings are the Schwann nuclei of the neurolemma, the sarcolemmal nuclei, and those associated with the nerve ending itself. The cytoplasm of the nerve ending is not clearly demonstrated, but the nuclei can be differentiated. The nuclei are larger and more oval than other nuclei in the region. They contain a granular chromatin material with a chromophilic nucleolus. The nuclei are dispersed among the plasmodesmata, and there are from 7 to 10 nuclei with each motor ending.

Phylogenetically, there is a steady progression from the primitive nerve ring found in the teleostomi, to the linear ending observed in amphibians (*terminaisons en ligne* ["linear ending"]), to the more complex nerve arrangement in the mammal (*terminaisons en plaque* ["plate endings"] and *terminaisons en grappe* ["bunched endings"]). The *terminaison en ligne* is associated with animals requiring quick, forceful muscle activity; this type of ending permits the greatest contact between nerve and muscle.

Studies in the rat have helped to clarify the subject of modifications in morphologic characteristics of the motor endings in muscles subserving different functions. Motor endings were classified as *terminaisons en grappe*, and *terminaisons en plaque*, the latter being subdivided into loose, compact, and multilemmal types. Because certain types of motor endings were associated with muscles having different basic functions, muscle groups could be identified by the morphology of the motor ending. Compact *terminaisons en plaque* were found on muscles with a high degree of coordination. Loose *en plaque* endings were found in the prime movers and the postural muscles. Multilemmal *en plaque* endings and the *terminaisons en grappe* were associated with specialized muscles such as the diaphragm.

W. V. COLE

References

Cole, W. V., "Motor endings in the striated muscles of vertebrates," J. Comp. Neurol., 102:671–716, 1955.

Cole, W. V., "Structural variations of nerve endings in the striated muscles of the rat," J. Comp. Neurol., 108:445–464, 1957.

Hines, M., "Nerve and muscle," Quart. Rev. Biol., 2: 149–180, 1927.

Hinsey, J. C., "The innervation of skeletal muscle," Physiol. Rev., 14:514–585, 1934.

Tiegs, O. W., "Innervation of voluntary muscle," Physiol. Rev. 33:90–144, 1953.

NERVOUS SYSTEM

Irritability and conductivity are fundamental properties of all PROTOPLASM, however, NERVE CELLS, which form the nervous system of all but the simplest metazoans, are highly specialized to subserve this function. The nervous system integrates and correlates information from the external and internal environment and delivers the resulting messages to the appropriate effectory apparatus. The degree to which the response to stimulation is modified and delayed is for the most part directly related to the complexity of the nervous system.

Within the protozoa a nervous system, as is usually recognized, does not exist, however, the protoplasm may differentiate into a highly complex excito-motor apparatus. Even in the amoeboid forms a physiologic gradient exists so that the ectoplasm is more responsive to stimulation and propagates the excitation wave. Within the Paramecium a central neuromotor mass is connected by fibrils to the base of each cilium forming a primitive coordinating center and conducting system. This arrangement reaches the highest stages of development in *Diplodinium ecaudatum* where a network of neuromotor strands connect a central neuromotor mass with retractor strands which in turn are continuous with adoral and oral CILIA and membranelles.

The first indication of cellular differentiation appears in the Porifera. In some of these forms a primitive ring of muscular tissue surrounds the osculum. Muscle then is the most ancient constituent of the neuromuscular mechanism.

The true nervous system has its origin in the Coelenterates. The first step is the differentiation of an epithelial cell into a neurosensory element. The surface of the cell serves as a receptor whereas nerve fibers ramify from the base to make contact with underlying muscle cells. True nerve cells having the primary function of conduction also appear in Coelenterates. These are interposed in a nerve net between receptive cells and muscle cells. This arrangement allows for the entire body to come under the influence of any one receptor.

The change from radial to bilateral symmetry allows

for further development of the nervous system. A simple plan of the latter is exemplified by the flatworm, Planaria. Bilateral ventral nerve cords extend the length of the animal. These are connected by numerous commissural bundles. Extending from the cords are fibers which communicate with a superficial nerve net. Nerve cells are sparse but scattered throughout the nerve cords, commissures and superficial nerve net. A number are concentrated in the head end to form small cephalic GANGLIA. Subsequent development of the nervous system in higher invertebrates occurs in the following ways: concentration of the nerve cells forming ganglionated nerve cords, fusion of the two longitudinal nerve cords, loss of the peripheral nerve net, development of nerves free of ganglion cells extending from the nerve cord to the muscle, development of specialized receptors and receptor organs concentrated in the head region, and progressive decrease in the number of ganglia by progressive fusion with an increase in size and importance of the head ganglion.

Within the segmented worms paired ganglionic chains extend from the head ganglion, which surrounds the esophagus, caudally to the last segment. A pair of ganglia are located in each segment. These send efferent fibers to the muscles and receive afferents from the receptors in the integument. In addition, they contain a large number of association neurons which interconnect the afferents and efferents and form contralateral intrasegmental and homolateral intersegmental connections. An extensive peripheral plexus containing ganglion cells, which have numerous connections with the central ganglia, is also present. A nervous system of this type allows for considerable correlation within a segment as well as between adjacent segments.

Within the molluscs, considerable transverse and longitudinal fusion of ganglia occurs. As an example, in the clam nerve cords extend back from paired cerebral ganglia to a pedal ganglion related to the foot. This is a purely motor ganglion sending efferents to the muscles of the foot and receiving afferents only from the cerebral ganglion. The cerebral ganglia are also connected with paired visceral ganglia. Subepidermal ganglion cells are few in number and confined to the foot. Facilitation and summation have been demonstrated in the nervous system of the clam. Furthermore, acetylcholine and cholinesterase activity has been noted. This is further indication of a high level of nervous integration.

In addition to the above, the squid presents a well developed giant fiber system. These result from a fusion of a number of individual axons whose cell bodies reside in the cerebral ganglia. They allow for a rapid conduction rate, and since they innervate the mantle, a massive, fast contraction of this structure is possible. This is particularly important in locomotion.

Considerable variation exists within the BRAIN of arthropods depending on the degree of fusion of ganglia, the importance of receptors in the head region and the importance of the head ganglia. In the extreme case, such as Sarcophaga carnaria, there is one head ganglion and all the abdominal and thoracic ganglia are fused into one. Segmented reflex behavior still occurs in arthropods, however, it is modified due to fusion of adjacent segmental ganglia and the dominance of the head ganglia. Within the latter considerable integration of sensory modalities occurs allowing for the complex instinctive behavior noted in some of the arthropods.

The chordates present a dorsal tubular nervous system as compared with the ventral laminar arrangement in the invertebrates. They are further characterized by progressive encephalization resulting in the increased dominance of the suprasegmental levels over the more primitive segmental portions of the central nervous system. The increase in the number of association neurons paralleling an increase in interconnections, allows for considerable integration and the ability to modify the behavioral response to any set of stimuli.

The central nervous system of all vertebrates presents the same basic architectural plan with all of the subdivisions relatively clearly delimited. There is, however, considerable variation in the degree of development of each subdivision and system. This depends in part on the animals position in phylogeny and also its habitat. The spinal cord probably undergoes the least amount of change. The basic function of the spinal cord is to integrate various sensorimotor mechanisms. With the progressive dominance of the brain various longitudinal pathways develop which serve to conduct sensory impressions directly to the brain. The brain in turn by direct descending pathways markedly modifies the basic spinal cord reflex mechanisms.

The vascularity of the central nervous system of the primitive chordate Amphioxus is extramedullary. This condition is also present in petromyzonts, however, in myxinoids and all other vertebrates it is intramedullary. Another feature common to both of these primitive forms is the separation of dorsal and ventral roots and alteration with the corresponding root of the opposite side. The roots fuse but remain alternate in myxinoids and plagiostomes. In all other forms they are fused and opposite. Bipolar dorsal root ganglion cells are characteristic of Amphioxus, cyclostomes, plagiostomes and most of the bony fish, although in the former they are entirely intramedullary. In Amphioxus and the cartilaginous fish the preganglionic autonomic fibers are primarily associated with the dorsal roots. Furthermore, the postganglionic cell bodies are widely scattered in relation to the organ innervated rather than aggregated in ganglia and chains as in other vertebrates.

Progressive differentiation of gray and white matter occurs in ascending the phylogenetic scale. This is associated with the development of appendages, discrete motor control and dominance of the brain. In Amphioxus no differentiation is apparent due to the absence of myelin, long conduction pathways and diffuse scattering of cell bodies throughout the spinal cord. Within the cyclostomes the cell bodies are located centrally and the axons peripherally, however, distinct dorsal and ventral horns and myelin do not appear until plagiostomes. A few distinct nuclear groups can be recognized in the gray matter of amphibians and reptiles. The number is increased somewhat in birds and reaches a maximum in mammals.

The first recognizable long conduction pathway appears in cyclostomes. It arises from the gray matter of the spinal cord and terminates in the brain stem reticular formation. In its manner and position of crossing and ascent in the cord, it resembles the spinothalamic tract of mammals. It may represent the first step in the phylogenetic development of this system which is concerned with pain and temperature. In plagiostomes some of the fibers of this tract reach the tectum of the midbrain, but only in mammals is the thalamus attained forming a true spinothalamic system. Nevertheless, the connections previously established with the brain stem and midbrain are retained.

Three other tracts make their appearance early in phylogeny. The reticulospinal, vestibulospinal and spinocerebellar tracts are present in plagiostomes. The latter is related to the development of the importance of the cerebellum in coordinating the muscular activity of this and subsequent forms. The vestibulospinal constitutes an important descending path from the vestibular nuclei for reflex maintainence of the body and its parts in space. The reticulospinal is an important descending path in all forms for the extrapyramidal system and cerebellum.

Three tracts which appear late phylogenetically are the rubrospinal, cerebellospinal and tectospinal. All three may be represented in reptiles but are clearly demonstrated only in birds and mammals. The former takes origin from the magnocellular portion of the red nucleus and constitutes another link between the cerebellum and lower motor neurons. The cerebellospinal takes origin largely from the vestibular portion of the cerebellum and is probably homologous to the fastigiospinal tract of mammals. The tectospinal serves as an efferent tract of the midbrain colliculi which are important optic and auditory reflex and integrating centers.

The dorsal columns, gracile and cuneate fasciculi, their corresponding medullary nuclei and the medial lemniscus which arises from these nuclei and terminates in the thalamus constitute a portion of the pathway to the conscious center for position and vibratory sense and two point discrimination. This pathway is, therefore, present in its entirety only in mammals. A small dorsal column can be recognized in the bony fish, amphibians, reptiles and birds. Most of these fibers terminate at various levels of the spinal cord. In the latter two classes a few reach brain stem nuclei. The "medial lemniscus" arising from these nuclei does not reach the thalamus but terminates in the brain stem and midbrain reticular formation.

The last important tract to be considered is the corticospinal. This is the voluntary motor pathway of neocortical origin. It is therefore present only in mammals. Since it is primarily concerned with the control of fine delicate movements, it is very small in all but the primates.

Most general somatic afferents from the head course with the trigeminal nerve. A few supplying the soma of the second through fifth branchial arches follow the facial, glossopharyngeal and vagus nerves. Their central connections, however, are with the trigeminal nerve. On the other hand visceral afferents including taste are related to the facial, glossopharyngeal and vagus nerves. A few visceral afferents have been described in the trigeminal nerve of cyclostomes.

A poorly developed descending root and nucleus of the trigeminal nerve can be recognized in cyclostomes. The cells which form a scattered discontinuous column project to the reticular formation. In all other vertebrates the tract and nucleus are well defined and continuous. In addition to the trigeminoreticular the nucleus also gives rise to a trigeminomesencephalic tract and in mammals a trigeminothalamic.

A few taste buds are found in the oral cavity and skin of the head in cyclostomes. They are more numerous in plagiostomes and reach a maximum in bony fish, being widely distributed over the body in some forms.

In the remaining vertebrates they are few in number and are confined to the oral cavity.

The visceral afferent fibers of cranial nerves VII, IX and X upon entering the brain stem form an ascending and descending solitary fasciculus. The ascending limb is absent in mammals. The fibers of the fasciculus terminate in facial and vagal nuclei, which in some forms are exceedingly enlarged to form lobes. The prevagal portion of the fasciculus associated with the facial nerve becomes progressively larger reaching a maximum size in bony fish. It then decreases in size and is smallest in mammals. The post vagal portion associated with the vagus is small in aquatic forms but increases in size to a maximum in mammals. This is related to the decline in importance of taste and the increased number of visceral afferents from the body cavities.

The vestibular, lateral line and cochlear systems are intimately related to each other centrally and peripherally. The vestibular receptors (semicircular canals, utricle and saccule) in the inner ear are poorly developed in cyclostomes but are essentially similar in all other vertebrates. The vestibular nerve is related to the ventral division of the auditory nerve. It terminates in the ventral vestibular nucleus of the brain stem of aquatic forms. This is probably homologous to the lateral vestibular nucleus of terrestial animals. Within the latter group additional primary vestibular nuclei are added up to five in mammals (lateral, tangential, superior, descending and medial). Direct and indirect connections, the latter by way of the vestibular nuclei, with cerebellum are present in all vertebrates. The other important secondary connection present in all vertebrates is with the lower motor nuclei of the brain stem and cord by way of the medial longitudinal fasciculus. The fibers are directed only caudally in this fasciculus in all forms except mammals where many reach the midbrain. Furthermore, a direct lateral vestibulospinal tract independent of the medial longitudinal fasciculus is present only in mammals. This is probably related to the reflex control of the extremities.

The function of the LATERAL LINE system is not fully understood. There is some evidence, particularly in deep sea forms, that it is a pressure receptor. The receptors also respond to low frequency vibrations of six per second. This may be important in avoidance reactions. It is also the basis for the theory that this system is analogous to the cochlear. The lateral line system is found only in aquatic forms reaching maximum development in bony fish. The receptor is a neuroepithelial hair cell around which a nerve fiber terminates. This is basically the same arrangement as found in the vestibular and cochlear systems. The anterior lateral line nerve from the receptors of the head enters the brain stem in association with the facial nerve. The posterior lateral line nerve enters with the glossopharyngeal or vagus. Both terminate in the dorsal and medial nuclei of the acousticolateral area. This also receives the dorsal division of the eighth nerve coming from the sacculus. Secondary connections are with the cerebellum, reticular formation, hypothalamus and caudal portion of the tectum.

Some aquatic forms respond to low frequency vibrations after sectioning the lateral line nerves. These receptors may be related to the sacculus. The dorsal division of the auditory nerve arises from these receptors and has the same central connections as the lateral line nerves. In terrestial forms the basal cochlear papilla may be the receptor for sound. A fully developed spiral cochlea is found only in mammals. The dorsal division of the auditory nerve is distributed to the papilla and terminates centrally in a dorsal magnocellular nucleus. Dorsal and ventral cochlear nuclei make their appearance in mammals. Secondary connections are with the caudal portion of the tectum of the midbrain by way of the bulbar lemniscus (lateral lemniscus of mammals). The nucleus laminaris and torus semicircularis of the tectum constitute the highest level of integration in cochlear system in submammalian forms. The inferior colliculus replaces these nuclei in mammals. It relays auditory modalities to the medial geniculate body of the thalamus which in turn projects to the auditory cortex located in the temporal lobe.

All vertebrates have a visual system although it is degenerate in myxinoids. Therefore, an optic nerve, chiasma and tract is always present. Crossing in the chiasma is complete in all but certain species of mammals which have binocular vision. The tectum of the midbrain, superior colliculus of mammals, and the dorsal portion of the lateral geniculate body of the thalamus constitute the central termination of the optic tract. The dorsal portion of the lateral geniculate body, which is concerned primarily with projections to the occipital cortex, is present only in birds and mammals.

The tectum constitutes the primary optic integrating and reflex center in most vertebrates. In addition to the optic tract it receives fibers from the auditory tectum and other afferent systems. Its efferent projections are to the motor nuclei of cranial and spinal nerves, cerebellum and reticular formation. Thalamic centers, lateral geniculate body and pulvinar, probably assume some of this function in birds and mammals. Cortical projections from the lateral geniculate body are also present in these two groups. In birds, however, the role of the cortex is probably dominated by subcortical centers, whereas in mammals the visual cortex is necessary for the complex behavioral patterns related to vision.

The skeletal musculature of the head is derived from two different sources, mesoderm of branchial arches and occipital somites. The somites give rise to the extraocular muscles innervated by cranial nerves III, IV and VI and the musculature of the tongue innervated by the hypoglossal nerve. These are referred to as general somatic efferent nerves. Branchial arch musculature is innervated by a special visceral efferent component. In aquatic forms these muscles are associated with the gills, however, they assume other functions in terrestial animals. The muscles of the first arch, which act primarily on the mandible, are supplied by the trigeminal nerve. The second arch gives origin to the facial muscles innervated by the seventh nerve, and the ninth and tenth nerves innervate the muscles of the larynx and pharynx which develop from the third, fourth and fifth arches. Portions of the sternocleidomastoid and trapezius innervated by the eleventh nerve are probably also of branchial arch origin.

In the brain stem of higher vertebrates the special visceral efferent column of cells is generally located in a ventrolateral position, whereas, the general somatic efferent column is dorsomedial. In the primitive forms all motor neurons are in the latter position. Phylogenetically, however, there occurs a progressive ventral and lateral migration of the visceral efferent group.

The cerebellum is primarily concerned with coordinating muscular activity. To perform this function it must receive messages from all motor centers and proprioceptors and project to these centers and lower motor neurons. The cerebellum arises phylogenetically as a plate of cells extending from the acoustico-lateral area over the fourth ventricle. This is the condition found in petromyzonts. No cerebellar tissue is recognized in myxinoids. Afferent connections are direct and indirect with the vestibular and lateral line systems, trigeminal and tectal nuclei and hypothalamus (lobocerebellar). All but the latter persist through the phylogenetic scale. The lobocerebellar tract is present only in aquatic forms. No basal nuclei are recognized in petromyzonts but efferent fibers project from diffuse cells to opposite brain stem and midbrain reticular formation.

The auricular lobe which is the homolog of the flocculus appears in plagiostomes. This is essentially the only part of the hemisphere present in submammalian forms. The extensive development of the hemisphere in mammals is associated with the development of the cerebral cortex. Spinocerebellar and accessory olivocerebellar tracts first appear in the cartilagenous fish. These are probably links in the proprioceptive paths to the cerebellum. Pontocerebellar and main olivocerebellar tracts are confined to mammals. They are probably related to the cerebral cortex and extrapyramidal system, respectively.

The basal nuclei give origin to most of the efferent projections of the cerebellum. One lateral nuclear mass can be identified in cartilagenous and bony fish. This assumes a medial position in amphibians. A medial and lateral nucleus is present in reptiles, birds and lower mammals. A third is present in most mammals and four in primates. The most medial, nucleus fastigii, is related largely to the vestibular system. It also projects to motor nuclei and reticular formation. The lateral nucleus, dentate of most mammals, gives rise to the brachium conjunctivum. This ascends and terminates in the opposite reticular formation and red nucleus of the midbrain. From these, descending tracts terminate in lower motor neurons. In mammals a portion of the brachium conjunctivum reaches the ventral nucleus of the thalamus. This, in turn, projects to the motor areas of the cortex thus completing a two way circuit between cerebral and cerebellar cortex.

The epithalamus consisting of the pineal body and habenular nuclei is relative stable phylogenetically. The former, whose function is unknown, is absent in cyclostomes and some reptiles. A true pineal eye is found in petromyzonts and lizards, whereas, a parapineal organ has been described in petromyzonts, plagiostomes and certain reptiles. The function of this structure is likewise unknown.

The habenular nuclei receive afferent fibers from olfactory areas and project to lower motor neurons via reticular formation. They are present throughout the vertebrate series varying only in size and symmetry. It is probably concerned with integrating olfactory impulses in activities such as feeding.

The hypothalmus is phylogenetically old and relatively stable in structure throughout the vertebrate classes. Although definite nuclei have been described, localization of function has been difficult to demonstrate. The cells of the supraoptic and periventricular nuclei, present in all vertebrates, are thought to be neurosecretory in function. Their axons extend by way of the supra-optico-hypophyseal tract to the neural lobe of the hypophysis. This system is concerned with water metabolism. A hormone may be the substance secreted by the neurons in the hypothalamus and stored in the neural lobe.

The principal afferent projections to the hypothalamus are from the olfactory areas and limbic lobe by way of the medial forebrain bundle, ascending fibers from the brain stem and thalamo-hypothalamic connections. The latter two are probably concerned with bringing various sensory modalities into the hypothalamus. Another important tract is the fornix arising from the hippo-campus. This has a diffuse termination, but with the appearance of a definitive mammillary body in reptiles, birds and mammals, it becomes primarily related to this structure.

Numerous descending fibers from the hypothalamus eventually reach the lower motor neurons of the autonomic and somatic nervous systems. Other afferent projections are the mammillothalamic tract to the anterior nucleus of the thalamus, which is a link in the projection to the limbic lobe, and ascending fibers in the medial forebrain bundle.

The saccus vasculosus is found only in aquatic forms. It is a convoluted extension of the third ventricle in the posterior hypothalamus. It has been suggested that it has a chemoreceptor function regulating the composition of the cerebrospinal fluid. Another possibility is a pressure receptor since it is well developed in deep sea forms.

The hypothalamus is an integrator of somatic, visceral and endocrine effectors in activities related to maintenance of the individual and species. It, therefore, has an important role in emotion, sexual behavior, feeding reactions, metabolism, maintenance of body temperature, etc.

The phylogenetically oldest portions of the thalamus are those related to subcortical structures. In all vertebrate forms the midline, reticular and dorsal portion of the lateral geniculate body can be recognized. The former are intimately related to the hypothalamus and visceral functions. The latter is associated with the optic tectum as a visual integrating and reflex center. The reticular nuclei are part of the general reticular activating system. The ventral nucleus appears in amphibia. This receives the termination of all ascending afferent systems but only in mammals does it serve to relay these modalities to the neocortex. In submammalian forms it is part of the thalamic complex associated with integrating afferent systems with the somatic and visceral efferent centers in the basal ganglia and hypothalamus.

With the development of the mammillary bodies and well developed limbic lobe in reptiles the anterior and medial nuclei appear. These are nuclei concerned with interconnecting hypothalamic and limbic areas. In higher mammals the medial nucleus also has strong connections with the prefrontal cortex.

Well developed cortical auditory and visual centers occur only in mammals. Therefore, the medial geniculate body and ventral portion of the lateral which are concerned with relaying these modalities are present only in these forms. The pulvinar and lateral nuclei are also confined to mammals. They have strong two way connections with the portions of the parietal, temporal and occipital lobes that are concerned with higher intellectual and psychic functions and are therefore best developed in man.

The telencephalon includes the cerebral cortex and the basal motor centers generally referred to as the basal ganglia. The latter centers are the highest level of somatic motor integration in submammalian forms. In mammals neocortical motor centers appear. The basal ganglia are still well developed but function in cooperation with the cortex. In aquatic forms these basal centers are referred to as the somatic area. It consists of large efferent type neurons which become progressively more well defined from cyclostomes to teleosts. Afferents come largely from the dorsal thalamic and olfactory areas. Efferents project to the hypothalamus and ventral thalmus. The pathway to the latter is the lateral forebrain bundle and is probably the homolog of the mammalian ansa lenticularis. From the latter two areas fibers descend to the reticular formation and thence by way of reticulospinal tracts to lower motor neurons. This is essentially the pattern in all vertebrates. The principal changes occur in the differentiation of the somatic area into a number of distinct nuclei and the presence in mammals of strong two way connections with the neocortical motor centers.

In amphibia a small celled neostriatum can be differentiated from the large celled paleostriatum. The former is probably homologous to the caudate and putamen of mammals and the latter to the globus pallidus. The exact function of the neostriatum is not known, but it is probably related to a terrestrial habitat since it is present only in these forms. The neostriatum receives most of the afferents and the paleostriatum gives rise to the efferents.

Differentiation of the striatal complex reaches its height in reptiles and birds where as many as four large circumscribed nuclei can be recognized. In the absence of a neocortex all the complex motor behavior of these forms is controlled by the striatum. This probably includes instinctive behavior.

The oldest portion of the telencephalon is that related to the olfactory system. Although not the same in structure, size and position all vertebrates possess an olfactory receptor, bulb, tract, anterior commissure and olfactory areas. The olfactory areas include the pirifrom cortex, lateral segment of the amygdaloid nucleus, nucleus of the stria terminalis and olfactory turbercle. These constitute the reflex and integrating centers of this system. Efferent projections from these areas are principally to the hypothalamus and tegmentum by way of the medial forebrain bundle and the habenular nuclei via the stria medullaris thalami.

Beginning with amphibia other structures appear which seem to be closely related to the olfactory system since they have strong connections with it. The first is the hippocampus and its efferent tract the fornix to the mammillary bodies. This is found in amphibia. Septal areas, which are probably homologous to the septum pellucidum, subcallosal gyrus and other frontal areas of mammals, make its appearance in reptiles. These are referred to as the mesocortex as opposed to the archicortex of the hippocampus and paleocortex of some of the olfactory areas. Other mesocortical areas appear in mammals, e.g. hippocampal gyrus, gyrus cinguli, paraolfactory areas and medial portion of the amygdaloid nucleus. These meso and archicortical areas have been collectively referred to as the limbic lobe. Recent evidence indicates that this lobe is concerned with the regulation of emotional tone. The effectors in emotional reactions are integrated at the hypothalamic level. Therefore, the limbic lobe has strong projections to the hypothalamus by way of the fornix and medial forebrain bundle. It also receives ascending afferent projections from the thalamus and hypothalamus and olfactory areas. In mammals there are also strong connections from the neocortex which can modify the stereotyped emotional behavior controlled by the limbic lobe.

The beginnings of the neocortex may appear in reptiles and birds, but it is generally considered exclusively a mammalian structure. Below primates most of the neocortex is occupied by primary sensory and motor areas. The former receive the general somesthetic projections from the ventral nucleus of the thalamus and the optic and auditory radiations from the lateral and medial geniculate bodies. The motor areas give rise to the pyramidal tract and project strongly to the basal ganglia. Strong interconnections exist and collectively they constitute a high level of sensorimotor integration.

In the higher mammals more and more discrete localization of the parts of the body occurs in these sensorimotor areas. In addition extensive areas appear which are not directly concerned with these primary functions. These have been referred to as silent areas. They are well developed only in primates and reach their zenith in man. They are concerned with the higher intellectual, abstract and psychic functions. They are located in man in the prefrontal lobes, posterior parietal area, lateral occipital lobe and inferior temporal areas. It is difficult to localize any specific intellectual function in any one of these areas. In general it may be stated that the prefrontal lobes allow man to delay reaction to an experience and project to the future. It also plays an important role in personality. Portions of the temporal, parietal and frontal lobes are concerned with motor speech and understanding the spoken and written word. The temporal lobe, no doubt in conjunction with other areas, plays an important role in memory. The parietooccipital cortex has been referred to as the gnostic center. This refers to the appreciation of symbolism, visual, auditory and tactile.

To speak of "centers' is a convenient way to discuss the nervous system. It should be emphasized, however, that all areas are either directly or indirectly connected to all others. Consequently, a "center" supposedly controlling a specific function is merely a portion of a complex circuit involving many other cortical and subcortical areas collectively subserving this function.

HOWARD A. MATZKE

References

Kappers, C. U. A., *et al.*, "The Comparative Anatomy of the Nervous System of Vertebrates, Including Man," 2 vols., New York, Macmillan, 1936.
Kuhlenbeck, H., "Central Nervous System of Vertebrates," 5 vols., New York, Academic Press, 1967.

NEST BUILDING

Nest building is such a familiar facet of bird study that the subject is usually treated as an exclusively avian habit, though it must be remembered the practice also occurs in other forms of animal life. The present treatment follows the accepted pattern and considers the art only as developed by the birds.

The avian habit of building nests appears to have had its origin in the gradual evolution of the earliest birds from their reptilian ancestors. Since the ancient reptiles buried their eggs in the sand or otherwise deposited them in some convenient place as in a hole in a stump or log or on the ground among decaying vegetation where the heat from the sun or the decomposing plant material could hatch them, it is likely that the egg-laying habits of the first birds paralleled those of the reptiles. But as the birds gradually evolved into warm-blooded creatures a change in their egg-laying habits became inevitable. Reptiles are coldblooded and are unaffected in health by marked temperature variations whereas the warm-blooded birds can suffer only slight changes in their normal blood temperature without harm. And what is true of the adult bird is equally true of the developing embryo within the egg, which will not only fail to develop but eventually die if a constant temperature is not maintained. Presumably the early birds discovered, as they evolved from the changeable to a constant temperature, that is, from cold-blooded to warm-blooded animals, that they could no longer employ the heat from the sun or decaying vegetation to hatch their eggs and had to find a substitute. They found a new source of heat in the warmth of their own bodies and thus arose the practice of incubation.

For the birds that laid their eggs in holes the matter of sitting on their eggs until they hatched was a simple one and posed no problems other than of learning how to excavate a cavity. Other birds, now faced with the necessity of incubating their eggs on the ground, were confronted with such hazards as floods, terrestrial enemies, and those offered by the ground itself as low temperatures and excessive moisture, which constantly threatened to, and often did, destroy their eggs or young. Eventually many of them found a way of circumventing such dangers by building a platform off the ground on which to incubate their eggs. Others, either unable to acquire the technique of building a raised platform or indifferent to the dangers attendant to incubating on the ground, continued to hatch them on the ground and still managed to survive. Even today there are many species that make no pretense at building a nest, as various shore birds that merely lay their eggs on the bare sand or in a gravel bed, the Whip-poor-wills that deposit their eggs on the litter of the forest floor, or the Nighthawks that make use of pavements, the flat-topped roofs of buildings, and similar locations. At the other extreme are the beautiful and often fantastic nests of some tropical birds. Between the two are all the gradations of nesting structures which may serve to indicate the probable steps through which the art of nest building has passed, in some instances beyond a purely utilitarian need.

The first step in the evolution of the nest building habit may possibly be illustrated by the scooped out hollows or depressions which birds like the Killdeer and Least Tern excavate in the ground to prevent their eggs from rolling. A slight step forward is provided by the Spotted Sandpiper and the Stilt which add a few grasses or weed stalks as a lining. Next in line would be the simple nests of the Gulls and Terns which fashion rather simple affairs of grasses or other vegetation in which they customarily deposit their eggs. Then would come the better made nests of the ground nesting Thrushes, Warblers, and Sparrows and lastly the domed or arched-over nest of the Meadowlark and the oven-like leafy nest of the Ovenbird, both of which provide a high degree of protection.

A similar parallel can be drawn among the birds that nest above the ground. The first nests were undoubtedly very crude affairs and even today many birds have failed to improve upon them. The Gallinules, for instance, construct a very simple platform of dead leaves and weed stalks and the Doves, Cuckoos, and Herons a similar structure of sticks, in some cases being so skimpily made that the eggs seem in constant danger of falling out. Other birds, as the Hawks, Crows, and Catbirds also use sticks but build more deeply hollowed nests which they line with softer materials. The nests of the Goldfinch, Yellow Warbler, and Redstart, which are made entirely of soft materials, represent a further advancement in nest building which reaches its climax in the exquisite structures of the Vireos, Orioles, and Weaver Birds.

The materials of which nests are made are essentially obtained from plants but hair, feathers, skin, shells and the skeletal fragments of animals are also frequently used. The materials gathered are generally those which are readily available but selection is governed by an innate urge to build a nest characteristic of the species. Thus, the Mourning Dove must have twigs, the Bluebird grasses, the Robin mud, and the Yellow Warbler cotton. The goldfinch characteristically uses thistledown, the Kingfisher bones and regurgitated scales, and the Barn Owl fur and bones of its prey. Frequently a bird is absent from a region where the traditional materials are lacking, as with the Cliff Swallow which is not to be found in an area where clay of the proper consistency is unobtainable. Usually other materials are substituted, however, provided they can be used to fashion the type of nest characteristic of the species. A Wren has been recorded as having built a nest entirely of hair pins and wire clippings instead of the customary twigs and a Robin has been observed to have used strips of paper instead of grasses. A Baltimore Oriole often substitutes pieces of string or yarn for plant fibers and the Crested Flycatcher more frequently uses cellophane in place of the traditional cast-off snake-skin.

Apart from a mere circular depression scooped out in the ground, nests are generally platform or cup-shaped structures. Variations of these two types occur, however, and a platform may be deeply hollowed or basket-like. Cup-shaped nests may also be slightly hollowed or deeply cupped and in some species are deep pensile baskets. A few species dome or arch-over their nests with the grasses in which they are built and still others construct a spherical or globular nest with an entrance in the side. In size they range from the tiny nest of the Ruby-throated Hummingbird to the large bulky nests of the predatory birds. A nest occupied by Bald Eagles over a period of thirty-five years and enlarged and rebuilt by them each year eventually crashed from the tremendous weight of an estimated two tons. Another nest of this same species and probably the largest ever known measured twenty feet deep and nine and a half feet wide.

The techniques followed in the construction of nests vary considerably. Many ingenious ways in which the bill and feet are employed in collecting, carrying, and weaving the materials together have been described. Cup-shaped cavities are fashioned by the rotary motion of the body; holes are drilled in decaying trees by bills admirably adapted for the purpose; tunnels are excavated in banks by the bill and the dirt kicked out by the feet; and in the weaving of a pensile nest both the bill and feet are brought into play.

Although the male selects the general nesting area or territory, the female usually decides upon the actual nesting site. In hole-nesting species, the male, however, often selects or at least designates suitable cavities and in the Cedar Waxwings site-hunting appears to be a cooperative effort.

Sites range from a chamber at the end of a burrow in the ground to the tallest treetop. On the ground the site may be on the bare sand of an open beach, among rocks of a rocky shore, on a ledge or a cliff, among herbaceous or shrubby plants, or in a cave. Above ground it may be in a vine, bush, or tree, either on a branch, crotch, or fork. Buildings provide a number of suitable locations and some birds find bridges, wharfs, culverts and similar situations to their liking. Hole nesters may either make use of a natural cavity or excavate one in a tree, telephone pole, fence post or appropriate tin cans, watering pots, pockets of scarecrows, flower pots, gutter pipes and the like. A few birds build floating nests on water; others suspend them above water in cattails, reeds, rushes, and water-loving bushes. Perhaps the most unusual sites are those selected by the Fairy Tern which balances its single egg on the horizontal branch of a tree without any material to hold it in place, and the Emperor Penquin which holds its egg on the web of the foot, keeping it warm with a fold of skin from the abdomen. Wrens often select the most unlikely places as the one that built her nest on the axle of an automobile kept in daily use, the bird accompanying the car wherever it was driven.

Birds generally conform to the species pattern of nest construction and follow much the same rule in selection of building sites. But as there is often a departure from the normal in nesting structure there is also at times a departure from the normal nesting site. Thus, many birds that normally nest on the ground may on occasion build an elevated structure and tree nesters may build on the ground and may build one type of nest in one situation and a different type in another.

The selection of a nesting site is largely governed by the need for concealment, protection against the elements, and the accessibility to a feeding area. Birds have learned how to circumvent their enemies by building their nests in inconspicuous or inaccessible places or by decorating them with materials, as the Ruby-throated Hummingbird which attractively camouflages its nest with lichens to simulate a knot on the branch on which it is built. Sometimes a change in environmental conditions necessitate a change of nesting site and birds that are not adaptable vanish whereas those that are able to cope with new conditions survive and often increase. The Phoebe and Barn Swallow formerly nested on cliffs but are now familiar birds about our dwellings and the Chimney Swift has practically abandoned hollow trees.

Typically the female does most of the building and in many species the male may assist in gathering materials. In a few cases it may appear to be a cooperative effort but whether the male actually assists in building the nest is open to question. Since it is the male's duty to keep the territory free of intruders it would appear he would have little time or inclination for nest building. The "dummy" nests built by male wrens, a common practice among these birds, is apparently a facet of the courtship performance as they are never used by the females.

The time required for nest building varies with the type of nest, the diligence of the workers, and the amount of time at their disposal. A Robin usually requires about six days but may take as many as twenty. Song Sparrows average five to six days but may extend their building operations to encompass a period of thirteen days. Field Sparrows generally complete their nests within three days. Goldfinches average thirteen days in July but can build them within five or six days in August. Northern birds, as a rule, require less time than southern birds since they are faced with a shorter breeding period. Arctic passerine birds for the same reason usually dispense with building new nests but reoccupy old ones. Of all birds the Hammerhead stork probably takes the longest—a period of four months. The nest once built is used year after year, however.

Nest building is usually not a continuous process and operations may be interupted by inclement weather or for other reasons as in the case of nests made of mud where the mud is allowed to settle before further work is done. A bird at first is not deeply attached to its nest and if frightened or disturbed may desert it and begin a new one elsewhere. Even if undisturbed the bird may suddenly decide on another site and may even leave her mate and territory. Generally the nest is completed the day before the first egg is laid and if destroyed while the eggs are being laid a new nest may be erected within a single day.

RICHARD HEADSTROM

References

Headstrom, Richard, "Birds' Nests," New York, Ives Washburn, 1949.
ibid., "Birds' Nests of the West," New York, Ives Washburn, 1951.
ibid., "The New and Revised Birds' Nests," New York, Ives Washburn, 1960.

NEUROGLIA

The neuroglia (Weigert 1885) is considered to be a connective tissue of the nervous system and of the retina. Two kinds of neuroglia have been distinguished —macroglia of ectodermal origin and mesenchematic microglia. Macroglia is composed of: (1) the ependyma which in the form of a ciliated epithelium covers the BRAIN ventricles and the central canal of the spinal cord (Fig. 1). Protoplasmatic astrocytes or glia—abundantly ramified, with a big easily stained nucleus and granulated protoplasm distributed in the entire cell. They are in close contact with neurocytes, and more rarely are terminated with pedicles on the blood vessels (Fig. 2). They are chiefly to be seen in the grey matter. Fibrous astrocytes predominate in the white matter; they differ from the preceding ones by very abundant, thin, feebly ramified fibers. The cytoplasm surrounds the fibrous astrocytes which, ramifying beyond the cell, form a dense net. Some of the fibers adhere to the blood vessels by dilated pedicles (Fig. 3). Oligogendrocytes or oligoglia, so called because of the possession of only a very few fibres, appear in the white matter along the nervous fibres and even accompany the peripheral nervous fibers in the form of lemmoblasts, which on the periphery are differentiated into Schwann's cells (Fig. 4).

The microglia is also known as the third element of the nervous tissue or "Rio-Hortega glia." Its cells have oval nuclei and their fibers have plumose ramifications (Fig. 5). The microglia is disseminated in the whole central nervous system and is considered to correspond to histiocytes of the CONNECTIVE TISSUE; in pathological states the microglia becomes phagocytic and is transformed into the so called Gluge's corpuscles.

The physiological role of neuroglia is not merely limited to its supporting function, but also plays an important role in the metabolism of the nerve cells. This is confirmed by the close contact between the fibrous glia, and—to a lesser degree—the protoplasmatic glia, and the blood vessels. As a rule the cells of the fibrous glia surround, or are connected by specific pedicles with the blood vessels and eventually come in touch with the internal surface of the pia mater. On the other hand some smaller protoplasmatic astrocytes adhering closely to the nervous cells produce satellite cells.

The neuroglial fibers of the fibrous astrocytes make a net-like web; in its meshes are placed neurons and other elements of the nervous tissue. Oligodendrocytes are characteristically found in the white matter but appear also in the vicinity of neurocytes forming satellite cells. Their function is probably limited to mediate or to create myeline.

Knowledge of the biological meaning of neuroglia is very sparse, just as, fifty years ago, were all notions concerning connective tissue. Most probably the functions of neuroglia are manifold. These cells are not only a support for neurons but play a definite part as isolating and nutritive agents in the metabolism of the nervous tissue, more particularly in the production of materials necessary for the normal functioning of the nervous tissue. Their importance in mammals seems proved by is indicated by the fact that all species of glia are restricted to this group. In lower animals—fishes, amphibians—the glial net is formed of ependyma and oligodendroglia, while other kinds do not appear, or are present only in very small quantities in reptiles and birds. Neuroglia plays the fundamental part in the process of healing lesions of the nervous system. The neurocytes have lost the capacity of mitosis and lesions caused by various agents are as a rule healed by the proliferating glial tissue.

S. BAGINSKI

References

De Robertis, E. D. P. and R. Carrea, "Biology of Neuroglia," New York, Am. Elsevier, 1965.
Nakai, J., ed., "Morphology of Neuroglia," Springfield, Ill., Thomas, 1963.

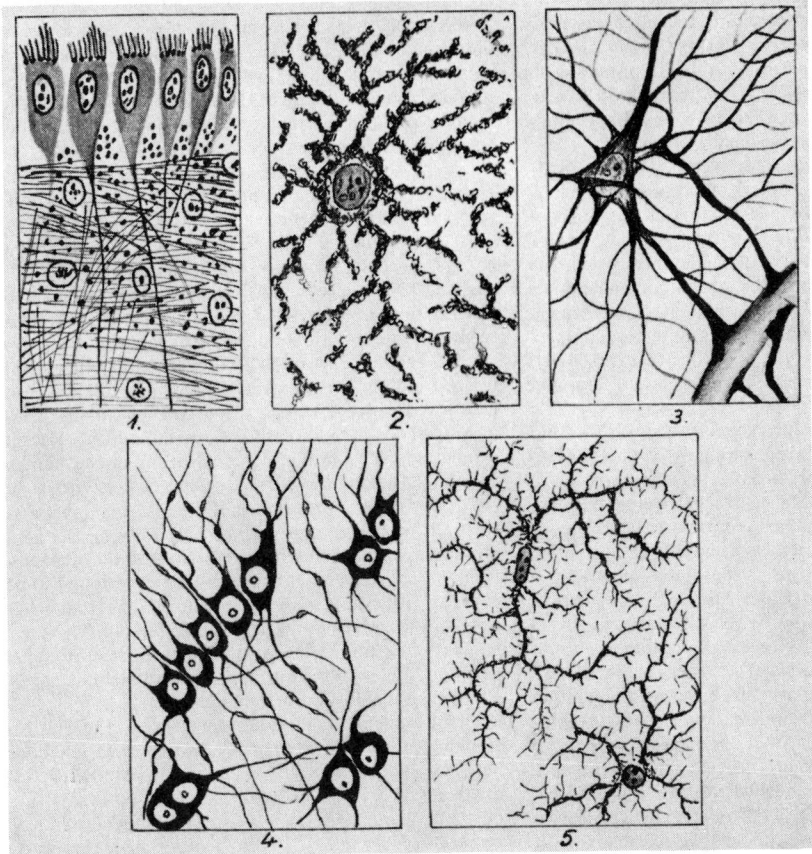

Figs. 1-5   1, ependyma; 2, protoplasmic astrocyte; 3, fibrous astrocyte; 4, oligodendrocytes; 5, microglia, mesoglia.

NEUROPHYSIOLOGY

Neurophysiology may be defined most simply as the physics and chemistry of the nervous system. It includes the study of nerve cells and their interrelations with all types of muscle and gland cells, the structural and functional relationships between nerve cells, the analysis of nerve networks, the biological bases of sensation (sensory input), coding and decoding of neural data, and control of motor and secretory activity (behavior). Involved in these processes are the phenomena of metabolism, growth, reflex activity, coordination, learning, emotion, memory, and abstract thought.

Neurophysiologists are concerned with the reactive properties of living organisms to external stimuli, starting with the simplest one-celled organisms whose "nervous system" is subcellular and part of the multifunctional cell unit, and, more often, with the highly specialized receptor, transmitter, integrative, and motor-control cells of more complex organisms; these cells are most often histologically categorized as *neurons*.

The term "physiology" although previously used, gained acceptance after Jean Fernel (1497–1558), Parisian physicist, astronomer, physician, and *physiologist* used it as a section heading in his famed treatise *Medicina*. Fernel might well be considered an early neurophysiologist, as he distinguished between voluntary and involuntary movements of muscle, and between sensory and motor nerves. It is not certain when the term "neurophysiology" first came into use—the ancients certainly pondered many of the problems attacked by modern scientists; and some of them made good guesses as to the autonomical locus of the control system. For example, Plato, in the fourth century B. C., placed his tripartite soul in the spinal column of the human, with the "lowest" functions lowest geographically, i.e. in the lumbar-sacral region. It has taken us some 23 centuries to adjust the principle loci of the control of sex, appetite, and emotion somewhat higher upstream, say in the region of the hypothalamus which lies in the diencephalon, or the most rostral part of the brain stem. Before the advent of the experimental method, not all guesses were as good as Plato's, and many considered the heart as the seat of emotion and of consciousness; in the 17th century, René Descartes located the soul in the pineal body, right in the middle of the brain. Despite the work of many brilliant pioneers prior to the 19th century, modern physiologists usually give the honorary title of "Father of Neurophysiology" to one or another 19th or 20th century figure—sometimes Ivan Sechenov (Russian, 1829–1905), sometimes Charles Sherrington (British, 1857–1952)—and occasionally others. Certainly, Sechenov was among the prominent investigators of the nature of the reflex arc, and laid the analytical groundwork for our very modern concept of the nervous system as maintaining a balance between excitatory and inhibitory influences; while the establishement of the neuron as the functional unit of the nervous system of complex organisms owed much to the extensive researches and theoretical analyses of both Sherrington and Santiago Ramón y Cajal (Spanish, 1852–1934).

In 1906, Cajal shared the Nobel prize with Camillo Golgi, another great neurohistologist, whose silver-staining techniques for demonstrating nerve fibers are still used. In their acceptance speeches, delivered from the same platform, Golgi "proved" the reticular theory of the nervous system, which argued that the entire nervous system is made of a continuous network of conducting fibers; while Cajal "proved" the neuron doctrine (earlier stated by Waldeyer-Hartz in 1891) which proposed that the system is discontinuous, made of discrete cellular units. Cajal's theory, while gaining wide acceptance largely on the basis of the unique chemical, temporal, and unidirectional characteristics of synaptic transmission, was not conclusively demonstrated until the advent of the electron microscope, whose magnification was needed to "see" the synapse or separation (in the order of 10^{-8} meter) between nerve cells. Today, we have come almost full circle again, with increasing evidence for instances of "tight junctions," where there is no discernible space between adjacent cell membranes, and with new electron microscope evidence of tiny (cytoplasmic?) bridges across the synaptic space and very recent experimental evidence for instances of two-way conduction of nerve impulses between adjacent neurons. Nevertheless, the neuron doctrine still stands, with overwhelming anatomical, chemical, electrophysiological, and behavioral evidence for the widely distributed, functioning, unidirectional synapses. A definitive history of neurophysiology is yet to be written; perhaps because it still is being enacted at such a rapid pace every day in thousands of laboratories all over the world. For the student entering this field, however, there are two delightful introductions; one, reproducing parts of selected classic papers (Fulton and Wilson, ref. 1) and the other, reviewing, in a short but scholarly fashion, the antecedents of the 20th century science (Brazier, ref. 2).

Some of the numerous disciplines contributing to, and in turn enriched by, neurophysiology, are outlined in Figure 1. Different investigators are trained in one or more of these fields and often specialize in a restricted area, but no one of these areas can be lost sight of if one is to gain insight into the remarkable functioning of this vastly complex system. Figure 2 is an oversimplified outline of the "system" studied by neurophysiologists. It could represent any of a wide range of species and omits reference to interactions between individuals except as inferred by the feedbacks to the external environment. Also omitted from the diagram

are many important sub-systems and epi-systems, such as those dealing with information storage and retrieval. Here again, many scientists have trained their sights on only a piece of this system—it is easy to spend a useful lifetime trying to discover the nature of but one of the many transducers in the sensory input of the machine, e.g. a visual receptor—but again the problems of the whole are so closely related to any one part that we can ignore the rest of the organism only at the peril of failing to understand the piece under inspection.

At first glance, Figure 2 seems quite simple, compared to a living organism. But, remembering the difficulty of Newton's 3-body problem, we can begin to appreciate the problems of trying to quantitatively describe a system with so many feedbacks and controls, each a variable function of all the others. It is pertinent to mention, in this respect, the number of units, or neurons, involved, leaving aside the enormous complexity (and individuality) of each neuron. Recent estimates place the number of neurons in the human cerebral cortex (only a part of the nervous system) in the order of 4×10^9. This does not include glial (supportive) cells of various types which play roles, at the very minimum, in the metabolic function of the neurons. Many of these cerebral neurons have been found to be covered with synaptic knobs, or connections from other cells, in the order of 10^2, or greater, per cell. Thus, the number of permutations of possible pathways through a single brain can be readily calculated to be of an astronomical size almost beyond imagination.

Despite such discouraging complexity, there is an amazing orderliness in the organization of this cellular network. Topographically and topologically, each individual's array and network is similar to that of the next individual, and even from species to species there is remarkable similarity in the anatomy, biochemistry, and biophysics of the system. Ontogenetically, each neuron from a specific locus in the central nervous system finds its way to the "correct" interneuron; the optic epithelial infolding in the embryo precisely meets the optic outfolding of the neural tube, the retina eventually is precisely hooked up through the proper way stations to the uniform (for all individuals) point-to-point localization in the visual cortex; each motor neuron sends its growing fiber coursing out from the developing spinal cord and makes contact with the "proper" muscle fiber in the "correct" muscle! Phylogenetically, the hypothalamus of the rat bears an amazing anatomical and functional resemblance to the hypothalamus of the human; the action potential of a squid axon looks and behaves very much like the action potentials of honey bees, frogs, and humans. Therefore the "system" can be studied, analyzed, dissected and statistically put together again so that the neurophysiologist begins to get insight into its mechanisms.

How does he study the nervous system? With all the tools of the disciplines outlined in Figure 1! Some "see" the nervous system through an optical microscope, some through an electron microscope, some through a spectrophotometer, some through an electrode; the information is read out on camera film, kymographs, polygraphs, oscilloscopes, computers, and other devices, both simple and complex.

In the 19th century and earlier, many contributions were made to knowledge of the nervous system by great clinicians with acute and discriminating powers of observation, by examination of patients and by clinical-

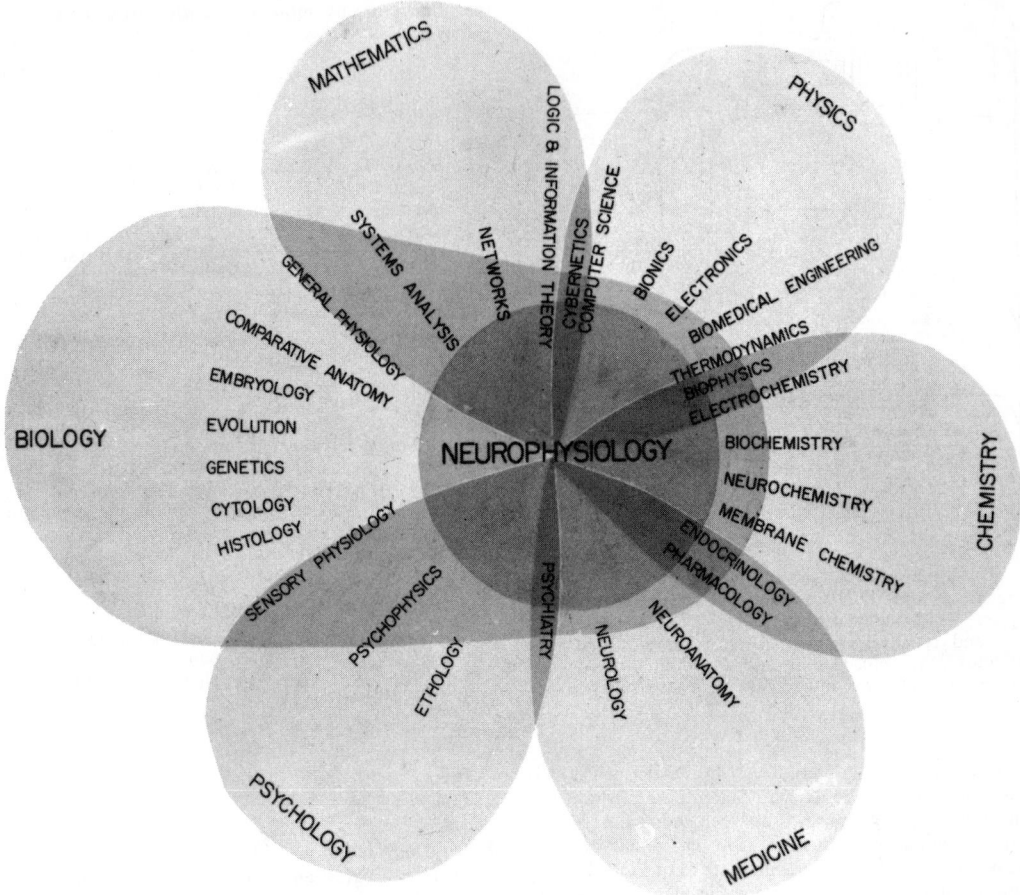

Fig. 1   Some of the scientific disciplines contributing to neurophysiology. In turn, each of these is enriched by the work of neurophysiologists.

anatomical correlations made upon post-mortem. To-day, the major part of our comprehension of human brain functions is based upon experimental work done in the laboratory upon other species. Behavioral and phys-iological effects of restricted lesions or ablations in various parts of the central nervous system are care-fully monitored and subjected to statistical analysis, while the anatomists obtain information on pathways by mapping the degeneration of nerve fibers after such lesions. Physical and chemical means of stimulation of specific loci are employed, both in acute experiments upon anesthetized animals, and in long-term ("chronic") studies of conscious animals with elec-trodes, thermodes, or cannulae surgically implanted. Electrical recordings from single neurons or from groups of cells can be visualized on an oscilloscope screen, listened to on audio amplifiers, placed on stor-age paper or magnetic tapes, and painstakingly ana-lyzed by inspection or via computerized mathematical procedures; behavioral responses are closely watched and recorded. Biophysical and biochemical properties of the isolated neuron and its components, *in vivo* and *in vitro*, are studied, as well as the properties of nerve cells with specialized functions, either alone or in the complex assemblage that exists in the intact animal.

Some of the basic problems and areas of study that occupy neurophysiologists today are: the nature of the transducer mechanism that transforms a mechanical, chemical, electrical, or electromagnetic signal or stimu-lus into a neuronal electrochemical signal or nerve im-pulse; the structure of and functional equations for nerve and muscle membranes; the qualitative and quantitative effects of ion concentrations, hormones, and other chemical influences upon neuron function; the physical-chemical equations for various types of synapses; the functions of specialized cells within the central nervous system, i.e. receptors, integrators, supportive cells; coding and decoding of information, for example the organization of "receptive fields" at various levels and in different degrees of complexity from the "external" receptors (e.g. in the retina) to the "internal" receptors (in this case, the visual areas of the cerebral cortex); the anatomical sites for specific sensory and motor functions of the central nervous sys-tem; the electrical, electromagnetic, and chemical manifestations of specific brain functions; the delicate

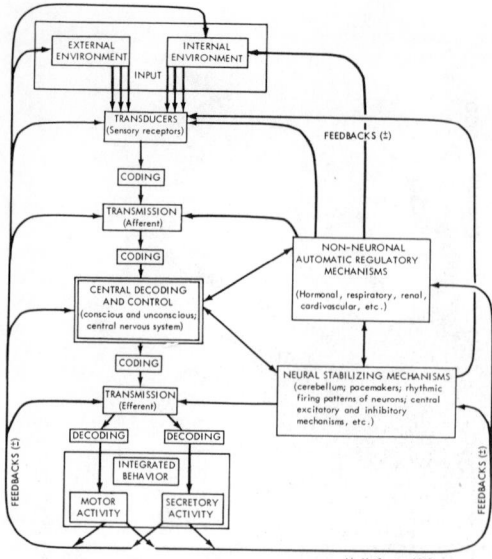

N. H. Spector 1968

Fig. 2 Neural data processing, a partial outline of the system studied by neurophysiologists. Omitted are storage and retrieval stations, filters, variable band-pass selectors, etc., etc. Coding and decoding stations generally indicate one or more synapses in a pathway.

the four journals most frequently cited by the respondents.

N. Herbert Spector

References

History
(1) Fulton, J. F. and L. G. Wilson, "Selected Readings in Physiology," (Chapts. I, IV, VI, VII, VIII and IX), 2nd ed., Springfield, Ill., Thomas, 1966.
(2) Brazier, M. A., "The Historical Development of Neurophysiology," *in* "Handbook of Physiology," Sect. 1, Vol. 1, Washington, D.C., 1959. Amer. Physiological Society.

Textbook
(3) Ruch, J. C., *et al.*, "Neurophysiology," 2nd ed., Philadelphia, Saunders, 1965.

Sources More Detailed than Textbooks
(4) Field, J. *et al.*, eds., "Handbook of Physiology," Sect. 1, Vols. 1–3, Washington, D.C., American Physiological Society, 1959–1960.
(5) "Progress in Brain Research," 27 vols., New York, Amer. Elsevier, various dates.
(6) "Physiological Reviews," 48 Vols., Washington, D.C., American Physiological Society, 1921.

Journals Giving Reports of Current Research
(7) Journal of Neurophysiology, Springfield, Ill., 1938-.
(8) Journal of Physiology, London, 1878-.
(9) Electroencephalography and Clinical Neurophysiology, Amsterdam, 1949–.
(10) Experimental Neurology, New York, 1959–.

balances between inhibition and excitation at the cellular level and in the over-all functioning of the organism; the interrelationships between endocrine concentrations in various tissues and the central nervous system. All of the foregoing contribute to our understanding of more general problems in neurophysiology, such as the nature of learning (acquisition and storage) and memory (retrieval), the neural bases of sensations, sleep, arousal, behavior, and abstract thought, as well as the etiology and treatment of neurological, neuroendocrine, and psychiatric diseases.

Today, neurophysiology is in its exponential phase of growth as a science. As Nobel laureate John C. Eccles recently pointed out, there are so many challenging and stimulating problems to be attacked, that no scientist in this field need worry about "competition" in his lifetime, nor ever lack for a new and exciting research project.

The above brief description can serve only as a somewhat expanded definition of "neurophysiology." For the student interested in delving further, the references below will serve as a more thorough introduction to the field. References 1 and 2, mentioned above, are excellent reviews of the history of the science; reference 3 is one of the better current textbooks available; the remainder are for the advanced student or investigator. There are many hundreds of journals that publish results of current research in neurophysiology and in such closely related topics as neurochemistry, neuropharmacology, biophysics, behavior, etc. In a questionnaire sent by this author to fifty neurophysiologists around the world, each was asked to list the five leading journals (continuing sources of new information) in the world. In 28 replies so far received, 27 journals were named; references 7, 8, 9, and 10 are

NEUROPTERA

An unspecialized order of holometabolous insects with biting mouth-parts, and with membranous, often net-veined wings, the front and back pairs usually similar, costal veinlets numerous except in groups with reduced venation, radial sector often pinnately branched. Larva campodeiform, either terrestrial or aquatic, in the latter case often with abdominal gills, which are probably homologous with segmental appendages; larval mouth-parts fitted for biting or secondarily modified for piercing and sucking. The Neuroptera are the most primitive of the insects with complete metamorphosis. They comprise two major groups, sometimes regarded as orders: the Megaloptera, with wing-veins not exceptionally multiplied in the adult and with biting mouth-parts in the larva, and the Planipennia, or Neuroptera proper, with wing-veins usually much multiplied by bifurcation towards the margin, and with suctorial mouth-parts in the larva. The Megaloptera contain three living families. The Corydalidae include giant insects, mostly subtropical or tropical, of which the North American hellgrammites or dobsonflies, *Corydalis*, are examples; the larvae are aquatic. The smaller alderflies, Sialidae, also have aquatic larvae. The Raphidioidea or snakeflies have the prothorax greatly elongated in the adult; the larva is terrestrial; some authorities place this family in a separate order.

The Planipennia consist mainly of smaller insects, with campodeiform, generally terrestrial, larvae. The Hemerobiidae and Chrysopidae are common and important families in temperate countries; both have predatory larvae that inhabit vegetation and eat aphids, mites and other small victims. Adult hemerobiids are generally brown and oval-winged; the eggs are un-stalked; the larvae are generally naked; there is a con-siderable development of flightless hemerobiids of bizarre form in the Hawaian Islands. Adult chrysopids are usually green; the eyes of the living insect often show golden reflections; some species have a disagree-able odor produced from paired glands on the protho-rax. The eggs of Chrysopidae are stalked, standing above the leaf-surfaces on which they are laid; the larvae are important predators of aphids. The Nemop-teridae have long, tail-like, fringed hind wings; they have a dancing, mayfly-like flight; the larva has the prothorax elongate; the pupa develops the hind wings in a rolled sheath. The Sisyridae are small Neurop-tera whose aquatic larvae feed on fresh-water sponges, piercing the cells with the suctorial mouthparts. The Myrmeleonidae or ant-lions are of above aver-age size, and are somewhat damselfly-like in gen-eral build. The larvae live singly at the bottom of fun-nel-shaped excavations in loose soil. They feed on ants and other insects trapped in these pits. Ascalaphidae are large tropical and subtropical insects, superficially somewhat resembling dragonflies, which some of the species resemble in habits as well as appearance. Ascalaphids typically have long, slender antennae, strongly clubbed at the tip. Mantispidae have raptorial fore legs in the adult. The larva of *Mantispa* is a hyper-metamorphic cuckoo-like parasite in the egg-cases of spiders, devouring the young spiders and being guarded by the mother. Coniopterygidae are minute psocid-like insects with reduced wing-venation; the larva is of the normal compodeiform type.

EUGENE G. MUNROE

NORDMANN, ALEXANDER VON (1805–1866)

The major contribution of this Finnish zoologist to science was his realization that the external shape of an animal gave few clues to its real affinities and, by ex-tension from this, that the environment played a major role in determining form. He established this view from a study of crustacea parasitic on fish which were the subject of his classic *Mikrographische Beitrage* which he wrote while a pupil of Rudolphi at Berlin. This work attracted a widespread attention and secured for him the chair of zoology at Odessa in South Russia. His excellent early monographs on the animals of this region secured for him in 1849 the chair in Helsingfors, where he speedily became known as much for his prankish eccentricities as for his zoology.

PETER GRAY

NUCLEIC ACIDS

The nucleic acids are of biological interest because they are the carriers and mediators of genetic information. They are of two types: deoxyribonucleic acid (DNA) and ribonucleic acid (RNA). In all organisms except a few viruses, DNA carries the genetic information, which according to present knowledge defines the

structures of all of the many proteins present in the organism, the amounts in which each is present, in which cells of a multicellular organism each is to be found, and the times during development at which each appears and disappears. RNA provides the link between the information contained by the DNA and the syn-thesis of the proteins.

The nucleic acids are long unbranched polymers. The backbone of the polymer is formed by alternate sugar and phosphate molecules which are joined by phospho-diester covalent bonds between the 3′ and 5′ hydroxyl groups of the sugars. Each sugar carries as a side group a nitrogenous base joined by a β-glycosidic bond. The sugar-phosphate-base unit is known as a nucleotide. In DNA the sugar is 2-deoxy-D-ribose, while in RNA it is D-ribose. In DNA there are four different bases, ade-nine (A), guanine (G), cytosine (C) and thymine (T). RNA contains the same bases with the exception that uracil (U) replaces thymine. In certain classes of RNA other rare bases are found, and in the DNA of some viruses and higher plants 5-methyl-cytosine replaces a portion of the cytosine. It is in the sequence in which the bases occur along the nucleic acid strand that in-formation is encoded. The complete sequence of bases has been determined for some species of RNA, for ex-ample, of certain of the transfer RNAs (see below).

The physical structure of DNA, which was first pro-posed by Watson and Crick in 1953, is that of a double-stranded helix. Two polynucleotide chains are held together by hydrogen bonds between the bases, which lie in planes perpendicular to the long axis of the mole-cule. The hydrogen bonding is highly specific. Adenine may bond only with thymine, and guanine only with cytosine. The chains of the helix are thus said to be complementary to one another in that the presence of a particular base at a position in one chain defines the base which occupies the same position in the opposite chain. In native double-stranded DNA, therefore, the content of adenine equals that of thymine while that of guanine equals that of cytosine. The ratio, $(A + T)/(G + C)$ is different for different species and charac-teristic for each species.

DNA molecules are the largest known natural poly-mers, with molecular weights of up to the order of 2×10^9 gm/mole. It is technically difficult to isolate such gigantic molecules intact since they are very sus-ceptible to physical shear. DNA as ordinarily isolated is therefore of smaller, although still great, size. RNA molecules are smaller, and except as they occur in the genetic material of some viruses, are always single-stranded. Even single-stranded RNA may possess con-siderable secondary structure due to intramolecular hydrogen bonding.

Almost the entire DNA complement of a cell is located in its nucleus as a major component of the chromosomes. Mitochondria and chloroplasts also contain small amounts of DNA which in fact carry the genetic information required for the function of these cell organelles. The amount of DNA doubles prior to cell division but is otherwise metabolically inactive. In contrast RNA is both synthesized and destroyed rapidly throughout the cell life cycle and is distributed throughout the cell. Typically half or more of the cell RNA is associated with the ribosomes which are the instruments of protein synthesis. The ribosomal RNA, as well as all other normal cellular RNA except the small amount contained in mitochondria and chloroplasts, is synthesized in the nucleus from whence it is distributed through the cell.

Direct evidence from studies of microorganisms and suggestive evidence from higher organisms show that DNA carries the genetic information. For example, the replication of viruses can occur by infection of the host cells by the viral DNA alone. Genetic characteristics of bacteria can be transferred from one strain to another by causing purified DNA of a donor strain to be taken up by cells of a receptor strain. The linkage of the genes of the bacterium E. coli can only be interpreted as a single circular genetic map. This corresponds to a single circular DNA molecule of giant size found in such bacterial cells. (Circular DNA molecules have also been obtained from mycoplasma, several species of virus, and from the mitochondria of higher organisms.)

In higher organisms the genes are organized in chromosomes which are composed mainly of DNA and protein. In some species, e.g., the salmon, the protein of chromosomes of sperm is different from that of somatic chromosomes, while the DNA (although halved in quantity as compared to somatic cells) remains otherwise similar. The amount of DNA in the somatic cells of a given species is constant and this amount is twice as much as in the haploid germ cells, and half as much as in tetraploid cells. The amount of DNA in each human cell, 5×10^{-12} gm, is approximately one thousand times greater than that in each bacterial cell and one million times greater than that in the simplest known viruses. Genes double or replicate before cell division. This is due to the replication of the DNA which composes the gene. In replication of the DNA helix the two complementary strands separate to serve as templates for the enzyme DNA polymerase. The enzyme lays down daughter complementary chains on each of the parent strands using nucleotide triphosphates as substrates. The specificity of hydrogen bonding between base pairs insures that the new daughter chain is the complement of the parent chain. Thus replication results in two new double-stranded helical DNA molecules, each identical to the original, and each containing one parental strand and one newly synthesized strand. The process is known as semi-conservative replication.

A gene is thought to be a discrete length of double-stranded DNA of characteristic base-pair sequence. The genes whose functions are best understood are those that serve as templates for the synthesis of RNA molecules. In this process, known as transcription, only one strand of DNA is copied, and an RNA molecule of base sequence complementary to that length of DNA which acted as template is produced. The process is carried out by the enzyme RNA polymerase.

Of the several known classes of RNA, three are involved in protein synthesis, messenger RNA (mRNA), ribosomal RNA (rRNA) and transfer RNA (tRNA). Each mRNA molecule of specific base sequence contains the information required to determine the amino acid sequence of a specific polypeptide. The information is written in the genetic code. The four kinds of bases can form 64 different sequences, each three bases long, and each called a codon. Sixty-one of these codons specify one or other of the 20 amino acids, of which proteins are made. An amino acid may be coded for by several codons, i.e., the code is degenerate. The mRNA molecule is therefore a sequence of codons which acts as an intermediate between the primary template,

which is the gene, and its final product, the polypeptide. Polypeptide synthesis is performed by the ribosomes which contain the species of mRNA as structural components. The ribosomes stabilize the mRNA and facilitate the translation of its sequence of codons by transfer RNA. The ribosome moves along the mRNA molecule progressing from codon to codon. As each codon is reached, a tRNA amino acid complex binds to it, and the amino acid is joined to the growing peptide chain. The peptide is held on the ribosome by the tRNA of the amino acid, which was the last to be attached. This "peptidyl tRNA" is displaced by the incoming amino acid, joined to its own tRNA. The latter then becomes the new peptidyl tRNA. The mRNA is read from its 5' to its 3' end, and the polypeptide grows from its -NH_2 terminal to its -COOH terminal. At least two enzymes, guanosine triphosphate, and a sulfhydryl group are required for the series of reactions. A single mRNA molecule may be read simultaneously by several ribosomes each one moving along the molecule and each producing one polypeptide chain. Such a complex is termed a polysome.

Artificial mRNA molecules such as polyuridylic acid, when supplied to an in vitro system containing ribosomes, charged tRNA and the other factors, effect protein synthesis, in the case of polyuridylic acid, the production of polyphenylalanine. This is because UUU is a codon for phenylalanine. This and other polymers have in fact been used to break the code. If a trinucleotide of known sequence is given to such an in vitro system, it causes binding of a type of charged tRNA, that which carries the amino acid corresponding to the codon of that sequence.

Since there are a large number of polypeptides in a cell, with a broad distribution of molecular weights, mRNA is extremely heterogeneous. In contrast there are only two types of ribosomal RNA of sedimentation coefficients 16S and 23S, and a small number of different tRNAs—probably one for each meaningful codon. tRNA molecules are very small polyribonucleotides, of molecular weight about 25,000, sedimentation coefficient 4S, and around 85 nucleotides in length.

The genetic phenomenon of mutation may be understood even at the molecular level. If a single base-pair of a gene is replaced by another, the codon which contains this base-pair is changed. This may lead to the incorporation of a different amino acid into the polypeptide which is the product of this gene. Thus a gene mutation causes alteration in the corresponding polypeptide. Polypeptides which differ by a single amino acid from the wild type have been isolated from many mutants. Other mutations have been found in which a sequence of amino acids at the N-terminal end identical to the wild type is followed by an entirely changed sequence. This is caused by insertion or deletion of a base-pair into the gene. This alters the reading of all the codons which follow it, since they are read sequentially. The sequence of mutations along a gene as determined by genetic mapping coincides exactly with the order of altered amino acids along the amino acid chain of the protein produced.

Two of the three codons which do not specify amino acids are chain termination or nonsense codons. These normally serve to end the growth of the amino acid chain of a protein when the correct size is attained. Mutation to a "nonsense" codon can, however, occur, and this blocks production of the protein normally coded for by that gene. Instead, incomplete amino acid chains are produced, their length increasing the further the nonsense mutation is located from the beginning of the gene.

Mutation can be induced by the use of chemical analogs of the four bases. These are incorporated into DNA and disturb specificity of replication. Bases already in the DNA molecule can be altered by chemical and physical methods, for example by the absorption of UV light or by treatment with nitrous acid. Acridine dyes distort the structure of the DNA double helix and lead to deletions or insertions on replication. Ionizing radiation, which causes breaks and hence deletions to appear in the DNA molecule is the most well known of mutagenic agents.

DAVID MCCONNELL
JAMES BONNER

References

Grossman, L. and K. Moldave, eds., "Methods in Enzymology," Vols. 12A and 12B, New York, Academic Press, 1967–1968.
Watson, J. D., "The Molecular Biology of the Gene," Benjamin, New York, 1965.

NUCLEUS

Nearly all CELLS have nuclei which are separated from the cytoplasm by a nuclear envelope. Typical nuclei are spherical bodies of about one-tenth the total volume of the cell, but great variations are found in special types of cells both in size and shape of nuclei. The nucleus typically has a refractive index different from the surrounding cytoplasm and is therefore visible in the living cell. In many cells the nuclei appear optically empty because of the absence of granules of different refractive index. However, most nuceli contain one or more dense bodies which are easily demonstrated in living cells, especially with the phase microscope; these are the nucleoli.

When the pH of the medium surrounding the cell is lowered, the nuclei become more conspicuous. When the cell is fixed, especially with acid fixatives, it stands out even more prominently. Basic dyes are readily bound by such nuclei even at the low pH. This is due to the presence of nucleic acids with the phosphoric acid groups available for dye binding. This material has usually been called chromatin because of this property. During the interphase stages between divisions the chromatin in fixed cells frequently appears to be a network of threads, sometimes with a few larger clumps. When cells are approaching division the chromatin becomes condensed into filaments or rods which are distinguishable as individual CHROMOSOMES as prophase advances. The chromosomes are usually visibly double as soon as they can be recognized at prophase. Although chromosomes are visible only during division stages, they are known to have existed in an intact but visually indistinguishable condition all during interphase. They usually are duplicated during middle interphase in rapidly dividing cells. Confidence that they persist intact from one division to the next comes from genetic evidence which shows that genetic loci (GENES) are arranged in a linear order which persists through numerous division cycles. In addition chromo-

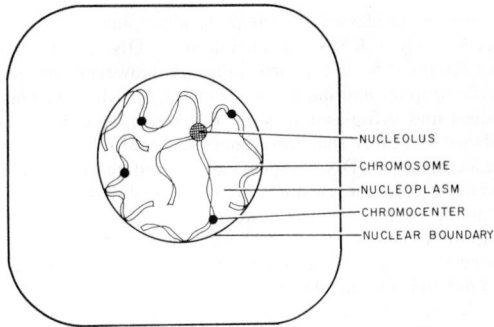

Fig. 1 Schematic representation of interphase nucleus. Note association of parts of the chromosomes with the nuclear boundary and nucleolus. (From Wilson and Morrison, "Cytology," 2nd ed., New York, Reinhold, 1966.)

somes with their DNA (see NUCLEIC ACIDS) labeled by radioactive hydrogen reproduce and pass through several divisions without losing their label or exchanging it with the newly-forming chrosome strands except by an occasional segmental exchange between sister chromatids (daughter chromosomes). This regular sorting of labeled DNA could only occur if the linear sub-units of the chromosome containing the DNA remains intact during interphase. Condensation must be accomplished by the folding and coiling of the strand or strands which make up the chromosome. The coiling has been followed by microscopic examination at successive stages of division in cells (see MITOSIS, MEIOSIS) with large chromosomes. Typically the long prophase chromosomes with many gyres of a coil of a small gyre diameter are transformed into relatively short thick rods with a few large gyres. The detailed changes are difficult to follow, but apparently are accomplished by kinking and formation of a regular coiled coil or by an increase in gyre diameter accompanied by the shortening of the long chromonema, either by coiling at the submicroscopic level or by folding at the molecular level. The coiling cycle is reversed after anaphase and two interphase nuclei are formed which are usually similar to the parental nucleus in appearance as well as genetic properties. Exceptions occur especially in differentiating tissues in which the appearance of the two daughter nuclei may differ from each other or from the parental nucleus. Whether they differ genetically is much more difficult to answer.   In most nuclei the nucleoli disappear during late prophase and reform during late telophase at specific sites on certain chromosomes referred to as nucleolar organizers.

Chemical Components of the Nucleus

The nucleus may contain many of the substances found in the cytoplasm, but relatively little is known of the low molecular weight components, since many of these may leak out or enter the nucleus during its isolation. *In situ* tests can sometimes be utilized, but most of these tests require treatments that would allow movement of small molecules. Nuclei usually do not contain secretory granules and reserves of stored food material, with the exception that some nuclei may contain glycogen.

All of the classes of large molecules, characteristic of cells, protein, RNA (ribonucleic acid), DNA, and lipids or lipoproteins, are present in nuclei. However, the specific proteins and nucleic acids characteristic of the nucleus and cytoplasm may be different. Available evidence indicates that differences do exist. With a few exceptions the DNA appears to be confined to the nucleus and to the chromosomes. Cells with DNA containing viruses or viroids are exceptions and a DNA-like material is found in the cytoplasm of some eggs where it may be a storage reserve of precursors for the rapid divisions during cleavage.

DNA is present in all types of nuclei so far investigated. The amount of chromosomal DNA present is a characteristic of the species and ordinarily increases only when the cell is about to divide or the cell is becoming polyploid. The increase in DNA is discontinuous even in rapidly dividing cells. Usually the cell spends only about one-third of the cell cycle in the replication of its DNA. Evidence indicates that each molecule replicates only once during each cell division cycle. Extra chromosomal DNA in viruses or other similar particles in cells are not under such rigid control, but frequently multiply out of sequence with the chromosomal complement.

Although the DNA per nucleus is relatively constant within the species, the variations between species is striking. The nuclei of amphibians and of higher plants of the lily family have the highest amounts of DNA. Man and other mammals have intermediate amounts, and fruit flies and fungi like Neurospora have one-hundredth or less than do those with the highest amounts per cell. One of the puzzling features is the significance of the great variations. The genetic information required for growth is probably not vastly different, yet the material for recording it is highly variable. Perhaps the same code is present many times in those with high amounts per cell. However, this would appear to present great problems in evolving changes in such a genetic code.

RNA is found in abundance in both nuclei and cytoplasm of rapidly growing cells, but may be much less abundant in certain non-growing cells. However, in most cells its concentration is low compared to proteins; 5–10 per cent of the dry weight would be considered a very high concentration. Analysis of the purine and pyrimidine bases in RNA from cytoplasm and nuclei indicate differences in some cells. However, there is considerable evidence that a part if not all of the RNA of the cytoplasm is derived from RNA synthesized in the nucleus. Cells with the nucleus removed do not synthesize additional RNA. RNA, labeled with a radioactive isotope in a nucleus transplanted to a homologous, unlabeled cell, moves into the cytoplasm after a few hours. Cells fed labeled intermediates in RNA formation, for example cytidine or uridine, incorporate these into the nuclear RNA first. Only after a considerable delay, dependent on the metabolic conditions of the cells, does the labeled RNA appear in the cytoplasm. To account for the differences in cytoplasmic and nuclear RNA, the various RNA's can be assumed to have different rates of destruction in the cytoplasm, be changed during the transfer, or else the nucleus contains RNA's with different metabolic characteristics and perhaps different functions. The evidence for RNA's of different metabolic properties has been found but the function of RNA is still obscure, although it is generally supposed to function in the synthesis of proteins. Certainly it is abundant in cells where protein synthesis is rapid and is a major component of the cytoplasmic particles, ribosomes, in which much of the protein synthesis appears to occur.

A characteristic class of basic proteins, the histones, are present in most metabolic nuclei; certain bacteria do not appear to have them. They may be present in sperm nuclei, but are sometimes replaced by a more basic class of protein designated protamines, for example in fish sperm. Present evidence indicates that the histones are intimately bound to the DNA of the chromosomes and proper isolation procedures yield nucleo-histone, a DNA-histone complex. The histones are present in amounts equivalent to the DNA in some nuclei. They appear to be synthesized simultaneously with DNA replication and to have a low rate of turnover. However, the ribosomes contain basic proteins with amino ratios similar to the histones. Histones have been reported to be present in nucleoli and to be transported to the cytoplasm. However, the cytochemical evidence on which this hypothesis is based has been questioned.

The most abundant proteins in nuclei of metabolically active cells are non-histones. The amount of histones and the DNA per nucleus is a species characteristic and does not vary except during the synthesis for cell division and during the formation of nuclei that are POLYPLOID (nucleic with multiple sets of chromosomes) or polytene (nuclei with giant multi-stranded chromosomes). However, the amount of the other proteins may vary greatly with the type of nucleus and its metabolic condition. They appear to be completely absent from sperm nuclei and very abundant in some nuclei of glandular tissues and developing oöcytes. Some of these proteins must be part of the enzymatic equipment of the nucleus, but perhaps a large fraction is synthesized and transferred to the cytoplasm. Proteins have also been shown to enter the nucleus from the cytoplasm.

The presence of high concentrations of proteins and nucleic acids give nuclei a high viscosity in most cells. The nucleoli are the densest structures in the cell. They have a high concentration of protein and relatively less RNA. Usually no DNA is present. Condensed chromosomes are also rather dense. They contain DNA, histone, some RNA and a residual protein (insoluble in salt solution). Most nuclei have in addition to the chromosomes and nucleoli a more fluid phase referred to as karyolymph or nuclear sap. Unlike the cytoplasm the nucleus is free of membranes except for the nuclear envelope which surrounds it. However, considerable spatial organization is indicated by the fact that chromosomes do not become excessively entangled. The chromosomes have been shown to retain the same position at prophase which they occupied at the preceding telophase.

In summary, then, the nucleus contains most of the cell's genetic information coded in molecules of DNA. This is translated by means of the RNA or perhaps by ribonucleoproteins which are released into the cytoplasm. Much of the protein synthesis and energy yielding metabolic processes occur in the cytoplasm. However, both protein synthesis and processes that yield ATP (adenosine triphosphate) occur also in the nucleus. Although cells soon die when deprived of a nucleus, nuclei are very dependent on the cytoplasm and cannot survive when freed of their cytoplasm. Contact with fluids other than homologous cytoplasm appears to result in a rapid deterioration of metabolic potential and almost instantaneous loss of ability to function as the genetic apparatus of another cell.

J. HERBERT TAYLOR

References

"Genetic Mechanisms: structure and function," Cold Spring Harbor Symposia, Vol. 21. Cold Spring Harbor, Carnegie Inst., 1956.

Bowen, V. T., ed., "The Chemistry and Physiology of the Nucleus," New York, Academic Press, 1952.

NUTRITION

From a primitive single-cell organism such as the amoeba, which derives its nutriture directly from its immediate environment, to higher-order animals which must search for highly selective diets, nutritional requirements reflect the evolutionary processes of adaptation and change. Consequently, they are dependent on acquired biochemical processes to convert available nutrients into those metabolites which are essential for the maintenance and growth of living cells. In essence, nutrition is a highly dynamic field of science and in order to comprehend its role in health and disease, an understanding of many related disciplines is mandatory. Since a comprehensive review is not feasible within this space, an attempt will be made to highlight some recent advances in nutritional research which may have a bearing on current problems.

The science of nutrition not only deals with the absolute and minimal intakes of nutrients but also requires knowledge of the relative interactions between foodstuffs. For example, if a compound is known to be required at a certain dietary level, it does not necessarily follow that higher levels are better or even advisable. Dietary excesses as well as deficiencies represent poor nutrition. It is interesting to note that during approximately the first half of this century the emphasis had been on nutritional requirements and deficiency diseases. With the discovery of the importance of the amino acids, the various minerals, and the vitamins, it was felt by many that the major contributions of nutrition to medicine had been accomplished. This attitude has now been dispelled by the emergence of a revitalized approach to the problems of over-nutrition. In this country, as well as many other prosperous societies where foods are readily available, diseases such as atherosclerosis and diabetes are believed to be related to diet. Although the underlying susceptibility in such conditions may be related to heredity and/or endocrine disorders, nevertheless the nutritional aspects of these diseases are quite impressive.

Basically, good nutrition demands that the caloric intake of foods be established in order for a person to maintain a desirable weight. What constitutes a desirable weight is dependent on height, age, sex, somatotype, and physical activity. For example, an active athlete may weigh more than a sedentary person of the same height and age but the weight difference may be largely due to muscle mass. However, it is recommended that with decreasing athletic activity, a concomitant program of reduced food intake be instituted to avoid the dangers of obesity which is common among many former athletes. Although a major health problem in the United States, the approach toward treating obesity has remained for the most part rather primitive. This lack of progress has been due largely to a tendency of oversimplifying the causes. Furthermore, there has been also a tendency of studying obesity as a group phenomenon. Fundamentally, most cases of obesity result from dietary excesses and lack of physical activity, but the problem is to explain why certain individuals are prone to obesity rather than vaguely attribute obesity to glandular disturbances. According to Mayer, obesity falls into two general categories: (a) "Regulatory" which denotes a faulty regulation of a central control of food intake, i.e., hypothalamic obesity; and (b) "Metabolic" which denotes altered metabolism of tissues. This hypothesis gives some measure of hope that continued research in this problem may reveal useful techniques for controlling excessive food intake.

The quacks and food faddists have done much to exploit this nation's idolism of the slim individual. Most food fads, however, have proven to be useless; many of them are actually hazardous to health and none have proven to be beneficial for maintaining a desirable weight. The only known effective reducing regimen is simply one that involves motivation, moderate exercise and sensible eating habits without excluding any essential nutrients. Exotic reducing regimens have little known scientific value. A reducing diet should be one that is low in calories but retains the required levels of good protein, vitamins, and minerals, with the caloric reduction being made at the expense of carbohydrate and fat. Essentially, one should continue to eat a well-balanced diet, but a certain degree of negative caloric balance should be maintained until a desirable weight is achieved. The caloric control period should not be the result of a crash program for it is neither desirable nor necessary that an obese individual lose a large amount of weight in a very short period of time; weight reduction should be gradual and the effectiveness of this program requires that a moderate amount of exercise be instituted along with intelligent eating habits which the patient is convinced will be helpful. On the other hand, there is a likelihood that some obese individuals may suffer from compulsive eating habits which reflect underlying emotional disturbances. In such cases, competent psychiatric assistance prior to undertaking any reducing program may prove useful.

Of the major foodstuffs, perhaps the most important are the proteins, the principal nitrogenous components of all plant and animal tissues. They constitute the essential chemical features of both the protoplasm and the nucleus of cells and, furthermore, they constitute enzymes, certain hormones, and antibodies; they are also important for maintaining osmotic relations between intracellular and extracellular fluids and for the actual stabilization of serum.

The PROTEINS are large, complex organic molecules consisting of amino acids linked together in peptide bonds. During the processes of digestion these large molecules are enzymatically reduced to their AMINO ACID components; these smaller moieties are then absorbed into the circulation. Following endogenous utilization they can be made available as precursors for body protein synthesis or are deaminated to other metabolites such as monosaccharides or converted to non-protein nitrogenous end products such as urea. The dietary amino acids are also available for the synthesis of related compounds such as heme, the PURINES, the PYRIMIDINES, etc. Eight of the many available amino acids are essential in the diet of man to maintain nitrogen balance. In other words, these eight cannot be synthesized *in vivo* and must be present in our food whereas the other required amino acids can be synthesized, for example, from related metabolites in the body. An approximation of the dietary requirements of these eight essential amino acids has been established for young adults.

In relation to amino acid intake, a newer concept has been proffered, i.e., amino acid imbalance. Studies with rats show that the excessive intake of certain amino acids will lead to tissue imbalances as evidenced by anorexia and ultimate weight loss. Such imbalance has only been shown at very low levels of protein intake. Therefore, not only are the total quantities of certain dietary amino acids important in the diet but at very low intakes the relative dietary proportions of these amino acids are also important.

In recent years, a disease related to protein deficiency and referred to as Kwashiorkor has been observed in many of the technically underdeveloped countries of the world. The clinical, chemical, and pathologic picture reveals retarded growth, edema, fatty infiltration of the liver, dispigmentation and dermatoses, gastrointestinal disorders with reduction of pancreatic and duodenal enzymes, irritability, follicular atrophy of the thyroid, atrophy of the adrenal cortex, hypoproteinemia, reduced concentration of several enzymes in the serum, reduction of several circulating vitamins, such as A and.E, and hypocholesteremia. If treated soon enough, the above changes can be dramatically corrected or reversed with a diet high in good quality protein. On the other hand, if such a condition is allowed to progress without treatment, death occurs.

The recommended protein allowance for adults has been set at approximately one gram of protein per kilogram of body weight per day; the allowance for children is higher. These recommended allowances have been arbitrarily set high to allow a sufficient safety margin; actually, it has been shown that levels of less than one-half gram per kilogram of body weight are sufficient *if* adequate calories are available and if the required level of essential amino acids are obtained. The latter depends on the biological value of the dietary protein. Perhaps the simplest procedure for assaying the biological value is based on the growth response of animals under a variety of feeding procedures. By chemical analysis, a protein may actually contain enough of a certain amino acid part of which may not be biologically available; thus only a fraction of the actual amount may be utilized by the body. On the other hand, a protein which is deficient in one or more amino acids can be readily supplemented by protein containing adequate levels of these amino acids or in some cases by the addition of the specific amino acid; for example the addition of lysine to wheat protein.

Atherosclerosis represents a classic condition related to dietary excesses and much of the current thinking on this problem stresses the role of dietary fats and/or excess total calories in the pathogenesis of this disease. About fifty years ago, it was observed that dietary cholesterol produces, in rabbits, atheromatous plaques. Since that time extensive research, particularly during the past decade, on the role of cholesterol metabolism in the atherogenic process has been carried out. The human intake of cholesterol is significant if a diet contains high levels of animal products. However, it has not been shown that in humans dietary cholesterol *per se* contributes to the atherosclerosis process.

A number of recent studies suggest that the FAT content of the diet may be important in producing a condition of hypercholesteremia which may lead to atherosclerotic plaque formation. Although one group of investigators believed that the total fat content of the diet may be responsible, others have now shown that it is the fatty acid composition which is more critical. For example, certain vegetable fats which are high in polyunsaturated fatty acids appear to protect against hypercholesteremia while certain animal fats or commercially hydrogenated fats favor higher serum cholesterol levels. At present, several suggestions have been offered on the cholesterol controlling properties of fats such as the (a) essential fatty acids (content of linoleic acid and arachidonic acid); (b) degree of unsaturation of fatty acids; (c) sterol content (phytosterols, cholesterol); (d) ratio of saturated versus unsaturated fatty acid content. In any event, there is little doubt that the lipid content of the diet contributes to the hypercholesteremia which is common among the more prosperous populations of the world but the most critical factor of fat intake has yet to be established. A reduction of the present very high fat intake in the United States to more reasonable levels is advisable, most importantly for the control of body weight. The intake of certain animal fats and hydrogenated fats is believed by some to be excessive and the replacing of such fats in the diet with those of vegetable origin has also been suggested as being very useful, especially in patients demonstrating high cholesterol levels and clinical manifestation of atherosclerosis. There is considerable experimental evidence that other dietary components including the proteins, carbohydrates, purines, pyrimidines, certain minerals and vitamins as being implicated in the disease. In general, these latter observations are based for the most part on specialized animal studies and therefore too preliminary in nature to be useful in the control of the human disease at this time. The experimental evidence strongly favors the theory that atherosclerosis is in large measure a dietary disease and the epidemiologic observations support this hypothesis.

Falling within the theme of overnutrition is the excessive intake of calcium among ulcer patients maintained on a diet high in dairy products. The symptoms are hypercalcemia and renal insufficiency which are reminiscent of those produced by hypervitaminosis D. There is some likelihood that calcium intakes among certain groups in the United States may be considered excessive. Since the requirements of calcium and vitamin D are generally met even when diets are relatively low in dairy products, adults should exercise a certain degree of moderation in the intake of foods high in calcium. Furthermore, experimental studies indicate that excessive levels of calcium and vitamin D lead to calcific deposits in many tissues, including the cardiovascular system.

Obviously, there exist real hazards to health when various nutrients are fed at excessively high levels. Another example is hypervitaminosis A. The symptome are dermatitis, cheilosis, epistaxis, gastrointestinal disturbances, anorexia, thinning of the hair, severe headaches, and increased fragility of the fingernails. Therefore, oversolicitous mothers should be cautioned against administering excessive doses of vitamin preparations to infants.

During the past two years, a renewal of interest in the essential fatty acids has been due in large part to the intensified effort in the fatty acid research as related to atherosclerosis. The term "essential fatty acids" includes linoleic and arachidonic acids, since these fatty acids have been shown to be necessary for good growth and for the maintenance of dermal integrity in rats. Skin changes in human infants maintained on a low fat formula have been corrected by feeding fats containing polyunsaturated fatty acids, and the present evidence suggests that in humans linoleic acid is an essential nutrient. However, the daily requirement for linoleic acid has not been established since the manifestations of the deficiency are difficult to evaluate on the basis of the present criteria. It is conceivable that subclinical changes due to diets poor in essential fatty acids may be present in large numbers of people. For this reason it is advisable that the selection of dietary fat should include those which contain significant levels of polyunsaturated fatty acids.

Extensive progress in vitamin E research has been made. Although the requirements for vitamin E in man have not been established, there has been an increasing awareness of the importance of this vitamin. A deficiency of vitamin E in animals results in hepatic necrosis, hemolysis of red cells, exudative diathesis, depressed oxidative phosphorylation, decreased lipogenesis, and altered RNA/DNA ratios in the liver; also, there is a loss in the antioxidative capacity of tissues resulting in decreased vitamin A storage, ceroid formation and brownish discoloration of the uterus. Muscular dystrophy and impaired reproductive capacity have been observed. Some of these changes are similar to those seen in animals also suffering from essential fatty acid deficiency. Furthermore, experimental evidence suggests that an interrelationship between vitamin E and the mineral selenium may exist since selenium protects against some of the above manifestations of vitamin E deficiency. Presumably, selenium substitutes for sulfur in the cystine moiety. Whether selenium exists in significant quantities in human tissue is not known. Selenium is present in grains and has been isolated from milk casein.

Attempts to treat human reproductive ailments with vitamin E have not been successful. Perhaps because of the widespread sources of this vitamin in plant and animal foods, clear-cut vitamin E deficiencies have not been delineated in man. Although Blanc et al. suggest that gastrointestinal tract lesions of patients with pancreatic cystic fibrosis represent the cumulative effects of avitaminosis E, such observations in humans require long-term assessment and vitamin treatment in order to establish clearly a causative relationship.

Several reports support evidence that naturally occurring goitrogens in certain plants constitute a serious public health problem in certain parts of the world. The occurrence of goitrogens in various foodstuffs, particularly in members of the Brassica (cabbage) family, is known. Recently, other plants such as the groundnut peanut, *Arachis hypogaea*, are believed to contain aromatic thyroid inhibitors. It has been suggested that phenolic metabolites of the peanut arachidoside are preferentially iodinated causing a depletion of iodine necessary for thyroid hormone synthesis. More recently, studies with rats have shown that uracil (the natural component of ribose nucleic acid) when consumed in excess, leads to changes similar to those produced with well known goitrogens such as thiouracil. Therefore, there is sufficient evidence in the literature to suggest further investigations in the goitrogenic properties of various natural foodstuffs. Although goiter is not a problem in areas of the world where iodine intake is high, nevertheless subclinical conditions of thyroid malfunction may exist which may not be related to iodine intake.

Another serious public health problem which still plagues certain areas of the world is lathyrism. Because of the chronic features of the disease, its symptoms have never been clearly defined. It is believed to be caused by the excessive ingestion of some toxic factor found in lathyrus pulses. Experimental studies with animals have suggested that the toxic factor is related to propionitrile compounds. However, there is still no evidence that this factor may be responsible in the human lathyrism which is widespread in India. The human disease is characterized by a crippling spastic paraplegia of the lower limbs. According to Filimonoff, the histological picture is reminiscent of a chronic myelitic process. In studies with animals, the excessive feeding of *Lathyrus odoratus*, for example, results in progressive skeletal changes and muscular paralysis, and in more prolonged experiments, dissecting aneurysms of the aorta have been observed. It is also interesting to note that a chemically related compound, the naturally occurring adenine, will also cause aneurysmal dilatations of the rat aorta, including medial necrosis and calcification of this vessel. Such studies serve to emphasize a real need for further research for the presence of food toxins which may have a bearing on human disease problems.

Of current interest is the elucidation of the nutritional importance of many less known trace minerals. For example, the work of Cotzias and his associates has shown that manganese is rapidly equilibrated between the blood and the cellular pool; that there is an active intracellular concentration of manganese within the mitochondria and that the metabolism of manganese is highly specific. Although a requirement for manganese has not been established for man, several animal studies demonstrate that it is an essential mineral since a dietary deficiency will produce a number of symptoms including retardation of growth, abnormal bone growth, nervous disorders, and reproductive failure. Similarly, the dietary requirements of magnesium are unknown for man although a deprivation of magnesium in animals leads to hyperirritability, vasodilation, convulsions, and death. Recently, a clear-cut magnesium deficiency was reported in man.

Epidemiologic, clinical, and experimental studies have conclusively shown that fluoride is desirable in man for the formation of a dental enamel with maximum resistance to decay. The natural source of fluorine is the fluoride usually found in drinking water. The best public health control of dental caries in many communities where the fluoride content of water is low is simply to supplement the water supply. This is a most desirable and intelligent public health procedure.

These comments should serve to emphasize that our knowledge of human dietary needs is far from complete and that it has a very important bearing on human health and disease. In the various areas of nutritional research new insights into the role of the minerals, vitamins, fats, carbohydrates and proteins are being continually revealed.

F. J. STARE

References

"Nutrition Reviews" published monthly by the Nutrition Foundation Inc., New York.

Chapters on Nutrition in "Annual Reviews of Biochemistry" published annually, Palo Alto, California.

Davidson, S. and R. Passmore, "Human Nutrition and Dietetics" 3rd. ed., Baltimore, Williams and Wilkins, 1966.

Stare, F. J., "Eating for Good Health" New York, Doubleday, 1964.

O

OCEAN

Sea water has its origin in incandescent gases of the primeval atmosphere. To some extent it is also juvenile water being gradually freed from magma. Water had remained in the original atmosphere until the temperature of the vapor dropped to the level of the critical temperature of water (374°C) before which the first condensation and the first rainfall took place. Salts contained in sea water are partly derived from the earth's interior, brought to the surface by the action of innumerable volcanic eruptions in the course of about two billion years, which is the age of the ocean, and partly from surface rocks weathered by streams ending in the sea. Anions present in sea water are mostly of volcanic origin while most of the cations come from land rocks. More then seven tenths of the earth's surface, i.e., 70.8%, are covered with sea water which occupies 60% of the Northern Hemisphere and 80% of the Southern Hemisphere. The Pacific is the largest, comprising about half of the total ocean. Next come the Atlantic, occupying about one quarter, and the Indian, occupying less than one quarter of the total ocean. The remaining one tenth of the total area belongs to adjacent and inland seas.

The average depth of the ocean is 4150 yds. while the average height of the land over sea level is about 915 yds. The maximum depth in the Indian Ocean amounts to 8120 yds. and is in the Sunda trench. The maximum depth in the Atlantic is off Puerto Rico 10,500 yds. The greatest depth ever found is the Vitez depth in the Marianas trench in the Pacific 12,000 yds. The ocean trenches are extended depths going beyond 6640 yds., often situated not far from a coast known for volcanic and seismic activities. The average ocean depths are the following: in the Pacific, 4560 yds.; in the Indian Ocean, 4300 yds.; in the Atlantic, 4210 yds.

The ocean has three distinctive parts, the continental shelf, the continental slope, and the abyss. The shelf zone lies around the land; it is shallow (less than 100 fathoms) and has a slight gradient of about 1:500. It is important for fisheries and sometimes also for the extraction of mineral oil from undersea deposits. The continental slope extends between the depths of 100 fathoms and 2 miles. It represents the real base of a continent. Its gradient is a steep one (1:20). Here lies the remotest area reached by fine sediment deriving from the land. The use of submarine detectors such as echo-sounding equipment has made it possible to ascertain that the ocean bottom is rather uneven, covered with valleys, ridges, fissures, and canyons. The sub-

marine canyons are often a mere continuation of river mouths, e.g. the Congo, the Hudson, etc. Their dimensions are considerable in many cases, and their origin has not yet been completely explained. The abyss begins at the depth of two miles. It is a region of gentle undulation with low temperatures, high pressure, permanent darkness, and slow water motion. Seven per cent of the total ocean area are occupied by the continental shelf, 15% by the continental slope, and 77% by the abyss. One per cent of the total belongs to deep ocean trenches.

Marine sediments are arranged in zones conforming in direction with the coast. The finer the sediment the farther away it is carried reaching greater depths. Coarse sediment is found in areas where significant water motion occurs at the bottom of the sea. The origin of the sediment is various: it may come from the land as a consequence of some mechanical or chemical action; it may be of volcanic or morainal character; it may be brought by a steady wind, e.g. the Saharan NE wind conveying sand to the Atlantic. An important part is also played by sediment of biogene origin. Plankton shells and skeletons are of mineral composition (silica, lime, etc.) and are deposited on the bottom when the organisms die. Shells belonging to globigerinae and other foraminifera, some mollusca and coccolithophoridae are made of lime. The skeletons of radiolaria and diatoms are made of silica. There is also sediment of chemical origin. Minute meteorites, coming from the space, volcanic dust, and sometimes shark teeth are also found. The red clay covering the deepest sea bottom seems to be of an inorganic origin. The accumulation of sediment in the open ocean is a slow process: 1 mm is added in 500 years; but its rate is much quicker in coastal waters: up to 30 cm. in 1000 years.

Owing to the unsymmetric structure of its molecules water possesses a number of important properties. It can dissolve various salts which are ionized. This makes a series of chemical reactions possible. Acid and alkaline compounds result from solutions occurring in water rendering it capable of dissolving further substances. Water molecules aggregate in 2, 4 and 8 molecules according to temperature. This again affects the water volume. The maximum density of water, therefore, occurs at 4°C, and, when water is turned into ice, the volume of the latter grows larger by 9% which makes it float in water. Life conditions in water are greatly affected by this phenomenon. Owing to the fact that three isotopes of oxygen and two isotopes of hydrogen occur in water (tritium is neglected here) there are nine kinds of water molecules. More than 0.25 per cent of ocean water is "heavy" water. The specific heat of water being a high one the storing up of heat is greatly favoured which is highly important in oceanography and climatology.

The total amount of salt present in ocean water is called salinity and is expressed in g/l. In contrast to freshwater, a remarkable steadiness is shown by ion ratios in sea water. On the other hand, the salinity rates of various seas differ widely. The salts found in sea water have the following percentage distribution.

Chloride (Cl⁻)............................ 55.20
Sodium (Na⁺)............................. 30.40
Sulfate (SO₄⁻⁻)........................... 7.70
Magnesium (Mg⁺⁺)....................... 3.70
Calcium (Ca⁺⁺)........................... 1.16
Potassium (K⁺⁺).......................... 1.10
Bicarbonate and Carbonate (HCO₃⁻ and
   CO₃⁻⁻).................................. 0.35
Bromide (Br⁻)............................ 0.19
Borate (BO₃⁻⁻⁻).......................... 0.07
Strontium (Sr⁺⁺)......................... 0.04

In addition to the above-mentioned ions, exceedingly small quantities of a great many chemical elements occur in sea water in salt form as minor constituents or trace elements. The upper limits of occurrence in sea water of some of them (mg/m³) are the following: Al 1900, F 1400, N 1000, P 60, Fe 60, J 50, Zn 21, Mo 16, V 7, Cu 5, U 2, Ag 0.3, Au 0.008.

The presence of more than fifty elements in sea water has been established. All the chemical elements are supposedly present in the ocean. The higher the atomic number of an element from a group of the Periodic System of elements, the lesser its occurence in sea water. The leading ten ions contained in larger quantities represent 99.5% of all the salts in the ocean. They also show the stability in their ratios of occurrence. The part played by trace elements is a very important one, particularly with regard to supporting life in the ocean. There is a considerable exchange of trace elements in the upper film of sediments on the ocean bed. The process of sedimentation prevails in oxidizing conditions while the process of solution is favored by reducing ones. The deep submarine volcanisms considerably contribute toward (1) the enrichment of sea water with trace elements and particularly with nutrient matter, (2) the maintenance of a reducing medium, and (3) the creation of an acid medium. Submarine volcanisms seem to produce a favorable effect by increasing the biological potential of the ocean. Harmful consequences of local extent inflicted on marine organisms by direct eruptions are neglected here. In the area of junction between the ocean and atmosphere there is a steady exchange, mostly consisting of salt transfer from the former to the latter. Foam and droplets, lifted from the sea surface by gales, are carried away into the atmosphere where, as "cyclic salts," they play an important part as nuclei of condensation in the process of precipitation. They also perform a geochemical function by intensifying the weathering of rocks thus contributing to soil formation.

Phosphates and N-compounds are nutrient salts of great importance and are extracted from sea water by phyto-plankton. They usually are the minimum factor of production in the low latitude ocean areas, but are rarely so in the areas of high latitudes since these substances are replenished from the deep layers which abound in them. The waters of the Antarctic Ocean are especially rich in these salts. Vertical components of the marine currents contribute to the enrichment of the upper layers with these salts and so does the inflow of freshwater and, indirectly, the submarine volcanisms as well.

Gases dissolved in sea water occur in the following average quantities: N₂, 12 ml/l; O₂, 6.4 ml/l; CO₃, 0.3 ml/l while the quantities occurring in the air are the following: N₂, 780 ml/l; O₂, 210 ml/l; CO₃, 0.3 ml/l. The higher the temperature the lesser the ability of sea water to dissolve these gases. The amount of oxygen occurring in the ocean ranges between 0 and 8 ml/l. It is abundant in the cold surface water, but the maximum occurs in the layers between 20 and 60 m which is explained by plant assimilation. The minimum O₂ is found in the ocean layers at the depth of about 3000 ft. This is the result of the activity of marine bacteria in burning organic compounds. The amount of oxygen grows again larger in the deeper layers. Special hydrographic conditions, however, prevailing in the Black Sea, in the Cariaco basin off Venezuelan coast, in some basins of the Indonesian mediterranean sea, in Norwegian fiords and Yugoslav bays, cause the amount of oxygen to grow smaller as we proceed toward the bottom and to disappear completely, being replaced by H₂S. The existence of higher forms of life is thereby excluded.

CO₂ is not only dissolved in sea water but it is chemically combined with the surplus of alkaline matter left over from neutralized strong gases. The higher the surplus of alkaline matter (excess base or alkalinity) the larger the quantity of CO₂ forming or being able to form carbonate and bicarbonate compounds in sea water. In the waters of the open ocean there are 2.38 milliequivalents of excess base. Their alkalinity amounts to 2,380. A smaller part of carbonic acid remains in free solution affecting the acid reaction of sea water (pH) which generally ranges between 8.00 and 8.20 pH. The waters of the ocean are, then, slightly alkaline. The hydrogen-ion concentration is intensified (pH is reduced) by the rise in temperature. Owing to their salinity the waters of the ocean possess some physical properties differing from those characteristic for freshwater. So the values of specific heat, thermal conductivity, and surface tension are much lower for sea water, and its freezing point lies lower than with freshwater. The sound travels faster through the sea than through the air.

The average salinity of the ocean waters is 35‰ Sal. The fluctuation of salinity from ocean to ocean ranges between 32‰ and over 37‰ Sal. The amount of salinity depends on the latitude and the meteorological factors determining the degree of evaporation and the formation of precipitations. A high salinity region is, therefore, to be found in the zones extending over 10°–30° of both latitudes. Low salinity often prevails in coastal regions while it is considerably higher in some mediterranean seas with a pronounced evaporation, e.g. 39‰ in the Mediterranean and 41‰ in the Red Sea. The fluctuation of salinity in the vertical sense occurring in the ocean is less than 1‰. The argentometric analysis is the most common method of measurng the degree of salinity. Among the factors influencing a number of phenomena in the ocean, temperature is one of the most important ones. It (1) affects the density of water and, consequently the dynamics of the sea; (2) contributes to the ability of sea water to dissolve gases and salts; (3) has an effect on the rate of chemical reactions influencing all the organic world of the sea.

The solar energy is the principal source of heat for the ocean waters. Other factors are of minor importance. They are: heat conduction from the earth's core, submarine volcanisms, friction of water particles as a result of tides, and the natural radio-activity of potassium, uranium, thorium, etc. The surface of the ocean, heated by day, cools at night, and the same process takes place in summer and winter respectively. The diurnal fluctuation of the ocean surface temperature ranges between 0.1°C and 0.6°C. The annual fluctuation amounts to about 2°C in the tropics, 8°C in the temperate zones, and 2°C in the regions of higher latitudes. About 53% of the total ocean area has a surface temperature above 20°C. The Red Sea is the warmest of the seas (35°C).

The deep sea water layers are not heated by direct insolation. The vertical water circulation caused by thermohaline convection and turbulence is responsible for the bottomward thermal transport. The water temperature falls by 20°C as we reach the first 1000 m but there is only a slight further change in larger depths. All the ocean waters lying deeper than 3000 ft represent an almost uniform mass of a low temperature. Cold water is also formed on the surface from where it sinks to the bottom. This particularly happens in the north, in the region around Greenland, and in the southern hemisphere around the border of the Antarctic Continent. This drop in the water temperature following the increasing depth is characteristic for the ocean. It is quite different in some landlocked seas, e.g., in the three mediterranean seas, where the temperature never falls beyond a certain depth under which uniform temperature prevails in a more or less thick water layer. In the Mediterranean, for example, the temperature found from the depth of 1200 ft downwards (i.e. no more than 4000 m) amounts to 13°C.

Sea water is light-absorbing owing to dispersed molecules and dissolved coloured matter that happens to be present in it. Optical measurements are made: (1) by lowering in the sea water a round white disc (Secchi disc); (2) by means of photographic plates; or (3) by means of photocells. The red part of the spectrum is absorbed by the upper layers while the green and blue parts penetrate to larger depths. The light absorption is greater in the coastal waters than in the ocean ones. The typical color of the warm and temperate parts of the ocean is blue. Light shades of green prevail near the coast. The occurrence of red tide along the coast is also known. The blue color of the sea is a property of sea water, not a mere reflection of the color of the sky.

If there is no change in the temperature, the specific gravity of sea water increases with the rise of salinity value. The higher the degree of salinity, the lower the temperature of maximum density. The highest densities are found in the seas of the frigid zones although they are diluted to a considerable extent. The lowest densities occur in the tropic seas with low salinity values. The density of sea water ranges between 1.0275 and 1.0290 (δt = 27.50–29.00). The dynamics of sea water bodies, the ecology of plankton, fish-eggs, etc, are highly affected by the degree of density.

Owing to its higher average temperature in comparison with the atmosphere, the ocean is a source of heat. By conveying some of its heat (10%) to the air the latter is directly heated by the sea. Additional heat is conveyed to the atmosphere by evaporation which is less pronounced in the equatorial region than in the zones of subtropical anticyclones (20°–30°). There the evaporation attains its maximum and it drops again in the regions of higher latitudes. The importance of evaporation is evident since the sea is the chief source of precipitations falling on the continents.

The factors which produce currents are external forces such as wind, alterations in the barometric pressure, tidal forces and internal forces which are caused by differences of specific gravity in the gravity field of the earth. So the factors producing the changes in density such as evaporation, precipitations, heating, cooling and other climatic factors may contribute to the formation of currents. They all produce changes in the ocean level thus causing displacements of water masses from a high level region to a low level one. The resulting differences are not great ones—they amount to a couple of inches only—but the part played by them is of paramount significance for the dynamics of ocean. Owing to the earth's rotation moving sea water turns right in the northern hemisphere and left in the southern hemisphere (Coriolis forces). The shape of a coast may also contribute to the forming of a current. Narrow passages increase the rate of motion of marine currents (e.g. the Antillean Passage, Florida Passage, etc.). Both the course and velocity of marine currents are rather fluctuating values. Even the largest among them such as Gulf Stream follow an irregular, meandering course. There are currents caused by the tides, too. They may be of considerable extent, and their characteristic consists in the reversal of their course following the tidal waves (Scylla and Charybdis in the Strait of Messina). The earth's rotation and trade winds are responsible for the rate of motion and course of large ocean currents. The surface currents usually involve a water layer in motion of a thickness not exceeding 600 ft.

Surface waves of the sea are caused by the action of winds, earthquakes, moon, etc. The simple wave motion occurring on the ocean surface has an $\sim$ shaped appearance (it is sinusoidal). The shape of the wind-swept waves does not conform to the above description but belongs to the trochoid type of curve: the crest of these waves is not rounded. Every wave has its length (i.e. the distance from crest to crest), its height (i.e. the vertical difference between its crest and trough), its amplitude (i.e. half the height), and its frequency (i.e. the number of oscillations in time). The development of waves depends on the force of the wind and the time and distance covered by it (the "fetch"). The greater the wave length the faster the wave moves. Ocean waves may reach a height of 50 ft and their length may amount to 300 ft. Instruments can predict by 24 hours the approach of swell produced by a gale raging in a distant region. Shock-waves are produced by submarine earthquakes. Their frequency is low and their wave length and velocity are large. The shock-waves frequent in the Pacific region, known as "tsunamis" may attain a length of ten miles or so and velocity of hundreds of knots. Their velocity dwindles in shallow water but the height increases up to 100 ft and their impact against the coast may be terrific causing great damage to coastal districts.

The tide is the periodic rise and fall of the sea caused by the attraction of the moon and sun (the attraction of the former is $2\frac{1}{2}$ times stronger than that of the latter). With the full and new moon the ocean waters are attracted in the same direction by the gravitational pull of both the moon and sun (syzygy) piling up water masses in two tidal waves. Tides with maximum range (spring tides) occur then. When the moon is at first or third quarter, its attraction, being at right angles to that of the sun (quadratures of the moon), is counteracted by the latter causing a high tide with small range (neap tide). The Antarctic Ocean, according to an opinion, is the most important region for the formation of the tidal wave. Every 24 h. 50 m. two tidal waves move there westwards around the globe unimpeded by land, affecting the three adjoining oceans. The tidal wave reaches the equator moving at a speed of 300 knots. Its height does not surpass 3 ft in the open ocean but it may grow to a considerable extent on approaching the coast (40 ft at St. Malo on the French coast, 55 ft in the Bay of Fundy in eastern Canada). Tidal streams in shallow waters are caused by the tide too. They usually run in one direction for a time ($6\frac{1}{4}$ or $12\frac{1}{2}$ hours) and then move in the opposite direction for the equal span of time. The rate of motion of such currents may sometimes reach considerable values (e.g. about 12 knots in the region of Orkney Islands).

The ocean is the largest biotope. The majority of animal classes live in it, and the development of biology has run concurrently with the extension, in the course of 19th century, of the knowledge of marine life. Marine plants assimilate and produce food eaten by all the marine organisms. The principal food supply comes from Phyto-plankton organisms present in the uppermost sea water layer (a hundred-odd meters thick). The most important areas for the food supply are chiefly the coastal ones, particularly where the upwelling takes place (the eastern coasts of the oceans, the Antarctic region, and the region of divergences of equatorial currents). No significant production takes place in the central areas of the oceans. The production of marine organisms is supposed to roughly match the entire organic production of the continents.

Studies of marine fishes are conducted for practical reasons in order to collect data on the condition of their life. Fish is caught along all the coasts. Considerable fishing industry has sprung up wherever large concentrations of fish occur, e.g. along the coast of California, in the North Sea, in the waters around Japan, etc. The catches are subject to substantial fluctuation, particularly as regards pelagic fish. The catches of the latter have considerably dwindled in recent years in the Mediterranean, in the English Channel, and in California. This fact has encouraged intensive oceanographic explorations of international character. The quantities of bottom-fishes caught on continental shelves grow smaller owing to intensive exploitation of fishing grounds. This is the consequence of modern fishing techniques. The problem of overfishing has been treated by international forums in cooperation with scientific and economic experts.

Marine organisms such as algae, worms, barnacles, shells and the like may become a nuisance to ships which have to be protected from them by means of antifoulings and special paints containing poisonous salts of heavy metals. One of the objectives of scientific research in this field is to provide a better remedy for this evil.

A liter of ocean water contains several thousand bacteria. Their number is immensely higher in the waters situated in the vicinity of the coast. MARINE BACTERIA generally are not of pathogenic character. Bacteria of the latter kind, brought into the sea by sewers, perish quickly owing to disinfectant properties of sea water. Fungi are often found in coastal waters.

Low temperature, icefields and icebergs render the navigation almost impossible in the region of the Canadian Archipelago, in the Arctic Ocean, and in a part of the Antarctic Ocean. A series of oceanographic explorations have been made in those regions in order to find out the maximum annual extent of their navigability. The efforts made in this field by the Ice Patrol Service are most important. Owing to similar efforts and explorations it has been possible to render the North-East Passage along the Siberian coast navigable for three to four months every year.

M. BULJAN

References

Buljan, M., "Deep submarine volcanisms and the chemistry of ocean," Bull. Volcanol. U. G. G. I. Série 2. 17: 1955.
Colman, J. S., "The sea and its mysteries," London, Bell, 1953.
Deacon, G. E. R., "Oceanography," London, Penguin, 1947.
Dietrich, G. and K. Kalle, "Algemeine Meereskunde," Berlin, Borntraeger, 1957.
Ercegović, A., "The life in the sea," Zagreb, Jugoslavenska Akademija, 1948.
Fairbridge, R. H., ed., "Encyclopedia of Oceanography," New York, Reinhold, 1966.
Harvey, H. W., "Chemistry and Fertility of sea water," Cambridge, University Press, 1957.
Kuenen, H., "Marine geology," New York, Wiley, 1950.

ODONATA

This primitive order of the INSECTA is generally placed between the EPHEMEROPTERA (Mayflies) and PLECOPTERA (Stoneflies). Its some 5,000 extant species are placed in approximately 500 Genera in two major suborders, the Zygoptera (Damselflies) and Anisoptera (Dragonflies). Fossil materials from Europe and America show abundant Odonata. Precursors (Protozygoptera), some with thirty-inch wingspread, appeared first in the Palaeozoic (Carboniferous and Permian). Small Mesozoic and sub-modern forms appeared in the Jurassic, and Coenozoic forms in the Eocene and Miocene.

The often beautifully colored adults range from less than one to over seven inches in wingspread. A freely movable head bears chewing mouth parts, bristle-like antennae, extremely large compound eyes, and three ocelli. The thorax is divisible into a small movable prothorax and a strongly fused meso-metathorax (pterothorax) that it tilted posteriorly, so shifting the wings that the mesothoracic pair may be over the metathoracic legs. The four richly veined membranous wings are about equal in size. Each usually has a *nodus* (notch) midway along its leading edge and a distal, pigmented *stigma*. Anisoptera have broader bases on the hind pair and the wings are held horizontal when at rest. Zygopteran wings are equal, narrowed basally and held vertically. The relatively unmodified legs are crowded forward, forming a basket useful in capturing prey in flight and for perching, but unsuited for walking. In the males the long abdomen terminates in unsegmented claspers used to grasp the female head or prothorax as she flies in tandem behind her mate. Sperm are transferred from terminally located vasa deferentia to the accessory male copulatory organs at the base of the abdomen, by his looping the abdomen forward. Copulation most often occurs at rest, the clasped female looping her abdomen forward to the male copulatory organs to effect sperm transfer.

All Zygoptera, and some Anisoptera, use well developed ovipositors to place elongate eggs in plant material (*endophytic*) in or above water. Others oviposit ovoid eggs in gelatinous envelopes or strings, on the water or in moist soil (*exophytic*). A male frequently accompanies the ovipositing female, even when she submerges. The number of eggs produced averages between 200 and 300. Hatching takes place in a few days or, in those developing in temporary waters, is delayed until spring.

Immatures live in all sorts of fresh water, though some occur in brackish and a few develop on sheltered moist tropical soil. Being hemimetabolous insects, the immature stages are referred to as naiads, larvae or, probably more correctly, nymphs. These campodeiform nymphs are insideous creatures, with a long prehensile labium that folds up against the antero-ventral part of the head. Spiracles are replaced by flattened terminal gills (Zygoptera) or rectal gills (Anisoptera). The latter nymph draws water into the rectum, which is surrounded by a tracheal network, and by forcible expulsion achieves additional locomotion. Nymphs are camouflaged by form and cryptic coloration, and are often encrusted with debris, algae or bryozoans. They burrow or settle into silt, or climb in vegetation. Development through 8 to 15 molts takes from a few months to as long as 5 years, transformation taking place above water on vegetation. Newly emerged adults are palid, weak *tenerals*. Until mature they hide in vegetation of fly far from water. Mature adults are vigorous, resplendent in color and true heliophiles, even a passing cloud causing them to seek shelter. Males, especially those which establish territories, exhibit a surprising curiosity, investigating every movement. Adults are unbelievably agile, attaining speeds to 60 miles per hour and, in some species, migrating long distances.

These superb predators are powerful determiners of the balance of insect life in all fresh waters. Nymphal food varies from Protozoa to tadpoles and fish fry, while adults capture any flying insect small enough to be overcome. Odonata in turn are preyed upon by other insects, spiders, frogs, fish, and aquatic birds. Their eggs are parasitized by curious swimming Hymenopterans, while adults bear Hydrachnid mites and Ceratopogonid flies. They serve also as second intermediate hosts of Amphibian and Avian flukes.

ROBERT W. ALRUTZ

References

Needham, James G. and H. B. Heywood, "A handbook of the dragonflies of North America," Springfield, Ill., Thomas, 1929.

Needham, James G. and M. V. Westfall, Jr., "A manual of the dragonflies of North America (Anisoptera)," Berkeley, University of California Press, 1955.

Walker, Edmund M., "The Odonata of Canada and Alaska. Vol. 1, Zygoptera. Vol. II, Anisoptera," Toronto, The University Press, 1953 and 1958.

OESTROUS CYCLE

Most mature female mammals experience recurring periods of sex desire commonly known as "heat" or *oestrus*. The period of the year during which the females of a species experience oestrus, and the males *rut*, is known as the *breeding season* (rutting season) of that species. The period of the year during which breeding, pregnancy, and lactation occur is the *reproductive season*. Under natural conditions the oestrous female has access to a male, so that *insemination*, *fertilization* and *pregnancy* ensue. The GESTATION period (pregnancy period) is terminated by *parturition* (birth). This is followed by a period of *lactation*, toward the end of which, in most species, a new oestrous occurs, thus marking the completion of a normal *reproductive* cycle or *pregnancy* cycle.

There are two common variations on this chain of events, neither of which is biologically ideal. The first follows a sterile mating in which either no fertilization occurs or the fertilized eggs die without establishing a placental relationship with the mother; the second variation occurs if the female has no access to a male during oestrus, a very common and usually planned situation in most domestic and laboratory animals.

Following sterile matings, females of many species enter upon a period of *pseudopregnancy*, during which they are in a physiological state somewhat resembling pregnancy.* This may last as long as a normal pregnancy as in the dog, but may also be much shorter, depending upon the species. Its end is signaled by the recurrence of oestrus. In other species a sterile mating does not result in a *pseudopregnancy cycle*, the reproductive sequelae being indistinguishable from those following oestrus without access to a male.

Commonly, when copulation is prevented, female mammals experience recurring periods of oestrus, usually at shorter intervals than would happen in the same species in the case of pregnancy or even of pseudopregnancy. Such cycles are the true *oestrous cycles*. When these have a periodicity of only a few days to three or four weeks, as in the rat (4 days) (fig. 2) and cow (21 days), the cycle consists of four major but integrading segments: 1. *oestrus*, the period during which positive external signs are generally most prominent, during which the female usually seeks out the male and is sought eagerly by the male, during which she accepts *copulation*, and during which she *ovulates*. 2. *metoestrus* (met = after), the "going out of heat,"

*Pseudopregnancy in some species is induced by sex play only, even if confined entirely to other females.

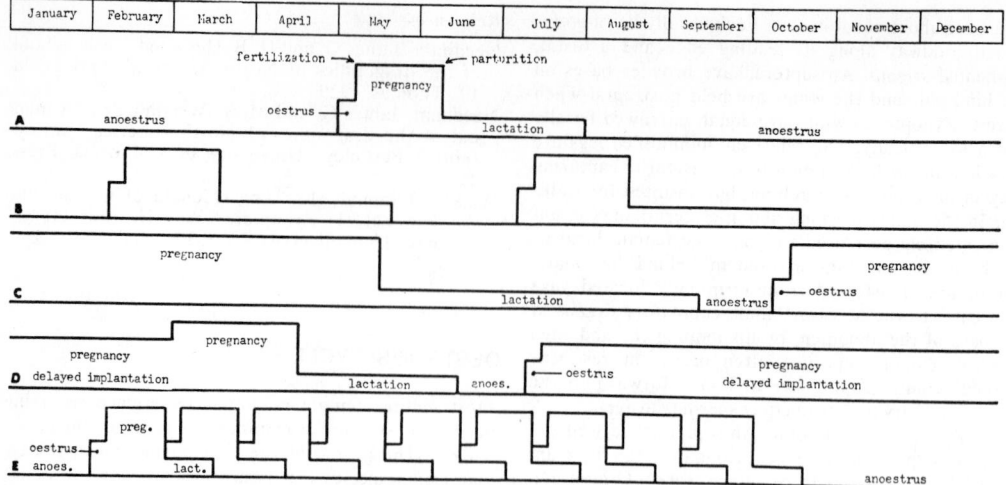

Fig. 1   Diagrams of some of the more common types of seasonal reproductive cycles. A, one short annual reproductive period; B, two short semiannual periods; C, one long annual period; D, one long annual period with very long phase of delayed implantation; E, one long annual period consisting of numerous repeated short pregnancy cycles. Based on the following examples: A, thirteen-striped ground squirrel, *Citellus tridecem-lineatus*, in southern Wisconsin; B, tree squirrels, *Sciurus* and *Tamiasciurus*, in southern Wisconsin; C, Canadian porcupine, *Erethizon dorsatus*, in northern Wisconsin; D, weasels, *Mustela frenata* and *erminea* in Montana; E, voles, *Microtus*, in north central United States.

a relatively brief period when behavioral, physiological and anatomical signs of oestrus are rapidly subsiding. 3. *dioestrus* (di = between), usually the longest segment, during which there is little or no manifestation of sexual activity, anatomically or functionally. This is the "resting" or "recuperative" phase. 4. *proestrus*, the "coming in heat," a short phase during which a set of eggs is rapidly maturing in the ovary, and the total pattern of full oestrus is obviously developing at a fast pace.

In some species, *monoestrous* forms, in contrast to the *polyoestrous* ones just described, only a single oestrus is experienced at one breeding season, and several weeks or months intervene before another occurs, even if no copulation has taken place. Dogs are commonly known examples of monoestrous animals. In these the proestrous, oestrous and metoestrous phases of the cycle are essentially the same as in polyoestrous forms, although usually each is somewhat more prolonged. The resting period is, however, far longer, and is usually called *anoestrous* (an = without) instead of dioestrus. During anoestrus the reproductive tract of these forms is truly quiescent for a long period of time, whereas in polyoestrous species dioestrus can best be considered a short reparative stage. A cycle including the long anoestrous period is often called an *anoestrous cycle*, in contrast to a *dioestrous cycle* characterized by only the short dioestrous period.

The distinction between polyoestrous and monoestrous forms is not as clear-cut as the above might imply. Many polyoestrous species, in fact probably most of them living under natural rather than domesticated conditions, have one or more periods during the year when their oestrous cycles cease for several weeks, i.e. they enter upon an anoestrous period. Wild mice (fig. 1) and rabbits are good examples of this group. Then too, many species which are usually considered

monoestrous, because they have a relatively restricted breeding season, are actually capable of experiencing two or three closely repeated oestrous cycles during any one breeding season, if they do not find a mate, or if a mating is infertile and pseudopregnancy does not result. The ground squirrels are examples of this group. Thus we have truly polyoestrous strains (possibly only under conditions of domestication); polyoestrous species most of which enter into an anoestrous period during at least a part of the year; monoestrous species which are capable of a few dioestrous cycles during the breeding season; and truly monoestrous mammals which have only anoestrous cycles.

The reproductive cycle in woman is basically the same as that of a polyoestrous mammal. However, the period of uterine hemorrhage, *menstruation*, is conspicuous, while there are usually no obvious physical signs of the period of oestrus. Therefore we usually reckon the sequences of the reproductive cycle in woman and the higher primates (which also have a conspicuous menstruation) in terms of the *menstrual cycle* rather than of the oestrous cycle. The menstrual period corresponds roughly to the late dioestrous period of an oestrous cycle. Ovulation in woman, therefore, normally occurs about midway in a 28-day menstrual cycle, but it is more accurate to say that ovulation occurs about 14 days before the onset of the next menstruation, even though the length of the menstrual cycle in the particular individual may be longer or shorter than 28 days.

It is obvious, as was stated above, that the most biologically desirable is not the oestrous or pseudopregnancy cycle, but the pregnancy cycle, and that the early recurrence of oestrus, if fertilization fails, is an adaptation which increases the possibility of successful breeding at a favorable time of year, i.e. during the normal *breeding season*. Definite seasonal breeding

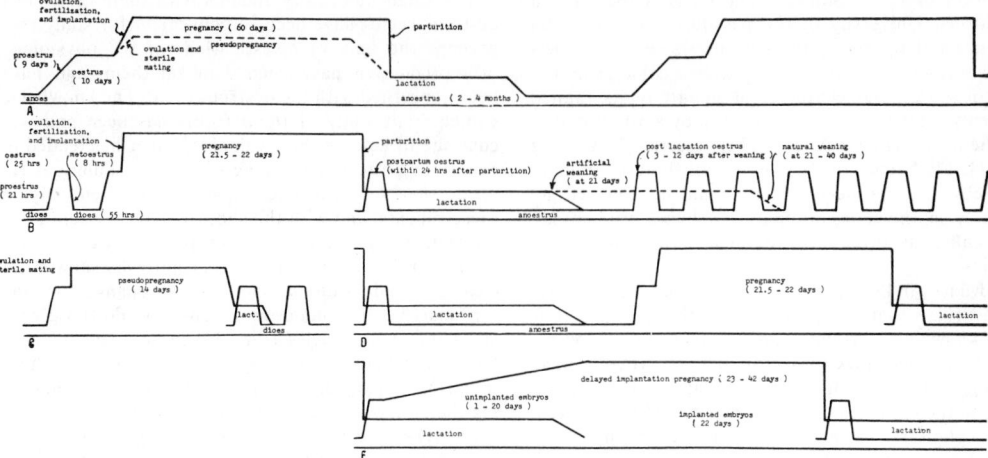

Fig. 2   Reproductive cycle of a monoestrous animal (dog) compared with that of a polyoestrous animal (rat). A, dog. Proestrus and oestrus are relatively long and pregnancy short. Pseudopregnancy is approximately as long as pregnancy. The next oestrus occurs at least two months after parturition, and lactation may continue to that time. If mating is prevented anoestrus lasts about 5 months. B, rat. Proestrus and oestrus are short and pregnancy relatively long. There is an oestrus immediately after parturition followed by an anoestrous phase during most of the lactation period. Toward the end of lactation, or soon after artificial weaning, oestrous cycles are resumed, and continue as long as age and nutritive conditions are favorable, or until another mating. C, rat. Sterile mating results in a period of pseudopregnancy which is shorter than pregnancy and is followed by a brief lactation, during which regular oestrous cycles are quickly resumed. D, rat. Poor nutritive or health conditions may result in an anoestrus of variable length, after which breeding may be again entirely normal. E, rat. Pregnancy resulting from breeding at the "postpartum" oestrus is prolonged by delay of implantation for from one to twenty days, the delay being roughly in direct proportion to the number of nursing young, and in indirect proportion to the general well-being and nutrition of the female.

periods during which the females experience oestrus and during which the males are in *rut*, are characteristic of mammals in all parts of the world, except where climatic conditions are very constant, as in some tropical and essentially "oceanic" climates. The time of year at which these breeding seasons occur is correlated less with favorable environmental conditions for breeding, than with a favorable season for birth and care of the young. For instance, in the northern hemisphere wild rats and mice litter throughout the spring and summer. Most of these have a gestation period of only three to four weeks, and the young grow rapidly and become self-sufficient within a month. However, one large rodent, the Canadian porcupine, has a seven month gestation. This species, like the northern deer, breeds in early autumn, apparently an adaptation to birth of its bulky and relatively mature young in the spring when succulent vegetation is available.

This apparent adaptation to favorable seasons has led to much speculation and investigation as to the possible control of breeding cycles by environment. It is now known that relative length of periods of light to darkness is a triggering factor in many species, but just as in plants where some flower with increasing light and others with decreasing light, so some mammals are spring breeders, some fall breeders, and a few breed at both periods (see PHOTOPERIODICITY). As examples of the latter are many North American tree squirrels which have one peak of breeding activity in January, February, and March, i.e. early in the period of increasing day lengths, and a second peak in

July and August, early in the period of decreasing day lengths. This situation, and many other similar ones, indicate that light factors are not themselves specific stimuli to breeding, but rather that through natural selection species have "used" light factors in "setting" their "physiological clocks" to the most favorable times for reproduction.

While many of the facts recounted above have been observed since prehistoric times, and are well-known among primitive cultures and among laymen today, relatively little progress in understanding or controlling the physiologic processes involved was made until this century. In fact the publication by Walter Heape in 1900 of an article entitled, The "Sexual Season" of Mammals and the Relation of the "Pro-oestrum" to Menstruation, gave the first great impetus to these studies, largely because it clearly defined a logical terminology of the subject, including for the first time the modern usage of the terms oestrus, metoestrus, dioestrus, anoestrus, and proestrus. Then in 1910 the first edition of Marshall's "The Physiology of Reproduction" appeared. This excellent compilation and synthesis of pertinent literature established the subject on a broad comparative basis. Also the importance of the endocrine system in reproduction was widely recognized during this decade, Fraenkel and Cohn (1903) and others having shown the *corpus luteum* of the ovary to be the "pregnancy gland."

As in all science, new advances usually follow upon the development of a new technique, so a second acceleration in investigations of reproductive physiology began in 1917 as the result of the relatively simple discovery by C. R. Stockard and G. N. Papanicolaou at Cornell University of the possibility of easily and accurately determining the various stages of the oestrous cycle of the guinea pig by microscopic examination of the cellular character of smears of its vaginal contents. This was followed quickly by similar studies on the mouse (Edgar Allen, 1922), and rat (J. A. Long and H. M. Evans, 1922), thus establishing techniques whereby the reproductive cycle could be followed accurately in the living animal, and hence studied scientifically, in three of the most useful laboratory species.

Advances in knowledge of female reproductive physiology since that time have been phenomenal. It is now known that visual stimuli affect the endocrine secretions (see HORMONE) of the PITUITARY gland through neurosecretory activity along the visual pathways in the *thalamus* of the forebrain, and that at least three pituitary hormones are carried in the blood stream to the ovaries where they (1) control growth and maturation of the eggs and their follicles (*follicle stimulating* hormone, or FSH), (2) stimulate formation of the corpus luteum from the cells of the ovulated follicles (*luteinizing* hormone, or LH), and (3) maintain the corpus luteum during the first stages of pregnancy (*luteotrophic* hormone, or LTH). However, LTH has been demonstrated only in the rat and mouse. It is also known that *oestrogenic* hormones produced by the ripening egg-containing ovarian follicles stimulate growth of the mucous lining of the uterus and bring about other physical as well as behavioral changes that characterize proestrus and oestrus. The cyclic nature of these phenomena is provided, in part at least, by the principle of reciprocal inhibition and stimulation, sometimes called the "push pull" or "feed-back" mechanism. For example, as the follicular hormone is secreted and increases in the circulating blood, it of course reaches the pituitary and hypothalamus, and at a certain level it begins to inhibit the secretion of the follicle stimulating hormone (FSH). Thus the production of ripening follicles in the ovary ceases for lack of stimulus and does not increase again until the ripe follicles have ovulated and ceased to provide FSH inhibiting oestrogens. Also, as the secretion of FSH decreases, the pituitary increases its output of luteinizing hormone (LH). This helps to cause the final maturation and the ovulation (rupture) of the ripening follicles, while after ovulation it is the chief stimulus to the formation of luteal gland (corpus luteum) tissue from the cells of the ovulated follicles. As these corpora lutea increase their output of *progesterone*, which prepares the uterus for reception and attachment of the embryos, this hormone at the same time inhibits LH production by the pituitary, just as oestrogens inhibit FSH production. If pregnancy does not ensue, the stage is now set for a renewal of the production of FSH by the pituitary, and thus the repetition of an oestrous cycle. If, instead, pregnancy does occur, then the physiological mechanisms become more complex, but need not be considered in detail here; yet a clue to their understanding lies in the fact that the placenta of some species, at least, takes over some of the endocrine functions of pituitary and ovaries, secreting both ovary stimulating hormones (*gonadotrophins*) resembling LH and LTH, and sex hormones similar to those formed in the ovary (progesterone and oestrogens).

The small laboratory rodents with their very short oestrous cycles have been so convenient for study that perhaps the bulk of investigations of the physiology of reproduction have been done on them, and have been concerned with the oestrous cycle. The knowledge gained from study of these forms has been and will continue to be extremely valuable, but it has tended to equate much of our thinking on the physiology of reproduction with the phenomena known well only in these rodents, and with the short oestrous cycle, rather than with the far more important pregnancy cycle. Animals with long anoestrous cycles (squirrels) show more clear-cut and profound anatomical changes in the reproductive and endocrine organs than do those with short dioestrous cycles (rat). Likewise, even in these latter, it is much easier to correlate anatomical changes in various organs during the pregnancy cycle than during their very short oestrous cycles when these changes are markedly compressed and overlapping. The same may also be said for the concomitant cyclic physiological changes. It seems certain that the common denominators or basic mechanisms in reproductive anatomy, physiology, and behavior will not be fully evident until many more data are obtained on a broad comparative spectrum of species.

What is the relation of all this to problems of human reproduction? This is certainly a question of paramount importance to the world. It is not just a matter of such immediate interests as the safe period, sterility, or sexual compatability; it relates directly to problems of far greater importance, such as the whole mechanism of sex determination and differentiation; environmental and psychological influences on normal sexual rhythms; the physiological well-being of mother and child during pregnancy and lactation; the chemistry of hormone production and of the action of hormones on the target organs; the mechanisms of physiologic integration of organ function; the evolution of human sex mores; and above all the problems of the means and motivation for human population control.

H. W. MOSSMAN

References

Asdell, S. A., "Patterns of Mammalian Reproduction," Ithaca Comstock Pub. Co., 1946.
Marshall, F. H. A., "Physiology of Reproduction," 3rd ed., 3 vols. edited by A. S. Parkes, Boston, Little Brown, 1956–1966.

OKEN, LORENZ (1779–1851)

This German naturalist, whose real name was Ockenfuss, was born on August 1, 1779, at Bohlsbach, Baden. He was educated at Würzburg and Göttingen and became Privatdozent at the latter University. In 1807 he began his appointment as professor extraordinarius of medical sciences at Jena with an inaugural address on the significance of the bones of the skull. In 1816 he began the publication at Weimar of the periodical *Isis, eine Encyclopädische Zeitschrift, vorzüglich für Naturgeschichte, vergleichende Anatomie und Physiologie.* His political writings in this periodical led to his resignation from Weimar, however, he continued to publish *Isis* at Rudolstadt until 1848.

Oken succeeded in establishing annual meetings of German naturalists and medical practitioners. The structure of his scientific society was used as the model for the British Association for the Advancement of Science.

The works of Oken are illustrations of his theory that the head is a modified vertebral column. He is remembered for his re-classification of the animal and mineral worlds.

In 1828 Oken returned to teaching at the University of Munich, where he soon became professor. In 1833 he was appointed professor of natural history at Zurich, where he remained until his death on August 11, 1851.

JAMES L. OSCHMAN

OLIGOCHAETA

A class of the Annelida defined as follows: Setiferous worms, lacking parapodia; with internal and external segmentation generally corresponding; hermaphroditic, the gonads of both sexes normally reduced to one or two pairs; special ducts are present to carry the sex products to the exterior; at full sexual maturity, a clitellum is present and this produces a capsule or öotheca in which the eggs undergo direct development without a free-living stage.

No significant geological record of the Oligochaeta is known and, although correlations with the Polychaeta and Hirudinea are indicated, both in embryonic development and adult characteristics, the phylogenetic line is not clear. Ordinal divisions have been erected as follows:

1. Plesiopora plesiotheca: with male pore opening in the segment immediately posterior to the testicular somite; the spermathecae are located in the same segment as the testes.
2. Plesiopora prosotheca: with male pore opening in the segment immediately posterior to testicular segment; the spermathecae are located several segments anterior to the one containing the testes.
3. Prosopora: with sperm duct not passing through posterior septum of segment containing sperm funnel and so male pore is located in testicular segment(s).
4. Opisthopora: with sperm duct opening some distance behind testicular segment or at least the length of a segment behind the male funnel.

Numerous qualifications are necessary to make these divisions include all known forms. Over 3,000 species of oligochaetes are grouped into 14 families; these family groups are also quite artificial and have been changed frequently. Many genera are poorly defined due mainly to the fact that characteristics such as the reproductive organs have, in the past, been presumed to be non-adaptive and hence were used extensively in generic diagnoses. Convergence has been shown for many of these structures thus reducing the efficiency of the present systems.

The Oligochaeta are nearly ubiquitous, occupying terrestrial and aquatic habitats on all continents and major island groups. With the exception of a very few littoral species, they have been unable to invade the oceanic realm. Earthworms are often considered rather sedentary but they do have substantial powers of dispersal and much of their widespread occurrence should be attributed to natural migration of the fauna rather than importation by man as has been so often stated. The nephridial-type excretory unit, together with the absence of a resistant exoskeleton, places even the hardiest soil species among the quasi-terrestrial animals which require that free water be available to them. The tubiculous habit is general for the group. Terrestrial worms excavate tunnels which can extend to a depth of several feet, or, in the case of inhabitants of surface leaf-litter, channels can be made from excreta and debris. A few species are known which live above the ground, usually associated with epiphytic plants. Aquatic oligochaetes burrow in bottom muds, build surface tubes of mucous and debris, or, smaller forms may wander freely over surfaces of aquatic vegetation.

Nutritional requirements of oligochaetes have been categorized generally as organic material of the soil and more specifically as plant residues. Precise trophic levels are unknown. Some earthworms may range through the soil mass, apparently surviving on consolidated organic elements; other species seem to use the burrow merely as a retreat from which they can extend their bodies over the soil surface to forage for plant remains. The nature, amount, and condition of the organic material, together with temperature and moisture factors, determine habitat occupancy at local, and probably regional, levels. Parasitism by oligochaetes is uncommon. The Branchiobdellidae are ectoparasites of crayfish. A number of commensals are known, primarily among the Naididae, most of the associations being with Porifera, Polyzoa, or Gastropoda.

Reproductive mechanisms are varied; in addition to the biparental sexual process, parthenogenesis is not unusual. Many of the aquatic species reproduce asexually, either through intercalation of new segments in the middle of the worm with subsequent fission, or by fragmentation followed by terminal regeneration.

WILLIAM R. MURCHIE

References

Stephenson, J., "The Oligochaeta," Oxford, Clarendon Press, 1930.
Lavarack, M. S., "The Physiology of Earthworms," New York, Macmillan, 1963.

ONYCHOPHORA

Onychophora comprise a small but unusually interesting group of animals of the phylum ARTHROPODA, at one time thought perhaps to be intermediate between the annelid worms and the arthropods. The Onychophora constitute a class of their own, and comprise few genera with but 100–200 species. There is little range of form within the Onychophora, they exhibit some primitive arthropodan features and some highly specialized ones, and their discontinuous distribution, scattered over all of the southern continents with a few species spreading into the Malay Archipelago

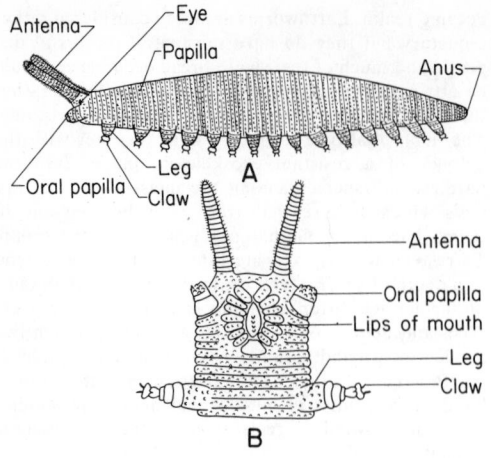

Fig. 1   The onychophoran *Peripatus*. A, lateral view of entire animal; B, ventral view of anterior end.

and India, is suggestive of survival of a once more widespread group of animals.

The fossil remains of arthropods are recognisable as far back in geological time as the Cambrian era, but even then various classes of marine Arthropoda were well differentiated. Many millions of years later the land became habitable to animals. An apparently marine deposit has yielded the middle Cambrian *Aysheaia*, a creature remarkably like the present day Onychophora in general body form, skin texture and limbs, but its head is ill preserved. It is thus possible that the Onychophora are a very ancient group. Comparative anatomy, embryology and experimental work, moreover, suggest that the Onychophora are persistent specialised relicts of an early land stock of animals which also gave rise to the modern pauropods, millipedes, centipedes, Symphyla and insects.

Peripatus is a typical onychophoran genus, but also a term in common use to signify any onychophoran species. In size the Onychophora range from one to several inches in length. They are elongated, with a dry velvety skin, and some 14–44 trunk segments each bear a pair of short legs, the number varying in the different species. The head is provided with: one pair of preoral antennae; a pair of jaws, each rather like a short wide leg with a much enlarged terminal paired claw; jaws and labrum are surrounded by a round lip; and a pair of short limbs, the oral papillae, are borne on the third head somite. They discharge a milky looking fluid as a jet which sets at once in contact with air. The glands supplying the fluid extend far back into the body, their branches reaching the posterior end. The ejection of "slime" represents the animal's only defence against predators; sizeable arthropods can be immobilised by the sticky threads. Onychophora are carnivorous, eating other small arthropods. No external demarcation of segmental boundaries is seen, because the cuticle is very thin, although sclerotized externally as in other arthropods. Bending can occur anywhere, and so no localised joints are needed, as found in the majority of arthropods bearing typical scutes.

The cuticle is furrowed and raised into innumerable papillae, each terminating in an elaborate sensory spine. The sub-ectodermal body wall is composed of a thick layer of connective tissue fibres; on to them are anchored the muscle fibres, circular, oblique, longitudinal, and deep dorsoventral, all of which are unstriated.

A typical arthropodan haemocoelic body cavity bathes the organs in blood, which is circulated by a dorsal heart situated above a pericardial floor.

Typical arthropodan "segmental organs" are present in most segments, consisting of an end-sac, representing the remains of the large segmental embryonic coelom which is like that of coelomate worms, and a duct passing to the exterior at the base of each leg. The segmental organs of the jaw segment are enlarged to form the salivary glands. In other arthropods, such organs are restricted in number, being present on some segments only.

Aerial respiratory organs are present in the form of unbranching tracheae, fine tubes carrying air from a number of spiracular depressions in the ectoderm of every segment to the various organs. In structure the tracheal system differs from that of other arthropods, but tracheae have undoubtedly be evolved several times in association with the land habit.

The alimentary canal is simple, with short fore- and hind-gut lined with chitin, and a long, ample mid-gut which serves for digestion, absorption of food and the daily excretion of uric acid crystals. A peritrophic membrane occurs as in some other arthropods, but its manner of formation and removal is unique.

The nervous system consists of paired ventral nerve cords enlarged in every segment to form a pair of ganglia from which segmental nerves pass out to the organs. At the head end a dorsal "brain" supplies the antennae and eyes and circumoesophageal commissures continue to the ventral chain. The paired eyes near the base of the antennae possess a lens and a simple retina; they are responsive to the changes in light intensity, but do not appear on record an acute image. Highly sensory taste spines lie on the lips and in the preoral cavity; the spines on the surface papillae, each with its own nerve and sense capsule, are responsive to air movements, vibrations and tactile stimuli.

The sexes are separate, and the organs paired, opening by a median pore near the posterior end of the body. Copulation occurs in the Australian species, spermatophores being deposited in the female genital tract. In the African species of *Peripatopsis*, spermatophores are deposited anywhere on the surface of the body, and sperms pass through the ectoderm and tissues to reach the ovary. Fertilisation is internal. A few species lay heavily yolked, shelled eggs, others retain a yolked or yolkless egg in the oviduct until the hatching of a miniature adult, pregnancy lasting 13 months. Growth is slow, four years or more being needed to reach full size in African species, and increase in size is intermittent, as in all arthropods, occurring just after the moult.

Onychophora are restricted to damp environments, where they find shelter in decaying logs, clefts in the ground, etc., but sodden conditions are intolerable. Onychophora cannot control water loss from their innumerable spiracles, and have no need to do so in their normal surroundings. Onychophora walk abroad at night, seeking their small prey, mates and fresh shelter, and can withstand a three months fast. The habit of paramount importance in the whole evolution of the Onychophora appears to be their extreme distortability of body shape, which enables them to squeeze through cracks, such as a small hole but 1/9th of their resting transverse sectional area. Thereby they can creep into places whose narrow access proves an insuperable barrier to carnivorous scute-bearing arthropods large enough to harm them. This extreme deformability of body far exceeds that of other arthropods and is made possible by: the furrowed flexible cuticle providing ample "slack" in all directions; the connective tissue "skeleton" which changes shape without stretching; by the unstriated muscles which allow great deformability, and by the compact jaws which slice backwards and not sideways, and by a lack of median tendon between adductor muscles so characteristic of many Arthropoda. All these features, at one time supposed to be annelidan characters, are clearly specialisations of the Onychophora essential for behaviour patterns of survival value.

Onychophora are slow in walking, and their limb movements display a simpler range of gaits than any other arthropod. These gaits, moreover, can be recognised in more specialised forms in the members of the several land classes, each of which has perfected more restricted types of gait in association with facilitating morphological features.

The embryonic development of the Onychophora shows many specialisations, but the basic pattern of development is that of an arthropod. The organisation of the head stands at a more primitive level than that of all other arthropods, but the head is clearly arthropodan in type and unlike that of annelids in both structure and development. There are many embryonic and other resemblances between Onychophora and PAUROPODA, DIPLOPODA, CHILOPODA and SYMPHYLA which suggest that the Onychophora, together with these classes and the insects, represent a monophyletic line of arthropod evolution quite separate from that of the primarily aquatic CRUSTACEA, Merostomata and TRILOBITA, but this does not imply a close unity between the last three classes.

S. M. MANTON

References

Bouvier, E. L. (1905, 1907), Monographie des Ony-chophores. *Ann. Sci. nat. Zool. Paris,* Ser. 9, **2** and **5**.
Zacker, F. (1933), Onychophora in Kükenthal und Krumbach, Handbuch der Zoologie, 3, 2, 4.
Manton, S. M. (1937), The Feeding, digestion, excretion and food storage of *Peripatopsis. Phil. Trans. B.,* **227**.
(1949) The early embryonic stages of *peripatopsis,* and some general considerations concerning the morphology and phylogeny of the Arthropoda. *Phil. Trans. B.,* **233**.
(1950) The locomotion of Peripatus. *J. Linn. Soc. (Zool.),* **41.**
(1958) Habits and evolution of body design in arthropods. *J. Linn. Soc. (Zool.),* **44.**
Tiegs, O. W. and S. M. Manton (1958) The evolution of the Arthropoda. *Biol. Rev.,* **33.**

OPHIDIA

(NL, pl. of *ophidium,* from Gr. ỏφιδιθγ, diminutive in form but not in sense of ὄφισ, a serpent.) a common but nowadays taxonomically inacceptable name for Serpentes, a suborder of the order Squamata of the class REPTILIA. Proposed in 1800 by Brongniart, the name has been replaced by common consent (no mandatory provisions existing in rules of nomenclature for names of taxa above the family level) with Linnaeus' name *Serpentes* of 1758.

Ophidians are related to and derived from lizards (suborder Sauria), first appearing in the upper Cretaceous period whereas lizards are known from as early as the upper Jurassic. Many lizards have paralleled snakes in departure from basic lizard structure, becoming for example limbless and earless, but snakes have acquired many more specializations than any one group of lizards. No one characteristic is unique to either snakes or lizards, but the following combination is unique to snakes (from Dowling): no pectoral girdle or limbs; no moveable eyelids; no external auditory meatus, middle ear cavity of tympanum; no upper temporal arch; an elongate, freely articulated supratemporal; numerous skull bones absent, *e.g.* squamosal, lacrimal, jugal, quadratojugal, epipterygoid, postfrontal; braincase completely enclosed by descending processes of frontals and parietals; vertebrae with zygosphenes and zygantra; halves of caudal hypapophyses disjunct midventrally; a ligamentous union of halves of lower jaw at chin.

Some authors recognize a third suborder of Squamata, the AMPHISBAENIA, a group otherwise placed with lizards in the suborder Sauria.

According to one view, primitive snakes were specialized burrowers, whereas according to the other view they were surface-dwelling or aquatic boa-like snakes from which specialized burrowers evolved in one line and fast moving terrestrial and arboreal types in another. Evidence at present at hand in not conclusive.

About 3000 species and subspecies of snakes range from southern Canada, northern Sweden and the Kamchatka Peninsula southward throughout Africa and most of the Americas, Australia, and temperate

Fig. 1   Skeleton of a snake (gaboon viper). (From Bogert.)

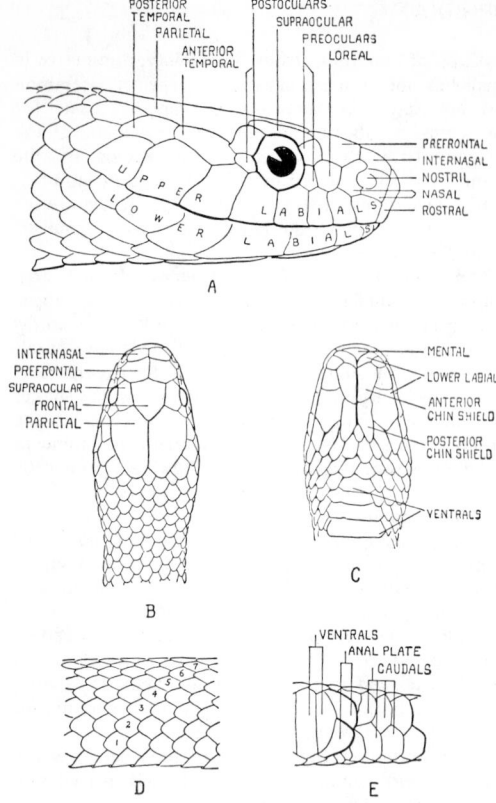

Fig. 2   Scale terminology applicable to most snakes, based upon the ringneck snake. (From Blanchard.)

and tropical islands except Polynesia. Only one tiny species occurs in Hawaii, none in Ireland or New Zealand. Islands have been populated by snakes from adjacent mainlands, either before separation from the mainland or by subsequent chance. Ireland lacks them because it was separated from continental contact before the last glacier departed, and no introduction has since occurred.

There are at least nine, and perhaps as many as fifteen families, the number variously estimated in accordance with degree of significance attached to specializations of diverse groups. Three super-families recognized are the **Leptotyphlopoidea**, containing the mostly tiny burrowing snakes of the families *Leptotyphlopidae, Typhlopidae* and *Anomalepidae*; the **Booidea**, containing the large constrictors (boas, pythons) of the family *Boidae*, as well as the smaller snakes of the families *Uropeltidae, Aniliidae* and *Xenopeltidae*; and the **Colubroidea**, including the family *Colubridae* which contains 75 per cent of the snakes of the world; the harmless Oriental aquatic trunk snakes of the family *Acrochordidae*; and the four families exclusively of venomous snakes. These four families are the *Hydrophiidae* (sea snakes), *Elapidae* (coral snakes, cobras, kraits, mambas, etc.), *Viperidae* (pitless vipers, strictly Old World: horned viper, rhinoceros viper, puff adder, Russell's viper, etc.); and *Crotalidae* (pit vipers, mostly New World:

rattlesnakes, cottonmouth, copperhead, fer-de-lance, bushmaster, etc.).

"Pit" vipers are members of the family Crotalidae, of both hemispheres but primarily American. This group includes all American venomous species except the coral snakes and Pacific sea snake. Snakes of this group have a deep facial pit, between eye and nostril, functioning as a delicate detector of heat. Some boas possess a series of pits, also heat detectors, along both upper and lower lips.

The rattle of rattlesnakes consists of several lobed and interlocked segments of thin keratin, one segment added at each moult. In rattlesnakes 2–4 moults occur each year, and a corresponding number of rattles is added. When first born the rattlesnake has a "button" at tip of tail, and moults a few days after birth, adding one segment to the rattle.

The most primitive families of snakes possess internal vestiges of the skeletal girdle of the hind limbs, and in some a vestige of the thigh (femur) remains, often visible externally as a small "spur" on either side of anal opening.

No snakes possess movable eyelids. The skin over the eye is in the form of a large scale flush with the rest of the cranial surface in the most primitive snakes. In more advanced forms the scale fits the eye perfectly, and is designated a "brille" or tertiary spectacle. Snakes sleep, but little outwardly visible evidence reveals when they are asleep.

The ears cannot directly receive air-borne sounds, since the external ear and middle ear cavity are missing. The middle ear ossicle or columella is present but is imbedded in muscle except where it fits into the *fenestra ovalis* of the capsule of the inner ear. Snakes can hear some ground-borne vibrations, since the inner ear is largely intact. They are deaf to music of snake-charmers and to their own noises.

The tongue in all snakes is long, slender and forked at the tip. It functions primarily in picking up air-borne particles that go into solution on its moist surface; by placing the tips of the tongue into the paired vomeronasal organs at the anterior tip of roof of mouth, the snake detects odors.

Snakes periodically moult the outer epidermal layers of the entire skin, including eye covering and tips of tongue. Frequency varies with species and health, is never less than once a year, and commonly is three times a year with more frequent occurrence after injury. The moult usually proceeds posteriorly starting at the snout and lower jaw where the skin is rubbed backward over the head, turning it wrong-side out and leaving it usually in one piece.

In preparation for the moult the basal layer of epidermis rapidly proliferates, producing a completely new outer epidermis, with cuboidal and squamous cells. Breakdown of the basal cuboidal cells of the older layer froms a milky fluid, giving a dull cast to the snake as a whole and a milky appearance to the eye; at this time snakes can see but dimly if at all, tend to remain inactive and hidden from view, and are likely to be on the defensive when disturbed. The fluid disappears by evaporation or absorption after about three days, and about two days thereafter the slough is shed.

Most snakes lay eggs. Some boas, most solenoglyph snakes and some common harmless varieties (*e.g.* watersnakes, gartersnakes) give birth to their young, numbering as many as 101. No parental care ordinarily

Fig. 3   A timber rattlesnake in the process of moulting. Note deep round pit in front of eye, rattle on tail, and keeled scales. (From a color painting in Life.)

occurs. The eggs are usually abandoned after deposition in damp sand, soil, sawdust, ground litter or rotten logs. They hatch in about three days to three months. Some viviparous types possess a placenta.

Like all other reptiles, snakes are ectothermic, their body temperatures being controlled largely by behavior and to only a very slight degree by intrinsic mechanisms. Air and substrate temperatures absolutely determine body temperatures except as (1) insolation or (2) reproductive hormones increase them, or (3) evaporation decreases them.

Many other general bodily adaptations to the mode

Fig. 4   A combat "dance" between two male copperheads. Note facial pit, large scales on belly, vertical pupil, and keeled scales. (From a color painting in Life.)

of life and limblessness of snakes occur, involving every system (see Bellairs).

Males court by repeatedly rubbing chin and body along length of the female's body. In rare circumstances the two animals intertwine the bodies and rise straight upward until only tail and rear part of body remain on the ground. Ordinarily the latter behavior is a "combat dance" between males, initiated by competition for mates. Territoriality is poorly developed. Home ranges vary in size more or less with size of snake.

Snakes, like lizards, possess two copulatory organs (hemipenes). Copulation occurs with either and with both structures, but not with both at once, and persists for long periods, at least ½ to 2 hours.

All snakes have numerous well-developed salivary glands the secretions of which have been shown experimentally (*fide Physalix*) to be powerfully toxic in almost all species. Yet the bites of most snakes produce no signs or symptoms of venom because of the absence of venom-conducting fangs.

"Non-poisonous" snakes, defined as species whose bites never evoke signs or symptoms of envenomation (despite proven presence of toxic saliva in many species), have only "solid" teeth, lacking a groove down one side. All "poisonous" species possess one to four pairs of fangs. Fangs are teeth specialized for injection of venom, having either a tubelike structure acting like a hypodermic needle, or a groove down one side conducting venom like a capillary tube when imbedded in a victim's flesh.

All snakes have teeth, usually all conical, slender, curved, of about the same size and shape. They are found on five paired jaw bones: the maxilla, palatine, pterygoid, dentary and rarely the premaxilla, and they vary in total number from about 20 to 200. Three sorts of fang modifications evolved. The least advanced type (*opisthoglyph*), unique to the family Colubridae, has one to four rear teeth on the maxilla enlarged and grooved medially or anteriorly. All opisthoglyphs or *rear-fanged* snakes are but very mildly poisonous to man, producing no reaction at all or only a bee-sting like reaction, except for a few deadly South African species such as the boomslang. In the *solenoglyph* families Crotalidae and Viperidae, with *movable front fangs*, two rear teeth on each maxilla were retained and became hollow; at the same time the maxilla lost its anterior parts and moved to the front of the mouth, becoming capable of rotating on a transverse axis so that the fangs could be directed either straight downward or straight back in the mouth. In another line of evolution, toward the *proteroglyph* families Elapidae and Hydrophiidae, with *fixed front fangs*, the maxilla become shortened from the rear instead of from the front but did not become mobile; it developed one pair of anterior hollow fangs (rarely followed by a few small solid teeth). In both types of front-fanged snakes usually but one fang is in position on each side at a time; the fangs are shed at frequent intervals, however, and the new fang always comes into position in the vacant fang socket; the old fang drops from its socket which remains vacant until the other fang on the same side of the mouth is to be replaced. A duct from the poison gland is connected with each fang socket so that venom injection is never impaired during fang replacement.

Snakes possess venoms containing numerous protein enzymes capable of promoting various types of tissue

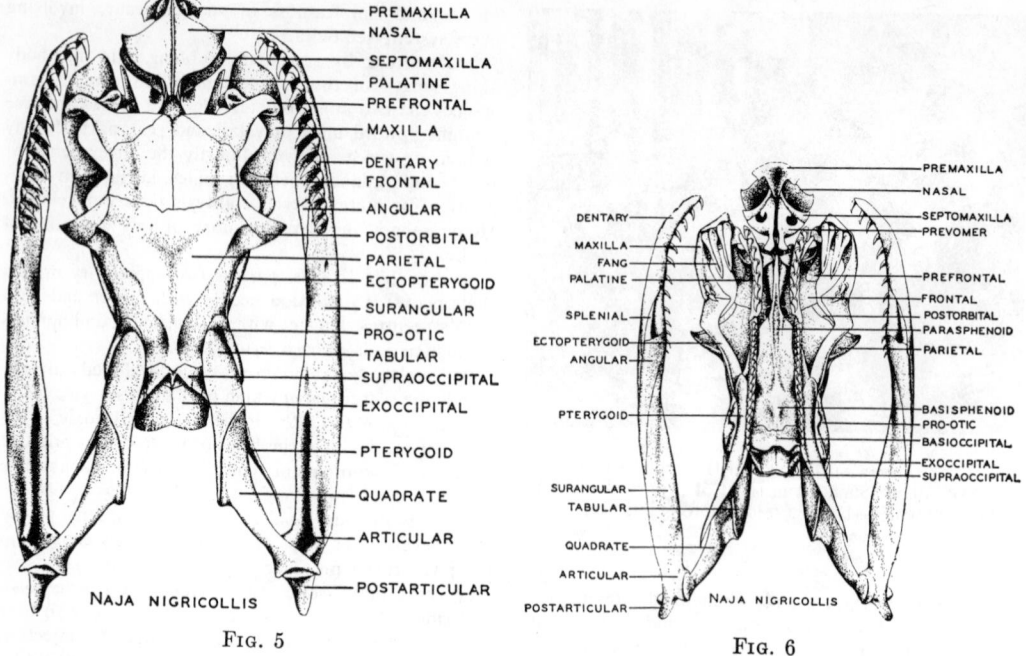

FIG. 5

FIG. 6

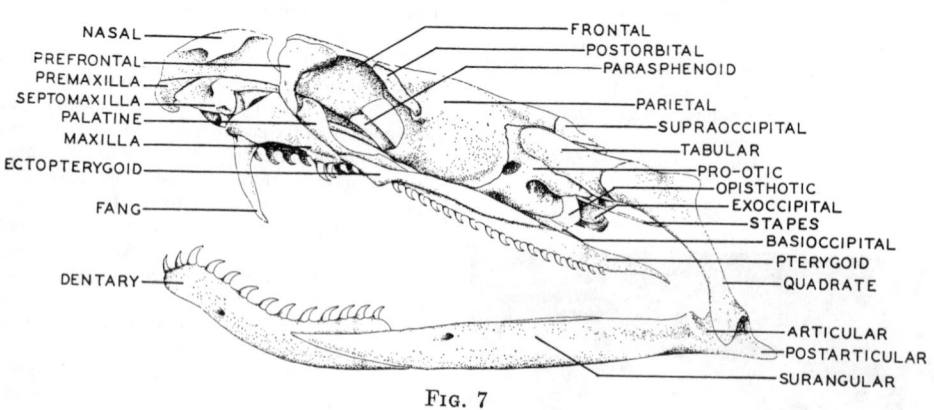

FIG. 7

Figs. 5–7 Skull of a cobra in dorsal, ventral and lateral views, showing fixed front fangs and identities of most bones. (From Bogert.)

degeneration. Those in which the enzymes affect mainly nervous tissues are *neurotoxic*, whereas those affecting primarily the circulatory system are *hemopathic*. No venoms are wholly neurotoxic or wholly hemopathic in action, yet most have a marked preponderance of one or the other type of action. Solenoglyphs are mostly hemopathic, proteroglyphs neurotoxic.

Death from snakebite in man may be virtually instantaneous or delayed two or three weeks. First aid treatment consists of use of a lymphatic tourniquet, incision in the form of short shallow longitudinal slits, and suction from the incisions. Medical treatment is of three sorts: *mechanical* treatment (tourniquet,

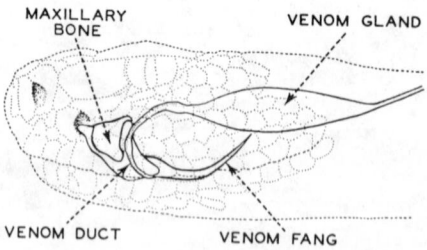

Fig. 8 Head of a rattlesnake in lateral view, showing poison apparatus in the perspective of external features, with mouth closed. (From Klauber.)

Fig. 9   Head of a striking rattlesnake, with fangs in "erect" position. (From Pope.)

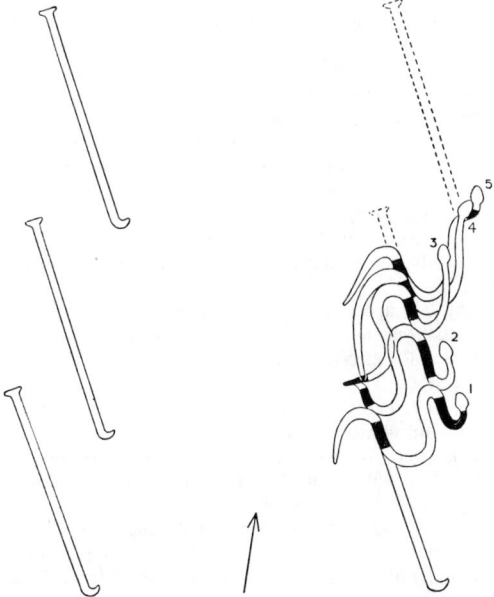

Fig. 10   Locomotion of a sidewinder rattlesnake. The track is shown at left, the direction of movement by the arrow, and at right a snake in the process of making tracks in successive stages 1–5. (From Oliver.)

incision, suction—an extension of first aid treatment); *antivenin*, prepared from blood serum of horses rendered immune to venoms of particular species of snakes; and *miscellaneous devices* such as calcium gluconate and local hypothermia (cryotherapy). Supportive measures are necessary in serious cases. Cryotherapy has been recommended by some for first aid use.

Snakes bite or envenomate their prey or enemies by *stabbing* (solenoglyphs), *snapping* (all snakes), *peck-*

ing (cobras, etc.), and by *squirting* (erroneously called spitting; some cobras).

Snakes feed only upon animals, never plant material. They often scavenge. Live food is killed by constriction or venom in some, but most species simply hold the prey in the mouth and swallow it as best they can. Constrictors seek to retain a hold tight enough to prevent breathing. Vipers usually strike, release the prey, and then seek the animal after it dies; other venomous snakes with shorter fangs retain their hold until the prey dies. All snakes usually swallow the food head first. The trachea can be extended forward below the food far enough to enable the snake to breathe while swallowing.

Four modes of locomotion are concertina, sidewinding, rectilinear and horizontal undulation.

HOBART M. SMITH

References

Bellairs, Angus d'A, "Reptiles," London, Hutchinsons Univ. Library, 195 pp., 12 figs, 1957.

Boys, Floyd and Hobart M. Smith, "Poisonous amphibians and reptiles," Springfield, Illinois, Thomas, 1959.

Pope, Clifford H., "Snakes alive and how they live," New York, Viking, 1937.

Schmidt, Karl P. and Robert F. Inger, "Living reptiles of the world," New York, Hanover House, 1957.

OPUNTIALES (CACTALES)

An order of DICOTYLEDONS placed in Engler's System between PARIETALES and MYRTALES consisting of a single family, the Cactaceae. The name Cactaceae, created by Lindley, comes from the Greek "Kaktos" meaning a spiny plant from Italy. Cacti are a group of plants which attract much attention by their odd spiny appearance and the rare beauty of their flowers. This family is supposed to be of recent origin, having evolved only ten or twenty thousand years ago. Restricted to America prior to 1492, with the exception of few *Rhipsalis* which were found in Madagascar and Africa having been imported as adherent berries by migratory birds.

The morphology of cacti, greatly varying even within a single species, shows a great adaptability to geological changes in the past and to present ecological conditions. Occurring ordinarily in arid regions, they may exist also in humid jungles, not only as epiphytes like *Rhipsalis*, but as tall shrubby columnars like *Cereus trigonodendron* from Peru. Cacti seedlings have a more or less strong primary root but in adult plants the roots may be fibrous, carrot-like, large tuberous, very long superficial, or whip-like, with or without, swellings. Adventitious roots are easily formed at any part of the succulent stems. The stem, though usually fleshy, is woody in *Pereskia*. The greatly developed cambium layer, which entirely surrounds the vascular bundles, makes grafting easy. The shape and ramification of the stem is very variable. It is globose with ribs in *Echinocactus, Echinopsis, Lobivia, Parodia,* etc., globose with tubercules in *Gymnocalycium, Neowerdermannia, Mammillaria,*

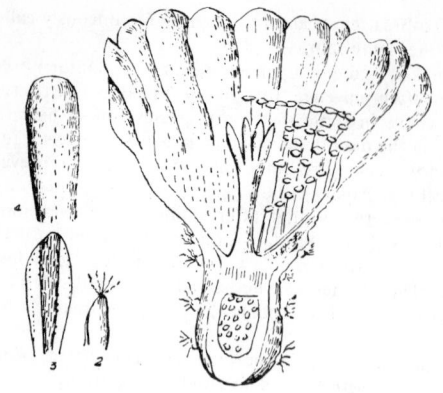

Fig. 1   Corryocactus charazanensis Cárd. 1, Flower; 2, 3, 4, outer, middle and inner perianth segments.

Coryphanta, etc., columnar with ribs in *Cereus, Trichocereus, Cephalocereus, Hylocereus, Cleistocactus*, etc., columnar tuberculate in *Cylindropuntia*, with flattened oval joints in *Platyopuntia*, bushy and much branched in *Pereskia* or reduced to a little disc in *Blossfeldia*. Most cacti are leafless, though there are well developed leaves in *Pereskia* and *Quiabentia* and rudimentary ones in *Opuntia*.

Two of the most characteristic organs of the cacti stem are the areoles and spines. Areoles are considered homologous to the axillary buds of other plants. Buds in the areoles are usually geminate, the upper one producing branches or flowers and the lower one, spines. Spines are considered as modified leaves and a proof of the similarity of spines and leaves, according to Prof. Buxbaum, is the existence of intermediate forms of both organs in *Pereskia* and *Opuntia*; these are the leaves with spiny tips. Wettewald also showed that spines enlarge like leaves with a basal meristem. Areoles are ordinarily felted. In *Neoraimondia* and

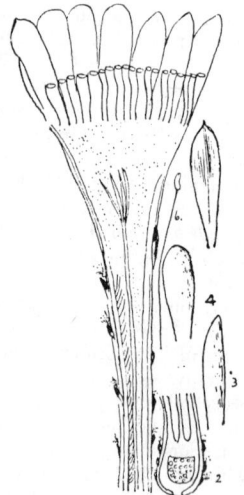

Fig. 2   Echinopsis pseudomammilosa Cárd.

Neocardenasia, areoles are very elongate and hard. Spines are subulate and strong in *Ferocactus, Oreocereus*, etc., acicular in *Cleistocactus*, bristle-like in *Rebutia*, feather-like in *Mammillaria plumosa*, very long and strong in *Neoraimondia* and *Neocardenasia* (30 cm) or only 1 to 2 mm long in *Trichocereus pachanoi* and *Echinopsis semidenudata*. In *Cephalocereus, Pilocereus* and *Vatricania*, there are spines reduced to wooly, silken or bristly hairs. The flowers of cacti are usually hermaphrodite, solitary or grouped in clusters or on a specialized region of stem called *cephalium* or *pseudocephalium* as in *Cephalocereus, Melocactus* and *Vatricania*.

Flowers are regular, rotate, urceolate, funnelform, tubulous or zygomorphic. The perianth phylla are indistinguishable as sepals and petals, and are usually designated as outer or inner segments. The color of the flowers may be white, yellow, purple or magenta. The ovary is inferior with the exception of *Pereskia* which, from this character, is considered the most primitive genus. The placenta is parietal but is not visible in *Opuntia*. Ovary cells are distinguishable only in *Pereskia*. Stamens are numerous, in most cases separated as lower and upper ones. The style has as many stigma lobes as the ovary has placentas. Pollination usually is performed by insects, although the flowers may open by day or by night. The fruits are one-celled, juicy, or dry, berries. They are juicy in *Corryocactus*, quite dry in *Oreocereus* and completely dry in *Pterocactus*. The fruit pulp may be white, greenish, magenta or reddish. The seeds are covered by a hard tegument (*esclerospermae*) in *Opuntia, Pterocactus, Quiabentia* or by a soft tegument (*malacospermae*) in *Pereskia, Maihuenia*, etc. The colour of the tegument may be whitish, brown or black with minute prominences or depressions (*puncticulate*).

The Cactaceae are arranged under various systems of classification. The most widely followed is the Britton and Rose system which embraces the following three tribes:

Tribe I. Pereskiae. Here are included woody trees, shrubs or climbing bushes with large areoles from which arise branches, spines and leaves. The flowers are rotate with reddish petals. They lack *glochids*. The tribe occurs in both hemispheres.

Tribe II. Opuntiae. The principal character of the Opuntiae is the presence of retrobarbelate bristles (*glochids*) filling the areoles at the base of the spines. There are conspicuous leaves in *Quiabentia* but only rudimentary ones on young shoots in the other genera. The flowers are sessile, rotate.

Tribe III. Cereae. This is the largest taxonomic group with leafless plants bearing rotate, tubulous, bell-shaped or funnel-shaped flowers. This tribe is divided in the following 8 subtribes: *Cereanal, Hylocereanae, Echinocereanae, Echinocactanal, Coctanae, Coryphantanae, Epiphyllanae* and *Rhipsalidanae*.

From the geographical point of view some genera are restricted to one of the hemispheres or some islands. Found in the Northern Hemisphere are *Wilcoxia, Echinocereus, Eriocarpus, Strombocactus, Lophophora, Astrophyton*, while *Lobivia, Echinopsis, Cymnocalycium, Corryocactus, Rebutia, Parodia, Frailea*, occur only in South America. *Brachycereus* and *Jasminocereus* are confined to the Galapagos Islands.

Fig. 3   Pereskia saipinensis Cárd.

The smallest known cactus is *Blossfeldia lilliputana* from Argentina and Bolivia, reduced to a disc 1–2 cm in diameter. The largest ones are the giant columnars: "Saguaro" (*Carnegia gigantea*) from Arizona and "Carapari" *(Neocardenasia Herzogiana)* from Boliva measuring 25 m and 18 m respectively.

The "Peyotl" (*Lophophora Williamsii*) from the Rio Grande area yields a narcotic drug used by Mexicans. The seeds of some South American upland Opuntias, contain a dye used to give a magenta colour to some beverages. The tall Cerei of Northern Argentina produce an exotic timber used in the manufacture of doors and furniture. Among the cacti fruits used as food, the "Carapari" fruit is the most delicious.

MARTIN CÁRDENAS

References

Britton, N. L. and J. N. Rose, "Studies in Cactaceae," Washington, Government Printing Office, 1913–1923.

Buxbaum, F., "Morphology of Cacti," Pasadena, Abbey Garden Press, 1950–1955.

Backeberg, C., "Die Cactaeae. Handbuch der Kakteen Kunde," Jena, Fischer, 1958–.

ORCHIDALES

The dominant feature of this order, which is regarded by many as the peak of plant development in the MONOCOTYLEDONS, is the more or less marked fusion of the reproductive organs—stamens and pistil—into a central organ, the column.

Herbs terrestrial (when rarely saprophytic) or epiphytic but never parasitic, very rarely subterranean (the Australian *Rhizanthella* and Cryptanthemis), cosmopolitan except for the polar regions, chiefly in the tropics of both hemispheres. Plants more or less erect, creeping or rarely climbing (*Vanilla, Galeola*), minute to 15 feet high or more, caespitose or with a creeping rhizome, bearing fibrous, tuberous, rarely coralloid (*Corallorrhiza, Hexatectris*), often fasciculate roots, the aerial ones (in epiphytes) coated with a layer of spongy cells (velamen) for the absorption of nutrients. Stems scapose or leafy, often thickened into corms or pseudobulbs (for water storage) bearing one to several leaves. Leaves simple, solitary to more or less numerous, alternate, frequently distichous or subopposite (*Scaphyglottis, Hexisea*), rarely whorled (*Isotria*), flat or equitant, membranaceous or fleshy, more or less sheathing below, often articulated at the base, parallel-veined, sometimes reduced to scales. Inflorescence terminal or lateral, one- to many-flowered, spicate, racemose or paniculate. Flowers minute to large and showy, of all colors, including white and black (*Coelogyne pandurata*), bracteate, bisexual, monoecious, dioecious, hermaphroditic, rarely polygamous, always zygomorphic. Perianth epigynous, consisting of six segments in two whorls; the outer whorl (sepals) free or more or less connate, calyx-like and rarely surrounded by a short, toothed crown or epicalyx (*Epistephium*); inner whorl (petals) corolla-like, with one segment (usually the lowermost resulting from a twist of the ovary) called the lip or labellum, commonly variously modified, simple or lobed and often adorned with keels or calli. Sometimes the perianth segments more or less united and frequently the lip, or rarely one or more sepals, produced at the base into a sac or often elongate spur which contains nectar (*Habenaria, Angraecum*, etc.) Stamens (one to three) and pistil more or less fused into a central organ (column or gynostemium). Anthers one to three, 1- or 2-celled, sometimes opening by a longitudinal slit. Stigmas three (*Apostasiaceae, Cypripediloideae*) or two *(Orchidaceae)* in which the third is often transformed into a sterile organ (rostellum) sometimes provided with a viscid disc or sticky mass to which the pollinia are attached either directly or by a caudicle or stipe. Pollen-grains minute, free (*Apostasiaceae, Cypripediloideae*) or more or less consolidated into masses, or bodies (pollinia) (Orchidaceae) which are either free or more or less united within the anther. Pollinia mealy, waxy or cartilaginous. Ovary inferior, one- or three-celled. Fruit generally a capsule, dehiscing by commonly three to six longitudinal slits. Seeds extremely numerous, minute, tunicate, often drawn out at each end.

The following key separates the components.

A. Perianth regular (actinomorphic), radiate; stamens and pistil distinct, although consolidated in the lower part.—*Apostasiaceae*
  I. Segments connivent, forming a tube; fertile stamens three.—*Neuwiedia*
  Ia. Segments spreading; fertile stamens two.
     1. Anthers versatile, one staminode present. —*Apostasia*
     1a. Anthers basifix, no staminode present.— *Adactylus*

A group apparently ancestral to the Orchidaceae, confined to the Asiatic tropics.

B. Perianth irregular (zygomorphic); stamens and pistil completely fused to form a more or less elongate column.

2. Fertile anthers two, representing the two lateral stamens of the inner whorl of the ancestral flower, the third transformed into a large staminode above and partially concealing the anthers; pollen granular, not consolidated into masses or distinct bodies; lip always slippershaped. Diandrae—Cypripedioloideae

   3. Ovary three-celled, with axila placentas; sepals valvate.

      4. Leaves plicate; perianth persistent; seeds subglobose.—*Selenipedium*

      4a. Leaves conduplicate; perianth deciduous; seeds fusiform.—*Phragmipedium*

   3a. Ovary one-celled, with parietal placentas; seeds fusiform.

      5. Leaves plicate; perianth persistent; sepals valvate.—*Cypripedium*

      5a. Leaves conduplicate; perianth deciduous; sepals imbricated.—*Paphiopedilum.*

2a. Fertile anther solitary, representing the upper member of the outer whorl of the ancestral flower. Pollen consolidated into masses or distinct bodies (pollinia). Lip very rarely slipper-shaped *(Calypso, Yoania.—* Monandrae.—*Archidaceae)*

   6. Caudicles and viscid disc arising from the base of the anther. Anther erect or more or less resupinate, very closely adnate to the column with a broad base, never deciduous after flowering. Pollinia always granular.

Division I. Basitonae, Tribe I. Ophrydoideae
(Ophrys, Orchis, Habenaria, Disa, etc.).

   6a. Caudicles and viscid disc arising from the appex of the anther. Anther erect or incumbent, attached by a short, thin, usually very narrow filament, commonly deciduous, more rarely persistent but soon drying up.

Division II. Acrotonae

   7. Pollinia granular, soft. Anther commonly persistent. Inflorescence always (normally) terminal—Tribe 2. Polychondreae.

(Pogonia, Vanilla, Sobralia, Spiranthes, Erythrodes, etc.).

   7a. Pollinia waxy or cartilaginous. Anther commonly soon deciduous. Inflorescence terminal or lateral.— Tribe 3. Kerosphaereae.

      8. Inflorescence normally terminal, sometimes by the abortion of the terminal inflorescence borne in the axils of the upper leaves. Series a. Acranthae.

(Masdevallia, Pleurothallis, Malaxis, Liparis, Coelogyne, Epidendrum, Cattleya, Dendrobium, etc.).

      8a. Inflorescence lateral, arising from near the base of the pseudobulbs or in the axils of the leaves or of the lower sheaths of the stems.

         Series b. Pleuranthae

   9. Plants forming a sympodium, i.e., with stems caespitose or superposed, manifestly terminated by apical, usually smaller leaves.

      Subseries a. Sympodiales.

(Balbophyllum [including *Cirrhopetalum*], *Eulophia, Cymbidium, Catasetum, Stanhopea, Lycaste, Zygopetalum, Maxillaria, Odontoglossum, Oncidium,* etc.).

   9a. Plants forming a monopodium, i.e., having stems with the apical growth unlimited and thus growing on indefinitely.

      Subseries b. Monopodiales.

(Dichaea, Aerides, Phalaenopsis, Vanda, Campylocentrum, Angraecum, Sarcanthus, Mystacidium, etc.).

The Orchidaceae, even without the traditional inclusion of the Cypripedium group, is surely one of the largest, if not the largest, of all plant families, with nearly 800 genera and about 30,000 species, depending upon the interpretation of the specialist. Without doubt, its members show the greatest diversity of vegetative and floral structure of any group in the plant world, and thus a more pleasing variety.

Yet the economic uses of this huge family (except for purely local applications) are surprisingly few. *Faham,* from the dried leaves of *Angraecum fragrans* (of Mauritius and Réunion), is used in France as a substitute for Chinese tea. *Salep,* from the dried tubers of some European species of *Orchis* or from the corms of several East Indian species of *Eulophia,* contains gum and starch, and is employed for food like tapioca and as a demulcent. The largest and most important economic product is Vanilla, a flavoring extract, discovered by the ancient Aztecs, derived from the dried pods ("beans") of *Vanilla planifolia* of the American tropics and now widely cultivated in various parts of the tropics of both hemispheres.

To the average layman, the outstanding importance of the Orchidaceae lies in the multimillion dollar industry that has been developed from its numerous striking members, such as *Cypripedium* (in its broad sense), *Cattleya, Dendrobium, Odontoglosum, Oncidium, Phalaenopsis* and *Vanda,* etc. Being a delight to the amateur hobbyist, this diverse family has given to the hybridist a practically unlimited field of experimentation and to the horticultural "trade" a heaven-sent opportunity to build up immense profits.

CHARLES SCHWEINFURTH
LESLIE A. GARAY

Reference

Withner, C. L., ed., "Orchids—a scientific survey," New York, Ronald, 1959.

ORGANELLE

An organelle (Gr. ergon-work + el-little) is to a cell as an organ is to a metazoan organism. It may be defined as a part of a cell possessing specific functions, distinctive chemical constituents, and characteristic morphology.

Cells perform a wide variety of functions, but are nevertheless composed of a few basic components which are similar throughout varying phyla and tissues. Functionally, an organelle usually performs circumscribed tasks with a degree of autonomy and can be regarded as a unit subsystem of the cell. Chemically, it has a distinctive composition closely related to its function. Morphologically, it has an identifying internal structure indicative of ordered molecular aggregation.

A given organelle may exist in great numbers in a particular cell, occasionally with arrangements into regular patterns. Little is known of the methods of duplication or multiplication of organelles since their size allows only detection and limited chemical analysis by light microscopy. Electron microscopy is necessary for study of their structure, and chemical analyses require large numbers of isolated organelles.

An organelle is characterized either by its function, chemical composition, or structure, or some combination of the three. For example, a cilium is usually an organelle of movement and is composed of a precise pattern of fibrils (see drawing). Cilia are present in almost all animal as well as many plant phyla. They may number from one to several hundred per cell and often are arranged in regular patterns (see electron micrograph).

Another organelle, the MITOCHONDRION, is present in large numbers in the cytoplasm of virtually all cells. Structurally, mitochondria are characterized by patterns of lamellae or tubules surrounded by a membrane, while chemically they almost invariably contain enzyme systems, e.g., Krebs cycle enzymes.

In the nucleus, one or two nucleoli, which lack obvious internal organization, are usually present. A nucleolus contains relatively high concentrations of ribose nucleic acid and is thus characterized on a chemical basis as an organelle.

Other examples of cellular components that can be strictly considered as organelles are the GOLGIBODY, centriole, endoplasmic reticulum, plastid, and CHROMOSOME.

The term has at times been used freely to denote almost any cellular inclusion, but the current tendency is toward a more restricted and confined usage. The concept that an organelle has a distinctive molecular array with a rather well defined function is implicit in most of the recent literature.

L. E. ROTH

References

Giese, A. C., "Cell Physiology," Philadelphia, Saunders, 1968.
Morrison, J. H., "Functional Organelles," New York, Reinhold, 1966.

ORGANIZER

The term "organizer" was originally applied to the dorsal lip of the blastopore of the Amphibian gastrula. Hans Spemann and his associates, in a masterly analysis of developmental mechanics emanating from the University of Freiburg in the two decades following the First World War, identified the germinal region anterior to the dorsal lip of the blastopore during gastrulation, destined to become notochord, as an area primarily possessing the ability, when transplanted to the ventral region of an early gastrula, of inducing the formation of a second embryonic axis on the host belly. Recently it has been shown that even the uncleaved egg protoplasm corresponding to this region can induce an axis. Such a secondary axis consists of nerve tube with associated sense organs; notochord, somites, pronephros, and even gut; host and donor cells participate side by side in the structure of any one of these organs; and typically the induced axis parallels that of the host, with brain vesicles, otocysts, etc. occupying identical levels.

The action of the organizer has, in principle, been analyzed into a series or network of simple *inductions*. In embryonic induction, contact or near contact between an active tissue (the inductor) and a receptive (competent) tissue will result in the latter's becoming *determined* or started in a causal pathway directed toward a specifically differentiated state. In normal development the primary action of the organizer is exerted on the ectoderm which it comes to underlie during the process of gastrulation; this ectoderm is induced to become neural plate whereas the organizer itself becomes archenteron roof. During neurulation the neural plate is subdivided into the organs of the central nervous system, while the adjacent ectoderm gives rise to sensory placodes and other accessory neural structures. The archenteron roof becomes subdivided into notochord, somites, nephros, and lateral plate. These progressive subdivisions demonstrably depend on further interactions between adjacent tissue and cell types.

The dorsal lip or organizer region is by no means homogenous in itself. Its normal behaviour is to roll gradually over the blastopore lip, growing forward to line the ectoderm above it. The material that is first invaginated thus advances farthest anteriorly, comes to underlie the anterior medullary plate (prospective forebrain). The last invaginated portion underlies spinal cord. Different levels of the organizer produce characteristically different levels of induction when brought into contact with standard receptive ectoderm. In practice, *archencephalic* (forebrain), *deuterencephalic* (mid-hindbrain) and *spino-caudal* (including mesodermal) levels have been distinguished.

Any ectoderm of the early gastrula is capable of responding to the stimulus of the organizer. As gastrulation proceeds, the ectoderm that has been induced becomes capable of neural differentiation autonomously, even if withdrawn from the inducing influence. Ectoderm not underlain by the organizer will form only non-neural epithelium resembling epidermis. If early gastrula ectoderm is isolated from organizer action, it gradually loses, with time, its capacity to respond. *Competence* for response is thus a condition of limited duration and unstable character. Amphibian species show marked variation with respect to the sensitivity of their competent ectoderm to experimental induction.

Once ectoderm has been induced in the neural direction, it itself acquires the capacity to perform neural inductions on sensitive tissues. Furthermore, although early gastrula ectoderm is not capable of acting as a neural inductor when alive, it acquires this capacity when killed by heat, by alcohol fixation, or in various other ways. Indeed, this is true of all early gastrula tissues, ectoderm or not, as well as of tissues from widely different animal and plant groups: killing releases inducing capacity in a great variety of tissues that have no normal relation to organizer activity.

The inducing activity of extracts not only of the organizer but of many adult tissues of widely different origin has been studied. Refinement of *in vitro* assay systems has given increased assurance of specificity in the response of competent ectoderm. The original näive search for an active chemical given off by the inductor, diffusing to the receptive tissue and there operating in an unspecified way to produce a visible result, has been replaced in recent years by a more realistic appraisal of the interaction involved.

The inductor. Various types of adult tissue have shown specific inducing activity for definite levels of the neural axis. For example, guinea pig liver typically produces archencephalic inductions, whereas kidney from the same animal induces deuterencephalic or spino-caudal levels, and bone marrow induces a meso-dermal response. Different protein fractions of chick embryos appear likewise to induce different levels of the Amphibian axis. Fractionation of embryonic Amphibian material has indicated that the microsomal component is the only one possessing inductive activity. Active material isolated from spinocaudal and meso-dermal inductors is of protein character. Denaturation evidently favors archencephalic induction.

Transport. Competent cells, in small groups, respond to inductor material in fluid culture conditions. This speaks for diffusion as a transport mechanism. Evidence of pinocytosis in the responding cells suggests the means by which the diffusing molecules may enter.

In experiments with larger numbers of cells, induction has been found to occur when the organizer is separated from the competent ectoderm by a filter not more than 20 μ thick, with pores 0.8 μ or more in diameter.

The reactive ectoderm. The responses of gastrula ectoderm are not understandable unless the ectoderm itself possesses regional organized properties. The cellular state that permits a gastrula ectoderm cell to become a neuroblast is a basic element in the response to the organizer, but very different mechanisms must be invoked to account for—to take one obvious example—the induction of a forebrain, with all that is implied in the way of developmental mechanics of the anterior brain vesicles and bilateral camera eyes. This so-called archencephalic induction is evidently a unitary and frequently obtained response of Amphibian gastrula ectoderm.

Interpretation. According to present views of cellular differentiation the stimulus of induction must operate somewhere in the sequence between gene and protein synthesis, switching the activity of a cell directly to neuroblast differentiation—or to that of some other element of the nervous system. It is clear that the *stimulus* lacks genetic specificity. Inductors need not originate from the same class or even phylum as that of the induced material. The *response* of the ectoderm, however, is species-specific. Competent ectoderm will form only structures within the genetic repertory of its species, even in response to a stimulus originating from an organism not possessing the particular structure that is being induced. It is easy to see how intracellular differentiation of specific character such as pig-

ment formation, or the lack of it, would follow this genetic restriction. To visualize a response involving groups of cells cooperating to form a species-specific organ is to beg the question of the whole problem of genic control of organized multicellular development.

The discovery of different sorts of inductors, inducing different levels of the axis, has encouraged visualizing the archenteron roof as possessing gradients of release of active substances. Perhaps the simplest hypothesis involves two such gradients, one antero-posterior ("neuralizing"), the other postero-anterior ("mesodermalizing"), with the assumption that varying proportions of the two inductors would produce various intermediate effects. Other hypotheses call attention to medio-lateral differences in the archenteron roof, as well as antero-posterior and medio-lateral differences in the overlying ectoderm.

Non-amphibian organizers. Analysis of the organizer and its normal relations in the Amphibia has led to similar investigations in other vetebrate groups. The precise and crucial experiment of removing prospective archenteron roof *alone*, placing it in association with indubitable prospective ventral epidermis, and being able to distinguish graft from host tissue cell for cell in the resulting induced axis—has for technical reasons not been fully achieved in any other vertebrate gastrula. Nevertheless, transplantation experiments in teleost fish, in birds, and even in mammals make it certain that the organization of the dorsal axis depends on a very small area corresponding to the dorsal blastoporal lip. When this area is transplanted to prospective ventral regions, a secondary axis results which in some cases is demonstrably formed partly at the expense of host tissues, i.e., by induction. Whether the induction is in all cases a lower-layer-to-upper-layer process, as in Amphibia, is not clear. It may be noted that in the Amphibian tail the invagination and hence the induction pattern is different from that in the head and trunk; it would hardly be surprising to find that in other vertebrates with different types of blastopore correspondingly different patterns of induction might prevail.

Various invertebrates (e.g., Echinoderms, Insects) display during their development *centers of differentiation*: i.e., localized areas necessary for important embryogenetic processes such as gastrulation or axis formation. Some investigators like to think of such centers as organizer regions; Spemann's own thinking was greatly influenced by the situation in the Echinoderm egg. The morphogenesis and mechanics in these cases, however, appear to be so different from the vertebrate situation that realistic comparison is difficult. The term organizer is probably best confined to vertebrate or chordate embryogenesis.

DOROTHEA RUDNICK

References

Saxén, L., and S. Toivonen, "Primary Embryonic Induction," Englewood Cliffs N.J., Prentice-Hall, 1962.

Spemann, Hans, "Embryonic Development and Induction," New Haven, Yale University Press, 1938.

Willier, B. H., P. Weiss, and V. Hamburger, eds., "Analysis of Development," Saunders, Philadelphia, 1955.

Yamada, T., "A Chemical Approach to the Problem of the Organizer," *in* "Advances in Morphogenesis" Vol. 1, New York, Academic Press, 1962.

ORGANOGENY

Organogeny, the process of organ-formation, has been studied largely in vertebrates where size, complexity of structure and developmental history, and the inevitable application to human medical problems have offered an irresistible challenge. Higher invertebrates—especially insects—have likewise furnished attractive material for investigation, and several useful concepts have been extended to, or borrowed from, simpler animals and plants which do not, strictly speaking, possess organs.

EMBRYOLOGY as a distinct discipline took origin from the germ layer concept. By studying closely spaced developmental stages of a single vertebrate species, the origin of individual organs was traced back to the original cellular sheets formed at the time of gastrulation. These germ layers, by folding, tube-formation, local thickening, apposition, separation of subsidiary layers, accumulation of fluid, or differential growth, were seen to give rise first to rudiments, then to the functional organs of the body, according to a pattern and sequence recognizably characteristic of vertebrates. During the middle of the nineteenth century the process of organogeny was considered to consist of such events as those cited above, supplemented by finer-scale cellular changes.

In the second half of the nineteenth century, Wilhelm His introduced a sophisticated and plastic concept of organ development, tracing all adult organs back, not only to visible embryonic rudiments, but to *germinal areas* in gastrulae or even younger stages. Differential growth would according to him be the main agent in organogeny, separating and molding the germinal areas, delimited originally only by their potentialities, into discrete rudiments, later into functional organs.

Origin of germinal areas. Some important germinal areas appear to be present in the uncleaved ovum. The amphibian organizer (future archenteron roof) is demonstrably in this category. Other areas, such as the medullary plate, arise by induction during the course of gastrulation and later stages. The potency to form nervous tissue, not present in ectoderm cells before gastrulation, is acquired by them as a result of contact with the archenteron roof. Before this event, the cells if transplanted would conform to whatever their new environment happened to be; after induction, they form nerve tube exclusively, even if such a structure is highly inappropriate to the new surroundings. Induction is not necessarily a single event. The ear vesicle for example depends first on underlying mesoderm, then on adjacent hindbrain. Even if the primary induction is by a single agent, normal morphogenesis and histogenesis of any organ depends on a series of relationships between the rudiment and the rest of the embryo, as well as on subordinate tissue and cell interactions within the rudiment itself. Many of these dependent relationships clearly belong to a much later phase of development than does induction.

Gradient properties. All organ-forming areas at first evidently possess, in some degree, gradient properties in their potency for independent differentiation. This is to say that the areas appear to be embryonic *fields*, in which there is a center or restricted area of very great tendency to differentiate in the given direction; this property diminshes peripherally from the center, until the area merges with a similar adjoining center. It must

be noted that the criterion of differentiation, experimentally, has in many cases been on the histogenetic rather than truly organogenetic level.

Axiation. The gradient or field properties of all organ-forming areas bear a precise relation to the main axes of the developing animal as a whole. Axial properties have been studied particularly in symmetrical organs such as limbs or ears, but also in medial organs such as nerve tube and heart. Whole organ areas, as early as they can be isolated and tested, appear to be irreversibly polarized as to their antero-posterior axis, even though regulation may occur in small parts of such areas if their axes are tampered with. If the limb field, in neurula stages, has its antero-posterior axis experimentally reversed, the limb will grow out according to its original, not its new, orientation. During this period, axes of symmetry can easily be reversed, and a right limb transformed to a left one by inverting its dorso-ventral axis. The latter axis becomes fixed much later, shortly before actual outgrowth of the bud begins. It is considerably later, during outgrowth of the bud, that the parts and tissues of the limb become determined. This takes place progressively through a complex series of interactions between the ectodermal and mesodermal components of the limb bud.

Guidance mechanisms. Organogeny frequently includes movement of the germinal area or rudiment itself with reference to the rest of the embryo, as well as immigration of additional cellular elements into the established rudiment. Blood vessels and nerves, with their mesenchymal accompaniments, grow into all rudiments soon after their delimitation. In addition, there are many cases of more specialized immigrations: pigment cells, ganglion cell bodies and chromaffin components, and primordial germ cells in the case of the gonad. We must thus include among principles of organogeny all guidance mechanisms by which cells or cell groups find their way around the embryonic framework or the original germ layer derivatives.

Growth. Increase in size of an organ rudiment during development is a fundamental organogenetic property as emphasized originally by His. Growth rates, especially differential ones (heterauxesis) are susceptible of mathematical treatment, and have furnished illuminating descriptions of development in stages after primary embryogenesis. Causal analysis of the mechanism of differential growth has not progressed nearly as far as the study of its results. Newer studies on differential nutritional requirements of various tissues point a valuable direction.

Cell differentiation and function. The functioning of an embryonic organ area includes its progressive changes. Specific cellular differentiation, implying specific adult function, although it may begin early, does not reach full expression until late in development. In some cases later changes have been shown to depend on the endocrine or other biochemical environment. Analysis of individual organs has shown many instances where two or more different cell types depend reciprocally on one another for maintenance and further differentiation. A classical example is the dependence of nerve cells on their terminal connections with other cells. For many, perhaps all, epithelial and glandular tissues, a relationship with underlying mesenchyme has been shown to be of crucial importance to differentiation and subsequent maintenance.

Organ function. Many organs and organ systems

develop without any clear relation to their prospective function. Thus a perfect neuromuscular system is formed in Amphibian embryos developing in anaesthetic solutions and hence not passing through the usual stages of embryonic motility. The skeletal system in amniotes is pre-adapted, in its finest details, to future mechanical function although the stresses in embryonic life are certainly very different from what will be encountered after birth. On the other hand, the heart shows a clear relation of function to form. In the absence or reduction of the blood stream, the heart tube, though fully competent as a muscle, does not acquire its normal shape and proportions. The general principles underlying the relation of embryonic to adult function have yet to be enunciated.

DOROTHEA RUDNICK

References

Willier, B. H. *et al.*, eds., "Analysis of Development," Philadelphia, Saunders, 1955.

Zwilling, E., "Limb Morphogenesis," *in* "Advances in Morphogenesis," Vol. 1, New York, Academic Press, 1961.

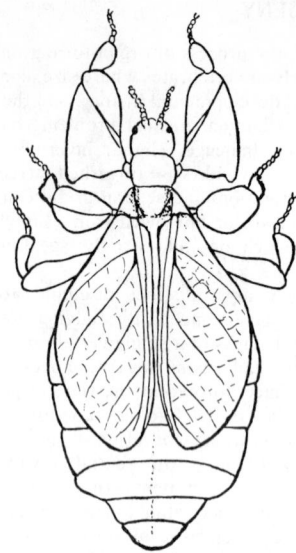

Fig. 1   Leaf-insect, *Phyllium*. Ceylon, length $3\frac{1}{2}$ inches.

ORTHOPTERA

Orthoptera are discussed here in a broad sense—that is, they are regarded as a single large order in the traditional way. However, there is a strong recent tendency to divide these insects into several orders, corresponding roughly to the suborders enumerated below, with Orthoptera restricted to the Saltatoria. For this reason, also see MANTODEA, for which a separate account has been prepared.

The 4 suborders of Orthoptera are:

Suborder Dictyoptera: Two superfamilies are included, Blattoidea (cockroaches, often abbreviated to "roaches") and Mantodea (mantids), each with several families. The great majority of cockroaches live in moist habitats of the tropics and subtropics, where they live primarily as scavengers. Some 3,500 species have been named, of which less than 75 occur in the United States, and the best known are about 10 cosmopolitan species carried by commerce to nearly every country. The nymphs of some species are born alive, that is, eggs hatch within a brood chamber in the female's body, but the majority hatch from eggs deposited in brownish packets called oothecae. Maturity is reached within a few weeks in some species, in others not for many months.

Mantids are most abundant in warm countries; only 21 species occur in the United States, 4 of them accidentally introduced from abroad. Most mantids spend their time climbing about and resting on weeds and shrubs, but some live on the ground, while others, like certain Malayan species, superficially resemble tiger beetles and actively pursue their prey on tree trunks. Among the remarkable types are species which in form and color resemble flowers; they lie amid leaves of shrubs and await their victims. The egg masses of a few species are familiar cocoon-like objects on weeds and shrubs during winter months.

Suborder Cheleutoptera: Because of their elongate bodies and adaptation for crawling instead of leaping, walking-sticks traditionally have been grouped near mantids and cockroaches, but studies of comparative anatomy indicate that they belong to a distinct orthopterous line. In the tropics, this group includes many oddities of the insect world, ranging from giants 13 inches in body length in the East Indies, and Oriental leaf-insects, to tiny species of *Timema* about ¾ inch long occurring on conifers and other trees in our Southwest. About 26 species inhabit the United States and serious defoliation of forest trees by *Diapheromera*

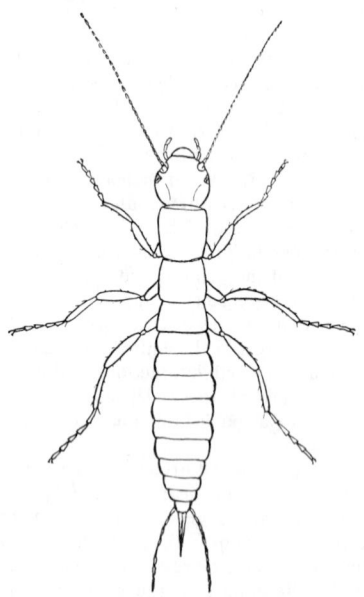

Fig. 2   Grylloblattid, *Grylloblatta*. U.S., length 1 inch.

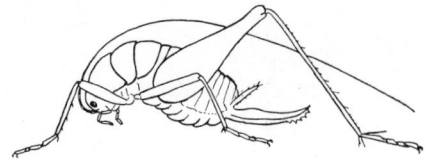

Fig. 3   Camel-cricket, *Ceuthophilus*. U.S., length 1¼ inches.

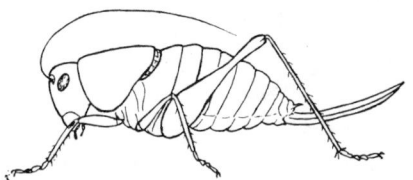

Fig. 4   Mormon-cricket, *Anabrus*. U.S., length 2 inches.

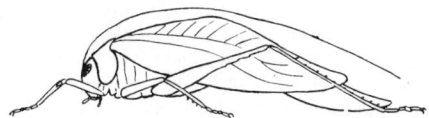

Fig. 5   Katydid, *Microcentrum*. U.S., length 3 inches.

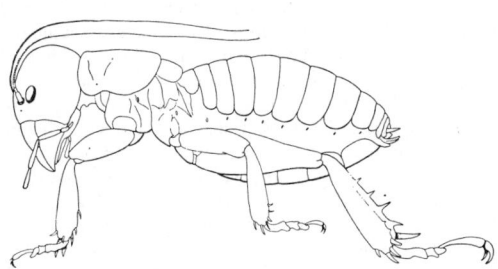

Fig. 6   Jerusalem-cricket, *Stenopelmatus*. U.S., length 1½ inches.

occasionally occurs over limited areas in the northeastern states. The genus *Anisomorpha* in the Southeast is able to eject a short stream of offensive fluid from pores in the thorax, which is very irritating to a person's eye.

Suborder Notoptera: There is a single family, the Grylloblattidae, containing 3 genera and about 9 species of these rare wingless insects. Females have ovipositors resembling those of crickets and katydids, but individuals run something like cockroaches, hence the name of the typical genus, *Grylloblatta*, which occurs in the mountains of western North America. The other genera are in Japan and eastern Siberia. Body length is about one inch. Food consists mainly of other insects, probably including some immobilized by the cold at the high altitudes where grylloblattids frequently live.

Suborder Saltatoria: These insects are the best known Orthoptera, in most areas are the most numerous both in species and individuals, and as a whole

are much the most important economically. The most distinctive character of the suborder is leaping hind legs, usually with the femur much enlarged. The principal superfamilies are:

Gryllacridoidea (camel-crickets, Jerusalem-crickets, cricket-locusts). In North America most of these insects are wingless, and non-noise making. *Ceuthophilus* is the dominant genus of camel-crickets, so-called because of the arched back. They live mainly as scavengers and frequent animal burrows, caves, and rocky areas. Jerusalem-crickets *(Stenopelmatus)* are distinctive, powerful insects in the western states.

Tettigonioidea (long-horned grasshoppers, katydids, shield-backed crickets, Mormon-crickets). These insects usually are winged, and the tegmina of males are equipped for stridulation, the "singing" of katydids famous in song and story. Stridulation results when the tegmina are shuffled, during which a row of tiny "teeth" on the lower side of the upper tegmen is drawn across a brittle scraper located at the inner edge of the lower tegmen. In the United States dominant groups are the green colored katydids and the shield-backed crickets, including Mormon-crickets. Typically, this superfamily differs from the Grylloidea, or true crickets, in having 4 instead of 3 tarsal segments, and in having the ovipositor sword-like, that is, flattened, rather than spear-like or round in cross-section.

Grylloidea (crickets, mole crickets). Crickets stridulate similarly to katydids, and the "songs" of field crickets and tree crickets are especially well known. Many minor pests are included, ranging from mole crickets which attack vegetables, and tree crickets which injure berry canes by ovipositing in them, to the house crickets, or "cricket on the hearth," which may chew household fabrics or infest city dumps. Mole crickets are a separate family, characterized by front legs equipped for digging burrows in moist soil.

Tridactyloidea (pigmy sand-crickets). Though widely distributed and often common along margins of streams and lakes, these insects are inconspicuous. Superficially, they seem to be miniature mole crickets, but the basic

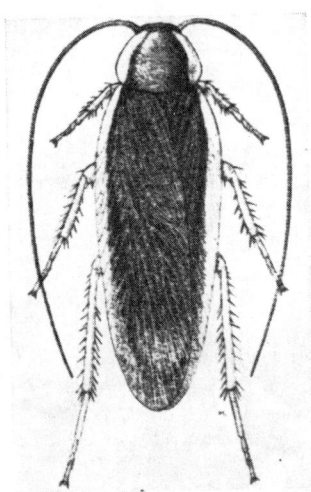

Fig. 7   Cockroach, *Parcoblatta*. U.S., length 1 inch.

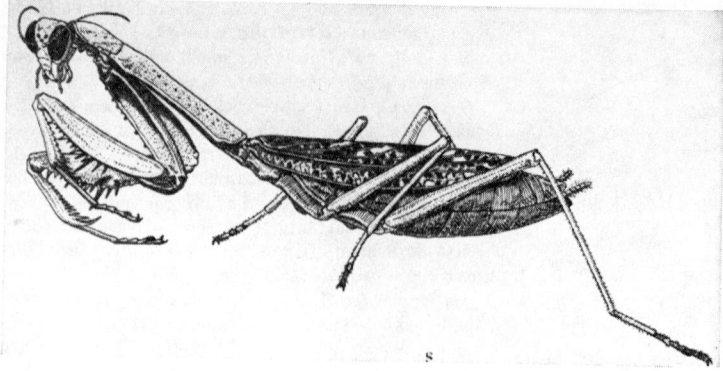

Fig. 8   Mantid, *Stagmomantis*. U.S., length 3 inches.

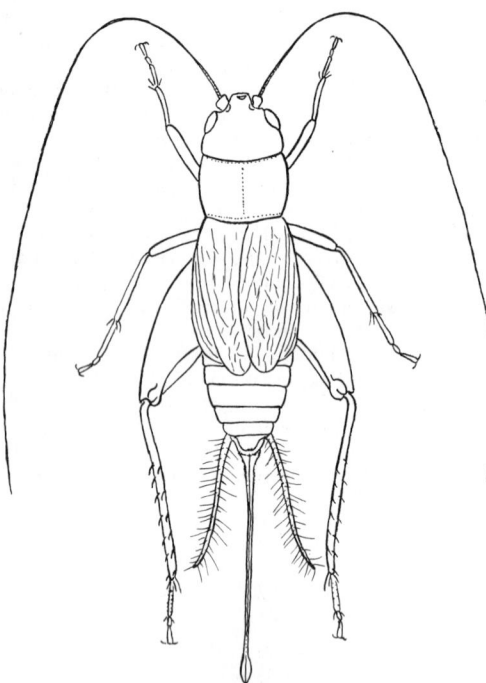

Fig. 9   Small field cricket, *Nemobius*. U.S., length ½ inch.

anatomy is distinctive. Only two species inhabit the United States; they live in tiny burrows in sandy soil.

Tetrigoidea (grouse-locusts, or pygmy-grasshoppers). Tetrigids resemble miniature grasshoppers with the pronotum greatly developed and covering much of the abdomen. Though all species are relatively small, and rarely exceed an inch in length, there are many bizarre tropical forms which have elaborate dorsal crests or clusters of spines.

Acridoidea (grasshoppers, locusts). Grasshoppers are the most numerous Orthoptera, with over 600 species in the United States. Tarsi have 3 segments, antennae are short, and the ovipositor consists of paired hook-like structures called valves at the apex of the female abdomen. Oviposition usually occurs in soil, where cylindrical clusters of eggs are laid, but in some species the ovipositor is variously modified for inserting the eggs into hard wood, soft plant tissue, or even for deposition on the leaves of aquatic plants. Grasshoppers are excellent indicators of ecological zones, and many species are extremely localized in distribution, often correlated with preferred host plants. Most grasshoppers are winged, and in some groups distinctive sounds are made by the wings when in flight, or by rubbing the hind femora against the folded tegmina when sitting on the ground or perched on vegetation. Many highly destructive species require control to protect crops and rangeland. About a dozen species, especially in Africa and Asia, are called locusts because of their strong migratory habits, on the ground as bands of marching nymphs, and as flying swarms when mature. Locusts differ from typical or "solitary"

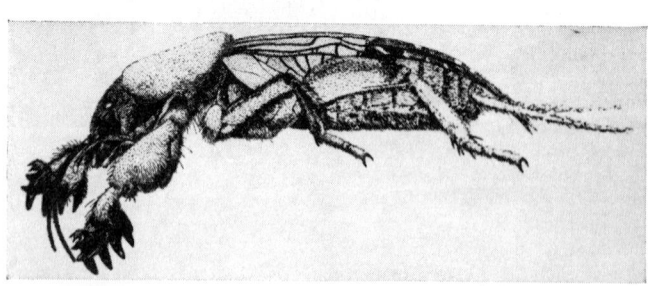

Fig. 10   Mole cricket, *Gryllotalpa*. U.S., length 1¼ inches.

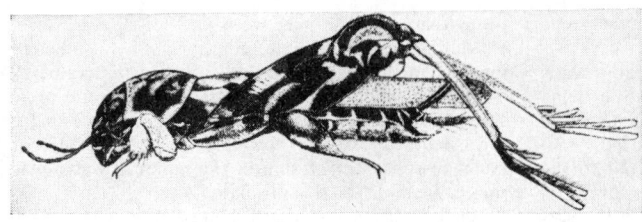

Fig. 11 Pygmy sand-cricket, *Tridactylus*. U.S., length, ½ inch.

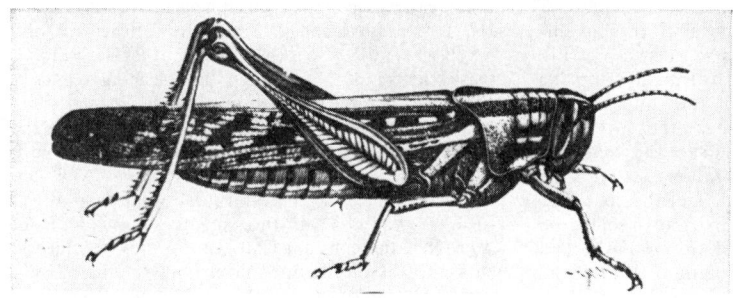

Fig. 12 Grasshopper, *Schistocerca*. U.S., length 3 inches.

Fig. 13 Grouse-locust, *Tetrix*. U.S., length ⅝ inch.

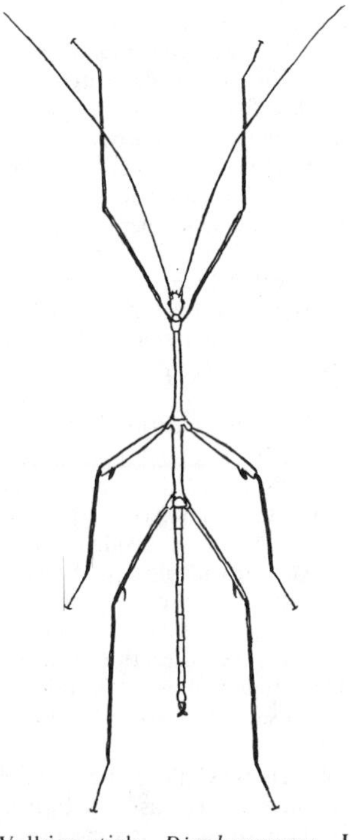

grasshoppers both physiologically and morphologically. They are not constantly in a migratory condition, but at intervals correlated with environmental conditions they develop the urge to migrate, and when in that condition, called the gregarious phase, the nymphs usually have darker colors, the adults usually have longer wings, and eggs usually are not ready for laying until after extensive flights have occurred.

ASHLEY B. GURNEY

References

Blatchley, W. S., "Orthoptera of Northeastern America," Indianapolis, Nature Publ. Co., 1920.
Hebard, M., "The Dermapatera and Orthoptera of Illinois," Bull. Illinois Nat. Hist. Surv., **20**: 125–279, 1934.
Uvarov, B., "Grasshoppers and locusts, a handbook of general acridology," Vol. 1, Cambridge, The University Press, 1966.

OSMOSIS

If, into the bottom of a jar containing water, a solution of cane sugar is introduced with care so as to avoid mixing, not only will the molecules of cane sugar diffuse into the water but the molecules of water will diffuse into the sugar solution. These processes will go on until

Fig. 14 Walking-stick, *Diapheromera*. U.S., length 3 inches.

the concentration of sugar, and water, is the same throughout.

If the solution is placed in a container, whose walls are relatively impermeable to the sugar while being permeable to the water, and the container is placed in water, the water will pass from the outside into the container. The term osmosis is usually restricted to the passage of water. If the influx of the water results in an overflow of solution to somewhere other than the surrounding water this overflow will continue until all the sugar is removed from the container. If the container is closed water will continue to enter until there is sufficient stress in the stretched walls to cause a pressure on the solution inside which will eventually stop the influx. Of course, if the walls of the container are not completely impermeable to sugar then the sugar will be escaping into the water outside the container and this will go on until the concentration of sugar is the same outside and inside. If the walls of the container were impermeable to water but permeable to solute the latter would escape. The cause of this osmosis, this "pushing," of water into the solution is that the tendency of the water molecules to escape from the pure water is greater than that of the water molecules in the solution. Consider water in contact with a limited volume of air. Of those molecules of water striking the surface some will have sufficient energy to escape into the air and this escape will result in net loss to the air which will continue until the concentration of water vapour molecules there is such that the rate of escape from the air (into the water) equals the rate of escape from the water (into the air). If the volume of the air space is fixed the pressure will rise. Just as the temperature of all bodies is the same when they are in thermal equilibrium, although their heat content per unit volume varies with their specific heat, so the escaping tendency of the water is the same in all systems when they are in aqueous equilibrium, whether the system is pure water, solution, gas phase, wettable solid, etc. The same concept can be applied to any substance, say mercury in pure mercury, in air containing mercury vapour, and in an amalgam with another metal such as zinc. The term osmosis is usually restricted to the passage of water from a solution where the escaping tendency is higher to a solution where it is lower. Moreover it is usually restricted to the passage through a solid or liquid barrier which prevents the solutions from rapidly mixing. It is not used for the passage of water in the form of vapour through the air from a dilute solution to a stronger solution in the same confined space, although the process is fundamentally the same. It is sometimes restricted to the case where the barrier is semipermeable, that is lets through water but not solute.

The escaping tendency of water is lowered by the addition of a solute. If the molecules of the solute have no other effect than to reduce the number of molecules of water in unit volume then the escaping tendency of the water will be reduced proportionately to the reduction in the mole fraction of water, N_1, the ratio of the moles of water to the sum of the moles of water and solute. Such is a "perfect" solution. If, however, there is some attraction between the solute and water molecules, a smaller fraction of the latter will have energy sufficient to escape—a "non-perfect" solution. The escaping tendency is increased by pressure. Hence a solution in which the water has a lower escaping tendency than it has in pure water at the same pressure, P^0, can be brought to water equilibrium by a sufficient increase in the pressure on the solution to a value P. This sufficient increase, $P - P^0$, is the osmotic pressure of the solution. In general we cannot state $P - P^0$, the osmotic pressure, knowing only N_2, the ratio of moles of solute to the sum of the moles of water plus solute, the mole fraction of solute ($N_2 = 1 - N_1$).

What we can say is, that if in a solution with a mole fraction N_2 of solute under a pressure P the water has the same escaping tendency as it has in pure water at the same temperature and at a pressure P^0 then $dP/dN_2 = A/B$ where dP/dN_2 is the increase of P relative to increase change of N_2 to keep the escaping tendency unchanged; A is the decrease of escaping tendency relative to increase of N_2 when P is unchanged; and B is the increase in escaping tendency relative to increase in P when N_2 is unchanged. For dilute solutions A/B approsimates to RT/V_1 and so $P - P^0$ approximates to N_2RT/V_1, where V_1 is the volume of one mole of water, R is the constant 82.07 cm.³ atmos per degree, and T the absolute temperature. For very dilute solutions N_2/V_1 approaches n^2/V, the number of moles of solute in a volume V of solution and $P - P^0 = n_2RT/V$ (van't Hoff's equation). This gives an osmotic pressure of 1 atmos for one mole of solute in 22.4 litres at 0°C. There is a departure from both these equations for stronger solutions. The fact that one mole of a perfect gas in 22.4 litres at 0°C. exerts a pressure of 1 atmos coupled with the above has led some to say that the osmotic pressure is the bombardment pressure of the solute molecules. It is correct to say that for very dilute solutions the osmotic pressure of a solution is equal in magnitude to the pressure the solute molecules would exert if they were alone in the same volume and behaved as a perfect gas; but that is another matter.

To measure the osmotic pressure a semi-permeable membrane must be prepared which itself can stand sufficient pressure or it must be deposited in the walls of a porous pot so that the pressure can be sustained. The solution being inside and water out pressure is applied to the former until there is no net movement of water.

Observations by Berkeley and Hartley showed that for 3.393 gms of cane sugar per 100 gms H_2O the osmotic pressure at 0°C. was 2.23 atmos while the van't Hoff equation gives 2.17 atmos since $n_2/V = 9.72 \times 10^{-5}$. If N_2/V_1 is used instead of n_2/V the value of 2.22 is obtained. With stronger solutions the measured osmotic pressure exceeds that calculated: with 33.945 gms of sugar 24.55 atmos was measured, while van't Hoff's equation gives 18.41 and the other 21.8 atmos.

The solutes in the vacuole of a plant cell are exposed to the inward pressure of the distended cell wall and that of the turgid surrounding cells. Water will pass into the cell vacuole from water outside as long as the total inward pressure on the vacuole falls short of the osmotic pressure of the solution in the vacuole. Passage of water into the vacuole dilutes the contents and lowers the osmotic pressure and increases the inward pressure by distension. The amount by which the inward pressure falls short of the osmotic pressure is called by some the suction pressure.

A substance such as cellulose or gelatin tends to take up water, the tendency decreasing with increase in water content until the stress in the substance causes a sufficient rise in the escaping tendency of the water in the substance. This process, which like osmosis is a movement from higher to lower escaping tendency, is called imbibition, and the pressure on the substance sufficient to stop the uptake is the imbibitional pressure. Hence, if a plant cell with a cellulose wall, after coming to equilibrium with a solution, is transferred to water the wall takes up water by imbibition and the vacuole by osmosis. The latter considers only the overall movement from outside to vacuole and does not consider the movement from cellulose to vacuole, a process the reverse of imbibition. A plant cell in equilibrium with a solution having an osmotic pressure of 25 atmos would also be in equilibrium with air about 98% saturated with water vapour. If transferred to a saturated atmosphere it would take up water. We lack precise terms for the passage of water from air into the cellulose and into the vacuole. Condensation, which might be used, ranges more widely.

The escaping tendency of water is affected by factors other than concentration of solute and pressure. Increase of temperature increases escaping tendency. This is a complex problem involving not only transfer of water but of heat also. To a minor extent the passage of water from pure water to a solution involves a heat transfer.

For many naturally occurring membranes which are not completely semi-permeable, i.e. let solute molecules through slowly, electro-osmosis is important. If the membrane tends to lose negative charges to, or take on negative charges from, water or solutions, then the water molecules, in the pores of the membrane, will tend to take on an opposite charge to the membrane. If there is a gradient of electric potential across the membrane the charged water will move in the appropriate direction. If the P.D. is established by the use of electrodes this is electro-osmosis.

With some membranes, particularly those containing protein, water may pass from a dilute solution on one side to water on the other: negative osmosis. Under such conditions if the solution was acid the membrane would be positively charged through the uptake of H^+ ions, leaving the water in the pores negative. The greater mobility of the H^+ ions relative to the anions will cause the side of the membrane towards the solution to be negative and so drive the negatively charged water in the pores across the membrane in the opposite direction to normal osmosis.

The rate of osmosis depends, not only on the excess of the escaping tendency of the water in the phase from which it moves over that in the phase to which it moves, but also upon the area of surface of interchange and the overall resistance experienced by the water. The rate of shrinkage of the vacuole of a plant cell when it is placed in a strong solution at first seems surprisingly high. When allowance is made for the fact that the ratio of surface to volume increases as the linear dimension is reduced then it is realised that when the vacuole of a spherical cell of radius 30 microns shrinks to half its volume in say 5 minutes the passage of water is only 1 ml. per 10,000 cm.² per minute although the thickness of the layer between vacuole and external solution is of the order of 1 micron in thickness. Under other circumstances this layer might be said to be relatively impermeable to water. It seems probable that much of the resistance resides not in the cellulose wall or cytoplasm but in the tonoplast which separates the latter from the vacuole.

G. E. BRIGGS

Reference

Potts, W. T. W., and G. Parry, "Osmotic and Ionic Regulation in Animals," New York, Pergamon, 1963.

OSTRACODA

An order of the Class CRUSTACEA comprising some 2,000 named species inhabiting both marine and freshwater environments. Closely related crustacean orders are the COPEPODA, CLADOCERA, and PHYLLOPODA. Ostracoda have a crustacean body and appendages enclosed within a more or less calcareous, bivalved shell. Superficially they appear like small clams but may be distinguished from the MOLLUSCA by the lack of concentric lines of growth in the valves and by the distinctly crustacean appearance of the internal anatomy. Certain concostracous Phyllopoda are also confused with ostracods, but this group has many more appendages. Likewise the internal anatomy of Cyprid larvae of barnacles will serve to distinguish the Ostracoda from the larval CIRRIPEDIA. Averaging about 1 millimeter in length, the Ostracoda commonly range in size from about 0.35 mm to 8.00 mm with one extreme marine example which reaches the size of a small cherry (24.00 mm).

A typical ostracod has an oval shaped shell, the surface of which may be either smooth or highly ornamented with sculptures, spines, pits, or tubercules, marine species being more commonly ornamented than freshwater forms. A hinge, similar to that found in the Mollusca, is present in most ostracods. An eye, either single or double is usually present. Seven pairs of appendages are found typically. There are, in order from anterior to posterior, the first antenna, second antenna, mandible, first maxilla, first thoracic leg, second thoracic leg, and third thoracic leg. The body terminates in a pair of caudal furca, typically armed with two claws. The first and second pairs of antenna are commonly provided with natatory setae, both pairs extending anteriorly through the valve aperture where they are vibrated rapidly back and forth in swimming. The mandible is a chitinous structure equipped with a heavy cutting edge, ordinarily divided into teeth. The first maxilla bears a respiratory plate which functions as in other crustaceans to create a current of water within the shell cavity. The first thoracic leg is usually poorly developed, although in the male of some forms it is often modified to form a prehensile palp for grasping the female during copulation. The second thoracic appendages are pediform and armed with a long tapering claw. The third thoracic legs are used in clearing the inside of the shell cavity of foreign matter. The abdomen, represented only by the caudal furca, is extremely variable and may be reduced to a pair of fine, tapering spines, or at the other extreme be greatly developed and appear similar to the furca of the Cladocera.

Ostracoda are omnivorous, feeding on all kinds of organic matter found in water environments. The mouth leads to a simple alimentary tract composed of a short oesophagus, a stomach and a short stomach-like intestine. Females may often be distinguished from males, especially in fresh-water species, by the large ovaries which shine through the valves in the posterior portion. The male organs are large, often coiled, masses of tubes which may also extend into the anterior part of the valve cavity. Spermatozoa in some species are longer than the body of the adult. Many Ostracoda are parthenogenetic, the males of some species being entirely unknown. The eggs develop into nauplii which somewhat resemble the adult shape and then pass through several molt stages during which the shape of the shell changes greatly. Externally the larval forms commonly resemble the Cyprid larvae of the Cirripedia in form and shape.

Ostracoda are found in salt and brackish water as well as in freshwater and in all intermediate concentrations. Many are very tolerant of wide ranges in salinity and almost all forms will withstand extreme conditions of pollution and water contamination. A few species prefer brackish water only, while the remainder are either distinctly marine or freshwater species. There are some pelagic groups among the marine Ostracoda, but the great majority of all ostracods live on or near the bottom, or in some cases in the upper few millimeters of the bottom oozes, through which they root and burrow after food. The nature of the substrate often determines the species which will be found at any given place, some forms preferring a hard sand bottom, others soft mud, clay or ooze. Still other species are always associated with algae to which they cling and which serve as a substratum independent of the bottom. Some ostracods prefer warm water while others have never been found except in cold water regions, either in the north or south, or at depths which have cold water the year around. Many ostracods, however, are seemingly unaffected by temperature and are found in waters of greatly varying temperatures. The Ostracoda are of economic importance because of their habits as scavengers, aiding other small forms in the restoration of polluted areas. They are known as fossils as far back as early Paleozoic times (Ordovician).

There are four suborders of the Order Ostracoda, as follows: (1) Suborder *Myodocopa.* Includes marine forms exclusively, which have a well marked aperture at the anterior portion of the valves through which the antennae may be extruded. Only the second antennae, however, are locomotory, and these have a greatly expanded basal portion which contains enormously developed muscles. One genus (Asterope) includes forms which alone among all the Ostracoda, possess gills. A well developed heart is present in the Order Myodocopa. (2) Suborder *Cladocopa.* This is another exclusively marine group composed of a small number of forms which lack an anterior shell aperture, and in which both antennae are natatory. Valves are of a more or less circular shape and can be closed tightly all around their edges. Eyes and a heart are completely lacking. (3) Suborder *Platycopa.* This small group includes a few marine species, usually grouped within a single genus, which exhibit strikingly different anatomical features. The appendages are much flattened, especially the second pair of antennae. There are no visual organs nor heart. (4) Suborder *Podocopa.* This suborder includes mostly fresh-water forms. The exopodite of the second antenna is either entirely rudimentary or persists as a long simple flagellum. The valves are without an anterior aperture and are commonly flattened on the ventral side. No heart is present and a simple ocellus, either single or paired, forms the visual apparatus. This suborder contains by far the greater bulk of all the known Ostracoda and is separated into two groups,

which are commonly given family rank. In the first group the internal structure is similar to that described for the typical ostracod, while in the latter, smaller group there are three similar pairs of thoracic appendages present and many species belonging to the second group are either brackish water forms or are marine.

WILLIS L. TRESSLER

References
Sars, Georg O., Ostracoda, in "The Crustacea of Norway," Vol. 9, Bergen, The Museum, 1928.
Tressler, Willis L., Ostracoda, in Ward and Whipple "Freshwater Biology," 2nd ed., Ed. by W. T. Edmondson. New York, Wiley, 1959.

OVARY, ANIMAL

Vertebrate Ovary

Origin of germ cells. The ovaries are the essential female reproductive organs and in them the eggs or ova develop. It is debated where the first primordial germ cells differentiate: some hold they arise from the germinal epithelium of the primordial ovary; others that they arise elsewhere (probably from the endoderm of the yolk sac) and subsequently migrate to the position of the genital ridges.

Histology of mammalian ovary. The paired ovaries are relatively solid ovoid bodies which are attached at their hila to the broad ligament by means of the mesovarium. The surface of an adult ovary is often scarred, pitted and uneven; this is due to having shed many ova, and having ova in the process of development. The ovary may be divided into a cortex (which contains the follicles) and a medulla which is composed primarily of connective tissue stroma, nerves and blood vessels. Covering the surface is the germinal epithelium and in the embryonic ovary it proliferates downward into the stroma, carrying with it the oögonia. Groups of cells are cut off from the surface and form primordial follicles. In the human ovaries there are said to be about 400,000 of these at birth, but only about 400 reach maturity; the others degenerate *in-situ* (*atretic follicles*). Beneath the germinal epithelium is a thickened sheath, the *tunica albugina.*

As development proceeds the follicles (Graffian) sink more deeply into the cortex and increase in number and in size. The maturing oocyte accumulates yolk and a *zona pellicuda* is formed. Its increase in size compresses and moulds the surrounding stroma into a fibrous membrane, the *theca folliculi,* which is further divisible into an outer dense fibrous layer *(theca externa)* which, according to some, gives rise to the interstitial cells. Concomitant with other changes in the ovary is the multiplication of the follicle cells. As they form several layers in thickness around the oögonium, small cavities (atria) appear among them; these fluid filled cavities eventually fuse, producing an enlarged antrum on one side of the oöcyte. The follicle cells now constitute the *stratum granulosum* which forms a lining for the follicular atrium and a mound or thickening of cells which project into the follicular cavity (*cumulus oöphorus*). The cells of the *cumulus oöphorus* which immediately surround the oöcyte, become columnar

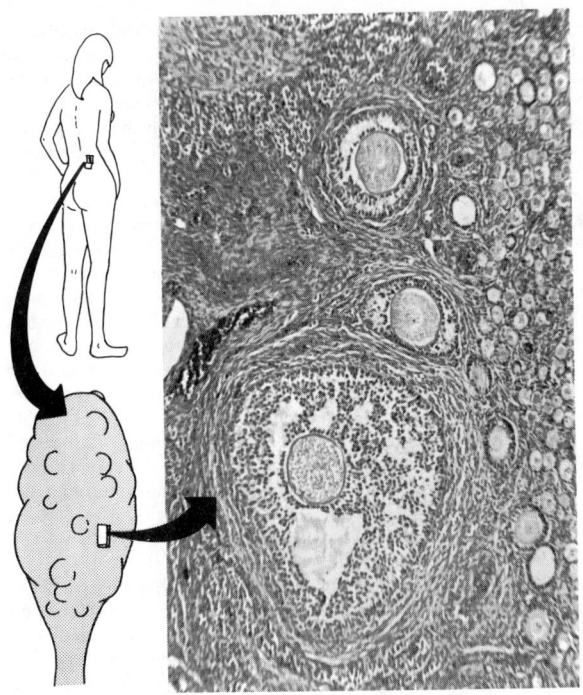

Fig. 1 The human ovary. The photomicrograph at the right shows a ripe egg in a follicle surrounded by nurse cells. The small cells at the right of the photomicrograph are oöcytes and all the intermediate stages can be seen.

and arrange themselves radially into a layer known as the *corona radiata*. By this time the oöcyte and follicle have greatly increased in size; for example, in man the former measures 100 to 150 μ in diameter. Both the growth of the egg and the follicle have been under the stimulus of the pituitary gonadotropic hormones (follicular stimulating and luteinizing). When a maximum size for the follicle has been reached, ovulation occurs.

Ovulation. The actual mechanics involved in the rupture of the follicle has not been clearly established; some suggest it is due to pressure caused by the secretion of follicular fluid by the follicle cells; others that it is due to contraction of smooth muscle fibers in the follicular wall and still others hold it is due to an enzymatic action along with the factors cited above. In any case, when the egg, surrounded by its cumulus cells, escapes from the ovary, it glides in a thin film of fluid to the mouth of the oviduct which is close at hand. It is directed into the oviduct by the current set up in the fluid by the beating of cilia in the oviduct. If sperm are present in the oviduct, they make their way between the cumulus cells (aided probably by the dissolving action of the hyaluronidase released from the acrosomes) until one fertilizes the egg and a zygote is formed.

Formation of corpus luteum. In placental mammals the ovary, in addition to forming the female germ cells, serves as a source of hormones necessary for the development of the embryo. In the ruptured follicle a blood clot is formed (*corpus hemorrhagium*) which is soon replaced by a group of relatively large, yellowish appearing glandular-like cells, the *corpus luteum*. The life of the *corpus luteum* depends upon whether or not the egg is fertilized; if it is fertilized, the *corpus luteum* persists as the *corpus luteum* of pregnancy; if unfertilized it remains for only about two weeks and is replaced by scar tissue (*corpus albicans*). The ovary secretes at least

two well defined sex hormones, namely, progesterone and estrogen. Progesterone from the *corpus luteum* causes the uterine glands to function, readies the uterine mucosa for reception of the developing zygote and inhibits spontaneous contraction of the uterine wall. Estrogen is produced by follicle cells and it stimulates growth and development of the female reproductive tract and mammary glands.

Ovaries of amphibia, fishes, reptiles and birds. The ovaries of amphibia, fish, reptiles and birds have much in common. Their eggs differ from mammals in possessing much yolk and their follicles do not possess fluid filled antra. The oögonia, which show well defined polarity, reveal also a well developed yolk nucleus. The follicles are made up of flattened granulosa or follicle cells and the *theca folliculi*. The mature eggs usually possess, in addition to the primary and secondary membranes of mammals, tertiary membranes formed by special regions in the oviduct, e.g., jelly of frog's egg, albumen and the shell membranes of amniots. The reproductive period is seasonal and unlike the condition in placental mammals, these do not display an estrous cycle.

Ovaries of Invertebrates

Protozoa. Although sex is manifested in the protozoa, the differentiation of well defined sex organs and germ cells does not occur, except in Volvox, where, through a division of labor, certain cells enlarge, become yolk-laden and function as ova. In Trichonympha, Cleveland has reported that asexual reproduction results in two unlike organisms or gametes, one of which is comparable to a sperm and the other to an egg.

Porifera. Sponges reproduce asexually by budding and gemule formation as well as sexually by the reproduction of eggs and sperm. The origin of the germ cells and ovaries is debated. Some think they arise from amoeboid wandering cells which have migrated into the mesoglea to a position below the collar cells where the ovary is formed. The egg is surrounded in its growing stages by food laden cells, the archeocytes; the oöcytes grow at the expense of these and in some forms eventually reach 160 to 175 μ in diameter. Numerous yolk granules are often seen in the cytoplasm of the egg.

Coelenterata; e.g., Hydra. In hydra the ovaries are temporary structures; the animal may reproduce either asexually by budding or sexually by the formation of eggs and sperm. The variation in method of reproduction seems to be seasonal; however, Loomis has shown it may be due also to the nature and amount of dissolved gases present in the culture. The ovary develops during the sexual phase by proliferation of interstitial cells between the mesoglea and ectoderm. As development proceeds, the egg enlarges, causing the epithelium to be reflected outward until eventually it disappears over the surface of the egg, leaving the ova naked except for its plasma and vitelline membranes; it remains attached to the animal by connections on its proximal side. Here the ova are fertilized and develop to the diploblastic stage when they become detached from the parent.

Platyhelminthes, e.g., Planaria. The tubellaria are hermaphroditic; the female reproductive system develops from the mesenchyme and is embedded in the parenchymatous tissue near the anterior end of the animal. The ova are extruded, fertilized and yolk cells from the yolk glands (*vitellaria*) added in the antrum; as they pass down the oviduct, a capsule is formed around them. Since many flatworms are parasitic, the chances for fertilization of the egg are reduced and the mortality of the young increased, so that the reproductive systems have compensated for this by becoming highly specialized in the production of large numbers of ova, as occurs in the tapeworms and flukes.

Nemathelminthes, e.g., Ascaris equorum. Ascaris possesses two cord like ovaries which arise slightly posterior to the middle of the body. Each possesses a central protoplasmic mass or rachis along which the developing ova are placed, and evidence indicates that there may be protoplasmic connections between the developing ova and rachis. The ova lie free in the oviduct where fertilization, maturation and the laying down of the shell takes place.

Ascaris equorum eggs are classical material for cytological studies; they possess only 2 to 4 large chromosomes and display chromosome diminution, a condition marking an early differentiation between the germ and somatic cells.

Mollusca, e.g., Crepidula. In Crepidula the primary gonad is ambisexual. It develops first as a functional testis but later transforms into a functional ovary. This process involves a degeneration and phagocytosis of the sperm and the development of ova within the tubules. The oöcytes develop in follicles and at the same time a transformation of the male genital ducts to female ducts occurs. The eggs are eventually ovulated into capsules composed of 100 to 200 each. Interested students of cytology and comparative embryology should study the classical papers by Conklin (J. Exp. Zool., 9; J. Acad. Natl. Sci. Philadelphia, 15) dealing with ascidian eggs, for here are accurately described and beautifully illustrated the organization of the egg, mitosis, maturation, cleavage, fertilization and cell lineage. Some molluscs are also interesting because of the relationship they show in the position of the first cleavage spindle to the direction of spiralling of the shell; others (Dentalium) also display polar lobe formation i.e., a condition where a lobe appears just before cleavage and if artificially removed, certain elements of the embryo are lacking.

Echinoderms, e.g., Arbacia punctulata. The gonads of Arbacia are suspended by folds of epithelium from the ambulaceral plates in the apical half of the body cavity. The immature gonads are indifferent to sex; however, those destined to become ovaries take on a reddish-brown color and in them are developed masses of large rounded follicles. According to Wilson, the ovary consists of the following five layers: (1) an outer ciliated epithelium; (2) collagenous and reticular connective tissue; (3) smooth muscle cells arranged in parallel bands over the surface of each acinus; (4) a second layer of connective tissue with scattered nerve cells and (5) an inner germinal epithelium containing developing ova and nutritive follicle cells. The eggs of echinoderms have contributed much to our knowledge of experimental embryology and physiology.

Annelida, e.g., earthworm. A single pair of ovaries is attached to the anterior septum of segment XIII, one on either side of the nerve cord. The ovary is an elongated sac-like structure, in the base of which immature eggs are found. The eggs mature as they move outward along the tapering or tail end of the ovary. Here the enlarged mature eggs are arranged in a single row, forming an egg string, each cell being surrounded by small follicle cells which probably function to supply nutrition and protection to the egg. The eggs are ovulated into the coelom, make their way to the oviduct and are stored in the ovisac until shed into the cocoon where they are fertilized. Finally, the cocoon is slipped off over the anterior end; the ends close and an egg capsule is formed.

Arthropoda, e.g., insects. There are usually two ovaries, each of which consists of from a few to many tubular ovarioles. In each of these, the oögonia are arranged in a single row with the more mature ones near the oviduct. The ovariole is surrounded by a thin wall composed of an inner coat of epithelium and an outer coat of connective tissue and muscle. Imms divides the ovariole into three sections, a terminal filament, germarium and vitellarium. Three types of ovarioles are recognized by him, depending upon the presence or absence of nutritive (nurse) cells and upon their location within the ovariole: (1) a panoistic type where nurse cells are wanting (e.g., Orthoptera); (2) polytropic type where nurse cells alternate with oöcytes (e.g., Neuroptera, Hymenoptera), and (3) the acrotropic type where nutritive cells are situated at the apices of the ovarioles (e.g., Hemiptera). The eggs of insects usually contain much yolk (*centrolecital*); they are surrounded by a vitelline membrane and a chorion; the latter may exhibit elaborate patterns and possesses one or more openings for the entrance of sperm (*micropiles*).

H. W. BEAMS

References

Andrew, W., "Textbook of Comparative Histology," New York, Oxford University Press, 1959.
Brambell, F. W. R., "Ovarian Changes," *in* Parkes, A. S. ed., "Marshall's Physiology of Reproduction," New York, Longmans, Green, 1956.

Cleveland, L. R., "Hormone induced sexual cycles of flagellates," J. Morph. **85**: 197, 1949.

Conklin, E. G., "The organization and cell-lineage of the ascidian egg," J. Acad. Natl. Sciences, ser. 2, **13**: 1, 1905.

Imms, A. D., "A General Textbook of Entomology," New York, Barnes and Noble, 1964.

Korschelt, E., and K. Heider, "Textbook of the Embryology of Invertebrates," New York, Sonnenschein, 1895.

Loomis, W. F., "The sex gas of Hydra," Sci. American, **200**: 145, 1959.

MacBride, E. W., "Textbook of Embryology," New York, Macmillan, 1914.

Nelson, O. E., "Comparative Embryology of Vertebrates," New York, Blakiston, 1953.

Pincus, G., "The Eggs of Mammals," New York, Macmillan, 1936.

Romanoff, A. L., "The Avian Embryo," New York, Macmillan, 1960.

Rugh, Robert, "The Frog: its reproduction and development," New York, McGraw-Hill, 1953.

OVULATION

Ovulation is the process by which ova, or eggs are released from the ovary. It should not be confused with oviposition, or their expulsion from the body.

Before ovulation can occur the ovum and the follicle containing it have to reach an appropriate degree of maturation. In those species in which hormonic influences have been shown to be at work (especially in birds and mammals) two anterior pituitary hormones are involved, viz., the follicle stimulator (FSH) and the ovulating or luteinizing hormone (LH). The first controls growth of the follicle and the second final maturation and ovum release. FSH secretion and follicle growth are influenced by several factors among which may be listed age of the animal, food supply and length of day. LH secretion and ovulation are also influenced by several factors among which diminishing light, low progesterone level and a neurogenic factor are important. The neurogenic factor manifests itself in those species in which ovulation does not normally occur until it has been provoked by the act of coitus or by some intense psychic sexual stimulus.

Among insects the tsetse fly (Glossina) is an example of coitus-induced ovulation. Probably there are others as many insects need this stimulus for oviposition. The condition may be common in birds. It is certainly present in pigeons though if the female sees her mirror image this is sufficient stimulus. In some other birds, in parrots for example, adequate stimulus is provided by preening or stroking the head and neck with the hand. Probably the courtship display is an ovulatory stimulus in some species. In mammals the condition is widespread but its incidence does not seem to follow any definite zoological pattern. It is found, for instance, in the rabbit and hare, in the cat, mink and ferret and probably in most carnivores except the Canidae. The ground squirrel and probably many other of the Sciuridae display it, and so do some strains of mice and microtines. Several insectivores display this pattern and so does the fruit bat, Plecotus.

In the rabbit ovulation follows coitus at an interval of about 10½ hours and, during the first two hours after coitus, sufficient LH has been released to allow it to occur. The psychic stimulus is central and affects the pituitary by way of the hypothalamus. The essential difference between spontaneous and provoked ovulation is that in the latter the neurogenic stimulus is needed to cause LH release; in the spontaneous ones this occurs as a result of an internal stimulus. Some workers have suggested that if a minute amount of progesterone is released from the growing follicle LH secretion follows. If it is not, the neurogenic stimulus is needed to produce the same effect.

The process of ovulation has been studied in detail in frogs, hens, rabbits and sheep. The follicle consists of three layers, the theca externa, theca interna and the granulosa cells. The latter, in mammals (except Ericulus) are separated in the mature follicles into two layers, a peri-follicular and a peri-ovan (cumulus) layer connected by a bridge of granulosa cells. Liquor folliculi occupies the intervening space and much of the final burst of follicular growth is caused by secretion of this fluid. In species in which it is present this rapid secretion has been widely regarded as the cause of ovulation, acting thus by increasing intrafollicular tension to the point where the free wall of the follicle gives way. The validity of this theory is doubtful since the follicle wall is not torn at the point of rupture and since in at least two mammalian species (cow and bat) intrafollicular pressure seems to fall during the final hours. In species with solid follicles ovum growth has been suggested as the cause of rupture.

In the hen the surface of the ripe follicle shows a long avascular region surrounded by a network of distended capillaries. In this area, the stigma, rupture occurs. It starts as a small triangular slit at one end and, as the ovum begins to erupt through the slit, a tear is produced that extends the length of the stigma. The extrusion of the ovum, which is almost ameboid in character, is believed to be caused by pressure from smooth muscle fibers in the follicle wall, but the nature of the initial slit does not fit the view that a tear due to pressure has opened the follicle in the first place.

In mammals a circular or oval avascular region, similar to the stigma and likewise surrounded by a ring of distended capillaries, develops on the exposed wall of the follicle. Within this region a small cone soon protrudes. This eventually gives way and the liquor folliculi, together with the ovum and some granulosa cells ooze through the opening. All accounts emphasize the slow nature of expulsion; it is not explosive. The rupture point does not resemble a tear as it has a smooth, blunt margin and is round or oval.

Besides the pressure theory, which has been discounted, several suggestions have been made to account for the rupture of the follicle. One of these is that smooth muscle cells contract radially thus causing a point of weakness which gives way. Another is that a necrotic area develops within the avascular region of the stigma and that proteolytic enzymes then erode it. Another theory holds that the rupture is caused in some undefined way by the direct action of gonadotrophic or ovarian hormones. None of these theories has been definitely proved and objections have been advanced to each of them. The main difficulty is to relate the possible causes in species with follicles that contain some liquor folliculi and those that do not.

In most mammalian species ovulation occurs towards the end of heat and, since the follicle is regarded as the source of the estrogens that cause sexual receptivity, it is easy to see that rupture should terminate heat. In the cow, however, ovulation follows the expiration of heat by about 12 hours, as a rule. In the dog and fox ovulation is near the beginning of heat but the first polar body has not yet been extruded from the ovum as it is before ovulation in all other species in which this has been studied.

In some species ovulation only occurs at a certain time of day. The rat ovulates soon after midnight and the necessary LH has been released about 10–12 hours previously. Reversal of the periods of light and darkness causes a corresponding shift in ovulation time. In the hen ovulation of the first egg of a clutch takes place early in the morning. The processes leading to oviposition require about 25–26 hours and a new ovulation follows oviposition in about 30 minutes. Thus, each egg is ovulated about 1–2 hours later than that of the previous day. In a few days the time for ovulation is delayed until 2–3 p.m. At this point the process is held up and the new ovulation does not occur until early next morning. Thus there is an interval of a day between clutches.

S. A. ASDELL

Reference

Greenblatt, R. B., ed., "Ovulation," Philadelphia, Lippincott, 1966.

OWEN, SIR RICHARD (1804–1892)

An English biologist, Sir Richard Owen was born on July 20, 1804, at Lancaster. After studying medicine at the University of Edinburgh, he was influenced by the famous surgeon, John Abernethy, to enter the field of scientific research. In 1856 he became superintendent of the Hunterian Museum.

When the Zoological Society began publication of scientific proceedings in 1831, Owen was the largest contributor of anatomical papers. His *Memoir on the Pearly Nautilus* (1832) was his first significant contribution and was soon recognized as a classic. For over fifty years he made contributions to every division of comparative anatomy and zoology. Among his more important discoveries was that of *Trichina spiralis*, the parasite involved in the disease trichinosis. He proposed the now accepted division of the Cephalopoda into the two orders of Dibranchiata and Tetrabranchiata.

Owen extended his studies to the remains of extinct groups and became a pioneer in vertebrate palaeontology. His work on extinct reptiles was reprinted in his *History of British Fossil Reptiles.* While most of his writings on mammals dealt with extinct forms, he did make important contributions in regard to the monotremes, marsupials, and the anthropoid apes.

JAMES L. OSCHMAN

OXIDATIVE METABOLISM AND ENERGETICS OF BACTERIA

Microorganisms, like other living matter, make use of the electron transport systems for the conversion of

energy into a chemically utilizable form. The "currency of energy" used for biosynthetic processes is in the form of adenosine triphosphate (ATP). The bacterial enzymes which participate in energy formation are found in highly organized structures which are biochemically analogous to mammalian mitochondria but differ in fine structure. The organelles which carry out the electron transport process in bacteria are associated with the cell membrane and are referred to as the electron transport particles. The structure of both the mammalian and bacterial organelles must be maintained intact in order to generate ATP from oxidative energy.

Biological oxidation. Biological oxidation consists of the stepwise catalytic transfer of hydrogen and electrons from an oxidizable compound to one that can be reduced. Under aerobic conditions, the final transfer of electrons involves the reduction of oxygen. Although the exact sequence of events and carriers is as yet unknown, a diagrammatic scheme for electron flow can be formulated on the basis of the known enzymes and coenzymes found in the isolated intracellular organelles which carry out these reactions, on the basis of the O/R potentials of the isolated coenzymes which act as electron carriers, and by spectrophotometric studies of the interaction of the respiratory chain components (Fig. 1). The sequence of electron flow in some bacterial systems is similar to that described for mammalian systems. It differs, however, from other schemes, in that an additional coenzyme, the naphthoquinone, may be substituted for the benzoquinone usually depicted in schemes of electron transport for mammalian systems (see Coenzyme Q). Both benzo- and naphthoquinones have been implicated as coenzymes in electron transport and in oxidative phosphorylation (see Bioenergetics below). In addition bacterial systems may contain a number of electron transport pathways not found in mammalian systems. These different electron transport pathways are a reflection of the diverse substrates which bacteria are capable of oxidizing. The scheme shown in Fig. 1 cannot be applied to all microorganisms since they differ considerably in the nature of their terminal respiratory chains. Microorganisms differ from one another in the nature and type of quinone(s) and in cytochrome composition.

The transfer of two electrons through this series of exergonic reactions usually results in the generation of three molecules of adenosine triphosphate. Although the mechanism of the coupling reactions between electron transport and the generation of phosphate bond energy remains obscure, the loci of the energy "transformers" in the electron transport chain have been described. Energy-rich phosphate bonds are generated between DPN and the quinone, between the quinone and cytochrome b (possibly at the naphthoquinone level), and at the cytochrome level of oxidation-reduction. All of the enzymes which carry out the reactions shown in Fig. 1 have been demonstrated in aerobic microorganisms. Such oxidations are generally referred to as complete oxidation since the end products are H_2O and CO_2 derived from decarboxylation at the substrate level. Some microorganisms, however, lack some of the enzymes necessary for complete oxidation and thus can carry out only a one- or two-step oxidation. This type of oxidation is referred to as incomplete and results in the accumulation of oxidized compounds.

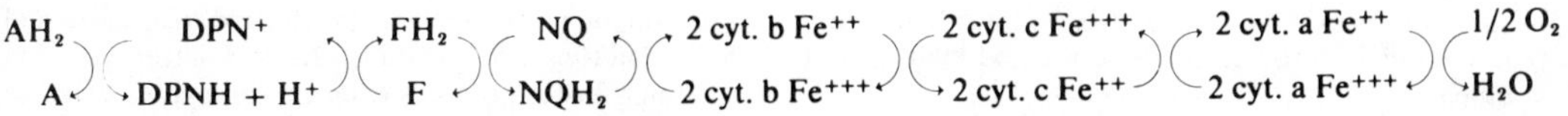

$$AH_2 \diagdown \diagup DPN^+ \diagdown \diagup FH_2 \diagdown \diagup NQ \diagdown \diagup 2\,cyt.\,b\,Fe^{++} \diagdown \diagup 2\,cyt.\,c\,Fe^{+++} \diagdown \diagup 2\,cyt.\,a\,Fe^{++} \diagdown \diagup 1/2\,O_2$$
$$A \diagup \diagdown DPNH + H^+ \diagup \diagdown F \diagup \diagdown NQH_2 \diagup \diagdown 2\,cyt.\,b\,Fe^{+++} \diagup \diagdown 2\,cyt.\,c\,Fe^{++} \diagup \diagdown 2\,cyt.\,a\,Fe^{+++} \diagup \diagdown H_2O$$

Fig. 1 The following abbreviations are used. A, AH_2 represents oxidized and reduced substrate; DPN and DPNH, oxidized and reduced diphosphopyridine nucleotide; F and FH_2, oxidized and reduced flavoprotein; NQ and NQH_2, oxidized and reduced naphthoquinone.

Citric acid cycle. The citric acid cycle appears to be the chief pathway for oxidative metabolism by most microorganisms and serves as the major route for energy formation, synthesis of amino acids and assimilation of carbon compounds. The existence of this pathway in bacteria was difficult to establish because of the lack of permeability to intermediates of the cycle by whole cells. Procedures used to establish the existence of this pathway in animal tissue could therefore not be used with bacteria.

Evidence for this oxidative pathway in bacteria was finally obtained by a number of different methods. The enzymes of the citric acid cycle were demonstrated in cell-free extracts. Analysis of the intracellular pools following the addition of C^{14}-labeled acetate showed that the C^{14} was equally labeled in all the intermediates of the cycle and the distribution of the label in the carbon atoms of the intermediates was that expected from the oxidative reactions of this cycle. Convincing evidence was obtained with mutants of bacteria which were blocked at the condensing enzyme but which retained the ability to form acetyl-CoA. These experiments showed that the condensing enzyme, a vital step in the citric acid cycle, is necessary for acetate oxidation by microorganism.

The individual enzymes found in this cycle have been isolated from bacteria and studied extensively. They are generally associated with intracellular organelles; however, some of these enzymes can be solubilized from the particles and purified by fractionation procedures. The necessity of intact particles for the coupling of phosphorylation to electron transport can be interpreted as a requirement for a specific spatial orientation of the enzymes and coenzymes involved in these reactions. The enzymes of the citric acid cycle and terminal respiratory chain are associated with the same intracellular structure.

Dehydrogenases. The electrons resulting from the dehydrogenation of isocitrate, α-ketoglutarate and malate are linked to oxygen by the terminal respiratory enzymes shown in Fig. 1. The dehydrogenase involved in the oxidation of malate is DPN^+-linked, whereas isocitric dehydrogenase is TPN^+-linked. Dehydrogenation of α-ketoglutarate differs however, in that it probably involves DPN and lipic acid. In addition to the DPN^+-linked malate dehydrogenase, some bacteria also oxidize malate by a DPN^+-independent pathway referred to as malate-vitamin K reductase. This flavin enzyme (FAD) links to the DPN^+ chain at the naphthoquinone level of oxidation-reduction.

Malate + $DPN^+ \rightarrow$ Oxalacetate + DPNH + H^+

   (Malic dehydrogenase)   (1)

(FAD)

Malate + NQ$\rightarrow$Oxalacetate + NQH_2

   (Malic-vitamin K reductase)   (2)

Succinate oxidation differs from DPN- and the malate-linked pathways in that it is mediated by a different flavoprotein and quinone and linked to the terminal respiratory chain at the cytochrome b level of oxidation-reduction.

TPNH oxidation is utilized by biosynthetic reactions and for ATP synthesis through the respiratory chain. An enzyme has been found in bacteria which transfers hydrogen from TPNH to DPN.

TPNH + H^+ + $DPN^+ \rightleftharpoons$ DPNH + H^+ + TPN^+ (3)

This enzyme, pyridine nucleotide transdydrogenase, may permit the oxidation of TPNH to occur via the DPN-linked electron transport pathway. The phosphorylation associated with TPNH oxidation arises from the DPN^+ pathway. The reversal of the transhydrogenase reaction (reduction of TPN from DPNH) was shown to require energy. The energy-dependent reversal of electron transport has been demonstrated in certain microorganisms.

Flavoproteins. There are many flavin enzymes which act as mediators in oxidative reactions. These enzymes differ from one another in the nature of their prosthetic group, in the type of compound oxidized and in their reactions with various electron acceptors. The prosthetic group of most flavoproteins is either riboflavin monophosphate or flavin adenine dinucleotide. A few electron transport enzymes are known which are activated by riboflavin.

Several flavin enzymes which promote the oxidation of reduced pyridine nucleotide coenzymes have been described. These enzymes differ with respect to the nature of the electron acceptor and the rate at which the acceptor can be reduced. The reaction carried out by each enzyme is generally specific with regard to the electron donor and acceptor. Cytochrome c, oxygen, glutathione, nitrate, menadione, vitamin K, and various dyes have all been described as electron acceptors for different flavin-linked pyridine nucleotide reductases. The pyridine nucleotide reductases are of particular importance since they form one of the essential links of the terminal respiratory pathway. Most of the reductases, however, utilize as electron acceptors compounds which are not found in biological systems. Some of the flavin enzymes contain metal ions in addition to the flavin prosthetic group. The metal ions are required for electron transfer to the oxidized acceptor.

Another class of flavoproteins catalyzes the direct transfer of electrons between glucose or amino acids and molecular oxygen. The flavins which react directly with molecular oxygen lead to the formation of hydrogen peroxide. The enzymes which carry out these oxidations are specific with respect to the configuration of the substrate. Thus, the D-glucopyranose configuration is necessary for activity with glucose oxidase and the amino acid oxidases are specific for either the D- or the L-amino acids. Xanthine oxidase differs from glucose or amino acid oxidase in that it is a molybdoflavoprotein like the TPNH-nitrate reductase and hydrogenase enzymes.

Nonheme iron. Although the electron transport particles contain large amounts of nonheme iron little is known concerning the participation and site(s) of action of this respiratory chain member. Nevertheless, iron chelating agents generally inhibit bacterial respiration. The reduction of the nonheme iron of the electron transport particles occurs with both succinate and DPN-linked substrates. Nonheme iron has been shown to participate in the succinoxidase pathway of *M. phlei*. This component interacts between the flavoprotein and cytochrome b. In addition the soluble malate-vitamin K reductase has been shown to contain nonheme iron which interacts between the quinone and dye electron acceptor.

Ferredoxin, an iron-containing protein found in certain microorganisms and plants, has been shown to mediate electron transfer to a flavoprotein which in turn reduces TPN+. In photosynthetic microorganisms

this compound is involved in electron transport and cyclic photophosphorylation. In the latter reaction electron transfer proceeds from chlorophyll *a*, activated by light, to ferredoxin which in turn reduces quinone. Ferredoxin contains two atoms of iron per molecule, is protein in nature and has a molecular weight of 12,000.

Quinones. A large variety of benzo- and naphthoquinones (Fig. 2) are found in microorganisms. The naturally occurring quinones differ with respect to the type of quinone nucleus and with the length and degree of saturation of the isoprenoid side chain. Quinones containing up to 5 saturated units in their side chain have geen found in nature. In addition, natural *cis/trans* geometric isomers of the natural quinones have been demonstrated in bacterial systems. These lipoidal compounds are localized in the subcellular organelles which carry out electron transport and are required for energy generation. The quinones have been shown to function as coenzymes in electron transport and are required for oxidative phosphorylation. Enzymes have been

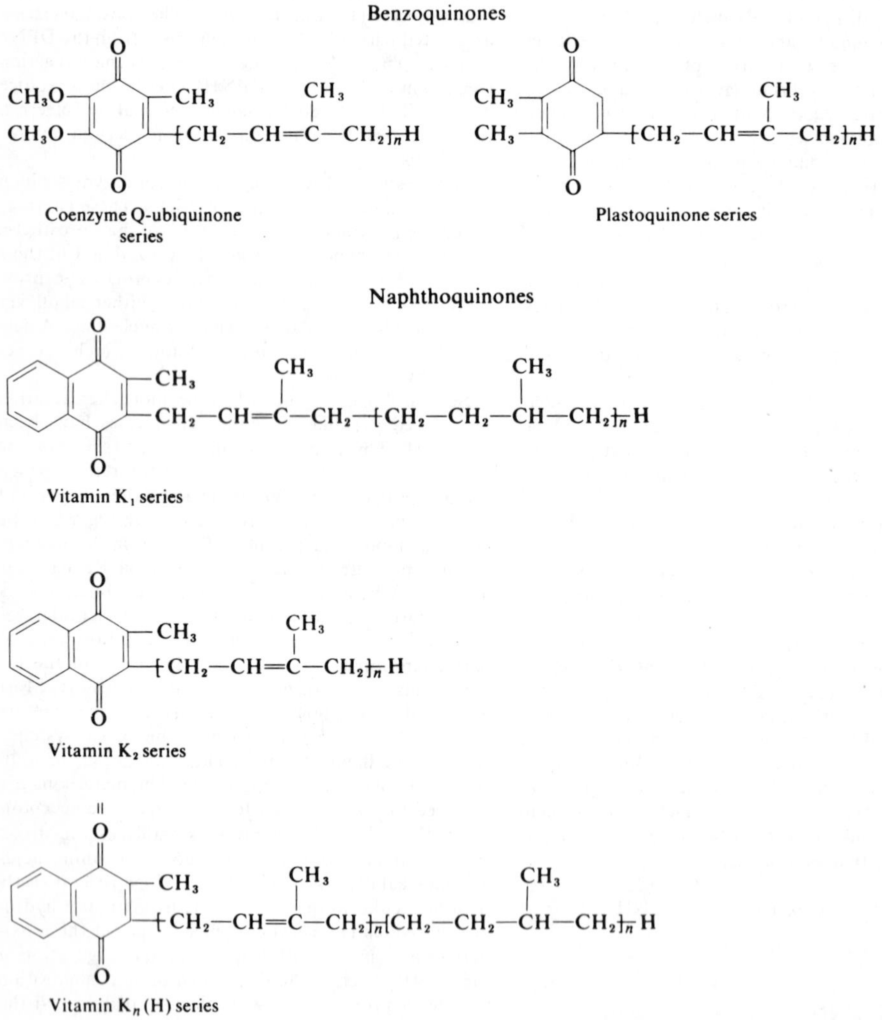

Fig. 2   Naturally occurring quinones.

isolated from bacteria which mediate the transfer of electrons from reduced DPN^+, malate (FAD) or succinate to certain specific quinones. Some microorganisms contain more than one type of quinone, and both appear to be functional in electron transport. For example, E. coli contains both a benzo- and napthoquinone; the napthoquinone serves as a coenzyme on the DPN^+-linked pathway whereas the benzoquinone mediates electron transport on the succinate-linked pathway. The dependence of electron transport and coupled phosphorylation on this lipoidal coenzyme was demonstrated following removal or destruction of the naturally occurring quinones from the subcellular organelles which carry out coupled phosphorylation. Both oxidation and phosphorylation are lost following such treatment and can be restored only by the addition of quinones with certain specific configurations. It is of interest to note that all of the naturally occurring quinones contain a long side chain adjacent to one of the carbonyl groups with a β-γ unsaturated position (Fig. 2). Quinones which restore both oxidation and phosphorylation contain the β-γ unsaturated side chain. Quinones which contain multiple saturated units in their side chain are active in restoring oxidative phosphorylation provided that the first unit adjacent to the ring is unsaturated. However, the cis naphthoquinone does not appear to be active in restoring phosphorylation. Quinone analogues are usually capable of restoring only oxidation. Quinones like menadione, which lack the side chain, restore oxidation by a bypass reaction transferring electrons from DPNH to oxygen or cytochrome b, whereas other quinones like lapachol, which contain substitutions in the 2-position of the naphthoquinone ring, transfer electrons by the same oxidative pathway as the natural quinone. Nevertheless, quinones belonging to the latter group do not restore phosphorylation.

It is interesting to note that certain bacteria have been shown to contain a multiplicity of related quinones. For example E. coli has been shown to contain 9 different benzoquinones ranging from coenzyme Q_1 to Q_9. The major quinone Q_8 has been shown to function in oxidative metabolism; however, a role of the other quinones has not been demonstrated. They may represent remnants of the biosynthesis of the major quinone.

The quinones mediate electron transport between the flavoprotein and cytochrome b on the DPN^+- and succinate-linked pathways. They have been shown to be involved in pyruvate, formate, and nitrate oxidation. Quinone-mediated oxidations are inhibited by dicumarol or other naphthoquinone analogues and following destruction of the natural quinone by irradiation with light at 360 mμ.

Terminal respiratory pigments. All aerobic forms of life possess iron porphyrin, proteins which link the respiratory enzymes to oxygen through a series of one-step electron transfers. Microorganisms, however, differ considerably from one another with regard to the nature of the terminal respiratory pigments. Some aerobic microorganisms contain cytochrome pigments which are similar in absorption characteristics to those found in yeast and mammalian tissues. These microorganisms contain cytochromes whose spectra resemble those described for cytochromes b, c_1, c, a, and a_3; however, they usually differ from those found in yeast and mammalian tissues in their sensitivity to Antimycin A, KCN, CO, and light reversibility. Cytochrome a is often missing in some aerobic and facultative microorganisms. This cytochrome, however, is usually replaced by cytochrome a_1 or a_2 or a mixture of the two. In certain facultative microorganisms cytochrome c and b appear to be replaced by one cytochrome pigment, b_1, whereas anaerobic bacteria with few exceptions are completely devoid of cytochromes. Variations in the type and content of the cytochrome pigments can occur by alteration of the degree of aeration, phase of growth, energy source, and iron content of the growth medium employed.

Although the spectral properties of cytochrome pigments among various bacterial species differ, these pigments, nevertheless, must assume the same physiological role in metabolism as that ascribed to the "typical" cytochromes found in yeast and mammalian tissues. Like mammalian tissues, a phosphorylative site has been demonstrated in the span from cytochrome c to oxygen in a few aerobic microorganisms. The differences between the various cytochromes in bacteria merely represent a variation on the same biological theme.

Bioenergetics. In contrast to animal systems which require organic compounds as a source of energy, some microorganisms are capable of utilizing a variety of other chemical substances for this purpose. The heterotrophic (organatrophic) bacterial forms require organic materials for foodstuff, whereas the autotrophic (lithotrophic) organisms can acquire energy from the oxidation of simple inorganic compounds. For example, organisms belonging to the genus Hydrogenomonos, an autotrophic microorganism, obtain energy by oxidation by molecular hydrogen. Other autotrophic microorganisms found in nature are capable of oxidizing hydrogen sulfide, sulfur, ammonia, nitrate, nitrite, or ferrous salts as their sole source of energy. Although the energy-forming systems utilize different types of substrates or electron donors (organic vs. inorganic compounds) and appear to involve different mechanisms for energy generation, they all have in common a series of oxidative-reductive reactions for the generation of energy-rich phosphate bonds (see Phosphate Bond Energies).

The process by which cells obtain energy is referred to as oxidative phosphorylation. This process differs from other energy-yielding reactions of the cell in that the substrates oxidized are not phosphorylated and that in the transfer of two electrons to oxygen 2 to 4 high-energy phosphate bonds are formed. In addition, oxidative phosphorylation, unlike substrate-level phosphorylation, does not occur under anaerobic conditions. The process of oxidative phosphorylation is measured in terms of the ratio of oxidation or two-electron transfer to ATP formation as an expression of the extent of high-energy phosphate formation. This ratio is called the P/O ratio. The mechanism by which inorganic phosphate is converted to a high-energy form has challenged scientists for over 35 years and remains unknown.

The process of oxidative phosphorylation may be explained with the mechanical energy generator shown in Fig. 3. The overall reaction may be represented by the formula:

$$AH_2 + 11/2 O_2 + P_i + ADP \rightarrow A + H_2O + ATP + CO_2$$

ENERGY GENERATOR

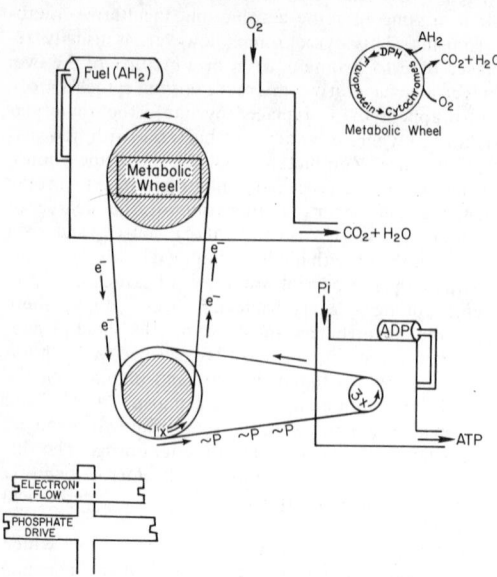

Fig. 3   Schematic representation of the process of oxidative phosphorylation.

Oxygen is utilized for the complete combustion of substrate to yield CO_2 and H_2O. The burning of fuel results in the removal of 2 electrons which in turn cause the metabolic wheel to move. The metabolic wheel is composed of the respiratory carriers, such as DPN^+, flavoprotein, nonheme iron, quinones, and the cytochromes, which are involved in the transfer of electrons and result in the reduction of oxygen to form H_2O. The metabolic wheel or the oxidative phase is coupled by a pulley to the energy phase. The ratio of the metabolic wheel to the phosphate drive is 1 to 3 so that for each turn of the metabolic wheel 3 ATP are formed. As can be seen, in the absence of oxygen or presence of inhibitors of respiration, phosphorylation cannot occur. Uncoupling agents, compounds which inhibit energy formation without interfering with oxidation, can be

viewed as slippage or a brake in the pulley. Under these conditions oxidation would continue, in fact with the loss of friction from the pulley, oxidation may be expected to occur at a faster rate.

A number of bacterial systems have been described which will carry out oxidative phosphorylation. In general these systems differ from intact mammalian mitochondrial systems in that the bacterial systems have lower P/O ratios and fail to exhibit respiratory control by ADP. Loss of sites of phosphorylation and nonphosphorylative electron transport reactions have been demonstrated in bacterial systems and may account for the lower P/O ratios. In only one bacterial system (*Mycobacterium phlei*) have all three phosphorylative sites been shown to be operative. Nevertheless, even with this system the P/O ratios are lower than those observed with mammalian mitochondria. Removal of nonphosphorylative oxidative enzymes results in higher P/O ratios.

Oxidative phosphorylation with most bacterial systems requires the presence of the highly organized electron transport particle and soluble protein components found in the supernatant fraction following removal of the particles from cell-free extracts. These soluble protein factors have been shown to be required for the coupling of phosphorylation to oxidation and are referred to as coupling proteins. Coupling proteins can also be obtained from mammalian mitochondria; however, with the mammalian system, the same coupling proteins are required for all three phosphorylative sites. Different coupling proteins for the three phosphorylative sites associated with oxidation have been demonstrated in a bacterial system indicating that at least one step in each of the phosphorylative sites may be different.

ARNOLD F. BRODIE

References

(1) Kaplan, N.O., Bacteriol. Rev., **19**: 234, 1952.
(2) Brodie, A. F., Federation Proc., **20**: 995, 1961.
(3) Dolin, M.I., "The Bacteria," vol. 2, p. 319, New York, Academic Press.
(4) Smith, L., Bacteriol. Rev., **18**: 106, 1954.
(5) Brodie, A.F. and T. Watanabe, "Vitamins and Hormones," **24**: 447, 1966.
(6) Pinchot, G.B., J. Biol. Chem. **229**: 1 (1957).

p

PAEDOGENESIS

This term was first used by von Baer in 1866 to mean precocious reproduction by a larva, as in some individuals of the gnat *Miastor* and the trematode *Polystomum*. Such precocity implies accelerated development of the reproductive organs relative to that of the whole body. Nevertheless the phenomenon is essentially the same as *neoteny*, a term introduced by Kollman in 1882 for retardation of bodily development, compared to that of the reproductive organs, as seen in those amphibians which breed while in the gill-bearing larval stage.

Bolk, in 1926, used "*foetalization*" to express the persistence of certain foetal or immature characters of an ancestor into the adult stages of a descendant. Man, for instance, shows several characteristics of foetal apes. Thus retardation of development may affect only a few characters instead of the entire body.

DeBeer (1958) favors the use of Garstang's word *paedomorphosis* to encompass all the modes by which larval or immature features of ancestors become adult characters of descendants, and he employs the still broader term *heterochrony* to designate any evolutionary changes in the relative rates of development of characters, reproductive or otherwise, during ontogeny. It is obvious that heterochrony reflects adaptively selected differences in the rate at which genes produce their ontogenetic effects, and that it is, therefore, a fruitful source of evolutionary novelties.

While developmental rates of different characters may be compared easily in a single species, the comparison tends to lose its meaning when made among distantly related animals, for there is no "constant rate" of development, by reference to which another rate may be considered accelerated or retarded. Animals do not go through their ontogenies in uniform periods of time. This is the reason for different interpretations of the broad idea of paedogenesis, as shown by the terms used above. Since comparisons lead to no real distinctions, and since no constant accomplishes this either, it is clear that the terms so far used are simply various designations for one broad group of phenomena, the *change in relative rates of development of characters* during phylogeny, and that this is one way of bringing about evolutionary changes in the nature of these characters. The obvious value of paedogenesis in many cases is a functional one, to facilitate reproduction; but this does not apply to neotenic salamanders, which do not reproduce more rapidly than their transforming relatives, nor to cases of partial neoteny or foetaliza-tion, which result in structural adaptations related to ecology but not to reproduction.

The concept of paedogenesis or neoteny has been used frequently in recent years to account for the origin of certain progressive groups of animals. For example, a larval millipede, halted in the multiplication of body segments, might serve as a structural ancestor of insects. Amphioxus has been interpreted as a specialized derivative of ancient larval cyclostomes. The Enteropneusta show a worm-like, burrowing adult habitus superimposed on an echinoderm type of larva.

In lower taxonomic categories we may also find cases to be explained in this manner. The pigmy sunfishes of the southeastern states (*Elassoma*) differ from other Centrarchidae in a series of characters which are essentially those of immature stages of normal sunfishes. The dwarf salamander, *Manculus*, differs from *Eurycea* by failure of the fifth toe to develop on the hind foot, perhaps in correlation with reduced size of the animal.

THEODORE H. EATON, JR.

Reference

DeBeer, G. R., "Embryos and Ancestors," 3rd ed., Oxford, Clarendon Press, 1958.

PALMALES

The Palmales (alternatively Principes) comprise the palms, a distinctive order and family (Palmae or Arecaceae) of about 230 genera and 2640 species of flowering plants found almost exclusively in the tropics of both hemispheres. They are of very ancient lineage, fossils of palms or palmlike plants dating from the Jurassic and Triassic. With the probable exception of the Cyclanthales, the order is without clear relationship among the Class MONOCOTYLEDONEAE.

Although a diverse group, the palms have in common a hard perennial stem sometimes of very large size and always of woody nature though lacking true wood, plicate leaf blades unique in their development by differential growth associated with splitting from a primordium, flowers with usually two regular whorls of floral envelopes (these sometimes reduced) and a superior gynoecium of uniovulate carpels variable in placentation, seeds with abundant hard endosperm and a small embryo with a single cotyledon.

The primary root of palms is early replaced by subterranean secondary roots and aerial roots are sometimes formed. Stems contain numerous bundles of conducting tissue scattered throughout a softer ground tissue and generally attain maximum girth before elongating significantly. They may creep underground, climb, or stand erect, and are solitary, caespitose, or colonial. Some are fiercely armed with prickles (as may be the leaves and inflorescences) or root-spines.

The leaf axis is divided into a basal sheath which partially or completely surrounds the stem, a petiole, and a blade. Blades are pinnate or pinnately nerved (bipinnate in *Caryota*), costapalmate or palmate. The plicate pinnae or segments are induplicate or reduplicate in vernation, 1-several-nerved, acuminate to obliquely toothed at the apex, and before unfolding are margined with prominent or vestigial strips (lorae or reins) that absciss with the blade apex (hook) and may remain attached to the lowermost pinnae at maturity.

Inflorescences are axillary, borne below (*infrafoliar*) or among the leaves (*interfoliar*) or terminal (*suprafoliar*) when usually very complex. They are paniculate, spicate or capitate on short to long 1-several-bracted peduncles.

Flowers are bisexual or more often unisexual and solitary, in lines (*acervulae*) or in clusters of two to three along the axes (*rachillae*). Bisexual flowers (in Coryphoideae, some Lepidocaryoideae, *Pseudophoenix*) have three imbricate or connate sepals, three imbricate, valvate or partially connate petals (rarely the perianth reduced), six or rarely more stamens, a gynoecium of three (rarely one) carpels separate or united. Remaining palms are monoecious or dioecious with dissimilar staminate and pistillate flowers. The flowers in monoecious genera are commonly in clusters or triads of two lateral staminate flowers and a central pistillate, at least at the base of the rachilla. In some cocoid palms, inflorescences may be entirely staminate or essentially pistillate.

Staminate flowers have three imbricate or connate sepals, three valvate petals (perianth rarely reduced), three to over 200 stamens and often a pistillode. Pollen is 1-sulcate, trichotomosulcate or bisulculate, ellipsoid to globose, with sexine as thick as nexine or thicker and reticulate, tegillate or sometimes warty, spiny or spinulose.

Pistillate flowers have three imbricate or connate sepals, three imbricate or very rarely valvate petals (perianth rarely reduced), six to no staminodes, and unilocular (by abortion) or trilocular ovary (or more in some Cocoideae and in Phytelephantoideae, tricarpellary in some *Chamaedorea* allies). With few exceptions, only one ovule develops into a seed.

Palm fruits are small to very large, 1-3-rarely 10-seeded, dry, fibrous, or fleshy with a smooth, scaly, prickly, or warty exocarp, fibrous or fleshy mesocarp, and thin to bony endocarp. The seed contains solid or hollow, homogeneous or ruminate endosperm and a subapical, lateral or basal embryo. Germination is of three types dependent on the amount of elongation of the cotyledon from the seed and the presence or absence of a ligule sheathing the first leaf. Haploid chromosome numbers of 6, 8, 12, 14, 16, and 18 are reported.

The diversity of characters in leaves, flowers and fruits is reflected in the current acceptance of nine sub

families. Four of these are characterized by the induplicate vernation of the pinnae or segments. These are the Coryphoideae, including about 33 genera and 330 species in both hemispheres, with plamate or costapalmate leaves and mostly bisexual flowers or these more rarely functionally unisexual but not strongly dissimilar; the dioecious Borassoideae, 7 genera and 42 species of Africa and Asia, with palmate or costapalmate leaves; the dioecious Phoenicoideae, one genus and about 12 species of Africa and Asia, with pinnate leaves, the lower pinnae modified into stout spines; and the Caryotoideae, three genera and about 38 species of the Old World, with pinnate, pinnately nerved or bipinnate leaves which are morphologically induplicate but anatomically reduplicate, and basipetally produced inflorescences.

The subfamilies with reduplicate leaves are the Lepidocaryoideae, about 24 genera and 500 species chiefly of the Old World but four genera in the New World, characterized by the imbricated scales that cover the fruit but diverse in floral morphology and with both pinnate and palmate or costapalmate leaves; the monoecious Cocoideae, 27 genera and about 610 species of the New World except two African genera and the pantropical *Cocos nucifera*, with pinnate leaves and fruit characteristically with bony endocarp marked by three pores; the Arecoideae, a somewhat heterogenous group at present comprising about 130 genera and 1100 species of monoecious, more rarely dioecious or hermaphrodite pinnate palms in both hemispheres differing from the Cocoideae in a usually thin endocarp lacking pores; the dioecious Phytelephantoideae, 4 genera and about 8 species in the New World, with pinnate leaves, massive staminate and capitate pistillate inflorescences, the perianth reduced; and the Nypoideae, with monotypic pinnate monoecious Old World genus unusual for its spicate staminate flowers on branches below a terminal capitulum of pistillate flowers, its reduced perianth, and the unique spiny bisulculate pollen.

Palms rank second only to the grasses in economic importance, furnishing indigenous peoples of tropical regions with everything from food to fuel. Commercially, the coconut (*Cocos nucifera*) and the African oil palm (*Elaeis guineensis*) are important sources of oil; coconuts and dates (*Phoenix dactylifera*) are important items of food.

HAROLD E. MOORE, JR.

References

Beccari, O. and R. E. G. Pichi-Sermolli, "Subfamiliae Arecoidearum Palmae Gerontogeae Tribuum et Generum Conspectus" Webbia, 11: 1–87, 1955.
Burret, M., "Systematische Übersicht über die Gruppen der Palmen," Willdenowia 1: 59–74, 1953.
Burret, M. and E. Potztal, "Systematische Übersicht über die Palmen," Willdenowia 1: 350–385, 1956.
Drude, O., "Palmae" *in* Engler, A. and H. Prantl, "Die natürlichen Pflanzenfamilien," 2(3): 1–93, Leipsig, Englemann, 1887.
Hooker, J. D., "Palmae" *in* Bentham, G. and J. D. Hooker, "Genera Plantarum," Vol. 3, London, Reeve, 1883.

PALPIGRADA

Definition. Almost microscopic ARACHNIDA of which the thin, elongated, segmented body ends in a many segmented flagella. The type (*Koenenia mirabilis* Grassi), discovered in 1885 in Sicily, has since been found in a variety of localities.

Anatomy. The body of Palpigrades is divided into three parts: prosoma, opisthosoma and flagella.

The prosoma is covered on the dorsal side by three unequal chitinous plates—a propeltidium covering the first 4 segments, a mesopeltidium, which is the rudimentary tergite of the fifth segment, and a metapeltidium, which is the tergite of the sixth segment. Ventrally, the body is covered with a series of ill defined sternal plates.

The opisthosoma consists of 11 segments with thin transparent tergites and sternites. The last three, greatly reduced, and diminishing progressively in size, form a sort of postabdomen, prolonged as a remarkable 15 jointed flagella which is mobile and almost as long as the rest of the body.

The Palpigrades possess the classical six pairs of prosomian appendages of the Arachnids. The chelicera are formed of three well developed joints. The mandible plays no role in feeding but functions as a walking leg. The first pair of legs, on the contrary, is not used for walking but is a sensory appendage which the animal holds up in front of itself. The internal organs are similar to, but more simple than those of the ordinary type of Arachnids. The mouth has retained its primary position in the cheliceral segment, and the intestinal diverticulae are still metameric. Since the tubes of Malpighi and nephrocytes are absent, excretion is conducted by a pair of well developed coxal glands which project into the opisthosoma. There is no trace of a specialized respiratory apparatus.

Biology. Little is known of the mode of life of these microscopic animals. The geographic range is very large including all tropical regions and hot temperate zones such as southern Europe, Africa, Madagascar, America from California to Chili, the extreme east of Asia and even Australia.

Palpigrades prefer stones half buried in the earth; they remain under these for so long as the ground retains that degree of humidity which suits them. After heavy rains they often walk on open ground. In dry periods, on the contrary, they dig themselves more and more deeply under the surface of the soil. They are extremely photophobic. We know practically nothing about their food, their sexual biology, their reproduction, or their embryonic development. Egg laying is not seasonal, but spread out through the whole year.

Systematics. The Palpigrades belong to only a single family (*Koeneniidae* Grassi), consisting of a score of species divided in several genera. Those characteristics which are peculiar to them, and which are difficult not to consider as primitive, are the segmentation of the sternum, the position of the mouth in the cheliceral segment, 3-jointed chelicerae, mandible not distinguished from walking legs, and the metameric segmentation of the opisthosoma. These made them in many respects the most archaic living Arachnids and necessitate maintaining them in a separate order.

J. MILLOT
(Trans. from French)

PALYNOLOGY

Palynology, or the study of spores and pollen grains, has found application in plant taxonomy, plant geography, climatology, geology, limnology, pedology, and archeology. The study of air-borne pollen and spores (aerobiology) has demonstrated a close relationship between specific pollen and spore types in the atmosphere and respiratory complaints, such as hayfever. The pollen and spores in honey (melito-palynology) can frequently be used to identify the plants from which the honey was made. Adulteration of commercial honey can be recognized by the presence of "foreign" pollen types.

An essential part of the reproductive process of most plants includes the production and dissemination of spores or pollen grains. Spores are characteristic of the lower plant groups (ALGAE, FUNGI), mosses (see MUSCI), and ferns (see FILICINEAE), whereas pollen grains are the male gametophytes of flowering plants (see SPERMATOPHYTA) and GYMNOSPERMS. Enormous quantities of these reproductive bodies are produced in a single season. Some figures on the production of pollen grains by individual flowers and inflorescences are given below: Austrian pine (*Pinus nigra*), 1,480,000 grains per strobilus, 22,550,000 grains per inflorescence; Durmast oak (*Quercus petraea*), 41,200 per flower, 555,000 per inflorescence; Norway maple (*Acer platanoides*), 8,000 per flower; Black alder (*Alnus glutinosa*), 4,445,000 per inflorescence. The spruce (*Picea* spp.) forests of southern and middle Sweden are estimated to produce 75,000 tons of pollen each spring.

Pollen and spores may be carried for long distances by wind because of their small size (10–250 microns) and light weight. Stray pollen grains of alder (*Alnus viridis*) have been found in air samples collected in the middle of the North Atlantic Ocean more than 400 miles from the nearest source. Although many plants are pollinated by insects, even these pollen types may be found in air samples.

The central, living portion of the spore or pollen grain is rich in fats (greater than 50 per cent). The intine, a thin membrane which surrounds the living cell, is composed of the same substances which form the bulk of ordinary plant cell walls. The outermost layer, or exine, is one of the most extraordinarily resistant organic materials known. The chemical composition of the exine has so far defied characterization, but the outer wall of pollen grains and spores will withstand treatment with concentrated mineral acids and alkalis at temperatures as high as 300°C. According to Zetsche (cf. Erdtman, 1954), a principal constituent of pollen and spore walls is the substance sporonine, which is related to terpenes and similar compounds. Experiments by Zetsche indicate that sporonine may undergo a spontaneous degradation by photochemical autooxidation, a fact which may account for the low incidence of pollen and spores in soil samples.

Pollen and spore morphology. The publications of G. Erdtman (see References below) are the standard works on the structure and organization of pollen and spores. In addition to bright-field microscopy, special techniques such as phase contrast microscopy, ultra-thin sectioning, and transmission as well as scanning electron microscopy are used to study the wall layers of spores and pollen grains.

Pollen grains and spores are usually the products of meiotic division in sporogenous tissues. With few exceptions, each of the meiotic products gives rise to a spore or pollen grain. Despite the fact that in most cases the grains are free from one another when mature, the tetrad stage is an important phase of their development, and the surface of each grain shows a distinct and definite relationship to the original orientation of the grain in the tetrad. The part of the pollen grain that is nearest the center of the tetrad is the proximal pole, and that furthest from the center of the tetrad is the distal pole. Pollen grains and spores are rotational ellipsoids about the polar axis and structures may be located on the surface of the grains by latitude and meridian.

The construction of the exine produces characteristic features useful in the identification of pollen and spores. Identification of plants to family, frequently to genus, and in some cases to species and even variety, is possible from the shape of the grain and the marking of the exine. In a few plant groups, the exine consists of a rather uniform, homogeneous sheath. More commonly, however, the exine is complex and consists of an outer ektexine and an inner endexine. The endexine is a smooth membrane which surrounds the intine. The ektexine may consist of small to large projections which may be isolated or clustered, separated or fused. In some groups of plants, the ektexine elements are clustered closely and are fused at the tips, forming a roof or tectum which may be perforated in characteristic ways. These features make it possible to distinguish between tectate and intectate pollen types, depending upon the structure of the ektexine. In addition, the ends of the ektexine elements may be sculptured in various ways, rounded, pointed, elongate, or arranged so as to form a reticulum.

Modifications of the ektexine structure include wings (vesiculate grains, Fig. 1, a), furrows (colpate grains, Fig. 1, b), and pores (porate grains, Fig. 1, c). Grains may have from none to many furrows or pores or both furrows and pores. Tricolporate (3 furrows, 3 pores) grains tend to be sub-triangular in equatorial section (Fig. 1, b) whereas monoporate (Fig. 1, d) and periporate (Fig. 1, c) grains tend to be sub-spherical in equatorial section. The polar axis of pollen and spores may be compressed (oblate) or elongated (prolate).

Pollen stratigraphy. The vast numbers of pollen grains and spores produced and released into the atmosphere during the flowering season eventually settle on lakes, bogs, oceans, and land surfaces. Most of the grains are destroyed, but in the anaerobic sediments of lakes and bogs, the pollen "rain" is incorporated in the

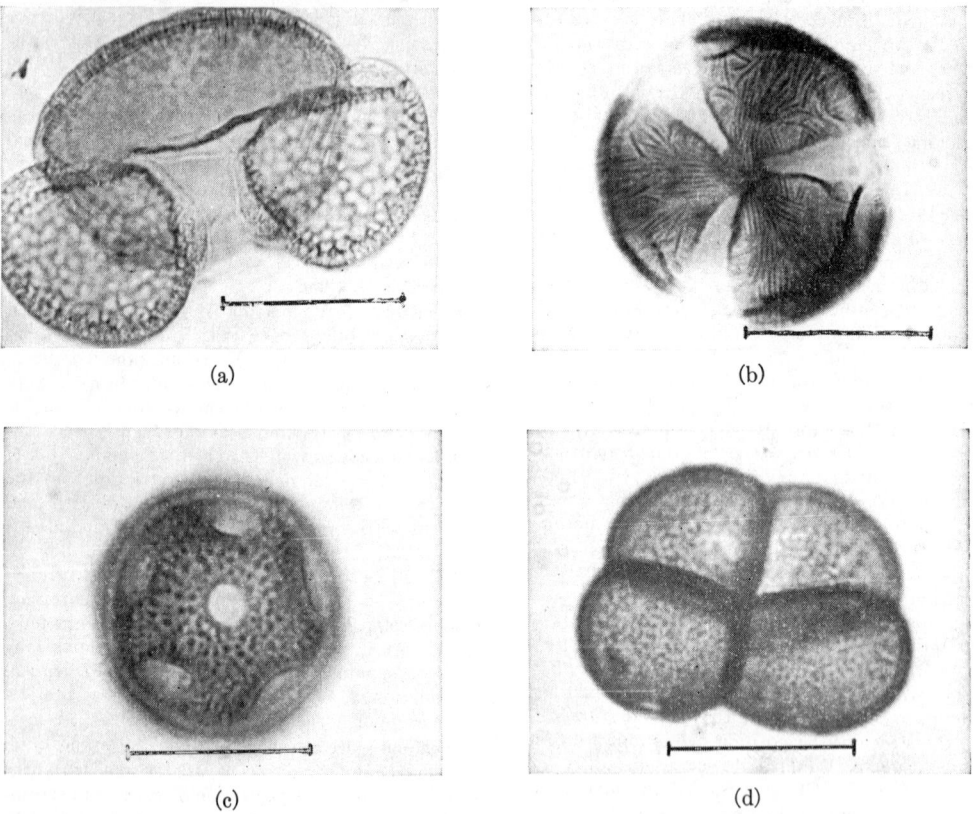

(a)   (b)

(c)   (d)

Fig. 1   Representative pollen types frequently found in lake and bog sediments. Length of the dark line in each microphotograph is 20 microns. (a), *Pinus resinosa*, vesiculate or winged grain; (b), *Menyanthes trifoliata*, tricolporate grain with striate sculpturing of ektexine; (c), *Arenaria groenlandica*, periporate; (d), *Typha latifolia*, tetrad of monoporate grains. Note reticulate marking of ektexine.

annual accumulation of sediment and may be successively buried and preserved. Borings taken through the successive layers of a sediment body permit the recovery of samples from which the pollen (and other micro-fossils) can be isolated, identified, and recorded to give statistical information about the frequency of different pollen types throughout the deposit. The results are usually presented in the form of a diagram which shows the percentage composition of the pollen "flora" at different levels. Climatic or physiographic changes in the deposit and the surrounding area may be inferred from changes in pollen types at different levels.

The effects of pre-Columbian and European forest clearance and agriculture can be recognized in pollen diagrams. Specific pollen types of agricultural weeds and European plants can be identified in deposits adjacent to habitation sites.

The accuracy of identification of pollen types and their parent plants decreases with increasing geological age. In deposits of Pleistocene age it is usually possible to relate most of the pollen types with present-day vegetation types. This relationship is progressively lost in older Cenozoic, Mesozoic, and Paleozoic deposits. Nonetheless, pollen and spore assemblages can be used as guide fossils for the recognition of strata of similar age even if the parent plants cannot be identified with certainty.

The method has limitations, however, due to problems of long-distance transport of "foreign" pollen types, to local over-representation or under-representation of some species, and to "contamination" due to secondary deposition of older or younger materials. In spite of these and other limitations, however, pollen analysis provides a much more representative picture of the flora of a region than do macrofossils, such as leaves, seeds, and wood.

Macrofossil deposits usually represent specialized environments such as swamp forests or river-bottom vegetation. Upland forest and vegetation types are usually poorly represented, if at all. Pollen diagrams, on the other hand, permit the qualitative reconstruction of the upland flora in addition to the vegetation surrounding the deposit.

J. GORDON OGDEN, III

References

Erdtman, G., "Pollen and Spore Morphology I: Angiosperms," New York, Ronald, 1952.
ibid, "An Introduction to Pollen Analysis," Waltham, Mass., Chronica Press, 1954.
ibid, "Pollen and Spore Morphology II: Gymnosperms, Pteridophytes, Bryophytes," New York, Ronald, 1957.
Faegri, K. and J. Iversen, "A Textbook of Pollen Analysis," 2nd ed., Copenhagen, Ejnar Munksgaard, 1965.

PANDANALES

Pandanales is the first order of the Monocotyledoneae in the Engler and Prantl system in the flowering plants. Trees, shrubs, vines, or herbs with long, linear leaves; plants monoecious or dioecious; flowers unisexual, in heads or spikes; perianth of bristles, or scales, or absent; pollen grains binucleate; seeds with endosperm; The order contains the following families:

Typhaceae. Perennial herbs of marshes or damp shores; rhizome horizontal, creeping, thick, crisp, starchy, with two rows of bracts; stem single, erect, as much as 8 ft. tall; leaves taller than the stem, strap-like, $\frac{1}{4}$ to 1 in. broad, thick and spongy; flowers unisexual, insect-pollinated, borne in cylindric spikes, the staminate one terminal, while below and touching or remote from it is the pistillate spike; staminate flowers of 2 to 5 stamens, these distinct or their filaments more or less united, the filaments borne on the spike axis and with several simple or forked long hairs attached near the base; anthers 2-loculed, linear-oblong, the connective often broadened and projecting at the apex; pistillate flowers with a long-stipitate, 1-celled ovary, the perianth of numerous bristles attached along the stalk. At maturity the pistillate spikes break, and the bristles spread, enabling the seeds to be dispersed by wind. Contains the single genus *Typha*, the "cat-tail," of about 25 species of temperate or tropical regions.

Pandanaceae. Trees, shrubs, or vines, mostly with aerial roots or prop roots; leaves firm, strap-shaped, mostly spiny on margins and midrib; flowers dioecious, in spikes or panicles; perianth of rudimentary scales or none; stamens many, free or connate; pistillate flowers of simple, 1-celled, or compound ovaries; fruits bony or drupe-like. The more primitive genus, *Sararanga*, of 2 species (Philippines, and Solomon Is.) has flowers in panicles, all with united perianth. *Freycinetia* (120 species) from tropical Asia to Polynesia has the spicate drupes many-seeded. They are high climbing vines of the rainforest. *Pandanus* (566 species) from tropical west and east Africa to Polynesia are mostly trees with prop roots. Their 1-seeded carpels are in simple or compound ovaries, in heads, spikes, or panicles. They produce edible seeds, and a few have sweet, tasty pulp. All have leaves valuable for thatch, mats, hats, etc. The common name is "screw pine."

Sparganiaceae. Submerged aquatic herbs, growing from a rhizome in the mud; leaves alternate, sessile, linear, 2-ranked; flowers wind pollinated, crowded in separate heads, the staminate ones above; staminate flowers with 3 or more stamens and bracts; pistillate flowers of a single carpel in a ring of bracts; styles simple or forked; fruits indehiscent; nutlets with a spongy exocarp. *Sparganium* (about 18 species), boreal and temperate regions of the Northern Hemisphere, and Australia and New Zealand. Commonly called "bur-reed."

HAROLD ST. JOHN

PARASITISM

Parasitism is a mode of life. A parasite is an organism which depends for some essential food factor which it must obtain from some other organism. This other organism is called the host and is larger than the parasite. The parasite must not kill the host, otherwise it would also destroy itself; consequently it is essentially nonpathogenic, causing disease only under exceptional circumstances. Parasitism differs from PREDATISM, in which the predator kills the prey although there is an intermediate relationship called parasitoidism, in which the host is ultimately killed; this is common among certain insects, particularly the "parasitic" hymenoptera. Parasitism differs also from saprobiosis, in which the organism feeds on material already dead; some saprobionts, however, can become parasites, and it is probable that parasitism evolved from some such primitive association.

Parasitism has frequently been classified as a series of phenomena, depending on the effect on the host. Thus commensalism has been used to describe cases in which there is not only no damage done to the host, there is no reaction to the presence of the parasite. Mutualism, in turn, has been used to express the situation in which the host derives a benefit from the presence of the parasite, while in SYMBIOSIS, the parasite is essential for the wellbeing or even the life of the host. In all cases, of course, the parasite always benefits, unless it causes serious disease or death. In practice, the effect on the host of the great majority of parasites is unknown and these terms can only be used to describe a minority of the cases.

Parasitism is accordingly a branch of ecology and a great many different organisms have acquired a parasitic mode of life. It is a phenomenon of long standing and during its evolution, parasites have become modified and specialized for this mode of life. They have lost organs which are no longer of value and they have hypertrophied old or developed new organs which increase the ability to exist as parasites. A parasite is not degenerate; it is specialized and some of the specialized modifications both of morphology and biology are of the greatest interest. Among the more important of these modifications are the physiological adaptation to the host and the complicated life cycles which ensure eventual infection of a new host.

The physiological adaptation to a host is shown in different degrees and has led to some degree of host specificity; that is, a parasite can only survive in certain kinds of hosts. Sometimes this specificity is very narrow and the parasite can utilize only a single species of host; sometimes it is wide and a large variety of organisms can be infected. It is always present to some degree, however, and in general the older the association, the more strict the host specificity. It follows that related species of hosts tend to have related parasites—a fact which is of value in phylogeny as well as in geographical distribution of both host and parasite. This rule, however, is not absolute.

As a parasite must pass from host to host, and as the transition must be full of hazards, there is usually modification of the reproductive system as well as specialization in the life history which will help to maintain the species. The number of young produced by parasites is often enormous and, as in nature the parasite problem must remain approximately constant (otherwise disease and final extermination of both host and parasite would result) the number of young produced gives some indication of the risks run in this transfer. This increase in progeny is accomplished by the development of asexual multiplication, of hermaphroditism, of colonization, and similar devices. To facilitate transfer from host to host, intermediate hosts are often employed—sometimes as many as four successive hosts are required to complete the life cycle. In addition, special taxes are often developed, especially when no inter-

mediate host is utilized, to assist in this transfer; these include thermotaxis, thigmotaxis, geotaxis, and so on.

The host in which a parasite becomes sexually mature is called the definitive host and that in which some essential developmental change takes place during the transfer, is called an intermediate host.

The distribution of a parasite is determined by the geographical distribution of the host or hosts and by the climatic conditions to which any free stages may be exposed. Few parasites accordingly are cosmopolitan. This distribution can be further controlled by food and other habits of the host.

Although parasites do not normally cause clinical disease, the most destructive diseases are infectious diseases produced by parasites. These diseases are in general, however, caused by some interference with the normal ecology of the parasite. Of all causes, the most important is segregation of the hosts in relatively confined spaces—this is true whether the hosts be human beings, domestic animals or wild animals in confinement. Here the natural destruction of young provided for in nature by the over-fecundity of the parasite, is prevented. Numbers of parasites increase and disease results. Disease may be caused by mechanical damage, by the production of toxic substances, or by the interference of some vital functions. The host often reacts to the presence of the parasite by the production of a temporary or permanent immunity. Disease is essentially due to numbers of parasites.

A great many different groups of organisms contain parasitic members. This is especially the case with the protista and such invertebrate animals as worms and arthropods.

Among the protista are the viruses, rickettsiae, bacteria, fungi, spirochaetes, and the protozoa. All of these have the ability to multiply asexually by simple division. Accordingly, unless held in check in a host by the development of some form of resistance, a single initial infection may cause serious disease.

Among the metazoan parasites, however, the situation is entirely different—especially with the internal parasites. This is because the metazoa have a highly evolved sexual life and often a very specialized obligate type of life history. A single exposure does not cause disease as one young metazoan parasite entering the host cannot multiply either asexually or sexually and its eggs or larvae must leave the host before they, in turn, can infect the same or another host. As disease is usually caused by numbers it results, therefore, in these cases from repeated exposure to the infective stages of the parasites—either free-living or in intermediate hosts. In a few cases, disease can be caused by an abnormal situation for the parasite or the presence of a parasite in an unusual host. It often happens that the larval stage of a parasite can damage the intermediate host, especially if it is an abnormal one.

A few metazoan parasites can multiply in or on the host and this is especially true of some external parasites such as mites and lice. Many ectoparasites can cause indirectly disease in small numbers, however, by the introduction of microorganisms which can multiply within susceptible animals.

While most groups of protista and of metazoa contain parasitic members, some are predominately parasitic while some have no free-living representatives.

The viruses and rickettsiae, for example, are exclusively parasitic, intra-cellular organisms, multiplying at the expense of the cell. So also are the sporozoa among the protozoa; they have a more complicated type of life cycle, involving not only a sexual phase, but a developmental stage outside of the body. Many of the other protozoa are also parasitic and some have actually become symbiotic in that, like the flagellates of termites, they are essential for the life of the host.

Among the metazoa, the trematodes and the cestodes are exclusively parasitic. The trematodes all require a molluscan intermediary in which a process of asexual multiplication takes place while some require a second or third intermediate host before reaching the definitive one. The trematodes possess a digestive tract unlike the cestodes which have none and which in consequence are parasitic in the digestive tract of their host. As a result of this they have in many cases become very host specific. The small group of possibly related thorny-headed worms, also are without a digestive tract at any stage of their development.

Several different groups of nematodes are exclusively parasitic but a great many more are free-living. Parasitism has occurred in several different evolutionary lines in this phylum and both plants and animals act as hosts.

This is the case also with the arthropods where parasitic habits have developed in several different groups of both arachnids, insects, and crustacea. Some of the most completely parasitic animals known are crustacea —especially among the barnacles, the copepods, and the isopods—where in many cases, modifications have become so extreme that only a larval stage will show its true relationship.

All other phyla, while predominately free-living, have a few parasitic members, the only known exception being the Echinoderma. All of these phyla have parasites, however, and it appears to be true that every species of animal is either a host or a parasite, so universal is the phenomonen.

THOMAS W. M. CAMERON

References

Croll, Neil A., "Ecology of Parasites," Cambridge, Harvard University Press, 1966.
Cameron, T. W. M., "Parasites and Parasitism," New York, Barnes and Noble, 1956.
Sprent, J. F. A., "Parasitism," Baltimore, Williams and Wilkins, 1964.

PARATHYROIDS

The parathyroid glands are small reddish-brown endocrine organs located in the neck adjacent to the THYROID from which they are distinct in structure and function. These glands are present in amphibia and all higher species but not in the teleostii or lower forms. They are four in number in man and most other mammals and the amphibia but only two in number in rats. They are derived embryologically from the third and fourth branchial pouches and histologically consists of bands of epithelial cells of two types, the oxyphil and chief cells. The latter are the predominant type. Histological and cytological evidence suggests that they are the cells which synthesize and secrete the parathyroid HORMONE.

The primary function of these glands, mediated by their secretion of the parathyroid hormone, is to maintain the concentration of calcium ions of plasma and extracellular fluids within narrow limits in spite of wide fluctuations in calcium intake, excretion and deposition in bone. Conversely the secretory function of these glands is controlled by the level of ionized calcium in the plasma of the organism. These two regulatory functions constitute a negative feedback mechanism, hormone being released in response to a lowering of the concentration of the calcium ions in plasma, the released hormone acting at peripheral sites to raise the concentration of calcium ions and thereby inhibit further production of hormone. There is no evidence that the parathyroids are under the control of a trophic hormone from the anterior pituitary.

The parathyroid hormone raises the concentration of calcium ion in plasma by increasing the rate of bone resorption, increasing the gastrointestinal absorption of calcium and increasing the reabsorption of calcium by the renal tubule. In addition, in increases the excretion of phosphate by the kidney. This latter effect tends to raise the calcium concentration indirectly due to the fact that these two ions are in equilibrium with the calcium phosphate salts (hydroxyapatite) in bone mineral ($Ca^{++} + HPO_4 \rightleftharpoons$ hydroxyapatite), and a lowering of phosphate concentration results in dissolution of hydroxyapatite and an elevation of calcium ion concentration in plasma.

The hormone has recently been purified from bovine glands and is a polypeptide having a molecular weight of approximately 9200. It contains all the common amino acids except cystine but the exact sequence of these in the molecule has not been determined. Mild oxidation results in inactivation of the hormone. Under the proper conditions activity can be regained by subsequent reduction.

Disease states associated with over- and underactivity of the gland are observed in man and dog. Hyperparathyroidism is glandular overactivity due to either adenoma or hyperplasia. It is characterized by hypercalcemia, hypophosphatemia, hypercalcuria and hyperphosphaturia associated with renal calculi and a bone disease, osteitis fibrosa cystica. The present treatment of the condition is surgical removal of the hyperfunctioning tissue.

Hyproparathyroidism is rare except due to inadvertent surgical removal of these glands during thyroid surgery. It is characterized by hypocalcemia, hypocalcuria, hyperphosphatemia and hypophosphaturia, leading to tetany, epilepsy, cataracts and calcification in the basal ganglia as well as trophic changes in the skin and nails. A high calcium, low phosphate diet supplemented by large doses of vitamin D is the method of current therapy since at present no preparations of the purified hormone are available.

HOWARD RASMUSSEN

References

Gaillard, P. J. ed., "Parathyroid Glands," Chicago, University Press, 1965.
Greep, R. O. and R. V. Talmage, "Parathyroids," Springfield, Ill., Thomas, 1961.

PARENCHYMA, PLANT

As the greek derivation of the word implies ("that which is poured in beside") the word parenchyma was originally intended as a name for soft or succulent tissues, or cell systems, which appeared to be filled in around fibrous, hardened or WOODY tissues; and it was used in contradistinction to this latter kind of tissue (later referred to as sclerenchyma). For example, the Asa Gray textbook of 1845 described parenchyma as the "Common cellular tissue, such as that which forms the pith of stems, the outer bark, and the green pulp of leaves. . .". As such, the original term seems to have been intended as a name for a type of tissue, in the sense in which tissues were conceived in a less sophisticated age. Later attempts were made by 19th and 20th century botanists, as refined methods and techniques of inquiry led to improved knowledge of the origin, structure and function of cells, to characterize the kind of cell making up this parenchyma. The term, however, encompassed such a wide variety of cellular types that any definitions at the cellular level were necessarily either broad and vague, or so precise as to exclude some cellular types which might have been included under the original meaning.

The present use of the term parenchyma is based on the fact that many tissues fitting the original meaning consist of (a) relatively thin walled, (b) many sided and possibly isodiametric cells, (c) which remain in a living condition (i.e. contain protoplasm) beyond the period of time for growth, and (d) which may (but not necessarily) be capable of renewed growth and division: Any cell fitting these criteria in the main, regardless of origin or physical location, may be called parenchymatous. It can readily be appreciated that the term is of general descriptive use only, since it lacks the specificity necessary for definitive studies of cells and tissues in all their myriad of ranges of form and function; nor does it serve a specific developmental purpose in describing the ontogenetic growth and differentiation of cells from their embryonic or meristematic stage to maturity.

ALBERT S. ROUFFA

PARIETALES

The Parietales is a large, artificial order of ANGIOSPERMS, including, according to Engler and Diels (1936), thirty-one families. Subsequent authors have not agreed with Engler and Diels, or with each other, for that matter, on the proper placement of the families of the Parietales. The order has been broken down into a number of smaller orders, and many of the families have been referred to other existing orders. The opinions of phylogenists are widely divided as to the present status of the order.

Those families referred to the Parietales by Engler and Diels have the following variety of characteristics: Flower biseriate, actinomorphic or zygomorphic, perfect, 5-merous, the 3 carpels apocarpous or syncarpous, epigynous or hypogynous, placentation parietal or sometimes axillary, endosperm present or absent, ovules usually numerous. The order was placed between the Malvales and Opuntiales by Engler and Diels. The majority of the families are tropical, only the Guttiferae (Hypericaceae) and the Violaceae being well represented in the North American flora. The size of the families is generally quite small, varying from a single species in the Medusagynaceae and Strausbergiaceae to about 900 in the Flacourtiaceae, Guttiferae, and Violaceae.

The phylogenetic (and hence taxonomic) status of the families of the Order Parietales is still in considerable doubt. Further phylogenetic evidences or more refined interpretations of presently available evidences will no doubt result in further realignments of the thirty-one families which have from time to time been included.

Several of the families are of particular interest. The Guttiferae includes the St. John's Worts (*Hypericum*), distinguished by the presence of many thin, translucent spots on the leaves. During the Middle Ages this plant was used as a medicant for puncture wounds—an excellent example of the Doctrine of Signatures. Among the dominant trees of southeastern Asia and the nearby Pacific Islands are many members of the Dipterocarpaceae, often reaching more than 100 feet in height. The Tamarisk or Salt Cedar (*Tamarix*-Tamaricaceae) is a familiar shrub in many desert areas, including the southwestern United States, where it is a serious agricultural problem, because of its propensity for absorbing great quantities of water from the soil. Among the violets (*Viola*-Violaceae) of Europe and North America are found some of the most complex patterns of variation and speciation in the Angiosperms. A number of genera (*Begonia*-Begoniaceae and *Passiflora*-Passifloraceae, for example) have beautiful flowers and are highly prized ornamentals.

NORMAN H. RUSSELL

PARTHENOGENESIS

Parthenogenesis ("virgin birth") is the reproduction of organisms from either male or female gametes without the concurrence of a gamete of opposite sex. It has appeared in organisms originally reproducing by *zygogenesis*, where male and female gametes and nuclei fuse. Reproduction by a male gamete, *male parthenogenesis*, occurs in some isogamous algae. Reproduction by an unfertilized ovum, *female parthenogenesis*, occurs in some plants and in nearly all animal classes: it is termed *obligatory* when it is the sole method of reproduction; *facultative* or *accidental* when ova can develop into adults whether fertilized or not and *rudimentary* when development ceases at a pre-adult stage. A *parthenogone* is a parthenogenetically-produced organism.

Female parthenogenesis may exhibit *arrhenotoky* —male production, *thelytoky*—female production, or *deuterotoky*—production of both sexes. The alternation of parthenogenetic and zygogenetic generations, e.g., in rotifers, cynipids, cladocerans and aphidids, is termed *heterogony*; in a specialized type, *paedogenesis*, e.g., in cecidomyids, it is some pre-adult stage that is parthenogenetic. Between zygogenesis and thelytoky is *pseudogamy (gynogenesis)* in which the spermatozoon enters and activates the egg but degenerates without its nucleus fusing with that of the egg, the parthenogones being female; examples are the plant *Atamosco mexicana* and some planarians and nematodes. In *geographical parthenogenesis* the parthenogenetic race or species occupies a different locality from that of its zygogenetic progenitor and is often polyploid; seen in many angiosperms and some crustacea and insects.

To appreciate the cytological adaptations involved in parthenogenesis normal zygogenesis must be understood. In zygogenetic animals where the male is the heterogametic sex, the zygoid (usually diploid) oogonium gives rise after meiosis to the functional hemizygoid (usually haploid) ovum and three hemizygoid polar bodies which degenerate. The zygoid spermatogonium also undergoes meiosis, forming four functional hemizygoid spermatozoa, two of which are male-determining and two female-determining. The spermatozoon activates the ovum and at syngamy the zygoidy and the nucleo-cytoplasmic (K/P) relationship essential for development are restored. The sex of the zygote depends upon which kind of sperm fertilizes the ovum.

From the cytological standpoint two types of parthenogenesis are distinguishable. In *hemizygoid (generative)* parthenogenesis the parthenogones arise from haploid (hemizygoid) eggs, whereas in *zygoid (somatic)* parthenogenesis the egg remains diploid (zygoid) or diploidy (zygoidy) is soon restored during development.

Hemizygoid parthenogenesis. Male and female gametes are parthenogenetic in some algae and fungi. Where arrhenotoky occurs (commonly in Rotifera, Hymenoptera, and Thysanoptera, and sporadically in Aleurodidae, Coccidae and Acari) the virgin (haploid) eggs produce males and the fertilized (diploid) eggs females. These males have been regarded as haploid and the females as diploid organisms, the chromosomal relationship being expressed as haploid-diploid or 1:2. Cytological and recent cytophotometric observations do not support this view because in both sexes polyploidization or polytenization occurs and produces increase in chromosome content in some tissues, the degree of this increase varying from tissue to tissue. However, the male germ tissue remains haploid, and males begin development as haploids; the 1:2 ratio may exist as regards some somatic tissues and, in general, the chromosome number and content of male adult tissues are less than those in corresponding female tissues.

The earlier view that male viability is consistent with haploidy may also require revision, since male viability may depend on the increase of chromosome content of tissues. Haploid and probably parthenogonic plants, e.g., *Datura, Nicotiana, Oenothera*, obtained in crossing experiments, have been recorded.

Sex determination under arrhenotoky involves the question of how males arise; for since virgin eggs have only one genome and their mothers two genomes, the balance between the sex genes and other genes is the same in both cases. By using mutant genes as markers, there has been developed a multiple-allele interpretation for *Habrobracon*. At least nine sex-gene alleles exist—*xa, xb*, etc., and sex depends upon which genes are present. Heterozygotes, e.g. *xa/xb*, are female. Their parthenogones, *xa* and *xb*, are hemizygous males. Zygogenetic offspring of *xa/xb* females and *xa* (or *xb*) males are *xa/xb* heterozygous diploid females and *xa/xa* (or *xb/xb*) homozygous diploid males. The latter for some reason are rare and arise after inbreeding.

Zygoid parthenogenesis is much more common than the hemizygoid type. The requisite zygoidy is assured (regulated) by diverse modes either *meiotic (automictic)* or *ameiotic (apomictic)*. The parthenogenesis is usually thelytokous but sometimes deuterotokous.

Meiotic parthenogenesis. Meiosis proceeds more or less normally. Zygoidy is restored during segmentation or somewhat later, e.g., in *Solenobia* moths and stick insects; or two of the four meiotic nuclei fuse as in *Drosophila parthenogenetica*; or one of the meiotic divisions is blocked and reduction in chromosome number thus prevented, e.g., in *Artemia* (Anostraca) and *Lecanium* (Coccidae); or the number of oogonial chromosomes is doubled by endomitosis and after normal meiosis the characteristic zygoid number is assured, e.g., in some planarians and earthworms.

Ameiotic parthenogenesis. This is the prevalent type of parthenogenesis found in angiosperms and animals and in the majority of organisms concerned there is seen the commonest of all modes of regulation. Instead of meiosis a single mitotic division occurs so that the ovum thereby remains zygoid. This is seen in many angiosperms, e.g., *Taraxacum, Euhieracium, Antennaria*, and in rotifers, nematodes, cladocerans, isopods, blattids, tettigonids, aphidids, coccids, cecidomyids, weevils, saw-flies and molluscs. Rarer modes of regulation cannot be described here.

Deuterotoky occurs in heterogonous organisms— cynipids, cladocera, aphidids, strongyloid nematodes, also in some cases, e.g., lepidoptera, where the female is the heterogametic sex. Life-cycles and cytology of heterogonous forms are diverse and complex; sex determination, except in cynipids, is independent of fertilization; the changeover from parthenogenesis to zygogenesis and the number of parthenogenetic generations are determined exclusively or partially by environmental factors, e.g., host, quality of food, light, temperature and crowding.

In lepidoptera the facultatively deuterotokous species, e.g., the Gypsy Moth *Lymantria*, males have a chromosome constitution $2A + ZZ$, females that of $2A + WZ$ (A representing autosomes, W and Z sex chromosomes). Eggs are therefore of two kinds, $A + W$ and $A + Z$, hence in regulation to diploidy a cytological difference is maintained between the organisms developing—those constituted $2A + WW$ are female and those $2A + ZZ$ are male.

Thelytoky. Sex determination here poses no problem since all individuals are female.

Artificial parthenogenesis. No artificially-induced parthenogenetic strain of animals has ever maintained itself permanently. A laboratory strain of the silk moth *Bombyx* has however maintained thelytoky for three generations after the ancestral virgin eggs had been subjected to increased temperature. The great majority of artificial parthenogones comprise some metamorphosed frogs and large numbers of pre-adult stages of polychaetes, gephyreans, starfish, sea urchins, molluscs and amphibia. They have developed after subjecting eggs to treatments involving, e.g., mechanical agitation, needle puncture, temperature change, acids, alkalis, salts (magnesium, sodium etc.), blood lymph, tissue extracts and sera. Rudimentary parthenogenesis has been induced in eggs of freshwater fish, e.g., perch and pike, by temperature change or distilled water.

Among mammals, seven parthenogonic female rabbits have been obtained: oocytes were immersed in hypertonic saline solution or blood-serum medium, and then developed in pseudo-pregnant does, or else were chilled for five minutes in the fallopian tube of the mother. Blastocysts of rabbit, rat and sheep have formed after the chilling of tubal eggs. The modes of regulation are similar to those of natural parthenogenesis.

Activation. The nature of activation of the egg is still unclear. In zygogenesis it begins at the egg cortex apparently through interaction between enzymes of the cortex and spermatozoon. But the fact that numerous and simple artificial agents can declench it shows activation to be polymorphic and basically due to the innate capacity of the egg. Sometimes the same agent is effective for a number of species and sometimes the same species reacts to a number of different agents. Cortical cytolysis occurs but is quickly corrected. The cortex is rendered permeable, probably through loss of salts (particularly calcium) thereby facilitating the internal reactions and re-arrangements upon which development depends. In fertilized eggs the spermatozoon introduces the functional centrosome; its appearance in unfertilized parthenogenetic eggs shows it to be an organelle of eggs also.

Intersexuality, as seen in certain parthenogenetic phasmids, reflects the unstable chromosome conditions existing in these insects. Intersexual offspring have resulted from Seiler's crossings between tetraploid parthenogenetic females and diploid normal males of a *Solenobia* moth. Each single brood comprised individuals among which a number of grades of intersexuality occurred and each individual displayed its own fixed grade in its organs throughout ontogeny. This result does not conform to Goldschmidt's classic time-law of intersexuality, based on the results of crossings between different Gypsy Moth races in which all the individuals of a brood possessed the same grade of intersexuality.

Genetics. Stereotypy in genetical constitution occurs under ameiotic parthenogenesis, chromosome pairing being absent and no genetic change possible except by mutation. This leads to the formation of clones, all individuals being of the same genotype. In meiotic parthenogenesis, however, if a chromosome pair be heterozygous for certain pairs of genes and if crossing-over occurs, the parthenogones will differ genetically from the parent.

Mutants have arisen under parthenogenesis, e.g., in cultures of thelytokous water-fleas and of the arrhenotokous chalcid *Habrobracon*. Local races of rotifers and water-fleas occur, thelytokous one-host aphidids and cynipids have arisen from heterogonous two-host species, and polyploid parthenogenetic plants and animals are the descendants of diploid zygogenetic species. Such novelties may be regarded as new species when they are unable to interbreed with the parent species.

Evolution. The descent of parthenogenetic forms from their zygogenetic ancestors, involving adaptations in activation, regulation of chromosome number and development, is not known in most cases. Judging from the successes obtained by using artificial agents zygogenetic forms seem as it were preadapted to parthenogenesis. Natural thelytoky could arise after a single mutation or after a series of mutations which enhanced viability and so promoted facultative parthenogenesis step by step from the rudimentary condition existing in many species. Mutation towards thelytoky has been witnessed, for in a culture of a zygogenetic nematode a pseudogamous strain appeared. Hybridization is regarded as responsible for parthenogenesis in some plants, e.g., hybrid roses and hawkweeds, and in a hybrid *Tephrosia* moth. However, rudimentary parthenogenesis is so widespread and so many new cases of it continue to be revealed that there is always the possibility that it already existed in the hybridizing species.

The evolution of arrhenotoky, suggested from observations on male coccids, has involved the gradual loss, through degeneration, of a genome from somatic and germ cells originally diploid. One result has been that spermatogenesis has become modified to avoid reduction below haploidy, the reduction division becoming abortive. The haploid spermatogonium thus produces only two haploid spermatozoa. In the honey bee the second division is unequal and only one functional spermatozoon forms.

In many plants and in some animals thelytokous forms are polyploid, exemplify geographical parthenogenesis, and their polyploidy has enabled them to colonize regions less hospitable than those occupied by their parent species.

The role of parthenogenesis in nature is severely limited, zygogenesis and other methods of reproduction being predominant. The reason seems to be that parthenogenetic species are not adaptable enough. In ameiotic parthenogenesis genetic variability, as explained above, is practically nil. In meiotic parthenogenesis, individuals tend to become homozygous thereby incurring the accompanying disadvantage of this condition. Moreover, variability is thus again restricted. As a rule, parthenogenetic lines are therefore shortlived, unless, as in heterogony and arrhenotoky, they combine the advantages of both sexual and parthenogenetic reproduction. The advantages of parthenogenesis are of several kinds. By arrhenotoky a large number of males is quickly assured. By thelytoky reproduction is speedier, lack of males allowing more females generation after generation. In ameiotic parthenogenesis—and this applies to homozygous genotypes in meiotic parthenogenesis—advantageous combinations of genes are maintained instead of being dispersed by meiosis as in zygogenetic species. Well adapted forms thus spread rapidly as long as the environmental conditions remain unchanged. Finally, the existence of hardy triploid (and aneuploid) organisms becomes possible in ameiotic parthenogenesis.

A. D. PEACOCK

References

Vandel, A., "La parthenogenèse," Paris, Doin, 1931.
Gustafsson, A., "Apomixis in Higher Plants," Act. Univ. Lund., **42**, 1946 and **43**, 1947.
Suomalainen, E., "Parthenogenesis in Animals," *in* Demerec, M., ed., "Advances in Genetics" Vol. 3, New York, Academic Press, 1950.
Schrader, F. and S. Hughes-Schrader, "Haploidy in Metazoa," Quart. Rev. Biol., **6**, 1931.
Whiting, P., "The Evolution of Male Haploidy," Quart. Rev. Biol., **20**, 1945.

PASSERIFORMES

The passerines or "perching birds" comprise the largest order of birds (see AVES), containing approximately 65 families and 5100 species, about three-fifths of all the living birds. The order is cosmopolitan in distribution and its fragmentary fossil record extends back into the early Tertiary. The Passeriformes are small to medium-sized land birds, the largest (Ravens, *Corvus* and Lyrebirds, *Menura*) about the size of chickens. This order is considered to be the most highly evolved of all orders of birds; passerines are numerically the dominant birds on all the continents except Antarctica.

Passerines have undergone extensive adaptive radiation complicated by considerable parallelism. Examples of the diversity and parallel development of feeding adaptations are as follows: swallows (Hirundinidae) and wood swallows (Artamidae) are aerial insect feeders; many tyrant flycatchers (Tyrannidae) and Old World flycatchers (Muscicapidae) "hawk" insects by flying out from a perch; vireos (Vireonidae) and many wood warblers (Parulidae) search foliage and small branches for insects; nuthatches (Sittidae), creepers (Certhiidae), and woodhewers (Furnariidae) search for insects in crevices in the bark of trees. Other feeding adaptations of passerines include aquatic feeding (dippers, Cinclidae), nectar feeding (sunbirds, Nectariniidae; honey-eaters, Meliphagidae; some wood warblers, Parulidae; some tanagers, Thraupidae; some Hawaiian honeycreepers, Drepanidae; some orioles, Icteridae), fish-eating (some tyrant flycatchers, Tyrannidae), hawk-like predation of insects and small birds and mammals (many shrikes, Laniidae), seed-eating (many finch-like birds, Fringillidae), fruit-eating (many tanagers, Thraupidae; waxwings, Bombycillidae; many thrushes, Turdidae; many orioles and blackbirds, Icteridae; etc.), and many others.

Nesting habits (see NEST BUILDING) are also diversified. Many cowbirds (Icteridae) and whydahs (Ploceidae) are "brood parasitic", i.e., lay their eggs in the nests of various host species which raise the "parasitic" young, often to the detriment of their own offspring. The Palm Chat (Dulidae) and some weavers (Ploceidae) have large communal nests; representatives of many families (blackbirds and allies, Icteridae; weavers, Ploceidae; swallows, Hirundinidae) have colonial nesting habits. Nest sites are diverse and include nests in holes in the ground, on the ground, in holes in trees, on cliff faces or ledges, and in a wide variety of locations in trees, shrubs, and other plants. Nest construction is usually elaborate and may involve weaving, knot-tying, sewing, or use of mud or clay.

Passerines lay from one to twelve or thirteen eggs; these are incubated from 11 to 21 days (35 to 40 days in the Lyrebird, Menuridae). The young at hatching are blind and helpless and may be naked (hole-nesting species), sparsely covered with down (most species), or densely covered with down (Menuridae). The young remain in the nest from eight days to four or five weeks but typically ten days to two weeks.

In COLORATION passerines vary from sombre browns and other plain colors to the elaborate and sometimes bizarre color patterns of tanagers (Thraupidae), birds of paradise (Paradiseidae), manakins (Pipridae), and others. Representatives of some families (Paradiseidae, Cotingidae, Bombycillidae, etc.) have highly modified feathers, some of which are celluloid-like or in part wax-like.

Courtship behavior of passerines is variable. Some manakins (Pipridae) and birds of paradise (Paradiseidae) have communal displays including dancing. Many bowerbirds (Ptilonorhynchidae) have an elaborate display involving the construction of a "bower" decorated with flowers, stones, and other objects of particular colors. In many species of many families courtship involves the establishment of a territory or defended area which may be advertised and often defended by means of song. Males may remain associated with their mates very briefly at the time of copulation, or until incubation begins, or most commonly throughout the nesting cycle. Most passerines are monogamous; a few are polygynous or polyandrous; some apparently pair for life.

Passerines as a group are noted for their singing abilities (see BIRD SONGS, BIOCOMMUNICATION); this is particularly true of the Suborder Passeres or "Oscines", the group which includes the thrushes (Turdidae), mimic thrushes (Mimidae), larks (Alaudidae), and other notable songsters. There is evidence that some species at hatching are endowed with a full complement of inherited songs and call notes; other species learn some or all of their songs and call notes. Representatives of many families are accomplished mimics.

The most obvious external characteristics of passerines are their perching feet: four unwebbed toes articulated with the distal end of the tarsometatarsus, three in front and one behind (the fourth never reversible).

On the basis of variations in the distribution and number of certain muscles of the trachea associated with the syrinx (the "lower larynx" or sound producing structure of birds), the Order Passeriformes is divided into four suborders:

Eurylaimi Broadbills; 14 species; tropical Africa and part of the Oriental Region.

Tyranni Contains two major groups or superfamilies: Furnarioidea.—Tapaculos, ant pipits, ant birds, ovenbirds, woodhewers; c. 520 species; Central and South America. Tyrannoidea.—Pittas and other Old World families; tyrant flycatchers, manakins, cotingas and other New World families; c. 550 species.

Menurae Lyrebirds and scrubbirds; 4 species; Australia.

Passeres (or "Oscines"; the aforementioned three suborders are often called collectively "suboscines"); true songbirds (including about 50 families); c. 4000 species; Cosmopolitan.

The passerines are comparable to teleost fishes and rodents in multiplicity of species, extent of adaptive radiation, and complexity of parallel evolution. Although the order can be divided into four quite distinct suborders, the largest of these assemblages, the Passeres, is itself quite homogeneous morphologically and is perhaps no more distinct than a family. The Passeres, however, have been subdivided into more than fifty families; the relationships within this large group of birds are poorly understood.

PHILIP S. HUMPHREY
(revised by MARY HEIMERDINGER CLENCH)

References

Austin, O. L., Jr., and A. Singer, "Birds of the World," New York, Golden Press, 1961.
Van Tyne, J. and A. J. Berger, "Fundamentals of Ornithology," New York, Wiley, 1959.

PASTEUR, LOUIS (1822–1895)

Pasteur's scientific achievements led not only to advances in biological theory but also to solving many practical agricultural problems which plagued France during the middle of the nineteenth century. His research was of a wide variety, from disease prevention in silk worms to a preventative treatment for rabies, indicating the versatility of Pasteur as a scientific investigator. His greatest contributions, however, were in the field of immunology and microbiology. Pasteur was born at Dole in east central France on December 27, 1822. He received his education at the Ecole Normale in Paris and also attended lectures on chemistry at the University of Paris. During this period he taught at the secondary schools. After his undergraduate work, Pasteur remained at the Ecole Normale where he received his doctorate in chemistry in 1848. His dissertation on stereoisomers of tartaric acid was widely recognized as a great pioneering work in stereochemistry. During the next six years Pasteur first taught physics at the Lycee at Dijon and then chemistry at the University of Strasburg where in 1849 he married Marie Laurient. In 1854 Pasteur went to Lille as the head of the science division. At this school, located in the center of the beer and wine industry, Pasteur began his research on fermentation. He reconfirmed that yeast was the cause of fermentation and conclusively proved that the yeast cells did not simply arise from fermenting liquid but came from pre-existing yeast cells. This work supported the earlier findings of Spallanzani and finally terminated the arguments in favor of spontaneous generation. In 1857, Pasteur returned to Ecole Normale as the director of the school. Between his teaching and administrative duties Pasteur devoted his time to finding the cause of wine spoilage. He was able to show that the spoiled wine contained a mixture of micro-organisms and identified specific organisms as the causative agents. He further demonstrated that the fermentation of wine by both normal and contaminating organisms could be inhibited by gently heating to a temperature of 60°C for thirty minutes, this process was later to become known as pasteurization.

At this time the silk industry of southern France was suffering heavy losses from a silk worm disease. Pasteur was able to show the farm managers the difference between infected and non-infected silk worms and instructed them only to use the healthy ones for silk production. So appreciative was the industry that in 1865 a statue was erected in his honor at Alais. During this period Pasteur suffered a disease which resulted in a permanent partial paralysis of his left hand and leg. He now turned his attention to diseases of other animals. While studying the micro-organisms that caused chicken cholera he accidently discovered that when the attenuated organism was injected into healthy chickens they became resistant to subsequent virulent doses of the organism. Pasteur further and more rigorously demonstrated the immunization theory on a larger scale when in 1881 he immunized sheep against the anthrax organism by injecting them with a weakened culture. The success of this public display led to a widespread use of Pasteur's methods of immunization against anthrax throughout France, saving the livestock industry from great financial losses and laying the foundation for modern methods of immunology. Pasteur next began investigating hydrophobia, at that time one of the most dreaded diseases of Europe. After long and laborious trial and error methods Pasteur was able to cultivate the rabies virus in his laboratory and was successful in his attempts to immunize dogs against the disease. He was reluctant to use his newly developed vaccine as a prophylactic measure with humans. But in 1885, he injected a boy who had been hopelessly bitten by a rabid dog. Immunity developed before the onset of the fatal disease, thus saving the boy's life. Hydrophobia had been conquered. Thus, at an age when most think of retirement, Pasteur, crippled by his earlier paralysis, achieved one of his most crowning triumphs. In 1888 as an honor to the man who had done so much for humanity, the Pasteur Institute was erected in his honor. He died in 1895.

RICHARD M. CRIBBS

Reference

Cuny, H., "Louis Pasteur," New York, Hill and Wang, 1965.

PAUROPODA

A class of progoneate ARTHROPODA. Small animals (Length < 1.5 mm), with head, 11 (rarely 12) segments and pygidium. 1st pair of legs reduced, 2nd to 9th pair of legs normal, 5-jointed. Body-segments partially fused, forming 4 to 5 diplo-somites (Fig. 2), covered dorsally by 6 corresponding diplotergites (Fig. 1, 2). This character probably is not to be regarded as a primitive structure tending to the real diplo-somites in DIPLOPODA; there seems to be no homology. It is highly probable that the partially fused somites in pauropoda are a special derivative character, already indicated in the class SYMPHYLA. Antennae divided in a characteristical manner: 4 basal joints bearing at the end a pair of styli, the anterior of them with 2 flagella, the posterior with 1 flagellum; between the styli 1 or more globulous sensory organs (Fig. 2, 3). Mouthparts consisting of 1 pair of mandibles and 2 pairs of maxillae. Cuticula smooth and soft in Pauropodidae (Fig. 1), strongly chitinized and sculptured in Eurypauropodidae (Fig. 3). Nervous system consisting of 13 ganglia, the first 3 ganglia fused, forming a large suboesophageal-ganglion, the following 10 connected by commisurae as in diplopoda. Tracheal and vascular system completely reduced. [1 single ovarium in females ventrally of intestine tract, vagina opening between the basis of 2nd pair of legs; 4 testicles in males dorsally of intestine tract, in connection with 2 penes behind coxae of 2nd pair of legs.] Ontogeny insufficiently known, smallest larvae with 3 pairs of legs, probably 4 larval stages.

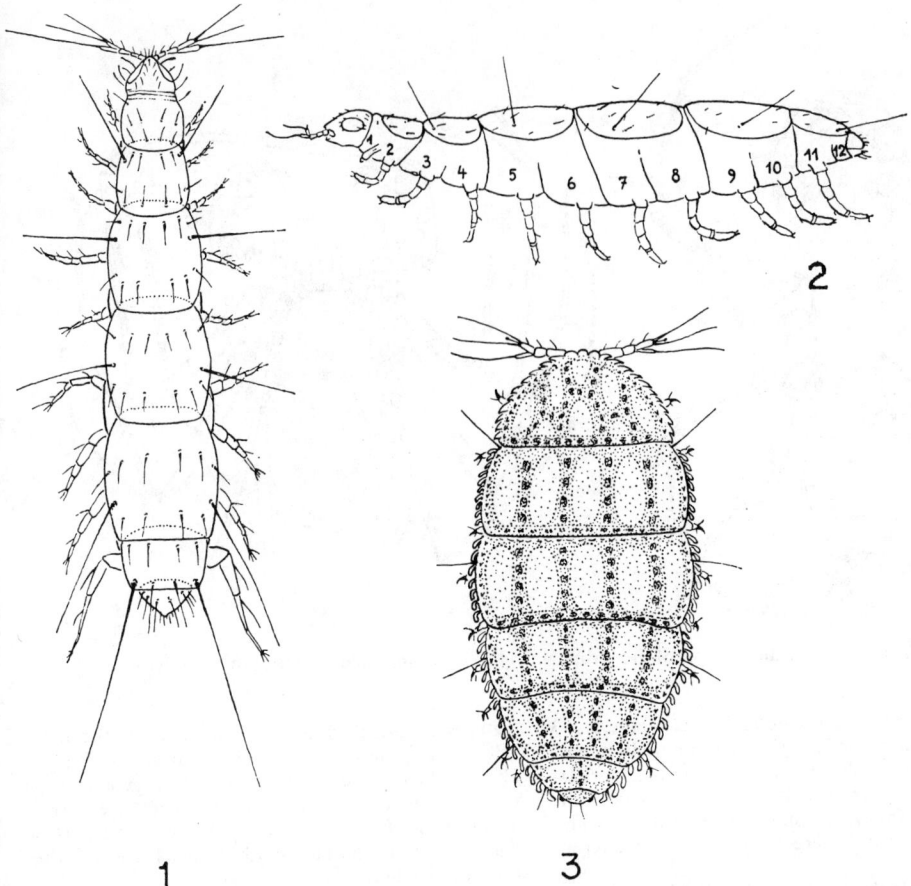

Fig. 1   *Pauropus Huxleyi* Lubbock (Europe), dorsal view.
Fig. 2   *Decapauropus cuenoti* Remy (Vosges), lateral view, segment-numbers indicated.
Fig. 3   *Eurypauropus ornatus* Latzel (Austria), dorsal view.

Occur under stones, dead leaves, rotten wood and bark. Distributed all over the world.

O. KRAUS

References

Attems, C., "Myriapoda" *in* Kükenthal, W. and T. Krumbach, "Handbuch der Zoologie" Vol. 4, Leipsig, de Gruyter, 1923–25.

Borner, C., "Beitrage zur Morphologie der Arthropoden. I. Ein Beitrag zur Kenntnis der Pedipalpen," Zoologica, **42**: 1–174, 1904.

Grassi, B., "Progenitori dei Miriapodi e degli Insetti. Mem. V.: Intorno ad un nuovo Aracnide artrogastro (*Koenenia mirabilis*) rappresentante di un nuovo ordine (*Microteliphonida*)," Bull Soc. Ent. Ital., **18**: 153–172, 1886.

Hansen, H. J. and W. Sorensen, "The order Palpigradi Thorell (*Koenenia mirabilis* Grassi) and its relationships to other Arachnida," Entom. Tidskrift, **18**: 223–240, 1897.

Kastner, A., "Palpigradi Thorell" *in* Kukenthal, W. and T. Krumbach, Eds., "Handbuch der Zoologie," Vol. 3, Leipsig, de Gruyter, 1923–25.

Millot, J., "Sur l'anatomie et l'histophysiologie de *Koenenia mirabilis* Grassi," Rev. franc. Entomol, **9**: 33–51, 1942.

ibid, "Notes complementaires sur l'anatomie, l'histologie et la réparition geographique en France de *Koenenia mirabilis*," Rev. franc. Entomol., **9**: 127–135, 1942.

Roewer, C. Fr., "Palpigradi" *in* Bronn, H. G. ed., "Klassen und Ordnungen des Tierreichs," Vol. 5, Leipsig, Winter, 1934.

Rucker, A., "A new *Koenenia* from Texas," Quart. Journ. Micr. Sci., **47**: 215–231, 1903. *ibid.,* "Further observations on *Koenenia*," Zool. Jahrb (Syst)., **18**: 401–434, 1903.

Wheeler, W., "A singular Arachnid (*Koenenia mirabilis* Grassi) occurring in Texas," Amer. Natural., **34**: 837–850, 1900.

Verhoeff, K. W., "Pauropoda" *in* Bronn, H. G. "Klassen und Ordnungen des Tierreichs," Vol. 5, Leipsig, Winter, 1934.

PEDIPALPI

The Pedipalpi ("whip-scorpions") are now regarded as two distinct orders of the class ARACHNIDA, the Uropygi and Amblypygi. The Uropygi are generally smaller,

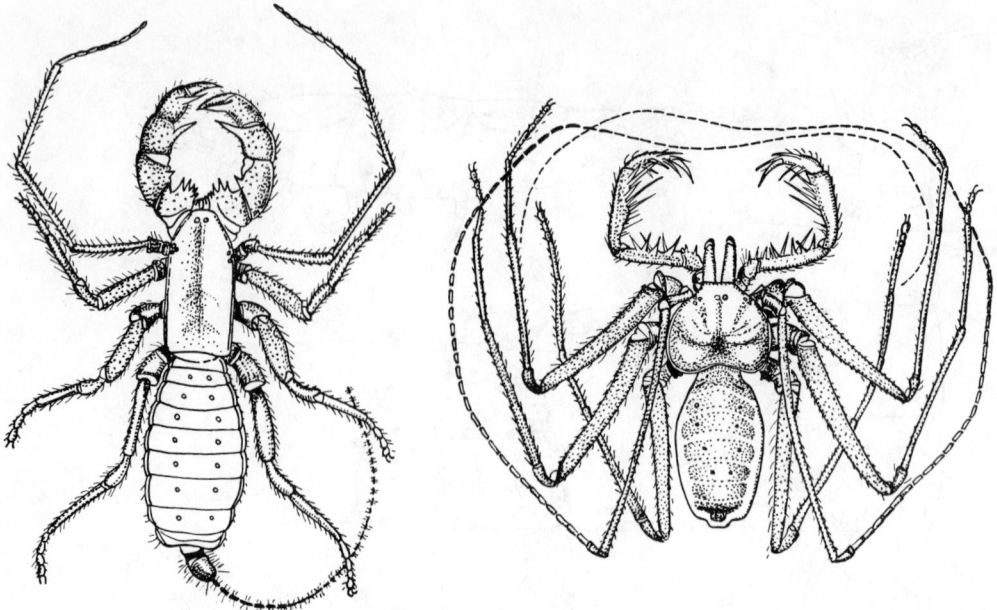

Fig. 1   Left, *Mastigoproctus*, an example of the Uropygi. Right, *Charinus*, an example of the Amblypygi.

the carapace longer than wide and divided, the last 3 abdominal segments forming a short or multi-articulate tail which probably has a tactile function, odoriferous glands being present in the anal region. The larger-bodied Amblypygi have the carapace wider than long and entire; there is no tail or odoriferous glands.

Appendages. In both orders the chelicerae have a large movable claw. In both the second pair of appendages (*pedipalps*) are large and prehensile, as in scorpions; the 6-jointed pedipalps of the Uropygi are simple and leg-like or extremely robust, with large apophyses and stout flexible claw (*Mastigoproctus*); in the Amblypygi they are elongate, with rows of powerful spines for impaling prey, the mobile tarsal claw forming a rudimentary pincer with the tibia. Walking legs with 7 segments, the first pair with elongate or annulate tarsi (Uropygi), modified as an exceedingly long lash-like tactile organ in the Amblypygi.

Sense organs. A pair of median eyes and a group of 3 minute lateral eyes on each side of carapace is present in the Amblypygi and larger Uropygi; vision however is poor and in the Schizopeltids (*Schizomus*) the eyes are atrophied; tactile hairs (*trichobothria*) are present on the first pair of legs; lyriform organs, chemoreceptors of taste or smell, similar to those of other Arachnids are present on the chelicerae.

Respiratory organs. The Uropygi have 1 or 2, the Amblypygi always 2 pairs of lung-books on the second and third abdominal sternites, resembling in structure those of the Scorpions and Spiders.

Reproduction and development. Secondary sex differences in the Uropygi are found on the last 3 abdominal segments (*Schizomus*) and in the curious deformities of the first tarsus of the female (*Thelyphonus*). There is an elaborate pattern of courtship, like that of Scorpions, only in the Uropygi. In both orders

the eggs are attached in a transparent sac to the under side of the abdomen, up to 35 (Uropygi) or 60 (Amblypygi) being laid; in the Uropygi a subterranean brood-chamber is made, while the eggs are carried about by the female in the Amblypygi; the emerging young climb on the mothers back but leave her after the second moult.

Ecology and distribution. The Uropygi are lucifugous, hygrophilous Arachnids, often found in damp humus soil; and Amblypygi rather less lucifugous and common in semi-desert regions, but some, like the Charontidae, living in the recesses of deep caves. Both orders inhabit the tropic belts expecting Australasia, some Amblypygi being partly domesticated. About 130 species, divided among 4 families, are known.

R. F. LAWRENCE

References

Hansen, H. J., Studies on Arthropoda, I: 5–55, Copenhagen, 1921.

Hansen, H. J. and W. Soerensen, "The Tartarides, a tribe of the order Pedipalpi," Arkiv. Zool., **2** (8): 1–78, 1905.

Millot, J., "Uropygi, Amblypygi," *in* Grassé, P.-P., ed., "Traité de Zoologie," vol. 6, pp. 535–88, Paris, Masson, 1949.

PELECANIFORMES

The Pelecaniformes include six comparatively small families of wide-ranging water birds (see AVES). All share the unique structural feature of a totipalmate foot, that is, all four toes, including the hind toe, are united in a web. The nostrils are rudimentary or entirely absent in the pelecaniform birds and they usually have a gular pouch, which reaches its highest development in the true pelicans.

There are also six families of fossil pelecaniform birds. Perhaps the most interesting of these is *Elopteryx*, a cormorant-like bird from the Cretaceous of Europe. It is among the oldest known fossil birds, indicating great antiquity for the cormorant stock.

The tropic-birds (Phaethontidae) include only three species (14 forms) that range over the tropical oceans of both the Old and New World. Their chief distinctions (apart from the rudimentary gular sac, small nostrils, and totipalmate foot) are their tern-like satiny plumage and a pair of attenuate median tail feathers often longer than the body. This gives them a very graceful appearance in flight. They feed on fish, squids, and crustaceans which they catch by hovering and dropping onto the surface of the water, in some cases submerging. They nest on islands, on rocky ledges, and lay a single egg on the bare rock without any attempt at nest building. The young are fed by regurgitation, get very fat and have to reduce before they can fly.

The pelicans (Pelecanidae) comprise six or eight species that are particularly noted for their large gular pouches and clumsy appearance on land. Pelicans are nearly world-wide in distribution, but avoid the polar extremes, and are absent from certain other regions, such as most of eastern North America and much of South America. Two well known species of quite different appearance and habits live in the New World. The White Pelican (*Pelecanus erythrorhynchus*) is chiefly western in distribution, breeding in the western states and Canadian provinces and wintering southward through Mexico into Central America. Like its close relative, the Brown Pelican (*Pelecanus occidentalis*), it is a grotesque-looking bird on land, with its huge bill, short legs and ponderous appearance, but once launched in the air it is graceful in flight, and large formations of a hundred or more soaring birds present a beautiful sight.

Pelicans are famous, and skillful, fishermen, but our two species use different fishing methods. The White Pelicans often hunt cooperatively in groups with a large driving line forming out in shallow water and then with raised wings to cast a shadow driving hapless schools of fish in toward shore where they are scooped up in the huge gular pouch. The pouch is used for scooping up fish but apparently not for transporting food to the young, which in the case of the pelicans nesting on Gunnison Island in Great Salt Lake (where there are no fish) may be 50 or more miles from the fishing grounds.

The Brown Pelican dives for its prey, often from considerable heights, hitting the water with such an impact that it is thought that the fish are stunned and thus easily captured. Subcutaneous air cells in the breast act as shock absorbers when the birds hit the water. These comical pelicans also often stand about piers and docks waiting for handouts from fishermen and tourists. In turn the pelicans often have their catches purloined by marauding gulls which hover around and actually steal fish out of the maws of the big birds as they rise from the water. Unfortunately, Brown Pelicans have suffered severe decline in the Gulf Coast states in recent years, probably due to contaminated prey in contaminated waters.

Pelicans nest in exposed places, either in trees or on the ground, and construct huge nests in which 1–4 (usually 2 or 3) young are raised. They feed by regurgitation, dribbling semisolid food into the mouths of the young until the latter are able to "fish" for themselves by thrusting their heads down into the huge mouth of the parents. The expansive gular membrane also serves as a respiratory membrane from which excess moisture can be evaporated in hot weather.

The family Sulidae includes two notorious types of birds, the boobies (6 species) and the gannets (3 species). The boobies are primarily tropical, but the North Atlantic Gannet (*Morus bassanus*) extends its range into far northern waters. Though the Sulidae resemble the pelicans in important structural details, they are less grotesque in appearance, the gannets in particular being graceful, well proportioned birds. Boobies and gannets have only a suggestion of the expansive gular pouch characteristic of the pelicans.

Boobies are much maligned birds of tropical seas. Their tameness led to the reputation for stupidity and some odd features, like the blue face and red feet of some, add to their queer appearance. Boobies nest on islands, in trees or on the ground, and usually lay two eggs. Oddly, in many cases, only one egg is hatched, or if both eggs hatch only one young is raised. This is thought to be an adaptation to the limited food supply in the vicinity of most nesting colonies, where only one young can be raised successfully. A West Indian species that nests near food-rich waters raises two or three young without difficulty.

The North Atlantic gannet is a fisherman par excellence. It dives for fish from great heights, up to 100 feet above the water, and strikes the surface with a great impact. As in the case of the Brown Pelican subcutaneous air cells tend to absorb the shock. Gannets nest in tremendous colonies, on Bird Rock, Bonaventure Island and on islands about the British Isles. Often the nesting colonies are so densely packed with birds that at a distance the ledges appear snow-covered. Possibly more film has been expended on gannet colonies than on any other bird, for excursion boats carry passengers to the Bird Rock and Bonaventure during the breeding season to witness the amazing spectacle. Gannets lay a single egg, which they guard carefully, as neighboring nests surround it only a bill's length away. That this compactness is due to sociability rather than lack of nest sites is indicated by the fact that there may be unused ledges on the outskirts of the colony.

Largest of the Pelecaniform families is the Phalacrocoracidae, which consists of about thirty species of cormorants or shags. Cormorants are also scattered over most of the world, but inhabit chiefly coastal areas in temperate or tropical waters. The Double-crested Cormorant (*Phalacrocorax auritus*) of North America inhabits inland waters, but the other five species in North America are chiefly coastal. Like most other pelecaniform birds, cormorants are excellent fishermen, but catch their prey by pursuing it swiftly under water.

The cormorant's fishing ability was capitalized on long ago by the Orientals who tamed the birds and employed them for fishing purposes. Trained birds were released from boats, with or without leashes, and made to bring back their catches to their masters. Collars fastened about the necks of the birds prevented them from swallowing their prey. Cormorants in this country, and in Canada, have often been accused of being detrimental to fishing interests, but studies have shown that they live mostly on coarse, non-game fishes and that in some cases at least their activities may be beneficial by culling out waste fish.

Perhaps the most famous of the cormorants is the Guanay Cormorant *(Phalacrocorax bougainvillii)* on the islands off the coast of Peru. Guano deposits that accumulate on the nesting islands, which are in an extremely dry climate and thus not leached out by rains, are worth millions of dollars annually and at one time defrayed the entire cost of the Peruvian Government. Now the guano is harvested more conservatively than in the past, so that this valuable resource may be perpetuated.

Another small family of pelecaniform birds is the snake-bird, water-turkey, or anhinga family (Anhingidae). This consists of one or two species of pantropical distribution which Peters divided into six forms. The anhingas, as one of their common names implies, are snake-like birds, often swimming about with only the head and long sinuous neck out of water. They differ from the cormorants by their elongated appearance, particularly of the long neck which has the eighth cervical vertebra articulated at an angle, making a conspicuous kink in the neck. The tail is also long, and decoratively fluted with cross striations in the male. The bill is long, sharp and serrate. Fish are impaled on the sharp bill when pursued under water.

The plumage of anhingas is not waterproof and they spend much time sunning and drying out after plunges into the water. The nesting habits of the New World species have been well documented by Brooks Meanley from a study in Arkansas.

Last and perhaps the most aberrant of the pelican-like birds is the frigate-bird or man-o'-war-bird family (Fregatidae). Some five species of these interesting birds inhabit the tropical waters of both the Old and New World. They have extremely long wings (7 foot wing spread), an inflatable pouch on the necks of the males, and a long hooked bill.

The chief distinction of the frigate-birds is their extremely buoyant flight. Though fairly large birds, three feet or more in length, their bones are pneumatic and the body very light in weight. Masters of the air, they rarely, if ever, voluntarily alight on the water. Their prey is secured by skimming the surface for floating animal matter, or by methodically stealing the catches of other birds at which they are very adept because of their powerful flight.

Frigate-birds nest in colonies on islands and incubation and brooding is done largely by the males which sit facing the wind with their colorful red pouches inflated. The ancient Polynesians employed frigate-birds to carry messages from island to island.

GEORGE J. WALLACE

References

Meanley, B., "Nesting of the Water-turkey in Eastern Arkansas," Wilson Bulletin, **66**: 81–88, 1954.
Murphy, R. C., "Oceanic Birds of South America," Vol. I, New York, Macmillan, 1936.
Wallace, G. J., "An Introduction to Ornithology," New York, Macmillan, 2nd ed., 1963.

PELECYPODA (BIVALVIA, LAMELLIBRANCHIA, ACEPHALA)

In this important class of the MOLLUSCA the mantle, and so the SHELL it forms, has been laterally extended and then compressed to enclose the animal. Compression is accompanied by subdivision of the mantle into lateral lobes and a mantle isthmus mid-dorsally, the former secrete the calcareous valves, the latter the ligament. Early marginal attachment of mantle to shell by orbicular muscles led, by compression, to formation of anterior and posterior adductor muscles which work in opposition to the opening thrust of the elastic ligament. The enclosed head is reduced to no more than the mouth. Feeding has been taken over by the enlarged ctenidia (gills) although probably following a stage, still indicated in the Nuculidae and Nuculanidae (Protobranchia) where extensions (proboscides) from the labial palps (lips) collect organic deposits from the soft substratum in which the animals live.

The shell (valves and ligament) consists of (1) superficial non-calcareous periostracum, (2) outer calcareous, or ligament, layer, (3) inner calcareous, or ligament, layer. The mantle margin bears three parallel folds, the outermost being concerned with secretion on its *inner* surface of (1) and on its *outer* surface of (2). The general mantle surface and the isthmus secrete (3).

The middle fold is sensory, taking over functions of the head; it usually bears tentacles (best developed around inhalant regions where water enters the mantle cavity) and sometimes eyes (e.g. *Pecten* and other scallops, also *Cardium*, the cockle). The innermost fold is muscular and often deep (e.g. the "velum" of *Pecten*, etc.). It controls the flow of water created by the lateral cilia on the gills into and out of the mantle cavity.

Primitively free, the mantle lobes may fuse posteriorly, ventrally and dorsally. This involves, always in this order, (1) innermost folds, (2) also middle folds, (3) also outermost folds, this involving union of the periostracal covering, as in deep burrowers, e.g. *Mya* (soft shell clam). Initial fusion is always posterior involving separation of exhalant (dorsal) from inhalant (ventral) openings, it is followed (in ontogeny *and* phylogeny) by ventral fusion separating the latter from a pedal opening. Dorsal fusion produces secondary extension of the ligament by formation of a fusion layer. Hypertrophy of the posterior mantle margins produces siphons of varying type according to the degree of fusion. Hence many bivalves can burrow, or bore, deeply while retaining contact with the water.

The body, initially little modified except by compression, retains bilateral symmetry except where a horizontal posture is assumed as in oysters, scallops, etc. This is usually associated with loss of the anterior adductor muscle (change from dimyarian to monomyarian condition), itself usually a result of attachment by byssus threads to a hard substratum. These protein fibers are formed in a gland at the base of the foot and especially characteristic of mussels (Mytilidae) though often present in early life and not infrequently throughout life.

In the Nuculidae (Protobranchia) the gill remains relatively small with primitively horizontal filaments but in the great majority of bivalves the gills extend almost to the mouth with a great increase in numbers and individual length of the filaments. These bend back, the two arms of each filament being united by tissue junctions, each gill (ctenidium) thus consisting of two plate-like half-gills (demibranchs). Adjacent filaments are united by cillary junctions (Filibranchia) or by tissue (Eulamellibranchia). The combined respiratory and feeding current is created by lateral cilia on the opposed faces of the filaments, food particles trapped on the surface are carried by frontal (originally cleansing) cilia to the margin or base of the demibranchs for transport to, and later selection by, the grooved, complexly ciliated, opposed faces of the labial palps. Further ciliary mechanisms deal with concentration of excess particles (pseudofaeces) which are rejected from the *inhalant* cavity by sudden contractions of the adductors which often contain areas of striated "quick" muscle. Swimming in scallops is an incidental consequence of possession of this muscle.

Most bivalves feed on suspended phytoplankton but the protobranch Nuculidae and Nuculanidae are deposit feeders (see above) and so, amongst the Eulamellibranchia, are the Tellinacea which draw in surface deposits through the separate inhalant siphon. The wood-boring Teredinidae (shipworms) digest cellulose. The Septibranchia, with ctenidia modified to form a muscular septum, are the only carnivorous bivalves. Except in this small group, the bivalve gut is modified for dealing with a constant supply of small particles of plant origin. The stomach is a ciliated sorting organ passing finer particles into the digestive diverticula where intracellular digestion occurs (also in phagocytic blood cells which enter the gut lumen). Some extracellular digestion follows liberation of carbohydrate and possibly fat-splitting enzymes from the crystalline style secreted in an elongated style-sac, sometimes in communication with the intestine, and consisting of a mucoid rod moulded and driven stomachward by cilia. Adsorbed enzymes are liberated when the style dissolves in the higher pH of the stomach. The style also mixes food in the stomach. It bears against a cuticular gastric shield which enlarges to cover much of the stomach wall where this serves as a gizzard, e.g. in the Nuculidae, Tellinacea and Septibranchia.

The foot is typically hatchet-shaped and adapted for movement through soft substrata. It is modified for rapid vertical burrowing (Solenidae, razor clams), for planting byssus threads (Mytilidae) or for sucker-like attachment in rock borers (Adesmacea; pholads and shipworms). The foot is lost in some, not all, cemented bivalves (e.g. Ostreaidae).

The nervous system consists of three pairs of ganglia, cerebropleural (largely pleural), pedal and visceral with connectives. Sense organs are reduced to statocysts and to tactile, olfactory and sometimes light perceptive organs in the middle mantle fold.

Fertilization is external or in the mantle cavity and the architecture of the reproductive ducts simple. Protandric hermaphroditism is common, alternating sex-change occurs in the Ostraeidae (oysters), some other bivalves are hermaphrodite with ova and sperm produced in different areas of the gonad. Incubation may occur, most strikingly in the genus *Ostrea* (flat oysters) where it takes place in the *inhalant* cavity, i.e. the eggs passing through the gills. Development is usually by way of a trochophore to a veliger larva with modifications in the freshwater Unionacea where there is a parasitic stage in development.

Invariably, because of their basic structure, aquatic, the Pelecypoda are immensely abundant and highly successful in both marine and freshwater environments occurring on all types of substrate. A few are commensal but only one parasitic species is recorded.

C. M. YONGE

PENTASTOMIDA

The Pentastomida are highly aberrant and entirely parasitic animals. They show definite affinities to the Arthropoda (Osche, 1963). Some workers mistakenly class them as Arachnida because primary larvae superficially resemble certain mites. Heymons (1935) considered them to have affinities with Annelida. It seems best to consider them an independent class of animals with remote relations to Arthropoda and Annelida, but more definitely to the former.

The adults vary in length up to four inches; are segmented and divided into head and abdominal regions; have an undifferentiated digestive tract; no trachea; nervous system with ganglia concentrated in the head; haemocoel; a cuticle which is molted; and separate sexes. Females are larger than males. The only appendages are two pairs of hooks in the head region of adults (Figs. 3 and 4) and two pairs of double hooks on parapodia-like appendages in primary larvae (Fig. 1). The hooks are equipped with strong muscles which make them useful in tissue migration and in anchoring them to the host.

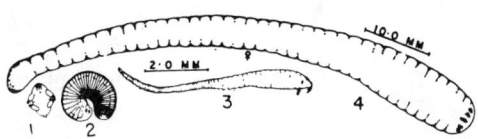

Fig. 1   1 and 2, first and third stage larvae of *Porocephalus clavitus* (modified from Heymans); 3, *Megadrepanoides solomonensis* adult; 4, *Porocephalus crotali.*

Adults are typically found in the respiratory tracts of their hosts. With few exceptions they occur in reptiles, particularly snakes and lizards. *Linguatula serrata* is unique in that the adults inhabit the nasal tracts of canines which are infected by eating larva-infected herbivorous mammals.

There are only two orders, the Porocephalida and Cephalobaenida. With few known exceptions adult Porocephalids live in the lungs of snakes, and their larvae (Fig. 2) parasitize mammals on which the snakes feed. Adult Cephalobaenida live in birds and reptiles and their larvae are known from both insects and their adult hosts.

J. TEAGUE SELF

References

Heymans, R., "Pentastomida," *in* "Bronn's Klassen u. Ordnungen d. Tierreichs," **5**: 1–268. Leipsig, Winter, 1935.

Osche, G., Die systematische Stellung und Phylogenie der Pentastomida. Embryologische und vergleich-end-anatomische Studien an Reighardia sternae. Zeitschrift für Morphologie und Ökologie der Tiere 52, 487–596, 1963.

PERISSODACTYLA

Perissodactyla—from the Greek, meaning "with an odd number of digits," is an order of hoofed four-footed herbivorous MAMMALS, including horses, rhinoceroses, and tapirs.

No perissodactyl is tiny and the shoulders of *Paraceratherium* (= "Baluchitherium") towered 17 feet above the ground. The enlarged third digit serves as axis of support and locomotion. In addition there is a third trochanter on the femur, the astragalus has flattened distal articular surfaces and a proximal gingly-mus, the premolar teeth tend to be molariform, and the molars are lophodont. The fox-sized *Hyracotherium* (Eohippus) of lower Eocene deposits in Europe and North America is the morphological ancestor for all perissodactyls and an annectent between perissodactyls and certain, probably ancestral, phenacodont condylarths. The front limbs of *Hyracotherium* were tetradactyl, the hind were tridactyl; and this is the primitive condition. The low-crowned bunodont-sublophondont upper molars show a cusp configuration from which the characterizing lophodont patterns of other groups of perissodactyls are derived (see Fig. 1).

Evolution of the horses, **Equidae**, centered in North America and dispersals penetrated all continents but

Australia and Antarctica; but the closely related Eocene and Oligocene palaeotheres were restricted to Eurasia. Wild populations of horses survive in Africa (zebras and asses) and in Asia (onagers and tarpans). *Equus,* the modern-type horse is a monodactyl unguligrade with high-crowned cheek teeth. Living species inhabit savannas and grasslands where sustained running gives escape from predators. **Brontotherioidea,** or "Titanotheres," largest members of the upper Eocene and lower Oligocene Holarctic terrestrial fauna, suffered an unexplainable extermination before Miocene time. The ponderous Oligocene titanotheres (*Teleodus, Brontops, Titanotherium, Embolotherium* and others) were characterized by massive supranarial bony "horns." All titanotheres had low-crowned teeth and probably were woodland or savanna browsers.

Chalicotherioidea, a unique group is first represented by lower Eocene *Paleomoropus* of Wyoming and *Lophiaspis* of France. Later chalicotheres, such as *Moropus* and *Ancylotherium* are clawed. Before complete skeletons were known, these claw bones were thought to represent monstrous pangolins or edentates. Chalicotheres are not comparable to carnivores, or diggers, or climbers; for the chalicothere limb bone is similar to that of other cursorial or ambulatory perissodactyls. The clawed foot may have been a means of pulling distant branches within reach of the cropping incisor teeth or a means for uprooting fleshy tubers. These probable woodland dwellers dispersed through the Northern Hemisphere and survived into the Pleistocene in Asia.

Tapiroidea, quadrupedal browsers, are represented by diverse forms in the Eocene of the Northern Hemisphere and survive in warm moist woodlands of southeast Asia and tropical America. The early Eocene

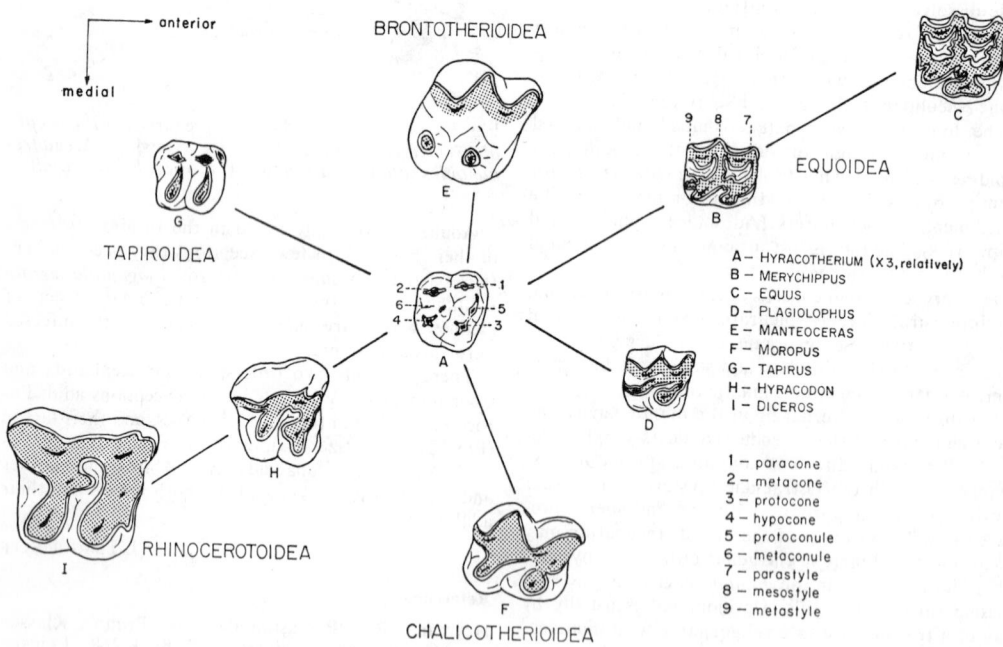

A – HYRACOTHERIUM (x3, relatively)
B – MERYCHIPPUS
C – EQUUS
D – PLAGIOLOPHUS
E – MANTEOCERAS
F – MOROPUS
G – TAPIRUS
H – HYRACODON
I – DICEROS

1 – paracone
2 – metacone
3 – protocone
4 – hypocone
5 – protoconule
6 – metaconule
7 – parastyle
8 – mesostyle
9 – metastyle

Fig. 1

Homogalax is almost inseparable from *Hyracotherium*. Tapiroids are primitive in their retention of 4 toes in the manus and 3 in the pes; and their teeth remain bunolophodont and unspecialized in the incisor-canine region. *Lophiodon* of the European late Eocene and *Megatapirus* of the Chinese Pleistocene were larger than the living *Tapirus*.

Rhinocerotoidea: These bizarre mammals probably originated in early Eocene tapiroid stock and subsequently dispersed through the Northern Hemisphere and into Africa. Rhinoceroses are a prominent element in the middle and late Cenozoic faunas. Most became heavy tridactyl browsing quadrupeds. Amynodontids retained the canine teeth as tusks, a singular modification among rhinoceroses. Teleoceratines became short-limbed and heavy-bodied and may have been amphibious browsers in riparian communities. Indricotheriines, culminating in *Paraceratherium*, were the mid-Oligocene to early Miocene colossi of the Eurasian land fauna. Elasmotherines, Pliocene and Pleistocene of Eurasia, developed high-crowned and complicated cheek teeth and were possibly steppe-land grazers. Among the rhinocerotines, specimens of *Coelodonta*, "woolly rhino" of northern Eurasia, have been found with hair preserved. The surviving groups of rhinoceroses *(Rhinoceros, Didermocerus, Ceratotherium,* and *Diceros)* live in brushlands in Africa and in certain high-grass country and bamboo-covered hillsides in tropical Asia.

D. E. SAVAGE

References

Colbert, E. H., "Evolution of the Vertebrates," New York, Wiley, 1955.
Viret, J., "Perissodactyla," *in* Piviteau, J., ed., "Traité de Palaeontologie," Paris, Masson, 1958.

PFEFFER, WILHELM (1845–1920)

This German botanist is far less known than he should be. Though very few of his actual writings have survived, his philosophy permeates contemporary thought in all branches of the biological sciences. While he espoused what is now known as the mechanistic viewpoint, he also insisted that involved physiological and biochemical processes were the cause, rather than the result, of mechanical processes. Thus he insisted that cell division in both plants and animals was triggered by the biochemical reactions within the cell itself, and not by a mere increase in size which produced a mechanical distention which thus necessitated cell division. He also insisted that any environmental effects to be observed in the growth of plants could only be produced if there was an internal interaction between the physiology of the cell and the environmental change. This is, in contemporary thinking, so obvious that the originator of this concept has tended to fade into obscurity. Yet almost all contemporary research in experimental cell physiology, and experimental morphology, both in plants and animals takes its origin from some of Pfeffer's ideas.

PETER GRAY

PHAEOPHYCEAE

Phaeophyceae or brown ALGAE form the only class of the division of Phaeophycophyta; they are sometimes united together with CHRYSOPHYCEAE, BACILLARIOPHYCEAE and XANTHOPHYCEAE in the larger phylum of Chromophycophyta.

Of a very different habit, structure and size, some of them are microscopic, others may be more than 30 m long, they are almost exclusively marine and widespread in all oceans but particularly numerous along the cold and temperate coasts of the Northern Hemisphere.

Phaeophyceae are mainly distinguished by their cytological and biochemical characters, as well as by their life-histories.

The cells of Phaeophyceae contain generally only one nucleus and one to many PLASTIDS of varying form and size (phaeoplastids) provided, in the less evolved forms, with an outer pyrenoid without starch-sheath.

Plastids contain Chlorophyll *a* and *b*; β carotin and special Xantophylls (see COLORATION, PLANT), the most characteristic of which is Fucoxanthin that give them an olive brown colour.

Starch is never produced by Phaeophyceae. The most important products of the photosynthesis are a polyholosid dissolved in vacuoles, laminarin and an alcohol, mannitol; fats are also present.

In the cytoplasm there are some refringent hyaline granules, physodes or fucosan-granules, which contain phenolic compounds giving the reactions of phloroglucol. Physodes are stained by vital stains such as cresyl-blue and neutral-red. They are most abundant near the nucleus.

The cell-wall contains cellulose in its inner part but it is mostly made of a pectic compound, *algin* (a calcium and magnesium salt of alginic acid). Alginates extracted from Phaeophyceae (mainly Laminariaceae), have many industrial uses.

Motile reproductive cells may be called zoids; some of them are asexual zoospores, others are gametes. Zoids are generally pear-shaped with two laterally inserted flagella going in opposite directions. The anterior flagellum, generally longer (except in Fucales), is provided with two opposite rows of mastigonema; the posterior flagellum is naked.

Unicellular forms and unbranched cellular filaments are unknown among Phaeophyceae. The simplest forms such as *Ectocarpus* have branched uniseriate filaments with intercalary growth, the branches often ending in pseudo-hairs.

In more evolved forms, intercalary growth may be restricted to a part of the filaments near the tips of the branches. In this part, by active division, the cells produce in the distal direction a single filament or a hair and, in the proximal direction, a vegetative axial filament with lateral branching (trichothallic growth).

Among forms of higher evolution (Sphacelariales, Dictyotales), growth may be terminal by means of a large initial apical cell. In some cases *(Padina, Zonaria)*, growth may be marginal by simultaneous division of many laterally coalescing initial cells.

In a large number of Phaeophyceae, branching filaments are always uniseriate *(Haplostichinae)*, among others there may be intercalary cell-multiplication by longitudinal cross-walls producing parenchymatous structure *(Polystichinae)*.

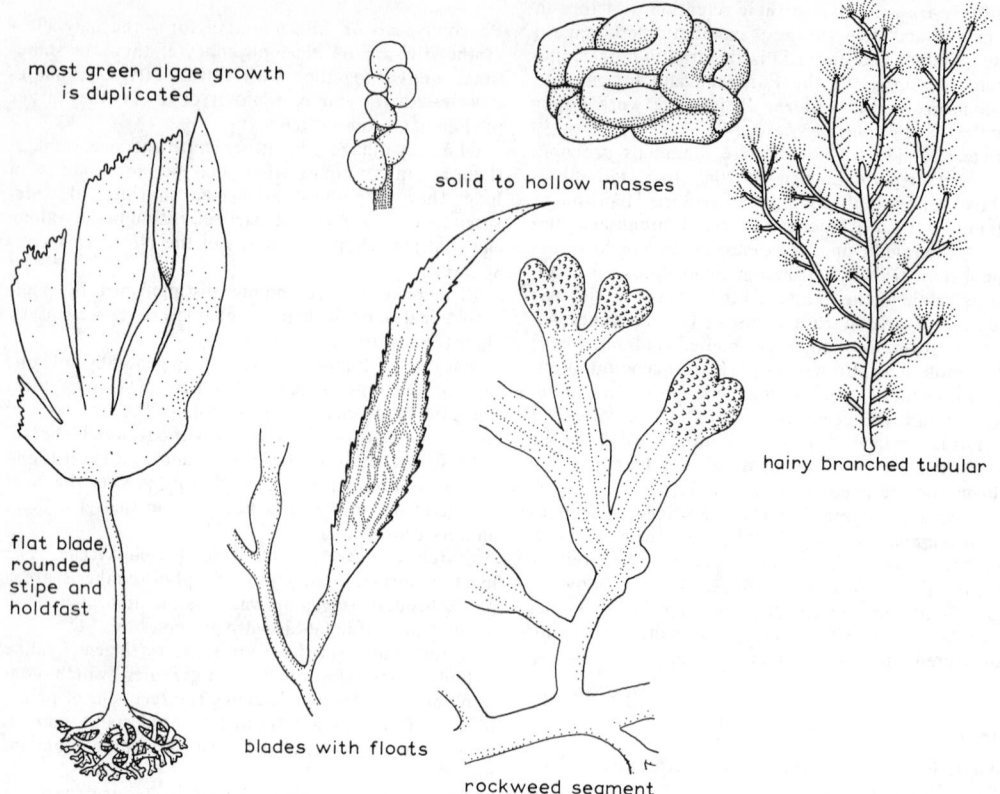

most green algae growth
is duplicated

solid to hollow masses

hairy branched tubular

flat blade,
rounded
stipe and
holdfast

blades with floats

rockweed segment

Fig. 1   Representative brown algae. Brown algae types include species that run the gamut of green algae growth forms. Most brown algae grow very large. (From Pimentel, "Natural History," New York, Reinhold, 1963.)

The greatest differentiation of tissues is attained in the order of Laminariales possessing in their blade a medulla provided with long conducting cells, connected together by pitted cross-walls, and a cortex with anastomosing mucilage-ducts.

Growth in Laminariales occurs by means of an intercalary meristem at the junction of the blade and the stipe (stipo-frondal zone). Among Fucales, growth is apical.

Some Fucales (e.g. *Sargassum*) show a great morphological complexity with perennial axis producing lateral branched ramuli bearing leaf-like expansions with a midrib, air-bladders and reproductive receptacles.

As to reproduction and life-cycle, Phaeophyceae may be divided into two major groups: *Phaeosporeae* showing an antithetic alternation of generations and *Cyclosporeae* without morphological alternation of generations.

Reproductive organs of Phaeosporeae are typically of two kinds: unilocular and plurilocular.

Unilocular ones develop from a single cell dividing its nucleus and protoplasm into numerous zoids filling the mother cell like nuts in a bag. Unilocular organs are borne on diploid plants: they are the seat of meiosis and they produce zoospores.

Plurilocular ones, on the contrary, develop generally from a series of cells partitioned into small compartments containing each a single zoid. Plurilocular organs are generally borne on haploid plants and they produce gametes.

Diploid plants may also bear plurilocular organs; in that case, the zoids issued never show sexuality but they develop without union into new diploid plants.

Reproduction among Phaeosporeae is isogamous, or more or less anisogamous, sometimes with relative sexuality and PARTHENOGENESIS. In most evolved forms, oogamy occurs between large motionless oospheres, produced singly in an oogonium, and very small antherozoids.

Reduction of the number of zoospores produced by unilocular organs may also occur: in Dictyotales they contain only four non motile tetraspores. Vegetative reproduction by means of propagules is frequent among Sphacelariales.

According to their life-cycle, Phaeosporeae are divided into two groups: Isogeneratae and Heterogeneratae. Among Isogeneratae (e.g. Ectocarpales, Dictyotales), an isomorphic alternation of generations takes place between haploid gametophytes and diploid sporophytes, both generations having the same shape, size and structure. Among Heterogeneratae, (e.g. Chordariales, Laminariales), alternation of generations is heteromorphic, the haploid gametophyte being of microscopic size and made of small branched uniseriate creeping filaments, sometimes laterally united in a disc. These gametophytes are *prothalli.*

In many Heterogeneratae, the development of zoo-spores or of zygotes may not give rise directly to the new generation. Sometimes, a *protonema* of creeping filaments is first formed on which buds give rise to new macroscopic plants.

In other cases, there is development of microscopic plants reproducing abundantly by means of asexual zoids born in plurilocular organs. These microscopic plants which look like prothalli but lack sexuality, are called *plethysmothalli.*

According to environmental conditions, numerous successive generations of microscopic plethysmothalli may develop before the appearance of the macroscopic sporophytic generation.

Some of the plethysmothalli, born of zoospores is-sued from unilocular organs are haploid; they may be parthenogenetic prothalli. Others, born of zoids pro-duced by plurilocular organs of diploid plants are also diploid and cannot be interpreted as prothalli.

The Cyclosporae (Fucales) occupy a rather isolated position among Phaeophyceae by their lifecycle with-out alternation of generations. Only sexual plants occur, producing large oospheres and very small an-therozoids. After fertilization, eggs give rise directly to new gametophytes. As in Metazoa, meiosis occurs during gametogenesis.

Classification of Phaeophyceae into twelve orders may be summarized as follows:

PHALANGIDA

Phalangida or Opiliones is an order of the class ARACHNIDA. The best known members of the order are the common daddy-long-legs or harvestmen. Phalangids are in part characterized by their posses-sion of four pairs of legs and a segmented abdomen broadly joined to the cephalothorax. The cephalo-thorax consists of nine modified segments separated from the abdomen by a groove. Located on the cara-pace of most species are two eyes usually situated on an eye tubercle. Near the anterior margin of the cara-pace are two openings which lead to a pair of odorif-erous glands. The cephalothorax bears six pairs of appendages, including the three-segmented chelate chelicerae, the six-segmented palpi, and the four pairs of seven-segmented legs. The genital opening is on the ventral side of the anterior part of the abdomen and is covered by a special plate, the genital operculum. In body size phalangids vary from one to twenty-two mm.

Three subgroups or suborders are recognized. The most primitive of these is the Cyphophthalmi. These are mite-like forms varying in length from one to three mm. In these the openings of the odoriferous glands of the cephalothorax are at the end of elevated tuber-cles. In the United States they are known from the Southeast and the Pacific Northwest. Another suborder is the Laniatores. Members of this group have well de-veloped palpi which often have long spines. A dorsal shield covers the cephalothorax and a portion of the abdomen. Frequently the abdominal portion of this

	Haplostichineae	*Ectocarpales*
Isogeneratae . . . . {		
	Polystichineae . . . . {	*Tilopteridales* *Sphacelariales* *Cutleriales* *Dictyotales* *Scytosiphonales*
	Haplostichineae . . . . {	isogamous or anisogamous . . . . *Chordariales*
		oogamous . . . . { *Sporochnales* *Desmarestiales*
Heterogeneratae . . . . {		
	Polystichineae . . . . {	isogamous or anisogamous . . . . *Dictyosiphonales*
		oogamous . . . . *Laminariales*
Cyclosporeae . . . . *Fucales*		

JEAN FELDMANN

References

Chadefaud, M., "Les végétaux et leur classification. Tome I Les végétaux non vasculaires (Crypto-games)," Paris, Masson, 1960.
Fritsch, F. E., "The structure and reproduction of the Algae," Vol. 2, Cambridge, The University Press, 1945.
Hamel, G., "Phéophycées de France," Paris, (pub-lished by the author) 1939.
Papenfuss, G. F. *in* Smith, G. M., ed., "Manual of Phycology," New York, Ronald, 1951.
Smith, G. M., "Cryptogamic Botany, 2nd ed., Vol. 1, New York, McGraw-Hill, 1955.

scute is brightly colored and bears spines and tubercles. Species of this suborder are found chiefly in tropical and subtropical regions of the world; though a number of species are found in the southern United States. Some laniatores live only in caves, and are white and without eyes. The most abundant suborder in the north and temperate regions of the world are the Palpatores. These include the daddy-long-legs, *Leio-bunum* sp., as well as a number of shorter-legged forms.

Phalangids are mostly carnivorous feeding on slow moving, soft-bodied insects or recently killed animals; however, they will eat bread, fungi, or other vegetable material. When feeding the chelicerae tear the food into small pieces.

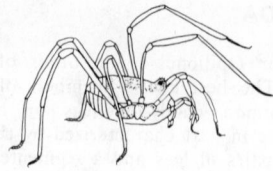

Fig. 1   An opilionid. (From Goodnight, Goodnight and Gray, "General Zoology," New York, Reinhold, 1964.)

No active courtship has been observed among most phalangids. Mature males and females mate briefly, separate, and continue their wandering. A short time later they may mate again with the same individual or with others. Eggs are laid at intervals throughout the summer and fall months, usually in rotting wood, crevices in the soil, or other moist places. The newly hatched young resemble the adult in general, but many special structures of the adult are lacking. These gradually appear during the later molts. In temperate regions, only a few species overwinter. Most pass the winter in the egg stage and hatch in the early spring.

Phalangids have few enemies, but are attacked by a number of parasites including mites, nematodes, and protozoans. In general, phalangids are of little economic importance to man. At times, in the dry Southwest, they become so abundant as to be a nuisance.

CLARENCE J. GOODNIGHT

References

Bishop, S. C., "The Phalangida (Opiliones) of New York," Proc. Rochester Acad. Sci. 9: 159–235, 1949.

Cloudsley-Thompson, J. L., "Spiders, Scorpions, Centipedes, and Mites," New York, Pergamon, 1958.

Roewer, C. Fr., "Die Weberknechte Der Erde," Jena, Fischer, 1923.

Savory, T. H., "Daddy Longlegs," Scientific American, 207:119–128, 1962.

PHARYNX

The pharynx, lying between the oral cavity and the esophagus, plays a dual role. It not only serves the DIGESTIVE SYSTEM as a passage way for food from the mouth to the esophagus but is also important in connection with the RESPIRATORY SYSTEM. In such a lower group as fishes it is provided with gills on each side, and through the gill-cleft it opens to outside. But it does not open to the nares (Fig. 1A). On both upper and lower sides of the pharynx of teleost fishes, there are pharyngeal teeth which are developed from the fifth epi- and ceratobranchial bones. In some aquatic, and all terrestrial animals the pharynx connects posteriorly with two tubular structures, an upper esophagus and a lower trachea which is a passage to the lung.

Among the vertebrates, the pharynx of the adult lamprey is attractive in its peculiar connection with the esophagus. The pharynx which bears 7 pairs of internal gill-slits, ends in a blind pouch, and lies below the esophagus. Anteriorly it opens into a buccal cavity, and there it simultaneously joins the origin of the esophagus.

The pharynx of humans and mammals is generally described as consisting of two parts: The upper portion above the plane of the soft palate is named the *nasopharynx*, and the part below the soft palate is termed as pharynx or *oral pharynx*. In these higher forms, the communication of the pharynx with the allied structures is more complicated than that in lower forms (Fig. 1B). The nasopharynx communicates anteriorly with the nares through the nasal passage, and dorsolaterally with the Eustachian tubes through which air reaches to the tympana. During the swallowing movement, the nasopharynx is transiently closed up from the pharynx by the soft palate pressed against the roof of the pharynx. Inferiorly, the pharynx is continuous to the larynx which leads to the trachea, bronchi, and lungs. The entrance of the larynx is guarded with platelike projection called epiglottis. And dorsally, the pharynx is contiguous to the esophagus.

The pharyngeal wall is principally composed of three layers; mucous epithelium, musculature, and fibrous tissue. The inner surface of the wall is lined with a stratified squamous epithelium provided with the mucous cells. The *tunica propria* is represented by elastic tissue. In the nasopharynx it occurs either stratified squamous or ciliated columnar epithelium. In addition the lymphoid tissues are abundant in the upper part of the nasopharynx. Successive layer forming the mesial wall of the pharynx is muscle fibers con-

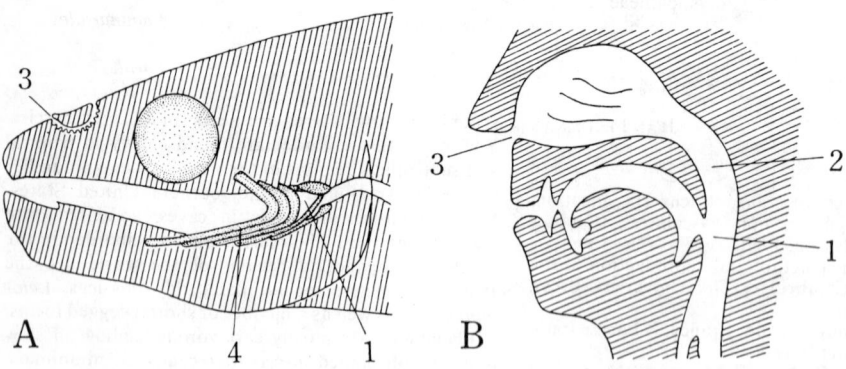

Fig. 1   Diagrammatic aspects of pharynx and its allied structures in teleost fish (A) and in human (B): 1, pharynx; 2, nasopharynx; 3, naris; 4, gill-arch.

stituting the constrictor muscles of the pharynx. The fibrous coat which lies beneath the muscular layer binds the pharynx to the surrounding structures.

Embryologically, the pharynx is formed partly from the endodermal portion of the gut and partly from the ectodermal stomodaeum. In the early embryonic stage, the pharyngeal pouches are derived from both sides of the anterior portion of the gut. The number of pharyngeal pouches is greater in lower vertebrates than in higher groups; the number being 7 in lamprey, 5–6 in most fishes, amphibian, and reptiles, 4–5 in aves and mammals. The pharyngeal pouches develop as a respiratory organ in lower aquatic animals, but they disappear in adult amniota which are equipped with well developed lungs derived from the pharyngeal floor.

TAMOTSU IWAI

PHENOCOPIES

Geneticists refer as "phenocopies" to non-hereditary modifications of form or function which arise in response to forces of the environment and which bear, in their expression, a distinct resemblance to the phenotype of known mutant conditions. The origin of phenocopies usually traces to events that had impinged upon embryonic, or larval and pupal, DEVELOPMENT, and that had in the sequence led to irreversible deviations from the normal course of events. Yet, within our definition, we must accept the temporary sun-tan of a light-complexioned, adult European as phenocopy of the skin pigmentation of a mulatto. There are many similar examples, especially among plants.

The term "phenocopy" was coined by R. B. Goldschmidt in 1935 as a designation for nonhereditary variants of DROSOPHILA imagines which he had obtained by treating definite larval stages with heat shocks and which mimicked mutant traits. The early work with physical agents (heat, X-rays) was followed more recently by experimentation with chemicals.

The more important information, gained from recent work with Drosophila and poultry, is summarized in the sequel. Factors which influence or determine the type of phenocopic response are: the developmental stage at time of treatment, the dosage or force of the external agent, the nature of the agent itself, and the genetic constitution of the treated organism. Identical treatments may in successive stages affect dissimilar parts and organs. The quantitative dosage-effect relations often show direct proportionality in regard to incidence and extent of modification, but a rising dosage frequently results in additional parts becoming involved. The chemical nature of the phenocopy-producing agent determines the type of metabolic event that will be interfered with, as was shown by the protective activity of specific metabolites if they are administered as supplements.

Genetic factors play an important and complex role in the origin of phenocopies, their nature and incidence. Different lines and stocks of one species are frequently quite dissimilar in their response to identical phenocopy-producing agents; such response differences are most commonly of a quantitative, but may even be of a qualitative nature. Plus and minus selection for response to specific agents is generally quite effective. The most interesting genetic evidence relating to the production of phenocopies is of three kinds: (1) the response to phenocopy-producing agents is often greatly enhanced in the presence of homologous hereditary (usually multifactorial) stock tendencies with low penetrance, i.e., the two sources of variation may become additive; (2) plus modifiers of homologous mutants may reduce and (3) heterozygosity for homologous, recessive mutants may potentiate the phenocopic response. The genetic evidence demonstrates, therefore, that phenocopies and homologous genetic traits may, and apparently often do, have certain developmental events in common.

Two principal genetic interpretations have been advanced to explain the nature of phenocopies. The first proposes that all phenocopies are due to a bringing into light of already present non-penetrant, subthreshold or isoallelic mutants; the second suggests that the responsible external agents intervene in gene-determined developmental events by preventing them at one point or another from accomplishing their appointed ends, leading thereby to mutant-like phenotypes.

WALTER LANDAUER

PHEROMONE

The term "pheromone" was introduced to biological literature in 1959 to replace the less apt terms, "exohormone" and "ectohormone," which had been used to designate substances acting as chemical communication signals between individuals of the same species. Subsequently, it was suggested that the term "pheromone" be restricted to communication substances which act biochemically, and "telomone" used to designate those acting on sensory receptors. This latter refinement has not been accepted by the majority of authors in the field.

In some respects, pheromones resemble hormones. They usually are produced in small amounts by glandular tissue, they may be stored in bound form subsequent to synthesis and prior to release, they may be under central neurosecretory or endocrine control, and they are active physiologically at extremely low concentrations. They differ from hormones in that, while their site of action is removed from the site of production, rather than acting on the producer, they act on a recipient of the same species either to release a directed behavior response, to initiate a developmental process, or to prime the recipient so it becomes liable to certain behavior if the requisite stimuli are subsequently encountered. In any case, the ultimate consequence is a specific, qualitative change in behavior.

Pheromones may be released into air (or water) and reach the recipient by diffusion; these chemical signals are detected by olfactory receptors. On the other hand, some pheromonal communication requires direct contact, with signal detection occurring orally. When the pheromonal secretion is a mixture of chemicals, the information communicated varies with context. For instance, the "queen substance" of honey bees acts olfactorily both as an attractant for drones in mating and for aggregation of workers in a swarm; orally, it acts to inhibit both ovarian development in worker bees and queen cell production.

Among the first pheromones to be identified were the insect sex attractants and the "queen substances" of Isoptera and Hymenoptera. In the past decade, alarm substances, aggression-promoting substances, territorial markers, trail substances, and caste- or group-recognition substances have been implicated in the regulation of behavior among social insects. The existence of chemical communication techniques in invertebrates other than insects and in some vertebrates has been demonstrated recently. In addition to the true pheromones, other chemical substances, including both repellants and attractants, transmit information between individuals of different species. Repellant substances often serve as defense mechanisms, while attractants include chemicals released by plants. Some of these compounds have pheromonal overtones. For example, the defense secretion produced by termites also serves as an alarm substance, alerting additional soldiers to the state of emergency. Then again, *trans*-2-hexenal derived from red-oak leaves acts on the antennal receptors of the female *Antheraea polyphemus* moth and release of the sex attractant pheromone is triggered. The pheromone can diffuse over a considerable distance and, when detected by the antennal receptors of a male, will elicit his upwind flight toward the female. Following mating, the female is in close proximity of host plants on which to deposit her eggs. Both the response of the female to the oak factor and the response of the male to the sex attractant are eliminated if the corpora cardiaca are ablated, suggesting that a neuroendocrine relay is involved in mediating the behavior of both male and female moths. In addition to the silk moth *Antheraea polyphemus*, release of sex pheromones by some cockroaches appears to be under neuroendocrine control. *Periplaneta americana* and *Byrsotria fumigata* females will release their pheromone only if the corpora allata are present. Furthermore, once mating has occurred, pheromonal production or release by the female of most insect species decreases markedly, suggesting an endocrine feedback mechanism of some sort.

The prediction on theoretical grounds that compounds employed as olfactory pheromones would be likely to have molecular weights of around 200 has been verified in the few which have been isolated. Such relatively small molecules can diffuse rapidly. Furthermore, storage reservoirs can contain more molecules per unit volume than if bulkier signal molecules were employed. A wide enough diversity of molecular structure is possible in organic compounds possessing more than 5 carbon atoms to account for the species specificity of pheromones. Olfactory pheromones include the sex attractants, trail substances, and alarm substances. The majority of the Lepidopteran sex attractants whose chemical structures have been defined are 12–16 carbon alcohols or alcohol derivatives, and valeric acid serves as a sex attractant for the click beetle, *Limonius californicus*. Various terpenoid aldehydes and ketones have been suggested as the alarm substances in a number of species of ants. No trail substance has been identified chemically.

Signal dissemination through air makes temporal or spatial patterning of stimulation improbable, since the time required to reach threshold concentration levels at a distance from the emitter will vary with conditions, and uncontrollable local concentrations of pheromone easily could occur. However, the threshold of antennal olfactory receptors of most insects is very low; *Bombyx mori* males, for example, require a concentration of only 200 molecules of sex attractant per cm^3 for stimulation. Furthermore, in many instances a change in information content is associated with a change in pheromone concentration. For example, sex pheromone at low concentrations (10^{-12} $\mu g/ml$) induces gradually increased excitation and finally, upwind flight in male *Bombyx*. Actual orientation due to intensity discrimination does not occur until the concentration of pheromone is 10^{-1} $\mu g/ml$. Also, in the ant *Pogonomyrmex badius* (Latreille), worker behavior changes from attraction to alarm and attack with a tenfold increase in the concentration of a pheromonal secretion of the mandibular glands.

Among the social insects (bees, ants, and termites), the caste system is maintained by inhibitory substances produced by either the queen or, in the case of termites, by both king and queen. These inhibitory pheromones are transmitted orally. The effect of the pheromone is to inhibit sexual development in members of the worker caste. There is evidence indicating that, in ants at least, the function of the corpora allata is disturbed at the last larval instar of female larvae; these larvae develop to sterile workers.

The use of chemical signals by animals other than insects has not yet been widely investigated. There is good evidence for the release of a sex attractant by the premolt female crab, *Portunus sanguinolentus*. Adult males exhibit typical sexual search and display behaviors either in the actual presence of premolt females or when they are exposed to water in which premolt females have been kept. Behavioral studies of the nematodes *Trichinella spiralis* and *Panagrolaimus rigidus* indicate the employment of chemical sex attractants by these species. In *Trichinella*, the male apparently releases the attractant substance. Finally, pheromonal substances appear to govern the social behavior of the visually deficient fish, *Ictalurus natalis*.

ELIZABETH J. ARTHUR

References

Novák, V. J. A., "Insect Hormones," New York, Methuen, 1966.
Haskell, P. T., ed., "Insect Behavior," Symposia of the Royal Entomological Society of London, vol. 3: 1966.

PHILOSOPHY OF BIOLOGY

Biology is a diffuse body of knowledge and methodology without a coherent philosophical tradition. Diversity marks the work of biologists even when the field is narrowly defined; their tasks are highly specific, and for the most part practical. Although philosophical presumptions may frame the biologist's work, there is little conscious recognition of philosophy-proper by biologists. Classification, for example, can be construed as an application of Platonic ideas or Aristotelian forms, but classification has been practiced quite independently of Greek philosophy. The biological work of Aristotle, his classification of animals for example, is still respected by biologists. Biological subject matter has been chosen by philosophers: Descartes, Schopenhauer, Dewey, and the philosophical writer Goethe. In Europe it has not been uncommon for biologists to become philosophers at the end of their careers, while American tradition has generally maintained a sharp and suspiciously regarded boundary between the two fields. Unless Bergson is considered a biologist, none of them seems to be regarded highly by philosophers, and biologists have paid scant heed to philosophers. Caspari (1) has written a lucid and knowledgeable account of the conceptual basis of biology and its relation to philosophic tradition.

In the diverse activity of biologists certain patterns of approach are evident; these are offered as their *working* philosophy. They are elucidated here under four categories:

I—Taxonomic Theory* (stressing forms of matter) and Ecological Theory (stressing energy)
II—Evolutionary Theory
III—A contemporary model of the biological system (stressing the transmission and accumulation of information)
IV—Contribution of biology toward a philosophy of man

I

Classification and identification have commanded much of the biologist's effort. The taxonomic approach has been justified on practical grounds (for example, knowing what to eat) or as an intellectual exercise in deriving rational ideas of natural order. Within the usual scope of the taxonomist, it is surprisingly inconsequential whether he believes that natural order means "God's plan" or "evolutionary arrangement." Furthermore, the taxonomist does not need a specific intellectual goal to justify his zeal: Sufficient is the vague belief that systemized knowledge shall be of use to mankind. Both the narrow focus of interest and the active inventing of new names have obscured the breadth of taxonomic activity throughout and beyond biology. There are about 1000 terms in the glossary of an *elementary* botany text. Most are *classes* of processes, structures, or environmental aspects, and thus are taxonomic in essence. It is simply by custom that the classifying of individuals is called taxonomy. In essence, taxonomy is an elaborated application of the logical devices of exclusion (*A* or not-*A*) and inclusion (all *B* is *A*). The device is used alike in the classical taxonomy of Aristotle or Linnaeus, or the new numerical taxonomy of Sokol and Sneath. (2)

The excluding relationship differentiates not only "objects" but any segments of experience from each other. Naming the segments enables communication of the perception: "Past" and "present" is a taxonomy of time. The inclusive relationship is an assertion of similarity among units which are differentiated by some

*Theory is used in the general sense of "idea," and "model" has a similar meaning. The somewhat ambiguous dual use is employed here because the separate terms are more familiar in their own historical contexts.

other means: Groups of individual (different) cells are "tissues." With this simply basic approach, the formal taxonomy (of individuals) has evolved two important elaborations: The *species*, a group of "similar" individuals, has been adopted as a basic unit, and the arrangement of classes has become hierarchical—all sub-units are nested within more and more inclusive units (species in genera, genera in families, etc.). Although both the species and the hierarchy are supposed to reflect nature, they may be regarded as devices for effective retrieval of information (or communication).

The ecological relationship makes the simple assertion that areas of experience may be related to each other. If taxonomy concentrates on objects or individual units and ignores background, ecology transfers the attention to the background, and in effect treats background as an object or structure. It is perhaps necessary (and certainly traditional) to exclude (separate, distinguish) a biological unit from its environment. Nevertheless unit boundaries must be open to some extent for "environment" to have any meaning, since it is not merely the surroundings, but *related* surroundings. Therefore, the testing of the ecological theory has demanded that the nature of the boundary relationship be specified. The changing means of specification have been the cause-effect relationship, flow of matter and energy, and flow of information.

Biologists have not challenged the validity of these specifications; they have accepted cause-effect and matter-energy, while information theory is new and has not yet been considered on a broad scale. If basic assumptions of the ecological theory were challenged, logical difficulties probably would appear. Where does a unit stop and environment begin? Where in space is the boundary of environment itself? The answers to both questions in actual biological studies have been practical: whatever answer is useful in the *particular* context of their problem. In the wider context of society, the identical questions are asked in terms of "loyalty." What is the context of a man's interest? Self? Family? Nation? Mankind? Life? It seems likely that his strong ego concept has caused him to harden the boundary about individuals and to be insensitive to his involvement in his environment.

II

In this analytical retrospect, evolutionary theories have been attempts to synthesize or unify the taxonomic and ecological theories by explaining both differences and similarities among biological units, and also the relationship of environment to unit structure. To achieve such a synthesis, the areas considered have required enormous expansion into the time dimension.

This cool analysis, however, does not suggest the philosophical storm which has raged since (in the words of Oscar Riddle), "the unleashing of evolutionary thought." To state it mildly, a biologically influenced philosophy *did* affect man's basic frame of reference. Certain aspects of the evolutionary theory have given it the specific form which non-biologists have variously embraced and denounced. "Natural selection" was Darwin's own nomenclature for the mechanism by which a population changes in time. Given only the assumption that variety *arises*, the mechanism largely suffices to synthesize taxonomy and ecology. Nevertheless, as an idea, natural selection has been subject to only limited logical criticism and empirical test. What are the alternates to "natural" selection? Selection by God? But, is God not "Nature's God" too? If the alternate to natural selection is artificial selection, why is man to be isolated taxonomically from the remainder of the environment?

A second major aspect of evolutionary theory, "adaptation," has become one of the most entrenched of biological dogmas and the routine explanation by biologists for any structure or behavior. The habit of language, at least, has spread in various directions from professional biology, for example, authorities on child development expect children to "adapt" to social situations. The assumption that organisms will adapt is virtually a universal article of faith. In the mouths of most users, the adaptive explanation is, in effect, teleological; the individual or population does what is best for it—at least according to the values which man imputes upon them.

There is a close relationship between the faith that the organism will somehow do what is best for itself under the circumstances and the faith in "survival of the fittest" (the statement was Herbert Spencer's, not Darwin's). The further belief in progress toward perfection has come easily from these ideas. The statements of these ideas have been the language in which both biologists and non-biologists have understood evolution: natural selection, survival of fittest, adaptation to environment, and progress to perfection. Biologists have scarcely questioned these canons of their faith. A conspicuous exception is Von Bertalanffy's slashing attack on "survival of the fittest" as being a "Tibetan prayer wheel" of logic (tautology): "What survives?" "The fittest." "How do you recognize the fittest?" "Observe what survives." The position of most biologists is fairly represented by Sir Julian Huxley who in his own right rather than by distinguished lineage is a preeminent philosopher of biology. Recently, he has yielded from "survival of the fittest" to "survival of the fitter," but this position is scarcely less vulnerable.

In fairness to the best biological tradition, it must be stated that they have gone far to supply tenable meanings for "fitness" and "adaptation." Henderson (3) emphasized minimum fitness, as opposed to perfection, while Dobzhansky and other geneticists have emphasized the opportunistic nature of selection, thus removing a good part of the guarantee of eventual perfection. Unfortunately, their qualifications have made little impression on lesser biologists or on laymen.

The impact of evolutionary philosophy beyond academic biology has been paradoxically divergent during "Darwin's century." On one hand, the optimism of 19th century social philosophy seemed to have good scientific grounds. The reformer was substantially inspired and encouraged by the beliefs that man was the supreme example of evolutionary perfection, and that progress of mankind was inevitable if its barriers were demolished. On the other hand, "social Darwinism" became a perversion. The intellectual error in interpreting Darwin's biology was a minor tragedy compared with the social crimes imposed upon the poor, the weak, and the young in good conscience—supposedly the fittest survived and adapted, so that the race was helped to perfection? As the century drew to a close, both of these

social extrapolations had substantially failed; such biological philosophy was not applicable to the world at large. All babies are unfit to survive alone; the less powerful individual workers organized revolutions; what was good for Standard Oil was at length deemed to be not necessarily good either for John Doe or for Uncle Sam. So strong had been the belief in the individual as a unit, that the social context had been ignored.

III

A more satisfactory philosophy of biology may be emerging now. At its center is a new concept or model of the living system itself. The implicit question which has called forth the new model as an answer is, "What is unique about life?" Although vitalism may persist strongly in habits of speech and thought, it has effectively been ignored as biologists have become more and more chemists, physicists, and mathematicians. Even where no question of vitalism is involved few biologists defend "natural history" as a serious pursuit. The uniqueness of the biological system is being reasserted in understanding it as a nesting hierarchy or organization of integration, with emergent properties at each level.

The thinking which has contributed to the new biological model is easier to identify in writings of non-biologists: a philosopher (Peirce), a statesman (Smuts), and a humanitarian social scientist (Kropotkin). The work of Gibbs in thermodynamics led to visualization of biological organization (negentropy) and disorganization (entropy) in meaningful terms. At a time when few biologists knew of Gibb's work Nabokov (4) opposed entropy and "revolution" as the choices of man in a literary work. Finally, it has been from communications technology that Information Theory has come. In general, Information Theory suggested the equivalence of information and organization, and, in specific practice, it has suggested the computer as a useful analog of biological systems.

Within biology the ideas have been integrated only very recently. Bernard's and Cannon's concept of homeostasis as a dynamic maintenance of order remained restricted in application. Morgan's (5) specific statement of the biological system being a hierarchy of organizational levels produced no immediate application. Over twenty years later (1945) a lively controversy between Novikoff (6) and Gerard and Emerson (7) on the nature of different levels simply died out from lack of a medium in which to test their views. Yet, within the last decade hierarchy of organization has become a popular format for elementary biology books. This is not to be taken as evidence that the new biological model is understood or even that it has fully formed from the chaos. Indeed, the model described here should be understood as a brief and tentative statement. Furthermore, the model does not presume to separate "living" from "non-living" systems with Aristotelian confidence; such precision is not characteristic of biological models.

The biological system is a complex pattern in space-time. It can be analyzed as a hierarchy of complexity with simpler units (necessarily also smaller) composing successively more complex units. Both the structure and behavior (changes) of more complex units are characterized in different terms from their less complex component unit. Changes (especially constructive and cyclical changes) occur in larger units over a longer time span (Rome was not built in a day, but many cells were).

Boundaries may be recognized between units of any size and environment, but "environment" has meaning only if the boundary is open to some extent. Consequently, boundaries must be defined in different terms applicable to the flow of matter and energy or to flow of information.

Both within units and at their boundaries, self-regulation is discernible, and self-regulation becomes a more complex activity with increasing structural complexity of a unit.

The result of regulatory activity is that the negative entropy (state of order) of the system can be maintained for an extent of time (condition of homeostasis), or can increase.

The general effect of the new model is to encompass previous theories, as increasingly general theories in physical science have done. In addition, a shift in the working philosophy of biologists is conceivable. The dignity of the taxonomist (anyone who classifies) may be reestablished when it is perceived that his task is to define meaningful boundaries whether he studies molecules, bacteria, or nations. He can outgrow his past roles: the naming of the different products of special creation and the forcing of all nature into an evolutionary tree. The ecologist is already learning the philosophical significance of what he is doing: discarding the cause-effect relationship between unit an environment, and reframing his questions in terms of flow, retention, and organization of matter and energy (or information) in chosen environments. Evolution theory should be applied less naively if "adaptation," "selection," and "survival," must be described with reference to more specific levels of organization (gene, cell, tissue, organism, community) and time span, whereas past attention has been concentrated at the level of individual and (more recently) population. It is even to be hoped that comprehension of the meaning of self-regulation may in one stroke banish the ghost of classical teleology from biology and heal the resentment of humanists against the "machine" concept of life.

IV

Only an optimistic visionary could foresee a biologically influenced philosophy as a major component of the philosophy of man (8); in the face of the immediately impending disaster of overpopulation, man seems unwilling to act as if he is biological. Nevertheless, the area of common concern is noteworthy, even startling. The old testament declaration, "The Lord is One!" has its modern affirmation, "All life is one!" from both the humanitarian Albert Schweitzer, and the biochemist, Albert Szyent-Gyorgyi. In 1954, Alfred Emerson, a respected ecologist, offered an assembly of churchmen, "Dynamic Homeostasis, a Unifying Principle in Organic, Social, and Ethical Evolution" (9). Recently, a governor publicly stated that his administration's primary consideration was the "total environment" of the state. How important is such evidence? In the very least it indicates that the philosophy of biology is being heard, even that it has achieved prestige. When cases of rhetoric and perversion are removed there remains quite ample content for a broad philosophy in the conclusions of serious scholars in their own areas of competence.

It is the biologist, after all, who should grasp the concept which draws the deepest awe of life. The maintenance and increase of order through the succor from a wider environment is a deep subject not only for scientific understanding and technological control, but also for broad philosophical searches which include at least the fields of moral philosophy and epistemology. Nevertheless, if the portent of biological philosophy is to be realized, the philosophy of biology may require the unifying infusion of Information Theory which has arisen from the pure technology of communications! Without it, the serious questions about the unity of life seem unresolvable. The environment of life may be more comprehensible to man as a system where information (even "order") moves, is accumulated, and is lost. Peirce or Dewey might have accepted successfully the opportunity to integrate information theory and the new biology into a philosophy of man, but they were too early. Among modern scholars, Ashby has made a worthy attempt to reach a new philosophy in his "Design for a Brain" (10). Yet, there has been too little effort in the difficult work of creating a sound, critically examined, Philosophy of Biology; the counterpart of *Principia Mathematica* has not been attempted.

HERMAN S. FOREST

References

(1) Caspari, E. W., "On the conceptual basis of biological sciences," *in* Colodny, R. C. ed., "Frontiers of Science and Philosophy," Pittsburgh, The University Press, 1962.
(2) Sokol, R. R. and P. H. A. Sneath, "Principles of numerical taxonomy," San Francisco, Freeman, 1963.
(3) Henderson, L. J., "The fitness of environment," New York, Macmillan,. 1913 [Reprint, Boston, Beacon Press, 1958.]
Ibid., "The order of nature," Cambridge, Harvard University Press, 1917.
(4) Nabokov, V. V., A book review in the *New York Times*, January 15, 1967 singles out this view for discussion. The review covered two works: V. V. Nabokov, "Speak memory. An autobiography revisited," New York, Putnam, 1966, and Page Stegner, "Escape into aesthetics. The art of Vladimir Nabokov," New York, Dial, 1966.
(5) Morgan, C. L. "Emergent evolution," New York, Holt, 1923.
(6) Novikoff, A. B., "The concept of integrative levels and biology," *Science*, 101: 209–215, 1945.
Ibid., "Continuity and discontinuity in evolution," *Science*, 102:405–406, 1945.
(7) Gerard, R. W. and A. E. Emerson, "Extrapolation from the biological to the social," *Science*, 101: 582–585, 1945 (Also involved, Joseph Needham, "A note on Dr. Novikoff's article," *Science*, 101: 582, 1945.
(8) In our conversation Dr. Caspari firmly declared that he *is* such an optimist, and, down deep, I will admit some such feelings myself.
(9) Emerson, A. E., "Dynamic homeostasis, a unifying principle in organic, social and ethical evolution," *Scientific Monthly*, 78: 67–85, 1954.
(10) Ashby, W. R., "Design for a brain," 2nd ed., New York, Wiley, 1962.

PHLOEM (BAST OR LEPTOM)

The phloem, a complex tissue, is the main food-conducting tissue of vascular plants. Rudimentary phloem occurs in some mosses. With the XYLEM, phloem forms the plant vascular system, primary phloem strands typically alternating with xylem in the root vascular cylinder, while stem primary phloem characteristically lies outside the xylem (sometimes inside it as well) in the vascular bundle. Secondary phloem is normally, but in some cases not exclusively, formed to the outside of the cambium; but definite phloem annual rings are not produced.

Historical

Although the *sieve element*, the essential phloem cell type, was identified by Hartig in 1837 and its general structure elucidated by Nägeli before 1880, the minute histology of the phloem, its phylogeny, and especially its food-transport mechanism, are still imperfectly known (cf. xylem). Phloem undergoes ontogenetic degeneration; its commercial uses do not demand structural knowledge; it fossilizes poorly.

Histology

Sieve elements may be *sieve cells* (Illustration, A), relatively little studied, which are characteristic of pteridophytes and gymnosperms, or *sieve-tube elements* (C) which are united end to end to form *sieve tubes* and which, with almost always associated *companion cells*, are confined to angiosperms. Other phloem cells are parenchyma (phloem parenchyma forming the vertical series, ray parenchyma the horizontal—B), fibers, sclereids and secretory cells—parenchyma only, in pteridophytes and many gymnosperms, parenchyma and fibers in some gymnosperms, various cell combinations in angiosperms. Further research on phloem parenchyma and its significances is urgently required. Fibers (hard bast) are conspicuously developed in some dicotyledons, and, late in phloem ontogeny, sclereids often form abundantly from parenchyma. In general, however, phloem cells do not develop rigid, persisting (secondary) walls as do xylem cells, and, after a short functional life, the phloem, if not crushed by axial thickening, becomes much modified, the sieve elements degenerating, sometimes even disappearing by absorption. In some dicotyledons, however, and especially in monocotyledons, phloem may remain active for years. In many woody plants periderm layers eventually form in the phloem, progressively destroying the tissue as new phloem develops.

Sieve elements are unique structurally, cytologically and functionally. They are cylindrical-elongate, living, cellulose-walled cells (clearly evolved from parenchyma cells) in which certain primary pit fields are metamorphosed into *sieve areas*—aggregations of fine pores through which cytoplasmic *connecting strands* usually pass. At functional maturity the element is *enucleate*.

Sieve cells are usually long and slender (sometimes branched or otherwise irregular, forming a phloem meshwork), with tapering ends tending to overlap, the sieve areas are relatively unspecialized, more or less alike, and scattered over the walls, especially end and radial walls. In some ferns they may be borne on cushion-like outgrowths.

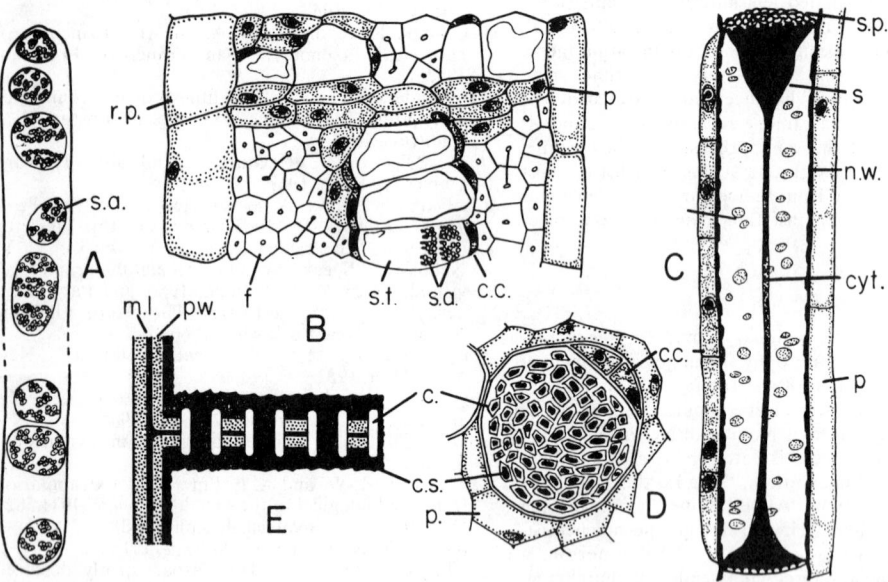

Fig. 1   A. Sieve cell (1/5 only shown) from secondary phloem of *Pinus radiata* × 25. B. T. S. of an angiospermic phloem (*Tilia europaea*) × 175. C. R. L. S. phloem of *Lagenaria vulgaris* showing sieve-tube element of advanced type × 55. D. *Lagenaria* phloem in T. S. showing a sieve plate × 100. E. Diagram (redrawn from Esau) of part of similar sieve plate in section. c, callose; c.c., companion cell; c.s., connective strand; cyt., cytoplasm; f, fiber; m.l., middle lamella; n.w., nacreous wall; p, parenchyma; p.w., primary wall; r.p., ray parenchyma; s, slime; s.a., sieve area; s.p., sieve plate; s.t., sieve tube.

Sieve-tube elements have certain highly specialized sieve areas known as *sieve plates* upon the end walls, these latter varying from very sloping, with typically *compound plates* (several sieve areas—B), to transverse, usually with a *simple plate* (one sieve area—D). In the first case the side walls may have numerous, relatively unspecialized sieve areas, even sometimes sieve plates, in the second, side wall areas are usually few, often vestigial (C).

Protoplast. The cytoplasm, of low metabolic activity and containing only atypical starch, is peripheral about a large, central vacuole. A vacuole membrane has lately been demonstrated. Nucleus loss is accompanied in many, especially herbaceous, dicotyledons by the accumulation in the vacuole of proteinaceous *slime*, derived by the dissolution of hitherto deep-staining *slime-bodies* in the cytoplasm. Slime has been detected in connecting strands.

Slime plugs (C) in sections are artifacts—for the system in active life is highly turgid.

Wall structure. "Nacreous" thickening (slight to very heavy, often crenulated—C) of uncertain, although cellulosic, nature, and unknown significance, develops in many plants. The thickening is not highly hydrated as formerly thought.

Sieve structure. The *pores* (from $< 1 \mu$ to c. 14μ in diameter, according to recent measurements) consist of cytoplasmic *connecting strand* and surrounding tubule of *callose* (D, E), a substance of uncertain nature. The tubule thickens as the element ages, continuity of the strands sometimes being broken, and eventually a callose cushion (*definitive callus*) may cover the once-depressed sieve area. This typically marks the death of the element and the callus itself eventually disappears, leaving a thin, openly-perforated sieve area: but in some plants the callus dissolves at the beginning of another season of phloem activity.

Companion cells (B, C, D), specialized parenchyma-type cells, arise by longitudinal division (C, D), or divisions (B), in the sieve-element mother cell. Subsequent transverse, and occasionally longitudinal (D), divisions in the c.c. initials may produce rows of c.c.'s C.c's have abundant nucleated cytoplasm, and are connected, it seems, by plasmodesmata to the sieve elements. They are thought to assist sieve-tube functioning: they die when the tubes die. Protophloem, however, usually lacks c.c.'s. and some primitive dicotyledons never develop them. The albuminous cells of conifers simulate c.c.'s.

Phylogeny

Sieve element and tracheid-vessel evolutions show parallelisms. The sieve element tends to become shorter (both through cambial initial shortening and secondary septation in element initials), its end wall less oblique, sieve areas reduced in number, pores larger, parenchyma more varied and abundant, c.c. divisions more prolonged.

Phloem Functions

The chief is conduction, both of carbohydrates (essentially sucrose) and organic nitrogen. Storage (parenchyma), support (fibers) and protection (fibers, sclereids) are secondary functions.

Economic Value

Fibers, tannin, gums, spices, drugs (e.g., cinnamon, quinine) and latex are phloem products.

L. H. Millener

References

Esau, K. *et al.*, "Physiology of phloem," Ann. Rev. Plant Phys., **8**: 349–74, 1957.
Zahur, M. S., "Comparative study of secondary phloem . . . of woody dicotyledons. . . ," Mem. Cornell Univ. Agr. Exp. Stn., No. 358: 1–160, 1959.

PHOLIDOTA (PANGOLINS)

These animals represent an order of MAMMALS. The Greek name Pholidota derives from *pholis* (scale), and *pholidota* means "the scaled." The phylogeny of the Pangolins is virtually unknown. The oldest bones, representing living forms, have been found in the Pleistocene of Southern India, therefore they have an age of only some hundred thousand years. Evidently the ancestors of the Scale Anteaters become separated during the Paleocene, that means 70–60 millions of years ago, from some early primitive stock of Placentalia (i.e. mammals with a placenta) and have evolved quite independently.

The Pangolin body size is between that of a marten and a badger, the total length being between 65 and 150 cm, the length of head and body between 30 and 75 cm, and the length of tail between 30 and 80 cm. In their general appearance the animals are cone shaped and are among the most curious living mammals of the world, as is indicated by the family name Manidae and the genus name Manis, since in Latin *manis—manes* means the spirits of the dead or ghosts.

The small, pointed, head bears only a narrow mouth, a fine muzzle and little squares between the nostrils, very small eyes, and rudimentary earconches. The head intergrades smoothly into the short neck, and the latter in the same manner into the roundish body. The short legs come down on the whole plantas of the feet which bear five toes with sharp claws, those on the fore feet longer than those on the hind, beyond which the second, third, and fourth are very long. Walking on the ground they are doubled under the fore feet. The claws of pollex and hallux are short, especially in the climbing species. The long, or extremely long, tail has the surface roundish above and flat below, a broad root, and tapers gradually. In *javanica* and *tricuspis* the under surface of the tip of the tail is naked, and may serve as an organ of touch for the climbing species. The under surface of the body, and the interior of the legs, is covered with coarse but not very dense blackish, brownish or whitish hairs, which develop relatively late. 1–4 hairs also stand behind each scale but are soon shed or rubbed off. The other parts of the body from front to tip of tail and down to toes including the under surface of the tail are covered with horny scales, tile shaped, arranged in regular overlapping series, tinged with a dark brown or more yellow-brown colour. They are built up by the epidermis, very strong, hard and sharp ridged, and may be spread at will by the animal. Each species has a fixed number within narrow limits.

In *tricuspis* f.i. there are 19–23 scales in each series, or crossrow, in *temmincki* 11–13. The skin bears no glands except a ring of perianal and circumanal ones around the anus, and the two lacteals with their axillary teats. Even eyelids are lacking glands, a feature unique among mammals. A broad band of skin muscles on each side of the backbone enables the Pangolins to roll themselves into a ball for sleeping or defense, exhibiting an enormous power, that defies any ordinary attempt to unroll them. A curious thing is the adhesive tongue with its abnormal great length, its roundish and at the tip more flattened shape, which can be stretched out at a great distance by the help of the complicated hyoid apparatus. It licks up ants and termites so quickly that man's eye is unable to follow. Therefore the salivary glands are of enormous extent. The insects licked up are devoured immediately and not chewed, because there are no teeth at all, not even a rudiment of them. The stomach is a simple sack, its inner wall covered with a horny epidermis, which in some species develops pyloric horn teeth to triturate the swallowed ants or termites. Only in the middle of the stomach is a field of glands secreting large quantities of gastric juice. In some species it is enlarged and folded up to a compact glandular corpus.

The intestine is a simple tube, its great gut or crassum showing no difference between colon and rectum, and there is not even a caecum. The deeply lobed liver includes a gall-bladder. The uterus bears two horns, the urethra entering the middle of the vagina so that the underpart of the latter exhibits a true urogenital tube. Anus and vulva are put together in an invagination forming a false cloaca. The testes are posed beneath the skin, no scrotum is to be found. The shape of the skull is that of a smooth and simple wedge, it bears no teeth, only incomplete zygomatic arches and a little jaw in form of a slim clasp only. The flat orbit and temporal grove are not separated each from the other, the tympanic bone is flat and ring-like, but the nasal cavity on the contrary is large and with well developed conchae. There are no clavicles. The sternum supports the muscles of the great tongue sack, which terminates behind in the form of a spade in the species of Asia, and as a very long double stick in the species of Africa. In the vertebral column there are 7 cervicals, 14–17 thoracics, 5–6 lumbars, 3–5 sacrals and 26–49 caudals. The end phalanges of the fingers are fissured to give an extreme stability to the roots of the claws while digging. The simple brain bears two very large olfactory lobes, the hemispheres with their indifferent sulcation are only

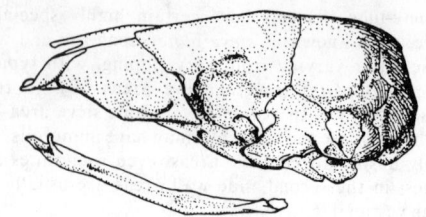

Fig. 2   Skull of a Pangolin in lateral view (*Paramanis javanita*), nearly two-thirds natural size.

small and the cerebellum uncovered. That means that the intellect of the Pangolins is not very important, but the animals have a good olfactory sense.

The habitat of the Scaled Anteaters is wooded, scrubbed, country in plains, hills, and mountains, *temmincki* only ranges steppes or semi-deserts. In daytime the solitary animals sleep curled up in burrows, several meters in depth (which they dig quickly or find abandoned) with a large chamber at the end. They block up the mouth of the hole on returning to it. The climbing species sleep in forked branches or tree holes. They are seldom abroad in the daytime. In searching for food they walk about and can travel when necessary at fair speed, displaying considerable agility. They often raise themselves on their haunches, and, owing to the balance of their tail, run on their hind feet. They are also able to swim, to climb on trees or on rocks to stir up the tree- and soil nests of the termites and ants, which they break asunder with their powerful claws. While climbing the tail is an important support.

Most of the biology of the abstruse mammals is unknown, especially about reproduction. The Pangolin produces only one young at a time, which is completely developed with soft scales growing hard the second day after birth. The mother has a very curious way of carrying her young. It scrambles on her tail and hangs on the root of the tail or tight on the back. In danger it remains in hiding on the belly where it is supported and protected by the mother's tail. A captured Pangolin does not show prolonged fear, but will soon uncurl and walk or climb about. The life span may reach a dozen of years or more, but no exact dates are known.

Scaled Anteaters are distributed in Asia and Africa. In Asia they settle in India and Ceylon and eastward to Fukien, Hainan, and Formosa, southward to Java, Su-

Fig. 1   The Giant Pangolin, erected, the long tongue stretched out.

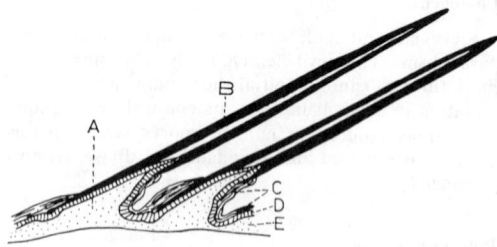

Fig. 3   Longitudinal schematic, up through two scales of a Pangolin (*Uromanis tricuspis*). A, perual papille; B, scale; C, horny stratum of the epidermis; D, muscle stratum of the epidemis; E, other part of the skin.

Fig. 4   A climbing Pangolin (*Uromanis triaispis*).

matra, Borneo and the Philippines, in Africa from the Sudan southward to the Cape.

The order Pholidota Weber, 1904 contains one family only, Manidae Gray, 1821, with 1 genus, 6 subgenera, and 7 species altogether.

THEODOR HALTENORTH

PHOSPHOLIPIDS

The phospholipids belong to the group known as "complex lipids." In addition to fatty acids they contain phosphoric acid, and further may contain glycerol, nitrogenous bases such as choline, ethanolamine, serine, sphingosine, and the sugar-like compound, inositol.

Although there is only one sphingosine containing phospholipid, sphingomyelin, there are several complex lipids which contain sphingomyelin but do not contain phosphoric acid. These are the cerebrosides and gangliosides, and will not be discussed here.

Typical formulas of phospholipids are shown below:

H_2—COCOR (saturated)

ROCOCH (unsaturated)

$$\begin{array}{c} O \\ \uparrow \\ C—OP—O— \quad \text{choline} \\ | \qquad\qquad \begin{bmatrix} \text{ethanolamine} \\ \text{serine} \\ \text{inositol} \end{bmatrix} \\ O— \end{array}$$

Phosphatidyl

H_2COCOR

HCOC=C—R

$$\begin{array}{c} O \\ \uparrow \\ COP— \quad O— \quad \text{choline} \\ | \qquad\qquad\qquad \text{(ethanolamine)} \\ O— \end{array}$$

Plasmalogen

$$CH_3(CH_2)_{12}CH=CH—CH—CH—CH_2OP—O$$

with the substituents: O, H; NH, COR; O; $(CH_2)_2N^+(CH_3)_3$ and O at top

Sphingomyelin

With the exception of the more complex sphingosine containing lipids or the polyglycerol or polyphosphoric acid derivatives, the phospholipids such as the phosphatidyl cholines, ethanolamines, serines, and inositols occur in both plant and animal tissues. Because of technical difficulties in their study, very little is known about either the function of these compounds or about differences in their metabolism. Unfortunately, for the most part, they have been studied and discussed simply as "phospholipids," although it is quite certain that each has its own particular metabolic pattern. It is hoped that with present methods of chromatographic separation, progress can be made in differentiating the metabolism of these important compounds.

The percentage phospholipid content of tissues varies but little under any physiological condition. They are, therefore, known as the *élemént constant* in contrast to the triglycerides which are known as the *élemént variable*.

In vitro studies with tissue extracts and subcellular particles indicate that phosphatidyl choline is derived from acyl coenzyme A and glycerol phosphate similarly to triglycerides. At the diglyceride stage, however, it is reacted with cytidine diphosphoryl choline to yield phosphatidyl choline and cytidine phosphate.

There is evidence, however, that phosphatidyl choline also arises from triglycerides, since the ingestion of glycerol- and fatty acid-labeled triglycerides results in glycerol- and fatty acid-labeled phospholipids in the liver, blood, and intestinal mucosa. Since, however, triglycerides are known to be in a dynamic state in which the fatty acids are constantly being removed and replaced, from the glycerol moieties, there could well be a diglyceride pool from both endogenous and preformed sources.

Phosphatidyl ethanolamine is probably a product of phosphatidyl choline, but is somewhat less dynamic than the phosphatidyl choline which, in turn, is less dynamic than triglycerides from which it is formed.

Although the functions of the phospholipids are somewhat obscure, they are probably multiple and different for each phospholipid. Except in pathological states, the sphingosine and carbohydrate containing lipids are found most concentrated in nervous tissues, although not exclusively so. Here they appear to serve some insulating function in the myelin sheath.

Since phospholipids have both hydrophilic and hydrophobic parts to the molecule, they may well function as emulsifying agents to maintain the proper colloidal state of protoplasm. Through their concentration in cell membranes they may somehow be involved in the transport of hydrophobic constituents into and out of cells. Since they are also amphoteric, it is likely that this property may be utilized in the maintenance of acid-base balance. The polar groups also give phospholipids the property of association with proteins in the physiologically important lipo-proteins.

In the avian egg the high phosphatidyl choline level

doubtlessly serves to maintain the proper colloidal state in the yolk and also as a source of phosphates for the developing embryo. In this instance, choline, which is a vitamin, is doubtlessly used for its vitamin function.

Phospholipids are thought to be involved in the transport of triglycerides through the liver, especially during mobilization from adipose tissue. Conditions which could be interpreted as interfering with phosphatidyl choline formation, such as a deficiency of choline or its precursors, results in a pronounced increase in liver triglycerides. Although some blood serum phospholipid may originate in the liver, postprandial serum phospholipid containing dietary fatty acids and glycerol, arises in extrahepatic tissue, probably the intestinal mucosa.

Phosphatidyl ethanolamine has been implicated in the mechanism of blood coagulation, probably as a part of the lipoprotein, thromboplastin.

The presence of phospholipids in mitochondria is probably not adventitious, but may well play a role in fatty acid oxidation.

RAYMOND REISER

References

Deuel, Harry J., "The Lipids," Vol. 1, 1951; Vol. 2, 1955; Vol. 3, 1957; New York, Interscience Publishers.

Hanahan, D. J., "Lipid Chemistry," New York, Wiley, 1960.

Wittcoff, Harold, "The Phosphatides," New York, Reinhold, 1951.

Holman, R. T., et al., "Progress in the Chemistry of Fats and Other Lipids," Vol. 1, 1951; Vol. 2, 1954; New York, Academic Press, Vol. 3, 1955; Vol. 4, 1957; Vol. 5, 1965; New York, Pergamon Press.

International Conferences (Colloquia) on the Biochemical Problems of the Lipids, 1948, Service des Publications du C.N.R.S., Paris; 1953, Paleis der Academien, Brussels; 1956, Butterworth's Scientific Publications, London; 1957, Butterworth's Scientific Publications, London (also Academic Press, New York); 1958, Pergamon Press, New York, (1962).

PHOTOPERIODISM, ANIMAL

The most conspicuous seasonal events in the lives of animals are migration of birds, reproduction, changes in pelage and plumage, and diapause (arrest of growth of developmental stages in the life cycle of insects and other animals, thereby resulting in seasonal forms such as egg, larva, or pupa). These events occur with marked regularity at a particular time each year and are closely correlated with seasonal changes in day length and temperature. By subjecting captive animals to artificial increases and decreases in day length, bird migration, reproductive activity, molt, and diapause have been induced out of season. It has been concluded, therefore, that in nature changes in day length determine the time when these annual or cyclic activities occur. Animals which respond in some way to changes in day length are said to exhibit photoperiodism.

Since many species live in environments with marked seasonal fluctuations in climate, photoperiodism enhances survival by inducing certain functions only during appropriate seasons. It also serves to synchronize events among the individuals of a population. It is, therefore, an important adaptation with high selective value. Photoperiodic control of annual events is undoubtedly widespread, but not universal among animals. It has been investigated most extensively in vertebrates and arthropods.

Studies of birds have been most numerous and have demonstrated that (a) vernal migration and concomitant changes in lipid metabolism and body weight, (b) reproduction and associated physiologic and behavioral activities, and (c) molt, are regulated by photoperiodism. The slate-colored junco, Junco hyemalis, a species of the North Temperate Zone, has been studied intensively and will serve as an example. Before vernal migration, there is a marked change in physiological state. This is shown by growth of the reproductive organs and deposition of large amounts of fat. By employing different schedules of day length at various seasons of the year, it has been demonstrated that the time of vernal migration and reproductive activity is regulated in two separate phases—a preparatory phase which occurs in autumn and is requisite to a subsequent progressive phase which occurs primarily in winter and spring. Both phases are regulated by day length. Short days, or more specifically, a daily period of darkness of 12–13 hours duration, is required for a number of weeks to complete the preparatory phase. The reproductive and metabolic responses occur during the subsequent progressive phase, but the rate and duration of these responses is a function of the daily photoperiod. The responses are rapid and of shorter duration under long days (20-hour photoperiods) and slower and of longer duration under short days (12-hour photoperiods). A photoperiod as short as 9 hours per day is effective. Constant day lengths are also effective, as are decreasing day lengths, provided that the daily photoperiod is stimulating. Figure 1 illustrates the changes which have been induced in the annual period of reproductive activity with various schedules of long and short days. Similar regulation by day length probably occurs in equatorial and transequatorial migrants.

Postnuptial molt normally follows a period of reproductive activity and can be induced by long days. In the willow ptarmigan, Lagopus lagopus, which is brown in summer and white in winter, molt into white plumage is induced by short days, that into brown plumage by long days; temperature is not a controlling factor.

The daily amounts of light and darkness regulate the events in the annual cycle through their control over certain nuclei in the hypothalamus. These nuclei contain neurons that synthesize neurosecretory material which is transported along their axons to the region of the median eminence where the material is stored. During the progressive phase, the daily photoperiod regulates the release of the neurosecretory material from the median eminence into the capillaries of the hypothalamo-hypophyseal portal system. When this material reaches the adenohypophysis, it secretes gonadotropic and other hormones which in turn induce the growth of the gonads and other physiological changes. The receptors for the light are in the retina or in the encephalic regions near the orbit, but how stimuli from the receptors reach the neurosecretory cells is not known. A new hypothesis suggests that the light and dark periods of a given day, or cycle, provide the bird with information that is interpreted as either a long or short day. The decision regarding the length of day is

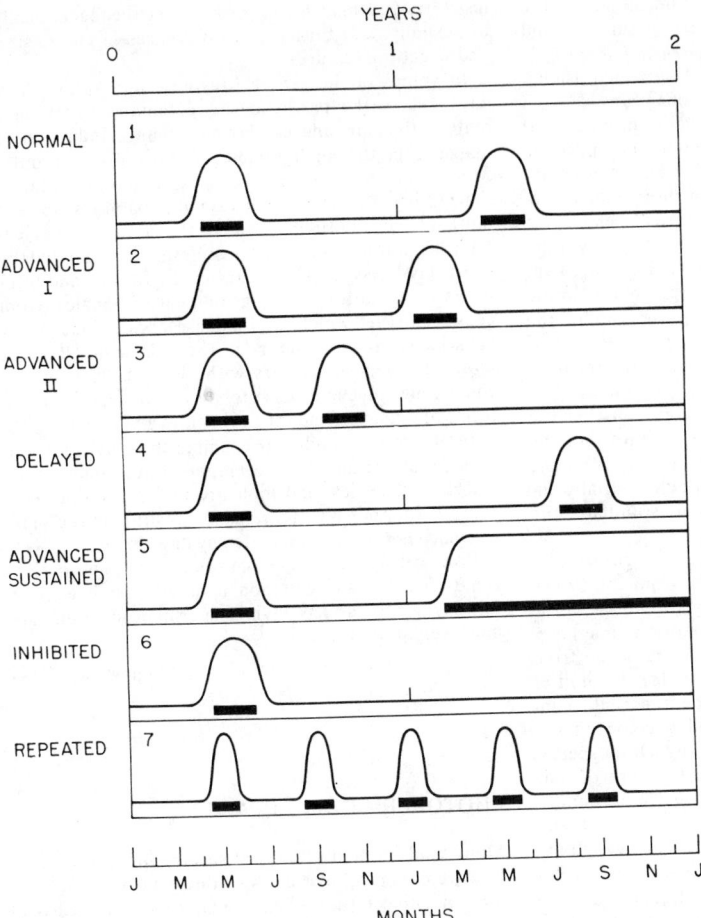

Fig. 1   Modifications in the reproductive cycle of the junco, *Junco hyemalis*. Treatment for each was as follows: *Advanced I*, long days in winter; *Advanced II*, short days in summer, long days in autumn; *Delayed*, short days from December to June, then long days; *Advanced-Sustained*, 12-hour days beginning in December; *Inhibited*, long days beginning in October which prevented completion of preparatory phase; *Repeated*, alternation of periods of long days and short days.

determined by the integration of the light-dark cycle with the circadian systems of the bird. Long-day and short-day decisions are then translated or manifested by the hypothalamo-hypophyseal system. A daily gonadotropic output that induces testicular growth reflects a long-day decision. Changes in the hypothalamus that prepare a bird for a subsequent gonadotropic response reflect a short-day decision. Patterns of motor activity show a clear-cut circadian rhythm and may play a role in the measurement of day length since they persist in darkness and show some correlation with the gonadal response. The measurement of day length, therefore, rather than the direct effect of light itself, could be the first critical step in the photoperiodic gonadal mechanism. Response to light can be modified by temperature, nutrition, behavior, and innate internal factors. The mechanisms for the metabolic changes which induce fat deposition and the postnuptial molt are not known, but the hypothalamus is probably involved.

Among the other classes of vertebrates, extensive experiments have been performed with fishes and mammals. In several species of fishes—three-spined stickleback, *Gasterosteus aculeatus*, banded sunfish, *Enneacanthus obesus*, minnow, *Phoxinus phoxinus*, and bridled shiner, *Notropis bifrenatus*, long days (15–17 hours) induce completion of gametogenesis, associated secondary sexual characteristics, and reproductive behavior and, hence, regulate the annual reproductive cycle. In the minnow, the photoperiodic mechanism functions only when the temperature is greater than 10° C, whereas in the three-spined stickleback, the photoperiodic responses occur over a wide range of temperature, with high temperatures augmenting the responses. Brook trout, *Salvelinus fontinalis*, which usually breed in fall, are stimulated by short days. The annual vernal migration of the stickleback from salt to fresh water, which usually occurs in late winter and spring, is induced by long days and augmented by high temperature. Preference for fresh water is controlled by increased secretion of thyrotropic hormone of the adenohypophysis. The reverse occurs in summer and fall. In juvenile Pacific salmon, *Oncorhynchus kisutch,* long days induce the seaward migration which usually occurs in spring when individuals are one year old.

Among mammals, reproductive cycles, molting, and many physiological rhythms are under photoperiodic control. In the female ferret, which has been studied intensively, long photoperiods in midwinter induce premature estrus, and their role, as in birds, seems to be regulation of the rate of response, since estrus will occur, but is delayed considerably, when animals are kept in constant darkness. Long days also regulate the timing of estrus in the raccoon, *Procyon lotor,* and in the domestic horse. In species which breed in the fall, for example deer, domestic goats, and several breeds of sheep, short days induce development of estrus. The organs involved in the photoperiodic response in the female ferret are similar to those in the bird, but the effect of the long photoperiods may be to depress the activity of a hypothalamic center which normally inhibits the neurosecretory cells that stimulate the adenohypophysis.

Molt of winter pelage and summer pelage are also under photoperiodic control. Change from the brown pelage of summer to the white pelage of winter in the shorttail weasel, *Mustela erminea,* and snowshoe hare, *Lepus americanus,* is regulated by the transition from long to short days. The opposite is true for the molt of the winter pelage. The similarity in photoperiodic control and adaptive significance of the white color in these mammals and the ptarmigan are striking. Other species in which a relation between molt and growth of hair and photoperiod has been demonstrated are the domestic sheep, domestic horse, cattle, marten, *Martes americana,* and mink, *Mustela vison.* Daily activity rhythms and other physiological rhythms, such as mitoses in cells and function of the adrenal gland, are also influenced by the daily periods of light and darkness. The relationship is primarily one of determining certain phase relations and not induction of periodicity.

When mammals and birds are transported from the southern to the northern hemisphere, or from the region of the equator to temperate latitudes, some species change their breeding season. Although day lengths are constant on the equator and change little in the tropics, there is some experimental evidence that day length influences the reproductive organs and plumage cycles of tropical birds.

Little work has been done with reptiles, but the data available for the turtle, *Pseudemys elegans,* the lizard, *Xantusia vigilis,* and a few other species suggest that day length may regulate the time of breeding. As in other cold-blooded vertebrates, temperature and thermoregulation are complicating factors. Several studies of different species of amphibians indicate that temperature and rainfall control the time of breeding and not photoperiod.

Among the invertebrates, the insects have been studied intensively. Their predominant photoperiodic responses fall into two categories: (a) the onset and termination of diapause of various stages of the life cycle, hence the control of growth or reproductive activity in species that hibernate or estivate, and (b) the control of differentiation in species which show seasonal differences in form and function, for example winged or wingless forms, forms which reproduce parthenogenetically or sexually, and forms with differences in color, size, and structural features.

In aphids of the genera Megoura and Aphis, short days induce the parthenogenetic females to produce male individuals and egg-laying females. Induction of diapause in the adult female of the red-spider mite, *Metatetranychus telarius,* occurs with days shorter than 12 hours at 15°C. Diapause of pupa of the noctuid moth, *Acronycta rumicis,* and of the larva of the oriental fruit moth, *Grapholitha molesta,* is regulated by short days, less than 15 hours in the former and 9–13 hours in the latter. The development of females (from eggs or larvae) which lay diapause eggs is induced by days longer than 15 hours in the silkworm, *Bombyx mori;* the opposite occurs with short days. Humoral controlling mechanisms probably participate in all forms of diapause, and the brain plays a key role.

In other arthropods, for example the crayfish, Cambarus, the ovarian cycle, development of secondary sexual characteristics, and molt are under photoperiodic control. In other invertebrates, oviposition in the snail, *Lymnaea palustris,* is regulated by day length.

Practical application of knowledge of photoperiodism is feasible and common in poultry management, for day length affects laying, sperm production, and body weight of fowl.

Albert Wolfson

PHOTOPERIODISM, PLANT

Many kinds of plants flower at specific times of year indicative of control by some variable environmental factor not subject to random fluctuation. The seasonal change in daylength is such a factor. Its controlling influence on plant growth and development was recognized first by W. W. Garner and H. A. Allard, of the United States Department of Agriculture, who named the phenomenon *photoperiodism* in their first publication on the subject in 1920.

Garner and Allard found that some plants, the so-called short-day ones, flower in nature only when days are short whereas others, the long-day ones, flower only when days are long. Still others, the indeterminate, or day-neutral, ones, flower when they reach proper size regardless of daylength. Examples of these three types are Biloxi soybean and Maryland Mammoth tobacco (short-day); sugar beet, wheat, and barley (long-day); and tomato and many bean varieties (day-neutral).

Many features of plant growth besides flowering are regulated by daylength. These include production of bulbs (onion), tubers (potato), and runners (strawberry), cessation of growth of woody plants in summer or fall accompanied by formation of winter buds and development of cold resistance and regulation of branching habit, leaf shape, leaf color, and leaf abscission. Artificial supplemental light at the beginning or end of short winter days substitutes effectively in causing plant response characteristic of summer daylengths.

The name *photoperiodism* suggests that daily duration of light controls flowering, but experiments show that daily duration of darkness actually does so. The time-measuring reactions of plants are dark reactions.

They are so sensitive that a minute or less of light near the middle of a long dark period prevents flowering of some short-day plants by effectively making two short dark periods out of a single long one. The same light treatment of a long-day plant on short days promotes flowering, showing that such plants fail to flower because the dark periods are long, not because the days are short.

The photoreaction that breaks the effect of a long dark period and thereby controls reproduction is now known to occur in all seed plants and in many of lower orders. It also influences many other kinds of plant responses including seed germination, lengthening of plant stems, and production of pigments. Even though these last three phenomena do not exhibit the time-measuring feature of photoperiodism, they are studied in conjunction with photoperiodism because they are now known to be controlled by the same photo-reaction. Although these several phenomena arise from a single causal photoreaction, the later courses of events leading to expression of particular phenomena are surely different. This discussion, therefore, attempts to present the current knowledge with minimum involvement with the details of flowering or of the various other phenomena that the photoreaction regulates.

Certain characteristic features of their action spectra clearly show that the photoreactions of these seemingly unrelated plant responses are identical. Action spectra as used here are quantitative measurements of the incident radiant energies required for a given response expressed as a function of wavelength of light. All these phenomena exhibit action maxima, one in the red part of the spectrum near 6500 Å and another in the far-red part near 7300 Å. The term "far red" is widely used in photoperiodic literature to designate a wavelength region extending roughly from 7000 to 8000 Å.

A second feature of the photoreaction is its photo-reversibility, red light driving it in one direction and far red driving it back. Reversibility of the reaction is revealed by study of any of several kinds of plant response. Cocklebur, soybean, or chrysanthemum plants, for example, flower if given several long (12-hour) daily dark periods but fail to flower if given a few minutes of red light near the middle of each dark period. Flowering is reinduced, however, if the plants are promptly treated with far red after the red. The far red reverses the flower-inhibiting reactions started by the red before they make measureable progress toward prevention of flowering. This illustrates the way in which one can examine the photoreaction without becoming involved with the many further reactions leading to flower formation about which little is now known.

Each of the several kinds of plant responses mentioned previously is repeatedly reversible without loss of effectiveness by successive alternating treatments with red and far red. This reversibility is one of the chief indications that the control results from a single reversible reaction rather than from two non-reversible ones having opposite effects. The reaction has a temperature coefficient of one in each direction, which indicates that both parts are truly photochemical and permits the use of first-order kinetics in further study of the reaction. Temperature coefficients were measured more satisfactorily in seed-germination than in flowering experiments, partly because of the greater accuracy with which temperatures of seeds instead of leaves can be controlled and partly because of the directness with which the seed-germination results can be given quantitative expression.

The absorption coefficients of the photoreversible pigment involved in photoperiodism were measured in the reaction that controls the length of bean stems. Values of about 10^4 obtained indicate a highly colored compound which is intense blue in its red-absorbing form.

The concentration of the pigment in the cells is so low, however, that it imparts no visible color to the plant. Albino seedlings of barley or corn, for example, appear completely colorless, but they exhibit the characteristic action spectrum determined by measurements of stem length induced by appropriate radiation treatments of the seedlings. The pigment, although present at such low concentration that it cannot be seen, is present in sufficient quantity to effect control of the plant. An estimate of the concentration of the absorbing substance is $< 10^{-7}$ molar, a value too low to permit its detection by ordinary laboratory spectrophotometers.

The pigment was detected *in vivo* in dark-grown corn seedlings by means of an especially sensitive spectro-photometer designed and built by electronic engineers in Agricultural Marketing Service at the Plant Industry Station, Beltsville, Maryland. The essential feature of the instrument is that it measures differences in total transmission of light through the tissue at two selected wavelengths (in this instance 6500 and 7300 Å) after the sample is first irradiated at one of these wavelengths with an energy much higher than that of the measuring beam. The instrument records the difference in transmission at the two wavelengths with a frequency of about 18 times per second. The sample is then irradiated at the other wavelength, which shifts the pigment from its previous absorbing form to the opposite one. The sample is then returned to the instrument and the transmission difference between the two wavelength forms is again measured. The difference between two successive measurements, one made after treatment with red and the other after treatment with far red, is a measure of the amount of reversibility exhibited by the sample.

The pigment was also detected in extracts obtained from corn seedlings and other plants. At low temperatures reversibility was unimpaired for many months but was quickly lost at 50°C. Reversibility was also retained upon dialysis at low temperature. These characteristics indicate that the active pigment is a protein. At the beginning of 1960 efforts to purify and identify the pigment were in progress but had not been completed, nor was the reaction that is catalyzed by the active pigment form known. Two years later the pigment was isolated at Beltsville and called a "cytochrome." At present (1968) several cytochromes have been isolated but neither their structure nor their mode of action is completely clear.

Knowledge concerning photoperiodism found practical use in improving plant-production procedures as soon as the discovery was reported. Agronomists and plant breeders began using artificial light in the early 1920's to extend natural winter days in the greenhouse to make possible the production of one or more extra generations a year. In small grains, for example, the practice permitted production of three generations in a single year thereby greatly accelerating the production of improved varieties.

Recognition of the photoperiodic requirements of

soybeans and other crops led to a more rational approach to varietal adaptation. In general, individual soybean varieties are grown in regions extending little more than 100 miles from north to south. Experience has shown that as the variety is grown farther south, its maturity date is too early and farther north too late. The principal reason is the delicacy with which the variety is adjusted to the seasonal shift in photoperiods. The response to daylength is obviously hereditary because varieties exhibiting specific adaptation to other areas are abundantly available. In fact, for any region from the Gulf of Mexico to the Canadian border a choice can be made among several available adapted varieties. In breeding new varieties, whether for improved quality, yield, or resistance to disease, incorporation of correct photoperiodic characteristics is a primary consideration.

Initially, supplemental light was used to extend the natural day, either morning or evening, but always without an interval of darkness because it was supposed that the continuity of light was an essential feature. With the discovery that the continuity of darkness was actually the controlling factor, plant breeders and others using light as an aid in plant culture quickly adopted the practice of interrupting the dark period. The change resulted in economies in the cost of electric current and gave equally effective control of flowering.

The chief application of light in actual production of crops in the United States is in the chrysanthemum industry. These flowers, originally available only in late summer and fall, are now produced throughout the year by growers who create the daylength sequence required by the plants for flowering at any given time. To do this the growers use artificial light during parts of the year when natural nights are long and they cover the plants with black cloth at other times to increase the length of short natural nights. Such a practice is practicable for crops that have a sufficiently high cash value per square foot of space but not for field crops.

H. A. Borthwick

Reference

Withrow, R. B., "Photoperiodism and Related Phenomena in Plants and Animals," Publication 55, A.A.A.S., Washington, D.C., 1959.

PHOTOSYNTHESIS

Photosynthesis is the utilization of light by plants for the conversion of carbon dioxide into organic matter. In green plants the overall chemical reaction may be represented by the balanced equation,

$$CO_2 + H_2O \rightarrow \{CH_2O\} + O_2$$

in which the newly formed organic mater is represented by {CH_2O}, a generalized formula for a carbohydrate.

The significance of this process lies in its conversion of energy from radiant into chemical form. Depending slightly on the particular carbohydrate formed, the amount of energy transformed is about 112 kcal. per mole of CO_2 fixed. Because of this large energy requirement, non-photochemical schemes for fixing CO_2 as organic matter are extremely rare. The chemical energy which a green plant stores by photosynthesis provides the total energy requirement of the plant, including that needed for the synthesis of the wide variety of its constituent organic substances. In addition, the green plants provide directly or indirectly the primary organic nutrient for most other living organisms.

Starting materials. Land plants take their carbon dioxide directly from the atmosphere. At low light the rate of photosynthesis may be independent of the CO_2 content; but in strong light the rate is linearly dependent on the carbon dioxide partial pressure at low CO_2 and reaches a limiting value at high CO_2. The average world CO_2 content of surface air, 0.03%, does not quite suffice for maximal rates in most land plants; but partial pressures above 0.2% are rarely required. Aquatic plants derive their requirements from dissolved CO_2. The other dissolved species, like carbonic acid, bicarbonate, and carbonate ions, act principally as reservoirs from which CO_2 itself can be drawn, but in some cases direct utilization of an ionic species may occur.

Water is not merely a convenient reaction medium nor an adjunct material for constituting the product carbohydrate; it is an essential substrate. Photosynthesis is essentially an oxidation-reduction reaction, the CO_2 being reduced to organic matter and the water being oxidized to molecular oxygen. It is known from tracer experiments, for example, that the atoms constituting the product oxygen arise from water and not from CO_2. In bacterial photosynthesis (see PHOTOSYNTHETIC BACTERIA) an exogenous reducing substrate replaces water as the ultimate reductant and oxygen is not evolved. A similar situation occurs in photoreduction, a process observed in several species of algae adapted anaerobically. Here molecular hydrogen acts as the reducing substrate.

In addition to the stoichiometric reactants, CO_2 and water, photosynthesis has a requirement for certain mineral ingredients, some in catalytic amounts. This category is hard to classify because of the problems of separating general growth requirements or ingredients for structural stability from photosynthetic essentials. Conclusive evidence from studies of requisite cofactors for individual steps in photosynthesis implicate phosphate in the carbon pathway, manganese in oxygen evolution, magnesium in the structure of chlorophyll and in photophosphorylation, and iron and copper as components of electron-transfer enzymes.

Products. Although 6-carbon sugars are the common carbohydrate products described in older textbooks, it is now generally recognized that the normal primary photosynthetically produced carbohydrate is phosphoglyceraldehyde, the phosphate ester of a 3-carbon sugar. The synthesis of hexoses and storage polysaccharides as well as the formation of fats, proteins, and other organic substances during periods of photosynthetic activity are attributed to secondary processes. Our understanding of the pathway of carbon in photosynthesis comes chiefly from labelling experiments with C^{14}-enriched CO_2.

The photosynthetic carbon cycle consists of the reactions concerned with the function and regeneration of the actual acceptor molecule reacting with CO_2, the assimilation step itself, and the photochemical reduction. The CO_2 acceptor, the 5-carbon ribulose diphosphate, is generated by condensations and hydrolyses of 3-, 4-, 6-, and 7-carbon carbohydrate phosphate esters; while part of the primary carbohydrates go directly to stored sugars or starch. The assimilation of CO_2 by ribulose diphosphate leads to two molecules of phosphoglycerate, which are reduced in the major energy-requiring step to phosphoglyceraldehyde; in some cases the 6-carbon assimilation complex may be reduced directly to form one molecule of a different 3-carbon compound.

The end-product oxygen is more unambiguously identified. It is not known at the time of this writing what the precursors of molecular oxygen are. Either peroxides of some sort (but not H_2O_2) or unstable high oxidation states of some metal co-factor are probably involved.

Gross kinetics. Photosynthesis consists of both light and dark reactions. In very weak light the rate measured either by CO_2 uptake or by O_2 production is proportional to the incident intensity. This weak light region is the experimental condition for the maximum utilization of absorbed light, that is, for optimum quantum efficiency. The wavelength of effective light may vary widely, from the near ultra-violet to the far red. Not all frequencies are absorbed equally, but on the basis of the light actually absorbed the efficiency varies by less than 30% between 400 and 700 nm for the green alga, *Chlorella*. Maximum efficiency is observed between 570 and 680 nm, where chlorophyll is the only significant absorber; and minimum efficiency occurs near 500 nm, where carotenoids and the accessory pigments are chiefly responsible for light absorption. Although monochromatic radiation of any wavelength in the 400 to 690 nm range may activate photosynthesis fairly well, the activity of light in the 690 to 750 nm range is enhanced by simultaneous illumination at lower wavelength.

At stronger illumination the rate increases less than linearly with increasing intensity and finally reaches a maximum saturation rate independent of further changes in intensity. For most plants the maximum is reached below the natural outside intensity on a bright sunny day. This high light region represents limitations by factors other than light availability. What was formerly known as the Blackman reaction is now known to consist of many different dark reactions, to be discussed in the following section. Many of these could limit the maximum rate of photosynthesis under different conditions. Flashing light with repeated dark-light cycles has been used to show the relatively slow process responsible for light saturation at optimal conditions, the approximately 10 millisecond turnover period of an essential intermediate step. The amount of endogenous substrate for this intermediate step is related to the maximum yield of photosynthetic product formed during a single short flash.

Mechanism. Hill made a significant discovery in 1940 that suspensions of chloroplasts with exogenous oxidants were able to promote the evolution of oxygen in the light at rates comparable to photosynthetic rates with whole cells of similar chlorophyll content. Since that time the reaction has been demonstrated with grana from chloroplast fragments. In its overall behavior the Hill reaction is similar to photosynthesis, reduction of the exogenous oxidant replacing the natural carbon cycle. The role of chlorophyll, the pro-

cesses of energy transfer, and the mechanism of oxygen evolution are presumed to be the same in the Hill reaction as in photosynthesis.

Hill's discovery opened the way to a more detailed study of the component reactions of photosynthesis. One of the significant results was the discovery that high-energy phosphate bonds can be formed in illuminated chloroplasts. The importance of phosphate chemistry had been anticipated from the nature of the carbon cycle. All the known carbon intermediates are phosphate esters, and the dark reactions associated with their interconversions could be related to known metabolic reactions by assuming the availability of high-energy phosphate compounds in amounts much larger than could be formed by fermentative or respiratory pathways. Particularly, the reduction of phosphoglycerate to triose and the conversion of ribulose phosphate to ribulose diphosphate are presumed to require high-energy phosphate. It is now well known not only that high-energy phosphate bonds are synthesized in the light at rates comparable to the Hill photoproduction of oxygen (approaching 2 phosphate bonds per O_2), but that the Hill reaction itself is accelerated if conditions for the simultaneous phosphorylations are optimized. The photophosphorylation includes both light and dark aspects. In the light a high-energy compound or state of the chloroplast is generated during electron transport. In a succeeding dark period the high-energy phosphate compounds are synthesized in the presence of ADP, Mg^{2+}, inorganic phosphate, and an endogenous protein coupling factor. Photophosphorylation can be decoupled from electron transport by certain substances, which need not be decouplers of oxidative phosphorylation.

The overall photosynthetic reaction, the reduction of CO_2 by water, can formally be described as an electron transfer over a large electrochemical potential gradient. Our understanding of the mechanism of the electron transport aspects received a great impetus in 1956 with the discovery by Emerson that there are two distinct photochemical events in the photosynthetic sequence, distinguishable not only in their chemical nature but also in the makeup of their associated pigment systems. These two photoreactions are believed to represent series events, each of which promotes the electron transfer over a part of the overall potential gradient. In the accepted nomenclature, photoreaction I is the formation of a very strong reductant, X^-, by the transfer of an electron from a donor of intermediate potential (about +0.4 volts on the conventional electrode potential scale) to an acceptor, X. Three possible donors for this reaction have been proposed, all of which have the same standard potential to within 0.06 volts: cytochrome f, the copper protein plastocyanine, and a special form of chlorophyll a designated as P700. The X-X^- couple has a standard potential of -0.6 volts or lower. The function of photoreaction II is to create a very strong oxidant, Y^+, by transfer of an electron from a donor substance, Y, to an intermediate-level acceptor, Q. The potential of the Y-Y^+ couple is at least +0.9 volts and of the Q-Q^- couple between 0.0 and +0.2 volts. Although there is no certainty at the time of this writing about the chemical identities of Q, X, and Y, something is known about their properties and functions. In overall steady-state photosynthesis the two photoreactions occur at equal rates. The strong reductant, X^- (or its conjugate acid, XH), is able to reduce the intermediates of CO_2 assimilation to the oxidation level of carbohydrate through a chain of dark processes including the reduction in turn of the non-heme iron protein ferredoxin, a flavoprotein, and NADP. The function of Y^+ is to oxidize water in some as yet unknown mechanism involving ultimately the cooperation of 4 one-electron oxidants in the production of one molecule of the 4-electron oxidation product, O_2. The role of Q is to restore electrons to the oxidized form of the primary donor of photoreaction I in a sequence of electron transfer steps.

Various parts of the above total scheme may be isolated by studying extracellular reactions or variants of whole-cell photosynthesis. In the Hill reaction, for example, the exogenous Hill reagent may replace ferredoxin in accepting electrons from X^-, and this is certainly the case for Hill oxidants of negative standard potential like the viologen dyes. Hill oxidants of higher standard potential, like ferricyanide and indophenol dyes, may accept electrons from members of the electron transport chain connecting the two photoreactions, and in so doing might completely bypass photoreaction I. Some exogenous electron donors of standard potential below +0.4, like reduced indophenol dyes or certain substituted hydroquinones, when coupled with Hill reagents for system I can completely bypass photoreaction II by supplying electrons to the oxidized donor of reaction I in competition with Q^-; oxygen is not evolved in such a case. In bacterial photosynthesis and in photoreduction only photosystem I operates, hydrogen or other exogenous donors replacing photoreaction II as the source of reducing power needed to restore the primary electron donor for photoreaction I to its reduced state. Other methods of isolating the photoreactions include the use of mutants lacking one photosystem or the other or of inhibitors specific for one of the photoreactions like chlorinated phenylureas which block photoreaction II. Photophosphorylation has been shown to occur in preparations where only photosystem I is operating.

Models for the carbon cycle of photosynthesis in partial systems also exist. Carbon dioxide assimilation itself is not a sufficient criterion for a photomimetic model; direct fixation into the terminal carbon of phosphoglycerate or of triose phosphate is a more rigid requisite. This type of fixation, so-called cell-free photosynthesis, occurs in illuminated chloroplasts containing as co-factors ADP, NADP, and one of a variety of carbohydrate initiators. In broken chloroplast suspensions the assimilation can occur when CO_2 is added in the dark to the colorless supernatant remaining after the previously illuminated grana are centrifuged out. Thus, the enzymes needed for the photosynthetic carbon cycle are localized in the aqueous fraction of the chloroplast, whereas the electron transport and the photophosphorylation are confined to the grana. The requirement for NADP is specific; NAD is not effective.

The photoreceptor. All photosynthetic organisms contain their light-sensitive pigments in more or less highly organized lamellar structures: chloroplasts, grana, or chromatophores. The ordered arrangement of the pigment molecules facilitates the transfer of excitation energy from one to another. The basic pigment for all non-bacterial photosynthetic cells is chlorophyll a, a magnesium-centered porphyrin containing as most distinctive substituents a hydrophilic 5-membered carbocyclic ring and a lipophilic phytol tail. In the lamellae the chlorophyll is bound in insoluble form. The spectroscopic properties are modified slightly with respect to those of extracted chlorophyll. Most green plants also contain chlorophyll b, which differs from a only in a minor substituent group. Carotenoids are present in most photosynthetic tissue. Another category of accessory pigments, the phycobilins, is found in the red and blue-green algae. All these pigments absorb incident light at their respective absorption regions and are at least partly effective in making radiant energy available for photosynthesis. Of all these compounds chlorophyll a absorbs at the longest wavelength. This property makes possible the transfer to it of energy from other pigments in its neighborhood by a resonance process. The chlorophyll a (or bacteriochlorophyll in the case of the photosynthetic bacteria), among all the pigments, thus has the unique role of the ultimate energy transducer which reacts with chemical substrates. In addition to their energy-collecting role the accessory pigments may have other functions as well. Carotenoids, for example, protect the plant from damaging photo-oxidations.

A major feature of the pigment structures is the cooperative action among several hundred pigment molecules and their associated structure comprising a "unit." This unit is a tribute to nature's economy in allowing a relatively small number of enzymes and substrates to be serviced by the energy absorbed over a much larger number of pigment molecules. Each unit may have only one reaction center at which the chlorophyll a excitation energy is transformed into chemical energy. The same resonance mechanism that allows the transfer of energy from an accessory pigment to chlorophyll a operates among identical chlorophyll a molecules as well. The ultimate trapping of the energy at that part of the unit that is provided with the enzymes and substrates needed for the primary photochemical act requires a special assumption. One possibility is that the chlorophyll at that spot, which may be called the trapping center, is in a slightly different environment because of its complexation with the reaction intermediates of the photosynthetic apparatus. If this modified milieu shifted the chlorophyll absorption spectrum slightly to the red, then energy from the whole unit would tend to migrate toward that point. Another view is that the pigment at the trapping site differs from the bulk of the pigment only in its capacity to perform the primary photochemistry and that the photochemical reaction is fast enough to be itself the trapping process, preventing a competing back-migration into the pigment bed of excitation energy which reaches the trap by a random migration process.

Corresponding to the two different photochemical reactions there are two different kinds of unit, one for pigment system I and one for II. These two systems were originally identified by their action spectra. The wavelength of maximum activity for photoreaction I is greater than for photoreaction II. Although normal chlorophyll a seems to be distributed in both pigment systems, chlorophyll b is preferentially concentrated in system II. In the red and blue-green algae, most of the chlorophyll a is in system I, while the phycobilins are in II. In all green plants system I has a preferential concentration of a small chlorophyll a component whose absorption maximum is shifted to about 700 nm, a slightly longer wavelength than for the bulk of the chlorophyll *a*. A characteristic property of system I is the ability of one of its components, P700, to undergo a reversible photobleaching at 700 nm. The extracted pigments from any photosynthetic tissue show only one spectroscopically distinct kind of chlorophyll *a*, so that the slight *in vivo* spectral differences must be due to second-order effects arising from interaction of pigment molecules with each other or with other substances in the environment. Fractional centrifugation of sonicated chloroplasts has yielded two kinds of particles which may be associated in pigment content and in photochemical function with systems I and II respectively. Kinetic data indicate that the unit for each pigment system has about 400 pigment molecules. This size has been confirmed for system I by the observation that in a short flash of saturating light the maximum amount of P700, cytochrome *f*, or plastocyanin undergoing oxidation is one molecule for several hundred total chlorophylls. Although each unit probably has only one molecule each of its primary donor and acceptor, the total pool of electron transport intermediates functionally connecting systems I and II is at least ten times the number of units of either kind. The unit of the photosynthetic bacteria, containing only about 40 pigment molecules, is much smaller than in the green plants.

At the time of writing this article it is not known whether the functional units are morphologically separated *in vivo* or whether the pigment arrays are much larger continua with an average ratio of total pigments to trapping centers corresponding to the unit sizes discussed above. The mapping of the mutual spatial arrangements of the various electron transport intermediates associated with the reaction centers remains for future research. Some indications of the intimacy of their positions are the rapidity of certain reactions, such as the oxidation of cytochrome *f* within 2 microseconds of a laser light pulse, and the persistence of some spectroscopically observed oxidation-reduction reactions down to 1° K.

The luminescence of chlorophyll is a useful property for study of the photoreceptor. Chlorophyll fluoresces with an efficiency dependent on the incident intensity as opposed to the case for chlorophyll *in vitro*. A unifying picture which has emerged from a study of fluorescence variations is that a photosynthetic unit can process only one quantum at a time and is able to process a succeeding quantum photochemically only after the substrates for the primary photochemical event are restored to their proper oxidation states. This processing time under optimal conditions may be the 10 millisecond period observed in flashing light experiments. Any quantum of excitation appearing within a unit during this working period would have to be dissipated by wasteful processes such as fluorescence or conversion into heat with some probability be made available to a neighboring unit whose reaction center is prepared to perform the photochemical act. In addition to fluorescence, which is an emission of light within 10^{-8} seconds of illumination, there is another dissipative process, a delayed emission of light which can be observed up to hours after the illumination of the photoreceptor. This delayed light, not observed in pigment extracts, depends on the organized structure of the lamella. The postponed emission is evidence for storage of a small fraction of the incident radiant energy, probably in the chemical form of metastable

trapped fragments resulting from chlorophyll photoionization.

Efficiency. One of the unique aspects of photosynthesis as opposed to all other known efficient photochemical processes is the accumulation of the effect of several light quanta for the production of a single stable product molecule. The mechanism discussed above implies that eight quanta must be processed by the photosynthetic machinery for the evolution of one oxygen molecule, four by system I and four by system II. The average experimental value obtained in several laboratories for the number of quanta absorbed per oxygen molecule evolved is about 10, indicating a relatively small wastage of whole quanta. The high efficiency of green plant photosynthesis can be appreciated by comparing the magnitudes of the incident light energy and the accumulated storage. Light of 680 nm, containing 42 kcal per mole of quanta, is as efficient as any other wavelength for photosynthesis. The *free* energy requirement of the overall process for the typical plant is about 116 kcal per mole of oxygen evolved. Thus, even a 10-quantum intake would correspond to a 28% energy efficiency.

J. L. ROSENBERG

References

Brookhaven Symposium in Biology No. 19, "Energy Conversion by the Photosynthetic Apparatus," Upton, N. Y., Brookhaven Lab., 1967.

Calvin, M. and J. A. Bassham, "The Photosynthesis of Carbon Compounds," New York, Benjamin, 1962.

Clayton, R. K., "Molecular Physics in Photosynthesis," New York, Blaisdell, 1965.

Gaffron, H., *in* Steward, F. C., ed., "Plant Physiology," Chap. 4, vol. 1B, New York, Academic Press, 1960.

Kamen, M. D., "Primary Processes in Photosynthesis," New York, Academic Press, 1963.

Rabinowitch, E. I., "Photosynthesis and Related Processes," 2 vols., New York, Wiley, 1945–1956.

Rosenberg, J. L., "Photosynthesis," New York, Holt, Rinehart, and Winston, 1965.

PHOTOSYNTHETIC BACTERIA

The photosynthetic BACTERIA are a morphologically and taxonomically diverse group which is characterized by the presence of a unique kind of photosynthetic metabolism. In contrast to the process in green plants, PHOTOSYNTHESIS in these bacteria proceeds without the liberation of molecular oxygen and requires the presence of an external hydrogen donor. The oxidation of the hydrogen donor is the equivalent of the release of oxygen in green plant photosynthesis.

Roughly, three major types of photosynthetic bacteria can be distinguished:

1) The green bacteria (Chlorobacteriaceae) use only reduced sulfur compounds as the external hydrogen donor; the green bacteria are all obligate anaerobes; they do not require organic growths factors.

2) The sulfur-purple bacteria (Thiorhodaceae) can use various organic compounds as the hydrogen donor as well as reduced sulfur compounds or hydrogen; they are obligate anaerobes and do not require organic growth factors.

3) The non-sulfur-purple bacteria (Athiorhodaceae)

differ from the other two groups in three respects: a) Organic compounds are used preferentially as the hydrogen donor, although some species can use reduced sulfur compounds. b) All species require one or more vitamins for growth. c) Many members of this group can grow in the presence of oxygen; these can grow aerobically in the *dark* using the same organic compounds as hydrogen donors for respiration as are used during phytosynthetic growth: during aerobic growth these bacteria appear to be ordinary heterotrophs; physiologically, respiration and photosynthesis are interchangeable.

Naturally, not all photosynthetic bacteria fall neatly into one of these groups; here only one example can be mentioned. *Rhodomicrobium vannielli* requires an organic compound as hydrogen donor but does not require any additional growth factors; it thus falls between the Athiorhodaceae and the Thiorhodaceae. Furthermore, it is morphologically unique. Instead of dividing by simple fission, the cells produce a thin tube-like outgrowth through a daughter nucleus migrates, the daughter cell is then formed at the end of the tube across the middle of which a cross-wall develops.

The photometabolism of the green and of the sulfur purple bacteria can be represented by the following equations

$$CO_2 + 2H_2A \qquad `CH_2O` + 2A + H_2O$$

Here, `CH_2O` represents cellular material (reduced carbon), and H_2A the external hydrogen donor. The similarity of this equation with that of green plants photosynthesis is apparent:

$$CO_2 + 2H_2O \qquad `CH_2O` + O_2 + H_2O.$$

The pathway of reduction of CO_2 in the photosynthetic bacteria is the same as in green plant photosynthesis, the first reaction being the formation of phosphoglyceric acid from ribulose-diphosphate and carbon dioxide.

In the case of the purple bacteria utilizing an organic compound as the external hydrogen donor the metabolism is more complex. For this compound serves not only as a source of hydrogen but can also serve as carbon source without having to be first oxidized to CO_2;

this is similar to the way in which the carbon source of an aerobic heterotroph is directly assimilated as partially oxidized intermediates. Two extreme cases can be recognized: (1) in which the organic compound serves as hydrogen donor *only*. The only known example being a strain which utilized *iso*-propanol, the alcohol is oxidized to acetone which is not further metabolized; here, all the carbon of the cell comes from CO_2 and *iso*-propanol is exactly analogous to a reduced sulfur compound. (2) The other extreme is represented by the photometabolism of acetate by the non-sulfur-purple bacterium *Rhodospirillum rubrum*: here one molecule of acetic acids provides the hydrogen to reduce eight other molecules of acetate to the level of hydroxybutyric acid. In this case none of the reduced cellular carbon is derived from CO_2.

As in green plants the *photosynthetic pigment* system of the photosynthetic bacteria is composed of a chlorophyll and various carotenoid pigments; both kinds of pigments show interesting differences from those of green plants. (table 2.)

Like chlorophyll, both bacteriochlorophyll and chlorobium chlorophyll are magnesium porphyrin compounds. The structure of the latter is unknown. Bacteriochlorophyll differs from chlorophyll a in two respects: (1) The substituent at position 2 on pyrrole ring I is acetyl instead of vinyl; and (2) pyrrole ring IV is reduced. It is this reduction which is responsible for the very marked shift in the position of the red adsorption peak in the spectrum of bacteriochlorophyll (see table 2). Not only do the bacterial chlorophylls absorb at longer wavelengths than green plant chlorophylls but also the bacterial carotenoids absorb at somewhat shorter wavelengths than their counterparts in green plants. The entire photosynthetic pigment system of these bacteria seems to be adapted to absorption of light which is not absorbed by green plants.

The photosynthetic pigments are associated with submicroscopic particles called *chromatophores*. The chromatophores have a diameter of about 50 mμ; so far as is known they do not have a lamellar structure, thus differing from the grana of green plants. The chromatophores are sites of numerous enzymes catalyzing reactions associated with photosynthesis: the light dependent production of adenosine triphosphate and reduction of pyridine nucleotide coenzymes, for ex-

Table 1. Some characteristics of the three major groups of photosynthetic bacteria

Group	Example	Pigment System		Nutrition	
		chlorophyll	carotenoid	hydrogen donor	growth factors
Green bacteria	Chlorobium	chlorobium chlorophyll	γ-carotene	reduced sulfur compounds	none
Sulfur purple bacteria	Chromatium	bacterio chlorophyll	acyclic carotenoids often methoxylated	reduced sulfur compounds: and organic compounds	none
Non-sulfur purple bacteria	Rhodospirillum			organic compounds, some species; reduced sulfur compounds	various vitamins

Table 2. Absorption maxima of bacterial and green plant chlorophylls

chlorophyll type	absorption maxima in ether	long wave length absorption maxima in vivo
	mμ	mμ
chlorophyll a	660, 614, 429	670–680
chlorobium chlorophyll	659, 431, 408	750
bacterio-chlorophyll	770, 574, 357	810; 850–870; 890

ample. The shift of the major absorption peak of bacteriochlorophyll from 775 mμ *in vitro* to between 800–900 mμ *in vivo* is due to the conjugation of the pigment with protein in the chromatophores.

In the non-sulfur purple bacteria which are capable of aerobic growth, pigment synthesis is totally inhibited at a sufficiently high oxygen tension, this inhibition is independent of light conditions; furthermore, light is not necessary for the synthesis of the photosynthetic pigments nor of the chromatophores, these syntheses occur only when the oxygen tension is reduced. In the absence of oxygen the concentration of photosynthetic pigments in the cell is dependent on the light intensity: high light intensities reduce the concentration of the pigments.

Many of the photosynthetic bacteria are motile and these exhibit a form of phototaxis. This response differs from that of an organism such as *Euglena* which moves towards the light source; the response in the photosynthetic bacteria is a reversal of swimming direction when a cell passes from a bright to a darker region. The result of this is that an initially uniform suspension of the bacteria becomes concentrated in a bright field which is surrounded by a dimmer background. No special phototactic pigment is present; the variation in the phototactic response with changing wavelengths of light is identical to the variation in the rate of photosynthesis.

W. R. Sistrom

References

Bergeron, J. A., "The Bacterial Chromatophore," *in* "Brookhaven Symposia in Biology, No. 11," U.S. Atom. Energ. Comm. Upton, N. Y., 1959.

Clayton, R. K., "Tactic responses and metabolic activities in *Rhodospirillum rubrum*," Arch. Mikrobiol. **22**: 204, 1955.

Gest, H. *et al.*, "Symposium on Bacterial Photosynthesis," Yellow Springs, Ohio, Antioch Press, 1963.

Larsen, H., "On the microbiology and biochemistry of the photosynthetic green sulfur bacteria," Kgl. Norske Videnskabers. Selskabs. Skrifter 1953 *Nr. 1.*

van Niel, C. B., "The culture, general physiology, morphology and classification of the non-sulfur purple bacteria," Bacteriological Reviews, **8**: 1, 1944.

van Neil, C. B., "The comparative biochemistry of photosynthesis," *in* Franck, J. and W. E. Loomis, eds., "Photosynthesis in Plants," Ames, Iowa State College Press, 1949.

Stanier, R. Y. and G. Cohen-Bazire, "The role of light in the microbial world: Some facts and speculations," *in* Williams, R. E. O. and C. C. Spicer, eds., "Microbial Ecology," Cambridge, The University Press, 1957.

Stanier, R. Y., "Formation and function of the photosynthetic pigment system in purple bacteria," *in* "Brookhaven Symposia in Biology, No. 11; U. S. Atom. Energ. Comm. Upton, N. Y., 1959.

PHYCOMYCETES

The Phycomycetes (*phykos*, alga + *mykes*, fungus) comprises a group of microscopic FUNGI of approximately 240 genera and a thousand or more species. These have been brought together on the basis of simplicity of body plan, coenocytic habit of the vegetative structures and production of a relatively indefinite number of non-sexual reproductive units termed "spores" borne in sporangia. The group should not be regarded as a natural assemblage of organisms but, rather a polyphyletic one composed of members of differing ancestry but with similar body plan.

These fungi occupy a wide variety of habitats. A great many are aquatic, amphibious or soil inhabitants and are found in fresh and marine waters as parasites and saprophytes on plants and animals and their remains and in "normal" soils. They are also present in such special habitats as sea strands and strongly alkaline semi-deserts. Numerous others are obligate parasites of vascular plants, often of considerable economic importance (viz. "Downy Mildews"), usually resisting all efforts to grow them apart from their hosts, whereas many, such as the familiar "Bread Molds" are found on stored foodstuffs, crop vegetables, etc. and are usually readily culturable. A few are parasites of animals, notably insects, and several are found on man.

The thallus or vegetative stage of a Phycomycete may consist of a single-celled multinucleate structure without special parts for absorption of food materials and which may be converted as a whole into a single reproductive structure, or at the other extreme, be composed of an extensive nonseptate, multinucleate system of much-branched nearly isodiametric threads of "hyphae," which may bear large numbers of reproductive structures. With rare exceptions, true septa are formed even in those types with an extensive *mycelium* (the collection of hyphae), only to delimit reproductive organs, special survival structures or injured areas.

Non-sexual reproduction (aside from mere multiplication by fragmentation, etc.) is a multiplying device designed to increase rapidly the numbers of individuals identical with the parent plant. In the aquatic, amphibious and soil-inhabiting members as well as many parasites of vascular plants, this is accomplished by the production in *sporangia* of naked, flagellated, free-swimming *zoospores*. These may have one posterior FLAGELLUM, or an anterior one, or two apically attached oppositely-directed, or two oppositely-directed, laterally attached, ones. Such spores are obviously well fitted for accomplishing the rapid and wide-spread dissemination of the species, for after their period of motility they come to rest, germinate, and form a complete individual like the parent. In strictly terrestrial types the spores are non-flagellated, walled, and function in the same manner as zoospores. There is perceptible in many groups of both aquatic and terrestrial types a graduated diminution in the number of spores formed in a sporangium, so that, in what might be termed the most highly specialized groups, the sporangium itself, often with the equivalent of only a single spore within it, dehisces and functions for reproduction. Such sporangia have often, and erroneously, been termed *conidia*.

Sexual reproduction is general among these fungi and is of a wide variety of types, the aquatic and semi-aquatic numbers exhibiting the greatest diversity. This process results in most Phycomycetes in the production of a thick-walled, one-celled extremely durable structure, the zygote, composed of the protoplasm and nuclei of the two fusing components or gametes. The exception to this is, so far as now known, in *Physoderma* of the primarily aquatic order Chytridiales, and in certain of the Blastocladiales. In the latter group it has been shown beyond question that there exists an alternation of an isomorphic 2-n spore-bearing generation with a 1-n gamete-bearing one, brought about by the immediate germination of the zygote without the intervention of meiosis. *Physoderma* is known to possess a morphological alternation of heteromorphic generations, but complete cytological details are lacking at the moment. Gametes fusing in sexual reproduction may be flagellated and isogamous or anisogamous. Fusion of a large non-flagellated egg or a female gamete, with a flagellated sperm are known and in two aquatic genera the zygote swims away propelled by the flagellum of the *male* gamete. Rhizoidal anastomosis of two plants or direct contact of two walled thalli, one of which contributes, the other of which receives, the gametic material, is also known. A greater number transmit the male gametic material to the large one or several-egged oogonium via a tube which penetrates the female. In all these instances the receptive structure becomes a durable resting zygote (*oospore*). In the Bread Molds (Mucorales) and their relatives (Entomophthorales, etc.) sexual reproduction is by some sort of conjugation of walled-off multinucleate hyphal tips which fuse distally and there produce the durable, thick-walled zygote (*zygospore*). Both oospores and zygospores may in a few forms be produced parthenogenetically. These resting structures unquestionably enable the fungus to withstand unfavorable conditions of life and, at germination, at which time in these sexually produced bodies meiosis is presumed to occur (cytological evidence is lacking in most instances), reestablish the plant.

A few Phycomycetes are, or were before their control, the causal agents of serious diseases of crop plants of economic importance. By far the most celebrated and significant of these was the "Late Blight" of Irish potato (produced by *Phytophthora infestans*) which because of its devastating destruction in the 1840's in successive years of the main staple of diet of that country caused not only tremendous loss of life but resulted in profound sociological and political upheavals in both Ireland and in the Atlantic seaboard cities of the United States. Attempts at control measures for this disease were widespread and elicited the attention not only of the great German mycologist deBary but of Charles Darwin, as well. Somewhat less important but no less devastating was the "Downy Mildew" of grape which although again of American origin and caused by a Phycomycete allied to *Phytophthora infestans*, produced its greatest depredations in French vineyards. This disease was ultimately controlled by the perfection by Millardet of the famous "Bordeaux mixture," the first of the genre of copper fungicidal sprays. It is interesting to note that one of man's most common frailties, thievery, contributed in large measure to this disease's control, for a mixture of copper sulfate and lime was commonly employed by vintners to dust grape leaves to prevent, by reason of the impalatable appearance, the stealing of the fruit. Millardet quickly noticed dusted plants suffered little from mildew and from this clue resulted the aforementioned fungicide.

F. K. SPARROW

References

Bessey, E. A., "Morphology and taxonomy of fungi," New York, Hafner, 1961.

Fitzpatrick, H. M., "The lower fungi. Phycomycetes," New York, McGraw-Hill, 1930.

Karling, J. S., "The simple holocarpic Phycomycetes," N. Y., Publ. by the author, 1942.

Sparrow, F. K., Jr., "Aquatic Phycomycetes," Ann Arbor, Univ. of Michigan Press, 1960.

PHYTOCHROME

Phytochrome is the name given to a plant pigment which is different from the chlorophylls, carotenoids, and the flavonoids. It is a blue-green protein which occurs in minute quantities in plant tissues. Phytochrome pigment is a photoreceptor for photomorphogenetic and photoperiodic responses in plants, and is found in many higher plants. It has been detected both in monocotyledonous plants, e.g., oats (*Avena sativa*), barley (*Hordeum vulgare*), corn (*Zea mays*), and the aquatic plant *Lemma*, and in dicotyledonous plants, e.g., sunflower (*Helianthus annus*), beans (*Phaseolus vulgaris*), peas (*Pisum sativum*), parsnip (*Pastinaca sativa*), soybean (*Glycine max*), lettuce (*Lactuca sativa*), radish (*Raphanus sativus*). It is present in aerial parts of plants (leaves, stems and buds, and inflorescences, as in the case of cauliflower, *Brassica oleracea* var. *botrytis*), as well as in roots. Phytochrome is present also in bryophytes, e.g., *Sphaerocarpus*, red algae, e.g., *Porphyra tenera*, green algae, e.g., *Mesotaenium* and *Mougeotia*.

Historically, lettuce seeds (*Lactuca sativa*) and cocklebur (*Xanthium*) were the first two plants in which phytochrome was detected. Before isolation, the existence of phytochrome as a specific pigment was implied by physiological work with red and far-red light. Red light (640–670 nm), which promotes germination in lettuce seeds is reversible by far-red (710–740 nm). In other words, far-red irradiation nullifies the promoting effect of the red light. Flint and McAlester (1935) were the first to find that red light stimulated the germination of lettuce seeds, and that light of somewhat longer wavelengths inhibited germination. No one paid much attention to these phenomena until 1952 when Hendricks, Borthwick and co-workers at the U.S. Department of Agriculture, Beltsville, Maryland, studied in detail the effects of red and far-red light on the germination of lettuce seeds. They found that there is a specific system sensitive to red and far-red, and the red-far-red reversibility could be repeated several times. If the last irradiation is red, regardless of how many red-far-red treatments previously given, the lettuce seeds will germinate, and if the last irradiation is far-red there is an inhibition of germination.

The second physiological response to red–far-red which was studied by the Beltsville group in 1952 was the flowering of short-day plants which are photoperiodically controlled. Cocklebur, *Xanthium pensylvanicum*, is a photoperiodically sensitive short-day plant. If given a long night it will start flowering. When the dark period is interrupted by red light, as short as one minute, flowering is inhibited. When far-red follows red, flowering is resumed (retained). This system of red–far-red which can be observed by the flowering response is a unique system in which a photoreceptor pigment must be involved.

Following 1952 many experiments were conducted on the flowering response, and it was found that long-day plants responded in an opposite way from that of short-day plants. That is to say, red light given in the middle of the dark period promoted flowering in long-day plants, while far-red reversed the reaction.

The photoreceptor pigment was detected in etiolated plants (grown in the dark). The etiolated plant parts, i.e., leaves, hypocotyls, mesocotyls, or coleoptiles, were found to show response to red–far-red. Green tissue cannot be used for the detection of phytochrome because chlorophyll has an absorption peak in the red which coincides with the peak of phytochrome. The action spectrum of seed germination and flowering corresponds to the absorption spectra of phytochrome. When the plants were irradiated with red, phytochrome shifted to a form in which the maximum absorption was in the 730 nm region. When far-red was

given phytochrome displayed a maximum absorption of 660 nm.

In order to clarify this point, the diagram of Fig. 1 will explain the phytochrome system. On the left side of the diagram, the normal state of phytochrome is P_r; when irradiated with red light the phytochrome will change into P_{fr}, which is the physiologically active form of phytochrome. The P_{fr} has an absorption maximum in the far-red region at 730 nm, P_r has an absorption maximum in the red region at 660 nm. Both P_r and P_{fr} have a small peak in the blue region at 400 nm. If P_{fr} is irradiated with far-red then it converts to P_r. The reversibility of P_r to P_{fr} is very rapid. Moreover, darkness will transform P_{fr} to P_r rather slowly at temperatures higher than 10°C, hence it is a thermal reaction. Some of the P_{fr} is consumed or destroyed and the other portion is reconverted into P_r in complete darkness. The $P_r \rightarrow P_{fr}$ is not a one-step reaction. Indications show that there are about five steps in this reaction; some are urea-sensitive and others are trypsin-sensitive. After irradiation with red light the phytochrome is present in an 80:20 ratio of P_{fr}:P_r due to transformation and/or the destruction of P_{fr}.

Some of the physiological reactions tied up with phytochrome are listed on the right side of the diagram. There are probably other morphological, physiological, and biochemical effects connected with phytochrome which are to be discovered. The effects of phytochrome could be classified into four major categories: (a) morphogenic effects, e.g., unfolding of the plumular hook, expansion of leaves, stem elongation, and germination of seeds; (b) photoperiodic effects, e.g., the flowering of short-day plants (even though photoperiodism is linked with light-dark durations within 24-hour cycles,

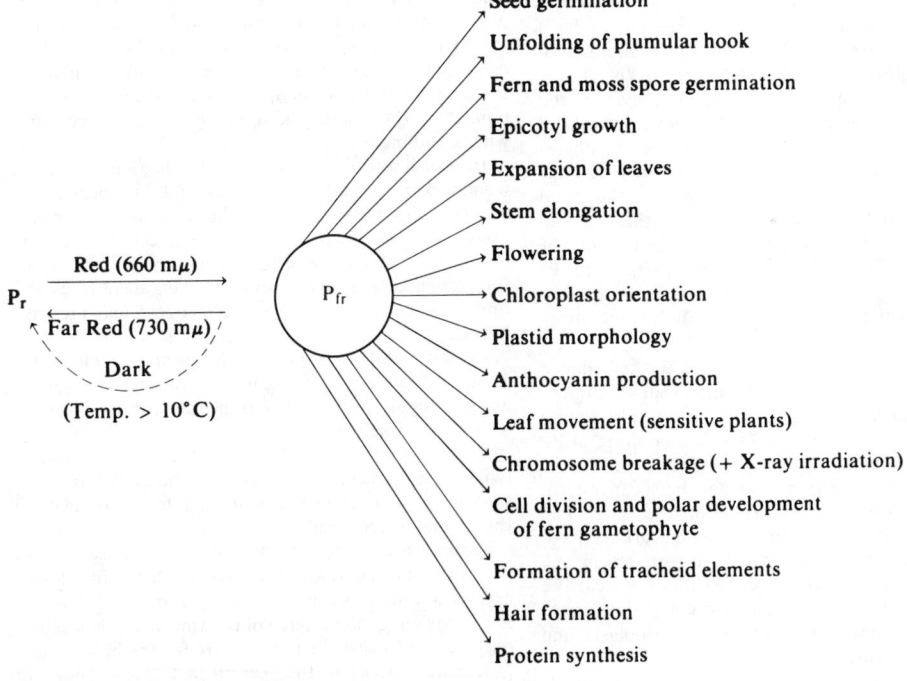

Fig. 1

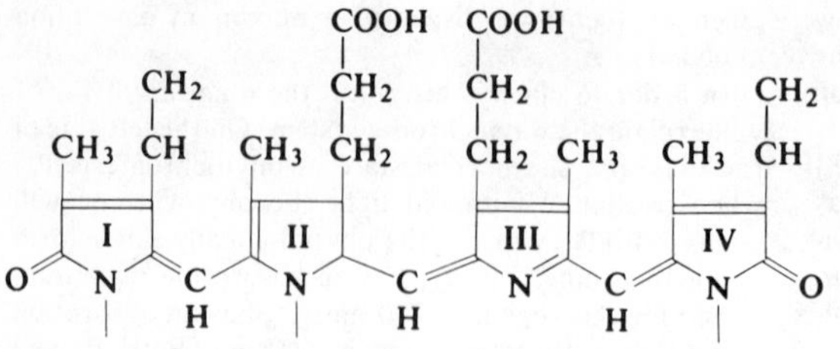

$$\text{COOH} \quad \text{COOH}$$

$$\text{CH}_2 \qquad \text{CH}_2 \quad \text{CH}_2 \qquad\qquad \text{CH}_2$$

$$\text{CH}_3 \ \text{CH} \quad \text{CH}_3 \ \text{CH}_2 \quad \text{CH}_2 \ \text{CH}_3 \quad \text{CH}_3 \ \text{CH}$$

I   II   III   IV

$$\text{O} \quad \text{N} \quad \text{C} \quad \text{N} \quad \text{C} \quad \text{N} \quad \text{C} \quad \text{N} \quad \text{O}$$

H   H   H

Fig. 2

phytochrome is critical in the formation of the flowering hormone); (c) biochemical effects, such as synthesis of anthocyanin, ascorbic acid, protein, and deaminase enzyme; and (d) movement, which includes movement of chloroplasts and movement of leaflets and whole leaves of *Mimosa*, the sensitive plant.

The detection of phytochrome can be made for either segments of etiolated plant organs, or the purified pigment by measuring the difference in the absorption of light using a special spectrophotometer. In this spectrophotometer, the optical density is measured at 660 and 730 nm. The sum of P_{660} (P_r) plus P_{730} (P_{fr}) is referred to as the total reversible phytochrome which can be indicated by the change in the optical density difference reading following irradiation of the sample with actinic sources of red and far-red light.

The phytochrome pigment was isolated and purified in 1959. The pigment was extracted from dark-green plant tissue with alkaline buffer of pH 7.8, then separated by column chromotography using dicalcium phosphate gel. The pigment is blue-green in color when irradiated with far-red, and blue when irradiated with red light. Structurally, phytochrome is a protein attached to a chromophore. Phycobiliprotein or phycobilins, which are common in algae, have protein molecules and chromophores. The molecular weight of the phytochrome protein is estimated to be 60,000. The chromophore, a bilitriene, is suggested to be a porphyrin molecule like the chlorophyll and hemoglobin molecules in general structure. It is closely related to the chromophores of C-phycocyanin and allophycocyanin. Figure 2 is the structural formula of bilitriene, which is thought to be similar to the phytochrome chromophore. It has been suggested that the two forms of P_r and P_{fr} are *trans* and *cis* forms of the chromophore bilitriene. Another hypothesis is that the change in the double bond of ring I and IV could be the reason for the difference between P_{fr} and P_r forms of phytochrome. The chromophore appears to be linked to the protein by a peptide linkage, or an ester linkage.

Phytochrome pigment is denatured in the presence of urea, pronase, and trypsin. It is thought that urea opens the protein molecule. High temperatures also cause denaturation of phytochrome. The chromophore of P_r and P_{fr} are stable in the pH range between 5 and 9.

One important question has not been answered: Where is phytochrome located inside the cells? The exact location is not yet known; some workers have shown evidence that phytochrome is present in the mitochondria. Other workers suggest that it is near the cell wall. Most likely, phytochrome could be located near the membranes of mitochondria, chloroplasts, and other cell membranes.

A. KARIM KHUDAIRI

Reference

"Symposium on Photomorphogenesis in Plants," The *Quarterly Review of Biology*, **39**: 1–34, 1964.

PICIFORMES

Piciformes is an order of birds (see AVES) that are distinguished from the PASSERIFORMES by the fact that they have ten feathers in the tail, and zygodactyl feet, that is with two toes in front and two in back. They nest in holes and have white, glossy eggs.

There are four families: Picidae, the woodpeckers and wrynecks; Capitonidae, the barbets; Galbulidae, the jacamars; and Rhamphastidae, the toucans.

Woodpeckers occur in both temperate and tropical regions of both the old and new worlds. Wrynecks occur in temperate regions of the old world. Barbets occur in tropical portions of Asia, Africa, and Central and South America. Jacamars and toucans occur in tropical regions of the new world.

Woodpeckers have the inner hind toe short, and in one group, known as three-toed woodpeckers, missing. The bills of woodpeckers are stout, straight and somewhat blunt at the end. The bill is used to drill into trees for borers and bark beetles, to dig out holes for their nests, and to make drumming sounds, which substitute for song. The tails of woodpeckers are composed of stiff, pointed feathers, and used to brace against a tree trunk when climbing.

In addition to insects obtained by digging into trees, woodpeckers eat ants and berries, and one species eats acorns. They are useful birds because of the harmful insects that they eat, but they require dead trees or branches in which to nest. Consequently the practice of foresters and tree surgeons in removing dead trees and limbs is likely to decrease woodpeckers and increase borers.

Certain small woodpeckers in tropical regions are known as piculets. They do not use the tail in climbing and are reported to climb with head up or down, somewhat as a nuthatch does.

Wrynecks are so called because they have a habit of twisting their necks. Their habits are much like those of woodpeckers. They feed largely on ants, and descend to the ground to get them.

Barbets are rather short birds, with more or less prominent bristles about the bases of their bills. There are a good many species, some Asiatic, some African, and some in Central and South America. They climb on trees and excavate holes in trees for their nests.

Jacamars occur in the American tropics. They are rather slender birds, with long sharp bills and brilliant colorings. They nest in holes in banks, rather after the manner of kingfishers.

Toucans occur in tropical America. They are distinguished by their enormous bills. These bills, however, are not as topheavy as they look, for they are thin, and not solid within. Both the bills and the plumages of these birds are brilliant in colors. There are smaller species of the family known as toucanettes. They feed on fruits, and some of the larger toucans are also predatory on the young of other, smaller birds, and have been known to break up their nests.

ARETAS A. SAUNDERS

Reference

Bent, A. C., "Life Histories of the North American Woodpeckers," U. S. Nat. Mus. Bull. 174, 1939.

PIGMENTATION

Pigmentation refers to the quality of color found in animals and plants. It may be caused either by the deposition of organic pigments (biochromes) in tissues, or by structural colors which are produced by optical effects of light rays. It also may be caused by a combination of the two methods. Pigmentation, which is almost universal in the animal and plant kingdom, may be permanent or changeable. It may perform important roles in the biological world, or it may be purely accidental or incidental to important physiological processes. It has many useful functions in the relationship of animals to their environment, such as protective coloration, mimicry, warning, and concealment; it aids in physiological processes, such as adjusting or regulating the light in organs of vision and in the formation of vitamins; in green plants a pigment (chlorophyll) is responsible for photosynthesis; and some pigments are waste substances discharged by the organism. Pigmentation is a valuable aid in distinguishing taxonomic units of plants and animals. Animal pigments may be derived from plant foods or they may be formed by biochemical processes in the metabolism of the animal; plant pigments are almost universally synthesized by the plants themselves.

In animal pigmentation, structural colors (schemochromes) are produced by physical surfaces which give the effect of various colors when light falls upon them in such a way as to disintegrate the spectrum. Thus, the white color of feathers and insects is often caused by the reflection and refraction of light; the blue and green colors of fishes and reptiles are due to a scattering of light rays upon a background of white and dark chromatophores; and the iridescent colors of the inner surface of mollusk shells are produced by interference of light. In many cases, a pigment or biochrome may modify a structural color. In contrast to organic pigments, no colored substance can be extracted from structural pigmentation, but it may be altered or destroyed mechanically.

True pigments (biochromes) exist in a variety of complex chemical compounds which are often difficult to analyze because of impurities and problems of isolation. Some have a wide distribution through the animal kingdom; others are restricted to a few animals. Among

the rarer pigments are the quinones (reddish cochineal of scale insects), flavones (yellow or red color of a few insects), and anthocyans (red-brown color of robber-flies). Other groups of pigments are more common, but impart little color to the animals involved, such as chromolipoids (yellow or darkish brown of ovaries and adrenal glands), and flavins (yellow-green fluorescence of many glandular tissues exposed to ultraviolet light). Another group of pigments (porphyrins) is widely distributed in nature. They are represented by the chlorophyll of green plants and by hemoglobin which contains the heme pigment. Hemoglobin and its derivatives (which are universal among vertebrates and found sporadically in many invertebrates) are of great importance as oxygen-carrying pigments and give red color to the blood. Three other respiratory blood pigments, chlorocruorin, hemocyanin, and hemerythrin, occur in the invertebrates. Chlorocruorin, probably a mutant form of hemoglobin, also contains iron in its porphyrin and is green in dilute solutions but reddish in concentrations. This pigment is found in the plasma of certain sessile marine annelids, one of which *(Serpula)* has both hemoglobin and chlorocruorin in its blood. Hemocyanin contains copper which in the presence of oxygen gives a blue color to the blood, but is colorless otherwise. This pigment is also dissolved in the plasma and occurs in certain mollusks and in some arthropods. Hemerythrin is present in the corpuscles (coelomic or blood) of a few annelids, *Sipunculus* and other sipunculid worms, and the brachiopod, *Lingula,* where it gives a purplish color with oxygen but is colorless in the absence of oxygen. Hemocyanin and hemerythrin contain no porphyrin groups. Three common animal pigments, carotenoid (carotene, xanthophyll), melanin and guanine are responsible for most of the striking surface pigmentation in animals. These pigments may be derived from food or from endogenous metabolism. Melanin, for example, may be produced by the interaction of an oxidase (tyrosinase) with tyrosine (an amino acid). These pigments all give characteristic absorption bands within the visible spectrum. Pigment granules may be scattered through ordinary tissue cells (e.g. epithelium), but often each kind is found in special branched pigment cells, such as xanthophores (lipophores) which are red or yellow; melanophores, blackish-brown; and guanophores, silvery white. Surface pigment may be found in both the epidermal and dermal layers of the skin. Pigment cells (melanoblasts of chromogens) originate in the neural crest and migrate to definitive positions where they synthesize melanin to become in succession melanocytes and melanophores (fishes, amphibians, reptiles) or only melanocytes (birds, mammals). Other chromoblasts (e.g. xanthophores) also originate in the neural crest.

In most animals, pigment cells are unicellular and color changes are produced when the pigment is concentrated in the center of the cell (paleness) or dispersed in the preformed branches (colored). In a few groups (e.g. crustaceans) the pigment unit is multicellular and may (polychromatic) or may not (monochromatic) bear different pigments. Another pigment unit is made up of a color cell with radially disposed muscle fibers which control the concentration and dispersal of the pigment (cephalopods). Color changes may be induced by stimuli, such as light and temperature, which act upon the nervous system. The nervous system may act directly upon the pigment units (muscular-pigment organs) or, more commonly, release neuro-humors which effect the color changes. Light may stimulate the chromatophores directly as in tanning. The skin color of man is due to (1) dendritic melanoblasts which are found in the base of the epidermis and which furnish melanin to the melanocytes and epidermal cells of the superficial layers of skin; (2) carotene in the outer layer of epidermis and subcutaneous fat; and (3) hemoglobin or oxyhemoglobin of the blood. Reddish portions of the body, such as lips, mucous membranes, and birthmarks, get their color from the propinquity of the hemoglobin in the blood. Poorly oxygenated blood gives a bluish or purplish tinge to the skin as in cyanosis. A yellowing of the skin is produced in jaundice by the deposition of bile pigments as does also the excessive consumption of carotene-containing fruits and vegetables. Hair color is mostly due to melanin in the cortical cells and to the concentration of air. Gray hair has almost no pigment as does hair produced by hereditary albinism. The dopa reaction selectively stains melanin-forming cells (melanoblasts). Melanomas are tumors which often contain excessive amounts of melanin.

In general, pigmentation in plants is caused by the same or closely related pigments as those found in animals. Many animals derive their pigments directly or indirectly from plants. The chlorophylls of green plants so important in photosynthesis contain the pyrrole nucleus which is also found in hemoglobin and bile. This important pigment exists in two forms, chlorophyll *a* and chlorophyll *b*, which are widely distributed in green leaves, bacteria and some marine organisms. Usually chlorophyll is formed only upon exposure to light, but there are some exceptions, (mosses, ferns). Chlorophyll does not mix with the other plant cell products, but is contained within chloroplasts. Chlorophyll is always accompanied by carotenoids, one of the most widely spread pigments in both the animal and plant kingdoms. The exact biological role of carotenoids is not clear, but they are known to synthesize vitamin A and may play some part in sexual reproduction in animals and plants and in the formation of chlorophyll. The principle carotenoids are orange-red carotenes and the yellowish xanthophylls. These pigments are found in chloroplasts along with chlorophylls, but they may also occur in non-green parts of the plant. They are responsible for the ripe colors of peppers, oranges, carrots and tomatoes as well as those of many flowers (zinnias, sunflowers, etc.). A third group of common plant pigments is the anthocyans which are usually formed in leaves late in the summer. They are very sensitive to acid-alkaline conditions, appearing red in acid, blue in alkaline, and violet in neutral solutions. These pigments produce the color of beets, purple cabbage, barberry and many others, and are dissolved in the cell sap especially in the epidermis of the leaf. Autumnal coloration is mostly due to the persistent carotenoid and anthocyan pigments which display a varied assortment of colors after the dominant chlorophyll decomposes and fades from the leaf. Weather conditions which promote the best fall coloration are low soil temperatures and warm, clear dry days alternating with low (not frosty) night temperatures. Such conditions favor the conversion of starch into sugars so necessary for anthocyan synthesis.

CLEVELAND P. HICKMAN

References

Fox, D. L., "Animal biochromes and structural colours," Cambridge, The University Press, 1953.

Fox, H. M. and G. Vevers, "The Nature of Animal Colours," London, Sidgwick and Jackson, 1960.

Parker, G. H., "Animal colour changes and their neurohumors," Cambridge, The University Press, 1948.

Mayer, F. and A. H. Cook, "The Chemistry of Natural Coloring Matters," New York, Reinhold, 1943.

McCormick, J., *The Colors of Autumn*, Natural History, October 1957, pp. 424–435.

Cott, H. B., "Adaptive Coloration in Animals," London, Methuen, 1957.

PINEAL (Epiphysis Cerebri; Corpus Pineale)

The pineal gland is a relatively small grayish-white structure on the upper surface of the diencephalon (see BRAIN). It develops as an evagination of the dorsal wall of the diencephalon to which it remains affixed by a short hollow stalk, the *pineal peduncle*. The cavity of the pineal peduncle constitutes the pineal recess of the third ventricle and is lined with *ependyma*. The upper aspect of the peduncle is attached to and continuous with the habenular commissure and the lower to the posterior commissure. From these commissures nerve fibers pass into the substance of the organ. In the adult mammal, the pineal gland lies immediately below the splenium of the corpus callosum and rests in the groove between the superior quadrigeminal bodies. The mammalian pineal gland varies considerably in size; its magnitude is apparently not a function of the animal's size, for it is small in the elephant and whale, and large in sheep and man. There is much variation in the shape of the organ; it is cylindrical-conical in cattle, oval in the horse and cone-shaped in sheep and man.

The pineal gland is invested by a relatively thick capsule provided by pia mater. From this capsule trabeculae containing blood vessels and nerve fibers penetrate deeply into the gland, thereby dividing it imperfectly into lobules.

In mammals the pineal gland consists of two types of cells (1) glial and (2) pinealocytes; the latter predominate. The pinealocytes vary in size, are usually stellate in appearance and are embedded in the dense network of glial fibers. With certain silver impregnation methods numerous long, thick or thin, processes can be demonstrated arising from the cell body and radiating in all directions, or they may be polarized in one, two or three directions only. The terminals of these processes end in bulblike enlargements either among the different cell types or on certain vascular elements. Each pinealocyte possesses a large nucleus whose perforated double membrane envelope limits a granular nucleoplasm in which is suspended a conspicuous nucleolus. The cytoplasm of pinealocytes contains numerous microtubules, mitochondria, Golgi complexes, endoplasmic reticulum of the rough and smooth variety and some dense-cored vesicles. Characteristic of the glial cytoplasm is a host of fine filaments and some single membrane-limited dense bodies of variable sizes that may be lysosomes. Histochemical studies by Wislocki and Dempsey have shown ribonucleoprotein, alkaline phosphatase and some glycogen in the cytoplasm of pinealocytes from the rhesus monkey.

One of the unique features of the pineal gland is the presence of certain concentrically lamellated concretions commonly referred to as "brain sand" *(corpora arenacea; acervulus cerebri)*. These bodies are thought to consist of calcium carbonate and calcium, magnesium and ammonium phosphates.

In mammals, some authorities have suggested that the pineal is a neuroendocrine gland, since it is related to the epithalamus through a system of nerve fibers called the epithelamico-epiphysial tract. Such a relationship appears to be analogous to a similar one found between the neurohypophysis and hypothalamus which are connected by a system of nerve fibers known as the hypothalamicohypophysial tract.

EVERETT ANDERSON

References

Anderson, E., "The anatomy of bovine and ovine pineals. Light and electron microscopical studies," J. Ultrastructure Research, Suppl. **8**: 1–80, 1965.
Del Rio-Hortega, P., "Pineal Gland," *in* Penfield, W. ed., "Cytology and Cellular Pathology of the Nervous System" vol. 2, New York, Hoeber, 1932.
Gladstone, R. J. and C. P. G. Wakeley, "The Pineal Organ," Baltimore, Williams and Wilkins, 1940.
Tilney, F. and L. Warren, "The Morphology and Evolutional Significance of the Pineal Body," [The American Anatomical Memoirs] Philadelphia, Wistar Inst., 1919.
Wislocki, G. B. and E. W. Dempsey, "The chemical histology and cytology of the pineal body and neurohypophysis," Endocrinology, **42**: 56–72, 1948.

PIPERALES

The Piperales is a natural order of DICOTYLEDONOUS flowering plants believed to represent an off-shoot of ranalian ancestry. The leaves are simple; flowers for the most part bisexual, very small, lacking perianth, each in the axil of a small bract, and arranged in spikes or racemes. It comprises the families Saururaceae, Piperaceae, Chloranthaceae, and Lacistemaceae. With the exception of the Saururaceae, of eastern Asia and North America, they are tropical and subtropical in distribution.

The Saururaceae is a small family of herbs represented in the northeastern United States by *Saururus cernuum*, the common Lizard's-Tail. The Chloranthaceae is a small family of trees, shrubs or herbs of the tropics and subtropics, the largest genus *Hedyosmum* comprises about 25 species in tropical America. The Lacismetaceae is a small tropical American family of trees and shrubs. None of these three families contains species of economic importance or is especially noteworthy.

The Piperaceae, the largest family of the order, is conservatively estimated to contain at least 2000 species. Members of the family are to be found throughout the tropics of both hemispheres with major concentrations and centers of dispersal occurring in Malasia and in Latin America which has the greatest representation of species of any area in the world. In the Americas they are to be found from Argentina northward to Mexico and the West Indies. A few species occur in southern Florida.

The family is subdivided into the two nearly equal-sized genera *Piper* and *Peperomia* plus a few small generic segregates. An anomalous feature in the anatomy of members of the family is the fact that the vascular bundles are arranged in two or more circles with the outer circle united, or are separated and scattered as are those of the monocots. The genus *Piper* includes woody or subwoody shrub, vine, or small tree-like species while those of *Peperomia* are small and herbaceous.

Economically, the family is of only minor importance, though certain members have been in use since time immemorial. The fruit of *Piper nigrum* is the source of our black and white peppers, the use of which dates from ancient times, the peppercorns having in the past been a very important trade item. The leaves, roots and fruits of certain species possess principles of a mildly narcotic or somewhat stimulative character and have been used for centuries to produce such effects. Two species, in particular, have since ancient times been used by the peoples of the orient and the Pacific area for this purpose. The leaf of *Piper betel* wrapped about a little lime and the seed of the betle-nut palm is extensively chewed by orientals and East Indians. The roots of *Piper methysticum*, when pulverized and prepared as a decoction, known variously as *ava, kava,* or *yagona,* provides a non-alcoholic but somewhat narcotic beverage used widely throughout Polynesia and other islands of the Pacific. *Piper cubeba* furnishes a drug used for asthmatic conditions. A few species provide a poison used by natives to stupefy fish. A number of species of Peperomia are used as ornamental house plants and some are reported as being used by native peoples as a salad-like food.

T. G. YUNCKER

References

Yuncker, T. G., "The Piperaceae: a family profile," Brittonia **10**: 1–7, 1958.
Trelease, W. and T. G. Yuncker, "Piperaceae of Northern South America," Urbana, Univ. Illinois Press, 1950.

PITH

Pith is a botanical term used to identify the tissue composed of loosely-arranged cells occupying the core of the vascular cylinder of most higher plants. This term should not be confused with the verb pith meaning to sever or destroy the spinal cord of vertebrate animals. The pith, also known as the medulla, occurs primarily in the stem. However, it also occurs in some roots, particularly in the monocotyledons. The predominant cells in the pith are polyhedral, thin-walled parenchyma. These cells exhibit the typical characteristics of parenchyma, i.e. they are undifferentiated, simple, living cells containing a protoplast which carries out the basic physiological activities associated with life (respiration, assimilation, storage, etc.). Parenchyma cells constitute the fundamental cell type of the plant body. Stem pith parenchyma is ontogenetically derived by periclinal divisions of mother cells laid down by the apical meristem (see figure). Root pith derives from potential vascular tissue that fails to differentiate. Upon maturation pith cells may have one of several fates: they may remain alive and contain starch-forming leucoplasts, crystals or tannins; they may die leaving cells devoid of contents; they may become torn apart leaving cavities as the later maturing surrounding tissues elongate; or, in some species they may differentiate to form thick-walled sclereids, lignified cells, laticifers, secretory canals or fibers. Several of these cell types may be present in the pith of a given species.

Between the vascular bundles of the stem, areas of parenchymatous cells called rays occur. These rays are identified by the tissue from which they derive. In those stems where the vascular cylinder is in the form of separate vascular bundles the rays are composed of parenchymatous extensions of the pith. These are called pith rays or medullary rays. In species where there is a continuous vascular cylinder in the stem the rays are not pith, but undifferentiated xylem and are called xylem rays or wood rays. In conifers where the primary vascular tissue consists of separate bundles and the secondary vascular tissue is continuous the primary rays are thought to be part of the pith, while the secondary rays are tracheary in origin. In most monocots the vascular bundles are evenly distributed throughout the stem. In them no pith as such is recognized. Instead, the basic parenchyma in which the bundles lie is called ground tissue.

Both the stem and root of primitive vascular plants are protostelic and lack a pith. The phylogenetic origin of the pith is accounted for by two theories. Shoute claims that the pith was derived from tracheary elements and cites as evidence species of the Paleozoic lycopod Lepidodendron which contained cores varying from a protostelic structure to a siphonostele. Jeffrey states, however, that the pith was derived from the cortex which intruded through the leaf gaps during evolution of the siphonostele. Evidence to support this view is derived from the fact that certain tissues cortical in origin are sometimes found between the vascular cylinder and the pith. The Osmundaceae lend evolutionary evidence to this theory in that Mesozoic species contain an extrasteler and an intrasteler endodermis while modern species lack the intrastelar endodermis. Thus, the cortical origin of the pith has been obscured by reduction through evolution. The origin of the pith in roots is more satisfactorily explained by the Shoute theory, but there is much evidence for the Jeffrey theory in stems, especially in ferns.

Until recently mature pith was thought to be a senescent tissue having little or no functional value. Experimental work has demonstrated that it has a regenerative capacity. *In vivo* experiments have shown that when vascular strands are severed or the shoot apex is surgically isolated on a plug of pith new vascular tissue develops in the pith as a response to hormonal stimuli and restores the vascular continuity. In sterile culture experiments living pith cells isolated from the inhibitory influence of the plant retain the power to enlarge, divide and, even, differentiate when grown on a medium supplying the tissue with nutrients and hormones. This capacity is apparently retained by the pith as long as the plant lives. Thus, pith is essentially a totipotent tissue whose potentiality is expressed only when the normal organismal relationships are disturbed.

MARY E. CLUTTER

References

Esau, Katherine, "Plant Anatomy," New York, Wiley 1965.
Wardlaw, C. W., "Phylogeny and Morphogenesis," London, Macmillan, 1952.

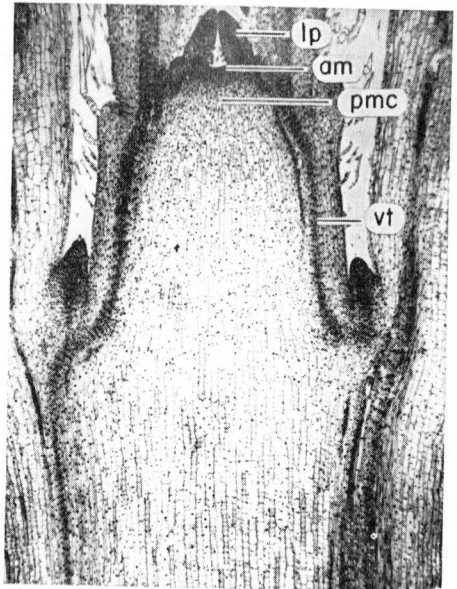

Fig. 1 Longitudinal section of shoot apex of *Helianthus tuberosus* illustrating leaf primordium (lp), apical meristem (am), pith mother cells (pmc), vascular tissue (vt) and pith (center) (×25).

PITUITARY

The pituitary (hypophysis) is an important gland of internal SECRETION (see also HORMONES) is divisible into (epithelial) *adenohypophysis* derived from Rathke's pouch, comprising pars tuberalis, pars distalis and pars intermedia, and (neural) *neurohypophysis* derived from the diencephalic floor and consisting of the median eminence (of the tuber cinereum), neural stalk and neural lobe. The *anterior lobe* is identical with pars distalis, *posterior lobe* with both epithelial and neural parts of the rest of the pituitary. The *pituitary stalk* is the combined neural stalk and pars tuberalis.

Development. Rathke's pouch invaginates dorsally just anterior to the bucco-pharyngeal membrane at the most dorsal point on its buccal side. This invagination becomes a pedunculated cyst; the dorsal part of the cyst becomes the pars tuberalis and pars intermedia and the ventral thickens and becomes the pars distalis, assuming the adult relationships as head flexure progresses. The peduncle may persist as the so-called cranio-pharyngeal canal. The dorsal portion of the cyst becomes wrapped around the infundibular process which grows downward from the diencephalon. The dorsal thickened end of the infundibular process develops into the neural lobe and the upper end develops into the median eminence. The narrow intermediate region forms the neural stalk.

Blood supply. In all vertebrates there is a common BLOOD supply which carries blood first to the median eminence and pituitary stalk and then to the pars distalis. In the tailless amphibians *(Anura)*, reptiles, birds and mammals, there are intermediate blood vessels, the hypophysio-portal vessels. The neural lobe has an independent blood supply in amphibians, reptiles, birds and mammals. In these forms the neural lobe seems to be specialized and to have developed *pari-passu* with independence from an aqueous habitat. Venous drainage in man is principally to the cavernous sinus, in other animals to derivatives of the primary head vein.

Innervation. The neurohypophysis is innervated by the hypothalamus. Axons from the supraoptic and paraventricular nuclei enter the neural stalk and are concerned with neurohypophysial secretions. Other fibers reach the neural stalk from neurons in the tuber cinereum (e.g., the lateral tuberal nuclei) and from the antero-lateral hypothalamic regions (e.g., medial forebrain bundle). Some of these fibers in the neural stalk send terminal branches toward the vessels of the pars tuberalis and zona tuberalis (a specialized functional region between pars distalis and pars tuberalis present in a few species). These vessels also receive autonomic fibers. The p. intermedia receives fibers from the neural lobe. These seem to be the only fibers in the adenohypophysis likely to be secretory ones. The consensus of opinion is that the pars distalis lacks a secretomotor nerve supply.

Phylogeny. The pituitary is present in all vertebrates and the characteristic location just rostral to the notochord, the epithelial and neural elements and the common blood supply is always present. The relative position of the lobes is somewhat variable. The pars intermedia is often absent (in many species of birds, for example). A specialized lobe associated with the pars intermedia (Wulzen's) lobe is present in some ungulates. In man the pars intermedia is represented by islands of pale cells and cystic remnants after childhood. In some species there is a dural cleft between the adenohypophysis and neurohypophysis, penetrated by the portal vessels (some reptiles, most birds, *cetacea*, etc.).

Cytology. Pars distalis contains three principal cell types: acidophils, basophils, and chromophobes. Acidophils: subdivisible into types with smaller granules electively stained by azocamine and with large granules stained by orange G. Basophils: many variations may be seen. Electron microscopy indicates variable and mixed granule populations. Some subdivide these cells into "thyrotrophs," "gonadotrophs," etc., depending on staining characteristics of predominant granule type and reaction to hormonal conditions. Chromophobes are (1) Primitive multinucleate so-called "stem cells" with small, pale nuclei, scanty cytoplasm; (2) True chromophobes, similar but with single nuclei and maturing to either acidophils or basophils; (a) false chromophobes (degranulated acidophils). The latter are large and contain large and darker nuclei, sometimes pycnotic.

The cells of the pars tuberalis are small, contain a variety of granules both acidophilic and basophilic and frequently much endoplasmic reticulum. The pars intermedia consists of small cells with poorly staining and very small basophilic granules. The cell margins are clearly defined since terminal bars and much intercellular cement is present. The neurohypophysis contains many types of parenchymatous cell, some derived from ependyma and others from glial elements.

Secretions. All the known pituitary secretions are protein like: of fairly large molecular weight in the pars distalis and low molecular weight in the neurohypophysis.

The pars distalis secretes: 1. *Growth hormone*, which probably influences both growth and carbohydrate metabolism. It is derived from the acidophils and its chemistry varies widely from one species to another; 2. *Adrenocorticotrophic hormone*, controlling the growth of the zona fasciculata and zona reticularis of the adrenal cortex and the production of glucocorticoids. It is probably secreted by the basophils; 3. *Thyrotrophic hormone*, also produced by the basophils, controls the growth and secretions of the thyroid hormone; 4. *The gonadotrophic hormones*, influencing the growth of ovarian follicles (follicle stimulating hormone), growth of the corpus luteum (luteinizing hormone), of the testis (interstitial cell stimulating hormone) probably identical with luteinizing hormone, of ovulation (follicle stimulation plus luteinizing hormones). These are believed by some to be derived only from the basophil. Others believe that the acidophils also play a role in their production; 5. *Prolactin*: a hormone playing an important role in lactation.

The pars tuberalis has not been shown to have important secretions so far. The pars intermedia secretes a melanophore expanding hormone active in some lower vertebrates.

The neurohypophysis secretes hormones which are probably polypeptides. Two have recently been synthesized. These hormones (1) arrest water diuresis (antidiuretic hormone), (2) cause uterine contraction and milk ejection in the lactating female (oxytocic hormone). Besides these effects, vasopressor, chloruretic, and increased peristaltic effects are produced. The two synthetic octapeptides have all these actions but one is chiefly antidiuretic and the other chiefly oxytocic. It is believed that these hormones are produced by the supraoptic and paraventricular neurons, either in their cell bodies with subsequent transport to the neural lobe by axoplasmic streaming or in graded fashion throughout the extent of the neurons with the chief production in the neural lobe. The once popular view that these hormones are produced by pituicytes (parenchymatous cells of the neurohypophysis) is now discounted. The supraoptic and paraventricular neurons control the liberation of the secretions as well as their formation. The activity of single cells in these nuclei has recently been studied and found to be quantitatively related to experimental changes in the tonicity of the arterial blood. This stimulus is known to regulate the release of antidiuretic hormone and to influence milk ejection.

The neurohypophysis may be selectively stained by special procedures and the intensity of the staining has been found to parallel the antidiuretic hormone content. The stained material is called neurosecretion.

Regulation of the hypophysis. External stimuli influence the secretion of most pituitary hormones. In the neurohypophysis the action of the central nervous system is direct. In the adenohypophysis gonadotrophins are believed controlled by a neurohormone produced in the neurohypophysis under nervous influences and carried to the pars distalis—the hypophysio-portal vessels. Currently there is still dispute about the mode of nervous regulation of other adenohypophysial hormones. It is claimed that a "corticoid releasing factor" can be isolated from the neuronhypophysis and this may also influence secretion of adrenocorticotrophin through the portal vessels. Evidence for neural control of thyrotrophin has been advanced and there is good evidence that prolactin is under nervous control. A little evidence for neural control of growth hormone also exists. On the other hand, a "feedback" of hormones from target organs also influences the secretion of adrenocorticotrophin and probably thyrotrophin. There is much evidence that the same peripheral and central nervous pathways are involved in the control of oxytocin and prolactin.

JOHN D. GREEN

References

Gorbman, A., ed., "Comparative Endocrinology," New York, Wiley, 1962.

Romeis, B. *in* Handb. mikr. Anat. Mensch. v. Möllendorf, Berlin, Springer, 1940.

Green, J. D., "The comparative anatomy of the hypophysis," Am. J. Anat. **88**: 225–312, 1951.

Harris, G. W. and B. T. Donovan, "Pituitary Gland," 3 vols. Berkeley, University of California Press, 1966.

Harris, G. W., "Neural control of the pituitary gland," London, Arnold, 1955.

Scharrer, E. and B. Scharrer, *in* Handb. mikr. Anat. Mensch, Berlin, Springer, 1954.

PLANKTON

Plankton, comprising passive plant and animal "drifters," contains some of the most widely distributed and abundant of all living organisms. Every major group of non-airbreathing animals is represented, some during the early stages only, others throughout their existence. Plankton abounds in the sea, brackish water and freshwater at all depths. For the most part these pelagic organisms are microscopic in size but some larger forms having feeble powers of locomotion, such as jellyfish and pyrosomes, are also included.

Importance. The importance of plankton cannot be overemphasized. It comprises the bulk of all aquatic living matter, and as on land, the plants (phytoplankton) are the producers and animals (zooplankton) the consumers. Directly or indirectly all high seas animal life is dependent on phytoplankton productivity in the upper layer illuminated by sunlight, because the larger sessile plants, also restricted to the photic zone, occur only as a relatively narrow fringe about the margins of continents and oceanic islands. In more turbid coastal waters photosynthesis may be limited to depths of fifty feet or less, but in the clear open ocean, sunlight at three hundred feet or more is sufficient for conversion of the carbon dioxide into organic matter.

Productivity. Phytoplankton require nutrients for successful growth similar to green plants on land. Some of these are almost always present in excess of requirements, but others like phosphorus and nitrogen are greatly depleted during periods of active algal growth which commonly occur in spring and autumn in northern open coastal waters. The plant population then declines and remains minimal until the nutrient supply is restored from deeper levels by vertical mixing and by decomposition of vegetation and animal remains. Light is a limiting factor to aquatic plant growth in high latitudes at certain seasons. Temperature exerts a selective influence on both plants and animals but does not in itself appear to limit the total quantity of plankton produced.

Local phytoplankton production determines the amount of zooplankton which can be supported at any particular time and place, but zooplankton response will depend to a considerable degree on physical environmental factors, differences in the breeding seasons of component species and the variable and relatively long developmental periods of the young, particularly in waters having marked regional and seasonal differences in temperature. Thus, although the quantitative distribution of phytoplankton provides a reasonably accurate index of local fertility, zooplankton abundance serves only as a measure of the result of production on different areas, and does not in itself necessarily indicate the points of origin. Under the influence of an active current system a relatively unproductive locality (like the Bay of Fundy) may at times harbor an appreciable zooplankton population consisting largely of immigrants from other breeding areas. Evaluation of animal productivity in any region thus involves a determination of the amount of actual successful production taking place there.

Composition of plankton populations. Phytoplankton is for the most part composed of a variety of unicelled yellow-green algae dominated by diatoms, armored dinoflagellates, naked flagellates and coccolithophorides, which together according to some authorities account for more than half of the organic production of the earth. The relative abundance of component groups varies regionally and seasonally. Diversity of species is greatest in tropical waters and least in boreo-arctic regions.

Diatoms, distinguishable by their siliceous shells, occur singly and in chains, have one or more chromatophores and during flowering seasons or "blooms" flourish in great abundance in coastal waters.

The dinoflagellates are more mobile, possess flagella and like diatoms at times dominate in the phytoplankton during bloom periods, particularly in warm neritic waters. At these times luminous species often cause brilliant bioluminescence. Some, possessing chlorophyll, are truly phytoplankton, but others are either parasitic, or feed like animals and are classed as zooplankton.

Little is understood about the naked flagellates because of their minute size and the difficulty in preserving them. However, they are now considered the principal food of most planktonic filter feeders and may play an equally important if not a more important role than diatoms, particularly in neritic environments. On occasions one or more species multiply so rapidly that the water becomes discolored and extensive mortality of fish and invertebrates may result. In the open ocean the naked flagellates tend to be augmented if not replaced by small calcareous coccolitho-phorides.

Other pelagic yellow-brown algal types at times abundant include the brownish colonial flagellate *Phaeocystis* which forms gelatinous clusters, particularly in northern coastal waters, and the globular green *Halosphaera*, widespread over the open ocean. A blue-green alga, *Trichodesmium*, appears at times as yellowish-brown patches on the surface in warm climates both in oceanic and neritic environments.

Unlike the pelagic algae which largely belong to a single major group, the plankton animal population is extremely diverse. Apart from coastal influence it consists almost entirely of endemic holoplanktonic species, nekton eggs and larvae (fish and squid) and occasional specimens of larval benthos. Euphausiids, because of their large size and abundance at the surface at night, are probably second only to copepods in importance in the natural economy of the open ocean. In neritic areas one finds in addition a much greater abundance and variety of eggs of nekton and young, eggs and larvae of benthonic invertebrates and at times swarms of adult benthos, particularly those species of annelids and crustacea which become pelagic during the breeding season.

Boreo-arctic zooplankton contrasted with that in more temperate and tropical regions is, like phyto-plankton, rich in individuals but relatively sparse in species. The contrast is further accentuated by the fact that, out of the total population, only a relatively small number of endemic species, usually copepods, comprise the greater part of the stock at all times. Other common holoplanktonic members of the northern community, occurring in variable abundance, include ctenophores, chaetognaths, pteropods, hyperiid amphipods, euphausiids, appendicularians, salps and medusae, etc. Most of these assume relative numerical importance only for a limited time following breeding periods.

Recent observations in the subtropical central Atlantic indicated the numerical abundance in the Sargasso Sea to average about one-third that of the colder northern waters and, like all tropical and subtropical ocean areas, it is characterized by very large numbers of species very few of which comprise an appreciable percentage of the population at any time. However, here again as in northern waters, copepods sometimes constitute the most stable and numerically dominant component, and variations in the total population tend to reflect fluctuations in this group. Other warm oceanic groups in order of abundance in weekly collections throughout the year were appendicularians, molluscs, ostracods, euphausiids and chaetognaths.

Seasonal distribution. The seasonal cycles in abundance of plankton are generally consistent in most areas and are indicative of successive augmentation and depletion of dominant components. In phyto-plankton these changes in neritic waters are most closely associated with light, temperature and nutrient supply, and are very similar in comparable thermal latitudes. In true arctic waters a single summer bloom period has been reported, but elsewhere the most luxuriant diatom growth does not occur in the warmest months. Progressing southward the typical vernal and autumnal maxima retreat farther and farther from the warmest season until in temperate neritic waters (south of Cape Cod and in the Adriatic) the major blooms

occur in mid-winter. In subtropical and tropical oceanic regions where the water masses are relatively stable phytoplankton growth is generally sparse.

Seasonal variations in numerical abundance of zoo-plankton in northern waters are predominantly influenced by a few dominant copepods, particularly *Calanus finmarchicus* in the north Atlantic. In winter when primary production is at a minimum these species hibernate at 600–1000 meters when depths permit. Following a vernal ascent of maturing individuals a new crop of young is produced and the parent stock dies off. The new generation descends to deeper water in late summer and autumn upon reaching late juvenile stages.

Farther south in both the eastern and western Atlantic there are three or more generations of the dominant copepods in coastal waters, the seasons starting earliest in the southern portion of the range and proceeding progressively northward. In the higher crustaceans and most non-crustaceans there appears to be but one annual brood, although a few (e.g. chaetognaths) have several.

Recent studies have revealed that in the central north Atlantic there is a valid annual biological cycle in the subtropical epizooplankton no less definite than that in temperate and boreo-arctic waters and, in spite of extreme short-period fluctuations in numerical abundance, there is a general uniformity in seasonal and yearly average composition. The principal difference in the subtropical cycle appears to be in the frequency of the short-period numerical fluctuations, explainable by the rapid growth rates characteristic of warmer seas. In this general region the zooplankton population shows remarkable agreement with the phytoplankton. The spring maximum and early winter minimum in plant growth is followed in one to two months by corresponding fluctuations in the zooplankton cycle.

Vertical distribution. Between the surface and the bottom in the open ocean there is a vertical series of life zones in which the plankton communities at different levels are in various ways adapted to their particular habitats.

Phytoplankton production extends downward to the lower limit of photosyntheses. Few pelagic plants thrive at the surface proper, however. Diatoms tend to concentrate at 10–25 meters in coastal areas and even deeper in clear mid-ocean waters where active reproduction may take place to depths of 100 meters or more. Some of the dinoflagellates favor bright sunlight but others comprise a "shade flora" frequently found most abundant in southern seas at depths of 100 meters where they have access to nutrients below the range of most active diatom growth. Here, where thermal stratification is permanent, fertilization of the photic zone is dependent on the nocturnal ascent of the animal plankton.

Except for the *Sargassum* community and a few unusual brilliantly colored but rarely abundant species, subtropical zooplankton is sparse at the surface during daylight hours. The subsurface zone extending to a depth of several hundred meters is characterized by more or less transparent species which make nocturnal upward migrations after sundown. Below the subsurface zone to a depth of approximately 1000 meters, the lower limit of visible light penetration and the depth beyond which migration to the surface does not appear to occur ordinarily, there is a relatively diverse and abundant population characterized by silvery species of fishes, and crustaceans commonly spotted with reddish chromatophores. The zone lacking sunlight with no seasonal environment changes, appears from approximately 1000 meters to the bottom. Within this zone there is also a varied population of pelagic animals with the greatest variety and smaller sizes in the upper part, and larger specimens of a smaller number of the same species declining in numbers to the bottom.

Diel migrations. Limited diel vertical migration of certain phytoplankton flagellates has been reported but the range is not great. Zooplankton inhabitants of the subsurface zone, however, characteristically rise to the surface at night. There they are joined by members of the transition zone, some of which are found concentrated as deep as 800 meters by day. Others like *Cyclothone* spp. (small luminous fishes) and alciopid worms never approach the surface. Light and temperature appear to be the major factors influencing these movements. Some copepods rise at more than a meter a minute and euphausiids at more than two meters per minute. As differential temperature tolerances may form a barrier to both upward and downward movement, not all species make the same nocturnal pilgrimages to and from the photic zone.

Estuarine zooplankton. Temperate-latitude estuarine zooplankton populations are composed primarily of copepods although planktonic larval stages, such as barnacle nauplii, are often important short-term contributors. In some estuaries, mysids may occupy a position comparable to that of the euphausiids of the open sea as food for nekton. Zooplankton populations in estuaries are usually characterized by the numerical dominance of relatively few species. In general, particular species will compose the bulk of the population at specific times and at specific locations. The composition of the population in a particular area at a given time is strongly influenced by the stage of the tide and the season of the year. Sharp vertical and horizontal gradients in population numbers occur, related to the distribution of the physical and chemical properties of the water mass within the estuary.

HERBERT C. CURL, JR.
HERBERT F. FROLANDER

References

Hardy, A. C., "The Open Ocean," London, Collins, 1956.
Sverdrup, H. W. *et al.,* "The Oceans," New York, Prentice-Hall, 1942.
Marshall, S. M. and A. P. Orr, "The Biology of a Marine Copepod," Edinburgh, Oliver and Boyd, 1955.
Murray, Sir. J. *et al.,* "The Depths of the Ocean," London, Macmillan, 1944.
Raymont, J. E. G., "Plankton and Productivity in the Oceans," New York, Macmillan, 1963.
Wimpenny, R. S., "Plankton of the Sea," New York, American Elsevier, 1966.

PLANTAGINALES

This order of plants belongs to the subclass Sympetalae of the class DICOTYLEDONEAE and is composed of a single family (Plantaginaceae). It has been thought to have arisen from the Primulales or Scrophulariaceae, but there is no unity of opinion. The plants are characterized by having basal leaves with conspicuous, apparently parallel venation and spicate or capitate inflorescences on wiry scapes. Flowering begins at the bottom of the scape and progresses upward at such a rate that mature seeds may be formed at the bottom while the upper flowers are still opening. The flowers are inconspicuous, 4-merous, mostly perfect, regular, with the corolla lobes scarious and the anthers on long, wiry filaments. The ovary is superior, 2-carpellate, with 1 style, and mostly develops into a circumscissile capsule with numerous seeds.

The order is composed of three genera (Plantago, Litorella, Bougueria) and around 200 species. Bougueria is a monotypic genus of the Andes. Litorella has 2 species, one, L. uniflora, an aquatic, occurring along the shores of lakes and ponds about the Great Lakes and St. Lawrence basins and the other in Europe and Antarctica.

Plantago is a cosmopolitan genus with 18 or 20 species occurring in the United States. These are mostly weedy plants. P. major is a common weed in lawns, golf greens, etc., that almost chokes out the grass at times. P. Psyllium is the one useful species. Its seed is the common "Psyllium Seed" whose coat becomes mucilaginous when wet, and is reputed to be of laxative value.

RAY J. DAVIS

Reference

Pilger, R., "Plantaginaceae" *in* Engler, A., "Das Pflanzenreich," 102: 1–466, Leipzig, Springer, 1937.

PLANT ANATOMY

Plant anatomy is one aspect of a broader field of study which is termed plant morphology. The latter embraces also the study of external morphology, *viz.,* the arrangement, form, and relationships of plant organs (e.g., STEMS, ROOTS, LEAVES). The study of internal structure, to include descriptions of cells, organization of tissues, and structural details of organs, constitutes plant anatomy. Plant cytology, another subdivision of morphology, is concerned with a study of the entities located in the protoplast—the living contents of a cell. To the anatomist and cytologist the light microscope is an indispensible instrument. In addition the electron microscope and the use of radioisotopes play important roles in modern research in cytology.

The attention of an anatomist, even though he may be concerned primarily with tissue organization, is drawn invariably to a consideration of individual cell structure. These minute components of plant parts were seen first by Robert Hooke, an English botanist, in the 17th century using the compound microscope. Hooke observed that cork from trees was made up of small cubicles or units, each surrounded by a wall. Later he recognized that certain other plant cells contained watery juices which today we term protoplasm —the living substance of life. With improvements in the microscope the various cellular components were described. For example, in 1831 Robert Brown described the occurrence of a spherical body in living cells and termed it the nucleus. We know today that the nucleus contains the hereditary units responsible for determining form, structure and physiologic specialization of an organism.

Cell types and classification of tissues. The various types of cells found in plant tissue have been described and a generally accepted classification system exists which encompasses all known types. For example, the PARENCHYMA cell, a very basic and fundamental cell in the plant body is variable in its size and form, but throughout its life it possesses a protoplast and may be active in PHOTOSYNTHESIS, e.g., a mesophyll cell of a leaf; or active in food storage, e.g., cell of a potato tuber. Not all cells retain a protoplast at the time when they are performing their principal functions. For example, the tracheid is the fundamental water conducting cell in vascular plants (plants possessing XYLEM— water conducting tissue and PHLOEM—food conducting tissue) and it is completely dead at functional maturity. Generally the cell is elongate, angular in outline and possesses a rather thick cellulose wall impregnated with lignin. Water moves through chains of such cells under the influence of the "transpiration pull" ini-

tiated at the surface of leaves by the process of evaporation. Water fills the cell cavity and movement to adjacent cells is facilitated by the presence of thin areas (pits) in the cell wall. A cavity of one tracheid is directly apposed to a cavity in an adjacent tracheid. Two such pits constitute a pit-pair. A cell type, phylogenetically derived from the tracheid, is the vessel member. The union of many of these cells into long chains constitute a so-called vessel. During ontogeny of each vessel member the end walls or portions of them are dissolved which results in the formation of actual perforations. The vessel, then, is a long tube made up of a series of superposed vessel members which are dead at maturity. Such a structure permits the more efficient movement of water through the plant. Vessels occur in some species of *Selaginella* and ferns, in certain gymnosperms, and in nearly all angiosperms. Space does not permit an extended discussion of this topic, but the reader is referred to the accompanying table which summarizes the main characteristics of recognized cell types (Table 1).

While purely descriptive accounts of individual cell types are sufficient in themselves, considerable difficulties are encountered when attempts are made to

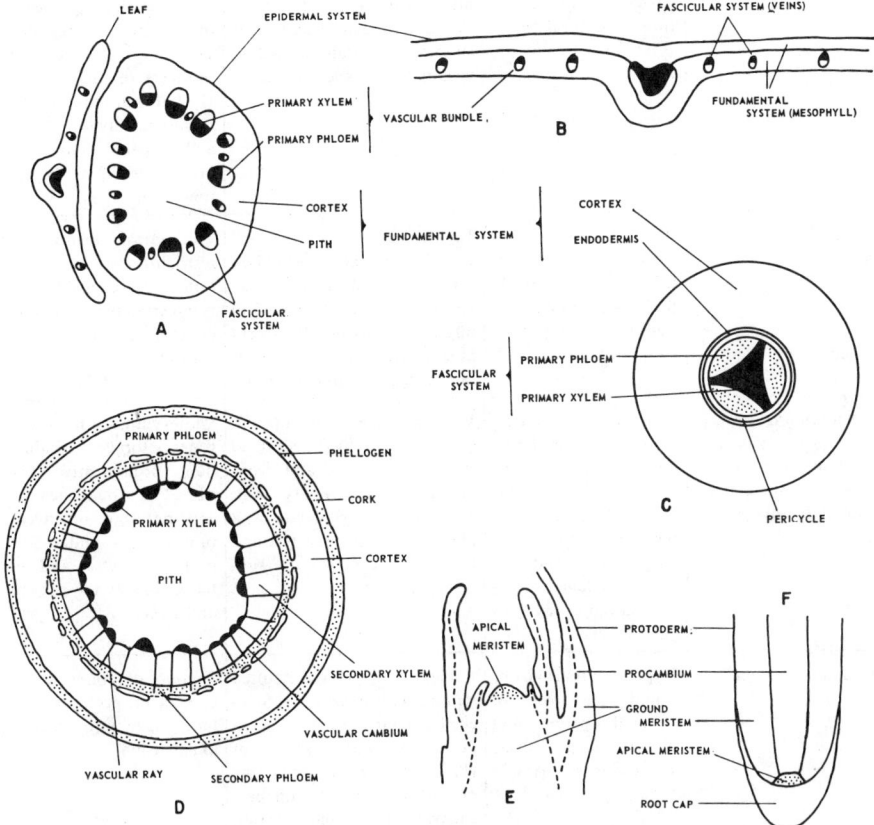

Fig. 1   Schematic representations of the anatomy of plant organs. A, transection of a young stem at level of mature primary tissues; B, section of a portion of a leaf showing one large midvein and smaller lateral veins; mesophyll differentiated into upper palisade and lower spongy parenchyma; C, root, at level of mature primary tissues; note presence of solid core of xylem; D, stem, in which some secondary growth has occurred; E, longitudinal view of a shoot tip and F, of a root, showing apical meristems and regions of primary meristematic tissues. See text for details.

Table I   Summary of Main Cell Types in Seed Plants

Cell Type	Origin	Location	Structural Characteristics	Functions
Apical Meristem	More or less direct, lineal descendents of cells of the embryo, except in case of adventitious roots and shoots	Apices of vegetative shoots, inflorescences, and developing flowers; in root, beneath inner edge of root cap	Polyhedral; primary wall thin or irregularly thickened; nucleus large, ovoid; cytoplasm vacuolated to varying degrees; mitochondria, plastids and storage products may be present.	Point of origin of the primary meristematic tissues, e.g. (protoderm, ground meristem and procambium) from which the "primary body" of shoot and root develop; in shoot the tissue from which foliar primordia originate
Vascular Cambium	From the procambium, from interfascicular areas in the vascular cylinder, and from reactivated parenchyma in cortex and phloem	*Lateral*, in stem and root, between secondary phloem and secondary xylem tissues	Two types of initials, viz.: *fusiform*, and *ray*. In both, cytoplasm highly vacuolated; storage products may be present; primary pit fields conspicuous	Produces secondary phloem and secondary xylem; results in growth in diameter of stems and roots
Phellogen (cork cambium)	In stems, phellogen arises in cortex, epidermis, or phloem	*Lateral*, in stem and root, between cork and subjacent phelloderm	Rectangular in shape, walls thin; cytoplasm vacuolated; may contain tannin	Produces phellem (cork) outwardly and phelloderm inwardly
Epidermal	Protoderm	Surface cells (including trichomes) of foliar organs, young stems, roots, and floral organs	Epidermal cells variable in shape; guard cells of stomata often reniform; outer walls overlaid by cuticle and often cutinized; protoplast active, vacuolated, plastids present	Mechanical protection; restriction of transpiration; aeration by means of stomata; photosynthesis; storage of water and products of metabolism; H_2O absorption in roots; retain capacity for growth and division; common point of origin of adventitious buds
Phellem (cork)	Phellogen or cork cambium	Peripheral regions of stems, roots and certain fruits; frequent in bud scales; often produced as a result of wounding	Tabular, compactly arranged without air spaces; secondary walls suberized or sometimes lignified and usually devoid of pits; protoplast usually absent at maturity; lumen often contains crystals, tannin, or resin	Mechanical protection; restriction of transpiration; cork tissue likewise is impervious to diffusion of gases except through lenticels
Parenchyma	Ground meristem, procambium, vascular cambium and, in the case of phelloderm, from the phellogen	Widely distributed throughout plant body; may constitute the dominant tissue in cortex, pith and mesophyll; present in secondary phloem and secondary xylem as vertical strands and as vascular rays	Vary in shape from approximately tetrakaidecahedral to stellately branched and cylindrical forms; primary wall with pit-fields, secondary wall present or absent; plastids and a wide range of ergastic substances present	Photosynthesis; food and water storage; conduction; prominently concerned in wound healing and the origin of adventitious roots and buds; protoplast active and retains capacity for growth and division
Collenchyma	Ground meristem	Hypodermal strands or cylinders in cortex of stems and petioles and ribs of foliage leaves; may occur in cortex of roots	± elongated, prismatic, often septate, with irregularly thickened primary walls, rich in pectin and with high % of H_2O; chloroplasts may be present; intergrades with parenchyma in form and structure	Support of young stems and leaves; protoplast retains capacity for growth and division
Sclereid	From protoderm (e.g., in developing seed coats), ground meristem, phellogen, vascular cambium and pro-	Diffused in cortex, phloem, pith and mesophyll as idioblasts or cell-clusters; in the laminae of some gen-	Polyhedral, columnar, or ± profusely branched; secondary wall massive, usually of lignified cellulose and often provided with ramiform pits;	Produces hard incompressible texture

Table I—(*Continued*)

Cell Type	Origin	Location	Structural Characteristics	Functions
	cambium; may arise by sclerosis of fully developed parenchyma cells	era terminal on the vein endings; may constitute large part of seed coat and pericarp of fruits; prominent in outer bark	protoplast may be retained at maturity	
Fiber	Protoderm, ground meristem, procambium and vascular cambium	Cortex; primary and secondary pholem, xylem; constitutes hypodermal strands or layers and the sclerenchymatous sheaths of vascular bundles in leaves; may occur as idioblast	Typical example of prosenchymatous cell, often attaining considerable length; secondary wall usually thick, often highly lignified; pits abundant or sparse; lumen continuous, ± occluded or septate; protoplast usually absent at maturity	Mechanical support
Tracheid	Procambium and vaslar cambium	Primary and secondary xylem, commonly formed in masses in callus tissues	Prosenchymatous without distinct end walls; imperforate at maturity; secondary wall of lignified cellulose, deposited as rings, spiral bands, transverse bars, or a reticulum, or is continuous except for bordered pits; protoplast usually absent at maturity	Conduction of water and certain solutes; mechanical support
Vessel member	Procambium and vascular cambium	Primary and secondary xylem of most dicotyledons; absent from xylem of gymnosperms except members of Gnetales; in certain monocotyledons, restricted to primary xylem of roots; occurring as ± extensive series of interconnected cells, each series termed a *vessel*	Prosenchymatous to cylindrical, with distinct perforated end walls; perforations either simple, scalariform or reticulate; secondary walls of lignified cellulose, with same range of pattern as in tracheid; protoplast absent at maturity	Conduction of water and certain solutes; and possibly as mechanical support
Sieve cell	Procambium and vascular cambium	Primary and secondary phloem of gymnosperms	Elongate in form with overlapping inclined or tapering ends; sieve areas numerous; each sieve area is a portion of primary wall traversed by connecting strands of cytoplasm enclosed by cylinders of callose	Conduction of organic solutes
Sieve-tube member	Procambium and vascular cambium	Primary and secondary phloem; usually in lateral connection with companion cells and occurring in vertical series, each series termed a *sieve-tube*	Prosenchymatous to cylindrical; end walls with highly specialized sieve areas termed sieve plates; lateral walls with less specialized sieve areas; protoplast at maturity enucleate	Conduction of organic solutes
Laticifers	Non-articulated type from initial cell which is devoid of septa, often ramifies throughout plant; articulated type from dissolution of end walls of a continuous series of cells	Cortex, phloem, xylem rays, pith, mesophyll; non-articulated type usually extensively branched tube; articulated type often interconnected by anastomoses thus forming a network	Contain latex and are multinucleate; in non-articulated type, may continue apical growth and branching throughout life of the plant	Probably excretory because of storage of nonfunctional metabolic products as rubber and resin

Adapted from "Practical Plant Anatomy" by A. S. Foster, D. Van Nostrand Company, Inc., New York.

classify aggregates of cells (tissues) on the basis of their morphologic characteristics and/or functions. Tissues have been classified as simple and complex. According to this concept a simple tissue is one in which the cells have had a common ontogenetic origin and possess a common function. Complex tissues are heterogeneous, consisting of several cell types, and may perform more than one function. Because so few tissues are simple and because of the occurrence of intergrading types of organization, the distinction is not well defined. A system based entirely upon function of the tissue, disregarding ontogeny, can be equally in error.

One system of tissue classification which merits attention because of its practicality and inviting simplicity and wide applicability is that proposed by Sachs (1875). For example, if a young DICOTYLEDONOUS stem is sectioned and the cut surface examined one can see that there are several well defined regions. On the outside is the epidermis, underlying this, the cortex, followed by a cylinder of vascular tissue. A PITH of varying proportions occupies the center of the stem. For these regions Sachs proposed the following classification system: for the outer layer or layers (if CORK is present) —Dermal tissue system; vascular tissues—Fascicular tissue system; cortex and pith—Fundamental tissue system (Fig. 1A). Each tissue system may be represented by various cell types. The fascicular system of the leaf ranges from one to several bundles of approximately equal size or a cylinder in the petiole, to a complex network of veins in the lamina or blade. In a root the fascicular system commonly forms a cylinder in the center of the organ. The roots of some species do have a parenchymatous central cylinder. The dermal system in the primary body generally is represented by an uniseriate layer of epidermis on leaves, stems, and roots. In leaves the photosynthetic mesophyll represents the fundamental tissue system. In most roots it is the cortex. The arrangement of these tissue systems does vary throughout vascular plants. For example, the stems of *Lycopodium* and most species of *Selaginella* possess a central cylinder of vascular tissue and hence lack a pith.

Apical meristems. Whatever the organization of tissues may be in vegetative organs, the patterns are established early near the tips of shoots and roots. Located at the terminus of the stem axis is a group of cells which are meristematic and are ultimately responsible for the formation of all cells of the shoot. These cells constitute the apical meristem. In many ferns and in some other lower vascular plants there is one conspicuously enlarged cell at the tip, termed the apical cell or apical initial. A group of apical initials characterizes other lower vascular plants and many gymnosperms. In angiosperms there are no individualized apical initials and furthermore there is stratification into an outer mantle of cells termed the tunica and an inner tissue termed the corpus. The term shoot apex, used in a very general way, is a convenient expression which may be used interchangeably with shoot apical meristem. Topographically, the shoot apex is often conical in shape (Fig. 1E). The immediate derivatives of the individualized apical initial(s) or uppermost cells are in turn meristematic, constituting the primary meristems, protoderm, ground meristem, and procambium. Through rapid cell division these meristems give rise to cells which will differentiate into

elements of the dermal, ground, and fascicular tissues, respectively (Fig. 1A). Leaves have their origin in cells on the flanks of the shoot apex and the three primary meristems also are present in the developing leaf.

An apical meristem is present also at the tip of a root but it differs from that of the shoot in being subterminal due to the presence of a root cap (Fig. 1F). Also the lateral appendages (lateral roots) arise endogenously from tissues some distance from the root tip. Structurally the root apical meristem may possess an enlarged initial cell (most ferns and some other lower vascular plants) or a distinction may exist between initials for the root cap and those for the main body of the root. In still other plants there may be a set of initials for root cap and future epidermis and variously stratified initials for the remainder of the root. As in the shoot, primary meristems continue meristematic growth proximal to the apical meristem.

Vascularization. A considerable volume of information has been accumulated concerning vascularization of shoots and roots. As can be inferred from the term shoot, the stem and leaves are closely related ontogenetically. This statement is particularly true in reference to vascularization. Procambium, the meristematic forerunner of vascular tissue, is present at the base of a newly initiated leaf. Procambium develops acropetally into leaves and is in continuity with the developing procambial strands of older leaves at lower levels in the stem. These connections are made in a definite pattern and the pattern is related to leaf arrangement on the stem. Differentiated procambial strands (vascular bundles) in the stem can be considered as simply the lower extensions of leaf vascular bundles or traces, so closely correlated is leaf and stem development.

A procambial strand consists of elongate meristematic cells. Associated with its occurrence in a young leaf, differentiation and maturation of procambial cells begins. Differentiation within procambial strands, as a shoot elongates, can be considered from two standpoints: longitudinal and radial differentiation. Under the first category it has been well established for seed plants that differentiation and maturation of the food conducting tissue, the phloem, proceeds acropetally and is continuous with the more mature phloem at lower levels in the stem. This is not the case for xylem —conducting tissue of water and inorganic salts. Xylem differentiation within a procambial strand is initiated near the base of a very young leaf and then proceeds acropetally into the leaf and basipetally into the stem where the xylem becomes joined with the older leaf traces.

In roots the course of longitudinal differentiation is somewhat easier to follow because the pattern is not disturbed by the presence of lateral organs near the apex. Differentiation of both xylem and phloem proceeds acropetally. The first mature sieve elements of the phloem invariably occur closer to the apex than do mature xylem elements.

The first formed xylem and phloem described above (for both shoots and roots) are termed protoxylem and protophloem and normally are formed during the initial growth in length of organs. Subsequent maturation of procambial cells occurs in a radial direction coordinated with the cessation of growth in length of an organ. In the stem of most seed plants radial maturation of xylem proceeds in a centrifugal direction resulting in the formation of metaxylem. Subsequent maturation of metaphloem is centripetal. The two components of the xylem constitute primary xylem, those of the phloem, primary phloem (Fig. 1A). In the root, maturation of metaxylem and metaphloem, although occurring on alternate radii, occurs in a centripetal direction (Fig. 1C).

Stem anatomy. It is convenient to describe STEM anatomy at the level of mature primary tissue (some time after growth in length has ceased). In many ancient vascular plants and in some living lower vascular plants there is a cylinder of primary xylem in the center of the stem ensheathed by a smooth cylinder of phloem or the phloem may be in the form of a cylinder of individualized strands. A stem which has a central core of xylem as just described is said to possess a protostele. By definition the stele comprises the xylem and phloem and pericycle (a layer or layers of parenchyma cells surrounding the phloem).

In the GYMNOSPERMAE and most dicotyledons vascular tissue is disposed in the form of a cylinder of individualized strands surrounding a pith (Fig. 1A). This orientation of vascular tissue constitutes a siphonostele. In many monocotyledons the vascular bundles appear to be scattered in disposition when the stem is viewed in transverse section. Commonly each vascular bundle consists of primary xylem toward the pith and primary phloem on the same radius but toward the periphery of the stem. This is a collateral bundle. In many herbaceous dicotyledons and monocotyledons each vascular bundle is capped by a conspicuous group of fibers which represents, developmentally, a part of the protophloem. The pericycle and endodermis, characteristically present in lower vascular plants, are variable in occurrence and morphology. Usually no pericycle, as a distinctive layer, is present. The endodermis may be represented physiologically by a limiting layer of cells which accumulate starch or possess other inclusions. Except for a few cases, endodermal layers with Casparian strips (see section on root) do not occur in the stems of seed plants. Vascular bundles are separated laterally by panels of parenchyma termed interfascicular regions. Although the bundles appear to be entirely separated from each other, they form an interconnected system when the extent of their longitudinal distribution is considered. Many dicotyledons and all gymnosperms develop into shrubs or trees and accordingly their stems increase in girth by a set of processes embraced by the expression—secondary growth. If secondary growth is characteristic of a plant, the undifferentiated procambial cells between the primary xylem and phloem divide longitudinally with the new walls being parallel with the surface of the stem. This layer of cells constitutes a lateral meristem, the vascular cambium, which produces new cells toward the center of the stem resulting in the formation of secondary xylem (wood); the contribution of cells toward the outside gives rise to secondary phloem. Sooner or later the vascular cambium assumes the shape of a complete cylinder by reactivation of parenchyma cells in the interfascicular regions (Fig. 1D). In latitudes where seasonal changes occur new bands of xylem and phloem are formed each growing season which results in the formation of visible growth rings or increments, readily visible in the xylem (wood). Ad-

ditional outer protective layers usually are formed by another lateral meristem, the cork cambium or phellogen which is initiated commonly in the outer cortical cells or in phloem parenchyma, or even in the epidermis itself. The cork cambium produces cork (compact layers of suberized cells) toward the outside and frequently some cells to the inside (phelloderm). The net result is the ultimate replacement of the epidermis (Fig. 1D). Subsequent cork cambia may be formed deeper in the cortex or in the secondary phloem with the result that the bark (all tissue outside the vascular cambium) is a highly complex tissue consisting of functional phloem near the vascular cambium and concentric cylinders (ringbark) or irregular scale-like patches of corky tissue (scalebark). The bark is sloughed away continually, being replaced by deeper lying cork layers.

Leaf anatomy. The term SHOOT implies ontogenetic and phylogenetic interrelationships between stem and leaf. For example, the petiole of the leaf often has a stem-like structure. Ontogenetically, the leaves arise from cells near the surface on the flanks of the shoot apical meristem. During early development the young leaf exhibits pronounced apical growth. Procambium develops acropetally into the primordium and the other primary meristems are in continuity with those of the stem axis. The leaf primordium, at first, is a peg-like structure, but very soon a lamina or blade is formed by a process termed marginal growth which results in the formation of mesophyll layers in addition to the upper and lower epidermal layers. Procambialization within the blade occurs early in development, resulting in the establishment of the basic pattern of venation. Depending on the species, apical growth ceases eventually and growth in length is intercalary. Initially the wave of differentiation, particularly of vascular tissue, is acropetal, but later maturation occurs basipetally. In mature leaves of the majority of dicotyledons, internal maturation results in the formation of a palisade layer(s) of columnar cells beneath the adaxial or upper epidermis. This tissue is well adapted for photosynthesis. A more loosely arranged tissue with a large air space system is present toward the abaxial side (lower side). In general, stomata are more numerous in the lower epidermis. Within vascular bundles of the major and minor veins the phloem is toward the abaxial side (Fig. 1B). Many monocotyledonous leaves (e.g. corn, wheat) have a generalized mesophyll made up of cells of rather uniform size. Many xerophytes form generous amounts of sclerenchyma in the leaves which render them coriaceous. In others, individualized cells may occur in the mesophyll such as laticiferous elements, secretory cells, and large ramified sclereids. In some plants leaf modification has resulted in the formation of spines or tendrils.

Root anatomy. There are several morphological differences between ROOTS and shoots, some of which have been mentioned earlier, e.g., the root has a cap over the apical meristem; roots are not segmented (a repetitive pattern such as nodes and internodes); lateral roots arise from deep lying tissue rather than from surface cells as do leaf primordia. Although not all vascular plants have roots, some primitive plants have underground stems which serve in anchorage and water absorption. Roots differ from the stem of the majority of seed plants in the presence of a central core of procambium rather than a cylinder of individualized pro-cambial strands which are related to leaves. Radial maturation within the procambial column results in the early blocking out of a distinctive pattern of vascular tissue in which strands of primary phloem and xylem lie on alternate radii (Fig. 1C). Continued centripetal maturation results in the formation of a solid central core of primary xylem. In some plants xylem maturation may not continue until a solid core is formed. In these cases a parenchymatous pith-like core is the result. If a solid core of primary xylem is characteristic it is usually in the form of a fluted or ribbed column. Primary phloem is present between the "arms" or ridges of xylem. Surrounding the vascular cylinder are two layers of parenchyma cells, the pericycle and endodermis. Cells of the latter have specialized bands of lignified or suberized cellulose (Casparian strips) on their end walls and radial walls which restricts water movement through cell walls. Water must move through protoplasts of endodermal cells and this appears to be a regulatory mechanism. In the primary state, most roots possess a rather broad cortex composed of parenchymatous cells. A limiting barrier, the epidermis, forms the outer layer. In a restricted region near the root tip certain epidermal cells are modified as water absorbing cells termed root hairs. Many dicotyledonous roots undergo secondary growth to varying degrees. A vascular cambium originates from arrested procambial cells lying between the primary xylem core and the strands of phloem. The pericycle plays an important role in secondary growth. The vascular cambium is made a complete cylinder by reactivation of pericycle cells opposite the xylem arms. Furthermore, the pericycle cells opposite the xylem become reactivated into a generalized meristematic tissue. With the onset of secondary growth the epidermis, cortex (including the endodermis) are sloughed off. Eventually a cork cambium is formed from the outer layers of the proliferated pericycle. Older roots resemble stems in their anatomy and often it is difficult to distinguish one from the other.

Floral anatomy. Morphologically, the flower is essentially a much shortened shoot consisting of an axis (the receptacle) and closely placed floral appendages. Sepals and petals exhibit many of the same anatomical features characteristic of leaves. Aside from color and form, the difference between sepals, petals, and leaves is often one of degree. Frequently the number of traces (vascular bundles) which establishes continuity with the stem is the same for leaves, carpels, and sepals. Petals and stamens usually have one vascular bundle at the level of attachment to the stem. The gynoecium may consist of a single carpel (single pistil, e.g., peach, almond, bean) or of several united carpels (compound pistil, e.g. apple, orange, lily). There are many concepts concerning the phylogenetic origin of the carpel. One commonly held view interprets the carpel as a modified leaf, that is, leaves and carpels are homologous (i.e., they have a common evolutionary ancestry).

ERNEST M. GIFFORD, JR.

References

Esau, K., "Anatomy of Seed Plants" 2nd ed., New York, Wiley, 1965.

Foster, A. S., "Practical Plant Anatomy" 2nd ed., New York, van Nostrand, 1949.

PLANT COLORATION

Nature affords no wider variety of colors than those found in the world of plants. The prevailing color of plants is the green of chlorophyll, often followed, in some parts of the world, by the yellow, orange and red of autumn foliage. The yellow colors of leaves are due principally to the group of fat-soluble plastid pigments known as the carotenoids.

The much more various colors of flowers and fruits range from dead white, through ivory, yellow, orange, red and finally violet and blue. Combinations of these can give rise to nearly the whole of the visible spectrum of color, and, in certain combinations and concentrations, the pigments responsible for the primary colors can produce brown and even what may appear to the eye as black colorations. Carotenoids, which will not be considered in detail in this section, are often involved in the pigmentation of yellow flowers and fruits; however, certain non-carotenoid yellow and orange pigments are often found to be the principal cause of yellow petal colors.

White flower petals are more often pale ivory than true white. The latter color is not common, and is the result of a total lack of pigmentation. Most white flowers contain one or more representatives of a class of pigments known as flavones (I). These compounds, usually bearing hydroxyl groups at three or more of the positions 3', 4', 3, 5, 6, 7, 8 (and most commonly at 3', 4', 5 and 7), are themselves pale yellow in color and usually impart little yellow color to the plant parts in which they occur. It is believed that in some cases their combination with metal ions in the cell sap may lead to flavone-metal complexes sufficiently deeper in color than the tree flavones to be capable of imparting yellow colors to what would otherwise be white petals. The presence of flavone derivatives in white petals can usually be demonstrated by treating the petal with an alkaline solution: the deeply yellow salt of the

I

II

III

IV

flavone can be recognized by the change in color of the petal.

Yellow flowers and fruits are nearly always carotenoid-pigmented. However, certain flowers (*Coreopsis, Antirrhinum, Oxalis* and some others) contain pigments belonging to the chalcone (II) and benzalcoumaranone (or aurone) (III) classes. The tri- and tetrahydroxy-II and -III are deep yellow to orange compounds and are capable of imparting lively yellow colors to flower petals. Exposure of a yellow *Coreopsis* petal to alkali causes the color to change to the orange-red color of the salts of chalcone and aurone pigments present in the flower.

Orange flowers may be pigmented with chalcones or aurones, but in most cases owe their colors to mixtures of yellow carotenoid and red anthocyanin pigments. The orange of snapdragons is the result of the presence of aurone and anthocyanin pigments, but this is an unusual example. Orange autumn leaves are pigmented with carotenoid and anthocyanin mixtures.

Red colors in flowers, fruits and leaves are nearly always due to a group of pyrilium salts called anthocyanins (IV), and pigments of this class are also responsible for the *blue* shades of color as well. By a surprisingly limited number of structural alterations, coupled with alteration of the cell-sap environment brought about by pH changes, the presence of metallic ions and other constituents, the extraordinary gamut of colors ranging from pale salmon, through pink, scarlet, lavender, crimson, violet, purple and finally to blue and blue-black are produced.

The fundamental structure IV appears in most of the natural anthocyanins in three modifications; all of them bear hydroxyl groups at 3, 5 and 7, and differ only in the presence of additional hydroxyl groups at 4'; 3', 4'; or 3', 4', 5'. These three anthocyanidins are called respectively, pelargonidin, cyanidin and delphinidin. In nature, they bear sugars, linked to form β-glycosides, at 3 or 5 or 3, 5. The glycosylated anthocyanidins are the anthocyanins. Pelargonidin glycosides are characteristic of scarlet, those of cyanidin of crimson to magenta, those of delphinidin of violet to blue flowers and fruits. In some cases one or more (usually 3' or 3', 5') of the hydroxyl groups may be methylated. Thus, a rather limited number of rather superficial structural changes can give rise to an almost endless number of subtle color changes in the mature plant tissues in which these pigments are found.

Most of the pigments mentioned in the above discussion occur, as do the anthocyanins, not as the free phenolic substances, but combined with sugars in the form of glyosides. The water solubility thus imparted to what would otherwise be rather insoluble compounds permits their occurrence in solution in the aqueous contents of the cell.

Recent studies have disclosed at least a part of the pathway by which compounds of the above classes (known collectively as flavonoid compounds) are formed in the plant. The basic carbon skeleton is derived by combination of the C_6–C_3 unit shown on the right in the formulas with three two-carbon ("acetate") fragments to give a precursor which, by eventual cyclization and alteration in the oxidation level of the C_3 portion of the molecule, leads to the final compounds. Little is known about the exact sequence of events by which two such compounds as quercetin (I, with OH at 3, 3', 4', 5, 7) and cyanidin (IV, with OH at 3, 3', 4', 5, 7) are linked. The relationship between quercetin and cyanidin is formally that of a single two-electron reduction of the flavone to the anthocyanidin, but existing evidence does not support the view that this process actually takes place in the cell.

T. A. GEISSMAN

PLANT CYTOTAXONOMY

Since the introduction into biology of the principle of Organic EVOLUTION, botanical taxonomy has sought to develop a system of classification that would best express GENETIC and phylogenetic relationships among plants. Although the principal taxonomic method for estimating genetic relationships remains that of comparative external morphology, data from other sources are becoming increasingly important in classification. Genetics and cytology, for example, have contributed greatly to our knowledge of how evolutionary changes occur, and today, in plants that have complex patterns of evolution, an adequate natural classification cannot be made without information from these fields. That branch of taxonomy that uses cytology as an aid in understanding natural relationships among species is cytotaxonomy.

The cytological structures of prime interest to the cytotaxonomist are the chromosomes. Each plant species has a characteristic karyotype which is determined by the number, size and shape of the chromosomes in its haploid complement, or genome. To study somatic karyotypes, one may cut meristematic tissue, such as a root tip, into thin sections or macerate it and squash it on a microscope slide. Exposing the cells to paradichlorobenzene or colchicine before killing, may cause long, tangled chromosomes to contract; such treatment, combined with techniques for flattening whole cells by squashing, greatly increases the ease and accuracy with which chromosome studies can be made.

Also important in cytotaxonomy is the behavior of CHROMOSOMES during MEIOSIS, the reductional cell divisions that precede spore formation. Meiosis is best studied in spore mother cells squashed from anthers or from the sporangia of ferns and mosses. The exact haploid number of chromosomes can be verified here and checked with the diploid number seen in somatic mitosis. More importantly, the synapsis of chromosomes early in meiosis allows one to observe similarities and differences in chromosome structure between related genomes. Irregularities in the normal two-by-two pairing of chromosomes and in their numerically equal segregation during meiotic anaphase are characteristic of interspecific hybrids or plants that are heterozygous for differences in chromosome structure. Hence, meiosis in wild plants or in experimentally cultivated hybrids can help in understanding natural genetic relationships. It can give a measure of the evolutionary differentiation of the chromosomes of related species. Although cytological characteristics are genetically more basic than features of external morphology, they cannot be the sole criteria for classification. Combined with traditional taxonomic methods, they clarify confusing evolutionary patterns and make classification more accurate.

Normally, all the members of a species are able to

interbreed freely, and all possess a similar karyotype. A difference between individuals in chromosome number or shape may thus imply that a barrier to free gene exchange exists. In the tarweed genus *Holocarpha*, two closely similar species were thought to be identical until it was discovered that one had four pairs of chromosomes and the other had six pairs; hybrids between these species are sterile. Even at the generic level the karyotype may be helpful in classification. The genera *Agave* and *Yucca* were once placed in different families because of a basic difference in floral structure. However, they have an identical and highly unusual karyotype, consisting of 10 large chromosomes and 50 much smaller ones; and this, together with other similarities, has caused their reclassification into a single family.

A major factor in plant evolution that is best approached by cytotaxonomic means is polyploidy. Evolution by multiplication of the whole genome produces related species whose chromosome numbers are multiples of one or more base numbers. Allopolyploid species, which arise by chromosome doubling in interspecific hybrids, are often difficult to recognize taxonomically. Being of hybrid origin, each is nearly intermediate between its two parental species, and the three together may appear to comprise a single variable species. Cytologically, however, an allopolyploid can be recognized as having the sum of the chromosome numbers of its parents. The marsh grass, *Spartina townsendii*, is polyploid with usually 122 chromosomes, while its parental species have 60 and 62 respectively. Although of hybrid origin, allopolyploids fully merit taxonomic recognition as species because they are genetically isolated yet are fertile and self-perpetuating.

More than simply indicating relationships, cytotaxonomy can often chart the direction of evolution in a group of species. Polyploidy is, with few exceptions, irreversible; when both diploid and polyploid species occur in a genus, the former probably antedate the latter. Where species evolution has proceeded by aneuploidy (the addition or subtraction of individual chromosomes), the direction of change may also be discernible. In *Crepis* the most primitive species have six pairs of chromosomes and evolution has been by stepwise reduction to species with five, four and three pairs.

Autopolyploidy, the doubling of chromosome number within a species, may produce plants that are genetically isolated but externally indistinguishable, posing a serious taxonomic problem. Most taxonomists hesitate to recognize such entities as separate species unless some consistent morphological difference can be found; sometimes the special category of "aggregate species" is used. Even more difficult are polyploid complexes, in which autopolyploidy and allopolyploidy combine to give a swarm of intergrading forms with varying chromosome numbers, and agamic complexes, which contain polyploids that are partially or wholly asexual. In each case, cytotaxonomy may help in understanding the source of the problem and in choosing the best criteria for classification.

Cytotaxonomic methods find increasing application in plant groups other than angiosperms. In lower plants with simpler body form, morphological differences are largely quantitative, and qualitative karyotypic differences are taxonomically very useful. POLY-PLOID series have been found in algae, and in mosses there is cytological variation similar to that in higher plants. Many species of pteridophytes are known cytologically, and some are remarkable for their high chromosome numbers. Polyploidy is absent in most gymnosperms, which are chromosomally less variable than other major groups of plants.

K. L. CHAMBERS

References

Clausen, J., "Stages in the Evolution of Plant Species," Ithaca, Cornell Univ. Press, 1951.
Davis, P. H. and V. H. Heywood, "Principles of Angiosperm Taxonomy," Edinburgh and London, Oliver & Boyd, 1963.

PLANT EMBRYOLOGY

Brief Historical Survey

With the rebirth of scientific botany during the earlier decades of the Nineteenth Century, and especially as a result of the emphasis placed by Schleiden, Naegeli, Von Mohl and others on the importance of the study of development in biology, it was only a matter of time till some attention would be given to the important phenomena of sexuality in plants, the process of fertilization, especially in the Flowering Plants, and the formation of the embryo. Thus, in 1847, Amici demonstrated in *Orchis* that a *germinal vesicle* was present in the ovule before the arrival of the pollen-tube, and that the vesicle in due course gave rise to the embryo. When, two years later, Wilhelm Hofmeister published an estensively illustrated monograph on the formation of embryos in phanerogams, the scientific study of plant embryology may be said to have been securely founded. In subsequent publications, which later appeared in English as the *Higher Cryptogamia* (1862), Hofmeister described and illustrated, with meticulous clarity and attention to detail, the embryological development, from the fertilized ovum onwards, in selected archegoniate plants. In 1870, Hanstein made detailed observations on the sequence of cell divisions in embryos of flowering plants, selecting *Capsella* and *Alisma* as exemplifying typical developments in dicotyledons and monocotyledons respectively. In 1879, Treub described the embryos in a number of orchids, with their unusual suspensor haustoria, and later contributed some remarkable and beautiful accounts of the varied embryological developments in different species of *Lycopodium*. In the same general field of work, reference must also be made to the very valuable contributions made by Bruchmann on embryogenesis in *Lycopodium, Selaginella, Ophioglossum* and *Botrychum*; and by Campbell (*Mosses and Ferns,* 1918) on primitive ferns. Since the beginning of the present century, a great many descriptive accounts of embryo development in gymnosperms and flowering plants have been published by investigators all over the world, Americans, French and Indians having excelled in this work. More recently, there has been a movement towards experimental investigations, the techniques of tissue culture and, more generally, of experimental morphology, having been introduced with advantage into embryological work. In this brief survey, the algae should not be forgotten; for, as will be seen, although the ova of algae are not encapsulated, and the group would, accordingly, be excluded by some investigators from the Embryophyta, the evident fact is that in some algae, e.g. *Fucus* and allied genera, there are embryological developments which in many respects correspond closely with those found in archegoniate and flowering plants. Lastly, it may be noted that botanical work on problems of embryogenesis has lagged considerably behind the zoological achievement. At the same time, the too-ready application of zoological concepts to plant embryos has evident dangers, since, whereas the animal embryo soon becomes definitively determined in its main organogenic developments, a majority of plants develop their axis and lateral members as a result of "continued embryogeny" at the shoot apical meristem.

Ovum and Zygote

Knowledge of the *specific organization* of the ovum and zygote in representative plant species is still quite inadequate. That such organization exists at the protoplasmic level in the zygote and at other levels during ontogenesis may be inferred from the orderly way in which embryogenesis normally proceeds. When the ovum is contained in an archegonium or embryo-sac it is a reasonable supposition, for which there is supporting evidence, that in addition to the genetically-determined specific properties of its protoplast, factors in the maternal environment, e.g. the position of the ovum, gradients of growth-regulating substances, nutrients, pressure of adjacent tissue, etc., will also affect its organization and subsequent development. The very early establishment of polarity (see below) almost invariably stands in some characteristic relationship to the enveloping gametophyte tissue. In free-floating eggs, like those of the Fucaceae, for example, there is also evidence of organization in the ovum and zygote; but in them it appears that external factors, such as gravity, light, etc., may play a greater part in some of the earlier embryogenic developments. In the ova and zygotes of some Bryophyta and Gymnosperms, visual evidence of differentiation in the protoplast is afforded by different degrees of vacuolation, concentration of granular matter, etc. at what will become the apical and basal poles of the young embryo.

A General Survey of Plant Embryos

In Figs. 1–8, stages in the embryology of an alga, a bryophyte, a lycopod, a fern, a gymnosperm, a dicotyledon and a monocotyledon are illustrated. With the exception of the fern in which the embryo passes through an approximately spherical phase in its development, of which more later, and of the gymnosperm in which there is an initial phase of free nuclear division, it will be noted that all of these embryological developments have evident features in common. In all of them, *polarity*, which may be regarded as the inception, indeed, the very foundation of differentiation, is established at a very early stage in the development of the zygote, if, in fact, it was not already determined in the still unfertilized ovum. A concomitant feature is that the first partition wall, which divides the zygote into two equal or unequal cells, is almost invariably laid down at right-angles to the polar axis. For, as the subsequent developments show, the very young embryo,

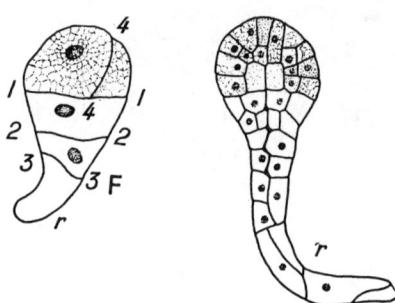

Fig. 1 *Fucus vesiculosus:* Young embryos showing the order in which the new cell walls were laid down.

whether it be evidently elongated or still spherical, has now undergone the first stage in differentiation, one of the cells, described as the distal cell, in due course giving rise to the apex of the nascent axis, while the other, the proximal cell, will give rise to the basal organs or tissues of the embryo, i.e. a suspensor or a foot, or to both, in different instances. At this early stage, too, evidence of differentiation can often be seen in the way in which the zygote divides: in some species it divides into two equivalent daughter cells, but in many the division is unequal, a typical result being that the distal cell is relatively small and densely protoplasmic, whereas the basal cell is larger and soon shows evident vacuolation. It is thus reasonable to infer that, at this initial stage, two rather different metabolic systems have been established at the embryo poles.

"The primitive spindle." In relation to these zygotic developments and the continued elongation of the embryo, with concomitant formation of further transverse walls, the young embryo typically becomes *filamentous*. Bower, in 1922, described this stage as "the primitive spindle," and pointed to it as a fundamental feature in the embryology of plants. To botanists interested in the evolution of the higher plants from a filamentous algal ancestry, the significance of this observation seemed evident. However, caution should be exercised in not pushing the phylogenetic inferences of the concept too far. In such developments the physico-chemical aspects must also be considered. The embryological development in leptosporangiate ferns, e.g. *Osmunda, Dryopteris* or *Adiantum* is not, in fact, in accord with the concept (see below), though

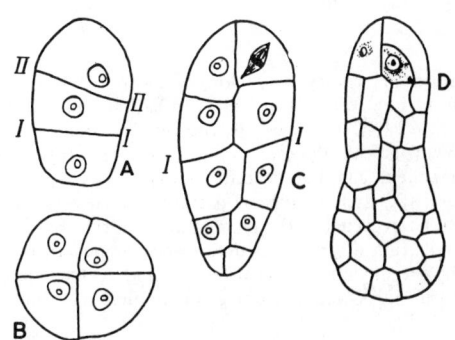

Fig. 2 *Fossombronia longiseta:* Young embryos showing order of cell divisions during development.

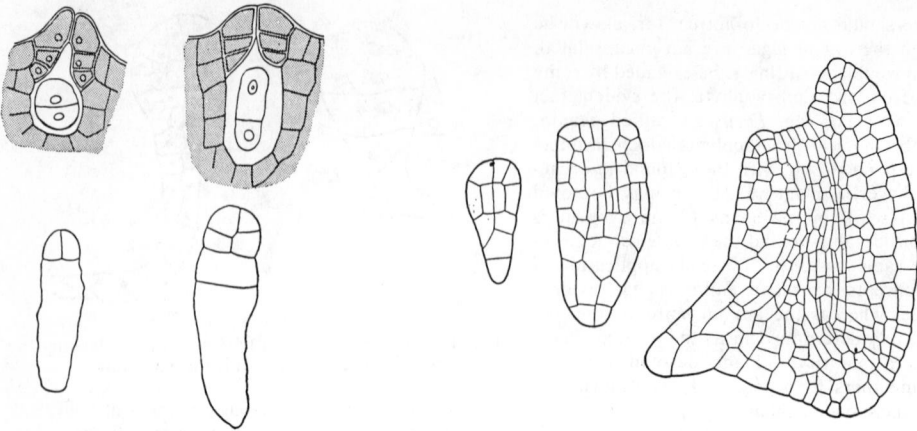

Fig. 3   Stages in the embryonic development of *Selaginella spinulosa* (× 300).

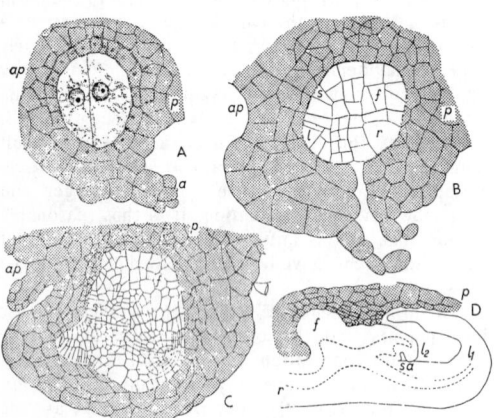

Fig. 4*   Embryogeny in leptosporangiate ferns. A, Pteris serratula. First division of zygote, as seen in l.s. of the archegonium (*a*); the smaller segment is towards the prothallus apex (*ap*); *p*, prothallus. B, C, Adiantum concinnum. B, Post-octant stage, as seen in a longitudinal median section of the prothallus and archegonium; *ap*, direction of apex of prothallus; *s*, shoot apex; *l*, first leaf; *f*, foot; *r*, first root. C, Older embryo, still enclosed in the prothallus; lettering as before. D, Onoclea sensibilis. Fully formed embryo, with first and second leaves, l_1 and l_2; *sa*, shoot apex. (A, B, × 250; C, × 135; after Atkinson; D, × 48; after Campbell.)

filamentous embryos are known in some of the eusporangiate ferns. What one can say at this point is that, in early embryogenesis, the vertical component of growth is considerably greater than the transverse component, and that, in accord with the laws of physical chemistry, a characteristic pattern of transverse walls is to be expected in the filamentous embryo. Since the distribution of growth is determined by genetical factors, the phyletic aspect of embryology cannot be excluded: it is, indeed, a legitimate field of enquiry,

*Figs. 4, 8, 9 and 12 from "Embryogenesis in Plants," C. W. Wardlaw, New York, John Wiley & Sons.

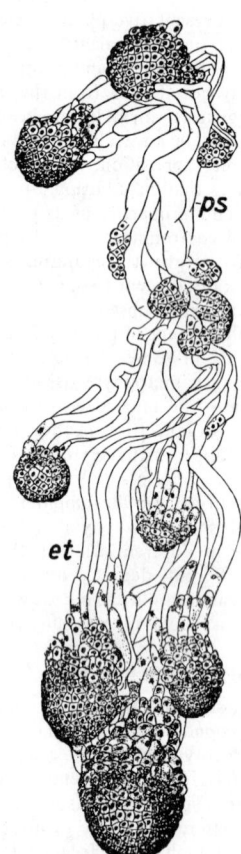

Fig. 5   *Torreya californica*. Complex embryo system dissected from a fully enlarged ovule; *ps*, prosuspensor cells; *et*, embryo tubes, from which the suspensor cells originate (× 37).

though so far it has not, with some exceptions, been of great use in the study of plant evolution.

The histological pattern. All young growing embryos, whether they be filamentous, spherical or ellipsoidal,

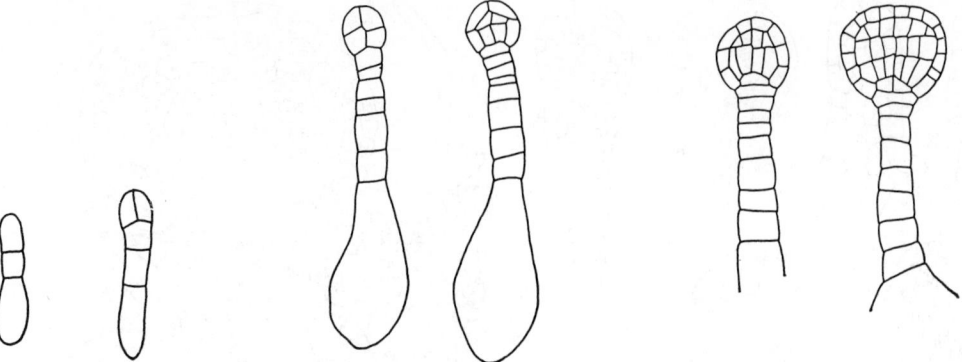

Fig. 6   Early stages in the embryogeny of a flowering plant, Shepherd's Purse, *Capsella bursa-pastoris.*

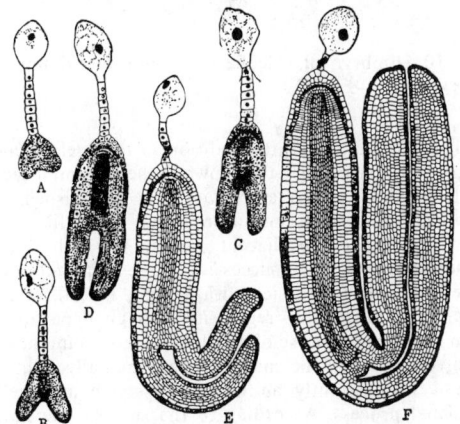

Fig. 7   Later stages in the embryogeny of *Capsella.* (After Schaffner, by permission from "An Introduction to the Embryology of Angiosperms" by P. Maheshwari; copyright 1950, McGraw-Hill Book Company Inc.)

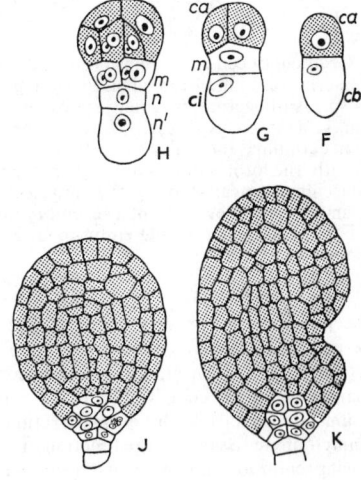

Fig. 8   *Luzula forsteri.*

are characterised by a sequence of definite and distinctive cellular patterns, i.e. almost as if some regular plan of development was being followed. This regularity and fidelity of histological development for the species was understandably seized upon by comparative morphologists in phyletic studies. However, as D'Arcy Thompson expounded in his remarkable book *On Growth and Form* (1917), these histological patterns in embryos and other growing tissues appear to exemplify in a high degree the law propounded by Errera in 1886: that enlarging cells, which can be regarded as physical systems, tend to divide into equal parts by walls of minimal area. This idea can now be extended by saying that the manner of division, and the position of the new partition wall, will also be affected by other factors in the developing organism. And here we should bear in mind that the embryo develops as a whole and, like any physical system, constantly tends towards a state of equilibrium. The histological pattern in many filamentous embryos, e.g. in bryophytes, is very much in accord with Errera's Law; and in the young spheroidal embryo of lepto-

sporangiate ferns the sequence of formation, and the positions, of the partition walls coincide in a most remarkable manner with the pattern which we should expect to find in an idealized physical system.

Gradient of cell size. With the establishment of polarity, and the distinction of apex and base, the filamentous embryo in all major groups soon affords evidence of metabolic, or chemical, differentiation. From the outset of the zygotic development, the apical pole becomes the principal seat, or locus, of the protein synthesis which we associate with embryonic and meristematic cells, whereas the basal pole becomes a region of conspicuous enlargement, vacuolation, and formation of parenchyma cells, from which it is inferred that the accumulation of osmotically active substances, including soluble carbohydrates is the characteristic metabolic activity. On further development, the cells lying between the poles often show a characteristic gradient in size, Fig. 9. Such observations suggest that, in the polarized embryo, various physiological gradients probably come into existence at an early stage. Actually, little biochemical or physiological work has

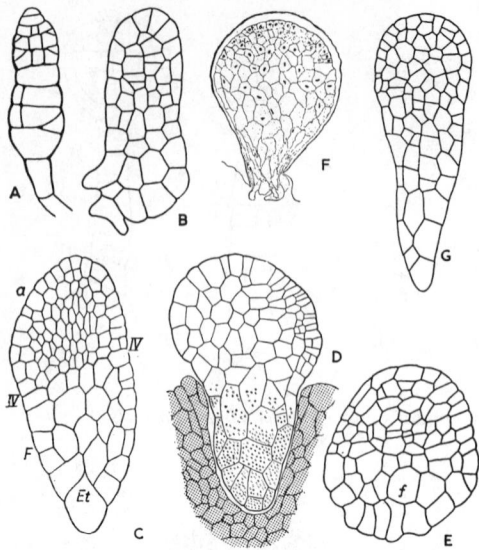

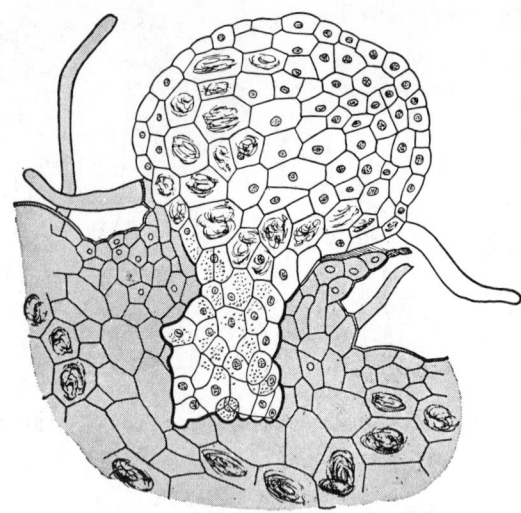

Fig. 10   Embryo of a leafless, rootless pteridophyte, *Tmesipteris*, (× c. 100).

Fig. 9   Gradient of cell size in embryonic development. A, Delesseria ruscifolia, red alga, germling. B, Notothylas sp., Anthocerotales, young sporophyte. C, Lycopodium selago; *et*, suspensor, *f*, foot, *a*, apex. D, Lycopodium cernuum, the enlarged cells of the foot in contact with the prothallus contain starch grains. E, Osmunda cinnamomea, showing the large cells of the foot, *f*, and the smaller cells of the embryonic distal region. F, Ginkgo biloba, young embryo. G, Zea mays, young embryo.

so far been carried out on very young plant embryos. In some embryos, e.g. *Lycopodium, Psilotum, Anthoceros, Sphagnum,* the basal, foot, region consists of large parenchymatous cells which make an almost haustorium-like contact with, or penetration of, the fleshy gametophyte tissue; i.e. the function of supplying the growing embryo with nutrients by active uptake from the gametophyte tissue pertains to the foot. In *Lycopodium cernuum* a gradient of starch grain density has been noted in the foot cells of some specimens. These developments of the foot could, however, also be interpreted as being a direct response of cells exposed to abundant soluble carbohydrates. In some embryos, the *suspensor* (*see* below), which initially constitutes the basal pole of the embryo, elongates conspicuously and thrusts the distal *embryonic cell* deeply into the nutritive tissues of the gametophyte. In flowering plants the embryonic cell is thrust more deeply into the embryo-sac. Remarkable examples of suspensor development, with dissolution of the adjacent gametophyte tissues, are known in some species of *Selaginella*, Fig. 3; but the phenomenon reaches its highest expression in the greatly elongated, compressed and contorted suspensors in some gymnosperms, Fig. 5. In some monocotyledons and dicotyledons, the basal cell of the suspensor may become conspicuously enlarged, as in the classic textbook illustrations of *Capsella*, or several suspensor cells may become conspicuously enlarged, Figs. 6, 7, whereas, in other species, no such enlargement takes place. The functional significance of these enlarged suspensors is not known. Curious developments of the suspensor, in-

cluding the multinucleate condition in the greatly enlarged cells, have long been known in the Leguminosae, while in species such as *Sedum acre* the suspensor may grow out basally into a branched, hypha-like organ which invades the adjacent tissues of the ovule.

Organogenesis. The successive stages in embryogenesis usually take place with a high, indeed, often remarkable degree of regularity, though departures from the normal course of development are sometimes observed and may be induced experimentally. Ontogenesis is apparently an inherently stable and well regulated process. According to the species, the first evidence of organ formation can usually be observed at a particular stage. In pteridophytes, such as *Lycopodium, Selaginella* and ferns, the organogenic activities include the organization of a shoot apical meristem, the inception of an associated first leaf, or leaves, the formation of a root apical initial cell and the organization of a root apical meristem, and the development of a parenchymatous foot. One or other of these organs may be relatively advanced or delayed in its inception and development, e.g. in some eusporangiate ferns the shoot apex is belatedly formed whereas root formation is relatively percocious and conspicuous. In different species of *Lycopodium* the foot may become very conspicuously enlarged, but sometimes, as in *L. cernuum*, it is the shoot region that becomes distended into a tuber-like body—the so-called "protocorm"; and in different species of *Selaginella* there may be notable differences in the juxtaposition of the foot and the suspensor. The ferns, e.g. *Dryopteris, Adiantum,* etc., are remarkable in that, presumably as a result of chemical differentiation, the four quadrants of the spherical embryo give rise, though not with absolute geometrical exactitude, to the shoot apex, the first leaf, the root, and the foot. These several nascent organs always originate in the same positions relative to the apex of the prothallus and to the enveloping archegonium. An auxin gradient in the gametophyte (or prothallus) in which the archegonium venter is completely immersed, and effects

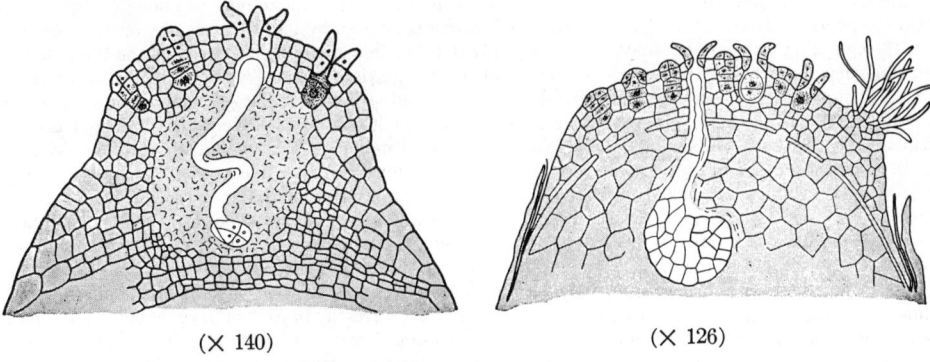

(× 140)   (× 126)

Fig. 11   Early stages in the embryogeny of *Selaginella galeottei* and *S. poulteri*.

proceeding from the archegonial wall and neck, appear to be among the factors to which these developments can be attributed. Such facts support the view that a patternised distribution of metabolic substances precedes, underlies and determines the visible organogenic and histogenic developments in embryos.

In the seed plants, although there may be considerable histological diversity in the earlier developments, the later stages of embryo formation are generally comparable in that the spheroidal, ellipsoidal or club-shaped mass of embryonic tissue becomes organized as an apical meristem and leaf primordia, a short axis (the hypocotyl) and the primary root. The distal *embryonic cell* in the young filamentous embryo in dicotyledons, for example, first enlarges and becomes a multicellular spherical structure; it then passes through "heart-shaped" and "torpedo" stages, in which the nascent cotyledons and apex and first root are being initiated; and later, the fully-formed embryo, a seedling in minature, fills the mature seed (together with such endosperm as may be present). The early developments in monocotyledons are closely comparable but, later, only one terminal, or seemingly terminal, cotyledon is formed. Some gymnosperms, on the other hand, may have many cotyledons, e.g. up to sixteen in some species of *Pinus*. In terms of organogenic developments, the Gramineae have the most complex embryos among plants (Figs. 6, 7, 8).

Not all flowering plant seeds contain, at maturity, fully developed embryos. In the seeds of many orchids, for example, the embryo may be a small gemma-like structure. In *Eranthis hiemalis*, the embryo remains at a club-like stage, and this has made possible some interesting experiments *in situ*, e.g. the inception of more than two cotyledons by chemical injections into the ovule.

The Nutrition of the Embryo

The algae apart, the entire development of the encapsulated embryo depends primarily on supplies of essential nutrients from the adjacent gametophyte and other tissues. The biological problem is the nutritional requirements of a plant in minature whilst coming into being.

In the flowering plants the formation and utilization of the endosperm are, of course, of major importance. By excising embryos at different stages and growing

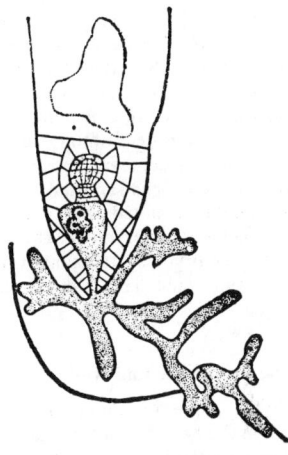

Fig. 12   Sedum acre. Apical portion of ovule, showing the embryo and suspensor; the basal region of the latter has grown out into long, branching haustorial processes.

them in aseptic media of known composition, it has been possible to obtain some information, though it is still scanty, regarding their nutritional requirements. Broadly speaking, minerals, nitrogen-containing substances, soluble carbohydrates and vitamins, hormonal substances, etc., are all required, the hormones being important in regulating the movement of supplies to the embryo. The morphological development of an embryo is undoubtedly affected in important ways by nutritional factors in its environment. The younger the embryo the more difficult is it to grow in culture. Embryos of *Capsella* at the globular stage have only recently been successfully cultured; those at the "heart-shaped" stage and older embryos can be grown with little difficulty in relatively simple media. Important ingredients of artificial media for embryo culture include sucrose and other sugars, caseinhydrolysate, glutamine and some other amino-acids, ammonium nitrate, various trace elements, the hormonal substances of coconut milk, etc. Not all sugars are equally effective sources of carbohydrate; some amino-acids limit or inhibit embryo growth; complete amino-acid mixtures are not necessarily more effective than individual amino-acids and may be less so; and ill-balanced media, especially in respect of the concentrations of growth-regulating substances, may result in poor or abnormal developments. Experiments on embryo culture have shown that some of the ingredients of coconut milk, indoleacetic acid, gibberellic acid, kinetin, adenine, etc., are very important in embryo development. The nutrition of embryos is, however, a very big subject and can only be touched on here. In some hybrids, in which the endosperm is apparently incompatible with the embryo, the latter may fail to develop fully or, if lethal factors are involved, it may become abortive at some stage. By excising and culturing such immature embryos, however, it has been possible to obtain plants of considerable scientific interest and practical value which otherwise would have been lost. Some important cytogenetical and physiological investigations of the relationships of embryo and endosperm have been carried out in recent years but the information is too complex to be considered here.

Studies of embryos in culture and of various hybrid embryos have shown that the "normal" course of development which, as we have seen, is usually followed with a high degree of constancy, is not the only possible course. "Modified ontogenies," which can usually be ascribed to departures from "normal" nutrition, have frequently been reported.

Lastly, it should be noted that many of the very curious phenomena of embryology, the suspensor, the endosperm, the embryo-sac tapetum, various haustorical developments, etc., have almost certainly a nutritional aspect.

Exoscopic and Endoscopic Embryos

In bryophytes, *Psilotum, Tmesipteris*, etc., the first partition wall in the zygote is at right-angles to the neck of the archegonium. The apex of the embryo grows towards the archegonium neck and the basal region is in contiguity with the gametophyte tissues from which nutrients will be drawn. Such embryos are said to be *exoscopic*. By contrast, in *Lycopodium, Selaginella*, some ferns, and seed plants, the apical pole of the embryo is directed inwards, i.e. towards the base of the archegonium. At the first division of the zygote, by a transverse wall, the inner daughter cell is recognised as the *embryonic cell* and the outer one as the *suspensor*, and such embryos are said to be *endoscopic*. On its further development, the enlarging embryo thus grows into the tissue of the gametopyte, subsequently curving upwards in response to the stimulus of gravity. Fern genera are known in which some of the species have a suspensor, whereas others have not; but all are endoscopic. While comparative morphologists have regarded the presence of a suspensor as evidence of ancient ancestry and of a primitive evolutionary condition, for which there is considerable evidence in the Pteridophyta, we may note the enigmatic fact that the embryogeny in gymnosperms and angiosperms is also endoscopic and that a suspensor is present. Causal as well as phyletic explanations of the two contrasted conditions are desirable.

Outlook: Embryology in Botanical Science

If our aim is to explain how specific organization in plants is brought about, then comprehensive studies of embryogenesis, including physiological-genetics, biochemistry, biophysics, etc., are obviously essential. Much fuller information is required on the constitution of the zygote, its further development, the assumption of form and structure in the growing embryo, and the functions of its parts. The zygote may be regarded as an organismal, physico-chemical reaction system which becomes progressively more complex during ontogenesis. As the factors in embryogenesis become more adequately understood, so may embryology gain in value for phylogenetic purposes. The descriptive aspect of embryology is now well advanced, though by no means exhausted, and from modest beginnings experimental investigations are now rapidly gathering momentum. Work on embryo culture, and on embryoids obtained by the culture of single cells and in other ways, is now being pursued with exciting results. The constancy and specificity of the embryonic developmental pattern, e.g. in flowering plants, and the occurrence of characteristic anomalous developments in certain hybrids suggest that further embryological investigations of related plants of known genetical constitution may yield information of both academic and practical value. There is also much scope for work on the enigmatic phenomena of parthenogenesis, apogamy, polyembryony, etc. Studies of the protein chemistry of embryos must eventually occupy a leading place in all investigations of a truly fundamental character.

C. W. WARDLAW

References

Johansen, D. A., "Plant embryology," Waltham, Mass., Chronica Botanica, 1950.
Maheshwari, P., "An introduction to the embryology of angiosperms," New York, McGraw-Hill, 1950.
Schnarf, K., "Embryologie der Angiospermen," in Linsbauer, K., ed., "Handbuch der Pflanzenanatomie," Berlin, Borntraeger, 1929.
Souèges, R., "L'embryologie végétale: résumé historique," Paris, Hermann, 1934.
ibid., "La segmentation," Paris, Hermann, 1936.
ibid., "La différenciation," Paris, Hermann, 1936.
ibid., "Les lois du développement," Paris, Hermann, 1937.
ibid., "Embryogénie et classification," Paris, Hermann, 1938–1951.
Steward, F. C., "Growth and Organization in Plants," Reading, Mass., Addison-Wesley, 1968.
Wardlaw, C. W., "Embryogenesis in plants," London, Methuen; New York, Wiley, 1955.
ibid., "Morphogenesis in Plants," London, Methuen, 1968.

PLANT GEOGRAPHY

Plant geography or phytogeography deals with the ranges and distributions of the different kinds of plants on the surface of the earth and with the underlying causes which have resulted in the existing diversification of floras and major vegetational communities. While it has its own aims and methods it is intimately linked with, and for facts partly dependent upon, many other branches of science both within and outside the realm of botany. It depends on taxonomy, that is on the accurate and workable classification and determination of the kinds of plants composing floras and communities. The second essential is a sound knowledge of geography, that is of the natural physical features of the earth's surface, and of the nomenclature employed to designate the various phenomena and divisions. When underlying causes of range and distribution are considered, data from meteorology, geology, ecology, pedology, and human history must be taken into account. There is no hard and fast line between plant geography and ecology, the former being concerned with ranges and distributions on a wide scale and the latter with the same subjects more intensively over smaller areas. As between plant geography, plant ecology, and plant taxonomy there is constant reciprocity, the more so that in their modern developments all three tend to become more synthetic.

The aims of plant geography are to provide a rational classification of floras based on the ranges of their constituent species and other taxa and as far as possible to correlate the results with the present and past environmental factors. The ranges of the plants themselves, whether considered as taxonomic units or as grouped into communities, give the basis and to determine these tabulations and range maps have to be prepared and summated. The primary data are obtained from herbaria and from published records which are considered reliable and filed on card indexes. The source of every datum must be kept for purposes of checking whenever this is necessary. Maps of great variety of projection, scale, and geographic content have been used. To a certain extent this variation in mapping is unavoidable but sometimes it makes comparison of phytogeographical maps difficult and standardization as far as possible is an outstanding need. There must be no attempt to support a hypothesis by manipulating data differently from one map (or tabulation) to another.

In causal plant geography explanation of ranges is sought in the correlations shown with factors of the environment. These are conveniently classified as climatic, edaphic, and biotic, with the proviso that past as well as present conditions have to be considered. This classification can be made to cover all known habitat factors, but it is sometimes convenient to separate certain factor combinations for treatment under special headings. Thus, altitude influences plant range through changes in climate, in soil, and in the influence of plant and animal communities on one another, but, in mountainous districts, altitudinal zonation of plant life is a striking feature and has to be considered as a whole. Evolution of species (and other taxa) and of plant communities has received much attention in recent years and this involves cytogenetics combined with studies of habitats. More data are needed from palaeobotany before the evolution of floras can be properly understood. While research on late Tertiary and Pleistocene fossil floras has progressed rapidly in recent years, especially through the study of fossil pollen, relatively little is known regarding the early evolution and spread of the Angiosperms in earlier times. Even such important hypotheses as continental drift are still in debate and the plant geographer is at times perplexed by contradictory statements by geologists and geophysicists. It follows that the division of the world into floristic and vegetational regions can only be partially explained by correlation with known environmental factors. More geological and palaeobotanical data are needed before

the past histories of floras can be traced and the relicts of such fully recognized in existing floras and plant communities.

Environmental factors are both complex and interact. The evidence clearly favours the view that on a large scale climate is the main factor controlling natural ranges of plants. Climate includes temperature, precipitation, humidity, and winds, and the maximum, minimum, mean, and seasonal distribution of these factors and their coincidence. Climate also acts indirectly in being one controlling factor in soil formation and, through natural selection, in the making and maintenance of biotic communities, while edaphic and biotic factors influence those of climate relatively slightly and locally. Both species ranges and the distribution of vegetational communities are certainly closely correlated with the major climatic divisions. Edaphic (soil) factors have importance on a smaller scale though often with striking results. Again, the factors are complex and involve the nature of the parent rock, the length of time of formation of soil under certain climatic and biotic conditions, water content, acidity or alkalinity, trace elements, etc. Biotic factors phytogeographically involve the action of animals and plants on plants. In the present state of knowledge they are less easy to evaluate than climatic and edaphic factors but attention must be called to the influence of man with his destructive and constructive activities and the effects of deforestation and of flocks and herds.

Phytogeographical groups are conveniently known as phytochoria, and, like systematic units, can be arranged hierarchically. Only the barest outline of the major phytochoria can be given here and the following classification is partly tentative.

I. *Northern Realm*

1. *Arctic and Subarctic Region.* This lies to the north of the closed coniferous forest. The flora is rich locally compared with the Antarctic and consists of herbs and low-growing shrubs, many of the species being more or less circumpolar. The climate is severe with a short growing season. The main plant communities are tundra, cold desert, strand, and aquatic. The woody plants are low bushes of willows, birches, junipers, and heaths. Amongst herbs are grasses, sedges, saxifrages, willow-herbs, crucifers, poppies, and members of the pink family.

2. *Boreal Region.* This is essentially the conifer belt, which extends across much of northern Eurasia, dominated by forests of spruce, pine, and larch, known as the taiga.

3. *Central European Region.* Before human exploitation this was, up to the altitudinal tree limit, occupied largely by broad-leaved forests of oaks, beech, ash, limes, maples, poplars, and elms, and, in wetter habitats, alders and willows. The western parts with a more equable climate are "oceanic" and the eastern with more extreme seasonal differences are "continental" and steppe communities become common. Locally soil and altitudinal factors result in special floras and communities.

4. *Mediterranean Region.* The vegetation is characterized by the dominance of evergreen hard-leaved plants correlated with the hot dry summer and mild wet winter. Aleppo pine, holm-oak, and macchia (maquis) brushwoods occur in the lower lands with woods of oaks, sweet chestnut, and conifers at higher altitudes. The flora contains many Tertiary relicts in the sense of

species and genera that were exterminated farther north by the oncoming of the Ice Age.

Macronesian Sub-region. The North Atlantic Islands of Madeira, Canaries, Azores, and Cape Verde Islands have a flora generically largely Mediterranean but with many endemic species and admixture of African and a few American elements.

5. *Central Asiatic Region.* This enormous region is being energetically studied by Russian botanists. It has often been recognized as a probable centre of development of many plants characteristic of the floras of the North Temperate Zone. Vegetational climaxes include coniferous forest, deciduous forest, grassland (steppes), and scrub. The area is rich in endemics—some of which are probably old relicts.

6. *The Regions of the Nearer East.* In the Nearer East there is some overlap of Mediterranean, Central European, and Central Asiatic regions with what have been called the Saharo-Sindian and Turanian regions. The Saharo-Sindian region is the great belt of desert and semi-desert lands from the Sahara across Egypt, Arabia, and southern Persia to the borders of India. It is the country of the date palm with a comparatively small flora adapted in various ways to survive drought. In saline areas members of the goosefoot family are common. Many of the species probably originated from members of the old Mediterranean flora. The Turanian flora awaits precise definition. It is assumed to be an extension south-westwards of the flora of southern Central Asia. The abundance of species of milk-vetches (*Astragalus*) is characteristic.

7. *Eastern Asiatic Region or Regions.* The enormous area of China has plant climax communities ranging from subalpine and subboreal forests to tropical and monsoon forests together with deserts, loess-steppe areas, and grasslands. The flora contains many ancient relicts including *Ginkgo* and *Metasequoia*. Much of Tibet remains unexplored botanically. The northern Japanese islands have deciduous forests of oak, birches, willows, etc. with some conifers. The central islands are characterized by forests comprised of numerous genera, with tall bamboos adding a special feature to the vegetation. The mountains show marked altitudinal zonation. Southern Japan is subtropical and the vegetation is composed of figs, oaks, laurels, woody climbers, orchards, ferns, and numerous other plants.

8. *North American Regions.* The North American continent has been intensively studied botanically. South of the Arctic and Subarctic Region is the wide belt of conifers. Farther south are the eastern and western forest regions separated by the prairies and semideserts, while in the extreme south the plant-life is subtropical. Of particular interest is the difference between the floras of the eastern and western parts and the East Asian affinities of the former.

II. *Palaeotropical Realm*

1. *Indian Regions.* The Himalaya are botanically divisible into an eastern and a western region, with well marked zonations. In Pakistan the flora contains many desert elements, while the eastern parts of the subcontinent botanically link on to China. The flora and vegetation of much of Peninsular India and Ceylon are tropical.

2. *Malaysian Region.* Floristically this is one of the richest in the world, the vegetation consisting mainly of dense tropical forests with many endemics. Dipterocarps and figs are characteristic.

3. *Tropical African Regions* include rain forests, dry forests, savannas, grasslands, and semi-deserts.

4. *Mascarene Region.* The flora of Madagascar contains a surprisingly large number of endemic genera and species.

5. *Pacific Region.* The volcanic islands have most often a rich forest vegetation with many endemics in Fiji, Samoa, New Caledonia, and Hawaii, while the low coral atolls have a poor native flora.

III. *Neotropical Realm*

1. *Tropical and subtropical American Regions* have four main groups of plant communities: forests, grasslands and savannas, deserts and semideserts, and montane. The Amazonian rain-forest is the most extensive in existence, with enormous numbers of species of many families. Lianes and epiphytes abound and palms are often common.

IV. *Southern Realm*

1. *South African Region,* with sclerophyllous brushwoods, forest, savannas, grasslands, and semideserts, is floristically rich and peculiar with Proteaceae, Restionaceae, heaths, mesembryanthema, spurges, and grasses, as prominent members.

2. *New Zealand Region* has over three-quarters of its flowering plants endemic. The vegetation ranges from kauri forests in the north to temperate rain forest in the south.

3. *Australian Region.* The Australian continent has an arid centre of desert, steppe and scrub with surrounding belts of savanna woodland, *Eucalyptus* forests, and, in the north east, rain forests. Floristically, the striking features are the genera *Eucalyptus* (and others of the myrtle family) and *Acacia* (and others of the Leguminosae) and the many members of the Proteaceae and Compositae. The native flora has, on the whole, long been isolated.

4. *Regions of Temperate South America.* The floristic richness decreases from north to south. Connections are mainly with the north but some Australasian and South African elements are of particular phytogeographical interest. In the southern parts *Nothofagus* forests are important.

5. *Antarctica.* Only two flowering plants have been recorded on the Antarctic continent: a grass (*Deschampsia antarctica*) and a member of the pink family (*Colobanthus crassifolius*). The marine algal flora is relatively rich.

W. B. TURRILL

References

Blake, S. F. and A. C. Atwood, "Geographical Guide to the Floras of the World," Washington, U.S. Dept. Agric., Misc. Pub., **401**, 1942.
Cain, S. A., "Foundations of Plant Geography," New York, Harper, 1944.
Dansereau, P., "Biogeography," New York, Ronald, 1957.
Good, R., "The Geography of the Flowering Plants," 3rd ed., New York, Wiley, 1964.
Richards, P. W., "The Tropical Rain Forest," Cambridge, The University Press, 1952.
Riley, D. R. and A. Young, "World Vegetation," Cambridge, The University Press, 1967.
Turrill, W. B., "Pioneer Plant Geography," The Hague, Nijhoff, 1954.
Turrill, W. B., "Plant Geography" *in* Turrill, W. B., ed., "Vistas in Botany," London, Pergamon, 1959–1964.

PLANT KINGDOM

How does a plant differ from an animal? Exactly, what makes a plant a plant and an animal an animal? There is no simple answer to this question. It is true that most plants are green while animals have no distinguishing color. This might constitute a valid difference were it not for the fact that one large group of plants, the fungi, are not green, and that certain microscopic organisms, the euglenoids, are green though in other respects they seem to resemble animals more than plants. The green color of plants is due to chlorophyll in their cells. This is the substance that enables plants to carry on photosynthesis, the process whereby the raw substances, carbon dioxide and water, are united chemically to produce glucose, a form of sugar. Though PHOTOSYNTHESIS takes place in green cells, it goes on only when these cells are exposed to light. The glucose made in the green cells then serves as the basic material for the synthesis of other more complex compounds such as sucrose, starches, fats, and proteins which are the principle food substances for both plants and animals. A by-product of photosynthesis is free oxygen, which is released from the carbon dioxide that is used. Since oxygen is taken in and carbon dioxide is released in the respiratory processes of most organisms, photosynthesis is instrumental in maintaining the normal oxygen content of the atmosphere. Because of their special ability to make food out of simple chemical compounds, green plants are the ultimate source of the food of all living things, and without them no animal life, including human life, could long exist. The ability to supply living things with food is a unique characteristic of green plants, and this probably constitutes the most significant difference between plants and animals.

Classification. The more than 350,000 kinds of plants can be classified in many ways. A relatively simple and semi-formal scheme that embodies some of the modern concepts of natural classification is as follows.

Non-vascular plants—the lower and intermediate plants

THALLOPHYTA—the simple plants

  ALGAE—thallophytes with chlorophyll

  FUNGI—thallophytes without chlorophyll

BRYOPHYTA—the intermediate plants

  HEPATICAE—liverworts and hornworts

  MUSCI—mosses

Vascular plants—the higher plants

SPORE BEARERS

  PSILOPSIDA—fossil *Psilophytales* and living *Psilotales*

  LYCOPSIDA—lycopods

  SPHENOPSIDA—scouring rushes and relatives

  FILICOPSIDA—ferns

SEED PLANTS

  CYCADOPSIDA—seed-ferns, cycads, and cycadeoids

  CONIFEROPSIDA—*Ginkgoales, Cordaitales, Coniferales,* and *Taxales*

  GENTOPSIDA—joint-firs and relatives

  ANGIOSPERMOPSIDA—flowering plants

The thallophytes. These are the algae and the fungi. They are often referred to as the simple plants because they are without true roots, stems, or leaves, and lack highly organized conducting systems. Although many of them reproduce sexually, their sex cells (gametes) are enclosed only within a cell wall, and where there is a distinct female organ, no embryo develops within it.

The thallophytes, however, are a large and complex assemblage of diverse organisms that exhibit a great variety of form and habit.

The algae grow mostly in water. They range from microscopic unicellular forms to giant seaweeds many feet long. The green color is often partially masked by other pigments—yellow, orange, red, brown, or even purple or blue. Pigmentation is usually constant within a group and it is widely used in classification. Thus there are the blue-green algae (*Cyanophyta*), the greens (*Chlorophyta*), the yellow-greens (*Chrysophyta*), the browns (*Phaeophyta*) and the reds (*Rhodophyta*).

The blue-greens are the simplest and most primitive. Their cells lack definite plastids and nuclei, and they multiply by fission. They are abundant in sewage and on wet soil. Some forms live in hot springs where the temperatures are as high as 80°C. Geologically they are the oldest plants, their remains having been found in rocks 2,000,000,000 years old.

The green algae are common in streams and ponds, though some live in the sea. Masses of their filaments often form floating scums. These scums are not poisonous, and have nothing to do with hydrophobia in dogs, as was formerly believed. In fact, they have a purifying effect on water because of the oxygen released during photosynthesis. The yellow-green group includes the diatoms, unique unicellular and colonial organisms with silicious walls. They form diatomaceous earth deposits which are the accumulations of the shells of dead diatoms.

The brown algae are the largest thallophytes. They are mostly marine plants and include the sea kelps. The reds are also sea plants.

Algae are one of the principal sources of food for fish which eat them directly or feed on other small animal forms that feed on algae. Certain members of the blue-greens, greens, and reds precipitate lime from water. These organisms are therefore important geological agents because of their role in limestone formation. Certain of the red algae, including *Lithothamnion*, are active in the building of coral reefs, actually contributing more lime than do the corals. The brown algae are the source of *algin*, a stabilizer now widely used in the manufacture of such commodities as ice cream, cosmetics, and paint. They were the original source of iodine, and near the coast they are sometimes used to fertilize the soil because of their high potash content. The red algae are the source of agar used for bacterial cultures.

Algae reproduce by several methods. Probably the most prevalent method is by fragmentation of the plant body. Portions of plants carried by moving water may establish them in new sites. Most algae, except the blue-greens, produce special reproductive cells. In the asexual methods the reproductive cells are motile (zoospores), and after a period of activity settle to the bottom and grow into new plants. Sexual reproduction involves fusion of *gametes*, which in some groups produce single-celled diploid resting spores (zygospores and oospores) but in others multicellular bodies.

The fungi resemble the algae except for their lack of chlorophyll. They use carbohydrates that are synthesized by green plants. Some are parasites, others are saprophytes.

The fungi include the BACTERIA and ACTINOMYCETES (*Schizophyta*), the slime molds (MYXOMYCOPHYTA), the algal fungi (PHYCOMYCETES, the sac fungi (ASCOMY-

CETES), and the club fungi (BASIDIOMYCETES). In addition there are the *Fungi Imperfecti* that cannot be properly classified because their life cycles are not completely known. The lichens are usually classified independently because they are plants composed of algae and fungi that grow in symbiotic relationship as single organisms.

The bacteria and viruses are the chief causes of infectious diseases in animals. Though they differ in certain respects, there are three main reasons why bacteria are regarded as fungi. They produce their own amino acids from inorganic nitrogen sources, their cells are surrounded by rather firm walls, and they produce spores that are typical of those of fungi.

The economic impact of the fungi is far-reaching. As scavengers, they prevent the endless accumulation of dead organic matter. Members of all of the groups cause plant diseases, and several, in addition to bacteria, infect animals. A well known example of a plant pathogen is the phycomycete *Phytophthora infestans* that causes potato blight. This disease caused famine conditions in Ireland during the last century which resulted in mass migration of Irish people to the United States. *Ceratostomella ulmi,* an ascomycete, is the cause of the Dutch elm disease which threatens the American elm with extinction. Cereal rusts and smuts and the white pine blister rust are caused by members of the basidiomycetes. Many of these parasites have complex life cycles that require more than one host plant for completion. Yeasts, also ascomycetes, are the principle agents in alcoholic fermentation. Species of *Aspergillus* are used in cheese manufacture, and certain strains of *Penicillium* and other genera of fungi that live in soil are the sources of antibiotic drugs. Some fleshy fungi are edible.

Fungi reproduce by spores that are formed both asexually and sexually. Since most fungi live out of water, in contrast to algae, their spores often have thick walls that resist drying. They are of many kinds—*zygospores, chlamydospores, conidiospores, ascospores, basidiospores, aeciospores, teliospores, urediniospores*—depending on how they are formed. They are often produced in prodigious numbers. The fruiting body of a large puffball has been estimated to yield 7,000,000,-000,000 basidiospores, probably the record number of reproductive cells of any organism.

The bryophytes. If the thallophytes are the simple plants, the bryophytes must be regarded as intermediate plants. They are above the algae in evolutionary development and bodily complexity, but because they lack vascular tissues they cannot be classified with the higher plants.

Bryophytes are small green plants that would generally escape notice were it not for their habit of forming green carpets on the ground in moist woods and on rocks and fallen tree trunks. Unlike the algae, they are mostly land dwellers though they all require considerable moisture for rapid growth and for the carrying out of certain stages in their life cycles.

The liverworts and hornworts grow more or less horizontally and are attached to the substrata by hair-like rhizoids that function much as roots. Most mosses are larger plants with upright or partially upright stems that are clothed with small leaves. They often grow in thick pads on the ground in forests, and are very effective in holding moisture and checking soil erosion.

The only bryophyte of any commercial importance is *Sphagnum,* the peat moss. It is gathered and baled, and sold as a soil conditioner for lawns and gardens. It has remarkable absorptive and antiseptic properties and has been used for surgical dressings when no other material was available. It is widely used for packing material where it is important to prevent drying during shipment.

Bryophytes have developed life cycles that are more complicated than those of algae. Their life cycles consist of two well defined phases, the *gametophyte* and the *sporophyte.* In the moss the gametophyte is the green leafy moss plant. Its body cells have the single, or haploid, chromosome number. It is also the sexual generation. The sex organs, the *antheridia* and *archegonia,* are borne in rosettes of leaves at the tips of the leafy stems. They are multicellular organs that differ rather markedly from the essentially single-celled structures that bear the sex cells in algae. The antheridia produce many motile sperm cells that escape when the moss plants are wet with dew or rain. These sperms either swim or are carried by splashing raindrops to the archegonia which are sometimes on separate but nearby plants. They make their way down the *neck canal* of the archegonium to the venter where the single *egg* is fertilized. This is the beginning of the sporophyte stage. The fertilized egg divides and forms a multicellular embryo within the venter of the archegonium. The embryo continues to grow and the basal part embeds itself in the tip of the leafy stalk. The upper part elongates into a slender naked stalk that in the largest mosses becomes several inches long. This stalk that grows from the fertilized egg is the sporophyte body which represents the diploid or asexual phase of the life cycle. Ultimately the tip of the sporophyte enlarges into a spherical or oval capsule within which spores develop. The spores result from the process of *reduction division* (MEIOSIS) and represent the first stage of the next gametophyte generation, and when they germinate they develop into new green moss plants. Although reduction division, spore formation, and fusion of gametes takes place in the life cycles of many of the algae and fungi, the sporophyte and gametophyte phases are not as distinctly expressed as in bryophytes. Life cycles that follow the same basic plan as that displayed in bryophytes occur in ferns, and in modified form in seed plants.

The vascular plants. The distinction between nonvascular and vascular plants is a precise one, there being no intermediate forms. Therefore, in modern botany the division of the plant kingdom into these two major categories is regarded as fundamental.

In the vascular plants the sporophyte generation constitutes the main plant. Typically it has roots, stems, and leaves. The very highly organized conducting system, the *vascular system,* lends these plants several advantages over those which lack it. The plants can become larger. The more expansive leaf surface can make more food, and more spores or seeds can be produced. The roots can penetrate deeply into dry soil so the plants can grow in drier places. Also roots, being below ground, will often keep the plant alive after aboveground parts have been burned or otherwise destroyed. The gametophyte, however, is as important as in lower plants for completion of the life cycle because it is the part that bears the sex organs, but it is very small as compared with the sporophyte.

The vascular plant complex is made up of at least

eight classes that represent developmental lines that are believed to have evolved quite independently of each other after having originated from some very remote common ancestor. To simplify characterization of these classes they may be divided into the *spore-bearers* (the pteridophytes of older classifications), and the *seed plants* (SPERMATOPHYTES). It is believed, however, that seeds have developed independently in some of the classes and that all seed plants are not close relatives of each other.

The PSILOPSIDA are mostly fossil. They include the rootless and leafless Psilophytales of the Devonian period and the living Psilotales, a group represented in today's flora by only two small tropical genera, *Psilotum* and *Tmesipteris*.

The LYCOPSIDA constitute an ancient class that during the Carboniferous period had members that were large trees. The living forms, on the other hand, are small plants with slender trailing or climbing stems covered with small spirally arranged simple leaves. Several species of *Lycopodium*, known as the club mosses and ground pines, grow abundantly in some forests. Their spores are borne in sporangia on specialized leaves called *sporophylls* which in most species are clustered into small compact cones, or *strobili*, at the tips of the stems. In another genus, *Selaginella*, two kinds of spores are produced, small *microspores* that develop into male gametophytes, and large *megaspores* that produce female gametophytes. This production of two kinds of spores is *heterospory*.

The SPHENOPSIDA are an ancient class of plants of which the only surviving genus is *Equisetum* with about 25 species that range from the arctic tundras to the tropics. These include the common scouring rushes and horsetails that are frequently seen in moist pastures and along railroad embankments. They are among the most peculiar of all plants. The green stems are jointed and lengthwise furrowed, and at each joint there is a circle of a dozen or more small pointed leaves joined together into a cuff at their bases but with free tips. They contain very little chlorophyll. Most of the photosynthesis goes on in the stem. The roots absorb silica from the soil which is precipitated in the walls of the epidermal cells, and this gives the stems a sandpapery feel. The spores, as in *Lycopodium*, are borne in sporangia in strobili at the tips of the stems.

The ferns (Filicopsida) (see FILICINEAE) constitute the largest group of spore-bearing vascular plants. They are old geologically, having first appeared during the Devonian period. They grew copiously in the expanded swamps of the Carboniferous period and were abundant throughout the Mesozoic era. Today there are about 200 genera and several thousand species. Many are cultivated as ornamental plants in temperate climates, and in the tropics the tall fibrous trunks of tree ferns are used for building huts.

In temperate climates ferns grow most abundantly in moist woods. Most of the common ones (for example the bracken fern) have horizontally growing underground stems, or rhizomes, that bear the roots and fronds (leaves). Some, however, such as the royal fern, have short upright stems bearing crowns of fronds at the top.

The frond is an elaborate organ that becomes several feet long in some ferns. It is commonly much divided, the main divisions being the *pinnae* and the smallest subdivisions the *pinnules*. In ferns with underground stems, the fronds may be the only visible parts of the plant.

The sporangia of ferns are rather specialized structures that are typically borne in small clusters (*sori*) on the lower surface of the pinnules. In many ferns the sorus is covered with a thin flap of tissue, the *indusium*, the characteristics of which are widely used in fern classification. Extending part of the way around the sporangium is a row or band of cells with unequally thickened walls, the *annulus*, that serves to open the sporangium when the spores are ripe.

Fern spores germinate on moist soil. They do not produce new plants directly, but develop into small, green, membranaceous gametophytes called *prothallia*. These bear the antheridia and archegonia on the lower surface among the rhizoids that attach the prothallia to the soil. When the spores escape from the antheridia they fertilize the egg cells in the archegonia, and new sporophyte plants develop. Thus the phenomenon of alternation of generations is more highly developed than in the mosses because the two generations are separate self-sustaining bodies.

Of all the spore-bearing vascular plants, the ferns are the closest to the seed plants. Both have similar vascular systems, and the ferns are connected to the more primitive of the living seed plants (the Cycadales and Ginkgoales) through the so-called seed-ferns, or *pteridosperms*, of the Carboniferous period. These seed-ferns looked much like ferns except that they bore seeds on their fronds in somewhat the same manner that ferns bear sori. Anatomically they resemble both groups.

The cycads are an ancient order of seed plants that are believed to have evolved from seed-fern ancestors during the latter part of the Paleozoic era. They spread widely over the earth during the Mesozoic era but began to decline in later times, and today only 9 genera remain and these are restricted to the tropics and subtropics. They are frequently cultivated in greenhouses in temperate climates where they are often mistaken for palms. The sago-palm (*Cycas circinalis*), for example, is a cycad. Most cycads have a short thick stem that bears a crown of large frondlike leaves at the top. In some species the stems are mostly beneath the ground, but others have trunks several feet tall. The trunk has a large pith which, in the sago-palm, is edible. The pith is enclosed by a cylinder of loosely arranged wood and a thick cortex.

In all genera of cycads except one (*Cycas*) the naked seeds are produced in rather solid cones borne at the stem apex in the center of the leaf crown. The pollen is produced in smaller cones on separate plants. It is carried by wind to the female plants and fertilization is accomplished by swimming sperms, a primitive method which in living seed plants is restricted to the Cycadales and Ginkgoales.

The cycadeoids are an extinct group of cycadlike plants that flourished during the Mesozoic era along with the dinosaurs. Instead of bearing cones, like true cycads, they had large flowerlike inflorescences that produced both pollen and seeds. They were worldwide in distribution, and during Jurassic and Early Cretaceous times they were the largest group of land plants.

The GINKGOALES are represented in the present era by a single species, *Ginko biloba*, the maiden-hair tree, but at least 20 genera are known to have thrived at various times during the past. The ginkgos reached their greatest development in the Jurassic period when they inhabited all continents.

The maidenhair tree is easily recognized by its simple bilobed fanshaped leaves that on old branches are borne in clusters on short shoots. The large fleshy seeds are produced on long pedicels among the leaf clusters. At present the tree is native only in remote parts of China where during ancient times it is believed to have been saved from extinction by cultivation in monastery gardens.

The Cordaitales were tall woody trees that grew in the Carboniferous coal swamp forests, and have long been extinct. They had long slender simple leaves and rather loosely constructed inflorescences that are believed to be the prototypes of the cones of pines and other modern conifers. Their winged seeds were scattered by the wind.

The conifers (CONIFERALES) include the pines, firs, spruces, junipers, and several other so-called "evergreens." To this group also belong the giant bigtrees and redwoods of the Pacific Coast region of western North America which are the largest of living trees and some of which are estimated to be at least 3000 years old. Like the cycads they bear their seeds and pollen in cones though the two kinds of cones may be on the same tree. When ripe, the seeds are scattered by the wind.

Most conifers retain their leaves throughout the year, though a few, such as the larches and swamp cypresses shed their leaves at the close of the summer season just as do most broad-leaved trees in temperate climates.

Though conifers grow in all climates except the coldest, they are best adapted to the cool temperate and boreal conditions of mountain systems and similar environments in middle and high northern and southern latitudes. Because of their small heavily cutinized leaves they transpire less water than broad-leaved trees and can grow better in cold soils where roots find it difficult to absorb enough water when evaporation is high.

The conifers are among our most valuable trees. Many species furnish lumber of high quality, and the tall spirelike trunks of spruces and firs furnish most of the poles for telephone and power lines. Smaller trees are important sources of paper pulp.

The taxads (Taxales) include the common yews. They resemble conifers in general appearance, but do not produce their seeds in cones.

The Gnetopsida are a small group of uncertain position in the plant kingdom, but because of their naked seeds they are usually classified near the cycads and conifers. There are only three genera. *Ephedra* is the most common, and is the source of the drug *ephedrin*.

The ANGIOSPERMS, or flowering plants (Angiospermopsida), have been the dominant land plants since the middle of the Cretaceous period. The group is represented today by about 300 families and 200,000 species. They are adapted to a wide range of environments and grow in swamps, deserts, frozen tundras, cultivated fields and gardens, vacant lots, through cracks in sidewalks, or almost anywhere that the roots can reach moisture. A few, like the mistletoe, are parasites; others, like the so-called Spanish moss, are epiphytes, growing suspended from tree branches and telephone wires, and absorbing water from the air. In size and bodily complexity flowering plants range from *Eucalyptus* trees 300 feet tall to minute duckweeds no larger than a pinhead. The furnish us either directly or indirectly with most of our food and clothing and much of our shelter. As far as human dependence is concerned, they rank next to the air we breathe and the water we drink.

There are two groups of angiosperms, MONOCOTS and DICOTS. The former include grasses, lilies, orchids, and palms. The latter include the "hardwoods" of our forests and thousands of species of shrubs, vines, and herbs.

Angiosperms are characterized by the production of flowers. These are specialized structures for the production of seeds. A typical flower consists of the stalk or *pedicel,* the *receptacle,* the *calyx* which is made up of *sepals,* the *corolla* consisting of *petals* which in many flowers are large and showy, the *stamens,* and the *pistil.* The stamens, which are usually stalklike, bear the pollen in *anthers* at their tips. The pistil is in the center of the flower with the other parts arranged around it. The enlarged pistil base is the *ovary* within which are the *ovules.* On top of the ovary is the columnlike *style* that is terminated by the flattened and often colored *stigma.* The special function of the stigma is for reception of the pollen which is brought to it by wind or insects. The stigmatic surface exudes a sticky sugary substance which stimulates germination of the pollen. A pollen tube forms and grows through the stigma and style. When it reaches the ovule two sperm cells are discharged, and one of these fertilizes the egg nucleus within the *embryo sac.* The other sperm fuses with two *polar nuclei* in a process called *triple fusion.* The fertilized egg develops into an embryo and the nucleus resulting from triple fusion produces *endosperm* tissue. While this is going on the ovule develops into the seed and the ovary becomes the *fruit* in which the seed or seeds are enclosed. In contrast, no fruits form after fertilization in the cone-bearing plants. The seeds are borne on the surfaces of the cone scales.

The fruits of flowering plants are primarily a means of seed dissemination. Whereas in cone-bearing plants seed scattering is largely a matter of chance and dependent mostly upon wind, fleshy fruits are eaten by animals and the seeds are carried as far as the animals may roam. Some seeds have become so dependent on animals that they will not germinate until digestive juices have softened their hard coats.

Plant and animal relationships are nowhere more strongly revealed than in insect pollination, and some of the modifications that have developed along with this relationship are among the most remarkable in the whole biological realm. Virtually all plants with showy flowers are insect pollinated, and the fact that angiosperms now dominate the plant world has without doubt come about by the liaison between plants and insects that developed at some time during the remote past.

CHESTER A. ARNOLD

References

Alexopoulos, C. J., "Introductory Mycology," New York, Wiley, 1962.

Black, C. A., "Soil-plant Relationships," New York, Wiley, 1957.

Campbell, D. H., "Structure and Development of Mosses and Ferns," New York, Macmillan, 1918.

Campbell, D. H., "The Evolution of Land Plants," Stanford Univ. Press, 1940.

Chamberlain, C. J., "Gymnosperms: Structure and

Evolution," Chicago, University of Chicago Press, 1935.

Chapman, V. J., "Introduction to the Study of Algae," Cambridge, Macmillan, 1941.

Daubenmire, R. F., "Plants and Environment," New York, Wiley, 1959.

Eames, A. J., "Morphology of Vascular Plants," New York, McGraw-Hill, 1936.

Foster, A. S. and E. M. Gifford, "Comparative Morphology of Vascular Plants," San Francisco, Freeman, 1959.

Gauman, E. A. and C. W. Dodge, "Comparative Morphology of Fungi," New York, McGraw-Hill, 1928.

Haupt, A. W., "Plant Morphology," New York, McGraw-Hill, 1953.

Lawrence, G. H. M., "Taxonomy of Vascular Plants," New York, Macmillan, 1951.

Tilden, J. E., "Algae and Their Life Relations; fundamentals of ecology," Minneapolis, Univ. of Minn., 1935.

Verdoorn, Fr. et al., "Manual of Pteridology," The Hague, Nijhoff 1938.

Wolf, F. A. and F. T. Wolf, "The Fungi," Vol. 1, New York, Wiley, 1947.

PLASMAGENE

It has been maintained by some investigators that, in addition to the ordinary GENES or Mendelian units of heredity which are located in the CHROMOSOMES of the NUCLEUS, there are also plasmagenes, non-chromosomal genes, or extra-nuclear genes, i.e., genes located outside of the nucleus in the cytoplasm. Like the chromosomal genes, plasmagenes are held to control hereditary traits, to be self-reproducing, and to be capable of mutating, i.e., to undergo changes which are thereafter reproduced as faithfully as the original type. In other respects, different conceptions of the plasmagene are held by different authors. The chief conceptions are: (1) The chromosomal genes produce copies of themselves which migrate into the cytoplasm where they become plasmagenes capable of (a) limited or (b) unlimited further reproduction; and to be dependent or independent of the presence of the gene from which they arose. (2) Plasmagenes are of independent origin, being perpetuated by their own reproduction. Most of the evidence adduced in support of the existence of plasmagenes comes from studies on microorganisms and plants. Except for one special kind of plasmagene, generally referred to as plastogenes, the very existence of plasmagenes is not generally agreed upon by geneticists. The exception—the plastogene—is associated with the plastids of plants and is responsible in part for traits of these plastids, including their capacity to form their characteristic pigments, such as chlorophyll. Other plasmagenes or independent genetic particles in the cytoplasm are often interpreted as viruses or symbiotic organisms. As a rule decisive evidence is lacking as to whether such particles are of extrinsic or intrinsic origin. If extrinsic they may more properly be considered symbionts; if intrinsic, plasmagenes. The noncommittal term, plasmid, has been proposed to include both, and to avoid the usually unresolvable question of ultimate origin. One of the most fully documented examples of a plasmid—probably of extrinsic origin—is the *kappa particle* (and others like it called *lambda, mu, pi, nu* and *sigma*, according to their properties and the traits controlled by them). These particles have been demonstrated to be bacteria in the cytoplasm of many strains of *Paramecium* and to confer upon the paramecia well defined hereditary traits. Among higher animals, a few well-established examples of the plasmid are known in insects, especially the fruit-fly, *Drosophila*, and the mosquito, *Anopheles*. Very extensive studies in some flowering plants such as the willow herb, *Epilobium*, have been held to warrant the interpretation that there exists a whole system of plasmagenes— together called the plasmon (in parallel with the term genome which signifies the totality of nuclear genes in a cell)—affecting many hereditary traits of the organism. In all fully studied examples of hereditary influences of the cytoplasm—whether due to plasmagenes or other physical bases—there is never a complete independence of cytoplasm and nucleus, but always a close integration. Ofen the same trait, e.g. the presence of certain active enzymes, may depend in the same organism both upon nuclear genes and upon cytoplasmic factors. Very little is known about plasmagenes or plasmids in comparison with the vast amount of information on nuclear genes.

In recent years, however, the existence of plasmagenes in plastids and mitochondria has been strongly supported by the finding of DNA (deoxyribonucleic acid) in plastids and mitochondria. Because DNA is the material of which genes in chromosomes are composed, it is reasonable to believe that mitochondria and plastids carry genes of their own. This is further supported by evidences of mitochondrial mutations in yeast and the bread mold (Neurospora) and of plastid mutations in many organisms. DNA has also been reported in other cellular organelles, but this is not yet firmly established. Finally, recent studies on a one-celled alga, Chlamydomonas, report a considerable number of traits held to be due to non-chromosomal genes. The whole subject of plasmagenes is now in a state of lively investigation.

T. M. SONNEBORN

PLASMODIUM

The genus *Plasmodium*, Marchiafava and Celli, 1885, is a moderately large one, containing slightly more than 60 known species; others almost certainly exist. These are the causative organisms of malaria, a disease of major importance in man, and probably of some importance in nature, where it may help in the maintenance of biological balances. The organisms reside in the red blood cells, in which they reproduce, destroying the cell. A dark pigment, haemozoin, is a characteristic by-product of haemoglobin metabolism. As far as is known, mosquitoes are always the transmitting agents (*vectors*).

The generic name *Laverania* for the malaria parasites of man and higher apes (chimpanzees and gorillas) producing elongate gametocytes (*Plasmodium falciparum* and *P. reichenowi*) is also sanctioned under the International Rules of Zoological Nomenclature, but is not generally used. At present there is difference of opinion about the best classification of the malaria parasites as a group; Garnham (1966) favors recognition of nine subgenera within the single genus *Plasmodium*, of which *Laverania* would be one.

Closely related to the genus are several others: *Haemoproteus*, only the gametocytes of which occur in red cells (in which they produce haemozoin), is very widely distributed in birds and even occurs in a few reptiles. *Leucocytozoon* is also very common in birds, but is unknown from other hosts. Its gametocytes occur in leucocytes, probably mainly the lymphocytes or, some think, in erythrocytes. Both genera reproduce only in the tissues. *Hepatocystis* (also producing haemozoin) has a life cycle somewhat resembling that of *Haemoproteus*, but is known only from mammalian hosts and from the Old World.

The host range of the malaria parasites is wide. They are extremely common in the passerine (perching) birds, and have been found in several hundred species; there is little doubt that they occur in many more. But other bird types, such as chickens, pigeons, ducks, turkeys, owls and hawks—and even penguins—have been found infected. Not only are birds frequent hosts, but more than 20 species of *Plasmodium* are known from them. Almost as many species occur in reptiles, mainly lizards, though such infection has recently been reliably reported from a snake. A related genus, *Dactylosoma*, occurs in frogs and toads.

Among mammals, rodents (especially rats) and primates are most often infected, but malaria is also found in a variety of other hosts, such as antelopes, water buffalo and bats. At least 24 species and subspecies occur in subhuman primates.

Man harbors four species of *Plasmodium*: *P. vivax* is the most widespread, being found in both temperate and tropical regions. *Plasmodium falciparum*, the most dangerous, is restricted to the warmer parts of the world. *Plasmodium malariae* (also a parasite of higher apes) has a distribution similar to that of *P. vivax*, but is relatively rare, while *P. ovale*, though widespread, is the rarest of the four. Man is also susceptible to several species of simian malaria, and partially so to several others.

The geographic distribution of most species of *Plasmodium* is also broad. Because of the migratory habits of most birds their plasmodia are likely to occur in many parts of the world, especially as most of the avian malarias have little host specificity. Certain kinds of avian *Plasmodium* even manage to perpetuate themselves in relatively cold climates: Infection has been found in some non-migratory species south of the Antarctic Circle and in the High Rockies. Reptilian malaria parasites, however, though widely distributed, often have a curiously localized occurrence even in a given area.

The life cycle of all species of *Plasmodium* is apparently very similar. In the vertebrate host the parasite lives in the erythrocytes (exceptionally in other blood and blood-forming cells). Here it grows and reproduces, often with a marked periodicity of the reproductive cycle. Some of the organisms, for reasons unknown, develop into sexual forms or gametocytes rather than multiplicative stages. Such parasites are incapable of further development unless picked up by susceptible mosquitoes of suitable species. On the whole, the avian malarias are transmitted by culicine mosquitoes and the mammalian malarias by anophelines, although some species of *Anopheles* may apparently be infected by avian plasmodia. In the gut of the mosquito (always a female) the gametocytes rapidly mature into gametes, with ensuing fertilization of the female cells. The fertilized cell (oökinete) penetrates the gut wall and develops within a few days (the time required varies with the temperature and species of parasite) into an oöcyst which protrudes like a tumor from the outer surface of the gut; from the oöcyst are soon released a myriad of the infective stages (sporozoites). These are minute filaments which make their way about in the body cavity of the mosquito, some of them eventually invading its salivary glands, to be injected into the victim when the insect bites. But here, instead of promptly penetrating the red cells, as was long believed, the organisms localize themselves in the tissues of the various organs, to which they are carried by the bloodstream. In mammals, the liver is the usual destination. After from one to several generations some of the offspring spill over into the blood, and in most plasmodial species enter the erythrocytes, thus completing the cycle. However, multiplication may continue in the tissues, and such foci of infection are thought to account for the relapses typical of the disease, which often lasts for years unless cut short by treatment. Quinine was for several centuries the only effective antimalarial, but a number of synthetic compounds are now known. To some of these the parasites have been able to develop resistance, however.

Many attempts have been made to cultivate the parasites, but so far with only limited success. The organisms may survive and multiply for several generations in the erythrocytes, when placed in suitable media, and Trager, at Rockefeller University, has even been able to secure growth and reproduction for a time of *Plasmodium lophurae* (an avian parasite) outside its host cells. The tissue or exoerythrocytic stages of some avian (but not mammalian) plasmodia may be quite easily grown in tissue culture. No one has yet been able to cultivate the stages occurring in the mosquito although Ball has recently reported notable progress in this direction.

The nutrient requirements of the parasites seem to be very similar to those of the host. Dextrose is the chief source of energy and the same amino acids are required. Within the red cell, haemoglobin (ingested, as is the food of an amoeba) is their chief food, and from this the characteristic haemozoin is derived. The globin fraction of the molecule furnishes the parasite most of its protein requirements. Apparently, however, haemoglobin S found in persons having the hereditary sickling trait is in some way unsuitable for the parasites, particularly *Plasmodium falciparum*. Such persons are partially protected against the often lethal consequences of falciparum malaria.

Although there are of course no fossil remains of malaria parasites, there is other evidence to indicate a probable origin of this group from the coccidia (a group of SPOROZOA). Most coccidia are resident in the tissues lining the alimentary tract of various vertebrates, and it seems possible that such parasites might on occasion have spilled over into the bloodstream, there to be picked up by blood-sucking insects such as mosquitoes. If a few of the more adaptable individuals were then able to establish themselves within the insect, the stage would be set for the evolution of a mosquito-vertebrate cycle. An alternative view is that insects, presumably mosquitoes or their ancestors, were the original hosts. Since a disproportionate number of species occur in Africa, the genus may have originated there.

REGINALD D. MANWELL

References

Garnham, P. C. C., "Malaria Parasites and Other Haemosporidia," Oxford, Blackwell Scientific Publications, 1966.

Hewitt, R., "Bird Malaria," Amer. J. Hyg., Monographic Series, No. 15, Baltimore, Johns Hopkins Press, 1940.

Manwell, R. D., "Some evolutionary possibilities in the history of the malaria parasites," Ind. Jour. Malariol., 9:247–53, 1955.

Moulton, F. R., ed., "A Symposium on Human Malaria," Pub. No. 15 of Amer. Ass. Adv. Sci., 1941.

Russell, P. F., "Man's Mastery of Malaria," New York, Oxford University Press, 1955.

Russell, P. F. et al., "Practical Malariology," New York, Oxford University Press, 1963 (Revised Edition).

Warshaw, L. J., "Malaria, the Biography of a Killer," New York, Rinehart, 1949.

PLASTID

A light microscopical study of the leaves of a plant reveals that all the green pigment, the chlorophyll, is contained in lens-shaped bodies within the cell. These are the *chloroplasts*, the most important category among the plastids. They convert light energy into chemical energy in the process called photosynthesis.

In plant cells several types of plastids are distinguished according to their morphology, structure, pigment content and function. The pigment-containing *chromoplasts* may be photosynthetic or without photosynthetic activity. An example from the latter group is the plastids of the carrot root. Here the carotene is stored in big crystals.

The yellow color of many flowers is due to chromoplasts. Those in the petals of the buttercup contain xanthophylls and carotene dissolved in lipid droplets. In the fruits of the red pepper and other red-colored fruits the carotenoids are associated with a fibrillar structure of the chromoplasts. The *leucoplasts* are plastids with little or no pigment. Some of these are specialized for storage of starch as found in the potato tuber, and they are then called *amyloplasts.* Other plastids are storage places for oil, the *elaioplasts.* Those accumulating protein are called *proteinoplasts.*

The photosynthetically active chromoplasts vary in size, shape and pigment content. In the algae they may be lens-shaped, bell-shaped, or they form big spiral bands or an irregular network. They are green, containing chlorophyll, carotene and xanthophylls; or they are brown as in the brown algae, where the chlorophyll is masked by carotenoids. In the red algae the red plastid color is due to phycobilin proteins, phycoerythrin and phycocyanin, which masks the chlorophyll color. All these different pigments in the plastids help the chlorophyll to absorb light for the photosynthetic process. Irrespective of their shape the chloroplasts of the green and brown algae (Fig. 1) consist of an envelope and a lamellar system embedded in

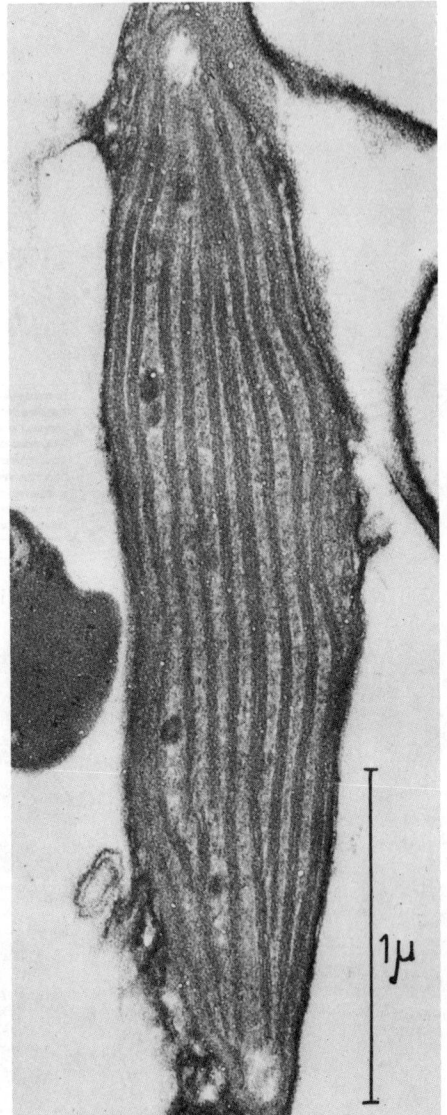

Fig. 1 Electron micrograph of cross section of plastid in the brown alga *Fucus.*

a granular matrix, the *stroma.* Generally, two or three membranous discs of the lamellar system, called *thylakoids,* are aggregated into the huge plates extending over the entire diameter of the plastid. These plates contain the chlorophyll and are homologous to the *grana* of the higher plant chloroplast. In the red algae the plates are formed by single discs with phycobilin-containing particles attached to their surface.

Each of the ten million cells in a leaf contains on an average 20 chloroplasts with a diameter of 5 μ and a thickness of 2–3 μ. About 30% of the dry weight of chloroplasts is lipid, 50% protein, 8% pigment and 1% nucleic acid. The two chlorophylls a and b occur in a ratio of 3:1, and there is as much carotenoid as

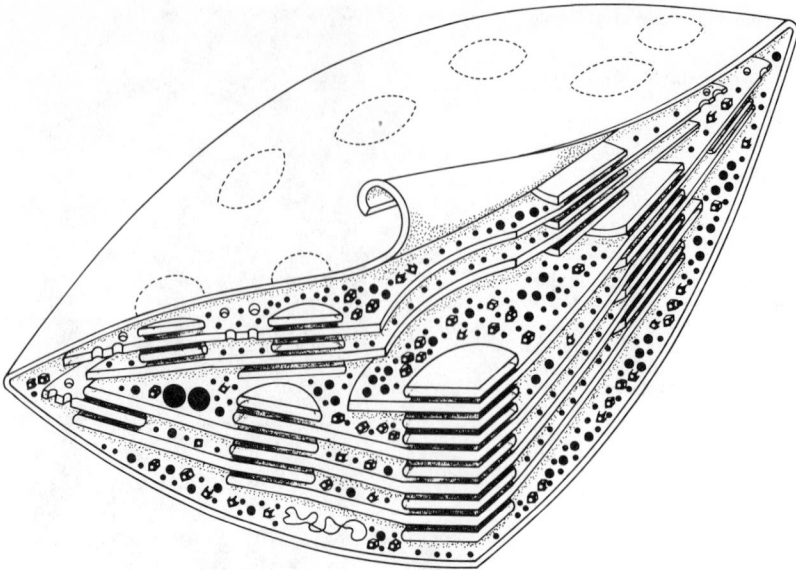

Fig. 2   Three-dimensional diagram of a chloroplast in higher plants.

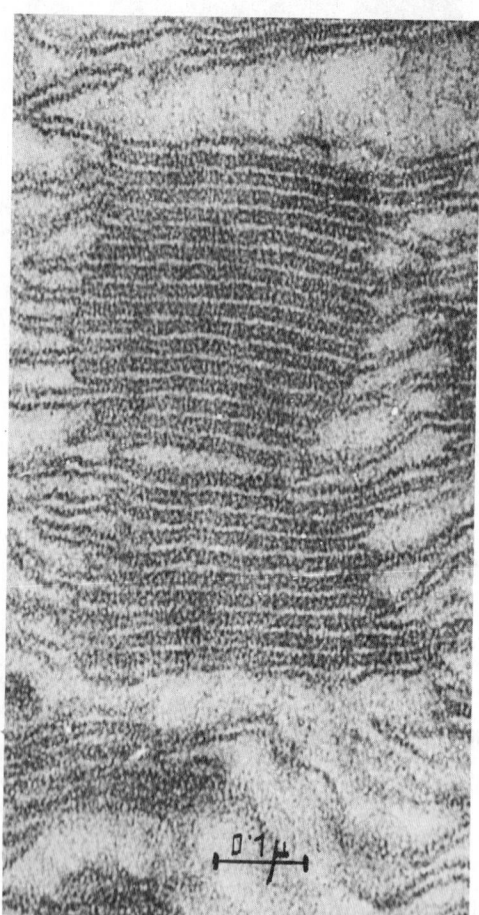

Fig. 3 Electron micrograph of part of a section through a chloroplast in higher plants.

there is chlorophyll *b*. The pigments are concentrated in the *grana*. These are made up of columns of round discs and are connected by large single discs into the lamellar system (Figs. 2, 3). The lipoprotein membranes of the chloroplast are about 70 Å thick.

Isolated chloroplasts have the ability to carry out complete photosynthesis. The photochemistry of this process is confined to the lamellar system, which contains the electron transport chain and can carry out O_2-evolution as well as cyclic photophosphorylation yielding adenosine triphosphate. Catalysts for non-cyclic photophosphorylation to yield further energy-rich phosphates are found in the stroma. These compounds provide the power to drive the dark reactions of CO_2-fixation and its reduction to sugars in the stroma. The sugars are usually polymerized into starch grains and stored there until transported away from the chloroplast during the night.

All plastids are differentiated during the ontogeny of the plant from small particles in the cytoplasm called *proplastids*. Proplastids and young chloroplasts can multiply by division. The lamellar system and stroma enzymes are synthesized in the greening plastid. For this purpose the stroma contains its own machinery to synthesize lipids and proteins. This is evidenced by the presence of chloroplast-specific ribosomes. Plastid development is controlled by a cooperation between nuclear genes and genes located in the chloroplasts themselves (the *plastome*). The genetic information contained in the chloroplast is coded in histon-free double-stranded deoxyribonucleic acid fibrils.

DITER VON WETTSTEIN

References

Kirk, J. T. O. and R. A. E. Tilney-Basset, "The plastids," San Francisco, Freeman, 1967.

Goodwin, T. W., ed., "Biochemistry of Chloroplasts," New York, Academic Press, 1966/67.

San Pietro, A., *et al.,* "Harvesting the Sun," New York, Academic Press, 1967.

PLATYHELMINTHES

The Platyhelminthes or flatworms comprise one of the major phyla of the ANIMAL KINGDOM. These animals manifest great diversity in structure and bionomic relations. They are included in a single taxonomic unit because of certain common features, *viz.*, bilateral symmetry, dorsoventral depression, a ciliated epidermis in free-living species and at some stage in most parasitic species, a solid body filled with loose connective tissue, an incomplete digestive system, and a protonephridial or flame-cell excretory system. One or more of these features may be reduced or lacking in parasitic species. The flatworms are characterized also by common negative features; the absence of metamerism, and of skeletal, circulatory, and respiratory systems. They were defined by Libbie Hyman as "acoelomate Bilateralia without a definite anus." As portrayed in most textbooks, the phylum contains three classes, the TURBELLARIA, TREMATODA, and CESTODA. However, two other groups, the MESOZOA and the NEMERTEA, are sometimes included as classes in the phylum.

The Mesozoa, often recognized as a separate and distinct phylum, are probably much simplified members of the Platyhelminthes. They are minute, ciliated, marine organisms, so named by E. van Beneden because he regarded them as intermediate between Protozoa and Metazoa. Morphologically, each consists of one or more germinal cells which are concerned with reproduction, and a covering of ciliated, somatic cells, typically constant in number for a given species. There are two subdivisions; the Orthonectida and the Dicyemida. Sexual stages of the orthonectids are free-swimming while asexual reproduction occurs in plasmodia in various invertebrates; turbellarians, nemerteans, annelids, ophiuroids, and bivalve mollusks. The life-cycle of the dicyemids is known only in part; both asexual and sexual stages occur in the renal organs of cephalopod mollusks. Infusoriform larvae leave the mollusk and other vermiform larvae infect young cephalopods. Presumably the vermiform larvae result from reproduction in another host-animal The simplicity of structure in the mesozoans can not be accepted as primitive; it is clearly secondary and the result of parasitic adaptation. In polyembryonic multiplication, cleavage patterns, and morphology of larvae, the mesozoans show marked agreement with the parasitic flatworms and in all probability should be included in the phylum Platyhelminthes.

The inclusion of the Nemertea as a class, the Rhynchocoela, is advocated by certain zoologists, especially German authors, but the majority of American writers recognize the nemerteans as a separate phylum. They differ from the rest of the flatworms in the presence of a complete alimentary tract, a vascular system, and an eversible proboscis enclosed in a dorsal tubular cavity, the rhynchocoel. All these characteristics, however, are foreshadowed in the Platyhelminthes. Many species of digenetic trematodes have communications between the digestive ceca and the outside of the body, others have a lymph-system which is almost certainly a precursor of the circulatory systems of higher animals and, given an invagination of the body-wall to form the rhynchocoel, the eversible pharynx of certain turbellarians suggests the proboscis of the Nemertea. Most of the nemerteans are marine, a few species occur in freshwater or terrestrial habitats, while the genus *Malacobdella* contains commensal or parasitic species.

Parasitism is widespread in the phylum, indeed, most of the members are parasitic, and this condition has profoundly influenced the character of the group. Within the phylum there are units which illustrate all stages in the gradual transition from a free-living existence to the extremes of dependency and morphological adaptation. With the adoption of a dependent condition, those structures which function most actively in the free-living stage undergo progressive reduction and may entirely disappear. Inactivation and atrophy of the sensory and locomotor organs is followed by degenerative changes in the entire nervous system. With the absorption of nutriment by the surface of the body, the alimentary tract undergoes regression and in the cestodes there is no trace of a digestive system and no endoderm formation in development. The overall rate of metabolism in parasitic species is probably not profoundly altered and the production of nitrogenous wastes requires an efficient excretory system. Living in the intestine and other locations where the amount of free oxygen is limited, many flatworms have developed an anaerobic type of respiration in which oxygen is obtained by the reduction of carbohydrates with the production of various higher fatty acids. The combination of these acids with other substances gives rise to the concretions so common in anaerobic species. The rich food supply and sedentary habits of parasitic species find expression in an enormously increased reproductive capacity. The adoption of the parasitic habit entails definite hazards on the part of any species which assumes this mode of life. Normally the parasite remains with and perishes with the host. For perpetuation of the species, progeny must escape from the first host and find new hosts, a precarious undertaking. The larval stages which accomplish such migrations are minute organisms, incapable of taking food, with limited locomotor ability and limited longevity, so the probability of completing the life-cycle is exceedingly remote. In certain species, no more than one in a million is successful. Survival of parasitic species is possible only because of the prodigious rate of reproduction.

Classification and Characteristics of Constituent Groups

Turbellaria. In general, Turbellaria are small, free-living carnivorous worms. Most of them are less than 5 mm. in length; the larger forms occur in the Tricladida and Polycladida; some of the terrestrial planarians attain a length of 50 cm. The turbellarians are oval to elongate in contour, flattened ventrally and somewhat convex dorsally; the body may bear tubercles or papillae and members of the Temnocephalida have anterior tenacles and a ventral, muscular, adhesive organ. The smaller species are translucent or white and present shades of gray or brown derived from the color of ingested food; the larger species often have pigment in or under the epidermis and may have striped, barred or blotched patterns in green, red, yellow, orange and black. The anterior end precedes in locomotion; it bears the sensory organs and contains the principal ganglionic mass of the nervous system. The body is clothed by a cellular or syncytial epidermis, usually ciliated and provided with minute, secreted, rhabdoid spicules. The worms are primarily hermaphroditic; the reproductive systems are often complex; internal fertilization is the rule, and the life-cycle is simple. The eggs usually contain large amounts of food material in the form of vitelline cells, but the Acoela and Polycladida lack vitelline glands and in some of the polyclads a free-swimming larva is produced. In these groups, cleavage is spiral and determinate, while in other orders it is usually irregular. Many of the Turbellaria live in commensal or parasitic association; the rhabdocoels in echinoderms and mollusks, the alloeocoels in crustaceans, the triclads in crustaceans, echinoderms, mollusks, chelicerate arthropods, and selachian fishes. Different authors classify the Turbellaria by details of the digestive or reproductive organs; the accepted system comprises 5 orders and is based on the form of the intestine.

Order Acoela.

Minute marine forms, ventral mouth, no distinct intestine, no delimited gonad, no excretory or reproductive ducts. Sometimes colored green or brown by symbiotic algae.

Order Rhabdocoela.

Small marine, freshwater, or terrestrial, free-living, commensal or parasitic worms, with simple pharynx, saccate intestine, compact gonads, few testes, definite reproductive ducts, and a relatively simple protonephridial excretory system.

Order Alloeocoela.

A variable group, representatives usually somewhat larger than the rhabdocoels, mostly marine, few freshwater, pharynx of variable types, lobate intestine, numerous testes, complex protonephridial excretory system.

Order Tricladida.

Often large, elongate worms with plicate pharynx, triclad intestine, single gonopore, complicated excretory system with numerous nephridiopores; 3 suborders, Paludicola in freshwater, Maricola in marine habitats, and Terricola in moist terrestrial locations.

Order Polycladida.

Typically marine, large, broad forms, often of striking shapes and coloration, with plicate pharynx, intestine much branched, gonads numerous and dispersed, no vitelline glands, common or separate male and female gonopores.

Trematoda. The trematodes are exclusively parasitic and comprise two distinct and distantly related subclasses: the Pectobothridia, primarily ectoparasites on the skin and gills of fishes and secondarily as endoparasites in the mouths and urinary bladders of amphibians and turtles; and the Malacobothridia, endoparasites of invertebrates and vertebrates, which begin their life-cycles as parasites of mollusks. The pectobothriid species are commonly known as polystomes and the malacobothriid species as distomes or monostomes. Van Beneden (1858) arranged the trematodes in two groups, Monogenea and Digenea, based on the type of life-history. Members of the Monogenea have a simple, one-host life-cycle and a single sexual generation; those of the Digenea have complex life-cycles with both invertebrate and vertebrate hosts and an alternation of asexual and sexual generations. The pectobothriid and monogenean groups are almost idential but exceptions invalidate the classification based on life-history. Members of the family Gyro-

dactylidae are viviparous with successive generations enclosed one within another in the embryo, and the polystomes of amphibians have an alternation of sexual generations, one on the gills of tadpoles and the other in the urinary bladder of mature animals. The Pectobothridia are typically ectoparasites of aquatic vertebrates, and Baer (1951) stated that at least 95 per cent of the genera have been reported from fishes, especially the elasmobranchs. Species occur also on the gills or in the mouths and urinary bladders of amphibians and reptiles, and one species occurs in the eye of the hippopotamus. Morphologically these trematodes are similar to rhabdocoele turbellarians of the family Graffillidae. In general, they are small worms which measure from 5 mm to 2.5 cm in length. The posterior portion of the body is modified to form a haptor, a disk-like structure which bears hooks or suckers or both and is a powerful adhesive organ. The animals maintain location by the attached haptor; the anterior end bears the mouth and pharynx. The intestine may be saccate or bifid and is often branched. The excretory system is duplex, with anterior, dorsolateral pores. The worms are hermaphroditic; the ovary is single; the testes one to many; and the eggs frequently bear long, terminal filaments. Typically, the life-cycle is simple and development is direct; the larva which hatches from the egg is ciliated; after a brief free-swimming period it attaches to a host and grows to the adult condition. The Pectobothridia are divided into two orders; the Monopisthocotylea which contains the Gyrodactyloidea with three families and the Capsaloidea with five families; and the Polyopisthocotylea which contains the Polystomatoidea with two families and the Diclidophoroidea with six families.

The Malacobothridia comprise two orders: the Aspidobothrea and the Digenea. Early authors recognized three major groups in the Trematoda: Pectobothrii (syn. Heterocotylea), Malacobothrii, and Aspidobothrii. This last group consists of a single family which, because of its multilocular ventral sucker, was included with the amphistomes in the Digenea. Although the aspidogastrids are morphologically similar to the Digenea, they are monogenetic. They are primarily parasites of mollusks, although they can persist for long periods in the digestive tracts of fishes and turtles that feed on mollusks. They do not have asexual multiplication although adults of *Stichocotyle* occur in the biliary ducts of rays and the larvae are encysted on the intestinal walls of lobsters. The aspidogastrids probably represent a primitive branch of the turbellarian stem that gave rise to the Digenea.

The Digenea are endoparasitic, with alternation of generations and hosts; the asexual generations are produced by polyembryony in mollusks, while the sexual generation occurs in vertebrates. The worms infect marine, freshwater, and terrestrial hosts, localizing in the digestive gland and gonads of mollusks and in the intestine, liver, lungs, pancreas, kidneys and blood-vessels of vertebrates. Some 40 different species have been recorded as human parasites and ten of them are of major medical importance. The digenetic trematodes normally maintain attachment by oral and ventral suckers, but in particular groups either or both may be lacking. The cuticula may be either smooth or spined; the body-wall consists of three layers of muscles; an external circular, a middle longitudinal, and an inner oblique layer. The mouth may be terminal or sub-terminal; in one group it is definitely ventral in location. A pharynx is usually present and the intestine is saccate or triclad, sometimes ramiform. The excretory vesicle is formed by fusion of the collecting ducts of the two sides of the body; the pore is single median, usually terminal and the flame-cell pattern is characteristic of taxonomic groups. The worms are hermaphroditic except for the schistosomes which live in the blood-vessels of birds and mammals. The testes are paired; their ducts usually unite to form a seminal vesicle, which may be enclosed in or followed by the copulatory organs. The ovary is single; the vitelline glands are paired; the oviduct frequently bears a diverticulum, the seminal receptacle; from which Laurer's canal extends to the dorsal surface of the body; the ootype is enclosed in the cells of Mehlis' gland; the uterus is long and the male and female ducts usually open into a common genital atrium. In different families, the eggs when laid may contain only the ovum and accompanying vitelline cells or a fully formed miracidial larva. The eggs may hatch in water or only after ingestion by a suitable intermediate host. In such a host, the miracidium loses its cilia and transforms into a sporocyst, in which germinal cells produce the next generation, which may be either sporocysts or rediae. Asexual reproduction may continue more or less indefinitely, but very soon a definitive larval type, the cercaria, is produced. As a rule, this larva leaves the mollusk and without further reproduction develops into the sexually mature adult in the final host. The cercariae may encyst as metacercariae in invertebrate or vertebrate animals which serve as transfer or paratenic hosts or on objects which are ingested by the definitive hosts. In certain strigeid species a mesocercarial stage precedes the metacercaria and a four-host cycle is established. The cercariae of the blood-flukes of fishes, reptiles, birds and mammals penetrate the skin of their hosts to reach the vascular system, where they become mature. The blood-flukes of the cold-blooded vertebrates are hermaphroditic and live in the arteries; those of birds and mammals are dioecious and live in the veins.

The classification of the Digenea is controversial; certain authors recognize as many as 70 families and 700 genera, whereas others join related genera in larger and more comprehensive family groups. In the recent "Systema Helminthum," Yamaguti recognized 54 families. Until recently, taxonomic work has been based almost entirely on the morphology of sexually mature worms, but increased knowledge of life-cycles and larval stages has provided the basis for a more adequate and realistic concept of these trematodes. La Rue (1957) has proposed a revision of the Digenea "based on life-history data and designed to show genetic relationships;" his system comprises two superorders, five orders, ten suborders, seventeen super-families, and eighty-four families.

Cestoda. The Cestoda comprise two groups, often listed as subclasses: one, the Cestodaria, is a small, relatively unimportant division containing monozoic parasites of archaic fishes; these worms have lycophoran larvae, each with ten hooks, but the life-history is incompletely known. The other division is composed of the Eucestoda or tapeworms, each of which typically consists of an adhesive organ or scolex, a proliferative zone adjacent to the scolex, and a strobila of one to hun-

dreds of proglottids. The merozoic cestodes are sometimes regarded as colonies, rather than individuals, since each proglottid contains complete sets of male and female organs and the terminal ones may detach and live for some time in the intestine of the host. The nervous, muscular and excretory systems, however, are continuous throughout the strobila and these features together with a lack of regenerative ability, strongly support the idea that the tapeworm is an individual. The scolex may have dorsal and ventral grooves termed bothria; sessile or stalked, simple, crenate or loculate leaf-like adhesive structures termed bothridia; or cup-shaped suckers termed acetabula. The bothridia may bear spines, hooks, or small acetabula. The excretory system is protonephridial with a pair of longitudinal vessels on either side of the body; one is dorsal, the other ventral and somewhat larger. The ventral vessels are usually connected by a transverse duct near the posterior end of each proglottid. Fluid passes forward in the dorsal vessels and backward in the ventral ones. There is a single bladder and excretory pore in the terminal proglottid but after it is shed, the vessels open separately. The reproductive systems are, in general, similar to those of the trematodes; in the Cyclophyllidea there is no uterine opening, the eggs accumulate in the uterus which ultimately fills the proglottid. Detached proglottids pass out of the body with fecal material. The cestodes infect all groups of vertebrates and localize in the alimentary tract. Their larvae occur in various invertebrate and vertebrate hosts which are ingested by the final vertebrate host. Eggs of cestodes contain onchosphere larvae, each with three pairs of hooks, and the life-cycle of the more primitive species involves three hosts, the first of which is typically an arthropod. In parasites of terrestrial hosts the life-cycle is abbreviated; only two hosts are required, and the first host may be some animal other than an arthropod. In the order Pseudophyllidea, the onchospheres are encased in a ciliated embryophore and hatch in water. They are eaten by crustaceans in whose bodies they become procercoids; the crustaceans are eaten by fishes and the parasites migrate to the muscles where they become plerocercoids; these larvae mature in the intestine of piscivorous hosts. In the Cyclophyllidea there are no free-living stages; the onchospheres emerge only after the eggs are ingested by a suitable host and develop into either solid-bodied cysticercoids in invertebrate hosts or bladder-like cysticerci in vertebrate hosts.

Different authors recognize from four to eleven orders. A conservative arrangement would include all families in five orders.

Order 1. Tetraphyllidea. Scolex usually complex with four bothridia, vitelline glands in two lateral bands; genital pores lateral; proglottids older and more mature posteriorly; uterus discharges embryonated eggs through ventral perforations of the body-wall; parasites of elasmobranch fishes.

Order 2. Trypanorhyncha. Scolex with two or four bothridia and four long, evertile and retractile proboscides armed with spiral rows of hooks or spines; vitelline glands form a continuous layer over entire proglottid; genital pores lateral; uterine opening ventral, eggs not embryonated; parasites of elasmobranch fishes.

Order 3. Pseudophyllidea. Scolex with bothria; proglottids mature simultaneously; genital pores midventral, vitelline glands in dorsal and ventral sheets; parasites of teleosts and terrestrial vertebrates.

Order 4. Proteocephala. Scolex small, very mobile, with four acetabula and occasionally a basal crumpled collar or an apical fifth sucker; reproductive organs as in the Tetraphyllidea; in freshwater fishes, amphibians and reptiles.

Order 5. Cyclophyllidea. Scolex with acetabula and often with an apical rostellum which may bear hooks or spines; proglottids older and more mature posteriorly; genital pores usually lateral; vitelline gland compact, posterior to the ovary; mostly in birds and mammals.

H. W. STUNKARD

References

Baer, J. G., "Ecology of Animal Parasites," Urbana, Univ. of Illinois Press, 1951.
Hyman, L. H., "The Invertebrates: Platyhelminthes and Rhyncocoela," Vol. 2, New York, McGraw-Hill, 1951.
La Rue, G. R., Exper. Parasitol., 6:306–344, 1957.
Stunkard, H. W., Quart. Rev. Biol., 37:23–34, 1962; 38:221–233, 1963.

PLECOPTERA

Stoneflies are a rather small order of aquatic INSECTS. Among winged insects they are rather primitive. Metamorphosis is hemimetabolous ("incomplete"). All the nymphs and most adults possess segmented cerci or "tails," as in THYSANURA, EPHEMEROPTERA, most ORTHOPTERA and ISOPTERA; on the other hand their winds fold backward when at rest, an advance over the strictly transverse hinging of ODONATA and EPHEMEROPTERA. Structurally the closest allies of Plecoptera are the Orthoptera; in flight they resemble cockroaches or termites.

The stonefly head has two eyes and two or three ocelli, and long antennae. The form of the mouth parts separates the two great divisions of the Order: the Holognatha with heavy mandibles and normal glossae, and the Systellognatha with thin mandibles and reduced glossae. The fore and hind wings are both membranous in texture and are similar in structure, except that the hind wings usually have a large anal fan, much as in grasshoppers. Radial veins are numerous; crossveins are numerous to few, being reduced in the more modern branches of each lineage. Some species, and some individuals of other species, have shortened non-functional wings, especially among the males; a few lack wings entirely. The abdomen always has ten segments; either the first or the last of these may be indistinguishable ventrally. Gills are sometimes present, usually on the thorax, less often on the head or abdomen.

In the male, the terminal (and some more anterior) tergites and sternites, the supra-anal process, the cerci and the paraprocts may all be diversely modified as part of the sexual apparatus. There may also be an eversible aedeagus, often armed with bristles, spines or sclerites. The female genital pore opens beneath the 8th sternite; this sternite and sometimes also the 7th is usually produced backward or otherwise modified.

The known world fauna of Plecoptera comprises approximately 1200 species; about 450 are North American. A conservative taxonomy recognizes nine families,

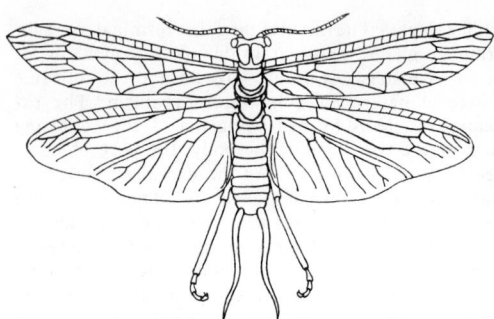

Fig. 1   Stonefly. (From Fox and Fox, "An Introduction to Comparative Entomology," New York, Reinhold, 1964.)

of which the more primitive mostly occupy the Australasian region and southern South America. Stoneflies are rare as fossils. They appear in the Permian along with the more numerous Protoperlaria; a few fragments, usually wings, are known from most of the succeeding periods, but only the middle Tertiary amber has yielded a real fauna, of 15 or more species.

Most stoneflies live in streams or rivers, less often along stony lakeshores, with a strong preference for cool water. The Perlidae, however, are characteristically tropical and warm-temperate, perhaps by virtue of their abundant, finely-dissected gills. The nymphs may develop for one, two or three years. They include carnivores, browsers and detritus feeders, and commonly comprise 5 to 25% of a river's invertebrate fauna, by weight. Adult life lasts up to a week or two. Most species emerge in spring, a few in summer or autumn, and a considerable group in winter. The latter sometimes attract attention as they crawl over snow to a convenient tree trunk, but most stoneflies shun the light and are rarely seen. Many species fast after emerging, others feed on algae, pollen, petals or tender young leaves. Copulation takes place near the water or in treetops not far away, and the females return to the stream to deposit their eggs.

W. E. RICKER

References

Needham, J. G. and P. W. Claassen, "The Plecoptera of North America," Washington, Entomological Society of America, 1925.
Frison, T. H., "The Plecoptera of Illinois," Illinois Nat. Hist. Survey Bull., 20: 281–471, 1935.
Ricker, W. E., "Systematic studies in Plecoptera," Bloomington, Indiana University Publications, Science Series, No. 18, 1952.
Hynes, H. B. N., "A key to the adults and nymphs of British stoneflies (Plecoptera)," London, British Freshwater Biological Association, Scientific Publication, No. 17, 1958.

PLINY THE ELDER (23 A.D.–79 A.D.)

The eruption of Vesuvius in 79 A.D. caused the death of Pliny the Elder, so well known for his 37-volume *Natural History*.

Born 23 A.D., in Transpadane Gaul, most of his life

was associated with the military, with whom he served in various parts of the world. His presence in the area at the time of Vesuvius' eruption was due to the fact that he had been appointed prefect of the Roman fleet at Misenum in Campania.

When not actively occupied with his various military and official duties, he led the life of a self-taught scientist, his study habits being well described by his nephew, Pliny the Younger.

Before daybreak he would call upon the Emperor Vespasian and after completing his official duties, would return home for a day of study. Upon finishing a light meal he would read, continually making notes, laboring under the idea that no book was so bad as not to contain something good. Then after a cold bath and short nap, he would resume reading until dinner, though this did not stop the process, for another book would be read during the meal and continued until he retired for the night.

Though mainly known for his *Natural History,* Pliny wrote widely in other fields, covering subjects ranging from the art of throwing the javelin from horseback to a work encompassing all the wars "waged between us and Germany." Of all the books in the *Natural History,* Books VII–XI, dealing with zoology are undoubtedly the most interesting. It is here that one finds such a fascinating collection of fact and fable. For example, coupled with fairly accurate data on Homo sapiens, one finds such creatures as the Mouthless Men, who subsisted upon the mere fragrance of flower and fruit, the Umbrella-foots, who used their extensive feet as parasols to protect them from the sun, or of men whose feet were turned the wrong way.

His treatises on animals other than man exhibit the same tendency, quotations from Aristotle being primarily responsible for such accurate scientific facts as were present.

It is impossible to read this highly interesting and entertaining work, whether it be Book II, which deals with a mathematico-physical description of the world and heavenly bodies, or Book XXXVII covering gems and precious stones, without being aware that here was a man who truly loved and was interested in all subjects.

LETITIA LANGORD

POGONOPHORA

The phylum Pogonophora—"beard wearing" (Greek, pogonos-beard) is the "youngest" of the phyla in the ANIMAL KINGDOM. It was established by the Russian zoologist A. V. Ivanov in 1955, but the first specimen was discovered by the French zoologist M. Gaullery in 1914 from the depths off the Malay Archipelago. The Soviet expeditions on the vessel "Vitjaz" in the Pacific and Indian Oceans produced much new material, as did also Soviet expeditions to the Antarctic seas and the North Atlantic.

Up to now 72 species are known, distributed among 23 genera, five families and two orders.

The Pogonophora are exclusively marine animals, living in tubes made of organic matter. The most specific external features are the vermiform, thread-like body and the complicated tentacle apparatus on the anterior end.

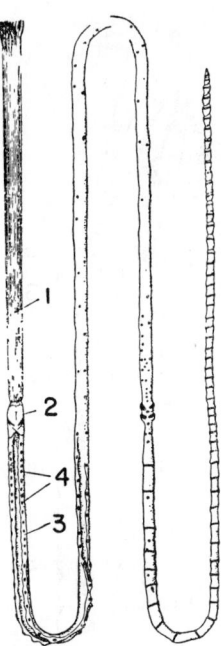

Fig. 1 Lamellisabella zachsi Uschakov—the general view. (After Ivanov, 1952). 1, tentacles; 2, protosome and mesosome; 3, metasome; 4, adhesive papillae.

The External Structure (Fig. 1)

The length of Pogonophora varies greatly. There are short specimens, only a few millimeters long, and others reaching several tens of centimeters. As a rule the tube is much longer than the animal itself. The length of the animal is many times the diameter, as in the genus Heterobrachia where the length is 270 times the diameter.

The trunk of Pogonophora has three sections, *protosome, mesosome* and *metasome.* The protosome bears the cephalic lobe on the dorsal side while the abdominal side carries the tentacular apparatus. The number of the tentacles varies greatly, from one in the genus Siboglinum to 223, in Polybrachia. The tentacles are more or less fused and form a cylinder (Lamellisabella) or a complicated spiral (Spirobrachia) with an internal cavity. Even when there is only a single tentacle, it twists to form a cavity inside itself. The external integument of the metasome is thickened in parts forming circular ridges of adhesive papillae, on which some species have cushions, provided with small chitin platelets. Apparently, these aid the animals to attach to, and move within, the tube. Sometimes the hinder end of the metasome shows pseudo-segmentation.

The Internal Structure (Fig. 2)

The Pogonophora are coelomate animals. There is one unpaired coelom in the protosome, from which caeca extend into the tentacles. In the mesosome and in the metasome there are paired coelomic sacs. *The complete absence of a digestive tract is one of the most distinctive features of the Pogonophora.* A. Ivanov (1955) has suggested that the nutrition of the Pogonophora is

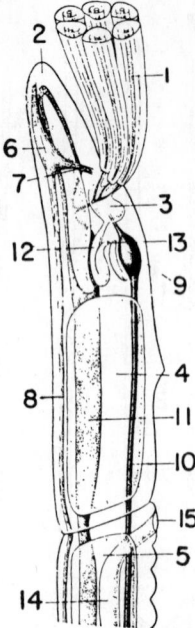

Fig. 2 Scheme of the internal structure of the Pogonophora; anterior end, looking from one side. (After Ivanov, 1955). 1, tentacles; 2, cephalic lobe; 3, protocoel; 4, mesocoel; 5, metacoel; 6, cephalic ganglion; 7, tentacular nerve; 8, dorsal nerve; 9, heart; 10, ventral blood vessel; 11, dorsal blood vessel; 12, coelomoduct; 13, its outlet; 14, gonad; 15, its outlet.

a function of the tentacles. As was mentioned above, there is a cavity inside the tentacular apparatus. The cilia, arranged on the tentacles within the cavity filter small particles from the surrounding water. In the epithelium of the tentacles there are glandular cells, which secrete digestive enzymes into the cavity. The digested food material is then absorbed by the epithelial cells of the tentacle, and is distributed through the body by the circulatory system. The Pogonophora thus exhibit extracellular digesting.

Respiratory organs are absent. Interchange of gases is carried out through the tentacles. A closed **circulatory system** consists of abdominal and dorsal blood vessels. The abdominal vessel at the roots of the tentacles forms a rather muscular bag (heart), which pushes the blood into the tentacles, from which it runs backwards through the dorsal vessel.

Excretion system. A pair of coelomoducts in the unpaired coelom of the protosome serves as the organs of excretion.

The nervous system is interaepidermal. There is a ganglionic mass in the cephalic lobe, from which arise a dorsal nervous trunk, and the nerves leading to the tentacles. Organs of sense are not found.

The reproductive system. In the Pogonophora the sexes are separate. The gonads are located in the coelomic sacs of the metasome. Rather complicated spermatophores are a characteristic feature, and various species show varying structures of the spermatophores.

Embryology

The female deposits eggs in the tube in which it lives, where the whole development takes place from fertilization on. There is no free-swimming larva.

The development has some unique features, which are probably a secondary character. These are: uneven segmentation, without any signs of radial or spiral cleavage; the poorly demarcated stages of segmentation and gastrulation. There is no real gastrula, therefore the blastocoel, the blastophore and archenteron are not formed. The coeloms are formed enterocoelously. The complete embryo has a ciliated band on the protosome, and the dorsal surface of the mesosome is partly covered by cilia. There are also ciliated bands on the metasome and two to four bundles of short, and rather thick bristles, two in each bundle. According to Ivanov the embryo's ciliated bands are a remnant of locomotor formations, retained apparently from free-swimming larvae, of ancestral forms. As to the bristles, Ivanov thinks that they are a temporary structure used by the larva for the support of the body in the tube.

The analysis of the structure and the development of Pogonophora permits the following conclusions about systematic position.

(1) The Pogonophora undoubtedly belong to the Deuterostomia as is shown by the enterocoelous formation of the coelom and by the position of the nervous trunk along dorsal surface of the whole body.

(2) Among the Deuterostomia the Pogonophora must be distinguished as a separate phylum. This is confirmed by such special features as the unique tentacular apparatus, the complete absence of the digestive tract, and the differentiation of the heart on the abdominal blood vessel.

It is not possible to place the Pogonophora in the

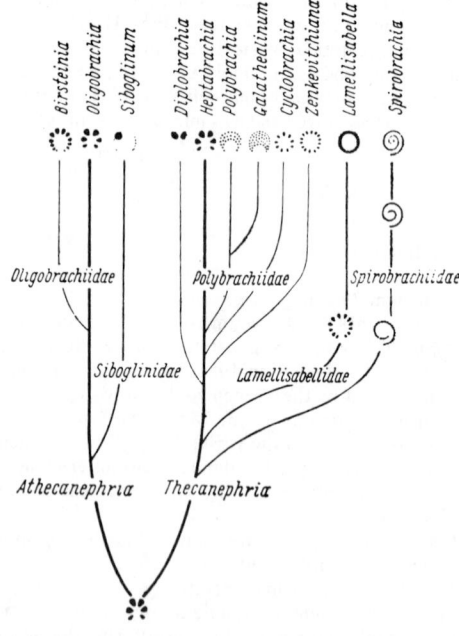

Fig. 3 Scheme of the phylogenetic relationships of the Pogonophora. (After Ivanov, 1960.)

phylum Hemichordata (together with the Enteropneusta and the Pterobranchia), as is suggested by some authors (V. N. Beklemishev, 1955), because they do not have such characteristic features of that phylum, as the stomatochordia and branchiate fissures. There is no basis for the suggestion that the Pogonophora are simply aberrant sedentary polychaetae (O. Hartman, 1954).

It can be said of the phylogenetic relations among Pogonophora that their evolution is connected with the progressive development of the tentacular apparatus, because of the physiological significance of this organ. Ivanov has developed a diagram of the phylogenetic relations of the Pogonophora (Fig. 3).

The Ecology and Geographical Distribution

The Pogonophora are exclusively marine animals. They were first discovered in abyssal depths which gave rise to the supposition that this was their sole habitat. The later discovery of some specimens in the middle, and even shallow, depths changed this view. Some Pogonophora are eurybathic; for example, *Siboglinum caulleryi* was found in the depths ranging from 23 to 8100 m.

Pogonophora are very widely distributed in the oceans of the world. Up to the present time they have been found all over the Pacific Ocean, including the Bering, Okhotsk and Japan Seas. They also occur in the Indian Ocean, the Antarctic, the Polar Basin, the Barents Sea, the North Atlantic, the Bay of Biscay, and the Skagerrack.

All this raises the question as to why Pogonophora were discovered so recently. This may be partly explained by the fact that the exploration of the abyssal depth of the World Oceans, where the greater part of the Pogonophora are found, was begun on the wide scale relatively recently. But the main reason is that the Pogonophora were simply overlooked. As has been pointed out, the animal itself occupies only a part of the tube, so these tubes often have the appearance of being empty, and apparently were considered as the empty tubes of some polychaetae. Besides, the very thin, thread-like, form of the animal was so similar to the scraps of the nets of the trawls and dredges, that it did not attract the notice of collectors.

G. G. ABRIKOSSOV

References

Hyman, L. H., "The Invertebrates" vol. 5, 208–227, New York, McGraw-Hill, 1959.

Ivanov, A. V., "Pogonosphora," New York, Consultants Bureau, 1963.

Southard, A. and E. Southard, "On some Pogonophora from the North-East Atlantic, including two new species," J. Mar. Biol. Assoc. U. K. **37**: 627–632, 1958.

Southward, E., "Two new species of Pogonophora from the North-East Atlantic," J. Mar. Biol. Assoc. U. K. **38**: 439–444, 1959.

POLYCHAETA

The Polychaeta constitute a class of ANNELIDA, sometimes considered, in contrast to the OLIGOCHAETA, as

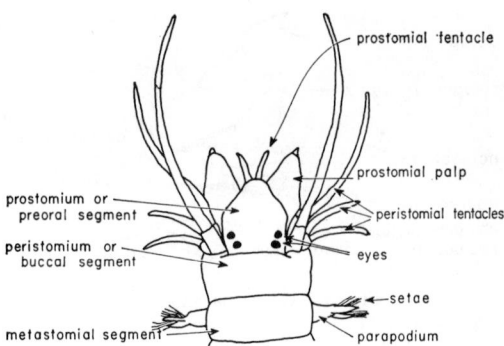

Fig. 1   Errant type polychaete, *Nereis*.

an order under the Chaetopoda. They are abundantly represented in the sea and estuaries, a few being fresh water.

Polychaete worms are generally elongate and cylindrical, sometimes flattened and compressed. The body may be separated into four divisions: a prostomium, provided variously with tentacles, palps, tentacular filaments, branchial feathery crown, eyespots, and ciliated pits; the eyes may be remarkably complex in the pelagic Alciopidae; a peristomium or buccal segment surrounding the mouth, sometimes provided with few to numerous tentacles; a metastomium consisting of a number of similar or dissimilar segments each with a pair of parapodia more or less developed; a posterior pygidium perforated by the anus, sometimes with anal cirri, disc or funnel. New segments are added in a growing zone just anterior to the pygidium. The parapodia, when well developed, are muscular lateral locomotor processes consisting of a basal portion, supported by internal spines (acicula) from which the parapodial muscles originate, and divided distally into a dorsal notopodium and a ventral neuropodium, both bearing numerous chitinoid fibrous setae of endless variety set in pockets in the integument; additional cirri, tongue-like lobes (ligules), lamellae, and gills may be present. The body wall consists of a thin, proteid, iridescent cuticle, epidermis, circular and longitudinal muscle bundles, and parietal peritoneum.

Some errant polychaetes, fitted for an active life, crawling, swimming, predatory, are equipped with sensory head appendages, powerful muscular eversible pharynx with jaws, and prominent parapodia, for example, the clam worm, *Nereis*, the blood worm, *Glycera*, and a pelagic form, *Tomopteris*. Other forms burrow in the bottom swallowing large quantities of soil by means of a saccular, papillate eversible pharynx,

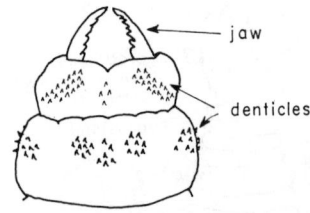

Fig. 2   *Nereis*, extended pharynx on proboscis.

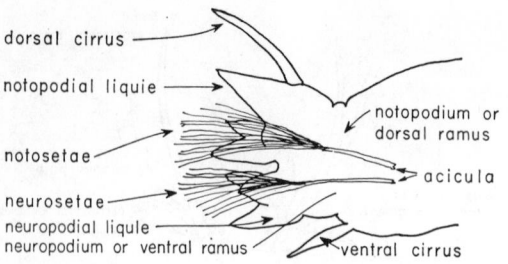

Fig. 3   *Nereis*, parapodium.

subsisting on the organic matter contained therein and giving off abundant castings or faeces, for example the lugworm, *Arenicola.* Some sedentary forms are tube-dwelling, feeding on microscopic food collected by ciliated mucus grooves, for example, *Amphitrite,* with numerous long extensible and contractile tentacular filaments, and the fan worms, as *Sabella* and *Serpula,* with a more rigid branchial plume; the parapodia are dissimilar, some reduced to fin-like extensions (pinnules) or to low ridges (tori) supporting the numerous small hooklike setae (uncini); undulatory pumping movements of the body wall bring about renewal of the water around the body for respiration; the tubes may be calcareous, of mucus or parchment-like material in which foreign material may be incorporated. Commensalism in varying degrees is frequent. Parasitism is rare.

Circulatory organs, when present, are closed, with some contractile vessels. The blood is red (hemoglobin in the plasma), green (chlorocruorin in the plasma), pink (hemerythrin in cells) or colorless. The coelom is a spacious perivisceral cavity, usually divided into incomplete chambers by intersegmental transverse septa, consisting of a double fold of peritoneum enclosing muscle fibers. The abundant coelomic fluid is corpusculated and may be red (hemoglobin in cells or in the fluid), with a ciliary circulation. It functions in respiration and in burrowing by acting as a hydraulic skeleton in maintaining turgor. The digestive tract consists of a mouth, buccal cavity, muscular pharynx, esophagus

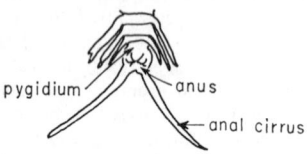

Fig. 4   *Nereis*, posterior end.

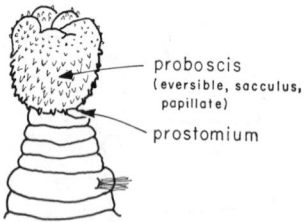

Fig. 5   Burrowing type polychaete, *Arenicola.*

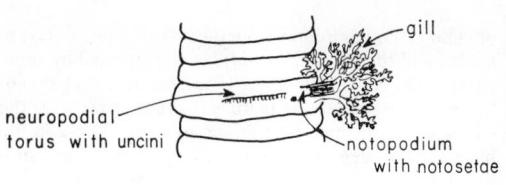

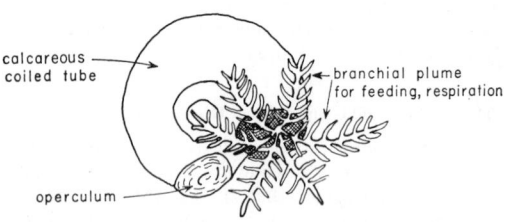

Fig. 7   Sedentary type polychaete, *Spirorbis.*

with a pair of caeca, simple or branched stomach-intestine, rectum and anus. The gut wall consists of visceral peritoneum of yellow cells, excretory in function, muscles, and mucosa, which is generally ciliated. The nervous system consists of dorsal bilobed cerebral ganglia and subpharyngeal ganglia with connectives and separate or closely applied ventral longitudinal ganglionated cords. Giant fibers may be present in the ventral cord, concerned with quick escape or withdrawal movements. A stomato-gastric system of nerves also occurs.

Primitively each segment is provided with a pair of genital tubes (coelomoducts) and a pair of excretory tubes (nephridia); these may be variously developed and united to form segmental organs. With few exceptions polychaetes are dioecious. Gonads develop by proliferation of certain parts of the coelomic peritoneum, usually throughout the greater part of their length, the developing ova and sperm causing distension of the coelom. Some polychaetes undergo remarkable transformations at the time of reproduction, with the acquisition of swimming setae and other structural changes associated with a pelagic existence, for example the heteronereis stage of *Nereis.* Certain regions containing the sexual elements may become modified and set free from the rest of the animal, as for example the famous Palolo worm. Fertilization is generally external, the sex products escaping through nephridiopores or by rupture of the body wall. The sexual adults often swarm on the bottom or swim toward the surface. Development is direct or by means of trochophore larvae and postlarval forms. Brood protection may occur; viviparity is rare. Asexual reproduction by fission or budding may occur. Regeneration of lost parts is common.

MARIAN H. PETTIBONE

References

Hartman, Olga, "Literature of the Polychaetous Annelids," vol. 1, Bibliography, Los Angeles, 1951.
Hartman, Olga, "Catalogue of the Polychaetous Annelids of the World, Pts. 1 and 2," Los Angeles, Allan Hancock Foundation Publications, 1959.

MacGinitie, G. E. and Nettie, "Natural History of Marine Animals," New York, McGraw-Hill, 1951.

POLYMORPHISM

While in the past under the head of polymorphism several aspects of VARIATION—such as geographical, seasonal and non-genetic variation, as well as mere recombinational variance, whether with or without previous HYBRIDISATION—were often loosely included, in recent years this term has been used in a more restricted and well defined sense. The definition generally accepted was given in 1940 by E. B. Ford: polymorphism is the occurrence together in the same habitat of two or more distinct forms of a species in such proportions that the rarest of them cannot be maintained by recurrent mutation. Seasonal and geographic variation is thus excluded for polymorphism concerns only those forms which occur together in the same habitat. Continuous variation—such as that falling within a single normal distribution—is not included in this definition in as much as polymorphism relates to relatively sharply contrasted differences which either do not overlap or else give rise to a bi-modal (or multi-modal) curve. When genetic variants coexist in temporary or permanent balance within a single interbreeding population in a single region and in frequencies higher than their respective mutation rates we speak of polymorphism. J. S. Huxley has proposed to substitute the term "morphism" (and its derivatives "morphic" and "morph") to the unnecessarily lengthy polymorphism.

Polymorphism is a very widespread phenomenon in the animal and plant kingdoms. Among animals it occurs in all classes of Vertebrates, and in the Urochorda, among Chordates; among Insects, in the Coleoptera, Diptera, Lepidoptera, Hymenoptera, Odonata, Dermaptera, Hemiptera, Orthoptera, and possibly Isoptera; among Aracnida, in Spiders; among Crustacea, in Copepoda, Isopoda and Decapoda; among Echinoderms, in Asteroidea, Holothuroidea and Ophiuroidea; among Molluscs, in Gastropoda and Lamellibranchia; among Coelentherates, in Anthozoa and possibly in Leptomedusae; among Sponges, in Demospongiae. Among plants it occurs in many orders of both Dicotyledons and Monocotyledons, and several groups of Fungi. The characters involved in polymorphism are quite varied. They include colour, pattern, structure, secondary sexual characters, caste in certain insects, essential oils in certain trees, reproductive incompatibility, clutch-size in birds, blood-groups, blood diseases, chromosome size, shape and number, chromosomal rearrangements, migratory behavior, heterokaryosis in fungi, sensory thresholds, general habit, temperature-tolerance, disease-resistance, cancer-proneness and any other conceivable character whether morphological, physiological or biochemical.

Polymorphism provides a method of intraspecific differentiation for adapting the species to sets of sharply distinct environmental conditions. If a new trait in an organism is originated by mutation and is disadvantageous to its survival, the mutant gene responsible will be subject to adverse selection and will remain always very rare in the population. Being constantly eliminated, it will be found in a population at frequencies corresponding to its mutation rate, i.e., very low. If, however, the new trait originated by mutation is advantageous, it will be favoured by the environment and its frequency will increase to values higher than mutation rates. If a genetically controlled variety is represented even by a few per cent of a population we can be fairly sure that it possesses some advantage. If this genetic variety is sharply enough distinct from the rest of the population we observe the occurrence of polymorphism. When an advantageous gene is in the process of spreading through a population we observe *transient* polymorphism. When a genetic condition is maintained at some fairly fixed level by a balance of selective forces we observe *balanced* polymorphism.

Transient polymorphism is maintained only during the time that a gene is spreading through a population and is in the process of replacing its allele. When it has reduced this to a very low frequency, though it was formerly the normal one, a new type has been reached and polymorphism no longer exists. Owing to the recurrent nature of mutation, transient polymorphism is likely to occur when the environment changes in such a way as to render the effects of a previously disadvantageous gene beneficial. Transient polymorphism has been observed in the spread of a blackish form of the hamster *Cricetus cricetus* L. in Central Russia; in the relative increase of the black form of the opossum *Trichosurus vulpecola* Kerr in Tasmania; etc. The best documented and most impressive case of transient polymorphism is that of the industrial melanism in moths. In certain industrial areas almost the whole population of moths, belonging to several species, has become black during the last hundred years, resulting in the most considerable evolutionary change which has ever been witnessed. The spread of the forms is due to the fact that mutants occurring as rarities in the ordinary populations of non-polluted areas have a selective advantage over the normal form when the environment becomes black because of smoke and soot.

Balanced polymorphism generally involves a high degree of permanence in the ratio of different forms. Variations in the relative frequencies of polymorphic forms may follow cycles, e.g., seasonal. Polymorphic forms may be distributed as clines, which may or may not lead from groups which are monomorphic. It has been shown in many cases that balanced polymorphism is the result of a selective advantage of the heterozygote—for one or more genes, or for more complex genetic conditions, such as chromosome rearrangements—over both homozygotes. The actual values of the observed ratios depend upon the relative selective values of the heterozygous and either homozygous forms. As a result of such type of heterosis lethal or highly deleterious recessive mutants can be maintained at relatively high frequencies in populations, even though the homozygotes are constantly eliminated.

Polymorphism is widely distributed in the human species. The most outstanding examples are: blood-groups, taste sensitivity to phenylthiocarbamide, thalassemia, sickle cell anemia and favism. In the case of sickle cell anemia the balanced polymorphism is probably maintained by the positive selective advantage of the heterozygote over the homozygous normal when infected by malaria.

A. A. BUZZATI-TRAVERSO

Reference

Ford, E. B., "Genetic Polymorphism," Cambridge, M.I.T. Press, 1965.

POLYPLOIDY

The presence, usually in the same genus, of related forms having CHROMOSOME numbers which are multiples of a basic number. Widespread in plants, though rare in some groups, such as fungi and gymnosperms; present in about ⅓ of the species of flowering plants (angiosperms). Rare in animals, the only clearly authenticated examples being in parthenogenetically reproducing forms, such as earthworms, *Artemia* (Crustacea), *Solenobia* (Lepidoptera), and some Curculionidae (Coleoptera); and in a few hermaphrodites, such as certain turbellarian flatworms. The rarity of polyploidy in animals is due partly to the fact that in many forms polypolidy disturbs the sex ratio; but the apparent absence of polyploidy in many groups of hermaphroditic animals, such as snails, and its presence in plant genera with separate sexes (*Salix*) suggests that other factors must also be considered. In urodele amphibia artificially produced tetraploids have impaired metabolism and inefficient nervous reactions, suggesting that physiological unbalance may hinder the establishment of polyploidy in some animals. Polyploids derived from interspecific hybrids with reduced chromosomal pairing (allopolyploids, see below) are very unlikely to occur in animals, partly because hybridization is rarer in animals than in plants, but chiefly because most species hybrids in animals have, in addition to chromosomal nonhomology, genic unbalance, leading to physiological abnormalities which are not corrected by polyploidy.

In flowering plants many genera contain polyploid series of chromosome numbers. An example is wheat (*Triticum*), with somatic numbers of 2n = 14, 28, and 42, the basic gametic number being x = 7. Such series are more common in perennial herbs than in annuals or woody plants, and are irregularly distributed among families of angiosperms. Frequencies are high in Rosaceae, Malvaceae, and Gramineae; intermediate in Ranunclaceae, Leguminosae, Cruciferae, Compositae, and Liliaceae; and low in Fagaceae, Moraceae, and Umbelliferae. Many examples exist of pairs of related genera, one with and one without polyploidy, such as *Salix* and *Populus, Betula* and *Alnus, Thalictrum* and *Aquilegia*.

Polyploids have been produced artificially in many species of crop plants and garden ornamentals, chiefly with the aid of the drug colchicine. Compared to their diploid progenitors, such plants have stouter stems, thicker leaves, larger flowers and seeds, slower growth, later flowering, and usually reduced fertility. Polyploids derived from a single ancestral species are termed *autopolyploids*. They are most often recognized by the formation of multivalent chromosomal associations and by complex tetrasomic segregation ratios. Other artificial polyploids have been produced from hybrids between widely different species, and are termed *allopolyploids*. If the chromosomes of the parental species are nonhomologous, pairing in the diploid interspecific hybrid is much reduced or absent, but since its allopolyploid derivative contains the complete somatic complement of both parental species, its meiotic cells exhibit pairing between homologous chromosomes derived from the same parent, and consequently its gametes contain the complete haploid complements or genomes of both parental species. This renders it both fertile and true breeding for the intermediate, hybrid condition.

Both auto- and allopolyploids are found in nature, but recent studies of some natural autopolyploids (*Dactylis, Medicago*) indicate that they have originated from hybrids between adaptively different ecotypes or subspecies of the same species. Since autopolyploids from a single strain usually have reduced vigor and fertility, most vigorous natural autopolyploids may have originated from intraspecific hybridization. Some natural autopolyploids resemble allopolyploids in having a low frequency or absence of multivalents. Natural polyploids derived from interspecific hybrids include not only typical allopolyploids, such as cultivated tobacco (*Nicotiana tabacum*) and the New World cottons (*Gossypium barbadense, G. hirsutum*), but also polyploids derived from hybrids between closely related species, such as *Zauschneria californica, Achillea collina,* and *Solanum nigrum.* These may have some characteristics of autopolyploids, such as multivalent chromosome association, and the presence of four sets of morphologically similar chromosomes. They often resemble one of their diploid ancestors closely enough so that taxonomic separation is difficult, and the two can be placed in the same species. Hence neither multivalent association, somatic chromosome morphology, nor taxonomic position provide by themselves reliable guides for distinguishing between auto- and allopolyploids. Some polyploids, like bread wheat (*Triticum aestivum*), have genomes sufficiently alike so that individual chromosomes derived from one ancestral diploid species can be substituted for the corresponding chromosome derived from a different species. Species complexes containing diploids and derived polyploids, known as *polyploid complexes,* are among the most difficult of plant groups for taxonomic analysis. Nevertheless, since polyploids are always derived from diploid ancestors and their origin can often be duplicated experimentally, such complexes provide a unidirectional phylogeny which is very useful in interpreting some aspects of phylogeny and plant geography.

The geographic distribution of polyploids usually differs from that of their diploid relatives. The floras of some northern regions (Iceland, Spitzbergen) contain very high percentages of polyploids, as do also Ceylon and New Zealand, which have milder, or even tropical climates. In the European Alps, the regions near the level of permanent snow and ice do not contain a higher percentage than the surrounding plains. Hence the incidence of polyploidy is imperfectly if at all correlated with increasing climatic severity. When diploids and polyploids of individual genera are compared (*Crepis, Antennaria, Achillea, Zauschneria, Eriogonum*), the polyploids are found to occupy more recently available habitats, indicating that habitat disturbance greatly aids the spread of polyploids.

Artificial polyploids of some cultivated plants have had moderate success as improved varieties, the most valuable being those of rye, sugar beets, oil seed rape, red clover, snapdragons, marigolds, various orchids, and some others, particularly garden ornamentals. Success is achieved only when intervarietal hybridization and selection have accompanied chromosomal doubling. Polyploids have also served as a medium for transferring by hybridization disease resistance to a cultivated species from a distantly related wild species; examples are rust resistance in bread wheat derived from *Aegilops umbellulata* (goat grass), and resistance to mosaic disease and black shank in tobacco, derived respectively from *Nicotiana glutinosa* and *N. plumbaginifolia*.

G. LEDYARD STEBBINS

Reference

Stebbins, G. L., "Variation and Evolution in Plants," New York, Columbia University Press, 1950.

POLYZOA see BRYOZOA

POPULATION GENETICS

Population genetics studies the consequences of mendelian inheritance on the populational level, in contradistinction to classical GENETICS which studies inheritance on the familial level. In the latter case we have one or more families and study the proportions or ratios of the various genotypes and phenotypes in the families; in the former case, analogously, we have one or more populations and study the proportions or ratios of the various genotypes and phenotypes in the populations.

Mendelian population, the basic unit of study in population genetics, is loosely defined as a group of individuals (plants, animals, or men) who interbreed and form a community by themselves. The group may be large or small. The aggregate of all the genes in a group is referred to as the "gene pool" of the population. Usually there is no clear cut boundary between two neighboring mendelian populations, because of the migrations of individuals between them. However, if the amount of migration is small, these partially isolated populations can still show considerable difference in their gene pools (genetic differentiation) if the environmental conditions for the two populations differ. Further, a population is immortal, as new generations are continuously replacing the old ones. Thus, population genetics studies not only the status quo of a population or the changes from place to place, but also the changes from generation to generation.

Gene frequency is the most important single index to characterize a mendelian population. Consider a group of 1000 diploid individuals and limit our attention to only one locus with two alleles, *A* and *a*. There are $2 \times 1000 = 2000$ genes for this locus. Suppose that 1200 of them are *A* and 800 *a*. We say that the gene frequency of allele *A* is $p = 1200/2000 = 0.60$ and that of *a* is $q = 800/2000 = 0.40$, so that $p + q = 1$. The gene frequency always refers to a certain locus. Its value varies from locus to locus and from population to population.

The knowledge of the gene frequency alone, however, does not tell us the number or proportion of the three genotypes in the population. Let D, H, R denote the *number* of *AA*, *Aa*, *aa* individuals respectively in the following populations:

Population I				Population II				Population III			
AA	*Aa*	*aa*	Total	*AA*	*Aa*	*aa*	Total	*AA*	*Aa*	*aa*	Total
D	*H*	*R*	*N*	*D*	*H*	*R*	*N*	*D*	*H*	*R*	*N*
360	480	160	1000	390	420	190	1000	600	0	400	1000

Each *AA* individual has two *A* genes; each *aa* has two *a*; each heterozygote has one *A* and one *a*. Thus, in each of the populations above, the gene frequencies are:

$$p = \frac{2D + H}{2N} = \frac{1200}{2000} = .60,$$

$$q = \frac{H + 2R}{2N} = \frac{800}{2000} = .40$$

For large populations, it is convenient to deal with the proportions rather than numbers of the various genotypes. Thus, $d = D/N$ is the proportion of *AA* individuals in a population and $d + h + r = 1$. The relationship between the gene frequencies (p, q) and the genotypic proportions (d, h, r) is determined by the *mating system* of the population.

Random mating (Panmixis) is a mating system in which any member of one sex is equally likely to mate with any member of the opposite sex. In other words, the mating of a male with a female is a chance event, occurring "at random." In population genetics, it is understood that random mating is relative to certain genetic characteristics of the mating individuals; it is not an absolute term. For example, in a human population the mating is at random with respect to the blood groups, but *not* at random with respect to height, education level, social and economic status. The non-randomness in one respect does not nullify the random nature in others.

Hardy-Weinberg law. If the mating is at random with respect to the genetic characteristics determined by the alleles *A* and *a*, the gene frequencies and genotypic proportions have a very simple relationship. *Random mating of individuals is equivalent to random union of gametes.* Thus we have the following situation:

		female gametes	
		$p(A)$	$q(a)$
male gametes	$p(A)$	$p^2(AA)$	$pq(Aa)$
	$q(a)$	$pq(Aa)$	$q^2(aa)$

The proportions of the genotypes in such a panmictic population are

$$d = p^2(AA), \quad h = 2pq(Aa), \quad r = q^2(aa)$$

e.g.

$$.36(AA), \quad .48(Aa), \quad .16(aa)$$

where $p = .6$ and $q = .4$ as in Population I shown previously.

In the absence of disturbing forces and differential fertility among the various types of families, the random mating population $(d, h, r) = (p^2, 2pq, q^2)$ remains the same generation after generation. We say that the population is in an *equilibrium* state. This is seen from the fact that the frequency of *A*-gametes produced by the population as a whole is $d + \frac{1}{2}h = p^2 + pq = p$. Weinberg (1908) gives a more detailed and instructive demonstration that the population $(p^2, 2pq, q^2)$ produces on random mating an identical offspring population by enumerating all types of families and their offspring. His table is essentially as the following one:

Type of mating	Frequency of mating	Offspring		
		AA	Aa	aa
$AA \times AA$	p^4	p^4	0	0
$AA \times Aa$	$4p^3q$	$2p^3q$	$2p^3q$	0
$Aa \times Aa$	$4p^2q^2$	p^2q^2	$2p^2q^2$	p^2q^2
$AA \times aa$	$2p^2q^2$	0	$2p^2q^2$	0
$Aa \times aa$	$4pq^3$	0	$2pq^3$	$2pq^3$
$aa \times aa$	q^4	0	0	q^4
Total	1.00	p^2	$2pq$	q^2

Hardy (1908) shows that the population is "stable" (i.e. remaining the same from generation to generation) on the condition $(\tfrac{1}{2})^2 = dr$, or $h^2 = 4dr$, which is equivalent to saying $(d, h, r) = (p^2, 2pq, q^2)$. Hence this is known as the Hardy-Weinberg equilibrium law for large random mating populations. Hardy has also proved that for any given arbitrary population, the equilibrium state is reached in one generation of random mating.   For most human populations it is safe to take the genotypic proportions to be $p^2, 2pq, q^2$.

For sex-linked genes, the genotypic proportions in the females (XX) are $(p^2, 2pq, q^2)$ and those in the males $(XY$ or $XO)$ are simply (p, q) in equilibrium condition. The proportions for the nine genotypes with respect to two pairs of genes are given by the terms of $(p_1A + q_1a)^2(p_2B + q_2b)^2$ in the equilibrium condition whether the two loci $(A, a$ and $B, b)$ are independent or linked. In these two cases, the equilibrium condition is not reached in one generation of random mating but is approached quite rapidly (in ten to fifteen generations). For multiple alleles, A_1, $A_2, \ldots, A_k$ with gene frequencies $p_1, p_2, \ldots, p_k$, the equilibrium genotypic proportions are given by the terms of $(p_1 + p_2 + \cdots + p_k)^2$ and this condition is reached in one generation of random mating.

Inbreeding population is one in which genetically related individuals mate more often than dictated by chance. Mating between relatives leads to correlation between the male and female gametes that unit to form the offspring. The situation may be represented as follows:

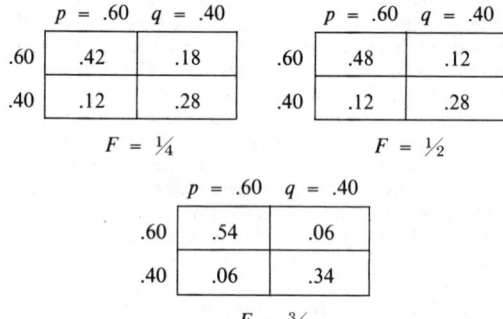

in which ϵ represents a positive fraction. Let F be the correlation coefficient between the uniting gametes. Calculation yields that $F = \epsilon/pq$ or $\epsilon = Fpq$. Wright (1921) defined F as the *inbreeding coefficient* of the population. Inbreeding increases the proportion of homozygotes at the expense of that of heterozygotes. The genotypic proportions with inbreeding coefficient F are:

$$AA: \ d = p^2 + Fpq = (1 - F)p^2 + Fp$$

$$Aa: \ h = 2pq - 2Fpq = 2(1 - F)pq$$
$$aa: \ r = q^2 + Fpq = (1 - F)q^2 + Fq$$

That inbreeding does not change the gene frequency but introduces association between the uniting gametes may be illustrated by the following two series of tables.

(i) gene frequency (p) increasing, but no inbreeding:

$p = .30 \quad q = .70$

.30	.09	.21
.70	.21	.49

$F = 0$

$p = .60 \quad q = .40$

.60	.36	.24
.40	.24	.16

$F = 0$

$p = .90 \quad q = .10$

.90	.81	.09
.10	.09	.01

$F = 0$

(ii) gene frequency (p) constant, but inbreeding increases:

$p = .60 \quad q = .40$

.60	.42	.18
.40	.12	.28

$F = \tfrac{1}{4}$

$p = .60 \quad q = .40$

.60	.48	.12
.40	.12	.28

$F = \tfrac{1}{2}$

$p = .60 \quad q = .40$

.60	.54	.06
.40	.06	.34

$F = \tfrac{3}{4}$

Continued systematic inbreeding. The most intensive form of inbreeding is *selfing* as practiced in many plants. The proportion of heterozygotes is halved in each generation of selfing, viz., $h = \tfrac{1}{2}h'$ where the prime indicates that of the preceding generation. The eventual result of continued selfing is complete homozygosity or "fixation" ($h = 0$, and $F = 1$) for all loci.

For bisexual organisms, the most intensive inbreeding system is continued *sib mating* (brother × sister). The proportion of heterozygosity decreases according to the rule $h = \tfrac{1}{2}h' + \tfrac{1}{4}h''$. The rate of decrease is slower than that of selfing but is still fairly rapid. After a number of generations, the heterozygosity proportion may be approximated by the relation $h = .809h'$, so that the decrease is 19.1% per generation. The ultimate result is also complete homozygosity.   The pure inbred strains of experimental animals are obtained essentially in this manner.

Many other inbreeding systems have been devised by animal breeders. The general conclusion is that continued mating between close relatives leads to complete fixation while mating between remote relatives has little effect on homozygosity.

Inbreeding in man. Marriage between very close relatives is forbidden and very rare; marriages between distant relatives has no genetic significance. The only kind of *consanguineous mating* that is of interest is marriage between first cousins. In a panmictic population such as that of man, the *children* of first-cousin-parents have an inbreeding coefficient $F = \frac{1}{16}$ so that the proportion of recessives (*aa*) among such children is

$$r_1 = Fq + (1 - F)q^2 = \frac{1}{16}q + \frac{15}{16}q^2.$$

This proportion is always larger than $r_0 = q^2$ for children of random parents. The ratio $r_1/r_0 = (1 + 15q)/16q$ is considerably higher than unity only when the frequency (q) of the recessive gene is low as shown in the following:

q	r_0	r_1	r_1/r_0
.20	.0400	.0500	1.25
.04	.0016	.0040	2.50
.01	.0001	.000719	7.19

If the genotype *aa* represents a *rare recessive* hereditary abnormality, its incidence among the children of first-cousin-parents will be higher than that among children from random parents. This is the so-called harmful effect of cousin marriages. The effect is important only with respect to very rare recessive hereditary diseases and negligible for common traits. The increase in incidence of *aa* (and *AA*) is entirely due to the correlation of the cousins' gametes; there is no increase in the recessive gene frequency among the cousins' children. In most human populations cousin marriages constitute only a small fraction of all marriages, and the increase in incidence of hereditary diseases due to such matings is, populationwise, negligible.

Assortative mating is a mating system in which individuals mate with preference on the basis of their phenotypes (where inbreeding is based on the relationship of the mates). If individuals of the same phenotype mate more often than by chance, it is positive assortative mating. If individuals of unlike phenotypes mate more often than by chance, it is negative assortative. The mating between a male and a female may be regarded as an extreme form of complete negative assortative mating, resulting in equal proportion of males and females. In general, assortative mating leads to a certain equilibrium condition different from that through inbreeding.

Forces that Change Gene Frequencies

Gene mutations occur at random in nature at a certain low rate. The causes for mutation remain unknown ("spontaneous"). Mutation rate varies from locus to locus. Let μ (e.g. 4×10^{-6}, 5×10^{-7}, etc.) be the mutation rate from allele A to a per generation. If the frequency of A is p_0 in any given generation, then $p_0 \cdot \mu$ of them will mutate to a in one generation, so that the frequency of A in the next generation will be $p_1 = p_0 - p_0 \cdot \mu = p_0(1 - \mu)$. After t generations, $p_t = p_0(1 - \mu)^t = p_0 e^{-t\mu}$; and $p_t \to 0$ as $t \to \infty$. This shows that an unopposed mutation can gradually change the gene frequency in a population.

Reverse mutations. Let ν be the mutation rate in the reverse direction, viz., from a to A. Then in each generation $q\nu$ of the a alleles will mutate to A, while $p\mu$ mutates from A to a. The simultaneous changes will cancel each other when $p\mu = q\nu$. Under this condition, there will be no change in gene frequency although the mutations are still occurring at the specified rates. The gene frequencies at this equilibrium state are $p:q = \nu:\mu$ or $p = \nu/(\mu + \nu)$ and $q = \mu/(\mu + \nu)$. If the opposing mutation rates are of the same order of magnitude, the equilibrium gene frequencies will be at intermediate value (around 0.50).

Genotypic selection (differential fertility, reproductive fitness). If not all genotypes produce the same average number of children, there will be a change in gene frequency in the offspring generation. Let $w_{11}:w_{12}:w_{22}$ denote the relative reproductive value of AA, Aa, aa, respectively. Since only the relative magnitudes of the w's affect the final result, we can always take one of them as unity, as exemplified in the following table. In this article the w's are assumed to be constants, and will be referred to as the "fitness" value for the sake of brevity.

In a random mating population the relative genotypic proportions after selection will be $p^2 w_{11}$, $2pq w_{12}$, $q^2 w_{22}$. The sum of these three quantities may be denoted by $\overline{w}$, the average fitness for the population. The new gene frequency after selection is $q' = (pq w_{12} + q^2 w_{22})/\overline{w}$; the amount of change is

$$\Delta q = q' - q = \frac{pq}{2\overline{w}} \frac{d\overline{w}}{dq}$$

In equilibrium condition, there is no change in gene frequency, $\Delta q = 0$ and thus $d\overline{w}/dq = 0$. The latter equation yields the solution

$$q = \frac{(w_{11} - w_{12})}{(w_{11} - w_{12}) + (w_{22} - w_{12})}$$

genotype	proportion before selection f	(general) fitness value w	(1) selection favors heterozygote w	(2) selection against heterozygote w	(3) selection against recessives w	(4) selection against dominants w
AA	$d = p^2$	w_{11}	.80	.90	1.00	0
Aa	$h = 2pq$	w_{12}	1.00	.70	1.00	.60
aa	$r = q^2$	w_{22}	.70	1.00	.75	1.00
Equilibrium remark	q see (formula)		$q = .40$ $0 \to .40 \leftarrow 1$ stable	$q = .40$ $0 \to .40 \leftarrow 1$ unstable	$q \to 0$ to be balanced	$q \to 1$ by mutation

In order that q be a positive fraction, the quantities $(w_{11} - w_{12})$ and $(w_{22} - w_{12})$ must be both negative as in case (1) of the table above, or both positive as in case (2). In other words, if q is to be an equilibrium gene frequency other than 0 and 1, the fitness of heterozygotes must be greater (example 1) or smaller (example 2) than those of *both* homozygotes.

Stability of an equilibrium. In example (1), if $q < .40$, it will gradually increase toward .40 as the limit. If $q > .40$, it will gradually decrease toward .40 as the limit. An equilibrium value with such properties is said to be *stable*. Only stable equilibria are expected to exist in nature. In example (2), the reverse is true: if q is smaller than .40, it will be smaller and smaller in subsequent generations; if q is larger than .40, it will increase in later generations. The eventual result is either $q = 0$ or $q = 1$. Such an equilibrium is said to be *unstable*. In example (3), q is always decreasing and approaches zero as limit. In example (4), q is always increasing and approaches unity as limit. There is no intermediate equilibrium value for these one-directional selection patterns.

Balance between selection and mutation. Selection against recessives (example 3) tends to diminish the frequency of the recessive gene, and yet the recessive deleterious gene persists, at a certain low level, in natural populations. This is due to the continuous occurrence of new mutations from A to a. Let this mutation rate be μ (e.g. 4×10^{-6}) per generation, and write the fitness of recessives $w_{22} = 1 - s$, where s (e.g., 0.25) is a positive fraction known as the *selection coefficient* against recessives. By the general method of finding the selection effect on q outlined previously, we obtain $\overline{w} = 1 - sq^2$ and $q' = (q - sq^2)/\overline{w}$ or $p' = p/\overline{w}$. The amount of loss in q due to selection is thus $\Delta q = q' - q = -spq^2/\overline{w}$, and the compensation gained by new mutations is $p'\mu = p\mu/\overline{w}$. Eventually a stable equilibrium condition will be reached, gain and loss balancing each other so that q remains the same. This obtains when $p\mu/\overline{w} = spq^2/\overline{w}$, or

$$\mu = sq^2, \quad q^2 = \mu/s, \quad q = \sqrt{\mu/s}$$

For example, with $\mu = 4 \times 10^{-6}$ and $s = 0.25$, the equilibrium condition will be $q^2 = .000,016$ and $q = .004$. The proportion of heterozygotes $h = 2(.996)(.004) = .007,968 \doteq .008 = 2q$, and there are $7968/16$ or approximately 498 or approximately 500 times as many heterozygotes as recessives. It is clear that most of the recessive deleterious genes are hidden in the heterozygotes, and selection acts upon only the 16×10^{-6} recessives. The small decrease in q inflicted by selection is compensated by new mutations. If the recessives are lethal ($w_{22} = 0$ or $s = 1$), the equilibrium value is $q = \sqrt{\mu}$ which is smaller than $\sqrt{\mu/s}$.

When the selection is against the dominants (example 4), the equilibrium frequency of the deleterious dominant gene is generally much smaller than indicated above for deleterious recessive genes. When AA is lethal and Aa has a fitness $w_{12} = 1 - s$, the p value is so small ($q \to 1$) that we may take $d = p^2 \doteq 0$, $h = 2pq \doteq 2p$, $r = q^2 \doteq 1 - 2p$. In each generation, $2ps$ dominant genes will be lost due to selection against the heterozygotes, and ν of the $2(1 - 2p)$ normal recessive genes will mutate to the dominant

condition. At equilibrium we have

$$ps = (1 - 2p)\nu = \nu \text{ approximately.}$$

The frequency $p = \nu/s$ is very small, of the order of the mutation rate. Many other types of selection have been investigated, but these examples should suffice to illustrate the principle of balance between selection and mutation. Deleterious genes are kept at a very low level by selection but are never completely eliminated from the population.

Migrations and intermixture. The introduction of individuals from one population to another changes, in general, the gene frequencies of both populations. Consider a population with gene frequency q. If a fraction m of this population consists of immigrants with gene frequency $\overline{q}$, then the new gene frequency of this population will be $q_1 = (1 - m)q + m\overline{q}$. The amount of change in gene frequency in one generation is $\Delta q = q_1 - q = -m(q - \overline{q})$, indicating that the change is proportional to the difference $(q - \overline{q})$ between the recipient population and the immigrants. Continued intermixture of neighboring populations will eventually make this difference disappear, and all subpopulations will have the same gene frequency in the absence of other factors. If the local partially isolated populations show genetic differentiation, there are probably local selection forces to counteract the homogenizing effect of migrations. Eventually, there will be an equilibrium condition between selection that tends to diminish the frequency of a certain gene and migration that prevents the complete elimination of that gene from the population. The principle is much like the balance between selection and mutation.

Random drift in small populations. Consider a population of three thousand adults, of whom two hundred become parents (the remaining ones childless). The genetic composition of the offspring generation is determined by these two hundred parents only. We say that the *breeding size* of the population is $N = 200$. Furthermore, if, in the idealized situation, there are 100 male and 100 female parents and they are random individuals of the parental generation with equal fertility, the gene frequency of the next generation will not be a constant but a random variable with mean q and variance $q(1 - q)/2N$. This kind of change in q is known as *random drift*, in contradistinction to directed changes induced by mutation, selection, migration. The smaller the population, the greater the random drift. Once a locus is fixed ($q = 0$ or 1), there will be no further random change due to sampling. Therefore random drift in small populations will eventually lead to complete fixation of all loci. The rate of fixation, known as the *rate of decay of variability*, is about $1/2N$ per generation, $1/4N$ reaching the terminal point $q = 0$ and $1/4N$ reaching $q = 1$. The theory of random fixation has been confirmed by observations in nature. Small isolated populations are usually homozygous for all loci and their differentiation is not necessarily correlated with local conditions. Unequal numbers of male and female parents, unequal size of families, and inbreeding all tend to increase the variance of gene frequency. In such cases, the breeding size $N = 200$ should be reduced to a smaller number N_e, known as the *effective size* of the population, in order to conform with the standard variance formula $q(1 - q)/2N_e$ or the standard rate of fixation $1/2N_e$.

The effective size could be much smaller than the breeding size in special cases.

Stationary distribution of gene frequencies. The systematic forces of selection, mutation and migration tend to determine an equilibrium value of q, while the process of random sampling of the parents in finite populations tend to scatter the q values away from it. The eventual result is a stationary distribution of gene frequency, a simplified form of which is

$$\phi(q) = C\overline{w}^{2N} q^{U-1} (1 - q)^{V-1}$$

where C is a constant, $\overline{w}$ is the average fitness, $U = 4N(\mu + m\overline{q})$, $V = 4N(\nu + m\overline{p})$, and N is the effective size of the population. Numerous special cases may be derived from this general expression. If, for instance, the distribution is essentially determined by selection pressure and random sampling, we may take U and V as zero and thus $\phi(q) = C\overline{w}^{2N}/q(1 - q)$. On the other hand, if selection and mutation pressures are negligible and migrations dominate the situation, the distribution becomes $\phi(q) = Cq^{U-1}(1 - q)^{V-1}$ where $U = 4Nm\overline{q}$ and $V = 4Nm\overline{p}$, and so on.

By and large, for very small populations, the distribution is U-shaped, indicating that most of the loci (or most of similar populations) are near fixation with only occasional migrations or mutations. For very large populations, the distribution centers around a certain mean value determined by the directed pressures with small variance. For intermediate populations, the distribution may have a peak at a certain q value and two asymptotic tails.

The most important single conclusion in the genetic studies of mendelian populations is that there is no one all-important factor in evolution. The theory of isolation, of natural selection, of migration, of hybridization, of mutation, etc., none of which is adequate by itself, are combined into one comprehensive theory which includes the additional factor of chance. Evolutionary changes depend on the interplay and balance of all factors.

C. C. LI

References

Fisher, R. A., "The genetical theory of natural selection," New York, Dover, 1958.

Haldane, J. B. S., "The causes of evolution," Harper, London, 1938.

Li, C. C., "Population genetics," Chicago, The University Press, 1955.

Wright, S., "Evolution in mendelian populations." Genetics **16**: 97–159, 1931.

Wright, S., "The general structure of populations," Ann. Eugenics **15**: 323–354, 1951.

POPULATIONS

The study of populations of organisms is concerned primarily with numbers of organisms and is necessarily to some extent a blend of biological and mathematical disciplines. The purely descriptive approach in terms of census data has obvious intrinsic value and is required for the study of the trend of numbers in time and the analysis of those forces which affect this trend. These forces, so basic to population studies, can be quite simply expressed by: (1) the positive force of natality; (2) the negative force of mortality; and (3) the force of dispersion—positive in the case of immigration or negative in the case of emigration.

In the widest sense, a population is any collection of things to which one wishes to apply the term, but more practically we seek to proscribe certain temporal and spatial restrictions and to force a degree of homogeneity —excluding those organisms which do not fit our criteria. It is often convenient to distinguish natural from artificial populations: a distinction reflecting the degree of control which the investigator exercises. Natural populations—those under little or no control— are often rather difficult to work with because of the numerous environmental variables involved. Through the use of artificial populations—in field or laboratory— the investigator seeks to simplify without excessively distorting the situation. The results of research utilizing artificial populations are subject to reservations that stem from manipulation of the conditions.

In general, the historical picture of population research shows observation of natural populations leading to speculation about the forces affecting numbers, and then the use of artificial populations to seek analytical, mechanistic explanations and to test hypotheses. The ultimate test, of course, is in the natural community; but much work in population methodology is required for the full development of this aspect.

The laboratory population is often regarded as an analog of a natural population and as such may consist of a pure culture of a single species or be a more complex system of two or more interacting species. Population studies on the process of interspecies competition have opened the field for analytical study of fundamental importance to ecology and evolution. Other notable studies conducted in controlled conditions afford a base for the study of predatory-prey and host-parasite relationships.

The biological properties of the elements of a population are fundamental but some of the major points of interest are strictly populational. An individual is born but once and hence cannot have a birth rate and although the individual may age irreversibly it lacks an age-structure. A survivorship curve is a good example of the distinction between individual and populational properties. By census and reporting of deaths by age one can construct a curve or a table which shows the proportion of organisms of a particular age which die during a given period of time. The age-specific mortality rates obtained from such a table may provide a good basis for predicting the number of deaths in the population but the table does not recognize a man on his death-bed as any different from the most vigorous fellow of similar age.

The study of a population must start with its definition. This definition may be in terms of a particular group of organisms, living, alike in some respect, located within specified spatial boundaries and so defined at a particular point in time. The population may include only a single species or several species. The most general statement about the population would tell how many individuals or elements are included in it.

Direct enumeration, the actual count of the number of individuals in a population, is of course the most precise measure. In many types of investigation it may be desirable to obtain this count according to various categories which reflect the heterogeneity of the population. The process of direct enumeration is straightforward; but in some cases the population may be too large or wide-spread, it may be very difficult to observe, or the population may be excessively disturbed by the methods which would be needed. The number of fish in a large body of water or the number of bacteria in a culture would be examples where a less direct method would be preferred.

Intercensus procedures afford a basis for estimating the population where periodic enumerations are made. An example of this would be the method of registration used in human populations where the machinery for accurate records exists. From a reliable census one would add the number of births and immigrants and subtract the number of deaths and emigrants. The ideal record system is rare but useful interpolations and extrapolations can often be obtained using various indices of population growth rates.

Sampling, by various methods, can afford good estimates where technical difficulties make direct enumeration impractical. A sampling procedure usually involves some sort of assumption about the spatial distribution of the organisms. The challenge which is posed to the investigator and the great importance in a practical sense has led to the development of sampling as a field of study in its own right.

Mark and recapture techniques are mostly applied in field work with animal populations. The method is applicable if, after capture, marking and release, the probability that an individual organism will be recaptured and detected as marked is exactly the same as the probability of capture of any other member of the population. This assumption has several important implications. The process of capture and marking must be harmless and must not modify the behavior of the organism. The mark must be permanent at least in so far as the duration of the study is concerned. Under these assumptions the total number of organisms in the population can be estimated by solution of the following equation:

$$\frac{\text{Total number of organisms in the population}}{\text{Total number of marked organisms in the population}} = \frac{\text{Number of organisms in the sample}}{\text{Number of marked organisms in the sample}}$$

Minor consistent deviations from the assumptions can be corrected for by arithmetic manipulation. This procedure has been widely used in fisheries, bird and game studies.

Distance to nearest neighbor is a measure which, when the distribution of organisms is random (Poisson), affords a basis for population estimation. The mean distance to the nearest neighbor is related to the population density as follows:

$$\bar{r} = \frac{1}{2}\sqrt{\rho}$$

where $\bar{r}$ is the mean distance to the nearest neighbor and rho is the density of the population. This procedure, although based in its theory on a random distribution, is not too sensitive to deviationa and may prove to be a useful technique in situations where the observer is

reasonably sure that all of the nearest neighbors are actually found.

All of the foregoing methods have to do with estimation of the number of individuals. Sometimes the investigator is primarily interested not in number but rather in mass. Studies concerned with productivity and community energetics would perhaps involve biomass and make use of appropriate modifications of census methods.

The interpretation of census data and particularly changes of numbers with time is facilitated by the use of various mathematical expressions. An equation, a mathematical model, which is derived from observational data is termed empirical or interpolary. The usual procedure here is to make empirical observations, then select or fabricate an equation. Next the model is tested, either by some goodness of fit test of the original data to the curve dictated by the equation or, preferably, to collect new data in a similar observational set-up and test the fit of this new data to the curve. In this type of model, it is generally preferred that the parameters of the model be associated with the biological properties of the organisms in a meaningful manner.

A second type of model might be referred to as a theoretical or *a priori* model. Observation of the biological properties of an organism or system of organisms, basic rates and various interactions lead to the synthesis of a mathematical model. Here the empirical test, fitting the data to the expected curve, constitutes a test of the assumptions underlying the model. If the data agree with the expectations, the worker tends to be satisfied that the mathematical expression represents in some sense a true statement concerning the processes going on within the population. The principal reservation needed here comes from recognition that in no case is this fit unique—there are always many other equations which would also produce as good a fit.

A mathematical population model may be of two basically different types. Let us first consider the historically older of the two, the deterministic model. Here the usual assumption is that there exists some one true curve and if the observations were to be repeated a sufficient number of times, the average of the points would give a better and better fit to the expected curve. An alternative statement is of particular relevance to theoretical models where the equations are derived from physiological parameters and the model itself represents a most probable curve. Knowing the limits of the physiological parameters and the empirical distribution of the various values, one can calculate the distribution of the various values for the model. In this line of argument such deterministic models may approach stochastic models in form with somewhat less complex formulation. A stochastic or probabilistic model is based on the underlying assumption that the various events in a population are matters of chance, that is fundamentally indeterministic.

One of the best known deterministic models is that of Verhulst and Pearl, the logistic growth curve. This can be written in the form:

$$\frac{dN}{dt} = rN\left(1 - \frac{N}{K}\right)$$

where N = number of individuals in the population
t = time
r = intrinsic rate of natural increase
K = environmental capacity in terms of number of animals

An arithmetic plot of N against time would show a sigmoid, symmetrical form increasing from zero to K. The rate of growth as represented by the slope of the curve would be greatest at one-half of the maximum. This general sort of pattern is demonstrated by a number of organisms and has been interpreted by Raymond Pearl and G. F. Gause in terms of the underlying biological and environmental factors. Those organisms for which the best fits to the logistic curve have been obtained were living under environmental conditions which changed rather radically during the course of the observations. The accumulation of waste products or depletion of food supply have a largely cumulative effect. The response cycle, organism acting on environment and environment affecting organism, can not be expected to be instantaneous; but such models do not take time lag into account. A model of this sort, while useful for prediction, must be viewed with reservation as an explanation of growth phenomena occuring in a population. With the introduction of more sophisticated computing methods, it is becoming feasible to make models more realistic and considerable progress can be anticipated in this field.

The problem of computation is even more difficult in stochastic models and most of the work thus far has been restricted to simplified situations. The use of probability generating functions facilitates the determination of an expectation curve and the distribution of the random variable, number of individuals, at various points in time. Again, as soon as one seeks to build in the various processes which are known to exist in a population, in particular the fact that the history of the population is not without influence, the mathematical problems become indeed very difficult. Two types of stochastic models are commonly used now; first, the model in the large—in which rather sweeping simplifications are made for the sake of generalization; second, the model in the small deals with a discrete process within the nexus of interaction in the population.

One aspect of population study which should be specifically mentioned is that concerned with community structure. This particular line of work has led to some of the major developments in field ecology and is of fundamental importance together with population genetics in the study of the dynamics of evolution.

EARL R. RICH

References

Andrewartha, H. G. and L. C. Birch, "The Distribution and Abundance of Animals," Chicago, Univ. of Chicago Press, 1954.
Gause, G. F., "The Struggle for Existence," Baltimore, Williams and Wilkins, 1934.
Volterra, V., "Variation and Fluctuations of the Number of Individuals in Animal Species Living Together," in Chapman, R. N. ed., "Animal Ecology," New York, McGraw-Hill, 1931.
Various Authors. Coldspring Harbor Symposia on Quantitative Biology. 1956.
Watt, K. E. F. Ed., "Systems Analysis in Ecology," New York, Academic Press, 1967.

PORIFERA

The Porifera, to which group belong the bath sponges, are multicellular organisms sufficiently distinct in their attributes from all other animals to be placed in a separate subkingdom of the ANIMAL KINGDOM. Technically, the correct term for the subkingdom is PARAZOA, but the term Porifera is more widely used.

The organization of sponges reflects, to a degree not apparent in any other filter-feeding organisms, their habit of passing through themselves water from which they extract oxygen and minute organic particles of food, and which carries away from them carbon dioxide, excretory products, spermatozoa, and developing larvae. Whereas in other filter-feeding organisms special portions of the body become hypertrophied to perform these functions, in sponges the whole mass of the organism is permeated by a filtering system. This *aquiferous system* is nonetheless not necessarily a well defined and permanent system, consisting as it does of an intricate and frequently changing complex of pores, tubes and irregular chambers. Water is drawn into the aquiferous system through small *ostia*, openings amongst the pinacocytes forming the outer cellular lining of the sponge. Passing through various lacunae and canals it reaches chambers lined with *choanocytes*: cells each bearing a large flagellum which is set within a porous cytoplasmic collar. The choanocyte flagella provide the force moving the water through the sponge, and in the turbulence around the choanocytes small food particles become trapped in the cytoplasmic collars. These particles are then ingested by the choanocytes and are transported from thence into the body of the sponge by other cells. Spermatozoa of the same species are captured and transported in similar fashion.

Choanocytes are frequently confined to special chambers, but in some forms they occur scattered over the surface of large areas of the aquiferous system. After passing through regions bounded by choanocytes the water moves along confluent canals and spaces of increasing diameter until it emerges through one or more large *oscula*.

Several attempts have been made to homologize the parts of sponges with tissues occurring in the Metazoa, but all such attempts have failed. While cytologically distinctive cells are always observable in any sponge, congregated in certain areas and performing distinct functions, e.g. choanocytes, pinacocytes, collencytes forming a meshwork in the body of the sponge, the origins of such cells, and the ease with which their cytological characteristics and their functions change, suggest that morphological integration in sponges is so flexible that the rigid concept of tissues—implying as it does achievement of a high degree of irreversible specialization of the majority of cells at an early embryonic stage—is misleading. The early embryo of sponges contains what are clearly choanocytes and collencytes, whose functions are already determined, but it also contains a large number of amoeboid cells, the *archaeocytes*, with a large nucleolus and much RNA. These archaeocytes persist, and are self-replicating throughout the life of the sponge, but they are also capable of giving rise to cells of all the other types found in the sponge. Thus they are the origin of the pinacocytes and the sclerocytes, which form the mineral portions of the skeleton, but they also give rise to additional choanocytes and collencytes. Besides forming a reservoir of cells for growth, replacement and regeneration, the archaeocytes, by their mobility and their ability for phagocytosis, carry out in their unaltered form a variety of functions essential to the well-being of the sponge.

The totipotency of the archaeocytes and the persistent dynamic intercellular balance which in sponges serves instead of the embryonic tissue determination found in the Metazoa is strikingly illustrated by the behavior of sponge cells dissociated by crushing through bolting silk or by chemical means. The scattered archaeocytes move around until they make contact with each other, when they adhere to form aggregates. These develop an envelope of pinacocytes, flatten, adhere to the substrate and reorganize to produce small, perfect, functional sponges. Because of their great mobility, archaeocytes may form a very great majority in such aggregates, and yet the final reconstituted sponge contains the usual cell complement.

The behavior of the archaeocytes in the reconstitution process has aroused much interest in recent years because of some similarities to the behavior of cancer cells. Researches initiated by this interest have cast much light on the nature and functions of the animal cell surface.

A generally acceptable terminology treats a sponge as delimited from the external environment by a layer of pinacocytes (the *pinacoderm*), except where the pinacocytes are replaced by choanocytes (the *choanoderm*). Distinctions are drawn between the pinacocytes next to the substrate (*basopinacoderm*), those lining the free surface of the sponge (*exopinacoderm*), and those lining any part of the aquiferous system (*endopinacoderm*). The rest of the sponge body is referred to as the *mesohyl*, through which ramify the aquiferous system and the *skeleton*, and in which lie the collencytes, archaeocytes, and other cells. Outer regions of the sponge, devoid of choanocytes, are referred to as *ectosome*, while those regions containing choanocytes are sometimes called *choanosome*. A portion of the ectosome distinguished by the presence of a special skeletal organization may be termed the *cortex*.

Some sponges contain contractile cells, deformations of which produce movement of relatively large portions of the sponge, e.g. oscular chimney, but despite several recent claims it appears highly unlikely that any sponges possess even the simplest form of nervous system.

Regeneration of injured parts and the total reorganization of fragments into fully functional sponges takes place readily. Closely adjacent sponges may grow towards each other and coalesce completely. Theoretically, sponges have the potential for almost one hundred percent regeneration, and also for unlimited coalescence with members of their own species. As a consequence it is impossible to agree on a suitable term to describe a single sponge. There is no such thing as an individual, except as a temporary phase, and the term "colony" is equally unsuitable. There is no guarantee that any one sponge under examination is or is not the product of the fertilization of a single ovum since embryos, larvae and postlarvae coalesce readily to produce one sponge that differs in no recognizable way from one produced from a single ovum, embryo, or larva.

These features are reflected in the reproduction, which is best treated under the two headings sexual and asexual.

So far as is known all sponges are hermaphrodite. Only once has claim been laid to the discovery of a male sponge, and that is somewhat questionable. Researches suggest that ova and sperms are derived from choanocytes which undergo modification, chiefly in size, at the onset of the breeding season. Choanocytes destined to form ova migrate into the choanosome and are there fed by nurse-cells. As they grow in size food granules appear in the cytoplasm. Maturation divisions are delayed until the advent of a spermatozoan.

Some species of sponge are oviparous and little is known of their further development. The majority are viviparous. In these, the fertilized ovum undergoes division within the mesohyl, a temporary capsule enclosing the embryo. Eventually, a free-swimming larva escapes by rupture of the capsule, passes into an exhalant canal and is ejected from an osculum by the exhalant current. Larvae of two types have been described; a hollow ovoid *amphiblastula* and a solid ovoid or flattened *parenchymella*. Larvae of both types are flagellated and capable of swimming or creeping for from a few hours to several days, after which this power of locomotion is gradually lost and the larvae settle to a suitable substrate and undergo metamorphosis. In the calcareous sponges this involves complete introversion of an amphiblastula; in other sponges metamorphosis is not so dramatic, although it involves thorough spatial reorganization of the cells. A minute functional sponge arises very shortly after the onset of metamorphosis.

Asexual reproduction may be by budding, fission or by asexually-formed embryos. In budding, groups of archaeocytes travel to the surface of the sponge, the pinacoderm bulges outwards to form pockets for their reception, and small functional sponges develop that are subsequently shed. A few species have been shown to multiply by fission, throwing off parts of the body, the fission being preceded by an hypertrophy over a limited area and the development of a line of weakness along which a split occurs separating the part from the parent body. In other sponges multiplication may arise by budding from stolons. In some species there is reason to believe that archaeocytes form asexually-produced embryos. These migrate from different points in the choanosome, assemble to form gemmules that behave subsequently like sexually-produced embryos, and like them eventually leave the parent as flagellated larvae indistinguishable from the product of a fertilised ovum.

Little is known of the longevity of sponges. Some littoral sponges may die off at the end of the second year, and the only other data are from the bath sponges which are believed to reach marketable size in 7 years. On this basis some commercial catches must include sponges perhaps 20 or more years old. Size is, however, no criterion. Abundance of food may produce sponges twenty or more times larger than normal for the species, and there is no evidence that they are of greater age than their smaller congeners. In addition, coalescence of neighbouring sponges may result in giants coating a surface of rock a metre or more square.

There are probably no more than 5,000 recent species distributed throughout the world, in every sea from mid-tide level to the deepest points from which dredgings have been made. Some marine species penetrate into estuarine waters of low salinity and species of one family (the Spongillidae) are distributed in rivers and lakes, streams and ponds, from sea-level to lakes in extinct volcanic craters 11,000 feet above sea-level.

Fossils indicate that sponges not markedly different from those of today except in details of their skeletons have existed since early Cambrian times at least.

The external form of sponges varies considerably and resembles more the patterns seen in the lower plants. The majority of sponges are of the form sometimes described as 'massive,' by other authors as 'amorphous.' That is, they are cushion-shaped or irregularly rounded masses. Others may be finger-shaped bushy, tree-like, cup- or funnel-shaped. On the whole, the greater the depth at which they grow the more symmetrical the form, but this is no inviolable rule. Deep-sea sponges also, growing on fine mud or ooze, tend to develop a stalk lifting the body up from the substratum, or to develop a raft of spicules by which the body floats on the surface of the mud or ooze.

Color is usually a monochrome. The commonest colors are browns, reds and purples, with green, yellow or cream, violet or lilac less common, and blue exceptional and usually due to the presence of bacteria. Sponges growing in caves or under overhanging ledges of rocks tend to be white or cream, and in some of these the color can be correlated with the presence or absence of light. Deep-sea sponges are usually white or cream. Freshwater sponges, and a few marine sponges, are green owing to the presence of symbiotic algae, and are consequently white or cream where shaded from light.

The external features such as shape and color are variable and only in a minority of species do they serve as a ready guide to identification. As a result, the characters of the skeleton are used almost exclusively for classification. Since these differ from class to class they are best discussed with the classification. It is, however, necessary to precede this with an explanation of the word *spicule* (i.e. little spike).

Spicules range in size from a few microns to upwards of several centimeters, are formed of colloidal silica or crystalline calcium carbonate and together with *spongin* fibrils (a form of "collagen") arranged in tracts form the structural units of the skeleton. In some very few forms spongin fibrils, not arranged in large tracts but scattered in the mesohyl, provide the only non-fluid skeletal units.

Spicules may be simple rods with variably shaped ends, or they may be complex. They may take the form of forks, anchors, rods, shovels, stars, plumes, and many other shapes. Some, larger than the others form the main skeleton and are called megascleres, the smaller spicules which occur interstitially to the main framework being called microscleres.

The sub-kingdom Parazoa (formerly known as Porifera or Spongiida) consists of two phyla: Nuda and Gelatinosa. In the first of these the network of cells forming the internal tissues is naked: in the second the interstices are filled with an inter-cellular matrix of fluids and fibrils.

The phylum Nuda contains a single class and order, the Hexactinellida. All spicules are formed on a six-rayed principle (hexaradiate) and in one sub-order (the Dictyonina) the main framework is strengthened by a secondary silicification to give a rigid and continuous skeleton. Hexactinellida are found at depths from 40 to several thousand fathoms, the majority being below the 100-fathom line. They are the most beautiful of all sponges and include the well-known Venus' Flower Basket (*Euplectella*), Glass-rope sponge (*Hyalonema*), and the Rat's Nest sponge (*Pheronema*).

The phylum Gelatinosa includes two classes: Calcarea and Demospongia. (By no means do all authorities accept this grouping, and the inclusion of the two classes in this one phylum and the merging of Tetraxonida and Keratosa into the single Class Demospongia are both matters of controversy.)

The Class Calcarea is characterized by the presence of calcareous spicules formed on a three-rayed (triradiate) or four-rayed (tetraradiate) plan. Calcareous sponges are mainly of small size, found especially between tidemarks or in shallow waters, but they also occur as deep as 1,000 meters. The Pharetronida, which are a largely extinct group, are usually included in the calcarea, despite their massive or spherulous skeleton.

The Class Demospongia is characterized by a skeleton of silicious spicules formed on a four-rayed (tetraradiate) plan and a highly developed system of spongin tracts. It includes the majority of marine sponges, distributed from the littoral to the greatest depths, and also the fresh-water sponges (Spongillidae) and the sponges of commerce.

Within the class the relative proportions of spicules and spongin in the skeleton vary continuously from the state in which the spongin tracts serve merely to cement the ends of spicules into a meshwork to the condition occurring in the bath sponges, in which there are no spicules, only a meshwork of spongin tracts. A few other forms contain neither thick spongin tracts nor spicules, while in yet others sand grains are bound into the spongin. An anomalous group, the Stromatoporoidea, with both calcareous and silicious skeletal elements, and until recently considered largely extinct, have been shown recently to be of great ecological importance in coral reefs.

Commercial sponges are distributed in all tropical and subtropical waters, down to moderate depths, but the greatest concentrations are in the Mediterranean, especially the eastern Mediterranean, and in the Gulf of Mexico, notably around the Bahamas and Florida. Elsewhere they occur sparsely, although in places they are fished and used locally. While the West Indian sponge fishery was opened up as recently as 1841, fishing for sponges in the Mediterranean was recorded in antiquity.

WILLIAM G. FRY

References

Fry, W. G., ed., "The Biology of the Porifera," Zool. Soc. London Symposia Vol. 25, 1969.
Hyman, L. H., "The Invertebrates: Protozoa through Ctenophora," New York, McGraw-Hill, 1940.

PORPHYRINS

The porphyrins do occur free in nature but have no known functions in normal tissues. In certain pathological conditions, such as the "cutaneous porphyrias" of humans, the high free porphyrin content of the tissues leads to photosensitivity of the skin. The biological importance of the porphyrins is due mainly to the manifold metabolic activities of their iron complexes, the hemes.

Porphyrins *occur* free in low concentrations in normal tissues, and in much higher concentrations in the tissues and excreta in certain pathological conditions.

Table 1   Order of Magnitude of Free Porphyrin Content of Tissues

	uropor- phyrin*	copropor- phyrin*	protopor- phyrin*
Normal human			
blood (μg/100 ml red cells)		0.5	30
faeces (mg/day)		0.5	0.6
urine (mg/day)	0.02	0.1	
Acute porphyria (human)			
faeces (mg/day)		2	3
urine (mg/day)	50	2	

*For structures, see Fig. 1 and Table 2. For references cf. ref. 1.

Faecal deuteroporphyrin results from bacterial degradation, in the intestine, of the heme of ingested blood or after intestinal hemeorrhage. Protoporphyrin is excreted by the Harderian glands of the rat, and occurs in the shells of eggs of hen and other birds and in the "red streak" of the earthworm *Lumbricus*. Uroporphyrin is found in the shells of certain molluscs, and in the bones of the grey squirrel. Free uro- and coproporphyrins are found in the root nodules of legumes, and uroporphyrin has been found in leaves. Porphyrins, usually coproporphyrin, are excreted by many bacteria.

The iron complex of protoporphyrin, heme, forms

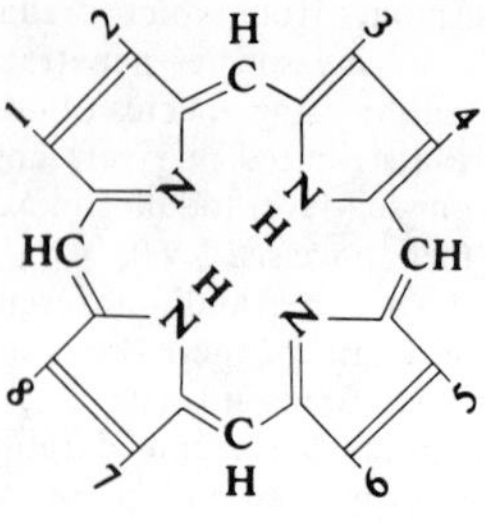

Fig. 1   Porphin. 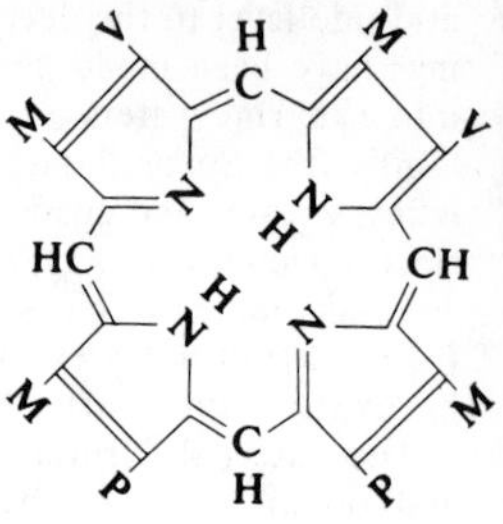   Fig. 2   Protoporphyrin.

the prosthetic group of the hemoproteins hemoglobin, myoglobin, catalases, peroxidases and the cytochromes *b*. The heme prosthetic group is modified in various ways in the cytochromes *c* and the cytochromes *a*.

In porphyrins, the H-atoms 1–8 of the parent compound porphin (1) are replaced by various side-chains (see Figs. 1 and 2 and Table 2).

The porphyrins have a planar, highly conjugated cyclic aromatic nucleus. They are stable to concentrated sulphuric acid, somewhat sensitive to light, and subject to oxidation by peroxides.

The systematic name of protoporphyrin (2) as an example is 1, 3, 5, 8-tetramethyl-2,4-divinyl-porphin-6, 7-dipropionic acid. The trival names are more convenient, and those of some common porphyrins are defined in Table 2. Porphyrins of isomer-types I and III only are known to occur in nature; coproporphyrins II and IV are included in Table 2 to illustrate the four possible isomers of this and similar porphyrins.

All naturally-occurring porphyrins have carboxylic acid side-chains (see Table 2), and are thus *ampholytes*; their isoelectric points are in the region of pH 3–5. Porphyrins are *extracted* readily from biological materials by acidified organic solvents. They are soluble in neutral ether; the variations in ether solubility and in basic dissociation induced by their different side-chains allow *separation* of different porphyrins by extraction from ether with increasing concentrations of aqueous

Table 2   Side-Chains and Properties of Some Important Porphyrins

		Positions of side-chains (cf. Fig. 1)								Soret band		
		1	2	3	4	5	6	7	8	λ max (mμ)	ε mM	HCl No.
Coproporphyrin I		M	P	M	P	M	P	M	P*			
"	II	M	P	P	M	M	P	P	M			
"	III	M	P	M	P	M	P	P	M	401	530	0.08
											(0.1N-HCl)	
"	IV	P	M	M	P	M	P	P	M			
Uroporphyrin I		A	P	A	P	A	P	A	P			†
"	III	A	P	A	P	A	P	P	A	406	539	†
											(2N-HCl)	
Protoporphyrin		M	V	M	V	M	P	P	M	408	275	2.5
											(2N-HCl)	
Deuteroporphyrin		M	H	M	H	M	P	P	M	404		0.3

*Abbreviations: A = —CH₂COOH; H = —H; M = —CH₃; P = —CH₂CH₂COOH; V = —CH=CH₂
†Uroporphyrins are soluble in water and insoluble in ether.

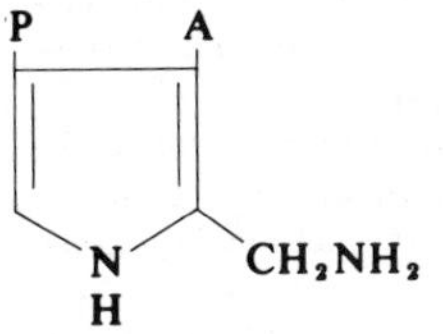

Fig. 3 Porphobilinogen. 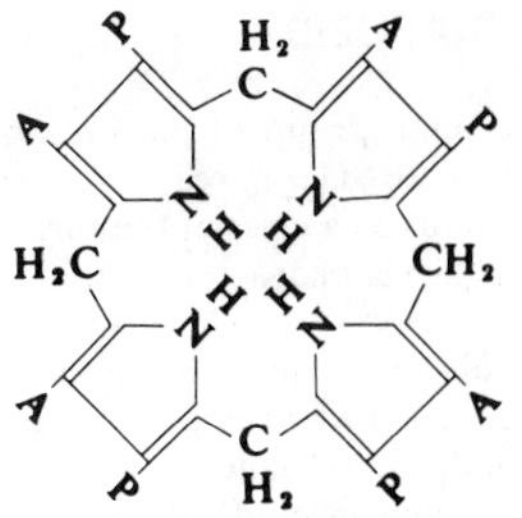   Fig. 4 Uroporphyrinogen.

HCl. The "HCl number"[1] of an individual porphyrin is fairly characteristic (see Table 2). The esters of porphyrins are much more soluble than the free acids in organic solvents.

Porphyrins have very intense, characteristic *absorption spectra*. In neutral or alkaline solvents there are four main bands in the visible region and a much more intense band, the "Soret" band, at about 4000 Å. The ratios of intensity of the visible bands change in a predictable manner with certain changes in substituents.[1] In acid solvents there are two main visible bands and the Soret band; the latter is commonly used for spectrophotometric determination (see Table 2).

The bright orange to red *fluorescence* of porphyrins under u. v. light (Woods glass) is characteristic and extremely intense. Less than 0.1 μg/ml is quite visible to the eye, and much lower concentrations are detectable by fluorimetry. The fluorescence is quenched by a variety of biological compounds and by some organic solvents; it is best observed in aqueous HCl solutions.

Paper and column *chromatographic* methods are available for the identification and separation of individual porphyrins.[2]

Many of the steps in the *biosynthesis* of protoporphyrin have been elucidated.[3] Glycine is condensed with succinyl coenzyme A to yield α-amino-β-ketoadipic acid (COOHCH$_2$CH$_2$COCH·(NH$_2$)COOH); this decarboxylates to δ-aminolevulic acid (COOHCH$_2$·CH$_2$COCH$_2$NH$_2$), two molecules of which condense (δ-aminolevulic acid dehydrase) with loss of 2H$_2$O, to the monopyrrole compound porphobilinogen (3). By a series of steps (porphobilinogen deaminase, porphobilinogen isomerase) four molecules of this condense to form uroporphyrinogen III (4), one of the pyrrole nuclei having been inverted. All the acetic acid side-chains of uroporphyrinogen III are decarboxylated (uroporphyrinogenase) to form coproporphyrinogen III. This in turn suffers oxidative decarboxylation of two only of its propionic acid side-chains (enzymes not yet identified) to yield protoporphyrin, oxidation of a porphyrinogen to the porphyrin level also having occurred.

J. E. FALK

References

1) Falk, J. E., "Porphyrins and Metalloporphyrins," New York, American Elsevier, 1964.
2) *ibid.*, Chromatographic Reviews, 4:137, 1962.
3) Granick, S. and Mauzerall, D., *in* Greenberg, D. M., ed., "Metabolic Pathways," New York, Academic Press, 1961.

PREDATION

The term "predation," derived from a word meaning "to catch," refers to the biological process resulting from the capture (and generally death) of members of a population. The individual that catches is the predator and the individual that is caught is the prey. The extent of predation describes the effect of the process on the population and for example may be "heavy" or "light." Usually the word is restricted to vertebrates and insects; for example the predation by foxes on mice and by wasps on caterpillars. However the term can properly be used for any species such as protozoa and pitcher plants. Recently the term has been extended to include the effect of parasites when the emphasis is on the result of deaths on the population rather than on the physiological adjustment of host to parasite. Actually these two processes intergrade at their extremes. Thus predation may include the effect on the mouse population of foxes, bacteria, and parasitic worms.

The study of predation has intrigued mathematicians as well as biologists for the last half-century since the predator-prey process lends itself to mathematical formulations. Assumptions about birth rates and movements are made and then the effect of varying the extent of predation is examined. From these results it is possible to calculate the rate of increase or decrease of predator and prey. Depending upon the coefficients, the result may be extinction of predator or of prey, equilibrium, or cyclic fluctuation. Unfortunately since natural systems are too complicated to fit the assumptions of the mathematical models, this approach has had little impact on the study of predation in nature.

The relation of predation to populations may be discussed under three aspects: natality, migrality, and mortality. Predation rarely directly affects natality (the number of young born) or migrality (the movement into or out of the population). However some diseases reduce the number of progeny and some predators frighten individuals away. Usually the effect is trivial.

Predation affects mortality directly and for most higher forms of animals a predator is the usual cause of death, even though starvation may predispose the individual. The extent of predation may vary on a wide spectrum from a level too low for the welfare of the species to a level too high for survival. For example, recent studies of fish populations show that, if predation is increased by introduction of predaceous fish or by increase of fishing intensity, then the population will be healthier and breed faster due to the reduction in competition. The removal of some fish results in better conditions for the survivors.

In other situations predators may have a negligible effect on the population. Although they may kill some individuals, these may be surplus, or diseased, or starved, and thus unimportant to the population. In other cases predators may remove a sufficient number of prey to reduce the population to a lower level, even to extinction. The particular relation that a predator has with its prey depends upon local circumstances. An example serves to illustrate this variability. A lake in central Montana served as a refuge for breeding waterfowl. In a year of low water the lake was essentially round, the ducks nested in weeds along the shore, and the skunks found few nests. In a year of high water, the contours of the land formed a peninsula. The skunks discovered every nest on the peninsula, but exerted only normal predation along the other shores. What is the effect of skunks on waterfowl? The only answer is "It depends upon the circumstances."

The results of an understanding of predation may be applied to the practical management of populations that are important to man-kind. One category of populations is the commercial fisheries and game populations. The objective is to decrease the natural predation which is caused by disease and enemies so that predation by humans for food may increase. First the extent of natural predation should be determined and then reduced if possible. However in several cases an increase in commercial catch resulted in an improvement in the yield over the years because the density of the population was decreased and hence reproduction improved. These effects have occurred primarily in commercial fisheries; as yet such spectacular results have not occurred in game populations although an adequate test is still in the future.

In another direction predation has practical importance. The control of agricultural pests depends upon an understanding of the role of predators. Insect pests are in nature controlled by environmental conditions and by predators, parasites, and disease. This latter group may be important and may be eliminated by chemical insecticides thereby benefiting the pest. Thus in apple orchards in Nova Scotia it was found that reduction and specialization in the use of insecticides resulted in less damage because the parasites and certain diseases were present to keep the pest in check. Thus predation can be beneficial when properly understood and utilized.

The process of predation is highly complex and may have paradoxical aspects. The story of the relation of mites to strawberries in California illustrates many aspects. One kind of mite damages the strawberry leaves while another kind preys upon the pest. When a farmer plants a new bed of strawberries both kinds are absent for the first year. In the second year the pest species appears, increases rapidly and then in the third year, damages the crop. By the time the predators enter the strawberry bed, the pests are far too abundant. In the fourth year the predators can reduce the pest but the damage has been done. However if in the first year some pests and some predators are introduced purposefully onto the strawberry plants, then the predators can survive by eating some pests and be ready in the second and third year to hold down the increase of the pest. Thus a balance is achieved without noticeable damage to the strawberry plants.

The principles of predation apply to diseases and parasites of humans. Indeed some of the best examples of predator-prey relations occur in measles, jungle yellow fever, and the encephalitides. However man can alter the course of the relationship either to his advantage or disadvantage according to his wisdom.

DAVID E. DAVIS

Reference

Errington, P. L., "Of Predation and Life," Ames, Iowa, State University Press, 1967.

PRIAPULIDA

A small group of marine animals, once placed in the obsolete "Gephyrea." Now regarded as a separate phylum, or as a class of the Phylum *Aschelminthes* because the larval characters are, in part, similar to those of the KINORHYNCHS. The body is divided into a protrusible proboscis, armed in front with teeth, and with 25 longitudinal rows of papillae; warty trunk; and in *Priapulus,* caudal appendages.

Found in muds of the floor of arctic and colder seas from the intertidal area to some 500 m.

A. C. STEPHEN

Reference

Hyman, L. H., "The Invertebrates" Vol. 3, New York, McGraw-Hill, 1951.

PRIMATES

Introduction

Since the publication of Linnaeus' "Systema naturae," man has been assigned to the mammalian order, Primates. From the first, the order has also included the anthropoid apes (Pongidae) and these, because of obvious similarities to man are now grouped with the Hominidae in a superfamily, Hominoidea. The monkeys of the New and Old Worlds (Ceboidea and Cercopithecoidea) are also closely related and together with the Hominoidea form the suborder, Anthropoidea.

Included in a second suborder (Prosimii) are the tree shrew and lemur (Lemuriformes), the loris, potto and galago (Lorisiformes) and the spectral tarsier (Tarsiiformes). These animals have few obvious resemblances to man, but in common with the Anthropoidea they display to a greater or lesser extent, a distinctive pattern of primitive mammalian attributes. These include a generalized pentadactyl pattern in the limbs with an opposable thumb or great toe, the presence of a collar bone, the orbits completely encircled by bone, three types of teeth (incisors, canines and molars), pectoral mammary glands and in the cerebral hemispheres, a posterior lobe which displays a calcarine fissure bounding an area associated with vision.

Prosimii

General characteristics. The three groups of Prosimii (Lemuriformes, Lorisiformes and Tarsiiformes) represent a grade of evolution believed to correspond broadly with that seen in generalized early mammals. But in a number of features, (e.g. various aspects of their cerebral development) they are somewhat more advanced than other mammalian groups, and certain species show anatomical peculiarities associated with differences in their way of life.

Prosimians are usually small animals, most of which live in trees and bushes, and while many are quadrupedal, some adopt specialized modes of locomotion (e.g. hopping).

Prosimians feed to a large extent on insects, although

vegetable and occasionally other animal matter is also included in their diet.

Subgroups and natural history. *Lemuriformes.* The Lemuriformes comprise first the tree shrews (Tupaioidea), second the lemurs (Lemuroidea) and third a group (Daubentonioidea) whose only known representative is the aye-aye.

The tree shrews *(Tupaia* and *Ptilocercus)* are found in South-East Asia and were formerly grouped with the Insectivora. They have a superficial resemblance to the squirrels, but the muzzle is longer and more pointed. They are quadrupedal animals and are diurnal in their habits.

The lemurs *(Lemur, Microcebus* etc.) are now found only on the island of Madagascar. Their size varies from that of a small rat to that of a fair-sized dog, and while many species are tree dwellers, others live on the ground. Most are true quadrupeds, but a few (e.g. *Indri)* move in long leaps from tree trunk to tree trunk.

The aye-aye *(Daubentonia)* is also confined to Madagascar. It is unusual, first in that practically all its digits bear claws rather than nails and second, in that its incisor teeth are big and rodent-like.

Lorisiformes. This group contains the loris *(Loris* and *Nycticebus)* from India and the Far East, which together with the potto *(Perodicticus)* from Equatorial Africa comprise a subfamily, Lorisinae. The galagos *(Galago* and *Euoticus)* from Equatorial and South Africa comprise a second subfamily, Galaginae. All are small, furry, nocturnal animals, but while the loris and potto move and climb slowly, not infrequently hanging by their limbs, the galago moves rapidly by hopping.

Tarsiiformes. The only living representative of the Tarsiiformes is the spectral tarsier *(Tarsius).* The trunk of this animal is not more than about six inches in length and the genus is found only in the East Indies and Phillipines. Its fur varies in colour from reddish-brown to grey, and the animal is characterized by enormous eyes that are set well towards the front of the head. It moves by hopping, and, in common with many other prosimians, is most active during darkness.

Anatomical and physiological features. *Skull.* The facial skeleton is big relative to the cranium and is set in front of, rather than below the anterior part of the brain-case. In *Tarsius,* the large orbits give a superficial impression that the snout is reduced.

In the Lemuriformes, the palatine and lacrimal bones adjoin in the medial orbital wall, but in other prosimians, they are separated by the ethmoid. In *Tarsius,* the orbits are more completely separated from the temporal fossae than in other prosimians.

In the Lemuriformes, the tympanic ring is partly enclosed by the prominent auditory bulla. In the Lorisiformes, it lies externally while in the Tarsiiformes, it is prolonged to form a short tubular meatus.

In the Lemuriformes and Lorisiformes the foramen magnum lies near the posterior end of the skull and faces backwards. In the Tarsiiformes, it lies more anteriorly and points more nearly downwards.

Teeth. In most prosimians, the dental formula is

$$\frac{2.1.3.3}{2.1.3.3},$$

but in the tree shrew there is an additional incisor in each half of the lower jaw, while in *Tarsius* the lower incisors are reduced to a single pair. In the Lemuriformes and Lorisiformes, the lower incisors and (except in the tree shrew) the lower canines lie almost horizontally to form a comb or scraper. The molar teeth show considerable variation in the arrangement of their cusps, and while in some (e.g. *Tarsius)* they conform to a relatively uncomplicated "tritubercular" pattern, in others (e.g. *Galago)* additional cusps are developed.

Limbs. The hind limb is the longer and while both retain many features of the primitive pentadactyl plan, the forelimb preserves considerably greater mobility. For instance, the bones of the forearm articulate to allow the movements of pronation and supination.

Except in the tree shrew and aye-aye, there is a flattened nail on each finger. The thumb is mobile, but it can be opposed to the other digits to only a limited extent.

In the potto and loris, the index finger is rudimentary and the thumb exceptionally widely abducted. In *Tarsius,* there are big, disc-like pads on the terminal phalanges.

In the leg, the tibia and fibula are joined by ligaments and cannot rotate, but sometimes an appreciable amount of lateral movement of the ankle is possible.

There are usually flattened nails on most pedal digits, but in each group a claw is retained on one or more toes. The great toe is well developed and can, to some extent, be opposed to the other digits.

In *Tarsius,* the tibia and fibula are fused in their distal half, and in both *Tarsius* and *Galago* certain ankle bones are markedly elongated.

Reproduction. The uterus is bicornuate, although in *Tarsius* the horns appear relatively shorter than in other prosimians. In the Lemuriformes (except the tree shrew) the placenta is of a relatively primitive epitheliochorial type, but in the Tarsiiformes, it is, as in the Anthropoidea, of the haemochorial variety.

Lemurs have a clearly-defined breeding season restricted to some four months of the year, but tree shrews breed throughout at least eight months. In the Lorisiformes, the Galago has a restricted breeding season but the Loris breeds throughout the year. *Tarsius* too, appears to be able to breed at all times.

Nursing behaviour varies somewhat in the three groups but in each, the infant normally clings to the fur of its mother's belly and is seldom handled by her.

Special senses. In the Lemuriformes and Lorisiformes there is a moist, naked rhinarium and the nasal cavity together with its enclosed turbinal bones is well developed. In *Tarsius,* there is no rhinarium and the nasal cavity is reduced in size.

In the Lemuriformes, the eyes are set towards the side of the head but in *Tarsius,* they point more directly forwards. It is, however, unlikely that any prosimian has stereoscopic vision.

In the Lemuriformes, facial vibrissae are well developed but they are reduced in the other prosimian groups.

Brain. Compared with other mammalian groups, the parts of the brain associated with olfaction, are reduced, while the visual and association areas are correspondingly increased. This tendency seems to have proceeded further in the lemurs than in the tree shrews and while in the latter, the surface of the cerebral hemispheres is practically smooth, in the former, a number of horizontally-disposed sulci are evident.

In the potto, there is a well-marked central sulcus delimiting the motor area.

In *Tarsius*, the occipital lobe, which includes the area associated with vision, is more prominent than in other prosimians.

Social habits and general behavior. Little is known of the social habits of prosimians, except that while tree shrews and lemurs are usually found in small bands, tarsiers appear to live in pairs.

Lemurs and the loris groom either their own or each others fur by combing or scraping with the procumbent lower incisors and by scratching with the claw on the second toe. *Tarsius,* on the other hand, grooms its fur by licking and by scratching with the claws on the second and third hind digits.

Facial musculature of the Lemuriformes and Lorisiformes is relatively undifferentiated and compared with monkeys and apes, the face is practically expressionless. In *Tarsius,* the facial musculature is somewhat more elaborate, and the facial expressions correspondingly more varied.

All three groups drink by lapping.

So far as can be judged from the very limited data that are available, the intelligence of prosimians is inferior to that of monkeys and apes.

Palaeontology. The first prosimians appeared in the Palaeocene, and by the Eocene, numerous genera were in existence. Few fossil remains have been found in later deposits.

Anthropoidea

General characteristics. Together with the apes (Pongidae) the monkeys of the New and Old Worlds (Ceboidea and Cercopithecoidea), while retaining many generalized features such as relatively unmodified and mobile forelimb, represent, in such features as the increasing elaboration of their brain, a stage of primate evolution more advanced than the Prosimii. Man too corresponds in his general anatomical features, but his cerebral development has proceeded much further than in any subhuman Primate.

In spite of an enormous variation in size, monkeys and apes almost invariably present a superficial resemblance to man. This results mainly from an expanded braincase and an abbreviated facial skeleton in which the eyes are set close together and face forwards. But while man is a biped, most monkeys are basically quadrupedal. Some "acrobatic" species, however, also swing from branch to branch by their arms, and when in the trees the apes move almost exclusively by this method (brachiation).

Monkeys and apes normally feed on fruit, leaves and roots, although some species also include insects, worms and small birds in their diet.

Subgroups and natural history. *New World monkeys* are found only in South and Central America, and are characterized by comma-shaped nostrils set diagonally on the snout, by series of three premolar teeth and by other peculiarities of their cranial structure. They are subdivided into two groups, of which the first (Callithricidae) contains the marmosets *(Callithrix)* and tamarins *(Leontocebus).* These are relatively small animals, little bigger than squirrels and are exceptional

in that there are only two molars in each half of each jaw, claws on most digits and few convolutions in the cerebral hemispheres.

The second group of New World monkeys (Cebidae) comprises six subfamilies of which the best known are the howler monkeys (Alouattinae), the spider monkeys (Atelinae) and the capuchins (Cebinae).

The howler monkeys *(Alouatta)* range from Central America to Bolivia and weigh up to twenty pounds. Their fur is black or reddish-brown and in common with many other New World monkeys they posess a prehensile tail. Their hyoid bone is enormously expanded and forms a resonating chamber for the voice. Howler monkeys are "acrobatic," living in the high canopy of the tropical forest.

The spider monkeys *(Ateles)* are found in Mexico and Brazil and are characterized, first by their exceptionally long prehensile tails and second by their bodily proportions, which approximate to those of the apes in spite of the fact that the spider monkey is a smaller animal. Spider monkeys are probably more completely acrobatic than any other New or Old World species and live in groups in the dense tree tops. Their fur is coarse, and in most species both it and the face are black.

The capuchins *(Cebus)* are small, robust animals weighing between two and four pounds and are found in an area extending from Colombia to Panama. Their fur varies in colour from species to species and on the head is arranged in the form of a cowl. The tail is to some extent prehensile, but the capuchin is less acrobatic than either the howler or spider monkey. Capuchins were, at one time, seen frequently as organgrinders' monkeys and are noted for their soft and pleasing voice. They have been used extensively in psychological studies.

Old World monkeys. The Old World monkeys are distributed throughout much of Africa and South East Asia. They are distinguished from New World species by such characters as their ischial callosities (sitting pads), by the lack of a prehensile tail and by having the premolar series reduced to two teeth. They comprise two subdivisions, of which the first (Colobinae) contains the colobus monkeys *(Colobus)* from Equatorial Africa, together with the langurs *(Presbytis)* from South-East Asia. Unlike other Old World types, these animals have no cheek pouches for storing food but their stomach is sacculated. Many species move by swinging from their arms more extensively than do other Old World monkeys.

The second group of Old World monkeys (Cercopithecinae) contains several genera of which the guenons *(Cercopithecus),* the macaques *(Macaca)* and the baboons *(Papio* and *Comopithecus)* are the best known.

The guenons *(Cercopithecus)* are distributed throughout much of Africa and comprise some twenty species. While some are no bigger than a squirrel, others are as large as a fox terrier. The colour of the fur varies widely from species to species but in the best known group *(Cercopithecus aethiops)* it is predominantly green. Although these animals normally live in the trees, they are almost completely quadrupedal.

The macaques *(Macaca)* are medium-sized monkeys found in India, Tibet, North China and Japan. They include the rhesus monkey *(Macaca mulatta)* which besides being another of the traditional organ-grinders' monkeys, is used extensively in biological and medical research. Another species *(Macaca nemestrina)* is exceptionally tractable and in some regions has been trained to collect fruit from trees. A further species *(Macaca sylvana)* is found in Morocco and Algeria, and a few specimens live wild on the rock of Gibraltar. It was this animal that was dissected by Galen in the second century A.D. as a basis for his descriptions of human anatomy.

Baboons *(Papio* and *Comopithecus)* are found in Africa and Arabia, and comprise some ten species. Adults grow as big as a large Alsatian hound, and most species live on the ground among rocky crags. They are characterized by a short tail and by an exceptionally well-developed muzzle, which, however, as in other monkeys, projects downwards rather than forwards.

Apes. The living apes differ from the monkeys in their exceptionally long arms, in having no tail, in a more elaborate configuration of their brain and in certain physiological features.

They comprise two subgroups, of which the first (Hylobatinae or lesser apes) contains the gibbons *(Hylobates)* and siamang *(Symphalangus)* from South-East Asia. Gibbons stand about three feet high and weight approximately one stone. The colour of their fur ranges from black, through brown to grey and white. The arms of the gibbon are exceptionally long and the technique of brachiation is more highly developed than in any other primate species.

The second subgroup of apes (Ponginae or great apes) contains first the chimpanzee *(Pan)* and gorilla *(Gorilla)* from Equatorial Africa, and second the orang-utan *(Pongo)* from Borneo and Sumatra.

The chimpanzee is the smallest member of the group weighing approximately 100 lbs. The fur is predominantly black but often shows patches of brown. When on the ground, the chimpanzee moves on all fours, but in the trees, it proceeds by brachiation. The chimpanzee is frequently exhibited in zoological gardens, and many aspects of its biology have been extensively studied, both in the wild and in the laboratory. It tractable nature has made it especially suitable as a subject in psychological experiments.

The gorilla is the biggest of the great apes and adults weigh up to 600 lbs. There are two varieties distinguishable by minor features, one living in the lowland forests of the Cameroons and Gaboon, and the other in the mountainous country to the West of Lake Kivu. The skull is often characterized by prominent bony crests for the attachment of the temporal and nuchal muscles, and although in common with other apes, the gorilla shows numerous anatomical adaptations to brachiation, the adult is so big that it spends much of its time on the ground.

The orang-utan is intermediate in size between the chimpanzee and gorilla form which it is also distinguished by its reddish-brown rather than black fur. In the male, there are conspicuous cheek pads, and in contrast to the African apes, the brow ridges are only feebly developed. The orang-utan lives mainly in the trees, but unlike the gibbon and to a lesser extent the chimpanzee, its movements are slow and deliberate.

Anatomical and physiological features. *Skull.* In monkeys and apes the braincase is relatively bigger than in prosimians. In addition, the facial skeleton projects less, partly as a result of a reduction in its size and partly as a result of the folding of the basicranial axis.

The ethmoid separates the palatine and lacrimal bones in the medial orbital wall, and in monkeys and apes, the orbit and temporal fossa are more completely separated by a bony plate than in any prosimian.

In New World monkeys, there is a prominent auditory bulla outside which is the tympanic ring. In Old World monkeys and apes, there is no bulla and the tympanic forms a tubular auditory meatus.

The occipital region of the skull is relatively well developed, but in monkeys and apes the foramen magnum does not lie nearly so far forward as in man.

The biggest endocranial capacity recorded even, for an adult gorilla is approximately 700 ccs. This is some 300 ccs below the minimum usually regarded as consistent with the social and intellectual behaviour characteristic of *Homo sapiens*.

Teeth. In Old World monkeys and apes, the dental formula is

$$\frac{2.1.2.3}{2.1.2.3},$$

but in New World Species, a third premolar is present. The incisors are relatively simple cutting teeth and the upper canines, which, especially in males, are often long and pointed, form a shear with the lower first premolar. The remaining premolars are simple biscuspid teeth. In the molar teeth of the Old World monkeys, opposite cusps are joined by prominent transverse bars of enamel, and a fifth cusp is usually found on only the third lower molar. In apes, opposite cusps are not joined, and each lower molar has five main cusps.

Limbs. In both monkeys and apes, the forelimb retains many primitive features such as the articulation of the radius and ulna which allows the movements of pronation and supination. In the quadrupedal monkeys, the legs are longer than the arms, but in the apes the arms are disproportionately developed, and in the great apes, the flexor tendons of the forearm are arranged so that when the wrist is extended, the fingers are bent into the form of a hook. The digits which, except in marmosets, have nails rather than claws, are freely mobile and while, in most quadrupedal types, the thumb is relatively well developed and to some degree opposable, in the apes and actobatic monkeys it is often much reduced.

In the hind limb, although the tibia and fibula are joined by ligaments, the ankle can be freely inverted and everted. The terminal phalanges bear nails rather than claws, and the great toe is opposable.

Blood. The plasma of monkeys and apes will, in contrast to that of prosimians, coagulate rabbit serum that has been immunised against human blood.

The human blood groups A and B are found in apes but not in monkeys or prosimians.

Reproduction. The uterus has no lateral horns, and the placenta is of a highly-differentiated haemochorial type.

Female monkeys and apes have a regularly-recurring menstrual cycle, but in New World species, the discharge at the end of the cycle is inconspicuous. In some species of Old World monkeys and apes, the skin around the external genitalia reddens and swells during the first phase of the menstrual cycle, the maximum swelling occurring at the time of ovulation.

The higher Primates breed throughout the year, but there is, in some species, a seasonal fluctuation in the numbers of births. The young are usually born singly and while in New World monkeys they often ride on the mother's back, in Old World types they cling to the fur of her belly, and are not infrequently handled.

Special senses. In monkeys and apes, there is no rhinarium and the system of turbinal bones in the nasal cavity is less extensive than in prosimians.

The eyes face forwards and monkeys and apes have both stereoscopic and colour vision.

Facial vibrissae, if present, are rudimentary.

Brain. The cerebral hemispheres become greatly expanded and are, especially in the apes, richly convoluted. The regions related to vision are, together with the association areas, more strongly developed than in prosimians, while those concerned with olfaction are much reduced.

The cerebellum is more prominent in monkeys and apes than in prosimians.

Social habits and general behavior. The basic social unit of monkeys and apes is the family party comprising a single adult male, one or more adult females and variable numbers of immature animals. As a result of, the regularly-recurring menstrual cycles of the females, the sexes associate throughout the year and social relationships are conditioned by a scale of dominance.

Monkeys and apes groom either themselves or each other, mainly by picking through the fur with their fingers. Sometimes the mouth and teeth are also brought into play.

As a result of their relatively highly-differentiated mimetic musculature, monkeys and apes are capable of a wide range of facial expressions. They normally drink by sucking rather than by lapping.

Monkeys and apes have been extensively used in behavioral experiments, and many species are attentive and relatively trainable. They appear to be more intelligent than lemurs but the data so far accumulated do not indicate any clear-cut differences in the mental capabilities of monkeys and apes.

Man, however, has a unique capacity for conceptual thought, and this, coupled with the freeing of the forelimb consequent upon the assumption of an upright posture, has led to patterns of behaviour that have resulted in his attaining biological dominance.

Palaeontology. The first monkeys and apes appeared in the Lower Oligocene and it is often thought that their ancestors were Eocene tarsiers. Man does not appear until the Pleistocene and it is uncertain when the line of descent leading to the Hominidae diverged from that leading to the extant apes.

E. H. Ashton

References

Clark, W. E. Le Gros, "The antecedents of man," Chicago, Quadrangle, 1960.

Forbes, H. O., "A handbook to the Primates," London, Allen, 1894.

Hooton, E., "Man's poor relations," New York, Doubleday, 1942.

Simpson, G. G., "The principles of classification and a classification of mammals," Bull. Amer. Mus. nat. Hist. **85**, 1, 1945.

Zuckerman, S., "Functional affinities of man, monkeys and apes," London, Kegan Paul, 1933.

PRIMULALES

The three families of this order resemble one another fairly closely in floral structure. They have generally 5 sepals, and 5 petals which are united at the base; there are 5 stamens which are opposite the petals and adhere to them; the ovary is superior and unilocular and contains many ovules, each of which has 2 integuments, and which are borne on a basal central placenta; the mature seeds are endospermic.

The Theophrastaceae is a small family (50 species) found mainly in tropical America, and consisting of trees or shrubs with simple leaves; the FLOWERS have 5 fertile and 5 infertile stamens (staminodes); and the FRUIT is generally stony (a drupe). The largest genus is *Clavija*, which has a palm-like appearance, with a slender trunk terminated by a rosette of leaves.

The Myrsinaceae is a big family (1000 species), widespread in the tropics. It is also woody, with simple leaves, but the leaves differ from those of the Theophrastaceae in being dotted with yellow glands from which resin is secreted. The flowers differ in having no staminodes; and the fruit may be a drupe or a berry. The largest genera are *Ardisia* and *Rapanea*. Two species of the genus *Aegiceras*, which occur in Indo-Malaya as part of the maritime mangrove vegetation, have seeds which germinate while the fruits are still attached to the tree, as in other mangroves (Rhizophoraceae), to which they are not related.

The Primulaceae (750 species) occurs mainly in north temperate regions, and is poorly represented both in the tropics and in the southern hemisphere. It differs from the other two families of the order in being herbaceous and in having a capsular fruit. Traces of staminodes may be seen in the flowers of some genera. Hutchinson groups the 3 families into two orders, Primulales and Myrsinales, which he does not consider to be closely related.

The largest and best known genus of the order is *Primula*, with about 500 species, most of which are found in the mountains of C. and E. Asia. The leaves are always in a basal rosette, and the flowers have a long corolla-tube and are visited by long-tongued insects. One of the most interesting sections of this genus is the Farinosae; its species form a polyploid series, with chromosome numbers ranging from 18 to 126. They are found throughout the northern hemisphere, and one, in southernmost S. America, is the only representative of the genus in the southern hemisphere. Many Primulas are grown in gardens, especially the polyanthus, which probably originated from hybrids between two European species, the primrose and the cowslip.

The flowers of the great majority of *Primula* species are dimorphic and heterostyled. As was first clearly shown by Darwin, crosses between long- and short-styled plants produce good seed, but self-pollination produces little or no seed; outcrossing is thus favoured and inbreeding discouraged. The differences between the two kinds of plant are controlled by a single gene. Other interesting genera of Primulaceae are *Hottonia* (aquatic and heterostyled) and *Cyclamen* (with coiled fruit-stalks and ant-dispersed seeds).

D. H. VALENTINE

References

Brunn, H. G., "Cytological studies in the genus *Primula*," Symbol. Bot. Upsal., **1**: 1939.

Ernst, A., "Stammgeschichtliche Untersuchungen zum Heterostylie—Problem," Arch. Jul-Klaus Stiftung, **34**: 58–191, 1959 (and many earlier papers in the same journal).

Thompson, H. J., "The biosystematics of *Dodecatheon*," Contrib. Dudley Herbarium, **4**: 73–154, 1953.

Wright Smith, W. and H. R. Fletcher, "Monograph (incomplete) of the genus *Primula*," Trans. and Proc. Roy. Soc. Edinburgh, 1941–49.

Wright Smith, W. and G. Forrest, "The sections of the genus *Primula*," Journ. Roy. Hort. Soc., **54**: 4–50, 1929.

PRINGSHEIM, NATHANAEL (1823–1894)

This German botanist was one of the leaders of the botanical renaissance of the 19th century, a fact which is the more remarkable in that he worked for almost all his life in a private laboratory of his own. Indeed, the only academic post he held was the chair of botany at Jena from 1864 to 1868. His most important works are a long series of studies on various groups of algae, in the course of which he showed for the first time that these forms have sexual reproduction. The most important of these works was *Pandorina* (1869) in which he showed for the first time the conjugation of zoo-spores and thus demonstrated the existence of iso-gametes. He also published important monographs on fungi and ferns but for the last 20 years of his life he concerned himself with the mechanism of photosynthesis, a subject on which, unfortunately, none of his views have stood up to subsequent experimental examination. He must also be remembered for having founded, in 1858, the important *Jährbuch fur Wissenschaftliche Botanik*. He was also the founder, and first president of, the *Deutsche Botanische Gesellschaft*.

PETER GRAY

PROTEIN BIOSYNTHESIS

A chronology of research developments in protein biosynthesis provides a very satisfactory framework for discussing the knowledge acquired in this area up to about 1960. Early in this century Fischer established that proteins were polymers, or more accurately, poly-peptides involving some twenty amino acids. In a formal sense the formation of a polypeptide from two amino acids could be visualized thus:

$$R_1-CH-COOH + NH_2-CH-COOH$$

$$\underbrace{\begin{array}{cc} | & | \\ NH_2 & R_2 \end{array}}$$

$$\text{synthesis} \quad \downarrow \uparrow \quad \text{hydrolysis}$$

$$R_1-CH-C\overset{O}{=}NH-CH-COOH + H_2O$$
$$\quad | \qquad\qquad\qquad | $$
$$\quad NH_2 \qquad\qquad R_2$$

Most living tissues and especially the digestive juices contain substantial amounts of proteolytic enzymes which catalyze the upward reaction of water and the peptide to give the free amino acids. In many instances it was found that attachment of appropriate substituents to the amino acids or working in solutions containing very little water permitted the downward or synthetic reaction. Consequently it was long believed that ubiquitous proteolytic enzymes were responsible for protein synthesis. In 1940 Borsook and Dubnoff (1) established by experiment and theory that the synthesis of the peptide bond was highly endergonic (energy consuming) under physiological conditions (500 to 4000 calories per bond). The synthesis of a peptide bond, therefore, had to be coupled with another reaction which had to be exergonic (energy yielding) under physiological conditions. Based on the evidence available in the 1940's Lipmann (2) and Chantrenne (3) postulated the intermediate existence of aminoacyl phosphate derivatives while Dounce suggested that the amino group could be activated by reaction with a phosphate group.

The other notable experimental observation in the pre-World War II period was from nutritional experiments. In order that experimental animals should thrive, it was necessary that all of the amino acids known to be essential should be available for protein synthesis at the same time. In other words, a lysine-deficient diet eaten at 9 a.m. followed by a leucine-deficient diet at 9 p.m. did not support growth while the complete diet given in a single meal sufficed. This was regarded as evidence that there was no storage mechanism for amino acids or, more precisely, that the organism is incapable of synthesizing any protein unless absolutely all of the component amino acids are simultaneously present (4).

By 1950 a number of new techniques and concepts did much to clarify the understanding of protein synthesis. Radioactive carbon (^{14}C) became available and was incorporated into synthetic amino acids such as alanine (5). The rate of protein synthesis is ordinarily so slow as to be immeasurable outside of a living organism, but the sensitivity of detectors for ^{14}C made it possible to measure the rate of synthesis of protein first in tissue slices and finally in cell-free homogenates. The first observations were that protein synthesis was extremely sensitive to a lack of oxygen or to various respiratory inhibitors such as dinitrophenol thus confirming the Borsook-Lipmann-Chantrenne idea that protein synthesis is endergonic and requires adenosine triphosphate or a similar energy source (6). Shortly thereafter it became possible to separate nuclei, mitochondria and microsomes from "cell sap" by successively centrifuging a tissue homogenate at 700 × g, 5000 × g and 100,000 × g respectively. In preliminary experiments it was found that nuclei contributed little or nothing to the cell's competence to synthesize protein (7). Mitochondria were necessary but could be dispensed with if adenosine triphosphate and an ATP generating system (PhosphoEnol Pyruvate and PEP kinase) were provided (8). Both the microsomes and the "cell sap" were necessary for protein synthesis. The newly synthesized protein was largely associated with the microsomes in small (150 A) particles of ribonucleoprotein (ribosomes) containing about equal amounts of ribonucleic acid and protein (see NUCLEIC ACIDS) (9). Subsequently newly synthesized protein appeared in the soluble or "cell sap" fraction. Much evidence indicates that cells that secrete protein transfer the protein from the ribonucleoprotein particles to the membranous part of the endoplasmic reticulum, thence into vesicles and finally into the extracellular space. Such ribonucleoprotein particles (or ribosomes) are found in all protein synthesizing cells, are always of about the same size and chemical composition, are generally nonspecific in that they will function equally well with other components derived from the same of different species and can always be dissociated reversibly into two unequal parts each containing ribonucleoprotein. Although much work has been done on the structure of ribosomes, they can best be described by their function which is to serve as the workbench on which enzymes, messenger-RNA, transfer-RNA and cofactors combine to assemble a protein molecule.

The composition of the soluble fraction of the cytoplasm or "cell sap" remains to be discussed. Very early on, it was discovered that treatment with anion exchange resins removed guanosine triphosphate (10) and that this nucleotide was essential for protein synthesis. Little more has been learned about this requirement except that it is universal and apparently involves the final step of peptide bond formation. Hoagland looked for some sort of amino acid activation enzyme in the cytoplasm. Following the lead of Lipmann's study of acetyl-CoA synthesis (11), he looked for an amino acid dependent exchange of pyrophosphate into adenosine triphosphate. He and other investigators quickly found and purified amino acid activating enzymes (now termed amino acid tRNA ligases) in almost every organism and generally quite specific for only one of the twenty natural L-amino acids (12). Not only did the enzymes catalyze an exchange of pyrophosphate into ATP but they catalyzed the reaction of ATP, amino acid and hydroxylamine to yield AMP, pyrophosphate and amino acyl hydroxamate, a formal prototype of a peptide synthesis. The first product of the reaction appeared to be an enzyme bound amino acyl adenylate ($E \cdot NH_2 - CH_2 - C \overset{O}{\diagdown}$. $-O-PO_3-RibAd$), which was remarkably stable considering the mixed acid anhydride character. Novelli, Meister, Berg and others synthesized such mixed amino acyl adenylic acid anhydrides and observed that in the presence of enzyme and pyrophosphate, ATP could be resynthesized. However, these mixed anhydrides did not polymerize to form protein even in the presence of microsomes.

In quick succession, Hoagland and Zamecnik, Hultin and Holley developed evidence that there existed a new kind of ribonucleic acid of low molecular weight that reacted with amino acid, enzyme and ATP to form AMP, pyrophosphate and an amino acyl ester of the RNA (amino acyl transfer RNA or AA~tRNA),

(13). Subsequent examination showed that there is at least one and sometimes as many as four distinct tRNA's in any one cell for any one amino acid. This amino acyl tRNA still possesses a reactive amino acyl group, its function being to carry the activated amino acid from the activating enzyme (in the cell sap) to the surface of the microsome where polymerization into polypeptide occurs. The Nobel Prize in Physiology in 1968 was shared by Robert Holley (14) who first isolated the alanine specific yeast tRNA and determined its sequence. Subsequently four other specific tRNA's have been isolated and had their sequences determined. In each case so far the esterified end has consisted of the sequence adenylic cytidylic cytidylic acid (15); in each case maximal hydrogen bonding between guanosine and cystoine, adenine and uridine has suggested a clover leaf structure; in each case a wide variety of unusual bases such as thiouracil, methylated purines, pseudo-uridine, reduced pyrimidines etc., have been located at specific loci; and in each case a suggestive "anticodon" (see below) has been identified in approximately the same part of the molecule about thirty nucleotides distant from the amino acid receptor site. (See typical structure and table of codons on opposite page.)

It was early realized that it would be difficult or impossible for the four bases of genetic material to "code" for the twenty amino acids of protein. It was immediately appreciated that two, three or more nucleotides would be necessary to specify a particular amino acid. Still no mechanism was apparent for some three nucleotides to select a unique amino acid such as leucine. The obvious credibility of the Watson-Crick-Wilkins structure of DNA (for which they received the Nobel Prize in 1963) and the evidence that amino acids were "activated" by attachment to unique nucleotides provided a possible missing link in this chain. Every amino acid was covalently bound to a specific RNA which in turn was very specifically hydrogen bonded to a specific portion of the RNA in the way that Watson and Crick had postulated. Thus each activated amino acid would be correctly positioned for polymerization.

As mentioned above, the ribosomal RNA appeared to be generally nondescript and accordingly was an unlikely candidate for this newly conceived coding action. At this point Astrakan and Volkin (16) observed that virus infection of bacteria completely changed the protein synthetic pattern and concommitantly generated a short-lived RNA whose composition reflected the composition of the infective DNA. Within months the concept developed that genetic DNA generated this new RNA (messenger RNA or mRNA) which interacted with ribosomes, aminoacyl~tRNA and cofactors to produce a specific protein. Working with this concept and cell-free microbial systems developed by Lamborg (17), Nirenberg (18) quickly established that the synthetic messenger RNA, polyuridylic acid, catalyzed the formation of polyphenylalanine from which he deduced that uridylic uridylic uridylic acid (UUU) must be the messenger code for phenylalanine. Quickly thereafter other synthetic mixed polynucleotides were tested and found to code for other amino acids. Finally Nirenberg (19) culminated his work by establishing that simple trinucleotides of known sequence stimulated the binding of unique amino acyl tRNA's to ribosomes. On the basis of the work that

```
                        ┌─────────┐
                        │ C—G—I   │
                        └─────────┘
                  MeI              U
                    ↘              U
                      G ·· C
                      G ·· C
                      G ·· C              DiH
                      A ·· U            U—G
                      G ·· C               G
                      A     G   DiMe  A     C
                      G     C—G—C—G          U
                      U     ┊ ┊ ┊ ┊        DiH
                      C     G—C—G—C—G
          C—ψ—T             GMe      U—A—G
        G    G—G—C—C—U
        A—U—U—C—C—G—G—A
                      U
                      C ·· G
                      U    U
                      C ·· G
                      U    G
                      C ·· G
                      C ·· G—P
                      A
                      C
                      C
                    A—OH
```

Yeast alanyl tRNA (14). The box indicates the possible anticodon which combines with the codons listed below. A = adenosine, U = uridine, G = guanosine, C = cytosine, ·· = hydrogen bonds, — = 3'–5' phosphodiester bonds, P = 5' terminal phosphate, —OH = 3' terminal hydroxyl, ψ = pseudouridine, diMeG = diMethylguanosine, MeG = methylguanosine, diH-U = dihydrouridine, I = inosine, MeI = methylinosine.

Probable Codons (E.coli (19), reticulocytes (23))

phenylalanine	UUU, UUC
leucine	UUA, UUG, CUU, CUA, CUG, CUC
isoleucine	AUU, AUC, AUA
valine	GCU, GCC, GCA
serine	UCU, UCC, UCA, UCG, AGU, AGC
proline	CCU, CCC, CCA, CCG
threonine	ACU, ACC, ACA, ACG
alanine	GCU, GCC, GCA, GCG
tyrosine	UAU, UAC
histidine	CAU, CAC
glutamine	CAA, CAG
asparagine	AAU, AAC
lysine	AAA, AAG
aspartic acid	GAU, GAC
glutamic acid	GAA, GAG
cysteine	UGU, UGC
tryptophan	UGG
arginine	CGU, CGC, CGA, CGG, AGA, AGG
glycine	GGU, GGC, GGA, GGG
chain	
initiation	GUG, AUG
chain ending	UAA, UAG, UGA

led to a probable genetic code, Nirenberg shared the Nobel Prize in 1968. (This probable code is summarized above.)

Still no well defined protein or protein-like material had been synthesized in a cell-free system. Using extraordinary skill in nucleotide chemistry, Khorana prepared a series of deoxyribonucleotides of established structure. Using these as primers in the Kornberg DNA-dependent DNA synthesis, Khorana (20) synthesized a number of DNA's of known structure and from these he prepared RNA's that would have been presumed to be mRNA's with specific coding properties. Happily, such mRNA's catalyzed the synthesis of polypeptides of exactly the sequence that would have been anticipated from Nirenberg's code. Khorana, accordingly, received the third part of the 1968 Nobel Prize.

Thus we have a picture in which genetic DNA begets more DNA; DNA produces complementary mRNA plus perhaps rRNA and tRNA; the mRNA uniquely defines the structure of the protein and correspondingly all or almost all phenotypic expressions of the genotype.

A few reservations or amplifications are necessary. 1) All eucaryotic cells appear to contain a full complement of genetic material, yet only a few proteins are being synthesized in a particular cell at a particular time. 2) At least two heat labile co-factors are required to assemble the protein on the microsome surface; very little is known of them. 3) The speed of protein synthesis is very great, perhaps 1/1000 to 1/10 second being required for the execution of the cumbersome reaction we have described. 4) The precision of protein synthesis vastly exceeds the precision of other chemical and biochemical reactions, the rate of error-making being well less than 1 in a thousand. This is not predictable from physio-chemical considerations and is not observed in the first steps of amino acid activation (21). 5) Even the first (activation) step of protein synthesis is not well understood. In many cases the corresponding tRNA is necessary or influences the reaction and it is possible that an amino acyl adenylate is not a true intermediate. 6) By the hypothesis outlined above every protein would appear first as a linear polypeptide. Ultimately the finished protein is folded, coiled, phosphorylated, acetylated, given secondary bonds such as disulfide bridges, amide bridges, degraded into several non-covalently organized chains etc. We know little of these processes. 7) Certain infective viruses contain too little DNA to account for the observed changes in the protein produced after infection. No explanation is wholly satisfying, but some of the influence of the input DNA must be more indirect than postulated. 8) Originally mRNA was recognized by its extremely rapid synthesis and degradation in virus infected bacteria. However, cells devoid of nuclei (such as reticulocytes and enucleate Acetabularia) continue to synthesize protein for long periods of time. Either the messenger is much more stable or other mechanisms obtain. 9) Formyl methionyl tRNA is widely regarded as the chain initiating amino acid at least in E. coli. However, under appropriate circumstances, a hexanucleotide of formyl methionine containing no anticodon will react with ribosomes and puromycin to simulate chain initiation (22).

Altogether, progress has been fantastic in the last twenty years but there are as many unanswered questions as answered.

The preparation of this article was supported by U.S.P.H. Grant CA 08000.

ROBERT B. LOFTFIELD

References

(1) Borsook, H. and J. W. Dubnoff, J. Biol. Chem., 132: 307, 1940.

(2) Lipmann, F., Advances in Enzymology 1: 100, 1941.

(3) Chantrenne, H., Biochim. Biophys. Acta, 2: 286, 1948.

(4) Cannon, P. R., Fed. Proc., 7: 391, 1948.

(5) Frantz, I. D., Jr., R. B. Loftfield, and W. W. Miller, Science, 106: 2762, 1947.

(6) Frantz, I. D., Jr., et al., J. Biol. Chem., 174: 773, 1948.

(7) Borsook, H., et al., J. Biol. Chem., 187: 839, 1950.

(8) Siekevitz, P. and P. C. Zamecnik, Fed. Proc., 10: 246, 1951.

(9) Littlefield, J. W., et al., J. Biol. Chem., 217: 111, 1955.

(10) Keller, E. B. and P. C. Zamecnik, J. Biol. Chem., 221: 45, 1956.

(11) Jones, M. E. et al., Biochim. Biophys. Acta, 12: 141, 1953.

(12) Hoagland, M. B., Biochim. Biophys. Acta, 16: 288, 1955.

(13) Hoagland, M. B., P. C. Zamecnik, and M. L. Stephenson, Biochim. Biophys. Acta, 24: 215, 1957.

(14) Holley, R. W., et al., Science, 147: 1462, 1965.

(15) Zamecnik, P. C., M. L. Stephenson, and L. I. Hecht, Proc. Natl. Acad. Sci., 44: 73, 1958.

(16) Volkin, E. and L. Astrakan, Virology, 2: 149, 1956.

(17) Lamborg, M. and P. C. Zamecnik, Biochem. Biophys. Acta, 42: 206, 1960.

(18) Matthaei, J. H. and M. W. Nirenberg, Biochem. Biophys. Res. Comm., 4: 404, 1961.

(19) Nirenberg, M. W., et al., Cold Spring Harbor Symp. XXXI, 11, 1966.

(20) Nishimura, S., et al., Fed. Proc., 24: 409, 1965.

(21) Loftfield, R. B., Biochemical J., 89: 82, 1963.

(22) Munro, R. E. and R. A. Marker, J. Mol. Biol., 25: 347, 1967.

(23) Gupta, N. K., J. Biol. Chem., 243: 4959, 1968.

PROTOPLASM

Protoplasm is the physical basis of life, in that it is the visible, organized substance with which we associate the sum total of chemical and physical transformations characteristic of life. In general, one considers the protoplasm, together with substances of known constitutions, such as the cell walls, vacuoles, starch grains, oil droplets, secretory colloids and similar materials as constituting the whole living unit, the cell. Conversely, the protoplasm may be defined as a cellular material of unknown constitution, in which occur reactions characteristic of life, many of which may be carried on outside of the cell but which, in the cell, are organized to produce the orderly seriation of events responsible for the living state.

Protoplasm is a heterogeneous mixture of materials. With the light microscope it presents a variety of appearances depending upon the particular cell or animal or plant form being examined. In its simplest form, i.e., a stamen hair cell, from a Tradescantia flower, it consists of a clear background substance, the hyaloplasm lining the cell wall, and transversing the central vacuole with large flowing strands. While not generally visible, membranes are thought to form interfaces between the hyaloplasm and the external environment, and between the hyaloplasm and the internal vacuole. Many small spherical granules, known as spherosomes, are visible in the strands of hyaloplasm.

With care, mitochondria may also be observed. Most plant cells contain some type of plastid. In the *Tradescantia* stamen hair cell these are colorless and are known as leucoplasts.

Mitochondria constitute another category of important cellular particulates. They are present in all cells critically examined for them, although they may be scarce or lacking from the photosynthetic tissue of leaves. They are small spheres or rods of variable shape about a micron in diameter. The electron microscope shows a bordering double membrane and an internal system of lamellae or tubules. The enzymes of the Krebs cycle are associated with mitochondria as are some other enzyme systems.

All living cells, with the known exceptions of mammalian red blood cells and the sieve tube elements of plant vascular tissue, contain one or more nuclei. The hyaloplasm, plastids, spherosomes, mitochondria and other extra-nuclear components of the protoplasm may be grouped together as the cytoplasm, which together with the nucleus constitutes the protoplasm.

The protoplasm of other kinds of cells varies according to its function or species from which the cell was derived. All green cells contain chloroplasts. Gland cells of animals are characterized by a peculiar area or zone, adjacent to the nucleus, which, under certain conditions, appears as a coarse net, a series of plates, or vacuoles with associated granules. It is in this region that the secretory granules first make their appearance and the region has been referred to as the Golgi Zone. In many cells appropriate techniques indicate areas of rather intense affinity for basic dyes, a phenomenon which has been referred to as basophilia. This region again seems to be protoplasmic area of rather active cellular transformations and the protoplasm comprising it has been called ergastoplasm.

The electron microscope shows, in protoplasm, a remarkable series of parallel lines. These have been interpreted to be lamellae, or tubes, the latter forming a reticulum and the lines are generally thought to have the properties of membranes. It would appear that they do present evidence of the existence within protoplasm of a system of orderly interfaces that may vary with function and consequently cell type, and with changes in the physiological state of the cell. Such regular lamellae are known to occur in chloroplasts, in the light receptive regions of the rod and cone cells of the retina of the eye. Concentric lamellae are found in the yolk nuclei of starfish eggs, in nerve fibers and in other cells. The Golgi Zone also many sometimes appear as groups of parallel double membranes with their edges on ends somewhat swollen.

The heavily basophilic areas of the cytoplasms show a series of interconnected lamellae or a network which seems to divide the protoplasm into two distinct regions. The lamellae adjacent to the nuclei seem to be connected to the nuclear membrane, which itself is double, in such a fashion that the interior of the lamellar system is separated from the interior of the nucleus by a single membrane about 40 A in thickness. The lamellar system is associated in a similar fashion with outer cytoplasmic membrane. This lamellar system has been called the ergastoplasm, endoplasmic reticulum or simply cytomembranes. Frequently associated with the outer surface of the double membranes are granules rich in ribose nucleic acid and therefore responsible for the basophilia.

Structurally the vegetative nucleus is not easy to interpret either with the light or electron microscope. It is bounded by a membrane, which at high magnification of the electron microscope appears to be double and provided with pores. It may also be associated with the endoplasmic reticulum. One to several dense spherical bodies of the nucleoli are present. They have a high content of ribose nucleic acid which is thought to be synthesized within them. Electron micrographs suggest that fine spiral fibrils and granules may constitute part of their fine structure. Optical studies indicate the chromosomes simply as numerous small dense bodies, the prochromosomes, or as an irregular system of fibrils thought to represent chromosomes. Nuclei of such tissues as liver may be isolated and broken apart to show fibrillar structures that have been interpreted to be chromosomes. Chromatin and nucleoli are embedded in a colorless matrix referred to as the karyolymph. Many of these details are shown in Fig. 1.

Chemically, protoplasm is exceedingly complex. A gross analysis of it contributes little toward its understanding. Literally the whole field of biochemistry, and much of organic chemistry, is simply the chemistry of protoplasm. Considering the complexity of its chemistry, the overall simplicity of its physical organization of protoplasm poses one of the toughest problems for those investigators who would find cellular

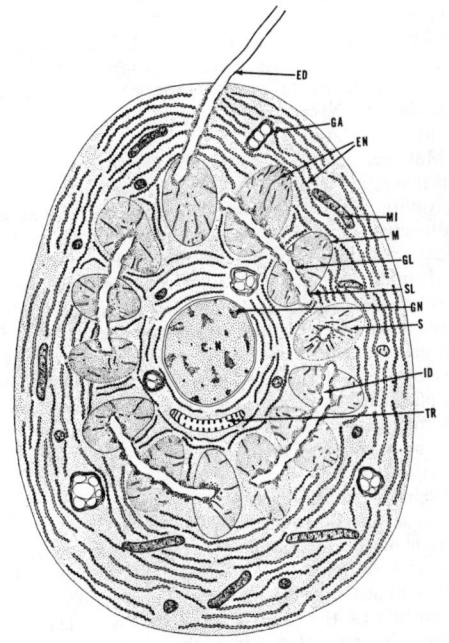

Fig. 1 Schematic drawing designed to show the various components of the nucleus and cytoplasm that can be demonstrated by various techniques. N, nucleus; ED, extracellular duct; ID, intracellular ductule; SL, inner spiral layer; GL, granular layer; M, outer membrane; S, secretion masses; EN, endoplasmic reticulum; MI, mitochondria; GA, Golgi material; GN, granular nucleoli; TR, tracheole. (From H. W. Beams, *et al.*, *Journal of Ultrastructure*, Vol. 3, 1959).

localizations for the multitude of reactions known to occur within the cell. There follow a few examples of technique's which are proving valuable in localizing specific reactions within the protoplasm.

One obvious method of localizing specific compounds within the cell is to use dyes having a special affinity for known cellular compounds. The most outstanding success in this area is the ability of aldehydes to regenerate Schiff's reagent, a solution of basic fuchsin decolorized by sulphur dioxide. Since many sugar molecules may be broken at the 2–4 linkage to form an aldehyde they may be localized by use of the Schiff reagent. Feulgen showed that the deoxyribose of the nucleic acid occurring within chromosomes and associated with the genetically active material may be broken by gentle hydrolysis in 1N HCl and subsequently stained with Schiff's reagent. Hotchkiss pointed out that periodic acid will similarly reveal aldehyde groups in numerous other carbohydrates that may be stained with Schiff's reagent. Since the uptake of the dye is directly related to the amount of aldehyde present a measurement of amount of dye absorbed will give a good indication of the amount of deoxyribose nucleic acid also present in parts of cells or chromosomes. For this work, known as microspectrophotometry, special instruments have been devised for measuring the dye absorption, or the ultraviolet absorption of portions of chromosomes or of whole cells. Cells may be broken apart and chloroplasts, mitochondria, and nuclei collected in test tubes. These isolated cellular organelles will continue to function, at least for part of their normal cellular activities. Steps of the Kreb cycle of respiration associated with mitochondria may be studied in test tubes or Warburg vessels and these experiments have yielded much information about cellular respiration.

Materials known to be normal precursors of certain substances synthesized by cells are needed for certain enzymatic reactions within cells may be labeled with radioactive atoms or withheld from the normal nutrient medium. When labeled materials are used such as thymidine labeled with H_3 which is a precursor for deoxyribose nucleic acid, sections of tissue are prepared and laid on photographic film. The active hydrogen acts upon the sensitive silver grains which may be developed. Ths tissue is stained and when matched with the blackened silver grains a very precise localization of old and newly synthesized deoxyribose nucleic acid may be made. This technique is known as autoradiography.

Depriving the fungus, *Neurospora*, of certain materials in its culture medium has yielded precise information of the succession of steps in the synthesis of numerous cellular materials used in the metabolism of this fungus. From these types of investigations it may be postulated that ribose nucleic acid is elaborated within the nucleolus under the control of the deoxyribose nucleic acid of the chromosomes and in collaboration with substances entering the nucleus from the cytoplasm. The newly formed ribose nucleic acid subsequently passes out of the nucleus into the cytoplasm where in conjunction with the substances of the cytoplasm the materials characteristic of the cell in question are elaborated. A feedback system may be considered as being operative in that the substances formed in the cytoplasm in cooperation with the ribosenucleic acid from the nucleolus also influence the formation of new ribosenucleic acid within the nucleolus.

T. Elliott Weier

References

Bourne, E. J. and J. F. Danielli, "The International Review of Cytology," New York, Academic Press.
Brachet, J. and A. E. Mirsky, "The Cell: Biochemistry, Physiology, Morphology" 6 vols., New York, Academic Press, 1959–1964.
Palay, S. E., "Frontiers in Cytology," New Haven, Yale University Press, 1958.
Brachet, J., "Biochemical Cytology," New York, Academic Press, 1958.

PROTOZOA

Protozoa are defined as primitive animals but, in current taxonomic practice, include certain plant-like organisms (e.g., plant-like flagellates, slime-molds). Colonies may contain differentiated reproductive and somatic zooids, but no tissues. Size ranges from about 2.0 μ to several centimeters (e.g., mycetozoan plasmodia). The largest are multinucleate; the smallest, uninucleate. Like bacteria, Protozoa live in fresh and salt water, in the soil, and as parasites in most animals and certain plants. Endoparasites of animals invade body cavities, tissues, or cells (e.g., malarial parasites (see PLASMODIUM), etc.). In freshwater species, hydrostatic regulation typically involves "contractile" vacuoles which accumulate water from the cytoplasm and later discharge it externally under pressure exerted by the cytoplasm. Vacuoles collapse during discharge and are later refilled.

Locomotor organelles include flagella, pseudopodia, cilia, and organelles composed of fused cilia. Ciliates possess cilia; other Protozoa, flagella or pseudopodia, or sometimes both or neither. Amoeboid movement involves protoplasmic flow. In the "hydraulic" types, according to one interpretation for *Amoeba proteus,* contraction of the outer cytoplasm (plasmagel) maintains pressure on the inner cytoplasm, the result being a bulge (pseudopodium) wherever the plasmagel is temporarily discontinuous. In the "sliding" method, as described for *Allogromia,* protoplasmic flow involves shearing forces between two distally continuous filaments composed of plasmagel alone (at least in small pseudopodia). Flow in opposite directions occurs on the two sides of a single pseudopodium. The contraction-hydraulic mechanism is attributed to amoebae, Arcellidae, Mycetozoa (see MYXOMYCETES); the sliding mechanism, to Foraminifera, Euglyphidae, Heliozoa, Radiolaria, and certain others.

A flagellum consists of a sheath enclosing a bundle of fibrils (1 pair central, 9 pairs peripheral) distinguishable in electron micrographs. Bases of the outer fibrils form a "granule" (*blepharoplast*) which may have genetic continuity. Except for trailing types and those of dinoflagellates, flagella typically extend anteriorly. The one to many flagella of flagellates supply the power for swimming or, rarely, for gliding. Flagellar activity, as characterized by Jahn, involves planar undulations (base to tip, *pushing,* posterior flagellum of dinoflagellates; or tip to base, *pulling,* trypanosomes) or helical waves (tip to base, *Peranema;*

base to tip, trichomonad undulating membrane). In some cases (e.g., *Rhabdomonas, Pyrsonympha*) the body acts as an inclined plane or Archimedean screw, turned spirally by flagellar activity.

A cilium, arising from a basal granule (*kinetosome*) homologous with a blepharoplast, parallels a flagellum in fine structure. Somatic distribution of cilia is a practical taxonomic criterion. Fused cilia form membranes (one or more longitudinal rows), membranelles (transverse rows), and cirri (tufts). Basal granules may persist in non-ciliated stages (e.g., adult Suctoria), and also have morphogenetic functions (origin of buccal organelles in fission, etc.).

A typical protozoan is covered with a pellicle, varying from a delicate layer in many amoebae to relatively firm layers in organisms with fairly constant shapes. A pellicle may be continuous or composed of articulating strips (e.g., certain Euglenidae). The body also may be covered by a cellulose-containing *theca* (phytomonads, dinoflagellates), or a test (or shell) in many Sarcodina and certain Mastigophora. Mostly inorganic (siliceous, calcareous) as a rule, tests may be secreted, or formed by cementing together foreign particles (sand grains, etc.). Foraminiferan tests are typically covered with cytoplasm and often consist of separate chambers formed consecutively during growth. Fossil tests help in identification of particular geological deposits. A lorica (conical, tubular), analogous to a test, occurs in certain flagellates and ciliates.

Endoskeletal structures include radiolarian skeletons (primarily siliceous), some of which are intricate latticeworks forming several concentric layers. Radiolarian skeletons have been preserved as fossils ("radiolarian ooze").

Nuclei of binucleate or multinucleate Protozoa are often homologous, but there are exceptions. Ciliates have two types (*macronucleus, micronucleus*), sometimes more than one of each. The polyploid macronucleus contains many chromatin granules and some nucleoli; the diploid micronucleus, fewer such granules. Both contain deoxyribonucleic acid and pentose-nucleic acid. The macronucleus, essential in reproduction and regeneration, is resorbed in conjugation and autogamy, and is replaced by a division-product of a synkaryon. Amicronucleate strains are well known (e.g., *Paramecium, Tetrahymena*). Such ciliates can reproduce indefinitely but cannot conjugate or undergo autogamy. Somewhat analogous "somatic" and "reproductive" nuclei are known in certain Foraminifera. Chromosome numbers range from two (probably haploid number) to an estimated 1600. Length ranges from less than 1.0 μ to about 50 μ. One or more nucleoli may be present. Division of the typical nucleus is mitotic; that of the macronucleus is not. In *Holomastigotoides,* a flagellate with very few large chromosomes, behavior of chromonemata and centromeres has been traced throughout mitosis. Information on small nuclei with more numerous chromosomes is incomplete. In different species, the spindle may lie outside or inside the nuclear membrane, which commonly persists through mitosis.

Reproduction involves division of the organism and is accompanied or preceded by nuclear division. Binary fission implies approximately equal division; plasmotomy, division of a multinucleate form into several multinucleate progeny; simple budding, division into two organisms differing in size; schizogony,

formation of several to many uninucleate buds from a multinucleate parent, typically leaving a cytoplasmic residue.

Sexual activity may or may not precede reproduction. Syngamy (fusion of two "gametes") occurs in certain Protozoa other than ciliates. Conjugation (involving exchange of haploid nuclei) occurs in ciliates and, with certain exceptions (e.g., various vorticellids), requires only temporary fusion of conjugants. In representative ciliates (e.g., *Paramecium, Tetrahymena, Stylonychia*) mating types are involved. A species includes "varieties" (*syngens*), each containing two or more mating types. Normally, conjugation involves any two mating types within one syngen. Intervarietal conjugations, if they occur, are often lethal. In autogamy, known in certain flagellates and ciliates, haploid "gametic" nuclei are formed and subsequently fuse into a synkaryon within one organism.

Life cycles often include only a trophic stage and a cyst, and the cyst may be lacking in certain parasites. Dimorphism in the trophic stage—flagellate/amoeba; amoeba/plasmodium; flagellate/non-flagellate; or ciliated/non-ciliated stage—is not uncommon. More than two trophic stages appear in the cycles of some species. Syngamy or other sexual activity may increase the complexity of a cycle. In malarial parasites, for example, an asexual phase in a vertebrate alternates with a sexual phase in a mosquito. Syngamy is often essential to completion of such two-host cycles.

Cysts may be primarily protective or reproductive. The organism is enclosed in a cyst wall, thick and often compound in protective cysts, thinner and usually single in reproductive cysts. Protective value may depend upon a layer of cellulose, protein, keratin-like, chitinous, or siliceous material. In certain species, cysts may protect the organism against desiccation for several months or more. In freshwater species and typical intestinal parasites, protective cysts are important in distribution. Encystment usually includes resorption of locomotor organelles and a change toward ovoid or spherical form. Such changes are reversed in excystment.

Dissolved foods may enter the surface by pinocytosis, perhaps by diffusion in some cases, or with the aid of enzymatic mechanisms. Particulate foods are ingested through a temporary or permanent "cytostome." Cytostomes of certain carnivores can be stretched to engulf large prey (e.g., *Didinium* eating *Paramecium*). Ingested particles are enclosed within a food vacuole (gastriole) formed at the base of the gullet. Digestion then occurs, usually within the gastriole, sometimes in small daughter vacuoles arising by pinocytosis. In suctorian ciliates, tentacles adhere to captured prey. Rupture of the prey's pellicle, apparently initiated by enzymes from the tentacles, is followed by a flow of protoplasm down the hollow tentacle into a developing food vacuole. The specific mechanism of ingestion is still uncertain. In any phagotroph, undigested materials are finally discharged from the gastriole to the outside.

Phytoflagellates often contain chromatophores and carry on photosynthesis. Other Protozoa obtain energy from oxidation of suitable organic foods. Unlike certain bacteria, Protozoa cannot use inorganic materials as a sole source of energy. Specific requirements are known for many species grown axenically in chemically defined media. Certain of these species are used in bioassay techniques (e.g., thiamin, vitamin B_{12}, nicotinamide, biotin, riboflavin).

Nitrogen requirements of certain, but not all, phytoflagellates may be satisfied by an ammonium salt or a nitrate (*Chlamydomonas*, etc.). Typical heterotrophs (e.g., ciliates) may need 10 or more amino acids, and often a purine or pyrimidine (or both). As major energy sources, fatty acids, alcohols, carbohydrates, or occasionally amino acids, are oxidized by particular heterotrophs. Carbon sources include CO_2 (probably used by all, but only by certain photosynthetic species as the sole carbon source), a variety of amino and other organic acids, alcohols, and carbohydrates. Suitability of a carbon source may vary with the species. In addition to minerals, other requirements often include growth factors: vitamin B_{12}, thiamine, or both in many but not all phytoflagellates, some of which may need biotin or (rarely) riboflavin in addition; three or four vitamins in some of the small amoebae; a larger number of vitamins (e.g., biotin, thiamine, riboflavin, nicotinic acid, folic acid, pyridoxine, thioctic acid, pantothenic acid, etc.) in typical heterotrophs, some of which also need a steroid (e.g., cholesterol, sitosterol) or a long-chain fatty acid.

Utilization of sugars (see CARBOHYDRATE METABOLISM) through the glycolytic (Embden-Meyerhof) pathway has been described in various species, although an aerobic monophosphate system may account for part of the sugar metabolism in some aerobes. Certain amino acids are deaminated aerobically, and ammonia seems to be the major nitrogenous waste. In addition, a few amino acids have been reported as excretory products of certain species. Comparatively little is yet known about metabolism of lipids, although most Protozoa apparently can synthesize and store them.

There may be as many parasitic as free-living species. Some parasites are harmful, often destroying tissues or individual cells. Others (*commensals*) seem to be essentially harmless in the normal host. A few are *symbiotes* in that they contribute to nutrition of the host. For example, certain intestinal flagellates of termites digest cellulose, whereas the wood-eating host produces no cellulase. Man harbors at least 25 species: mouth, 2; small intestine, 2; large intestine, at least 10; urogenital tract, 1; blood and other tissues, at least 10. In terms of potential damage, man's important parasites are: *Entamoeba histolytica*, causing amoebiasis; *Leishmania donovani*, kala-azar (visceral leishmaniasis); *Trypanosoma gambiense, T. rhodesiense*, African sleeping sickness; *Trypanosoma cruzi*, Chagas' disease; and above all, malarial parasites (*Plasmodium falciparum, P. vivax, P. malariae, P. ovale*).

Taxonomically, Protozoa are grouped into Mastigophora, Sarcodina, Sporozoa, Cnidospora, and Ciliophora.

Mastigophora include Phytomastigophorea and Zoomastigophorea. Many phytoflagellates contain chlorophyll and accessory pigments. The latter vary qualitatively and quantitatively, so that chromatophores may be green, yellow, red, brown, or blue in different species or larger groups. A number of phytoflagellates lack chromatophores. Food requirements are comparatively simple. Phytomastigophorea include the orders Chryosomonadida (*Ochromonas*, etc.), Cryptomonadida (*Chilomonas, Cryptomonas*, etc.), Dinoflagellida (*Ceratium, Gonyaulax*, etc.), Volvicida

(= Phytomonadida) (*Volvox, Chlamydomonas*, etc.), and Euglenida (*Euglena, Peranema*, etc.), among others. The Zoomastigophorea include the orders Kinetoplastida (trypanosomes, etc.), Trichomonadida (trichomonads), Hypermastigida (*Trichonympha*, etc.), and others.

Sarcodina, which move by protoplasmic flow, are subdivided on the basis of pseudopodial structure and other features, but a completely satisfactory classification is yet to be reached. The group has been split into Rhizopodea and Actinopodea, but the slender pseudopodia in some members of each class show "sliding" protoplasmic flow. Also, certain Actinopodea possess "axopodia" (slender pseudopodia with axial "filaments") while others have slender pseudopodia without the filaments. Sarcodina include amoebae, Mycetozoa, shelled rhizopods (e.g., *Arcella, Euglypha*), Radiolaria, Heliozoa, and Foraminifera, among others.

Sporozoa and Cnidospora are parasitic. Sporozoa include malarial parasites, Coccidia (e.g., *Eimeria, Isospora*), and gregarines (*Monocystis, Gregarina*, etc.). Malarial parasites live in both vertebrate and invertebrate hosts; Coccidia, in either, depending upon the parasite. Gregarines, on the other hand, are parasites of invertebrates. Cnidospora typically produce cysts ("spores") containing one or more coiled "filaments" which, in at least certain species, form a tube through which the parasite apparently passes from the ingested cyst to the gut wall. Cnidospora are parasites of fishes, sometimes of Amphibia and reptiles, and of invertebrates (especially annelids and arthropods).

Ciliophora are subdivided mainly on the basis of type and distribution of ciliary organelles. Holotrichia include species without cirri or prominent membranelle zones—e.g., *Paramecium, Tetrahymena*. Cilia of adult Peritrichia typically are limited to the peristomial region, as in *Vorticella, Epistylis*, and *Carchesium*. Reproduction may yield a motile *telotroch* stage in addition to a typical form; in such cases the motile daughter, equipped with extra cilia, eventually changes into a typical adult. Suctoria, as adults, possess tentacles but no cilia; budding results in a ciliated "larva" which, after a period of swimming, discards its cilia and develops tentacles characteristic of the adult. Spirotrichia have a well developed zone of membranelles leading clockwise toward the cytostome. Representatives include *Blepharisma, Bursaria, Euplotes, Stylonychia*, typical "rumen ciliates" (e.g., *Entodinium, Diplodinium*) of herbivores, and also a large group of loricate marine ciliates (*Tintinnus*, etc.).

R. P. HALL

References

Corliss, J. O., "The ciliated Protozoa," London, Pergamon, 1961.

Grassé, P.-P., "Traité de Zoologie," vol. 1, Paris, Masson, 1952.

Hall, R. P., "Protozoology," Englewood Cliffs, N.J., Prentice-Hall, 1953.

Ibid. "Protozoan Nutrition," New York, Blaisdell, 1965.

Hutner, S. H., ed., "Biochemistry and Physiology of the Protozoa," vol. 3, New York, Academic Press, 1964.

Hutner, S. H. and A. Lwoff, eds., "Biochemistry and Physiology of the Protozoa," vol. 2, New York, Academic Press, 1955.

Jahn, T. L. and F. F. Jahn, "How to Know the Protozoa," Dubuque, Iowa, Brown, 1949.

Kudo, R. R., "Protozoology," Springfield, Thomas, 1966.

Lwoff, A., ed., "Biochemistry and Physiology of the Protozoa," vol. 1, New York, Academic Press, 1951.

Manwell, R. D., "An Introduction to Protozoology," New York, St. Martin's Press, 1961.

PSAMMON

The interstitial fauna and flora of BACTERIA, diatoms (see BACILLARIACEAE) PROTOZOA and micrometazoa that lives between the sand grains of marine, brackish and freshwater beaches is known collectively as *Psammon*, a term introduced by Sassuchin, Kabanov, and Neizwestnova in 1927. The biotope, a variable mixture of sand grains, air, water and detritus, inhabited by this group, has been called by Wiszniewski the *Psammolittoral* (an area composed of the unsubmerged sands of the littoral zone); it has been referred to by Remane as the *Lückensystem*, and as the *Système Lacunaire* by Fauré-Fremiet. Remane preferred the term *Mesopsammon*, while Nicholls in 1935 described psammozoa as the *interstitial fauna* and the biotope in which they lived as the *interstitial milieu*. Thalassopsammon is now accepted as a term describing the marine flora and also the fauna is less than 2 mm long, inhabiting the lacunar spaces of interstidal and subtidal sands, while Mesopsammon is more or less restricted to freshwater forms.

It is not very long ago that beach sands, submerged or not, were considered to be biologically poor, a concept very likely originating in an unconscious linkage with the idea of a desert. The history of the investigation of the freshwater interstitial micrometazoa starts with the work of Sassuchin in 1926, who called attention to the protozoa in a sandy beach bordering the Olga River in Russia. This was followed the next year by Sassuchin, Kabanov, and Neizwestnova's paper above; and later on by the work of Wiszniewski (1932), Varja (1938), Myers (1936), Pennak (1939), and Neal (1948). A bulletin *Psammonalia*, is now available to investigators working on all aspects of the biology of interstitial fauna and flora.

Marine beach psammon investigations started with the work of Girard (1904) and of N. A. Cobb (1914), who indicated that the intertidal zone of marine beaches was inhabited by large populations of harpacticoid copepods and nematodes. Their work was largely ignored until the early nineteen thirties when the study of this milieu was again undertaken, this time on a broader scale. Zoologists, especially taxonomists and ecologists, became actively interested in Great Britain, Germany, South Africa, Western Australia, France, Italy, Madagascar, India, Canada, Brazil, Peru, and the United States. Several psammologists as Remane and Pennak, have worked both with the freshwater and marine forms. Remane has achieved distinction with investigations of the ecologically difficult brackish water psammon. The four centers of psammon research today are Kiel on the Baltic Sea, Banyuls and Roscoff on the Mediterranean coast of France and the Marine Biological Laboratory in Woods Hole, Mass.

The passages between the sand grains form a micro-labyrinth. The water in the pores of the deeper sand of fresh water beaches often becomes oxygen poor to the point of anaerobiosis either because of becoming covered with detritus, or because of the presence of hydrogen sulfide-rich decaying organic material. The interstitial fauna can ordinarily withstand a moderate oxygen deficiency; moreover, in the upper layers of marine beaches, renewal of oxygenated interstitial water is accomplished cyclically by tides. Disturbance of the sand on a large scale in the surf zone, dilution of the interstitial water by rain, and extremes of temperature apparently do not disturb the psammon. The psammon, by creeping or gliding, migrate vertically or laterally with apparent ease in escaping from these disturbing elements. Other unstable environmental factors that must be met are dilution by subterranean land drainage water (dependent on topography), gross changes in the amount of capillary water because of evaporation, variations in concentrations of available food, manifold effects of storms, and the fact that even under the most favorable circumstances, light penetrates to no more than 15 mm. below the sand surface. The nature of the sand itself modifies the impact of these seemingly grossly harmful environmental exposures.

The substratum of sand grains in freshwater and marine beaches in temperate latitudes are rounded to angular particles of quartz and biotite, plus a small amount of additional hard material accumulating from the weathering of local rocks. Compression will not decrease the size of the interstitial milieu, the diameter of whose passages depends on the sorting, size, and angularity of the sand grains. Interstitial space as measured by retention of greater volumes of water has been shown to become larger with increasing fineness of graded sands. Extremely fine sand beaches (grain size less than 0.047 mm.) have a negligible microfauna because of the ultraconfining nature of the micropassages. The same situation obtains in tropic marine beaches that contain large amounts of the disintegration products of coral that block the passages, or large amounts of fractured mollusk shells whose compacted flat surfaces practically obliterate the interstices. In typical temperate marine and freshwater beaches, the largest percentage of sand particles is between 0.3 and 0.6 mm., the total sand substratum when saturated occupying above 60% of the aggregate volume. Recent papers on various aspects of the granulometry of beaches by geologists at the Woods Hole Oceanographic Institution have been very helpful to investigators of the ecology of both mesopsammon and thalossopsammon.

Investigation of the chemistry of interstitial fresh water (initiated by Stangenberg in 1934), reveals a content of about 54% more dissolved inorganic material, 42% more dissolved salts, and relatively large quantities of plant nutrient materials as nitrates, phosphates, iron, and sulphates than the bordering body of water. This makes the interstitial milieu a decidedly eutrophic habitat capable of supporting a large population. The hydrogen-ion concentration in the top layers of sand is close to that of the adjacent body of water, but below 15 cm., it tends to become more acid. Similar microstratification, greater in fresh water than in marine beaches, occurs with oxygen, free carbon dioxide, and other dissolved materials.

A relatively tremendous population both in numbers of species and individuals has been found in the interstitial milieu of the psammolittoral biocoenose. Pennak found in the 2 to 3 cubic centimeters of water from an average 10 cubic centimeter sample of freshwater beach sand, 150 centimeters from the water's edge, a biota containing 4,000,000 bacteria, 8,000 protozoans, and 400 varied micrometazoa. Deboutteville (1953), Remane (1951), Zinn (1942), and others have demonstrated that the psammon of marine beaches shows even greater diversity, more complex relationships and a plainly evident vertical microstratification. Classified according to eating habits, there are carnivores, herbivores, omnivores, detritus feeders, scavengers, and browsers. Important primary food sources are the usually plentiful bacteria and detritus and algal populations which may reach 325,000 cells per cubic centimeter in the top layers of sand (diatoms, peridinians, green or blue-green algae). These figures must be downgraded for marine beaches. The total population supported is apparently a direct function of the total area of exposed surfaces on which the organisms may crawl, creep, glide, or become attached in

with only minor population changes during the winter. In intertidal beaches, copepods, nematodes, and tardigrades, and stragglers from other groups, have been found in depths over 30 cm., but in freshwater beaches, where more than 95 per cent of the population occurs in the top 6 cm., very few descend as far as 11 cm. Horizontally, marine psammon is found much further from the water's edge than the 350 to 375 cm. limit of freshwater forms. Many species show definite horizontal and vertical distribution profiles, patterns largely dependent on the slope of the beach. Unique morphological adaptations to this micro-labyrinthine environment are small size; linear, slender, wormlike bodies; flexibility; increase of tactile hairs or setae; spines used in creeping; ovisacs pressed against the body; and adhesive organs.

Phylogenetically, more groups are represented in the microfauna of marine beaches than in the interstices of freshwater beaches although total numbers of organisms are usually similar. Their distribution may be summarized as follows:

Animal Groups with Species Peculiar to Marine Psammon	Animal Groups Found Elsewhere but with Unique Representation in the Marine Psammon	Animal Groups with Species Found Abundantly in the Freshwater Psammon

Turbellaria:
    Otoplanidae
    Monocelididae
Gnathostomulida
    Gnathorhynchidae
    Cicerinidae
Gastrotricha:
    Macrodasyoidea
    Chaetonotoidea
Archiannelida:
    Protodorilidae
    Polygordiidae
    Nerillidae
    Dinophilidae
    Pisionidae
Insecta (Collembola)
Crustacea:
    Mystacocarida (subclass)
    Ostraocoda (Cladocopa)
Cnidaria
    Actinulida
	Ciliata
	Foraminifera
	Cnidaria:
	    Hydrozoa
	Turbellaria:
	    Acoela
	    Polycladida
	Nematoda
	Ostracoda:
	    Cytheridae
	Copepoda (Harpacticoida)
	Isopoda
	Tardigrada
	Acarina
	Ectoprocta
	Polychaeta:
	    Syllidae
	    Hesionidae
	Nemertinea
	Gastropoda;
	    Opiertobranchia
	    Philinoglossidae
	    Hedylidae
	Echinodermata
	    Holothuroidea
	Chordata:
	    Tunicata
		Ciliata
		Nematoda
		Rotatoria
		Tardigrada
		Copepoda (Harpacticoida)
		    (very few species)

the aquatic surface film of the sand grains. Interstitial protozoa and micrometazoa exhibit a regular seasonal succession, particularly in fresh water beaches, disappearing extensively during the cold months, reappearing during spring, and attaining a maximum during August. The marine microfauna is far more stable

The intertidal zone of marine beaches and the area above the water line of brackish and fresh water beaches, contain a productive, dense, complex population of bacteria, algae, protozoa and micrometazoa inhabiting the aqueous interstitial labyrinth between the sand grains; collectively known as thalassopsammon.

The unusual range of environmental conditions in this biocoenose gives a unique aspect to the ecology of the psammon.

DONALD J. ZINN

References

Deboatteville, C. D., "Biologie des Eaux Souterraines Littorales et Continentales," Paris, Hermann, 1960.
Swedmark, B., "The interstitial fauna of marine sand," Biol. Rev., **39**: 1–42, 1964.
Zinn, D. J., "Between grains of sand," Underwater Naturalist 4(2): 4–10, 1900.

PSEUDOSCORPIONIDA

The order Pseudoscorpionida or Chelonethida of the class ARACHNIDA includes animals commonly called pseudoscorpions. Except for the body being smaller, usually between 1 and 6 mm. in length, and the opisthosoma or abdomen not being specialized to form a sting-bearing metasoma or post-abdomen, the pseudoscorpions are very similar to SCORPIONS in general appearance. The anterior part of the body, the *prosoma* or cephalothorax, is covered dorsally by a *carapace*. Each finger of the chelate, two-segmented chelicera has a comblike *serrula*. Near the distal end of the movable finger is often found a styliform or branched *galea* or spinneret. The chelate, six-segmented pedipalps are large and conspicuous. The chelal fingers bear marginal teeth and have specialized tactile setae, of which typically four occur on the movable and eight on the fixed finger. One or both of the fingers may have a venom apparatus and a venom tooth. All the legs may be similar in general structure or, especially because of specializations of the femur, the legs of the first and second pairs may differ conspicuously from the legs of the third and fourth pairs. The number of tarsal segments serves as the basis for division into three suborders. In the Monosphyronida all pedal tarsi have one segment; in the Diplosphyronida each tarsus consists of two segments; in the Heterosphyronida the tarsi of only the third and fourth legs are divided. Two pairs of spiracles and the genital opening occur towards the anterior end of the ventral surface of the abdomen. Except for a few species with obligate parthenogenesis, reproduction is bisexual. In at least many species, spermatophores are deposited by the male and picked up by the female during a mating dance. Eggs and embryos are carried on the ventral surface of the abdomen, where the young are nourished by a secretion from the mother. Subsequent growth usually includes three nymphal instars.

Pseudoscorpions are found in virtually all parts of the world. Species are especially numerous in moist tropical and subtropical regions. While many species are adapted to microhabitats in semiarid areas, few tolerate the low and prolonged temperatures of high latitudes. Pseudoscorpions are commonly associated with woody plants and usually remain well hidden in crevices of bark, in woody debris, and in leaf litter. Some species live in the nests of rodents, birds, and social insects. Phoresy is fairly common. Little is known about the economic significance of pseudoscorpions, but some species no doubt prey on microarthropods important as parasites and vectors. Others certainly play a part in the control of plant-damaging arachnids and insects. A very common and virtually cosmopolitan species, *Chelifer cancroides* (Linnaeus), is associated with man. This and a few other species occur frequently in barns and chicken houses and are found occasionally in human dwellings.

C. CLAYTON HOFF

References

Beier, Max., 1932a. Pseudoscorpionidea. I. Subord. Chthoniinea et Neobisiinea. Das Tierreich, vol. 57, pp. 1–258. 1932b. Pseudoscorpionidea. II. Subord. Cheliferinea. Das Tierreich, vol. 58, pp. 1–294.
Chamberlin, Joseph C., "The arachnid order Chelonethida," Stanford Univ. Publs., Biol Sci., vol. 7: 1931.

PSILOPHYTINA (PSILOPSIDA)

The naturalness of this subdivision, formerly composed of two orders, the Psilophytales of Silurian and Devonian age and the Psilotales represented only by two extant genera in a single family, has been questioned for some time (see 1st edition of this volume). In the interim between the publication of the two editions of the Encyclopedia, research has uncovered a number of significant new facts which have forced a withdrawal of recognition of the subdivision as well as its two included orders. The fossil forms have been segregated and the component groups have been elevated to three new subdivisions: RHYNIOPHYTINA, ZOSTEROPHYLLOPHYTINA, and TRIMEROPHYTINA.

The extant Psilotales have been taxonomically reassigned to a position close to the filicalean families Stromatopteridaceae (segregated from the Gleicheniaceae), Gleicheniaceae and Schizaeaceae. The new interpretations are based on the discovery of psilotoid gametophytes in *Stromatopteris* and *Actinostachys* (*Schizaeaceae*), psilotoid embryology in *Actinostachys*, and close structural and developmental conformance of nearly all entities within the life cycles of the Psilotaceae and the Stromatopteridaceae.

DAVID W. BIERHORST
HARLAN P. BANKS

References

Bierhorst, D. W., Bull. Torrey Bot. Club, **92**: 475–488, 1965.

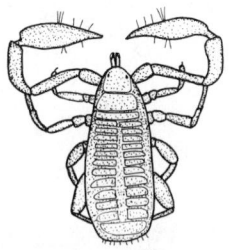

Fig. 1 A pseudoscorpion. (From Goodnight, Goodnight and Gray, "General Zoology," New York, Reinhold, 1964.)

ibid. Amer. Jour. Bot., **53:** 123–133, 1966.
ibid. Amer. Jour. Bot. 54: 651.
ibid. Amer. Jour. Bot., 1968 (in press).
ibid. Phytomorphology, 1968 (in press).

PSOCOPTERA

Psocoptera (Corrodentia; Copeognatha) are small IN-SECTS, rarely as much as one-half inch long, usually less than one-fourth inch. The majority have 4 functional wings, with distinctive venation of several types, but some have very short non-functional wings or are wingless. Tarsi are either 2- or 3-segmented, cerci are absent, mouthparts are for chewing, and the clypeus usually is much enlarged. A very distinctive structure is the lacinia of the maxilla, usually elongate and chisel-like. Antennae comprise 13 or more segments. Life history stages include egg, usually 6 nymphal instars, and the adult, so that metamorphosis is gradual. A general common name is psocid, pronounced so-sid (plural, psocids).

The best known psocids are pale wingless types usually called book-lice, which occur among books and papers. Several species with fully developed wings also occur indoors, frequently on the walls of new houses and in cereals and various other stored products. Most of the same species, as well as many others, occur outdoors. There, the majority live on the foliage or bark of

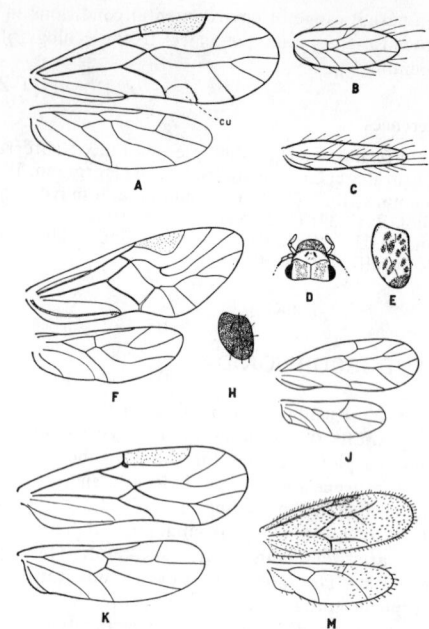

Fig. 2   Psocids. (A) wings of *Lachesilla pedicularia,* (B) wing of *Psocathropos,* (C) wing of *Dorypteryx,* (D) head of *Myopsocnema,* (E) wing of *Myopsocnema,* (F) wings of *Psocus venosus,* (H) wing of *Lepinotus reticulatus,* (J) wings of *Psyllipsocus,* (K) wings of *Ectopsocus pumilis,* (M) wings of *Archipsocus.* cu, cubital cell. (Drawings by A. B. Gurney, reproduced by permission of National Pest Control Association).

trees, and hence are called bark-lice, though others live under dead bark, on or beneath stones, in leaf mold, or in the nests of vertebrate animals. Among species living on tree trunks, gregarious nymphs hatching from a single egg mass sometimes gather as a conspicuous cluster, then scatter when mature.

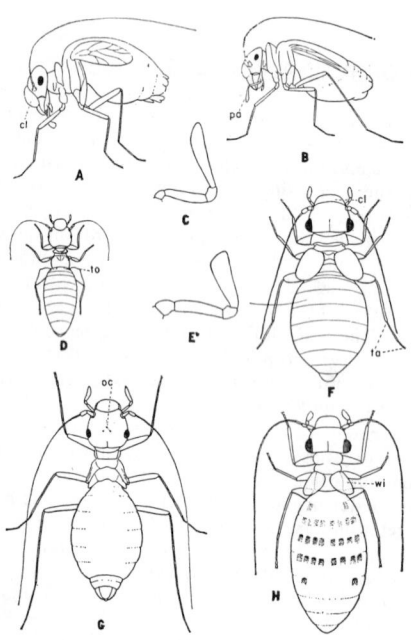

Fig. 1   Psocids. (A) *Psocathropos,* (B) *Dorypteryx,* (C) palpus of *Dorypteryx,* (D) *Liposcelis,* (E) palpus of *Psocathropos,* (F) *Lepinotus,* (G) *Psyllipsocus* (short-winged adult), (H) *Trogium.* cl, clypeus; oc, ocelli; pa, palpus; ta, tarsus; to, femoral tooth; wi, wing. (Drawings by A. B. Gurney, reproduced by permission of National Pest Control Association).

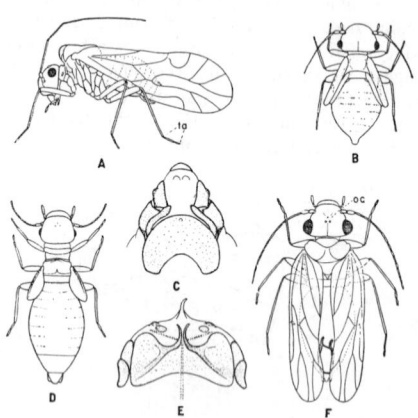

Fig. 3   Psocids. (A) *Lachesilla,* (B) *Ectopsocus* (nymph), (C) *Lachesilla pedicularia,* external genitalia of female, (D) *Archipsocus* (short-winged adult), (E) *Lachesilla pedicularia,* external genitalia of male, (F) *Ectopsocus.* oc, ocelli; ta, tarsus.

Lichens and molds are common foods of psocids, also cereals, dead insects, and organic matter of various kinds. Psocids often create a nuisance by infesting houses, museum collections of insects, granaries, straw packing, and other plant products, but control measures are necessary only in exceptional cases.

Although a dozen or so of the species usually found indoors occur nearly worldwide, most of the strictly outdoor species conform to usual zoogeographic patterns. Of about 1400 known species, about 155, in 17 families, have been found in the United States, and new species are being found continually. Psocids have not been a popular group with insect collectors because they are neither showy nor large, and only a few taxonomists have studied them consistently.

A. B. GURNEY

Reference

Smithers, C. N., "A bibliography of the Psocoptera (Insecta)," Australian Zoologist, **13**: 137–209, 1965.

PURINES

Purine compounds are found in nature especially in plants, to a lesser degree in animals. They are derived from purine, which may be called 7-imidazo-(4.5d)-pyrimidine, imidazolo-4'.5':4.5-pyrimidine, or 1.3.4.6-tetraazindene. Emil Fischer, who gave this ring system the name "purin" (from purum and uricum), introduced the following schematic representation and numbering,

$$N_1=_6C-H$$
$$H-C_2 \quad {}_5C-_7N \diagdown$$
$$\qquad\qquad\qquad {}_8C-H$$
$$N_3-_4C-_9N \diagup$$

and this symbol is still commonly used.

Purine ($C_5H_4N_4$, mol. wt. 120.06, m.p. 217°C) forms colorless crystals, soluble in water with neutral reaction to litmus or turmeric. Several syntheses are available, and many of its salts have been prepared. It is mainly used for organic syntheses, and in studies of metabolism.

The purine derivative of greatest biological importance is uric acid ($C_5H_4N_4O_3$), a white tasteless and odorless crystalline compound. It is only slightly soluble in water, and decomposes above 250°C without melting. Uric acid is very widely distributed in animals, plants, and in bacteria, usually in form of its salts. It is the main nitrogenous end product of protein metabolism in reptilian and avian excrements (guano). Small amounts occur in normal human urine, blood, saliva, and cerebrospinal fluid; under pathological conditions the amount is increased, e.g. in gout, when uric acid crystals are deposited in the joints. Uric acid has been synthesized in a number of ways. Shaking with phosphorus oxychloride under pressure gives 2,6,8-trichloropurine (old name): The 3 chlorine atoms may be catalytically removed, and replaced by hydrogen, to give purine. The reactivity of the chlorine atoms de-

$$N=C-Cl$$
$$\qquad\qquad\qquad H$$
$$Cl-C \quad C-N \diagdown$$
$$\qquad\qquad\qquad\qquad C-Cl$$
$$N-C-N \diagup$$

creases from position 6, the most reactive, to 2, and finally 8; this behavior is most important for the synthesis of various purine derivatives.

Hypoxanthine (6-hydroxypurine, $C_5H_4H_4O$) is widely distributed in plants and animals, and is obtained by hydrolysis of nucleic acids.

Xanthine (2,6-dihydroxypurine, $C_5H_4N_4O_2$) occurs in tea leaves, in beet juice, and in the actively growing tissues of sprouting lupine seedlings; it is also found in urine, blood, liver tissue, and in urinary calculi.

Adenine (6-aminopurine, $C_5H_5N_5 \cdot 3H_2O$) is present in tea leaves and in beet juice. It is a constituent of the nucleic acids, from which it may be prepared by hydrolysis with mineral acids.

Guanine (2-amino-6-hydroxypurine, $C_5H_5N_5O$) is found in guano, fish scales, leguminous plants, sprouting lupine seedlings, and in various animal tissues. Its white crystals have a pearly luster, and are used in making artificial pearls. Guanine forms a white deposit in the tissues of swine affected with a king of gout.

The guanine isomer, 2-hydroxy-6-aminopurine, (isoguanine) has been found in certain seeds and in insects.

The nucleic acids from yeast, and those from thymus tissue, contain the residues of the two aminopurines, adenine and guanine. 1-Methylxanthine, 7-methylxanthine, and 1.7-dimethylxanthine (paraxanthine) occur in urine. A number of alkaloids are purine derivates.

Caffeine (theine, caffeine) (1,3,7-trimethyl-2.6-diketopurine, $C_8H_{10}N_4O_2$)

$$CH_3-N-C=O$$
$$\qquad\qquad\qquad\qquad CH_3$$
$$O=C \quad C-N \diagdown$$
$$\qquad\qquad\qquad\qquad C-H$$
$$CH_3-N=C-N \diagup$$

is the most widely used purine derivative. It is the main alkaloid in coffee beans (1 to 1.75%, Phillipine coffee 1.6 to 2.4%), tea leaves (1 to 4.8%), cola nuts (2.7 to 3.6%), Maté (Paraguay Tea, 1.25 to 2%), guarana paste (2.7 to 5.1%). Small amounts of it are present in cocoa. Caffeine is a stimulant for the central nervous systems, heart and muscle activity, and has diuretic effect.

Theobromine (3.7-dimethyl-2,6-diketopurine, $C_7H_8N_4O_2$) is the main alkaloid of the Cocoa bean (unroasted 1.5 to 1.8%, roasted 0.6 to 1.4%); small amounts of it are found in tea leaves and in cola nuts. Its effect on the central nervous system is less than that of caffeine, but it is a stronger diuretic. Pharmaceutically its sodium compound is usually applied in form of double salts with sodium acetate, lactate, salicylate, or other salts.

Theophylline (1.3-dimethyl-2.6-diketopurine, $C_7H_8N_4O_2$) isomeric with theobromine, occurs in small quantities in tea leaves. Its physiological effects are similar to those of caffeine. Since it is an excellent diuretic, it is synthesized in relatively large quantity.

Finally, it should be mentioned that a number of purines, singly or in combination with pyrimidine derivatives, have been found to be important in the nutrition of bacteria. They act either as growth stimuli, or are essential growth factors.

PAUL ROTHEMUND

PURKINJE, JOHANNES EVANGELISTA (1787–1869)

Purkinje coupled with a genius for biology a fierce Czech nationalism which was to hamper him throughout his career. He learned German at a theological college but spoke it all his life with a strong Czech accent which was always ridiculed by his detractors as an affectation. Fortunately for science, he early attracted the attention of Goethe, who was able to procure for him, purely on the basis of political influence, the chair of physiology in Breslau in 1823. The manner in which his appointment had been secured outraged his colleagues and it was not until 17 years later that he was able to found the separate physiological institute on which he had set his heart. In the course of this 17 years, however, he ranged over an extraordinarily wide field of biological discovery. Much of his early work was on the physiology of sight and his investigations into the electrical stimulation of the retina led him close to the discovery of the mechanism of nerve transmission. He passed from this to a study of the anatomy of the brain and discovered both the existence of the axis cylinder of nerves and the large tree-like cells in the cerebellum which bear his name. He elucidated the structure of cartilage and of the sweat glands and discovered the function of the germinal vesicle of the hen's egg some years before von Baer demonstrated the existence of mammalian eggs. He also demonstrated the existence of ciliated epithelium in vertebrates, thus demonstrating that cilia were not, as had previously been thought, confined to the invertebrates. His experiments on the autonomous behavior of ciliated epithelium led him to postulate the existence of a cell even though the latter had not yet been described as a living concept.

However, after the brilliance of these discoveries had led to the establishment of his institute, he permitted his fierce nationalism to override his interest in science, resigned from the directorship of the institute, and returned to Prague where he devoted the rest of his life to the cause of the political freedom of his country. Indeed, in Czechoslovakia he is today far better known as a patriot who called himself Jan Purkyne than as the internationally famous biologist, Johannes Purkinje.

PETER GRAY

PYCNOGONIDA

Pycnogonida (or Pantopoda) is a subphylum of Arthropoda of exclusively marine habit, ranking with the Chelicerata and Mandibulata but more closely affiliated with the former in lacking antennae, in having the second pair of appendages (chelifores) chelicerate and with a patellar segment in the uniramous legs. They differ from the other two Recent subphyla in having multiple gonopores in an essentially thoracic position, and the abdomen reduced to a brief process supporting the anus. The most notable anatomic peculiarities of the pycnogonida are the reduction of the body in many species to a central tube-like structure with lateral processes supporting the legs (in some species these may be close together, producing a sort of central disc), and a large proboscis-like structure at the anterior end which is a sucking and filtering apparatus. The pycnogonids, like chelicerates, are primarily feeders on fluids or juices and use the chelifores to attach themselves to their prey.

Pycnogonids lack respiratory and excretory systems (although cement-producing glands of various types, serially repeated on the legs of the males, may be derived from some precursory segmental organs), have the usual ladder-like ventral nervous system, circumesophogeal ring and dorsal ganglion of arthropods. Above the dorsal ganglion are the four simple eyes, mounted in a dorsal process or tubercle (there is no well defined head region); eyes may be lacking in deepsea species. The circulatory system is essentially a dorsal tubular heart. Reproductive and digestive systems extend as diverticula far out into the very long legs; the gonopores are on the second segment ("trochanter") of the legs. There are usually four pairs of legs, but several species (including a Jurassic fossil) having five pairs, and two having six pairs of legs are known. Two 12-legged species are so far known, one from only two specimens and the other from six; both are Antarctic species. The significance of these extra-legged forms, which are otherwise no different from certain "normal" species of closely related genera, is unknown, but certainly does not represent a primitive condition. This phenomenon may be associated with polyploidy, but as yet nothing is known about the chromosomes of pycnogonids. In addition to having a pair of chelicerate appendages and several pairs of walking legs, pycnogonids possess sensory palpi, and in the males and some females there is a ventral accessory pair of appendages (ovigers) which in the male are used for carrying eggs. These ovigers are peculiar in having intratarsal muscles in their terminal segments, a character of unique but uncertain significance among the Arthropoda. They have a peculiar larval stage, the protonymphon (Fig. 1), bearing three

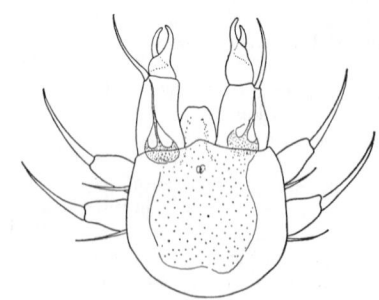

Fig. 1

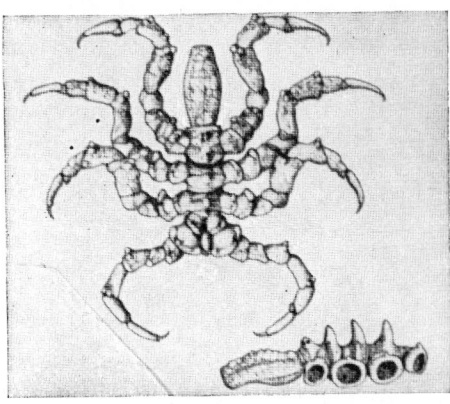

Fig. 2

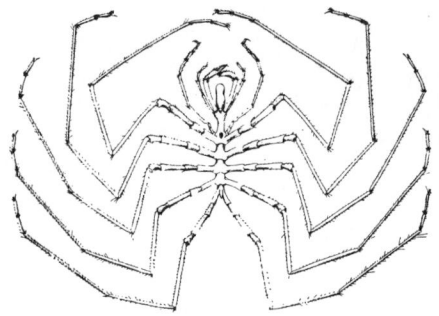

Fig. 3

pairs of appendages and thus superficially resembling the Crustacean nauplius, but the resemblance goes no further since the characteristic pycnogonid proboscis is already well developed, and the homologies of the appendages are arachnoid rather than crustacean. Sexes are usually separate but one hermaphroditic species has been described.

There are perhaps 600 species of pycnogonids, in perhaps 50 genera in eight well defined families. Classification is based primarily on the presence or absence of various anterior appendages. In *Pycnogonum* (Fig. 2), for example, there are no chelifores or palpi and only the males have ovigers, whereas all the appendages occur in both sexes of such genera as *Nymphon* and *Achelia* (Fig. 3). There are so many transitional combinations of characters that ordinal divisions cannot be maintained within the group, as recognized by D'Arcy Thompson half a century ago; Recent species are assigned therefore to the order Pantopoda. The Devonian, Palaeopantopus and Palaeoisopus and a pentamerous Jurassic form are placed in the Order Palaeopantopoda. Pycnogonids are common but inconspicuous members of the intertidal fauna; the larger species occur in deeper water, and some are found at depths beyond 6000 meters. One species seems to be bathypelagic, and another lives in coarse sand. Many of the shore and shallow-water forms pass their early stages in hydroids, octocorals, etc., as parasites; the natural history of the deep-water forms is unknown, and for some of them even eggs have never been observed. Some species are parasitic in clams and mussels and a few are ectoparasites of sea anemones, nudibranchs and other soft-bodied organisms. No economic significance has yet been demonstrated for these animals, but they sometimes occur by the thousands in fouling accumulations on buoys and in similar aggregations on the bottom. Since they seem to prey upon filter feeders and small-particle feeders for the most part, they could, when abundant, be significant links in the transfer of radioactive materials.

Important contributions to our knowledge of these animals have been made by Anton Dohrn, G. O. Sars, E. B. Wilson, T. H. Morgan and Vladimir Schimkevich, as well as a host of less eminent zoologists including the late Errol Flynn's father; the best general treatments are those of Fage in Traite de Zoologie, D'Arcy Thompson in Cambridge Natural History and Snodgrass in Arthropod Anatomy.

JOEL W. HEDGPETH

References

Hedgpeth, J. W., "On the phylogeny of the pycnogonida," Acta Zool., **35**: 193–213, 1954.
Helfer, H. and E. Schlottke, "Pantapoda" *in* Bronn, Y. G. ed., "Klassen und Ordnungen des Tierreichs" vol. 5, Leipzig, Akad. Verlags. 1935.

r

RABL, CARL (1853–1917)

This anatomist was born on May 2, 1853 at Wels, Austria and studied under Haeckel, Leuckhart, and Brücke. He then served as Professor of Anatomy at the Universities of Vienna (1885), Prague (1886), and Leipzig (1904) and lectured on normal and topographical anatomy and embryology. He published numerous papers on comparative anatomy and embryology occupied mostly with fundamental problems of morphology, the origins of the mesoderm and joints, and the development of the eye and head. Perhaps his greatest contribution to science was the detection of the constancy of the chromosome number in the developing egg. Rabl's contributions in anatomy and embryology were also of the highest value. He died at Lipsia on December 24, 1917.

Martin J. Nathan

RADIATION EFFECTS

Introduction

The various ionizing radiations to which all living matter (including man) is constantly subject are basically similar in their physical make-up and in the biological responses which they elicit. In this article the discussions will be limited to ionizing radiations which may be incorporated in the more general term *atomic radiations*. These radiations are produced when the electrical balance of atoms, which comprise matter, is disrupted.

While the biological effect of such radiation depends upon the energy absorbed* from it by the tissues concerned, there are physical differences between the various types of radiation, particularly with respect to the density of the ionizations and their penetrability into or through tissue. A very brief description of the variety of commonly encountered radiations might clarify this picture.

Sources of Ionizing Radiations

There are three major sources of ionizing radiations:
1. **Natural.** This includes the cosmic, terrestrial,

*Absorption refers to assimilation within the atom or molecule of energy from an electromagnetic or particulate radiation.

and atmospheric radiations, the latter two due to elements such as K-40, C-14, radon, thoron, and radium.
2. **Man-Made.** This includes x-rays and isotopes used in medicine, research and industry.
3. **Environmental contamination.** This includes fall-out and radioisotope disposal, luminous dials, some television sets and shoe-fitting fluoroscopes.

From any of these sources various radiations are physically similar, although the intensity or ion density may vary. Ionizing radiations are not to be confused with ultraviolet or infrared or even the longer radio waves where there is no free charged particle of molecular, atomic or subatomic size. An ion, in the usage here, may be a free electron, free proton, atom or even a molecule carrying an electric charge. It must not be confused with pH ions of dissociated molecules.

Types of Ionizing Radiations

What is ionization? A fast moving particle (concentration of energy) such as a photon, which penetrates matter, can be likened to a bowling ball accomplishing a strike in bowling pins. There is a disturbance in the balance (organization) of the atomic structure and this electrical imbalance results in ionization. In addition to ionization by impact, there may be ionizations by secondary means where uncharged neutrons can excite an atomic nucleus to cause it to emit a charged particle such as a proton. This charged particle then causes ionization by impact. There is also secondary ionization where a liberated electron dislodges other electrons from their orbits. This may even be a continuing process. The major difference relates to whether the ionizations are by particles or by electromagnetic forces.

For our purposes here, a brief summary statement will be made about the several more common sources and types of radiation.

Alpha particles or rays. These are helium nuclei emitted by the nuclei of an isotope (e.g., radon or polonium) during disintegration, or they may be produced by high energy accelerators. They consist of two protons and two neutrons propelled from the disintegrating nucleus. They are charged (2+), 8,000 times as heavy as an electron, have small velocity and from natural sources have low penetrance (max. 0.1 mm in tissue) and an energy range from 4 to 10 million electron volts with high specific ionization. Their biological hazard comes from inhalation or ingestion and alphas from artificial sources can affect the skin. Due to the high specific ionization its energy may be concentrated in a single cell, doing its damage there. It is generally considered to be biologically more hazardous than beta or gamma radiations.

Beta particles. These are high-speed negative electron beams from certain radioactive isotopes (disintegrating nuclei) or from certain accelerators such as the Betatron. They are charged particles which produce ionizations as they traverse tissue, and can create as many as 200 ion pairs per centimeter of track in air. Their ionizations are less dense than those from alpha particles, and their tissue penetrance is greater than alphas but from natural sources never exceeds 8 cm in tissue. From artificial sources penetration can be greater. In deceleration through tissue they can cause the production of x-radiation which is then an additional hazard.

Gamma rays. There are uncharged electromagnetic radiations from nuclei of certain radioactive isotopes, with extremely short wave length and high frequencies. They ionize matter by the ejection of high speed electrons from the absorbing material, but penetration through tissue is considerable because the attenuation of the primary gamma rays is small. High energy have greater penetrance than low energy rays, the latter being absorbed more readily by the heavy elements. Gamma rays can destroy tissue and cause burns rapidly, reacting with matter much in the same way as do x-rays. The charged particles produced may have sufficient energy to further excite and ionize biological material.

X-Rays. These are very similar to gamma rays, are electromagnetic radiations which interact with matter, but the emission process is extra-nuclear in contrast with that from gamma rays. X-rays may be produced by retardation of accelerated electrons at the anode of an x-ray tube, and are most commonly used to reveal internal (skeletal) structure. The degree of penetration is related to the voltage available to the tube. In general use, they are less penetrating than gamma rays but they can be made even more penetrating by increasing the voltage and by filtering out the soft (long) rays. Ionizations from x-rays and gamma rays are produced in matter by photoelectric effect, Compton effect, or by pair production.

Neutrons. These are uncharged constituents of atomic nuclei which may be ejected during the process of fission. They may be obtained from a cyclotron, an atomic pile or bomb, or from a van de Graaff generator. Their range is limited only by the density of the material traversed. *Fast neutrons* do not produce ionizations directly but give rise to charged, recoiling nuclei (usually protons) in the matter they traverse. These (secondary means) ionizations which are more dense than from Beta particles, are concentrated along the recoil track. They have great powers of penetration and are best impeded or absorbed by paraffin rather than lead because of the high concentration of hydrogen in paraffin. *Slow neutrons* do not produce recoils of significant energy but may be involved in capture by atomic nuclei to produce new nuclei which may be radioactive. They react particularly with hydrogen which is, of course, abundant in protoplasm. These unstable elements may then emit Beta or gamma rays, and are known as radioactive isotopes. In fact this is the common method of isotope production, by bombardment of elements with neutrons.

Cosmic rays. These are mixed, highly penetrating radiations arising from both light and heavy particles (with varying specific ionizations) in outer space, but they have been recorded deep below the earth's surface. Most of the radiations are absorbed by the atmosphere before reaching the earth's surface. It is estimated that the inescapable exposure of animals and plants on the earth's surface from cosmic rays will average about 0.03 rads per year. Even high altitude fliers are not likely to receive serious exposure because the ionization is limited to a small number of the body cells. Secondary cosmic radiations consist of a shower of particles produced from explosive collisions resulting in secondary and in tertiary particles, but these constitute little or no hazard to man. It is estimated that on earth there are at least a thousand such collisions between body nuclei and cosmic particles every second, without noticeable effects.

The biological effect of any of the above mentioned radiations is due to the absorption of energy by protoplasm. This is not heat energy, which is negligible. It is the kinetic energy of an ejected electron and the attendant ionization, or excitation, which brings about the biological effect. The physical imbalance, due to ionization, leads to a chemical change and a consequent biological adjustment. The physical and chemical changes may be immediate, but the biological adjustment may take from minutes to decades.

Major Terms and Concepts in Radiobiology

While radiobiology is a very young discipline, there are already a number of concepts and terms relating to it which should be briefly defined in order to facilitate an understanding of the ensuing discussions.

Radiation vs. irradiation. These terms are often confused and while there are a few situations in which they can be used interchangeably, they never mean the same thing. The sun gives off rays so that it radiates light, heat, U.V. etc. but to get a sunburn the skin must be irradiated by those rays. Radiant energy is emitted by the sun just as gamma rays are emitted by certain isotopes, or x-rays by an x-ray tube. Irradiation is the act or process of exposing an object to such rays, and the radiologist irradiates a patient or a tumor. Irradiation may be prescribed so that a tumor is treated by 1,000 r x-radiation.

Radiosensitivity vs. radioresistance. Ionizing radiations are not selective with regard to their targets in any biological system so that any differences which exist in the response (by damage or death) to equivalent ionizing radiations between species, organs, tissues, or cells may be attributed to inherent resistance or sensitivity. Further, there are differences in the response during the maturation process of any particular cell, or during the development of any organ, or the life of any organism. The precursor of the neuron, the embryonic neurectoderm, is about 10 times as resistant to radiation damage as is the transforming neuroblast, and the adult neuron may be 25 times as radioresistant as is the neuroblast from which it develops. It requires 1,000 times as much absorbed ionization to kill an adult Drosophila as it does the newly fertilized egg from which it is derived. Thus, while ionizing radiations can affect any biological system functionally and/or structurally, the dose required to disturb or to kill may vary with the maturity or the mitotic activity of the system. Certain animals are more resistant to whole body irradiation than are others (e.g., monkey more than guinea pig), some tissues more than others (e.g., bone more than epithelium and neurons more than germ cells). Some investigators have arbitrarily set a figure, such as 5,000 r, and regard failure to exhibit structural changes at this level to evidence of radio-resistance. At the other extreme is the geneticist who finds the gene extremely radiosensitive, so that these two terms have a relative value. These differences are unexplained, but may be related to degree of hydration and to genotype.

Dose of radiation. The dose (exposure) of radiation is the quantity of radiant energy delivered and may be expressed as total dose or in terms of dose rate with respect to time. A chronic exposure is that which is delivered over a long period, at a low level and continuously. A fractionated exposure is discontinuous. An acute exposure is generally a high level dose delivered in a few minutes or less. In any case, reference is made to the energy delivered. In biological studies it is not the dose delivered but the dose absorbed which is of importance. Radiation dose is expressed in several ways as follows:

The Roentgen: This is the unit of exposure to x- or gamma rays based upon ionizations produced, namely 1 esu of either charge in 1 cc of air under standard conditions. It is physically equivalent to 1.61×10^{12} ion pairs or 83 ergs per gram of material. It does not refer to absorbed dose, only to the delivered dose, and is limited to x- and gamma rays only.

The Rad: (radiation absorbed dose). This is the new unit, referring to absorbed energy equivalent to 100 ergs per gram of irradiated material at the point of interest, and applies to any ionizing radiations in any medium, therefore of more universal use. Since in biological studies it is the absorbed and not the delivered dose which is effective, the term "rad" is more frequently used.

The Rep: (roentgen equivalent physical). This refers to ionization in soft tissue equivalent to that of the roentgen in air. The energy dissipation for 1 rep of particulate radiation is generally regarded as 93 ergs/gram. Tissue absorption of energy depends upon its composition and the energy of radiation. The denser the tissue the greater the ionization produced and the greater the energy absorption. This term is useful only in soft tissue studies.

Permissible dose. This is the maximum amount of radiation which one could presumably be exposed to, within a specified time, without incurring permanent ill effects. Naturally, permissible levels for chronic low level exposures, or fractionated exposures, are greater than for acute exposures (see "Recovery" below). The value of the permissible dose has been successively reduced with further knowledge of the biological effects of ionizing radiations. It is currently less than 0.3 rad/week of chronic exposure or 25 rads in a single acute, emergency exposure. At these levels there appears to be no permanent, observable damage to somatic tissues.

Tolerance dose. This is regarded as a safe dose with respect to survival and may be many times the permissible dose. Exposure to this level may have drastic effects, involving radiation sickness and certain sequelae, but it does not kill (i.e., is tolerated). The term is used in civil defense directives, setting maximum limits in an emergency.

Radioprotection. Ever since the discovery that ioniz-

ing radiations can be harmful, there have been investigations to discover some means of protecting the individual against either the radiations themselves, or against the consequences of irradiation. If survival probability is increased by some drug, homogenate, or substance interposed, then it is said that there is some protection. Protection can include (1) Interposing distance from the source of the radiations, since the inverse square law is applicable, (2) Interposing an absorbing medium between the source and object such as lead, concrete, or paraffin, (3) Injecting into the body some chemical prior to exposure (e.g., cysteinamine), (4) Placing the organism in an hypoxic or anoxic environment during exposure, or (5) Injecting cell homogenates such as spleen (I.P.) or bone marrow (I.V.) from newborn within several hours after exposure. In each of these instances "protection" does not mean the prevention of irradiation sequelae (see below) but rather the providing of a better chance for survival. It really means reduction in lethality or the prolongation of the latent period for the biological effect(s). Thus far there is nothing that completely protects against the deleterious effects of ionizing radiations except being completely out of range of the exposure.

LD/50/30. This is the lethal dose for 50% of the exposed organisms which occurs within 30 days subsequent to exposure. For instance, this value for the adult guinea pig is about 225 r, and for man about 400 r, and for the chicken about 1300 r, etc. It should always be specific in including the age, sex, and weight of the organism concerned as these alter the LD/50/30 value. This is a convenient figure to use in studying the effectiveness of the so-called "protective" agents, substances, or devices. The MLD is the median lethal dose of radiation which results in the death of half of the exposed animals in any specified time. It cannot be used with animals of short lives.

Recovery. This is a term that must be clearly defined in radiobiology whenever it is used. The resilience of living systems is well known, and the ability to repair wounds or "recover" from sunburn is established. Skin exposed to radiation may exhibit erythema, like sunburn, which later disappears but in which there *may* develop, some 20 or more years later, skin cancer. If "recovery" means a return to the former state, it is questionable whether recovery from irradiation damage ever occurs since chromosomes are involved. But the term "recovery" is often used in the sense that there has been repair, regeneration, or replacement of tissue in a manner similar to that after surgery. In radiobiology where there may be fractionation of exposure over a period of days or longer, the tissue or organism can accept a much larger dose without the expected effects so that there has arisen the concept of "recovery" occurring between the fractionated exposures. However, on a cellular basis it is believed that changes brought about by ionizing radiations are irreversible and irrevocable. Since it is agreed that a single ionization may cause a mutation which, by definition, is a permanent change, it is doubtful that any cell structurally changed by ionizing radiations could in fact "recover" (return to) the prior state. The word has been used most often in connection with radiation sterilization of the testes which may be followed by a resumption of fertility. Fertility is indeed "recovered" but no one would argue that the spermatozoa produced are unaffected by irradiation during early maturation stages, but rather that the more resistant and unkilled precursors of the spermatozoa are eventually able to repopulate the testes with functional spermatozoa, by maturation. But certainly these surviving cells will carry their full quota of mutations.

Radiation sequelae. The full consequences of exposure to ionizing radiations may not be realized for decades. It is now well known that an animal may survive an LD/50 exposure for thirty days only to die later from other and less acutely lethal consequences of the same irradiation. Some of these consequences may be shortening of life, reduced fertility, skin lesions, cataracts, nephrosclerosis, reduced resistance to disease, effects on the blood components, blood forming organs and blood pressure, and various types of cancer such as epidermoid carcinoma, basal cell sarcoma, fibrosarcoma, osteogenic sarcoma, and myelogenous leukemia. When one includes the genetic effects of ionizing radiations, the consequences of one roentgen to the germ cells may be expressed in a long line of future generations even though such consequences may not be entirely obvious.

Half-life. Radioactive isotopes* disintegrate at a rate proportional to the number of atoms present although the fraction decaying in any unit of time is characteristic of each isotope. If the arbitrarily chosen fraction is 50%, then for each isotope there is a specific period of time for 50% of the nuclei to disintegrate, known as the "half-life" of that isotope. For each succeeding similar period of time the amount of radioactivity is further reduced by 50%. The half-life of radium C' is a fraction of a second, carbon 14 is 5,600 years and that of uranium 238 is more than a thousand millions of years.

Biological half-life. The quantity of any administered isotope present at any given time in any organism depends upon the physical half-life of the element and the rate or degree of elimination. This varies with the element, certain organs, and with the organism, because it is dependent upon affinity and excretion.

R.B.E. (Relative biological effect): This implies different biological reactions to different types of radiations, due largely to the linear energy transfer along the path of the ionizing particle. It is also due to the part of the body (e.g., tissue) irradiated. But it must be remembered that the rate and level of exposure are involved. When all variables except the biological response are equated, then any differences in response may be attributed to type of radiation. For each biological effect and each type of ionizing radiation there is an RBE factor. For instance, if the RBE for an alpha source is given as 10 for a particular biological reaction, this generally means that for the particular process only 1 rad of alpha radiation is biologically equivalent in effect to 10 rads of an x- or gamma ray source. The term REM (roentgent equivalent man) is

*An isotope is an atom with different atomic weight than other atoms of the same element, caused by an excess or deficiency of neutrons in the nucleus. Radioactive isotopes are therefore unstable or disintegrating isotopes. This instability results in emission of radiation. The new term radionuclide refers to any radioactive element, which may be found in the form of various isotopes.

sometimes used to incorporate RBE values as, for instance, the Dose in REM = Dose in RAD × RBE.

Fractionation. The rate of exposure is to be distinguished from the total dose, but when the rate is discontinuous (interrupted) it is designated as fractionated exposure, even though the total delivered dose over the extended period may be the same. For example, 600 r whole body exposure to man is 100% lethal within 30 days, but if this is fractionated into 150 r/week for the same 30 days, it is not 100% lethal and may be tolerated by many. Should such a man be exposed to 10 r/week for 60 weeks the total exposure would still be 600 r but since it was fractionated he might show no ill effects whatsoever. This better tolerance of fractionated exposures is believed to be based upon some degree of "recovery" between the exposures.

Fallout. This refers to the contaminating radiations from atomic or hydrogen bombs which consist of various isotopes with varying half-lives that actually are dispelled from or by the bomb. It has been estimated that the average exposure of plants, animals and humans at the present rate of production of fallout contamination is about 0.1 rad in 30 years, or less than from the luminous dials of watches.

Background. This refers to the radiations found at any one time in the environment, and naturally varies with the environment and time. It includes cosmic, local gamma rays, and radon from the atmosphere. It also includes radiations from internal sources such as K^{40}, and C^{14} and radon products. The total exposure of animals and plants on the earth's surface to background radiations is estimated at an average of about 4 rads in 30 years. Background would include "scattering" which is a process by which the direction of an incident beam of radiation is altered. The most common usage of the word "scattering" refers to the back-scattering which occurs when the radiation is affected so as to alter the delivered dose. There is scatter from one animal to another, or from one organ to another, all contributing to background radiation.

Latent period. Ionizations may be instantaneous but the biological consequences may be delayed. Lymphocytes respond within 2 hours to very low levels of exposure (25 r or less) while cancer of the skin may not develop for 17–35 years after contamination. These differences in response vary with different cells and tissues, and, to some extent, with variations in the imposed physical environment. Even with hyper-acute exposures, animals rarely die in less than a few days, and with low level but lethal exposures death may occur at three different peaks during the thirty day period. These peaks represent death due to different causes such as hemorrhage, intestinal malfunction, anemia and secondary infections. Genetic consequences may not be revealed for generations.

Biological Effects of Ionizing Radiations

Ionizing radiations are probably the most potent of all physical variables to which biological systems are subject. A single ionization may cause the mutation of a gene or the death of a virus and an exposure of 1,000 roentgens will kill almost every mammal known even though it means the ionization of but one molecule in every ten million. The un-ionized 9,999,999 molecules are unable to survive the presence of the single ionized molecule. An exposure at the rate of 1 r per second continuously for 24 hours each day would require 500 years before three quarters of the molecules would be ionized.

Even though any and every biological system can be affected by ionizing radiations, resulting in functional and/or structural changes, little is known about the mechanism of such effects. Most experiments have involved a single exposure and a subsequent examination of the changes brought about in the chromosomes, cells, tissues, organs or organisms. There is a great unknown chasm between the exposure, which may be instantaneous, and the biological manifestations which may either be long delayed or develop in a variety of ways over a long period. It is generally believed that the physical exposure brings about a chemical change which leads to the biologic adjustments. But the adjustments may vary with the total dose, the dose rate, with ion density, and with the particular reacting system. There is no general rule regarding the biological effects of ionizing radiations except that rarely do two apparently identical cells or organisms react in the same exact way to the same exposure of ionizing radiations. Whether this is due to transient variables or to inherent genetic differences has not been established. It points up the importance of statistical data in radiobiology with the establishment of probabilities.

Theories of Biological Effects of Ionizing Radiations

From time to time various theories have been advanced regarding the mechansim of radiation effects. While none has proven to be satisfactory and most are descriptive rather than explanatory, it is of interest to briefly define these in order to point up the problem.

Fluid flow theory (Failla). This theory emphasizes the fact that following irradiation there is a freer flow of water across living membranes, and cells and their contents tend to swell.

Activated water theory. Ionizing radiations produce short-lived peroxides in the exposed water, and since protoplasm invariably contains water it is believed that these brief exposures to toxic peroxides may explain the effects of radiations.

Poison theory. There are generally toxic by-products in any biological system exposed to ionizing radiations, aside from peroxides, which may arise from dying or dead protoplasm, but which could presumably affect healthy tissue.

Enzyme inactivation theory. It is known that enzymes, under certain conditions, are quite radiosensitive. Since enzymes are inevitably found in protoplasm, and are necessary for its survival and reproduction, they could represent the radiosensitive element in living matter.

Protein denaturation theory. Protein denaturation does occur but it may be the end product of other biological effects rather than the cause. This would include the enzymes.

Point heat theory. The heat equivalent of energy transferred to water by 1,000 rads is only about 0.002 calorie per gram for x- and gamma rays of high energy, and yet there are those who believe that there may be sub-microscopic points where ionizations occur accompanied by great heat concentration, sufficient to bring about a biological change. This is doubtful.

Point hit theory. This is also known as the target theory. It emphasizes the discontinuous nature of radiation absorption, and envisions irradiation as a bombardment with ions and extremely submicroscopic points which are more likely to be hit with greater bombardment, but that a single hit can be effective. An effect is seen when such a critical point is directly ionized by a quantum hit of radiant energy. These points, whatever they are, are presumed to be of unusual importance in the survival of the protoplasm. This theory has considerable merit and is supported by some data.

Mitotic inhibition theory. It is true that certain phases of mitosis are more radiosensitive than are others. It is also true that tissue can be killed by impeding mitosis, which is normally a never-ending process in living matter. While there is some relationship between radiosensitivity and cell proliferation (although the relationship is not inevitable), it may be that direct chromosome damage is the prime factor in radiation effects.

Somatic and/or genetic mutation theory. It has long been known that ionizing radiations are the most effective means of bringing about mutations in the germ cells. It is therefore possible that similar mutations may also be brought about in the formed somatic cells and these genic changes would be evident in the function or in the progeny of the somatic cells. It is logical that this should be the basis of some theories as to the origin of radiation-induced cancerous cells.

Since none of these theories has been widely accepted, it is obvious that little is really known of the mechanism which brings about the biological effects of ionizing radiations.

General Considerations

Gross or whole body effects of ionizing radiations. Radiation affects the structure of the electrical charges of the atoms of irradiated material. If an animal is subjected to whole body ionizing radiations at an LD/50% level there are certain major changes that can be expected such as (a) immediate reduction in circulating lymphocytes, (b) immediate rise in granulocytes, (c) reduction in platelets, (d) loss of appetite, weight, and activity, (e) gradual fall in erythrocytes, (f) diarrhea, watery to bloody, due to damage to gut epithelium, (g) emaciation, prostration, sepsis, fever, and death within two to three weeks. Radiation sickness is similar to many other illnesses except that ionizing radiations can penetrate to any of the cells of the body so that the effects may be more widespread. In postmortem studies of such irradiated animals there is gross hemorrhage, sepsis, ulceration of the small intestines, splenic atrophy, and depletion of the bone marrow. There are subjective symptoms such as headaches, vertigo, debility, abnormal sensations of taste and smell, anorexia, nausea, vomiting, tachycardia, fall in blood pressure, shortness of breath, increased irritability, insomnia and fear.

Those animals which survive the acute exposure may later manifest certain delayed effects such as reduction in life span, high death level due to infections, cataracts, alteration of media of arterioles, nephrosclerosis, arterial hypertension, cardiac atrophy, cerebral hemorrhage, general cachexia, late epilations, malignancy, splenomegaly, anemia, and higher incidence of developing cancer.

There are certain variables that may contribute to the survival of animals exposed to an LD/50/30 level such as (1) *Sex:* It is known that females are more radioresistant than are the males, and female hormones injected into castrate males give them added radioresistance. (2) *Age:* The embryo is the most sensitive stage in ontogeny, the newborn surprisingly resistant, the sexually mature adult the most resistant, and the senile again radiosensitive. (3) *Basal metabolic rate:* Hyperactivity or oxygenation, during or after exposure, renders the animal more sensitive to ionizing radiations, so that reduced activity and anoxia seem to increase radioresistance. (4) *Infection:* Changes in the blood constituents reduce the resistance to secondary infections which certainly contribute to post-irradiation deaths. (5) *Hydration:* The most hydrated tissues are the most radiosensitive so that it is possible that variations in total water content may alter radiosensitivity. (6) *Protective agents:* (see above) There are protective agents or conditions which will affect survival.

Organ and tissue responses to ionizing radiations. When an animal is subject to whole body ionizing radiations certain of the tissues suffer first and most, and are often responsible for death, while other tissues may appear to be unaffected. Listing of such tissues in the order of decreasing radiosensitivity would be as follows:

1. Lymphoid tissue: Some lymph nodes respond within 2 hours to exposures as low as 25 r, lymphocytes, spleen, nodules of the G.I. tract, and thymus.

2. Germinal epithelium: The maturation stages in the testes are so sensitive that the degree of their depletion has been regarded by some as an indication of the degree of radiation exposure. Ovaries of the adult are less responsive but, in contrast with the testes, there is generally no "recovery" following sterilization doses of 100 r for mice or 500 r for the human. In the embryo the presumptive testis is more radiosensitive than is the presumptive ovary.

3. Formative cells: The erythroblast and myelocyte of the bone marrow are quite radiosensitive.

4. Epithelium: The skin and hair follicles and the lining of the gastrointestinal tract are next.

5. Mature structures: Blood constituents (except lymphocytes), pancreas, liver, lungs, kidneys, and endocrine organs.

6. Mesodermal derivatives: Connective tissue, muscle and bone.

7. Ectodermal derivations: Nervous system, peripheral and central.

There is a 10,000 fold range in sensitivity among the various tissues of the adult, even though all may be equally ionized. The sequence of effects is generally an altering of the function, an altering of the structure, and finally necrosis or death.

If one were to cite the single most sensitive tissue it would be the hematopoietic. This tissue is rather widespread in the body and any exposure of a part will generally involve some of this tissue. The blood forming organs of children are more radiosensitive than those of the adults, and there may well be some causal relationship between exposure to ionizing radiations and the incidence of leukemia. Of course the skin is always involved in irradiations and in earlier days skin erythema was regarded as a good radiation dosimeter. Skin exposure is always a problem in radiation therapy since there are often sequelae such as changes in pigmentation, epilation, severe necrosis and ulceration.

When one visualizes the adult organism as a mosaic of a variety of tissues, each with specific and different reactions to ionizing radiations, it is obvious that an organism will exhibit a mosaic of reactions with some tissues responding quickly and drastically while others react either not at all or not for many years. If reacting tissues are vital to the organism's survival, then it may succumb, taking along many apparently undamaged tissues.

Radiobiologists have long been in search of a built-in dosimeter. As previously mentioned, the testes reduction in weight and skin erythema are fair indications of the degree of radiation. The lymphocyte count (lymphopenia) also shows a rapid response, and is easily determined. Certain bi-lobed lymphocytes have appeared in the blood of cyclotron workers and it was thought that their presence might be a certain proof of radiation exposure, but such cells are found in other abnormal conditions. There is such a variety among hybrid or heterogenous human beings, added to variations in tissue response, that it is unlikely that any reliable living dosimeter will be found.

Cytoplasmic and nuclear effects of ionizing radiations. Cytological effects of irradiation concern particularly the nucleus, although the protoplasm outside of the nucleus does show some effects such as vacuolization, degeneration of mitochondria and plastids, change in degree of basophilia, fragmentation and swelling of the Golgi apparatus, increased permeability of both the nuclear and cell membranes, decrease in protoplasmic movements such as streaming, pseudopod formation or ciliary movements, decrease in secretions, and changes in the pH, stickiness, and a drop in viscosity. There has been reported a change in cell turgor due to protoplasmic contraction. Since all protoplasm is aqueous, at least to a degree, there is ionization of the water molecule with subsequent (but instantaneous) production of hydrogen peroxide which is itself presumed to be toxic and a causative factor in biological reactions.

It has long been known that the nucleus of any cell is more sensitive to irradiation damage than is either the cytoplasm or the cell wall. There is a swelling of nuclear volume in intact nuclei. The cytoplasmic nuclear volume relations are often disturbed. Giant cells, resulting from failure of cytoplasmic division, or bilobed or polyploid nuclei may appear.

Effects on the nuclei concern particularly the chromosomes and changes in the chromatic material although there may also be nuclear vacuolization. There appears to be about a 60 fold range of difference in sensitivity of the chromatic material with the most sensitive at the diplotene and the least sensitive at the interphase stage. The order of decreasing sensitivity seems to be next diakenesis, metaphase, anaphase, pachytene, leptotene, telophase, and finally the interphase. Meiotic cells seem to be more sensitive than somatic cells.

Much emphasis has been placed upon the presumed hyperradiosensitivity of nuclear DNA as opposed to cytoplasmic and nucleolar RNA, and the possible mechanism of mitotic effects through upsetting the DNA/RNA ratio which must remain within certain limits for normal cell life and growth. In fact there are some who believe that the ratio of DNA/RNA may be an indication of nuclear radiosensitivity. However, chromosomes may be damaged even after DNA synthesis is completed, when the DNA is presumed to be less radiosensitive.

Regarding the chromosomes themselves, following irradiation they may become fluid and sticky so that they clump together, resulting in the pyknotic condition so often seen in irradiated cells. At certain stages *karyorrhexis* (chromatin fragmentation) is a result. There are chromosomes, chromatid, and isochromatid breaks, chromosomal and chromatid interchanges, and, as a result of the stickiness, anaphase movements are affected so that bridges appear between separating chromosomes. Chromosomes may be broken by direct effect of ionizing radiations, the particles passing through or in the immediate vicinity of the chromosome where the break occurs. It may be that a single ionization is insufficient to cause a break so that ionizations of greater linear ion density seem to be more effective in producing breaks than, for instance, neutrons. Following a chromosome break there may be a refusion of the separated parts, or a re-arrangement of the segments including inversions, so that with respect to the linear order of the genes, complete "recovery" from chromosome fragmentation is very unlikely.

There is some evidence that nuclear sensitivity is mediated in part through the associated cytoplasm since nuclei, removed from all cytoplasm, appear to be more radioresistant than when in contact with even a very small amount of cytoplasm. Investigations along this line should be encouraged, particularly with a variety of cell types.

Irradiation causes the greatest delay in cleavage or mitosis if it occurs during the late prophase when the chromosomes are discrete, but before the breakdown of the nuclear membrane. Exposures in the range of 10–20 r can delay cleavage in certain cells. Cells which have entered upon the active movements of mitosis are apparently less sensitive, and may not be delayed in the process or impeded even with higher doses. If delay is caused by irradiation it certainly implies some change (i.e., damage) to the cleavage process which may well have a permanent effect and from which complete recovery is impossible.

One might arrange various mammalian cells in the order of diminishing radiosensitivity thus: germ cells, lymphocytes, granulocytes, basal cells, alveolar lung cells, bile duct cells, kidney tubule cells, endothelial cells, connective tissue cells, muscle, bone and nerve cells.

Effects of ionizing radiations on the embryo and fetus. Probably the most sensitive stage in all development is immediately after fertilization at the time the two pronuclei are fusing. If mouse eggs at this stage are exposed to as little as 5 roentgens there is an increase by 10% of intra-uterine deaths and resorptions and an exposure of 50 r causes 42% deaths. As the embryo develops and organogenesis is completed, resistance is increased both with respect to survival and the production of congenital anomalies. However, it should be pointed out that during organogenesis the gonad primordia appear and they contain cells which will be the precursors of all future germ cells, so that ionizing radiations will then have genetic as well as congenital anomaly effects.

It has been shown that there are certain stages in development when irradiation is more likely to produce certain specific anomalies, and higher levels of exposure produce not only a greater incidence of these anomalies but additional defects which were not caused by the lower exposures. It is in error, however, to assume that there is any limited "critical stage," only during which certain anomalies can be caused by ionizing radiations. It is known that if the embryo is irradiated at the time of most active neuroblast differentiation that central nervous system anomalies are most likely to occur. This probably applies in a similar manner to differentiating cells of all the tissues. However, it has also been found that irradiation of the fertilized but uncleaved egg will cause the same congenital anomalies of the central nervous system so that the so-called "critical period" may begin with fertilization and extend through the completion of organogenesis.

The dose levels which cause congenital anomalies of a gross structural nature are higher than those generally used in radio-diagnosis, but not in radiotherapy. A severe central nervous system anomaly known as exencephalia or herniated brain has been produced in mice with exposures of 15 r x-rays, particularly during pre-differentiation stages. While the incidence of this anomaly rises with higher exposures, even at 15 r it is not presumed that litter mates which appear to be normal are in fact undamaged by the exposure. It is a general finding that all irradiated embryos give rise to stunted fetuses, often with microcephaly. Thus, x-irradiation of the embryo or fetus should be avoided as well as any exposure of the germ cells of fetus or of those of reproductive age.

But one must be practical and weigh the benefits against the possible hazards. It is not so much in diagnostic but in therapeutic radiology that there may be dangerous exposure of the embryo, fetus, or germ cells (gonads). To shield the germ cells from all irradiations (cosmic, natural, etc.) is impossible and may not even be wise, but there is no justification in adding to the inherited debilities or deficiencies of the human stock by unnecessary exposure of the germ cells of people who might subsequently reproduce. Radiology, or the use of ionizing radiations in medical practice, is absolutely essential to proper diagnosis and often in therapy, saving the lives of many and alleviating pain in others. But, it is becoming increasingly obvious that there are two places where extreme caution should be exercised, namely in possible exposures of the *embryo* (fetus) (causing congenital anomalies and mutations) or of the *germ cells* of potential parents (causing mutations which alter their hereditary contributions to their progeny).

General Summary

While this is a chapter on radiation effects, it should be obvious that the effects of ionizing radiations on biological systems are many and varied, that responses are not of the all-or-none category, and that in all probability we have but scratched the surface in exposing the consequences of irradiation. Radiation effects may be direct or indirect, localized within a cell nucleus or transmitted to remote body cells. The biological responses may be immediate or delayed, or both. Radiation produces both functional and structural changes and the effectiveness depends first upon the amount of ionization and second upon the reacting system involved. Functional changes may be reversible, but structural changes appear to be irreversible.

Radiobiology has become of universal interest and concern, not so much because of the hysteria occasioned by ignorance of irradiation hazards but because it is possible that medical radiology, research and industrial uses of ionizing radiations could pose a contamination hazard that could get out of control. Radiations are poisons, useful under controlled conditions as are poisonous drugs, but dangerous when not understood and controlled.

It is impossible to gather experimental data from human exposures, and at the present rate of accumulation of data from medical practice and accidental exposures, statistically reliable data may never be available. For this reason experiments with other organisms, from viruses to mammals, are important. It is impossible and unwise to extrapolate data from any other form to man, and yet, if we are never to have reliable data from the human, we must recognize the basic and fundamental similarities in biological responses in all mammals. It seems wise to derive some tentative convictions from animal experiments for the direction of human conduct. The excuse that genetic experiments deal "only with the fruitfly" and hence have no value for human inheritance predictions is indefensible. When it has been found that the mouse, a mammal, and therefore presumably closer to the human genotype, responds to ionizing radiations by a 14 times greater sensitivity than does Drosophila, we must admit the possibility that the human genotype may be even more radiosensitive than that of the mouse. Thus, whatever can be learned from any biological reaction does have some value for the human and must be accepted as suggestive until such time as there are adequate data from direct human studies.

Ionizing radiations probably preceded life on this earth, and will probably be with us as far as we can project our imagination into the future. They are exceedingly useful in medicine, research, and increasingly so in civilized living. Yet they are poisons which must be controlled in order to avoid destruction of life. This is indeed possible and must be insured.

ROBERTS RUGH

RADIOLARIA see ACTINOPODA

RAMÓN Y CAJAL, SANTIAGO (1852–1934)

This Spanish histologist was born at Petila de Aragon on May 1, 1852. He was an impetuous youth and was backward in his studies, so his father, a country surgeon, apprenticed him first to a barber then to a carpenter. Eventually permitted to enter the medical school of Zaragoza University, he obtained his license to practice medicine in 1873. He then served for one year as an army surgeon in Cuba. In 1875 he was appointed to the medical faculty of Zaragoza and devoted himself to the study of anatomy. He was soon promoted to extraordinary professor and then to director of the medical museum of Zaragoza University. In 1883 he joined the faculty of the University of Valencia as Professor of Descriptive Anatomy and in 1887 he was appointed Professor of Histology and Pathological Anatomy at the University of Barcelona. He held the chair of Histology and Pathological Anatomy at the University of Madrid from 1892 until his retirement in 1922. Using his own modification of the Golgi silver stain he explored the hitherto unknown worlds of the cerebrum and the cerebellum. Perhaps his most fundamental achievement was the establishment of the neuron as the basic unit of the nervous system. He shared the 1906 Nobel Prize for Medicine with Camillo Golgi in recognition of their work on the structure of the nervous system. His best known work is "Textura del Sistema Nervioso del Hombre y de los Vertebrados" (1899). Santiago Ramón y Cajal, recognized as Spain's greatest scientist, died at Madrid on October 17, 1934.

MARTIN J. NATHAN

RANALES

The origin of the ANGIOSPERMS was termed "an abominable mystery" by the father of evolutionary thought, Charles Darwin. This statement was occasioned by the fact the fossil record was of little or no value in indicating which group of flowering plants was the first to have arisen from some ancient group of GYMNOSPERMS. Now, 100 years after Darwin's irritable pronouncement on the subject, paleobotanists are still far from providing a solution to this mystery. Therefore, in the absence of concrete evidence (which *could* be provided by a continuous fossil plant record), botanists must rely on salient morphological features of extant angiosperms in order to delineate their probable phylogeny.

Although innumerable systems of putatively phylogenetic classification have been proposed in the past, whenever these are carefully examined, it is discovered that they are usually based on two opposing theories of which group best represents the most primitive angiosperms: 1) the "Amentiferae" (willows, oaks, birches, etc.), whose flowers are unisexual, apetalous, and form in aments or catkins; or 2) the "Ranales" (magnolias, buttercups, annonas, etc.), whose flowers are relatively large, solitary, bisexual, and possess a definite perianth. Therefore, without consciously taking sides in this argument, the present note is intended to outline the origin and manifold concepts of the taxon Ranales.

In 1830 John Lindley (1799–1865), professor of botany at the University of London, published his "Introduction to the Natural Orders of Plants." This system, based essentially on the "anatomical" classification of the Swiss botanist, A. P. de Candolle (1778–1841), was widely accepted in all English-speaking countries. Lindley's revision, "Vegetable Kingdom," published in 1846, although claiming more originality than his initial effort, was a far cry from a truly natural system. Nevertheless, Lindley should be remembered by both current and future generations of botanists for his introduction of the taxon RANALES into the world's botanical literature. Lindley established the Ranales as the "first alliance" under the first subclass (Polypetalae) of de Candolle's "Exogens" (essentially dicotyledons).

This concept of the Ranales included the Ranunculaceae, Podophylleae, Papaveraceae, Fumariaceae, Nymphaeacea, and Nelumbonea. This alliance was circumscribed as follows: herbaceous; carpels single, or if several, they are separable; if united, they have partitions (parietal) arising from the placenta. Lindley's "second alliance" (Anonales) was deemed closely related to the Ranales, and included three apocarpous woody families—Magnoliaceae, Wintereae, and Anonaceae.

Quite understandably, as elements of the plant kingdom became better known through extensive and intensive botanical researches, Lindley's original concept of the Ranales became gradually modified. Accordingly, when one refers to the Ranales today, one must append a systematist's name, in order to indicate exactly to *which* concept of the Ranales he refers. In this connection, the uses of the taxon Ranales in the most important systems of classification are briefly outlined in the following paragraphs.

Bentham and Hooker (1862–1893) adopted the Ranales as their first dicotyledonous cohort (a taxonomic category intermediate between that of a class and an order). Like Lindley, Bentham and Hooker relied heavily on the earlier system of de Candolle. However, Bentham and Hooker broadened Lindley's concept of the Ranales in that they included both woody and herbaceous taxa as follows: Ranunculaceae, Dilleniaceae, Calycanthaceae, Magnoliaceae, Anonaceae, Menispermaceae, Berberideae, and Nymphaeaceae. These authors listed as significant distinguishing features of the Ranales: usually two whorls of perianth parts; superior ovary; stamens and perianth hypogynous; stamens rarely definite in number; carpels free or immersed in the torus, very rarely united; micropyle usually inferior; embryo minute in fleshy endosperm. Although essentially non-Darwinian in concept, the Bentham and Hooker system is still used today in the great British herbaria, e. g., Kew and the British Museum of Natural History.

In 1892 Adolf Engler, professor of botany at the University of Berlin and director of the Berlin Botanical Garden, inaugurated a new system of classification which is still in general use in herbaria throughout the world. The widespread adoption of this system is due more to its completeness than to the general agreement of systematists that this classification is the most natural one proposed. According to Engler and his coauthor K. Prantl, the "Amentiferae" represent the most primitive dicotyledons and the members of the Ranales are distant and fairly highly-evolved derivatives of these amentiferous forms. The Ranales is listed as the 18th order of Engler & Prantl's Archichlamydeae and is further subdivided into four suborders: Nymphaeineae, Trochodendrineae, Ranunculineae, and Magnoliineae. The complete list of families is as follows:

Order 18. Ranales. Herbs or woody plants with spiral, spirocyclic, or cyclic flowers; perianth usually present; stamens indefinite in number; carpels many to one, rarely united.

Suborder I. Nymphaeineae
1. Nymphaeaceae
2. Ceratophyllaceae
Suborder II. Trochodendrineae
3. Trochodendraceae
4. Cercidiphyllaceae

Suborder III. Ranunculineae
5. Ranunculaceae
6. Lardizabalaceae
7. Berberidaceae
8. Menispermaceae
Suborder IV. Magnoliineae
9. Magnoliaceae
10. Calycanthaceae
11. Lactoridaceae
12. Anonaceae
13. Eupomatiaceae
14. Myristicaceae
15. Gomortegaceae
16. Monimiaceae
17. Lauraceae
18. Hernandiaceae

Hans Hallier (1905) rejected the Englerian concept of the primitive flower, adopting instead the strobiloid type. In his system (never fully delineated) woody plants containing oil cells are listed as his first order Polycarpicae; this taxon included the Magnoliaceae, Canellaceae, Anonaceae, Myristicaceae, Calycanthaceae, Monimiaceae, and Lauraceae. His second order was the Ranales and was characterized by the absence of oil cells, and essentially herbaceous, but with some woody shrubs and climbers; included were: Berberidaceae, Menispermaceae, Ranunculaceae, Nymphaeaceae, and Ceratophyllaceae.

Charles Bessey (1915) agreed with Hallier that the Ranales was a primitive group; however, he included more families (24) in this order than anyone ever had previously, which fact probably lessens the naturalness of the taxon. Bessey's ranalian families included (although not necessarily in order of increasing complexity): Magnoliaceae, Calycanthaceae, Monimiaceae, Cercidiphyllaceae, Trochodendraceae, Leitneriaceae, Anonaceae, Lactoridaceae, Gomortegaceae, Myristicaceae, Saururaceae, Piperaceae, Lacistemaceae, Chloranthaceae, Ranunculaceae, Lardizabalaceae, Berberidaceae, Menispermaceae, Lauraceae, Nelumbaceae, Cabombaceae, Ceratophyllaceae, Dilleniaceae, and Canellaceae.

Alfred Rendle (1925 and 1930) followed the basic tenets of the Englerian system, but altered some of the names of the higher taxa. He placed at the base of his system 13 orders of Monochlamydeae (amentiferous and related forms); under his second group (Diapetalae) he lists the Ranales as the first order, but recognizes only the following 12 families: Magnoliaceae, Anonaceae, Myristicaceae, Calycanthaceae, Monimiaceae, Lauraceae, Ranunculaceae, Berberidaceae, Lardizabalaceae, Menispermaceae, Nymphaceaceae, and Ceratophyllaceae.

Warming and Mobius (1929) follow Rendle's concept of the Ranales, with the exception that the 12 families are treated in a different sequence. The Ranunculaceae and Nymphaeaceae are listed as their first two families, followed by the Anonaceae, Magnoliaceae, Calycanthaceae, Monimiaceae, Berberidaceae, Menispermaceae, Lardizabalaceae, Lauraceae, Myristicaceae, and Ceratophyllaceae.

John Hutchinson (1926 and 1959) followed Hallier and Bessey in the sense that the strobiloid flower was regarded as primitive, but departed from all previous systems by the separation of woody and herbaceous families of dicots. In his second edition he makes the

Ranales the first order of his division Herbaceae and includes the following families: Paeoniaceae, Helleboraceae, Ranunculaceae, Nymphaeaceae, Podophyllaceae, Ceratophyllaceae, and Cabombaceae. His allied primitive order for woody dicots (division Lignosae) is the Magnoliales and includes nine families. His second and third woody orders, Annonales and Laurales, include nine other families which are normally included in the Ranales (*sensu* Engler).

By now it should be quite obvious that the order Ranales has multiple meanings, varying considerably according to the interpretations of various systematists —at one time or another the Ranales has been regarded as including either the most primitive or else relatively advanced DICOTYLEDONS; also, as comprising either strictly herbaceous or else both woody and herbaceous groups combined.

JAMES E. CANRIGHT

Reference

Hutchinson, J., "The families of flowering plants," 2nd ed., Oxford, Clarendon Press, 1959.

RATITES

Former classifications of birds (see AVES) have often recognized two main divisions, a primitive *ratite* group, composed of flightless, ostrich-like types with a flat or "raft-shaped" sternum with no keel for the attachment of the muscles of flight, and a *carinate* group, composed of some primitive birds (tinamous and penguins) and the more modern birds with a keeled sternum for the attachment of the pectoral muscles. Later classifications utilized the structure of the palate and divided birds into *palaeognathous* (old palate) types and *neognathous* (new palate) types. This newer arrangement has also run into difficulties and included orders have recently been subject to considerable juggling and combining. But the families of ratite or struthious birds have been rearranged very little of late; these are discussed in their apparent phylogenetic sequence below.

The ratite birds include the ostriches, rheas, cassowaries, emus, the extinct moas and elephant birds, and the peculiar kiwis of New Zealand. All are primitive flightless birds with poorly developed wings and unkeeled (raft-shafted) sterna. All have stout or well developed legs; strong, clawed toes; and are or were cursorial in habit. The prevailing belief with regard to the origin of flightless stock is that all are descended from flying ancestors, and arrived at their flightless condition by degeneration of the wings. Some scientists, however, have proposed that these types were always flightless, and were derived from cursorial, non-flying ancestors. This presupposes an improbable diphyletic origin of birds, with separate ancestral lines for the flightless and flying forms.

Best known, to the layman at least, of the ratite or struthious birds is the Ostrich (Struthionidae). A single species (*Struthio camelus*), divided into six subspecies or geographic races, prevails over the arid and semiarid areas of Africa. Formerly a North African form extended into Arabia and Syria, but it is doubtful if it still exists there. An ostrich-type from the Eocene of Switzerland and fossils from Asia, including Mongolia, suggest that in former times (and in warmer climates) the ostriches were more widely distributed than now.

By weight the Ostrich is by far the world's largest living bird. Males reach a maximum of 300 pounds and stand eight feet high. The scanty plumage is black, with white wing and tail feathers, in the male, and grayish in the female; both have the long neck and thighs largely bare. The long plumes from the wing and tail are so valuable in the millinery industry that ostrich-farming, at least in the past, has been a flourishing and profitable industry. The demand for ostriches in zoos throughout the world has also enhanced their commercial value.

Ostriches are noted for their great speed. The legs are long and strong, the toes reduced to one main middle toe and a smaller lateral digit, and once momentum is gained they can take strides measuring 15 feet. They can readily outdistance a horse; yet sometimes they are run down and captured by horsemen who change to fresh steeds during the chase and cut corners off the Ostrich's persistent circular path.

Ostriches are polygamous, the males usually having one "major hen" and two or more "minor hens." The females all contribute to a common nest which may then contain a clutch of two dozen extremely large eggs. The males assume the main responsibility for care of the eggs and young, keeping the eggs covered with warm sand by day and uncovering them and incubating them at night. Sometimes the females uncover the eggs for short spells of day-time brooding. The males are very pugnacious and dangerous during the breeding season; a single strike from the clawed middle toe can disembowel a man. Ostriches sometimes rest by lying with their long necks stretched out on the ground, which may have given rise to the fallacious idea that they bury their heads in the sand.

The rheas (family Rheidae) include two species restricted to the grasslands and open brush country of South America. Their isolation from their nearest relatives in the Old World raises the question of their place of origin and how they got to South America. Fossil evidence from Switzerland and Asia suggests the traditional explanation of the invasion of the New World by Old World forms spreading across the Alaskan land corridor at a time when northern climates were comparatively mild.

Rheas differ from the true Ostriches by their much smaller size (males attain a weight of about 40 pounds compared to 300 in the Ostrich), fully feathered neck, and a three-toed foot. Rheas also have no tail feathers. They are primarily grazing animals but capture some animal life. Their nesting habits are similar to those described for the Ostrich—several females contribute to a large clutch of eggs (up to 30) which are cared for by the male. Rhea skins for rugs and feathers for dusters have been in such demand in the past that these birds have become quite scarce. They are also hunted to some extent for sport, chiefly by gauchos who pursue them on horseback and lasso them by throwing bolas whose weighted ends entangle the neck and legs of the fleeing birds.

The cassowaries (Casuariidae) consist of three or more species which in turn are composed of about 30 recognized forms distributed over New Guinea, north

Queensland in Australia, and nearby islands. Their peculiarities include highly colored wattles and caruncles about the head and neck, a bony casque on the crown, and a coarse hair-like plumage with the remiges (wing feathers) reduced to mere quills. Unlike the closely allied ostriches, rheas and emus, the cassowaries are forest birds, very skillful at manoeuvering their large bodies through dense underbrush by using their helmeted head and rapier-like wing quills to part resisting vegetation.

Cassowaries are omnivorous in diet and very belligerent in disposition. Many natives have lost their lives in encounters with the fierce birds. They have little commercial value, though their eggs are edible and their skin and feathers are sometimes utilized. Cassowaries are monogamous, rather than polygamous; the female deposits 6-12 eggs in a hollow on the ground and these are incubated by the male.

The emus (Dromiceiidae) of Australia are now reduced to a single species. Another species survived until recently and several other emus are known only as fossils. The surviving Emu (*Dromiceius novaehollandiae*) is the world's second largest (heaviest) living bird and stands about five feet high. It differs from the cassowaries (its closest relatives) mainly in the absence of the prominent head and neck ornaments. Emus are also friendly and curious, not belligerent. They inhabit open country, often travel in flocks, and graze so heavily on the scanty vegetation in arid regions that Australian ranchmen consider them enemies. In addition to competing with domestic stock for forage, emus are destructive to fences, pushing their way through these barriers and breaking them down. Their nesting habits resemble those of the cassowaries; a ground nest, scantily lined with leaves and twigs, houses 7-12 dark green (sometimes black) eggs which are incubated mainly by the male.

Apparently closely allied to the living ratites, and included with them in current classifications, are two interesting groups (orders or families) of extinct birds known only from archaeological and fossil evidence. These are the elephant-birds (Aepyornithidae) of Madagascar and the moas (Dinornithidae) of New Zealand.

The elephant-birds, including a dozen or so species, are known only from eggs, egg fragments, and bones discovered by early settlers and traders in Madagascar. The initial discovery, interestingly enough, was made when natives took two huge eggshells to the island of Mauritius and wanted them filled with rum. The eggshells and a piece of metatarsal bone were sent to Paris for identification, which resulted in the description of a new species (*Aepyornis maximus*). Later, additional eggs and bones were found. Though elephant-birds were large, probably standing 7-8 feet high, with immense legs, they were exceeded in size by some of the largest moas. Their eggs, however, are the largest eggs known, as well as the largest single animal cell. They measured 13 x 9½ inches, and had a capacity of about two gallons, or the equivalent of about 12 dozen hen's eggs.

Even more interesting are the extinct moas. About 30 species, ranging from the size of a turkey to *Dinornis maximus* which stood 10-12 feet high, lived in New Zealand within recent times. Eggshells and piles of bones recovered from campsites of the early Maoris and their predecessors indicate that the birds and their eggs were used for food. This may well have been a factor in their extinction, as the birds were flightless and easily dispatched with clubs. Additional clues to their extermination, however, are found in huge graveyards of bones in certain depressions on the island, suggesting that some natural calamity, such as fire, herded the birds together in some swamp or lake bottom where they perished en masse.

Last of the ratite types is *Apteryx* (Apterygidae), which includes three species of kiwis also confined to New Zealand. Although much reduced in numbers, and clinging to a precarious existence in spite of complete protection, moderate numbers of this unique bird survive. Kiwis are among the most peculiar of living birds. A long curved bill with the nostrils near the top instead of at the base; long, hair-like drooping feathers almost completely covering the body; no tail and no discernible wings; and stout legs and feet terminating with sharp claws are among the peculiarities of these strange birds. They are largely nocturnal, have poor vision but apparently a keen sense of smell, and grope about in the darkness for worms, insects, berries and succulent plant material. Tests with earthworms buried in buckets of sand seem to indicate that kiwis locate the worms by some olfactory sense. Tactile bristles at the base of the beak may also aid them in locating prey.

Kiwis nest in burrows where the female lays one or two extremely large eggs. A kiwi egg is about five inches in length and equal to one fourth the weight of the parent. The incubation period is one of the longest known among birds; it takes the male, unassisted by the female, about 75 days to hatch the bulky egg. Kiwis were formerly hunted for food, and their leg bones used for pipe stems, but now the remaining wild birds are carefully protected—the proud symbol of the New Zealanders.

GEORGE J. WALLACE

References

Austin, O. L., Jr., "Birds of the World," New York, Golden Press, 1961.
McDowell, S., "The Bony Palate of Birds. Part I: The Palaeognathae," Auk, **65**: 520-549, 1948.
Sauer, E. G. F., and E. M. Sauer, "The Behavior and Ecology of the South African Ostrich," The Living Bird, Ithaca, N.Y., 1966.
Wallace, G. J., "An Introduction to Ornithology," New York, Macmillan, 1955.

REDI, FRANCESCO (1626-1698)

This Florentine physician deserves mention as one of the first recorded exponents of the method of controlled experimentation. He exposed putrified meat and observed that it developed maggots. However, he exposed at the same time similar putrified meat protected by a thin cloth and showed that though flies' eggs were laid on the cloth, they did not turn into maggots unless they touched the meat. Moreover, like all good experimenters, he was moderate in his claims and considered this result to have shown solely that maggots did not result from the putrefaction of meat. He drew from these experiments no general conclusions on the subject of spontaneous generation.

PETER GRAY

REGENERATION, INVERTEBRATE

For the invertebrates it is best to define regeneration as any process restoring a complete, functioning individual from some part of the body. This includes a variety of types of asexual reproduction as well as very extensive powers to make good accidental damage and loss. Many Protozoa, Porifera, Coelenterata, planarians and Nemertinea can regenerate from fragments as small as 0.5% of the body, and under special conditions from considerably smaller pieces. Frequently special *autotomy* mechanisms have been evolved to facilitate a rapid and neat casting of parts seized by predators, while many Holothuria when irritated eject most of their viscera as a placebo. There is usually very good regeneration of the parts shed. Sometimes, as in the *heterochelic* change in decapod Crustacea, regeneration is indirect or *allotopic* (compensatory hypertrophy). Again, many sponges and coelenterates can reconstitute a new individual by the reaggregation of isolated cells.

Coelenterates, planarians, nemertines, Polyzoa, phoronids, pterobranchs and tunicates are able to degrow and dedifferentiate, under starvation and other adverse conditions, sometimes to minute gastrula-like bodies, and to regenerate when fed. Many colonial hydroids cast their hydranths under such conditions, a curiously wasteful process, and regenerate from the hydrocaulus. Some Ectoprocta and tunicates regenerate from "reduction bodies" while some sponges and Ectoprocta form special asexual reproductive bodies, "gemmules" and "statoblasts," in anticipation of the adverse season. More spontaneous regenerative processes include the fission of Protozoa, some Coelenterata, Turbellaria, Nemertinea, Annelida and Hemichordata, and the multiple fragmentation of coelenterates and worms. In some, regeneration follows fission (*architomy*) but in others it is virtually complete before separation of the daughters (*neotomy*), and then is effectively a budding process, often producing a regular pattern of buds. The Coelenterata, in particular, also have a great variety of less regular processes of asexual reproduction by *frustules*, buds, pedal-laceration, etc. In *Hydra* there is a continuous turnover of cells—proliferation at the apex and resorption at the base of the peduncle (Brien).

In general, regenerative capacity progressively declines with the degree of differentiation and the grade of evolution of animals. The examples above are mainly from the lower invertebrates, with high capacities. Some starfish can regenerate from a single arm and some holothurians from an empty body wall, but some echinoderms show limited powers and the arthropods and molluscs regenerate at most only appendages and some internal organs. The evolutionary-taxonomic trend is not simple however; in most phyla some groups have poor capacity. Most small, prolific, short-lived animals, e.g. rotifers, copepods, insects, and many larvae, do not regenerate. The capacity is small in nematodes, leeches, and other types with very "determinate" development, and in most parasites, notwithstanding the great powers of asexual reproduction and growth by many of them; otherwise animals which reproduce asexually usually also regenerate well, whereas sexual reproductive activity tends to be inversely correlated with regeneration.

The explanation of this distribution of regenerative capacity is not always clear. There are often sharp contrasts between the capacities of related genera and even species. Types with mosaic, non-regulative eggs (ctenophores and tunicates) or with very determinate cleavage (annelids) often regenerate well. There is no very simple correlation with habitat but the capacity is usually adaptive, and is correlated with the incidence of losses in the field. Its variations within the body show a similar correlation. For evident reasons, therefore, motile animals usually best regenerate their hind end and appendages, and sedentary types their heads and hydranths: isolated hydranths often cannot regenerate a stolon posteriorly, simple structure as it is, and reciprocally many worms show very limited regeneration from anterad surfaces. The capacity to regenerate at a surface of amputation usually decreases progressively with distance from the end of the body, where injury is most common.

Protozoa can regenerate completely in a few hours and the lower Metazoa in a few days: the rate is related to the extent of regeneration in the different animals and is a second measure of capacity. Rate is affected by temperature, ionizing radiations and other factors which promote or inhibit growth in general. The major internal factors controlling rate are the local nerve supply, hormones and nutrients. Hormonal control has been detected in most phyla possessing circulatory systems.

Abnormal regenerates are absolutely, if not relatively, common in invertebrates and are readily induced experimentally. Most common are supernumerary growths, often associated with subdivision of the wound surface, or of the nerve supplying it, and deficiencies, due to various inhibitory factors, of which premature wound closure may be one, as in the vertebrates. There also occur mosaic deficiencies, of sharply delimited regions of the regenerate. Also common are qualitative abnormalities, or *heteromorphs,* in particular the mirror type, e.g. biheads, and *homoeotic* replacement, by a serial homologue. Regenerates sometimes have atavistic characters. The explanation of some of these abnormalities is still uncertain. Regeneration commonly recapitulates embryogenesis and may become arrested before completion; these and other abnormalities usually become more normal in time, particularly if reamputated.

Wound closure is by cellular contraction or by blood clotting, or both, and later by cell migration and proliferation. There is local inflammation and tissue dedifferentiation but possibly less than in vertebrates. Epithelial tissues usually regenerate in continuity and mesodermal tissues from a *blastema* of undifferentiated, migratory cells (*neoblasts*) and local, dedifferentiated cells. The various tissues show some independence in phasing and in other respects. The degree of *metaplasia,* or change in cell type, during regeneration is still uncertain. As in embryogenesis, there is a *mosaic* stage when re-regeneration cannot be induced.

Epimorphic regeneration from a blastema is often supplemented, or even replaced, by *morphallaxis,* a remodelling of the stock portion, which is rarely significant in vertebrate regeneration. It involves (1) concurrent de- and re-differentiation, (2) the destruction of some and the migration of other cells, with resulting movements of whole organs and (3) disseminated foci of epimorphic neoformation, in which alone there is much cell proliferation. Reconstitution from tissue minces is entirely morphallactic and differentiation therefore precedes all growth, in contrast to the sequence in epimorphosis.

The causal analysis of the process is progressing. A wound is the prime initiator of regeneration and may initiate even without any loss. Other types of irritation also can do this. The wound factor evokes the immigration of neoblasts, to form the blastema. The polarity of the regenerate depends largely on its nerve supply. This affects other qualitative features only indirectly, through the polar organization. The head end of the body, with the brain, regenerates most rapidly and acts as an *organizer,* preventing lower levels from producing similar structures and ultimately ensuring that they produce structures appropriate to their level (Child, Brøndsted). A diffusing *organisin* has been recognized in planarians: It is produced by the brain and is not species-specific. Control is serially deputed as in embryogenesis, lower-level organizers becoming in turn independent of higher levels (Wolff: *cascade determination*). Organization becomes multipartite, therefore. Control is essentially inhibitory (Rose). Both newly regenerated and parental head regions are able to organize both parental and regenerating tissues in the rest of the body, but it is difficult to change either polarity or specific fate of a region except through the natural system. Direct electric current is the most effective experimental agent, indicating that an intrinsic electrical field may be the basis of the observed polar organization (Moment, Becker).

Metabolic properties are often distributed more according to two countercurrent gradients than to a single anteroposterior system, and there is growing evidence of such a morphogenetic system (Tucker, Smith). Ciliate Protozoa have the indications of such a system. They resemble the Metazoa in many more respects (Tartar) than might have been anticipated.

A. E. NEEDHAM

References

Abeloos, M., "La Régénération et les Problemes de la Morphogenèse," Paris, Gauthier-Villars, 1932.
Hay, E. D., "Regeneration," New York, Holt, Reinhart & Winston, 1966.
Korschelt, E., "Regeneration und Transplantation," Berlin, Borntrager, 1927. Band I.
Needham, A. E., "Regeneration and Wound-Healing," London, Methuen, 1962.
Vorontsova, M. A. and L. D. Liozner, "Asexual Propagation and Regeneration," Trans. P. M. Allen, Oxford, Pergamon, 1960.

REGENERATION, VERTEBRATE

Regeneration is the replacement of lost parts of an organism at any time in its life cycle. The process is as highly varied as the replacement of a damaged nerve axon in a vertebrate to the completion of a new individual from a fragment of an old one in some invertebrates. Regeneration is widely found among animals and plants but in the present discussion space allows brief treatment of only one group—the vertebrates. Among the vertebrates it is primarily the lower forms which exhibit the most striking examples of regenerative capacity. In the reptiles regeneration is confined

largely to the replacement of the tail and scales, and to the process of wound healing; in birds to replacement of feathers, claws and spurs; in mammals to tissue repair, wound healing, and, in deer, replacement of antlers. Since the attention of a majority of investigators has centered on the regeneration of organs of Amphibia, with consequent enrichment of our knowledge of the phenomenon in this group particularly, the chief emphasis here will be on organ regeneration of Amphibia.

The Amphibian Limb. The limbs of larval and adult salamanders regenerate completely, the process proceeding much faster in the larva than in the adult. Thus in *Amblystoma punctatum* larvae wound healing of the amputation surface of the limb stump is completed in 4 to 6 hours by migration of the surrounding epidermal cells of the stump. Following a two-day period of phagocytosis of tissue debris resulting from the trauma of amputation, there occurs a period, lasting 4 to 6 days, of "tissue dedifferentiation" during which the injured muscle, skeletal and connective tissues undergo partial disintegration and liberate cells consisting of a nucleus and scanty cytoplasm. These cells accumulate at the tip of the limb stump beneath a thickened, richly innervated epidermal cap. This aggregation of mesenchymatous cells is known as the blastema. Continued growth of the blastema is a function of the mitotic proliferation of the blastemal cells. The blastema at 10 days resembles to a remarkable degree the embryonic limb bud, and indeed, tissue differentiation proceeds within it much as it does in the limb bud.

The origin of the blastemal cells has been the subject of much controversy. Among the theories to account for blastemal cell origin which have been proposed are the following: (a) Hematogenic origin. Leucocytes entering the limb stump by means of the blood vessels were thought to become extravasated and transformed into blastemal cells. This theory became untenable when blastemal cells were proved to be of purely local origin. Thus a haploid limb regenerating on a diploid host produces haploid blastemal cells. Regeneration after exposure of all but a short middle section of a limb to X-rays, which inhibit regeneration, followed by amputation through the shielded section, pointed even more rigorously to a highly restricted origin of the blastemal cells. (b) Mesodermal tissue origin. Unspecialized reserve cells contained in the various connective tissues of the stump have been suggested as the source of blastemal cells. However, no distinctive reserve cell has yet been discovered in the amphibian limb, and tissue culture techniques have failed to disclose any potential reserve cells even under a variety of culture environments. It is generally conceded that all stump tissues injured at the time of amputation contribute cells to the blastema. Traditional histological techniques as well as the newer techniques of histochemistry and electronmicroscopy have provided abundant evidence of the migration of cells freed from the "dedifferentiating" tissues of the stump into the blastema. A central problem remains: Do cells derived from one type of stump tissue participate in the differentiation of another type of tissue in the developing blastema, or do they differentiate only into the tissue of their origin? The chief difficulty in answering this question has been an inability to identify regeneration cells at all stages of the process. Recently, however, it has been possible to label *triploid* axolotl limb cells with ^{3}H-thymidine, transplant them to the limbs of *diploid* axolotls and follow those cells throughout the process of regeneration. Doubly labeled limb cartilage cells are found only in cartilage cells of the regenerate. Doubly labeled limb musculature cells are found in musculature of the regenerate and also in cartilage of the regenerate. It should be noted, however, that musculature contains connective tissue cells, and until pure clones of myoblasts can be cultured, the problem of possible muscle metaplasia remains open. Transplanted nuclei of regeneration blastemata in enucleate frog eggs fail to support development, indicating restriction in the developmental capacity of blastemal nuclei. Thus, although the question of metaplasia in regeneration is not finally answered, the evidence at hand does not encourage a belief in metaplasia.

Although it is clear that the majority of the stump tissues undergo "dedifferentiation" during early regeneration stages, two tissues do not, but rather restore themselves by direct growth from their remnants in the stump. These tissues are the skin and the nerves, and they have a peculiarly significant effect on regeneration. Thus if a limb is denervated and amputated, regeneration fails. Denervation of the limb stump after a blastema has formed, however, does not prevent continued regeneration. Nerves, therefore, are required only for those processes of regeneration which produce the blastema. Whatever the nerve contribution to regeneration may be it is a quantitative one since above a certain threshold ratio of number of nerve fibers to wound area regeneration will always occur when the limb is challenged by amputation. Below the threshold ratio regeneration fails. Motor, sensory or even intra-central nerve fibers are capable of supporting regeneration so that no correlation between impulse conduction and the nerve influence on regeneration obtains. Attempts to discover a chemical basis for this "trophic" nerve influence have failed. It is interesting, however, that if a limb develops from early embryonic stages without nerves it is able to regenerate normally in later larval stages. Such aneurogenic limbs differ, therefore, in their response to amputation as compared with denervated limbs—limbs deprived of nerves previously present. Although the mechanism is not known the limb tissues thus appear to become "addicted" to the neural influence during development. It has been proposed that the trophic substance is produced in great abundance in the neuron primarily to maintain its great mass of active neuroplasm, but that significant amounts spill over onto other tissues which thus depend on the nerve for their own regenerative activity. Nerve fibers of large diameter convey larger amounts of trophic substance and thus fewer of them are needed as compared to fibers of small diameter. Since aneurogenic limbs possess no nerve-derived trophic substance and since normally innervated amphibian limbs require few or no nerves after homoplastic transplantation, it has been suggested that under certain circumstances non-nervous tissues can be caused to produce trophic substance needed for limb regeneration.

The skin also strongly influences regeneration. Thus if the fresh amputation surface of a limb is sealed by skin taken from the body, regeneration is inhibited. Regenerative failure is not due to mechanical restriction of blastemal outgrowth since a wound epithelium contributed by grafted head or back skin also fails to stimulate regeneration. Skin from the hind limb or tail will, when grafted to the limb, provide a wound epi-

thelium which supports regeneration. The evidence thus allows the interpretation that wound skin of the limb or tail possesses unique properties for stimulating the onset of regeneration. Attention in recent years has been directed to the apical cap, a thickening of the wound epidermis which develops in correlation with a massive invasion of this epithelium by regenerating nerve fibers. Blastemal cells aggregate beneath the cap even when it is directed, by experiment, to form at one side of the limb stump. Such apical cap asymmetry is correlated with a corresponding asymmetry of the blastema and, later, in the completed regenerate which projects at an acute angle from the limb stump. It has been proposed that the apical cap somehow influences the aggregation of blastemal cells. It has also been proposed that the apical cap is primarily concerned with initiating the dedifferentiation of the stump tissues which provides the cells of regeneration. Perhaps the skin, particularly the epidermis, also produces trophic substance. For example, when skin is interchanged between normally innervated and aneurogenic limbs, regeneration is inhibited in the aneurogenic limbs. In these cases the grafted skin is in a denervated condition, although the aneurogenic mesodermal tissues are capable of regenerating under the appropriate stimulus. Muscle and cartilage will readily take part in limb regeneration when these tissues replace the mesodermal tissues of an aneurogenic limb. Thus the aneurogenic skin, with its apical cap, can stimulate mesoderm tissues, which are in a denervated condition, to form blastema cells. Thus, in these experiments the skin, and not the mesodermal tissues, acts in a trophic capacity.

Stimulation of limb regeneration in amphibians has also been attributed to hormonal action. Pituitary hormones, particularly growth hormone, ACTH and prolactin, have been implicated in the stimulation of the growth of the blastema. Thyroxin given during the growth phase, on the other hand, retards limb regeneration in newts. Since there is some evidence of an antagonism between prolactin and thyroxine, it is possible that thyroxin treatment may act by repressing prolactin production. The mechanism of hormone action on limb regeneration is not known. Increase in survival of well nourished, hypophysectomized newts which then are able to regenerate their limbs points to the likelihood that the pituitary hormones have a general and not a specific action on regeneration.

The Amphibian Tail. The morphogenetic mechanisms underlying the regeneration of the amphibian tail have in general proved to be more directly available to systematic analysis than have those of the limb. The spinal cord is of central importance to tail regeneration, just as it is in embryonic development, and is the first structure to show outgrowth after tail amputation, maintaining this regenerative lead throughout the subsequent morphogenesis of the regenerate. A blastema, morphologically similar to that found in limb regeneration, develops under the wound epithelium a few days after amputation. There is considerable experimental evidence that the majority of the blastemal cells are derived from the cut axial muscles which undergo partial disintegration. During subsequent morphogenesis the blastemal cells differentiate primarily into the cartilaginous vertebrae, a process seemingly involving metaplasia. New muscle first appears closely associated with the former muscle of the stump, although the origin of these muscle cells is uncertain. Differentiation of the cartilaginous vertebrae from blastemal cells is the result of an inductive action of the motor half of the spinal cord and is transmissible through an intervening mesenchymal matrix. Finally, the spinal cord has been shown to direct the regenerative outgrowth of the tail. For example, a segment of spinal cord which is deflected dorsally into the tail fin induces a supernumerary tail projecting at an acute angle with the long axis of the host.

The Amphibian Eye. Regeneration of parts of the eye (retina, lens and iris) is unique in that a specialized cell (the retina pigment epithelial cell) has been clearly traced through a morphological dedifferentiation to a redifferentiation into quite other eye structure such as lens or neural retina. It should be noted, however, that all the structures involved are derivatives of one germ layer—the ectoderm.

Lens regeneration, by a budding process of the epithelium along the free margin of the dorsal iris, has been known for many years. No injury to the dorsal iris is necessary to initiate lens regeneration; merely removing the lens serves as sufficient stimulus. The presence of a normal, living lens, even though it is derived from the eye of another species of salamander, specifically inhibits lens regeneration from the dorsal iris. Glass beads, wax lenses and spherical masses of tissue exert no inhibiting effect on lens regeneration when inserted in the eye, indicating that ordinary mechanical forces apparently play no role in the inhibition. When a section of the dorsal iris is isolated from the normal lens, and the aqueous humor bathing it, by means of an impermeable plastic membrane a new lens regenerates from the dorsal iris just as well as in complete lentectomy. When, however, aqueous humor from eyes containing a normal lens is injected daily into lentectomized eyes lens regeneration is inhibited. Lens regeneration resumes when the injections are stopped. Daily injections of Ringer's solution are without effect. It is important to note that the inhibitory substance is limited to the eye—it is not circulated in the blood. It is also to be noted that lens regeneration is limited to the family Salamandridae.

Considerable interest accrues to the fact that the presence of a neural retina is necessary for lens regeneration from the dorsal iris. Since in embryonic development the neural retina is formed from an evagination of part of the brain, there is the possibility that lens regeneration is dependent on a neural influence— a phenomenon which might indicate a common property of regenerative mechanisms in lens, limb and tail reconstitution. It should be borne in mind, however, that there are also basic differences in the regenerative phenomena of these three organs. In the case of lens regeneration, no blastema is formed, nor are pituitary, adrenal or thyroid hormones important inhibitors of the process.

Regeneration in Fish. Regeneration studies in fish have been concerned largely with the reconstitution of the fins, and the taste barbel of the catfish. The fin consists of skin, connective tissue and bony fin rays. After amputation through the fin rays a terminal blastema forms beneath an innervated, thickened epidermis. Osteoblasts appear first beneath the epidermal basement membrane where they initiate deposition of dermal bone. Bony fin rays are not regenerated from

stumps from which they have previously been removed; nor does fin regeneration occur after denervation of the stump.

Of particular interest are the results of hormone studies of fin regeneration. In Platypoecilus the immature anal fin and the anal fin of the female are capable of complete regeneration after amputation, but the anal fin of the adult male, modified by androgen to a copulatory organ, regenerates abortively and atypically. Castration of adult males does not restore regenerative ability, while treatment of females with methyl testosterone previous to fin amputation results, even though androgen treatment is discontinued, in abortive, male-type anal fin regenerates. Androgen, through its morphogenetic effect on the anal fin, is regarded primarily as the precipitating agent in the loss of regenerative ability. This situation is reminiscent of that found in the anuran limb where prior experience with thyroxin inhibits regeneration.

Like the fin, the taste barbel of Ameiurus is a structure with striking regenerative capabilities. It is composed of an outer layer of skin, connective tissue, a central rod of cartilage with a thick perichondrium, and is extensively innervated. After amputation, nerves penetrate the wound epithelium which becomes thickened. A blastema appears, its cells derived from the perichondrium of the cartilaginous rod. Continuation of barbel regeneration is reportedly dependent on a mutual interaction between the regenerating cartilaginous rod and the wound skin. Removal of the central rod of cartilage inhibits regeneration of the barbel by removing the source of supply of blastemal cells. Similarly, denervation of the barbel inhibits regeneration and, indeed, induces an extensive regression similar to that observed in denervated, amputated limbs of salamander larvae.

Regeneration in Reptiles. The dramatic, but limited, powers of regeneration of lizards are well known. Two vertebrae behind the pelvic girdle possess, in their mid-section, special articular surfaces which can be separated by strong muscular contraction with consequent loss of the distal part of the tail. Externally, the regenerated tail looks fairly normal, but internally there is regeneration only of an unsegmented cartilaginous rod, which gradually ossifies, and a spinal cord which is only about one-quarter the diameter of the normal cord although it contains a central canal with ependyma. No gray matter can be seen in the regenerated cord and there are no spinal ganglia. As with amphibian limbs and tails, fish fins and barbels, destruction of the neural component, the spinal cord, inhibits regeneration. It is now known that the ependyma alone can support tail regeneration in the lizard, indicating that in this case a specific segment of the central nervous system is of particular importance to regeneration. Reptiles generally are unable to regenerate limbs, although some cases of abortive limb regeneration have been reported.

C. S. THORNTON

References

Hay, E. D., "Regeneration," New York, Holt, Rinehart and Winston, 1966.

Kiortsis, V. and H. A. L. Trampusch eds., "Regeneration in Animals and Related Problems," Amsterdam, North-Holland Publishing Co., 1965.

REPRODUCTION

Reproduction is basically the capacity of living systems to give rise to new systems identical with themselves. For extant organisms and proto-organisms, this function implies the ability to take up raw materials from the environment, to convert such raw materials into more substances having the exact physico-chemical constitution of the parent substances, and to transmit to the offspring a code which will enable it to reproduce itself with precision in its turn.

The ability to take up raw materials which are reorganized into material like that of the active agent probably antedates life itself as we now define it. It is generally believed that complex organic systems arose spontaneously under conditions existing at an early stage in earth history. The systems that survived were those in which relations became organized in such a way as to lead to self-augmentation. The formation of a nucleic acid in which one or more items of transmissible information could be invested was probably an essential step in the evolution of systems that could be called organisms.

Of critical importance in the cycle of reproductive events as we know it today is the division of the NUCLEUS. This ORGANELLE, which is a conspicuous feature of all cells except bacteria, contains the units of heredity, the GENES, that subsequently direct the growth and development of the cells that are newly formed from the division of the parent cell. Each nucleus contains numerous CHROMOSOMES, the full or diploid number being characteristic of the species. All the chromosomes occur in matched pairs except the sex chromosomes, which may be of two different forms, or may occur singly. It is known that the genes, though not discrete units, do exist in linear order on the chromosomes. As a cell prepares to divide, its chromosomes split lengthwise, with the apparent result that each gene is duplicated. The division therefore gives each daughter cell a set of chromosomes having a point-by-point identity with those of the parental set. The splitting of the chromosomes can be seen by microscopic examination, and the distribution of genes can be established by experiments on heredity. In addition, the reduplication of the chromosomal material has been verified by measuring the amount of deoxyribonucleic acid (DNA), which appears to be the information-bearing component of the chromosome; after division, each new cell contains the same amount of DNA as the parent cell had.

Reproduction in its simplest form involves the splitting of the whole organism into equal or unequal parts. Among the bacteria, algae, and protozoa generally, binary fission is accomplished by division in either the transverse or the longitudinal plane of the body. In yeasts the parent cell ordinarily produces one or more small cells on its surface by a process called budding. The cytoplasmic division is in most cases accompanied by visible reduplication of the chromosomes. The absence of a defined nucleus in bacteria makes it difficult to determine what happens to the genetic material in these forms when division occurs, but it has been observed that some bacteria contain DNA-rich granules that are duplicated at division.

Even among the unicellular organisms, however, reproduction is not exclusively of the asexual type just described. Many protozoa undergo a pairing, or conjugation, in which nuclear material is exchanged between the temporarily attached cells, each of which later divides by fission. Other types, including particularly the parasitic Sporozoa, produce gametes that combine to form the spores from which the new animals develop. The male gametes are flagellated, and so able to swim. Under appropriate circumstances, certain yeasts form four or eight ascospores subsequent to the uniting of two cells and the fusion of their nuclei. The spore consists of a diploid nucleus surrounded by cytoplasm and invested with a spore wall.

Even among the bacteria, which were long thought to be entirely asexual, genetic interactions have now been identified (see BACTERIAL GENETICS). The most conventional of these is *recombination*, in which cells of opposite mating types pair. There is however no exchange of DNA-containing material, but rather a unidirectional transmission from donor to recipient, with the recombinant then developing within the recipient cell. The other sexual processes of bacterial are even less like those of higher forms. In *transformation* (which is not known to occur outside the laboratory), one cell may incorporate DNA from another, and thereafter display certain characteristics of the donor cell. *Transduction* depends on the ability of bacteriophage to incorporate heritable characteristics of parasitized cells into their own genetic apparatus, and then to confer these characteristics on other cells that are subsequently invaded.

In higher organisms, the primitive mode of halving of the cytoplasm, accompanied by precise mitotic division of the nucleus, remains the standard method of cell multiplication. The division of labor that is necessitated by increasing complexity, however, has led to the evolution of special mechanisms for reproducing the organism as a whole. In the great majority of cases, these are sexual mechanisms, involving the production of gametes of opposite sex. The phenomenon of sexuality is not an innovation of multicellular organisms, but as has just been shown, is clearly foreshadowed at the unicellular level. The gametes produced by the multicellular forms are usually of unequal size, the EGGS being fixed (in plants) or immotile and heavy with incorporated food material (in animals), the smaller male elements being motile (see SPERMATOZOA). Despite the differences in their appearance, the gametes are alike in that each has in its nucleus a haploid set of chromosomes, that is, one member of each pair of chromosomes characteristic of the species. The gamete nuclei may however differ in the distribution of the sex chromosomes, since these are usually not a matched pair. In human beings, for example, in which the nuclei of females bear two "X" chromosomes whereas those of males have an "XY" pair, the sperms will have either an "X" or a "Y" chromosome. Since every egg has one "X" chromosome, the sex of the offspring is determined by the character of the sex chromosome of the sperm that effects fertilization. The union of two gametes of opposite sex restores the full complement of chromosomes to the zygote that will differentiate into the new organism. Although the development of the egg is normally dependent on its penetration by a sperm, ripe eggs are in many species able to develop in the absence of sperms. Such parthenogenetic development may be activated experimentally, and occurs in nature among both plants (e.g., dandelion) and animals (e.g., social insects).

Reproduction in multicellular plants proceeds by an alternation of gametophyte and sporophyte generations. Among lower plants, like the mosses, the plant itself is the gamete-producing generation. Male plants bear ANTHERIDIA, in which the sperms are produced, and female plants bear ARCHEGONIA, in which the eggs are formed. When moisture is present, the sperms enter the archegonia to fertilize the eggs. The resulting sporophyte gives rise to numerous spores. It continues to live as a parasite on the gametophyte until it liberates the spores, each of which has the capacity to develop into a new moss plant. This pattern is foreshadowed in some of the more complex fungi.

The vascular plants on the other hand are sporophytes. Their reproductive structures are the FLOWERS, which contain an egg-producing ovary and pollen-bearing anthers; the sexes are sometimes separate, eggs and pollen being formed in different flowers, or even on different plants, but in most species both types of reproductive products are borne on a single flower. At the onset of fertilization, the pollen grain, which is the male gametophyte, divides into two cells, one of which gives rise to the two haploid male nuclei. One of these nuclei enters the egg cell to become the parent of the future offspring. The other enters the endosperm surrounding the egg, and participates in the formation of the stored food with which the seed is supplied. In bisexual forms, the pollen may effect fertilization in the same flower, but frequently there are mechanisms to prevent self-pollination. Generally pollen is blown about by the wind, or is carried from flower to flower by nectar-seeking insects. By the time the seed is separated from the parent plant, it has accumulated a quantity of food, and contains an embryo that has begun to develop from the fertilized egg.

Most plants have the capacity to reproduce asexually, from a piece of stem, root, or even a leaf. Much used in horiculture, this phenomenon is of considerable importance in nature. Grasses spread by vegetative reproduction, and some trees, like the sassafras, multiply by the growth of suckers from the roots of the parent plant. Numbers of plants have evolved structures especially adapted to vegetative propagation; the potato tuber, and the runners of the strawberry, are examples.

Among the metazoa, asexual reproduction is significant only in the sponges, the simpler coelenterates such as hydra, and the flatworms and nemerteans. Almost any piece broken from a living sponge can become a complete new individual, and similarly a piece of a nemertean remaining after an attack by an enemy organism can become a whole new worm. Hydras produce on their body surface buds that develop into miniature hydras and then pinch off to pursue an independent life. Flatworms sometimes separate spontaneously into anterior and posterior segments, each of which reforms itself into a complete individual. These animals can also reproduce sexually. In the more highly organized invertebrates, as well as in all the vertebrates, reproduction is solely by the production and union of eggs and sperms, the only exception being the previously-mentioned parthenogenetic activation of the eggs from which some classes of social insects arise.

The development of the eggs and sperm is carried out in more or less specialized organs known as ovaries or testes. Occasionally the situation may be simpler, as in some of the segmented marine worms, in which the sex cells are budded directly off the lining of the coelom. Hermaphroditism is not uncommon, particularly among the flatworms, annelids, and gastropods, but among the more complex forms the sexes are invariably separate. Even where hermaphroditism occurs, however, self-fertilization is rare.

In its simplest form, FERTILIZATION requires only the discharge of the male and female products into the ambient water. Even in this case, it is essential that members of both sexes be assembled in a limited area, with discharge occurring during a limited period of time. Although some chemical attraction may be exerted, effective contact depends largely on chance, and is assured by the production of tens or hundreds of thousands of eggs. (Sperms are always produced in enormous numbers.) Considering the apparent inefficiency of this method, it is not surprising that members of almost all animal phyla above the Coelenterates and Ctenophores have evolved mechanisms to promote successful fertilization. Copulation, in which sperms are introduced directly into the body of the female, is the most common of such mechanisms, and is obligatory in land-living forms. The sex act may however be effected in other ways. In lobsters, for example, the male places sperms near the female pores, so that the eggs are fertilized as they emerge; whereas the male salamander deposits sperms in packets that the female draws into her own body.

In addition to mechanisms to insure fertilization, animals have also evolved a great variety of means for fostering the development of the zygote. These are of two major types: provision of a food supply, and elaboration of devices to protect the embryo during its formative stages. Food (yolk) is generally built directly into the egg cell, so that among both vertebrates and invertebrates, ripe eggs may be of macroscopic size. The flatworms however form food cells that are external to the egg, but are included with it in a cocoon. Some of the more complex animals also undertake to feed the young after hatching or birth. The social insects as well as the higher vertebrates present food directly to the young. But sometimes the provision is made in a rather indirect way. The wasp Habrobracon for example lays her eggs in the body of a caterpillar which she has paralyzed but not killed; the larvae, on emerging from the eggs, feed themselves on the living flesh of the host.

Devices for protecting the embryo until it is able to fend for itself are very varied. Some of these devices are formed within the body of the mother. Thus the eggs may be invested in a tough, unpalatable capsule (squid), may be incorporated in a mass of jelly that can be anchored to a branch in the pond (various amphibians), may be enclosed in a hard shell that prevents drying (birds). Frequently the eggs are laid under a stone to keep them from being washed away (lamprey), or in a crude nest in the water (stickleback), or in a shallow pit dug in the ground (many insects). A few invertebrates and many vertebrates carry the fertilized eggs in or on the body of the mother, or, occasionally, of the father. The female lobster, for example, carries her young attached to her swimmerets. Members of several vertebrate classes have acquired the ability to retain the developing egg within the maternal oviduct. The effect of all these techniques is to insure adequate replacement of a species from small numbers of eggs.

Although reproductive mechanisms have been subject to continued experimentation and improvement in both plants and animals of all kinds, it is among the vertebrates that reproduction has been made a decisive instrument in progress toward greater independence of, and control over, the environment. Probably the most important innovations that have appeared among the vertebrates are the production of large closed eggs containing a supply of water as well as food; the maintenance of the embryo within a uterus that supplies it with food and oxygen, as well as protecting it; and the care and tutelage of the emerged young, including the feeding of milk.

The large egg enclosed in a leathery shell enabled the reptiles to become completely independent of water, in contrast to the amphibians, which must lay their eggs in water, even though the adults may be able to survive under rather dry conditions. The generous provision of raw materials made it possible for the young reptile, and later the bird, to emerge not as a helplessly small larva, but as a good-sized replica of its parents. Among birds, reproductive efficiency is further enhanced by the habit of placing the eggs in a carefully constructed nest in which they can be incubated. After hatching the young are fed and tended until they are able to fly. Only some of the higher mammals have developed parental care to a greater extent than have the birds.

The maintenance of the developing young within the maternal oviducts, though exploited most sucessfully by the mammals, has appeared also among the fishes and reptiles. These lower forms may be merely ovoviviparous, which is to say that the oviduct simply provides safe housing for a closed egg, but many of them are truly vivaparous, in the sense that the embryo actually receives oxygen and water from the maternal system, and eliminates waste into it. Only the mammals however produce small eggs, almost free of yolk, which must derive their nutriments as well as gas and water from the maternal blood stream. This supply function is performed by the placenta, a complex structure that the mammals have evolved out of the extraembryonic membranes by which reptiles (and birds) carry on gas exchange with the atmosphere outside the shell.

The evolution of increasingly elaborate patterns of reproductive activity has depended to an important extent on the functioning of the ENDOCRINE SYSTEM. The role of the endocrines in reproduction is best understood in the vertebrates, although it is clear that hormonal activity is involved in gamete production and mating in the higher invertebrates. The key to the vertebrate system is the hypothalamus, a part of the brain that mediates between nervous and endocrine systems, and thus makes possible the seasonal control of reproductive activity that is essential in most species if the young are to be raised under optimal conditions. Under the direction of the hypothalamus, the anterior lobe of the pituitary (adenohypophysis) liberates gonadotrophins that stimulate the activity of the gonads. The gonads in turn release estrogens and androgens that maintain the reproductive tracts and promote the differentiation of secondary sexual characteristics—body size and form, color, plumage or hair pattern, horns and antlers, etc.—that may play an important role in attracting or commanding mates. A critical, and little understood, function of the sex hormones is their influence on behavior. Sex drive and its expression in patterns of courtship and of the sex act itself are ultimately dependent on some effect of hormones on the nervous system, and this is true of maternal care too.

The hormonal control of reproduction has been elaborated to its greatest extent in female mammals. In these, cyclical changes in the uterus must be synchronized with the release of eggs from the ovary, so that the uterus is prepared to receive the eggs when fertilization occurs. Proper timing is effected by the fact that the gonadotrophin-stimulated phases of egg production in the ovary are accompanied by the serial production of two ovarian hormones that prepare the uterus for the reception of the developing embryo. Once conception has occurred, further hormone release from the pituitary, the ovaries, and the placenta is required for continued growth of the uterus, and for maintenance of favorable conditions for fetal development. In addition, these hormones promote the enlargement of the mammary gland, and, after birth has occurred, they bring about the production of milk.

Although asexual or vegetative reproduction, as well as PARTHENOGENESIS, occur among both plants and animals, it is obvious that the vast majority of new organisms, at least above the unicellular level, are formed from the union of gametes of opposite sexes. Moreover, in those species in which single individuals produce gametes of both sexes, measures have in most cases been evolved that prevent self-fertilization. This all but universal adoption of bisexual mechanisms clearly demonstrates that sexuality has some substantial advantage over the asexual condition. This advantage is generally conceded to be the facilitating of evolutionary change by permitting continuous recombination of the units of heredity. Fertilization ordinarily brings together two sets of chromosomes from different individuals. The new assortments of genes made up in this way may possess qualities not manifested by either of the parent sets. Further, the combining of chromosomal material from separate individuals allows a new mutant gene to be tested in a wide variety of genetic situations, instead of remaining limited by the situation in which it arose. Finally, the mechanism by which haploid gametes are formed (MEIOSIS) in many species involves breakage of chromosomes and recombination of parts of different chromosomes. Thus an individual may pass on to his offspring new chromosomes which were not represented in either of the sets he received from his two parents. Taken together, these possibilities explain why life, which is characterized by the ability to change and adapt, should reproduce by a mechanism that favors change and adaptation.

FLORENCE MOOG

REPRODUCTIVE SYSTEM

Components of the Reproductive System. The primary raison d'être of all living things is to reproduce their kind. To accomplish this mission a basic reproductive system has been evolved which consists of three components. The gonads produce the gametes, ova or sperm. The duct system (which in mammals consists of the vagina, cervix, uterus and oviducts) conducts the gametes to their point of union within the body in mammals, birds, reptiles, and some fishes. In some amphibia and fishes it conducts the gametes outside the body where fertilization occurs.

In viviparous fishes and reptiles the duct system provides an attachment site and shelter for the developing zygotes and young which, however, derive all the nourishment they need for prenatal development from the yolk supply of the egg from which they originate. Eggs of mammals being devoid of yolk, the mammalian duct system is modified to provide not only shelter for the embryo, but also to make it possible for the developing young to parasitize the mother and derive all the nourishment from the maternal host by means of the placenta through which the exchange of nutrients occurs.

The male system consists of paired testes and a duct system which conducts the sperm cells to the ejaculatory organ. The sperm cells become motile when they become suspended in the secretions of the male accessory glands, (seminal vesicles, prostate and bulbourethral glands), which pour their products into the duct system. Ejaculation occurs into the posterior port of the vagina.

The testes of most mammals are sheltered in the scrotum which, because of its temperature-sensitive musculature, can regulate the distance of the testes from the body wall and thus provide the testes with an environmental temperature which is 1–8° F. lower than body temperature. If the thermo-regulatory function of the scrotum is impaired mechanically or physiologically, complete sterility will result because spermatogenesis is impossible at body temperature. Exceptions to this rule are pachyderms and aquatic mammals which have abdominal testes and no scrota but which are said to have normally a lower body temperature than do terrestrial mammals. Birds also have abdominal testes but a body temperature which is significantly higher than that of mammals. In them spermatogenesis is said to occur only in the early morning hours when their body temperature is at its lowest.

In those species in which the male is seasonally breeding, the testes migrate from the scrotum into the body cavity as the amount of day light decreases and they return to the scrotum in the spring. This is known to be due to waning and waxing rates of secretion and androgen which in turn is controlled by light. In all animals the testes originate in the kidney region and migrate into the scrotum during late embryonic or early post-embryonic development. Occasionally their descent is prevented by lack of proper hormones or by anatomical obstruction of the inguinal canal. If this condition (which in some species is hereditary) is not corrected sufficiently early, such cryptorchids will be sterile although their sexual drives and instincts will not be impaired since high body temperature has no effect on normal function of Leydig cells and hence androgen secretion.

The third component of the reproductive system is the neuro-endocrine synchronizing system which insures that the shedding of gametes, FERTILIZATION, GESTATION, parturition and LACTATION—all events which require relatively accurate timing—occur in such a way that the young life depending on them has a good chance of surviving. The neuro-endocrine system consists of the anterior and posterior lobes of the PITUITARY gland (or hypophysis), the hypothalamus, and the nerves which connect parts of the reproductive system with the hypothalamus. The hypothalamic nuclei communicate with the posterior lobe by means of nerve fibers. The anterior lobe of the pituitary gland is not innervated and there is a hypophyseal-hypothalamic portal system which conducts humoral products from the hypothalamic nuclei to the anterior lobe.

Gametogenesis. Anterior pituitary glands of both sexes secrete at least 6 different protein HORMONES, 3 or 4 of which are concerned with reproduction. In the female (see OESTRUS CYCLE) the follicle stimulating hormone (FSH) causes growth of the ova-containing follicles, and the luteinizing hormone (LH) causes the rupture (ovulation) of mature follicles. In the male, FSH causes growth of the seminiferous tubules which produce sperm, while LH (also called ICSH for interstitial cell stimulating hormone) acts on the Leydig cells of the testicular interstitial tissue causing them to secrete the male sex hormone (androgen). Recent evidence suggests that both FSH and LH must interact in order to produce the effects noted and that neither hormone is ever secreted by the anterior lobe to the exclusion of the other. It is probable that the ratio between the amounts of FSH and LH poured out by the pituitary gland changes during the different stages of the normal reproductive cycle of the female. While it is important for the specialist to remember that the two hormones have distinctly different physiological properties, even within the same sex, it is often more convenient to think of an FSH-LH mixture and to speak of the gonadotrophic complex, especially in view of the complementary action of FSH and LH in their stimulatory action on the gonads.

Regulation of hormone flow. The rate of flow of gonadotrophic hormones from the pituitary glands is subject to macro and micro regulation. While the micro regulation is accomplished by a feed-back mechanism which depends on the internal hormonal environment of individuals, macro regulation is largely under the control of the external environment. Chief among the forces of the external environment is light. Light, impinging on the retina of the eyeballs, causes a neural signal to be conducted via the eye nerve to the hypothalamic nuclei which respond by release of unidentified humors (FSH and LH precursors?), which in turn cause the hypophysis to produce or to release gonadotrophic hormones. Many animals are seasonal breeders because in them the pituitary gland produces gonadotrophins only in response to increasing light in spring and summer. Only during these seasons do the gonads get adequate hormonal stimulation for full development. With the seasonal decrease in day length the pituitary cells secreting gonadotrophins become quiescent and the gonads, in the absence of hormonal stimuli, regress. Such seasonal breeding behavior is typical of all non-domesticated birds and many small and large wild-living mammals. By genetic selection seasonal breeders can be converted into continuous breeders although some latent dependence on macro regulation of reproduction by light persists even after many generations of selection. Even in such domesticated mammals as man and cattle, who originally were probably seasonal breeders, the influence of light on reproductive patterns can be seen by comparing birth rates of their young with light changes at different latitudes.

This effect of light has been put to practical use by chicken raisers who have learned that additional light during winter months will prevent the normally occurring decline in egg production. Neither blindness nor even enucleation impares the action of light on the hypothalamus so long as the eye nerve remains intact. In some animals the nerve endings in the eyelids or the thin skin around the eyes are light sensitive and can transmit light stimuli (whatever their nature may be) to the eye nerve and through it to the hypothalamus.

Quantity of light, as the macro-regulator of hypophyseal gonadotrophic function, is permissive of optimal gonadal development during the breeding season. The micro-regulatory mechanism, which controls the finer adjustment and timing of the rate of hypophyseal, and hence, gonadal function, consists of the gonadal sex hormones and the neuro-humoral system governed by the hypothalamus.

The quality and quantity of the gonadotrophic complex is profoundly changed by the physiological activity of the endorgans (gonads) it affects. Castration of males or females and menopause in women, increase the amount of the complex produced and secreted into the blood stream by the pituitary gland. From this observation it was concluded that the male and female sex hormones normally participate in regulating the rate of secretion of the gonadotrophic complex and play an important role in the auto-regulatory feedback systems which exist between gonads and hypophysis. In the female, hypophyseal gonadotrophins cause growth of ovarian follicles which, as they increase in size, secrete more and more female sex hormones (estrogen). As the titer of the latter rises in the blood stream it inhibits the rate at which gonadotrophic hormone is released from the pituitary gland. Diminishing stimulation from the pituitary gland and a change in the FSH/LH ratio in the complex, causes ovulation. With ovulation the titer of estrogen drops, the inhibition of the pituitary gland is thus removed, and the gonadotrophic complex once again flows initiating a new phase of follicular growth, completing the cycle. This is the basis of the cyclic reproductive behavior of female mammals.

Sex cycles. Non-primate mammalian females have an OESTROUS CYCLE which consists of a relatively long anestrous period followed by a short period of sexual receptivity which is called heat or oestrus (literally: frenzy). Copulation is tolerated or encouraged only during oestrus. Heat coincides with maximal follicular development and is caused by estrogen which, when injected even into castrated females, causes psychological heat which persists for the duration of treatment with estrogen. As a rule follicles ovulate shortly before or shortly after the end of heat. The estrous cycle thus extends from the end of one heat to the end of the next heat, and includes an anestrous period, a heat period and culminates in ovulation. In group-living females with regularly recurring estrous cycles, heat is the only time when copulation is permitted, and since the time of ovulation is related to heat, the likelihood of the presence of living sperm and eggs is maximized. In some females living in diaspora (rabbits, cats, mink etc.), chance encounters between females in heat and males were apparently too uncertain to permit highest reproductive efficiency. Here selective advantage was in favor of females which solved the problem of synchronizing reproductive events differently. In these species the females are in heat almost continuously or at least for prolonged periods of time, and ovulation is induced, e.g. it occurs only following

mating. During copulation the penis stimulates the nerve endings in the cervix and vagina and these impulses are instantly transmitted through the nervous system to the hypothalamus. The latter decodes the signal, translates it into an as yet unidentified humoral substance which it secretes into the hypothalamo-hypophyseal portal system, through which it reaches the anterior lobe of the pituitary gland which responds by the release of LH into the systemic circulation. When LH reaches the ovarian follicles the latter respond by ovulation just as they do in spontaneous ovulators in which LH is released periodically.

In some fish and amphibians ovulation is also neuro-humorally controlled although in them the release of LH is not due to actual copulatory stimulus but depends on the courtship maneuvers in which ripe males and females indulge prior to the shedding of ova and sperm directly into the water.

In primates peaks of sexual receptivity are either completely absent or are very indistinct. In the absence of a mechanism synchronizing mating and ovulation and barring daily copulation, the statistical chances of intercourse taking place around the time of ovulation, are relatively small. Furthermore, in women at least, the time of ovulation is not nearly as predictable or standard as it is in non-primates. For these and other reasons to be discussed later, fertility in animals with long periods of sexual receptivity is lower than it is in females with short, clear-cut heat periods. In women sexual receptivity lasts about 28 days and ovulation occurs midway between menstrual periods. In mares heat lasts 4 to 7 days and they ovulate 1 day before or one day after the end of heat. Rats have a 4 to 5 day cycle and a heat period lasting about 14 hours. In sheep one oestrous cycle lasts 16 days and heat 34 hours, while in cows and sows cycles are 21 days long and heat lasts 15 hours in the former and 2 to 3 days in the latter.

Lack of well defined heat periods in primates, and prolonged heat periods in such animals as mares and pigs, affect fertility in yet another way. Mating may occur long before or long after ovulation. In the first case sperm cells age while waiting for the arrival of a newly ovulated egg while in the second case the eggs age prior to the arrival of the freshly ejaculated sperm cells in the oviduct where fertilization takes place. In either case the available evidence shows that zygotes resulting from the union of two germ cells, one of which is senile, are less viable or give rise of less viable or abnormal young than do zygotes produced by young vigorous germ cells. Such aging of germ cells also leads to greater embryonal mortality.

Effect of sex hormones on uterus. After ovulation follicles rapidly transform into corpora lutea which secrete still another steroidal female sex hormone, called progestin. (The synthetic chemical compound which has progestinlike properties is called progesterone). In the absence of pregnancy a corpus luteum remains functional only during about the first half of the succeeding cycle. The uterus during the cycle is thus influenced by two female sex hormones: first by estrogen of follicular origin, then by progestin from the corpus luteum. Progestin causes the uterine glandular layer (endometrium) to thicken because it is able to induce extreme convolutions of the uterine glands which, under estrogen stimulation, were straight. These coiled glands secrete glycogen-containing uterine milk which is essential for the early nutrition (embryothrophe) of the zygote before it becomes implanted and establishes a permanent connection with the maternal blood supply (hemotrophy). The endometrium is maintained in its proliferated state only so long as it is stimulated by progestin. As the corpus luetum wanes, the rate of progestin secretion drops and the endometrium is sloughed off. This sloughing of the endometrium is observed in all mammalian females but only in primates does the thinning of the endometrium lead to bleeding (menstruation) due to the exposure of the spiral arteries of the vascularis layer of the uterus. Following sloughing, the endometrium is repaired largely under the influence of estrogen secreted by the growing follicles of the next cycle.

Menstruation. The menstrual cycle extends from the end of one menstrual episode to the end of the next. Menstruation itself is due to the absence of hormones and is specifically brought on by the drop in the blood progestin titer which accompanies the functional death of the corpus luteum. In women ovulation occurs most commonly on days 13 to 16 after the previous menstruation although considerable individual variation is observed, thus making application of the so-called "safe period" rather hazardous. The menstrual flow itself lasts 3 days in most women but here again considerable variation is encountered. While complete absence of menstruation may indicate nonfunctional ovaries, amenorrhea is known in women with normally functioning ovaries who also show normal fertility. Amenorrhea in these cases is due to the fact that the blood hormone titers do not fall low enough to permit denudation of the endometrium to the point where bleeding results.

With approaching menopause (when the ovaries become non-functional) menstrual cycles become irregular and menstruation may become very frequent and even continuous. This condition can be easily controlled by the oral administration of mixtures of estrogen and progesterone which control the bleeding and improve the mental state of women who are frequently adversely affected by the absence of their own ovarian hormones.

Reproduction in birds. Birds face a problem that is essentially different from that of mammals. Since the entire embryonal development takes place outside the body of the mother, the young must be provided with all the food it will require during the incubation period. This problem was solved by providing large stores of lipo-protein in the yolk, protein and water in the albumen, and minerals (Ca and P) in the egg shell which also protects its contents. Because of the large size of the yolk and the large amount of the other ingredients of the whole egg, maturation and ovulation of eggs has to be timed in such a way that not more than one ovum is ovulated at one time, that this ovum be allowed sufficient time to acquire the albumen coat, the calciferous shell and that it be laid before the next ovum is released from the ovary. Basically the same neurohumoral mechanism operates in birds as the one described for mammals. The gonadotrophic complex stimulates follicular growth and causes ovulation. The ovulated ovum is engulfed by the oviduct and is invested with albumen as it slowly progresses through the oviduct. The latter is differentiated into three specialized glandular portions, one of which secretes albumen, another the soft shell membrane, and the third portion secretes the calcareous shell. In the chicken the whole process from ovulation to the laying of the finished egg requires 25 to 28 hours. The secretory activity of oviducal glands is controlled by joint action of sex hormones (estrogen and androgen) which are produced by the ovary.

While an ovum is traveling through the oviduct, nerve endings in this duct are stimulated sending signals that the oviduct is occupied and that ovulation (e.g. LH release) should be postponed. As the egg approaches the time of expulsion (induced by oxytocin), the signals stop, LH is released and ovulation of the next ovum can occur.

Hormones play an important role in activities of birds which are related to reproduction. Nuptial plumages, head adornments where present, colors of bills etc. are all controlled by androgen. Amounts of androgen secreted also establish peck orders in flock-living birds of both sexes, and androgen causes males to establish and to defend nesting territories. Nest building itself is hormonally controlled although it is not clearly established which hormone is directly responsible. Normally birds will lay the number of eggs typical of the species and then begin incubation. If eggs are removed from the nest before the full contingent has been laid, laying will continue as the female attempts to meet her quota of eggs. By this device flickers and sparrows have been caused to lay up to 250 eggs in succession. This is another example of the involvement of the nervous system in the reproductive mechanism. If the female is allowed to lay her normal quota of eggs, the ovary degenerates rapidly and incubation of eggs begins. The urge to incubate and the subsequent urge to feed, protect and guide the young is controlled by prolactin (the hormone which causes initiation of lactation in mammals). In such specialized birds as pigeons and doves prolactin causes the glandular crop to secrete crop or "pigeon milk" which is regurgitated into the gullets of the young and serves as their only food until they leave the nest.

Pregnancy. If the mammalian female becomes pregnant, the endometrium is not allowed to degenerate and the corpus luteum is maintained thus insuring continued progestin secretion. It is almost certain that the presence of the zygote, and later the embryo, in the uterus, causes a neurally transmitted signal to go from the uterus to the pituitary gland which responds by releasing a hormone called luteotrophin which is responsible for the maintenance of the corpus luteum throughout all or part of pregnancy. In all mammals malfunction or surgical removal of the corpus luteum early in pregnancy (prior to implantation) invariably leads to abortion or resorbtion of the conceptus. Females of some species (women, cows, mares etc.) can tolerate castration after the first $\frac{1}{3}$ or $\frac{1}{2}$ of pregnancy because their placental membranes secrete progestin and estrogen which are essential for maintenance of the proper uterine environment in pregnancy. In other species (rats, rabbits, sows) castration is never tolerated presumably because in them the corpus luteum remains the only source of progestin throughout gestation.

Parturition. Gestation culminates in parturition. The exact endocrine mechanisms have not been worked out but it seems probable that the following events take place. It is known that gestation can be terminated by a decrease in progestin levels or prolonged beyond normal time limits by the injection of exogenous progesterone. Thus, decrease of progestin levels appears to be an essential prerequisite for the onset of the birth process. There is no evidence for such a decrease at parturition and it seems most probable that the pregnancy maintaining properties of progestin may be counteracted by estrogen which has a well known ability to antagonize the physiological effects of progestin. If the progestin level remains unchanged during the last part of pregnancy, and if the estrogen level continues to rise, a ratio of these two hormones will be reached when progestin is completely antagonized by rising estrogen levels. In effect, then, the uterus will be dominated by estrogen a condition essential for the expulsion of the fetus. The contractions of the uterine muscle resulting in "labor pains" are brought about by the action of the posterior pituitary gland hormone, oxytocin, on the smooth musculature when it is predominantly under control of estrogen.

Older theories which proposed that parturition was precipitated by the accumulation of fetal waste products or by the size of the fetus which made its expulsion essential, are now completely discredited. In most species male conceptuses are normally carried longer and are heavier at birth than female fetuses. In women and in cows possessing certain genetic traits parturition may be significantly delayed beyond the normal time. In cows this is known to be caused by gene combinations which presumably act by preventing the shifts in the hormone combinations compatible with the onset of labor. Thus in these cows and in diabetic human mothers who, as a rule, produce heavier than normal babies at the expected time, fetal size can not be the cause of onset of labor.

Lactation. As noted earlier the placenta secretes estrogen and progestin during gestation. While there are some species differences, these two hormones in most mammals are responsible for building the mammary gland during the last stages of gestation. Estrogen is primarily responsible for building the duct system while progestin prepares the intricate loboalveoler system in which milk is formed. Shortly before impending birth the building of the breast is essentially completed and the gland secretes a substance resembling milk (colostrum). Postnatally milk secretion is initiated by the action of hypophyseal hormones on the breast which was anatomically prepared for secretion by estrogen and progestin. Which pituitary hormone is most responsible for precipitating lactation is not completely clear although it is probable that prolactin, growth hormone, the adrenocorticotrophic and thyrotropic hormones all play a role in the formation and in the mobilization of precursors for milk and milk fat.

Yet another neuro-humoral mechanism is involved in the release ("let-down") of the accumulated milk from the breast. Stimulation of the nipple by the suckling young causes a nerve impulse from the nipple to be transmitted via the hypothalamus directly to the innervated posterior lobe of the pituitary gland. The posterior lobe responds by the instant release of oxytocin into the systemic circulation which carries this hormone to the breast and causes there the constriction of the myoepithelium which surrounds the individual alveoli. The milk is squeezed out into the minor and major ducts from which it can be sucked out by the nursing young.

In lactating females the nursing stimulus is not absolutely essential and the chain of events leading to oxytocin release can also be initiated by such stimuli as the voice or sight of the baby, and, in the case of dairy animals, by the sounds of milking machines. Conversely, once milk letdown has been initiated by one of these stimuli, it can be interrupted by such events as embarrassment, fright or stress of nursing mothers. Since oxytocin affects not only the breast but also the uterine musculature, uterine contractions occur during nursing, or conversely milk may gush from the nipples of lactating females during mating. The ability of oxytocin to cause contractions of the uterus and oviducts may be also responsible for the extremely rapid transportation of sperm through the whole duct system. Semen has been shown to reach the oviducts within a few minutes after ejaculation into the posterior portion of the vagina.

A. V. NALBANDOV

References

Nalbandov, A. V., "Reproductive Physiology," San Francisco, Freeman, 1964.
Parkes, A. S., ed., "Marshall's Physiology of Reproduction," 3 vols. New York, Longmans, Green, 1956–1966.

REPTILIA

Introduction. Reptiles were recognized as a group of poikilothermal tetrapods that hatch from an amniote egg and have a scaly, non-glandular skin, a single occipital condyle, and a three-chambered heart as early as 1816, but the class Reptilia was not distinguished from the AMPHIBIA in these terms until 1895.

This definition, derived from study of living animals, is based chiefly upon characters of physiology and soft anatomy that are not preserved in the fossil record. Increased knowledge of fossils reveals that living reptiles are relicts of a large and heterogeneous group. Study of fossil reptiles has led to greater use of osteology in general classification, and subdivision of the Reptilia depends more upon trends of skeletal evolution than upon classic characters.

In revealing animals annectant between tetrapod classes, the fossil record has forced recognition of the arbitrariness of any classification dealing with such forms. Late Paleozoic reptiles and amphibians have many characters in common; some reflect common inheritance from fish ancestors and others, common tendencies toward a terrestrial habit. A few characters may indicate a reptilian evolutionary level because they appear consistently in later animals of closer relationship to living reptiles.

Basic structure. *Subclasses Anapsida and Synapsida.* The surface of the primitive tetrapod skull consists of a continuous sheet of dermal bones, perforated only by orbits, external nares, and parietal foramen.

Masticatory muscles originate inside the skull roof behind the orbit and pass downward to insert on the lower jaw. In more advanced reptiles the bony covering of the cheek region becomes variously fenestrated to allow for expansion of these muscles. Temporal fenestration appears to correlate with consistent trends of other features in a number of reptilian lines, and provides a basis for establishment of subclasses.

The subclass Anapsida is defined primarily by the lack of any consistent opening in the temporal region. Anapsids tend to be conservative, retaining a primitive build of body and limbs even in such a specialized order as the turtles.

The order Cotylosauria (lower Permian through Triassic) includes some of the earliest reptiles; it is based upon postcranial characters which also define primitive reptilian structure. The pleurocentrum is large and discoidal while the intercentrum is small and wedge-shaped; the amphibian intercentrum equals or exceeds the pleurocentrum in size. Neural arches of dorsal vertebrae are wider and present a characteristically massive appearance; neural spines tend to be low. The ilium is larger in reptiles than in amphibians and articulates with two vertebrae instead of one. The pectoral girdle includes a separate coracoid and a stem on the interclavicle. An entepicondylar foramen perforates the humerus, and the feet show less reduction of toe and phalangeal number than is characteristic of amphibians.

Differences between primitive amphibian and reptilian skull structure are illustrated by characters separating cotylosaur suborders. The amphibian-like seymouriamorph pattern contrasts with the reptilian captorhinomorph pattern primarily in jaw suspension and middle ear structure, which are intimately related throughout tetrapod evolution. The seymouriamorph quadrate slants forward and upward from the jaw articulation; above it the posterior edge of the skull roof is deeply emarginated to form an otic notch in which lay a superficial ear-drum. A small stapes runs from the ear-drum to the otic capsule, which houses the neurosensory part of the ear. The captorhinomorph quadrate stands vertically; its upper end has migrated posteriorly and closed the otic notch, which is marked by a zone of weakness in primitive forms. The massive stapes passes distally to the quadrate; the ear-drum is not superficial.

The captorhinomorph palatal pattern, which remains remarkably stable throughout the Reptilia, is distinctive in its transversely-arched structure, larger internal nares and masticatory muscle passages, and prominent transverse pterygoid flanges.

The seymouriamorph skull retains a primitive number and pattern of dermal bones; the temporal series is reduced in captorhinomorphs.

The third cotylosaurian suborder, Diadectomorpha, probably arose from the amphibians independently of the Captorhinomorpha. The quadrate stands vertically, apparently because its lower end moved forward, which resulted in exaggeration of a seymouriamorph otic notch rather than closure as in captorhinomorphs. The quadrate is concave posteriorly and the tympanum remains superficial; the short stapes has no ossified process to the quadrate. The temporal series of dermal bones is complete. In the palate the basipterygoid articulation, movable in other cotylosaurs, is fused, and the pterygoid flanges are poorly developed. In the family Diadectidae a turtle-like secondary palate is foreshadowed.

The anapsid order CHELONIA, turtles (Triassic to Recent), is marked by its unique modification of the axial skeleton, origin of which is clearly referable to no other group. The chelonian skull, however, recalls the diadectomorph skull in unfenestrated temporal region, posteriorly-concave quadrate, and palatal structure.

The subclass Synapsida is distinguished by a single temporal fenestra below the squamosal and postorbital bones, and tendencies toward dental specialization and enlargement of the dentary at the expense of posterior lower jaw bones. The synapsid order Pelycosauria lived contemporaneously with the Cotylosauria. The most primitive pelycosaurs are very similar to primitive captorhinomorphs in general shape of skull, dermal bone pattern, ear, palate, and most postcranial features; ultimately they are distinguishable only on the arbitrary basis of termporal fenestration. Advanced pelycosaurs display subclass characters more clearly and, in the postcranium, tend toward reduction of the swollen zygapophysial supports and increase in height of neural spines.

The advanced synapsid order Therapsida (middle Permian to lower Jurassic) stems from one suborder of pelycosaurs, and its earliest members are very pelycosaur-like. "Mammal-like" tendencies (reduction of quadrate and stapes with increasing mobility of quadrate; increasing heterodonty; development of secondary palate, rhinarium, and turbinate bones; increasing complete covering of the sides of the braincase by dermal bones) appear at different times in different suborders, and evolve toward a mammalian structural level at different rates.

Despite rapid development of one or more mammalian characters, most lines retain a characteristic therapsid aspect. Dermal bones, excepting those around the jaw joint, are not reduced in number or size. The most important postcranial changes involve alteration of a sprawling reptilian gait to a "legs-under-body" mammalian gait, but this does not produce much loss of elements beyond reduction of the phalangeal formula to the mammalian 23333 pattern in advanced forms. A tendency toward differentiation of thoracic and lumbar vertebrae appears, but even advanced therapsids tend to retain ribs on the pre-sacral vertebrae, reptile-fashion.

Central reptiles. *Subclasses Lepidosauria and Archosauria.* Most Reptilia belong to the subclasses Lepidosauria and Archosauria, characterized by two temporal fenestra separated by the squamosal and postorbital bones, and by a posteriorly-concave quadrate and slender stapes. Similarity of the middle ear to that of the Diadectomorpha has been used to brigade these two subclasses with the diadectomorphs and turtles as Reptilia Sauropsida, in contradistinction to the Reptilia Theropsida, which includes the Synapsida, captorhinomorph cotylosaurs, and minor subclasses. But the peculiarities of the "sauropsid" middle ear seem to be adaptive toward improved hearing of air-borne sound, and could have arisen more than once in animals trending toward a terrestrial habit. Since the earliest synapsids and the captorhinomorphs are primitive groups, it is likely that their common type of middle ear is a mark of that primitiveness rather than a reliable taxonomic criterion. The lower Permian forms and some of the minor subclasses are probably related, but most of the latter are modified for marine life, which might be expected to produce parallel changes in the middle ear. The lepidosaur skull is derived from a captorhinomorph pattern more readily than from a diadectomorph pattern in all features except the middle ear. The sauropsidtheropsid concept probably reflects parallel adaptive modifications at different levels of evolution rather than true lines of descent.

The Lepidosauria are more conservative than the Archosauria, with no particular postcranial tendency toward bipedalism; the skull tends to remain of primitive aspect. Typical members of the order Eosuchia (middle Permian to Eocene) are unarmored and lizard-like and provide good morphological ancestors for the other two orders, which differ from them in dermal bone reduction and in acrodont or pleurodont dentition. The Squamata (Jurassic to Recent) differ from the Rhynchocephalia (lower Triassic to Recent) chiefly in their movable quadrate and kinetic skull, features that are carried to an extreme in snakes. The Squamata have evolved hemipenes that are lacking in *Sphenodon,* the only living rhynchocephalian.

The most primitive Archosaurian order, Thecodontia (Triassic), shows characteristic specializations of the subclass, i.e., antorbital and mandibular fenestrae, implantation of teeth in sockets, tendency toward bipedalism, and body armor. The skulls of some small thecodonts are bird-like.

The order CROCODILIA (lower Jurassic to Recent) consists of archosaurs which early assumed an aquatic, fish-eating habit, as reflected by their distinctive morphology (long snout, complete secondary palate, and laterally flattened, sculling tail). Little trace of bipedalism remains except for the slightly greater length of hind legs. Some crocodilians, the only archosaurs to invade the sea successfully, achieved large size.

The orders Saurischia (upper Triassic through Cretaceous) and Ornithischia (middle Jurassic through Cretaceous) were formerly brigaded as dinosaurs, but it is now accepted that each took independent origin from thecodonts. In addition to different structure of the pelvis (from which the ordinal names are derived), and the lack of any forms more advanced than thecodonts annectant between them, these orders are separated on the basis of characteristic trends. Some of the Saurischia early became quadrupedal, attained large size, and assumed an herbivorous diet; those that remained a bipedal also remained carnivorous. The Ornithischia produced no carnivores, although most of them remained bipedal; characteristically they tended to evolve a grinding dentition and beak-like anterior jaw structure. The quadrupedal ornithischians tended to retain and to elaborate the ancestral body-armor.

A fifth order, the Pterosauria (lower Jurassic through Cretaceous), consists of forms which are classified as archosaurs on the basis of skull and jaw fenestrae and implantation of teeth, but which at their first appearance are already highly specialized for flight and bear little postcranial resemblance to the rest of the subclass. In loss of tail and teeth they parallel the birds.

Minor subclasses. The remaining Reptilia are divided into three subclasses of uncertain relationship.

The subclass Ichthyopterygia contains the single order Ichthyosauria (middle Triassic through Cretaceous), whose members are highly specialized for a marine existence. Advanced forms have evolved a fish-like habitus; the tail is the main propulsive organ and bears an adipose fin. A dorsal fin is also present and the limbs are modified into control planes. There is a single dorsal temporal fenestra, the stapes is massive, and the snout is drawn out into a slender rostrum, at the base of which the nares have coalesced to form a blow-hole.

The subclass Euryapsida (Synaptosauria) is characterized by a dorsal temporal fenestra and by rib articulation different from both the primitive tetrapods and the lepidosaur-archosaur group. The Protorosauria (lower Permian through Triassic) consists of lightly-built animals of lizard-like appearance. In some the neck is peculiarly elongate by reason of lengthening of individual vertebrae rather than by addition of segments. The other euryapsid order. the Sauropterygia (lower Triassic through Cretaceous) is marked by specializations for an aquatic or marine existence. In the nothosaurs, marine modifications are limited to head and dentition; limbs are little changed from a primitive terrestrial pattern. More advanced forms, the plesiosaurs, show a characteristic modification of the limbs into rowing paddles. The body is stout, the tail short, and the neck long; the plesiosaurs have been said to look like "a snake strung through a turtle shell." A third group, the placodonts, was probably derived independently from nothosaur stock; the limbs are not particularly paddle-like and the tail is longer than in the plesiosaurs, but the skull, jaws and dentition are radically modified toward a mollusc-eating habit.

The lower Permian order Mesosauria, of uncertain subclass assignment, consists of one genus of small reptiles distinguished by such aquatic specializations as an isodont, fish-catching dentition and a long, laterally-flattened, sculling tail. As might be expected from its antiquity, it retains many primitive features (e.g., large stapes and expanded zygapophyses) reminiscent of the Cotylosauria. There is a lateral temporal fenestra, and the group may represent an aquatic offshoot from a pelycosaur stock.

NICHOLAS HOTTON III

References

Gadow, H., "Amphibia and reptiles," London, Macmillan, 1901.
Goodrich, E. S., "Studies on the structure and development of vertebrates," 2 vols., New York, Dover, 1958.
Romer, A. S., "Osteology of the reptiles," Chicago, University of Chicago Press, 1956.

RESPIRATION

Respiration is primarily an energy yielding (see ENERGY PRODUCTION) dissimilation process and a phenomenon that is exhibited by all living cells. In this process, high energy containing substances generally carbohydrates, e.g. starch, glycogen, sucrose, glucose, or lipids are broken down in a stepwise manner, under enzymatic control, to simpler substances of lower energy content. Chemical or free energy is liberated at certain specific stages in the form of high energy phosphate bonds

($\sim$ P) which are trapped by the adenylic acceptor system (adenosine diphosphate, ADP) and stored in the pyrophosphate bonds of adenosine triphosphate (ATP). The over-all thermodynamic efficiency is estimated at 60–70% which is high in comparison to man-made machines (steam or combustion engines etc).

Respiration may be either aerobic or anaerobic (FERMENTATIVE) depending on the organism or tissue concerned, and also on the oxygen tension in the atmosphere or at the sites of respiration.

Aerobic respiration is essential for the survival of the vast majority of terrestrial and aquatic organisms, although some fungi (yeasts) and, for example, the enteric bacteria are facultative anaerobes and other bacteria (Clostridia) are obligate anaerobes. In animals, fishes and insects, aerobic respiration includes breathing (pumping air in or out of air sacs and lungs or water over gills). This is referred to as external respiration (*vide infra*) in contrast to internal or cellular (tissue) respiration.

Respiration is a catabolic process and as well as being the least complex it is also the best understood of all physiological processes. However, all other synthetic (anabolic) processes which are essential for growth and reproduction of organisms are in one way or another completely dependent on its normal operation.

Thus in obligately aerobic organisms, if respiration is inhibited by heavy metal poisons (e.g. cyanide and carbon monoxide) that interfere with the iron porphyrin oxygen acceptor systems, death ensues rapidly. Similarly in organisms that "breathe" in their overall aerobic respiration, considerable discomfort and death may be experienced if the oxygen tension of the atmosphere (21%) falls below a critical point (14%).

Aerobic Cellular Respiration

The aerobic respiration of glucose involves the consumption of oxygen, the release of an equivalent volume of carbon dioxide, the liberation of energy and the formation of water according to the following over-all equation:

$$C_6H_{12}O_6 + 6O_2 \rightarrow 6CO_2 + 6H_2O$$

$$\Delta F: -688 \text{ kg. cal/mole.}$$

The volume ratio of CO_2/O_2 is known as the respiratory quotient (R.Q.) and as indicative of the substrate being respired. Thus the R.Q. is unity for hexoses and varies to <1 for fats and to >1 when organic acids are being preferentially respired.

The equation for aerobic respiration does little to indicate either its mechanism or its special relationship with fermentation or anaerobic respiration. It is now well established that the aerobic respiration of carbohydrates takes place in three main phases of which the first two are shared with fermentation.

Phase I. Mobilization of reserve carbohydrates. In phase I, the reserve carbohydrates are mobilized by phosphorylations involving inorganic phosphate and ATP to glucose-6-phosphate and fructose-1, 6-diphosphate in which the esterfied phosphate linkages are "energy poor" ($-$P ca. 3 kg. cal/mole) in comparison with the "energy rich" pyrophosphate bonds of ATP ($\sim$ P ca. 12 kg. cal/mole). No oxidation occurs in this phase and energy is lost from the ATP reservoir.

Phase II. The Embden-Meyerhof-Parnas sequence of glycolysis. In Phase II the process of glycolysis occurs in which this hexosediphosphate is cleaved into two triosephosphates. One of these, 3-phosphoglyceraldehyde undergoes a coupled oxidation and substrate phosphorylation involving inorganic phosphate and diphosphopyridine nucleotide (DPN, Coenzyme I, Vitamin B group). Sufficient energy is liberated by the oxidation of the carbonyl to the carboxyl group to reduce DPN and to form an unstable $\sim$ P in 1,3-diphosphoglyceric acid which is trapped by the adenylic acceptor system and stored as a pyrophosphate bond in ATP. A second $\sim$ P is formed in the intramolecular transfer of the low energy ester phosphate of 3-phosphoglyceric acid in the 2-position accompanied by a dehydration with the formation of phosphoenolpyruvic acid. This phosphate is transferred to ATP with the formation of pyruvic acid.

The second triosephosphate arising from hexosediphosphate is converted to 3-phosphoglyceraldehyde and is carried through the same series of reactions. Thus, per mole of hexosediphosphate, two moles of pyruvic acid are formed, two moles of DPN are reduced and four moles of ATP are synthesized from ADP according to the following equation:

$$C_6H_{12}O_{6-}(2P) + 2DPN + 4ADP \rightarrow$$

$$2CH_3COCOOH + 2DPNH_2 + 4ATP$$

and this ends the phase of glycolysis.

Phase III. The Krebs tricarboxylic acid cycle. By far the major part of the energy of the glucose molecule is liberated in the steps that lead to the ultimate aerobic fate of pyruvic acid into carbon dioxide and water. The ENZYMES of the KREBS CYCLE in contrast to glycolytic enzymes are bound up in discrete cellular organelles called MITOCHONDRIA. Entrance to the cycle is gained by acetylcoenzyme A which arises from the oxidative decarboxylation of pyruvic acid involving coenzyme A (CoA Vitamin B group) and the reduction of DPN. Acetyl CoA condenses with one molecule oxaloacetic acid to form citric acid which undergoes successive decarboxylations and oxidations involving primarily DPN, but also TPN (triphosphopyridine nucleotide, coenzyme II, Vitamin B group) and also coenzyme A to form another molecule of oxaloacetic acid which condenses with acetyl CoA arising from the second molecule of pyruvic acid. Thus, for two complete revolutions of the cycle, the carbon skeleton of glucose is entirely released as carbon dioxide in these decarboxylations.

The remaining available energy of the glucose molecule withheld in two molecules of pyruvic acid is liberated by processes of oxidative phosphorylation which accompany the passage of H^+ and electrons along hydrogen and electron carrier systems involving the pyridine nucleotides, flavoproteins, cytochromes and cytochrome oxidase to molecular oxygen.

Inorganic phosphate is trapped and stored in the pyrophosphate bonds of ATP by the energy liberated at the successive steps in the transfer of H^+ and electrons to oxygen where they react to form water. Similarly reduced DPN arising from triosephosphate dehydrogenation in Phase II (glycolysis) would also release its energy by oxidative phosphorylations of this type. It is estimated that per mole of hexose diphosphate between 30 and 40 pyrophosphate bonds are synthesized by the steps of oxidative phosphorylation in which oxygen is an essential factor as compared to the four that are produced by substrate phosphorylations in anaerobic glycolysis.

The primary function of respiration is to release chemical energy, but this would be to no avail if there were no energy conserving systems such as the adenylic acceptor system. Likewise if ATP were not utilized for synthetic reactions in growth the adenylic acceptor system would become saturated and would operate as a feedback control that would regulate the rate of respiration. Thus the highest respiration rates are found at the sites of most active synthesis.

The Pentose Phosphate Pathway

Many, if not all organisms have an alternative aerobic pathway for the oxidation of glucose and one of widespread occurrence has been variously referred to as the "hexosemonophosphate shunt," the "direct oxidation pathway" or the "pentose phosphate cycle." The starting point is glucose-6-phosphate which undergoes two successive dehydrogenations involving TPN followed by a decarboxylation of the No. 1 carbon to pentose phosphate. The subsequent interconversions involving tetrosephosphate and sedoheptulose etc. are complex and result in the formation of 3-phosphoglyceraldehyde and the regeneration of glucose-6-phosphate. The net effect of one revolution of the cycle which involves three glucose-6-phosphate molecules and in which two are regenerated is as follows:

Glucose-6-phosphate →

3-phosphoglyceraldehyde + 3 CO_2

3-Phosphoglyceraldehyde does not accumulate and may be metabolized in the glycolytic sequence (Phase II) or a more subtle channel may be the condensation of 2 molecules of 3-phosphoglyceraldehyde by a reversal of glycolysis to glucose-6-phosphate. Such a cycle could effect a complete breakdown of glucose-6-phosphate by the successive action of only two dehydrogenation steps involving TPN and one decarboxylation step.

In some organisms the pentose phosphate pathway may represent a major channel of glucose breakdown and energy release to ATP by oxidative phosphorylations accompanying the transfer of H^+ and electrons from reduced TPN to the flavoprotein cytochrome-oxygen carrier systems. However, it is more probable that this pathway represents a primary source of pentose-phosphate for photosynthesis in green plants and the synthesis of nucleic acids in organisms generally.

Anaerobic Cellular Respiration

Anaerobic respiration is synonymous with FERMENTATION, the latter is often the preferred term and connotes in Pasteur's original definition "Life without oxygen," or more specifically the degradation of carbohydrate into two or more simpler molecules by processes not requiring molecular oxygen. In fermentations the carbon skeleton of glucose is never completely released as carbon dioxide and in some it may not appear at all. The term aerobic fermentation may be used to designate fermentations occurring in aerobic atmospheres along with a component of aerobic respiration.

The classical fermentation is the alcoholic fermentation of glucose by yeasts according to the equation:

$$C_6H_{12}O_6 \rightarrow 2CO_2 + 2C_2H_5OH$$

$$\Delta F = -56.1 \text{ kg. cal/mole.}$$

Since one of the products of this fermentation cannot escape from the cells it eventually becomes toxic and yeasts vary widely in their capacity to withstand the accumulation of alcohol. Thus Baker's yeast which has been selected for the rapid production of carbon dioxide can only withstand a concentration of 3–4% alcohol. Brewer's yeasts ferment glucose more slowly and can produce as much as 10% alcohol and wine yeasts can withstand concentrations of 14–16% alcohol. When yeasts are transferred to an aerobic environment they continue to ferment (aerobic fermentation), but also carry out a small aerobic respiration.

Certain body tissues e.g. muscle and liver when respiring may support components of both aerobic respiration and a lactic acid fermentation according to the equation:

$$C_6H_{12}O_6 \rightarrow 2C_3H_6O_3 \ \Delta F = -47.4 \text{ kg. cal/mole.}$$

The relative contribution of each process to the total respiration is determined by the supply of oxygen to the muscle tissue. Thus lactic acid fermentation is augmented by anoxia brought about by heavy exercise and the accumulation of lactic acid can cause stiffening of the muscles.

In the tissues and organs of green plants, aerobic respiration is the normal process, although some organs, e.g. apple fruits etc., can endure long exposure (100 hours or more) of anaerobic conditions without harmful effects. In such cases fermentations take place exclusively. In between a completely anaerobic environment and one that contains a critical volume of oxygen (ca. 5%) components of both aerobic respiration and aerobic fermentation contribute to the overall process. This critical oxygen tension is known as the "extinction point" of aerobic fermentation, above which aerobic respiration is exclusively dominant. When fermentations occur in plants the products are generally alcohol and carbon dioxide as in yeast fermentation e.g. apples, carrot roots, grapes, pea seedlings, but in potato tubers and other tissues, lactic acid and carbon dioxide are the main products.

Many microorganisms ferment carbohydrates to a variety of products including formic, acetic, propionic, butyric and succinic acids etc.

Both alcoholic and lactic acid fermentations share a common pathway with aerobic respiration leading to the production of pyruvate in glycolysis (Phase I and II). The primary function of these fermentative processes is to release energy in the form of ~ P which is trapped and stored in the pyrophosphate bonds of ATP for subsequent use in synthetic growth processes. Pasteur recognised that they could only be substitute energy yielding processes in normally aerobic organisms, but a comparison of the free energy changes (ΔF) indicates that they would have to utilize about 12 times as much glucose per unit time in order to yeild the same amount of energy as aerobic respiration and this is never achieved in nature.

In many normally aerobic animal and plant tissues the actual rate of carbohydrate destruction increases on exposure to anaerobic conditions, but this rarely exceeds a two to four fold increase. Conversely on transfer from anaerobic to aerobic conditions the rate of carbohydrate utilization decreases. It is quite obvious why aerobic respiratory processes do not occur in the absence of oxygen, but it is not so clear why fermentations are extinguished in an oxygen atmosphere above a critical oxygen tension (extinction point). Oxygen, therefore, plays a positive role in suppressing or decreasing fermentation and the products of anaerobic metabolism. Much more subtle to the organism is the action of oxygen in diminishing carbohydrate destruction, in other words oxygen conserves carbohydrates and prevents the inefficient drain on foodstuffs that would occur if fermentations operated independently and unlimited. This is known as the PASTEUR effect which is considered to be operating when it can be shown that the rate of carbohydrate utilisation under anaerobic conditions exceeds that under aerobic conditions at oxygen tensions above the extinction point of anaerobic respiration. The mechanism of the "Pasteur effect" and the role of oxygen has been the subject of considerable investigation and much controversy.

The lower efficiency of alcoholic and lactic acid fermentations as compared to aerobic respiration can be readily explained by the fact that in the absence of oxygen the processes of oxidative phosphorylation which release the major part of the energy of the glucose molecule and which accompany the transfer of H^+ and electrons to molecular oxygen are inoperative in the absence of oxygen. Since these processes are irrevocably linked with the Krebs tricarboxylic acid cycle which is responsible for the aerobic degradation of pyruvate, this substance is forced into anaerobic channels of metabolism where it is reduced directly or indirectly by reduced DPN produced in the oxidation of 3-phosphoglyceraldehyde in glycolysis (Phase II).

In alcoholic fermentation pyruvate is decarboxylated to acetaldehyde and carbon dioxide by a powerful carboxylase involving a coenzyme (cocarboxylase; thiamine pyrophosphate; Vitamin B group). In turn, acetaldehyde, in the presence of an alcohol dehydrogenase, is reduced to ethanol by $DPNH_2$ arising from the oxidation of 3-phosphoglyceraldehyde in Phase II. The oxidized DPN is regenerated which can participate in the oxidation of another molecule of 3-phosphoglyceraldehyde and the process is cyclical until the substrate is exhausted or the ethanol becomes toxic.

$$CH_3CO \cdot COOH \rightarrow CH_3CHO + CO_2$$

$$CH_3CHO + DPNH_2 \rightleftharpoons C_2H_5OH + DPN.$$

In muscle tissue in which lactic acid fermentation is predominant, carboxylase is lacking and pyruvate is reduced to lactic acid directly by $DPNH_2$ in the presence of a lactic dehydrogenase, without the evolution of CO_2,

$$CH_3CO \cdot COOH + DPNH_2 \rightleftharpoons$$

$$CH_3CHOH \cdot COOH + DPN.$$

External Respiration

The maintenance of respiration within cells demands an external system to facilitate gaseous exchange. In microorganisms aqueous diffusion is sufficient, but in multicellular organisms, more elaborate mechanisms are required to maintain a suitable gaseous environment at the cellular level. In land plants this takes the form of gaseous diffusion paths in the intercellular spaces, controlled at the epidermis, of LEAVES, for example, by STOMATA. Insects also rely upon diffusion and air sacs, but in higher animals where much of the actively respiring tissue is far from the surface, a more complex two stage system is found which involves the mechanical pumping (breathing) of air or water (in fishes) into organs that effect the attainment of gaseous equilibria (lungs or gills) and the transport of respiratory gases in a circulatory system.

External respiration involves expiration (breathing out) of carbon dioxide and inspiration (breathing in) oxygen of the air or water transported to and from the lungs or gills by hemoglobin of the venous and arterial blood. Breathing is involuntary, but may be controlled voluntarily. The rate of involuntary breathing is adjusted by nerve centres of the brain so that 20 liters of fresh air are inspired for every liter of carbon dioxide expired. Normally a man at rest breathes 7 liters of air per minute. When oxygen is in short supply the need for oxygen takes precedence over the elimination of carbon dioxide. The most important organ of respiration in higher vertebrates is the lung where the exchange of gases occurs. The LUNG has no muscle tissue, but during inspiration the diaphragm contracts and flattens extending the chest cavity downwards and the outer muscles of the ribs etc., raise the ribs and increase the diameter of the chest. This enlargement decreases the lung air pressure resulting in a flow of air through the mouth, trachea, bronchus and bronchioles. Expiration is a passive movement in which the reverse process takes place. The bronchioles end in thin walled sacs (alveoli) which allow oxygen and carbon dioxide to diffuse through. It is here that the exchange of gases occurs. The ability of blood to transport large volumes of oxygen and carbon dioxide depends on the special properties of hemoglobin which is found inside the red blood cell membrane. Hemoglobin containing iron in the ferrous state (Fe^{++}) is reversibly oxygenated without valence change by oxygen in the lungs at a partial pressure of 100 mm Hg. The passage of this arterial blood to tissues having partial pressure of oxygen about or less than 40 mm Hg results in a dissociation of the loose complex and at the same time about as much carbon dioxide is gained by the hemoglobin. Whereas the oxygen pressure falls about 60 mm Hg there is a gain of only 6 mm Hg due to carbon dioxide association and this complex is now transported back by the venous blood to the alveoli where it gives up its carbon dioxide and is replenished by oxygen to produce arterial blood. The transport of oxygen and carbon dioxide by blood affords the body tissue with ample oxygen to carry out the processes of aerobic cellular respiration with the removal of carbon dioxide. However, under conditions of anoxia the fermentative mechanisms, of muscle tissue for example, provide a substitute source of energy.

E. R. WAYGOOD

References

Spector, W. S. ed., "Handbook of Biological Data," Philadelphia, Saunders, 1956.

Krebs, H. A. and H. L. Kornberg, "Energy Transformations in Living Matter," New York, Springer, 1964.

RHAMNALES

An order of DICOTYLEDONOUS seed plants, consisting of the two families Rhamnaceae and Vitaceae. The most noteworthy characteristic is the single whorl of stamens, borne opposite the petals. A disk surrounds the ovary. All members of the order are woody plants, being trees, shrubs, or lianas. In the Rhamnaceae, the fruit is a caspule, samara, or drupe, while in the Vitaceae, the fruit is a berry.

The Rhamnaceae, or Buckthorn Family, includes 45 genera, with about 550 species, widely distributed throughout the world. The flowers are inconspicuous, perfect or imperfect, regular, sometimes apetalous. The flower parts are in fours or fives. Several species are of economic value, as a source of drugs, edible fruits, timber, or for ornamental planting. *Paliurus* ranges from the Mediterranean region to eastern Asia; *P. spina-Christi*, Christ's thorn, bears stipular spines. *Zizyphus* includes 60 species of warm regions of both hemispheres; *Z. jujuba*, jujubes, is cultivated for its edible fruits. *Rhamnus*, with about 100 species, is mostly restricted to the north temperate zone; several species yield medicinal substances. *R. cathartica*, purging buckthorn, of Eurasia, has purgative berries; *R. frangula*, alder buckthorn, of Europe, produces purgative frangula bark, while *R. purshiana*, of western North America, is cultivated for the bark, which yields cascara sagrada. *Ceanothus* includes 50 species of North America, particularly common in the western part, and especially in the California chaparral; *C. americanus*, of eastern North America, is New Jersey Tea. *Krugiodendron ferreum*, leadwood or black ironwood, of Florida and the West Indies, produces one of the heaviest of timbers, with a specific gravity of 1.3.

The Vitaceae, or Grape Family, includes mostly climbing shrubs (lianas). The flowers are cymose, small, regular, perfect or imperfect, with the flower parts in fours or fives. The petals are often united at the tips and fall off together, like a little hood. The family includes 11 genera, with 600 species, widely distributed in tropical regions, but extending into north temperate regions. The stems often have terminal growing points that develop into what appear to be lateral tendrils, as they are subordinated by the more rapid growth of the axillary branch from the next lower leaf axil. Thus the tendrils appear opposite the leaves. Economically the family is of great importance, being cultivated for the fruits, which are used as food and for making wine. *Vitis* is the principal genus, with 60 species, mostly in the northern hemisphere. *V. vinifera*, the European grape, is one of the oldest and most valuable of cultivated fruits, and one of the chief sources of wine; the dried fruits are raisins; currants (named from Corinth, an ancient production center) are the dried fruits of a Greek variety. Several American species also have commercial value, particularly *V. labrusca*, of the eastern United States, the source of the Concord and other popular horticultural varieties. *V. rotundifolia*, muscadine, is the source of the scuppernong grape. *Parthenocissus*, with 15 species, occurs in temperate Asia and North America. *P. tricuspidata*, of China and Japan, is an ornamental wall vine, known as Boston ivy. *P. quinquefolia*, of eastern North America, is Virginia creeper, also sometimes grown as an ornamental. *Ampelopsis* includes 20 species of Asia and North America; *A. arborea*, pepper-vine, is native to the southern United States. The genus *Leea*, here included in the Vitaceae, is sometimes separated as the only genus of the family Leeaceae.

EARL L. CORE

Reference

Suessenguth, V., "Rhamnaceae, Vitaceae, Leeaceae" *in* Engler, A. and H. Prantl, eds., "Die natürlichen Pflanzenfamilien" 2nd ed., vol. 20., Leipsig, Engelmann, 1953.

RHIZOME

Rhizomes are elongated plant stems which grow more or less horizontally below the surface of the ground. Sometimes they are only partially buried as, for example, in certain species of *Iris*. They generally bear roots at the nodes but differ from roots in their internal structure, in the absence of a root-cap and in the possession of scale leaves. Rhizomes are produced by many herbaceous flowering plants and pteridophytes. In plants where the aerial parts die in winter, rhizomes, like certain other underground organs, can serve as a means of perennation. Often they contain large amounts of stored food. As they elongate, subterranean rhizomes push their way through the soil but the delicate terminal growing-point is usually protected from injury by a hard, pointed scale leaf which is folded over the apex.

In spring, aerial leaves may develop in place of scale leaves, and aerial shoots which give rise to foliage and later to flowers and seeds may arise from the terminal bud. Axillary buds can give rise to underground branches of the rhizome. Rhizomes vary in length from a few cm. up to 100 cm. or more. By traversing the soil in various directions in a horizontal plane they enable the plant to spread rapidly. Later, the aerial portions become separated from one another by the death and disintegration of the older regions of the rhizome. Some grasses develop an extensive system of underground rhizomes and the network of rhizomes and roots formed by certain species plays an important part in the building up of sand dunes and in aiding the consolidation of soft mud. The eradication of weeds possessing rhizomes is often difficult since portions of the rhizome which becomes broken off and left in the soil may give rise to new plants.

The behavior of underground rhizomes raises some interesting problems. Their horizontal growth is undoubtedly a response to a gravitational stimulus, but whereas positive and negative geotropic responses can be attributed to an accumulation on the lower side of an organ of a hormone which promotes or inhibits the rate of growth, a more complex system is required to explain a diageotropic or a plagiotropic response. This is especially the case where pieces of rhizome which have been detached from the plant and have no visible structural dorsiventrality are found to curve downward in darkness when placed so as to point obliquely upward, and to curve upwards when pointed obliquely downward.

By means of time-lapse photography using infrared radiation Bennet-Clark and Ball (1951) recorded the elongation and curvature of portions of the rhizome of *Aegopodium podagraria* while growing in darkness in a humid atmosphere. These rhizomes are extremely sensitive to light. Even 30 secs. exposure to a dim red light causes them to turn down. This is not a phototropic response but a light-induced change in their reaction to gravity. After about 6 hours a reversal of this curvature commences and by the end of about 24 hours in darkness the tips have returned to an approximately horizontal position. This responsiveness to light accounts for the downward bending of rhizomes which may happen to come near the surface when growing horizontally through uneven ground.

When the position of a horizontal rhizome was altered either by turning it through 180° so as to place it upside down, or by displacing it through any smaller angle, it was found that the side which was originally lowermost elongated less rapidly than the opposite one and therefore became concave. After 1–2 hours this side began to grow faster and straightening took place, followed usually by bending in the opposite direction. Additional up and down movements with decreasing amplitude often occurred, but ultimately the tip of the rhizome returned to a more or less horizontal position. While the details are still far from clear it seems probable that the final orientation in a vertical plane of rhizomes growing in darkness below the surface of the soil results from the establishment of an equilibrium between opposing hormones; either two growth accelerators on opposite sides, or a growth accelerator and a growth inhibitor on the same side. It may be postulated that any change of position leads to a disturbance of this equilibrium, perhaps by the displacement of these hormones at unequal rates. Up and down curvatures then occur with decreasing amplitude as the equilibrium becomes re-established. Immersion of similar rhizomes in solutions of the growth-promoting substances indole-3-acetic acid or 2, 4-dichlorophenoxyacetic acid, in addition to increasing their rate of elongation, causes them to curve upwards (Ball, 1953). This suggests that these substances become more concentrated on the lower side and thus disturb a pre-existing equilibrium.

While diageotropism, or plagiotropism, usually determines the angle to the vertical at which the growth of rhizomes takes place, some additional factor or factors must regulate their depth below the surface of the soil. As mentioned above, rhizomes which approach near enough to the surface to become illuminated will turn down, but something else is required to prevent them from going too deep. About 50 years ago Raunkiaer suggested that there was some kind of vital activity which enabled the plant to measure the distance between the rhizome and the part of the aerial shoot that was illuminated. A more plausible explanation can be based on an experiment of Bennet-Clark and Ball (1951) who showed that 5 per cent CO_2 modified the diageotropic response of the rhizome of *Aegopodium podagraria* so that it was induced to curve upwards. An increasing concentration of CO_2 therefore probably prevents a rhizome from going far below the surface, as might otherwise happen if it were growing hori-zontally into a sloping bank of earth. Finally, it is worth nothing that the responses to change of position of the diageotropic aerial branches of *Asparagus plumosus*, investigated by Rawitscher (1930), show a good deal of similarity to those of the underground rhizomes of *Aegopodium*.

NIGEL G. BALL

References

Bennet-Clark, T. A. and N. G. Ball, J. Exp. Bot. **2**: 169, 1951.
Ball, N. G., J. Exp. Bot. **4**: 349, 1953.
Rawitscher, F., Zeit. f. Bot. **23**: 537, 1930.

RHIZOPODA

The Rhizopoda are a group of PROTOZOA which, in the narrowest sense, contain three orders: Amoeba, Testacea, and Foraminifera. Common to all is the capacity to form from the protoplasm, processes which are designated as pseudopods, and which serve for locomotion and absorption of food. While both orders occur in fresh water, the FORAMENIFERA are limited to the sea and for that reason are treated separately.

In all species the surface of the body is naked. In the simplest case (*Amoeba*) the animal consists of a mass of protoplasm, which is surrounded by a plasmatic covering (*pellicle*). The more developed order of Testacea is distinguished by the formation of shells, which in the various species exhibit a high degree of differentiation. All imaginable forms of transition will be observed: from bowl-shaped hemispheres of pseudochitin in which the protoplasmic body is fastened to the groove by extensions of the cytoplasm, up to highly complicated shells which almost completely cover the plasma body and allow only one or two openings to be left. From this the pseudopods emerge.

In a few species the shell is of pure organic origin and therefore hyaline (*Hyalosphenia*). In other cases embedded in this covering are endogenous pieces of silicic acid or limestone, which are formed differently being elliptical in the case of *Nebela*, quadrangular in the case of *Quadrula* or rod-like in the case of *Lesquereusia*. Denticulate or spined sections are also frequently encountered (*Euglypha*).

Species which attach to their shells foreign material (sand grains, quartz, diatomic shells) are common. The fastening of this stratified foreign material is with the aid of a deposited cement.

The shells are rigid and growth has not been observed. On the other hand shell molting is known to occur with subsequent production of a new shell.

The development of the pseudostomate is of taxonomic significance. The mouth can be circular, oval, denticulated or lobed, or even slit-like.

The cell body itself consists of a lump of protoplasm, which usually shows an outer thinner layer of viscous consistency (*ectoplasm*), while the inner watery part is granular and contains many inclusions (*endoplasm*). Both phases are different colloidal states but a firm boundary does not exist between them.

Among cytoplasmic inclusions are the contractile vacuoles and food vacuoles. The former regulate the os-

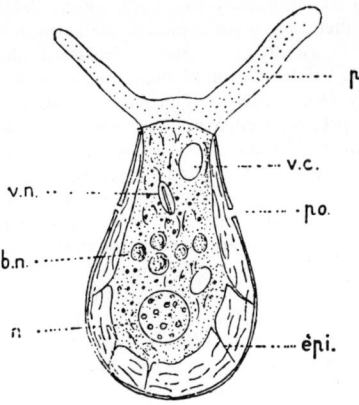

Fig. 1 Schematic survey concerning the organization of a Testacea. (Penard.) n, nucleus; vc, contractile vacuole; p, pseudopods; po, pores; vn, food vacuole.

motic pressure in the interior of the cell. Their number is variable. In the case of the amoeba the contractile vacuoles occur in different parts of the body. In the case of the Testacea, they are mainly found in the boundary zone between the plasmatic body and pseudopods and in the vicinity of the nucleus.

The food particles in the food vacuoles become surrounded by the pseudopods and thus absorbed and digested. The indigestible residue is transported to the surface and eliminated there. Moreover various metabolic products are found in the plasma, for unknown reasons, probably connected with reserve material or non-utilizable material.

Many Rhizopoda contain apparently symbiotic Cyanophycea or Zoochlorella, but in other species they are apparently only occasionally symbiotic.

The diagnostic character of the Rhizopods is the root-like processes of the plasma (*pseudopods*), whose form and development differs in a taxonomically significant manner. Three types may be distinguished:

1. *Lobopodia* of broad, lobed or digital form, with rounded ends. The proportion of ectoplasm and endoplasm changes according to size.

2. *Filopodia* of slender, filamentous form with clearly cuspidate ends. They consist only of ectoplasm and are consequently very transparent. Their number is usually large, occasionally they are arranged in clusters. Their motion is much faster than that of the Lobopodia.

3. Reticular pseudopods are a transition between these types and are digital, but exhibit at the ends distinct tips.

4. A broader character in the case of Radiolarians and Foraminiferans is the loosely reticulated pseudopods, also called *rhizopods*, which also appear in a few Rhizopods. Their form is linear, but they ramify and anastomose easily, forming net-like complexes. The most important distinction is the presence of a cytoplasmic spindle, which lays down a liquid margin, in which a distinct flowing cell follows.

The different formations of pseudopods depend also on a divergent kind of locomotion. The typical flowing "amoeboid" movement is characteristic of naked amoebae. In this case the large body is able to flow in a forward direction or it will form principal pseudopods, which always form more. In this case, forms with thicker pellicles will develop from the flowing typical of a rolling locomotion. While for the most part, the plasmatic body lies on the substratum, single species exhibit a striding movement, by which the pseudopods are alternately attaching firmly and then contracted or lengthened.

All species equipped with filopods attach the ends of expanded pseudopods to the substrate and then shorten them, thus applying traction to the body.

The rate of locomotion is a function of the temperature and is only known in the case of a few species. It fluctuates between 0.5 and 3 ( )/sec.

Free living Rhizopods are holozoic, and can therefore absorb and digest large organisms or particles. The indigestible remainder is subsequently excreted. The absorption of food can occur in different ways: through flowing (around), through import, through circulation or through invagination. In the case of species with rhizopods, the digestion is external to the plasmatic body, since the powerful anastomozing pseudopods encircle the food.

By far the most frequent kind of asexual reproduction is binary fission in which, after a preliminary nuclear division, the plasmatic body divides. In the case of the Thekamoebae, development is by means of daughter capsules; in this case previously collected reserve nutrients are laid down in forms which are typical for the individual species. In this method, nuclear division proceeds according to the structure of daughter capsules. Concerning sexual reproduction, there are only a few contradictory observations.

As in other Protozoa, many Rhizopods exhibit encystinent, in which the activity of the contractile vacu-

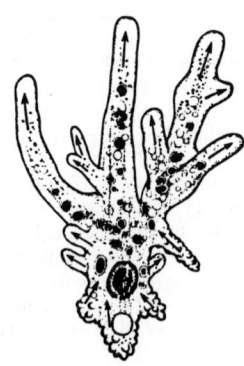

Fig. 2 Locomotion of an amoeba with lobopodia. (From Kühn.)

g

Fig. 3 Striding locomotion. (From Grell.)

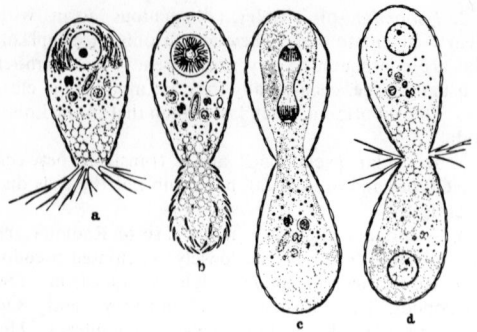

Fig. 4 Division of *Euglypha alveolata* in various phases. (According to Schewiakoff from Grell.)

oles concentrates the plasma to a viscous state and thus permits the individual to survive drought.

There are parasitic species especially in the order Amoebae. Numerous species of the genera Entamoeba, Endamoeba, and Malpighiella are parasitic in the intestinal canal of different animals and of men. While many are harmless, other species are known pathogens (i.e. *Entamoeba hystolytica,* the cause of amoebic dysentery).

TH. GROSPIETSCH

Reference

Hyman, L., "The Invertebrates. Vol. 2. Protozoa through Ctenophora," New York, McGraw-Hill, 1940.

RHODOPHYCEAE

The red ALGAE owe their name to the presence in their cells of a water-soluble proteinaceous red PIGMENT, r-phycoerythrin. Most species also contain a water-soluble blue phycobilin, r-phycocyanin, which is formed more readily than the red pigment in regions of intense illumination so that most fresh-water forms and marine species of the upper intertidal zone lack the reddish hues characteristic of the group. Wherever the blue rays of the visible spectrum penetrate marine waters the r-phycoerythrin is an efficient photosynthetic pigment and typical red forms thrive. Along with the two pigments already mentioned red algae ordinarily possess chlorophyll a, chlorophyll d, alpha and beta carotene and the xanthophyll lutein. The chief food reserve is floridean starch which occurs as cytoplasmic granules and gives a red-violet rather than a blue-violet reaction to the iodine-starch test.

About 3000 species of red algae are recognized at the present time; only 200 of these are found in fresh-water and they are usually restricted to clear, well aerated streams. The phylum is represented in all marine communities, but most species occur in the sublittoral and intertidal zones of tropical and subtropical seas. In those regions red algae have been dredged from a depth of 500 feet.

The plants vary in size from single cells (*Porphy-*

ridium) to large leafy thalli (*Porphyra, Iridophycus*) several feet long; none approach the kelps in size. The thallus is usually a laterally compacted filamentous structure. A single axial filament may give off filaments laterally (*Rhodochorton*), or on all sides (*Lemanea*), and a pseudoparenchymatous structure may result (*Gelidium*). The multiaxial thallus may have distinct filaments arranged around a central hollow space (*Cumagloia*), or be closely compacted (*Furcellaria*); in both cases lateral branches of limited growth are apt to produce a pseudo-parenchymatous cortex. While the monaxial thallus is most common, neither type is uniquely associated with advanced reproductive structures. The thalli of coralline red algae are heavily calcified as are certain other genera of warm seas and these plants have been shown to be of greater importance than the corals themselves in the building of coral reefs.

The cell wall of the red algae is ordinarily differentiated into an outer pectic and an inner cellulosic portion. Cells of most species have a large central vacuole with a thin layer of cytoplasm next to the cellulose. Strands of cytoplasm appear to connect sister cells through pore-like openings (pit-connections) in the largest subclass of the group. In the cells of primitive species (*Porphyra*) there is a single stellate plastid at the center of which lies a dense proteinaceous body, the pyrenoid. These bodies lack the encircling sheath of starch grains, typical of the pyrenoids of green algae. Pyrenoids do not occur in the many bandshaped or discoid plastids found in the cells of more advanced species. Vegetative cells of lower forms are uninucleate but upwards of 4000 nuclei have been counted in the vegetative coenocytic structures common in higher species. Although the nuclei themselves are very small, they possess distinct membranes and unusually large nucleoli. It is especially noteworthy that distinct centrosomes are present during mitosis in most red algae, but the spindle appears to be intranuclear in origin where it has been studied.

Reproductive structures of the Rhodophycophyta are so different from those of other algae that special terms are used for them. Sexual reproduction is always oögamous with quite distinctive features. The male sex organ is a *spermatangium (antheridium)* and produces a single colorless, nonflagellated, uninucleate male gamete, the *spermatium.* The spermatium is passively borne to the uninucleate female sex organ, the *carpogonium* which usually has a distal receptive process, the *trichogyne,* against which the spermatium lodges and into which the male nucleus enters and migrates to the carpogonial base. There it fuses with the single egg nucleus to form the zygote. Either directly or indirectly nonflagellated *carpospores* are produced from the zygote.

Vegetative reproduction by thallus fragmentation is infrequent. Asexual reproduction is accomplished by one or more kinds of nonflagellated spores which are usually capable of amoeboid movement. These include *neutral spores, monospores, bispores, tetraspores, polyspores,* and *paraspores.* The exact role which some of these asexual structures play in life-cycles is not yet clear.

All of the red algae are placed in the single class, *Rhodophyceae,* which is divided into two subclasses, *Bangiophycidae* and *Floridiophycidae.* The more primitive and smaller (15 genera, 60 species) group, the Bangiophycidae, is characterized by (a) a diffuse type of growth, (b) no pit-connections, and (c) the direct division of the zygote into carpospores. *Bangia* is represented by both freshwater and marine species and the latter are fairly common in the upper intertidal zone on both coasts of North America. The black to purplish uniseriate and multiseriate filaments grow to be several inches long and form smooth plaques on rocks and pilings. *Bangia fuscopurpurea* has recently been shown to have a juvenile *Conchocelis*-phase, similar to that in *Porphyra*, and is reminiscent of the *Chantransia*-phase of higher forms.

In the Floridiophycidae mitosis is limited to apical or marginal cells; pit connections are well developed; and the carpospores are always formed on *gonimoblast filaments*. These filaments grow directly from the base of the carpogonium in primitive species, but in higher groups fertilization is followed by the growth of oöblast filaments from the carpogonium to another cell of the plant (auxiliary cell). After the zygote nucleus has migrated to it, the auxiliary cell produces gonimoblast filaments. Whether directly or indirectly developing from the carpogonium the mass of gonimoblast filaments and the carposporangia they bear constitute the carposporophyte (*cystocarp*), an asexual generation which always grows parasitically on the female gametophyte. When the carposporophyte is surrounded by protective gametophytic tissue the whole structure is known as a gonimocarp.

In lower forms the zygote divides meiotically, but reduction division is delayed in more specialized species. Liberated, unreduced carpospores germinate and develop into free-living diploid tetrasporophytes. The latter plants are usually the same size and have the same vegetative structure as the gametophytes of the same species. After reduction division produces the tetraspores in the tetrasporangium they are liberated and develop into gametophytes to complete the life-cycle.

The structure and development of the carposporophyte along with the presence or absence of a tetrasporophyte constitute the natural basis for the classification of the Floridiophycidae into six orders. The *Nemalionales* (8 families; 35 genera) have no tetrasporophyte and haploid carpospores are generally produced after the direct development of gonimoblast filaments from the carpogonium. The *Gelidiales* (1 family; 6 genera), like all the remaining orders, have a tetrasporophytic generation but lack auxiliary cells. *Cryptonemiales* (14 families; 85 genera) have auxiliary cells borne on special filaments which may be adjacent to or remote from the supporting cell of a carpogonial filament. *Gigartinales* (21 families; 65 genera) form the only order in which the auxiliary cell is an intercalary cell of the gametophyte. The *Rhodymeniales* (2 families; 25 genera) have an auxiliary cell which is the outer cell of a two-cell filament that develops from the supporting cell of a carpogonial filament. The auxiliary cell is cut off *before* fertilization in this order whereas in the *Ceramiales* (4 families; 160 genera) the auxiliary cells are formed directly from the supporting cell of a carpogonial filament *after* fertilization.

JOHN H. MULLAHY, S.J.

References

Drew, K. M., "The Rhodophyta," *in* Smith, G. ed., "Manual of Phycology," New York, Ronald, 1951.

Fritsch, F. E., "The Structure and Reproduction of the Algae," Vol. 2, Cambridge, The University Press, 1952.

Kylin, Harald, "Die Gattungen Der Rhodophyceen," *in* Lund, CWK Gleerups Forlag, 1956.

Papenfuss, G. F., "Classification of the Algae" *in* "A century of progress in the natural sciences 1853–1953," San Francisco, Cal. Acad. Sci., 1955.

RHOEDALES

An order of about seven flowering plant families, although only four are of broad interest. They are mostly herbaceous, the flowers of all are bisexual and the gynoecium consists of two to several united carpels. The placentation is principally parietal but this may often be a derived condition from the axile type.

The Papaveraceae or Poppy Family includes mostly herbs with milky or colored sap, estipulate, entire to variously divided leaves. The flowers are regular, bisexual and often attractive; the two to three sepals usually fall quickly, the four to twelve petals are borne in one to three whorls; the filaments of the many stamens, arranged in several whorls, are often winged and petal-like. The single pistil is composed of several carpels but is typically unilocular; numerous ovules are produced on each of the parietal placentas. The fruit is a capsule which opens by pores or valves to release many small seeds. The principal economic importance of the family is in the opium obtained from the immature capsules; in addition, several species are grown as ornamentals.

The Fumariaceae, or Bleeding-heart Family, is similar to the poppies but the sap is watery, the flowers are irregular, the stamens are only six; the pistil is bicarpellate, unilocular, and two to many-ovulate. The fruit is a capsule which dehisces by valves or it is an indehiscent nut. The only economic value of the family is in the few ornamental species cultivated.

The Caper Family, the Capparidaceae, is easily recognized by the long stalk which typically supports the gynoecium. In addition, the flowers are generally irregular and four-parted; the fruit is usually a few to many-seeded capsule which opens by valves. Several species are cultivated ornamentals and the condiment known as capers is the flower buds of a Mediterranean species.

The Mustard or Cress Family, the Cruciferae, includes almost exclusively herbs with forked or star-like hairs, when pubescence is present. The leaves are simple and without stipules but frequently much-divided. The flowers are bisexual with the perianth four-parted; both the petals and sepals are arranged in the form of a cross (hence the family name). The tetramerous perianth coupled with the six stamens of most species is adequate for recognition of the family. The pistil and fruit too are characteristic; the ovary is typically separated into two locules by a false wall and the few to many ovules are borne on its wall. The fruit opens by two valves and is short and thick or slender and elongate. The family is very important economically as the source of many food plants (cabbage and its near relatives, radishes, horseradish, mustard and water cress), weeds (the weed mustards, pepper-grass), and ornamentals (candytuft, alyssum, rockcress).

R. S. COWAN

References

Engler, A. and K. Prantl, "Die natürlichen Pflanzen-familien," vol. 3, Leipzig, Engleman, 1891.
Hutchinson, J., "The Families of Flowering Plants," 2nd ed., Oxford, Clarendon Press, 1959.

RHYNIOPHYTINA

These vascular plants are known only as fossils from the uppermost Silurian and the Devonian Periods. They are characterized by leafless stems, terminal sporangia and terete, primary xylem strands whose maturation is centrarch. *Cooksonia* from Upper Silurian of Bohemia is the earliest representative and *Rhynia, Horneophyton, Hicklingia* and *Taeniocrada* are four of the Devonian descendants. *Hedeia* and

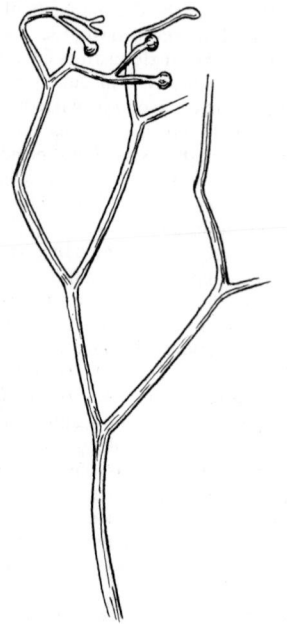

Fig. 1 *Cooksonia* cf. *hemispherica*. (Drawn from Fig. 2 of Obrhel, 1962.)

Yarravia illustrate probable stages in the aggregation of sporangia in the terminal position. They may have evolved in the direction of *Trimerophytina*, thence into ferns and gymnosperms.

H. P BANKS

Reference

Banks, H. P., "The early history of land plants," *in* Drake, E. T., ed., "Evolution and Environment," New Haven, Yale University Press, 1968.

RICKETTSIA

The generic name of rickettsia is used for a group of minute organisms possessing many features in common. The rickettsiae, about the size of small bacteria, usually appear microscopically in stained smears as pleomorphic coccobacillary organisms; they are obligate intracellular parasites, and on the biologic scale lie between the *viruses* and the *bacteria*. The genus *Rickettsia* was later established by da Rocha-Lima in honor of Dr. H. T. Ricketts who died of typhus fever during the course of his investigation of the disease. The rickettsiae are usually found as commensals in arthropods, and some are transmitted from one arthropod generation to the next through the egg. Although over forty species of them are apparently non-pathogenic for mammals, some rickettsiae are incidentally capable of causing serious disease in man. Some rickettsiae have intermediate hosts, such as rodents, that serve as reservoirs in nature. These rickettsiae are, with very few exceptions, transmitted by the bite of the arthropod vector. That is to say, these vectors are true hosts of the rickettsiae in nature.

Morphology and cultivation. The rickettsiae usually appear as small coccobacilli or diplobacilli about 0.6μ in length and 0.3μ in width, but under certain conditions they are pleomorphic in culture. They are nonmotile, and form no spores. When stained by Gram's method they are found to be gram-negative but almost invisible so they are usually stained by Giemsa's or Macchiavello's stain. Detailed structure of the rickettsial body, elucidated by ELECTRON MICROSCOPE, is almost similar to bacterial one. However, the rickettsiae can multiply only in the susceptible, living cells. Then their own enzyme systems are incomplete in comparison with bacterial ones, but not quite so restricted as those of the viruses. Their oxygen needs, pH and osmotic pressure requirements are the same as those of their host cells. The embryonated hen's eggs are a suitable culture medium. Almost all the rickettsiae pathogenic to man grow in abundance in cells which line the yolk sac. These cultured rickettsiae are available in relative abundance for antigenic studies and for the preparation of vaccines. On the other hand the mouse and guinea pig are usually employed as susceptible hosts of rickettsiae in the laboratory.

Other properties. In general the rickettsiae are easily destroyed by heat, drying and chemical agents, but under certain exceptional conditions they may survive in nature for more than one year. Especially *Coxiella burnetii* is more resistant than the other rickettsiae and can be transmitted directly from man to man. Para-aminobenzoic acid effects a marked inhibition of growth, but the broad spectrum antibiotics, such as tetracyclines and chloramphenicol, are more rickettsicidal than PABA and now provide excellent specific therapy for human infections caused by pathogenic rickettsiae. Rickettsial toxin within the bodies of the rickettsiae to which mice succumb within a few hours (to large doses of either *Rickettsia prowazekii* or *R. mooseri*), is species specific and is neutralizable by specific antiserum. However, this toxic factor is intimately associated only with living richettsiae, and has not yet been separated from them.

Serologic reactions. By the use of antigens derived

from rickettsiae cultivated in the yolk sac of the embryonated egg, specific complement fixation effects have been demonstrated in almost all rickettsiosis. On the other hand all of the rickettsiae except *C. burnetii* and *R. akari* share a common antigen with one or another of certain strains of *Proteus vulgaris.* Then the serum of patients infected with these rickettsiae may agglutinate one or another of the *Proteus* strains such as *Proteus OX₁₉, OX₂* or *OXK* (Weil-Felix Reaction). The antibodies, however, induced by these common antigens afford no protection against rickettsial infections.

Rickettsiosis. Rickettsial diseases of man are widespread and probably exist anywhere in the world where men, rodents, and arthropods live in intimate or occasional contact with each other. There is a known rickettsial infection of animals: a tick-borne fever of sheep, cattle, and goats which occurs in Africa and is known as heartwater disease. The rickettsiosis of man may be arbitrarily and broadly divided into four main groups on the basis of their clinical features, their epidemiologic aspects and their immunologic characteristics: (1) Typhus fever group—Epidemic (louse-borne) typhus (etiologic agent—*Rickettsia prowazekii*), Brill's disease (*R. prowazekii*), Murine (flea-borne) typhus (*R. mooseri*); (2) Spotted fever group—Rocky Mountain spotted fever (*R. rickettsii*), Mediterranean fever (*R. conorii*), other tick-borne typhus fevers (*R. pijperi*, etc.), Rickettsial-pox (*R. akari*); (3) Scrub typhus (Tsutsugamushi disease) (*R. orientalis*), (4) Q fever (*Coxiella burnetii*). Trench fever may be presumed to be rickettsial in etiology (*R. quintana*).

AKIYOSHI KAWAMURA, JR.

Reference

Horsfall, F. L. and I. Tamm, "Viral and Rickettsial Infections of Man," 4th ed., Philadelphia, Lippencott, 1965.

RODENTIA

Rodents are placental MAMMALS whose unique characters are the single pair of upper and lower incisors, the masseteric musculature, and zygomatic and mandibular architecture associated with gnawing. In adapting to gnawing the masseter muscle of rodents has specialized, with concomitant modification of the skeletal parts to which it attaches. It is divided into two major portions, the lateral and internal, which in turn may be subdivided. In most rodents one or both of these divisions have had their origin translocated anteriorly. The lateral masseter extends forward onto the zygomatic process of the maxilla, which is usually broadened into a plate of considerable dimensions, and may be horizontal or tilted strongly upward. In advanced forms the muscle may spread onto the side of the rostrum. The internal masseter may retain its attachment to the horizontal arm of the zygoma, or transmit part or all of its bulk through the enlarged infraorbital aperture to the side of the rostrum. This musculature elevates the mandible and draws it obliquely forward as well. Since the mandibular con-

dyle is knob-shaped and the mandibular fossa is a longitudinal groove, the lower jaw is drawn strongly fore-and-aft in mastication. The cheek-teeth, which are in partial or complete opposition on right and left sides simultaneously, are ground to and fro on one another like millstones. The upper and lower incisors grow constantly from persistent pulp, and consist of dentine faced anteriorly with a plate of enamel. Unequal wear of these components imparts a chisel-edge to the teeth. Gnawing is accomplished by the lower incisors, the uppers serving merely as holdfasts. The upper incisors are rooted in the enlarged premaxillae. The mandibular incisors extend posteriorly at least to the level of the last molar and often into the condylar process. Canine teeth are absent in all rodents as are some or all premolars, but the molar number is 3/3 except in the Australasian subfamily Hydromyinae, which may have the molars reduced to 1/1. Two genera possess 3/3 premolars, but usually do not have all six cheekteeth in place simultaneously. Most rodents which have premolars have them numbering only 2/1 or 1/1. Approximately half of all species lack premolars entirely. The total number of teeth therefore varies from 16 to 28, with a majority having only 16 and few indeed with more than 22. Other rodent characteristics included the lack of folding in the cerebral cortex, the failure of the fibula to articulate with the calcaneum, the retention in most forms of the clavicle, and the unguiculate condition of the toes. Pentadactyly is normal, but a few forms lack a pollex, and reduction of hind toes sometimes occurs in saltatorial species.

The order is ancient, being known from the North American Palaeocene and the Eocene of North America and Europe. It is represented in post-Eocene deposits of all continents, except Australia which lacks pre-Pleistocene placental mammals. South American rodents of pre-Pleistocene age stem from a single stock apparently isolated there since the Oligocene. Europe and North America have provided the greatest variety of fossil rodents, yet withal the incompleteness of the early Tertiary record and repeated cases of evolutionary convergence, divergence, and parallelism make it impossible at present either to trace with certainty the descent of modern groups or to relate them to one another with confidence. This is especially true with supra-familial categories. At least nine major classifications have been proposed, none of which have proven universally acceptable. Rodent families are fairly well established, but opinions vary on whether certain groups should rank as families or subfamilies. Ellerman (1940) recognized 23 families, Miller and Gidley (1918) and Simpson (1945) recognized 32. There are approximately 350 genera, and between 6,000 and 7,000 named forms. They are presently the most actively speciating of all mammals.

The Ethiopian Region is currently the most richly endowed with rodents. In named forms it is matched by other zoogeographic regions, but more higher categories occur in Africa than elsewhere. Numerous families are exclusively African. The Neotropical Region rivals Africa in named forms, and in endemic families, but is inferior in higher categories; the result of evolution in isolation. The Nearctic and Palaearctic Regions each contain named forms comparable to Africa and South America, but are poorer in the higher categories, and much poorer in endemic families. The two show many close relationships, even to the sharing of numerous genera, but each manifests certain unique elements, such as the Nearctic geomyoids, the Palaearctic spalacids, etc. The Indomalayan Region is rich in named forms but depauperate in families, containing only six or seven, none of which are completely endemic. The Australasian Region contains only some 240 named forms of the single family Muridae whose range includes all major areas of the old world.

Adaptive diversity among rodents exceeds that of any other order, similar adaptations often appearing independently in distantly related families. The more extreme adaptations are the aquatic, four families; arboreal volant, two families; arboreal scansorial, seven families; fossorial, six or more families; cursorial, two families; echinate hair, six families. In at least three families hibernation has developed to a state of perfection unequalled in other orders. Body temperature, respiratory rate, and heartbeat drop to the minimum necessary to sustain life, and may persist at that rate with only intermittent fluctuation for as much as eight months.

Most rodents are relatively short-lived, have a high reproductive potential, short gestation period, rapid growth, and early maturity. Many species reproduce before becoming full-grown. Numerous litters per year are usual. Populations therefore react promptly to favorable and unfavorable environmental conditions. Cyclic fluctuations in numbers are a widespread phenomenon. When population outbreaks occur in agricultural or urban areas they are of considerable economic detriment. In all natural communities rodents play a vital part in food interactions as converters of vegetable matter into animal protein.

BRYAN P. GLASS

References

Ellerman, J. R., "The families and genera of living rodents" Vols. 1 and 2, London, Brit. Mus., 1940.
Miller, G. S. and J. W. Gidley, "Synopsis of the supergeneric groups of rodents," Jour. Wash. Acad. Sci., 8: 431–448, 1918.
Simpson, G. G., "The principles of classification and a classification of mammals," Bull. Amer. Mus. Nat. Hist. 85: 1–350, 1945.

ROOT

In the evolution of Tracheophytes, the root was apparently the last organ of the "big three" (roots, STEMS, and LEAVES) to develop. Stems in the form of RHIZOMES were the anchoring organs and rhizoids arising from the epidermal cells of these rhizomes provided the necessary absorptive function for the earliest of the vascular plants, the Psilophytales (see PSILOPSIDA), and for a few species of the Psilotales still living today. Development of the primary axis resulted in a differentiation of root and stem and both organs assumed an independent division of labor as well as conspicuous morphological differences. However, the root has always remained a comparatively simple structure. This has generally been attributed to the greater uni-

formity of the underground habitat, as contrasted with the aerial, where greater modifications in tissue and organ development have led to more highly specialized structures in stems and leaves.

Ontogenetically roots usually arise endogenously from fairly deep-seated tissues. In primary growth root tissues develop from an apical meristem which is surrounded by and protected by a cellular and often mucilaginous root cap. Early in its development the root differentiates a typically uniseriate epidermis although the air roots of some epiphytes, such as tropical orchids, develop a multiseriate epidermis termed the *velamen*. Lying beneath the epidermis is the *cortex*, a tissue that is made up of a number of layers of thin-walled storage cells. The innermost cortical layer in the roots of seed plants differentiates into an endodermis with thickened walls which usually become suberized. Roots of most plants have more or less extensive areas in which the endodermis is partially suberized and other areas where it may be so completely. The pattern of suberin formation shows a relation to the proximity of phloem which often lies within a cell or two's width of the endodermal ring. As the root matures, suberization spreads laterally to form a continuous ring of suberized endodermal cells around the vascular tissue.

The central portion of a root is comprised of thick-walled xylem cells and some associated parenchyma (Fig. 1). In dicotyledonous plants the next tissue surrounding the xylem is cambium. It is the last tissue to differentiate from the ground meristem but soon after its differentiation it begins to form new xylem from its inner half and when it again reaches full size may divide to form another inner xylem cell or an outer

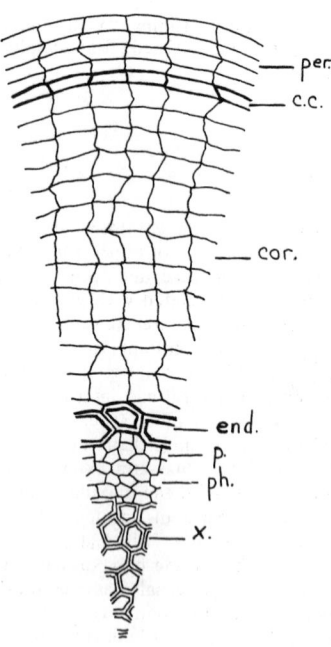

Fig. 1   A diagram of a section through a root showing: per, periderm; cc, cork cambium; cor, cortex; end, endodermis; p, pericycle; ph, phloem; x, xylem.

phloem cell. The continuous cylinder of cambium thus contributes many layers of cells to the diameter of the root. Outside of the phloem is a layer of pericycle, then a suberized endodermal layer. The endodermis is usually surrounded by many layers of cortical cells which in young roots are covered by an epidermis. Older roots often develop a cork cambium within the outer layers of the cortex which then divides as does the vascular cambium. However, the cork cambium gives rise mostly to periderm cells towards the outside of the root, ultimately forming bark as in the stem.

The significance of cambial activity, in increasing the conductive capacity of the gymnosperm and dicotyledonous root, is soon apparent with the formation of secondary xylem and phloem from this lateral meristem. Those plants with tap root systems have no other way of increasing their bulk and conductive elements. Plants such as monocotyledons, which form fibrous root systems, may have dozens of roots arising from the base of the plant. Their conductive capacity is increased by the continuous production of new roots and in some dicotyledonous plants cambial activity may also increase markedly the vascular tissues in these roots.

Branch roots in both gymnosperms and angiosperms usually arise by meristematic activity in the pericycle; those of the pteridophytes more commonly originate in the endodermis. In the higher plant, after the primary root has completed its tissue differentiation, but usually before cambial activity, several pericylic cells opposite the xylem points begin division in a tangential plane. Early in its development and before the new root has completely penetrated the tissues of the parent root four distinct layers are formed. The *calyptrogen-dermatogen* is formed from the outside layer, the second layer becomes *periblem*, and the two inside layers become *plerome*. At this stage, when the rootlet is four cells in length, the endodermis can be observed in a crushed condition. Increase in root length is mostly the result of a very active plerome which divides rapidly by both periclinal and anticlinal divisions. When the secondary root has forced its way through the cortex to the epidermis periclinal divisions occur in the single calyptrogen-dermatogen layer to initiate the appearance of a root cap. Periblem cells divide periclinally, forming two layers, and in the plerome provascular tissue increases to a width of six cells, at which stage the pericyclic layer becomes distinct. After a secondary root has increased to a length of several hundred microns there is a differentiation of xylem and phloem, each connecting with the corresponding tissue of the primary root.

There is no distinct pattern of secondary root origin although they usually arise at the points of the xylem, or at a tangent from them. Some plants have roots which branch only into the secondary division, but more commonly those of the secondary division also give rise to tertiary roots. Occasionally those of a quarternary category are formed but only rarely, and this in large woody plants, is there branching beyond the fourth division.

Environmental factors which have a demonstrable effect on the branching of roots include water, aeration, soil texture, nutrition, and competition. Whether one or the other exerts a greater stimulus to the form a root may take depends largely on the interaction of the other factors concerned. The writer has experi-

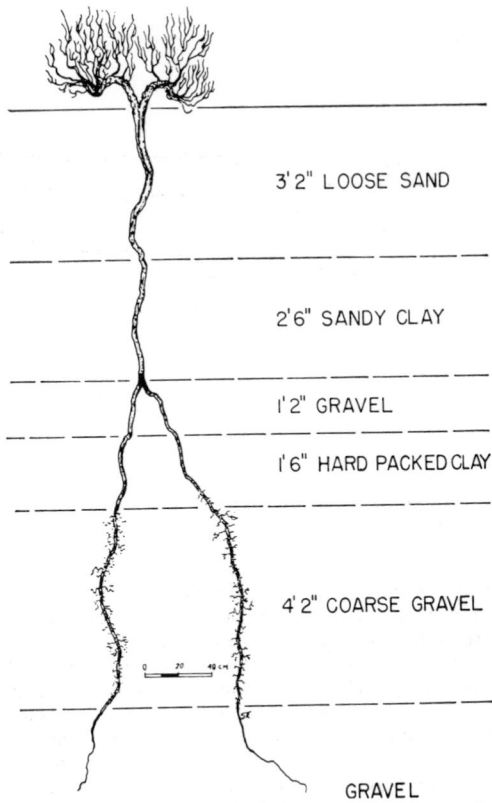

3'2" LOOSE SAND

2'6" SANDY CLAY

1'2" GRAVEL

1'6" HARD PACKED CLAY

4'2" COARSE GRAVEL

GRAVEL

Fig. 2   The deep penetrating tap root system of *Chrysothamnus nauseosus*. Note the prolific branching of fine roots in the coarse gravel stratum which carried more water than the other strata.

mentally shown that certain grass plants, when grown well apart from other plants of the same or different species, branch into the quarternary division, but when planted close together branch only into the tertiary category. Weaver and others have shown that roots of many native plants of the plains area can penetrate a hardpan layer, and often form the bulk of their root system beneath this layer, while those of cultivated plants fail to grow into and through such layers.

In general roots have a tendency to branch much more prolifically in soil strata carrying greater amounts of water than in those layers above or below which may be drier. Obviously this may be also controlled by the amount of aeration. Roots of most terrestrial plants cannot grow continuously in a complete watery medium but where the amount fluctuates and free oxygen is adequately supplied better root growth will be observed than in soils where water is at a minimum. After learning of the tremendous number of roots and root hairs produced by a winter rye plant, Kramer concluded that enough new roots were produced each day to handle the day to day water needs of the plant, even though the remainder of the root system may no longer act as a functional absorptive organ. The number and size of root hairs varies tremendously from species

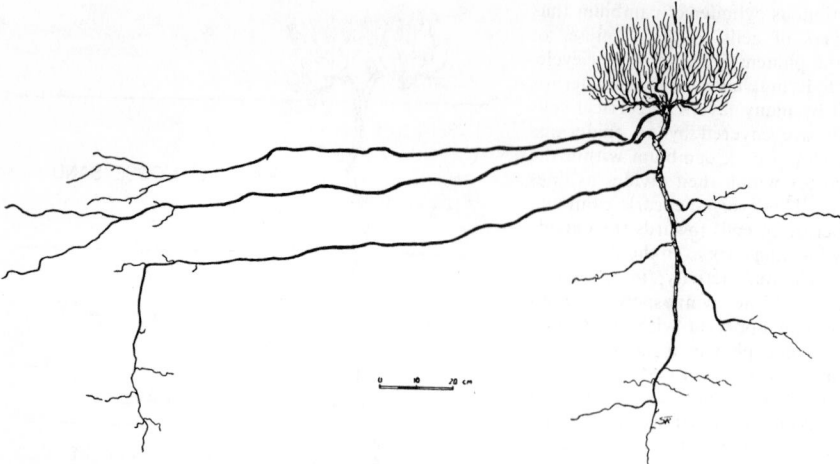

Fig. 3 *Dalea scoparia*. Although essentially a tap root system, very prominent surface lateral roots develop in this plant.

to species but to most plants the *root hair* is the most intimate organ of contact the plant has with the soil.

Growth of root hairs is by an apical or tip extension and not by a general wall elongation. This manner of growth is typical of most plant cell growth, not only in tissues of higher plants but also in lower forms such as fungal hyphae (see FUNGI), in which there is a migration of protoplasm from the older portions of the cell to the tip thereby providing food for the formation of new cell wall as the tip moves forward. Recently it was shown that the cell wall consists of cellulose and pectic substances, and that the cellulose layer is extremely thin at the tip.

Through the aid of the electron microscope Daws and Bowler have determined that the cell wall of the radish root-hair, at least, consists of two layers of microfibrils. The outer layer originates at the tip and has a reticulate pattern, but the inner microfibrils are parallel and longitudinal and arise farther back from the tip. Possibly due to the expansion of the hair tube, the outer reticulate pattern thins or spreads out and more pectic matrix is deposited between the cellulose meshwork. Minute thin areas in the cell wall, possibly developing pits, are evident. Such root hair walls are composed of an outer layer of mucilage surrounding a very delicate cuticle. Just beneath the cuticle lies an amorphous matrix which both covers and impregnates the meshes of the cellulose microfibrils described above.

The significance of root hairs in imbibing and absorbing substances from their environmental medium is greatly enhanced by the mucilaginous layer. It also makes the root parts more efficient in binding soil and enables the microflora and microfauna to make close contact with the root. Microorganisms in the soil contribute materially in making available to the plant organic and inorganic substances not readily soluble.

Root systems are generally classified as tap or fibrous; however, if a number of the roots arising from the base of the plant show considerable secondary thickening, as in dahlias or sweet potatoes, the system is termed fascicled. Roots are also classified on the basis

of their origin. Seminal roots are those which arise from the embryo in germination of the seed; adventitious roots are those which arise from part of the plant other than the seed. Usually these roots arise from the stem or leaves. In the germination of corn seed, one or more seminal roots emerge and become established in the soil. Before the corn plants become very large new roots begin arising from the base of the stem. These are of adventitious origin and before the corn plant has reached maturity will make up almost 100% of the entire root system. Although the initial roots of the corn plant were seminal and those arising later were of adventitious origin, we classify the root system of corn as fibrous because of the general nature of the numerous fine roots which make up the ultimate system as opposed to the single large root (and its branches) which would be considered a tap root system.

Undoubtedly the most important factor in root form is heredity. Although there are many individual differences which exist between roots of different species, and even between those within the same species, one who is familiar with a particular plant species root system can distinguish it from all others even though the edaphic factors under which it was grown were different. Inheritance has a powerful influence and the type of root system typical of the species will remain essentially the same under whatever conditions it may be found.

Plants are classified as xerophytes, mesophytes, or hydrophytes, depending largely on the amount of moisture available to them. Most investigators, when describing the morphological differences exhibited by plants of these diverse environments, confine their descriptions to the aerial portions. Certainly the atmospheric factors of light, temperature, and humidity are more variable than the edaphic factors, but the roots do occupy an environment that is always changing, both chemically and physically, and to which it must adapt itself in form and structure. We do not, however, find the degree of modification in roots that we do in the aerial plant parts. Apparently plants have been more successful in adapting the above ground

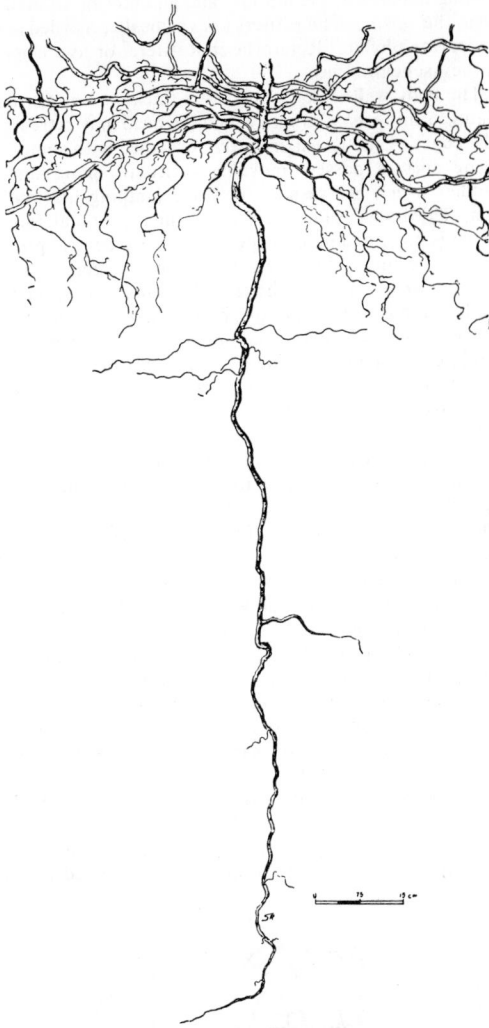

Fig. 4 *Franseria acanthicarpa*. This plant combines a profusely branched fibrous root system in the upper soil layers with a very deep penetrating tap root.

parts to control the water needs than modifications of the subterranean structures. We do not, for example, find more extensive root systems on desert plants than we do on those of moister habitats. In fact the opposite condition is more common. Roots develop more extensively in soils of greater water content and have a much larger surface area exposed in the soil for absorption, in proportion to the surface area of the tops, than those of desert plants. Investigations concerning the top-root surface ratios are continuing and should disclose some interesting facts about these plant parts in different ecological situations.

HOWARD J. DITTMER

References

Esau, K., "Plant Anatomy," New York, Wiley, 1965.
Weaver, J. E. and F. E. Clements, "Plant Ecology," 2nd ed., New York, McGraw-Hill, 1938.

ROSALES

The second largest order of flowering plants, consisting of at least 17 families of great structural diversity. About half of the families produce stipules; the flowers are usually complete, typically five-parted or the parts are in multiples of five, the stamens often are numerous and the one to several carpels are free or variously united, the styles usually distinct. Almost every known type of fruit is produced by members of the order, but achenes and follicles are perhaps the most frequent. The most important families in the order are described below.

The Stonecrop Family (Crassulaceae) consists mostly of annual or perennial herbs or shrubs which are found in the arid parts of the western United States, the Mediterranean Region, South Africa, and south-central Asia. The stems and leaves are usually succulent, simple, entire, and estipulate. The flowers characteristically have as many carpels as petals and stamens; the carpels are free from each other and a scale-like gland subtends each. The only economic importance of the family is the use of many of the species as ornamentals.

The Saxifrage Family, Saxifragaceae, are herbs, shrubs, or woody climbers; the leaves are mostly alternate and estipulate. The flowers are usually complete with as many stamens as petals or twice as many and one to five, free or united carpels composing the pistil. The fruit is usuallly a berry or a capsule and the seeds are abundantly endospermous. The family is difficult to separate from related families, especially from the Rose Family; the presence of abundant endosperm in the seeds, generally estipulate leaves, and the constantly few stamens and carpels separate most of the genera from those in the Rosaceae. A large number of ornamentals belong to this family (Philadelphus, Heuchera, Hydrangea, Deutzia, etc.), as well as small fruits (currants and gooseberries).

The Witch-hazel Family (Hamamelidaceae) is characterized by deciduous or evergreen trees and shrubs which, if pubescent, bear stellate hairs. The leaves are alternate, stipulate and toothed to lobed. The perianth consists of four to five parts or is absent, the stamens two to eight in one cycle, the ovary bicarpellate and inferior, and the fruit is a dry capsule. Besides Witch-hazel (Hamamelis) the family also includes the Sweet Gum (Liquidambar) and several ornamental shrubs.

The Sycamore or Plane-tree Family (Platanaceae) includes a single genus of trees. The trunk and branches lose patches of the older bark periodically and a mottled appearance results. The simple, palmately lobed leaves are alternate, the petiole encloses the axillary bud, and pubescence, when present, is stellate. The flowers are very small and arranged in dense, globose, pendulous heads. Most of the species are cultivated for ornament.

The Rose Family is commonly divided into six subfamilies based principally on the morphology of the gynoecium and the fruit; although all six have been considered as distinct families, most botanists do not accept such fragmentation. The family is characterized by stipulate, alternate leaves, mostly five-parted flowers (except in the gynoecium), the development of an hypanthium, and seeds almost entirely lacking endosperm. In the temperate zone the family is of considerable economic value. Several of the species produce edible fruits (apple, pear, cherry, plum, peach, black-berry, strawberry, etc.), while others are cultivated ornamentals (hawthorn, spiraea, mountain ash, fire-thorn, the roses, etc.).

The Bean or Pea Family (Legumonosae) is one of the three largest of the flowering plant families and is ex-ceedingly diverse morphologically. It is by most au-thorities subdivided into three sub-families on floral morphology: (1) Mimosa-like plants, (2) Cassia-like plants, and (3) Bean-like plants. In all three groups, the habit ranges from herbs to large trees; the leaves are usually compound, twice to thrice divided in the mimo-soids, and stipules are characteristic; the floral mor-phology is exceptionally variable and, although, by definition, the family is characterized by a legume for the fruit, actually almost all fruit types known occur in some form within the family. Economically it is one of the most important families: (1) some members pro-duce foods for Man (peas, beans, peanuts, lentils, soy beans); (2) others are valuable forage crops (alfalfa, clover, vetch); and numerous ornamentals are included in the family (redbud, poinciana, acacia, lupine, wis-teria, sweet peas, broom).

R. S. COWAN

References

Engler, A. and K. Prantl, "Die natürlichen Pflanzen-familien," vol. 3, Leipsig, Engelmann, 1888–1894.
Hutchinson, J., "The Families of Flowering Plants," 2nd ed., Oxford, Clarendon Press, 1959.

ROTIFERA

A group of minute (mostly $50–500\mu$, up to about 2500μ) aquatic animals distinguished by a complex feeding apparatus, the *mastax,* and an anterior ciliated area, the *corona.* The rotifers are variously regarded as a separate phylum, Rotatoria or Rotifera, or as a class of the Aschelminthes.

The body wall in some species is thickened into stiff plates forming a *lorica* into which the head may retreat. The edges of the lorica are often furnished with spines. Some species possess movable cuticular spines and paddles which are responsible for a jumping locomo-tion. The body usually terminates in a postanal foot which is furnished in most species with two conical pro-cesses, the toes. The foot contains cement glands the ducts of which open on the toes. In planktonic species the foot tends to be small or absent. The principle tac-tile sense organs are a dorsal antenna on the head and two lateral antennae on the posterior part of the body. In addition various setigerous papillae may exist on the foot and corona.

The corona, which in its simplest development is an elongated ciliated area around the mouth, varies greatly within the group in relation to the swimming and feed-ing habits. In some there is a bare area in the ciliated field near the anterior end. The bare area may be very large, the area posterior to the mouth reduced, and the corona then consists of a strongly ciliated groove around the edge of the head which may be thrown into large lobes. In some rotifers, the ciliated area is greatly reduced and the edge of the corona produced as a bowl-shaped structure that is edged with very fine fila-ments that form a trap for smaller animals.

A ciliated gullet leads from the mouth to the muscular pharynx (mastax). Highly characteristic of the rotifers is the development of the cuticular lining of the mastax to form specialized feeding structures, the trophi. The trophi usually consist of seven basic pieces, supple-mented in some rotifers by accessory pieces. Several major types exist, the variations being associated with

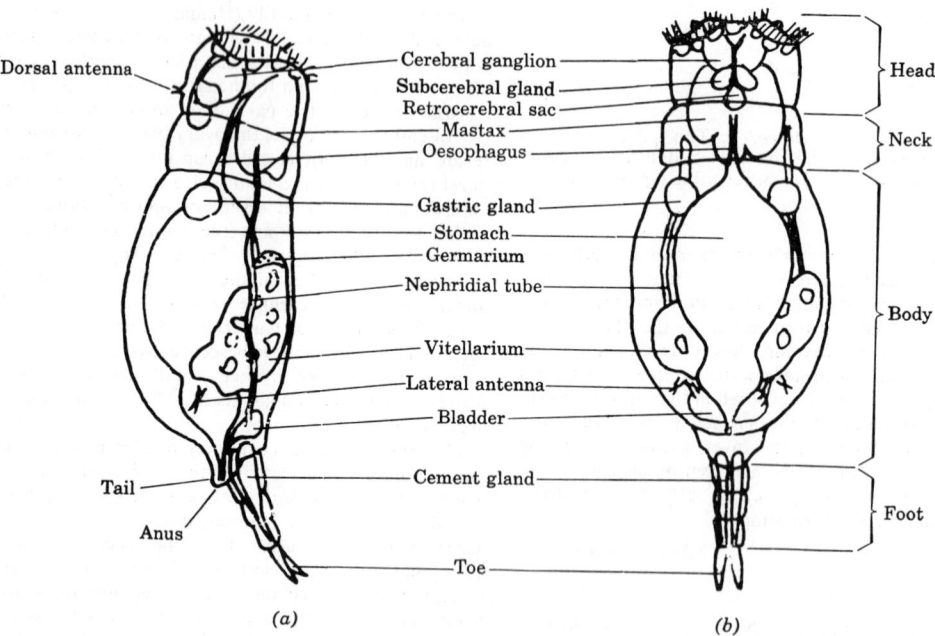

Fig. 1   *(Courtesy John Wiley & Sons.)*

differences in food and feeding methods. Some types function primarily to grind food particles, others function to pump out the contents of prey organisms, and still others may be thrust from the mouth to grasp prey.

The body musculature consists of longitudinal and transverse muscle bands and strands; there is no thick layer of muscle associated with the body wall. The nervous system consists of three major ganglia with associated nerves. Photoreceptors, when present, are closely associated with the cerebral ganglion. The excretory system has two protonephridial tubes, with flame bulbs, which empty directly into the cloaca or into a special bladder. The digestive system posterior to the mastax usually has an oesophagus, stomach furnished with gastric glands, and intestine, but in a few genera the stomach ends blindly.

The female reproductive system has one or two ovaries, each consisting of a yolk gland and a germarium in most families. Most reproduction is parthogenetic, and in one group, Bdelloidea, males are unknown. In most, the haploid males are produced only occasionally. The fertilized eggs generally have a thick shell and may withstand severe conditions. There is a high degree of sexual dimorphism.

The rotifers inhabit waters of almost all kinds, but most species are limited to fresh or slightly brackish water. Many have rather distinct ranges of tolerance for certain environmental variables and occur only in rather special types of habitats. Most species appear to be widely distributed, but some are strikingly limited.

W. T. EDMONDSON

References

Edmondson, W. T., "Rotifera" *in* Ward, H. B. and C. C. Whipple, "Fresh-water Biology," 2nd ed., New York, Wiley, 1959.
Hyman, L. H., "The Invertebrates," Vol. 3, New York, McGraw-Hill, 1951.

ROUX, PIERRE PAUL EMILE (1853–1933)

This French bacteriologist and student of Louis Pasteur was born on December 17, 1853, at Confolens, Charente. A student of medicine, he held a position on the faculty of medicine in Paris from 1874 to 1878. He then worked in Louis Pasteur's laboratory until his appointment to a post in the Pasteur Institute in 1888. He directed the Pasteur Institute from 1904 to 1918.

Roux collaborated with Pasteur in the study of hydrophobia. He also worked with Pasteur and Charles Edouard Chamberland on the production of anthrax vaccines. With the Swiss bacteriologist Alexandre Yersin, he studied the diptheria bacillus and its toxins. In conjunction with Emil von Behring, he introduced an antitoxin for both prevention and treatment of diptheria. The value of the serum was shown clearly in 1894, when the death rate in Paris was cut in half. For this work, Roux and Behring were awarded the Nobel Prize in Medicine in 1901.

Roux died in Paris on November 3, 1933.

JAMES L. OSCHMAN

RUBIALES

One of the higher orders of DICOTYLEDONOUS flowering plants, comprising the families Rubiaceae, Caprifoliaceae, Valerianaceae, and Dipsacaceae. This group forms a major component of many tropical floras, is much less important in most temperate regions, and has a rather minor status in high latitudes. The Rubiaceae, largely tropical, are mostly shrubs and trees, the Caprifoliaceae, largely temperate, are almost entirely woody, the Valerianaceae, largely temperate and Andean, are mostly herbs but some Andean ones are shrubs, and the Dipsacaceae, mostly Mediterranean, are all herbs, or half-shrubs.

The important gross-morphological features shared by these families are opposite or verticellate simple leaves, stipules, if any, usually interpetiolar, cymose or thyrsoid inflorescences, cyclic arrangements of floral parts, petals united, at least at base, single whorl of stamens united to corolla tube equal in number to or fewer than corolla lobes and alternating with them, another 2-celled and dehiscing longitudinally, ovary inferior, mostly 2-celled but many 3- or even more, Dipsacaceae 1-celled, ovules many to one, anatropous, placentation axile, basal, or apical, style mostly one, stigmas 1-several, seeds mostly with endosperm, embryo mostly small, straight or curved, integuments one or none. There are exceptions, usually secondary or of a derived nature, to most of these characteristics. For example, a few Rubiaceae have a superior ovary (e.g. *Pagamea*), some have spicate inflorescences (e.g. *Alseis*), a few have apparently alternate leaves by reduction of one of each pair (e.g. *Didymochlamys*), *Sambucus,* usually included in Caprifoliaceae, has apparently compound leaves, and *Dialypetalanthus* (here included in the Rubiaceae but by some made a separate family) has truly separate petals and supernumerary stamens inserted on the epigynous disk (all three of these characters in this case regarded as primitive in the order and not derived).

Nectaries form an epigynous cushion in the Rubiaceae and in most Caprifoliaceae, some of the latter have a patch of glandular hairs near the base of the corolla tube, a tendency leading to the Valerianaceae which have the basal part of the corolla nectariferous. Secretory hairs are common inside the stipules of the Rubiaceae. Domatia are common in leaf axils, especially in Rubiaceae, and some genera have nitrogen-fixing bacteria in leaf nodules. Pollen morphology differs strongly from family to family and even within families so that no ordinal characteristics can be pointed out, although there seems to be a tendency toward colpate grains.

Anatomical features vary so widely that few generalizations can be made. No features are known to be peculiar to this order nor are any generally characteristic of it. The wood is regarded as rather unspecialized, as none of the families have storied structure, only the Caprifoliaceae have scalariform perforation plates, all have apotracheal parenchyma, all have fiber tracheids, some Rubiaceae and Caprifoliaceae have libriform fibers.

The chromosome numbers vary widely, but for the majority of genera for which sufficient counts are available the basic haploid numbers are 8, 9, 10, and 11, with the Rubiaceae having higher numbers, mostly 11, the other families mostly 7 to 9.

In spite of the fact that the revised Hutchinson classification scatters the families included here into several widely separated orders, they do not seem to differ very fundamentally and are probably derived from a common stock. It is, indeed, difficult to find features that will separate them very sharply. The Rubiaceae and Caprifoliaceae are very close, even the traditional key difference, the presence or lack of stipules, breaking down in *Viburnum* (Caprifoliaceae) of which some species do possess stipules. Certain species of *Sambucus* also have stipules, but it is here suggested that the genera *Sambucus* and *Adoxa*, usually placed in the Caprifoliaceae (or *Adoxa* separated as a monotypic family), should be excluded from this order. This suggestion is made on the basis of a number of vegetative characters, such as compound leaves which resemble nothing else anywhere near this relationship, crassicaulous branchlets in *Sambucus*, otherwise unknown in anything close, as well as on the rather different gynaecium structure.

The derivation of the Rubiales is not clear. Most authors have suggested that they may come from the UMBELLALES, but that order as usually constituted is a rather unnatural aggregation and certainly does not seem close to what are certainly the primitive Rubiales, e.g. *Dialypetalanthus*. It may be that the Rubiaceae and Loganiaceae arose from common ancestors, as the general habit, usually interpetiolar stipules, and the placentation are common features. Several genera (e.g. *Gaertnera*) have been shifted back and forth between the two families. There is considerable similarity to the Cornaceae and the Oleaceae, which may have also diverged from this same stock. One tribe, the Henriqueziae, have been shifted back and forth between the Rubiaceae and Bignoniaceae, and it is entirely likely that the Bignoniaceae also may derive from the same common ancestral stock.

Within the Rubiales the Rubiaceae seem to be the most primitive family. Major trends may be noted in the order toward loss of stipules, toward reduction of stamens, fertile carpels, ovary cells, and ovules, from woody to herbaceous habit, toward condensation of inflorescences into heads with development of showy involucral bracts, toward fusion of ovaries into fleshy compound fruits, and in floral symmetry from radial to bilateral or irregular. These tendencies are variously repeated in many different lines, but generally run from the Rubiaceae through the Caprifoliaceae, and Valerianaceae, to the Dipsacaceae. A tendency to enlargement of one or all calyx lobes occurs sporadically in the Rubiaceae and Caprifoliaceae, and more frequently in the Valerianaceae where in some cases it suggests the pappus of the Compositae. Certain of the tendencies noted above seem to lead easily from the Rubiales into the Asterales (Calyceraceae and Compositae).

Economically this order is relatively unimportant.

The Rubiaceae furnish coffee, Cinchona alkaloids, a few minor edible fruits, and several not widely known but excellent timber trees. Madder dye was formerly produced from the roots of *Rubia tinctoria* and other related Rubiaceae. A few Caprifoliaceae yield small edible fruits. Some members of each of the principal families are reputed to have medicinal properties. Many members of the Caprifoliaceae and a few of each of the other families are grown as ornamentals, and *Dipsacus* furnishes the fuller's teasel, used in carding wool.

F. R. FOSBERG

References

Hutchinson, J., "The Dicotyledons," London, Oxford Univ. Press, 1959.

Lawrence, G. H. M., "Taxonomy of Vascular Plants," New York, Macmillan, 1951.

Engler, A. and E. Gilg, "Syllabus der Pflanzenfamilien," Berlin, de Gruyter, 1924.

Hutchinson, J., "The Families of Flowering Plants," Vol. 1, Oxford, Clarendon, 1959.

Utzschneider, R., "Zur Abstammung der Rubiaceen," Mitt. bot. Staatssammlung Munchen 3: 96–98, 1961.

RUDOLPHI, CARL ASMUND (1771–1832)

Carl Asmund Rudolphi was born in 1771 in Stockholm of German parents. He studied medicine at Greifsward and attended the German university of Swedish Pomerania where he later became professor of anatomy. He then went to Berlin as professor of anatomy and remained there until his death in 1832.

Among achievements, Rudolphi founded the Berlin zoological museum and was successful in educational work. Though he was highly critical of others and himself, he labored hard so that his writings give the impression of solid reality. He contributed to three important branches of biology which are: (1) parasitic research in which he investigated numerous germ-carrying animals and gave detailed accounts of the appearance and conditions of life of the parasites existing in them, (2) comparative anatomy which is described in his collection of short essays including comparative microscopical studies of the intestinal villi in different vertebrates and study of cerebral cavities, (3) physiology of which *Grundriss der Physiologie* is his most important work. Even though this work was unfinished at his death, it testifies that exact research had progressed immensely during the first decades of the nineteenth century.

BETTY WALL

S

SACHS, JULIUS (1832–1897)

The founder of experimental plant biology was born in Breslau in conditions of utter poverty, alleviated only by the kindness shown him by Purkinje whom he followed to Prague. Though the research of Sachs roamed over a wide area of botany, his two most important discoveries were in the field of the irritability of plants, and in photosynthesis. In the former capacity, he may be regarded as the originator of the theory of tropisms and his brilliant demonstration of geotropism by fastening a bean to the hour-axis of a clock is still used as a demonstration to the present day. In the field of photosynthesis, he was the first to demonstrate clearly that starch is the end product of assimilated carbon dioxide and that sunlight is necessary for this reaction to take place. Many of his early experiments on the growth of plants in variously colored light are being repeated, admittedly under less crude conditions, at the present time. Sachs was one of the first botanists to embrace the theories of Darwin in their entirety and their spread in continental Europe is in no small measure due to his brilliance as a writer of text books.

Sachs old age was as embittered as was his youth for his arrogant refusal to consider any views but his own left him in an intellectual isolation which he blamed on the jealousy of his contemporaries.

PETER GRAY

SAINT-HILAIRE, AUGUSTIN FRANCOIS CESAR (1799–1853)

This French botanist, who is in no way related to the subject of the next entry, was one of the earliest botanical explorers of South America. He published ten major contributions to the flora of that region and wrote one of the earliest texts in plant morphology.

PETER GRAY

SAINT-HILAIRE, ETIENNE GEOFFROY DE (1772–1844)

Saint-Hilaire was very much a product of his times in that he regarded philosophical speculation as of more value than accurate observation. Thus the comparative anatomy in the study of which he vied, in his day, with Cuvier, postulates such imbecilities as a sternum in fish and a vertebrate affinity for sepia, the "shell" of which he presumed to be a set of fused ribs. In spite of all this, however, he made some very valuable contributions to comparative anatomy through the pursuit of his fundamental idea that there must be somewhere a common type from which all others were derived. That this idea led to the exaggerations listed did not prevent him, for example, from accurately drawing attention to the homologies between the branchial arches of many different groups of vertebrates. He was an indefatigable collector and took advantage of Napoleon's numerous expeditions of conquest to stock the museum in Paris with what speedily became the largest osteological collection in the world.

The subject of this biography should properly be called, and indexed as, "Geoffroy." He is, however, inserted at this point in the encyclopedia to diminish any chance of perpetuating the usual confusion with the previous entry.

PETER GRAY

SALICALES

Salicales is an order of DICOTYLEDONOUS plants which is composed of the Willows and Poplars. Because of the simplicity of the flowers, earlier workers, as well as some present-day ones, considered the order a very primitive one but this simplicity is without much doubt the result of much specialization and reduction. Their long geologic history, judging from the fossil record, is thought to be more a result of their habitat preferences in moist to wet areas where preservation of structure by fossilization is more likely, than evidence of their primitive nature. There is some evidence that the single family which makes up this order may have been derived, evolutionarily, from several advanced families.

The Salicaceae is primarily a family of trees and shrubs, although some of the arctic species are only sub-shrubby. The leaves are deciduous, simple, almost invariably alternate, and generally with stipules which typically fall soon after expansion. The flowers are unisexual and the staminate and pistillate ones are borne on separate plants; flowers of both sexes are borne in dense, finger-like clusters called catkins. Recognizable sepals and petals are lacking in flowers of both sexes but vestigial in the form of small bracts and either a cup-shaped disc or one to two nectar-producing glands at the base of each flower. In the staminate flowers two or more stamens are produced and in the pistillate a single pistil. The ovary is one-celled but 2–4-carpellate, the two to four placentae are parietal. The fruit is a capsule which dehisces by two to four valves to release numerous hairy seeds. The latter characteristic along with the dioecious habit and the nature of the vestigial perianth serve to distinguish plants of this group from other catkin-bearing plants.

The only economic uses for the plants are the ornamental culture of some species, others furnish wood for charcoal, and the twigs of others provide basketry materials.

RICHARD S. COWAN

References

Hutchinson, J., "The Dicotyledons," London, Oxford Univ. Press, 1959.

Lawrence, G. H. M., "Taxonomy of Vascular Plants," New York, Macmillan, 1951.

SALT MARSH

Salt marshes are areas of land adjacent to the sea which have a cover of vegetation and generally subject to periodic inundation by the tide. There are also inland areas associated with saline springs or with salt lakes that may bear a similar type of vegetation, though they are not subject to regular inundation by salt water. When very moist these inland areas also can be regarded as salt marsh, but as they become drier they grade gradually into salt steppe and salt desert. Maritime salt marshes originate on stable, emerging or submerging shore lines, though in the last-named case the rate of sedimentation must be greater than the rate of subsidence. Salt marsh will only develop where there is active sedimentation and protection. The sediment may come from local cliff erosion, e.g., Bay of Fundy marshes, or it may be brought down by rivers. Adequate protection is provided by estuaries, enclosed bays, spits and bars. Sometimes salt marsh appears to develop on an open coast, but in such cases it will be found that either there is no extensive body of water in front, or else the sea is very shallow so that in neither case do large destructive waves arise. Maritime salt marsh originates as mud or sand flats which become colonized first by algae and then by phanerogams. The advent of the higher plants aids the rate of sedimentation as the plants help to trap the silt burden of the water. The marshes therefore steadily increase in height, and this leads to a gradation in height between the seaward and the landward parts of the marsh. Associated with this gradation in height there is a zoning of the vegetation. Similar zones of vegetation can often be observed in inland saline areas with increasing distance from the salt spring or the salt lake.

The plants that grow on salt marshes are known as halophytes. These plants are commonly characterized by one or more of the following features: succulence, glabrousness, salt excreting glands, gray-green foliage. The vegetational similarity of maritime and inland salt areas indicates that there must be a common determining factor. The principal common factor is the sodium ion, which in the exchangeable state in the soil causes the soil colloids to become very highly dispersed. This results in a number of secondary effects, e.g., poor aeration, retention of water, impeded water movement, all of which are adverse to plant growth. The soil water of salt marshes contains an excess of alkali salts: in maritime marshes it is sodium chloride; in inland haline areas it is sodium chloride, sodium carbonate, sodium sulphate or magnesium sulphate.

Physiographic features of maritime salt marshes are the meandering creeks and salt pans. Creek building is a complex phenomenon associated with sedimentation and erosion, both of which can vary with age of the creek. In some marshes the creek system is extensive, in others it is not so. On the east coast of North American marshes ditches have been dug with a view to reducing the mosquito nuisance. This ditching has resulted in changing the marsh vegetation and also in altering the animal population. It has subsequently been suggested that the ditching is not really necessary. Pans are bare areas, somewhat below the level of the surrounding vegetation, in which water stands for much of the year. These pans can develop through failure of the vegetation to colonize the original bare flat (primary pan), by subdivision of a primary pan (residual pan), by blocking off the end of a creek (creek pan), by death and decay of vegetation from trash or other cause ('rotten spot').

Maritime salt marshes are essentially features of mid and high latitudes, their place in the tropics and subtropics being taken by mangrove swamps. On the basis of vegetation, associated in some cases with geomorphological features, it is believed that these marshes can be divided into the following major groups: North European, Arctic, Mediterranean, Eastern North American, Western North American, Sino-Japanese, South American, Australasian and Tropical (in places these last can develop on the landward side of mangroves). Insofar as the vegetation of the maritime salt marshes changes as the marsh level is elevated, the zoning must be regarded as dynamic. In nearly all parts of the world there is evidence that given the right physiographic conditions, salt marsh vegetation will eventually progress through brackish to fresh water swamp and thence to the regional climax. Just as the phanerogamic vegetation is so characteristic of salt marsh, the associated algal vegetation is equally characteristic, a number of species and genera being widely distributed. On the Atlantic-North Sea salt marshes there are very interesting fucoids to be found.

Considerable attention has been paid to the environmental factors. On the maritime salt marshes tidal phenomena are obviously of great significance and marshes can be divided into upper and lower relative to submergence and exposure. Periods of continuous exposure are particularly important in relation to salinity changes and seedling germination. Movements of the water table are related to distance from creeks or ditches, and the distribution of some phanerogams is related to this drainage. One of the most important features of maritime marshes is the existence of an aerated layer in the upper soil layers even when the marsh is flooded. The proportion of exchangeable sodium and soluble chloride vary seasonally, and in a number of salt marshes it has been found that a lowering of the salinity takes place in the spring at a time when seeds germinate. It has also been shown that a number of salt marsh plants will not germinate at high salinities. Although the high salinity has been regarded as making the salt marshes 'physiologically dry' for plants, the halophytes appear quite capable of absorbing water from the highly concentrated soil solutions. Economically salt marshes represent valuable agricultural land. In many parts of the world maritime marshes are reclaimed by building enclosing sea walls with one-way sluices for drainage. Elsewhere animals graze on the wild marsh, the grass is cut for hay and the turf is used for golf greens. Such marshes are also a haunt for wild fowl and are therefore of considerable interest to the sportsman.

V. J. CHAPMAN

Reference

Chapman, V. J., "Salt Marshes and Salt Deserts of the World," London, Leon Hill, 1960.

SANTALALES

This order of plants belongs to the Apetalae, one of 3 subdivisions of the DICOTYLEDONEAE. Some authorities feel that it cannot be a primitive group of plants, where it is usually considered, because of the presence of parasitic members. It is thought to be closely allied to the Proteales, except that it has mostly an inferior ovary.

The Santalales are shrubs, small trees or herbs, many of them partial root or tree parasites, or completely so. The leaves are simple, or some of the species leafless. The flowers are inconspicuous, perfect or imperfect, mostly without petals, regular, cyclic. The stamens are mostly as many as the perianth segments and opposite them. There are usually 2-3 united carpels or rarely 1. The fruit may be a drupe, nut, or berry, the seeds mostly with endosperm.

The order is usually considered to consist of eight families: Balanophoraceae, Grubbiaceae, Loranthaceae, Myzodendraceae, Octoknemaceae, Olacaceae, Opiliaceae, Santalaceae. They are mainly found in tropical or subtropical regions, but plants belonging to three of them are found in the United States (Loranthaceae, Olacaceae, Santalaceae).

The Balanophoraceae (Balanophora family) is a tropical family of annual or perennial herbs parasitic on the roots of trees and shrubs. They are fleshy and yellowish or reddish, with tuberous, branching rhizomes. They attach themselves to the roots of their host plants by means of haustorial suckers. The flowering stem is short, erect, and at first has scale-like leaves which soon wither and fall off. The small flowers are borne in dense spikes or heads, or in one genus in a panicle. There is evidence that in some cases the inflorescence develops inside the rhizome, then pushes to the outside and above ground.

The flowers are mostly unisexual, the staminate usually having a 3-4 or 8-lobed, partly united calyx and as many stamens as perianth lobes. The pistillate flowers usually lack a perianth and have 1-3 carpels, forming a 1-chambered ovary with 1-3 nutlets.

There are fifteen genera and about forty species in the family, the main genus being Balanophora, with about a dozen species, in Asia and Australia.

The Grubbiaceae (Grubbia family) is composed of the single genus Grubbia, containing three species, occurring in South Africa. They are shrubs with flowers possessing 4 sepals, 8 stamens, and a 1-chambered, inferior ovary of 2 carpels, with 2 pendulous ovules. The fruit is a drupe.

The Loranthaceae (Mistletoe family) are usually small shrubs, completely or partially parasitic on the branches of trees, or rarely they are erect trees (Nuystsia of Australia). They attach themselves to their hosts by modified roots (haustoria) and gain all or part of their food through these structures. The stems are usually brittle, with opposite leathery leaves, or these reduced to scales. Two subfamilies are generally recognized: the Loranthoideae, with a calyculus at the base of the flower, and the Viscoideae, lacking this structure.

The flowers are bisexual or unisexual, in panicles, racemes, spikes, or solitary. Some are green, but in the tropics many are brightly colored. They are usually small, with a cup-shaped receptacle and separate or united perianth parts. The stamens usually equal in number the perianth lobes and borne on them or at their base. There is one pistil, the ovary inferior, with 3-4 carpels and one style or sessile stigma. This develops into a 2-3-seeded berry or a drupe, which is often viscid.

The family is primarily tropical, but some members extend into the temperate zones of both hemispheres. It is often divided into about 30 genera and 1000 species. The largest genus, Loranthus, mainly in Africa, has about 500 species. The members of this family in North America belong to the subfamily Viscoideae, and to the two genera, Phoradendron and Arceuthobium.

Phoradendron includes the common, leafy, green mistletoe of America, with more than 100 species. They are partly parasitic, particularly on oaks, cottonwoods, alders, and some evergreens. The mistletoe of American commerce is *Phoradendron flavescens*, while that of Europe is *Viscum album*, a plant closely related to Phoradendron. It was used as a ceremonial plant by the Druids and other early Europeans, and from this arose the custom of kissing under the mistletoe at Christmas time.

Arceuthobium includes the yellowish, leafless mistletoes, parasitic on conifers. Some species are restricted to a particular species of evergreen, while others may attack several species. They produce swellings and usually rapid proliferation of branches on the host tree near the point of entrance of the haustoria. This results in the "witches brooms" so injurious to our conifers. These become so plentiful in some places that large areas of trees are cut to check the spread of the parasite.

The fruit is mucilaginous and birds carry the seeds on their beaks and feet to new host trees.

The Myzodendraceae (Myzodendron family) is a small family consisting of a single genus (Myzodendron). The plants are shrubby, semi-parasites growing on beech trees, and occurring in the temperate regions of South America. The leaves are scale-like and alternate. The plants are dioecious, the staminate flowers with discs but without a perianth. The pistillate flowers have a partly inferior ovary, a 3-angled, nut-like fruit, with 3 feathery brushes that aid in its dispersal.

The Octoknemaceae (Oktoknema family) is composed of a single genus of shrubs or trees occurring in West Africa. The plants have simple, alternate leaves and unisexual flowers, in racemes. These have 5 sepals, 3–4 carpels, with 1 style and 1 locule. The placenta is thread-like and extends from the base to the top of the chamber. The fruit is a drupe.

The Olacaceae (Olax family) has about 25 genera and 150 species which are found in the tropics around the world. Two genera (Schoepfia, Ximenia) occur in Florida, but are of no economic importance. This family can usually be distinguished from the others of the order because of having twice as many stamens as petals, or more, and a superior ovary with the ovules pendulous from the placenta.

The plants usually have alternate, simple, entire leaves and bisexual flowers. The sepals and petals are 4–6 each, the parts distinct or variously united. There are 4–12 stamens, the anthers opening by apical, pore-like slits. There is 1 pistil with 3–4 carpels and locules, bearing 1 ovule in each. The fruit is a berry or drupe, often surrounded by the enlarged, persistent calyx. Some members of this family are semi-parasitic.

The Santalaceae (Sandalwood family) are herbs, shrubs or trees, with simple, mostly alternate, entire leaves, or these reduced to scales. They are widely distributed in the tropics and temperate regions. A few of the members are semi-parasitic on the roots of the host plants.

The flowers are small, regular, and the 4–5 sepals sometimes petaloid. There are 4–5 stamens, opposite the perianth parts. There is 1 ovary, often inferior and embedded in the resceptacular tissue, composed of 3–4 carpels, with 1–5 ovules, but often only 1 maturing. The fruit is an achene or drupe without a seed coat.

There are 26 genera in the family and about 400 species. The family in general is of little economic importance, but the sweet scented sandalwood (Santalum album) is used extensively in perfumes and in cabinet making. The genera Pyrularia and Buckleya are occasionally cultivated as novelties. Thesium, largely in Africa, is the largest genus, having about 220 species. Comandra (false toadflax) is wide-spread in North America, being a semi-parasite on the roots of trees and shrubs. Buckleya, Geocaulon and Nestronia grow in Eastern United States. The latter is a monotypic genus of semi-parasites growing on the roots of trees, occurring along streams. Pyrularia, known as buffalo or oil nut, is a small shrub found from Pennsylvania to Florida.

The Opiliaceae (Opilia family) is a small family found in the tropics of South America, Africa and Asia. They may be trees, shrubs, or vines, with simple, alternate leaves. The flowers are usually bisexual, the ovary being partly inferior. There are 4–5 sepals and 4–5 petals, with 4–5 stamens opposite the petals and sometimes adnate to their bases. There are 4 carpels and 1 style, with 3 locules, each with 1 ovule. The fruit is a drupe. The family is of no economic importance.

RAY J. DAVIS

References

Rao, L. N., "Studies in the Stantalaceae," Ann. Bot. NS 6: 151–175, 1942.

Trelease, W., "The Genus Phorandendron," Urbana, Univ. Illinois Press, 1916.

SAPINDALES

The Sapindales are DICOTYLEDONOUS woody plants of wide distribution throughout the temperate and tropical regions of the world. The chief families included in the order by recent authors are the Melianthaceae, Sapindaceae, Sabiaceae, Anacardiaceae, Aceraceae and Hippocastanaceae. Economic species include many ornamental trees and shrubs, such as maples, horsechestnuts and smoke tree (Cotinus coggyria); they also yield such products as cashew and pistachio nuts (Anacardium occidentale and Pistacia vera, both in Anacardiaceae), fine hardwood and sugar (Acer saccharum) and lacquer (Toxicodendron vernicifera). The Anacardiaceae are notorious for such poisonous plants as poison ivy and poison sumac (Rhus spp.).

The Sapindales have probably evolved from the Rutales, and are closely related to the Meliales and Geraniales. Various and conflicting hypotheses have been advanced as to their phylogeny. The chief obstacle to more certain knowledge lies in the paucity of fossil evidence; but more needs to be known, also, about the present-day plants. A major evolutionary trend has been the reduction of the flowers from ancestral forms that were pentamerous in all whorls, with 3 whorls of stamens. The Rutales include species having completely pentamerous flowers, with 2 whorls of stamens, and with obdiplostemony suggesting that there were primitively 3 stamen whorls.

The genus Acer illustrates many steps in the reduction in number of floral appendages from the basic pentamerous plan; similar reduction series are found in other families of the order. The sepals usually number 5, although in some species often only 4 are present. There are usually 5 petals, though the flowers of some species lack petals. Stamens vary from 10 to 7 or 8 (one complete and one partial whorl), to a single whorl of 5 or 4; or in some cases are absent in female flowers. The carpels are mostly 2 in number, yet in some species flowers having 3, or even 4 or 5 carpels, are occasionally found. Rarely flowers have a single carpel, while carpels are absent in male flowers of strictly dioecious species.

Reduction of stamens and carpels has resulted in curious mechanisms that insure outbreeding. In some Acer species the structurally perfect flowers become functionally male or female by the abortion of, respectively, the carpels or stamens. The first flowers to mature on one tree will be female, those on another, male. Later flowers on the first tree will be male, on the latter, female. Thus each tree functions as both male and female parent, but only as one sex at a given time; and the species becomes, in function at least, dioecious. This abortion of stamens and carpels is a step, evidently, in the evolution of strictly unisexual flowers, such as those of A. negundo.

Reduction appears to have occurred also in the inflorescences, chiefly through the loss of branches and of bracts.

Because they vary so widely in floral as well as in

other characteristics, these families furnish a rich field for further taxonomic and morphological investigation.

BENEDICT A. HALL

References

Engler, A. and H. Prantl, "Die natürlichen Pflanzen-familien," 2nd ed., Leipzig, Engelmann 1924–1942 (incomplete).
Hutchinson, J., "The families of flowering plants" 2nd ed., Oxford, Clarendon Press, 1959.
Lawrence, G. H. M., "Taxonomy of vascular plants," New York, Macmillan, 1951.

SARRACENIALES

A small order of DICOTYLEDONOUS flowering plants, comprising 3 families, 8 genera and some 165 species.

I. Sarraceniaceae: 1. *Darlingtonia*. 1 species. Northern California, southern Oregon. 2. *Heliamphora*. 5 species. Northern South America. 3. *Sarracenia*. 9 species. Eastern North America. **II. Nepenthaceae:** 4. *Nepenthes*. 60 species. Eastern tropics, chiefly Borneo. **III. Droseraceae:** 5. *Aldrovanda*. 1 species. Widespread throughout Old World. 6. *Dionaea*. 1 species. Southeastern United States. 7. *Drosera*. 90 species. Almost world-wide; most abundant in Australia. 8. *Drosophyllum*. 1 species. Morocco, Spain, Portugal.

FLOWERS in the Sarraceniales are actinomorphic, with cyclic parts and uniseriate or biseriate perianth; bisexual, except in *Nepenthes*; pistil syncarpous, ovary superior (sometimes half-inferior in *Drosera*); placentation axile or parietal; embryo usually straight. Plants herbaceous or sometimes shrubby; leaves insectivorous.

The major systems of classification differ as to the position and phylogenetic relationships of these families. Authors derive them from the Saxifragales or from the Helleboreae in the RANALES, and place the Nepenthaceae with the other two families or separately in the Aristolochiales; two recent authors have placed the Drosehaceae in the PARIETALES, close to Violaceae.

Because of their insectivorous habit, these plants are of great biological interest. The leaves are modified, structurally and functionally, into several types of traps, with a variety of devices for attracting and capturing prey. In most of the genera the prey are digested by proteolytic enzymes secreted by the plant (*Heliamphora* is an exception in which the animals are digested by bacterial action), and the nitrogenous products of digestion are absorbed.

Drosera has contributed significantly to current knowledge of evolutionary mechanisms. *D. longifolia* was the first plant demonstrated to be a natural allopolyploid. The genus is now known to contain species having diploid CHROMOSOME numbers of 20, 28, 30, 40, 60 and 80. Also *Drosera* and *Drosophyllum* furnish an extreme example of variation in chromosome size within a family, the chromosomes of the latter being 1000 times as large as those of the former.

If insectivorous plants have turned the tables on the animals, reversing the more usual dependence of animals on plants for their energy foods, some animal species have succeeded in returning the tables. Many animals, chiefly arthropods, find suitable living conditions in the pitcher leaves, getting their food from the plant's prey or eating the tissues of the plant itself. Some insect species are found, at least in the larval stage, only in these pitchers. The hollow tendril, near the pitcher, in one species of *Nepenthes*, serves as a formicary, in a manner similar to the thorns of *Acacia*.

Although much work has been done on this order of plants, they remain a remarkably fertile source of biological problems. Much more needs to be known about the homologies of the highly modified leaves, the physiology of digestion and absorption of animal prey, the delicate responses involved in the capturing of prey, and the ecology of the fauna of the pitchers.

BENEDICT A. HALL

References

Darwin, Charles, "Insectivorous Plants," London, Appleton, 1915.
Lloyd, Francis Ernest, "The Carnivorous Plants," New York, Ronald, 1942.

SCALES

Animal scales are flattened structures which range from millimeters in diameter, in certain insects, to inches in diameter in certain REPTILES and armadillos. Scales may cover all, or part, of the body. Moths and butterflies have the wings and some other parts of the body covered with minute, flattened scales that usually contain an iridescent pigment, giving the wing a metallic or glossy color because of rays of light which they reflect. Insect scale structures are probably modified SETAE.

Some minute, sucking, insects which live on plants, have the whole body covered by a single scale, usually characteristic of the species, and are thus called scale insects. Some are of economic importance in the production of cochineal shellac, and china wax. On the negative economic side are the purple scale of citrus fruits, oyster scale and the San Jose scale.

There are at least four different kinds of fish scales, the placoid, ganoid, cycloid and the ctenoid. Placoid scales are found on the ELASMOBRANCHS, in some parts of the body scattered over the surface, but for the most part arranged in the form of a mosaic. This scale consists of a bony plate, buried beneath the surface of the skin with an exposed layer of enamel covering the spine. Under the enamel is a layer of *dentine* covering a center of *pulp*. The placoid scale is in the shape of a curved spine partially exposed to the surface and is often known as a dermal denticle. This scale is supposed to be the basis from which vertebrate teeth originated.

The ganoid scale is found on ganoid fishes like Lepidosteus (garpike) or Amia (bowfin). These scales are arranged in a mosaic over the body on Lepidosteus but imbricated on Amia. The basal part is bony, covered by some modified dentine-like material and the exposed surface overlayed with a peculiar enamel called ganoin. This is sometimes regarded as a modification of the placoid scale.

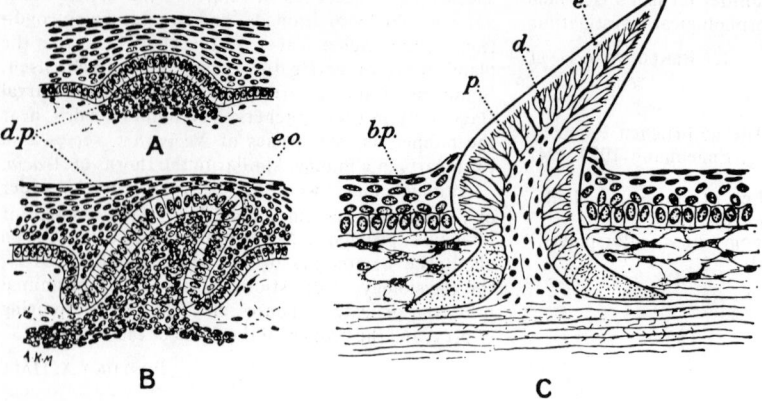

Fig. 1   Placoid scale, as seen in longitudinal section. A, B, early stages; C, fully-developed scale. *b.p.*, basal plate; *d*, dentine; *d.p.*, dermal papilla; *e*, enamel; *e.o.*, enamel-organ; *p*, pulp.

The cycloid scale is a rather large scale found on carp, buffalo and similar fish. On its free (external) surface it is covered with concentric growth rings. Since growth is interrupted during the winter, the years of age of the fish can be obtained by counting the rings. The attached edge of the scale is scalloped in shape while the free edge is rounded. They are imbricated, like shingles on a roof, in their position on the surface of the skin.

The ctenoid scale is found on the perch, sunfishes and in other representatives of that group. They are similar to cycloid scales except that, at the free edge, there are comb-like rows of teeth or spines. These scales are also imbricated over the body, and have apparent growth-rings.

Among Amphibia only the APODA have rudimentary scales imbedded under the skin.

In reptiles and birds the scales are similar. In lizards, snakes and crocodiles the scales are in the form of *corneoscutes* of varying size. In turtles the carapace and plastron, which are sectioned into scales, have a bony base. The scales which cover the shank and foot of the bird are enlarged corneoscutes, and are composed chiefly of keratin.

The scales of mammals differ from those of any other group. The tails of rodents, for example, have a scaly surface produced from modified skin, hardened and thickened by the presence of keratin. Horns, and hoofs are composed largely of the same substance. Armadillos and pangolins have an almost complete cover of hard, bony-based scales over the body. Spiny anteaters have a body covered with scales modified into spines.

All types of scales are functionally effective. In insects they provide not only protection but also smooth, firm planing surfaces for flight. In other groups protection is the primary function.

GEORGE E. POTTER

SCAPHOPODA

The Scaphopoda are a small, but very distinct class of marine mollusks (see MOLLUSCA). About 350 living species are known, and approximately 400 extinct species have been described. Geologically the scaphopods date from the Devonian, but representatives of the class are not common in the fossil record until the Cretaceous.

The shell is a slightly to moderately curved, tapered, non-chambered aragonitic tube that is open at both ends. The tube is commonly attenuated posteriorly to resemble an elephant tusk or a canine tooth of a carnivore, thus giving rise to the common names: tusk or tooth shells. The tube is commonly sculptured with longitudinal or annular ribs, or, rarely, lacks surface ornamentation. The body closely fits the tube to which it is attached along the concave (dorsal) side by the columellar muscles.

The character of the muscular foot serves to divide the class into two families. In the Dentaliidae, the foot is pointed with the epipodial collar interrupted dorsally to give a trifid appearance, whereas in the family Siphonodentaliidae the subterminal epipodial ridge is not slit dorsally and terminates in a crenulated disk. A head is not developed, the mouth being located at the base of the foot in a projection of the pharynx; a mandible and a simple radula are present. Eyes are lacking, but at the base of the snout arise two tassels composed of numerous prehensile filamentous tentacles that terminate in minute food-gathering capsules. The oral cavity opens into a V-shaped intestine, the posterior part of which contains a small pyloric caecum into which the liver ducts open. The two lobes of the liver are symmetrical in the Dentaliidae, asymmetrical in the Siphonodentaliidae. The intestine forms several loops and terminates at the anal opening near the visceral GANGLION. The heart is rudimentary and lacks auricles. The blood system consists of five major sinuses. RESPIRATION occurs by an exchange of gases through the wall of the mantle. The NERVOUS SYSTEM is typically molluscan and is well developed. Paired NEPHRIDIA open into the mantle cavity below the retractor muscles. The sexes are separate and the sexual products are discharged through the right nephridium. The anatomical characters are shown in Fig. 1.

These animals are largely restricted to subtidal depths. The shell is held in an oblique position in the

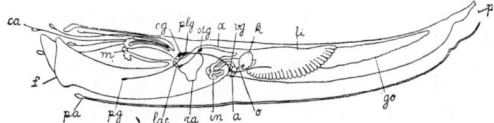

Fig. 1 Diagram of the organization of *Dentalium*, left-side view. *a*, anus; *ca*, captacula; *c.g.* cerebral ganglion; *f*, foot; *go*, gonad; *in*, intestine; *k*, left kidney; *la.c,* labial commissure; *li*, liver; *m*, mouth; *o*, orifice leading into the perianal sinus; *oe*, oesophagus; *pa*, mantle; *p.g.*, pedal ganglion, with otocyst; *pl. g*, pleural ganglion; *po*, posterior orifice of the mantle; *ra*, radular sac; *st.g*, stomatogastric ganglion. (After Pelseneer, P. in Lankaster, E. R. ed., 1906.)

substrate, with the foot extended out of the larger opening of the tube into the bottom sediments and with the posterior third of the tube exposed as the surface of the bottom to permit the circulation of water. The expended contractile tentacles select minute organisms, commonly FORAMINIFERA, from the substrate as the animal burrows into the sediment.

WILLIAM K. EMERSON

Reference

Pelseneer, P. *in* Lankaster, E. R., ed., "A Treatise on Zoology," Vol. 5, London, Black, 1906.

SCHLEIDEN, MATTHIAS JACOB (1804–1881)

Matthias Jacob Schleiden was born in Hamburg in 1804 and in his lifetime he became a strange scientific personality. He studied jurisprudence, became a doctor of law, and took up a practice as a barrister in his native town. After an attempted suicide because of his increased melancholy disposition in this field, he devoted himself to natural science. In obtaining his degree as a doctor of philosophy and medicine, he later gained a great reputation through his writings and became professor at Jena. He stayed twelve years there before resigning and being appointed professor at Dorpat. After less than a year there, he left and led a life of wandering until his death in 1881. Thus his life was evident of a soul without balance which is also true of his scientific work.

In his work Schleiden was able to recognize the importance of the cell nucleus and developed Robert Brown's views on the nucleus. He studied the embryonic cells of phanerogams and discovered the nucleolus. The significant part of his work was that he insisted upon the independence of the cell as the essential unit of the living organism.

His impetuous temper and arrogant character led him into many errors, but he eagerly applied himself to the microscopic analysis of structure and growth. Published in 1842 was his textbook on botany which received both favorable and unfavorable criticism. In the introduction of his book, Schleiden presents his philosophical view of nature in saying that the aim of natural science is "to relate all physical theories to mathematical grounds of explanation." The

valuable aspect of his work lies in his method of presentation in the section explaining cytology, morphology, and physiology of plants, and not in the actual contents.

BETTY WALL

SCHWANN, THEODOR (1810–1882)

Theodor Schwann, German physiologist, was born at Neuss in Rhenish Prussia on December 7, 1810. He studies at Cologne, Bonn, Worzburg, and graduated in medicine at Berlin in 1834. While in Berlin he assisted James Müller in experimental work on physiology. In 1838 Schwann was called to the chair of anatomy at the Roman Catholic university of Louvain, and in 1847 he went to Liège as professor, staying there until shortly before his death on January 11, 1882.

Schwann was gentle, reserved, easy-going, and because of this disposition, his scientific activities fall within the period which he worked with Müller. Working with Müller, Schwann was concerned with the physico-chemical basis of life. Respiration of the chick embryo was presented in his doctor's dissertation, and in 1836 he demonstrated the influence of organisms and lower fungi in production of fermentation and putrefaction which led him to disprove spontaneous generation. He also demonstrated (1) the organic nature of yeast and (2) the necessity of the presence of a ferment pepsin in digestion. Besides investigation of laws of muscular contraction he also discovered the striped muscle in the upper part of the esophagus and the envelope of nerve fibers which bears his name.

His publication: *Microscopic Investigations on the Accordance in the Structure and Growth of Plants and Animals* in 1839 led to the cell theory discovery. A general statement of his belief as to the cellular origin and structure of animals and plant lies in this conclusion: The entire animal or plant is composed either of cells or of substances thrown off by cells; cells have a life that is to some extent independent, and this individual life of all the cells is subject to that of the organism as a whole. His general attitude is still valid.

BETTY WALL

SCORPIONES

Scorpions are easily distinguished from other orders of ARACHNIDA by a combination of characters never absent: the large pedipalps always held forward; the small anterior segment of the abdomen (pedicel) joining it to the carapace, is lacking; the presence of comb-like appendages on the third abdominal segment; the sharp division of the abdomen into a wide anterior portion and a much more slender, clearly segmented tail.

Scorpions are of great antiquity and the earliest Arachnida to have appeared on land; *Palaeophonus* from the Silurian deposits of both N. America and Europe differed little in structure from modern representatives of the order.

External structure. Large Arachnida (the largest African Scorpion, *Pandinus imperator,* is about 7″ long), the body consisting of two regions, a prosoma and opisthosoma, the former entire, the latter divided into a wider abdomen with 7 and a narrow tail with 5 segments, ending in a sting containing the poison glands.

The appendages consist of small pincer-like chelicerae at the anterior apex of the head followed by powerful pedipalps in which the last segment is usually enlarged and provided with pincers like the claws of a lobster. The four walking legs have 7 segments and two claws; behind the last pair on the ventral surface are the pectines, peculiar modified appendages with numerous comb-like teeth (3–40); between them lies the sex opening.

Sense organs. The cephalothorax bears a pair of median eyes and at each anterior angle 2–5 much smaller ones; they are of a simple type; vision is feeble and compound eyes are absent; numerous special tactile hairs, the trichobothria, are present on the femur, tibia and pincers of the pedipalps. The function of the pectines, the teeth of which contain numerous sense cells, is, according to Alexander's researches, the selection and exploration of a suitable mating site by the male.

Internal structure. The oesophagus and intestine are narrow, the latter having large diverticula buried in the lobes of the digestive gland which occupies most of the abdomen. The large tube-like heart has 7 pairs of osteoles and is associated with a well developed system of large arteries and veins. Respiration is carried out by 4 pairs of lung-books (the largest number in the Arachnida) on the third to the sixth abdominal sternites, the air circulating between numerous flat lamellae like the pages of a book; the intake of air is slow and feeble, respiratory movements being imperceptible.

Poison weapon. The poison vesicle contains a pair of large poison glands each connected by a duct with a minute orifice at the apex of the sharp sting. The poison glands are much larger in the African Buthidae than in the Scorpionidae, the yield of poison being three times as great; the venom is also far more toxic in the former and may occasionally cause death in humans.

Development. There is a complicated courtship which however does not end in copulation; the male produces a spermatophore by means of which sperms are introduced into the vulva of the female. Scorpions are ovo-viviparous, the period of gestation long, viz: 6–16 months. In the Scorpionidae the embryo is nourished internally with maternal secretions through a tube attached to the diverticulum from the ovary in which it lies.

Ecology and distribution. Scorpions are solitary nocturnal animals living under stones or in burrows in the soil, feeding on any smaller arthropod which they can overcome. They are found in all arid regions of the world, being especially abundant in deserts, but are very poorly represented in the temperate zones. About 600 species are divided among 6 families.

R. F. LAWRENCE

Reference

Alexander, A. J., "Courtship and mating in the buthid scorpions," Proc. Zool. Soc. London, **133**: 145.

SCYPHOZOA

The class Scyphozoa (Hyman, 1940; M. E. Thiel, 1936–1962), with the Hydrozoa (→) and the Anthozoa (→), forms the phylum Cnidaria. Together with the class Ctenophora (→) (Phylum: Acnidaria), they are grouped into the Coelenterata.

Within the Scyphozoa, which has altogether about 180 species, there are five present-day genera, four of which, the Rhizostomae, the Semaeostomae, the Coronatae and the Cubomedusae, are distinguished by a metagenesis between the sessile polyp stage and the pelagic medusa stage, while the Stauromedusae are purely polypoid. The structure of the polyps and medusae can be reduced to a quadriradiate symmetry.

Morphology. (Figures 1–6; from Hj. Thiel, 1966.) In the four genera which show metagenesis, the polyps or scyphistomes of the Cubomedusae, the Semaeostomae and the Rhizostomae are small and so similar to one another that they can not be differentiated from one another. While the polyps of the genus Coronate are surrounded by a peridermal sheath that seems to be species-characteristic, the scyphistomes of the remaining genera seldom form one, and then only very slightly, in the region of the pedal disc. The scyphistomes consist of three parts: the trunk with its broadened pedal disc, the cup-shaped middle part extending above it and terminal the oral disc. The four-cornered manubrium, whose edges mark the perradii, is located centrally on the oral disc. Between the perradii lie the interradii with their funnel-shaped indentations, the septal funnels. They invaginate into four gastric septa;

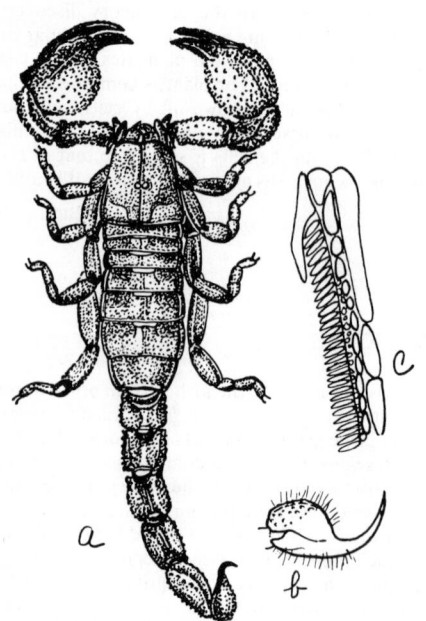

Fig. 1 *a,* The Mediterranean scorpion, *Scorpio*; *b* and *c,* the sting and pectinal comb enlarged.

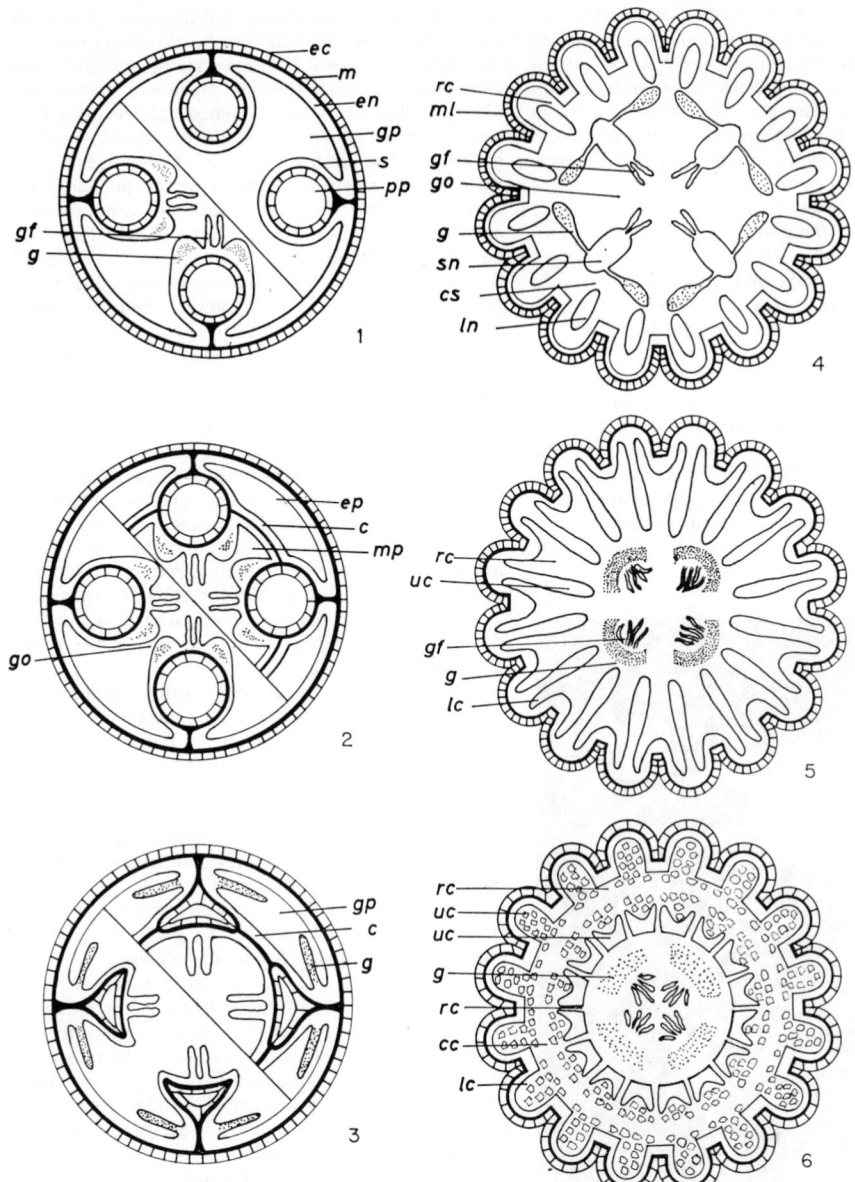

Figs. 1–6 Diagrammatic cross cection through various Scyphozoa. *c*, claustrum; *cc*, coronal canal; *cs*, coronal stomach; *ec*, ectoderm; *en*, entoderm; *ep*, exogon pocket; *g*, gonad; *gf*, gastric filament; *go*, gastric ostium; *gp*, gastric pocket; *lc*, lappet canal; *ln*, lappet node; *m*, mesoglea; *ml*, marginal lappet; *mp*, mesogon pocket; *pp*, peristomial pit; *rc*, radial canal; *s*, septum; *sn*, septal node; *uc*, umbrellar connection.

the muscles which effect the contraction of the polyp also proceed in the gastric septa. Numerous contractile tentacles are carried on the periphery of the oral disc. The gastric cavity, arranged in one middle piece and four gastric pouches that are separated by septa, is connected to the manubrium. The polyps of the Coronatae are distinguished from the other genera not only by the peridermal sheath, but also by the special characteristic possession of a ring canal and four radial canals in the perradii.

The asexually reproducing polyps of the four genera named above show alternation of generations (are in metagenesis) with sexually reproducing medusae. The Stauromedusae, on the other hand, are distinguished by a permanently sessile way of life; they develop gonads and reproduce sexually. Like the scyphistomes of the other genera, the Stauromedusae are fastened to the substrate by a pedal disc. The wide calyx above the trunk has its external layer designated as the exumbrella, as is the case in the medusae. The oral disc or subumbrella carries a short, four-cornered manubrium and is invaginated in the per- and interradi. Eight adradial lobes are formed; these carry tufts of capitate tentacles. The rhopaliods, which are homologous with tentacles and secrete a sticky secretion, lie in the per- and interradial scalloped edges. From the oral disc also, four funnels invaginate into the gastric septa, on which gastric cirri are located and in whose ectoderm the gonads lie. The septa are partially connected by a claustrum which the gastric pockets organize into an inner and an outer part. The musculature is arranged as a ring in the subumbrella, and in four double strands in the septa. The nerve plexus shows slight concentrations in the tentacle area. Sense organs occur only as ocelli in some juvenile stages.

The polypoid *Conulariida* (Fig. 7), which are found only as fossils, are also included at the present time in the Scyphozoa. They possessed in part four-edged peridermal sheaths with four three-cornered occluding lids, and in part, round or oval thecae and no lids.

In all metagenetic species, the medusae are larger than the polyps. Individual species of jellyfish reach a diameter of one meter. In medusae, the upper side of

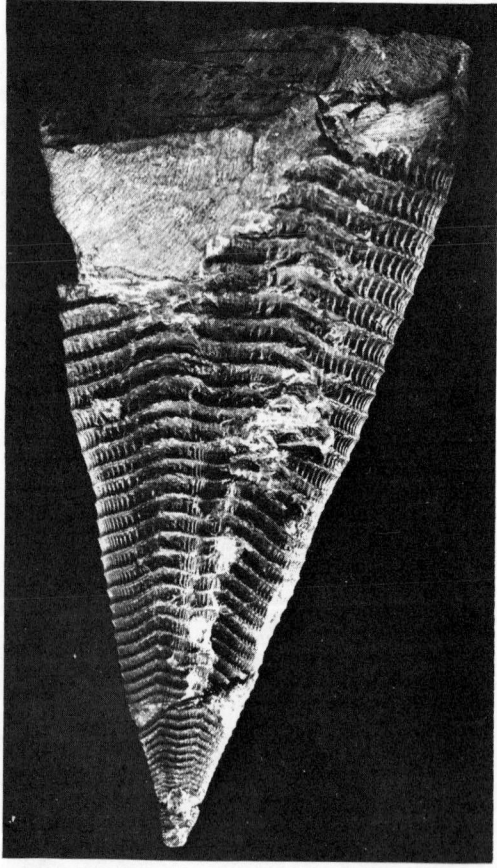

Fig. 7   Conularia spec., an extraordinarily well prepared specimen, which shows the outer sculpture to be very similar to that of present-day Stephanoscyphus. Found near Hamburg, Germany. (Collection Kausch, Phot. Werner, Original.)

the umbrella or exumbrella is distinguished from the lower side of the umbrella or subumbrella, in which lie the intrinsic muscles. In the Semaeostomae and some Coronatae, the umbrella is shallowly arched and disc-shaped, while in the Cubomedusae, the other Coronatae and the Rhizostomae, it is bell-shaped or hemispherical. The umbrella of the Coronatae is subdivided by a ring-furrow into a central portion and a peripheral ring. Through gelatinous concentrations and broad and articulated appendages in the interradii, on which one tentacle or tentacular tuft is located, the umbrella of the Cubomedusae is four-sided. In the perradii of this genus, one always finds a sensory body or rhopalium, in a deep sensory niche. In the other genera, tentacular organs are numerous: Tentacles and rhopalia exist on the umbrellar margin in the Coronatae and most of the Semaeostomae in species- and genus-specific numerical orders, while in other Semaeostomae, numerous tentacles may hang between each two sensory knobs on the umbrellar margin, or in bunches on the subumbrella, or no tentacles may be present at all; this latter condition is the case in all Rhizostomae. Between these marginal organs, lappets are formed on the umbrellar margin. In the case of the Coronatae, these, like the tentacle organs, are located on pedalia that are radially separated from one another. An ectodermal velum is found in the Cubomedusae.

The four-sided manubrium, which encloses the mouth opening, hangs down from the subumbrella in the centrum of the medusae. In the Cubomedusae and the Coronatae it is small; in the Rhizostomae and Semaeostomae, it is pulled out to long arm folds. In the genus Rhizostomae, the originally four arms always are cleft into two parts, and are secondarily grown together to a strong pendulum whose edges are highly folded. It is thickly covered with small, knobby tentacles, the digitelles. The coalesced arm folds are penetrated by fine canals which end externally through pores and substitute for the mouth opening.

In all medusae, the gastric cavity is divided in a radially symmetrical manner. While in the Cubomedusae and the Coronatae the original organization is always formed by four septa, as it is in the polyps, no homologs of the septa are formed in the Semaeostomae and the Rhizostomae. The septa of the Coronatae are detached from the margin and are designated as septal nodes, peripherally connected to the stomach. In the Semaeostomae and Rhizostomae, radial canals are formed by entodermal and mesogloeal fusion of the inner walls of sub- and exumbrellas; these canals form a more-or-less branching and anastomosing canal system which is attached to the central stomach. In the Semaeostomae the canal system ends with a ring canal on the umbrella margin, while the canal net in the Rhizostomae can expand between the ring canal and umbrellar margin and into the marginal lappets. Eight gonads and four groups of gastric filaments are found in the Cubomedusae on the four septa and in the Coronatae, on the septal nodes. Since the septa are not present in the Semaeostomae and the Rhizostomae, these organs lie in the entoderm of the subumbrella in a homologous position. Besides, two gonads always are fused into a single organ, so that only four are present.

The sensory knolbs or rhopalia are the tentacle homologs and similar to them in their external form. They are covered with sensory epithelium and lie protected in a niche lined with sensory epithelium. Crystals of calcium carbonate are found in the entoderm cells on the tips of the rhopalia, thus conferring on these organs their pendulum function and permitting them, with the niche, to form a functional unit. Frequently, one or two ocelli are found on the rhopalia; these serve as light sensory organs. The so-called outer sense groove, to which taste function has been ascribed, is found in different species in the cover plate over the rhopalium. Two nerve nets (plexuses) can be demonstrated in the medusae. The giant fiber net regulates the beat of the umbrella, while the diffuse nerve net subserves all other functions. Both systems are connected with each other through the marginal ganglia Horridge (1956).

Up to the present time, three types of nematocysts have been found in the Scyphozoa; however, the cnidae of the Cubomedusae and many species of other genera have not yet been investigated. Only the micro-basic eurytelae occur in all four of the studied genera. In the Stauromedusae, the Semaeostomae and the Rhizostomae, atrich haplonema also are present; in the Coronatae and the Semaeostomae, holotrich haplonema are present in addition to these (Werner, 1965).

The Tetraplatidae (Fig. 8), having two species, are considered to be aberrant medusae. These were represented by Ralph (1960) as a family of the Coronatae. The cone-shaped exumbrella is opposed by a second, inverted cone which is formed of subumbrella and manubrium. A coronal groove runs between the cones.

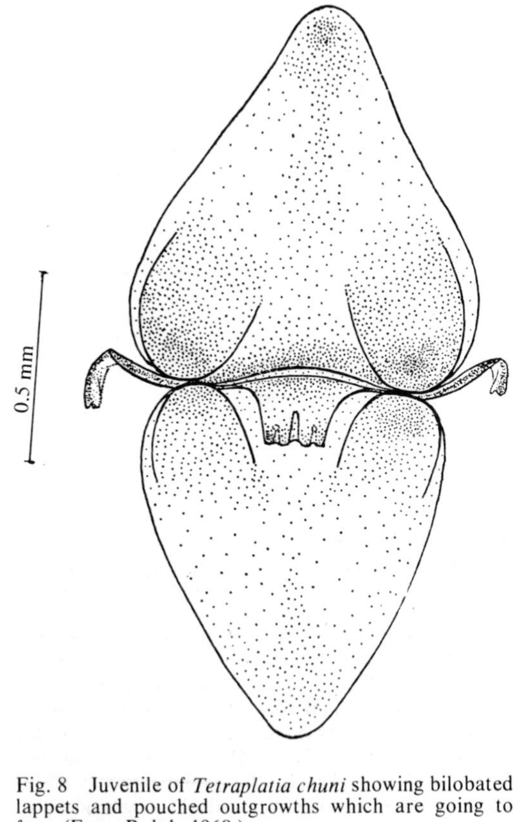

Fig. 8   Juvenile of *Tetraplatia chuni* showing bilobated lappets and pouched outgrowths which are going to fuse. (From Ralph, 1960.)

On this, an ephyroid pair of lappets that carry two rhopalia is prominent on each of the four radii. Between the lappets, pockets of the ex- and the subumbrella grow opposed to one another, fuse, and thus span the coronal groove.

Development. Mature sex products travel through the opened entoderm into the gastric cavity and are released through the mouth opening into the water. Fertilization and development occur mainly in the water, although in some species fertilization can have occurred already within the gastric cavity, and development can begin and be carried on within the arm folds of the manubrium on certain filaments (Rhizostomae). In *Aurelia aurita,* special pockets are formed in the arm folds to hold the developing stages.

Cleavage is total equal in the Stauromedusae and the Cubomedusae, and total unequal in the other three genera. Gastrulation by unipolar migration occurs in the Stauromedusae, Cubomedusae and Coronatae and by invagination in the Semaeostomae and Rhizostomae. Development continues to a planula larva stage. In the Stauromedusae, this is not ciliated but it can creep on the substrate; the planulae of the other genera are ciliated, with the cilia subserving swimming. The planula larva fastens itself to a solid substrate and from it the young polyp, which initially has four tentacles, develops first. After a certain size is attained and the tentacle number has increased, asexual reproduction begins. In Figure 9 (from Hj. Thiel, 1963), the different reproduction possibilities are represented, as they might be found in the Semaeostomae (at the time, only for *Aurelia aurita*): From the planula (1), the young (2) and the older (3) polyp develops. Strobilation, the transverse partitioning of the scyphistome into discs, begins (9) under certain external influences. Species that always produce only one disc have a monodisc strobile; in species that show simultaneous development of more than one disc, one speaks of polydisc strobiles. Normally, development leads to the appearance of the young medusa, the ephyra (16), which grows up to the adult medusa (18). After the ephyra separate, a rest-polyp (17) remains. This can begin the course of development anew. Development as it is represented by (1), (2), (3), (9), (15), (16), and (18) shows the alternation between the polyp and the medusa generations, metagenesis, which, with the exception of the hypogenetic Stauromedusae, is typical of the Scyphozoa. Two Semaeostomae also are exceptions: (a) *Pelagia noctiluca,* which lives on the high seas, shows direct development from the planula to the ephyra; and (b) the viviparous species, *Stygiomedusa fabulosa* (Russell and Rees, 1960), in which a greatly altered polyp generation develops in a chorion of the "mother" medusa and produces a single medusa.

In addition to strobilation, in the normal course of development other forms of asexual reproduction are encountered which start out from the polyp (3): polyp production by way of the lateral bud (4), by the stolon bud (5), by hydratype budding (6), by the podocyst (7), and by the pseudoplanula (8). All these pathways lead back to the polyp (3). However, a polyp (3) can be formed by asexual reproduction if strobilation is disturbed: re-development of the polyp from the strobile (10), the detachment of a disc and its development into a polyp (11), the detachment of a disc by means of a stolon (12), the transformation of a disc to a polyp head and its division into two polyps (13), the breaking

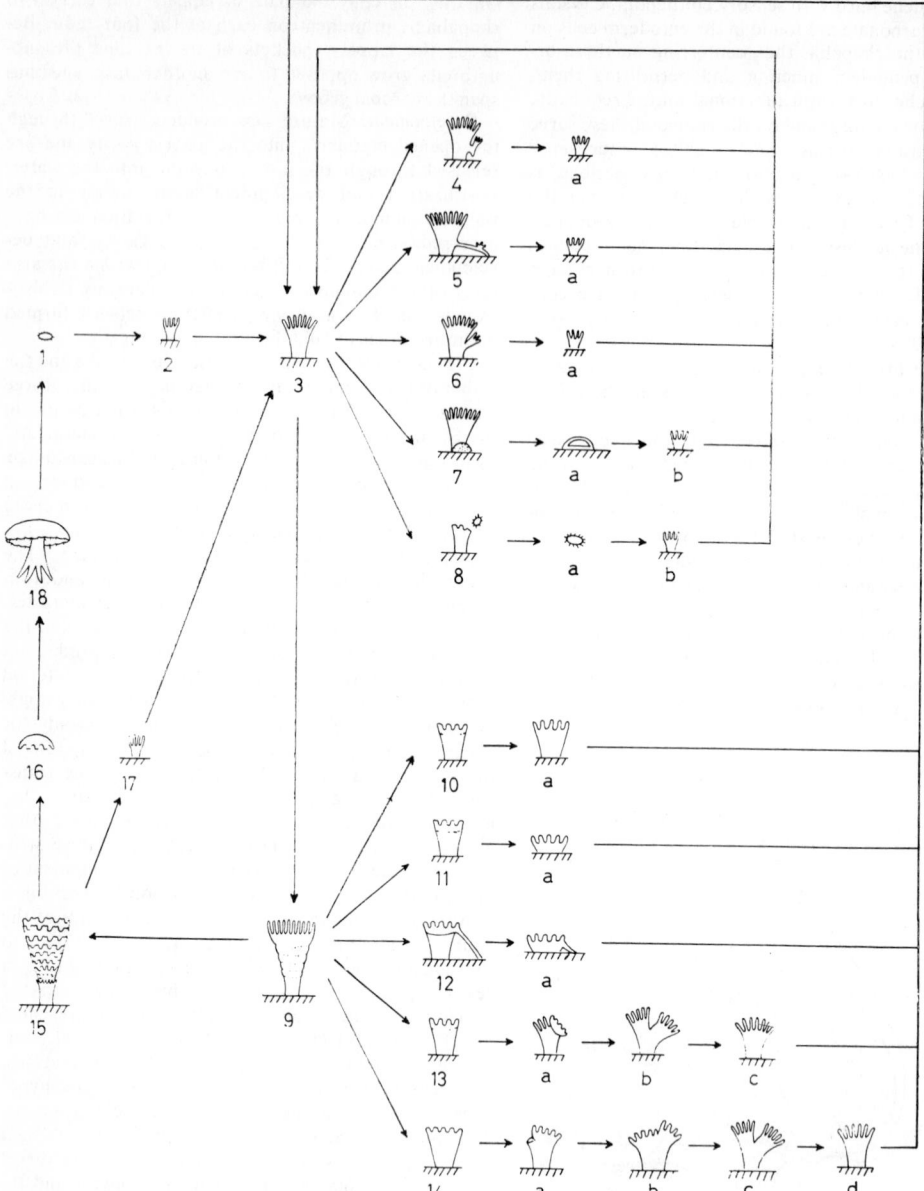

Fig. 9

through of the rest-polyp near the disc, the formation of two polyp heads and their longitudinal division (14).

Evolution. While, because of their similarity, the scyphistomes of the different Scyphozoan genera provide little insight into their evolution, the medusae indicate a close relation between the Coronatae, Semaeostomae and Rhizostomae, as opposed to the Stauro- and Cubomedusae. This relationship is expressed in the phylogenetic schemes of several authors, and is also expressed by Hj. Thiel (1966), who arranged the Scyphozoa on the basis of a sexually reproducing polyp, which in its other characteristics correspond to the present-day scyphistomes of the

Semaeostomae and Rhizostomae. Werner's studies (1966 and 1967) on various species of *Stephanoscyphus*, the polyps of Coronatae, showed that this genus corresponds to the fossil Conulariida, and stands particularly close to the original Scyphozoa. For the stem form of the Scyphozoa, one must therefore assume: (1) at least four tentacles, (2) a four-sided and four-lappet manubrium, (3) four taenioli, (4) four gastric pouches, (5) four or twice four gonads in each of the (6) four septa, (7) a ring canal, (8) four radial canals and (9) a periderm tube. In the evolution to present-day Scyphozoa, one can postulate the origin of metagenesis as the change between polyp and medusa

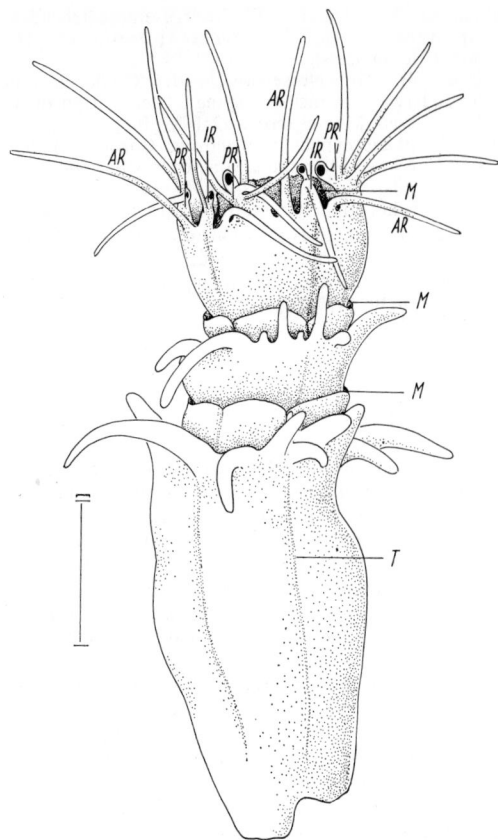

Fig. 10 Polyp-strobila. The terminal polyp carries the original tentacles, the statocysts grow out, the manubrium (*M*) is to be seen, and the septa (*T*) shine through the body wall. (From Hj. Thiel, 1963a.) *Ar*, adradii; *IR*, interradii; *PR*, perradii. Scale: 2.4 cm = 1 mm.

through the "polyp-strobile" (Fig. 10), described by Hj. Thiel (1963). This particular form of strobile does not consist of scyphistome and ephyrae, but of several polyps located one upon the other, as they can be observed in the disturbed strobilation of *Aurelia aurita*. In the polyp-strobile, therefore, equivalent polyps resulted from a particular type of asexual reproduction, the transverse partition of polyps. These equivalent polyps also became sexually mature and could reproduce. One can postulate that as the next step, the basal polyp lost the ability for sexual reproduction, and sexual reproduction was therefore carried on only just by the dividing polyps. Metagenesis, according to this study, was in its first step an alternation of asexually and sexually reproducing polyps. Later, the medusa generation developed from the sexually reproducing polyp generation and, with this, metagenesis between polyp and medusa. Present-day Scyphozoan genera have evolved according to this scheme of development with the exception of Stauromedusae.

Before the evolution of present-day Coronatae, the Cubomedusae, having polyploid symmetry in their medusae and the sessile Stauromedusae each followed their own evolutionary lines which, in their earliest stages, may have been the same. The divergence from the common line of the Semaeostomae and Rhizostomae, and the further development of the Coronatae can be postulated to have evolved from an ephyroid intermediate form. The morphological differences between all groups appear in the sexual generation, since the polyps of the Cubomedusae, Semaeostomae and Rhizostomae do not show any difference with regard to the periderm sheath and the ring canal. The primordia of the organs, nevertheless, are potentially present in them, and always appear in the medusae.

Ecology. The sessile species and developmental stages of the Scyphozoa belong to the benthos; their free-living stages belong to the plankton, as do also the large scyphomedusae whose active movement is less than their drifting because of currents. They represent an entirely marine group and are present in all oceans. Some species penetrate into brackish water, up to a salinity of 15% or less; *Aurelia aurita* is found in the Baltic Sea as far as the Finish coast, where the full life cycle is covered within a water of 6% salinity. The medusae of the Coronatae are found, with few exceptions, in the deep waters of the high seas. The dark red color of many species is typical of occurrence in the deep sea. The polyps of this genus, also easily recognized through their peridermal sheath, are frequently brought up by trawls from 5000 meters and even greater depths. The polyps and medusae of the other genera live predominantly in the shelf region. Generally, only the medusae are found, and the scyphistomes are overlooked since they are insignificant and contract when disturbed. As exceptions already mentioned in the discussion of life cycles, the Semaeostomae *Stygiomedusa fabulosa* and *Pelagia noctiluca* live in the deep sea and on the high seas respectively.

The Scyphozoa primarily capture prey by using their tentacles which, with the aid of the cnidae, attach to nekton and plankton organisms, as well as to detritus. Rhizostomae, which lack tentacles around their margin, capture their prey with the digitelles, past which water and particles are carried with every movement of the umbrella (M.E. Thiel, 1964). If fine particles come into contact with a jellyfish, these will be surrounded by mucus and transported to the mouth opening by the activity of cilia.

Currents, temperature and food are to be listed as important environmental factors for the Scyphozoa. Essential changes in location of medusae occur because of currents, not because of intrinsic motile ability. Because of wind-driven currents, medusae frequently are driven onto the beaches in large numbers. Various authors have described strobilation as being temperature-dependent, but the qualitative and quantitative composition of food likewise plays an essential role. According to Hj. Thiel (1962), the polyps of *Aurelia aurita* in the Baltic Sea can produce ephyrae from two up to four times in one year, in each case under different temperature conditions. Under natural conditions, scyphistomae fed additional food form strobiles earlier than unfed animals. As a function of temperature and food, the life cycle of *Aurelia aurita* of the eastern Baltic compared to that of the same species in the western Baltic is shifted in time and shortened. This is manifested particularly in smaller size and later appearance of sexual medusae in the eastern Baltic.

About 55 species of fish stay in the shelter of the umbrella and tentacles of scyphomedusae and sometimes seek refuge in the subgenital pits or other hollows. (Mansueti, 1963 and Thiel, M. E., 1970). Amphipods of the genus Hyperia live on the medusae and in their gastric cavities, primarily as food parasites. However, they also can eat the tentacles and gastric filaments of their host. *Hyperia galba* is found in the Baltic in dense congestions of polyps of Aurelia aurita. The animals attach themselves in the dorsal position by their pereiopods to several polyps, seize the tentacles of the polyps with their gnathopods, and eat these up from the tip (unpublished observation). The fishes and amphipods, whose sizes correspond completely to the food of the scyphomedusae, are not paralyzed by the nematocyst poison.

The jellyfish are not economically useful. Only one group of Australian natives eats certain species. Massive appearances of jellyfish can damage the fishing industry severly, since they fill the nets, thereby reducing the fish catch, they make hauling in difficult, and they frequently tear the nets because of too heavy a load.

The nematocyst effect of jellyfish on man generally remains slight, according to Stadel (1965), and without medical treatment, does not even appear for some hours. More severe stinging which occurs particularly if a swimmer gets into a swarm of jellyfish can lead to severe health disturbances and even to death. Some Cubomedusae of the Australian waters seem to be especially dangerous to man, despite their small size, since contact with them can lead to death through respiratory and heart failure within a few minutes.

Hjalmar Thiel

(Trans. from German)

References

Horridge, G. A. "The nervous system of ephyra larvae of Aurelia aurita," Quart. Journ. Microsc. Sci., **97**: 59–74, 1956.

Hyman, L. H., "The Invertebrates: Protozoa through Ctenophora," New York, McGraw-Hill, 1940.

Mansueti, R., "Symbiotic behavior between small fishes and jellyfishes, with new data on that between the Stromateid, Peprilus alepidotus and the Scyphomedusa, Chrysaora quinquecirrha (Dactylometra q.)," Copeia, 40–80, 1963.

Ralph, P. M., "Tetraplatia, a coronate scyphomedusan, Proc. Roy. Soc., B, **152**:263–281, 1960.

Russell, F. S. and W. J. Rees, The viviparous Scyphomedusa *Stygiomedusa fabulosa* Russell, J. mar. biol. Ass. U. K., **39**: 303–317, 1960.

Stadel, O., Über die Nesselwirkung der Quallen auf den Menschen und ihre medizinische Behandlung, Abhandl. Verhandl. Naturwiss. Ver. Hamburg, N. F., **9**:61–80, 1965.

Thiel, Hj., Untersuchungen über die Strobilisation von Aurelia aurita LAM. an einer Population der Kieler Förde. Kieler Meeresforsch, **18**:198–230, 1962.

ibid., Untersuchungen über die Entstehung abnormer Scyphistomae, Strobilae und Ephyrae von Aurelia aurita LAM und ihre theoretische Bedeutung, Zool. Jb. Anat., **81**:311–358, 1963.

ibid., The evolution of syphozoa, a review, *in:* "The Cnidaria and their evolution," ed. by W. J. Rees, Symp. Zool. Soc. London, **16**: 77–117, 1966.

Thiel, M. E., Scypomedusae (Stauromedusae through Semeostomae, Rhizostomae in preparation) *in:* Bronns K1. Ordn. Tierreichs 2 (Lieferungen 1–7):1–1308, 1936–1962.

ibid., Untersuchungen über die Ernährungsweise und den Nahrungskreislauf bei Rhizostoma octopus LAG, Mitt. hamb. zool. Mus. Inst. Kosswig-Festschrift: 247–269, 1964.

Thiel, M. E., "Die mit Fischen zusammenlebenden Rhizostomae," Dt. wiss. Komm. Meeresforsch., **21**: 30p, 1970 (in press).

Werner, B., Die Nesselkapseln der Cnidaria, mit besonderer Berücksichtigung der Hydroida. Helgol. wiss. Meeresunters, **12**:1–39, 1965.

ibid., Stephanoscyphus (Scyphozoa, Coronatae) und seine direkte Abstammung von den fossilen Conulata. Helgol. wiss. Meeresunters, **13**:317–347, 1966.

ibid., Morphologie, Systematik und Lebensgeschichte von Stephanoscyphus (Scyphozoa, Coronatae) sowie seine Bedeutung für die Evolution der Scyphozoa. Verh. dt. Zool. Ges., Göttingen 1966, Zool. Anz. Suppl., **30**:297–319, 1967.

SECRETION

Secretion is classically defined as the action of a gland in separating certain matters from the blood and elaborating from them a particular substance, either to fulfill some function within the body or to undergo excretion as waste. The word is derived from the Latin *secretionem*, which itself was formed on the verb *secernere* —to separate (from *se,* aside, and *cerno,* to distinguish), whence the German *sezernieren* and the obsolete English, to *secern*. We may note that the adjective 'secret' is derived from the same root and is essentially an extension of the sense of separation. As we shall see, the term 'secretion' has come to mean more than the act of separation from blood (or sap in the case of plants), and now it does not necessarily imply the 'elaboration of a particular substance.' Furthermore, the word denotes not only the act, but also the product of this activity; thus the saliva is formed by a *process* of secretion, and is described as the *secretion* of the salivary glands.

Glands. Many of the secretions of the complex organism are obvious, not only in so far as the secreted materials are concerned but also in so far as the site at which the process occurs. Thus the saliva appears as a mucinous fluid in the mouth and may be seen emerging from ducts opening into this cavity and leading from organized arrangements of secretory cells—the *salivary glands.* Other examples are found in the tears, secreted by the lacrimal gland in the upper eyelid and orbit, the sweat secreted by glands in the dermis and opening by ducts into the pores of the skin, and so on. These glands open on to the surface of the body; not so obvious are those that open on to certain internal epithelial surfaces; for example, the pancreas, opening on to the duodenum and supplying enzymes required for digestion; nevertheless they are exactly analogous in that they produce a characteristic fluid, or secretion, which is carried away from the cells that produce it by means of a duct. The glands that are the seats of this secretory activity may be characterized, generally, as organized arrangements of specialised epithelial cells which have invaginated into a primary epithelial surface, as illustrated in Fig. 1; it is clear from this diagram that many degrees of complexity can be achieved as developments of this basic pattern of closely packed secretory cells coming into relationship on the one hand with the blood system and on the other with a cavity—acinus or alveolus—whose wall is formed by the cells. The secretory activity, on this basis, is the summated

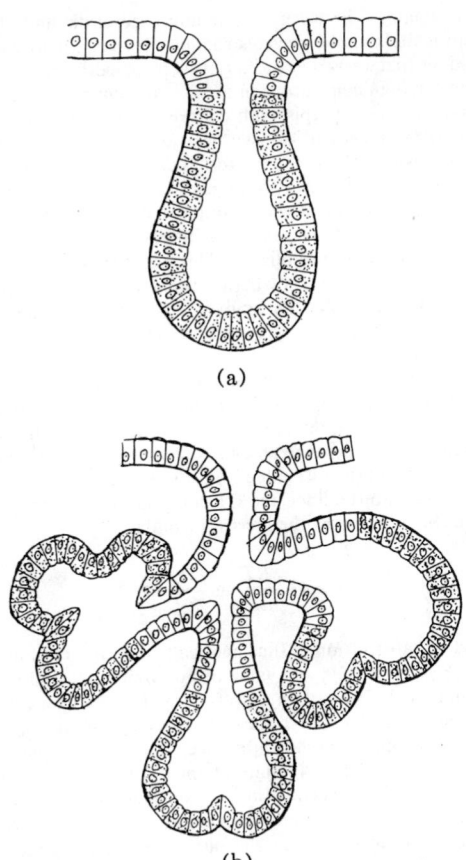

(a)

(b)

Fig. 1 Illustrating invagination of epithelium to form (a) a simple cavity and (b) a more complex arrangement of cavities lined with specialized secretory cells. (After Maximow & Bloom).

Secretion. The products of the exocrine secretory glands have been submitted to chemical analysis, whilst the individual cells have been examined histologically in an effort to discover what visible changes take place in them during the formation of the secretion. The most obvious histological feature in certain cells is the appearance of *secretory granules* within the cytoplasm, usually close to the Golgi apparatus; during the phase of activity these granules multiply and grow and finally appear to empty their contents into the cavity of the acinus or duct constituted by the cells. These granules appear to be the storage form of some essential constituents of the secretion, since high-speed centrifugation of homogenates of pancreatic cells has permitted their separation which on analysis may be shown to contain the characteristic pancreatic enzymes. Fig. 2 is an electron micrograph of the cytoplasm of a pancreatic exocrine cell in the region of the Golgi apparatus; the bodies marked S are secretory granules.

Chemical analysis. With many secretions, however, unequivocal histological evidence of activity has not been demonstrated, and we must rely on less direct, but often more revealing, studies of the process of secretion. The first step is clearly a chemical analysis of the secretion; this tells us in what way it differs from the blood plasma—the fluid from which all secretions are ultimately derived. In general, such analyses have revealed the presence of one or more specific substances, synthesized by the secretory cells from mate-

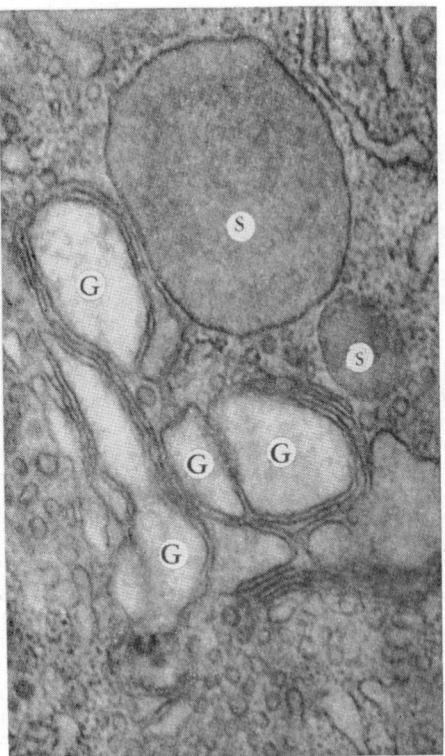

Fig. 2 Electron micrograph of part of an exocrine pancreatic cell illustrating Golgi apparatus and secretory granules. S, secretory granules; G, Golgi vacuoles. × 45,000. (Courtesy J. D. Robertson).

activity of the individual cells; if these are similar, the secretion is homogeneous, but it may happen that several different types of cell line the walls, as with the gastric glands of the stomach which contain four different types responsible for producing the mucus, the pepsin and the acid of the mixed secretion. With certain glands, or parts of glands, the secretory cells are not arranged so that their products lead out to the surface of the epithelium; instead, their secretions pass back into the bloodstream; they are called *endocrine* glands by contrast with the *exocrine* glands so far described. The secretions of the endocrine glands conform with our classical definition in so far as they 'fulfill some function'; in this case they are concerned with the control of the animal's activity, being described as *hormones* which circulate in the blood and influence a variety of centres of activity; e.g. adrenaline, the secretion of the suprarenal medulla, which causes acceleration of the heart, constriction of the blood vessels, dilatation of the pupil, and so on. The pancreas is both endocrine and exocrine, containing endocrine cells that secrete the hormone, *secretin*, and exocrine cells that elaborate the fluid secretion, containing digestive enzymes, that is emptied into the duodenum.

rials derived from the blood plasma, together with most of the constituents of the blood plasma, namely sodium, chloride, potassium, glucose, etc., but these are often in markedly different concentration. Thus saliva from the parotid gland contains an enzyme, amylase, synthesized by the cells of the gland, whilst the concentration of sodium is only of the order of 5 meq. per liter comparing with about 150 meq. in plasma. Because of this deficiency of sodium the fluid has a lower osmolarity than that of plasma. Again, the gastric secretion is a mixture of separate secretions from different types of cell; by indirect methods it has been deduced that the parietal cell secretes what is essentially a solution of hydrochloric acid; i.e., this cell does not, apparently, synthesize any new substance, merely withdrawing H^+ and Cl^- ions from the plasma and secreting them as a solution of HCl.

Plasma filtrates. With some body fluids the concentrations of ions and nonelectrolytes are very similar to those of blood plasma, with the exception that the large protein molecules—serum albumin and globulin and fibrinogen—are absent or present only in traces. Examples are the pericardial and ascitic fluids. Analysis of these suggests that they could have been formed by a process of *ultrafiltration* from plasma, i.e., if the capillary membrane behaves as a porous filter, holding back the plasma proteins by virtue of their large size, it is possible that these fluids are formed by leakage of filtered plasma from these vessels The composition of such an ultrafiltrate can be predicted on thermodynamic grounds and conforms to the Gibbs-Donnan equilibrium. For example the ratio: Concn. in Fluid/Concn. in Plasma will be about 0.96 for Na^+ and K^+ and the reciprocal of this, 1.04, for anions such as Cl^- and HCO_3^-.

Studies of the ionic composition of pericardial and ascitic fluids strongly suggest that this is how they are formed, i.e. no special cells are required to elaborate them; they are indeed *separated* from plasma, but mechanically by filtration; and physiologists distinguish sharply between this process and secretion. Work is required to produce both types of fluid but in the case of filtration it is purely mechanical and the energy is derived from the contraction of the heart; in the case of a secretion work is required, both for synthesis of specific molecules and for the establishment of gradients of concentration of ions and molecules, but the energy cannot be provided mechanically; instead we must invoke the intervention of certain energy-giving chemical reactions. It is by considering the processes whereby cells establish and maintain different concentrations of molecules and ions between the plasma, on the one hand, and a secretion or other fluid on the other, that the real significance of the word secretion, as used today, can be grasped.

Active transport. Let us consider the relative concentrations of ions inside and outside the muscle fiber. The concentration of K^+ is approximately fifty times higher inside than outside, while the reverse is the case with Na^+ and Cl^-. The fiber contains a high concentration of organic anions, represented by A^-, which are not present in the outside fluid and have been mainly synthesized within the cell. We may regard the fluid within the cell, then, as a secretion in so far as the cell has synthesized specific molecules from materials it has ultimately derived from the blood plasma. Moreover, the ionic composition is greatly different from plasma so

the analogy between a glandular secretion and this intracellular fluid is even stronger. However, before deciding that *secretory activity* was involved in establishing the high concentration of K^+, for example, we must enquire whether this high concentration might not be the consequence of the synthesis of the organic anions *per se*; similarly, the low internal concentrations of Na^+ and Cl^- might have been caused in the same way. Moreover, we must ask whether energy is required to *maintain* the existing state of affairs or whether it is only required to establish it. This last question is essentially a question as to whether the system is *in equilibrium* or not, so that the study of ionic equilibria is vital for determining the existence and nature of secretory processes.

There is a difference of potential across the fiber membrane, the inside being negative; this is the resting potential and amounts to about 100 mV. If the muscle membrane is impermeable to Na^+ and A^-, we can compute that there will, indeed, be a potential accelerating K^+ ions into the fiber and Cl^- ions out of it, its magnitude being given by the Nernst formula:

$$E = \frac{RT}{zF} \ln \frac{[K]_{in}}{[K]_{out}}$$

and by substitution of the chemically determined values of $[K]_{in}$ and $[K]_{out}$ a value of approximately 100 mV is obtained. The system therefore appears to be in equilibrium because the observed difference of potential matches the observed difference of concentrations of K^+ and Cl^-. Thus, in spite of the existence of a large difference of concentration of ions, no secretory activity is necessary to maintain the state of affairs. If the concentration of K^+ in the outside medium is raised K^+ passes into the fiber against a gradient of concentration to establish a new equilibrium distribution. Thus, it is not sufficient to state that secretion is a matter of causing ions to move up a gradient of concentration, and we must define more strictly just what is meant by secretion and active transport. If we indicate the difference of potential across the membrane by $\pi_{in} - \pi_{out}$ we have as a condition for equilibrium:

$$\pi_{in} - \pi_{out} = \frac{RT}{zF} \ln \frac{[K]_{in}}{[K]_{out}}$$

which gives:

$$\pi_{in} ZF + RT \ln [K]_{in} = \pi_{out} ZF + RT \ln [K]_{out}.$$

If we define as the *electrochemical potential* the sum of the two terms: $\pi ZF + RT\ln[K]$, we can say that the condition for equilibrium is that the electrochemical potential of a given ion be the same inside and outside the fibre. If equilibrium is disturbed, by adding K^+ to the outside medium as above, the electrochemical potential of K^+ outside has been raised above that inside, and K^+ moves *down a gradient of electrochemical potential* although it is *up a gradient of concentration*. In general terms, then, if we know the p.d. across the boundary between two compartments, and the distribution of a given ion, we can compute whether it is in equilibrium with respect to these compartments, and therefore whether or not active transport mechanisms must be invoked to explain the relative concentrations. This begs the question, of course, as to the origin of the potential, which may require work to establish it and also to maintain it. In the case of the muscle fiber, if our initial postulate is true, namely that the membrane is impermeable to Na^+ and A^-, the system is in equilibrium and the potential requires no energy to maintain it; to establish it, of course, involved the synthesis of anions which required energy. Modern work with isotopes, however, has shown that the membrane is *not* impermeable to Na^+ which can pass back and forth into and out of the fiber. The system is clearly not in equilibrium with respect to Na^+, since the p.d. is such as to assist movement inwards. Thus, on simple thermodynamic grounds Na^+ should diffuse into the cell, down its gradient of concentration and potential, leading to a final condition in which the ratio: $[Na]_{in}$ / $[Na]_{out}$ is equal to $[K]_{in}$ /$[K]_{out}$. In effect, this does not happen, and this can only mean that the Na^+ entering the fibre is 'pumped' out as fast as it enters, i.e. secretory activity, or *active transport*, is directed towards moving Na^+ continuously uphill against a gradient of electrochemical potential. The system as a whole is therefore not in equilibrium; but if we suppose that there is an active transport of $Na\Delta$ maintaining an 'effective impermeability' to this ion, then the behaviour of K^+ and Cl^- can be described without invoking active processes directed towards them specifically.

To summarize, we may say that 'secretion' meant originally the formation of a fluid by a process of separation from the blood; as our knowledge of the chemical composition of secretions increased the term implied also the synthesis of new matter and the directed movement of molecules and ions against gradients of electrochemical potential. It is this last aspect—active transport—that has come to dominate thought in this connection so that 'secretory activity' is now almost synonymous with 'active transport,' although strictly speaking the former is the more general term.

HUGH DAVSON

References

Brown, R. and J. F. Danielli, eds., "Active transport and secretion," Symposium Soc. Exp. Biol., No. 8, New York, 1954.

Harris, E. J., "Transport and accumulation in biological systems," London, Butterworths, 1956.

Turner, C. D., "General Endocrinology," Philadelphia, Saunders, 1966.

SEDIMENTATION*

For the purification, concentration and characterization of biological materials, such as cell components or viruses, the ultracentrifuge is an indispensable tool. Crude suspensions of biological material may be prepared for analytical study by employing a procedure of differential sedimentation, involving alternate cycles of high and low speed centrifugation. For example in the purification of a virus, by a judicious choice of speed and time, one can eliminate both molecules smaller than the virus, which are characteristic of the

*Publication No. 79 from the Dept. of Biophysics, University of Pittsburgh.

host cell's cytoplasm, and larger particles which are in the main cellular debris resulting from rupture of the cells during release of the virus.

After sufficiently concentrating the purified material, be it microsomal particles, nucleic acid strands, an enzyme or an as yet unidentified nucleoprotein, an estimate of its size and shape can frequently be obtained by studying its sedimentation behavior in the analytical ultracentrifuge. Under the influence of a sufficiently high centrifugal field, such molecules, when suspended in a medium of lower density, will migrate peripherally toward the "bottom" of the cell (Figure 1). If all the particles move at the same rate, a boundary is established between the trailing edge of the suspension and the resulting solvent. Figure 2 shows a series of exposures, at given time intervals, demonstrating a moving boundary as revealed by ultraviolet light. It can be seen that the biological material to the right of the boundary absorbs readily in the ultraviolet, whereas the solvent to the left of the boundary is relatively transparent to light of this wavelength. This boundary, or concentration gradient, since it represents a refractive index gradient too, can be made visible by employing such other optical systems as the Schlieren Cylindrical Lens Method or the interferometric technique. An example of a boundary as viewed with the Schlieren optical system is shown in Figure 3.

The ultracentrifuge can offer some indication of the homogeneity of a macromolecular suspension of particles. For a solution containing one solute component of constant size, only one gradient should be apparent in the ultracentrifuge, while its rate of spreading to a first approximation should be characteristic of the diffusion coefficient of the particle. If, however, more than one peak is observed it can be assumed that either additional components are present or the material is inhomogeneous with respect to size, shape or density. Contamination, unless it is homogeneous within itself and of sufficient concentration to be resolved by the optical system may not be observed, nor may very large particles such as bacteria which sediment too rapidly for observation.

In sedimentation velocity experiments, one measures the rate at which this boundary moves. The velocity of the particle, dx/dt, depends upon the angular velocity of the centrifuge, ω, where $\omega = 2 \cdot \pi \cdot$ revolutions/sec, the distance of the particle from the axis of rotation, x, the density of the suspending medium, d, and the mass, m, frictional coefficient, f, and partial specific volume, V, of the particle, respectively, according to the relationship $m(1 - Vd)\omega^2x/f$. The velocity per unit field, $(1/\omega^2x)$ (dx/dt), has been defined as the sedimentation coefficient, S, and generally is expressed in "Svedbergs," as $cm/sec/unit$ field x 10^{13}.

Sedimentation is the result of the two forces which act upon a particle in the centrifuge; F_c, the force due to the centrifugal field and, F_r, the force due to the frictional resistance to movement of the particle as it migrates through the suspending medium. When the two forces become equal, the net force on the particle is zero and it moves with approximately constant velocity (for small increments of distance) away from the axis of rotation. This situation, which is basic to all sedimentation velocity measurements, can be described mathematically as $F_c = m(1 - Vd)\omega^2x = F_r = f (dx/dt)$.

The sedimentation coefficient, S, can be evaluated by

Fig. 1 Analytical rotor capable of operating at 60,000 rpm. Cell counterbalance, cell housing, centerpiece and sector cups holding quartz windows. (Courtesy Specialized Instruments Corporation.)

plotting the natural logarithm of the distance, ln x, as a function of time. Dividing the slope, $S\omega^2$, by ω^2 yields the sedimentation coefficient.

In order that particles studied under different conditions may be compared as to sedimentation rate, the measured value of the sedimentation coefficient, S_t, is corrected to the sedimentation rate which the particle would possess in a medium having the viscosity and density of water at 20° C, designated S_{20}^w, by the equation

$$S_{20}^w = S_t \frac{(1 - V_{20}^w d_{20}^w)}{(1 - V_t d_t)} \frac{\eta t}{\eta_{20}^w}$$

Here η_{20}^w and d_{20}^w represent the viscosity and density, respectively, of water at 20° C, and η_t and d_t denote the viscosity and density of the solution at the temperature of the experiment.

Since the sedimentation rate of a biological macromolecule is proportional to the difference in density between it and the surrounding medium, one can estimate

Fig. 2 Sedimentation of the DNA from T2 bacteriophage as seen by ultraviolet light absorption. The corrected sedimentation coefficient is 33S. Rotor speed 35,600 rpm. Time between exposures 8 minutes.

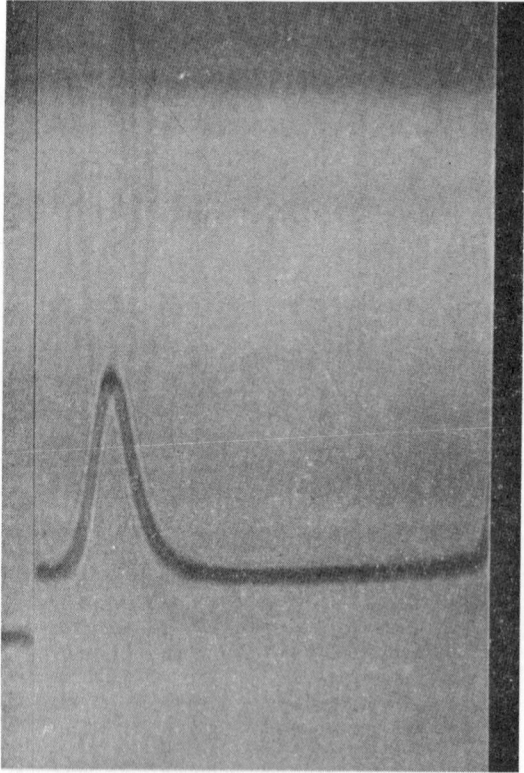

Fig. 3 Sedimentation of a 1.03% solution of bovine plasma albumin. Schlieren optics, bar angle 60°. Rotor speed 59,780 rpm.

the density of the particle in solution by increasing the density of the medium until there is zero sedimentation. Materials such as sucrose, glycerine, serum albumin and deuterium oxide have been used for this purpose. The "hydrated" density of several viruses have been determined in this manner.

The basic sedimentation velocity equation indicates that the sedimentation rate is inversely proportional to the frictional coefficient of the particle, and that if the latter value is known the mass or molecular weight of the particle can be determined. The frictional coefficient can be obtained from a diffusion experiment. Thus, by combining sedimentation and diffusion data one can arrive at the molecular weight of a given substance (see DIFFUSION TECHNIQUES). The hidden assumption in this procedure is that the frictional coefficient for the particle during sedimentation is exactly equivalent to the frictional coefficient for the particle during diffusion. This implies no orientation of the particles under the influence of a centrifugal field.

The sedimentation coefficient itself is sufficient to yield information concerning the mass of a spherical particle of known density. For non-spherical particles, however, one must combine sedimentation measurements with either diffusion or intrinsic viscosity results to obtain the molecular weight. Theoretical interpretations of the sedimentation coefficient, as it relates to the size, shape and hydration of macromolecules in solution, as well as comprehensive discussions of ultracentrifugation in general, are available in the references cited.

Molecular weights can be determined directly from the ultracentrifuge by the method of sedimentation equilibrium. With this technique, by operating the ultracentrifuge at a comparatively low speed, there is set up at equilibrium a distribution of particles resulting from the combined effect of simultaneous sedimentation and diffusion. Analogous to the distribution with height of gas molecules in the atmosphere, particles in solution under the influence of a centrifugal field will distribute themselves in the sedimentation cell when equilibrium is achieved in accordance with the equation

$$ M = \frac{2RT \ln (c_2/c_1)}{(1 - Vd)\omega^2(x_2^2 - x_1^2)}, $$

where M is the molecular weight, R is the universal gas constant, $(1 - Vd)$ is the buoyancy correction, and c_1 and c_2 are the concentrations at distances x_1 and x_2 from the axis of rotation.

A major disadvantage of the equilibrium method is that the equation is valid only after complete equilibrium has been attained, which may require several days if extremely small particles such as protein molecules are being studied. For such molecules, which are too small to arrive at equilibrium within a reasonable time, a technique intermediate between sedimentation velocity and sedimentation equilibrium, an approach to sedimentation equilibrium, can be employed. The applicable equation for this technique is

$$ M = \frac{RT[1/x_b (dc/dx)_b - 1/x_m (dc/dx)_m]}{(1 - Vd)\omega^2(c_b - c_m)} $$

where $(dc/dx)_m$ and $(dc/dx)_b$ are the concentration gradients at the meniscus and bottom of the cell, and c_b and c_m are the concentrations at positions x_b and x_m, respectively.

A highly concentrated solution of a salt such as cesium chloride establishes a density gradient in a centrifugal field. If material is sedimented in such a medium, at equilibrium the macromolecules are confined to a narrow band of the density gradient, which approximates the solvated density of the biological material. Besides, the width of the band, being a function of the diffusion coefficient of the particle, is related to the particle's molecular weight, M, by the equation

$$ M = \frac{RT}{V(d\rho/dx)_{x_0} \omega^2 x_0 \sigma^2} $$

where $d\rho/dx$ is the density gradient, x_0 is the radius of the band center, and σ is the standard deviation of the normal curve. As an indication of the resolving power of this method, deoxyribonucleic acid containing N^{14} has been separated from deoxyribonucleic acid containing the heavy isotope N^{15}.

IRWIN BENDET

References

Svedberg, T. and K. O. Peterson, "The Ultracentrifuge," Oxford, Clarendon, 1940.

Schachman, H. K., "Ultracentrifugation in Biochemistry," New York, Academic Press, 1959.

SEED

A seed is a ripe ovule which typically consists of one or more seed coats, an embryo, and stored food. The seed habit is known to have evolved independently at least four times in the plant kingdom (in the Mazocarpales, Lepidocarpales, and Miadesmiales of the LYCOPSIDA; and in the Marattiales of the PTEROPSIDA), and exists today in the familiar and numerous descendants of the marattiaceous line—the GYMNOSPERMS (which bear "naked" ovules on the surface of sporophylls) and ANGIOSPERMS (which bear ovules enclosed in folded sporophylls or carpels, comprising the ovary).

The protective seed coats (integuments) are derived from outgrowths of the upper (inner) surface of the sporophyll. There may be none (Symplocarpus) to several (Xylopia), but typically one surrounds the gymnosperm seed and two envelop the angiosperm seed—an outer layer or testa and an inner, often papery layer. There is a small pore—the micropyle—which is classically described as the entrance point of the pollen tube, although pollen tubes may penetrate the ovule at any point along its surface. In the angiosperms, a scar called the hilum is to be seen near the micropyle. This marks the site of attachment to the funiculus or ovarian stalk. In some cases, accessory structures, e.g. the aril of Taxus, Araucaria, Myristica and the caruncle of Ricinus develop on or around the seed.

The embryo or young sporophyte is formed from the fertilized egg cell, develops to varying degrees in different species, and typically becomes dormant. Usually, one or more cotyledons (typically one in the monocots, two in the dicots, and several in the gymnosperms), an epicotyl (shoot apex), hypocotyl (embryonic stem),

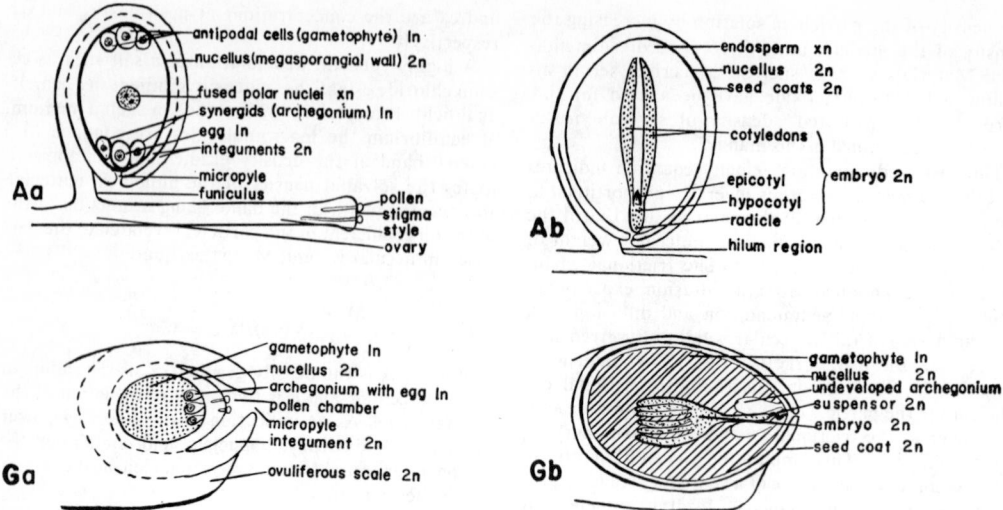

Fig. 1 Diagrammatic embryo sacs (a) and maturing seeds (b) of a gymnosperm (G) and a dicot angiosperm (A). 1n = monoploid; 2n = diploid; xn = variable ploidy.

and *radicle* may be distinguished. The bulk of the seed consists of a mass of stored food (see figure). In the gymnosperms, the nutritive tissue is the monoploid female gametophyte. In the angiosperms, a POLYPLOID, nutritive endosperm tissue proliferates as the result of a fusion of the second sperm nucleus (two are produced in each pollen tube) with one or more polar nuclei of the embryo sac (female gametophyte). It may remain separate from the embryo (albuminous seeds) or become incorporated into the cotyledons (exalbuminous seeds). In a few angiosperms (*Castalia, Coffea*) some nutritive tissue (*perisperm*) develops from the diploid *nucellus* or megasporangial wall. In any case, the stored food is a rich supply of carbohydrates in the form of sugars, starches, and hemicelluloses (see CARBOHYDRATE STORAGE); fats in the form of oils; proteins usually as aleurone grains; and various vitamins and growth factors. Some authorities have advocated the use of fortified cereal grain (seed) proteins (with lysine, tryptophan and threonine added if necessary) to supplement animal sources in population dense, underdeveloped areas of the world.

Many species regularly produce normal looking seeds which contain no embryo (*Ginkgo, Citrus*) or seeds which are polyembryonic (*Alnus rugosa*), and in some cases (*Taraxicum, Crepis*) seeds may form apomictically—asexually. Parthenogenesis is but one popularly cited example of APOMIXIS.

Since the FRUITS and seeds of the higher plants furnish the principal means of dispersal for the species, interesting adaptations may be cited for disseminating these structures by wind, water, and animals. Adapted for wind dispersal are the winged seeds (*Pinus, Catalpa*) and fruits (*Acer, Ulmus*) which are produced by many trees, the small, dust-like seeds produced by the orchids, and the feathery, parachute seeds (*Asclepias*) and fruits (*Taraxicum*) borne by many weed plants. Buoyant seed and fruit coats are found in such water dispersed species as *Castalia* and *Cocos*. Some plants, like *Impatiens (Touch-me-not), Ecballium elaterium* (squirting cucumber), and *Hamamelis* (witch hazel)

forcibly eject their seeds as turgor pressure builds up during fruit ripening. Plants such as *Bidens, Zanthium,* and *Phoradendron* (mistletoe) produce barbed, hooked, and sticky fruits respectively. These adhere to the animal body and may thus be transported to new locations. Most animal dispersed seeds are encapsulated in attractively colored, fleshy fruits which are eaten by birds or other animals. The seeds withstand enzymatic activity and are passed out of the digestive tract unharmed. There is evidence that birds may retain viable seeds for periods of 8 to 340 hours (the latter recorded for *Rhus glabra* seeds in sandpipers). Dispersal distances of up to several thousand miles may thus be provided by migratory, seed retaining birds.

The physiology of seeds has been of interest both commercially and scientifically. The auxin (see HORMONE, PLANT) content of angiosperm seeds at varying developmental stages indicates that the growth of the pollen tubes down the floral styles initiates free auxin production by the ovary wall. After FERTILIZATION, the endosperm produces auxin and finally the embryo itself becomes the major synthetic site of growth hormones. This auxin supply emanating from the seed prevents abscission of the flower and promotes the development of the ovary and accessory structures into the fruit. *Parthenocarpic* (seedless) fruits are variously induced by alien pollen, pollination not followed by fertilization, insect infestation, application of auxins to the pistils, and other irregular stimuli. Recently, excised embryos and young, fertilized ovules have been successfully grown in sterile culture on controlled media.

Although some cases of direct germination of seeds are known (mangrove and corn seeds often germinate within the fruit while still attached to the parent plant), most seeds require a period of dormancy before germination is possible. This is, of course, desirable for species which must overwinter in unfavorable climates or for desert species which must adapt their life cycles to limited periods of moisture. There are five principal factors which influence dormancy:

1. immaturity of the embryo (*Ginkgo, Cocos*)

2. chemical immaturity requiring a period of after-ripening (peaches, plums, apples, etc.)

3. hard seed coats (*Alisma, Chenopodium*)

4. impermeability of seeds to water or oxygen (*Juniperus, Zanthium*)

5. presence of inhibitory substances in the fruit (tomato, *Iris*)

These factors are overcome in time, prompted by external stimuli such as successive freezing and thawing, drying and moistening, or partial digestion within the intestine of an animal. Specific periods of dry and cold storage, scarification of seed coats, etc. are employed to break the dormancy of seeds of commercial interest. Biennials or winter annuals such as winter cereals and lettuce are routinely subjected to low temperature vernalization. The seeds are soaked and then maintained at temperatures just above the freezing point for several weeks. This shortens the vegetative growth stage and insures that flowers will form and fruit will set in one season, instead of the two normally required by these plants.

Viability is the capacity of a seed to germinate (resume active growth and produce a young plant). Longevity is the length of time a seed may remain dormant and still be viable. These are variable characteristics. The Indian Lotus (*Nelumbo nucifera*) holds the record for maximum longevity. There are authenticated reports of seeds now estimated to be about 1000 years old (by radio-carbon measurements) which were discovered in a drained lake bed in Manchuria and successfully germinated. By contrast, the seeds of *Acer saccharinum* have a life span of only a few weeks. In general, low relative humidity and low temperatures (—4°C.) are ideal storage conditions for seeds, although *Citrus* and coffee are notable exceptions requiring high moisture storage.

Germination of a viable seed will occur provided there are present water, oxygen, a suitable temperature range (usually 25–30°C), and light (this last is not required by many seeds). Water is imbibed and the seed coats either soften by absorption or burst by the swelling of the interior. This is now accompanied by heightened physiological activity. The respiratory rate goes up (as evidenced by heat generated and CO_2 liberated), ENZYMES digest the stored foods making them available to the embryo, and auxins are transferred to the embryo. Ultimately the radicle and cotyledons emerge from the seed coat, the hypocotyl or epicotyl elongates rapidly, and the seedling appears.

JOAN EIGER GOTTLIEB

References

Crocker, W. and L. V. Barton, "Physiology of Seeds," Waltham, Mass., Chronica Botanica, 1953.
Meyer, B. S., D. B. Anderson and R. H. Böhning, "Introduction to Plant Physiology," Princeton, N.J., Van Nostrand, 1960.

SENSE ORGANS

A sense organ is a part of an organism that is especially sensitive to a particular form of energy and relatively insensitive to others. The simplest sense "organs" are those that consist of the histologically unspecialized final terminations of a single afferent nerve fiber. Aggregations of such individual receptors comprise the sensory surface of more complex sense organs. Accessory structures which serve to transmit the stimulus to the neural receptor are common for both single receptor organs and complex ones.

The receptor cells of all invertebrate sense organs are primary sensory neurons and, therefore, may consist entirely of the dendritic termination(s) of the cell. Many of these, especially in the Arthropoda, end in association with a sensillum; that is, an architectural modification of the cuticle with its associated epidermal cells. Many of these are multiply innervated and approach the more elaborate vertebrate organs in structural complexity. Homologous primary sensory cells also occur in the vertebrates but the receptor surface of most vertebrate sense organs is composed of secondary sensory cells; that is, modified epithelial cells which receive the stimulus and transmit the response to associated primary sensory neurons.

One method of classifying sense organs is based on the source of the stimulus and the anatomical location of the receptors. *Exteroceptors* are those located near the surface of the body and which provide information about the immediate external environment, whereas *interoceptors* are located in visceral organs and give information about these structures. *Proprioceptors* are found in muscles, tendons and joints and thus signal movements and position of the body, or parts of the body. *Teloceptors,* such as the eye and ear, are located at the surface of the body but respond to stimuli that normally originate at some distance from the body.

Another method of sense organ classification uses the kind of stimulus to which the receptors are most sensitive as a basis. Thus are distinguished *photoreceptors* stimulated by light, *chemoreceptors* sensitive to chemical change, *thermoreceptors* sensitive to temperature and *mechano-receptors* which respond to physical forces. Subdivisions of these categories are based upon further restrictions on the nature of the most adequate stimulus. Thus, for *mechanoreceptors* are distinguished, among others, *phonoreceptors* normally stimulated by vibratory motion, *stretch receptors* affected by a stretch of the peripheral endings and *touch* or *pressure receptors* usually affected by compression forces. Similarly among *chemoreceptors* there are distinguished, in some organisms, olfactory receptors normally stimulated by molecules which arrive at the receptor surface in a gaseous state, and *taste receptors* stimulated by chemical solutions. However, this method of distinguishing between olfaction and gustation does not pertain to aquatic organisms in which the two modalities can sometimes be assumed on other, though related, grounds.

Neither of these methods of classification, nor any other single method, can adequately describe all of the different kinds of sense organs found in the animal kingdom. Often, an adequate description of the function of a given sense organ is given only by a precise description of the adequate stimulus, the means by which this stimulus is caused to act on the receptor surface and the relation the sense organ has to the rest of the body. For example, the invertebrate statocyst serves as a *gravity* or *equilibrium receptor* because it is so structured and located that it is stimulated mechanically by a solid body, the statolith, which acts upon different areas of the sensory epithelium when the organism changes position relative to gravity. Thus the statocyst is at once a gravity receptor, a proprioceptor and a mechanoreceptor.

The only way that sensation can be studied directly is by correlating the introspective report of a human subject with experimentally controlled stimuli applied to a specific sense organ. However, considerable information about the functions of sense organs, apart from the sensations they cause, may be obtained in nonhuman subjects by studying either the behavior of an organism in response to controlled stimulation or the electrical response of the receptor or its sensory nerve fibers, or both. The former method has yielded much valuable information in the past, and will undoubtedly continue to do so, but the method that presents final proof of function, eliminates much of the subjective judgement necessary in behavioral observations and provides more direct quantitative data is the electrophysiological one.

It is known that sensory neurons signal receptor response by generating repetitive, self-propagating, all-or-none electrical variations called action potentials or spike potentials. This method of response is, of course, the common one for all neurons, motor and internuncial, as well as sensory.

In primary sensory neurons the action potential response is not complicated by the transmission of the excitation from other cells to the sensory neuron ending. In many sense organs it is possible to record action potentials from a single sensory unit. Ideally this consists of a single sense cell which connects with a single afferent fiber, although in some sense organs more complex arrangements of sense cells and fibers may respond as a unit. Unitary sensory responses are recognized by the regular repetitive nature of like action potential discharges that are detected when the end organ is stimulated. This was first accomplished from the frog muscle spindle by Adrian and Zotterman. Many properties of sense organ response have since been ascertained by this method.

Since action potentials, and thus nerve impulses, which arise from different sense organs are qualitatively similar, the different sensations they produce must be due to their different central terminations. This is direct evidence for the *doctrine of specific nerve energies* which states that the sensation produced is a function of the sense organ stimulated regardless of the method of stimulation.

Stimulus intensity, and, therefore, response intensity, is signalled by variations in action potential frequency. Weak stimulation of single sensory units results in a low frequency of action potential discharge and stronger stimulation yields higher discharge frequencies. In some sense organs, it has been shown that within a limited range of stimulus intensities the frequency of discharge is a linear function of the logarithm of the stimulus. Within these ranges then the spike discharges, if they are substituted for intensity of sensation, follow the Weber-Fechner Law, which states that sensation intensity is a linear function of the logarithm of the stimulus intensity. Sensation intensity, like spike frequency, follows this law only over a limited range.

An additional neural correlate of stimulus intensity is the number of end-organs discharging. The weakest stimuli applied to a population of like receptors activate those with the lowest threshold. Stronger stimuli not only increase the discharge frequency of those organs already activated but also initiate discharges in organs with higher thresholds. Maximal stimulation is reached when all receptors are responding at maximal frequency.

Spike discharge frequency, usually not constant during the action of one stimulus, is influenced by the length of time a stimulus is allowed to act and the rapidity with which it is applied. Typically the frequency is initially high and then becomes progressively less until either a plateau rate is achieved or the rate drops to zero. The decrease in frequency from the initial rate is called adaptation. Complete adaptation is achieved when the discharge rate reaches zero. Adaptation rates vary with different sense organs from incompletely adapting organs to rapidly adapting receptors. Among incompletely, or very slowly adapting, receptors are some vertebrate proprioceptors, thermoreceptors and pain receptors as well as some invertebrate proprioceptors. Tough receptors such as those associated with hairs in mammals, the encapsulated Pacinian corpuscle as well as analogous structures in the invertebrates are among the rapidly adapting receptors. An extreme example of these is the touch receptor which adapts to zero after one impulse. Receptors which adapt completely but intermediate in time between the slow and fast adaptors include chemoreceptors and photoreceptors. Some receptors show a continuous background discharge frequency even in the absence of any apparent stimulation. These receptors respond to stimulation by an increased discharge frequency above the basal level and adapt completely by returning to the basal frequency. Among these are mammalian phonoreceptors and receptors of the fish lateral line organ.

Another kind of electrical potential is known to occur, consequent to stimulation, in sense organs. These potentials are slow, graded (not all-or none) and non-propagating (confined to the region of the sense organ). They are called receptor potentials or generator potentials. It is now widely believed that these potentials are the immediate cause of impulse initiation in the afferent nerve fibers of sense organs. The nature of receptor potentials varies widely with the sense organ studied. For complex organs like the mammalian eye and ear, they are complex wave forms which are given special names such as electroretinogram and cochlear microphonices respectively. For single receptors the potential takes the form of a slowly developed and long-lasting negativity of the actual receptor region as against the more centripetal region of the nerve ending. Among single receptors from which generator potentials have been recorded are Pacinian corpuscles and muscle spindles in vertebrates as well as certain invertebrate stretch receptors and photoreceptors. Current notions on the origin of the generator potential and the mechanism by which it causes repetitive spike potential activity in the sensory nerve ending are still speculative and controversial.

In the invertebrates relatively few receptors have been identified with reasonable certainty by careful physiological experimentation. Tentative identifications in the older literature without adequate experimental observations are to be viewed with skepticism.

Among invertebrate chemoreceptors that have been identified are the auricular organs of turbellarian Platyhelminthes, certain hair cells in the epithelium of molluscs, sensilla basiconicae, trichodeae and placodeae in insects, certain spines on the coxae of *Limulus* legs and hair sensilla on the appendages of decapod Crustacea. In the vertebrates the taste buds of the mammalian tongue and a restricted area of the nasal epithelium have been so identified. Homologous structures in the lower vertebrates are also known.

Statocysts are known in the coelenterates and crustaceans and their analogous structures, the membranous labyrinth of the inner ear, are known in the vertebrates. Other kinds of proprioceptors have been identified in the invertebrates only in the arthropods. These include primary sense cells whose dendrites end among epithelial cells of soft leg-joint cuticle in *Limulus* and Crustacea, hair and campaniform sensilla of insects, slit sensilla of Arachnida, muscle receptors in insects and Crustacea and many chordotonal organs of insects. In the vertebrates muscle and tendon receptors are conspicuous proprioceptors.

In the vertebrates certain nerve endings on the tongue, skin and cornea have been identified as either warm or cold receptors. More specialized receptors so identified are the ampullae of Lorenzini in elasmobranch fishes and, among snakes, the facial pits of pit vipers.

Invertebrate tactile receptors are known in the arthropods, especially insects, as long movable hairs with joints in the skeleton. Vertebrate tactile receptors are known as undifferentiated nerve endings in the skin, mucous membranes and internally, as well as encapsulated endings with a wide bodily distribution. Some of the vertebrate undifferentiated endings may also be specific for noxious stimulation of any kind and thus subserve the modality of pain.

A variety of vibration receptors and phonoreceptors are known in the arthropods. These are the subgenual organs and tympanal organs of insects, hair sensilla of insects and spiders and slit sensilla of spiders. In higher vertebrates the cochlea of the inner ear contains phonoreceptors while in amphibians the lagena, an evolutionary precursor to the cochlea, serves this function. In fishes the sacculus, utriculus and lagena of the inner ear may serve as phonoreceptors. The lateral line organs of fishes are also vibration receptors.

Photoreceptors are easily identified throughout the animal kingdom. Ocelli are known in coelenterates, flatworms and annelids. Molluscs have more complex eyes, especially the cephalopods in which the eye resembles that of the vertebrates. Both ocelli, and compound eyes are common in arthropods. The sixth abdominal ganglion of the crayfish is also a photoreceptor. In the vertebrates the eye is conspicuous, easily identified and similar throughout the group.

SAUL B. BARBER

References

Prosser, C. L. and F. A. Brown, "Comparative Animal Physiology," Philadelphia, Saunders, 1961.
Granit, R., "Receptors and Sensory Perception," New Haven, Yale University Press, 1955.

SETA

Seta (pleural, setae) is a rod-, needle-, or hair-like structure growing within and/or projecting from the body wall of certain invertebrates. The structures are usually associated with the ANNELID worms and certain ARTHROPODS, although other forms possess them.

Marine annelids possess setae in great numbers, frequently several millimeters in length. They extend laterally from the segments, occurring in bundles, and functioning principally in locomotory and respiratory movements. The aquatic (primitive) oligochaeges also have bundles of setae.

The terricolous oligochaetes (earthworms) possess setae which occur singly. As few as eight per segment occur in the Lumbricidae, or as many as one hundred per segment in the Megascolecidae. These usually are about one millimeter in length. Muscles are attached to these setae so that they can be extended, withdrawn, or moved oar-like for locomotion. They may also be accessory to reproduction, skeletal support, and excretion.

The chemical composition of setae is not well known, but it does vary among the species. Those of *Enchytraeus* are not chitinous since they are insoluble in hydrochloric acid. Those of *Lumbricus* are chitinous, or at least partly, since they are insoluble in potassium hydroxide and partly soluble in hydrochloric acid.

Setae are apparently formed from a single cell. When a seta is lost, a new generative cell produces a new seta, beginning with the distal end.

Setae in the insects are also hair-like processes of unicellular origin, which may constitute the chief covering of the exoskeleton. Some setae are spine-like, others branched or plumose, and others flat and squamous-scaled. The core of the fully formed seta shrinks and withdraws, leaving a cavity as in the Oligochaeta. There is even an open pore at the distal end of the setae in the alder flea beetle.

Scale setae are best known in the Lepidoptera. Here setae are formed from the wing epidermis, and by a series of changes of shape become flattened into a scale. Pigmentation of the scale is said to be formed by blood corpuscles entering the fully formed scale after retraction of the primary scale. Iridescence so characteristic of these setae is the result of surface sculpturing.

Poison setae in certain lepidopterous larvae are also known. The venom is formed by a special poison gland cell associated with the formative cell. The poison issues from the seta when its tip is broken off. Annually, many people are "stung" by these caterpillars simply by brushing against them. Notodontidae, Liparidae, Megalopygidae, Arctiidae, Noctuidae, Eucleidae, Saturniidae, and Nymphalidae all have larvae equipped with such setae.

Aquatic crustaceans have setae that function in a variety of ways. Some are sensory or tactile, others are branched and serve to increase surface area, thereby increasing buoyancy. Still others are so situated and shaped that only ornamentation can be attributed to them.

Students of the arthropods lack uniformity in the application of the word seta. Usually "small" projections of the exoskeleton are termed seta and "larger" ones are termed spurs, spines, or bristles.

WALTER J. HARMAN

References

Richards, A. G., "The Integument of Arthropods," Minneapolis, University of Minnesota Press, 1951.
Stephenson, John, "The Oligochaeta," Oxford, Clarendon Press, 1930.

SEX

As a biologic concept, sex is a comparison of contrasting characteristics in organisms that produce either eggs or sperm. It is not a force to produce these contrasts, but simply the aggregate of genetic, anatomic, physiologic and psychologic qualities that we recognize as maleness or femaleness. Egg and sperm may be produced by a single individual; such cases are known as hermaphrodites. Since some lower animal forms are naturally hermaphroditic, sexuality may relate only to the genetic characteristics of the gametes and omit completely any consideration of somatic variations among individuals. But for most animals, including all the protochordates and chordates sexuality refers to the composite of genetic, somatic and behavioral contrasts between males and females. The somatic differences involve mostly the reproductive system and various accessory structures, frequently ornamental, that are related somehow to the process of reproduction. Even though sex-specific characteristics may exist that have no apparent association with reproduction, these usually result from the action of hormones of the reproductive system. Examples would include sex-differences in the normal concentration of certain blood constituents or the fact that tolerance to drugs may differ between the two sexes of a single species.

Sex determination. *Genetic* sex is established at fertilization and until differentiation of the gonads during embryonic development of the individual, chromosomal constitution may be considered the sole manifestation of sex. The term *sex determination* is often applied to the establishment of chromosome constitution at the time of fusion of egg and sperm. With each gamete contributing a haploid compliment of chromosomes, the fertilization process establishes the diploid chromosome number, characteristic of the species and found in the nuclei of all somatic cells. In the human, for example, the diploid number of 46 chromosomes has now been amply verified. In males and females, 44 chromosomes may be classified in 22 pairs, termed autosomes. The remaining two of females are paired, also, and are known as the sex chromosomes symbolized as XX. Males, however, have sex chromosomes that are morphologically distinguishable; one resembles the X chromosomes of females and the other is the much smaller Y chromosome. The sex chromosome constitution of the male, therefore, is XY and it is referred to as the heterogametic sex. Consequently, when chromosomal reduction to the haploid number occurs in the process of sperm formation (spermatogenesis) half of the sperm are endowed with a set of autosomes and an X chromosome, while the other half contain the 22 autosomes and a Y chromosome. The former may be termed gynosperm and the latter androsperm. Obviously, the female produces only one kind of egg, all carrying the X chromosome; this is the homogametic sex. It is assumed that sex is genetically determined on the basis of a quantitative balance between male-determining and female-determining genes. Although the sex chromosomes are the primary vehicles for sex-determining genes, in certain animal forms autosomal factors may contribute to the quantitative balance that decides sexuality. The "balance" system works so that two doses of the female determiners on the X-chromosomes (XX) overcome the male determiners outside of the X,

but one dose (XY) is insufficient for the dominance of female factors. The quantitative balance theory of sex determination emphasizes that male and female sex determiners are present in both sexes, accounting for the fact that in each sex the potentiality for the other sex is present.

Heterogamety and heterochromosy. Considering the distribution of chromosomes during the genesis of gametes, it is apparent that in species with male heterogamety, as in the human, the chromosomal constitution of the sperm is of major importance in determining sex at fertilization. Efforts have been made to separate androsperm and gynosperm on the basis of supposed differences in physical-chemical properties. Such separation, followed by artificial insemination could permit the control of sex of offspring. To date, however, confirmed and definitive results have not been achieved. Promising investigations are continuing, utilizing the techniques of electrophoresis, immunology and differential centrifugation. It has been claimed by Shettles that androsperm and gynosperm of the human can be identified by morphologic characteristics evident under the phase microscope. Future confirmation will be required to establish the validity of this observation. Within the last decade Barr discovered a previously ignored sexual dimorphism in somatic cell nuclei that permits the identification of male or female tissue. It is possible to identify female tissue by the presence of a characteristic chromatin mass, which is absent or only occasionally seen in male cells. This body, the *sex chromatin mass*, is probably a manifestation of the intermitotic phase of the X chromosomes. There remains some controversy as to its precise origin. However, the study of sex-chromatin has become of considerable importance in diagnosing and classifying cases of human intersexuality. For this purpose, the terms chromatin-positive and chromatin-negative have been introduced to refer to females and males, respectively.

Primary sex differentiation. The sex glands serve the dual function of gamete production and hormone secretion. They originate in both males and females as similar undifferentiated primordia consisting of a cortical and medullary region. The cortex is an inductor of female differentiation and the medulla an inductor of male differentiation. In the event of male development the embryonic medulla predominates to form the major part of the testis and the cortex regresses. Female development is marked by the dominance of the cortex which gives rise to an ovary. Normally, the pathway of differentiation to form testes or ovaries, is determined by the genetic sex established at fertilization. Nevertheless, this process can be influenced by non-genetic factors. It has been demonstrated in amphibia, birds and mammals that testes may very well develop in genetic females, if the proper hormonal or physical influence is exerted while the embryonic gonad is still in the indifferent state. Conversely, ovaries may be induced to differentiate in genetic males. The factors that decide the alternatives of testicular or ovarian differentiation may be placed in three groups: *genetic, environmental and localized internal agents*. As Witschi has shown in amphibia, *environment conditions* such as extreme temperatures or delayed fertilization can partly or completely reverse the genetic determination. In nature, a striking example of the importance of environmental conditions during differentiation of the gonads is the case of the eel. Sex ratios of eel populations differ greatly in different geographic locales, in spite of the fact that the genetic mechanism for sex inheritance should result in a 50:50 sex ratio. Among certain European eels, females are more prevalent in the higher reaches of rivers while in the estuaries and coastal sea-waters, males are more frequent. Also, in the farthest regions of the geographic area of distribution of a species, females tend to predominate. The accumulated facts and supporting experimental evidence suggest that environmental conditions such as temperature and crowding can overcome genetic sex determination and exert a primary influence over sex differentiation. External factors do not interfere by changing the genetic constitution. They affect adversely the cortical or medullary inductor systems that operate in the differentiation of the gonad. Among the invertebrates, the influence of external environmental factors on sex differentiation is noted in many species. In *Bonellia*, a gephyrean worm studied by Baltzer, the males are minute and undergo sexual differentiation, while parasitically attached to the larger female. Most of the free-swimming larvae which come to rest upon the proboscides of the females differentiate in the male direction; those which settle on the sea bottom transform into females. Extracts of the proboscis, added to the water in which sexually indifferent larvae are maintained, are effective in encouraging the production of males. In another annelid worm, *Ophryotrocha*, all young individuals are functional males, but become females under favorable environmental conditions. Adverse conditions, such as inanition, lack of oxygen, accumulated excretions resulting from overcrowding, cause the female to revert to the male phase. These examples serve to demonstrate that many of the invertebrates have genetic mechanisms so labile that natural or experimental alterations of the environment may suffice to produce sex reversals.

The role of localized internal factors in sex differentiation becomes evident through an analysis of experimental and naturally-occurring sex reversal in vertebrates. Lillie's descriptions of different sexed cattle twins illustrate this concept. When the fetal membranes of male and female calf embryos become united in the uterus so that a common blood circulation develops, the female embryo is modified in the male direction, forming the so-called *free-martin*. So complete is this transformation that the female develops a sterile testis instead of an ovary. Absolute sex-reversal, to the extent of transforming genetic males or females into reproductively functional individuals of the opposite sex has been achieved experimentally by the administration of hormonal substances before the embryonic gonad is differentiated. An example is the sex-reversal of *Xenopus* larvae, by Chang and Witschi. If young tadpoles are reared in aquarium water containing small quantities of estrogenic hormones, the expected 50:50 sex ratio does not prevail; instead all larvae become ovary-bearing females. That genetic males are functioning as reproductive females *without alteration of the original male genetic constitution* can be proven by breeding experiments between sex-reversed males and normal males. Since in this species the male is homogametic (XX), 100% of the offspring of such a mating are male. Many species of amphibia have been completely sex-reversed in this fashion. Functional sex reversal may be produced, also, in birds.

It is entirely reasonable that this could occur in higher vertebrates although it has not yet been achieved experimentally or proven to have occurred spontaneously.

Secondary sex differentiation. Differentiation of the accessory sex structures follows the primary differentiation of the gonad and almost without exception, throughout the vertebrate class, these organs are not under direct genetic control, but under the influence of secretions from the newly-formed gonads. In the embryo during the indifferent stage, the oviducts (ducts of Müller) and mesonephric ducts (of Wolff) appear in both sexes. The oviduct is the primordium of the Fallopian tubes, uterus and upper vagina. The mesonephric duct gives rise to the epididymis, vas deferens and seminal vesicle. In the male, the oviduct degenerates during the process of secondary sex differentiation, while in the female it is the mesonephric duct that does not persist. External genitalia of each sex develop from bipotential primordia, the urogenital sinus and the genital tubercle, which have the capacity to develop along masculine or feminine lines. *Thus, each embryo possesses the potentiality to develop internal and external genital organs of either sex.* Recent experiments, employing the technique of fetal castration have elucidated the influence of the fetal gonad on the differentiation of the sex accessory organs. In the absence of gonads, mammalian fetuses will develop female accessory structures, regardless of whether the castrated individual be genetic male or genetic female. If genetic males are deprived of their fetal gonads after the masculinization of the upper portion of the genital duct system has begun, then only the lower regions and the external genitalia are of the feminine type. In this fashion, pseudohermaphrodites can be experimentally produced which possess both male and female genital structures. By unilateral castration, it can be demonstrated that the morphologic inductive capacity of the mammalian fetal testis acts locally to suppress the oviduct and activate the mesonephric duct. Removal of a single fetal testis results in the appearance of a female duct system on the operated side, while male development proceeds on the unmolested side. These experiments of Jost, with rabbit fetuses, have contributed greatly toward the understanding of human pseudohermaphroditism and intersexuality. Parallel experiments of Wolff in birds, reveal the interesting fact that in this group of vertebrates, it is the *ovary* that must be present to suppress the tendency for masculine differentiation to occur. The principles regarding the inductive role of the gonad in the differentiation of the accessory sex organs are identical. Why the agonadal condition results in masculinization in one group of vertebrates and in another brings about feminization is problematical. In all likelihood, the explanation is linked to the fact that females are the homogametic sex in mammals while male birds are homogametic. How this genetic factor acts in secondary sex differentiation is unknown.

Control of adult secondary sex characteristics. The establishment and maintenance of sexual patterns, both morphologic and behavioral, usually involve coordinated hormonal interactions. Some sexual dimorphisms, particularly among birds, are not hormonally controlled, but are determined by genetic constitution directly. This is true or normal sex differences in plumage of a number of avian species. In the English sparrow, neither removal of the gonad nor hormone injection has any noticeable effect on plumage dimorphism. An intermediate type is represented by the pheasant in which full development of sexually characteristic plumage is dependent upon simultaneous actions of both genetic and hormonal factors.

In contrast to those instances, mainly in birds, of direct genetic control, is the vast majority of behavioral and morphologic expressions of sexuality which are established and controlled by hormonal mechanisms. The range of hormonally controlled sexual characteristics in vertebrates extends from exotic courtship rites of salamanders to such majestic ornaments as the antlers of the deer or the mane of the lion. Almost all of these sex-specific features are under the influence of steroid hormones, produced by the sex glands. However, exceptional situations have been described in which there is a direct influence of protein hormones from the pituitary on secondary sex characteristics. In several genera of African finches, the male bird assumes a bright nuptial plumage at the onset of the breeding season. He maintains this adornment for two or three months and then, after molting, dons the hentype plumage, a constant characteristic of the female. This plumage change coincides with the cyclic change of the gonads from the quiescent to the breeding stage. Castrated males, however, continue to develop the colored plumage rhythmically. This indicates that feather pigmentation during the male phase of the plumage cycle is not controlled by the gonad. It has been established that a pituitary hormone, lutenizing hormone, controls directly this secondary sex characteristic. This represents an unusual case, perhaps an evolutionary transition, for almost always pituitary hormones direct their action toward endocrine glands and it is the secretions of the target glands that influence the rest of the body soma.

As each vertebrate organism approaches the stage of gonadal maturation, greater and greater contrast between male and female becomes apparent. Male guppies develop a gonopod; thumbpads appear on the digits of male frogs. In the male turtle, *Pseudemys elegans,* the three middle foreclaws, which are used to stimulate the female during courtship, begin to elongate. Voice changes, not unlike those of young boys approaching puberty, become apparent in such diversified vertebrates as the leopard frog, tree toad, prairie chicken, domestic duck and the male mink. With the onset of gonadal function in the male Virginia deer, antler growth begins. By the time these appendages are needed for fighting during courtship, they have shed the velvet and grown hard in response to increased production of testis hormones. The boar's tusks, the bull's horn and crest, the goat's odor gland, the ram's horns and the rooster's comb and spurs are all well-known secondary sex characteristics that respond to the action of testicular hormones. Females are equally dependent upon hormonal stimuli from the gonads for the development of sex-contrasting characteristics. The thread like oviducts of the female frog enlarge to fill most of the abdominal cavity, as the first breeding season approaches. The female opossum's vicious resentment of the male's advances is replaced by docile acceptance as ovarian function becomes established. The female clawed-toad, *Xenopus,* undistinguishable from males as young juveniles, responds to ovarian hormone production by a typically feminine growth pattern just as the awakened ovary stimulates the developing of feminine contours in the human female at puberty.

Sex cycles: neural influence on sexuality. In many vertebrates, the majority of sex-specific characters appear and regress seasonally at the time of the breeding season, while others, including all primates studied, maintain most sexual features from the time of puberty until the onset of gonadal senility. These differences depend on events controlling the function of the gonads. All lower vertebrates, most birds and some wild mammalian species, are seasonal breeders. Ovaries and testes regress each year at the end of the breeding season. Consequently, the hormone-sensitive secondary-sex characters become quiescent, ready to be reawakened the following Spring. The cottontail rabbit is representative of this group. The testes, for example, return to a completely immature condition each Autumn. In mid-winter, they begin to grow and in early spring reach fifty to a hundred times their weight in the inactive phase. Correspondingly, sex structures such as the seminal vesicles and prostate glands, as well as sexual behavior patterns, regress and awaken each year, under the influence of the falling and rising tide of testicular hormones. Seasonal breeders exhibit an annual ovarian quiescence, also. The ovary of starlings increases nearly thirty-fold in weight from early winter to spring. Ovarian hormones are not produced during the inactive phase, but gradually reappear each year as ovarian growth is initiated. Consequently, oviduct development and other factors necessary for successful reproductive function are provided with the required hormonal stimulus in time for the breeding season.

The rhythmic character of gonadal function in both sexes of seasonal breeders is the result of concomitant waves of anterior pituitary production of gonad-stimulating-hormones (gonadotrophins). In the male cottontail rabbit whose testicular history was described above, the gonadotrophin content of the pituitary increases over the inactive phase by 600% at the opening of the breeding season. The importance of seasonal changes in length of daylight in controlling these cyclic peaks of gonadotrophin production has been well-established in many species. Brook trout, for example, which usually spawn in December will produce offspring in August if the light-dark rhythm is appropriately manipulated. Similarly, ovarian development, consequent to the awakening of gonadotrophin production, may be stimulated in female sparrows during the winter by increasing the light ratio. Nevertheless, light is not the only cause for the cyclic nature of the pituitary; all factors are not completely understood. It is apparent, however, that these factors, including light, operate through the neural system—the hypothalamus or higher brain centers—in exerting their influence of the pituitary. Thus, the control of gonadotrophin production and release exemplifies a neuro-endocrine mechanism which links the two important coordinating systems of the body. It is evident that differences in sex cycles among mammalian females simply reflect different degrees of spontaneity in the function of neuro-endocrine pathways controlling the production and release of gonadotrophic hormones. Monestrus, seasonal breeders like the cottontail rabbit or ferret, in response to photostimulation in the spring, elaborate the gonadotrophic hormones which develop ovarian follicles and stimulate the ovary to produce

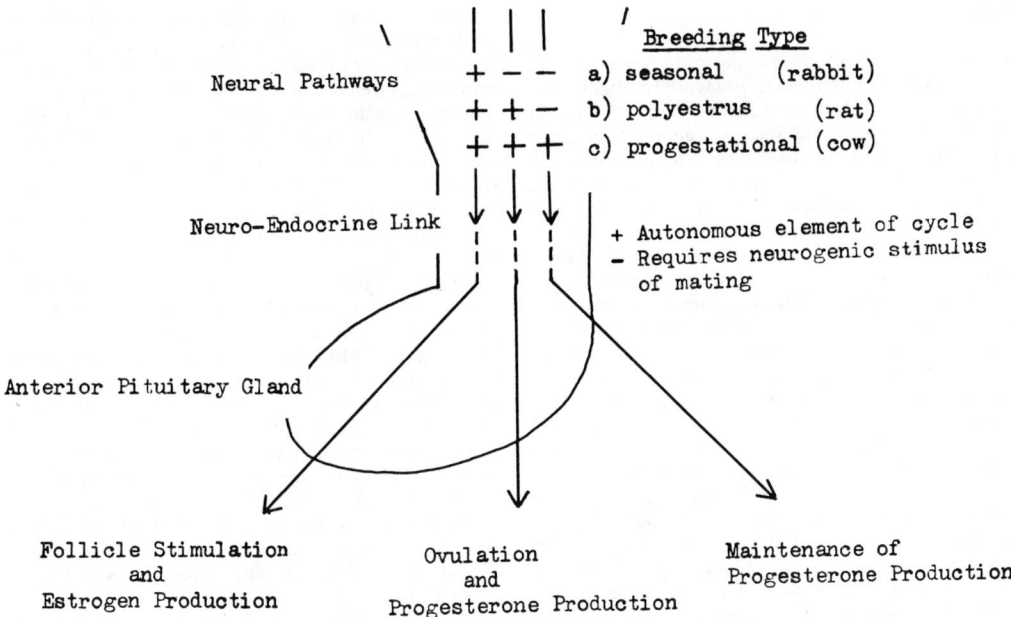

Fig. 1 Spontaneous release of gonadotrophins by female mammals.

estrogenic hormones. The ovary may remain in this condition for the remainder of the breeding season. No further gonadotrophin-induced events occur spontaneously. When the neural stimulus initiated by the mating act reaches the pituitary, the necessary gonadotrophins for ovulation are released, and the ovary responds to this hormonal stimulus by releasing eggs from the matured follicles and by switching its hormone production from the sexual-preparedness type (estrogens, mainly) to the pregnancy-preparedness (progesterone, mainly) type. The mating stimulus activates, also, the release of the gonadotrophic hormone required to maintain the ovary as a progesterone-producer. This sequence is the basic pattern of gonadotrophin production for the complete manifestation of gonad-controlled sexuality in all mammalian females. Differences in sex-cycle type reflect differences in the extent to which neural stimuli for the release of the various gonadotrophins occur spontaneously, without the requirement of the neurogenic influence of mating. The neuro-endocrine pathways of polyestrus breeders, like the rat or mouse, bring about the release of both follicle-stimulating and ovulation-inducing gonadotrophins in repetitive sequence. Thus, spontaneous ovulation becomes a regular event in the reproductive cycle of these animals. However, like the rabbit, they require the stimulus of mating to activate the neural pathways that control the release of gonadotrophins to maintain progesterone production by the ovary. Progestational breeders (cows and guinea pigs) and menstrual breeders (humans and other primates) include even the latter event among the spontaneous occurrences of the reproduction cycle, accounting for the prolonged, post-ovulatory phase of their cycles. This comparison of breeding types is illustrated in the following diagram:

Neural control of gonadotrophin release may be considered the key mechanism through which "male-ness" and "femaleness" become contrasted in the sex hormone activity of individuals of many species. Except for seasonal breeders, males tend to release at a steady rate the full complement of gonadotrophins necessary for complete gonad function—both gamete and hormone production. Even seasonal breeders follow this rule during the reproduction season each year. Females of all species, however, exhibit a cyclic gonadotrophin activity which controls the observed temporal variations in ovarian function and almost all secondary sex characteristics, including sexual receptivity. In many ungulates, a parallelism exists between the duration of heat and the relative preponderance of follicle-stimulating-hormone in the pituitary. Normally this favors a desirable synchronization between the time of maximal sexual receptivity and the spontaneous release of ovulation-inducing gonadotrophins.

Although it is clear that neural control of gonadotrophin release as described for lower mammals applies, also, to the human, almost nothing is known of the mechanisms through which psychological factors influence sexuality. It would be tempting to suppose that the gonadotrophin-release mechanism of the neural system is ultimately the basic *modus operandi*, but this explanation fits only some of the observed facts and not others. Psychological pseudo-pregnancy in the human could be explained on this basis, but in psychic infertility, a far more prevalent condition, gonadotrophin function appears to be completely normal. Similarly, abnormal conditions of sexual performance, including frigidity, impotence and nymphomania do not appear to be correlated with abnormal gonadotrophin function. It may be pointed out, however, that our present tests for gonadotrophin function are far from precise and with the development of better methods, our current concepts may undergo considerable change.

SHELDON J. SEGAL

SHELL

The external calcified covering of most MOLLUSKS. In rare instances, this structure is merely a small plate and may be internal, or even absent, in this phylum of animals. In general, the shell is composed of three layers. The outer or surface layer, the *periostracum,* is composed of an organic nitrogenous substance allied to chitin, called *conchiolin.* The *ostracum,* the second and third layers together, is composed of a framework of conchiolin in which is deposited mineral salts, mainly calcium carbonate with traces of calcium phosphate and magnesium carbonate. The ostracum is composed of two quite different layers. The first layer, developed below the periostracum, consists of prismatic crystals arranged at right angles to the surface of the shell. The inner or laminated layer, which may or may not be nacreous, consists of plates of calcium which parallel the outer surface layer and are produced at right angles to the prismatic layer. This inner laminated layer is sometimes referred to as the *hypostracum.* The periostracum and prismatic layers are produced by the marginal cells of the epithelium of the mantle. The inner laminated layer is produced by submarginal cells as well as the entire surface epithelium of the mantle. In certain families of mollusks, particularly the freshwater clams (Unionidae), the mantle surface produces not only the framework of conchiolin but can produce relatively thick layers of conchiolin below any area of the outer surface which has become corroded or eroded. The organic periostracum serves as a protective layer against any acids which may develop in the substrate in or on which the mollusk is living.

Mother-of-pearl, the nacreous laminated layers, is produced by exceedingly thin and very short cells of calcium, their junctions forming minute prisms. Only mollusks which produce the nacreous type of shell can produce commercial pearls. The nucleus of most pearls is generally a parasite which has invaded the thin epithelial tissue of the mantle. The mollusk, in order to protect itself, surrounds the parasite with laminated layers of shell and thus the pearl is formed. In time, the pearl is slowly forced through the mantle tissue and, in the case of the pearl "oyster," which is a marine bivalve, the pearl drops into the mantle cavity and may remain there until the death of the mollusk, or, if very small, may be forced out by water currents created by sudden contractions of the two valves. In certain freshwater mussels, irregular pearls are frequently produced in the soft anatomy of the dorsal area, usually above the large adductor muscles. Commercial pearls are occasionally produced, but most are lost through the ventral opening when these pearls are set free in the mantle cavity. The irregular pearls formed in the dorsal area probably remain in the tissues of the mollusk during the life of the animal.

Various color PIGMENTS are found in all layers but are most highly developed in the prismatic portion of the shell. These various colors give rise to many distinctive patterns among several groups of marine and even terrestrial mollusks, particularly in the tropical parts of the world. In many areas of the tropics, land snails have become tree living and, with few exceptions, are highly colored, the various colors forming intricate patterns which may be in the form of spiral bands, axial streaks and spots, all existing as single characters or in any combination.

Many mollusks produce elaborate structures on the outer surface of the shell, such as spines and ridges. Among the rock whelks, in the family Muricidae, these structures become exceedingly complex in their multiple branching. The same holds true for the spiny "oysters" in the family Spondylidae. These bivalves are sessile and are generally attached to rocks, ledges or coral. They produce long, pointed or flattened spines up to two and three inches long. Such structures offer considerable protection against predatory fish.

The commercial value of shells is quite important in various parts of the world. Oyster shells are used extensively for road beds in many areas in the southern United States. Mother-of-pearl, from the pearl oyster, is used in jewelry, inlay, and other ornamental work. In the study of prehistoric man, shells play an important part in working out trade routes. That is, shells obtained from burial sites indicate that some sort of trade existed, sometimes over considerable distances. For example, certain burial sites in inland Asia Minor, hundreds of miles from the sea, contained shells which indicate that trade existed between these areas and coastal areas of the Mediterranean, as well as the Persian Gulf.

Fossil shells play a very important part in Geology. Many genera and several species are known as "index fossils" and their presence in a given horizon is indicative of a given period of time. Also, by knowing the ecology of our Recent species, we can give a basic interpretation of conditions existing in the past; that is, these fossils may indicate marine, freshwater, temperate or tropical conditions.

WILLIAM J. CLENCH

References

Abbott, R. T., "American Sea Shells," New York, Van Nostrand, 1954.

Clench, W. J., *et al.,* "Johnsonia, Monographs of the Marine Mollusks of the Western Atlantic," Vols. 1–4, Cambridge, Harvard University Press, 1941–1960.

Cooke, A. H., "Mollusca," *in* Harmer, S. E. and A. E. Shipley, eds., "Cambridge Natural History," London, Macmillan, 1895.

Morton, J. E., "Molluscs," London, Hutchinson, 1958.

Pilsbry, H. A., "Land Mollusca of North America," Vols. 1 and 2, Philadelphia Acad. Nat. Sciences, 1939–1948.

Tryon, G. W., (continued by) H. A. Pilsbry, "Manual of Conchology," 28 vols., Philadelphia, Academy Nat. Sciences, 1885–1935.

SHERRINGTON, SIR CHARLES (1857–1952)

This English physician was primarily concerned with the workings of the nervous system, i.e. neurophysiology. In 1894, Sherrington showed that nerves going to muscles were not all concerned with stimulating muscle contraction. From one-third to one-half of these nerves were sensory, carrying sensations to the brain. Thus, the brain was aware of the tensions upon muscles and joints and therefore possessed a sense of position and equilibrium. Sherrington's work helped explain why certain nervous disorders resulted in loss of muscular coordination. In 1906, he proposed a theory of reflex behavior of antagonistic muscles; which helped to explain how the body, under the coordinating guidance of the nervous system, functioned as a unit. He also determined, with greater accuracy than any previous work, the motor areas of the cerebral cortex, determining which regions governed the motion of the various parts of the body. For his work with the nervous system, Sherrington received a share of the 1932 Nobel Prize in medicine and physiology.

DOUGLAS G. MADIGAN

SHOOT

The term *shoot* designates a stem together with the leaves and buds that it bears. It is recognized as a single structural unit with no clearly defined boundaries between the STEM axis and the associated appendages.

The general form of a shoot is determined in large measure by the degree of elongation of the stem. *Long shoots* possess internodes which separate the successive nodes and leaves of the axis, whereas *short shoots* exhibit little or no internodal growth and bear leaves in close succession upon a short compact stem. The two contrasting habits are established early in the course of development. In plants which possess long shoots the formation of internodes takes place coincident with the expansion and maturation of the leaves, but in short shoots leaf development occurs without appreciable internodal growth.

Most vascular plants possess long shoots. Short shoots, however, are not uncommon, for they appear among the ferns, in certain of the cone-bearing trees and in cycads and *Ginkgo* as well as in such angiosperms as *Cercidiphyllum,* the palms, agaves, and yuccas. Further, the rosette stage of biennials and the spurs of fruit trees are examples of short shoots. Although the pattern of shoot development is usually rather stable, it may vary during ontogeny. For example, in *Larix* and *Ginkgo,* an established long shoot may spontaneously change to the short shoot habit, or, similarly, a short shoot may alter its pattern of development to form a long shoot. Biennials characteristically produce short shoots in the first season of growth, but in the second year, as they enter the reproductive phase, they form an inflorescence at the end of a long shoot. The shoots of ANGIOSPEARMS in the vegetative stage may be either long or short, but the flowering axis is always short.

The structure of LEAVES and the manner of their arrangement on the stem also contribute significantly to the form of a shoot. For each species leaf distribution along the axis is usually constant with one, two, or several leaves at each node, in arrangements termed *alternate, opposite,* and *whorled,* respectively. Leaf position is determined at the apical meristem at the time that leaf primordia are initiated. Development at the apex may be so orderly in the progressive appearance of new leaf primordia as to be interpreted by mathematical formulation. Such values, representing the fraction of the stem circumference between two adjacent leaves, are commonly ½, ⅓, ⅖, ⅜, fractions of the Fibonacci series. In some plants, as in *Helianthus,* leaf arrangement changes during development, gradually shifting from one pattern at the first-formed nodes to others at the later-formed nodes, in seeming correlation with an increasing diameter of the apical meristem.

The growth of a woody shoot takes place in a well-defined pattern. At the beginning of the growing season, after the bud of a shoot opens, the leaves at successive nodes gradually and progressively enlarge and mature. Concomitant with such development, the stem passes through a period of differentiation, with the formation of internodes in long shooots, and with little or no internodal growth in short shoots. The entire complement of foiliage leaves for the season's growth may be contained in the winter buds, or additional leaf and stem tissue may be initiated and attain maturity within the same growing season. Growth of a shoot is completed with the formation of a terminal bud or it ceases with the death of the tip. Terminal buds form during the growing period in some plants, as in the conifers and in many angiosperms, as for example in *Acer, Quercus,* and *Carya.* Continued development of the shoot in the following year takes place from these buds. In most angiosperms, however, as in *Ulmus, Betula, Salix,* and *Syringa,* the tip of the shoot normally dies during the growing period. In *Syringa,* the aborted tip consists of four or five pairs of young leaves on a short unextended axis; after growth in the terminal parts ceases, the foliage leaves gradually become yellow, and finally the entire tip dries up and dies. The shriveled tip sometimes persists, or at other times it falls off leaving a scar at the end of the shoot. In *Syringa* abortion of the tip may be prevented and additional growth and development in the apical part can be induced by the destruction of the uppermost pair of axillary buds.

Cessation of shoot growth may also be due to other factors. In *Robinia* and *Philadelphus,* frost halts development. In *Aesculus, Betula,* and other genera, the formation of an inflorescene brings shoot extension to an end. This change to the flowering pattern in many plants may be induced by the appropriate exposure to light and dark conditions. Thus there are short-day plants and long-day plants.

A terminal bud of a woody plant is initiated in the late spring or early summer after the foliage leaves have expanded, and it continues development during the remainder of the growing season. Axillary buds are usually present as a part of a shoot. Each lateral bud has its origin in the apical meristem and is initiated at the same time as the subtending leaf. The scales and leaves of a lateral bud are subsequently formed in orderly progression during the growing season.

Two distinctive types of branching occur among shoot systems. *Dichotomous branching* is characterized by a forked pattern. It is derived from the subdivision of the apical meristem, each half becoming independent and producing a single shoot. Dichotomy is a primitive form of branching and appears among the early vascular plants. *Monopodial branching* is a system in which the main axis dominates the laterals. Each year growth from the terminal bud adds to the stature of a plant, thereby maintaining the identity of the main stem. *Sympodial branching,* a modification of a monopodium, results when the main axis ceases growth or exhibits limited growth and the axillary buds near the tip assume the major role in shoot development. In alternate-leaved plants in which shoot tip abortion occurs, the uppermost axillary bud may shift its position slightly and in its development appear to extend the main axis as in a monopodium. In opposite-leaved plants a sympodium may superficially resemble a dichotomy if the stem tip ceases growth and the uppermost pair of axillary buds produces shoots of the same size and vigor. Examples of plants with monopodial branching include the conifers; among those with sympodial branching are *Betula, Syringa,* and *Alnus.*

RHODA GARRISON

References

Wetmore, R. H., "Growth and development in the shoot systems of plants," *In* Fourteenth Symposium of Soc. for Study of Development and Growth, Chap. 8, Cellular Mechanisms in Differentiation and Growth, Princeton, N. J., Princeton Univ. Press, pp. 173–90, 1956.

SIPHONAPTERA

This order of an inexplicible origin, but supposedly derived from a MERCOPTERAN stock, comprises about 1000 species. It includes fleas which are small, bilaterally compressed, wingless holometabolous insects whose adults are blood-sucking and parasitic on mammals (including man) and birds. A flea can glide swiftly through the pelt and plumage of its hosts as the numerous bristles on its body are backwardly directed. While retaining the fundamental characteristics of a Pterygote, the flea's anatomy is highly specialized and remarkably different from that of other insects.

The lateral sockets concealing structurally peculiar antennae, the lower cranial border fringed in some fleas with a genal comb and reposing over the anteriorly projected propleura, an interantennal falx and laterally displaced ocelli (absent in some fleas) are some of the unusual features of the head, which is closely approximated to the thorax.

The piercing stylets lodges within appressed labial palpi, a very short hypopharynx, and a pair of voluminous maxillary lobes atypically attached to the lateral borders of the perioral ring are exceptional characteristics of the mouth parts.

The anomalous shape and disposition of the propleura, presence of pronotal comb, and practically detached mesopleural ridge in many fleas, and the first abdominal spiracle eccentrically borne by the capacious metapleuron have all deformed the thorax into an entomological oddity. The powerful legs are adapted for jumping, and the overdeveloped coxae and femora accomodate additional muscles for unexpected leaps.

The abdomen is characterized by the presence of segmental ctenidia, pygidium, and pygidial bristles. The atypical ninth sternum of the male bears claspers and the intromittent organ exhibits the greatest complexity. The proctiger in the female bears a pair of setose stylets.

The otherwise undiversified internal anatomy includes the nervous system of a primitive type and a proventriculus bearing an armature of spines.

The eggs laid in the pelts of hosts fall off and hatch into apodous eyeless larvae resembling those of some Nematocera (Mycetophilidae). They feed on organic debris containing remnants of blood defaecated by the adults, and pupate within cocoons. The usual period of life cycle (three weeks) may be shorter among tropical fleas under optimum conditions with a temperature of 23°C and humidity of 80% to 90%.

Fleas being born and bred under unhygienic conditions act as vectors of many human diseases and many are potential reservoirs of Protozoal and Spirochaetal infections. The common rat flea (*Xenopsylla cheopis*) is an effective vector of bubonic plague and murine (endemic) typhus. The ingested plague bacilli, which often block its proventriculus, are regurgitated and inoculated into the body of the victim. Some fleas parasitic on cats, dogs and man are probably intermediate hosts of certain helminths affecting man. Haemorrhagic septicaemia of cattle is probably transmitted by the cat flea. The female of the 'Chigoe flea' (*Tunga penetrans*) bores through the skin of man and later oviposits, infecting him occasionally with tetanus and gangrene.

In the light of the recent decisions of the International Science Congress of Zoology, G. H. E. Hopkins has reviewed the family-group names which have been applied to this Order and has given a detailed list of the same.

D. M. MUNSHI

References

Hopkins, G. H. E. and Mariam Rothschild, Illustrated, "Catalogue of the Rothschild Collection of Fleas," London, British Museum, 1953.
Snodgrass, R. E., "The skeletal anatomy of fleas," Smithson. Misc. Coll., **104**, 1946.
Hopkins, G. H. E., "Order-group and family-group names for the fleas," Ann. and Mag. Nat. Hist. I(7), 1958.

SIPHONOPHORA

The Siphonophora, of which there are presently fewer than 300 species, referred to three suborders, are essentially colonial HYDROZOA. Of these only one or two conspicuous genera, *Velella* ('Jack Sail-by-the-wind') of the rather aberrant Sub-Order Chronodrophorae and *Physalia* ('Portuguese-man-of-war') an atypical representative of the Sub-Order Physophorae, are familiar to seafarers and scientists alike in warmer water throughout the world. *Physalia* have been noisome to bathers and admirers of their beautiful bluish floats, because the batteries of nematocysts on their long streaming tentacles cause reactions in man almost as severe as some representatives of the Carybdeidae (seawasps) of the Scyphomedusae. Representatives of both genera are, however, less numerous and taken less frequently in plankton samples than other members of the Physophorae and the Calycophorae.

The Physophorae most regularly taken in plankton tows have a relatively tough, stocky stem, which may be considered as the pipe or support for the gastrovas-cular cavity common to all members of the colony. It is capped by a small float (*pneumatophore*). Just below this lies a compact whorl of gelatinous swimming bells (*nectophores*), all of similar shape but with the younger and smaller individuals situated nearest the float. Lower on the stem, there may be a compact solid whorl of bracts with an orderly arrangement as in *Agalma okeni*. In other species, such as *Forskalia edwardsi*, the bracts are more loosely associated and extend for a greater distance down the stem. Bracts in this sub-order are more or less rigid gelatinous leaflets with little internal structure, but with a characteristic outline. They afford protection for the more delicate medusoid and polypoid individuals, the *gonophores* (which are essentially hydromedusae with a gono-style bearing the sexual products), the tentacles (*ten-tilla*), palpons, siphons (*gastrozoids*), etc. Together these individuals form a cluster known as a *cor-midium*.

The Calycophorae are perhaps the best known sub-order among students of this group. They usually have two, sometimes more, nectophores, either arranged in tandem as in the Diphyidae, similar and opposite to one another as in the Prayidae or with roughly eight almost identical nectophores in an alternating series or in whorls extending down the stem as in the Hip-popodidae. In the Diphyidae and in the Prayidae one of the nectophores may be missing in certain genera (*Muggiaea* and *Nectopyramis* respectively). In the Diphyidae, the two nectophores are rather similar in outline (Diphyinae, Galettinae, Chumiphyinae) except in the Abylinae. In all the latter but for *Ceratocymba sagittata*, the superior (anterior) nectophore is cubical (Fig. 1). It nevertheless contains the typical internal structures of the family: a *nectosac* with a radial and four lateral canals similar to those in simple hydro-medusans, a *somatocyst* (which corresponds possibly to the float of the Physophorae) and a *hydroecium* (a de-pression or hollow) within which the apex of the inferior (posterior) nectophore fits and to which the stem is attached. The inferior (posterior) nectophore (Fig. 2) likewise has a hydroecium running its entire length and formed by two ventral wings to protect the stem which runs down through it. There is also a nectosac not unlike that of the superior nectophore. The external ridges and teeth on both nectophores are

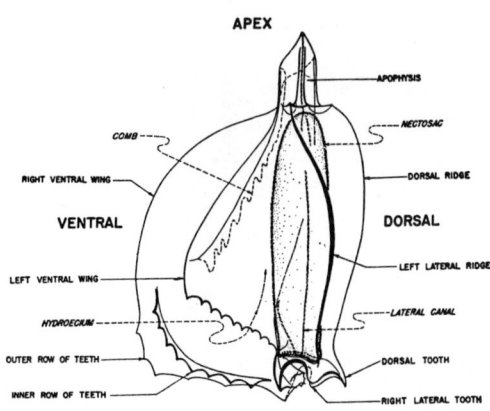

Fig. 2 Diagrammatic drawing of the inferior nec-tophore of *Abyla tottoni*. Left lateral view.

the chief characteristics used in identification of the individual species. The cormidia on the stem are rather similar to those of the Physophorae. However, the older ones in most species break off and are known as free-swimming *eudoxids*. In the Calycophorae, in con-trast to the Physophorae, the bracts (Fig. 3) have well developed and characteristic internal structure: a so-matocyst and a more or less well-defined hydroecium. They are readily distinguished from nectophores by the absence of the nectosac. These bracts are usually found with little more than the gonophores (Fig. 4) attached. Hence, there is here an obvious alternation of generations with the sexual stage (eudoxid) being fragmented off the asexual stage. To the eudoxids in the genus *Diphyes* there are also added so-called special nectophores.

This simple generalized description is perhaps mis-leading because the individuals taken in plankton nets are almost always so badly fragmented and often so poorly preserved—especially representatives of the Physophorae—that a complete colony is almost never found. Therefore it has been difficult to establish the identity and relationships of the various parts. Thus, particularly in early studies, the two generations of one species were frequently given entirely different generic and specific names. Thus, *Enneagonum hyalinum*, the nectophore and "parent" of the eudoxid, *Cuboides*

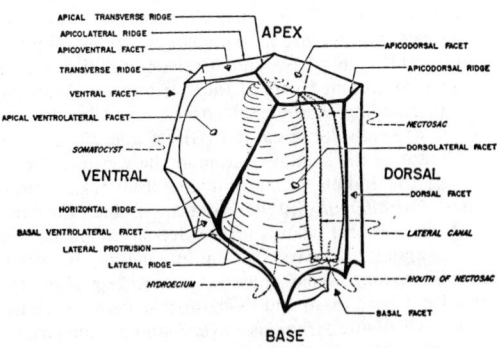

Fig. 1 Diagrammatic dorsolateral view of the supe-rior nectophore of *Albyla haeckeli*.

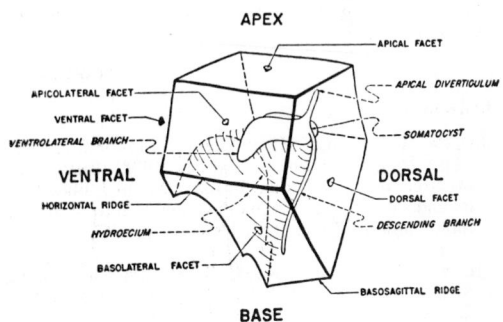

Fig. 3 Diagrammatic dorsolateral view of a bract of *Abylopsis eschscholtzii*.

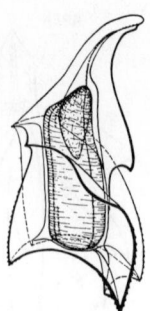

Fig. 4 Laterial view of a gonophore of *Abylopsis eschscholtzii.*

vitreus, was first described by Quoy & Gaimard (1827). The selection of the former name for this monotypic genus was only recently agreed to and it merely depends on page priority! Totton (1932) has clarified certain relationships by abolishing Sphaeronectidae Huxley 1859 (the Monophyidae of Chun). He has subdivided the genus *Diphyes* into four: *Diphyes, Eudoxoides, Chelophyes* and *Lensia* (see also Bigelow and Sears, 1937). On the other hand, there is still disagreement about the appropriate name for certain of the Galettinae. Many of the Physophorae are but poorly known. Similarly, throughout the literature, the terminology used for the anatomy of the siphonophores is inconsistent and undefined. Hence, it is difficult for the novice to compare descriptions of the same parts or even species when written by different authors. One further word of caution is needed, namely, that siphonophores although presumably radially symmetrical, are usually described as if they were bilateral. To complicate matters further, Totton (1932; 1954) uses one system for this and Bigelow and Sears (1937) and Sears (1953 p. 6) another.

Siphonophores are widely distributed throughout the oceans of the world, often in considerable numbers —up to 35,000 colonies in an hour's tow with a meter net (i.e., roughly 6 per cubic meter of water). They live at all levels down to considerable depths (Leloup and Hentschel, 1935; Bigelow and Sears, 1937, Fig. 83), certain forms possibly as deep as 4000 meters. From closing net samples taken by the "Meteor," they appear to inhabit waters over wide temperature ranges (Leloup and Hentschel, 1935; Bigelow and Sears, 1937; Fig. 81) with some representatives in the Arctic and Antarctic regions.

MARY SEARS

References

Bigelow, H. B. and Mary Sears, "Siphonophorae," Rept. Danish Oceanogr. Exped., 1908–1910 to the Mediterranean and adjacent seas, 2 (Biol.): 144, 1937.
Garstang, W., "The morphology and relations of the Siphonophora," Quart. J. Micros. Sci. 87: 103–193, 1946.
Moore, H. B., "Plankton of the Florida Current. II. Siphonophora," Bull. Mar. Sci., Gulf & Caribbean, 2: 559–573, 1953.
Sears, M., "Notes on siphonophores. 2. A revision of the Abylinae," Bull. Mus. Comp. Zool., Harvard, 109: 1–119, 1953.

Totton, A. K., "Siphonophora of the Indian Ocean together with systematic and biological notes from other oceans," Discovery Repts., 27: 1–162, 1954.

SIPUNCULIDA

Once included in the obsolete "Gephyrea," now regarded as a separate phylum with affinities to the ANNELIDS. Exclusively marine animals, inhabiting all seas from the polar regions to the equator from the intertidal area to the ocean depths.

Some 250 species have been described, classified in about 13 genera, some with only a few species, *Golfingia* the largest having about 100.

The body is unsegmented and in two parts; the thin retractile introvert and the stouter trunk. The introvert may have a series of hooks anteriorly and may bear papillae. The trunk may be almost smooth, or covered with papillae. Two genera found in corals have horny or calcareous shields.

The body-wall is made up of several layers, and in some genera the longitudinal mucles are gathered into strands. Internally, the chief muscles are the important retractors 1–6 in number which control the extrusion and contraction of the introvert, and the spindle muscle.

The gut is coiled and doubled on itself so that the anus lies at the anterior end of the body or on the introvert. One or two segmental organs present. The nervous system is of the annelid type. No definite circulatory system has been developed. The sexes are alike externally but separate.

A. C. STEPHEN

Reference

Hyman, L. H., "The Invertebrates," Vol. 5, pp. 610–696, New York, McGraw-Hill, 1959.

SIRENIA

The sirenian or sea cow is a little known, retiring, almost extinct, herbiverous mammal in which the vernacular name "sea cow" conveys no suggestion of relationship to other mammals nor to the appearance of the members of the order Sirenia. There are two living genera: *Trichechus* (manatee), which inhabits shallow salt water bays and fresh or brackish waters of sluggish coastal rivers of the Caribbean islands and the surrounding continental coasts, and the Atlantic coasts of South America and Africa; and *Dugong*, an inhabitant of the shallow waters of the western Pacific islands, the Indian Ocean, and the Red Sea. A third genus *Hydrodamalis* (Steller's sea cow), which formerly lived near several islands in the Bering Strait became extinct about 200 years ago due to whaling operations in this region. Dense massive bones, especially the ribs, from both fossil and living forms characterize the order. All of the sirenians have a similar appearance due, in part, to their complete adaptation to water. Body hair is essentially lacking and the thick skin is described as naked. The bulbous seal-shaped body tapers gradually back to a broad, round paddlelike fin in the case of *Trichechus* and to a flattened, forked tail in *Dugong* and *Hydrodamalis*. The tail fin is horizontally flattened unlike the vertical flattening in fishes, and serves as the chief organ in swimming. When at rest the two living genera have a strange vertical posture of back humped up or rounded with head and tail dangling, the tail curled under the body so that its dorsal surface rests upon or points to the bottom. The anterior appendages are modified into paddles for maneuvering, grasping the young, and manipulating food. Posterior appendages are vestigial or lacking in recent species. A round, elongate head with a blunt snout merges with the body through a short neck. In relation to the large size of the animal the ears and eyes are minute. The cleft upper lip covered with stiff bristles is used for grasping succulent water vegetation and the slow manner of grazing is the basis for the name "sea cow." Horny plates are present on the front part of the jaws behind which lie 4 or 5 peglike (*Dugong*) or 7 or 8 similar lophodont molariform (*Trichechus*) teeth which serve as the chief masticating surfaces. *Hydrodamalis* is edentulous and retained only horny plates throughout its mouth. Although some individuals fall outside this growth range the usual adult *Trichechus* has a length of 7–10 feet, *Dugong* 8–12 feet and *Hydrodamalis* is estimated to have been 20–25 feet. It is a strange paradox that these ugly creatures may have formed the basis for the legends of sirens or mermaids.

Sirenians are descended from quadrupedal terrestrial mammals in early Eocene or earlier time and the fossil record indicates a close relationship with the proboscideans. The most ancient forms date back to the Middle Eocene of Egypt, Florida, and Jamaica, 50 million years ago, but even at that time the Sirenia were well adapted to an aquatic life and would have been unable to move on land. The order is subdivided into two families, the Dugongidae and Trichechidae, of which the former has a geologic range from Middle Eocene to Recent, a moderately documented history with fossil representatives from all continents but is well known only from North Africa, Europe, and the United States. Other than the living *Trichechus* the Trichechidae are known from relatively rare specimens from the Miocene of South America and the Gulf coast Pleistocene of the United States. Though they have achieved a worldwide distribution sea cows apparently have always been relatively limited in numbers. With the exception of the recently extinct *Hydrodamalis* both living and fossil forms have been confined to warm waters. A climax in number and variety was reached during the middle and late Miocene epoch.

ROY H. REINHART

References

Moore, J., "The Status of the Manatee in the Everglades National Park, with Notes on its Natural History," Jour. Mammal., 32: 22–36, 1951.
Reinhart, R. H., "A Review of the Sirenia and Desmostylia," Univ. of Calif. Publ. Geol. Sci., 36: 1–146, 1959.
Sickenberg, O., "Beitrage zur Kenntnis tertiaren Sirenen," Mem. Mus. Hist. nat. Belg., 53: 1–352, 1934.
Simpson, G. G., "Fossil Sirenia of Florida and the Evolution of the Sirenia," Bull. Amer. Mus. Nat. Hist., 59: 419–503, 1932.

SKELETON, INVERTEBRATE

If we define the term "skeleton" as a rigid supporting or protective structure or assemblage of structures, the invertebrates exhibit a variety of skeletal devices and patterns. These may be grouped for purposes of discussion into endoskeletal and exoskeletal categories, although this distinction is artificial in many cases (e.g., in Foraminifera).

Endoskeleton. The most obvious endoskeletal structures are those which are more or less permanent, usually being composed of organic fibrils or of crystalline inorganic compounds. Among the Protozoa, in which the endoskeleton is necessarily intracellular, the *axostyle* of polymastigid flagellates is an example of the fibrillar (proteinaceous) type, while the *tests* of radiolarians and foraminiferans illustrate the crystalline type. Foraminiferan tests are usually composed of calcium carbonate but they may be "chitinous," arenaceous, or siliceous. Radiolarian tests are siliceous except for certain ones of strontium sulfate.

Spicules are the characteristic skeletal structures of sponges, but the Demospongiae exhibit a fibrous protein (*spongin*) skeleton in addition to, or, in some cases, instead of, spicules. Spicules are formed by deposition of calcium carbonate or silica upon initially intracellular fibrils. A simple monaxon spicule is formed by a pair of cells, but more complex spicules are formed by the cooperative activity of larger groups of secretory cells. The endoskeleton of alcyonarians is also comprised of calcareous spicules, which in

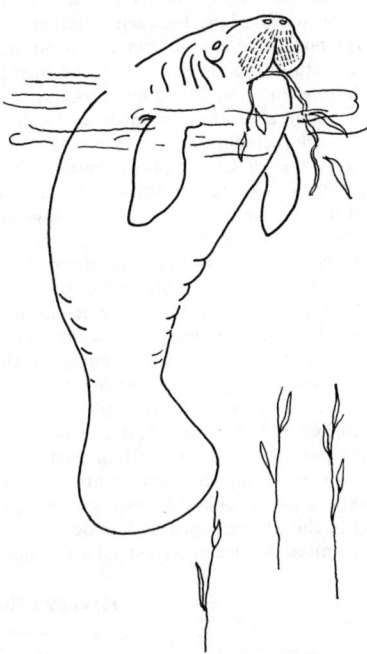

Fig. 1 *Trichechus* from Florida.

some cases are fused to form such structures as precious coral. In the articulate brachiopods, a calcareous endoskeleton associated with the dorsal valve supports the lophophore. Among the dibranchiate cephalopod mollusks, the skeleton produced by the mantle is internal, if present at all; in the cuttlefish it is calcareous, but in the squid it is horny (chitinous). In addition, however, the cephalopods also possess a cartilage-like tissue which forms a "skull" and supports for the muscular fins and tentacles. The skeletal structures of the protochordates—the notochord, gill bar rods, and rays—are composed of still different, somewhat cartilage-like tissues. Some arthropods have a well-developed endoskeleton formed of the *apodemes* (hollow) and *apophyses* (solid) which project inward and serve for muscular attachment. These projections are essentially invaginations of the integument.

The mesodermal *ossicles* of echinoderms are remarkable calcite crystals which contain appreciable quantities of $MgCO_3$ and $CaSO_4 + 2H_2O$. These ossicles may serve merely to toughen the skin or they may form armor plate, or movable, articulated skeletal structures with more or less intricate muscular attachments. In the arms of Ophiuroidea, the ossicles serve a function strikingly like that of vertebrate caudal vertebrae. Ossicles are also reminiscent of vertebrate bone in the relationship between organic and inorganic components.

Less obvious as an endoskeletal feature, but of major importance, is *water*, acting through turgor or hydraulic pressure to provide the rigidity essential for support and locomotion in such diverse groups as amebas, cnidarian polyps, nemerteans, nematodes, rotifers, acanthocephalans, sipunculids, priapulids, echiuroids, annelids, mollusks, softbodied arthropods, certain chordates, and echinoderms. Since such liquids as sea water, coelomic fluid, blood, etc. are virtually incompressible, hydrostatic and hydraulic "skeletons" are quite practical and highly adaptable. Antagonistic sets of muscles (e.g., circular vs. longitudinal) can function as successfully as with any other type of skeleton. Although no sharp distinction can be drawn between the two, we may consider liquid under pressure which serves merely to provide rigidity as constituting a *hydrostatic* skeleton; when moving liquid is involved, as in the extension of a tentacle, the skeleton may better be termed *hydraulic*.

Exoskeleton. Exoskeletal structures are found in many phyla, and range in complexity from a thickened cell wall or pellicle to such elaborate physico-chemical productions as a butterfly wing or an ammonite shell. The advent of the reflection scanning electron microscope offers an expansion of resolving power (useful magnification) for observation of minute surface structural details comparable to that provided by the conventional electron microscope for sectioned material.

Among the Protozoa, exoskeletons are prominent in the armored dinoflagellates, euglenoids, silicoflagellates, testaceans, foraminiferans, and tintinnids. The *theca* of such forms as phytomonads and dinoflagellates is in direct contact with the cell membrane, and is the homolog of the cell wall of higher plants, being composed of cellulose and pectins. Impregnation with inorganic salts often increases the rigidity. Less snugly-fitting coverings are designated by such terms as *lorica, test,* or *shell.* These may be composed solely of secreted material or may be formed of such objects as sand grains, diatom shells, or sponge spicules cemented together.

The massive calcareous exoskeletons of stony corals are feasible only for sessile existence, and the skeletons of barnacles, encrusting bryozoa, brachiopods, and diverse mollusks allow for but little more movement. Flexibility is essential to mobility, and is achieved (in organisms with exoskeletons) either through overall flexibility of the *cuticle* as in nematomorphs and many nematodes, or through flexible regions between rigid sections. The latter plan, present in rotifers and miscellaneous other groups, reaches its culmination in the arthropods, most of whose evolutionary advances are related to the nature of the cuticle.

The arthropod cuticle, secreted by the cellular epidermis, consists of a thin outer *epicuticle*, 0.1–1 micron thick, and a thicker *endocuticle*. The epicuticle is usually a wax, and is largely responsible for the impermeability of the cuticle to water. The endocuticle typically consists of a chitin-protein complex which varies considerably in its precise composition. Chitin, isolated from the naturally-occurring complex, is a high molecular weight polymer of anhydro-N-acetyl-glucosamine residues joined by ether linkages of the β-glycosidic type between carbon atoms 1 and 4 of adjacent residues. These form a chain of at least several hundred residues linked into one long molecule (perhaps as much as a micron in length). The complex is tough but pliable. Rigidity of portions of the cuticle (*sclerites*) is due to the presence of additional substances, including scleroproteins, waxes, and often inorganic compounds such as calcium salts. This hardening or sclerotization is often accompanied by pigmentation, and involves both the inner portion of the epicuticle and the outer portion of the endocuticle (*exocuticle*). The flexible portions of the cuticle are termed *membranes*; these are present wherever there are movable articulations between adjacent sclerites.

Since the non-living cuticle cannot expand gradually with the growth of the animal, it must be shed periodically (*molting* or *ecdysis*). This process, which is under hormonal control, also involves shedding the linings of such ectodermal derivatives as the fore- and hind-guts, the tracheae, and the reproductive ducts. A new cuticle is secreted beneath the old while the bulk of the old endocuticle is being digested (by enzymes in the *molting fluid*) and its components salvaged for re-utilization in the deposition of the new endocuticle. After the new cuticle has been secreted, the animal ruptures the old skin, typically along the mid-dorsal line, and emerges—then, in an enlarged or swollen condition, awaits the hardening of the new cuticle. The swelling by which the old cuticle is ruptured and the new kept in shape until it has hardened is accomplished in part by forcing excess blood into one or another part of the body, and in part by overall swelling through taking in water or air. The problem of adequate support following ecdysis and prior to hardening of the cuticle is probably a significant factor which has limited the size of terrestrial arthropods.

HARLEY P. BROWN

References

Hyman, L. H., "The invertebrates," Vols. 1–6, New York, McGraw-Hill. 1940–1967.

Richards. A. G., "The integument of arthropods, Minneapolis, University of Minnesota Press, 1951.

SKELETON, VERTEBRATE

The VERTEBRATES are characterized by the possession of a well-developed internal (endoskeleton) consisting of cartilage and/or BONE plus an external (exoskeleton) of HAIR, FEATHERS, SCALES, horns, antlers, hoofs, nails, claws, and ossicles in the skin (sturgeons, crocodiles, alligators, and armadillos). Bone is classified into two types depending on its embryonic origin. One type is called endochondral or replacing bone because it is first formed in cartilage which is later converted into bone. The second type is called membrane, dermal, or investing bone because it ossifies directly from a membrane and is not pre-formed in cartilage. This classification refers only to the way the bone originates and not to its appearance or physical characteristics.

The skeleton of CYCLOSTOMES and ELASMOBRANCHS is entirely cartilaginous throughout life although calcium salts may be deposited in the skeleton of the large sharks so that it becomes quite hard. The skeleton of all other adult vertebrates has bone, the amount of which varies from approximately half bone and half cartilage in some of the lower fishes to almost completely bone in mammals. The possession of bone distinguishes vertebrates from all other animals.

The endoskeleton is often classified into an *axial skeleton*, consisting of the SKULL, vertebral column, ribs, sternum, and an *appendicular skeleton*, composed of the bones of the pectoral and pelvic girdles and of the appendages. The latter may take the form of fins, flippers, wings, legs or arms. Membrane bones are restricted to certain parts of the skull and the pectoral girdle, the rest of the skeleton being composed of cartilage bones.

Vertebrates are also distinguished from other animals by the possession of a spinal column (vertebral column) composed of individual elements called vertebrae. The presence of the vertebral column is the basis for the name "vertebrate" or "animals with backbones." A typical vertebra (Fig. 1) consists of a centrum (body) with a neural arch which surrounds the neural canal containing the spinal cord lying above the vertebral column, and a haemal arch surrounding the caudal artery and vein lying below the vertebral column. The number of individual vertebrae composing the vertebral column, which forms the main axial support of the body, varies from as few as six in some frogs to nearly five hundred in some legless lizards and snakes.

The vertebral column is preceded in embryonic development by a continuous, unsegmented rod (the notochord, composed of a special non-cartilaginous and non-bony type of tissue). The notochord persists as a well-developed structure throughout life in Cyclostomes and the more primitive Elasmobranchs (sharks, rays), Chondrostei (sturgeons), and Dipnoi (lung fish) in which the vertebral column consists of only neural and haemal arches without any centrum. In other vertebrates the notochord is replaced in adult life by the centrum of the vertebrae. However, remnants of the

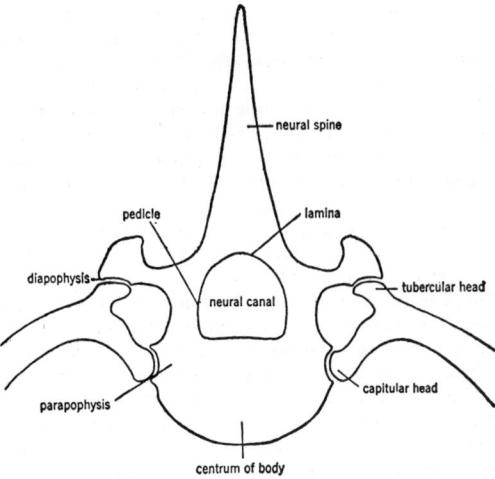

Fig. 1   Diagram of a typical thoracic vertebra. (After Adams, Introduction to the Vertebrates, 2nd ed., Wiley, 1938).

notochord may persist between or within the bodies of the vertebrae (sharks and some amphibia).

The neural arch typically possesses a spinous process and two transverse processes (*diapophyses*) which serve for the attachment of muscles, ligaments, and ribs. Articular processes (*zygapophyses*) are also present for articulation with adjacent vertebrae. The anterior pair (*prezygapophyses*) have their articular surfaces facing in the direction of the spinous process while the articular surfaces of the posterior pair (*postzygapophyses*) are directed toward the centrum of the vertebrae. Other articular processes (*zygosphenes* and *zygantra*) are present on the neural arches in some REPTILES (e.g. snakes), while certain fossil reptiles (diadectid cotylosaurs, placodonts, and dinosaurs) have additional articular processes called hyposphenes and hypantra which are located below the level of the zygopotheses. The hyposphene and the zygosphene are projecting processes which fit into corresponding cavities, the zygantrum and hypantrum. The zygosphene is at the anterior end of the neural arch above and between the prezygopophyses while the hyposphene is a projection below and between the inner ends of the postzygopophyses. The later pterodactyles (extinct flying reptiles) had additional articulating processes, exapophyses, at each end of the centrum of the cervical vertebrae. The haemal arch is restricted to the tail region and in higher vertebrates is replaced by small v-shaped "chevron" bones.

Some living fish and certain fossil amphibians (Embolomeri) have two centra per muscle segment while other fossil amphibia (Rhachitomi) have a centrum composed of three elements per segment. However, the great majority of vertebrates, both fossil and living, have a centrum consisting of a single element.

The anterior and posterior surfaces of the centrum vary considerably in different vertebrate groups. In the most primitive vertebrates the centra are bi-concave (*amphicoelous*). However, in some fish and many reptiles the centra are concave in front and convex behind (*procoelous*), while in many other forms the reverse condition (*opisthocoelous*) occurs. Except for the cervical vertebrae of ARTIODACTYLES (e.g. cow), which are opisthocoelous, the centra of mammals have both surfaces flat (*amphyplatyan*). The centra of all freely movable vertebrae in mammals, except for the first two in the neck, are separated by intervertebral discs. In tetrapods the centra often have a process (*parapophysis*) for articulation with one of the heads of the ribs.

The vertebral column exhibits considerable regional differentiation in different vertebrate groups. Only two regions, trunk and caudal (tail), are present in fish. The trunk vertebrae have haemal ribs while the caudal vertebrae have haemal arches, at least anteriorly. Living AMPHIBIA usually have four regions—cervical, trunk, sacral, and caudal—in the vertebral column. In the amniotes (reptiles, birds, and mammals) five regions—cervical, thoracic, lumbar, sacral, and caudal—are typically present. A similar condition existed in some fossil amphibia (Labyrinthodonts).

The number of cervical (neck) vertebrae varies from one in living amphibia to as many as seventy six in the plesiosaur *Elasmosaurus*, a fossil marine reptile. In the amniotes and labyrinthodonts the first two cervical vertebrae are specialized for support and movement of the head. When the head is turned, the first cervical vertebra (the *atlas*) moves with the skull around the second cervical vertebra (the *axis*). Movements can also occur between the skull and the atlas.

Cervical vertebrae usually have no ribs or only very short ones. In mammals the transverse process of the cervical vertebrae is typically perforated by a foramen (*foramen transversarium*) for the passage of the vertebral blood vessels and a nerve plexus. The members of the camel family, the great anteater (*Myrmecophaga*) and *Macauchenia* (a fossil South American mammal) differ from other mammals in having the vertebral vessels passing through the pedicle of the cervical vertebrae rather than through a foramen in the transverse process. The number of cervical vertebrae in reptiles and birds varies with the length of the neck. Mammals, living and fossil, typically have seven neck vertebrae, the only exceptions being the manatee (*Trichechus*) and the two-toed sloth (*Cholaepus*) which have six and the three-toed sloth (*Bradypus*) which has nine. Except for the first one, all trunk vertebrae in living amphibia have short ribs but in the amniotes well developed ribs are limited to the thoracic vertebrae. The number of thoracic vertebrae varies in different groups of vertebrates but is usually fairly constant for a particular group. In turtles the ten thoracic vertebrae are fused with the ribs and bony plates to form the carapace. Birds have four to six thoracic vertebrae while mammals have ten to twenty such vertebrae. In the glyptodonts (fossil South American mammals) most of the thoracic vertebrae were fused with a heavy turtle-like carapace composed of a mosaic of many small polygonal bony plates.

Lumbar vertebrae are located in the region between the thorax (chest) and the pelvis. Typically, they either lack ribs or have them indistinguishably fused with the transverse process. Reptiles usually have a very poorly defined lumbar region of the vertebral column with no very definite lumbar vertebrae. In birds the lumbar vertebrae generally are fused with the thoracic, sacral, and urosacral vertebrae (ten to twenty three

bones in all) to form the synsacrum. The number of lumbar vertebrae in mammals varies from three to as many as twenty four or more in some of the true Dolphins. Except in legless forms the sacral vertebrae of tetrapods are intimately attached to the pelvis via sacral ribs which may be fused with the sacral vertebrae. Living amphibia have only one sacral vertebra although two were present in some of the fossil forms. Only two sacral vertebrae are present in living reptiles but some of the ceratopsian dinosaurs and the pterosaurs (fossil flying forms) have as many as ten. Birds usually have two sacral vertebrae while the number in mammals varies from one (SIRENIA) to ten (armadillos). The sacral vertebrae, especially in birds and mammals, are usually fused to form the sacrum but they may remain separate. There is no sacrum in whales (CETACEA) and sea cows (Sirenia).

True sacral vertebrae have sacral ribs which articulate with the pelvis. In many urodeles (salamanders), crocodilians and turtles the sacral ribs remain separate but in most tetrapods they fuse with the sacral vertebrae. Additional vertebrae (pseudosacral) may join the pelvis by their transverse process, since they have no ribs, to give rise to a structure called a synsacrum. In birds the synsacrum may consist of the two true sacral vertebrae plus as many as eighteen pseudosacral vertebrae.

The number of caudal vertebrae varies tremendously depending on the length of the tail. In frogs and toads an unknown number of caudal vertebrae are fused into a single bone, the urostyle. Living birds have four to six caudal vertebrae which are fused into a single bone (*pygostyle, urostyle* or "ploughshare bone"). Some fossil birds (*Archaeopteryx*) had a long tail with as many as twenty vertebrae in it.

Two kinds of ribs, haemal and pleural, occur in vertebrates. Haemal ribs lie inside of the muscles forming the body wall while the pleural (true) ribs are in the horizontal septum, separating the body musculature into dorsal (*epaxial*) and ventral (*hypaxial*) divisions at its points of junction with the myosepta between the muscle segments. Haemal ribs are the "ribs" of most fish although in some species both types of ribs are present on the same vertebra. Pleural ribs are the only kind present in tetrapods (land vertebrates). In the higher vertebrates the ribs typically have two heads (Fig. 1). One head (*tuberculum*) articulates with the transverse process (*diapophysis*) of the neural arch while the other head (*capitulum*) articulates with the parapophysis, a process on the centrum.

Living amphibia have short straight ribs on all the trunk vertebrae except the first but long, curved ribs are restricted to the thoracic region of the vertebral column in amniotes. Typically most of the ribs in amniotes curve around the body so as to join with a sternum ("breastbone") in the ventral midline. The ribs which articulate directly with the sternum are called the "true" ribs while those which reach the sternum indirectly via costal (rib) cartilages are "false" ribs. Ribs which do not reach the sternum at all are the "floating" ribs.

In turtles the ribs are expanded and fused with the corresponding vertebrae and a series of lateral (*costal*) and median (*neural*) bony plates to form the carapace or upper shell. These plates are covered superficially with corresponding horny scales.

In the Crocodilia, *Sphenodon* (a very primitive

lizard-like living reptile) and many fossil reptiles a series of dermal rib-like bones (*gastralia, parasternalia* or "abdominal ribs") are found in the ventral abdominal wall between the last true rib and the pelvis. The number of gastralia varies from one pair per body segment in the Crocodilia to two in *Sphenodon* and six or more in some fossil reptiles.

The sternum is considered to be restricted to the tetrapods although in certain sharks there is a medial cartilage which is thought to be homologous with the presternum of mammals. In the amphibia the sternum is largely cartilaginous although it may consist of several parts (frog). The sternum of amniotes which is calcified cartilage or bony, consists of several parts, and is usually associated with the pectoral girdle and the ribs. In legless lizards the sternum is reduced or absent. Birds have a well developed sternum which has a prominent median keel in flying birds and also in penguins. Except in penguins, flightless birds have no keel on the sternum. In the amniotes the sternum arises embryologically by a fusion of cartilage in the ventral midline between the ends of the ribs.

In contrast to the vertebral column ribs and sternum, which are composed entirely of cartilage bones, the skull of a vertebrate (Fig. 2) consists of both cartilage and dermal bones. For descriptive purposes the skull may be divided into a neurocranium, consisting of the parts surrounding the brain and a branchiocranium (*splanchocranium* or *viscerocranium*) which is associated with the visceral or branchial arches (jaws, gills, and their derivatives). In the elasmobranchs, because of its cartilaginous condition, the entire neurocranium is frequently called the chondrocranium.

The neurocranium consists of three pairs of capsules associated with the organs for the sense of smell, sight, balance and hearing. These capsules are arranged in anteroposterior order, the olfactory capsules being first and the otic capsules last. The olfactory capsules, which develop around the olfactory lobes of the brain, and the otic (auditory) capsules around the ear fuse with the rest of the neurocranium. However, the optic capsules which develop around the eyes do not fuse with the rest of the cranium because if they did the animal would be unable to move its eyes. The cyclostomes are unique among the living vertebrates in having only one olfactory capsule rather than the usual two. Since the neurocranium encloses the brain it is necessary to have foramina to provide for the passage of the cranial nerves and of blood vessels. The floor of the neurocranium is formed by cartilage bones while the side walls and the roof consist of dermal (membrane) bones. Thus, the adult skull is essentially a box of cartilage bones more or less surrounded by a second box of membrane bones.

The branchiocranium (*viscerocranium*) consists of the branchial or visceral arches. The first (*mandibular*) arch forms the jaws, the second (*hyoid*) arch forms the hyoid apparatus while the remaining arches (three through seven) constitute the gills. The first and second arches have both cartilage and dermal (membrane) bones but the gill arches consist entirely of cartilage bones.

Except in the cyclostomes (lampreys and hagfish) and the elasmobranchs (sharks and rays), the jaws of all other living vertebrates consist of an inner (primary) and an outer (secondary) set of bones. The inner jaws are composed of cartilage bones developed from the

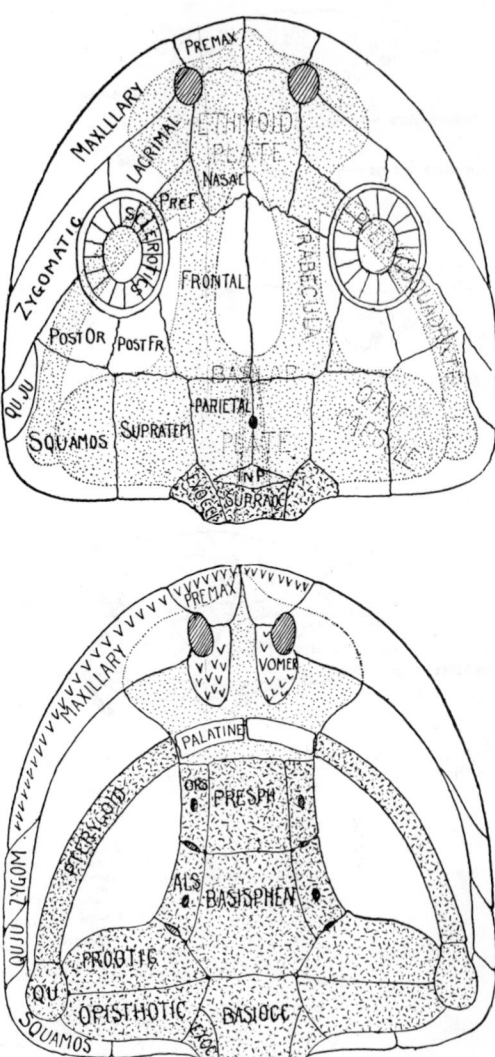

Fig. 2   Diagram of a tetrapod skull. Upper dorsal view; Lower ventral view. Chondrocranium shaded; cartilage stippled; cartilage bone with lines and dots; membrane bones outlined. (After Kingsley, The Vertebrate Skeleton, Blakiston, 1925.)

palato-quadrate (*pterygoquadrate*) cartilage, forming the upper jaw, and mandibular (Meckel's) cartilage, forming the lower jaw. Around these bones are developed the membrane bones forming the outer jaws.

The jaws of most vertebrates have teeth of various sizes and types. Generally the teeth are confined to the membrane bones along the margins of the jaws but some fish and the more primitive fossil tetrapods have teeth on the roof of the mouth. Certain fish also have teeth on some of the gill arches. In the elasmobranchs the teeth are located on the palato-quadrate and Meckel's cartilage since no bones are present. Teeth are absent in turtles, living birds, and certain highly specialized mammals (e.g., the greater anteater and baleen or whalebone whales). Some of the lower

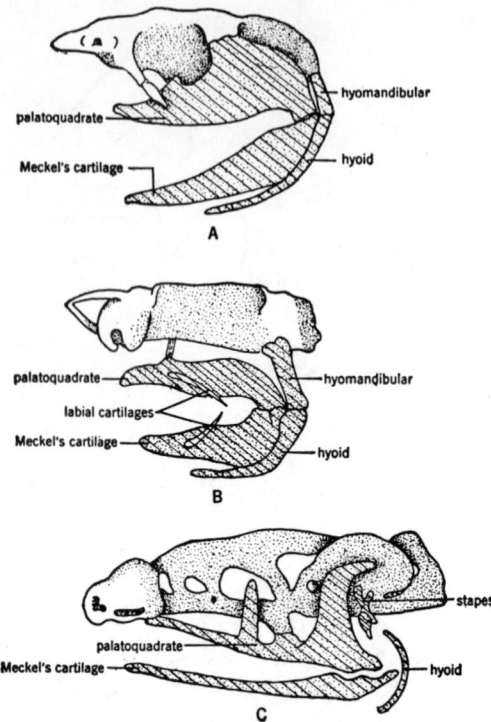

A

B

C

Fig. 3 (After Goodrich, From Adams, Introduction to the Vertebrates, 2nd ed., Wiley, 1938.)

vertebrates (sharks, snakes) have several sets of teeth during their life but mammals usually have just two, deciduous or milk teeth and permanent teeth. The type and number of teeth varies greatly in the different vertebrate groups. In mammals the number and arrangement of the cusps on the cheek teeth is very important in their classification. This is especially true of fossil mammals since the teeth, because of their hardness, are often the only parts preserved.

The hyoid (second visceral) arch of the lower vertebrates is involved in various ways with the support of the jaws (Fig. 3). In the amniotes it contributes to the formation of the hyoid apparatus for support and movements of the tongue. The hyoid bone in adult man arises from a union of the hyoid and the third visceral arch elements.

The remaining branchial arches form the cartilages (sharks and rays) or bones (other fish and certain amphibia) of the functional gills. In the amniotes, which never have functional gills at any time, these elements either disappear completely or contribute to the formation of the laryngeal and tracheal (windpipe) cartilages.

The number of individual bones in the skull of adult vertebrates of different classes varies tremendously. Thus, most bony fishes (Fig. 4) have over one hundred bones, while in the living amphibia (frogs, toads, salamanders, caecilians) the number varies from fifteen to nineteen (Fig. 5). Reptiles may have from forty to seventy individual bones in the skull which in certain groups has one or two openings in the region behind the eyes. Birds (Fig. 6) may have only twenty bones,

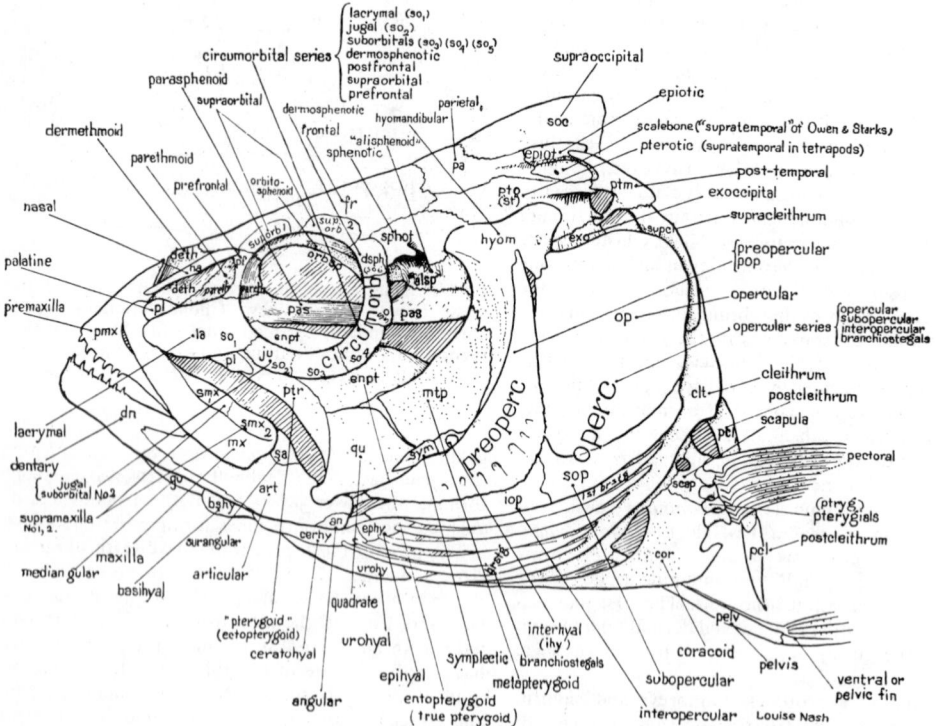

(Fig. 4 Composite diagram of the skull of a bony (teleost) fish. Based on skulls of fifty species. (After Gregory, Fish Skulls: A Study of the Evolution of a Natural Mechanism. Transact., Amer. Phil. Soc. N.S., Vol. 23, Pt. II, 1933.)

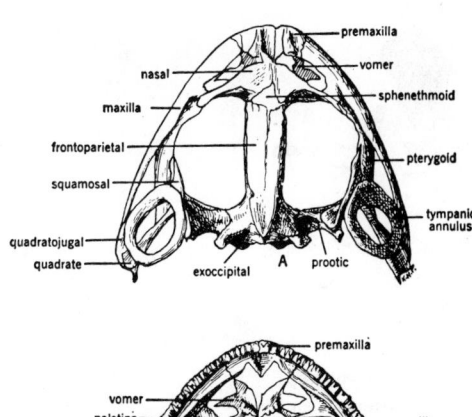

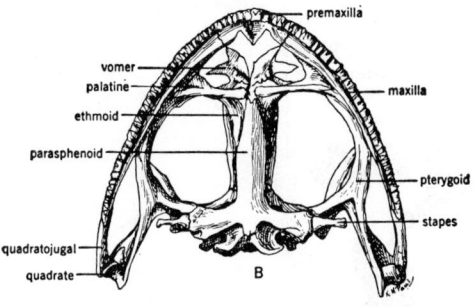

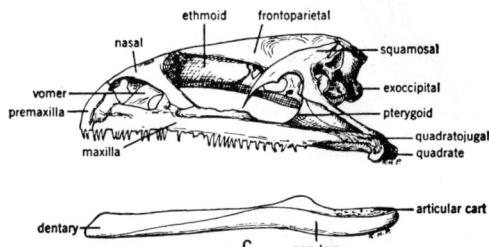

Fig. 5   Skull of bullfrog. (After Adams, Introduction to the Vertebrates, 2nd ed., Wiley, 1938.)

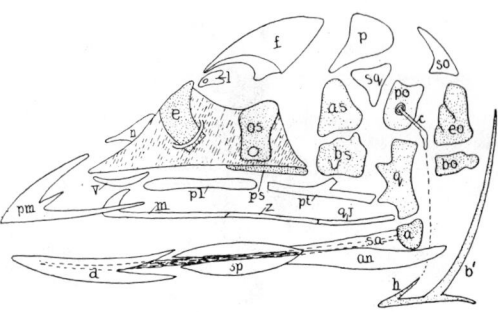

Fig. 6   Skull of a bird. Cartilage bones dotted, cartilage lined obliquely. a, articular; an, angular; as, alipshenoid; b¹, 1st branchial; bo, basioccipital; bs, basisphenoid; c, columella; d, dentary; e, ethmoid; eo, exoccipital; f, frontal; h, hyoid; l, lacrimal; m, maxilla; n, nasal; os, orbitosphenoid; p, parietal; pl, palatine; pm, premaxilla; ps, rostrum (?presphenoid); pt, pterygoid; q, quadrate; qj, quadrato-jugal; so, supraoccipital; sp, splenial; sq, squamosal; v, vomer; z, zygomatic. (After Kingsley, The Vertebrate Skeleton. Blakiston, 1925.)

most of which, especially in the neurocranium, are completely fused. Birds also resemble some of the fossil reptiles and living fish in having a circle of sclerotic bones in the eyeball. The number of individual bones in the skull of an adult mammal (Fig. 7) also varies. In a primitive mammal such as *Didelphis* (Virginia opossum) there are only forty two elements whereas an adult chimpanzee has the number reduced to twenty two. In the human skull there are approximately twenty seven bones. Thus, in contrast to the popular belief that evolution always progresses from the simple to the complex, the reverse has occured with respect to the number of individual elements in the vertebrate skull.

The reduction in the number of skull bones during the evolution of the vertebrates from fish to mammals is the result of the loss of individual bones or their fusion with other bones. The operculum, a bony flap covering the gills, and most of the gills, were lost when the land vertebrates arose but many bones of the neurocranium which are separate in the lower tetrapods have fused in mammals.

In the Crossopterygians (*Eusthenopteron*), the group believed to be ancestral to the tetrapods, there were as many as sixteen individual elements in the primary

(inner) upper jaw whereas this number has been reduced to three or four in mammals. No reduction has occurred in the number of the individual elements of the inner lower jaw which is two throughout the vertebrates. The number of dermal bones forming the secondary (outer) upper jaw has been reduced from four in the primitive fish to only two in mammals. A tremendous reduction has occurred in the number of elements forming the secondary (outer) lower jaw. In *Eusthenopteron* there were sixteen such dermal bones but mammals typically have only one in each half of the jaw, and in man they fuse to form a single bone called the mandible.

Mammals differ from all other animals in the possession of three little bones (Fig. 8) in the middle ear cavity on each side. One of these bones, the *stapes*, is considered to be homologous with the hyomandibular (part of the second visceral arch) of fish. The second bone, the *incus*, is considered to be homologous with the quadrate, a bone of the primary (inner) lower jaw of other vertebrates while the third bone, the *malleus*, is considered to be homologous with the articular, a bone of the primary (inner) upper jaw of other vertebrates.

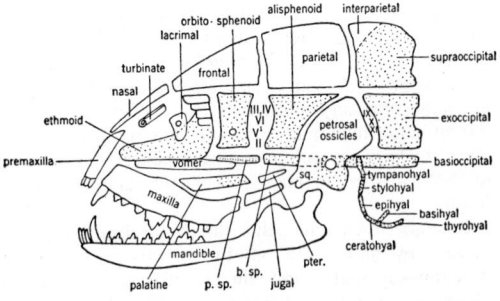

Fig. 7   Diagram of a mammal skull. Cartilage bones stippled, dermal bones white. (From Adams, Introduction to the Vertebrates, 2nd ed., Wiley, 1938.)

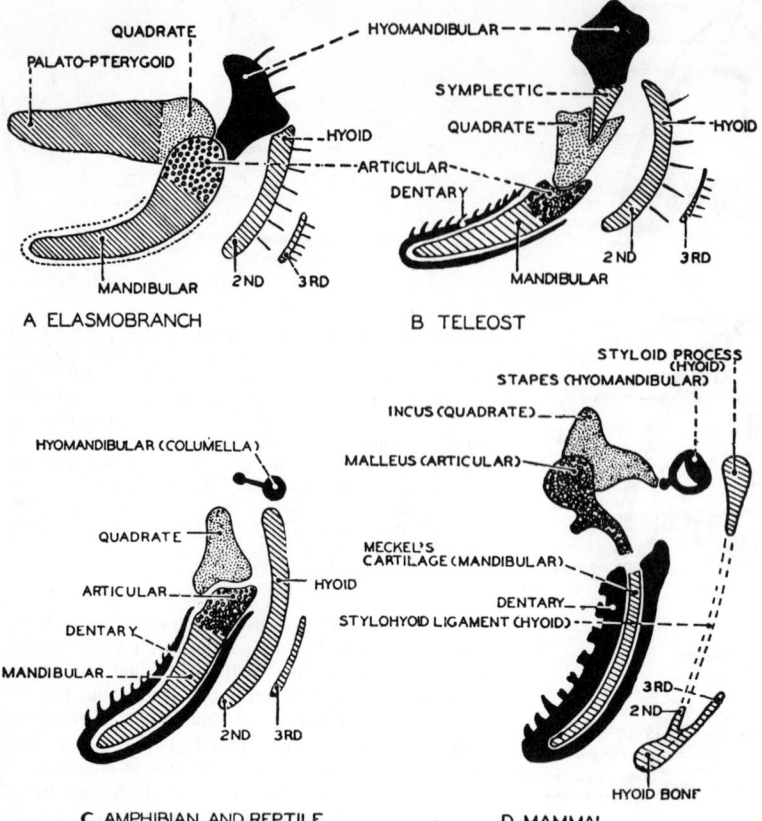

Fig. 8   Diagrams of the first and second visceral arches in A, Elasmobranch; B, of the hinge of the jaw of lower vertebrates into the malleus and incus of the mammal. The third earbone, the stapes, comes from the hyomandibular. (From Neal and Rand, Chordate Anatomy, Blakiston, 1939.)

The pectoral girdle of vertebrates also consists of cartilage bones (the scapula and coracoid) which in the lower tetrapods and in fish is covered by a series of membrane bones. In fact, in bony fishes (Fig. 4) the pectoral girdle is intimately associated with the skull and appears to be a part of it. However, with the rise of the land vertebrates the pectoral girdle becomes separated from the skull and a neck develops. The parts of the pectoral girdle in a typical land vertebrate (tetrapod) are diagrammatically illustrated in Fig. 9A. In the early, primitive tetrapods the girdle consisted of a *scapulo-coracoid* (cartilage bone) more or less covered by several membrane bones. However, during the course of evolution the membrane bones have been progressively reduced in size and lost until in typical mammals the only one left is the clavicle. It is even lost in some mammals (e.g. the horse). In mammals the clavicle is the only membrane bone found outside of the skull and in man it is partly cartilaginous in origin.

In contrast with the pelvis the pectoral girdle is held in place by muscles. In only two groups of vertebrates (the rays or skates and the pterodactyls) is there a direct cartilaginous or bony connection between the pectoral girdle and the vertebral column.

The pelvic girdle is usually associated with the out-

lets of the reproductive, excretory, and digestive systems, and hence located near the end of the body. However, in some of the bony fishes the pelvic girdle is located in the "throat" region and may actually be in front of the pectoral girdle (Fig. 4). In typical tetrapods, the pelvic girdle, (Fig. 9B) is intimately associated with the vertebral column via sacral ribs and vertebrae. In legless forms the pelvic girdle may be entirely absent. As far as known the pelvic girdle has never had any membrane bones contributing to it.

The free limbs of tetrapods (Fig. 9) typically consist of three segments, a proximal segment (arm or thigh), an intermediate segment (forearm or leg), a distal segment (hand or foot), each with its own skeletal elements.

The main variations found in the bones of the arm and thigh consist of shortening of the *humerus* and *femur* which in some of the marine forms (dolphin, ichthyosaur) are only slightly longer than they are wide. The long bones of all tetrapods are usually hollow but those of birds are especially light and thin. As a further adaptation to flight, air sacs from the lungs extend into the humerus. In the forearm the ulna may be reduced to a splint (cow) or fused with the radius (frog). The fibula is splint-like in cows and fused with the tibia in frogs and living birds.

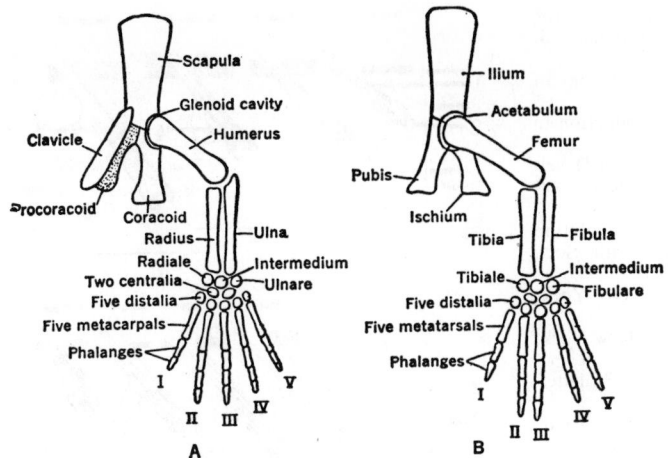

Fig. 9 Diagram of the plan of structure of a tetrapod limb. A, pectoral girdle and fore limb; B, pelvic girdle and hind limb. (From Messer, An Introduction to Vertebrate Anatomy, The Macmillan Co., 1938.)

Considerable variation occurs in the number of carpal (wrist) and tarsal (ankle region of foot) bones. Some of the primitive fossil amphibia had twelve or thirteen separate bones in these regions but in living amphibia the number varies from three to nine for the carpus and three to twelve for the tarsus. Certain living salamanders (*Siren*) have lost the hind legs while the caecilians are entirely legless.

The number of bones (phalanges) in each digit is fairly constant for different vertebrate groups. This number, expressed in order from the first to the last digit is called the phalangeal formula. The phalangeal formula in living amphibia, which usually have only four well developed digits, is 2–2–3–2 in the fore-foot of salamanders (e.g. *Necturus*), and 2–2–3–3 in the frog. Both forms have six tarsals but two of them in the frog are elongated, a specialization for hopping. Some of the primitive fossil amphibia had twelve or thirteen bones in the carpus and tarsus, respectively. The phalangeal formula in a lizard (*Iguana*) is 2–3–4–5–3 in the fore-foot and 2–3–4–5–4 in the hind foot. This is the normal condition in reptiles although some variation does occur. In the fossil ichthyosaurs and plesiosaurs (marine reptiles) the limbs were paddle-like with the bones reduced to polygonal or rectangular pieces. There was also an increase in the number of phalanges per digits. Ichthyosaurs had three to eight digits.

The wing of living birds has three digits, a reduced number of phalanges, and fused metacarpals. In the foot the proximal tarsal bones fuse with the tibia to form the tibio-tarsus. Only four toes are present in most birds, the fifth always being absent. The phalangeal formula is 2–3–4–5–0.

Five is the primitive number of digits in mammals but many living forms have fewer digits. Tapirs have four toes on the front feet and three on the hind feet. Rhinos have three toes on both feet but horses only have one toe on each foot. Other hoofed mammals have two toes on each foot. Whales and porpoises (Cetacea), sea cows (Sirenia) and marine carnivores (seals, sea lions, and walruses) have the feet modified into flippers which usually have five toes and the typical mammalian phalangeal formula of 2–3–3–3–3. In whales the second and third digits have more than three phalanges. The wing of bats has elongated metacarpals and phalanges. Five fingers are present but the number of phalanges varies in different families.

The skeleton of the paired fins of fishes is quite different from that of tetrapods with the exception of the pectoral fin of the Crossopterygii, the group of fishes believed to be ancestral to the tetrapods. In these fish the arrangement of the skeletal parts in the pectoral fin was somewhat similar (Fig. 10) to that in the front legs of the primitive fossil amphibia. This resemblance to the primitive tetrapod pectoral limb is one of the reasons for believing the crossopterygian fishes gave rise to the amphibia and other tetrapods.

F. GAYNOR EVANS

References

Adams, L. A., "Introduction to the Vertebrates," 2nd ed., New York, 1938.

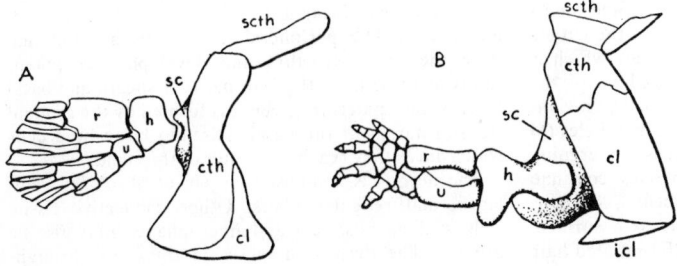

Fig. 10 Comparison of A, the pectoral girdle and fin of a crossopterygian fish (*Sauripterus*), with B, a diagram of the pectoral girdle and fore leg of a primitive fossil tetrapod. cl, clavicle; ch, cleithrum; h, humerus; icl, interclavicle; r, radius; sc, scapulocoracoid; scth, supracleithrum; u, ulna. (From Romer, Vertebrate Paleontology, 2nd ed., The University of Chicago Press, 1945.)

Evans, F. Gaynor, "The Morphological Status of the Modern Amphibia Among the Tetrapoda," J. Morph., **74**:43–100, 1944.

Flower, W. H., "An Introduction to the Osteology of the Mammalia," London, Macmillan, 1870.

Flower, W. H. and R. Lydekker, "An Introduction to the Study of Mammals," London, Black, 1891.

Goodrich, E. S., "Studies on the Structure and Development of Vertebrates," London, Macmillan, 1930 [reprint, New York, Dover, 1958].

Gregory, W. K., "Fish Skulls: A Study of the Evolution of Natural Mechanisms," Trans. Amer. Phil. Soc., N.S., **23**: i–vii, 75–481, 1933.

Jollie, M., "Chordate Morphology," New York, Reinhold, 1962.

Kent, J. C., "Anatomy of the Vertebrates," New York, Mosby, 1967.

Kingsley, J. S., "The Vertebrate Skeleton," Philadelphia, Blakiston, 1925.

Messer, H. M., "An Introduction to Vertebrate Anatomy," London, Macmillan, 1938.

Reynolds, S. H., "The Vertebrate Skeleton," Cambridge, The University Press, 1897.

Romer, A. S., "Vertebrate Paleontology" 2nd ed., Chicago, University of Chicago Press, 1945.

ibid, "The Vertebrate Body," Philadelphia, Saunders, 1962.

ibid, "Osteology of the Reptiles," Chicago, University of Chicago Press, 1956.

Van Tyne, J. and A. J. Berger, "Fundamentals of Ornithology," New York, Wiley, 1959.

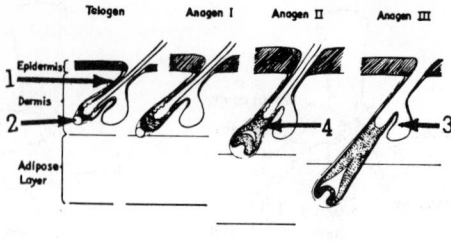

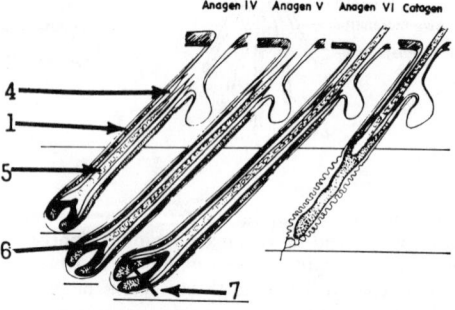

Fig. 1   (After Montagna, "The Structure and Function of Skin," Academic Press Inc., N. Y.)

SKIN

Function of skin is to produce surface keratin, HAIR, sebrum, and in certain mammals, sweat. These processes are integrated and can best be understood by their relationship to the hair growth cycle. All mammals possess hair growth cycles which are divided into 3 stages, a resting stage (*telogen*), an active stage (*anagen*), and a transitional stage between these two (*catagen*). Only the details of the cycles vary. The description which follows pertains largely to mice and rats, but its relationship to other mammals also will be given.

Resting hair follicles (telogen) are composed of a single epithelial sheath, the external root sheath (1), and an inactive dermal papilla (2). Growth of resting follicles may be initiated spontaneously or by plucking their hairs. Growing (anagen) stage begins with a burst of mitotic activity in the base of the resting hair follicles and activation of the dermal papilla. The cells of the external sheath grow into the subcutis, envelop the dermal papilla, and form the internal sheath (4) which signifies the beginning of keratinization. Internal sheath is composed of 3 layers. Henle's layer, next to the external root sheath, Huxley's layer, and innermost, the cuticle, which interdigitates with the cuticle of the hair shaft (5). The cortex of the hair which is inside the hair cuticle, surrounds the medulla of the hair. If the skin is of the pigmented variety, melanocytes become active in the bulb of the hair follicles (6) and the pigment is transferred to cells undergoing keratinization. The differentiating follicles continue to elongate and extend deep into the subcutis, and concomitantly, their hair shafts reach the skin surface. During the next 10 days the already differentiated hair

follicles primarily produce hair. When a sufficient quantity of hair has been formed the growing hair follicles enter the transitional phase, catagen. Hair production ceases, the lower halves of the follicles degenerate, the dermal papillae move up and the follicles enter the resting phase again.

The epidermis is composed of a basal layer, stratum germinativum. The stratum spinosum is the next layer, and it is usually in turn covered by a layer of cells containing granular cytoplasm, stratum granulosum. The keratinized surface covering forms the stratum corneum. To what extent, if at all, each of these layers will be present is dependent on the species of mammal and the phase of the hair cycle under consideration.

Keratinization in the epidermis is accomplished through cell proliferation in the basal cell layer and subsequent orderly suicide of many of the cells as they move upwards to the skin surface. Hair production is basically a similar process with cell proliferation occurring in the matrix (7) of the follicle and the dying cells ordered longitudinally to form a hair. Chemically, keratin production, in either case, is largely a transformation of SH containing protein to S-S-protein within the dying cells.

The sebaceous glands (3) are multiple acinar glands. Their changes during the hair growth cycle are not understood. The peripheral gland cells are flat and have the same structure and developmental potentiality as the cells of the external root sheath and basal layer of the epidermis. Sebum is formed by the gradual transformation of the basal cells into large lipid cells which die as they reach the center of the gland.

The dermis is composed largely of fibroblasts, collagen, and reticular fibers. Other connective tissue cells, such as mast cells and macrophages, may also be present. The dermis increases in thickness throughout the growth phase of the hair cycle. To what extent this represents an increase of protoplasmic mass is unknown. The subcutis is made up largely of fat cells, and it similarly enlarges during the growth phase. To some extent this may be due to an increase in lipid content. Associated with these increases in the size of the dermis and subcutis is an increase in blood flow. The skin muscle, panniculus carnosus, just beneath the subcutis, is typical skeletal muscle. It does not change during the hair cycle.

In mammals with seasonal shedding, new hair growth occurs after each shedding while in man and guinea pig each hair follicle has its own hair cycle. In these mammals, the basic changes of the hair follicles during the hair cycle are the same, however, it is not known to what extent the rest of the skin components undergo changes during the hair cycle.

Sweat glands are simple tubular coiled glands of two kinds, eccrine or apocrine. They are best seen in primates. Their relation to the hair cycle is unknown. Eccrine glands are distributed generally. They originate from the epidermis and extend into the dermis and subcutis. The apocrine glands have a limited distribution. They originate from hair follicles and also extend into the subcutis. Both kinds have an upper duct lined by 2 cell layers, and a lower secretory portion lined by one cell layer. Eccrine glands produce typical sweat, whereas apocrine glands produce a milky fluid.

THOMAS S. ARGYRIS

References

Montagna, W., "The structure and Function of Skin," New York, Academic Press, 1962.

Montagna, W. and R. A. Ellis, eds., "The Biology of Hair Growth," New York, Academic Press, 1958.

Chase, H. B., "Growth of Hair," Physiological Reviews **34**: 113–126, 1954.

Rothman, S., "Physiology and Biochemistry of the skin," Chicago, Univ. Press. 1954.

SKULL

This survey will chiefly deal with the ontogenetical and phyletical aspects of the development of the skull. It cannot be extended to include a review of the morphology of the skull in the different groups of vertebrates, as the diversity is so great that a short presentation would limit itself to generalities of little interest and value. Thus it will give a general idea of some of the major problems concerning the morphology of the skull or serve as a starting point for the prospective student of the subject.

The skull, and the vertebral column, form together the axial skeleton of the vertebrates (see SKELETON, VERTEBRATES). It consists in principle of two morphologically different parts, the neural skull, *cranium cerebrale* or *neurocranium,* and the visceral skull, *cranium viscerale* or *splanchnocranium.* The neural skull surrounds the brain and the sense organs, the visceral skull the anterior gut. The two elements are most easily discernible in the gill-breathing vertebrates.

In the fully formed neural skull are distinguished 1) the nasal or ethmoidal region, *pars ethmoidalis,* 2) the orbital region, *pars orbito-temporalis,* 3) the otical region, *pars otica,* and 4) the occipital region, *pars occipitalis.*

The visceral skull consists orginally of a series of arches, the visceral arches, and stretches as far back as the gill gut. The first arch is the mandibular arch, the second the hyoid (hyal or lingual) arch, and the following are the branchial arches.

In the ontogeny the first primordia appear as blastemas. After having passed a protochondral and a prochondral stage they develop into cartilage that ossifies in most vertebrates.

The first elements visible in the developing neural skull are the notochord, the trabecles (*trabecula cranii*), the polar bodies or cartilages, the parachordals and, in fishes, the basiotic laminas. The notochord develops first. Its anterior tip never reaches further rostrad than the place of the future hypophysis. During the ontogeny the anterior end reduces gradually, and in higher vertebrates it does not take any part in the formation of the skull.

The trabecles are two rod-like structures developing on each side in the prechordal region. In mammal embryos they appear as a single central stem. They are formed chiefly by ectomesenchyme derived from the neural crest, the most caudal portion probably also containing endomesenchymatic material.

The polar cartilages are rounded bodies lying immediately caudal to the trabecles. They exist as separate elements during the ontogeny only in some primitive fishes, and are of endomesoblastic origin.

The parachordals have sclerotomic origin and form as a ledge on each side of the anterior end of the notochord. In primitive fishes they show a tendency to be segmented in three portions (cf. table 1).

Table 1   Head segmentation

		somites	somitic material	visceral arches	ventral roots	dorsal roots
prootic somites		1. premandibular (preoral)	part of trabecles	(premandibubula)	III	V_1
		2. mandibular	polar cartil.	mandibula	IV	$V_2 - V_3$
		3. hyal	parachordals (lamina basiotica)	hyal arch	VI	VII
metotic somites		4. —	—	branchial 1	—	IX
		5. —	—	" 2	—	X_1
		6. —	—	" 3	hypoglossal	X_2
		7. —	—	" 4		X_3
		8. —	—	" 5		X_4

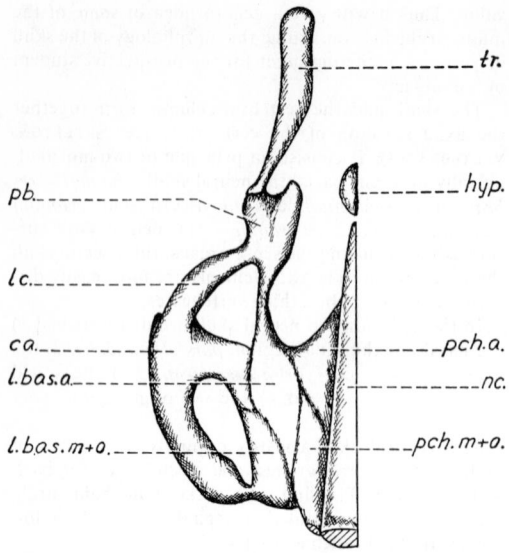

Fig. 1 *Amia calva*, 8 mm. Reconstruction of right half of skull in ventral view. 25 × 1. (Partly from G. Bertmar.) *ca.*, otic cartilage; *hyp.*, hypophysis; *l. bas. a.*, lamina basiotica metotica and occipitalis; *l.c.*, lateral cartilage; *nc.*, notochord; *p.b.*, polar body; *pch. a.*, anterior parachordal; *pch. m. + o*, metotic and occipital parachordal; *tr.*, trabecle.

The basiotic laminas develop in fishes as a product of the dermo-myotome portion of the somites nos. 3–5 on each side (fig. 1). Later they produce the ventro-medial parts of the labyrinth capsules.

An original segmentation of the head is shown in embryos of cyclostomes, as well as in several higher vertebrates, by the formation of more or less rudimentary somites as a continuation rostrad of the somitic metamerism of the trunk. In this way the head is segmented as far rostrad as the tip of the notochord. In gnathostomes the series of head somites is disturbed by the development of the comparatively spacious otic capsules. The result is a partial suppression of one or several somites. Three somites, the prootic somites, develop in front of the otic capsules followed by a number of metotic somites.

As each somite corresponds to a complete segment it is accompanied by a nerve with a dorsal and a ventral root. The ventral roots of the three prootic somites innervate the somitic muscles of the head, i.e. the intrinsic eye-muscles. They are the nerves nos. III, IV and VI. The dorsal roots of the same segments are respectively the *ramus ophtalmicus profundus* (V_1) belonging to somite no. 1, the rest of the trigeminus nerve (V_2 and V_3) to somite no. 2 and the *nervus facialis* to somite no. 3. In the two following somites the ventral roots are suppressed, the dorsal roots being represented by *nervous glossopharyngeus* (somite no. 4) and the following head segments by branches of *nervus vagus*. The ventral roots of the segments nos. 6–8 are branches of the hypoglossal nerve.

The dorsal roots innervate the visceral muscles of the head. As the trigeminal nerve (V_2 and V_3) pertains to the mandibular arch, the segment where this nerve originates is called the mandibular. Hence, the segment

in front, innervated by the profundus nerve, is called the premandibular. The third (last prootic) segment sends the glossopharyngeal nerve to the hyal arch, and is thus the hyal segment. The following segments are branchial (see table).

The somites deliver material for the development of the neural skull as well as of the meninges which take part in the later forming of the skull, when they act as the perichondrium and the periost respectively of the growing skull.

The visceral skull is essentially formed by ecto-mesenchymatic material originating from the neural crest. As this is the case also with the main part of the trabecles they are considered by several investigators to belong originally to the splanchnocranium and to represent a premandibular arch existing as such in some of the fossil agnathous fishes but missing in all gnathostomes.

The primary segmentation of the head is best seen in *Agnatha* and in selachians. It is indicated in the early ontogeny of actinopterygians, reptiles and some mammals.

During the following development the skull is built up of the material accounted for above. The trabecles fuse caudally with the polar cartilages which join the parachordals and basiotic laminas. As a rule the parachordals are fused with the elements in fromt and behind already from the blastematic stage. The tips of the trabecles connect rostrally either by means of a cartilage bridge, the trabecular commissure, or by means of their rostral portions meeting in the mid-line. In the first case the developing skull is broad (platybasic), in the second narrow (tropibasic). The former type is considered to be more primitive. Between the posterior parts of the trabecles in either case an opening forms, the *fenestra basicranialis* or *f. hypophyseos* gradually diminishing to the hypophyseal foramen. In front the joined trabecles form the ethmoidal plate, the prospective building material of the nasal capsule, and the septum nasi.

Laterally on each side the auditory sacs become surrounded by cartilage forming the otic capsules, the lateral walls of which in selachians are partly formed by tissue derived from the neural crest. In front of the otic capsules arise the pleurosphenoid or sphenolateral cartilages. They connect on each side with the growing nasal capsule by means of a chondral bridge, the *taeniae marginales,* and caudally with the otic cartilages. In the occipital region there is formed a basal plate in which the parachordals take part enclosing the anterior end of the shrinking notochord. The primordium of the skull terminates with the two *pila occipitalia* from which develops the main part of the occipital region. Morphologically they correspond to neural arches.

The skull, forming in this way, has in all vertebrates except the *Agnatha* assimilated, in the occipital region, a varying number of vertebrae, as is shown by the nerve roots leaving the occipital region. The end of the primary skull is indicated in cyclostomes by the exit of the vagal nerve. The two added cranial nerves (nos. XI and XII) are the ventral roots belonging to assimilated vertebrae.

There is no special cranio-vertebral joint in fishes except in rays and holocephalians. In all tetrapodes the basioccipital articulates with the first vertebra. Amphibians have paired occipital condyles. In reptiles and birds the condyle is single, formed by the two lateral

SKULL   867

occipitals (exocipitals) joined ventrally and medially by a process from the basioccipital. The mammal skull has two condyles formed only by the exocipitals. In the mammalian ancestral reptile group, the permian therapsides, the condyle is divided in the middle, and in *Echidna* a chondral strip connects the two condyles. These conditions are to be considered intermediary between the reptile and the mammalian type.

The result of an abrupt turn ventrad at an early stage, followed by a straightening out, is that a depression, the *sella turcica,* is formed at the base of the skull, immediately behind the hypophysis in the basisphenoid portion of the skull. It is limited caudally by a sharp edge, the *dorsum sellae.*

The roofing of the skull in fishes begins with the formation of an anterior connection between the two *taeniae,* the epiphyseal bridge, and a similar connection between the otic capsules, the *tectum synoticum,* also developing in tetrapodes where there is as well a *tectum posterius.*

By means of peripheral as well as interstitial growth of the chondral elements a more or less closed cartilaginous capsule is formed where foramina remain for blood vessels and nerves. The skull so formed, the primordial neural skull, is best developed in selachians. In all other vertebrates it is either less complete, especially as far as the roofing is concerned, or partly reduced by resorption of cartilage.

Also the visceral skeleton passes through a cartilaginous stage persisting only in cyclostomes and *Chondrichthyes* (sharks, rays and holocephalians), but otherwise ossified.

The original function of the visceral arches as supports of the gills is indicated by a rudimentay gill slit, the spiracle, between the mandibular and the hyal arches in lower fishes (selachians, crossopterygians, polypterides, palaeoniscids and sturgeons). Ostracoderms had a gill slit also between the premandibular and the mandibular arches.

The mandibular arch consists of two parts, one dorsal, the palato-quadrate, and one ventral, the meckelian cartilage. The hyal arch is usually composed of three parts. The most dorsal, the hyomandibula, is in fishes as a rule engaged as the suspensorium of the jaw. The branchial arches consist of five elements each. This original segmentation also of the first two visceral arches can be traced in fishes, indicating the primary function as branchial arches.

The visceral skeleton is greatly reduced in lung-breathing tetrapodes. Instead it takes part in various ways in the forming of the hyoid apparatus and the larynx.

In urodeles the original elements are still discernible, but, in anurans, functional demands have produced the development of a broad plate supporting the tongue, and formed chiefly by the hyal arch. Elements of the fifth, possibly also of the sixth branchial arch, take part in the formation of the larynx.

In reptiles, as in birds, the first branchial arch is well developed, and functions as the real *cornu hyale.* The skeleton of the larynx is a product of the fifth branchial arch.

The *os hyoides* in mammals has a pair of anterior horns formed by parts of the hyal arch, and a pair of long posterior horns originating from the branchial arch. The branchial arches nos. 2 and 3 form the *cartilago thyreoidea* of the larynx.

The cartilaginous skull is called the *chondrocranium,* or the primordial skull, although it has been preceded by an embryonic stage when the different parts are preformed as blastematic primordia. In cyclostomes parts of the skull are preformed by a special kind of cartilage, the mucocartilage.

In all vertebrates except the cyclostomes and the *Chondrichthyes* the primordial skull is partly, or entirely, replaced by bone forming the *osteocranium* (fig. 2). It consists partly of the more or less ossified primordial skull, the *endocranium,* partly of covering bones of dermal origin, the *exocranium.* Earlier the chondrocranium was considered to be phyletically older, hence

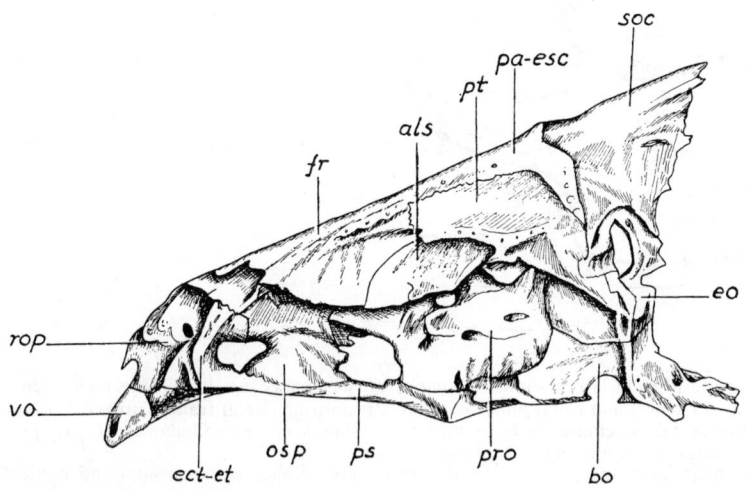

Fig. 2 *Abramis brama.* Neural skull in lateral view. (Partly from Holmgren-Stensiö.) Chondral bones: als, alisphenoid; *bo,* basioccipital; *ect-et,* ectethmoid; *eo,* exocipital; *osp,* orbitosphenoid; *pro,* prootic; *pt,* pterotic; *soc,* supraoccipital. Dermal bones:*fr,* frontal; *pa-esc,* parietal-extracapular; *ps,* parasphenoid; *rop,* rostral-postrostral; *vo,* vomer.

the name primordial skull. The oldest known vertebrates, however, the ordovician ostracoderms as well as the devonian gnathostomes, had an ossified skull. On the other hand no remains of geologically older chondral predecessors of osseous fishes can be expected to have been preserved, especially as the first real vertebrates were certainly jawless, and probably also toothless beings. It is then reasonable that even if the oldest known vertebrates had an ossified skeleton they might have been preceded by fish-like animals equipped with a skeleton of soft tissues.

The osseous tissues in the ossifying skull form chondral or replacing bones. The endoskeleton so formed is surrounded by the exoskeleton, consisting of dermal or membrane bones. Bones may also be composed of both elements, the prefix auto- and dermo- respectively then being added to the name of the bone in order to indicate when they are derived from replacing or membrane bones only.

The endoskeleton of the skull is mostly derived from the sclerotomes of the somites (exception e.g. the trabecles, cf. Fig. 1). The exoskeleton on the other hand has its source exclusively in ectomesenchyme. In fishes three types of dermal bones may be distinguished. One type develops in all bony fishes originally as a protection of the sense organs in the lateral line system in the head (fig. 3). In a primitive fish like *Polypterus* one bone primordium develops in relation to each sense organ. During the ontogeny they often fuse and form larger entities. The innervation of the sense organs is

very constant, and the homologies of the bones, therefore, comparatively reliably established. As a sensory line system enclosed in bones is known only in fishes, and in extinct labyrinthodonts, homologisation with the conditions in tetrapodes is generally difficult, but the discovery of transitory stages between crossopterygians and the most primitive labyrinthodonts has facilitated the solution of the problem. Latero-sensory bones are e.g. nasals and frontals.

The two other types of dermal bones in fishes arise without connection with the lateral line system. Some of these bones carry teeth, and their development is induced by the presence of tooth primordia. The homologies can be followed from fishes to tetrapodes because of the position in the skull even if in some groups as chelonians, and all recent birds, they are toothless. Such bones are e.g. the dentary, the maxillary, the premaxillary and in the mouth the vomers, the palatines and the parasphenoid (parabasale). In connection with the development of a secondary palate in crocodiles, birds and mammals, the vomers become excluded from the roof of the mouth. The parasphenoid, which is a big and important bone in the roof of the mouth of lower vertebrates, has disappeared entirely in mammals.

The third type of dermal bones develops independently of the lateralis system and never carries teeth. Most of them fill the spaces between the other dermal bones and are called anamestic, space-filling bones. Their homologies are often difficult or impos-

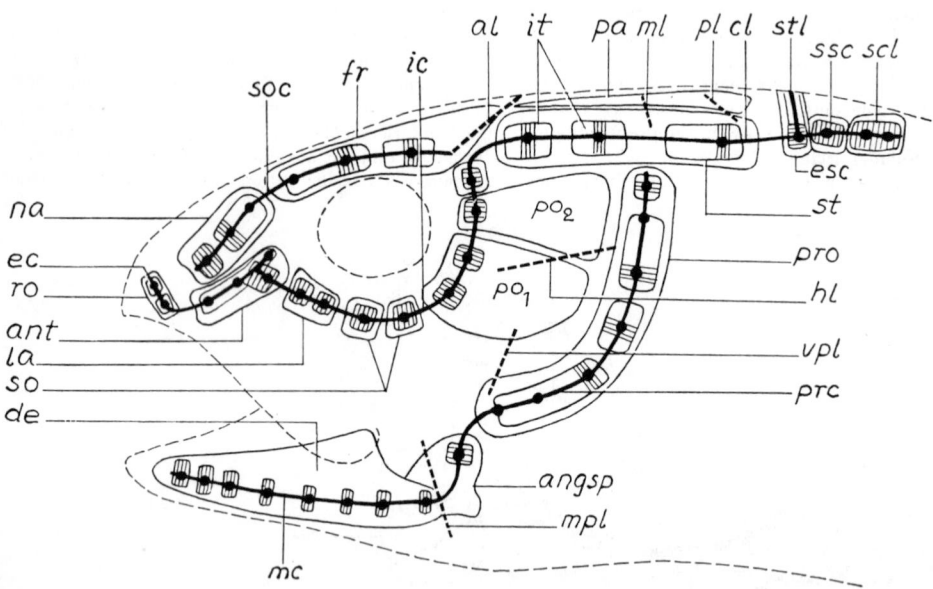

Fig. 3 *Amia calva.* Diagram of the head in lateral view showing sensory canal system (thick lines) with sense organs (dots), pit organ lines (short dashes) and canal bones with bone primordia (small framed areas). *Sensory canals: cl,* cephalic part; *ec,* ethmoidal cross commissure; *ic,* infraorbital canal; *mc,* mandibular canal; *mc,* preopercular canal; *soc,* supraorbital canal; *stl,* supratemporal cross commissure.
Pit organ lines: *al, ml, pl,* anterior, middle and posterior head lines of pits; *hl* and *vpl,* horizontal and vertical cheek lines; *mpl,* mandibular line.
Bones: angsp., angular with splenial primordium; *ant,* antorbital; *de,* dentary with eight splenial primordia; *esc,* extrascapular; *fr,* frontal; *it* and *st,* inter- and supra-temporal premordia of supratemporo-intertemporal; *la,* lacrymal; *na,* nasal; *pa,* parietal; *po₁* and *po₂,* postorbitals; *pro,* preopercular; *ro,* rostral; *scl,* supracleithral; *so,* suborbital; *ssc,* suprascapular; *st,* supratemporal primordium.

sible to settle as they may vary as to size, localisation, and number, even in different individuals of the same species.

Already the oldest known vertebrates, the jawless pteraspid ostracoderms had a well developed dermal skull. It consisted of a bony shield enclosing the head, and formed by a varying number of plates. Their endo-cranium is not known, and was probably formed by soft tissues. In the cephalaspid ostracoderms the exo- and often also the endo-skeleton was entirely ossified and both fused with each other and with the dorsal elements of the splanchnocranium.

Also the earliest gnathostomians, e.g. acanthodians and arthrodires, had a well ossified skull, in the fishes mentioned forming one single piece. The exocranium consisted of a shield of bony plates, not homologous with the dermal bones in higher bony fishes, the *Osteichthyes*.

This group was represented, in upper devonian, by crossopterygians, dipnoans and actinopterygians. Also in these the neural skull was ossified in toto, only that in crossopterygians the skull was divided in an anterior ethmo-sphenoid portion, and a posterior otico-occipital portion, allowing a dorso-ventral articulation as is the case in the recent *Latimeria*. In this fish the skull consists, however, partly of cartilage.

In the palaeozoic dipnoans as well as in the contemporary actinopterygians, the palaeoniscids, the skull was ossified in one piece, while in younger representatives of these groups separate ossifications were usual. This gradual reduction of bone is especially striking in dipnoans where, in the recent species, the neural skull is entirely chondral.

The dermal skull in the palaeozoic *Osteichthyes* is built mainly on the pattern common to all modern representatives of the group. Thus, the different bones belonging to the categories indicated above can be comparatively easily homologized in the different groups.

As mentioned above the primordial skull is con-sidered to be best developed in elasmobranchs. Already the palaeozoic members of the group had to all appearance an entirely cartilaginous skull, even if the cartilage, as in recent selachians, was calcified, the only hard elements of the skull being the teeth. For various reasons it is considered, however, that the skull in selachians, already when the first representatives of the group appeared in upper devonian, had undergone the phase of osseous degeneration indicated in the *Osteichthyes*. They should accordingly be descendants of fishes with an originally ossified neural skull, probably covered with an exoskeleton of the same type as in some placoderms. Mesenchymatic fields in the head of shark embryos are considered to be remnants of earlier dermal ossifications possible to homologize with dermal bones in arthrodires.

The separate bone elements which in geologically younger vertebrates substitute the primary homogenous neural bone skull appear mostly in fixed places, and are hence considered to be homologous. The naming of the bones, chondral as well as dermal, is originally derived from human anatomy, hence it is in many cases grossly mis-leading. Comparatively constant through the series of vertebrates are, in the occipital region, the basi-, supra- and exoccipitals, in the otic region the epi-, pter-, sphen-, pro- and epiotics, in the orbitotemporal region the ali-, latero-, orbito- and basi-sphenoids, and in the ethmoidal region the ethmoids. Comparatively constant in most vertebrates are the dermal bones nasals, frontals, parietals, squamosals, naxillaries, vomers, palatines, pterygoids, the basisphenoid, and, in the lower jaw, the dentaries.

The palato-quadrate ossifies postero-ventrally as the quadrate bone, dorsally as the metapterygoid, and anteriorly as the auto-palatine. In the posterior end of the meckelian cartilage the articular ossifies, forming together with the quadrate the jaw-joint in all gnathostomes except the mammals.

The variations in the shaping of the osseous skull are

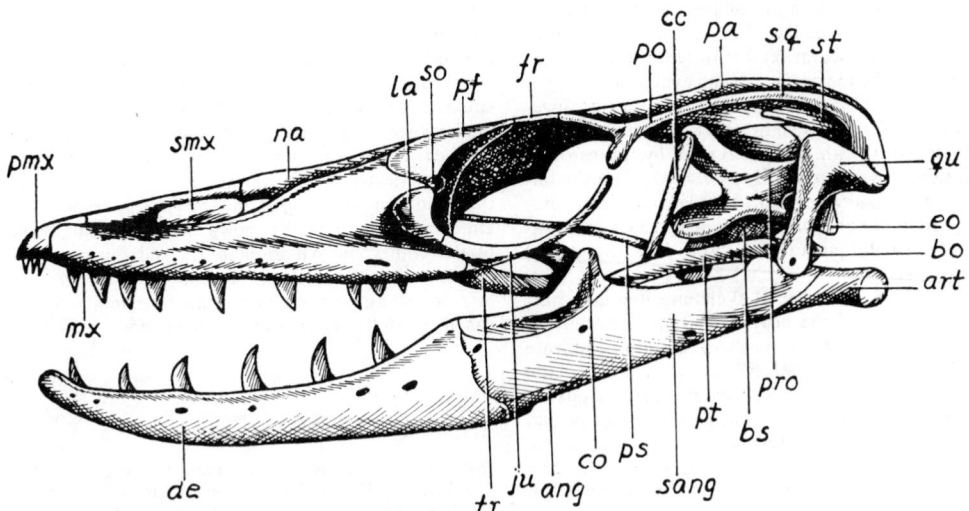

Fig. 4 *Varanus salvator*. Skull in lateral view. *ang*, angular; *art*, articular; *bo*, basioccipital; *bs*, basisphenoid; *cc*, columella cranii (epipterygoid); *co*, coronoid; *de*, dentary; *eo*, exoccipital; *fr*, frontal; *ju*, jugal; *la*, lacrymal; *mx*, maxillary; *na*, nasal; *pa*, parietal; *pf*, postfrontal; *pmx*, premaxillary; *po*, postorbital; *pro*, prootic; *ps*, parasphenoid; *pt*, pterygoid; *qu*, quadrate; *sang*, supraangular (surangular); *smx*, supramaxillary; *so*, supraorbital; *sq*, squamosal; *st*, supratemporal; *tr*, transverse.

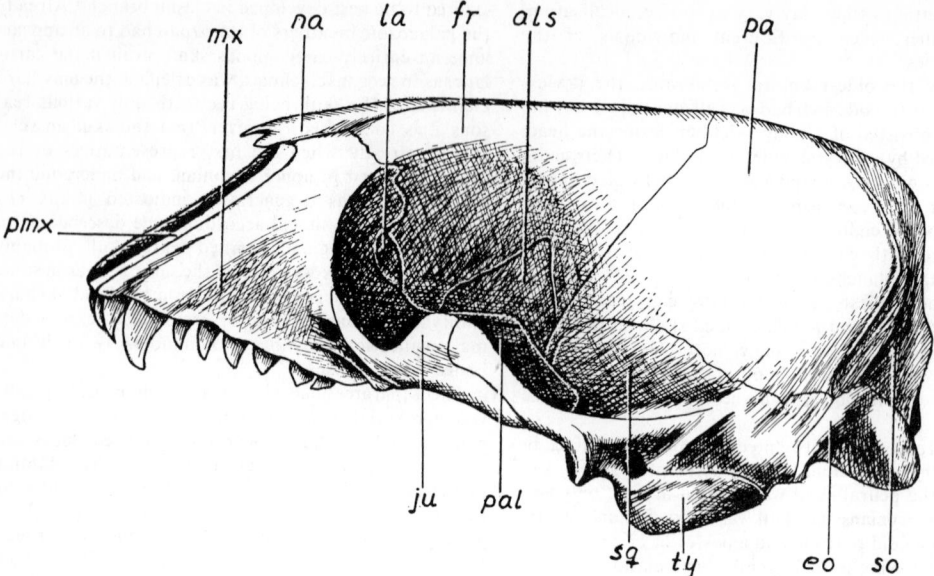

Fig. 5 *Halichoerus grypus.* Lateral view. *als,* alisphenoid; *eo,* exoccipital; *fr,* frontal; *ju,* jugal; *la,* lacrymal; *mx,* maxillary; *na,* nasal; *pa,* parietal; *pal,* palatine; *pmx,* premaxillary; *so,* supraoccipital; *sq,* squamosal; *y,* tympanic.

exceedingly numerous. Bones may be missing entirely, or may fuse with other components. In most vertebrates with an osseous skull the anterior end of the upper jaw is formed by premaxillaries and maxillaries, but these bones are missing in e.g. dipnoans, and amongst the teleosts, in the muraenas. In man premaxillary and maxillary are fused, and form one bone on each side.

The development of the splanchnocranium and its relations to the neural skull also vary. In the elasmobranch fishes the palato-quadrate acts as the upper jaw, the posterior portion articulating with the meckelian cartilage, the lower jaw. The palato-quadrate may articulate with the neural skull in different ways, but as a rule in fishes the top portion of the hyal arch, the hyomandibula, acts as an intermediary part, an arrangement called hyostyly. In some sharks the palato-quadrate is connected with the neural skull by means of an articulation between *processus postorbitalis* on the latter and a *processus oticus* on the palato-quadrate, amphistyly. In holocephalians and dipnoans, as in all tetrapodes, the palato-quadrate is fused with the neural skull, autostyly.

Elements from the visceral skeleton often take part in the formation of the neural skull. In crossopterygians and in the lung-fish *Neoceratodus* the quadrate part of the palato-quadrate has a dorsal process, the *processus ascendens.* It is well developed in urodeles, but in anurans it appears only during the ontogeny, and disappears later. In most reptiles, fossil as well as recent, it exists as a process from the *pars quadrata,* dorsally connected with the neural skull. In lacertilians it forms a special bone, the epipterygoid or *columella cranii* (fig. 4). It is missing in snakes but developed in the therapsids, the reptile ancestors of mammals. In these animals it has lost the connection with the visceral parts of the skull. Instead it forms the lateral portion of the alisphenoid, the *ala temporalis,* and has thus

entered as an element of the cranial wall in the temporal region (fig. 5). Also the teleost neural skull has a bone called alisphenoid (fig. 2). It forms, however, as an ossification in the primordial skull and is, thus, not homologous with the mammalian bone with that name.

As mentioned above the hyomandibula acts as the suspensorium of the lower jaw in all gnathostome fishes (except where there is autostyly). In all tetrapodes on the other hand it is seemingly missing. The reason is that it has changed into an ear ossicle, the *stapes.* This bone was developed already in the lower permian labyrinthodonts, probably connected with an eardrum laterally and medially with the *fenestra ovalis* in the labyrinth capsule. In mammals the jaw-joint is no longer situated between the quadrate and the articular bone but between the dentary and the squamosal—both originally dermal bones—in the temporal region. The quadrate and the articular have instead joined the *stapes* in the middle ear as the *incus* and the *malleus* respectively. Ontogenetically it can be shown that the proximal ends of the embryonic palato-quadrate and the meckelian cartilage detach and form two ear ossicles. This process as well as the formation of a new mandibular joint can also be followed in permian therapsids.

TORSTEN PEHRSON

References

Bolk, L. *et al.* eds., "Handbuch der vergleichenden Anatomie, Bd. IV," Berlin und Wien, Urban, 1936.
De Beer, G. R., "The development of the vertebrate skull," Oxford, The University Press, 1937.
Goodrich, E. S., "Studies on the structure and development of vertebrates," London, Constable, 1930 (*Reprinted* New York, Dover, 1958).
Grassé, P.-P., "Traité de Zoologie," Tomes XII–XVI, Paris, Masson 1950–58.

SOCIAL INSECTS

The usual definition of these insects is that the young remain with their parents forming a large family group, the young in turn aiding in the rearing of yet more insects. Under this definition only ants, some bees and some wasps (all HYMENOPTERA) and termites (ISOPTERA) are social insects. However, attempts have been made to include all insects which show behavioral ties; thus locusts (SEE ORTHOPTERA) have been suggested as social insects. Such behaviour as locusts show e.g. turning towards the swarm when an individual is tending to head out of it, is minor compared with the number and strength of the behavioral inter-relations in an insect society.

Various degrees of social development can be found extending from bugs (HEMIPTERA) which protect their eggs, Earwigs (DERMAPTERA) which protect the neonates as well as the eggs, through various wasps and bees whose few daughters remain with and aid the parents in feeding the young, to the highly developed societies of honey-bees, wasps and most ants in which the queen has become an egg-laying machine taking no part in foraging for food and rarely, if ever, leaving the colony. She is fed and cleaned entirely by workers. Exceptions to this are found in the Ponerine ants, generally considered the most primitive of the Formicidae; the queen forages as well as the workers, each obtaining enough food for itself but probably none for the other members of the group.

It is characteristic of social insects mutually to exchange food among the members of the hive or nest; this takes place not only among the adults but also between adult and larva. This relationship is trophallaxis. Food-soliciting and food-begging behavior by workers are equally important in worker honeybees. There is doubt, however, whether worker wasps, such as *Vespa germanica,* exchange food with each other: they undoubtedly feed the larvae but again whether they in turn obtain secretions from the larvae is still uncertain. Where it exists, the mutual exchange system provides the means whereby the PHEROMONES (ectohormones, sociohormones) can be passed through the colony and the odor of the individuals, apparently derived from the food being taken by the foragers, may be made more uniform and unique to the particular colony. Such a colony odor is an important distinguishing mark by which the individual is recognized by its fellow members. Controlling, for example, the admission of foragers at the entrance of the bee-hive, robber bees of unfamiliar odor being excluded.

There are three main castes within hymenopteran colonies; queens (reproductive females), males and workers (females, usually infertile, though in ants these may take on an egg-laying role, the eggs giving rise only to males in most cases). The workers may be subdivided in the ants into soldiers of various size groups, whose function is to guard the nest or marching columns, and workers, again often of various size groups, who forage for food, care for larvae, construct the nest and so forth. Since the larvae are immobile, and sometimes immured in cells, the work of the colony is carried out by the adults. In termites, on the other hand, much of the work is carried out by the active nymphs which are relatively undifferentiated and therefore form a pool of potential adult workers, soldiers or reproductives.

The determination of caste may be by at least two means, genetic and trophogenic. Sex determination is essentially genetic, fertilized eggs giving rise to females, unfertilized to males, but the subsequent development is influenced by other factors. Nutrition of honeybee female larvae seems to determine whether the consequent adult will be a worker or a queen. Future queens are raised in cells of a shape different from the rest and the formation of these cells is suppressed so long as "queen substance" (pheromone) licked from the queen's body is circulating in the trophallactic current through the colony. Similar substances appear to be responsible for the control of production of the different castes in a termite colony though their determination seems to be genetic in some species.

Often the castes are distinguished by morphological as well as behavioral differences. Worker honeybees possess pollen-collecting apparatus, not found on the limbs of queens, for example, and soldiers of both ants and termites may be widely different from the workers e.g. the nasute termite soldier has a head which is prolonged into a projection through which irritant fluid can be shot and rudimentary mouthparts.

Since the essence of an insect society is its high reproductive capacity, it is not surprising to find that the bringing of food essential to the colony, and particularly the young, is highly organized. Insurance of the return of a successful forager with its load of food is of overriding importance. Landmarks along the route may be learned having been recognised visually or chemically. Flying insects, such as bees, utilize various aspects of the sun as a guide, reacting to ultra-violet rays and to the polarized light from the sky reflected into the eyes from the background. Ants and termites, being earthbound, use mainly chemical clues for orientation, though a number of ants certainly recognize visual landmarks. Termites are blind and often forage only underground or through covered ways should they have to emerge into the light.

The diet of social insects is varied, there being examples of both carnivorous and herbivorous insects among them. Wasps are mainly carnivorous as are Ponerine ants, though whereas wasp larvae are fed on nectar and juices, the very active larvae of Ponerines are given insect carcases to dismember. Bees are generally nectivorous. Most ants have a mixed diet but feed their larvae on regurgitated juices. Some are seed eaters. Plant tissue may be stored in the nest and infected with fungi which then grow upon it. The leafcutter ants (*Atta*) do this, and then feed upon the fructifications of the fungi. Among the grass-eating termites, such a method of dealing with the otherwise indigestible cellulose is common and the chambers containing the fungi seem also to be humidity and temperature controlled nurseries for the youngest nymphs. The wood-eating termites digest their cellulose food by means of protozoa in the gut.

A variety of materials is used for the construction of the nests of social insects. These may be simple tunnels and chambers under the ground or in earth piles on the surface, as are the nests of most ants. Those of termites, which are a feature of the landscape in Africa and Australia, are made of triturated earth cemented with secretions to form a material with the firmness of concrete, rising to twenty feet above the ground. Within, these large nests often have an elaborate system of air-conditioning tunnels which serve to control the nest temperature and humidity.

Bees use wax secreted in plates from abdominal glands to construct their hexagonal cells; wasps and some ants use a wood pulp made by chewing wood from posts and so forth to construct their carton nests. Plants may also be used. Certain ants live within the tissues of plants, whose response to their presence is to produce large gall-like excrescences which become penetrated by the galleries of the ants. The sewing ant (*Oecophylla*) joins leaves together to form a nest.

Colony foundation may occur by a female alone, or by a male and a female. The pattern in ants and bees is of foundation by a fertilized female, the male dying. Termite colonies are founded by a royal pair, the male continuing to live beside the queen. Some species of ant are parasitic on others; in these the queen invades a nest of the host species and kills the host queen.

Into the colonies are introduced other insects which either share the life of the owners or prey upon them. Sometimes this has important economic effects, as does, for example, the care given to plant lice and coccids by ants. These yield secretions to their keepers. The protection afforded by the nest permits the pest to avoid predators and to evade insecticidal sprays, thus they form a reservoir from which the pest population outside can be renewed.

Other important economic effects of social insects come from the production of honey by bees, used as a source of high grade sugar all over the world and their function as pollinators. But on the other hand, the damage done by termites to cane sugar crops in one Indian province has been estimated at $4,500,000 in one year, and repairs to government buildings in British West Africa in 1955 cost nearly $1,000,000.

J. D. CARTHY

References

Brian, M. V., "Social Insect Populations," New York, Academic Press, 1965.
Goetsch, W., "Vergleichende Biologie der Insekten-Staaten," Leipzig, Acad. Verlags Gesellschaft, 1953.
Richards, O. W., "The Social Insects," London, Macdonald, 1953.
Wheeler, W. M., "Social Life among the Insects," New York, Harcourt-Brace, 1923.

SOIL MICROORGANISMS

The science of soil microbiology has a relatively brief history beginning about the 1890's. Its basic aims are to study the minute forms living in soil, particularly as they influence fertility. Because of the extremely complex nature of soil relative to the three major disciplines involved, physics, chemistry and biology, and because of the involved interactions occurring both within and between disciplines, it has been a field fraught with many difficulties for the investigator. To comprehend better the functions and interactions of various soil microorganisms, some knowledge of the way in which environmental conditions affect these organisms is necessary. In addition to the need of an adequate supply of food, three additional factors are important, namely, oxygen supply, moisture level,

and soil pH. Oxygen supply is important because it influences the relative activity of three groups of microorganisms: 1. the aerobes which need free oxygen to carry on their normal activities, 2. the anaerobes which only develop in the absence of oxygen, and 3. the facultative forms which can adapt themselves to either situation.

Aerobes tend to bring about complete oxidation of organic matter, the ultimate products being primarily carbon dioxide and water, while anaerobes, developing in the absence of oxygen cannot completely oxidize the organic matter upon which they feed, and consequently produce many partially decomposed products uncluding humic acids and other components of the soil humus. Since anaerobes can be isolated easily from soil, it follows that oxygen depletion must occur, at least in isolated pockets in the soil, often enough for them to maintain themselves without difficulty.

The presence of moisture in soil is essential for the development of microorganisms. Studies of microbial respiration indicate that these organisms can remain dormant in dry soil for many years, and, in a matter of minutes after the soil is remoistened, many can immediately resume their metabolic processes. With but few exceptions, soil organisms cannot feed on substances insoluble in water, and in order to utilize such materials they excrete enzymes capable of partially degrading particulate food materials into water soluble forms available to the cell. Apart from the direct effect of soil moisture on the microbial population there is an indirect effect based on the fact that soil aeration is related inversely to the amount of water present in soil. Most mineral soils display optimum microbial activity at moisture levels close to 60% saturation.

In addition to the effect soil pH has on the availability of many nutrients (through its influence over their solubilities) pH is extremely important because it exerts control over many of the microbial groups in soil. Thus, in most soils of the temperate region, nitrogen fixation by aerobic organisms is limited to pH levels between 6 and 7.5, while fixation by anaerobic organisms proceeds at levels as low as 3.5. It is a well known fact too that fungi, being much more tolerant of acid conditions than are bacteria, constitute a greater proportion of the total population in acid than in neutral soils.

Microbial Population

The soil population is composed of representatives of many different groups. Many forms possess unusual nutritional requirements, often satisfied by the excretions from other microbial groups. Many representatives produce substances antagonistic to other types of organisms; in fact most of the antibiotics in current use are produced by organisms originally isolated from soil.

The enumeration of the soil population presents many problems, some still unsolved. Many of the groups are present in numbers corresponding to millions per gram of soil, but since figures representing numbers of such variously sized organisms might be deceiving, it seems more pertinent to give estimates of their live weights in the plough depth (6–7 inches) of an acre of fertile agricultural soil. The microbial population consists primarily of seven groups.

1. Molds (2000 lb per acre)—all are aerobic, and many are acid tolerant. As a group they possess potent degradative powers.

2. BACTERIA (1000 lb per acre)-this group contains many types ranging from strict aerobes to strict anaerobes, (SEE ANAEROBIC BACTERIA) and having growth requirements that vary from the extremely simple, developing in inorganic mineral salts solution (autotrophs), to the very complex, requiring a suitable organic energy source, minerals, amino acids, VITAMINS, and sometimes unknown constituents from soil extract (heterotrophs). Special groups can obtain energy from bizarre sources such as chitin, paraffinic hydrocarbons, sulfur and even phenol. In general bacteria do less well under acid conditions than molds. The group includes most of the nitrogen fixing organisms. (SEE BACTERIAL NUTRITION)

3. Actinomycetes (1000 lb per acre)—the normal soil inhabitants are aerobic and most prefer neutral to alkaline pH for best development; they can survive very low nutrient levels. Thermophilic forms are active in the rapid high-temperature decomposition of compost and manure.

4. PROTOZOA (200 lb per acre)—these constitute the major groups representing the soil microfauna; they consist of various species, all aerobic. While some feed saprophytically upon portions of the soil organic matter, many devour soil bacteria and consequently tend to vary inversely with the bacterial population.

5. ALGAE (100 lb per acre)— this group is of little significance in agricultural soils except in the paddy culture of rice where certain blue-green forms capable of fixing nitrogen are important. Because of their photosynthetic (SEE PHOTOSYNTHESIS) power the algae are peculiarly adapted to colonize barren rock surfaces and to initiate soil formation.

6. YEASTS (100 lb per acre)—Aside from wine yeasts which temporarily dwell in vineyard soils, only a few species regularly inhabit soil. They are limited in number, probably unimportant, and little is known about them.

7. PHAGE—specific phage (VIRUS diseases) capable of attacking various species of bacteria and actinomycetes are normal soil inhabitants. While present in astronomical numbers in soil, their ecological importance has not been assessed and their contribution on a weight basis will be insignificant.

In total, these microorganisms comprise over two tons of living cells per acre, an amount which might be expected to exert a considerable effect upon the development and nature of soil. Other forms of life in soil include plant roots, mites, insects and earth worms; these, while important components of the biological phase of the soil, are not properly considered under the heading of soil microbiology. The principle functions of the soil microbic population may be discussed under the following headings.

Decomposition of plant and animal residues. The microorganisms prevent the smothering of soil by deposits of organic matter through their ability to decompose almost all types of animal and plant residues. In so doing they contribute greatly to the formation of humus which improves soil in many ways: through the increase of drought resistance, base exchange capacity, the improvement of structure and tilling characteristics of soil, and finally, the protection of the microbial population against harmful effects arising from the use of chemical pesticides.

Nitrogen fixation. The process of nitrogen fixation has, throughout the brief history of soil microbiology, held the attention of a large number of workers. It is now known that nitrogen may be fixed by many different organisms working independently of or in association with plants. Because fixation of nitrogen requires much energy, the organisms concerned require large supplies of food to function effectively. The most important microorganisms capable of fixing nitrogen independently are now classified in three genera, two of which are aerobes. The latter include the Azotobacters, a group having world wide distribution and capable of fixing nitrogen in a pH range near neutrality, but ineffective at a pH below 6.0. The second group have been placed in the genus *Beijerinckia*, which so far appears to be restricted to tropical regions; these organisms are acid tolerant and are active at a pH as low as 4.0. The third group consists of anaerobic organisms belonging to the genus *Clostridium*. They are active over a wide pH range from above neutrality to 3.5; these anaerobic organisms can use only a fraction of the energy contained in organic matter and therefore fix nitrogen less efficiently than aerobic groups. Most estimates of nitrogen fixed by these free living organisms range from 12-40 lbs. per acre per year.

From the agricultural viewpoint, the most important nitrogen fixation is that brought about through the symbiotic relationship between members of the genus *Rhizobium* and legume plants. By this process fixation of as much as 250 lbs. per acre per year has been reported. The relationship between the rhizobia and legumes is very complex, different strains of host and of rhizobia display complex interactions that influence the quantity of nitrogen fixed. The rhizobia exist in a wide range of forms varying from those capable of very effective fixation, ineffective, and very rarely even truly parasitic forms. The effective relationship results in the production symbiotically of the pigment leghemoglobin, which gives nodules a pinkish colour and constitutes the only known production of a type of hemoglobin in the plant kingdom. Its connection with the process of symbiotic fixation, while not understood, is apparently very close. Nitrogen fixation has also been demonstrated to occur symbiotically between an unknown root nodule organism and the alder.

Nitrification. Before fixed nitrogen can be made available for plants other than the microorganism or the legume host, the nitrogen combined in organic compounds must be broken down. Most microorganisms are capable of releasing ammonia from the nitrogen compounds in microbial and plant cells. While many or possibly all plants can utilize ammonia-nitrogen directly, there are some which seem to prefer nitrate-nitrogen. Members of two autotrophic genera of bacteria together accomplish this transformation: *Nitrosomonas* species oxidize ammonia to nitrite and *Nitrobacter* species oxidize nitrite to nitrate. The activities of these bacteria are limited to a pH range of about 5.0 to 8.0, therefore in strongly acid soil little or no nitrification occurs.

Denitrification. This process, favoured by anaerobic conditions, can cause the loss of much soil nitrogen obtained either through fixation or fertilization, by converting nitrite or nitrate to gaseous nitrogen.

Though occurring to some degree aerobically, it is most rapid under conditions of low oxygen tension: in wet or waterlogged soil or in pockets poorly supplied with oxygen in well aerated soil. It may proceed rapidly under suitable conditions, and is retarded by management practices that improve soil aeration.

The dissolving of soil minerals. The gradual release, in an available form, of nutrient elements locked up in insoluble soil minerals is another important function of soil microorganisms. They do this through the excretion of a number of organic and inorganic acids. Although produced in small amounts, these acids slowly dissolve the minerals, thereby replenishing the supply of available plant nutrients.

Soil Structure. Soil microorganisms contribute significantly to the improvement of soil structure. This comes about not only as an indirect benefit from the formation of humus in soil, but as a direct result of the evolution of polysaccharide gums which tend to cement the mineral particles together into aggregates, and also by the production of stable aggregates held together by the mechanical binding action of the mycelial filaments of mold and actinomycete colonies.

F. E. Chase

References

Alexander, M., "Introduction to Soil Microbiology," New York, Wiley, 1961.
Bollen, W. B., "Micro-organisms and soil fertility," Oregon State Monographs, No. 1, 1959.
Russell, E. J., "Soil Conditions and Plant Growth," 9th ed., New York, Wiley 1961.
Waksman, S. A., "Soil Microbiology," New York, Wiley, 1952.

SOLIFUGAE (Sun-spiders)

The order is distinguished from other ARACHNIDA by the very large carapace consisting of the head and first thoracic segment, the powerful 2-jointed chelicerae, the large sensory pedipalpi and the peculiar stalked leaf-like organs (malleoli) of the fourth pair of legs. Size from 10 to 70 mm.

External structure. The body is divided into a large segmented prosoma and an opisthosoma or abdomen of 11 segments. The appendages consist of enormous 2-jointed chelicerae, both jaws of which are armed with large crushing teeth, the ventral jaw being movable; the pedipalps have 7 segments, the last one bearing an adhesive sucker but no claw; of the four pairs of legs the first is weak and usually clawless, the remainder robust, with two claws.

Sense organs. Vision is good, the median pair of eyes of the cephalothorax being large; there are also 1 or 2 rudimentary lateral pairs. The body and appendages are remarkable for the large number of very long, fine tactile hairs. Five pairs racquet-shaped organs on the fourth pair of legs are peculiar, being strongly innervated and containing organs of sense with unknown functions. Modified hairs on the inner surface of the chelicerae are perhaps organs of taste, others on the pedipalps having olfactory and tactile functions.

Respiration. The order is remarkable for its highly

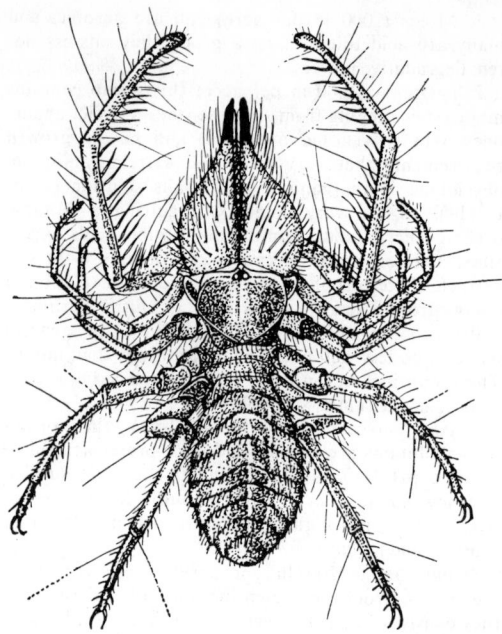

Fig. 1   Galeades, a Salifugid. (After Millot.)

developed tracheal system, the tracheae being numerous and much larger than in other Arachnid orders, opening by 3 paired ventral stigmata and a single median spiracle on the fifth sternite. In association with the rapid movements and intense activity of the animal, breathing movements like those of insects are practised.

Reproduction and development. The male sex organ is located on the dorsal jaw; it may be leaf or cup-shaped or simply spine-like, sometimes assuming fantastic shapes; it assists the introduction of the spermatophore into the vulva of the female; Solifugae are viviparous, 50–200 eggs being laid in a tunnel underground; the young, growing progressively larger after successive moults, soon leave the mother and disperse.

Ecology and distribution. Solifuges are active non-social predators, some being nocturnal, others diurnal in habit; some short-legged forms like *Chelypus* are subterranean in habit, feeding on termites. The larger species are extremely savage, attacking other arthropods or even small vertebrates. The 700 species are divided among 10 families, and like the scorpions, live in hot arid parts of both hemispheres except Australasia, the larger oceanic islands, and Madagascar.

R. F. Lawrence

SOMATIC CELL GENETICS

Before the developments of microbial genetics, the studies of heredity were confined to species with sexual reproduction which is mediated by specialized germ cells. Sexual reproduction involves all the processes on which genetic analysis depends: fusion, segregation and recombination of structures responsible for the genetic constitution. Microbial genetics stimulated speculations as to possible counterparts of these processes in somatic cells.

The potential advantages of using somatic cells for the genetic analysis of multicellular organisms are obvious:

(1) Mapping of chromosomes, detection of heterozygotes for recessive genes and other genetic problems, which are hardly accessible in species with a long generation time or without a possibility of experimental breeding, would become approachable when the block of sexual reproduction could be by-passed. The studies on the mutation rates also would be greatly facilitated when the populations of individuals could be replaced by cell populations whose size is much less a limiting factor.

(2) It would be possible to experimentally approach such physiological or pathological processes as differentiation, antibody formation or neoplasia whose mechanisms seem to involve genetic changes at the somatic level and which are thus, via sexual reproduction, inaccessible.

Let us thus consider the available evidence on the existence in the somatic cells of processes equivalent to the basic components of the sexual cycle. It is quite obvious that such processes must be very rare in the somatic cells because permanent segregation of homozygous cell variants should otherwise occur in individuals heterozygous for many genetic loci. Nevertheless, some phenomena are occasionally observed that can be ascribed to such rare events at the somatic level. For example, the occurrence of twin spots of two recessive phenotypes, aa and bb on Drosophilas, double heterozygotes Ab/aB, were interpreted as being due to mitotic crossing over between two of the four chromatids of the respective homologous chromosomes.[1] Reports on similar phenomena, scattered in the literature, seem to make a strong case for the occurrence, however rare, of mitotic crossing over also in the laboratory rodents.[2] A great deal of experimental evidence in favor of mitotic crossing over is based on immunoselection of homozygous cell variants arising in mouse tumors heterozygous for the histocompatibility locus H–2.[3] One of the main problems concerning the somatic cell variants is the difficulty in eliminating the possibility that the observed change occured only at the phenotypic level since the variants cannot be simply submitted to progeny test. Reported findings on quadriradical configurations in human cells dividing *in vitro*[4] may, however, be interpreted as cytological evidence that processes which can result, under appropriate conditions, in gene recombination do occasionally occur during mitosis.

Spontaneous mitotic crossing over is not the single potential mechanism to give rise to somatic recombination. Under experimental conditions crossing over might take place, in principle, between a somatic cell and a naked gene, isolated chromosome, transplanted nucleus or even another cell, if some of these structures tend to be taken up by the somatic cell. There are indications, more or less strong, that all these possibilities exist. Virus-induced tumorigenesis of somatic cells might be the best analogy of the transduction phenomenon outside bacteria. Mammalian chromosomes isolated from cells of one type can be taken up by cells of a different type, and although the uptake is restricted to the cytoplasm during the early phase, penetration into the nucleus seems to occur later.[5] Transplantation of the cell nuclei into enucleated cells of amphibian embryos represents a promising approach to studies of cellular differentiation. Finally, fusions of somatic cells have been recently repeatedly demonstrated with tumor or normal cells of the same species or of two different animal species in tissue cultures *in vitro*.[6,7]

Therefore, potential conditions for the first essential step of genetic analysis at the somatic level are available. The second and third steps (segregation and recombination) require some segregation mechanism. This is no problem when mitotic crossing over takes place in metaphase and the recombinants, i.e. the two complementary daughter cells, are separated in telophase. A double complementary mitotic non-disjunction (which is probably extremely rare) would have a similar result, except that the recombination would concern the whole chromosome while only a part of a linkage group would be involved in mitotic crossing over. Fusion of two diploid somatic cells, however, results in tetraploidy, and no regular mechanism of mitotic segregation follows. In tissue culture cells,[8] similarly as in filamentous fungi,[9] cell fusion may be followed by gradual, probably accidental loss of chromosomes. The intermediary aneuploid cell stages would have a very low chance of survival in multicellular organisms where, nevertheless, a mitotic segregation seems to occur occasionally as indicated by some cytological and genetical findings. Alternative mechanisms, tetrapolar mitosis or two partially overlapping bipolar mitoses, were suggested to account for the occurrence of diploid somatic cell recombinants, originating most likely from fusion of two different cells.[10]

The evidence thus exists, even though it is partially indirect, that the separate processes, on which the possibility of genetic analysis critically depends, do occur in some form in the somatic cells. The fact that they are not regular processes but rather rare mitotic accidents is then of minor significance. The frequency of mitotic accidents is probably proportional to the mitotic rate which can be experimentally manipulated. In addition, the frequency of mitotic crossing over can be increased by means of several factors and the frequency of cell fusion by other agents which facilitate cell contact, e.g. a hemagglutinating virus, etc.[11]

A further requirement is the availability of suitable genetic markers, i.e. stable, heritable characters detectable in alternative forms at the cellular level (preferentially such which are subject to known selective pressures). The choice of such markers is still limited: cellular alloantigens (which can be submitted to immunoselection), resistance versus sensitivity to various drugs, different nutritional requirements, different forms of hemoglobin, marker chromosomes and a few others.

The available experimental systems are not confined to tissue cultures *in vitro* or to the search for spontaneous homozygous cell variants in heterozygotes which might be at a selective disadvantage and therefore difficult to reveal. Spontaneous and particularly artificial chimeras (e.g. radiation chimeras) can be created so that two different homozygous cell types are made to coexist within one animal where heterozygous cell variants may arise by cell fusion and segregation and various efficient selective pressures can be further applied.[12] Several techniques which have been successfully used in bacterial genetics are in principle also available with the somatic cells.

So, although the obstacles may be formidable, the outlook that Mendelian analysis at the somatic level might be feasible is by no means less promising than it was with regard to bacteria or Aspergillus some twenty years ago.

ALENA LENGEROVÁ

References

1 Stern, C., Genetics, 21:625, 1936.
2 Grüneberg, H., Genet. Res., Camb., 7:58, 1966.
3 Klein, G., p. 407 in Burdette, W. ed., "Methodology in mammalian genetics," San Francisco, Holden-Day, 1963.
4 German, J., Science, 144:298, 1964.
5 Chorazy, M., et al., J. Cell Biol., 19:71, 1963.
6 Barski, G., et al., Comptes rendus, 251:1825, 1960.
7 Sorieul, S. and B. Ephrussi, Nature, 190: 653, 1961.
8 Ephrussi, B. and S. Sorieul, in Merchant, D. J. and J. V. Neel eds., "Approaches to the genetic analysis of mammalian cells," Ann Arbor, The University of Michigan Press, 1962.
9 Pontecorvo, G. and E. Käfer, Advances Genet., 9:71, 1958.
10 Ohno, S., p. 46 in Goldstein M. N. ed., "Phenotypic expression," Baltimore, Williams and Wilkins, 1966.
11 Harris, H., Proc. Roy. Soc. B, 166:358, 1966.
12 Lengerová, A., Science, 155:529, 1967.

SPACE BIOLOGY

With the earth as a point of reference, *space biology* is the science of living things beyond the atmosphere. Here the subject is treated in its broadest sense and is meant to include and to some extent be synonymous with a number of currently used terms including space medicine, bioastronautics, space bioscience, and exobiology. Thus its scope ranges from the influence of the space environment on fundamental life processes through the application of biomedical knowledge to sustain human life in space, to questions of extraterrestrial life.

Recorded evidence of the first fairly serious thinking about subjects closely related to space biology appeared in Johannes Keppler's "Somnium" published about 1634 and dealing with a fictional voyage to the moon during which it was observed that the lunar inhabitants were equipped by nature with a special thick fur which, when exposed to the sun's radiation would form into a hard, heat-insulating shell thus permitting survival in the environmental extremes of the moon. In the ensuing two and one-half centuries several popular science fiction-type works appeared, dealing with trips to the moon and the planets and including some very shrewd speculations on space biological matters, principally having to do with the effects of lunar and planetary environments on the nature of living things.

The first living organisms intentionally used to test the effects of a sustained ascent into the atmosphere were the chicken, duck, and sheep used in 1783 by the Montgolfier brothers in testing their newly invented hot air balloons. In the early part of the 19th Century, Claude Ruggieri (Paris) gave public demonstrations

with rockets that carried small animals aloft and brought them safely to earth by parachute; however, this was for profit rather than for science. Hermann Ganswindt designed a space ship in the latter part of the 19th Century in which he proposed to counteract reduced gravity effects by causing the cabin to revolve. Another pioneer space ship designer of the same period, Konstantin Ziolkovsky, considered the need for a sealed cabin, oxygen stores, and air purification. Even as recently as the 1920's a presumably authoritative medical criticism of Hermann Oberth's classic scientific contribution, "The Rocket into Planetary Space," alleged that human space travel would be out of the question because, once beyond the earth's atmosphere the human body would collapse from the powerful gravity of the sun.

The forerunner of modern space biology is the broad field of aviation medicine and its fundamental science, aviation physiology, which originated not long after a particularly harrowing balloon flight in 1862. During the flight, British aeronauts Glaisher and Coxwell reached about 29,000 feet altitude but were very nearly fatally incapacitated by acute oxygen deprivation, or hypoxia. Reports of this experience stimulated French physiologist Paul Bert to investigate the physiological effects of decreased and increased barometric pressure as well as to conduct extensive researches in respiratory physiology. Bert's was the first low-pressure chamber for human physiological studies and his work laid the foundation for the science and the art of aviation medicine. Impetus was given to the advancement of this field both in the United States and abroad when military flying began to become important during the First World War. New vigor was added during the 1930's and subsequently by the original investigations of a young medical officer serving with the Army Air Corps, Captain Harry G. Armstrong, well known today as a pioneer leader in modern aviation medicine research. Important milestones in the evolution of space biology included, for example, the remarkable balloon ascent to 34,120 feet in 1932 by A. Piccard of Belgium and the long-time record balloon ascent to 72,395 feet in 1935 by Stevens and Anderson of the USA. World War II saw the build-up among the Western Allied Nations, as well as in Germany, of an impressive array of aeromedical research and development programs. These programs have flourished since World War II, with a natural but limited extension of interest beyond the earth's atmosphere into the exciting and challenging field of space biology. Space biology sprang into international prominence along with other areas of space science and technology shortly after Sputnik I, October 4, 1957, but especially after Sputnik II, November 3, 1957, with its famous canine passenger, Laika.

However, investigations involving space biology had been going on for several years prior to these spectacular events. A department of Space Medicine was established at the U.S. Air Force School of Aviation Medicine in 1949. The Air Force aeromedical Aerobee rocket flight, 1951, in which two monkeys and two mice survived without ill effects a flight to about 37 miles altitude was a milestone in space biology achievement. Other significant pre-Sputnik work included the Air Force and Navy studies on the biologic effects of primary cosmic particles; Professor H. Strughold's series of studies on space medicine and related subjects including ecological studies of the planetary atmospheres,

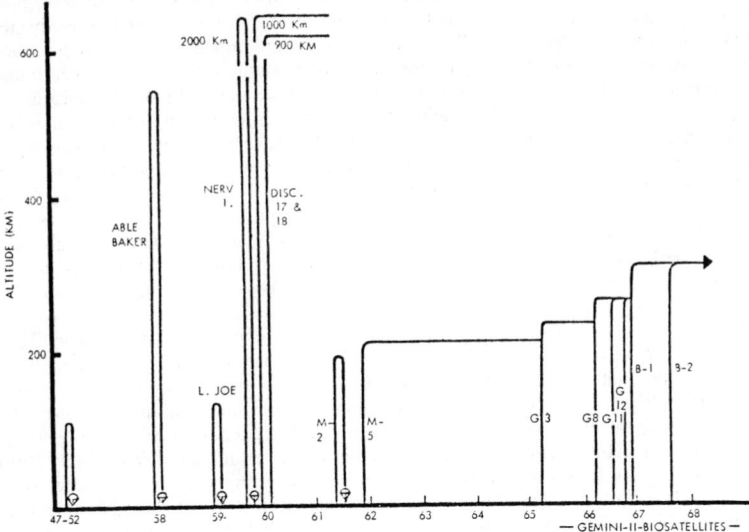

Fig. 1   United States biological space flights.

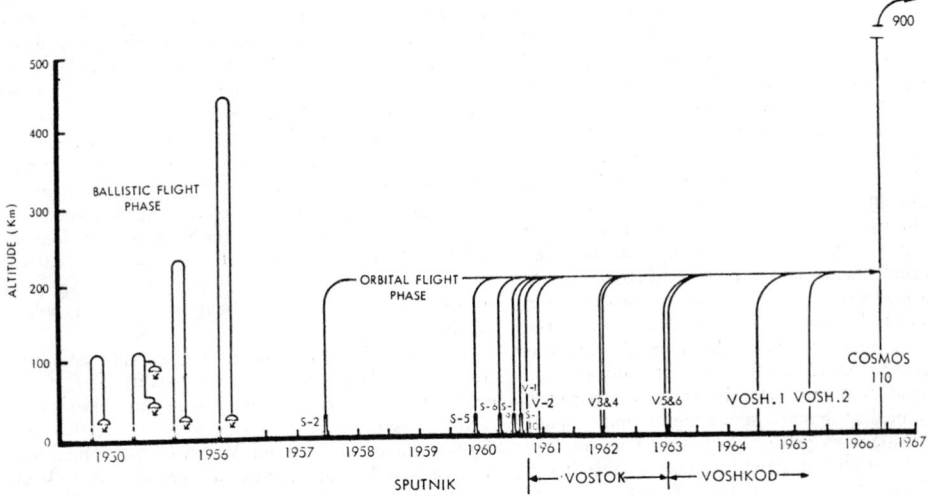

Fig. 2   Russian biological space flights.

extra-terrestrial life, space-equivalent levels of the atmosphere, and requirements for sustaining life in sealed cabins; the U.S. Air Force investigations of short-term weightlessness; and the record-manned balloon ascent to 102,000 feet by Major David Simons in 1957. Since Sputnik II, Soviet and American space biological research and development have advanced rapidly. Figures 1 and 2 present the space flights of the U.S. and USSR containing payloads of biological interest.

Space

An absolute definition of where the earth's atmosphere stops and space begins is not entirely feasible. For example, for aerodynamic lift 120 miles altitude represents the approximate upper limit of the atmosphere. Exposure of unacclimatized persons to altitudes in excess of 10,000 feet without supplemental oxygen leads to the deleterious effects of hypoxia. Even with pressure breathing apparatus, oxygen alone suffices only to about 47,000–50,000 feet and for limited periods of time because of the extra physical effort of pressure breathing and the marginal oxygen saturation of the blood under these conditions. At about 63,000 feet in the unprotected human being the blood freely liberates its dissolved nitrogen and severe aeroembolic damage results. For practical purposes a protective artificial pressure environment is required at and above 45,000 feet in order that life-sustaining respiration and circulation may proceed unimpaired. For the radiobiologist on the other hand, space in terms of possible important radiation hazards appears to start in the vicinity of 90,000 feet, an altitude to which heavy primary cosmic particles penetrate without much atmospheric interference. Thus it appears that for space biologists in general, space begins between 45,000 and 100,000 feet in that many of the environmental hazards of space which are important to man are encountered at and above these altitudes.

Problems in space biology. The immediately practical problems in space biology concern potential adverse effects of the space environment on the human traveller as well as the stresses which will be imposed by his space vehicle. A related field of study involves the use of the space environment as a tool for the study of the properties and processes of earth organisms (e.g. weightlessness in the study of response to gravity). Of both theoretical and practical interest would be knowledge of the possible existence in deep space of bioorganic molecules and living spores (see ARRHENIUS); discovery and study of extraterrestrial life forms; the ability of some earth life forms to adapt to the environments of other planets and their natural satellites; and the possibility that the introduction of earth forms of life might cause irreversible and widespread changes in the ecological balance of other worlds or the corollary possibility of back-contamination of the earth from extraterrestrial bodies.

Weightlessness. Weightlessness, which is often referred to as "zero-G," is not caused by the absence of gravity but rather by a dynamic condition in which the centripetal acceleration of the earth's gravitational field is exactly balanced by the opposing centrifugal acceleration of an orbiting body. Another way of stating this is that weightlessness is one of the characteristics of a body in a free fall. An orbiting satellite, for example, is in a free-fall mode with respect to the earth, but it remains in orbit because the earth's surface continuously curves away from the satellite at the same rate at which it is falling. In this situation, the satellite and its contents are said to be weightless and things that are not tied down are free to float about within the open spaces of the vehicle. It is noteworthy that the estimated "equilibrium point" between the gravitational fields of the earth and the sun is at about 162,000 miles distance from the earth. Before any experience had been gained in manned space flights, many space medical investigators predicted that serious adverse effects would result from living for hours or days in the weightless state. In particular, the heart, kidneys, organs of balance, and musculo-skeletal system were thought to be vulnerable. The aggregate human experience from the Mercury, Gemini, Apollo, and Soviet space flights has shown that man can tolerate weightlessness for up to 14 days without prohibitively detrimental effects. Very sensitive post-flight tests of astronauts have shown a temporary interference with autonomic control of arterial blood pressure, and measurable decreases in bone density. It remains to be determined whether or not these changes were caused by weightlessness, *per se*, or perhaps by a combination of factors including weightlessness, hampered body movement in space suits and cramped space cabins, imperfect body water balance, variations in nutrition during flights, and the consummate fatigue of all the mental and physical demands experienced by the astronauts. Even more important may be the question of whether or not longer flights will worsen these effects. Space biologists look forward to investigating man's ability to remain in the weightless state for many weeks

or months. Whether or not there may be biological limits for exposure to zero-G for man and other species will be determined in future long-duration space flights inasmuch as no suitable method has been found of simulating weightlessness in earthbound laboratories.

Radiation. With the exception of Apollo 8 manned space flights thus far have been below the Van Allen Belts; the associated radiation exposure has been negligible. The hazards to the space explorer of the various types of radiation and energetic particles that exist in and beyond the Van Allen Belts are only partially assessed. We are concerned with very high energy particles such as protons and the nuclei of atoms of greater mass extending up in the periodic table to iron or higher; and with both ionizing and non-ionizing electromagnetic radiations such as solar x-rays, ultraviolet, thermal, visible, and microwaves. Sources of extraterrestrial radiations besides our sun include our own galaxy, and other galaxies. Although experiments to date have not disclosed serious biological effects from exposure to the heavy primary cosmic particles such as the iron nucleus, it is theoretically possible that direct hits on certain body tissues such as brain regulatory center neurons and the lens of the eye could result in irreversible damage and functional impairment. Effective shielding against such atomic nuclei, travelling with velocities equivalent to billions of electron volts of energy, appears impractical within the limits of current knowledge. The Van Allen radiation belts are zones of increasing radiation intensity between roughly 200 and 1000 miles altitude, with a marked reduction in intensity beyond 17,000 miles. These belts contain electrons and protons which move at high velocities under the influence of the earth's magnetic fields. Although more knowledge is needed about the nature, distribution, and intensities of the ionizing radiation in the Van Allen Belts, it is estimated that radiation doses to astronauts passing through these regions on, for instance, a flight to the moon, will be acceptable.

Another radiation problem of great interest results from solar flares and their attendant solar proton events. For the Apollo manned lunar mission totalling 7 to 8 days in space, it has been estimated that total doses to the crew, including Van Allen Belt and a solar flare of 1 to 2+ magnitude would not exceed 25 RAD. Similarly, the solar protons from a Class III flare could elevate the total dose to 50 RAD. Doses of this size are considered operationally acceptable and would not interfere with crew effectiveness and mission completion. There is optimistic hope that a dependable method of predicting solar flares will be developed so that space flights may be planned for times of low solar flare activity. Beyond earth's atmosphere, the intensities of ultraviolet, visible, and thermal radiations from the sun represent potential biological hazards; however, effective protection by shielding and filtering techniques is practical.

Psychological aspects. The splendid records of performance of the crews in the Mercury, Gemini, Apollo and Soviet programs give reassurance that the highly trained, very select astronauts were successfully in control of essentially all aspects of their space flights. Terrestrial experiences had suggested that some serious psychological problems might be expected. For instance, it has been shown that sensory deprivation by the artificial reduction of normal physical stimuli may cause profound psychological and physiological changes including hallucinations, which not only reduce human effectiveness but also may extend into temporary psychotic states. Similar results have happened in the case of castaways, explorers, and others in isolated situations. The so-called break-off phenomenon that has been described by aviators as a strange feeling of withdrawal from earthly realities, especially in cases of single-place jets on high-altitude missions suggests a type of reaction which might have been expected in the astronaut. Experiments on the effects of confinement, as in a space cabin simulator, have shown a wide variation in human adaptability and tolerance, some subjects remaining effective for many days, others demanding to terminate the experiment after only a few hours. These sorts of problems have not been observed in the astronauts. Nevertheless, it would appear worthwhile to develop techniques to screen out the unadaptable as well as to select the most promising; and for various artificial aids and psychological techniques to help the space traveller remain mentally effective. The broad picture of the psychology of space travel will come into focus with more experience and as much longer space voyages are accomplished. Until then, the most important contributions of the psychologist appear to be in selection and training of astronauts and in improving the equipment they use and the space cabin environment they inhabit.

Bioengineering of environment systems. Of great interest to space biologists is the matter of life-sustaining and protecting techniques for the space traveller. The requirements and their solutions constitute the vital field of life support. Life support is the provision of nutrition and a suitable artificial environment including oxygen, an inert gas, water vapor, and trace elements; the removal of carbon dioxide and other body wastes and their regeneration to useful products; and the control of such factors as temperature, pressure, humidity, illumination, odors, noise, noxious substances, and hazardous radiations. In addition, weightlessness may require that normal gravity be simulated. Other practical needs include emergency escape systems and design to protect against the various patterns of acceleration throughout a trip including take-off, in-flight maneuvers, atmospheric reentry, powered landings, and parachute-capsule landings. For long space voyages of a year or more such as a 420-day round trip to Mars, the ultimate aim of life support is the development of a closed ecological system which would salvage and regenerate to food, water, breathing oxygen, and other needed products, the human wastes and other wastes from the operation of the space vehicle. It would, in a small space, duplicate the essentials of man's ecological system on earth. Ideally, not a single atom would be unrecoverable. Excellent progress has been made in methods for recovering oxygen and water from respiratory CO_2 and water vapor, urine, and used wash water. However, regeneration of foods is a formidable problem. When one considers that the feasibility of chemical synthesis of carbohydrates, fats, and proteins for human nutrition has not been demonstrated and that an adequate diet for man includes no less than 15 atomic species, 11 amino acids, 9 or more mineral salts, 20 vitamins, and carbohydrates and fats, the complexity of the problem is impressive. Nevertheless, there is optimism that techniques for combining the use of growing organisms such as algae and bacteria with chemical syn-

thesis may prove feasible. Oxygen production by such methods as electrolysis of water or the high-temperature thermolysis of carbon dioxide is considered operationally feasible. Pure water may be recovered from waste water and urine by any of several proved techniques, the most promising of which is hot-air evaporation.

Because the exchange of heat between a space vehicle and the vacuum of space cannot be accomplished, without loss of mass, by any means other than radiation, the removal of heat from the cabin atmosphere is expected to be a difficult engineering problem for the long-range space ship or manned satellite. The design of radiating systems to accomplish this essential operation, compact enough to be carried, ingenious enough to permit automatic operation in flight, and efficient enough to work with the limited power available will tax the state of the engineering arts.

The photosynthetic gas exchanger. One of the most thoroughly studied approaches to solving the very complex problems of an artificial ecological system is based upon the use of PHOTOSYNTHESIS for the removal of carbon dioxide from the sealed atmosphere, and its conversion to free oxygen and to carbon for organic synthesis. In the laboratory it has been possible to meet the carbon dioxide absorption and the breathing oxygen requirements of human test subjects for periods of many days using a suspension of green ALGAE with artificial illumination. Experiments are in progress to extend this approach by various means; for instance, small, fast-growing fish may be added to the system. The fish eat the algae; the man eats both the fish and some nutrients prepared from the algae, and portions of the human and fish waste products are cycled back into the system for conversion to useful products. An interesting question pertaining to photosynthetic systems involves the possibility of spontaneous or induced mutations of the algal cultures with resulting changes in performance as a gas exchanger and the possible production of toxic by-products and ultimate failure of the culture. Another concerns the problem of protective environmental control for the algal suspension in order to prevent damage to it by the various physical stresses of space flight and by infectious agents. The search for suitable green plants has included many varieties. Laboratory studies have shown that the sweet potato, for example, offers excellent possibilities provided that its growth and development will not prove susceptible to the weightless state.

Exobiology

The search for and study of extra-terrestrial life is a field of great fundamental significance in biology. Its close relation to theories of the origin of life and the range of adaptation to environment of which living organisms are capable produce a broad area of commonality with terrestrially oriented biological study. Laboratory studies in many countries have shown that virtually all of the simple organic compounds of which terrestrial life is composed are produced by introduction of energy into mixtures of reducing gasses such as are present in the primordeal atmospheres of planets. When the large number of stars in the universe supporting planetary systems is considered ($\approx 10^{21}$), and assuming the history of many of these planets must have been similar to that of earth, it seems most improbable that the earth is the only planet upon which life has arisen. However, only the planets of the solar system are accessible to direct examination by us under any presently known technology of space travel, so the nearby bodies, the moon, Mars, and Venus are the principal targets of search. Remote observation of these bodies by passing and orbiting spacecraft and limited landed payloads on the moon and Venus have done much to characterize the environments of these bodies.

In the near future samples of lunar material returned to earth and landed experiments on Mars and Venus can be expected to produce direct evidence bearing on the origin and evolution of life in these extra-terrestrial environments.

Environmental Biology In Space Flight

In recent years both U.S. and USSR have developed the capability of automating biological experimentation to a degree that allows the conduct of experiments in space flight which approach in sophistication those conducted in the earth laboratory. Recent examples of such developments are found in the Soviet Cosmos 110 and the U.S. Biosatellite II. In the latter mission the payload contained experiments aimed at studying the biological effectiveness of ionizing radiation, plant growth, and cellular growth and reproductive processes in the weightless environment. Principal findings were as follows: Some factor in space flight (presumptively weightlessness) has a significant effect on the action of ionizing radiation on many types of cellular processes. Plants display disturbed orientation in growth and motion of a type predicted by plant behavior in the gravity-compensation device termed the clinostat, thereby confirming a part of a body of theory developed over many decades of laboratory study.

Future studies of the new environments available to us through the technology of space flight may be expected to produce both new information and confirmation of prior theoretical work as a contribution to and profit from the adventure of exploration of space.

O. E. REYNOLDS
JOHN M. TALBOT

References

Armstrong, H. G., "Aerospace Medicine," Baltimore, Williams and Wilkins, 1961.

Chase, H. B. and J. S. Post, "Damage and Repair in Mammalian Tissues Exposed to Cosmic Ray Heavy Nuclei," J. Aviation Med., 27: 533–540, 1956.

Clark, B. and A. Graybiel, "The Break-off Phenomenon," J. Aviation Medicine, 28: 121–126, 1957.

Henry, J. P., "Flight Above 50,000 Feet: A Problem in Control of the Environment," Astronautics, 1: 12–19, 1954.

Hessberg, R., "Accelerative Forces Associated with Leaving and Re-entering the Earth's Gravitational Field," J. Astronautics, 4: 6–8, 1957.

Hitchcock, F. A., "Present Status of Space Medicine," J. Astronautics, 3: 41–42, 51–52, 1956.

Isakov, P. K., "Life in Sputnik," Astronautics, 3: 38–39, 49–50, 1958.

Konecci, E. B., "Human Factors and Space Cabins," Astronautics, 3: 42–43, 71–73, 1958.

Ley, Willie, "Rockets, Missiles, and Space Travel," New York, Viking, 1957.

Mayo, A. M., "Environmental Considerations of Space Travel from the Engineering Viewpoint," J. Aviation Med., 27: 379–389, 1956.

Schaeffer, H. J., "Definition of a Permissible Dose for Primary Cosmic Radiation," J. Aviation Med., 25: 392–398, 411, 1954.

Strughold, H., "Atmospheric Space Equivalence," J. Aviation Med., 25: 420–424, 1954.

Strughold, H., "The Green and Red Planet, A Physiological Study of the Possibility of Life on Mars," Albuquerque, Univ. of New Mexico Press, 1953.

White, C. S. and O. O. Benson, Jr., eds., "Physics and Medicine of the Upper Atmosphere, A Study of the Aeropause," Albuquerque, Univ. of Mexico Press, 1952.

Wilcox, E. J., "Psychological Consequences of Space Travel," J. British Interplanetary Society, 16: 7–10, 1957.

"Aviation Medicine on the Threshold of Space: A Symposium," J. Aviation Medicine, 29: 485–540, July, 1958.

"Behind the Sputniks: A Survey of Soviet Space Science," Washington, Public Affairs Press, 1958.

"Space Travel: A Symposium," J. Aviation Med., 28: 479–512, 1957.

Bedwell, T. C. and H. Strughold, eds., "Bioastronautics and the Exploration of Space," 1965, U. S. Government (Chief, Input Section), CFSTI, Sills Bldg, 5285 Port Royal Road, Springfield, Virginia 22151.

Burns, N. M., R. M. Chambers and E. Hendler, "Unusual Environments and Human Behavior," London, Collier-MacMillan, 1963.

"A Review of Space Research," National Academy of Sciences, National Research Council Publication, 1079, 1962.

Fox, S. W. Ed., "The Origin of Prebiological Systems and Their Molecular Matrices," New York, Academic Press, 1965.

Miller, S. L. and H. C. Urey, "Organic Compound Synthesis on the Primitive Earth," Science, 130, p. 245, 1959.

"Physiology in the Space Environment" Vol II, Respiration, National Academy of Sciences, National Research Council, Washington, D. C., 1967.

Morowitz, H. and C. Sagan, "Life in the Clouds of Venus," Nature, 215: p. 1259–1260, 1967.

"Radiobiological Factors in Manned Space Flight," National Academy of Sciences, National Research Council, Washington, D. C., Printing and Publishing Office, NAS-NRC, 2101 Constitution Ave, N.W., 1967.

Henry, J. P., "Biomedical Aspects of Space Flight," New York, Holt, Rinehart and Winston, 1966.

Foster, J. F., "Life Support Systems and Outer Space," January 1966 Battelle Technical Review, Battelle Memorial Institute, 505 King Avenue, Columbus, Ohio 43201.

Berry, C. A., "Space Medicine in Perspective," JAMA, 201, No. 4, 24 July 1967.

Pittendrigh, C. S., et al., "Biology and the Exploration of Mars," Publication 1296, National Academy of Sciences, National Research Council, 1966.

Haldane, et al., "Extraterrestrial Life: Anthology and Bibliography," National Academy of Sciences National Research Council, 1966.

"Symposium on the Biosatellite II Experiments—Preliminary Results," Bioscience, 18, June 1968.

SPALLANZANI, LAZARO (1729–1799)

A master of scientific thinking with firm adherence to the truth Spallanzani was born at Scandiano in northern Italy on January 10, 1729. The son of a lawyer, he began his education at the college of Reggio. Here under the influence of his cousin Laura Brassi, a professor at the college, he studied mathematics, philosophy and the languages. He became a priest in order to help support himself by saying Masses. His reputation as a scholar became well known and in 1754 he became a professor of logic, metaphysics and Greek at the University of Bologna. A man with an inquisitive mind, Spallanzani spent his leisure time investigating the laws of nature. The ingenuity and complete thoroughness with which he conducted his experiments along with his successful teaching led Maria Theresa, in 1768, to offer him a chair of natural history and the headship of the museum at the University of Pavia. At Pavia his abilities not only made him popular with his students but he also gained recognition from the nobility as a leading authority in the field of science.

His years at the university were occupied with teaching, securing collections for the museum and advancing the logic of scientific experimentation. He died on February 12, 1799 at Pavia.

The theory of spontaneous generation advanced by J. T. Needham of England and G. L. Buffon of France as the origin of the animalicules recently discovered by Leeuwenhoek was largely disproved by Spallanzani. The results of his rigidly controlled infusion experiments demonstrated that the minute organisms could only arise from pre-existing organisms and not spontaneously as popularly believed. In 1765, he published his results in *Saggio di osservazioni microsopiche concernenti il sistema della generazione de' signori de Needham e Buffon*. In 1780, his article *Dissertationi de fisica animale e vegetale* pointed out that digestion was a chemical process rather than a physical grinding and that saliva had a preliminary chemical action on food. Spallanzani came to these conclusions by swallowing small hollow blocks of wood with meat embedded on the inside, then regurgitating them and examining the contents. In the field of embryology he followed the ovist theory of preformation and ascertained, in lower animals, that some material part of the semen was necessary for fertilization and not a spermatic vapor. He was also successful in artificial insemination of a dog, and investigated other aspects of fertilization and reproduction. His interests were not only in the field of biology but also extended into meteorology and vulcanology. Spallanzani's investigations gave important principles to the infant science of biology, but even more important was the scientific methodology he contributed to research.

RICHARD M. CRIBBS

SPECIATION

To understand speciation, the process by which new species are formed, it is essential to define what is meant by a species. To the taxonomist or paleontologist studying two groups of dead specimens, the criteria for deciding whether they should be called separate species are morphological discontinuities. The biologist likewise does this, but because he studies living organisms he is enabled in many cases to apply a more clear cut definition based on GENETICS. To him, a species is any group of sexually reproducing individuals which does not and can not interchange genes through hybridization with any other group. Nevertheless, the possibility to test the interchangeability of GENES through experiment or observation often does not exist and then even the biologist falls back on the criteria of the paleontologist and taxonomist. It is, however, a striking fact that the species pronouncements of a competent taxonomist are almost always confirmed when subjected to the genetical criteria of the biologist. It is therefore clear that morphological divergence of groups is a reflection of fundamental genetic differences between them and although it is impossible to put fossil species to the biological test, their designation as such is certainly correct in most if not all cases.

Now morphological discontinuity can arise in two fundamentally different ways, one based simply on the passage of time (*allochronic speciation*) and the other on geographic fragmentation combined with passage of time (*allopatric speciation*). In allochronic speciation, new species arise along direct lines of descent because natural selection molds each species by maintaining it in a high degree of adaption to each phase of the continually changing environment. The problem in this type of speciation is where to draw the line between old and new species. This is fortunately resolved by the fossil record which is represented by samples discontinuous with respect to time that are for this reason morphologically different enough to be confidently called at least separate species. It should be clear that it would be impossible to draw the lines if a continuum existed.

Allopatric speciation occurs when an original species becomes geographically fragmented into two or more groups which are then unable to breed with one another because of their spatial isolation. Because it is very improbable that the two areas in which the isolates occur will be similar in all respects, genetic divergence by means of natural selection will occur and as in the instance of the allochronic process, the greater the lapse of time, the greater this will be. Subsequently, the geographic barriers between the two may break down with the result that the two divergent stocks meet again. Provided that the isolation has been long enough to allow sufficient differences to develop, they will be incapable of exchanging genes with one another and as such each will be a valid species.

Before going on to discuss the fate of these new species as well as the complex cases where some interbreeding occurs, the relative importance of allochronic and allopatric speciation in evolution should be considered. In the examples just discussed, allopatric speciation in evolution should be considered. In the examples just discussed, allopatric speciation resulted in the formation of two genetically divergent species from one, whereas in allochronic speciation there can be no such dichotomization because by the nature of the process there has to be genetic continuity between each successive generation. While it is true that within a group of allopatric species, each must bear an allochronic relationship to the single parent species of the group (and in this sense all species are of allochronic descent from one) it should be evident that the *diversity* of species present at any one time on the earth since the first sexually reproducing organisms is a product of allopatric and not allochronic speciation.

What happens when two species which arise from geographic isolation come to overlap one another. Because of their recent descent from a common ancestor, they will be similar with regard to morphology, physiology, behavior, *etc.*, and as a consequence also in their ecological requirements such as food and habitat. As a result, ecological competition will occur. Mathematical considerations show that it is so unlikely that the two will be equally adapted in utilizing the same limited environmental resources that one or the other of two alternatives has to take place. Either the competitively superior one will win out and the inferior one become extinct, or fairly rapid divergence will occur to reduce the competition and allow their coexistence. This latter phenomenon is particularly well shown by *character displacement* in which differences between two species are exaggerated by natural selection in the areas where they overlap but not in areas where each exists alone. Another competitive relationship that can occur is interspecific sexual behavior. Even though they do not hybridize, they may respond sexually towards one another because of their similarity, and this may also result in character displacement. It is of course possible that the divergence in isolation could have been sufficient to allow them to be completely free from both ecological as well as sexual competition upon their arrival in each other's range. At the opposite extreme, the divergence may have been so slight that upon overlapping the two freely interbreed and as time passes remerge completely with one another. In this situation, speciation has not occurred, but it is interesting to note that the intercrossing may result in a single species which is intermediate between the two previous isolates and adaptively superior to both of them.

Although there are these two discrete situations, one in which no interbreeding occurs, and the other where they freely interbreed and merge back into a single species, there can be an entire spectrum between the two. It is for example possible that they interbreed when first overlapping but the hybrid offspring exhibit genetic imbalances such that they are sterile, inviable, or in some other way competitively inferior to the parental species. In this instance, natural selection will tend to favor any mechanisms which reduce or prevent the tendency for hybridization including differences in courtship behavior, time of occurrence, habitat, and many others. Or again, it is possible that there will be a partial transference of genetic characters from one to the other during the initial overlap period before the build-up of complete barriers to crossing. This latter phenomenon, known as *introgressive hybridization,* is especially important in the higher plants and is also known to occur in animals. (Some botanists maintain that certain plant species are derived from hybridization and are able to survive as genetical units because of their superior adaptation to particular environments which neither of the parental species is able to tolerate.)

To this point we have been considering that complete geographic isolation between the two groups is a necessary prerequisite to speciation and it may be asked, is sympatric speciation possible? Commonly in plants, there is a certain hereditary phenomenon (POLYPLOIDY) which allows this to occur, but its importance in evolution is probably small compared to speciation through geographic isolation, and in animals this is certainly so. Nevertheless, there is a pattern of geographic distribution found in both plants and animals where a species occurs over a very wide area that encompasses a broad range of climatic conditions which are often graded, as for example in a species occurring from high

to low latitudes. If individuals of such a species are collected at various points along the distribution, it is found that a gradient of change in morphological and other characters occurs between the extremes, that is to say, it has a distribution known as a *cline.* To the taxonomist who sees specimens only from the two ends of the cline, the differences are often sufficient to call them separate species. And yet because of the cline, the two are connected by a continuity of interbreeding populations. When these are tested biologically by laboratory breeding, it is sometimes found that the extremes are genetically incompatible while in other cases this is not so, and in fact a spectrum again exists between these two alternatives. The question then is should they be called separate species or not? Fortunately in several instances nature has performed an experiment for us known as *ring speciation.* Here the cline is extensive, but curves around on itself so that the extremes overlap and interbreeding does not occur. In these examples it is likely that interbreeding occurred when the two ends first met, but that barriers to hybridization were built up by natural selection due to the inferiority of the hybrids. Where natural distributional rings do not occur, it is really not possible to say whether the extremes should be called separate species. The main point is that speciation is a continuous and dynamic process and by observing a great number of species at one time we are bound to see all stages represented. Occasionally breeding occurs in some areas of overlap, but not in others. These situations deserve far more study than they have been given, not only from the point of view of further elucidating the speciation process but also as a means of studying rates of evolution which are probably greater than anticipated to the present.

LINCOLN P. BROWER

References

Dobzhansky, Th., "Genetics and the origin of species" 3rd ed., New York, Columbia University Press, 1951.
Roe, Anne and G. G. Simpson, eds., "Behavior and evolution," New Haven, Yale University Press, 1958.

SPERMATOPHYTA

1. Introduction. Spermatophyta literally means seed plants. Since the botanical name is not implicitly legitimate under the Code of Botanical Nomenclature, whose rule of priority does not extend beyond the rank of an order, Spermatophyta is a name which stands or falls with the definition of the word seed.

Like regarding so many othher plant organs, such as STEM, LEAF, ROOT, etc., which have of old been used in daily speech and have only later on been more or less standardized as botanical terms, the original definition of SEED referred to the Phanerogams and particularly to the Angiosperms.

2. Definition of the term seed. (Fig. 1) In the Angiosperms a seed has developed from an ovule (cf Fig. 3) and it mostly remains for some time in a state of dormancy, particularly in regions with a seasonal periodicity in climate. Its most significant part is the embryo of a young sporophyte, and mostly this is, together with some other tissues (perisperm, endosperm), enveloped

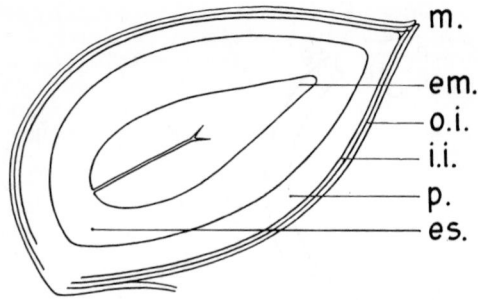

Fig. 1 "Ideal" seed of a dicotyledonous plant, longitudinal section. m., micropyle; o.i., outer integument, i.i., inner integument, p., perisperm, all diploid and belonging to mother sporophyte; es., endosperm (triploid); em., embryo (daughter sporophyte). The gametophyte has disappeared.

by some special protective layers, the integuments (cf Fig. 1), which have ultimately formed the seed coat. In addition the seed as a whole is protected by some other organ(s), the carpels or pseudocarpels which together may form the fruit.

While the perisperm, if any, is a remainder of the original tissue of the ovule, that is, the nucellus, both embryo and (secondary) endosperm are products of the process of double fertilization: one of the male generative NUCLEI from the pollen tube fuse with the haploid egg cell, thus forming the diploid embryo, the other male nucleus unites with the diploid secondary embryosac nucleus (originated from the fusion of the (mostly) two haploid polar nuclei), thus giving rise to the growth of a—mostly triploid—endosperm (albumen) with a nutritional function.

Although such an endosperm is, with very few exceptions, characteristic for all Angiosperms, it is by no means found in all mature seeds. From this it may be inferred that the endosperm either fails to develop to some considerable size (e.g. in Orchids) or is consumed at an early stage by the growing embryo, or again that both processes take place more or less simultaneously.

Anyway, the result is that in mature seeds there may either be no endosperm at all (left), in which case the nutritive substances are mostly found stored in the embryo, notably the cotyledons (e.g. in beans and peas); or that an endosperm is extant, and in all possible gradations, from a scanty film around the basal part of the embryo to a voluminous tissue much larger than the embryo itself and sometimes enveloping it (e.g. palms, Cyperaceae) or adnate to it laterally (e.g. grasses).

Such being a brief and more or less diagrammatic description of the angiospermous seed, the question arises whether or not it is compatible with what is usually called a seed in Gymnosperms.

Before entering upon this subject, it is, for a better understanding, indispensable to discuss some fundamental points pertaining to those groups of plants which are generally considered to produce seeds. To this end it is first of all necessary to recall the process of alternation of generations or, in short, of the various types of life cycles.

3. Life cycles in principle. In all plants the life cycle runs as follows: *spore → gametophyte* (the haploid prothallus, n-phase) *→ gametangia* (sexual organs: arche-

gonia in ♀, antheridia in ♂) *→ gametes* (sexual cells: egg cells or egg nuclei in ♀, spermatozoids, sperm cells, or generative nuclei in ♂) *→ fertilization → zygote* (first diploid cell, beginning of 2n-phase) → embryo (if distinguishable as such) *→ sporophyte → sporangia → reduction division* (each spore mother cell produces a tetrad, i.e. four haploid daughter cells) *→ spore* (first haploid cell).

4. Iso- and heterospory. In the lower groups the SPORES are all of one size (iso- or homospory), but in the higher ones there are two types: megaspores and microspores, the latter always smaller and more numerous than the former. Accordingly, in such heterosporous groups there are two types of sporangia: megasporangia and microsporangia.

5. Types of life cycles. In the Thallophyta the pattern of this process is often still more or less irregular and to various extent subject to environmental influences which can even temporarily suppress one of the generations and favor the duration of the other.

In the Cormophyta, however, the cycle, though not absolutely rigid, is more strictly fixed. Within the taxon a series is to be distinguished which, among the Thallophyta, has a striking parallel in the Phaeophyta. This series, the sequence of which is determined by particulars of both generations, can be subdivided into three parts (Figs 2 and 3): I. the sporophyte parasitizes on the dominating gametophyte: BRYOPSIDA; II. the two generations are physiologically independent: PSILOPSIDA, LYCOPSIDA, SPHENOPSIDA, PTEROPSIDA (A. isosporous, B. heterosporous); III. the gametophyte parasitizes on the dominating sporophyte: Gymnosperms and Angiosperms.

This picture, however, is extremely simplified and accordingly not entirely correct. For, since the life cycle always runs: spore—gametophyte—gametes (fertilization)—sporophyte (reduction division)—spores, and the first diploid cell, the zygote, never possesses the means for physiologically independent life, the sporophyte in principle always parasitizes on the gametophyte. What is significant is, that in group I the sporophyte is a lifelong parasite, that in group II it only parasitizes in its earliest stages and soon becomes physiologically dependent from the ever decreasing gametophyte, and that in group III the gametophyte has been reduced so much that it cannot possibly lead an independent existence. On looking at the representation of Fig. 2, it seems, indeed, as if the process of sexuality, first allotted to the gametophyte alone, gradually encroaches upon the originally asexual sporophyte (Fig. 2, IV), a process which in some plants, may lead to dioecy in the diploid generation (Fig. 2, V).

6. "Lines" and "levels." Although the characters of the gametophyte in the three groups just mentioned are fairly well correlated with characters of the sporophyte, it seems as though they are, to a lesser degree than the sporophyte is, based on "taxonomic relationship:" the gametophyte (originally a water plant) is of a clearly more functional, the sporophyte (essentially a land plant) of a more constitutive nature. Nevertheless, the gametophyte shows many signs of stubborn resistance against reduction and of great conservatism regarding its most essential parts.

What exactly is meant by these expressions, can, in this context, be best explained by discussing the case of heterospory, which has a direct bearing on seed-formation.

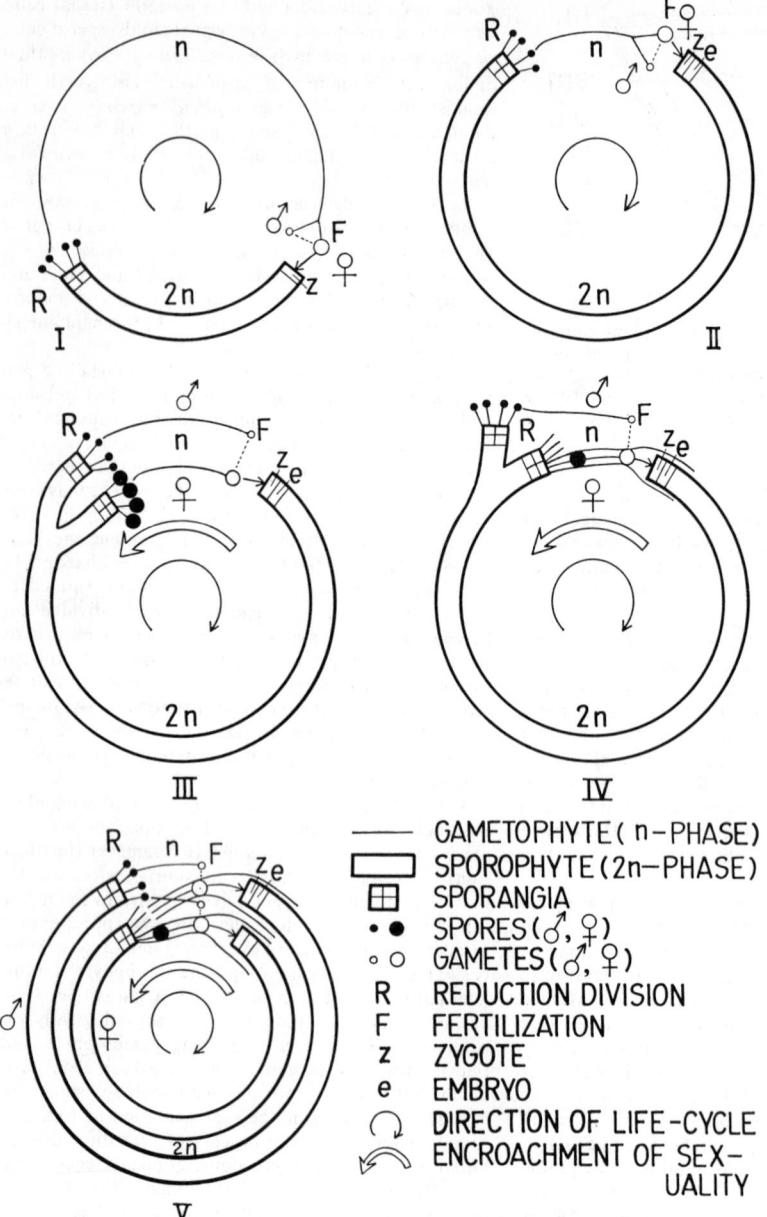

——	GAMETOPHYTE (n−PHASE)
▭	SPOROPHYTE (2n−PHASE)
⊞	SPORANGIA
• ●	SPORES (♂, ♀)
○ ○	GAMETES (♂, ♀)
R	REDUCTION DIVISION
F	FERTILIZATION
z	ZYGOTE
e	EMBRYO
↻	DIRECTION OF LIFE-CYCLE
↙	ENCROACHMENT OF SEX- UALITY

Fig. 2   Patterns of life cycles in the Cormophyta. I. Astelocormophyta—Bryopsida. II. Stelocormophyta, isosporous—Lycopsida (Lycopodium), Sphenopsida (most), Pteropsida (Filices). III. Stelocormophyta, hetero-sporous—Lycopsida (Selaginella, Isoëtes), Sphenopsida (some), Pteropsida (Hydropteridales). IV–V. Stelocor-mophyta, heterosporous—Pteropsida V. Gymnospermae, Angiospermae—IV. monoecious, V. dioeceous. V. (Pteridospermales).

During the process mentioned above, heterospory has apparently originated independently in various groups. This, of course, is a thesis which cannot be proven, although being of theoretically great impor-tance, since it touches upon the problems of analogy and homology.

We cannot dwell upon this vast subject here; suffice it to state that homology is referring to genealogical relationship and constant characters, analogy to adap-tation by means of variable features. Thus, if we look at a phylogenetical chart like that of Fig. 5, we perceive mainly more or less vertical lines, which represent lines of development of genetically isolated taxa during the geological eras. These lines have been constructed more or less intuitively, but their factual basis is formed by the apparently less speculative evolution of certain characters. Of these, heterospory in various groups seems, originally at least, a "level" or developmental phase rather than a "line," an analogy rather than an homology.

Our point is that the phenomenon is found scattered among lower groups of very different systematic position. A number of carboniferous Lycopsids (e.g. Lepidocarpon) were heterosporous; among recent ones only Isoëtes and Selaginella are heterosporous, Lycopodium is isosporous. It depends on our opinion regarding the grade of relationship between these genera, whether we should here speak of a single, double, or treble appearance of heterospory.

Heterospory seems to have been an occasional condition in Sphenopsida (Sphenophyllum, Calamites) but its only surviving member, Equisetum, shows some interesting particulars: at first sight it seems isosporous, but a statistical investigation discloses that, in species with separate fertile and sterile stems, such as in *E. arvense*, there are actually two types of spores, whose frequency curves overlap. The smaller ones are pale green and give rise to small male prothalli, the larger ones are brighter green and produce bisexual prothalli. Both types may be found in the same tetrad. This condition could be explained as an incipient heterospory which has hardly risen above the physiological level.

In the Pteropsids, heterospory has apparently originated at least twice, once in the Hydropteridales which are clearly related to the leptosporangiate ferns, and once in the extinct Pteridospermales. As the last-named group is likely to have included the ancestral stock of most (if not all) higher classes (GYMNOSPERMS and ANGIOSPERMS), heterospory has obviously reached the stage of being a constant character, thus entering in the sphere of homology, and presenting an example to the thesis that analogy is, in principle, an incipient homology.

The opinion that analogy and homology are quantitative concepts which are connected by a great number of transitions and that neither is possible in an absolute sense, renders it extremely difficult and often impossible to distinguish between "lines" and "levels." Accordingly, the decision is often subjective. There are, however, indications that not only heterospory and seed-information but dioecy, gymnospermy, angiospermy and the development of many protective structures are to be considered analoga. The difficulty is that such analoga apparently gradually turn into homologa, when functional characters become part of the homozygotous genome. Still deeper these ideas rest on the fundamental homology of sterile and fertile, and of ♂ and ♀, for which there are many indications. In interpreting terata, these ideas may sometimes present a simple though never conclusive explanation.

7. The principle of protection. Essentially the same has apparently happened to the seed, which yields a good example of the principle of protection of vital parts in which comparative morphology of both plants and animals abound. This is not an expression of finality; it is a very real concept involving both the strongest and the most subtle intervention of natural selection.

We will have to refer to examples of this principle repeatedly in what follows, and in order to understand how a seed may have originated during the process of evolution it is appropriate to consider somewhat more in detail the gradual and, as it were, irresistible reduction of the gametophyte, which we briefly mentioned above.

8. Reduction of the gametophyte. (Fig. 3) In the lower groups the haploid generation is, as we saw, still physiologically independent, and the principle of protection is particularly discernible in the more and more sheltered position of the gametangia. As soon, however, as heterospory sets in, the female gametophyte enjoys an increasingly better protection, while its male counterpart is, so to speak, largely left to look after itself.

2. The first "seed." The first sign of this principle is to be observed in the carboniferous Lepidospermales (e.g. Lepidocarpon) in which the megaspore (often only one of a tetrad) remained inside the sporangium which, in its turn long remained attached to the stegophyll (cf Fig. 4) to which it was adnate. This megaspore developed a gametophyte, the gametophyte one or more archegonia (probably only one) and this archegonium contained an egg cell, the latter being, of course, the most vital cell of the whole structure as the procreation was allotted to it. In addition, the lateral wings of the stegophyll were folded upward, thus enveloping the megasporangium and rendering another protection to the vital cell. This condition has led to various interpretations. Indeed, a cross-section (Fig. 4, I at b, and II) deceivingly suggests something like what we know as an integument (cf Fig. 1) in higher groups, but it is quite a different thing; neither is there any reason to compare it with the indusium of the leptosporangiate ferns.

An embryo has never been found inside the gametophyte and as the whole structure loosened itself at the base of the stegophyll (Fig. 4, I at a) and thus fell to the ground, it is likely that fertilization happened there and that what has been termed "seed" in this case is nothing but a megaprothallus enclosed in its megaspore, in its various envelopments.

Thus, in Lepidocarpon and related extinct plants we meet with an example of high standard protection of the future generation with the means provided by the possibilities in the group and, remarkably enough, foreshadowing, as it were, what would happen later on in the Gymnosperms. Incidentally, it is noteworthy that among the recent Selaginellas which usually shed their megaspores, there are some species in which these spores occasionally remain inside their sporangia and germinate there, so that young sporophytes develop in the axils of the stegophylls: a sort of attempt at seed-formation which supports the assumed common ancestry of Selaginella and the Lepidospermales.

Anyway, if we should include these "seed-bearing" plants in the "taxon" Spermatophyta, we should be aware that this taxon cannot be considered a natural one since it would comprise two "lines" (Lycopsida and "Gymnospermae"), albeit on the same or a similar "level."

10. Gymnosperms. What is true for the term Spermatophyta, to some degree also holds for Gymnospermae, an artificial group comprising some "lines" which are all characterized by having their "seeds" less well protected than e.g. the Angiosperms. In this respect, however, the delimitation is far from sharp, and this applies also to the seeds. Gymnospermy, again, is a "level," rather than a line.

There is some difference of opinion whether or not the so-called seed-ferns (Pteridospermales) which are

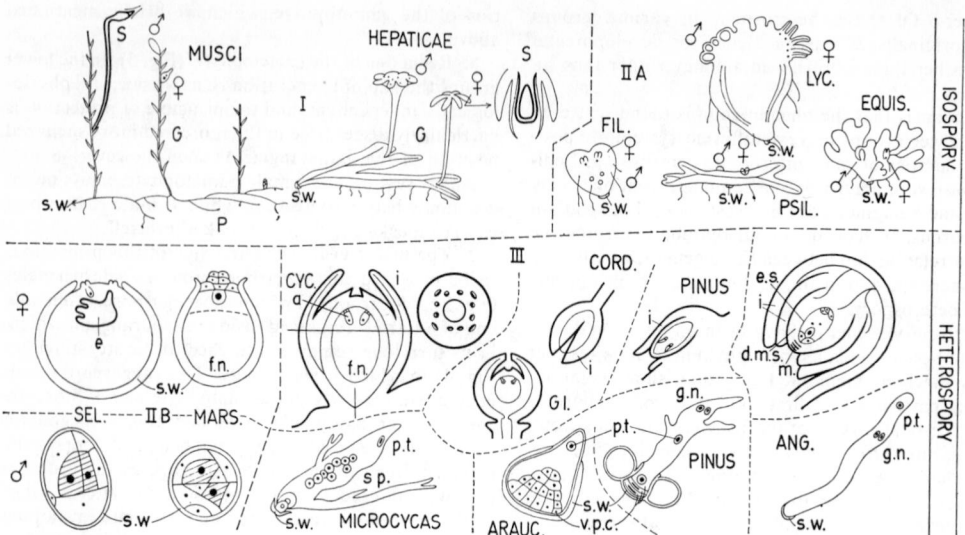

Fig. 3 Reduction of the Gametophyte in Cormophyta (gametophyte, thin line; sporophyte, heavy line).
I. Gametophyte (G) dominates, with sporophyte (S) parasitizing. P. Protonema—Bryopsida. II. A. Both generations independent but for initial parasitism of isosporous sporophyte (not pictured)—Pictured are gametophytes
of a fern, Lycopodium, Psilotum, and Equisetum. B. Same as in A but heterosporous. Pictured are Selaginella and
a waterfern (Marsilea)—e. embryo; f.n. free nuclear stage; hatched, antheridial cells. III. Sporophyte (not pic-
red) dominates and enclosed gametophyte. Pictured are gametophytes, the ♀ one in its megasporangium (ovule),
of: Arucaria (♂), Cordaianthus† (♀), Cycas (♀, with cross-section), Ginkyo (♀), Microcycas (♂), Pinus (♂, ♀),
and an Angiosperm (♂, ♀). a., archegonium; d.m., degenerated megaspores; e.s., embryo sac; f.n., free nuclear
stage; g.n., generative nuclei; i., integument; m.s., megasporangium (= nucellus); p.t., pollen tube nucleus; sp.,
spermatogene nuclei (spermatocytes); s.w., spore wall; v.p.c., vegetative prothallus cells.

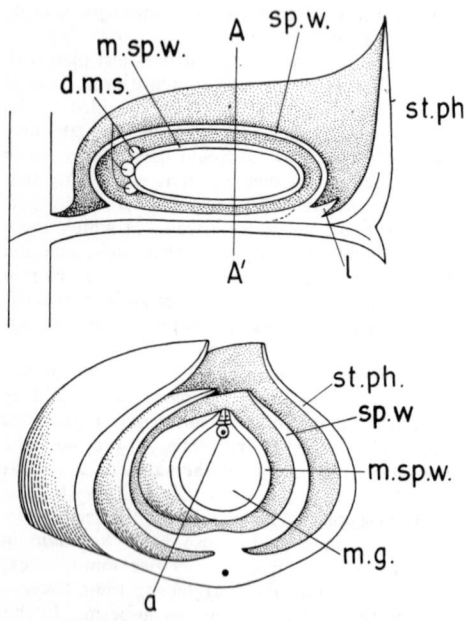

Fig. 4 "Seed" of a Lepidosperm, longitudinal section (top) and cross section along A-A' (bottom) with
proximal part. a., archegonium; d.m.s., degenerated
megaspores; l., ligule; m.g., megagametophyte;
m.sp.w., megaspore wall; sp.w., megasporangium wall;
st.ph., stegophyll.

now all extinct, should be considered Gymnosperms
with fernlike foliage or Pteropsids with secondary wood
and "seeds." Since, however, the Pteridosperms are undoubtedly related to several groups of Phanerogams (cf
§ 15, Phylogeny), the question is of little consequence,
and whatever the differences of the various gymnospermous groups, it must be stated that all of them, like
the Angiosperms possess an "ovule."

Before discussing this interesting organ, we have to
complete our survey of the gametophyte in Phanerogams.

11. Gametophytes of the Phanerogams. (Fig. 3) In the
Gymnosperms the megagametophyte remains comparatively large and, accordingly, the megasporangium
(now called nucellus) is sometimes of considerable size
(Cycas, Ginkyo), with a large (basal) part in a free nuclear stage, in other cases much smaller (Taxales, Coniferales, Chlamydospermae). It is called "primary endosperm," "primary" so as to distinguish it from the "secondary endosperm" of the Angiosperms, which is
formed after fertilization.

In this megagametophyte a number of archegonia is
formed, in some cases as many as 200 or more (Microcycas in the Cycads, Widdringtonia in the Conifers),
but in most cases it is reduced to two which are situated
near the micropyle; as a rule, only one of these is fertilized. The archegonia themselves are also much reduced
and entirely embedded in the vegetative tissue.

There is, however, a sharp drop, as far as the megagametophyte is concerned, between the Gymnosperms
and the Angiosperms. In the latter the mature megagametophyte, called embryosac, is hardly more than a

large cell with 16, 8, or 4 nuclei, or a tissue with so many uninucleate cells. It is tempting to recognize in this organ remnants of one or more archegonia, as was suggested by Porsch as early as 1907. And though there are a number of arguments against this obvious interpretation and some authors maintain that in the Angiosperms and in some Chlamydosperms archegonia are essentially missing (cf Maheshwari, p. 412 ff.), in the opinion of the writer of these lines Porsch's suggestion has not only the advantage of simplicity but it is also supported by the general phytogenetic trend of thought: what happened to the archegonium which is present in all lower groups and so far has shown to be a highly conservative organ? It therefore seems more logical to assume that also in these groups archegonia have been there (as they are in Ephedra) and that they have lost their recognizable structure in some of the higher groups.

In this context it should be recalled that in the higher groups, the megagametophyte is parasitizing on the sporophyte and it is a well-known fact that parasitizing organisms often suffer a considerable reduction (dedifferentiation) of organs and tissues which have lost their function (have become rudimentary).

A few words are due to the male gametophyte, which, of course, like the female one, exists only as such in heterosporous plants. Its original form is already strongly reduced and this is comprehensible in connection with its function: the aerial transport of the male gametes to the megagamete. Those more primitive microgametophytes (Selaginella, Hydropteridales) contain, in a mature state, some 10 to 20 cells and the 2 or 4 antheridial cells among them produce a great number of spermatozoids. In higher groups all these numbers gradually decrease while the spermatozoids ultimately loose their motility. Among the Gymnosperms only the Cycads and Ginkyo have motile (multiciliate) spermatozoids, the other groups, as well as the Angiosperms have generative nuclei, which are passively transported; in the last-named group the whole microgametophyte consists of a pollentube with three nuclei, of which one is vegetative and two generative. A distinct antheridium is nowhere distinguishable; it has probably got lost as such at the sudden decrease in size of the microgametophyte as heterospory appeared.

In lower groups there is no distinct correlation with other characters regarding the number of cilia on the spermatozoids. In Bryopsida, Lycopodium, and Selaginella there are two of them, the other groups (fossils unknown) are multiciliate, a type which, amongst Thallophyta, is only known in the Oedogoniales.

Whereas the microgametophyte necessarily remained fit for aerial transportation, the female one is, as it were, swallowed down by, or incorporated in the sporophyte (Fig. 2, IV–V). What it looses in independence, it gains in ever more perfect protection, and in doing so it caters for providing the liquid medium (pollination drop and dissolving of cell membranes in the female gametophyte) to the degree necessary for bringing the gametes together, thus replacing the outside water which served as such in lower groups.

12. Ovules. The definition of ovule is: a megasporangium with a special protection by one of more integuments. It is the first superprotection of the gametophyte which is entirely enclosed in it, having originated from the only functional megaspore.

The ultimate reduction in number is, that of the only

tetrad only one megaspore matures. This is the usual condition in Phanerogams, the functional megaspore always being the basal one in Gymnosperms, more rarely so in Angiosperms, in which several types are known. In some of these the megaspore mother-cell (megasporocyte) directly forms the embryosac while in reduction division.

In the Lycopsida this process of reduction is paralleled by some fossil Lepidospermales (cf Fig. 4) and the taxon which has proceeded farthest in this respect is Fucus (Phaeophyta) in which the spores of both sexes at the same time function as gametes, the "gametophyte" being altogether eliminated. This is the only example in the vegetable kingdom of a condition which is the usual one in animals, obviously another analogy.

A beautiful example of a pseudo-ovule is presented by Azolla, in which the megasporocarp contains a single megasporocyte which produces only one functional megaspore and three abortive ones. The indusium strikingly resembles an integument. Again a remarkable case of analogy.

In recent times some palaeobotanical evidence has become known as to the possible origin of these integuments. The fact that, in many cases, they are provided with vascular bundles suggests that they are built up of telomes or reduced syntelomes; in fact, several reconstructions have been proposed. There are, however, from the outset, apparently at least two types of integuments, viz the type with many bundles in a circle (cross-section) which are the result of repeated dichotomies (Pteridosperms, CYCADS, some Angiosperms) and the type with only two bundles (Ginkyo) (see GINKOALES) or entirely devoid of vascular supply (Cordaitales [bivalved], CONIFERALES).

In the interpretation of these facts the old question arises whether simple is the same as primitive. For instance, is the Ginkyo type reduced from the Cycad one or is it original and is the single evascular (bivalved) integument of the Cordaitales and the integument of the related Coniferales to be derived from it or a separate type?

Too little is known as yet regarding the correlation of these characters with others but in any case these structures may be interpreted as sterile protective organs surrounding the most vital part of the sporophyte, the megasporangium. Integuments may be laciniate (the Pteridosperm cupula) or entire, free or adnate, absent (obviously by reduction) or single, or there may even be three or four of them; structures of evidently the same status are arilli (e.g. in Taxus).

13. The seed in Gymnosperms and Angiosperms. In the Gymnosperms there has long been uncertainty as to where the limit between ovule and seed should be drawn and only recently it has appeared that what has long been regarded as a seed, need not contain an embryo at all. In this respect Emberger (from 1942 onward) distinguishes two groups: Praephanerogamae and Phanerogamae. The latter are characterized by having a mature, mostly dormant embryo in the seed when shed, in the former such an embryo is not necessarily extant and is, indeed, often missing on dissemination. In fact, what in the Cycads and in Ginkyo looks like a seed, in reality is mostly only an ovule containing a mature megagametophyte. This ovule is often brightly coloured and does not increase in size as the embryo develops. It may be fertilized while still on the tree, or when already fallen to the ground, and the process of maturing of the embryo is independent of the sporophytic nutritional sources, which are provided for in the primary endosperm. In addition, there is no marked dormancy of the seed. According to Emberger Praephanerogams are the fossil classes of Pteridospermales and Cordaitales, and the surviving ones of Cycadales and Ginkyoales. On the other hand true Phanerogams should be the fossil Cycadeoidales, Coniferales, Taxales, Chlamydospermae and Angiospermae.

If the characters of seed-formation are, in this way, used for a taxonomical subdivision, closely related groups are severed, notably Cycadales from Cycadeoidales, Cordaitales and Ginkyoales from Coniferales and Taxales. The conclusion must be—in full accordance with what we have said above—that seed-formation is a matter of "level" (analogy) rather than of "line" (homology) and quite correctly Martens (1951) has refuted Emberger's opinion of Praephanerogams being a taxon of any distinct circumscription, pointing out the very gradual nature and the variable limits of the development, even within a single genus or species, between a (whether or not fertilized) megaprothallus and an embryo.

14. The sporophyte of Phanerograms. So far most attention has been given to the gametophyte, since the nature and evolution of the seed could not be well explained without some wider knowledge of the sexual generation. We have now to turn to the sporophyte, in which tendencies appear, similar to those in the gametophyte; notably, the principle of protection is even more strikingly manifest here. The favored organ, then, is the sporangium, and particularly again, the female one.

Basing himself on Zimmermann's telome theory, the present writer distinguishes two fundamental types of sporangial arrangement, the stachyosporous type and the phyllosporous one. In both cases the sporangia are essentially homologous with sterile telomes, but in the stachyosporous groups the sporangia long remain independent axial structures and only secondarily get some protection by means of special bifurcations of which one part—usually the upper or adaxial one—is fertile, i.e. is a sporangium or a sporangial truss, and the lower or abaxial one sterile (stegophyll or false sporophyll), whereas in the phyllosporous groups the sporangia are from a very early stage, incorporated in a system of ramification in which sterile telomes abound and are, at least in principle, in the majority. Such "mixed syntelomes" are called (true) sporophylls.

We may recall here that "leaves" in this category obviously have originated by the process of overtopping, special systems of originally dichotomous telomes being pushed aside. These developed into flattened photosynthetic organs, the telome leaves, the more primitive types of which are still to be observed in ferns (Adiantum!) and Ginkyo. The origin of the "microphylls" is more obscure. In the Sphenopsida they are obviously reduced telome leaves, but in other groups they are sometimes regarded as having originated as epidermal products (enation leaves). We cannot dwell upon problems of ramification and phyllotaxis, but must restrict ourselves to express the opinion that all types of ramification in the Cormophyta are to be derived from the dichotomous one; this applies both to stems and nerves. Also in the vegetative parts

the principle of protection is manifest, since the leaf (bract) of Phanerogams and its axillary bud are to be considered a bifurcation of a protective nature.

The distinction in stachyosporous and phyllosporous types is never sharp. In accordance with the telome theory, stachyospory is the primitive condition, and apparently in some groups this has been more or less purely conserved while in others there has been a strong tendency towards phyllospory. And though perhaps initially an analogy, the latter type has developed sufficiently strong to allow a taxonomical distinction (homology).

The distinction between the two types is most striking in the lower groups. Stachyosporous are the (Bryopsida), Psilopsida, Lycopsida, and Sphenopsida, phyllosporous (after a stachyosporous start) the Pteropsida (ferns). The distinction is not correlated with iso- or heterospory.

In the higher groups the picture is somewhat blurred. We know not enough of the Pteridosperms to arrive at a definite conclusion, but they may have been a group of some considerable variability in this respect. Preponderantly phyllosporous are the Cycadales, the Cycadeoidales being apparently only so in the male sex. Preponderantly stachyosporous are the Coniferopsida (most ♂ Conifers are weakly phyllosporous), and the Chlamydospermae. As to the Angiosperms there is much uncertainty and the present writer holds that the two types are recognizable here and that there are—perhaps next to mixed types—on the one hand plants with true sporophylls (carpels) with ovules on leaf margins (or leaf blade?) and stamens which are homologous with leaves; and, on the other hand, plants with false carpels (pseudocarpels, stegophylls) with axillary ovules or ovular trusses of an axial nature, and axial ("epipetalous") stamens situated in the axil of a "leaf." All this is still an open problem and the discussion about it is well under way.

As to the principle of protection, this is, as we said, most striking regarding the megasporangium (ovule). In the Pteridosperms Calathospermum is known to have developed a "supercupula," enclosing a number of stalked ovules, and also in the Gymnosperms that organ may be perfectly protected, e.g. in Conifers, where there is only a short period in which a connection with the outer world is open on behalf of the pollination. In the Angiosperms it is generally (but not exclusively) still more perfect and more permanent, the protection being offered by either a sporophyll or a stegophyll and the connection with the outer world being provided for by the pollentube of the male gametophyte. An extreme form of protection is manifest in inferior ovaries. It may be noted here that there is quite some difference of opinion regarding the morphological status of the flower parts, particularly those which bear the sporangia. Some authors still adhere to the classical theory that they are all homologa of "leaves," others hold that some of them are "axes," and it has even been suggested that they are neither the one nor the other, but rather some mixed form (stachyophyll) or, which is still more non-committal, organs "sui-generis."

The only character which perfectly, or almost perfectly, characterizes the Angiosperms, seems to be the double fertilization. All other differential characters are vague and overlapping.

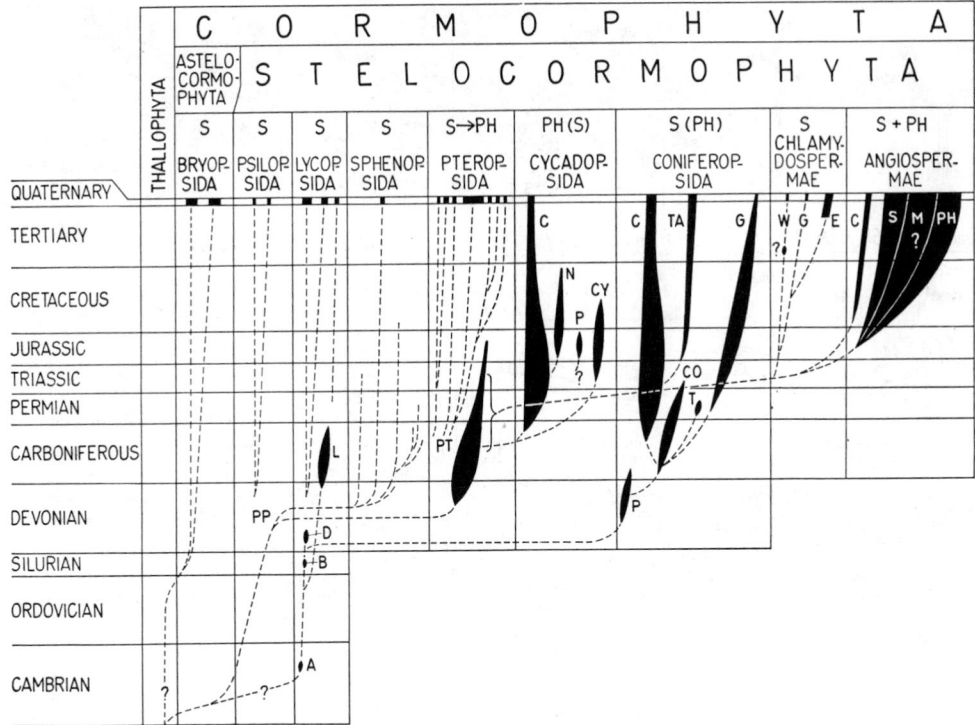

Fig. 5 Phylogeny of Spermatophytes. S., Stachyosporous; PH., Phyllosporous; M., Mixed. Psilopsida: PP., Psilophytales. Lycopsida: A., Aldanophyton; B., Baragwanathia; D., Drepanophycus; L., Lepidospermales. Pteropsida: PT., Pteridospermales. Cycadopsida: C., Cycadales; CY., Cycadeoidales; N., Nilssoniales; P., Pentoxylales. Coniferopsida: C., Coniferales; CO., Cordaitales; G., Ginkyoales; P., Pityales; TA., Taxales; T., Trichopityales. Chlamydospermae: E., Ephedra; G., Gnetum; W., Welwitschia. Angiospermae: C., Casuarina.

In all Phanerogams the embryo is endoscopic, with a suspensor at the micropylar pole of the ovule.

15. Phylogeny of Phanerograms. Group phylogeny is always a matter of speculation and its pictorial representation is always suspect of being greatly over-simplified. Yet, if we can recognize taxa as more or less distinct units in the present time, the question naturally arises: what did their ancestors look like and in what period may they be supposed to be connected. What we believe may have some sense, is based on fossil evidence and, particularly, on the chronological development of separate characters of fossil plants, and if these satisfactorily correlate we may have some confidence that they were incorporated in what might be called a genorheithrum (Lam), a "stream of potentialities" which keep together without a possibility of exchange with other taxa of a similar rank, in short, a phylad, i.e. the hologeny (Zimmermann) of a taxon.

Conclusions in this matter largely rest on in how far we consider a taxon a natural one and consequently such conclusions are likely to be very subjective. Accordingly, a representation like that of Fig. 5 is little more than an "aide-mémoire," a personal impression, gained by reflections during a life-time. In our chart we have only considered those groups which have been mentioned in the above as having seeds or seed-like structures, the remaining related groups having been indicated in a more simplified way. On studying the picture it is obvious that the Lepidospermales are

taxonomically far apart from what we call Spermatophytes or Phanerogams. Yet, it must be noted that the carboniferous Lepidosperms belong to the Lycopsids, a class which is suspect to bear some relationship to the Coniferopsida (cf Fig. 6). Nevertheless the analogy of the Lepidosperm-"seed" with that of Phanerogams is clearly expressed in Fig. 4.

As to the "true" Spermatophytes, there are four main groups, the Cycadopsida, Coniferopsida, Chlamydospermae, and Angiospermae, which in spite of some difference of opinion as to the taxonomic position of subclasses (including fossil ones) and the interpretation of some of their organs, give the impression of being quite natural taxa.

The Cycads have long been linked with the Angiosperms, notably the (phyllosporous) Magnoliales. On the other hand it has been suggested that there is also a distinct connection between the (stachyosporous) Chlamydospermae and the Angiosperms, notably the (stachyosporous) Monochlamydeae.

As the double fertilization, or rather the development of a secondary endosperm, seems to be the only truly differential character of the Angiosperms, which occurs both in stachyosporous and phyllosporous groups—the exact features of correlation are still unknown—, the conclusion must be that the endosperm-development has arisen twice "independently," since this character is obviously much younger than that of stachyo- and phyllospory. However, so long as we

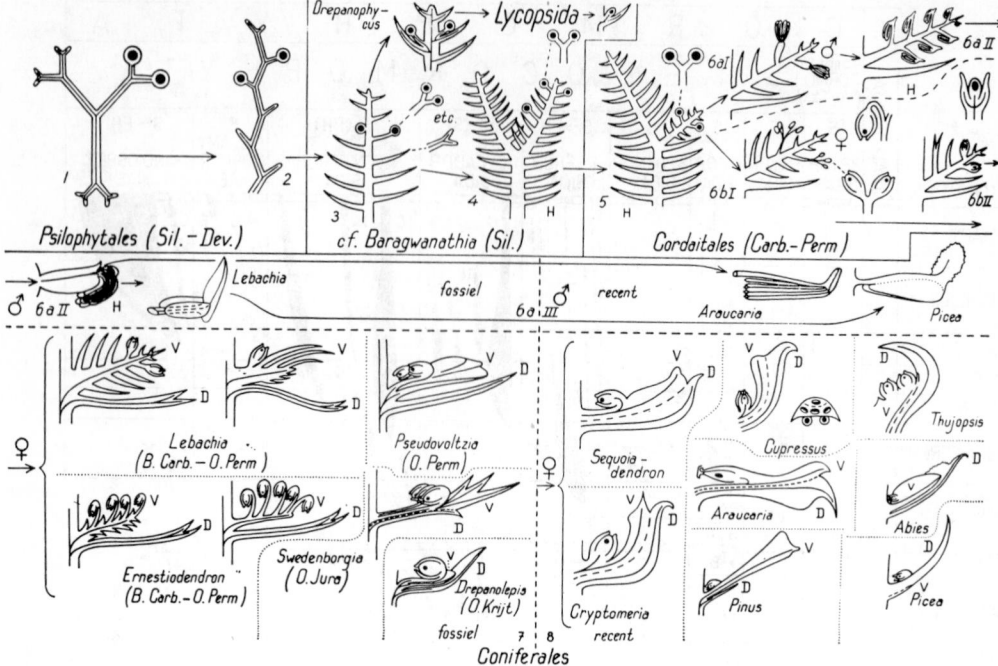

Fig. 6   Possible evolution of the male and female cones of the Coniferales from Psilophytales through Lycopsida and Cordaitales. fossiel = fossil; B. (Carb.) = Upper; O. (Perm.) = Lower (Permian); Jura = Jurassic; Krijt = Cretaceous.—H. = Hypothetical transitionary stage; V—Ovuliferous scale (axillary axis); D—bract.

know so little of the significance and distribution of the last-named characters in Angiosperms and we may suspect that the two are, in some groups, mixed up, the above conclusion need not alarm us too much, and the result must be that for the time being the ancestors of the three groups mentioned (Cycadopsida, Chlamydospermae, en Angiospermae), are to be looked for in the Pteridosperms, provided that these have included a wide range of potentialities. The "biphyly" of the Angiosperms need then be of not too radical a nature.

There remains the Coniferopsida, whose ancestry is more obscure. Their subclasses are undoubtedly closely akin; the Conifers clearly root in the older Cordaitales (Fig. 5) and these are probably related to the Pityales of which only the wood is known. We will not altogether exclude the possibility of a pteridospermic ancestry, but the facts so far known rather point to a descent from some much older group, related to the Lower Devonian Lycopsids. Fig. 6 shows in which way the process may have evolved mainly as far as the sporangia are concerned. At the same time it shows the probable evolution of the female coniferous cone, so brilliantly explained by the Swedish palaeobotanist Florin.

H. J. LAM

References

Turrill, W. B. ed., "Vistas in Botany," London, Pergamon Press, 1959.
Lam, H. J., "Some Fundamental Considerations on the 'New Morphology'," Trans Bot. Soc. Edinb., 38: 100–134, 1959.

SPERMATOZOA

Spermatozoa were first observed by a medical student Johan Hamm who saw them in human semen; his observations were communicated to the Royal Society in London by Leeuwenhoek in 1677. Their role in fertilization was not clearly understood until about 200 years later when Van Benedin and Hertwig gave the first clear description of the fusion of the nuclei of ovum and spermatozoon, although Prevost and Dumas had previously shown that spermatozoa were necessary for FERTILIZATION. Towards the end of the 19th century there was intense interest among cytologists in the development and morphology of spermatozoa and the classical work of Retzius and Ballowitz belongs to this period. More recently interest in spermatozoa has been of two kinds: first, the introduction of new microscopical techniques has made possible an extension of the investigations of the classical cytologists and, second, interest in problems of fertility in animals and man and particularly in the problems of artificial insemination of farm animals has led to a search for information on many aspects of the physiology of spermatozoa.

The spermatozoa are formed within the testes by the multiplication of the parent spermatogonia which form the basal layers of the tubular epithelium. Division begins in a special class of spermatogonia—the 'stem cells.' There are species differences in the exact pattern of development but in all, after a varying number of mitoses, some spermatogonia enter a resting stage and become new 'stem cells'; the rest continue to divide and eventually become transformed into primary spermatocytes. Each primary spermatocyte divides meiotically to produce two daughter secondary spermatocytes from which four spermatids are produced. The mature spermatozoa are shed from the tubule and are discharged via the efferent tubules of the testis into the epididymis. In most mammalian species the duration of spermatogenesis is about 30 days. In the ram, labelled spermatozoa arrear in the proximal part of the epididymis about 30 days after injection of P^{32}, and the ejaculate contains labelled spermatozoa after a further 13–15 days.

The following morphological description of the mammalian spermatozoon is based largely on ungulate spermatozoa. Although great variations in morphological features exist between different mammalian species, especially as regards head shape, the features described here are probably common to all (Fig. 1). The *head* is shown here as a paddle shaped structure which is attached at the neck to a long flagellum, the *tail*. The tail has two main components, the *middle-piece* and the main *tail-piece*. The head is covered in front by a cytoplasmic cap-like structure which is here referred to as

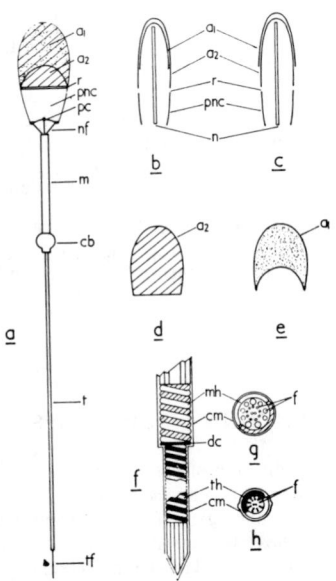

Fig. 1 a. Diagram of the main components of the mammalian (ungulate) spermatozoon. b. Longitudinal vertical section of the head of the boar spermatozoon (Hancock, 1957). c. Longitudinal vertical section of the head of the ram spermatozoon (after Randall and Friedlaender, 1950). d. Plan view of the larger component of the acrosome (a_2). e. Plan view of the smaller, crescentic component of the acrosome (a_1); this is the *galea capitis* of Blom (1945). F. Diagrammatic reconstruction of the partly dissected middle-piece and tail (after Challice, 1952, and Bradfield, 1953, 1955). g. Transverse section of middle-piece (Bradfield, 1955). h. Transverse section of main tail-piece (Bradfield, 1955). a_1, smaller component of acrosome (*galea capitis*); a_2, larger component of acrosome; r, nuclear ring; pnc, post-nuclear cap; pc, proximal centriole; nf, neck fibers; m, middle-piece; n, nucleus; cb, cytoplasmic bead or droplet; t, main tail-piece; tf, terminal filament; mh, mitochondrial helix; cm, cell membrane; f, fibrils or middle-piece and tail; dc, distal centriole; th, tail helix.

the *acrosome*. There is some evidence that the acrosome consists of two components, one of which may be identical with the structure known as the *galea capitis*. The *perforatorium* is most readily demonstrated in certain rodent spermatozoa where it forms a rod-like support for the overlying acrosome. It is said to be a modified area of the nuclear membrane. The surface of the area of the head caudal to the acrosome is readily impregnated with silver and this area is usually referred to as the *post-nuclear* cap. The common boundary of the acrosome and post-nuclear cap forms the *nuclear ring*. The *equatorial segment* is shown in Fig. 1 as a segment-shaped area in front of the nuclear ring. Its precise identity is still controversial but differences in the appearance of this feature are probably related to differences in the structure of the acrosome before and after death of the cell and to species differences in the structure of the acrosome.

Little can be seen with the light microscope of the structure of the neck but in suitably-stained preparations three *neck granules* are visible; the central granule is probably the *proximal centriole*. In certain circumstances the two lateral granules only can be seen. There is some evidence that the component fibrils of the tail are inserted into the head in three bundles; the points of insertion may correspond to the neck granules seen in stained preparations.

With the light microscope, the middle-piece normally shows no differentiation except for the presence in some spermatozoa of a *bead* or *droplet* of cytoplasm. Investigations with the electron microscope have shown that the tail is composed of a number of fibrils. In the middle-piece there are two sets, each of nine fibrils, arranged concentrically; a central pair of smaller fibrils run the whole length of the flagellum. In the main tail-piece the double set is reduced to a single set of nine double fibrils surrounding the central pair.

In the middle-piece the fibrils are surrounded by a double helix of mitochondria. A separate double helix of some fibrous protein surrounds the fibrils of the main tail-piece. These features are illustrated in Figs. f, g, h. The helix does not extend to the tip of the tail—the naked *terminal filament* can be distinguished from the rest of the tail, even with the light microscope.

The nucleus which comprises most of the head carries the material which at fertilization forms the paternal genetic contribution to the zygote. About 43% of the nuclear chromatin is deoxyribonucleic acid (DNA) (see NUCLEIC ACIDS), the DNA content of spermatozoa is about half that of a somatic nucleus. Genetically, two kinds of spermatozoa can be distinguished according to whether they contain the 'X' or the 'Y' sex CHROMOSOME.

Until relatively recently the acrosome was believed to function as a mechanical device for effecting sperm penetration; this view has largely been abandoned. It seems more likely that the role of the acrosome at fertilization is concerned with the release of necessary enzymes. In some species it has been shown that spermatozoa are capable of fertilization only after a stay of several hours in the female genital tract; in these species morphological changes occur in the acrosome during this period of *capacitation*. The acrosome reaction which is observed at the time of fertilization in certain invertebrate species has not been shown to occur in mammalian spermatozoa.

The tail is the motor organ of the spermatozoon. The middlepiece is the power house where energy derived from glycolysis is transferred to high energy phosphate compounds. In swimming, a two dimensional wave is propagated along the length of the tail. The speed of the bull spermatozoon has been estimated at 100μ per second.

A feature of active ram and bull semen is the so called 'wave motion' which is evident on microscopical examination. This wave formation is apparently due to a periodic aggregation of spermatozoa, the sperm tails beating synchronously in these aggregations. The 'wave motion' is the source of the changes in electrical impedance which can be detected in active ram and bull semen.

The cytoplasmic bead is the residual cytoplasm of the spermatid. It is present in virtually all spermatozoa in the epididymis but is lost at or about the time of ejaculation. In the proximal part of the caput epididymis the bead is located at the neck but it migrates distally during the course of the journey through the epididymis and in spermatozoa from the cauda epididymis it is found at the junction of the middle-piece and the tail-piece. The function of the bead is unknown. It has been suggested that the bead is necessary for normal locomotion but this is certainly not true; confusion about the significance of this structure might well be traced to the readiness with which it disintegrates after death of the cell.

Little is known with certainty about the physical properties of the cell surface but it has been suggested that the electrical charges on the head and tail differ. Of great practical interest is the claim that X and Y bearing spermatozoa can be separated electrophoretically.

It has been shown that in several species, contrary to expectation, the transport of spermatozoa from the site of deposition to the site of fertilization is effected by contractions of the tubal genitalia rather than by sperm motility. The proportion of spermatozoa which reach the site of fertilization is small relative to the number deposited in the female tract. For example, a single boar ejaculate may contain 50 thousand million spermatozoa but usually not more than a few thousand can be found in the sow's Fallopian tube at the time of fertilization; in the rat and rabbit, only a few hundred spermatozoa reach the Fallopian tube. The length of survival of spermatozoa in the female tract varies with the species. In most domestic mammals, fertilizing capacity is retained for only a few hours or at most a few days. It is retained for longer periods in birds and some bats are notable in that their spermatozoa may retain their fertilizing capacity for several months. It is of interest that spermatozoa stored in the epididymis retain their fertilizing capacity for up to two months even in species where it is lost rapidly in the female tract.

The maintenance of fertilizing capacity of spermatozoa stored *in vitro* is of great practical importance in animal husbandry. With bull semen stored at $5°C$ there is a small but progressive decline in fertility with the age of semen over the first three days of storage and fertility declines rapidly thereafter, but bull spermatozoa stored at $-79°C$ retain their fertilizing capacity for several years.

The spermatozoon must traverse three barriers before penetration of the ovum is effected so that fertilization can take place. The first is the loose mass of cells forming the *cumulus oophorus*. These are embedded in a cement substance which has a high content of hyaluronic acid. Dispersion of the *cumulus oophorus* can be effected *in vitro* by treatment with the enzyme hyaluronidase. Although complete dispersal of the cumulus is unnecessary for sperm penetration and fertilization, there is good reason to believe that the hyaluronidase which is associated with the spermatozoa in many species, is concerned with sperm penetration. Penetration of the zona pellucida apparently takes place with the aid of other, as yet unidentified enzymes—spermatozoa which have not undergone capacitation fail to penetrate the zona pellucida. Passage of the third barrier, the vitelline membrane and entry of the spermatozoon into the cytoplasm is followed by the changes which characterize fertilization. Among these changes are those which prevent entry of further spermatozoa "the block to polyspermy." After penetration the head of the penetrating spermatozoon rapidly increases in size and forms the male pro-nucleus; this comes into contact with the recently formed female pronucleus and syngamy ensues.

J. L. HANCOCK

References

Bishop, M. W. H. and A. Walton, "Spermatogenesis and the structure of the mammalian spermatozoa," *in* Parkes, A. S. ed., "Marshall's Physiology of Reproduction," 3rd Ed. Vol. 1, London, Longman, 1960.
Bishop, M. W. H. and A. Walton, "Metabolism and motility of mammalian spermatozoa," *loc. cit. supra.*
Mann, T., "The biochemistry of semen," London, Methuen, 1964.

SPHENOPHYTINA (= SPHENOPSIDA)

This is a subdivision of vascular plants under the division Tracheophyta in recent systems of classification. Its known chronological range is from Devonian to Recent. In most systems it consists of a single class: Equisetinae (= Articulatae) and several orders: Hyeniales, Pseudoborniales, Cheirostrobales, Sphenophyllales, Calamitales, and Equisetales. Of these only the Equisetales is represented by extant species of *Equisetum*. The group is usually defined in terms of the jointed nature of the stems and the whorled arrangement of leaves and branches. Dichotomous branching, however, is not unknown, e.g., in *Hyenia* and even rarely in the living *Equisetum*. Lateral branches in *Equisetum* and some species of *Calamites* are located at the nodes, but are not axillary. They alternate in position with the leaves. In some species of *Calamites* lateral branches tend to be some distance above the nodes.

Within the group there are trees (*Calamites*), creeping herbs (*Sphenophyllum*), and upright herbs (*Equisetum*). The stems are often ridged (*Sphenophyllum, Calamites,* and *Equisetum*) and possess either protosteles (*Sphenophyllum*) or a type of siphonostele unlike those found among Pteropsids (*Calamites, Equisetum*). Primary xylem development in the stems is either exarch (*Sphenophyllum*) or weakly mesarch

(*Equisetum*) with the possibility of some endarchy existing in the genus *Equisetum*. Endarchy is attributed to *Calamites*. Secondary xylem is either present (*Sphenophyllum, Calamites*) or absent (*Equisetum*).

The leaves range from small tooth-like, unbranched structures with a single midvein (*Equisetum, Calamites*) to wedge-shaped with dichotomous venation (*Sphenophyllum*) to dichotomously branched (*Hyenia, Asterocalamites*). The morphology of Devonian forms as well as the fact that in *Hyeniopsis* a more obvious branch system may replace a leaf in a given whorl suggest that the sphenopsid leaf represents phylogenetically a modified branch system and not a microphyll or enation.

The growing points, both shoot and root, of *Equisetum* possess a single apical cell from which the tissues are derived. The apical cell in the shoot apex is so large and conspicuous that it is often used to demonstrate the single initial cell to beginning students.

Roots are present in the sphenopsids and are not strikingly peculiar; they can usually, however, be identified to order on the basis of certain characteristic features.

Sporangia are typically borne on recurved tips of over-topped branches. Such a branch system is represented by *Calamophyton* (Fig. 1). In *Equisetum*, the fertile branch system is modified into a peltate sporan- giophore without a subtending bract. The true nature of the sporangiophore in *Equisetum* is made clear not only by comparison with less specialized members of the class, but also by its ontogeny and vascularization at maturity (Fig. 2, 3). The sporangiophore of *Calamites* is similar to that of *Equisetum*, but possesses a subtending bract, this being the major difference between the two orders. The fertile branches are fused to subtending bracts in some species of *Sphenophyllum*.

Gametophytes are known only for the extant species of *Equisetum*. The unicellular stage, i.e., the spore, is spherical without a triradiate marking. Most of the fossil forms do show triradiate markings on the cuticular remains of their spores. Attached to the outer wall of the spore in *Equisetum* are four strap-shaped non-cellular elaters, which are technically not a part of the spore since they are of tapetal origin. The spores contain well developed chloroplasts. The spores of *Equisetum* are all of one morphological kind, i.e., the genus is homosporous. Heterospory did exist in some calamites and sphenophylls. Spores of *Equisetum* are easy to germinate in the laboratory merely by placing them on wet filter paper or porous stone. Germination is rapid and immediate; two-celled gametophytes may be seen in a very few days. Germinating spores of *Equisetum* are excellent material for demonstrating dividing chloroplasts. Young gametophytes tend to be

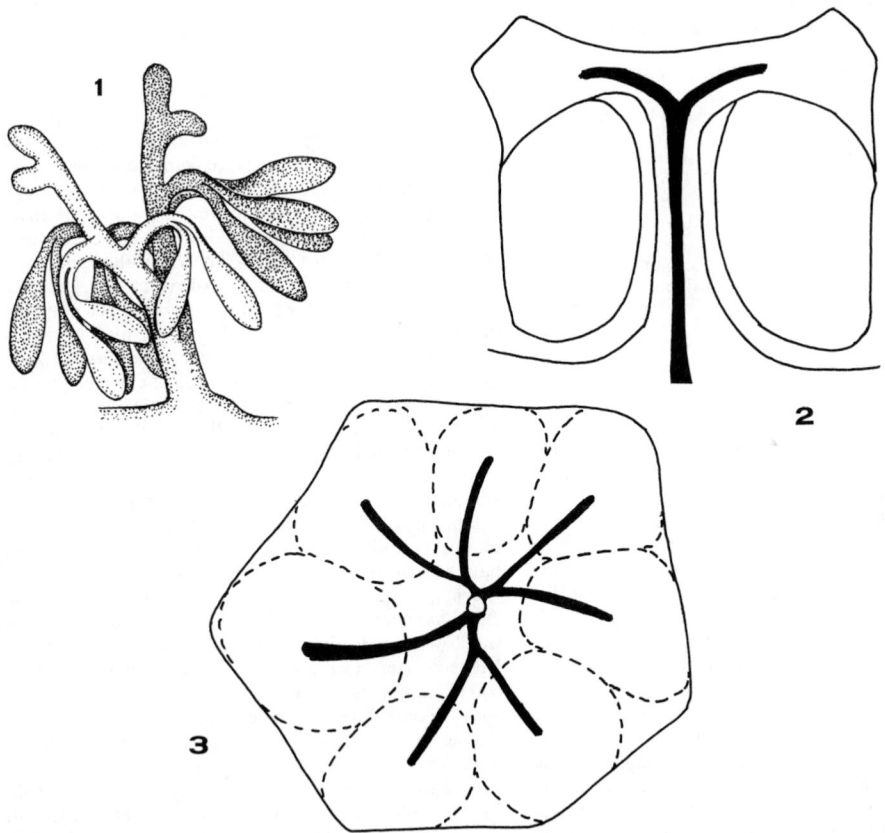

Figs. 1–3   1, *Calamophyton bicephalum*, a single sporangiophore. (Redrawn from Leclercq and Andrews, text fig. 8, 1960.) 2, Longitudinal section of same in extant *Equisetum* with vascular tissue blackened. 3, Top view of same with outlines of sporangia shown.

somewhat filamentous, with broad cells, but soon the body becomes more fleshy with a basal pad of tissue and upright filamentous to spatulate lobes and with fine rhizoids penetrating the substratum. Antheridia are large, semi-sunken, and situated at or near the tips of the upright lobes. Archegonia are mostly sunken and located mostly on the basal pad between the upright lobes. A unique feature of the archegonium is the arrangement of the two neck canal cells which lie side by side and not in a row corresponding to the axis of the canal. The gametophyte is annual in colder climates and may reach an overall size of nearly 3 cm in diameter. Gametophytes often tend to bear only antheridia or only archegonia. This led to the now discarded belief that *Equisetum* was heterothallic. There still remains the possibility that genetically different gametophytes may respond differently in terms of sex expression to the same environment. If this is so, then in *Equisetum* there may be the beginnings of heterothallism in association with homospory, a condition which is considered not to exist in vascular plants.

DAVID W. BIERHORST
HARLAN P. BANKS

References

Bierhorst, D. W., Bull. Torrey Bot. Club, 85:416–433, 1958.
Foster, A. S. and E. M. Gifford, "Comparative Morphology of Vascular Plants," Los Angeles, Freeman, 1959.
Hirmer, M., "Handbuch der Paläobotanik" Vol. 1, Munich, Oldenbourg, 1927.
Leclercq, S. and H. N. Andrews, Ann. Missouri Bot. Gard., 47:1–23, 1960.
Leclercq, S. and H. J. Schweitzer, Bull. Acad. Roy. Belgique (Sciences), 51:1395–1403, 1965.
Manton, I., "Problems of Cytology and Evolution in the Pteridophyta," Cambridge, University Press, 1950.
Walton, J., "An Introduction to the Study of Fossil Plants" 2nd ed., New York, Macmillan, 1953.

SPLEEN

The spleen is a distinctive purplish-red body situated in the dorsal mesentery to the left of the stomach or intestine. Among the vertebrates it is lacking only in cyclostomes and lungfishes. Throughout the group, lymphoid tissue forms the main splenic substance, or pulp, but this differs widely in structural organization from one species to another. Its structure is most distinctive in mammals.

The living spleen in most mammals is shaped like one segment of an orange (bilobate in Marsupalia; trilobate in Echidna) and is situated in the dorsal mesogastrium with its concave surface applied to the greater curvature of the stomach. A fibro-elastic capsule containing varying amounts of smooth muscle surrounds the organ. The capsule gives off trabeculae which penetrate inward, branch repeatedly and interlace to form a coarse three dimensional net-like framework; its interstices contain the splenic pulp.

Seen grossly in a slice of the fresh organ, the pulp is a thick pasty substance in which small circumscribed areas of grayish white color (white pulp) contrast with larger areas of red color (red pulp). Microscopically, each island of white pulp is a nodular accumulation of small lymphocytes (Malpighian corpuscle). The nodules may have germinal centers containing earlier cells of the lymphoblastic series. White pulp grades into red pulp; the two are intermixed in varying proportions in different animals. In certain mammals, and in man, a marginal zone of mixed red and white pulp occurs; in others the pulp zones are more sharply defined. Erythrocytes predominate in the red pulp surrounding the nodules but other cells representing all types found in the circulating blood are present and arranged as cords (cords of Bilroth). The cords occupy intervals between wide, blood filled, venous channels which in some mammals (e.g. rat, guinea pig, man) have highly specialized sinusal walls constructed of elongated reticular cells oriented longitudinally and held in place by ring-like condensations of reticular fibers; the arrangement is likened to a barrel with reticular-fiber hoops and cellular staves. In others (e.g. cat, mouse) the channels are non-sinusal and exist as simple, dilated endothelial tubes.

A delicate network of fine argyrophylic reticular fibers extends throughout the pulp enclosing the cells and vascular structures in its meshes. Phagocytic reticular cells occur among the reticular fibers. The reticulum of the spleen is identical in character to that found in other lymphatic structures such as bone marrow, lymph node and thymus.

The organ receives blood through one or more branches of the abdominal aorta which enter its hilus. Its venous blood passes to the liver through tributaries of the portal vein. Afferent lymphatics are absent. Efferent lymph vessels are found in the capsule and trabeculae and have been traced into the white pulp in several mammals (mouse, guinea pig, horse, monkey). The main arteries send branches into the trabeculae. These branch repeatedly to follow the trabecular pattern. Small branches leave the trabeculae and each traverses, in succession, an area of red pulp and a white pulp nodule. The artery then re-enters the red pulp and divides to form a spray of straight arterioles termed a "penicillus." Capillary-sized continuations of the penicilli acquire sheaths (ellipsoid or Schweiger Seidel sheath) formed of densely packed reticular cells. Beyond the sheathed segments the capillaries continue through the red pulp and come into intimate relationship with the venous sinuses. The precise nature of the vascular connections at this point is in doubt. Two views are proposed: (1) an "open circulation in which post-ellipsoid capillaries funnel out to blend with the red pulp reticulum permitting blood to percolate through the pulp and to enter the sinus through the "barrel stave" arrangement of its wall; (2) a "closed circulation" with unbroken endothelial continuity existing between arterial capillary and sinus; the blood is everywhere confined, as in other organs, to vessels lined with endothelium. The characteristic presence of masses of erythrocytes lying free in the pulp in histological preparations of the spleen is believed to result from agonal changes in permeability of the sinus wall. Species differences appear to exist. Further, agreement on the details of structure within the species has not been reached. Some results of current research with the electron microscope indicate that the Bilroth cords in the rabbit spleen are venous sinuses with collapsed walls—the absence of visible lumina accounting for their appearance as solid cellular formations or cords. Other studies which employ similar techniques have not confirmed this concept. Further study is required to elucidate the finer structure of the splenic pulp and to correlate this with the organ's several known functions.

Certain functions of the spleen are related to particular structural components; the contractile and expansible capsule and trabeculae and its unusual vasculature favor a reservoir function, the heavy muscled spleens of ungulates and of some other mammals enlarge to accommodate an increased volume of blood which may then be forced into the circulation; the nodules of the white pulp continually produce masses of lymphocytes; the reticulum is a component of the reticuloendothelial system, its phagocytic reticular cells engulf and destroy bacteria and other foreign organisms and matter. Presumed "aged" or otherwise damaged erythrocytes undergo phagocytosis by the macrophages of the reticulum which ingest their iron-containing pigments and transport them to the liver to be released and used in the production of hemoglobin and bilirubin. The spleen is the original vertebrate blood forming organ. In fish and urodel amphibians hematopoiesis in the spleen continues throughout life. In anuran amphibians and in other vertebrates with true bone marrow the production of erythrocytes and myelocytes by the spleen diminishes. In mammals this function is limited to the embryonic period (man), or, may continue into young adulthood (opposum, rat, mouse). Other functions attributed to the spleen include its role in the production of antibodies. Total splenic function is incompletely assessed. The production of a hormonal substance "splenin" which acts to depress hematopoiesis in the bone marrow has been postulated. The close evolutionary and embryological association of the spleen with the digestive tract suggests that it may perform an as yet unknown digestive function.

The forerunner of the spleen appears in cyclostomes. Blood forming tissue resembling bone marrow is scattered in the submucosa of the hagfish intestine; a more organized mass of similar tissue forms the core of the typholosole of the lamprey. Except for the Dipnoi where "spleen" tissue is again found in stomach wall, the organ is isolated from the digestive tract and has only vascular connections. Its anlage forms in the dorsal mesentery from cells of the coelomic epithelium to which mesenchymal cells are added. The embryonic rotation of the digestive tube carries the organ to the left side of the body except in birds where it remains to the right of the stomach and applied to the right-sided avian aorta. The entire dorsal mesentery has the phylogenetic capacity to form spleen tissue. In primitive reptiles (Rhynchocephalia) and amphibians (Siren lacertina) the spleen extends the length of the mesentery, i.e. from foregut to hindgut, probably representing the ancestral type. Among other vertebrates it is more compact and lies opposite the hindgut (frog, toad, turtle) or opposite the foregut (snake, alligator bird, mammal).

ROBERT C. MURPHY

References

Björkman, S. E., "The Splenic Circulation," Acta med. Scandinav., supp. 191, pp. 1–89, 1947.
Blaustein, A. W., "The Spleen," New York, Mc-Graw-Hill, 1963.

SPORANGIOPHORE

A sporangiophore is an aerial, branched or unbranched HYPHA which produces one or several sporangia containing many spores or sporangioles with few spores. Sporangia and sporangioles may be formed on the same sporangiophore. Sporangiophores are initiated by vegetative hyphae and are asexual reproductive structures characteristic for most FUNGI belonging to the Mucorales. In *Pilobolus* they originate from a greatly enlarged cell which forms a basal swelling (trophocyst). Sporangiophore length varies in different species from microscopic size to over 20 cm. The longest sporangiophores are unbranched and are produced by *Pilaira, Pilobolus* and *Phycomyces*. Several types of branching occur, ranging from simple dichotomous to complex, repeated branching. In some sporangiophores the main axis forms a terminal enlarged vesicle with many short stalks bearing sporangioles. Remarkably complex structures result by the formation of secondary vesicles each with clusters of stalked sporangioles, as in *Radiomyces* Embree. There may or may not be septa in the sporangiophores, or between sporangiophores and vegetative mycelium. Sporangia are separated from the sporangiophores by septa, or commonly by a columella. The columella is laid down in the cytoplasm of the young sporangium and forms an apical, often enlarged extension of the sporangiophore which is surrounded by spores originating by cleavage of the sporangium contents. Sporangiophores of *Phycomyces* and probably other fungi have a thin primary cell wall in the growing region and a secondary wall layer is deposited in the older portions which stop growing. The cell wall contains chitin and is covered by a waxy cuticle.

In *Phycomyces* the sporangiophore first consists of an upright hypha, about 100 μ or more in diameter, with an apical growth zone extending to 1–2 mm below the tip. After some time elongation stops, and the apex bulges out forming the sporangium which attains full size within a few hours. Sporangiophore elongation then resumes and continues for many hours. The sporangium darkens due to spore formation at the beginning of this stage. The growth zone extends from very near the sporangium base to about 2.5 mm below. A remarkable feature of growth in this stage is the rotational component or spiral growth. The rotation is counterclockwise as seen from above immediately after resumption of elongation, but about 1–2 hours later the direction of rotation changes to clockwise and continues in this fashion.

Sporangiophores are sensitive to external physical agents and asymmetrical stimulation induces curvatures (tropisms). Many Mucorales are phototropic, and *Phycomyces* and *Pilobolus* give positive curvatures towards a unilateral source of blue light. *Phycomyces* also gives negative curvatures away from an ultraviolet source (280 mμ), but red or green light are phototropically inactive. Symmetrical illumination with blue or ultraviolet promotes growth, at least temporarily, and this response partly explains the negative curvatures in the ultraviolet. Since the sporangiophore growth zone is opaque to this region of the spectrum growth promotion is limited to the irradiated side. However, the mechanism of phototropic curving in blue or ultraviolet is not yet fully understood. Light also influences the morphology of some sporangiophores, for instance in *Pilobolus*. Sporangiophores are negatively geotropic, and curve in response to temperature and humidity gradients.

Certain sporangiophores show highly developed morphological and physiological adaptations for spore dispersal. *Pilobolus* forms a large vesicle filled with vacular sap just below the sporangium, and, eventually, high internal pressure ruptures the sub-sporangial vesicle along a line of dehiscence close to the sporangium. The vesicle contents are violently ejected and project the entire sporangium to distances of up to about 200 cm. The sporangium adheres to substrates by a ring of mucilaginous material.

HANS E. GRUEN

References

Castle, E. S., "Spiral growth and reversal of spiraling in Phycomyces, and their bearing on primary wall structure," Amer. J. Botany, **29**: 664–672, 1942.
Curry, G. M. and H. E. Gruen, "Action spectra for the positive and negative phototropism of Phycomyces sporangiophores," Proc. National Acad. Sciences, **45**: 797–804, 1959.
Grehn, J., "Untersuchungen über Gestalt und Funktion der Sporangienträger bei den Mucorineen," Jahrb. wissenschaf. Bot., **76**: 93–165, 167–207, 1932.
Ingold, C. T., "Dispersal in Fungi," Oxford, Clarendon Press, 1953.
ibid., "Spore liberation," Oxford, Clarendon Press, 1965.

SPOROZOA

The Sporozoa are often classified as one of the four subphyla of the PROTOZOA. All Sporozoa are parasitic. Their hosts are widely distributed in the animal kingdom, from the Protozoa to the Chordata. The life cycle is complicated and in many species involves an alternation of a sexual with an asexual phase in reproduction. Numerous species are transmitted from one host to another by means of spores. A spore is an infective cell or group of cells (*sporozoites*) capable of producing a new infection, surrounded by a resistant membrane, and formed at a definite stage in the life history of the species. The spore membrane permits the sporozoite to withstand considerable variation in chemical and physical conditions. In those species which are transmitted directly from one host to another, no spore membrane is present, and the sporozoites are naked. Some spores contain polar filaments (i.e. Class Cnidosporidea) others do not (i.e. Classes Telosporidea and Acnidsporidea).

In the class Telosporidea, the spore lacks a polar filament or a polar capsule. The Telosporidea is divided into three subclasses: Gregarinida, Coccidia and Haemosporidia.

In the gregarines the development is largely or completely extracellular, while in the other two subclasses development is chiefly intracellular. The gregarines are chiefly parasites of the digestive tract and body cavities of invertebrates, especially arthropods and annelids. *Schizogony* (multiple fission) occurs in the order Schizogregarinida, while those gregarines without schizogony belong to the order Eugregarinida. In the latter, the body of the adult may be separated by

a septum into two parts, an anterior protomerite, and the posterior portion, known as the deutomerite. These septated individuals belong to the suborder Cephalina. Non-septate forms belong to the suborder Acephalina. Many gregarines are solitary; others are found in an endwise association of two or more individuals. This condition is known as *syzygy*. The anterior individual is the primite; the posterior, the satellite.

In a typical cephaline gregarine, the spore is ingested by the host; the sporozoites are released in the gut-lumen and penetrate the epithelial cells of the digestive tract. In the early stages of development the trophozoite grows intracellularly. As growth proceeds the trophozoite migrates to the free surface of the host cell. Here the trophozoite extends into the lumen remaining attached to the epithelial cell by means of an organelle known as the epimerite. As development proceeds, the gregarine becomes detached from the host epithelium, the epimerite is lost, and the sporont moves freely about in the gut-lumen. Syzygy occurs, and a gametocyst is formed. Multiple fission occurs within the gametocyst to form gametes which fuse to form many zygotes. Each zygote becomes barrel-shaped and secretes a membrane. This is an oocyst. Within each oocyst are formed eight sporozoites. Thousands of sporozoites may be formed in one gametocyst.

The coccidians are telosporidian parasites of both invertebrates and vertebrates. These organisms cause severe cases of diarrhea and dysentery in vertebrates which sometimes prove fatal to the vertebrate host. Coccidia in general are intracellular parasites of the gut-epithelium. The mature trophozoite is intracellular and small, whereas in the gregarines it is extracellular and large. The zygote is non-motile in both gregarines and coccidia. In some coccidia syzygy occurs, and in other species it is absent. Some species produce only a few microgametes, while others produce many microgametes. *Eimeria tenella* is the causative organism of acute coccidiosis in chickens. Many different species of *Eimeria* occur in pigeons, turkeys, pheasants, ducks, geese, sheep, goats, cats, dogs and several species of wild animals. The genera *Haemogregarina* and *Hepatozoon* are found in the blood of vertebrates and show an intermediate relationship between the Coccidia and the Haemosporidia. *Isospora hominis* is the only coccidian parasite of man.

The development of the Haemosporidia is similar to that of the Coccidia. Both groups undergo asexual reproduction resulting in sporozoite formation. In the Haemosporidia, schizogony occurs in the blood of vertebrates, while in the Coccidia it occurs in the alimentary canal of some bloodsucking invertebrate. The Haemosporidia are divided into two orders; Plasmodiida and Babesiida. The order Plasmodiida is divided into two families, the Haemoproteidae and the Plasmodiidae. The genera *Haemoproteus* and *Leucocytozoon* include forms that are blood parasites of birds and reptiles. The family, Plasmodiidae consists of a single genus, *Plasmodium*.

The genus PLAMODIUM is widespread both taxonomically and geographically. Many species are found in reptiles, birds and mammals, including man. Those species parasitizing birds are transmitted by mosquitoes belonging to the genera *Culex*, *Aedes* and *Theobaldia*. Four species of Plasmodia parasitize man and are transmitted by various species of female mosquitoes belonging to the genus Anopheles. *Plasmodium vivax* is the most common cause of benign tertian malaria. Other species causing malaria in man are *Plasmodium ovale, P. falciparum* and *P. malariae. P. falciparum* is the causative organism of malignant tertian malaria and is the most deadly of the malarial diseases. In *P. vivax* a paroxysm occurs every 48 hours, while in *Plasmodium malariae* (the causative organism of quartan malaria) fever and chills occur every 72 hours. Human malaria can be treated successfully by several drugs, such as quinine, atabrine, chloroquine and others. In the past, and at the present time, malaria is the most important protozoan disease of man. There has been a constant decline in the number of cases during the past 15 years.

The genus *Babesia* contains the organism causing "Redwater fever" in cattle. The causative organism is *Babesia bigemina.* The discovery of the method of transmission from one cow to another by means of the tick, *Margaropus annulatus*, was an important discovery in protozoology. Theobald Smith and F. L. Kilbourne demonstrated this fact in 1893, thus showing that an arthropod was the transmitting agent of a protozoan disease.

The class Cnidosporidea is distinguished from the Telosporidea by the spore containing one to four polar filaments. This class is divided into four orders: Myxosporida, Microsporida, Actinimyxida and Helicosporida. The Myxosporidians are primarily parasites of fishes, but a few are found in amphibia and reptiles. The parasite may infect various tissues and organs, and one species is the causative organism of "twist disease" in salmonid fish. In the Actinomyxida, the spore contains three valves, while the Myxosporidans contain only two. Species in this order have been found only in fresh or salt water annelids. The Microsporidian spore contains no separated valves as in the two previous orders. The pebrine disease of the silkworm and Nosema disease of honey bees are caused by organisms that are members of this order. The order Helicosporida contains a single genus *Helicosporidium*, and the complete life history is still unknown.

The class Acnidosporidea consists of organisms that do not produce spores containing polar filaments. Their life cycles fail to suggest any close relationship to the Telosporidea. The genera *Haplosporidium* and *Sarcocystis* are included within this class. The taxonomy of the Acnidosporidea is in a confused state at the present time due to the lack of information concerning development and known life cycles of representative forms within this class.

CHARLES F. ALLEGRE

References

(1) Hall, R. P., "Protozoology," Prentice-Hall, New York, 1953.
(2) Jahn, T. L. and F. F. Jahn, "How to Know the Protozoa," Dubuque, Iowa, 1949.
(3) Kudo, R. R., "Protozoology," 5th ed., Springfield, Illinois, Brown, Thomas, 1966.

STEM

A *stem* may be defined as an axis that bears leaves and buds. It is characterized by the possession of *nodes*,
the points of attachment of the appendages, and *internodes*, the regions between two successive nodes. In the course of development as well as at maturity there exists a close relationship between a stem and its leaves, and hence the two structures are commonly treated as a single unit, the SHOOT.

The structure of a stem forms the basis for the distinction between two widely recognized types of shoots. In stems in which significant internodal growth takes place, as in most vascular plants, *long shoots* are produced. By contrast, in *short shoots* there is little or no internodal growth, with the result that the leaves are arranged in close succession upon the axis. Examples of the short shoot habit include the spurs of fruit trees, the stems of rosette plants, certain of the branches of such trees as *Larix, Gingko,* and *Cercidiphyllum,* and such large leaved plants as tree ferns, cycads, and palms.

In addition to a basic difference in the extent of internodal growth, stems exhibit other variations in form and behavior. *Woody stems*, characteristic of trees and shrubs, live for many years. In each growing season, they increase in length and become larger in diameter. The epidermis, the outermost tissue of young stems, is early replaced by cork, which produces a hard bark with its rough exterior surface. *Herbaceous stems* form little woody tissue and have little or no increase in diameter. Consequently, such a stem remains relatively soft in texture and slender in form. The stems of most herbaceous plants live but a single growing season; even among the perennials, certain plants have herbaceous stems and die back at the end of the growing period to a more permanent root or modified stem system, as for example larkspur, rhubarb, dahlias, and tulips.

The form of a stem may be highly modified from the usual concept of an aerial axis bearing foliage LEAVES. Among the variations in structure are underground stems. These often are little more than horizontal stems bearing the appendages, or they may be thick and fleshy with a major function of storage of food reserves. Such stems or *rhizomes* are the main axes of many ferns, grasses, and some herbaceous perennials. Adventitious roots develop at intervals from the lower surface of the rhizome, usually in association with the nodes, and leaves and reproductive axes extend above ground. Another modification is a *bulb*, a short upright stem bearing a terminal bud or a flower primordium, the entire structure being enclosed by fleshy leaf bases and covered by dry scale leaves. Plants of this type include tulip and narcissus. *Corms* differ from bulbs in that the bulk of the short vertical axis is stem tissue which is surrounded by dry, fibrous leaf bases. Freesia, Gladiolus, and Crocus are examples.

Development of the stem. The stem is initiated during the development of the embryo when the hypocotyl takes form as the first stem axis. Subsequently, in addition to the hypocotyl, cotyledons, and root primordium, a mature embryo possesses a shoot primordium. The degree of differentiation of the shoot varies from species to species, and may, on the one hand, be represented by only an apical meristem, or, on the other hand, by an axis bearing one or more leaf primordia. Later, at the time of seed germination, the embryonic shoot resumes growth.

As a shoot develops, leaf primordia are successively produced at the flanks of the apical meristem in accordance with the phyllotactic pattern characteristic of the species. Concomitant with the increase in size of each leaf, the subjacent stem tissue may enlarge and elongate with the resultant formation of an internode. An internodal region begins to grow slowly, the rate of elongation then increases and continues for an extended period at a uniform rate before finally tapering off. Growth takes place first at the base of an internode,

then at progressively higher levels until it finally becomes localized in the upper end of a maturing internode. Analyses of the pith have shown that increase in both cell number and in cell length contribute to internodal growth; in some species cell division is the dominant process, in others cell elongation is of greater relative importance.

During the period of internodal elongation, the

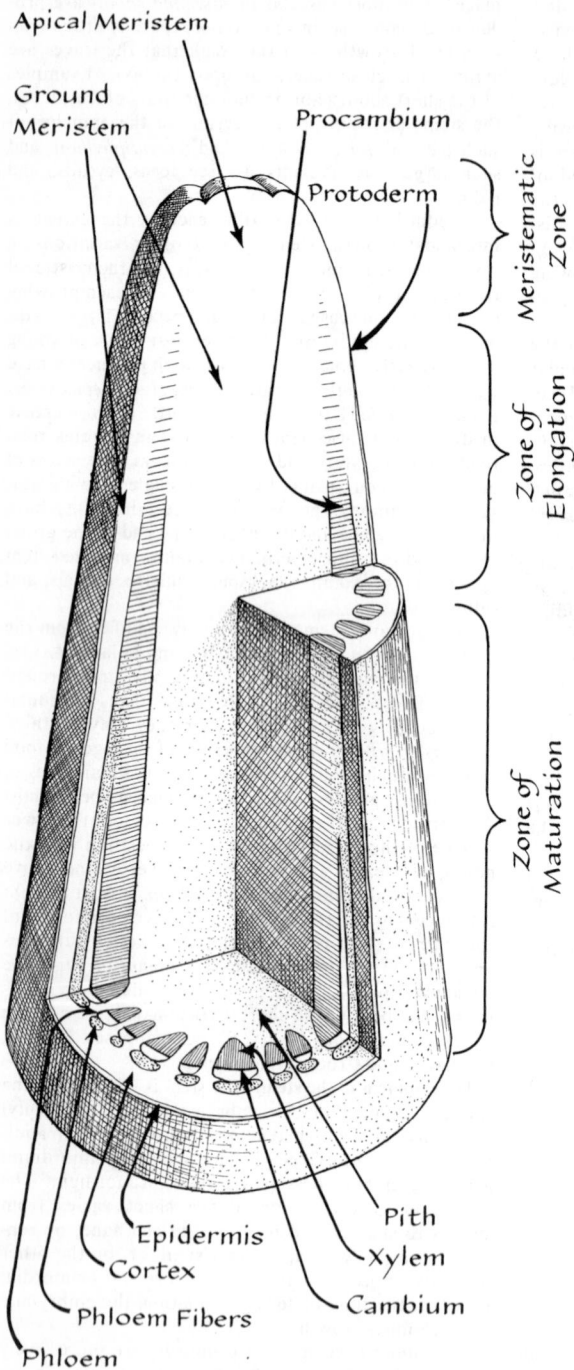

Fig. 1 Development of primary tissue in a dicot stem. (From Platt and Reid, "Bioscience," New York, Reinhold, 1967.)

supra-adjacent node does not change appreciably in size. However, as an internode approaches maturity, the node above enters a period of expansion during which it adds to its own stature as well as to the internodal regions on each side.

The timing of the development of successive internodes varies. In some species it is sequential, each internode approaching maturity before the next one above enters its most rapid phase of growth. By contrast, in other species several internodes develop more or less simultaneously, though at any one time the lower ones are in progressively more advanced stages. In addition to these two distinctive growth patterns, transitional forms also appear.

Although internodal elongation occurs in most stems as a result of a wave of differentiation proceeding from the base to the upper end, in some plants increase in length is governed by the activity of intercalary meristems, which though not permanent, persist for some time. These meristematic regions are usually located at the base of internodes. In the earliest stages of development, such an internode is composed only of meristematic tissue, but in the course of differentiation the intercalary meristem becomes set off as the adjacent cells gradually develop into permanent cell types. Subsequently, after the internode ceases elongation, the intercalary meristem remains inactive though it maintains for some time the capacity for further growth. Eventually, however, the meristem loses its identity as it undergoes cellular changes and becomes mature tissue. Intercalary meristems have been intensively studied in the internodes of the Gramineae and Equisetaceae.

The center for the initiation of leaf primordia and stem tissue is at the apex of the shoot. As the cells derived from the apical meristem enlarge and become vacuolated and thus begin to be differentiated, the tissue systems of the stem become blocked out. From the outermost layer of the apical meristem the precursor of the epidermis, the protoderm, takes form. Adjacent to the protoderm as well as in the central part of most stems, ground meristem originates, and from this tissue the cortex and pith are subsequently derived. Maturation of the epidermis, cortex, and pith occurs gradually in the successive nodal and internodal regions.

As the protoderm and ground meristem become delimited, strands of elongate meristematic cells, comprising the procambium, gradually arise in the region just below the apical meristem. In all vascular plants it appears that the procambium differentiates progressively and continuously in an acropetal direction through the stem axis and into the leaves. Likewise, in vascular plants generally PHLOEM is formed at the outer periphery of the procambium and forms centripetally within the strand as well as acropetally and continuously in the longitudinal path of the procambium. XYLEM varies in its orderly development in the different groups. In all, it develops acropetally in the procambium. However, its timing with respect to that of the phloem varies. In the Lycopsida, the xylem precedes the phloem in its acropetal progress. In the Sphenopsida and the Pteropsida, the xylem characteristically lags behind the phloem in its apical development. Interestingly, in Angiosperms and in a few ferns, when the advancing phloem reaches a leaf base, there is promptly induced a second locus of xylem formation at the leaf base. This xylem development proceeds both acropetally and basipetally, in this direction meeting the advancing xylem from below.

Within the procambium of most Pteropsida—not true for the ferns—the initiation of xylem and its subsequent differentiation in a radial direction takes place from the inner edge of the procambium centrifugally, i.e. toward the center of the strand in a pattern described as *endarch*. However, in stems of primitive vascular plants—the Psilopsida and Lycopsida—the direction of xylem maturation is from the outside of a

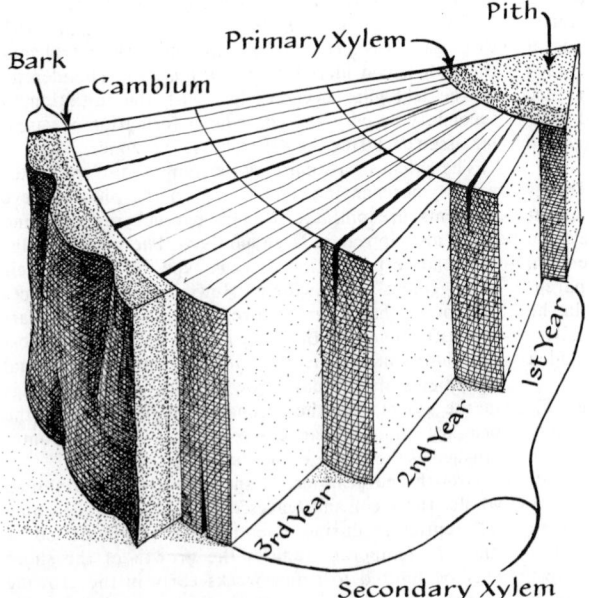

Fig. 2   Principal tissues in a three-year-old stem. (From Platt and Reid, "Bioscience," New York, Reinhold, 1967.)

strand inward (*exarch*), and in ferns from one or more centers in each bundle in all directions more or less haphazardly (*mesarch*). The tracheary elements that are formed during the elongation of the stem are annular or helical, types that are less resistant to expansion than are the scalariform and pitted types which develop after the stem has completed elongation.

The above description of stem growth does not consider in any detail the diversity of developmental patterns in the stem. Variables include the relative amount of vascular tissue in relation to parenchymatous tissue, the form that the primary vascular tissue or stele assumes, and the rates of development of the several tissues.

With the maturation of the epidermis, ground parenchyma, and vascular systems, stem growth in the lower vascular plants, the herbaceous angiosperms, and in most monocotyledons ceases. All of the above tissues had their origin in the apical meristem, and no additional tissues are formed. A shoot system of this type, complete with the above organization is said to possess *primary tissues* only. By contrast, among the gymnosperms and many dicotyledons, other tissues arise secondarily and are not associated in origin with the apical meristem. Such derivatives are designated *secondary tissues*, and together they constitute the secondary body. The vascular cambium, derived from residual procambial cells, develops between the primary xylem and primary phloem, and by continued mitotic divisions during the growing season, produces secondary xylem and secondary phloem. In addition, another secondary meristem, the cork cambium, originates in the epidermis, cortex, or phloem, and by its activity forms cork on its outer surface and may produce phelloderm on its inner surface.

Structure of mature stems. The epidermis is the outermost layer of the mature primary body. It is chiefly composed of tabular-shaped cells, termed epidermal cells, arranged in such a manner that there are essentially no intercellular spaces. The protoplasts of these cells characteristically produce cutin which is deposited in their walls and further in a more or less continuous layer over the entire stem surface. Interspersed amongst the epidermal cells of the stems of Sphenopsida and of Angiosperms are stomata, each bounded by a pair of guard cells. No stomata are present in the stems of Lycopsida or of ferns and gymnosperms. In many plants, epidermal appendages in the form of hairs or scales are also present.

The fundamental tissue system likewise exhibits variety in cell structure. Topographically, there are at least two distinct regions: the cortex located between the epidermis and the vascular system; and the pith, forming the central core of cells. In both parts, parenchyma constitutes the basic cell type. However, in the outer cortical regions there may be zones of collenchyma or sclerenchyma cells which provide added flexibility and mechanical support to the stem.

Of the three basic systems, by far the greatest diversity is displayed by the vascular tissues. Already early in the development of the stem the form of the primary vascular tissue is outlined by the position of the procambium.

The entire complex of vascular tissue together with the pith and pericycle, when present, are considered to form a unified whole, designated the *stele*. The most primitive type of stele, the *protostele*, is essentially a solid column of vascular tissue with xylem occupying the center and phloem forming the surrounding tissue. This kind of structure appears among the Psilopsida and Lycopsida. An alternate type is a *siphonostele*, of which there are several modifications. Basically, a siphonostele possesses a core of pith which is enclosed by vascular tissue. Variations of this fundamental form include: the occurrence of phloem on both sides of the xylem (*amphiphloic siphonostele* or *solenostele*), as is characteristic of certain ferns; phloem on only the outer periphery of the xylem (*ectophloic siphonostele*), a type found among many seed plants; and a form similar to the latter but dissected vertically by radial bands of parenchyma (*eustele*), as in numerous dicotyledons. The vascular system of monocotyledons is composed of strands which are scattered through the stem (*atactostele*) rather than arranged in a circle, as in those dicotyledons which possess separate vascular bundles.

The organization of the stele within the nodes is distinctive, for in these regions strands of vascular tissue are diverted into the leaves. The deflection of these vascular bundles, or *leaf traces*, may be gradual or abrupt, and the number of traces for each leaf, though specific for each kind of plant, may range from one to several. A zone of parenchyma is located immediately above the place where a leaf trace separates as a distinct unit from the main vascular system. Such a parenchymatous region, a *leaf gap*, appears in median longitudinal section to interrupt the continuity of the vascular tissue, but vascular elements extend laterally and continuously in an acropetal direction around the gap. Leaf gaps develop with developing leaves. When leaves are destroyed as they appear, leaf gaps are not formed. Leaf gaps may be short or long in vertical extent. In a eustele, the parenchyma of a leaf gap may merge with interfascicular parenchyma, thereby making the limits of each indistinguishable. Leaf gaps develop in most Pteropsida, but they are absent in the Psilopsida, Lycopsida, and Sphenopsida.

In the primary body the structure of a node is characterized by the number of gaps. When there is one leaf trace, the node is described as *unilacunar*, with three gaps it is termed *trilacunar*, and with several gaps the node is considered to be *multilacunar*. The identity of the leaf gaps becomes lost with the initiation of secondary activity and consequent formation of vascular tissue enclosing the parenchymatous zone.

The xylem of a stem may be composed of tracheids, vessels, parenchyma, and fibers. In the phloem sieve elements predominate, though parenchyma cells and fibers also appear in large numbers. The kinds of cells, the relative proportion of each type, as well as their arrangement, vary widely. However, the differences are not of a haphazard nature, rather, they are so characteristic of species that they may be used as a diagnostic feature in the identification of plants. The annual increments of secondary growth appear as growth rings in the xylem, but the secondary phloem, just as the primary phloem earlier, sooner or later becomes crushed as the stem expands laterally.

Growth and behavior. A shoot develops as a unified whole, the stem and leaves enlarging and undergoing differentiation during essentially the same period of time. In temperate regions the growth of the shoot may be limited to a few weeks early in the growing season, as in some trees and shrubs, or the growing period may be of much longer duration and continue for several months. Nodes and internodes cease to form after an inflorescence or a terminal bud develops at the end of the shoot. Growth in length may also become limited by the abortion of the young leaves and stem at the tip of the shoot, as in *Ulmus, Syringa, Betula,* and many other trees and shrubs.

The formation of either long shoots or short shoots is not always a constant character. In *Ginkgo,* for example, a branch that has exhibited the long shoot habit for several years may spontaneously become a short shoot. A similar reversal in behavior is often exhibited by short shoots. Even though most angiosperms form long shoots in the vegetative stage, their reproductive structures or flowers are always short shoots. Further, the stem of biennials in the first year of growth is a short shoot, yet in the second year, with the development of an inflorescence the axis characteristically becomes a long shoot.

Stems, just as leaves and roots, have the potential for regeneration. In fact, such a growth phenomenon forms the basis for the horticultural uses of cuttings and grafting.

Stem cuttings are made from both woody and herbaceous plants. After excision of a shoot from the parent plant, parenchyma cells adjacent to the vascular system dedifferentiate. By subsequent mitotic divisions and gradual orderly differentiation, an adventitious root becomes organized. As the new root continues to develop, it grows at right angles to the stem, extending through the cortex and finally emerging from the epidermis or cork. The vascular tissue at the basal end of the adventitious root becomes continuous with the vascular system of the stem cutting. In the period during which roots are initiated, callus tissue sometimes develops at the base of the cutting.

Regeneration of stem tissue is also essential for success in grafting. In this practice, a union between stock and scion is effected through the development of a callus. A vascular cambium, xylem, and phloem develop within the callus and become continuous with the vascular tissues of the stock and scion.

RHODA GARRISON

Reference

Esau, K., "Plant Anatomy," New York, Wiley, 1965.

STEMS, MODIFIED

In the development of the STEM a number of modifications from the usual cylindrical organ may be observed. These modified structures are essentially storage organs used by the plant for vegetative propagation. Man has recognized the potentialities of these structures not only in the food he obtains from them but also in that they hasten growth of the plant and may often save a year's time in crop production. Often the modification is so different from the usual stem type that close anatomical examination is necessary to ascertain that the structure is not a root or a leaf. Proof that the structure is a stem is usually obtained by examining the internal tissues and determining that it has the siphonostele or dictyosteles of stems, and externally produces buds and leaves characteristic of a stem structure.

Rhizome. The rhizome consists of a more or less elongated underground stem which arises from a lateral bud near the base of the main stem axis and extends horizontally through the soil. Externally a rhizome has nodes and internodes with small leaves arising from the nodes and with axillary buds produced within the leaf axils. Aerial shoots frequently develop from these buds, making possible the continued production of these rhizomatous plants by vegetative propagation. Adventitious roots are usually produced by the rhizomes in large numbers.

Closely related to the rhizome in its general habit of growth is the stolon or runner. Probably the only significant difference is that a stolon is an above ground structure, such as the runners of a strawberry, and the rhizome is a subterranean structure, such as the rhizome of Kentucky bluegrass or Iris. Some plants, for example Bermuda grass, produce both stolons and rhizomes. (See also RHIZOME)

Tuber. A tuber is really a horizontal rhizome that is considerably limited in length but enormously increased in diameter. Best known of all tubers is the white or Irish potato (Fig. 1) which ranks with the top ten food plants for man and is actually about fourth in this important list. It literally stands between man and starvation in many countries. A native of the New World, the white potato probably originated in the mountains of south or central America, probably Peru.

As the subterranean parts of the white potato develop slender rhizomes branching out from the base of the plant enlarge terminally to form the thickened structures we call tubers. Externally there are a number of compound lateral buds (*eyes*) arising in a spiral pattern which agrees with the phyllotaxic pattern typical of the vegetative stem. Each eye actually consists of a series of small scale leaves and several rudimentary buds. Below each eye is a ridge or leaf scar marking this area as a node on the stem. When prepared for planting the potato is cut into several pieces, leaving at least one eye on each piece. One or more aerial shoots develop from an eye in the process of growth.

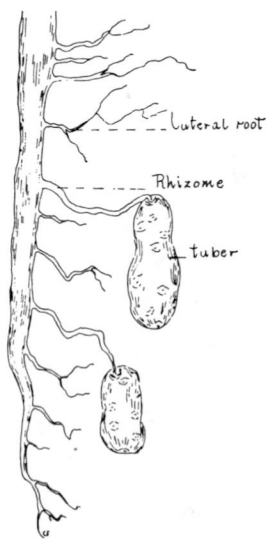

Fig. 1   Tubers of Irish potato.

Potatoes also demonstrate a high degree of apical dominance in the process of shoot development. If a whole potato is planted usually only the apical bud develops into an aerial shoot, the other buds are inhibited in growth by a hormone liberated from the terminal bud at the apical end of the tuber.

Internally tuberization is marked by a very active growth of storage tissues which form the bulk of the mature tuber. It is fairly well agreed by those who have studied the internal anatomy of the tuber that there are three very active PARENCHYMATOUS regions; (1) the medulla, including the central pith, (2) the procambial area; and just outside the PHLOEM, (3) the cortex. The usual stem tissues are present in the young potato tuber but as it matures the endodermis finally disappears.

The epidermis of the potato tuber is a short-lived tissue chiefly because of the activity of the cells which divide rapidly to compensate for increasing size of the tuber. These epidermal cells gradually build up an initial periderm of several cells thickness but, with increase in size, a hypodermal layer begins to function as a phellogen and ultimately builds a periderm up to fifteen cell layers in width. The phellogen remains active throughout the growth of the tuber and new phellem cells are formed to replace those that are torn apart and disintegrate as a result of enlargement. The individual cells are brick-shaped, have thin suberized walls, and act as an excellent protective bark. Potatoes owe their excellent keeping qualities to this corky periderm which forms over the moist parenchymatous cells below.

Bulb. A bulb is essentially a lateral bud which consists of a small stem tip at its base and a number of fleshy leaves arising from and usually surrounding the stem tip. Typical of all buds the bulb produces additional axillary buds in the axils of each fleshy leaf. These buds are the forerunners of the next year's bulbs or may produce aerial shoots, usually considered the function of the centrally located terminal bud, within the bulb.

Most bulbs are produced by plants in the *monocoty-ledonous* group including onions, garlic, lily, and tulip. The onion (Fig. 2) is undoubtedly the bulb of greatest

economic importance. In addition to the large bulb associated with the onion the mature plant also produces a large number of bulblets on the flower stalks well above the ground. In the onion these tiny bulbs are termed "sets" and are used by most gardeners for planting next year's onion crop.

Anatomically the onion bulb consists of a short stem which bears a series of fleshy leaves above and around the stem. Arising from the base of the dome-shaped stem are numerous adventitious roots. The arrangement of the leaves is concentric so that the older ones surround the younger with the apical meristem lying in the center of the entire structure. Usually the outer leaves of the bulb are green, well-developed aerially, and photosynthetic but many of the inner leaves do not develop the typical elongated blade. As a bulb matures the outermost leaves become dry and break off so that only the sheaths remain. These protect the fleshy storage leaves and enable the bulb to keep for long periods without drying out.

In spring many axillary buds begin to develop into new bulbs from the old structure. The youngest leaves in the center of the old bulb, as well as the storage leaves outside them, elongate, become green, and provide food for the new bulbs now forming.

One characteristic of the onion and garlic bulbs is the possession of lactiferous cells within the succulent portions of the leaves. The substance formed within these cells is a pale milky fluid which has been hydrolyzed to allyl sulphide. It is an excretion product, remains localized and is not associated with the vascular tissues.

The essential anatomical differences between onion and garlic bulbs is that in the former the entire bulb is essentially a mass of fleshy leaves with very small axillary buds. In garlic the axillary buds are very prominent (and become the garlic cloves of commerce) while the fleshy leaves are reduced in size and number and become chiefly scale leaves around the developing cloves.

Corm. A corm is a type of underground stem with a short fleshy vertical axis. The interior is quite compact with a highly vascular central zone surrounded by a large cortical region in which the vascular bundles are more or less dispersed typical of most monocotyledons. Large numbers of adventitious roots arise from the base of the corm and at its apex tufts of leaves develop.

Externally the corm is surrounded by scale leaves which arise from concentric nodal rings. Lateral buds develop in the axils of these scale leaves and the following year produce new shoots and ultimately new corms in a cluster around the old corm which then dries up and ceases to function. In contrast with bulbs, corms have much more stem tissue, fewer scale leaves, and longer internodes. Corms differ from tubers in that the former is the enlarged base of the stem rather than the swollen tip of a rhizome.

The *Gladiolus* (Fig. 3) is one of our most common corm producing genera and along with *Crocus* is a very desirable garden plant. Corms are also produced by a number of species in the family Orchidaceae.

Cladophyll. Those modified stems which superficially most closely resemble leaves are cladophylls. For the most part they are green, photosynthetic in function, and are often readily mistaken for leaves. The cladophylls of cacti are the large fleshy "hands" or

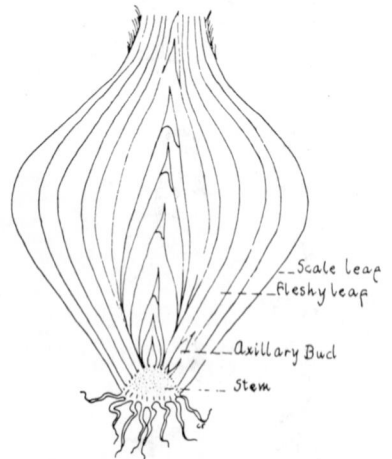

Fig. 2   Longitudinal section through bulb of onion.

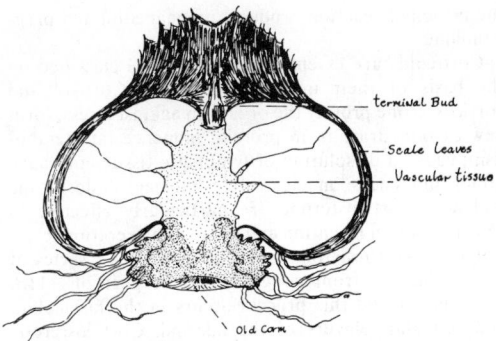

Fig. 3   Longitudinal section through corm *Gladiolus*.

"pads" which form the major portion of the aerial growth. Close inspection reveals that they have distinct nodes and internodes. When leaves are formed on these cacti pads they are usually small, often scale-like, and absciss from the stem soon after they appear. For the most part cactus cladophylls have clusters of tiny spines at the nodes. These spines were at one time considered modified leaves but Boke, following his morphological examinations, concluded they were analogous to bud scales. Some cacti also have tiny spines scattered along the entire surface of the cladophyll. These are of epidermal origin and are therefore more like trichomes or epidermal hairs.

Least conspicuous of the cladophylls are those of *Asparagus* and *Ruscus* (Fig. 4). Those of the former are narrowly cylindrical and look more like ordinary short stems; those of the latter are quite leaf-like with tiny leaves and flowers developing in their axils.

Thorns. Usually we think of spines as superficial structures, either of epidermal origin with no vascular connection to the main axis of the plant or as highly modified leaves or stipules of exogenous origin but with very little connection to the underlying tissues. Thorns, on the other hand, are usually much larger, arise in the axils of leaves, as ordinary branches do and even may bear leaves to further the evidence that they are indeed stems. The honey locust tree produces large branched thorns, often 18 inches long; those of *Pyracantha* are much shorter but still quite formidable structures. The large branched thorns of the crown-of-thorns plant, a native of our southwestern deserts, produce short but numerous leaves during the rainy season.

These are but a few of the numerous plants that produce thorns. From an evolutionary standpoint they may be looked upon as organs of survival. Over the years they were formed and persisted. Because of spines and thorns the plants were not eliminated by herbivorous animals. Whether or not thorns really played a part in this survival picture is purely conjectural but for the fanciful they provide an affirmative argument. Anatomically thorns have less transpiring surface than leaves and certainly they can withstand the buffeting of hot desiccating winds better but even here it takes a good deal of imagination to argue that these plants owe their survival to the fact that the main stems produced thorns instead of broad, flat leaves.

HOWARD J. DITTMER

References

Hayward, H. E., "The Structure of Economic Plants," New York, Macmillan, 1938 [reprint, New York, Steehert, 1968].

"Crops in Peace and War," 1950–51, Year Book of Agriculture, U.S. Department of Agriculture, Washington.

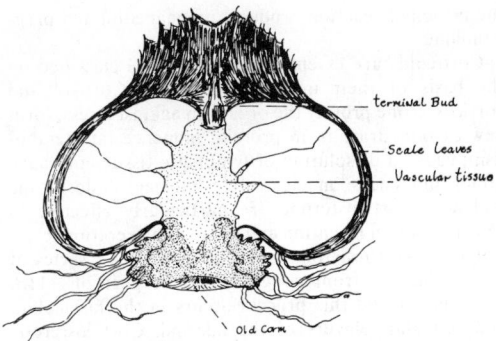

Fig. 4   *Ruscus*. The broad leaflike appendages are cladophylls. The actual leaves are the tiny bracts sticking up from the cladophyll surface. In the axis of the tiny leaves are the small flowers.

STEROID HORMONES

Steroid hormones are a class of organic compounds containing C, H and O, possessing the cyclopentanoperhydrophenanthrene nucleus, are produced in endocrine glands (see HORMONE, ANIMAL) such as the adrenal, gonads and placenta, and exert their action at a distant site. The steroid hormones may be subdivided into androgens, estrogens, progestational substances and corticoids on the basis of their physiological activity.

Androgens are produced in all of the steroid producing tissues and have the specific function of maintaining the male sex characters of the mammal such as the prostate and seminal vesicles. In man, other secondary sex characters as facial hair and pitch of voice are controlled by androgens. In the fowl, androgens maintain the comb, wattles and spurs. In addition to sex specific functions, androgens influence nitrogen metabolism, fat distribution, and to a limited extent exert control over electrolyte balance. Testosterone is the most active naturally occurring androgen. Other less active androgens include Δ^4-androstene-3, 17-dione, dehydroepiandrosterone, and 11β-hydroxy-Δ^4-androstene-3, 17-dione. Structurally, androgens possess 19 carbon atoms with oxygen substituents at carbons 3, 11 and 17.

The formation of androgens in the gonads is controlled by the pituitary gonadotropic hormones and the general sequence of biosynthesis in these tissues, the placenta and to some extent the adrenal, is cholesterol → pregnenolone → progesterone → 17α-hydroxyprogesterone → Δ⁴-androstene-3, 17-dione ⇌ testosterone. Only the testis produces significant amounts of testosterone. Adrenal androgens are formed from two additional pathways: cholesterol → pregnenolone → 17α-hydroxypregnenolone → dehydroepiandrosterone → Δ⁴-androstene-3, 17-dione and cortisol → 11β-hydroxy-Δ⁴-androstene-3, 17-dione. Androsterone and etiocholanolone are the two principal catabolites of testosterone, Δ⁴-androstene-3, 17-dione and dehydroepiandrosterone. Four 11-oxygenated 17-ketosteroids, 11β-hydroxyandrosterone, 11-ketoandrosterone 11β-hydroxyetiocholanolone and 11-ketoetiocholanolone are the principal catabolites of 11β-hydroxy-Δ⁴- androstene-3, 17-dione.

Estrogens are a class of C_{18} phenolic steroids where either one or two of the four rings in the structure is aromatic. Estrogens are produced in all steroid producing tissues. The primary physiological action is maintenance of the female sex characters including growth of the vagina, uterus, mammary glands and fallopian tubes. Other effects include fat distribution, influence on electrolyte balance, calcium metabolism, and blood clotting.

Estradiol-17β is the principal and most active estrogen and together with estrone, the corresponding 17-ketosteroid, is widely distributed in body fluids. Equilin and equilenin are representative ring B unsaturated estrogens and are associated mainly with the pregnant mare.

Estrogen production in the gonads is controlled by pituitary gonadotropic hormones, while adrenal estrogen formation is regulated by the pituitary adrenocorticotropic hormone. The pathway testosterone → 19-hydroxytestosterone → estradiol-17β probably accounts for 95% or more of the estrogens biosynthesized. Catabolic reactions lead to the formation of a dozen or more products of which estradiol-17β, estrone, and estriol may be considered to be major metabolites. Other catabolites include 16α, 16β, and 16-keto derivatives of estrone and estradiol-17β.

Progestational hormones include progesterone and its 20-reduced derivatives 20β-hydroxy-Δ⁴-pregnene-3, 20-dione and 20α-hydroxy-Δ⁴-pregnene-3, 20-dione, of which progesterone possesses the highest order of physiological activity. These compounds are produced in all steroid producing tissues and especially in the corpus luteum and placenta. Progesterone has a specific function on the vaginal and uterine epithelium and mammary glands and in concert with estrogens has a role in the maintenance of the female sexual cycle whether estrus in the rodents or menstrual cycle in the primate. Progestational substances have a unique role in the maintenance of pregnancy. A second primary role of progesterone is to serve as an intermediate in the biosynthesis of all classes of steroid hormones.

The biosynthesis of progesterone in gonadal tissue is regulated by pituitary gonadotropins and by adrenocorticotropin in the adrenal. The pathway of formation is cholesterol → pregnenolone → progesterone. Progesterone undergoes reductive catabolic reactions of the ketone groups and the nuclear double bond forming the principal reaction products pregnanediol and pregnanolone.

Corticoids are essential to life and are classified on the basis of their metabolic function. Cortisol and corticosterone protect the organism against stress, form new carbohydrate from protein, influence fat metabolism, cause a dissolution of lymphatic tissue, and have minor influences on electrolyte balance. Aldosterone and deoxycorticosterone are particularly effective in causing sodium retention and potassium excretion.

The biosynthesis of cortisol proceeds by a series of hydroxylations from progesterone at positions 11β, 17α, and 21 and this process occurs in the fasciculata and reticularis layers of the adrenal. Corticosterone is produced in all three zones of the adrenal and is formed by 11β and 21 hydroxylation. Aldosterone is formed exclusively in the glomerulosa and requires the hydroxylation of progesterone at positions 11β, 18, and 21. The 18-hydroxy group is oxidized to an aldehyde function.

The principal but not the only catabolites of cortisol include cortisone, tetrahydrocortisol (3α, 5β), tetrahydrocortisone (3α,5β), tetrahydroallocortisol (3α, 5α) and the 17-ketosteroids: 11β-hydroxyandrosterone, 11-ketoandrosterone, 11β-hydroxyetiocholanolone, and 11-ketoetiocholanolone. The catabolites of aldosterone include various reorg A tetrahydro derivatives.

RALPH I. DORFMAN

STEROLS, BIOGENESIS OF

It has now been firmly established that all sterols are derived from the triterpenes squalene and lanosterol. The biogenesis of sterols may, therefore, be logically discussed in two phases: (1) the formation of these two triterpenes, and (2) their further transformation to cholesterol and other sterols.

Early work by organic chemists has resulted in the recognition that a larger number of natural products (the terpenes, rubber, etc.) may be visualized as condensation products of the branched-chain hydrocarbon isoprene. Until 1956, the nature of the biological equivalent of isoprene and the mode of its condensation in living organisms were unknown, although it has been long recognized that this "biological isoprene unit" is derived from acetate. Following the discovery of mevalonic acid as an intermediate in cholesterol biogenesis in 1956, both the nature of this "biological isoprene unit" and its mode of polymerization have been elucidated. These reactions are summarized in Figure 1. The chemical acid-catalyzed polymerization of isoprene is also shown here to illustrate the similarity between the two processes.

Certain aspects of this series of reactions deserve additional comment.

(1) More than half of the reactions are irreversible, thus furnishing the driving force for an efficient synthesis of sterols from acetate.

(2) In the many reactions requiring a pyridine coenzyme, TPNH, and not DPNH, is utilized. This serves as another example of the current theory that the *in vivo* functions of these two reduced coenzymes are dif-

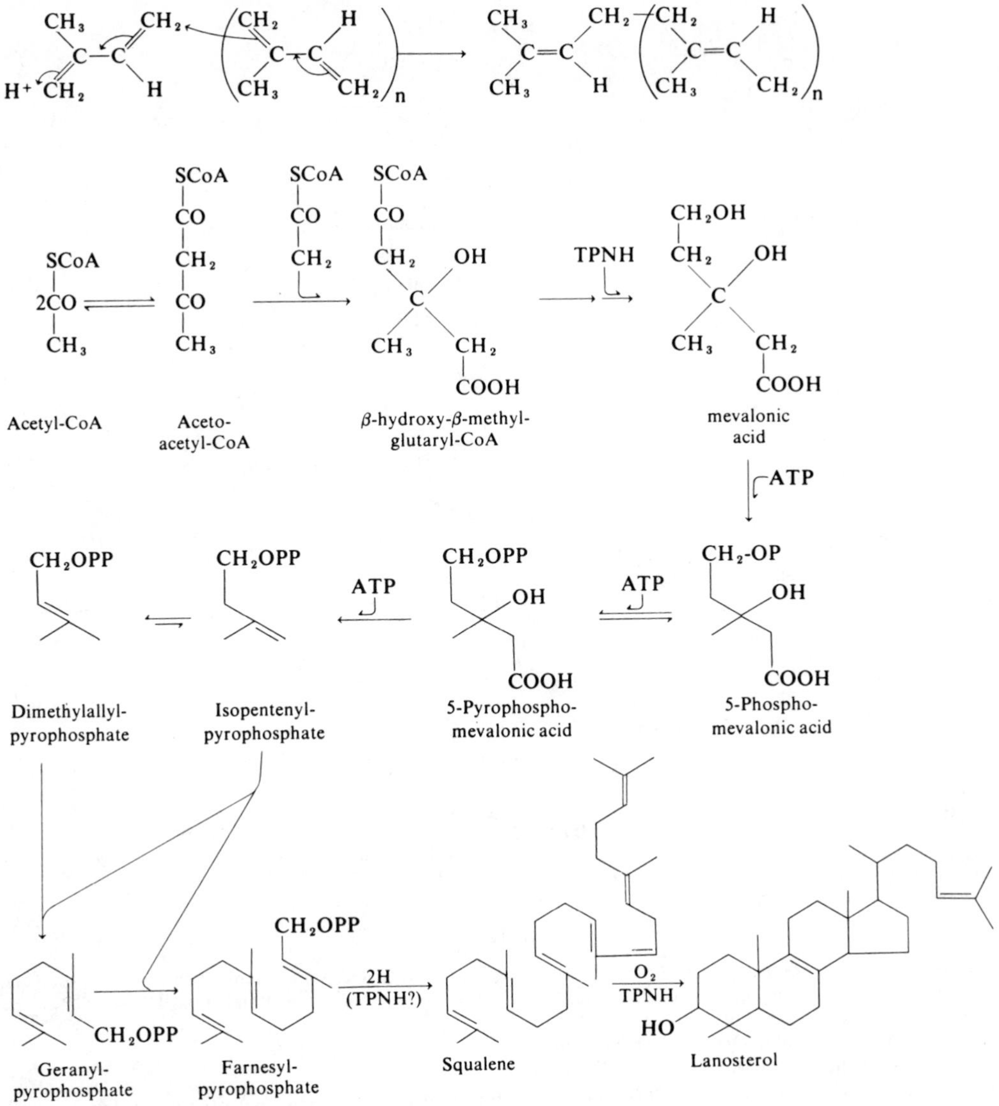

Acetyl-CoA

Aceto-
acetyl-CoA

β-hydroxy-β-methyl-
glutaryl-CoA

mevalonic
acid

Dimethylallyl-
pyrophosphate

Isopentenyl-
pyrophosphate

5-Pyrophospho-
mevalonic acid

5-Phospho-
mevalonic acid

Geranyl-
pyrophosphate

Farnesyl-
pyrophosphate

Squalene

Lanosterol

Fig. 1   Acid-catalyzed polymerization of isoprene and enzymatic formation polyisoprenoids.

ferent, TPNH being the one most utilized for synthetic purposes.

(3) These reactions represent a general mechanism for the biogenesis of the carbon skeleton of terpenes. Of particular interest is the fact that the biogenesis of a large number of tetra- and penta-cyclic triterpenes can all be rationalized as derived from a one step cyclization of squalene.

(4) The subcellular loci where these reactions take place have been studied by the use of fractions of homogenate. With the reasonable assumption that these results do correspond to the true distribution *in vivo*, it may be said that mitochondria are not directly involved in sterol biogenesis. The early steps (formation of mevalonic acid) and the late steps (after farnesyl pyrophosphate) take place in (or on) the microsomes while the steps in between occur in the soluble portion of cytoplasm.

(5) It has also been shown that the rate-limiting step in this process is the reduction of β-hydroxy-β-methylglutaryl-CoA to mevalonic acid.

The conversion of lanosterol to cholesterol is illustrated in Figure 2. It is apparent that our knowledge here is very limited. The three carbon atoms which are lost during this process are probably oxidized to carboxyl groups and subsequently decarboxylated. In the later steps, desmosterol, Δ^7-cholesterol and 7-dehydrocholesterol (not shown in Figure 2) have all been postulated as intermediates. Since these compounds do not fit into one pathway, two pathways are shown between zymosterol and cholesterol.

Plants and some lower forms of animals, contain

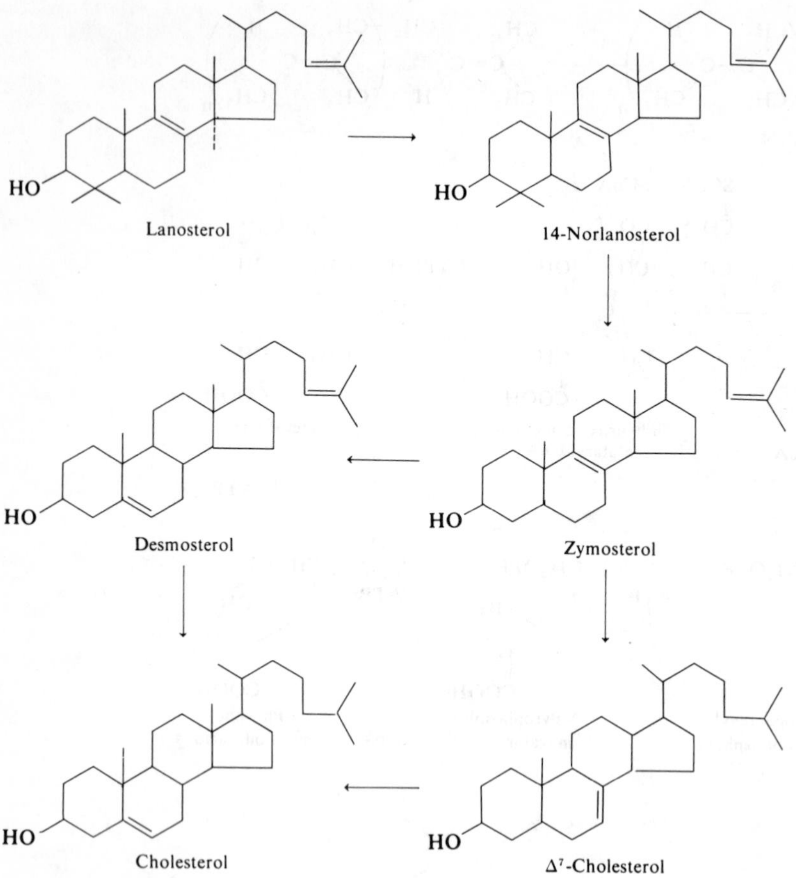

Fig. 2 Conversion of lanosterol to cholesterol.

sterols with one or two additional carbons attached to C-24 of the side chain (ergosterol, sitosterol, etc.). It has been shown that these carbon atoms are added to the molecule after the sterol nucleus is already formed. Unlike the rest of the carbon atoms, these additional ones are not derived from acetate. Where only one additional carbon is attached to C-24, this carbon is probably derived from active methionine (S-adenosyl methionine).

In the higher animals, the highest concentration of sterol is found in the nervous tissue where it is present almost exclusively as cholesterol. This cholesterol is synthesized during the growth of the tissue (young or regenerating) and laid down as structural units. Mature nerve tissue does not synthesize cholesterol, nor does its cholesterol "turn over" at any appreciable rate. In the other tissues where cholesterol biogenesis has been demonstrated, cholesterol is continuously synthesized to be secreted (skin, perhaps intestines), or converted to other physiologically important derivatives (adrenals, sex organs), or both (liver).

T. T. Tchen

References

Cook, R. P. ed., "Cholesterol," New York, Academic Press, 1958.

Ciba Foundation Symposium, "Biosynthesis or Terpenes and Sterols," Boston, Little, Brown, 1959.

STOMA (pl. Stomata; from Gr. στόμα, στόματα)

In botany, a stoma consists usually of a pair of specialized bean-shaped epidermal cells (guard cells), on a LEAF or other organ. These can become more curved, due to increase in their turgor pressure relative to that of the adjacent epidermal cells (subsidiary cells), and so widen the lenticular slit or pore between them (stomatal pore) which communicates with the intercellular space system within the leaf. Such turgor movements are largely due to the distribution of thickening of the guard cell walls; in a simple type the (ventral) walls bordering the pore are thickened above and below, making them relatively inextensible, but the thin dorsal walls extend with turgor increase, lengthening and curving the cell and so opening the pore. Also the cellulose chains run transversely to the long axis of the guard cell which thus extends longitudinally rather than laterally.

Stomata respond to many different stimuli; especially they tend to open with increased light intensity or reduced atmospheric carbon dioxide concentration and close with water deficit. *Rapid* wilting, however, causes transient wide opening, due to loss of turgor in the subsidiary cells, followed by closure as these withdraw water from the guard cells; rapid recovery from wilting causes the converse changes. Wilting causes massive increases in guard cell starch content, implying condensation from sugar and fall of osmotic potential—these changes are the converse of those in mesophyll cells. How light and carbon dioxide produce changes in guard cell turgor is as yet uncertain. Observed falls in guard cell vacuolar pH with darkness suggest "dark fixation" with organic acid formation at increased carbon dioxide concentrations; it has been demonstrated with radioactive carbon that guard cell chloroplasts can fix carbon dioxide in the dark and the same reactions may proceed at high concentrations in light. Even so, the link with guard cell turgor is unknown. Atmospheric carbon dioxide operates from the substomatal cavity and not from the exterior, at least in some species, for with completely closed stomata external concentration changes are without effect, though carbon dioxide-free air can cause wide opening in darkness of stomata not initially closed.

Light apparently affects stomatal movement by at least three separate mechanisms. The first operates through reduction of carbon dioxide concentration in the guard cells, by photosynthesis both in the mesophyll and in the guard cells themselves. Low carbon dioxide concentration plus high humidity tend to reduce guard cell starch content; part of this first light effect may thus operate indirectly by hydrolyzing starch to sugar, raising the osmotic potential and hence turgor. Responses to light can, however, be independent of changes in starch content. Production of osmotically active substances directly by photosynthesis in the guard cells is not of primary importance, for at high carbon dioxide concentration when photosynthesis should be most active the stomata tend to close. The second light effect is independent of carbon dioxide concentration and is especially associated with blue light. Stomata have been found to open wider in blue than in red light even when the concentration of carbon dioxide in the intercellular spaces is higher in the blue light. There is some evidence that blue light, even at very low intensity, causes decrease of guard cell starch and increase in osmotic potential; this may be the mechanism of the blue light effect. Red light causes an increase in guard cell starch even at the light compensation point when there can be no net gain of photosynthates. The third light effect controls the phase of endogenous cyclic rhythms of stomatal opening and closing tendencies. The rhythm in darkness is timed by the change from light to dark or from high to low light intensity; in low light it proceeds more slowly than in dark so that the phase is shifted and this can be effected by as little as 10 lux or 100 erg cm^{-2} sec^{-1}; red light (700 nm) is the most effective. The phase of the dark rhythm greatly affects the rate of opening on bright illumination. Similar rhythms in light have been less studied. Endogenous diurnal rhythms of guard cell starch content (and by implication sugar) are also found and may be causal.

There are up to about 500 stomatal pores per mm^2 of leaf surface, each of the order of 10 μ deep by 10 μ long by 0–5 μ wide, accounting generally for less than 2% of the total area. Such small pores are very efficient and through this minute fraction diffuse nearly all the carbon dioxide assimilated and water vapor transpired. Complete stomatal closure almost stops both processes; partial closure probably causes a relatively greater reduction in transpiration than assimilation, because of the added resistance for carbon dioxide diffusion into the cells, and is especially important in wind when the risk of desiccation can be great. This theory is involved in the proposed use in agriculture

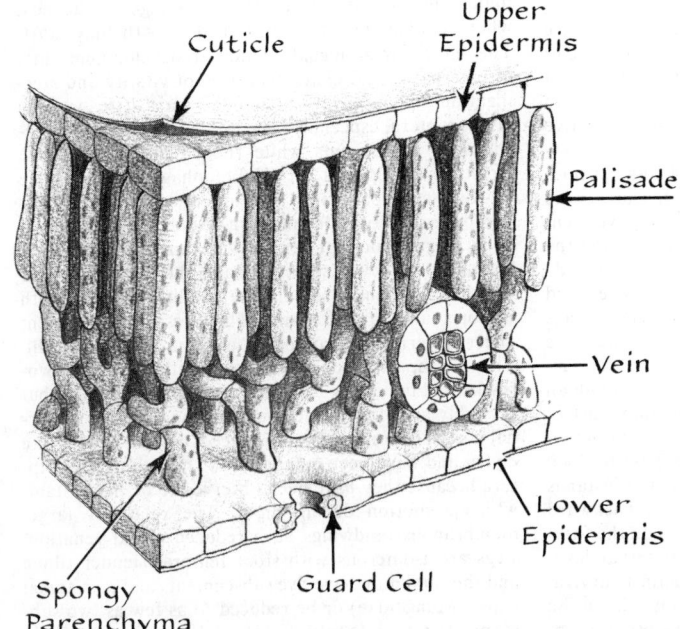

Fig. 1 Diagram, showing details of leaf. (From Platt and Reid, "Bioscience," New York, Reinhold, 1967.)

of chemicals as antitranspirants intended to reduce stomatal apertures without decreasing photosynthesis unduly.

O. V. S. Heath

References

Heath, O. V. S., *in* Steward, F. C. ed., "Plant Physiology," Vol. 2, New York, Academic Press, 1959.
Heath, O. V. S., *in* Ruhland, W. ed., "Encyclopedia of Plant Physiology," Vol. 17/1, Berlin, Springer, 1959.
Heath, O. V. S. and T. A. Mansfield, *in* Wilkins, M.B. ed., "The Physiology of Plant Growth and Development," London, McGraw-Hill, 1969.
Meidner, H. and T. A. Mansfield, Biol. Rev., **40**: 483–509, 1965.
ibid., "Physiology of Stomata," London, McGraw-Hill, 1968.
Penman, H. L. and R. K. Schofield, Symp. of Soc. for Exp. Biol., **5**: 115–129, 1951.

Fig. 1   (From Fox and Fox, "An Introduction to Comparative Entomology," New York, Reinhold, 1964.)

STREPSIPTERA

The Strepsiptera have led a checkered taxonomic career ever since 1793, when Petrus Rossius described *Xenos vesparum* as a member of the Hymenoptera. Latreille in 1809 transferred the species to the order Diptera. Kirby in 1816 established the order Strepsiptera or "twisted-winged insects," then comprising only two genera, *Xenos* Rossius and *Stylops* Kirby. During subsequent years these insects were assigned to various orders and by many workers were treated as family Stylopidae of the Coleoptera. Comprehensive studies by W. D. Pierce from 1908 to 1918 unequivocally established the Strepsiptera as a separate order, an arrangement now used by all students of these parasites (although Zoological Record as recently as 1966 treated them as a family in the order Coleoptera). At present about 250 to 300 species are known.

The life histories of those Strepsiptera that have been fully studied are among the most complex cycles known among the insects, involving a host-seeking triungulin larva, almost always a neotenic female and sometimes different host species for the respective sexes.

The family Mengeidae, known from a fossil species in Tertiary amber and a score of living species, is the most primitive. Both males and females have pentamerous clawed tarsi, are free-living as adults and the females of some species are fully developed and winged.

In the other five families the male tarsus is reduced and clawless and the neotenic larviform females are permanently endoparasitic. Adult winged males live for only about five hours, during which time they actively seek a mate. Fertilization takes place through an opening of the female cephalothorax, the only part of her body not buried in the host. Sperm pass through a brood canal into the body cavity where 2000 to 7000 fertilized eggs develop larviparously. The first instar is a triungulin larva with sharply pointed head, backward pointing bristles and slightly adhesive pads on the foretarsi. They emerge forcibly from the maternal host, then must overcome tremendous odds against survival by finding their own host insects, which must be in the larval stage. Those that parasitize Hymenoptera hitch-

hike on the maternal host to her nest. Those that parasitize insects such as Homoptera or Orthoptera drop to the ground or are brushed off on flowers. Those successful in finding a proper host immediately burrow in between abdominal segments, where subsequent instars are apodous and endoparasitic. Metamorphosis is similar for males and females during the first three instars, but divergent thereafter. Females become reproductive in the last larval instar, never passing through the pupal instar or becoming true morphological adults. Males complete normal holometabolous development, the seventh or pupal instar being enclosed by the sloughed skin of the sixth instar, and the fully developed adult leaves the maternal host immediately.

Parasitism by Strepsiptera is not usually lethal to the host insect but results in abnormalities known collectively as "stylopidization" from the large and well-known genus *Stylops*. The females are extremely successful parasites, never directly destroying any of the host organs, but an infestation can have damaging results by attrition. The forcible emergence of the adult males always leaves some tissue damage to the host where a focal infection or fungal growth may start. The effects may include retarded development, loss of external sexual characters, loss of vitality and sterilization. The family Stylopidae parasitize various Hymenoptera, especially bees. Males of family Myrmecolacidae attack ants while their females parasitize Orthoptera. Species of Halictophagidae parasitize crickets, cockroaches or Homoptera, and the Elenchidae parasitize only Homoptera. Adult males range in body length from 0.1 mm to 4.0 mm. Antennae vary from four to seven segments, and the third always is flabellate; the entire appendage usually is covered with sensoria and minute hairs. The eyes are prominent and raspberry-like, composed of from 20 to 50 ocelli. The mouthparts, reduced to the mandibles and the two-segmented maxillary palpi, are not used for feeding but probably are of assistance in emerging from the maternal host. The forewings are reduced, twisted paddle-like elytra and have been compared to the halteres of Diptera because they have been observed to be in constant, whirling motion during flight. The relatively large, membranous hindwings have reduced radial venation. Legs are isomerous with stout femora, slender tibiae and the tarsi may have five subsegments and post-tarsal claws (Mengeidae) or be reduced to as few as two subsegments (other families). The abdomen is apparently

10-segmented, the tenth forming a dorsal flap over the ninth, in which the aedeagus is pocketed. The aedeagus varies with the species and is an important diagnostic character.

Identification of the female is based mainly on the formation of the cephalothorax, the number of spiracles, the shape of the opening of the brood canal and the number of genital pores, but larviform females unassociated with males are now generally regarded as of doubtful taxonomic value. Males are identified by means of the more distinctive adult characters present. Triungulin larvae may be distinguished by the tarsi, segmentation and arrangement of bristles and ocelli.

Some study has been made on the role of the Strepsiptera in biological control of their hosts, but the relative scarcity of the parasites and the fact that few of the host species are pests suggests that, on the whole, the twisted-winged insects have little or no economic importance. The only possibility appears to be in effecting some deterrence to certain leafhoppers pestiferous on various grain crops (Graminaceae) and sugar cane.

JEAN W. FOX
RICHARD M. FOX

SUCCESSION

That a sequence of changes of vegetation occurs on any area which has been disturbed by agencies such as fire, lumbering, or agriculture is apparent even to a casual observer and was noted in various writings hundreds of years ago. There are several detailed 18th century accounts of vegetational changes associated with bogs forming from lakes, but the writers saw no more than a special situation. As early as 1792, the botanist Willdenow described precisely the colonization of bare rock by pioneer plants, their influence on the habitat, especially in building soil, and the resulting slowly successive, predictable invasion by other species which could go on until the community was finally similar to the vegetation of the region. However, the implications of these observations were ignored. Recognition of progressive and orderly change in all communities of organisms did not come until much later.

Nineteenth century literature frequently refers to *succession* but in an incidental fashion which shows no real appreciation of the universality of the phenomenon. Even when Kerner (1863), describing the vegetation of the Danube Basin, did, for the first time, recognize the significance of succession in all communities, his ideas were passed over. It was the work of Cowles (1899, 1901) on the Lake Michigan sand dunes that clarified the meanings and implications of succession and stimulated interest in its study. At about the same time Clements, working in Nebraska, was formulating ideas that culminated in his classic publication (Plant Succession, 1916) which has influenced concepts and thinking to the present day.

Because plants are more obvious than animals and because the latter are commonly dependent on the vegetation, succession is usually considered in terms of plants. Yet animals of many kinds and sizes are invariably associated in definite relationships with the plants. Thus, vegetational change is paralleled by biotic changes.

The concept of primary succession assumes first that in any area, there are, or once were, unpopulated habitats which, for the general climate, range from relatively dry (*xeric*) to wet (*hydric*) moisture conditions. As extreme as these habitats may be, there are invariably some pioneer organisms especially adapted to invading the bare areas and becoming established in them. Mosses and lichens initiate succession on bare rock and submerged vascular plants appear first in aquatic habitats.

Once established, the pioneers react upon the habitat to make it less extreme. Isolated plants on rock increase in numbers, hold dust and washed down mineral material, and slowly build mats of which their organic remains form a substantial part. Similarly, in the pond or lake, the hydrophytes check currents, hold sediments, and help build up a substrate as their partially decomposed remains accumulate. Such habitat changes accomplished by the pioneers eventually create conditions favorable to other plants and animals. When they come in, as they invariably do, they gradually exclude the pioneers by competition and a second stage in succession has been accomplished.

Biotic succession of this type occurs everywhere at differing rates and with a variety of stages possible. In habitats less extreme than bare rock or open water, the invaders are less specialized and the rate of succession is more rapid.

The general trend of succession is invariably toward relative mesophytism among the organisms as the habitat becomes more mesic. Thus the moisture extremes of both hydric and xeric environments tend to be reduced and the kinds of organisms in these sites become more similar until, in late stages of succession, when theoretically at least, the habitats might become very similar, the species would also tend to become identical. Similarity of successional trends and their tendency to converge toward a particular type of community is characteristic of every climatic area.

This relatively mesophytic community toward which all successional trends lead is termed the CLIMAX, partially because it terminates succession and partially because it is related to climate. It is terminal because there are no other species available which can replace it and because its species can reproduce and maintain themselves in competition with each other.

Succession is often interrupted, sometimes dramatically as by fire, lumbering, or agriculture. If then there is no further disturbance, the subsequent recovery of the vegetation is again successional (secondary succession). Although numerous species unrelated to primary succession may appear immediately after the disturbance, they are soon superseded and the regular sequence of succession is reactivated. In fact, if the disturbance has not been too radical, the resulting succession toward climax may proceed more rapidly than it would have without the interruption.

In a given climatic area the tendency for all successional trends to converge toward a single climax community is real. Nevertheless, many localized habitats, because of parent material, soil, or physiography, may have a greater controlling influence on vegetation than the general climate. Here, succession may proceed to a community with the stability or equilibrium of climax but made up of quite different species which can maintain themselves indefinitely. To apply the monoclimax interpretation, special terminology is used to recognize these localized situations (postclimax, pre-climax, subclimax, etc.). Increasingly though, a poly-climax interpretation is applied and these communities are referred to directly as climaxes although often with a descriptive term (*physiographic, edaphic, pyric*) related to the major controls.

The classic concept of succession is one in which biological influences, reaction of organisms on the habitat, result in environmental changes paralleled by changes in the community occupying a site. But, environmental modification also comes with physiographic change as when a stream cuts more deeply into a flood plain or a bluff erodes. Then too adjustments in the biota occur and the progressive changes constitute succession. Finally, it is well known that climates have changed in the past and that the vegetation of extensive regions become modified accordingly. Such climatically controlled changes, slow and scarcely perceptible, are nevertheless also involved in succession at any time and undoubtedly are going on today.

Succession, then, is universal, implemented always by biotic reaction, often influenced by physiographic change, and generally controlled by climate which itself may be changing.

HENRY J. OOSTING

References

Clements, F. E., "Plant Succession: An Analysis of the Development of Vegetation," Washington, Carnegie Inst. Wash. Publ. No. 242, 1916.
Cowles, H. C., "The ecological relations of the vegetation on the sand dunes of Lake Michigan," Bot. Gaz., 27: 95–117, 167–202, 281–308, 361–391, 1899.
Oosting, H. J., "Study of Plant Communities," 2nd ed., San Francisco, Freeman, 1956.

SUCTORIA

The class Suctoria was originally established to include those PROTOZOA which are non-ciliated and without locomotor organelles in the adult, being attached to the substrate either directly or by means of a stalk. Although the majority of these Suctoria are free-living, some have been reported as being endoparasitic (e.g. *Sphaerophrya* in *Paramecium* and *Stentor*) and others (*Ophryodendron faurei* on *Psamathe longicaudata*) as ectoparasitic. Their close affinity to the CILIATA is indicated in several ways: the possession of dimorphic nuclei, nuclear exchange by means of conjugation, and the presence of cilia in the embryo. The exact relationship is still in doubt: in primitive genera, such as *Podophrya*, the cilia are arranged in longitudinal rows similar to those of the Holotrich ciliates, while the presence of cilia arranged in circlets (characteristic of most genera) indicates a closer affinity to the Peritrich ciliates.

In form the Suctoria may be rounded, pyramidal, or branched. There is no mouth; food is captured and ingested by means of knobbed or pointed tentacles. In most species, each tentacle is made up of two parts: an outer contractile sheath and a stiff, needle-like tube. The type and arrangement of the tentacles varies: in some (e.g. *Acineta*) they are arranged in terminal groups, while in *Paracineta* and *Podophrya* they are distributed over the distal surface.

When one of these tentacles comes in contact with a suitable food organism (*Paramecium, Colpidium*) the tip of the tentacle adheres firmly. A single individual of *Tokophrya infusionum* has been observed to feed simultaneously on ten or more ciliates, each of which was completely devoured in 25 minutes. Within a few minutes the cilia of the prey (in the region of contact) stop

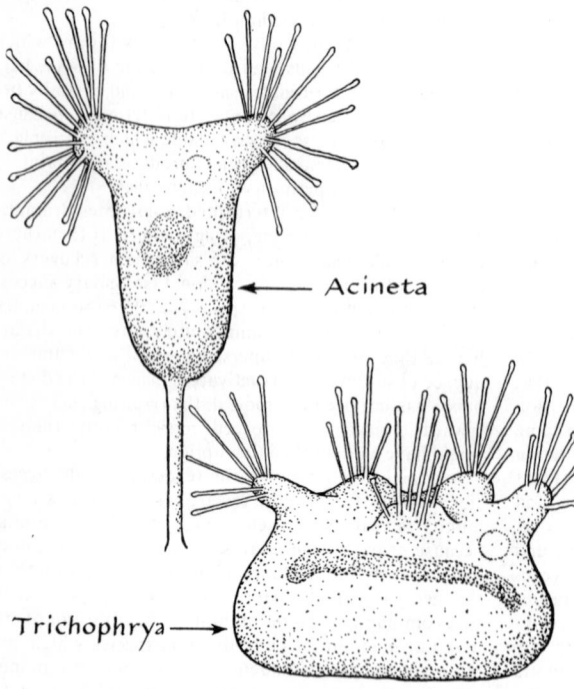

Acineta

Trichophrya →

Fig. 1 Representative suctorians. (From Platt and Reid, "Bioscience," New York, Reinhold, 1967.)

beating. This paralysis of the cilia spreads gradually over the surface of the prey, indicating the secretion of a toxic substance. The protoplasm of the food organism is then broken down in the area of contact, and is drawn up thru the hollow portion of the tentacle. This process suggests that digestion is, in part at least, extracellular.

Reproduction is by means of budding, either endogenous or exogenous. In exogenous budding, the outer surface of the Suctorian invaginates, forming a pouch which soon becomes ciliated on its inner surface. At the same time, the single elongated or branched macronucleus undergoes division, and several new vacuoles are formed. Following this, the pouch evaginates (the cilia are now external) and one of the newly formed macronuclei, along with one of the micronuclei, vacuoles, and protoplasm, flows into the "bud." This bud then pinches off and swims away as the ciliated embryo. In endogenous budding, the bud is produced within a temporary vacuole, the "brood pouch." This first appears as a large vacuole in the apical region of the body. After a short interval, the vacuole encompasses one of the micronuclei, together with some cytoplasm and a small portion of the macronucleus. The embryo thus formed becomes ciliated, and is ejected to the exterior. After a variable period of free-swimming activity, the embryo becomes attached, loses its cilia, and reaches the adult stage in a few minutes.

JAMES M. CRIBBINS

References

Canella, M. F., "Studi e recherche sui Tentaculiferi (Suctoria) nel quadro della biologica generale," Ann. Univ. Ferrara, 1: 259–716, 1957.

Guilcher, Y., "Morphogénèse chez l'Acinétien, *Ephelota gemmiparva*," Bull. Soc. Zool. France, 73: 24–27, 1948.

ibid., "Sur quelques Acinétiens nouveaux ectoparasites de Copepodes Harpacticides," Notes et Rev. Arch. Zool. Exptl. et Gén., 87: 24–30, 1950.

Kitching, J. A., "The physiology of contractile vacuoles," J. Exp. Biol., 28: 203–214, 1951.

Rudzinska, M., "The influence of amounts of food on the reproductive rate of *Tokophyra infusionum*," Science, 113: 10–11.

SULFUR BACTERIA

The sulfur BACTERIA do not form a homogeneous morphologic or physiologic group of microorganisms. Morphologically, the sulfur bacteria are placed in four Families of the Schizomycetes Class. Three Families, Beggiatoaceae, Thiorhodaceae, and Chlorobacteriaceae, are comprised of filamentous forms which closely resemble the lower fungi, and are referred to as the "higher bacteria." The organisms are heterotrophic, using both organic and inorganic materials as sources of carbon. Sulfur granules usually accumulate within the cells as a result of metabolic processes. The Thiorhodaceae and Chlorobacteriaceae Families are endowed with pigments, making possible a photosynthetic metabolism. The fourth Family, Thiobacteriaceae, morphologically indistinguishable from the usual bacterial forms, are chemoautotrophic bacteria. They derive energy from the oxidation of sulfur or sulfur compounds, and carbon from carbon dioxide only.

The Beggiatoaceae Family is composed of four Genera separated on a morphological basis: Beggiatoa, Thiospirillopsis, Thioploca, and Thiothrix. Beggiatoa is the best established Genus, and probably is the most prevalent of the filamentous sulfur bacteria. Species are found in both fresh-water and marine environments containing hydrogen sulfide and soluble sulfides. The cells occur in unattached, motile, segmented trichomes. The trichomes move over solid substrates in a slow, gliding, undulating motion. Though occurring in white to cream-colored masses, the trichomes retain their individuality. In Thiospirillopsis, the trichomes are wound spirally, and move with a slow corkscrew-like motion. The trichomes of Thioploca are in braided bundles enclosed in a wide slime-sheath, and move with the characteristic gliding motion. The Thiothrix genus is non-motile, and the trichomes are differentiated into base and tip. These structures are attached to solid surfaces by gelatinous hold-fasts.

The Thiorhodaceae Family consist of microorganisms that produce a pigment system composed of green bacteriochlorophyll and yellow and red carotenoids. The pigments make possible a photo-synthetic metabolism. Under anaerobic or microaerophilic conditions carbon dioxide is reduced without the liberation of molecular oxygen. Hydrogen sulfide usually serves as the hydrogen donor, but may be replaced by many organic substances. Sulfur accumulates as droplets within the cells. Single cells, unless very large, appear colorless, but aggregates of cells appear bluish-violet, pale purple, and brownish to deep red. The size and shape of the aggregates separates the Family into thirteen Genera.

The Chlorobacteriaceae Family is comprised of microorganisms referred to as the green bacteria because they contain green pigments. The pigments are of a chlorophyllus nature, though not identical with plant chlorophylls nor with bacteriochlorophylls. The organisms are capable of photosynthesis in the presence of hydrogen sulfide, but do not liberate oxygen. Individual cells are small and occur singly, or in cell masses of various shapes and sizes. Sulfur is not contained by the cells, but is deposited in the elemental state outside the cells. Owing to their symbiotic relationship with other microorganisms, the validity of the classification of this Family is doubtful.

The Thiobacteriaceae Family consists of five genera separated on a morphological basis. The cells are coccoid, straight, or curved rods, but are never filamentous. These are colorless sulfur bacteria, and are sometimes embedded in gelatinous pellicles, or in bladder-like colonies. Sulfur and sulfur compounds are oxidized for energy. In some genera, sulfur granules are deposited inside or outside the cells. All of the genera inhabit places where hydrogen sulfide occurs and may oxidize elemental sulfur, polythionates, and related compounds.

Thiobacillus, the best established Genus of the Thiobacteriaceae Family, is a small, rod-shaped cell. Energy is derived from the oxidation of elemental sulfur and thiosulfate. In some cases, energy is obtained from sulfide, sulfite, and polythionates. The principal product of the oxidation is sulfate, but sometimes sulfur is formed. Growth of individual species occurs under acid or alkaline conditions. Some species are facultatively autotrophic, using carbon dioxide or bicarbonate as sources of carbon. Other species, *Thiobacillus thioparus, T. ferrooxidans, T. concretivorus,* and *T. thiooxidans,* are strictly autotrophic, utilizing carbon dioxide as the sole source of carbon. Sulfur or sulfur compounds are oxidized to sulfate for energy. *T. thiooxidans* lives and grows in an environment more acid than can be tolerated by any other microorganism.

Bacteria which oxidize elemental sulfur in addition to iron are designated as *Ferrobacillus sulfooxidans,* while those which oxidize thiosulfate as well as iron have been placed in the genus *Thiobacillus.*

WILLIAM W. LEATHEN

References

Breed, Robert S., *et al.,* "Bergey's Manual of Determinative Bacteriology," 7th ed., Baltimore, Williams and Wilkins, 1957.

Ellis, D., "Sulfur Bacteria," New York, Longmans, Green, 1932.

Kinsel, N. A., "New Sulfur Oxidizing Iron Bacterium: Ferrobacillus Sulfooxidans Sp. N.," J. Bacteriol., **80:** 628–32, 1960.

van Neil, C. B., "The Bacterial Photosyntheses and Their Importance for the General Problem of Photosynthesis," Adv. Enzymology, **1:** 263–328, 1941.

van Neil, C. B., "The Chemoautotrophic and Photosynthetic Bacteria," Ann. Rev. of Microbiol., **8:** 105–132, 1954.

Vishniac, W. and M. Santer, "The Thiobacilli," Bacteriol. Rev., **21:** 195–213, 1957.

SWAMMERDAM, JAN (1637–1680)

A Dutch naturalist and pioneer microscopist, Jan Swammerdam was born on February 12, 1637, at Amsterdam. His father was an apothecary and naturalist. He obtained a medical degree at Leyden in 1667, but disregarded his practice because of his preference for microscopic investigation.

Swammerdam made extensive studies of insects, classifying them according to their type of metamorphosis. He examined the internal structure of the mayfly and bee with great detail. He described the cleavage of the frog's egg and its subsequent development. In 1672 he observed the ovarian follicles of mammals. He was the first to illustrate fern sporangia. He is also credited with the discovery of the valves of the lymphatic system and the first description of the red blood corpuscles. He devised an experiment which showed that contracting muscle tissue changes in shape but not in size, contrary to the popular theory that movement is caused by the flow of fluid through the nerves.

His unstable personality led Swammerdam to turn against his father and become a temporary member of Antoinette Bourignon's religious reformation movement. After a period of severe physical and spiritual misery, he died in Amsterdam on February 15, 1680.

JAMES L. OSCHMAN

SYMBIOSIS

Symbiosis is a mode of living, characterized by intimate and constant association or close union of two dissimilar organisms. Although the definition does not specify whether the results should be beneficial, indifferent, or harmful to either partner, authors have tended to restrict the word to only mutually beneficial associations. Paradoxically, the estimation of mutualism is often based merely on the criterion of constant association. Truly mutual physiological dependence is sometimes difficult to prove. Etymologically, the word "symbiote" is derived from the Greek "symbiotes" and is correct, while "symbiont" has no original and its use should be discontinued.

Organisms living in symbiosis cover a wide range of plant-plant, animal-animal, or plant-animal associations. The classic is algae within a fungus producing a lichen. This terminology is exceptional because most partnerships are not designated by a new name. Usually if one organism is microscopic, the partnership is known by the larger of the two, the host or macrosymbiote.

Students of lichens have proposed the term *consortium* to emphasize partnership. Actually, the ideal of mutual benefit is subject to question. While the fungus provides moisture and protection for the alga, and the alga synthesizes carbohydrates for absorption by the fungus; nevertheless it has been demonstrated that hyphae penetrate algal cells to the detriment of the latter. Thus the partnership may be "controlled parasitism."

The majority of lichens are ascomycetes combined with Chlorophyceae; but three tropical genera are basidiomycetes with Myxophyceae (Cyanophyceae). The fungal components of lichens are found only in lichens but the algae are the same as free-living aerial ones. *Protococcus* is the most common unicellular green alga in lichens.

Mycorhizae, found in practically all bryophytes and vascular plants, are homologous with lichens. The mycelia of basidiomycetes or phycomycetes form either a dense mantle over the surface of the roots or inhabit the protoplasts, in either case with hyphae extending outward into the soil. There is a deficiency of at least one nutrient in most natural soils. In these the short roots are invaded by fungi which take over the absorptive functions. This stimulates the roots to repeated branching and thus the mycorhizae become efficient absorptive organs.

Various species and strains of *Rhizobium* bacteria enter the roots via root hairs of most legumes as well as a number of non-leguminous plants. Nodules are formed around the bacteria, which then convert atmospheric nitrogen into amino acids utilized by the plant host. This is obviously a facultative mutualism, because under anaerobic conditions the bacteria are parasitic; and in the presence of available nitrogen, nodule formation is weak. Nitrogen-fixing bacteria also form colonies in the leaves of numerous non-leguminous plants.

Soil microorganisms exert much influence over one another antibiotically and symbiotically, benefits accruing from the performance by different species of various steps in the nutritional chain.

Most animal symbioses are between marine invertebrates. Whether both animals are locomotory or one is sessile, the establishment of each partnership invokes highly evolved and adaptive behavior. Most of the symbiotes live at least part of their lives independently. Of some species, only a fraction of the individuals ever become symbiotic; so research emphasis has been placed on the behavior of the animals rather than on the consequences of the association. Since in the majority of cases, the ultimate partners are born in spatial segregation, the first step toward a relationship is searching behavior. The microsymbiote may search for its host or the host may select the appropriate microsymbiote from among similar ones. Sometimes three different species unite. Initially significant is the recognition of the habitat environment of the sought-for partner. Chemical attractants as well as physical stimuli, including thigmo-, photo-, and geotaxes, elicit highly evolved responses. Such behavior patterns have been likened to the releasing mechanisms of social animals.

The degree of physical contact of the symbiotes varies widely. In the most intimate, polychaete annelids or nemertean worms enter and feed in the gastric cavities of starfish or anemones without being digested; worms dwell in the ambulacral grooves of starfish; while peacrabs live in the tubes of annelids, the mantle cavity of mollusks, or the cloaca of sea-cucumbers. Some worms steal food from the mouths of commensal crabs without being captured. Crabs arm themselves by carrying anemones in their claws or on their backs. At critical times in the life-cycle the relationship is interrupted and must be re-established. Whenever the hermit crab changes to a new snail shell after outgrowing the old one, the anemones must be transferred to the new shell.

Certain fishes live unharmed among the lethal nematocyst-bearing tentacles of anemones or jelly-fishes; while other fishes stand on their heads among the spines of sea-urchins. In such situations the more active animal catches food, dropped particles feeding the sedentary partner; while in return protection from predators is gained from living among the tentacles, etc.

Hazardous conditions are survived because of immunity to attack. Immunity from stinging coelenterates or pinching crabs seems to result from specific avoiding behavior. Moribund or dead animals are immediately attacked. How immunity to digestive juices is achieved is unknown. In this connection, the well-known role of protozoa in the digestion of ruminants is merely mentioned. Flagellates in the intestines of insects are discussed below for convenience.

The third and most versatile group of symbiotes is comprised of algae, yeast or bacteria in combination with animals. Algae were observed a century ago to be responsible for the green or yellow color of marine and freshwater protozoa, especially amoebae and ciliates, and certain coelenterates, turbellaria, and sponges. Early investigations on these organisms established the fact that there is no "animal chlorophyll" and, as a result of observing the method of transmission through egg-infection, laid the basis for studying symbiosis with yeasts and bacteria.

Most algae in protozoa are of the Chlorella type, but there are some blue-greens scattered throughout the flagellates, ciliates, and sarcodina. The associations are seasonal; so that the animals may be green, yellow or colorless at various times of the year. The particular tissues in which the algae are located vary with different species.

A lamellibranch mollusk, *Tridacna,* exemplifies algal symbiosis. The edge of the mantle, so thickened and enlarged that the shell can no longer close, lies in folds over the edge of the shell. As long as the mantle is exposed to light, lacunae in the vascular system contain masses of brown zooxanthellae, each individual enclosed in a much expanded amoebocyte. The zooxanthellae possess only a very thin membrane and in their plasma, oil droplets and much starch. The excess algae are taken to phagocytes accumulated between the intestinal diverticulae and the intestinal canal, where the algae are dissolved.

In cases where animal nutrition is changed from carnivorous to algal symbiosis, life habits must be changed. The animal must live on the edges of coral reefs or in such locations as to permit adequate illumination to reach it while anatomical modifications must insure exposure of the algae. Experimentally starved host-animals can survive longer in light than in dark. Likewise, the release of oxygen within the tissues of an aquatic animal allows it to live in shallow, stagnant water that would be an unfit habitat without algae.

Of 128 species of photogenic cephalopods, only Myopsid squid of the family Sepiolidae possess demonstrable luminescent bacteria. These are located in accessory sex glands or in these glands as well as in light organs. Bacteria of some squids which have been cultured without difficulty represent several kinds of bacilli, vibrios, cocci, and coccobacilli. There is considerable confusion of these bacteria with external luminous contaminants. Whether or not the true symbiotes are transmitted from generation to generation through the egg is questionable. Luminous bacteria of accessory gland secretions cover the eggs when they are laid, but the bacteria are not found in the fluids or tissues of the embryos. In some juvenile animals the tubes of the accessory glands apparently become infected by bacteria accompanying detritus of sea-water.

Although many cartilaginous and bony fishes are luminous, only six or eight families of teleosts owe their photogenic properties to symbiotic bacteria. Beneath the eyes of *Anomalops* and *Photoblepharon,* primitive, dark brown fish inhabiting medium deep seas, there are large light organs. The ducts of the organs contain motile, rod- or granular-shaped bacteria, which are often united in spirillum-like chains. The continuous light given off by the bacteria can be voluntarily shut off by the fish which either rotate the light organ into the eye socket or cover it with a dark fold of skin. The eggs are not luminous and the means of infection of the embryo is unknown. The bacteria have been cultured on peptone-agar but did not produce light.

Bacterial luciferin and luciferase, evidently chemically different from those in other systems, have been demonstrated only recently. The reported glow of many animals is caused by infections with pathogenic bacteria or by saprophytes growing on decaying flesh.

There are many variations among insect symbiotes. Species of ants, termites, and wood-boring beetles, with complex behavior and probably the ability to excrete antibiotic substances, inoculate and maintain pure cultures of particular fungi in their nests or tunnels. The fungi are eaten but never depleted. Microorganisms harbored internally are differentiated on the basis of extracellular or intracellular locations. Extracellular organisms are in the lumen of the gut or Malpighian tubules, or in pouches or diverticulae of the gut. Intracellular organisms are in gut epithelium, gonads, or highly differentiated cells (mycetocytes) or organs (mycetomes) in the fat body or hemocoele. Intracellular bacteria are most common, found in about 25% of the orders of insects; but scattered throughout so that one can not discern an evolutionary trend. Intracellular yeasts, less numerous, occur in some of the same orders and families. Bacteria and yeasts are abundant in Coleoptera and Homoptera.

In general, intracellular symbiotes enter the ovaries (or testes in rare cases) and infect the embryos, although less frequently special vaginal pouches simply smear symbiotes over the surface of eggs as they are laid. In the latter case, hatching larvae become infected while gnawing through the egg shells. When the symbiotes are in the lumen of the gut, infection of the next generation is accomplished by feeding on fecal pellets or contaminated detritus.

Intracellular microorganisms appear comparable to parasitic and free-living species; but most have not been identified because of the difficulties of culturing them. They are usually non-motile and are engulfed and carried to the definitive site by mycetocytes in both embryos and adults.

One generalization can be made about insect symbiosis. Nearly all symbiotic insects feed on restricted and incomplete diets, such as hair, feathers, wood, plant sap, dried grains, or vertebrate blood. Insects that are omnivorous, carnivorous, leaf-feeders, or blood-suckers only in the adult stages lack symbiotes. It seems, therefore, that symbiotes perform a nutritional function. Yet, three groups of insects—cockroaches, ants, and the primitive termites—are general feeders and possess elaborate symbiote systems. Nevertheless, the nutritional theory is upheld in experimental work on cockroaches.

Feeding antibiotics or diets deficient in trace minerals, or simply surface-sterilizing eggs, have been fruitful methods of experimentally eliminating symbiotes. Nearly all of the aposymbiotic insects resulting need dietary supplements to survive, let alone grow. The reproductive ability is also impaired; speculatively, this may reflect a hormone imbalance.

The flagellates of wood-roaches and termites break down cellulose to utilizable sugar for their hosts; the bacteria of the blood-sucking bug, *Rhodnius*, synthesizes B-vitamins as do the yeasts of Anobiid beetles; and there is evidence that the symbiotes of aphids fix atmospheric nitrogen in nitrogen-free culture media or otherwise utilize excretory nitrogen. The bacteria-like symbiotes of cockroaches probably are involved in synthetic metabolism of nutritional intermediates and mobilization of excretory wastes.

The ultimate goal is to assess the consequences of eliminating symbiotes, then re-establish the normal functions by re-infection with the cultured (and identified) microorganisms.

MARION A. BROOKS

Reference

Henry, S. M. ed., "Symbiosis," 2 vols., New York, Academic Press, 1966–1967.

SYMMETRY

In living beings symmetry defines the geometric relationships that exist in their structures. It indicates the disposition and harmony of the parts. Living things, animal or vegetable, always present a certain regularity of form and, therefore, a certain amount of symmetry.

Various types of symmetry are distinguished in the living. In spherical symmetry of the body there is a symmetrical center; every plane that passes through the center divides of body into symmetrical halves. Some Protozoa approach this type of symmetry. More frequently, as in the case of the Coelenterata, there is a symmetrical axis and through it pass a variable number of planes, each of which divides the body into symmetrical halves (radial symmetry). In doubled, bilateral symmetry there are two symmetrical planes, the one perpendicular to the other. The Ctenophora are an example of this type. In simple, bilateral symmetry (called more precisely "bilateral symmetry") there is only one symmetrical plane that divides the body into symmetrical halves (antimeres). The Bilateria consist of animals of the primary, bilateral type of symmetry. Such symmetry in some cases is lost, or substituted by another type of symmetry. To the Bilateria belong all of the animals superior to the Coelenterata and the Ctenophora: they are animals that move themselves actively. Finally, there are always some asymmetrical organisms.

The most interesting type of symmetry, most important, and also most diffuse, is bilateral symmetry. This type is the one with which this article shall deal.

Any given figure and its reflected image in a flat mirror are in bilateral symmetry (one says also that they are in mirrored symmetry). If the given figure is absolutely asymmetrical, that is if no plane exists that divides it into symmetrical halves, then the figure itself and its reflected image in the mirror are enantiomorphic (two bodies that are not superimposable). The term enantiomorphism, in its strictest sense, signifies contrary form. Looking at the right hand in the mirror, one sees a left hand. The right hand and the left (the hand is completely asymmetrical; one cannot superimpose the right hand on the left) give a typical example of enantiomorphism. In the bilaterally symmetrical organisms, there are paired organs, like the limb and the ear, that are asymmetrical in themselves and are mirrored in the corresponding organs of the other side. They are enantiomorphic. One might say the same thing for the two halves of the body. Thus, one usually speaks of bilateral, or mirrored, symmetry in the sense of enantiomorphism.

Pasteur (1884) discovered optical isomerism and noted that it was a question of the forms that were located between them, like the right hand and the left. They are called optical opposites, or enantiomorphic forms, and they have equal and contrary optical activity.

An examination of some research projects on symmetry and asymmetry which have deepened our knowledge of the subject proves valuable.

The origin of symmetry and asymmetry has not yet had a clear explanation. It has been observed that in some cases its physiological base is already present in the egg before fertilisation. Sometimes in the unfertilised egg the bilateral symmetry is recognizable because in it there is an oblong, pigmented part, but at this plane of symmetry one may superimpose another one of them determined by the point of penetration of the sperm. In other cases (*e.g.*, the frog) the bilateral symmetry is established with the fertilisation (the symmetry being normally dependent on the point of en-

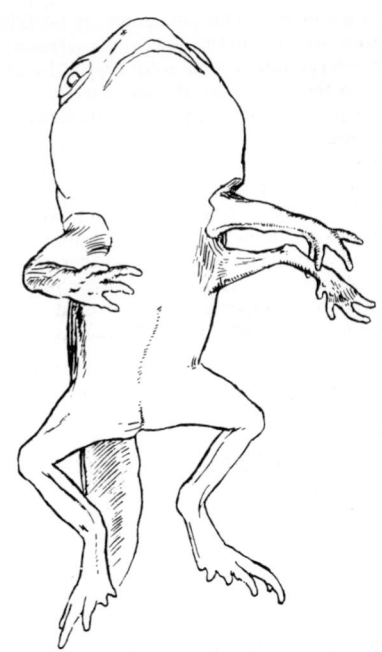

Fig. 1 *Bufo vulgaris*. The induction of an anterior supernumerary right limb on the left side of the body. Mirrored symmetry between the normal limb and the supernumerary one. (Design) (From Perri, Experientia xii, 1956.)

trance of the sperm) and becomes more stable with the appearance of the grey cresent.

In gynandromorphism, asymmetry between the halves of the body finds its cause (as generally thought) in the irregular transmission of the sexual chromosomes in the first two blastomeres. We find an example of predetermination in the water snail, *Limnea peregra*. In this species there are sinistral and dextral individuals. Dextrality appears to behave as dominant to sinistrality. If a dextral individual is homozygous in regard to the recessive gene of sinistrality, the one produces all sinistral offspring. In this case the maternal genes determine, before maturation, the type of asymmetry that will begin to appear during the early cleavage divisions of eggs. This case and that of gynandromorphism are examples of the genetic control of asymmetry.

All of the Vertebrates have symmetrical and asymmetrical organs. Along the latter, for example, are the heart, the intestinal tract, and the pancreas. They are always of the same asymmetry and only rarely individuals with inverted asymmetry occur. In the stage of young neurula in amphibians, Spemann removed a piece of presuntive neural tube material together with the underlying gutroof and replanted it after having rotated its anteroposterior axis 180°. In this manner the orginal right part of the piece was found on the left. The embryo developed normally but presented *situs inversus*. In other words, the heart, the intestine etc. were in abnormal asymmetry. In the only removal of the piece, that is to say without successive replanting, Spemann did not obtain any other asymmetrical inver-

sions. Similarly, if only the presumptive material of the neural tube turned, that is to say without the underlying gut-roof turning, the normal asymmetry remained. Therefore, the agents that control asymmetry are distributed in the archenteron.

In the embryos at the tail-bud stage, the rudiment of the heart can be divided into halves. The left half of the rudiment forms a heart with normal asymmetry. Instead, the right half develops itself with inverted asymmetry. It is notable that Elkman's experiment demonstrated that the *situs inversus* develops only in the right half.

In the blastula stage of amphibians (Spemann and Falkemberg) with constriction in the plane of bilateral symmetry there were cases of *duplicitas anterior*. In the left member there was normal asymmetry, while in the right there was usually *situs inversus*. When the constriction was complete (Ruud and Spemann) the left half had developed into an embryo with normal asymmetry while that originating from the right half presented in half of the cases *situs inversus*. From all of these experiments of constriction and from those on the division of the heart it follows that the left half has its laterality stably determined while the right half has it still undetermined. In other words, with the asymmetrical inversion in the right member, there is established a mirrored symmetry (between the asymmetrical organs) between one member and another. And, therefore, this inversion is also a result of the tendency toward mirrored symmetry.

If the constriction is effected at the two-cell stage (Mangold) it has no asymmetrical inversion. That can depend on the regulative conditions.

Johannson described a cow with *duplicitas anterior* that arrived behind the anterior limbs, in which there was clear symmetry in the coloration of the skin.

Let us examine some experiments on the paired organs that were conducted on the limbs, which lend themselves to experimentation better than other subjects. The hands, or the feet, as previously stated, are enantiomorphic. In a limb three axes of polarity are distinguished: anteoposterior, dorsiventral, and mediolateral, that in a given moment, are parallel to the axes of the embryo. Harrison followed his experiments on urodele *Amblystoma* at the tail-bud stage. He transplanted the rudiment of the limb with one or more inverted axes, either on the same side or on the opposite one. Granted that Harrison calls a limb harmonic when it is of the same laterality with the side of the body on which it lies (for example, a right limb on the right side of the body); further, he calls a limb disharmonic when it is laterally opposed to the side of the body on which it lies. (for example, a right limb on the left side of the body) Following this train of thought, in the limb transplants without inversion of the anteoposterior axis, he had harmonic limbs. When, instead, he inverted the anteoposterior axis, he obtained disharmonic limbs. The cause is that the anteoposterior axis is already fixed, while the other two axes are not yet fixed and are imposed by the embryo. The anteoposterior axis becomes determined rather quickly. Detwiler at the middle gastrula stage in *Amblystoma* removed the presumptive limb area and transplanted it on a neurula with the anteoposterior axis inverted. A limb was developed with inverted laterality demonstrating in this manner that the determination of this axis is contemporary with the determination of the limb. The other two axes of the limb are fixed during the advanced embryonic life. When all three axes are definitely fixed, no other inversion of laterality is possible. In order to impress on the limb the dorsiventral axis and the mediolateral axis, the tissues that surround the limb-bud are important, and not the whole embryo. In other words, in the transplanting of the limb-bud in the somite zone, there are no other inversions of the limbs and the limbs develop according to their origin and thus demonstrate that the dorsiventral and mediolateral axes already have a weak fixation.

Whether transplanting a limb-bud in the *Anura* or in the *Urudela*, one has often the development of two limbs. These are in mirrored symmetry. Similarly, following an amputation of a limb, sometimes with regeneration of two, one finds these limbs always in mirrored symmetry with one another.

Bateson (1894) examined a large number of cases (in Vertebrates and Invertebrates) with supernumerary appendages (palps, antennae, limbs). He observed that the supernumerary appendages are usually two. They are in mirrored symmetry with one another and this, he says, depends on the fact that there are a pair of appendages. Of the two supernumerary appendages, that closest to the normal appendage is also in perfect mirrored symmetry with it. These are, basically, Bateson's rules of mirrored symmetry. He observes that it does not verify itself mathematically but in its essence. Bateson's rules were confirmed anew by Przibram and others. The exceptions are most rare.

Bateson's rules of mirrored symmetry have been verified largely in the research on supernumerary limbs by abnormal induction. Balinsky, in experiments involving the transplanting of otocysts in *Triton*, had several cases of the formation of well-formed supernumerary limbs. The somatopleure stimulated by this transplant gave origin to such limbs. The same results, always in *Urodela*, were had with the transplanting of other rudiments. In the *Anura* Perri obtained in various species (above all in *Bufo vulgaris*) a large number of cases with supernumerary limbs. He used the transplanting of some cephalic parts of embryos, strongly X-rayed, on the flank of a normal embryo in the tail-bud stage. These buds, because of the radiation undergone, go into necrosis and free cytolitic substances. These are the ones that, existing in substantial amounts, stimulate the somatopleure and thus, conforming to the competence of this last, has given the skeleton one or more supernumerary limbs. Well-shaped limbs are often developed with very clear laterality and thus lend themselves readily to the study of symmetry. When one has developed a single supernumerary limb, one knows that it is disharmonic (Fig. 1). When there are two (or one with duplication) with laterality easily recognizable in both, one finds them, without exception, in mirrored symmetry with one another. The limb closest to the normal limb (to the anterior if the supernumerary limbs are anterior, posterior if posterior) is in mirrored symmetry with it also. If, for example, the operation has been performed on the right side of the body, one has the following succession of limbs: right normal, left supernumerary, right supernumerary. These cases have been interpreted considering that in the normal limb zone, that undoubtedly has a high physiological capacity, the normal limb imposes on the neighboring supernumerary limb an opposite laterality (that is,

mirrored symmetry). This latter limb, in turn, imparts an opposite laterality to the other supernumerary limb. Such an interpretation is in accord with the concepts of Child on the physiological dominance of one organ over another, tending to impose on the latter a mirrored symmetry.

Various interpretations have been formulated in order to explain, in a concise manner, the facts of symmetry and asymmetry. One speaks of a gradient right-left, of agents of symmetry, of facts that are very analogous with the crystals, taking into consideration, above all, the molecular arrangement. At present one can only confirm the great importance of mirrored symmetry, even if its intimate significance still eludes us.

TEODORO PERRI
Translated by V. Moorefield and U. Pittola

Reference

Abuladze, K. S., "Functioning of Paired Organs," New York, Pergamon, 1963.

SYMPHYLA

This is a class of many-legged tracheate ARTHROPODS of which the largest representatives are less than 1 cm. in length. The integument of these creatures is thin, lacks pigment and never contains calcium or other hardening salts. They live under logs, leaves and stones, where their white bodies contrast sharply with the dark earth or loam. While occasionally devouring dead insects, they otherwise subsist wholly upon decaying vegetable matter or the more delicate rootlets and tissues of living plants, one species in particular (*Scutigerells immaculata*) being known at times to do extensive damage to such crops as celery, lettuce, asparagus and beets. Hence, this form is often spoken of as the "garden centiped," although the absence of poison jaws and other basic differences widely separate it from true centipeds.

As in the other so-called myriapods (see CHILOPODA and DIPLOPODA), the head is distinct. The body forms a continuous region composed of fourteen segments, bearing fifteen, or, rarely 22 tergites of which some segments may bear two, and having 12 pairs of legs, one pair to each of the first twelve segments. The head, which is shaped much like that of the insects termed Thysanura, bears a Y-shaped epicranial suture like that of insects in general, lacks eyes, and bears a pair of long antennae composed of many segments. The mouth parts include anteriorly a pair of mandibles, followed by a pair of maxillulae *(paragnatha)*, and two pairs of maxillae of which the second pair combine to form a labium homologous with that of insects and centipeds. With the exception of the reduced first pair, each leg consists of five segments of which the terminal one bears two claws and the first one usually a slender process (parapodium) suggestive of the styli of the THYSANURA. Only one pair of respiratory spiracles are present, these opening on the head near the bases of the antennae. The paired reproductive organs open by a single pore lying between the third and fourth pairs of legs, a character which has led most students to place the members of the class with millipeds and pauropods in a group designated Progoneata. Female symphylids lay their eggs in clusters in the earth or loam. As with the millipeds, the young hatch out with three pairs of legs, intermediate stages between this and adults appearing with from three to eleven additional pairs of legs.

The symphylids have a world-wide distribution. Because of neglect by most collectors only about one hundred species have been recorded. These have been placed in three families—the Scolopendrillidae, Scutigerellidae and Geophillidae. It was the fact that these animals show obviously annectant characters with the other classes of tracheate arthropods, those with insects having been repeatedly pointed out, that the name Symphyla was chosen to designate them.

H. J. Hansen's work is still the best for entering into a study of this group. Since that publication many shorter and more restricted papers have appeared, such as those by Bagnall in England, Silvestri in Italy, Hilton in America and Remy in France. In the United States various agricultural experiment stations have issued reports on *Scolopendrella* from the economic point of view.

R. V. CHAMBERLIN

Reference

Hansen, H. J., "Genera and species of the Symphyla," Quart. J. Microsc. Sci., **53**, 1901.

t

TARDIGRADA

A group of about 350 aquatic and semiaquatic species of microscopic animals, variously considered as a phylum, an order of ARACHNIDS closely related to the mites, or as a class of ARTHROPODS. They are often called "water bears" because of their fancied resemblance to microscopic, bearlike creatures, chiefly on the basis of the stout cylindrical trunk and stumpy legs.

Some common species are found in fresh waters, especially on aquatic vegetation or debris and algal growths in the shallows or in puddles. A few occur in similar marine habitats, and some forms are found in the interstitial spaces of fresh-water and marine sandy beaches. Much more typically, however, tardigrades occur on LICHENS, liverworts (see HEPATICAE), and mosses (see MUSCI), sometimes in enormous numbers, especially if such plants are occasionally wetted or splashed with water. Tardigrades do not swim.

Most mature tardigrades range between 100 and 800 microns in length. The external surface is a cuticle which may be embellished with spines, plates, papillae, or granules. The four pairs of legs terminate in minute claws which aid in clinging to the substrate as the animals clamber about in an awkward, deliberate manner. The head is not well defined but is merely the cylindrical anterior portion of the trunk.

The mouth is surrounded by cuticular rings and opens into a tubular pharynx. The tubular pharynx, in turn, is terminated by a thick, oval, muscular pharynx having small internal sclerotized pieces. The tubular pharynx is supplied with a pair of minute, needle-like stylets which can be protruded through the mouth and withdrawn by means of a set of muscles. The stylets are used for piercing the cells of algae, mosses, and even the bodies of other microscopic animals. By means of the muscular pharynx the fluid contents of the food organisms are sucked into the tardigrade digestive tract. Two salivary glands empty into the tubular pharynx. A short esophagus connects the muscular pharynx with the large stomach, and the latter tapers to a short rectum and cloacal aperture located ventrally between the bases of the last two legs. Typically, three small glands empty into the posterior part of the digestive tract. Two long ventral glands (*Malpighian tubules*) are thought to have an excretory function. The function of the unpaired dorsal gland is not yet established, although it is also thought to be excretory.

The brain is a large, lobed, dorsal mass surrounding and obscuring much of the anterior end of the pharynx. It is connected with a chain of four ventral ganglia. Two simple eyespots are usually imbedded in the brain.

Thin muscle fibers are found attached at either end along the inner surface of the thin body wall.

There is no special circulatory system, but the extensive body cavity is filled with a fluid containing reserve food particles and colorless corpuscles. Tardigrades are variously colored, depending on the body fluid, contents of the digestive tract, and nature of the cuticle.

Female tardigrades are much more common than males, and in some species males are unknown. Parthenogenesis is thus common. The testis or ovary is a large median dorsal sac, and sperm or eggs leave by way of the cloacal aperture. Some tardigrade eggs are fascinating objects, especially since the thick shells may be highly sculptured or covered with a wide variety of spines or other types of protuberances. One to several eggs are often seen in the castoff cuticle of the female.

There is no special larval stage, the newly-hatched miniature tardigrade having most of the adult characters. The exoskeleton is shed periodically and the available evidence indicates four to six such ecdyses (five to seven instars) in the life history. An unusual feature is the fact that egg production may begin as early as the third instar.

The phylogenetic affinities of the Tardigrada are highly controversial. Since the main body cavity is probably a hemocoel, they are often considered close relatives of the mites, crustaceans, or ONYCHOPHORA. Other specialists believe that they have NEMATODE, CHAETOGNATH, or ANNELID affinities. The early embryology is said to show a prostomium and five segments.

One of the most remarkable features of tardigrade biology is the ability of many species to survive desiccation. When the habitat dries up, the tardigrade contracts and shrivels into a wrinkled spherical mass which may persist in a state of suspended animation for many years, in spite of exceptionally high or low temperatures and an absence of water. When ecological conditions again are suitable, the animal becomes turgid and resumes its normal activities in a few minutes to a few hours.

Tardigrades are found everywhere, from the tropics to arctic areas. Less than ten per cent of the known species are marine. They are most easily collected from soft bits of moss in which the cell walls are not especially thick. If the moss is wet, simply rinse it out in water and examine the washings. If the moss is dry, it should be soaked for 30 minutes to several days so that the tardigrades can emerge from the anabiotic state. *Macrobiotus* and *Echiniscus* are common genera.

ROBERT W. PENNAK

References

Marcus, E., "Tardigrada," *in* Edmondson, W. T. ed., "Freshwater Biology," New York, Wiley, 1959.

ibid., "Tardigrada" *in* Bronn, H. G., ed., "Klassen und Ordungen des Tierreichs," Vol. 66, Leipzig, 1936.

Pennak, R. W., "Freshwater Invertebrates of the United States," New York, Ronald, 1953.

TAXIS

A movement of an animal oriented with respect to a source of stimulation is called a taxis. Most frequently this term is applied to movements directly away from (negative) or towards (positive) the stimulus source, but has been used to refer to movement at a fixed angle to the stimulus e.g. light-compass reaction or *menotaxis*. Originally particulary used of movements with respect to light sources(*phototaxis*), it has been extended to include movements with respect to chemical stimuli (*chemotaxis*), the force of gravity (*geotaxis*), tactile stimuli (*thigmotaxis*), currents in water (*rheotaxis*), currents in air (*anemotaxis*) and so forth. The term taxis should be reserved for descriptions of the movements of whole animals while the word *tropism,* sometimes also used for such movements, should be reserved for use in the botanical sense to describe movements of parts of plants, e.g. root tips or shoots produced by growth.

The classification of taxes devised by Fraenkel and Gunn subdivides this group of reactions into *klinotaxis, tropotaxis* and *telotaxis,* the intention being to apply these terms to reactions to any form of stimulation, though the concepts were mainly derived from studies of light reactions.

In *klinotaxis* the organism moves its head, bearing symmetrically placed receptors, from side to side as it travels forward. Since it will move equidistant between two stimuli sources of equal strength, it appears that the direction is determined by equal stimulation of the receptors on either side of the head. Moreover this is supported by its behaviour when confronted by two unequal sources for then it will veer towards the stronger if it is an animal which is reacting positively and towards the weaker source if it is reacting negatively. The head movements would, in this case, be for the purpose of receiving *successively* the stimuli reaching the two sides of the animal. If a receptor is eliminated on one side, e.g. an eye is blackened with opaque varnish, the animal will circle towards the operated side if it is a negative animal and to the unoperated side if positive. A typical example is the movement of blowfly larvae away from a light source (negative photo-klinotaxis). The light sensitive cells embedded in the head skeleton are exposed on alternate sides by the whole anterior end of the body being swung over and the head extended to that side. If the larva is lit from one side only, that is, say, whenever it swings over to the left, but not when it swings to the right, it will crawl in a circle away from the stimulated side. The movement of a planarian worm (SEE TURBELLARIA) toward food is an example of positive chemo-klinotaxis. A swing in the direction of the food produces greater stimulation by the substances diffusing from it than does a swing away from it. The worm turns towards the more stimulated

side until it is heading towards the food when the receptors will be stimulated equally whichever side they face.

When showing *tropotaxis,* the animal moves forward without bending from side to side, but reacts to two sources in the same way as an animal said to be displaying klinotaxis. Again when one of the two symmetrically placed receptors is eliminated, circling towards or away from the intact side occurs as before. Here then, comparison between the stimulation on the two sides seems to be a *simultaneous* one. Examples of this are the movement of honeybees (SEE HYMENOPTERA) towards a light source (positive photo-tropotaxis) and the crawling of planarians into a current of water (positive rheotropotaxis).

An animal capable of moving toward or away from the stimulus source with only one receptor intact is described as showing *telotaxis.* Thus the animal does not circle when one of two receptors is eliminated and will head towards one of two sources of equal strength instead of steering between them. If one source is stronger than the other, positive animals will move to this one and negative animals to the weaker. A fixation area in the anterior part of the eye of a drone fly (*Eristalis*) (SEE DIPTERA) has been postulated to explain the ability of the fly to move towards the light with one eye blackened. Light falling on ommatidia lateral to this causes the animal to turn towards the light (i.e. righthandedly when the right eye is stimulated in this way); while the few ommatidia medial to the area cause the fly to turn away from the light (i.e. left-handedly when stimulated in the right eye). Both these actions will cause the light to be brought to bear on the fixation area. This variation in response according to the part of the eye which is stimulated is a central process and is not the result of unique properties of the various ommatidia. Thus inversion of the head causes a complete antero-posterior reversal of the pattern of reactions to light, a result which would not be expected if the seat of the differentiation of reaction were peripheral.

Many other reactions particularly of invertebrates have been given names involving a taxis e.g. *photohorotaxis* describing the way in which stick insect nymphs (SEE ORTHOPTERA) align themselves with the edges of black stripes painted on a white background. It must be stressed that all of these terms are merely descriptions of types of reactions convenient for classifying them and give no more than a general indication of the neural processes involved. Thus the tactic movements described above are all reflex-like in character but *pharotaxis,* the recognition and use of landmarks for navigation, involves learning. The wider use of the word taxis as a suffix in this way would seem to weaken its meaning.

The light compass reaction (menotaxis) by which an animal moves at a fixed angle to a source of light, e.g. the sun, similarly involves learning, at least for short periods. The angle for the return journey of a homing insect may be acquired during the outward journey as it will be the reverse of the one used then. Here again there is an acquired component in another of the reactions classified as a taxis. The dorsal light reaction (and ventral light reaction) of fish and insects, for example, can be considered as a special case of menotaxis, for the animals are maintaining their longitudinal axes at right angles to the light source. Learning is not necessarily involved here.

The sign of the reaction may vary according to the

physiological state of the animal. Thus *Paramecium* (SEE CILIATA) under certain conditions is positively geotactic in light but negatively so in darkness. Indifferent *Daphnia* (SEE CLADOCERA) become strongly positively phototactic when the concentration of carbon dioxide in the water is increased to above the normal. Planarian worms about to lay eggs are positively rheotactic and head upstream, but as soon as they have shed their eggs they become negative and turn down stream.

A taxis component may be part of innate behaviour directing the instinctive act in space; thus inherited rigid automatisms can be qualified in accordance with the stimuli impinging on the animal at the moment. The egg-retrieving instinctive act of geese (SEE ANSERIFORMES) involves a rigid motor pattern of arching the neck while pulling the egg towards the nest with the underside of the beak. However, since the egg is not a perfect cylinder, it will tend to roll sideways, in which case the beak must be turned to counteract this movement. Such regulation—which is adaptive—is the role of the taxis component.

J. D. CARTHY

References

Fraenkel, G. and D. L. Gunn, "The Orientation of Animans," London, 1940 [reprint, New York, Dover, 1961].

Kuhn, A., "Der Orientierung der Tiere im Raum," Jena, 1919.

Lorenz, K. and N. Tinbergen, "Taxis und Instinkthandlung in der Eirolbewegung der Graugans. I," Z. Tierpsychol., **2**: 1–29, 1938.

Precht, H., "Das Taxisproblem in der Zoologie," Zeit. f. wiss. Zool., **156**: 1–128, 1942.

TAXONOMY

Taxonomy (from τάξις, arrangement, and νόμος, law) is a science of classification. From the point of view of biology, taxonomy is the science of classification of organisms including the principles, procedures, and rules for nomenclature. The term taxonomy was proposed by the botanist De Candolle for the theory of plant classification. The term systematics may be used more or less interchangeably with taxonomy, but may imply a much broader study, including the kinds and diversity of organisms and all the various relationships existing among them. With this wider context, biosystematics, overlapping broadly with evolutionary and population biology, is concerned with the taxonomic implications of the genetic, geographic, and ecologic aspects of species and species formation.

Objectives of Taxonomy

The principal objectives of taxonomy are (1) to discriminate among organisms and to provide means for the subsequent recognition (identification) of the discriminated entities (taxa), (2) to develop a suitable procedure (nomenclature) for designating taxa for reference purposes, and (3) to devise and perfect a scheme of classification in which the named taxa can be arranged. The functions of the classification are (1) to provide a means for the communication and retrieval of information concerning each of the 1.5 million different kinds of organisms, (2) to facilitate the gathering of new information by permitting the prediction of characters in unfamiliar organisms, and (3) to demonstrate at once the unity and diversity of organic life by expressing the evolutionary origins of the various taxa.

Materials of Taxonomy

The basic raw materials of taxonomy are the individual organisms which make up the species, subspecies, and other population units which are found in nature. Nearly half a million plant species are now known, yet new ones are being named at the rate of approximately 5000 per year. The number of known species of animals is over one million and to these approximately 10,000 new discoveries each year are still being added. When one considers the infinite number of genetically and phenotypically diverse intraspecific populations represented by this vast array of species, it would appear that the materials of taxonomy are practically inexhaustible.

History of Taxonomy

From earliest times man has sought knowledge of animals and plants, distinguishing among them the sources of food and potion, clothing and shelter, and danger to his life and well being. However, the written record of his biological knowledge is sparse before the works of the natural philosophers Democritus and Hippocrates, and the accumulated knowledge remained largely unsystematized until it was organized by Aristotle (384–322 B.C.) and Theophrastus (370–285 B.C.), the respective fathers of zoological and botanical science. The classifications of these authors were in part natural, in part artificial, and for 2000 years were scarcely improved upon, although Renaissance botanists such as Brunfels (1488–1534), Tragus (1498–1554), Cesalpino (1519–1603), Bauhin (1550–1624), and Tournefort (1656–1708) and zoologists such as Gesner (1516–1565), Aldrovandi (1522–1605), Rondelet (1507–1556), and Belon (1517–1564) greatly expanded biological knowledge and contributed something to classification. However, taxonomic progress lagged for want of a suitable system, and it was not until John Ray (1628–1705) and Carolus Linnaeus (1707–1778), building on the work of their predecessors and crystallizing emerging philosophical concepts and techniques of presentation, published their great works (Ray, *Methodus plantarum;* Linnaeus, *Critica botanica, Philosophia botanica, Species plantarum, Systema Naturae,* etc.) that the foundation was laid for modern biological taxonomy. In the Linnaean system, species were characterized and recognized as the basic biological taxa, a binomial system of nomenclature was utilized to permit their scientific designation, and a hierarchy of taxonomic categories was utilized for their grouping according to the degrees of similarity and difference exhibited by selected organs. The simplicity and usefulness of this system and the prestige of its author brought immediate and widespread acceptance and stimulated the study of local faunas and floras, which were described and classified by this method. The system of Linnaeus became known as the natural system, and dominated taxonomic philosophy for a century. A contemporary, Michel Adanson (1727–1806), proposed a different scheme in his *Familles des Plants* based on many equally weighted characters. His theory of classification was largely overlooked until the recent interest in phenetic classification.

The publication in 1859 of Darwin's "The origin of species by means of natural selection" opened up new philosophical vistas in biology and caused much speculation on evolutionary history (phylogeny) and the significance of higher categories, stimulated in part by the tree-like phylogenetic diagrams invented by Haeckel (1866).

The emergence and growth of the science of genetics in the twentieth century provided a better understanding of individual variation and focused the attention of taxonomists on species and their subdivisions as natural populations. This dynamic (biological) concept places emphasis on geographical distribution, ecological requirements, genetic mechanisms, and degrees of reproductive isolation. This concept has proven extremely useful in the theoretical interpretation of the species of sexually reproducing organisms, but the key factors are often difficult to determine in nature.

Taxonomic Activities

An essential phase of taxonomic activity is the assembling of taxonomic materials. Among living species this involves the sampling of populations and preservation of the samples for subsequent study, procedures for which vary from group to group. In the case of higher plants and animals the harder (skeletal) parts are usually preserved in dry form, the softer parts and the whole of soft-bodied creatures in liquid or, after appropriate treatment, on microscope slides. Such materials form the basis of most collections, whether maintained by amateur collectors and individual scientists or by the large herbaria and zoological museums of the world. These collections are essential sources of material for comparative morphological purposes, for certain kinds of ecological data (e.g. seasonal and geographical distribution) and as repositories for the type specimens which serve as points of reference for biological nomenclature. However, preserved materials cannot be used for experimental purposes nor, except in a limited manner, as sources of genetic or physiological taxonomic data. For these purposes living samples must be maintained in cultures, zoological and botanical gardens, arboreta, vivaria, aquaria, insectaries, nature preserves, and so forth. The organized maintenance of living cultures for taxonomic purposes is most highly developed in those fields of microbiology in which the species classification is based largely upon physiological characters. However, experimental taxonomists in all fields assemble living materials for comparative study.

A second taxonomic activity involves discrimination among assembled taxonomic materials. The principal objective of this discrimination is the recognition of natural populations of organisms (species). Discrimination is practiced by comparing population samples by means of taxonomic characters. Appropriate statistical techniques are often utilized in these comparisons. If the samples consist of dead material, or dry preparations made from freshly killed organisms, the taxonomic characters available are largely morphological, whether at the level of the cell (cytological, karyological) or of the gross external anatomy. Living organisms may be compared on the basis of physiological characters (serological reactions, metabolic factors, body secretions, genetic makeup, means of reproduction, biochemical properties). Field studies provide a source of ecological characters (habitat requirements, geographical distribution, limiting factors) and behavioral characters (orientation, movements). The number of taxonomic characters available from these and other sources may be very large, and species in the same genus may differ by several hundred characters, although most of these will be difficult to demonstrate and the exact total in a given instance will probably never be determined.

A third activity which involves the taxonomist is the maintenance and improvement of the generally accepted nomenclature for designating taxa. From the set of rules and aphorisms proposed by Linneaus in his *Critica Botanica* international codes of nomenclature have been developed which govern the application of scientific names in bacteriology, botany, and zoology. These rules are under constant study and are subject to modification and change by international congresses of the sciences concerned.

A fourth taxonomic activity involves the grouping of species into progressively larger taxa, forming the hierarchy of genera, families, orders, classes, phyla, and kingdoms. This is discussed in detail in the next section.

Higher Classification

In view of the several functions of a classification (information retrieval, prediction of unobserved characters, and expression of evolutionary relationships), much interest now surrounds the methods by which the higher classification is formed. Although a single general classification is now in use, it is clear that the various sections have been constructed with different theoretical viewpoints in mind, depending on the knowledge and attitude of the taxonomist responsible. For information retrieval these differing viewpoints are of little consequence since the hierarchic structure and rules of nomenclature provide a unique name for each of the 1.5 million kinds of plants and animals. It is unlikely that a different classificatory scheme would be as effective in ordinary use, but for electronic systems of retrieval other schemes with numerical designations would be more efficient.

If the biologist is concerned with prediction or evolutionary lineages then the theoretical bases of the classification are important. Generally, as a consequence of the evolutionary process, taxa that are similar are closely related and those that are different are more distantly related. But this axiom is not always true. Evolution does not necessarily proceed at the same rate and its course is often obscured by extinction of lines and confused by parallelism or convergence. Fossil evidence of the evolutionary lineages is often lacking and the probable relationships must be reconstructed using the comparative morphology, physiology, embryology, and ecology of living forms. When a newly discovered organism is placed in the classification the evidence is usually morphological. The predictive value of the classification, however, permits the biologist to anticipate the physiological, ecological, and other characteristics of the new taxon with some degree of confidence. As additional information is gathered the classification is revised and refined, sometimes with major changes, to improve the expression of evolutionary relationships.

The construction of the higher classification involves the evaluation of two distinctive sources of evidence: (1) the similarity among the different organisms and (2) the clues to the relative recency of their common ancestry. In practice both kinds of evidence usually support the same classificatory arrangement, but in theory two separate approaches to classification can be recognized: the phenetic and the phyletic. The phenetic approach places emphasis on the similarities and differences existing among the taxa as judged by their phenotypes. The phyletic approach emphasizes the reconstruction of ancestral lineages.

Degree of similarity is determined by the number of taxonomic characters which species have in common. For this purpose, characters may be drawn from the same sources (comparative morphology, comparative physiology, comparative ecology) as those utilized for taxonomic discrimination. A rigorous form of phenetic classification, called numerical taxonomy (Sokal and Sneath), seeks repeatability and objectivity of the classification by using large numbers of characters, often in excess of 100, and by giving each equal weight. A character may exist in two or more conditions or states, e.g., different numbers of leaves or teeth. Each of the initial taxa (operational taxonomic units or OTU's) is examined for each character and the state of the character recorded. Numerical methods are used to evaluate the degree to which each OTU is similar to each of the other OTU's under study. The extensive computations usually require the services of an electronic computer. The result of this analysis, called the similarity matrix, contains the most information about how similar or dissimilar an OTU is to each of the other OTU's. The matrix, however, does not resemble the familiar classificatory hierarchy. Again using numerical methods, the OTU's may be grouped or clustered together, beginning with those having the highest index of similarity and progressively grouping those less similar into a hierarchic scheme. The classification may be graphically presented as a tree-like diagram, called a dendrogram or phenogram, in which the "twigs" represent the initial OTU's and the branches and trunk represent the successive groupings of the OTU's at lower levels of similarity. The vertical dimension of the tree expresses the relative similarity. Some information on the phenetic relationships of the OTU's is lost when the similarity matrix is translated into a phenogram. This loss results because the matrix contains information which is geometrically multidimensional while the phenogram contains only two dimensions. The tree-like nature of the diagram should not be confused with a phylogenetic diagram because the phenogram does not imply ancestral lineages. The strictly phenetic classification is well-suited for prediction because of its statistical basis, but may fail to reflect the evolutionary origins of the taxa.

In contrast, the phyletic approach is concerned with deducing the phylogeny or ancestral lineages of the taxa. A phylogeny is traditionally represented by a branching tree with ancestral types represented by the trunk, later forms as branches, and modern taxa as the twigs and leaves. The vertical dimension of a phylogenetic tree is time or relative advancement. Only rarely (horses, elephants, redwoods) is the fossil evidence adequate to justify this truly phylogenetic representation. Usually the positions of the trunks and branches can only be inferred from the degrees of relationship seen in present-day forms.

The phylogenetic method depends on the recognition and use of homologous characters. Thus the wings of birds and bats, both derived from fore limbs and therefore homologous, show the relationship of these vertebrate types whereas the wings of insects, not being derived from limbs, are analogous and do not reveal relationship. There are many pitfalls in the interpretation of homology but it is the essence of the phylogenetic method. Inherent in the method also is the actual or implied recognition of "archetypes" or ancestral types. Even when available, fossils rarely represent the precise ancestral types that are assumed to have given rise to two branches of a phyletic line. Therefore the phylogenetic taxonomist seeks to distinguish "primitive" or "ancestral" characters in contrast to "specialized" or "derived" characters. This may be done by comparison with related forms and the selection of those characters that are relatively simple or generalized. For example, the bill of a sparrow is obviously less specialized than that of a woodpecker. Specialization is often by adaptive modification, like the web-feet of aquatic birds. In other cases derived types are the result of reduction, such as loss of wings in ectoparasites. The survival of relics is often helpful to the phylogenist. Such living fossils as *Lingula, Neopilina, Latimeria,* and *Metasequoia* are near enough to hypothetical ancestors of their respective groups to serve as model "archetypes."

Evolution is usually thought to be irreversible and unique; i.e., a change from a complex character A to character B occurs only once and B cannot return to the condition A. If two taxa each have the character B, the assumption is that they have a common ancestor. Following these end similar arguments an exacting approach has been formulated by Hennig. Many characters, drawn from all aspects of the organism, are considered. A phylogenetic evaluation is made for each character of a given taxon with reference to the other taxa under study. A single character may be judged primitive (pleisiomorphic) when the taxon is compared with a second taxon, but the same character also may be considered specialized (apomorphic) when compared with a third taxon. For example, the possession of hair would be a pleisiomorphic character if only mammals were being compared with each other, but a mammal's hair would be an apomorphic character when compared with the scales of a reptile. Two taxa possessing an apomorphic character in common exhibit the crucial evidence of common ancestry and form a "sister group," while the sharing of a pleisiomorphic character provides no evidence of phylogenetic value. The classification is constructed by grouping the taxa according to the relative recency of common ancestry, based on the apomorphic characters. Of all the characters initially considered, these receive special weight. The strictly phylogenetic classification is intended to express only the lines of evolution, irrespective of the relative similarity of the taxa.

The differences between a phenetic and a phyletic classification of the same organisms may be minor and of little significance for the average user. In fact, much taxonomic work is an intuitive blend of phenetic and phyletic reasoning, i.e., an evaluation of similarity with special emphasis placed on those characters of phylogenetic importance.

Species. The species, though elemental and basic in classification, remains a subject of controversy. To some it is a building block for classification, analogous to the cell in an organism or to an atom in a physical system, representing a significant level of integration in our universe. To others the species is an arbitrary stage in evolution, a fixed point to attach a name or a convenient pigeon-hole for filing information. As understood by Ray and Linnaeus, species were distinct entities in the local flora and fauna of England and Sweden. Members of a species showed individual variation, males differed from females, and some showed seasonal changes, but hybrids were unknown or rare. This gave rise to the static species concept and was in keeping with the prevailing idea of special creation.

One hundred years later Darwin provided a theoretical basis for the evolutionary species concept. He visualized species as arbitrary divisions of the continuous and ever-changing series of individuals found in nature. This is the view of many paleontologists who visualize species as an arbitrary stage in the divergence of a phyletic line.

Early in the twentieth century genetics provided a new concept of species based on the idea of genetic compatibility within the species and reproductive isolation between different species. Species, then, were all individuals that could potentially or actually share the same gene pool, the populations varying in space and time but always in dynamic equilibrium with their environment.

The species may be defined strictly on morphological grounds as a group of individuals that resemble each other in most of their visible characters. However, a more satisfactory definition on theoretical grounds is the so-called "Biological Species Definition": Species are groups of actually (or potentially) interbreeding natural populations which are reproductively isolated from other such groups (Mayr).

Plant species consist of numerous local forms or ecotypes which differ morphologically and usually also genetically from adjacent types. These arise through the selective influence of local climatic or soil conditions. In animals local populations are likewise different and these geographical races are commonly named as subspecies. The degree of difference that is commonly recognized as namable is a debatable point but the seventy-five per cent rule is one attempt at an arbitrary criterion. If seventy-five per cent of the individuals within a given population can be distinguished from all those of an adjacent population the subspecies may be named. Species that break up into recognizable subspecies are called "polytypic species." Subspecies are allopatric, living in adjacent areas and showing more or less complete geographical replacement. Hybridization is common at the zones of contact, if any, and, in plants, may result in polyploidy, or duplication of chromosomes. Intermediate zones may be as narrow as a few feet or may extend for many miles. Some characters (color, body weight) within large populations may change gradually from one extreme to another over thousands of miles. These character gradients are called clines (Huxley) and are not to be confused with taxonomic categories.

Extensive study of many characters over a wide geographic range sometimes reveals a complex pattern of variation which cannot be easily expressed by the designation of subspecies. The complexity and lack of congruent patterns in geographic variation have led some taxonomists to abandon the subspecies concept in favor of an accurate analysis of the variation and the underlying causes.

Species names are binominal, consisting of a generic designation (*Homo*) and a specific epithet (*sapiens*). Subspecies are trinominal, (*Ondatra zibethica alba*, a northern population of muskrat), with the "nominate" or originally described form designated by doubling the specific epithet (*Ondatra zibethica zibethica*, the muskrat of the Eastern United States).

A different species concept is necessary for asexual and parthenogenetic forms. The strains of bacteria and viruses and the clones, if plants, are reproductively isolated. There is no fusion of gametes and no mixing of genetic factors. The taxonomic treatment of asexual types varies but forms that differ in constant morphological characters or in their reactions to various culture media are generally named. Others that differ in degree of virulence as determined by the reaction of the host are referred to as types or strains.

Individual variation is a characteristic of organisms and is conspicuous in bisexual plants and animals. In its simplest form, variation reflects the mutations in genes and such variants segregate according to the Mendelian laws of heredity. However, in the museum or herbarium extreme variants are not easy to distinguish from species.

Taxonomic Methods

Descriptive method. The descriptive method was established by Linnaeus as part of his nomenclatural system, and was designed to provide a point of reference for zoological and botanical names. The taxonomic description is still an essential part of the nomenclatural system, and a description must accompany the proposal of names for previously unrecognized taxa. However, the original description accompanying the proposal of a new scientific name, especially at the species level, is rarely sufficiently detailed to permit its use as a standard for comparison. As a result, species and other taxa must be redescribed to take into account increasing knowledge. The redescription thus becomes an important method of communication among taxonomists. Its preparation requires a thorough knowledge of the group of organisms involved as well as familiarity with, and ability to effectively utilize descriptive terminology.

The type method. The shortcomings of the descriptive method are partially offset by the type method, a taxonomic and nomenclatural device for providing a fixed reference point for the application of scientific names to taxa. As biological nomenclature has expanded since its establishment by Linnaeus, the need for fixed points of reference has become more urgent. The first application of the type concept involved the names of species, the point of reference (type) being the material which the author had before him when he described a new species, and often, also, such material as he subsequently added. However, greater precision became essential, and as type specimens have been more rigidly defined more than 100 kinds have been differentiated. These fall into three principal classes: (1) *primary types* (including holotypes, allotypes, paratypes, syntypes, and lectotypes), the original specimens used by an author in describing or illustrating a new species, (2) *supplementary types* (including neotypes and plesiotypes), the specimens serving as a basis for descriptions or illustrations which supplement or correct knowledge of a previously described species, and (3) *typical specimens* (including topotypes, metatypes, and homotypes), specimens of varying degrees of authenticity which have not been used in connection with published descriptions or illustrations.

As classifications were revised and the number of named taxa grew, points of reference for the names of collective (group) categories became increasingly necessary. This was particularly true of generic names, since these often provide roots for the names of higher taxa. As a result the type concept was extended to collective categories, the type of a genus being a species, the type of a higher category being a genus. With the application of the type concept to collective categories, precise rules have been developed in the nomenclatural codes governing type designation and type selection.

Publication forms. Taxonomic information is presented in the form of isolated descriptions, revisions of existing classifications, or comprehensive monographs. Catalogues provide lists of species, usually with citations of literature and synonymy. Field handbooks are prepared for purposes of identification and faunal works and "Floras" treat in a technical way the animals and plants of particular regions.

Taxonomic publications consist of descriptions, diagnoses, keys, synonymies, and diagrammatic representations of the relationships of taxa. Keys present a series of choices, usually dichotomous, leading to the correct name of a species. Key characters are those attributes which readily distinguish between species or higher taxa. Synonymies are lists of names that have been applied to a particular taxon. The names are usually arranged in chronological order and preferably with citation of the bibliographical reference to the original publication of each. Synonyms are different names applied to the same taxon and homonyms are identical names applied to different taxa. In either case, the later name (or names) is suppressed in favor of the earlier one.

Diagrams are often used to represent the author's views of relationships. They may be in the form of a two-dimensional "tree," called a dendrogram, or they may employ perspective and thus represent the facts in three dimensions. In some groups such as the horses, elephants, and sequoias, fossil evidence is sufficient to provide a "phylogenetic" tree, showing the time of origin and extinction of groups.

Status of Taxonomy

Taxonomy dominated the field in biological sciences up to a century ago. This was a period of great exploration and the discovery of exciting new plants and animals. Then, with the rise of physiology, genetics, and other experimental sciences, taxonomy was looked upon more as a source of scientific names than as a science in its own right. Gradually, because of the economic importance of plants and animals and the

necessity for accurate identifications this situation has changed. Also it has come to be recognized that taxonomy is a discrete science and has as its primary purpose the development of a sound and meaningful classification.

Unfortunately, there are relatively few positions in the world devoted to pure taxonomy. Most of the 11,-000 taxonomists are employed as curators, teachers, or researchers in applied fields and many are students. Hence, taxonomy still lacks the regular formal support that is needed for most efficient use of its limited manpower for accomplishment of its enormous tasks. At present, the coverage of groups of plants and animals by specialists is very uneven and fortuitous. There are 2000 plant taxonomists working on 413,570 species and 9000 taxonomists covering over a million species of animals. In zoology seventy-five per cent of taxonomists specialize on invertebrates and twenty-five per cent on vertebrates. Twenty per cent are paleontologists and one-third are entomologists (Blackwelder).

Substantial progress has been made in the organization and support of Taxonomy during the past decade. There is now an "International Association for Plant Taxonomy and Nomenclature" of the "International Association for Plant Taxonomy" and a "Society for Systematic Zoology." Nomenclature has been coordinated at least partially at international conferences of microbiologists, botanists and zoologists. Also, the section on Systematic Biology of the National Science Foundation has administered a far-sighted and broadly based program for the support of sound taxonomic work. This has resulted in more and better taxonomic studies and increased recognition of Taxonomy by its sister disciplines.

The challenge of taxonomic research. At the present time it can be said that the broad outlines of biological classification have been worked out. In most groups, the pioneer stage is over and taxonomists are no longer preoccupied with mere naming of new forms, though much of this still remains to be done.

For the future, exciting new vistas are seen. Experimental taxonomy is already far advanced in both botany and zoology (especially entomology) and is contributing major evidence to evolutionary and population biology. Cytological, serological, biochemical, mathematical, and other techniques are adding to the precision of taxonomic interpretation. At the synthetic level, taxonomic research is approaching closer to a picture of the course of evolution, with the mass of new evidence from paleontology, experimental embryology, comparative biochemistry, comparative ecology, and, of course, zoogeography. As a result, taxonomy stands at mid-twentieth century as one of the most challenging fields of biology.

HOWELL V. DALY AND E. GORTON LINSLEY

References

Blackwelder, R. E., "Taxonomy, a text and reference book," New York, Wiley, 1967.
Davis, P. H. and V. H. Heywood, "Principles of Angiosperm Taxonomy," Edinburgh, Oliver and Boyd, 1963.
Hennig, W., "Phylogenetic Systematics," Urbana, University of Illinois Press, 1966.
Heywood, V. H. and J. McNeill, "Phenetic and Phylo-

genetic Classification," London, The Systematics Association, 1964.

Mayr, E., "Animal Species and Evolution," Cambridge, Harvard University Press, 1963.

ibid., "Principles of Systematic Zoology," New York, McGraw-Hill, 1969.

Simpson, G. G., "Principles of Animal Taxonomy," New York, Columbia University Press, 1961.

Sokal, R. R. and P. H. A. Sneath, "Numerical Taxonomy," San Francisco, Freeman, 1963.

TELEOSTEI

The Teleostei may be regarded as a superorder of modern bony fishes, in which the vast majority, more than 20,000 species, are included. They first appeared in the early Jurassic, and they differ from their holostean ancestors in the absence of a spiracle, absence of microscopic canals in the scales, and presence of a homocercal tail (the vertebral column does not turn toward the upper lobe). The skeleton in postlarval stages is almost completely bony, except for partial retention of cartilage in a few, such as *Mola*, the ocean sunfish.

Teleostei are the most progressive division of the subclass Actinopteri, fishes with separate rays supporting membranous fins. They share with other members of the class Osteichthyes a hinged, bony operculum which covers the gills and acts as a valve in rhythmic breathing movements. It is probable that teleost fishes originated from two or three different sources among Holostei.

Certain adaptive trends can be seen in the structure of teleosts on comparing the more primitive with the advanced orders. In the Isospondyli (Clupeiformes), a stem order, the finrays are soft, flexible, minutely jointed, but most members of the advanced order Acanthopterygii are spiny-rayed, with the more anterior rays of the dorsal and anal fins hard and sharp. Pelvic fins in lower teleosts are located on the abdomen, but in advanced groups they become thoracic (under the pectorals) or jugular (in front of the pectorals). Again, the SCALES in more primitive teleosts are generally cycloid, that is, round or oval without toothed edges, whereas the spiny-rayed fishes have ctenoid scales, which are toothed or comblike along the exposed margins. Isospondyli and other primitive orders retain an open duct from the esophagus to the airbladder (the physostomous condition); Acanthopterygii are without such a duct except in early development (physoclistous). Another special feature is the development of a mechanism for protracting the bones of the upper jaw, used in nibbling or snapping at food; this is not seen in more primitive teleosts but occurs in at least four or five different patterns among the advanced orders. The combination of these several features shown by any particular fish can serve as a rule-of-thumb for judging its status among the teleosts, if due allowance is made for the highly complicated branching of the evolutionary tree.

Although the early teleosts were largely marine, their more remote ancestors among the Actinopteri were not. During teleost history many independent invasions of fresh water have occurred, some of them so long ago that there is no trace of a connection with any particular marine group. Ecologically, long-standing groups that are unable to survive at all in salt water are called primary freshwater groups; those in which at least some members extend into brackish or sea water are called secondary. For instance, although minnows, suckers, characins and most catfishes are considered primary, a couple of families of catfishes became marine, and some members of them have again entered rivers. Some other groups, such as the cyprinodonts (killifishes) and the perchlike cichlids, contain members that live in brackish water. Shad, several species of salmon and trout, striped bass, and others, have seasonal spawning runs from the sea into freshwater streams; this behavior is anadromous, in contrast to the downstream, catadromous migration of American and European eels, which spawn in mid-Atlantic.

Many marine teleosts produce vast numbers of minute, transparent eggs (over 6 million in cod), externally fertilized, and usually with the same specific gravity as the water. These develop rapidly into translucent larvae, and thus both eggs and larvae constitute an important part of the plankton, consumed by numerous plankton-feeding fishes and other animals. Spawning in more restricted habitats is accompanied by reduction in number but increase in size of eggs. Generally, freshwater and shore fishes do not have planktonic eggs or larvae; the eggs may be attached to objects, concealed or attended by the adult. Elaborate nest-building, nursing or guarding adaptations are seen in many teleosts, for instance the artificial nests of sticklebacks, the broodpouches of seahorses and the "mouth-breeding" habits of *Tilapia*, a cichlid. Finally, internal fertilization and viviparity occur in a few families, notably the cyprinodonts (guppies, etc.), embiotocids (surfperches) and certain blennies.

Besides furnishing food and sport to man from prehistoric times, teleosts in both sea and fresh water constitute important links in the food chains and chemical cycles involving aquatic life. Plankton feeders such as herring and menhaden are tremendously prolific, and they serve as food for numerous predatory fishes (bluefish, tuna, etc.) as well as for sea birds and aquatic mammals. Although remarkable advances have been made in recent years in the techniques of commercial fisheries, we may expect to realize even greater advantages from exploitation of new areas in the sea and consumption of species not yet fully utilized. Fish farming in both fresh and salt water, especially in tropical countries with overcrowded human populations, is already helping to relieve food shortages. But careful conservation is necessary because of the ever-present danger of extinction of rare species.

THEODORE H. EATON

References

Berg, L. S., "Classification of fishes, both recent and fossil," Ann Arbor, Edwards, 1947.

Brown, M. E., ed, "The physiology of fishes," 2 vols., New York, Academic Press, 1957.

Greenwood, P. H., *et al.*, "Phyletic studies of teleostean fishes, with a provisional classification of living forms," Bull. Amer. Mus. Nat. Hist., **131**:341–455, 1966.

Norman, J. R., "A history of fishes," 2nd ed., New York, Hill and Wang, 1958.

THEOPHRASTUS (c. 370 B.C.–287 B.C.)

Greek philosopher and student of both Plato and Aristotle, Theophrastus was born at Eresus on the island of Lesbos. At an early age he went to Athens and became a pupil of Plato. From Plato Theophrastus learned the importance of the principle of classification, which he applied for the first time to the plant kingdom. When Plato died Theophrastus went to the Peripatos and became head of that school when Aristotle retired to Chalcis. When Aristotle died at age 63, Theophrastus inherited his manuscripts and botanical garden. It was in this garden that the first systematic botanist made many of his observations. Theophrastus also obtained botanical data from the scientifically trained observers who accompanied Alexander of Macedon on his marches to the East. It is believed that his accounts of the flora of regions nearer home were not first hand —that a number of traveling students were employed to make observations and collect facts in adjoining lands.

Of the few works of Theophrastus which have been preserved, the most important are his nine books of the *Enquiry into Plants*, and six books *On the Causes of Plants*. His famous *Characters*, containing brief portraits of moral types, was the basis for Jean de La Bruyére's well known satire of the same title. His *Ethics*, which upheld Aristotle's principle of a multitude of virtues with their corresponding vices, is famous because of the attacks made on it by the Stoics. His *Doctrines of the Natural Scientists* is the basic source for the philosophy of antiquity.

Theophrastus died at the age of 85. We are told that he complained that "we die just when we begin to live." However, his life must have been a very full one as he was the personal friend of both Plato and Aristotle and witnessed the careers of Philip and Alexander of Macedon.

JAMES L. OSCHMAN

Reference

Stratton, G. M., "Theophrastus," Chicago, Argonaut, 1967.

THERMODYNAMICS OF CELLS

Within the short space available for an article such as the present one it is impossible to do more than briefly cover some of these aspects of thermodynamics which the author feels are most relevant to the understanding of cellular processes. Nor will the literature drawn upon be adequately cited. The interested reader should consult the longer article on these and other such matters by Best and Hearon (1960) which cites more references to relevant books and the original literature. Chapter 9 of Fruton and Simmonds (1958), Podolsky (1959), Hutchens (1951), and Lumry (1959) should also be consulted as guides to this general area.

The *universe* can be partitioned into two primary regions, the *system* of interest and the *environment*. Dividing the system from the environment will be some real, or conceptual, closed surface called the *boundary* of the system, across which any exchanges of matter and energy between system and environment

must occur. The system may do work upon, or have work done upon it by, the environment; it may exchange heat or matter with the environment; and it may undergo changes in its internal composition or configuration. Work done by the system upon the environment is conventionally designated as positive. Work done by the environment on the system is negative. The word "system" is used here in the broadest sense and is meant to include living as well as non-living systems.

A system which can exchange neither work, heat, nor matter with its environment is *isolated*. If it cannot exchange matter with its environment, it is said to be *closed*. If it can exchange matter with its environment, it is *open*. A process in which no work is done is called *isochoric*; one in which the pressure is held constant, *isobaric*; one in which the temperature is held constant, *isothermal*.

The *first law of thermodynamics* asserts the conservation of energy and the equivalence of heat and work. Denoting the energy of the system as E, the infinitesimal contribution to the energy of the system from both heat transfer and exchange of matter across the boundary as $d\Phi$, and the infinitesimal work done by the system as dW, one can write the differential change, dE, in the energy as

$$dE = d\Phi - dW \tag{1}$$

This is the assertion of the first law valid for open systems. In the event that the system is closed, $d\Phi$ reduces to dQ, the contribution to the energy from heat transfer only. Heat transfer from environment to system is positive.

The first law places no restriction upon the interconversion of one form of energy into another. Such a limitation is, however, imposed by the *second law of thermodynamics* upon the conversion of heat into work. Of the various completely equivalent statements of this law, one of the more useful ones is that

$$TdS \geq dQ \text{ for a closed system,} \tag{2}$$

where dS is the differential change in the *entropy* of a closed system and T its absolute temperature. The equality sign holds for *reversible* processes and the inequality for *irreversible* ones. If the system is isolated then $dQ = 0$ so

$$(dS)_{isolated} \geq 0, \tag{3}$$

from which it is clear that the entropy of an isolated system can never decrease. An isolated system will therefore be most stable when its entropy is a maximum.

Because living organisms can decrease their entropy it is sometimes asserted that they act in violation of the second law. This conclusion is incorrectly drawn. Such organisms are not closed, much less isolated, systems. A decrease in their entropy need not, therefore, entail any paradox in regard to the second law.

Certain thermodynamic variables, e.g. pressure, P, temperature, T, volume, V, number of moles of the kth constituent, n_k, characterize the system and have a definite value when the system is in a particular state. When certain subsets of these descriptive variables are assigned unique values, the state of the system is defined. The infinitesimal increment of change in a variable of state is a perfect differential, e.g. dE, dS. Other quantities of interest such as, Q and W do not have a unique correspondence to the state of the system, their change depending rather on the process, or "path," by which the system is moved from one state to another. The infinitesimal increment of change in such quantities is not a perfect differential and, to indicate this, is denoted by đ instead of d, e.g. đΦ, đW, đQ.

The question of whether a system, biological or organic, is governed by the laws of thermodynamics does not depend upon its being at equilibrium as some writers assert, but rather upon whether it can be described in terms of thermodynamic variables having unique values associated with the various states of the system. Thus, for example, there may exist a temperature gradient within the system so that a single temperature cannot be assigned to the entire system. In a large proportion of such instances, however, the temperature within an infinitesimal element of volume will be uniquely defined at a given time. The laws of thermodynamics will, in such instances, be applicable to each infinitesimal element of volume and by appropriate summation conclusions concerning the behavior of the entire system can be obtained. By such straightforward extensions, which will not, however, be discussed here, it would appear that virtually all biological systems are at least in principle amenable to thermodynamic analysis.

The systems in which one is interested are seldom isolated but frequently under conditions of constant temperature and pressure. A more useful function than S therefore is the Gibb's *free energy*, F, defined by

$$F = H - TS \qquad (4)$$

where

$$H = E + PV \qquad (5)$$

is the *enthalpy* or "*heat content.*" The work can be written as the sum of pressure-volume work and work of other kinds, dW', thus

$$đW = đW' + PdV. \qquad (6)$$

It can be demonstrated (e.g., cf. Best and Hearon) that

$$dF_{T,P} \leq -dW' \qquad (7)$$

for a closed system at constant T and P (indicated by the subscript position of these quantities by the differential) where the equality sign holds for reversible processes and the inequality for irreversible ones. Thus, the free energy of a closed isobaric isothermal system can never increase unless non-PV work is done on it.

Living cells can, in the vast majority of cases, be regarded as isothermal, isobaric systems. In some cases they will be isochoric but in many instances of interest, e.g., muscle fibers, kidney tubules, electric organs, the non-PV work terms will be of primary interest.

Consider a system comprised of materials $1, \ldots, C$. Let n_k denote the number of moles of the kth material and μ_k its chemical potential defined as

$$(\partial F/\partial n_k)_{T,P,n_{j \neq k}} = \mu_k$$

i.e., the free energy change per mole of the kth material added to the system keeping T, P and the amounts of the other constituents constant. In the case of an ideal solute, μ_k can be expressed in the form

$$\mu_k = \mu^0_k + RT \ln C_k \qquad (8)$$

where R is the gas constant and μ^0_k and C_k are the standard chemical potential and concentration of the kth constituent. Suppose that these materials participate in a reaction as either reactants or products. The change in n_k due to the reaction will be

$$dn_k = m_k \, d\xi \qquad (9)$$

where n_k is the stoichiometric coefficient of the kth substance and ξ is the *extent of the reaction*. If k is a product, m_k is positive; if a reactant, m_k is negative.

The free energy change per unit change in the extent of the reaction will be

$$dF/d\xi = \sum_{k=1}^{C} m_k \mu_k$$

If this is the only reaction taking place in the system and the system is closed, isothermal, isobaric, and dW' = 0, it is clear from equation (7) that the reaction can proceed only if $dF/d\xi < 0$. The quantity α, called the *affinity* of the reaction is defined as

$$\alpha = -dF/d\xi$$

If the extent of the reaction is independent of that of any other reaction in the system then

$$\alpha \, d\xi > 0$$

i.e. the reaction will proceed only in that direction for which its affinity is positive. Consider a system with two reactions of affinities, α_1 and α_2 and extents ξ_1 and ξ_2. If a change in ξ_1 entails an obligatory change in ξ_2, the two reactions are said to be "*coupled.*" The observation that some reaction 1 in an isobaric isothermal system, e.g. a tissue, cell, or reconstituted system, proceeds in such a direction that

$$\alpha_1 \, d\xi_1 < 0$$

is sufficient grounds to infer the existence of a second reaction, to which 1 is coupled, such that

$$\alpha_1 \, d\xi_1 + \alpha_2 \, d\xi_2 > 0.$$

Such coupling is both prevalent and important in the metabolic reactions of living cells.

It is sometimes thought that a certain amount of energy is made available to a cell by its food stuffs and that this energy can either be used for work or wasted as heat. This is not correct. The heat absorbed by a reaction at constant T and P is the change in enthalpy $dH/d\xi$ (usually denoted as ΔH). The thermodynamic "driving force" of the reaction is α or $dF/d\xi$ (usually denoted as ΔF). Depending upon the magnitude of the term $T\Delta S$, ΔH can be either positive or negative for a spontaneously occurring reaction. The non-pressure volume work which can be done by coupling of a reaction is determined by ΔF, not ΔH.

In a homogeneous system the free energy change at constant temperature and pressure will be

$$dF_{T,P} = \Sigma \mu_k dn_k$$

Consider the special case of such a system in which the changes in the amount of the constituents are the results of a single reaction of extent ξ which is electrochemically coupled to an external potentiometer circuit, i.e. the reaction can proceed only by donating or accepting electrons from an electrode. The potential of the electrode is adjusted to a point at which the current flow is just zero. At equilibrium with the external circuit

$$-dW' = \Sigma \mu_k m_k d\xi.$$

The external non-PV work will be the electrical work $-\epsilon \mathfrak{F} dz$ where ϵ is the electromotive force, $\mathfrak{F}$ the Faraday constant, i.e. the number of coulombs per chemical equivalent, and dz the number of equivalents of electrons transferred. Then

$$\epsilon = -\frac{1}{\mathfrak{F}} \frac{d\xi}{dz} \Sigma \mu_k m_k = \frac{\alpha}{\mathfrak{F}} \frac{d\xi}{dz} \qquad (10)$$

This relation between the affinity (or free energy) of an oxidation-reduction reaction and the electromotive force is of interest in that frequently the affinities of the various electron donor acceptor systems involved in the Embden Meyerhoff, Krebs cycle, flavoprotein, and cytochrome systems are expressed in terms of ϵ and measured in volts.

The previous discussion regarding coupling can be extended. If two reactions are coupled in such a way that the "forward" reaction of one necessitates that the other move "backward" or vice versa then

$$d\xi_1 = -d\xi_2 = d\xi.$$

Since

$$0 < \alpha_1 d\xi_1 + \alpha_2 d\xi_2 = (\alpha_1 - \alpha_2)d\xi \qquad (11)$$

$d\xi$ must be positive if $\alpha_1 - \alpha_2$ is positive and negative if $\alpha_1 - \alpha_2$ is negative. This means that the reaction of greater affinity will drive the other of lesser affinity backwards. With $d\xi/dz$ and $\mathfrak{F}$ the same for the two reactions, it is clear, in the light of relations (10) and (11), that the reaction of greater electromotive force will drive the reaction of lesser electromotive force backward. In the case of two hydrogen donors, the donor with the greater affinity in the direction of being oxidized will become oxidized, driving the other in the direction of the reduced state.

In a living cell, or reconstituted enzyme system, many reactions are proceeding simultaneously. The preceding discussion can be extended as follows. Let ξ_j be the degree of advancement of the jth reaction and m_{kj} the stoichiometric coefficient of the kth substance in the jth reaction. The differential change in amount of the kth constituent due to the reactions transpiring within the system will be

$$dn_k = \Sigma m_{kj} d\xi_j. \qquad (12)$$

The change due to addition of k from the environment (as contrasted to the addition of k by conversion of some other material inside the cellular system) will be indicated as $d_e n_k$. The free energy change of this open system will then be

$$dF = \sum_k \sum_j \mu_k m_{kj} d\xi_j + \sum_k \mu_k d_e n_k \qquad (13)$$

The simplest state in which to conceive a cell, consistent with reality, is the steady state. A steady state system may have reactions proceeding inside of it but its composition remains unchanging with time. Each of its constituents will be used up as fast as they are formed or supplied from the outside, each reaction product consumed or transported out as fast as they are produced. If any of its reaction rates are non-zero, it is, of necessity, open. The free energy of such a system will remain constant while that of the environment undergoes dissipation due to the processes occurring within the system. Thus, for a steady state system of the type to which equation (13) is applicable, one would have

$$(dF/dt)_{system} = 0$$

$$(dF/dt)_{environment} \leq \sum_k \sum_j \mu_k m_{kj} \frac{d\xi_j}{dt} < 0$$

$$d_e n_k / dt = -\sum_j m_{kj} \frac{d\xi_j}{dt} j$$

In all this discussion it should be pointed out that homogeneity of the system has been assumed. No account was taken of spatial nonuniformities in the concentrations or chemical potentials of the constituents nor of the free energy dissipation due to flows across barriers such as the membrane surrounding the cell surface or its nucleus.

A necessary free energy dissipation is entailed by the movement of metabolites into and out of the cell. Just as the free energy change serves as a criterion of spontaneity for a process in which materials are undergoing transformation in a chemical reaction, so does it serve for one in which materials are redistributing themselves between phases or different spatial regions. Denote the chemical potential of the kth substance at regions 1 and 2 as μ_{k1} and μ_{k2} respectively. The free energy change per mole of k moving from 1 to 2 will then be $\mu_{k2} - \mu_{k1}$. This will be negative only if $\mu_{k2} < \mu_{k1}$ so it follows that such a transition can occur, in the absence of some kind of coupling, only from regions of higher to regions of lower chemical potential.

A material diffusing into a cell will dissipate free energy in this way. If A is the cell area, C_1 and C_2 the concentrations outside and inside respectively, h the permeability coefficient of the cell membrane, and μ_1 and μ_2 the respective chemical potentials, the rate of free energy dissipation will be

$$(dF/dt)_{membrane} = Ah(C_1 - C_2)(\mu_1 - \mu_2) \qquad (14)$$

If, in addition, it is assumed that the solute is perfect inside and out and the standard chemical potential inside and out is the same, then

$$(dF/dt)_{membrane} = AhRT(C_1 - C_2) \ln C_1/C_2 \qquad (15)$$

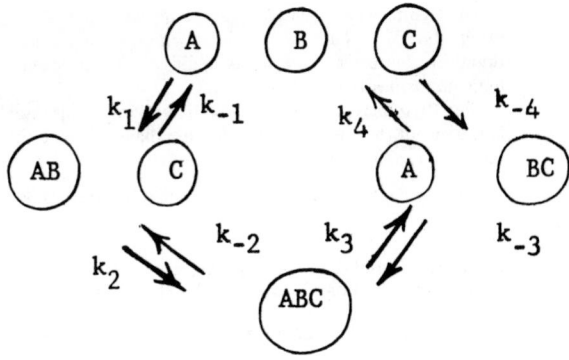

Fig. 1

The free energy cost per mole of such material taken into the cell will be

$$(dF/dn) = RT \ln \frac{C_1}{C_2} = RT \ln \frac{C_1}{C_1 - r/Ah} \qquad (16)$$

where r is the rate of entry. It can be seen from expression (16) that, for more rapid rates of entry, not only will the free energy dissipation per unit time be greater, but also the dissipation per mole. Thermodynamics and kinetic properties exhibit such an interplay in many cellular processes.

In addition to the first and second laws of thermodynamics, there is a third principle with wide implications for cellular processes. This is the *principle of microscopic reversibility* or its corollary, the *principle of detailed balancing*. To illustrate this principle consider the alternate pathways of reaction by which three materials A, B, and C may react to give the compound ABC. Excluding trimolecular reactions as sufficiently improbable to be ignored one could have A first reacting with B to give the intermediate AB, or one could have B reacting with C to give the intermediate BC. These are indicated in Figure 1. The rate constants $k_1, \ldots, k_{-4}$ refer to the process indicated by the arrow they stand beside. The principle of detailed balancing asserts that at equilibrium the forward and reverse rate of each reaction step must be equal. This imposes the relation

$$k_1 k_2 k_3 k_4 = k_{-1} k_{-2} k_{-3} k_{-4} \qquad (17)$$

upon the rate constants. To indicate the relation between this equation and the thermodynamic properties, equation (17) can be rearranged into the form

$$\frac{k_1}{k_{-1}} \frac{k_2}{k_{-2}} \frac{k_3}{k_{-3}} \frac{k_4}{k_{-4}} = K_1 K_2 K_3 K_4 = 1$$

where $K_1, \ldots, K_4$ are the equilibrium constants for the various steps. Inasmuch as the relation $\Delta F_0 = -RT\ln K$ exists between the equilibrium constant, K, of a reaction and its standard free energy ΔF_0, the relationship is evident.

J. BOYD BEST

References

Best, J. B. and J. Z. Hearon, *in* Bronner, F. and C. Comar eds., "Mineral Metabolism," Vol. 1, New York. Academic Press, 1960.

Fruton, J. S. and S. Simmonds, "General Biochemistry," 2nd ed., New York, Wiley, 1958.

Hutchens, J. O., Federation Proc., **10**: 622, 1951.

Lumry, R. *in* Boyer, P. D. *et al.* eds., "The Enzymes," Vol. 1, New York, Academic Press, 1959.

Podolsky, R. J., Ann. N.Y. Acad. Sci., **72**: 522, 1959.

THYMUS

The thymuses are two bilaterally paired organs which develop from the embryonic pharynx. They are present in all animals from the bony fish to mammals. In fish, the left and right thymus lie widely separated from each other at the lateral extreme of the body, just under the gill cover. In reptiles and amphibians, the thymuses are less widely separated from each other. In birds, the two thymuses lie in the neck, each one consisting of an irregular string of lobes. In some mammals like the guinea pig, the two thymuses persist in the neck, but in the majority of mammals, they lie in the chest, just above the heart. Because the two thymuses lie close to each other in mammals, they had been mistakenly regarded as being fused in the midline, and were erroneously regarded as "lobes" of a single midline organ, "the thymus." This old error in gross anatomy still persists widely today. The correct gross anatomy of the thymuses is shown in Figure 1.

Each thymus is partially divided into many lobules, consisting of an inner medulla, and an outer cortex. The medulla contains many epithelial cells, some of which form Hassall's corpuscles. These cyst-like epithelial structures may possibly play a role in eliminating unwanted cells within the thymuses. Cells have been shown to enter these cysts and to be digested therein. The cortex may be regarded as a loose sponge-like mesh of epithelial cells, which serve as a matrix to store lymphocytes. As lymphocytes accumulate within this storage area, the cortex expands and the thymuses enlarge. Since the number of epithelial cells does not increase appreciably after birth, the size of the thymuses thus depends solely on the number of stored lymphocytes, and not on intrinsic thymic growth.

The thymic lymphocytes are extremely sensitive to adrenal-mediated general body stress. Following stress of any kind, the adrenal glands secrete cortisone into

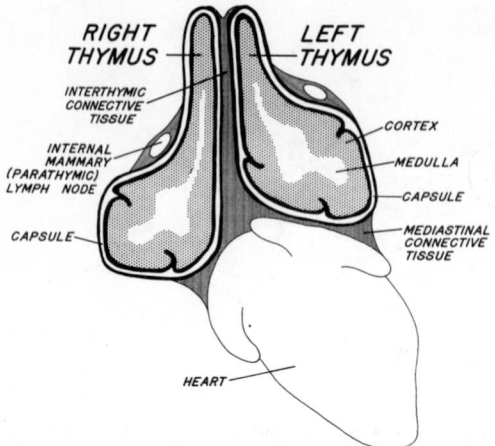

Fig. 1 Diagrammatic representation of the anatomic relations of the left and right thymus in the anterior mediastinum of most mammals.

the bloodstream. The thymuses respond to this cortisone by rapidly lysing or discharging their stored lymphocytes. After lysis and discharge of the cortical lymphocytes, the thymuses shrink in size and become quite small. After a period of time, however, regeneration of the lymphocytes occurs, and the thymuses re-expand. This rapid *physiologic* lymphocyte depletion was formerly termed "accidental involution" because of its relation to severe trauma of accidents. After puberty, the thymuses store fewer and fewer lymphocytes, until in the adult, the thymuses appear as two thin strap-like cords deeply embedded in the fat of the anterior mediastinum.

Formerly, thymuses were studied only in humans, who usually had died in hospitals after prolonged stress, and had small, lymphocyte-depleted thymuses. When, by chance, a child died suddenly, with no previous history of illness, a normal-sized thymus was found, which was incorrectly thought to be *enlarged*. Because earlier anatomists had no idea of the normal size of the thymuses, they established a disease entity for sudden death in which "a large thymus" was found at autopsy—the so-called "status thymico-lymphaticus." Many children had their chests irradiated when a normal thymic shadow was visualized by X-ray, in the erroneous belief that they had "an enlarged thymus." Many of these children have developed cancer of the thyroid gland as a result of this radiation.

The function of the thymuses is at present only partly understood. The relation of the thymuses to immunity was established only recently, when it was found that removal of the thymuses in newborn mice made them unusually susceptible to infection, which led to wasting and death. It is not yet certain whether the thymuses exert their effect in initiating and promoting the immune response by way of a hormone, by dissemination of thymic lymphocytes, or by both methods.

It is generally agreed that the thymuses exert some primary role in regulating the immune defenses of the body, but the details of this mechanism are not yet known.

The thymuses are related in some obscure manner to other diseases. Tumors of the thymuses have been found in some cases of myasthenia gravis, immunologic abnormalities, and anemias.

The thymuses are the site of origin for lymphatic leukemia in the mouse. Whether they have this role in human lymphatic leukemia is not known.

RICHARD SIEGLER

References

Good, R. A. and A. E. Gabrielsen, eds., "The Thymus in Immunobiology," New York, Harper and Row, 1964.
Metcalf, D., "The Thymus," New York, Springer, 1966.
Rich, M. A., ed., "Experimental Leukemia," New York, Appleton-Century-Crofts, 1968.

THYROID

Morphology. The thyroid gland, found in all vertebrates from the cyclostomes up, is a structure ranging from a shield-like (Greek, Thyreos) appearance to a completely separated pair of ovoids, lying below the cricoid cartilage, near or even adherent to the trachea. It receives an exceedingly rich blood supply, a fact which caused the ancients to consider that it had an important function in regulation of blood supply to the head.

Microscopic examination of the thyroid reveals a mass of follicles varying in size from 15 to 100 μ in diameter. Each follicle is normally a spheroid shell of thin flattened epithelium, one cell thick, enclosing a central mass of proteinaceous colloid, the *thyroglobulin.* When the gland becomes hyperactive, the cells increase in height and in number. At the same time, there is depletion of colloid, producing a characteristically altered histological appearance, as seen in Fig. 1. These changes occur whenever there is demand for thyroid hormone in excess of the "resting" thyroid cells to produce it and is seen in chronic iodine deficiency, treatment with goitrogenic chemicals or the human clinical syndrome of thyrotoxicosis, involving hyperthyroidism.

Function. The thyroid is uniquely able to concentrate iodide from the blood plasma and to activate it in such a way that it adds onto the amino acid tyrosine, present in the protein of the colloid. Mono- and diiodotyrosine make up half to two-thirds of the organic iodine of the thyroid, most of the remainder being thyroxine formed by the combination of two diiodotyrosines. The discovery in 1952 of small amounts of 3,5,3'-triiodothyronine stimulated a successful search for the other partially iodinated thyronines, 3,3'-diiodothyronine and 3,3',5'-triiodothyronine. No thyronine or 3,5-diiodothyronine has been encountered, suggesting some specific requirement for at least one iodine on each of the two tyrosyl moieties being considered.

Proteolysis of thyroglobulin results in the production of free amino acids including monoiodotyrosine, diiodotyrosine and thyroxine. The first two of these are rapidly deiodinated, so that the principal substance liberated into the blood stream is thyroxine. Some, 3,5,3'-triiodothyronine is often found in the blood,

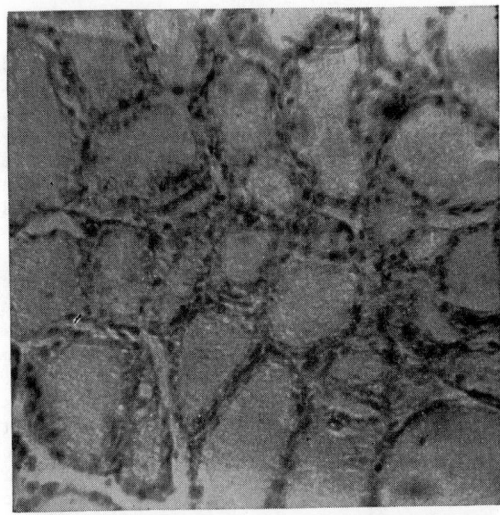

Fig. 1   Photomicrographs of rat thyroids: (above) normal, (below) thyrotrophin-stimulated.

and extremely small quantities of the 3,3'-diiodo- and 3,3',5'-triiodothyronines mentioned above have been reported.

Transport. Thyroxine is carried in the blood plasma

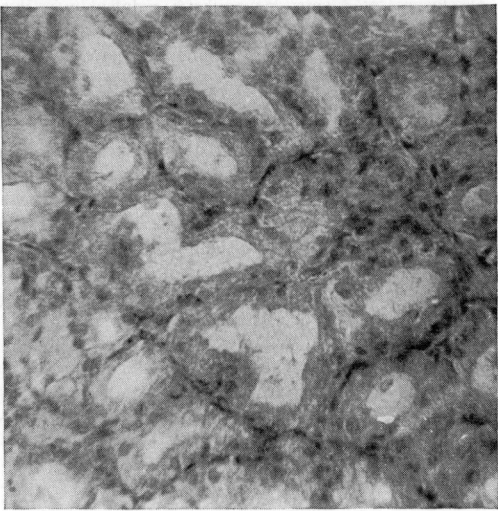

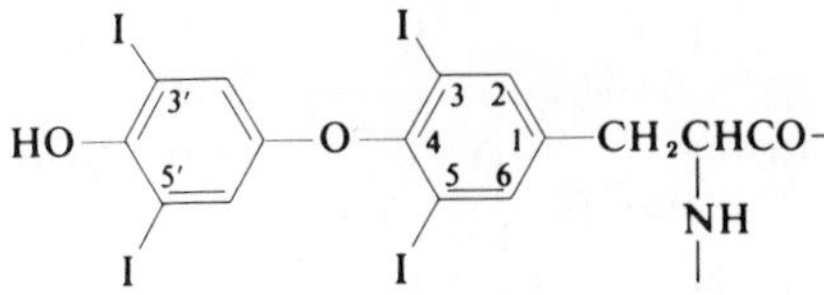

Fig. 3

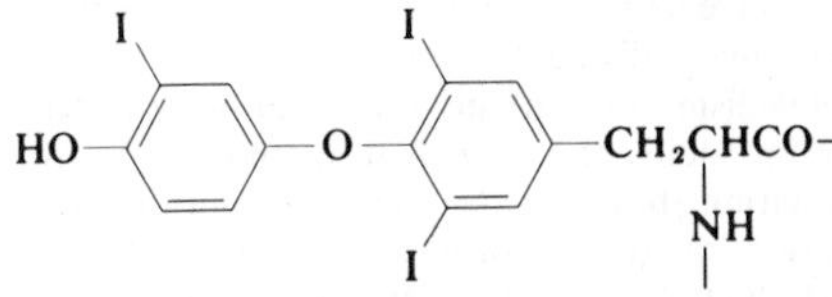

Fig. 4

in loose combination with a protein which migrates electrophoretically as an α-globulin in veronal buffer and as a "pre-albumin" in tris buffer. Triiodothyronine, when present, is much less specifically attached to any protein.

Target cell activity. It has long been known that thyroxine is involved in a highly diverse series of biological reactions: metamorphosis, growth and development, metabolic rate, heart rate, neuromuscular activity, reproduction, cerebral activity, fluid and electrolyte balance, etc. It is undoubtedly far too much of an oversimplification to state that control of metabolic activity is the basis of all these. Discrepancies have been noted specifically between metabolic stimulation and both amphibian metamorphosis and lowering of serum cholesterol values in the case of certain analogs with fatty acid side-chains. In other instances, a close relationship between energy metabolism and cerebral function or ovarian activity has never been demonstrated.

Nonetheless, the general proposal that thyroxine action is somehow related to its fundamental control of metabolic activity remains most attractive. Many theories are current as to how thyroxine acts metabolically, including dissociation of oxidation and phosphorylation ("uncoupling"), production of alterations in mitochrondrial morphology and even direct participation in enzyme systems. At present, it is safe to say that no single proposal is fully satisfactory, although it seems probable that the hormone is in some way responsible for the transformation of some of the large

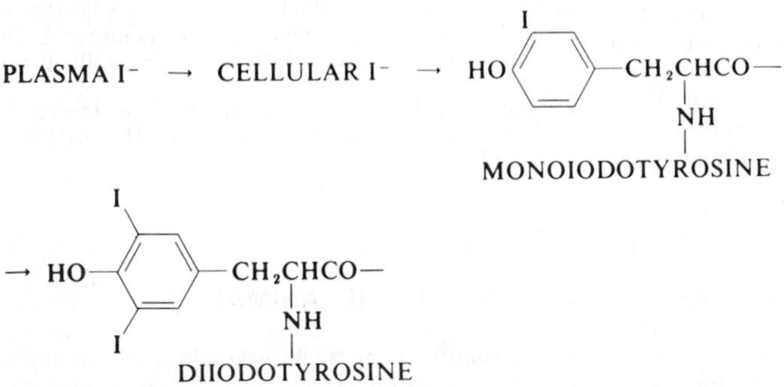

PLASMA I⁻ → CELLULAR I⁻ →

MONOIODOTYROSINE

DIIODOTYROSINE

Fig. 2

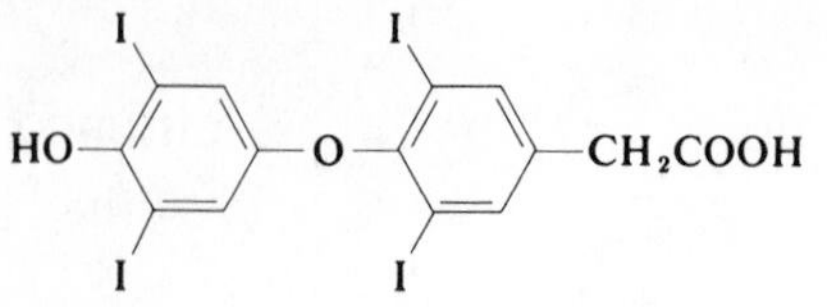

Fig. 5

amount of enzymatically inert protein present in cells into enzymatically active forms.

Metabolism of thyroxine. As would be expected from an amino acid, the principal metabolic pathway is deamination followed by decarboxylation with the formation of tetraiodothyroacetic acid a derivative having about 25% of the metabolic activity of thyroxine.

Small amounts of triiodothyronine may also be formed by partial deiodination of thyroxine. This compound is more potent than thyroxine, possibly because of more rapid penetration. Triiodothyroacetic acid is encountered as a product of deamination and decarboxylation of triiodothyronine.

At least two types of conjugates of thyroxine and related compounds have been identified, those with glucuronic and with sulfuric acids.

Attempts have been made to ascribe special activity to one or another of the derivatives of thyroxine mentioned. It is an intriguing thought that the appreciable latent period always found for thyroxine is due to formation of an active derivative, but scattered evidence for such a substance has thus far yielded only discrepancies.

S. B. Barker

References

Pitt-Rivers, R. and J. R. Tata, "The Thyroid Hormones," New York, Pergamon Press, 1959.
Pitt-Rivers, R. and W. R. Trotter, "Thyroid Gland," 2 vols., New York, Plenum, 1964.

THYSANOPTERA

The Thysanoptera, commonly called thrips, is one of the smaller orders of INSECTS. It contains nearly 5000 species whose members average about 2 mm. in size (extremes 0.6 to 14 mm.). Thrips belong to the Corrodentia-Hemiptera phyletic line, presumably derived from pre-hemipterous ancestors.

Based on fossils from Ural shale, the order has existed since Permian times. Other fossils, nearly 100 species, have been found in Cretaceous and Eocene amber, and in Tertiary shales, nodules, and peat. Modern Thysanoptera are abundantly represented throughout most of the world, especially in the tropics and warm temperate regions.

Thrips occur in most major habitats. A large number feed on juices of living plants. They cause damage to grass (*Poa*), gladiolus, corn, oats, cotton, tobacco, onions, *Ficus,* rice, banana, privet hedge, lilac, and many fruits, herbs, and flowers. Besides living plants, thrips dwell and feed under bark of dead trees, in standing dead leaves, in ground litter of forests, in

polypore fungi, and in crowns of grass clumps. A few thrips attack other insects and mites.

Reproduction is principally by means of bisexual union, although unisexual reproduction (parthenogenesis) often occurs. Larvae hatch from deposited eggs, except in several species purported to bear the young alive. There are two larval instars, and one to three pupal instars. Some spend the duration of the pupae in cocoons, some in earthen cells, but the majority spend the pupal stage uncovered above ground.

The Thysanoptera is divided into two suborders. The suborder Terebrantia, the group in which the females bear sawlike ovipositors, contains five families: Aeolothripidae, Heterothripidae, Merothripidae, Uzelothripidae, and Thripidae. The second suborder, the Tubulifera, contains only the family Phlaeothripidae.

Thrips may be distinguished from all other insects by the characteristic of the absence of the right mandible. The head is opisthgnathus, that is, the base of the mouth parts is directed posteriorly. Each antenna is composed of four to nine segments and often these segments support sensoria. Adults have compound eyes of closely set facets; larvae have reduced eyes with only a few facets. Ocelli are usually present in adults, never in larvae. The single mandible and the lacinae of the maxillae are produced into stylets. Maxillary and labial palps are present, although each is usually reduced to a few segments. The wings when fully developed are two paired, ordinarily slender, and fringed with long setae. Wing veins are reduced to a few major veins or are absent. The legs often display spurs, toothlike projections, and tactile setae. In all cases the tarsi, which are one or two segmented, bear claws and protrusible bladders. The abdomen contains ten distinct segments, many of which may have wing-holding setae. Cerci are absent. Three or four pairs of spiracles are present. Frequently males, and sometimes females, exhibit abdominal sternal glands. Often parts of the body are sculptured with hexagonal or striate designs.

Major and minor forms exist, in addition to apterous, brachypterous, micropterous, and macropterous forms, even within one species.

Lewis J. Stannard

References

Bailey, S. F., "The Thrips of California, Part I: Suborder Terebrantia," Berkeley, Univ. of Calif. Press, 1957.
Cott, H. E., "Systematics of the suborder Tubulifera (Thysanoptera) in California," Univ. of Calif. Pub. in Ent., **13:** 1–216, 1956.
Priesner, H., "Genera Thysanopterorum," Bull. Soc. Fouad 1er Entom., **33:** 31–157, 1949.
Stannard, L. J., "The Phylogeny and Classification of the North American genera of the suborder Tubulifera (Thysanoptera)," Urbana, Univ. of Ill. Press, 1957.
Zimmerman, E. C., "Thysanoptera," *in* Insects of Hawaii, Vol. 2, Honolulu, Univ. of Hawaii Press: 387–475, 1948.

TISSUE CULTURE, ANIMAL

Tissue culture is the name generally given to any of several methods for keeping small pieces of animal or plant tissues alive outside of the body. The methods consist primarily of removing a bit of living tissue (with a minimum of mechanical damage and with protection from bacterial contamination) and placing it in a sterile, suitable medium in an appropriate chamber at the body temperature of the donor. This suffices for short-time observations: a day or two for warm-blooded animals to a week or more for cold-blooded animals, and longer for plants. To keep cells alive and normally active for extended periods, there must be intermittent or continuous change of the medium to remove the waste products of their metabolism and to supply nutriment. If there is rapid cell division, cell density must be controlled by subculturing or by inhibiting cell division.

The methods of tissue culture have been applied most successfully to problems of cytology, histology, embryology, physiology, immunology, and pathology. Plant tissue culture is discussed under a separate entry. Here, the discussion is limited to the culture of animal tissues, and of these, vertebrate tissues have been used most successfully. Tissue culture offers the investigator the theoretical advantage of isolating a group of cells from physical, neural, and chemical influences of the body, and so maintaining the cells under known and controllable conditions. It theoretical disadvantage is that the cells may be exposed to artificial and unphysiological media or to foreign bodies that invoke inflammatory reactions, wound-healing, and regenerative responses. Often it is difficult to interpret the results of tissue culture experiments or to reconcile the results with *in vivo* observations.

Yet there are many biological problems that defied solution until the methods of tissue culture were applied to them. For example, such methods have made it possible to show that nerve cells develop axons independently of adjacent cells, that many kinds of cells may continue to divide and grow for a period longer than the life span of the donor; that cells generally maintain their type in tissue culture but also may mutate to other relatively stable types; that most cell structures seen in fixed and stained histological preparations are valid and not artifacts; that the promordia of embryos may undergo self-differentiation; that certain embryonic structures influence the differentiation of certain other structures. Cells in tissue culture can be shown to respond to inflammatory stimuli and to undergo characteristic changes; virus particles can be seen to enter cells and increase in number; cells can be observed to change from normal to malignant types and to retain their new characteristics. Adult cells can be shown to continue their adult functions when appropriate conditions are present, etc. Some of these points will be discussed in more detail after the general techniques are described. The use of tissue culture methods in applied biology is noteworthy, for example, in the production of vaccines (e.g., poliomyelitis) and in the study of agents that change normal cells into cancer cells or destroy cancer cells.

The simplest method of tissue culture is the slide culture method. A tissue fragment, about 1 mm. in diameter, is placed (explanted) into a drop of clotting lymph or plasma on a coverslip; then a 1 × 3 inch depression slide is inverted over the coverslip and sealed with vaseline and paraffin. Such a preparation lasts two or three days (for mammalian and avian tissue at 38°C.) to several weeks (for cold-blooded animals at

room temperatures). Cells wander about phagocytize, divide and differentiate, depending upon the type of cell and the circumstances. The cells die within a few days unless the medium is changed.

To keep the preparation alive longer, it must be subcultured or the medium renewed. To subculture, the fragment may be excised with a sharp fine knife, divided into two or more pieces, and each of these explanted into fresh drops of medium on clean coverslips. An alternate method is to make the original preparation in such a way that the culture-bearing coverslip adheres to a larger (40 × 40 mm.) coverslip by means of a drop of water. The pair of coverslips is then covered with a 2 × 3 inch depression slide. After two or three days, instead of subculturing, and thereby disturbing the cells in their matrix, the cell metabolites are washed out by transferring the culture-bearing coverslip to an appropriate physiological saline solution for about 15 minutes. The bathed culture is then mounted on a fresh 40 × 40 mm. coverslip and covered as before. Such slide cultures readily permit microscopic examination. In a more refined method, known as perfusion, the tissue is mounted on a coverslip as before, but the coverslip is placed in a chamber which has a second coverslip held a mm. or two below the culture-bearing coverslip. Between the two is a ring with two openings through which a stream of nutrient medium is passed. In perfused cultures cells may be kept alive and functioning for long periods and under continuous observation. The medium may be constantly renewed, may be altered from time to time or chemical substances may be added or removed.

Larger numbers of cells in larger volumes of medium are kept in tubes, flasks, or bottles. Tube cultures are made by streaking the sides of test tubes (15 × 150 mm. is standard) with tissue fragments, then adding medium and stoppering. Every few days the medium is changed. If the tubes are placed in a drum and the drum rotated, cells may cover the walls of the culture tubes. This method is useful for maintaining large numbers of cell strains, or for performing experiments requiring many easily handled replicates. It has the disadvantage of limiting observation to low magnifications, although special flat-sided tubes overcome this defect to some extent. When enormous numbers of cells or special conditions are required, such as in certain virus cultures or in obtaining a sufficient quantity of some metabolite for further analysis, bottles of various shapes and sizes, even up to five gallons capacity, may be used for growing massive cultures.

Subculturing from tubes, flasks, and bottles may be accomplished by removing the cells from the walls of the container mechanically, as by scraping or shaking, or chemically, as by digesting away the adhesive substance with trypsin. The cells are concentrated by centrifuging, washed free of old medium, and replanted in fresh bottles with new medium.

Cells grown in suspension, by keeping the medium constantly agitated, are useful for studies in which estimates of the cell population are necessary. Cells may also be grown in a matrix of perforated cellophane, glass fibers, glass cloth, gelatin sponge, cellulose sponge, or other inert surface. Such dense populations of tissue cells are useful in biochemical studies and for problems requiring tissue structure that is more organized than that of the flat sheets of cells produced on glass surfaces.

Media for tissue culture vary with the problem at hand, but generally consist of an aqueous solution of inorganic ions, glucose, aqueous extracts of embryonic tissues, amino acids, vitamins, and other metabolic substances. Cells will survive (but will not grow) in isotonic salt solutions containing sodium, potassium, calcium, magnesium, chloride, bicarbonate, and phosphate ions and also glucose. Cells will survive, grow and divide when placed in serum and a saline extract of embryonic tissue. A complex mixture of proteoses and enzymatic hydrolysates may replace nearly all, but not all, of the tissue extract, because cells can utilize large fragments of proteins but not as their only source of nitrogen. A mixture of the "essential" amino acids may serve as the source of nitrogen, but growth is not as rapid as it is when some of the complex nitrogen compounds are present as well. Cells of different kinds and from different species differ with respect to their nutritive requirements, so that no one formula may be used to support maximum growth of all cells. Furthermore, our scanty knowledge of these differences does not yet permit us to prescribe formulas either for maximum rate of cell division or for maximum survival time without cell division. Fortunately, the study of cell nutrition is one of the most active fields in cell physiology. In addition to providing the compounds already mentioned, it is necessary to adjust the pH of the medium to a proper physiological level (usually 7.4 to 7.6) and to maintain oxygen and carbon dioxide tensions at physiological levels appropriate to the cells studied.

Though the methods of animal tissue culture have contributed importantly to many fields of biological research, only a few can be mentioned here. *Cytology.* Mitosis and meiosis have been studied in great detail and under phase contrast cinephotomicrography. Much evidence from in vivo studies has been confirmed and greatly elaborated. The behavior of normal and malignant cells, under various experimental conditions, such as after radiation and after treatment with various chemical substances, has been carefully followed. Many so-called artifacts in sections and stained preparations have been shown to be normal cell structures, clearly visible in unstained living cells. Single cells have been isolated and grown into cell strains with well-defined characteristics. *Histology.* The behavior of cells in tissue culture indicates that scattered cells are able to reform epithelial sheets, reorganize into kidney tubules, and form other structures such as secretory glands. Hemopoesis can be observed in tissue culture, and bone seen to form. *Embryology.* Almost all embryonic precursors continue to differentiate, and the influence of one type of tissue in the differentiation of another has been extensively studied. *Physiology.* Biochemical studies have been made of a number of links in the metabolic sequences of glucose metabolism, nitrogen metabolism, and the formation of hormones. The effects of hormones and other substances have been followed. *Immunology and virology.* Most significant here are studies on the effects of virus particles on cells and the production of additional virus particles; and considerable work has been done on the relation between macrophages and antibody formation. Much is known about the phagocytosis of bacteria *in vitro*. *Pathology.* Malignant cells have been studied and compared with normal cells under various experimental conditions. Effects of external radiation and of

radioactive isotopes in the medium have been the subject of much work. Bacteria, animal parasites, and toxic agents have been added to cultures. Aging in relation to cell behavior is being studied. Inflammation has been the subject of a great many tissue culture investigations.

As mentioned earlier, the cells in tissue cultures are exposed to glass and to other "foreign bodies," therefore much of the "normal behavior" of cells *in vitro* must be regarded as inflammatory behavior. Despite the inherent limitations, the methods of tissue culture are indispensable for attacking any biological problem in which it becomes necessary to observe living cells continuously under high magnifications, or to isolate living cells from the more complex environmental influences of the whole organism in which they are found.

RALPH BUCHSBAUM

References

Murray, M. P. ed., "A Bibliography of Research in Tissue Culture," New York, Academic Press, 1953.

White, P. R., "The Cultivation of Animal and Plant Cells," New York, Ronald, 1954.

Willmer, E. N., "Tissue Culture," 2nd ed., New York, Wiley, 1954.

Fell, H. B., "Methods for study of organized growth in vitro," Methods in Med. Res., **4**: 233–237, Chicago, Year Book Pub., 1951.

Evans, V. I., *et al.*, "The preparation and handling of replicate tissue cultures for quantitative studies," J. Nat. Cancer Inst., **11**:907–927, 1951.

Harris, M., "Cell Culture and Somatic Variation," New York, Holt, Rinehart and Winston, 1964.

Paul, J., "Cell and Tissue Culture," Baltimore, Williams and Wilkins, 1960.

Wilmer, E. N. ed., "Cells and Tissues in Culture," 2 vols., New York, Academic Press, 1965.

TISSUE CULTURE, PLANT

Plant tissue culture provides a method by which living plant tissues may be maintained for prolonged periods outside the organism. The excised tissue is sustained by a nutrient medium which may also support additional growth and proliferation. When all the nutritional requirements of the tissue are met by the culture medium it is possible, by the use of appropriate subculture techniques, to maintain continuous growth through potentially unlimited periods of time. Precautions must be observed to free the tissue, the culture medium, and the various manipulative procedures from contamination by microorganisms; cultured tissues are, therefore, sterile. However, various pathological plant tissues may be grown in culture, and previously sterile cultured tissues may be purposely infected.

Although the definition emphasizes the culture of plant tissues, the techniques have proven equally applicable, on the one hand, to the culture of excised organ fragments, whole organs and intact organisms and, on the other, to the culture of cell populations and single isolated cells. Some workers, as alternatives for the name tissue culture, have used such designations as sterile culture, *in vitro* culture, callus culture, organ culture, or cell culture to describe the culture of various plant parts.

The particular advantages of the tissue culture method derive from the ability to place the excised tissue in a chemical and physical environment which can be more accurately specified than the environment of the same tissue forming part of the organism. Thus the normal, but largely unknown, interactions of the organism are eliminated, chemically defined substances can be supplied quantitatively, and the physical environment can be critically controlled.

History. The first systematic attempt at plant tissue culture was undertaken by the German botanist, G. Haberlandt concerning the changes in zygotic totipotence and the physiological interactions between cells during growth. His attempts in 1902 to maintain living plant cells and tissues outside the organism met with failure. However, within a few years, in the hands of zoologists the methods suggested by Haberlandt's work were successful and it became possible to grow animal tissues in culture. It was not until 1939 that plant tissue cultures capable of unlimited growth were first achieved, and only in 1956 were single isolated plant cells successfully grown in culture.

Despite the difficulty experienced in obtaining plant tissue cultures progress had been made in another direction, the growth of excised organs. In 1934 tomato roots were grown through successive subcultures without any diminution in the rate of growth, and by 1937 a chemically defined medium, which would support their continuous growth, had been developed. Since that time a large number of different organs and plant parts has been grown in culture. Among these are stem tips, leaves, flowers, stamens, ovaries, ovules, gametophytes, endosperm and embryos in addition to tissues from several kinds of normal organs, pathological tissues of crown-gall, insect, or viral origin, and spontaneous, genetic, and chemically induced tumors and galls.

Most attention has been directed to the culture of tissues and roots and, in all, tissues from about 100 species and roots of about 35 species of plants have been grown in culture. These, however, have been derived from a very restricted range of plants. Root cultures have been derived almost exclusively from herbaceous dicotyledonous species, and tissue cultures principally from dicotyledons, although several gymnosperm tissue cultures have been established in recent years. The monocotyledons and the pteridophytes are markedly under-represented in root and tissue culture studies, but gametophytes of numerous pteridophytes, and stem tips and excised leaves of several species of ferns have been grown in culture. The gametophytes of some bryophytes have been grown in culture, but the basic technique, that of growing a detached portion of an organism in isolation, has not been generally adopted by cryptogamic botanists. The pure culture techniques of microbiologists were developed independently of those in plant tissue culture, but the recently developed free cell cultures, in which higher plant tissues are dissociated into cell populations, should permit the use of microbiological techniques in approaching certain problems in the vascular plants.

The earlier preoccupation with the largely technical problem of obtaining rapid, unlimited proliferation of cultured plant tissues as an end in itself has given way in the past few years to a situation where plant tissue and organ culture has emerged as a general tool of research which is increasingly being employed in studies of nutrition, metabolism, auxin physiology, histogenesis, MORPHOGENESIS, pathology, cytology, RADIATION EFFECTS, tumorization and many more specialized areas of botany.

Establishment and maintenance of cultures. Cultures are established by placing a suitable piece of plant material on or in a suitable nutrient medium in a closed container under aseptic conditions. Considerable success has resulted from the use of plant parts already in an active state of division at the time of excision from the plant. For this reason root tips and stem tips from newly germinated seeds are favored material. Similarly, fragments from the dividing cambial zone (see CAMBIUM) of woody plants, or the immature elongating zone of the stem usually grow well in culture. However, mature PARENCHYMA from fleshy organs such as the tuber of potato, the root of carrot, or from tobacco stem pith may undergo dedifferentiation and resume meristematic growth.

The plant material must either be initially free of microorganisms, as are most internal tissues and many stem tips, or must first be rendered sterile by brief immersion in ethyl alcohol, dilute chlorine or bromine solutions, or occasionally by exposure to high temperature, ultraviolet light, or antibiotics.

The composition of the nutrient medium is of great importance and in several laboratories considerable effort has been devoted to obtaining chemically defined media which will support continuous tissue growth. Such media usually consist of the following fractions: (a) a carbon-energy source, (b) a mixture of major inorganic salts, (c) a trace element mixture, (d) various growth supplements e.g. B-vitamins, hormones. Frequently, pieces of tissue excised from a plant will undergo some growth on a plain agar medium, but as the stored nutrients are depleted further growth is dependent on those supplied in the culture medium. When new tissue cultures, whose nutritional requirements are unknown, are being established some workers prefer to make unit additions in the above 4 fractions until a suitable synthetic medium is devised. Others prefer to replace or to supplement fraction (d) with various mixtures of unknown chemical composition which may initiate or sustain growth. These include coconut milk, horse chestnut milk, yeast extract, malt extract, hydrolysed casein, and occasionally tomato juice or watermelon juice. The active component of these additives may later be established by substitution in the medium.

Four different procedures have been commonly used for growing plant parts in culture (a) agar solidified medium, used for growth of many tissues and organs (b) stationary liquid medium, used principally in root culture (c) shaking liquid medium, used for growth of cell populations (d) single drop media, used in single cell isolations.

When planted on an agar medium explanted tissues may begin to proliferate within a few days and in a few weeks may have given rise to voluminous parenchymatous callus outgrowths. The explant may not proliferate uniformly. Stem pieces, which are usually implanted in an inverted position, proliferate most at the exposed morphological base, and cambial explants proliferate most extensively from the face consisting of the youngest cells (Fig. 1a). Proliferation of the explant ceases

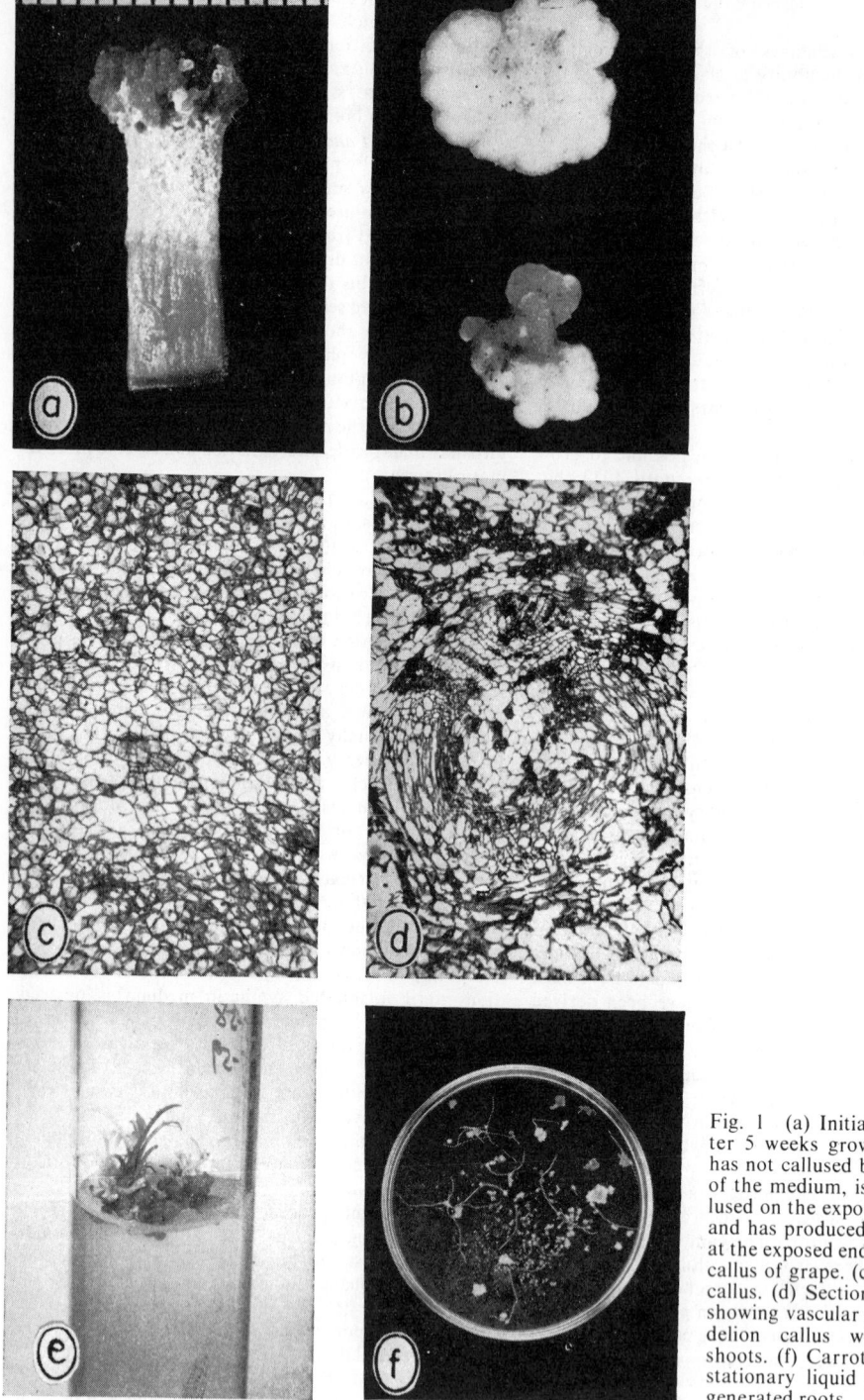

Fig. 1   (a) Initial explant of lilac ter 5 weeks growth. The explant has not callused below the surface of the medium, is moderately callused on the exposed cambial face, and has produced a massive callus at the exposed end. (b) Subcultured callus of grape. (c) Section of lilac callus. (d) Section of grape callus showing vascular nodule. (e) Dandelion callus with regenerated shoots. (f) Carrot callus grown in stationary liquid culture with regenerated roots.

in a few weeks but may be prolonged by subculturing fragments from the callus to fresh medium. The rate of growth in successive passages may decrease to zero due to dilution of substances not replaced by the culture medium. If, however, all necessary growth factors are supplied by the culture medium a uniform rate of growth occurs in successive passages.

Organ cultures are established on agar or in stationary liquid medium in a generally similar manner. Sterilized seeds provide sterile root and shoot tips on germination. Tightly compacted buds may be surface sterilized and used to obtain leaf primordia and floral organs which usually do not require further sterilization. Root and stem tip subcultures may be made from the main axis tip or from lateral tips in species where branching is prevalent.

Recently cultures of free cells (Fig. 2) have been obtained by placing a piece of a plant organ or a fragment from an established tissue culture into a shaking liquid culture. Such cultures may be continued indefinitely by subculturing a few drops of the cell suspension into fresh medium. Single cells removed from such cultures to a single drop of nutrient medium on a cover-glass have been used to establish cell clones in a few cases. Considerable difficulty has been experienced in inducing division in such cells, but division is sometimes stimulated if a "nurse-piece" of the original tissue is included.

Morphology. Tissue cultures possess a characteristic morphology which is different from that of tissues within the organism. On agar the tissue forms a more or less coherent mass (Fig. 1b) which may be firm e.g. lilac, or friable, falling spontaneously into smaller nodules e.g. carrot. Color is characteristic, usually being pale. Chlorophyll is absent or present in reduced amounts, even in tissue cultures derived from chlorophyllous organs. Lilac tissue is pale green, that of tobacco and grape is white, and carrot root tissue is yellow-green becoming scarlet in later stages of a subculture. Cultured tissues from different strains or varieties of the same species may appear different, but in many plants tissues derived from different organs resemble each other in appearance. From several plants, e.g. grape, Jerusalem artichoke, it has been possible to obtain "sets" of tissues of distinctive origin. These include normal tissue, tumor tissue of bacterial or viral origin, and habituated tissue, a spontaneous development from normal tissue growing in culture. These 3 kinds of tissues possess distinctive physiological properties but may differ morphologically in only minor ways. It is usually impossible, therefore, to categorize an unknown tissue by its morphology, and prior knowledge of its origin is necessary.

Histology. Only occasionally do cultured plant tissues consist of a single cell type. In a few cases cultures are entirely parenchymatous, but even here a degree of heterogeneity may exist in that certain cells contain tannins e.g. grape, or are enlarged e.g. lilac (Fig. 1c). Most cultured tissues are heterogeneous consisting, in addition to the parenchyma ground tissue, of xylem and phloem elements, and occasionally of periderm elements and laticifers also. Vascular elements are usually produced in nodules (Fig. 1d) which are formed by the activity of isolated cambia. There is no vascular continuity through the tissue. Some parenchyma cells may also become lignified forming occasional xylem elements. The degree of internal differentiation may undergo change in successive subcultures. Early subcultures of willow, pine, and grape produce vascular nodules which are not formed in later subcultures, while in lilac and Parthenocissus the early subcultures are entirely parenchymatous, but after several passages vascularization of the tissue occurs.

Experimental control of histogenesis has been made difficult because of the natural heterogeneity of most cultured tissues. However, differences in the proportion of lignified cells have frequently been induced. The addition of various auxins e.g. indoleacetic acid, or of lignin precursors e.g. coniferyl alcohol, to the culture medium has resulted in an increase in the number of lignified cells in many tissues and organs. Recently it has been found that the addition of dilute auxin solutions directly into cultured tissues through incisions or implanted pipettes also induces vascularization of the tissue.

Morphogenesis. Many tissues produce shoots e.g. Sequoia, or roots e.g. willow, some carrot clones, or both kinds of organs e.g. tobacco, some carrot clones, when growing in culture (Fig. 1e, f). The capacity for morphogenesis may remain constant through successive passages e.g. carrot, may diminish gradually e.g. certain tobacco tissues produce shoots and roots only in early subcultures, may increase e.g. Salvia tumor tissue produced roots only after prolonged culture, or may change with time e.g. elm callus in early passages regenerates shoots and in later passages roots are formed. Buds regenerated from cultured tissues have been excised and grown into adult plants.

The chemical basis of morphogenesis has been intensively studied in tobacco tissue cultures. Two groups of substances have been shown to be important, the auxins and substances which are components of nucleic acids. High levels of sugar, phosphate, various purines, and pyrimidines in the presence of low auxin levels stimulate bud initiation, but high levels of auxin inhibit buds. More recently the balance between kinetin (6-furfurylaminopurine) and indoleacetic acid has been shown to be critical. A low kinetin/IAA ratio favors root initiation, a balanced ratio promotes undifferentiated tissue growth, and a high kinetin/IAA ratio results in shoot initiation.

Tissue or organ cultures of several species are now known to regenerate entire plants by a developmental sequence that more or less resembles the steps in normal embryogeny. Vegetative *embryoids* have been obtained from *Apium, Citrus, Cuscuta, Datura, Daucus* (carrot), *Nicotiana* (tobacco), and *Ranunculus* among others. The development of carrot embryoids has been studied critically, and the origin can be traced to one or a few differentiated cells which divide to form a small globular mass similar to the globular proembryo of the sexually produced embryo, and a tail of larger, vacuolated cells similar to the suspensor of the normal embryo. Cotyledons, shoot and root apices, and internal procambium are then differentiated, and the fully developed embryoid closely resembles the normal embryo in size and morphology. The carrot embryoids continue to develop in tissue culture, and may then be planted out into soil where they grow into mature plants of normal appearance which flower, set fruit, and produce normal seed.

The vegetatively produced embryoids have significance for several different areas of plant research. In studies of development their origin from single differentiated cells of the tissue culture demonstrates that all the genes present in the zygote are present in each somatic cell, and that differentiation is not, therefore, caused by or accompanied by irreversible gene mutation or loss. In studies of plant propagation large numbers of genetically identical individuals could be produced quite rapidly by the embryoid technique. For example, it has been estimated that over 100,000 carrot embryoids developed in a single Petri dish on an agar surface of about 7 sq. in. Rapid propagation

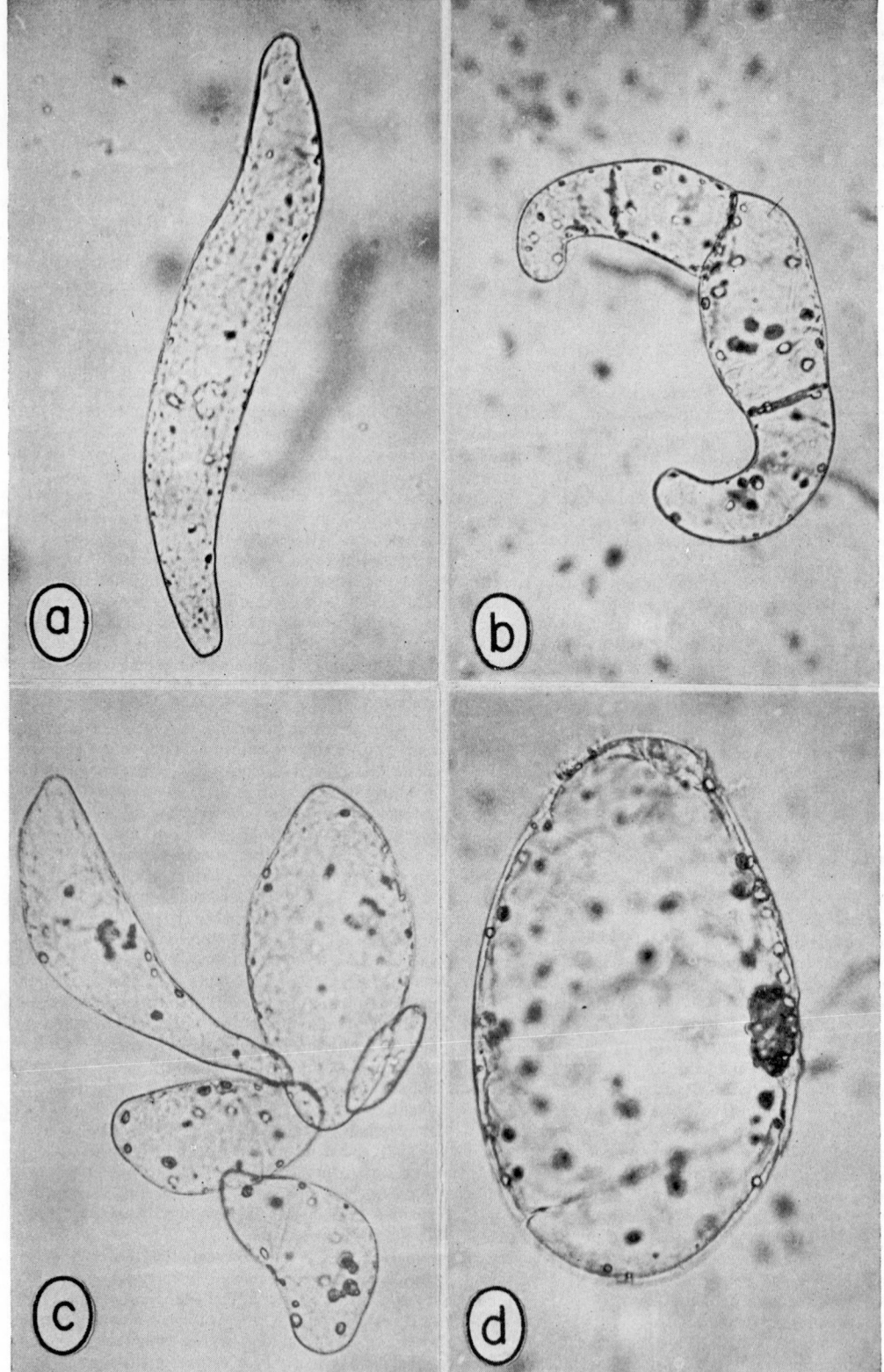

Fig. 2   Living cells from free cell cultures (a–c) carrot cells, (d) tobacco cell. All × 325. (a) Single cell. (b) Divided cell. (c) Cell cluster. (d) Single cell with nucleus at right center obscured by leucoplasts.

of desirable clones of crop plants or forest trees for breeding purposes might be economically possible using this technique. In studies of biochemical genetics the recent discovery that haploid embryoids of *Datura* and tobacco can be produced from mature pollen and microspores opens the way for the examination of gene controlled metabolic pathways using induced mutant individuals, which has previously been possible only in microbial systems.

Physiology. Tissue and organ cultures have been used extensively in the study of nutrition. The essentiality of various major elements, trace elements and vitamins for continued growth has been demonstrated. Considerable effort has been devoted to the importance of carbohydrates and auxins in tissue growth. In culture all tissues become heterotrophic for organic carbon. Even if the tissue is grown in light and was derived from a chlorophyll-rich organ it must be supplied with exogenous carbon. Normal tissues usually require also an exogenous supply of growth hormone for prolonged growth. Various natural auxins or synthetic hormones can maintain growth, as can complex substances such as coconut milk. Synergistic action between pairs of growth regulators occurs in some tissues e.g. potato callus. Certain tissues and particularly crown gall e.g. sunflower, and genetic tumors e.g. Nicotiana, are autotrophic for auxins and will grow indefinitely on an auxin-free medium. Nor does exogenous auxin stimulate their growth. Habituated tissue is auxin autotrophic, but exogenous auxin is beneficial, causing stimulation of growth. The exact nature of the tumorous transformation in crown gall or habituated tissue is not fully understood, nor are the occasional recoveries which have occurred in culture.

Little physiological work has yet been carried out on free cell cultures (Fig. 2), but they undoubtedly represent an important means of approaching several fundamental problems at a cellular, rather than at a tissue or organ, level.

IAN M. SUSSEX

References

Steward, F. G. *et al.*, "Growth and Development of Cultured Plant Cells," Science, **143**: 20–27, 1964.
White, P. R., "The cultivation of animal and plant cells," New York, Ronald Press, 1954.
ibid. ed., "Decennial review conference on tissue culture," Jour. Natl. Cancer Inst., **19**: 467–843, 1957.
White, P. R. and A. R. Grove, "Plant Tissue Culture," Berkeley, Cal., McCutchan, 1965.

TOOTH

Teeth vary in form from scale-like protrusions or bony plates in fishes to tusks or involuted molars in mammals. Fish and reptiles have large numbers and sometimes many successions of homodont teeth. In mammals the maximum number of 44 teeth is found only in dogs and pigs. They are replaced only once.

In *Carnivora* all teeth have narrow or cutting points and edges with no occlusal or grinding surfaces. In *Omnivora* the anterior teeth or incisors are wedge-shaped and the posterior molar teeth have broad, cusped and grooved surfaces. *Ungulata* have wide molars wearing into multiridged patterns. *Rodentia*

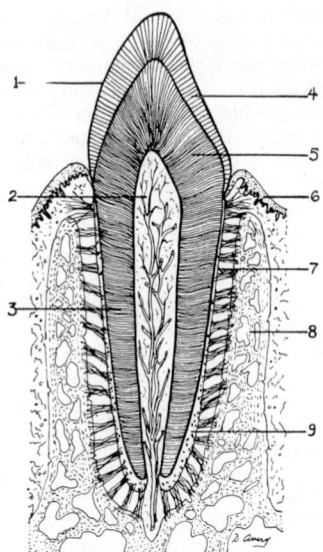

Fig. 1   Cross section of cuspid tooth. 1, enamel; 2, pulp; 3, root dentin; 4, enamel rods (diagrammatic enlargement); 5, crown dentin; 6, gingival epithelium; 7, periodontal fibers; 8, alveolar bone; 9, cementum.

have continuously erupting and self-wearing anterior teeth, and varied molars with compensatory eruption in some species.

Man has 20 deciduous teeth—2 incisors, 1 canine or cuspid, and 2 molars—and 32 permanent teeth, made up of 2 incisors, 1 cuspid, 2 premolars and 3 molars, in each quadrant. Eruption ages for deciduous teeth are, (± 4 months): Central Incisor 8, Lateral Incisor 9, Cuspid 16, 1st Molar 12, 2nd Molar 20 months; for permanent teeth, ($\pm 1\frac{1}{2}$ years): Central Incisor 7, Lateral Incisor 8, Cuspid 11, 1st Premolar 10, 2nd Premolar 11, 1st Molar 7, 2nd Molar 13, 3rd Molar 21 years. Lower teeth tend to appear slightly before upper, and girls' teeth approximately 6 months earlier than boys'.

In man, incisors are somewhat shovel-shaped, with a narrow cutting edge; cuspids are pointed with a narrow edge; bicuspids present cusps separated by an anterior-posterior groove, and have one or two roots; the molars, except the lower first and the third molars, present four cusps separated by anterior-posterior and buccal-lingual grooves or fissures; the lower first molar has five cusps. Upper first and second molars have three roots and the lowers two. The crowns and roots of third molars are more variable.

The crown, or exposed portion, of the tooth is covered by enamel and the root by cementum into which the periodontal fibers supporting the tooth are attached. Dentin makes up the bulk of the underlying tissue and surrounds the pulp chamber.

The enamel is mainly tricalcium phosphate. This exists in plate-like apatite crystallites up to several hundred angstroms in length which unite in parallel orientation to hexagonal to arch-like rods or prisms approximately four microns in diameter. Mucopolysaccharides and soluble proteins have also been described. Isotope studies show that the enamel is essentially inert, having only about 10% of the isotope exchange of the dentin, and 1% or less of that of the alveolar bone. Almost all of the isotope exchange is at the enamel surface supporting the concept that after eruption the enamel surface reacts with its salivary environment, losing carbon dioxide, but picking up organic material and trace elements such as fluorine and lead, thus contributing to the greater resistance of older teeth to dental decay.

Dentin contains approximately seventy per cent of inorganic material and thirty per cent organic material. The inorganic material is a calcium phosphate as unoriented apatite crystallites of smaller dimensions than in the enamel. The organic material is a matrix of collagen and the contents of tubules which extend from the odontoblasts of the underlying pulp throughout the whole width of the dentin. These extensions of the odontoblasts seem to be responsible for conducting sensation, but definite demonstration of nerve fibrils is lacking. The dentin shows greater isotope exchange than the enamel, which is in keeping with histologically demonstrable reactions in the form of transparent dentin or secondary dentin resulting from the stimulus of caries or tooth wear.

The cementum which covers the root closely resembles bone histologically and chemically. In it are embedded the periodontal fibers by which the tooth is attached to the bony alveolar plate or lamina dura of the alveolar tooth socket. This attachment is somewhat flexible allowing for tooth movement and is associated with a sensitive propriosensory mechanism which influences jaw movement in the final stages of tooth occlusion and mastication.

A cuff of epithelium, the epithelial attachment, extends from the gingival epithelium over the attaching fibers of the tooth socket to the tooth. In young teeth this fuses with the enamel cuticle which is the epithelial residue of the enamel organ, but when teeth erupt to compensate for occlusal wear, or when the gingival margin recedes, this epithelial attachment invests the cementum and fuses with cuticular material which forms thereon.

The tooth pulp which occupies the central portion of the tooth is connective tissue containing blood vessels and nerve fibers as they pass from the apical foramen of the tooth root to the region of the odontoblasts which make up the outer layer of the pulp subjacent to the dentin. Fibers or processes from the odontoblasts enter tubules in the dentin and continue across its width to the base of the enamel. Irritation of these processes causes odontoblastic reaction and sealing off of the dentinal tubules, and, if severe, a patch of secondary dentin at the periphery of the pulp.

Teeth are formed from an invagination of the alveolar epithelium differentiating into an enamel organ with an internal layer of ameloblasts which deposit the enamel. A corresponding mesodermal differentiation produces odontoblasts which form the dentin.

Destruction of the teeth results from dental caries which is most active in youth. While alternative theories are offered, the weight of evidence indicates that caries is primarily a dissolution of enamel by acids formed on the tooth surface by bacterial fermentation of carbohydrate foodstuffs, followed by decalcification and proteolysis of the dentin. Fluorine in the drinking water, or applied externally to the teeth, slows down the caries attack and reduces tooth destruction by

from 50%–70%. In adults tooth loss is principally caused by periodontal disease.

BASIL G. BIBBY

References

Jenkins, G. N., "The Physiology of the Mouth," Philadelphia, Davis, 1966.
Kronfeld, R. and P. E. Boyle, "Histopathology of the Teeth," Philadelphia, Lea and Febiger, 1949.
Scott, J. H. and N. B. B. Symonds, "Introduction to Dental Anatomy," Baltimore, Williams and Wilkins, 1958.

TOXINS

Toxins may be defined as high molecular weight poisons, usually protein in nature, that are produced by certain species of animals, plants or bacteria and which are highly toxic for certain other species. Most toxins are antigenic and when injected into animals in sublethal doses, or in a non-toxic form (toxoid), they induce the formation of antitoxins which specifically combine with the toxin in question and usually neutralize its toxicity.

As examples of toxins produced by animals, the snake venoms may be cited. Crotoxin, a neurotoxic protein, has been crystallized from the venom of the rattlesnake, *Crotalus terrificus* and highly toxid lecithinases have been isolated in crystalline form from the venoms of *Naja tripudians* and *Bungarus fasciatus*. Toxins are produced by certain insects, by scorpions and by other species.

Among the plant toxins, ricin has been isolated as a crystalline hemagglutinating protein from seeds of the castor bean, *Ricinus communis,* and the toxins abrin and crotin are found in the seeds of *Abrus precatorus* and *Croton Tiglium.*

The most studied toxins are those produced by certain pathogenic bacteria such as *Clostridium botulinum,* Type A, *Clostridium tetani* and *Corynebacterium diphtheriae.* The toxins produced by these three organisms have each been isolated in highly purified form, as heat-labile crystalline proteins the composition of which has been completely accounted for in terms of the known amino acids. They are among the most potent of all known poisons and almost all of the signs and symptoms of the diseases botulism, tetanus and diphtheria can be accounted for in terms of the corresponding bacterial toxin. The toxicity of botulinus, Type A and tetanus toxins is such that it has been calculated that 1 mg would be sufficient to kill some 1200 tons of susceptible animal (guinea pig). Although crystalline botulinus toxin (Type A) has a molecular weight of nearly 10^6, recent work suggests that it may be an aggregate built up from many toxic sub-units with molecular weights as low as 12,000.

When botulinus, tetanus or diphtheria toxins are treated with certain reagents, such as with dilute formalin at slightly alkaline pH and ordinary temperatures, they are converted to toxoids; they become completely detoxified without losing either their capacity to combine with antitoxin or their ability to induce the

Table 1   Properties of Bacterial Toxins

	Diphtheria	Tetanus	Botulinus Type A
Toxicity* for guinea pig	8×10^{11}	8×10^{13}	1.1×10^{15}
Toxicity* for mouse	8×10^{8}	1.3×10^{13}	0.5×10^{15}
Molecular weight	66,000	67,000	900,000
Sedimentation Constant	4.6S	4.5S	17.3S
Isoelectric point	pH 4.1	pH 5.1	pH 5.6

*Toxicity expressed as LD50 per kilo animal per mole of toxin

formation of specific neutralizing antitoxin upon injection into animals or man. Immunization of children with formol diphtheria and tetanus toxoids was first introduced in 1923 by Ramon. The diseases diphtheria and tetanus have virtually disappeared from those countries where universal immunization with toxoids has been introduced.

The mode of action of most toxins is not understood. Some of the toxins of snake venoms are hemolytic esterases attacking lecithin, and others have been shown to be enzymes attacking nucleotides. Many toxins produced by bacteria are cytolytic and cause lysis of red blood cells, leucocytes, etc., presumably by destroying essential structures on the cell membrane. Included in this group are toxins produced by hemolytic streptococci, staphylococci and many of the *Clostridia*. The hemolytic α-toxin produced by *Clostridium welchii*, for example, is an enzyme hydrolyzing lecithin to phosphoryl choline and a diglyceride. Botulinus, tetanus and Shiga dysentery toxins act specifically on nervous tissue. Little is known of their mode of action, but it has been shown recently that tetanus toxin combines specifically with certain gangliosides isolated from brain tissue. The action of botulinus toxin is restricted to the terminal twigs of the myoneural junctions, at a point proximal to the site of acetylcholine release. Because it has been most studied, diphtheria toxin is worthy of particular mention. This toxin is produced only by diphtheria bacilli infected by a particular temperate bacteriophage under conditions in which the iron supply becomes limiting and the organisms are becoming depleted in cytochromes. Recent studies with labelled toxin indicate that relatively few molecules, fixed to cell membranes of cultured cells derived from susceptible animal species, completely and irreversibly block protein synthesis within 1–2 hours. The mechanism by which protein synthesis is inhibited is now known. Diphtheria toxin catalyzes the rapid transfer of ADP-ribose from NAD (nicotinamide adenine dinucleotide) to a point at or near the active site of transferase II, one of the essential enzymes required for transfer of amino acids from aminoacyl-tRNA's to the growing peptide chains on the ribosomes. For each molecule of transferase II converted to its inactive ADP-ribosyl derivative, one molecule of nicotinamide is released.

A. M. PAPPENHEIMER, JR.

References

Gill, D.M., A.M. Pappenheimer, Jr., R. Brown and J. M. Kurnick, J. Exp. Med. **129:** 1, 1969.

Van Heyningen, W. E. and S. N. Arseculeratne, Ann. Rev. Microbiol. **18:**195, 1964.

Howie, J. W. and A. J. O'Hea, eds., "Mechanisms of Microbial Pathogenicity," Cambridge, The University Press, 1955.

Welsh, J. H., "Composition and Mode of Action of Invertebrate Toxins," Ann. Rev. Pharmacol. **4:**293, 1964.

Buckley, E. E. and N. Porges, eds., "Venoms," Publication #44, Washington, D. C., AAAS.

TRACE ELEMENTS

A century ago Sachs and Knop, by growing plants in solutions of known composition, claimed that along with carbon obtained from the carbon dioxide of the air and hydrogen and oxygen derived from water, plants could grow and complete their life-cycle if provided with salts containing altogether only seven elements which were nitrogen, sulphur, phosphorus, potassium, calcium, magnesium and iron. This view of the essential elements for plant growth was generally accepted for nearly half a century until in 1905 Bertrand showed that a small quantity of manganese was necessary for the successful growth of the oat plant, an observation which has since been extended to many other species. In subsequent years other elements were shown to be necessary for plant growth and by 1939 it was recognized that in addition to manganese plants required small quantities of boron, zinc, copper and molybdenum. These elements are required by plants in very small quantity and for this reason they are known as trace elements, minor elements or micro-nutrients. In larger quantity they generally act harmfully. Since 1939 some evidence has been produced that for some species other elements may be necessary, as for example, sodium for some marine and bluegreen algae, silicon for sunflower, maize, barley, rice and millet, aluminum for maize and millet, chlorine for sugar beet and some other plants and cobalt for some blue-green algae. However, it is still generally considered that for the vast majority of higher plants the five essential trace elements are manganese, boron, zinc, copper and molybdenum.

Shortage of any one of the trace elements may result in reduced growth and in a pathological condition which may end in the death of the affected plant. With shortage of manganese the first symptoms usually take the form of an intervenal chlorosis which with parallel-leaved monocotyledons results in the appearance of yellow or grey streaks between the veins and with net-veined dictoyledons of a mottling. Some of these conditions are frequent enough to have received names; such are the gray-streak of gray-stripe of oats, barley, wheat and rye, the Pahala blight of sugar-cane, speckled yellows of sugar-beet and the frenching of tung and mu-oil trees. With peas manganese deficiency results in the condition called marsh-spot in which brown or black spots or cavities appear on the inner surface of the cotyledons.

Boron deficiency generally leads to the death of the

stem apex and subsequent death of the apices of the side shoots which develop after the death of the apex of the main stem. Very generally there is disintegration of thin-walled tissues such as cambium, phloem and ground parenchyma. Diseases resulting from boron deficiency are heart-rot of mangold and sugar-beet, canker and internal black spot of red beet and brown heart of swede and turnip, all of which are characterised by a breakdown of the root tissues.

A pathological condition resulting from a deficiency of zinc has been particularly observed in trees where an intervenal chlorosis is followed by failure of the stem to grow in length and of leaves to grow to their proper size with the result that instead of elongated shoots bearing normal leaves the trees bear rosettes of small leaves. The conditions known as pecan rosette, little-leaf or rosette of apples, pears, peaches and other deciduous fruit trees, mottle-leaf or little-leaf of *Citrus* and rosette of *Pinus radiata* are all attributable to zinc deficiency.

Copper deficiency has also been observed with fruit trees including both *Citrus* and deciduous trees such as apples, pears and plums, the resulting condition being known as exanthema or die-back. Small swellings may develop on the young shoots, but particularly characteristic is the dying of the shoots backward from the apex. The reclamation disease of various crop plants such as cereals, beet and leguminous plants, occurring on reclaimed land in various parts of Europe and characterized by chlorosis of the leaf tips and failure to set seed, is also attributable to shortage of copper.

Deficiency of molybdenum first results in a partial chlorosis of the leaves which may be followed by death of the leaf margins, while seed formation may fail. The condition of cauliflowers and other *Brassica* crop plants known as whiptail in which most of the lamina of the leaves of the middle region of the stem dies so that the leaves consist of little more than their midribs is due to molybdenum deficiency.

Various functions have been ascribed to the trace elements in plants. It is clear that at least one reason for the necessity of the four metallic trace elements is that they form essential constituents of certain enzymes. A number of enzymes concerned in the oxidation of carbohydrates in respiration are activated by manganese and for one of them, oxalosuccininc decarboxylase, it may be essential although for others activation can be effected also by other metallic ions, as for example, magnesium. Zinc is a constituent of carbonic anhydrase which catalyzes the decomposition of carbonic acid to carbon dioxide and water, of alcohol dehydrogenase in yeast and of hexokinase in *Neurospora*. Several oxidizing enzymes contain copper or require it for their action. Such are polyphenol oxidase, tyrosinase, luccase and ascorbic acid oxidase. Molybdenum is intimately connected with nitrogen metabolism and is contained in the nitrate reductase system. It appears likely that some, at least, of these trace elements may have other functions but although much work has been done on the question no definite conclusions can as yet be drawn from it. Similarly, as regards boron, although a number of functions have been ascribed to it, no general picture of its mode of action can so far be discerned.

In animals, as well as plants, trace elements may be required. Manganese is widely distributed throughout the animal kingdom and may possibly be generally essential for the utilisation of vitamin B_1. Deficiency of

it in chicks results in deformity of bones, particularly those of the leg. Shortage of iodine in mammals is generally considered to be the cause of goitre; this element is sometimes described as a semi-trace element. The best known instance of a trace element in animals is cobalt, a deficiency of which causes a disease of sheep and cattle recognized as occurring in many parts of the world where the pasture plants contain an exceptionally low content of cobalt. This disease, known as pining, enzootic marasmus, bush sickness or Morton Mains disease, involves progressive debility, anaemia and emaciation of the affected animals and generally ends in their death. The importance of cobalt may lie in the fact that not only is in an activator of a number of enzymes but is a constituent of the vitamin B_{12}.

WALTER STILES

References

Borwen, H. J. M., "Trace Elements in Biochemistry," New York, Academic Press, 1967.

Stiles, W., "Trace Elements in Plants," 3rd ed., Cambridge, The University Press, 1961.

TRANSLOCATION

The evolution of land plants that began in the Devonian era was faced not only with the problem of the mechanical support of the plant body, but also with the distribution of water, mineral nutrients and photosynthetic products. The two centers of supply, namely the roots which take up moisture and minerals from the soil, and the green aerial organs, which carry out photosynthesis, are localized apart from each other in these plants, but connected by the so-called vascular system. This consists of two "channels," the XYLEM and the PHLOEM. In MONOCOTYLEDONS (grasses, palms, etc.) and herbaceous plants, strands of xylem and phloem are joined into conducting bundles which are distributed throughout the stem cross section. In GYMNOSPERMS and DICOTYLEDONS with secondary growth (increase in stem diameter), the two channels are in the form of cylinders, the wood (xylem) and the inner bark (phloem). The two systems work in essentially opposite directions: water and assimilated soil minerals move upward from roots to leaves in the xylem (transpiration stream), organic solutes move down from leaves to stem and roots in the phloem (assimilate stream). Toward growing centers, such as shoot tips, flowers and fruits, movement is unidirectional in the two systems.

The transpiration stream moves from the roots, up through the xylem into leaves (*fascicular path*), leaves the xylem there and spreads out into the parenchyma-cell walls (*extrafascicular path*). Water evaporates from the cell walls into the intercellular spaces which communicate with the open through the stomata of the leaf. Water penetrates the whole xylem-and-cell-wall system (*apoplast*) and thus surrounds all the living cells. Movement occurs within the apoplast presumably in the directions of hydrostatic gradients, which are brought about—in addition to various minor factors—mainly by transpiration. This is obvious from the following findings: In the morning of an average summer day, movement of water begins earlier in the twigs than in the stem. In later afternoon when transpiration ceases, the water in the twigs comes to a stand-still while there is still water movement from roots to stem. The ratio of velocity in twigs to velocity in stem thus varies within a 24-hour period. It is higher before than after the midday transpiration maximum. This finding also demonstrates the elasticity of the plant body, even a tree trunk. It has indeed been found that the diameter of trees varies within 24 hours, that is, it becomes minimal after the midday maximum of transpiration rate. During normal plant activity the xylem is under less than atmospheric pressure. Positive pressures are found only under conditions favorable for water uptake and unfavorable for transpiration, for instance early in spring before leaf expansion. They are brought about by the metabolic activity of roots and stem, and in some cases perhaps also by bubble formation upon freezing of the xylem water. Velocities in a system of various capillary sizes vary between zero at the edge of the capillaries and maximal values in the center of the capillaries with the largest diameter. The maximal midday velocities in higher plants range from 1.2 meters per hour in conifers to 10–60 m./hr. in herbaceous plants, and up to 150 m./hr. in lianas. How this "ascent of sap" occurs in tall plants is one of the oldest problems in plant physiology. The most widely accepted theory, the cohesion-transpirational pull theory, was put forth by Böhm (1893), Dixon & Joly (1894) and others. It postulates that adhesion forces between cell walls and water, and cohesion forces within the water columns enables the water to be pulled above barometric height. The tensile strength of water is one of the points of controversy today. Theoretical considerations (calculations from surface tension relationships, heats of vaporization, etc.) yield values of the order of 15,000 atm. Experimental measurements with various methods range from 0.05 atm. to 300 atm. An estimated minimal tensile strength of 20 to 30 atm. is required to lift the water into the tops of the tallest trees. Large vessel diameter and great vessel length are favorable for water transport by causing least resistance to flow. Phylogenetic development, indeed, points in this direction. However, the same trend increases the risk of inactivation of the vessels by air bubble formation. Significantly, trees with the widest vessels lose them as water-conducting elements often within a year after formation, whereas the xylem of more primitive trees that contains only tracheids may continue to conduct for many years. Wide and long vessels, for the same reason, make a tree more susceptible to certain pathological disturbances (chestnut blight, Dutch elm disease, etc.).

Movement of soil nutrients into the xylem of the roots requires metabolic energy; the mechanism of this is still insufficiently known. Movement to aerial organs takes place in the water of the transpiration stream. Most mineral nutrients, particularly nitrogenous compounds, are transformed to organic form in the roots before their ascent in the xylem. The solute concentration of xylem sap is usually quite low (less than 1% solids), but does reach 5% in exceptional cases (mainly sugars, in the sugar maple, and only in early spring during hydrolysis of reserve starch in the wood before the leaves emerge).

Export of photosynthates from mature leaves to places of consumption takes place in living tissues, the sieve tubes of the phloem. In contrast to xylem transport, phloem transport can be stopped completely by killing the tissue with a fine jet of steam (*steam girdling*). 90% or more of the organic material moves out of the mature leaves in the form of carbohydrates, and in many plants sucrose is the only carbohydrate form translocated. However, there are plant families in which raffinose, stachyose and verbascose (consisting of sucrose with one, two, and three galactose units attached, respectively) are equally or even more important translocation sugars. In some plants (e.g. ash [*Fraxinus*] and lilac [*Syringa*] of the Oleaceae) D-mannitol plays also an important role; in the giant brown alga *Macrocystis* it is the major translocation substance. Sorbitol, another sugar alcohol, is translocated together with sucrose in some members of the Rosaceae (cherry, apple, etc.). Besides carbohydrates there are other organic as well as inorganic substances translocated through the phloem, some of them particularly at certain times. For instance, one finds nitrogen- and phosphorus-containing compounds moving out of aging leaves and flower petals prior to their falling off. The concentration of transported material in the phloem is much higher than in the xylem (10 to 30% dry weight of solutes), and translocation velocities are lower (around 100 cm. per hour). There is good evidence that transport in the sieve tubes occurs *en masse*. Incisions into the inner bark of many dicotyledonous woody plants result in an exudation from the sieve tubes. Exudation also occurs from mouth parts of experimentally cut-off aphids which were feeding on these translocation channels, and from punctures left by withdrawn aphid stylets. The rates of flow from aphid stylet bundles may be 5 mm.[3] per hour or more. The sieve element in which the stylet bundle tip is located is therefore refilled 3 to 10 times per second from an axial direction, a phenomenon which strongly suggests a mass flow. The pressure drop along the stylet bundle has been calculated to be 20 to 40 atm., which is of the order of magnitude of osmotic pressures within the sieve tubes. Phloem translocation is many thousand times faster than pure diffusion could account for. At the present time it looks as though the whole sieve-tube system contains a pool which is moved from places of production to places of consumption. The driving force of the movement is the plant's metabolism, but the question whether metabolic forces are directly or indirectly involved has not yet been answered. Movement from leaf mesophyll, where photosynthesis takes place, into the sieve tubes of the phloem, where long distance transport begins, takes place against a concentration gradient and requires metabolic energy. Removal of substances from the sieve tubes is also a metabolic process, possibly hormonally controlled by the leaves. At places of solute entry, the turgor of the sieve tubes is being built up, at places of solute exit (or cell expansion by rapid growth) the turgor drops. Passive flow could occur along such a turgor gradient. This theory, in its original form including all living cells, was first put forth by Münch in 1930. It is correct (1) if the side-wall cytoplasm of the sieve tubes is semipermeable and (2) if the transverse walls (sieve plates) which separate individual sieve elements, are permeable to the moving solution. In addition one should be able to find osmotic gradients in the sieve tubes along the direction of transport, except toward shoots and fruits where growth could cause a turgor drop. Such gradients have, indeed, been found along tree trunks, and more than that, upon defoliation the total molar gradient disappeared, while some of the individual gradients became inverted. Since sieve tubes remain turgescent after defoliation, their side-wall cytoplasm must be regarded as semipermeable. The most critical point is the permeability of the sieve plates. The observed osmotic gradients are sufficiently great to force the solution through the lumen of the sieve elements and across the plates, if the sieve pores were "open." However, pores appear to contain protoplasmic material; they have therefore been regarded as the site of enormous resistance to flow. Exudation from aphid stylets, on the other hand, indicates that flow across the plates does take place easily. It has recently been suggested that electroosmosis may cause the solution to cross the sieve plates and would thus represent the major driving force of translocation.

Distribution of soil nutrients and photosynthates is accomplished even in the largest trees by the combination of phloem and xylem transport. Minerals that ascend into mature leaves and are not utilized there may be returned via the phloem to other plant parts. Subsequent transfer from phloem to xylem can lead to a circulation within the plant. Not all the minerals are equally phloem-mobile; calcium, for instance, is regarded as highly phloem-immobile. It is easily trapped in various cells in the form of oxalate crystals.

There are other mechanisms of translocation in plants than the two described above. Of great interest is the transport of phytohormones. Although these substances can be translocated in the phloem and in exceptional (mainly artificial) cases even in the xylem, their movement seems to go ordinarily through other tissues. Transport velocities are very small, that of indole acetic acid in intact *Avena* coleoptiles for instance being only 15 mm. per hour, and they are independent of the transport distance (see HORMONES, PLANT). This as well as other findings show that it is not a simple diffusion process. One of the most striking features of hormone transport is its apex-to-base polarity which is largely responsible for the polarity of shoots, particularly in regard to bud development and root formation. Hormone transport is one of the processes that are described by the term "secretion," the metabolic accumulation of substances that can occur against a concentration gradient across membranes. Secretion processes are typical for living cells. They transfer, for instance, soil minerals into the xylem of the roots, allow algae to grow in very dilute nutrient solutions; they introduce sugars into the sieve tubes of the phloem and "pump" sugars out of the plant body in floral and extrafloral nectaries, etc. (see SECRETION).

MARTIN H. ZIMMERMANN

References

Scholander, P. F. et al., "Cohesive lift of sap in the rattan vine," Science, **134:** 1835–1838, 1961.
Scholander, P. F. et al., "Sap pressure in vascular plants," Science, **148:** 339–346, 1965.
Bollard, E. G., "Transport in the xylem," Ann. Rev. Plant Physiology, **11:** 141–166, 1960.
Zimmermann, M. H., "Translocation of nutrients in plants," and Goldsmith, M. H., "Transport of plant growth substances," *in:* M. B. Wilkins, ed., "The physiology of plant growth, development and responses," London, McGraw-Hill, 1969.

TRANSPIRATION (EVAPOTRANSPIRATION)

Transpiration is the process by which water vapor is lost to the atmosphere from living plants. Evapotranspiration describes water losses from soil by evaporation and transpiration by plants. It is synonymous with consumptive use in denoting total water loss. In most experimental studies the separation of soil evaporation and plant transpiration is difficult and often impracticable. Hence evapotranspiration is the more accurate description of water loss from a vegetated area than is transpiration, even though transpiration is generally the larger and most important fraction.

Transpiration by plants occurs in response to the same energy sources as does evaporation from a water or moist soil surface. The response, however, is modified by physical character of the soil, and physiological characteristics of the plant, the effects of which are only imperfectly understood. The transpired vapor escapes principally from the leaves of the plant through their stomates, which occur in greater numbers on the lower surfaces of the leaves. Transpiration is the end process in the circulation of water from the roots through the stem to the leaves, and then to the atmosphere. For a given climatic condition, with water non-limiting, the rate of transpiration depends on the species, its cover density and plant size, stage of maturity, and its tolerance to mineral salts in the soil and water. For a given plant species, the rate is affected by climatic conditions such as temperature, wind movement, humidity, solar radiation, and growing season period.

Plant growth, that is the addition of new cells, tissues, and organs to the plant body, requires among other things a nearly constant supply of water otherwise metabolic processes are altered and growth is retarded. Most of the water absorbed by plants is lost by transpiration and only a small part is utilized in the formation of plant tissues. Small quantities are combined with carbon dioxide from the air to form sugar which is the basic food from which starch, fats, amino acids, and proteins are manufactured in plant cells.

Experiments have shown that the ratio of plant tissue produced to the total water use—evaporation plus transpiration—ranges from about one hundred to one to over one thousand to one, for different species of plants under different climatic conditions. Further discussion of the role of water in plants may be found in Meyer and Anderson (1952).

Differences in losses per unit of leaf area for different species, and differences in total leaf area in plants of equal size, account for the variation between species. As an example, the loss per unit area of saltcedar (Tomanek)[2] is less than for either cottonwood or willow; but because of its much greater leaf area the transpiration loss is greater for saltcedar than for either of the other two. Other influences being equal, the difference in transpiration between species is the resultant of unit loss and total leaf area.

Some species of xerophytic desert plants are adapted to an extreme economy of water and, during prolonged dry periods, maintain themselves in a nearly dormant condition (Meinzer)[3] during which transpiration is at a minimum. When the water supply becomes adequate the transpiration rate returns to normal and the plant to full vigor.

For some plants the quality of the water supply is

known to affect the rate of transpiration. As the dissolved-solids concentration increases the rate of transpiration decreases. This was observed for seepwillow (*Baccharis*) in the Safford Valley, Ariz., (Gatewood and others)[4] and determined experimentally for cultivated plants at the U.S. Salinity Laboratory (U. S. Dept of Agriculture).[5]

An individual plant growing in an isolated position transpires more water than one of equal size growing in a closed stand. The increase results from a more favorable micro-climate. The isolated plant receives more sunshine, is subject to higher temperatures, lower humidity, and greater wind movement than in a closed stand; all of which combine to increase transpiration.

Most of the elements of climate affect the rate of transpiration. Sunlight is necessary in the process of photosynthesis, essential for plant growth. Because photosynthesis is the basic process in the stomatal mechanism, light intensity has a basic effect on the rate of transpiration. The elements of temperature, humidity, wind movement, rainfall, length of growing season, and daytime hours influence transpiration.

Of these the greatest effects are due to temperature and to light. Temperature controls the length of the growing season and affects the rate during the growing season. In the Safford Valley, Ariz., (Gatewood and others) it was found that changes in transpiration paralleled changes in temperature both seasonal and daily, increasing as the temperature increased and decreasing as it decreased. Light, the principal factor controlling the operation of the stomates, is of prime importance in controlling transpiration.[6] The nightly decline of transpiration is the result of the closing of stomates in the absence of light. Even when temperatures remain high, transpiration rates at night drop markedly. In some species there is appreciable transpiration at night, in others hardly any. Experiments in the Safford Valley, Ariz., (Gatewood and others) indicated that for saltcedar, on the average, 80 percent of the daily transpiration occurred during the day and 20 percent at night. For seepwillow about 93 percent occurred during the day and 7 percent at night; but for cottonwood there was no measurable transpiration at night.

The effect of humidity is the reverse of temperature. Transpiration rate increases as the humidity decreases and decreases as it increases. Low humidity combined with high temperature is conducive to a high rate of transpiration. The combination of these two elements and their effect on the water level in a shallow well, in an area of transpiring saltcedar, is illustrated in Fig. 1. Rainfall affects transpiration in much the same manner as high humidity. Wind movement increases transpiration by removing air made humid by transpiration next to the foilage and replacing it with less humid air.

Annual evapotranspiration rates for several species of phreatophytes, plants that have a perennial and adequate water supply, growing in different parts of the western States are given in Table 1. The variation in evapotranspiration rates illustrates the differences be-

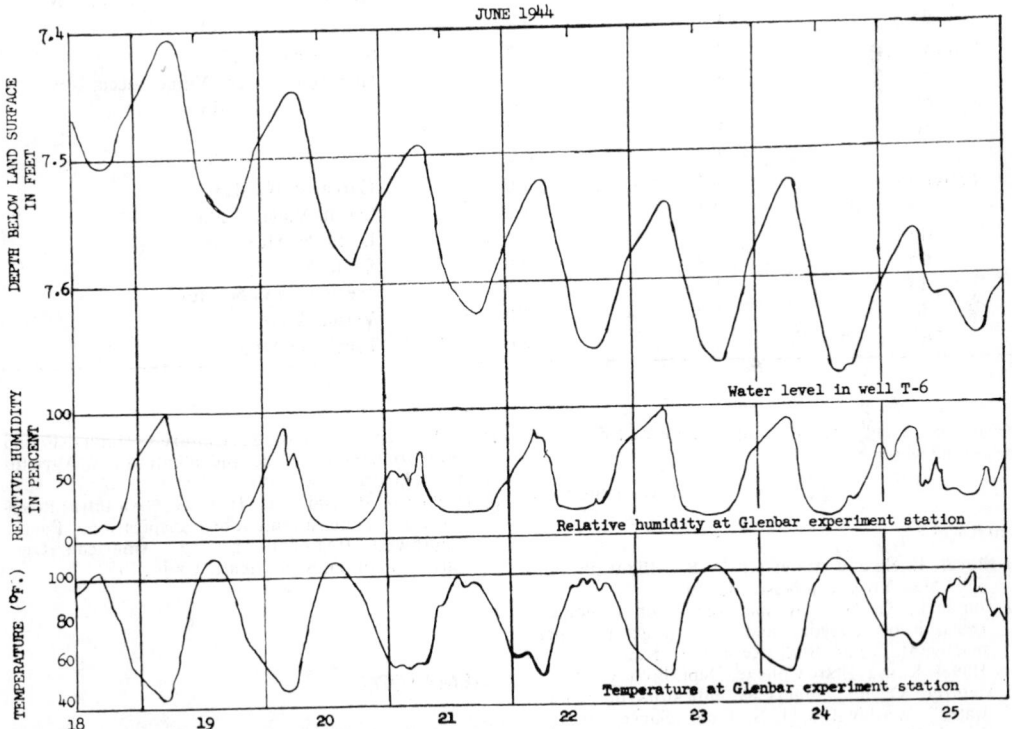

Fig. 1 Relation of fluctuations of the water table due to transpiration by saltcedar to fluctuation of relative humidity and temperature, Safford Valley, Arizona. The ground water level falls as transpiration increases and rises as it decreases. (After Gatewood, Robinson, *et al.*, 1950 Fig. 41.)

Table 1

Species	Water use, as depth of water over a unit area	Depth to water level below land surface	Locality
	Feet	Feet	
Alder	5.3	—	Santa Ana River, Calif.
Arrowweed	8.7	5.6	Colorado River, Yuma, Ariz.
Baccharis	4.6	5.9	Gila River, Safford, Ariz.
do	6.1	4.0	do
Cottonwood	6.0	6.0	do
Cottonwood and willow	7.7	3.0	San Luis Rey River, Calif.
do	5.4	4.0	do
Greasewood	2.1	2.0	Escalante Valley, Utah
do	1.7	5.0	Humboldt River, Winnemucca, Nev.
do	1.4	6.0	do
do	1.2	7.5	do
Mesquite	3.3	9.9	Gila River, Safford, Ariz.
Sacaton	4.0	—	Pecos River, N. Mexico
Rabbitbrush	1.6	5.0	Humboldt River, Winnemucca, Nev.
Saltcedar	9.2	4.0	Gila River, Safford, Ariz.
do	7.3	6.9	do
do	4.8	4.0	Pecos River, New Mex.
do	5.5	2.0	do
Wildrose	2.1	4.0	Humboldt River, Winnemucca, Nev.
do	1.6	6.0	do
Willow	4.4	2.0	Santa Ana, Calif.
do	3.5	4.0	Humboldt River, Winnemucca, Nev.
do	2.4	5.0	do
do	2.1	5.7	do
do	1.6	6.0	do
Saltgrass	4.5	2.0	Carlsbad, N. Mex.
do	4.1	1.5	Owens Valley, Calif.
do	2.6	.65	Isleta, N. Mex.
do	2.9	2.0	Santa Ana, Calif.
do	1.9	2.2	Mesilla Dam, N. Mex.
do	2.0	2.0	Vernal, Utah
do	1.6	2.6	Escalante, Utah

tween species and the effect of climate and depth to the ground water.

T. W. ROBINSON

References

1) Meyer, B. S., and Anderson, D. B., "Plant physiology" New York, van Nostrand, 1952.
2) Tomanek, G. W., "Annual report on ecological research of saltcedar and other vegetation primarily at Cedar Bluff Reservoir, Kansas" Fort Hayes, Kansas State College, Dept. Botany, 1957
3) Meinzer, O. E., "Plants as indicators of ground water" Washington, U.S. Geol. Survey Water-Supply Paper 577, 1927
4) Gatewood, J. S., T. W. Robinson, et al., "Use of water by bottom-land vegetation in lower Safford Valley, Ariz," Washington, U.S. Geol. Survey Water-Supply Paper 1103, 1950
5) U. S. Department of Agriculture, "Diagnosis and improvement of saline and alkali soils" Agriculture Handbook 60, 1954
6) Decker, J. P., and Janet D. Wien, "An infrared apparatus for measurement of transpiration;" Pacific Southwest Regional Meeting, American Geophysical Union, Sacramento, Calif., 1957.

TREMATODA

The trematoda or flukes constituting a second great subdivision of PLATYHELMINTHES are unsegmented, commonly hermaphroditic animals possessing a digestive tract. They are completely adapted to their parasitic life on or in the body of the host animals—the more the danger of their existence increases, the more eggs or the more larval generations are produced. With some exceptions of Digenea found in invertebrate hosts they are parasitic exclusively on or in vertebrates, some on the surface of the host body, in the mouth or branchial chamber, others in the gastro-intestinal tract or particular locations such as body cavity, liver, lung, kidney, heart or blood vessel, etc., intruding from the exterior into the interior further and further, even into the central nervous system, though the skeletal system and certain viscera are free from the invasion.

On the whole the adult tremadodes differ from the first subdivision of Platyhelminthes, TURBELLARIA, in lacking integumentary cilia and possessing a protective cuticle.

Trematodes are classified according to their life histories into two orders, Monogenea and Digenea, while the third order, Aspidocotylea, is monogenetic in development but digenetic in general anatomy. The most outstanding features of these three orders may be tabulated as follows:

ulate or non-loculate disc, muscular or not (Capsaloidea), or may bear one pair or two or anchors usually supported by transverse bars and marginal larval hooklets (Gyrodactyloidea), with or without additional adhesive plaques such as squamodisc or sucker, or three pairs of anchor complexes (*Megaloncus arelisci* Yamaguti, 1958, Fig. 1). In this type of haptors, especially in Gyrodactyloidea symmetrical groups of cement glands are developed in the posterior part of the body proper or at the base of the cotylophore, with their ducts opening at the base of the anchors. In the disc-shaped haptor as in *Udonella* the paired ducts of the cement glands are united to form a median duct which opens in the center of the disc.

In Polyopisthocotylea in which no cement glands are known, the opisthohaptor is more complex in structure than that of Monopisthocotylea; it consists of one pair or three to four pairs of sessile or pedunculate suckers with or without additional paired anchors, or a variable number of valvate, sessile or pedunculate clamps on each lateral edge of the cotylophore with or

	Monogenea	Aspidocotylea	Digenea
Prohaptor	Head organs, intra-or extra-buccal suctorial organs, almost always in pair	None, or oral sucker	Rhynchus or oral sucker, with or without accessory adhesive organs
Opisthohaptor	Single or multiple suckers or clamps, with or without anchors	Adhesive disc loculate or not	Acetabulum with or without accessory adhesive organs, occasionally absent
Excretory pore	Paired, anterior	Unpaired, posterior	Unpaired, posterior
Genito-intestinal canal	Present or absent	Absent	Absent
Laurer's canal	Absent	Homologous canal present	Present or absent
Vagina	Present or absent	Absent	Absent
Development	Direct	Direct	Indirect
Mode of life	Ectoparasitic	Endoparasitic	Endoparasitic

For further classification see Fuhrmann (1928) and Yamaguti (1958).

The cuticle, as to the origin of which different opinions exist, is commonly smooth in Monogenea and Aspidocotylea, but in Digenea it is often armed with spines. Below the cuticle is a basement membrane beneath which are layers of circular and longitudinal muscles, and then a layer of subcuticular cells. The interior of the body is occupied by the parenchyma, which consists of vacuolated, granular, branched cells and in which are embedded various internal organs and muscle fibers or bundles.

The adhesive organs are divided into a *prohaptor* and an *opisthohaptor* according to their position. The prohaptor of Monogenea consists of so-called head organs comprising a pair of adhesive or sticky gland complexes or a pair of suckers or suctorial grooves or lobes; these are independent of the buccal cavity in Monopisthocotylea, whereas in Polyopisthocotylea the small, often partitioned, paired suckers are enclosed in the buccal cavity and associated with the pharynx directly following them. In the onococtylid or hexabothriid Monogenea (Polyopisthocotylea) the prohaptor is represented by an oral sucker. The opisthohaptor of Monopisthocotylea may be a simple or armed, loc-

without paired caudal anchors; these clamps may be limited on one side of the cotylophore or more numerous on one side than on the other, thus presenting a marked asymmetry of the haptor. Of the above mentioned suckers each may retain a larval hooklet, the remaining hooklets being left outside the suckers (*Polystomum*), or some may be provided each with a chitinous supporting hook (*Erpocotyle*), or one pair is more strongly developed than the remaining pairs (*Anthocotyle*) or reduced on the contrary (*Pedocotyle*). The valvate clamps are supported by a chitinous framework consisting of an unpaired median piece and paired lateral pieces, one of the quadrants sometimes being modified into a sucking pad or sucker to which a clamp fixer is attached; the inner surface of the clamp may be provided with spines or hooks to aid in fixation. In *Osphyobothrus parapercis* there is an accessory adhesive organ at the posterior end of the body in the form of an ovoid contractile muscle bulb opening dorsally and containing transversely striated, longitudinal muscle bands.

In Aspidocotylea the prophaptor is usually represented by a weakly developed oral sucker; the opisthohaptor has the form of a single or multiple, shield or sucker occupying entire or greater part of the ventral surface of the body; it may be subdivided by transverse and longitudinal ridges into a number of areolae usually provided with retractile sense organs.

In the gastrostomatous Digenea the prohaptor is represented by a *rhynchus*, the shape and structure of which are variable in different genera. In the prosostomatous Digenea the prohaptor is replaced by a circumoral sucker usually without projections or with such (*Bunodera, Liliatrema, Crepidostomum,* etc.) or with spined collar (*Echinostoma*) or spines (*Acanthostomum*) and the opisthohaptor by an acetabulum, but the oral sucker may be poorly developed or lacking in the flukes parasitic in the circulatory system or air sac where no need for adhesion is required (Schistosomatidae, Cyclocoelidae). On the other hand accessory adhesive organs are developed in those flukes, in which the oral sucker and acetabulum are reduced in development, viz., additional ventral acetabulum in *Cadenatella*, additional dorsal acetabulum in *Polycotyle*, ventral glandular pits in *Notocotylus*, glandular papillae in *Homalogaster*, head collar and ventral glands in *Adenogaster*, ventral pouch in *Gastrothyrax*, pseudosucker and tribocytic organ in Diplostomidae and Strigeidae in which the forebody is modified into a scoop or cup which also aids in adhesion. The paired proboscides of *Rhopalias* serve as the most powerful accessory fixer; the lateral body folds in *Lomasoma, Oistosomum*, posterior sleeve-like body folds in *Trigonotrema*, wing-like lateral expansions of the forebody in *Bianium* or *Scaphanocephalus*, etc. belong in this category.

Development. Each egg of Monogenea, when laid, contains an unsegmented ovum or an embryo. In *Gyrodactylus* the first embryo in the uterus of the adult contains a second embryo within itself. In other Monogenea, however, the embryo is produced after the egg has been laid on the body of the host, and the hatched embryo provided with larval adhesive organs stays *in situ* and grows up into an adult without intervention of asexual generation. In *Polystomum*, parasitic in the urinary bladder of frogs, the eggs are laid in the water, and when hatched the oculate larva swims in the water by means of cilia, seeking for a young tadpole; the posterior end of the body is produced into a caudal disc armed with 16 hooklets, but without suckers. Arrived on the gill of a tadpole the larva undergoes a gradual metamorphosis—the cilia disappear, suckers appear in pairs on the caudal disc, each being formed around a larval hooklet, and 10 hooklets remaining outside the suckers. As the metamorphosis of the tadpole proceeds the larva leaves the gill and passing through the pharynx and esophagus comes to lie in the intestine without being affected by the digestive secretion of the tadpole. Upon the formation of the cloacal bladder the young *Polystomum* enters it and attains maturity. If this larva has attacked a young tadpole in which the external gills still persist, it grows more rapidly and produces eggs in only five weeks. This so-called "gill parasite" differs in several points from the "cloacal form," notably in the absence of a vagina.

In Aspidocotylea development is direct as in Monogenea; there is no intermediate host. On the other hand in Digenea development is indirect; alternation of generations and hosts occur, and large numbers of new flukes are produced from one egg in strong contrast with Monogenea. The fertilized ovum gives rise to a

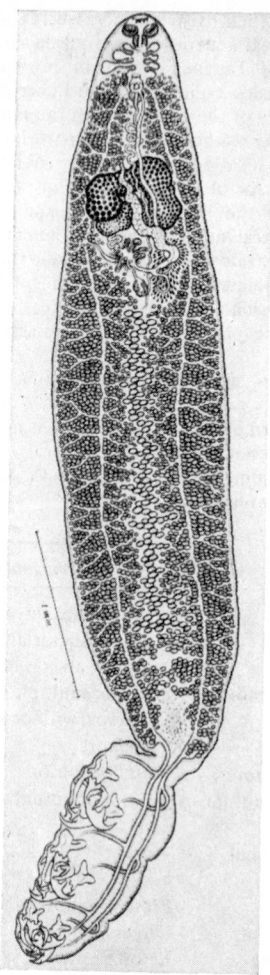

Fig. 1   *Megaloncus arelisci.* Yamaguti, 1958.

miracidium *in utero* or in the water after having passed out of the host body.

The free-swimming miracidium is covered with cilia all over except for the anterior end or on certain areas alone. The flat epidermal cells of the miracidium are usually arranged in five zones; the anterior end devoid of cilia is differentiated into a snout or terebratorium used for boring into the intermediate host, at the apex of which opens the apical or cephalic gland regarded formerly as "enteron." Immediately behind this gland usually lie two lensed eyes close together, resting on the cephalic ganglion. On each side of the head region is an elongate cervical gland whose duct apparently opens on the terebratorium. Laterally between the first and the second epidermal zone is a sensory papilla. The excretory system has made its appearance at so early a stage in form of a pair (or two) of flame cells, the capillary of which opens outside laterally between the fourth and the fifth epidermal zone. The germinal cells lie as undifferentiated blastomeres in the posterior "coelomic" cavity of the miracidium. Arrived on the surface of an appropriate molluscan host the miracidium bores its way into the body of the host, eventually into the liver where abundant nutriment is supplied for the growing larval form. Here it loses its cilia and the integumentary cell layer degenerates, whereas the germinal cells divide to form so-called *germ cell balls*. This first motionless *sporocyst* stage (mother sporocyst) gives rise to a number of new sporocysts (daughter sporocysts) or *rediae* by a series of changes from the germ cell balls. The rediae capable of migrating through the tissue of the host are characterized, when fully developed, by the possession of a pharynx, intestine, cervical collar, feet and birth pore. They escape from the sporocyst and feed on the tissue of the liver and gonad of the host and produce in their body cavity a number of new generations, daughter rediae or cercariae, in the same way as they themselves were produced.

The cercariae, especially of the furcocercous type, are produced directly in sporocysts in Schistosomatidae, Cyathocotylidae, Strigeidae, Leucochloridiidae, Brucephalidae, etc., but those of other types develop generally in rediae. They escape from the sporocyst by rupture of the wall of the latter, and from the redia through its birth pore, usually into the gill chamber of the host, eventually into the water, but in the microcercous cercariae (*Paragonimus*) or cercariaea without tail (*Leucochloridium*) they remain in the body of their matrix until they are devoured by the next host. The cercaria possesses all the organs of the young fluke in a rudimentary condition, including the anlagen of the genitalia and the originally paired excretory system with a definitive flame cell formula characteristic of the species. In addition there are a tail of varying size and structure, cystogenous cells, mucoid cells, penetration glands, and in some cases eyes, stylets, virgula organ, pigment cells, etc.—organs used only during their brief larval life.

The cercaria swimming freely in the water by the movement of the tail attaches itself to aquatic vegetation or anything submerged in the water, even the body surface of its own host, or the wall of the container in which the host is placed (*Fasciola*), or sinks to the bottom substratum (*Diplodiscus*) and is enclosed in a cyst formed by the secretion of the cystogenous cells, or directly invades the skin of the definitive host (*Schistosoma*), but in the majority of cases it swims about in search of a second intermediate host, into which it enters with the aid of the secretion of the penetration gland and its accessory apparatus (stylet, spinelets) and encysts, occasionally remains free, in the tissue of the host skin, scales, fins, subcutaneous tissue or muscle, body cavity, etc., the tail having been dropped off in the meantime. In this encysted or unencysted metacercarial stage the parasite awaits the chance of being devoured by the final host. In general the metacercaria undergoes further development without attaining complete sexual maturity which usually occurs in the body of the definitive host, but occasionally a neoteny or progenesis has been observed in the second intermediate host; sometimes a mesocercarial stage intervenes as in *Alaria* when a natural or experimental intermediate host is devoured by a host in which the metacercaria encysts again.

Diplostomulum and *Tetracotyle* are the terms used for those advanced larval forms of Diplostomidae and Strigeidae respectively which are characterized by their body distinctly divided into two body regions and the forebody being expanded in the form of a scoop and often provided with a pseudosucker on each side of the oral sucker and the anlagen of the tribocytic organ behind the acetabulum.

SATYU YAMAGUTI

References

Benham, W. B., "Platyhelmia, Mesozoa, and Nemertini" *in* Lankester, E. R., ed., "A Treatise on Zoology," Pt. 4, London, Black, 1901.
Fuhrmann, O., Zweite "Klasse des Cladus Plathelminthes, Trematoda" *in* Kükenthal and Krumbach, "Handbuch der Zoologie" Bd. 2, Leipzig, Springer, 1928.
Yamaguti, S., "Systema Helminthum" 5 vols. New York, Wiley, 1958–1963.

TRICHOPTERA

An order of holometabolous insects with mouth-parts of biting type, often functionless in the adult, with two pairs of membranous wings with primitive venation, the hind wing with expanded anal area or, in small forms, broadly fringed with hairs; surface of wings generally densely hairy or scaly. Larva generally aquatic, campodeiform or eruciform, usually living either in a net-like silken nest or more commonly in a portable case of pebbles, leaves, wood-chips or other debris. The Trichoptera or caddisflies are closely related to the Lepidoptera, to which they are ancestral; they differ in usually having the wings hairy rather than scaly and in never having the body scaled, in not having the maxillae developed into a suctorial tube (a rudimentary tubular development of the maxillae reputedly occurs in one rare genus) and in the nature of the wing-coupling apparatus. Trichoptera are numerous in both temperate and tropical countries. About 1,000 species are known from North America. They are associated with all types of fresh water. A few species live in brackish water or in wet moss. The adults form mating swarms like those of Nematocera; sometimes enormous numbers occur over rivers and lakes; these aggregations may be a pest in residential or resort areas, especially when attracted to artificial light. Trichoptera are recorded as causing allergic reactions in some individuals. Caddisflies are useful as an important natural food of fish and as significant predators of black-fly larvae. Adult caddisflies are rather uniform in appearance; they are mostly moth-like insects of dull color. The Hydroptilidae and some Psychomyiidae are minute, resembling Microlepidoptera; some Phryganeidae and Limnephilidae exceed three inches in wing-span. A few species are ornate, such as the striped species of *Macronemum*, and the white, speculum-bearing species of *Leptocella*.

The Rhyacophilidae are the most primitive family, with wing-venation of almost diagrammatic completeness and simplicity. This family and the related but more specialized Hydropsychidae, Philopotamidae and Psychomyiidae have campodeiform larvae living in nests under stones, mainly in rapid streams; hydropsychid larvae have prominent tracheal gills, lacking in the other larvae of this section of the order. Aside from these families, the majority of Trichoptera have case-making larvae. Larvae of the minute Hydroptilidae are transitional, being campodeiform and naked when young, but building cases in later instars. The larvae of the remaining families are mostly eruciform, and are often highly adapted for life in their portable cases. The cases show great variety of form and materials. Many Limnephilidae construct rough, cylindrical cases of pebbles, chips, fibers or leaves. Phryganeidae have characteristic tapering cases made of spirally arranged pieces of grass. The sand-grain cases of Leptoceridae may be tusk-shaped, or may be flattened and provided with lateral flanges; the latter type of case reappears in Molannidae and is an adaptation to life on sandy lake-bottoms. Brachycentrid larvae construct a chimney-like, tapering, twig case of square section; a cylindrical case of Helicopsychidae is shaped exactly like a small snail shell. The eggs of caddisflies may be laid singly or in masses of jelly, sometimes formed into spiral ropes. The pupae are aquatic; those of lacustrine forms usually swim to the surface before emergence of the adult. The adults are mainly crepuscular and nocturnal; they may fly long distances from water.

EUGENE MUNROE

TRILOBITA

To the biologist the trilobites offer particular interest primarily because they represent the oldest, and in many ways the most generalized, ARTHROPODS known. The trilobites are unique in having all the postoral limbs practically alike and a distinct dorsal segmentation of the larva. To the stratigraphers the trilobite species serve as excellent guide fossils on account of their wide horizontal and limited vertical development in the stratigraphic column.

The trilobites played an important role in marine faunas from the beginning of the Paleozoic. Their major development took place in the Cambrian, after which the group gradually became less important and apparently died out in the Permian. The great importance of the Class Trilobita is illustrated by the huge number of species described—about 10,000 up to the present.

The exoskeleton, which is solid and mineralized, covers the broad body and is inflected along the margin forming a narrow doubloure on the ventral side (Fig. 1). In front a median plate, the hypostoma or upper lip, is also mineralized and often preserved together with the rest of the exoskeleton. The remaining ventral surface as well as the delicate ventral appendages were less mineralized and are only preserved under special conditions.

The dorsal skeleton is divided lengthwise by two furrows into an elevated median axis and two flattened lateral areas. The principal body cavity and the appendages belong to the axis, the lateral areas are merely broad and thin lateral extensions serving as a protection for the appendages below. The anterior segments of the trilobite body are fused into a semicircular shield, the cephalon or head. Remnants of the original segmentation are indicated in the lobes of the axis (Fig. 1 and 2a, c, d), particularly in larval stages (Fig. 3a–d).

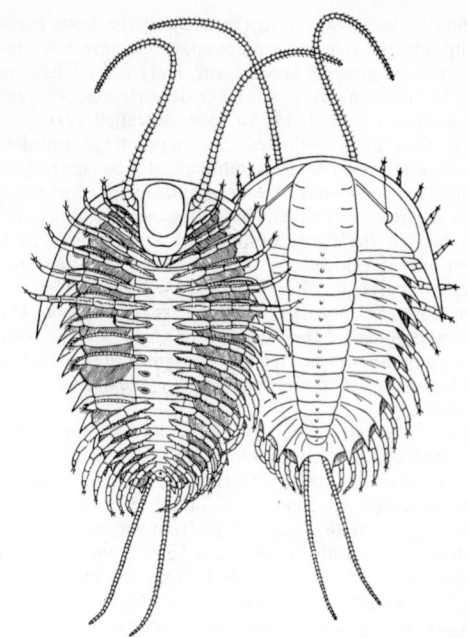

Fig. 1   Reconstruction of the ventral and dorsal surface of the trilobite *Olenoides serratus* (Rominger) from the Middle Cambrian, British Columbia, Canada, 1 ×.

The head may be provided with long spines or horns (Fig. 2a, c, g). A pair of beanshaped compound eyes, sometimes stalked (Fig. 2g) is situated on the cheeks on either side of the axis. The visual surface has numerous facets appearing in cross sections as lenticular bodies or prisma. Certain forms had no eyes (Fig. 2b), but certain "facetted" *maculae* on the hypostoma may possibly represent visual organs. The exoskeleton of the head has distinct lines of weakness, facial sutures, which facilitate the moulting.

The abdomen is divided into a thorax with a number of free segments ranging from 2 to more than 40, and a tail-shield, the *pygidium*, composed of 1 to 30 fused segments. The tergites of the thorax are movable on each other permitting the body to be rolled up for protection. Thy pygidium varies greatly in size, in early Cambrian forms (Fig. 2a) it may be very small, in later forms exceeding the size of the cephalon (Fig. 2d, e). In certain forms the axis is provided with long spines directed backwards, the last one serving as a telson spine (Fig. 2c).

The ventral appendages are fairly well known thanks to a few famous finds of well-preserved forms from the Cambrian, Ordovician and Devonian of North America and Europe. Fig. 1 presents a reconstruction of a Middle Cambrian species found by Ch. D. Walcott in British Columbia. On the ventral surface a pair of long flexible antennae is attached in front on either side of the hypostoma which covers the mouth below; a small plate, a metastoma, is situated near the hind margin of the hypostoma. Behind the antennae the 4 pairs of cephalic appendages and all the appendages of the abdomen (except the caudal filaments which are only known in this form) are mutually alike. This fea-

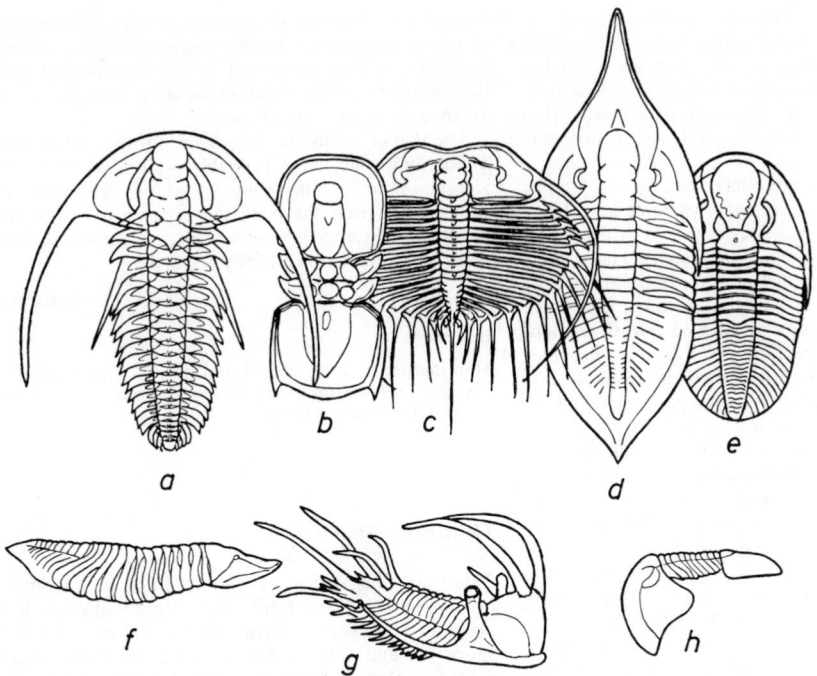

Fig. 2   Trilobites from the Lower Cambrian to the Middle Permian. *a, Fallotaspis typica* Hupé, Low. Cambrian, Morocco, 1 ×; *b, Peronopsis fallax* (Linnarsson), Mid. Cambrian, Norway, 1 ×; *c, Ctenopyge pecten* (Salter); Upp. Cambrian, England, Sweden 3 ×; *d, Megistaspis acuticauda* (Angelin), Low. Ordovician, Norway, 0.25 ×; *e, Pseudophillipsia sumatrensis* (Roemer), Mid. Permian, Sumatra, 1.3 ×; *f, Trimerus delphinocephalus* (Green), Silurian, New York, U.S.A., 0.17 ×; *g, Ceratarges armalus* (Goldfuss), Mid. Devonian, Germany, 2 ×; *h, Illaenus sinuatus* Holm, Ordovician, Esthonia, 2 ×.

ture, characteristic of all trilobites known, represents a very generalized structure unknown in other arthropods.

The nature and position of the antennae suggest that they correspond to the antennules and antennae of CRUSTACEANS and INSECTS, appendages which are enerved from the deutocerebrum and are primarily preoral. The postoral appendages in trilobites have a very characteristic structure. Each appendage is composed of a walking leg or telopodite, with a gill-bearing branch or pre-epipodite (or epipodite), attached to its base. The telopodite has 8 or 9 joints (9 if the narrow basal portion is interpreted as a separate precoxa). The coxa is developed as a powerful podite with a slight median extension (endite). Those of the head-appendages are situated so far apart that the short spineferous endites could not have served as jaws. The lateral branch of the limb is attached to the base of the

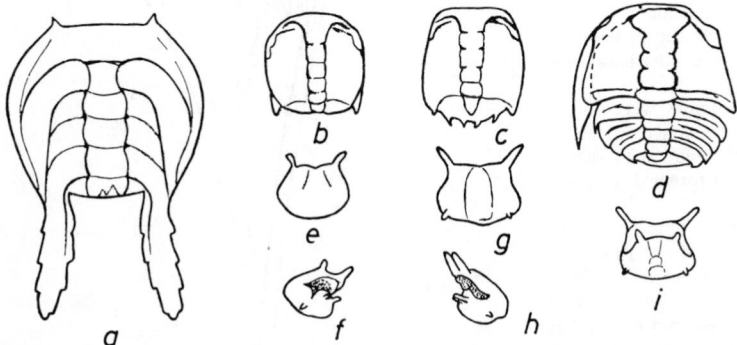

Fig. 3   Larval stages of trilobites. *a, Olenellus gilberti* Meek, Low. Cambrian, Nevada U.S.A., 43 ×. Early larval head showing dorsal segmentation. *b–d, Sao hirsuta* Barrande Mid. Cambrian. Bohemia, 30 ×. *b–c*, protaspids; *d*, first post-protaspid stage. *e–i, Menoparia genalunata* Ross, Low. Ordovician, Utah, U.S.A., 20 ×. *e, f*, small protaspid in dorsal and anterioventral view, *g–i*, larger protaspid in dorsal, lateral and ventral view. Specimens unrolled.

telopodite. This means that the branch represents a pre-epipodite (if attached to a precoxa) or epipodite (if attached to the coxa), rather than an exopodite which is fastened to the end of a usually jointed sympod. The lateral branch is composed of a shaft with a broad fringe of narrow blade-shaped filaments probably serving as gills.

Larval stages of trilobites have been found in fine grained sediments. In recent years valuable knowledge on the larval structures has been obtained by dissolving limestones containing silicified specimens. The earliest stages, measuring less than 1 mm in length, are usually strongly convex with a circular outline. Cambrian species in particular, show a marked dorsal segmentation (fig. 3a-d). The portion which later on becomes the head has an axis with five rings or segments corresponding to the antennae and four pair of legs on the ventral side. If one regards the preoral antennae (deuto-cerebral) segments as belonging to the acron then the structure suggests a segmentation also of the preoral portion in the trilobite (Lower Cambrian forms from Morocco may suggest even three segments in the acron). In certain Lower Cambrian species the lateral areas are also segmented (fig. 3a). It is possible that the head segments were formed more or less simultaneously while the rest were added one by one by teloblastic growth in front of a telopore. Three succeeding stages of *Sao* (fig. 3b-d) illustrate the gradual increase of segments. The first larval stages, those in which the body is covered by one continuous shield, is called the protaspis. At a certain growth stage a transverse suture is formed between the head and the abdominal portion. Later on new sutures are formed beyond the first one, releasing thoracic segments migrating forward from the telopore. The first transverse suture seems to appear very early in Lower Cambrian forms (In Fig. 3a a small "pygidium" is probably lost), while in Ordovician species it may appear after a considerable number of abdominal segments were developed (Fig. 3e-i).

The trilobites were marine animals. Certain species (olenids) probably survived in less aerated waters with a low oxygen content. Most trilobites were bottom dwellers, but were evidently able to swim by means of their "walking legs." It is difficult, however, to understand how the pre-epipodites (or epipodites) could have been used in swimming. While in *Limulus* the abdominal foot has a free hind margin, that of the trilobite fringe is covered by the following appendage, a fact that would prevent a free movement of the pre-epipodite (fig. 1). Smaller trilobites with long horizontal spines may have been pelagic.

The trilobites, with their lack of jaws, were probably mud feeders. Since the mouth was concealed below a large hypostoma or upper lip the mud with the food particles was possibly transported forward by a water current created by the appendages.

The relationships of the trilobites have been subject to much discussion. The structure of the exoskeleton and limbs shows that the trilobites belong to the Arthropoda. Because of the antennae and "biramous limbs" the group was generally referred to the Crustacea. A more detailed knowledge of the appendages, however, has as mentioned above, raised strong doubts as to the homology of the Crustacean and trilobite limbs. The trilobite limb seems to have more in common with the limbs of *Limulus* which also resembles the trilobites in the general morphology of the

exoskeleton. The lack of antennae in the Chelicerata is of minor importance in this connection since indications of antennae glomeruli in *Limulus* suggest that the ancestors of the Chelicerata may have possessed antennae in front of the chelicerae.

The Class Trilobita belonging to the subphylum Trilobitomorpha, were probably related to the Chelicerata. Whether this common group developed independently from the Annelida or were linked to the Crustacea, Insecta and others in a common arthropod stem, cannot at present by decided.

LEIF STØRMER

Reference

Harrington, J. J. *et al.*, "Trilobita" *in* Moore, R. C., ed., "Treatise on Invertebrate Paleontology," Kansas, University of Kansas Press, 1959.

TRIMEROPHYTINA

These early vascular plants appeared first in the middle of Lower Devonian time, after Rhyniophytina had become established. Their leafless stems, terminal sporangia and centrarch maturation resembled those of rhyniophytes. However, their lateral branch systems were considerably advanced over the simple axes of rhyniophytes. They were terminated by large clusters of fusiform, longitudinally dehiscent sporangia and branched first by double dichotomies then by simple dichotomies (Fig. 1). Their vascular strand was more massive than that of rhyniophytes although still centrarch in maturation. The main axis produced

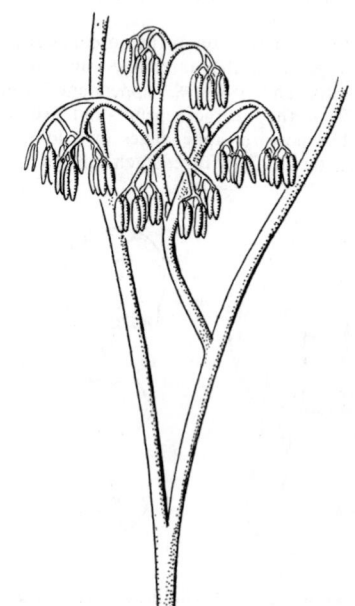

Fig. 1 *Psilophyton.* (Drawing supplied by F. M. Hueber based on a new species of *Psilophyton* currently under study.)

lateral branches pseudomonopodially rather than dichotomously and this difference was reflected in the internal structure of the stem. Tracheids were basically scalariform but showed peculiar thin spots in the wall between successive bars. This condition might, in the course of evolution, have given rise to the purely scalariform pattern found in many ferns or to the pitted condition characteristic of many gymnosperms. The trimerophytes appear to be precursors of both ferns and progymnosperms and some related groups which appeared later in geologic time.

H. P. BANKS

References

Banks, H. P., "The early history of land plants," *in* Drake, E. T., ed., "Evolution and Environment," New Haven and London, Yale University Press, 1968.

Hueber, F. M. and H. P. Banks, "*Psilophyton princeps:* The search for organic connection," Taxon, **16**:81–85, 1967.

TRYPANOSOMA

The Trypanosomatidae constitute a family of obligate PROTOZOAN parasites which have been found in plants, arthropods, nematodes, rotifers, mollusks and leeches. Six to eight genera are included. Most species are polymorphic. The body varies from minute globose to elongate spindle-shaped, sinuous or falciform, and is enclosed in a definite periplast. The flagellum marks the anterior pole. In addition to the rounded nucleus with its large central karyosome there occurs a conspicuously staining basophilic kinetoplast. At the base of the flagellum lies the dot-like basal body or rhizoplast, visible only in special preparations. These structures normally lie in close apposition, but preceding cell division they appear to bracket a small vacuole. From the basal body an axoneme passes to the surface where it becomes the core of the marginal filament or free flagellum when one is present. The different body forms are characterized by the location of the kinetoplast. In the globular leishmania nucleus and kinetoplast are in close proximity and a flagellum is lacking. In the leptomonas the kinetoplast lies near the anterior end and produces a terminal free flagellum. The crithidia has the kinetoplast just anterior to the nucleus, and the flagellum supports a short undulating membrane before becoming free at the anterior end. In the trypanosome the kinetoplast is posterior to the nucleus, the undulating membrane fringes the convex border of the body for the greater part of its length, and a free flagellum may be present or lacking.

Electron microscopy reveals a subpellicular layer of hollow fibrils passing obliquely around the body. The basal body consists of a nine-fibered cylinder which merges with the axoneme of the flagellum. The flagellum arises within a spacious pocket or reservoir, lined with pellicle, which is reflected at the bottom to form the flagellar sheath; it contains the usual nine and one arrangement of double fibrils. The undulating membrane is formed by the simple application of the flagellum and its sheath to the pellicle along one side of the body.

The kinetoplast is a disk-shaped modified mitochondrion, containing a fibrous mat of DNA, sandwiched between two layers of lighter amorphous matrix. Arising from one edge of the kinetoplast in blood forms, a cord-like mitochondrion passes forward and loops in the anterior region of the body. In culture or insect gut forms a posterior elongate mitochondrion also arises from the kinetoplast. Trypanosome strains formerly thought to lack a kinetoplast are now known to lack only the DNA component.

The leishmania shows essentially the same ultrastructure as the trypanosome including the reservoir and the stump of the flagellum. It has a double-layered pellicle, but the leptomonas has only a single-layered membrane. In the leishmania the kinetoplast is compact, and peripheral mitochondria may be present. In the leptomonas the kinetoplast again branches to form mitochondria which extend into the posterior region.

Multiplication is ordinarily by longitudinal binary fission. The kinetoplast divides first. A new flagellum arises from the daughter kinetoplast, the nucleus divides endomitotically, and finally the organism splits from anterior to posterior. In most species division may occur in any stage. In some forms multiple division or schizogony from plasmodial forms occurs. Sexual processes have not been demonstrated.

The trypanosomes appear to have evolved as intestinal parasites of insects, from which they have spread to other hosts with the development of the hemophagous habit. The genera *Leptomonas, Crithidia* and *Herpetomonas* have only an invertebrate host. The actively multiplying flagellate stages occur free in the intestinal lumen, attached to the epithelial cells, or actively invade these cells. As they migrate posteriorly they diminish in size, round up, and pass in the feces as leishmanial cysts. These infect other insects when eaten. The genus *Phytomonas* has a leptomonas parasitic in the latex vessels of euphorbiaceous plants and is transmitted by Hemiptera.

The genera *Trypanosoma* and *Leishmania,* the "hemoflagellates," resemble *Herpetomonas* and *Leptomonas* respectively, but differ in having both vertebrate and invertebrate hosts. Trypanosomes are predominantly extracellular parasites of the blood of all classes of vertebrates but pathogenic forms invade the lymph and tissue fluids also. The largest species are upwards of 70 microns in length. *Endotrypanum schaudinni* is an intracellular parasite of the erythrocytes of the two-toed sloth. The trypanosomes of fishes, amphibia and turtles are transmitted by leeches, those of terrestrial vertebrates by blood-sucking insects. Transmission may be direct, involving only the immediate transfer of organisms on the contaminated mouthparts of a vector, or cyclic, involving multiplication and metamorphosis in the gut of the vector. In cyclic development the trypanosomes or crithidia in the gut of the insect are not infective for the vertebrate host. After several weeks the parasites migrate either to the mouthparts and salivary glands (anterior station) or to the rectum (posterior station). These metacyclic forms are now infective. Many variations of this scheme occur, but the pattern is characteristic for the species. Anterior station forms are transmitted by bite (inoculation). Posterior station forms are transmitted by contamination, the vector defecating the infective forms upon the skin at the time of feeding. These may invade the puncture wound, or penetrate through the mucous

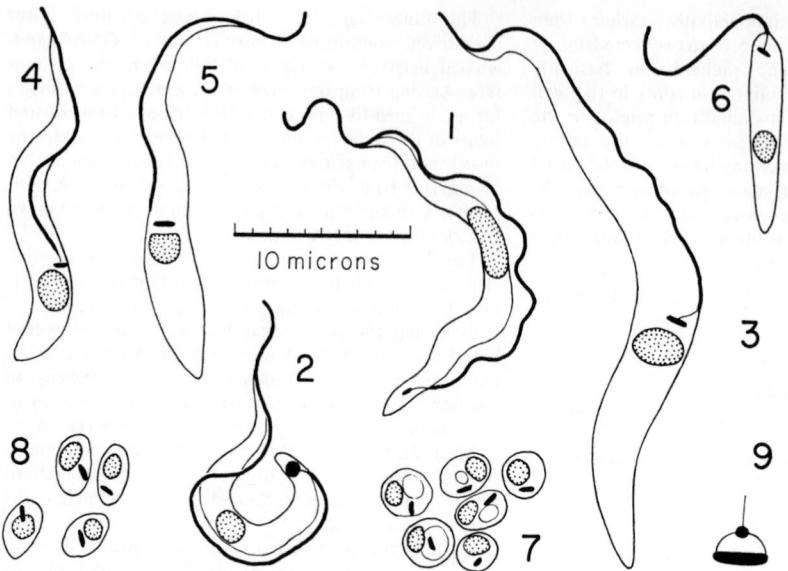

Fig. 1   Various Human Trypanosomatidae. 1, *Trypanosoma rhodesiense*, in blood film. 2, *T. cruzi*, in blood film; 3, 4, *T. cruzi*, crithidial forms from gut of triatomid bug. 5, do., from culture; 6, *Leishmania donovani*, leptomonas, from culture; 7, *L. donovani*, in spleen smear, several showing vacuoles; 8, *L. braziliensis*, in smear from bed of lesion; 9, detail of kinetoplast, as seen in disintegrating forms, or prior to division. All of above as seen by light microscopy with conventional blood stains.

membrane when the animal bites or licks the area. *T. equiperdum* is a monomorphic species which has dispensed with the insect vector and is transmitted by coitus.

Trypanosomes are usually non-pathogenic for their normal hosts, but a number cause serious disease in man and domestic animals. Pathogenicity is regarded as evidence of a new host relationship which has not yet attained equilibrium. African sleeping sickness, cause by *T. gambiense* and *T. rhodesiense;* and nagana of horses and cattle, caused by *T. brucei, T. congolense, T. vivax* and others, effectively prevent development of approximately 4 million square miles of Africa. The African trypanosomes have reservoirs in antelopes and big game, and are transmitted cyclically by tsetse flies (*Glossina*), and mechanically by tsetse, tabanid, and *Stomoxys* flies. The widespread surra, and South American mal de caderas, caused by *T. evansi* and *T. equinum* respectively, are equine diseases transmitted mechanically by tabanids; while *T. equiperdum* causes a cosmopolitan venereal disease of horses, known as dourine.

American trypanosomiasis, or Chagas's disease of man, caused by *T. cruzi*, occurs in South, Central, and North America, including the U.S., with reservoirs in small mammals. It is transmitted contaminatively by blood-sucking triatomid bugs. Both trypanosome and leishmania forms occur in the vertebrate host. The leishmanias invade cells of various tissues, particularly myocardium and brain. Damage to the autonomic nervous system of various muscular hollow organs often leads to megaoesophagus, megacolon, megaureter, etc.

In the leishmaniases the parasite occurs as a leishmania in man and various reservoir hosts, mainly dogs and rodents, and as a leptomonas in the gut and mouthparts of the sandfly vector, *Phlebotomus,* whence it is inoculated at the time of biting. The leishmanias are intracellular parasites of the reticulo-endothelial system. The tiny parasites occur in the macrophages in enormous numbers and lead to the destruction and compensatory proliferation of these cells. The infection assumes a generalized or visceral form in kala azar, caused by *L. donovani;* and is limited to localized cutaneous lesions in Oriental sore, caused by *L. tropica;* or muco-cutaneous lesions in espundia, caused by *L. braziliensis.* Kala azar is pantropical. Oriental sore is essentially a Mediterranean disease, while espundia is limited to Central and South America. One species, *L. enriettae* is a natural parasite of guinea pigs in Brazil.

Trypanosomes and leishmanias are fairly resistant organisms. They are not particularly host specific, and can for the most part be readily adapted to other animals. They can be maintained by syringe passage indefinitely, frequently thereby losing their ability to infect the natural vector. They can be readily cultured on non-cellular media at room temperature, in such cultures taking the form found in the invertebrate vector. The African trypanosomes require tsetse fly extracts and explants for such cultures to develop infectivity. *T. lewisi* of rats and fleas, *T. equiperdum,* and *L. donovani* have been extensively studied from the standpoint of physiology and immunity. Trypanosomal and leishmanial diseases are treated with arsenical and antimonial drugs or complex organic compounds. But in many cases results are disappointing.

JUSTUS F. MUELLER

References

Grassé, P-P., "Trypanosoma" *in* Grassé, P-P. ed., "Traité de Zoologie" Vol. 1, Paris, Masson, 1952.

Rudzinska, M. A., *et al.*, "The fine structure of *Leishmania donovani* and the role of the kinetoplast in the leishmania-leptomonad transformation," J. Protozool., **11**: 166–191, 1964.

Sanyal, A. B. and P. C. Sen Gupta, "Fine structure of *Leishmania* in dermal leishmanoid," Trans. Roy. Soc. Trop. Med. Hyg., **61**: 211–216, 1967.

Vickerman, K., "The mechanism of cyclical development in trypanosomes of the *Trypanosoma brucei* sub-group: an hypothesis based on ultrastructural observations," Trans. Roy. Soc. Trop. Med. Hyg., **56**: 487–495, 1962.

TUBIFLORAE

A group of tetracyclic sympetalous dicotyledonous flowering plants defined by Eichler in his classification of 1875 with five main constituent families: Convolvulaceae, Polemoniaceae, Hydrophyllaceae, Boraginaceae and Solanaceae. It was taken over by Engler, given the rank of order and extended to include Eichler's next group Labiatiflorae. This is the sense in which the group is now most often used. It is then divided into 8 suborders:—(i) Convolvulineae (Convulvulaceae, Polemoniaceae, Fouquieriaceae); (ii) Lennoineae (Lennoaceae); (iii) Boragineae (Boraginaceae, Hydrophyllaceae); (iv) Verbenineae (Verbenaceae, Labiatae); (v) Solanineae (Solanaceae, Nolanaceae, Scrophulariaceae, Orobanchaceae, Gesneriaceae, Bignoniaceae, Pedaliaceae, Lentibulariaceae, Martyniaceae, Columelliaceae, Globulariaceae); (vi) Acanthineae (Acanthaceae); (vii) Myoporineae (Myoporaceae); (viii) Phrymineae (Phrymaceae).

Of the above families, Fouquieriaceae has numerous stamens and its inclusion here is doubtful; Columelliaceae may be better placed near Saxifragales; Lentibulariaceae has the free central placentation characteristic of Primulales and is said to share certain serological features with that order. Lennoaceae are sometimes placed near Monotropaceae in Ericales, but they lack characteristic morphological and embryological features of that well-marked order and their inclusion in Tubiflorae is justified. Buddleiaceae is a group of some 8 genera of woody plants and has been placed in the order Contortae (in or near Loganiaceae) but is sometimes included in Scrophulariaceae. Rendle regarded Convolvulaceae as a separate order between Tubiflorae and Contortae. The difficulties of placing the Convolvulus and Buddleia families indicate the lack of any clear and satisfactory line of demarcation between Contortae and Tubiflorae.

Within Tubiflorae, in Engler's sense, the first three suborders and part of the family Solanaceae tend to have 5 fertile stamens and actinomorphic flowers (Echium and some other genera in Boraginaceae are exceptions) while the remainder of the group have largely zygomorphic flowers and four or two fertile stamens. Engler's view of a close affinity between Solanaceae and Scrophulariaceae, however, is not endorsed by modern studies which emphasize the importance of the oblique position of the ovary cells in Solanaceae and tend to regard the nearly actinomorphic Scrophulariaceae (Verbascum and Celsia) as having attained that condition secondarily. Solanaceae are also sharply distinguished by the possession of intra-xylary phloem.

Variation in ovary characters can be arranged in two main series, though these may not be evolutionary lines. In one, reduction in the number of ovules and, finally, lobing of the ovary leads to a 4-nutletted fruit in most Boraginaceae and Verbenaceae and in all Labiatae. In the other there is an increase in the number of seeds, often correlated with elongation of the ovary; this series reaches its climax in the pod-like fruits of some herbaceous Gesneriaceae and woody Bignoniaceae. In Orobanchaceae too there is a very marked increase in the number of seeds, made possible here by reduction in their size and increase of the placental surfaces. Orobanchaceae are parasites on the roots of other dicotyledons and a high seed production is vital to this way of life. In Scrophulariaceae there are a large number of semi-parasites (the greater part of the tribes Buchnereae and Euphrasieae), from which there seems little doubt that Orobanchaceae is derived. The genus Lathraea is somewhat intermediate between the two families and authorities differ as to its placing.

Tubiflorae are one of the most highly evolved groups of dicotyledons. Although congested inflorescences do occur, the emphasis in floral organisation is characteristically on the individual flower which is often highly specialised for pollination by insects or birds and usually produces numerous seeds. This is in contrast to the other peak of dicotyledonous floral evolution in Compositae and similar types. In these there is an aggregation of numerous small single-seeded flowers into a dense head. Just as all such plants bearing capitula of flowers are not closely related, so it seems likely that the order Tubiflorae in the broad sense includes the end-points of a number of different lines of evolution and that the group will disappear from future classifications, as it has already done from Hutchinson's (1926, 1959). It is not possible, however, to follow Hutchinson's latest scheme and regard the woody and herbaceous families of Tubiflorae as having evolved independently from primitive dicotyledons. It is a notable feature of the group that, though predominantly herbaceous, woody plants are widely scattered through it and occur in all the larger families.

Economically by far the most important family is Solanaceae containing the potato (Solanum tuberosum), egg plant or brinjal (Solanum melongena), tomato (Lycopersicum esculentum), chili (Capsicum spp.), tobacco (Nicotiana spp.), as well as important drug plants such as henbane (Hyoscyamus niger), deadly nightshade (Atropa belladonna) and thorn apple (Datura stramonium). Convolvulaceae includes the sweet-potato (Ipomoea batatas): Verbenaceae one of the most important of timbers, the teak (Tectona grandis), a native of Burma now cultivated in some other tropical and subtropical areas.

Labiatae are noteworthy for their essential oils and the leaves of nearly all species bear sessile glands: lavender (Lavandula), marjoram (Origanum), rosemary (Rosmarinus), patchouli (Pogostemon), mint (Mentha), thyme (Thymus), basil (Ocimum) are amongst the most important. In nearly every family the bright flowers, attractive to pollinators, result in many species being cultivated as ornamentals.

B. L. BURTT

TUBULIDENTATA

These animals represent an order of MAMMALS. Their latin name Tubulidentata derives from *tubulus* (small tube), and *dens* (tooth), so that *tubulidentata* means mammals with tube like teeth. They were given the name "Aardvark" by the first pioneers of the Boers, when beginning their settlement in Cape Province, in view of the piglike environment and behavior. The word means a pig like animal burrowing in the ground.

The fossil history of the Tubulidentates is for the most part unknown. Our knowledge does not extend with certainty beyond the Miocene epoch. Fossils are found in pliocene and miocene sediments of India, Greece, France and Algeria. They are so similar to modern forms, that we must put them into the genus *Orycteropus* of the recent living representatives. Studying the structure of anteater's skeleton feet and brain there is to be found a resemblance with primitive hoofed mammals or ungulates of the earliest tertiary epochs, the Paleocene and Eocene, some 70–60 millions of years ago, called Condylarthra. The peculiar structure of the teeth, with the vasodentin as a building material, suggests a relationship with tapirs and SIRENIA, the latter also a tribe arisen from primitive ungulates. During their evolution the primitive hoofed aardvarks adapted completely to a termite and ant diet, simplifying skull and dentition, enlarging tongue and salivary glands, varying the stomach for the new demands, and learning to dig and burrow.

The aardvark is a sturdy animal, about the size of a pig, with a pecular habitus. The long conical head bears small eyes and large ears, each of the latter moving independently. The short neck leads to a round high-backed rump with a long and stout tail and short stout legs. The total length of the animal varies between 170 and 200 cm, the length of head and body between 110 and 145 cm, the length of ear between 15 and 25 cm, the length of tail between 60 and 75 cm, the height of trunk between 50 and 60 cm, and the weight between 70 and 80 kg.

The snout terminates with a kind of rounded disk—resembling the trunk-shaped one of the pig—and bears in the middle two round nostrils capable of being closed, surrounded by four little clusters of stiff hairs. The mouth is narrow and allows only the long, strap like, adhesive tongue to pass, which can be stretched out to a distance of 45 cm. The small eyes in the middle of the head are surrounded by fine stiff hairs also. The long and pointed ears have a more tube like base and can detect sounds at a great distance. The conical tail begins with a broad root and tapers gradually. The fore feet show four fingers only (the first or thumb vanished completely), beyond which the second is the longest, the fifth the shortest, all bearing long hooflike claws. Between the fingers a webbed skin is stretched as a help in ejecting soil while burrowing. The hind feet have five toes, with broad claws and webbed between the second and fourth. The legs stand on the hole plantas of the feet. Therefore the animal is able to erect the fore body or to sit down in erected position supported by the hind feet and the stout tail only. Head and ears are nearly naked, the other parts of the body are covered with coarse to bristle like hairs sparsely, a little elongated on belly and backside of fore legs. The colour differs between a yellow-gray and a brown, especially the legs show often a darker

head with vibrissae

above: hand in lateral view
below: foot

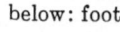

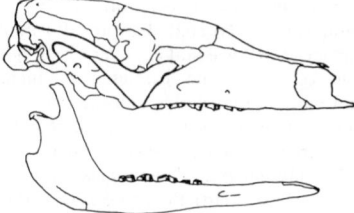

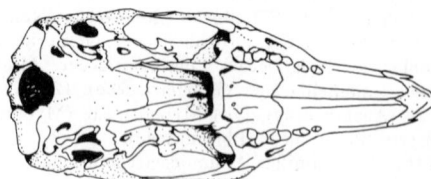

above: skull in lateral view
below: skull underpart

habitus

Fig. 1 Head, hand, foot, skull, and habitus of the ardvark. (After Grasse, 1955).

brown or even a black. Besides for scattered glands of the alveolar and tubulus type the skin bears a small compact aggregation on the elbow, the hip, the root of penis, and the border of vulva. Anal glands are lacking. There are four lacteals with two ventral and two inguinal teets. The salivary glands reach to an enormous extent to provide the richly papilla bespangled long tongue with an adhesive film. The simple stomach shows a crop like sacculation behind the entrance receiving and storing the swallowed humps of termites.

The pylorus is distinct and is provided with a thick layer of musculature to triturate the food before it enters the gut. The latter reaches a length of nearly 11 m, divided into the small intestine 9 m long, and the large intestine 2 m long. On the border between the two enters a short ampulla-like coecum. The lungs in the left part consist of two lobes, in the right of four. The liver includes a gall-bladder. The kidney surface is smooth, the pelvis has only one papilla. The uterus bears two horns, each entering separately. The placenta has a zonal form. The male shows an uterus masculinus but externally no scrotum, because the testes are posed beneath the skin. The vesiculae seminales are of enormous size. The penis is small. The simple brain shows two very great olfactory lobes, demonstrating the power of smelling as the principal sense. The hemispheres with their simple and shallow sulcation are only small and leave the cerebellum uncovered. This means that the aardvark's intellect is not very important.

The skeleton shows the keybones well developed, the great bones of arms and legs short and stout, the vertebras strong (with a number of 7 cervicles, 13 thoracics, 8 lumbars, 6 sacrals, and 25 caudals), and the pelvis with a long pubical symphysis. The smooth and nearly tube like skull has a long rostrum, small orbits, contiguous and weak zygomaticarches and a long low jaw. The teeth are unique among mammals. Their number in each side of the upper jaw is 8 in younger animals and 5 in older ones, in the lower jaw 6 respectively 4. The teeth have the form of vertical cylinders without roots and grow continuously. Each cylinder consists of a bundle of many (up to 1500) fine dentine tubes open at the root but filled with marrow. The dentine lies in prismatic layers around each tube penetrated by very fine transverse channels. All the cylinders are cemented together. There is no enamel. Beneath the small milk teeth—never perforating the gums—the hindmost is greater, furnished with two roots and a knobbed crown, thus demonstrating the form of the ancestral teeth.

Anteaters are found in Africa south of the Sahara including Ethiopia, but are nowhere common, in certain countries even rare, most common in those grasslands which have numerous termite hills. There is only one family (Orycteropidae) with one genus (*Orycteropus*) and with one species (*afer*, Pallas 1766). A greater number of subspecies throughout the continent have been described, but the individual variation is not important. Thus the descriptions, mostly based on a single specimen, are without adequate foundation.

The aardvark is solitary. In the daytime it sleeps curled up like a dog in one of its burrows—often beneath a termite hill—dug out to two or three meters in depth and mostly furnished with several entrances. It is seldom abroad in the daytime, but moves almost entirely at night leaving its burrow at dusk, seeking termite hills and destroying them to reach the interior chambers and tunnels alive with insects which it rapidly licks up in lumps with the long adhesive tongue. The ability to dig is astonishing, since, within a period of seconds, the large animal passes completely below the surface. While burrowing, clumps and particles of soil fly like bullets behind it, enveloping the digger in a great cloud of dust, and preventing any enemy from attacking it. The Aardvark buries its droppings like a cat. Though mostly silent, the animal can grunt like a hippopotamus. The breeding season, in the northern and southern ranges of distribution (Sudan—Ethiopia, South Africa) is chiefly in December and January. In the tropical parts of Africa the breeding season is not known, but is probably seasonally unfixed. The period of gestation is about seven months, the period of parturition in the north and south is chiefly in May, June and July. The young are born fully developed with open eyes, and able to walk about some hours later. The life span is at least ten years (as shown by captive specimens), but may probably extend to fifteen years or more. In captivity aardvarks soon become accustomed to the keeper but do not show great intelligence. It is not very difficult to feed them with milk, eggs, chopped mince-meat, ants, termites, etc.

THEODOR HALTENORTH

TUNDRA

Tundra is the term commonly applied to the 'naturally' treeless areas of land lying north of the tree-limit in the northern hemisphere and south of the tree-limit in the southern hemisphere. It is often used, too, for locally treeless tracts lying within or seaward of the boreal forests, and for those above the tree-limit on mountains and high plateaus—but not normally for other treeless areas such as cleared ones or most deserts or semi-deserts. Thus although, in the original Russian (or Lappish?) sense of the term, reference was to more or less marshy, treeless plains lying beyond the forest, lately this has been widely and very loosely extended to apply also to many other types of vegetation, or even of terrain regardless of vegetation. Since such wide use is apt to embrace a great diversity of vegetational types and sometimes virtually plantless land, yet implies a degree of uniformity even beyond that implicit in the lack of trees (though without including by any means all treeless areas), it is considered best for scientific purposes to restrict the use of the term 'tundra' to the more or less sedgy-grassy types of vegetation that are at least half-closed and widely characterize undisturbed lowland areas of low- and middle-arctic regions—together with their counterparts in treeless tracts within the boreal forests, in the southern hemisphere, and on mountains and high plateaus elsewhere. Scrub and heathlands, fell-fields and barrens, seaside and other local types, and seral eca, are thus (or incidentally) excluded, together of course with those of fresh and salt waters and the cryophytic communities of snow and ice, all being treated, at least in outline for the Arctic, in the separate article on ARCTIC VEGETATION. The component flora of all types of arctic vegetation is dealt with (under the various systematic group headings) in a special article on ARCTIC FLORA.

Even in this restricted sense of at least half-closed and predominantly herbaceous or cryptogamic vegetation in cold regions, tundras include a considerable diversity of vegetation-types situated in almost all major portions of the globe, and far too numerous to describe in detail here. Accordingly in this account the arctic tundras will be the ones mainly described, with supplementary reference to those of other regions.

Mesic tundra. The general run of tundra which

covers a large proportion of the lowland plains and some less extensive upland areas of most low-arctic and many middle-arctic regions, is commonly a rather thin 'grassy' sward dominated by mesophytic sedges such as the Rigid Sedge (*Carex bigelowii* agg.) and grasses such as the Arctic Meadow-grass (*Poa arctica* s.l.), with various associated forbs (i.e. herbs of other than grass habit) and undershrubs including dwarf willows. The whole forms a continuous if often poor sward commonly 15–35 cm. high, in which a mixture of various bryophytes and lichens usually forms a rather poorly-marked second layer a very few centimeters high. Such tundra is commonly a mosaic made up of faciations having each some lesser number of the total association dominants, and including consociations having only one of these. The areas of the component eca are often small and the variations from spot to spot accordingly considerable. In addition there are often local societies dominated by species other than the association dominants. The (sometimes unaccountable) mixing and even intergradation of all these communities is often intricate and may be indicative of their relative youth, many having apparently failed to come to a state even approaching equilibrium with the environment since emergence from glaciation or other extreme disturbance. Moreover, especially in many of the damper areas, the persistent frost activity is such that the vegetation can probably never stabilize the surface properly, the main tendency being towards repeated readjustment to almost perpetual disturbance. This is particularly noticeable in areas of 'polygonal' or other 'patterned' soils.

In the middle-arctic belt the mesic tundra is generally poorer both floristically and vegetationally than to the south, although the dominants are commonly the same, while in high-arctic regions still further depauperation is general, and indeed only limited and relatively few areas are here sufficiently vegetated to be properly designated as tundra in the present sense, and these are usually of hydric type (*see* below). Tracts of mesic tundra in the far north are usually very limited in extent, and, though sometimes characterized by grasses or, particularly, Wood-rushes (*Luzula* spp.) and Polar Willow (*Salix polaris* agg.), are commonly dominated by mosses and/or lichens.

Xeric tundra. On raised areas or well-drained surface material the tundra tends to be relatively poor and thin. Such xeric types are composed of an extremely various array of more or less xerophilous sedges (such as the Rock Sedge, *Carex rupestris*, and the Nard Sedge, *C. nardina* s.l.), willows (particularly the Arctic Willow, *Salix arctica* s.l.), grasses (such as Alpine Holygrass, *Hierochloe alpina*), Northern Wood-rush (*Luzula confusa* agg.), and various forbs (such as Viviparous Knotweed, *Polygonum viviparum*), also Mountain and Arctic Avens (*Dryas* spp.) which are somewhat woody. Scattered heathy plants may also occur, especially in low-arctic areas, as may lichen-rich tracts. The vegetation is usually rather poor, often barely covering the ground in spite of a plentiful admixture of lichens and sometimes also of bryophytes, while on limestone or porous sandy substrata it may be relatively sparse, although the component flora particularly in calcareous areas may be very various.

In the middle-arctic belt the xeric tundra tends to be similar but poorer than in low-arctic regions. Thus although the dominants are often much the same, they

are usually of poorer growth, while some of the associates which were important in low-arctic areas are absent. In the fartherest north lands the dry types of tundra (if such they can be called) tend to be especially thin and limited in extent, dominated largely by lichens (especially species of *Cetraria* and *Alectoria*), and much interrupted by rocks and bare patches.

Hydric tundra. With the generally poor drainage resulting from the soil being permanently frozen to not far beneath the surface, damper depressions or marshy tracts tend to be plentiful in the Arctic though often of quite limited extent, being indeed rarely absent except in regions of porous substrata. In the low-arctic they are commonly rather luxuriantly vegetated, the sward often being taller than in drier areas. Such hydric tundra is usually dominated by Cotton-grasses (*Eriophorum* spp.) and relatively hygrophytic sedges such as the marshland ecads of the Water Sedge (*Carex aquatilis* agg.), and by grasses such as the Arctagrostis (*Arctagrostis latifolia* s.l.), with a few hygrophilous willows (such as the Arctic Marsh Willow, *Salix arctophila*) or other ground-shrubs and many hygrophilous or ubiquitous forbs such as Viviparous Knotweed and Yellow Marsh Saxifrage (*Saxifraga hirculus* agg.). The fairly luxuriant cryptogamic layer is largely composed of mosses and helps to consolidate the sward.

In the middle-arctic belt, as with other types of tundra, the same dominants and most of the same associates persist—though in less luxuriance and with some losses at least among the latter. And although the tendency to lower and poorer development and to restriction in area becomes more and more marked with increasing latitude in the Arctic, marshy tracts may still support a fair hydric tundra even in high-arctic regions. In the far north such areas are limited in extent but usually still dominated by sedges, cotton-grasses, and grasses—often of the same species as in the south, and including similar associated forbs and consolidating mosses, though woody plants apart from tiny willows are usually absent. In some places characteristic moss mats occur, and in others there is marked disturbance and even interruption by polygon formation.

Practically throughout the Arctic, marshy and other humid tracts may be beset with small hummocks commonly about 20 cm. high; such areas are usually referred to as 'hillock tundra,' although the hummocks and their vegetation are of various types. The hummocks are commonly peaty and introduce a range of different microhabitats, including dryish tops which are often populated largely by lichens or xeromorphic ground-shrubs, damper sides supporting sedges and grasses, and intervening dark boglets or puddles with more hygrophilous herbs (often including rushes, *Juncus* spp.), mosses, or even algae.

Forest-tundra. Outside the open, park-like subarctic 'taiga' of scattered trees occurring towards the northern limit of arborescent growth, is commonly found a belt within which, in damp depressions and especially along the courses of rivers, there occur faciations approaching the ordinary northern coniferous forest of the region which in Siberia is itself called 'taiga'. They may project as timbered tongues containing well-grown trees, or even form outliers in the tundra—in either case characterizing the so-called 'forest-tundra.' Such better growth often seems to be correlated with better aeration of the roots where there is active drainage of water. Elsewhere around the tree-line, with inimical conditions of soil and, particularly, exposure, tree growth tends to be extremely poor and slow, yielding gnarled or stick-like dwarfs often of great age. Indeed in many such regions trees fail to return after cutting or burning, so that the tree-line is likely to be man-made.

Alpine tundras. The tree-line on mountains appears to be determined mainly by local exposure, though often to some extent also by soil conditions or by grazing etc. In any case the timber-line 'Krummholz' of stunted, twisted, and sometimes more or less prostrate 'trees' tends to be even more grotesque on mountains than on the borders of the Arctic. Apart from 'outliers' in sheltered situations, the alpine tundras begin where the trees leave off—at altitudes which tend to get lower and lower as the poles are approached, though in this there may be considerable variation with local conditions even on the selfsame parallel of latitude.

In temperate regions the vegetation of tundra etc. just above the timber-line is in many ways comparable with that of low-arctic regions near sea-level, while higher up in both instances a similar sequence of altitudinal climaxes prevails. Thus tundras of various types roughly comparable with those of the Arctic are found above the highest 'elfin' forest, with often extensive tracts of scrub where conditions are suitable and pasturing is not too severe. Above this may stretch increasingly limited tracts of 'alpine meadow' consisting of grasses, sedges, forbs, and often undershrubs—also constituting a form of alpine tundra, but interrupted by heathlands in suitable situations and by fell-fields or barrens in detrital or exposed areas. Mossy mats are especially characteristic of run-off areas below snow-banks hereabouts—especially on the side facing away from the Equator. Elsewhere, with the decreased precipitation at very high altitudes and especially in the tropics, lichens are often the chief plants, the vascular species of such thin tundras and fell-fields as persist tending to be xeromorphic, reduced in stature, and often very hairy and tussocky.

Antarctic tundras. These are in general comparable with the arctic types but tend often to be more tussocky and to include fewer woody plants. Also, with the prevailingly exposed and otherwise inimical conditions of the much more widely scattered southern lands, they persist in most sectors to considerably lower latitudes, and exhibit, floristically, a much higher degree of endemism.

On the Antarctic Continent itself, extremely few and rare vascular plants are found, such very limited tracts of thin and more or less closed 'tundra' as occur in a few places being composed of mosses, lichens, and algae. The most luxuriant vegetation on the Continent is found in particularly favorable seaside areas that are manured by penguins etc., and consists mostly of mosses which form an almost continuous investment very locally.

The islands scattered in the wide seas surrounding the Continent, such as Kerguelen, are also exposed and in general poorly vegetated, though in the less desert-like situations, and especially away from the prevailing winds, the tussocks of such plants as *Azorella selago*, with associated *Acaena adscendens* and grasses and other vascular plants, may form an almost continuous tundra, consolidated by various lichens and mosses. The slightly more southerly island of South Georgia, in spite of its poverty in species and position within the pack-ice, supports relatively luxuriant vegetation near the shore and in sheltered valleys—particularly of *Acaena* and the tall tussocky grass *Poa flabellata*. In some places an unbroken grassy tundra may extend to an altitude of 200 or even 300 meters on sheltered north-facing slopes. The Falkland Islands, near southern South America, support in places quite large bushes, besides grasses which may approach the height of a man and grow so closely together as virtually to exclude other plants.

NICHOLAS POLUNIN

References

"Great Soviet Encyclopedia"; Polunin, N., "Botany of the Canadian Eastern Arctic. III: Vegetation and Ecology," Bull. Nat. Mus. Canada, **104**, 1948. *Idem,* "Introduction to Plant Geography and Some Related Sciences," Chapter 13, London, Longmans, and New York, McGraw-Hill, copyright 1960.

TURBELLARIA

This class includes all of the free living as well as a few commensal and parasitic species of the phylum PLATYHELMINTHES. For many years five orders were recognized: Acoela (220 species); Rhabdocoela (1200 species); Alloeocoela (400 species); Tricladida (1100 species); and Polycladida (800 species), but recently the number of orders has been increased to eleven by dividing the Rhabdocoela into Catenulida, Macrostomida and Neorhabdocoela; the Alloeocoela into Archoophora, Lecithoepitheliata, Holocoela, and Seriata, and creating a new order, Gnathostomulida, for two aberrant species. Turbellaria are of little economic importance and do not even form an important link in food chains, since they are rarely eaten except by other Turbellaria. Biologically, however, they are of considerable importance because they are the most primitive Metazoa to possess well-developed bilateral symmetry, cephalization, a central nervous system, endo-mesoderm, an excretory and hydrostatic system, organ muscles, and complex reproductive systems with accessory organs.

The epidermis may be cellular or syncitial, is generally at least partly ciliated, and secretes mucus. It often also produces rhabdoids which occur in two forms, the short, thick rhabdites and the less common, long, often sinuous rhammites. A basement membrane usually lies immediately beneath the epidermis, followed in order by an outer circular and an inner longitudinal muscle layer, the parenchyma and, except in Acoeles, the gut wall. The mouth, located medially on the ventral surface, opens into the glandular pharynx or into the pharyngeal cavity. Three types of pharynges occur: *simple*, formed by an inturning of the epidermis, not separated from the mesenchyme, and without special muscles; *bulbous*, completely separated by a membrane from the mesenchyme and quite muscular; and *plicate*, also quite muscular and separated from the mesenchyme by a pharyngeal sheath except at its base. Except in Acoeles the pharynx opens directly or by way of a short esophagus into the gut cavity which may be simple and sac like, diverticulated or three or many branched. Acoeles

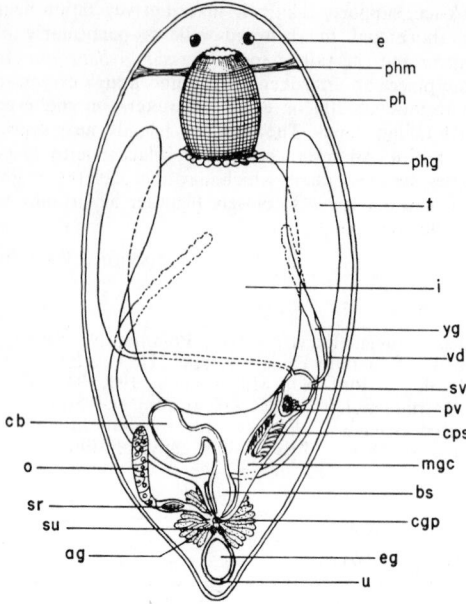

Fig. 1   *Microdalyellia rossi* × 65, showing basic structure of a Turbellarian. (After Graff 1911.) ag, atrial glands; bs, bursa stalk; cb, copulatory bursa; cgp, common genital pore; cps, cuticular penis stilet; e, eye; eg, egg in uterus; i, intestine; mgc, male genital canal; o, ovary; ph, pharynx; phg, pharyngeal glands; phm, pharyngeal muscle; pv, prostate vesicle; sr, seminal receptacle; su, sphincter of uterus; sv, seminal vesicle; t, testis; u, uterus; vd, vas deferens; yg, yolk gland.

have no gut cavity and food passes from the pharynx into the central parenchyma. The protonephridial system consists of one or more tubules with attached flame bulbs and one to several external openings. Primitively, the nervous system is made up of a number of longitudinal fibers lying in the epidermis and connected by commissures and a cerebral nerve ring. In higher forms the system sinks inward into the parenchyma; the number of longitudinal nerves is reduced, leaving only one to three pair, and cerebral ganglia are well developed. Sense organs may include eyes; statocysts; tactile bristles and hairs; and chemo-receptors such as ciliated pits, grooves or rings.

Asexual reproduction by fission or by fragmentation occurs, especially in the Catenulida and Tricladida, but for most species sexual reproduction is the rule. The gonads may be paired or single, and compact or follicular. Most Turbellaria produce ectolecithal eggs with yolk deposited around the outside of the egg, a condition seldom encountered except in the Platyhelminthes. The female gonad may be paired or single and is generally differentiated into egg and yolk producing sections which are sometimes combined as a germovitellarium but usually separated into distinct ovary and vitellarium, either or both of which may be paired. Accessory reproductive organs, which are generally well developed, may include such structures as copulatory bursa, seminal receptacle, uterus, vagina, spermatic duct and oviduct for the female system, and vas

deferens, seminal vesicle, prostate glands and vesicle, and an armed or unarmed penis for the male. The two systems may have a common gonopore or may be entirely separate with two or even three gonopores. The male gonopore is sometimes combined with the mouth. With the exception of Müller's larva, which occurs in some Polyclads, there is no larval stage. Many Turbellaria possess great power of regeneration and they have been much used for the study of problems of regeneration, differentiation and metabolic axial gradients.

Turbellaria are common in fresh and brackish water as well as in marine and terrestrial habitats but are easily overlooked because of their secretive habits, small size and inconspicuous coloring. Their usual length is 0.5 to 20 mm. but in some land planaria it may exceed 50 centimeters. Most species are colorless or almost transparent but some of the larger species are brightly colored and others contain black, brown or gray pigment; eat colored food; or contain zoochlorellae.

There are two chief theories as to the evolutionary origin of the Turbellaria. The Ctenophore-Polyclad theory postulates the origin of the Polyclads from the creeping Ctenophores (Platyctenea) on the basis of such common attributes as possession of similar shape, two dorsal tentacles, a mouth in the center of the ventral surface, from which the branches of the gut radiate; a radiating nervous system, determinate cleavage and a creeping habit. This theory presupposes the evolution of the other orders from the Polyclads, but the Polyclads are less primitive than the Acoeles and there is more evidence of relationship between the Acoeles and each of the other orders of Turbellaria than between the Polyclads and the other orders. The second theory derives the Turbellaria from a planuloid ancestor by way of the Acoeles and seems to be supported by stronger evidence. Primitive characteristics which are found in at least some species of the order Acoela include intraepidermal nerve plexus with little differentiation of a brain; statocyst comparable to that of coelenterate medusae; primitive reproductive system with germ cells differentiating from parenchyma cells; no real gonads and comparatively little development of ducts or accessory organs; epidermis sometimes syncitial and containing muscle fibers; remarkable powers of regeneration so that even a very small piece from almost any section of the body may regenerate an entire worm. Many of these conditions are also found in Coelenterates. In addition, like the planula larva, the Acoela have no digestive cavity.

The Nemertina obviously arose from the Turbellaria, but differ in the possession of an anus, a circulatory system and an eversible probascis. The proboscis probably represents a much elaborated form of that seen in the Kalyptorynchia among the Neorhabdocoela. The spiral, determinate type of cleavage which is characteristic of many Turbellaria indicates a relationship with the Annelid-Mollusc or protostome group of the animal kingdom. However, the presence of large masses of yolk in the ectolecithal eggs often profoundly modifies development. All Turbellaria appear to produce a stereogastrula which resembles the planula larva of coelenterates in lacking an archenteron.

E. RUFFIN JONES

References

Bresslau, E., "Turbellaria" *in* Kukenthal, W. and T. Krumback eds., "Handbuch der Zoologie," Vol. 2, Berlin, de Gruyter, 1933.

Hyman, L. H., "The Invertebrates," Vol. 2, New York, McGraw-Hill, 1951.

TWINNING

The litter size of each mammalian species is relatively constant. Even when the typical species number of young at a birth is one, as in all primates except the marmoset, exceptional births occur in which several offspring are delivered.

Litter size is roughly correlated with six other characters. (1) *Body size:* the larger the adult animal the fewer the young. (2) *Duration of gestation:* animals with a period of pregnancy normally exceeding 150 days rarely bear more than one at a birth. (3) *Age of pubescence:* mammals which become pubescent after the age of five usually have one offspring. (4) *Life span:* a mammalian species with a life span of more than 20 years is commonly uniparous, while the animal which ordinarily dies before the age of 10 is likely to be multiparous. (5) *Type of uterus:* the simplex or unicornuate uterus ordinarily harbors a single embryo. (6) *Number of breasts:* two breasts are associated with uniparity.

On the basis of the above characters, humans ordinarily should and do bear single young. However, on the basis of 87,876,745 consecutive United States births, twin deliveries occur at the rate of 10.9 per 1000 births (proportion of 1:90.3 singletons) and triplets 0.105 per 1000 (proportion of $1:98 \times 10^3$ singletons).

In 1895 Hellin stated that twins occur once in 89 births, triplets in 89^2 and quadruplets in 89^3. In 1920, Zeleny modified this by claiming that when $1/N$ expresses the relative frequency of twins, $1/N^2$ and $1/N^3$ expresses the frequency of triplets and quadruplets respectively. Zeleny's hypothesis takes into account the fact that the incidence of nultiple births varies in different population samples. When tested against a large number of births Zeleny's formulation is more correct than Hellin's, but nevertheless is only an inexact biologic approximation. However, it is true that when an ethnic group has frequent or infrequent twin births it has relatively frequent or infrequent triplet and quadruplet births.

This is clearly demonstrated by U.S. white and nonwhite births. Among whites, twins were born once in 93.3 and triplets once in 10,200 births; while in U.S. colored twins occurred once in 73.3 deliveries and triplets once in 6,200. It has been clearly demonstrated that multiple births are most common among the "colored," intermediately common among the "white" and least common among the "yellow" races. According to the Japanese Bureau of Statistics, in 10,427,779 births (1951–55) twins occurred once in 155 deliveries and triplets once in 17,710.

There are two biological varieties of twins, those arising from the fertilization of a single zygote (monozygotic) and those originating from two eggs (dizygotic). In monozygotic twinning a single SPERMATOZOON fertilizes a single EGG and either during the morula stage the early cell mass divides into two cell masses or after the blastocyst is formed the embryonic area either divides or buds. In either variety, morula- or blastocyst-twinning, the embryos have the same identical genetic material and therefore must be of the same sex, have the same blood groups, same coloration, etc. If there is any doubt about the uniovular origin of a pair of twins, skin transplants between the two offer incontrovertible proof since tissue transplants between monozygotic twins behave like autografts instead of homografts.

In morula-monozygotic twinning which occurs in approximately 30 per cent of one-egg twins, they either have separate placentas, since the division occurs before the trophoblastic anlagen of the placenta is formed, or a fused single placenta with a partition between the two sacs consisting of four membranes: amnion-chorion-chorion-amnion. In other words, their placental relations would be wholly indistinguishable from ordinary two-egg twins. When such placental relations are observed, in 90 per cent two eggs are involved, but in 10 per cent, it is associated with a pair of one-egg morula-twins. In blastocyst-monozygotic twinning, one always finds a single placenta with only a two layer partition between the sacs, amnion-amnion; therefore whenever such a placenta is encountered, one may be sure that the twins are from one egg.

Conjoined (or siamese) twins are one-egg blastocyst twins in which the embryonic area failed to divide completely. The location in the embryonic area where division failed to occur determines which tissues the twins will have in common and their point of union.

The etiology of one-egg twinning is obscure but there is experimental evidence that it may be due to the developing ovum's temporary deprivation of oxygen in the very early phases of embryogenesis. Two-egg twinning is simply a litter of two, two eggs fertilized by two separate spermatozooa within the same ovulation cycle. Monozygotic twinning is uninfluenced in its incidence by race, age, number of previous children, previous twinning or heredity. On the other hand, the occurrence of dizygotic twinning is influenced by many factors. Dizygotic twinning is most frequent in colored races and least common among yellow ethnic groups. It is most common in women between the ages of 35 to 40 and least common in women below 20 and above 45. Two-egg twinning is more frequent in mothers who have previously had a pair of two-egg twins and in families where there is a strong family history of multiple twin births.

ALAN F. GUTTMACHER

References

Guttmacher, A. F., "The incidence of multiple births in man and some other unipara," Obstet. and Gynecol., **2:** 22, 1953.

Waterhouse, J. A., "Twinning in twin pedigrees," Brit J. Soc. Med., **4:** 197, 1950.

Newman, H. H., "The Biology of Twins," Chicago, The University Press, 1917.

ibid., "The Physiology of Twinning," Chicago, The University Press, 1923.

U

ULTRASONICS

Sound waves that exceed the audible limit of 20,000 vibrations per second are defined as *ultrasound*. However, the term ultrasonics is generally applied to frequencies of 500,000 vibrations per second and higher. Most of the work in which animals and plants have been treated with ultrasonics has been of such high frequencies.

The first ultrasonic waves were produced over forty years ago by means of the electric spark and the singing arc. A dependable source of ultrasonic waves of controllable frequency and intensity stems from Langevin's work in underwater signalling around 1912. Although Langevin hinted at the possibilities of the piezo-electric effects of quartz on biological materials nothing was published until 1926. The range of ultrasonic wave lengths in liquids is from 6 cm to approximately 2×10^{-3} cm. These extremely short wave lengths have made possible the discovery of many interesting phenomena that are not ordinarily observed in the acoustics range.

In the course of fifty years this physical force, ultrasonic vibrations, has been investigated and has been applied to practically every branch of science. Ultrasonic irradiation was introduced as an investigative tool to the field of biology by Wood and Loomis in 1927. At this time the destructive force of ultrasonic energy upon biological organisms and living tissues was discovered.

Ultrasonics exerts its effects by pressure alternations and cavitations. In addition to these mechanical effects there have also been reported various thermal effects largely due to the adsorption of the sound energy by living tissues. Either or both may cause death to an organism. Mechanical death may be due to the explosive violence with which cavitations occur thereby resulting in the dissociation of molecules whereas thermal death is the result of the high level of heat produced.

Ultrasonic energy applied in large dosages is an injurious agent like x-rays and atomic radiations, but if used in smaller dosages, like the latter, it has proved to be a therapeutic agent in the treatment of certain diseases such as arthritis. As a method for investigating the mechanisms of biological processes it has raised questions of fundamental biological importance. The general pathology of the mechanical, thermal, and chemical effects of ultrasonic energy in large doses is easily defined, but the use of small dosages has resulted in difficulty in determining the subtle changes produced in living tissues. That changes do occur is evident but the nature of the changes is more difficult to define. In some instances they are beneficial and in others they are not. The minimum ultrasonic dosage is not easily defined at present nor is it presently possible to correlate a definite type of tissue damage with a universally standardized ultrasonic dosage.

Ultrasonic irradiation has been used with some success in sterilizing bacterial cultures, in altering the permeability of living membranes, in experimental neurology, in inhibiting types of sarcomas, hemolysing blood, and on chromosome behavior.

DALE C. BRAUNGART

References

Dognon, A. and E. Biancani, "Ultrasonics and their Biological Action," Radiobiologica, **3**: 40–54, 1958.

El'pener, I. E., "Ultrasound: Its Physical, Chemical, and Biological Effects," New York, Consultants Bureau, 1964.

Kelly, E., "Ultrasonic Energy; Biological Investigations and Medical Applications," Urbana, University of Illinois Press, 1965.

ULTRASTRUCTURE

Ultrastructure refers to the structure of matter at a level below the dimensions accessible for light microscopical examination. In Biology today Ultrastructure deals with the structure of macromolecules and of supramolecular components consisting of a few types of molecules organized into membranous, filamentous, or particulate components which constitute the biologically functional structural units.

Methods of analysis. Ultrastructure is studied by means of electron microscopy, X-ray diffraction, polarization optical analysis, and various physical chemical methods. Electron microscopy allows direct visualization of the structural components, which makes it possible to measure directly the dimensions of these components and to observe their shape and the variation of shape and dimensions. The range of resolution of electron microscopy extends from 6 Å to that of low resolution light microscopy. Light microscopy and electron microscopy therefore overlap from about 0.2μ upwards. The resolving power of the electron microscope can rarely be exploited completely, mainly due to unfavorable contrast conditions. 10–15 Å can be considered the limit when working on biological material.

X-ray diffraction studies reveal periodically repeated structural details in an indirect way through reflections of a narrow X-ray beam. The periodicity of regularly repeated patterns can be calculated, and through trial and error models can be designed which, according to calculations, would give rise to the observed diffraction patterns. Applied to biological material X-ray diffraction analysis is in most cases greatly limited because this material yields few reflections of the X-ray beam due to a low degree of order. This means a restricted amount of information and great uncertainty regarding the reliability of the deduced models. X-ray diffraction reveals the average periodicity in a repeated pattern, but not the range of variability. The range of dimensions to which X-ray diffraction is applicable is from 1–2 Å to over 1000 Å. There is therefore a considerable overlap with electron microscopy as far as dimensions are concerned.

Polarization optical studies allow observing anisodiametrical particles, that is, either more or less rod-shaped or plate-shaped particles, which are small in relation to the wave length of light, when these particles show some preferred orientation. Such particles give rise to so-called *form birefringence*. The interpretation of form birefringence is based on the application of the Wiener's theory for form birefringence. Polarization optical analysis also discloses asymmetries in molecules, *intrinsic birefringence*, and in the arrangement of anisodiametric molecules.

Information which is collected with these methods may be interpreted to reveal the arrangement of the atoms in molecules and of the molecules in supramolecular structural units. A certain knowledge of the chemistry of the supramolecular units, for instance the relative proportions of lipids and proteins, is important for these deductions. All information brought together allows the proposal of models describing the intramolecular structure (X-ray diffraction) as well as the molecular architecture of various supramolecular components (X-ray diffraction, polarization optical analysis, and electron microscopy). The reliability of these models depends on the amount of information available. Since for most biological material this is rather limited, the models frequently represent crude approximations.

The ultrastructure of molecules is discussed under such headings as Proteins, Nucleic Acids, etc. Some basic supramolecular structural units of cells and tissues will be discussed here.

Some basic supramolecular components of cells and tissues. Cells are bounded by a membrane, the cell membrane or plasma membrane, which regulates the exchange between the cells and their surrounding medium. In the part of the cell body which extends between the plasma membrane and the cell nucleus, the cytoplasm, several organized bodies or cell organelles are present. They are, in most cases, characterized by the presence of membranes. In fact, the organization of organic molecules into membranes appears as a most basic principle of organization of living matter, and the formation of a bounding membrane separating a small amount of water containing complex organic molecules from the surrounding medium certainly represents one of the basic steps in the development of life.

Available chemical data show that these membranes consist largely of protein and lipid molecules, and possibly carbohydrates. Studies of the electrical impedance of the cell membrane, its mechanical and optical properties and its chemistry, have contributed indirect evidence that a continuous layer of lipids, presumably a double layer sandwiched between two thin layers of protein, is a basic component of the plasma membrane. Electron microscopic observations support this model and indicate furthermore that the membrane is asymmetric; the layer of protein molecules covering the cytoplasmic surface, that is, the inner surface of the membrane, is thicker than that forming the outer surface of the cell membrane. Fig. 1 illustrates a model incorporating the available information relating to the molecular architecture of the plasma membrane. This seems to be a basic pattern for most membrane components in cells. The thicker layer of protein molecules on one side of the lipoprotein complex can vary

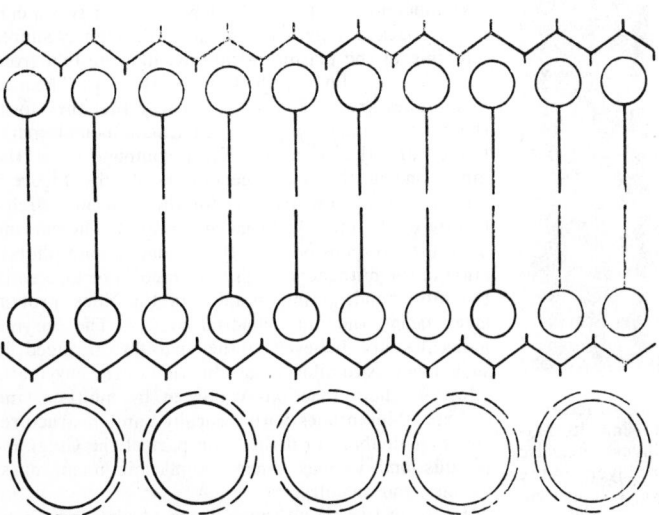

Fig. 1 Molecular model for the plasma membrane. The small circles with tails represent lipid molecules with the hydrophilic end indicated by the small circles. The zigzag lines indicate protein molecules with stretched peptide chains. Large circles indicate an additional layer of protein molecules which has been proposed to consist of globular proteins.

in thickness. In certain compound membranes there is no obvious difference in thickness between the two protein layers sandwiching the lipid layers.

In certain specialized cases such as the lipoprotein sheath enveloping the myelinated nerve fibers, the so-called myelin sheath, X-ray diffraction studies have allowed the interpretation that the lipid and protein molecules form alternating layers, with the lipid molecules oriented at right angles to thin layers of protein. The lipid molecules appear to be arranged in double layers measuring about 60 Å in thickness. Polarization optical data can also be interpreted as demonstrating a layered arrangement, but do not allow any conclusions regarding the actual dimensions of the layers. The electron microscope reveals a layered structure, and the dimensions of the layers can be measured directly (Fig. 2). The layers are of molecular dimensions, and the structural pattern may be interpreted as reflecting an ordered arrangement of molecules. A reasonable model for the molecular architecture of the myelin sheath is shown in Fig. 3. Pairs of double layers of lipid molecules sandwiched between

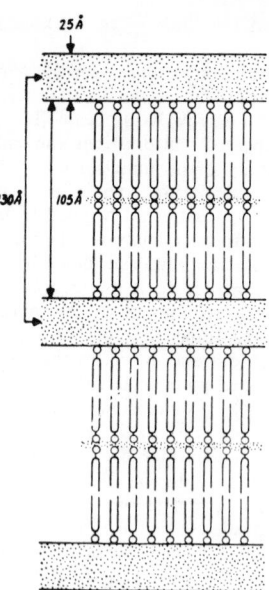

Fig. 3 Molecular model describing the molecular structure of the myelin sheath. The stippled layers indicate the position of proteins, which are stained black in Fig. 2. The small circles with tails indicate double layers of oriented lipid molecules.

protein layers form the fundamental unit of the sheath, and the difference in thickness of the sheath in different nerve fibers depends on the number of such units. The layered structure of the myelin sheath is continuous with the plasma membrane of the Schwann cells surrounding the nerve fibers as rows of satellite cells (Fig. 4). The relation between the structural pattern of myelin sheath and that of the plasma membrane supports the interpretation that the plasma membrane consists of one double layer of lipid molecules sandwiched between two layers of protein or between one layer of protein and another layer containing carbohydrates.

Compound membranes, which consist of two membranes packed together, can arise through a simple infolding of the plasma membrane like the fold from which the myelin sheath develops (Fig. 4). The two membranes are packed in such a way that they form the mirror image of each other. Compound membranes are the basic structural components of the mitochondria, the energy generators of cells. Figure 5 illustrates a proposed model for the molecular architecture of the mitochondrial membranes. The enzyme molecules responsible for the oxidative phosphorylation in the mitochondria are assumed to be located in · the mitochondrial membranes, in the thick protein layer facing the mitochondrial matrix. The enzyme molecules are believed to be spatially arranged in such a way as to allow a coordination of the enzymatic effect in chain reactions catalyzed by multi-enzyme systems. Membranes with basically similar structure, single or double, form the main part of the Golgi apparatus and various other cytoplasmic membranes, vesicles, and vacuoles.

In connective tissue the collagen fibrils are respon-

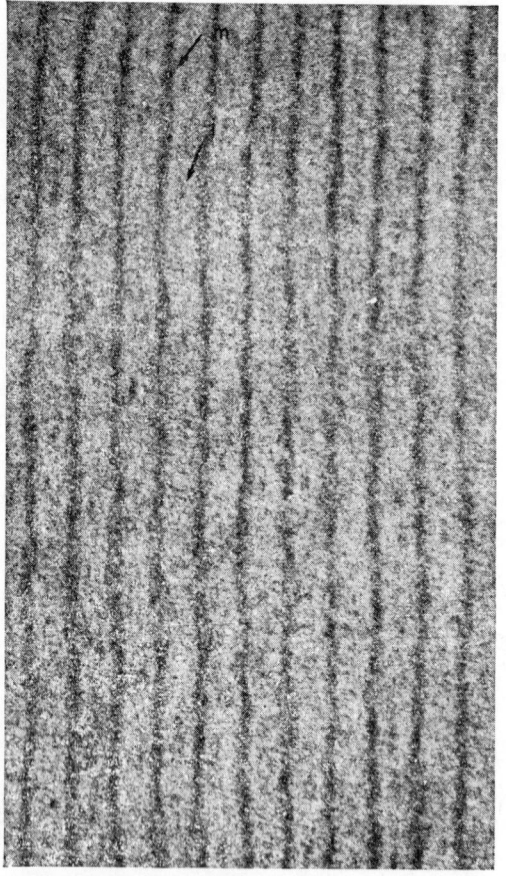

Fig. 2   Electron micrograph of the layered structure of the myelin sheath of peripheral nerve showing alternating dark and light layers. The dark layers measure about 30 Å in thickness. Between the dark layers, in the middle of the light layers, a thin dark layer is indicated. × 5000,000.

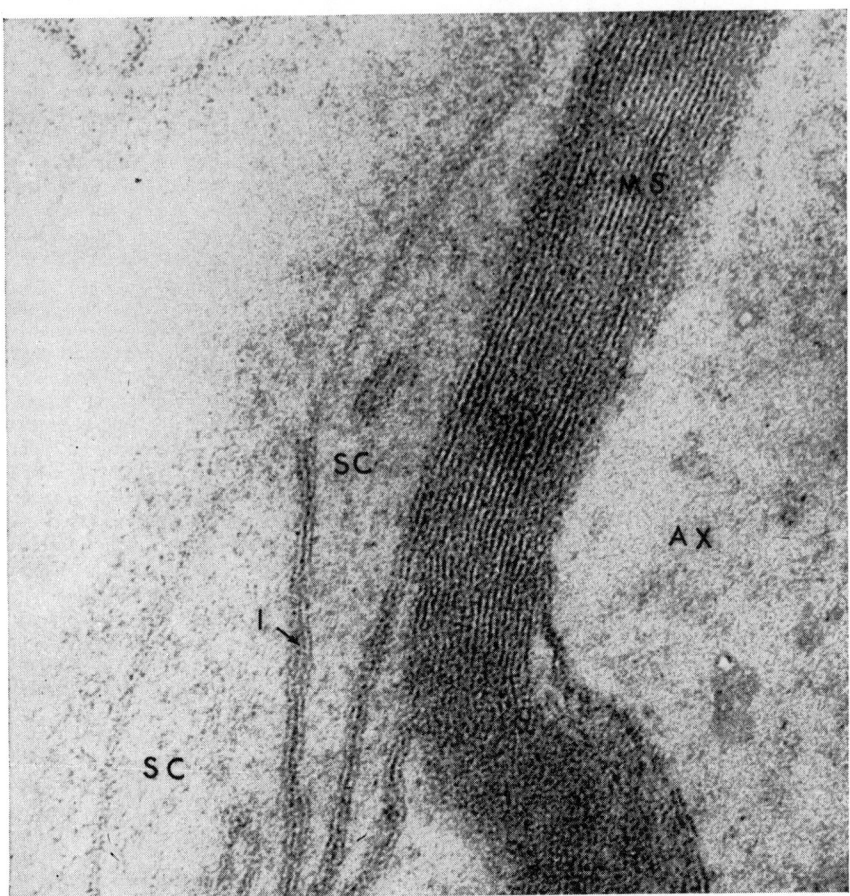

Fig. 4 Part of the myelin sheath of a myelinated peripheral nerve fiber showing the connection between the myelin sheath (MS) structure and the plasma membrane of the Schwann cell (SC) surrounding the nerve axon (AX). The connection is seen at I as an invagination of the Schwann cell plasma membrane, which forms a multi-layered spiral around the axon. × 192,000.

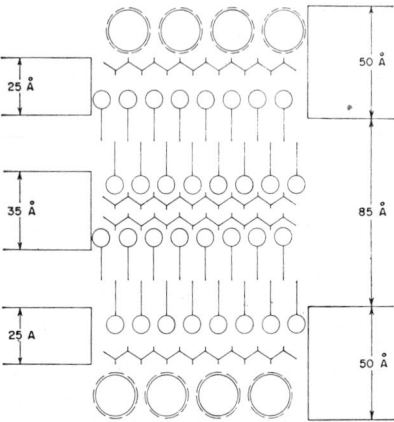

Fig. 5 Model describing molecular structure of composite membranes as, for instance, the membranes in mitochondria.

sible for the tensile strength of the tissue, which in the tendons reaches its highest values. The high tensile strength is assumed to be due to a regular staggered packing in thin collagen filaments of about 3000 Å-long protein molecules, which in turn consist of three peptide chains twisted together to form a three-stranded helix. Within each molecule a high tensile strength results from the strong bonds between carbon atoms and between carbon and nitrogen atoms along the peptide chains. Because of the staggered packing of the collagen molecules, adjacent molecules can be bound together by a large number of cross links between the amino acid side chains of adjacent molecules. The regular packing of the collagen molecules has been assumed because collagen filaments show a periodic and very regular cross banding in the electron microscope with a length of the period of 640 Å. This cross banding depends on the ability of certain amino acids in the collagen molecule to bind the stains used. X-ray diffraction analysis has made possible the deduction of a model describing the three-stranded helical structure of the collagen molecule as one possible model. Electron microscopy has made it possible to deduce the length and shape of the collagen molecule and to study the packing of the molecules in supramolecular units, the collagen filaments. The dimensions of the collagen molecule have been confirmed by physical chemical methods.

FRITIOF S. SJØSTRAND

References

Kurtz, S. M., "Electron Microscopic Anatomy," New York, Academic Press, 1964.

Sjostrand, F. S., "Electron Microscopy of Cells and Tissues," 2 vols., New York, Academic Press, 1966–1968.

UMBELLALES

The order comprises moderately specialized DICOTYLEDONOUS flowering plants possessing small actinomorphic FLOWERS in frequently involucrate massed INFLORESCENCES, a reduced calyx, predominantly 4- or 5-merous choripetalous corolla, usually isomerous stamens, few carpels, pendent anatropous ovules, and few-seeded fleshy drupaceous or baccate, or dry schizocarpous fruits. Six families are included: *Cornaceae* (11 genera, 90 species), temperate zones and Old World tropics; *Nyssaceae* (3 genera, 12 species), eastern Asia and North America; *Alangiaceae* (1 genus, 22 species), Old World tropics; *Garryaceae* (1 genus, 15 species), southwestern North America; *Araliaceae* (65 genera, 750 species), chiefly pan-tropical; *Umbelliferae* (200 genera, 3,000 species), cosmopolitan.

Nyssaceae, Alangiaceae, Garryaceae, and most Cornaceae and Araliaceae are trees or shrubs; Umbelliferae are predominantly herbaceous, of diverse habit, but a number are woody. Wood is diffuse-porous in Nyssaceae, Garryaceae, and most Cornaceae and Araliaceae, and tends to be ring-porous in Alangiaceae, many Araliaceae, and Umbelliferae. Vessel segments range from slender, very long to medium-sized, with oblique end walls in Cornaceae, Nyssaceae, Alangiaceae, Garryaceae, and some Araliaceae, to extremely short and transverse in more advanced Umbelliferae. Perforations are scalariform in Cornaceae, Nyssaceae, and Garryaceae, scalariform to simple in Alangiaceae and Araliaceae, and consistently simple in Umbelliferae. Vascular rays are mostly heterogeneous of types I, IIA, and IIB, but with some homogeneous types I and II in Araliaceae. Arrangement of vascular parenchyma varies from diffuse in Alangiaceae and Garryaceae, to paratracheal in individual genera of Nyssaceae and Cornaceae and most Araliaceae and Umbelliferae.

Simple exstipulate leaves characterize Cornaceae, Nyssaceae, Garryaceae, and Alangiaceae. Those of Araliaceae and Umbelliferae are commonly divided or compound, and a sheathing petiolar base and/or stipular structures are usually prominent. The vascular supply is trilacunar in Cornaceae, Nyssaceae, Alangiaceae, and Garryaceae, and multilacunar in Araliaceae and Umbelliferae.

Congested inflorescences are characteristic of all but Alangiaceae, which have conspicuously larger flowers borne on articulated pedicles in axillary cymes. Garryaceae, apparently specialized for wind pollination, bear densely flowered, pendent, ament-like racemes, which has frequently led to association of the family with the artificial Englerian "Amentiferae." Hutchinson has recently concluded that the flowers of Garryaceae are epigynous, as in other Umbellales. Aestivation is valvate in Alangiaceae and Garryaceae, mostly so in Cornaceae and Araliaceae, and usually imbricate in Nyssaceae and Umbelliferae. Alangiaceae have bisporangiate flowers only, Garryaceae are dioecious, and other families show some unisexuality. The stamens are attached to a prominent disc surmounting the ovary; the anthers dehisce longitudinally. The ovary contains usually 1–5 locules, and the ovules are solitary or paired (Garryaceae) in each locule. Placentation is axile except in Garryaceae and *Aucuba* (Cornaceae). The ovule, which has a single integument, varies from crassi- to tenuinucellate, but constantly lacks a parietal cell only in Umbelliferae. The massed nectar-secreting flowers are probably attractive to small insects, and the common proterandry is believed to promote cross-pollination. Pollen grains are usually tricolporate or tricolpate. Embryo sacs mostly undergo the Normal type of development, although other types occur in Umbelliferae. Endosperm development is cellular in Cornaceae and Nyssaceae, and nuclear in Garryaceae, Araliaceae, and Umbelliferae. The embryo is typically small and embedded in endosperm.

The fossil record reveals the circumboreal occurrence of Nyssaceae, Cornaceae, and Araliaceae almost throughout the Tertiary, of Alangiaceae in Mid-Tertiary, of Umbelliferae only in Mid- and later Tertiary, and of Garryaceae from uppermost Tertiary, and then only in western North America.

Basic gametic chromosome numbers would appear to be: *Alangiaceae*, 8, 11; *Nyssaceae*, 11; *Garryaceae*, 11; *Cornaceae*, 8, 9, 10, 11; *Araliaceae*, 12; *Umbelliferae*, 5, 6, 7, 8, 9, 10, 11, 12 (with 8 and 11 most frequent). Polyploidy is apparently present in Nyssaceae, Cornaceae, Araliaceae, and Umbelliferae.

The order consists of two major groups: (1) the Araliaceae-Umbelliferae alliance, with primary stems strengthened by peripheral collenchyma, containing abundant aqueous tissue, and bearing predominantly divided or compound leaves supplied by multilacunar nodes, tissues permeated by secretory canals, and possessing widely spaced multiseriate vascular rays and paratracheal parenchyma; and (2) the Cornaceae-Nyssaceae alliance, with primary stems supported by pericyclic sclerenchyma, lacking aqueous tissue, and bearing simple leaves attached by trilacunar nodes, tissues lacking secretory canals, but having both uniseriate and multiseriate vascular rays and diffuse parenchyma. Garryaceae and Alangiaceae are families of moderate advancement which show similarities to Cornaceae-Nyssaceae in their wood and pollen. The two alliances are linked by the polymorphic families Cornaceae and Araliaceae. Comparative anatomy fails to support separation of Araliaceae and Umbelliferae into separate orders; indeed, no sharp boundary exists between Araliaceae and Umbelliferae. Extra-ordinal affinities may be with Hydrangeaceae ("Saxifragaceae") and Caprifoliaceae (*Sambucus, Viburnum*).

Economic value is slight, comprising chiefly vegetable crops (carrot, celery, parsely, parsnip) and condiments (anise, caraway, coriander, dill) from Umbelliferae. Some genera of each family are used as ornamentals.

An exhaustive bibliography is contained in the reference given.

LINCOLN CONSTANCE

Reference

Rodriguez, R. L., "Systematic anatomical studies on Myrrhidendron and other woody Umbellales," Univ. Calif. Publ. Bot., **29**: 145–318, 1957.

UROCHORDATA

One of three major divisions of the animal phylum Chordata (subphyla Urochordata or Tunicata, Cephalochordata, and Vertebrata or Craniata). Urochordates differ from other chordates in the following features: (1) the notochord is confined to the tail and posterior trunk region of the larva in most species, and is usually resorbed along with the dorsal nerve cord during metamorphosis; (2) the mesoderm forms no somites during development, thus typical chordate metamerism is absent; (3) a typical coelom is lacking, and the main body cavity is haemocoelic; however, two thin-walled sacs, the epicardia, develop from the posterior pharynx in most species and may represent enterocoelic pouches; the pericardial cavity is sometimes considered a coelomic space; (4) an atrial cavity, formed from one or two dorsal or dorsolateral epidermal invaginations, surrounds the pharynx in most species; (5) the body is encased in a secreted test or tunic containing living cells and resembling an external connective tissue rather than a cuticle; the tunic usually contains fibers of cellulose or a similar polysaccharide, and sometimes calcareous or organic spicules, blood vessels, and nerves; (6) the heart periodically reverses the direction of its beat; (7) all species are hermaphroditic; and (8) cleavage is determinate.

About 2000 species of urochordates are known, all from marine environments. Three classes are recognized: Ascidiacea, Thaliacea, and Larvacea. The Ascidiacea, the largest and most varied class, represent sedentary forms with the pharynx modified in connection with ciliary-mucoid filter feeding on small plankton and detritus; the gut is strongly recurved (due to expansion of the morphologically ventral surface) and the oral and atrial apertures, sometimes borne on tubular siphons, are thus generally directed away from the substrate. The oral aperture admits a steady inflow of water laden with food and oxygen, while the atrial aperture expels water carrying metabolic and fecal wastes. Muscles in the body wall permit body contractions which may forcibly eject materials from both apertures. The pharynx is often greatly enlarged; its entrance is guarded by a ring of tentacles which prohibit ingestion of large objects, and its walls are pierced by numerous elongate, curved, or spiral stigmata which probably represent specialized and subdivided gill-slits. Cilia lining the stigmata create the current of water flowing through the siphons. The endostyle, a longitudinal glandular furrow in the pharyngeal floor (and an evolutionary forerunner of the vertebrate thyroid), secretes strands or porous sheets of mucus which are carried upward on either side, propelled by cilia on the pharyngeal walls and by ciliated peripharyngeal grooves anteriorly. Trapping food particles as it moves past the stigmata, the mucus is carried to the dorsal midline of the pharynx where it is rolled into a rope by

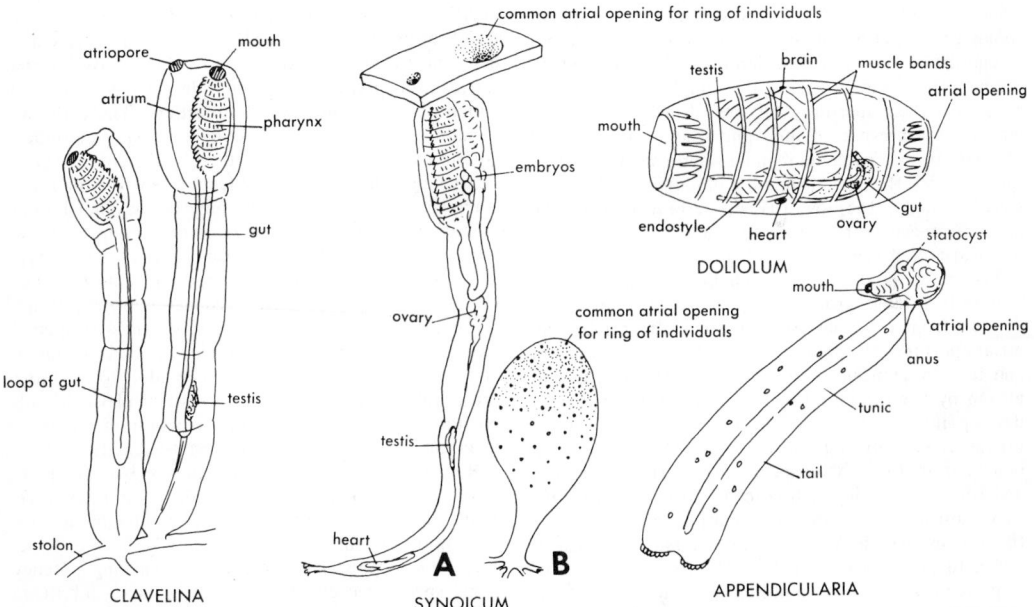

Fig. 1   Basic kinds of tunicates: solitary (*Clavelina*), colonial (*Synoicum*), and pelagic (*Doliolum*, *Appendicularia*). A shows individual removed from colony, B. (From Jollie, "Chordate Morphology," New York, Reinhold, 1962.)

a row of tentacles or a ciliated membrane and is passed posteriorly. The remainder of the gut is ciliated internally and is differentiated into esophagus, stomach, and intestine, the latter often showing specialized regions. The gut bears a pyloric gland and sometimes a liver. Digestion is wholly extracellular. Fecal pellets pass from anus to atrium for expulsion. Nitrogenous wastes are eliminated as ammonia in some species; in others specialized blood cells (nephrocytes) or renal vesicles representing modified epicardia appear to function as storage excretory organs. The tubular heart bears pacemakers at each end which alternate in their control of the beat, resulting in periodic reversal of blood flow throughout the body. Formed vessels may occur in the pharynx and tunic; elsewhere the blood follows haemocoelic channels, and capillaries are absent. Blood cells of several types are known. A single solid ganglion, located dorsally between the apertures, serves as an adult brain, sending nerves to the rest of the body. The neural gland lies near the dorsal ganglion and communicates with the pharynx by a duct; it contains gonadotropic substances and is sometimes considered homologous with part of the vertebrate pituitary. Gonads are borne within, or posterior to, the intestinal loop, or are attached to the atrial wall laterally. Hermaphroditism prevails, though some species are markedly protandric. Fertilization may be internal or external, some forms being oviparous, others viviparous. Some species show a high degree of self-sterility. Cleavage is bilateral and determinate. In most species a tadpole larva is produced, equipped with a muscular tail, notochord, dorsal nerve cord, brain, sense organs, and structures for attachment. The larva swims briefly, settles, attaches, undergoes metamorphosis, and grows to the adult condition. Some ascidians remain solitary, their ovoid or stalked bodies attached to hard bottoms or partially buried in mud or sand. In other species asexual reproduction results in clones or colonies which may appear as groups of zooids connected basally or may grow as massive lobes or encrusting sheets of common test material in which the numerous minute zooids are completely embedded except for their apertures. Of special interest in ascidians are the presence of phosphocreatine, the storage of quantities of vanadium and sometimes niobium in some species, the organic binding of iodine by the endostyle in some, the low pH of certain tissues, the marked capacity for regeneration, and the mosaic eggs which are often used in embryological experiments.

The much smaller classes Thaliacea and Larvacea represent pelagic groups. The thaliaceans, probably derived from a sessile ascidian stock, bear oral and atrial apertures at opposite ends of the body and swim and feed by pumping water through the pharynx and atrium by contraction of body muscle bands. The eggs develop into motile oozooids which in turn produce groups of sexually reproducing blastozooids by means of a budding stolon. Most species are nearly transparent and some are highly luminescent. The Larvacea are small forms, probably representing paedomorphic thaliaceans, which retain the tail, dorsal nerve cord, and notochord throughout life. No atrium is present. The gelatinous tunic contains polysaccharides but no cellulose, and in some species forms an elaborate apparatus capable of concentrating nannoplankton for food.

D. P. Abbott

References

Berrill, N. J., "The Tunicata," London, Ray Society, 1950.
Van Name, W. G., "The North and South American Ascidians," Bull. Amer. Mus. Nat. Hist., **84,** 1945.
Thompson, H., "Pelagic Tunicates of Australia," Commonwealth Counc. Sci. Industr. Res. Org., Melbourne, 1948.

URODELA

Salamanders. Amphibians, with head, trunk, and tail. Usually tetrapods but with reduced limbs. The skin is scaleless and protective embryonic membranes are absent.

The urodela are closely related to the living APODA, which are limbless with minute scales and reduced eyes and which lack a tail and have sensory tentacles. They are also closely related to the fossil Aistopoda, Nectridia and Microsaura which lived in the Carboniferous and Permian eras. The earliest recorded urodele is from the Lower Cretaceous. These groups constitute the Subclass Lepospondyli which are those amphibians where the centrum forms directly by the deposition of bone around the embryonic notochord. The other subclass of AMPHIBIA (Apsidospondyli) include the fossil Labyrinthodonts and the living Salientia (Frogs and Toads) and are known from the Devonian to the recent.

The urodela are characterized by skeletal degeneracy and the loss of the tympanum. The limb girdles are cartilaginous. The vertebra are numerous and the branchial arch skeleton is well developed. Teeth are present in both jaws. Sexual dimorphism is not great, the females are usually slightly larger than the males, and occasional differences are reported in body proportions, glands, teeth and color. Breeding is aquatic in all forms except the plethodontids. Primitively urodeles have external fertilization but most have internal fertilization preceded by a complex courtship pattern which differs in the several suborders. The courtship includes the deposition by the male of a spermatophore (sperm packet) which is picked up by the cloacal lips of the female. The eggs number from a few to a thousand. The eggs are holoblastic and may or may not be pigmented. Cleavage is complete and development follows the usual amphibian pattern resulting in (1) a pond type larva with long gills, well developed fins and balancers (2) a stream type larvae with reduced fins and gills and lacking balancers or (3) a terrestrial form with specialized gills and a very abbreviated larval development and which hatches directly into the adult form. Lungs are developed in all but the terrestrial plethodontids. Neotony (larval reproduction) is common in many forms and some forms are permanently aquatic.

Regeneration of lost parts is characteristic of aquatic larvae and regeneration of the tail is common in the adult terrestrial plethodontids. The integument contains both mucous and granular glands, both of which may secrete poisonous alkaloids of varying potency. Some specialized glands are involved in sexual attraction of the female. The skin is highly vascularized in most forms and is a major respiratory structure in the lungless plethodontids. Experimental work suggests that the skin is a major osmoregulatory organ in the salamander. Lipophores, guanophores and melanophores are present in most forms and a vast array of colors and patterns are found. Cave living species are usually colorless.

The heart is amphibian-like except in the terrestrial lungless plethodontids and in the permanently aquatic forms where the auricular septum and spiral valves are reduced or lost. The digestive system is the usual vertebrate form; the choice of food is apparently opportunistic and without great selectivity and consists mainly of insects, worms and occasionally larger animals.

The segmental nature of the body musculature is apparent in the costal grooves which can be seen externally. The limb musculature is greatly reduced. The branchial musculature is particularly well developed in aquatic forms.

Many salamanders have been reported to be capable of long migrations and homing has been reported in several species. Hibernation and aestivation are common habits of these animals. Frequently the salamander will live most of its life in the terrestrial habitat where shelter and food are sought in rotten logs, leaf litter and under rocks, boards, trash, and in mammal nests. In apparent response to rising temperatures and increased humidity they migrate to the breeding site. The seasonal nature of reproduction has allowed the establishment of age groups in the young of some species. In captivity the larger forms have been reported to have very long life spans (up to 52 years) but field studies indicate that the life span of an individual that reaches sexual maturity may be five or six years at the most.

About 250 species are known among the 8 families and 5 suborders recognized. The suborder Cryptobranchoidea includes the families Hynobiidae of eastern Asia and Cryptobranchidae in eastern Asia and eastern United States. These are the most primitive urodeles with a generalized skeleton and with external fertilization. The Hynobiidae are land forms with aquatic reproduction while the Cryptobranchidae are semilarval and spend their entire life in water. The suborder Proteida (family Proteidae) includes the European blind salamander and the mudpuppy, *Necturus* common in eastern North America. They are permanently larval bottom walkers with internal fertilization. The Suborder Meantes (family Sirenidae) are aquatic eellike salamanders which lack hind limbs and are limited to eastern North America. The Suborder Ambystomoidea (family Ambystomidae) is limited to North America. The adults are typically terrestrial but reproduction is aquatic with internal fertilization. The Suborder Salamandroidea includes three families (Amphiumidae, Salamandridae and Plethodontidae.) They all have internal fertilization. The Amphiumidae are eellike permanently aquatic forms with reduced limbs and are known only in the southeastern United States. The Salamandridae are commonly known as newts and are found throughout the northern hemispheres. They may be either terrestrial or aquatic. The Plethodontidae are known from southern Europe, South America and North America. They are lungless forms, many of which have direct development.

D. L. JAMESON

Reference

Bishop, S., "Handbook of Salamanders," Ithaca, Cornell University Press, 1943.

URTICALES

An order of at least three flowering plant families which is variously interpreted evolutionarily as a primitive or an advanced group. The flower structure is relatively simple but this may reflect primitive simplicity or it may be an example of simplification by reduction; the latter is the more current view. The order is distinguished from related orders by its uniovulate, unilocular, superior bicarpellate ovary and its few to several stamens.

The Ulmaceae or Elm Family is a family of woody plants whose leaves are often obliquely inequilateral at the base. The flowers are mostly unisexual, although the elms are bisexual. The perianth consists of four to eight sepals which are more or less joined basally in a cup. The stamens are erect in the bud and are usually borne opposite the sepals; pollination is by wind-borne pollen. The fruit is winged or it is a fleshy drupe. There is little economic importance for the family aside from the use of several species as specimen trees in landscape planting, and the lumber produced by some species.

The Moraceae (Mulberry Family) is with very few exceptions a family of trees and shrubs with milky sap. The leaves often have three to five principal veins which originate from the apex of the petiole. The stipules are often joined around the bud in a cap which leaves a circular scar when it falls. The flowers are unisexual and often produced in much-condensed inflorescences. The fruit is an aggregate of small drupelets which may or may not incorporate fleshy perianth parts or inflorescence axes. The family is of considerable economic importance because of the several edible fruits it produces; figs, mulberries, and breadfruit belong here. The drug marijuana is produced from the staminate flowers of the hemp and its stem-fibers are important as a source of cordage. Several species are cultivated as ornamentals.

The Nettle Family (Urticaceae) is predominately herbaceous, although many woody forms occur in the tropics. Several genera produce stinging hairs and the leaves often have crystals of characteristic shapes in the epidermal cells. The flowers are unisexual and are either naked or the perianth is in two undifferentiated series. The stamens are usually four, opposite the perianth segments, and infolded; at anthesis the filaments reflex explosively, scattering the pollen. The economic value of the family is slight, aside from a few cultivated, ornamental herbs.

R. S. COWAN

References

Engler, A. and K. Prantl, "Die naturlichen Pflanzenfamilien," vol. 3, Leipsig, Engelmann, 1889.
Hutchinson, J., "The Families of Flowering Plants," 2nd ed., Oxford, Clarendon Press, 1959.

V

VARIATION

Variation between individual organisms, with respect to a given character, results from either or both of two influences, GENETIC and environmental. In other words, the total phenotypic variance between individuals is the sum of the genotypic and environmental variances. Characters acquired during the lifetime of the organism are not inherited. Biological variation is best studied at the level of the species population.

In the experimental separation of the influences of heredity and environment in variation, two general methods are emphasized. Thus, one may attempt to minimize either the hereditary or environmental contribution to variation. Organisms possessing identical genotypes can be obtained and the variance between the individuals measured. Variations arising may, in most instances, be safely ascribed to the effects of the environment. This method was first approximated by Johannsen, who through inbreeding, was able to reduce the genetic variance in garden beans to the point where he could observe directly the effect of the environment on plants which had quasi-identical genotypes. Reduction of genetic variation through inbreeding and selection over a number of generations has, since Johannsen's time, been perhaps the most widely-used method for achieving low genetic variance in sexually-reproducing organisms.

The complete exhaustion of genetic variation by inbreeding, however, is very difficult to achieve in actual practice. Almost all sexually-reproducing organisms appear to harbor great amounts of minor genetic difference which, under normal cross-breeding make it very difficult, if not in many cases impossible, to obtain a group of organisms which are genotypically identical at all the thousands of gene loci which exist on the chromosomes of that species. One must, in most cases, resort to the use of fortuituously produced conditions of genotypic identity, such as identical (monozygotic) TWINS. Such individuals, especially when reared in different environments, provide an important source of data on the effect of the environment. In certain higher plants and in some microorganisms, use of zygotic identity is possible because the individual organism arising from a single zygote (genotype) can be divided and propagated clonally. The replicates may then be reared in experimental plots under diverse conditions, and direct measurements of the effects of these different environments made.

The fruitless debates of the past over whether heredity is generally more important than environment, or vice versa, have been abandoned in favor of a much more sensitive and sensible view. Extensive data are now available which indicate that both influences are so deeply involved in so many characters of biological significance that the characters themselves can only be viewed as resulting from the interaction of the two components.

Such a formulation of the problem has no lack of scientific rigor if each character is considered separately. Thus, an attempt is made, for a particular character, to measure the relative contributions of heredity and environment to the expression of that character. Of particular significance for human traits are the extensive studies of twins. Data are obtained for a series of pairs of identical (monozygotic) and fraternal (dizygotic) twins of like sex. A twin pair is said to be concordant when both members show the character and discordant when the two members of the twin pair differ with respect to the character. An illustrative example will make clear the way in which this method may be applied to many important physical, mental and behavioristic traits. Thus, we may collect all cases of twins in which at least one member of the pair has developed, for example, diabetes mellitus. Among *identical* twins, the data show that in 84% of such twin pairs *both* members develop the disease (i.e. are concordant). Now, standing alone, these data do not tell us much about the heredity and environment problem because identical twins are ordinarily reared in the same home and thus are subject to very similar environmental conditions. The data acquire meaning when we compare nonidentical twins of the same sex in the same way for the same character. In diabetes, such pairs show only 37% concordance. Evidently the genotype plays an important role in the development of this disease. At the same time, we observe that concordance in identical twins is not absolute, pointing to a role of the environment in the development of the same character. Independent evidence indicates a simple genic basis for diabetes mellitus and the high concordance even among non-identical twins is also attributable to this fact. Nevertheless this "inborn error of metabolism" does not always result in a diseased condition unless the right environmental conditions prevail, in this case, a high carbohydrate intake.

The great value of studies of concordance in twins is that it is possible to assess accurately the roles of heredity and environment in a wide variety of characters which undoubtedly have a complex genetic basis and are refractory to the usual method of family analysis. Thus strong hereditary ingredients may be recognized in such diverse conditions as, for example, tuberculosis (65% concordance in identicals; 25% concordance in non-identicals), rickets (88%–22%), beginning of walking (76%–30%), any type of tumor (61%–44%) and schizophrenia (68%–11%). These data show that certain genotypes predispose their carriers to the development of the trait which may even be the onset of an infectious disease, providing that the particular disease-producing agent comes along. The traditional "cause" of tuberculosis is a bacterium but the higher concordance in identical twins shows that this organism succeeds in producing disease more easily in certain genotypes than in others.

Certain traits, of course, like eye color and blood type are little affected by the environment. Even here, however, the character we observe results from the interaction between gene and environment; the environment being, in such cases, largely internal to the organism, that is embryological or cytoplasmic.

The ultimate source of genetic variability is the process of gene mutation. For operational reasons, geneticists have emphasized those gene mutations which tend to have large, conspicuous and easily recognized phenotypic effects (oligogenes). Most mutations, however, tend rather to have small effects even when homozygous and alter the phenotype very little. Recent studies of the inheritance of the so-called "quantitative characters" in man and other organisms (stature, weight, yield of seeds, milk, eggs, beef, etc.) have shown that these characters are influenced by the separate contributions of a great many separate gene loci (polygenes). The genetic effects of any single locus are small and the general effect of polygenes is additive. Frequency distribution curves of such characters show that the variability generally follows the shape of a normal bell-shaped curve which is smoothed both by the multiple contributions of different genes and by the effects of the environment, which are usually great on such characters.

Most sexually-reproducing populations of higher animals and plants which have been tested harbor an enormous amount of minor concealed variability, components of which may affect almost every character. This is usually in the form of recessive genes carried in the heterozygous conditions. The internal genetic processes connected with sexual reproduction (synapsis of homologous chromosomes, crossing over, random segregation of cross-over products into the gametes and chance recombination of gametes at fertilization) result in the production of an almost incalculably great variety of genotypes. It is probably safe to say that in cross-fertilizing, sexually-reproducing organisms no two individuals have the same total genotype nor is there finite probability that the same genotype will ever occur again. Indeed, recombination of existing genes in populations is the major source of individual genetic variability. The process could theoretically be effective for a long period even in the absence of new variations arising through mutation.

From sexual populations which are widely distributed geographically, processes of isolation and selection produce separate gene pools, in other words, species populations which may become genetically quite different from one another. Such divergence is not always accompanied by complete reproductive isolation between the members of the two groups. Thus, if the members of two such entities come in contact, interspecific hybridization may occur. In a number of plants, such hybrids are fertile and may cross among themselves or may backcross to one parental form in particular (introgressive hybridization). The result is extensive genetic recombination and the generation thereby of large amounts of genetic variability, ordinarily much greater than that observed when recombination is wholly intraspecific.

Because of the variety of genotypes within populations, it if often necessary, from a practical point of view, to attempt to study variations using organisms which are not genotypically identical. A series of genotypes may be profitably compared by rearing them under as closely uniform environmental conditions as can be attained in the laboratory or in the experimental plot. Indeed, this is the method most widely employed by the plant and animal breeder or the geneticist. The subtleties of environmental influence on variation, however, are so all-pervading as to render interindividual comparisons less useful than those in which essential genotypic uniformity has been achieved. In practice, this does not invalidate the work of the practical breeder, however, who does not necessarily have to distinguish the genetic from the environmental components. Successful artificial selection has been practiced since the dawn of human history in the absence of rigorous analysis of the variations being selected. Parents may thus be selected on a scale of phenotypic merit in the hope that one may continue to accumulate, generation by generation, genes which lead to increased merit. The confusion of the psychologist and physician in the nature-nurture area is understandable in the light of the above facts. Normally his human patients are not only genotypically very diverse but have been subjected to widely diverse environments as well.

A number of recent studies have shown that outbred organisms (hybrids, or the results of matings between relatively unrelated parents) generally are vigorous, symmetrical and show low environmental variation. Conversely, inbred organisms frequently show the reverse tendencies. There is considerable evidence which relates hybrid vigor to a condition of extensive heterozygosity at many gene loci. Accordingly, hybrid vigor is often referred to as heterosis. In discussing variation, it is important to note that the heterotic individual appears to be better buffered developmentally against the onslaughts of the environment. Thus, if a series of highly heterozygous individuals (hybrids) are subjected to a series of environmental shocks during their development, they withstand these shocks and develop into normal, uniform, relatively invariable organisms. They may be said to be well canalized or buffered (homeostatic) in their developmental and physiological pathways. On the other hand, homozygotes (inbreds) show a much lower degree of developmental homeostasis or buffering and are thus more subject to variations inflicted by the environmental shocks in which they have been tested. These findings are of far-reaching importance in the understanding of the relation of the genetic material to the general health of the individual. In many population contexts, natural selection appears to have favored heterozygous genotypes because of their better general homeostatic properties.

The generation of genetic variability in natural populations depends on the mutation rate and the rate of recombination of these genes. All degrees of the latter are found among organisms in nature. Recombination tends to be retarded by low chromosome number, low crossing over, and inbreeding or self-fertilization. Organisms showing these characteristics generally show low variation compared to those showing converse properties. The intrinsic genetic mechanism thus generates high or low genetic variability, depending on the species. Evolution, descent with change, is characteristically associated with populations which generate high individual variability through mutation and recombination. Natural or artificial selection, operating in populations of critical sizes must have genetic variability to work on or else the result is merely descent without change, as in asexual organisms.

HAMPTON L. CARSON

References

Stern, C., "Principles of Human Genetics," 2nd ed., San Francisco, Freeman, 1960.
Stebbins, G. L., Jr., "Variation and Evolution in Plants," New York, Columbia Univ. Press, 1950.
Mather, K., "Biometrical Genetics," New York, Dover, 1949.
Lerner I. M., "Population genetics and animal improvement," Cambridge, University Press, 1950.
Cold Spring Harbor Symposia on Quantitative Biology. XX. "Population genetics: The nature and causes of genetic variability in Populations," Cold Spring Harbor, New York, Carnegie Corporation, 1955.
Clausen, J. et al., "Experimental studies on the nature of species," Washington, Carnegie Inst. Publs. No. 520 (1940), 564 (1945) 581 (1948).
Darlington, C. D., "The Evolution of genetic systems," New York, Basic Books, 1958.

VENOMOUS FISHES

Fishes that are toxic to man are of two general types, those which are poisonous to eat, and those which sting. Stinging or venomous fishes inflict their injuries with the use of stings comprised of dentinal spines to which are attached venom glands. The venomous properties of a fish have no relationship to their edibility. Some species which are extremely venomous are also valuable food fishes. The venom organs, or stings, of most fishes are generally situated on either the gill covers or on the dorsal, pectoral, ventral, or anal fins. Fish venoms are collectively termed *ichthyoacanthotoxins*, from the Greek, *ichthyos*, a fish; *akantha*, a thorn or prickle, *toxikon*, poison. For the most part little is known regarding either the pharmacology or chemistry of ichthyoancanthotoxins. Venomous fishes are widely distributed throughout the piscine group and are found in both fresh and salt water throughout most parts of the world. However, venomous fishes are found in greatest abundance in tropical and temperate waters. The following constitute some of the more important groups of venomous fishes:

Horned or spiny sharks. There are several species of horned sharks, or spiny dogfish, which range throughout temperate and tropical seas. They are largely inhabitants of shallow, protected, muddy or sandy bays. The venom apparatus consists of the dorsal stings which are located along the anterior margins of each of the two dorsal fins. The venom gland is a glistening whitish mass of epithelial tissue situated in a shallow groove along the back of each dorsal spine.

Stingrays. There are numerous species of sting-rays, all of which are members of seven different families, inhabiting tropical and warm temperate seas, with the exception of one family which is confined to freshwater. Stingrays are most common in shallow, sheltered bays, river mouths, or in sandy areas. They have the habit of covering themselves with sand or mud with the use of their pectoral fins. The venom apparatus of stingrays is an integral part of the caudal appendage and varies in structure according to the species of stingray. There are four general types of stingray venom organs: 1) *Gymnurid type*, having a short poorly developed tail, with a small sting situated near the base, thus making it a feeble striking organ. This type is found in the butterfly rays. 2) *Myliobatid type*, having a long whip-like tail, with a well-developed sting situated near the base of the tail—an effective stinging device. This type is found in the bat, or eagle rays. 3) *Dasyatid type*, similar to the Myliobatid type, but the sting is generally situated further out from the base of the tail making it a more effective striking organ. The tail terminates in a long whip-like appendage. This type is found among the stingrays proper which are some of the most dangerous kinds known. 4) *Urolophid type*, having a short, muscular caudal appendage, and a well-developed sting situated near the terminal end of the rounded compressed tail. This type constitutes a powerful defensive weapon and is found in the round stingrays.

The stingray sting consists of a retrorse serrated vasodentinal spine which is enveloped by a sheath of skin, the integumentary sheath. The ventrolateral-glandular grooves are situated along either edge on the underside of the spine. Situated within these grooves is a strip of soft, spongy, grayish tissue, which is the main venom-producing area of the sting, although lesser amounts are produced in other areas of the integumentary sheath. The grooves tend to protect the delicate glandular tissue within them even though the remainder of the integumentary sheath is worn away. Withdrawal of the sting from the wound generally results in further damage to the surrounding tissues by the recurved spines.

Chimaeras or ratfishes. These are a group of cartilaginous fishes having a rounded or cone-shape snout, a single external gill opening on either side, a laterally compressed body, tapering posteriorly to a slender tail. They prefer cooler waters and have a depth range down to 1400 fathoms or more. The venom apparatus consist of the single dorsal sting which is situated along the anterior margin of the first dorsal fin. The venom gland appears as a strip of soft grayish tissue lying in a shallow depression on the back side of the dorsal spine.

Catfishes. They vary greatly in size and shape, ranging from short to greatly elongate, or even eel-like. Their lips are usually equipped with barbels. There is said to be about one thousand species, most of which are tropical freshwater, although a few are marine or brackish water species. The venom apparatus of catfishes usually consists of a single, sharp, stout sting immediately in front of the soft-rayed portion of the dorsal and pectoral fins. The sting is comprised of a spine enveloped by a thin layer of skin, the integumentary sheath, which is continuous with the soft-rayed portion of the fin. There is no external evidence of the venom glands, which consist of a series of glandular cells lying along the anterolateral and posterolateral margins within the epidermal layer of the integumentary sheath. The fin spines of some catfishes have a series of sharp retrorse teeth similar to those found on stingray spines and are capable of producing a severe laceration of the victim's flesh.

Weeverfishes. These are small marine fishes which inhabit flat, sandy or muddy bays of the temperate Atlantic Ocean and Mediterranean Sea. They frequently bury themselves in the soft sand or mud with only the head exposed. The venom apparatus of weeverfishes consists of the dorsal and opercular spines and their associated venom glands. The venom glands are enclosed within a thin-walled integumentary sheath. If the integumentary sheath is removed the venom glands appear as a thin, elongate, fusiform strip of whitish spongy tissue lying within the grooves near the tips of each fin spine. The opercular glands are pear-shaped. Weever venom is said to have both neurotoxic and hemotoxic components similar to that of certain snake venoms. Weeverfish stings are said to be extremely painful.

Stargazers. They are bottom-dwelling marine fishes having a cuboid head, an almost vertical mouth with fringed lips, and eyes on the flat upper surface of the head. A large part of their time is spent buried in the sand with only their eys and a portion of the mouth protruding. The venom apparatus of star-gazers consists of two double-grooved shoulder spines, one on either side, surrounded by an integumentary sheath. The venom glands are attached to these spines.

Dragonets. These are small scaleless marine fishes having flat heads that are armed with a strong preopercular spine which is said to be venomous. Some of the species are found in deep water, whereas others inhabit shallow bays and reefs. Little is known regarding the structure of the venom organs, or the nature of the venom.

Scorpionfishes. The scorpionfishes range throughout all tropical and temperate seas, and a few species are found in arctic waters. Scorpionfishes are divided into three groups on the basis of the morphology of their venom organs, viz: 1) *Zebrafish type.* These are ornate, beautiful, coral reef fishes having plume-like dorsal and pectoral fins. Hidden within the dorsal, anal and pelvic fins are grooved, elongate, needle-like spines to which are attached venom glands. 2) *Scorpionfish type.* The scorpionfishes proper are largely shallow water, bottom dwellers, and are found in bays, along sandy beaches, rocky coastal areas, or coral reefs. They have the habit of concealing themselves in coral, seaweed, rocks, or other debris, which they closely resemble in their coloration. The venomous stings are found in the dorsal, anal and pelvic fins, but are shorter and heavier than those found in the zebrafishes. 3) *Stonefish type.* Stonefishes have a thick warty skin which gives them the appearance of a rock or clump of mud. They are most commonly encountered in tidepools and shoal reef areas, lying motionless under rocks, in crevices, or buried in the sand or mud. Their stings are situated on the dorsal, anal and pelvic fins, and are comprised of short, heavy bodied, grooved spines, associated with very large venom glands. The venom organs of stonefishes are among the most highly developed of any fishes. Their stings are extremely painful and may be fatal.

Toadfishes. These are small, bottom, marine fishes which inhabit the warmer waters along the coasts of America, Europe, Africa and India. Toadfishes have broad, depressed heads, large mouths, and are somewhat repulsive in their appearance. They hide in crevices, in burrows, under rocks, seaweed, debris, or lie buried in the sand and mud. The venom apparatus consists of two dorsal fin spines, two opercular spines, and their associated venom glands. The spines are slender, hollow and needle sharp. The stings operate somewhat like a hypodermic needle, thus making this a highly-developed venom apparatus. Stings from toadfishes are said to be similar to a scorpion sting. However, no fatalities from toadfish stings have been reported.

Surgeonfishes. These are shore fishes characterized by the presence of a sharp, lance-like, movable spine which rests in a shallow groove on either side of the caudal peduncle. The point of the spine is directed forward, but when the fish becomes excited the spine is extended at right angles from the body. When the spine is extended an agitated surgeonfish can inflict a deep wound with the lashing of its tail. The peduncle spines of some species of surgeonfishes are believed to be venomous.

Rabbitfishes. These are a group of reef fishes which resemble the surgeonfishes, but differ from all other fishes in that the first and last rays of the pelvic fins are spinous. Rabbitfishes range from Polynesia to the Red Sea. The venom apparatus of rabbitfishes consists of the dorsal, pelvic, and anal spines and their associated venom glands.

Pompanos. These are small to large-sized, metallic, silvery or golden-colored fishes which are usually adapted for rapid swimming. They are inhabitants of oceanic and coastal areas in all warm and some temperate seas. The anal spines of some species are said to be venomous.

Clinical characteristics of venomous fish stings. Injuries resulting from venomous fishes vary according to the species of the fish, the area of the body involved, the physical condition of the victim, mechanical trauma produced, the nature and amount of venom introduced into the wound. Most fish stings are of the puncture-wound variety. Wounds caused by stingrays and catfishes may result in severe tissue lacerations because of the retrorse teeth. The character and intensity of the pain varies greatly. In some instances the pain may amount to only a mild prick or stinging sensation whereas in others it may be very severe, resulting in loss of consciousness of the individual. Stingray, catfish, scorpionfish, zebrafish, weeverfish, and stonefish stings are said to be especially severe. Swelling, redness, and cyanosis in the area about the wound are commonly observed. Tingling, followed by numbness in the vicinity of the wound, frequently occurs. Shock and secondary infections are not uncommon complications.

Treatment of venomous fish stings. The treatment of venomous fish stings is concerned with the alleviation of the pain, combating the effects of the venom, and the prevention of secondary infection. Whenever possible bleeding should be encouraged, and particularly so in puncture wounds. It is desirable to irrigate the wound when possible to do so. Most physicians recommend soaking the injured member in hot water for 30 minutes to one hour. The water should be maintained at as high a temperature as the victim can tolerate without injury. Infiltration of the tissues about the wound with 0.5 to 2% procaine has been used with good results. Demerol may be required to control the pain. Supportive therapy, antibiotics, and a course of tetanus antitoxin may be required.

An antivenin for stonefish stings has recently been developed at the Commonwealth Serum Laboratories, Department of Health, Parkville, Victoria, Australia. This is the only fish antivenin that has been developed to date, and is reputed to be highly efficacious in treating stonefish stings.

BRUCE W. HALSTEAD

VENOMOUS PLANTS

Poisonous plants contain or produce substances capable of harming or killing animal life. The number of toxic species runs into many thousands. At least 400 species in 68 families are listed for the United States; tropical floras are very much richer. Occurring throughout the Plant Kingdom, they are most frequent in fungi and the Apocynaceae, Anacardiaceae, Compositae, Ericaceae, Euphorbiaceae, Leguminosae, Loganiaceae, Menispermaceae, Moraceae, Solanaceae, Umbelliferae.

Plant poisons may be classified botanically, chemically or according to their physiological activity. In this last case, there are 5 main groups; blood poisons; nerve; muscular; surface or contact; irritant. Toxic substances are very diverse: the commonest are alkaloids; saponine, cyanogenic, solanine and mustard oil glucosides; resinoids; volatile oils; tannins; oxalic acid; phytotoxins or toxalbumins; selenium.

Surface or contact poisons cause dermatitis, ranging from itching rashes to painful blisters of short or long duration. Many plants may induce dermatitis in susceptible individuals or only at certain seasons. The commonest belong to the Anacardiaceae, including sundry species of *Rhus*: *R. Toxicodendron*, poison ivy, and *R. Vernix*, sumac (N. Am.); *R. juglandifolia* (S. Am.). Dermatitis is an occupational disease of collectors and artisans working with the Asiatic lacquer tree, *R. vernicilflua*, and its product. The active principle seems to be catechol compounds with unsaturated side chains. *Metopium* (N. Am.) and *Comocladia* (W. Ind.) cause even severer dermatitis, while many genera, like *Mangifera* and *Anacardium*, are only mildly irritant. The tropical euphorbiaceous trees, *Hippomane mancinella* and *Hura crepitans* possess very caustic latex.

Blood poisons act on corpuscles, hemoglobin or plasma, producing cyanosis or ecchymosis. Toxalbumins such as ricine (*Ricinus communis*), abrine (*Abrus precatorius*), croton oil (*Croton tiglium*) and robine (*Robinia pseudo-acacia*) and cyanogenic glucosides like amygdaline (*Prunus*) are important in this category.

Nerve poisons induce either spasms and convulsions by overstimulation of nerves or depression or paralysis. Most are alkaloidal, such as *Atropa, Datura, Hyoscyamus* (atropine, hyoscyamine); *Strychnos Nuxvomica* (strychnine); *Aconitum* (aconitine, aconine);

Papaver somniferum (morphine, codeine, narcotine, etc.). Some narcotics fall into this group.

Muscular poisons, usually ,containing alkaloids or glucosides, may cause tremours, convulsions or paralysis by direct action on muscles. Examples are *Veratrum* (alkaloids) and *Digitalis* and *Strophanthus* (glucosides), the last two affecting heart muscles.

Irritant poisons act, sometimes fatally, on the gastrointestinal tract through their essential oils, saponines or resins. This group is characterized by *Brassica* (mustard oil glucosides) and *Podophyllum peltatum* (podophyllin, a resin). Some are employed medicinally as purgatives.

Primitive man early acquired discerning knowledge of poisonous plants, using them in witchcraft, medicine, hunting, fishing, administering justice, killing enemies and the infirm. Civilization has long found them useful as medicines, insecticides, rodenticides, and new uses are constantly arising.

An ancient use of poisons surviving in Madagascar and Africa is determination of guilt or innocence through ordeal. Poisons are administered in the belief that their good spirits will punish guilt by death, sparing the innocent. The important ordeal poisons are Leguminosae: bark of species of *Erythrophleum, Parkia, Detarium;* seeds of *Physostigma venenosum. Erythrophleum* contains a cardiac glucoside, erythrophlein; *Physostigma*, best known ordeal poison, contains an alkaloid, physostigmine, a sedative of the spinal cord causing paralysis of the limbs, loss of muscular control and death by asphyxiation; physostigmine is valuable in a modern ophthalmology. Minor ordeal poisons belong to the Apocynaceae, Asclepiadaceae, Euphorbiaceae, Loganiaceae, Sapotaceae.

Fish poisoning is world wide. At least 154 species in 68 families are employed, most (109) in South America. The practice is ancient; some species are known only as cultigens. Piscicides owe their activity to a variety of principles, especially to alkaloids and saponine and cyanogenic glucosides. Usually crushed leaves, stems or roots are thrown into still or sluggish water. Few piscicides kill fish, merely stupefying, usually by interfering with respiration. Scattered widely, they are known from the Acanthaceae, Amaryllidaceae, Araceae, Ebenaceae, Flacourtiaceae, Lecythidaceae, Passifloraceae, Solanaceae, Taxacaeae; but the Compositae, Euphorbiaceae, Leguminosae, Sapindaceae are the most important. *Derris* (S.E. Asia, Austr.) and *Lonchocarpus* (S. Am.) are outstanding. Both genera contain a ketone, rotenone, which rapidly stupefies fish; it has become the most effective of modern contact insecticides. The insecticidal properties were discovered in 1848 when a solution of *Derris elliptica* was sprayed on nutmeg trees, but as a commercial item rotenone was almost unknown before 1930. Sundry species of the legume *Tephrosia*, native to both hemispheres and owing its effects to the alkaloid tephrosine, are employed in North and South America, Africa, Asia, Australia, an example of independent invention. Many species of *Serjania* and *Paullinia* (Sapindaceae), acting through saponines, are used locally throughout the tropics. *Clibadium* (Compositae) and *Phyllanthus* (Euphorbiaceae) are important South American piscicides. Fish poisons have long been used in Europe, especially *Verbascum, Cyclamen, Taxus*. An important North American piscicide was the saponine-rich *Yucca*.

Few plants, other than *Derris* and *Lochocarpus*, have found use as insecticides. The most notable are tobacco (*Nicotiana*), hellebore (*Helleborus*), pyrethrum flowers (*Chrysanthemum*). Fly agaric (*Amanita muscaria*), a dangerous nerve poison to man, is said to have had folk-use in Europe as a fly-killer. The lily, *Urginea maritima*, employed as a rodenticide in the Mediterranean since ancient times, became important commercially in the 1930's only to be displaced by more efficient synthetic products.

The use of arrow poisons, obviously ancient, is nearly world wide. Our words *toxic, toxin*, etc., coming from the Greek τοξογ (arrow), indicate great age for the practice in the classical world. Many families enter into their preparation: Araceae, Dioscoriaceae (E. Ind.); Amaryllidaceae, Leguminosae, Rubiaceae (Africa); Ranunculaceae (Europe, India); Annonaceae (S. Am.); Celastraceae, Rutaceae (Philippines); Sapindaceae (W. Ind.); Asclepiadaceae (N. Am.). The notable families are Apocynaceae, Loganiaceae, Menispermaceae, Moraceae. The active principles are primarily alkaloids and glucosides. Historically interesting is *Antiaris toxicaria*, upas tree, basis of arrow poisons of southeastern Asia. The latex contains two similar glucosides for which good antidotes are not known. First reported in Europe by Odoric in 1300, this poison wrought great slaughter amongst the Portuguese in the taking of Malacca in 1511. Rumpf discovered the source plant in 1750. *Strophanthus* is a major arrow poison in Africa. The greatest number of species is found in South America where the poison is called *curare*. The usual but meaningless classification of curares according to containers used (pot-, tube-, gourd-curare) should be replaced with a botanical classification. Although many plants enter into the complex formulas, varying from tribe to tribe, most curares owe their activity either to loganiaceous (many species of *Strychnos*) or menispermaceous (*Abuta, Chondrodendron, Sciadotenia*) plants. Rarely are both families mixed in one curare. Bark, leaves roots are boiled to a syrup and sun-dried to a paste. About 42 alkaloids enter into South American curares. One, curarine, is employed therapeutically as a muscle relaxant in shock therapy. Research on arrow poisons still promises many new discoveries.

Plants capable of absorbing selenium from certain soils cause "alkali disease" or "blind staggers" in livestock. Common in western North America where they are called locoweeds, they include *Aster, Atriplex, Zygadenus* and, especially, certain legumes such as *Astragalus*. Wheat may be deleterious to animals if grown on selenium soils.

Food plants may have toxic substances which must be removed. An outstanding example is tapioca root (*Manihot*) from which a cyanogenic glucoside must be leached before utilizing the starch. Amazon Indians employ rubber (*Hevea*) seeds as food after similar detoxication. Many plants containing crystals or similar mechanical irritants (*Rumex*, certain aroids) may be eaten safely after treatment.

One danger from poisoning lies in unfamiliar, introduced ornamentals. This is especially true of tropical ornamentals, but temperate and warm areas have species which, when eaten, are suspect: *Aconitum, Colchium, Delphinium, Lupinus, Hedera, Narcissus, Taxus, Nerium*, etc.

A few photodynamic plants sensitize domestic animals to light when ingested. Examples are *Hypericum, Fagopyrum* (buckwheat), *Agave, Trifolium, Medicago*.

Of the hundreds of poisonous species of native floras, only a few, important historically, culturally or economically, may be named. Ergot, *Claviceps purpurea*, a fungal parasite on rye and containing several alkaloids, occasionally causes fatal mass poisoning in European cities when infected grains accidently pass through the mill to contaminate bread flour. Superstitions have often been associated with toxic plants; the frenzy of witches in the Middle Ages was sometimes due to unguents of mandrake and other solanacous poisons. Poison hemlock, *Conium maculatum* (Umbelliferae) of Europe, now widely naturalized, was given as a death-potion to Socrates; its alkaloids, chiefly coniine, act of the sensory and phrenic nerves, killing by asphyxiation resulting from paralysis of the diaphragm. Water hemlock, *Cicuta*, has similar, often fatal effects, due to a resinoid, cicutoxin. *Equisetum*, mixed with hay in cutting, may slowly weaken cattle through a nerve poison. The wood, bark, needles and fruits of *Taxus* contain taxine, poisonous to man and browsing animals. *Sorghum*, producing a glucoside, dhurrin, hydrolyzing to hydrocyanic acid, may poison livestock. Berries of mistletoe (*Phoradendron*) and pokeweed (*Phytolacca*) occasionally poison children. Some of our commonest temperate plants may have toxic effects with certain susceptible individuals at certain seasons: oaks (*Quercus*), lady slipper (*Cypripedium*), a number of grasses, holly (*Ilex*), mountain laurel (*Kalmia*) and other ericaceous species, milkweeds (*Asclepias*), etc. Toxic plants are abundant in the Compositae, especially in warmer, dryer areas; and many legumes common in both temperate and tropical climates are suspect.

Poison plants have been employed to administer capital punishment and have entered into political intrigue, often influencing the course of history. Undoubtedly the most intensive and cleverest use of poisonous plants grew up during the Middle Ages in Europe, when the Italians developed the study of criminal and political poisoning into a cult; murders were commonly committed for fees, and elaborate formulas for slow or fast poisons were created. The name of the Borgia family is inextricably linked with this phase of European culture. The *hashshashin* of ancient Asia Minor were political murderers, excited to their nefarious work by eating hasheesh (*Cannabis sativa*); from this term comes our word *assassin*. It has been suggested that the Viking *berserker*, who went on periodic frenzies of lust and murder, worked themselves to an uncontrollable state of madness by eating *Amanita muscaria*. The use of toxic plants has influenced affairs in many cultures and lands far beyond the extent which is usually suspected.

RICHARD EVANS SCHULTES

VENOMOUS REPTILES

Of the 5 broad categories of living reptiles—the snakes, lizards, crocodilians, turtles, and beakheads (*Sphenodon*)—only the first two include venomous species.

Lizards

Of the world-wide lizard order Sauria, only a single North American genus (*Heloderma*) is venomous. There are two species: the Gila monster (*H. suspectum*), of sw. U. S. and nw. Mexico, and the Mexican beaded lizard (*H. horridum*) of w. Mexico. The biting mechanism is inefficient, involving grooved fangs in the lower jaw and requiring persistent chewing to inject venom. So the bite has variable effects, but the venom is powerful and painful, and human fatalities have been recorded.

Snakes

When considering venomous animals, snakes (order Serpentes) come first to mind. But, although this order contains many species highly dangerous to man, in most areas (except Australia) harmless snakes outnumber the venomous kinds, both in number of species and individuals.

Poisonous snakes are widespread throughout the world; they occur in every continent and in almost every area wherein the summer season lasts long enough to permit them to live and reproduce. However, no venomous land snakes occur in several large islands, including New Zealand, Madagascar, Ireland, Iceland, the Azores, Canaries, and Hawaii.

Classes of venomous snakes. Snakes are divided into four categories having different biting mechanisms:

(1) The aglyphs, with solid conical teeth, and without differentiated venom glands.

(2) The opisthoglyphs, with enlarged grooved teeth at the rear of the maxillary series. Some secrete venom in effective amounts.

(3) The proteroglyphs, with permanently erect, hollow fangs preceding the maxillary series. They have venom glands and ducts for its discharge, through the fangs, into the victim.

(4) The solenoglyphs with cannulated fangs attached to rotatable maxillaries. Except while biting, the fangs are folded back against the roof of the mouth; they are much longer than would be practicable if permanently erect.

Almost all venomous snakes dangerous to man belong to classes (3) and (4). The rearward fangs of an opisthoglyph can be imbedded only in slender objects, so most are dangerous only to their small prey. However, two opisthoglyphs, the African boomslang (*Dispholidus typus*) and the African vine snake (*Thelotornis kirtlandii*) have caused human fatalities, and there may be others equally dangerous.

Factors affecting the gravity of snake bite. There are so many variable conditions affecting the gravity of snake bite that no one can hazard a dependable prognosis in any specific case. Some factors relate to the snake, such as its species, affecting venom toxicity and quantity; its size, influencing venom quantity and depth of injection; its condition, affecting the fullness of the glands and state of the fangs; the degree of anger or fear that motivates the snake, determining the quantity of venom discharged. Certain factors depend on the victim, *e.g.* his age, size, and vigor, and his susceptibility to protein poisoning. Other circumstances vary with each particular accident, such as the site of the bite, whether on the trunk or extremities; whether the venom is injected into a blood vessel, into muscle, or in fat; also the protection afforded by clothing. Usually many important factors are indeterminate, leading to uncertainties on prognosis and treatment, and to unexpectedly mild or, alternatively, dangerous cases.

Snake venoms, which are primarily proteins, differ greatly both in toxicity and physiological effects. They fall into several categories such as hemotoxins, neurotoxins, and cardiotoxins. However, these distinctions are not inflexible; the venoms of some species exhibit multiple effects. There may be extensive intrageneric differences in venom toxicites (*e.g.* among rattlesnakes), and intraspecific differences also occur.

Despite extensive researches on the chemical analysis of venoms, many uncertainties remain, particularly regarding their enzymatic activities, and the generation of autogenous toxins in their victims. Whereas the fixed-fanged snakes are usually supplied with neurotoxic, and the folding-fanged snakes with hemotoxic venoms, this allocation is not universally accurate and the differences are not sharply defined.

Although venom, as a protection against enemies, undoubtedly has survival value, this is not its primary function, which is to aid in securing prey. Venomous snakes invariably bite their prey, which, being small, quickly succumb to venom quantities hardly dangerous to man. The venom immobilizes prey that might otherwise escape; it prevents retaliation injurious to the snake; and its introduction into the circulation of the victim initiates digestion, important to a creature swallowing its prey whole, without dismemberment or mastication.

As stated, with few exceptions snakes dangerous to man fall into two categories—those with fixed anterior fangs (*proteroglyphs*), and those with folding anterior fangs (*solenoglyphs*). These two are each divided into two families having the following names and continental distributions:

Proteroglyphs: Elapidae (elapids). Asia, Africa, Australia, North and South America
Hydrophiidae (hydrophids). Tropical seas (except the Atlantic)
Solenoglyphs: Viperidae (viperids). Asia, Africa, Europe.
Crotalidae (crotalids). Europe (extreme se.), Asia, North and South America

These four families differ in certain important characteristics besides fang arrangement.

The elapids are living proofs of the inaccuracy of the often-heard statement that venomous snakes are distinguished by their broad, triangular heads and stout bodies. For most elapids are slim, without especially distinct heads, yet this group contains such dangerous kinds as the cobras, kraits, coral snakes, and the Australian black and tiger snakes. Indeed, the elapids include the most dangerous snakes in the world: the king cobra (*Ophiophagus hannah*) of southeastern Asia; the mambas (*Dendroaspis*) of Africa south of the Sahara; and the taipan (*Oxyuranus scutellatus*) of New Guinea and northern Australia. Many elapid species have extremely powerful neurotoxic venoms, more than compensating for their short fangs. Elapids probably cause more human fatalities than any other group. This is because they are alert and agile, and the natives, in many areas where they are prevalent, go about barelegged, and have religious scruples against the killing of any creature, however dangerous.

Most hydrophids are true dwellers of the sea, although concentrating along the coast, especially near river mouths. They are quite helpless if cast ashore. All sea snakes are venomous; indeed they are sea-going elapids feeding on fish. They have flat, paddle-like tails, and valved, dorsal nostrils. Although much smaller than the fabled sea serpents, some kinds do reach 10 feet, but most are shorter. Having extremely toxic venoms, they constitute a distinct hazard to fishermen but fortunately are rarely aggressive, and many fishermen habitually remove them from their nets with bare hands.

Sea snakes frequent the tropical waters of the Indian and western Pacific oceans. One species (*Pelamis platurus*) ranges from the east coast of Africa to the tropical west coast of America, the greatest range of any snake.

The viperids: These are the true or typical vipers; they are exclusively Eurasian and African. They are usually broad-headed and stout-bodied, thus resembling the popular conception of a venomous snake. The group includes such characteristic and pathologically important forms as the European viper or adder (*Vipera berus*), the venomous snake most often mentioned in western literature; extremely stout-bodied African vipers of the genus *Bitis*; Russell's viper of southern Asia; and various small but dangerous vipers inhabiting the deserts of north Africa and southwest Asia.

The crotalids: These are the pit vipers, distinguished from the viperids by a special sense organ evident as a deep depression, or pit, on each side of the head between nostril and eye. The pit, a heat-sensitive receptor, permits detection of prey, such as small mammals, having temperatures above their surroundings. In the Old World, crotalids range from West-Caspian Europe across Asia to the Pacific, including such large islands as Japan, Taiwan, the Philippines, Borneo, Sumatra, and Java. In the New World crotalids constitute the most important venomous snakes, including such dangerous forms as the rattlesnakes, fer-de-lance, bushmaster, and other Neotropical pit vipers.

Venomous snakes, although not so numerous as their harmless relatives, nonetheless are greatly diversified morphologically, and have become adapted to every kind of terrain and climate entailing a summer season long enough to permit reproduction, for these ectotherms are the slaves of temperature.

The longest venomous snake is the king cobra, occasionally exceeding 18 feet. The bulkiest are several African vipers, genus *Bitis*, and the eastern diamondback rattlesnakes of se. U. S. Certain dangerous desert forms rarely exceed 1½ feet. Some are arboreal, such as the mamba, the viperid genus *Atheris* of Africa, and several species of the crotalid genera *Trimeresurus* of Asia and *Bothrops* from Mexico to Argentina. There are burrowers (*Atractaspis*) of Africa, and fresh-water snakes, such as the African water cobras (*Boulengerina*). Virtually every ecological niche, from prarie to mountain, from desert to rain forest, has its venomous tenants.

Snake-bite hazard and treatment. The severity of the snake-bite hazard depends on a number of local factors: on the relative populations of dangerous snakes and people; on agricultural methods; on the nature and extent of clothing worn, particularly on the legs; taboos against killing snakes; and on the availability and dissemination of effective methods of treatment.

Accurate statistics on fatalities from snake bite affect the adequacy of institutional support for research on curative methods. Unfortunately snake-bite statistics are relatively inconclusive and inaccurate for two reasons: (1) the difficulty of securing accurate reports from primitive or illiterate rural populations; (2) the present method of classifying deaths under an international system, which, unfortunately, combines, in a single category, deaths from the bites and stings of all venomous animals. In several areas where dangerously venomous snakes abound, the stings or bites of bees, wasps, spiders, scorpions, etc. often cause more deaths than snakes.

The following approximate mortality statistics indicate the variability of the snake-bite hazard as influenced by human and ophidian factors:

In the United States the death rate is about 0.008 per 100,000 of population per annum. The mortality rate of those bitten by venomous snakes is about 0.2 per cent. Agriculture is extensively mechanized, protective clothing is worn, and antivenin and skilled treatment are widely available. The most dangerous snakes are rattlesnakes of the genera *Crotalus* and *Sistrurus*, and the moccasins, *Agkistrodon*. They are slow-moving and rarely aggressive. The most dangerous rattlers are the larger species, including the eastern diamondback (*C. adamanteus*), western diamondback (*C. atrox*), the timber rattler (*C. horridus*), the Mojave rattlesnake (*C. scutulatus*), and the western rattlesnake (*C. viridis*). Altogether there are 30 subspecies of rattlers in the U. S. out of the 65 known subspecies, all restricted to the New World. Although coral snakes of the genus *Micrurus*, having powerful neurotoxic venoms, occur in se. U. S., accidents from them are rare, as they are secretive and inoffensive.

In Australia the annual mortality is 0.07 per 100,000 population. Although there are many dangerous elapids, some with extremely toxic venoms, there is a low death rate because the rural population is sparse and well shod, and antivenin is available.

These rates are to be contrasted with 15.4 per 100,000 in Burma, 5.4 in India, and 4.2 in Ceylon. These countries not only have many dangerous snakes, both elapids and viperids, but dense populations of ill-shod people engaged in manual agriculture. About 15,000 people are said to die annually from snake bite in India. The principal offending snakes are cobras (*Naja*), kraits (*Bungarus*), Russell's viper (*Vipera russelli*), and, in the west, the saw-scaled viper (*Echis carinatus*).

The best treatment for snake bite varies from place to place depending on the snakes encountered. Some treatments may be recommended almost universally, such as the use of a ligature as a first-aid measure. Incision and suction to drain venom from the wound are sometimes useful, particularly for viperine or crotaline venoms having hemotoxic effects but only if applied promptly. Cryotherapy has been proposed, but requires further validation. Supportive measures, such as transfusions, shock preventives, and antibiotics are important and should be widely used. But primary dependence should be placed on antivenin, prepared from the lyophilized sera of immunized horses. The victim must be tested for horse-serum sensitivity. No antivenin is universally polyvalent; most are polyvalent for the snakes of a particular area. Often there are separate antivenoms for the elapids and the viperids in a district.

LAURENCE M. KLAUBER

References

Buckley, E. E. and N. Porges, eds., "Venoms," Washington, American Association for Advancement of Science, 1956.

Klauber, L. M., "Rattlesnakes," 2 vols., Berkeley, University of California Press, 1956.

Klemmer, K., "Liste der rezenten Giftschlangen" *in* Die Giftschlangen der Erde, pp. 255–464, 1963.

Parrish, H. M., "Incidence of Treated Snakebites in the U.S.," Public Health Reports, vol. 81, no. 3, pp. 269–276, 1966.

VERTEBRATA see CHORDATA

VERTICILLATAE

Casuarinales (Verticillatae), unifamilial and unigeneric (*Casuarina*), consist of forty to fifty species distributed chiefly in Indo-Malayan, Australian, and East Indies regions. Certain species are cultivated in the subtropical United States. Familiar species are Beefwood, Australian Pine, Ironwood, and Sheoak. "She . . . " refers to the whisper of the wind in the branches. *Casuarina* are trees and shrubs with long, slender, jointed, erect branches. The leaves are inconspicuous, evergreen, scalelike, and occur in alternating whorls of four to twelve. Since the leaves at a node are fused basally to form a sheath, only the distal tips are free. Many species grow under alkaline conditions and all have xeric modifications, as: reduced leaves, stomata sunken in internodal furrows with insulating hairs, considerable extraxylary sclerenchyma throughout, heavy cuticle. Photosynthesis occurs in chlorenchymatous cortex to each side of the furrows.

The flowers are reduced and unisexual. Male flowers occur in whorls in the leaf axils forming catkinlike spikes at the ends of the finer branches. Female flowers are axillary and capitately arranged on the tips of lateral branches. The plants are monoecious or dioecious, and pollen is wind-borne.

Each male flower, surrounded by pair of lateral bracteoles and a pair of median ones, consists of one stamen with a bilobed, basifixed anther with the conventional four microsporangia. The bracteoles enclose the stamen until anthesis. Dehiscence is longitudinal.

Each female flower has a single median bicarpellary, bilocular pistil, with a short style and two filiform stigmas, and is subtended by two lateral bracteoles. The stigmas mature before ovary or ovules have begun to develop. The posterior locule aborts before anthesis: the other bears two parietal, ascending, orthotropous, bitegumented ovules. One ovule receives no pollen tube and aborts. Several megasporocytes, hence several megaspores and several eight-celled embryosacs, with chalazal caecae, develop. The megagametophyte is the normal monosporic, *Polygonum* type. The pollen tube is intercellular and traverses successively: stigma, style, a bridge of tissue between the stylar base and the raphe, and raphe. Further growth is chalazogamic, the tube finally entering the micropylar end of one embryosac. Other embryosacs degenerate.

The fruit is an indehiscent samara subtended by two sclerenchymatous, capsulelike bracteoles. The fruits, bracteoles, and bracts form a conelike, multiple fruit.

According to one theory of Angiosperm origin, amentiferous, achlamydeous families, as *Casuarina* and Willows, are primitive Angiosperms. *Casuarina* is considered the best example of a transitional group between Gnetales (especially *Ephedra*) and Angiosperms. Gross structural resemblances between strobili of Gnetales and inflorescences of *Casuarina* and other "Amentiferae" provide the principal evidence. Flowers of achlamydeous families supposedly arose by compacting of gnetalean strobili with loss of bracts and bracteoles, except those bracteoles immediately subtending the ovules. The latter bracteoles enclosed the ovules evolving carpels. Families with perianth-bearing flowers evolved subsequently from amentiferous ancestral families. Unbiased evidence, however, from wood, floral, and phloem anatomical research and other botanical disciplines precludes this concept of Angiosperm origin in amentiferous families. The evidence indicates, rather, Angiosperm origin in woody ranalian groups and a derived nature for "Amentiferae."

M. F. MOSELEY

References

Moseley, M. F., "Comparative anatomy and phylogeny of the Casuarinaceae," Bot. Gaz., **110**: 231–280, 1948.

Swamy, B. G. L., "A contribution to the life history of *Casuarina*," Proc. Amer. Acad. Arts Sci., 77: 1–32, 1948.

VESALIUS, ANDREAS (1514–1564)

This sixteenth century Flemish anatomist gained fame for his dissections. Finding it difficult to carry out his work in northern Europe, Vesalius traveled to Italy where there was more intellectual freedom. He taught anatomy at the Universities of Pavia, Pisa and Bologna where he personally supervised anatomical demonstrations. Vesalius gained recognition for a simple demonstration in which he showed that men and women have the same number of ribs. The product of his scholarship is one of the great books of scientific history, *De Corporis Humani Fabrica* (*On Structure of the Human Body*). This book was the first accurate work on human anatomy. Its great strength is found in the illustrations done by Jan Stephen van Calcar, the artist-pupil of Titan. Van Calcar showed the human body in its natural positions with muscle illustrations so exact that nothing since has surpassed them. This work marked the beginning of modern anatomy. While Vesalius was an accurate anatomist, he still clung to the old physiological ideas of Galen. However, Vesalius' work together with the celestial work of Copernicus (both published in the same year, 1543) marked the birth of the Scientific Revolution.

DOUGLAS G. MADIGAN

Reference

O'Malley, C. D., "Andreas Vesalius of Brussels," Berkeley, University of California Press.

VIRCHOW, RUDOLF LUDWIG CARL (1832–1902)

Though Virchow was primarily a pathologist, his contributions to the theory of cellular biology make it imperative that he should be included in a list of any great biologists. He saw in his life time the death of the theory of spontaneous generation, but carried theory one stage further to the point of stating that nothing could arise within a cell save by the action of the cell itself. He therefore made it possible for contemporary biologists to accept the theory that the cell was an independent life unit and that an organ was no more than an assemblage of cells. Moreover, when this organ reached a pathological condition, the pathology was the result of the cell structure and not of the whole organ. His espousal of this very important concept was to some extent marred by his refusal to accept any other view so that he could never permit himself to be convinced that bacteria, arising external to the cell, could cause a pathological condition. Apart from this concept, he must always be remembered by his contributions to pathology and, possibly even more important his contributions to the social development of Germany.

PETER GRAY

VIRUSES

The first scientific evidence for the existence of such sub-microscopic agents as viruses was given in 1892 by Iwanowski, a Russian botanist, who showed that the infective principle which caused the mosaic disease of tobacco could pass through a bacteria-proof filter without losing infectivity. Shortly after this, Loeffler and Frosch demonstrated that the foot-and-mouth disease of cattle was caused by a similar type of agent. These discoveries laid the foundation for the development of the science of *virology*, the study of viruses. All types of living organism, from bacteria to man, are susceptible to these agents; more than 300 different viruses have been recorded from plants alone, and those attacking animals of all kinds, including man, are probably no fewer in number.

Virus diseases of plants include tobacco mosaic, beet yellows, beet curly-top, turnip yellow mosaic, potato leaf-roll and many others. Insects are susceptible to the polyhedroses and granuloses, so called because the virus particles are occluded in many-sided (polyhedral) protein crystals and capsules. Several different viruses, known as bacteriophages (bacteria-eaters), or 'phages, attack bacteria and recently certain algae have been shown to be susceptible to a virus similar in some respects to a bacterial virus (Fig. 1). Many familiar diseases of animals are caused by viruses such as foot-and-mouth disease of cattle, dog distemper, rabies, and the tumor in fowls known as the Rous sarcoma, discovered in 1911 by Peyton Rous and recently shown capable of causing tumors in primates.

Tumor viruses are common in birds, especially fowls, and in some mammals. In mice the viruses of breast cancer and the polyoma virus are cases in point. So far no virus has been definitely associated with cancer of

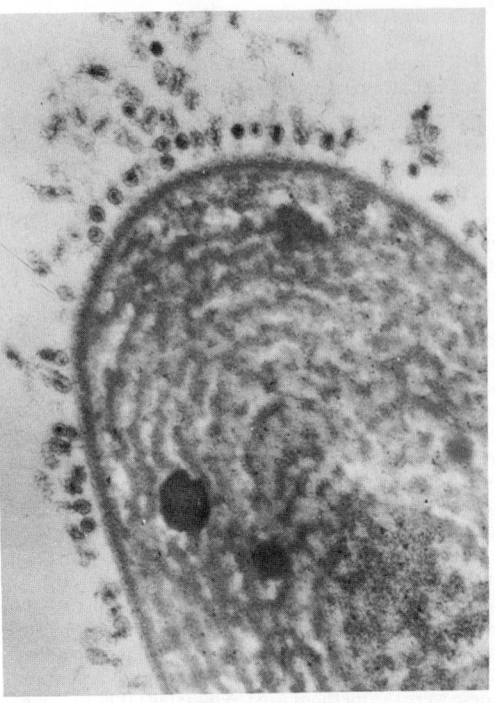

Fig. 1 Section through part of a cell of the blue-green alga, *Plectonema boryanum*, infected with a virus: note the virus particles attached to the outer cell wall by their "tails." × 54,000. (K. M. Smith and R. Malcolm Brown.)

man, though virus-like particles have been observed in patients suffering from leukemia. These resemble the particles which have been proved to cause leukemia in mice.

Man is susceptible to influenza, poliomyelitis, measles, mumps, chicken pox, smallpox, yellow fever, herpes, the common cold, which itself may be due to one or other of numerous viruses, and many other diseases, all caused by viruses.

All these agents, the causes of so many diseases in such a wide range of hosts, are viruses, and, though different, have many fundamental characters in common. A virus is a virus regardless of the host from which it comes, and a virus from a vertebrate animal may be superficially similar in its morphology to a virus from a plant.

Definition of a virus. Many attempts have been made to formulate a definition which would set forth the main characteristics of viruses, and one of the most recent is quoted here from Lwoff and Tournier (1)

(a) Viruses possess only one type of nucleic acid, either deoxyribonucleic acid (DNA) or ribonucleic acid (RNA); other agents possess both types.

(b) Viruses are reproduced from their sole nucleic acid, whereas other agents are reproduced from the integrated sum of their constituents.

(c) Viruses are unable to grow or to undergo binary fission.

(d) Viruses lack the genetic information for the synthesis of the Lipman System, the system responsible for the production of energy with high potential.

(e) Viruses make use of the ribosomes of their host cells; this is defined as absolute parasitism.

These features, being absent in other agents, are characteristic of viruses; they are present in all viruses, but absent in all nonviruses such as bacteria and including the agents of psittacosis, protozoa, etc. Bawden (2), however, considers this list as too exclusive and demanding more knowledge than there is about the constitution and behavior of most plant viruses. He suggests as a definition instead, the following: "viruses are submicroscopic, infective entities that multiply only intracellularly and are potentially pathogenic."

Isolating the virus. With the advent of the electron microscope it is now possible for the first time to visualize and characterize the virus particle itself. However, before that can be done, the virus must be freed from the constituents of the host cell and obtained in a pure state; otherwise it would be difficult to differentiate between the virus particle and similar-sized cell constituents.

There are many ways of purifying viruses; for plant viruses these may be chemical methods whereby the virus can be precipitated out of the extracted sap with such agents as ammonium sulphate or ethanol. This has to be done several times and the final purification is carried out by means of the high-speed centrifuge. These were the methods used by Stanley (3) in 1935 who was the first to isolate a virus, that of tobacco mosaic, in pure form. Animal viruses are best purified from tissue cultures and here again ultracentrifugation must be employed. In some cases where the virus occurs in high concentration in the host, centrifugation alone suffices. For example insect larvae infected with the *Tipula* iridescent virus have 25% of their dry weight as virus and it is necessary only to grind up the diseased larvae in distilled water, filter off the debris and then to spin the virus down on the centrifuge to obtain a pellet of pure virus. This, however, is exceptional and viruses, as a rule, are more difficult to isolate; indeed the majority of plant viruses still await isolation and characterization on the electron microscope. Several of the small plant viruses, and also the virus of poliomyelitis, have been obtained in crystalline form. The first virus to be crystallized was tobacco mosaic virus (TMV) (3) but as this virus is rod-shaped, only a 2-dimensional crystal, or paracrystal, is obtained. However, the plant viruses, tomato bushy stunt, tobacco necrosis, turnip yellow mosaic among others, as well as poliovirus, have all been obtained in 3-dimensional crystals. It is, of course, necessary to have the virus in crystalline form in order to undertake X-ray diffraction studies.

Size, shape and chemical nature of viruses. Recent improvements in the electron microscope and the development of electron "stains," which are not stains in the usual sense but are substances which enhance the contrast between the dark and the light, have enabled the size, morphology and some of the ultrastructure of virus particles to be elucidated.

A typical virus particle consists of a protein coat, or more accurately a protein framework in which the nucleic acid, the essential part of all viruses, is embedded. The nomenclature suggested for the virus particle and its components is as follows; the whole particle is called

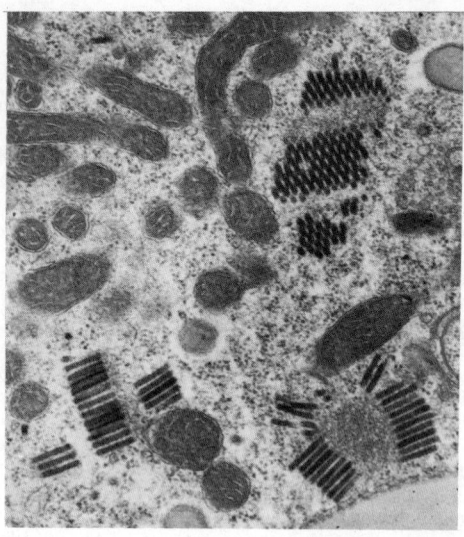

Fig. 2   Section through part of a fatbody cell from a larva of the Indian meal moth (*Plodia interpunctella*) infected with a granulosis virus: note the rod-shaped virus particles in regular array. × 50,000. (H. J. Arnott and K. M. Smith.)

a "virion," the protein framework is the "capsid" and the individual protein subunits are known as "capsomeres." The structure composed of the nucleic acid contained in its protein framework is the "nucleocapsid."

By means of negative staining and the use of ultrathin sectioning much of the structure of virus particles can be visualized on the electron microscope, but to elucidate the complete build-up, together with the position of the nucleic acid in the protein framework, X-ray diffraction studies are necessary.

Viruses differ widely in size and shape. They are, broadly speaking, of two kinds: *anisometric* and *isometric*. The first group may be subdivided into bacillus-like bodies, as in some plant viruses, rigid rods (Fig. 2), long flexible threads or brick-like shapes. The isometric particles, which are themselves a crystalline form that has three equal axes at right angles to one another, appear spherical but are in fact polyhedral (Fig. 4). Actually, the very small plant viruses and some of the small animal viruses are *icosahedra*, a figure with twenty sides. A technique known as "shadowing" allows for a 3-dimensional characterization of a virus particle in the electron microscope. A vapor of metal is thrown at an angle, in a high vacuum, over the virus particle. The area in the background obscured by the particle does not become covered with the electron dense film of metal and this is the "shadow." By casting a shadow from two different angles it was shown that the *Tipula* iridescent virus was an icosahedron, because this is the only shape that throws a blunt-ended and a sharp-ended shadow (Figs. 3, 4). Some of the bacterial viruses are unique in being "tadpole-like" with a "head" and a "tail."

X-ray diffraction studies (4) and electron microscopy (5) have shown the virus particle of tobacco mosaic (TMV) to be a rigid rod measuring 300 millimicrons in

Fig. 3   A cardboard model of an icosahedron, shadowed from two directions: note one blunt- and one sharp-ended shadow.

length and 15 millimicrons in diameter (one millimicron (1 mμ) equals one millionth of a millimeter). The protein framework consists of 2120 subunits, set in a helical array of pitch 23 Å [one angstrom (Å) equals a tenth of a millimicron] containing a single chain of ribonucleic acid (RNA) which follows the same basic helix at a radius of 40 Å (Fig. 5). There are 49 protein subunits in three turns of the helix. The virus particle

Fig. 4   A frozen-dried particle of the *Tipula* iridescent virus, shadowed similarly to Fig. 3: note identical shadows. × 105,000. (Figs. 3, 4, R. C. Williams and K. M. Smith.)

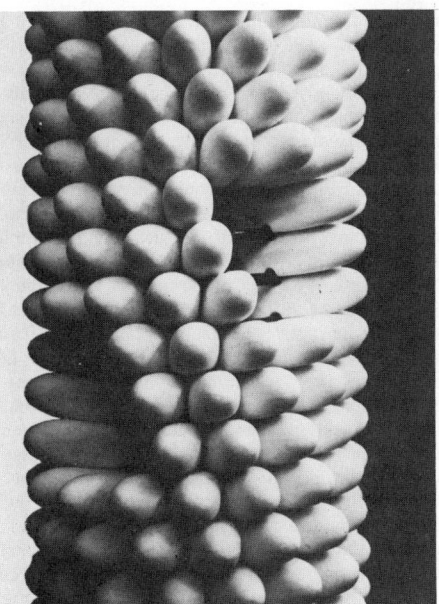

Fig. 5   Model of a tobacco mosaic virus particle, showing the protein subunits set in a helix; some of the subunits have been removed to show the strand of ribonucleic acid. (R. E. Franklin and A. Klug.)

has a hole of diameter 40 Å extending along the axis and the cylindrically averaged diameter of the particle is close to 150 Å.

Tobacco mosaic virus contains 95% protein and 5% ribonucleic acid, and in the small plant viruses there is no other constituent except protein and nucleic acid. All viruses attacking the higher plants, so far studied, contain ribonucleic acid (RNA). Among the different bacteriophages some contain RNA and some deoxyribonucleic acid (DNA); similarly with the viruses of the higher animals, but, as already mentioned, no virus contains both types of nucleic acid.

The structure and morphology of a small isometric virus which causes turnip yellow mosaic (TYMV) have also been studied in great detail by electron microscopy and X-ray diffraction (6, 7). Earlier X-ray studies and also electron microscopy (8,9) have shown that the protein framework of TYMV has a surface structure of 32 knobs; these are known as "morphological units." X-ray diffraction studies have now shown that there are actually 180 units, called "structural units." The arrangement of these units is fairly uniform with no clumping into hexameters and pentameters. In the images of the virus particles there is additional density linking the centers of the units into the 32 large morphological units. This additional density is attributed to the presence of RNA in the virus particles, so that the RNA must be distributed in such a way that local concentrations occur in the regions of the 32 surface lattice points. Photographs from crystals with a well-defined lattice show that the gross distribution of the RNA has the same icosahedral symmetry as the protein.

A significant proportion of the RNA is deeply embedded within the protein shell, and the mode of winding of the single RNA chain must be such that large segments of it are intimately associated with the rings of 6- and 5-protein structure units which make up the protein shell. It is the presence of the RNA in and about these positions that enhances the appearance of 32 morphological units in the electron micrographs.

Virus replication. One of the characteristics of viruses is that they cannot grow or divide by binary fission like a microorganism. Therefore the method of replication must be different; it is actually a building-up or biosynthesis of the new virus particles. In 1956 the important discovery was made almost simultaneously in Germany and California (10, 11) that the nucleic acid of the TMV particle is by itself capable of replication. This has been confirmed with other RNA and DNA viruses. Thus, the nucleic acid is the bearer of the genetic information. It is generally accepted that when a virus particle enters a susceptible cell, the first event to occur is the release of the nucleic acid and the removal of the protein framework. It must be remembered that in the synthesis of viruses two kinds of fundamentally different processes are involved (a) "reading" the viral genetic information and carrying into effect the messages which it contains and (b) "copying" this information in order to reproduce identically the nucleic acid moiety of the infecting particle. These processes can be accomplished only if the protein is removed from the nucleic acid (12).

When the virus takes charge of the metabolism of the cell and compels it to produce viral nucleic acid and viral protein certain other phenomena may occur. In the cells of plants infected with some viruses, what are known as "intracellular inclusions" can be observed with the optical microscope. These inclusions are found only in virus-diseased plants and they take several forms: amorphous inclusions or X-bodies, crystalline inclusions and peculiar formations resembling "pinwheels" or "cat-o'-nine-tails" (Fig. 6). Some of these inclusions are undoubtedly agglomerates of the virus particles, but the nature of the "pinwheels" is not certainly known; they may possibly consist of virus protein only.

Insect viruses. Certain virus diseases of insects are known as "inclusion body" diseases; these are of two main kinds, the "polyhedroses" in which large numbers of rod-like virus particles are occluded in many-sided protein crystals, the "polyhedra," and the "granuloses" in which a single rod-like particle is occluded in a very small protein crystal or "capsule." A section through a typical capsule is shown in Fig. 7; note the virus rod with its surrounding membranes and the paracrystalline lattice of the crystal. The function of these crystals is not known but it is thought that the virus incites and controls the crystallization process. It seems that sometimes the mechanism controlling the crystallization is faulty, and the process gets out of hand. When this happens abnormal "capsules" are formed; these may be "giants," elongated crystals with a central channel occasionally with a right-angled turn (Fig. 8), or bizarre agglomerates of crystals often with no occluded virus particle.

Incomplete viruses. Sometimes the mechanism of virus replication itself may go astray and give rise to viruses which do not function properly.

Such "incomplete" viruses can be of two kinds, those caused experimentally and those which occur naturally. In the first category are certain mutants of

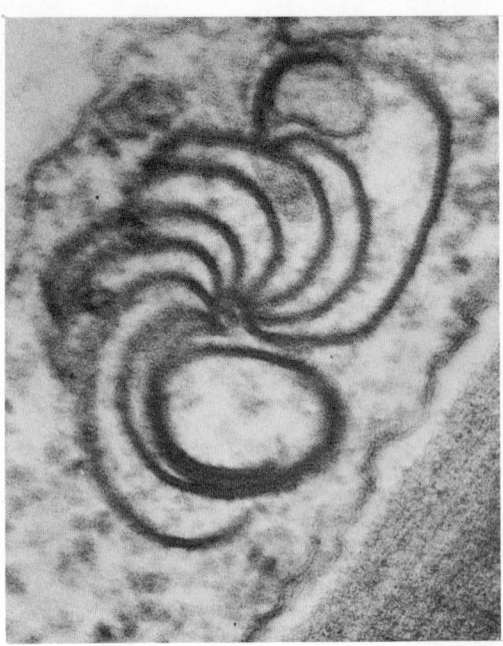

Fig. 6 Section through part of a cell from a virus-diseased sunflower (*Helianthus annuum*) showing one of the "pinwheel" inclusions. × 200,000. (H. J. Arnott and K. M. Smith.)

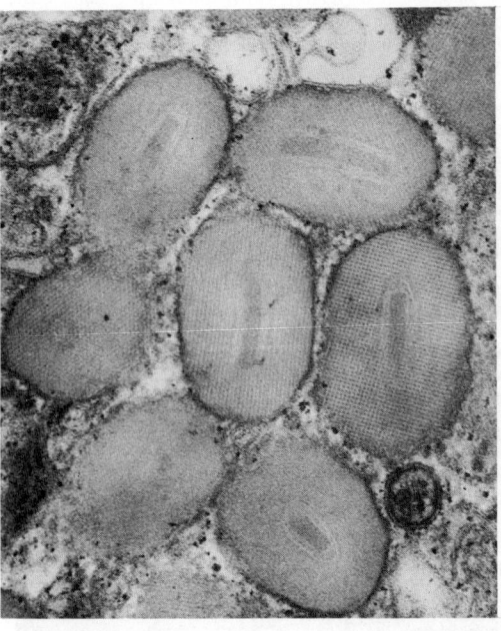

Fig. 7 Sections of the crystalline capsules in a fatbody cell of the Indian meal moth (*Plodia interpunctella*) affected with a granulosis virus: note the virus particle, with its membranes, inside the capsule and the crystalline lattice of the capsule. × 99,000. (H. J. Arnott and K. M. Smith.)

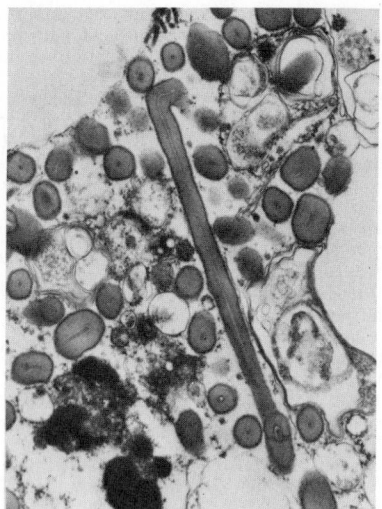

Fig. 8 Similar section to that in Fig. 7, showing one of the long aberrant capsules. × 37,700. (H. J. Arnott and K. M. Smith.)

tobacco mosaic virus which arise after treatment with a mutagenic agent such as nitrous acid. Three types of defective mutants of TMV have been isolated in this manner; plants infected with the first type contain a virus-like protein which is defective in the sense that it will not reconstitute with TMV nucleic acid to form intact, infectious rods. The second type of defective strain does not induce the formation of virus-like protein which can aggregate into rods but the protein is present in a disaggregated form. In the third type of mutant infected leaves contain small quantities of non-infectious rod-like material. It is interesting to note that these mutants are unable to move through the plant via the phloem but are restricted to a slow cell-to-cell movement, possibly because the naked RNA, unprotected by its protein framework, is subject to enzyme action (13).

The naturally occurring defective viruses lack the capacity to replicate by themselves and need the assistance of another virus. The former are known as "satellite" viruses and the latter as "helper" viruses, and they are found in both plants and animals; the helper virus is said to "activate" the satellite virus. Plants infected with tobacco necrosis (TNV) may contain two types of particles, one much larger than the other. The larger particle (TNV) is able to multiply indefinitely in the plant without the presence of the small particle, the satellite virus (SV). The latter is very small, measuring only 17 mμ in diameter, and cannot replicate by itself. It is, however, a complete and stable virus and what it obtains from the activating virus to enable it to multiply is not at present known (14).

A similar phenomenon has been observed in the Rous sarcoma virus (RSV) from fowls; the Bryan high-titre strain of this virus is always accompanied by another virus, known as the Rous-associated virus (RAV). Apparently RSV is unable to supply its own protein coat and depends on RAV or any one of several avian leukosis (tumor) viruses to fulfill this need (14).

Virus transmission. Viruses spread from one host to another by various means; animal viruses may spread by contact or, especially in man, by the respiratory tract. Some, like the viruses of yellow fever and equine encephalomyelitis, are carried by mosquitoes. Plant viruses, more than any other type of pathogen, are dependent upon insects, and other organisms, which prey upon plants, for their dissemination in nature (15).

Sap-sucking insects belonging to the order Hemiptera are the most effective carriers, or "vectors" as they are called, of plant viruses. One or two are transmitted by leaf-biting insects such as beetles. There are three types of relationships visualized between a plant virus and its insect vector: (a) *Stylet-borne,* a more or less mechanical contamination of the mouth-parts by the virus; (b) *Circulative,* here the virus is swallowed by the insect and has to travel via the gut to the salivary glands before being ejected into the plant; (c) *Propagative,* in this relationship the virus multiplies inside the insect vector. This is interesting because it illustrates the multiplication of a plant virus inside an animal and also raises the question as to whether the virus is primarily a plant or an insect virus.

The majority of plant viruses are spread by aphids, and one species, *Myzus persicae* Sulz., is implicated in the spread, in different parts of the world, of 50 different viruses. Most aphid-spread viruses are of the stylet-borne or circulative types while the propagative viruses are mainly carried by leaf-hoppers (Jassidae). Several of these viruses have been visualized by means of the electron microscope in various organs of the vector insect; most of the plant viruses which multiply in their leaf-hopper vectors are transmitted through the egg for many generations.

There are other types of vectors besides insects; several plant viruses are transmitted by mites, and a large group of soil-borne viruses are spread by root-feeding nematode worms. It has recently been shown that some viruses are transmitted by root-infesting fungi; the virus is carried inside the spores and not just as a casual contamination of the outside of the spore. The fungus is therefore a true virus vector (16).

Bacterial Viruses. The discovery in 1915 by Twort, and a little later by d'Herelle, of the bacterial viruses, to which d'Herelle gave the name "bacteriophages," led to a series of investigations almost unparalleled in virus research being concerned mainly with the genetics of the phages. This was largely because the bacterium is an ideal subject for studying a virus. It is a single-celled organism, it can be grown easily under controlled conditions, in small or large quantities, and a culture can be obtained from one organism, thus giving identical progeny. Moreover the life cycle of the host is short, the bacterium *Escherichia coli* divides every 20–30 minutes and the virus multiplies by a factor of 100 in 15–20 minutes (17).

Two interesting and important phenomena were discovered concerning the bacteriophages. One was the existence of an extreme form of latency, known as "lysogeny." The DNA of the virus is incorporated in the DNA of the host cell; such bacteria do not lyse, or destroy, other bacteria unless the lysogenic bacterium is subjected to some shock such as ultra-violet irradiation. The phage from these lysed bacteria is then capable of lysing other bacteria.

The other phenomenon is known as "transduction," in which some of the host cell DNA becomes combined with the viral DNA of the phage. When infecting new cells these fragments may be added to the bacterial chromosome so altering in some manner the genetic "make-up" of the bacterium.

Virus vaccines. Viruses cannot multiply outside a living susceptible cell. But cells can be grown, detached from the parent organism, in various kinds of nutrient media, and viruses can be cultured in these cells. This has given rise to the technique of "tissue-culture" from which much knowledge concerning the relationships between the virus and the cell has been derived. Furthermore the tissue-culture of viruses has resulted in the production of "vaccines" against some of the most serious virus diseases of man. Such vaccines can be of two kinds, the "dead" virus which has been inactivated in such a way as to invoke the production of antibodies without causing the disease. The Salk vaccine against poliomyelitis consisted of a virulent virus inactivated with formaldehyde and is an example of the first kind of vaccine. The second kind of vaccine consists of a "living" virus in an attenuated form; examples of this type are the smallpox, yellow fever and Sabin poliomyelitis vaccines.

Slow Viruses. One of the unsolved problems of the moment is the existence of the "slow viruses" so called because of the extremely long incubation period of the disease. The agent causing the "scrapie" disease of sheep, a name derived from the animal's habit of scraping and scratching the skin, has the attributes of a virus but is unusually resistant. Thus it withstands boiling at 100°C for 30 minutes or longer. It retains activity in 0.35% formalin for long periods of time and ultra-violet light does not appear to inactivate it. The virus is transmissible from sheep to goats and also to rats and mice; but it has not been isolated or characterized on the electron microscope. The possibility of the relevance of scrapie to certain human diseases such as affections of the central nervous system has been suggested (18).

Further information on viruses is available in the journal *Virology*, in *Advances in Virus Research* Volumes 1–13 edited by K. M. Smith and M. A. Lauffer and in *The Viruses*, Volumes, 1, 2 and 3 edited by F. M. Burnet and W. M. Stanley; all published by Academic Press, New York. For a more elementary survey of the whole subject the reader can refer to this contributor's *The Viruses*, Cambridge University Press (1962), and *Biology of Viruses*, Oxford University Press (1965).

KENNETH M. SMITH

References

(1) Lwoff, A. and P. Tournier, "The classification of viruses," Ann. Rev. Microbiol., **20**: 46–74, 1966.
(2) Bawden, F. C., "Plant Viruses and Virus Diseases," 4th ed., New York, Ronald, 1964.
(3) Stanley, W. M., "Isolation of a crystalline protein possessing the properties of tobacco mosaic virus," Science N.S., **81**: 644–645, 1935.
(4) Klug, A. and D. L. D. Caspar, "The structure of small viruses," Adv. Vir. Res., 7: 225, 1960.
(5) Finch, J. T., "Resolution of the substructure of tobacco mosaic virus in the electron microscope," J. Mol. Biol., **8**: 872–874, 1964.
(6) Finch, J. T. and A. Klug, "Arrangement of protein subunits and the distribution of nucleic acid in turnip yellow mosaic virus. II. Electron microscope studies," J. Mol. Biol., **15**: 344–364, 1966.
(7) Klug, A., *et al.*, "Arrangement of protein subunits and the distribution of nucleic acid in turnip yellow mosaic virus. I. X-ray diffraction studies," J. Mol. Biol., **15**: 315–343, 1966.
(8) Nixon, H. L. and A. J. Gibbs, "Electron microscope observations on the structure of turnip yellow mosaic virus," J. Mol. Biol., **2**: 197–200, 1960.
(9) Huxley, H. E. and G. Zubay, "The structure of the protein shell of turnip yellow mosaic virus," J. Mol. Biol., **2**: 189–196, 1960.
(10) Gierer, A. and G. Schramm, "Infectivity of Ribonucleic acid from tobacco mosaic virus," Nature, **177**: 702, 1956.
(11) Fraenkel-Conrat, H., "The role of the nucleic acid in the reconstitution of active tobacco mosaic virus," J. Amer. Chem. Soc., **78**: 882, 1956.
(12) Mundry, K. W., "Plant virus-host cell relations," Ann. Rev. Phytopathol., **1**: 173–196, 1963.
(13) Siegel, A., *et al.*, "The isolation of defective tobacco mosaic virus strains," Proc. Natl. Acad. Sci. U.S., **48**: 1845, 1962.
(14) Kassanis, B., "Satellitism and related phenomena in plant and animal viruses," Adv. Vir. Res. **13**: 147–180, 1968.
(15) Smith, K. M., "Plant virus-vector relations," Adv. Vir. Res., **11**: 61–96, 1965.
(16) Grogan, R. C., and R. N. Campbell, "Fungi as vectors and hosts of viruses," Ann. Rev. Phytopathol., **4**: 29–52, 1966.
(17) Fraser, D., "Viruses and Molecular Biology," New York, Macmillan, 1967.
(18) Stamp, J. T., "Scrapie and its wider implications" *in* "Aspects of Medical Virology," London, The British Council, 1967.

VITAMINS, FAT-SOLUBLE

During the first part of the 20th Century, medical interest began to shift from infections to metabolic, nutritional and endocrine diseases. In 1912 the word "vitamine" was first employed to describe substances necessary for the prevention of deficiency diseases in animals. Shortly thereafter two groups of investigators working independently found that *something* in fats was necessary for the growth of animals. Soon it was learned that cod liver oil contained a high concentration of unknown substances which would cure rickets, and relieve xerophthalmic diseases of the eyes. For a time these effects were ascribed to a single fat-soluble substance. Soon E. V. McCollum furnished proof that two vitamins existed in cod liver oil. An unstable factor became known as vitamin A, a stable one as vitamin D.

Vitamin A. Vitamin A is an essential dietary factor for man and most animals. It is necessary for the normal function of epithelial tissues, the retina and for growth. There are three forms of vitamin A, retinol, retinal and retinoic acid; and several precursors called carotenes which are abundant in such plants as green leafy vegetables, peas, carrots, sweet potatoes, apricots and peaches. Active vitamin A substances are present in many foods of animal origin including milk, butter, cheese, eggs, and liver, but the amount varies seasonally and to some extent is dependent upon the vitamin content of the diet of the animals.

A deficiency of vitamin A, rare in the United States, may be encountered in other countries. The manifestations of human vitamin A deficiency include faulty growth, and disorders of the epithelial tissue. At first there is poor accommodation of vision in darkness. This results from lack of retinal rhodopsin or visual purple. Later the eyes may become dry with thickening of the conjunctiva and loss of the clear white appearance of the sclera. Finally the cornea becomes opaque, the whole eye becomes inflamed, ulcerations occur and vision is lost completely. A characteristic change of the skin is follicular hyperkeratosis resembling "goose flesh." It must be differentiated, however, from folliculitis of adolescent acne vulgaris or that caused by irritants or by atrophy of skin in any form of severe malnutrition. Another change in the skin has been called xerosis or dry crinkled skin. Another result of deficiency is the sicca syndrome, a loss of secretion of the lachrymal and salivary glands.

Carotenes which occur in so many foods are not destroyed by usual methods of cooking and are absorbed readily through the gastrointestinal tract. Rarely in conditions such as diarrhea, or serious disease of the liver or gallbladder there may be inadequate absorption of vitamin A or its precursors. In healthy people the blood contains detectable amounts of carotenoids. Both vitamin A and its precursors are stored in the liver. In myxedema, diabetes, and a few other diseases the utilization of carotene is slowed and higher concentrations may accumulate. This results in a peculiar yellow discoloration of the skin. Indeed the yellow color of human and animal fat is largely the result of carotenes. With such plentiful storage of these substances in the liver and in the body fat, deficiency of vitamin A develops slowly over a period of months or years.

The exact amount of vitamin A necessary for health is difficult to determine. About 20 international units of the preformed vitamin or 40 units of carotene per kg. of body weight are adequate under most circumstances. In some diseases or with increased metabolic demands greater amounts may be desirable. The recommended daily allowance is 5,000 international units per day. This unit is determined by the effect of the vitamin upon the growth rate of young rats under deficient conditions.

Normal persons obtain an abundance of vitamin A from their diet since most milk, butter and margarine in this country are fortified with it. The carotenes in vegetables are absorbed and utilized less efficiently, but contribute significantly to the human supply, especially if foods are cooked.

Excessive amounts of vitamin A can produce serious toxic reactions. These do not occur in persons eating a normal diet but only after ingestion of excessive amounts of vitamin concentrates for a prolonged period of time. The clinical features are headache, itching of the skin, folliculitis and xerosis, scaliness and an increase of dandruff, irritation and dryness of the lips, nausea, loss of appetite, and alternate diarrhea and constipation. Dizziness and generalized muscular weakness with fatigue and aching of the extremities may occur. In children typical changes in the x-ray appearance of the bones also occur.

Vitamin D. Although rickets has been recognized for centuries it was not considered to be a disease of deficiency until 1906 when Hopkins proposed this explanation. By 1919 Mellanby induced rickets in young dogs and demonstrated that certain fats would cure it. After McCollum demonstrated that the antirachitic factor differed from the antixerophthalmic factor the presence of a separate vitamin was established.

Many substances possess antirachitic activity. A number of provitamins such as the ergosterol in plants and other foods can be irradiated with actinic rays to produce vitamin D_2 (calciferol, viosterol). Cod liver oil has long been a rich source of vitamin D, the USP reference unit of which is determined by assay in chicks. The several vitamins D are sterols which are distributed throughout the body, especially the liver, adipose tissues, and central nervous system.

The skin of animals and of man contains sterols which can be converted into active vitamin D by the action of ultraviolet rays. During summer these may contribute small but significant increments to the human source. In winter, especially in northern regions, the amount so obtained is probably negligible.

Rickets is primarily a disease of children less than three years of age, although it may occur later. Premature infants are most susceptible. Clinical symptoms include a poor rate of growth and development and a characteristic change of growing bones. Cartilage is formed normally, but fails to become calcified, particularly in the region of rapidly growing epiphyses. This results in bowing and knobby deformities of the bones of the legs and of the lower portion of the chest cage. The skull may become flattened by the weight of its contents. Inadequate deposition of calcium phosphate seems to be the chief defect. Fractures are common in rachitic children. Administration of vitamin D to a rachitic child results in better absorption of calcium from the gastrointestinal tract leading to gradual healing of the defect. Permanent deformities may persist. Some rachitic children develop tetany from an inadequate concentration of calcium in the blood; others may have normal concentrations. Vitamin D is absorbed from the gastrointestinal tract of normal individuals but diseases characterized by malabsorption of fats impair absorption and utilization.

The clinical use of vitamin D is limited largely to the prevention or treatment of rickets. In adults a deficiency of vitamin D produces less spectacular diseases, but may result in widespread demineralization of bones. Fractures may occur. Large quantities of vitamin D may be given in the treatment of a deficiency of parathyroid hormone to facilitate absorption of calcium, and to raise the concentration of calcium in the blood. Deficiency of vitamin D may impair the growth and development of the teeth and result in faulty development of the enamel leading to caries. Excessive amounts of vitamin D will not prevent caries in people whose supplies are adequate.

The common foods which supply vitamin D are liver, milk and eggs. Milk, in this country, is fortified with enough vitamin D to prevent rickets in most people. Most vegetable foods and meats contain insignificant amounts of vitamin D.

The amount of vitamin D recommended for the average child is 400 units per day. Adult requirements may be similar. Excessive quantities may produce serious or even fatal toxic reactions. Unfortunately some proprietary vitamin preparations contain large amounts of vitamin D. Ill-advised self-medication has been the most common cause of vitamin D poisoning. Some years ago large quantities of vitamin D were recommended in the treatment of rheumatoid arthritis, hay fever, asthma and psoriasis. Sometimes doses as high as 100,000 to 200,000 units per day were given. This resulted in a rise in the serum calcium accompanied by loss of appetite, nausea, vomiting, diarrhea, excessive thirst, and urinary frequency. In some instances disastrous calcification occurred in the kidneys, the heart and blood vessels and in muscular tissues. Recent experimental work in animals indicates that even brief periods of excessive vitamin D therapy may result in premature arteriosclerotic changes.

Vitamin D is available in its natural form in cod liver oil and as the irradiated sterols (vitamin D_2), which will mix readily with milk. Vitamin D_2 (calciferol or viosterol) is employed most frequently for the enrichment of milk and for therapeutic administration.

Vitamin K. This vitamin was discovered accidentally by H. Dam while feeding a highly purified diet to fowl which developed a hemorrhagic tendency and anemia. Later Horvath found that the blood of chickens fed sprouted soy beans did not clot after standing overnight. Other workers encountered similar findings and learned that fatty substances extracted from rice, corn, and sunflower seeds could correct the illness. An active material also was found in green leafy plants. Chemists who attempted to isolate and synthesize the substance discovered that there were two active compounds, both naphthoquinones. Later a synthetic substance called Menadione (2 methyl-1, 4 naphthoquinone) was found to be even more effective. Vitamin K enables a healthy liver to form the prothrombin necessary for normal clotting of blood. A deficiency of prothrombin may develop not only from absence of vitamin K, but also from diseases of the liver and biliary tract. Vitamin K is widely distributed in nature and is present in adequate quantities in most diets. It is readily absorbed with bile in the gastrointestinal tract. Some is stored, presumably in the liver or fatty tissues. Even without a dietary source the normal bacteria of the intestinal tract may produce enough vitamin K to meet the requirements of man and animals.

A deficiency of vitamin K does not develop unless production or absorption is impaired. In newborn infants the intestinal tract is sterile and the amount provided in a formula may be negligible. Hypoprothrombinemia may develop and result in hemorrhagic disease of the newborn. In adults defective absorption of vitamin K from the intestine may occur in the absence of bile. Defective utilization of vitamin K may occur in the presence of parenchymal disease of the liver. Sometimes prolonged treatment with antibiotics will impair presynthesis of vitamin K. In any of these events the prothrombin is impaired and hemorrhage may develop.

An antagonist to vitamin K was discovered through "Sweet Clover Disease" which occurred in cattle after they had eaten spoiled hay. Excessive bleeding followed any form of trauma and many animals died. In 1929 L. M. Roderick demonstrated that the bleeding was due to a deficiency of prothrombin and suggested that the toxic substance might be due to decomposition of coumarin. Later this substance was found to be bishydroxycoumarin or dicumarol which antagonized vitamin K causing a deficiency of prothrombin. Administration of vitamin K in sufficient quantities overcomes the lack of prothrombin and the resulting hemorrhage. In recent years dicumarol has been used widely in the treatment of coronary artery disease and other conditions in which reduced blood clotting is desired.

The clinical uses of vitamin K are limited to situations which interfere with production or absorption or for correction of hypoprothrombinemia induced by drugs. Excessive doses have been reported to cause jaundice in premature infants. Prophylactic administration of the synthetic substance, menadione, late in pregnancy has been said to prevent hemorrhagic disease of the newborn but recent evidence has cast doubt on this belief. The only known pathologic effect of vitamin K deficiency is that of hypoprothrombinemia with resultant hemorrhage. There is no generally accepted figure for the daily human requirement for vitamin K. Although little is known about it, apparently vitamin K is stored rather poorly.

Vitamin E: (The tocopherols). In 1920 Matill and Conklin found evidence of a factor necessary for growth of rats. Later Evans demonstrated that it was not vitamins A, B, C, or D. Eventually the tocopherols were identified as naturally occurring oily substances and characterized as alpha, beta and gamma forms. They have biological activity in descending order. Vitamin E is necessary for the normal growth of animals. Without it they develop infertility, abnormalities of the central nervous system and myopathies involving both skeletal and cardiac muscle. The tocopherols exert an antioxidant effect chemically, the magnitude of which is in reverse order to that of its vitamin activity. Muscular tissue taken from a deficient animal has an increased rate of oxygen utilization. The tocopherols are so widely distributed in natural foods that a spontaneous deficiency does not occur unless diseases of the gastrointestinal or biliary systems hinder absorption. Substitution of polyunsaturated fats for saturated fats in the diet may increase the need for Vitamin E.

The evidences for such deficiency have rested largely upon the findings of a low level of tocopherol in the serum and upon an abnormal hemolysis of erythrocytes by hydrogen peroxide.

Although nearly every vitamin has been used unwisely in the treatment of human diseases, perhaps no other substance has aroused a greater degree of controversy among clinicians than vitamin E. Because deficient animals develop a form of myopathy it was natural to test the therapeutic efficacy of vitamin E in various forms of progressive muscular dystrophy, and in diseases of the reproductive system. Many enthusiastic claims have been refuted by investigators whose methods were meticulous and objective. At present there is no recognized indication for the administration of vitamin E except in patients with a malabsorption syndrome.

Despite the lack of therapeutic indication many pharmaceutical preparations contain the tocopherols. As yet there is no evidence that an excess of vitamin E produces toxic reactions. On the basis of animal studies it has been estimated that normal persons require approximately 30 mg. of naturally occurring tocopherols daily, but more than this amount is supplied by the average diet. Vitamin E is essential for man yet much remains to be learned about it.

ROBERT E. HODGES

References

Jolliffe, N. *et al.,* eds., "Clinical Nutrition," New York, Hoeber, 1950.

Clark, G. W., "A vitamin digest," Springfield, Ill., Thomas, 1953.

VITAMINS, WATER-SOLUBLE

Water-soluble vitamins are a group of organic compounds, more or less soluble in water, which are either synthesized or obtained exogenously by all organisms. They are active at extremely low concentrations and their usual metabolic role is as a cofactor or prosthetic group of an enzyme.

Ascorbic acid. (Vitamin C); $C_6H_8O_6$; 1-Threo-2,4,5, 6-pentoxyhexen-2-carboxylic acid lactone. A white crystalline material which melts at 192°C.; it is soluble in water and has a specific rotation in water of $[\alpha]_D + 23°$. It is a powerful reducing substance. Ascorbic acid is widely distributed in plants and animals; the richest sources are paprika plant, rose hips, and West Indian cherry; it is also made synthetically.

Ascorbic acid is required by man (1.2 mg./kg. of body weight for adult male), other primates, and the guinea pig. It is synthesized by rat and higher plants. Deficiency results in scurvy in man indicated by a loss of appetite and physical energy, muscle soreness, poor wound healing, follicular keratosis, swollen gums and loosening teeth, capillary fragility, and failure of normal collagen synthesis.

The strong reducing power of this compound and its general distribution in organisms suggest that it has a role in many oxidation-reduction reactions. It participates in electron transfer reactions *in vitro* and may participate in the formation of Fe-containing enzymes. It is a hydrogen donor for many enzyme systems, including polyphenol oxidase, peroxidase, cytochrome, and a specific ascorbic acid oxidase. It is also involved in the oxidative catabolism of tyrosine and the synthesis of tetrahydro derivatives of pteroylglutamic acid.

Biotin. $C_{10}H_{16}O_3N_2S$; 2-keto-3, 4-imidazolido-2'-tetrahydrothiophene-n-valeric acid. It forms crystals which melt with some decomposition at 230–232°C.; it is soluble in dilute alkali and hot water and sparingly soluble in dilute acid and cold water. Widely distributed in nature, the richest sources are egg-yolk, liver, and yeast. The vitamin is made synthetically.

Biotin is required by man, vertebrates (except those supplied by their intestinal flora), insects, and some microorganisms. It is synthesized by higher plants. Deficiency is indicated by dermatitis, alopecia in some mammals, atrophy of lingual papillae, electrocardiographic changes, anorexia, lassitude, sleeplessness, muscle pain (man), spasticity, and paralysis of hind quarters (rat).

Metabolic role in metabolism not certain. There is evidence for its participation in aspartic acid synthesis, decarboxylation of some acids, oleic acid synthesis, CO_2 fixation, and fat synthesis.

Vitamin B$_{12}$. (Cobalamin, Extrinsic factor); C_{63}·$H_{88}O_{14}N_{14}PCo$. B_{12} forms dark red crystals which are soluble in water and stable to heat and light. It has absorption maxima at 278,361, and 550 mμ. It is widely distributed in animals and lower plants but has not been found in higher plants. A good natural source is liver.

B_{12} is required by man, all vertebrates studied, and some microorganisms. It is synthesized by many microorganisms; bacteria are the present industrial source. Anemia, degenerative changes in the spinal cord, and glossitis follow from a lack of B_{12}.

The cofactor form or precise metabolic role of B_{12} is not yet known but it appears to effect methyl group synthesis, deoxyriboside synthesis, activation of sulphhydryl enzymes, and protein synthesis. B_{12} is the extrinsic factor necessary to alleviate pernicious anemia.

Carnitine. $C_{17}H_{15}O_3N$; trimethylbetaine of β-OH-γ-aminobutyric acid. Required for the metamorphosis of the mealworm, *Tenebrio,* and several other insects.

Choline. $C_5H_{15}O_2N$; β-hydroxyethyltrimethyl ammonium hydroxide. It is a colorless crystal which acts as a strong base. It decomposes on heating. Dilute water solutions are heat stable. It is soluble in water, formaldehyde, and in absolute methanol and ethanol. It is found in plant and animal lecithins, acetylcholine, and phospholipids.

It is not required by man but is required by other vertebrates, some insects, and a few microorganisms. Deficiency results in renal lesions, fatty and cirrhotic liver, hypertension, perosis, anemia, ulcers, and nerve degeneration. Chronic alcoholism leads to cirrhosis due to an induced choline deficiency.

Choline serves as a major source of methyl groups in one carbon metabolism. It is a constituent of acetylcholine and of phospholipids. In an ill-defined manner it affects the deposition of fat in the liver.

Folic acid group. Pteroylglutamic acid; $C_{19}H_{19}$·O_6N_7; N-[4-{[(2-amino-4-hydroxy-6-pteridyl) methyl]-amino}-benzoyl]glutamic acid. This group includes folinic acid (Citrovorum factor) tetrahydrofolic acid, p-aminobenzoic acid, biopterin, and their several analogs. Pteroylglutamic acid forms yellow, spear-shaped leaflets which char around 250°C. It is very sparingly soluble in water and somewhat more soluble in alkaline 20% ethanol-water. In 0.1 N NaOH it has the following absorption maxima: 256, 282, 365 mμ. It is stable to heating for a short time but is unstable in light. This vitamin group is widely distributed among plants and animals; a good natural source is liver. It is made synthetically.

The folic acid group of vitamins is required by all vertebrates (except those whose needs are met by their intestinal flora), insects, and some microorganisms. Deficiency manifests itself in retardation of growth, sprue, anemia, diarrhea, perosis, and oral lesions. Analogs of this vitamin have been extensively studied for their antitumor possibilities.

This vitamin in a form, as yet not definitely known (perhaps tetrahydrofolic acid), is a cofactor for the synthesis of purines, pyrimidines, and several amino acids as well as one carbon metabolism involving both formate and methyl. It is thus vital to nucleic acid and protein synthesis.

Inositol. $C_6H_{12}O_6$; Hexahydroxycyclohexane; 9 possible isomers. D-Inositol is a white crystal melting at 249–250°C; it is non-reducing and soluble in water and shows optical rotation, $[\alpha]_D^{25} = +65.0°$ in water.

It is a constituent of the lipids and phosphatides of most, if not all, organisms.

It is not required by man and is required by mouse, cotton rat and hamster, and a few microorganisms.

Deficiency results in alopecia in mouse and dermatitis.

α-Lipoic acid. (DL-Thioctic acid, protogen); $C_8H_{14}O_2S_2$; d-5(dithiolane-3)-pentanoic acid. This is a crystalline material melting at 50–61°C. and boiling at 160–165°C.; it is soluble in fat solvents but barely soluble in water. The vertebrate requirement for lipoic acid is uncertain. It is required by a ciliate protozoan and replaces acetate for a lactobacillus. It is involved in oxidative decarboxylation.

Nicotinic acid. (Niacin); $C_6H_5O_2N$; pyridine-3-carboxylic acid and nicotinic acid amide (Niacinamide); $C_6H_6ON_2$; pyridine-3-carboxylic acid amide. A white crystalline material which melts at 234–237°C.; it is soluble in water and ethanol. It forms quaternary ammonium compounds and salts. It has an absorption maximum at 262 mμ; this does not vary with pH. It is made synthetically.

Nicotinic acid is required by man (0.25 mg./kg. of body weight for adult male), vertebrates (except those supplied by their intestinal flora), insects, and some microorganisms. It is made by higher plants. Deficiency is shown in man (pellagra) by dermatitis, diarrhea, and delirium and death and in dog (black tongue) by oral lesions, oral odor, diarrhea, emaciation; other symptoms are degeneration of nervous system, anemia, and perosis.

Nicotinic acid as its amide in diphosphopyridine nucleotide (coenzyme I) or triphosphopyridine nucleotide (coenzyme II) is a cofactor in hydrogen transfer in numerous enzyme reaction in all organisms studied.

Pantothenic acid. $C_9H_{17}O_5N$; α, γ-dihydroxy-β,β-dimethylbutyryl-β'-alanine. It is a yellow viscous oil which forms crystalline salts. It is soluble in water and ethanol; neutral solutions are stable but hot acid or alkaline solutions are not. Free pantothenic acid has following rotation $[\alpha]_D^{26} +37.5$. It is made synthetically; the richest natural source is royal jelly.

Pantothenic acid is required by man, most vertebrates, insects, and some microorganisms. Deficiency is characterized by retarded growth, abnormal gait, ataxia, diarrhea, anemia, fatty liver, dermatitis of chicken and rat, graying in several mammals, and burning sensation of hands and feet in man.

As coenzyme A, pantothenic acid is essential for the normal interchange of fats, proteins, and carbohydrates in metabolism. It is a cofactor for the metabolism of acetyl and acyl, succinyl, and propionyl groups.

Pyriodoxine. (Vitamin B_6); $C_8H_{11}O_3N$; 3-hydroxy-4, 5-dihydroxymethyl-2-methylpyridine. Also, Pyridoxal; $C_8H_9O_3N$; 3-hydroxy-4-formyl-5-hydroxymethyl-2-methylpyridine and Pyridoxamine; $C_8H_{12}\cdot O_2N_2$; 3-hydroxy-4-aminomethyl-5-hydroxymethyl-2-methylpyridine. Pyridoxine as the hydrochloride occurs as white platelets melting at 204–206° C.; it is freely soluble in water and is destroyed by light in neutral or alkaline solution. The molecule is tautomeric as shown by its absorption pattern with maxima at 292 mμ at pH 2, 328 mμ at pH 4.5, and 328 and 256 mμ at pH 6.8. It is made synthetically.

B_6 is required by man (15–31 μg./kg. of body weight for adult male), most vertebrates, insects, and some microorganisms; it is synthesized by higher plants. Deficiency is characterized by retarded growth, anemia, degeneration of myelin sheath, convulsions, lesions about eyes, nose and mouth, loss of hair from paws, snout and eartips, and insomnia. In higher animals tryptophan metabolism is abnormal and there is less kynurenine and nicotinic acid derivatives and more xanthurenic acid excreted in the urine in B_6 deficiency.

B_6 as pyridoxal phosphate plays an essential role in amino acid metabolism since it is required for the decarboxylation of amino acids and for transamination; it is also part of muscle phosphorylase.

Riboflavin. (Vitamin B_2); $C_{17}H_{20}N_4O_6$; 6,7-dimethyl-9-(D-1'-ribityl)isoalloxazine. Orange-yellow needles which decompose at 278–282°C. They are sparingly soluble in water or ethanol and very soluble but unstable in alkali. Neutral solution shows absorption maxima at 475, 445, 359–372, 268, and 223 mμ and a yellowish-green fluorescence with maximum at 565 mμ which is destroyed by acid or alkali. Solutions of this vitamin are heat stable and light unstable. It is made synthetically and also by several fungi such as *Eremothecium ashbyii* and *Ashbya gossypii*.

Riboflavin is required by man (25 μg./kg. of body weight for adult male), most vertebrates, some insects, and several microorganisms. It is synthesized by higher plants. Deficiency results in the cessation of growth, dermatitis, epidermal atrophy, neuritis, light sensitivity, itchy eyes and lids, and muscular weakness.

Riboflavin as riboflavin-5-phosphate and flavin adenine dinucleotide is a cofactor in respiration, acting as a hydrogen acceptor and donor. It is a prosthetic group in flavoprotein enzymes such as yellow enzyme, xanthine oxidase, diaphorase, TPNH-cytochrome C reductase. It also plays an important role in bioluminescence (constituent of bacterial luciferin) and may be involved in the reaction of plants to light.

Thiamine. (Vitamin B_1, Aneurin); $C_{12}H_{17}N_4OS\cdot$ClHCl; 3-(4-amino-2-methylpyrimidyl-5-methyl)-4-methyl)-5-β-hydroethylthiazolium chloride hydrochloride. Colorless needles which melt at 250°C. It is soluble in water and shows absorption bands at 235 and 267 mμ at acid pH; it is unstable at pH 7 or higher. This vitamin is widely distributed in plants and animals; it is made synthetically, a rich natural source, however, is yeast.

Thiamine is required by man (25 μg./kg. of body weight for adult male), vertebrates (except for those which obtain it by way of their intestinal flora), insects, and several microorganisms. Deficiency results in beriberi in man characterized by retardation of growth, loss of weight, anorexia, neuron degeneration, convulsions, and neurasthenia. Some mammals show myocardial lesions, heart dilatation, bradycardia, and edema. Pyruvic acid accumulates in the blood and tissues and there is a drop in the products of the tricarboxylic acid cycle.

Thiamine as thiamine pyrophosphate functions as a coenzyme in the key reaction connecting glycolysis with the tricarboxylic acid cycle as well as many other decarboxylation reactions.

S. AARONSON

References

Sebrell, W. H. Jr. and R. S. Harris, eds., "The Vitamins," 6 vols., New York, Academic Press, 1954–1965.

Harris, R. S., G. F. Marrian and R. V. Thimann, eds., "Vitamins and Hormones," New York, Academic Press, (annual review).

Luck, J. M., F. W. Allen and G. Mackinney, eds., "Annual Review of Biochemistry," Palo Alto, California, Annual Reviews, Inc.

Robinson, F. A., "The Vitamin B Complex," New York, Wiley, 1951.

WAGNER, RUDOLF (1805–1864)

Rudolf Wagner was a German physiologist who studied medicine and obtained his degree at Worzburg. He worked under Cuvier, a comparative anatomist in Paris, and later was appointed the successor to Blumenbach, the comparative anatomist at Gottingen. As an investigator and teacher he led the way for German zoologist Leuckart and philosopher Lotze.

His work included investigations into spermato- and ovogenesis, and into the corpuscles. Under Blumenbach's influence he was an anthropologist. Along this line Wagner gave a lecture at a scientific meeting at Gottingen in which he discussed the question of the origin of man from one single pair in accordance with the church's doctrines of creation. Thus he had an opportunity to make a violent attack upon the materialistic soul-theories of the time, which he inveighed against from the point of view of both science and morality. He worked out a theory of the soul as a kind of ethereal substance which leaves the body at death and imparts itself to the children that are born. Though neither Wagner nor his contemporary antagonists realized the importance of evolution, their service to natural science deserves recognition.

BETTY WALL

WALLACE, ALFRED RUSSELL (1823–1913)

Unlike many of his contemporaries Wallace received little formal education. Born at Usk, in Wales in 1823 he left home when he was fourteen years of age. In the following six years he worked as a land surveyor with his brother. It was during this period that he became interested in the Nature he was later to write about. Returning to London in 1843, Wallace taught for a year at the Collegiate School at Leicester. He spent many hours at the library reading the works of Malthus and Humboldt and it was here that he met H. W. BATES, an entomologist with whom he became fast friends. In 1848 Wallace and Bates left England and sailed to the Amazon. The next four years were spent in observing the flora and fauna of the Amazon Valley. In 1852, because of poor health, Wallace returned alone to England, and reported his observations in *Travels on the Amazon and Rio Negro* which was published in 1853. In 1854, he left on an expedition to the Malay Archipelago where he remained for eight years collecting specimens for the University of Oxford and the British Museum and making notes for his later publications. In 1869, he published the results of his comprehensive studies in the book, *Malay Archipelago.*

During his travels in the Archipelago, Wallace was aware of the great differences that existed between animals inhabiting different islands. He records that the fauna of Borneo and Bali are quite divergent from the fauna of Celebes and Lombok, yet these two pairs of islands are separated by only a narrow strip of sea, later named Wallace's Line, in honor of the naturalist. His first publication from Malay, *On the Law which Has Regulated the Introduction of New Species,* appeared in the Annals and Magazine of Natural History in 1855. But it was the manuscript he forwarded to his friend, Charles Darwin, in 1858 that was to have far greater consequence. In his letter, Wallace gave the essential basis for the theory of evolution by selection, conclusions at which he had independently arrived while isolated in Ternate, a small island near New Guinea. This correspondence contained an almost exact duplication of the theories Darwin had been formulating since 1844. Giving equal credit therefore, to both men Charles Lyell and Joseph Hooker simultaneously delivered both papers before a meeting of the Linnean Society on July 1, 1858. After his return from the Archipelago, Wallace compiled a two volume work entitled, *The Geographical Distribution of Animals.* In later years his interests turned to anthropology and the evolution of man, although he never published any of his ideas of this subject. He died in 1913.

RICHARD M. CRIBBS

WATER METABOLISM OF VERTEBRATES

Water metabolism concerns the balancing of water loss (by evaporation from SKIN and LUNGS, in urine and feces, by OSMOSIS through skin) against water gain (by drinking, in food, from metabolism of food, osmotic uptake through the skin). Balance is necessary for osmotic homeostasis, maintenance of adequate circulatory volume, and also temperature regulation (evaporative dissipation of heat).

Vertebrates fall into three groups according to how they maintain water balance:

(1) Fishes and AMPHIBIANS in fresh water are very hypertonic to their medium and must counteract a con-

989

tinual dilution of their body fluids. Water influx is reduced by the relative impermeability of the skin, and is balanced by diuresis. Electrolytes lost in this urine are replaced in food eaten and by absorption through gill surfaces (fishes) or skin (amphibians).

(2) Marine ELASMOBRANCHS are unique in maintaining themselves slightly hypertonic to sea water by retention of urea (2±%) in their body fluids, making its osmotic pressure more than twice that of marine teleosts.

(3) For marine TELEOSTS and terrestrial tetrapods the continual problem is to counteract desiccation due to osmotic loss to a hypertonic medium or to evaporation.

(a) Marine teleosts counteract osmotic dehydration by drinking large amounts of water and forming little urine. Nitrogenous wastes and surplus salt gotten by drinking sea water are secreted by the gills.

Terrestrial tetrapods adjust by: avoidance of evaporative stress, reduction of evaporative and urinary water losses, and temporary toleration of hyperthermia or hypernatremia. Antidiuretic hormone (ADH) from the neurohypophysis is very important in enhancing uptake of water through the skin (amphibians), reduction in glomerular filtration (amphibians, reptiles, birds), and increase in tubular reabsorption of water (mammals).

Water balance processes are best developed in species inhabiting deserts, where little drinking water is available and climatic conditions accentuate evaporation.

(b) Certain toads and frogs survive in deserts, needing open water only for breeding, largely by remaining dormant during dry periods. Evaporation is greatly retarded in a cool damp burrow, and urine volume is reduced by 98–99% (filtration antidiuresis), but urine remains hypotonic. Urinary water may be recycled through the body by reabsorption from the bladder. Dormant animals tolerate a loss of 50–60% of their body water. They emerge during rains, and in their dehydrated state quickly reabsorb water through the skin.

(c) Terrestrial reptiles also avoid considerable evaporation by being quiescent in burrows much of the time. Also, their skin is more impermeable than amphibians', though water is still lost in expired air. Hydrated lizards have low urine filtration rate (urine always hypotonic), and may become almost anuric when dehydrated. During dehydration electrolyte wastes are retained in the body and tolerated in concentrations fatal to birds and mammals, until water is available for their excretion. A carnivorous diet (70±% water) provides adequate water intake, while food is available. Water can be reabsorbed osmotically from the cloaca, reabsorption being particularly effective because of the nature of the principal nitrogenous waste, uric acid, which has a very low solubility. As uric acid precipitates in the cloaca, its osmotic effect is removed, and further water can then be absorbed by osmosis. This is probably the major value of uric acid excretion. Precipitated wastes are excreted en masse, with very little fluid loss.

(d) Birds, being homeothermic, cannot reduce their evaporative loss by becoming dormant. Being diurnally active and exposed to radiant energy, they must often expend water for cooling, by panting. Consequently, in arid regions the distribution of birds is limited to areas within flying distance of water. Some water expenditure is avoided by allowing hyperthermia (up to 3°C.) in the daytime.

Bird urine (chicken, cormorant, house finch) has a maximum concentration equal to 0.3 M NaCl, which is on a par with most mammals. Ability to form hypertonic urines may depend on the loop of Henle component of nephrons (see EXCRETORY SYSTEM), present only in birds and mammals. Urine water is conserved in dehydrated chickens by filtration antidiuresis while uric acid, the main nitrogenous waste, continues to be independently secreted. Uric acid becomes supersaturated in kidney tubules (150 to 21600 mg%) where it gelates and then precipitates. Branches of the ureter are continuous with the tubules (there is no renal pelvis as in mammals) and strong peristaltic contractions milk the viscous, lumpy urine into the cloaca. Cloacal osmotic reabsorption of water may occur, but is not as important as tubular reabsorption.

(e) Desert rodents lead the most water-independent life of all vertebrates. Kangaroo rats can so reduce their evaporation that they are able to maintain water balance on only metabolic water. Other species survive on only metabolic water plus free water in air-dry seeds. Respiratory water loss is reduced by cool nasal mucosal surfaces, which condense water from warm air coming from the lungs, before it can be expired. Skin impermeability involves a physical vapor barrier in the epidermis, plus unknown physiological factors.

Only larger mammals are exposed to daytime radiant energy and need to dissipate heat by sweating, panting or wetting themselves with saliva (marsupials). These water expenditures must be balanced periodically by drinking. A dehydrated camel is particularly physiologically adapted to store heat (rather than dissipate it by evaporation), undergoing a temperature rise of up to 6°C. in the daytime.

MAMMALS are the most effective of vertebrates in conserving urine water, by concentrating the urine, which is achieved by reabsorption of water in the kidney tubules. Kangaroo rats attain maximum osmotic concentrations equal to 1.1 M NaCl plus 3.84 M urea (a highly soluble nitrogenous waste, in contrast to uric acid, which can be excreted in solution from the renal pelvis through the ureter). Hypertonicity of the urine is achieved in the collecting ducts of nephrons. In the presence of ADH, which makes the tubule membranes highly water permeable, water passes by osmosis from the tubules into the hypertonic tissue of kidney medulla and papilla. Kangaroo rats, which attain the highest concentrations, have a chronic high plasma ADH concentration (19 milliunits/ml. versus 0.0 mU for normal white rats and a maximum of 6 mU for dehydrated ones).

ROBERT M. CHEW

References

Chew, R. M., "Water metabolism of mammals," in "Physiological Mammalogy," Mayer, W. V. and R. G. VanGelder, eds., New York, Academic Press, 1963.
Jones, I. C. and P. Eckstein, eds., "Hormonal control of water and salt-electrolyte metabolism in vertebrates," Cambridge, Univ. Press, 1956.
Prosser, C. L., and F. A. Brown, "Comparative animal physiology," Philadelphia, Saunders, 1961.

WEB BUILDING

The origin of web building is seen either in the trailing drag-lines of primitive spiders or in the building of egg cocoons. As a next step silk-lined tubes with diverging threads at the entrance constitute first traps. The number of threads is then further increased to form a sheet leading to the tube. The sheet can possess a superstructure of tangled threads causing insects to fall. By a process of simplification, sheet developed into orb, the latter covering a comparatively large area with little silk. Building of two-dimensional orb-webs has been studied extensively while influences which govern construction of other spider-webs (irregular or three dimensional, sheet, funnel, triangular webs or combinations thereof) and their sequence of construction are practically unexplored. Different types of orbs are: complete orbs (e.g. *Araneus diadematus*), asymmetric orbs (e.g. *Metepeira labyrinthea*), orbs without hubs (e.g. *Theridiosoma gemmosum*), orbs with free sector (e.g. *Zillax-notata*), and orbs with hackled bands or stabilimentum (e.g. *Argiope aurantia*). All these are built in the same phasic sequence. Web-building is an innate pattern of behavior which is executed by a spider in a species specific manner almost daily throughout its life; the preferred web-building time is early in the morning before sunrise. The main purpose of web-building is to trap animals for food. The behavior pattern is considered innate because it is executed independently of learning. In evidence it was observed that young spiders built their first webs in final form without ever having seen another individual's web, and the web changed with age independent of routine; e.g. the free sector of a Zilla web is built only in later life and is preceded by full orbs. When Zilla was kept from building webs until adulthood, it immediately constructed a free sector. Web-building is best understood if it is realized that spiders let out a thread wherever they run. Phases of web-building are (for names of parts see half-schematic figure of unfinished Zilla-web): Phase I: construction of a bridge connecting two points horizontally. A bridge can be built by letting a thread drift freely until it catches a branch and then pulling it straight, or in stretching a thread straight across two arms of an angle. A Y-structure is achieved by the spider's letting itself drop vertically from the middle of the bridge. Phase II: Each radius is built together with a frame thread or auxiliary frame thread as a Y. All radii meet in almost one point in the closely knit hub area. Phase III: Wide turns are constructed from the hub towards the periphery as temporary fixation (Provisional spiral). Phase IV: The fine thread of the catching spiral is built by the spider, working inwards from the frame and running in loops or full circles towards the free zone which surrounds the hub. Simultaneously, the provisional spiral is eliminated. Phase V: The spider eats the white bundle of threads in the hub. It now settles down, keeping in touch with the completed web via radii or signal thread. Only the shape of the frame depends on the surroundings. Frequency of web-building and size of catching area are determined by outside stimuli, e.g. length of day, humidity, barometric pressure, abundance of food, and inner factors such as hunger, age, and sex. Temperature minimum before sunrise is known to be among stimuli releasing web-building in predisposed spiders. Interference experiments have shown that animals try to complete one phase before they proceed to the next, e.g. up to 150% of radii were built if some were destroyed during construction until this phase was finally exhausted, whereupon construction of a spiral was started on an incomplete basic structure. A sense of gravity regulates the relationship between the horizontal and vertical diameters of a web; e.g. turning an oval web 90° during its construction results in the building of an added area which re-establishes the diameters in the originally planned ratio. Blinded animals have built normal webs, hence vision seems unimportant in web construction. The following procedures have been described as ancillary techniques which spiders utilize during web building: probing for empty spaces, turning their bodies, pulling threads to measure tension or vibration frequency, estimating distances by spreading their legs or running a definite number of steps, and dropping the perpendicular onto a thread to establish the shortest distance. Material for web-silk is a scleroprotein which is secreted into and stored in glands in viscous liquid form. It hardens by mechanical stretching when drawn out of the spinnerets. Threads are simple or composite, of extraordinary strength and elasticity, and between 0.03 and 0.1 micron thick. Radii and frame threads come from the aciniform and ampullate glands and are thicker and less elastic than the spiral threads. Attachment discs (composed of a large number of minute loops) are produced by pyriform

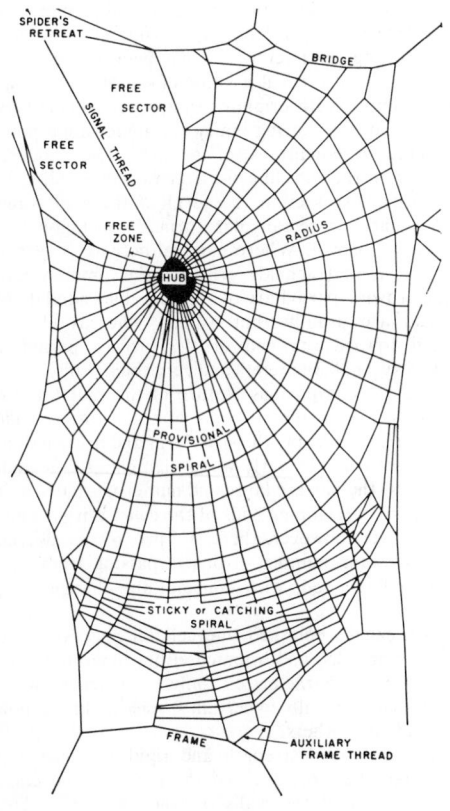

Fig. 1

glands. The spiral thread is coated with sticky fluid from aggregate glands. It is pulled out slowly by the fourth pair of legs and released with a jerk which results in formation of sticky globules along the spiral. Silk-glands are transformed coxal glands which fill a large part of an orb-weaver's abdomen. Spinnerets are vestiges of ancestral abdominal limbs of the fourth and fifth abdominal somites.

PETER N. WITT

References

Gertsch, W. J., "American spiders," New York, 1949.
Savory, T. H., "The spider's web," New York, Warne, 1952.
Tilquin, A., "La toile géométrique des araignées," Paris, 1942.
Wiehle, W. J., *in* "Die Tierwelt Deutschlands," vol. 23, Jena, 1931.
Witt, P. N., "Die Wirkung von Substanzen auf den Netzbau der Spinne als biologischer Test," Berlin, 1956.

WEISMANN, AUGUST (1834–1914)

Weismann developed one of the first tenable, though erroneous, theories of heredity which would account for the evolutionary facts established by Darwin. This theory is the "germinal plasm" in which it is stated that all heritable characters are carried in the plasm of the sperm and egg so that the physical make-up of the offspring is determined by the fusion of these. Since this theory involved a disbelief in the inheritance of acquired characters, he devoted much time to a series of experiments in which he endeavored by all possible means to induce a change in offspring through the mutilation of the parents. His failure to do this confirmed him in his views. His logical mind led him to inquire in what part of the cell the germ plasm might be localized if it were not the entire plasm itself, and he was therefore led to postulate—in the absence of any evidence whatever—that the chromosomes of sex cells might be considered the site of the germ plasm. The fact that he subsequently had a number of ideas of considerable less accuracy in this field does not deprive him of the claim to be the ancestor, if not the father, of contemporary genetic theory.

PETER GRAY

WOHLER, FRIEDRICH (1800–1882)

This German chemist made only one contribution to biology but it was of profound significance. In 1828 he synthesized urea from inorganic compounds and thus destroyed what was at that time an article of faith —that organic compounds could only be produced by the action of "vital force." The significance of this breakthrough does not appear to have impressed him and he devoted the rest of his life to studies in inorganic chemistry. A fuller account of the man and his work will, therefore, be found in the Clark-Hawley "Encyclopedia of Chemistry." (New York, Reinhold, 1957)

PETER GRAY

WOOD

Introduction. To the botanist "wood" is a collective term, synonymous with secondary XYLEM, which comprises cellular elements derived from the activity of the vascular CAMBIUM. These elements function jointly in conduction and storage, and in imparting strength and rigidity to the plant.

The bulk of the cellular components of wood are aligned with their long axes roughly parallel to that of the stem or branch in which they occur. However, permeating the wood are ribbon-like aggregates of cells termed xylem rays, which are aligned transversely.

Both living and dead elements occur in wood. Living or parenchymatous cells are found in the xylem rays, and occur also in the form of short, vertically-aligned series of elements, termed xylem parenchyma. However, the bulk of the secondary xylem comprises dead elements collectively termed tracheary cells which, despite a common origin, manifest considerable structural heterogeneity.

Coniferous woods consist largely of tracheids: elongate, imperforate elements with tapering end walls, which function jointly in conduction and support. Tracheid length ranges from approximately one millimeter to one centimeter, and has been found to vary with lateral and vertical position of sample in tree, with age of sample, with origin or sample (clear bole versus branch wood), following injury and under abnormal or unfavourable conditions of growth. The tangential diameter of tracheids likewise may vary markedly.

In transverse section the form of tracheids ranges from square or rectangular through pentagonal and hexagonal to oval. Where cambial activity is seasonal or intermittent, tracheids formed at the commencement of the growing period are of considerable radial diameter and are usually thin walled, whereas elements formed toward the end of the growth period are marked by small diameters and very thick walls.

Pits are a conspicuous feature of the radial and occasionally the tangential walls of tracheids, considerable diversity marking the size, form and distribution of these structures. In early-wood elements (large radial diameters) the pits are large and numerous, particularly toward the overlapping ends of the cells. Within a given growth layer, however, there is a progressive decrease in the size and number of pits in passing from early-wood to late-wood, with concomitant changes in pit structure.

Resin canals, enlarged intercellular spaces jacketed by secreting cells, and oriented either longitudinally or transversely, characterize some coniferous woods. These occur naturally in certain species and in response to wounding in others.

The mechanical strength and rigidity of wood derives from the structure and the physical and chemical properties of the walls of tracheary cells. These walls comprise a multi-layer structure consisting of an outer tenuous primary wall, and an inner three or more-layered secondary wall.

The primary wall of coniferous tracheids comprises a coherent porous matrix of cellulose microfibrils of varying diameter and orientation. This matrix is heavily impregnated with lignin, pectic substances and hemicelluloses. The several layers of the secondary wall are also characterized by an anisotropic cellulose matrix. Within a given layer the microfibrils are arranged parallel to one another in preferred orientations. The fibrils of the tenuous inner and outer layer are arranged approximately transversely to the long axis of the cell, whereas those of the prominent central layer show longitudinal orientation.

Further structural heterogeneity derives from the non-uniform distribution of cellulose in the central layer of the secondary wall. Variation in the density or porosity of microfibrillar aggregation leads to the formation of concentric lamellae or radial patterns of varying complexity. The interstices comprise a microcapillary network oriented parallel to the microfibrils, in which lignin, hemicelluloses and other complex organic material are deposited.

Dicotyledonous woods are structurally complex, and in the following paragraphs appeal is made to evolutionary sequences in treating this complexity.

In dicotyledonous woods, marked structural differences distinguish supporting from conducting elements. Units especially modified for conduction are called vessels. These cells consist of vertical series of segments manifesting considerable variation in length, diameter, form and wall sculpture. In cross section, vessel segments range from angular to circular with diameters of 50 to 400 μ. Laterally they vary from drum or barrel-shaped to linear.

Vessel members, in contrast to tracheids from which they are phylogenetically derived, are perforate elements. In primitive woods adjacent members are small in diameter with extensively overlapping end walls, and scalariform (ladder-like) perforation plates with numerous bars. In more advanced woods the amount of overlapping is less, the number of bars reduced, the structural characteristics of the perforation areas modified and the diameter of vessel segments increased. The trend culminates in short, barrel-shaped segments with almost transverse end walls and a single large perforation. Concomitant changes in the types of lateral pitting in vessel members accompany the foregoing trend.

DICOTYLEDONOUS woods of temperate climates are divided into ring porous and diffuse porous types on the basis of vessel distribution. In the ring porous structure the early-wood is marked by the presence of vessel members of extremely large diameter with a pronounced decrease in the diameter of later formed vessel segments. In diffuse porous woods, however, the vessel segments within a specific growth layer are of approximately uniform diameter, vessels occurring singly or in clusters.

Concomitant changes have marked the evolutionary development of fibrous, imperforate, supporting elements in dicotyledonous woods. These modifications include a gradual increase in wall thickness, a progressive decrease in cell diameter and pronounced changes in pitting. In many cases these supporting elements retain their protoplasts at maturity and assume both a storage and supporting function.

The cell walls of vessel members and fibrous elements are basically similar to the walls of coniferous tracheids. Minor differences include deviations from the normal three-layered structure, and variations in the distribution and amounts of non-cellulosic materials in the secondary wall.

In dicotyledonous woods living cells are found in xylem rays and in longitudinal parenchyma strands. The evolutionary sequence in parenchyma distribution is from diffuse in primitive woods through aggregate apotracheal to paratracheal in specialized forms. Xylem ray structure and distribution is extremely variable and complex, with evolutionary sequences again being evident.

K. N. H. GREENIDGE

References

Bailey, I. W., "Contributions to Plant Anatomy," Chron. Bot., 15: 1–262, 1954.

Brown, H. P., et al., "Textbook of Wood Technology," 1st ed., Vol. 1, New York, McGraw-Hill, 1952.

Metcalfe, C. R. and L. Chalk, "Anatomy of the Dicotyledons," 2 vols., Oxford, Clarendon Press, 1950.

XIPHOSURA

The Xiphosura (king crabs) are marine arthropods of large size (up to 24 inches) which are placed, together with the Arachnida and the Gigantostraca, in the subphylum Chelicerata. They are distinguished by possessing a single pair of preoral appendages, the chelicera, and by the fact that the anterior region of the body, or prosoma, is produced by the fusion of the preoral segments with at least five posterior segments. This region is separated from the posterior part of the body, or opisthosoma, by a transverse articulated groove. The body ends with a projecting caudal spine.

The body which is entirely covered with a thick coat of olive-green chitin is thus formed of 15 segments, more or less fused together, and of a caudal spine projecting from the last segment. The prosoma bears, in addition to the chilicerae, five pairs of walking feet. All the appendages terminate with a chela though in this respect the fifth pair are incomplete. The opisthosoma bears severn appendages: the chilaria, which lie on the posterior margin of the mouth; the genital operculum, a membranous disc which lies over the two sexual apertures and five pairs of legs carrying gills on their posterior surface.

The nervous system is remarkable for its extreme condensation and by the great development of stalked bodies, as well as by the fact that it lies for the most part enclosed within the arterial system. There are two lateral eyes and one pair of median ocellae.

The mouth, which lies unusually far back, opens into the esophagus which leads directly to a gizzard which lies anterior to it, and in which the food is ground up. The inedible large fragments are pushed out of the mouth while the digestible paste passes into the stomach and thence to the intestine. A large pair of branched hepatic sacs open by four holes into the anterior region of the intestine.

The dorsal heart is furnished with eight pairs of ostioles provided with valves. The indigo blue blood, after having bathed the tissues, reaches a system of lucunae which take it back to the gills. From these, it reenters the pericardium and returns to the heart.

The sexes are separate. In the breeding season, males and females come together in shallow water. The males climb on the back of the females and hold themselves there with the aid of the hooks on the first pair of feet. The female then digs a hole in the sand and deposits its eggs in this hole where they are subsequently moistened with sperm by the male. The eggs measure two to three millimeters in diameter and two to three hundred are shed at one time. Many such nests are thus left in the intertidal zone but, since the eggs are buried in the sand, they remain constantly moist.

The newly hatched larva is about one centimeter long and has the appearance of a trilobite. It swims on its back, as does the adult, and is capable of burying itself in the sand. After the first molt, the telson becomes elongated and the abdominal feet are fully formed. In each successive molt, the animal develops more and more the shape of the adult. There is no doubt that the Xiphosura are related to the trilobites in spite of the presence of antennules and of biramous appendages in the latter; on the other hand, they may be classified with the Gigantostraca in a single class, the Me-

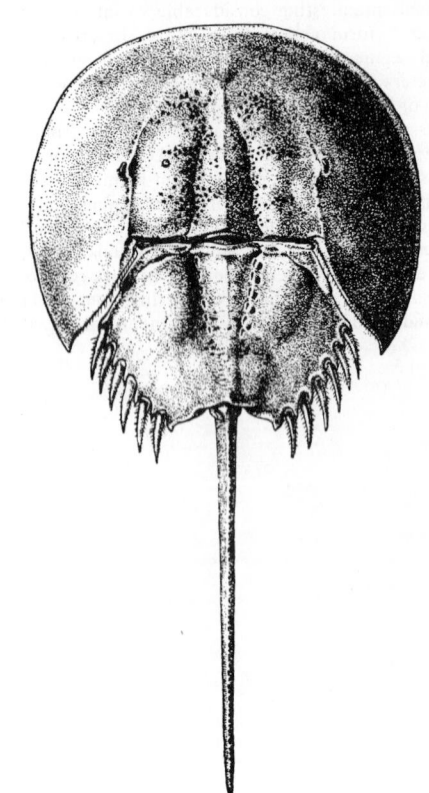

Fig. 1 *Xiphosura polyphemus* (L.) (Male). Dorsal side. (After Van der Hoeven.)

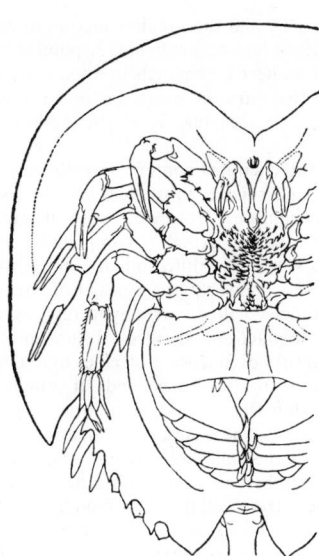

Fig. 2 *Xiphosura polyphemus* (L.) (female). Ventral side. (After Van der Hoeven, slightly modified.)

rostomacea. Nevertheless, even if it is admitted that these two groups have developed on independent lines from the main trunk of the trilobites, their evolution has been very different. Whilst the Xiphosura are extant and have remained marine, the Gigantostraca gave rise to terrestrial forms, happily no longer living, but in which the ancestors of the arachnida can be recognized.

Present day Xiphosura are allied species, of littoral habitat, and essentially fossorial by nature. They swim upside down—that is, belly upwards—with the aid of their abdominal appendages. All are carnivorous, existing principally on mollusca.

The five species alive at the present time belong to three genera: the genus Xiphosura with a single species (*X. polyphemus*) is found on the Atlantic American coast from Nova Scotia to Yucatan and in the Gulf of Mexico and the Caribbean. The genus *Tachypleus* has three species: *T. gigas* occurs in southeast Asia, in the Torres Strait, at Singapore, and in the Gulf of Siam; *T. tridentatus* is widely distributed over the Sea of China as far as the coast of Japan; *T. hovena* is confined to the Moluccas. Finally there is the genus *Carcinoscorpius* with a single species found in the Gulf of Bengal, the Gulf of Siam, southeast Asia and the Philippines.

Thus the existing Xiphosura are confined in two isolated regions: the North American Atlantic on the one hand and the Far East Indo-Pacific area on the other. But the suborder is found in the Devonian and the fossil remains indicate clearly that they were once dispersed over the whole of the North Atlantic and Eurasiatic coast.

LOUIS FAGE
(trans. from French)

References

L. Störmer, "On the Relationships and Phylogeny of fossil and recent Arachnomorpha," Norsk, Vidensk. Akad. Oslo; 1944, **1**, 158 p.

XYLEM

The term xylem is from the Greek *xylon* meaning WOOD. In the plant body, xylem comprises the chief mechanical and conductive tissues, and functions to a lesser extent in food storage. The xylem is a "complex tissue" and certain of its cells are modified in form relative to their function(s). At maturity, that is, in the functional condition, most cells of the xylem are dead, and all that remains is the rigid cell wall. Xylem study, therefore, is largely the investigation of the non-living walls of woody cells—form, function, chemical composition, and physical and mechanical properties. Specialized studies of these aspects may be considered under separate topics, in order: wood anatomy, plant physiology, wood chemistry and wood technology.

Wood anatomy, the study of form in wood, is concerned primarily with the microscopic structure of the cells and tissues which comprise wood, and to a somewhat lesser extent with the origin of wood in the development of the plant. Wood or xylem is produced by two kinds of generative regions in the plant, one located at the tips of root and shoot, and the other ensheathing the stem and root between the bark and wood. These generative regions are termed meristems, and also produce tissues other than xylem. The meristems at the tips of the plant body are the apical meristems. They are responsible for the growth in length of the plant, and produce primary xylem. The ensheathing meristem referred to above is the vascular cambium. It is responsible for growth in diameter of the plant, and produces secondary xylem. Xylem is found in both herbaceous and woody plants, the former ordinarily being characterized by the presence of primary xylem, the latter by the presence of both primary and secondary xylem.

Most of the volume of STEMS and ROOTS in woody plants, i.e., trees and shrubs, is occupied by xylem, and the great preponderance of this xylem is secondary xylem produced by the vascular cambium. In herbaceous plants, by and large, probably less than half of the volume of stems and roots is occupied by xylem, and this xylem is mostly primary xylem produced by the apical meristems. The xylem body of plants consists of two systems, an axial system oriented parallel to the axis of plant organs, and a radial system oriented perpendicularly with the axis of plant organs. The axial system consists of cells which function in the conduction of water and dissolved minerals, cells which are largely mechanical (supportive) in function, and cells which serve in the storage and translocation of elaborated foodstuffs. Cells of the radial system are concerned only with the storage and translocation of food.

Cells functional in conduction. Vessel cells (often called vessel elements or vessel members) are part of the axial system and are elongate in shape, possess pitted, more or less thickened walls, are non-living at maturity, exhibit variously specialized, but always open, end walls (i.e., cells are perforate), and are united end-to-end with similar cells to form a continuous cylinder designated a vessel. Vessels characterize the conductive systems of flowering plants (ANGIOSPERMS).

Cells functional in conduction and support. Tracheids are elongated cells which are part of the axial xylem system, possess more or less thickened, pitted walls, are non-living at maturity and differ sharply from vessel cells in that they do not possess open end walls (i.e., cells are imperforate) and are never united end-to-end with other cells. Tracheids are usually the sole conducting cells of cone-bearing plants, ferns, horsetails, club mosses and other lower vascular plants (i.e., GYMNOSPERMS and PTERIDOPHYTES). Tracheids ordinarily serve two functions in contrast to vessel cells: conduction and support.

Cells functional in support. Cells in this category are part of the axial xylem system and commonly show thick, pitted walls, are fusiform (spindle shaped) in shape, dead at maturity, do not possess open end walls (i.e., cells are imperforate) and, together with the tracheids in gymnosperms, comprise the major part of the mass in secondary xylem. Cells in this grouping are commonly referred to as "fibers," a term of convenience which includes all imperforate cells of the xylem, and sometimes even vessel cells.

Cells functional in food storage and translocation. Cells of this kind may be part of the axial xylem system or may form the radial system. Collectively they are called parenchyma cells. Parenchyma cells differ markedly from cell types mentioned previously in that they are living at maturity and are roughly isodiametric in shape (not elongated). Their walls may or may not be thickened, but are mostly pitted. Parenchyma cells which form the radial xylem system are aggregated into tissues known as vascular rays. Parenchyma cells which are oriented parallel to the axis of the plant are united into tissues termed axial parenchyma. Vascular rays occur in the secondary xylem of angiosperms and gymnosperms. Axial parenchyma, although a prominent feature in the xylem of most woody angiosperms is a minor constituent in the wood of gymnosperms and indeed is commonly lacking entirely in these plants.

An idealized section of the secondary xylem in an angiosperm would show that the bulk was composed of imperforate cells, or "fibers," which taken together make up the main supportive tissue. "Embedded" in this groundmass would be vessels and axial parenchyma arranged in a manner parallel to the imperforate cells, and radiating from the center of the plant organ, the vascular rays composed of ray parenchyma cells. The picture would be much the same in the secondary xylem of a gymnosperm, with exceptions noted in the total absence of vessels, and the great reduction in amount, or lack, of axial xylem parenchyma. A point which might be emphasized here is that cells in wood are not isolated units, and they are in more or less direct contact through pits, or thin places, in the walls of adjacent cells.

Most of the chemical content of walls in woody cells is cellulose and related substances which together are called holocellulose. However, were it not for the presence of a second group of compounds, the lignins, woody cells would not differ much from cotton fibers. Holocellulose is made up of a number of chemically related substances called carbohydrates, examples of which include glucose and starch. In its simplest form, cellulose consists of glucose molecules united by oxygen "bridges" to form a chain. Strands of cellulose are aggregated within the cell walls to form larger strands called micelles. These micelles make up the basic framework of the cell wall. Lignin, which is a generic term for a number of similar substances, is chemically quite different from cellulose. Although it contains carbon, hydrogen and oxygen as do the carbohydrates in cellulose, the arrangement into molecules and the proportions of these elements make lignin a peculiar substance still incompletely understood. In any event, it is the presence of lignin which imparts to wood its special character. The lignin of the cell wall is very intimately related to the cellulose, but it is not chemically combined with it in any manner. There may be a number of other lesser substances in the cell wall, chief among which is perhaps silicon. Wood also may contain other minor components not directly associated with the cell wall; nevertheless, these often give distinct characteristics to the wood: oleoresins, terpenes, fats and tannins along with certain dyestuffs as haematoxylin, brazilin and lapachol may be present.

WILLIAM L. STERN

References

Brown, H. P., *et al.,* "Textbook of wood technology," 2 vols., New York, McGraw-Hill, 1949–52.
Esau, K., "Plant anatomy," New York, Wiley, 1965.
Jane, F. W., "The structure of wood," New York, Macmillan, 1956.
Metcalfe, C. R., and L. Chalk, "Anatomy of the dicotyledons," 2 vols., Oxford, Clarendon Press, 1950.
Wise, L. E. and E. C. Jahn, eds., "Wood chemistry," 2 vols., New York, Reinhold, 1952.

Z

ZINGIBERALES

This order (which corresponds with the Scitamineae of many authors) consists of six families (78 genera, 1400 spp.) of tropical and sub-tropical perennial plants, mostly herbaceous but sometimes somewhat woody, ranging in size from small to gigantic. Members are rhizomatous and have sheathing leaves, the sheaths usually open, either distichous or spiral in disposition. Flowers are of advanced structure, homo- or heterochlamydeous, usually strongly zygomorphic, the androecium often petaloid or reduced from the basic 3 + 3; they are generally aggregated in infloresences and subtended by more or less elaborate (often conspicuous) bracts. The ovary is inferior, trilocular; the fruit 1–many seeded, dry or fleshy, dehiscent or not, the seeds commonly rather large and hard, sometimes arillate.

There is some disagreement among authors as to the limits and affinities of the six families of Zingiberales; thus Lane and others unite Strelitziaceae with Musaceae. One view of the group is represented by the following key (condensed from Simmonds):

1a. Androecium not petaloid, stamens 5–6; 2a. Leaves spiral, fruit many seeded, baccate—Musaceae. 2b. Leaves primarily distichous, fruit dry or, if fleshy, few-seeded; 3a. Venation parallel, sepals not tubular—Strelitziaceae. 3b. Venation reticulate, sepals tubular—Lowiaceae. 1b. Androecium petaloid, stamen solitary. 4a. Plants aromtaic, leaves ligulate, spiral or distichous—Zingiberaceae. 4b. Not aromatic, leaves not ligulate, distichous. 5a. Leaves pulvinate, ovules solitary—Marantaceae. 5b. Leaves not pulvinate, ovules numerous—Cannaceae.

Musaceae is distributed in the Old World from West Africa to mid-Pacific and contains two genera (*Musa, Ensete*) and about 40 spp.—the bananas, wild and cultivated[2]. The major group of edible bananas originated by the evolution of vegetative parthenocarpy and polyploidy from *Musa acuminata* and *M. balbisiana* (section Eumusa) in South-East Asia. The Fe'i bananas of the Pacific islands originated independently, probably in the New Guinea area, from one or more species of the section Australimusa. *Musa textilis* yields the Manila hemp of commerce. *Ensete ventricosum* is a minor food and fibre plant of upland Abyssinia. Strelitziaceae (4 genera, 160 spp.) is distributed in warm countries of both hemispheres and contains a few ornamentals (*Ravenala, Strelitzia*) but no significant economic plants. Lowiaceae (only genus *Orchidantha* (2spp.)) is native to Malaya and Borneo. Zingiberaceae is distributed in both hemispheres but is predominantly Asiatic; it contains about 45 genera (800 spp.) which yield many ornamental and spices (e.g. *Elettaria, Aframomum, Zingiber, Curcuma,* etc.). Marantaceae (about 25 genera, 300 spp.) is likewise distributed in both hemispheres but is predominantly American; it yields two minor root crops (*Maranta arundinacea*—Arrowroot; *Calathea allouia*—Topi tamboo) and some ornamentals. Cannaceae contains but one genus (*Canna*, 60 spp.) distributed in both hemispheres; it yields many ornamentals (the gaily coloured "petals" of *Canna* are androecial in origin) and a minor root crop (*Canna edulis*).

N. W. SIMMONDS

References

Lane, I. E., Mitt. bot. Staats. München, **13**: 114–31, 1955.
Simmonds, N. W., "Bananas," New York, Humanities, 1966.
Moore, H. E., Baileya," **5**: 166–94, 1957.

ZOOGEOGRAPHY

Zoogeography, or animal geography, is a field, scarcely a separate science, that synthesizes data drawn from various branches of knowledge for the purpose of explaining the facts of animal distribution. From the earliest hunter many millennia ago to the most sedentary urban dweller of today there has been common recognition that different places have different faunas. It is generally recognized, also, that the faunas of today differ from those of the past. No one expects to see a live whale on a mountain top, nor a browsing giraffe in Alaska, but whale bones protrude from terrestrial cliffs and rhinoceros carcasses lie frozen in Siberian muskeg.

Such facts are easier to observe than to explain. Efforts to do so have been largely associated with three approaches: *descriptive* (an aspect of general natural history); *historical* (a mixture of geology, paleontology and evolution); and *interpretive* (ecology on a broad scale). Since about 1942, when Ernst Mayr's *Systematics and the Origin of Species* appeared, the earlier approaches have become so merged and supplemented that recent contributions have become essentially *interdisciplinary* or *multidisciplinary.*

The classical approach to zoogeography was purely descriptive. From Aristotle on many writers contributed information on what animals occurred where until by the mid-nineteenth century the essential distribution of conspicuous species was known and generalizations were being attempted. In 1858 terrestrial regions based upon the known facts of bird distribution were mapped by P. L. Sclater, and closely similar regions derived from reptile distribution were delimited by Alfred Günther. Two mutually helpful developments then occurred: the publication, in 1859, of Darwin's *Origin of Species*, which greatly stimulated intensive biological investigation of the nooks and crannies of the world; and the studies of Alfred Russell Wallace, culminating in the publication, in 1876, of his two-volume compendium, *The Geographical Distribution of Animals*, followed in 1880 by *Island Life*, in which the useful differentiation of "continental" and "oceanic" islands was proposed. The time of Wallace was the heyday of descriptive zoogeography, when subdivision of the earth's surface into faunal realms, regions, subregions, and provinces led to heated controversies over the relationships of such constructs and the placement of dividing lines. Three realms were commonly recognized: *Arctogaea*, encompassing the Palearctic, Nearctic, Oriental, and Ethiopian regions; *Neogaea*, including only the Neotropical Region; and *Notogaea*, including the Australian Region. Heilprin subsequently proposed the convenient term "Holarctic" for the combined Palearctic and Nearctic regions. The overconcentration of many workers upon regional classification appears in retrospect to have been largely sterile, and certainly unwarranted in its exclusive emphasis upon existing faunas without consideration of their origins and dispersal. (It should not be inferred, however, that biological exploration is no longer needed. The faunas of *most* terrestrial areas are still imperfectly known, many invertebrate groups are critically in need of taxonomic investigation, and real exploration of the ocean depths is just beginning.)

Darwin and Wallace were far more scientific in one matter than the following generation of zoogeographers. They believed firmly in the stability of continents whereas from 1880 to 1915 it became the fashion for nearly every writer to postulate a former land connection every time related creatures were found to be separated by a water barrier. Geologists used "facts" of animal distribution to buoy up hypothetical continents and zoologists utilized geological assertions as piers for their land bridges. So many land connections were proposed during this period that it has been computed that scarcely a cubic mile of existing ocean escaped displacement. Arguments raged over Gondwana Land, Archhelenis, Pacelia, Atlantis, and many others; fortunately, most of these have been submerged in the literature of science although Atlantis occasionally bobs up in Sunday supplements. A few breaks and reconnections in Central America, at Bering Strait, and along the island chain from Malaya to Australia suffice for most modern zoogeographers and few, if any, would countenance any land bridge not solidly supported by sound geology.

Historical zoogeography began in 1896 with the publication of *A Geographical History of Mammals*, by R. Lydekker, who fully understood that present distribution could be explained only in the light of past distribution and evolutionary history. Osborn's *The Age of Mammals*, in 1910, benefitted from rapidly accumulating paleontological data, but it remained for William Diller Matthew to lay the unshakable foundation for historical zoogeography with the publication of *Climate and Evolution* in 1915. Matthew's detailed knowledge of fossil mammals enabled him to demonstrate that mammals, at least, needed no trans-Atlantic or trans-Pacific bridges to reach their present habitations. Periodic uplift of present continents with associated climatic changes, followed by extensive erosion, base-leveling, invasion by shallow seas, and climatic geniality, provided an evolutionary pulse; most major "new models" of mammals originated in the Northern Hemisphere (Holarctic Region) and spread east or west across a Bering Strait bridge whenever it was in existence and southward along the great continental superhighways; old styles, resistant to change, moved as the climate moved or were shunted into peripheral areas or habitats by rapidly evolving improved models.

Climate and Evolution alone might not have initiated a new era in zoogeography in spite of its global sweep and scholarly perspective of animal dispersal during a long span of geological time. Matthew's influence upon his associates was equally potent and many American zoogeographers have been his students or disciples. Much additional work by Romer, Camp, Noble, Ruthven, Schmidt, Dunn, Myers, Simpson and others has served to correct inaccuracies and to ratify the basic concepts. Darlington favored the tropics as the centers of dispersal for many cold-blooded vertebrates without challenging the main thesis: others assert that the Holarctic itself was subtropical when many groups arose.

George Gaylord Simpson, the most productive Matthewsian of all, with an incomparably broad command of the now far more extensive data of mammalian paleontology, has both enriched and enlivened the literature of zoogeography with apt terms and unforgettable cartoons. His numerous writings on routes of faunal interchange have treated and illustrated *corridors*, "along which the spread of many or most of the animals of one region to another is probable;" *filters*, "across which spread of some animals is fairly probable but spread of others is definitely improbable;" and *sweepstakes*, "across which spread is highly improbable for most or all animals, but does occur for some." Simpson has computed the odds for colonization in the Hawaiian Sweepstakes, for example, as 1 to 200,000 per year for land snails, 1 to 20,000 per year for insects. His Condon Lectures, published in 1953 under the title *Evolution and Geography* (from which the above definitions were quoted), is the best brief introduction to the historical geography of mammals.

The very year that Matthew revivified historical zoogeography, an appealing hypothesis created a diversion that led many astray. Alfred Wegener, impressed by the similarities in the continental outlines of eastern South America and western Africa, offered as an alternative to a trans-Atlantic bridge the hypothesis of continental drift, expanded, in 1924, in *The Origin of Continents and Oceans*. The hypothesis was espoused by those ungrounded in Matthew, and the literature on the subject is now tremendous. It is not yet possible to prove that never in earth history did a huge, combined continental mass occur, break into continents and slowly separate, but there is abundant evidence that drifting back and forth accordion fashion (a later modification), could not have happened since the Triassic. The controversy will eventually be resolved by geologists; meanwhile, there is growing evidence that Europe-Africa and the Americas have been moved slowly apart by the upwelling of magma along the mid-Atlantic Ridge.

The interpretive or ecological approach to problems of animal distribution was foreshadowed in Ludwig Schmarda's *Die geographische Verbreitung der Thiere* (1853), and ecological plant geography was launched as early as 1898 by A. F. W. Schimper in *Pflanzengeographie auf physiologischer Grundlage*. The first comprehensive treatment of ecological zoogeography was not published until 1924, when Richard Hesse's *Tiergeographie auf oekologischer Grundlage* appeared. This epochal work, issued in English in 1937 (revised edition, 1951) as *Ecological Animal Geography*, by Richard Hesse, W. C. Allee and Karl P. Schmidt, remains the standard reference for all interested in relating animal distribution patterns to the broad biotic areas determined by vegetation—forest, tundra, desert, etc.

For a more adequate review of this subject reference should be made to "Animal Geography," by Karl P. Schmidt, in *A Century of Progress in the Natural Sciences—1853-1953.*

Nineteenth-century zoogeographers, hampered by limited techniques and inadequate data, exercised great ingenuity in making sweeping inferences. Now, isotope dating, ocean-bottom core sampling, soil washing for complete recovery of fossil assemblages, and many other new research tools are being utilized. Highly pertinent studies of speciation, faunal interchange, refugia and redispersals, isolating mechanisms, thermal regulation, photoperiodism, dendrochronology, and climatic shifts are abundantly available. Pleistocene studies have multiplied, pollen stratigraphy has flourished, and prehistory has been correlated with biogeography. An increasing number of workers blame man, and concommittant increase in fire frequency, for the relatively recent and biologically abrupt extinction of many large vertebrates.

M. GRAHAM NETTING

References

Darlington, P. J., "Zoogeography," New York, Wiley, 1957.
Hubbs, C. L., ed., "Zoogeography," Washington, Amer. Assoc. Advan. Sci., 1958.

ZOOLOGICAL GARDENS

Historically, wild animals have been kept in captivity, privately and publicly, since the earliest days of recorded history.

Formally organized zoos, so-called *living museums*, identified as zoological gardens, zoological parks, menageries and aquariums, have been in operation for 150 years. Corporate non-profit organizations associated with zoos and identified as Zoological Societies have made zoo operation a scientific program. Notable and representative are the Zoological Societies of Antwerp, Chicago, London, New York, Philadelphia and San Diego. There are many other active societies in other parts of the world.

Official organizations. There is an International Union of Directors of Zoological Gardens. The Secretary, Dr. Wilhelm Windecker, is Director of the Cologne Zoological Garden, Cologne, Germany.

The American Association of Zoological Parks and Aquariums, affiliated with the American Institute of Park Executives, has a central office at Oglebay Park, Wheeling, West Virginia. All those associated with the management of zoological parks and aquariums, public and private, in the Americas are eligible for membership. For the most part membership consists of those active in the field in the United States and Canada.

Publications. A small publication, privately printed, known as "International Zoo News" is published by G. Th. Van Dam, Zoo-Centrum, Zeist, Holland, and is an excellent source of zoo activities world-wide. The American Association of Zoological Parks and Aquariums publishes papers in the Journal of Parks and Recreation, an organ of the American Institute of Park Executives, and the A.A.Z.P.A. separately publishes a monthly Newsletter. Zoological societies and many zoos distribute both scientific and popular publications. Representative are,

"Proceedings of the Zoological Society of London" and "The Transactions," Zoological Society of London, Regent's Park, London, N. W. 1, England.

"Zoologica," New York Zoological Society, 185th Street & Southern Boulevard, New York 60, N.Y.

"Der Zoologische Garten," official organ of the International Union of Directors of Zoological Gardens, Steindamm 9, Hamburg 1, Germany.

"Artis," Koninklijk Zoologisch Genootschap, Natura Artis Magistra, Plantage, Kerklaan 40, Amsterdam, Holland.

"Zoo" Societé Royale de Zoologie D'Anvers, 26, Place Reine Astrid, Antwerp, Belgium.

Some popular publications are:

"Animal Life," published by the Zoological Society of London, Regent's Park, London, N. W. 1, England.

"Animal Kingdom," published by the New York Zoological Society, 185th Street and Southern Boulevard, New York 60, N.Y.

"Zoonooz," published by the Zoological Society of San Diego, P.O. Box 551, San Diego 12, California.

"Animals and Zoo," published by Tokyo Zoological Park Society, Ueno Zoological Gardens, Tokyo, Japan.

"I Giardino Zoologico," Roma, Viale del Giardino Zoologico 20, Rome, Italy.

"Freunde Des Kolner Zoo," Zoological Garden, Köln, Germany.

"Ziva," Zoological Garden of Prague, Prague, Czechoslovakia.

and there are many others.

Most zoos publish guide books which quite adequately describe and illustrate the animal collection.

In 1960 the Zoological Society of London, Regent's Park, London, N. W. 1, England, published an International Zoo Year Book which included the official name and address of zoos throughout the world. It also includes the name of the director, governing body, source of financial support, admission price, number of visitors annually, physical size, number of employees, number of species and specimens in each animal class, number and kinds of buildings, whether the zoo has animal rides, an aquarium, a botanical section, a children's zoo, a children's club; something about its educational facilities; whether it has a film unit; whether they produce films; whether there is a pathology department, a library, a radio or TV program, a veterinarian and veterinary section, a research division, and other miscellaneous activities. Also included are the names of the zoo's publications and frequency of appearance; successful breeding of rare species; the names of the senior staff, appointments, retirements and deaths.

The American Association of Zoological Parks and Aquariums biannually publishes a roster of zoos and aquariums in the United States and Canada. This small, factual manual quite well covers the Americas.

Exhibits. Zoos, for the most part, are maintained for public recreation, and secondly for educational and scientific purposes. Most zoos exhibit representative collections of both mammals and birds; many include reptiles and amphibians; and some have aquariums and insectariums associated with, or a part of, the zoo. The largest zoos exhibit approximately 3000 specimens, often over 4000, exclusive of fish and insects, representing 800 to 1200 species. There is no set pattern for *Class* representation. In some zoos 50% of the total collection is birds; in others half the total may be mammals. Some excel in large animals while others have large collections of very small mammals, like rodents. The reptile collections are usually smaller in both species and number.

Most zoos make great effort to have world-wide representation and also show great rarities, such as okapis, Indian and White rhinoceroses, platypuses, kiwis, Komodo dragons, Mountain gorillas, koalas, Giant pandas, and Emperor penguins. However, all actively engage in conservation programs, and some make an effort to reproduce in captivity those species facing extinction in the wild.

The largest animal collections have a total value of about half a million dollars. The most valuable animals are the Giant panda, the Komodo dragon, the Indian and White rhinoceroses. Many other rarities cannot have a value placed upon them, such as kiwis, koalas and platypuses.

Housing. Revolutionary changes are taking place in the design of structures for the exhibition of wild animals. Every effort is being made to permit visitors an unobstructed view of animals leading to the elimination of bars and wire; to bring the specimens up forward for close observation; to associate the animal with plantings typical of its native habitat; to use structural material impervious to urine and feces; to minimize unpleasant odors by the use of ceramic tiles, structural glasses, ceramic coated metals, stainless steels and anodized aluminum. Plastic sheets and plastic coatings are being used extensively and successfully. It has been shown that radiant floor heat, by means of warm gas, hot water or other liquids, plastic or lead covered electric cables, or electric heated copper tubing, is the ideal, the purpose being to avoid animal heat loss by conduction, convection and radiation. This calls for the use of styrofoam sheets or aluminum foil built in as vapor barriers to permit floors, walls, and ceiling to reflect the temperature of the room or enclosure, and not present cold surfaces. Of equal importance is the avoidance of draft or chimney effect in the holding or exhibit quarters.

Size. Zoos vary widely in size from the largest 650 acre Whipsnade Zoo, forty miles from London, operated by the Zoological Society of London, to the very smallest roadside zoos of less than an acre. So-

called children's zoos, for the most part associated with larger institutions, cover areas from one-third to approximately ten acres. The average major zoo varies in size between forty and one hundred and fifty acres.

The visitor. Zoos of average size throughout the world cater to from 500,000 to a million visitors, but the largest zoos have two or more million patrons annually. Greatest attendance is in the summer period of July and August in the Northern Hemisphere, and January and February in the Southern Hemisphere. The largest zoos have from twenty to fifty thousand visitors per day as a peak. The present greatest problem for all zoos, world-wide, is transportation—means of carrying the visitor to the zoo, accommodating his car, and transporting him about the zoo. Communicating the story about the animals to the visitor in an entertaining, effortless manner calls for new and improved techniques.

Feeds and feeding. Many industrial nutrition laboratories, in addition to those in academic institutions, are actively engaged in studying the nutritional needs of all classes of animals.

Commercially prepared rations for domestic poultry, swine, cattle, dogs and fur bearers are being successfully fed to captive wild animals. Special balanced monkey rations are sold commercially as compressed pellets. Alfalfa and other hays and special grain formulas are prepared as pellets and are accepted by most herbivores.

Food costs of the largest collections approximate one hundred thousand dollars annually.

Research and health. Comparative research is gaining momentum, especially in the fields of behavior, anatomy, histology, protozoology, mycology, bacteriology, virology, pathology, hematology, serology, parasitology and biochemistry. The study of infectious disease, especially host resistance and susceptibility, is attracting the attention of national and international health bodies, including the investigation of the relationship of diseases of captive wild animals to the diseases of man. Reference to the international medical literature in this field can be found in Index Veterinarius, Commonwealth Agricultural Bureaux, Farnham Royal, England. Another excellent source is a bibliography of references to diseases in wild mammals and birds compiled by Patricia O'Connor Halloran, available through the American Veterinary Medical Association, 600 S. Michigan Avenue, Chicago 5, Illinois.

Perhaps of greatest interest is the accumulation of overwhelming evidence pointing to the relationship of stresses and strains to infectious disease. Although the terminal cause of death may be apparent, the contributing causes, such as insecurity, irritations, chilling excessive heat or cold, nutritional deficiencies and many bad management practices must be considered. The belief that wild animals are unhappy in their captive state is a great fallacy. If given comfortable quarters with minimal disturbance and good nutrition, with the predator threat removed, and security absolutely assured, wild animals, birds, mammals, or reptiles live for long periods and are obviously relaxed and content.

Fiscal. Operating budgets, exclusive of animal purchases and new capital structures, exceed four hundred thousand dollars annually for very modest zoos and exceed two million dollars for the larger zoos.

The operating staff of a major zoo may be as large as 400, but averages about 250 in busy seasons. The scientific (curatorial) staff and animal keepers may represent a small part of the total staff, the greater number devoted to serving the public.

Capital expenditures are sizeable. Starting a new zoo in a new area, covering thirty to fifty acres, requires a minimum expenditure of four million dollars. In those zoos being built in temperate climates, requiring central heating, single buildings may have construction costs exceeding one million dollars. Depending on terrain and climatic conditions a simple bear or large cat moat may cost fifty thousand dollars.

Most zoos stress the importance of appropriate plantings and maintenance of an arboretum to increase the attractiveness of the zoo and furnish better backgrounds for the display of animals.

Many zoos are dependent for funds for animal purchases on income derived from concessionaires, or for food, souvenir, and camera units operated by their own organizations. Income of public zoos is, of course, derived from taxes. Many zoos, like privately operated zoos, are seeking new sources of income by way of gate admissions, parking fees and guided motorized tours.

In the U.S.A. special bond issues have made it possible to build entirely new zoos or to rebuild obsolete structures. Income from endowments, although sizeable, is small compared to income from other sources.

World-wide interest in zoos is rapidly increasing. The public demand is being satisfied by the development and construction of zoos and menageries, in the modern manner, in even the smallest communities throughout the world.

C. R. SCHROEDER

References: see in text.

ZOSTEROPHYLLOPHYTINA

In earliest Devonian time simple vascular plants bearing lateral, globose to reniform sporangia appeared alongside the Rhyniophytina. Dehiscence of the sporangia was distal. The vascular strand was elliptical in transection and its maturation was exarch. Leaves and roots were lacking but unicellular spines, multicellular spines, or multicellular teeth arranged in opposite rows along the stem, all indicate variation from the

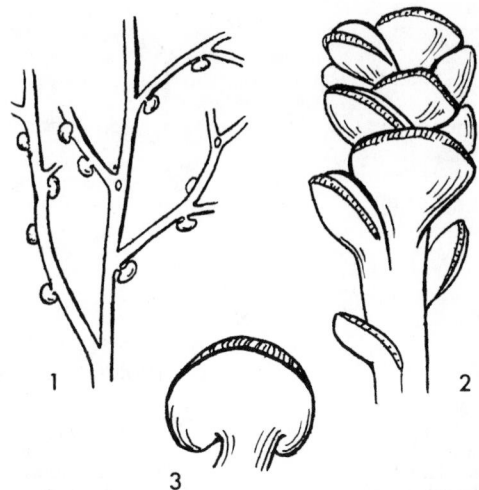

Figs. 1–3 1, *Gosslingia breconensis*. Redrawn from text fig. 86 of Croft and Lang, 1942. 2, *Zosterophyllum australianum*. One "spike" of sporangia. Drawn from Lang and Cookson Pl. 12, fig. 16 (1930). 3, *Zosterophyllum australianum*. A single sporangium showing its reniform shape and distal line of dehiscence. Drawn from Lang and Cookson Pl. 12, fig. 29 (1930).

naked stems of *Zosterophyllum*. Sporangia were aggregated into spike-like clusters (*Zosterophyllum, Bucheria*) or were scattered along the stem (*Gosslingia*, and the ornatum type of *Psilophyton* for which a new generic name is needed).

Knowledge of this group is expanding rapidly, with several contributions now nearly ready for publication. The group seems likely to have been the evolutionary precursor of the lycopods which appear first in the middle of Lower Devonian time, after the zosterophyllophytes became established.

H. P. BANKS

References

Banks, H. P., "The early history of land plants," *in* Drake, E. T. ed., "Evolution and Environment," New Haven, Yale University Press, 1968.

Croft, W. N. and A. H. Lang, Phil. Trans. Roy. Soc. London. B., **231**: 131–163, 1942.

Lang, W. H. and I. C. Cookson, Phil. Trans. Roy. Soc. London. B., **219**: 133–163, 1930.

INDEX

(Boldface numbers refer to main articles)

Aardvark, 956
A-bands, 582
Abbe, E., 555
Abdominal ribs,
Abduction, 586
Aberrations, 555
ABIOGENESIS, 1
Abomasum, 63
Abortion, 374
Abrine, 974
Abscisic acid, 433
Abyss, 629
Acantharia, 8
Acanthinae, 955
ACANTHOCEPHALA, 2, 460
  larva, 482
Acanthopterygii, 925
Acanthor, 482
Acaridida, 4
ACARINA, 3
Accipitridae, 328
Acephala, 680
Acetaldehyde, 140
Acetyl choline, 399
Acetyl-CoA, 146, 332, 806, 905
Acetylglucosamine, 171
Acid fermentations, 332
Acid mucopolysaccharides see AMP
Acidophil cells, 713
Acnidosporidea, 896
Acoela, 742, 959
ACOELOMATA 4, 31
Aconitase, 473
Aconitine, 974
Acorn shells, 189
Acorn worm, 401, 460
Acoustic stimulators, 280
Acridoidea, 652
Acrosome, 891
Acrothoracica, 189
ACTH, 292, 426, 713
Actin, 585
Actinaria, 42
Actinedida, 4
Actinomycetales, 5, 73
ACTINOMYCETES, 5
  in soil, 873
Actinophrydia, 8
ACTINOPODA, 7
Actinopterygii, 178
Actinotroch larva, 482
Actinotrochida, 4
Actinozoa larvae, 482
Actinula larva, 441, 482
Active transport, 838, 908
  in excretion, 313
Aculeata, 444
ADANSON, M., 10, 920
ADAPTATION, 11
Adders, 640

Adduction, 586
Adenine, 621, 779
Adenocorticotrophin, 478
Adenohypophysis, 712
  hormones of, 426
Adenophorea, 601
Adenosine diphosphate see ADP
Adenosine triphosphate see ATP
Adenyl-oxyluciferin, 109
Adephaga, 197
ADH, 291, 310, 427, 713
Adipose tissue, 329
Adjustor neurones, 127
Adonic acid fermentation, 333
ADP, 106
  glucose metabolism, 140
  in cell, 152
  respiration, 806
ADRENAL, 13, 291, 294
  hormones of, 427
  in hibernators, 412
Adrenal androgens, 904
Adrenaline, as synaptic inhibitor, 452
Adrenocorticotrophic hormone see ACTH
Adventitious embryogeny, 46
Aepyornithidae, 792
Aerobic respiration, 806
Aerobiology, 667
Agamidae, 477
Agamospermae, 46
Agar-agar, 18
AGASSIZ, A. E., 14
AGASSIZ, J. L. R., 14
Aggregate fruit, 346
AGGRESSION, 15
Aggression releasers, 101
Aggressive behavior, 101
Aglossa, 45
Aglyph snakes, 976
Agnatha, 176, 271
Ahermatypic corals, 222
Air sacs, 67
Aizoaceae, 155
Akaniaceae, 370
Akinetes, 16, 172
Alae, 598
Alangiaceae, 966
Alar plate, 126
Alaudidae, 675
Albinism, of troglobionts, 150
Albumen, egg, 271
Alcae, 163
Alcedinidae, 220
Alcohol, 377
Alcoholic fermentation, 331, 807
Alcyonaria, 41, 221
Alder, 326
Alder flies, 620
Aldonic acid fermentation, 333
Aldosterone, 14, 294, 424

ALDROVANI, U., 15, 920
Alfalfa, 820
ALGAE, 16, 733
  alternation of generations, 20
  Antheridium of, 39
  arctic, 53, 59
  benthonic, 93
  blue-green, 230
  calcareous, 221
  cell wall, 154
  chlorophyll in, 171
  chloroplasts in, 173
  conjugation in, 212
  in soil, 873
  in space flight, 879
  planktonic, 715
  red, 812
  symbiotic, 913
Algin, 683
Alima larva, 483
Alismaceae, 593
Alisphenoid, 940
Alkaloids, 974
Alleles, 243, 363
ALLELISM, 19
Allergy, 448
Alligators, 178, 224
Allochronic speciation, 881
Alloeocoela, 742, 959
Allogromiina, 343
Allopatric speciation, 881
Allopolyploids, 750
Allotheria, 522
Allotype, 924
Aloe, 572
Alpha ketoglutarate oxidase, 474
Alpha particle, 782
Alpine tundra, 959
Alternation of generations, animal, 742
ALTERNATION OF GENERATIONS, PLANT,
  20, 46
Alveoli, 515
Amanita muscaria, as drug, 975
Amaranthaceae, 155
Amaryllidaceae, 501, 571
"Amber" imitation, 362
Amblypygi, 677
Ambulacrum, 465
Ambystomoidea 22, 969
Amebocytes, sponge, 31
Amentiferae, 21, 245, 789
Amera, 598
Amici, G. B., 724
Amino acid hormones, 424
Amino acids, fermentation, 333
  in food, 626
  in mesoglea, 545
  origin of, 500
  secretion, 308
Aminoacyl phosphate, 767

1002

Birches, 326, 789
Bird communities, 208
BIRD SONGS, 10, **114**
Birds (*see also* Aves), 66, 179
  arctic, 52
  beak, 91
  colonial, 199
  coloration, 201
  flight, 338
  group names, 594
  longevity, 514
  malaria, 738
  migration, 564
  parasitic, 675
  photoperiodism, 696
  reproduction, 802
Bird's egg, hormonal control, 802
  largest, 792
Bird's nests, 614
Birds of Paradise, 675
Bispores, 812
Biting lice, 520
Bivalent chromosomes, 540
Bivalvia, 680
Blackbud, 675
Blackfish, 159
Blackflies, 254
BLAINVILLE, H. M. D. DE, **116**
Blastemas, of skull elements, 865
Blastocyst, 270
Blastopore, dorsal lip, 576, 647
Blastula, 270
  gradient in, 383
Blastulation, 288, 289
Blattoidea, 650
Bleeding heart, 813
Blepharoplast, 772
Blood, 186
  annelid, 36
  development of, 403
  hormonal control of, 426, 427
  INVERTEBRATE, **116**
  polychaetae, 768
  VERTEBRATE, **117**
Blood count, 118
Blood poisons, plant, 974
Blood sugar, hormonal control, 427
Bloom, water, 16
Blow fly, 254
Blubber, 159
Bluebell, 572
Blueberries, 304
Blue-green algae, 16
Boa constrictors, 640
Bojanus organ, 314
Bolus, 63
Bombacaceae, 521
Bombicylidae, 675
BONE, **121**, 213, 416
Bone formation, effect of parathyroid, 293
Bony fishes, 177
Boobies, 679
Booidea, 640
Book-lice, 778
Book-lungs, 48
Boomslang, 967
Bopyridea, 462
Boraginae, 955
Boron deficiency in plants, 942
BOSE, J. C., **123**
Boston ivy, 809
Botryoidal tissue, 413
Botulism, 28, 254, 940
Botworm, 254
Boveri, 235
Bowerbirds, 675
Bowman's capsule, 319
BRACHIOPODA, 32, **123**, 460
  larvae, 482
Brachypteraciidae, 220
Bract (plant), 240

Bract (Siphoniphoran), 853
BRAIN, **125**, 417, 612
  crustacean, 227
  divisions of, 129
  of anthropoids, 42
  of Periplaneta, 429
Brain sand, 711
Branchial arches, 859
  myxinoidea, 234
Branchial pouches, 234, 671
Branching, of shoots, 851
Branchiocranium, 860
Branchiomeric muscles, 589
BRANCHIOPODA (*see also* Cladocera), **130**, 227
Branchiostoma, 24
Brassi, L., 881
Bread mold, 706
Breastbone, 858
Breeding season, 633
BREHM, A. E., **131**
Briar, 304
Brille, 640
Bristlecone pine, 392
Brittle stars, 264
Brome, 375
Bromeliaceae, 571
Bronchi, 515
Bronthotheroidea, 682
BROWN, R., **131**, 716, 829
Brown algae, 16, 683
Brownian motion, 241
Bruche, H., 275
Brucke, E. von, 111
Brunfels, O., 920
Brunner's glands, 000
BRUNO, G., **132**
Brush border, 303
Bryales, 582
BRYOPHYTA, **132**, 733
  alternation of generations in, 20
  antheridium, 40
  archegonium, 50
  arctic, 55
  cell wall, 154
Bryopsida, 405
BRYOZOA, **134**
  eggs, 271
Bubonic plague, 852
Buccal skeleton, 234
Bucerotidae, 220
Buckbean, 368
Buckthorns, 809
Budding bacteria, 74
BUFFON, G. L. L., **137**, 419, 881
Bufonidae, 45
Bugs, 409
Bulbils, of mosses, 582
Bulbs, 897, 902
Bur-reed, 669
Bursa of Fabricius, 578
Burseraceae, 370
Bush-babies, 491
Bushmaster, 640, 977
Bustards, 390
Butomaceae, 571, 593
BUTSCHLI, O., **137**
Buttercup, 155, 245, 571, 789
Butterflies, 492
  coloration of, 201
Button quail, 389
Butylene glycol fermentation, 332
Butyric acid fermentation, 332

Cabbage, 246, 813
Cacti (*see also* Opuntiales), 242
Caddisflies, 949
Caecilians, 21
Caffeine, 779
Caiman, 224
Cairinini, 38
Cajal, R. y, 618

Calanoida, 219
Calcarea, 759
Calciferol, 985
Calcification, 416
Calcite, in branchiopod shells, 123
  in molluscan shells, 569
Calcitonin, 427
Calcium, excretion, 310
  hormonal control of, 671
  in diet, 626
Calcium phosphate, in bone, 121
Caligoida, 219
Calla lily, 49
Callitrichaceae, 371
Calobryales, 407
Caltrops, 369
Calyciferae, 571
Calycophorae, 000
Calyptrogen, 817
Calyx, 340
CAMBIUM, **138**
  of root, 817
  origin, 544
Cambombaceae, 571
Camel crickets, 651
Camerarius, R. J., 436
Camerostome, 465
de Candolle, 245
de Candolle, A. P., 789
Canidae, 143
Cannaceae, 997
Cannon, W. B., 421
Canoidea, 143
Capelin, 508
Capers, 813
Capillaries, 187
  bacterial, 77
  moss, 582
Capitonidae, 708
Capitulum, floral, 209
Capparidaceae, 813
Caprellidea, 26
Caprifoliaceae, 821
Capsid, 980
Capsomere, 980
Capsule, 346
Capuchins, 764
Carapace, 164
Carapari, 645
Carbohydrate, aerobic oxidation, 472
  source of, 139
  synthesis, 141
CARBOHYDRATE METABOLISM, **139**
  hormonal control of, 425
Carbohydrate reserve, of fruit, 347
Carbon cycle, photosynthetic, 700
Carbon dating, 141
Carbon-14, 141, 142
Carbon sources, bacterial use, 85
β-Carboxylation, 475
Cardiac, cycle, 186
  muscle, 417
  output, 187
  rhythm, of insects, 431
Cardueae, 209
Cariamidae, 389
Caridea, 238
Carinini, 33
Carminic acid, 203
Carnassial tooth, 143
Carnitine, 987
CARNIVORA, 143
  dormancy in, 411
  energy relations, 208
Carotenes, 145
  as vitamin precurser, 985
  in bacteria, 704
CAROTENOIDS, **145**, 202, 710, 723
Carpal bones, 863
Carpel, 340
Carpet beetle, 455
Carpet weed, 155